Completely Revised Edition 2021-22

NEET

Objective CHEMISTRY

Volume 02

Completely Revised Edition 2021-22

NEET

Objective CHEMISTRY

Volume 02

Dr. RK GUPTA

ARIHANT PRAKASHAN (Series), MEERUT

ARIHANT PRAKASHAN (Series), MEERUT

卐 **Administrative & Production Offices**

Regd. Office

'Ramchhaya' 4577/15, Agarwal Road, Darya Ganj, New Delhi -110002
Tele: 011- 47630600, 43518550

卐 **Head Office**

Kalindi, TP Nagar, Meerut (UP) - 250002, Tel: 0121-7156203, 7156204

卐 **Sales & Support Offices**

Agra, Ahmedabad, Bengaluru, Bareilly, Chennai, Delhi, Guwahati, Hyderabad, Jaipur, Jhansi, Kolkata, Lucknow, Nagpur & Pune.

卐 **ISBN** 978-93-25299-42-9

卐 **PRICE** ₹705.00

PO No : TXT-XX-XXXXXXX-X-XX

Published by Arihant Publications (India) Ltd.

PRODUCTION TEAM

Publishing Managers	:	Mahendra Singh Rawat, Keshav Mohan	Inner Designer	:	Ankit Saini
Project Head	:	Sorabh Chaudhary	Page Layouting	:	Hament Kumar, Shishpal
Cover Designer	:	Shanu Mansoori	Proof Readers	:	Jyoti, Salman

For further information about the books published by Arihant, log on to www.arihantbooks.com or e-mail at info@arihantbooks.com

Follow us on   

PREFACE

Medical offers the most exciting and fulfilling of careers. As a doctor you can find satisfaction from curing other persons. Although the number of medical colleges imparting quality education and training has significantly increased after independence in the country but due to the simultaneous increase in the number of serious aspirants, the competition is no longer easy for a seat in a prestigious medical college today.

For success, you require an objective approach to your study in the test subjects. This does not mean you 'prepare' yourself for just 'objective questions'. Objective Approach means more than that. It could be defined as that approach through which a student is able to master the concepts of the subject and also the skills required to tackle the questions asked in different formats in entrances such as NEET (National Eligibility cum Entrance Test). These two-volume books, Objective Chemistry (Vol.1 & 2) are borne out of my experience of teaching Chemistry to medical aspirants, fill the needs of such books in the market.

The plan of the presentation of the subject matter in the books is as follows

- The whole chapter has been divided under logical topic heads to cover the whole syllabi of NEET developing the concepts in an easy going manner, taking the help of suitable examples.
- Important points of the topics have been highlighted in the text under notes, some extra points regarding the topics have been given in Notes to enrich the students.
- The **Solved Examples** given with different concepts of the chapter make the students learn the basic problem solving skills in Chemistry. It has been ensured that given examples cover all aspects of a concepts comprehensively.
- **Check Point** Exercises given in between the text of all chapter help the readers to remain linked with the text given as they provide them an opportunity to assess themselves while studying the text.
- Chapter Exercises at the end of the chapters have been divided into three parts:
  Part A- Topical Exercise has Objective Questions arranged in topically as per the flow of concepts of the chapter.
  Part B- Miscellaneous Questions has objective questions arranged as per the level of difficulty, Matching and Statement type questions are also covered here.
  Part C- 'Medical Entrances gallery' covering all questions asked in last 10 years' (2020-2011) in NEET & Other Medical Entrances.
- The answers / solutions to all the questions given in different exercises have been provided.

At the end I would like to say that suggestions from the respected Teachers & Students for the further improvement of the book will be welcomed open heartedly.

Huge efforts have been made from my side to keep this book error free, but inspite of that if any error or whatsoever is skipped in the book then that is purely incidental, apology for the same, please write to me about that so that it can be corrected in the further edition of the book.

Dr. RK Gupta
Email: raghuveshg@gmail.com

CONTENTS

SYLLABUS

UNIT I Solid State

Classification of solids based on different binding forces, molecular, ionic covalent and metallic solids, amorphous and crystalline solids (elementary idea), unit cell in two dimensional and three dimensional lattices, calculation of density of unit cell, packing in solids, packing efficiency, voids, number of atoms per unit cell in a cubic unit cell, point defects, electrical and magnetic properties, Band theory of metals, conductors, semiconductors and insulators.

UNIT II Solutions

Types of solutions, expression of concentration of solutions of solids in liquids, solubility of gases in liquids, solid solutions, colligative properties- relative lowering of vapour pressure, Raoult's law, elevation of boiling point, depression of freezing point, osmotic pressure, determination of molecular masses using colligative properties abnormal molecular mass. van Hoff factor.

UNIT III Electrochemistry

Redox reactions, conductance in electrolytic solutions, specific and molar conductivity variation of conductivity with concentration, kohlrausch's Law, electrolysis and Laws of electrolysis (elementary idea), dry cell-electrolytic cells and Galvanic cells; lead accumulator, EMF of a cell, standard electrode potential, Relation between Gibbs' energy change and EMF of a cell, fuel cells, corrosion.

UNIT IV Chemical Kinetics

Rate of a reaction (average and instantaneous), factors affecting rates of reaction, concentration, temperature, catalyst, order and molecularity of a reaction, rate law and specific rate constant, integrated rate equations and half-life (only for zero and first order reactions), concept of collision theory (elementary idea, no mathematical treatment). Activation energy, Arrhenious equation.

UNIT V Surface Chemistry

Adsorption physisorption and chemisorption, factors affecting adsorption of gases on solids, catalysis homogeneous and heterogeneous, activity and selectivity, enzyme catalysis, colloidal state, distinction between true solutions, colloids and suspensions, lyophillic, lyophobic multimolecular and macromolecular colloids, properties of colloids, Tyndall effect, Brownian movement, electrophoresis, coagulation, emulsions- types of emulsions.

UNIT VI General Principles and Processes of Isolation of Elements

Principles and methods of extraction concentration, oxidation, reduction electrolytic method and refining, occurrence and principles of extraction of aluminium, copper, zinc and iron.

UNIT VII *p*- Block Elements

Group 15 elements General introduction, electronic configuration, occurrence, oxidation states, trends in physical and chemical properties, preparation and properties of ammonia and nitric acid, oxides of nitrogen (structure only), **Phosphorous** allotropic forms, compounds of phosphorous, preparation and properties of phosphine, halides (PCl_3, PCl_5) and oxoacids (elementary idea only).

Group 16 elements General introduction, electronic configuration, oxidation states, occurrence, trends in physical and chemical properties, dioxygen, preparation, properties and uses, classification of oxides, ozone. **Sulphur** allotropic forms, compounds of sulphur, preparation, preparation, properties and uses of sulphur dioxide, sulphuric acid, industrial process of manufacture, properties and uses, oxoacids of sulphur (structures only).

Group 17 elements General introduction, electronic configuration, oxidation states, occurrence, trends in physical and chemical properties, compounds of halogens, preparation, properties and uses of chlorine and hydrochloric acid, interhalogen compounds oxoacids of halogens (structures only).

Group 18 elements General introduction, electronic configuration, occurrence, trends in physical and chemical properties, uses.

UNIT VIII *d*-and *f*-Block Elements

General introduction, electronic configuration, characteristics of transition metals, general trends in properties of the first row transition metals metallic character, ionization enthalpy, oxidation states, ionic radii, colour, catalytic property, magnetic properties, interstitial compounds, alloy formation. Preparation and properties of $K_2Cr_2O_7$ and $KMnO_4$. **Lanthanoids** electronic configuration, oxidation states, chemical reactivity and lanthanoid contraction and its consequences. **Actinoids** Electronic configuration, oxidation states and comparison with lanthanoids.

UNIT IX Coordination Compounds

Coordination compounds Introduction, ligands, coordination number, colour, magnetic properties and shapes, IUPAC nomenclature of mononuclear coordination compounds, isomerism (structural and stereo) bonding, Werner's theory VBT,CFT, importance of coordination compounds (in qualitative analysis, biological systems).

UNIT X Haloalkanes and Haloarenes

Haloalkanes Nomenclature, nature of C –*X* bond, physical and chemical properties, mechanism of substitution reactions. Optical rotation.

Haloarenes Nature of C–*X* bond, substitution reactions (directive influence of halogen for monosubstituted compounds only).

Uses and environment effects of – dichloromethane, trichloromethane, tetrachloromethane, iodoform, freons, DDT.

UNIT XI Alcohols, Phenols and Ethers

Alcohols Nomenclature, methods of preparation, physical and chemical properties (of primary alcohols only), identification of primary, secondary and tertiary alcohols, mechanism of dehydration, uses with special reference to methanol and ethanol. Phenols, Nomenclature, methods of preparation, physical and chemical properties, acidic nature of phenol, electrophillic substitution reactions, uses of phenols. Ethers, Nomenclature, methods of preparation, physical and chemical properties uses.

UNIT XII Aldehydes, Ketones and Carboxylic Acids

Aldehydes and Ketones Nomenclature, nature of carbonyl group, methods of preparation, physical and chemical properties, and mechanism of nucleophilic addition, reactivity of alpha hydrogen in aldehydes, uses. **Carboxylic Acids** Nomenclature, acidic nature, methods of preparation, physical and chemical properties, uses.

UNIT XIII Organic Compounds Containing Nitrogen

Amines Nomenclature, classification, structure, methods of preparation, physical and chemical properties, uses, identification of primary secondary and tertiary amines. **Cyanides and Isocyanides** will be mentioned at relevant places. **Diazonium salts** Preparation, chemical reactions and importance in synthetic organic chemistry.

UNIT XIV Biomolecules

Carbohydrates Classification (aldoses and ketoses), monosaccharide (glucose and fructose), D.L. configuration, oligosaccharides (sucrose, lactose, maltose), polysaccharides (starch, cellulose, glycogen): importance. Proteins Elementary idea of – amino acids, peptide bond, polypeptides, proteins, primary structure, secondary structure, tertiary structure and quaternary structure (qualitative idea only), denaturation of proteins, enzymes.

Hormones- Elementary idea (excluding structure). **Vitamins** Classification and function. **Nucleic Acids** DNA and RNA

UNIT XV Polymers

Classification Natural and synthetic, methods of polymerisation (addition and condensation), copolymerization.

Some important polymers natural and synthetic like polyesters, bakelite, rubber, Biodegradable and non-biodegradable polymers.

UNIT XVI Chemistry in Everyday Life

Chemicals in medicines analgesics, tranquilizers, antiseptics, disinfectants, antimicrobials, antifertility drugs, antibiotics, antacids, antihistamines.

Chemicals in food preservatives, artificial sweetening agents, elementary idea of antioxidants. **Cleansing agents** soaps and detergents, cleansing action.

CHAPTER 01

Solid State

Solid state is among the prominent states in which matter occurs. For different applications, we need solids with widely different properties, which depend upon the nature of constituent particles and the binding forces operating among them.

Solids differ from liquids and gases in the fact that gases and liquids possess fluidity, i.e. they can flow. Solids do not possess fluidity, instead they possess rigidity.

The solids possess following general characteristics

(*i*) They have definite mass, volume and shape.

(*ii*) They have strong inter-particle forces.

(*iii*) They are incompressible and rigid.

(*iv*) They have short inter-particle distances.

(*v*) Their constituent particles (atoms, molecules or ions) have fixed position and can only oscillate about their mean position.

Inside

CLASSIFICATION OF SOLIDS

Depending upon the arrangement of the building constituents, solids are classified into two types, as given below

Crystalline solids

The solids having definite geometrical shape are called **crystalline solids**.

Characteristics

(*i*) They are anisotropic, i.e. physical properties are different in different directions. It is due to different arrangement of particles along different directions, as shown in Fig 1.1.

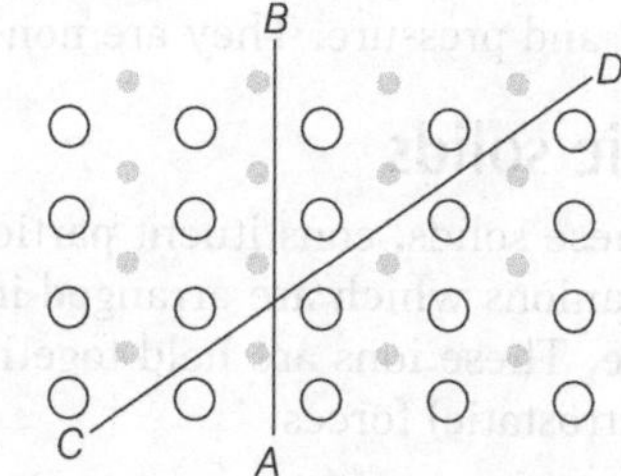

Fig. 1.1 Anisotropy in crystalline solid

(*ii*) They have long range order arrangement in constituent particles, e.g. sodium chloride (NaCl), Quartz (as shown in Fig 1.2).

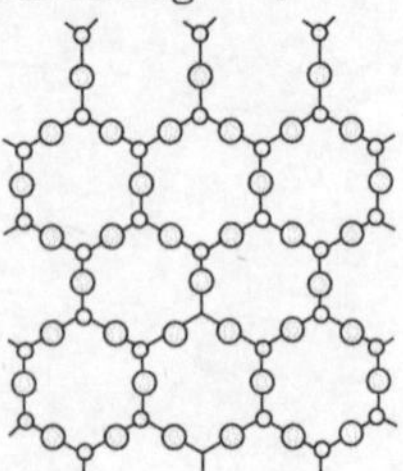

Fig. 1.2 Long range order (Quartz)

Amorphous solids

The solids that do not have definite geometrical shape are known as **amorphous solids.**

Characteristics

(*i*) They are isotropic in nature, i.e. the properties such as electrical conductivity, thermal conductivity, mechanical strength and refractive index are same in all directions.

(*ii*) They have short range order arrangement in constituent particles, e.g. glass, plastic, rubber, wood etc. (as shown in Fig 1.3)

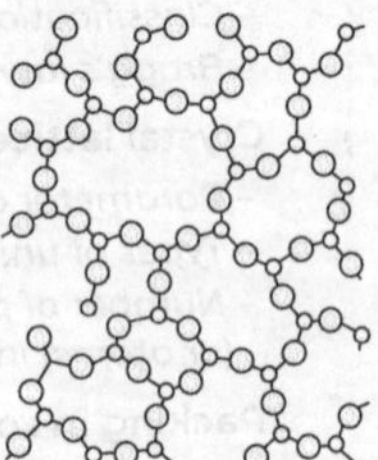

Fig. 1.3 Short range order (Glass)

Table 1.1 Differences between crystalline and amorphous solids

Basis of difference	Crystalline solids	Amorphous solids
Shape	They have definite geometrical shape.	They have irregular shape.
Melting point	They melt at a sharp and characteristic temperature.	They melt over a range of temperature.
Cleavage property	On cutting with sharp edged tool, they split into two pieces and the newly generated surfaces are plain and smooth.	They give irregular surfaces.
Nature	They are anisotropic in nature and called true solids.	They are isotropic in nature and called pseudo-solids or supercooled liquids.
Order in arrangement of constituent particles	They have long range order. e.g. quartz	They have short range order. e.g. glass.
Compressibility	They are rigid and incompressible.	They are significantly compressible.
Heat of fusion	They have definite heat of fusion.	They do not have definite heat of fusion.

Classification of crystalline solids

Depending upon the binding force operating between constituent particles, the crystalline solids are classified into four types, as given below

Molecular solids

These solids have molecules as constituent particles which are bonded by van der Waals' forces, e.g. I_2, dry ice, solid CO_2, solid CH_4, ice etc.

These solids are further classified on the basis of types of existing forces between the molecules as given

(*i*) **Non-polar molecular solids** The molecular solids which consist of either atoms, or the molecules formed by non-polar covalent bonds, are called non-polar molecular solids. In this type of solids, the atoms or molecules are held together by a weak dispersion forces or London forces, e.g. Ar, He, H_2, Cl_2, CCl_4, CO_2 etc. They are non-conductors of electricity.

(*ii*) **Polar molecular solids** These solids consist of molecules having polar covalent bond, i.e. molecules are held together by relatively stronger dipole-dipole interactions, e.g. solid NH_3, HCl, SO_2 etc. They are non-conductors of electricity.

(*iii*) **Hydrogen bonded molecular solids** These solids consist of molecules containing polar covalent bonds between H and F, O or N atoms. Strong hydrogen bonding binds the molecules of such solids like H_2O (ice). They are volatile liquids at room temperature and pressure. They are non-condutors of electricity.

Ionic solids

In these solids, constituent particles are ions, i.e. cations and anions which are arranged in three-dimensional space. These ions are held together by strong coulombic (electrostatic) forces,

e.g. KCl, NaCl, $CuSO_4$, CaF_2, CsCl, ZnS, MgO, etc.

They are electrical insulator in the solid state. However, in the molten state or in aqueous state, the ions become free to move and conducts electricity.

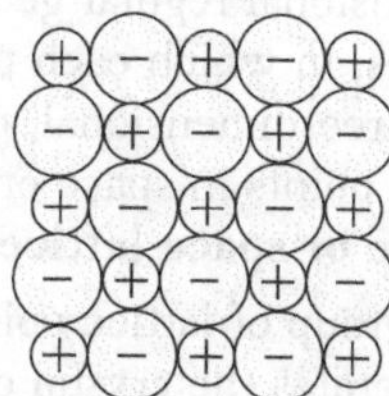

Fig. 1.4 General structure of ionic solid

Metallic solids

Metallic solids are solids composed of metal atoms/ions that are held together by metallic bonds. In these type of solids, constituent particles are positively charged metal ions known as **Kernels** and are surrounded by free electrons. These free electrons can easily flow throughout the metal crystal like water in the sea hence they are called sea of free electrons. They are responsible for high electrical and thermal conductivity of metals (Fig 1.5). e.g. Fe, Cu, Mg, Co, Al, Au, Ag, etc.

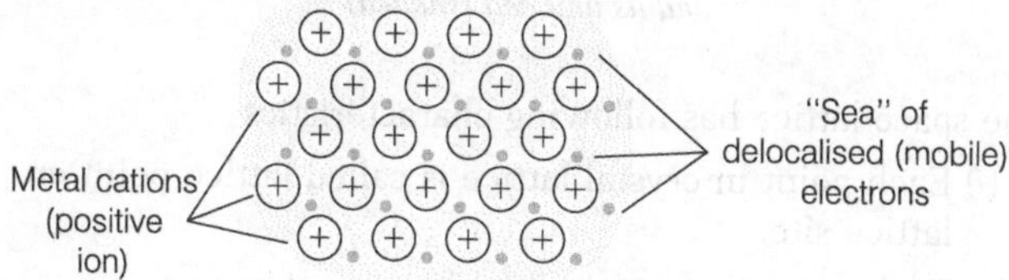

Fig. 1.5 Metallic solid

Covalent or network solids

A wide variety of crystalline solids of non-metals result from the formation of covalent bonds between adjacent atoms throughout the crystal. They are also called **giant molecules.** In these solids, atoms are strongly held by covalent bond at their positions (Fig 1.6), e.g. diamond, graphite, silicon carbide (SiC), silica, AlN, etc.

They are hard in nature and are bad conductor of electricity (except graphite).

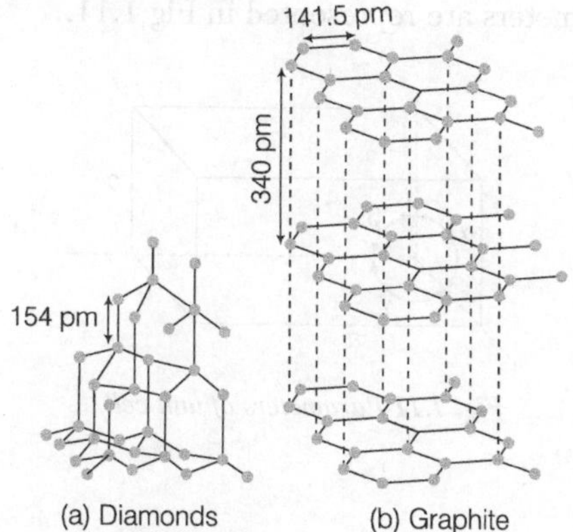

Fig. 1.6 Covalent or network solids

External features of crystalline solids

(i) **Faces** The crystals are bounded by surfaces which are usually planar and arranged in definite pattern. These surfaces are called 'faces'. These are of two types, like faces (e.g. alums, fluorspar etc) and unlike faces (e.g. glue).

(ii) **Forms** All the faces corresponding to a crystal are said to constitute a form. They may be simple (with like faces) or in combination (with unlike faces.)

(iii) **Edge** When two adjacent faces intersect then we get an edge.

(iv) **Solid angle** When three or more edges intersect then we get a solid angle.

(v) **Interfacial angle** The angle between the normals to the two intersecting faces are called interfacial angle.

Note *The faces, edges and interfacial angles are related as,*

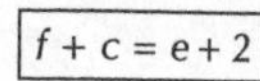

$$f + c = e + 2$$

f = number of faces
e = number of edges
c = number of interfacial angles

Elements of symmetry in crystalline solids

The following three are called elements of symmetry in crystal

(i) **Centre of symmetry** An imaginary point within a crystal through which any line drawn, intersects the surface of crystal at equal distance in all directions, is called the centre of symmetry.

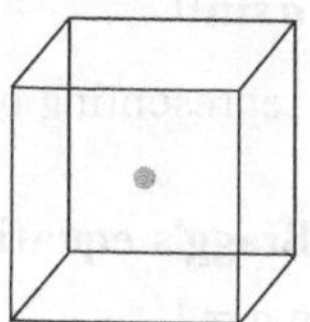

Fig. 1.7 Centre of symmetry

(ii) **Plane of symmetry** It is an imaginary plane passing through the crystal which can divide crystal into equal halves such that one half is the mirror image of the other half. In a crystal, two different planes are present namely rectangular (i.e. horizontal and vertical) and diagonal planes.

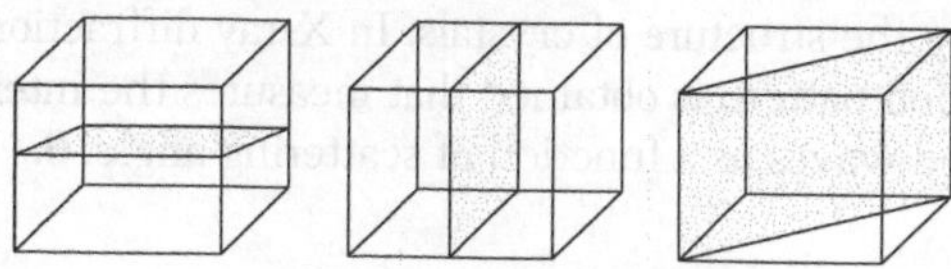

Fig. 1.8 Plane of symmetry

(*iii*) **Axis of symmetry** It is an imaginary line passing through the crystal, along which if a crystal is rotated, it presents the same appearance more than once in one complete rotation (360°).

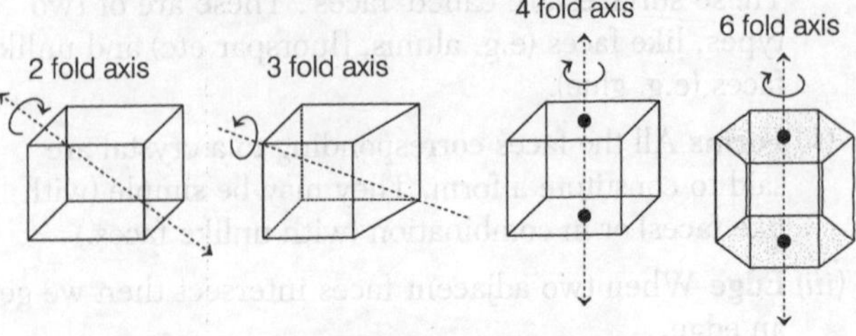

Fig. 1.9 Axis of symmetry

e.g. The cube has the greatest symmetry (23 elements of symmetry)

Axes of symmetry : Four - fold (3), three-fold (4), two-fold (6)

Plane of symmetry : 9

Centre of symmetry : 1

Bragg's law

Bragg's law was proposed by **W H Bragg.** According to this, diffraction of electromagnetic radiation occurs when a beam is scattered by an object containing regularly spaced lines (such as those in a diffraction grating) or points (such as the atoms in a crystal). Diffraction is due to interference between two waves passing through the same region of space at the same time.

The path difference between the waves is given as $2a\sin\theta$ where, θ is the scattering angle and '*a*' is the interplanar distance. Therefore,

$$\boxed{n\lambda = 2a\sin\theta}$$

where, '*n*' is an integer representing order of diffraction, and 'λ' is wavelength.

This equation is called **Bragg's equation.**

For first order reflection $n = 1$

and if $\theta = 30°$, $a = \dfrac{\lambda}{2\sin\theta} = \lambda$

if $\theta = 90°$, $a = \dfrac{\lambda}{2\sin\theta} = \dfrac{\lambda}{2}$

Application of Bragg's law

Bragg's law can be applied to both neutron diffraction and electron diffraction. Electron diffraction provides a base for studying the structure of crystals. In X-ray diffraction, a diffraction pattern is obtained that measures the intensity of scattered waves as a function of scattering angle, θ.

CRYSTAL LATTICE AND UNIT CELL

In a crystalline solid, constituent particles are arranged in a definite, three-dimensional regular geometrical order along all the three axes, in which each particle is depicted as a lattice point. A three-dimensional, regular arrangement of lattice points in space or in a crystal is called a **crystal lattice or space lattice** (Fig 1.10).

A three-dimensional group of lattice points which when repeated in space generates the crystal called **unit cell.** Refer to the shaded portion of Fig 1.10.

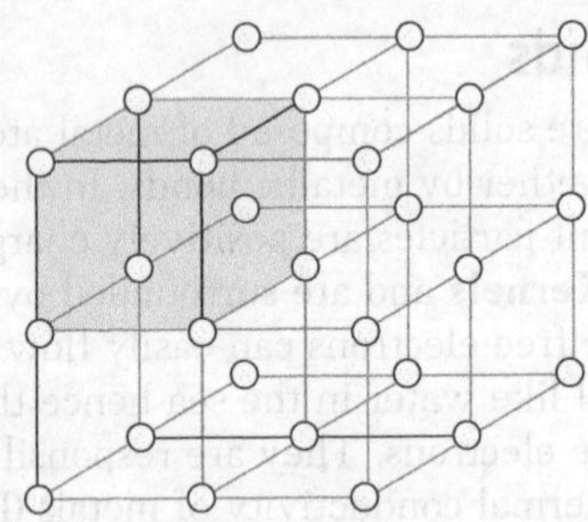

Fig 1.10 A portion of a three-dimensional cubic space lattice and its unit cell (shaded)

The space lattice has following characteristics

(*i*) Each point in crystal lattice is called lattice point or lattice site.

(*ii*) Each point in crystal lattice represents one constituent particle which may be an atom, a molecule or an ion.

(*iii*) Lattice points are joined by straight lines to bring out geometry of the lattice.

(*iv*) A lattice is made up of various unit cells.

Parameters of unit cell

The unit cell is described by the length of its edges *a*, *b*, *c* and the angles between the edges α, β, γ. These are called parameters of unit cell.

These parameters are represented in Fig 1.11.

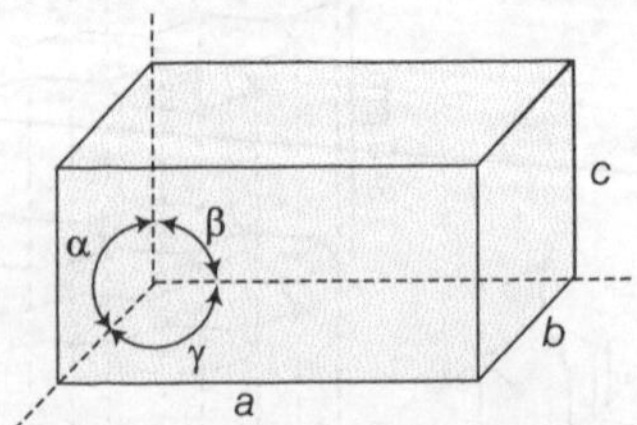

Fig. 1.11 Parameters of unit cell

Types of unit cell

On the basis of constituent particles present in unit cell, these are two types of unit cells

1. Simple or primitive unit cell

When constituent particles are present only at the corner of the unit cell then it is called **simple or primitive unit cells.** They are of seven types given in table 1.2.

2. Centred or non-primitive unit cell

When a unit cell contains one or more constituent particles present at positions other than corners in addition to those at corners are known as **centred unit cell.**

These are of three types

(*i*) **Body centred unit cell** In addition to the particles at the corners, there is one particle present at the centre of body of the unit cell.

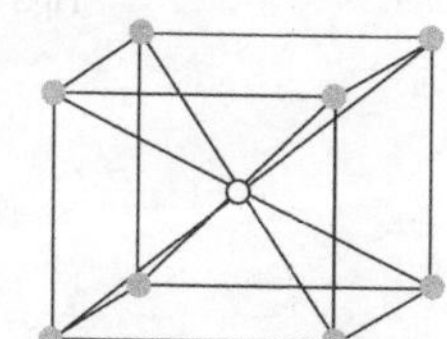

Fig. 1.12 Body centred cubic

(*ii*) **Face centred unit cell** In addition to the particles at the corners, there is one particle present at the centre of each face.

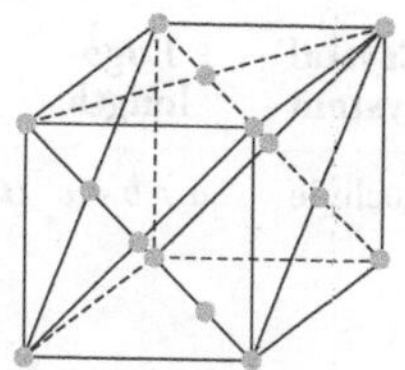

Fig. 1.13 Face centred cubic

(*iii*) **End centred unit cell** In addition to the particles at the corners, there is one constituent particles at the centre of any two opposite faces.

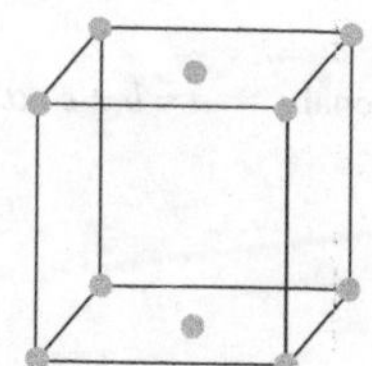

Fig. 1.14 End centred cubic

Crystal systems and Bravais lattice

On the basis of the six parameters of unit cell, they are divided into seven types of crystal systems. By bringing out possible variations in each of the crystal system, they are further classified into 14 types of lattices called **Bravais lattice.** Their distinguishing parameters and examples are listed in Table 1.2.

Table 1.2 Their distinguishing parameters and examples

Crystal system	Edge length	Axial angles	Bravais lattice	Examples
Cubic	$a = b = c$	$\alpha = \beta = \gamma = 90°$	Primitive, Face centred, Body centred	NaCl, zinc blende and Cu
Tetragonal	$a = b \neq c$	$\alpha = \beta = \gamma = 90°$	Primitive, Body centred	White tin, SnO_2, TiO_2 and $CaSO_4$
Orthorhombic	$a \neq b \neq c$	$\alpha = \beta = \gamma = 90°$	Primitive, Face centred, Body centred, End centred	Rhombic sulphur, KNO_3 and $BaSO_4$, $MgSO_4 \cdot 7H_2O$ (epsomite)

Crystal system	Edge length	Axial angles	Bravais lattice	Examples
Monoclinic	$a \neq b \neq c$	$\alpha = \gamma = 90°, \beta \neq 90°$	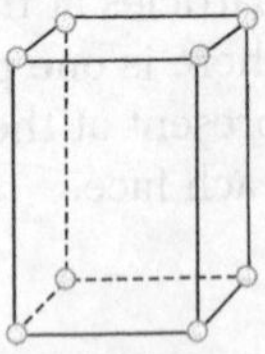 Primitive, 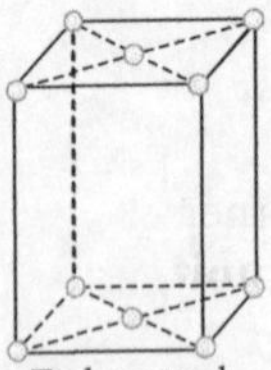 End centred	Monoclinic sulphur and $Na_2SO_4 \cdot 10\,H_2O$, $CaSO_4 \cdot 2H_2O$
Hexagonal	$a = b \neq c$	$\alpha = \beta = 90°\ \gamma = 120°$	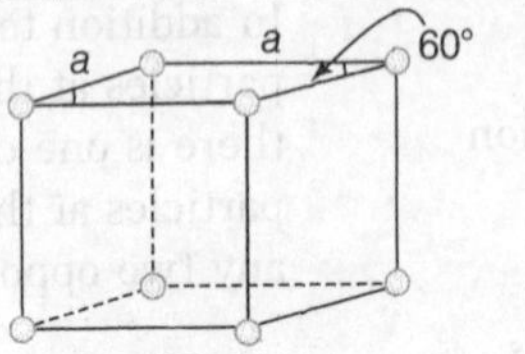 Primitive	Graphite, ZnO, CdS, SiO_2
Rhombohedral or trigonal	$a = b = c$	$\alpha = \beta = \gamma \neq 90°$	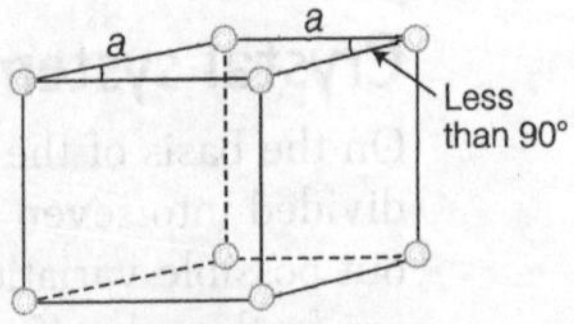 Primitive	Calcite ($CaCO_3$), HgS (cinnabar)
Triclinic	$a \neq b \neq c$	$\alpha \neq \beta \neq \gamma = 90°$	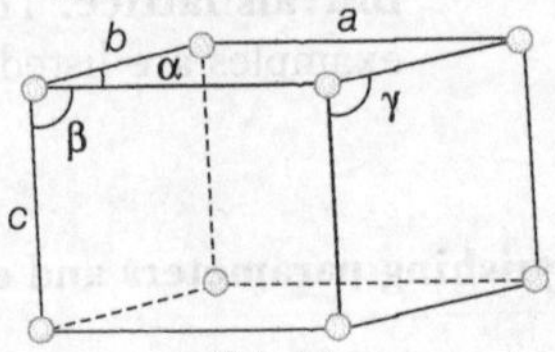 Primitive	$CuSO_4 \cdot 5H_2O$, H_3BO_3, $K_2Cr_2O_7$

Number of particles (or atoms) in a unit cell

Each type of crystal system has characteristic number of particles per unit cell. These particles can be located at the corner, body centre and face centre of the unit cell. Those particles present at a position other than body centre, are shared by other adjacent unit cells as well. So, we consider only a fraction of such particle belonging to a particular unit cell. The actual number of particle (sum of all fractions) is represented by Z or Z_{eff}.

To calculate the number of particles in a unit cell, follow the rules listed below

(*i*) A point that lies at the corner of a unit cell is shared among eight unit cells and therefore, only one-eighth (1/8th) of each such point lies within the given unit cell.

(*ii*) A point along an edge is shared by four unit cells and only one-fourth (1/4th) of it lies within the given cell.

(*iii*) A face centred point is shared by two unit cells and only one-half (1/2) of it is present in a given unit cell.

(*iv*) A body centred point lies entirely (1 particle) within the unit cell and contributes one complete point to the cell.

Table 1.3 Contribution of different atoms per unit cell

Type of crystal	Type of lattice point	Contribution to one unit cell (in cubic unit cell)
Cubic	Corner	1/8
	Edge	1/4
	Face centre	1/2
	Body centre	1
Hexagonal	Corner	1/6
	Face centre	1/2
	Body centre	3

Calculation of number of particles in cubic crystal system

As, in cubic crystal, all three axis have equal length ($a = b = c$) and all axial angles are of 90°. Therefore, in cubic crystal system, there are three types of unit cell.

(*i*) **Simple cubic unit cell** In this unit cell, constituent particles are situated at all eight corners of the cube, therefore, contribution to the unit cell is 1/8.

$\therefore$ Total particles in an unit cell, $(Z) = \frac{1}{8} \times 8 = 1$

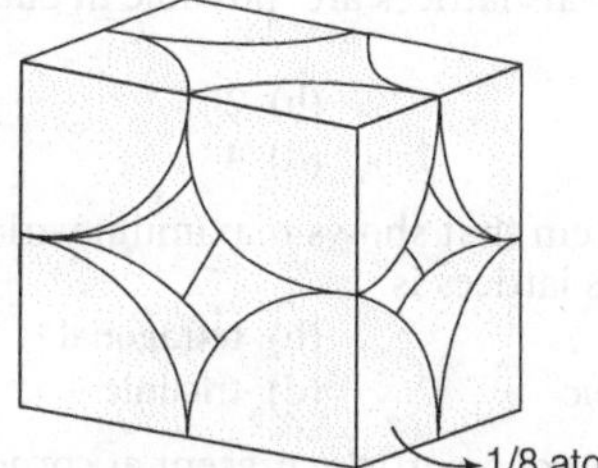

Fig. 1.15 Simple cubic cell

(*ii*) **Body centred cubic unit cell** In this unit cell, constituent particles are present at all eight corners as well as at the body centre.

$\therefore$ Contribution to the unit cell is '1' and therefore, total number of particles in an unit cell

$$(Z) = \frac{1}{8} \times 8 + 1 = 2$$

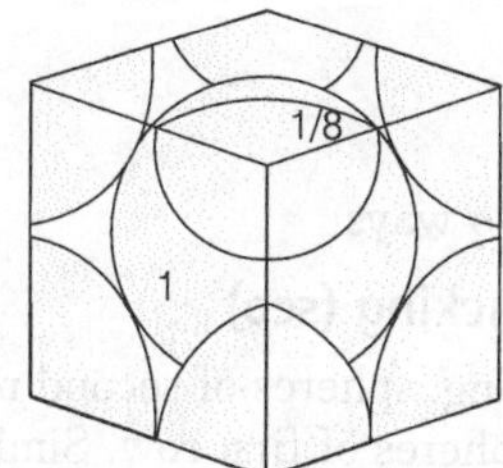

Fig. 1.16 Body centred cubic cell

(*iii*) **Face centred cubic unit cell** In this unit cell, constituent particles are present at the corners as well as at six faces.

Hence, contribution to the unit cell.

$$= \frac{1}{8} \text{ (at corner) and } \frac{1}{2} \text{ (at face centre)}$$

$\therefore$ Total number of particles in fcc.

$$= \frac{1}{8} \times 8 + 6 \times \frac{1}{2} = 4$$

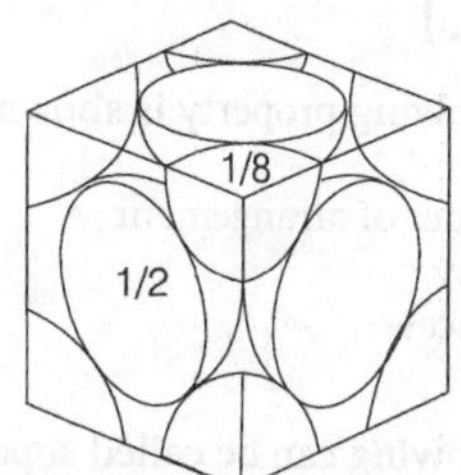

Fig. 1.17 Face centred cubic cell

Z can be determined by the simplest relation

$$Z = \frac{n_c}{8} + \frac{n_f}{2} + \frac{n_i}{1}$$

S.No.	Cubic unit cell	n_c	n_f	n_i	Total atoms in unit cell (Z)
1.	Simple cubic	8	0	0	1
2.	Body centred cubic	8	0	1	2
3.	Face centred cubic	8	6	0	4

where, n_c = number of atoms at the corners of the cube = 8

n_f = number of atoms at six faces of the cube = 6

n_i = number of atoms inside the cube = 1

Example 1.1 *A compound formed by elements 'X' and 'Y' has a cubic structure in which X-atoms are at the corner of the cube and Y-atoms are at the face centres. One atom 'X' is missing from the corner. Find out the formula of the compound.*

Sol. In a cube, there are 8-atoms. Since, one atom 'X' is missing from the corner. The number of X-atoms at the corners = 7

$$\therefore \quad X_{\text{eff}} = 7 \times \frac{1}{8} = \frac{7}{8} \text{ atom/unit cell}$$

$$Y_{\text{eff}} = 6 \times \frac{1}{2} = 3 \text{ atoms/unit cell}$$

Thus, formula of the compound is

$$X_{7/8}\,Y_3 = X_7\,Y_{24}$$

CHECK POINT 1.1

1. Which of the following property is shown by a crystalline solid ?
   (a) Long range order of arrangement
   (b) Compressibility
   (c) Irregular surfaces
   (d) Isotropy
2. Which of the following can be called supercooled liquid ?
   (a) Table salt (b) Quartz (c) Glass (d) Iron bar
3. In which of the following, Coulombic force is responsible for the substance being in solid state ?
   (a) HCl (b) NaCl (c) H_2 (d) CCl_4
4. Which of the following is not a network solid ?
   (a) SiO_2 (b) SiC (c) AlN (d) H_2O
5. Which of the following crystal systems has the edge length parameters as $a \neq b \neq c$ and angle parameters as $\alpha \neq \beta \neq \gamma \neq 90°$?
   (a) Tetragonal (b) Trigonal
   (c) Monoclinic (d) Triclinic
6. How many Bravais lattices are possible in cubic crystal system?
   (a) 1 (b) 2
   (c) 3 (d) 4
7. The crystal system that shows maximum variations to produce bravais lattices is
   (a) cubic (b) tetragonal
   (c) orthorhombic (d) triclinic
8. The contribution of a particle, present at corner, to the unit cell, is
   (a) 1/2 (b) 1/3 (c) 1/4 (d) 1/8
9. In face centred cubic unit cell, total number of particles contributed by the particles present at face is
   (a) 1 (b) 2 (c) 3 (d) 4
10. Lowest number of particle per unit cell is in
   (a) simple cubic unit cell
   (b) body centred cubic unit cell
   (c) face centred cubic unit cell
   (d) hexagonal unit cell

PACKING IN SOLIDS, DENSITY AND RADIUS RATIO

The packing of the constituent particles in a crystal is defined as the packing of the spheres in such a way that it occupies the maximum available space and there is minimum empty space left. Thus, it forms close-packed structures.

The number of nearest neighbours of any constituent particle in a crystal is called the **coordination number.**

The steps in which constituent particles build up the three dimensional structures are given below

One dimensional close packing

In one dimensional close packed structure, lattice points are arranged in a row and touching each other. In this arrangement, each sphere is in contact with two other neighbouring spheres.

Thus, in one dimensional close-packed arrangement, the **coordination number is 2.**

Fig. 1.18 Close packing of spheres in 1D

Two dimensional close packing

When the rows are stacked over one an other, a two-dimensional close packed structure (known as crystal plane) is produced.

It can be done in two ways

1. Square close packing (scp)

In this type of packing, spheres of second row are placed exactly above the spheres of first row. Similarly, spheres of third are just above second row and so on.

This is called *AAA* ... type of arrangement. Centres of four immediate neighbouring spheres form a square. Hence, the packing is **called square close packing (scp) in two dimension.**

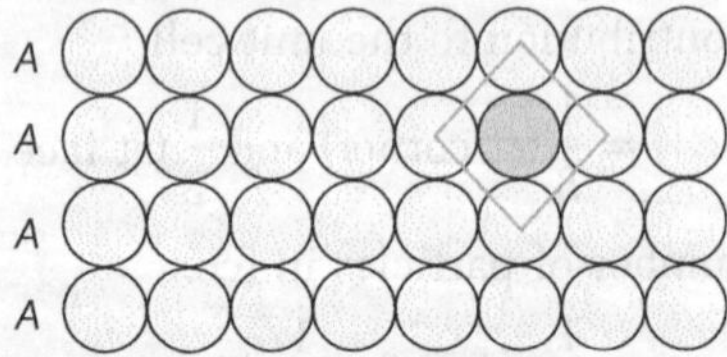

Fig. 1.19 Square close packing of spheres in 2-D

Each sphere is in contact with four other spheres, so **coordination number is 4.** Atom (◉) is in contact with other four atoms (○) as shown in Fig. 1.19.

2. Hexagonal close packing (hcp)

In this type of close packing, spheres of second row fit in the depression of the first row spheres and those of third row in the depression of the spheres of second row and so on. The arrangement in each alternate row is similar, that is, first row is similar to the third row and second row is similar to the fourth row and so on.

If we call the first row as *A*-type and second row as *B*-type, thus, this arrangement will be called *ABAB*...... type.

In this arrangement, each sphere is in contact with six of its neighbours, hence its **coordination number is 6.**

The centres of these six spheres are at the corners of a regular hexagonal, hence this packing is called **two dimensional hexagonal close packing** (hcp).

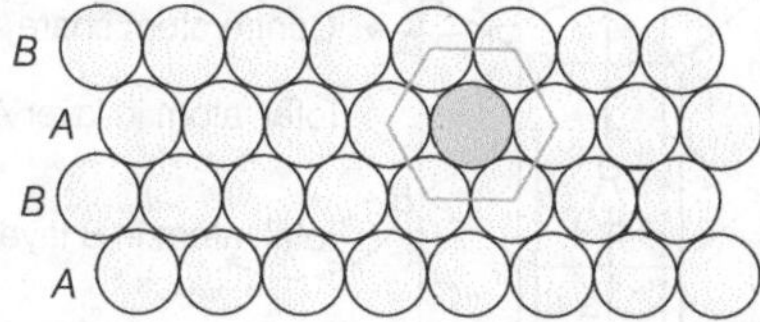

Fig. 1.20 Hexagonal close packing of spheres in 2D.

Atom (◎) is in contact with other six atoms (○) as shown in Fig. 1.20.

Three dimensional close packing

The geometry of original crystals are three dimensional.

There are two main types of three dimensional packing as described below

1. Three dimensional close packing from two dimensional square close packed layers

These can further be classified into two groups, depending upon how the second layer is placed over the first one.

(*i*) ***AAAA*-type or simple cubic close (scc) packing** It is formed when spheres of each layer is placed over spheres of layer below and not in the depressions of the layer below. It gives rise to **simple cubic close (scc)** packing as shown in Fig 1.21. Here, **the coordination number is 6.** There is no known natural crystal with this type of packing.

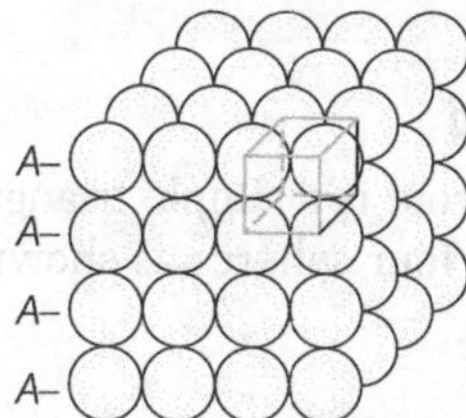

Fig. 1.21 AAAA-type simple cubic packing in 3D

(*ii*) ***ABAB*-type or body centred cubic (bcc) packing** It is formed when spheres of next layer is placed in the depressions in the layer below. It gives rise to **body centred cubic (bcc)** packing. In this case coordination number is 8. Metals like Li, Na, K, Rb and Cs crystallises in bcc structure.

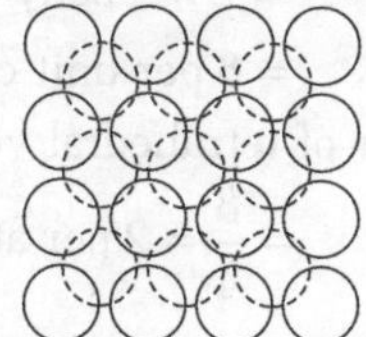

Fig. 1.22 bcc packing of sphere

2. Three dimensional packing from two dimensional hexagonal closed packed layers

This three dimensional packing is made by placing layers of spheres one over the other as illustrated below

(*i*) **Placing second layer over the first layer** There are triangular voids in the first layer '*A*' of hexagonal close packing. Spheres of the second layer are placed on triangular voids (or in the depressions of the first layer). From the Fig. (1.23) given below, it can be observed that not all the triangular voids of the first layer are covered by the spheres of the second layer. This gives rise to different arrangements.

Whenever, a sphere of the second layer is placed above the void of the first layer (or *vice-versa*), a **tetrahedral void** is formed. At other places, where the triangular voids of second layer are above the triangular voids of first layer, a new type of voids are formed by the combination of two tetrahedral voids. These voids are called **octahedral voids.**

The number of these two types of voids depend upon the number of close packed spheres.

If close packed spheres = N

then octahedral voids = N

and tetrahedral voids = $2N$

These voids are discussed further in this chapter.

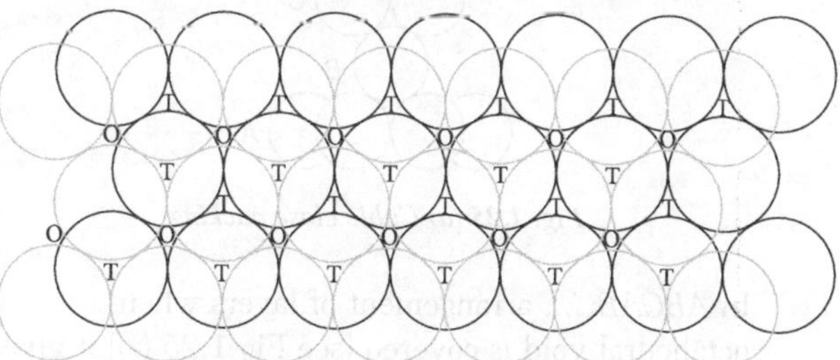

Fig 1.23 Voids between hcp layer

(*ii*) **Placing third layer over the second layer** When third layer is placed over the second, there are two possibilities.

- **Covering tetrahedral voids** Spheres of the third layer may cover tetrahedral voids of the second layer. In this case, the spheres of third layer are exactly aligned with those of the first and fourth layer with the second and so on. Thus, the pattern of spheres is repeated in alternate layers. This pattern is called *ABAB*...... pattern.

  This structure is called **hexagonal close packed** (hcp) **structure**. Here, **coordination number is 12.**

Metals like zinc, magnesium crystallises in hcp structure.

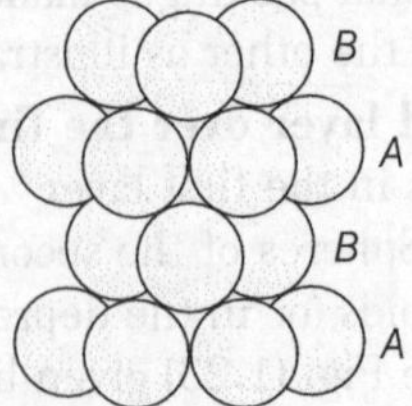

Fig. 1.24 Hexagonal close packing

- **Covering octahedral voids** The third layer may be placed above the second layer in such a manner that its spheres may cover octahedral voids of second layer. In this case, the spheres of the third layer are not aligned with those of either the first or the second layer.

  This layer is different from layers *A* and *B*. Let us call it as layer *C*. Continuing further, a packing is obtained where spheres in every fourth layer will be vertically aligned with those of the first layer. This pattern is called *ABCABC......* pattern.

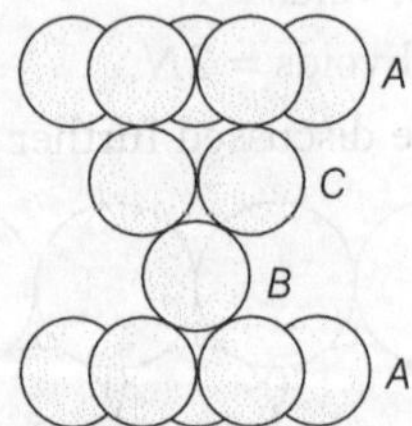

Fig. 1.25 (a) Cubic close packing

In *ABCABC*.... arrangement of layers when octahedral void is covered [see Fig 1.25 (a)] it gives rise to a structure called **cubic closed packed (ccp) or face centred cubic (fcc)** structure [see Fig 1.25 (b)]. Here, like **hcp, coordination number is 12.**

Metal like copper, silver etc. crystallise in ccp structure.

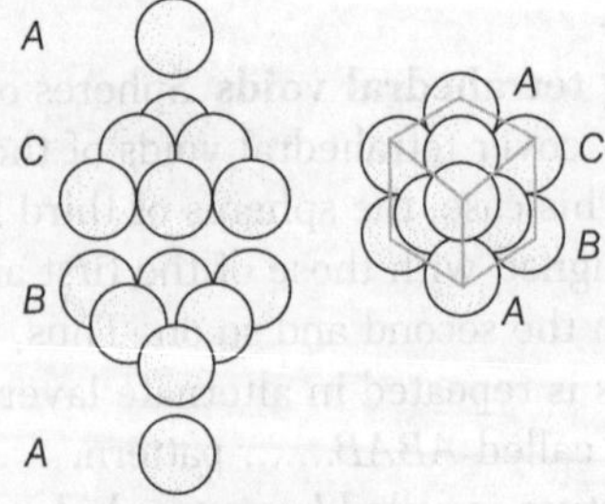

Fig. 1.25 (b)

Calculation of number of particles in hexagonal crystal system

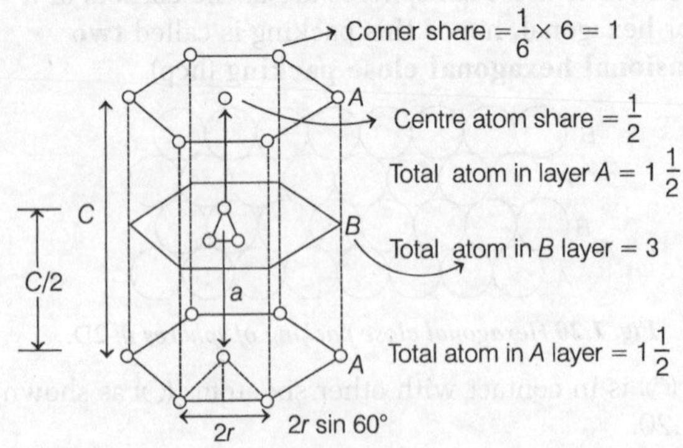

Fig. 1.26

$$\therefore \quad Z_{\text{eff}} = 1\frac{1}{2} + 1\frac{1}{2} + 3 = 6$$

Interstitial holes or voids

In closely packed structures, the empty space is called 'void'. In case of two dimensional packing it is called **trigonal void.**

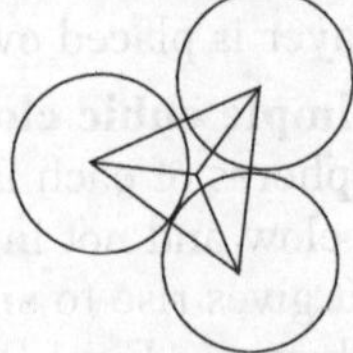

In three dimensional closed packing patterns, the voids are classified as

1. Tetrahedral void

- A tetrahedral void is a simple triangular space surrounded by four spheres as shown below

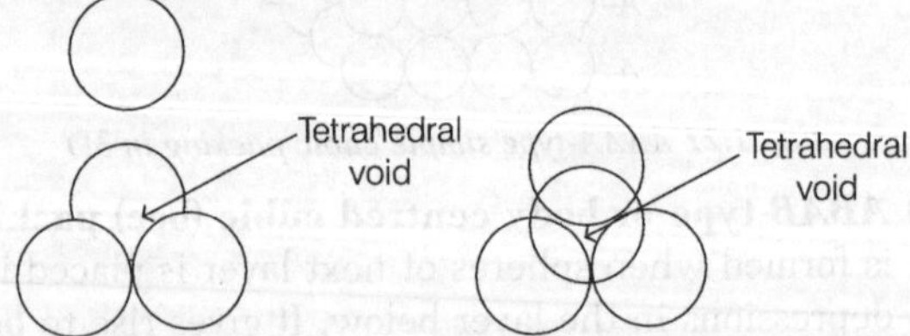

- 2 tetrahedral voids are present on each body diagonal of a cube. Therefore, number of tetrahedral voids

$$= 2 \times 4 \text{ body diagonal}$$

$$= 8 \text{ per unit cell}$$

Thus, the number of tetrahedral voids

$$= \frac{8}{4} = 2 \text{ per atom}$$

- Coordination number of tetrahedral void = 4
- Relation between radius of the tetrahedral voids (r) and radius of sphere (R) is

$$r = 0.225\,R$$

2. Octahedral void

An octahedral void is a double triangular void surrounded by six spheres as shown below

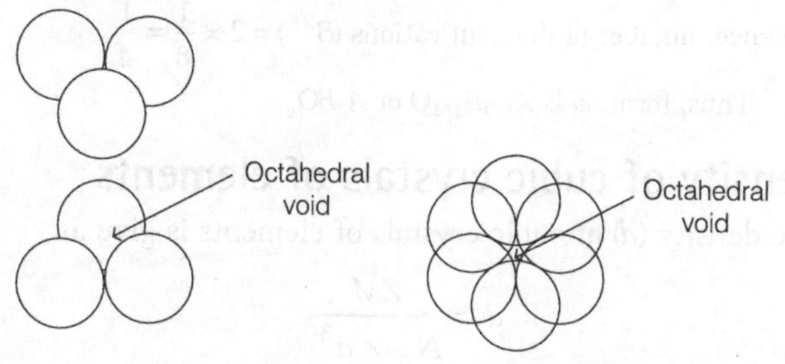

Number of octahedral void

$= 12$ edge centres $\times \frac{1}{4}$ per edge centre share $+1$ body centre

$= 3 + 1 = 4$ per unit cell $= \frac{4}{4} = 1$ **per atom**

- Coordination number of octahedral voids = 6
- Relation between radius of octahedral void (r) and radius of sphere (R) is

$$r = 0.414\,R$$

Note
- *Tetrahedral void = 2 × number of lattice point*
- *Octahedral void = Number of lattice point*
- *Tetrahedral void = 2 × octahedral void*
- *Size of voids can be arranged in increasing order as, trigonal void < tetrahedral void < octahedral void*

Example 1.2 *A compound is formed by two elements X or Y. Atoms of the element Y (as anion) make ccp and those of the element X (as cations) occupy all the octahedral voids. What is the formula of the compound?* (NCERT)

Sol. If number of atoms $= N$

then octahedral voids $= N$

and tetrahedral voids $= 2N$

Thus, **anions** in ccp $= N = Y$

cations in octahedral voids $= N = X$

$\therefore$ Formula : XY

Example 1.3 *Atoms of elememt B form hcp lattice and those of the element A occupy 2/3rd of the tetrahedral voids. What is the formula of the compound formed by the elements A and B?*

Sol. If atoms $= N$

then, element B at hcp lattice $= N$

Octahedral voids $= N$

Tetrahedral voids $= 2N$

Element A with $\frac{2}{3}$ rd of tetrahedral sites $= 2 \times \frac{2}{3} = \frac{4}{3}N$

Thus, $A : B = \frac{4}{3}N : N = 4N : 3N$

Thus, A_4B_3

Packing efficiency

It is the ratio of total volume of spheres to the volume of the unit cell. It can also be defined as the percentage of total space filled by particles.

$$\text{Packing efficiency} = \frac{\text{total volume of spheres}}{\text{volume of the unit cell}}$$

$$\%\ \text{packing efficiency} = \text{packing efficiency} \times 100$$

$$\text{or}\ \%\ \text{packing efficiency} = \frac{\frac{4}{3}\pi r^3 \cdot Z}{a^3} \times 100$$

where, r = radius

Z = number of atoms in a unit cell

a = edge length of the unit cell

Different unit cells have different percentage of packing efficiency. This is summarised in the Table 1.4 below.

Table 1.4 Characteristics of cubic unit cell

Unit cell	Z_{eff}	Coordination number	Packing efficiency	Radius (r)	Neighbouring distance
scc	1	6	52.4%	$r = \frac{a}{2}$	$d = a$
bcc	2	8	68%	$r = \frac{\sqrt{3}a}{4}$	$d = \frac{\sqrt{3}a}{2}$
fcc	4	12	74%	$r = \frac{a}{2\sqrt{2}}$	$d = \frac{a}{\sqrt{2}}$
hcp	6	12	74%	as above	as above

Example 1.4 *Potassium crystallises in a bcc lattice as shown in figure.*

(a) What is the distance between nearest neighbours ?

(b) What is the distance between next nearest neighbours ?

(c) How many nearest neighbour does each K atom have ?

(d) How many next nearest neighbours does not each K atom have ?

Sol. (a) The body diagonal $= \sqrt{3}a = \sqrt{3} \times 5.20$ Å $= 9.01$ Å

The nearest neighbours along that diagonal are at the half distance apart $= \frac{9.01}{2} = 4.50$ Å

(b) Distance between the next nearest neighbours = 5.20 Å along the cell edge.

(c) Nearest neighbours having surrounded by each K atom = 8 (atom at the centre is surrounded by eight corners at the nearest)

(d) Next nearest neighbours having surrounded by each K atom= 6 (CN of each K at corner is six).

Example 1.5 *The length of the unit cell edge of a bcc lattice metal is* 352 pm. *Calculate radius of the atom of the metal.*

Sol. In a 'bcc' structure, the atoms touch each other along the body diagonal.

Let, radius of the atom = R

then body diagonal = $R + 2R + R = 4R = \sqrt{3}a$

$$\therefore \quad R = \frac{\sqrt{3}}{4}a = \frac{\sqrt{3}}{4} \times 352 = 152.4 \text{ pm}$$

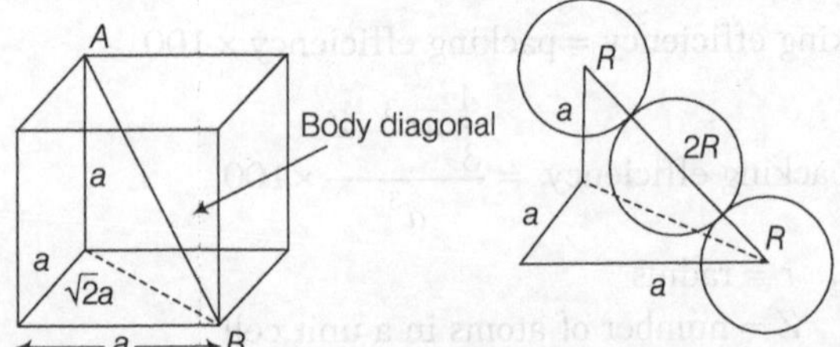

Example 1.6 *Silver metal crystallises in a cubic close packed arrangement with the edge of the unit cell having a length* a = 407 pm. *What is the radius of silver atom?*

Sol. $AC^2 + AB^2 = BC^2$

Here, $AC = AB = a$

$BC = 4r$

$a^2 + a^2 = (4r)^2$

$2a^2 = 16r^2$

$$\therefore \quad r^2 = \frac{a^2}{8}$$

$$\therefore \quad r = \frac{a}{2\sqrt{2}} = \frac{407}{2\sqrt{2}} = 143.9 \text{ pm}$$

Example 1.7 *Aluminium crystallises in a cubic close-packed structure. Its metallic radius is* 125 pm.

(i) What is the length of the side of the unit cell?

(ii) How many unit cells are there in 1.00 cm^3 *of aluminium?*

Sol. (i) For cubic close-packed structure,

$$r = \frac{a}{2\sqrt{2}}$$

$$a = 2\sqrt{2}\,r = 2\sqrt{2} \times 125 \text{ pm} = 353.6 \text{ pm}$$

(ii) Volume of face centred cube (cubic-close packed)

$= a^3$

$= (353.6 \times 10^{-10} \text{ cm})^3$

$= 4.42 \times 10^{-23} \text{ cm}^3$

Thus, 4.42×10^{-23} cm^3 = 1 unit

$$\therefore \quad 1 \text{ cm}^3 = \frac{1}{4.42 \times 10^{-23}} \text{ units} = 2.26 \times 10^{22} \text{ units}$$

Example 1.8 In a CPS (close *packed structure) of mixed oxides, it is found that lattice has* O^{2-} *(oxide ions), and one-half of octahedral voids are occupied by trivalent cations* (A^{3+}) *and 1/8th of tetrahedral voids are occupied by divalent cations* (B^{2+}). *Derive formula of the mixed oxide.*

Sol. Number of octahedral voids per ion in lattice = 1

hence, number of trivalent cations $(A^{3+}) = 1 \times \frac{1}{2} = \frac{1}{2}$

number of tetrahedral voids per ion in lattice = 2

hence, number of divalent cations $(B^{2+}) = 2 \times \frac{1}{8} = \frac{1}{4}$

Thus, formula is $A_{1/2}B_{1/4}O$ or A_2BO_4.

Density of cubic crystals of elements

The density (d) of cubic crystals of elements is give as

$$d = \frac{ZM}{N_A \times a^3}$$

where, d = density of crystal

Z = number of atoms per unit cell

M = molar mass,

a^3 = volume of unit cell

N_A = Avogadro's number $= 6.022 \times 10^{23}$

The SI unit of d is kgm^{-3}.

Example 1.9 *Titanium metal has a density of* 4.54 g/cm^3 *and an edge length of* 412.6 pm. *In what cubic unit cell does titanium crystallise?* (Ti= 48)

Sol. Density $d = \frac{MZ}{N_0 a^3}$, $d = 4.54$ g / cm^3, $M = 48$ g mol^{-1},

$Z = ?$

$N_0 = 6.023 \times 10^{23}$ mol^{-1}

Edge length, $a = 412.6$ pm $= 412.6 \times 10^{-10}$ cm

If value of Z is known, structure can be decided,

$$Z = \frac{dN_0 a^3}{M} = \frac{4.54 \times 6.023 \times 10^{23} \times (412.6 \times 10^{-10})^3}{48} = 4$$

Thus, titanium has **face centred cubic** structure.

Radius ratio in ionic crystal

Ionic crystals are made by cations and anions. These ions are arranged in three-dimensional array to form an aggregate of the type $(A^+B^-)_n$. Since, the Coulombic forces are non-directional, hence, the structures of such crystals are mainly governed by the ratio of the radius of cation (r_+) to that of anion (r_-). The ratio r_+ to r_- (r_+/r_-) is called radius ratio.

$$\text{Radius ratio} = \frac{\text{radius of positive ion}}{\text{radius of negative ion}} = \frac{r_+}{r_-}$$

Greater is the radius ratio, the larger is the size of the cation and therefore, greater is its coordination number. The relationship can be shown as,

Table 1.5 Coordination number and shape from radius ratio

Limiting radius ratio (r_+/r_-)	Coordination number	Shape	Structure type	Example
Less than 0.225	2 or 3	Linear for 2, triangular planar for 3	Linear (2) or triangular (3)	B_2O_3
0.225 - 0.414	4	Tetrahedral	Sphalerite (ZnS)	CuCl, CuBr, CuI, HgS, BaS
0.414 - 0.732	4 or 6	Square planar for 4 and octahedral for 6	NaCl	MgO, NaBr, CaS, MnO, KBr, CaO, AgCl
0.732 or more	8	Cube (Body centered)	CsCl	CsI, CsBr, TlBr, NH_4Br

Significance of radius ratio in the stability of ionic crystals

On increasing the radius of positive ion, radius ratio and coordination number of positive ion increases and therefore stability of ionic crystal also increases.

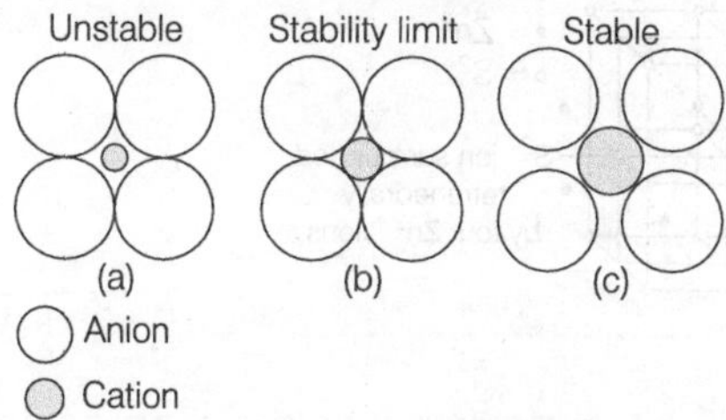

Fig. 1.27 Variation of coordination number with radius ratio

Example 1.10 *Radii of A^+, X^- and Y^- have been given as:*

A^+	1.00 pm
X^-	1.00 pm
Y^-	2.00 pm

Assign structure of AX and AY crystals and also determine volume of the unit cell of AX and AY crystals.

Sol. For AX, $\frac{r_{A^+}}{r_{X^-}} = \frac{1}{1} = 1.0$

Since, $\frac{r_+}{r_-} = 1.0$ hence, AX has **bcc structure.**

For bcc structure, edge length a is related to r_+ and r_- by

$$a\sqrt{3} = 2(r_{A^+} + r_{X^-})$$

$$a = \frac{2}{\sqrt{3}}[1.00 + 1.00] = \frac{4}{\sqrt{3}}$$

$$\therefore \quad \text{Volume} = a^3 = \left(\frac{4}{\sqrt{3}}\right)^3 = \frac{64}{3\sqrt{3}}$$

$$= 12.32 \text{ pm}^3$$

For AY, $\frac{r_{A^+}}{r_{Y^-}} = \frac{1}{2} = 0.5$

Since, $\frac{r_+}{r_-}$ of AY is 0.5. AY has **square planar or octahedral structure**

$$\therefore \quad r = \frac{a}{2}$$

$$a = 2r = 2(r_+ + r_-) = 6 \text{ pm}$$

$$\therefore \quad \text{Volume} = a^3 = 216 \text{ pm}^3$$

STRUCTURE OF IONIC COMPOUNDS

Simple ionic compounds are of two types, i.e. AB or AB_2 where, A and B are the positively and negatively charged ions respectively.

Structure of type AB and AB_2

In AB type of ionic compounds, the positively charged ion 'A' and negatively charged ion 'B' are present in the ratio 1 : 1. These compounds can have the three types of structures.

In AB_2 type of ionic compounds, the cation (A) and anion (B) are present in the ratio 1 : 2.

Different ionic solids having AB, AB_2 and A_2B type structures are summarised in the Table 1.6.

Table 1.6 Main characteristics and examples of some ionic solids

Crystal structure	Description	Coordination number	Numbers of ions per unit cell	Structure	Example
		Type *AB*			
(i) Rock salt type (NaCl)	It has fcc arrangement in which anions (i.e. Cl^-) occupy the corners and cations (i.e. Na^+) are present at body centre, edge centre and occupy all the octahedral voids.	$Na^+ = 6$ $Cl^- = 6$	4	• = Na^+; ∘ = Cl^-; Cl^- ion surrounded octahedrally by six Na^+ ions; Na^+ ion surrounded octahedrally by six Cl^- ions	NaI, KCl, RbI, FeO, AgCl, MgO etc.
(ii) Caesium chloride structure (CsCl)	It has bcc arrangement with anions (i.e. Cl^-) at corners and cations (i.e. Cs^+) at body centre.	$Cs^+ = 8$ $Cl^- = 8$	2	Cs^+ ion surrounded by 8Cl^- ions; Cl^- ion surrounded by 8Cs^+ ions; • = Cs^+; ∘ = Cs^-; (a) 1 unit cell; (b) 8 units cells	CsBr, CsI, TlCl, TlBr, etc.
(iii) Zinc sulphide structure (ZnS)	Anions are present at face centre and cations (i.e. Zn^{2+}) at alternate tetrahedral voids.	$Zn^{2+} = 4$ $S^{2-} = 4$	4	Zn^{2+} ion surrounded tetrahedrally by four S^{2-} ions; • = Zn^{2+}; ∘ = S^{2-}; S^{2-} ion surrounded tetrahedrally by four Zn^{2+} ions	CuCl, CuBr, CuI, AgI etc.
		Type *AB*$_2$			
Fluorite structure (CaF_2)	It has ccp arrangement in which Ca^{2+} (i.e. cations) occupy fcc lattice and F^- (i.e. anions) occupy all the tetrahedral voids.	$Ca^{2+} = 8$ $F^- = 4$	4	Ca^{2+} ∘; F^- ◯	BaF_2, $BaCl_2$, CdF_2, ThO_2, SrF_2 etc.
		Type *A*$_2$*B*			
Anti fluorite structure (Na_2O)	Anions are arranged in ccp form and cations occupy all the tetrahedral voids.	$Na^+ = 4$ $O^{2-} = 8$	4	∘ O^{2-}; • Na^+	K_2O, Li_2O etc.

Effect of temperature and pressure on ionic solids

On increasing pressure, coordination number increases whereas on increasing temperature coordination number decreases.

e.g. $\underset{CN=6:6}{\text{NaCl}}$ type structure $\underset{\text{High 'T' (760 K)}}{\overset{\text{High 'p'}}{\rightleftharpoons}}$ $\underset{CN=8:8}{\text{CsCl}}$ type structure

Example 1.11 *The radius of calcium ion is 94 pm and of an oxide ion is 146 pm. Predict the crystal structure of calcium oxide.*

Sol. The ratio $\frac{r_+}{r_-} = \frac{94}{146} = 0.644$

The prediction is an octahedral arrangement of the oxide ions around the calcium. Because the ions have equal but opposite charges, there must also be an octahedral arrangement of calcium ions around oxide ions. Thus, CaO structure is similar to **rock salt** (NaCl) structure.

Example 1.12 *CsCl crystallises in a cubic lattice that has a Cl^- at each corner and Cs^+ at the centre of the unit cell. If $(r_{Cs^+}) = 1.69$ Å and $(r_{Cl^-}) = 1.81$ Å, what is the value of edge length (a) of the cube?*

Sol. We assume that, the closest Cs^+ to Cl^- distance is the sum of the ionic radii of Cs^+ and Cl^-

$$= 1.69 + 1.81 = 3.50 \text{ Å}$$

This distance is one-half of the cubic diagonal $= a\sqrt{3}/2$

$$\therefore \frac{a\sqrt{3}}{2} = 3.50 \text{ Å}, \quad a = 4.04 \text{ Å}$$

Example 1.13 *The coordination number of the barium ion Ba^{2+}, in BaF_2 is 8. What must be the coordination number of F^- ion?*

Sol. Coordination number of barium ion tells us that it is surrounded by eight fluoride ions (charge $8\times(-1) = -8$). In order to balance out the eight negative charges, we need four barium ions (charge $4\times(+2) = +8$). Hence, the coordination number of F^- ions must be + 4.

Example 1.14 *The metal ion-halide ion distances in several alkali metal halides are given below.*

	Distance, nm	
	I^-	Cl^-
Li^+	0.310	0.249
Na^+	0.317	0.279
K^+	0.352	0.314
Rb^+	0.367	0.329

Suggest why there is such a small difference in internuclear distance between LiI and NaI compared to that between LiCl and NaCl.

Sol. Difference of internuclear distance between NaI and LiI

$$= 0.317 - 0.310 = 0.007 \text{ nm}$$

Difference of internuclear distance between NaCl and LiCl

$$= 0.279 - 0.249 = 0.03 \text{ nm}$$

$$0.03 \text{ nm} > 0.007 \text{ nm}$$

Iodide ion-iodide ion contact in LiI and NaI makes closer approach impossible.

Example 1.15 *Predict the coordination number (CN) of the cation in crystals of each of the following compounds.*

Compounds	Radius	
	Cation	Anion
MgO	0.65 Å	1.40 Å
MgS	0.65 Å	1.84 Å
CsCl	1.69 Å	1.81 Å

Sol.

Compounds	Radius ratio(r_+/r_-)	CN
MgO	0.46	6
MgS	0.35	4
CsCl	0.934	8

Spinels

The compounds which have two different type of cations, i.e. A^{2+} and B^{3+} and oxide (O^{2-}) as anion are called spinels. They have general formula $A^{2+}, B_2^{3+}, O_4^{2-}$. These structures can be further classified into two types

Normal spinel structure The structures in which O^{2-} anions form fcc packing and divalent metal ion A^{2+} occupies 1/8th of the tetrahedral voids and trivalent metal ion B^{3+} occupies half of the octahedral voids are known as normal spinel structures, e.g. $MgAl_2O_4$, $ZnFe_2O_4$ (zinc ferrite) etc.

Inverse spinel structure The structures in which O^{2-} anions form fcc packing and divalent metal ion A^{2+} occupies 1/8th of the tetrahedral voids whereas trivalent metal ions B^{3+} are present in 1/8th of the tetrahedral voids and 1/4th of the octahedral voids. e.g. Fe_3O_4 (magnetite), $MgFe_2O_4$, etc.

CHECK POINT 1.2

1. The maximum number of particles that can surround a particle in square closed packing in two dimension is
   (a) 3 (b) 4 (c) 6 (d) 8
2. When 6 particles surround a certain particle, with each particle touching the central particle leaving no space between two surrounding particle, this type of packing in two dimension is called
   (a) scp (b) hcp (c) ccp (d) fcc
3. The *AAAA* type three-dimensional packing, forms from two-dimensional closed packing is called
   (a) scc (b) hcp (c) bcc (d) fcc
4. When trigonal void of an hcp layers lies over trigonal void of another hcp layer beneath, the new type of void formed is
   (a) tetrahedral (b) inverted tetrahedral
   (c) octahedral (d) Both (a) and (c)
5. Which of the following metals crystallises in hcp form where each third layer is placed to cover tetrahedral voids?
   (a) Na (b) Zn (c) Mg (d) Both Zn and Mg
6. The ratio of radius of cation and anion of tetrahedral void is
   (a) 0.225 (b) 0.375
   (c) 0.415 (d) 0.667
7. In a crystal system, the ratio of tetrahedral voids and octahedral voids is usually
   (a) 1 : 1 (b) 1 : 2 (c) 2 : 1 (d) 1 : 3
8. Sizes of trigonal (T_g), tetrahedral (T_d) and octahedral (O_h) voids increase in the order
   (a) $T_g < O_h < T_d$ (b) $T_g < T_d < O_h$
   (c) $O_h < T_d < T_g$ (d) $T_d < T_g < O_h$
9. Which of the following types of structures has highest coordination number ?
   (a) Triangular (b) ZnS-type
   (c) NaCl-type (d) CsCl-type
10. Which of the following crystal has highest limiting radius ratio?
    (a) B_2O_3 (b) CuCl (c) CsCl (d) NaCl

IMPERFECTIONS OR DEFECTS IN SOLIDS

Any deviation from perfectly ordered arrangement of particles in a crystal is called **'crystal defect'.** These defects change the properties of crystals.

Point defect

When a defect occurs due to irregularities or deviations from the ideal arrangement around a point (or a particle), it is called **point defect.**

Line defect

When a defect occurs due to irregularities or deviation from ideal arrangement of entire rows of lattice points, it is called **line defect.** This defect deal during advance studies of crystallography.

Types of point defects

Some types of point defects are as follows

1. Stoichiometric defects

These are the point defects that do not disturb the stoichiometry of a solid, which are also known as **intrinsic** or **thermodynamic defect.**

These defects are of two types

Vacancy defect

When some of the lattice sites are empty, the crystal is said to have vacancy defect. This defect arises when substance is heated, due to which density of the substances decreases.

Interstitial defect

When some constituent particles occupy an interstitial site, the crystal is said to have **'Interstitial defect'.** In this defect, density of substance increases.

Rather than showing simple vacancy or interstitial defects, most of the ionic solids show different kinds of defects to maintain electrical neutrality.

(*i*) **Schottky defect** This defect arises when equal number of cations and anions are missing from the lattice to maintain the electrical neutrality.

This defect is shown by those ionic substances in which cations and anions are of almost **similar sizes.** Due to missing of ions from the lattice, density decreases, e.g. CsCl, NaCl, KCl, KBr, etc.

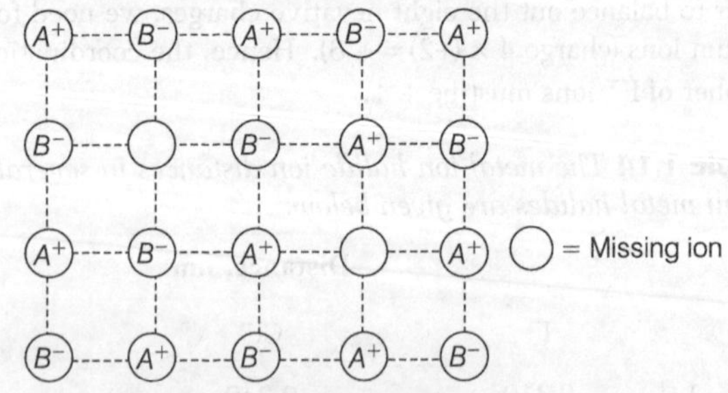

Fig. 1.28 Schottky defect

(*ii*) **Frenkel defects** This defect arises when smaller ion is dislocated from its original site to an interstitial site.

This defect is found in ionic crystals where size of anion is much larger than that of cation.

e.g. AgBr, AgCl, AgI, ZnS etc.
In this defect, density remains the same.
Since, small positive ions are highly polarising and large negative ions are readily polarised, these compounds have some covalent character.
The defect can be understood by the Fig 1.29 given below.

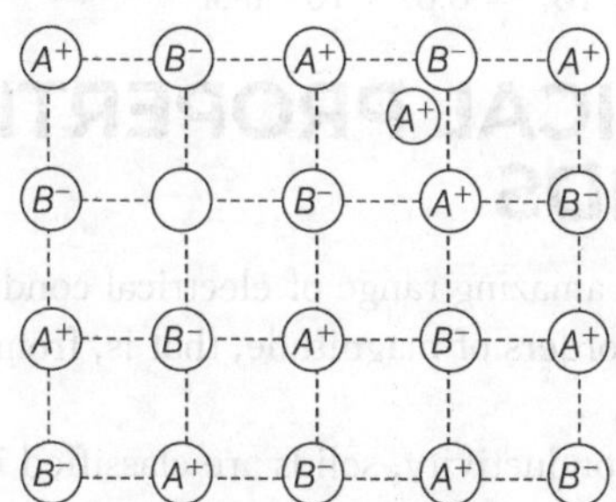

Fig. 1.29 Frenkel defect

Note *A compound, AgBr shows both Schottky and Frenkel defects because AgBr is highly ionic and there is large difference in size of Ag^+ and Br^-.*

Table 1.7 Differences between Frenkel and Schottky defects

Schottky defect	Frenkel defect
It is a vacancy defect.	It is an interstitial defect.
Positive as well as negative ions have almost equal size.	Positive ions are smaller but negative ions are bigger in size.
Ions are removed from crystal lattice	Ions penetrates into interstitial points.
Density decreases	Density does not alter
Value of dielectric constant remains unchanged	Value of dielectric constant increases
This defect found in crystal of high coordination number.	This defect found in crystal having lower coordination number.
In this defect, stability and electrical conductivity of crystal increases.	In this defect, stability of crystal decreases but conductivity increases.

2. Non-stoichiometric defects

Those imperfections in the crystals, which lead to change in the composition of solids are called **non-stoichiometric defects.** They contain constituent elements in non-stoichiometric ratio due to defects in their crystal structures, e.g. $Fe_{0.93}O$ and $Ni_{0.94}O_{1.00}$.

These defects are of two types which are discussed below

Metal excess defect

Metal excess defect are of two types which are discussed below

(*i*) **Metal excess defect due to anionic vacancies** These types of defects are produced when metal halide is heated in metal vapours, e.g. in alkali halide such as NaCl, KCl etc.

***F*-centre** A negative ion may be missing from its lattice site leaving a hole which is occupied by an extra electron to maintain electrical neutrality.

This anionic site is called ***F*-centre,** i.e. colour centre, as these are responsible for imparting colour to the crystals.

This defect is similar to *Schottky defects,* depicted in the Fig. 1.30 given below

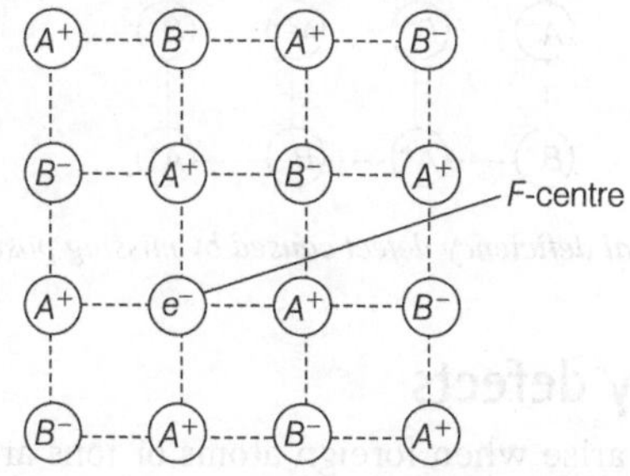

Fig. 1.30 Metal excess defect because of absent anion

(*ii*) **Metal excess defect due to interstitial cation** In this defect, an extra positive ion occupies interstitial position in the lattice and the free electron is trapped in the neighbouring interstitial site, the crystal remains neutral as depicted in the Fig. 1.31.

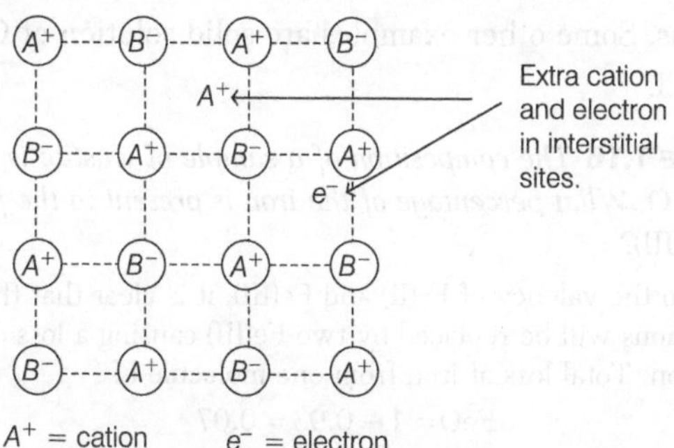

Fig. 1.31 Metal excess defect due to presence of interstitial cation

e.g. zinc oxide which is white in colour on heating loses oxygen and turns yellow.

$$ZnO \xrightarrow{\text{Heating}} Zn^{2+} + \frac{1}{2}O_2 + 2e^-$$

Now, there is excess of zinc in the crystal. Thus, formula becomes $Zn_{1+x}O$. These excess ions occupy interstitial sites. This defect is similar to Frenkel defect.

Metal deficiency defect

This defect arises due to the missing of a cation from its lattice site and the presence of the cation having higher charge in the adjacent lattice site. This defect occurs where the metal shows variable valency, i.e. in transition metals, e.g. FeO, FeS, NiO etc.

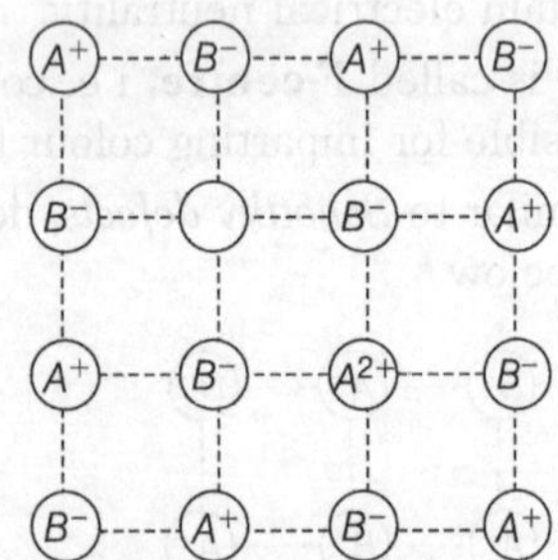

Fig. 1.32 Metal deficiency defect caused by missing positive ion.

3. Impurity defects

These defects arise when foreign atoms or ions are present as impurity in the lattice site.

e.g. If molten NaCl having some amount of $SrCl_2$ is crystallised, some of the sites of Na^+ ions are occupied by Sr^{2+} ions. Each Sr^{2+} ion will replace two Na^+ ions (as charge on Sr is +2). Therefore, it occupies the site of one ion and the other site remains vacant. Hence, this crystal lattice has cationic vacancies which are same in number of Sr^{2+} ions. Some other examples are solid solution of $CdCl_2$ and AgCl.

Example 1.16 *The composition of a sample of wustite is* $Fe_{0.93}O$. *What percentage of the iron is present in the form of* Fe(III)?

Sol. From the valency of Fe(II) and Fe(III), it is clear that three Fe(II) ions will be replaced by two Fe(III) causing a loss of one iron ion. Total loss of iron from one molecule of

$$FeO = 1 - 0.93 = 0.07$$

Hence, total Fe(III) present in one molecule of

$$FeO = 2 \times 0.07 = 0.14$$

Total number of Fe(II) and Fe(III) present in one molecule of FeO = 0.93

Hence, percentage of Fe(III) $= \frac{0.14}{0.93} \times 100 = 15.05\%$

Example 1.17 *If NaCl is doped with* 10^{-3} mol% *of* $SrCl_2$, *what is the concentration of cation vacancies?*

Sol. Due to addition of $SrCl_2$, each Sr^{2+} ion replaces two Na^+ ions,but occupies only one Na^+ lattice point. This makes one cationic vacancy. 100 moles of NaCl contains cation vacancies $= 10^{-3}$ mol.

Thus, 1 mole contains cation vacancies $= 10^{-5}$ mol

Thus, total cationic vacancies $= N_0 \times 10^{-5}$

$= 6.02 \times 10^{23} \times 10^{-5} = 6.02 \times 10^{18}$ mol^{-1}

ELECTRICAL PROPERTIES OF SOLIDS

Solids exhibit an amazing range of electrical conductivities ranging over 27 orders of magnitude, that is, from 10^{-20} to 10^7 ohm^{-1} m^{-1}.

On the basis of conductivity, solids are classified into three categories

Conductors

Solid substance which conduct electricity is known as conductor. Conductors have range of conductivity from 10^4 to $10^7 \, \Omega^{-1} m^{-1}$

On the basis chemical composition of substances, conductors are of two types.

(*i*) **Metallic conductors** In metallic conductors, the flow of electricity is due to the free electrons. In these conductors, after increasing temperature, conductivity of metal decreases, e.g. Al, Cu, Ag, Au etc.

(*ii*) **Electrolytic conductors** In these conductors, flow of electricity is carried out by the movement of ions. On increasing temperature, conductivity of these conductors increases. e.g. NaCl, KCl etc.

Insulators

The solid, which does not conduct electricity is called **'Insulator'**. In insulator, range of conductivity is between 10^{-20} to $10^{-10} \, \Omega^{-1} m^{-1}$.

e.g. phosphorus (P), wood rubber, sulphur (S) etc.

Semiconductors

The solid, whose conductivity lies between conductors and insulators is known as **semiconductors**. In these conductors, conductivity range is 10^{-8} to $10^{-4} \, \Omega^{-1} m^{-1}$, but at absolute temperature, its conductivity becomes zero.

There are two types of semiconductors

1. Intrinsic semiconductors

The substance, which functions as semiconductor in its purest state is known as intrinsic semiconductors, e.g. Si, Ge, etc.

- Conduction follows *via* two routes in this semiconductor
  (*i*) Displacement of electron from valence band to conduction band.
  (*ii*) By the locomotion of valence band.
- On the other hand, at absolute temperature (0 K), semiconductors behave as insulators.
- Electrical conduction in this semiconductor can be figured as,

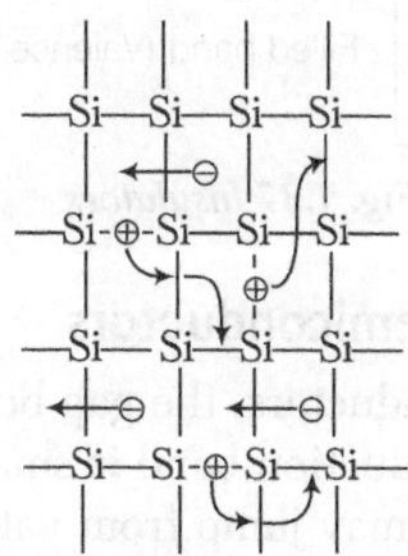

Fig. 1.33 Electrical conduction in an intrinsic semiconductor

2. Extrinsic semiconductor

For practical purposes the conductivity of pure Si and Ge is very low at room temperature. In order to increase their conductance, the pure substances are carefully doped (introduced with small amount of impurities in the form of elements of group 13 and 15 of the periodic table to increase their conductivity). Such conductance is called **extrinsic conductance.** Depending upon the nature of the elements used for doping, the extrinsic semiconductor can be of two types

(*i*) ***n*-type semiconductors** Silicon and germanium belong to group 14 and have four valence electrons each. In their crystals, each atom forms four covalent bonds with its neighbours shown in Fig. 1.34.

When silicon or germanium crystal is doped with a group 15 elements like P or As, which contains five valence electrons, they occupy some of the lattice sites in silicon or germanium crystal [Fig. 1.34]. Four out of five electrons are used in the formation of four covalent bonds with the four neighbouring silicon atoms. The fifth electron is not used in bonding thus, it is considered as extra and hence, becomes delocalised. This delocalised (or mobile) electron increases the conductivity of doped silicon or germanium.

Here, the increase in conductivity is due to negatively charged electron, hence silicon doped with electron rich impurity is called *n*-type semiconductor.

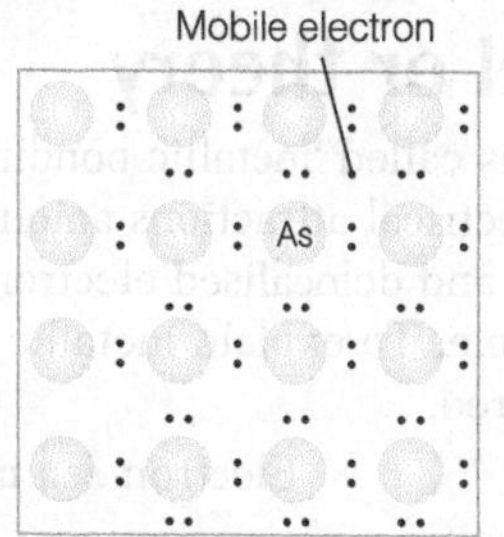

Fig. 1.34 n-type semiconductors

(*ii*) ***p*-type semiconductor** When silicon or germanium is doped with group 13 elements like B or Al, which contains only 3 valence electrons in the place of fourth electron, a hole is created. This is called **electron hole** or **electron vacancy.**

An electron from a neighbouring atom can come and fill the electron hole, but in doing so it would leave an electron hole at its original position. In this way, this hole moves through the crystal like positive charge giving rise to electrical conductivity.

When an electric field is applied, the electrons move toward positively charged plate and electron holes move towards the negatively charged plate. Hence, silicon and germanium doped with electron deficit impurities are called ***p*-type semiconductors.**

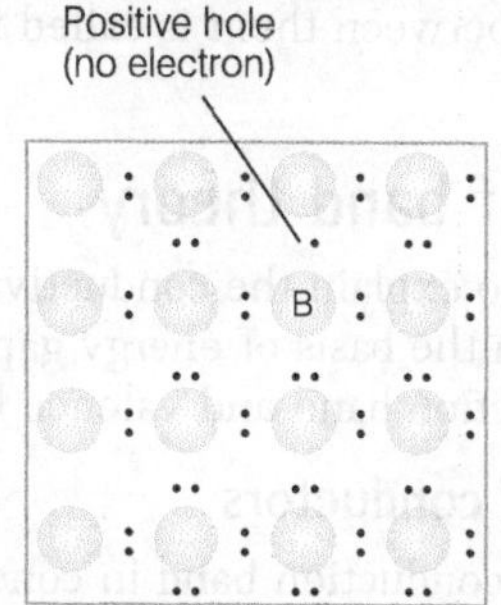

Fig. 1.35 p-type semiconductors

Applications

(*i*) **Diode** It is a combination of *n*-type and *p*-type semiconductors and is used as a rectifier.

(*ii*) ***n-p-n* and *p-n-p* type transistors** Transistors are made by sandwiching a layer of one type of semiconductor between two layers of the other type of semiconductors. These are used to detect or amplify radio or audio signals.

(*iii*) **Photodiode** These are diodes which are capable of converting light energy into electrical energy and are used in solar cells. Amorphous silica can be used for this purpose.

Band model or theory

Bonding in metals is called 'metallic bonding', which results from the electrical attractions among positively charged metal ions and delocalised electrons belonging to the crystal as a whole. To explain metallic bonding, two models are considered.

- Band model
- Electron-sea model

Note *As per the syllabus is concerned, only band model or band theory has been discussed in this chapter.*

Characteristics of band model

(*i*) Atomic orbitals of valence shell overlap with each other forming molecular orbitals whose energy difference is very low and they form a continuous band known as **energy band.**

(*ii*) The energy band in which valence electron is present, known as **valence band.**

(*iii*) If valence band is half-filled and overlaps with a higher energy vacant band then electron easily flows to this band. This band is called **conduction band.**

(*iv*) Combination of **valence** and **conduction band** depends upon structure of atoms and symmetry of crystal lattice.

(*v*) When these bands do not combine mutually then the gap arise between them is called **forbidden energy.**

Significance of band theory

Band theory helps to explain the conductivity of solids. It can be explained on the basis of energy gap present between the conduction band and valence band.

1. Explanation of conductors

When valence and conduction band in conductors combine, electrons flow from valence band to conduction band due to low excitation of electrons, that's why electrical conductivity is high in conductors.

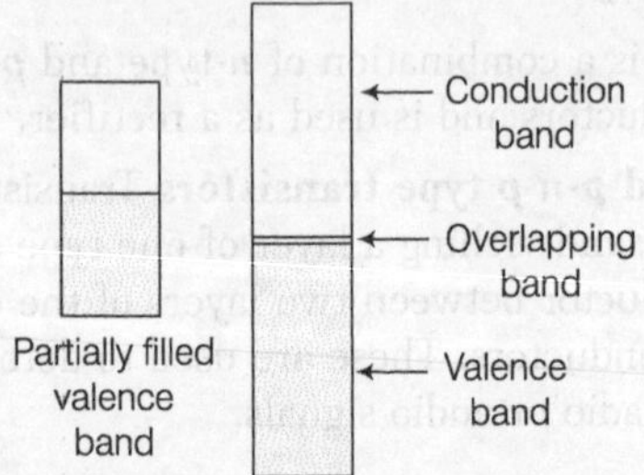

Fig. 1.36 Conductor

2. Explanation of insulators

Due to higher difference in energy of conduction band and valence band in insulator, electrons cannot enter into conduction band and hence there is no electrical conductivity.

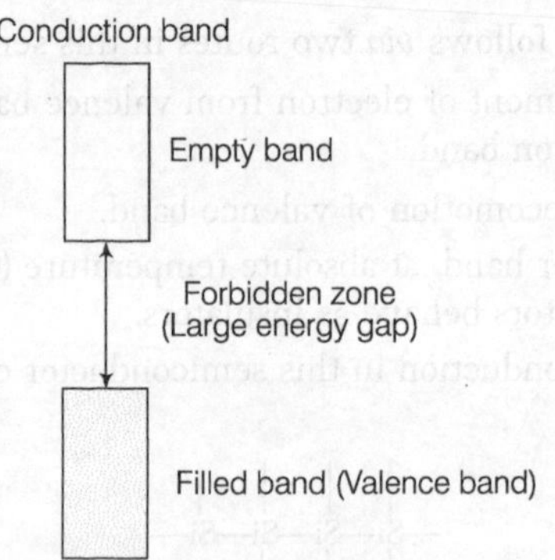

Fig. 1.37 Insulators

3. Explanation of semiconductors

In the case of semiconductors, the gap between valence band and conduction band is small and therefore, some of the electrons may jump from valence band to conduction band and some conductivity is observed.

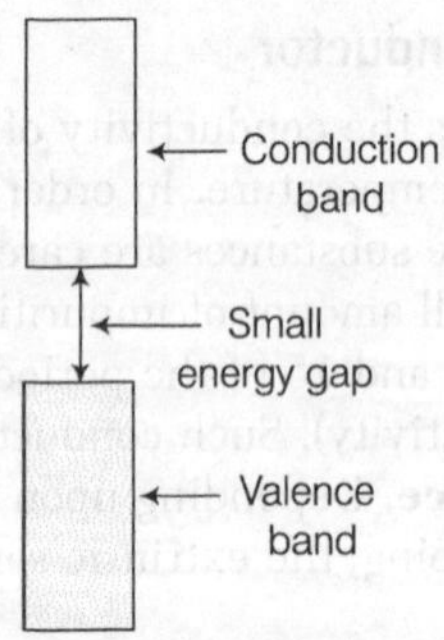

Fig. 1.38 Semiconductor

The conductivity increases with increase in temperature. The reason for such an increase is lowering of energy gap as shown in the figure. 1.38.

Super conductivity

The phenomenon where electricity passess through a material without resistance is called **super conductivity.** The material showing this property is said to be **super conductor.** Kammerlingh Onnes first reported the case of Hg as superconductor at a very low temperature of 4.2 K. Afterwards, some metal alloys and certain organic solids have been reported to be superconducting but at very low temperatures. Now a days, super conductors are also reported at comparatively higher temperatures.

e.g. $YBa_2Cu_3O_7$ at 90 K, $Bi_2Ca_2Sr_2Cu_3O_{10}$ at 105 K and $Tl_2Ca_2Ba_2Cu_3O_{10}$ at 125 K, etc.

MAGNETIC AND DIELECTRIC PROPERTIES

Magnetic properties

Each substance has magnetic and dielectric properties for which electron is responsible. Each electron behaves like a tiny magnet. Magnetic moment of each electron originates from two types of motions

(*i*) its orbital motion around the nucleus [Fig.1.39 (a)] and

(*ii*) its spin around its own axis [Fig. 1.39 (b)].

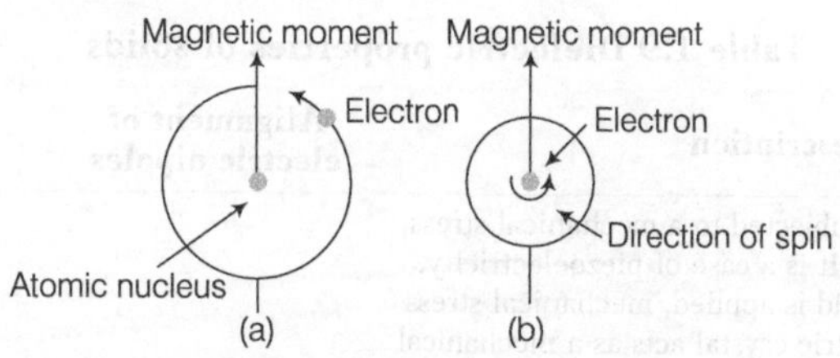

***Fig. 1.39** Magnetic moment associated with (a) orbit motion of the electron. (b) spin motion of the electron.*

As electron is a charged particle, thus its motion can be considered as a small loop of current which possesses a magnetic moment. Hence, each electron has a permanent spin and an orbital magnetic moment is associated with it. Magnitude of this magnetic moment is very small, its unit is **Bohr magneton μ_B**, which is equal to 9.27×10^{-24} Am^2.

Based on the behaviour of substances in the presence of magnetic field, these are classified into five classes and can be summarised in Table 1.8.

Table 1.8 Magnetic properties of solids

S.No.	Properties	Description	Alignment of magnetic dipoles	Examples	Applications
1.	Diamagnetic	Feebly repelled by the magnetic fields. Non-metallic elements (except O_2, S), inert gases and species with paired electrons are diamagnetic.	All paired electrons ↑↓ ↑↓ ↑↓	TiO_2, V_2O_5, NaCl, C_6H_6 (benzene)	Insulator
2.	Paramagnetic	Attracted by the magnetic field due to the presence of permanent magnetic dipoles (unpaired electrons). In magnetic field, these tend to orient themselves parallel to the direction of the field and thus, produce magnetism in the substance.	↑↓ ↑ ↑ At least one unpaired electron	O_2, Cu^{2+}, Fe^{3+}, TiO, Ti_2O_3, VO, VO_2, CuO	Electronic appliances
3.	Ferromagnetic	Permanent magnetism even in the absence of magnetic field. Above the **Curie temperature**, there is no ferromagnetism.	Dipoles are aligned in the same direction ↑↑↑↑↑	Fe, Ni, Co, CrO_2	CrO_2 is used in audio and video tapes
4.	Antiferromagnetic	This arises when net dipole alignment is zero due to equal and opposite alignment.	↑↓↑↓↑↓	MnO, MnO_2, Mn_2O, FeO, Fe_2O_3; NiO, Cr_2O_3, CoO, Co_3O_4,	–
5.	Ferrimagnetic	This arises when there is net dipole moment.	↑↓↓↓↑↓	Fe_3O_4, ferrites	–

Effect of temperature on magnetism

On increasing temperature, spin of electron get disordered and thus, at high temperature ferromagnetic, ferrimagnetic and anti-ferromagnetic solids get converted into paramagnetic solids, e.g. Fe_3O_4, (NiO, V_2O_3) are converted into paramagnetic solid at 850 K and 395 K respectively.

Dielectric properties

When a non-conducting material is placed in an electrical field, the electrons and the nuclei in the atoms or molecule of that material are pulled in the opposite directions, then negative and positive charges are separated and dipoles are generated. In an electric field,

- these dipoles may align themselves in the same direction, so that there is net dipole moment in the crystal.
- these dipoles may align themselves in such a manner that the net dipole moment in the crystal is zero.
- based on these facts, dielectric properties of crystals are summarised as

Table 1.9 Dielectric properties of solids

S. No.	Property	Description	Alignment of electric dipoles	Examples	Applications
1.	Piezoelectricity	When polar crystal is subjected to a mechanical stress, electricity is produced. It is a case of piezoelectricity. Reversely if electric field is applied, mechanical stress is developed. Piezoelectric crystal acts as a mechanical electrical transducer.	—		Record players
		• Piezoelectric crystals with permanent dipoles are said to have **ferroelectricity.**	↑ ↑ ↑ ↑ ↑	$BaTiO_3$, KH_2PO_4, etc,	
		• Piezoelectric crystals with zero dipole are said to have **anti-ferroelectricity.**	↑ ↓ ↑ ↓ ↑ ↓	Rochelle salt, $PbZrO_3$ etc.	
2.	Pyroelectricity	Small electric current is produced due to heating of some of polar crystal.	—	Perovskite, ionic crystals like GaN, $CsNO_3$ etc.	Power generation

CHECK POINT 1.3

1. Ionic solids show
(a) Schottky defect (b) Frenkel defect
(c) Interstitial defect (d) Both (a) and (b)

2. Density remains same in
(a) Vacancy defect (b) Interstitial defect
(c) Schottky defect (d) Frenkel defect

3. In which of the following defects, stoichiometry of the crystal is not maintained ?
(a) Interstitial defect (b) Metal excess defect
(c) Schottky defect (d) Frenkel defect

4. Slight yellowish colour of NaCl crystal is due to
(a) Vacancy defect (b) Interstitial defect
(c) Metal excess defect (d) Metal deficiency defect

5. Zinc oxide crystal when heated turns yellow, from white, after losing oxygen. The change in colour occurs due to development of
(a) Interstitial defect
(b) Metal excess defect due to interstitial anion
(c) Metal excess defect due to interstitial cation
(d) Metal deficiency defect

6. A substance has conductivity nearly equal to 10^7 S. The substance may be
(a) conductors (b) insulators
(c) semiconductors (d) cannot be

7. The electrical conductivity of silicon is improved by doping it with aluminium, it forms
(a) *n*-type semiconductors (b) *p*-type semiconductors
(c) intrinsic semiconductors (d) diode

8. According to band model of conductivity, semiconductor have
(a) partially filled band (b) small energy gap
(c) overlapped bands (d) large energy gap

9. In which of the following all the dipole moments are aligned in the same direction ?
(a) Diamagnetic (b) Paramagnetic
(c) Ferromagnetic (d) Anti-ferromagnetic

10. The compound which has at least one unpaired electron behaves as
(a) paramagnetic (b) diamagnetic
(c) ferromagnetic (d) anti-ferromagnetic

Chapter Exercises

(A) Topical exercise

I. Classification of solids

1 Which of the following is not a characteristic of a solid ?
(a) Rigidity (b) Fluidity
(c) High density (d) Definite volume

2 The binding force operating in molecular solids is/are
(a) van der Waals' forces (b) hydrogen bond
(c) London forces (d) All of these

3 Amorphous materials are infact considered as
(a) supercooled liquids (b) supercooled solids
(c) covalent network (d) molecular crystals

4 The nature of cohesive forces in water is
(a) van der Waals' (b) covalent network
(c) H-bonding (d) None of these

5 Glasses and plastics are
(a) amorphous solids (b) supercooled liquids
(c) Both (a) and (b) (d) crystalline

6 Silica in amorphous form is used in
(a) photovoltaic cell
(b) conductivity cell
(c) standard hydrogen electrode
(d) None of the above

7 Which of the following is not a characteristic of a crystalline solid ?
(a) Definite and characteristic heat of fusion.
(b) Isotropic nature.
(c) A regular periodically repeated pattern of arrangement of constituent particles in the entire crystal.
(d) A true solid.

8 Amorphous solids
(a) do not have sharp melting points.
(b) are isotropic.
(c) have same mechanical strength, electrical and optical properties in all directions.
(d) all the facts given in (a), (b) and (c) are correct about them.

9 Which of the following conditions favours the existence of a substance in the solid state ? [NCERT Exemplar]
(a) High temperature
(b) Low temperature
(c) High thermal energy
(d) Weak cohesive forces

10 Which of the following is an amorphous solid ? [NCERT Exemplar]
(a) Graphite (C)
(b) Quartz glass (SiO_2)
(c) Chrome alum
(d) Silicon carbide (SiC)

11 At very low temperature, oxygen O_2, freezes and forms a crystalline solid. Which term best describes the solid ?
(a) Covalent network (b) Molecular crystals
(c) Metallic (d) Ionic

12 As it cools, olive oil slowly solidifies and forms a solid over a wide range of temperature. Which term best describes the solid?
(a) Ionic (b) Covalent network
(c) Metallic (d) Molecular crystal

13 Which of the following statements is not true about amorphous solids?
(a) On heating they may become crystalline at certain temperature.
(b) They may become crystalline on keeping for long time.
(c) Amorphous solids can be moulded by heating.
(d) They are anisotropic in nature.

14 The sharp melting point of crystalline solids is due to ....
(a) a regular arrangement of constituent particles observed over a short distance in the crystal lattice.
(b) a regular arrangement of constituent particles observed over a long distance in the crystal lattice.
(c) same arrangement of constituent particles in different directions.
(d) Different arrangement of constituent particles in different directions.

15 Iodine molecules are held in the crystals lattice by .... [NCERT Exemplar]
(a) London forces
(b) dipole-dipole interactions
(c) covalent bonds
(d) coulombic forces

16 Which of the following is true about the value of refractive index of quartz glass?
(a) Same in all directions
(b) Different in different directions
(c) Cannot be measured
(d) Always zero

17 Which of the following is not true about the ionic solids?
(a) Bigger ions form the close packed structure.
(b) Smaller ions occupy either the tetrahedral or the octahedral voids depending upon their size.
(c) Occupation of all the voids is not necessary.
(d) The fraction of octahedral or tetrahedral voids occupied depends upon the radii of the ions occupying the voids.

18 Select the correct statement (s)..
(a) Crystalline solids are anisotropic..
(b) Amorphous solids are isotropic..
(c) Both (a) and (b)
(d) Amorphous solid has sharp melting point

19 The melting point of RbBr is 682°C while that of NaF is 988°C. The principal reason that the melting point of NaF is much higher than that of RbBr is that
(a) the two crystals are not isomorphous.
(b) the molar mass of NaF is smaller than that of RbBr.
(c) the bond in RbBr has more covalent character than the bond in NaF.
(d) the internuclear distance is greater for RbBr than for NaF.

20 Column I describes nature of bonding and Column II the solid having that type of bonding.

	Column I (Nature of bonding)		Column II (Solid)
A.	van der Waals'	1.	Al, Cd
B.	Ionic	2.	CO_2, H_2
C.	Metallic	3.	Si, diamond
D.	Covalent	4.	MgO, NaCl

Correct matching of I and II is in alternatives is

	A	B	C	D		A	B	C	D
(a)	1	2	3	4	(b)	2	4	1	3
(c)	2	1	3	4	(d)	4	1	2	3

II. Crystal lattice and unit cell & their parameters

21 The number of parameters that defines a unit cell is
(a) 2 (b) 3
(c) 4 (d) 6

22 When constituent particles are present only at the corner of a unit cell, it is called?
(a) Primitive (b) Body centred
(c) Face centred (d) End centred

23 In which of the following crystal systems all the axial angles are not of 90°?
(a) Cubic (b) Trigonal
(c) Tetragonal (d) Orthorhombic

24 End centred lattices can be found in which of the following?
(a) Monoclinic (b) Hexagonal
(c) Rhombohedral (d) Triclinic

25 Total number of particles in a body centred cubic unit cell is
(a) 1 (b) 2
(c) 3 (d) 4

26 Column I describes nature of bonding and Column II describes no. of units in that type of bonding in unit cell.

Column I (Nature of bonding)	Column II (Unit)
A. Simple cubic	1. 4
B. Body centred cubic	2. 1
C. Face centred cubic	3. 2

	A	B	C		A	B	C
(a)	2	1	3	(b)	3	2	1
(c)	1	3	2	(d)	2	3	1

27 A crystal is a repetition of a pattern called
(a) motif (b) basis
(c) Both (a) and (b) (d) None of these

28 The lattice site in a pure crystal cannot be occupied by .... **[NCERT Exemplar]**
(a) molecule (b) ion
(c) electron (d) atom

29 An fcc crystal contains how many atoms in each unit cell?
(a) 4 (b) 6
(c) 3 (d) 8

30 bcc type structure is for
(a) NaCl (b) CsCl
(c) Both (a) and (b) (d) None of these

31 fcc type structure is for
(a) CsCl (b) AgCl
(c) MgO (d) Both (b) and (c)

32 The mass of a unit cell of NaCl corresponds to
(a) 1 Na^+ and 6 Cl^- (b) 1 Cl^- and 6 Na^+
(c) 1 Na^+ and 1 Cl^- (d) 4 Cl^- and 4 Na^+

33 **Statement I** The total number of atoms present in a simple cubic unit cell is one.
Statement II Simple cubic unit cell has atoms at its corners, each of which is shared between eight adjacent unit cells.
(a) Only statement I is true
(b) Only statement II is true
(c) Both statements I and II are true
(d) Neither statement I nor statement II is true

34 Of the following crystal lattices, the one that has the maximum number of atoms in unit cell is
(a) bcc (b) simple cubic
(c) simple tetragonal (d) fcc

35 It is stated that uncharged atoms or molecules never crystallise in simple cubic structures because
(a) they are water insoluble
(b) they are water soluble
(c) they pack more efficiently in closest packed structure
(d) None of the above

36 If the three interaxial angles defining the unit cell are all equal in magnitude, the crystal cannot belong to the
(a) orthorhombic system
(b) cubic system
(c) hexagonal system
(d) tetragonal system

37 Numbers of formula units in unit cells of MgO (rock salt), ZnS (zinc blende) and Pt (fcc) respectively are
(a) 4, 3, 2 (b) 4, 3, 4
(c) 4, 4, 4 (d) 4, 3, 1

38 Monoclinic crystal has dimension
(a) $a \neq b \neq c$; $\alpha = \gamma = 90°$, $\beta \neq 90°$
(b) $a = b = c$; $\alpha = \beta = \gamma = 90°$
(c) $a = b = c$; $\alpha = \beta = 90°$, $\gamma = 120°$
(d) $a \neq b = c$; $\alpha = \beta = \gamma = 120°$

39 Match the type of unit cell given in Column I with the features given in Column II. **[NCERT Exemplar]**

Column I	Column II
A. Primitive cubic unit cell	1. Each of the three perpendicular edges compulsorily have the different edge length, i.e. $a \neq b \neq c$
B. Body centred cubic unit cell	2. Number of atoms per unit cell is one
C. Face centred cubic unit cell	3. Each of the three perpendicular edges compulsorily have the same edge length, i.e. $a = b = c$
D. End centred orthorhombic unit cell	4. In addition to the contribution from the corner atoms the number of atoms present in a unit cell is one
	5. In addition to the contribution from the corner atoms the number of atoms present in a unit cell is three

	A	B	C	D
(a)	2, 3	3,5	3, 4	1, 4
(b)	1, 2	3,5	4, 4	2, 3
(c)	2, 3	3, 4	3, 5	1, 4
(d)	4, 5	2, 3	1, 2	3, 4

40 A compound contains two types of atoms, *X* and *Y*. Its crystal structure is a cubic lattice with *X*-atoms at the corners of the unit cells and *Y*-atoms at the body centres. The simplest formula is
(a) X_2Y (b) XY (c) XY_2 (d) XY_3

41 In a solid *AB* having the NaCl structure, '*A*'-atom occupies the corners of the cubic unit cell. If all the face centred atom along one of the axes are removed, then the resultant stoichiometry of the solid is
(a) AB_2 (b) A_2B (c) A_4B_3 (d) A_3B_4

III. Packing in solids

42 The coordination number of crystal structure in which potassium metal crystallises is
(a) 4 (b) 6 (c) 8 (d) 12

43 *ABCABC* pattern of arrangement is found in
(a) scc (b) bcc (c) hcp (d) ccp

44 The metal that crystallises in ccp is
(a) Rb (b) Cs (c) Ag (d) Zn

45 Percent of space present as void in a bcc structure is
(a) 52% (b) 48%
(c) 68% (d) 32%

46 **Statement I** The packing efficienty is maximum for the fcc structure.
Statement II The coordination number is 12 in fcc structures.
(a) Only statement I is true
(b) Only statement II is true
(c) Both statements I and II are true
(d) Neither statement I nor statement II is true

47 Of the following crystal lattices, the one that has the largest packing fraction is
(a) bcc (b) fcc
(c) simple cubic (d) simple tetragonal

48 The coordination number of ccp structure is
(a) 16 (b) 12 (c) 8 (d) 4

49 Which has the least packing efficiency?
(a) Simple cubic
(b) Body centred cubic
(c) Hexagonal closesd packed
(d) Cubic closesd packed

50 In hcp structure, the packing fraction is
(a) 0.68 (b) 0.74 (c) 0.50 (d) 0.54

51 The hcp arrangement of ions is described by
(a) *ABAB* (b) *ABABAC*
(c) *ABCABC* (d) *ABCABCABC*

52 The total number of tetrahedral voids in the face centred unit cell is .... **[NCERT Exemplar]**
(a) 6 (b) 8 (c) 10 (d) 12

53 Match the elements given in Column I with the shape of the crystal given in Column II.

Column I	Column II
A. Be	1. Body-centred cubic
B. Ca	2. Simple cubic
C. Ba	3. Face centred cubic
D. Mn	4. Hexagonal close-packed

	A	B	C	D		A	B	C	D
(a)	1	4	3	2	(b)	4	3	1	2
(c)	3	2	1	4	(d)	4	1	3	2

54 **Statement I** Total number of octahedral voids present in unit cell of cubic close packing including the one that is present at the body centre, is four.

Statement II Besides the body centre there is one octahedral voids present at the centre of which is shared between two adjacent unit cells.

(a) Only statement I is true
(b) Only statement II is true
(c) Both statements I and II are true
(d) Neither statement I nor statement II is true

55 In which pair most efficient packing is present? **[NCERT Exemplar]**

(a) hcp and bcc (b) hcp and ccp
(c) bcc and ccp (d) bcc and simple cubic cell

56 A mineral having the formula AB_2 crystallises in the cubic close packed lattice, with the A-atoms occupying the lattice points. Hence, coordination numbers of A and B-atoms are

(a) 4, 8 (b) 4, 4
(c) 8, 8 (d) 8, 4

57 A metallic element crystallises into a lattice containing a sequence of layers of $ABABAB$....... Any packing of spheres leaves out voids in the lattice. Percentage of empty space (by volume) is

(a) 52% (b) 48% (c) 50% (d) 74%

58 An element crystallises in a bcc lattice. The coordination number of nearest neighbours and next nearest neighbours of the element are respectively

(a) 8, 8 (b) 8, 6 (c) 6, 8 (d) 6, 6

59 The coordination number of a M^{2+} in MX_2 is 8. Hence, coordination number of X^- is

(a) 8 (b) 6 (c) 4 (d) 2

60 In the cubic close packing, the unit cell has ....

(a) 4 tetrahedral voids each of which is shared by four adjacent unit cells
(b) 4 tetrahedral voids within the unit cell
(c) 8 tetrahedral voids each of which is shared by four adjacent unit cells
(d) 8 tetrahedral voids within the unit cells

61 Lithium selenide can be described as a closest packed array of selenide ions with lithium ions in all of the tetrahedral holes. Formula of lithium selenide is

(a) Li_2Se (b) LiSe
(c) $LiSe_2$ (d) Li_3Se

62 Match the type of packing given in Column I with the items given in Column II. **[NCERT Exemplar]**

Column I	Column II
A. Square close packing in two dimensions	1. Triangular voids
B. Hexagonal close packing in two dimensions	2. Pattern of spheres is repeated in every fourth layer
C. Hexagonal close packing in three dimensions	3. Coordination number = 4
D. Cubic close packing in three dimensions	4. Pattern of sphere is repeated in alternate layers

	A	B	C	D		A	B	C	D
(a)	3	1	4	2	(b)	4	3	2	1
(c)	3	4	1	2	(d)	2	1	3	4

63 The correct order of the packing efficiency in different types of unit cells is ....

(a) fcc < bcc < simple cubic
(b) fcc > bcc > simple cubic
(c) fcc < bcc > simple cubic
(d) bcc < fcc > simple cubic

64 Hexagonal close packed structure and cubic close packed structure for a given element would be expected to have the same density because of

(a) same molar mass
(b) same coordination number
(c) same packing fraction
(d) Both (b) and (c)

65 The two types of holes which occur in any close packed structures are

(a) tetrahedral, octahedral
(b) trigonal, octahedral
(c) trigonal, tetrahedral
(d) Only octahedral

66 Zinc blende structure is obtained when Zn^{2+} occupies

(a) all tetrahedral sites (b) half tetrahedral sites
(c) all octahedral sites (d) half octahedral sites

67 The intermetallic compound LiAg crystallises in cubic lattice in which both lithium and silver have coordination number of eight. The crystal class is

(a) simple cube (b) body centred cube
(c) face centred cube (d) None of these

IV. Density and radius ratio

68 The following crystallographic data were obtained for a protein.
Volume of unit cube $= 1.50 \times 10^{-19}$ cm^3
Density $= 1.35$ gcm^3
$Z = 4$ and protein fraction $= 0.75$. Thus, molar mass of protein is
(a) 2.3×10^4 g mol^{-1} (b) 3.048×10^4 g mol^{-1}
(c) 1.725×10^4 g mol^{-1} (d) None of these

69 The length of the unit cell edge of a bcc lattice metal is 352 pm. Thus, volume of atoms in one mole of the metal is
(a) 8.93 cm^3 (b) 7.25 cm^3 (c) 1.21 cm^3 (d) 4.00 cm^3

70 In octahedral void and tetrahedral void, radius ratio is (in closest packing)
(a) $\frac{0.414}{0.225}$ (b) $\frac{0.225}{0.414}$
(c) 0.225×0.414 (d) None of these

71 A salt AB crystallises in the CsCl structure. The anions at the corners touch each other and the cation in the centre. Hence, radius ratio r_+/r_- is
(a) 0.225 (b) 0.441
(c) 0.625 (d) 0.732

72 The radius of a divalent cation M^{2+} is 94 pm and of divalent anion X^{2-} is 146 pm. Thus, $M^{2+}X^{2-}$ has
(a) rock salt (NaCl) structure
(b) zinc blende structure
(c) anti-fluorite structure
(d) bcc (CsCl) structure

73 If the edge length of NaH unit cell is 488 pm, what is the length of Na—H bond if it crystallises as fcc structure?
(a) 488 pm (b) 122 pm (c) 244 pm (d) 976 pm

74 The edge length of the NaCl unit cell is ... if radius of Na^+ is 97 pm and radius of Cl^- is 181 pm.
(a) 556 pm (b) 278 pm (c) 139 pm (d) 120 pm

75 Match the limiting radius ratio given in Column I with the shape of the crystal given in Column II.

Column I	Column II
A. < 0.155	1. Octahedral
B. 0.155-0.225	2. Body-centred cubic
C. 0.225-0.414	3. Planar triangle
D. 0.414-0.732	4. Linear
E. 0.732-0.999	5. Tetrahedral

	A	B	C	D	E		A	B	C	D	E
(a)	2	3	4	5	1	(b)	4	5	3	1	2
(c)	4	3	5	1	2	(d)	5	4	2	3	1

76 Density of a crystal is (where, a is the edge length, Z is the number of atoms in the unit cell and M is the molecular weight).
(a) $\frac{a^3 \times N_0}{Z \times M}$ (b) $\frac{Z \times M}{a^3 \times N_0}$
(c) $\frac{Z \times M}{a^3}$ (d) $\frac{M}{a^3 N_0}$

77 If the radius of an octahedral void is r and radius of the atom in close packing is R, then
(a) $r = 0.732\,R$ (b) $R = 0.414\,r$
(c) $r = 0.414\,R$ (d) $R = 0.732\,r$

78 Which of the following statement is true for the radius ratio of $\frac{r^+}{r^-}$?
(a) Range 0.732 to 0.414 has CN is 6 and is octahedral
(b) Radius ratio above 0.732, CN is 8, and is cubic
(c) Radius ratio 0.225 to 0.155, CN is 3, and is triangular
(d) All of the above

79 An element (atomic mass = 100) having bcc structure has unit cell edge 400 pm. Then, density of the element is
(a) 10.376 g/cc (b) 5.19818 g/cc
(c) 7.289 g/cc (d) 2.144 g/cc

80 Ice crystallises in a hexagonal lattice having volume of the unit cell as 132×10^{-24} cm^3. If density is 0.92 g cm^{-3} at a given temperature, then number of H_2O molecules per unit cell is
(a) 1 (b) 2 (c) 3 (d) 4

81 In the structure of NaCl given below, ratio r_{Na^+}/r_{Cl^-} is

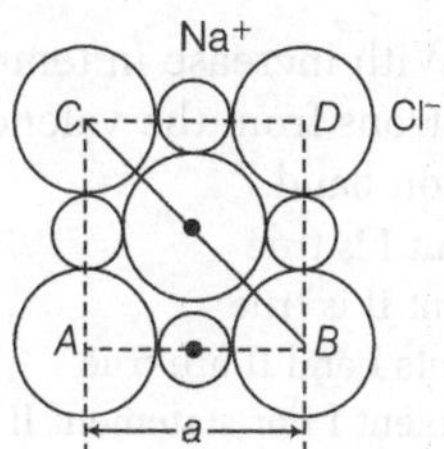

(a) 0.414 (b) 0.681 (c) 0.745 (d) 0.520

82 CsCl crystallises in body centred cubic system. If a is its edge length then which of the following expressions is correct ?
(a) $r_{Cs^+} + r_{Cl^-} = 3a$
(b) $r_{Cs^+} + r_{Cl^-} = \frac{3a}{2}$
(c) $r_{Cs^+} + r_{Cl^-} = \frac{\sqrt{3}a}{2}$
(d) $r_{Cs^+} + r_{Cl^-} = \sqrt{3}a$

V. Structure of ionic compounds

83 In which of the following structures coordination number for cations and anions in the packed structure will be same ?
(a) Cl^- ions form fcc lattice and Na^+ ions occupy all octahedral voids of the unit cell
(b) Ca^{2+} ions form fcc lattice and F^- ions occupy all the eight tetrahedral voids of the unit cell
(c) O^{2-} ions form fcc lattice and Na^+ ions occupy all the eight tetrahedral voids of the unit cell
(d) S^{2-} ions form fcc lattice and Zn^{2+} ions go into alternate tetrahedral voids of the unit cell

84 6 : 6 of NaCl coordination changes to 8 : 8 coordination on
(a) applying high pressure (b) increase in temperature
(c) Both (a) and (b) (d) no effect on coordination

85 8 : 8 coordination of CsCl is found to change into 6 : 6 on
(a) applying high p (b) increase in temperature
(c) Both (a) and (b) (d) None of these

86 For which of the following crystals would you expect the assumption of anion-anion contact to be valid ?
(a) CsBr (b) NaI
(c) NaF (d) KCl

87 In crystals of which of the following ionic compounds would you expect maximum distance between centres of cations and anions?
(a) LiF (b) CsF
(c) CsI (d) LiI

VI. Imperfections or defects in solids

88 **Statement I** Electrical conductivity of semiconductor increases with increasing temperature.
Statement II With increase in temperature, large number of electrons from the valence band can jump to the conduction band.
(a) Only statement I is true
(b) Only statement II is true
(c) Both statements I and II are true
(d) Neither statement I nor statement II is true

89 The presence of *F*-centres in a crystal makes it
(a) conducting (b) colourless
(c) coloured (d) non-conducting

90 The *p*-type semiconductor is obtained when Si is doped with
(a) P (b) Ge (c) Ga (d) As

91 The point defect which lowers the density of the material is
(a) Schottky (b) Frenkel
(c) Both (a) and (b) (d) Interstitial defect

92 Select the correct statement(s) about non-stoichiometric compounds.
(a) They are called Berthollide compounds
(b) They do not obey the law of constant composition
(c) Electrical neutrality is maintained either by having extra electrons in the structure or changing the charge on some of the metal ions
(d) All of the above

93 Schottky defect in crystals is observed, when
(a) equal number of cations and anions are missing from the lattice
(b) density of crystals is increased
(c) an ion leaves its normal site and occupies an interstitial site
(d) unequal number of cations and anions are missing from the lattice

94 Match the defects given in Column I with the statements in given Column II. **[NCERT Exemplar]**

Column I	Column II
A. Simple vacancy defect	1. Shown by non-ionic solids and increases density of the solid.
B. Simple interstitial defect	2. Shown by ionic solids and decreases density of the solid.
C. Frenkel defect	3. Shown by non-ionic solids and density of the solid decreases.
D. Schottky defect	4. Shown by ionic solids and density of the solid remains the same.

	A	B	C	D
(a)	3	2	1	4
(b)	2	4	3	1
(c)	3	1	4	2
(d)	1	3	4	2

95 Frenkel defect is the
(a) Schottky defect (b) Interstitial defect
(c) combination of (a) and (b) (d) Metal deficiency defect

96 Calcium crystallises in a fcc unit cell with $a = 0.556$ nm. It contained 0.1 % Frenkel defect and 0.1 % Schottky defect. Density with Frenkel defect $= d_f$ and density with Schottky defect $= d_s$, then
(a) $d_f = d_s$
(b) $d_f > d_s$
(c) $d_f < d_s$
(d) density is not concerned with defect

97 Cations are present in the interstitial sites in .... **[NCERT Exemplar]**
(a) Frenkel defect (b) Schottky defect
(c) Vacancy defect (d) Metal deficiency defect

98 Which of the following defects is also known as dislocation defect? **[NCERT Exemplar]**
(a) Frenkel defect (b) Schottky defect
(c) Non-stoichiometric defect (d) Simple interstitial defect

99 Which kind of defects are introduced by doping? **[NCERT Exemplar]**

(a) Dislocation defect (b) Schottky defect
(c) Frenkel defect (d) Electronic defect

100 Match the types of defect given in Column I with the statement given in Column II. **[NCERT Exemplar]**

Column I	Column II
A. Impurity defect	1. NaCl with anionic sites called *F*-centres
B. Metal excess defect	2. FeO with Fe^{3+}
C. Metal deficiency defect	3. NaCl with Sr^{2+} and some cationic sites vacant

	A	B	C		A	B	C
(a)	2	3	1	(b)	3	1	2
(c)	1	3	1	(d)	3	2	1

101 Structure shown here represents

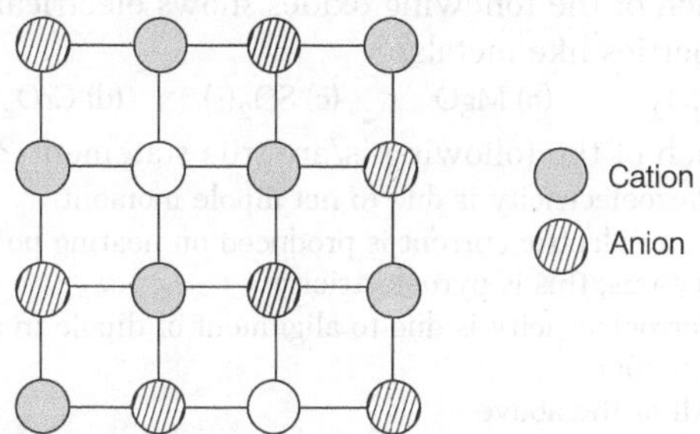

(a) Schottky defect
(b) Frenkel defect
(c) Metal excess defect
(d) None of the above

102 Consider the following statements

I. If an Fe^{2+} is missing from its lattice site in FeO, then there must be two Fe^{3+} ions somewhere in the lattice to balance the electrical charges.
II. Crystals with metal deficiency defects are called super-conductors.
III. Crystals with metal deficiency defects are called semiconductors.

Select the correct alternate.

(a) Both I and II (b) Both I and III
(c) Only II (d) Only I

103 Superconductivity was first found by

(a) Debye-Huckel (b) Kammerling Onnes
(c) Both (a) and (b) (d) None of these

104 The substance whose resistance gets reduced to virtually zero at very low temperature is called

(a) electrical conductor (b) hyper conductor
(c) semiconductor (d) super conductor

105 Which of the following is true about the charge acquired by *p*-type semiconductors? **[NCERT Exemplar]**

(a) Positive
(b) Neutral
(c) Negative
(d) Depends on concentration of *p* impurity

106 Graphite is a good conductor of electricity due to the presence of ....

(a) lone pair of electrons
(b) free valence electrons
(c) cations
(d) anions

107 Statement I Group 13 doped crystals of silicon are called *p*-type semiconductors.

Statement II Holes (positive in charge) appear to be responsible for the semiconducting properties.

(a) Only statement I is true
(b) Only statement II is true
(c) Both statements I and II are true
(d) Neither statement I nor statement II is true

108 Match the items given in Column I with the items given in Column II. **[NCERT Exemplar]**

Column I	Column II
A. Mg in solid state	1. *p*-type semiconductor
B. $MgCl_2$ in molten state	2. *n*-type semiconductor
C. Silicon with phosphorus	3. Electrolytic conductors
D. Germanium with boron	4. Electronic conductors

	A	B	C	D		A	B	C	D
(a)	4	1	3	2	(b)	1	3	2	4
(c)	3	2	4	1	(d)	4	3	2	1

109 Which has least energy gap between conduction band and valence band ?

(a) Carbon (diamond) (b) Silicon
(c) Germanium (d) Tin

110 Which has maximum band-gap energy ?

(a) ZnSe (b) GaAs
(c) Ge (d) Equal

111 A group of 13 elements if added in small amounts to Ge, ... is formed.

(a) *n*-type semiconductor
(b) *p*-type semiconductor
(c) insulator
(d) rectifier

112 Select the correct statement(s).

(a) Highest filled bonding orbitals are called valence band
(b) Lowest unfilled anti-bonding orbitals are called the conduction band
(c) Both (a) and (b)
(d) Highest filled bonding orbitals are called conduction bond.

113 Statement I Semiconductors are solids with conductivities in the intermediate range from 10^{-6} - 10^{4} ohm^{-1} m^{-1}.

Statement II Intermediate conductivity in semiconductor is due to partially filled valence band.

(a) Only statement I is true
(b) Only statement II is true
(c) Both statements I and II are true
(d) Neither statement I nor statement II is true

114 M^+ forms $C_{60}M_3$ where, C_{60}^{3-} is fulleride (a superconductor). It forms octahedral holes. If radius of C_{60}^{3-} is 500 pm, then the radius of M^+ is

(a) 207 pm (b) 307 pm (c) 400 pm (d) 500 pm

115 Which of the following oxides behaves as conductor or insulator depending upon temperature ?

(a) TiO (b) SiO_2
(c) TiO_3 (d) MgO

116 To get a *n*-type semiconductor from silicon, it should be doped with a substance with valency ....

(a) 2 (b) 1
(c) 3 (d) 5

117 Which of the following solids is/are not an electrical conductor? **[NCERT Exemplar]**

1. Mg (*s*) 2. TiO (*s*)
3. I_2 (*s*) 4. H_2O (*s*)

(a) Only 1 (b) Only 2
(c) 3 and 4 (d) 2, 3 and 4

118 Which of the following point defects are shown by AgBr(*s*) crystals?

1. Schottky defect 2. Frenkel defect
3. Metal excess defect 4. Metal deficiency defect

(a) 1 and 2 (b) 3 and 4
(c) 1 and 3 (d) 2 and 4

VII. Magnetic and dielectric properties

119 A ferromagnetic substance becomes a permanent magnet when it is placed in a magnetic field because ....

(a) all the domains get oriented in the direction of magnetic field
(b) all the domains get oriented in the direction opposite to the direction of magnetic field
(c) domains get oriented randomly
(d) domains are not affected by magnetic field

120 Ferrimagnetism is in

(a) ↑↑↑↑↑ (b) ↑↓↑↓
(c) ↑↑↑↓↓ (d) None of these

121 Ferromagnetism is in

(a) ↑↑↑↑↑ (b) ↑↓↑↓
(c) ↑↑↑↓↓ (d) None of these

122 Which of the following arrangements shows schematic alignment of magnetic moments of anti-ferromagnetic substances? **[NCERT Exemplar]**

(a) ↑ ↑ ↑ ↑ ↑ ↑
(b) ↓ ↓ ↓ ↓ ↓ ↓
(c) ↑ ↑ ↑ ↑ ↑ ↓
(d) ↑ ↓ ↑ ↓ ↑ ↓

123 Which is true about Rochelle's salt and lead zirconate?

(a) Rochelle's salt is ferroelectric while lead zirconate is anti-ferroelectric
(b) Rochelle's salt is anti-ferroelectric and lead zirconate is ferroelectric
(c) Both are ferroelectric
(d) Both are anti-ferroelectric

124 Which of the following oxides shows electrical properties like metals?

(a) SiO_2 (b) MgO (c) $SO_2(s)$ (d) CrO_2

125 Which of the following is/are true statements?

(a) Piezoelectricity is due to net dipole moment
(b) Some electric current is produced on heating polar crystals, this is pyroelectricity
(c) Ferroelectricity is due to alignment of dipole in same direction
(d) All of the above

126 Which of the following statements is not true?

(a) Paramagnetic substances are weakly attracted by magnetic field
(b) Ferromagnetic substances cannot be magnetised permanently
(c) The domains in anti-ferromagnetic substances are oppositely oriented with respect to each other
(d) Pairing of electrons cancels their magnetic moment in the diamagnetic substances

(B) Miscellaneous questions

1 Which of the following has definite heat of fusion ?

(a) Wood (b) Glass (c) Rubber (d) Sodium chloride

2 The example of attraction between positively charged kernels and sea of electron, in solid state of substance is found in

(a) Ne (b) H_2O (c) Fe (d) CaF_2

3 Which of the following is not an NaCl-type of crystal?

(a) KBr (b) CuCl (c) AgCl (d) MgO

4 The *n*-type semiconductor is obtained when Si is doped with

(a) Al (b) Ge (c) B (d) As

5 Statement I Group 15 doped crystals of silicon are called a *n*-type semiconductor.

Statement II Neutrons, are responsible for the semiconducting properties.

(a) Only statement I is true
(b) Only statement II is true
(c) Both statements I and II are true
(d) Neither statement I nor statement II is true

6 Match the cubic structure in Column I with packing density in Column II.

Column I	Column II
A. Simple cubic	1. 0.68
B. fcc	2. 0.52
C. bcc	3. 0.74

	A	B	C		A	B	C
(a)	3	2	1	(b)	2	3	1
(c)	2	1	3	(d)	1	2	3

7 Ferrimagnetic is converted into ferromagnetic at
(a) 300 K (b) 400 K
(c) 600 K (d) 850 K

8 An electron trapped in an anion vacancy within the crystal is called
(a) *n*-type conductor (b) *p*-type conductor
(c) *F*-centre (d) insulator

9 Number of atoms in different cubic unit cells of monoatomic substance is
(a) simple cubic-1 (b) bcc-2
(c) fcc-4 (d) All are correct

10 Silicon doped with electron rich impurity forms....
(a) *p*-type semiconductor
(b) *n*-type semiconductor
(c) intrinsic semiconductor
(d) insulator

11 Structure shown here represents

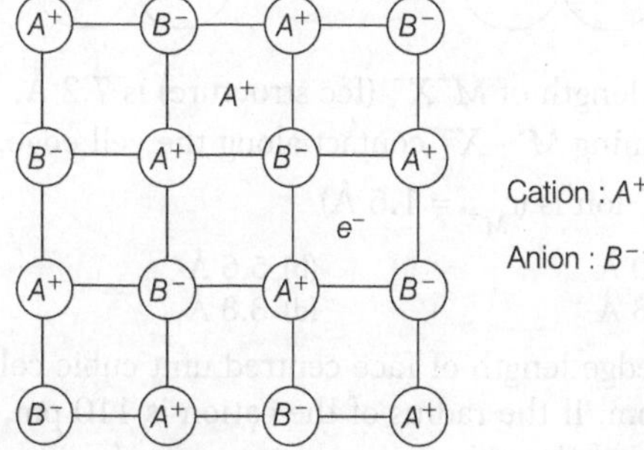

(a) Schottky defect
(b) Frenkel defect
(c) Metal excess defect because of interstitial cation
(d) Metal excess defect because of excess cation

12 Statement I Graphite is a good conductor of electricity however diamond belongs to the category of insulators.

Statement II Graphite is soft in nature on the other hand diamond is very hard and brittle.

(a) Only statement I is true
(b) Only statement II is true
(c) Both statements I and II are true
(d) Neither statement I nor statement II is true

13 Match the shape of the molecule in column I with coordination number in column II.

Column I	Column II
A. Linear	1. 6
B. Planar triangle	2. 4
C. Square planar	3. 8
D. Octahedral	4. 2
E. Body-centred cubic	5. 3

	A	B	C	D	E		A	B	C	D	E
(a)	4	5	1	2	3	(b)	5	1	3	4	2
(c)	4	5	2	1	3	(d)	3	4	5	2	1

14 Which of the following statement is not true about the hexagonal close packing?
(a) The coordination number is 12
(b) It has 74% packing efficiency
(c) Tetrahedral voids of the second layer are covered by the spheres of the third layer
(d) In this arrangement spheres of the fourth layer are exactly aligned with those of the first layer

15 Which of the following represents correct order of conductivity in solids?
(a) $\kappa_{\text{metals}} >> \kappa_{\text{insulators}} < \kappa_{\text{semiconductors}}$
(b) $\kappa_{\text{metals}} << \kappa_{\text{insulators}} < \kappa_{\text{semiconductors}}$
(c) $\kappa_{\text{metals}} > \kappa_{\text{semiconductors}} > \kappa_{\text{insulators}} = \text{zero}$
(d) $\kappa_{\text{metals}} < \kappa_{\text{semiconductors}} > \kappa_{\text{insulators}} \neq \text{zero}$

16 Silica (SiO_2) can be crystalline as well as amorphous, with following properties.

I. Crystalline
II. The SiO_4^{4-} tetrahedra are randomly joined giving rise to polymeric chains or three-dimensional sheets.
III. Have high and sharp melting point.

Which of I, II and III is/are matched with amorphous solids?
(a) I and II (b) I and III
(c) II and III (d) Only II

17 What is the coordination number in a square close packed structure in two dimensions? **[NCERT Exemplar]**
(a) 2 (b) 3
(c) 4 (d) 6

18 Anti-fluorite structure is for
(a) ThO_2 (b) Na_2O
(c) Both (a) and (b) (d) None of these

19 Which of the following is a network solid? **[NCERT Exemplar]**
(a) SO_2 (solid) (b) I_2
(c) Diamond (d) H_2O (ice)

20 Graphite cannot be classified as .... **[NCERT Exemplar]**
(a) conducting solid (b) network solid
(c) covalent solid (d) ionic solid

21 Select the correct statement(s).
(a) The non-stoichiometric form of NaCl is yellow and that of KCl is blue-lilac
(b) Solids containing *F*-centres (Farbe) are paramagnetic
(c) Conduction by electrons is called *n*-type semiconductor
(d) All of the above

22 Which of the following is not the characteristic of ionic solids?
(a) Very low value of electrical conductivity in the molten state
(b) Brittle nature
(c) Very strong forces of interactions
(d) Anisotropic nature

23 *ABCABC* type of arrangement of different layers, in a closest packed structure is called
(a) hcp (b) bcc
(c) fcc (d) cubic close packed (ccp)

24 Which set forms covalent crystals involving sp^3-hybrid orbitals?
(a) C, Si, Ge, Sn (b) B, Al, Ga, In
(c) C, B, N, P (d) Li, Na, K, Rb

25 It is stated that ZnS does not crystallise in the NaCl structure. It is due to
(a) the r_+ / r_- ratio is 0.402, too low to avoid anion-anion contact in the NaCl structure
(b) ZnS is water insoluble, NaCl is water soluble
(c) ZnS is water soluble, NaCl is water insoluble
(d) Zn belongs to *d*-block, Na belongs to *s*-block

26 If we mix a pentavalent impurity in a crystal lattice of germanium, what type of semiconductor will be formed?
(a) *p*-type semiconductor (b) *n*-type semiconductor
(c) Both (a) and (b) (d) Intrinsic semiconductor

27 AgI crystallises in the cubic close packed zinc blende structure. Assuming that the iodide ions occupy the lattice points, fraction of the tetrahedral sites occupied by silver ions is about
(a) 50% (b) 75%
(c) 100% (d) 33%

28 Match the crystal system/unit cells mentioned in Column I with their characteristic features mentioned in Column II.

Column I	Column II
A. Simple cubic and face-centred cubic	1. Have these cell parameters $a = b = c$ and $\alpha = \beta = \gamma$
B. Cubic and rhombohedral	2. are two crystal systems
C. Cubic and tetragonal	3. Have only two crystallographic angles of 90°
D. Hexagonal and monoclinic	4. Belong to same crystal system

	A	B	C	D
(a)	1,4	1,2	2	2,3
(b)	1,2	1,3	4	2
(c)	1,3	1,4	2	2,3
(d)	1,4	2	1,2	2,3

29 In the following NaCl type structure.

Salt	Anion-anion distance	Cation-anion distance
MT	2.40 Å	1.70 Å
MQ	1.63 Å	1.15 Å
RT	2.66 Å	1.88 Å
RQ	2.09 Å	1.48 Å

MT is represented as,

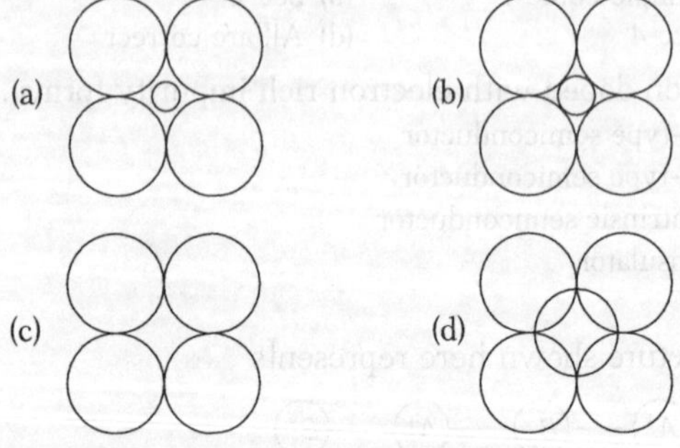

30 Edge length of M^+X^- (fcc structure) is 7.2 Å. Assuming $M^+ – X^-$ contact along the cell edge, radius of X^- ion is (r_{M^+} = 1.6 Å)
(a) 2.0 Å (b) 5.6 Å
(c) 2.8 Å (d) 3.8 Å

31 The edge length of face centred unit cubic cell is 508 pm. If the radius of the cation is 110 pm, the radius of the anion is
(a) 288 pm (b) 398 pm
(c) 144 pm (d) 618 pm

32 A metallic element has a cubic lattice. Each edge of the unit cell is 2 Å. The density of the metal is 2.5 g cm^{-3}. The unit cells in 200 g of the metal are
(a) 1×10^{24} (b) 1×10^{20}
(c) 1×10^{22} (d) 1×10^{25}

33 Each rubidium halide crystallising in the NaCl-type lattice has a unit cell length 0.30 Å greater than that for corresponding potassium salt ($r_{(K^+)}$ = 1.33 Å) of the same halogen. Hence, ionic radius of Rb^+ is
(a) 1.18 Å (b) 1.48 Å (c) 1.63 Å (d) 1.03 Å

34 A compound alloy of gold and copper crystallises in a cubic lattice in which the gold atoms occupy the lattice points at the corners of a cube and the copper atoms occupy the centres of each of the cube faces. Hence, compound alloy has formula
(a) AuCu (b) Au_3Cu (c) Au_2Cu (d) $AuCu_3$

35. The edge lengths of the unit cells in terms of the radius of spheres constituting fcc, bcc and simple cubic unit cells are respectively ....
(a) $2\sqrt{2}r, \frac{4r}{\sqrt{3}}, 2r$ (b) $\frac{4r}{\sqrt{3}}, 2\sqrt{2}r, 2r$
(c) $2r, 2\sqrt{2}r, \frac{4r}{\sqrt{3}}$ (d) $2r, \frac{4r}{\sqrt{3}}, 2\sqrt{2}r$

36. A crystal is formed by two elements *P* and *Q*. The element *Q* forms cubic close packing and atoms of *P* occupy one-third of tetrahedral voids. The compound is thus represented as,
(a) PQ_3 (b) P_3Q (c) P_2Q_3 (d) P_3Q_2

(C) Medical entrances' gallery

1 An element has a body centered cubic (bcc) structure with a cell edge of 288 pm. The atomic radius is **[NEET 2020]**
(a) $\frac{\sqrt{2}}{4}\times 288$ pm (b) $\frac{4}{\sqrt{3}}\times 288$ pm
(c) $\frac{4}{\sqrt{2}}\times 288$ pm (d) $\frac{\sqrt{3}}{4}\times 288$ pm

2 Which one of the following compounds show both, Frenkel as well as Schottky defects? **[NEET 2020]**
(a) AgBr (b) AgI
(c) NaCl (d) ZnS

3 Formula of nickel oxide with metal deficiency defect in its crystal is $Ni_{0.98}O$. The crystal contains Ni^{2+} and Ni^{3+} ions. The fraction of nickel existing as Ni^{2+} ions in the crystal is **[NEET (Odisha) 2019]**
(a) 0.96 (b) 0.04
(c) 0.50 (d) 0.31

4 A compound is formed by cation *C* and anion *A*. The anions form hexagonal close packed (hcp) lattice and the cations occupy 75% of octahedral voids. The formula of the compound is **[NEET 2019]**
(a) C_3A_2 (b) C_3A_4
(c) C_4A_3 (d) C_2A_3

5 Silicon carbide is an example of **[JIPMER 2019]**
(a) molecular solid (b) metallic solid
(c) ionic crystal (d) covalent solid

6 Quartz is crystalline silica and silica glass is
(a) amorphous (b) crystalline **[AIIMS 2019]**
(c) Both (a) and (b) (d) None of these

7 Which type of defect present in NaCl? **[JIPMER 2019]**
(a) Frenkel defect
(b) Schottky defect
(c) *F*-centre defect
(d) None of the above

8 Frenkel defect is present in which of the following?
(a) NaOH (b) NaI **[JIPMER 2019]**
(c) AgBr (d) None of these

9 **Assertion** (A) ZnO becomes yellow when it is heated.
Reason (R) The anionic sites is (*F*-centres) occupied by unpaired electrons. **[AIIMS 2019]**
(a) Both A and R are correct and R is the correct explanation of A
(b) Both A and R are correct but R is not the correct explanation of A
(c) A is correct but R is incorrect
(d) Both A and R are incorrect

10 The tetrahedral voids present in 0.5 mole of hcp crystal structure are **[AIIMS 2019]**
(a) 3.6×10^{23} (b) 9×10^{23}
(c) 3.6×10^{24} (d) 6.02×10^{23}

11 Which of the following is an example of network solid ?
(a) SO_2 (solid) (b) I_2 **[AIIMS 2018]**
(c) Diamond (d) H_2O (ice)

12 Niobium crystallises in body centred cubic structure. If density is 8.55 g cm^{-3}, then the atomic radius of niobium is (atomic mass of niobium = 93u) **[AIIMS 2018]**
(a) 163 pm (b) 143 pm
(c) 182 Å (d) 152 Å

13 An atom forms face centred cubic crystal with density d = 8.92 g / mL and edge length $a = 3.6\times10^{-8}$ cm. The molecular mass of atom in amu is **[JIPMER 2018]**
(a) 98 (b) 63
(c) 32 (d) 93

14 In a crystalline solid, having formula XY_2O_4, oxide ions are arranged in cubic close packed lattice while, cations X are present in tetrahedral voids and cations Y are present in octahedral voids. The percentage of tetrahedral voids occupied by X is **[JIPMER 2018]**

(a) 12.5% (b) 25% (c) 50% (d) 75%

15 A solid has a structure in which W atoms are located at the corners of a cubic lattice. O atoms at the centre of edges and Na atom at centre of the cube. The formula for the compound is **[AIIMS 2018]**

(a) $NaWO_2$ (b) $NaWO_3$ (c) Na_2WO_3 (d) $NaWO_4$

16 Which is the incorrect statement? **[NEET 2017]**

(a) Frenkel defect is favoured in those ionic compounds in which sizes of cation and anion are almost equal.
(b) $FeO_{0.98}$ has non-stoichiometric metal deficiency defect.
(c) Density decreases in case of crystals with Schottky's defect.
(d) NaCl(s) is insulator, silicon is semiconductor, silver is conductor, quartz is piezoelectric crystal.

17 Lithium has a bcc structure. Its density is 530 kg m^{-3} and its atomic mass is 6.94 g mol^{-1}. Calculate the edge length of a unit cell of lithium metal.

($N_A = 6.02 \times 10^{23}$ mol^{-1}) **[NEET (I) 2016]**

(a) 352 pm (b) 527 pm
(c) 264 pm (d) 154 pm

18 The ionic radii of A^+ and B^- ions are 0.98×10^{-10} m and 1.81×10^{-10} m. The coordination number of each ion in AB is **[NEET (I) 2016]**

(a) 1 (b) 8 (c) 2 (d) 6

19 In calcium fluoride, having the fluorite structure, the coordination numbers for calcium ion (Ca^{2+}) and fluoride ion (F^-) are **[NEET (II) 2016]**

(a) 8 and 4 (b) 4 and 8 (c) 4 and 2 (d) 6 and 6

20 A given metal crystallises out with a cubic structure having edge length of 361 pm. If there are four metal atoms in one unit cell, what is the radius of one atom? **[CBSE AIPMT 2015]**

(a) 40 pm (b) 127 pm (c) 80 pm (d) 108 pm

21 The vacant space in bcc lattice cell is **[CBSE AIPMT 2015]**

(a) 26% (b) 48% (c) 23% (d) 32%

22 The correct statement regarding defects in the crystalline solid is **[CBSE AIPMT (Re-Exam) 2015]**

(a) Schottky defects have no effect on the density of crystalline solids
(b) Frenkel defects decreases the density of crystalline solids
(c) Frenkel defect is a dislocation defect
(d) Frenkel defect is found in halides of alkaline metals

23 Match the following columns. **[Kerala CEE 2015]**

	Compound		Magnetic property
(A)	NaCl	(i)	Ferrimagnetic
(B)	MnO	(ii)	Paramagnetic
(C)	$CrCl_3$	(iii)	Ferromagnetic
(D)	CrO_2	(iv)	Diamagnetic
(E)	$MgFe_2O_4$	(v)	Anti-ferromagnetic

	A	B	C	D	E
(a)	(i)	(iii)	(ii)	(v)	(iv)
(b)	(v)	(ii)	(iii)	(i)	(iv)
(c)	(iii)	(v)	(ii)	(i)	(iv)
(d)	(iv)	(v)	(ii)	(iii)	(i)
(e)	(iv)	(iii)	(v)	(ii)	(i)

24 A unit cell with edge length $a \neq b \neq c$ and axial angles $\alpha = \beta = \gamma = 90°$ is called **[J&K CET 2015]**

(a) cubic (b) tetragonal
(c) orthorhombic (d) hexagonal

25 Which of the following defect is seen in FeO? **[Guj. CET 2015]**

(a) Metal deficiency defect
(b) Metal excess defect
(c) Displacement defect
(d) Impurity defect

26 Which of the following substances possesses anti-ferromagnetic property? **[Guj. CET 2015]**

(a) CrO_2 (b) Fe_3O_4
(c) H_2O (d) MnO

27 Pure silicon doped with phosphorus is **[EAMCET 2015]**

(a) amorphous
(b) p-type semiconductor
(c) n-type semiconductor
(d) insulator

28 If a is the length of the side of a cube, the distance between the body centred atom and one corner atom in the cube will be **[CBSE AIPMT 2014]**

(a) $\frac{2}{\sqrt{3}}a$ (b) $\frac{4}{\sqrt{3}}a$ (c) $\frac{\sqrt{3}}{4}a$ (d) $\frac{\sqrt{3}}{2}a$

29 Copper crystallises in a face centred cubic lattice with a unit cell length of 361 pm. What is the radius of copper atom in pm? **[Manipal 2014]**

(a) 157 (b) 181
(c) 108 (d) 128

30 The number of tetrahedral and octahedral voids in a CCP array of 100 atoms are respectively **[Kerala CEE 2014]**

(a) 200 and 100 (b) 100 and 200
(c) 200 and 200 (d) 100 and 100
(e) 50 and 50

31 Which one of the following is used as a piezoelectric material? **[Kerala CEE 2014]**

(a) Silicones (b) Silica gel
(c) Kieselguhr (d) Quartz

32 A solid AB has NaCl structure. If the radius of cation A^+ is 170 pm. Calculate the maximum possible radius of the anion B^-. **[WB JEE 2014]**

(a) 210.3 pm (b) 397.4 pm
(c) 410.6 pm (d) 347.9 pm

33 The diffraction of a crystal of barium with X-rays of wavelength 2.29 Å gives a first order reflection at 27°8′. What is the distance between the diffracting planes? [sin 27°8′ = 0.4561] **[WB JEE 2014]**

(a) 1.46 Å (b) 1.59 Å
(c) 2.51 Å (d) 5.46 Å

34 The appearance of colour in solid alkali metal halides is generally due to **[UK PMT 2014]**

(a) Frenkel defect (b) interstitial positions
(c) F-centres (d) Schottky defect

35 The percentages of void space for simple cubic, body centred cubic and hexagonal close packed arrangement respectively are **[EAMCET 2014]**

(a) 48, 32, 26 (b) 48, 26, 32
(c) 26, 48, 32 (d) 32, 48, 26

36 Volume occupied by single CsCl ion pair in a crystal is 7.014×10^{-23} cm^3. The smallest Cs—Cs internuclear distance is equal to length of the side of the cube corresponding to volume of one CsCl ion pair. The smallest Cs to Cs internuclear distance is nearly **[KCET 2014]**

(a) 4.4 Å (b) 4.3 Å
(c) 4 Å (d) 4.5 Å

37 The metal has a fcc lattice. The edge length of the unit cell is 404 pm. The density of the metal is 2.72 g/cm^3. The molar mass of the metal is (N_A = Avogadro's constant = 6.02×10^{23} mol^{-1})

(a) 40 g/mol (b) 30 g/mol **[NEET 2013]**
(c) 27 g/mol (d) 20 g/mol

38 The number of carbon atoms per unit cell of the diamond unit cell is **[NEET 2013]**

(a) 4 (b) 8 (c) 6 (d) 1

39 Which one of the following defects in the crystal lowers its density? **[KCET 2013]**

(a) F-centres
(b) Interstitial defect
(c) Frenkel defect
(d) Schottky defect

40 The metal crystallises with a face-centred cubic lattice. The edge of the unit cell is 408 pm. The diameter of the metal atom is **[CBSE AIPMT 2012]**

(a) 288 pm (b) 408 pm
(c) 144 pm (d) 204 pm

41 The number of octahedral void (s) per atom present in a cubic close packed structure is **[CBSE AIPMT 2012]**

(a) 1 (b) 3
(c) 2 (d) 4

42 Structure of a mixed oxide is cubic close packed (ccp). The cubic unit cell of mixed oxide is composed of oxide ions. One fourth of the tetrahedral voids are occupied by divalent metal A and the octahedral voids are occupied by a monovalent metal B. The formula of oxide is **[CBSE AIPMT 2012]**

(a) ABO_2 (b) A_2BO_2
(c) $A_2B_3O_4$ (d) AB_2O_2

43 Which one of the following is a covalent crystal?

(a) Rock salt (b) Ice **[KCET 2011]**
(c) Quartz (d) Dry ice

44 If in a crystal lattice of a compound, each corner of a cube is enjoyed by sodium, each edge of a cube has oxygen and centre of cube is enjoyed by tungsten (W), then give its formula **[Guj CET 2011]**

(a) Na_2WO_4 (b) $NaWO_3$
(c) Na_3WO_3 (d) Na_2WO_3

45 The first order reflection (n = 1) from a crystal of the X-ray from a copper anode tube (λ = 1.54 Å) occurs at an angle of 45°. What is the distance between the set of plane causing the diffraction? **[Guj CET 2011]**

(a) 0.1089 nm
(b) 0.1089 m
(c) 1.089 Å
(d) 1.089×10^{-9} m

ANSWERS

Check point 1.1

1. (a) 2. (c) 3. (b) 4. (d) 5. (d) 6. (c) 7. (c) 8. (d) 9. (c) 10. (a)

Check point 1.2

1. (b) 2. (b) 3. (a) 4. (c) 5. (d) 6. (a) 7. (c) 8. (b) 9. (d) 10. (c)

Check point 1.3

1. (d) 2. (d) 3. (b) 4. (c) 5. (c) 6. (a) 7. (b) 8. (b) 9. (c) 10. (a)

(A) Topical exercise

1. (b) 2. (d) 3. (a) 4. (c) 5. (c) 6. (a) 7. (b) 8. (d) 9. (b) 10. (b)
11. (b) 12. (d) 13. (d) 14. (b) 15. (a) 16. (a) 17. (d) 18. (c) 19. (d) 20. (b)
21. (d) 22. (a) 23. (b) 24. (a) 25. (b) 26. (d) 27. (c) 28. (c) 29. (a) 30. (b)
31. (d) 32. (d) 33. (c) 34. (d) 35. (c) 36. (c) 37. (c) 38. (a) 39. (c) 40. (b)
41. (a) 42. (c) 43. (d) 44. (c) 45. (d) 46. (c) 47. (b) 48. (b) 49. (a) 50. (b)
51. (a) 52. (b) 53. (b) 54. (a) 55. (b) 56. (d) 57. (b) 58. (b) 59. (c) 60. (d)
61. (a) 62. (a) 63. (b) 64. (d) 65. (a) 66. (b) 67. (b) 68. (a) 69. (a) 70. (a)
71. (d) 72. (d) 73. (c) 74. (a) 75. (c) 76. (b) 77. (c) 78. (c) 79. (b) 80. (d)
81. (a) 82. (c) 83. (a) 84. (a) 85. (b) 86. (b) 87. (c) 88. (c) 89. (c) 90. (c)
91. (a) 92. (d) 93. (a) 94. (c) 95. (b) 96. (b) 97. (a) 98. (a) 99. (d) 100. (b)
101. (a) 102. (d) 103. (b) 104. (d) 105. (b) 106. (b) 107. (c) 108. (d) 109. (d) 110. (a)
111. (b) 112. (c) 113. (a) 114. (a) 115. (c) 116. (d) 117. (c) 118. (a) 119. (a) 120. (c)
121. (a) 122. (d) 123. (a) 124. (d) 125. (b) 126. (b)

(B) Miscellaneous questions

1. (d) 2. (c) 3. (b) 4. (d) 5. (a) 6. (b) 7. (d) 8. (c) 9. (d) 10. (b)
11. (c) 12. (c) 13. (c) 14. (d) 15. (a) 16. (d) 17. (c) 18. (c) 19. (c) 20. (d)
21. (b) 22. (a) 23. (d) 24. (a) 25. (a) 26. (b) 27. (a) 28. (a) 29. (a) 30. (a)
31. (c) 32. (d) 33. (b) 34. (d) 35. (a) 36. (c)

(C) Medical entrances' gallery

1. (d) 2. (a) 3. (a) 4. (b) 5. (d) 6. (a) 7. (b) 8. (c) 9. (a) 10. (d)
11. (c) 12. (b) 13. (b) 14. (a) 15. (b) 16. (a) 17. (a) 18. (d) 19. (a) 20. (b)
21. (d) 22. (c) 23. (d) 24. (c) 25. (a) 26. (d) 27. (c) 28. (d) 29. (d) 30. (a)
31. (d) 32. (c) 33. (c) 34. (c) 35. (a) 36. (c) 37. (c) 38. (b) 39. (d) 40. (a)
41. (a) 42. (d) 43. (c) 44. (b) 45. (c)

Hints & Explanations

(A) Topical exercise

7 (*b*) Crystalline solid is anisotropic in nature as this solid shows different physical properties such as electrical resistance, refractive index in different directions.

Isotropic and anisotropic properties are related to amorphous solid and crystalline solid respectively, which can be clearly understood as

Property	Types of solid	Physical properties
Isotropy	Amorphous solid	Same value
Anisotropy	Crystalline solid	Different values

9 (*b*) At low temperature existence of a substance in solid state is due to

(a) slow molecular motion and

(b) strong cohesive forces

These two forces hold the constituent particles together thus causes existence of substance in solid state.

10 (*b*) • Quartz glass (SiO_2) is an amorphous solid due to its short range order of constituent particles.

• Quartz is a crystalline solid while quartz glass is an amorphous solid.

13 (*d*) Amorphous solids are isotropic in nature because it has no long range order and any physical property will be same on all direction. On the other hand, anisotropic nature is a characteristic feature of crystalline solid.

14 (*b*) Crystalline solid has regular arrangement of constituent particles observed over a long distance in the crystal lattice.

Due to this crystalline solids having regular arrangement have sharp melting point.

15 (*a*) Iodine molecules are a class of non-polar molecular solid in which constituent molecules are held together by London or dispersion forces. These solids are soft and non-conductors of electricity.

16 (*a*) Since, quartz glass is an amorphous solid having short range order of constituents. Hence, value of refractive index is same in all directions, can be measured and not be equal to zero always.

17 (*d*) The fraction of octahedral or tetrahedral voids occupied depends upon the radii of the ions present at the lattice points. As we know the radii of octahedral or tetrahedral void is related to radii of atoms (r) as

Radius of octahedral void $(R_o) = 0.414\, r$

Radius of tetrahedral void $(R_t) = 0.225\, r$

where, $r =$ radius of bigger atom involved.

28 (*c*) Each point in a lattice is known as lattice point which are either atom or molecule or ion joined together by a straight line to bring out geometry of lattice in pure crystal.

Here, constituents are arranged in fixed stoichiometric ratio. Hence, existence of free electrons are not possible, it is possible in case of imperfection in solid.

32 (*d*) Every unit cell of NaCl has $4Cl^-$ and $4\ Na^+$ ions.

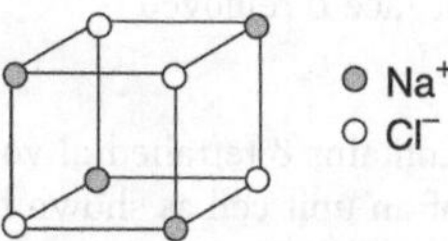

39 (*c*) A. For primitive unit cell, $a = b = c$

Total number of atoms per unit cell $= 1/8 \times 8 = 1$

Here, 1/8 is due to contribution of each atom present at corner.

B. For body centred cubic unit cell, $a = b = c$

This lattice contain atoms at corner as well as body centre.

Contribution due to atoms at corner $= 1/8 \times 8 = 1$

Contribution due to atoms at body centre $= 8$

C. For face centred unit cell, $a = b = c$

Total constituent ions per unit cell present at corners

$$= \frac{1}{8} \times 8 = 1$$

Total constituent ions per unit cell present at face centre

$$= \frac{1}{2} \times 6 = 3$$

D. For end centered orthorhombic unit cell, $a \neq b \neq c$

Total contribution of atoms present at corner $= \frac{1}{8} \times 8 = 1$

Total contribution of atoms present at end centre

$$= \frac{1}{2} \times 2 = 1$$

Hence, other than corner it contain total one atom per unit cell.

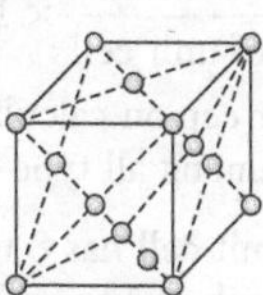
Face centred

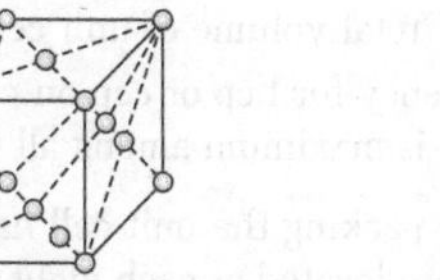
End centred

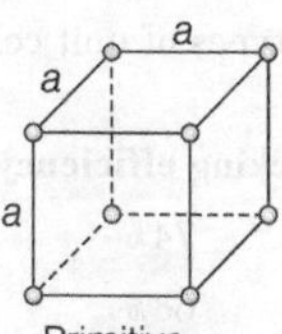

Primitive (or simple)

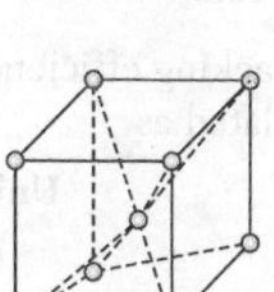
Body centred

40 (*b*) Cubic lattice is bcc type.

Number of X atoms at the corner $= 8/8 = 1$

Number of Y atoms at the body centre $= 1/1 = 1$

Thus, ratio of $X : Y = 1 : 1$

Thus, formula is XY.

41 (*a*) AB has fcc type structure. Atoms at corner $(A) = 8/8 = 1$

Atoms at faces $= \frac{4}{2} = 2$

Since, atoms B at face B removed

Hence, AB_2.

52 (*b*) fcc unit cell contains 8 tetrahedral voids at centre of each 8 smaller cube of an unit cell as shown below

Eight tetrahedral voids per fcc unit cell

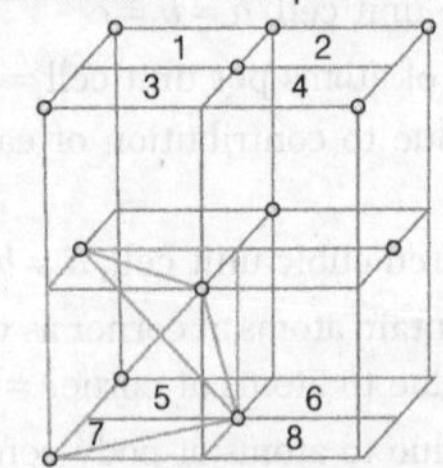

Each cube represented by numeric 1, 2, 3, 4, 5, 6, 7, 8 contain one tetrahedral voids

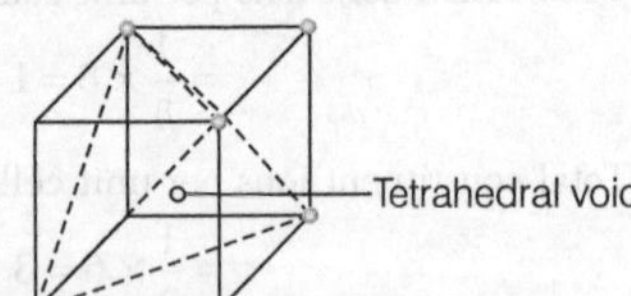

Each cube contain one tetrahedral void at its body centre as shown above

Number of atoms in fcc unit cell = 4

Number of octahedral voids = 4

Number of tetrahedral voids = 8

55 (*b*) Packing efficiency is the percentage of total filled space by particles and it can be calculated as

$$\text{packing efficiency} = \frac{\text{volume occupied by four spheres in the unit cell}}{\text{total volume of unit cell}} \times 100$$

Since, packing efficiency for hcp or ccp on calculation comes out to be 74% which is maximum among all type of crystals.

60 (*d*) In the cubic close packing the unit cell has 8 tetrahedral voids within it and are located at each eight smaller cube of unit cell.

63 (*b*) Packing efficiency in different types of unit cells can be tabulated as

Unit cell	Packing efficiency
fcc	74%
bcc	68%
Simple cubic	52%

Hence, correct order is fcc (74%) > bcc (68%) > simple cubic (52%).

68 (*a*) Density $(d) = \frac{MZ}{N_0 a^3}$

$$\therefore \quad M = \frac{dN_0 a^3}{Z}$$

$$= 3.048 \times 10^4 \text{ g mol}^{-1}$$

Since, protein fraction is 0.75, hence

$$M\text{(protein)} = 3.048 \times 10^4 \times 0.75$$

$$= 2.286 \times 10^4$$

$$= 2.3 \times 10^4 \text{ g mol}^{-1}$$

69 (*a*) In a bcc structure, the atoms touch each other along the body diagonal.

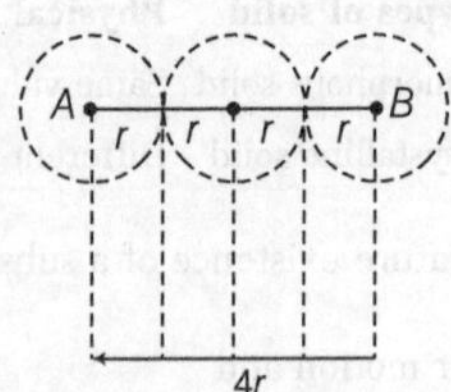

Radius of the atom $= r$

$$AB^2 = BC^2 + AC^2$$

$$BC^2 = BD^2 + CD^2$$

$$= a^2 + a^2 = 2a^2$$

$$\therefore \quad AB^2 = 2a^2 + a^2 = 3a^2$$

$$\therefore \quad AB = \sqrt{3}a$$

But $\quad AB = 4r$

$$\therefore \quad 4r = \sqrt{3}a$$

$$\Rightarrow \quad r = \sqrt{3}a/4$$

$$r = \sqrt{3} \times \frac{352}{4} \text{ pm}$$

$$= \sqrt{3} \times \frac{352}{4} \times 10^{-12} \text{ m}$$

$$= \sqrt{3} \times \frac{352}{4} \times 10^{-10} \text{ cm}$$

$$= 152.42 \times 10^{-10} \text{ cm}$$

Volume of one atom $= \frac{4}{3}\pi r^3 = \frac{4}{3}\pi(152.42 \times 10^{-10})^3 \text{cm}^3$

$$= 1.48 \times 10^{-23} \text{cm}^3$$

Volume of N_0 atoms in one mole of the metal

$$= 6.02 \times 10^{23} \times 1.48 \times 10^{-23} \text{cm}^3$$

$$= 8.93 \text{ cm}^3$$

73 (*c*) $2(r_{Na^+} + r_{H^-}) = 488$ pm

$$\therefore \quad r_{Na^+} + r_{H^-} = 244 \text{ pm}$$

74 (*a*)

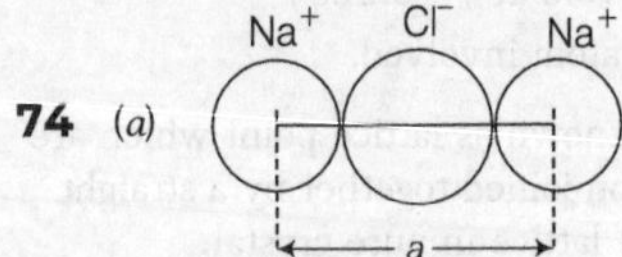

$$a = 2r_{Na^+} + 2r_{Cl^-} = 2(97 + 181) = 556 \text{ pm}$$

80 (*d*) Density, $d = \frac{MZ}{N_0 a^3}$

where

M = molar mass of H_2O = 18 g mol^{-1}

N_0 = Avogadro's number = 6.02×10^{23}

Volume $(= a^3) = 132 \times 10^{-24}$ cm^3

$$\therefore \quad 0.92 = \frac{18 \times Z}{6.02 \times 10^{23} \times 132 \times 10^{-24}}$$

$$\therefore \quad Z = 4.06 \Rightarrow Z = 4$$

81 (*a*) Along the edge, $AB = AC = 2(r_{Cl^-} + r_{Na^+}) = a$

$$\therefore \quad r_{Na^+} = \left(\frac{a}{2} - r_{Cl^-}\right)$$

Also, $AC^2 + AB^2 = BC^2$

$$\Rightarrow \quad a^2 + a^2 = (4r_{Cl^-})^2$$

$$\therefore \quad a = 2\sqrt{2}r_{Cl^-}$$

$$\therefore \quad r_{Na^+} = \left(\frac{2\sqrt{2}r_{Cl^-}}{2} - r_{Cl^-}\right)$$

$$= r_{Cl^-}(\sqrt{2} - 1) = r_{Cl^-}(1.414 - 1)$$

$$\therefore \quad \frac{r_{Na^+}}{r_{Cl^-}} = 0.414$$

82 (*c*) In bcc structure, Cl^- ions lie at the corners of simple cubic and Cs^+ at the body centre. Hence, along the body diagonal, Cs^+ and Cl^- touch each other.

$$r_{Cs^+} + r_{Cl^-} = 2r$$

$$\text{But, } r = \frac{\sqrt{3}}{4}a$$

$$\therefore \quad r_{Cs^+} + r_{Cl^-} = 2\frac{\sqrt{3}}{4}a = \frac{\sqrt{3}}{2}a$$

83 (*a*) NaCl crystal have rock salt structure having fcc lattice in which Cl^- ions are present at fcc lattice points and Na^+ occupies all the octahedral void of given unit cell.

Where, coordination number of Na^+ = 6

coordination number of Cl^- = 6

96 (*b*) Frenkel defects do not change the value of the theoretical density because the atom is only occupying an interstitial position instead of lattice position. But in case of Schottky defect,

$$Z = 4\left(1 - \frac{0.1}{100}\right) = 3.996$$

$$d(\text{density}) = \frac{MZ}{N_0 a^3}$$

$\frac{MZ}{N_0 a^3}$ is constant, thus $d \propto Z$

$$Z(\text{Frenkel}) = 4$$

$$Z(\text{Schottky}) = 3.996$$

Thus, $d_f > d_s$

97 (*a*) When smaller ion (usually cation) is dislocated from its normal site in crystal and move to interstitial site is known as Frenkel defect as shown below :

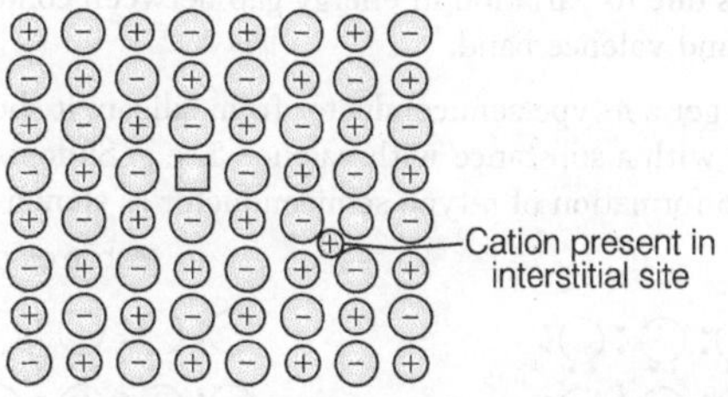

Frenkel defect

98 (*a*) Frenkel defect is also known as dislocation defect because in Frenkel defect atoms present in crystal lattice is dislocated to interstitial site.

99 (*d*) When electron rich or electron deficient impurity is added to a perfect crystal it introduces electronic defect in them.

105 (*b*) When group 14 elements are doped with group 13 elements, it creates a hole in a molecule but the molecule as a whole remain neutral. Thus, *p* -type semiconductor is neutral in nature.

Positive hole (no electron)

B

Perfect crystal

106 (*b*) Graphite is a good conductor of electricity due to presence of free valence electrons. In graphite, each carbon is sp^2-hybridised having one free electron which makes graphite a good conductor of electricity.

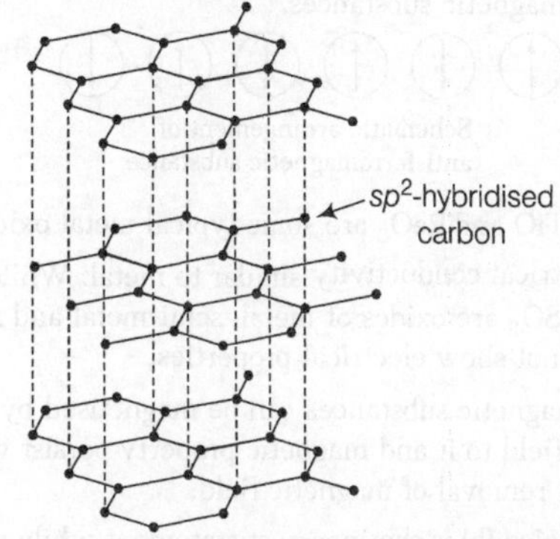

111 (*b*) Ge is tetravalent while group 13 element has three electrons in valence shell, thus there is positive charge (*p*), or hole hence *p*-type semiconductor.

114 (*a*) For octahedral holes,

$$\frac{r_{M^+}}{r_-} = 0.414 \Rightarrow r_{M^+} = 0.414 \times r_-$$

$$= 0.414 \times 500 = 207 \text{ pm}$$

115 (*c*) Certain metal oxides like VO_2, VO, VO_3 and TiO_3 show metallic or insulating property depending upon temperature. As temperature varies metallic or insulating property varies. This is due to variation in energy gap between conduction band and valence band.

116 (*d*) To get a *n*-type semiconductor from silicon, it should be doped with a substance with valence 5. e.g. Si doped with P lead to formation of *n*-type semiconductor as shown below

Mobile electron

Doping with P (valency= 5)

Perfect crystal *n*-type

117 (*c*) Iodine is a non-polar molecular solid in which iodine molecules are held together by London force or dispersion force. This is soft and non-conductor of electricity.

Water is a hydrogen bonded molecular solid in which H and O are held together by polar covalent bond and each water molecule held together by hydrogen bonding. Due to non-ionic nature, they are not electrical conductor.

118 (*a*) AgBr shows both Schottky as well as Frenkel defect. In AgBr, both Ag^+ and Br^- ions are absent from the lattice causing Schottky defect. However, Ag^+ ions are mobile so they have a tendency to move aside the lattice and trapped in interstitial site. Hence, causes Frenkel defect.

119 (*a*) When a ferromagnetic substance is placed in a magnetic field it becomes a permanent magnet because all the domains get oriented in the direction of magnetic field even after removal of applied magnetic field.

122 (*d*) The substances in which domain structure are oppositely oriented and cancel each other's magnetic field are known as anti-ferromagnetic substances.

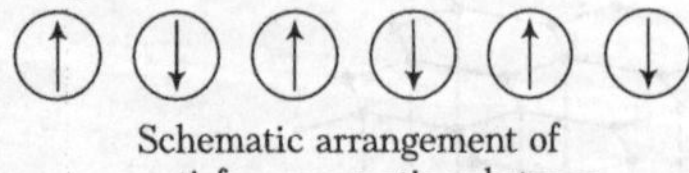

Schematic arrangement of anti-ferromagnetic substance

124 (*d*) CrO_2, TiO and ReO_3 are some typical metal oxides which show electrical conductivity similar to metal. While SiO_2, MgO and SO_2 are oxides of metal, semi-metal and non-metal which do not show electrical properties.

126 (*b*) Ferromagnetic substances can be magnetised by applying magnetic field to it and magnetic property persist within it even after removal of magnetic field.

Hence, choice (b) is the incorrect statement while other three choices are correct statements.

(B) Miscellaneous questions

4 (*d*) *n*-type semiconductor (negative) is formed when there is extra electron available for conduction. Si has four electrons in valence shell, while As has five. Thus, extra one electron makes *n*-type semiconductor on doping Si with As.

10 (*b*) Silicon doped with electron rich impurity such as phosphorus forms a *n*-type semiconductor. This is due to presence of mobile electron.

14 (*d*) Hexagonal close packing can be formed by arranging two layers *A* and *B* one over another which can be diagramatically represented as,

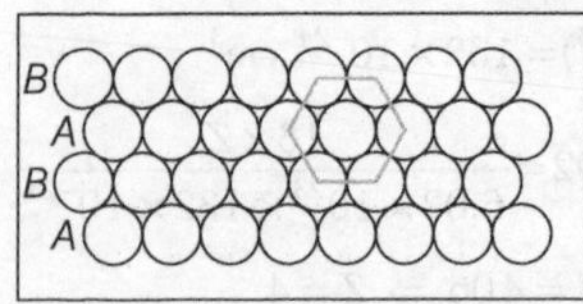

Here, we can see easily that 1st layer and 4th layer are not exactly aligned.

Thus, statement (d) is not correct while other statements (a), (b) and (c) are true.

15 (*a*) Conductivity of metal, insulator and semiconductors can be represented in the term of κ (Kappa) which depends upon energy gap between valence band and conduction band.

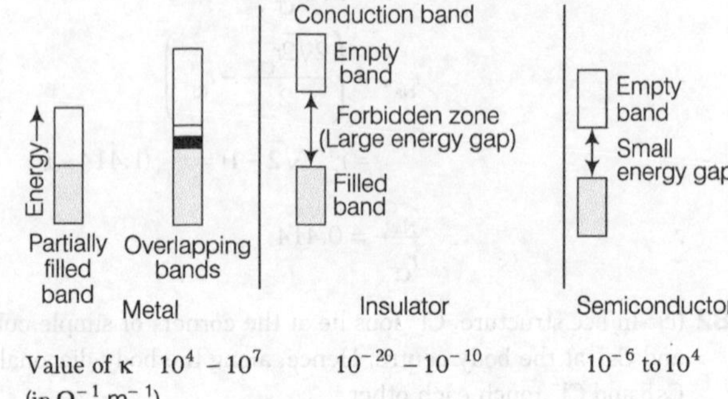

Value of κ (in $\Omega^{-1}\,m^{-1}$) $10^4 - 10^7$ $10^{-20} - 10^{-10}$ 10^{-6} to 10^4

Hence, correct order is

$$\kappa_{metals} >> \kappa_{insulators} < \kappa_{semiconductors}$$

17 (*c*) Coordination number in a square closed packed structure in two dimensions is equal to 4 as shown below

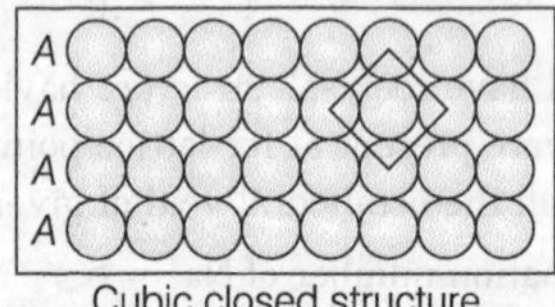

Cubic closed structure

19 (*c*) Diamond is a giant molecule in which constituent atoms are held together by covalent bond. Hence, this is a network solid.

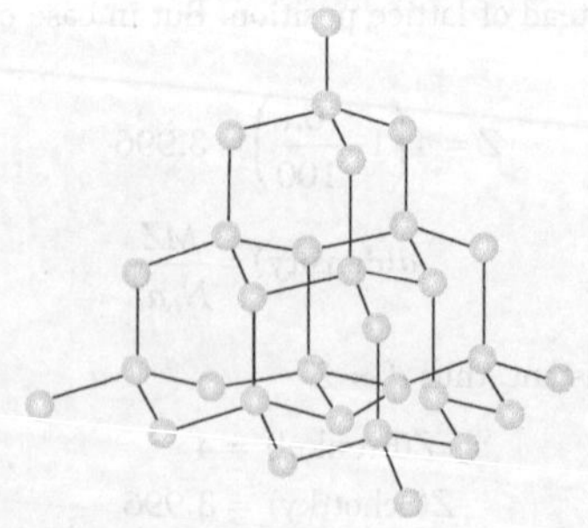

20 (*d*) Graphite can't be classified as ionic solid as graphite is not made up of ions. It is made up of carbon atoms covalently bonded to three carbon atoms so, it is a covalent solid.

Since, the formation of covalent bond occurs throughout the crystal therefore, it is a type of network solid. Due to presence of free electron, graphite is also classified as conductor solid.

22 (*a*) Ionic solids easily dissociates into its ions in molten state and show high electrical conductivity. So, statement (a) is incorrect while ionic solids are anisotropic and brittle linked with very strong force of interactions.

30 (*a*)

X^- M^+ X^- A B C D E

$$AE = 7.2\ \text{Å} = AB + BC + CD + DE$$

$$= r_{X^-} + r_{M^+} + r_{M^+} + r_{X^-}$$

$$= 2(r_{X^-} + r_{M^+})$$

$$3.6 = r_{X^-} + r_{M^+} \quad (\text{Given } r_{M^+} = 1.6\ \text{Å})$$

$$\therefore \quad r_{X^-} = 2.0\ \text{Å}$$

33 (*b*) There is increase in 0.30 Å unit length. This is due to two atoms at the corners.

Thus, $r_{\text{Rb}^+} = 1.33\ \text{Å} + \dfrac{0.30\ \text{Å}}{2} = 1.48\ \text{Å}$

34 (*d*) Number of Au atoms at the corners $= \dfrac{8}{8} = 1$

Number of Cu atoms at the faces $= \dfrac{6}{2} = 3$

Hence, formula is $AuCu_3$.

35 (*a*) Edge length for different types of unit cells can be tabulated as,

Types of unit cell	Edge length
fcc	$2\sqrt{2}r$
bcc	$\dfrac{4}{\sqrt{3}}r$
scc	$2r$

36 (*c*) Let the number of atoms in cubic close packing be x thus, number of tetrahedral voids $= 2x$

Number of atoms of $Q = x$ (due to cubic close packing)

P atoms occupy one-third of tetrahedral voids, thus number of

P atoms $= \dfrac{2}{3}x$

Thus, $\dfrac{P \text{ atoms}}{Q \text{ atoms}} = \dfrac{\frac{2x}{3}}{x} = \dfrac{2}{3}$

Thus, compound is P_2Q_3.

(C) Medical entrances' gallery

1 (*d*) In bcc crystal,

$$\because \quad 4r = \sqrt{3}a$$

Given, cell edge $(a) = 288$ pm

$$r = \frac{\sqrt{3}a}{4} = \frac{\sqrt{3} \times 288}{4}\ \text{pm}$$

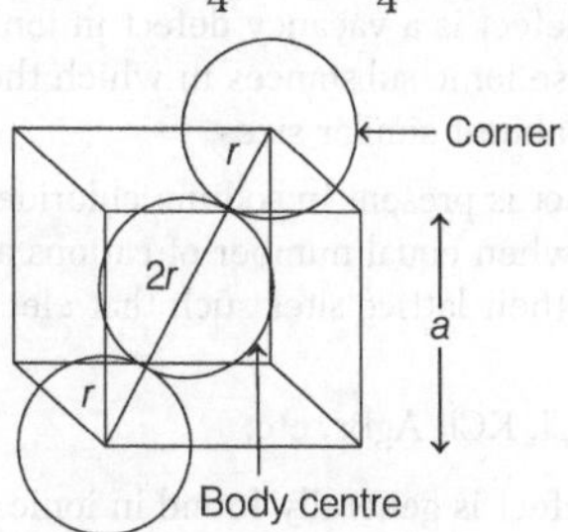

where, r = radius of atoms,

a = edge length of the unit cell.

2 (*a*) Frenkel defect is shown by ionic substances in which there is a large difference in size of ions.

e.g. AgBr, AgI, ZnS

Schottky defect is shown by ionic substances in which the cation and anion are of almost similar sizes.

e.g. AgBr, NaCl, KCl, CsCl

So, AgBr shows both, Frenkel as well as Schottky defects.

3 (*a*) Let, in the given crystal $Ni_{0.98}O$

$Ni^{2+} = x$ and $Ni^{3+} = 0.98 - x$

Total charge on M^{2+} and M^{3+}

$$= (+2)x + (+3)(0.98 - x)$$

$$= 2x + 2.94 - 3x = 2.94 - x$$

As metal oxide is neutral. Therefore,

total charge on cations = total charge on anions.

$$2.94 - x = 2 \quad [\because \text{ Charge of oxygen atom} = -2]$$

$$x = 2.94 - 2 = 0.94$$

So, the fraction of Ni^{2+} ions in the crystal

$$= \frac{0.94}{0.98} = 0.959 \sim 0.96$$

4 (*b*) Anions (A) form hexagonal close packed (hcp) lattice, so number of anions $(A) = 6$

number of octahedral voids = number of atoms in the close packed structure = 6.

Cations (C) occupy 75% of octahedral voids, so number of cations (C)

$$= 6 \times \frac{75}{100} = 6 \times 3/4 = 9/2$$

$\therefore$ The formula of compound $= C_{9/2}A_6 = C_9A_{12} = C_3A_4$

Thus, option (b) is correct.

5 (*d*) Silicon carbide (SiC) is an example of covalent solids.

Covalent or network solids are crystalline solids of non-metals that result from the formation of covalent bonds between adjacent atoms throughout the crystal. They are insulators and do not conduct electricity.

6 (*a*) Silica and silica glass is an amorphous solid.

7 (*b*) Schottky defect is a vacancy defect in ionic solids. It is shown by those ionic substances in which the cations and anions are of almost similar sizes.

Schottky defect is present in sodium chloride (NaCl). This defect arises when equal number of cations and anions are missing from their lattice sites such that electrical neutrality is maintained.

e.g. CsCl, NaCl, KCl, AgBr, etc.

8 (*c*) Frenkel defect is generally found in ionic crystals where anion is much larger in size than the cation. It is also known as dislocation defect.

Frenkel defect is present in AgBr. This defect arises when the smaller ion (usually cation) is dislocated from its normal site. It creates a vacancy defect at its original site and an interstitial defect at its new location.

e.g. AgBr, AgCl, AgI, ZnS.

9 (*a*) Zinc oxide which is white in colour on heating loses oxygen and turns yellow.

$$\underset{}{ZnO} \xrightarrow{\text{Heating}} \underset{\text{Yellow}}{Zn^{2+}} + \frac{1}{2}O_2 + 2e^-$$

It is due to metal excess defect in which excess ions (cation and electron) occupy interstitial site.

A negative ion missing from its lattice site leaves a hole which is occupied by an extra electron to maintain electrical neutrality. Thus, anionic site is called *F*-centre.

Hence, both Assertion and Reason are correct and Reason is the correct explanation of Assertion.

10 (*d*) Number of particles or atoms present in 1 mole

$= 6.022 \times 10^{23}$

$\therefore$ Number of particles or atoms present in 0.5 mole

$= 0.5 \times 6.022 \times 10^{23} = 3.011 \times 10^{23}$

Number of tetrahedral voids

$= 2 \times$ number of atoms in close packaging

$= 2 \times 3.011 \times 10^{23} = 6.022 \times 10^{23}$

$= 6.02 \times 10^{23}$ voids

11 (*c*) Diamond is a giant molecule in which constituent atoms are held together by covalent bonds. Hence, it is a network solid.

Note SO_2 (solid), H_2O (ice) and I_2 are examples of molecular solid.

12 (*b*) Given, ρ (density) = 8.55 g cm^{-3}

Atomic mass of niobium = 93u

As we know that,

$$\text{density } (\rho) = \frac{MZ}{a^3 N_A} \quad \text{...(i)}$$

where, M = atomic mass

Z = number of unit cell

a = edge length

N_A = Avogadro's number

For bcc, $Z = 2$

By putting the given values in equation (i), we get

$$8.55 = \frac{93 \times 2}{a^3 \times 6.02 \times 10^{23}}$$

or, $a^3 = 3.6137 \times 10^{-23} \text{ cm}^3$

or, $a = 3.306 \times 10^{-8} \text{ cm}$

For body centred cube, radius $= \frac{\sqrt{3}}{4} a$

or, $r_{bcc} = \frac{\sqrt{3}}{4} \times 3.306 \times 10^{-8}$

or, $r_{bcc} = 1.43 \times 10^{-8} \text{ cm} = 1.43 \times 10^{-10} \text{ m} = 143 \text{ pm}$

13 (*b*) Given

Density (*d*) = 8.92 g/mL

Edge length (*a*) = 3.6×10^{-8} cm

$Z = 4$ (fcc)

The molecular mass of atom in a lattice can be caluculated by using the formula,

$$d = \frac{Z \times M}{N_A a^3}$$

$$\therefore \quad M = \frac{d N_A a^3}{Z}$$

$$M = \frac{8.92 \times 6.023 \times 10^{23} \times (3.6)^3 \times (10^{-8})^3}{4}$$

$M = 62.66 \approx 63$ amu

14 (*a*) In a cubic close packed lattice of oxide ions, there would be two tetrahedral and one octahedral void per oxide ion.

Since, the formula shows the presence of 4 oxide ions, the number of tetrahedral voids is eight and that of octahedral voids is four. Out of the eight tetrahedral voids, one is occupied by *X*.

$\therefore$ Percentage of tetrahedral voids occupied $= \frac{1}{8} \times 100 = 12.5\%$

15 (*b*) In a unit cell,

W atoms at the corner $= \frac{1}{8} \times 8 = 1$

O atoms at the centre of edges $= \frac{1}{4} \times 12 = 3$

Na atoms at the centre of the cube = 1

$\therefore$ W : O : Na = 1 : 3 : 1

Hence, formula of compound is $NaWO_3$.

16 (*a*) Size of cation must be smaller than that of anion.

17 (*a*) Given, Li has a bcc structure.

Density (ρ) = 530 kg m^{-3}

Atomic mass (*M*) = 6.94 g mol^{-1}

Avogadro's number (N_A) = 6.02×10^{23} mol^{-1}

We know that, number of atoms per unit cell in bcc (Z) = 2

$\therefore$ We have the formula for density, $(\rho) = \dfrac{ZM}{N_A a^3}$

where, a = edge length of a unit cell

$$\text{or } a = \sqrt[3]{\frac{ZM}{\rho N_A}} = \sqrt[3]{\frac{2 \times 6.94 \text{ g mol}^{-1}}{0.53\text{g cm}^{-3} \times 6.02 \times 10^{23} \text{ mol}^{-1}}}$$

$$= \sqrt[3]{4.35 \times 10^{-23} \text{ cm}^{-3}} = 3.52 \times 10^{-8} \text{cm}$$

$$a = 352 \text{ pm}$$

18 (*d*) Given, ionic radius of cation (A^+) = 0.98×10^{-10} m

Ionic radius of anion (B^-) = 1.81×10^{-10} m

$\therefore$ Coordination number of each ion in AB = ?

Now, we have

$$\text{Radius ratio} = \frac{\text{radius of cation}}{\text{radius of anion}} = \frac{0.98 \times 10^{-10} \text{m}}{1.81 \times 10^{-10} \text{ m}} = 0.541$$

If radius ratio range is in between 0.441– 0.732, ion would have octahedral structure with coordination number 'six'.

19 (*a*) In CaF_2, the coordination numbers for $Ca^{+2} = 8$, $F^- = 4$

20 (*b*) Given, edge length = 361pm

Four metal atoms crystallises in one unit cell

i.e. effective number in unit cell (Z) = 4 (given)

$\therefore$ Face diagonal = $4r$

$$\sqrt{2}\, a = 4r$$

$$r = \frac{\sqrt{2} \times 361}{4} = 127.63 \approx 127 \text{ pm}$$

21 (*d*) $\because$ Packing efficiency in bcc lattice = 68%

$\therefore$ Vacant space in bcc lattice = 100 − 68 = 32%

22 (*c*) In Frenkel defect, ions in solid dislocate from their positions. Hence, Frenkel defect is a dislocation defect.

23 (*d*)

Compound		Magnetic properties
NaCl	–	Diamagnetic
MnO	–	Anti-ferromagnetic
$CrCl_3$	–	Paramagnetic
CrO_2	–	Ferromagnetic
$MgFe_2O_4$	–	Ferrimagnetic

24 (*c*) A unit cell with edge length $a \neq b \neq c$ and axial angles $\alpha = \beta = \gamma = 90°$ is called orthorhombic, e.g. rhombic sulphur, KNO_3 and $BaSO_4$.

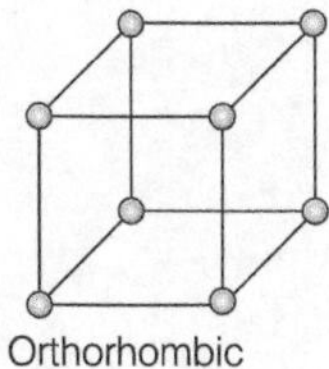

Orthorhombic

25 (*a*) FeO shows metal deficiency defect. In crystal of FeO, some Fe^{2+} cations are missing and the loss of positive charge is made up by the presence of required number of Fe^{3+} ions.

26 (*d*) MnO shows anti-ferromagnetism having domain structure similar to ferromagnetic substance, but their domains are opposite oriented and cancel out each other's magnetic moment.

↑↓↑↓↑↓↑

27 (*c*) Doping means incorporating small amount of another element. Doping of group 14 elements such as silicon with group 15 elements such as phosphorus gives rise to excess electrons.

This type of doped silicon is referred as negative type or *n*-type silicon and it is called *n*-type semiconductor.

28 (*d*) In bcc, 2 atoms are present. One atom lie at the centre of cube while, other lies at the corner of the cube. Hence, the distance between the body centred and one corner atom is half of the body diagonal, i.e. $\dfrac{\sqrt{3}}{2}a$.

29 (*d*) For fcc, $$r = \frac{a}{2\sqrt{2}}$$

$$a = 361 \text{pm (given)}$$

$$r = \frac{361}{2\sqrt{2}} = 127.6 \text{ pm} \approx 128 \text{ pm}$$

30 (*a*) The close packed structure have both octahedral and tetrahedral voids. For *n* atoms, there are 2*n* tetrahedral voids and *n* octahedral voids.

For 100 atoms, there will be 200 tetrahedral voids and 100 octahedral voids.

31 (*d*) Piezoelectric effect is the ability of some materials to create an electric potential in response to applied mechanical stress. The commonly used piezoelectric materials are quartz (SiO_2), zinc oxide etc.

32 (*c*) For NaCl type structure,

$$\frac{r_{A^+}}{r_{B^-}} = 0.414 - 0.737$$

For maximum possible radius of B^-,

$$\frac{r_{A^+}}{r_{B^-}} = 0.414 \Rightarrow r_{B^-} = \frac{170}{0.414} = 410.6 \text{ pm}$$

33 (*c*) $\because$ $$n\lambda = 2a \sin\theta$$

$$1 \times 2.29 = 2d \times 0.4561$$

$$a = 2.51 \text{ Å}$$

34 (*c*) In metal excess defect, *F*-centres are produced in an ionic crystal which are responsible for the appearance of colour in solid alkali metal halides.

35 (*a*)

	Packing fraction	Void Space
Simple cubic unit cell	52%	48%
Body centred cubic unit cell	68%	32%
Face centred cubic unit cell	74%	26%

36 (*c*) $\because$ Volume,

$$V = 7.014 \times 10^{-23}\ cm^3$$

$\therefore$ $$a^3 = 7.014 \times 10^{-23}\ cm^3$$

or $$a = \sqrt[3]{7.014 \times 10^{-23}\ cm^3}$$

$$= \sqrt[3]{70.14 \times 10^{-24}\ cm^3}$$

$$= 4.125 \times 10^{-8}\ cm^3$$

$$= 4.12\ Å \approx 4\ Å$$

37 (*c*) Given cell is fcc. So, $Z = 4$

Edge length, $a = 404 pm = 4.04 \times 10^{-8}$ cm

Density of metal $d = 2.72\ g / cm^3$

$N_A = 6.02 \times 10^{23}\ mol^{-1}$

Molar mass of the metal, $M = ?$

We know that, density

$$d = \frac{Z \times M}{a^3 \cdot N_A}$$

$$\Rightarrow \quad M = \frac{d \cdot a^3 N_A}{Z}$$

$$= \frac{2.72 \times (4.04 \times 10^{-8})^3 \times 6.02 \times 10^{23}}{4} = 27\ g / mol$$

38 (*b*) Carbon as diamond crystallises in a face centred cubic structure and possesses four more carbon atoms that are found to be present in the alternate tetrahedral voids.

Thus, Z_{eff} [number of C-atoms] in diamond

$= Z_{eff}$ in fcc unit cell + Z_{eff} in alternate tetrahedral voids

$$\therefore Z_{eff} = \underbrace{\left[\frac{1}{8} \times 8 + \frac{1}{2} \times 6\right]}_{\text{fcc unit cell}} + \underbrace{[1 \times 4]}_{\text{Alternate tetrahedral voids}} = 8$$

Hence, the number of carbon atoms per unit cell of the diamond unit cell is 8.

39 (*d*) Schottky defect lowers the density of a crystal. It is the vacancy defect in ionic solids. This defect arises when equal number of cations and anions are missing from the lattice and thus, electrical neutrality is maintained, e.g. CsCl, NaCl, KCl, AgBr in which cations and anions are of almost same size.

40 (*a*) For fcc lattice, $4r = \sqrt{2}a$

$$r = \frac{\sqrt{2}a}{4} = \frac{a}{2\sqrt{2}}$$

$$= \frac{408}{2\sqrt{2}} = 144\ pm$$

Diameter, $$d = 2r = 2 \times 144\ pm$$

$$= 288 = pm$$

41 (*a*) Number of octahedral voids = Number of atoms in the close packed structure

Since, number of atom = 1

So, number of octahedral void = 1

42 (*d*) From ccp arrangement, number of O^{2-} ions = 4

Number of octahedral voids = 4

Thus, number of tetrahedral voids = 4 × 2 = 8

Number of A^{2+} ions $= \frac{1}{4} \times 8 = 2$

Again, B^+ ions occupies all octahedral voids

$\therefore$ Number of B ions = 4

$$A : B : O = 2 : 4 : 4 = 1 : 2 : 2$$

$\therefore$ Structure of oxide $= AB_2O_2$

43 (*c*) Rock salt (NaCl) is ionic solid. Ice and dry ice are molecular solids and quartz (SiO_2) is covalent or network solids.

44 (*b*) 8 corners $\times \frac{1}{8}$ per corner sodium atom

$$= 8 \text{ corners} \times \frac{1}{8} \text{per corner}$$

$$= 8 \times \frac{1}{8} = 1$$

12 edges $\times \frac{1}{4}$ per edge Oxygen atom $= 12 \times \frac{1}{4} = 3$

Tungsten atom in the centre of cube = 1

Hence, chemical formula = $NaWO_3$.

45 (*c*) $n\lambda = 2a \sin \theta$

$$1 \times 1.54 = 2\,a \sin 45°$$

$$1 \times 1.54 = 2\,a \times \frac{1}{\sqrt{2}}$$

$$a = 1.089\ Å$$

CHAPTER 02

Solution

A solution or a true solution is a homogeneous mixture of two or more substances, whose composition can be varied within certain limits. A homogeneous mixture means that its composition and properties are uniform throughout the mixture.

In a solution, generally, the component that is present in the largest quantity is known as **solvent**. One or more components present in the solution other than solvent are called **solutes.** Solvent determines the physical state in which solution exists. Lemon water, air, etc. are the examples of solution.

Important characteristics of true solution are as follows

(*i*) A solution is a single phase system.

(*ii*) A solution does not allow a beam of light to scatter and the particles of solute in solution cannot be seen by naked eye.

(*iii*) A solution is stable and the solute from the solution cannot be separated by filtration process.

(*iv*) Solution is completely uniform, hence its properties are similar, e.g. density, refractive index.

(*v*) The composition of solution is variable and can be changed within certain limits of temperature and pressure.

Inside

- **Types and concentration of solution**
  - *– Concentration of solution*
- **Solubility**
  - *– Solubility of gases in liquids*
  - *– Henry's law*
  - *– Solubility of solids in liquids*
- **Vapour pressure & Raoult's law**
  - *– Ideal solution*
  - *– Non-ideal solution*
- **Colligative properties of solutions**
  - *– Relative lowerting of vapour pressure*
  - *– Elevation in boiling point*
  - *– Depression in freezing point*
  - *– Osmotic pressure*
  - *– Abnormal molar mass*
  - *– van't Hoff factor*

TYPES AND CONCENTRATION OF SOLUTION

Types of solutions

Solutions are classified on following two basis

I. On the basis of quantity of components, solutions are classified in the following way

(*i*) **Saturated solution** It is the solution containing the maximum possible quantity of solute dissolved in a solvent at a certain temperature.

(*ii*) **Unsaturated solution** Lower quantity of solute than saturated solution is dissolved in this solution. It can be further divided into two types :

- **Dilute solution** Solution containing lower quantity of solute.
- **Concentrated solution** Solution containing higher quantity of solute but still lower than the quantity of a saturated solution.

(*iii*) **Supersaturated solution** At constant temperature, quantity of solute is greater than that in saturated solution is known as supersaturated solution.

II. On the basis of physical states of components, there are three types of solutions which are tabulated below

Table 2.1 Different types of solution

Types of solutions	Solute	Solvent	Examples
Gaseous solutions	Gas	Gas	Mixture of nitrogen and oxygen gases
	Liquid	Gas	Chloroform mixed with nitrogen gas, fog
	Solid	Gas	Camphor in nitrogen gas, dust or smoke
Liquid solutions	Gas	Liquid	Oxygen dissolved in water
	Liquid	Liquid	Ethanol dissolved in water
	Solid	Liquid	Glucose dissolved in water
Solid solutions	Gas	Solid	Solution of hydrogen in palladium
	Liquid	Solid	Amalgam of mercury with sodium
	Solid	Solid	Copper and gold alloy

Concentration of solution

The concentration of a solution is defined as the relative amount of a solute present in a solution. It is generally expressed as the amount of solute present in a unit volume of solution.

Various expression for the percentage concentrations of solutions can be summarised as

1. Mass percentage (*w/W*)

The mass percentage of a component in a given solution is the mass of the component per 100 g of the solution.

$$\text{Mass \% of a component} = \frac{\text{mass of the component in the solution}}{\text{total mass of the solution}} \times 100$$

e.g. A 10% glucose solution in water means that 10 g of glucose is dissolved in 90 g of water resulting in a 100 g solution.

Mass percentage is used in industrial chemical applications.

2. Volume percentage (*v/V*)

The volume percentage is the volume of the component per 100 parts by volume of the solution.

$$\text{Volume \% of a component} = \frac{\text{volume of the component}}{\text{total volume of solution}} \times 100$$

e.g. 10% by volume of ethanol solution means that 10 mL of ethanol is dissolved in enough water, so that the total volume of the solution is 100 mL.

Solutions containing liquids are commonly expressed in volume percentage.

3. Mass by volume percentage (*w/v*)

It is the mass of solute dissolved in 100 mL of the solution.

e.g. A 10% solution of sodium chloride (*w/V*) means that 10 g of sodium chloride is dissolved in 100 mL of solution. This unit is commonly used in medicine and pharmacy.

4. Parts per million (ppm)

When a solute is present in trace quantities, the concentration is expressed in **parts per million** (ppm)

$$\text{ppm} = \frac{\text{number of parts of the component}}{\left[\begin{array}{c}\text{total number of parts of all}\\ \text{components of the solution}\end{array}\right]} \times 10^6$$

Concentration in parts per million can also be expressed as mass to mass, volume to volume and mass to volume. The concentration of pollutants in water or atmosphere is often expressed in terms of $\mu g\ mL^{-1}$ or ppm.

5. Parts per billion (ppb)

It is the amount of solute, which is dissolved in 10^9 parts of a solution.

$$\text{ppb} = \frac{\text{mass of a solute particles}}{\text{mass of a solution}} \times 10^9$$

Example 2.1 *Calculate the mass percentage of benzene and carbon tetrachloride, if* 22 g *of benzene is dissolved in* 122 g *of* CCl_4.

Sol. Total mass of solution

= mass of benzene in solution + mass of carbon tetrachloride in solution

$= (22 + 122)\ g = 144\ g$

$$\text{Mass \% of benzene} = \frac{\text{mass of benzene in solution}}{\text{total mass of solution}} \times 100$$

$$= \frac{22}{144} \times 100 = 15.28\%$$

$$\text{Mass \% of } CCl_4 = \frac{\text{mass of } CCl_4 \text{ in solution}}{\text{total mass of solution}} \times 100$$

$$= \frac{122}{144} \times 100 = 84.72\%$$

Example 2.2 *Glucose solution is 20 % by weight of solution. What is its percentage concentration by weight of solvent (water) ?*

Sol. *w*% percentage concentration by weight of solution means 100 g solution has *w* g solute.

Thus, glucose = 20 g

water + glucose (solution) = 100 g

∴ water = 80 g

Percentage concentration by weight of solvent means how much solute is present in 100 g solvent.
Thus, percentage concentration by weight of solvent

$$= \frac{20}{80} \times 100 = 25\ \%$$

6. Concentration or strength in g/L

Strength in g/L represents the number of grams present in 1L of solution.

$$\text{Strength of solution} = \frac{\text{mass of solute (in g)}}{\text{volume of solution (in L)}}$$

7. Mole fraction (χ)

It is the ratio of the number of moles (n) of one component (either solute or solvent) to the total number of moles of solute and solvent.

$$\boxed{\chi_{\text{solute}} = \frac{n_{\text{solute}}}{n_{\text{solute}} + n_{\text{solvent}}}}$$

$$\boxed{\chi_{\text{solvent}} = \frac{n_{\text{solvent}}}{n_{\text{solute}} + n_{\text{solvent}}}}$$

Sum of mole fractions of solute and solvent is always unity

$$\boxed{\chi_{\text{solute}} + \chi_{\text{solvent}} = 1}$$

8. Molality (*m*)

It is the number of moles of the solute dissolved in 1000 g of the solvent.

$$\text{Molality } (m) = \frac{\text{number of moles of solute}}{\text{mass of solvent (in g)}} \times 1000$$

$$\boxed{m = \frac{\text{mass of solute (in g)} \times 1000}{\text{molecular weight of solute} \times \text{mass of solvent (in g)}}}$$

(*i*) Molality is independent of temperature

(*ii*) Unit of molality = mol/kg of solvent

9. Normality (*N*)

Normality is the number of gram equivalents of substance dissolved per litre of the solution.

$$\text{Normality } (N) = \frac{\text{number of g-equivalent of solute}}{\text{volume of solution (in L)}}$$

$$= \frac{\text{mass of solute (in g)}}{\text{equivalent weight} \times \text{volume (in L)}}$$

$$N = \frac{\text{mass of solute } (w) \text{ (in g)}}{\text{equivalent weight } (E) \times \text{volume } (V) \text{ (in mL)}} \times 1000$$

or $\boxed{N = \frac{w \times 1000}{E \times V}}$

(*i*) Normality changes with temperature since it involves volume.

(*ii*) Unit of normality = gram equivalents per litre.

Predicting normality of different solutions (Dilution law)

If normality and volume of solution are changed from N_1, V_1 to N_2, V_2 then

$$\boxed{N_1V_1 = N_2V_2} \quad \text{(Normality equation)}$$

If n_1 moles of reactant '1' react with n_2 moles of reactant '2', then

$$\frac{N_1V_1}{n_1} = \frac{N_2V_2}{n_2}$$

If two solutions of the same solute are mixed then normality of the resulting solution

$$N = \frac{N_1V_1 + N_2V_2}{V_1 + V_2}$$

Volume of water added to get a solution of normality N_2 from V_1 mL of normality N_1 is

$$\boxed{V_2 - V_1 = \left(\frac{N_1 - N_2}{N_2}\right) V_1}$$

Example 2.3 *Calculate the amount of* H_2SO_4 *in* 150 mL *of* $N/7\ H_2SO_4$ *solution.*

Sol. Equivalent weight (E) of $H_2SO_4 = \frac{98}{2} = 49$

Amount of H_2SO_4,

$$w = \frac{N \times E \times V}{1000} = \frac{1}{7} \times \frac{49 \times 150}{1000};\ w = 1.05 \text{ g}$$

Example 2.4 *Calculate the normality of a solution obtained by mixing* 0.45 N *and* 0.60 N NaOH *in the ratio* 2 : 1 *by volume?*

Sol. Gram-equivalent weight $= 0.45 \times 2V + 0.60 \times V$

$\because$ Total volume $= 2V + V = 3V$

$\Rightarrow$ Gram-equivalent weight of NaOH solution required

$$= N \times 3V$$

$$N \times 3V = 0.45 \times 2V + 0.6V$$

$$3NV = 0.9V + 0.6V$$

$$N = \frac{1.5}{3} = 0.5$$

or Alternatively, $N_{\text{mix}} = \frac{N_1V_1 + N_2V_2}{V_1 + V_2}$

$$= \frac{0.45 \times 2 + 0.60 \times 1}{3} = 0.5$$

10. Molarity (*M*)

It is the number of moles of solute present in one litre of the solution.

$$\text{Molarity } (M) = \frac{\text{number of moles of solute}}{\text{volume of solution (in L)}}$$

$$= \frac{\text{mass of solute (in g)}}{\text{molecular weight of solute} \times \text{volume of solution (in L)}}$$

$$= \frac{\text{number of millimoles of solute}}{\text{volume of solution (in mL)}}$$

$$= \frac{\text{per cent of solute} \times 10}{\text{molecular weight of solute}}$$

$$= \frac{\text{weight \% of solute} \times \text{density of the solution} \times 10}{\text{molecular weight of the solute}}$$

- Molarity varies with temperature due to accompanied change in volume of the solution.
- Unit of molarity = g-mol/L

Predicting molarity of different solutions

If molarity and volume of solution are changed from M_1, V_1 to M_2, V_2 then,

$$M_1V_1 = M_2V_2 \quad \text{(Molarity equation)}$$

If n_1 moles of reactant '1' react with n_2 moles of reactant '2' then,

$$\frac{M_1V_1}{n_1} = \frac{M_2V_2}{n_2}$$

If two solutions of the same solute are mixed then molarity of the resulting solution,

$$M = \frac{M_1V_1 + M_2V_2}{V_1 + V_2}$$

Volume of water added to get a solution of molarity M_2 from V_1 mL of molarity M_1 is

$$\boxed{V_2 - V_1 = \left(\frac{M_1 - M_2}{M_2}\right) V_1}$$

Example 2.5 *Calculate the molarity of solution of* $CaCl_2$, *if on chemical analysis it is found that* 200 mL *of* $CaCl_2$ *contains* 3.01×10^{22} Cl^- *ions.*

Sol. $\underset{1 \text{ mol}}{CaCl_2} \longrightarrow Ca^{2+} + \underset{2 \times 6.023 \times 10^{23} \text{ ions}}{2Cl^-}$

$\because$ $2 \times 6.023 \times 10^{23}$ Cl^- ions are present in 1 mol of $CaCl_2$

$\therefore$ 3.01×10^{22} Cl^- ions are present in

$$= \frac{1 \times 3.01 \times 10^{22}}{2 \times 6.02 \times 10^{23}} \text{ mol of } CaCl_2$$

$$= 0.025 \text{ mol of } CaCl_2$$

$$M = \frac{\text{number of moles}}{\text{volume of solution (in L)}}$$

$$= \frac{0.025}{200} \times 1000 = 0.125 \text{ M}$$

11. Formality (*F*)

This is the concentration unit for ionic compounds which dissolve in a polar solvent to give pair of ions. This represents number of gram-formula weight of the substance dissolved per litre of the solution. It is almost the same as molarity.

$$\text{Formality} = \frac{\text{moles of substance added to solution}}{\text{volume of solution (in L)}}$$

(*i*) If formula weight is equal to molecular weight, formality and molarity will be same.

(*ii*) Formality is also **dependent on temperature.**

Example 2.6 *Calculate the formality of solution of* 58.5 g NaCl *in* 1 *litre of solution.*

Sol. Formula mass of NaCl = 23 + 35.5 = 58.5

Given mass = 58.5 gm

$$\text{Formality} = \frac{\text{mass of NaCl}}{\text{formula mass} \times \text{volume of solution (in L)}}$$

$$= \frac{58.5}{58.5} \times 1 = 1 \text{F}$$

Inter-relation between different concentration of solution

Various expressions for concentration of solutions can be related as given below

1. **Relation between molality** (*m*) **and weight percentage** (*W*)

$$= \frac{1000 \times \text{weight \% of solute } (x)}{(100 - x) \times \text{molecular weight of solute}}$$

2. **Relation between molality** (*m*) **and mole fraction** (χ)

$$\chi = \frac{m M_1}{(1000 + m M_1)}$$

or

$$m = \frac{1000\chi}{(1-\chi)M_1}$$

Where, M_1 = molecular mass of the solvent

3. **Relation between molality** (*m*) **and molarity** (*M*)

$$\text{Molality } (m) = \frac{M}{d - MM_2}$$

(If, d in kg L^{-1} and M_2 in kg mol^{-1})

$$\text{Molality } (m) = \frac{M}{1000d - MM_2} \times 1000$$

(If, d in g mL^{-1} and M_2 in g mol^{-1})

$$\text{Molarity } (M) = \frac{m \times d}{1 + \frac{m \times M_2}{1000}}$$

Where, M_2 = molar mass of the solute
d = density of the solution

4. **Relation between normality (N) and strength in gL^{-1}**

$$N = \frac{x \times d \times 10}{E}$$

Where, x = % of the solute by mass
d = density of solution in $g\,mL^{-1}$
E = equivalent mass of solute

5. **Relation between molarity and strength in g/L**

$$\text{Molarity } (M) = \frac{\text{weight of solute (in g)}}{\text{molecular weight of solute} \times \text{volume of solution (in L)}}$$

$$\text{Molarity } (M) = \frac{\text{strength in g L}^{-1}}{\text{molecular weight of solute}}$$

6. **Relation between normality and molarity**

$$\frac{\text{Normality}}{\text{Molarity}} = \frac{\text{molecular weight}}{\text{equivalent weight}}$$

For an acid, $\frac{\text{molecular weight}}{\text{equivalent weight}} = \text{basicity}$

$\therefore$Normality of an acid = molarity × basicity

For a base, $\frac{\text{molecular weight}}{\text{equivalent weight}} = \text{acidity}$

$\therefore$Normality of a base = molarity × acidity

7. **Relation between molarity (M) and mole fraction of the solute (χ)**

$$\chi = \frac{MM_1}{M(M_1 - M_2) + 1000d}$$

Where,
d = density of the solution $g\,mL^{-1}$
M_1 = molar mass of solvent
M_2 = molar mass of solute
M = molarity of solution

Example 2.7 *What mass (in grams) of a* 0.500 *molal solution of sodium acetate in water would you use to obtain* 0.150 *mol of sodium acetate?*

Sol. CH_3COONa solution is 0.5 molal, it means 0.5 mol CH_3COONa (molar mass = 82 g mol^{-1}) or 41 g are in 1000 g water.

or 0.5 mol CH_3COONa are in = (1000 + 41)
= 1041 g solution

Hence, 0.15 mol CH_3COONa are in $= \frac{1041}{0.5} \times 0.15$
= 312.3 g solution

Example 2.8 *Molarity of* H_2SO_4 *is* 18 M. *If density d is* 1.8 g cm^{-3}, *calculate its concentration in* mol kg^{-1} *solvent* (H_2O), *i.e. molality.*

Sol. Molality (m) is related to molarity M by equation,

$$m = \frac{1000\,M}{(1000d - MM_2)}$$

Where, m_1 is the molar mass of H_2SO_4

$$m = \frac{1000 \times 18}{(1000 \times 1.8 - 18 \times 98)}$$
$$= 500 \text{ molal}$$

Example 2.9 *Mole fraction of ethanol in ethanol-water system is* 0.25. *What is the per cent concentration of ethanol by weight of solution?*

Sol. Mole fraction of ethanol (χ_1) = 0.25

$\therefore$ Mole fraction of H_2O (χ_2) = 0.75

$$\chi_1 = \frac{n_1}{n_1 + n_2},$$

$$\chi_2 = \frac{n_2}{n_1 + n_2}$$

$$\therefore \quad \frac{\chi_1}{\chi_2} = \frac{n_1}{n_2}$$

$$\frac{w_1}{m_1}\frac{m_2}{w_2} = \frac{1}{3}$$

$$\frac{w_1}{46} \times \frac{18}{w_2} = \frac{1}{3}$$

$$\therefore \quad \frac{w_1}{w_2} = \frac{46}{54}$$

Thus, if w_1 (ethanol) is 46 g then w_2 (water) is 54 g in ($w_1 + w_2$) = 100 g solution.

Per cent concentration of ethanol in solution $= \frac{46}{100} \times 100 = 46\%$

Example 2.10 *An 'm' molal solution of a compound in benzene has mole fraction of solute equal to* 0.2. *Calculate the value of 'm'.*

Sol. Mole fraction of solute is given by,

$$\chi = \frac{mM_1}{1000 + mM_1}$$

$$\Rightarrow \quad 0.2 = \frac{m \times 78}{1000 + m \times 78}$$

$$(0.2 \times 1000) + (0.2 \times 78m) = 78\,m$$
$$62.4\,m = 200$$
$$m = \frac{200}{62.4} = 3.205$$

CHECK POINT 2.1

1. Which of the following statements is incorrect about a true solution?
(a) A solution is a multiphase system
(b) A solution is a homogeneous mixture
(c) Solute cannot be separated from solvent by simple filtration
(d) A solution does not scatter the beam of light

2. The correct order of concentration of different solutions, *viz*. dilute (*A*), concentrated (*B*), saturated (*C*) and supersaturated (*D*) will be
(a) $A > B > C > D$ (b) $A > D > C > B$
(c) $C > B > D > A$ (d) $D > C > B > A$

3. Which of the following is an example of liquid in gas solution ?
(a) Clean air (b) Fog (c) Smoke (d) Dust

4. Salt water is an example of solution of type
(a) gas in liquid (b) liquid in liquid
(c) solid in liquid (d) liquid in solid

5. 10 g of table salt is dissolved 90 g of water what is the mass percentage of the salt in the solution?
(a) 5% (b) 10% (c) 11.11% (d) 20%

6. Glucose solution is 25% by weight of solution. Its per cent concentration by weight of solvent is
(a) 20% (b) 25%
(c) 33.3% (d) 16.66%

7. The unit in which parts per million (ppm) can be expressed is
(a) $g\ L^{-1}$ (b) $mg\ mL^{-1}$
(c) $\mu g\ L^{-1}$ (d) $\mu g\ mL^{-1}$

8. What mass of solution of 0.50 molal glucose solution is required to obtain 0.15 mol glucose?
(a) 327 g (b) 300 g (c) 273 g (d) 373 g

9. Sum of mole fraction of all the solutes and that of a solvent in a solution is always
(a) 0 (b) 1
(c) 100 (d) ∞

10. Which of the following concentration term does not change with change in temperature?
(a) Molality (b) Molarity
(c) Formality (d) Normality

SOLUBILITY

The maximum amount of a solute that can be dissolved in a given amount of solvent (generally 100 g) at a given temperature is termed as **solubility** at that temperature.

Solubility of gases in liquids

All gases are soluble in water as well as in other liquids to a greater or smaller extent, e.g. gases like H_2, O_2, N_2 etc. dissolves in water to a small extent whereas the gases like NH_3, HCl, SO_2 etc. are highly soluble in water.

Factors affecting solubility of gases

(*i*) **Nature of the gas** The gases, which can be easily liquified are more soluble in common solvents, e.g. CO_2 is more soluble than H_2 and O_2 in water.

(*ii*) **Nature of the solvent** The gases which are capable of forming ions in aqueous solutions are much more soluble in water than in other solvents, e.g. gases like HCl and NH_3 are highly soluble in water but not in organic solvents in which they do not ionise.

(*iii*) **Temperature** The solubility of most gases in liquids decreases with increase of temperature. When a solution of a gas is heated, the gas is usually expelled. However, some gases are more soluble at higher temperature than at lower temperature.

(*iv*) **Pressure** The solubility of a gases increase with increase in pressure. As pressure increases, gaseous molecules come closer and get easily solubilised in solvent. Solubility of gas in liquid varying with change in pressure is governed by the Henry's law.

Henry's law

Henry was the first one to give a quantitative relation between **pressure and solubility** of a gas in a solvent and accordingly gave a law, which is called **Henry's law**.

It states that, "At a constant temperature, the solubility of a gas in a liquid is directly proportional to the partial pressure of the gas present above the surface of liquid or solution".

$$m \propto p \text{ or } m = K'p$$

Where, m = dissolved quantity of gas
p = pressure of gas at equilibrium
K' = Henry's law constant

In other words, "The partial pressure of the gas in vapour phase (p) is proportional to the mole fraction of the gas (χ) in the solution".

$$p \propto \chi \quad \text{or} \quad p = K_H \cdot \chi$$

Where, K_H is called Henry's law constant.

Hence, $K_H = \frac{1}{K'}$

Value of K_H depends on the nature of the gas at constant temperature. If, Henry's law is true then graph between partial pressure and mole fraction is a straight line with slope K_H.

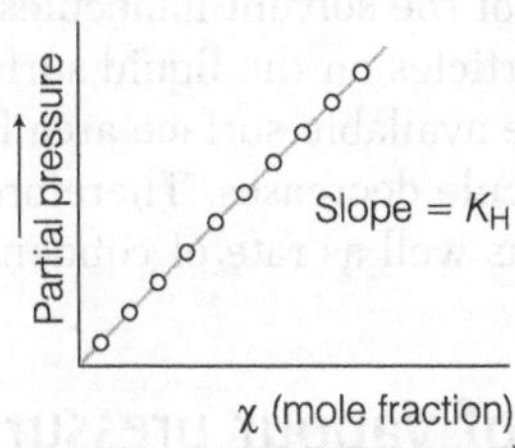

Fig. 2.1 Graph between partial pressure and mole fraction

- To increase the solubility of CO_2 in soft drinks and soda water, the bottle is sealed under high pressure.
- Higher the value of K_H at a given pressure, the lower is the solubility of the gas in the liquid.

Different gases have different K_H values at the given temperature. Values of K_H (Henry's law constant) have been given for some gases in water.

Gas	Temp/K	K_H (kbar)
He	293	145.0
H_2	293	69.0
N_2	293	76.5
N_2	303	88.8
O_2	293	35.0
O_2	303	47.0

From the above values, it is clear that K_H values for both N_2 and O_2 increases with increase of temperature and thus, solubility of gases decreases with increase in temperature. Thus, aquatic species are more comfortable in cold water rather than in warm waters.

Limitations

Although, it is an important law of solubility of gases but it is applied only, when

(*i*) pressure is not in excess.

(*ii*) temperature is not at the lowest.

(*iii*) gases are feebly soluble in solvent.

(*iv*) the atomic state of gases is similar in both the liquid and gaseous states.

Applications

(*i*) **In the production of carbonated beverages** To increase the solubility of CO_2 in soft drinks and soda water, the bottle is sealed under high pressure.

(*ii*) **In the deep sea diving** Scuba divers (deep sea divers) breathe air at high pressure under water. Increased pressure, increases the solubility of atmospheric gases in blood. When divers come towards surface, the pressure gradually decreases resulting in the release of dissolved gases. This leads to the formation of bubbles of nitrogen in the blood, which blocks capillaries and gives rise to disease called **bends** or **decompression sickness.** To minimise this effect, O_2 diluted with less soluble the gas is used as breathing gas.

(*iii*) **Respiratory problem at high altitudes** At high altitudes, the partial pressure of oxygen is less than that at the ground level. This leads to low concentration of oxygen in the blood and tissues of people living at high altitudes or climbers. Consequently, they feel weak both physically and mentally. This medical condition is known as **anoxia.**

Example 2.11 For a solution of acetone in chloroform, Henry's law constant is 150 torr at a temperature of 300 K. Calculate the vapour pressure of acetone when the mole fraction is 0.12.

Sol. The given values are $K_H = 150$ torr

$$\chi_{acetone} = 0.12$$

We have the Henry's law equation,

$$p = K_H \times \chi$$

$$p = 150 \text{ torr} \times 0.12 = 18.0 \text{ torr}$$

Solubility of solids in liquids

Some solids are readily soluble in a liquid while others are least soluble. When solute dissolves and its concentration increases in solution, the process is known as **dissolution.** When solute particles in solution colloide with the solid particles and get separated out of solution, the process is called **crystallisation.**

Factors affecting the solubility of solids in liquids

(*i*) **Nature of solute and solvent** In general, a solute dissolves in a solvent, if the intermolecular interactions are similar in the two then it said to be 'like dissolves like'.

It means that, polar (i.e. ionic) compounds (like NaCl) dissolve in polar solvents (like water) and almost insoluble in non-polar solvents (like benzene). Similarly, non-polar (i.e. covalent or organic) compounds (like naphthalene, anthracene) are soluble in non-polar solvents (like benzene, ether) and are very less soluble in water.

Solubility is due to solute-solute, solute-solvent and solvent-solvent interactions. When these three kinds of interactions are similar, solution is formed.

(*ii*) **Effect of temperature** The solubility of a solid in a liquid is significantly affected by temperature change. In general, if in a nearly saturated solution, the dissolution process is endothermic, i.e. $\Delta_{sol}H > 0$, the solubility should increase with the rise in temperature and if it is exothermic, i.e. $\Delta_{sol}H < 0$, the solubility should decrease in accordance with Le-Chatelier's principle.

(*iii*) **Effect of pressure** Pressure does not have any significant effect on solubility of solids in liquids because solids and liquids are highly incompressible and practically remain unaffected by the pressure change.

VAPOUR PRESSURE AND RAOULT'S LAW

The pressure exerted by the vapour molecules above the liquid surface in equilibrium with the liquid at a given temperature is called **vapour pressure.**

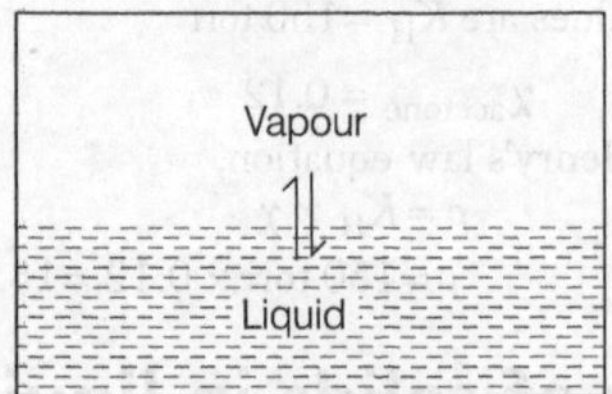

Fig. 2.2 Vapour pressure of a liquid

$$H_2O(l) \rightleftharpoons H_2O(g); \ \Delta H > 0$$

Above equilibrium is endothermic process. Hence, increase in temperature increases vapour pressure. Temperature at which vapour pressure of a liquid is equal to atmospheric pressure is called **boiling point** of liquid. Liquid solutions are formed when solvent is a liquid. The solute can be a gas or a liquid or a solid. In the solutions of liquids and solids in a liquid, generally, the liquid solvent is volatile. The solute may or may not be volatile.

Factors affecting vapour pressure

Vapour pressure gets affected by following factors

(*i*) **Purity of the liquid** Pure liquid always has a vapour pressure higher than its solution.

(*ii*) **Nature of the liquid** Liquids, which have weak intermolecular forces are volatile and have greater vapour pressure.

(*iii*) **Temperature** The vapour pressure of a liquid increases with increase in temperature. On increasing the temperature the kinetic energy of molecules increases and more number of molecules of the liquid go into vapour phase.

(*iv*) **Effect of adding solute** When a liquid contains a solute, some of the solvent molecules are replaced by the solute particles on the liquid surface and therefore, the available surface area for the escape of solvent molecule decreases. Therefore, rate of evaporation as well as rate of condensation both decrease.

Composition of vapour pressure

If the mole fraction of component A and B in vapour state are χ_A and χ_B.

Then, $$p_i = \chi_i p_T \quad \text{(by Dalton's law)}$$

Where, p_i = vapour phase in ith state.
χ_i = mole fraction of component in ith state
p_T = total pressure

Vapour pressure of liquid-liquid solution

In a binary solution of two **volatile** liquids in a closed vessel both the components evaporated and finally a state of equilibrium is reached between the vapour and liquid phase.

According to the Dalton's law of partial pressure, the total vapour pressure (p_T) in this case is equal to the sum of the partial pressures (p_A and p_B) of each of the two components.

i.e. $$p_T = p_A + p_B$$
$$p_A = \chi_A p_T$$
$$p_B = \chi_B p_T$$

Raoult's law

The French chemist, **Francois Marte Raoult** gave a quantitative relationship between the partial pressures and the mole fractions of two components. This relationship is known as Raoult's law which states that, "At a given temperature, for a solution of volatile liquids, the partial vapour pressure of each component of the solution is directly proportional to its mole fraction present in solution".

For a solution containing two liquids A and B, the partial vapour pressure of liquid A is

$$p_A \propto \chi_A \text{ or } p_A = K\chi_A$$

Where, χ_A = mole fraction of liquid A

or $$\chi_A = \frac{n_A}{n_A + n_B}$$

Where,

n_A and n_B = number of moles of liquid A and B respectively.

If $\chi_A = 1$, then $p_A = p_A^\circ$

(vapour pressure of pure solvent A)

Thus, $K = p_A^\circ$

$\therefore \quad p_A = p_A^\circ \times \chi_A$

Similarly, $p_B = p_B^\circ \chi_B$

From Dalton's law of partial pressure,

$$p_T = p_A^\circ \chi_A + p_B^\circ \chi_B$$

$$p_T = p_A^\circ + (p_B^\circ - p_A^\circ)\chi_B \qquad [\because \chi_A + \chi_B = 1]$$

Mole fraction of A in vapour phase,

$$\chi'_A = \frac{p_A}{p_{\text{total}}} = \frac{p_A^\circ \chi_A}{p_A^\circ \chi_A + p_B^\circ \chi_B}$$

Mole fraction of B in vapour phase,

$$\chi'_B = \frac{p_B}{p_{\text{total}}} = \frac{p_B^\circ \chi_B}{p_A^\circ \chi_A + p_B^\circ \chi_B} = (1 - \chi'_A)$$

Raoult's law as a special case of Henry's law

Raoult's law becomes a special case of Henry's law in which the partial pressure of the volatile component is directly proportional to its mole fraction in the solution where, K_H becomes equal to p_i°.

$$p = K_H \chi \qquad \text{(Henry's law)}$$

$$p_i = \chi_i p_i^\circ \qquad \text{(Raoult's law)}$$

Limitations

(*i*) It is applicable only to very dilute solutions.

(*ii*) It is applicable only to solutions containing non-volatile and non-electrolytic solutes, which exist as a single molecule.

(*iii*) It is not applicable to solutes, which dissociate or associate in the particular solvent.

On the basis of Raoult's law, liquid-liquid solutions can be divided into two type of solution, i.e. ideal and non-ideal solution, as discussed below.

Ideal solution

Ideal solutions are those which obey Raoult's law over all concentration ranges at a given temperature. The total vapour pressure of an ideal solution containing components A and B is represented as

$$p_T = p_A + p_B \quad \text{or} \quad p_T = p_A^\circ \chi_A + p_B^\circ \chi_B$$

e.g. Benzene-toluene, n-hexane-n-heptane, ethyliodide-ethylbromide, chlorobenzene -bromobenzene.

Ideal solutions can be represented graphically as

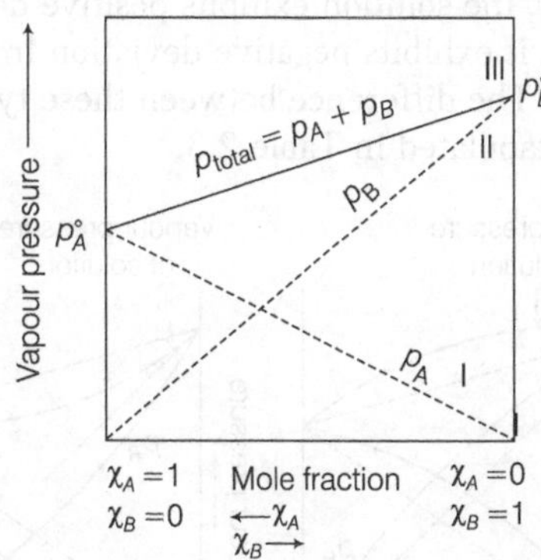

***Fig. 2.3** Plot of vapour pressure and mole fraction of an ideal solution at constant temperature*

Curve I Variations of vapour pressure of liquid A with mole fraction.

Curve II Variations of vapour pressure of liquid B with mole fraction.

Curve III Straight line representing $p_T = p_A^\circ \chi_A + p_B^\circ \chi_B$

Non-ideal solution

Solutions, which do not obey Raoult's law are called **non-ideal solutions.** Hence, for such solutions $p_A \neq p_A^\circ \chi_A$ and $p_B \neq p_B^\circ \chi_B$.

e.g. chloroform – benzene, phenol – aniline

Table 2.2 Difference between ideal and non-ideal solutions

Ideal solution	Non-ideal solution
1. Each component obeys Raoult's law at all temperature and concentrations, i.e. $p_A = p_A^\circ \chi_A$ and $p_B = p_B^\circ \chi_B$	Their components do not obey Raoult's law. They show positive and negative deviations from Raoult's law, i.e. $p_A \neq p_A^\circ \chi_A$ and $p_B \neq p_A^\circ \chi_B$
2. On mixing, there is no enthalpy change, $\Delta_{\text{mixing}} H = 0$	On mixing, there is an enthalpy change, $\Delta_{\text{mixing}} H \neq 0$
3. On mixing, there is also no change in volume, $\Delta_{\text{mixing}} V = 0$	On mixing, there is volume change also, $\Delta_{\text{mixing}} V \neq 0$
4. Interactions between components are similar to those in pure components.	Interactions between components are different from those of pure components.

Deviation shown by solutions from Raoult's law

Since, the vapour pressures of non-ideal solutions are either higher or lower than that predicted by Raoult's law. If it is higher, the solution exhibits positive deviation and if it is lower, it exhibits negative deviation from Raoult's law. The difference between these type of solution are tabulated in Table 2.3.

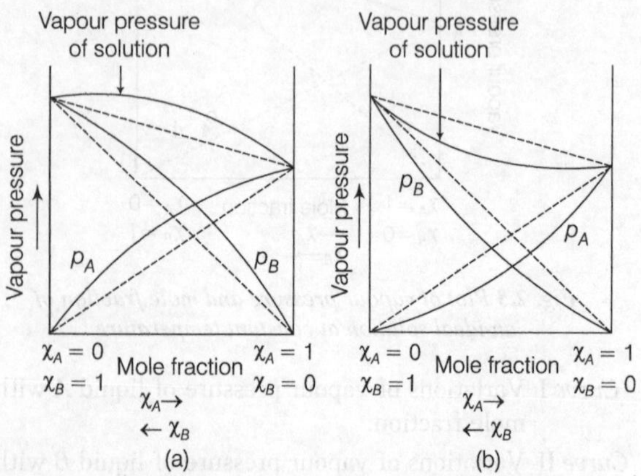

Fig. 2.4 The vapour pressures of two components systems as a function of composition
(a) a solution that shows positive deviation from Raoult's law
(b) a solution that shows negative deviation from Raoult's law

Table 2.3 Difference between solutions having positive and negative deviation from ideal behaviour

S. No.	Solutions with positive deviation	Solutions with negative deviation
1.	The intermolecular interactions, i.e. A—B interactions are weaker than in pure components, i.e. between A—A or B—B.	The intermolecular interactions, i.e. A—B interactions are stronger than in pure components, i.e. between A—A or B—B.
2.	$p_A > p_A^\circ \chi_A$ and $p_B > p_B^\circ \chi_B$	$p_A < p_A^\circ \chi_A$ and $p_B < p_B^\circ \chi_B$
3.	$\Delta_{mix} H$ = positive	$\Delta_{mix} H$ = negative
4.	$\Delta_{mix} V$ = positive	$\Delta_{mix} V$ = negative
5.	Heating increases solubility	Heating decreases solubility
6.	**Examples** Ethanol-water, ether-acetone, acetone-benzene, carbon disulphide-acetone, carbon tetrachloride-methanol, etc.	**Examples** Acetic acid-methanol, acetic acid-pyridine, acetone-aniline, etc.

Example 2.12 *The vapour pressure of a pure liquid A is* 50 mmHg *at* 310 K. *The vapour pressure of this liquid in a solution with liquid B is* 30 mmHg. *If the solution obeys Raoult's law then calculate the mole fraction of A in the solution.*

Sol. Given, vapour pressure of pure A, $p_A^\circ = 50$ mmHg

Vapour pressure of A in solution, $p_A = 30$ mmHg

$\therefore$ From Raoult's law $p_A = p_A^\circ \chi_A$

$$\Rightarrow \quad \chi_A = \frac{p_A}{p_A^\circ} = \frac{30 \text{ mmHg}}{50 \text{ mmHg}} = 0.6$$

Example 2.13 *At* 90°C, *the vapour pressure of toluene is* 400 torr *and that of o-xylene is* 150 torr. *What is the composition of the liquid mixture that boils at* 90°C *when the pressure is* 0.50 atm? *What is the composition of the vapour produced?*

Sol. Toluene and *o*-xylene both are volatile liquids and the mixture boils at 90°C, when total pressure is 0.50 atm (= 380 torr)

Let mole fraction of toluene in the mixture be $= \chi_T$

Its vapour pressure in mixture $p_T = p^\circ{}_T\, \chi_T = 400\chi_T$

and vapour pressure of *o*-xylene $p^\circ = 150(1 - \chi_T)$

$$\therefore \quad 400\chi_T + 150(1 - \chi_T) = 380$$

$$\therefore \quad \chi_T = 0.92$$

Hence, mixture contains 92% mol of toluene

mole fraction of toluene in vapour

$$\chi_T' = \frac{p^\circ{}_T\, \chi_T}{p_{total}} = \frac{400 \times 0.92}{380} = 0.968$$

Hence, mixture contains 96.8 mol% toluene in vapour.

Example 2.14 *Heptane and octane form an ideal solution. At* 373 K, *the vapour pressures of the two liquid components are* 105.2 kPa *and* 46.8 kPa *respectively. What will be the vapour pressure of a mixture of* 25.0 g *of heptane and* 35 g *of octane?*

Sol. Moles of heptane $(C_7H_{16}) = \frac{25}{100} = 0.25$

Moles of octane $(C_8H_{18}) = \frac{35}{114} = 0.307$

Total moles = 0.557

$\chi_{heptane} = 0.4488$

$\chi_{octane} = 0.5512$

$$p_{total} = \chi_H\, p_H^\circ + \chi_O\, p_O^\circ$$

$$= 0.4488 \times 105.2 + 0.5512 \times 46.8$$

$$= 73.01 \text{ kPa}$$

Azeotropic mixture or constant boiling mixture

The binary liquid mixtures having the same composition in liquid and vapour phase and boil at constant temperature are called **azeotropic mixtures** or **azeotropes**. In such cases, it is not possible to separate the components by fractional distillation.

There are two types of azeotropes called minimum boiling azeotrope and maximum boiling azeotropes.

- **Minimum boiling azeotrope** The solutions, which show a large positive deviation from Raoult's law form minimum boiling azeotrope at specific composition. e.g. Ethanol-water mixture on fractional distillation gives a solution containing approximately 95% by volume of ethanol. After achieving this composition, no further separation occurs as the liquid and vapour have the same composition.

- **Maximum boiling azeotrope** The solutions, which show a large negative deviation from Raoult's law form maximum boiling azeotrope at a specific compositions. e.g. Azeotrope formed from nitric acid and water has the approximate composition of 68% nitric acid and 32% water by mass with a boiling point of 393.5 K.

CHECK POINT 2.2

1. Solubility of a gas in liquids depends upon
   (a) nature of the gas (b) temperature
   (c) pressure (d) All of these
2. Mass of a certain gas dissolved in a liquid is directly proportional to
   (a) temperature
   (b) pressure
   (c) Both (a) and (b)
   (d) inversely proportional to both (a) and (b)
3. The solubility of $N_2(g)$ in water exposed to the atmosphere, when the partial pressure is 593 mm is 5.3×10^{-4} M. Its solubility at 760 mm and at the same temperature is
   (a) 4.1×10^{-4} M (b) 6.8×10^{-4} M
   (c) 1500 M (d) 2400 M
4. In case of true solution of a solid in liquid the interactions among solute particles (say A) and solvent particles (say B) should be like
   (a) $A—A = B—B > A—B$
   (b) $A—A = B—B < A—B$
   (c) $A—A = B—B = A—B$
   (d) $A—A > B—B > A—B$
5. Which of the following will have least vapour pressure?
   (a) Salt water solution
   (b) Pure ethanol
   (c) Ethanol water solution
   (d) All of the above have same vapour pressure
6. According to Raoult's law, at a given temperature for a solution of volatile liquids, the partial vapour pressure of each component of the solution is directly proportional to its in solution
   (a) mole fraction (b) mole number
   (c) molarity (d) normality
7. The vapour pressure of a liquid in pure state is 50 mmHg while that in solution state is 40 mmHg. Find the mole fraction of that liquid in solution state.
   (a) 0.40 (b) 0.50 (c) 0.60 (d) 0.80
8. Which of the following is not true about the Raoult's law?
   (a) It is applicable to only very dilute solutions
   (b) It is applicable to solution containing non-volatile solute
   (c) It is applicable to solution containing electrolytic solute
   (d) All of the above statements are true
9. Which of the following solution obey Raoult's law at almost all concentration ranges?
   (a) Ideal solution
   (b) Non-ideal solution with positive deviation
   (c) Non-ideal solution with negative deviation
   (d) All of the above solution
10. Which of the following is/are true for an ideal solution
    (a) $\Delta H_{mix} = 0$
    (b) $\Delta V_{mix} = 0$
    (c) $p_A = p_B$
    (d) Both (a) and (b)

COLLIGATIVE PROPERTIES OF SOLUTIONS

The properties of a dilute solution, which depend only upon the number of particles present in a solution and their concentration are called **colligative properties.** They depend on the dative fraction of solute and solvent particles but not on the chemical nature of solute, i.e. these are proportional to the number of solute particles in the solution.

There are following four types of colligative properties

1. Relative lowering of vapour pressure

When a non-volatile solute is added to a solvent, the vapour pressure is lowered. So, vapour pressure of a solution (p) is always less than the vapour pressure of pure solvent ($p°$) at that temperature. Therefore, ($p° - p$) is called the **lowering in vapour pressure** and the ratio $\left(\frac{p° - p}{p°}\right)$ is known as the **relative lowering in vapour pressure.**

Raoult's law for non-volatile solutes

According to Raoult's law, relative lowering in vapour pressure of an ideal solution containing the non-volatile solute (*B*) is equal to the mole fraction of the solute at a given temperature.

$$\frac{p^\circ - p}{p^\circ} = \chi_B = \frac{n_B}{n_A + n_B}$$

Where, χ_B = mole fraction of non-volatile solute (*B*).

Raoult's law for dilute solution

According to Raoult's law,

$$\boxed{\frac{p^\circ - p}{p^\circ} = \frac{n_A}{n_A + n_B}}$$

Where, n_A = moles of solute, n_B = moles of solvent

For dilute solution,

$\therefore \quad n_A <<< n_B$

$$\therefore \quad \frac{p^\circ - p}{p^\circ} = \frac{n_A}{n_B}$$

Determination of molecular weight of solute

Generally, on the basis of Raoult's law, molecular weight of solute in dilute solution can be formulated as

$$\boxed{\frac{p^\circ - p}{p^\circ} = \frac{n_2}{n_1 + n_2} \approx \frac{n_2}{n_1} = \frac{w_2\, m_1}{m_2\, w_1}}$$

where w_2 = mass of solute
m_2 = molecular mass of solute
w_1 = mass of solvent
m_1 = molecular mass of solvent

Example 2.15 *The vapour pressure of* 100 g *water reduces from* 17.53 mm *to* 17.22 mm *when* 17.10 g *of substance X is dissolved in it. Identify the substance.*

Sol. We have the relation between relative lowering of vapour pressure and molar mass.

$$\therefore \quad \frac{p^\circ - p}{p^\circ} = \frac{w_2\, m_1}{m_2\, w_2}$$

$$\frac{17.53 - 17.22}{17.53} = \frac{17.10}{m_2} \times \frac{18}{100}$$

$$m_2 = \frac{17.10 \times 18 \times 17.53}{0.31 \times 100} = 174.05$$

Since, 174 is nearest to the molecular weight of glucose, thus the substance *X* can be glucose.

Example 2.16 *The vapour pressure of water is* 12.3 kPa *at* 300 K. *Calculate the vapour pressure of* 1 molal *solution of a non-volatile solute in it.*

Sol. Solution is 1 molal,

$\therefore$ Solute = 1 mol

$$\text{Water} = 1000 \text{ g} = \frac{1000}{18} \text{ mol} = 55.55$$

Total moles = 56.55

$$\text{Mole fraction of water} = \frac{55.55}{56.55}$$

$$\therefore \quad p_{\text{solution}} = p^\circ_{\text{solvent}} \times \text{mole fraction of water}$$

$$= 12.3 \times \frac{55.55}{56.55}$$

$$= 12.08 \text{ kPa}$$

Example 2.17 *Calculate the mass of a non-volatile solute* (molar 40 g mol^{-1}) *which should be dissolved in* 114 g *octane to reduce its vapour pressure to 80%.*

Sol. If, $p^\circ_{\text{solvent}} = 100$ units then $p_{\text{solution}} = 80$ units

$$\text{Moles of octane} = \frac{114}{114} = 1, \text{ moles of solute} = \frac{w}{40}$$

$$\frac{p^\circ_{\text{solvent}} - p_{\text{solution}}}{p^\circ_{\text{solvent}}} = \chi_{\text{solute}}$$

$$0.2 = \frac{\frac{w}{40}}{\frac{w}{40} + 1} \Rightarrow 0.2 = \frac{w}{w + 40}$$

$$\therefore \quad w = 10 \text{ g}$$

Example 2.18 *A solution containing* 30 g *of non-volatile solute exactly in* 90 g *of water has a vapour pressure of* 2.8 kPa *at* 298 K. *Further,* 18 g *of water is then added to the solution and the new vapour pressure becomes* 2.9 kPa *at* 298 K. *Calculate the*

(i) molar mass of the solute

(ii) vapour pressure of water at 298 K

Sol. By Raoult's law, relative decrease in vapour pressure is equal to the mole fraction of solute.

Let molar mass of solute = *m*

and vapour pressure of solvent = p°

$$p_{\text{solution}} = p^\circ_{\text{solvent}} \times \text{mole fraction of solvent}$$

$$\textit{Case I} \quad 2.8 = p^\circ_{\text{solvent}} \times \frac{5}{\left(5 + \frac{30}{m}\right)}$$

$$2.8 = p^\circ_{\text{solvent}} \left(\frac{m}{m + 6}\right)$$

$$\textit{Case II} \quad 2.9 = p^\circ_{\text{solvent}} \left(\frac{m}{m + 5}\right)$$

$$\therefore \quad \frac{2.9}{2.8} = \frac{m + 6}{m + 5} \quad \therefore \; m = 23 \text{ g mol}^{-1}$$

$$\therefore \quad p^\circ_{\text{solvent}} = 3.53 \text{ kPa}$$

2. Elevation in boiling point

As the vapour pressure of a solution is decreased due to the presence of a non-volatile solute, the solution boils at a higher temperature as compared to the pure solvent.

The boiling point of the solution is therefore higher than the boiling point of the pure solvent. If $T_b°$ is the boiling point of pure liquid and T_b is the boiling point of solution then, elevation in boiling point,

$$\Delta T_b = T_b - T_b°$$

(*i*) The elevation in boiling point is proportional to the molality of the solution, i.e.

$$\Delta T_b \propto m \quad \text{or} \quad \Delta T_b = K_b m$$

Where,

K_b = boiling point elevation constant or molal elevation constant or ebullioscopic constant

$$\text{Since, molality} = \frac{\text{moles of solute}}{\text{kg of solvent}}$$

$$\therefore \quad \Delta T_b = \frac{1000 w_2 K_b}{m_2 w_1}$$

Where, $\frac{w_2}{m_2}$ = moles of solute

$\frac{w_1}{1000}$ = kg of solvent

(*ii*) **Unit of K_b, K kg mol^{-1}** The variation of vapour pressure of the pure solvent and solution as a function of temperature can be depicted as below.

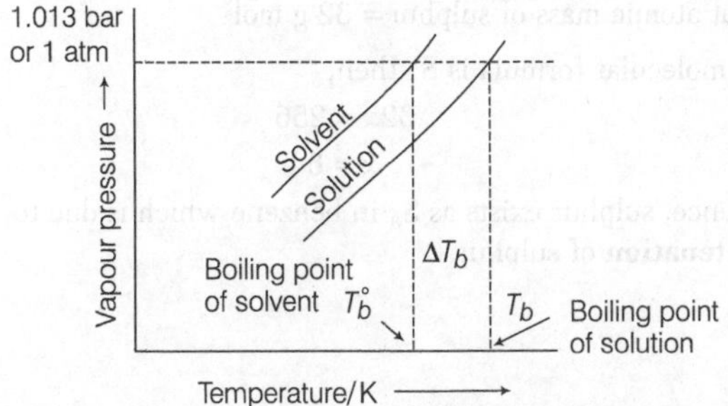

Fig. 2.5 The vapour pressure curve for solution lies below the curve for pure water. The figure shows that ΔT_b denotes the elevation in boiling point of a solvent in solution

Determination of molar mass of solute

We can calculate molar mass of solute from the relation of elevation in boiling point.

$$m_2 = \frac{K_b \times w_2 \times 1000}{\Delta T_b \times w_1}$$

where, m_2 = molar mass of solute, w_2 = weight of solute
w_1 = weight of solvent

Some important relationships of K_b

$$K_b = \frac{m_1 R (T_b°)^2}{\Delta H_V \times 1000}$$

Where, R = gas constant

ΔH_V = molar enthalpy of vaporisation of the solvent

T_b° = normal boiling point of pure solvent

T_b = boiling point of solvent

m_1 = molecular weight of solvent

If, ΔH_V is given in heat units mol^{-1} divide it by m_1.

Example 2.19 *A solution containing* 0.456 g *of camphor (molar mass* = 152) *dissolved in* 3.14 g *of acetone* (bp = 56.30°C), *if the molal elevation of acetone is* 17.2°C kg mol^{-1}, *then calculate the boiling point of the solution.*

Sol. We have, $\Delta T_b = \frac{1000\, K_b \times w_2}{w_1 \times m_2} = \frac{1000 \times 17.2 \times 0.456}{31.4 \times 152}$

$= 0.16°\text{ C}$

$\therefore$ Boiling point of solution $(T_b) = T_b° + \Delta T_b$

$= 56.30°\text{ C} + 0.16°\text{ C}$

$= 56.46°\text{ C}$

3. Depression in freezing point

The freezing point is that temperature, at which the solid and the liquid states of the substance have the same vapour pressure. A solution will freeze when its vapour pressure will be equal to the vapour pressure of the pure solid solvent.

When a non-volatile solute is added to a solvent, the freezing point of the solution is always lower than that of pure solvent as the vapour pressure of the solvent decreases in the presence of non-volatile solute.

This difference in freezing point is known as **depression of freezing point.** This is shown by the following figure.

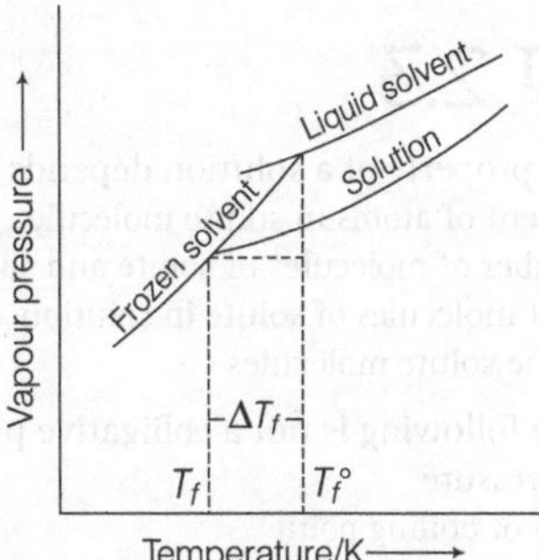

Fig. 2.6 ΔT_f, depression of the freezing point of a solvent in a solution

Depression of freezing point (ΔT_f) is given by

$$\Delta T_f = T_f^\circ - T_f$$

Where, T_f° = freezing point temperature of pure solvent
T_f = freezing point temperature of the solvent when non-volatile solute is dissolved in it

For dilute solution, $\Delta T_f \propto m$

$$\Delta T_f = K_f m$$

Where, K_f = molal depression constant or cryoscopic constant
m = molal concentration of the solution (molality)

Unit of K_f, **K kg mol^{-1}**

Since, $m = \dfrac{1000 w_2}{m_2 w_1}$

$\therefore \quad \Delta T_f = \dfrac{1000 K_f w_2}{m_2 w_1}$

Where, $\dfrac{w_1}{1000}$ = kg of solvent, $\dfrac{w_2}{m_2}$ = moles of solute

Calculation of molar mass of solute

$$M_2 = \frac{K_f \times w_2 \times 1000}{\Delta T_f \times w_1}$$

Where, m_2 = molar mass of solute, w_2 = weight of solute
w_1 = weight of solvent

Important relationships of K_f

(*i*) Depression in freezing point (ΔT_f) is directly proportional to vapour pressure.

$$\Delta T_f \propto p^\circ - p$$

Where, K_f = molecular depression constant

(*ii*) $K_f = \dfrac{m_1 R (T_f^\circ)^2}{\Delta H_f \times 1000}$

where, ΔH_f = molar heat of freezing
T_f° = freezing point of solvent

Relation between K_b and K_f

$$\frac{K_b}{K_f} = \frac{\Delta T_b}{\Delta T_f}$$

for water, $K_f > K_b$

Example 2.20 *The molal freezing point constant for water is* 1.86 °C kg mol^{-1}. *If* 342 g *of cane sugar is dissolved in* 1000g *of water, then calculate the temperature at which the solution will freeze.*

Sol. Given, $K_f = 1.86°\text{C kg mol}^{-1}$, $m_2 = 342$ g, $w_1 = 1000$ g

Then, we have the formula

$$\Delta T_f = \frac{1000 K_f w_2}{m_2 w_1} = \frac{1000 \times 1.86 \times 342}{342 \times 1000} = 1.86°\text{ C}$$

$$\therefore \quad T_f = T - \Delta T_f = 0 - 1.86 = -1.86°\text{ C}$$

Example 2.21 25.6 g *of sulphur in* 100 g *benzene shows depression in freezing point of* 5.12K. *If* K_f *of benzene is* 5.12 K kg mol^{-1} *then what is the molecular formula of sulphur in benzene?*

Sol. $\Delta T_f = \dfrac{1000 K_f w_2}{m_2 w_1}$

$$m_2 \text{ (sulphur in benzene)} = \frac{1000 K_f w_2}{\Delta T_f w_1} = \frac{1000 \times 5.12 \times 25.6}{5.12 \times 100} = 256 \text{ g mol}^{-1}$$

But atomic mass of sulphur = 32 g mol^{-1}

If molecular formula is S_x then,

$$32x = 256$$
$$x = 8$$

Hence, sulphur exists as S_8 in benzene which is due to **catenation** of sulphur.

CHECK POINT 2.3

1. A colligative property of a solution depends on the
(a) arrangement of atoms in solute molecule
(b) total number of molecules of solute and solvent
(c) number of molecules of solute in solution
(d) mass of the solute molecules

2. Which of the following is not a colligative property?
(a) Vapour pressure
(b) Elevation of boiling point
(c) Lowering of freezing point
(d) Osmotic pressure

3. If vapour pressure of a pure solvent is p° and that of its solution, with a non-volatile solute is p. Then which of the following relationship is correct
(a) $p > p^\circ$ (b) $p < p^\circ$
(c) $p = p^\circ$ (d) cannot be predicted

4. At boiling point, under normal conditions, vapour pressure a liquid is
(a) greater than atmospheric pressure
(b) smaller than atmospheric pressure
(c) equal to the atmospheric pressure
(d) double of the atmospheric pressure

5. When certain amount of an organic compound is dissolved in acetone, its boiling point increase by 0.34°C. If K_b for acetone is 17.0°C kg mol^{-1}, then find the molality of the solution
(a) 0.1 molal (b) 0.2 molal
(c) 0.01 molal (d) 0.02 molal

6. When 40 g of substance is dissolved in 1000 g of water, its freezing point is depressed by 1.86°C. K_f for water is 1.86°C mol^{-1} then find the molar mass of the solute
(a) 4 (b) 10 (c) 40 (d) 400

7. Which of the following relation is true for water
(a) $K_f = K_b$
(b) $K_f > K_b$
(c) $K_f < K_b$
(d) None of the above

8. The order of boiling points of four equimolar aqueous solutions is $C < B < A < D$. The correct order of their freezing points is
(a) $D < C < B < A$ (b) $D < C < B < A$
(c) $D < B > A < C$ (d) $D > A > B > C$

4. Osmotic pressure

Before discussing about osmotic pressure, let us first discuss osmosis. When two solutions of different concentrations are separated by a semipermeable membrane, then solvent from the solution of lower concentration, flows to the solution of higher concentration through the semipermeable membrane, until concentrations on both sides become equal. This spontaneous flow of solvent through the semipermeable membrane is called **osmosis.**

Semipermeable membranes

These are thin membranes, that allow the passage of only selective solvent molecules through them, e.g. egg membrane, goat's bladder, cell membrane are natural semipermeable membrane. While gelatinous inorganic substances such as calcium phosphate, copper ferrocyanide, freshly prepared silicates of iron, cobalt, nickel, cellophane and parchment papers etc., are the artificial ones.

Difference between osmosis and diffusion

(i) Diffusion is intermingling of molecules or ions etc., resulting from the random thermal agitation. Thus, there is no restriction to motion of any kind of particles. While in osmosis only solvent molecules which can flow through a semipermeable membrane.

(ii) Osmosis takes place through a semipermeable membrane. While for diffusion such things are not required.

(iii) A homogeneous mixture is obtained as a result of diffusion. On the other hand no mixing is possible in osmosis.

(iv) Diffusion cannot be prevented or stopped. While osmosis can be stopped, by applying pressure to the solution.

The external pressure which must be applied on the solution in order to stop the flow of the solvent into the solution through semipermeable membrane is termed as **osmotic pressure.**

The hydrostatic pressure developed as a result of osmosis is a measure of osmotic pressure of the solution and is defined as the pressure built up on the solution which just stops thc osmosis.

From the figure 2.7,

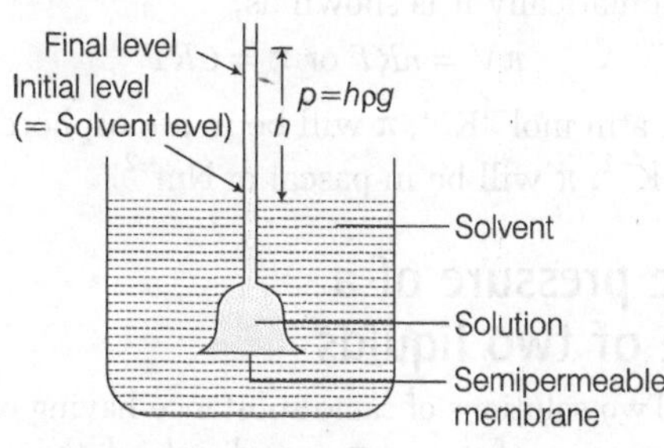

Fig. 2.7 Process of osmosis

Osmotic pressure (π) = $h\rho g$ (approximately)

Where,

h = increase in level in the tube of unit cross-section

ρ = density of solution

g = acceleration on due to gravity

Laws of osmotic pressure

van't Hoff realised that an analogy exists between the gases and solutions provided the osmotic pressure of solutions is used in place of ordinary gas pressure. He showed that for solutions of non-electrolytes following laws hold good.

(i) **Boyle-van't Hoff law** Which states that the osmotic pressure (π) of a solution is directly proportional to its concentration (C), when the temperature is kept constant.

$$\pi \propto C \quad \text{(Temperature, } T \text{ is constant)}$$

If, V litres of solution contain n moles of solute then

$$C = \frac{n}{V} \text{ or } \pi \propto \frac{1}{V}$$

$$\pi V = \text{constant}$$

(ii) **Charles van't Hoff law** (Gay-Lussac law) According to which the osmotic pressure of a dilute solution is directly proportional to its absolute temperature (T), when the concentration of solution is kept constant.

$$\pi \propto T \quad \text{(concentration, } C \text{ is constant)}$$

or $$\frac{\pi}{T} = \text{constant}$$

(*iii*) **Avogadro-van't Hoff law** This law states that equal volumes of dilute solutions of different solutes, having the same temperature and osmotic pressure contain equal number of molecules, i.e. isotonic solutions at a given temperature have same molar concentration.

$$\pi \propto n$$

Here,

n = number of moles of solute per litres of solutions

(*iv*) **General equation for dilute solutions** It is obtained by combining the above three laws.

Mathematically it is shown as,

$$\pi V = nRT \text{ or } \pi = CRT$$

If R is in L atm $mol^{-1}K^{-1}$, π will be in atmosphere. If R is in J mol^{-1} K^{-1}, π will be in pascal or Nm^{-2}.

Osmotic pressure of a mixture of two liquids

Case I Two solutions of same substance having osmotic pressure of π_1 and π_2 are dissolved, then

$$\pi_1 V_1 = \pi_2 V_2 = \pi_R (V_1 + V_2)$$

Case II If number of moles of two solutes are n_1 and n_2, which are present in volume V_1 and V_2 respectively then, $\pi = \pi_1 + \pi_2 = \left(\frac{n_1 + n_2}{V_1 + V_2}\right) RT$

Calculation of molecular weight

Calculation of molecular weight of solute since, osmotic pressure is proportional to the concentration (C) of the solution at a given temperature (T).

$$\pi \propto C \Rightarrow \pi = CRT = \frac{w_2}{m_2}\frac{RT}{V}$$

On the basis of above formula, we can calculate molecular mass of solute as $m_2 = \frac{w_2 RT}{V\pi}$

Where, m_2 = molecular mass of solute,

w_2 = mass of solute, V = volume of solution (in litre)

π = osmotic pressure

Relation between osmotic pressure and other colligative properties

(*i*) Osmotic pressure is related with lowering of vapour pressure as given below, (assume dilute solution)

$$\pi = \left(\frac{p^\circ - p}{p^\circ}\right)\frac{RTd}{m_1}$$

Where, d = density of solution at T temperature

m_1 = molecular weight of solvent

p° = vapour pressure of solvent

p = vapour pressure of solution

(*ii*) Osmotic pressure is related with elevation in boiling point

$$\pi = \Delta T_b \times \frac{dRT}{1000 \times K_b}$$

(*iii*) Osmotic pressure is related with depression in freezing point

$$\pi = \Delta T_f \times \frac{dRT}{1000 \times K_f}$$

In all of the above cases

Weight of solution ≈ weight of solvent = Vd

Example 2.22 *How many gram of glucose is present in* 0.5 L *of solution for its osmotic pressure to be the same as that of* 9.2 g *of glucose per litre?*

Sol. For isotonic solution, $\pi_1 = \pi_2$

$$\frac{w_1}{m_1 V_1} = \frac{w_2}{m_2 V_2}$$

$$w_1 = \frac{w_2 m_1 V_1}{m_2 V_2} = \frac{9.2 \times 180 \times 0.5}{180 \times 1}$$

$$w_1 = 4.60 \text{ g}$$

Important terminology related to osmosis

(*i*) **Isotonic solution** Two solutions having same osmotic pressure at a given temperature are called isotonic solutions. There is no osmosis between these two solution.

$$\pi_1 = \pi_2$$

$$\frac{n_1}{V_1}RT_1 = \frac{n_2}{V_2}RT_2$$

if $T_1 = T_2$ and $V_1 = V_2$

then $n_1 = n_2$

Thus, equal volume of isotonic solutions will have equal mole of solute under isothermal conditions.

$$\frac{w_1}{m_1} = \frac{w_2}{m_2}$$

If w_1, m_1, w_2 are known than m_2 can be calculated.

(*ii*) **Hypertonic solution** A solution having more osmotic pressure than the other solution is called hypertonic solution.

(*iii*) **Hypotonic solution** A solution having less osmotic pressure than the other solution is called hypotonic solution.

(*iv*) **Reverse osmosis** When a pressure more than the osmotic pressure is applied to the solution, the solvent pass from solution of higher concentration to the solution of lower concentration, through the semipermeable membrane. This type of osmosis is known as **reverse osmosis.** It is used for the **desalination of sea water.**

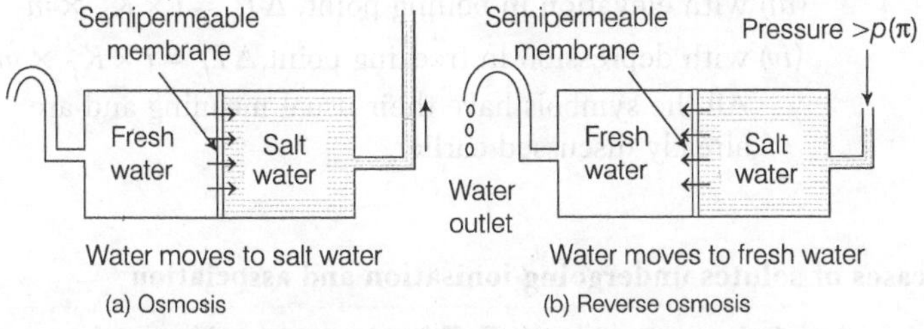

Fig. 2.8 A schematic set up for the reverse osmosis used for desalination of sea water

(v) **Isobaric solutions** At same temperature, the solutions containing equal vapour pressure are known as **isobaric solutions.**

Abnormal molar mass

Ionic compounds when dissolved in water dissociates into cations and anions, e.g. one mole of KCl (74.5 g) in water dissociates into K^+ and Cl^- ions. While the molecules of few organic compounds like acetic acid or benzoic acid dimerise (associate) in benzene due to H-bonding.

These association or dissociation leads to new apparent molar mass. Since, a particle upon dissociation acts like two while two particles upon association act like one. This change in molar mass is called **abnormal molar mass.**

Various relations of colligative properties hold good in dilute solutions only, when there is no change in molecular state of solute. In case the total number of particles of solute changes in solutions the colligative properties also change accordingly. This change can be estimated using the van't Hoff factor, as discussed below.

van't Hoff factor

It is the ratio of normal molar mass to the abnormal molar mass and is represented by '*i*'. It also gives relation between observed (abnormal) colligative property and calculated (normal) colligative property.

The factor '*i*' is defined as

$$i = \frac{\text{normal molecular mass}}{\text{abnormal molecular mass}}$$

or, $$i = \frac{\text{(abnormal) observed colligative property}}{\text{(normal) calculated colligative property}}$$

$$= \frac{\text{actual number of particles in solution}}{\text{number of particles taken}}$$

van't Hoff factor can be applied in case of dissociation and association as follows

(i) **In case of dissociation** When substance is an electrolyte, the number of particles increases in solution and the colligative property increases accordingly.

Suppose 1 of solute A dissociates to give n ions and α is the degree of dissociation.

	$A \rightleftharpoons$	nB
Initially	1 mol	0
After dissociation	$(1-\alpha)$ mol	$n\alpha$

Total number of moles present in solution

$$= (1-\alpha) + n\alpha$$
$$= 1 + (n-1)\,\alpha$$

van't Hoff factor, $i = 1 + (n-1)\,\alpha > 1$ if $n \geq 2$

and $$\alpha = \frac{(i-1)}{(n-1)}$$

(ii) **In case of association** When substance undergoes association, the number of particles decreases in solution and consequently the value of colligative property decreases.

e.g. molecules of acetic acid dimerise in benzene due to hydrogen bonding.

$$2CH_3COOH \rightleftharpoons (CH_3COOH)_2$$

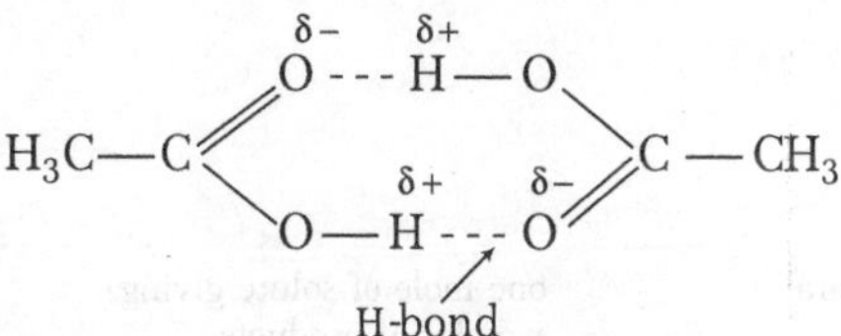

Suppose n molecules associate to form one giant molecule and α is the degree of association.

	$nA \rightleftharpoons$	$(A)_n$
Initially	1 mol	0
At equilibrium	$(1-\alpha)$ mol	$\frac{\alpha}{n}$

Total number of moles in solution

$$= (1-\alpha) + \frac{\alpha}{n} = 1 + \left(\frac{1}{n} - 1\right)\alpha$$

$$\text{van't Hoff factor} = 1 + \left(\frac{1}{n} - 1\right)\alpha < 1$$

if, $n \geq 2$ and $$\alpha = \frac{(i-1)}{\frac{1}{n} - 1}$$

Colligative properties for solution of electrolytes

Relation between van't Hoff factor and different colligative properties is established as following

(*i*) with osmotic pressure,

$$\pi = \frac{n_2}{V} RTi = \frac{w_2}{m_2 V} RTi$$

(*ii*) with relative lowering of vapour pressure,

$$\frac{p^\circ - p}{p^\circ} = i \frac{w_2 m_1}{m_2 w_1} i$$

(*iii*) with elevation in boiling point, $\Delta T_b = i \times K_b \times m$

(*iv*) with depression in freezing point $\Delta T_f = i \times K_f \times m$

All the symbols have their usual meaning and are already discussed earlier.

Table 2.4 van't Hoff factor '*i*' for different cases of solutes undergoing ionisation and association

Solute	Example	Association/dissociation (x = degree)	y^*	van't Hoff factor $i = [1 + (y^* - 1)\,x]$	Abnormal molecular weight (m_2')
Non-electrolyte	urea-glucose, sucrose, etc.	none	1	1	normal molecular weight (m_2)
Electrolyte of type A^+B^-	NaCl, KCl, HCl, CH_3COOH, $FeSO_4$, etc.	$\underset{1-x}{AB} \rightleftharpoons \underset{x}{A^+} + \underset{x}{B^-}$	2	$(1+x)$	$\frac{m_2}{(1+x)}$
Electrolyte of type A_2B, AB_2	K_2SO_4, $BaCl_2$, $K_2[PtCl_6]$, etc.	$\underset{1-x}{A_2B} \rightleftharpoons \underset{2x}{2A^+} + \underset{x}{B^{2-}}$ $\underset{1-x}{AB_2} \rightleftharpoons \underset{x}{A^{2+}} + \underset{2x}{2B^-}$	3	$(1+2x)$	$\frac{m_2}{(1+2x)}$
Electrolyte of type A_3B, AB_3	Na_3PO_4, $AlCl_3$, $K_3[Fe(CN)_6]$, $FeCl_3$	$\underset{1-x}{A_3B} \rightleftharpoons \underset{3x}{3A^+} + \underset{x}{B^{3-}}$ $\underset{1-x}{AB_3} \rightleftharpoons \underset{x}{A^{3+}} + \underset{3x}{3B^-}$	4	$(1+3x)$	$\frac{m_2}{(1+3x)}$
Associated solute	benzoic acid in benzene forming dimer, any solute forming polymer A_n	$2A \rightleftharpoons A_2$ $\underset{(1-x)}{A} \rightleftharpoons \underset{x/2}{\frac{1}{2}A_2}$ $nA \rightleftharpoons A_n$ $\underset{(1-x)}{A} \rightleftharpoons \underset{\left(\frac{x}{n}\right)}{\frac{1}{n}A_n}$	$\frac{1}{2}$ $\frac{1}{n}$	$\left(1-\frac{x}{2}\right) = \left(\frac{2-x}{2}\right)$ $\left[1+\left(\frac{1}{n}-1\right)x\right]$	$\frac{2m_2}{(2-x)}$ $\frac{m_2}{1+\left(\frac{1}{n}-1\right)x}$
General	one mole of solute giving y moles of products	$A \rightleftharpoons yB$	y	$[1+(y-1)\,x]$	$\frac{m_2}{[1+(y-1)\,x]}$

y^* = Total mol of product formed after dissociation/association of one mole of reactant.

Example 2.23 *Calculate the van't Hoff factor in each of the following cases?*

(a) Aq. $K_4[Fe(CN)_6]$ *which is 20% ionised*

(b) PCl_5 *which is 50% dissociated*

(c) Benzoic acid which forms dimer to the extent of 50% in benzene

(d) Glucose solution

Sol. (*a*) $\underset{\substack{1\\(1-x)}}{K_4[Fe(CN)_6]} \rightleftharpoons \underset{\substack{0\\4x}}{4K^+} + \underset{\substack{0\\x}}{[Fe(CN)_6]^{4-}}$

$$y = 5$$

$$i \text{ (van't Hoff factor)} = [1 + (y-1)\,x]$$
$$= 1 + 4x = 1 + 4 \times 0.2 = 1.8$$

(b) $\underset{\substack{1\\(1-x)}}{PCl_5(g)} \rightleftharpoons \underset{\substack{0\\x}}{PCl_3(g)} + \underset{\substack{0\\x}}{Cl_2(g)}$

$$y = 2$$

$$i = [1 + (y-1)\,x] = 1 + x = 1 + 0.5 = 1.5$$

(c) $2C_6H_5COOH \rightleftharpoons (C_6H_5COOH)_2$

$$C_6H_5COOH \rightleftharpoons \frac{1}{2}(C_6H_5COOH)_2$$

$$y = \frac{1}{2}$$

Dimer formation is due to intermolecular H-bonding.

$$i = [1 + (y-1)\,x]$$
$$= 1 - \frac{x}{2} = 1 - \frac{0.5}{2} = 0.75$$

(d) There is no change in the molecular state of glucose being a non-electrolyte.
Hence, $y = 1$

$$i = [1 + (y - 1)x] = 1$$

Example 2.24 *A solution of* 2.8 g *of* CdI_2 *(molar mass* $= 364$ g mol^{-1}*) in* 20 g *water has elevation in boiling point of* 0.20 K. *Determine the molecular state of* CdI_2 *in aqueous solution?*

$$[K_b(H_2O) = 0.52 \text{ K mol}^{-1} \text{ kg}]$$

Sol. To decide molecular state, determine value of i or its molar mass.

(a) If, $i = 1$ and molar mass = calculated molar mass then solute remains without change.
(b) If, $i > 1$ and molar mass < calculated molar mass then solute is ionised.
(c) If, $i < 1$ and molar mass > calculated molar mass then solute is polymerised (dimer, trimer, etc.)

$$\Delta T_b = \frac{1000\, K_b\, w_2}{m_2\, w_1}(i)$$

$$i = \frac{\Delta T_b\, m_2\, w_1}{1000\, K_b\, w_2} = \frac{0.20 \times 364 \times 20}{1000 \times 0.52 \times 2.8} = 1.0$$

Thus, CdI_2 **remains as** CdI_2 (without ionisation) in aqueous solution.

Example 2.25 Urea solution is 20% *by weight of solution. Calculate its boiling point? When water boils at* 1 atm *pressure and* 1 molal *glucose solution boils at* 100.52°C.

Sol. Urea and glucose are non-electrolytes hence, $i = 1$

Water boils at 1 atm pressure, it means boiling point of water is 100°C.

ΔT_b (glucose solution) $= 100.52 - 100.00 = 0.52$°C
ΔT_b (glucose solution) $= mK_b i$

$$0.52 = 1 \times K_b \times 1$$

$$\therefore \quad K_b = 0.52° \text{C mol}^{-1} \text{ kg}$$

urea solution is 20% by weight of solution, it means 100 g solution (solute + solvent) has 20 g solute and 80 g solvent

$$\Delta T_b \text{ (urea solution)} = \frac{1000\, K_b\, w_2 i}{m_2\, w_1}$$

$$= \frac{1000 \times 0.52 \times 20 \times 1}{60 \times 80} = 2.17° \text{C}$$

$\therefore$ Boiling point of urea solution $= 100.00 + 2.17° = 102.17$°C

Example 2.26 *The freezing point of a solution prepared by dissolving* 1.00 mole *of* HF *in* 500 g *water is* −3.8°C, *but the freezing point of a solution prepared by dissolving* 1.00 mole *of* HCl *in* 500 g *water is* −7.4°C. *Justify.*
$K_f(H_2O) = 1.86°$ mol^{-1} kg.

Sol. $\Delta T_f = mK_f i$

For HF, $$m = \frac{1}{0.5} = 2 \text{ mol kg}^{-1}$$

$$\Delta T_f = 0 - (-3.8) = 3.8° \text{C}$$

$$\therefore \quad 3.8 = 2 \times 1.86 \times i$$

$$i = \frac{3.8}{2 \times 1.86} = 1.02 \approx 1$$

For HCl, $$m = \frac{1}{0.5} = 2 \text{ mol kg}^{-1}$$

$$\Delta T_f = 0 - (-7.4) = 7.4°\text{C}$$

$$i = \frac{\Delta T_f}{mK_f} = \frac{7.4}{2 \times 1.86} = 1.99 \approx 2$$

Example 2.27 *Phenol* (C_6H_5OH) *is found to exist as polymer. If there is* 100% *polymerisation, determine number of phenol molecules polymerised in aqueous solution if* 9.4 g *aqueous solution in* 100 g *water freezes at* −0.93°C.

$$K_f(H_2O) = 1.86° \text{ mol}^{-1} \text{ kg}$$

Sol. Observed $(\Delta T)_f = 0.93$°C,

$$\text{Calculated, } \Delta T_f = 1000\, K_f \frac{w_2}{m_2\, w_1}$$

$$= 1000 \times \frac{1.86 \times 9.4}{94 \times 100} = 1.86° \text{C}$$

$$i = \frac{\text{observed } (\Delta T)_f}{\text{calculated } (\Delta T)_f} = \frac{0.93}{1.86} = \frac{1}{2}$$

$$\therefore \quad \frac{1}{n} = \frac{1}{2}$$

$$\Rightarrow \quad n = 2$$

CHECK POINT 2.4

1. The passage of a solvent across a semipermeable membrane because of concentration differences is called

(a) evaporation (b) diffusion
(c) hydration (d) osmosis

2. Osmosis involved in

(a) excretion of urine
(b) interchange of nutrients and waste products between tissue cells and their surroundings
(c) in both cases
(d) evaporation of liquid

3. If m_1 is the normal molar mass and m_2 is abnormal molar mass of an electrolyte in solution state. Then van't Hoff factor (i) is given as

(a) $i = \frac{m_1}{m_2}$ (b) $i = \frac{m_2}{m_1}$
(c) $i = m_1 \cdot m_2$ (d) $i = m_1 + m_2$

4. Osmotic pressure method is used to determine molar mass of protein, macromolecules like polymers, biomolecules etc., because

(a) osmotic pressure is measured at room temperature
(b) osmotic pressure depends upon molality
(c) these are very big molecules and have poor solubility
(d) All of the above

5. Osmotic pressure present in the fluid inside the blood cell is equivalent to

(a) 0.9% (m/V) NaCl solution
(b) less than 0.9 % (m/V) NaCl solution
(c) more than 0.9 % (m/V) NaCl solution
(d) 0.9% (m/V) Na_2SO_4 solution

6. The degree of dissociation (α), if a weak electrolyte, A_xB_y is related to van't Hoff factor (i) by the expression

(a) $\alpha = \frac{i-1}{(x+y-1)}$ (b) $\alpha = \frac{i-1}{x+y+1}$
(c) $\alpha = \frac{x+y-1}{i-1}$ (d) $\alpha = \frac{x+y+1}{i-1}$

7. The van't Hoff factor for 0.1 M $Ba(NO_3)$ solution is 2.74. The degree of dissociation is

(a) 91.3% (b) 87%
(c) 100% (d) 74%

8. The elevation in boiling point of a solution containing 13.44 g of $CuCl_2$ in 1 kg of water using the following information will be (molecular weight of $CuCl_2$ = 134.4 and $K_b = 0.52\ Km^{-1}$).

(a) 0.16°C (b) 0.05°C
(c) 0.1°C (d) 0.2°C

Chapter Exercises

(A) Topical exercise

I. Types and concentration of solution

1 Which of the following units is useful in relating concentration of solution with its vapour pressure? **[NCERT Exemplar]**

(a) Mole fraction (b) Parts per million
(c) Mass percentage (d) Molality

2 The concentration of pollutant in ppm, that has been measured at 450 mg per 150 kg of sample is

(a) 3 ppm (b) 6 ppm
(c) 3000 ppm (d) 330 ppm

3 Persons are medically considered to have lead poisoning if they have a concentration of greater than 10 μg of lead per decilitre of blood. Concentration in parts per billion is

(a) 1000 (b) 100 (c) 10 (d) 1

4 Which solution is the most concentrated? (Assume same solvent and solute in each case)

(a) 2.4 g solute in 2 mL solution
(b) 2.4 g solute in 5 mL solution
(c) 20 g solute in 50 mL solution
(d) 30 g solute in 150 mL solution

5 How many moles of HCl are present in 35.5 mL of a 0.20 M solution?

(a) 0.236 mol (b) 0.178 mol
(c) 0.974 mol (d) 0.0071 mol

6 Match the terms given in Column I with expressions given in Column II.

Column I		Column II
A. Mass percentage	1.	$\frac{\text{Number of moles of the solute component}}{\text{Volume of solution in litres}}$
B. Volume percentage	2.	$\frac{\text{Number of moles of a component}}{\text{Total number of moles of all the components}}$
C. Mole fraction	3.	$\frac{\text{Volume of the solute component in solution}}{\text{Total volume of solution}} \times 100$
D. Molality	4.	$\frac{\text{Mass of the solute component in solution}}{\text{Total mass of the solution}} \times 100$
E. Molarity	5.	$\frac{\text{Number of moles of the solute components}}{\text{Mass of solvent in kilograms}}$

Codes

	A	B	C	D	E
(a)	4	5	3	2	1
(b)	5	3	4	1	2
(c)	4	3	2	5	1
(d)	1	4	5	3	2

7 Consider the following statements.

Statement I Molarity of a solution in liquid state changes with temperature.

Statement II The volume of a solution changes with change in temperature.

(a) Only statement I is correct
(b) Only statement II is correct
(c) Both statements I and II are correct
(d) Both statements I and II are incorrect

8 A beaker contains a solution of substance *A*. Precipitation of substance *A* takes place when small amount of *A* is added to the solution. The solution is .... **[NCERT Exemplar]**

(a) saturated (b) supersaturated
(c) unsaturated (d) concentrated

9 On dissolving sugar in water at room temperature solution feels cool to touch. Under which of the following cases dissolution of sugar will be most rapid? **[NCERT Exemplar]**

(a) Sugar crystals in cold water
(b) Sugar crystals in hot water
(c) Powdered sugar in cold water
(d) Powdered sugar in hot water

10 Maximum amount of a solid solute that can be dissolved in a specified amount of a given liquid solvent does not depend upon .... **[NCERT Exemplar]**

(a) temperature (b) nature of solute
(c) pressure (d) nature of solvent

11 4 L of 0.02 M aqueous solution of NaCl was diluted by adding 1 L of water. The molality of the resultant solution is .... **[NCERT Exemplar]**

(a) 0.004 M (b) 0.008 M
(c) 0.012 M (d) 0.016 M

12 What is the normality of 0.30 M H_3PO_4 (a tribasic acid) in the following reaction?

$$H_3PO_4 + 2OH^- \longrightarrow HPO_4^{2-} + 2H_2O$$

(a) 0.30 N (b) 0.60 N (c) 0.90 N (d) 0.15 N

13 Mole fraction of the toluene in the vapour phase which is in equilibrium with a solution of benzene ($p° = 120$ torr) and toluene ($p° = 80$ torr) having 2.0 moles of each is

(a) 0.50 (b) 0.25 (c) 0.60 (d) 0.40

14 All of the water in a 0.20 M solution of NaCl was evaporated and a 0.150 mol of NaCl was obtained. What was the original volume of the sample?

(a) 30 mL (b) 333 mL (c) 750 mL (d) 1000 mL

15 A 20.0 mL sample of $CuSO_4$ solution was evaporated to dryness, leaving 0.967 g of residue. What was the molarity of the original solution? (Cu = 63.5)

(a) 48.4 M (b) 0.0207 M
(c) 0.0484 M (d) 0.303 M

16 When 25 g of Na_2SO_4 is dissolved in 10^3 kg of solution its concentration will be

(a) 2.5 ppm (b) 25 ppm
(c) 250 ppm (d) 100 ppm

17 There are some of the characteristics of the supersaturated solution.

I. Equilibrium exists between solutions and solute.
II. If a crystal of solute is added to supersaturated solution, crystallisation occurs rapidly.
III. Supersaturated solutions contain more solute than they should at a particular temperature.

Correct characteristics of supersaturated solutions are

(a) I, II and III (b) Both II and III
(c) Both I and III (d) Both I and II

18 During dissolution of a solute in a solvent, following are the three operative interactions.

I. Solvent-solvent
II. Solute-solvent
III. Solute -solute

Which of the above takes place with the absorption of heat? Choose the correct option.

(a) Both I and II (b) Both II and III
(c) I, II and III (d) Both I and III

19 Which concentration can be calculated if the mole fraction and density of an aqueous solution of HCl are known?

I. Molality II. Molarity
III. Per cent by mass

Choose the correct option.

(a) Only I (b) Only III
(c) Both I and II (d) I, II and III

20 Which of the following solutions represent a metastable system?

(a) Supersaturated (b) Staurated
(c) Unsaturated (d) Concentrated

II. Solubility

21 The solubility of gases in liquids

(a) increases with increase in pressure and temperature
(b) decreases with increase in pressure and temperature
(c) increases with increase in pressure and decrease in temperature
(d) decreases with increase in pressure and increase in temperature

22 At equilibrium the rate of dissolution of a solid solute in a volatile liquid solvent is .... **[NCERT Exemplar]**

(a) less than the rate of crystallisation
(b) greater than the rate of crystallisation
(c) equal to the rate of crystallisation
(d) zero

23 Naphthalene dissolves in ether or benzene due to

(a) similar dipole-dipole attraction
(b) similar London dispersion forces
(c) H-bonding
(d) ion-ion attractions

24 Which statement best explains the meaning of the phrases "like dissolves like"?

(a) A solute will easily dissolve a solute of similar mass
(b) A solvent and solute with similar intermolecular forces will readily form a solution
(c) The only true solutions are formed when water dissolves a non-polar solute
(d) The only true solutions are formed when water dissolves a polar solute

25 Match the items given in Column I with the type of solutions given in Column II.

	Column I		Column II
A.	Soda water	1.	A solution of gas in solid
B.	Sugar solution	2.	A solution of gas in gas
C.	German silver	3.	A solution of solid in liquid
D.	Air	4.	A solution of solid in solid
E.	Hydrogen gas in palladium	5.	A solution of gas in liquid
		6.	A solution of liquid in solid

Codes

	A	B	C	D	E
(a)	3	5	1	4	2
(b)	5	3	4	2	1
(c)	2	4	3	1	5
(d)	5	2	3	4	1

26 Value of Henry's constant K_H for N_2 and O_2 **[NCERT Exemplar]**

(a) increases with increase in temperature
(b) decreases with increase in temperature
(c) remains constant
(d) first increases then decreases

27 The value of Henry's constant, K_H is .... **[NCERT Exemplar]**

(a) greater for gases with higher solubility
(b) greater for gases with lower solubility
(c) constant for all gases
(d) not related to the solubility of gases

28 Partial pressure needed to dissolve 21 mg of CO_2 in 100 g of water at 298 K (Henry's law constant for CO_2 is 2.937 kPa m^3 mol^{-1}) is
(a) 14.0 kPa (b) 7.0 kPa
(c) 121.0 kPa (d) 79.0 kPa

29 Select the incorrect statement.
(a) Higher the value of K_H (Henry's law constant) at a given pressure, the lower is the solubility of the gas in the liquid
(b) Solubility of a gas in a liquid decreases with increase in temperature and pressure
(c) To minimise the painful effects accompanying the decomposition of deep sea divers, O_2 diluted with less soluble He gas is used as breathing gas
(d) The solubility of a gas in a liquid is governed by Henry's law

30 K_H value for Ar(*g*), $CO_2(g)$, HCHO (*g*) and $CH_4(g)$ are 40.39, 1.67, 1.83×10^{-5} and 0.413 respectively. Arrange these gases in the order of their increasing solubility. **[NCERT Exemplar]**
(a) HCHO < CH_4 < CO_2 < Ar
(b) HCHO < CO_2 < CH_4 < Ar
(c) Ar < CO_2 < CH_4 < HCHO
(d) Ar < CH_4 < CO_2 < HCHO

III. Vapour pressure and Raoult's law

31 Which of the following is true about enthalpy of solution containing benzene and toluene?
(a) $\Delta H_{solution} < 0$
(b) $\Delta H_{solution} = 0$
(c) $\Delta H_{solution} > 0$
(d) $\Delta H_{solution}$ may be zero or greater than zero

32 Which of the following interactions are endothermic?
(a) NaCl with H_2O (b) CH_3CH_2OH with H_2O
(c) Both (a) and (b) (d) H_2O with H_2O

33 If p_A is the vapour pressure of a pure liquid *A* and the mole fraction of *A* in the mixture of two liquids *A* and *B* is χ, the partial vapour pressure of *A* is
(a) $(1-\chi)p_A$ (b) χp_A
(c) $\frac{\chi}{(1-\chi)} p_A$ (d) $\frac{(1-\chi)}{\chi} p_A$

34 Total vapour pressure of mixture of 1 mole of volatile component *A* ($p°_A$ = 100 mmHg) and 3 moles of volatile component *B* ($p°_B$ = 60 mmHg) is 75 mm Hg. For such case
(a) there is positive deviation from Raoult's law
(b) boiling point has been lowered
(c) force of attraction between *A* and *B* is smaller than that between *A* and *A* or between *B* and *B*
(d) All the above statements are correct

35 Consider the following statements.
Statement I If one component obeyed Raoult's law over a certain range of composition, the other component would not obey Henry's law in that range.
Statement II Raoult's law is a special case of Henry's law.
(a) Only statement I is correct
(b) Only statement II is correct
(c) Both statements I and II are correct
(d) Both statements I and II are incorrect

36 A mixture contains 1 mole volatile liquid *A* ($p°_A$ = 100 mmHg) and 3 moles volatile liquid *B* ($p°_B$ = 80 mmHg). If solution behave ideally, total vapour pressure of the distillate is approximately
(a) 85 mmHg (b) 86 mmHg
(c) 90 mmHg (d) 92 mmHg

37 Relative decrease in vapour pressure of an aqueous NaCl is 0.167. Number of moles of NaCl present in 180 g of H_2O is
(a) 2 mol (b) 1 mol (c) 3 mol (d) 4 mol

38 Which of the following statements comparing solutions with pure solvent is not correct?
(a) A solution containing a non-volatile solute has a lower vapour pressure than pure solvent
(b) A solution containing a non-volatile solute has a lower boiling point than pure solvent
(c) A solution containing a non-volatile solute has a lower freezing point than pure solvent
(d) A solution will have a greater mass than an equal volume of pure solvent if the solute has a molar mass greater than the solvent

39 The most likely of the following mixtures to be an ideal solution is
(a) NaCl—H_2O (b) C_2H_5OH—C_6H_6
(c) C_7H_{16}—H_2O (d) C_7H_{16}—C_8H_{18}

40 An azeotropic solution of two liquids has a boiling point lower than either of them when it
(a) shows negative deviation from Raoult's law
(b) shows positive deviation from Raoult's law
(c) shows ideal behaviour
(d) is saturated

41 At 25°C, the vapour pressure of pure methyl alcohol is 92.0 torr. Mole fraction of CH_3OH in a solution which vapour pressure of CH_3OH is 23.0 torr at 25°C is
(a) 0.25 (b) 0.75 (c) 0.50 (d) 0.66

42 Total vapour pressure of mixture of 1 mol *A* ($p°_A$ = 150 torr) and 2 mol *B* ($p°_B$ = 240 torr) is 200 mm. In this case
(a) there is positive deviation from Raoult's law
(b) there is negative deviation from Raoult's law
(c) there is no deviation from Raoult's law
(d) molecular masses of *A* and *B* are also required

43 Mole fraction of the component A in vapour phase is χ_1 and mole fraction of the component A in liquid mixture is χ_2 ($p°_A$ = vapour pressure of pure A, $p°_B$ = vapour pressure of pure B) then total vapour pressure of the liquid mixture is

(a) $\frac{p°_A \chi_2}{\chi_1}$ (b) $\frac{p°_A \chi_1}{\chi_2}$ (c) $\frac{p°_B \chi_1}{\chi_2}$ (d) $\frac{p°_B \chi_2}{\chi_1}$

44 An ideal solution is made from 5 moles of benzene and 3.25 moles of toluene. At 298 K, the vapour pressure of pure benzene and pure toluene are 96.4 torr and 28.9 torr. Vapour phase first appears at

(a) 760 torr (b) 96.4 torr
(c) 28.9 torr (d) 69.8 torr

45 At 100°C and 1 atm of the density of liquid water is 1.0 g cm^{-3} and that of water vapour is 0.0006 g cm^{-3}, then the volume occupied by water molecules in 1 L of steam at that temperature is

(a) 6 cm^3 (b) 60 cm^3
(c) 0.6 cm^3 (d) 0.06 cm^3

46 An ideal solution of benzene and toluene boils at 1 atm pressure and 90°C temperature. At 90°C, benzene has a vapour pressure of 1022 mm and toluene has a vapour pressure of 406 mm. The mole fraction of benzene in the solution is

(a) 0.284 (b) 0.364 (c) 0.574 (d) 0.475

47 At 40°C, the vapour pressure in torr of methanol and ethanol solution is $p = 119x + 135$ where, x is the mole fraction of methanol. Hence,

(a) vapour pressure of pure methanol is 119 torr
(b) vapour pressure of pure ethanol is 135 torr
(c) vapour pressure of equimolar mixture of each is 127 torr
(d) mixture is completely immiscible

48 What is the vapour pressure if a solution with a benzene to octane molar ratio of 2 : 1, at 50°C vapour pressure of benzene is 280 mmHg and that of octane is 400 mmHg?

(a) 120 mmHg (b) 360 mmHg
(c) 680 mmHg (d) 320 mmHg

49 The vapour pressure of a pure liquid A is 40 mmHg at 310 K. The vapour pressure of this liquid in a solution with liquid B is 32 mmHg. Mole fraction of A in the solution, if it obeys Raoult's law is

(a) 0.8 (b) 0.5
(c) 0.2 (d) 0.4

50 Considering the formation, breaking and strength of hydrogen bond predict which of the following mixtures will show a positive deviation from Raoult's law? **[NCERT Exemplar]**

(a) Methanol and acetone (b) Chloroform and acetone
(c) Nitric acid and water (d) Phenol and aniline

51 If two liquids A and B form minimum boiling azeotrope at some specific composition then **[NCERT Exemplar]**

(a) A—B interactions are stronger than those between A—A or B—B
(b) vapour pressure of solution increases because more number of molecules of liquids A and B can escape from the solution
(c) vapour pressure of solution decreases because less number of molecules of only one of the liquids escape from the solution
(d) A—B interactions are weaker than those between A—A or B—B

52 At -47°C, the vapour pressure of pure ethyl bromide (C_2H_5Br) is 10.0 torr and that of pure ethyl chloride (C_2H_5Cl) is 40.0 torr. Assuming ideal behaviour, mole fraction of C_2H_5Cl in the vapour phase is 0.80. Hence, mole fraction of C_2H_5Cl in the liquid state is

(a) 0.20 (b) 0.50 (c) 0.75 (d) 0.25

53 The vapour pressure of water at 20°C is 17.54 mmHg. What will be the vapour pressure of the water in the apparatus shown after the piston is lowered, decreasing the volume of the gas above the liquid to one half of its initial volume? (assume temperature is constant)

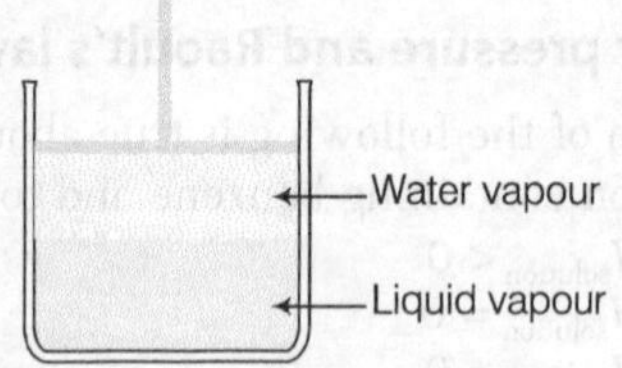

(a) 8.77 mmHg
(b) 17.54 mmHg
(c) 35.08 mmHg
(d) between 8.77 and 17.54 mmHg

54 At a given temperature, total vapour pressure in torr of a mixture of volatile components A and B is given by

$$p = 120 - 75\chi_B$$

hence, vapour pressure of pure A and B respectively (in torr) are

(a) 120, 75 (b) 120, 195
(c) 120, 45 (d) 75, 45

55 In the following aqueous solutions

(A) 1 m sucrose
(B) 1 m potassium ferricyanide and
(C) 1 m potassium sulphate

maximum value of vapour pressure of solution is that of

(a) Only A (b) Only B
(c) Only C (d) equal

56 If two substances A and B have $p_A^\circ : p_B^\circ = 1:2$ and mole fraction in solution $= 1:2$, then mole fraction of A in vapours is

(a) 0.33 (b) 0.25 (c) 0.52 (d) 0.2

57 At 300 K two pure liquids A and B have 150 mm Hg and 100 mm Hg vapour pressures, respectively. In an equimolar liquid mixture of A and B, the mole fraction of B in the vapour mixture at this temperature is

(a) 0.6 (b) 0.5
(c) 0.8 (d) 0.4

IV. Colligative properties-I

(Relative lowering vapour pressure, Elevation in boiling point and depression in freezing point)

58 $K_f m$ is expressed in the unit of

(a) molar mass (b) mass
(c) temperature (degree) (d) molality

59 $\Delta T_f / K_f$ is expressed in the unit of

(a) degree (b) degree mol^{-1} kg
(c) degree mol kg^{-1} (d) mol kg^{-1}

60 Which has the maximum freezing point?

(a) 6 g urea solution in 100 g H_2O
(b) 6 g acetic solution in 100 g H_2O
(c) 6 g sodium chloride in 100 g H_2O
(d) All have equal freezing point

61 Vapour pressure of pure water is 40 mm. If a non-volatile solute is added to it vapour pressure falls by 4 mm. Hence, molality of solution is

(a) 6.173 m (b) 3.0864 m
(c) 1.543 m (d) 0.772 m

62 Consider the following statements.

Statement I The vapour pressure of 0.1 M sugar solution is more than that of 0.1 M potassium chloride solution.

Statement II Lowering of vapour pressure is directly proportional to the number of species present in the solution.

(a) Only statement I is correct
(b) Only statement II is correct
(c) Both statements I and II are correct
(d) Both statements I and II are incorrect

63 If relative decrease in vapour pressure is 0.4 for a solution containing 1 mole NaCl in 3 moles H_2O, NaCl is ... % ionised.

(a) 60% (b) 50% (c) 100% (d) 40%

64 Lowering of vapour pressure due to a solute in 1 molal aqueous solution at 100°C is

(a) 13.44 torr (b) 14.12 torr
(c) 312 torr (d) 352 torr

65 Vapour pressure of a solution falls in presence of solute. Best explanation can be given by

(a) solute-solvent interaction
(b) entropy effect
(c) hydrogen bonding
(d) dipole-dipole interactions

66 Vapour pressure of CCl_4 at 25°C is 143 mmHg. 0.5 g of a non-volatile solute (molecular weight = 65) is dissolved in 100 mL CCl_4. Find the vapour pressure of the solution (Density of CCl_4 = 1.58 g / cm^3)

(a) 141.9 mm (b) 142.5 mm
(c) 140.0 mm (d) 144.0 mm

67 Two beakers of capacity 500 mL were taken. One of these beakers, labelled as A was filled with 400 mL water whereas the beaker labelled B was filled with 400 mL of 2 M solution of NaCl. At the same temperature both the beakers were placed in closed containers of same material and same capacity as shown in figure. **[NCERT Exemplar]**

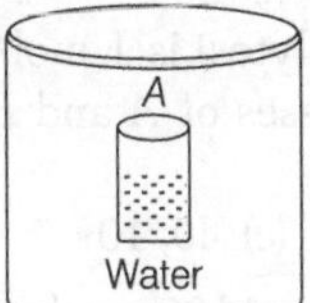

At a given temperature, which of the following statements is correct about the vapour pressure of pure water and that of NaCl solution?

(a) Vapour pressure in container (A) is more than that in container (B)
(b) Vapour pressure in container (A) is less than that in container (B)
(c) Vapour pressure is equal in both the containers
(d) Vapour pressure in container (B) is twice the vapour pressure in container (A)

68 The vapour pressure of pure liquid solvent A is 0.80 atm. When a non-volatile substance B is added to the solvent, its vapour pressure drops to 0.60 atm, mole fraction of the component B in the solution is

(a) 0.50 (b) 0.25
(c) 0.75 (d) 0.40

69 Select the correct statement(s).

(a) The fundamental cause of all colligative properties is the higher entropy of the solution relative to that of the pure solvent
(b) The freezing point of hydrofluoride solution is larger than that of equimolal hydrogen chloride solution
(c) 1 M glucose solution and 0.5 M NaCl solution are isotonic at a given temperature
(d) All the above are correct statements

70 Elevation in bp of a molar glucose solution ($d = 1.2$ g mL^{-1}) is

(a) $0.98\,K_b$ (b) K_b (c) $1.20\,K_b$ (d) $1.02K_b$

71 Select the correct statement(s).

(a) Boiling point of 1 molal NaCl solution is twice that 1 molal sucrose solution
(b) Boiling point elevation of 1 molal glucose solution is half of the 1 molal KCl solution
(c) Boiling point is a colligative property
(d) All of the above

72 Consider the following statements.

Satement I When NaCl is added to water a depression in freezing point is observed.

Statement II The lowering of vapour pressure of a solution causes depression in the freezing point.

(a) Only statement I is correct
(b) Only statement II is correct
(c) Both statements I and II are correct
(d) Both statements I and II are incorrect

73 Ratio of $\Delta T_b / K_b$ of 6% AB_2 and 9% A_2B (AB_2 and A_2B both are non-electrolytes) is 1 mol/kg in both cases. Hence, atomic masses of A and B are respectively

(a) 60, 90 (b) 40, 40 (c) 40, 10 (d) 10, 40

74 60 g of urea is dissolved in 1100 g solution. To keep $\Delta T / K_f$ as 1 mol/kg, water separated in the form of ice is

(a) 40 g (b) 60 g (c) 100 g (d) 200 g

75 2.56 g of sulphur in 100 g of CS_2 has depression in freezing point of 0.010°, $K_f = 0.1°$ (molal)$^{-1}$. Hence, atomicity of sulphur in CS_2 is

(a) 2 (b) 4 (c) 6 (d) 8

76 A solution contains 0.1 mole acetamide in 1 L of glacial acetic acid. When the solution is cooled, the first crystal that appeared at the freezing point contains the molecules of

(a) acetamide only
(b) acetic acid only
(c) Both acetamide and acetic acid
(d) either acetamide and acetic acid depending upon the conditions of the experiment

77 Density of 1 M solution of a non-electrolyte $C_6H_{12}O_6$ is 1.18 g/mL. If K_f (H_2O) is 1.86°C mol^{-1} kg, solution freezes at

(a) −1.58°C (b) −1.86°C
(c) −3.16°C (d) 1.86°C

78 Mole fraction of a non-electrolyte in aqueous solution is 0.07. If K_f is 1.86°C mol^{-1} kg, depression in fp, ΔT_f is

(a) 0.26°C (b) 1.86°C (c) 0.13°C (d) 7.78°C

79 The fundamental cause of ΔT(depression) is

(a) higher entropy of the solution relative to that of pure solvent
(b) lower entropy of the solution relative to that of pure solvent
(c) higher enthalpy of the solution relative to that of pure solvent
(d) lower enthalpy of the solution relative to that of pure solvent

80 250 g of an organic solvent contains x g of a non-electrolyte of molar mass m. The cryoscopic constant of the solvent is K_f degree mol^{-1} kg. Thus, ΔT_f (depression in freezing point) is

(a) $\dfrac{K_f x}{m}$ (b) $\dfrac{4K_f x}{m}$

(c) $\dfrac{K_f x}{4m}$ (d) $\dfrac{K_f x}{250\,m}$

81 Which of the following statements is false? **[NCERT Exemplar]**

(a) Two different solutions of sucrose of same molality prepared in different solvents will have the same depression in freezing point.
(b) The osmotic pressure of a solution is given by the equation $\pi = CRT$ (where, C is the molarity of the solution)
(c) Decreasing order of osmotic pressure for 0.01 M aqueous solutions of barium chloride, potassium chloride, acetic acid and sucrose is
$BaCl_2 > KCl > CH_3COOH >$ sucrose
(d) According to Raoult's law, the vapour pressure exerted by a volatile component of a solution is directly proportional to its mole fraction in the solution

V. Colligative properties-II
(Osmotic pressure, Abnomal molar mass, van't Hoff factor)

82 Desalinating of sea water is now done by using

(a) reverse osmosis (b) osmosis
(c) filtration (d) evaporation

83 A process analogous of osmosis is

(a) peptisation (b) coagulation
(c) distillation (d) dialysis

84 Osmosis results from

(a) an increase in entropy
(b) a decrease in entropy
(c) a decrease in enthalpy
(d) a decrease in internal energy

85 Important component in medical technologies used in kidney dialysis is

(a) semipermeable membrane
(b) salt-bridge
(c) quinhydrone electrode
(d) calomel electrode

86 pH of 1M HA (weak acid) is 2. Hence, van't Hoff factor is
(a) 1.2 (b) 1.02 (c) 1.1 (d) 1.01

87 In which case van't Hoff factor is maximum ?
(a) KCl 50% ionised (b) K_2SO_4 40% ionised
(c) $FeCl_3$ 30% ionised (d) $SnCl_4$ 20% ionised

88 Consider the following statements.
Statement I When a solution is separated from the pure solvent by a semipermeable membrane, the solvent molecules pass through it from pure solvent side to the solution side.
Statement II Osmosis of solvent occurs from a region of high concentration solution to a region of low concentration solution.
(a) Only statement I is correct
(b) Only statement II is correct
(c) Both statements I and II are correct
(d) Both statements I and II are incorrect

89 Select the correct statement(s).
(a) Osmosis, like all colligative properties results from an increase in entropy as pure solvent passes through the membrane and mixes with the solution
(b) Desalination of sea-water is done by reverse osmosis
(c) Both (a) and (b) are correct statements
(d) Osmosis is a non-spontaneous process

90 At 25°C, the highest osmotic pressure is exhibited by 0.1 M solution of
(a) $CaCl_2$ (b) KCl
(c) glucose (d) urea

91 Cellulose acetate is used as semipermeable membrane in desalination of sea water (reverse osmosis). It is because it is
(a) permeable to water but impermeable to impurities and ions
(b) impermeable to water but permeable to impurities and ions
(c) permeable to water impurities and ions
(d) impermeable to water impurities and ions

92 An unripe mango placed in a concentrated salt solution to prepare pickle shrivels because .... **[NCERT Exemplar]**
(a) it gains water due to osmosis
(b) it loses water due to reverse osmosis
(c) it gains water due to reverse osmosis
(d) it loses water due to osmosis

93 At a given temperature, osmotic pressure of a concentrated solution of a substance .... **[NCERT Exemplar]**
(a) is higher than that of a dilute solution
(b) is lower than that of a dilute solution
(c) is same as that of a dilute solution
(d) cannot be compared with osmotic pressure of dilute solution

94 If A contains 2% NaCl and is separated by a semi permeable membrane from B, which contains 10% NaCl then which of the following event will occur?
(a) NaCl will flow from A to B
(b) NaCl will flow from B to A
(c) Water will flow from A to B
(d) Water will flow from B to A

95 What is osmolarity of a 0.20 M KCl solution?
(a) 0.10 osmol (b) 0.20 osmol
(c) 0.30 osmol (d) 0.40 osmol

96 Red blood cells are placed in a solution and neither haemolysis nor crenation occurs. Therefore, the solution is
(a) hypertonic (b) hypotonic
(c) isotonic (d) isotopic

97 Osmotic pressure of blood is 7.40 atm at 27°C. Number of moles of glucose to be used per litre for an intravenous injection that is to have the same osmotic pressure as blood is
(a) 0.3 mol L^{-1} (b) 0.2 mol L^{-1}
(c) 0.1 mol L^{-1} (d) 0.4 mol L^{-1}

98 Sea water is found to contain NaCl and $MgCl_2$. If NaCl is 80% ionised and $MgCl_2$ is 50% ionised then van't Hoff factor in this case is
(a) 1.3 (b) 5.0 (c) 3.3 (d) 3.8

99 The relationship between osmotic pressure at 273 K when 10 g glucose (p_1), 10 g urea (p_2) and 10 g sucrose (p_3) are dissolved in 240 mL of water is
(a) $p_1 > p_2 > p_3$ (b) $p_3 > p_1 > p_2$
(c) $p_2 > p_1 > p_3$ (d) $p_2 > p_3 > p_1$

100 Which of the following salt has the same value of van't Hoff factor i as that of $K_3[Fe(CN)_6]$? (If all are 100% ionised)
(a) $Al_2(SO_4)_3$ (b) NaCl (c) Na_2SO_4 (d) $Al(NO_3)_3$

101 Which of the following has the maximum value of van't Hoff factor?
(a) KO_2 (b) NH_2CONH_2
(c) $C_6H_{12}O_6$ (d) $C_{12}H_{22}O_{11}$

102 In the following equilibrium

$$N_2O_4(g) \rightleftharpoons 2NO_2(g)$$

NO_2 is 50% of the total volume. Hence, degree of dissociation (x) and van't Hoff factor (i) respectively are
(a) 0.5, 1.5 (b) 0.25, 1.25
(c) 0.33, 1.33 (d) 0.66, 1.66

103 Consider the following statements.
I. Isotonic solution have the same molar concentration at a given temperature.
II. The molal elevation constant K_b is characteristic of a solvent and is independent of the solute added.

III. The freezing point of 0.1 M KCl solution is more than that of 0.1 M aqueous $AlCl_3$ solution.

Which of the above statements are correct? Choose the correct option accordingly.

(a) Both I and II (b) Both II and III
(c) Both I and III (d) I, II and III

104 A complex containing K^+, Pt (IV) and Cl^- is 100% ionised giving $i = 3$. Thus, complex is

(a) $K_2[PtCl_4]$ (b) $K_2[PtCl_6]$
(c) $K_3[PtCl_5]$ (d) $K[PtCl_3]$

105 In which of the following case, van't Hoff factor i remains unchanged

(a) $PtCl_4$ reacts with KCl
(b) *aq* .$ZnCl_2$ reacts with *aq.* NH_3
(c) *aq.* $FeCl_3$ reacts with *aq.* $K_4[Fe(CN)_6]$
(d) $KMnO_4$ reduced to MnO_2 in alkaline medium

106 van't Hoff factors are x, y, z in the case of association, dissociation and no change respectively. Increasing order is

(a) $x < y < z$ (b) $x = y = z$
(c) $y < x < z$ (d) $x < z < y$

107 Vapour density of $PCl_5(g)$ dissociating into $PCl_3(g)$ and $Cl_2(g)$ is 100. Hence, van't Hoff factor for the case

$$PCl_5(g) \rightleftharpoons PCl_3(g) + Cl_2(g) \text{ is}$$

(a) 1.85 (b) 3.70 (c) 1.085 (d) 1.0425

108 Aluminium phosphate is 100% ionised in 0.01 molal aqueous solution. Hence, $\Delta T_b / K_b$ is

(a) 0.01 (b) 0.015
(c) 0.0175 (d) 0.02

109 Aqueous solution of barium phosphate which is 100% ionised has $\Delta T_f / K_f$ as 0.05. Hence, given solution is

(a) 0.01 molal (b) 0.02 molal
(c) 0.04 molal (d) 0.05 molal

110 Maximum freezing point will be for 1 molal of solution (assuming equal ionisation in each case) is

(a) $[Fe(H_2O)_6]Cl_3$
(b) $[Fe(H_2O)_5Cl]Cl_2 \cdot H_2O$
(c) $[Fe(H_2O)_4Cl_2]Cl \cdot 2H_2O$
(d) $[Fe(H_2O)_3Cl_3] \cdot 3H_2O$

111 An aqueous solution of a solute *AB* has boiling point of 101.08°C (*AB* is 100% ionised in boiling point of the solution) and freezes at –1.80°C.

$$(K_b / K_f = 0.3)$$

Hence, *AB*

(a) is 100% ionised at the freezing point of the solution
(b) behaves as non-electrolyte at the freezing point of the solution
(c) forms dimer
(d) None of the above

112 Consider the following statements.

Statement I The boiling point of 0.1 M urea solution is less than that of 0.1 M KCl solution.

Statement II Elevation of boiling point is directly proportional to the number of species present in the solution.

(a) Only statement I is correct
(b) Only statement II is correct
(c) Both statements I and II are correct
(d) Both statements I and II are incorrect

113 Depression of freezing point of 0.01 molal *aq.* CH_3COOH solution is 0.02046°. 1 molal urea solution freezes at –1.86° C. Assuming molality equal to molarity, pH of CH_3COOH solution is

(a) 2 (b) 3
(c) 3.2 (d) 4.2

114 12.2 g of benzoic acid ($M = 122$) in 100 g benzene has depression in freezing point 2.6°C, $K_f = 5.2°$ C kg / mol. If there is 100% polymerisation then number of molecules of benzoic acid in associated state is

(a) 1 (b) 2
(c) 3 (d) 4

115 In comparison to a 0.01 M solution of glucose, the depression in freezing point of a 0.01 M $MgCl_2$ solution is .... **[NCERT Exemplar]**

(a) the same (b) about twice
(c) about three times (d) about six times

116 Match the terms given in Column I with its explanation given in Column II.

	Column I		Column II
A.	Upper critical solution temperature	1.	At which a gas cannot be liquefied
B.	Isotonic solutions	2.	Solutions have same osmotic pressure
C.	Cryoscopic constant	3.	Solutions have same viscosity
		4.	ΔT produced by 1 g of a solute in a litre
		5.	ΔT per mole of solute/ 1000 g of solvent and above
		6.	At and above which two immiscible liquids become miscible

Codes

	A	B	C		A	B	C
(a)	6	4	3	(b)	1	6	5
(c)	3	4	6	(d)	6	2	5

117 If $pK_a = -\log K_a = 4$ and $K_a = Cx^2$ and C = 0.01M then van't Hoff factor for weak monobasic acid is

(a) 1.01 (b) 1.02 (c) 1.10 (d) 1.20

118 $FeCl_3$ on reaction with $K_4[Fe(CN)_6]$ in aqueous solution gives blue colour. These are separated by a semi permeable membrane *AB* as shown.

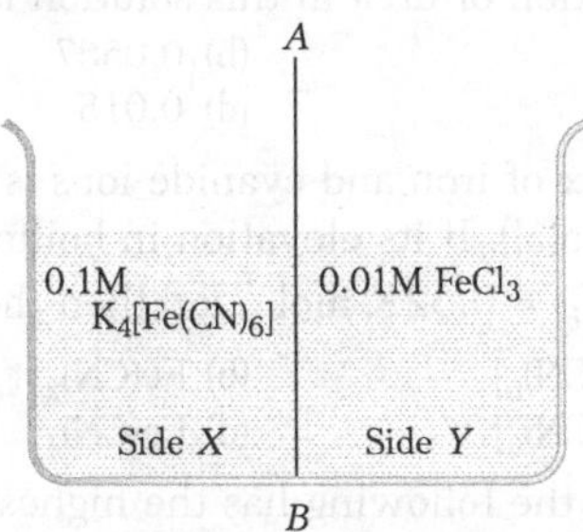

Due to osmosis there is

(a) blue colour formation in side *X*
(b) blue colour formation in side *Y*
(c) blue colour formation in both of the sides *X* and *Y*
(d) no blue colour formation

119 Match the statement (in Column I) with the single term of the statement (in Column II).

	Column I		Column II
A.	Hypotonic	1.	A solution in which more solute can be dissolved.
B.	Haemolysis	2.	Refers to two solutions having the same osmolarity.
C.	Hypertonic	3.	The swelling and bursting of a red blood cell when placed in a solution of lower osmolarity than the cell itself.
D.	Crenation	4.	The shrinking of a red blood cell when placed in a solution of greater osmolarity than the cell itself.
E.	Hygroscopic	5.	Refers to a solution having greater osmolarity than 0.30 osmol, the osmolarity of normal red blood cells.
F.	Isotonic	6.	An ionic compound which attracts water molecules from the atmosphere.
G.	Unsaturated	7.	Refers to a solution having a lower osmolarity than 30.0 osmol, the osmolarity of blood serum.

Codes

	A	B	C	D	E	F	G
(a)	7	3	5	4	6	2	1
(b)	5	7	4	3	2	1	6
(c)	7	5	4	6	3	2	1
(d)	3	4	5	7	6	1	2

120 Human blood gives rise to an osmotic pressure of approximately 7.65 atm at body temperature, 37°C. Hence, molarity of an intravenous glucose solution be to have the same osmotic pressure as blood is

(a) 0.30 M
(b) 0.20 M
(c) 0.10 M
(d) 0.50 M

121 Osmotic pressure of insulin solution at 298 K is found to be 0.0072 atm. Hence, height of water column due to this pressure is [given d(Hg) = 13.6 g /mL]

(a) 0.76 cm (b) 0.70 cm
(c) 7.4 cm (d) 76 cm

122 We have 200 mL of 6% M/V urea solution at 27°C. Osmotic pressure due to osmosis is ($R = 0.0821$ L atm mol^{-1} K^{-1})

(a) 24.63 atm (b) 12.315 atm
(c) 4.926 atm (d) 6.157 atm

123 Sea water is approximately 0.70 M in NaCl. It can be desalinated by reverse osmosis. For 50% recovery additional osmotic pressure required (on sea-water side) at 298 K is

(a) 30 atm (b) 45 atm
(c) 68 atm (d) 90 atm

124 What will be the ratio of the masses of formalin (HCHO) and glucose ($C_6H_{12}O_6$) contained in equal volumes of solutions having the same osmotic pressure at the given temperature?

(a) 1 : 1 (b) 1 : 2
(c) 1 : 3 (d) 1 : 6

125 Consider the following figure :

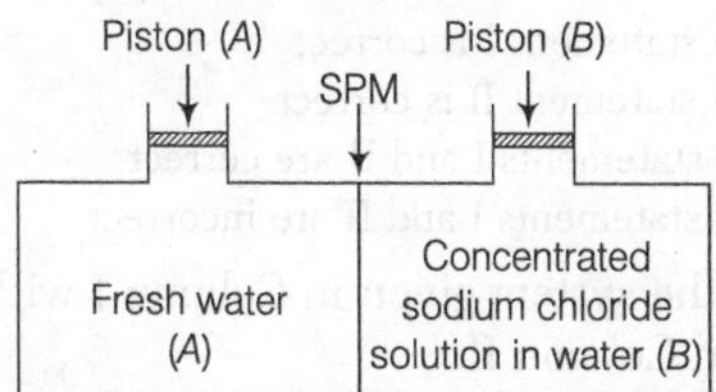

Mark the correct option. **[NCERT Exemplar]**

(a) Water will move from side (*A*) to side (*B*) if a pressure lower osmotic pressure is applied on piston (*B*)
(b) Water will move from side (*B*) to side (*A*) if a pressure greater than osmotic pressure is applied on piston (*B*)
(c) Water will move from side (*B*) to side (*A*) if a pressure equal to osmotic pressure is applied on piston (*B*)
(d) Water will move from side (*A*) to side (*B*) if pressure equal to osmotic pressure is applied on piston (*A*)

(B) Miscellaneous questions

1 The unit of ebullioscopic constant is **[NCERT Exemplar]**
(a) K kg mol^{-1} or K $(molality)^{-1}$
(b) mol kg K^{-1} or K^{-1} (molality)
(c) kg mol^{-1} K^{-1} or K^{-1} $(molality)^{-1}$
(d) K mol kg^{-1} or K (molality)

2 What quantity (in mL) of a 45% acid solution of a mono-protic strong acid must be mixed with a 20% solution of the same acid to produce 800 mL of 29.875% acid solution?
(a) 320 (b) 325 (c) 316 (d) 330

3 Which of the following has the maximum osmotic pressure at temperature T?
(a) 100 mL of 1 M urea solution
(b) 300 mL of 1 M glucose solution
(c) Mixture of 100 mL of 1 M urea solution and 300 mL of 1 M glucose solution
(d) All the above are isotonic

4 When mercuric iodide is added to the aqueous solution of potassium iodide, then
(a) freezing point is raised
(b) freezing point is lowered
(c) freezing point does not change
(d) boiling point does not change

5 We have 100 mL of 0.1 M KCl solution. To make it 0.2 M
(a) evaporate 50 mL water (b) evaporate 50 mL solution
(c) add 0.1 mol KCl (d) add 0.01 mol KCl

6 **Statement I** When methyl alcohol is added to water, boiling point of water increases.
Statement II When a volatile solute is added to a volatile solvent elevation in boiling point is observed.
(a) Only statement I is correct
(b) Only statement II is correct
(c) Both statements I and II are correct
(d) Both statements I and II are incorrect

7 Match the system given in Column I with its type given in Column II.

Column I	Column II
A. Ethyl alcohol-water system	1. Immiscible liquid mixture
B. Phenol-water system	2. Azeotropic mixture
C. Water-nitrobenzene system	3. Partially miscible liquid mixture
	4. Ternary mixture

Codes

	A	B	C
(a)	2	4	1
(b)	3	1	4
(c)	2	3	1
(d)	4	2	3

8 Elevation in boiling point of an aqueous urea solution is 0.52 K. (K_b = 0.52 K mol^{-1} kg). Hence, mole fraction of urea in this solution is
(a) 0.982 (b) 0.0567
(c) 0.943 (d) 0.018

9 A complex of iron and cyanide ions is 100% ionised at 1 m (molal). If its elevation in boiling point is 2.08K. (K_b = 0.52K mol^{-1} kg) then the complex is
(a) $K_3[Fe(CN)_6]$ (b) $Fe(CN)_2$
(c) $K_4[Fe(CN)_6]$ (d) $Fe(CN)_4$

10 Which of the following has the highest boiling point?
(a) 0.1 M Na_2SO_4
(b) 0.1 M $C_6H_{12}O_6$ (glucose)
(c) 0.1 M $MgCl_2$
(d) 0.1 M $Al(NO_3)_3$

11 Select the correct statement(s).
(a) Heats of vaporisation for a pure solvent and for a solution are similar because similar intermolecular forces between solvent molecules must be overcome in both cases.
(b) Entropy change between solution and vapour is smaller than the entropy change between pure solvent and vapour.
(c) Boiling point of the solution is larger than that of the pure solvent.
(d) All the above are correct statements.

12 The dissolving process is exothermic when
(a) the energy released in solvation exceeds the energy used in breaking up solute-solute and solvent-solvent interactions
(b) the energy used in solvation exceeds the energy released in breaking up solute-solute and solvent-solvent interactions
(c) the energy released in solvation is about the same as the energy used in breaking up solute-solute and solvent-solvent interactions
(d) the energy used in solvation is about the same as the energy used in breaking up solute-solute and solvent-solvent interactions

13 Low concentration of oxygen in the blood and tissues of people living at high altitude is due to ……… **[NCERT Exemplar]**
(a) low temperature
(b) low atmospheric pressure
(c) high atmospheric pressure
(d) Both low temperature and high atmospheric pressure

14 Mole fraction of glycine molecule in 0.140 molal aqueous solution is
(a) 1.4×10^{-1}
(b) 8.6×10^{-1}
(c) 2.51×10^{-3}
(d) 9.9748×10^{-1}

15 Select the correct statement(s).

(a) When solid $CaCl_2$ is added to liquid water, the temperature rises
(b) When solid $CaCl_2$ is added to ice at 0°C, the freezing point falls
(c) Both (a) and (b)
(d) Adding $CaCl_2$ to water has no effect

16 An ionic compound that attracts atmospheric water so strongly that a hydrate is formed is said to be

(a) dilute (b) hygroscopic
(c) immiscible (d) miscible

17 pH of a 0.1 M monobasic acid is found to be 2. Hence, its osmotic pressure at a given temperature T K is

(a) 0.1 RT (b) 0.11 RT
(c) 1.1 RT (d) 0.01 RT

18 Acetic acid in benzene solution forms dimer due to intermolecular H-bonding. For this case van't Hoff factor is

(a) $i=1$ (b) $i>1$
(c) $i<1$ (d) inclusive

19 Consider the following statements.

Statement I Acetic acid has a molecular weight of 120 in benzene solution.

Statement II There is dimer formation by H-bonding.

(a) Only statement I is correct
(b) Only statement II is correct
(c) Both statements I and II are correct
(d) Both statements I and II are incorrect

20 On the basis of information given below mark the correct option.

Information

On adding acetone to methanol some of the hydrogen bonds between methanol molecules break. **[NCERT Exemplar]**

(a) At specific composition methanol acetone mixture will form minimum boiling azeotrope and will show positive deviation from Raoult's law
(b) At specific composition methanol-acetone mixture will form maximum boiling azeotrope and will show positive deviation from Raoult's law
(c) At specific composition methanol-acetone mixture will form minimum boiling azeotrope and will show negative deviation from Raoult's law
(d) At specific composition methanol-acetone mixture will form maximum boiling azeotrope and will show negative deviation from Raoult's law

21 At 35°C, the vapour pressure of CS_2 is 512 mmHg and that of acetone is 344 mm Hg. A solution of CS_2 and acetone in which the mole fraction of CS_2 is 0.25 has a total vapour pressure of 600 mm Hg. Which of the following statements is/are correct?

(a) A mixture of 100 mL of acetone and 100 mL of CS_2 has a volume of 200 mL
(b) When acetone and CS_2 are mixed at 35°C, heat must be absorbed in order to produce a solution at 35°C
(c) Both (a) and (b)
(d) None of the above

22 At a given temperature, the vapour pressures of pure A and B are 108 and 36 torr respectively. What will be the mole fraction of B in the vapour phase which is in equilibrium with a solution containing equimolar fraction of A and B under ideal behaviour conditions?

(a) 0.25 (b) 0.33 (c) 0.50 (d) 0.60

23 Relative decrease in vapour pressure of an aqueous solution containing 2 moles of $[Cu(NH_3)_3Cl]Cl$ in 3 moles of H_2O is 0.50. On reaction with $AgNO_3$, this solution will form

(a) 1 mol AgCl (b) 0.25 mol AgCl
(c) 2 mol AgCl (d) 0.40 mol AgCl

24 Consider the following cases.

I. 2 M CH_3COOH solution in benzene at 27°C where there is dimer formation to the extent of 100%.
II. 0.5 M KCl *aq.* solution at 27°C, which ionises 100%; which of the above is/are true statement(s)?

Choose the correct alternative accordingly.

(a) Both are isotonic (b) I is hypertonic
(c) II is hypertonic (d) None of these

25 Consider the following four liquids.

1. Water : highly polar, H-bonding
2. Hexanol : slightly polar, some H-bonding
3. Chloroform : slightly polar, no H-bonding
4. Octane : non-polar, no H-bonding

Which of the following pair of liquids is immiscible?

(a) Water and octane
(b) Water and hexanol
(c) Hexanol and chloroform
(d) Chloroform and octane

26 Which of the following azeotropic solutions has the boiling point less than boiling point of the constituents A and B?

(a) $CHCl_3$ and CH_3COCH_3
(b) CS_2 and CH_3COCH_3
(c) CH_3CH_2OH and CH_3COCH_3
(d) CH_3CHO and CS_2

27 The vapour pressure of a solvent is decreased by 10 mm of mercury when a non-volatile solute was added to the solvent. The mole fraction of the solute in the solution is 0.2. What should be the mole fraction of the solvent if the decrease in the vapour pressure is to be 20 mm of mercury?

(a) 0.8 (b) 0.6
(c) 0.4 (d) 0.2

28 The vapour pressure of chloromethane at various mole fraction (χ) in a mixture at 298 K was found to be as follows :

χ	0.005	0.009	0.019	0.024
***p*/torr**	205	363	756	946

Thus, Henry's law constant is
(a) 4×10^4 torr (b) 2×10^4 torr
(c) 2.5×10^{-5} torr^{-1} (d) 5×10^{-5} torr^{-1}

29 Mole fraction of vapours A above the solution in the mixture of A and B ($\chi_A = 0.4$) will be
$[p^\circ_A = 100$ mm, $p^\circ_B = 200$ mm$]$
(a) 0.4 (b) 0.8
(c) 0.25 (d) None of these

30 At what temperature does an aqueous solution containing 3×10^{23} molecules of a non-electrolyte in 250 g H_2O freeze? (Molal elevation constant of H_2O is 1.86 K mol^{-1} kg)
(a) 273 K (b) 269.28 K (c) 271.14 K (d) 270 K

31 van't Hoff factors of aqueous solutions of X, Y, Z are 1.8, 0.8 and 2.5. Hence, their
(a) boiling point : $Z<X<Y$
(b) freezing point : $Z<X<Y$
(c) osmotic pressure : $X=Y=Z$
(d) vapour pressure : $Y<X<Z$

32 An aqueous solution of 0.01 M CH_3COOH has van't Hoff factor of 1.01. If pH = – log [H^+], pH of 0.01 M CH_3COOH solution would be
(a) 2 (b) 3 (c) 4 (d) 5

33 Decimolar solution of potassium ferricyanide, $K_3[Fe(CN)_6]$ has osmotic pressure of 3.94 atm at 27°C. Hence, per cent ionisation of the solute is
(a) 10% (b) 20% (c) 30% (d) 40%

34 The degree of dissociation 'α' of a weak electrolyte is (where, N is the number of ions given by 1 mole of the electrolyte.)
(a) $\frac{i-1}{N+1}$ (b) $\frac{i-1}{N-1}$ (c) $\frac{N-1}{i-1}$ (d) $\frac{N+1}{i-1}$

35 Select the correct statement(s).
(a) Solution has more molecular randomness than a pure solvent has, the entropy change between solution and solid is larger than the entropy change between pure solvent and solid
(b) Heats of fusion of solution and solvent are similar since, similar forces of intermolecular forces are involved
(c) Sugar containing solution freezes at a lower temperature than pure water
(d) All the above are correct statements

36 Which of the following statements is false? **[NCERT Exemplar]**
(a) Units of atmospheric pressure and osmotic pressure are the same
(b) In reverse osmosis, solvent molecules move through a semipermeable membrane from a region of lower concentration of solute to a region of higher concentration
(c) The value of molal depression constant depends on nature of solvent
(d) Relative lowering of vapour pressure is a dimensionless quantity

37 Which of the following aqueous solutions should have the highest boiling point? **[NCERT Exemplar]**
(a) 1.0 M NaOH (b) 1.0 M Na_2SO_4
(c) 1.0 M NH_4NO_3 (d) 1.0 M KNO_3

38 The values of van't Hoff factors for KCl, NaCl and K_2SO_4 respectively are .... **[NCERT Exemplar]**
(a) 2, 2 and 2 (b) 2, 2 and 3
(c) 1, 1 and 2 (d) 1, 1 and 1

39 We have three aqueous solutions of NaCl labelled as A, B and C with concentrations 0.1 M, 0.01 M and 0.001 M, respectively. The value of van't Hoff factor for these solutions will be in the order .... **[NCERT Exemplar]**
(a) $i_A < i_B < i_C$ (b) $i_A > i_B > i_C$
(c) $i_A = i_B = i_C$ (d) $i_A < i_B > i_C$

40 The vapour pressure of pure benzene, C_6H_6 at 50°C is 268 torr. How many moles of non-volatile solute per mol of benzene is required to prepare a solution of benzene having a vapour pressure of 167 torr at 50°C?
(a) 0.377 (b) 0.605 (c) 0.623 (d) 0.395

41 The boiling point elevation constant for toluene is 3.32 K kg mol^{-1}, the normal boiling point of toluene is 110.7°C. The enthalpy of vaporisation of toluene would be nearly
(a) 17.0 kJ mol^{-1}
(b) 34.0 kJ mol^{-1}
(c) 51.0 kJ mol^{-1}
(d) 68.0 kJ mol^{-1}

42 Consider following two solutions
I. 0.500 M KCl or 0.500 wt. % KCl in water.
II. 1.75 M glucose or 1.75 m glucose in water.
Which of the following solutions is more concentrated in each pair?

	I	II
(a)	0.500 M KCl	1.75 M glucose
(b)	0.500 wt. % KCl	1.75 M glucose
(c)	0.500 M KCl	1.75 m glucose
(d)	0.500 wt. % KCl	1.75 m glucose

43 Colligative properties have many practical uses, some of them may be
I. melting of snow by salt
II. desalination of sea water
III. determination of molar mass
IV. determination of melting point and boiling point of solvent

Actual practical uses are
(a) Both I and II (b) Both III and IV
(c) I, II and III (d) II, III and IV

44 Consider following terms.
I. mK_b II. mK_f
III. $\frac{\Delta T_b}{i}$ IV. K_b

Terms which can be expressed in degree (temperature) are
(a) Both III and IV (b) Both I and II
(c) I, II and III (d) Both I and III

45 The freezing point of benzene decreases by 0.45° C, when 0.2 g of acetic acid is added to 20 g of benzene. If acetic acid associates to form a dimer in benzene, percentage association of acetic acid in benzene will be
(K_f for benzene = 5.12 K kg mol^{-1})
(a) 64.6 % (b) 80.4 %
(c) 74.6 % (d) 94.6 %

46 Relative decrease in vapour pressure of an aqueous glucose dilute solution is found to be 0.018. Hence, elevation in boiling point is (It is given 1 molal *aq.* urea solution boils at 100.54°C at 1 atm pressure)
(a) 0.018° (b) 0.18°
(c) 0.54° (d) 0.03°

47 Match the items given in Column I and Column II.

	Column I		Column II
A.	Saturated solution	1.	Solution having same osmotic pressure at a given temperature as that of given solution.
B.	Binary solution	2.	A solution whose osmotic pressure is less than that of another.
C.	Isotonic solution	3.	Solution with two components.
D.	Hypotonic solution	4.	A solution which contains maximum amount of solute that can be dissolved in a given amount of solvent at a given temperature.
E.	Solid solution	5.	A solution whose osmotic pressure is more than that of another.
F.	Hypertonic solution	6.	A solution in solid phase.

Codes

	A	B	C	D	E	F
(a)	5	2	4	3	6	1
(b)	4	3	1	2	6	5
(c)	6	4	3	2	5	1
(d)	4	1	5	2	3	4

48 Vapour pressure of pure A, $p_A^\circ = 100$ mm Hg
Vapour pressure of pure B, $p_B^\circ = 150$ mm Hg
Distillate of vapours of a solution containing 2 moles A and 3 moles B will have total vapour pressure, approximately, (on condensation)
(a) 135 mm (b) 130 mm (c) 140 mm (d) 145 mm

49 1 mole each of the following solutes are taken in 5 moles of water,
A. NaCl B. K_2SO_4
C. Na_3PO_4 D. glucose

Assuming 100% ionisation of the electrolyte, relative decrease in vapour pressure will be in the order
(a) A < B < C < D (b) D < C < B < A
(c) D < A < B < C (d) equal

50 Which combination of vapour pressure, intermolecular forces and ΔH_{vap} (latent heat of vaporisation) is matched correctly?

	Vapour pressure	Intermolecular forces	ΔH_{vap}
(a)	High	Weak	Small
(b)	High	Strong	Large
(c)	Low	Weak	Large
(d)	Low	Strong	Small

51 Consider the following solutions
I. 1 M *aq.* glucose
II. 1 M *aq.* sodium chloride
III. 1 M benzoic acid in benzene
IV. 1 M ammonium phosphate

Select the incorrect statement(s).
(a) All are isotonic solutions
(b) III is hypotonic of I, II, IV
(c) I, II, IV are hypertonic of III
(d) IV is hypertonic of I, II III

52 Match the laws given in Column I with expressions given in Column II.

	Column I		Column II
A.	Raoult's law	1.	$\Delta T_f = K_f m$
B.	Henry's law	2.	$\pi = CRT$
C.	Elevation of boiling point	3.	$p = \chi_1 p_1^\circ + \chi_2 p_2^\circ$
D.	Depression in freezing point	4.	$\Delta T_b = K_b m$
E.	Osmotic pressure	5.	$p = K_H \cdot \chi$

Codes

	A	B	C	D	E
(a)	3	1	4	5	2
(b)	5	3	1	2	4
(c)	4	2	5	1	3
(d)	3	5	4	1	2

53 The phase diagrams for the pure solvent (solid lines) and the solution (non-volatile solute, dashed line) are recorded below

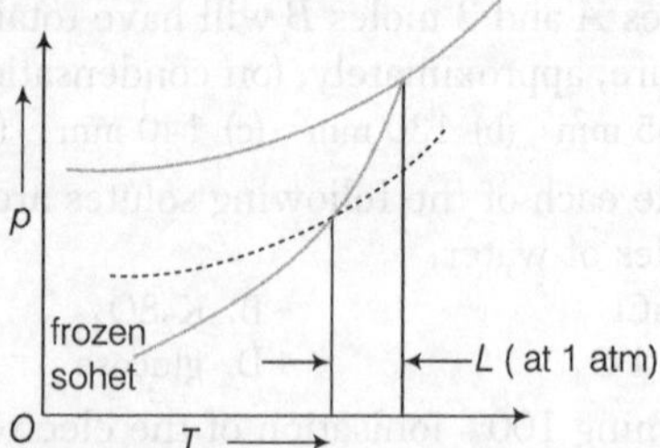

The quantity indicated by L in the figure is (m = molality)

(a) Δp (b) ΔT_f (c) $K_b m$ (d) $K_f m$

54 Some of the following gases are soluble in water due to formation of their ions

I. CO_2 II. NH_3
III. HCl IV. CH_4
V. H_2

Water insoluble gases can be

(a) I, IV and V (b) Both I and V
(c) I, II and III (d) Both IV and V

55 Consider the following statements.

I. An ideal solution is one in which both components obey Raoult's law.
II. A colligative property is a property that depends on the number of solute particles, not their chemical identity; they arise from the effect of a solute on the entropy of solution.
III. An azeotrope is a mixture that vaporises and condenses without a change of composition.

Select the correct statement(s) and mark the correct option.

(a) Both I and II (b) Both I and III
(c) Both II and III (d) I, II and III

56 **Statement I** The freezing point of 0.05 M urea solution is different from that of 0.05 M sodium chloride solution.

Statement II The depression in freezing point is directly proportional to the number of species present in the solution.

(a) Only statement I is correct
(b) Only statement II is correct
(c) Both statements I and II are correct
(d) Both statements I and II are incorrect

57 Some entropy change are represented in figure.

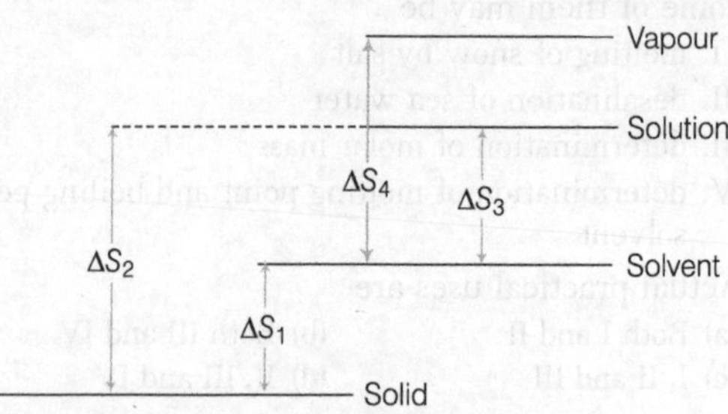

Select the correct entropy change

(a) $\Delta S_1, \Delta S_2, \Delta S_3$
(b) $\Delta S_1, \Delta S_2, \Delta S_4$
(c) $\Delta S_1, \Delta S_2, \Delta S_3, \Delta S_4$
(d) ΔS_2 and ΔS_4

58 5 g of Na_2SO_4 was dissolved in x g of H_2O. The change in freezing point was found to be 3.82°C. If Na_2SO_4 is 81.5% ionised, the value of x (K_f for water = 186°C kg mol^{-1}) is approximately. (molar mass of S = 32 g mol^{-1} and that of Na = 23 g mol^{-1}

(a) 15 g (b) 25 g
(c) 45 g (d) 65 g

59 Based on the given diagram

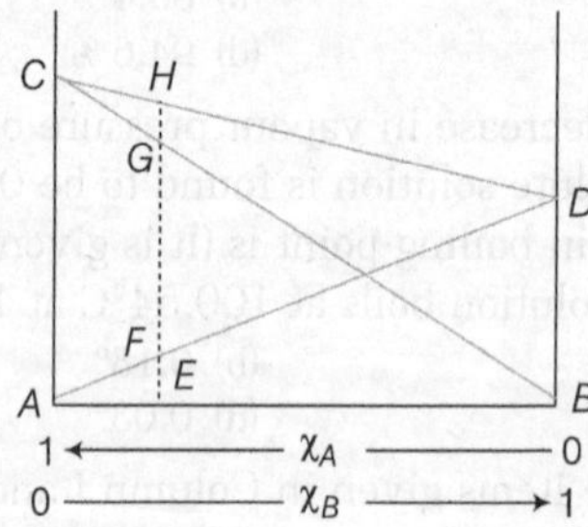

Which of the following statements regarding the solutions of two miscible volatile liquids are correct?

I. Plots AD and BC show that Raoult's law is obeyed for the solution in which B is a solvent and A is the solute and as well as for that in which A is solvent and B is solute
II. Plots CD shows that Dalton's law of partial pressures is observed by the binary solutions of components A and B
III. AC and BD correspond to the vapour pressure of the pure solvents A and B respectively.

Select the correct answer using the codes given below.

Codes

(a) Both I and II (b) Both II and III
(c) Both I and III (d) I, II and III

(C) Medical entrances' gallery

1 The mixture which shows positive deviation from Raoult's law is **[NEET 2020]**
(a) benzene + toluene
(b) acetone + chloroform
(c) chloroethane + bromoethane
(d) ethanol + acetone

2 The freezing point depression constant (K_f) of benzene is 5.12 K kg mol^{-1}. The freezing point depression for the solution of molality 0.078 m containing a non-electrolyte solute in benzene is (rounded off upto two decimal places) **[NEET 2020]**
(a) 0.80 K (b) 0.40 K
(c) 0.60 K (d) 0.20 K

3 Isotonic solution have same **[NEET 2020]**
(a) vapour pressure
(b) freezing temperature
(c) osmotic pressure
(d) boiling temperature

4 Which of the following statements is correct regarding a solution of two compounds *A* and *B* exhibiting positive deviation from ideal behaviour? **[NEET (Odisha) 2019]**
(a) Intermolecular attractive forces between *A*—*A* and *B*—*B* are stronger than those between *A*—*B*.
(b) $\Delta_{mix} H = 0$ at constant *T* and *p*
(c) $\Delta_{mix} V = 0$ at constant *T* and *p*
(d) Intermolecular attractive forces between *A*—*A* and *B*—*B* are equal to those between *A*—*B*.

5 For an ideal solution, the correct option is **[NEET 2019]**
(a) $\Delta_{mix} V \neq 0$ at constant *T* and *p*
(b) $\Delta_{mix} H = 0$ at constant *T* and *p*
(c) $\Delta_{mix} G = 0$ at constant *T* and *p*
(d) $\Delta_{mix} S = 0$ at constant *T* and *p*

6 The mixture that forms maximum boiling azeotrope is **[NEET 2019]**
(a) ethanol + water
(b) acetone + carbon disulphide
(c) heptane + octane
(d) water + nitric acid

7 Benzoic acid undergoes dimerisation in benzene solution. The van't Hoff factor (*i*) is related to the degree of association '*x*' of the acid as **[NEET 2018]**
(a) $i = (1 - x)$ (b) $i = (1 + x)$
(c) $i = \left(1 - \frac{x}{2}\right)$ (d) $i = \left(1 + \frac{x}{2}\right)$

8 If molality of the dilute solution is doubled, the value of molal depression constant (K_f) will be **[NEET 2017]**
(a) unchanged (b) doubled
(c) halved (d) tripled

9 Which of the following is dependent on temperature? **[NEET 2017]**
(a) Weight percentage
(b) Molality
(c) Molarity
(d) Mole fraction

10 At 100°C, the vapour pressure of a solution of 6.5 g of a solute in 100 g water is 732 mm. If, $K_b = 0.52°$ mol^{-1} kg the boiling point of this solution will be **[NEET (I) 2016]**
(a) 100°C (b) 102°C
(c) 103°C (d) 101°C

11 Which of the following statements about the composition of the vapour over an ideal 1 : 1 molar mixture of benzene and toluene is correct? Assume that the temperature is constant at 25°C. (Given, vapour pressure data at 25°C, benzene = 12.8 kPa, toluene = 3.85 kPa) **[NEET (I) 2016]**
(a) The vapour will contain a higher percentage of toluene
(b) The vapour will contain equal amounts of benzene and toluene
(c) Not enough information is given to make a prediction
(d) The vapour will contain a higher percentage of benzene

12 The van't Hoff factor (*i*) for a dilute aqueous solution of the strong electrolyte barium hydroxide is **[NEET (II) 2016]**
(a) 2 (b) 3
(c) 0 (d) 1

13 Which one of the following is incorrect for ideal solution? **[NEET (II) 2016]**
(a) $\Delta p = p_{obs} - p_{\text{calculated by Raoult's law}} = 0$
(b) $\Delta G_{mix} = 0$
(c) $\Delta H_{mix} = 0$
(d) $\Delta U_{mix} = 0$

14 Which one is not equal to zero for an ideal solution? **[CBSE AIPMT 2015]**
(a) ΔH_{mix} (b) ΔS_{mix}
(c) ΔV_{mix} (d) $\Delta p = p_{observed} - p_{Raoult}$

15 The boiling point of 0.2 mol kg^{-1} solution of *X* in water is greater than equimolal solution of *Y* in water. Which one of the following statements is true in this case? **[CBSE AIPMT 2015]**
(a) *X* is undergoing dissociation in water
(b) Molecular mass of *X* is greater than the molecular mass of *Y*
(c) Molecular mass of *X* less than the molecular mass of *Y*
(d) *Y* is undergoing dissociation in water while *X* undergoes no change

16 Which one of the following electrolytes has the same value of van't Hoff's factor (*i*) as that of $Al_2(SO_4)_3$ (if all are 100% ionised)? **[Kerala (Medical) 2015]**

(a) K_2SO_4 (b) $K_3[Fe(CN)_6]$
(c) $Al(NO_3)_3$ (d) $K_4[Fe(CN)_6]$

17 The vapour pressure of a solvent at 293 K is 100 mmHg. Then the vapour pressure of a solution containing 1 mole of a strong electrolyte (AB_2) in 99 moles of the solvent at 293 K is (assume complete dissociation of solute) **[Kerala (Medical) 2015]**

(a) 103 mmHg (b) 99 mmHg
(c) 97 mmHg (d) 101 mmHg
(e) 98 mmHg

18 The correct order of osmotic pressure of 0.01 M aqueous solution of the following is **[Kerala (Medical) 2015]**

(a) sucrose > CH_3COOH > KCl
(b) CH_3COOH > sucrose > KCl
(c) sucrose > KCl > CH_3COOH
(d) KCl > sucrose > CH_3COOH
(e) KCl > CH_3COOH > sucrose

19 0.5 molal solution of a solute in benzene shows a depression in freezing point equal to 2 K. Molal depression constant for benzene is 5 K kg mol^{-1}. If the solute forms dimer in benzene, what is the % association? **[J & K 2015]**

(a) 40 (b) 50 (c) 60 (d) 80

20 The vapour pressure of pure benzene at certain temperature is 1 bar. A non-volatile, non-electrolyte solid weighing 2 g when added to 39g of benzene (molar mass 78g mol^{-1}) yields solution of vapour pressure of 0.8 bar. The molar mass of solid substance is **[J & K 2015]**

(a) 32 (b) 16 (c) 64 (d) 48

21 Which option is inconsistent for Raoult's law? **[Guj. CET 2015]**

(a) The change in heat of dilution for solution = 0
(b) Volume of liquid solvent + volume of liquid solute = volume of solution
(c) Solute does not undergo association in solution
(d) Solute undergoes dissociation in solution

22 Which colligative property is more useful to determine the molecular weight of the substances like proteins and polymers? **[Guj. CET 2015]**

(a) Elevation in boiling point
(b) Lowering of vapour pressure
(c) Depression of freezing point
(d) Osmotic pressure

23 18 g of glucose is dissolved in 90 g of water. The relative lowering of vapour pressure of the solution is equal to **[EAMCET 2015]**

(a) 6 (b) 0.2 (c) 5.1 (d) 0.02

24 A gas *X* is dissolved in water at 2 bar pressure. Its mole fraction is 0.02 in solution. The mole fraction of water when the pressure of gas is doubled at the same temperature is **[EAM CET 2015]**

(a) 0.04 (b) 0.98
(c) 0.96 (d) 0.02

25 Of the following 0.10 M aqueous solutions, which one will exhibit the largest freezing point depression? **[CBSE AIPMT 2014]**

(a) KCl (b) $C_6H_{12}O_6$
(c) $Al_2(SO_4)_3$ (d) K_2SO_4

26 What is the molality of a solution containing 200 mg of urea (molar mass 60 g mol^{-1}) dissolved in 40 g of water? **[MHT CET 2014]**

(a) 0.0833 (b) 0.825
(c) 0.498 (d) 0.0013

27 The volume of ethyl alcohol (density 1.15 g/cc) that has to be added to prepare 100 cc of 0.5 M ethyl alcohol solution in water is **[WB JEE 2014]**

(a) 1.15 cc (b) 2 cc
(c) 2.15 cc (d) 2.30 cc

28 To observe an elevation of boiling point of 0.05°C, the amount of a solute (molecular weight = 100) to be added to 100 g of water (K_b = 0.5) is **[WB JEE 2014]**

(a) 2 g (b) 0.05 g
(c) 1 g (d) 0.75 g

29 The unit of ebullioscopic constant is **[UK PMT 2014]**

(a) K kg mol^{-1} (b) mol kg K^{-1}
(c) kg mol^{-1} K^{-1} (d) K mol kg^{-1}

30 If α is the degree of dissociation of Na_2SO_4 the van't Hoff factor (*i*) used for calculating the molecular mass is **[UK PMT 2014]**

(a) $1+\alpha$ (b) $1-\alpha$ (c) $1+2\alpha$ (d) $1-2\alpha$

31 van't Hoff factor, (i) of a 0.5% (*w/W*) aqueous solution of KCl which freezes at – 0.24°C is (K_f of water = 1.86 K kg mol^{-1}, molecular weight of KCl = 74.5) **[EAMCET 2014]**

(a) 1.52 (b) 2.32 (c) 1.92 (d) 1.32

32 A solution of 1.25 of *P* in 50 g of water lowers freezing point by 0.3°C. Molar mass of *P* is 94. $K_{f(\text{water})}$ = 1.86 K kg mol^{-1}. The degree of association of *P* in water is **[KCET 2014]**

(a) 80% (b) 60% (c) 65% (d) 75%

33 16.02×10^{-20} molecules of urea are present in 100 mL of its solution. The concentration of solution is **[NEET 2013]**

(a) 0.02 M (b) 0.01 M
(c) 0.001 M (d) 0.1 M

34 How many gram of concentrated nitric acid solution should be used to prepare 250 mL of 2.0 M HNO_3. The concentrated acid is 70% HNO_3? **[NEET 2013]**
(a) 45.0 g conc. HNO_3
(b) 90.0 g conc. HNO_3
(c) 70.0 g conc. HNO_3
(d) 50.0 g conc. HNO_3

35 The normality of orthophosphoric acid having purity of 70% by weight and specific gravity 1.54 is **[AIIMS 2013]**
(a) 11 N (b) 22 N
(c) 33 N (d) 44 N

36 The molarity of H_2SO_4 solution, which has a density 1.84 g/cc at 35°C and contains 98% by weight is **[AIIMS 2013]**
(a) 1.84 M (b) 18.4 M
(c) 20.6 M (d) 24.5 M

37 The colligative property is not represented by **[AIIMS 2013]**
(a) elevation in boiling point
(b) osmotic pressure
(c) optical activity
(d) relative lowering of vapour pressure

38 Which one of the statements given below covering properties of solution, describes a colligative effect? **[AIIMS 2013]**
(a) Boiling point of pure water decreases by the addition of ethanol
(b) Vapour pressure of pure water decreases by the addition of nitric acid
(c) Vapour pressure of pure benzene decreases by the addition of naphthalene
(d) Boiling point of pure benzene increases by the addition of toluene

39 The osmotic pressure of 12% solution of cane sugar at 17°C is **[UP CPMT 2013]**
(a) 2.42 atm (b) 4.33 atm
(c) 8.35 atm (d) 16.30 atm

40 Which of the following is not a colligative property? **[Manipal 2013]**
(a) Depression in freezing point
(b) Osmotic pressure
(c) Elevation in boiling point
(d) Freezing point

41 Distribution law was given by **[OJEE 2013]**
(a) Ostwald (b) Nernst
(c) Henry (d) van't Hoff

42 p_A and p_B are the vapour pressure of pure liquid components A and B respectively of an ideal binary solution. If χ_A represents the mole fraction of component A, the total pressure of the solution will be **[CBSE AIPMT 2012]**
(a) $p_A + \chi_A (p_B - p_A)$ (b) $p_A + \chi_A (p_A - p_B)$
(c) $p_B + \chi_A (p_B - p_A)$ (d) $p_B + \chi_A (p_A - p_B)$

43 Vapour pressure of chloroform ($CHCl_3$) and dichloromethane (CH_2Cl_2) at 25°C are 200 mmHg and 41.5 mmHg respectively. Vapour pressure of the solution obtained by mixing 25.5 g of $CHCl_3$ and 40 g of CH_2Cl_2 at the same temperature will be (Molecular mass of $CHCl_3$ = 119.5 u and molecular mass of CH_2Cl_2 = 85 u) **[CBSE AIPMT 2012]**
(a) 173.9 mmHg (b) 615.0 mmHg
(c) 347.9 mmHg (d) 90.63 mmHg

44 In which case Raoult's law is not applicable? **[AIIMS 2012]**
(a) 1 M NaCl (b) 1 M urea
(c) 1 M glucose (d) 1 M sucrose

45 Which of the following will show a negative deviation from Raoult's law? **[WB JEE 2012]**
(a) Acetone-benzene (b) Acetone-ethanol
(c) Benzene-methanol (d) Acetone-chloroform

46 The vapour pressure lowering caused by the addition of 100 g of sucrose (molecular mass = 342) to 1000 g of water, if the vapour pressure of water at 25°C is 23.8 mm Hg is **[AFMC 2012]**
(a) 0.12 mmHg (b) 0.125 mmHg
(c) 1.15 mmHg (d) 1.25 mmHg

47 The freezing point of one molal NaCl assuming NaCl to be 100% dissociated in water is (molal depression constant is 1.86). **[Manipal 2012]**
(a) −2.72°C (b) −3.72°C
(c) 2.72°C (d) 3.72°C

48 A solution has higher osmotic pressure than its standard solution. Which of the following term will be used for this solution? **[OJEE 2012]**
(a) Isotonic (b) Hypertonic
(c) Dilute (d) Hypotonic

49 58.5 g of NaCl and 180 g of glucose were separately dissolved in 1000 mL of water. Identify the correct statement regarding the elevation of boiling point of the resultiag solution **[WB JEE 2012]**
(a) NaCl solution will show higher elevation of bp
(b) glucose solution will show higher elevation of bp
(c) Both the solution will show equal elevation of bp
(d) the bp elevation will be shown by neither of the solutions

50 The mass of non-volatile solute of molar mass 40 g/mol that should be dissovled in 114 g of octane to lower its vapour pressure by 20% is **[KCET 2012]**
(a) 11.4 g (b) 9.8 g (c) 12.8 g (d) 10 g

51 Molarity of a given orthophosphoric acid solution is 3 M. Its normality is **[UP CPMT 2012]**
(a) 9 N (b) 0.3 N
(c) 3 N (d) 1 N

52 138 g ethyl alcohol is mixed with 72 g of water. The ratio of mole fraction of alcohol to water is **[Manipal 2012]**

(a) 3 : 4 (b) 1 : 2
(c) 1 : 4 (d) 1 : 1

53 Mole fraction of the solute in a 1.00 molal aqueous solution is **[CBSE AIPMT 2011]**

(a) 1.7700 (b) 0.1770
(c) 0.0177 (d) 0.0344

54 Which one of the following gases has the lowest value of Henry's law constant? **[Kerala CEE 2011]**

(a) N_2 (b) He
(c) CO_2 (d) O_2

55 Which one of the following statements is false? **[Kerala CEE 2011]**

(a) Raoult's law states that the vapour pressure of a component over a binary solution of volatile liquids is directly proportional to its mole fraction.
(b) Two sucrose solutions of the same molality prepared in different solvents will have the same depression of freezing point.
(c) The correct order of osmotic pressures of 0.01 M solution of each compound is $BaCl_2 > KCl > CH_3COOH >$ glucose.
(d) In the equation osmotic pressure is the molarity of the solution.
(e) The molecular weight of NaCl determined by colligative property measurement is lass than its theoretical molecular weight.

56 The system that forms maximum boiling azeotrope is **[Kerala CEE 2011]**

(a) carbon disulphide — acetone
(b) benzene — toluene
(c) acetone — chloroform
(d) *n*-hexane — *n*-heptane
(e) ethanol — acetone

57 How many grams of sulphuric acid is to be dissolved to prepare 200 mL aqueous solution having concentration of $[H_3O^+]$ ions 1 M at 25°C temperature?
[H = 1, O = 16, S = 32 g - mol^{-1}] **[Guj. CET 2011]**

(a) 4.9 g (b) 19.6 g
(c) 9.8 g (d) 0.98 g

58 Choose the correct statement. **[Guj. CET 2011]**

When concentration of a salt solution is increased

(a) boiling point increases while vapour pressure decreases.
(b) boiling point decreases while vapour pressure increases.
(c) freezing point decreases while vapour pressure increases.
(d) freezing point increases while vapour pressure decreases.

59 A solution of two liquids boils at a temperature more than the boiling point of either them. Hence, the binary solution shows **[KCET 2011]**

(a) negative deviation from Raoult's law
(b) positive deviation from Raoult's law
(c) no deviation from Raoult's law
(d) positive or negative deviation from Raoult's law depending upon the composition

60 The empirical formula of a non-electrolyte is CH_2O. A solution containing 3 g of the compound exerts the same osmotic pressure as that of 0.05 M glucose solution. The molecular formula of the compound is

(a) CH_2O
(b) $C_2H_4O_2$ **[KCET 2011]**
(c) $C_4H_8O_4$
(d) $C_3H_6O_3$

61 1×10^{-3} m solution of $Pt(NH_3)_4Cl_4$ in H_2O shows depression in freezing point by 0.0054°C. The structure of the compound will be [given $K_f(H_2O) = 1.860$ Km^{-1})] **[DUMET 2011]**

(a) $[Pt(NH_3)_4]Cl_4$
(b) $[Pt(NH_3)_3Cl]Cl_3$
(c) $[Pt(NH_3)_2Cl_2]Cl_2$
(d) $[Pt(NH_3)Cl_3]Cl$

ANSWERS

Check point 2.1

1. (a)	2. (d)	3. (b)	4. (c)	5. (b)	6. (c)	7. (d)	8. (a)	9. (b)	10. (a)

Check point 2.2

1. (d)	2. (b)	3. (b)	4. (c)	5. (a)	6. (a)	7. (d)	8. (c)	9. (a)	10. (d)

Check point 2.3

1. (c)	2. (a)	3. (b)	4. (c)	5. (d)	6. (c)	7. (b)	8. (c)

Check point 2.4

1. (d)	2. (c)	3. (a)	4. (c)	5. (a)	6. (a)	7. (b)	8. (a)

(A) Topical exercise

1. (a)	2. (a)	3. (b)	4. (a)	5. (d)	6. (c)	7. (c)	8. (b)	9. (d)	10. (c)
11. (d)	12. (b)	13. (d)	14. (c)	15. (d)	16. (b)	17. (b)	18. (a)	19. (d)	20. (a)
21. (c)	22. (c)	23. (b)	24. (b)	25. (b)	26. (a)	27. (b)	28. (a)	29. (b)	30. (c)
31. (b)	32. (c)	33. (b)	34. (d)	35. (c)	36. (b)	37. (b)	38. (b)	39. (d)	40. (b)
41. (a)	42. (b)	43. (a)	44. (d)	45. (c)	46. (c)	47. (b)	48. (d)	49. (a)	50. (a)
51. (d)	52. (b)	53. (b)	54. (d)	55. (a)	56. (d)	57. (d)	58. (c)	59. (d)	60. (a)
61. (a)	62. (c)	63. (c)	64. (a)	65. (b)	66. (a)	67. (a)	68. (b)	69. (d)	70. (a)
71. (b)	72. (c)	73. (c)	74. (a)	75. (d)	76. (b)	77. (b)	78. (d)	79. (a)	80. (b)
81. (a)	82. (a)	83. (d)	84. (b)	85. (a)	86. (d)	87. (c)	88. (a)	89. (c)	90. (a)
91. (a)	92. (d)	93. (a)	94. (c)	95. (d)	96. (c)	97. (a)	98. (d)	99. (c)	100. (d)
101. (a)	102. (c)	103. (d)	104. (b)	105. (b)	106. (d)	107. (d)	108. (d)	109. (a)	110. (d)
111. (b)	112. (c)	113. (b)	114. (b)	115. (c)	116. (d)	117. (c)	118. (d)	119. (a)	120. (a)
121. (c)	122. (a)	123. (c)	124. (d)	125. (b)					

(B) Miscellaneous exercise

1. (a)	2. (c)	3. (d)	4. (b)	5. (d)	6. (b)	7. (c)	8. (d)	9. (a)	10. (d)
11. (d)	12. (a)	13. (b)	14. (c)	15. (c)	16. (b)	17. (b)	18. (c)	19. (c)	20. (a)
21. (b)	22. (a)	23. (a)	24. (a)	25. (a)	26. (c)	27. (b)	28. (a)	29. (c)	30. (b)
31. (b)	32. (c)	33. (b)	34. (b)	35. (d)	36. (b)	37. (b)	38. (b)	39. (c)	40. (b)
41. (b)	42. (a)	43. (c)	44. (c)	45. (d)	46. (c)	47. (b)	48. (c)	49. (c)	50. (a)
51. (a)	52. (d)	53. (b)	54. (d)	55. (d)	56. (c)	57. (c)	58. (c)	59. (d)	

(C) Medical entrances' gallery

1. (d)	2. (b)	3. (c)	4. (a)	5. (b)	6. (d)	7. (c)	8. (a)	9. (c)	10. (d)
11. (d)	12. (b)	13. (b)	14. (b)	15. (a)	16. (d)	17. (c)	18. (e)	19. (a)	20. (b)
21. (d)	22. (d)	23. (d)	24. (c)	25. (c)	26. (a)	27. (b)	28. (c)	29. (a)	30. (c)
31. (c)	32. (a)	33. (b)	34. (a)	35. (c)	36. (b)	37. (c)	38. (c)	39. (c)	40. (d)
41. (b)	42. (d)	43. (d)	44. (a)	45. (d)	46. (b)	47. (b)	48. (b)	49. (a)	50. (d)
51. (a)	52. (a)	53. (c)	54. (c)	55. (b)	56. (c)	57. (c)	58. (a)	59. (a)	60. (b)
61. (c)									

Hints & Explanations

(A) Topical exercise

1 (*a*) According to Henry's law, partial pressure of gas in the solution is proportional to the mole fraction of gas in the solution.

$$p = K_H \chi$$

Where, K_H = Henry's constant

Hence, (a) mole fraction is the correct choice.

2 (*a*) 150 kg = 15×10^4 g sample

$$450 \text{ mg} = 450 \times 10^{-3} \text{ g pollutant}$$

Thus, $$\text{ppm} = \frac{450 \times 10^{-3}}{15 \times 10^4} \times 10^6$$

$$= 3 \text{ ppm}$$

3 (*b*) Lead = 10 μg per decilitre or in 0.1 L = 10 μg in 100 mL

Thus, 100 mL blood has lead = $10 \times 10^{-6} = 10^{-5}$ g

$$10^9 \text{ mL blood has lead} = \frac{10^{-5} \times 10^9}{100} = 100 \text{ ppb}$$

5 (*d*) 1000 mL solution contains = 0.2 mol HCl

$$1 \text{ mL solution contains} = \frac{0.2 \text{ mol HCl}}{1000}$$

$$\therefore \text{ 35.5 mL solution contains} = \frac{0.2}{1000} \times 35.5 \text{ mol} = 0.0071 \text{ mol}$$

6 (*c*)

Column I (Concentration terms)	Column II (Mathematical formula)
A. Mass percentage	$\frac{\text{Mass of the solute component in solution}}{\text{Total mass of the solution}} \times 100$
B. Volume percentage	$\frac{\text{Volume of the solute component in solution}}{\text{Total volume of solution}} \times 100$
C. Mole fraction	$\frac{\text{Number of moles of a component}}{\text{Total number of moles of all the components}}$
D. Molality	$\frac{\text{Number of moles of the solute components}}{\text{Mass of solvent in kilograms}}$
E. Molarity	$\frac{\text{Number of moles of the solute component}}{\text{Volume of solution in litres}}$

8 (*b*) When solute is added to the solution three cases may arise

(i) It dissolves into solution then solution is unsaturated.

(ii) It does not dissolve in the solution then solution is known as saturated.

(iii) When solute get precipitated, solution is known as supersaturated solution.

9 (*d*) Dissolution of sugar in water will be most rapid, when powdered sugar is dissolved in hot water because powder form can easily insert in the vacancies of liquid particles.

Further dissolution of sugar in water is an endothermic process. Hence, high temperature will favour the dissolution of sugar in water.

10 (*c*) Maximum amount of solid that can be dissolved in a specified amount of a given solvent does not depend upon pressure. This is because solid and liquid are highly incompressible and practically remain unaffected by change in pressure.

11 (*d*) Calculate required parameter using formula,

$$M_1V_1 = M_2V_2$$

Where, V_1 = volume of solution before dilution

V_2 = volume of solution after dilution

M_1 = strength of solution before dilution

M_2 = strength of solution after dilution

Given, $M_1 = 0.02$ M, $V_1 = 4$ L, $M_2 = ?$, $V_2 = 5$ L

As we know, $M_1V_1 = M_2V_2$

$$0.02 \times 4 \text{ L} = M_2 \times 5 \text{ L}$$

$$M_2 = \frac{0.08}{5} = 0.016 \text{M}$$

12 (*b*) $H_3PO_4 \equiv 2OH^-$

$$\frac{H_3PO_4}{2} = OH^-$$

$$\therefore \quad N = 2 \times M = 0.60 \text{N}$$

13 (*d*) Mole fraction of toluene vapour $= \frac{p_T^\circ \chi_T}{p_T^\circ \chi_T + p_B^\circ \chi_B}$

χ_T = mole fraction of toluene in solution = 0.5

χ_B = mole fraction of benzene = 0.5

p_T° = vapour phase of pure toluene = 80 mm

p_T° = vapour phase of pure benzene = 120 mm

$$= \frac{0.5 \times 80}{80 \times 0.5 + 0.5 \times 120} = \frac{0.5(80)}{0.5(80 + 120)} = \frac{80}{80 + 120} = 0.40$$

14 (*c*) $$\text{Mol} = \frac{\text{molarity} \times \text{volume (mL)}}{1000}$$

$$0.15 = \frac{0.2 \times V}{1000}$$

$$V = 750 \text{ mL}$$

15 (*d*) M (molarity) = mol L^{-1}

$$= \frac{1000\, w_2}{m_2 V} = \frac{1000 \times 0.967}{159.5 \times 20}$$

$$= 0.303 \text{M}$$

20 (*a*) Supersaturated solution represent a metastable system.

22 (*c*) At equilibrium, the rate of dissolution of solid in a volatile liquid solvent is equal to the rate of crystallisation.

25 (*b*) A. **Soda water** A solution of gas in liquid. e.g. CO_2 in soft drinks.

B. **Sugar solution** A solution of solid in liquid in which sugar particles (solid) are dissolved in water (liquid).

C. **German silver** German silver is an alloy which is a solid solution of solid in solid. It is an alloy of Cu, Zn and Ni.

D. **Air** Air is a mixture of various gases. It is example of a solution of gas in gas.

E. **Hydrogen gas in palladium** is an example of solution of gas in solid. This is used as an reducing agent.

26 (*a*) Value of Henry's constant (K_H) increases with increase in temperature representing the decrease in solubility for O_2 and N_2.

27 (*b*) According to Henry's law,

$$p \propto \chi \Rightarrow p = K_H \chi$$

As value of K_H rises solubility of gases decreases.

28 (*a*) $21 \text{ mg} = 21 \times 10^{-3}$ g CO_2 in 100 g H_2O

$$= \frac{21 \times 10^{-3}}{44} \text{ mol } CO_2 \text{ in 100 mL } H_2O \ (d = 1 \text{ g / mL})$$

$$= \frac{21 \times 10^{-3} \times 10}{44} \text{ mol / dm}^3 = \frac{21 \times 10}{44} \text{ mol / m}^3$$

$$p_{CO_2} = K_H \times [CO_2]$$

$$= 2.937 \text{ kPa m}^3 \text{ mol}^{-1} \times \frac{21 \times 10}{44} \text{ mol m}^{-3} = 14.0 \text{ kPa}$$

30 (*c*) Value of K_H depends upon nature of gases dissolved in water.

Gas	Temperature (K)	K_H /kbar
Ar	298 K	40.3
CO_2	298 K	1.67
CH_4	298 K	0.413
HCHO	298 K	1.83×10^{-5}

So, the order of their increasing solubility is Ar < CO_2 < CH_4 < HCHO.

34 (*d*) $p = p_A^\circ \chi_A + p_B^\circ \chi_B = \frac{100}{4} + \frac{60 \times 3}{4} = 70$ mm Hg

70 mm (calculated) < 75 mm Hg (experimental)

Thus, there is positive deviation (a) true mixture is more volatile due to decrease in bp. Thus, (b) is true; also force of attraction is decreased thus (c) is true.

36 (*b*) $p_{\text{total}} = p_A^\circ \chi_A + p_B^\circ \chi_B = 100 \times \frac{1}{4} + 80 \times \frac{3}{4} = 85$

$$\chi_A{}'(\text{vapours}) = \frac{p_A^\circ \chi_A}{p_{\text{total}}} = \frac{25}{85} = \frac{5}{17}$$

$$\chi_B{}'(\text{vapours}) = \frac{12}{17}$$

$$p_{\text{total (distillate)}} = p_A^\circ \chi_A' + p_B^\circ \chi_B'$$

$$= 100 \times \frac{5}{17} + 80 \times \frac{12}{17} = 85.88$$

$$\simeq 86 \text{ mmHg}.$$

37 (*b*) $\frac{\Delta p}{p^\circ} \frac{n_2 i}{n_2 i + n_1}$

n_2 = moles of solute (NaCl)

n_1 = moles of solvent (H_2O) $= \frac{180}{18} = 10$

$i = 2$

$$\therefore \quad 0.167 = \frac{2n_2}{2n_2 + 10}$$

$$n_2 = 1 \text{ mol}$$

39 (*d*) Both are non-polar solution and do not change the strength of inter-molecular force of attraction on mixing.

40 (*b*) When boiling point is lower, it means vaporisation is increased, hence vapour phase is greater than theoretical value, hence, positive deviation.

41 (*a*) $p_{CH_3OH} = p^\circ_{CH_3OH} \chi_{CH_3OH}$

$$\therefore \quad \chi_{CH_3OH} = \frac{23}{92} = \frac{1}{4} = 0.25$$

42 (*b*) $p_{\text{total}} = p_A^\circ \chi_A + p_B^\circ \chi_B = 150 \times \frac{1}{3} + 240 \times \frac{2}{3}$

= 210 mm > 200 mm (given)

Observed value of total vapour pressure is decreased hence, there is negative deviation.

43 (*a*) χ_1 = mole fraction of A in vapour phase $= \frac{\chi_2 p_A^\circ}{p_{\text{total}}}$

$$\therefore p_{\text{total}} = \frac{p_A^\circ \chi_2}{\chi_1}$$

44 (*d*) $p_{\text{total}} = p_B^\circ \chi_B + p_T^\circ \chi_T$

$$= \frac{96.4 \times 5}{8.25} + \frac{28.9 \times 3.25}{8.25} = 69.8 \text{ torr}$$

45 (*c*) Density of water vapour = 0.0006 g / cm^3

Hence, weight of water vapour in the steam

$= 0.0006 \times 1000$ g = 0.6 g

Also, density of liquid water = 1.0 g / cm^3

$\therefore$ Volume of liquid water in 0.6 g steam

$$= \frac{0.6 \text{ g}}{1.0 \text{ g / cm}^3} = 0.6 \text{ cm}^3$$

46 (*c*) $p_B^\circ = 1022$ mm, $p_T^\circ = 406$ mm

$$p_B^\circ \chi_B + p_T^\circ \chi_T = p_{\text{total}}$$

$$1022\chi_B + 406(1 - \chi_B) = 1 \text{ atm} = 760 \text{ mm}$$

$$1022 \cdot \chi_B + 406 - 406\chi_B = 760$$

$$1022 \cdot \chi_B - 406\chi_B = 760 - 406$$

$$616\chi_B = 354$$

$$\chi_B = \frac{354}{616}$$

$$\Rightarrow \quad \chi_B = 0.574$$

47 (*b*) $p_{\text{total}} = p_m^\circ \chi_m + p_e^\circ \chi_e = p_m^\circ \chi + p_e^\circ (1-x)$

$$= p_m^\circ \chi + p_e^\circ - p_e^\circ \chi = (p_m^\circ - p_e^\circ)\chi + p_e^\circ = 119x + 135$$

$\therefore \quad p_e^\circ = 135$ torr

50 (*a*) In pure methanol, molecules are hydrogen bonded. On adding acetone, its molecules get in between the host molecules and break some of the hydrogen bonds between them. Therefore, the intermolecular attractive forces between the solute-solvent molecules are weaker than those between the solute-solute and solvent-solvent molecules.
On the other hand, other three remaining options will show negative deviation from Raoult's law where the intermolecular attractive forces between the solute-solvent molecules are stronger than those between the solute-solute and solvent-solvent molecules.

51 (*d*) If two liquids *A* and *B* form minimum boiling azeotrope at some specific composition then *A*—*B* interactions are weaker than those of *A*—*A* and *B*—*B*. It is due to the fact that in case of positive deviation, we get minimum boiling azeotropes whereas in case of negative deviation we get maximum boiling azeotropes.

52 (*b*) $p_{\text{total}} = p^\circ_{C_2H_5Br}\chi_{C_2H_5Br} + p^\circ_{C_2H_5Cl}\chi_{C_2H_5Cl}$

$$\chi'_{C_2H_5Cl} \text{ (vapour phase)} = \frac{p^\circ_{C_2H_5Cl}\chi_{C_2H_5Cl}}{p_{\text{total}}}$$

$$0.80 = \frac{40\chi_{C_2H_5Cl} \text{ (liquid phase)}}{40\chi_{C_2H_5Cl} + 10(1-\chi_{C_2H_5Cl})} = \frac{4\chi_{C_2H_5Cl}}{4\chi_{C_2H_5Cl} + 1 - \chi_{C_2H_5Cl}}$$

$\therefore \chi_{C_2H_5Cl} = 0.5$

54 (*d*) $p = p_A^\circ \chi_A + p_B^\circ \chi_B = p_A^\circ (1-\chi_B) + p_B^\circ \chi_B$

$$= p_A^\circ - (p_A^\circ - p_B^\circ)\chi_B = 120 - 75\chi_B$$

Thus, $\quad p_A^\circ = 120$ torr

$\Rightarrow p_A^\circ - p_B^\circ = 75$ torr

$\therefore \quad p_B^\circ = 45$ torr

55 (*a*) Sucrose is non-electrolyte.
Hence, it will not dissociate into ions and as a result there will be no lowering of vapour pressure.
Alternatively,

Vapour pressure $\propto \frac{1}{i}$

For non-electrolyte solution (sucrose) $i = 1$

For $K_3[Fe(CN)_6]$

$$K_3[Fe(CN)_6] \longrightarrow 3K^+ + [Fe(CN)_6]^{3-}, i = 4$$

For K_2SO_4

$$K_2SO_4 \longrightarrow 2K^+ + SO_4^{2-}, \ i = 3$$

60 (*a*) $\Delta T_f = K_f mi$

$$m \text{ (molality)} = \frac{1000 \times w_2}{m_2 w_1} = 1 \text{ in all cases}$$

Freezing point $= (T_0 - \Delta T_f)$

Smaller the value of ΔT_f, larger the freezing point

Urea remains undissociated, thus $i = 1$

61 (*a*) $\frac{\Delta p}{p^\circ} = \chi_{\text{solute}} = \frac{n_2}{n_1 + n_2}$

$$\frac{4}{40} = \frac{n_2}{n_1 + n_2}$$

$$n_1 = 9n_2, \ \frac{w_1}{m_1} = 9n_2$$

$\therefore \quad w_1 \text{ (solvent)} = 9n_2 m_1$

$$= 9n_2 \times 18 = 162n_2$$

$$\text{Molality} = \frac{\text{moles of solute}}{\text{kg solvent}} = \frac{1000\, n_2}{162 n_2} = 6.173 \text{ molal}$$

63 (*c*) NaCl is ionised in aqueous solution

$$\therefore \quad \frac{\Delta p}{p^\circ} = \frac{(1+x)n_2}{(1+x)n_2 + n_1} = \frac{1+x}{1+x+3} = 0.4$$

$$1 + x = 0.4 + 0.4x + 1.2$$
$$x - 0.4x = 0.4 + 1.2 - 1$$
$$0.6x = 0.4 + 0.2 \Rightarrow 0.6x = 0.6$$
$$\therefore \quad x = 1$$

Thus, % of NaCl ionised is 100.

64 (*a*) 1 molal aqueous solution means 1 mol solute in 1000 g H_2O.

$$\therefore \quad \chi_{\text{solute}} = \frac{1}{1 + \left(\frac{1000}{18}\right)} = \frac{18}{1018}$$

$$\therefore \quad \frac{\Delta p}{p^\circ} = \frac{18}{1018}$$

$\therefore \Delta p$ (lowering of vapour phase) $= p^\circ \chi_{\text{solute}}$

$$= 760 \times \frac{18}{1018} = 13.44 \text{ torr}$$

66 (*a*) Vapour pressure of solution $= p \cdot \chi_{\text{solvent}}$

$$n_2 \text{ (solute)} = \frac{0.5}{65} = \frac{1}{130}$$

$$n_1 \text{ (solvent)} = \frac{100 \times 1.58}{154} = 1.026$$

$$X_1 = \frac{1.026}{\left(\frac{1}{130} + 1.026\right)} = 0.9926$$

$\therefore$ Vapour pressure of solution $= 143 \times 0.9926 = 141.93$ mm.

67 (*a*) When salt is added to water to make the solution, the vapour pressure of solution get decreases. This is due to decrease in surface covered by solvent molecule, which leads to decrease in number of solvent molecule escaping from the surface corresponding to pure solvent.
Hence, vapour pressure also get reduced.

68 (*b*) By Raoult's law, relative decrease in vapour pressure = mole fraction of solute *B*

$$\frac{0.80 - 0.60}{0.80} = \chi_B = 0.25$$

70 (*a*) 1 molar glucose solution means 1L solution has solute (since, $d = 1.2$ g mL^{-1})

1020 g solvent (water) has solute = 180 g glucose

$$\therefore \quad \Delta T_b = \frac{1000\,K_b\,w_2}{m_2\,w_1} = \frac{1000 \times K_b \times 180}{180 \times 1020} = 0.98\,K_b$$

73 (c)
$$\Delta T = \frac{1000 K_b w_2}{m_2 w_1}$$

$$\frac{\Delta T}{K_b}(AB_2) = \frac{1000 \times 6}{m_2 \times 100} = 1$$

$$\therefore \quad m_2(AB_2) = 60 = A + 2B \quad \text{...(i)}$$

$$\frac{\Delta T}{K_b}(A_2B) = \frac{1000 \times 9}{m_2 \times 100} = 1$$

$$\therefore \quad m_2(A_2B) = 90 = 2A + B \quad \text{...(ii)}$$

By solving equation (i) and (ii), we get the value of A and B.

$$\therefore \quad A = 40, \; B = 10$$

74 (a) 1100 g solution has 60 g urea

Thus, water will be = 1040 g

$$\Delta T_f = \frac{1000 K_f w_2}{m_2 w_1}$$

$$\frac{\Delta T_f}{K_f} = \frac{1000 \times 60}{60 \times w_1} = 1$$

$$\therefore \quad w_1 = 1000 \text{ g}$$

Thus, ice formed = 1040 − 1000 = 40 g

75 (d) Let atomicity be n, then it exists as $(S)_n$

molar mass = $(32n)$

$$\Delta T_f = \frac{1000 w_2 K_f}{m_2 w_1}$$

$$\therefore \quad m_2\,(\text{sulphur}) = \frac{1000 \times 2.56 \times 0.1}{0.01 \times 100} = 256$$

$$\therefore \quad 32n = 256$$

$$\therefore \quad n = 8$$

77 (b) 1 M $C_6H_{12}O_6$ (molar mass = 180 g mol^{-1})

1000 mL solution has = 180 g solute

1180 g solution has = 180 g solute

1000 g solvent has = 180 g solute

Thus, molality = 1 molal

$$\therefore \quad \Delta T_f = K_f \times \text{molality} = 1.86 \times 1 = 1.86°$$

$\therefore$ Freezing point = − 1.86° C

81 (a) According to definition of depression in freezing point,

$$\Delta T_f = K_f m$$

Where, K_f = freezing point depression constant, value of K_f depends upon nature of solvent. This is why although the solution have same molality two different solutions of sucrose of same molality prepared in different solvents will have different depression in freezing point.

86 (d) pH = 2

$$\therefore \quad [H^+] = 10^{-2}\text{ M} = C\,x = x \quad \text{(by Ostwald's dilution law)}$$

$$\therefore \quad i = (1 + x) = 1 + 0.01 = 1.01$$

87 (c)

	Solute	y	x	$i = [1 + (y-1)x]$
(a)	KCl	2	0.5	1.5
(b)	K_2SO_4	3	0.4	1.8
(c)	$FeCl_3$	4	0.3	1.9
(d)	$SnCl_4$	5	0.2	1.8

Hence, in case of $FeCl_3$ van't Hoff factor is maximum.

90 (a) $\pi = CRT(i)$

Solutions are isotonic thus,

$$\pi \propto i$$

For $CaCl_2$ [$i = (1 + 2x)$] is maximum.

92 (d) When an unripe mango is placed in a concentrated salt solution to prepare pickle then mango loose water due to osmosis and get shrivel.

93 (a) According to definition of osmotic pressure we know that $\pi = CRT$. For concentrated solution C has higher value than dilute solution. Hence, as concentration of solution increases osmotic pressure will also increase.

95 (d)
$$KCl \rightleftharpoons K^+ + Cl^-$$

Osmosis is a colligative property

Osmolarity = molarity × number of particles of that solute dissociates into

Osmolarity of 0.2 M KCl = 0.2 × 2 = 0.400 osmol

97 (a) $\pi = \frac{n}{V}RT = CRT$

$$\therefore \quad C(\text{mol L}^{-1}) = \frac{\pi}{RT} = \frac{7.4}{0.0821 \times 300} = 0.3 \text{ mol L}^{-1}$$

98 (d) van't Hoff factor for NaCl = 1 + x = 1 + 0.8 = 1.8

van't Hoff factor for $MgCl_2$ = 1 + 2x = 1 + 2 × 0.5 = 2.0

$\therefore$ Total van't Hoff factor = 3.8

99 (c) $p = CRT(i) = \frac{w}{mV}RT(i)$

All are non-electrolyte $i = 1$

$$\frac{w}{V} \text{ same}$$

Thus,
$$p \propto \frac{1}{m\,(\text{molar mass})}$$

m(urea = 60) < m (glucose = 180) < m (sucrose = 342)

Thus, p_2 (urea) > p_1 (glucose) > p_3 (sucrose)

100 (d) $K_3[Fe(CN)_6] \rightleftharpoons \underbrace{3K^+ + [Fe(CN)_6]^{3-}}_{y=4}$

$$i = 1 + (y-1)x = 1 + 3x$$

		i
$Al_2(SO_4)_3 \rightleftharpoons 2Al^{3+} + 3SO_4^{2-}$	$y = 5$	$(1 + 4x)$
$NaCl \rightleftharpoons Na^+ + Cl^-$	$y = 2$	$(1 + x)$
$Na_2SO_4 \rightleftharpoons 2Na^+ + SO_4^{2-}$	$y = 3$	$(1 + 2x)$
$Al(NO_3)_3 \rightleftharpoons Al^{3+} + 3NO_3^-$	$y = 4$	$(1 + 3x)$

Hence, $Al(NO_3)_3$ has the same value of van't Hoff factor (i) as that of $K_3[Fe(CN)_6]$.

102 (c) $N_2O_4 \rightleftharpoons 2NO_2$, $y = 2$

$\underset{(1-x)}{1} \qquad \underset{2x}{0}$

Since, $V \propto n$

Hence $\left(\dfrac{2x}{1+x}\right) = 0.5$

$x = 0.33$, $i = 1 + x = 1.33$

104 (b) i = van't Hoff factor $= 1 + (y-1)x$

(y = number of ions, x = degree of ionisation) = 1, because given complexis 100% ionised

		y	$i = 1 + (y-1)x$
(a)	$K_2PtCl_4 \longrightarrow 2K^+ + PtCl_4^{2-}$	3	$1 + 2 = 3$
(b)	$K_2PtCl_6 \longrightarrow 2K^+ + PtCl_6^{2-}$	3	3

In (a), oxidation number of Pt = 2

In (b), oxidation number of Pt = 4

105 (b) i remains unchanged when number of ions before and after complex formation remains constant.

Solute	y	Complex	y
$ZnCl_2$	3	$[Zn(NH_3)_4]Cl_2$	3

107 (d) $PCl_5 \rightleftharpoons PCl_3 + Cl_2$

$i = 1 + (y-1)x = (1 + x)$

$x = \dfrac{D-d}{d} = \dfrac{104.25 - 100}{100} = \dfrac{4.25}{100} = 0.0425$

$\therefore \quad i = 1 + x = 1.0425$

Where, D = theoretical vapour density $= \dfrac{\text{molar mass}}{2}$

d = experimental vapour density.

108 (d) $AlPO_4 \rightleftharpoons Al^{3+} + PO_4^{3-}$ $y = 2$

$i = [1 + (y-1)x]$

$i = 1 + x = 2$

$\Delta T_b = \text{molality} \times K_b \times i = 0.01 \times K_b \times 2$

$\therefore \quad \dfrac{\Delta T_b}{K_b} = 0.02$

109 (a) $Ba_3(PO_4)_2 \rightleftharpoons 3Ba^{2+} + 2PO_4^{3-}$, $y = 5$

$x = 1$

[as $Ba_3(PO_4)_2$ solution is 100% ionised]

$i = 1 + (y-1)x$

$= 1 + (5-1)x$

$= 1 + 4 = 5$

$\therefore \quad i = 5$

$\Delta T_f / K_b = mi$

$\therefore \quad m = 0.01$ molal

110 (d)

(a) $[Fe(H_2O)_6]Cl_3 \rightleftharpoons \underbrace{3Cl^- + [Fe(H_2O)_6]^{3+}}_{4\text{ ions}}$, $i = (1 + 3x)$

(b) $[Fe(H_2O)_5Cl]Cl_2 \rightleftharpoons \underbrace{2Cl^- + [Fe(H_2O)_5Cl]^{2+}}_{3\text{ ions}}$, $i = 1 + 2x$

(c) $[Fe(H_2O)_4Cl_2]Cl \rightleftharpoons \underbrace{Cl^- + [Fe(H_2O)_4Cl_2]^+}_{2\text{ ions}}$, $i = 1 + x$

(d) $[Fe(H_2O)_3Cl_3] \rightleftharpoons$ no ionisation $\quad i = 1$

$\Delta T_f \propto i$

and freezing point $= 0 - \Delta T_f$

Thus, greater the value of ΔT_f, smaller the freezing point hence, maximum freezing point is of (d).

111 (b) $i = 2$ for AB under boiling condition

$\Delta T_b = mK_b i$

$1.08 = mK_b \times 2$

$\Delta T_f = mK_f i_f$

$1.80 = mK_f i_f$

$\therefore \quad \dfrac{1.80}{1.08} = \dfrac{mK_f i_f}{2mK_b} = \dfrac{(i)_f}{2 \times 0.3}$

$i_f = \dfrac{5}{3} \times 0.6 = 1$

Thus, no change in molecular state of AB.

113 (b) $(\Delta T)_{CH_3COOH} = K_f \times 0.01 \times i = 0.02046$ (given)

$(\Delta T)_{urea} = K_f \times 1 = 1.86$ (given)

$\therefore \quad K_f = 1.86° \text{ mol}^{-1}\text{ kg}$

$\therefore \quad i = 1.1 = (1 + x)$

$\therefore \quad x = 0.1$

$[H^+]$ in $CH_3COOH = Cx = 1 \times 10^{-3}$ M

$\therefore \quad pH = 3$

114 (b) $\Delta T_f = \dfrac{1000 K_f w_2}{m_2 w_1}$

$\therefore$ m_2 (benzoic acid in benzene) $= \dfrac{1000 K_f w_2}{\Delta T_f w_1}$

$= \dfrac{1000 \times 5.2 \times 12.2}{2.6 \times 100}$

$= 244 \text{ g mol}^{-1}$

But theoretical value $= 122 \text{ g mol}^{-1}$

Hence, there is dimer formation (by H-bonding) then number of molecules of benzoic acid in associated state is 2.

115 (c) As we know depression in freezing point is directly related to van't Hoff factor (i) according to which greater the value of i greater will be the depression in freezing point.

Solution	i
0.01 M glucose	1
0.01 M $MgCl_2$	3

Hence, depression in freezing point of $MgCl_2$ is about 3 times that of glucose.

117 (c) $HA \rightleftharpoons H^+ + A^-$

$i = [1 + (y-1)x] = 1 + x$

$pK_a = 4 = -\log K_a$

$\therefore \quad K_a = 10^{-4} = Cx^2$

$$1 \times 10^{-4} = 0.01 \times x^2$$
$$x = 0.1$$
$$\therefore \quad i = 1 + x = 1.1$$

118 (*d*) Osmosis of the solvent (H_2O) and not of ions takes place. Hence, no colour change appears.

120 (*a*) $\pi = CRT$

$$7.65 = M \times 0.0821 \times 310$$
$$\therefore \quad M = 0.30\,\text{M}$$

121 (*c*) $p = 0.0072$ atm $= 0.0072 \times 76$ cm of Hg

$= 0.0072 \times 76 \times 13.6$ cm of mercury column

Also, $p = h \times 1$cm of water column

$h \times 1 = 0.0072 \times 76 \times 13.6$ cm

$\therefore \quad h = 0.0072 \times 76 \times 13.6\text{ cm} = 7.4$ cm

122 (*a*) Given, urea solution is 6%, that means 6 g urea present in 100 mL of solution.

Temperature = 27° C = 300 K

$R = 0.0821\,\text{L atm K}^{-1}\,\text{mol}^{-1}$

$H_2N\overset{\overset{\displaystyle O}{\|}}{C}—NH_2$

$(M = 12 + 28 + 16 + 4 = 60)$

We know that,

$$C = \frac{\text{mole}}{\text{volume of solution}} = \frac{6}{60} \times \frac{1000}{100}$$

Osmotic pressue $(\pi) = iCRT$

$$= 1 \times \frac{6}{60} \times \frac{1000}{100} \times 0.0821 \times 300 = 24.63$$

123 (*c*) $\pi = CRT(i) = 0.7 \times 0.0821 \times 298 \times 2 = 34.25$ atm

For 50% recovery pressure is nearly double of required to prevent osmosis. Thus, required pressure is 68 atm.

124 (*d*) Solutions are isotonic, hence, number of moles are equal

$$\therefore \quad n_1\,(\text{HCHO}) = n_2\,(C_6H_{12}O_6)$$
$$\frac{w_1}{m_1} = \frac{w_2}{m_2} \Rightarrow \frac{w_1}{w_2} = \frac{m_1}{m_2} = \frac{30}{180} = 1:6$$

125 (*b*) We know that, if a pressure higher than the osmotic pressure is applied on the solution, the solvent will flow from the solution into the pure solvent through the semipermeable membrane. This process is called reverse osmosis.

Thus, in this case, water will move from side (*B*) to side (*A*), if a pressure greater than osmotic pressure is applied on piston (*B*).

(B) Miscellaneous questions

1 (*a*) As we know from elevation in boiling point that,

$$\Delta T_b = K_b m \Rightarrow K_b = \frac{\Delta T_b}{m}$$

$$\text{Unit of } K_b = \frac{\text{unit of } \Delta T_b}{\text{unit of } m} = \frac{K}{\text{molality}}$$

$$= \frac{K}{\text{mol kg}^{-1}} = \text{K kg mol}^{-1}$$

2 (*c*) $X_{\text{final}} = \dfrac{X_1V_1 + X_2V_2}{V_1 + V_2}$

$$29.875 = \frac{45V + 20(800 - V)}{800}$$
$$45V + 20\,(800 - V) = 800 \times 29.875$$
$$45V + 16000 - 20V = 23900$$
$$25V = 23900 - 16000 = 7900$$
$$V = \frac{7900}{26}$$
$$\therefore \quad V = 316\text{ mL}$$

3 (*d*) $\pi = CRT$, osmotic pressure depends on molarity which is constant in all cases.

4 (*b*) $\underbrace{HgI_2}_{\text{Insoluble}} + 2KI \longrightarrow K_2HgI_4$

$$K_2HgI_4 \rightleftharpoons 2K^+ + HgI_4^{2-}] \; i = 3$$
$$KI \rightleftharpoons K^+ + I^-] \; i = 2$$

'i' increases, hence ΔT_f increases.

Hence, freezing point is lowered.

5 (*d*) 1000 mL solution contains = 0.1 mol KCl

100 mL solution contains = 0.01 mol KCl

Required solution

100 mL solution contains = 0.2 mol KCl

$\therefore$ 100 mL solution contains = 0.02 mol KCl

Thus, KCl required= 0.02 − 0.01 = 0.01 mol KCl

8 (*d*) $\Delta T_b = \text{molality} \times K_b$

$$0.52 = m \times 0.52$$
$$\text{Molality} = 1\text{ mol kg}^{-1}$$
$$\therefore \quad \text{Urea} = 1\text{ mol}$$
$$\text{Water} = \frac{1000}{18} = 55.55$$
$$\text{Mole fraction of urea} = \frac{1}{56.55} = 0.018$$

9 (*a*) $\Delta T_b = mK_b i$

$$2.08 = 1 \times 0.52\,i$$
$$i = 4$$
$$i = [1 + (y - 1)x]$$
$$4 = 1 + (y - 1)$$
$$\therefore \quad y = 4 \quad \text{(means there are total 4 ions)}$$
$$K_3[Fe(CN)_6] \equiv 3K^+ + [Fe(CN)_6]^{4-}$$

13 (*b*) Low concentration of oxygen in the blood and tissues of people living at high altitude is due to low atmospheric pressure. Because at high altitude, the partial pressure of oxygen is less than that at the ground level. This decreases atmospheric pressure and causes release of oxygen from blood.

14 (*c*) Moles of glycine = 0.140

$$\text{Moles of } H_2O = \frac{1000}{18} = 55.56$$

Total moles = 0.14 + 55.56 = 55.70

$$X_{glycine} = \frac{n\,(\text{glycine})}{n\,(\text{total})} = \frac{0.14}{55.70} = 2.51 \times 10^{-3}$$

17 (*b*) pH = 2

$[H^+] = 10^{-2}M$

$\therefore \quad Cx = 0.1 \times x = 0.01$

$\therefore \quad x = 0.1$

$\therefore$ *i* for monobasic acid $= 1 + x = 1 + 0.1 = 1.1$

$\therefore \quad \pi = CRT\, i = 0.1(RT)\,(1.1) = 0.11RT$

20 (*a*) At specific composition methanol acetone mixture will form minimum boiling azeotrope and will show positive deviation. This is due to weaker *A*—*B* interaction than *A*—*A* or *B*—*B* interaction, i.e. *A*–*B* < *A*–*A* and *B*–*B*.

21 (*b*) $p_{total} = p^{\circ}_{CS_2}\chi_{CS_2} + p^{\circ}_{CH_3COCH_3}\chi_{CH_3COCH_3}$

$= 512 \times 0.25 + 344 \times 0.75 = 386$ mm Hg

Experimental $p_{total} = 600$ mm Hg > 386 mm Hg

Thus, positive deviation, in such case

(i) $\Delta V \neq 0$ (ii) $\Delta H = +ve$

22 (*a*) χ_B' (mole fraction of *B* in vapour phase) $= \dfrac{p^{\circ}_B\chi_B}{p^{\circ}_A\chi_A + p^{\circ}_B\chi_B}$

$$= \frac{36 \times 0.5}{108 \times 0.5 + 36 \times 0.5} = 0.25$$

23 (*a*) $[Cu(NH_3)_3Cl]Cl \rightleftharpoons [Cu(NH_3)_3Cl]^+ + Cl^-$

Complex ionises in aqueous solution and Ag^+ reacts with free Cl^- ion.

Degree of ionisation = *x*

$i = (1 + x)$

$$\frac{\Delta p}{p^{\circ}} = \frac{n_2(1+x)}{n_2(1+x) + n_1} = \frac{2(1+x)}{2(1+x) + 3} = 0.5$$

$\therefore x = 0.5$

hence, complex is 50% ionised

$\therefore \quad Cl^- = 2 \times 0.5 = 1$ mol

$Cl^- + Ag^+ \longrightarrow AgCl\downarrow$

$\therefore \quad AgCl = 1$ mol

24 (*a*) I. Due to dimer formation of CH_3COOH in benzene, it is therefore, 1 M.

$$\underset{2\,M}{2CH_3COOH} \rightleftharpoons \underset{1\,M}{(CH_3COOH)_2}$$

II. 0.5 M KCl ionises in aqueous solution hence, it is 1 M.

$$\underset{0.5\,M}{KCl} \rightleftharpoons \underset{1\,M}{K^+ + Cl^-}$$

Thus, both are isotonic (equal osmotic pressure).

27 (*b*) By Raoult's law,

relative decrease in vapour pressure = mole fraction of solute

$$\frac{\Delta p}{p^{\circ}} = \chi_1 \Rightarrow \frac{10}{p^{\circ}} = 0.2$$

In second case, $\dfrac{20}{p^{\circ}} = \chi_1'$

$$\therefore \quad \frac{20}{10} = \frac{\chi_1'}{0.2} \Rightarrow \chi_1' = 0.4$$

$\therefore$ Mole fraction of solvent $= 1 - 0.4 = 0.6$

28 (*a*) $p_{CM} = K_H\chi_{CM}$

$K_H = \dfrac{p_{CM}}{\chi_{CM}}$ (CM = chloromethane)

$= \dfrac{205}{0.005} = 4 \times 10^4$ torr

take other cases similarly.

29 (*c*) Mole fraction of *A* in vapour phase

$$= \frac{p^{\circ}_A\chi_A}{p_{Total}} = \frac{100 \times 0.4}{100 \times 0.4 + 200 \times 0.6} = 0.25$$

30 (*b*) Moles of solute $= \dfrac{3 \times 10^{23}}{6 \times 10^{23}} = 0.5 = \dfrac{w_2}{m_2}$

$$\Delta T_f = \frac{1000 \times w_2 \times K_f}{m_2 \times w_1} = \frac{1000 \times 0.5 \times 1.86}{250}$$

$= 3.72$ K

$\therefore$ Freezing point $= 273 - 3.72 = 269.28$ K

32 (*c*)

$$\underset{\substack{C \\ (C - Cx)}}{CH_3COOH} \rightleftharpoons \underset{\substack{0 \\ Cx}}{CH_3COO^-} + \underset{\substack{0 \\ Cx}}{H^+}$$

$i = [1 + (y - 1)x] = 1 + x = 1.01 \Rightarrow x = 0.01$

$[H^+] = Cx = 0.01 \times 0.01 = 1 \times 10^{-4}$ M

$pH = -\log H^+$

$\therefore \quad pH = 4$

33 (*b*) $\pi = MRTi$

$3.94 = 0.1 \times 0.0821 \times 300 \times i \Rightarrow i = 1.6$

$K_3[Fe(CN)_6] \equiv 3K^+ + [Fe(CN)_6]^{3-}$

$y = 4$

$\therefore \quad i = (1 + 3x) \Rightarrow x = 0.2$

In terms of %, it is = 20%

34 (*b*) $i = 1 + (N - 1)\alpha$

$$\therefore \quad \alpha = \frac{i - 1}{(N - 1)}$$

36 (*b*) In reverse osmosis, solvent molecules move through a semipermeable membrane from a region of higher concentration of solute to lower concentration.

37 (*b*) As we know greater the value of van't Hoff factor higher will be the elevation in boiling point and hence higher will be the boiling point of solution.

Solution	van't Hoff factor (*i*)
1.0 M NaOH	2
1.0 M Na_2SO_4	3
1.0 M NH_4NO_3	2
1.0 M KNO_3	2

Hence, 1.0 M Na_2SO_4 has highest value of boiling point.

38 (*b*) Number of total ions present in the solution is known as van't Hoff factors (i).

Substances	van't Hoff factor (i)
For KCl	2
For NaCl	2
For K_2SO_4	3

Hence, correct choice is (b).

39 (*c*) van't Hoff factor is the measurement of total number of ions present in the solution.

$$i = [1 + (y+1)x] = (1+x) \text{ for NaCl}$$

Being strong electrolyte, $x = 1$

in all cases. Thus, $i = 2$ (equal)

40 (*b*) By Raoult's law,

relative decrease in vapour density = mole fraction of solute

$$\therefore \quad \frac{268-167}{268} = \chi_{\text{solute}} = \frac{n}{1+n} \quad (n = \text{mole of solute})$$

$$\therefore \quad n = 0.605$$

41 (*b*) $K_b = \dfrac{RT_0^2}{1000\,\Delta H_V}$

Note ΔH_V is in energy unit per gram.

$$\therefore \quad \Delta H_V = \frac{RT_0^2}{1000K_b} = \frac{8.314\times10^{-3}\times(383.7)^2}{1000\times3.32}$$

$$= 0.3687 \text{ kJ g}^{-1} = 0.3687\times92 \text{ kJ mol}^{-1}$$

$$= 33.9 \text{ kJ mol}^{-1} \simeq 34.0 \text{ kJ mol}^{-1}$$

45 (*d*) $\Delta T_f = 0.45°\text{ C}$

$$2B \rightleftharpoons B_2$$

$$B \rightleftharpoons \frac{1}{2}B_2$$

$$y = \frac{1}{2} \text{ [moles of product from 1 mole } B \text{ (benzoic acid)]}$$

$$i = [1 + (y-1)x]$$

$$= 1 + \left(\frac{1}{2} - 1\right)x = \left(1 - \frac{x}{2}\right)$$

$$\therefore \quad \Delta T_f = \frac{1000\,w_2\,K_f(i)}{m_2 w_1} \Rightarrow 0.45 = \frac{1000\times0.2\times5.12}{60\times20}\left(1-\frac{x}{2}\right)$$

$$\left(1-\frac{x}{2}\right) = 0.527$$

$$\frac{x}{2} = 0.473 \Rightarrow x = 0.946$$

$$\therefore \quad 94.6\%$$

46 (*c*) $(\Delta T)_b$ for urea solution $= 0.54°\text{ C} = K_f \times \text{molality} = K_f \times 1$

$$\therefore \quad K_f = 0.54°\text{ C kg mol}^{-1}$$

$$\frac{\Delta p}{p°} = \frac{w_2\,m_1}{m_2\,w_1}$$

$$\frac{w_2}{m_2 w_1} = \frac{\Delta p}{p°\,m_1} = \frac{0.018}{18}$$

$$\Delta T_b(\text{glucose}) = \frac{1000K_f w_2}{m_2 w_1}$$

$$= 1000\times0.54\times\frac{0.018}{18} = 0.54°\text{ C}$$

47 (*b*)

A. **Saturated solution** A solution which contains maximum amount of solute that can be dissolved in a given amount of solvent at a given temperature.

B. **Binary solution** A solution with two components is known as binary solution.

C. **Isotonic solution** A solution having same osmotic pressure is less than another is known as isotonic solution.

D. **Hypotonic solution** A solution whose osmotic pressure is less than another is known as hypotonic solution.

E. **Solid solution** A solution in solid phase is known as solid solution.

F. **Hypertonic solution** A solution whose osmotic pressure is greater than that of another is known as hypertonic solution.

48 (*a*) Vapours are condensed hence, to determine vapour density of the distillate, we take mole fractions of vapours above liquid mixture.

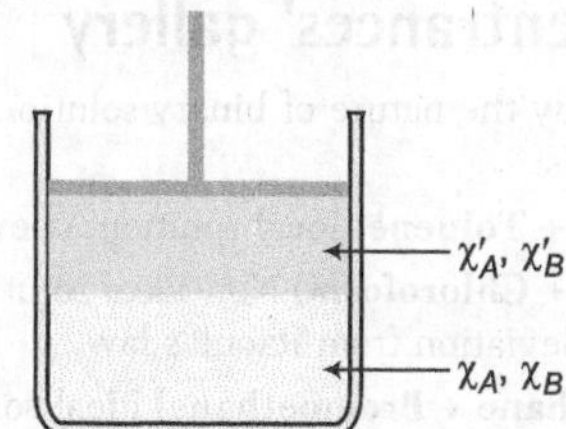

$$\chi_A' = \frac{p_A}{p_{\text{total}}} = \frac{p_A^\circ\chi_A}{p_A^\circ\chi_A + p_B^\circ\chi_B}$$

$$= \frac{100\times0.4}{100\times0.4 + 150\times0.6} = \frac{4}{13}$$

$$\chi_B' = \frac{9}{13}$$

$$\therefore p'_{\text{total}} \text{ (distillate)} = p_A^\circ\chi_A' + p_B^\circ\chi_B' = 100\times\frac{4}{13} + 150\times\frac{9}{13}$$

$$= \frac{1750}{13} = 135 \text{ mm}$$

49 (*c*) Relative decrease in vapour pressure × mole fraction of solute water = 5 mol

Solute	Mole in solution from 1 mol solute	Mole fraction of solute
A (NaCl)	2	$\frac{2}{7} = 0.29$
B (K_2SO_4)	3	$\frac{3}{8} = 0.38$
C (Na_3PO_4)	4	$\frac{4}{9} = 0.44$
D (Glucose)	1	$\frac{1}{6} = 0.17$

Hence, $D < A < B < C$.

51 (*a*) I. 1 M *aq.* glucose $i = 1$

II. 1 M *aq.* NaCl $i = 2$

III. 1 M benzoic acid $i = 0.5$

IV. 1 M $(NH_4)_3PO_4$ $i = 4$

Thus, osmotic pressure of solution increases in order as

$$\underset{\text{Hypotonic}}{\text{III}} < \text{I} < \text{II} < \underset{\text{Hypertonic}}{\text{IV}}$$

52 (*d*)

A. **Raoult's law** Mathematical representation of Raoult's law $p = \chi_1 p_1^\circ + \chi_2 p_2^\circ$

B. **Henry's law** Its mathematical representation is, $p = K_H \cdot \chi$

C. **Elevation of boiling point** Mathematical representation, $\Delta T_b = K_b \cdot m$

D. **Depression in freezing point** Mathematical representation, $\Delta T_f = K_f \cdot m$

E. **Osmotic pressure** Mathematical representation, $\pi = CRT$.

58. (*c*) $\Delta T_f = \dfrac{1000\, K_f w_2}{m_2 w_1}(1 + 2\alpha)$

$w_1 = x = 45$ g

(C) Medical entrances' gallery

1 (*d*) Let us study the nature of binary solutions made of two volatile liquids:

(a) **(Benzene + Toluene)** Ideal solution, obeys Raoult's law.

(b) **(Acetone + Chloroform)** Non-ideal solution, shows negative deviation from Raoult's law.

(c) **(Chloroethane + Bromoethane)** Ideal solution, obeys Raoult's law.

(d) **(Ethanol + Acetone)** Non-ideal solution, shows positive deviation from Raoult's law.

2 (*b*) The expression of freezing point depression is

$K_f = 5.12$ K kg mol^{-1}

Molality $(m) = 0.078$ molal

$i = 1$, for non-electrolyte solute

$$\Delta T_f = K_f \times m \times i$$
$$= 5.12 \times 0.078 \times 1 = 0.39936 \simeq 0.40 \text{ K}$$

3 (*c*) Isotonic solutions have same osmotic pressure (π) at a given temperature.

$$\pi = CRT$$

When, two solution have same molar concentration (C) their osmotic pressure will be equal.

4 (*a*) Solution exhibits positive deviation from ideal behaviour if the intermolecular interactions, i.e. *A*—*B* interactions are weaker than in pure components, i.e. between *A*—*A* or *B*— *B*. Thus. option (a) is correct. The remaining options are valid only for ideal solutions.

5 (*b*) Ideal solutions are those which obey Raoult's law over all concentration ranges at a given temperature, e.g. benzene-toluene, *n*-hexane-*n*-heptane, etc. For an ideal solution,

$$\Delta V_{mix} = 0, \Delta H_{mix} = 0$$
$$\Delta G_{mix} < 0, \Delta S_{mix} > 0$$

Hence, option (b) is correct.

6 (*d*) The binary liquid mixtures having the same composition in liquid and vapour phase and boil at constant temperature are called azeotropic mixtures or azeotropes.

The solutions that show large negative deviation from Raoult's law forms maximum boiling azeotrope. For, e.g. nitric acid and water. The remaining option containing different mixtures forms minimum boiling azeotrope.

7 (*c*)

	$2C_6H_5COOH \longrightarrow$	$(C_6H_5COOH)_2$
Before association	2	–
	$2-2x$	x
After association	$1-x$	$x/2$

$$\text{Total} = 1 - x + \frac{x}{2} = 1 - \frac{x}{2}$$

$$\therefore \quad i = \frac{1 - x/2}{1} \text{ or } i = 1 - \frac{x}{2}$$

8 (*a*) K_f is property of solvent and is independent of concentration of solution.

9 (*c*) Molarity is volume dependent, hence affected by change in temperature.

10 (*d*) From Raoult's law of partial pressure,

$$\frac{p_1^\circ - p_s}{p_s} = \frac{n_2}{n_1} \Rightarrow \frac{760 - 732}{732} = \frac{w_2 \times m_1}{m_2 \times w_1}$$

$$\Rightarrow \frac{28}{732} = \frac{w_2 \times 18}{m_2 \times w_1}$$

$$\frac{w_2}{m_2 \times w_1} = \frac{2.8}{(732 \times 18)}$$

$$\Delta T_b = \frac{1000 K_b w_2}{m_2 w_1} = \frac{1000 \times 0.52 \times 28}{732 \times 18}$$
$$= 1.1$$

$\therefore$ Boiling point $= 100 + 1.10 = 101.1^\circ\text{C} \approx 101^\circ\text{C}$

(1 is solvent, 2 is solute).

11 (*d*) Since, component having higher vapour pressure will have higher percentage in vapour phase benzene has vapour pressure 12.8 kPa which is greater than of toluene 3.85 kPa. Therefore, the vapour will contain a higher percentage of benzene.

12 (*b*) $Ba(OH)_2$ is strong electrolyte so its 100% dissociation occurs in solution

$$Ba(OH)_2 \longrightarrow Ba^{+2}(aq) + 2OH^-(aq)$$

van't Hoff factor = total number of ions present in solution

$\therefore \quad i = 3$

13 (*b*) For an ideal solution, $\Delta H_{mix} = 0$

$$\Delta U_{mix} = 0$$
$$\Delta S_{mix} \neq 0$$

According to $\Delta G_{mix} = \Delta H_{mix} - T\Delta S_{mix} \Rightarrow \Delta G_{mix} \neq 0$

Incorrect answer is $\Delta G_{mix} = 0$

14 (*b*)

(i) There will no change in volume on mixing the two components, i.e. $\Delta V_{mixing} = 0$

(ii) There will no change in volume on $\Delta H_{mixing} = 0$

So, $\Delta S_{mix} \neq 0$ for an ideal solution.

15 (*a*) Molality of solution X = molality of solution Y = 0.2 mol/kg

we know that elevation in the boiling point (ΔT_b) of a solution is proportional to the molal concentration of the solution, i.e.

$$\Delta T_b \propto m, \Delta T_b = K_b m$$

Where, m is the molality of the solution and K_b is molal boiling point constant or ebullioscopic constant.

$\therefore$ By elevation in boiling point relation

$$\Delta T_b = iK_b m, \ \Delta T_b \propto i$$

where, i is van't Hoff factor

Since, ΔT_b of solution X is greater than ΔT_b of solution Y (observed colligative property is greater than normol colligative property)

$\therefore$ i of solution $X > i$ of solution Y

$\therefore$ Solution X undergoing dissociation.

16 (*d*) $Al_2(SO_4)_3 \rightleftharpoons Al^{3+} + 3SO_4^{2-}$

Value of van't Hoff's factor (i) = 5

(a) $K_2SO_4 \rightleftharpoons 2K^+ + SO_4^{2-}$ ($i = 3$)

(b) $K_3[Fe(CN)_6] \rightleftharpoons 3K^+ + [Fe(CN)_6]^{3-}$ ($i = 4$)

(c) $Al(NO_3)_3 \rightleftharpoons Al^{3+} + 3NO_3^-$ ($i = 4$)

(d) $K_4[Fe(CN)_6] \rightleftharpoons 4K^+ + [Fe(CN)_6]^{4-}$ ($i = 5$)

Therefore, $K_4[Fe(CN)_6]$ has same value of i that of $Al_2(SO_4)_3$, i.e. $i = 5$.

17 (*c*) $\because \dfrac{p^\circ - p_s}{p^\circ} = i\dfrac{n_1 i}{n_1 i + n_2}$

$$\therefore \quad \frac{100 - p_s}{100} = 3 \times \frac{1}{3 + 99}$$

$$\therefore \quad p_s = 97 \text{ mmHg}$$

18 (*e*) $\because \pi \propto i$

(i) The molecular state of sucrose remains unchanged in solution, hence $i = 1$.

(ii) KCl, CH_3COOH undergo dissociation in solution, hence $i > 1$ but KCl is strong electrolyte thus, order or osmotic pressure will be KCl > CH_3COOH > sucrose.

19 (*a*) $2B \rightleftharpoons B_2$

$$B \rightleftharpoons \frac{1}{2}B_2 \qquad (B \text{ is benzoic acid})$$

$$\begin{matrix} 1 & 0 \\ (1 - x) & x/2 \end{matrix}$$

$$\therefore \quad i = \left(1 - \frac{x}{2}\right)$$

$$\Delta T_f = K_f m i$$

$$2 = 5 \times 0.5\left(1 - \frac{x}{2}\right)$$

$$x = 0.4$$

Thus, 40% .

20 (*b*) *Step* I Calculation of number of moles

$$p_{solution} = p_{benzene} \times \chi_{benzene}$$

$$p_{solution} = 0.8 \text{ bar}$$

$$p_{C_6H_6} = 1 \text{ bar}$$

$$\chi_{C_6H_6} = \frac{p_{solution}}{p_{C_6H_6}} = \frac{0.8 \text{ bar}}{1 \text{bar}}$$

$$\chi_{C_6H_6} = 0.8$$

$$\chi_{C_6H_6} = \frac{\text{moles of } C_6H_6}{\text{moles of } C_6H_6 + \text{moles of solute}}$$

$$\chi_{C_6H_6} = \frac{(39/78) \text{ g mol}^{-1}}{(39/78) \text{ g mol}^{-1} + n_{solute}}$$

$$\frac{\Delta p}{p^\circ} = \frac{n_2}{n_1 + n_2}$$

$$\frac{0.2}{1} = \frac{n_2}{0.5 + n_2}$$

$$\therefore \quad n_2 = \frac{1}{8}$$

$$\therefore \quad \frac{w_2}{m_2} = \frac{1}{8}$$

$$\therefore \quad m_2 = w_2 \times 8 = 2 \times 8$$

$$= 16 \text{ g mol}^{-1}$$

21 (*d*) Raoult's law is not applicable to solutes which dissociate or associate in the particular solutions.

22 (*d*) Osmotic pressure is widely used to determine molar masses of proteins, polymers and other macromolecules.

$$M_2 = \frac{w_2 RT}{\pi V}$$

Where, π = osmotic pressure

R = gas constant

V = volume of the solution (in L)

M_2 = molar mass,

w_2 = weight of solute

23 (*d*) From Raoult's law for non-volatile solute.

We know that,

$$\frac{p^\circ - p_s}{p^\circ} = \frac{n_2}{n_1 + n_2}$$

For dilute solution, it becomes $\dfrac{p^\circ - p_s}{p^\circ} = \dfrac{n_2}{n_1} = \dfrac{w_2}{m_2} \times \dfrac{m_1}{w_1}$

Given that, $w_2 = 18$ g $w_1 = 90$ g

Molecular mass of glucose

$$(C_6H_{12}O_6) = 12 \times 6 + 12 \times 1 + 6 \times 16$$

$$= 180 \text{ g mol}^{-1}$$

Molecular mass of $(H_2O) = 2 + 16 = 18$ g mol^{-1}

Now, putting the values, we get

$$\frac{p^\circ - p_s}{p^\circ} = \frac{18}{180} \times \frac{18}{90} = 0.02$$

24 (c) We know that $\frac{p_1}{p_2} = \frac{\chi_1}{\chi_2}$

Where, p_1 = intial pressure

p_2 = final pressure

χ = mole fraction of solute

Given that, $p_1 = 2$ bar, $p_2 = 4$ bar. $\chi_1 = 0.02$, $\chi_2 = ?$

Now, putting the values in the equation

$$\frac{2}{4} = \frac{0.02}{\chi_2} \Rightarrow \chi_2 = 0.04$$

In a solution, $\chi_{solute} + \chi_{solvent} = 1$

or $0.04 + \chi_{solute} = 1$

or $\chi_{solvet} = 1 - 0.04 = 0.96$

25 (c) Freezing point depression (ΔT_f) is a colligative property and depends upon the van't Hoff factor (i), i.e. number of ions given by the electrolyte in aqueous soluton.

$$\Delta T_f = i \times K_f \times m$$

Where, K_f = molal freezing point depression constant

m = molality of the solution

$\therefore \quad \Delta T_f \propto i$

(a) $KCl\,(aq) \rightleftharpoons K^+(aq) + Cl^-(aq)$ (Total ions = 2 thus, $i = 2$)

(b) $C_6H_{12}O_6 \rightleftharpoons$ no ions [$i = 1$]

(c) $Al_2(SO_4)_3(aq) \rightleftharpoons 2Al^{3+} + 3SO_4^{2-}$ (Total ions = 5, thus, $i = 5$)

(d) $K_2SO_4\,(aq) \rightleftharpoons 2K^+ + SO_4^{2-}$ (Total ions = 3, thus, $i = 3$)

Hence, $Al_2(SO_4)_3$ will exhibit largest freezing point depression due to the highest value of i.

26 (a) Molality (m)

$$= \frac{\text{mass of solute (in g)} \times 1000}{\text{molecular weight of solute} \times \text{mass of solvent (in g)}}$$

$$= \frac{200 \times 10^{-3} \times 1000}{60 \times 40} = 0.0833$$

27 (b) Molarity (M)

$$= \frac{\text{mass of solute} \times 1000}{\text{molecular weight of solute} \times \text{volume of solution}}$$

Molecular weight of ethyl alcohol,

$C_2H_5OH = 12 \times 2 + 5 + 16 + 1$

$= 24 + 22 = 46$ g mol^{-1}

On substituting the values, we get

$$0.5\,M = \frac{\text{mass of ethyl alcohol} \times 1000}{46 \times 100}$$

$$\text{Mass of ethyl alcohol} = \frac{0.5 \times 46}{10} = 2.3 \text{ g}$$

$\therefore$ Volume of ethyl alcohol required,

$$V = \frac{\text{mass of ethyl alcohol}}{\text{density of ethyl alcohol}} = \frac{2.3 \text{ g}}{1.15 \text{ g/cc}} = 2 \text{ cc}$$

28 (c) Elevation of boiling point, $\Delta T_b = \frac{w \times K_b \times 1000}{M \times W\,(g)}$

(Here, w and W = weights of solute and solvent respectively)

M = molecular weight of solute and K_b = constant.

On substituting values, we get

$$0.05 = \frac{w \times 0.5 \times 1000}{100 \times 100} \text{ or } w = \frac{0.05 \times 100 \times 100}{0.5 \times 1000} = 1 \text{ g}$$

29 (a) Ebullioscopic constant, $K_b = \frac{\Delta T_b}{\text{molality}}$

Units will be K kg mol^{-1}

30 (c) Degree of dissociation (α) $= \frac{i-1}{n-1}$

n = number of particles per molecule

$n = 3$ for Na_2SO_4 $\alpha = \frac{i-1}{3-1} \Rightarrow i = 1 + 2\alpha$

31 (c) Depression in freezing point $\Delta T_f = K_f \times \text{molality} \times i$

$$i = \frac{\Delta T_f}{K_f \times \text{molality}}$$

$$\text{Molality} = \frac{0.5 \times 1000}{74.5 \times 100} = 0.067 \text{ m}$$

$$\text{van't Hoff factor } (i) = \frac{0.24}{1.86 \times 0.067} = 1.92$$

32 (a) ***Step* I** Calculation of van't Hoff factor of P

$$\Delta T_f = i \times K_f \times m = i \times K_f \times \frac{w_1 \times 1000}{w_2 \times M_1}$$

$$0.3 = i \times 1.86 \times \frac{1.25 \times 1000}{50 \times 94}$$

$$\Rightarrow \quad i = \frac{0.3 \times 50 \times 94}{1.86 \times 1.25 \times 1000} = 0.6064$$

***Step* II** Calculation of degree of association of P

Degree of association

$$\alpha = \frac{i-1}{\frac{1}{n}-1} = \frac{0.6064-1}{\frac{1}{2}-1} = \frac{-0.3936}{-0.5} = 0.7872 = 78.72\% \approx 80\%$$

Also refer Q.8.

33 (b) Given number of molecules of urea $= 6.02 \times 10^{20}$

$$\therefore \text{Number of moles} = \frac{6.02 \times 10^{20}}{6.2 \times 10^{23}} = 1 \times 10^{-3} \text{ mol}$$

$$\text{Volume of the solution} = 100 \text{ mL} = \frac{100}{1000} \text{ L} = 0.1\text{L}$$

Concentration of urea solution (in mol L^{-1})

$$= \frac{1 \times 10^{-3}}{0.1} = 1 \times 10^{-2} \text{ mol L}^{-1} = 0.01 \text{ mol/L or M}$$

34 (a) Given molarity of solution = 2

$$\text{Volume of solution} = 250 \text{ mL} = \frac{250}{1000} = 0.25\text{L}$$

Molar mass of $HNO_3 = 1 + 14 + 3 \times 16 = 63$ g mol^{-1}

$$\text{Molarity} = \frac{(\text{weight of } HNO_3)}{\text{molecular mass of } HNO_3 \times \text{volume of solution (L)}}$$

Weight of HNO_3 = molarity × molecular mass × volume (L)

$$= 2 \times 63 \times 0.25 \text{ g} = 31.5 \text{ g}$$

It is weight of 100% HNO_3.

But the given acid is 70% HNO_3.

$$\therefore \text{ Its weight} = 31.5 \times \frac{100}{70} \text{ g} = 45 \text{ g}$$

35 (*c*) 100 g H_3PO_4 solution has H_3PO_4 = 70 g

$$\frac{100}{1.54} \text{ mL} \left(= \frac{1}{15.4} \text{ L}\right) \text{ solution has } H_3PO_4 = 70 \text{ g}$$

$$\therefore \quad \text{Molarity} = \frac{\text{mol } (70/98)}{\text{volume } 1/15.4} = \frac{70}{98} \times 15.4 = 11 \text{ M}$$

Since $H_3PO_4 \equiv 3H^+$

$\therefore$ Normality = 33 N

36 (*b*) 98% by mass means 98 g H_2SO_4 is present in 100 g of acid

$$M = \frac{\text{given mass} \times 1000}{\text{molecular mass} \times V_{cc}}$$

Given, mass = 98 g, molecular mass = 98 g

$$V = \frac{\text{mass}}{\text{density}} = \frac{100}{1.84} \quad (\because \text{ density of } H_2SO_4 \text{ is } 1.84)$$

$$\text{Putting the values } M = \frac{98 \times 1000 \times 1.84}{98 \times 100} = 18.4 \text{ M}$$

37 (*c*) Colligative properties are defined as properties of dilute solutions of non-volatile solutes whose values depend upon the concentration and number of solute particles in the solution but not on the individual identity of the solute.

So, optical activity is not a colligative property.

38 (*c*) The colligative properties are the properties of dilute solution of non-volatile solutes whose value depend on concentration of solute particles in the solution. Naphthalene is a non-volatile solid, hence on addition in benzene it decreases its vapour pressure.

39 (*c*) Use the following formula to find osmotic pressure, $\pi V = RT$.

where, V = volume of solution containing 1 mole of solute.

Given, $T = 17° \text{C} = 17 + 273 = 290$ K

$R = 0.082 \text{ L atm K}^{-1} \text{ mol}^{-1}$; C = 12%

$\because$ 12 g of sugar is dissolved in 100 mL water

$$\therefore \text{ 342 g of sugar is dissolved in } \frac{100}{12} \times 342 \text{ m/water}$$

$$= 2850 \text{ mL } V = 2.85 \text{ L}$$

$$\pi V = RT$$

$$\pi = \frac{0.082 \times 290}{2.85} = 8.35 \text{ atm}$$

40 (*d*) Colligative properties are those properties which depend on the number of solute particles and not on the nature of solute. The colligative properties are depression in freezing point, elevation in boiling point, osmotic pressure and relative lowering of vapour pressure.

41 (*b*) Nernst gave the distribution law. According to this law, 'if a solute X distributes itself between non-miscible solvents A and B, at constant temperature and X is in the same molecular condition in both the solvents, then

$$\frac{\text{concentration of } X \text{ in } A}{\text{concentration of } X \text{ in } B} = k_d\text{'}$$

Where, k_d is called distribution coefficient.

42 (*d*) Total pressure, $p_T = p'_A + p'_B$ ...(i)

We, know that $p'_A = p_A \chi_A$ and $p'_B = p_B \chi_B$

Substituting the values of p'_A and p'_B in Eq. (i)

$$p_T = p_A \chi_A + p_B \chi_B \quad [\chi_A + \chi_B = 1]$$

$$\Rightarrow \quad \chi_A = 1 - \chi_B \text{ or } \chi_B = 1 - \chi_A$$

$$p_T = p_A \chi_A + p_B (1 - \chi_A) = p_A \chi_A + p_B - p_B \chi_A$$

$$p_T = p_B + \chi_A (p_A - p_B)$$

43 (*d*) Number of moles of $CHCl_3$, $n_A = \frac{w}{M} = \frac{25.5}{119.5} = 0.213$

Number of moles of CH_2Cl_2, $n_B = \frac{40}{85} = 0.47$

Mole fraction of $CHCl_3$, $\chi_A = \frac{n_A}{n_A + n_B} = \frac{0.213}{0.684} = 0.31$

Mole fraction of CH_2Cl_2, $\chi_B = 1 - x_A = 1 - 0.31 = 0.69$

$$p_{\text{total}} = p_A \chi_A + p_B \chi_B$$

$$200 \times 0.31 + 41.5 \times 0.69 = 62 \times 28.63 = 90.63 \text{ mmHg}$$

44 (*a*) Raoult's law is not applicable, if the total number of particles of solute changes in the solution due to association or dissociation. Among the given compounds NaCl undergoes dissociation and forms Na^+ and Cl^- ions is not applicable to NaCl.

45 (*d*) $CH_3 - \overset{\overset{O}{\|}}{C} - CH_3$ (Acetone), $CHCl_3$ (Chloroform)

Acetone and chloroform will show a negative deviation due to their association after mixing.

46 (*b*) Number of moles of $C_{12}H_{22}O_{11}$, $n = \frac{100}{342} = 0.292$ mol

Number of moles of H_2O, $N = \frac{1000}{18} = 55.5$ mol

Vapour pressure of pure water $p° = 23.8$ mmHg

According to Raoult's law, $\frac{\Delta p}{p°} = \frac{n}{n + N}$

$$\Rightarrow \quad \frac{\Delta p}{23.8} = \frac{0.292}{0.292 + 55.5}$$

$$\Delta p = \frac{23.8 \times 0.292}{55.79} = 0.125 \text{ mmHg}$$

47 (*b*) For NaCl, $i = 2$

$$\Delta T_f = i K_f m = 2 \times 1.86 \times 1 = 3.72$$

$$T_{\text{solution}} = \Delta T_f^{\circ} - \Delta T_f = 0 - 3.72° \text{C} = -3.72° \text{C}$$

48 (*b*) A solution having lower or higher osmotic pressure than the other is said to be hypotonic or hypertonic respectively.

49 (*a*) Elevation in boiling point $\Delta T_b = i \times K_b \times m$

$$\text{Molality of NaCl solution} = \frac{n}{w} \times 1000 = \frac{\frac{58.5}{58.5}}{w_{H_2O}} \times 1000 = \frac{1000}{w_{H_2O}}$$

$$\text{Molality of } C_6H_{12}O_6 \text{ solution} = \frac{\frac{180}{180} \times 1000}{w_{H_2O}} = \frac{1000}{w_{H_2O}}$$

Both solution have same molality but values of i, van't Hoff factor, for NaCl and glucose are 2 and 1 respectively. Hence, NaCl will show higher elevation.

50 (*d*) Given, $p° = 100$, $p = 100 - 20 = 80$, $M_2 = 40 \text{ g mol}^{-1}$,

$M_1 \text{ (octane)} = C_8H_{18} = 12 \times 8 + 18 = 114$

$w_2 = ?$ and $w_1\text{(octane)} = 114 \text{ g}$

$$\frac{p° - p}{p} = \frac{w_2}{M_2} \times \frac{M_1}{w_1}$$

$$\frac{100 - 80}{80} = \frac{w_2}{40} \times \frac{114}{114}$$

or $$\frac{20}{80} = \frac{w_2}{40} = 10 \text{ g}$$

51 (*a*) Orthophosphoric acid is H_3PO_4. It is a tribasic acid

Normality = Molarity × basicity = 3×3 = 9 N

52 (*a*) Number of moles of $C_2H_5OH = \frac{138}{46} = 3$

$[M_{C_2H_5OH} = 46 \text{ g/mol}]$

Number of moles of $H_2O = \frac{72}{18} = 4$ $\quad [\because M_{H_2O} = 18 \text{ g/mol}]$

$$\text{Mole fraction, } \chi_{C_2H_5OH} = \frac{3}{3+4} = \frac{3}{7}$$

$$\chi_{H_2O} = \frac{4}{3+4} = \frac{4}{7}$$

$$\Rightarrow \quad \frac{\chi_{C_2H_5OH}}{\chi_{H_2O}} = \frac{3/7}{4/7} = \frac{3}{4} = 3:4$$

53 (*c*) Mole of solute = 1

$$\text{Moles of water } \frac{1000}{18} = 55.55$$

$$\text{Mole fraction of solute} = \frac{1}{56.55} = 0.0177$$

54 (*c*) As the solubility increases K_H decreases. Among the given set of gases, CO_2 is the most soluble in water.

55 (*b*) Two sucrose solutions of the same molality prepared in different solvents will have the different depression of freezing point.

$$\Delta T_f = K_f \times m$$

K_f values are different for different solvents.

56 (*c*) Solutions which show negative deviation from Raoult's law are also called maximum boiling azeotrope.

Chloretone molecules are joined by H-bonding.

$$CHCl_3 + CH_3COCH_3 \longrightarrow Cl_3C\text{—}C(OH)(CH_3)\text{—}CH_3$$

Chloretone

Hence, acetone-chloroform forms maximum boiling azeotrope.

57 (*c*) $2M[H_3O^+]$ ions are obtained from 1 M H_2SO_4

$1M[H_3O^+]$ ions are obtained from 0.5 M H_2SO_4

Molarity,

$$M = \frac{w(H_2SO_4) \times 1000}{M(\text{molar mass}) \text{ of } H_2SO_4 \times \text{volume of solution (mL)}}$$

$$0.5 = \frac{w \times 1000}{98 \times 200}$$

$$w = \frac{0.5 \times 98 \times 200}{1000} = 9.8 \text{ g}$$

58 (*a*) When concentration of a salt solution increases, vapour pressure decreases and boiling point increases.

59 (*a*) The vapour pressure of the solution is lower than that of either of two liquids. Hence, the solution shows negative deviation from Raoult's law.

60 (*b*) For isotonic solution, $\pi_1 = \pi_2$.

$$\text{So,} \quad \frac{w_1}{M_1} = \frac{w_2}{M_2}$$

Mass of 0.05 M of glucose = 0.05 × 180 = 9 g

$$\frac{3}{M_1} = \frac{9}{180}$$

$$\Rightarrow \quad M_1 = 60 \text{ g}$$

$$n = \frac{\text{molecular weight}}{\text{empirical formula weight}} = \frac{60}{30} = 2$$

∴ The value of $n = 0$

The empirical formula of the given non-electrolyte = CH_2O

The molecular formula of compound will be = $2 \times (CH_2O)$

$= C_2H_4O_2$

61 (*c*) $\Delta T_f = iK_f\, m$

$$0.0054 = i \times 1.86 \times 0.001$$

$$i = \frac{5.4}{1.86} = 3$$

$$[Pt(NH_3)_2Cl_2]Cl_2 \rightleftharpoons [Pt(NH_3)_2\ Cl_2] + 2Cl^-$$

So, the structure of the compound will be $[Pt(NH_3)_2Cl_2]Cl_2$.

CHAPTER 3

Electrochemistry

Electrochemistry is a branch of chemistry that deals with the study of relationship between electron-flow and redox reactions. It is related with electrical conductance in a solution and associated chemical changes. We can simply say, it is change of electrical energy into chemical energy and *vice-versa.* As these chemical changes involve production or consumption of electricity, hence these are called as **electrochemical changes.**

The device inside which these changes are carried out are known as **cells.** These cells have commercial importance in our daily life.

In this chapter we will study about different aspects of electrochemistry one-by-one.

Inside

CONDUCTORS

Substances which allow electric current to flow through them are known as **conductors**, e.g. metals like Fe, Ag, Cu. Whereas, the substance which do not allow the passage of electricity through them are called insulators, e.g. plastics.

Types of conductors

Metallic conductors or electronic conductors

These are metallic substances that conduct electric current due to the mobility of free electrons, without undergoing any chemical change, e.g. metals like Cu, Ag, Fe, non-metals like graphite, some minerals etc.

Note *Cs, Hg and Ga are metallic conductors but are liquids at room temperature.*

Electrolytic conductors (or electrolytes)

These are the substances which allow the electricity to pass through them in molten or fused state and in their solutions. They also undergo chemical decomposition during the process. The conduction of current through electrolyte is due to the movement of ions present in them.

On the basis of dissociation in aqueous solution, electrolytic conductors are further classified into two types

(*i*) **Strong electrolytes** The electrolytes which dissociate completely into ions are called strong electrolytes,
e.g. NaOH, NaCl, HCl, KCl, HNO_3, H_2SO_4 etc.

(*ii*) **Weak electrolytes** The electrolytes which do not ionise completely in aqueous as well as in molten state are called weak electrolytes, e.g. H_2CO_3, CH_3COOH, $ZnCl_2$, $HgCl_2$ etc.

CONDUCTANCE AND CONDUCTIVITY OF ELECTROLYTIC SOLUTIONS

Just like metallic conductors, electrolytic solutions also obey Ohm's law, which states that,

"The current flowing through a conductor is directly proportional to the potential difference across it."

Resistance (*R*)

The hinderance provided by the medium in the flow of current is called **resistance.**

From Ohm's law, $V \propto I$

$$\boxed{V = RI} \quad \text{(where, } R = \text{resistance)}$$

Unit of resistance is Ω (ohm).

Resistivity or specific resistance (ρ)

Resistance provided by unit volume of the medium is called, resistivity.

For any medium, Resistance ∝ length

Also, $\text{resistance} \propto \frac{1}{\text{area}}$

So, $R \propto \frac{\text{length}}{\text{area}}$

$$\boxed{R = \rho \frac{l}{a}}$$

where, ρ is called **resistivity** or **specific resistance**

If $l = 1\,\text{cm}$, $a = 1\,\text{cm}^2$, then $R = \rho$, its unit is ohm-cm.

Cell constant (G^*)

It is the ratio of distance between electrode (l) and total surface area (a) of the electrodes.

$$\boxed{G^* = \frac{l}{a}}$$

Unit of cell constant = m^{-1}, or cm^{-1}

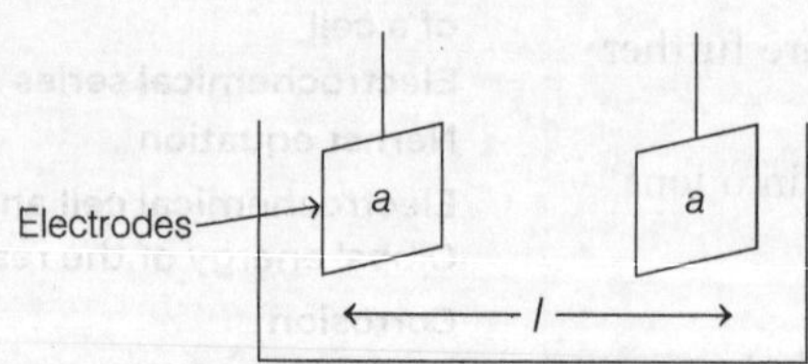

Fig. 3.1 Electrolytic cell

This cell constant is a function of the electrode areas, the distance between the electrodes and the electrical field pattern between the electrodes.

$$\boxed{\text{Cell constant} = \frac{\text{Resistance}}{\text{Resistivity}}}$$

Conductance (*G*)

The property of a medium which tells about ease with which current flows through the medium is called conductance.

By definition, conductance (G) $= \frac{1}{\text{resistance } (R)}$

Unit of conductance (G) = ohm^{-1}, mho, Siemens(S)

Conductivity or specific conductance (κ)

The reciprocal of specific resistance is called specific conductance, i.e. the conductance of a solution of 1 cm length and having $1\,\text{cm}^2$ area of cross section. It is denoted by κ (kappa).

$$\text{Conductivity} = \frac{1}{\text{resistivity}}$$

$$\boxed{\kappa = \frac{1}{\rho}}$$

Unit of κ is $\text{ohm}^{-1}\text{cm}^{-1}$ or Siemens cm^{-1} (S cm^{-1}) or mho cm^{-1}.

$$1\,\text{ohm}^{-1}\text{cm}^{-1} = 100\,\text{ohm}^{-1}\text{m}^{-1}$$

Example 3.1 *When a certain conductivity cell was filled with* 0.020 M KCl *which has specific conductivity of* $0.2768\,\Omega^{-1}\,\text{m}^{-1}$, *it had a resistance of* 82.40 Ω *at* 25° C. *With* 0.0025 M K_2SO_4, *it has a resistance of* 326.0 Ω. *Calculate*

(a) the cells constant. (b) conductivity of K_2SO_4 *solution.*

Sol. (a) Calibration of the cell has been done using 0.020 M KCl solution.

$$\text{specific conductance} = \frac{\text{cell constant}}{\text{resistance } (R)}$$

∴ cell constant = specific conductance × resistance

$$= 0.2768 \times 82.4 = 22.81\,\text{m}^{-1}$$

(b) Thus, specific conductance of K_2SO_4 solution.

$$= \frac{22.81}{326} = 0.07\,\Omega^{-1}\,\text{m}^{-1}\,(\text{S cm}^{-1})$$

Equivalent conductivity (Λ_{eq})

The conductance of all the ions produced from gram equivalent of the electrolyte such that electrolyte is kept between two electrodes at 1 cm apart.

$$\text{Equivalent conductivity} = \frac{\text{conductance}}{\text{gram equivalent of electrolyte}}$$

$$\Lambda_{eq} = \frac{G}{\text{g.eq}} = \frac{\kappa(a/l)}{N\frac{V(\text{cc})}{1000}}$$

$$= \frac{\kappa(a/l)}{N(a.l)} \times 1000$$

$$= \frac{\kappa}{Nl^2} \times 1000$$

As $l = 1$, cm then $\boxed{\Lambda_{eq} = \frac{\kappa \times 1000}{N}}$

Its unit is mho-cm^2- eq^{-1}. (S cm^2 eq^{-1})

1 S cm^2 eq^{-1} = 10^{-4} S m^2 eq^{-1}

Example 3.2 *The resistance of a decinormal solution of a salt occupying a volume between two platinum electrodes* 1.80 cm *apart and* 5.4 cm^2 *in area was found to be* 50 Ω. *Calculate the equivalent conductance of the solution.*

Sol. We know equivalent conductance,

$$\Lambda_{eq} = \frac{1000 \times \text{conductance} \times \text{cell constant}}{\text{normality}}$$

$$\text{Given, conductance} = \frac{1}{\text{resistance}} = \frac{1}{50}\Omega^{-1} = 0.02\text{ S}$$

$$\text{Cell constant} = \frac{1}{a} = \frac{\text{distance}}{\text{area}} = \frac{1.80}{5.40} = \frac{1}{3}\text{ cm}^{-1}$$

Normality = 0.1 N

$$\therefore \quad \Lambda_{eq} = \frac{1000 \times \text{conductance} \times \text{cell constant}}{\text{concentration (normality)}}$$

$$= \frac{1000 \times 0.02}{0.1} \times \frac{1}{3}$$

$$= 66.66\text{ S cm}^2\text{ equivalent}^{-1}$$

Molar conductivity (Λ_m)

Molar conductivity of a solution is the conductance provided by 1 mole of electrolyte such that electrolyte is kept between two electrode at 1 cm apart.

$$\Lambda_m = \frac{\text{conductance}}{\text{mole of electrolyte}} = \frac{G}{n} = \frac{\kappa(a/l)}{M \cdot \frac{V(\text{cc})}{1000}}$$

[where, M is molarity and V is volume]

$$= \frac{\kappa(a/l)}{M \cdot (a.l)} \times 1000 = \frac{\kappa \times 1000}{M.l^2}$$

When $l = 1$cm, then $\boxed{\Lambda_m = \frac{\kappa \times 1000}{M}}$

Unit of Λ_m is mho-cm^2-mol^{-1} (S cm^2 mol^{-1})

Relation between molar conductivity and equivalent conductivity

Molar conductivity (Λ_m) and equivalent conductivity (Λ_{eq}) can be related as,

$$\Lambda_m = n \times \Lambda_{eq}$$

where, n = valence factor and can be calculated as,

$$n = \frac{\text{atomic weight}}{\text{equivalent weight}}$$

Example 3.3 *Equivalent conductance of* 0.01 N *solution of* $CaCl_2$ *is* 120.36 mho cm^2 equiv^{-1}. *Convert this value to the molar conductance in units of* S m^2 mol^{-1}.

Sol. mho (S) cm^2 equiv^{-1} is to be changed into S m^2 mol^{-1} for $CaCl_2$

$$\Lambda_{eq} = 120.36\text{ mho cm}^2\text{ equiv}^{-1}$$

$$= \frac{120.36\text{ S}\left(\frac{1}{100}\text{ m}\right)^2}{\left(\frac{\text{mol}}{2}\right)} = 2 \times 120.36\text{ S }(10^{-4})\text{ m}^2\text{ mol}^{-1}$$

$$= 240.72 \times 10^{-4}\text{ S m}^2\text{ mol}^{-1} = 2.4072 \times 10^{-2}\text{ S m}^2\text{ mol}^{-1}$$

Note 1 mol $CaCl_2$ = 2 equivalent $CaCl_2$

100 cm = 1 m ; 1 cm = 10^{-2} m ; 1 cm^2 = 10^{-4} m^2; $\Lambda_m = 2\Lambda_{eq}$ for $CaCl_2$

Example 3.4 *A* 0.05 M KOH *solution offered as resistance of* 31.6 Ω *in a conductivity cell of cell constant* 0.367 cm^{-1} *at* 298 K. *Calculate the molar conductance of* KOH *solution?*

Sol.

$$\Lambda_m = \frac{1000 \times \text{conductance} \times \text{cell constant}}{C_m\text{ (mol L}^{-1})}$$

$$= 1000 \times \frac{1}{31.6} \times \frac{0.367}{0.05} = 232.24\text{ S cm}^2\text{ mol}^{-1}$$

Variation of conductivity with dilution

Depending upon the values of molar conductivity, the electrolytes can be divided into two groups namely strong electrolytes and weak electrolytes.

For strong electrolyte

Strong electrolytes are completely ionised in all concentrations. For strong electrolytes, Λ_m increases slowly with dilution. It is found to vary with concentration according to the Debye-Huckel-Onsager equation.

$$\Lambda_m = \Lambda_m^\infty - [A + B\Lambda_m^\infty]\sqrt{C}$$

where, A and B are called Debye-Huckel constants.

$$A = \frac{82.4}{(DT)^{1/2}\eta}, \quad B = \frac{8.20 \times 10^5}{(DT)^{3/2}}$$

D and η are the dielectric constant and coefficient of viscosity of the medium respectively at temperature T(K) and C is the molar concentration (mol L^{-1}).

Λ°_m is the molar conductivity at infinite dilution called as **limiting molar conductivity**.

Interpretation of limiting molar conductivity

This equation found to hold good at low concentrations. Thus, if we plot molar conductivity (Λ_m) against $\sqrt{C}$, a linear graph is obtained with negative slope. Whereas, it is not linear for higher concentrations as shown in Fig. 3.2 in which KCl has been taken as an example of a strong electrolyte.

This decrease of the slope is not due to a change in the number of ions with concentration, but due to change in the interionic attractions. At high concentrations, there are greater number of ions per unit volume which increases attraction between cation (K^+) and anion (Cl^-). Attraction between ions reduces the conducting ability of the ions hence, Λ_m falls with increasing concentration.

If solution is diluted (concentration is decreased), there is decrease in ionic attractions hence, molar conductivity increases with decrease in concentration.

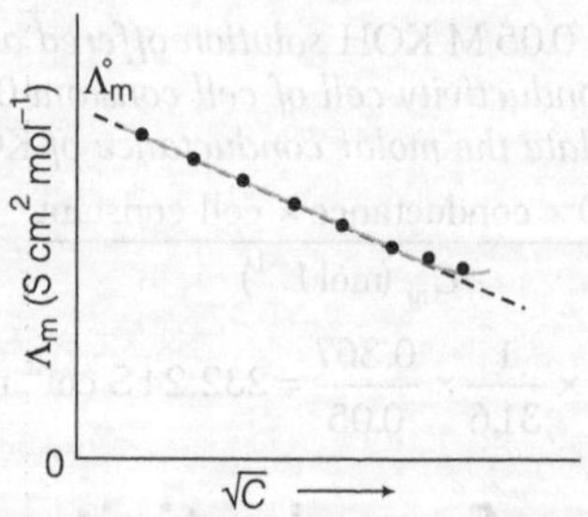

Fig. 3.2 Variation of Λ_m with $\sqrt{C}$ for strong electrolytes (KCl)

The value of limiting molar conductivity can be obtained by extrapolating the curve to $\sqrt{C} = 0$.

For weak electrolyte

In case of weak electrolytes (as CH_3COOH, NH_4OH, AgCl etc.), ionisation is very small as compared to strong electrolytes, hence, Λ_m (or Λ_{eq}) of a weak electrolyte is low.

In very low concentration range, the degree of ionisation rises sharply hence Λ_m increases sharply when the solution is diluted in lower concentration range.

In this region, the graph of weak electrolyte becomes parallel to the Y-axis. So, the limiting molar conductivity of a weak electrolyte cannot be determined by the extrapolation of the plot. It is rather determined by applying the **Kohlrausch law** of independent migration of ions.

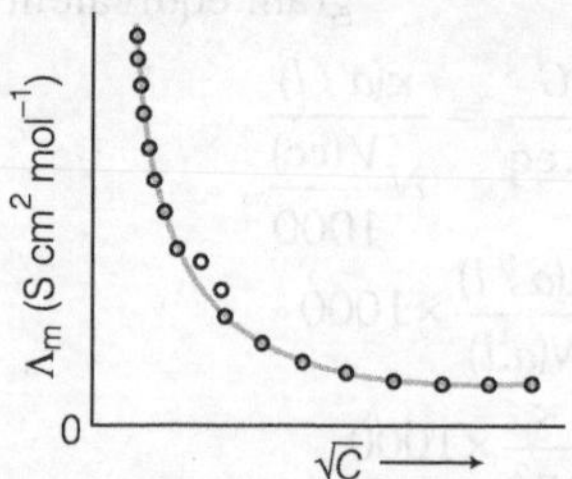

Fig. 3.3 Variation of Λ_m with $\sqrt{C}$ for weakly electrolyte (CH_3COOH)

Factors affecting conductivity

The conductivity of electrolytes can be affected by the following factors.

(*i*) **Interionic attraction** If interionic attraction between solute molecules is higher than conductivity becomes lower.

(*ii*) **Polarity of solvent** If solvent is more polar than its ionisation and conductivity will be greater.

(*iii*) **Viscosity of medium** On increasing viscosity of medium, conductivity decreases.

(*iv*) **Temperature** On increasing the temperature of electrolytic solution, conductivity increases.

(*v*) **Concentration of the solution** All types of conductivities are affected by changing the concentration. Their respective variations with concentration are discussed below,

- **Effect of concentration on specific conductivity** Conductivity always decreases with decrease in concentration (i.e. with dilution) of both the strong and weak electrolytes. This is due to the fact that the number of ions that carry current in a unit volume of solution always decreases with decrease in concentration.
- **Effect of concentration on molar and equivalent conductivity** On decreasing the concentration of electrolytic solution, molar conductivity and equivalent conductivity increases. It is so because on progressively decreasing the concentration the dissociation of electrolytes goes on increasing. Thus, the number of ions goes on increasing.

Limiting conductivity and Kohlrausch's law

At infinite dilution each ion makes a definite contribution towards molar conductivity (or equivalent conductivity) of the electrolyte irrespective of the nature of the other ions present.

In other words, we can say that the **limiting molar conductivity** of an electrolyte is the sum of the limiting molar conductivities of the cation and the anion each multiplied with the number of ions present in one formula unit of the electrolyte.

Mathematically, Λ°_m for $A_xB_y = x\lambda^\circ_{(A)^{y+}} + y\lambda^\circ_{(B)^{x-}}$

where, Λ°_m = limiting molar conductivity of electrolyte

$\lambda^\circ_{(A)^{y+}}$ and $\lambda^\circ_{(B)^{x-}}$ are the limiting molar conductivities of cation (A^{y+}) and anion (B^{x-}) respectively.

The multiplication of number of ions with molar conductivity of a particular ion is called **molar ionic conductance** of that particular ion.

Some example of electrolyte with their number of ions per formula unit are given below

Electrolyte	HCl	K_2SO_4	$BaCl_2$	$AlCl_3$	Na_3PO_4	$Fe_2(SO_4)_3$
x	1	2	1	1	3	2
y	1	1	2	3	1	3

In terms of equivalent conductivity Kohlrausch's law is defined as "The limiting equivalent conductivity (or equivalent conductivity at infinite dilution) is the sum of the equivalent conductivities of cations and anions", i.e.

$$\Lambda^\circ_{eq} = \lambda^\circ_a + \lambda^\circ_c$$

where λ°_c and λ°_a are called limiting ionic conductivities for cation and the anion respectively.

e.g. Molar conductance of $CaCl_2$

If equivalent conductance of

$$Ca^{2+} = x \text{ S cm}^2\text{eq}^{-1}$$

$$Cl^{-1} = y \text{ S cm}^2\text{eq}^{-1}$$

at infinite dilution,

then equivalent conductance of $CaCl_2$ will be

$$\Lambda^\infty_{eq} = \lambda^\infty_{Ca^{2+}} + \lambda^\infty_{Cl^-}$$

$$= (x + y)$$

Molar conductance $\quad \Lambda^\infty_m = 2\Lambda^\infty_{eq}$

$$= 2(\lambda^\infty_{Ca^{2+}} + \lambda^\infty_{Cl^-})$$

$$= 2(x + y)$$

or $\quad \lambda^\infty_{Ca^{2+}} \text{(molar)} = 2\lambda^\infty_{Ca^{2+}} \text{(equi)}$

$$\lambda^\infty_{Cl^-} \text{(molar)} = \lambda^\infty_{Cl^-} \text{(equi)}$$

$\therefore \quad \Lambda^\infty_m(CaCl_2) = \lambda^\infty_{Ca^{2+}} \text{(molar)} + 2\lambda^\infty_{Cl^-} \text{(molar)}$

$$= 2x + 2y = 2(x + y)$$

Example 3.5 *For strong electrolytes the values of molar conductivity at infinite dilution (or limiting molar conductivity) are given below.*

Electrolyte	Λ^∞_m (S m^2 mol^{-1})
$BaCl_2$	280×10^{-4}
NaCl	126.5×10^{-4}
NaOH	248×10^{-4}

Calculate $\Lambda^\infty_{[Ba(OH)_2]}$.

Sol. $BaCl_2 + 2NaOH \longrightarrow Ba(OH)_2 + 2NaCl$

Thus, $\Lambda^\infty_{[Ba(OH)_2]} = \Lambda^\infty_{(BaCl_2)} + 2\Lambda^\infty_{(NaOH)} - 2\Lambda^\infty_{(NaCl)}$

$$= 280 \times 10^{-4} + 2 \times 248 \times 10^{-4} - 2 \times 126.5 \times 10^{-4}$$

$$= 523.0 \times 10^{-4} \text{ S m}^2 \text{ mol}^{-1}$$

Applications of conductance measurement and Kohlrausch's law

Some important applications of Kohlrausch's law are as follows :

(*i*) **Determination of limiting molar conductivity (Λ°_m) for weak electrolyte**

This law is helpful to calculate molar conductance at infinite dilution for weak electrolytes. Suppose, we have to calculate the limiting conductance for CH_3COOH. Then, according to Kohlrausch's law

$$\Lambda^\circ_{(CH_3COOH)} = \Lambda^\circ_{(CH_3COO^-)} + \Lambda^\circ_{(H^+)} \quad ...(i)$$

This equation can be obtained by the knowledge of molar conductance at infinite dilution for some strong electrolytes, i.e.,

$$\Lambda^\circ_{(CH_3COO^-Na^+)} = \Lambda^\circ_{(CH_3COO^-)} + \Lambda^\infty_{(Na^+)} \quad ...(ii)$$

$$\Lambda^\circ_{(HCl)} = \Lambda^\circ_{(H^+)} + \Lambda^\circ_{(Cl^-)} \quad ...(iii)$$

$$\Lambda^\circ_{(NaCl)} = \Lambda^\circ_{(Na^+)} + \Lambda^\circ_{(Cl^-)} \quad ...(iv)$$

The above Eq. (i) can be obtained by adding Eqs. (ii) and (iii) and subtracting Eq. (iv) from the sum, i.e.

$$\Lambda^\circ_{(CH_3COOH)} = \Lambda^\circ_{(CH_3COONa)} + \Lambda^\circ_{(HCl)} - \Lambda^\circ_{(NaCl)}$$

(*ii*) **In the calculation of degree of dissociation** For calculating degree of dissociation (α) of weak electrolyte, formula used is

$$\alpha = \frac{\Lambda_m}{\Lambda^\circ_m}$$

where, α = degree of dissociation

Λ^c_m = molar conductance at any concentration

Λ°_m = limiting molar conductance

(*iii*) **In the calculation of ionisation or dissociation constant (K_a) for weak electrolyte** For a weak electrolyte, dissociation constant can be formulated as,

$$K_a = \frac{C\alpha^2}{1-\alpha} = \frac{C\Lambda_m^2}{\Lambda_m^\circ(\Lambda_m^\circ - \Lambda_m)}$$

where, C = concentration of electrolyte
Λ_m° = limiting molar conductivity
Λ_m = molar conductivity

(*iv*) **In the calculation of solubility of sparingly soluble salts** If solute is sparingly soluble in a given solvent, its concentration is taken as its solubility in the saturated solution.

[Also $\lim_{c\to 0} \Lambda_m^c = \Lambda_m^\circ$ or Λ_m^∞]

Thus, $\Lambda_m^c = \Lambda_m^\infty$

or $\Lambda_m^\infty = \dfrac{1000 \times \text{specific conductance}}{C}$

Λ_m° can be computed by use of Kohlrausch's law

$$\Lambda_m^\infty \text{ or } \Lambda_m^\circ = \lambda^\circ_{\text{cation}} + \lambda^\circ_{\text{anion}}$$

Thus, C (which is also the solubility of the sparingly soluble salt) and hence, K_{sp} is known.

A_xB_y $K_{sp} = x^x y^y (S)^{x+y}$

S is the solubility in mol L^{-1}

Solute	K
AgCl	$K_{sp} = [Ag^+][Cl^-] = S^2$
PbI_2	$K_{sp} = [Pb^{2+}][I^-]^2 = 4S^3$
$Al(OH)_3$	$K_{sp} = [Al^{3+}][OH^-]^3 = 27S^4$

Note *If equivalent conductance is given, then solubility is in eq L^{-1}. Convert it into mol L^{-1} for K_{sp} value.*

(*v*) **Conductometric titration** Titration in which end point can be determined by measuring conductance at each addition of the titre solution (from burette) is called conductometric titration. We take the fact that conductance of non-electrolyte or insoluble solute is lower than that of the electrolyte or its ion.

Titration of HCl (*aq*) with NaOH (*aq*)
(Strong base *vs* strong acid)

$$HCl(aq) + NaOH(aq) \rightleftharpoons NaCl(aq) + H_2O$$
$$H^+(aq) + OH^-(aq) \rightleftharpoons H_2O$$

Initially, there is maximum conductance due to HCl (free H^+). As NaOH (OH^-) is added from burette, equivalent

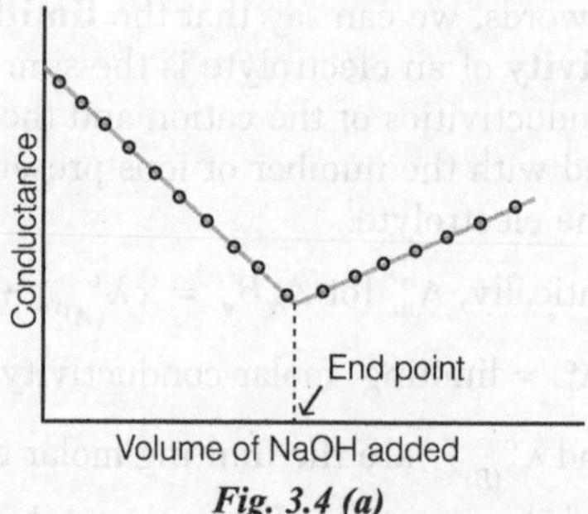

Fig. 3.4 (a)

H^+ ion is neutralised forming weak electrolyte H_2O. Thus, conductance falls. This decrease appears till all the HCl is neutralised. After complete neutralisation, further addition of NaOH will result in the increase in conductance.

A graph between conductance and volume of NaOH added from burette is of the type as shown in Fig. 3.4(a). End point of the titration is where two linear curves intersect.

Titration of CH_3COOH (*aq*) with NaOH (*aq*)
(Weak acid vs strong base)

$$CH_3COOH(aq) + OH^-(aq) \rightleftharpoons CH_3COO^- + H_2O$$

In this case every addition of OH^- ion from burette into CH_3COOH solution will form more conducting CH_3COO^- ion, thus, conductance starts increasing slightly. After complete neutralisation of CH_3COOH, further addition of OH^- ion will make rapid increase in conductance (Fig. 3.4(b)). End point is that where two linear curves intersect.

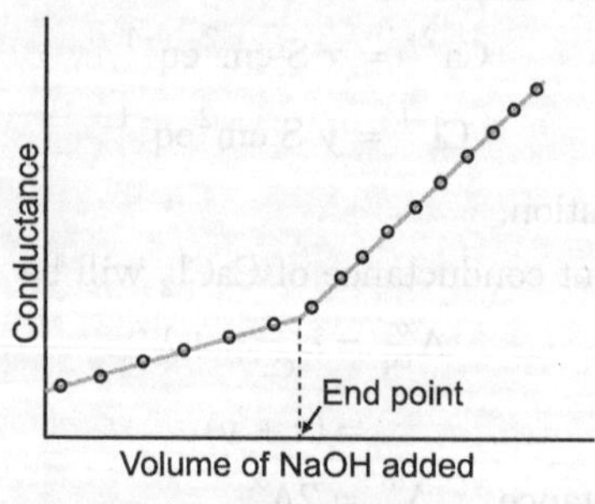

Fig. 3.4 (b)

Transport number

Transport number is defined as the fraction of the current carried by anion.

$$\text{Transport number} = \frac{\text{current carried by ion}}{\text{total current carried}}$$

Transport number of cation + transport number of anion = 1

Example 3.6 *Λ°_m for NaCl, HCl and NaAc are 120, 400.2 and 80.5 S cm^2 mol^{-1} respectively. Calculate Λ°_m for HAc.*

Sol. $\Lambda^\circ_{m(HAc)} = \Lambda^\circ_{m(HCl)} + \Lambda^\circ_{m(NaAc)} - \Lambda^\circ_{m(NaCl)}$

$= (400.2 + 80.5 - 120)$ S cm^2 mol^{-1}

$= 360.7$ S cm^2 mol^{-1}

Example 3.7 *The molar conductivity of 0.25 mol L^{-1} formic acid is 40.1 S cm^2 mol^{-1}. Calculate its dissociation constant. [Assuming, $\Lambda^\circ_{m(H^+)} = 349.6$ S cm^2 mol^{-1} $\Lambda^\circ_{m(HCOO^-)} = 54.6$ S cm^2 mol^{-1}]*

Sol. $\Lambda^c_m = 40.1$ S cm^2 mol^{-1}

$$\Lambda^\circ_m = \Lambda^\circ_{m(HCOO^-)} + \Lambda^\circ_{m(H^+)}$$

$= (54.6 + 349.6)$ S cm^2 mol^{-1}

$= 404.2$ S cm^2 mol^{-1}

$$\alpha = \frac{\Lambda^c_m}{\Lambda^\circ_m} = \frac{40.1}{404.2} = 0.0992$$

Now, $HCOOH \underset{}{\overset{H_2O}{\rightleftharpoons}} HCOO^-(aq) + H^+(aq)$

Initial conc.	C	0	0
Equi conc.	$C(1-\alpha)$	$C\alpha$	$C\alpha$

$$K_a = \frac{C\alpha^2}{1-\alpha} = \frac{0.25 \times (0.1140)^2}{(1 - 0.1140)}$$

$= 2.73 \times 10^{-3}$ mol L^{-1}

Example 3.8 *For a saturated solution of AgCl at 25° C, specific conductance is 3.41×10^{-6} Ω^{-1} cm^{-1} and that of water used for preparing the solution was 1.60×10^{-6} Ω^{-1} cm^{-1}. Calculate the solubility product of AgCl?*

$\Lambda^\infty_{eq}(AgCl) = 138.3\,\Omega^{-1}$ cm^2 equiv^{-1}

Sol. For AgCl, molarity = normality

Actual specific conductance

= (specific conductance of AgCl − specific conductance of water)

$= (3.41 - 1.60) \times 10^{-6}$

$= 1.81 \times 10^{-6}\,\Omega^{-1}$ cm^{-1}

For saturated solution of sparingly soluble salt

$\lambda_{eq} = \lambda^\infty_{eq}$, and solubility = concentration

$$\lambda^\infty_{eq} = \frac{1000 \times \text{specific conductance}}{\text{solubility}}$$

$$\therefore \quad 138.3 = \frac{1000 \times 1.81 \times 10^{-6}}{S}$$

$$S\,(\text{mol L}^{-1}) = \frac{1000 \times 1.81 \times 10^{-6}}{138.3}$$

$= 1.31 \times 10^{-5}$ mol L^{-1}

$$AgCl(s) \rightleftharpoons Ag^+(aq) + Cl^-(aq)$$

$\therefore \quad K_{sp} = [Ag^+][Cl^-] = S^2$

$= (1.31 \times 10^{-5})^2 = 1.72 \times 10^{-10}$ mol^2 L^{-2}

CHECK POINT 3.1

1. Which of the following is not a strong electrolyte?
(a) KNO_3 (b) NaCl
(c) NH_4OH (d) $FeSO_4$

2. Resistivity of a material is equal to resistance, when
(a) cell constant = 1 cm^{-1} (b) cell constant = 0.1 cm^{-1}
(c) cell constant = 10 cm^{-1} (d) cell constant = 100 cm^{-1}

3. Reciprocal of Siemen is
(a) volt (b) ohm (c) ampere (d) coulomb

4. Specific conductance has unit
(a) Ω-cm (b) Ω^{-1} cm (c) Ω^{-1}cm^{-1} (d) Ω cm^{-1}

5. Which of the following has maximum conductivity?
(a) Purified drinking water (b) Distilled water
(c) Rain water (d) River water

6. Which of the following relation is correct?
(a) Conductance = specific conductance × cell constant
(b) Cell constant = $\dfrac{\text{conductivity}}{\text{conductance}}$
(c) Equivalent conductance = $\dfrac{\text{conductivity}}{\text{volume}}$
(d) Molar conductance = conductivity × concentration

7. Which of the following statements is incorrect.
(a) Equivalent conductance increases with dilution
(b) Specific conductance increases with dilution
(c) Molar conductance of the solution increase with dilution
(d) At infinite dilution each ion (cation or anion) plays a definite role towards electrical conductance.

8. Which factor(s) is/are responsible for decrease in electrolytic conduction?
(a) Dilution
(b) Increase in temperature
(c) Interionic attraction in higher concentration range
(d) None of the above

9. In the variation of Λ_m with $\sqrt{C}$, $\Lambda_m = \Lambda^\infty_m - (A + B\Lambda^\infty_m)\sqrt{C}$, A and B are called
(a) van der Waals' constant (b) Debye-Huckel constant
(c) Onsager constants (d) Critical constants

10. From the equation in above question (9), it is evident that
(a) $\Lambda_m > \Lambda^\infty_m$
(b) $\Lambda_m = \Lambda^\infty_m$ as $\sqrt{C} \to 0$
(c) A and B both are dependent on temperature T and dielectric constant D of the medium
(d) All the above are correct statements

CELL

Cells are the devices in which interconversion of electrical energy and chemical energy takes place.

A cell is made up of various components, i.e. electrolytic salt-bridge, electrodes and solution.

Salt-bridge

It is a U-shaped glass tube which contains agar-agar paste with NH_4NO_3, KNO_3 or KCl as conducting electrolytes. **KCl is not used when electrode is made of Ag.** This salt-bridge is only used in electrochemical cell.

Functions of salt-bridge

(*i*) The function of salt-bridge is to allow the movement of ions from one solution to other without mixing of the two solutions.

(*ii*) As electrons flow in the outer circuit, in the wire the inner circuit is completed by the flow of ions from one solution to the other through the salt-bridge.

(*iii*) It helps to maintain the electrical neutrality of the solution in the half cells.

(*iv*) It also prevents the liquid-liquid junction potential that arises between two solutions.

Electrodes

These are the materials that conduct electricity to or from the cell due to movement of electrons. It may be taken in any form like wire, rod, sheet etc. Electrodes may be further classified in following ways.

(*i*) **Anode and cathode** Electrode at which oxidation takes place is called as **anode** whereas the electrode at which reduction takes place is called as **cathode.**

(*ii*) **Inert and attackable electrode**

- **Inert electrode** Electrode which only conducts electricity but does not participate in the chemical reactions are called inert electrodes.
- **Attackable electrode** The electrode which conduct electricity and also participate in the chemical reactions are called attackable electrodes.

(*iii*) **Reversible electrode and irreversible electrodes**

- **Irreversible electrode** At this electrode, reaction occurs only in one direction.
- **Reversible electrode** At this electrode, reduction as well as oxidation will takes place.

Some examples of reversible electrodes are as follow

(a) **Metal-metal salt electrode** When metal electrode immersed in its salt solution a reversible reaction will take place between metal atoms and its metal ions.

$$M(s) \rightleftharpoons M^{n+}(aq) + ne^-$$

The metal atom have tendency to move in the solution as metal ion. Whereas, the metal ions from the solution have tendency to get deposited at the metal surface.

(b) **Metal insoluble metal salt electrode** In this type of electrodes, the metal surface of the electrode is coated with some of its insoluble salt. This coated electrode is now immersed in aqueous solution of some salt that have common anion with metal electrode; as AgCl on Ag electrode.

(c) **Gaseous electrode** In this type of electrode gases (like H_2) are taken which are kept in contact with the solution of its corresponding ion. e.g. hydrogen electrode, oxygen electrodes.

Note *In this case Pt-wire is normally used to conduct electricity.*

(d) **Redox electrodes** The electrode in which electrolytic solution contains the same element in two different oxidation state. In this case also Pt-wire is used to conduct electricity in the solution.

e.g. $$Sn^{2+}(aq) \rightleftharpoons Sn^{4+}(aq) + 2e^-$$

Electrode potential

When a metallic electrode is placed in its salt solution, exchange of electrons between the atoms of electrodes and cation in its salt solution will take place.

On account of this, a definite potential difference is developed between the electrode and its salt solution. This potential is called **electrode potential** (or half-cell electrode potential).

In other words, it is the measure of the tendency of an electrolyte to gain or lose electrons in a half-cell. It is denoted by ***E***.

Standard electrode potential

In the standard state, when pressure is 1 bar, temperature is fixed (usually 298 K or 25°C) and concentration is 1M electrode potential is called as standard electrode potential. It is denoted as ***E***°.

Type of electrode potential is tabulated below.

Half-cell	Reaction	Electrode potential
$Zn \mid Zn^{2+}(aq)$	$Zn \rightarrow Zn^{2+}(aq) + 2e^-$	E_{ox} or $E_{Zn \mid Zn^{2+}}$ represents oxidation electrode potential in the given state and E°_{ox} or $E^\circ_{Zn \mid Zn^{2+}}$ in the standard state.
$Cu^{2+}(aq) \mid Cu$	$Cu^{2+}(aq) + 2e^- \rightarrow Cu$	E_{red} or $E_{Cu^{2+} \mid Cu}$ represents reduction electrode potential in the given state and E°_{red} or $E^\circ_{Cu^{2+} \mid Cu}$ in the standard state.

Generally, the half-cell reactions are written as reduction half reactions and their potential are represented as reduction potential as given below

$$E_{\text{oxd}} = -E_{\text{red}}$$

Factors affecting the magnitude of electrode potential

The magnitude of electrode potential depends upon

(*i*) Nature of the metal (electrode)

(*ii*) Concentration of metal ions in solution

(*iii*) Temperature

Measurement of electrode potential

The electrode potential of an electrode is always determined with respect to a standard electrode or reference electrode. Because neither oxidation nor reduction takes place independently that is, no half-cell work independently.

Thus for indirect measurement of potential of an electrode, we couple the electrode with reference electrode (as its potential is known). The emf (potential difference when no current is drawn from the system) of resulting cell is measured experimentally. Now, as we know the emf as well as the potential of reference electrode, the potential of other electrode can be determined. One of the important reference electrodes i.e. SHE is discussed below.

Standard hydrogen electrode (SHE)

SHE consist of Pt-wire sealed in a glass tube and has a platinum foil attached to it. The foil is coated with finely divided platinum and acts as platinum electrode.

It is dipped into an acid solution containing H^+ ions in 1 M concentration (1M HCl). Hydrogen gas is bubbled into the electrode through inlet as shown in figure. The pressure of hydrogen gas is maintained at 1 atm.

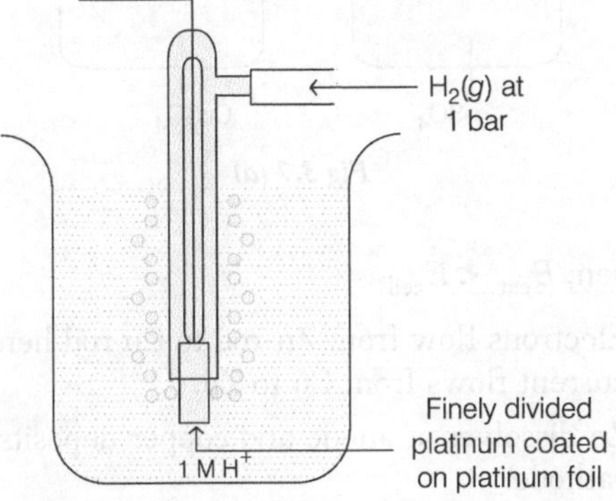

Fig. 3.5 Standard hydrogen electrode (SHE)

H_2 gas gets absorbed on the finely divided platinum. Thus, hydrogen is in contact with H^+ ions of HCl solution at the electrode surface. The assembly so obtained is standard hydrogen electrode.

The following reaction occurs in this half-cell depending upon whether it acts as anode or as a cathode

- If SHE acts as anode

$$H_2(g) \longrightarrow 2H^+ + 2e^-$$

- If SHE acts as cathode

$$2H^+ + 2e^- \longrightarrow H_2(g)$$

- The SHE potential is taken as zero. The electrode potential of a metal electrode is determined with respect to a standard (or normal hydrogen electrode) is called standard electrode potential of that electrode.

Note *Pt foil is used in SHE because it is a very good absorber of* H_2 *gas.*

Types of cell

Cells are of two basic types as discussed below

(*i*) **Electrochemical cell** In this cell chemical reaction (redox reactions) are carried out that result in production of electricity. It is also called voltaic or galvanic cell. One of the important example of this type of cells is Daniell cell.

And other some cells are classified into two more categories, primary cell and secondary cell.

(*ii*) **Electrolytic cell** In this cell electricity is used to produce non-spontaneous chemical changes. As there occurs, breaking down of a molecule with the help of electricity, the process is called **Electrolysis**.

How to apply the terms cathode and anode to an electrolytic cell ?

- Regardless whether a cell is a voltaic or an electrolytic cell
- The anode is the electrode at which oxidation occurs
- The cathode is the electrode at which reduction occurs

	Electrochemical cell	Electrolytic cell
Anode	Oxidation, negative (-) terminal	Oxidation positive (+) terminal
Cathode	Reduction positive (+) terminal	Reduction negative (-) terminal

We will explore furthur descriptions of cell types, after study of conductance and conductivity of electrolytic solutions.

ELECTROCHEMICAL CELL

As discussed, electrochemical cell (or voltaic or galvanic cell) is a device that produce electricity as a result of chemical reactions (redox reactions). In other words, in this device the Gibbs energy of the spontaneous redox reaction is converted into electrical energy. e.g. Daniell cell.

When a spontaneous reaction takes place in a galvanic cell, electrons are deposited in one electrode (the site of oxidation– anode) and collected from another (the site of reduction–the cathode). So, there is net flow of current which can be used to do work.

Note *In a cell, if two electrolytic solutions of different concentrations are in contact with each other, a potential difference develops across the boundary of the two solutions. It is called* ***liquid-junction potential*** *or the diffusion potential across the interface of two electrolytes. Electrolyte concentration cells always have liquid junction.*

Daniell cell

(*i*) In this cell, anode is made up of Zn-rod which is placed in zinc sulphate solution (oxidation half-cell). Similarly, a Cu-rod is placed in copper sulphate solution taken in another beaker (reduction half-cell). In general, electrochemical cell involves the chemical reactions that takes place in two vessels each of which is called **half-cell.**

(*ii*) The two electrodes are connected by a wire and the two solutions are connected by a **salt-bridge.**

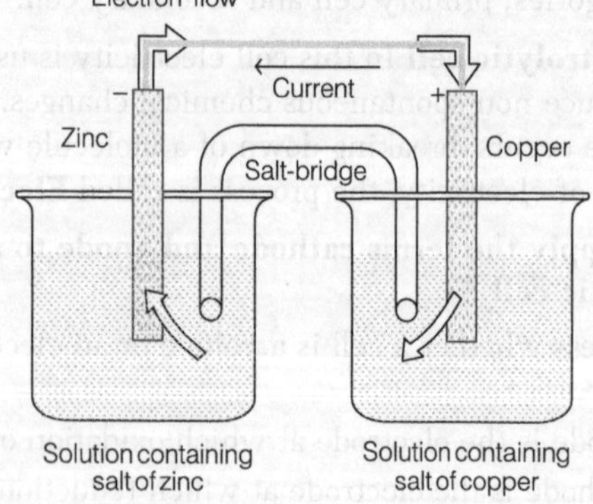

Fig. 3.6 Daniell cell having electrodes of zinc and copper dipping in the solutions of their respective salts

When the Zn and Cu electrodes are joined by wire, following observations are noted.

- Reactions involve in two half-cells are

At cathode $Cu^{2+} + 2e^- \longrightarrow Cu(s)$ (Reduction half reaction)

At anode $Zn(s) \longrightarrow Zn^{2+} + 2e^-$ (Oxidation half reaction)

- Since, electrons are produced at anode, this electrode is rich in electrons thus also called as **negative electrode**. Whereas, cathode electrode (i.e. Cu-electrode) needs electron for reduction therefore acts as **positive electrode.**

Note *The ⊕ sign of the cathode can be interpreted as indicating the electrode at which electrons enter the cell and the ⊖ sign of the anode is where the electrons leave the cell.*

- There is a flow of electric current in the external circuit from Cu electrode to Zn electrode, i.e. + ve terminal to –ve terminal.
- Zn-rod loses its mass, while Cu-rod gains its mass. Conversely, the concentration of $ZnSO_4$ solution increases while that of $CuSO_4$ solution decreases.
- To maintain the electrical neutrality the cations and anions move through the salt-bridge, this helps to complete the inner circuit.

Functioning of Daniell cell when external voltage (E_{ext}) is applied

Following three cases are possible when external voltage (E_{ext}) opposing the cell potential is applied :

(*i*) The chemical reactions continue to take place as long as external opposing potential (E_{ext}) is less than the potential of cell (E_{cell}). (Fig 3.7 (a))

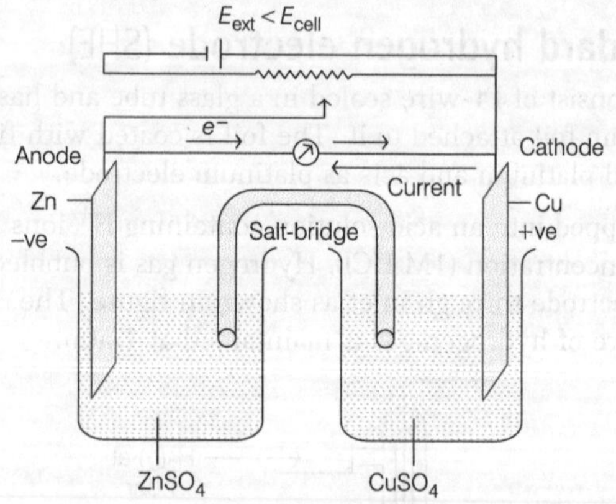

Fig 3.7 (a)

When, $E_{ext} < E_{cell}$

- Electrons flow from Zn-rod to Cu rod hence, current flows from Cu to Zn.
- Zn dissolves at anode and copper deposits at cathode.

(*ii*) When $E_{ext} = E_{cell}$, the reaction will stop and current becomes zero. (Fig 3.7 (b))

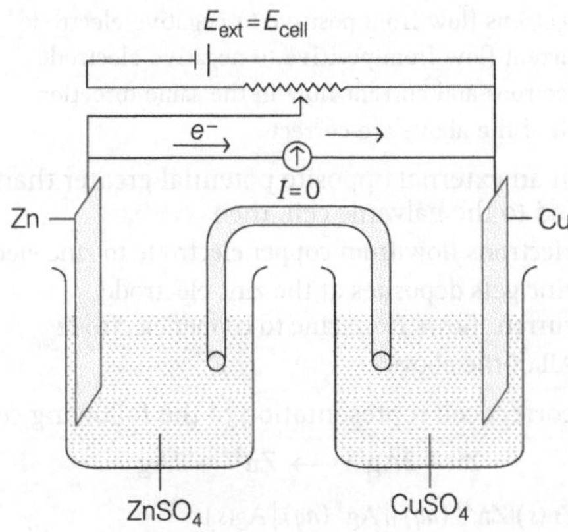

Fig 3.7 (b)

(*iii*) When E_{ext} increases, then current flows in the opposite direction and chemical reaction of the cell is reversed. (Fig 3.7 (c))

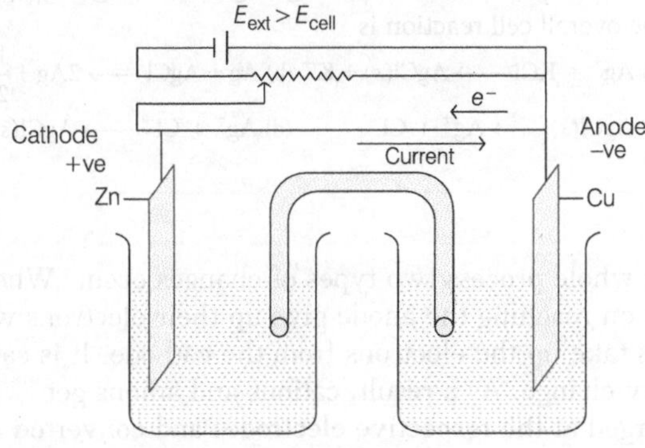

Fig. 3.7 (c)

When $E_{ext} > E_{cell}$

- Electrons flow from Cu to Zn and current flows from Zn to Cu.
- Zinc is deposited at the zinc electrode and copper dissolves copper electrode.

Representation of electrochemical cell

The following conventions are used in representing an electrochemical cell.

(*i*) The electrode on which oxidation takes place is written on the left hand side. While the electron on which reduction takes place is written on the right hand side.

(*ii*) The electrode on the left hand side is written by writing the symbol of the metal (or the gas) first followed by the symbol of the ion with its concentration in brackets.

(*iii*) The electrode on the right hand side, is represented by writing the electrolyte first (anion) and then the metal or solid phase.

(*iv*) Single vertical lines represent the base boundaries of the electrodes.

(*v*) Double vertical line represents the salt-bridge.

(*vi*) C_1 and C_2 (or a_1 and a_2) represents the concentrations that are written within the brackets after the cation.

To sum up,

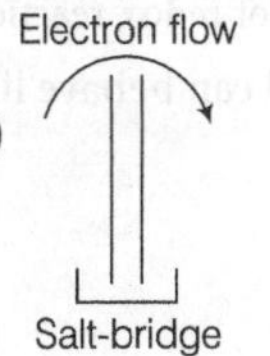

Left electrode	**Right electrode**
• Metal/metal (conc.)	Metal ion (conc.)/metal
• Oxidation occurs	Reduction occurs
• Negative pole	Positive pole

Thus, a complete Daniell cell can be represented as,

$$Zn|Zn^{2+}(1\ M)||Cu^{2+}(1\ M)|Cu$$

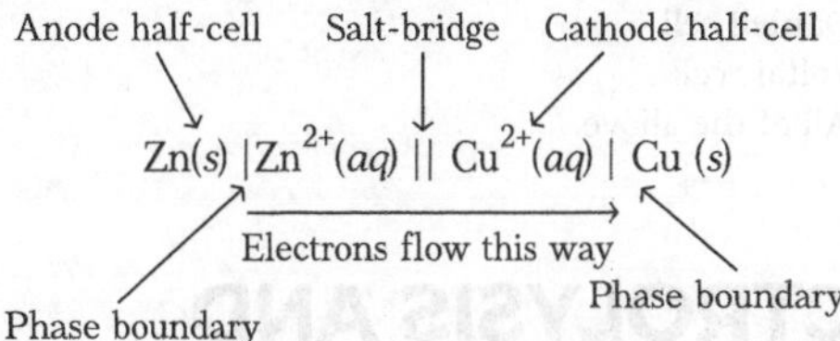

Similarly, Cu-$AgNO_3$ cell may be represented as,

$$Cu|Cu^{2+}(1M)||Ag^{+}(1M)|Ag$$

Table 3.1 Difference between electrochemical cell and electrolytic cell

Electrochemical cell	Electrolytic cell
This cell converts chemical energy into electrical energy	This cell converts electrical energy into chemical energy.
It is based upon the redox reaction which is spontaneous.	The redox reaction is non-spontaneous and takes place only when electrical energy is supplied.
Two different electrodes with two different electrolytes are usually set up in two separate beakers.	Both the electrodes are suspended in electrolyte solution (or melt of electrolyte) in the same beaker.
To set up this cell, a salt-bridge is used.	No such salt-bridge is used in this case.
In this case, anode act as negative pole while cathode act as positive pole.	In this case, anode act as positive pole while cathode act as negative pole.

CHECK POINT 3.2

1. Which of the following statements is correct?
(a) At anode oxidation will takes place
(b) At cathode reduction will takes place
(c) Standard hydrogen electrode is reference electrode
(d) All of the above statements are correct

2. Which of the following statements is not correct about an inert electrode in a cell?
(a) It does not participate in the cell reaction
(b) It provides surface either for oxidation or for reduction reaction
(c) It provides surface for conduction of electrons
(d) It provides surface for redox reaction

3. An electrochemical cell can behave like an electrolytic cell when ....
(a) $E_{cell} = 0$
(b) $E_{cell} > E_{ext}$
(c) $E_{ext} > E_{cell}$
(d) $E_{cell} = E_{ext}$

4. $Zn(s) + Cu^{2+}(aq) \longrightarrow Zn^{2+}(aq) + Cu(s)$

The above redox reaction is used in
(a) Galvanic cell
(b) Daniell cell
(c) Voltaic cell
(d) All of the above

5. Generally, in a galvanic cell
(a) electrons flow from positive to negative electrode
(b) current flow from positive to negative electrode
(c) electrons and current flow in the same direction
(d) All of the above are correct

6. When an external opposite potential greater than 1.1 V is applied to the galvanic cell, then
(a) electrons flow from copper electrode to zinc electrode
(b) zinc gets deposites at the zinc electrode
(c) current flows from zinc to copper electrode
(d) All of the above

7. The correct cell representation of the following cell is

$$Zn + 2Ag^{+} \longrightarrow Zn^{2+} + 2Ag$$

(a) $Zn(s)|Zn^{2+}(aq)\,||\,Ag^{+}(aq)\,|\,Ag(s)$
(b) $Zn^{2+}\ |Zn\ ||Ag|Ag^{+}$
(c) $Zn(aq)|\,Zn^{2+}(s)||Ag^{+}(s)|Ag(aq)$
(d) $Zn(s)|Ag^{+}(aq)||Zn^{2+}(aq) + Ag(s)$

8. For the electrochemical cell, $Ag^{-}|AgCl|KCl\ ||AgNO_3|Ag^{+}$, the overall cell reaction is
(a) $Ag^{+} + KCl \longrightarrow AgCl(s) + K^{+}$ (b) $Ag + AgCl \longrightarrow 2Ag + \frac{1}{2}Cl_2$
(c) $AgCl(s) \longrightarrow Ag^{+} + Cl^{-}$ (d) $Ag^{+} + Cl^{-} \longrightarrow AgCl(s)$

ELECTROLYSIS AND ELECTROLYTIC CELL

Electrolysis is a process of decomposition of an electrolyte by the passage of electricity through its aqueous solution or molten state. As a result, cations move toward cathode while anions toward anode. This movement of ions in solution produce electrical conductance and constitutes flow of current through the solution.

Electrolysis occurs with various components, these tend to set up a system, this system is called as electrolytic cell. (Fig. 3.8).

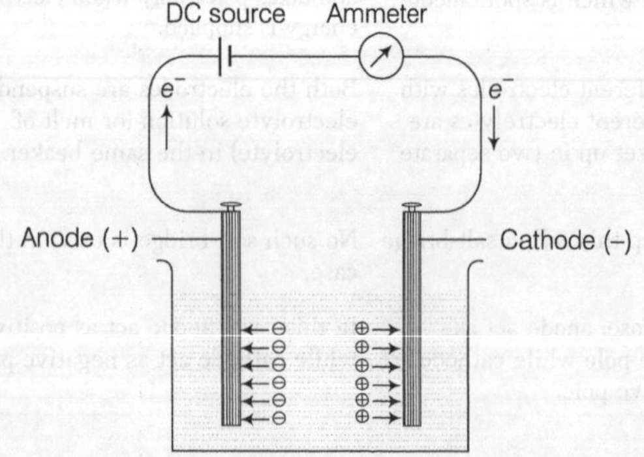

Fig. 3.8 Electrolytic cell

In this whole process two types of changes occur. When anions on reaching the anode give up their electrons while cations take up the electrons from the cathode. It is called primary change. As a result, cations and anions get discharged at the respective electrodes and converted into neutral species.

In simple words, the reduction and oxidation changes which occur at respective electrodes during electrolysis are called primary changes (or primary reactions). When products of primary change (anion and cation) undergo further changes to form molecules, the changes are called secondary changes.

$$AB \rightleftharpoons \underset{\text{Cation}}{A^{+}} + \underset{\text{Anion}}{B^{-}}$$

At anode (Oxidation) $\quad B^{-} \longrightarrow B + e$

At cathode (Reduction) $\quad A^{+} + e^{-} \longrightarrow A$

Discharge potential and preferential discharge theory

An electrolytic solution usually consists of more than two ions. When electrolysis is done then all the ions are not discharged at the electrodes simultaneously, certain ions are liberated at the electrodes in preference to other. Which of them will discharge depends upon their discharge potential. **Discharge potential** is define as the potential at which discharge of a cation occurs at cathode or that of anion occurs at anode. The above phenomenon is explained by the **preferential discharge theory**.

It states that if more than one type of ions are attracted toward a particular electrode, the ions which discharge first is the one which requires least energy.

(*i*) **For cation** In general, cations which is stronger oxidising agent (have high value of standard reduction potential) is discharged first at the cathode. The increasing order of decomposition of few cations is

$$\xrightarrow[\text{Increasing order of decomposition}]{K^+, Ca^{2+}, Na^+, Mg^{2+}, Al^{3+}, Zn^{2+}, Fe^{2+}, H^+, Cu^{2+}, Ag^+, Au^{3+}}$$

(*ii*) **For anion** The anion which is stronger reducing agent (have lower value of standard reduction potential) is liberated first at anode. The increasing order of discharge of few anions is

$$\xrightarrow[\text{Increasing order of discharge}]{SO_4^{2-}, NO_3^{2-}, OH, Cl^-, Br^-, I^-}$$

Note *The +ve or –ve terminal will depends on the direction of the electrons and ions flow during the process. In electrolytic cell cathode is negatively charged whereas anode is positively charged.*

Examples of electrolysis

Some examples of electrolysis are discussed below

1. Electrolysis of molten NaCl

$$NaCl(\text{molten}) \longrightarrow Na^+ + Cl^-$$

At anode, $2Cl^- + 2e^- \longrightarrow Cl_2 \uparrow$

At cathode, $2Na^+ + 2e^- \longrightarrow 2Na\ (l)$

Thus, Cl_2 and Na are the products of electrolysis of molten NaCl.

2. Electrolysis of *aq.* NaCl

In case of aqueous NaCl, H_2O is also present alongwith Na^+ and Cl^- ions. On passing electricity, the following reactions take place.

At anode $2Cl^-(aq) \longrightarrow Cl_2(g) + 2e^-$; $E° = -1.36$ V

$2H_2O(l) \longrightarrow O_2(g) + 4H^+(aq) + 4e^-$; $E° = -1.23$ V

If the solution of NaCl is concentrated, Cl_2 is the main product. However, if it is quite dilute, the product is O_2 instead of Cl_2. This is because of overvoltage.

At cathode $2Na^+ + 2e^- \longrightarrow 2Na$; $E° = -2.71$ V

$2H_2O + 2e^- \longrightarrow H_2 + 2OH^-$; $E° = -0.83$ V

Similar to above, in case of dilute NaCl solution, here also, the product is H_2 instead of Na due to overvoltage.

However, if cathode is made up of mercury, Na-Hg is the main product as Na is soluble in Hg and the over potential of hydrogen is higher on mercury. In such case, on hydrolysis NaOH is obtained as main product.

$$2Na\text{-}Hg + 2H_2O \longrightarrow 2NaOH + 2Hg + H_2$$

3. Electrolysis of water (acidified)

Dil. H_2SO_4 is added to water to increase conductance.

At anode, $H_2O(l) \longrightarrow 4H^+(aq) + O_2(g) + 4e^-$

At cathode, $4H_2O(l) + 4e^- \longrightarrow 4OH^-(aq) + 2H_2(g)$

SO_4^{2-} ions are resistant to oxidation and are not discharged at the cathode.

4. Electrolysis of aqueous $CuSO_4$ solution

(*i*) With inert electrode (platinum)

At anode, $2H_2O(l) \longrightarrow O_2(g) + 4H^+(aq) + 4e^-$;

$E° = -1.23$ V

At cathode,

$2Cu^{2+}(aq) + 4e^- \longrightarrow 2Cu(s)$; $E° = +0.337$ V

Net reaction is,

$2Cu^{2+}(aq) + 2H_2O(l) \longrightarrow 2Cu(s) + 4H^+(aq) + O_2(g)$;

$E_{cell} = -0.89$ V

This is the basic principle of electroplating. Copper metal is liberated at the cathode.

Due to electrolysis, blue colour of the solution gradually fades.

(*ii*) With copper electrodes,

At anode,

$Cu(s) \longrightarrow Cu^{2+}(aq) + 2e^-$; $E° = -0.337$ V

At cathode,

$Cu^{2+}(aq) + 2e^- \longrightarrow Cu(s)$; $E° = +0.337$ V

The net reaction is,

$$\underset{\text{Anode}}{Cu(s)} \longrightarrow \underset{\text{Cathode}}{Cu(s)}; \quad E°_{cell} = 0$$

- Only a very slight voltage is required for this electrolysis. For every Cu atom that enters the solution at the anode. A Cu^{2+} ion deposit as Cu atom at the cathode.
- All that happens is that copper is transferred from the anode to the cathode (as Cu^{2+} through the solution).
  The solution concentration remains unchanged.
- This makes the basis of purification of metals called **refining** (by electrolysis).
- In many cases the voltage necessary to bring about a particular electrode reaction may exceed what we calculate theoretically. Interactions called

polarisation may occur between the electrode surface and the species involved in an electrode reaction. An **overpotential** is the voltage in excess of that calculated to electrolyse.

Example 3.9 *List the anode and cathode half-reactions that might occur when an aqueous solution of $MgCl_2$ is electrolysed in a cell having inert electrodes. Predict which half-reactions will occur and justify your answers?*

Sol. At anode, $2Cl^-(aq) \longrightarrow Cl_2(g) + 2e^-$

$2H_2O(l) \longrightarrow O_2(g) + 4H^+ + 4e^-$

At cathode, $2H_2O(l) + 2e^- \longrightarrow 2OH^-(aq) + H_2(g)$

$Mg^{2+}(aq) + 2e^- \longrightarrow Mg(s)$

Actual reactions

At anode, $2Cl^-(aq) \longrightarrow Cl_2(g) + 2e^-$

At cathode, $2H_2O(l) + 2e^- \longrightarrow H_2(g) + 2OH^-(aq)$

This anode reaction takes place because of a high overvoltage for formation of gaseous O_2. This cathode reaction takes place because H_2O is easier to reduce than Mg^{2+}.

Quantitative aspects of electrolysis

Michael Faraday showed (in 1830) that the amount of substance produced at electrode by electrolysis depends on the quantity of charge passed through the cell. For example, passage of 1 mole of electrons through a cell yields 1 mole (23.0 g) of sodium at the cathode

$$\underset{1\text{ mol}}{Na^+(l) + e^-} \longrightarrow \underset{1\text{ mol}}{Na(l)}$$

Quantitative aspects of electrolysis are governed by **Faraday's law of electrolysis.**

1. First law of electrolysis

It states that "the amount of chemical change produced is proportional to the quantity of electric charge passing through an electrolytic cell."

Greater the number of moles of electrons transferred, the greater the mass of the products.

[current (C / s)× time (s) = charge (C)]

$$w = zit = zC$$

where, w is the mass of the metal deposited/gas liberated due to passage of current i ampere for t seconds, z is the **electrochemical equivalent.** It is the amount of the substance which is deposited when 1A of current is passed. Its unit is $g\,C^{-1}$.

Mathematically,

$$z = \frac{\text{atomic mass}}{\text{electrons exchanged} \times 96500} \quad [1\text{ F} = 96500\text{ C}]$$

e.g. $Ag^+ + e^- \longrightarrow Ag$

$Cu^{2+} + 2e^- \longrightarrow Cu$

$Al^{3+} + 3e^- \longrightarrow Al$

Thus, 1 Faraday (96500 C) of electricity will produce 1g equivalent each of Ag, Cu and Al at cathode or 1 mole of Ag, 0.5 mole of Cu and 0.33 mole of Al.

2. Second law of electrolysis

A given quantity of electricity produces the same number of equivalents of any substance on electrolysis.

$$\left(\frac{\text{mass of metal }(a)}{\text{mass of metal }(b)} = \frac{\text{equivalent weight of }a}{\text{equivalent weight of }b}\right)$$

Example 3.10 *From the definition of a Faraday, calculate charge on one electron.*

Sol. 1 Faraday = 96500 C mol^{-1} electrons.

Thus, 1 mol of electron (= 6.02×10^{23}) has charge = 96500 C

Hence, charge on one electron $= \dfrac{96500\text{ C}}{6.02\times 10^{23}}$

$= 1.603\times 10^{-19}$ C / electron

Example 3.11 *How many electrons per second pass through a cross-section of a copper wire carrying* 10^{-16} *A?*

Sol. 10^{-16} A $= 10^{-16}$ Coulomb / second

Charge on one electron = 1.603×10^{-19} coulomb / electron

Thus, $\text{rate} = \dfrac{10^{-16}\text{ coulomb / second}}{1.603\times 10^{-19}\text{ coulomb / electron}}$

= 624 electron / second

Example 3.12 *A solution of copper (II) sulphate is electrolysed between copper electrodes by a current of* 10.0 A *for exactly* 1 h. *What changes occur at the electrodes and in the solution?*

Sol. Reactions at,

Anode, $Cu \longrightarrow Cu^{2+}(aq) + 2e^-$

Cathode, $Cu^{2+}(aq) + 2e^- \longrightarrow Cu$

By Faraday's first law, amount of copper deposited or dissolved

$$w = zit = \frac{63.5}{2\times 96500} \times 10\times 60\times 60\text{ g}$$

$$\text{Moles of Cu} = \frac{10\times 60\times 60}{2\times 96500}\text{ mol} = 0.18653\text{ mol}$$

Example 3.13 *In an aqueous solution* $AgNO_3$ *and* $CuSO_4$ *are connected in series. If Ag deposited at cathode is* 1.08 g *then calculate the amount of Cu deposited.*

Sol. Eq. wt. of Ag = Eq. wt. of Cu

$$\frac{1.08}{1.08} = \frac{W_{Cu}}{\frac{63.5}{2}} \Rightarrow W_{Cu} = \frac{63.5\times 1.08}{2\times 108} = 0.3175\text{ g}$$

CHECK POINT 3.3

1. What will be the weight of silver deposited, if 95.5 A of current is passed into aqueous solution of $AgNO_3$ for 100 s?
(a) 1.08 g (b) 10.8 g
(c) 108 g (d) 1080 g

2. How many Coulombs are required in order to reduce 12.3 g of nitrobenzene to aniline?
(a) 579 C
(b) 5790 C
(c) 57900 C
(d) 579000 C

3. When aqueous sodium chloride solution is electrolysed
(a) at cathode H^+ is reduced into H_2 instead of Na^+
(b) at cathode Na^+ is reduced to Na
(c) Cl^- is oxidised into Cl_2 at cathode
(d) Both (b) and (c)

4. Which of the following statements is incorrect about electrolytic cell ?
(a) Cations are reduced at anode and anions are oxidised at cathode
(b) Cations move towards cathode and anions move towards anode
(c) Electrons leave through anode
(d) Electrons enter through cathode

5. If an aqueous NaCl solution is electrolysed, which phenomenon is not observed ?
(a) The concentration of Na^+ ion decreases
(b) $Cl_2(g)$ is produced at the anode
(c) The solution become more basic
(d) $H_2(g)$ is produced at the cathode

6. Which of the following statements is/are correct?
(a) The amount of chemical change produced is proportional to the quantity of electric charge passing through an electrolysis cell.
(b) A given quantity of electricity produces the same number of equivalents of any substance or electrolysis
(c) None of the above are correct
(d) Both (a) and (b) statement are correct

7. Electrolysis of acidified water gives following at anode
(a) O_2 (b) H_2
(c) $H_2S_2O_3$ (d) None of these

8. Electrochemical equivalent of Cu in the reaction
$$Cu^{2+} + 2e^- \longrightarrow Cu \text{ is}$$
(a) $\frac{63.5}{96500}$ (b) $\frac{63.5}{96500 \times 2}$ (c) $\frac{63.5 \times 2}{96500}$ (d) $\frac{96500}{63.5 \times 2}$

ELECTROMOTIVE FORCE (EMF) OF A CELL

Electromotive force (emf) is the potential difference, when no current is drawn from the system. It is measured in volt (V).

Mathematically it is expressed as,

$$E^\circ_{cell} = \begin{bmatrix}\text{Standard reduction}\\ \text{potential of cathode}\end{bmatrix} - \begin{bmatrix}\text{Standard reduction}\\ \text{potential of anode}\end{bmatrix}$$

or $$E^\circ_{cell} = E^\circ_{cathode} - E^\circ_{anode} \quad (\text{values of } E^\circ_{red})$$

$$E^\circ_{cell} = E^\circ_{red} + E^\circ_{ox} \quad (\text{values as reaction takes place})$$

or $$E^\circ_{cell} = E^\circ_{right} - E^\circ_{left}$$

Note *Standard reduction potential of an electrode is simply called its standard potential, i.e. E°_{red} of cathode = $E^\circ_{cathode}$ and E°_{red} of anode = E°_{anode}*

Let us understand the different expression of EMF of a cell by considering the example of Daniell cell.

In Daniell cell, the reaction takes place as

$$Cu^{2+}(aq) + Zn(s) \longrightarrow Zn^{2+}(aq) + Cu(s)$$

At anode, the oxidation of zinc takes place,

$$Zn(s) \longrightarrow Zn^{2+}(aq) + 2e^-; E^\circ_{Zn/Zn^{2+}} = 0.76 \text{ V}$$

At cathode, the reduction of cupric ion takes place,

$$Cu^{2+}(aq) + 2e^- \longrightarrow Cu(s); E^\circ_{Cu^{2+}/Cu} = 0.34 \text{ V}$$

Before we proceed, it is important to note that,

$$E^\circ_{Zn/Zn^{2+}} (\text{i.e. } E^\circ_{ox} \text{ of Zn}) = -E^\circ_{Zn^{2+}/Zn}$$
$$(\text{i.e. } E^\circ_{red} \text{ of zinc ion}) = -E^\circ_{anode} = -E^\circ_{left}$$

Similarly, $$E^\circ_{Cu^{2+}/Cu} (\text{i.e. } E^\circ_{red} \text{ of } Cu^{2+} \text{ ion}) = -E^\circ_{Cu/Cu^{2+}}$$
$$(\text{i.e. } E^\circ_{ox} \text{ of Cu}) = E^\circ_{cathode} = E^\circ_{right}$$

Verifying E°_{cell} using different relationships

(*i*) $E^\circ_{cell} = E^\circ_{cathode} - E^\circ_{anode}$ (E°_{red} values)
$= 0.34 - (-0.76)$ $[E^\circ_{anode} = -E^\circ_{Zn/Zn^{2+}} = -0.76 \text{ V}]$
$= 1.10\text{V}$

(*ii*) $E^\circ_{cell} = E^\circ_{red} + E^\circ_{ox} = E^\circ_{Cu^{2+}/Cu} + E^\circ_{Zn/Zn^{2+}}$ (as reaction takes place)
$= 0.34 + 0.76$
$= 1.10\text{V}$

(*iii*) $E^\circ_{cell} = E^\circ_{right} - E^\circ_{left}$
$= 0.34 - (-0.76)$ $[E^\circ_{left} = -E^\circ_{Zn/Zn^{2+}} = -0.76 \text{ V}]$
$= 1.10 \text{ V}$

As value of E°_{cell} is same in all the case, hence, any of these relationships can be used to calculate EMF of a cell.

Example 3.14 *The following half-cell reactions are given:*

I. $Cu^{2+}(aq) + e^- \longrightarrow Cu^+(aq)$; $E° = 0.15$ V

II. $2H^+(aq) + 2e^- \longrightarrow H_2(g)$; $E° = 0.00$ V

III. $CuI(s) + e^- \longrightarrow Cu(s) + I^-(aq)$; $E° = -0.17$ V

IV. $Zn^{2+}(aq) + 2e^- \longrightarrow Zn(s)$; $E° = -0.76$ V

Calculate the voltage of the cell

$$\underset{\ominus}{Zn} \mid Zn^{2+}\ (1\text{ M}) \parallel I^-\ (1\text{ M}) \mid CuI \mid \underset{\oplus}{Cu}$$

Sol. Oxidation at LHS

$$Zn \longrightarrow Zn^{2+} + 2e^- \;;\; E°_{ox} = 0.76 \text{ V (from IV)}$$

Reduction at RHS

$$\left.\begin{array}{ll} & CuI + e^- \longrightarrow Cu + I^- \\ \text{or} & 2CuI + 2e^- \longrightarrow 2Cu + 2I^- \end{array}\right] E°_{red} = -0.17 \text{ V (from III)}$$

Overall reaction

$$Zn + 2CuI \longrightarrow Zn^{2+} + 2Cu + 2I^-$$

$$E° = E°_{ox} + E°_{red} = 0.59 \text{ V}$$

Note Values given in I and II are not required. Do not double $E°$ values when you double the number of moles in a half-reaction.

Example 3.15 *Given the following half-cell reactions and corresponding standard (reduction) electrode potentials :*

I. $A + e^- \longrightarrow B^-$ $E° = -0.24$ V

II. $B^- + e^- \longrightarrow B^{2-}$ $E° = +1.25$ V

III. $C^- + 2e^- \longrightarrow C^{3-}$ $E° = -1.25$ V

IV. $D + 2e^- \longrightarrow D^{2-}$ $E° = +0.68$ V

V. $E + 4e^- \longrightarrow E^{4-}$ $E° = +0.38$ V

Which combination of two half-cells would result in a cell with the largest potential?

Sol. If half-cell III is made the anode, and II is made the cathode then

Oxidation $C^{3-} \longrightarrow C^- + 2e^-$; $E°_{ox} = 1.25$ V

Reduction $B^- + e^- \longrightarrow B^{2-} \times 2$; $E°_{red} = 1.25$ V

Overall $C^{3-} + 2B^- \longrightarrow C^- + 2B^{2-}$

$$E°_{cell} = E°_{ox} + E°_{red} = 2.50 \text{ V}$$

Note The two half-cells with the greatest difference in potential yield the cell with the largest potential.

Example 3.16 *For the cell reaction,*

$$2Ce^{4+} + Co \longrightarrow 2Ce^{3+} + Co^{3+}$$

$E°_{cell}$ *is 1.89 V. If* $E_{Co^{2+}/Co}$ *is* -0.28 V, *what is the value of* $E°_{Ce^{4+}/Ce^{3+}}$?

Sol. $E°_{cell} = E°_{ox} + E°_{red} = E°_{Co/Co^{2+}} + E°_{Ce^{4+}/Ce^{3+}}$

$$E°_{cell} = -E°_{Co^{2+}/Co} + E°_{Ce^{4+}/Ce^{3+}}$$

$$1.89 = -(-0.28) + E°_{Ce^{4+}/Ce^{3+}} = 0.28 + E°_{Ce^{4+}/Ce^{3+}}$$

$$\therefore E°_{Ce^{4+}/Ce^{3+}} = 1.89 - 0.28 = 1.61 \text{ V}$$

ELECTROCHEMICAL SERIES

After measuring the potential of various electrode *versus* hydrogen electrode, a series of standard electrode potential has been established which is known as **electrochemical series** or **activity series** or **electromotive series.** In this series, the reduction potentials of various electrodes have been arranged in the increasing order. See Table 3.2

Table 3.2 Electrochemical series

Element	Electrode reaction (Reduction)	Standard reduction potential
Li	$Li^+ + e^- \rightarrow Li$	−3.05
K	$K^+ + e^- \rightarrow K$	−2.925
Ca	$Ca^{2+} + 2e^- \rightarrow Ca$	−2.87
Na	$Na^+ + e^- \rightarrow Na$	−2.714
Mg	$Mg^{+2} + 2e^- \rightarrow Mg$	−2.37
Al	$Al^{+3} + 3e^- \rightarrow Al$	−1.66
Zn	$Zn^{+2} + 2e^- \rightarrow Zn$	−0.7628
Cr	$Cr^{+3} + 3e^- \rightarrow Cr$	−0.74
Fe	$Fe^{+2} + 2e^- \rightarrow Fe$	−0.44
Cd	$Cd^{+2} + 2e^- \rightarrow Ni$	−0.403
Ni	$Ni^{+2} + 2e^- \rightarrow Ni$	−0.25
Sn	$Sn^{+2} + 2e^- \rightarrow Sn$	−0.14
H_2	$2H^+ + 2e \rightarrow H_2$	0.00
Cu	$Cu^{+2} + e^- \rightarrow Cu$	+0.337
I_2	$I_2 + 2e^- \rightarrow 2I^-$	+0.535
Ag	$Ag^+ + e^- \rightarrow Ag$	+0.799
Hg	$Hg^{+2} + 2e^- \rightarrow Hg$	+0.885
Br_2	$Br_2 + 2e^- \rightarrow 2Br^-$	+1.08
Cl_2	$Cl_2 + 3e^- \rightarrow 2Cl^-$	+1.36
Au	$Au^{+3} + 3e^- \rightarrow Au$	+1.50
F_2	$F_2 + 2e^- \rightarrow 2F^-$	+2.87

Increasing tendency to accept electrons (↓)

Increasing tendency to lose electrons (↑)

Activity decreases (↓)

Characteristics

(*i*) The substances which are stronger reducing agents than hydrogen are placed above hydrogen in series.

(*ii*) While the substances which are stronger oxidising agents than hydrogen, are placed below hydrogen in the series.

(*iii*) Metal present on the top of the series are **active metals.**

(*iv*) Negative sign of standard reduction potential (SRP) indicates that an electrode when joined with standard hydrogen electrode (SHE) act as anode and oxidation occurs on this electrode.

(*v*) Positive sign of SRP indicates that an electrode when joined with SHE acts as cathode and reduction occurs on this electrode.

Applications

The important applications of electrochemical series are

(*i*) **To compare relative strength of oxidising and reducing agent** Greater is the reduction potential, more easily is the substance (element or ion) reduced and stronger oxidising agent it is. e.g. F_2 has the highest reduction potential and Li^+ ions the lowest reduction potential this means that F_2 is reduced most easily whereas Li^+ ions are reduced with great difficulty.

Thus, F_2 is the strongest oxidising agent and Li^+ ions are the weakest oxidising agents.

(*ii*) **To compare the relative activities of metals** A metal with greater oxidation potential can displace metals with lower oxidation potential from their salt solution. In other words, a metal with high negative reduction potential can displace metals with less negative reduction potentials from their salt solution.

This concept is very useful in displacement reactions, e.g. oxidation potentials of Al, Mn, Zn, Cr and Fe are in the order

$$Al > Mn > Zn > Cr > Fe$$

Thus, their reactivity is same as of the above order.

(*iii*) **To calculate the emf of the cell** The emf of the cell can be determined by knowing the electrode potential from the series as

$$E^\circ_{cell} = E^\circ_{right} - E^\circ_{left}$$

(If standard reduction potential are given)

(*iv*) **To predict the product of electrolysis** The ion which is stronger oxidising agent is discharged first at cathode.

(*v*) **Help in extraction of metals from their ores.**

(*vi*) **To predict whether a metal can liberate hydrogen from acid or not.** The metals which have negative reduction potentials (or present above hydrogen) can only liberate hydrogen from the acids.

(*vii*) **To predict the feasibility of a redox reaction** In general a redox reaction is feasible only if the species which has higher reduction potential is reduced, while the species which has lower reduction potential is oxidised, otherwise redox reaction is not feasible.

Example 3.17 *Predict what will happen if molecular bromine* (Br_2) *is added to solution containing* NaCl *and* NaI *at* 25°C. *Assume all species are in their standard states.*

$$I_2(s) + 2e^- \longrightarrow 2I^-(1M); \quad E^\circ = 0.53\ V$$
$$Br_2(l) + 2e^- \longrightarrow 2Br^-\ (1\ M); \quad E^\circ = 1.07\ V$$
$$Cl_2(1\ bar) + 2e^- \longrightarrow 2Cl^-\ (1\ M); \quad E^\circ = 1.36\ V$$

Sol. Standard reduction potentials have been given. Based on these values oxidising nature of Br_2 relative to Cl_2 and I_2 can be compared.

E°_{red} of $I_2 < Br_2 < Cl_2$. Hence, oxidising power of $Cl_2 > Br_2 > I_2$.

Thus, $Br_2(l)$ can oxidise I^- to I_2 but cannot oxidise Cl^- to Cl_2,

i.e. $$2I^-(aq) + Br_2(l) \longrightarrow 2Br^-(aq) + I_2(s)$$

is spontaneous, but

$$2Cl^-(aq) + Br_2(l) \longrightarrow 2Br^-(aq) + Cl_2(g)$$

is not spontaneous.

Note *Determine* E°_{cell}, *if* $E^\circ_{cell} > 0$ *then cell reaction is spontaneous.*

Example 3.18 *Arrange the following oxidising agent in order of increasing strength under standard-state conditions* $Br_2(l)$, $Fe^{3+}(aq)$, $Cr_2O_7^{2-}(aq)$ *given.*

$$Cr_2O_7^{2-} + 14H^+ + 6e^- \longrightarrow 2Cr^{3+} + 7H_2O, \quad E^\circ_1 = 1.33\ V$$
$$Br_2(l) + 2e^- \longrightarrow 2Br^-(aq), \quad E^\circ_2 = 1.06\ V$$
$$Fe^{3+}(aq) + e^- \longrightarrow Fe^{2+}(aq), \quad E^\circ_3 = 0.77\ V$$

Sol. E° values given are standard reduction potentials. We find that $E^\circ_1 > E^\circ_2 > E^\circ_3$.

Thus, $Cr_2O_7^{2-}$ has the maximum tendency to be reduced, hence, it is the best oxidising agent, order being $Fe^{3+} < Br_2 < Cr_2O_7^{2-}$.

Note *More positive value of the* E°_{red} *implies its greater tendency to be reduced hence, bettter oxidising agent.*

Example 3.19 *Given the following half-reactions and* E° *values*

$$Mn^{3+}(aq) + e^- \longrightarrow Mn^{2+}(aq); \quad E^\circ = 1.54\ V$$
$$MnO_2(s) + 4H^+(aq) + e^- \longrightarrow Mn^3(aq) + 2H_2O(l); E^\circ = 0.95V$$

Does Mn^{3+} *(aq) undergo disproportionation?*

Sol. Mn^{3+} undergoes disproportionation if $E^\circ_{cell} = +$ ve or E°_{red} is more positive.

$$Mn^{3+}(aq) + e \xrightarrow{Red} Mn^{2+}(aq); \quad E^\circ = 1.54\ V$$
$$Mn^{3+}(aq) + 2H_2O(l) \xrightarrow{Oxid} MnO_2(s) + 4H^+(aq) + e^-;$$
$$E^\circ = -0.95\ V$$
$$E^\circ_{cell} = 1.54 - 0.95 = +0.59\ V$$

Since, E°_{cell} is positive hence, Mn^{3+} undergoes disproportionation.

Example 3.20 *If* $Co^{3+} + e^- \longrightarrow Co^{2+}; E° = 1.82$ V

$$2H_2O \longrightarrow O_2(g) + 4H^+ + 4e^-; E° = -1.23 \text{ V}$$

Explain Co^{3+} *is not stable in aqueous solution.*

Sol. $4 \times [Co^{3+} + e^- \longrightarrow Co^{2+}]; \quad E° = 1.82$ V

$$2H_2O \longrightarrow O_2(g) + 4H^+ + 4e^-; \quad E° = -1.23 \text{ V}$$

$$\text{Net}: 4Co^{3+} + 2H_2O \longrightarrow 4Co^{2+} + O_2(g) + 4H^+; \; E° = 0.59 \text{ V}$$

Cell reaction is spontaneous ($E°$ = +ve), hence Co^{3+} is reduced to Co^{2+} thus, unstable in aqueous solution.

Example 3.21 *Zinc dissolves in dil.* HCl *to liberate* $H_2(g)$, *but from concentrated* H_2SO_4, SO_2 *is evolved. Explain.*

Sol. I. $Zn + 2H^+ \underset{\text{Dil.}}{\longrightarrow} Zn^{2+} + H_2; \; E° = 0.76$ V

II. $Zn + SO_4^{2-} + 4H^+ \underset{\text{Conc.}}{\longrightarrow} 2H_2O + SO_2 + Zn^{2+};$

$$E° = 1.93 \text{ V}$$

SO_4^{2-} is a better oxidising agent than H^+, hence with concentrated H_2SO_4 reaction II predominates in which SO_2 is evolved.

CHECK POINT 3.4

1. The difference between the electrode potentials of two electrodes when no current is drawn through the cell is called.........

(a) cell potential (b) cell emf
(c) potential difference (d) cell voltage

2. Consider the following sets

Zn → Ag → Fe →
$CuSO_4$ (blue) I $CuSO_4$ II $CuSO_4$ III

Blue colour solution changes to colourless (or fades) in

(a) I, II and III (b) Both I and II
(c) Both II and III (d) Both I and III

3. Using the data given below find out the strongest reducing agent.

$$E^\circ_{Cr_2O_7^{2-}/Cr^{3+}} = 1.33\text{V}; \; E^\circ_{Cl_2/Cl^-} = 1.36 \text{ V}$$

$$E^\circ_{MnO_4^-/Mn^{2+}} = 1.51\text{V}; \; E^\circ_{Cr^{3+}/Cr} = -0.74 \text{ V}$$

(a) Cl^- (b) Cr (c) Cr^{3+} (d) Mn^{2+}

4. Use the data given in Q. 3 and find out which of the following is the strongest oxidising agent?

(a) Cl^- (b) Mn^{2+} (c) MnO_4^- (d) Cr^{3+}

5. EMF of the cell reaction,

$$2Ag^+ + Cu \longrightarrow 2Ag + Cu^{2+}$$

is 0.46 V. If $E^\circ_{Cu^{2+}/Cu}$ is + 0.34 V, $E^\circ_{Ag^+/Ag}$ is

(a) 0.80 V (b) 0.12 V
(c) 0.40 V (d) 1.60 V

6. Which of the following oxidise fluoride ions?

(a) O_3 (b) Cl_2
(c) Br_2 (d) None of these

7. Fluorine is best oxidising agent because it has

(a) highest E°_{red}
(b) highest electron affinity
(c) highest E°_{oxi}
(d) lowest electron affinity

8. Compound *XA*, *XB*, *XC* are added into separate test tubes containing *A*, *B*, *C* solutions. *XB* react with *A* and *C*. *XA* does not react with any of these. *XC* reacts with *A*.

Arrange the anion in the decreasing order of their oxidation.

(a) A^-, B^-, C^- (b) B^-, C^-, A^-
(c) C^-, A^-, B^- (d) B^-, A^-, C^-

NERNST EQUATION

The electrode potential of an electrode as well as EMF of cell is determined with the help of an equation known as **Nernst equation.**

This equation was given by **W Nernst** in order to relate reduction potential of an electrochemical reaction to standard electrode potential, temperature and concentrations (activities of electrolyte). Nernst equations for different conditions are discussed below.

Nernst equation for electrode potential

For an hypothetical reaction,

$$M^{n+} + ne^- \longrightarrow M$$

Reaction quotient, $Q = \dfrac{[M]}{[M^{n+}]}$

By thermodynamics Gibbs equation,

$$\Delta G = \Delta G° + 2.303 \, RT \log Q$$

where,

ΔG = Gibbs free energy change under given condition
$= -nFE_{M^{n+}/M}$

$\Delta G°$ = Standard Gibbs free energy change
$= -nFE°_{M^{n+}/M}$

n = number of electrons

R = gas constant (= 8.314 J K^{-1} mol^{-1})

F = Faradays of electrolytes (1F = 96500 coulombs)

$$\therefore \quad E_{M^{n+}/M} = E^\circ_{M^{n+}/M} - \frac{2.303RT}{nF}\log\frac{[M]}{[M^{n+}]}$$

$$= E^\circ_{M^{n+}/M} + \frac{2.303RT}{nF}\log[M^{n+}]$$

Active mass, $M = 1.0$ for an element (or any solid)

At 298 K,

$$\frac{2.303RT}{F} = 0.0591\,\text{V}$$

$$\therefore \quad E_{M^{n+}/M} = E^\circ_{M^{n+}/M} + \frac{0.0591}{n}\log[M^{n+}]$$

This is called Nernst equation for electrode potential.

Thus for,

$$Zn^{2+} + 2e^- \longrightarrow Zn$$

$$E_{\text{red}} = E_{Zn^{2+}/Zn} = E^\circ_{Zn^{2+}/Zn} + \frac{0.0591}{2}\log[Zn^{2+}]$$

and for $Zn \longrightarrow Zn^{2+} + 2e^-$

$$E_{\text{ox}} = E_{Zn/Zn^{2+}} = E^\circ_{Zn/Zn^{2+}} - \frac{0.0591}{2}\log[Zn^{2+}]$$

Thus $E_{\text{ox}} = -E_{\text{red}}$

(*ii*) $Cl_2 + 2e^- \longrightarrow 2Cl^-$

$$E_{Cl_2/Cl^-} = E^\circ_{Cl_2/Cl^-} - \frac{0.0591}{2}\log\frac{[Cl^-]^2}{p_{Cl_2}}$$

where p_{Cl_2} is partial pressure of $Cl_2(g)$, $[Cl^-]$ is molar concentration of Cl^-.

Nernst equation for EMF of a cell

Let us consider a hypothetical redox reaction.

$$A(s) + B^{n+}(aq) \longrightarrow A^{n+}(aq) + B(s)$$

For oxidation half-cell reaction,

$$A(s) \longrightarrow A^{n+}(aq) + ne^-$$

$$Q = \frac{[A^{n+}]}{[A]}$$

$$E_{A/A^{n+}} = E^\circ_{A/A^{n+}} - \frac{2.303RT}{nF}\log[A^{n+}]$$

For reduction half-cell reaction,

$$B^{n+}(aq) + ne^- \longrightarrow B(s)$$

$$Q = \frac{[B]}{[B^{n+}]} = \frac{1}{[B^{n+}]}$$

$$\therefore \quad E_{B^{n+}/B} = E^\circ_{B^{n+}/B} - \frac{2.303RT}{nF}\log\frac{1}{[B^{n+}]}$$

Since $E_{\text{cell}} = E_{\text{ox}} + E_{\text{red}}$ and $E^\circ_{\text{cell}} = E^\circ_{\text{ox}} + E^\circ_{\text{red}}$

or $(= E^\circ_R - E^\circ_{\text{left}})$

$$E_{\text{cell}} = E_{A/A^{n+}} + E_{B^{n+}/B}$$

$$= E^\circ_{A/A^{n+}} - \frac{2.303\,RT}{nF}\log[A^{n+}] + E^\circ_{B^{n+}/B} - \frac{2.303RT}{nF}\log\frac{1}{[B^{n+}]}$$

$$\therefore \quad E^\circ_{\text{cell}} = \left[E^\circ_{A/A^{n+}} + E^\circ_{B^{n+}/B}\right] - \frac{2.303RT}{nF}\log\frac{[A^{n+}]}{[B^{n+}]}$$

$$= E^\circ_{\text{cell}} - \frac{0.0591}{n}\log\frac{[A^{n+}]}{[B^{n+}]}$$

In general,

$$\boxed{E_{\text{cell}} = E^\circ_{\text{cell}} - \frac{2.303RT}{nF}\log\frac{[\text{Products}]}{[\text{Reactants}]}}$$

e.g. for Daniell cell,

$$Zn(s) + Cu^{2+}(aq) \longrightarrow Zn^{2+}(aq) + Cu(s)$$

$$E_{\text{cell}} = E^\circ_{\text{cell}} - \frac{0.0591}{2}\log\frac{[Zn^{2+}][Cu(s)]}{[Cu^{2+}][Zn(s)]}$$

$$= E^\circ_{\text{cell}} - \frac{0.0591}{2}\log\frac{[Zn^{2+}]}{[Cu^{2+}]} \quad [\because [Cu(s)] = 1, [Zn(s)] = 1]$$

Example 3.22 *Graph between E_{cell} and $\log\frac{[Zn^{2+}]}{[Cu^{2+}]}$ is linear with intercept on E°_{cell} axis 1.10 V. Calculate the E_{cell} for*

$$Zn|Zn^{2+}(0.1\text{ M})||Cu^{2+}(0.01\text{ M})|Cu$$

Sol. Nernst equation $E = E^\circ - \frac{0.0591}{n}\log Q$ can be compared with $y = c - mx$ (an equation of straight line) where c is intercept and m is slope of the line. Thus, E° is known as E_{cell} and is calculated using Nernst equation.

Given cell reaction can be

$$Zn + Cu^{2+} \longrightarrow Cu + Zn^{2+}$$

$$\therefore \quad Q = \frac{[Zn^{2+}]}{[Cu^{2+}]} = \frac{0.1}{0.01} = 10$$

$$\therefore \quad E_{\text{cell}} = E^\circ_{\text{cell}} - \frac{0.0591}{2}\log\frac{[Zn^{2+}]}{[Cu^{2+}]} \quad \text{...(i)}$$

We can compare Eq. (i) with $y = c - mx$ with intercept c.

Thus,

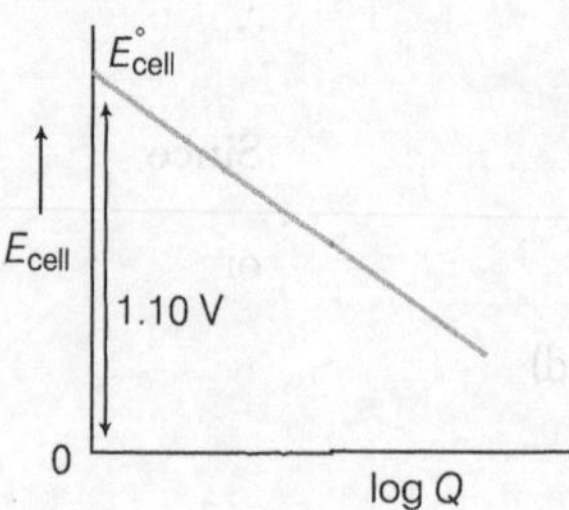

$$c = E^{\circ}_{cell} = 1.10 \text{ V}.$$

$$\therefore \quad E_{cell} = 1.10 - \frac{0.0591}{2} \log 10 = 1.10 - 0.0295 = 1.0705 \text{ V}$$

Various types of half-cells

Various types of half-cells have been used, some as **reference half-cells,** to make complete cell with spontaneous reaction in forward direction. These half-cells have been summarised in Table 3.3.

Table 3.3 Various types of half-cell

Type	Example	Half-cell reaction	Q =	Reversible to	Electrode potential (oxidation), E =
1. Gas-ion half-cell	$Pt(H_2) \mid H^+(aq)$	$\frac{1}{2}H_2(g) \longrightarrow H^+(aq) + e^-$	$[H^+]$	H^+ ion	$E^{\circ} - 0.0591 \log [H^+]$
	$Pt(Cl_2) \mid Cl^-(aq)$	$Cl^-(aq) \longrightarrow \frac{1}{2}Cl_2(g) + e^-$	$\frac{1}{[Cl^-]}$	Cl^- ion	$E^{\circ} + 0.0591 \log [Cl^-]$
2. Metal-metal ion half-cell	$Ag \mid Ag^+(aq)$	$Ag(s) \longrightarrow Ag^+(aq) + e^-$	$[Ag^+]$	Ag^+ ion	$E^{\circ} - 0.0591 \log [Ag^+]$
3. Metal-insoluble salt anion half-cell	$Ag, AgCl \mid Cl^-(aq)$	$Ag(s) + Cl^-(aq) \longrightarrow AgCl(s) + e^-$	$\frac{1}{[Cl^-]}$	Cl^- ion	$E^{\circ} + 0.0591 \log [Cl^-]$
4. Calomel electrode	$Hg, Hg_2Cl_2 \mid Cl^-(aq)$	$2Hg(l) + 2Cl^-(aq) \longrightarrow Hg_2Cl_2(s) + 2e^-$	$\frac{1}{[Cl^-]^2}$	Cl^- ion	$E^{\circ} + 0.0591 \log [Cl^-]$
5. Metal-metal oxide hydroxide half-cell	$Hg, HgO \mid OH^-(aq)$	$Hg(l) + 2OH^-(aq) \longrightarrow HgO(s) + H_2O(l) + 2e^-$	$\frac{1}{[OH^-]^2}$	OH^- ion	$E^{\circ} + 0.0591 \log [OH^-]$
6. Oxidation-reduction half-cell	$Pt \mid Fe^{2+}_{(aq)}, Fe^{3+}_{(aq)}$	$Fe^{2+}(aq) \longrightarrow Fe^{3+}(aq) + e^-$	$\frac{[Fe^{3+}]}{[Fe^{2+}]}$	Fe^{2+}, Fe^{3+} ions	$E^{\circ} - 0.0591 \log \frac{[Fe^{3+}]}{[Fe^{2+}]}$
7. Mercury-mercury sulphate half-cell	$Hg, HgSO_4 \mid SO_4^{2-}(aq)$	$SO_4^{2-}(aq) + Hg(l) \longrightarrow HgSO_4(s) + 2e^-$	$\frac{1}{[SO_4^{2-}]}$	SO_4^{2-}	$E^{\circ} + \frac{0.0591}{2} \log [SO_4^{2-}]$
8. Quinhydrone half-cell	$Pt \mid Quinhydrone \mid H^+(aq)$	Hydroquinone (OH, OH) $\longrightarrow$ Quinone (O, O) $+ 2H^+ + 2e^-$	$[H^+]^2$	H^+	$E^{\circ} - 0.0591 \log [H^+]$

- **In gas-ion half-cell,** gas at 1 bar is passed into aqueous solution of the ion of 1 M concentration having Pt electrode dipped into it.

$$\underset{1\text{ bar}}{Pt(Cl_2)} \mid Cl^-(1\text{ M})$$

For such cases, $E_{Pt(Cl_2)/Cl^-} = E^\circ_{Pt(Cl_2)/Cl^-}$

- Quinhydrone is a mixture of quinone and hydroquinone in 1:1 molar ratio; they are joined by H-bonding.

δ– δ+ δ–
O–H-------O
O–H-------O
δ– δ+ δ–

In aqueous solution, following equilibrium is set-up

$$\underset{(A)}{\text{OH–C}_6\text{H}_4\text{–OH}} \rightleftharpoons \underset{(B)}{\text{O=C}_6\text{H}_4\text{=O}} + 2H^+ + 2e^-$$

$$Q = [H^+]^2 \qquad \text{(since, } [A] = [B])$$

Note *In all cases electrode potential depends on concentration of ion e.g.* $[Ag^+]$, $[Cl^-]$, $[H^+]$

Nernst equation for concentration cells

When same metal electrodes are dipped in same electrolyte having different concentration in two half-cells, such a cell is called as **concentration cell.**

For concentration cell,

$$E^\circ_{cell} = 0 \text{ but } E_{cell} \neq 0$$

The EMF of a concentration cell is

$$E_{cell} = \frac{2.303RT}{nF} \log \frac{C_2}{C_1}$$

where C_1 is concentration of electrolyte in which anode is dipped, C_2 is concentrations of electrolyte in which cathode is dipped.

1. Concentration cell in which electrode is reversible with respect to cation

- $Zn \mid Zn^{2+}(C_1) \parallel Zn^{2+}(C_2) \mid Zn$

Half-cell,

At LHS $\quad Zn \longrightarrow Zn^{2+}(C_1) + 2e^-; \quad E^\circ_L = 0.76\text{ V}$

At RHS $\quad Zn^{2+}(C_2) + 2e^- \longrightarrow Zn; \quad E^\circ_R = -0.76\text{ V}$

Net $\quad Zn^{2+}(C_2) \longrightarrow Zn^{2+}(C_1); \quad E^\circ_{cell} = 0.00\text{ V}$

$$E_{cell} = E^\circ_{cell} - \frac{0.0591}{2} \log \frac{C_1}{C_2}$$

$$E_{cell} = \frac{0.0591}{2} \log \frac{C_2}{C_1}$$

- Reaction will be spontaneous, when $E_{cell} > 0$, i.e. $C_2 > C_1$

Cell reaction is spontaneous in forward direction, if

$$[Zn^{2+}]_{RHS}, (C_2) > [Zn^{2+}]_{LHS}, (C_1)$$

- $\underset{1\text{ atm}}{Pt(H_2)} \mid HCl(C_1) \parallel HCl(C_2) \mid \underset{1\text{ atm}}{Pt(H_2)}$

$$E_{cell} = E^\circ_{cell} - \frac{0.0591}{1} \log \frac{C_1}{C_2} = 0.0591 \log \frac{C_2}{C_1}$$

2. Concentration cell in which electrode is reversible with respect to anion

$$\underset{1\text{ atm}}{Pt(Cl_2)} \mid Cl^-(C_1) \parallel Cl^-(C_2) \mid \underset{1\text{ atm}}{Pt(Cl_2)}$$

$$E_{cell} = \frac{0.0591}{1} \log \frac{C_1}{C_2}$$

For such cases, cell reaction is spontaneous in forward reaction if $C_1 > C_2$.

3. Concentration cell having electrodes at different concentration dipped into same electrolyte

$$\underset{\text{at } p_1}{Pt(H_2)} \mid \underset{1\text{ M}}{HCl} \mid \underset{\text{at } p_2}{Pt(H_2)}$$

Two half-cells are joined by two salt-bridges since they have common electrolyte.

LHS half-cell $\quad H_2(p_1) \longrightarrow \underset{1\text{ M}}{2H^+} + 2e^-$

RHS half-cell $\quad \underset{1\text{ M}}{2H^+} + 2e^- \longrightarrow H_2(p_2)$

Net reaction $\quad H_2(p_1) \longrightarrow H_2(p_2) \quad E^\circ_{cell} = 0.00\text{ V}$

$$E_{cell} = E^\circ_{cell} - \frac{0.0591}{2} \log \frac{p_2}{p_1} = \frac{0.0591}{2} \log \frac{p_1}{p_2}$$

Reaction will be spontaneous, when $E_{cell} > 0$, i.e. $C_2 > C_1$

Cell reaction is spontaneous in forward direction, if

$$p_1(\text{LHS}) > p_2(\text{RHS}) \text{ or } (C_1 = C_2)$$

Example 3.23 *For the cells given below E°_{cell} have been given.*

(a) Pt; H_2(g,1 bar) | H^+(aq, 1 M) || Cu^{2+}(aq, 1 M) | Cu, $E^\circ_{cell} = +0.34$ V

(b) Zn | Zn^{2+}(aq,1 M)|| H^+(aq,1 M)|| H_2(g,1 bar), Pt $E^\circ_{cell} = 0.76$ V

What are the values of $E^\circ_{Cu^{2+}/Cu}$ and $E^\circ_{Zn^{2+}/Zn}$?

Sol. (a) For the given cell,

$$E^\circ_{cell} = E^\circ_{SHE} + E^\circ_{Zn^{2+}/Cu}$$

$$0.34 = 0.00 + E^\circ_{Cu^{2+}/Cu}$$

$$\therefore \quad E^\circ_{Cu^{2+}/Cu} = 0.34 \text{ V}$$

(b) For this cell, $E^\circ_{cell} = E^\circ_{SHE} + E^\circ_{Zn/Zn^{2+}}$; $0.76 = 0.00 + E^\circ_{Zn/Zn^{2+}}$

$$\therefore \quad E^\circ_{Zn/Zn^{2+}} = 0.76 \text{ V}$$

$$\therefore \quad E^\circ_{Zn^{2+}/Zn} = -0.76 \text{ V}$$

Example 3.24 *A certain electrode has a reduction potential of 1.40 V when measured against a standard calomel electrode. Calculate its potential versus a SHE.* $E^\circ_{calomel}$ *(oxidation)* $= -0.244$ V

Sol. $M^{n+} + ne^- \longrightarrow M,\ E^\circ = x$

Reaction at calomel electrode

$$2Hg + 2Cl^- \longrightarrow Hg_2Cl_2 + 2e^- \ ; \ E^\circ = -0.244 \text{ V}$$

$$E^\circ_{M^{n+}/M} + E^\circ_{cal} = 0.140 \text{ V}$$

$$E^\circ_{M^{n+}/M} - 0.244 \text{ V} = 0.140 \text{ V}$$

$$E^\circ_{M^{n+}/M} = 0.140 + 0.244 = 0.384 \text{ V}$$

Since, $E^\circ_{SHE} = 0.00$ V

$$\therefore \quad E^\circ_{cell}(\text{wrt SHE}) = E^\circ_{SHE} + E^\circ_{M^{n+}/M} = 0.00 + 0.384 = 0.384 \text{ V}$$

Note Calomel electrode is Hg, $Hg_2Cl_2(s)$ / Cl^-.

Equilibrium constant from Nernst equation

Nernst equation may also be used to calculate the equilibrium constant for a cell reaction. Let us consider an hypothetical equation,

$$aA + bB \rightarrow cC + dD$$

The Nernst equation can be written as,

$$E_{cell} = E^\circ_{cell} - \frac{RT}{nF}\ln\frac{[C]^c[D]^d}{[A]^a[B]^b}$$

$$= E^\circ_{cell} + \frac{RT}{nF}\ln\frac{[A]^a[B]^b}{[C]^c[D]^d} \quad ...(i)$$

At equilibrium, $E_{cell} = 0$

and $$\frac{[C]^c[D]^d}{[A]^a[B]^b} = K_C = K_{eq}$$

(equilibrium constant)

Thus, equation (i) becomes

$$0 = E^\circ_{cell} + \frac{RT}{nF}\ln K_{eq}$$

or $$E^\circ_{cell} = \frac{2.303RT}{nF}\log K_{eq}$$

On putting the value of given constants (F, R and T)

or $$E^\circ_{cell} = \frac{0.0591}{n}\log K_{eq}$$

K_{eq} is actually thermodynamic equilibrium constant.

Note Nernst equation helps in the determination of pH

Formula used, $\text{pH} = \dfrac{E_{cell} - E^\circ_{cell}}{0.0591}$

$$\text{pH} = \frac{E_{cell} - E^\circ_{reference}}{0.0591}$$

CHECK POINT 3.5

1. For the cell Mg | Mg^{2+} (0.01 M) || pH = 1,H^+ | Pt(H_2), $E^\circ_{cell} = +2.37$ V, hence

(a) $E_{cell} = 2.37 + \frac{0.0591}{2}$ V
(b) $E_{cell} = 2.37 + 0.0591$ V
(c) $E_{cell} = 2.37$ V
(d) None of the above

2. For the cell,

(a) $E^\circ_{cell} = E^\circ_{cathode} + E^\circ_{anode}$
(b) $E^\circ_{cell} = E^\circ_{cathode} - E^\circ_{anode}$
(c) $E^\circ_{cell} = E^\circ_{anode} - E^\circ_{cathode}$
(d) None of the above

3. Select the most appropriate Nernst equation for Daniell cell containing Cu^{2+} and Zn^{2+} ions.

(a) $E_{(Cu^{2+}/Cu)} = E^\circ_{(Cu^{2+}/Cu)} - \frac{RT}{F}\ln\frac{1}{[Cu^{2+}(aq)]}$
(b) $E_{(Zn^{2+}/Zn)} = E^\circ_{(Zn^{2+}/Zn)} - \frac{RT}{2F}\ln\frac{1}{[Zn^{2+}(aq)]}$
(c) $E_{cell} = E^\circ_{cell} - \frac{RT}{2F}\ln\frac{[Zn^{2+}]}{[Cu^{2+}]}$
(d) All of the above

4. Given, $E^\circ_{Ni^{2+}/Ni} = 0.25$ V, $E^\circ_{Cu^{2+}/Cu} = 0.34$ V

$E^\circ_{Ag^+/Ag} = 0.80$V and $E^\circ_{Zn^{2+}/Zn} = -0.76$V

Which of the following reactions under standard condition will not take place in the specified direction ?

(a) $Ni^{2+}(aq) + Cu(s) \longrightarrow Ni(s) + Cu^{2+}(aq)$

(b) $Cu(s) + 2Ag^+(aq) \longrightarrow Cu^{2+}(aq) + 2Ag(s)$

(c) $Zn(s) + 2H^+(aq) \longrightarrow Zn^{2+}(aq) + 3H_2(g)$

(d) All of the above

5. In the given reaction,

$$2Cu^+(aq) \rightleftharpoons Cu^{2+}(aq) + Cu(s)$$

$E^\circ_{Cu^+/Cu} = 0.6$ V and $E^\circ_{Cu^{2+}/Cu} = 0.41$ V

The equilibrium constant for this reaction will be

(a) 276×10^2 (b) 276×10^4

(c) 276×10^6 (d) 276×10^8

6. $E^\circ_{cell} = 1.1$ V for Daniell cell. Which of the following expressions are correct description of state of equilibrium in this cell ?

(a) $1.1 = K_C$ (b) $\frac{2.303RT}{2F}\log K_C = 1.1$

(c) $\log K_C = \frac{3.4}{0.059}$ (d) $\log K_C = 1.1$

7. If $E^\circ(Zn^{2+}/Zn) = -0.763$ V and $E^\circ(Fe^{2+}/Fe) = -0.44$ V, then the emf of the cell $Zn|Zn^{2+}(a = 0.001)||Fe^{2+}(a = 0.005)|Fe$ is

(a) equal to 0.323 V

(b) less than 0.323 V

(c) greater than 0.323 V

(d) equal to 1.103 V

8. For the cell, $Zn(s)|Zn^{2+}(aq)||M^{x+}(aq)|M(s)$, different half cells and their standard electrode potentials are given below.

$M^{x+}(aq)/M(s)$	$Au^{3+}(aq)/Au(s)$	$Ag^+(aq)/Ag(s)$	$Fe^{3+}(aq)/Fe^{2+}(aq)$	$Fe^{2+}(aq)/Fe(s)$
$E^\circ_{M^{x+}/M}$/V	1.40	0.80	0.77	− 0.44

If $E^\circ_{Zn^{2+}/Zn} = -0.76$ V, which cathode will give a maximum value of E°_{cell} per electron transferred?

(a) $\frac{Ag^+}{Ag}$ (b) $\frac{Fe^{2+}}{Fe}$

(c) $\frac{Au^{3+}}{Au}$ (d) $\frac{Fe^{3+}}{Fe^{2+}}$

ELECTROCHEMICAL CELL AND GIBBS' ENERGY OF THE REACTION

The relationship between potential difference (volt), energy (Joule) and electrical charge (Coulomb) is given by

$$1J = 1C \times 1V$$

where, 1 Coulomb is the amount of charge transferred when a current of 1 ampere (A) flows for 1 second (s). When 1 Coulomb of charge moves between two electrodes that differ in electrical potential by 1V, 1 Joule of energy is released by the cell and can be used to do electrical work. Thus, quantitative measures of the driving force of a chemical reactions are :

(a) The cell potential E (an electrochemical quantity)

(b) The free-energy change ΔG (a thermochemical quantity).

ΔG represents the maximum amount of useful work at constant-pressure process that a system can perform.

$$\Delta G = -W_{max} \text{ (useful)}$$

For a spontaneous change $\Delta G < 0$ and work done by the system is negative. When a reaction occurs in a voltaic-cell it does work (electrical work).

$$W_{electrical} = -nFE_{cell}$$

$$\Delta G = -nFE_{cell}$$

$$\Delta G = -2.303RT\log K_{eq}$$

where, n is the number of moles of electrons transferred in the reaction and F is the Faraday, the electric charge on 1 mole of electrons (96500/mol e^-). If E is in volt then ΔG is in Joule.

- If the reactants and products are in their standard states

$$\Delta G^\circ = -nFE^\circ_{cell}$$

- If E_{cell} is **positive,** reaction occurs spontaneously in the forward direction.
- If E_{cell} is **negative,** the reaction occurs spontaneously in the reverse direction.
- If a cell reaction is reversed, E_{cell} **changes sign.**

Relation between standard potentials of half-cells containing a metal in different oxidation state

If two half cell reactions having potentials E_1° and E_2° are combined to give a third-half cell reaction having a potential E_3° then,

$$\Delta G_3^\circ = \Delta G_1^\circ + \Delta G_2^\circ$$

$$-n_3FE_3^\circ = -n_1FE_1^\circ - n_2FE_2^\circ$$

$$n_3E_3^\circ = n_1E_1^\circ + n_2E_2^\circ$$

$$\Rightarrow \quad E_3^\circ = \frac{n_1E_1^\circ + n_2E_2^\circ}{n_3}$$

Note *While using this equation, take values of E° according to reaction. If equal number of electrons are involved, then $E_3^\circ = E_1^\circ + E_2^\circ$*

Determination of thermodynamic data

(*i*) Using Gibbs-Helmholtz equation, temperature coefficient of emf of the cell can be determined.

$$\left[\frac{dE_{cell}}{dT}\right]_p = \frac{\Delta H}{nFT} + \frac{E_{cell}}{T}$$

(*ii*) Enthalpy change can be determined as,

$$\Delta S = nF\left(\frac{dE_{cell}}{dT}\right)_p.$$

Example 3.25 *If a reaction has an equilibrium constant $K < 1$, then $E°$ is positive or negative? What is the value of K when $E° = 0$ V?*

Sol. $E° = \frac{0.0591}{n}\log K$

If $K < 1$ (say 0.01) then

$$E° = \frac{0.0591}{n}\log 0.01 = \frac{-2\times 0.0591}{n}$$

$E°$ is thus negative.

If $E° = 0$ V then $K = 1$

Example 3.26 *Is* 1.0 M H^+ *solution under* H_2SO_4 *at* 1.0 bar *capable of oxidising silver metal in the presence of* 1.0 M Ag^+ *ion?*

$$E°_{Ag^+/Ag} = 0.80\text{ V}$$

$$E°_{H^+/H_2(Pt)} = 0.00\text{ V}$$

Sol. Ag is to be oxidised to Ag^+ by H^+ which can be reduced to H_2.

$$2Ag + 2H^+ \longrightarrow 2Ag^+ + H_2$$

$$E°_{cell} = E°_{Ag/Ag^+} + E°_{H^+/H_2(Pt)}$$

$$= -E°_{Ag^+/Ag} + E°_{H^+/H_2(Pt)}$$

$$= -0.80 + 0 = -0.80\text{ V}$$

Thus, $\Delta G° = +$ ve hence, cell reaction is non-spontaneous and H^+ solution does not oxidise silver metal under these conditions.

Example 3.27 *Given,*

$$H_2O_2 \longrightarrow O_2 + 2H^+ + 2e^-;\ E° = -0.69\text{ V}$$

$$H_2O_2 + 2H^+ + 2e^- \longrightarrow 2H_2O;\ E° = 1.77\text{ V}$$

$$2I^- \longrightarrow I_2 + 2e^-;\ E° = -0.535\text{ V}$$

Predict whether H_2O_2 *behaves as oxidant or reductant for* I_2/I^-

Sol. **H_2O_2 is a reducing agent (if it is oxidised):**

$$H_2O_2 \longrightarrow O_2 + 2H^+ + 2e^-;\ E° = -0.69\text{ V}$$

$$I_2 + 2e^- \longrightarrow 2I^-;\qquad E° = +0.535\text{ V}$$

$$\text{Overall}\quad H_2O_2 + I_2 \longrightarrow 2I^- + O_2 + 2H^+;\ E° = -0.155\text{ V}$$

Since, $E° = -$ ve hence, this cell reaction is not spontaneous, hence H_2O_2 is not a reducing agent.

H_2O_2 is an oxidising agent (if it is reduced)

$$H_2O_2 + 2H^+ + 2e^- \longrightarrow 2H_2O;\quad E° = 1.77\text{ V}$$

$$2I^- \longrightarrow I_2 + 2e^-;\quad E° = -0.535\text{ V}$$

$$\text{Overall}\ H_2O_2 + 2H^+ + 2I^- \longrightarrow 2H_2O + I_2;\ E° = 1.235\text{ V}$$

Since, $E°$ is positive, hence, cell reaction is spontaneous and H_2O_2 acts as an oxidising agent for I^-.

Example 3.28 *For the half-cell reactions*

$$Cu^{2+} + e^- \longrightarrow Cu^+;\quad E° = 0.15\text{ V}$$

$$Cu^{2+} + 2e^- \longrightarrow Cu;\quad E° = 0.34\text{ V}$$

calculate $E°$ of the half-cell reaction

$$Cu^+ + e^- \longrightarrow Cu$$

Sol. $Cu^{2+} + e^- \longrightarrow Cu^+,\ E° = 0.15$ V

I. $Cu^+ \longrightarrow Cu^{2+} + e^-;\ E_1° = -0.15\text{ V},\ n_1 = 1$

II. $Cu^{2+} + 2e^- \longrightarrow Cu;\ E_2° = 0.34\text{ V},\ n_2 = 2$

(I + II) gives

$$Cu^+ + e^- \longrightarrow Cu,\ E_3° = ?,\ n_3 = 1$$

$$\therefore\quad \Delta G_3° = \Delta G_1° + \Delta G_2°$$

$$-n_3E_3° = -n_1E_1° - n_2E_2° \Rightarrow E_3° = \frac{n_1E_1° + n_2E_2°}{n_3}$$

$$= \frac{1\times(-0.15) + 2\times(0.34)}{1}$$

$$= -0.15 + 0.68 = +0.53\text{ V}$$

Example 3.29 *From the values of example 3.28, predict whether* Cu^+ *undergoes disproportionation or not.*

Sol.

$$2Cu^+ \longrightarrow Cu^{2+} + Cu$$

(Oxidation: $Cu^+ \to Cu^{2+}$; Reduction: $Cu^+ \to Cu$)

$$Cu^+ \longrightarrow Cu^{2+} + e^-;\quad E° = -0.15\text{ V (given)}$$

$$Cu^+ + e^- \longrightarrow Cu;\quad E° = 0.53\text{ V}$$

(calculated in example 3.28)

$$\text{Overall}\quad 2Cu^+ \longrightarrow Cu + Cu^{2+};\quad E° = 0.53 + (-0.15) = 0.38\text{ V}$$

Since $E°$ is positive hence, Cu^+ disproportionates to Cu^{2+} (oxidation) and Cu (reduction).

Example 3.30 *The Edison storage cell is represented as*

$$Fe(s)|FeO(s)|KOH(aq)|Ni_2O_3(s)|Ni(s)$$

The half-cell reactions are,

$$Ni_2O_3(s) + H_2O(l) + 2e^- \longrightarrow 2NiO(s) + 2OH^-(aq);\ E° = 0.40\text{ V}$$

$$FeO(s) + H_2O(l) + 2e^- \longrightarrow Fe(s) + 2OH^-(aq);\ E° = -0.87\text{ V}$$

(*i*) *What is the cell reaction?*

(*ii*) *What is the cell emf? How does it depend on the concentration of KOH ?*

(*iii*) *What is the maximum amount of electrical energy that can be obtained from one mole of* Ni_2O_3*?*

Sol. (i) Given cell is $Fe(s)|FeO(s)|KOH(aq)|Ni_2O_3(s)|Ni(s)$

Half-cell reaction

LHS $\quad Fe(s) + 2OH^-(aq) \longrightarrow FeO(s) + H_2O(l) + 2e^-$;

$$E^\circ_{ox} = 0.87\ V$$

RHS $\quad Ni_2O_3(s) + H_2O(l) + 2e^- \longrightarrow 2NiO(s) + 2OH^-(aq)$;

$$E^\circ_{red} = 0.40\ V$$

Net $\quad Fe(s) + Ni_2O_3(s) \longrightarrow FeO(s) + 2NiO(s)$;

$$E^\circ_{cell} = 1.27\ V$$

(ii) $E^\circ_{cell} = E^\circ_{ox} + E^\circ_{red} = 1.27\ V$

Since net reaction does not contain OH^-, hence cell emf does not depend on the concentration of KOH.

Note *$E°$ values of E°_{ox} and E°_{red} are as per reaction at half-cell.*

(iii) Maximum amount of electrical energy

$$\Delta G^\circ = -nFE^\circ_{cell} = -2 \times 96500 \times 1.27$$
$$= -2.45 \times 10^5\ J\ mol^{-1}$$
$$= -2.45 \times 10^2\ kJ\ mol^{-1}$$

CHECK POINT 3.6

1. Following is the concentration cell reversible to anion

$$\underset{1\ bar}{Pt(Cl_2)}|Cl^-(C_1)||Cl^-(C_2)|\underset{1\ bar}{Pt(Cl_2)}$$

cell reaction is spontaneous, if

(a) $C_1 > C_2$ (b) $C_1 < C_2$ (c) $C_1 = C_2$ (d) $C_1 = 0$

2. For the cell given below $Zn \mid Zn^{2+} \parallel Cu^{2+} \mid Cu$, $(E_{cell} - E^\circ_{cell})$ is – 0.12 V. It will be when

(a) $[Zn^{2+}]/[Cu^{2+}] = 10^2$ (b) $[Zn^{2+}]/[Cu^{2+}] = 10^{-2}$

(c) $[Zn^{2+}]/[Cu^{2+}] = 10^4$ (d) $[Zn^{2+}]/[Cu^{2+}] = 10^{-4}$

3. The standard Gibbs free energy of

$Zn(s) + Cu^2 + (aq) \longrightarrow Zn^{2+}(aq) + Cu(s)$ is

(a) $-91\ FE_{cell}$ (b) $-2\ FE_{cell}$ (c) $-3\ FE_{cell}$ (d) $-4\ FE_{cell}$

4. Consider the reduction half-reactions as

Cathode $\quad A^{x+} + xe^- \longrightarrow A$

Anode $\quad B^{y+} + ye^- \longrightarrow B$

then

(a) net reaction is $yA^{x+} + xB \longrightarrow yA + xB^{y+}$

(b) number of electrons transferred per reaction is xy

(c) free energy change is $\Delta G^\circ = y\Delta G^\circ_{cathode} - x\Delta G^\circ_{anode}$

(d) All of the above are correct statements

5. Given below are the half-cell reactions,

$Mn^{2+} + 2e^- \longrightarrow Mn$; $E^\circ = -1.18\ V$

$2(Mn^{3+} + e^- \longrightarrow Mn^{2+})$; $E^\circ = +1.51\ V$

The E° for $3Mn^{2+} \longrightarrow Mn + 2Mn^{3+}$ will be

(a) –2.69 V, the reaction will not occur

(b) –2.69 V, the reaction will occur

(c) –0.33 V, the reaction will not occur

(d) –0.33 V, the reaction will occur

6. At equilibrium

(a) $E^\circ_{cell} = 0, \Delta G^\circ = 0$

(b) $E_{cell} = 0, \Delta G = 0$

(c) Both (a) and (b) are correct

(d) None of the above is correct

7. For electrosynthesis of a substance

(a) $\Delta G = +$ ve and $E_{cell} = -$ve (b) $\Delta G = -$ ve and $E_{cell} = +$ve

(c) $\Delta G = -$ ve and $E_{cell} = -$ve (d) $\Delta G = +$ ve and $E_{cell} = +$ve

8. Which of the following statements about a reaction occurring in galvanic cell is true?

(a) If $E_{cell} > 0, \Delta G < 0$ (b) If $E^\circ_{cell} < 0, \Delta G < 0$

(c) If $E^\circ_{cell} < 0, K_{eq} > 1$ (d) If $E^\circ_{cell} < 0, K_{eq} > 1$

9. $\Delta G = \Delta H - T\Delta S$ and $\Delta G = \Delta H + T\left[\dfrac{d(\Delta G)}{dT}\right]_p$ then $\left(\dfrac{dE_{cell}}{dT}\right)$ is

(a) $\Delta S/nF$ (b) $nE/\Delta S$

(c) $-nFE_{cell}$ (d) $+nFE_{cell}$

CORROSION

Corrosion is the oxidative deterioration of a metal, such as conversion of iron to rust $[Fe_2O_3 \cdot xH_2O]$, tarnishing of silver (due to formation of Ag_2O), development of a green coating on copper and bronze.

Corrosion is basically an electrochemical phenomenon. A metal is oxidised by loss of electrons to oxygen and forms metal oxide.

Mechanism of corrosion

Let us consider the example of rusting of iron. The spot of an iron object, where oxidation occurs is called the anode.

At anode, $Fe(s) \longrightarrow Fe^{2+}(aq) + 2e^-$; $E^\circ = 0.44\ V$

Electrons from the anode move through the metal and goes to another spot on the metal and reduce O_2 in presence of H^+ ions. This spot where reduction takes place is called cathode.

At cathode,

$$O_2(s) + 4H^+(aq) + 4e^- \longrightarrow 2H_2O(l), E^\circ = 1.23\ V$$

Note *The H^+ ion is produced when CO_2 gets dissolved in water forming H_2CO_3, which releases H^+ ions,*

$$CO_2 + H_2O \longrightarrow H_2CO_3 \longrightarrow H^+ + HCO_3^-.$$

Net reaction,

$$2Fe(s) + O_2(g) + 4H^+(aq) \longrightarrow 2Fe^{2+}(aq) + 2H_2O(l);$$
$$E^\circ_{cell} = 1.67\ V$$

Fe^{2+} ions move through water on the surface of the iron object. If water present is saline, it will help more in carrying the current in the electrochemical cell formed. Fe^{2+} ions are further oxidised by atmospheric O_2 to Fe^{3+} (as Fe_2O_3) and comes out as rust in the form of a hydrated iron (III) oxide represented as $Fe_2O_3 \cdot xH_2O$

$$4Fe^{2+}(aq) + O_2(g) + 4H_2O(l) \longrightarrow 2Fe_2O_3(s) + 8H^+$$

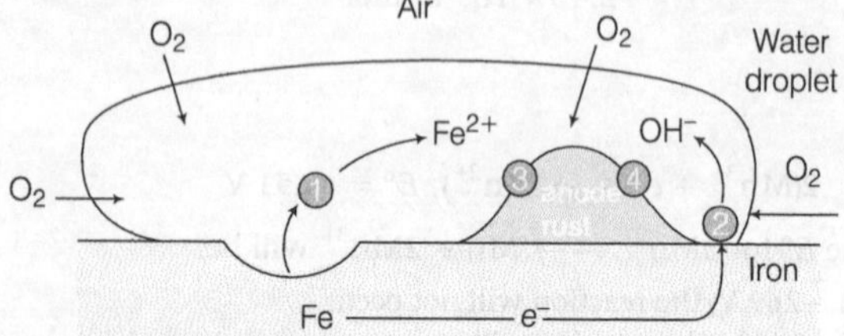

Fig. 3.9 (a) A water droplet on the surface of iron forms an electrochemical cell which leads to rusting. Atmospheric oxygen

① $Fe(s) \longrightarrow Fe^{2+}(aq) + 2e^-$ (Oxidation)

② $O_2(g) + 2H_2O(l) + 4e^- \longrightarrow 4OH^-(aq)$ (Reduction)

③ $Fe^{2+}(aq) \xrightarrow{O_2} Fe^{3+}(aq)$

④ $Fe^{3+}(aq) \xrightarrow{OH^-(aq) \cdot H_2O} Fe_2O_3 \cdot xH_2O$ (Rust)

H^+ ions produced in this reaction help further in rusting.

$$Fe + 2H^+ \longrightarrow Fe^{2+} + H_2$$

Impurities present in iron also enhance rusting by setting a number of miniature cells. Rusting of very pure iron takes place very slowly.

Prevention of corrosion

Corrosion of iron can be prevented or at least minimised by shielding the metal surface from oxygen and moisture. It can be done by using following methods.

(*i*) **Barrier protection** A coat of paint, thin flim of oil, grease etc., are very effective for a while, but rust begins to form as soon as the coating is broken or scratched or chipped.

(*ii*) **Electrical protection or cathodic protection** Metals such as Cr, Sn, or Zn afford a more durable surface coating for Fe. The steel used in making automobiles is coated by dipping into a bath of molten zinc, a process known as galvanisation. Since

$$E^\circ_{Fe^{2+}/Fe} = -0.44 \text{ V},$$

$$E^\circ_{Zn^{2+}/Zn} = -0.76 \text{ V}$$

Hence, zinc is oxidised more readily than iron, and it, in turn, reduces Fe^{2+} (formed during oxidation rusting) to Fe.

$$Fe^{2+} + Zn \longrightarrow Zn^{2+} + Fe, \quad E^\circ_{cell} = 1.20 \text{ V}$$

This reaction is spontaneous in forward direction hence, Zn can reduce Fe^{2+} to Fe. As long as the zinc and iron are in contact, the zinc protects the iron from oxidation even if the zinc layer becomes scratched.

The technique of protecting a metal from corrosion by connecting it to a second metal that is more easily oxidised is called **cathodic protection.**

(*iii*) **Sacrificial protection or sacrified anode** Underground steel pipeline is protected by connecting it through an insulated wire to magnesium (Fig. 3.9(b)) which acts as a **sacrified anode** and corrodes instead of iron. In this, easily oxidised Mg acts as anode, pipeline behaves as cathode, the moist soil is the electrolyte.

At anode, $Mg \longrightarrow Mg^{2+}(aq) + 2e^-$; $E^\circ = 2.37$ V

At cathode, $O_2(g) + 4H^+(aq) + 4e^- \longrightarrow 2H_2O(l)$;

$E^\circ = 1.23$V

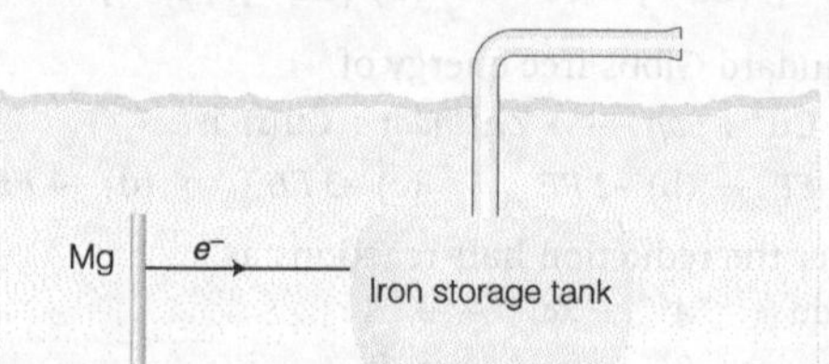

Fig. 3.9 (b) Sacrificial protection of an iron storage tank (cathode) by magnesium, a more electropositive metal

COMMERCIAL CELLS OR BATTERIES

Any cell or batteries that we employ as source of electrical energy is basically an electrochemical cell. If a number of cells are arranged in a series then this arrangement is collectively called as **battery.** In this, the anode of each cell connected to the cathode of the adjacent cell and the voltage provided by the battery is the sum of the individual cell voltages. The electrochemical cells can be broadly classified into following two categories.

1. Primary cell

In these type of cell, redox reactions occur only once and the cell becomes dead after sometime. They cannot be recharged or used again, e.g. dry cell like Leclanche cell, mercury cell.

(i) Leclanche cell

It is called dry cell as it has constituents in dry powdered or paste form and not in solution form. The components of this cell are

At anode zinc metal container.

At cathode inert graphite rod surrounded by a paste of MnO_2 and carbon black.

Electrolyte moist paste of NH_4Cl and $ZnCl_2$ in starch.

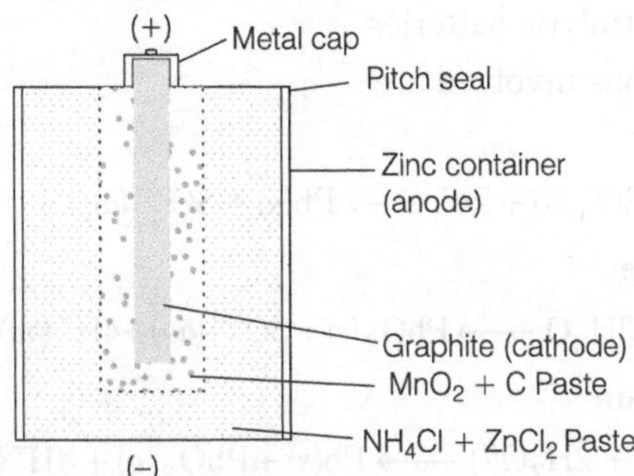

Fig. 3.10 The dry cell

Reactions involved

At anode $Zn(s) \longrightarrow Zn^{2+}(aq) + 2e^-$

At cathode $2MnO_2(s) + 2NH_4^+(aq) + 2e^- \longrightarrow Mn_2O_3(s) + 2NH_3(g) + H_2O(l)$

Note *Oxidation state of Mn changes from + 4 in* MnO_2 *to +3 in* Mn_2O_3.

Net reaction

$$Zn(s) + 2NH_4^+ + 2MnO_2 \longrightarrow [Zn(NH_3)_2Cl_2] + Mn_2O_3 + H_2O$$

Alkaline dry cell is a modified version of the Leclanche cell in which the acidic electrolyte NH_4Cl is replaced by an alkaline (basic) electrolyte (NaOH or KOH). It involves reactions.

At anode,

$$Zn(s) + 2OH^-(aq) \longrightarrow ZnO(s) + H_2O(l) + 2e^-$$

At cathode,

$$2MnO_2(s) + H_2O(l) + 2e^- \longrightarrow Mn_2O_3(s) + 2OH^-(aq)$$

Net reaction

$$Zn(s) + 2MnO_2(s) \longrightarrow ZnO(s) + Mn_2O_3(s)$$

- Thus, cell potential is independent of concentration of OH^- ions.
- Corrosion in alkaline medium takes place slowly as compared to acidic medium hence, alkaline batteries have longer life.
- Alkaline cell also produces higher power, more stable current and voltage because of more efficient ion transport in the alkaline electrolyte.

(ii) Mercury cell

It has

Anode	Zinc and mercury amalgam
Cathode	Paste of carbon and HgO
Electrolyte	Moist HgO mixed with KOH

Reactions involved

Anode,

$$Zn(s) + 2OH^-(aq) \longrightarrow ZnO(s) + H_2O(l) + 2e^-$$

Cathode,

$$HgO(s) + H_2O(l) + 2e^- \longrightarrow Hg(l) + 2OH^-(aq)$$

Net reaction

$$Zn(s) + HgO(s) \longrightarrow ZnO(s) + Hg(l)$$

- Net reaction does not involve any ion, i.e. OH^- hence, cell potential is independent of OH^- concentration. Therefore, the cell gives a constant potential of 1.35V.

2. Secondary cells

- It is a chemical device in which energy may be stored in the form of chemical energy and it may be drawn out when required.
- It can be recharged by passing an electric current through them.
- It can be used over and over again.
- It acts as an accumulator or storage cell.
- It shows reversible reactions.

(i) Nickel-cadmium batteries

Nickel-cadmium or 'ni-cad' batteries are rechargeable storage batteries. They have longer life but are more expensive.

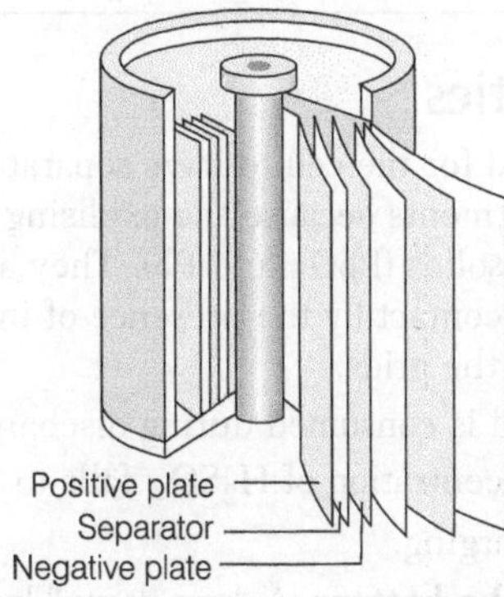

Fig. 3.11 Nickel-cadmium batteries

Anode cadmium metal,

Cathode nickel (III) compound like NiO(OH) supported on nikel.

Electrolyte KOH

Reactions involved during discharge

At anode $Cd(s) + 2OH^-(aq) \rightleftharpoons Cd(OH)_2(s) + 2e^-$

At cathode

$$NiO(OH)(s) + H_2O(l) + e^- \longrightarrow Ni(OH)_2(s) + OH^-(aq)$$

Net reaction

$$Cd(s) + 2NiO(OH)(s) + 2H_2O(l) \underset{\text{Recharging}}{\overset{\text{Discharging}}{\rightleftharpoons}} Cd(OH)_2(s) + 2Ni(OH)_2(s)$$

When cell is recharging, the reactions are reversed.

(ii) Lead-storage batteries

A typical 12-V battery consists of six cells connected in series, each cell providing a potential of about 2 V.

Anode A series of lead grids packed with spongy lead

Cathode A series of grids packed with lead dioxide.

Electrolyte An aqueous solution of H_2SO_4, 38% by weight.

- **Discharge of battery** When the cell gets discharged, (providing current), it acts as an electrochemical cell. The reactions involved are

At anode $Pb(s) + HSO_4^-(aq) \longrightarrow PbSO_4(s) + H^+ + 2e^-, E° = 0.296$ V

At cathode

$$PbO_2(s) + 3H^+(aq) + HSO_4^-(aq) + 2e^- \longrightarrow PbSO_4(s) + 2H_2O(l), E° = 1.628 \text{ V}$$

Net reaction

$$Pb(s) + PbO_2(s) + 2H^+ + 2HSO_4^-(aq) \longrightarrow 2PbSO_4(s) + 2H_2O(l), \quad E° = 1.924 \text{ V}$$

Characteristics

There is no need for the cell to have separate anode and cathode compartments because the oxidising and reducing agents are both solids (PbO_2 and Pb). They are kept away from coming in contact by the presence of insulating spaces between the grids.

- Sulphuric acid is consumed during discharging and when the concentration of H_2SO_4 falls to 1.20 g L^{-1} it requires recharging.
- **Recharging the battery** A "run-down" lead storage battery can be recharged by using an external source of direct current to drive the cell reaction in the reverse non-spontaneous direction.
Actually, it forces the electrons to flow in opposite direction resulting in the deposition of Pb on the anode and PbO_2 on the cathode. In this case battery behaves like electrolytic batteries.

The reactions involved are

At anode

$$PbSO_4(s) + 2e^- \longrightarrow Pb(s) + SO_4^{2-}(aq)$$

At cathode

$$PbSO_4(s) + 2H_2O \longrightarrow PbO_2(s) + SO_4^{2-}(aq) + 4H^+(aq) + 2e^-$$

Net reaction

$$2PbSO_4(s) + 2H_2O(l) \longrightarrow Pb(s) + PbO_2(g) + 4H^+(aq)$$

The recharging process becomes possible because $PbSO_4$ formed during process is solid and stick to the electrons. Therefore, it can easily lose or gain electrons during electrolysis.

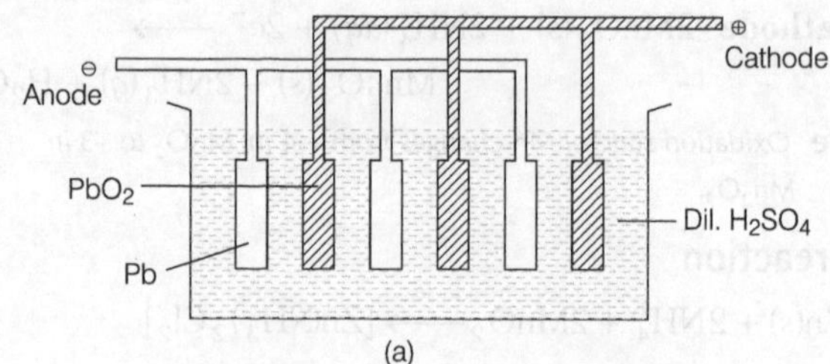

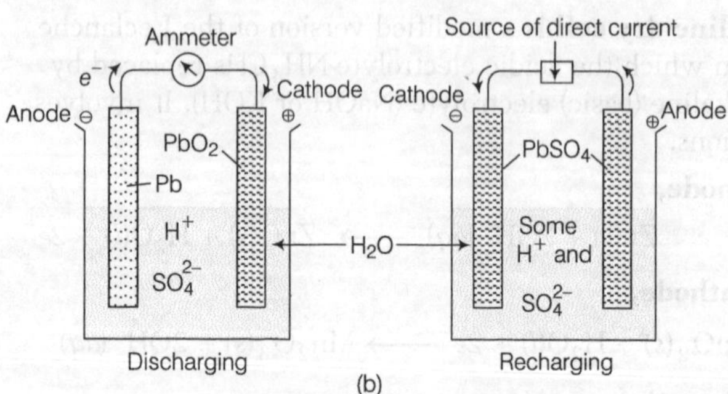

Fig. 3.12 Lead storage battery
(a) Construction of a lead storage battery
(b) Changes occurring during discharging and recharging of

(iii) Fuel cells

A fuel cell is a **galvanic cell** in which one of the reactants is a traditional fuel such as CH_4 or H_2 or CO.

Reactants are not contained within the cell but instead are continuously supplied from an external reservoir.

- **Electrodes** are made of porous carbon impregnated with metallic catalysts (Pt, Ag or metal oxide).

- **Electrolyte** Concentrated KOH or NaOH
- **Catalysts** Generally Pt or Pd or Ag

Reactions involved The fuel (gaseous H_2) and the oxidising agent (gaseous O_2) do not react directly but instead of flow into separate cell compartments where H_2 is oxidised at the anode and O_2 is reduced at the cathode.

At anode $2H_2 \longrightarrow 4H^+ + 4e^-$

$$4H^+ + 4OH^- \rightleftharpoons 4H_2O$$

$$2H_2(g) + 4OH^-(aq) \longrightarrow 4H_2O(l) + 4e^-$$

At cathode $O_2(g) + 2H_2O(l) + 4e^- \longrightarrow 4OH^-(aq)$

Net reaction $2H_2(g) + O_2(g) \longrightarrow 2H_2O(l)$

Catalyst used in this cell, increases the rate of electrode reactions.

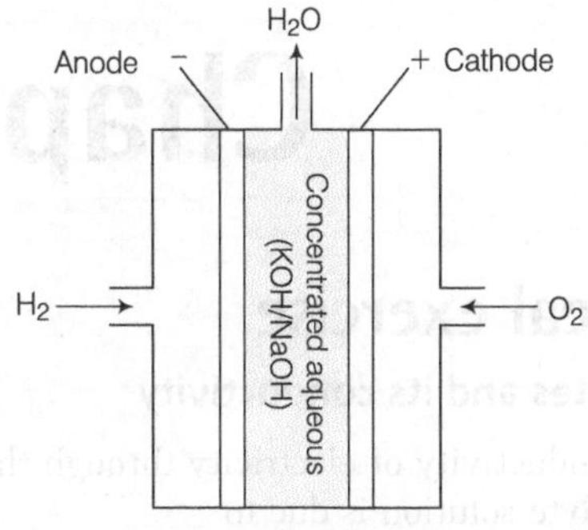

Fig. 3.13 Fuel cell

Advantages

(i) Fuel cells are pollution free cell, light in weight and have high reliability.

(ii) Efficiency of fuel cell is very high (about 70%).

Disadvantages

(i) It is very difficult to handle gaseous fuels.

(ii) Catalyst used are very costly.

(iii) Electrode preparation is very difficult and time taking.

Uses

The primary use of this kind of cells is in space craft, where by-product (H_2O) can be used for other purposes as well. These cells are also used in other vehicles like electric cars, submarines etc.

CHECK POINT 3.7

1. The corrosion of iron object is favoured by
(a) presence of moisture in air
(b) presence of H^+ ions
(c) presence of CO_2 in air
(d) All of the above

2. Rusting on the surface of iron involves
(a) $4Fe^{2+}(aq) + O_2(g) + 4H_2O(l) \longrightarrow 2Fe_2O_3(s) + 8H^+$
(b) $O_2(g) + 4H^+(aq) + 4e^- \longrightarrow 2H_2O(l)$ (at cathodic site)
(c) $Fe(s) \longrightarrow Fe^{2+}(aq) + 2e^-$ (at anodic site)
(d) All of the above

3. The technique of protecting a metal from corrosion by connecting it to a second metal (that is more easily oxidised) is called
(a) cathodic protection
(b) anodic protection
(c) galvanisation
(d) annealing

4. Metal can be prevented from rusting by
(a) connecting iron to more electropositive metal—a case of cathodic protection
(b) connecting iron to more electropositive metal—a case of anodic protection
(c) connecting iron to less electropositive metal—a case of anodic protection
(d) connecting iron to less electropositive metal—a case of cathodic protection

5. Which of the following statements is incorrect about dry cell ?
(a) It is also called Leclanche cell
(b) It is also called Daniell cell
(c) Electrolyte is a moist paste of NH_4Cl and $ZnCl_2$ in starch
(d) Cathodic reaction is
$$2MnO_2(s) + 2NH_4^+(aq) + 2e^- \longrightarrow Mn_2O_3(s) + 2NH_3(g) + H_2O(l)$$

6. Anode in the Leclanche cell is
(a) zinc container (b) graphite electrode
(c) carbon (d) $MnO_2 + C$

7. Which is/are rechargable batteries?
(a) Nickel-cadmium (b) Lead-storage
(c) Both (a) and (b) (d) None of these

8. Which of the following about batteries is/are correct ?
(a) Batteries are classified as primary, secondary and flow batteries.
(b) Leclanche cell is a primary battery.
(c) Lead acid (storage) battery is a secondary cell.
(d) All of the above

9. In which reactants are not contained within the cell but are continuously supplied from external source?
(a) Lithium battery
(b) Fuel cell
(c) Dry cell
(d) Lead-storage battery

Chapter Exercises

(A) Topical exercise

I. Electrolytes and its conductivity

1 The conductivity of electricity through the electrolyte solution is due to
(a) movement of molecules of electrolyte
(b) movement of ions of electrolyte
(c) movement of particles of the solvent
(d) movement of separate atoms

2 Which of the statements about solutions of electrolytes is not correct? **[NCERT Exemplar]**
(a) Conductivity of solution depends upon size of ions
(b) Conductivity depends upon viscosity of solution
(c) Conductivity does not depend upon solvation of ions present in solution
(d) Conductivity of solution increases with temperature

3 The cell constant of a conductivity cell .... **[NCERT Exemplar]**
(a) changes with change of electrolyte
(b) changes with change of concentration of electrolyte
(c) changes with temperature of electrolyte
(d) remains constant for a cell

4 Cell constant has unit
(a) cm (b) cm^{-1}
(c) cm^2 (d) $mol\ L^{-1}$

5 Electrical potential of a cell is
(a) an intensive property
(b) an extensive property
(c) an isothermal property
(d) an isobaric property

6 Consider the following statements.
Statement I Cell constant is the emf of a cell.
Statement II Cell constant is determined by using saturated KCl solution.
(a) Both statements I and II are correct
(b) Only statement I is correct
(c) Only statement II is corect
(d) Both statements I and II are incorrect

7 Cell constant is maximum in case of a
(a) wire of length 100 cm and area 100 cm^2
(b) wire of length 10 cm and area 10 cm^2
(c) one centimeter cube of a material
(d) equal in all cases

8 Variation of resistance (R) with increase in cell constant (l/a) gives graph of the type

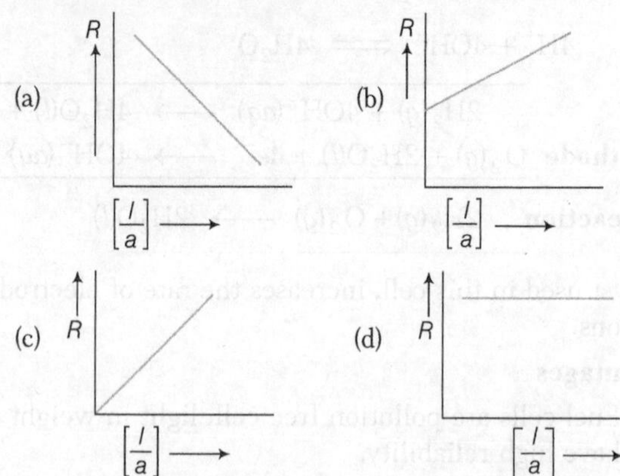

9 The specific conductance of 0.1 N KCl solution at 23°C is 0.012 $\Omega^{-1}\ cm^{-1}$. The resistance of cell containing the solution at the same temperature was found to be 55 Ω. The cell constant will be
(a) 0.142 cm^{-1} (b) 0.66 cm^{-1}
(c) 0.918 cm^{-1} (d) 1.12 cm^{-1}

10 When a certain conductivity cell was filled with 0.01 M solution of KCl, it had a resistance of 160 Ω at 25°C and when filled with 0.005 M NaOH, it had a resistance of 190 Ω. If specific resistance of KCl solution is 700 Ω-cm, specific conductance of NaOH solution is
(a) 0.00120 $\Omega^{-1}\ cm^{-1}$ (b) 0.00170 $\Omega^{-1}\ cm^{-1}$
(c) 0.00180 $\Omega^{-1}\ cm^{-1}$ (d) 0.00190 $\Omega^{-1}\ cm^{-1}$

11 500 mL of an aqueous solution contains 0.1 mole of the solute AB, if its specific conductance is x S cm^{-1}, its molar conductance will be .... S $cm^2\ mol^{-1}$.
(a) $10x$ (b) $5000x$ (c) x (d) $200x$

12 0.1 M aqueous solution of a dibasic acid is diluted such that its equivalent conductance is five times. Hence, final concentration is
(a) 0.01 N (b) 0.02 N (c) 0.04 N (d) 0.055 N

13 The resistance of 0.1 N solution of salt is found to be 2.5×10^3 Ω. The equivalent conductance of solution (cell constant = 1.15 cm^{-1}) in $\Omega^{-1}\ cm^2\ equiv^{-1}$ is
(a) 3.8 (b) 4.6 (c) 6.4 (d) 7.6

14 Statement I Equivalent conductance increases with dilution for an electrolyte solution.

Statement II The number of ions in one litre of electrolyte increases with dilution.

Consider the following statements.
(a) Both statements I and II are correct
(b) Only statement I is correct
(c) Only statement II is corect
(d) Both statements I and II are incorrect

15 The molar conductivity of strong electrolytes
(a) does not change on dilution
(b) increases on dilution slightly
(c) decreases on dilution
(d) depends on density of electrolyte itself

16 0.1 M H_2SO_4 is diluted to 0.01 M H_2SO_4. Hence, its molar conductance will be
(a) 10 times (b) $\frac{1}{10}$th (c) 100 times (d) 10000 times

17 For pure water degree of dissociation of water is 1.9×10^{-9}

$\Lambda_m^\infty (H^+) = 350 \text{ S cm}^2 \text{ mol}^{-1}$

$\Lambda_m^\infty (OH^-) = 200 \text{ S cm}^2 \text{ mol}^{-1}$

Hence, molar conductance of water is
(a) 1.045×10^{-6} S cm^2 mol^{-1} (b) 1.045×10^{-9} S cm^2 mol^{-1}
(c) 1.045×10^{-10} S cm^2 mol^{-1} (d) 1.04×10^{-14} S cm^2 mol^{-1}

18 Select the correct statement(s).
(a) Λ_m^∞ of strong electrolytes as well as weak electrolytes can be obtained by Kohlrausch's law as
$\Lambda_m^\infty = x_+ \lambda_+^\infty + y_- \lambda_-^\infty$
(b) Λ_m^∞ of weak electrolytes as well as strong electrolytes is obtained by extrapolation to $\sqrt{C} \to 0$.
(c) Both (a) and (b)
(d) None of the above

19 Ionic conductances of H^+ and SO_4^{2-} at infinite dilution are x and y S cm^2 equiv^{-1}. Hence, equivalent conductance of H_2SO_4 at infinite dilution is
(a) $x + y$ (b) $2(x + y)$ (c) $2x + y$ (d) $x + 2y$

20 Which of the following facts about electrolytic conduction is correct?
(a) Λ_m decreases with increase of $\sqrt{C}$ due to increase in interionic attraction between ions.
(b) For weak electrolytes, there is rapid increase in the degree of ionisation with dilution hence, interionic attraction decreases and Λ_m or Λ_{eq} increases rapidly in lower concentration range.
(c) Λ_m^∞ can be obtained by extrapolation to $\sqrt{C} = 0$.
(d) All of the above

21 Match the physical term in Column I with the unit in Column II.

	Column I		Column II
A.	Conductance	1.	Siemen m^2 mol^{-1}
B.	Resistivity	2.	m^{-1}
C.	Conductivity	3.	volt per ampere
D.	Molar conductivity	4.	Siemen
E.	Cell constant	5.	Siemen m^{-1}
F.	Resistance	6.	Ohm metre

Choose the correct match.

	A	B	C	D	E	F
(a)	6	5	4	3	2	1
(b)	5	6	1	2	3	4
(c)	4	6	5	1	2	3
(d)	3	4	6	5	1	2

22. Statement I Conductivity of all electrolytes decreases on dilution. **[NCERT Exemplar]**

Statement II On dilution number of ions per unit volume decreases.

Consider the following statements.
(a) Both statements I and II are correct
(b) Only statement I is correct
(c) Only statement II is corect
(d) Both statements I and II are incorrect

23 Statement I Λ_m for weak electrolytes shows a sharp increase when the electrolytic solution is diluted.

Statement II For weak electrolytes degree of dissociation increases with dilution of solution. **[NCERT Exemplar]**

Consider the following statements.
(a) Both statements I and II are correct
(b) Only statement I is correct
(c) Only statement II is corect
(d) Neither statement I nor statement II is correct

24 Match the terms in Column I with corresponding units Column II.

	Column I		Column II
A.	Transport number	1.	cm^2 V^{-1} s^{-1}
B.	Ionic mobility	2.	Ω^{-1} cm^2 mol^{-1}
C.	Ionic conductivity	3.	Ω^{-1} cm^{-1}
D.	Cell constant	4.	unit-less
		5.	cm^{-1}

Choose the correct match.

	A	B	C	D
(a)	5	1	2	3
(b)	2	1	5	3
(c)	3	5	4	2
(d)	4	1	2	5

25 Λ^{∞}_{AgCl} can be obtained
(a) by extrapolation of the graph Λ and $\sqrt{C}$ to zero concentration
(b) by known values of Λ^{∞} of $AgNO_3$, HCl and HNO_3
(c) Both (a) and (b)
(d) None of the above

26 Which of the following graph is/are correct?

(a) 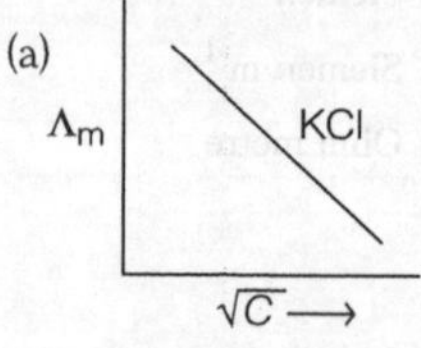

(b) 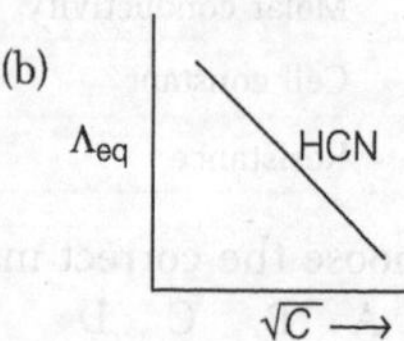

(c) Both (a) and (b) are correct (d) None of these

27 Λ^{∞}_m ($BaSO_4$) is equal to
(a) Λ^{∞}_m of ($BaCl_2 + H_2SO_4 - HCl$)
(b) Λ^{∞}_m of ($BaCl_2 + H_2SO_4 - 2HCl$)
(c) limiting molar conductance when graph between Λ_m and $\sqrt{C}$ is extrapolated to zero
(d) None of the above

28 $\Lambda^{\circ}_{m\,(NH_4OH)}$ is equal to ……… . **[NCERT Exemplar]**
(a) $\Lambda^{\circ}_{m\,(NH_4OH)} + \Lambda^{\circ}_{m\,(NH_4Cl)} - \Lambda^{\circ}_{(HCl)}$
(b) $\Lambda^{\circ}_{m\,(NH_4Cl)} + \Lambda^{\circ}_{m\,(NaOH)} - \Lambda^{\circ}_{(NaCl)}$
(c) $\Lambda^{\circ}_{m\,(NH_4Cl)} + \Lambda^{\circ}_{m\,(NaCl)} - \Lambda^{\circ}_{(NaOH)}$
(d) $\Lambda^{\circ}_{m\,(NaOH)} + \Lambda^{\circ}_{m\,(NaCl)} - \Lambda^{\circ}_{(NH_4Cl)}$

29 Equivalent conductance at infinite dilution of $BaCl_2$, H_2SO_4 and HCl *aq* solutions are x_1, x_2 and x_3 respectively. Equivalent conductance of $BaSO_4$ solution is
(a) $x_1 + x_2 - x_3$ (b) $x_1 - x_2 - x_3$
(c) $x_1 + x_2 - 2x_3$ (d) $x_1 - 2x_2 + x_3$

30 Equivalent conductance of 1 M CH_3COOH is 10 Ω^{-1} cm^2 $equiv^{-1}$ and at infinite dilution 200 Ω^{-1} cm^2 $equiv^{-1}$. Hence, percentage ionisation of CH_3COOH is
(a) 5% (b) 2%
(c) 4% (d) 1%

31 The equivalent conductance of Ba^{2+} and Cl^- are respectively 127 and 76 Ω^{-1} cm^2 $equiv^{-1}$ at infinite dilution. The equivalent conductance (in Ω^{-1} cm^2 $equiv^{-1}$) of $BaCl_2$ at infinite dilution will be
(a) 139.5 (b) 203
(c) 279 (d) 101.5

32 Equivalent conductance of NaCl, HCl and C_2H_5COONa at infinite dilution are 126.45, 426.16 and 91 Ω^{-1} $equiv^{-1}cm^2$, respectively. The equivalent conductance of C_2H_5COOH is
(a) 201.28 Ω^{-1} $equiv^{-1}$ cm^2
(b) 390.71 Ω^{-1} $equiv^{-1}$ cm^2
(c) 698.28 Ω^{-1} $equiv^{-1}$ cm^2
(d) 540.48 Ω^{-1} $equiv^{-1}$ cm^2

33 Equivalent conductance of $BaCl_2$, H_2SO_4 and HCl are x_1, x_2 and x_3 S cm^2 $equiv^{-1}$ at infinite dilution. If specific conductance of saturated $BaSO_4$ solution is of y S cm^{-1}, then K_{sp} of $BaSO_4$ is
(a) $\dfrac{10^3\,y}{2(x_1 + x_2 - 2x_3)}$ (b) $\dfrac{10^6\,y^2}{(x_1 + x_2 - x_3)^2}$
(c) $\dfrac{10^6\,y^2}{4\,(x_1 + x_2 - x_3)^2}$ (d) $\dfrac{x_1 + x_2 - 2x_3}{10^6\,y^2}$

34 Which has maximum value of Λ_{eq} at constant temperature assuming 100% ionisation of each electrolyte ?
(a) 0.1 M HCl (b) 0.1 M H_2SO_4
(c) 0.1 M H_3PO_4 (d) Equal

35 Which has maximum value of Λ_{eq} at constant temperature assuming 100% ionisation of each electrolyte?
(a) 0.1 M H_2SO_4 (b) 0.1 M H_3PO_3
(c) 0.1 M $H_2PO_4^-$ (d) Equal

36 Ionisation constant of a weak acid (*HA*) in terms of Λ^{∞}_m and Λ_m is
(a) $K_a = \dfrac{C\Lambda^{\infty}_m}{(\Lambda_m - \Lambda^{\infty})}$ (b) $K_a = \dfrac{C\Lambda^2_m}{\Lambda^{\infty}_m\,(\Lambda^{\infty}_m - \Lambda_m)}$
(c) $K_a = \dfrac{C\,(\Lambda^{\infty}_m)^2}{\Lambda^{\infty}_m\,(\Lambda^{\infty}_m - \Lambda_m)}$ (d) None of these

37 $E^{\circ}_{Al^{3+}/Al} = -1.66$ V and K_{sp} of $Al(OH)_3 = 1.0 \times 10^{-33}$. Reduction potential of the above couple at pH = 14 is
(a) − 2.31 V (b) + 2.31 V (c) +1.01 V (d) +1.01 V

38 For the following cell reaction,
$$Pb(s) + Hg_2SO_4\,(s) \rightleftharpoons PbSO_4\,(s) + 2Hg(l)$$
$$E^{\circ}_{cell} = 0.92 \text{ V}$$
$K_{sp}(PbSO_4) = 2 \times 10^{-8}$, $K_{sp}(Hg_2SO_4) = 1 \times 10^{-6}$;
Hence, E_{cell} is
(a) 0.92 V (b) 0.89 V (c) 1.04 V (d) 0.95 V

39 Reduction potential of half-cell
AgI(*aq*) saturated | Ag, ($E^\circ_{Ag^+/Ag} = 0.8$ V, K_{sp} of AgI is 10^{-16}) is V
(a) 0.328 (b) 1.272
(c) −0.328 (d) − 0.146

40 For a reaction, $A + B^{2+} \longrightarrow B + A^{2+}$ at 25°C, $E^\circ = 0.2955$ V. Hence, K_{eq} is
(a) 10 (b) 10^{10}
(c) − 10 (d) 10^{-10}

II. Cell and electrochemical cell

41 Salt-bridge contains
(a) calomel (b) corrosive sublimate
(c) H_2O (d) *agar-agar* paste

42 Which of the following is/are function(s) of salt-bridge?
(a) It completes the electrical circuit with electrons flowing from one electrode to other through external wires and a flow of ions between the two compartments through salt-bridge.
(b) It prevents the accumulation of the ions.
(c) Both (a) and (b)
(d) None of the above

43 Following behaves as SHE at a pressure of
Pt, $H_2 | H_2O$
(a) 1 bar (b) 10^{-14} bar
(c) 10^7 bar (d) 10^{14} bar

44 Select the correct statement (s).
(a) In a galvanic cell, a spontaneous chemical reaction generates an electric current.
(b) In an electrolytic cell, an electric current drives a non-spontaneous reaction.
(c) In a galvanic cell, cell reaction is exothermic.
(d) All the above statements are correct.

45 Which of the following statements is correct about galvanic cell.
(a) Cathode is positive terminal
(b) If $E_{cell} < 0$, it is a spontaneous reaction
(c) Current flows from anode to cathode
(d) All are correct

46 Consider the following statements about a galvanic cell.
I. Electrons flow through H_2O (medium) from anode to cathode.
II. Electrons flow through wire from anode to cathode.
III. Salt-bridge is necessary to complete the electrical circuit.
Select the correct alternate(s).
(a) I, II and III (b) Both I and III
(c) Both II and III (d) Both I and II

47 $Cu^+ + e^- \longrightarrow Cu, \quad E^\circ = x_1$ V
$Cu^{2+} + 2e^- \longrightarrow Cu, \quad E^\circ = x_2$ V, then for
$Cu^{2+} + e^- \longrightarrow Cu^+$, E° (volt) will be
(a) $x_1 - 2x_2$ (b) $x_1 + 2x_2$ (c) $x_1 - x_2$ (d) $2x_2 - x_1$

48 For the reactions,
I. $MnO_4^- + 8H^+ + 5e^- \longrightarrow Mn^{2+} + 4H_2O$, $E^\circ = 1.51$V
II. $MnO_2 + 4H^+ + 2e^- \longrightarrow Mn^{2+} + 2H_2O$, $E^\circ = 1.23$ V
then for the reaction,
III. $MnO_4^- + 4H^+ + 3e^- \longrightarrow MnO_2 + 2H_2O$, E° is
(a) 1.70 V (b) 5.09 V (c) 0.28 V (d) 0.84 V

49 Electrochemical cell stops working after sometime because
(a) electrode potential of both the electrodes becomes zero
(b) the reaction reverts its direction
(c) one of the electrode completely vanishes
(d) electrode potential of both the electrodes equalise

50 For the cell reaction,
$$Cu(s) + Cl_2(aq) \longrightarrow Cu^{2+}(aq) + 2Cl^-(aq)$$
Cell notation is
(a) $Cu(s)|Cu^{2+}(aq)||Cl^-(aq)|Cl_2(g)|C(s)$
(b) $C(s)|Cl^-(aq)|Cl_2(g)||Cu^{2+}(aq)|Cu(s)$
(c) $Cu(s)|Cu^{2+}(aq)||Cl^-(aq)|Cl_2(g)$
(d) $Cu(s)|Cu^{2+}(aq)||Cl_2(aq)|Cl^-(aq)$

51 For the electrochemical cell,
$\underset{\ominus}{Ag} \mid AgCl(s) \mid KCl(aq) \mid\mid AgNO_3(aq) \mid \underset{\oplus}{Ag}$
The overall cell reaction is
(a) $Ag^+ + KCl \longrightarrow AgCl + K^+$
(b) $Ag^+ + Cl^-(aq) \longrightarrow Ag + \frac{1}{2}Cl_2$
(c) $AgCl(s) \longrightarrow Ag^+ + Cl^-$
(d) $Ag^+ + Cl^- \longrightarrow AgCl$

52 If, $Sn^{2+} + 2e^- \longrightarrow Sn \quad E^\circ = -0.14$ V
$Sn^{4+} + 2e^- \longrightarrow Sn^{2+} \quad E^\circ = 0.13$ V
then,
(a) Sn^{2+} is unstable and disproportionates to Sn^{4+} and Sn
(b) Sn^{2+} is stable and disproportionation reaction is not spontaneous
(c) Sn^{4+} is easily reduced to Sn
(d) None of the above is correct

53 In which of the following cells will the emf be independent of the activity of the chloride ion?
(a) $Zn|ZnCl_2(aq)||KCl(aq)|Pt(Cl_2)$
(b) $Zn|ZnCl_2(aq)||KCl(aq)|AgCl(s), Ag$
(c) Ag, AgCl (s)| KCl (aq)| $Pt(Cl_2)$
(d) Hg, $Hg_2Cl_2(s)|KCl(aq)||AgNO_3(aq)|Ag$

54 In the galvanic cell,

$$Fe(s) + 2Fe^{3+}(aq) \longrightarrow 3Fe^{2+}(aq)$$

(a) $Fe(s)$ is the cathode because it is in contact with $Fe^{2+}(aq)$
(b) $Fe(s)$ is the cathode because it is in contact with $Fe^{3+}(aq)$
(c) $Fe(s)$ cannot be used as cathode since it would directly react with Fe^{3+}, and short-circuit the cell
(d) salt-bridge contains Fe^{2+} and Fe^{3+} ions

55 In the Daniell cell constructed in the laboratory, the voltage observed was 0.9 V instead of 1.1 V of the standard cell. A possible explanation is

(a) $[Zn^{2+}] > [Cu^{2+}]$
(b) $[Zn^{2-}] < [Cu^{2-}]$
(c) the Zn electrode has twice the surface of Cu electrode
(d) mol ratio of $Zn^{2+} : Cu^{2+}$ is 2:1

56 An electrochemical cell is set up as

$Pt(H_2) | HCl(0.1\ M) || CH_3COOH(0.1\ M) | Pt(H_2)$

The emf of this cell will not be zero, because

(a) the temperature is constant
(b) pH of 0.1 M HCl and 0.1 M CH_3COOH is not same
(c) acids used in two compartments are different
(d) emf depends on molarities of acids used

57 **Statement I** When an aqueous solution of $NaNO_3$ is electrolysed, sodium is liberated at the cathode.

Statement II Na^+ ions are discharged at cathode and NO_3^- at the anode.

Consider the following statements.

(a) Both statements I and II are correct
(b) Only statement I is correct
(c) Only statement II is corect
(d) Both statements I and II are incorrect

58 Match electrode in Column I with its presentation in Column II.

Column I	Column II
A. Electrode reversible with respect to cation	1. $Pt/Fe^{2+}, Fe^{3+}$
B. Electrode reversible with respect to anion	2. $Pt, H_2 \mid H^+ (a=1)$ 1 bar
C. Redox electrode	3. $Ag \mid AgCl(s), HCl(aq)$
D. Reference electrode	4. $Ag \mid AgNO_3$

Choose the correct match.

	A	B	C	D		A	B	C	D
(a)	4	3	1	2	(b)	3	2	4	1
(c)	2	4	1	3	(d)	4	1	3	2

III. Electrolysis

59 During the electrolysis of aqueous zinc nitrate

(a) zinc plates out at the cathode
(b) zinc plates out at the anode
(c) hydrogen gas H_2 is evolved at the anode
(d) oxygen gas O_2 is evolved at the anode

60 When a dilute aqueous Li_2SO_4 solution is electrolysed, the products formed at the anode and cathode, respectively are

(a) S and Li (b) O_2 and Li
(c) SO_2 and H_2 (d) O_2 and H_2

61 For reducing 1 mole of $Cr_2O_7^{2-}$ to Cr^{3+} the charge required is

(a) 3×96500 C (b) 6×96500 C
(c) 0.3 F (d) 0.6 F

62 **Statement I** At the end of electrolysis using platinum electrodes, an aqueous solution of $CuSO_4$ turns colourless.

Statement II $CuSO_4$ changes to $Cu(OH)_2$ during electrolysis.

Consider the following statements.

(a) Both statements I and II are correct
(b) Only statement I is correct
(c) Only statement II is corect
(d) Both statements I and II are incorrect

63 When $Na_2SO_4(aq)$ is electrolysed between Pt electrodes, the reaction that occurs at cathode is

(a) $Na^+ + e^- \longrightarrow Na$
(b) $H^+ + e^- \longrightarrow \frac{1}{2}H_2$
(c) $2OH^- \longrightarrow \frac{1}{2}O_2 + H_2O + 2e^-$
(d) $2H_2O + 2e^- \longrightarrow H_2 + 2OH^-$

64 The quantity of charge required to obtain one mole of aluminium from Al_2O_3 is .... **[NCERT Exemplar]**

(a) 1 F (b) 6 F (c) 3 F (d) 2 F

65 1 mole each of $AgNO_3$, $CuSO_4$ and $AlCl_3$ is electrolysed. Number of Faradays required are in the ratio of

(a) 1:1:1 (b) 1:2:3 (c) 3:2:1 (d) 1:3:1

66 How many Faradays are required to reduce one mole of MnO_4^- to Mn^{2+}?

(a) 1 (b) 5 (c) 3 (d) 2

67 A current of 5.00 A flowing for 30.00 min deposits 3.048 g of zinc at the cathode. Hence, equivalent weight of zinc is

(a) 65.4 g (b) 130.8 g (c) 32.7 g (d) 16.85 g

68 96500 Coulomb deposits 107.9 g of silver. If electronic charge is 1.6×10^{-19} C, value of Avogadro's number is

(a) 6.02×10^{-23} mol^{-1} (b) 6.02×10^{23} mol^{-1}
(c) 6.02×10^{19} mol^{-1} (d) 6.02×10^{-19} mol^{-1}

69 A certain current liberates 0.500 g of H_2 in 2.00 h. How many gram of oxygen can be liberated by the same current in the same time?

(a) 0.500 g (b) 8.00 g (c) 4.00 g (d) 16.00 g

70 How many coulombs are required for the oxidation of 1 mole H_2O to O_2 ?

(a) 3.86×10^5 C (b) 9.65×10^4 C
(c) 1.93×10^5 C (d) 4.825×10^4 C

71 In a process of electrolysis, MnO_4^- is converted into MnO_2. By 1 Faraday, one mole of MnO_4^- will give

(a) 1 mol of MnO_2 (b) 0.5 mol of MnO_2
(c) 0.66 mol of MnO_2 (d) 0.33 mol of MnO_2

72 Which has maximum potential (numerical value) for the half-cell reaction ?

$$2H^+ + 2e^- \longrightarrow H_2$$

(a) 1.0 M HCl (b) A solution having pH 4
(c) Pure water (d) 1.0 M NaOH

73 Number of electrons lost during electrolysis of 0.355 g of Cl^- is

(a) 0.01 (b) 0.01 N_0 (c) 0.02 N_0 (d) 0.01/2 N_0

74 A steady current of 30.0 A for 70.2 min corresponds to a passage of

(a) 1.26×10^5 electrons (b) 1.31 Coulombs
(c) 1.26×10^5 Faradays (d) 1.31 Faradays

75 A 0.200 M KOH solution is electrolysed for 1.5 h using a current of 8.00 A. How many moles of O_2 were produced at the anode?

(a) 0.48 mol (b) 0.224 mol
(c) 0.112 mol (d) 2.24×10^{-2} mol

76 A conducting wire carries a current of 0.965 A. Rate of flow of electrons per second at a given point is

(a) $1 \times 10^{-5} N_0$ (b) N_0 (c) $0.965\, N_0$ (d) $\dfrac{N_0}{0.965}$

77 For $M^{2+} + 2e^- \longrightarrow M$, 0.275 g of metal M is deposited at the cathode due to passage of 1 A of current for 965 s. Hence, atomic weight of the metal M is

(a) 27.5 g (b) 55.0 g (c) 110.0 g (d) 13.75 g

78 The current in a given wire is 1.8 A. In a 1.36 min the number of coulombs transferred in wire is

(a) 136.44 C (b) 138.88 C (c) 146.88 C (d) 150.66 C

79 1 A of current is passed to supply 72000 C of charge in

(a) 10 h (b) 5 h (c) 10 h (d) 20 h

80 Consider the following statements.

Statement I 1 Faraday of electricity deposits 1 g equivalent of Ag or Cu or Al.

Statement II 1 mole electrons are required to reduce 1 mole of Ag^+ or Cu^{2+} or Al^{3+} ion.

(a) Both statements I and II are correct
(b) Only statement I is correct
(c) Only statement II is corect
(d) Both statements I and II are incorrect

81 Assume that, during electrolysis of $AgNO_3$, only H_2O is electrolysed and O_2 is formed as,

$$2H_2O \longrightarrow 4H^+ + O_2 + 4e^-$$

O_2 formed at NTP due to passage of 2 A of current for 965 s is

(a) 0.112 L (b) 0.224 L (c) 11.2 L (d) 22.4 L

82 If a steady current of 15.0 A is passed through an aqueous solution of $CuSO_4$, how many minutes will it take to deposit 0.250 mol of Cu at the cathode, assuming 100% efficiency?

(a) 3.217×10^3 (b) 1.613×10^3
(c) 53.62 (d) 0.893

83 Same quantity of current is passed through molten NaCl and molten cryolite containing Al_2O_3. If 4.6 g of sodium were liberated in one cell, the mass of aluminium liberated in other cell was

(a) 0.9 g (b) 2.7 g (c) 1.8 g (d) 3.6 g

84 The electric charge for electrode deposition of gram equivalent of a substance is

(a) 1 A per second
(b) 96500 C per second
(c) 1 A for 1 h
(d) charge on one mol of electrons

85 1 g equivalent of Na metal is formed from electrolysis of fused NaCl. Number of moles of Al formed by the same current from fused Na_3AlF_6 will be

(a) 1 (b) 3
(c) 1 / 3 (d) 2

86 Consider the following electrolytes,

1. $AgNO_3$ 2. $CuSO_4$ 3. $AlCl_3$ 4. $Bi_2(SO_4)_3$

The quantity of electricity needed to electrolyte separately 1 M solutions of these electrolytes will be (F is Faraday)

(a) 1 F, 2 F, 3 F, 6 F (b) 2 F, 3 F, 6 F, 1 F
(c) 3 F, 6 F, 2 F, 1 F (d) 6 F, 3 F, 2 F, 1 F

IV. Electromotive force of a cell (emf) and electrochemical series

87 In the electrolysis of aqueous sodium chloride solution which of the half-cell reaction will occur at anode ?

(a) $Na^+ (aq) + e^- \longrightarrow Na (s); E^{\ominus}_{cell} = -2.71$ V

(b) $2H_2O (l) \longrightarrow O_2 (g) + 4H^+ (aq) + 4e^-; E^{\ominus}_{cell} - 1.23$ V

(c) $H^+ (aq) + e^- \longrightarrow \frac{1}{2} H_2 (g); E_{cell} = 0.00$ V

(d) $Cl^- (aq) \longrightarrow \frac{1}{2} Cl_2 (g) + e^-; E^{\ominus}_{cell} = 1.36$ V

88 Consider the given data

Half-cell reaction $E°$

1. $Cr_2O_7^{2-} + 14H^+ + 6e^- \rightleftharpoons 2Cr^{3+} + 7H_2O$; +1.33 V
2. $Cr^{3+} + 3e^- \rightleftharpoons Cr$; - 0.74 V
3. $Cu^+ + e^- \rightleftharpoons Cu$; + 0.52 V
4. $Cu^{2+} + 2e^- \rightleftharpoons Cu$; + 0.34 V

$E°$ for the reaction,

$2Cr(s) + 3Cu^{2+} (aq) \rightleftharpoons 2Cr^{3+} + 3Cu$ is

(a) -1.08 V (b) -0.70 V (c) 1.08 V (d) 2.50 V

89 Statement I E_{cell} should have a positive value for the cell to function. **[NCERT Exemplar]**

Statement II $E_{cathode} < E_{anode}$

Consider the following statements.

(a) Both statements I and II are correct
(b) Only statement I is correct
(c) Only statement II is corect
(d) Both statements I and II are incorrect

90 For the cell ,

$Hg| Hg_2Cl_2 | Cl^- (0.1 M)|| Cl^- (0.01 M)| Pt(Cl_2), E°_{cell}$ is 1.10 V. Hence, E_{cell} is

(a) 1.1591 V (b) −1.1591V (c) 1.0409 V (d) −1.0409 V

91 $2Ag^+ (aq) + Cu(s) \longrightarrow Cu^{2+} (aq) + 2Ag(s)$

The standard potential for this reaction is 0.46 V. Which change will increase the standard potential the most?

(a) Doubling the $[Ag^+]$
(b) Halving the $[Cu^{2+}]$
(c) Doubling the size of the Cu(s) electrode
(d) Standard potential is considered only for 1 M solutions

92 For the half-cells,

Cu^+ / Cu $E° = +0.52$ V ; Cu^{2+} / Cu^+ $E° = +0.16$ Vhence, $E°$ for the disproportionation reaction

$2Cu^+ \longrightarrow Cu^{2+} + Cu$ would be

(a) 0.36 V (b) 0.78 V
(c) 0.84 V (d) − 0.36 V

93 In the following electrochemical cell

$Zn | Zn^{2+} || H^+ | Pt(H_2)$
$\ominus$ $\oplus$

$E°_{cell} = E_{cell}$. This will be when

(a) $[Zn^{2+}] = [H^+] = 1$ M and $p_{H_2} = 1$ atm
(b) $[Zn^{2+}] = 0.01$ M, $[H^+] = 0.1$ M and $p_{H_2} = 1$ atm
(c) $[Zn^{2+}] = 1$ M, $[H^+] = 0.1$ M and $p_{H_2} = 0.01$ atm
(d) All of the above

94 Consider the following statements.

Statement I Cu is less reactive than hydrogen.

Statement II $E^{\ominus}_{Cu^{2+}/Cu}$ is negative. **[NCERT Exemplar]**

(a) Both statements I and II are correct
(b) Only statement I is correct
(c) Only statement II is corect
(d) Both statements I and II are incorrect

95 Consider the following statements.

Statement I Copper sulphate can be stored in zinc vessel. **[NCERT Exemplar]**

Statement II Zinc is less reactive than copper.

(a) Both statements I and II are true
(b) Only statement I is true
(c) Only statement II is true
(d) Neither statement I nor statement II is true

96 It is not advisable to

(a) stir copper sulphate solution with a zinc spoon
(b) stir silver nitrate solution with a copper spoon
(c) Both (a) and (b)
(d) stir iron sulphate solution with a copper spoon

97 Consider the following sets

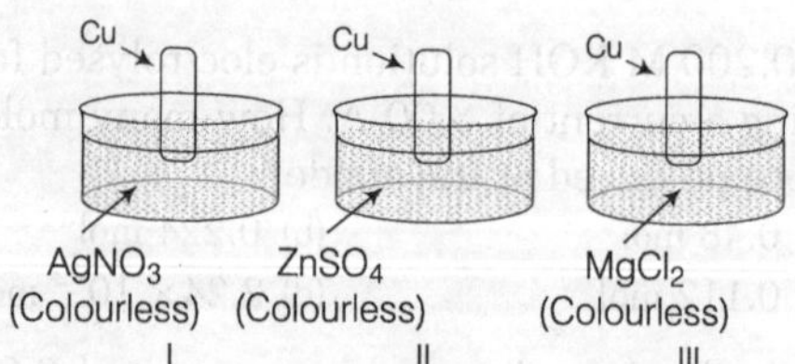

Colourless solution changes to blue coloured solution in

(a) Only I (b) Only II (c) Only III (d) Only I, III

98 Given, $E°_{Ag^+/Ag} = 0.80$ V, $E°_{Mg^{2+}/Mg} = -2.37$ V, $E°_{Cu^{2+}/Cu} = 0.34$ V, $E°_{Hg^{2+}/Hg} = 0.79$ V

Which of the following statements is/are correct?

(a) $AgNO_3$ can be stored in copper vessel
(b) $Cu(NO_3)_2$ can be stored in magnesium vessel
(c) $CuCl_2$ can be stored in silver vessel
(d) $HgCl_2$ can be stored in copper vessel

99 Consider the following statements.

Statement I When acidified zinc sulphate solution is electrolysed between zinc electrodes, it is zinc that is deposited at the cathode and hydrogen evolution does not take place.

Statement II The electrode potential of zinc is more negative than hydrogen as the over-voltage for hydrogen evolution on zinc is quite large.

(a) Both statements I and II are correct
(b) Only statement I is correct
(c) Only statement II is corect
(d) Both statements I and II are incorrect

100 The $E^\circ_{M^{3+}/M^{2+}}$ values for Cr, Mn, Fe and Co are −0.41 V, +1.57 V , +0.77 V and 1.97 V respectively. For which one of these metals the change in oxidation state from +2 to +3 is easiest?

(a) Cr (b) Mn
(c) Fe (d) Co

101 The reaction,

$Cu^{2+}(aq) + 2Cl^-(aq) \longrightarrow Cu(s) + Cl_2(g)$ has $E^\circ_{cell} = -1.02$ V. This reaction

(a) can be made to produce electricity in a voltaic cell
(b) can be made to occur in an electrolysis cell
(c) occurs whenever Cu^{2+} and Cl^- are brought together in an aqueous solution
(d) can occur in acidic but not in basic solution

102 Match the items of Column I and Column II on the basis of data given below **[NCERT Exemplar]**

$$E^\ominus_{F_2/F^-} = 2.87\text{V}, E^\ominus_{Li^+/Li} = -3.5\text{V},$$
$$E^\ominus_{Au^{3+}/Au} = 1.4\text{V}, E^\ominus_{Br_2/Br^-} = 1.09\text{V}$$

Column I	Column II
A. F_2	1. Metal is the strongest reducing agent
B. Li	2. Metal ion which is the weakest oxidising agent
C. Au^{3+}	3. Non-metal which is the best oxidising agent
D. Br^-	4. Unreactive metal
E. Au	5. Anion that can be oxidised by Au^{3+}
F. Li^+	6. Anion which is the weakest reducing agent
G. F^-	7. Metal ion which is an oxidising agent

Choose the correct match.

	A	B	C	D	E	F	G
(a)	2	5	4	3	7	1	6
(b)	3	1	7	5	4	2	6
(c)	4	6	2	7	2	1	5
(d)	7	4	1	6	3	5	2

V. Nernst equation

103 For the half-cell, $Cl^- / Pt(Cl_2)$, $(E - E^\circ)$

(a) increases as $[Cl^-]$ increases
(b) decreases as $[Cl^-]$ increases
(c) remains constant as $[Cl^-]$ increases
(d) cannot be predicted

104 For $Pt(H_2)|H_2O$, electrode potential at 298 K and 1 atm is

(a) −0.2364 V (b) −0.4137 V (c) 0.4137 V (d) 0.00 V

105 $Zn|\ Zn^{2+}(10^{-4}\ M)||\ H_2O|Pt(H_2)_{1\ bar}$, $E^\circ_{cell} = 0.76$ V at 298 K. Hence, E_{cell} is

(a) 0.93 V (b) 0.23 V (c) 0.46 V (d) 0.64 V

106 100 mL of a buffer of 1 M NH_3 (*aq*) and 1 M NH_4^+ (*aq*) are placed in two volataic cells separately. A current of 1.5 A is passed through both cells for 20 min. If electrolysis of water only takes place

$2H_2O + O_2 + 4e^- \longrightarrow 4OH^-$(RHS)

$2H_2O \longrightarrow 4H^+ + O_2 + 4e^-$ (LHS)

then pH of the

(a) LHS half-cell will increase
(b) RHS half-cell will increase
(c) Both half-cells will increase
(d) Both half-cells will decrease

107 Nernst equation is useful in determining

(a) emf of a cell
(b) equilibrium constant
(c) spontaneity of a cell reaction
(d) All of the above

108 A solution containing one mole per litre each of $Cu(NO_3)_2$, $AgNO_3$, $Hg_2(NO_3)_2$ and $Mg(NO_3)_2$ is being electrolysed using inert electrodes. The values of standard reduction potentials are

$$E^\circ_{Ag^+/Ag} = +0.80\text{ V}, \quad E^\circ_{Hg_2^{2+}/Hg} = +0.79\text{ V}$$
$$E^\circ_{Cu^{2+}/Cu} = +0.34\text{ V}, \quad E^\circ_{Mg^{2+}/Mg} = -2.37\text{ V}$$

with increasing voltage, the sequence of deposition of metals on the cathode will be

(a) Ag, Hg, Cu, Mg (b) Mg, Cu, Hg, Ag
(c) Ag, Hg, Cu (d) Cu, Hg, Ag

109 Which cell will measure standard electrode potential of copper electrode?

(a) Pt (*s*) $|H_2(g, 0.1\text{ bar})|\ H^+(aq, 1\text{ M})\ ||\ Cu^{2+}(aq, 1\text{ M})|$ Cu
(b) Pt (*s*) $|H_2(g, 1\text{ bar})|\ H^+(aq, 1\text{ M})\ ||\ Cu^{2+}(aq, 2\text{ M})|$ Cu
(c) Pt (*s*) $|H_2(g, 1\text{ bar})|\ H^+(aq, 1\text{ M})\ ||\ Cu^{2+}(aq, 1\text{ M})|$ Cu
(d) Pt (*s*) $|H_2(g, 0.1\text{ bar})|\ H^+(aq, 0.1\text{ M})\ ||\ Cu^{2+}(aq, 1\text{ M})|$ Cu

110 If for the half-cell reactions E° values are known

$$Cu^{2+} + e^- \longrightarrow Cu^+, \quad E^\circ = 0.15\ V$$

$$Cu^{2+} + 2e^- \longrightarrow Cu, \quad E^\circ = 0.34\ V$$

Predict which exists in aqueous solution?

(a) Copper (I) sulphate (b) Copper (II) sulphate
(c) Both (a) and (b) (d) None of the above

111 For the following half-cell reactions, E° values are also given

$$Mn^{2+} + 2H_2O \longrightarrow MnO_2 + 4H^+ + 2e^-, \quad E^\circ = -1.23\ V$$

$$MnO_4^- + 4H^+ + 3e^- \longrightarrow MnO_2 + 2H_2O, \quad E^\circ = +1.70\ V$$

Select the correct statement(s).

(a) Mn^{2+} reacts with MnO_4^- in acid solution to form MnO_2
(b) $Mn(MnO_4)_2$ is stable in acid solution
(c) MnO_2 disproportionates to Mn^{2+} and MnO_4^- in acid solution
(d) None of the above is correct

112 Consider the following half-cell reactions,

$$Cu^{2+} + 2e^- \longrightarrow Cu; \quad E^\circ = 0.34\ V$$

$$Zn^{2+} + 2e^- \longrightarrow Zn; \quad E^\circ = -0.76\ V$$

$$SO_4^{2-} + 4H^+ + 2e^- \longrightarrow SO_2 + 2H_2O; \quad E^\circ = 1.17\ V$$

Hence, select the correct statement(s).

(a) Cu reacts with dil. H_2SO_4 forming H_2 and SO_2
(b) Cu reacts with concentrated H_2SO_4 forming SO_2
(c) Zn reacts with concentrated H_2SO_4 forming SO_2
(d) Both (b) and (c)

113 Electrode potential for Mg electrode varies according to the equation **[NCERT Exemplar]**

$$E_{Mg^{2+}/Mg} = E^\ominus_{Mg^{2+}/Mg} - \frac{0.0591}{2}\log\frac{1}{[Mg^{2+}]}$$

The graph of $E_{Mg^{2+}/Mg}$ *vs* $\log[Mg^{2+}]$ is

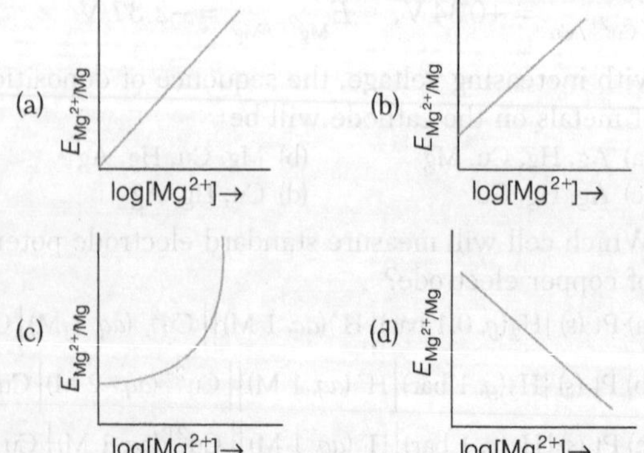

114 Consider the half-cell reaction,

$$O_2(g) + 4H^+(aq) + 4e^- \longrightarrow 2H_2O$$

By what factor are *n*, *Q*, *E* and E° changed, if all the stoichiometric coefficients are multiplied by the factor two?

	n	*Q*	*E*	E°
(a)	doubled	squared	no change	no change
(b)	doubled	no change	doubled	no change
(c)	doubled	doubled	doubled	no change
(d)	doubled	doubled	doubled	doubled

115 The standard reduction potentials of Cu^{2+}/Cu and Cu^+/Cu are 0.339 V and 0.518 V respectively. The standard electrode potential of Cu^{2+}/Cu^+ half-cell is

(a) 0.16 V (b) − 0.179 V
(c) − 0.16 V (d) +0.179 V

116 Which of the following changes will increase the emf of the cell?

$$Co(s) \mid CoCl_2(M_1) \parallel HCl(M_2) \mid Pt(H_2, g)$$

(a) Increase the volume of $CoCl_2$ solution from 100 mL to 200 mL
(b) Increase M_2 from 0.01 M to 0.50 M
(c) Increase the pressure of the $H_2(g)$ from 1.00 to 2.00 atm
(d) Increase M_1 from 0.01 M to 0.50 M

117 For the calomel half-cell, Hg, $Hg_2Cl_2 \mid Cl^-(aq)$ values of electrode potentials are plotted at different $\log[Cl^-]$ variation is represented by

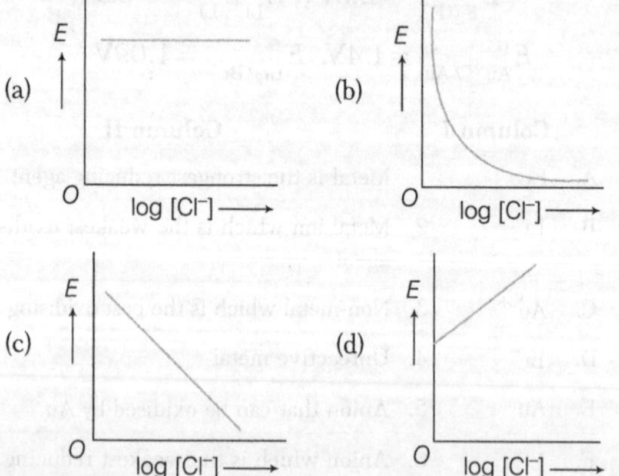

118 For the given cell,

$$Ag \mid AgNO_3\ (1\ M) \parallel AgNO_3\ (2\ M) \mid Ag,\ E_{cell}\ \text{is}$$

(a) 0.0591V
(b) (0.0591 × 0.3010) V
(c) 0.0591 × 2 V
(d) zero

119 For the half-cell

OH OH $\rightleftharpoons$ O O $+ 2H^+ + 2e^-$; $E° = 1.30$ V

At pH = 2, electrode potential is
(a) 1.36 V (b) 1.30 V
(c) 1.42 V (d) 1.20 V

120 For the cell, $Zn \mid Zn^{2+}(1M) \| Cu^{2+}(1\ M) \mid Cu$, $E°_{cell}$ is 1.10 V, $E°_{Cu^{2+}/Cu} = 0.34$ V and for the cell $Cu \mid Cu^{2+}(1M) \| Ag^+(1M) \mid Ag$, $E°_{cell} = 0.46$ V hence, $E°_{cell}$ of the cell
$Zn \mid Zn^{2+}(1M) \| Ag^+(1M) \mid Ag$ is
(a) − 0.04 V (b) + 0.04 V (c) + 0.30 V (d) +1.56 V

121 $Cu^{2+} + 2e^- \longrightarrow Cu$, For this between log ($Cu^{2+}$) *versus* E_{red} is a straight line of intercept 0.34 V then electrode potential of the half-cell Cu/Cu^{2+} (0.1 M) will be
(a) $0.34 + \frac{0.0591}{2}$ (b) $-0.34 - \frac{0.0591}{2}$
(c) 0.34 (d) $-0.34 + \frac{0.0591}{2}$

122 Half-cell reaction for a half-cell,
$Hg(l) + 2OH^-(aq) \longrightarrow HgO(s) + H_2O(l) + 2e^-$
This half-cell is reversible to
(a) $H_2O(l)$ (b) $HgO(s)$
(c) $OH^-(aq)$ (d) All are correct

123 For the half-cell given below
$Pt(H_2) \mid pH = 2$
1 atm
Half-cell potential is
(a) 0.0591 V (b) 0.0295 V
(c) 0.1182 V (d) 0.00 V

124 Quinhydrone half-cell is reversible to
(a) H^+ (b) quinone
(c) quinol (d) None of these

125 A certain electrode has a reduction potential of 0.100 V when measured against a standard calomel electrode (anode); $E°_{calomel} = -0.244$ V. Potential *versus* SHE is
(a) − 0.344 V (b) − 0.144 V
(c) + 0.144 V (d) + 0.344 V

126 Indicator electrode is present in
(a) glass electrode (b) SHE
(c) quinhydrone electrode (d) None of these

127 Match the salt in Column I with corresponding use in Column II.

Column I	Column II
A. Hg_2Cl_2	1. Used in salt-bridge
B. Quinhydrone	2. Measurement of cell constant
C. Lead acetate	3. Redox electrode
D. NH_4NO_3	4. Calomel electrode
E. 0.1 N KCl	5. Used in platinising solution

Choose the correct match.

	A	B	C	D	E
(a)	2	4	1	5	3
(b)	3	5	4	2	1
(c)	4	3	5	1	2
(d)	5	1	2	3	4

128 Match the electrode in Column I with its general name in Column II.

Column I	Column II
A. Calomel	1. Reference
B. Glass	2. Redox
C. Hydrogen	3. Membrane
D. Quinhydrone	4. Gas

Choose the correct match.

	A	B	C	D
(a)	1	3	4	2
(b)	2	4	1	3
(c)	3	2	4	1
(d)	4	1	3	2

VI. Gibbs's energy of electrochemical reactions

129 Temperature coefficient of the emf of the cell in terms of entropy change is
(a) $\frac{\Delta S}{nF}$ (b) $\frac{\Delta S}{T}$ (c) $-\frac{\Delta S}{nF}$ (d) $\frac{\Delta S}{n}$

130 Which of the following statements is correct ? **[NCERT Exemplar]**
(a) E_{cell} and $\Delta_r G$ of cell reaction both are extensive properties.
(b) E_{cell} and $\Delta_r G$ of cell reaction both are intensive properties.
(c) E_{cell} is an intensive property while $\Delta_r G$ of cell reaction is an extensive property.
(d) E_{cell} is an extensive property while $\Delta_r G$ of cell reaction is an intensive property.

131 For $Pt(H_2) \mid H^+(1\ M) \mid Pt(H_2)$ (where p_1 and p_2 are pressures) cell reaction will be spontaneous, if

(a) $p_1 = p_2$ (b) $p_1 > p_2$
(c) $p_2 > p_1$ (d) $p_1 = 1$ atm

132 For the fuel-cell reaction, $2H_2 + O_2 \longrightarrow 2H_2O$, $\Delta G° = -475$ kJ. Hence, $E°_{cell}$ is

(a) 1.23 V (b) 2.46 V
(c) 0.615 V (d) 0.31 V

133 The emf of a galvanic cell for spontaneous cell reaction with $E°_{Al^{3+}/Al} = -1.66$ V and $E°_{Mg^{2+}/Mg} = -2.37$ V and that is

(a) − 0.71 V (b) + 1.42 V
(c) 0.71 V (d) − 1.42 V

134 For the electrochemical cell, $(M/M^+) \mid\mid (X^-/X)$, $E°(M^+/M) = 0.44$ V and $E°_{X^-/X} = 0.33$ V

From this data one can conclude that

(a) $M + X \longrightarrow M^+ + X^-$ is the spontaneous reaction
(b) $M^+ + X^- \longrightarrow M + X$ is the spontaneous reaction
(c) $E°_{cell} = 0.77$ V
(d) $E°_{cell} = -0.77$ V

135 The rusting of iron takes place as follows

$$2H^+ + 2e^- + \frac{1}{2}O_2 \longrightarrow H_2O(l);\quad E° = +1.23\ V$$

$$Fe^{2+}(aq) + 2e^- \longrightarrow Fe(s);\quad E° = -0.44\ V$$

Calculate $\Delta G°$ for the net process giving oxidation of Fe is

(a) − 322 kJ mol^{-1} (b) −161 kJ mol^{-1}
(c) − 152 kJ mol^{-1} (d) − 76 kJ mol^{-1}

136 A galvanic cell with standard oxidation electrode potential of zinc = + 0.76 V and of copper = − 0.34 V. The emf with $\Delta G°$ negative is

(a) + 076 V (b) + 0.34 V (c) − 1.1V (d) +1.1 V

137 Consider certain observations about electrolysis.

I. Electric current is used to drive a non-spontaneous reaction.
II. E_{cell} is positive and ΔG is negative.
III. Cations and anions are discharged at the cathode and anode respectively.
IV. Over-voltage is responsible that a particular reaction takes place.

Correct observations are

(a) I, II and III (b) II, III and IV
(c) I, III and IV (d) I, II and IV

138 For the net cell reaction of the cell $Zn(s) \mid Zn^{2+} \mid Cd^{2+} \mid Cd(s)$, $\Delta G°$ in kilo Joules at 25°C is ($E°_{cell} = 0.360$ V)

(a) 112.5 (b) 69.48 (c) -34.73 (d) -69.48

139 For the cell reaction,

$$Cu^{2+}(C_1) + Zn(s) \rightleftharpoons Zn^{2+}(C_2) + Cu(s)$$

of the electrochemical cell, the change in free energy, at a given temperature is function of

(a) $\ln(C_1)$ (b) $\ln(C_2/C_1)$ (c) $\ln(C_2)$ (d) $\ln(C_1 + C_2)$

140 For the following cell-reaction, $Cu(s) + 2Ag^+(aq) \longrightarrow 2Ag(s) + Cu^{2+}(aq)$, $E°_{cell}$ is 0.46 V at 300 K and 0.48 V at 310 K. Hence, entropy change is

(a) 193 JK^{-1} (b) 386 JK^{-1} (c) 579 JK^{-1} (d) − 193 JK^{-1}

141 For the following cell-reaction,

$$2Ag^+ + Cu \longrightarrow Cu^{2+} + 2Ag$$

temperature coefficient is 2×10^{-3} V / K. Hence, entropy change is

(a) − 3.86 J / K (b) 3.86 J/K
(c) − 386 J/K (d) 386 J/K

142 $Al^{3+}(aq) + 3e^- \longrightarrow Al(s);\quad E° = -1.66$ V

$Cu^{2+}(aq) + 2e^- \longrightarrow Cu(s);\quad E° = +0.34$ V

What voltage is produced under standard conditions to give a spontaneous reactions by combination of these two half-cells?

(a) 1.32 V (b) -1.32 V (c) 2.00 V (d) -2.00 V

143 $E°_{red}$ (standard reduction electrode potentials) of different half-cells are given

$$E°_{Cu^{2+}/Cu} = 0.34\ V;\quad E°_{Zn^{2+}/Zn} = -0.76\ V$$

$$E°_{Ag^+/Ag} = 0.80\ V;\quad E°_{Mg^{2+}/Mg} = -2.37\ V$$

In which cell is $\Delta G°$ most negative?

(a) $Zn \mid Zn^{2+}$ (1 M)$\mid\mid Mg^{2+}$(1 M)$\mid Mg$
(b) $Zn \mid Zn^{2+}$(1 M)$\mid\mid Ag^+$(1 M)$\mid Ag$
(c) $Cu \mid Cu^{2+}$(1 M)$\mid\mid Ag^+$(1 M)$\mid Ag$
(d) $Ag \mid Ag^+$(1 M)$\mid\mid Mg^{2+}$ (1 M)$\mid Mg$

144 Consider the following statements.

Statement I For three half-cell reactions involving different number of electrons

$$E_3 = E_1 + E_2$$

Statement II For three half-cells, $\Delta G_3 = \Delta G_1 + \Delta G_2$

(a) Both statements I and II are correct
(b) Only statement I is correct
(c) Only statement II is corect
(d) Both statements I and II are incorrect

VII. Corrosion and commercial cells of batteries

145 In corrosion of iron

(a) an electrochemical (galvanic cell) is formed in which Fe acts as anode and cathode where O_2 is reduced
(b) electrons flow from anode to cathode through the metal while ions flow through the water droplets
(c) dissolved O_2 oxidises Fe^{2+} to Fe^{3+} before it is deposited as rust ($Fe_2O_3 \cdot H_2O$).
(d) All of the above take place

146 The formation of rust on the surface of iron occurs through the reactions

(a) $Fe(s) \longrightarrow Fe^{2+}(aq) + 2e^-$ at anode
(b) $O_2(g) + 4H^+(aq) + 4e^- \longrightarrow 2H_2O(l)$ at cathode
(c) $Fe^{2+}(aq) + O_2(g) + 4H_2O(l) \longrightarrow 2Fe_2O_3(s) + 8H^+$
(d) All of the above take place

147 In mercury battery, cathode is made up of

(a) Zn
(b) ZnO in contact with steel
(c) steel in contact with HgO
(d) Zn in contact with HgO

148 While charging the lead storage battery .... **[NCERT Exemplar]**

(a) $PbSO_4$ anode is reduced to Pb
(b) $PbSO_4$ cathode is reduced to Pb
(c) $PbSO_4$ cathode is oxidised to Pb
(d) $PbSO_4$ anode is oxidised to PbO_2

149 Some of the batteries have alkaline electrolytes

I. Mercury battery
II. Nickel-cadmium
III. Modified Leclanche cell

Cell potential is independent of $[OH^-]$ in

(a) I, II and III (b) I and II
(c) II and III (d) Only III

150 During discharging of lead-storage acid battery following reaction takes place

$$Pb(s) + PbO_2(s) + 2H_2SO_4 \longrightarrow 2PbSO_4(s) + 2H_2O$$

If 2.5 A of current is drawn for 965 min H_2SO_4 consumed is

(a) 0.75 mol (b) 3.00 mol
(c) 1.50 mol (d) 4.50 mol

151 For hydrogen-oxygen fuel cell with reaction

$$H_2(g) + \frac{1}{2}O_2(g) \longrightarrow H_2O(l)$$

$$\Delta G_f^\circ(H_2O) = -237.2 \text{ kJ mol}^{-1}$$

Hence, emf of the fuel cell is

(a) +2.46 V (b) − 2.46 V
(c) + 1.23 V (d) − 1.23 V

152 When a lead-storage battery is recharged

(a) Pb is formed
(b) H_2SO_4 is formed
(c) H_2SO_4 is consumed
(d) Both (a) and (b) both will occur

153 Select the correct statement(s).

(a) In lead-storage battery, galvanic cells are linked in series.
(b) Cathode and anode compartments are not separated in a battery as oxidising agents and reducing agents both are solids.
(c) Recharging of a storage battery is a non-spontaneous process.
(d) All the above are correct statements

154 "Maintenance-free" batteries now is use in place of common batteries have

(a) electrodes made of lead-lead oxide
(b) electrodes made of calcium-containing lead alloy
(c) non-aqueous solvents as medium
(d) platinum electrodes

155 When a lead storage-battery is discharged

(a) SO_2 is evolved
(b) lead is formed
(c) lead sulphate is formed
(d) sulphuric acid is consumed

156 Aluminium-air battery is also called

(a) flow-battery
(b) Leclanche cell
(c) dry cell
(d) fuel cell

157 For a Ag-Zn button cell, net reaction

$$Zn(s) + Ag_2O(s) \longrightarrow ZnO(s) + 2Ag(s)$$

$$\Delta G_f^\circ(Ag_2O) = -11.21 \text{ kJ mol}^{-1}$$

$$\Delta G_f^\circ(ZnO) = -318.3 \text{ kJ mol}^{-1}$$

Hence, E°_{cell} of the button cell is

(a) 3.182 V (b) 1.71 V
(c) -1.591 V (d) 1.591 V

158 Consider the following batteries/cells.

I. Dry-cell battery
II. Lead-storage (acid) battery
III. Nickel-cadmium battery
IV. Lithium battery
V. Fuel cell

Choose the batteries that are rechargable and select the correct option

(a) I, II and III
(b) II, III and IV
(c) III, IV and V
(d) II, III and V

159 Match the items of Column I and Column II. **[NCERT Exemplar]**

Column I	Column II
A. Lead storage battery	1. Maximum efficiency
B. Mercury cell	2. Prevented by galvanisation
C. Fuel cell	3. Gives steady potential
D. Rusting	4. Pb is anode, PbO_2 is cathode

Choose the correct match.

	A	B	C	D		A	B	C	D
(a)	1	2	4	3	(b)	2	4	1	3
(c)	3	1	4	2	(d)	4	3	1	2

160 Match the items of Column I and Column II. **[NCERT Exemplar]**

Column I	Column II
A. Leclanche cell	1. Cell reaction : $2H_2 + O_2 \longrightarrow 2H_2O$
B. Ni-Cd cell	2. Does not involve any ion in solution and is used in hearing aids
C. Fuel cell	3. Rechargeable
D. Mercury cell	4. Reaction at anode, $Zn \longrightarrow Zn^{2+} + 2e^-$
	5. Converts energy of combustion into electrical energy

Choose the correct match.

	A	B	C	D		A	B	C	D
(a)	5	1	3	4	(b)	2	4	5	3
(c)	3	5	2	1	(d)	4	3	1	2

161 Statement I Mercury cell does not give steady potential. **[NCERT Exemplar]**

Statement II In the cell reaction, ions are not involved in solution.

Consider the following statements.

(a) Both statements I and II are correct
(b) Only statement I is correct
(c) Only statement II is corect
(d) Both statements I and II are incorrect

(B) Miscellaneous questions

1 One mole of Ag_2CO_3 is strongly heated in an open vessel. Residue will be

(a) 1 mole of Ag_2O
(b) 2 moles of Ag
(c) 1 mole of Ag_2O and 1 mole of CO_2
(d) 2 moles of Ag, 0.5 mole of O_2 and 1 mole of CO_2

2 Consider the following statements.

Statement I The mobility of sodium ion is lower than that of potassium ion.

Statement II The ionic mobilities depend upon the effective radius of the ion.

(a) Both statements I and II are correct
(b) Only statement I is correct
(c) Only statement II is corect
(d) Both statements I and II are incorrect

3 If all species are in their standard states, which of the following is the strongest oxidising agent?

(a) Br^- (b) Zn^{2+} (c) Co^{3+} (d) Fe^{2+}

4 If $E^\circ_{Fe^{2+}/Fe}$ is x_1, $E^\circ_{Fe^{3+}/Fe}$ is x_2 then $E^\circ_{Fe^{3+}/Fe^{2+}}$ will be

(a) $3x_2 - 2x_1$ (b) $x_2 - x_1$
(c) $x_2 + x_1$ (d) $2x_1 + 3x_2$

5 For this cell $Zn\,|\,Zn^{2+}(C_1)\,||\,Zn^{2+}(C_2)\,|\,Zn$ ΔG is negative, if

(a) $C_1 = C_2$ (b) $C_1 > C_2$
(c) $C_2 > C_1$ (d) None of these

6 Match the terms in Column I with the terms in Column II.

Column I	Column II
A. Weak electrolyte	1. Copper
B. Strong electrolyte	2. Quinhydrone
C. Insulator	3. Sodium chloride
D. Electronic conductor	4. Buffer
E. Indicator electrode	5. Acetic acid

Choose the correct match.

	A	B	C	D	E
(a)	5	3	4	1	2
(b)	4	5	2	3	1
(c)	3	1	2	5	4
(d)	2	4	5	1	3

7 $\underset{x \text{ atm}}{Pt(H_2)}\,|\,0.01\text{ M }H^+\,||\,0.1\text{ M }H^+\,|\,\underset{y \text{ atm}}{Pt(H_2)}$.

If $E_{cell} = 0.00$ V then $\frac{x}{y}$ is

(a) 100 (b) 10 (c) 0.01 (d) 0.1

8 For the following cell $Zn\,|\,Zn^{2+}\,||\,Cd^{2+}\,|\,Cd$, $E_{cell} = 0.30$ V, $E^\circ_{cell} = 0.36$ V, then value of $\frac{[Cd^{2+}]}{[Zn^{2+}]}$ is

(a) 10 (b) 0.01 (c) 0.1 (d) 100

9 One litre of 1 M $CuSO_4$ solution is electrolysed. After passing 2 F of electricity, molarity of $CuSO_4$ solution will be

(a) M/2 (b) M/4 (c) M (d) 0

10 $2Hg(l) + 2Cl^-(aq) \longrightarrow Hg_2Cl_2(s) + 2e^-$, half-cell will be

(a) Hg, Hg_2Cl_2 | Cl^- (b) Cl^- | Hg, Hg_2Cl_2
(c) Hg | Cl^- (d) None of these

11 $2Ce^{4+} + Co \longrightarrow 2Ce^{3+} + Co^{2+}$; $E^\circ_{cell} = 1.89$ V, $E^\circ_{Co^{2+}|Co} = -0.28$ V, hence, $E^\circ_{Ce^{4+}/Ce^{3+}}$ is

(a) 0.805 V (b) 1.61 V
(c) −0.805 V (d) −1.61 V

12 Equivalent conductance of saturated $BaSO_4$ is 400 Ω^{-1} cm^2 equiv^{-1} and specific conductance is $8 \times 10^{-5}\ \Omega^{-1}$ cm^{-1}. Hence, K_{sp} of $BaSO_4$ is

(a) 4×10^{-8} M^2 (b) 1×10^{-8} M^2
(c) 2×10^{-4} M^2 (d) 1×10^{-4} M^2

13 Specific conductance of 0.01 N KCl solution is $x\ \Omega^{-1}$ cm^{-1} having conductance $y\ \Omega^{-1}$. Specific conductance of 0.01 N NaCl having conductance $z\ \Omega^{-1}$ is

(a) $\frac{yx}{z}$ (b) $\frac{zx}{y}$
(c) $\frac{z}{yx}$ (d) None of these

14 What is the value of reaction quotient, Q for the cell Ni(s) | Ni^{2+} (0.190 M) || Cl^- (0.40 M) |Pt(s) Cl_2(g, 1.0 atm)

(a) 3.1×10^{-1} (b) 1.3×10^{-1}
(c) 8.0×10^{-2} (d) 3.04×10^{-2}

15 18 g water is to be decomposed electrolytically into H_2 and O_2 by a current of 3 A. Time required is approximately

(a) 17.87 h (b) 35.94 h (c) 8.94 h (d) 26.80 h

16 A 5 A current is passed through a solution of zinc sulphate for 40 min. Find the amount of zinc deposited at the cathode

(a) 40.41 g (b) 4.041 g (c) 0.4041 g (d) 65.04 g

17 0.04 N solution of a weak acid has specific conductance 4.23×10^{-4} mho cm^{-1} and degree of dissociation is 0.0612. The equivalent conductance (ohm^{-1} cm^2 equiv^{-1}) of weak acid at infinite dilution is

(a) 1.72 (b) 17.29 (c) 142.27 (d) 172.79

18 Through a solution of $CuSO_4$ a current of 3 A was passed for 2 h. At cathode 3 g of Cu^{2+} ions were discharged. The current efficiency is [Atomic weight of Cu = 63.5]

(a) 33% (b) 42.2% (c) 48.7% (d) 54.4%

19 Consider the following statements.

Statement I Electrolysis of NaCl solution gives chlorine at anode instead of O_2. **[NCERT Exemplar]**

Statement II Formation of oxygen at anode requires over voltage.

(a) Both statements I and II are correct
(b) Only statement I is correct
(c) Only statement II is corect
(d) Both statements I and II are incorrect

20 If the following half-cells have the E° values as

$Fe^{3+} + e^- \longrightarrow Fe^{2+}$; $E = +0.77$ V

and $Fe^{2+} + 2e^- \longrightarrow Fe$; $E^\circ = -0.44$ V,

the E° of the half-cell $Fe^{3+} + 3e^- \longrightarrow Fe$ will be

(a) 0.605 V (b) 0.33 V
(c) − 1.21 V (d) − 0.037 V

21 For the half-cells Tl^{3+} / Tl^+, Tl^+ / Tl, E° values are 1.26 V and − 0.336 V respectively. Hence, for the half-cell Tl^{3+} / Tl, E° is

(a) 0.728 V (b) 2.184 V
(c) 0.924 V (d) 1.596 V

22 The ionic conductance of H^+ and SO_4^{2-} are 350 and 80 S cm^2 equiv^{-1}, hence equivalent conductance (S cm^2 equivalent^{-1}) and molar conductance (S cm^2 mol^{-1}) of H_2SO_4 will be

(a) 430, 430 (b) 860, 430
(c) 215, 430 (d) 430, 860

23 Consider the following half-cell reactions

$Fe^{2+} \longrightarrow Fe^{3+} + e^-$ $E^\circ = -0.77$ V

$O_2 + 4H^+ + 4e^- \longrightarrow 2H_2O$ $E^\circ = 1.229$ V

$Fe \longrightarrow Fe^{2+} + 2e^-$ $E^\circ = 0.44$ V

Select the correct statement(s).

(a) Fe^{2+} solutions are not stable in air
(b) Fe nails can prevent oxidation of Fe^{2+} to Fe^{3+} by air
(c) Corrosion of Fe takes place by air
(d) All of the above

24 The value of the reaction quotient, Q, for the cell Zn(s)| Zn^{2+} (0.01 M)|| Ag^+ (1.25 M)| Ag(s) is

(a) 156 (b) 125
(c) 1.25×10^{-2} (d) 6.40×10^{-3}

25 What products are formed during the electrolysis of a concentrated aqueous solution of sodium chloride?

I. $Cl_2(g)$ II. NaOH(aq) III. $H_2(g)$

(a) Only I (b) Both I and II
(c) Both I and III (d) I, II and III

26 Which of the following statements about the spontaneous reaction occurring in a galvanic cell is always true?

(a) $E^\circ_{cell} > 0, \Delta G^\circ < 0$ and $Q < K_{eq}$

(b) $E^\circ_{cell} > 0, \Delta G^\circ < 0$ and $Q > K_{eq}$

(c) $E^\circ_{cell} > 0, \Delta G^\circ > 0$ and $Q > K_{eq}$

(d) $E_{cell} > 0, \Delta G < 0$ and $Q < K_{eq}$

27 100 g Al is to be deposited on the cathode of the electrolytic cell containing Al_2O_3 in the solvent Na_3AlF_6 by a current of 125 A. If at the cathode, reaction is

$$Al^{3+} + 3e^- \longrightarrow Al$$

then time required for the passage of current is (Al = 27)

(a) 143 min (b) 429 min
(c) 48 min (d) 286 min

28 When Al_2O_3 is electrolysed, aluminium is produced at one electrode and oxygen gas at the other electrode. For a given quantity of electricity, ratio of number of moles of aluminium to the number of moles of oxygen gas is

(a) 1:1 (b) 2:1 (c) 2:3 (d) 4:3

29 1 Faraday of electricity is passed through the solutions containing 1 mole each of $CuSO_4$, $AlCl_3$ and $SiCl_4$. Number of moles of Cu, Al and Si formed will be in the ratio

(a) 1:1:1 (b) 2:3:4 (c) 6:4:3 (d) 3:4:6

30 If the Pb^{2+} concentration is maintained at 1.0 M. What is the $[Cu^{2+}]$ when the cell potential drops to zero? $E^\circ_{cell} = 0.473$ V,

$Pb | Pb^{2+}(1.0\ M) || Cu^{2+}(1.0 \times 10^{-4}\ M) | Cu(s)$

(a) 1×10^{-16} M (b) 1×10^{16} M
(c) 1.0×10^{-14} M (d) 1.0×10^{14} M

31 pH of the solution in the anode compartment of the following cell at 25°C is x when

$E_{cell} - E^\circ_{cell} = 0.0591$ V,

$\underset{1\ atm}{Pt(H_2)} | pH = x || Ni^{2+}(1\ M) | Ni$, x is

(a) 4 (b) 3 (c) 2 (d) 1

32 For the half-cell reaction,

$2H_2O + 2e^- \longrightarrow H_2 + 2OH^-$,

$E^\circ = -0.8277$ V at 298 K. Hence, autoprotolysis constant for water is

(a) 1×10^{-10} (b) 1×10^{-12}
(c) 1×10^{-13} (d) 1×10^{-14}

33 Consider the following statements.

Statement I For measuring resistance of an ionic solution an AC source is used. **[NCERT Exemplar]**

Statement II Concentration of ionic solution will change if DC source is used.

(a) Both statements I and II are correct
(b) Only statement I is correct
(c) Only statement II is corect
(d) Both statements I and II are incorrect

34 Which statement is true and about a galvanic cell employing Pb, Cu, Pb^{2+} and Cu^+?

$E^\circ_{Pb^{2+}/Pb} = -0.127$ V; $E^\circ_{Cu^+/Cu} = +0.518$ V

(a) Spontaneous cell-reaction will be in the cell $Pb | Pb^{2+} || Cu^+ | Cu$
(b) $E^\circ_{cell} = 0.645$ V
(c) Both (a) and (b) are correct
(d) None of the above is correct

35 Consider the following cell reaction,

$2Fe(s) + O_2(g) + 4H^+(aq) \longrightarrow 2Fe^{2+}(aq) + 2H_2O(l)$, $E^\circ = 1.67$ V

At $[Fe^{2+}] = 1 \times 10^{-3}$ M, $p_{O_2} = 0.1$ atm and pH = 3, the cell potential at 298 K is

(a) 1.47 V (b) 1.77 V
(c) 1.87 V (d) 1.57 V

36 Consider the following equations for a cell reaction

$A + B \rightleftharpoons C + D$ $E^\circ = x$ volt, $K_{eq} = K_1$

$2A + 2B \rightleftharpoons 2C + 2D$ $E^\circ = y$ volt, $K_{eq} = K_2$

then

(a) $x = y, K_1 = K_2$ (b) $x = 2y, K_1 = 2K_2$
(c) $x = y, K_1^2 = K_2$ (d) $x^2 = y, K_1^2 = K_2$

37 $10Cl^-(aq) + 2MnO_4^-(aq) + 16H^+(aq) \rightleftharpoons 5Cl_2(g) + 2Mn^{2+}(aq) + 8H_2O(l)$

For this reaction, $E^\circ = 0.15$ V at 298 K. Hence, equilibrium constant is

(a) 2.4×10^{25} (b) 4.9×10^{12}
(c) 1.2×10^5 (d) 3.4×10^2

38 $Ag | \underset{1\ L\ solution}{Ag^+(1\ M)} || \underset{1\ L\ solution}{Ag^+(2M)} | Ag$, 0.5 F of electricity in the LHS (anode) and 1 F electricity in the RHS (cathode) is first passed making them independent electrolytic cells at 298 K. EMF of the cell after electrolysis will be

(a) increased
(b) decreased
(c) no change
(d) time is also required

39 In acidic medium, MnO_4^- is an oxidising agent.
$MnO_4^- + 8H^+ + 5e^- \longrightarrow Mn^{2+} + 4H_2O$.

If H^+ ion concentration is doubled, electrode potential of the half-cell MnO_4^-, Mn^{2+}/Pt will

(a) increase by 28.46 mV (b) decrease by 28.46 mV
(c) increase by 14.23 mV (d) decrease by 142.30 mV

40 During electrolysis of acidified water, O_2 gas is formed at the anode. To produce O_2 gas at the anode at the rate of 0.224 cc per second at STP, current passed is

(a) 0.224 A (b) 2.24 A
(c) 9.64 A (d) 3.86 A

41 Consider the cell,

$$Cd(s) \mid Cd^{2+}(1.0\text{ M}) \parallel Cu^{2+}(1.0\text{ M}) \mid Cu(s).$$

If we want to make a cell with a more positive voltage using the same substances, we should

(a) increase both the $[Cd^{2+}]$ and $[Cu^{2+}]$ to 2.00 M
(b) increase only the $[Cd^{2+}]$ to 2.00 M
(c) decrease both the $[Cd^{2+}]$ and $[Cu^{2+}]$ to 0.10 M
(d) decrease only the $[Cd^{2+}]$ to 0.100 M

42 The standard electrode potentials of Fe^{2+}/Fe and Sn^{2+}/Sn electrodes are - 0.44 and -0.14 V respectively. For the reaction,

$Fe + Sn^{2+} \longrightarrow Fe^{2+} + Sn$, the standard emf is

(a) -0.58 V (b) -0.30 V
(c) + 0.30 V (d) + 0.58 V

43 $E^\circ_{Zn^{2+}/Zn}$ and $E^\circ_{Cl_2/Cl^-}$ are -0.76 V and 1.36 V respectively. the emf of the cell

$$Zn \mid Zn^{2+}\ (1\text{ M}) \parallel Cl^-\ (1\text{ M}) \mid \underset{(1\text{ atm})}{Cl_2(g)}, Pt \text{ is}$$

(a) 0.6 V (b) + 2.12 V
(c) -2.12 V (d) -0.6 V

44 Standard reduction potentials at 25°C or $Li^+ \mid Li$, $Ba^{2+} \mid Ba$, $Na^+ \mid Na$ and $Mg^{2+} \mid Mg$ are -3.05, -2.90, -2.71, and -2.37 V respectively. Which one of the following is the strongest oxidising agent?

(a) Na^+ (b) Li^+ (c) Ba^{2+} (d) Mg^{2+}

45 Oxidation potential for the following half-cell reactions are

$Zn \rightleftharpoons Zn^{2+} + 2e^-$; $E^\circ = 0.76$ V

$Fe \rightleftharpoons Fe^{2+} + 2e^-$; $E^\circ = +0.44$ V

The emf for the cell reaction,
$Fe^{2+}(aq) + Zn \longrightarrow Zn^{2+}(aq) + Fe$ will be

(a) -120 V (b) + 0.32 V
(c) - 0.32 V (d) -1.20 V

46 E° for the cell, $Zn \mid Zn^{2+}(aq) \parallel Cu^{2+}(aq) \mid Cu$ is 1.10 V at 25°C. The equilibrium constant for the reaction,

$$Zn(s) + Cu^{2+}(aq) \rightleftharpoons Cu(s) + Zn^{2+}(aq)$$

is of the order

(a) 10^{37} (b) 10^{28} (c) 10^{18} (d) 10^{17}

47 Consider half-cell reactions,

$$H_2O_2 \longrightarrow O_2 + 2H^+ + 2e^-; E^\circ = -0.69 \text{ V}$$
$$H_2O_2 + 2H^+ + 2e^- \longrightarrow 2H_2O; \quad E^\circ = 1.77 \text{ V}$$
$$Fe^{2+} \longrightarrow Fe^{3+} + e^-; \quad E^\circ = 0.771 \text{ V}$$
$$Fe \longrightarrow Fe^{2+} + 2e^-; E^\circ = 0.44 \text{ V}$$

If H_2O_2 is mixed with Fe^{2+}, which reaction is more likely?

(a) Oxidation of Fe^{2+} to Fe^{3+}
(b) Reduction of Fe^{2+} to Fe
(c) Both of the reactions
(d) None of the above reactions

48 Rusting of iron takes place in

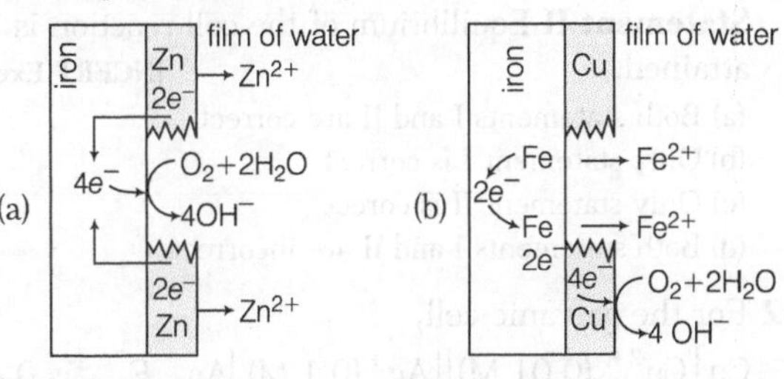

(c) Both (a) and (b) (d) None of these

49 $Zn + Cu^{2+}(aq) \rightleftharpoons Cu + Zn^{2+}(aq)$ reaction quotient is $Q = \frac{[Zn^{2+}]}{[Cu^{2+}]}$. Variation of E_{cell} with log Q is of the type with $OA = 1.10$ V. E_{cell} will be 1.1591 V when

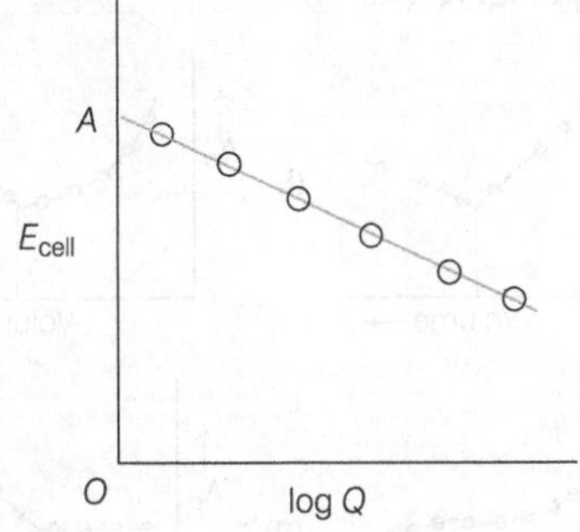

(a) $\frac{[Cu^{2+}]}{[Zn^{2+}]} = 0.01$ (b) $\frac{[Zn^{2+}]}{[Cu^{2+}]} = 0.01$
(c) $\frac{[Zn^{2+}]}{[Cu^{2+}]} = 0.1$ (d) $\frac{[Zn^{2+}]}{[Cu^{2+}]} = 1$

50 Match the items of Column I and Column II. **[NCERT Exemplar]**

Column I	Column II
A. κ	1. $i \times t$
B. Λ_m	2. $\Lambda_m / \Lambda_m^\circ$
C. α	3. $\dfrac{\kappa}{C}$
D. Q	4. $\dfrac{G^*}{R}$

Choose the correct match.

	A	B	C	D
(a)	4	3	2	1
(b)	3	4	1	2
(c)	2	1	4	3
(d)	3	2	1	4

51 Consider the following statements.

Statement I Current stops flowing, when $E_{cell} = 0$.

Statement II Equilibrium of the cell reaction is attained. **[NCERT Exemplar]**

(a) Both statements I and II are correct
(b) Only statement I is correct
(c) Only statement II is corect
(d) Both statements I and II are incorrect

52 For the galvanic cell,

$\underset{\ominus}{Cu}|Cu^{2+}$ (0.01 M)$||Ag^+$(0.1 M) $|\underset{\oplus}{Ag}$, E°_{cell} is 0.46 V

at 298 K. Hence, E_{cell} is

(a) 0.460 V (b) 0.519 V
(c) 0.401 V (d) 0.430 V

53 $AgNO_3$ (aq) was added to an *aq.* KCl solution gradually and the conductivity of the solution was measured. The plot of conductance (Λ_C) *vs* the volume of $AgNO_3$ is

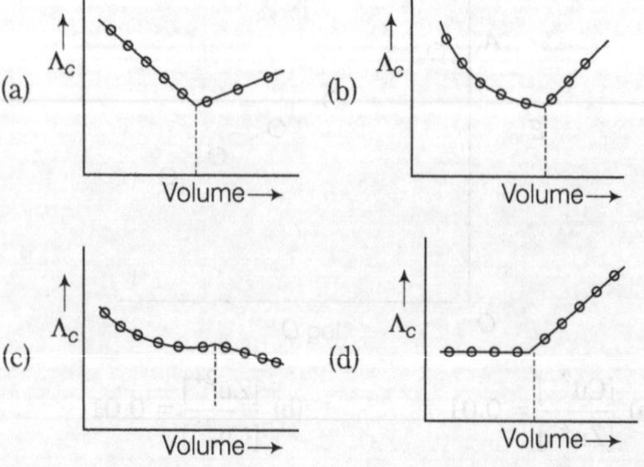

54 Match the terms given in Column I with the items given in Column II. **[NCERT Exemplar]**

Column I	Column II
A. Λ_m	1. Intensive property
B. $E^\ominus_{cell}$	2. Depends on number of ions/volume
C. κ	3. Extensive property
D. $\Delta_r G_{cell}$	4. Increases with dilution

Choose the correct match.

	A	B	C	D
(a)	1	4	2	3
(b)	2	3	4	1
(c)	3	4	1	2
(d)	4	1	2	3

(C) Medical entrances' gallery

1 On electrolysis of dilute sulphuric acid using platinum (Pt) electrode, the product obtained at anode will be **[NEET 2020]**

(a) oxygen gas (b) H_2S gas
(c) SO_2 gas (d) hydrogen gas

2 The number of Faradays (F) required to produce 20 g of calcium from molten $CaCl_2$ (Atomic mass of Ca = 40 g mol^{-1}) is **[NEET 2020]**

(a) 2 (b) 3 (c) 4 (d) 1

3 Identify the reaction from following having top position in EMF series (Standard reduction potential) according to their electrode potential at 298 K. **[NEET 2020]**

(a) $Mg^{2+} + 2e^- \longrightarrow Mg(s)$ (b) $Fe^{2+} + 2e^- \longrightarrow Fe(s)$
(c) $Au^{3+} + 3e^- \longrightarrow Au(s)$ (d) $K^+ + 1e^- \longrightarrow K(s)$

4 For the cell reaction,

$2Fe^{3+}(aq) + 2I^-(aq) \longrightarrow 2Fe^{2+}(aq) + I_2(aq)$

$E^\ominus_{cell} = 0.24$ V at 298 K. The standard Gibb's energy ($\Delta_r G^\ominus$) of the cell reaction is **[NEET 2019]**

[Given that, Faraday constant $F = 96500$ C mol^{-1}]

(a) -23.16 kJ mol^{-1} (b) 46.32 kJ mol^{-1}
(c) 23.16 kJ mol^{-1} (d) -46.32 kJ mol^{-1}

5 For a cell involving one electron, $E^\circ_{cell} = 0.59$ V at 298 K, the equilibrium constant for the cell reaction is **[NEET 2019]**

$$\left[\text{Given that, } \frac{2.303\,RT}{F} = 0.059 \text{ V at } T = 298 \text{ K}\right]$$

(a) 1.0×10^5 (b) 1.0×10^{10}
(c) 1.0×10^{30} (d) 1.0×10^2

6 Cell notation, $M|\underset{0.01}{M^{2+}}||\underset{0.0001}{M^{2+}}|M$. If value of $E°_{cell}$ is 4 volt (Given, $\frac{RT}{F}\log 10 = 0.06$) **[AIIMS 2019]**

(a) 3.94 V (b) 4.06 V (c) 2.03 V (d) 8.18 V

7 The standard electrode potential (E^{-}) values of, (Al^{3+} / Al, Ag^{+} / Ag, K^{+} /K and Cr^{3+} / Cr are − 166 V, 0.80V, 2.93 V and − 0.74 V, respectively. The correct decreasing order of reducing power of the metal is **[NEET (Odisha) 2019]**

(a) Ag > Cr > Al > K (b) K > Al > Cr > Ag
(c) K > Al > Ag > Cr (d) Al > K > Ag > Cr

8 Given, $E^{\circ}_{Cr^{3+}/Cr} = -0.72$ V, $E^{\circ}_{Fe^{2+}/Fe} = -0.42$ V, The potential for the cell $Cr|Cr^{3+}$ (0.1 M) || Fe^{2+} (0.01 M) | Fe is **[JIPMER 2019]**

(a) − 0.339 V (b) − 0.26 V
(c) 0.26 V (d) 0.3 V

9 Following limiting molar conductivities are given as,

$\Lambda^{\circ}_{m(H_2SO_4)} = x$ S $cm^2 mol^{-1}$ **[NEET Odisha 2019]**

$\Lambda^{\circ}_{m(K_2SO_4)} = y$ S $cm^2 mol^{-1}$

$\Lambda^{\circ}_{m(CH_3COOK)} = z$ S $cm^2 mol^{-1}$

$\Lambda^{\circ}_{m\,(in\ S\ cm^2\ mol^{-1})}$ for CH_3COOH will be

(a) $x - y + 2z$ (b) $x + y - z$
(c) $x - y + z$ (d) $\frac{(x-y)}{2} + z$

10 For a Ag-Zn button cell, net reaction is

$Zn(s) + Ag_2O(s) \longrightarrow ZnO(s) + 2Ag(s)$

$\Delta G^{\circ}_f (Ag_2O) = -11.21$ kJ mol^{-1}

$\Delta G^{\circ}_f (ZnO) = -318.3$ kJ mol^{-1}

Then, $E°_{cell}$ of the button cell is **[AIIMS 2018]**

(a) 3.182 V (b) −1.621 V (c) 1.591 V (d) −1.591 V

11 The emf of a Daniell cell at 298 K is E_1,

$Zn \mid \underset{(0.01M)}{ZnSO_4} \mid\mid \underset{(1.0M)}{CuSO_4} \mid Cu$

When the concentration of $ZnSO_4$ is 1.0 M and that of $CuSO_4$ is 0.01 M, the emf changed to E_2. What is the relationship between E_1 and E_2?

(a) $E_1 = E_2$ (b) $E_1 < E_2$
(c) $E_2 = 0, E_1 \neq 0$ (d) $E_1 > E_2$

12 If, $E°(Zn^{2+}, Zn) = -0.763$ V and $E°(Fe^{2+}, Fe) = -0.44$ V, then the emf of the cell Zn| Zn^{2+} (a = 0.001)|| Fe^{2+} (a = 0.005)| Fe is **[JIPMER 2018]**

(a) equal to 0.323 V (b) less than 0.323 V
(c) greater than 0.323 V (d) equal to 1.103 V

13 For the electrochemical cell, $Ag^{+}|AgCl|KCl\ ||AgNO_3|Ag^{+}$, the overall cell reaction is **[JIPMER 2018]**

(a) $Ag^{+} + KCl \longrightarrow AgCl(s) + K^{+}$

(b) $Ag + AgCl \longrightarrow 2Ag + \frac{1}{2}Cl_2$

(c) $AgCl(s) \longrightarrow Ag^{+} + Cl^{-}$

(d) $Ag^{+} + Cl^{-} \longrightarrow AgCl(s)$

14 At 25°C, the molar conductance at infinite dilution for the strong electrolytes NaOH, NaCl and $BaCl_2$ are 248×10^{-4}, 126×10^{-4} and 280×10^{-4} $Sm^2 mol^{-1}$ respectively. λ°_m $Ba(OH)_2$ in $Sm^2 mol^{-1}$ is **[AIIMS 2018]**

(a) 362×10^{-4} (b) 402×10^{-4}
(c) 524×10^{-4} (d) 568×10^{-4}

15 If molar conductivity of Ca^{2+} and Cl^{-} ions are 119 and 71 S $cm^2 mol^{-1}$ respectively, then the molar conductivity of $CaCl_2$ at infinite dilution is **[JIPMER 2018]**

(a) 215 S $cm^2 mol^{-1}$ (b) 340 S $cm^2 mol^{-1}$
(c) 126 S $cm^2 mol^{-1}$ (d) 261 S $cm^2 mol^{-1}$

16 In the electrochemical cell :

Zn| $ZnSO_4$ (0.01 M)|| $CuSO_4$ (1.0M)| Cu, the emf of this Daniell cell is E_1. When the concentration of $ZnSO_4$ is changed to 1.0 M and that of $CuSO_4$ changed to 0.01 M, the emf change to E_2. From the following, which one is the relationship between E_1 and E_2 ? $\left(\text{Given, } \frac{RT}{E} = 0.059\right)$ **[NEET 2017]**

(a) $E_2 = 0 \neq E_1$ (b) $E_1 = E_2$
(c) $E_1 < E_2$ (d) $E_1 > E_2$

17 The pressure of H_2 required to make the potential of H_2-electrode zero in pure water at 298 K is **[NEET (I) 2016]**

(a) 10^{-12} atm (b) 10^{-10} atm
(c) 10^{-4} atm (d) 10^{-14} atm

18 The molar conductivity of a 0.5 mol/dm^3 solution of $AgNO_3$ with electrolytic conductivity of 5.76×10^{-3} S cm^{-1} at 298 K is **[NEET (II) 2016]**

(a) 0.086 S cm^2 / mol (b) 28.8 S cm^2 /mol
(c) 2.88 S cm^2 /mol (d) 11.52 S cm^2 /mol

19 During the electrolysis of molten sodium chloride, the time required to produce 0.10 mol of chlorine gas using a current of 3 amperes is **[NEET (II) 2016]**

(a) 220 minutes (b) 330 minutes
(c) 55 minutes (d) 110 minutes

20 If the E°_{cell} for a given reaction has a negative value, which of the following gives the correct relationships for the values of ΔG° and K_{eq} ? **[NEET (II) 2016]**

(a) $\Delta G^\circ < 0; K_{eq} > 1$ (b) $\Delta G^\circ < 0; K_{eq} < 1$
(c) $\Delta G^\circ > 0; K_{eq} < 1$ (d) $\Delta G^\circ > 0; K_{eq} > 1$

21 The number of electrons delivered at the cathode during electrolysis by a current of 1 ampere in 60 seconds is (charge on electron $= 1.60 \times 10^{-19}$ C) **[NEET (II) 2016]**

(a) 3.75×10^{20} (b) 7.48×10^{23}
(c) 6×10^{23} (d) 6×10^{20}

22 A device that converts energy of combustion of fuels like hydrogen and methane, directly into electrical energy is known as **[CBSE AIPMT 2015]**

(a) fuel cell (b) electrolytic cell
(c) dynmo (d) Ni-Cd cell

23 Standard electrode potentials of three metals X, Y and Z are + 0.52 V − 2.87 V and −0.44 V respectively. The reduction power of these metals are **[Kerala (Medical) 2015]**

(a) $X > Y > Z$ (b) $X > Z > Y$
(c) $Y > X > Z$ (d) $Z > X > Y$
(e) $Y > Z > X$

24 Calculate ΔG° for the following cell reaction,

$$Zn(S) + Ag_2O(s) + H_2O(l) \longrightarrow Zn^{2+}(aq) + 2Ag(s) + 2OH^-(aq)$$

$E_{Ag^+/Ag} = +0.80$ V and $E_{Zn^{2+}/Zn} = -0.76$ V **[EAMCET 2015]**

(a) −305 kJ/mol (b) −301 kJ/mol
(c) −305 kJ/mol (d) 301 kJ/mol

25 The value of E°_{red} for metal A, B and C are 0.34 V, −0.80 V and −0.46 V respectively. State the correct order for their ability to act as reducing agent. **[Guj. CET 2015]**

(a) $A < B > C$ (b) $C < B > A$ (c) $B > C > A$ (d) $C < A > B$

26 Two electrolytic cells containing molten solution of nickel chlorine and aluminium chloride are connected in series. If same amount of electric current is passed through them, what will be the weight of nickel obtained when 18 g of a aluminium is obtained? [Al = 27 g/mol, Ni = 58.5 g/mol^{-1}] **[Guj. CET 2015]**

(a) 117 g (b) 58.5 g (c) 29.25 g (d) 5.85 g

27 For the reaction,

$A(s) + 2B^+(aq) \longrightarrow A^{2+}(aq) + 2B(s)$; the E is 1.18 V. Then, the equilibrium constant for the reaction is **[J & K 2015]**

(a) 10^{10} (b) 10^{20} (c) 10^{40} (d) 10^{60}

28 If $E_{M^+/M} = 1.2$ V, $E^\circ_{X_2/X^-} = -1.1$ V and $E^\circ_{O_2/H_2O} = 1.23$ V, then on electrolysis of aqueous solution of salt MX, the products obtained are **[J & K 2015]**

(a) M, X_2 (b) H_2, X_2 (c) H_2O_2 (d) M, O_2

29 When 0.1 mol MnO_4^{2-} is oxidised, the quantity of electricity required to completely oxidise MnO_4^{2-} to MnO_4^- is **[CBSE AIPMT 2014]**

(a) 96500 C (b) 2 × 96500 C (c) 9650 C (d) 96.50 C

30 The weight of silver (atomic weight = 108) displaced by a quantity of electricity which displaces 5600 mL of O_2 at STP will be **[AFMC 2014]**

(a) 5.4 g (b) 10.8 g (c) 54.0 g (d) 108.0 g

31 The quantity of electricity needed to separately electrolyse 1 M solution of $ZnSO_4$, $AlCl_3$ and $AgNO_3$ completely is in the ratio of **[WB JEE 2014]**

(a) 2 : 3 : 1 (b) 2 : 1 : 1 (c) 2 : 1 : 3 (d) 2 : 2 : 1

32 When molten magnesium oxide was electrolysed for a certain period, 150 mg of Mg was deposited on the cathode. The volume of oxygen gas in cm^3 at STP conditions liberated at the anode during the same period is [atomic mass of Mg = 24 g mol^{-1}] **[Kerala CEE 2014]**

(a) 140 cm^3 (b) 280 cm^3
(c) 70 cm^3 (d) 120 cm^3

33 In the electrolysis of aqueous solution of copper sulphate using copper strips as anode and cathode, the anode reaction is **[Kerala CEE 2014]**

(a) $Cu^{2+} + 2e^- \longrightarrow Cu$
(b) $Cu \longrightarrow Cu^{2+} + 2e^-$
(c) $2HO^- \longrightarrow H_2 + \frac{1}{2}O_2 + 2e^-$
(d) $SO_4^{2-} \longrightarrow SO_4 + 2e^-$
(e) $2HSO_4^- \longrightarrow H_2S_2O_8 + 2e^-$

34 Impure copper containing Fe, Au, Ag as impurities is electrolytically refined. A current of 140 A for 482.5 s decreased the mass of the anode by 22.26 g and increased the mass of cathode by 22.011 g. Percentage of iron in impure copper is [Given molar mass of Fe = 55.5 g mol^{-1}, molar mass of Cu = 63.54 g mol^{-1}] **[KCET 2014]**

(a) 0.95 (b) 0.85 (c) 0.97 (d) 0.90

35 For hydrogen-oxygen fuel cell at 1 atm and 298 K

$$H_2(g) + \frac{1}{2}O_2(g) \longrightarrow H_2O(l), \ \Delta G^\circ = -240 \text{ kJ}$$

E° for the cell is approximately,
[Given F = 96500 C] **[KCET 2014]**

(a) 2.48 V (b) 1.25 V (c) 2.5 V (d) 1.26 V

36 How is electrical conductance of a conductor related with length and area of cross-section of the conductor ? **[MHT CET 2014]**

(a) $G = l \cdot a \cdot \kappa^{-1}$ (b) $G = \kappa \cdot l \cdot a^{-1}$

(c) $G = \kappa \cdot a \cdot l^{-1}$ (d) $G = \kappa \cdot l \cdot a^{-2}$

37 At 25° C, the molar conductance of 0.007 M hydrofluoric acid is 150 mho cm^2 mol^{-1} and its Λ°_m = 500 mho $cm^2 mol^{-1}$. The value of the dissociation constant of the acid at the given concentration at 25° C is **[WB JEE 2014]**

(a) 7×10^{-4} M

(b) 7×10^{-5} M

(c) 9×10^{-3} M

(d) 9×10^{-4} M

38 Conductivity of a saturated solution of a sparingly soluble salt AB at 298K is 1.85×10^{-6} Sm^{-1}. Solubility product of the salt AB at 298 K is [given, $\Lambda^\circ_m (AB) = 140 \times 10^{-4}$ Sm^{-2} mol^{-1}] **[KCET 2014]**

(a) $5.7 \times 10^{-12} mol^2 L^{-2}$ (b) $1.32 \times 10^{-12} mol^2 L^{-2}$

(c) $7.5 \times 10^{-12} mol^2 L^{-2}$ (d) $1.74 \ mol^2 L^{-2}$

39 The formal potential of Fe^{3+} / Fe^{2+} in a sulphuric acid and phosphoric acid mixture ($E^\circ = + 0.16$ V) is much lower than the standard potential ($E^\circ = + 0.77$ V). This is due to **[Manipal 2014]**

(a) formation of the species $[FeHPO_4]^+$

(b) lowering of potential upon complexation

(c) formation of the species $[FeSO_4]^+$

(d) high acidity of the medium

40 Given, $E^\circ_{Cr^{3+}/Cr} = - 0.72$ V $E^\circ_{Fe^{2+}/Fe} = - 0.42$ V.

The potential for the cell **[UK PMT 2014]**

Cr / Cr^{3+} (0.1 M) | | Fe^{2+} (0.01 M) 1 Fe is

(a) – 0.339 V (b) – 0.26 V

(c) 0.26 V (d) 0.3 V

41 For $Cr_2O_7^{2-} + 14H^+ + 6e^- \longrightarrow 2Cr^{3+} + 7H_2O$; $E^\circ = +1.33$ V. At $[Cr_2O_7^{2-}]$ = 4.5 millimole, $[Cr^{3+}]$ = 15 millimole, E is 1.067 V. The pH of the solution is nearly equal to **[KCET 2014]**

(a) 2 (b) 3 (c) 5 (d) 4

42 The emf of a galvanic cell with electrode potential of zinc = – 0.76 V and that of copper = 0.34 V is **[KCET 2014]**

(a) 0.76 V (b) 1.1 V

(c) – 1.1 V (d) 1.6 V

43 Standard electrode potential data are useful for understanding the suitability of an oxidant in a redox titration. Some half-cell reactions and their standard potentials are given below

$$MnO_4^-(aq) + 8H^+(aq) + 5e^- \longrightarrow Mn^{2+}(aq) + 4H_2O(l); \ E^\circ = 1.51 \text{ V}$$

$$Cr_2O_7^{2-}(aq) + 14H^+(aq) + 6e^- \longrightarrow 2Cr^{3+}(aq) + 7H_2O(l); \ E^\circ = 1.38 \text{ V}$$

$$Fe^{3+}(aq) + e^- \longrightarrow Fe^{2+}(aq); \ E^\circ = 0.77 \text{ V}$$

$$Cl_2(g) + 2e^- \longrightarrow 2Cl^-(aq); \ E^\circ = 1.4 \text{ V}$$

Identify the only incorrect statement regarding the quantitative estimation of aqueous $Fe(NO_3)_2$ **[WB JEE 2014]**

(a) MnO_4^- can be used in aqueous HCl

(b) $Cr_2O_7^{2-}$ can be used in aqueous HCl

(c) MnO_4^- can be used in aqueous H_2SO_4

(d) $Cr_2O_7^{2-}$ can be used in aqueous H_2SO_4

44 At 25° C, molar conductance of 0.1 molar aqueous solution of ammonium hydroxide is 9.54 Ω^{-1} cm^2 mol^{-1} and at infinite dilution, its molar conductance is 238 $ohm^{-1} cm^2 mol^{-1}$. The degree of ionisation of ammonium hydroxide at the same concentration and temperature is **[NEET 2013]**

(a) 2.080% (b) 20.800% (c) 4.008% (d) 40.80%

45 A hydrogen gas electrode is made by dipping platinum wire in a solution of HCl and pH = 10 and by passing hydrogen gas around the platinum wire at 1 atm pressure. The oxidation potential of electrode would be **[NEET 2013]**

(a) 0.059 V (b) 0.59 V (c) 0.118 V (d) 1.18 V

46 A button cell used in watches, functions as following

$$Zn(s) + Ag_2O(s) + H_2O(l) \rightleftharpoons 2Ag(s) + Zn^{2+}(aq) + 2OH^-(aq)$$

If half-cell potentials are :

$$Zn^{2+}(aq) + 2e^- \longrightarrow Zn(s); E^\circ = -0.76 \text{ V}$$

$$Ag_2O(s) + H_2O(l) + 2e^- \longrightarrow 2Ag(s) + 2OH^-; \ E^\circ = 0.34 \text{ V}$$

The cell potential will be **[NEET 2013]**

(a) 1.1 V (b) 0.42 V (c) 0.84 V (d) 1.34 V

47 Aluminium displaces hydrogen from acids but copper does not. A galvanic cell prepared by combining Cu / Cu^{2+} and Al / Al^{3+} has an emf of 2.0 V at 298 K. If the potential of copper electrode is 0.34 V that of aluminium electrode is **[KCET 2013]**

(a) – 2.3 V (b) + 2.34 V

(c) – 1.66 V (d) + 1.66 V

48 Which of the following reaction is used to make a fuel cell ? **[AIIMS 2013]**

(a) $Cd(s) + 2Ni(OH)_3(s) \longrightarrow CdO(s) + 2Ni(OH)_2(s) + H_2O(l)$

(b) $Pb(s) + PbO_2(s) + 2H_2SO_4(aq) \longrightarrow 2PbSO_4(s) + 2H_2O(l)$

(c) $2H_2(g) + O_2(g) \longrightarrow 2H_2O(l)$

(d) $2Fe(s) + O_2(g) + 4H^+(aq) \longrightarrow 2Fe^{2+}(aq) + 2H_2O(l)$

49 Which of the following reactions takes place at the cathode in the electrolytic cell used for extraction of aluminium from alumina? **[WB JEE 2013]**

(a) $12F^- \longrightarrow 12F + 12e^-$

(b) $Al^{3+} + 3e^- \longrightarrow Al$

(c) $2C(s) + O_2 \longrightarrow 2CO(g)$

(d) $2Al_2O_3 + 12F \longrightarrow 4Al^{3+} + 3O_2 + 12F$

50 When 9.65 Coulomb of electricity is passed through a solution of silver nitrate (atomic mass of Ag = 108 mol^{-1}). The amount of silver deposited is **[KCET 2013]**

(a) 16.2 mg (b) 21.2 mg (c) 10.8 mg (d) 6.4 mg

51 Molar conductivities (Λ°_m) at infinite dilution of NaCl, HCl and CH_3COONa are 126.4, 425.9 and 91.0 S cm^2 mol^{-1} respectively. Λ°_m for CH_3COOH will be **[CBSE AIPMT 2012]**

(a) 425.5 S $cm^2 mol^{-1}$ (b) 180.5 S $cm^2 mol^{-1}$ (c) 290.85 S $cm^2 mol^{-1}$ (d) 390.5 S $cm^2 mol^{-1}$

52 Read the following statements and predict the corresponding law. "At infinite dilution, when dissociation is complete, each ion makes a definite contribution towards total equivalent conductance of the electrolyte irrespective of the nature of the ion." **[OJEE 2012]**

(a) Ostwald's dilution law (b) Kohlrausch's law (c) Nernst equation (d) Ohm's law

53 The Gibbs energy for the decomposition of Al_2O_3 at 500° C is as follows

$$\frac{2}{3}Al_2O_3 \longrightarrow \frac{4}{3}Al + O_2;\ \Delta_r G = +960 \text{ kJ mol}^{-1}$$

The potential difference needed for the electrolytic reduction of aluminium oxide (Al_2O_3) at 500° C is at least **[CBSE AIPMT 2012]**

(a) 4.5 V (b) 3.0 V (c) 2.5 V (d) 5.0 V

54 The standard emf for the given cell reaction, $Zn + Cu^{2+} \longrightarrow Cu + Zn^{2+}$ is 1.10 V at 25°C. The emf for the cell reaction, when 0.1 M Cu^{2+} and 0.1 M Zn^{2+} solutions are used at 25° C is **[AIIMS 2012]**

(a) 1.10 V (b) – 1.10 V (c) 2.20 V (d) – 2.20 V

55 The equilibrium constant of the reaction, $Cu(s) + 2Ag^+(aq) \longrightarrow Cu^{2+}(aq) + 2Ag(s)$ $E^\circ = 0.46$ V at 298 K is **[AFMC 2012]**

(a) 2.4×10^{10} (b) 2.0×10^{10} (c) 4.0×10^{10} (d) 4.0×10^{15}

56 Li occupies higher position in the electrochemical series of metals as compared to Cu since **[WB JEE 2012]**

(a) the standard reduction potential of Li^+/Li is lower than that of Cu^{2+}/Cu

(b) the standard reduction potential of Cu^{2+}/Cu is lower than that of Li^+/Li

(c) the standard oxidation potential of Li/Li^+ is lower than that of Cu/Cu^{2+}

(d) Li is smaller in size as compared to Cu

57 What is the time required (in seconds) for depositing all the silver present in 125 mL of 1M $AgNO_3$ solution by passing a current of 241.25 A? [1F = 96500 C] **[UP CPMT 2012]**

(a) 10 s (b) 50 s (c) 1000 s (d) 100 s

58 On passing 3 A of electricity for 50 min, 1.8 g metal is deposited. The equivalent mass of metal is **[Manipal 2012]**

(a) 9.3 (b) 19.3 (c) 38.3 (d) 39.9

59 Saturated solution of KNO_3 is used to make salt-bridge because **[Manipal 2012]**

(a) velocity of K^+ is greater than that of NO_3^-

(b) velocity of NO_3^- is greater than that of K^+

(c) velocity of both K^+ and NO_3^- are nearly same

(d) KNO_3 is soluble in water

60 Which of the following used widely in the manufacture of lead storage battery? **[Manipal 2012]**

(a) Arsenic (b) Lithium (c) Bismuth (d) Antimony

61 The approximate time duration in hour to electroplate 30 g of calcium from molten calcium chloride using a current of 5 A is [Atomic mass of Ca = 40] **[KCET 2012]**

(a) 80 h (b) 10 h (c) 16 h (d) 8 h

62 The electrode potentials for $Cu^{2+}(aq) + e^- \longrightarrow Cu^+(aq)$ and $Cu^+(aq) + e^- \longrightarrow Cu(s)$ are + 0.15 V and + 0.50 V respectively. The value of $E^\circ_{Cu^{2+}/Cu}$ will be **[CBSE AIPMT 2011]**

(a) 0.150 V (b) 0.500 V (c) 0.325 V (d) 0.650 V

63 Standard reduction electrode potential of three metals X, Y and Z are -1.2 V, + 0.5V and - 3.0 V respectively. The reducing power of these metals will be **[CBSE AIPMT 2011]**

(a) $X > Y > Z$ (b) $Y > Z > X$
(c) $Y > X > Z$ (d) $Z > X > Y$

64 Standard electrode potential for Sn^{4+} / Sn^{2+} couple is + 0.15 V and that for the Cr^{3+} / Cr couple is − 0.74 V. These two couples in their standard state are connected to make a cell. The cell potential will be **[CBSE AIPMT 2011]**

(a) + 1.83 V (b) + 1.19 V
(c) + 0.89 V (d) + 0.18 V

65 Which pair of electrolytes could not be distinguished by the products of electrolysis using inert electrodes? **[Kerala CEE 2011]**

(a) 1 M $CuSO_4$ solution, 1 M $CuCl_2$ solution
(b) 1 M KCl solution, 1 M KI solution
(c) 1 M $AgNO_3$ solution, 1 M $Cu(NO_3)_2$ solution
(d) 1 M KCl solution, 1 M NaCl solution
(e) 1 M $CuBr_2$ solution, 1 M $CuSO_4$ solution

66 Which one of the following has the highest molar conductivity? **[Kerala CEE 2011]**

(a) Diamminedichloroplatinum(II)
(b) Tetraamminedichlorocobalt(III) chloride
(c) Potassium hexacyanoferrate(II)
(d) Hexaaquachromium(III) chloride
(e) Potassium hexacyano ferrate(III)

67 A current is passed through two cells connected in series. The first cell contains $X(NO_3)_3(aq)$ and the second cell contains $Y(NO_3)_2(aq)$. The relative atomic mases of X and Y are in the ratio 1 : 2. What is the ratio of liberated mass of X to that of Y? **[Kerala CEE 2011]**

(a) 3:2 (b) 1:2 (c) 1:3 (d) 3:14
(e) 2:1

68 Given the following reactions involving A, B, C and D

(i) $C + B^+ \longrightarrow C^+ + B$
(ii) $A^- + D \longrightarrow$ No reaction
(iii) $C^+ + A \longrightarrow$ No reaction
(iv) $D + B^+ \longrightarrow D^+ + B$

The correct arrangement of A, B, C, D in order of their decreasing ability as reducing agent **[DPMT 2011]**

(a) $D > B > C > A$ (b) $A > C > D > B$
(c) $C > A > B > D$ (d) $C > A > D > B$

69 9.65 C of electric current is passed through fused anhydrous $MgCl_2$. The magnesium metal thus obtained is completely converted into a Grignard reagent. The number of moles of Grignard reagent obtained is **[KCET 2011]**

(a) 5×10^{-4} (b) 1×10^{-4}
(c) 5×10^{-5} (d) 1×10^{-5}

70 The standard emf of a galvanic cell involving 2 moles of electrons in its redox reaction is 0.59 V. The equilibrium constant for the redox reaction of the cell is **[KCET 2011]**

(a) 10^{20} (b) 10^5 (c) 10 (d) 10^{10}

71 E_1, E_2 and E_3 are the emf of the following three galvanic cells respectively

(i) $Zn(s) \mid Zn^{2+}$ (0.1 M) || Cu^{2+} (1 M) | $Cu(s)$
(ii) $Zn(s) \mid Zn^{2+}$ (1 M) || Cu^{2+} (1 M) | $Cu(s)$
(iii) $Zn(s) \mid Zn^{2+}$ (1 M) || Cu^{2+} (0.1 M) | $Cu(s)$

Which one of the following is true? **[KCET 2011]**

(a) $E_2 > E_1 > E_3$ (b) $E_1 > E_2 > E_3$
(c) $E_3 > E_1 > E_2$ (d) $E_3 > E_2 > E_1$

72 Two different electrolytic cells filled with molten $Cu(NO_3)_2$ and molten $Al(NO_3)_3$ respectively are connected in series. When electricity is passed 2.7 g Al is deposited on electrode. Calculate the weight of Cu deposited on cathode.
[Cu= 63.5; Al= 27.0 g mol^{-1}] **[Guj. CET 2011]**

(a) 190.5 g (b) 9.525g (c) 63.5 g (d) 31.75 g

73 Which of the following reactions is correct for a given electrochemical cell at 25°C? **[Guj. CET 2011]**

$$Pt / Br_2(g)/Br^-(aq) // Cl^-(aq)/ Cl_2(g)/Pt$$

(a) $2Br^-(aq) + Cl_2(g) \longrightarrow 2Cl^-(aq) + Br_2(g)$
(b) $Br_2(g) + 2Cl^-(aq) \longrightarrow 2Br^-(aq) + Cl_2(g)$
(c) $Br_2(g) + Cl_2(g) \longrightarrow 2Br^-(aq) + 2Cl^-(aq)$
(d) $2Br^-(aq) + 2Cl^-(aq) \longrightarrow Br_2(g) + Cl_2(g)$

ANSWERS

Check Point 3.1

1. (c) 2. (a) 3. (b) 4. (c) 5. (d) 6. (b) 7. (b) 8. (c) 9. (b) 10. (d)

Check Point 3.2

1. (b) 2. (d) 3. (b) 4. (c) 5. (b) 6. (d) 7. (a) 8. (c)

Check Point 3.3

1. (b) 2. (c) 3. (a) 4. (a) 5. (a) 6. (d) 7. (a) 8. (b)

Check Point 3.4

1. (b) 2. (d) 3. (b) 4. (c) 5. (a) 6. (d) 7. (a) 8. (b)

Check Point 3.5

1. (c) 2. (b) 3. (c) 4. (a) 5. (c) 6. (b) 7. (c) 8. (a)

Check Point 3.6

1. (a) 2. (c) 3. (b) 4. (d) 5. (a) 6. (c) 7. (a) 8. (a) 9. (a)

Check Point 3.7

1. (d) 2. (d) 3. (a) 4. (a) 5. (b) 6. (a) 7. (a) 8. (d) 9. (b)

(A) Topical exercise

1. (b) 2. (c) 3. (d) 4. (b) 5. (a) 6. (c) 7. (d) 8. (c) 9. (b) 10. (a)
11. (b) 12. (c) 13. (b) 14. (a) 15. (b) 16. (a) 17. (a) 18. (a) 19. (a) 20. (d)
21. (c) 22. (a) 23. (a) 24. (d) 25. (b) 26. (a) 27. (b) 28. (b) 29. (a) 30. (a)
31. (b) 32. (b) 33. (c) 34. (a) 35. (d) 36. (b) 37. (a) 38. (d) 39. (d) 40. (b)
41. (d) 42. (c) 43. (b) 44. (d) 45. (a) 46. (c) 47. (d) 48. (a) 49. (d) 50. (a)
51. (d) 52. (b) 53. (c) 54. (c) 55. (a) 56. (b) 57. (c) 58. (a) 59. (d) 60. (d)
61. (b) 62. (b) 63. (d) 64. (c) 65. (b) 66. (b) 67. (c) 68. (b) 69. (c) 70. (c)
71. (d) 72. (d) 73. (b) 74. (d) 75. (c) 76. (a) 77. (b) 78. (c) 79. (d) 80. (b)
81. (a) 82. (c) 83. (c) 84. (d) 85. (c) 86. (a) 87. (d) 88. (c) 89. (b) 90. (a)
91. (d) 92. (a) 93. (b) 94. (b) 95. (d) 96. (c) 97. (a) 98. (c) 99. (a) 100. (a)
101. (b) 102. (b) 103. (b) 104. (c) 105. (c) 106. (b) 107. (d) 108. (c) 109. (c) 110. (b)
111. (a) 112. (d) 113. (b) 114. (d) 115. (a) 116. (b) 117. (d) 118. (b) 119. (c) 120. (d)
121. (d) 122. (c) 123. (c) 124. (a) 125. (d) 126. (c) 127. (c) 128. (a) 129. (a) 130. (c)
131. (b) 132. (a) 133. (c) 134. (b) 135. (a) 136. (d) 137. (c) 138. (d) 139. (b) 140. (b)
141. (d) 142. (c) 143. (b) 144. (c) 145. (d) 146. (d) 147. (c) 148. (a) 149. (a) 150. (c)
151. (c) 152. (d) 153. (d) 154. (b) 155. (c) 156. (d) 157. (d) 158. (b) 159. (d) 160. (d)
161. (c)

(B) Miscellaneous questions

1. (d) 2. (a) 3. (c) 4. (a) 5. (c) 6. (a) 7. (c) 8. (b) 9. (d) 10. (a)
11. (b) 12. (b) 13. (b) 14. (d) 15. (a) 16. (b) 17. (d) 18. (b) 19. (a) 20. (d)
21. (a) 22. (d) 23. (d) 24. (d) 25. (d) 26. (d) 27. (a) 28. (d) 29. (c) 30. (a)
31. (d) 32. (d) 33. (a) 34. (c) 35. (d) 36. (c) 37. (a) 38. (c) 39. (a) 40. (d)
41. (d) 42. (c) 43. (b) 44. (d) 45. (b) 46. (a) 47. (a) 48. (b) 49. (b) 50. (a)
51. (a) 52. (a) 53. (d) 54. (d)

(C) Medical entrances' gallery

1. (a) 2. (d) 3. (c) 4. (d) 5. (b) 6. (b) 7. (b) 8. (c) 9. (d) 10. (c)
11. (d) 12. (c) 13. (c) 14. (c) 15. (d) 16. (d) 17. (d) 18. (d) 19. (d) 20. (c)
21. (a) 22. (a) 23. (e) 24. (b) 25. (c) 26. (b) 27. (c) 28. (b) 29. (c) 30. (d)
31. (a) 32. (c) 33. (b) 34. (d) 35. (b) 36. (c) 37. (d) 38. (d) 39. (a) 40. (c)
41. (a) 42. (b) 43. (a) 44. (c) 45. (b) 46. (a) 47. (c) 48. (c) 49. (b) 50. (c)
51. (d) 52. (b) 53. (c) 54. (a) 55. (d) 56. (a) 57. (b) 58. (b) 59. (c) 60. (d)
61. (d) 62. (c) 63. (d) 64. (c) 65. (d) 66. (c) 67. (c) 68. (d) 69. (c) 70. (a)
71. (b) 72. (b) 73. (a)

Hints & Explanations

(A) Topical exercise

2 (*c*) Solution consisting of electrolytes is known as electrolytic solution and conductivity of electrolytic solution depends upon the following factors.

(a) **Size of ions** As ion size increases, ion mobility decreases and conductivity decreases.

(b) **Viscosity of solution** Greater the viscosity of the solvent lesser will be the conductivity of the solution.

(c) **Solvation of ions** Greater the solvation of ions of an electrolyte lesser will be the electrical conductivity of the solution.

(d) **Temperature of medium** Conductivity of solution increases with increase in temperature.

3 (*d*) Cell constant is defined as the ratio of length of object and area of cross section.

$$G^* = \frac{l}{A}$$

Since, l and a remain constant for any particular object, hence value of cell constant always remains constant.

9 (*b*) Specific conductance = conductance × cell constant

$$0.012 = \frac{1}{55} \times \text{cell constant}$$

$\therefore$ Cell constant = 0.66 cm^{-1}

10 (*a*) Resistance = specific resistance × cell constant

$$(R)_{KCl} = (S)_{KCl} \times (\text{cell constant})$$

$$(R)_{NaOH} = (S)_{NaOH} \times (\text{cell constant})$$

$$\frac{(R)_{KCl}}{(R)_{NaOH}} = \frac{(S)_{KCl}}{(S)_{NaOH}} \Rightarrow \frac{160}{190} = \frac{700}{(S)_{NaOH}}$$

$\therefore$ $(S)_{NaOH} = 831.25$ ohm-cm

$\therefore$ Specific conductance $(\kappa)_{NaOH} = 1.20 \times 10^{-3}\ \Omega^{-1}\ cm^{-1}$

11 (*b*) Concentration of solute = 0.1mole in 0.5 L = 0.2 M

$$\text{Molar conductance} = \frac{1000 \times \text{specific conductance}}{\text{molarity}}$$

$$= \frac{1000 \times x}{0.2} = 5000x \text{ S cm}^2\text{mol}^{-1}$$

12 (*c*) Equivalence conductance is five times, hence, dilution is also **five times**

$\therefore$ $0.1\text{M} \times V \equiv M' \times 5\ V$

$\therefore$ M' (final) = 0.02 M

= 0.04 N (due to dibasic acid)

13 (*b*) $\text{Equivalent conductance} = \frac{1000 \times \text{conductance} \times \text{cell constant}}{\text{concentration}}$

$$= 1000 \times \frac{1}{2.5 \times 10^3} \times \frac{1.15}{0.1} = 4.6$$

16 (*a*) If dilution is made V-times molar and specific conductance becomes V-times 0.1 M H_2SO_4 changes to 0.01 M H_2SO_4 (10 times)

Thus, dilution is 10 times.

17 (*a*) $\overset{\circ}{\Lambda}_{H_2O} = \overset{\circ}{\Lambda}_{H^+} + \overset{\circ}{\Lambda}_{OH^-} = 550 \text{ S cm}^2 \text{ mol}^{-1}$

$$x = \frac{\Lambda_C}{\Lambda_0}$$

$$\therefore \quad \Lambda_C = x \times \Lambda_0$$

$$= 1.9 \times 10^{-9} \times 550$$

$$= 1.045 \times 10^{-6} \text{ S cm}^2 \text{ mol}^{-1}$$

19 (*a*) $\overset{\circ}{\Lambda}_{eq} = \lambda^{\infty}_{H^+} + \lambda^{\infty}_{SO_4^{2-}} = (x + y)$

22 (*a*) Both statements are correct.

Since, conductivity depends upon number of ions per unit volume. Therefore, the conductivity of all electrolytes decreases on dilution due to decrease in number of ions per unit volume.

23 (*a*) Both statements are true.

Molar conductivity of weak electrolytic solution increases on dilution, because as we add excess water to increase the dilution degree of dissociation increases which lead to increase in number of ions in the solution. Thus, Λ_m shows a very sharp increase.

28 (*b*)

$$\overset{\circ}{\Lambda}_{m\,(NH_4Cl)} = \overset{\circ}{\Lambda}_{m\,(NH_4^+)} + \overset{\circ}{\Lambda}_{m\,(Cl^-)}$$

$$\overset{\circ}{\Lambda}_{m\,(NaOH)} = \overset{\circ}{\Lambda}_{m\,(Na^+)} + \overset{\circ}{\Lambda}_{m\,(OH^-)}$$

$$\overset{\circ}{\Lambda}_{m\,(NaCl)} = \overset{\circ}{\Lambda}_{m\,(Na^+)} + \overset{\circ}{\Lambda}_{m\,(Cl^-)}$$

$$\overset{\circ}{\Lambda}_{m\,(NH_4Cl)} + \overset{\circ}{\Lambda}_{m\,(NaOH)} - \overset{\circ}{\Lambda}_{m\,(NaCl)} = \overset{\circ}{\Lambda}_{m\,(NH_4OH)}$$

Hence, option (b) is correct.

30 (*a*) $x = \frac{\Lambda_c}{\Lambda_0} = \frac{10}{200} = 0.05$, % ionisation = 5%

31 (*b*) $\Lambda^{\infty}_{BaCl_2} = \Lambda^{\infty}_{Ba^{2+}} + \Lambda^{\infty}_{Cl^-}$ (in terms of equivalent conductances)

$$= 127 + 26 = 203$$

$$\Lambda^{\infty}_{molar} = \text{valency} \times \Lambda^{\infty}_{eq}$$

$\therefore$ $\Lambda^{\infty}_m(Ba^{2+}) = 2 \times 127$

$$\Lambda^{\infty}_m(Cl^-) = 1 \times 76$$

$$\Lambda^{\infty}_m(BaCl_2) = 254 + 2 \times 76 = 406$$

$$\Lambda^{\infty}_{eq}(BaCl_2) = \frac{406}{2} = 203\ \Omega^{-1}\text{ cm}^2\text{ equiv}^{-1}$$

32 (*b*) $\Lambda^{\infty}_{CH_3CH_2COOH} = \Lambda^{\infty}_{CH_3CH_2COONa} + \Lambda^{\infty}_{HCl} - \Lambda^{\infty}_{NaCl}$

$$= 390.71\ \Omega^{-1}\text{ equiv}^{-1}\text{ cm}^2$$

33 (*c*) $\Lambda^{\circ}_{BaSO_4} = \Lambda^{\circ}_{BaCl_2} + \Lambda^{\circ}_{H_2SO_4} - \Lambda^{\circ}_{HCl} = (x_1 + x_2 - x_3)$

Equivalent conductance of $BaSO_4$

$$\Lambda^{\circ}_{BaSO_4} = \frac{1000 \times \text{specific conductance}}{\text{solubility (in saturated solution)}}$$

$$(x_1 + x_2 - x_3) = \frac{1000\,y}{\text{Solubility}}$$

$$\therefore \quad \text{Solubility } (BaSO_4) = \frac{1000\,y}{(x_1 + x_2 - x_3)} \text{ N}$$

$$= \frac{1000\,y}{2\,(x_1 + x_2 - x_3)} \text{ M}$$

$$BaSO_4 \rightleftharpoons Ba^{2+} + SO_4^{2-}$$

$$K_{sp}(BaSO_4) = [Ba^{2+}][SO_4^{2-}]\text{ M}^2$$

$$= \frac{10^6\,y^2}{4(x_1 + x_2 - x_3)^2}$$

37 (*a*) pH = 14, pOH = 0

$$\therefore \quad [OH^-] = 1 \text{ M}$$

$$K_{sp}[Al(OH)_3] = 1.0 \times 10^{-33}$$

$$[Al^{3+}][OH^-]^3 = 1 \times 10^{-33}$$

$$\therefore \quad [Al^{3+}] = 1 \times 10^{-33} \text{ M}$$

$$Al^{3+}(aq) + 3e^- \longrightarrow Al(s)$$

$$K = \frac{1}{[Al^{3+}]} = 1 \times 10^{33}$$

$$\therefore \quad E = E^{\circ} - \frac{0.0591}{3}\log K$$

$$= -1.66 - \frac{0.0591}{3}\log 1 \times 10^{33}$$

$$= -2.31 \text{ V}$$

38 (*d*)

$$Pb(s) \rightleftharpoons Pb^{2+}(aq) + 2e^-$$

$$Hg_2^{2+}(aq) + 2e^- \longrightarrow 2Hg(l)$$

$$Hg_2^{2+}(aq) + Pb(s) \longrightarrow Pb^{2+}(aq) + 2Hg(l)$$

$$\therefore \quad K = \frac{[Pb^{2+}]}{[Hg_2^{2+}]}$$

$[Pb^{2+}]$ in LHS is from $PbSO_4$

$$\therefore \quad [Pb^{2+}] = \sqrt{K_{sp}\,(PbSO_4)} = \sqrt{2 \times 10^{-8}}$$

$$= 1.41 \times 10^{-4} \text{ M}$$

$[Hg_2^{2+}]$ in RHS is from Hg_2SO_4

$$\therefore \quad [Hg_2^{2+}] = \sqrt{K_{sp}\,(Hg_2SO_4)}$$

$$= \sqrt{1 \times 10^{-6}} = 1 \times 10^{-3} \text{ M}$$

$$\therefore \quad K = \frac{1.41 \times 10^{-4}}{1 \times 10^{-3}}$$

$$= 1.41 \times 10^{-1} \text{ M}$$

$$\therefore \quad E = E^{\circ} - \frac{0.0591}{2}\log 1.41 \times 10^{-1}$$

$$= 0.92 + 0.025 = 0.945 \text{ V} \approx 0.95\text{V}$$

39 (*d*)

$$AgI(s) + e^- \longrightarrow Ag(s) + I^-(aq) \qquad E^{\circ}_{red} = x\,\text{V}$$

$$Ag(s) \longrightarrow Ag^+(aq) + e^- \qquad E^{\circ}_{ox} = -0.80 \text{ V}$$

$$AgI(s) \longrightarrow Ag^+(aq) + I^-(aq) \qquad E^{\circ}_{cell} = (x - 0.80)\,\text{V}$$

$$K_{eq} = [Ag^+][I^-] = K_{sp} = 1 \times 10^{-16}$$

$$E_{cell} = E^{\circ}_{cell} - \frac{0.051}{1}\log 1 \times 10^{-16}$$

At equilibrium, $E_{cell} = 0.80$ V

$$\therefore \quad E^{\circ}_{cell} = 0.0591 \log 10^{-16}$$

$$(x - 0.80) = -0.9456 \text{ V}$$

$$\therefore \quad x = -0.146 \text{ V}$$

40 (*b*)

$$E^{\circ} = \frac{2.303\,RT}{nF}\log K_{eq}$$

$$0.2955 = \frac{0.0591}{2}\log K_{eq}$$

$$\therefore \quad K_{eq} = 10^{10}$$

43 (*b*) Instead of H^+ (1 M), H_2O has been taken

$$Pt, H_2(g) \mid H_2O\,[H^+] = 10^{-7}\text{ M}$$

$\ominus$

Oxidation takes place being at anode

$$H_2(g) \longrightarrow 2H^+(aq) + 2e^-$$

$$E = E^{\circ} - \frac{0.0591}{2}\log\frac{[H^+]^2}{p_{H_2}}$$

$$0 = 0 - \frac{0.0591}{2}\log\frac{(10^{-7})^2}{p_{H_2}}$$

$$\frac{(10^{-7})^2}{(p_{H_2})} = 1$$

$$\therefore \quad p_{H_2} = 10^{-14} \text{ bar}$$

48 (*a*) For reaction involving different electrons

$$\Delta G_3 = \Delta G_1 - \Delta G_2$$

$$-n_3FE_3{}^{\circ} = n_1FE^{\circ}{}_1 - (-n_2FE^{\circ}{}_2)$$

$$n_3E^{\circ}{}_3 = n_1E^{\circ}{}_1 - n_2E^{\circ}{}_2$$

$$E^{\circ}{}_3 = \frac{5 \times 1.51 - 2 \times 1.23}{3}$$

$$= 1.70 \text{ V}$$

52. (*b*) A cell reaction is spontaneous, if $E^{\circ}_{cell} = +\text{ve}$

$$Sn^{2+} \longrightarrow Sn^{4+} + 2e^- \qquad E^{\circ} = -0.13\,\text{V}$$

$$Sn^{2+} + 2e^- \longrightarrow Sn \qquad E^{\circ} = -0.14\,\text{V}$$

$$2\,Sn^{2+} \longrightarrow Sn^{4+} + Sn \qquad E^{\circ}_{cell} = -0.27\,\text{V}$$

Thus, (a) is not possible.

53 (*c*)

$$Ag + Cl^- \longrightarrow AgCl + e^-$$

$$\frac{1}{2}Cl_2 + e^- \longrightarrow Cl^-$$

$$Ag + \frac{1}{2}Cl_2 \longrightarrow AgCl$$

55 (a) $E_{cell} = E^\circ_{cell} - \dfrac{0.0591}{2} \log \dfrac{[Zn^{2+}]}{[Cu^{2+}]}$

$= 1.1 - \dfrac{0.0591}{2} \log \dfrac{[Zn^{2+}]}{[Cu^{2+}]}$

$E_{cell} < 1.1$ V if, $\dfrac{[Zn^{2+}]}{[Cu^{2+}]} > 1$

56 (b) Since, CH_3COOH is a weak acid and emf is dependent on pH, also pH of 0.1 M HCl < pH of 0.1 M CH_3COOH.

61 (b) $\underset{6\text{ mol}}{Cr_2O_7^{2-} + 6e^-} \longrightarrow 2Cr^{3+}$

$Q = 6 \times 96500$ C

64 (c) The quantity of charge required to obtain one mole of aluminium from Al_2O_3 is equal to number of electron required to convert Al_2O_3 to Al.

$Al^{3+}(aq) \xrightarrow{+3e^-} Al(s)$

Hence, total 3F is required.

65 (b) $Ag^+ + e^- \longrightarrow Ag$

$Cu^{2+} + 2e^- \longrightarrow Cu$

$Al^{3+} + 3^- \longrightarrow Al$

Thus, number of Faradays required by 1 mole each of Ag^+, Cu^{2+} and Al^{3+} = 1 : 2 : 3

66 (b) $MnO_4^- \xrightarrow{+5e^-} Mn^{2+}$

67 (c) Amount of zinc = 3.048 g

$Zn^{2+} + 2e^- \longrightarrow Zn(s)$

$w = zit$

where, z = electrochemical equivalent $\left(= \dfrac{\text{atomic weight}}{\text{electrons} \times 96500} \right)$

i = current in amperes

t = time in seconds

$3.048 = \dfrac{A}{2 \times 96500} \times 5 \times 30 \times 60$

$A = 65.36 \text{ g mol}^{-1}$

$\therefore$ Equivalent weight $= \dfrac{A}{2} = 32.68 \text{ g equivalent}^{-1}$

$\approx 32.7 \text{ g equivalent}^{-1}$

68 (b) Number of moles of Ag $= \dfrac{107.9}{107.9} = 1$

Number of Ag atoms $= 1 \times N_0$ (where, N_0 is the Avogadro's number)

Total charge $= N_0 \times 1.6 \times 10^{-19}$ C

$\therefore \quad N_0 \times 1.6 \times 10^{-19} = 96500$

(charge carried by 1 mole of electrons)

$\therefore \quad N_0 = \dfrac{96500}{1.6 \times 10^{-19}} = 6.02 \times 10^{23} \text{ mol}^{-1}$

69 (c) $\underset{\substack{1\text{ mol}\\0.5\text{ g}\\=0.25\text{ mol}}}{H_2O(l) \longrightarrow H_2(g)} + \underset{\substack{0.5\text{ mol}\\0.125\text{ mol}\\=4\text{ g}}}{\dfrac{1}{2}O_2(g)}$

Alternately

$2H^+ + 2e^- \longrightarrow H_2(g)$

Equivalent weight of $H_2 = \dfrac{2}{2} = 1$

$2O^{2-} \longrightarrow O_2 + 4e^-$

Equivalent weight of $O_2 = \dfrac{32}{4} = 8$

By Faraday's second law of electrolysis,

$\dfrac{\text{weight of } O_2}{\text{weight of } H_2} = \dfrac{\text{equivalent weight of } O_2}{\text{equivalent weight of } H_2}$

weight of $O_2 = \dfrac{8 \times 0.5}{1} = 4$ g

Note *Equivalent weight of $O_2 = 8 \text{ g equiv}^{-1}$*

70 (c) $\underset{1\text{ mol}}{H_2O(l)} \longrightarrow \dfrac{1}{2}O_2(g) + 2H^+ + 2e^-$

Moles of electrons exchanged = 2

Charge carried by 1 mole of electrons = 96500 C

Thus, charge carried by 2 moles of electrons $= 2 \times 96500$ C

$= 1.93 \times 10^5$ C

71 (d) $\underset{+7}{MnO_4^-} \longrightarrow \underset{+4}{MnO_2}$

Thus, electrons gained = 3

Thus, 1 mole of MnO_4^- gives $\dfrac{1}{3}$ mol MnO_2 = 0.33 mol

72 (d) $2H^+ + 2e^- \longrightarrow H_2$; $K = \dfrac{1}{[H^+]^2}$

$E = E^\circ + \dfrac{0.0591}{2} \log [H^+]^2$

$= E^\circ + 0.0591 \log [H^+]$

$= E^\circ - 0.0591 \text{ pH} = -0.0591 \text{ pH}$

(a) $E = 0.0591 \log 1 = 0$

(b) $E = -0.0591 \times 4 = -0.2364$ V

(c) $E = +0.0591 \times \log 10^{-7} = -0.4137$ V

(d) 1.0 M NaOH = 1.0 M $[OH^-] = 10^{-14}$ M $[H^+]$

$\therefore \quad E = 0.0591 \log 10^{-14} = -0.8274$ V

73 (b) $Cl^- \longrightarrow \dfrac{1}{2}Cl_2 + e^-$

0.355 g Cl^- = 0.01 mol of Cl^- = $0.01N_0$

74 (d) $q = it = 30 \times 70.2 \times 60$ C = 126360 C

$= \dfrac{126360}{96500}$ C = 1.31 F

75 (c) Electrolysis of H_2O takes place,

At anode, $2H_2O \longrightarrow 4H^+ + O_2 + 4e^-$

$w = zit = \dfrac{32}{4 \times 96500} \times 8 \times 1.5 \times 3600$ g = 0.112 mol

76 (a) Rate of flow of electrons per second $= \dfrac{0.965}{96500} N_0$

$= 1 \times 10^{-5} N_0$

77 (*b*) $w = \dfrac{\text{atomic weight}}{96500 \times \text{electron}} \times i \times t$

Atomic weight = 55 g

78 (*c*) Quantity of electricity $= it = 1.8 \times 1.36 \times 60\ C = 146.88\ C$

79 (*d*) Quantity of electricity $= it$

$$72000 = 1 \times t$$
$$t = 72000\ s = 20\ h$$

81 (*a*) $w(O_2) = zit = \dfrac{32 \times 2 \times 965}{4 \times 96500}$

$= 0.16\ g = 0.005\ mol$

$= 0.112\ L$ at NTP

83 (*c*) $\dfrac{w(Al)}{w(Na)} = \dfrac{\text{equivalent weight of Al}}{\text{equivalent weight of Na}}$

$\dfrac{w(Al)}{4.6} = \dfrac{9}{23}$

$\therefore \quad w = 1.8\ g$

87 (*d*) In case of electrolysis of aqueous NaCl oxidation reaction occurs at anode as follows,

$$Cl^- (aq) \longrightarrow \frac{1}{2} Cl_2(g) + e^-; \qquad E° = 1.36\ V$$
$$2H_2O(l) \longrightarrow O_2(g) + 4H^+(aq) + 4e^-; \quad E^°_{cell} = -1.23\ V$$

But due to lower $E^°_{cell}$ value, water get oxidised in preference of Cl^- (*aq*). However, the actual reaction taking place in the concentrated solution of NaCl is (d) and not (b), i.e. Cl_2 is produced and not O_2.

This unexpected result is explained on the basis of the concept of 'overvoltage', i.e. water needs greater voltage for oxidation to O_2 (as it is kinetically slow process) than that needed for oxidation of Cl^- ions to Cl_2. Thus, the correct option is (d) not (b).

88 (*c*) Half-cell reactions in (1) and (3) are of no use; from (2) and (4)

$3 \times (4) - 2(2)$ we get required equation

$\therefore \quad E^°_{cell} = 0.34 + 0.74 = 1.08\ V$

$E°$ values are additive, when equal number of electrons are involved.

89 (*b*) Statement I is true but the statement II is false.

Feasibility of chemical reaction depends on Gibbs' free energy which is related to $E^°_{cell}$ as,

$$\Delta G° = -nFE^°_{cell}$$

When value of $E^{\ominus}_{cell}$ is positive then $\Delta G^{\ominus}$ becomes negative. Hence, reaction becomes feasible.

Correct statement II is $E_{cathode} > E_{anode}$.

90 (*a*) $2Hg(l) + \underset{L}{2Cl^-(aq)} \longrightarrow Hg_2Cl_2(s) + 2e^-$

$$Cl_2(g) + 2e^- \longrightarrow \underset{R}{2Cl^-(aq)}$$

$$\underset{(L)}{2Cl^-(aq)} \longrightarrow \underset{(R)}{2Cl^-(aq)}$$

$$K = \frac{[Cl^-]^2_R}{[Cl^-]^2_L} = \frac{(0.01)^2}{(0.1)^2} = 0.01$$

$$E_{cell} = E°_{cell} - \frac{0.0591}{2} \log K$$
$$= 1.10 - \frac{0.0591}{2} \log(0.01)$$
$$= 1.1591\ V$$

92 (*a*)

I.	$Cu^+ + e^- \longrightarrow Cu;$	$E^°_1 = 0.52\ V$
II.	$Cu^{2+} + e^- \longrightarrow Cu^+;$	$E^°_2 = 0.16\ V$
III.	$2Cu^+ \longrightarrow Cu^{2+} + Cu;$	$E^°_3 = ?$

Reverse the equation (II) and then add both the reaction.

$$E^°_3 = E^°_1 + E^°_2 \Rightarrow E^°_3 = 0.52 + (-0.16)$$
$$\therefore \quad E^°_3 = 0.36\ V$$

93 (*b*) $Zn + 2H^+ \longrightarrow Zn^{2+} + H_2$ hence, $(E - E°) = 0$

$[H^+] = 0.1\ M$

$[Zn^{2+}] = 0.01\ M \Rightarrow K = \dfrac{[Zn^{2+}]}{[H^+]^2} = 1$

Hence, $(E - E°) = 0$

94 (*b*) Statement I is true but the statement II is false.

Electrode potential of Cu^{2+}/Cu is + 0.34 V and electrode potential of $2H^+/H_2$ is 0.00 V.

Hence, correct reason is due to positive value of Cu^{2+}/Cu it looses electron to H^+ and get reduces, while H_2 gas evolves out.

95 (*d*) Both statement I and statement II are false.

Copper sulphate can not be stored in zinc vessel as zinc is more reactive than copper due to negative value of standard reduction potential of Zn.

98 (*c*) Reducing nature varies as Mg > Cu > Hg > Ag, hence $CuCl_2$ can be stored in silver-vessel.

101 (*b*) When $E^°_{cell} < 0$ electrolysis takes place.

102 (*b*)

A. F_2 is a non-metal and best oxidising agent because SRP of F_2 is + 2.87 V.

B. Li is a metal and strongest reducing agent because SRP of Li is – 3.05 V.

C. Au^{3+} is a metal ion which is an oxidising agent as SRP of Au^{3+} is +1.40 V.

D. Br^- is an anion that can be oxidised by Au^{3+} as Au^{3+} ($E^° = 1.40$) is greater than Br^- ($E^° = 1.09$ V).

E. Au is an unreactive metal.

F. Li^+ is a metal ion having least value of SRP (– 3.05 V), hence it is the weakest oxidising agent.

G. F^- is an anion which is the weakest reducing agent as F^-/F_2 has low oxidation potential (–2.87 V).

103 (*b*) $Cl_2 + 2e^- \longrightarrow 2Cl^-$

$$K = [Cl^-]^2$$

$$\therefore \quad E = E^\circ - \frac{0.0591}{2}\log[Cl^-]^2$$
$$(E - E^\circ) = -0.0591 \log [Cl^-]$$
$$(E - E^\circ) = 0.0591 \log\left[\frac{1}{Cl^-}\right]$$

Thus, as $[Cl^-]$ increases, $(E - E^\circ)$ decreases.

104 (*c*) $H_2 \longrightarrow 2H^+ + 2e^-$

$Q = [H^+]^2 = 10^{-14}$ M in H_2O

$$E = E^\circ - \frac{0.0591}{2}\log 10^{-14} = 0 + 0.0591 \times 7 = 0.4137 \text{ V}$$

105 (*c*)
$$Zn \longrightarrow Zn^{2+} + 2e^-$$
$$2H^+ + 2e^- \longrightarrow H_2$$
$$\text{overall,} \quad Zn + 2H^+ \longrightarrow Zn^{2+} + H_2$$
$$K = \frac{[Zn^{2+}][H_2]}{[Zn][H^+]^2}$$

In H_2O,

$[H^+] = 1 \times 10^{-7}$ M under standard condition (298 K, 1 bar)

$$K = \frac{(10^{-4})(1)}{(1)(10^{-7})^2} = 10^{10}$$
$$\therefore \quad E = E^\circ - \frac{0.0591}{2}\log K$$
$$= 0.76 - \frac{0.0591}{2}\log 10^{10} = 0.4645 \text{ V}$$

106 (*b*) At LHS there is formation of H^+ ion and in RHS OH^- ion

At LHS $\quad NH_4OH + H^+ \longrightarrow NH_4^+ + H_2O$

At RHS $\quad NH_4^+ + OH^- \longrightarrow NH_4OH$

At LHS concentration of NH_4OH decreases and that of NH_4^+ increases and reverse at RHS

$$pOH = pK_b + \log\frac{[NH_4^+]}{[NH_4OH]}$$

Since, pOH at LHS increase hence pH decreases and that of RHS increases.

108 (*c*) Reducing power of Mg > Cu > Hg > Ag

Thus, Cu^{2+}, Hg^{2+} and Ag^+ are reduced to metals and Mg remain as oxidised part.

109 (*c*) Standard electrode potential of copper electrode can be calculated by constructing a concentration cell composed of two half-cell reactions in which concentration of species on left hand and right hand side are unity. In such case, cell potential is equal to standard electrode potential.

$$\underbrace{Pt(s)|H_2(g, 1\text{bar})|H^+(aq, 1\text{M})}_{\text{Oxidation half cell reaction}} \,\Big|\Big|\, \underbrace{Cu^{2+}(aq, 1\text{M})|Cu}_{\text{Reduction half cell reaction}}$$

110 (*b*)
$$2 \times [Cu^+ \longrightarrow Cu^{2+} + e^-] \quad E^\circ = -0.15 \text{ V}$$
$$Cu^{2+} + 2e^- \longrightarrow Cu \quad E^\circ = 0.34 \text{ V}$$
$$Cu^+ \longrightarrow Cu^{2+} + Cu \quad E^\circ = 0.19 \text{ V}$$

In aqueous solution, Cu^+ disproportionates to Cu^{2+} and Cu. Thus, Cu^+ is not stable but Cu^{2+} is stable.

111 (*a*) $2MnO_4^- + 3Mn^{2+} + 2H_2O \longrightarrow 5MnO_2 + 4H^+$;

$E^\circ = 0.47$ V

Thus, cell reaction in which Mn^{2+} on reaction with MnO_4^- forms MnO_2 in acidic medium.

112 (*d*) $Cu + 2H^+ (\text{dil. } H_2SO_4) \longrightarrow Cu^{2+} + H_2$

$$E^\circ_{cell} = E^\circ_{Cu/Cu^{2+}} + E^\circ_{SHE} = -0.34 \text{ V}$$

Thus, (a) is not feasible.

$$Cu + SO_4^{2-} + 4H^+ \longrightarrow Cu^{2+} + SO_2 + 2H_2O$$
$$E^\circ_{cell} = -0.34 + 1.17 = 0.83 \text{ V}$$
$$E^\circ_{cell} > 0$$

Thus, (b) is feasible.

Similarly, (c) is feasible.

113 (*b*) Electrode potential for Mg electrode varies according to the equation,

$$E_{Mg^{2+}/Mg} = E^\circ_{Mg^{2+}/Mg} - \frac{0.059}{2}\log\frac{1}{[Mg^{2+}]}$$
$$E_{Mg^{2+}/Mg} = E^\circ_{Mg^{2+}/Mg} + \frac{0.059}{2}\log[Mg^{2+}]$$
$$E_{Mg^{2+}/Mg} = \frac{0.059}{2}\log[Mg^{2+}] + E^\circ_{Mg^{2+}/Mg}$$

This equation represents equation of straight line. It can be correlated as,

$$\underset{Y}{E_{Mg^{2+}/Mg}} = \underset{M}{\left(\frac{0.059}{2}\right)} \underset{X}{\log[Mg^{2+}]} + \underset{+C}{E^\circ_{Mg^{2+}/Mg}}$$

So, intercept $(C) = E^\circ_{Mg^{2+}/Mg}$.

Thus, equation can be diagramatically represented as,

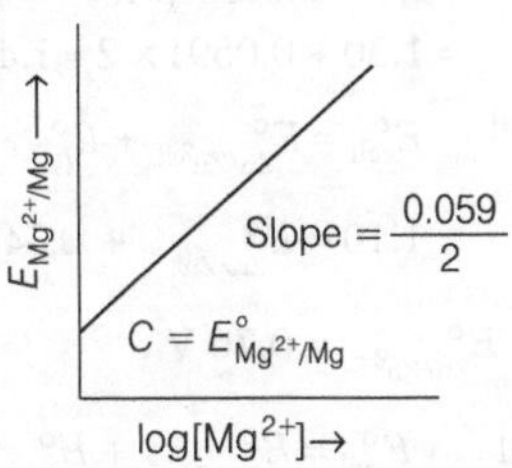

115 (*a*) I. $Cu^{2+} + 2e^- \longrightarrow Cu \quad E^\circ_1 = 0.339$ V, $n_1 = 2$

II. $Cu^+ + e^- \longrightarrow Cu \quad E^\circ_2 = 0.518$ V, $n_2 = 1$

III. $Cu^{2+} + e^- \longrightarrow Cu^+ \quad E^\circ_3 = ? \quad n_3 = 1$

I - II = III

$$\therefore \quad \Delta G^\circ_{III} = \Delta G^\circ_I - \Delta G^\circ_{II}$$
$$-n_3FE_3^\circ = -n_1FE_1^\circ + n_2FE_2^\circ$$
$$\therefore \quad E_3^\circ = \frac{n_1E_1^\circ - n_2E_2^\circ}{n_3}$$
$$= \frac{0.678 - 0.518}{1} = 0.160 \text{ V}$$

116 (*b*) $Co + 2H^+ \longrightarrow Co^{2+} + H_2$

$$Q = \frac{[Co^{2+}]}{[H^+]^2}$$

$$E = E^\circ - \frac{0.0591}{2}\log\frac{[Co^{2+}]}{[H^+]^2}$$

E will increase, when $[H^+]$ increases or $[Co^{2+}]$ decreases.

117 (*d*) Half-cell reaction is

$$2Hg(l) + 2Cl^-(aq) \longrightarrow Hg_2Cl_2(s) + 2e^-$$

$$\therefore \quad K = \frac{1}{[Cl^-]^2}$$

$$E = E^\circ - \frac{0.0591}{2}\log\frac{1}{[Cl^-]^2}$$

$$E = E^\circ + 0.0591 \log [Cl^-]$$

Thus, E increases as $[Cl^-]$ increases.

119 (*c*) In this electrode,

$[A] = [B]$ in quinhydrone electrode

OH ... OH $\rightleftharpoons$ O ... O $+ 2H^+ + 2e^-$

(*A*) (*B*)

Hence, $Q = [H^+]^2$

$$E = E^\circ - \frac{0.0591}{2}\log [H^+]^2$$

$$= E^\circ - 0.0591 \log [H^+]$$

$$= E^\circ + 0.0591 \text{ pH}$$

$$= 1.30 + 0.0591 \times 2 \approx 1.42 \text{ V}$$

120 (*d*) For first cell $\quad E^\circ_{cell} = E^\circ_{Zn/Zn^{2+}} + E^\circ_{Cu^{2+}/Cu}$

$$1.10 = E^\circ_{Zn/Zn^{2+}} + 0.34$$

$$\therefore \quad E^\circ_{Zn/Zn^{2+}} = 0.76 \text{ V}$$

For second cell $\quad E^\circ_{cell} = E^\circ_{Cu/Cu^{2+}} + E^\circ_{Ag^+/Ag}$

$$0.46 = -0.34 + E^\circ_{Ag^+/Ag}$$

$$E^\circ_{Ag^+/Ag} = 0.80 \text{ V}$$

For third cell,

$$E^\circ_{cell} = E^\circ_{Zn/Zn^{2+}} + E^\circ_{Ag^+/Ag}$$

$$= 0.76 + 0.80 = 1.56 \text{ V}$$

123 (*c*) $Pt(H_2) \mid pH = 2$

$$H_2(g) \rightleftharpoons 2H^+ + 2e^-$$

$$K = [H^+]^2 = (10^{-2})^2 = 10^{-4} \text{ M}$$

$$E = E^\circ - \frac{0.0591}{2}\log 10^{-4} = 0.1182 \text{ V}$$

124 (*a*) Quinhydrone contains quinone (*A*) and quinol (*B*) in 1 : 1 ratio

O ... O $+ 2H^+ + 2e^- \longrightarrow$ OH ... OH

Quinone (*A*) (*B*)

Thus, $\quad K = \frac{1}{[H^+]^2}$

Thus, E depends on $[H^+]$.

125 (*d*) $E^\circ_{calomel} + E^\circ_X = 0.10$

$$E^\circ_X = 0.344 \text{ V}$$

Since, $\quad E^\circ_{SHE} = 0.00 \text{ V}$

Hence, $\quad E^\circ_X \,(wrt) = 0.344 \text{ V}$

129 (*a*) $\left(\frac{dE}{dT}\right)_p = \frac{\Delta S}{nF}$

130 (*c*) E_{cell} is an intensive property as it does not depend upon mass of species (number of particles) but $\Delta_r G$ of the cell reaction is an extensive property because this depends upon mass of species (number of particles).

132 (*a*) $2H_2 + O_2 \longrightarrow 2H_2O$

Total electrons involved = 4

$$\therefore \quad \Delta G^\circ = -nFE^\circ$$

$$-475 \times 1000 = -4 \times 96500\, E^\circ$$

$$\therefore \quad E^\circ = 1.23 \text{ V}$$

133 (*c*) $2 \times [Al^{3+} + 3e^- \longrightarrow Al] \qquad E^\circ = -1.66 \text{ V}$

$3 \times [Mg \longrightarrow Mg^{2+} + 2e^-] \qquad E^\circ = +2.37 \text{ V}$

$2Al^{3+} + 3Mg \longrightarrow 3Mg^{2+} + 2Al \qquad E^\circ_{cell} = 0.71 \text{ V}$

140 (*b*) By Gibbs Helmholtz equation,

$$\Delta G = \Delta H + T\left[\frac{d(\Delta G)}{dT}\right]$$

Also, $\quad \Delta G = \Delta H - T\Delta S$

$$\therefore \quad \Delta S = -\left[\frac{d(\Delta G)}{dT}\right] = \frac{-d(-nFE)}{dT} = nF\left(\frac{dE}{dT}\right)$$

$$= 2 \times 96500 \times \frac{(0.48 - 0.46)}{310 - 300}$$

$$= 386 \text{ JK}^{-1}$$

141 (*d*) Temperature coefficient $= \frac{\Delta S}{nF}$

$$2 \times 10^{-3} = \frac{\Delta S}{2 \times 96500}$$

$$\Delta S = 386 \text{ JK}^{-1}$$

142 (*c*) Reducing power of Al is larger than that of Cu.

$$\begin{array}{lr} & E^\circ \\ [Al \longrightarrow Al^{3+} + 3e^-] \times 2 & 1.66\text{ V} \\ [Cu^{2+} + 2e^- \longrightarrow Cu] \times 3 & 0.34\text{ V} \\ \hline \therefore \quad 2Al + 3Cu^{2+} \longrightarrow 3Cu + 2Al^{3+} & 2.00\text{ V} \end{array}$$

148 (*a*) While charging the lead storage battery the reaction occurring on cell is reversed and $PbSO_4$ (*s*) on anode and cathode is converted into PbO_2 respectively.

Hence, option (a) is the correct choice.

The electrode reactions are as follows

At cathode,

$$PbSO_4(s) + 2e^- \longrightarrow Pb(s) + SO_4^{2-}(aq) \quad \text{(Reduction)}$$

At anode,

$$PbSO_4(s) + 2H_2O \longrightarrow PbO_2(s) + SO_4^{2-} + 4H^+ + 2e^- \quad \text{(Oxidation)}$$

Net reaction

$$2PbSO_4(s) + 2H_2O \longrightarrow Pb(s) + PbO_2(s) + 4H^+(aq) + 2SO_4^{2-}(aq)$$

150 (*c*) **Oxidation**

$$Pb(s) + SO_4^{2-}(aq) \longrightarrow PbSO_4(s) + 2e^-$$

Reduction

$$\begin{array}{l} PbO_2(s) + 4H^+(aq) + SO_4^{2-}(aq + 2e^- \longrightarrow PbSO_4(s) + 2H_2O(l) \\ \hline Pb(s) + PbO_2(s) + 2H_2SO_4(aq) \longrightarrow 2PbSO_4(s) + 2H_2O(l) \end{array}$$

Thus, 2 mol H_2SO_4 are consumed by exchange of 2 mol of electrons.

$$\therefore \quad z = \frac{\text{mol of } H_2SO_4}{\text{mol of electrons} \times 96500} = \frac{2 \times 98}{2 \times 96500}$$

$$w = zit$$

$$= \frac{2 \times 98 \times 2.5 \times 965 \times 60}{2 \times 96500}$$

$$= 147\text{ g} = 1.5\text{ mol}$$

157 (*d*) $\Delta G^\circ = \Delta G_f(ZnO) - \Delta G_f(Ag_2O)$

$$= -318.30 + 11.21 = -307.09\text{ kJ}$$

$$= -307.09 \times 10^3\text{ J}$$

$$\Delta G^\circ = -nFE^\circ_{cell}$$

$$\therefore \quad -307.09 \times 10^3 = -2 \times 96500 \times E^\circ_{cell}$$

$$\therefore \quad E^\circ_{cell} = 1.591\text{ V}$$

159 (*d*)

A. Chemical reaction occurring on lead storage battery can be represented as

At anode, $Pb(s) + SO_4^{2-}(aq) \longrightarrow PbSO_4(s) + 2e^-$

At cathode, $PbO_2(s) + SO_4^{2-}(aq) + 4H^+(aq) \xrightarrow{+2e^-} 2PbSO_4(s) + 2H_2O(l)$

Thus, Pb is anode and PbO_2 is cathode.

B. Mercury cell does not include ions during their function hence produce steady current.

C. Fuel cell has maximum efficiency as they produce energy due to combustion reaction of fuel.

D. Rusting is prevented by corrosion.

160 (*d*)

A. **Leclanche cell** The electrode reaction occurs at Leclanche cell are

At anode, $Zn(s) \longrightarrow Zn^2 + 2e^-$

At cathode, $MnO_2 + NH_4^+ + e^- \longrightarrow MnO(OH) + NH_3$

B. Ni-Cd cell is rechargeable. So, it has more life time.

C. Fuel cell produces energy due to combustion. So, fuel cell converts energy of combustion into electrical energy, e.g.

$$2H_2 + O_2 \longrightarrow 2H_2O$$

D. Mercury cell does not involve any ion in solution and is used in hearing aids.

161 (*c*) Statement I is false but statement II is true.

Correct statement I is mercury cell gives steady potential.

Statement II is correct as ions are not involved in cell reaction.

(B) Miscellaneous questions

1 (*d*) $Ag_2CO_3 \rightarrow 2Ag + CO_2 + \frac{1}{2}O_2$

4 (*a*)

I : $Fe^{2+} + 2e^- \longrightarrow Fe \qquad E^\circ_{Fe^{2+}/Fe} = x_1, \quad n_1 = 2$

II : $Fe^{3+} + 3e^- \longrightarrow Fe \qquad E^\circ_{Fe^{3+}/Fe} = x_2, \quad n_2 = 3$

III : $Fe^{3+} + e^- \longrightarrow Fe^{2+} \qquad E^\circ_{Fe^{3+}/Fe^{2+}} = ?, \quad n_3 = 1$

II - I = III

$$\Delta G_3 = \Delta G_2 - \Delta G_1$$

$$-n_3FE^\circ_3 = -n_2FE^\circ_2 + n_1FE^\circ_1$$

$$E^\circ_3 = \frac{n_2E^\circ_2 - n_1E^\circ_1}{n_3} = 3x_2 - 2x_1$$

5 (*c*) $E = E^\circ - \frac{0.0591}{2}\log\left(\frac{C_1}{C_2}\right) = \frac{0.0591}{2}\log\frac{C_2}{C_1}$

To make $\Delta G = -$ve, $E = +$ve

Hence, $C_2 > C_1$

7 (*c*)

$$\begin{array}{lccc} \text{LHS} & \underset{x\text{ atm}}{H_2(g)} & \rightleftharpoons & \underset{0.01\text{ M}}{2H^+(g)} + 2e^- \\ \text{RHS} & \underset{0.1\text{ M}}{2H^+(aq)} + 2e^- & \longrightarrow & \underset{y\text{ atm}}{H_2(g)} \\ \hline & \underset{x}{H_2(g)} + \underset{0.1\text{ M}}{2H^+(aq)} & \rightleftharpoons & \underset{0.01\text{ M}}{2H^+} + \underset{y\text{ atm}}{H_2(g)} \end{array}$$

$$K = \frac{(0.01)^2 y}{(0.1)^2 x} = 0.01\left(\frac{y}{x}\right)$$

$$E_{cell} = E^\circ_{cell} - \frac{0.0591}{2}\log(0.01)\left(\frac{y}{x}\right)$$

$$0 = 0 + 0.0591\log 10 - \frac{0.0591}{2}\log\frac{y}{x}$$

$$\log 10 = \log\left(\frac{y}{x}\right)^{1/2}$$

$$\therefore \quad 10 = \left(\frac{y}{x}\right)^{1/2};\ \left(\frac{x}{y}\right)^{1/2} = \frac{1}{10}$$

$$\therefore \quad \frac{x}{y} = 0.01$$

8 (*b*) $E = E^\circ_{cell} - \frac{0.0591}{2}\log\frac{[Zn^{2+}]}{[Cd^{2+}]}$

$$0.30 = 0.36 - \frac{0.0591}{2}\log\frac{[Zn^{2+}]}{[Cd^{2+}]}$$

$$\log\frac{[Zn^{2+}]}{[Cd^{2+}]} = \frac{0.06\times 2}{0.0591} \approx 2$$

$$\therefore \quad \frac{[Zn^{2+}]}{[Cd^{2+}]} = 10^2$$

$$\therefore \quad \frac{[Cd^{2+}]}{[Zn^{+2}]} = 0.01$$

9 (*d*) 1 L of 1 M $CuSO_4$ = 1 mol of Cu^{2+}

$$Cu^{2+} + 2e^- \longrightarrow Cu(s)$$

2 F of electricity will convert dil. Cu^{2+} into Cu, hence molarity of $[CuSO_4] = 0$

12 (*b*) $\lambda_0(BaSO_4) = \frac{1000 \times \text{specific conductance}}{\text{conc. (normality)}}$

$$\text{Normality} = \frac{1000\times 8\times 10^{-5}}{400}$$

$$\therefore \quad \text{Molarity} = \frac{\text{normality}}{2} = 1\times 10^{-4}\text{ M}$$

$$\therefore \quad K_{sp} = S^2 = 1\times 10^{-8}\text{ M}^2$$

Note *In K_{sp}, S is solubility in mol L^{-1} (molarity).*

13 (*b*) Specific conductance = conductance × cell constant

$\kappa_{KCl} = (\text{Conductance})_{KCl} \times$ cell constant

$\kappa_{NaCl} = (\text{Conductance})_{NaCl} \times$ cell constant

$$\therefore \quad x = y \times \text{cell constant}$$

$$\kappa_{NaCl} = z \times \text{cell constant}$$

$$\therefore \quad \kappa_{NaCl} = \frac{xz}{y}$$

14 (*d*) Cell reaction is

$$Ni(s) + Cl_2(g) \longrightarrow Ni^{2+}(aq) + 2Cl^-(aq)$$

$$Q = \frac{[Ni^{2+}][Cl^-]^2}{p_{Cl_2}}$$

$$= \frac{(0.19)(0.40)^2}{1} = 3.04 \times 10^{-2}$$

15 (*a*)

$$\underset{18\text{ g} = 1\text{ mol}}{H_2O} \longrightarrow \underset{1\text{ mol}}{H_2} + \underset{0.5\text{ mol}}{\frac{1}{2}O_2}$$

$$w = zit$$

$$2\text{ g } H_2 = \frac{2}{96500\times 2}\times 3\times t$$

When z is electrochemical equivalent = $\frac{\text{molar mass of } H_2}{\text{electron}\times 96500}$

$$\therefore \quad t = 64333.33\text{ s}$$

$$\therefore \quad t = 17.87\text{ h}$$

16 (*b*) $Zn^{2+} + 2e^- \longrightarrow Zn$

$$z\text{ (electrochemical equivalent)} = \frac{65}{2\times 96500}$$

$$i\text{ (current)} = 5\text{ A}$$

$$t\text{ (time)} = 40\text{ min} = 2400\text{ s}$$

$$\therefore \quad w = zit = \frac{65\times 5\times 2400}{2\times 96500} = 4.041\text{ g}$$

17 (*d*) Equivalent conductance of 0.04 N weak acid

$$= \frac{1000\times 4.23\times 10^{-4}}{0.04}$$

$$= 10.575\ \Omega^{-1}\text{ cm}^2\text{ equiv}^{-1}$$

Also, x (degree of ionisation) $= \frac{\Lambda_c}{\Lambda_\infty}$

$$0.0612 = \frac{10.575}{\Lambda_\infty}$$

$$\Lambda_\infty = 172.79\ \Omega^{-1}\text{ cm}^2\text{ equiv}^{-1}$$

18 (*b*) Theoretical value of Cu^{2+} discharged at cathode = zit

$$= \frac{63.5\times 3\times 2\times 3600}{2\times 96500} = 7.107\text{ g}$$

Actual weight of Cu^{2+} discharged = 3.0 g

Thus, % efficiency $= \frac{3.0}{7.107} = 42.2\%$

19 (*a*) Both Statement I and Statement II are correct.

Electrolysis of NaCl is represented by following chemical reactions.

At cathode, $H^+(aq) + e^- \longrightarrow \frac{1}{2}H_2(g)$

At anode $\quad Cl^-(aq) \longrightarrow \frac{1}{2}Cl_2 + e^-;\quad E^\circ_{cell} = -1.36\text{V}$

$$2H_2O(aq) \longrightarrow O_2(g) + 4H^+(aq) + 4e^-;$$

$$E^\circ_{cell} = -1.23\text{ V}$$

E°_{cell} for this reaction has lower value but formation of oxygen at anode requires over potential.

20 (*d*) $\Delta G_3^\circ = \Delta G_1^\circ + \Delta G_2^\circ$

$$-n_3FE_3^\circ = -n_1FE_1^\circ - n_2FE_2^\circ$$

$$\therefore \quad E_3^\circ = \frac{n_1E_1^\circ + n_2E_2^\circ}{n_3}$$

$$= \frac{0.77 + 2\times(-0.44)}{3} = -0.037\text{ V}$$

21 (*a*)

$$Tl^{3+} + 2e^- \longrightarrow Tl^+;\quad E_1^\circ = 1.26\text{ V}$$

$$Tl^+ + e^- \longrightarrow Tl\quad E_2^\circ = -0.336\text{ V}$$

$$\therefore \quad Tl^{3+} + 3e^- \longrightarrow Tl\quad E_3^\circ = ?$$

$$\Delta G_3 = \Delta G_1^\circ + \Delta G_2^\circ$$
$$-n_3 E_3^\circ = -n_1 E_1^\circ - n_2 E_2^\circ$$
$$+3E_3^\circ = 2\times 126 - 1\times 0.336$$
$$\therefore \quad E_3^\circ = 0.728 \text{ V}$$

22 (*d*) $\Lambda_{eq} = \lambda^\infty(H^+) + \lambda^\infty(SO_4^{2-})$

$= 80 + 350$

$= 430 \text{ S cm}^2 \text{ equivalent}^{-1}$

Since, molar mass of $H_2SO_4 = 98 = 2\times$ equivalent mass

$\therefore \quad \Lambda_m = 2\times 430 = 860 \text{ S cm}^2 \text{ mol}^{-1}$

24 (*d*) $Zn + 2Ag^+ \rightleftharpoons 2Ag + Zn^{2+}$

$$Q = \frac{[Zn^{2+}]}{[Ag^+]^2} = 6.40\times 10^{-3}$$

27 (*a*) $Al^{3+} + 3e^- \longrightarrow Al$

$$w = zit$$
$$100 = \frac{27\times 125\times t}{3\times 96500}$$
$$t = 8578 \text{ s} = 143 \text{ min}$$

28 (*d*) $Al_2O_3 \longrightarrow 2Al + \frac{3}{2}O_2$

Ratio of Al to $O_2 = 2 : \frac{3}{2} = 4 : 3$

29 (*c*) **Reaction** **Number of moles formed by 1F**

Reaction	Number of moles formed by 1F
$Cu^{2+} + 2e^- \longrightarrow Cu$	1/2
$Al^{3+} + 3e^- \longrightarrow Al$	1/3
$Si^{4+} + 4e^- \longrightarrow Si$	1/4
Ratio	$\frac{1}{2} : \frac{1}{3} : \frac{1}{4}$
	$6 : 4 : 3$

30 (*a*) $Pb(s) \rightleftharpoons Pb^{2+}(aq) + 2e^-$

$Cu^{2+}(aq) + 2e^- \rightleftharpoons Cu(s)$

$Cu^{2+}(aq) + Pb(s) \rightleftharpoons Pb^{2+}(aq) + Cu(s)$

$$K = \frac{[Pb^{2+}]}{[Cu^{2+}]} = \frac{1}{[Cu^{2+}]}$$
$$\therefore \quad E = E^\circ - \frac{0.0591}{2}\log K = E^\circ + \frac{0.0591}{2}\log[Cu^{2+}]$$
$$0 = 0.473 + \frac{0.0591}{2}\log[Cu^{2+}]$$
$$\log[Cu^{2+}] = -16$$
$$\therefore \quad [Cu^{2+}] = 1\times 10^{-16} \text{ M}$$

31 (*d*) $Ni^{2+}(aq) + H_2(g) \rightleftharpoons Ni(s) + 2H^+(aq)$

$$K = \frac{[H^+]^2}{[Ni^{2+}]} = [H^+]^2$$
$$\therefore \quad E_{cell} = E_{cell}^\circ - \frac{0.0591}{2}\log[H^+]^2$$
$$(E_{cell} - E_{cell}^\circ) = 0.0591 \text{ pH}$$
$$0.0591 = 0.0591 \text{ pH}$$
$$\therefore \quad \text{pH} = 1$$

32 (*d*) $2H_2O + 2e^- \longrightarrow H_2 + 2OH^- \qquad E^\circ = -0.8277 \text{ V}$

$H_2 \longrightarrow 2H^+ + 2e^- \qquad E^\circ = 0.00 \text{ V}$

$\therefore \quad 2H_2O \rightleftharpoons 2H^+ + 2OH^- \qquad E^\circ = -0.8277 \text{ V}$

$$E = E^\circ - \frac{0.0591}{2}\log[H^+]^2[OH^-]^2$$

At equilibrium, $E = 0$

$$\therefore \quad E^\circ = \frac{0.0591}{2}\log[H^+]^2[OH^-]^2$$
$$-0.8277 = 0.0591\log[H^+][OH^-]$$
$$-0.8277 = 0.0591\log K_w$$
$$\therefore \quad K_w = 1\times 10^{-14}$$

33 (*a*) Both statements are correct.

Concentration of ionic solution changes on using DC current as a source of energy while on passing AC current, concentration does not change. Hence, AC source is used for measuring resistance.

34 (*c*)

	E°
$Pb \longrightarrow Pb^{2+} + 2e^-$	+ 0.127 V
$2Cu^+ + 2e^- \longrightarrow Cu$	+ 0.518 V
$2Cu^+ + Pb \longrightarrow Pb^{2+} + Cu$	+ 0.645 V

Thus, (a) and (b) are true.

35 (*d*) $Q = \dfrac{[Fe^{2+}]^2}{[H^+]^4 p O_2} = \dfrac{(1\times 10^{-3})^2}{(1\times 10^{-3})^4(0.1)} = 1\times 10^7$

$$\therefore \quad E = E_{cell}^\circ - \frac{0.0591}{4} = \log_{10^7}$$
$$= 1.67 - \frac{0.0591\times 7}{4} = 1.57 \text{ V}$$

36 (*c*) $A + B \rightleftharpoons C + D$

$$K_1 = K_{eq} = \frac{[C][D]}{[A][B]}$$

$E^\circ = x$ volt

$$2A + 2B \rightleftharpoons 2C + 2D$$
$$K_2 = K_{eq} = \frac{[C]^2[D]^2}{[A]^2[B]^2} = K_1^2$$

$E^\circ = y$ volt

Since, E° is an intensive property, hence $x = y$ and $K_2 = K_1^2$

37 (*a*) $E^\circ = \dfrac{0.0591}{n}\log K$

$$0.15 = \frac{0.0591}{10}\log K$$
$$K = 2.4\times 10^{25}$$

38 (c) $E = E^\circ + \frac{0.0591}{1}\log\frac{[Ag^+]_R}{[Ag^+]_L}$

$= 0 + 0.0591 \log 2 = 0.0591 \times 0.301\,V$

After current is passed $[Ag^+]_R = 1\,M$

$[Ag^+]_L = 0.5\,M$

Hence, no change in emf takes place.

39 (a) $E = E^\circ + \frac{0.0591}{5}\log(2)^8$

$E - E^\circ = \frac{0.0591}{5} \times 8 \log 2$

$= 0.02846\,V = 28.46\,mV$

40 (d) $0.224\text{ cc } O_2 = 3.2 \times 10^{-4}\,g$

$z = \frac{32}{4 \times 96500}$

$i = \frac{w}{zt} = \frac{3.2 \times 10^{-4} \times 96500 \times 4}{32 \times 1} = 3.86\,A$

44 (d) More negative values of $E^\circ_{M^{n+}/M}$ represents better reducing agent and more positive values of $E^\circ_{M^{n+}/M}$ represents better oxidising agent.

45 (b)

$$Zn \longrightarrow Zn^{2+} + 2e^- ; \quad E^\circ = 0.76\,V$$
$$Fe^{2+} + 2e^- \longrightarrow Fe ; \quad E^\circ = -0.44\,V$$
$$Fe^{2+} + Zn \longrightarrow Zn^{2+} + Fe \quad E^\circ = 0.32\,V$$

46 (a) $E^\circ = \frac{0.0591}{n}\log K_{eq}$

$1.10 = \frac{0.0591}{2}\log K_{eq}$

$\therefore \quad K_{eq} = 37.22$

$K_{eq} = 1.68 \times 10^{37}$

47 (a) $H_2O_2 + 2H^+ + 2Fe^{2+} \longrightarrow 2Fe^{3+} + 2H_2O$

$E^\circ_{cell} = 1.77 + 0.771 = 2.541\,V$

49 (b) $E_{cell} = E^\circ_{cell} - \frac{0.0591}{2}\log\frac{[Zn^{2+}]}{[Cu^{2+}]}$

From graph, $OA = E^\circ_{cell} = 1.1\,V$

If $\frac{[Zn^{2+}]}{[Cu^{2+}]} = 10^{-2}\,M$, then $E_{cell} = 1.1591\,V$

50 (a)

A. Conductivity $(\kappa) = \frac{G^*}{R}$

B. Molar conductivity $(\Lambda_m) = \frac{\kappa}{C}$

C. Degree of dissociation $(\alpha) = \frac{\Lambda_m}{\Lambda^\circ_m}$

D. Charge $Q = I \times t$

where, Q is the quantity of charge in Coulomb when I ampere of current is passed through an electrolyte for t second.

51 (a) Both Statement I and Statement II are correct and current stop flowing, when $E_{cell} = 0$

At $E_{cell} = 0$, reaction reaches the equilibrium.

52 (a) $Cu + 2Ag^+ \longrightarrow 2Ag + Cu^{2+}$

$K = \frac{[Cu^{2+}]}{[Ag^+]^2} = \frac{0.01}{(0.1)^2} = 1$

$\therefore \quad E = E^\circ - \frac{0.0591}{2}\log K$

$= 0.46 - \frac{0.0591}{2}\log 1 = 0.46\,V$

54 (d)

A. Λ_m (molar conductivity) is the conductivity due to number of ions furnished by one mole of electrolyte. As dilution increases number of ions present in the solution increases hence molar conductivity increases.

B. E°_{cell} of any atom/ion does not depend upon number of atom/ion, hence E°_{cell} of any atom/ion is an intensive properties.

C. κ represents specific conductivity which depends upon number of ions present in per unit volume.

D. $\Delta_r G_{cell}$ is an extensive property as it depends upon number of particles(species).

(C) Medical entrances' gallery

1 (a) Dilute H_2SO_4 contains H_2SO_4 and H_2O and they show electrolytic dissociation as,

$$\underset{\text{Strong electrolyte as it is dilute}}{H_2SO_4} \longrightarrow 2H^+ + SO_4^{2-}$$

$$H_2O \rightleftharpoons H^+ + OH^-$$

Reaction at cathode (Pt)

$$2H^+ + 2e^- \longrightarrow H_2\uparrow$$

Reaction at anode (Pt)

$$4OH + 4e^- \longrightarrow O_2\uparrow + 2H_2O$$

We get, O_2 gas at anode.

2 (d) $\underset{\text{Molten}}{CaCl_2} \rightleftharpoons Ca^{2+} + 2Cl^-$

At cathode (Reduction) $Ca^{2+} + 2e^- \longrightarrow Ca$

According to Faraday's first law,

Charge passed in Faraday = Gram equivalent of product

$= \frac{\text{Given mass}}{\text{Atomic mass}} \times \text{no. of } e^- \text{ released}$

$= \frac{20}{40} \times 2 = 1F$

So, one Faraday (F) is required for the process.

3 (*c*) Let us consider the segment of EMF series (standard reduction potential) of the given reduction half-cell reactions:

Reduction half reaction	E/V at 298K
(a) $Mg^{2+} + 2e \longrightarrow Mg$	− 2.36
(b) $Fe^{2+} + 2e \longrightarrow Fe$	− 0.44
(c) $Au^{3+} + 3e \longrightarrow Au$	+ 1.40
(d) $K^{+} + e \longrightarrow K$	− 2.93

Hence, option (c) is correct.

4 (*d*) The relation between E°_{cell} and $\Delta_r G^{\circ}$ is as follows

$$\Delta_r G^{\circ} = -nFE^{\circ}_{cell}$$

For the cell reaction,

$$2Fe^{3+}(aq) + 2I^{-}(aq) \longrightarrow 2Fe^{2+}(aq) + I_2(aq),$$

$$n = 2$$

Given, $F = 96500 \text{ C mol}^{-1}$, $E^{\circ}_{cell} = 0.24 \text{ V}$

Now, we know that

$$\Delta_r G^{\circ} = -n\,FE^{\circ}_{cell}$$

On substituting the given values in above equation,

$$\Delta_r G^{\circ} = -2 \text{ mol} \times 96500 \text{ C mol}^{-1} \times 0.24 \text{ J mol}^{-1}$$

$$= -46320 \text{ J mol}^{-1}$$

$$= -46.32 \text{ kJ mol}^{-1}$$

Hence, option (d) is correct.

5 (*b*) We know that,

$$E_{cell} = E^{\circ}_{cell} - \frac{2.303\,RT}{nF}\log Q$$

At equilibirum,

$$Q = K_{eq} \text{ and } E_{cell} = 0$$

$$\therefore \quad E^{\circ}_{cell} = \frac{2.303RT}{nF}\log K_{eq} \quad \text{... (i)}$$

Given, $E^{\circ}_{cell} = 0.59 \text{ V}$

Also, at 298 K, $\frac{2.303RT}{F} = 0.059 \text{ V}$,

$$T = 298\text{K}, n = 1$$

On substituting the given values in Eq. (i)

$$0.59 = \frac{0.059}{1}\log K_{eq}$$

$$\therefore \quad \log K_{eq} = \frac{0.59}{0.059} = 10$$

$$K_{eq} = 10^{10} = 1 \times 10^{10}$$

Thus, option (b) is correct.

6 (*b*) According to Nernst equation,

$$E_{cell} = E^{\circ}_{cell} - \frac{2.303}{nF}RT\log\frac{[M_p]^{n+}}{[M_r]^{n+}}$$

where, $[M_p]^{n+}$ and $[M_r]^{n+}$ are the concentrations of products and reactants respectively.

For the given cell, emf can be calculated as follows :

$$E_{emf} = E^{\circ}_{cell} - \frac{2.303RT}{nF}\log\frac{[M_p]^{n+}_{(aq)}}{[M_r]^{n+}_{(aq)}}$$

For the reaction at 25° C,

$$E_{emf} = E^{\circ}_{cell} - \frac{0.059}{n}\log\frac{[M_p^{n-1}]}{[M_r^{n+}]}$$

$$\therefore \quad E^{\circ}_{cell} = 4 \text{ V}$$

(Given) and $n = 2$

$$E_{emf} = 4 - \frac{0.059}{2}\log\frac{10^{-4}}{10^{-2}}$$

$$E_{emf} = 4 - 0.03\log 10^{-2}$$

$$E_{emf} = 4 + 0.06 = 4.06 \text{ V}$$

Hence, option (b) is correct.

7 (*b*) More negative the value of standard reduction potential, higher is the reduction power.

i.e. Reducing power $\propto \frac{1}{\text{standard reduction potential}}$

Thus, the correct decreasing order of reducing power of the metal is

K > Al >

$(E^{\circ}_{K^+/K = -2.93V})\,(E^{\circ}_{Al^{3+}/Al = -1.66V})$

Cr > Ag

$(E^{\circ}_{Cr^{3+}/Cr = -0.74V})\,(E^{\circ}_{Ag^+/Ag = 0.80V})$

So, the correct decreasing order of reducing power of the metal is,

$$K > Al > Cr > Ag$$

Hence, option (b) is correct.

8 (*c*) $E^{\circ}_{cell} = E_{cathode} - E_{anode}$

$$= -0.42 - (-0.72) = 0.3 \text{ V}$$

At anode $Cr \longrightarrow Cr^{3+} + 3e^{-}] \times 2$

At cathode $Fe^{2+} + 2e^{-} \longrightarrow Fe] \times 3$

$$2Cr + 3Fe^{2+} \longrightarrow 2Cr^{3+} + 3Fe$$

$$\therefore \quad E = E^{\circ}_{cell} - \frac{0.0591}{6}\log\frac{[Cr^{3+}]^2}{[Fe^{2+}]^3}$$

$$= 0.3 - \frac{0.0591}{6}\log\frac{(0.1)^2}{(0.01)^3} = 0.26 \text{ V}$$

9 (*d*) According to Kohlrausch's law,

Λ°_m for $A_xB_y = x\Lambda^{\circ}_{(A)^{y+}} + y\Lambda^{\circ}_{(B)^{x-}}$

where, Λ°_m = limiting molar conductivity of electrolyte and $\Lambda^{\circ}_{(A)^{y+}}$ and $\Lambda^{\circ}_{(B)^{x-}}$ are the limiting molar conductivities of cation (A^{y+}) and anion (B^{x-}), respectively.

Given $\Lambda^{\circ}_{m(H_2SO_4)} = 2\Lambda^{\circ}_{H^+} + \Lambda^{\circ}_{SO_4^{2-}}$

$$= x \text{ S cm}^2\text{mol}^{-1} \quad \text{...(i)}$$

$$\Lambda^{\circ}_{m(K_2SO_4)} = 2\Lambda^{\circ}_{K^+} + \Lambda^{\circ}_{SO_4^{2-}} = y \text{ S cm}^2\text{mol}^{-1} \quad \text{...(ii)}$$

$$\Lambda^{\circ}_{m}(CH_3COOK) = \Lambda^{\circ}_{CH_3COO^-} + \Lambda^{\circ}_{K^+} = z \text{ S cm}^2\text{mol}^{-1} \quad \text{...(iii)}$$

To find :

$$\Lambda^\circ_{m(CH_3COOH)} = \Lambda^\circ_{CH_3COO^-} + \Lambda^\circ_{H^+}$$

The above equation can be obtained by

$$\text{Eq. (iii)} + \frac{\text{Eq. (i)}}{2} - \frac{\text{Eq. (ii)}}{2}$$

$$\therefore \Lambda^\circ_{m(CH_3COOH)} = \Lambda^\circ_{m(CH_3COOK)} + \frac{\Lambda^\circ_{m(H_2SO_4)}}{2} - \frac{\Lambda^\circ_{m(K_2SO_4)}}{2}$$

$$= \left(z + \frac{x}{2} - \frac{y}{2}\right) \text{S cm}^2 \text{ mol}^{-1}$$

$$= \left(\frac{x-y}{2} + z\right) \text{S cm}^2 \text{ mol}^{-1}$$

10 (c) Given,

$$\Delta G_f^\circ (Ag_2O) = -11.21 \text{ kJ mol}^{-1}$$

$$\Delta G_f^\circ (ZnO) = -318.3 \text{ kJ mol}^{-1}$$

As we know that,

$$\Delta G^\circ = \Delta G_f^\circ (ZnO) - \Delta G_f^\circ (Ag_2O)$$

Putting the values in the above given equation,

$$\therefore \quad \Delta G^\circ = (-318.30 + 11.21) \text{ kJ mol}^{-1}$$

$$= -307.09 \text{ kJ} = -307.09 \times 10^3 \text{ J}$$

$$\because \quad \Delta G^\circ = -nFE^\circ_{cell}$$

$$\therefore \quad -307.09 \times 10^3 = -2 \times 96500 \times E^\circ_{cell}$$

or $\quad E^\circ_{cell} = 1.591 \text{ V}$

11 (d) $E_1 = E_0 - \frac{0.0591}{n} \log \frac{[Zn^{2+}]}{[Cu^{2+}]}$

$$= E_0 + \frac{0.0591}{2} \log \frac{0.01}{1}$$

$$E_2 = E_0 - \frac{0.0591}{n} \log \frac{[Zn^{2+}]}{[Cu^{2+}]}$$

$$= E_0 - \frac{0.0591}{2} \log \frac{1}{0.01}$$

$\therefore E_1 > E_2$

12 (c) The cell reaction is

$$Zn + Fe^{2+} \longrightarrow Zn^{2+} + Fe$$

From Nernst equation,

$$E_{cell} = E^\circ_{cell} - \frac{0.0591}{n} \log \frac{a_{Zn^{2+}}}{a_{Fe^{2+}}}$$

$$= (0.763 - 0.44) - \frac{0.0591}{2} \log \frac{0.001}{0.005} = 0.344 \text{ V}$$

13 (c) For the electrochemical cell,

$Ag^+ | AgCl | KCl || AgNO_3 | Ag^+$

The cell reaction is,

$$AgCl\,(s) + e^- \longrightarrow Ag + Cl^-(aq)$$

$$Ag \longrightarrow Ag^+(aq) + e^-$$

Overall reaction, $AgCl(s) \longrightarrow Ag^+ + Cl^-$

14 (c) Given, $\Lambda^\circ_m (NaOH) = 248 \times 10^{-4} \text{Sm}^2\text{mol}^{-1}$

$$\Lambda^\circ_m (NaCl) = 126 \times 10^{-4} \text{Sm}^2\text{mol}^{-1}$$

$$\Lambda^\circ_m (BaCl_2) = 280 \times 10^{-4} \text{Sm}^2\text{mol}^{-1}$$

The reaction for the formation of $Ba(OH)_2$ can be written as,

$$BaCl_2 + 2NaOH \longrightarrow Ba(OH)_2 + 2NaCl$$

$$\because \quad \Lambda^\circ_{m\,Ba(OH)_2} = \Lambda^\circ_{m\,BaCl_2} + 2\Lambda^\circ_{m\,NaOH} - 2\Lambda^\circ_{m\,NaCl}$$

$$\therefore \quad \Lambda^\circ_{m\,Ba(OH)_2} = (280 \times 10^{-4}) + (2 \times 248 \times 10^{-4}) - (2 \times 126 \times 10^{-4})$$

$$= (280 + 496 - 252) \times 10^{-4}$$

$$= 524 \times 10^{-4} \text{ Sm}^2\text{mol}^{-1}$$

15 (d) Given, $\Lambda^\circ_{Ca^{2+}} = 119 \text{ S cm}^2 \text{ mol}^{-1}$

$$\Lambda^\circ_{Cl^-} = 71 \text{ S cm}^2 \text{ mol}^{-1}$$

As we know that,

$$\Lambda^\circ_{CaCl_2} = \Lambda^\circ_{Ca^{2+}} + 2 \times \Lambda^\circ_{Cl^-}$$

$$\therefore \quad \Lambda^\circ_{CaCl_2} = 119 + 2 \times 71$$

$$\Lambda^\circ_{CaCl_2} = 261 \text{ S cm}^2 \text{ mol}^{-1}$$

16 (d) $Zn + Cu^{2+} \longrightarrow Cu + Zn^{2+}$

$$\therefore \quad K = \frac{[Zn^{2+}]}{[Cu^{2+}]}$$

$$E_1 = E^\circ - \frac{0.0591}{2} \log \frac{0.01}{100}$$

$$= (E^\circ + 0.0591)$$

$$E_2 = E^\circ - \frac{0.0591}{2} \log \frac{1.00}{0.01}$$

$$= E^\circ - 0.0591$$

Thus, $\quad E_1 > E_2$

17 (d) From the question, we have an equation

$$2H^+ + 2e^- \longrightarrow H_2(g)$$

According to Nernst equation,

$$E = E^\circ = \frac{0.0591}{2} \log \frac{p_{H_2}}{[H^+]^2}$$

$$= 0 - \frac{0.0591}{2} \log \frac{p_{H_2}}{(10^{-7})^2} \quad [\because [H^+] = 10^{-7}]$$

$\therefore$ For potential of H_2 electrode to be zero, p_{H_2} should be equal to $[H_2]^2$, i.e. 10^{-14} atm.

$$\therefore \quad \log \frac{10^{-14}}{(10^{-7})^2} = 0$$

18 (d) $C = 0.5 \text{ mol/dm}^3$, $\kappa = 5.76 \times 10^{10^{-3}} \text{ S cm}^{-1}$, $T = 298 \text{ K}$

$$\lambda_m = \frac{\kappa \times 100}{M} = \frac{5.76 \times 10^{-3}}{0.5}$$

$$= 11.52 \text{ S cm}^2/\text{mol}$$

19 (d) $2Cl^- \longrightarrow Cl_2(g) + 2e^-$

$$w = zit$$

$$z = \frac{E}{96500}$$

$$W = \frac{E}{96500} \times it$$

$$0.1 \times 71 = \frac{35.5}{96500} \times 3 \times t\,(\text{sec})$$

$$t\,(s) = 6433.33 \text{ sec}$$

$$t\,(\text{min}) = 107.22 \text{ min} \approx 110 \text{ min.}$$

20 (c) $\because E^\circ_{cell} = -\text{ve}$

$\therefore \quad \Delta G^\circ = -nF\,E^\circ_{cell}$

$\therefore \quad \Delta G^\circ = +\text{ve}$

$\Rightarrow \quad \Delta G^\circ > 0$

$\because \quad \Delta G^\circ = -2.303\,RT \log K_{eq}$

$\therefore \quad K_{eq} < 1$

21 (a) $Q = ne$

$i.t. = n.e$

$$n = \frac{1 \times 60}{1.6 \times 10^{-19}} = 3.75 \times 10^{20} \text{ electrons}$$

22 (a) Fuel cell is a device that converts energy of combustion of fuels like hydrogen and methane, directly into electrical energy. Electrolytic cell converts electrical energy into chemical energy. Dynamo its an electrical generator that produces direct current with the use of a commulator.

Ni-Cd cell is a type of rechargable battery which constants of a cadmium anode and a metal grid containing NiO_2 acting as a cathode.

23 (e) The reducing power will be higher when reduction potential is lower.

Thus, the correct order of reduction power is $Y > Z > X$.

24 (b) Given that, $E^\circ_{Ag^+/Ag} = +0.80$ V

$E^\circ_{Zn^{2+}/Zn} = 0.76$ V

Cell is $Zn|Zn^{2+}||Ag^+|Ag$

$$E^\circ_{cell} = E^\circ_R - E^\circ_L$$

$$= 0.80 - (-0.76) = 1.56 \text{ V}$$

or $\quad E^\circ_{cell} = E^\circ_{ox} + E^\circ_{red}$

From the equation, $\Delta G^\circ = -nFE^\circ$

(where, $F = 96500$ C, $n = 2$)

$$= -2 \times 96500 \times 1.56$$

$$= -301080 \text{ J/mol}$$

$$= -301.080 \text{ kJ/mol} = -301 \text{ kJ/mol}$$

25 (c)

Metal	E°_{cell}
A	0.34 V
B	−0.80 V
C	−0.46 V

A negative E° means that the redox couple is a stronger reducing agent that the H^+/H_2 couple while a positive E° means that the redox couple is a weaker reducing agent than that H^+/H_2 couple, so correct order will be $B > C > A$.

26 (b) Since, $\quad w = Zit$ or $\frac{w}{Z}$ = constant

$$\frac{(W)_{Al}}{(Z)_{Al}} = \frac{(W)_{Ni}}{(Z)_{Ni}} \Rightarrow \frac{\frac{18}{27}}{3} = \frac{\frac{(W)_{Ni}}{58.5}}{2}$$

$$\frac{18}{9} = \frac{2 \times (W)_{Ni}}{58.5}$$

$\Rightarrow \quad (W)_{Ni} = 58.5$ g

27 (c) $A\,(s) + 2B^+\,(aq) \longrightarrow A^{2+}\,(aq) + 2B(s)$

Oxidation $\quad A\,(s) \longrightarrow A^{2+}\,(aq) + 2e^-$

Reduction $2B^+\,(aq) + 2e^- \longrightarrow A^{2+}\,(aq) + 2B(s)$

Here, $n = 2$, $E^\circ = 1.18$ V

$K_{eq} = ?$

$\because \quad E^\circ_{cell} = \frac{2.303\,RT}{nF} \log K_{eq}$

$$= \frac{0.0591}{n} \log K_{eq} \qquad \left[\frac{F}{RT} = \frac{1}{0.059}\right]$$

$\therefore \quad \log K_{eq} = \frac{2 \times 1.18}{0.059}$

$\therefore \quad \log K_{eq} = 40$

or $\quad K_{eq} = 10^{40}$

28 (b) Given, $\quad E^\circ_{M^+/M} = 1.2$ V

$E^\circ_{X_2/X^-} = -1.1$ V

$E^\circ_{O_2/H_2O} = 1.23$ V

On electrolysis of aqueous solution of salt MX, the products obtained depend on the given E° values, i.e. cathodic reaction will be one with higher E°_{cell} value. Anodic reaction will be one with higher E°_{cell} value or lower E°_{red} value. Hence, the products are H_2 and X_2.

29 (c) $\quad \underset{(+6)}{MnO_4^{2-}} \xrightleftharpoons{1 \text{ mole } e^- \,[=1F]} \underset{(+7)}{MnO_4^-}$

As per the equation, for 1 mole of MnO_4^{2-}, 1 F of electricity is required. Thus, for 0.1 mole of MnO_4^{2-}, 0.1 F of electricity is required.

Since, $\quad 1 \text{ F} = 96500 \text{ C}$

$\therefore \quad 0.1 \text{ F} = 0.1 \times 96500 \text{ C} = 9650 \text{ C}$

Hence, 9650 C of electricity is required to completely oxidise MnO_4^{2-} to MnO_4^-.

30 (d) Since, 22400 mL volume is occupied by 1 mole of O_2 at STP.

Thus, 5600 mL O_2 means $= \frac{5600}{22400}$ mol O_2

$$= \frac{1}{4} \text{ mol } O_2$$

$\therefore \quad$ Weight of $O_2 = \frac{1}{4} \times 32 = 8$ g

According to problem,

Equivalents of Ag = Equivalents of O_2

$$\Rightarrow \frac{\text{weight of Ag}}{\text{equivalent weight of Ag}} = \frac{W_{O_2}}{\text{equivalent weight of } O_2}$$

$$\Rightarrow \frac{W_{Ag}}{\frac{M_{Ag}}{1}} = \frac{W_{O_2}}{\frac{M_{O_2}}{4}}$$

$$\therefore \frac{W_{Ag}}{108} \times 1 = \frac{8}{32} \times 4$$

$$[\because 2H_2O \longrightarrow O_2 + 4H^+ + 4e^-]$$

$$\Rightarrow W_{Ag} = 108 \text{ g}$$

31 (*a*) In $ZnSO_4$, $\underset{1 \text{ mol}}{Zn^{2+}} + \underset{\substack{2 \text{ mol or} \\ 2 \text{ Faraday}}}{2e^-} \longrightarrow Zn$

In $AlCl_3$, $\underset{1 \text{ mol}}{Al^{3+}} + \underset{3 \text{ F}}{3e^-} \longrightarrow Al$

In $AgNO_3$, $\underset{1 \text{ mol}}{Ag^+} + \underset{1 \text{ F}}{e^-} \longrightarrow Ag$

Thus, the ratio of amounts of electricity required is 2 : 3 : 1.

32 (*c*) Mg = 150 mg = 0.150 g

At cathode $Mg^{2+} + 2e^- \longrightarrow Mg$; $E_{Mg} = \frac{24}{2} = 12$

At anode $2O^{2-} \longrightarrow O_2 + 4e^-$; $E_{O_2} = \frac{32}{4} = 8$

$$\frac{w(O_2)}{w(Mg)} = \frac{E_{O_2}}{E_{Mg}}$$

$$W(O_2) = 0.150 \times \frac{8}{12}$$

$$= 0.100 \text{ g} = \frac{0.100}{32} \text{ mol}$$

$$\therefore O_2 \text{ at STP} = \frac{0.100}{32} \times 22400 \text{ cm}^3$$

$$= 70 \text{ cm}^3$$

33 (*b*) In the electrolysis of aqueous solution of copper sulphate using copper strips as anode and cathode, the anode reaction is

$$Cu \longrightarrow Cu^{2+} + 2e^-$$

34 (*d*) Amount of impurity

= decreased mass of anode – increased mass of cathode

= 22.26 – 22.011 = 0.249 g

Amount of pure Cu deposited,

$$w = zit = \frac{E}{96500} \times it$$

$$= \frac{63.54}{2 \times 96500} \times 140 \times 482.5 = 22.239 \text{ g}$$

But increased mass of cathode = 22.011 g

∴ Amount of impurity (Fe)

$$= 22.239 - 22.011 = 0.228 \text{ g}$$

Now, from Faraday's second law of electrolysis

$$\frac{\text{weight of Fe deposited}}{\text{weight of Cu deposited}} = \frac{\text{equivalent weight of Fe}}{\text{equivalent weight of Cu}}$$

(≈ amount of impurity)

$$\therefore \frac{\text{weight of Fe deposited}}{0.228} = \frac{27.55}{31.77}$$

$$\therefore \text{Weight of Fe} = \frac{27.75 \times 0.228}{31.77} = 0.199$$

$$\therefore \text{\% of iron in impure copper} = \frac{0.199}{22.26} \times 100$$

$$= 0.89 \approx 0.90$$

35 (*b*) From $\Delta G^\circ = -nFE^\circ$

where, n = number of moles of electrons transferred = 2

$$\Rightarrow F = 96500 \text{ C}$$

$$\Delta G^\circ = -240 \text{ kJ} = -240 \times 1000 \text{ J}$$

$$\therefore E^\circ = \frac{-240000}{-2 \times 96500} = 1.25 \text{ V}$$

36 (*c*) Specific conductance, $\kappa = G \times \frac{l}{a}$

where, G = electrical conductance or $G = \kappa \times \frac{a}{l}$

thus, electrical conductance is given by,

$$G = \kappa \cdot a \cdot l^{-1}$$

37 (*d*) Degree of dissociation, $\alpha = \frac{\lambda^c}{\lambda^\infty} = \frac{150}{500} = 0.3$

Given, $C = 0.007$ M

Hydrofluoric acid dissociates in the following manner

	HF ⟶	H^+	+ F^-
Initial	C	0	0
At time, t	$C(1-\alpha)$	$C\alpha$	$C\alpha$

Dissociation constant,

$$K_a = \frac{[H^+][F^-]}{[HF]} = \frac{C\alpha \cdot C\alpha}{C(1-\alpha)} = \frac{C\alpha^2}{(1-\alpha)}$$

On substituting values, we get

$$K_a = \frac{0.007 \times (0.3)^2}{(1-0.3)}$$

$$= \frac{63 \times 10^{-3} \times 10^{-2}}{0.7}$$

$$= 9 \times 10^{-4} \text{ M}$$

38 (*d*) $\Lambda_m^\circ = \frac{\kappa \times 1000}{S}$

$$S = \frac{\kappa \times 1000}{\Lambda_m^\circ} = \frac{1.85 \times 10^{-6} \times 1000}{140 \times 10^{-4}} = 1.321\text{M}$$

$$AB \rightleftharpoons A^+ + B^-$$

$$K_{sp} = [A^+][B^-] = S^2 = (1.321)^2$$

$$= 1.74 \text{ mol}^2 \text{ L}^{-2}$$

39 (*a*) Since, H_2SO_4 and H_3PO_4 both are strong acids so, their presence makes the medium more acidic. The Fe^{3+} ions form complex, $[FeHPO_4]^+$ with these acids and hence, the concentration of Fe^{3+} ion reduces which results in lowering in the formal reduction potential.

40 (c) $E^\circ_{cell} = E_{cathode} - E_{anode}$

$= -0.42 - (-0.72)$

$= -0.42 + 0.72 = 0.3$

At anode $Cr \longrightarrow Cr^{3+} + 3e^-] \times 2$

At cathode $Fe^{2+} + 2e^- \longrightarrow Fe] \times 3$

$$2Cr + 3Fe^{2+} \longrightarrow 2Cr^{3+} + 3Fe$$

$$\therefore \quad E = E^\circ_{cell} - \frac{0.0591}{6}\log\frac{[Cr^{3+}]^2}{[Fe^{2+}]^3}$$

$$= 0.3 - \frac{0.0591}{6}\log\frac{(0.1)^2}{(0.01)^3}$$

$$= 0.26\text{V}$$

41 (a) $Cr_2O_7^{2-} + 14H^+ + 6e^- \longrightarrow 2Cr^{3+} + 7H_2O$

$$E = E^\circ - \frac{0.059}{n}\log\frac{[Cr^{3+}]^2[H_2O]^7}{[Cr_2O_7^{2-}][H^+]^{14}}$$

$$1.067 = 1.33 - \frac{0.059}{6}\log\frac{(15\times10^{-3})^2(1)^7}{(4.5\times10^{-3})[H^+]^{14}}$$

$$\frac{0.059}{6}\log\left[\frac{225\times10^{-6}}{(4.5\times10^{-3})\times[H^+]^{14}}\right] = 0.263$$

$$\Rightarrow \quad \log\left[\frac{225\times10^{-6}}{(4.5\times10^{-3})\times[H^+]^{14}}\right] = \frac{0.263\times6}{0.059}$$

$$\log\left[\frac{50\times10^{-3}}{[H^+]^{14}}\right] = 26.74$$

$$\log(50\times10^{-3}) - \log(H^+)^{14} = 26.74$$

$$-1.3010 - 14\log[H^+] = 26.74$$

$$-14\log[H^+] = 26.74 + 1.3010$$

or $$+14\,\text{pH} = 28.041 \quad [\because \text{pH} = -\log[H^+]]$$

$$\text{pH} = \frac{28.041}{14} = 2$$

42 (b) The emf of a galvanic cell = electrode potential of zinc − electrode potential of copper

$= 0.34 - (-0.76) = 1.1$ V

43 (a) MnO_4^- will oxidise Cl^- ion according the following equation,

$$2MnO_4^- + 16H^+ + 10Cl^- \longrightarrow 2Mn^{2+} + 8H_2O + 5Cl_2$$

The cell corresponding to this reaction is as follows,

Pt, Cl_2 (1 atm) | Cl^- || MnO_4^-, Mn^{2+}, H^+ | Pt

$$E^\circ_{cell} = 1.51 - 1.40 = 0.11\text{ V}$$

As E°_{cell} is positive, the above reaction is feasible. Thus, MnO_4^- will not oxidise Fe^{2+} ion but oxidise Cl^- ion.

44 (c) Given, molar conductance at 0.1 M concentration,

$$\Lambda_c = 9.54\ \Omega^{-1}\text{cm}^2\text{mol}^{-1}$$

Molar conductance at infinite dilution, $\Lambda_\infty = 238\ \Omega^{-1}\text{cm}^2\text{mol}^{-1}$

We know that, degree of ionisation,

$$\alpha = \frac{\Lambda_c}{\Lambda_\infty}\times100 = \frac{9.54}{238}\times100 = 4.008\%$$

45 (b) For hydrogen electrode, oxidation half-reaction is

$$\underset{(1\text{ atm})}{H_2} \longrightarrow \underset{(\text{at pH }10)}{2H^+} + 2e^-$$

If pH = 10, $H^+ = 1\times10^{-pH} = 1\times10^{-10}$

From Nernst equation, $E = E^\circ - \frac{0.0591}{2}\log\frac{[H^+]^2}{p_{H_2}}$

For hydrogen electrode, $E^\circ = 0$

$$E = -\frac{0.0591}{2}\log\frac{(10^{-10})^2}{1}$$

$$= \frac{0.0591\times2}{2}\log(10)$$

$$= 0.0591\times\log10^{10} = 0.59\text{ V}$$

Note *Direct* $E = 0.0591\,\text{pH}$

46 (a) Anode is always site of oxidation and cathode is a site of reduction. Thus, anode half-cell is

$$Zn^{2+}(aq) + 2e^- \longrightarrow Zn(s);\quad E^\circ = -0.76\text{ V}$$

Cathode half-cell is

$$Ag_2O(s) + H_2O(l) + 2e^- \longrightarrow 2Ag(s) + 2OH^-(aq);\quad E^\circ = 0.34\text{ V}$$

$$E^\circ_{cell} = E^\circ_{cathode} - E^\circ_{anode}$$

$$= 0.34 - (-0.76) = 1.1\text{ V}$$

47 (c) $\because\ E_R = Cu/Cu^{2+}$

$E_L = Al/Al^{3+}$

We know that, $E_{cell} = E_R - E_L$

$$2 = +0.34 - x$$

$$x = 0.34 - 2 = -1.66\text{ V}$$

48 (c) Cell which converts chemical energy of a fuel directly into electrical energy is called fuel cell, e.g. H_2—O_2 fuel cell also called Bacon cell. The net reaction taking place in Bacon cell is

$$2H_2(g) + O_2(g) \longrightarrow 2H_2O(l)$$

Other fuel cells are

Propane oxygen fuel cell

$$C_3H_8 + 5O_2 \longrightarrow 3CO_2 + 4H_2O$$

Methane oxygen fuel cell

$$CH_4 + 2O_2 \longrightarrow CO_2 + 2H_2O$$

Methanol oxygen fuel cell

$$2CH_3OH + 3O_2 \longrightarrow 2CO_2 + 4H_2O$$

Note *Efficiency of fuel cell is 70-75% which is much higher than connected cells (40% only).*

49 (b) At cathode, reduction takes place is

$$Al^{3+} + 3e^- \longrightarrow Al$$

50 (c) Electricity passed = 9.65 Coulomb

The amount of silver deposited = ?

We know that, 96500 coulomb electricity deposits

$$Ag = 108\text{ mol}^{-1}$$

$$\therefore \text{ 9.65 coulomb electricity deposits Ag} = \frac{9.65}{96500}\times108$$

$$= 108\times10^{-4}\text{g} = 10.8\text{ mg}$$

51 (*d*) $CH_3COONa + HCl \longrightarrow NaCl + CH_3COOH$

$$\Lambda^\circ_{m(CH_3COOH)} = \Lambda^\circ_{m(CH_3COONa)} + \Lambda^\circ_{m(HCl)} - \Lambda^\circ_{m(NaCl)}$$

$$\Lambda^\circ_{m(CH_3COOH)} = (91.0 + 425.9 - 126.4)\ S\ cm^2mol^{-1}$$

$$\Lambda^\circ_{m(CH_3COOH)} = 390.5\ S\ cm^2mol^{-1}$$

52 (*b*) The given statement is termed as Kohlrausch's law.

$$K_C = 3.89 \times 10^{15} \approx 4 \times 10^{15}$$

53 (*c*) $4Al^{3+} \longrightarrow 4Al + 12e^-$

$$12e^- + 6O^{2-} \longrightarrow 3O_2$$

or $\frac{4}{3}Al^{3+} \longrightarrow \frac{4}{3}Al + 4e^-$

$$4e^- + 2O^{2-} \longrightarrow O_2$$

$$\Delta G^\circ = -nFE^\circ$$

$$\Delta G = +960\ kJ\ mol^{-1}$$

$$= 960 \times 1000\ J\ mol^{-1}$$

$$nF = 4F = 4 \times 96500\ C$$

$$\therefore\ 960 \times 1000 = -4 \times 96500 \times E^\circ$$

$$E^\circ = \frac{960000}{4 \times 96500}$$

$$= -2.48\ V$$

Potential difference ≈ 2.5 V

54 (*a*) $E_{cell} = E^\circ_{cell} - \frac{0.059}{2}\log\frac{[Zn^{2+}]}{[Cu^{2+}]}$

$$= 1.10 - \frac{0.059}{2}\log\frac{0.1}{0.1}$$

$$E_{cell} = 1.10\ V$$

55. (*d*) $Cu(s) + 2Ag^+(aq) \longrightarrow Cu^{2+}(aq) + 2Ag(s)$

$$E^\circ = 0.46\ V\ \text{at}\ 298\ K$$

$$E^\circ = \frac{0.059}{2}\log K_C$$

$$0.46 = \frac{0.059}{2}\log K_C$$

$$\log K_C = 15.59$$

$$K_C = \text{antilog}\ 15.59$$

$$= 3.91 \times 10^{15}$$

$$\approx 4.0 \times 10^{15}$$

57 (*b*) 1000 mL of $AgNO_3$ solution contains 108 g Ag^+ = 1 mol Ag^+

Ag^+ in 125 mL

$$[Ag^+] = \frac{125 \times 1}{1000} = 0.125\ mol$$

$$= 0.125 \times 108\ g$$

$$w = zit$$

$$0.125 \times 108 = \frac{108 \times 241.25t}{1 \times 96500}$$

$$t = 50\ s$$

58 (*b*) $w_{metal} = \frac{Eit}{96500} = \frac{E \times 3 \times 50 \times 60}{96500}$

$$E = \frac{96500 \times w}{3 \times 50 \times 60}$$

$$\Rightarrow\ E = \frac{96500 \times 1.8}{3 \times 50 \times 60} = 19.3\ g\ equiv^{-1}$$

59 (*c*) Saturated solution of KNO_3 is used to make salt-bridge because they have same ionic mobility.

60 (*d*) Antimony is used in making lead storage batteries because lead containing antimony is harder and more resistant towards the action of acids than ordinary lead.

61 (*d*) $Ca^{2+} + 2e^- \longrightarrow Ca$ (at cathode)

$$w = zit$$

$$30 = \frac{40 \times 5 \times t}{2 \times 96500} \Rightarrow t = 28950\ s = 8.04h$$

62 (*c*)

	E°	n
$Cu^{2+} + e^- \longrightarrow Cu^+$	0.15 V	1
$Cu^+ + e^- \longrightarrow Cu$	0.50	1
Thus, $Cu^{2+} + 2e^- \longrightarrow Cu$	?	2

$$\Delta G_3 = \Delta G_1 + \Delta G_2$$

$$\Rightarrow\ -2FE^\circ = -1F \times 0.15 + (-1\ F \times 0.50)$$

$$\Rightarrow\ -2FE^\circ = -0.15\ F - 0.50\ F$$

$$\Rightarrow\ -2FE^\circ = -F\ (0.15 + 0.50)$$

$$\therefore\ E^\circ = \frac{0.65}{2} = 0.325\ V$$

63 (*d*) $Z > X > Y$ more negative the reduction potential, higher the reducing power.

64 (*c*) $E^\circ_{cell} = E^\circ_{cathode\ (RP)} - E^\circ_{anode\ (RP)}$

$$= 0.15 - (-0.74) = +0.89\ V$$

65 (*d*) Electrolysis of both KCl and NaCl liberates H_2 at the cathode and Cl_2 at the anode. The resulting solutions will be different.

66 (*c*) Maximum number of ions (5) is obtained from potassium hexacyanoferrate(II), i.e. $K_4[Fe(CN)_6]$, thus maximum molar conductivity.

67 (*c*)

	Atomic mass	Equivalent mass
$X^{3+} + 3e^- \longrightarrow X$	A_X	$E_X = \frac{A_X}{3}$
$Y^{2+} + 2e^- \longrightarrow Y$	B_Y	$E_Y = \frac{A_Y}{2}$

Mass is deposited in the ratio of equivalent mass

$$\frac{E_X}{E_Y} = \frac{2A_X}{3A_Y} = \frac{2}{3} \times \frac{1}{2} = \frac{1}{3}$$

68 (*d*) $C + B^+ \longrightarrow C^+ + B$ So, $C > B$

$A^+ + D \longrightarrow$ No reaction So, $A > D$

$C^+ + A \longrightarrow$ No reaction So, $C > A$

$D + B^+ \longrightarrow D^+ + B$ So, $D > B$

Hence, $C > A > D > B$.

69 (*c*) $Mg^{2+} + 2e^- \longrightarrow Mg(s)$

Equivalent mass of Mg $= \frac{24}{2} = 12$

Mg formed = Grignard reagent formed

$$w = zit = \frac{12 \times 9.65}{96500} = 12 \times 10^{-4}\ \text{g} = \frac{12 \times 10^{-24}}{24}\ \text{mol}$$

$$= 0.5 \times 10^{-4}\ \text{mol}$$

$$= 5 \times 10^{-5}\ \text{mol}$$

$\therefore$ Grignard reagent formed $= 5 \times 10^{-5}$ mol

70 (*a*) $nFE° = 2.303RT \log K_p$

$$E° = \frac{2.303RT}{nF} \log K$$

$$= \frac{0.0591}{n} \log K$$

$$\log K = \frac{0.59 \times 2}{0.059} = 20$$

$$K = \text{antilog } 20.0 = 1 \times 10^{20}$$

71 (*b*) $E_{cell} = E°_{cell} + \frac{0.0591}{n} \log \frac{[Cu^{2+}]}{[Zn^{2+}]}$

$$E_1 = E°_{cell} + \frac{0.0591}{2} \log \frac{1}{0.1};\quad E_1 = E°_{cell} + 0.02955$$

$$E_2 = E°_{cell} + \frac{0.0591}{2} \log \frac{1}{1};\quad E_2 = E°_{cell}$$

$$E_3 = E°_{cell} + \frac{0.0591}{2} \log \frac{0.1}{1};\quad E_3 = E°_{cell} - 0.02955$$

$\therefore \quad E_1 > E_2 > E_3.$

72 (*b*) According to Faraday second law,

$$Al^{3+} + 3e^- \longrightarrow Al;\quad E(Al) = \frac{27}{3} = 9$$

$$Cu^{2+} + 2e^- \longrightarrow Cu;\quad E(Cu) = \frac{63.5}{2} = 31.75$$

By Faraday second law,

$$\frac{w(Cu)}{w(Al)} = \frac{E(Cu)}{E(Al)}$$

$$w(Cu) = 2.7 \times \frac{31.75}{9}$$

$$= 9.525\ \text{g}$$

73 (*a*) Correct reaction for a given electrochemical cell is

$$2Br^-(aq) + Cl_2(g) \longrightarrow 2Cl^-(aq) + Br_2(g)$$

CHAPTER 04

Chemical Kinetics

The feasibility of a reaction can be predicted with the help of thermodynamic principle but it does not deal with the rate at which chemical reaction occurs. The branch of chemistry which deals with the study of rate of reactions, factor affecting the rate of reaction and mechanism of the reaction is known as **chemical kinetics.**

The word **kinetics** is derived from the Greek word *kinesis* which means movement. Therefore, chemical kinetics explains the movement of reacting particles.

In this chapter, we shall be dealing with above mentioned aspects of chemical reactions.

CHEMICAL REACTIONS

When one or more substances undergo a change which result in the formation of a new product, it is called **chemical reactions.**

Types of chemical reactions

Depending upon reaction rates or speeds, chemical reactions can be classified in the following ways,

(i) **Instantaneous or fast reactions** These reactions proceed at very fast speed. They occur as soon as the reactants are brought together. So, it is practically impossible to measure the speed of such reactions.
e.g. neutralisation reactions, precipitation reactions, explosive reactions of oxygen with hydrogen and hydrocarbons.

(ii) **Extremely slow reaction** These type of reactions proceed at very low speed. So, it is also very difficult to measure the speed of these reactions. e.g. rusting of iron, etc.

(iii) **Moderately slow reactions** These type of reactions proceed with a measurable speed at room temperature. Thus, these reactions are utilised in the study of chemical kinetics. e.g. inversion of cane sugar, decomposition of ammonium nitrate etc.

Depending upon number of steps, chemical reactions are classified in the following ways,

(i) **Elementary reactions** A balanced chemical equation never gives us a true picture of how a reaction takes place. It can be completed in one step or in more than one steps. The reactions taking place in one step are called elementary reactions.

Inside

(*ii*) **Complex reactions** The reactions which do not take place in one step are called complex reactions. In these reactions, all the atoms, ions or molecules as represented in the balanced chemical equation may not come together to collide simultaneously. Such reactions occur in a sequence of a number of steps.

e.g. $$4HBr + O_2 \longrightarrow 2H_2O + 2Br_2$$

Step I $$HBr + O_2 \xrightarrow{\text{Slow}} HOOBr$$

Step II $$HOOBr + HBr \xrightarrow{\text{Fast}} 2HOBr$$

Step III $$[HOBr + HBr \xrightarrow{\text{Fast}} H_2O + Br_2] \times 2$$

$$4HBr + O_2 \longrightarrow 2H_2O + 2Br_2$$

RATE, RATE LAW EXPRESSION AND RATE CONSTANT

Rate of a chemical reaction

The rate (or speed or velocity) of a reaction is defined as the change in the concentration of any one of the reactants or products per unit time.

Consider a hypothetical reaction,

$$nA \longrightarrow mB$$

When $t = t_1$ $\quad [A_{t_1}] \quad [B_{t_1}]$

When $t = t_2$ $\quad [A_{t_2}] \quad [B_{t_2}]$

Rate of consumption of $A = -\dfrac{\Delta[A]}{\Delta t}$

Rate of formation of $B = \dfrac{\Delta[B]}{dt}$

Rate of a reaction

$$= \frac{1}{n}\left[\frac{\text{decrease in the concentration of reactants}}{\text{time taken}}\right]$$

$$\text{or} \quad \frac{1}{n}\left[\frac{\text{increase in the concentration of product}}{\text{time taken}}\right]$$

$$\text{Rate of reaction} = -\frac{1}{n}\frac{[A_{t_2}] - [A_{t_1}]}{[t_2 - t_1]} = -\frac{1}{n}\frac{\Delta[A]}{\Delta t}$$

$$= \frac{1}{m}\frac{[B_{t_2}] - [B_{t_1}]}{[t_2 - t_1]} = +\frac{1}{m}\frac{\Delta[B]}{\Delta t}$$

where, n and m are the stoichiometric coefficient of A and B respectively.

Sign convention

The negative sign indicates that the concentration of reactants decreases with time whereas the positive sign indicates that the concentration of products increases with time. The concentration change may be positive or negative but the rate of reaction is always positive.

Average and instantaneous rate

The rate of reaction as defined above is the average rate of reaction for a chemical reaction. The average rate depends on the values of t_2 and t_1, i.e. time interval chosen graphically. It is shown in fig. (a) or (b) both.

To know the rate of reaction at any instant of time during the course of a reaction, a term instantaneous rate of reaction is introduced. It is defined as the rate of change of concentration of any one of the reactants or products at that particular instant of time. In other words, it is the average rate during the smallest possible interval of time (i.e. when Δt approaches zero.)

Suppose the small change in concentration is dx in small interval of time dt.

Then, rate of reaction or instantaneous rate $(r_{\text{inst}}) = \dfrac{dx}{dt}$

For given hypothetical reaction,

$$r_{\text{(inst)}} = \frac{-1}{n}\frac{d[A]}{dt} = \frac{1}{m}\frac{d[B]}{dt}$$

Thus, for the reaction

$$N_2(g) + 3H_2(g) \longrightarrow 2NH_3(g)$$

$$-\frac{d[N_2]}{dt} = -\frac{1}{3}\frac{d[H_2]}{dt} = +\frac{1}{2}\frac{d[NH_3]}{dt}$$

Graphical interpretation of rate

Graphically rate can be determined by drawing a tangent at time (t) on either curves for concentration of reactant or product *vs* time and calculating its slope as shown in Fig. 4.1 (a) or (b) both.

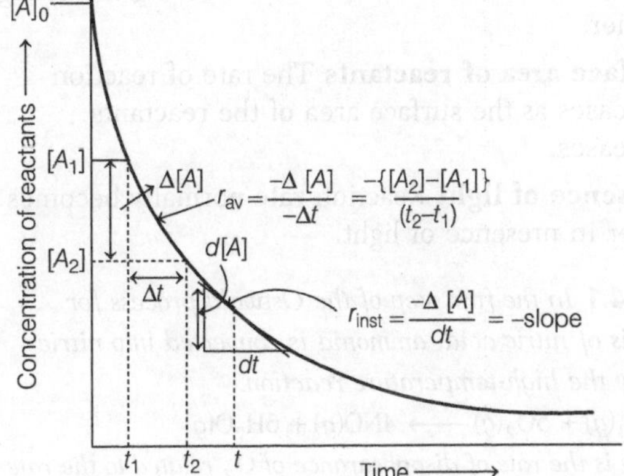

Fig. 4.1 (a)

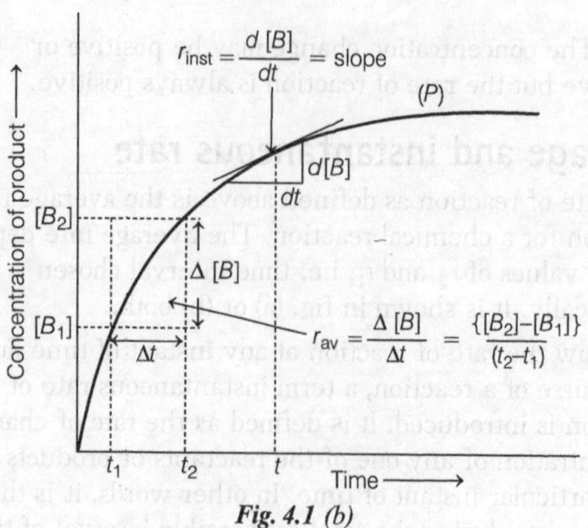

Fig. 4.1 (b)

Instantaneous and average rate of a reaction.

Units of the rate of reaction

$$\text{Rate of reaction} = \frac{dx}{dt} = \frac{\text{mol L}^{-1}}{\text{s}}$$

$$\boxed{\text{Rate} = \text{mol L}^{-1}\text{s}^{-1}}$$

Factors affecting rate of reaction

(*i*) **Concentration of reactants** An increase in concentration of reactants, brings more frequent collisions between the reacting molecules. That increases the rate of reaction.

(*ii*) **Nature of reactants** The nature of bonds in a molecule influences the rate of reaction at which it changes into products.

(*iii*) **Temperature** Reaction rates are normally favoured by increase of temperature. Increase in temperature causes a greater fraction of the molecules to have an energy at least equal to the threshold energy.

(*iv*) **Presence of catalyst** Presence of catalyst increases the rate of reaction by lowering the energy barrier.

(*v*) **Surface area of reactants** The rate of reaction increases as the surface area of the reactants increases.

(*vi*) **Presence of light** Reaction rate normally becomes faster in presence of light.

Example 4.1 *In the first step of the Ostwald process for synthesis of nitric acid, ammonia is converted into nitric oxide by the high-temperature reaction.*

$$4NH_3(g) + 5O_2(g) \longrightarrow 4NO(g) + 6H_2O(g)$$

(*a*) *How is the rate of disappearance of* O_2 *related to the rate of disappearance of* NH_3*?*

(*b*) *How are the rates of appearance of* NO *and* H_2O *related to the rate of disappearance of* NH_3*?*

Sol. (a) $-\frac{1}{4}\frac{d[NH_3]}{dt} = -\frac{1}{5}\frac{d[O_2]}{dt}$

$$\therefore \quad \frac{d[O_2]}{dt} = 1.25\frac{d[NH_3]}{dt}$$

(b) $\frac{1}{4}\frac{d[NO]}{dt} = \frac{1}{6}\frac{d[H_2O]}{dt} = -\frac{1}{4}\frac{d[NH_3]}{dt}$

$$\therefore \quad \frac{d[NO]}{dt} = -\frac{d[NH_3]}{dt}$$

$$\frac{d[H_2O]}{dt} = -1.5\frac{d[NH_3]}{dt}$$

Example 4.2 *In the following reaction,*

$$2H_2O_2 \longrightarrow 2H_2O + O_2$$

rate of formation of O_2 *is* 36 g min^{-1}.

(*a*) *What is the rate of formation of* H_2O ?

(*b*) *What is the rate of disappearance of* H_2O_2?

Sol. Caution Take amount in mol or mol L^{-1}

When 2 moles H_2O_2 (68 g) decomposes, 2 moles H_2O (36 g) and 1 mole O_2 (32 g) are formed.

$$36 \text{ g min}^{-1}\ O_2 = \frac{36}{32} = 1.125 \text{ mol min}^{-1}$$

$$-\frac{1}{2}\frac{d[H_2O_2]}{dt} = \frac{1}{2}\frac{d[H_2O]}{dt} = \frac{d[O_2]}{dt}$$

$$-\frac{d[H_2O_2]}{dt} = 1.25 \times 2 = 2.5 \text{ mol/min}$$

Rate law expression

It is defined as a mathematical expression in which reaction rate is given in terms of molar concentration of reactants with each term raised to some power.

Consider a hypothetical expression,

$$xX + yY \longrightarrow aA + bB$$

where, *x*, *y*, *a* and *b* are the stoichiometric coefficients of the reactants and products.

The rate expression for this reaction is

$$\text{Rate} \propto [X]^c[Y]^d \quad \text{or} \quad = k\,[X]^c[Y]^d \qquad \text{...(i)}$$

(this is also known as differential rate equation)

where exponents *c* and *d* may or may not be equal to the stoichiometric coefficients (*x* and *y*) of the reactants.

Characteristics

(*i*) It must be noted that the form of the rate law cannot be deduced from the equation for a given reaction. It is found by experiment only.

(*ii*) The rate law may not have a simple relationship for the stoichiometric equation.

(*iii*) The rate law may not depend on the concentration of every reactant or product of the reaction.

(*iv*) The rate law may depend on the concentration of species (e.g. catalyst) which do not appear in the equation for the overall reaction.

Rate constant or specific reaction rate

In the rate law expression (i) k is the proportionality constant called rate constant. It is defined as "A coefficient of proportionality relating the rate of a chemical reaction to the concentration of reacting species on which the rate of reaction actually depends.

If all concentration in expression (i) are taken as unity, i.e. $[X]$ and $[Y] = 1\,\text{mol/L}$.

then $\boxed{\text{Rate} = k}$

Thus, **rate constant is also defined as the rate of the reaction when the concentration of each reactant is taken as unity. Therefore, it is also known as specific reaction rate.**

Characteristics

(*i*) Rate constant is a measure of the rate of reaction. Greater the value of the rate constant, faster is the reaction.

(*ii*) Each reaction has a definite value of rate constant at a particular temperature and its value for the same reaction change with change in the temperature.

(*iii*) Its value does not depend upon the concentration of the reactants.

(*iv*) The unit of rate constant depends upon the order of reaction.

Unit

Since, the dimensions of rate are concentration/time and the SI units for concentration are mol m^{-3}, mol dm^{-3}, or mol L^{-1} and second(s) for time.

Let consider a hypothetical reaction.

$$nA \longrightarrow \text{Product}$$

$$\text{Rate} = k[A]^n;\; k = \frac{\text{Rate}}{[A]^n}$$

Thus, the dimension of k is equal to

$$\frac{\text{concentration / time}}{[\text{concentration}]^n} = [\text{concentration}]^{1-n}[\text{time}]^{-1}$$

$$= (\text{mol})^{1-n}\,\text{L}^{n-1}\text{s}^{-1} \text{ or } \left[\frac{\text{mol}}{\text{L}}\right]^{1-n}\text{s}^{-1}$$

where, n = order of reaction

For gaseous reactions, concentrations are expressed in terms of pressure with unit of atmosphere. Thus, the rate has the unit of atm per second.

Difference between rate of reaction and rate constant

Following are the differences between rate of reaction and rate constant

Rate of reaction	Rate constant
1. It is the speed with which reactants are converted into products.	It is the proportionality constant in the rate law.
2. It is measured as the rate of decrease in the concentration of reactants or increase in the products with time.	It is equal to the rate of reaction when the concentration of reactants is unity.
3. It depends upon the initial concentration of reactants or product.	It is independent of the initial concentration of reactants or product.
4. Its concentration unit are always mol L^{-1} time $^{-1}$	Its unit depend upon the order of reaction. $(\text{conc.})^{1-n}$ time^{-1} or $(\text{mol/L})^{1-n}$ time^{-1}

CHECK POINT 4.1

1. A graph of volume of hydrogen released *vs* time for the reaction between zinc and dil. HCl is given in figure. On the basis of this mark the correct option. **[NCERT Exemplar]**

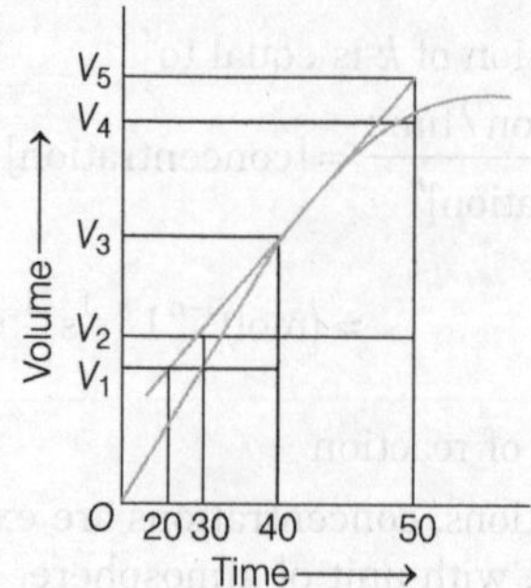

(a) Average rate upto 40 s is $\frac{V_3 - V_2}{40}$

(b) Average rate upto 40 s is $\frac{V_3 - V_2}{40 - 30}$

(c) Average rate upto 40 s is $\frac{V_3}{40}$

(d) Average rate upto 40 s is $\frac{V_3 - V_1}{40 - 20}$

2. While considering the graph given in (Qus 1), which of the following options does not show instantaneous rate of reaction at 40 s ?

(a) $\frac{V_5 - V_2}{50 - 30}$ (b) $\frac{V_4 - V_2}{50 - 30}$

(c) $\frac{V_3 - V_2}{40 - 30}$ (d) $\frac{V_3 - V_1}{40 - 20}$

3. The following reaction is first-order *wrt*. *A* and *B* both

$$A + B \longrightarrow \text{Product}$$

Rate $= k[A][B]$

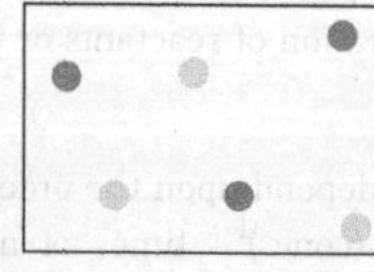

I

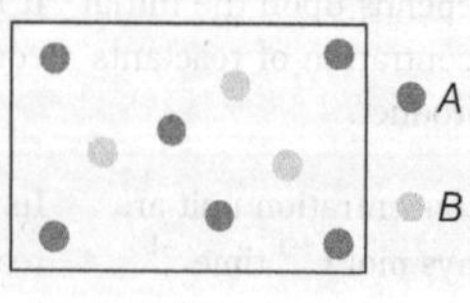

II

Ratio of relative rate of this reaction in vessels I and II of equal volume is

(a) 1 : 1 (b) 1 : 2 (c) 2 : 1 (d) 1 : 4

4. For the reaction, $A \longrightarrow B$; $-\frac{\Delta[A]}{dt} = 2\frac{\Delta[B]}{dt}$

then rate law is

(a) $-\frac{\Delta[A]}{\Delta t} = k[A]^2$ (b) $\frac{\Delta[B]}{\Delta t} = k[A]$

(c) $\frac{\Delta[A]}{\Delta t} = k[A]^2$ (d) $-\frac{\Delta[B]}{\Delta t} = k[B]^2$

5. The rate of the reaction, $A + B_2 \longrightarrow AB + B$ is directly proportional to the concentration of *A* and independent of concentration of B_2, hence, rate law is

(a) $k[A][B_2]$ (b) $k[A]^2[B_2]$ (c) $k[A]$ (d) $k[B_2]$

6. Rate law for the reaction, $A + 2B \longrightarrow C$ is found to be

Rate $= k[A][B]$

Concentration of reactant '*B*' is doubled, keeping the concentration of '*A*' constant, the value of rate constant will be.......... .

(a) the same (b) doubled

(c) quadrupled (d) halved

7. For the reaction, $2NH_3 \longrightarrow N_2 + 3H_2$

$$-\frac{\Delta[NH_3]}{\Delta t} = k_1[NH_3]; \ \frac{\Delta[N_2]}{\Delta t} = k_2[NH_3]; \ \frac{\Delta[H_2]}{\Delta t} = k_3[NH_3]$$

Then relation between k_1, k_2 and k_3 is

(a) $1.5k_1 = 3k_2 = k_3$ (b) $2k_1 = k_2 = 3k_3$

(c) $k_1 = k_2 = k_3$ (d) $k_1 = 3k_2 = 2k_3$

8. For $xA + yB \longrightarrow zC$, $-\frac{d[A]}{dt} = -\frac{d[B]}{dt} = 1.5\frac{d[C]}{dt}$

then *x*, *y* and *z* are

(a) 1, 1, 1 (b) 3, 2, 3

(c) 3, 3, 2 (d) 2, 2, 3

9. For a hypothetical reaction, $aA + bB \longrightarrow P$, the rate law is rate $= k[A]^x[B]^y$, then

(a) $(a + b)$ may be equal to $(x + y)$

(b) $(a + b)$ may be greater than $(x + y)$

(c) $(a + b)$ may be lesser than $(x + y)$

(d) All of the above case are correct

10. Rate constant of the nth order reaction can be expressed as,

(a) $(\text{mol})^{n-1}\text{L}^{n-1}\text{s}^{-1}$ (b) $(\text{mol})^{n-1}\text{L}^{-n}\text{s}^{-1}$

(c) $(\text{mol})^{n}\text{L}^{n-1}\text{s}^{-1}$ (d) $(\text{mol})^{1-n}\text{L}^{n-1}\text{s}^{-1}$

MOLECULARITY AND ORDER OF A REACTION

Molecularity

It can be defined as the number of reacting species (atoms, ions or molecules) taking part in an elementary reaction which must collide simultaneously in order to bring about a chemical reaction.

In other words, the type of molecule to which the concentration terms in the rate law expression are raised to express the observed rate of reaction is called molecularity. e.g.

(*i*) If one reactant is involved, the reaction is called **unimolecular reactions.**

e.g. $$NH_4NO_2 \longrightarrow N_2 + 2H_2O$$

(*ii*) Reaction involving simultaneous collision between two species are called **bimolecular reactions.**

e.g. $$2HI \longrightarrow H_2 + I_2$$

(*iii*) Reactions involving simultaneous collision between three reacting species are called **trimolecular reaction.**

e.g. $$2FeCl_3 + SnCl_2 \longrightarrow 2FeCl_2 + SnCl_4$$

The probability that more than three molecules can collide and react simultaneously is very small. Thus, the molecularity greater than three is not observed.

Characteristics

(*i*) It cannot be zero, fractional or more than three.

(*ii*) In case of complex reactions, the atoms, molecules or ions participating in rate determining step decide the molecularity. Among several steps, the slowest step is the rate determining step, e.g. if two molecules participate in the rate determining step, then molecularity is two.

(*iii*) It does not provide the information about mechanism of the reaction.

Order

The order of reaction is defined as "the sum of the power to which the concentration terms in the rate law equation are raised to express the observed rate of the reaction."

$$\text{For the rate law} = k\,[A]^c[B]^d \qquad \ldots\text{(i)}$$

Thus, for the rate expression (i) the sum of $c + d$ is the order of the taken reaction.

Characteristics

(*i*) Order of reaction is purely experimental quantity. It may be whole number, zero or in fractions.

(*ii*) Depending upon whether $(c + d)$ in rate law (eq. (i)) is equal to zero, 1, 2 or 3, the reactions are said to be zero order, first order, second order or third order respectively.

Table 4.2 Difference between molecularity and order of reactions

S.No.	Molecularity	Order
1.	The number of reacting species which must collide simultaneously in order to bring about a chemical reaction is called molecularity of a reaction.	The sum of powers of the concentrations of the reactants in the rate law expression is called the order of the chemical reaction.
2.	Molecularity is always a whole number value.	Order of reaction may be zero, whole number or fractional values, even negative.
3.	Molecularity is a theoretical concept.	Order of reaction is determined experimentally.
4.	Molecularity is applicable to elementary reaction only. For complex reactions, it has no meaning.	Order to reaction is applicable to elementary as well as complex reactions.
5.	For simple reactions, molecularity can be obtained from the stoichiometry of the equations.	For simple reactions, order of reaction may not be equal to the number of molecules of the reactants as seen from the balanced equation.

The overall rate of the reaction is controlled by the slowest step in a reaction called the rate determining step.

Decomposition of H_2O_2 is catalysed by I^-.

$$2H_2O_2 \xrightarrow[\text{Alkaline medium}]{I^-} 2H_2O + O_2$$

$$-\frac{d[H_2O_2]}{dt} = k_2[H_2O_2][I^-]$$

The reaction is first order *wrt* H_2O_2 as well as I^-.

Thus, rate determining step can be bimolecular

$$H_2O_2 + I^- \xrightarrow{\text{slow}} H_2O + IO^-$$

In the following bimolecular step IO^- disappears

$$H_2O_2 + IO^- \xrightarrow{\text{fast}} H_2O + I^- + O_2$$

IO^- is the intermediate of this reaction. In this reaction, rate of formation of intermediate determines rate of the reaction.

Example 4.3 *The gas-phase reaction of nitric oxide and bromine yield nitrosyl bromide.*

$$2NO(g) + Br_2(g) \longrightarrow 2NOBr(g)$$

The rate law is, rate = k $[NO]^2[Br_2]$

Calculate the reaction order with respect to each of the reactants and what is the overall reaction order?

Sol Rate law is, rate = $k[NO]^2[Br_2]$

hence, *wrt* NO, order = 2 and *wrt* Br_2, order = 1

Overall reaction order = 2 + 1 = 3

Different order of reactions

1. Zero order reaction

The reaction in which rate of reaction is directly proportional to the zero power of product of concentrations of reactants, is known as zero order reaction.

For the reaction, $A \longrightarrow$ Product

	A	Product
At $t = 0$	a	0
At $t = dt$	a	x

$$\text{Rate} = -\frac{d[A]}{dt} = k[A]^0 = k$$

or

$$\frac{dx}{dt} = k$$

(*i*) Since, the rate is independent of concentration, the plot of rate *versus* concentration will be a straight line parallel to the concentration axis.

(*ii*) Integrated

$$\frac{d[A]}{dt} = k[A]^0 = k$$

$$-\int d[A] = k\int -dt$$

Assume $[A] = [A]_0$ at $t = 0$

and $[A] = [A]_t$ at time t

$\therefore$ $-[A]_t = kt + I$ ...(i)

I = Integration constant

If $t = 0$, $[A]_t = [A]_0$

$\therefore$ $-[A]_0 = I$

$\therefore$ $-[A]_t = kt - [A]_0$

$\therefore$ $[A]_0 - [A]_t = kt$

$\therefore$ For zero order reaction,

$$[A]_t = [A]_0 - kt \quad \text{...(ii)}$$

Since, $[A]_0 - [A]_t = x$

$$x = kt$$

$$k = \frac{x}{t}$$

[Graph: $[A]_0$ on vertical axis, time t on horizontal axis, straight line decreasing from $[A]_0$]

Graphically, equation (ii) is represented as given

slope = k = rate constant

(*iii*) Some examples of zero order reactions are

$$2NH_3(g) \xrightarrow[\Delta]{Pt} N_2(g) + 3H_2(g)$$

In this reaction, Pt is a catalyst. At very high pressure, the metal surface gets saturated with gas molecules and any further change in experimental conditions has no effect on NH_3 concentration on the surface and rate remains independent of $[NH_3]$.

$$H_2(g) + Cl_2(g) \xrightarrow{hv} 2HCl$$

$$2HI \xrightarrow{Gold} H_2 + I_2$$

(*iv*) Unit of rate constant $k = \left(\frac{mol}{L}\right)\frac{1}{s} = mol\,L^{-1}s^{-1}$

(*v*) If A_1 = concentration of A at $(t = t_1)$

and A_2 = concentration of A at $(t = t_2)$

then, $$k = \frac{A_2 - A_1}{(t_2 - t_1)}$$

(*vi*) **Half-life time period ($t_{1/2}$)**

The time required to reduce the concentration of the reactant to half of its initial value is called half-life time period. It is denoted by $(t_{1/2})$.

For zero order reaction,

if concentration of product is $[A_0]$ at time t and $\frac{[A_0]}{2}$ at time t_2 then

$$k = -\frac{\left[\frac{A_0}{2}\right] - [A_0]}{t_2 - t_1} = \frac{[A_0] - \left[\frac{A_0}{2}\right]}{t_{1/2}}$$

$$\boxed{t_{1/2} = \frac{[A_0]}{2k}} \quad \text{...(v)}$$

Therefore, $t_{1/2}$ for zero order reaction is directly proportional to the initial concentration of the reactant.

(*vii*) **Quarter life time** (Time of 75% reaction)

The time required to reduce the concentration of the reactant to one fourth of its initial value is called quarter life time.

$\therefore$ For zero order reactions, $t_{1/4} = \frac{3}{4}\frac{[A_0]}{k}$;

(*vii*) Relation between $t_{1/4}$ and $t_{1/2}$

$$\frac{t_{1/4}}{t_{1/2}} = \frac{\frac{3}{4}\frac{[A_0]}{k}}{\frac{[A_0]}{2k}} = \frac{3}{2}$$

$$\therefore \quad \boxed{t_{1/4} = \frac{3}{2}t_{1/2}}$$

Note *$t_{1/4}$ is also written as t_{75}.*

Average life is defined as reciprocal of the rate constant.

$$t_{av} = \frac{1}{k}$$

For 1st order reaction, $t_{av} = 1.44 t_{1/2}$

Example 4.4 *If half-life period for zero order reaction is 50min, calculate the quarter life time in second.*

Sol. Given $t_{1/2} = 50$ min

We know that, $t_{1/4} = \frac{3}{2} \times t_{1/2} = \frac{3}{2} \times 50 \text{ min}$

$$= 75 \text{ min} = 75 \times 60 \text{ s}$$

$$t_{1/4} = 4500 \text{ s}$$

2. First order reaction

A reaction is said to be of first order if its rate is determined by the change of one concentration term only. It means rate is directly proportional to the concentration of one reactant raised to first order.

Consider a hypothetical equation,

	$A \longrightarrow$	Product
Concentration at $t = 0$	$a = [A]_0$	0
Concentration at $t = t$	$(a - x) = [A]_t$	x

where, a or $[A]_0$ is the concentration of 'A' at initial time. $(a - x)$ or $[A]_t$ is the concentration of A after time t.

∴ Different rate equation can be written as,

$$\therefore \quad \frac{dx}{dt} = k(a - x)$$

$$\Rightarrow \text{ or } \quad -\frac{d[A]}{dt} = k[A]_t \quad \text{...(i)}$$

Some example of first order reaction

$$2N_2O_5 \longrightarrow 4NO_2 + O_2$$

$$SO_2Cl_2 \longrightarrow SO_2 + Cl_2$$

(*i*) Integrated equation is

$$k = \frac{2.303}{t} \log_{10} \frac{a}{(a - x)}$$

or

$$t = \frac{2.303}{k} \log_{10} \frac{a}{(a - x)}$$

or

$$k = \frac{2.303}{t} \log \frac{[A]_0}{[A]_t} \quad \text{...(ii)}$$

(*ii*) Unit of rate constant for first order reaction is **time^{-1}**, i.e. **s^{-1}**.

(*iii*) Since, the rate directly depends upon the concentration, the plot of rate *vs* concentration will be a straight line passing through the origin.

(*iv*) If $\log_{10}\left(\frac{a}{a - x}\right)$ is plotted against time t, a straight line is obtained which is passing through the origin with slope $\frac{k}{2.303}$.

(*v*) Exponential expression for first order reaction

$$kt = \log_e \frac{[A]_0}{[A]_t}$$

$$-kt = \log_e \frac{[A]_t}{[A]_0}$$

$$[A]_t = [A_0]\, e^{-kt}$$

The concentration of reactant in first order reaction decreases exponentially with time.

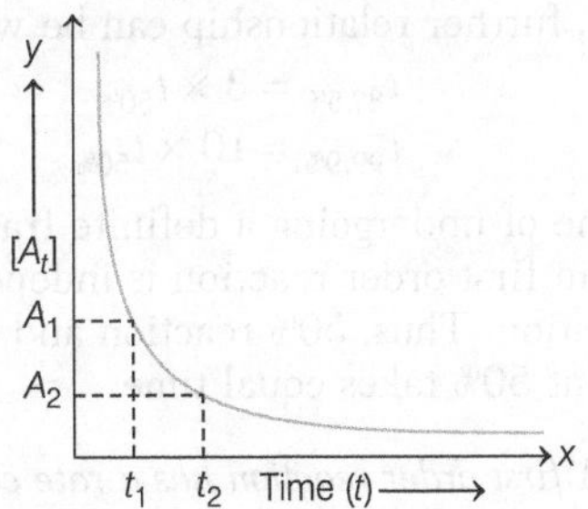

(*vi*) If initial concentration of the reactant is not known, rate constant k can be derived by taking its concentration at two time intervals.

Let $[A]_1 = (a - x_1)$ = concentration of A at time t_1

$[A]_2 = (a - x_2)$ = concentration of A at time t_2

$$\therefore \quad \frac{kt_2}{2.303} = \log_{10} a - \log_{10}(a - x_1)$$

$$\frac{kt_1}{2.303} = \log_{10} a - \log_{10}(a - x_1)$$

$$\therefore \quad k = \frac{2.303}{(t_2 - t_1)} \log \frac{(a - x_1)}{(a - x_2)}$$

$$k = \frac{2.303}{(t_2 - t_1)} \log \frac{[A]_1}{[A]_2}$$

(*vii*) Half-life period ($t_{1/2}$)

From the definition of half-life period, $t = t_{1/2}$, then $(a - x) = \frac{a}{2}$, putting this value in rate equation

For first order, we get

$$t_{1/2} = \frac{2.303}{k} \log\left(\frac{a}{a - (a/2)}\right) = \frac{2.303}{k} \log 2$$

$$t_{1/2} = \frac{0.693}{k}$$

(viii) **Quarter life time**

From the definition of quarter life time,

when $a - x = \frac{a}{4}$, $t = t_{1/4}$,

then rate constant $(k) = \frac{2.303}{t_{1/4}} \log \frac{a}{a - x}$

or,
$$t_{1/4} = \frac{2.303}{k} \log 4$$
$$= 2\left(\frac{2.303 \log 2}{k}\right)$$
$$= 2 \times \left(\frac{0.693}{k}\right)$$
$$t_{1/4} = 2 \times t_{1/2}$$

Similarly, further relationship can be written as,

$$t_{87.5\%} = 3 \times t_{50\%}$$
$$\Rightarrow \quad t_{99.9\%} = 10 \times t_{50\%}$$

Thus, time of undergoing a definite fraction of reactant in first order reaction is independent of the concentration. Thus, 50% reaction and each subsequent 50% takes equal time.

Example 4.5 *A first order reaction has a rate constant of* $15 \times 10^{-3} \, s^{-1}$. *Calculate the time taken by* 5.0 g *of reactants to reduce to* 3.0 g.

Sol. From first order equation ,

$$t = \frac{2.303}{k} \log \frac{a}{a - x}$$

Given, $k = 15 \times 10^{-3} \, s^{-1}$, $a = 5g$, $(a - x) = 3 \, g$

Given,
$$t = \frac{2.303}{15 \times 10^{-3} s^{-1}} \log \frac{5}{3}$$
$$t = 34.06 \, s$$

Example 4.6 *The half-life of a first order reaction is* 60 min. *Calculate the time taken by reactant to consume 90% of itself ?*

Sol. For the first order reaction,

$$k = \frac{0.693}{t_{1/2}} = \frac{0.693}{60 \text{min}} = 11.55 \times 10^{-3} \, \text{min}^{-1}.$$

We have, the equation

$$t = \frac{2.303}{k} \log \frac{a}{(a - x)}$$
$$t = \frac{2.303}{11.55 \times 10^{-3} \, \text{min}^{-1}} \log \frac{100}{10}$$
$$t = 199.40 \, \text{min}$$

3. Second order reaction

A reaction is said to be second order, if its reaction rate is determined by the variation of two concentration terms.

The kinetics of second order reactions are given as follows

(i) When both reactants have same concentration.

	$2A \longrightarrow$	Products or	A	$+ B \longrightarrow$	Products
At $t = 0$,	a	0 or	a	a	0
At $t = t$,	$(a - x)$	x or	$(a - x)$	$(b - x)$	x

where, a is the concentration of A at $t = 0$ and $(a - x)$ is at $t = t \, [(b - x)$ is concentration of B which is equal to $(a - x)]$

$\therefore$ Differential rate equation can be written as,

$$\frac{dx}{dt} = k(a - x)^2$$

After integrating and rearranging the above equation, we get integrated rate equation, i.e.

$$k = \frac{1}{t} \frac{x}{a(a - x)}$$

or
$$k = \frac{1}{t}\left[\frac{1}{[A]_t} - \frac{1}{[A_0]}\right]$$

[where, $[A_0]$ = initial concentration and $[A]_t$ = concentration at time t.]

(ii) When the two reactants have different concentrations.

	A	$+$	$B \longrightarrow$	Products
At $t = 0$	a		b	0
At $t = t$	$(a - x)$		$(b - x)$	x

$\therefore$ Differential rate equation,

$$\frac{dx}{dt} = k(a - x)(b - x)$$

On integrating and rearranging the above equation, we get integrated rate equation

$$k = \frac{1}{t} \frac{2.303}{(a - b)} \log \frac{b(a - x)}{a(b - x)}$$

If one of the reactant is in excess say $a >> b$, then $(a - x) \approx a$, $(a - b) \approx a$

then,
$$k = \frac{2.303}{ta} \log\left(\frac{b}{b - x}\right)$$

or
$$k' = ka = \frac{2.303}{t} \log\left(\frac{b}{b - x}\right)$$

(k' is pseudo molecular rate constant)

Some exmaples of second order reactions are

$CH_3COOH + C_2H_5OH \longrightarrow CH_3COOC_2H_5 + H_2O$

- $NO + O_3 \longrightarrow NO_2 + O_2$
- As concentration of reactant increases 'n' times, rate of reaction increases 'n^2' times.
- Unit of rate constant = $mol^{-1} \, L \, s^{-1}$.

- $t_{1/2}$ for 2nd order reaction is shown as,

$$t_{1/2} = \frac{1}{ak}$$

- $t_{1/4}$ for 2nd order reaction is shown as,

$$t_{1/4} = \frac{3}{ak}$$

$$\boxed{t_{1/4} = 3 \times t_{1/2}}$$

Example 4.7 *The reaction, $P \longrightarrow Q$ is a second order reaction. If concentration of reactant increases five times, what will be the effect on rate of the reaction ?*

Sol. For the second order reaction ,

$$P \longrightarrow Q$$

$$r_1 = k[P]^2 \Rightarrow r_1 = ka^2 \quad \text{...(i)}$$

$$[\because [P] = a \text{ mol L}^{-1}]$$

If $[P]$ increases 5 times,

then $r_2 = k(5a)^2 = 25\, ka^2$

$\therefore$ $\frac{r_1}{r_2} = \frac{ka^2}{25\, ka^2}$

$\therefore$ $r_2 = 25r_1$

$\therefore$ On increasing the concentration of reactant 'P' five times, the rate becomes 25 times more.

4. Third order reaction

A reaction is said to be third order if its rate is determined by the variation of three concentration terms.

When the concentration of all the three reactants is same or three molecules of the same reactant are involved, the rate expression is written as

$$3A \longrightarrow \text{Products}$$

or $A + B + C \longrightarrow \text{Products}$

$\therefore$ **Differential rate equation** can be written as,

$$\frac{dx}{dt} = k(a - x)^3 = k[A]^3 \quad \text{...(i)}$$

On integration and solving Eq.(i), we get

$$k = \frac{1}{2t}\left[\frac{1}{(a-x)^2} - \frac{1}{a^2}\right]$$

or

$$k = \frac{1}{2t}\left[\frac{1}{[A]^2} - \frac{1}{[A_0]^2}\right]$$

Some examples of third order reactions are

$$2NO + O_2 \longrightarrow 2NO_2$$

$$2NO + Cl_2 \longrightarrow 2NOCl$$

- Unit of rate constant (k) = $\text{mol}^{-2}\ \text{L}^2\text{s}^{-1}$.
- Half-life time ($t_{1/2}$) for third order reaction is $\frac{3}{2a^2k}$.
- Quarter life time ($t_{1/4}$) for third order reaction is $\frac{15}{2a^2k}$.

$\therefore$ $t_{1/4} = 5t_{1/2}$

Note *Reactions of 3rd or higher order are rare. This is because the probability of trimolecular or more molecular collisions is much less as compared to uni or bi-molecular reactions.*

Fractional order reactions

In these type of reactions, rate is expressed by fractions of concentration terms.

e.g. $CO + Cl_2 \longrightarrow COCl_2$

Rate = $k[CO]^2[Cl_2]^{1/2}$

Order = 2.5

Pseudo first order reactions

As discussed for an elementary chemical reactions, the order is same as molecularity. But this is not so when one of the reactants is present in excess. In that case the molecularity is two, whereas its order is one. So, such reactions in which the molecularity is two but they confirm to the first order are known as pseudo first order reaction or pseudo unimolecular reactions.

e.g. Suppose 0.1M solutions of $CH_3COOC_2H_5$ undergoes complete hydrolysis.

	$CH_3COOC_2H_5$ +	$H_2O \xrightarrow{H^+}$	CH_3COOH +	C_2H_5OH
Moles before hydrolysis	0.1	$\frac{1000}{18} = 55.5$	0	0
Moles after hydrolysis	0	55.40	0.1	0.1

The concentration of water remains constant. It does not influence the reaction rate.

Thus, rate = $k'[CH_3COOC_2H_5]$

$= k[CH_3COOC_2H_5][H_2O]$

Pseudo molecular rate constant

$$k' = k[H_2O]$$

$\therefore$ $k = \frac{k'}{[H_2O]} = \frac{k'}{55.56}$

Plots for different order reaction

For zero order	For first order	For second order	For third order

(a) Plot of rate of reaction *vs* concentration

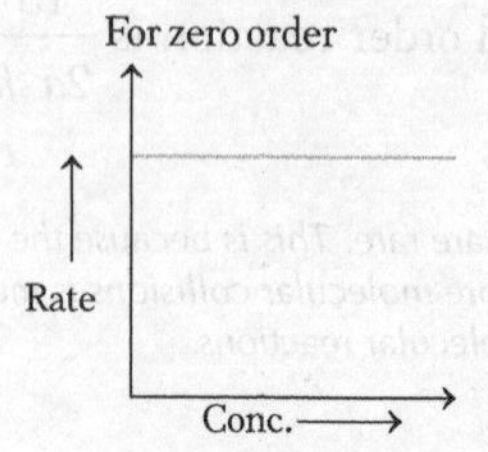

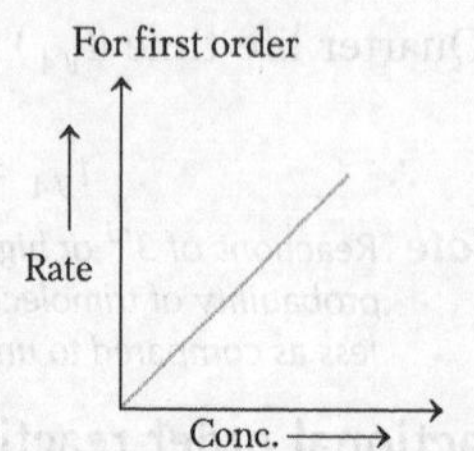

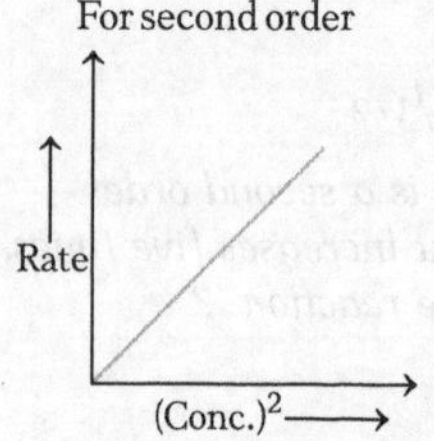

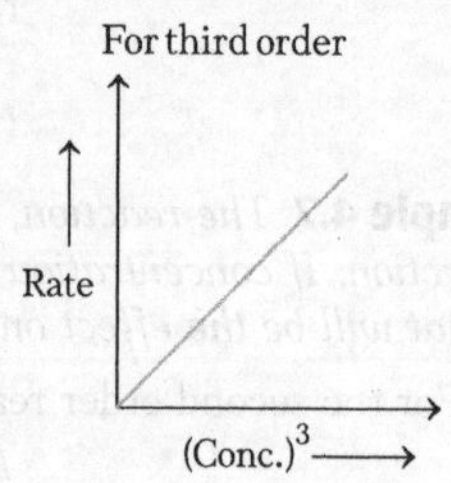

(b) Plot from integrated rate equations

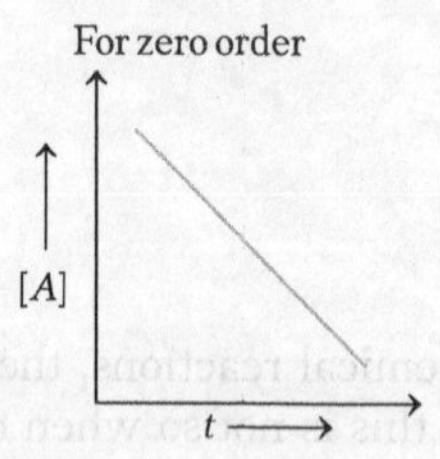

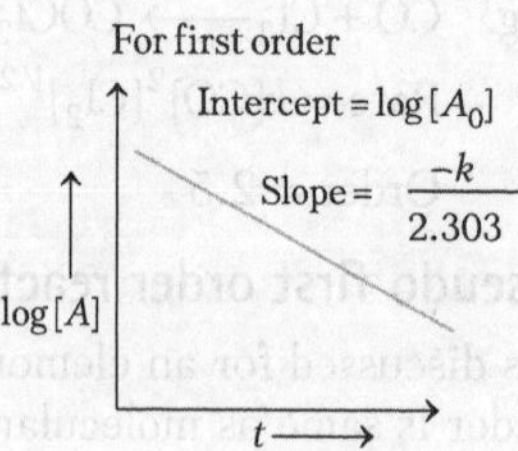

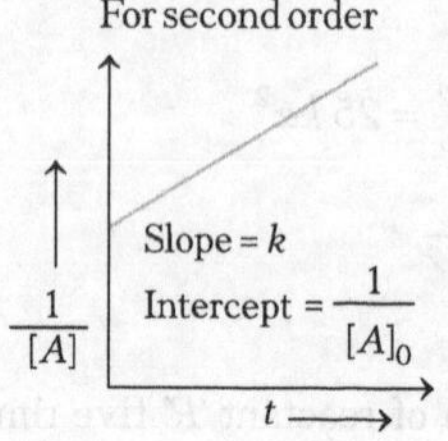

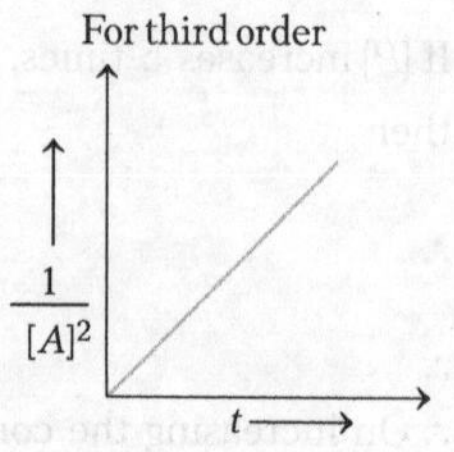

(c) Plot of half-life *vs* concentration

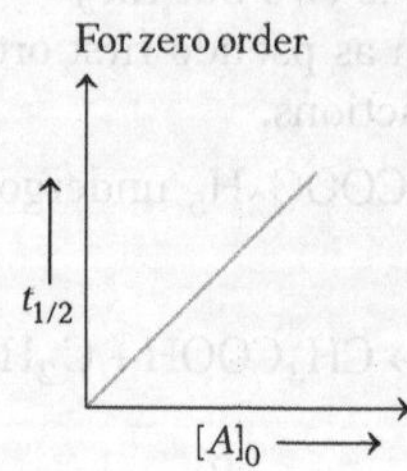

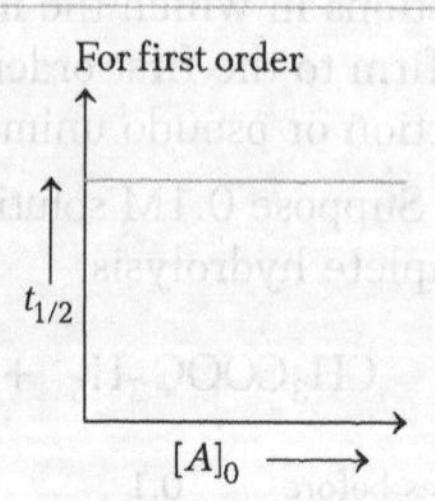

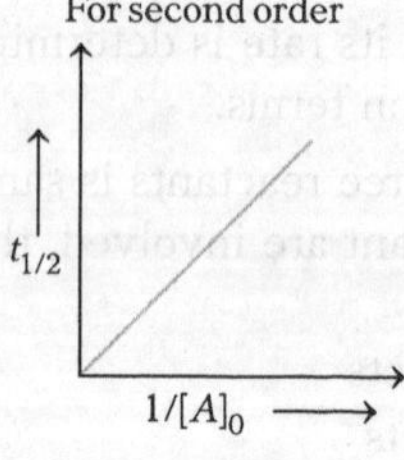

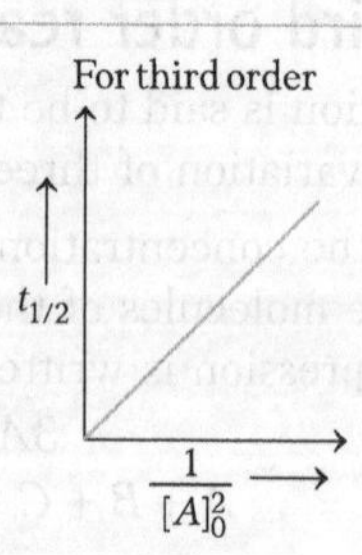

CHECK POINT 4.2

1. Half-life of first order and zeroth order reactions are same. Ratio of rates at the start of reaction is

(a) 0.693 (b) $\frac{1}{0.693}$ (c) 2×0.693 (d) $\frac{2}{0.693}$

2. Which is not true for a second order reaction ?

(a) It can have rate constant 1×10^{-2} L mol^{-1} s^{-1}

(b) Its half-life is inversely proportional to its initial concentration

(c) Time to complete 75% reaction is twice of half-life

(d) $T_{50} = \frac{1}{ka}$

3. For a second order reaction,

$$2A \longrightarrow \text{Product}$$

a straight line is obtained, if we plot

(a) concentration *vs* time (b) log (conc.) *vs* time

(c) log (conc.) *vs* time^{-1} (d) (conc.)$^{-1}$ *vs* time

4. For the second order reaction $A \longrightarrow$ Product

$[A]_0$ = concentration of the reactant A at $t = 0$

$[A]$ = concentration of the reactant at time t, then

(a) $\frac{1}{[A]} - \frac{1}{[A]_0} = kt$ (b) $[A] = \frac{[A]_0}{1 + kt[A]_0}$

(c) $t_{50} = \frac{1}{k[A]_0}$ (d) All are correct relations

5. $A \rightarrow$ Product, $\frac{dx}{dt} = k[A]^3$ If a is the initial concentration and $(a - x)$ is the concentration of A after time t, then rate constant is given by

(a) $k = \frac{1}{t}\left[\frac{1}{(a-x)} - \frac{1}{a}\right]$ (b) $k = \frac{1}{2t}\left[\frac{1}{(a-x)^2} - \frac{1}{a^2}\right]$

(c) $k = \frac{2.303}{t}\log\left(\frac{a}{a-x}\right)$ (d) $k = \frac{x}{t}$

6. For a third order reaction, $2NO + O_2 \longrightarrow 2NO_2$

(a) $+\frac{d[NO]}{dt} = k[NO]^2[O_2]$

(b) $-\frac{d[NO]}{dt} = k[NO]^2[O_2]$

(c) $-\frac{d[NO]}{2dt} = -\frac{d[O_2]}{dt} = +\frac{d[NO_2]}{2dt}$

(d) Both (b) and (c)

7. Half-life of a third order reaction is

(a) $T_{50} = \frac{0.693}{k}$ (b) $T_{50} = \frac{1}{ka}$

(c) $T_{50} = \frac{a}{2k}$ (d) $T_{50} = \frac{1.5}{ka^2}$

8. The value of rate constant of a pseudo first order reaction......

(a) depends on the concentration of reactants present in small amount

(b) depends on the concentration of reactants present in excess

(c) is independent of the concentration of reactants

(d) depends only on temperature

9. The esterification of acetic anhydride (A) with ethyl alcohol (B) can be shown as,

$$\underset{A}{(CH_3CO)_2O} + \underset{B}{C_2H_5OH} \longrightarrow CH_3COOC_2H_5 + CH_3COOH$$

Select the correct statement.

(a) In dilute hexane solution, rate is $k[A][B]$

(b) In ethyl alcohol as solvent, rate is $k[A]$

(c) In ethyl alcohol, reaction is pseudo first order

(d) All of the above are correct statements

10. In the following second order reaction,

$$A + B \longrightarrow \text{Product}$$

$$\text{rate constant } k = \frac{2.303}{t(a-b)} \log_{10} \frac{b(a-x)}{a(b-x)}$$

(where, a and b are the initial concentrations of A and B). This changes to first order rate constant, when

(a) $a >> b$ (b) $t \to \infty$

(c) $t \to 0$ (d) $x \to a$

Note *For the reactions of n order, $nA \longrightarrow$ Products*

The following rate equation is obtained.

$$k_n = \frac{1}{t(n-1)}\left[\frac{1}{(a-x)^{n-1}} - \frac{1}{a^{n-1}}\right]$$

This equation is applicable for all orders $n > 1$.

METHODS FOR DETERMINATION OF ORDER OF A REACTION

Order of the reaction is the most important part of the kinetics to study the rate law. Order is generally deduced from the rate determining step which is generally the slowest step. Following are some of the methods adopted to determine order of the given reaction.

Integration method

(Hit and Trial method)

In this method, the data is substituted in the integrated rate equation for different order reactions. In other words, the values of a, $(a-x)$, and x at different interval of time t are fitted into the equation of rate constants one by one as given in Table 4.1. The equation which gives almost a constant value of k decides the order of the reaction.

Table 4.1 Integrate rate equation for different order of a reaction

Type of reaction	Equation corresponding to $y = mx + c$	Linear plot	Slope	Intercept
Zero order (rate = k)	$[A] = -kt + [A]_0$	$[A]$ *vs* t	$-k$	$[A]_0$
First order (rate = $k[A]$)	$\log[A] = -\frac{kt}{2.303} + \log[A]_0$	$\log[A]$ *vs* t	$-\frac{k}{2.303}$	$\log[A]_0$
Second order (rate = $k[A]^2$)	$\frac{1}{[A]} = kt + \frac{1}{[A]_0}$	$\frac{1}{[A]}$ *vs* t	k	$\frac{1}{[A]_0}$

Half-life method

This method is employed when the rate law involves only one concentration term.

As already discussed for zero order, $t_{1/2} \propto [A_0]$

for first order, $t_{1/2} \propto [A_0]^0$

Similarly, for second order $t_{1/2} \propto [A_0]^{-1}$ and so on.

In general, for a reaction of nth level.

$$t_{1/2} \propto \frac{1}{[A_0]^{n-1}}$$

Starting with two different initial concentrations, i.e. $[A_0]$, $[A_0]_2$ and finding their half-lives, the order of reaction (n) can be calculated as follows

$$\frac{(t_{1/2})_1}{(t_{1/2})_2} = \frac{[A_0]_2^{n-1}}{[A_0]_1^{n-1}} = \left\{\frac{[A_0]_2}{[A_0]_1}\right\}^{n-1}$$

By taking log on both side,

$$\therefore \quad \log(t_{1/2})_1 - \log(t_{1/2})_2 = (n-1)\{\log[A_0]_2 - \log[A_0]_1\}$$

$$n = 1 + \frac{\log(t_{1/2})_1 - \log(t_{1/2})_2}{\log[A_0]_2 - \log[A_0]_1}$$

Graphical method

(Using differential rate law equations)

This method requires less time and avoid tedious calculations. The rate of the reaction is plotted against the concentration, i.e. $(a-x)$, $(a-x)^2$ and so on.

The rate constant of the reaction can be calculated from the slope of the resultant straight line plot.

(*i*) If the plot of rate *vs* molar concentration of the reactant is constant, then rate is independent of the concentration of the reactant, i.e.

$$\text{Rate} = k[a-x]^0$$

or $$k[R]^0 = k \text{ and}$$

The given reaction is zero order

(*ii*) If the plot of rate *vs* concentration of reactant $(a-x)$ gives a straight line, then the rate is directly proportional to the concentration of reactant.

$$\text{Rate} = k[R] \text{ or } k[a-x]$$

Therefore, the reaction is of first order.

(*iii*) If plot of $(a-x)^2$ *vs* rate obtained is straight line, then rate $= k[a-x]^2$

Therefore, the reaction is of second order.

(*iv*) If plot of $(\text{concentration})^3$, i.e. $[a-x]^3$ *vs* rate is a straight line then, rate $= k[a-x]^3$

Therefore, the reaction is third order.

Initial rate method

In this method, the order of a reaction is determined by varying the concentration of one of the reactants while others are kept constant.

The initial rate is determined either numerically or from the slope of the curve *vs* time at $t = 0$.

Consider a hypothetical reaction,

$$n_1A + n_2B + n_2C \longrightarrow \text{Products}$$

If order with respect to *A*, *B* and *C* are *a*, *b* and *c* then we take four sets of rate *R* to determine order *a*, *b*, *c* starting with two different initial concentrations of *A*, keeping concentrations of *B* and *C* constant. The initial rate of the reaction are determined, as

$$(r_0)_1 = k[A_0]_1^a[B]^b[C]^c \quad \text{...(i)}$$

$$(r_0)_2 = k[A_0]_2^a[B]^b[C]^c \quad \text{...(ii)}$$

On dividing equation (i) by equation (ii), we get

$$\therefore \quad \frac{(r_0)_1}{(r_0)_2} = \frac{[A_0]_1^a}{[A_0]_2^a} = \left\{\frac{[A_0]_1}{[A_0]_2}\right\}^a$$

By taking log on both side, we get

or $$\log\frac{[r_0]_1}{[r_0]_2} = a\log\frac{[A_0]_1}{[A_0]_2}$$

$$\boxed{a = \frac{\log[(r_0)_1/(r_0)_2]}{\log[[A_0]_1/[A_0]_2]}}$$

This is the order with respect to *A*. Similarly, order with respect to *B* and *C* can be determined. By adding the value of *a*, *b* and *c* we get the overall order of the reaction.

Note *If concentration is made m times of the nth order reaction, new rate is m^n times.*

Example 4.8 *At 600°C, acetone* (CH_3COCH_3) *decomposes to ketene* ($CH_2{=}C{=}O$) *and various hydrocarbons. The observations are tabulated below*

Experiment	Initial $[CH_3COCH_3]$	Rate Ms^{-1}
1.	6.0×10^{-3} M	5.2×10^{-5}
2.	9.0×10^{-3} M	7.8×10^{-5}
3.	1.8×10^{-3} M	?

(*a*) *What is the order?*
(*b*) *Write rate law.*
(*c*) *Calculate rate constant.*
(*d*) *Calculate the rate of decomposition when the acetone concentration is* 1.8×10^{-3}M.

Sol. (a) $CH_3COCH_3 \xrightarrow{\Delta} CH_2{=}C{=}O + \ldots$

By general rule, for *nth* order reaction, if concentration is made *m* times, rate becomes m^n times. On comparison of experiment values in (1) and (2),

$$\frac{7.8\times10^{-5}}{5.2\times10^{-5}} = \left[\frac{9.0\times10^{-3}}{6.0\times10^{-3}}\right]^n$$

$$\Rightarrow \quad (1.5)^n = 1.5$$

thus, $$n = 1$$

(b) $$\frac{dx}{dt} = k[CH_3COCH_3]$$

(c) $$5.2\times10^{-5} = k[6\times10^{-3}]$$

$$\therefore \quad k = \frac{5.2\times10^{-5}\text{Ms}^{-1}}{6\times10^{-3}\text{M}} = 8.67\times10^{-3}\ \text{s}^{-1}$$

(d) Concentration has been made 0.3 times *wrt* (1), hence, rate becomes 0.3 times for first order reaction.

$$\therefore \quad \left(\frac{dx}{dt}\right) = 1.56 \times 10^{-5} \text{ M s}^{-1}$$

Example 4.9 *Initial rate at 25°C are listed in the table for the reaction,*

$$NH_4^+(aq) + NO_2^-(aq) \longrightarrow N_2(g) + 2H_2O(l)$$

Experiment	Initial $[NH_4^+]$	Initial $[NO_2^-]$	Initial rate of consumption of $[NH_4^+]$, Ms^{-1}
1.	0.24	0.10	7.2×10^{-6}
2.	0.12	0.10	3.6×10^{-6}
3.	0.12	0.15	5.4×10^{-6}

(a) *What is the rate law?*
(b) *What is the value of the rate constant?*

Sol. (a) Let rate law be

$$\left(\frac{dx}{dt}\right) = k[NH_4^+]^x[NO_2^-]^y$$

By observation of (1) and (2) experimental values, when concentration of NH_4^+ has been halved at constant $[NO_2^-]$, rate is also halved, hence, $x = 1$.

From (2) and (3), when concentration of NO_2^- has been made 1.5 times at constant $[NH_4^+]$, rate is also 1.5 times.

Hence, $y = 1$.

Hence, rate law $= [NH_4^+][NO_2^-]$

(b) From any value out of (1), (2) and (3), say (2)

$$-\frac{d[NH_4^+]}{dt} = k[NH_4^+][NO_2^-]$$

$$3.6 \times 10^{-6} = k(0.12)(0.10)$$

$$\therefore \quad k = 3.00 \times 10^{-4} \text{ M}^{-1}\text{s}^{-1}.$$

CHECK POINT 4.3

1. For a given reaction, $A \to$ Product, rate is 1×10^{-4} M s^{-1} when $[A] = 0.01$ M and rate is 1.41×10^{-4} M s^{-1} when $[A] = 0.02$ M. Hence, rate law is

(a) $-\frac{d[A]}{dt} = k[A]^2$

(b) $\frac{-d[A]}{dt} = k[A]$

(c) $\frac{-d[A]}{dt} = \frac{k}{4}[A]$

(d) $\frac{-d[A]}{dt} = k[A]^{1/2}$

2. For the first order reaction T_{av} (average life), T_{50} and T_{75} in the increasing order are

(a) $T_{50} < T_{av} < T_{75}$ (b) $T_{50} < T_{75} < T_{av}$
(c) $T_{av} < T_{50} < T_{75}$ (d) $T_{av} = T_{50} < T_{75}$

3. For n^{th} order reaction,

$$\left(\frac{dx}{dt}\right) = \text{Rate} = k[A]_0^n$$

Graph between log (rate) against log $[A]_0$ is of the type

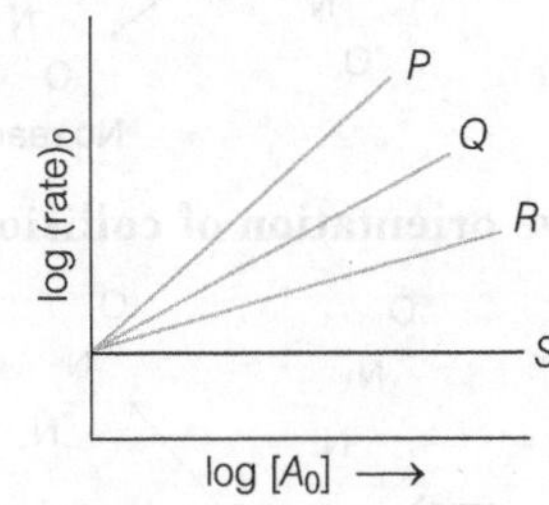

Lines *P, Q, R, S* are for the order

	P	*Q*	*R*	*S*		*P*	*Q*	*R*	*S*
(a)	0	1	2	3	(b)	3	2	1	0
(c)	1	2	3	0	(d)	0	3	2	1

4. $A + B \longrightarrow$ Product, $\frac{dx}{dt} = k[A]^a[B]^b$

If $\left(\frac{dx}{dt}\right) = k$, then order is

(a) 4 (b) 2 (c) 1 (d) 0

5. Concentration of the reactant *A* in the reaction ($A \to B$) at different times are given below.

Time (minutes)	*[A]*
0	0.069 M
17	0.052 M
34	0.035 M
51	0.018 M

Hence, order and rate constant are respectively

(a) zero, 0.001 M min^{-1} (b) 1, 0.001 min^{-1}
(c) 2, 0.001 M^{-1} min^{-1} (d) 1, 0.01 min^{-1}

6. For the reaction,

$$2A + 3B \longrightarrow \text{Product}$$

A is in excess and on changing the concentration of *B* from 0.1 M to 0.4 M, rate becomes doubled. Thus, the order of reaction is

(a) 1 (b) 2 (c) 3 (d) 1/2

7. For the reaction,

$$aA + bB \longrightarrow \text{Product}$$

$$\frac{dx}{dt} = k[A]^a[B]^b$$

If concentration of *A* is doubled, rate is four times. If concentration of *B* is made four times, rate is doubled. Hence, the order of the reaction will be

(a) 0 (b) 1 (c) 2 (d) 5/2

8. For the reaction,

$$A + B \longrightarrow C$$

S.No.	$[A]$	$[B]$	$-\frac{d[A]}{dt}$
1.	1.0 M	1.0 M	0.25 M min^{-1}
2.	2.0 M	1.0 M	0.50 M min^{-1}
3.	1.0 M	2.0 M	0.25 M min^{-1}

Hence, rate law is

(a) $k[A][B]$ (b) $k[A][B]^{1/2}$ (c) $k[A]$ (d) $k[B]$

9. The data for the reaction, $A + B \longrightarrow C$, is

Exp.	$[A]_0$	$[B]_0$	Initial rate
1	0.012	0.035	0.10
2	0.024	0.070	0.80
3	0.024	0.035	0.10
4	0.012	0.070	0.80

The rate law corresponds to the above data is

(a) rate = $k[B]^3$

(b) rate = $k[B]^4$

(c) rate = $k[A][B]^3$

(d) $k[A]^2[B]^2$

10. The experimental data for the reaction $2A + B_2 \longrightarrow 2AB$ is

Experiment	$[A]$	$[B_2]$	Rate (Ms^{-1})
1	0.50 M	0.50 M	1.6×10^{-4}
2	0.50 M	1.00 M	3.2×10^{-4}
3	1.00 M	1.00 M	3.2×10^{-4}

The rate equation for the above data is

(a) rate = $k[B_2]$

(b) rate = $k[B_2]^2$

(c) rate = $k[A]^2[B]^2$

(d) rate = $k[A]^2[B]$

THEORIES OF REACTION RATE

Following theories are given to explain the rate of a reaction.

Collision theory

The fundamental aspect of the collision theory of reaction rates is that for reaction to occur, reaction species like, molecules, atoms or ions must collide first. Increased concentrations of reacting species result in greater number of collisions per unit time. However, not all collisions result in reaction, i.e. not all collisions are effective collisions.

Factors affecting effective collisions,

(*i*) For a collision to be effective, the reacting species must

(*a*) **possess at least a certain minimum energy, i.e. threshold energy** which is defined as the minimum amount of energy which the colliding molecules must possess so as to make the chemical reaction to take place.

(*b*) **have proper orientation toward each other at the time of collision** If colliding molecules have improper orientations, they do not react even though they may possess sufficient energy.

We know that the average kinetic energy of a collection of molecules is proportional to the absolute temperature. At high temperatures, more molecules possess sufficient energy to react.

Reaction, $NO + N_2O \longrightarrow NO_2 + N_2$

- **Effective orientation of collision**

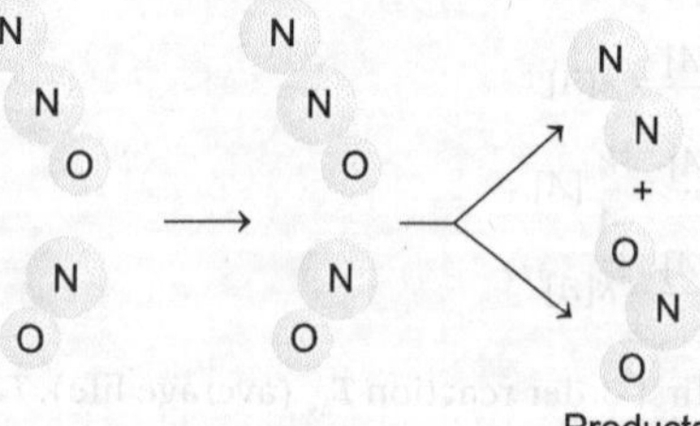

- **Ineffective orientation of collision**

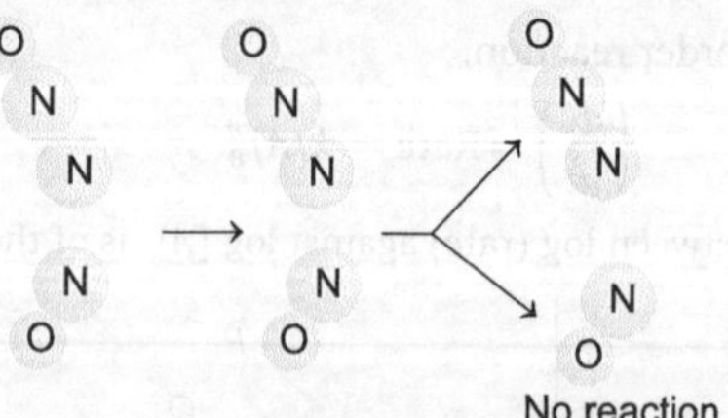

- **Ineffective orientation of collision**

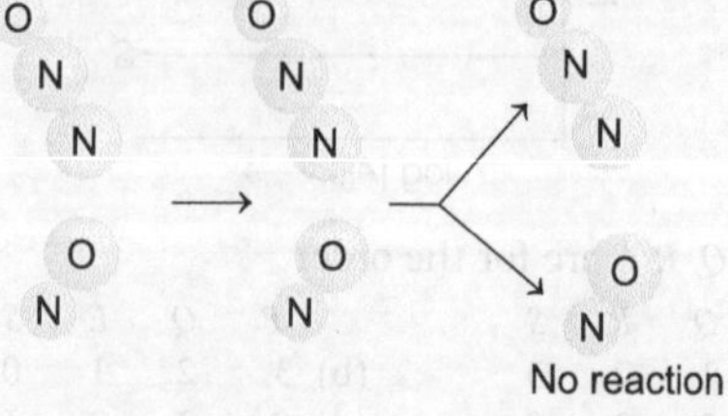

(*ii*) The minimum additional energy required by the molecule to attain threshold energy is called **activation energy**, which is required by the species during collision as a result of interchange of energies.

$E_a = E_{(\text{threshold energy})} - E_{(\text{reactants})}$

- The fraction of collisions (f) with an energy equal to or greater than the activation energy E_a is given as,

$$f = e^{-E_a/RT}$$

where, $e^{-E_a/RT}$ called **Boltzmann factor.**

The fraction of collisions having proper orientation for conversion of reactants to products is called the **steric factor,** P.

For the reaction, $A + BC \longrightarrow AB + C$, the value of P is expected to be about 0.5 because A has a nearly 50 : 50 probability of colliding with each of B and C, of BC.

Collision rate = $Z_{AB}[A][BC]$

where, Z_{AB} is a constant related to **collision frequency** (in the unit of $M^{-1}\ s^{-1}$)

and reaction rate = $P \times f \times$ collision rate

$= PfZ_{AB}[A][B]$

Thus, rate constant $k = PfZ_{AB} = PZ_{AB}e^{-E_a/RT}$

If we compare this equation with Arrhenius equation,

$$k = Ae^{-E_a/RT}$$

then $PZ = A$

Transition state theory

For reaction to occur, some covalent bonds break so that others can be formed. This can occur only if the molecules collide with enough kinetic energy to overcome the potential energy stabilisation of the bond.

The process of conversion of reactant into product is shown in fig 4.2.

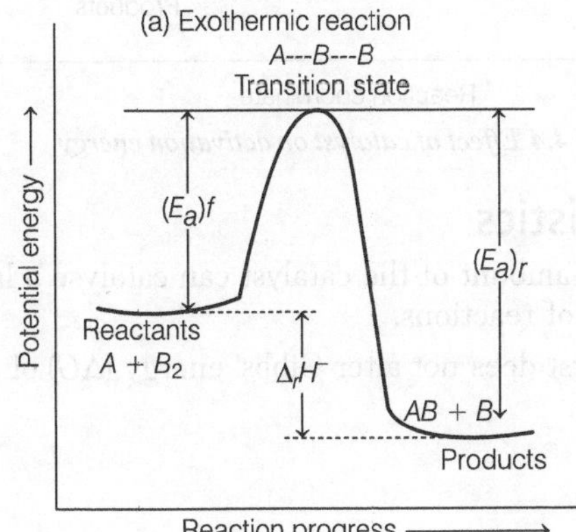

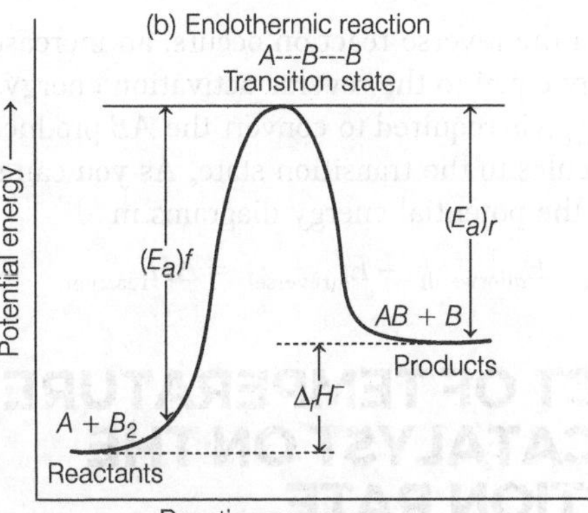

Fig. 4.2 Potential energy profile for the reaction $A + B_2 \rightarrow AB + B$ showing the energy barrier between the reactants and products. As the reaction progresses, kinetic energy of the reactants is first converted into potential energy of the transition state and is then transformed into kinetic energy of the products. At each point along the profile, the total energy is conserved.

According to the transition state theory,

(*i*) The reactants pass through a short-lived, high-energy intermediate state, called a **transition state**, before the products are formed.

$$\underset{\substack{\text{Reactants}\\A+B_2}}{A+B-B} \longrightarrow \underset{\substack{\text{Transition state}\\AB_2}}{A\cdots B\cdots B} \longrightarrow \underset{\substack{\text{Products}\\AB+B}}{A-B+B}$$

(*ii*) The activation energy E_a is the additional energy that must be absorbed by the reactants in their ground states to allow them to reach the transition state.

(*iii*) The activated complex is supposed to be in equilibrium with the reactant molecules. This complex though unstable, has a transient existence and its potential energy is maximum. Finally, it decomposes into products.

(*iv*) If A and B_2 molecules do not possess the necessary amount of energy E_a, above their ground states when they collide reaction cannot occur.

If they possess sufficient energy to "climb the energy barrier" (to reach the transition state), the reaction can proceed. When the atoms go from the transition state arrangement to the product molecules, energy is released. This type of reaction is called **exothermic reaction.**

(*v*) If the reaction results in a net absorption of energy the amount less than E_a is given off when the transition state is converted to products and the reaction is endothermic.

(vi) When the reverse reaction occurs, an increase in energy equal to the reverse activation energy, $E_{a(\text{reverse})}$ is required to convert the AB product molecules to the transition state. As you can see from the potential energy diagrams in

$$E_{a(\text{forward})} - E_{a(\text{reverse})} = \Delta H_{\text{reaction}}$$

EFFECT OF TEMPERATURE AND CATALYST ON THE REACTION RATE

Effect of temperature

Most of the chemical reactions are accelerated by increase in temperature. It has been found that for a chemical reaction with rise in temperature by 10°, the rate constant is nearly doubled. The effect of temperature is usually expressed in terms of temperature coefficient.

Mathematically,

$$\text{Temperature coefficient } (n) = \frac{\text{rate constant at } (T+10)^\circ}{\text{rate constant at } T^\circ}$$

Explanation

At a particular temperature T, all the molecules of reactants do not have the same kinetic energy. Only the fraction of molecules having particular kinetic energies at a particular temperature remains constant. The distribution of kinetic energy may be described by ploting the fraction of molecules (N_E / N_T) with a given kinetic energy (E) *versus* kinetic energy. This is called Maxwell's distributions of energies. Here, N_E is the number of molecules with energy E and N_T is the total number of molecules.

The peak of the curve represents the kinetic energy possessed by the large fraction of molecules and is known as **most probable** kinetic energy. Small number of molecules possess either very low or very high kinetic energy. Most of the molecules have intermediate kinetic energies between the highest kinetic energy and the lowest kinetic energy.

On raising temperature the maximum of the curve shifts to higher kinetic energy value and the curves spreads to the right such that there is a greater proportion of molecules with much higher energies. However, the total area under the curve must be constant, since total probability must be one at all times.

We can determine E_a (activation energy) with the Maxwell Boltzmann distribution curve.

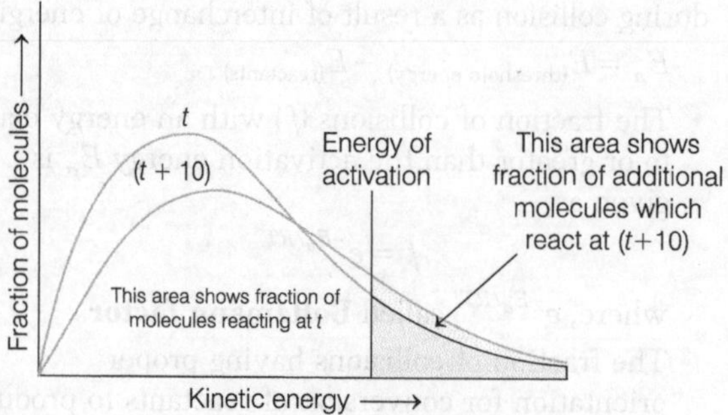

Fig 4.3 Distribution curve showing temperature dependence of rate of a reaction

Effect of catalyst

Catalyst are substances that can be added to reacting species to increase the rate of reaction. A catalyst does take part in the reaction, but all of it is regenerated in later steps.

Thus, a catalyst does not appear in the balanced equation for the reaction. They allow reactions to occur *via* alternative pathways that increases the reaction rate by lowering activation energy as depicted in Fig. 4.4.

For constant T and the same concentrations, if E_a decreases than rate of reaction increases

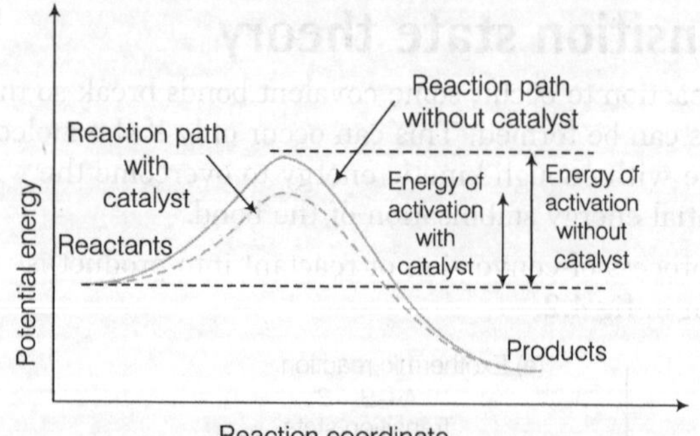

Fig 4.4 Effect of catalyst on activation energy

Characteristics

(i) A small amount of the catalyst can catalyse a large amount of reactions.

(ii) A catalyst does not alter Gibbs' energy (ΔG) of a reaction.

(*iii*) It does not change the equilibrium constant rather it helps in attaining the equilibrium faster.

(*iv*) A catalyst catalyses the spontaneous reactions but does not catalyse non-spontaneous reactions.

Arrhenius equation

Variation of rate constant k with temperature T (K) is given by Arrhenius equation.

It is based on the two theories of reaction rates, i.e. collision theory and transition state theory.

$$k = Ae^{-E_a/RT} \quad \text{...(i)}$$

where, A is called the frequency factor or the pre-exponential factor.

E_a is activation energy and

$e^{-E_a/RT}$ is referred as Boltzmann factor.

'A, has same unit as the rate constant. Collectively A and E_a are called the Arrhenius parameters of the reaction.

Calculation of activation energy

(i) Graphical method

By taking logarithm of Eq. (i) on both sides,

$$\log k = \log_e A - \frac{E_a}{RT}$$

$$\log_{10} k = \log_{10} A - \frac{E_a}{2.303RT} \quad \text{...(ii)}$$

Comparing it with $y = c - mx$

A plot Fig. 4.5 between $\log_{10} k$ and $(1/T)$ is a straight line slope is given by

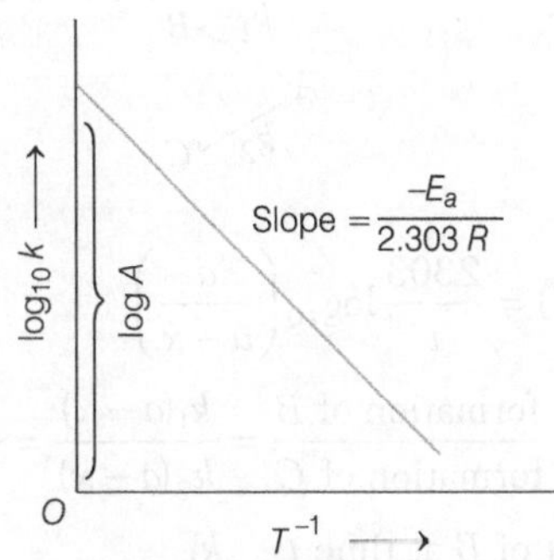

Fig. 4.5 Plot of $\log_{10} k$ *vs* $1/T$

$$\text{Slope} = -\frac{E_a}{2.303R} \text{ (slope has negative value)}$$

The intercept of the line will give the value of log A.

(ii) Experimental method

Now at temperature T_1 and T_2 rate constant are k_1 and k_2 respectively, hence from Eq (i) we get

$$\log_{10} k_2 = \log_{10} A - \frac{E_a}{2.303RT_2} \quad \text{...(iii)}$$

and $$\log_{10} k_1 = \log_{10} A - \frac{E_a}{2.303RT_1} \quad \text{...(iv)}$$

By substracting Eq (iv) from Eq. (iii) we get,

$$\therefore \quad \log_{10}\left(\frac{k_2}{k_1}\right) = \frac{E_a}{2.303R}\left[\frac{1}{T_1} - \frac{1}{T_2}\right] \quad [\text{Here, } T_2 > T_1]$$

This relation is used when rate constant at two different temperature are given.

Important points related to Arrhenius equation

(*i*) The temperature dependence on the rate of a chemical reaction can be accurately explained by this equation.

If T increases ⇒ E_a/RT decreases ⇒ $-E_a/RT$ increases ⇒ $e^{-E_a/RT}$ increases ⇒ k increases ⇒ Reaction speeds up

(*ii*) When $T \to \infty$, $K = Ae^{\circ}$ or rate constant becomes equal to Arrhenius or frequency factor.

(*iii*) Variation of k leads to variation of rate of reaction of the same degree since for a rate of reaction = $K[\text{conc.}]^n$.

(*iv*) Larger is the activation energy, smaller is the value of rate constant.

If E_a increases ⇒ E_a/RT increases ⇒ $-E_a/RT$ decreases ⇒ $e^{-E_a/RT}$ decreases ⇒ k decreases

(*v*) Lager the activation energy, greater the effect of a given temperature rise on rate constant (k).

Example 4.10 *Calculate the activation energy for a reaction when a change of temperature from 20° C to 30° C exactly triples the reaction rate constant?*

Sol. Given, $T_1 = 273 + 20 = 293$ K; $k_1 = k$ (say)

$T_2 = 273 + 30 = 303$ K; $k_2 = 3k$

On substituting these values in the equation,

$$\log\frac{k_2}{k_1} = \frac{E_a}{2.303R}\left(\frac{T_2 - T_1}{T_1T_2}\right), \text{ we get}$$

$$\log\frac{3k}{k} = \frac{E_a}{2.303 \times 8.314}\left(\frac{303 - 293}{303 \times 293}\right)$$

$$\Rightarrow \quad E_a = 81100 \text{ J mol}^{-1} \text{ or } E_a = 81.100 \text{ kJ mol}^{-1}$$

Example 4.11 *The rate constant for the decomposition of a certain reaction is described by the equation*

$$\log_{10} k(\text{s}^{-1}) = 14 - \frac{1.25 \times 10^4 \text{K}}{T}$$

(a) *Calculate pre-exponential factor?*

(b) *Calculate energy of activation (in cal)?*

(c) *At what temperature, rate constant is equal to pre-exponential factor?*

Sol. We start with Arrhenius equation in logarithm form.

$$\log_{10} k = \log_{10} A - \frac{E_0}{2.303RT} \quad ...(i)$$

Given $$\log_{10} k(s^{-1}) = 14 - \frac{1.25 \times 10^4 K}{T} \quad ...(ii)$$

(a) From Eqs. (i) and (ii),

$$\log_{10} A = 14$$

$\therefore$ A (pre-exponential factor) $= 10^{14}\ s^{-1}$

(b) $$\frac{E}{2.303R} = 1.25 \times 10^4$$

$$E_a = 2.303R \times 1.25 \times 10^4$$

$$= 2.303 \times 2 \times 1.25 \times 10^4 \text{ cal} = 57575 \text{ cal}$$

(c) $\log_{10} k = \log_{10} A$ if $T \to \infty$ or $k = A$ if $T \to \infty$

Example 4.12 *The rate constant for the first order gas phase decomposition of ethyl-iodide is $1.60 \times 10^{-5}\ s^{-1}$ at 600 K and $6.36 \times 10^{-3} s^{-1}$ at 700 K. Calculate the energy of activation for this reaction.*

Sol. Given

$$k_1 = 1.60 \times 10^{-5} s^{-1};\ T_1 = 600 \text{ K}$$

$$k_2 = 6.36 \times 10^{-3} s^{-1};\ T_2 = 700 \text{ K}$$

we have, $$\log_{10}\left(\frac{k_2}{k_1}\right) = \frac{E_a}{2.303\,R}\left[\frac{1}{T_1} - \frac{1}{T_2}\right] \quad ...(i)$$

On substituting the given values in equation (i), we get

$$\log_{10}\left(\frac{6.36 \times 10^{-3} s^{-1}}{1.60 \times 10^{-5} s^{-1}}\right) = \frac{E_a}{2.303 \times 8.314}\left[\frac{1}{600} - \frac{1}{700}\right]$$

$$\Rightarrow \log(3.98 \times 10^2) = \frac{E_a}{19.15 \times 4200}$$

$$\therefore \quad E_a = 209 \text{ kJ mol}^{-1}.$$

MECHANISM OF CHEMICAL REACTIONS

A series of elementary reactions proposed to account for the overall reaction is called the mechanism of the reaction. The writing of elementary reactions is usually based upon experimental evidence. In some reactions, intermediates are formed, which accumulate during the early period of the reaction. This reaches the maximum concentration and then reacts to give final products. The outline of mechanism of some of the simple reactions are written below.

1. Reaction involving two first order consecutive reactions

Such reactions involve two consecutive first order steps. In first step, an intermediate is formed which is consumed in the second step.

Thus, intermediate accumulate and reaches to maximum after which it decays to zero concentration.

e.g. $$\underset{\text{[Reactant]}}{R} \xrightarrow{k_1} \underset{\text{[Intermediate]}}{I} \xrightarrow{k_2} \underset{\text{Final product}}{P}$$

In this example, the reaction takes place in two steps both of which are first order, i.e.

Step I. $R \xrightarrow{k_1} I$ Step II. $I \xrightarrow{k_2} P$

The concentration profile of reactant, intermediate and products is shown in Fig. 4.6

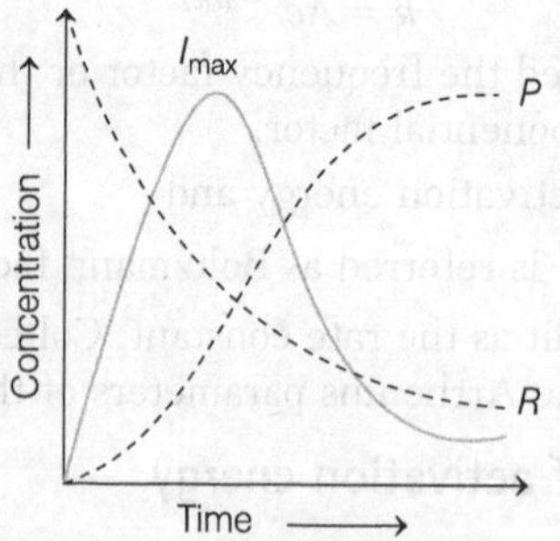

Fig. 4.6 Graph of concentration vs time

Concentration of I is maximum at time t

$$\ln\frac{k_2}{k_1} = (k_2 - k_1)t$$

$$\therefore \quad t_{max} = \frac{\ln k_2 - \ln k_1}{k_2 - k_1}$$

For parallel reactions

$$A \begin{cases} \xrightarrow{k_1} B \\ \xrightarrow{k_2} C \end{cases}$$

$$(k_1 + k_2) = \frac{2.303}{t}\log_{10}\left(\frac{a}{a-x}\right)$$

Also, $$\frac{\text{rate of formation of } B}{\text{rate of formation of } C} = \frac{k_1(a-x)}{k_2(a-x)} = \frac{k_1}{k_2}$$

and $$\frac{\text{amount of } B \text{ at time } t}{\text{amount of } C \text{ at time } t} = \frac{k_1}{k_2}$$

For opposing reactions

$$\underset{(a-x)}{A} \underset{k_2}{\overset{k_1}{\rightleftharpoons}} \underset{x}{B}$$

$$k_1 = \frac{2.303 x_e}{at}\log_{10}\frac{x_e}{x_e - x}$$

where, x_e is the equilibrium concentration.

2. Reaction involving slow step

Those reactions which occur by a sequence of steps, and among them the slowest step is considered as the rate determining step of that reaction.

e.g. $$NO_2 + CO \xrightarrow{500K} CO_2 + NO$$

Mechanism

$$\textit{Step I}\ NO_2 + NO_2 \xrightarrow{Slow} NO + NO_3$$

$$\textit{Step II}\ NO_3 + CO \xrightarrow{Fast} CO_2 + NO_2$$

Step I is the slowest step and also the rate determining step of the reaction, thus rate $= k[NO_2]^2$, i.e. the rate of reaction does not depend upon the concentration of CO but depends upon $[NO_2]^2$.

CHECK POINT 4.4

1. The temperature coefficient of a certain reaction is found to be 3. If temperature changes from 25° to 55°, the new rate of reaction will be

(a) 9 times (b) 27 times (c) 16 times (d) 8 times

2. Select the correct statement.

(a) A collision between reactant molecules must occur with a certain minimum energy before it is effective in yielding product molecules

(b) $\frac{d(\log_{10} k)}{dt} = \frac{E_a}{2.303\, RT^2}$ is called Arrhenius equation.

(c) Both (a) and (b)

(d) None of the above

3. The role of a catalyst is to change ....

(a) Gibbs energy of reaction

(b) enthalpy of reaction

(c) activation energy of reaction

(d) equilibrium constant

4. Consider the hypothetical reaction, $A + 2B \longrightarrow AB_2$

It follows rate law, rate $= k[A][B]$

hence, rate determining step is

(a) $A + B \longrightarrow AB$ (b) $A + 2B \longrightarrow AB_2$

(c) $AB + B \longrightarrow AB_2$ (d) $AB + A \longrightarrow A_2 + B$

5. For the following reaction,

$(CH_3)_3CCl + H_2O \longrightarrow (CH_3)_3COH + HCl$

$\left(\frac{dx}{dt}\right) = k[(CH_3)_3CCl]$, hence, rate determining step is

(a) $(CH_3)_3CCl \longrightarrow (CH_3)_3C^{\oplus} + Cl^{\ominus}$

(b) $(CH_3)_3CCl + H_2O \longrightarrow (CH_3)_3COH + HCl$

(c) $(CH_3)_3C^{\oplus} + H_2O \longrightarrow (CH_3)_3COH + H^+$

(d) $H^+ + Cl^- \longrightarrow HCl$

6. A two-step mechanism has been suggested for the reaction of nitric oxide and bromine

$$NO(g) + Br_2(g) \xrightarrow{k_1} NOBr_2(g)$$

$$NOBr_2(g) + NO(g) \xrightarrow{k_2} 2NOBr(g)$$

Observed rate law is, rate $= k[NO]^2[Br_2]$

hence, rate determining step is

(a) $NO(g) + Br_2(g) \longrightarrow NOBr_2(g)$

(b) $NOBr_2(g) + NO(g) \longrightarrow 2NOBr(g)$

(c) $2NO(g) + Br_2(g) \longrightarrow 2NOBr(g)$

(d) None of the above

7. The reaction, $H_2(g) + 2ICl(g) \longrightarrow I_2(g) + 2HCl(g)$ has a second order rate law, rate $= k[H_2][ICl]$. Hence, rate-determining step is

(a) $H_2(g) + ICl(g) \longrightarrow HCl(g) + HI(g)$

(b) $HI(g) + ICl(g) \longrightarrow HCl(g) + I_2(g)$

(c) Both (a) and (b)

(d) None of the above

8. For the reaction,

$$2Fe^{2+} + H_2O_2 \longrightarrow 2Fe^{3+} + 2OH^-$$

$$\frac{d[Fe^{3+}]}{dt} = k[Fe^{2+}][H_2O_2]$$

hence, slow step is

(a) $Fe^{2+} + H_2O_2 \longrightarrow Fe^{3+} + \dot{O}H^- + OH$

(b) $Fe^{2+} + OH \longrightarrow Fe^{3+} + \dot{O}H^-$

(c) $2Fe^{2+} + H_2O_2 \longrightarrow 2Fe^{3+} + 2OH^-$

(d) $Fe^{2+} \longrightarrow Fe^{3+} + e^-$

9. The acid catalysed bromination of acetone in aqueous solution

$$CH_3COCH_3 + Br_2 \xrightarrow{H^+} CH_3COCH_2Br + H^+ + Br^-$$

is found to obey the following rate law

$$\text{rate} = k[CH_3COCH_3][H^+][Br_2]^0$$

Thus, rate determining step is

(a) $CH_3COCH_3 + H^+ \longrightarrow CH_3\overset{\oplus}{C}(OH)CH_3$

(b) $(CH_3)_2\overset{\oplus}{C}OH + Br_2 \longrightarrow CH_3COCH_2Br + 2H^+ + Br^-$

(c) Both (a) and (b)

(d) None of the above

10. For the complex reaction,

$$Ag^+ + 2NH_3 \rightleftharpoons [Ag(NH_3)_2^+]$$

$$\left(\frac{dx}{dt}\right) = 2 \times 10^7\ L^2\ mol^{-2}s^{-1}\ [Ag^+][NH_3]^2 - 1 \times 10^{-2}\ s^{-1}\ [Ag(NH_3)_2^+]$$

Hence, ratio of rate constants of the forward and backward reaction is

(a) $2 \times 10^7\ L^2\ mol^{-2}$ (b) $2 \times 10^9\ L^2\ mol^{-2}$

(c) $1 \times 10^{-2}\ L^2\ mol^{-2}$ (d) $0.5 \times 10^{-9}\ L^2\ mol^2$

Chapter Exercises

(A) Topical exercise

I. Rate, rate law expression and rate constant

1 For the reaction, $2A + B \longrightarrow 3C + D$, which of the following does not express the reaction rate ?

(a) $\frac{-\Delta[A]}{2\Delta t}$ (b) $\frac{\Delta[D]}{\Delta t}$ (c) $\frac{-\Delta[C]}{\Delta t}$ (d) $\frac{-\Delta[B]}{\Delta t}$

2 For the reaction, $2A + 3B \longrightarrow 4C$ the rate of reaction may be represented as,

(a) $r = -2\frac{\Delta[A]}{\Delta t} = \frac{-3\Delta[B]}{\Delta t} = \frac{4\Delta[C]}{\Delta t}$

(b) $r = -6\frac{\Delta[A]}{\Delta t} = \frac{-4\Delta[B]}{\Delta t} = \frac{3\Delta[C]}{\Delta t}$

(c) $r = -\frac{1}{2}\frac{\Delta[A]}{\Delta t} = \frac{1}{3}\frac{\Delta[B]}{\Delta t} = \frac{1}{4}\frac{\Delta[C]}{\Delta t}$

(d) $r = -\frac{1}{2}\frac{\Delta[A]}{\Delta t} = -\frac{1}{3}\frac{\Delta[B]}{\Delta t} = \frac{1}{4}\frac{\Delta[C]}{\Delta t}$

3 Which of the following expressions is correct for the rate of reaction given below?

$$5Br^-(aq) + BrO_3^-(aq) + 6H^+(aq) \longrightarrow 3Br_2(aq) + 3H_2O(l)$$

(a) $\frac{\Delta[Br^-]}{\Delta t} = 5\frac{\Delta[H^+]}{\Delta t}$ (b) $\frac{\Delta[Br^-]}{\Delta t} = \frac{6}{5}\frac{\Delta[H^+]}{\Delta t}$

(c) $\frac{\Delta[Br^-]}{\Delta t} = \frac{5}{6}\frac{\Delta[H^+]}{\Delta t}$ (d) $\frac{\Delta[Br^-]}{\Delta t} = 6\frac{\Delta[H^+]}{\Delta t}$

4 A reaction follows the given concentration-time graph. The rate for this reaction at 20 s will be

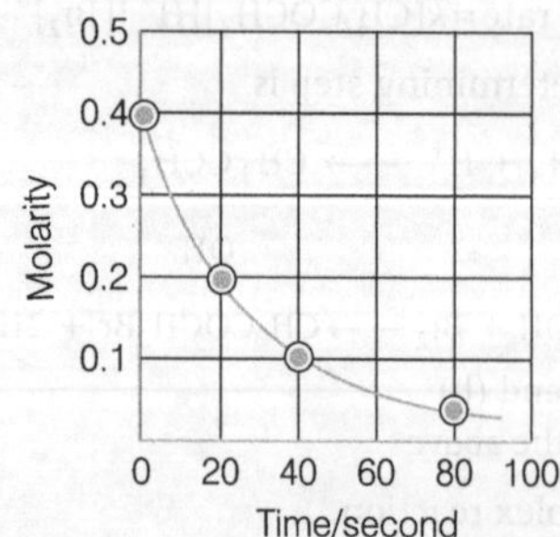

(a) 4×10^{-3} M s^{-1} (b) 8×10^{-3} M s^{-1}

(c) 2×10^{-2} M s^{-1} (d) 1×10^{-2} M s^{-1}

5 Which of the following affect the rate of reaction?

(a) Nature of reactants
(b) Temperature
(c) Presence of catalyst
(d) All of the above

6 For zeroth order reaction

(a) E_a (energy of activation) is zero
(b) heat of reaction is zero
(c) rate of disappearance is zero
(d) rate of disappearance of reactant is independent of concentration

7 Which of the following statements is correct? **[NCERT Exemplar]**

(a) The rate of a reaction decreases with passage of time as the concentration of reactants decreases.
(b) The rate of a reaction is same at any time during the reaction.
(c) The rate of a reaction is independent of temperature change.
(d) The rate of a reaction decreases with increase in concentration of reactant (s).

8 Which of these factors affect the value of the specific rate constant for the reaction?

$$2A(g) \longrightarrow B(g)$$

(a) Temperature only
(b) Temperature and concentration
(c) Pressure only
(d) Pressure and concentration

9 Match the items of Column I and Column II. **[NCERT Exemplar]**

Column I	Column II
A. Diamond to graphite	1. Short interval of time
B. Instantaneous rate	2. Ordinarily rate of conversion is imperceptible
C. Average rate	3. Long duration of time

Codes

	A	B	C		A	B	C
(a)	2	1	3	(b)	1	2	3
(c)	3	2	1	(d)	1	3	2

10 For the reaction, $A(g) + 2B(g) \longrightarrow C(g) + D(g)$

$$\frac{dx}{dt} = k[A][B]^2$$

Initial pressure of A and B are respectively 0.60 atm and 0.80 atm. At a time when pressure of C is 0.20 atm, rate of the reaction, relative to the initial value is

(a) $\frac{1}{6}$ (b) $\frac{1}{48}$

(c) $\frac{1}{4}$ (d) $\frac{1}{24}$

11 Consider the reaction $A \longrightarrow B$. The concentration of both the reactants and the products varies exponentially with time. Which of the following graph correctly describes the change in concentration of reactants and products with time? **[NCERT Exemplar]**

(a)
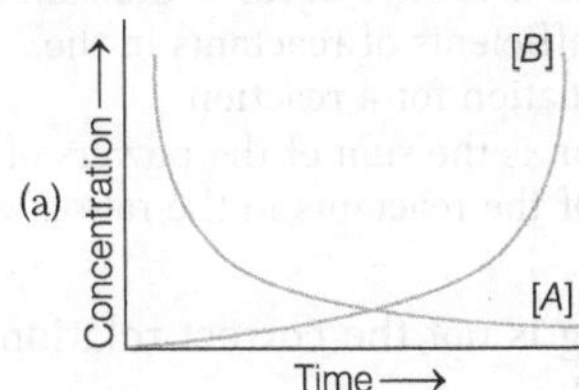

(b)
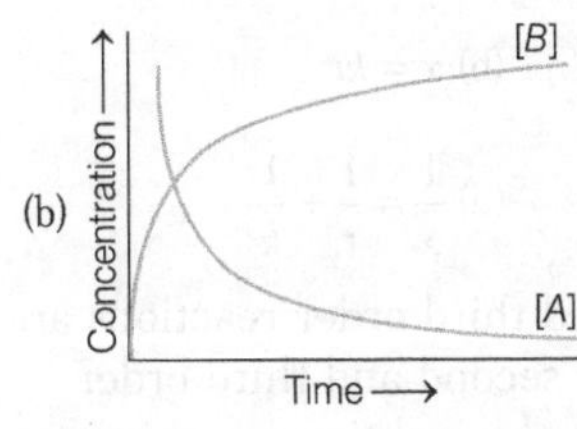

(c)
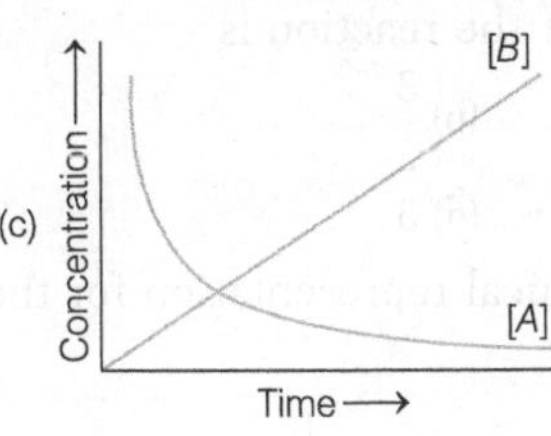

(d)
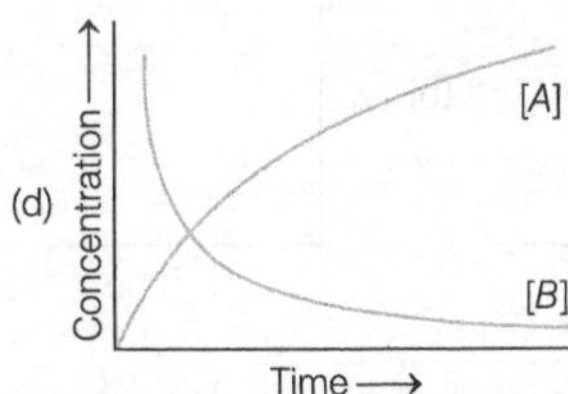

12 Consider the following statements.

Statement I In rate laws, the exponents for concentration do not necessarily match the stoichiometric coefficients.

Statement II It is the mechanism and not the balanced chemical equation for the overall change that governs the reaction rate.

Choose the correct option.

(a) Only statement I is correct.
(b) Only statement II is correct.
(c) Both the statements are correct.
(d) None of above statement are correct.

13 Rate of the reaction in terms of the pressure of the reactant (assume rate as an extensive property) is

$CH_3CHO \longrightarrow CO + CH_4$

(a) $-\frac{V}{RT}\frac{dp_{CH_3CHO}}{dt}$ (b) $-\frac{1}{RT}\frac{dp_{CH_3CHO}}{dt}$

(c) $-\frac{1}{RTV}\frac{dp_{CH_3CHO}}{dt}$ (d) $\frac{n_{CH_3CHO}}{V}\frac{dp_{CH_3CHO}}{dt}$

14 The rate law for the dimerisation of NO_2 into N_2O_4 is

$$-\frac{d[NO_2]}{dt} = k[NO_2]^2$$

Which of the following changes will change the value of the specific rate constant k?

(a) Doubling the total pressure
(b) Decreasing the pressure
(c) Changing the volume of the flask
(d) Changing the temperature

15 In a first order reaction, $A \longrightarrow P$, the ratio of $a/(a-x)$ was found to be 8 after 60 min. If the concentration is 0.1 M then the rate of reaction is

(a) 2.226×10^{-3} mol L^{-1} min^{-1}
(b) 3.466×10^{-3} mol L^{-1} min^{-1}
(c) 4.455×10^{-3} mol L^{-1} min^{-1}
(d) 5.532×10^{-3} mol L^{-1} min^{-1}

16 In a first order reaction the concentration of reactant decreases from 800 mol/dm^3 to 50 mol / dm^3 in 2×10^4 s. The rate constant of reaction is

(a) 2×10^4 s^{-1} (b) 3.45×10^{-5} s^{-1}
(c) 1.386×10^{-4} s^{-1} (d) 3×10^{-4} s^{-1}

17 Rate of formation of SO_3 in the following reaction

$$2SO_2 + O_2 \longrightarrow 2SO_3$$

is 100 g min^{-1}. Hence, rate of disappearance of O_2 is

(a) 50 g min^{-1} (b) 100 g min^{-1}
(c) 200 g min^{-1} (d) 20 g min^{-1}

18 $2N_2O_5 \longrightarrow 4NO_2 + O_2$

If $-\frac{\Delta[N_2O_5]}{\Delta t} = k_1[N_2O_5]$, $\frac{\Delta[NO_2]}{\Delta t} = k_2[N_2O_5]$, $\frac{\Delta[O_2]}{\Delta t} = k_3[N_2O_5]$, then

(a) $k_1 = k_2 = k_3$ (b) $2k_1 = k_2 = 4k_3$
(c) $2k_1 = 4k_2 = k_3$ (d) None of these

19 Which factors would increase the rate of a chemical reaction?

I. Increasing the temperature.
II. Removing products as they are formed.
III. Adding a catalyst.

(a) Both I and II (b) Both II and III
(c) I, II and III (d) Only I

20 In the complex reaction,

$$N_2 + 3H_2 \rightleftharpoons 2NH_3$$

$$\left(\frac{dx}{dt}\right) = 1\times 10^2 M^{-3}[N_2][H_2]^3 - 1\times 10^{-3} M^{-1}[NH_3]^2$$

$\frac{[N_2][H_2]^3}{[NH_3]^2} = 10^{-5}\ M^{-2}$ if $\left(\frac{dx}{dt}\right)$ is

(a) 0 (b) 1×10^5 (c) 1×10^{-5} (d) 1×10^{-3}

21 For the reaction,

$$3BrO^- \longrightarrow BrO_3^- + 2Br^-$$

in alkaline aqueous solution, the value of the second order (in BrO^-) rate constant at 80°C in the rate law for $-\Delta[BrO^-]/\Delta t$ was found to be 0.056 L $mol^{-1}s^{-1}$. Thus, select correct alternate

(a) rate constant is 0.019 L mol^{-1} s^{-1} when rate law is $\frac{\Delta[BrO_3^-]}{\Delta t}$

(b) rate constant is 0.037 L mol^{-1} s^{-1} when rate law is $\frac{\Delta[Br^-]}{\Delta t}$

(c) rate constant is 0.037 L mol^{-1} s^{-1} when rate law is $\frac{\Delta[BrO_3^-]}{\Delta t}$

(d) Both (a) and (b) are correct

22 For the reaction, $A \longrightarrow$ Products, $-\frac{d[A]}{dt} = k$ and at different time interval, [A] values are

Time	0	5 min	10 min	15 min
[A]	20 mol	18 mol	16 mol	14 mol

At 20 min, rate will be

(a) 12 mol/min (b) 10 mol/min
(c) 8 mol/min (d) 0.4 mol/min

23 For the given reaction, $A \longrightarrow$ Product $\frac{dx}{dt} = k[A]^2$

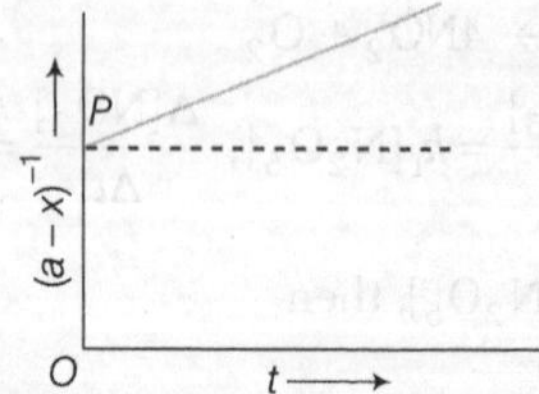

If $k = 1.0 \times 10^{-2}$ L mol^{-1} min^{-1} and $OP = 0.10$ L mol^{-1} then rate at the start of the reaction is

(a) 1.0×10^{-2} mol L^{-1} min^{-1}
(b) 1.0 mol L^{-1} min^{-1}
(c) 1.0×10^{2} mol L^{-1} min^{-1}
(d) 1.0×10^{-4} mol L^{-1} min^{-1}

II. Molecularity and order of reaction

24 Which of the following statements is not correct about molecularity of a reaction ?

(a) It can be obtained from balanced chemical equation.
(b) It can be more than three.
(c) It cannot be fractional.
(d) It may or may not be equal to the order of reaction.

25 Which of the following statements is not correct about order of a reaction? **[NCERT Exemplar]**

(a) The order of a reaction can be a fractional number.
(b) Order of a reaction is experimentally determined quantity.
(c) The order of a reaction is always equal to the sum of the stoichiometric coefficients of reactants in the balanced chemical equation for a reaction.
(d) The order of a reaction is the sum of the powers of molar concentration of the reactants in the rate law expression.

26 Which of the following is not the correct relation for zeroth order reaction ?

(a) $\left(\frac{dx}{dt}\right) = k[A]^0$ (b) $x = kt$

(c) $\log x = \log t + \log k$ (d) $\frac{1}{x} = \frac{1}{t} + \frac{1}{k}$

27 Half-lives of second and third order reactions are equal. Ratio of rates of second and third order reactions at the start of the reaction is

(a) $\frac{2}{3}$ (b) $\frac{3}{2}$
(c) 0.693 (d) 3

28 Which is not the graphical representation for the zeroth order reaction ?

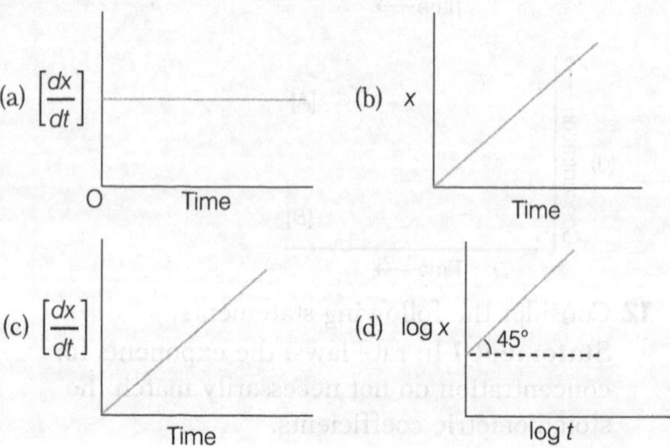

29 Match the order of reaction (in Column I) with its property (in Column II).

	Column I (Order)		Column II (Property)
A.	Zeroth	1.	Half-life $\propto \frac{1}{a^2}$
B.	First	2.	Half-life $\propto \frac{1}{a}$
C.	Second	3.	Half-life is doubled on doubling the initial concentration.
D.	Third	4.	50% reaction takes at same time even if concentration is halved or doubled.

Codes

	A	B	C	D		A	B	C	D
(a)	3	4	2	1	(b)	3	4	1	2
(c)	4	3	2	1	(d)	4	3	1	2

30 What function of $[x]$, plotted against time, will give a straight line for a second-order reaction ?

(a) $[x]$ (b) $[x]^2$

(c) $\log[x]$ (d) $\frac{1}{[x]}$

31 For the first order reaction ($A = A_0 e^{-k_1 t}$) and $T_{av} = k_1^{-1}$. After two average lives, concentration of the reactant is reduced to

(a) 25% (b) 75% (c) $\frac{100}{e}$ % (d) $\frac{100}{e^2}$ %

32 A first order reaction is 50% completed in 1.26×10^{14} s. How much time would it take for 100% completion? **[NCERT Exemplar]**

(a) 1.26×10^{15} s (b) 2.52×10^{14} s

(c) 2.52×10^{28} s (d) Infinite

33 Match order of the reaction (in Column I) with the unit of rate constant (in Column II).

Column I (Order)		Column II (Unit)
A. Zeroth	1.	$L^2 \, mol^{-2} \, s^{-1}$
B. First	2.	$L \, mol^{-1} \, s^{-1}$
C. Second	3.	$mol \, L^{-1} \, s^{-1}$
D. Third	4.	s^{-1}

Codes

	A	B	C	D		A	B	C	D
(a)	3	4	2	1	(b)	3	4	1	2
(c)	4	3	2	1	(d)	4	3	1	2

34 A reaction of first order completed 90% in 90 min, hence, it is completed 50% in approximately

(a) 50 min (b) 54 min (c) 27 min (d) 62 min

35 The rate constant for the reaction,

$$2N_2O_5 \longrightarrow 4NO_2 + O_2$$

is $3 \times 0 \times 10^{-5}$ s^{-1}. If the rate at a given time is 2.40×10^{-5} mol L^{-1} s^{-1}, then concentration of N_2O_5 at that time is

(a) 1.4 (b) 1.2 (c) 0.04 (d) 0.8

36 Rate constant of a first order reaction is 0.0693 min^{-1}. If we start with 20 mol L^{-1}, it is reduced to 2.5 mol L^{-1} in

(a) 10 min (b) 20 min

(c) 30 min (d) 40 min

37 For the reaction, $A(g) \longrightarrow B(g) + C(g)$

rate law is $-\frac{d[A]}{dt} = k[A]$

At the start pressure is 100 mm and after 10 min, pressure is 120 mm. Hence, rate constant (min^{-1}) is

(a) $\frac{2.303}{10}\log\frac{6}{5}$ (b) $\frac{2.303}{10}\log 5$

(c) $\frac{2.303}{10}\log\frac{5}{4}$ (d) $\frac{2.303}{10}\log\frac{5}{6}$

38 T_{50} (Half-life period) of first order reaction is 10 min. Starting with 10 mol L^{-1}, rate after 20 min is

(a) 0.0693 min^{-1} (b) $0.0693 \times 2.5 \text{ min}^{-1}$

(c) $0.0693 \times 5 \text{ min}^{-1}$ (d) $0.0693 \times 10 \text{ min}^{-1}$

39 Rate constant of a reaction with a virus is 3.3×10^{-4} s^{-1}. Time required for the virus to become 75% inactivated is

(a) 35 min (b) 70 min

(c) 105 min (d) 17.5 min

40 Match order of the reaction (in Column I) with the corresponding rate constant (in Column II).

Column I (Order)		Column II (Rate constant)
A. Zeroth	1.	$k = \frac{1}{2t}\left[\frac{1}{(a-x)^2} - \frac{1}{a^2}\right]$
B. First	2.	$k = \frac{1}{t}\left[\frac{1}{(a-x)} - \frac{1}{a}\right]$
C. Second	3.	$k = \frac{x}{t}$
D. Third	4.	$k = \frac{1}{t}\log_e\left(\frac{a}{a-x}\right)$

Codes

	A	B	C	D
(a)	3	4	2	1
(b)	3	4	1	2
(c)	4	3	2	1
(d)	4	3	1	2

41 Consider the following statements.

Statement I Order and molecularity are same.

Statement II Order is determined experimentally and molecularity is the sum of the stoichiometric coefficient of rate determining elementary step.

Choose the correct option.

(a) Only statement I is correct

(b) Only statement II is correct

(c) Both the statements are correct

(d) None of above statement are correct

42 $A \longrightarrow$ Product, $[A]_0 = 2$ M. After 10 min, reaction is 10% completed. If $-\dfrac{d[A]}{dt} = k[A]$ then T_{50} is approximately

(a) 0.693 min (b) 69.3 min
(c) 66.0 min (d) 0.0693 min

43 Half-life period of the first order reaction is 100 days. After 144.3 days, concentration of the reactant is reduced to what times of the original value ?

(a) $\dfrac{1}{e}$ (b) $\dfrac{1}{e^2}$ (c) 10% (d) 20%

44 Number of average life times (T_{av}) required for a first order reaction to achieve 99.9% level of completion is

(a) 2.3 (b) 6.9 (c) 9.2 (d) 0.105

45 For the second order reaction,

$$2A \longrightarrow \text{Product}$$

the initial concentration of A is 0.1 M and rate constant is 2×10^{-3} dm^3 mol^{-1} s^{-1}. The half-life period of the reaction is

(a) 5 s (b) 50 s
(c) 500 s (d) 5000 s

46 Radioactive material A has decay constant '8λ' and material B has decay constant 'λ'. Initially they have same number of nuclei. After what time the ratio of number of nuclei of material B to that A will be e ?

(a) $\dfrac{1}{9\lambda}$ (b) $\dfrac{1}{\lambda}$ (c) $\dfrac{1}{7\lambda}$ (d) $\dfrac{1}{8\lambda}$

47 Consider a first order reaction,

$$(A) \longrightarrow \text{Product}$$

Concentration of A, changes from 0.1 M to 0.025 M in 40 min. Find the rate of reaction, when concentration of A is 0.01 M

(a) 3.47×10^{-4} M min^{-1} (b) 3.47×10^{-5} M min^{-1}
(c) 1.73×10^{-4} M min^{-1} (d) 1.73×10^{-5} M min^{-1}

48 A certain zeroth order reaction has $k = 0.025$ M s^{-1} for the disappearance of A. What will be the concentration of A after 15 s if the initial concentration is 0.50 M ?

(a) 0.50 M (b) 0.375 M
(c) 0.125 M (d) 0.060 M

49 For the zeroth order reaction, sets I and II are given,

I $t = 0$ → $t = 2$ min

II $t = 0$ → $t = x$ min

hence, x is

(a) 2 min (b) 3 min (c) 4 min (d) 6 min

50 In the question 49, in set II final stage as shown is attained after

(a) 4 min (b) 3 min
(c) 2 min (d) 1 min

51 If a is the initial concentration of the reactant and $(a - x)$ is the concentration at time t for the first order reaction (rate constant k_1), then

I. $x = a(1 - e^{-k_1 t})$

II. $(a - x) = a\left(\dfrac{1}{2}\right)^y$ where, $y = \dfrac{t}{T_{50}}$

III. $k_1 = \dfrac{2.303}{t} \log_{10}\left(\dfrac{a}{a - x}\right)$

Choose the correct option.

(a) Both I and II (b) Both II and III
(c) Both I and III (d) I, II and III

52 Consider a first order gas phase decomposition reaction given below

$$A(g) \longrightarrow B(g) + C(g)$$

The initial pressure of the system before decomposition of A was p_i. After lapse of time 't' total pressure of the system increased by x units and became 'p_t'. The rate constant k for the reaction is given as .... **[NCERT Exemplar]**

(a) $k = \dfrac{2.303}{t} \log \dfrac{p_i}{p_i - x}$ (b) $k = \dfrac{2.303}{t} \log \dfrac{p_i}{2p_i - p_t}$

(c) $k = \dfrac{2.303}{t} \log \dfrac{p_i}{2p_i + p_t}$ (d) $k = \dfrac{2.303}{t} \log \dfrac{p_i}{p_i + x}$

III. Determination of order of a reaction

53 Which of the following reaction is not of first order ?

(a) $2H_2O_2 \longrightarrow 2H_2O + O_2$

(b) C_6H_5–$N_2Cl \longrightarrow C_6H_5$–$Cl + N_2$

(c) $CH_3COOC_2H_5 + H_2O \longrightarrow CH_3COOH + C_2H_5OH$

(d) $CH_3COOC_2H_5 + NaOH \longrightarrow CH_3COONa + C_2H_5OH$

54 Which of the following graph is incorrect for zero order reaction?

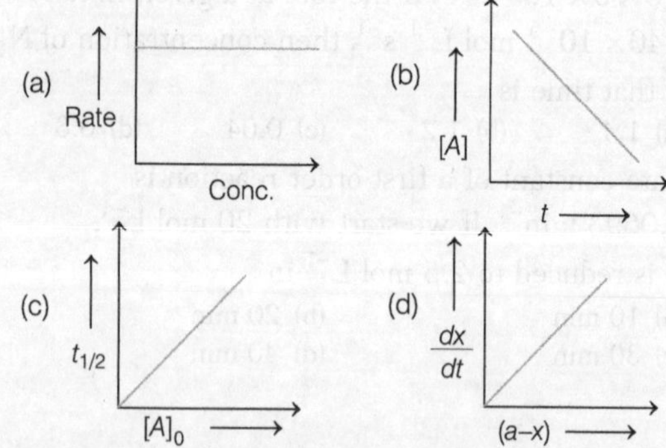

55 Which of the following graphs represent relation between initial concentration of reactants and half-life for third order reaction ?

(a) 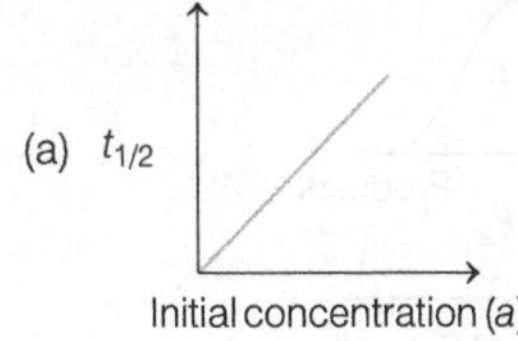

(b) 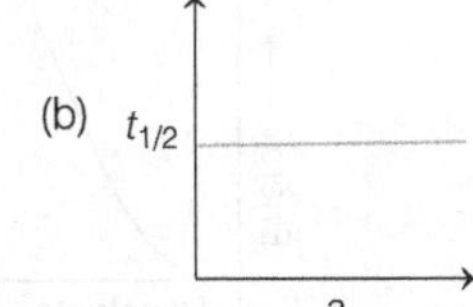

(c) 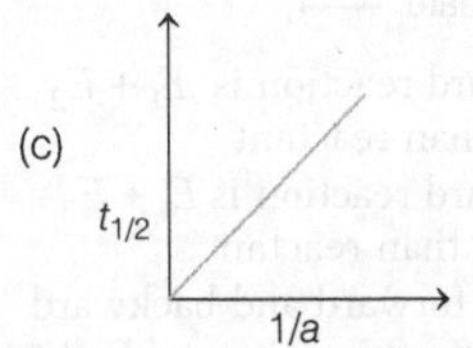

(d) 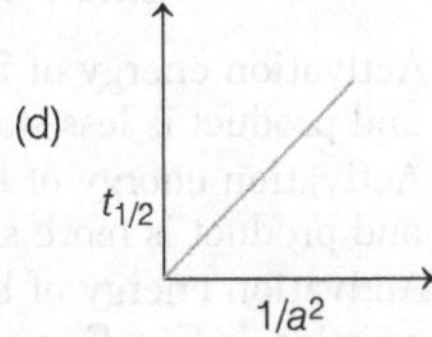

56 For a reaction, graph between log T_{50} (y-axis) and log a (x-axis) gives a straight line parallel to x-axis. Hence, order is

(a) 0 (b) 1 (c) 2 (d) 3

57 Time of 75% reaction is thrice of time of 50% reaction. Hence, order is

(a) 0 (b) 1 (c) 2 (d) 3

58 If in the fermentation of sugar in an enzymatic solution that is initially 0.12 M the concentration of the sugar is reduced to 0.06 M in 10 h and to 0.03 M in 20 h, then, order of the reaction is

(a) 0 (b) 1
(c) 2 (d) 3

59 Consider a reaction, $aG + bH \longrightarrow$ Products.
When concentration of both the reactants G and H is doubled, the rate increases by eight times. However, when concentration of G is doubled keeping the concentration of H fixed, the rate is doubled. The overall order of the reaction is

(a) 0 (b) 1
(c) 2 (d) 3

60 The rate constant for the reaction in gaseous phase

$$O + O_3 \longrightarrow 2O_2$$

is 8.0×10^{-15} cm^3 molecule^{-1} s^{-1} at 298 K. Corresponding value in dm^3 mol^{-1} s^{-1} is

(a) 4.8×10^6 dm^3 mol^{-1} s^{-1}
(b) 1.33×10^{-6} dm^3 mol^{-1} s^{-1}
(c) 4.8×10^8 dm^3 mol^{-1} s^{-1}
(d) 1.33×10^6 dm^3 mol^{-1} s^{-1}

61 If $\left(\frac{dx}{dt}\right) = k\,[H^+]^n$ and rate becomes 100 times when pH changes from 2 to 1. Hence, order is

(a) 1 (b) 2
(c) 3 (d) 0

62 Which represents first order reaction out of I, II and III ?

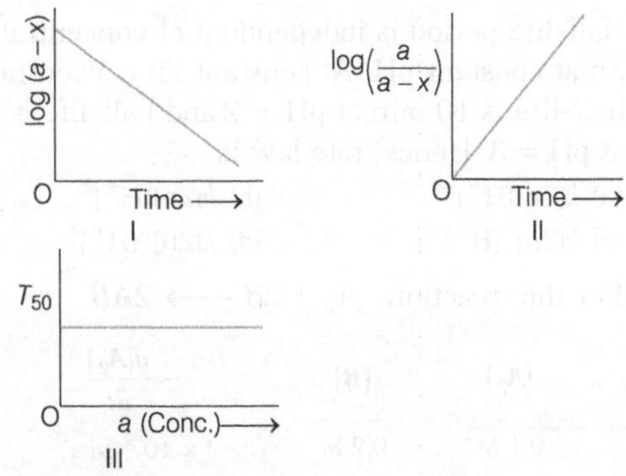

Choose the correct option.

(a) I, II and III
(b) Both I and II
(c) Both II and III
(d) Both I and III

63 $A \rightarrow$ Product and $\left(\frac{dx}{dt}\right) = k[A]^2$, If $\log\left(\frac{dx}{dt}\right)$ is plotted against log [A], then graph is of the type

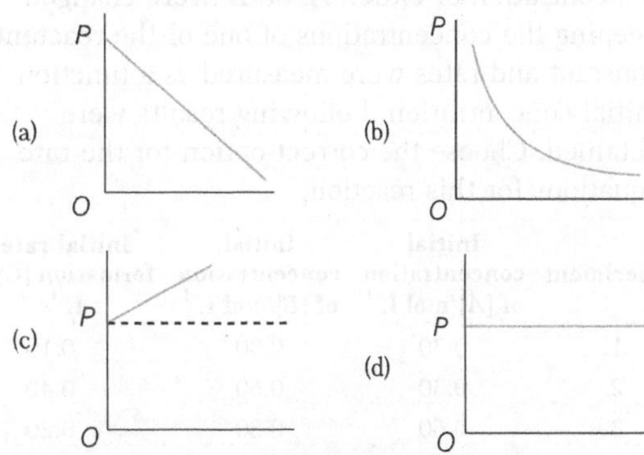

64 Slope of the line of the above question (Q-63) is

(a) 1 (b) 2
(c) 3 (d) 4

65 Match the graphical study with their order of the reactions

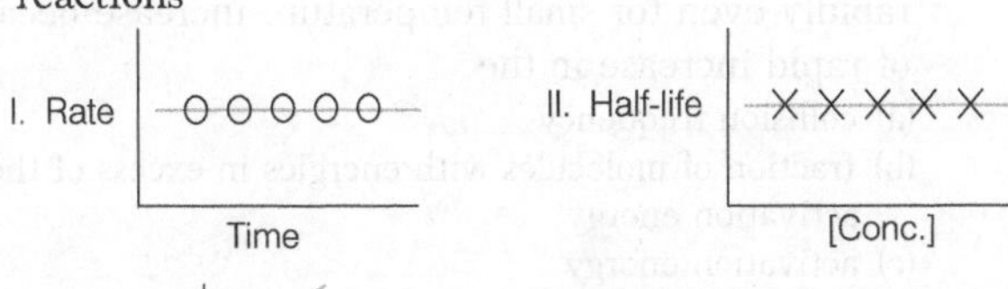

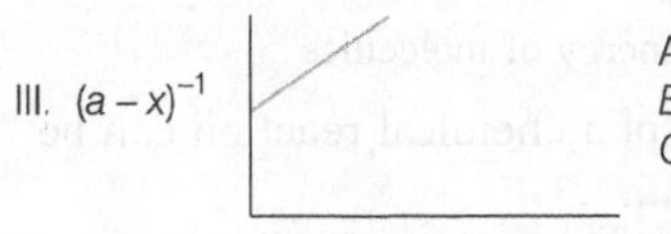

A : First
B : Second
C : Zero

	I	II	III		I	II	III
(a)	A	B	C	(b)	B	C	A
(c)	C	B	A	(d)	C	A	B

66 Consider the following reaction,

$$Zn + 2H^+ \longrightarrow Zn^{2+} + H_2$$

Half-life period is independent of concentration of Zn at constant pH. At constant Zn concentration, half-life is 10 min at pH = 2 and half-life is 100 min at pH = 3. Hence, rate law is

(a) $k[Zn][H^+]$ (b) $k[Zn][H^+]^2$
(c) $k[Zn]^0[H^+]$ (d) $k[Zn]^0[H^+]^2$

67 For the reaction, $A_2 + 2B \longrightarrow 2AB$

$[A_2]$	$[B]$	$-\frac{d[A_2]}{dt}$
0.1 M	0.2 M	1×10^{-2} M s^{-1}
0.2 M	0.2 M	2×10^{-2} M s^{-1}
0.2 M	0.4 M	8×10^{-2} M s^{-1}

Order of reaction *wrt* A_2 and B are respectively

(a) 1 and 2 (b) 2 and 1 (c) 1 and 1 (d) 2 and 2

68 Compounds '*A*' and '*B*' react according to the following chemical equation. **[NCERT Exemplar]**

$$A(g) + 2B(g) \longrightarrow 2C(g)$$

Concentration of either '*A*' or '*B*' were changed keeping the concentrations of one of the reactants constant and rates were measured as a function of initial concentration. Following results were obtained. Choose the correct option for the rate equations for this reaction.

Experiment	Initial concentration of [*A*]/mol L^{-1}	Initial concentration of [*B*]/mol L^{-1}	Initial rate of formation [*C*]/mol L^{-1}s^{-1}
1.	0.30	0.30	0.10
2.	0.30	0.60	0.40
3.	0.60	0.30	0.20

(a) Rate = $k[A]^2[B]$ (b) Rate = $k[A][B]^2$
(c) Rate = $k[A][B]$ (d) Rate = $k[A]^2[B]^0$

IV. Theories of reaction rate

69. The rate of a chemical reaction generally increases rapidly even for small temperature increase because of rapid increase in the

(a) collision frequency
(b) fraction of molecules with energies in excess of the activation energy
(c) activation energy
(d) average kinetic energy of molecules

70. Activation energy of a chemical reaction can be determined by ....

(a) determining the rate constant at standard temperature
(b) determining the rate constant at two temperatures
(c) determining probability of collision
(d) using catalyst

71. Consider the figure and mark the correct option. **[NCERT Exemplar]**

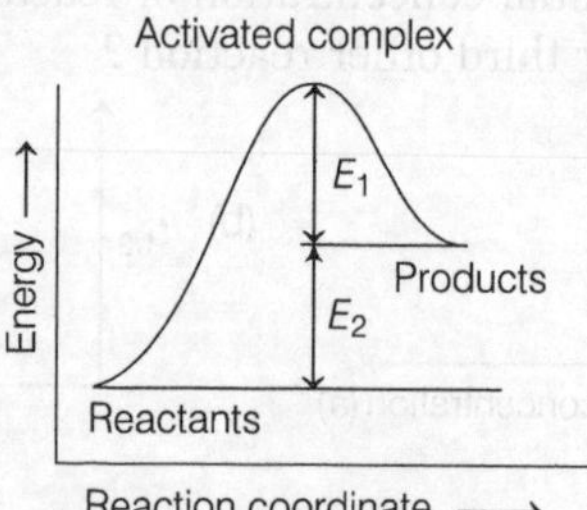

(a) Activation energy of forward reaction is $E_1 + E_2$ and product is less stable than reactant
(b) Activation energy of forward reaction is $E_1 + E_2$ and product is more stable than reactant
(c) Activation energy of both forward and backward reaction is $E_1 + E_2$ and reactant is more stable than product
(d) Activation energy of backward reaction is E_1 and product is more stable than reactant

72 According to Arrhenius equation rate constant k is equal to $A\,e^{-E_a/RT}$. Which of the following options represents the graph of $\ln k$ *vs* $\frac{1}{T}$?

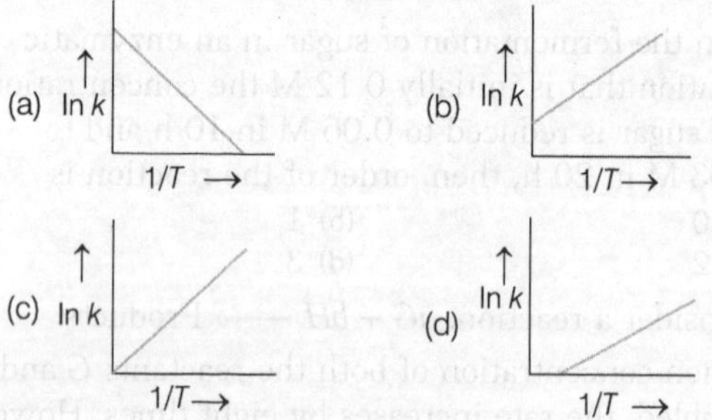

73. Select the incorrect statement.

(a) Larger the value of E_a, smaller the value of rate constant (k)
(b) Larger the value of E_a, greater is the effect on the value of k for a given temperature change
(c) At lower temperature, increase in temperature causes more change in the value of k than at higher temperature
(d) k is affected by change in concentration

74 Given the following diagram for the reaction,

$$A + B \longrightarrow C + D$$

Energy ↑
$A + B$
x
y
$C + D$
z
Progress of reaction ⟶

The enthalpy change and activation energy for the reverse reaction, $C + D \longrightarrow A + B$ are respectively

(a) x, y (b) $x, x + y$

(c) $y, x + y$ (d) $y, y + z$

75 Rate constant k of a reaction is dependent on temperature

$k = Ae^{-E_a/RT}$. k has the least value at

(a) high T and high E_a (b) high T and small E_a

(c) low T and low E_a (d) low T and high E_a

76 The activation energy of a reaction is zero. At 280 K rate constant is $1.6 \times 10^{-6}\ s^{-1}$. The rate constant at 300 K is

(a) zero (b) $3.2 \times 10^{-6}\ s^{-1}$

(c) $1.6 \times 10^{-5}\ s^{-1}$ (d) $1.6 \times 10^{-6}\ s^{-1}$

77 The correct expression for Arrhenius equation (where, k is rate constant)

(a) $k = e^{-\Delta H/RT} A$

(b) $\ln k = \dfrac{\Delta E_a}{RT} + \ln A$

(c) $\ln k = \ln A + \dfrac{E_a}{RT}$

(d) $\ln\left(\dfrac{A}{k}\right) = \dfrac{E_a}{RT}$

78 In Arrhenius equation $K = Ae^{-E_a/RT}$. The 'A' is

(a) enthalpy of reaction

(b) activation of energy

(c) free energy of reaciton

(d) rate constant of the reaction of infinite temperature

79 In the presence of a catalyst the heat evolved or absorbed during the reaction .... **[NCERT Exemplar]**

(a) increases (b) decreases

(c) remains unchanged (d) may increase or decrease

80 For a chemical reaction to occur, all of the following must happen except

(a) a large enough number of collisions must occur

(b) chemical bonds in the reactants must break

(c) reactant particles must collide with enough energy for change to occur

(d) reactant particles must collide with correct orientation

81 Reaction rates are determined by all of the following factors except

(a) the number of collisions between molecules

(b) the force of collisions between molecules

(c) the orientation of collisions between molecules

(d) the spontaneity of the reaction

82 For a reaction, $A \longrightarrow B, E_a = 10\ \text{kJ mol}^{-1}$ and $\Delta H = 5\ \text{kJ mol}^{-1}$. Thus, potential energy profile for this reaction is

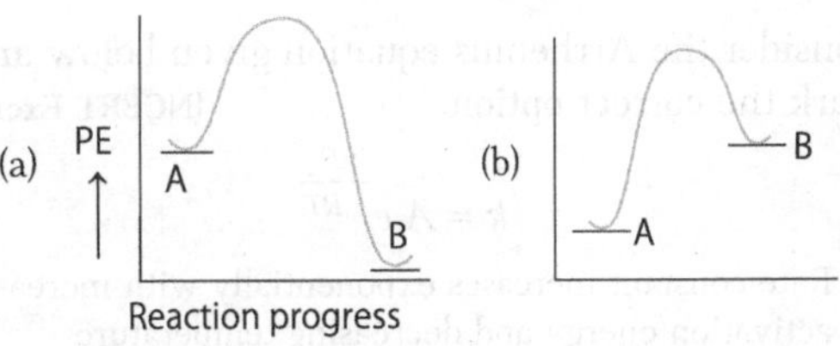

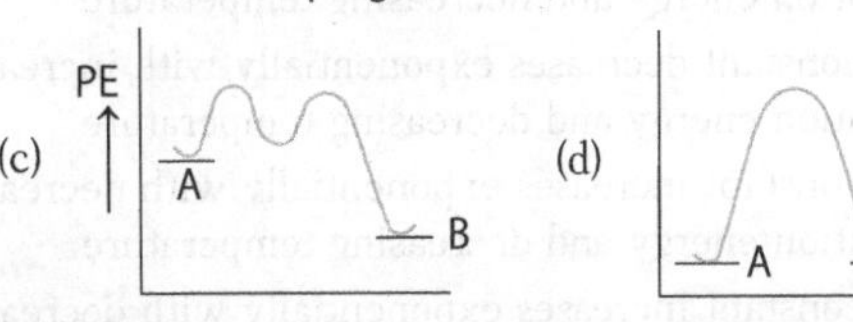

83 Energy of activation of $B \longrightarrow A$ in Q. (82) is

(a) $25\ \text{kJ mol}^{-1}$ (b) $10\ \text{kJ mol}^{-1}$

(c) $5\ \text{kJ mol}^{-1}$ (d) $-25\ \text{kJ mol}^{-1}$

84 Consider the following statements.

Statement I The rate of a chemical reaction increases as the temperature is increased.

Statement II The collision frequency between the reactant molecules increases on increasing temperature.

Choose the correct option.

(a) Only statements I is correct.

(b) Only statement II is correct.

(c) Both the statements are correct.

(d) None of above statement are correct.

85 For the following reaction,

$$A \rightleftharpoons B + 10\ \text{kcal}$$

increase in temperature

(a) increases rate of the forward as well as backward reaction

(b) increases rate of the forward reaction but decreases that of backward reaction

(c) decreases rate of the forward reaction but increases that of the backward reaction

(d) decreases rate of forward as well as backward reaction

86 When temperature of a reaction is changed from T_1 and T_2, half-life is found to decrease. Thus,

(a) $T_1 > T_2$ and reaction is endothermic

(b) $T_2 > T_1$ and reaction is exothermic

(c) $T_1 > T_2$ and reaction is exothermic

(d) $T_2 > T_1$ and reaction can be exothermic or endothermic

87 Two reactions with different activation energies have the same rate at room temperature. Which statement correctly describes the rates of these two reactions at the same higher temperature?

(a) The reaction with the greater activation energy will be faster

(b) The reaction with the smaller activation energy will be faster

(c) The two reactions will have the same rate

(d) Temperature range is also required

88 Consider the Arrhenius equation given below and mark the correct option. **[NCERT Exemplar]**

$$k = A e^{-\frac{E_a}{RT}}$$

(a) Rate constant increases exponentially with increasing activation energy and decreasing temperature
(b) Rate constant decreases exponentially with increasing activation energy and decreasing temperature
(c) Rate constant increases exponentially with decreasing activation energy and decreasing temperature
(d) Rate constant increases exponentially with decreasing activation energy and increasing temperature

89 Which of the following statements is not correct for the catalyst ? **[NCERT Exemplar]**
(a) It catalyses the forward and backward reactions to the same extent.
(b) It alters ΔG of the reaction.
(c) It is a substance that does not change the equilibrium constant of a reaction.
(d) It provides an alternate mechanism by reducing activation energy between reactants and products.

90 Which of the following statements is incorrect about the collision theory of chemical reaction ? **[NCERT Exemplar]**
(a) It considers reacting molecules or atoms to be hard spheres and ignores their structural features
(b) Number of effective collisions determines the rate of reaction
(c) Collision of atoms or molecules possessing sufficient threshold energy results into the product formation
(d) Molecules should collide with sufficient threshold energy and proper orientation for the collision to be effective

91 Consider the following statements.
Statement I A catalyst enhances the rate of a reaction.
Statement II The energy of activation of the reaction is lowered in presence of a catalyst.
Choose the correct option.
(a) Only statement I is correct.
(b) Only statement II is correct.
(c) Both the statements are correct.
(d) None of above statement are correct.

92 The rate of a reaction A doubles on increasing the temperature from 300 to 310 K. By how much, the temperature of reaction B should be increased from 300 K so that rate doubles if activation energy of the reaction B is twice to that of reaction A.
(a) 9.84 K (b) 4.92 K
(c) 2.45 K (d) 19.67 K

93 The decomposition of N_2O into N_2 and O in the presence of gaseous argon follows second order kinetics with $k = (5.0 \times 10^{11}\ \text{L mol}^{-1}\text{s}^{-1})\, e^{-29000(K)/T}$
Hence, E_a (energy of activation) is
(a) 5.0×10^{11} kcal (b) 29000 kcal
(c) 58 kcal (d) −29000 cal

94 At a certain temperature, the first order rate constant k_1 is found to be smaller than the second order rate constant k_2. If the E_a (1) of the first order reaction is greater than E_a(2) of the second order reaction, then as temperature is raised,
(a) k_2 will increase faster than k_1
(b) k_1 will increase faster than k_2 but will always remain less than k_2
(c) k_1 will increase faster than k_2 and become equal to k_2
(d) k_1 will increase faster than k_2 and become greater than k_2.

95 Consider the following statements.
Statement I The enthalpy of reaction remains constant in the presence of a catalyst.
Statement II A catalyst participating in the reaction forms different activated complex and lowers down the activation energy but the difference in energy of reactant and product remains the same.
Choose the correct option
(a) Only statement I is correct.
(b) Only statement II is correct.
(c) Both the statements are correct.
(d) None of above statement are correct.

96 Graph between log k and $\frac{1}{T}$ (k is rate constant in s^{-1} and T is the temperature in K) is a straight line.
If $OX = 5$ and slope of the line $= -\frac{1}{2.303}$ then E_a is

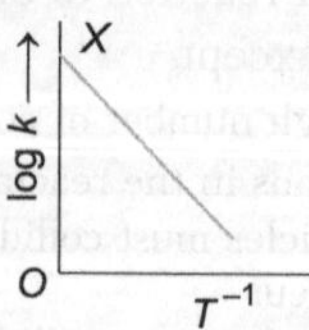

(a) 2.303×2 cal (b) $\frac{2}{2.303}$ cal
(c) 2 cal (d) None of these

97 $A \longrightarrow B, \Delta H = -10\ \text{kJ mol}^{-1}, E_a = 50\ \text{kJ mol}^{-1}$ then E_a of $B \longrightarrow A$ will be
(a) 40 kJ mol^{-1} (b) 50 kJ mol^{-1}
(c) −50 kJ mol^{-1} (d) 60 kJ mol^{-1}

98 Rate constant $k = 1.2 \times 10^3 \text{ mol}^{-1}\text{L s}^{-1}$ and $E_a = 2.0 \times 10^2 \text{ kJ mol}^{-1}$. When $T \to \infty$

(a) $A = 2.0 \times 10^2 \text{ kJ mol}^{-1}$

(b) $A = 1.2 \times 10^3 \text{ mol}^{-1} \text{ L s}^{-1}$

(c) $A = 1.2 \times 10^3 \text{ mol L}^{-1} \text{ s}^{-1}$

(d) $A = 2.4 \times 10^3 \text{ kJ mol}^{-1} \text{ s}^{-1}$

99 Consider the following statements.

Statement I Rate constant determined from Arrhenius equation are fairly accurate for simple as well as complex molecules.

Statement II Reactant molecules undergo chemical change irrespective of their orientation during collision.

Choose the correct option

(a) Only statement I is correct.

(b) Only statement II is correct.

(c) Both the statements are correct.

(d) None of above statement are correct.

100 Consider the following reaction,

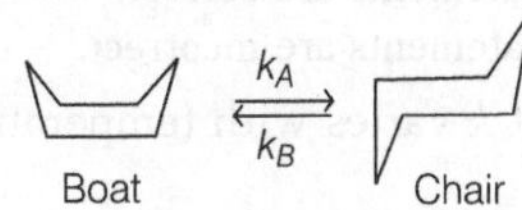

The reaction is of first order in each direction, with an equilibrium constant of 10^4. For the conversion of chair form to boat form $e^{-Ea/RT} = 4.35 \times 10^{-8}$ at 298 K with pre-exponential factor of 10^{12} s^{-1}. Apparent rate constant ($= k_A + k_B$) at 298 K is

(a) $4.35 \times 10^4 \text{ s}^{-1}$ (b) $4.35 \times 10^8 \text{ s}^{-1}$

(c) $4.35 \times 10^{-8} \text{ s}^{-1}$ (d) $4.35 \times 10^{12} \text{ s}^{-1}$

101 Milk turns sour at 40°C three times as faster as 0°C. Hence, E_a in cal of turning of milk sour is

(a) $\dfrac{2.303 \times 2 \times 313 \times 273}{40} \log 3$

(b) $\dfrac{2.303 \times 2 \times 313 \times 273}{40} \log\left(\dfrac{1}{3}\right)$

(c) $\dfrac{2.303 \times 2 \times 40}{273 \times 313} \log 3$

(d) None of the above

102 Consider the following statements.

Statement I All collision of reactant molecules lead to product formation.

Statement II Only those collisions in which molecules have correct orientation and sufficient kinetic energy lead to compound formation.

Choose the correct option.

(a) Only statement I is correct.

(b) Only statement II is correct.

(c) Both the statements are correct.

(d) None of above statement are correct.

(B) Miscellaneous questions

1 For the reaction of the zeroth order, variation of $\log T_{50}$ with $\log a$ (where, T_{50} is half-life period and a is the initial concentration) is given by

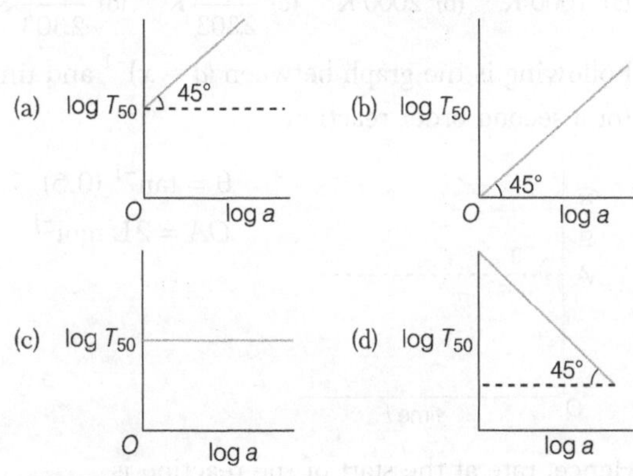

2 Graph between concentration x of the product and time of the reaction, $A \longrightarrow B$ is of the type graph I. Hence, graph between $-\dfrac{d[A]}{dt}$ and time will be of the type

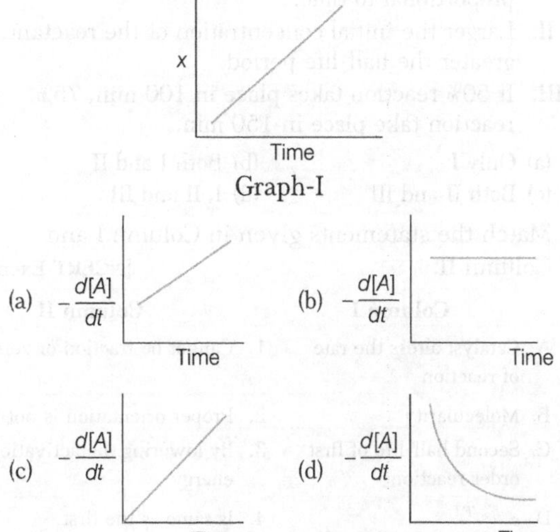

3. Consider the following statements.

Statement I The rate of a reaction normally increases by a factor of 2 or 3 for every 10° rise in temperature.

Statement II Increase in temperature increases the number of collisions.

Choose the correct option.

(a) Only statement I is correct.

(b) Only statement II is correct.

(c) Both the statements are correct.

(d) Both the statements are incorrect.

4 $A \longrightarrow B \; k_A = 10^{15} e^{-2000/T}$
$C \longrightarrow D \; k_C = 10^{14} e^{-2000/T}$
Temperature T / K at which $(k_A = k_C)$ is

(a) 1000 K (b) 2000 K (c) $\frac{2000}{2.303}$ K (d) $\frac{1000}{2.303}$ K

5 Following is the graph between $(a - x)^{-1}$ and time t for a second order reaction

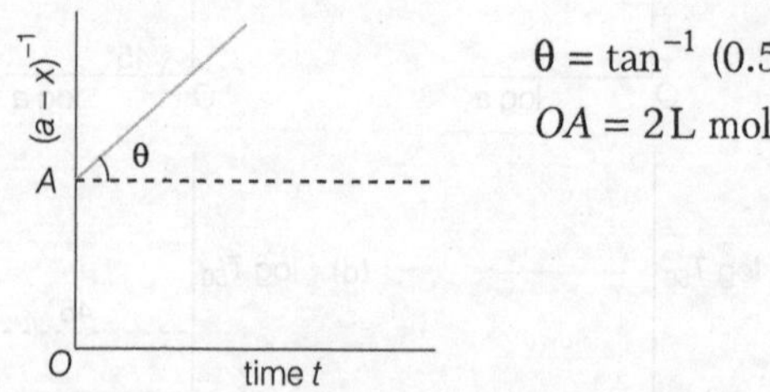

$\theta = \tan^{-1}(0.5)$
$OA = 2 \text{L mol}^{-1}$

Hence, rate at the start of the reaction is

(a) 1.25 mol^{-1} min^{-1} (b) 0.5 mol L^{-1} min^{-1}
(c) 0.125 mol L^{-1} min^{-1} (d) 1.25 mol L^{-1} min^{-1}

6 Select the correct statements out of I, II and III for zeroth order reaction.

I. Quantity of the product formed is directly proportional to time.
II. Larger the initial concentration of the reactant, greater the half-life period.
III. If 50% reaction takes place in 100 min, 75% reaction take place in 150 min.

(a) Only I (b) Both I and II
(c) Both II and III (d) I, II and III

7 Match the statements given in Column I and Column II. **[NCERT Exemplar]**

Column I	Column II
A. Catalyst alters the rate of reaction	1. Cannot be fraction or zero
B. Molecularity	2. Proper orientation is not there
C. Second half-life of first order reaction	3. By lowering the activation energy
D. $e^{-E_a/RT}$	4. Is same as the first
E. Energetically favourable reactions are sometimes slow	5. Total probability is one
F. Area under the Maxwell-Boltzmann curve is constant	6. Refers to the fraction of molecules with energy equal to or greater then activation energy.

Codes

	A	B	C	D	E	F
(a)	3	1	4	6	2	5
(b)	3	1	4	6	5	2
(c)	3	1	4	2	5	6
(d)	3	1	6	4	2	5

8 Two reactions R_1 and R_2 have identical pre-exponential factors. Activation energy of R_1 exceeds that of R_2 by 10 kJ mol^{-1}. If k_1 and k_2 are rate constants for reactions R_1 and R_2 respectively at 300 K, then $\ln\left(\frac{k_2}{k_1}\right)$ is equal to

(R = 8.314 J mol^{-1} K^{-1})

(a) 8 (b) 12
(c) 6 (d) 4

9 Consider the following statements.

Statement I The Arrhenius equation explains the temperature dependence of rate of a chemical reaction.

Statement II Plots of $\log k$ vs $\frac{1}{T}$ are linear and the energy of activation is obtained from such plots.

Choose the correct option.

(a) Only statement I is correct.
(b) Only statement II is correct.
(c) Both the statements are correct.
(d) Both the statements are incorrect.

10 Rate constant k varies with temperature as given by equation

$$\log k\,(\text{min}^{-1}) = 5 - \frac{2000 \text{ K}}{T}$$

Consider the following about this equation.

I. Pre-exponential factor is 10^5.
II. E_a is 9.212 kcal.
III. Variation of $\log k$ with $1/T$ is linear.

Choose the correct option.

(a) I, II and III (b) Both I and III
(c) Both II and III (d) Both I and II

11 For the second order reaction, concentration (A) of the reactant at time t starting with initial concentration $[A_0]$ is

(a) $\frac{[A_0]}{kt[A_0 + 1]}$ (b) $\frac{k + [A_0]^2}{1 + kt}$
(c) $\frac{1 + kt[A_0]^2}{k + [A_0]^2}$ (d) None of these

12 Consider the following statements.

Statement I Order of the reaction can be zero or fractional.

Statement II We cannot determine order from balanced chemical equation.

Choose the correct option

(a) Only statement I is correct.
(b) Only statement II is correct.
(c) Both the statements are correct.
(d) Both the statements are incorrect.

(d) Both the statements are incorrect.

13 Saponification of ethyl acetate ($CH_3COOC_2H_5$) by NaOH is studied by titration of the reaction mixture have 1 : 1 molar ratio of the reactants. If 10 mL of 1N HCl is required by 5 mL of the solution at the start and 8 mL of 1N HCl is required by another 5 mL after 10 min, then rate constant is

(a) $k = \frac{2.303}{10} \log \frac{10}{8}$ (b) $k = \frac{2.303}{10} \log \frac{10}{2}$

(c) $k = \frac{1}{10}\left[\frac{1}{8} - \frac{1}{10}\right]$ (d) $k = \frac{1}{10}\left[\frac{1}{2} - \frac{1}{10}\right]$

14 Which of the following graphs represents exothermic reaction ? **[NCERT Exemplar]**

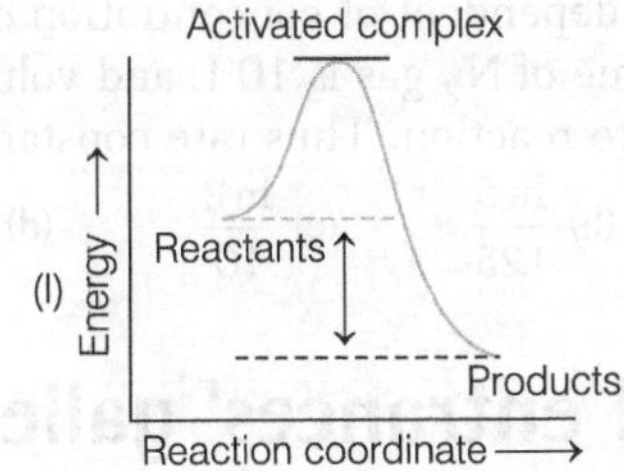

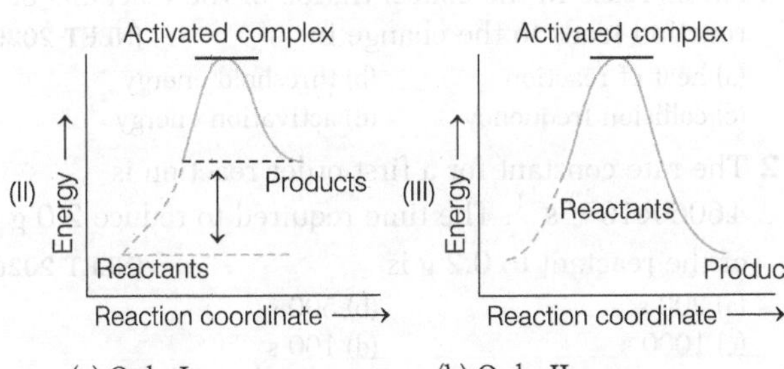

(a) Only I (b) Only II
(c) Only III (d) Both I and II

15 A reaction takes place by the following mechanism

$$A + BC \longrightarrow AC + B$$
$$AC + D \longrightarrow A + CD$$

The potential energy profile for this is shown below

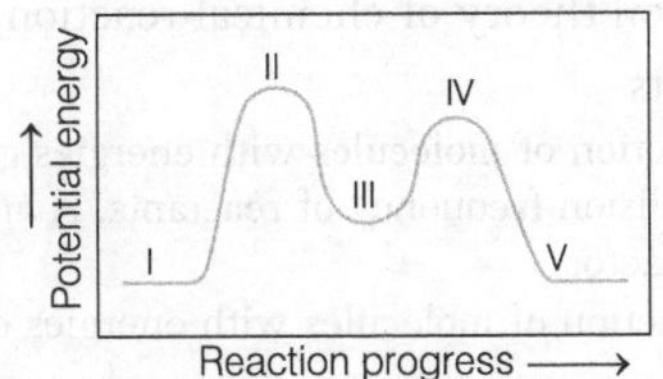

Transition states are shown by

(a) Both I and V (b) Both II and IV
(c) II, III and IV (d) Only III

16 Consider the following statements.

Statement I Order can be different from molecularity of a reaction.

Statement II (R) Slowest step is the rate determining step and may involve lesser number of reactants.

Choose the correct option.

(a) Only statement I is correct.
(b) Only statement II is correct.
(c) Both the statements are correct.
(d) Both the statements are incorrect.

17 Match the items of Column I and Column II. **[NCERT Exemplar]**

Column I	Column II
A. Mathematical expression for rate of reaction	1. Rate constant
B. Rate of reaction for zero order reaction is equal to	2. Rate law
C. Unit of rate constant for zero order reaction is same as that of	3. Order of slowest step
D. Order of a complex reaction is determined by	4. Rate of reaction

Codes

	A	B	C	D		A	B	C	D
(a)	2	1	4	3	(b)	2	1	3	4
(c)	1	2	4	3	(d)	1	2	3	4

18 In presence of catalyst E_a is lowered by 2 kcal at 27°C. Hence, rate will be

(a) 20 times (b) 14 times
(c) 28 times (d) 2 times

19 A G.M. Counter is used to study the process of first order of radioactive disintegration. In the absence of radioactive substance *A*, it counts 3 dps (disintegration per second). When *A* is placed in G.M Counter it records 23 dps at the start and 13 dps after 10 min. It records *x* dps after next 10 min. Half-life period of *A* is *y* min. Thus, *x* and *y* are

(a) 8 dps, 10 min (b) 5 dps, 10 min
(c) 5 dps, 20 min (d) 5 dps, 5 min

20 Inversion of sucrose ($C_{12}H_{22}O_{11}$) is a first order reaction and is studied by measuring angle of rotation at different interval of time

$$\underset{\substack{\text{Sucrose}\\ d\text{-}}}{C_{12}H_{22}O_{11}} + H_2O \xrightarrow{H^+} \underset{\substack{\text{Glucose}\\ d\text{-}}}{C_6H_{12}O_6} + \underset{\substack{\text{Fructose}\\ l\text{-}}}{C_6H_{12}O_6}$$

r_0 = angle of rotation at the start,
r_t = angle of rotation at time t
r_∞ = angle of rotation at the complete reaction

There is 50% inversion, when

(a) $r_0 = 2r_t - r_\infty$ (b) $r_0 = r_t - r_\infty$
(c) $r_0 = r_t - 2r_\infty$ (d) $r_0 = r_t + r_\infty$

21 Match the graph given in Column I with the order of reaction given in Column II. More than one item in Column I may link to the same item of Column II. **[NCERT Exemplar]**

Column I	Column II
A. 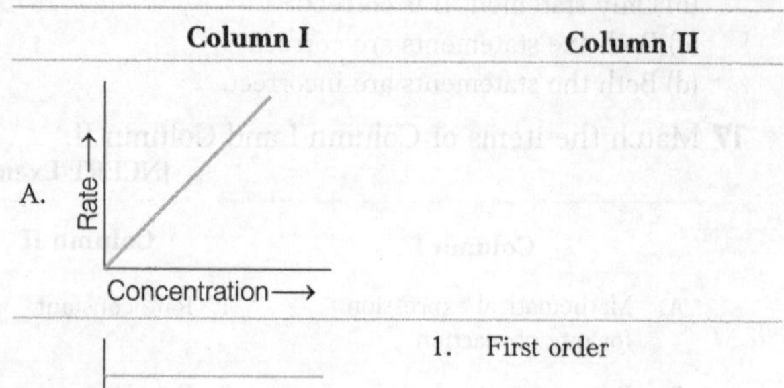	1. First order
B.	2. Zero order
C. 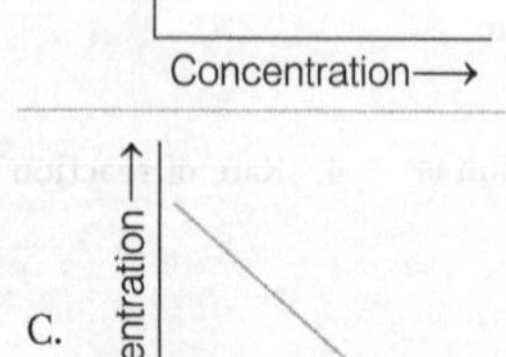	
D. 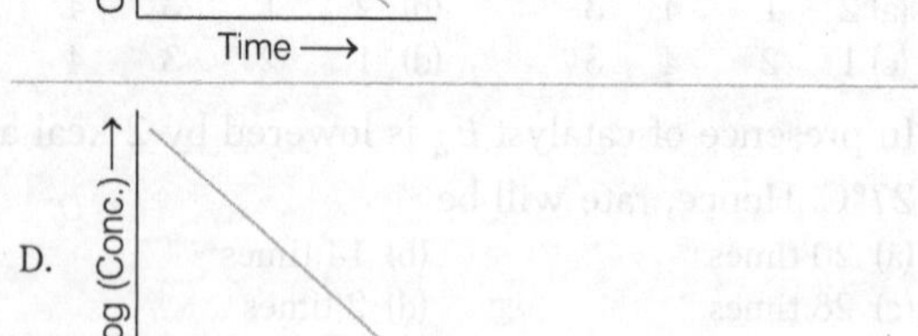	

Codes

	A	B	C	D		A	B	C	D
(a)	1	2	2	1	(b)	1	2	1	2
(c)	2	1	2	1	(d)	2	1	1	2

22 Following radioactive disintegration follows first order kinetics

$$_{Z}^{M}A \longrightarrow {}_{Z-4}^{M-8}B + \alpha\text{-particles}$$

Starting with 1 mol of A in a 1L flask at 300 K, pressure set up after two half-lifes is approximately

(a) 25 atm (b) 12 atm (c) 37 atm (d) 40 atm

23 The rate of a first order reaction is 0.04 mol L^{-1} s^{-1} at 10 min and 0.03 mol L^{-1} s^{-1} at 20 min. Thus, half-life of the reaction is

(a) 24.1 min (b) 34.8 min
(c) 50.1 min (d) 48.2 min

24 For a reaction of the type $X \underset{k_2}{\overset{k_1}{\rightleftharpoons}} Y$,

$[X]_0$ is the concentration of X at time $t = 0$ and $[X]$ is the concentration of X at time $t = t$

Thus, correct rate expression is

(a) $\frac{-d[X]}{dt} = k_1[X]_0 - (k_1 + k_2)[X]$

(b) $\frac{-d[X]}{dt} = (k_1 + k_2)[X] - k_2[X]_0$

(c) $\frac{-d[X]}{dt} = (k_1 + k_2)[X]_0 - k_2[X]$

(d) $\frac{-d[X]}{dt} = (k_1 - k_2)[X] - k_2[X]_0$

25 For the given reaction,

$$C_6H_5-N_2Cl \xrightarrow{Cu,\ \Delta} C_6H_5-Cl + N_2\uparrow$$
(A)

half-life is independent of concentration of A. After 10 min volume of N_2 gas is 10 L and volume is 50 L after complete reaction. Thus rate constant is

(a) $\frac{\ln 1.25}{10}$ (b) $\frac{\ln 5}{1.25}$ (c) $\frac{\ln 2}{10}$ (d) $\frac{\ln 2}{5}$

(C) Medical entrances' gallery

1 An increase in the concentration of the reactants of a reaction leads to the change in **[NEET 2020]**

(a) heat of reaction (b) threshold energy
(c) collision frequency (d) activation energy

2 The rate constant for a first order reaction is $4.606 \times 10^{-3}\ s^{-1}$. The time required to reduce 2.0 g of the reactant to 0.2 g is **[NEET 2020]**

(a) 200 s (b) 500 s
(c) 1000 s (d) 100 s

3 The half-life for a zero order reaction having 0.02 M initial concentration of reactant is 100 s. The rate constant (in mol L^{-1} s^{-1}) for the reaction is **[NEET 2020]**

(a) 1.0×10^{-4} (b) 2.0×10^{-4}
(c) 2.0×10^{-3} (d) 1.0×10^{-2}

4 In collision theory of chemical reaction, Z_{AB} represents **[NEET 2020]**

(a) the fraction of molecules with energies greater than E_a
(b) the collision frequency of reactants, A and B
(c) steric factor
(d) the fraction of molecules with energies equal to E_a

5 For the chemical reaction,

$$N_2(g) + 3H_2(g) \rightleftharpoons 2NH_3(g)$$

The correct option is **[NEET 2019]**

(a) $-\frac{d[N_2]}{dt} = 2\frac{d[NH_3]}{dt}$ (b) $-\frac{d[N_2]}{dt} = \frac{1}{2}\frac{d[NH_3]}{dt}$

(c) $3\frac{d[H_2]}{dt} = 2\frac{d[NH_3]}{dt}$ (d) $-\frac{1}{3}\frac{d[H_2]}{dt} = -\frac{1}{2}\frac{d[NH_3]}{dt}$

6 A first order reaction has a rate constant of 2.303×10^{-3} s^{-1}. The time required for 40 g of this reactant to reduce to 10 g will be
[Given that, $\log_{10} 2 = 0.3010$]
(a) 230.3 s (b) 301 s **[NEET (Odisha) 2019]**
(c) 2000 s (d) 602 s

7 If the rate constant for a first order reaction is k, the time (t) required for the completion of 99% of the reaction is given by **[NEET 2019]**
(a) $t = 6.909/k$ (b) $t = 4.606/k$
(c) $t = 2.303/k$ (d) $t = 0.693/k$

8 For a reaction, activation energy $E_a = 0$ and the rate constant at 200 K is $1.6 \times 10^6 s^{-1}$. The rate constant at 400 K will be
[Given that gas constant $R = 8.314 JK^{-1}mol^{-1}$] **[NEET 2019]**
(a) $3.2 \times 10^4 s^{-1}$ (b) $1.6 \times 10^6 s^{-1}$
(c) $1.6 \times 10^3 s^{-1}$ (d) $3.2 \times 10^6 s^{-1}$

9 What is the activation energy (kJ/mol) for a reaction, if its rate constant doubles when the temperature is raised from 300 K to 400 K ($R = 8.314 Jmol^{-1}K^{-1}$) **[AIIMS 2019]**
(a) 68.8 kJ mol^{-1} (b) 6.88 kJ mol^{-1}
(c) 34.4 kJ mol^{-1} (d) 3.44 kJ mol^{-1}

10 Formula for half- life of a zero order reaction is **[JIPMER 2019]**
(a) $\frac{C_0}{k}$ (b) $\frac{C_0}{2k}$ (c) $\frac{2C_0}{k}$ (d) $\frac{2C_0}{2k}$

11 If half-life of a substance is 36 minutes. Find amount left after 2 hrs. Initial amount is 10 g **[JIPMER 2019]**
(a) 1 g (b) 2 g (c) 3 g (d) 4 g

12 Find out time period of Ist order reaction when reaction complete 2/3 rd, if the value of rate constant is 4.3×10^{-4} s^{-1} **[JIPMER 2019]**
(a) 0.0025×10^3 s (b) 0.25×10^3 s
(c) 0.025×10^3 s (d) 2.5×10^3 s

13 The correct difference between first and second order reactions is that **[NEET 2018]**
(a) a first-order reaction can be catalysed; a second-order reaction cannot be catalysed
(b) the half-life of a first-order reaction does not depend on $[A]_0$; the half-life of a second-order reaction does depend on $[A]_0$
(c) the rate of a first-order reaction does not depend on reactant concentrations; the rate of a second-order reaction does depend on reactant concentrations
(d) the rate of a first-order reaction does depend on reactant concentrations; the rate of a second-order reaction does not depend on reactant concentrations

14 When initial concentration of the reactant is doubled, the half-life period of a zero order reaction **[NEET 2018]**
(a) is tripled (b) is doubled
(c) is halved (d) remains unchanged

15 Among the following statements, the correct statement about the half-life period for a first order reaction is
(a) independent of concentration **[AIIMS 2018]**
(b) proportional to concentration
(c) inversely proportional to concentration
(d) inversely proportional to the square of the concentration

16 In the reversible reaction, $2NO_2 \underset{k_2}{\overset{k_1}{\rightleftharpoons}} N_2O_4$ the rate of disappearance of NO_2 is equal to **[JIPMER 2018]**
(a) $\frac{2k_1}{k_2}[NO_2]^2$ (b) $2k_1[NO_2]^2 - 2k_2[N_2O_4]$
(c) $2k_1[NO_2]^2 - k_2[N_2O_4]$ (d) $(2k_1 - k_2)[NO_2]$

17 A chemical reaction was carried out at 300 K and 280 K. The rate constant were found to be k_1 and k_2 respectively, then **[JIPMER 2018]**
(a) $k_2 = 4k_1$ (b) $k_2 = 0.25k_1$
(c) $k_2 = 2k_1$ (d) $k_2 = 0.5k_1$

18 The rate constant for the first order decomposition of a certain reaction is described by the equation

$$\ln k (s^{-1}) = 14.34 - \frac{1.25 \times 10^4 \text{ K}}{T}.$$

The energy of activation for this reaction is **[AIIMS 2018]**
(a) 1.26×10^4 cal mol^{-1}
(b) 4.29×10^4 cal mol^{-1}
(c) 3.12×10^4 cal mol^{-1}
(d) 2.50×10^4 cal mol^{-1}

19 **Assertion** (A) The reaction
$2NO + O_2 \longrightarrow 2NO_2$ and $2CO + O_2 \longrightarrow 2CO_2$
proceeds at the same rate because they are similar.
Reason (R) Both the reactions have same activation energy. **[AIIMS 2018]**
(a) Both (A) and (R) are true and (R) is the correct explanation of (A).
(b) Both (A) and (R) is true but (R) is not the correct explanation of (A).
(c) (A) is true but (R) is false.
(d) Both (A) and (R) are false.

20 A first order reaction has a specific reaction rate of 10^{-2} s^{-1}. How much time will take for 20 g of the reactant to reduce to 5 g? **[NEET 2017]**
(a) 693.0 s (b) 238.6 s (c) 138.6 s (d) 346.5 s

21 Mechanism of a hypothetical reaction $X_2 + Y_2 \longrightarrow 2XY$ is given below **[NEET 2017]**

(i) $X_2 \rightleftharpoons X + X$(fast)

(ii) $X + Y_2 \longrightarrow XY + Y$(slow)

(iii) $X + Y \longrightarrow XY$(fast)

The overall order of the reaction will be

(a) 1.5 (b) 1 (c) 2 (d) 0

22 The addition of a catalyst during a chemical reaction alters which of the following quantities? **[NEET (I) 2016]**

(a) Internal energy (b) Enthalpy
(c) Activation energy (d) Entropy

23 The rate of a first order reaction is 0.04 mol L^{-1} s^{-1} at 10 s and 0.03 mol L^{-1} s^{-1} at 20 s after initation of the reaction. The half-life period of the reaction is **[NEET (I) 2016]**

(a) 34.1 s (b) 44.1 s (c) 54.1 s (d) 24.1 s

24 The activation energy of a reaction can be determined from the slope of which of the following graphs? **[CBSE AIPMT 2015]**

(a) $\ln k$ vs T (b) $\dfrac{\ln k}{T}$ vs T (c) $\ln k$ vs $\dfrac{1}{T}$ (d) $\dfrac{T}{\ln K}$ vs $\dfrac{1}{T}$

25 When initial concentration of a reactant is doubled in a reaction, its half-life period is not affected. The order of the reaction is **[CBSE AIPMT 2015]**

(a) zero (b) first (c) second
(d) more that zero but less than first

26 The rate constant of the reaction $A \rightarrow B$ is 0.6×10^{-3} mL^{-1} per second. If the concentration of A is 5 M then concentration of B after 20 min is **[CBSE AIPMT 2015]**

(a) 1.08 M (b) 3.60 M
(c) 0.36 M (d) 0.72 M

27 The decomposition of N_2O_5 is CCl_4 at 318 K is studied by monitoring the concentration of N_2O_5 is 2.4 mol L^{-1} and after 200 min, it is reduced to 2.0 mol L^{-1}. What is the rate of production of NO_2 during this period in min^{-1}? **[Kerala (Medical) 2015]**

(a) 4×10^{-3} (b) 2×10^{-3} (c) 1×10^{-3} (d) 2×10^{-4}
(e) 5×10^{-3}

28 In a first reaction, the concentration of the reactant decreases from 0.6 M to 0.3 M in 30 min. The time taken for the concentration to change from 0.1 M to 0.025 M is **[Kerala (Medical) 2015]**

(a) 60 min (b) 30 min
(c) 15 min (d) 50 min
(e) 90 min

29 The time required for a first order reaction to complete 90% is t. What is the time required to complete 99% of the same reaction? **[EAM CET 2015]**

(a) $2t$ (b) $3t$ (c) t (d) $4t$

30 Total order of reaction $X + Y \longrightarrow XY$ is 3. The order of reaction with respect to X is 2. State the differential rate equation for the reaction. **[Guj. CET 2015]**

(a) $-\dfrac{d[X]}{dt} = k[X]^0[Y]^3$ (b) $-\dfrac{d[X]}{dt} = k[X]^3[Y]^0$

(c) $-\dfrac{d[X]}{dt} = k[X]^2[Y]$ (d) $-\dfrac{d[X]}{dt} = k[X][Y]^2$

31 $X \xrightarrow{\text{Step I}} Y \xrightarrow{\text{Step II}} Z$, is a complex reaction. Total order of reaction is 2 and step II is slow step. What is molecularity of step II ? **[Guj. CET 2015]**

(a) 2 (b) 1
(c) 3 (d) 4

32 Reaction, $3ClO^- \longrightarrow ClO_3^- + 2Cl^-$ occurs in following two steps. **[Guj. CET 2015]**

I. $ClO^- + Cl^- \xrightarrow{k_1} ClO_2^- + Cl^-$ (slow step)

II. $ClO_2^- + ClO^- \xrightarrow{k_2} ClO_3^- + Cl^-$ (fast step)

Then, the rate of given reaction = ...

(a) $k_1[ClO^-]$ (b) $k_1[ClO^-]^2$
(c) $k_2[ClO^-][ClO^-]$ (d) $k_2[ClO^-]^3$

33 For a reaction $2A \longrightarrow 3B$, if the rate of formation of B is x mol/L, the rate of consumption of A is **[J&K 2015]**

(a) x (b) $3x/2$ (c) $2x/3$ (d) $3x$

34 Energy of activation of forward reaction for an endothermic process is 90 kJ. If enthalpy change for the reaction is 50 kJ then, activation energy for backward reaction will be **[J&K 2015]**

(a) 40 kJ (b) 140 kJ
(c) 90 kJ (d) 50 kJ

35 The potenital energy diagram for a reaction $R \rightarrow P$ is given below, **[AIIMS 2014]**

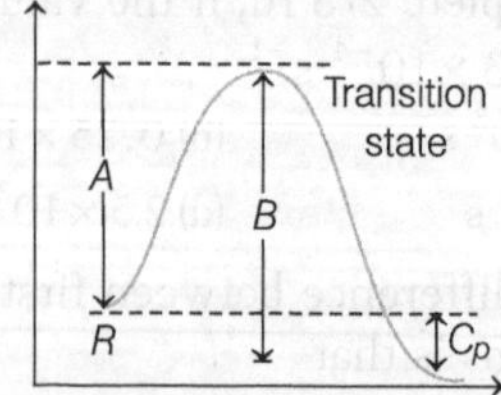

ΔH° of the reaction corresponding to the energy

(a) A (b) B (c) C (d) $A + B$

36 A plot of $\dfrac{1}{T}$ vs $\ln k$ for a reaction gives the slope -1×10^4 K. The energy of activation for the reaction is (Given, R = 8.314 K^{-1} mol^{-1}) **[KCET 2014]**

(a) 8314 J mol^{-1} (b) 1.202 kJ mol^{-1}
(c) 12.02 J mol^{-1} (d) 83.14 kJ mol^{-1}

37 The correct statement regarding the following energy diagrams. **[WB JEE 2014]**

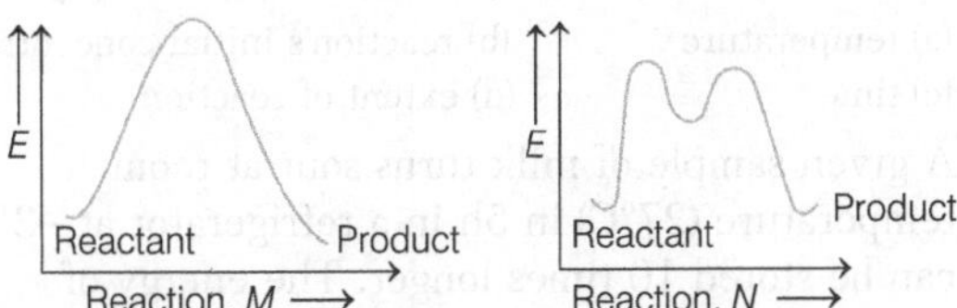

(a) Reaction M is faster and less exothermic that reaction N
(b) Reaction M is slower and less exothermic that reaction N
(c) Reaction M is faster and more exothermic than reaction N
(d) Reaction M is slower and more exothermic that reaction N

38 According to Arrhenius equation, the slope of $\log k \longrightarrow \frac{1}{T}$ plot is **[GCET 2014]**

(a) $\frac{-E_a}{2.303R}$ (b) $\frac{-E_a}{2.303}$
(c) $\frac{-E_a}{2.303RT}$ (d) $\frac{-E_a}{2.303RT}$

39 Rate law for the reaction, $A + 2B \longrightarrow C$ is found to be, rate $= k[A][B]$. **[UK PMT 2014]**
Concentration of reactant 'B' is doubled, keeping the concentration of 'A' constant, the value of rate constant will be

(a) same (b) doubled
(c) quadrupled (d) halved

40 Rate law for the reaction, $A + B \longrightarrow$ product, is rate $= k[A]^2[B]$. What is the rate constant, if rate of reaction at a given temperature is $0.22\ \text{Ms}^{-1}$, when $[A] = 1\ \text{M}$ and $[B] = 0.25\ \text{M}$? **[MHT CET 2014]**

(a) $3.52\ \text{M}^{-2}\text{s}^{-1}$ (b) $0.88\ \text{M}^{-2}\text{s}^{-1}$
(c) $1.136\ \text{M}^{-2}\text{s}^{-1}$ (d) $0.05\ \text{M}^{-2}\text{s}^{-1}$

41 For reaction, $2N_2O_5 \longrightarrow 4NO_2 + O_2$, the rate is directly proportional to $[N_2O_5]$. At 45°C, 90% of the N_2O_5 reacts in 3600 seconds. The value of the rate constant is **[Manipal 2014]**

(a) $3.2 \times 10^{-4}\text{s}^{-1}$ (b) $6.4 \times 10^{-4}\ \text{s}^{-1}$
(c) $8.5 \times 10^{-4}\text{s}^{-1}$ (d) $12.8 \times 10^{-4}\text{s}^{-1}$

42 The $t_{1/2}$ for a zero order reaction at the initial concentration of 6×10^{-3} M is one minute at 27°C. The rate constant at 27°C in mol dm^{-3} s^{-1} is **[Kerala (Medical) 2014]**

(a) 3×10^{-4} (b) 6×10^{-4}
(c) 5×10^{-5} (d) 5×10^{-4}
(e) 3×10^{-5}

43 The inversion of cane sugar is first order and proceeds with half-life of 600 minutes at pH = 4 for a given concentration of sugar. However, if pH = 5, the half-life changes to 60 minutes. The rate law expression for the sugar inversion can be written as **[Kerala (Medical) 2014]**

(a) rate $= k[\text{sugar}]^1[\text{H}^+]^2$ (b) rate $= k[\text{sugar}]^1[\text{H}^+]^1$
(c) rate $= k[\text{sugar}]^1[\text{H}^+]^4$ (d) rate $= k[\text{sugar}]^1[\text{H}^+]^0$
(e) rate $= k[\text{sugar}]^1[\text{H}^+]^5$

44 In an attempt to compare the half-life of two radioactive elements A and B, a scientist set aside 400 g of each. After 3 months, the scientist found 25 g of A and 200 g of B. Which one of the following statement is correct? **[Kerala (Medical) 2014]**

(a) Half-life of B is twice that of A
(b) Half-life of B is four times that of A
(c) Half-life of A is four times that of B
(d) Half-life of B is eight time that of A

45 The rate of a certain reaction is given by rate $= k[\text{H}^+]^n$ The rate increases 100 times when the pH changes from 3 to 1. The order (n) of the reaction is **[WB JEE 2014]**

(a) 2 (b) 0
(c) (d) 1.5

46 If 60% of first order reaction was completed in 60 min, 50% of the same reaction would be completed in approximately **[UK PMT 2014]**

(a) 45 min (b) 60 min
(c) 40 min (d) 50 min

47 $A(g) \xrightarrow{\Delta} P(g) + Q(g) + R(g)$, follow first order kinetics with a half-life of 69.3 s at 500 °C. Starting from the gas 'A' enclosed in a container at 500 °C and at a pressure of 0.4 atm, the total pressure of the system after 230 s will be **[KCET 2014]**

(a) 1.15 atm (b) 1.32 atm
(c) 1.22 atm (d) 1.12 atm

48 The half-life period for a first order reaction is **[GCET 2014]**

(a) independent of concentration
(b) proportional to concentration
(c) inversely proportional to concentration
(d) inversely proportional to the square of the concentration

49 The value of rate constant for a first order reaction is $2.303 \times 10^{-2}\ \text{s}^{-1}$. What will be the time required to reduce the concentration to $\frac{1}{10}$th of its initial concentration? **[GCET 2014]**

(a) 100 s (b) 10 s
(c) 2.303 s (d) 23.03 s

50 What is the activation energy for a reaction if its rate doubles when the temperature is raised from 20° C to 35°C ($R = 8.314 \text{ J mol}^{-1} \text{ K}^{-1}$) **[NEET 2013]**

(a) 342 kJ mol^{-1} (b) 269 kJ mol^{-1}
(c) 34.7 kJ mol^{-1} (d) 51.1 kJ mol^{-1}

51 A reaction having equal energies of activation for forward and reverse reaction has **[NEET 2013]**

(a) $\Delta S = 0$ (b) $\Delta G = 0$
(c) $\Delta H = 0$ (d) $\Delta H = \Delta G = \Delta S = 0$

52 75% of a first order reaction is completed in 30 minutes, what is the time required for 93.75% of the reaction (in minutes)? **[KCET 2013]**

(a) 45 (b) 120
(c) 90 (d) 60

53 For a given reaction, $t_{1/2} = \frac{1}{ka}$, the order of the reaction is

(a) 1 (b) 0
(c) 3 (d) 2

54 According to Arrhenius hypothesis, rate of reaction increases with **[Manipal 2013]**

(a) rise in temperature (b) decrease in temperature
(c) rise in pressure (d) decrease in pressure

55 A chemical reaction was carried out at 300 K, and 280 K. The rate constant were found to be k_1 and k_2 respectively, then **[WB JEE 2013]**

(a) $k_2 = 4k_1$ (b) $k_2 = 2k_1$
(c) $k_2 = 0.25k_1$ (d) $k_2 = 0.5k_1$

56 In a reaction, $A + B \rightarrow$ product, rate is doubled when the concentration of B is doubled, and rate increases by a factor of 8 when the concentrations of both the reactants (A and B) are doubled. Rate law for the reaction can be written as **[CBSE AIPMT 2012]**

(a) Rate $= k[A][B]^2$ (b) Rate $= k[A]^2[B]^2$
(c) Rate $= k[A][B]$ (d) Rate $= k[A]^2[B]$

57 In a zero order reaction for every 10° rise of temperature, the rate is doubled. If the temperature is increased from 10°C to 100°C, the rate of reaction will become **[CBSE AIPMT 2012]**

(a) 256 times (b) 512 times
(c) 64 times (d) 128 times

58 Activation energy (E_a) and rate constant (k_1 and k_2) of a chemical reaction at two different temperatures (T_1 and T_2) are related by **[CBSE AIPMT 2012]**

(a) $\ln \frac{k_2}{k_1} = -\frac{E_a}{R}\left(\frac{1}{T_1} - \frac{1}{T_2}\right)$ (b) $\ln \frac{k_2}{k_1} = -\frac{E_a}{R}\left(\frac{1}{T_2} - \frac{1}{T_1}\right)$

(c) $\ln \frac{k_2}{k_1} = -\frac{E_a}{R}\left(\frac{1}{T_2} + \frac{1}{T_1}\right)$ (d) $\ln \frac{k_2}{k_1} = \frac{E_a}{R}\left(\frac{1}{T_1} - \frac{1}{T_2}\right)$

59 For the first order reaction rate constants depend upon **[OJEE 2012]**

(a) temperature (b) reaction's initial concentration
(c) time (d) extent of reaction

60 A given sample of milk turns sour at room temperature (27°C) in 5h in a refrigerator at −3° C. It can be stored 10 times longer. The energy of activation for the souring of milk is **[KCET 2012]**

(a) $2.303 \times 5R \text{ kJ mol}^{-1}$ (b) $2.303 \times 3R \text{ kJ mol}^{-1}$
(c) $2.303 \times 2.7R \text{ kJ mol}^{-1}$ (d) $2.303 \times 10R \text{ kJ mol}^{-1}$

61 **Assertion** (A) The order of reaction may be defined as the sum of the powers which the concentration terms are raised in order to determine the rate of reaction gives the total order of reacting.

Reason (R) The number of molecules whose concentrations determine the ratio of reaction at a given temperature is called order of reaction. **[AIIMS 2012]**

(a) Both A and R are correct and the R is the correct explaination of the A
(b) Both A and R are correct but the R is not the correct explanation of the A
(c) A is correct but R is incorrect
(d) Both A and R are incorrect

62 **Assertion** (A) Acid catalysed hydrolysis of ethyl acetate is a first order reaction. **[AIIMS 2012]**

Reason (R) Water does not take part in the reaction.

(a) Both A and R are correct and the R is the correct explanation of the A
(b) Both A and R are correct but the R is not the correct explanation of the A
(c) A is correct but R is incorrect
(d) Both A and R are incorrect

63 Half-life of a reaction is found to be inversly proportional to the cube of initial concentration. The order of reaction is **[AFMC 2012]**

(a) 5 (b) 2 (c) 4 (d) 3

64 Units of specific reaction rate for second order reaction **[UP CPMT 2012]**

(a) s^{-1} (b) $\text{mol L}^{-1}\text{s}^{-1}$
(c) $\text{L}^2 \text{ mol}^{-2}\text{s}^{-1}$ (d) $\text{L mol}^{-1} \text{ s}^{-1}$

65 For which order half-life period is independent of initial concentration? **[Manipal 2012]**

(a) Zero (b) First (c) Second (d) Third

66 For a reaction, $A + B \rightarrow$ products the rate of the reaction at various concentrations are given below.

Exp. No	[A]	[B]	Rate (mol $\text{dm}^{-3}\text{s}^{-1}$)
1	0.2	0.2	2
2	0.2	0.4	4
3	0.6	0.4	36

The rate law for the above reaction is **[KCET 2012]**

(a) $r = k[A][B]^2$ (b) $r = k[A]^3[B]$
(c) $r = k[A]^2[B]^2$ (d) $r = k[A]^2[B]$

67 Which one of the following statements for the order of a reaction is incorrect? **[CBSE AIPMT 2011]**

(a) Order of reaction is always whole number.
(b) Order can be determined only experimentally.
(c) Order is not influenced by stoichiometric coefficient of the reactants.
(d) Order of reaction is sum of power to the concentration terms of reactants to express the rate of reaction.

68 A chemical reaction proceeds into the following steps

Step I, $2A \rightleftharpoons X$ fast
Step II, $X + B \rightleftharpoons Y$ slow
Step III, $Y + B \rightleftharpoons$ Product fast

The law for the overall reaction is **[DUMET 2011]**

(a) rate $= k[A]^2$
(b) rate $= k[B]^2$
(c) rate $= k[A][B]$
(d) rate $= k[A]^2[B]$

69 The time required for 100% completion of a zero order reaction **[KCET 2011]**

(a) ak (b) $\frac{a}{2k}$ (c) $\frac{a}{k}$ (d) $\frac{2k}{a}$

70 The following data were obtained during the first order decomposition of $2A(g) \longrightarrow B(g) + C(s)$ at a constant volume and at a particular temperature.

S.No.	Time	Total pressure in Pascal
1.	At the end of 10 min	300
2.	After completion	200

The rate constant in min^{-1} is **[KCET 2011]**

(a) 0.0693 (b) 69.3
(c) 6.93 (d) 6.93×10^{-4}

71 The rate law for the reaction,

$$xA \longrightarrow mP + nQ$$

is rate $= k[A]^c[B]^d$. What is the total order of the reaction ? **[Kerala CEE 2011]**

(a) $(x + y)$ (b) $(m + n)$ (c) $(c + d)$ (d) $\frac{x}{y}$
(e) Two

72 The initial rates of reaction,

$$3A + 2B + C \longrightarrow \text{Products}$$

at different initial concentration are given below.

S.No.	Initial rate, Ms^{-1}	$[A]_0, M$	$[B]_0, M$	$[C]_0, M$
1.	5.0×10^{-3}	0.010	0.005	0.010
2.	5.0×10^{-3}	0.010	0.005	0.015
3.	10×10^{-3}	0.010	0.010	0.010
4.	1.25×10^{-3}	0.005	0.005	0.010

The order with respect to the reactants A, B and C are respectively **[Kerala CEE 2011]**

(a) 3, 2, 0 (b) 3, 2, 1
(c) 2, 2, 0 (d) 2, 2, 1
(e) 2, 1, 0

73 For a first order reaction, the time taken to reduce the initial concentration by a factor of 1/4 is 20 min. The time required to reduce initial concentration by a factor of 1/16 is **[Kerala CEE 2011]**

(a) 20 min (b) 10 min
(c) 80 min (d) 40 min
(e) 5 min

74 Give the relation between half reaction time ($t_{1/2}$) and initial concentration of reactant for $(n-1)$ order reaction **[Guj. CET 2011]**

(a) $t_{1/2} \propto [R_0]$ (b) $t_{1/2} \propto [R_0]^{2-n}$
(c) $t_{1/2} \propto [R_0]^{n+1}$ (d) $t_{1/2} \propto [R_0]^{n-2}$

75 For a first order reaction, the initial concentration of a reactant is 0.05 M. After 45 min, it is decreased by 0.015 M. Calculate half reaction time ($t_{1/2}$). **[Guj. CET 2011]**

(a) 87.39 min (b) 25.90 min
(c) 78.72 (d) 77.20 min

76 The activation energy of a reaction at a given temperature is found to be 2.303 RT J mol^{-1}. The ratio of rate constant to the Arrhenius factor is **[KCET 2011]**

(a) 0.01 (b) 0.1
(c) 0.02 (d) 0.001

ANSWERS

Check point 4.1

1. (c)	2. (b)	3. (b)	4. (a)	5. (c)	6. (a)	7. (a)	8. (c)	9. (d)	10. (d)

Check point 4.2

1. (c)	2. (c)	3. (d)	4. (b)	5. (b)	6. (d)	7. (d)	8. (a)	9. (a)	10. (a)

Check point 4.3

1. (d)	2. (a)	3. (b)	4. (d)	5. (a)	6. (d)	7. (d)	8. (c)	9. (a)	10. (a)

Check point 4.4

1. (b)	2. (c)	3. (c)	4. (a)	5. (a)	6. (c)	7. (a)	8. (a)	9. (a)	10. (b)

(A) Topical exercise

1. (c)	2. (d)	3. (c)	4. (d)	5. (d)	6. (d)	7. (a)	8. (a)	9. (a)	10. (a)
11. (b)	12. (c)	13. (a)	14. (d)	15. (b)	16. (c)	17. (d)	18. (b)	19. (c)	20. (a)
21. (c)	22. (d)	23. (b)	24. (b)	25. (c)	26. (d)	27. (a)	28. (c)	29. (a)	30. (d)
31. (d)	32. (d)	33. (a)	34. (c)	35. (b)	36. (c)	37. (c)	38. (b)	39. (b)	40. (a)
41. (c)	42. (c)	43. (a)	44. (b)	45. (d)	46. (c)	47. (a)	48. (c)	49. (d)	50. (a)
51. (c)	52. (b)	53. (d)	54. (d)	55. (d)	56. (b)	57. (c)	58. (b)	59. (d)	60. (a)
61. (b)	62. (a)	63. (c)	64. (b)	65. (d)	66. (b)	67. (a)	68. (b)	69. (b)	70. (b)
71. (a)	72. (a)	73. (d)	74. (c)	75. (d)	76. (d)	77. (d)	78. (d)	79. (c)	80. (a)
81. (d)	82. (b)	83. (c)	84. (c)	85. (a)	86. (d)	87. (a)	88. (d)	89. (b)	90. (c)
91. (c)	92. (b)	93. (c)	94. (a)	95. (c)	96. (c)	97. (d)	98. (b)	99. (c)	100. (b)
101. (a)	102. (b)								

(B) Miscellaneous questions

1. (a)	2. (b)	3. (c)	4. (d)	5. (c)	6. (b)	7. (a)	8. (d)	9. (a)	10. (a)
11. (a)	12. (c)	13. (c)	14. (a)	15. (b)	16. (c)	17. (a)	18. (c)	19. (a)	20. (a)
21. (a)	22. (c)	23. (a)	24. (b)	25. (a)					

(C) Medical entrances' gallery

1. (c)	2. (b)	3. (a)	4. (b)	5. (b)	6. (d)	7. (b)	8. (b)	9. (b)	10. (b)
11. (a)	12. (d)	13. (b)	14. (b)	15. (a)	16. (b)	17. (b)	18. (d)	19. (d)	20. (c)
21. (a)	22. (c)	23. (d)	24. (c)	25. (b)	26. (d)	27. (a)	28. (a)	29. (a)	30. (c)
31. (a)	32. (b)	33. (c)	34. (a)	35. (c)	36. (d)	37. (c)	38. (a)	39. (a)	40. (b)
41. (b)	42. (c)	43. (d)	44. (b)	45. (c)	46. (a)	47. (d)	48. (a)	49. (a)	50. (c)
51. (c)	52. (d)	53. (d)	54. (a)	55. (c)	56. (d)	57. (b)	58. (b,d)	59. (a)	60. (c)
61. (b)	62. (d)	63. (c)	64. (d)	65. (b)	66. (d)	67. (a)	68. (d)	69. (c)	70. (a)
71. (c)	72. (e)	73. (d)	74. (b)	75. (a)	76. (b)				

Hints & Explanations

Check point 4.1

1 (*c*) $Zn + Dil.\ HCl \longrightarrow ZnCl_2 + H_2 \uparrow$

$$\text{Average rate of reaction} = \frac{\text{change in concentration of } H_2}{\text{change in time}}$$

$$= \frac{V_3 - 0}{40 - 0} = \frac{V_3}{40}.$$

2 (*b*) Reaction occurring at smallest time interval is known as instantaneous rate of reaction, e.g. instantaneous rate of reaction at 40 s is the rate of reaction during a small interval of time close to 40 s. Volume change during a small time interval close to 40 s, i.e. $40 \rightarrow 30$ s, $50 \rightarrow 40$ s, $50 \rightarrow 30$ s, $40 \rightarrow 20$ s.

$$\text{Instantaneous rate of reaction} = \frac{\text{change in volume}}{\text{time interval close to 40 s}}$$

(a) $r_{inst}\ (20\ s) = \dfrac{V_5 - V_2}{50 - 30}$ correct

(b) $r_{inst}\ (20\ s) = \dfrac{V_4 - V_2}{50 - 30}$ incorrect, correct is $\dfrac{V_5 - V_3}{50 - 30}$

(c) $r_{inst}\ (10\ s) = \dfrac{V_3 - V_2}{40 - 30}$ correct

(d) $r_{inst}\ (20\ s) = \dfrac{V_3 - V_1}{40 - 20}$ correct

6 (*a*) Rate constant is affected only by change in temperature.

7 (*a*) $2NH_3 \longrightarrow N_2 + 3H_2$

$$\frac{-d[NH_3]}{2dt} = \frac{+d[N_2]}{dt} = \frac{1}{3}\frac{d[H_2]}{dt}$$

$$\frac{k_1[NH_3]}{2} = k_2[NH_3] = \frac{k_3[NH_3]}{3}$$

$$1.5k_1 = 3k_2 = k_3$$

Check point 4.2

2 (*c*) $T_{50}(\text{zeroth}) = \dfrac{a}{2k_0}$

$$T_{50}(\text{first}) = \frac{0.693}{k_1}$$

$$\frac{a}{2k_0} = \frac{0.693}{k_1}$$

$$\therefore \quad \frac{k_1}{k_0} = \frac{2 \times 0.693}{a}$$

$$\frac{\left(\dfrac{dx}{dt}\right)_1}{\left(\dfrac{dx}{dt}\right)_0} = \frac{k_1 a}{k_0} = 2 \times 0.693$$

Hence, option (c) is correct.

4 (*b*) For second order reaction,

$$k = \frac{1}{t}\left[\frac{1}{[A]} - \frac{1}{[A]_0}\right]$$

$$\therefore \quad kt = \frac{1}{[A]} - \frac{1}{[A]_0}$$

$$kt + \frac{1}{[A]_0} = \frac{1}{[A]}$$

$$\frac{kt[A]_0 + 1}{[A]_0} = \frac{1}{[A]}$$

$$\therefore \quad [A] = \frac{[A]_0}{kt[A]_0 + 1}$$

Check point 4.3

1 (*d*) $\dfrac{-d[A]}{dt} = k[A]^n$

$$1 \times 10^{-4} = k\,[0.01]^n$$

$$1.41 \times 10^{-4} = k\,[0.02]^n$$

$$(1.41) = (2)^n$$

$$(2)^{1/2} = (2)^n$$

$$\therefore \quad n = \frac{1}{2}$$

$$\therefore \quad \frac{-d[A]}{dt} = k[A]^{1/2}$$

5 (*a*) We can apply **hit** and **trial** method. First we use integration equation of zeroth order.

t	$[A]$		x or $= [[A_0] - [A]_t]$	$k_0 = \dfrac{x}{t} = \dfrac{[[A_0] - [A]_t]}{t}$
0	0.069 M = a		0	
17	0.052 M		0.017 M	$\dfrac{0.017}{17} = 1 \times 10^{-3}$ M min^{-1}
34	0.035 M	$(a - x)$ or $[A]_t$	0.034 M	$\dfrac{0.034}{34} = 1 \times 10^{-3}$ M min^{-1}
51	0.018 M		0.051 M	$\dfrac{0.051}{51} = 1 \times 10^{-3}$ M min^{-1}

Since, for every set of values k_0 comes out to be constant hence, zeroth order reaction.

$$k_0 = 1 \times 10^{-3}\ \text{M min}^{-1}$$

6 (*d*) A is in excess, hence, rate is not affected by change in concentration of A.

$$R = k[0.1]^b \quad \text{...(i)}$$

$$2k = k(0.4)^b \quad \text{...(ii)}$$

On dividing by Eq. (ii) by (i), we get

$$2 = (4)^b \Rightarrow (4)^{1/2} = (4)^b$$

$$\therefore \quad b = \frac{1}{2}$$

7 (*d*)
$$\left(\frac{dx}{dt}\right) = k[A]^a[B]^b$$

$$4\left(\frac{dx}{dt}\right) = k[2A]^a[B]^b \Rightarrow 2\left(\frac{dx}{dt}\right) = k[A]^a[4B]^b$$

$$\therefore \quad 4 = (2)^a \Rightarrow a = 2 \Rightarrow 2 = (4)^b$$

or
$$(4)^{1/2} = (4)^b \Rightarrow b = \frac{1}{2}$$

$$\text{Order} = 2 + \frac{1}{2} = \frac{5}{2}$$

8. (*c*)
$$\frac{dx}{dt} \text{ or rate} = k[A]^a[B]^b$$

$$0.25 = k\,[1.0]^a[1.0]^b \quad \text{...(i)}$$

$$0.50 = k\,[2.0]^a[1.0]^b \quad \text{...(ii)}$$

$$0.25 = k\,[1.0]^a[2.0]^b \quad \text{...(iii)}$$

∴ Divide values of Eq. (ii) by that of Eq. (i) and (iii) by that of Eq. (i).

$$\therefore \quad 2 = (2)^a \Rightarrow a = 1$$

$$1 = (2)^b$$

$$2^0 = (2)^b \Rightarrow b = 0$$

Thus,
$$\frac{dx}{dt} = k[A]$$

9 (*a*) Assume rate law as

$$\left(\frac{dx}{dt}\right) = k[A]^a[B]^b$$

By general rule, if order $= n$

Multiple of concentration $= m$

then rate becomes $= m^n$ times

On comparison of values of sets (1) and (4).

Concentration of B has been doubled keeping $[A]$ constant, and rate becomes 8 times.

Thus, $(2)^b = 8 = 2^3$

$$\therefore \quad b = 3$$

On comparison of values of sets (1) and (3).

$$(2)^a = 1 = (2)^0$$

$$\therefore \quad a = 0$$

Thus, rate law $= k[B]^3$

10 (*a*)
$$\left(\frac{dx}{dt}\right) = k[A]^a[B_2]^b$$

Comparison of values in (1) and (2).

On doubling concentration of B_2, rate is doubled, i.e.

$$(2)^b = (2) \Rightarrow b = 1$$

Comparison of values in (2) and (3).

$$(2)^a = 1 = 2^0 \Rightarrow a = 0$$

Thus, rate $= k[A]^0[B_2] = k[B_2]$

Check point 4.4

3 (*c*) The role of a catalyst is to change the activation energy of reaction. This is done by either increasing or decreasing activation energy of molecule as catalyst are mainly of two types; +ve catalyst and –ve catalyst.

Note Catalyst are of two types one is positive catalyst which increases rate of reaction by decreasing activation energy and another is negative catalyst which decreases rate of reaction by increasing energy of activation.

9 (*a*) By rate law, reaction is first order in acetone, first order in H^+.

10 (*b*)
$$Ag^+ + 2NH_3 \underset{k_b}{\overset{k_f}{\rightleftharpoons}} Ag(NH_3)_2^+$$

$$\left(\frac{dx}{dt}\right)_{net} = k_f[Ag]\,[NH_3]^2 - k_b[Ag[NH_3]_2^+]$$

On comparison,
$$k_f = 2 \times 10^7 \text{ L}^2 \text{ m}^{-2}\text{s}^{-1}$$

$$k_b = 1 \times 10^{-2} \text{ s}^{-1}$$

$$\therefore \quad \frac{k_f}{k_b} = \frac{2 \times 10^7 \text{ L}^2 \text{ mol}^{-2} \text{ s}^{-1}}{1 \times 10^{-2} \text{ s}^{-1}}$$

$$= 2 \times 10^9 \text{ L}^2 \text{ mol}^{-2}$$

(A) Topical exercise

3 (*c*) Given, chemical reaction is

$$5\,Br^-(aq) + BrO_3^-(aq) + 6H^+\,(aq) \longrightarrow 3Br_2(aq) + 3H_2O(l)$$

Rate law expression for the above equation can be written as,

$$-\frac{1}{5}\frac{\Delta[Br^-]}{\Delta t} = -\frac{\Delta[BrO_3^-]}{\Delta t}$$

$$= \frac{-1}{6}\frac{\Delta[H^+]}{\Delta t}$$

$$= \frac{+1}{3}\frac{\Delta[Br_2]}{\Delta t}$$

$$\Rightarrow \quad -\frac{\Delta[Br^-]}{\Delta t} = -\frac{5\,\Delta[BrO_3^-]}{\Delta t}$$

$$= \frac{-5}{6}\frac{\Delta[H^+]}{\Delta t}$$

$$\Rightarrow \quad \frac{\Delta[Br^-]}{\Delta t} = \frac{5}{6}\frac{\Delta[H^+]}{\Delta t}$$

7 (*a*) Rate of reaction is defined as rate of decrease of concentration of any one of reactant with passage of time.

$$\text{Rate of reaction} = \frac{\text{rate of disappearance of reactant}}{\text{time taken}}$$

Thus, as the concentration of reactant decreases with passage of time, rate of reaction decreases.

10 (*a*) $\left(\frac{dx}{dt}\right)_0 = k(0.60)(0.80)^2 = 0.384\,k$

$$\begin{array}{ccccc} A(g) & + & 2B(g) & \longrightarrow & C(g) + D(g) \\ 0.60 & & 0.80 & & 0 \quad 0 \\ (0.60-x) & & (0.80-2x) & & x \quad x \end{array}$$

$$x = 0.2 \text{ atm}$$

$\therefore$ $[A] = 0.4$ atm

$[B] = 0.4$ atm

$\therefore$ $\left(\frac{dx}{dt}\right)_t = k\,(0.4)(0.4)^2 = 0.064\,k$

$\therefore$ $\left(\frac{dx}{dt}\right)_t \Big/ \left(\frac{dx}{dt}\right)_0 = \frac{0.064k}{0.384k} = \frac{1}{6}$

11 (*b*) $A \longrightarrow B$

Concentration of reactants and products varies exponentially *wrt* time.

(i) Concentration of reactant (here, A) decreases exponentially respect to with respect to time.

(ii) Concentration of product (here, B) increases exponentially with respect to time.

13 (*a*) $pV = nRT$

$$p_{CH_3CHO} = \frac{n}{V}RT$$

$\therefore$ $-\left(\frac{dp_{CH_3CHO}}{dt}\right) = \left(\frac{dn}{dt}\right)\frac{RT}{V}$

$\therefore$ $\frac{dn}{dt} = -\frac{V}{RT}\frac{dp_{CH_3CHO}}{dt}$

15 (*b*) $k = \frac{2.303}{t}\log\left(\frac{a}{a-x}\right)$

$$= \frac{2.303}{60}\log 8 = 0.03466 \text{ min}^{-1}$$

$$\left(\frac{dx}{dt}\right) = k[A] = 0.03466 \times 0.1 = 3.466 \times 10^{-3} \text{ mol L}^{-1} \text{ min}^{-1}.$$

16 (*c*) $k = \frac{2.303}{2\times 10^4}\log\frac{800}{50} = 1.386 \times 10^{-4} \text{ s}^{-1}$.

18 (*b*) Based on the given reaction,

$$-\frac{1}{2}\frac{\Delta[N_2O_5]}{\Delta t} = \frac{1}{4}\frac{\Delta[NO_2]}{\Delta t} = \frac{\Delta[O_2]}{\Delta t}$$

$$\frac{k_1}{2}[N_2O_5] = \frac{k_2}{4}[N_2O_5] = k_3[N_2O_5]$$

$\therefore$ $2k_1 = k_2 = 4k_3$

Hence, option (b) is correct.

20 (*a*) $N_2 + 3H_2 \underset{k_b}{\overset{k_f}{\rightleftharpoons}} 2NH_3$

$$\left(\frac{dx}{dt}\right)_{net} = k_f[N_2][H_2]^3 - k_b[NH_3]^2$$

Also, $\frac{[NH_3]^2}{[N_2][H_2]^3} = \frac{k_f}{k_b}$

or $\frac{[N_2][H_2]^3}{[NH_3]^2} = \frac{k_b}{k_f} = \frac{1\times 10^{-3}}{1\times 10^2} = 1\times 10^{-5}$

$$\left(\frac{dx}{dt}\right) = 1\times 10^2 M^{-3} \times 10^{-5}[NH_3]^2 - 1\times 10^{-3}\,[NH_3]^2$$

Thus, $\left(\frac{dx}{dt}\right) = 0$

Hence, option (a) is correct.

21 (*c*) $-\frac{1}{3}\frac{\Delta[BrO^-]}{\Delta t} = \frac{\Delta[BrO_3^-]}{\Delta t} = \frac{1}{2}\frac{\Delta[Br^-]}{\Delta t}$

$\therefore$ Rate constant of $\frac{\Delta[BrO_3^-]}{\Delta t} = \frac{0.056}{3} = 0.019 \text{ L mol}^{-1}\text{ s}^{-1}$

and rate constant of $\frac{\Delta[Br^-]}{\Delta t} = \frac{2}{3} \times 0.056$

$$= 0.037 \text{ L mol}^{-1}\text{ s}^{-1}$$

22 (*d*) $\frac{dx}{dt} = k$

Rate is constant but concentration of the reactant decreases linearly.

At 20 min $(A) = 12$ mol

Thus, $\Delta x = 20 - 12 = 8$ mol

$\Delta t = 20 - 0 = 20$ min

$\therefore$ $\frac{\Delta x}{\Delta t} = \frac{8}{20} = 0.4$ mol / min

23 (*b*) $\left(\frac{dx}{dt}\right) = k[A]^2$

It represents second order reaction

$$k = \frac{1}{t}\left[\frac{1}{a-x} - \frac{1}{a}\right] \Rightarrow \frac{1}{(a-x)} = kt + \frac{1}{a}$$

Thus, graph between $(a-x)^{-1}$ and time t is a straight line

$(y = mx + c)$

$\therefore$ $OP = \frac{1}{a} = 0.10 \text{ L mol}^{-1}$

$\therefore$ $a = 10 \text{ mol L}^{-1}$

$\therefore$ $\left(\frac{dx}{dt}\right) = 1\times 10^{-2} \text{ L mol}^{-1}\text{ min}^{-1} \times (10 \text{ mol L}^{-1})^2$

$$= 1.0 \text{ mol L}^{-1}\text{ min}^{-1}$$

30 (*d*) For second order reaction,

$$k = \frac{1}{t}\left(\frac{1}{a-x} - \frac{1}{a}\right)$$

$\therefore$ $\left(\frac{1}{a-x}\right) = kt + \frac{1}{a}$

Graph between $\frac{1}{a-x}$ and t is a straight line.

31 (*d*) $A = A_0\, e^{-k_1 t}$

After $t = \frac{2}{k_1}$ ($\frac{1}{k_1}$ is called average life)

$$A = A_0\, e^{-2} = \frac{A_0}{e^2} = \frac{100}{e^2}\,\%$$

32 (*d*) The time taken for half the reaction to complete, i.e. the time in which the concentration of a reactant is reduced to half of its original value is called half-life period of the reaction.

But it is impossible to perform 100% of the reaction. Whole of the substance never reacts because in every half-life , 50% of the substance reacts. Hence, time taken for 100% completion of a reaction is infinite.

34 (*c*) $k = \frac{2.303}{t}\log\left(\frac{a}{a-x}\right)$

$$k = \frac{2.303}{90}\log\left(\frac{100}{100-90}\right) \quad \text{...(i)}$$

$$k = \frac{2.303}{90}\log 10 \Rightarrow \frac{2.303}{90}\text{ min}^{-1} \quad \text{...(ii)}$$

Also, $k = \frac{2.303}{t_{50}}\log\frac{100}{100-50} = \frac{2.303}{t_{50}}\log 2$ ...(iii)

From Eqs. (ii) and (iii),

$$\therefore \quad \frac{2.303\log 2}{t_{50}} = \frac{2.303}{90}$$

$$\therefore \quad t_{50} = \log 2 \times 90 = 0.3 \times 90 = 27 \text{ min}$$

36 (*c*) $T_{50} = \frac{0.693}{0.0693} = 10$ min

$$C = C_0\left(\frac{1}{2}\right)^y$$

$$2.5 = 20\left(\frac{1}{2}\right)^y$$

$$y = \frac{\text{Total time}}{t_{50}}$$

$$\therefore \quad \left(\frac{1}{8}\right) = \left(\frac{1}{2}\right)^y \Rightarrow \left(\frac{1}{2}\right)^3 = \left(\frac{1}{2}\right)^y$$

$$\therefore \quad y = 3 = \frac{t}{t_{50}}$$

$\therefore$ t = 30 min

37 (*c*) Reactants and products are in gaseous state thus,

$$A(g) \longrightarrow B(g) + C(g)$$

	$A(g)$	$B(g)$	$C(g)$
At $t = 0$	p_i	0	0
At $t = t$	$(p_i - x)$	x	x

Total pressure at time $t = p_t = p_i - x + x + x = p_i + x$

$$\therefore \quad x = (p_t - p_i)$$

and $(a - x) = p_i - (p_t - p_i) = 2p_i - p_t$

Given, $p_i = a = 100$ mm

$p_t = 120$ mm

$$\therefore \quad (a - x) = 2p_i - p_t = 80 \text{ mm}$$

$$\therefore \quad k = \frac{2.303}{10}\log\frac{100}{80} \text{ or } \frac{2.303}{10}\log\frac{5}{4}$$

Note If 1 mole of gaseous species gives y moles y gaseous products then $(a-x) = \frac{yp_i - p_t}{(y-1)}$

38 (*b*) $T_{50} = 10$ min

$$\therefore \quad k = \frac{0.693}{10} = 0.0693 \text{ min}^{-1}$$

conc. at start = 10 M

conc. after 20 min (two half-lifes) = 2.5 M

$$\therefore \quad \left(\frac{dx}{dt}\right) = k[A]$$

$$= 0.0693 \times 2.5 \text{ M min}^{-1}$$

39 (*b*) $k = 3.3 \times 10^{-4}\text{ s}^{-1}$

Unit indicates that it is a first order reaction.

$$\because \quad k = \frac{2.303}{t}\log\frac{a}{a-x}$$

$$3.3 \times 10^{-4} = \frac{2.303}{t}\log\left(\frac{100}{25}\right)$$

$$t = 4201.6 \text{ s} = \frac{4201.6}{60}\text{ min}$$

$$t = 70 \text{ min}$$

42 (*c*) $k = \frac{2.303}{10}\log\frac{100}{90} = 0.0105\text{ min}^{-1}$

$$\therefore \; T_{50} = \frac{0.693}{0.0105} = 66 \text{ min}$$

43 (*a*) $T_{50} = 100$ days

$$\therefore \quad k \text{ (first order)} = \frac{0.693}{T_{50}} = \frac{0.693}{100}\text{ day}^{-1}$$

$$A = A_0 e^{-kt}$$

$$= A_0(e^{-\frac{0.693}{100}\times 144.3}) = A_0 e^{-1} = \frac{A_0}{e},$$

44 (*b*) Natural life times $= \frac{1}{k} = T_{av}$

$$T_{99.9} = 10 \times T_{50} = \frac{10 \times 0.693}{k} = \frac{6.93}{k}$$

$\therefore$ Number of natural life times $= \frac{6.93}{k}\Big/\frac{1}{k} = 6.93$

45 (*d*) T_{50} (second order) $= \frac{1}{ka} = \frac{1}{2\times 10^{-3} \times 0.1} = 5000$ s

46 (*c*) $N_A = N_0 e^{-8\lambda t}$

$$N_B = N_0 e^{-\lambda t}$$

Given $\frac{N_B}{N_A} = e = \frac{e^{-\lambda t}}{e^{-8\lambda t}} = e^{7\lambda t}$

$$7\lambda t = 1$$

$$\therefore \quad t = \frac{1}{7\lambda}$$

47 (*a*) $k = \frac{2.303}{40}\log\frac{0.1}{0.025} = 0.03466\text{ min}^{-1}$

$$\frac{dx}{dt} = k[A] = 0.03466 \times 0.01 = 3.47 \times 10^{-4}\text{ M min}^{-1}$$

48 (*c*) $\frac{dx}{dt} = k = 0.025 \text{ M s}^{-1}$

or $\frac{x}{t} = k = 0.025 \text{ M s}^{-1}$

$x = 15 \times 0.025 = 0.375$ M

Thus, $[A] = 0.50 - 0.375 = 0.125$ M

49 (*d*) By set I, half-life is 2 min. In set II, number of moles have been doubled thus, half-life is also doubled, i.e. now it is 4 min.

Thus, 8 mol change to 4 mol in 4 min

$\therefore$ 4 mol change to 2 mol in 2 min

Thus, total time = 6 min

52 (*b*)

	$A(g) \longrightarrow$	$B(g)$ +	$C(g)$
Initially p_i		0	0
At time t	$p_i - x$	x	x

$$p_t = p_i - x + x + x = p_i + x$$

For first order reaction,

$$x = p_t - p_i$$

$$k = \frac{2.303}{t} \log \frac{p_i}{p_i - x}$$

$$= \frac{2.303}{t} \log \frac{p_i}{p_i - (p_t - p_i)}$$

$$= \frac{2.303}{t} \log \frac{p_i}{2p_i - p_t}$$

56 (*b*) $T_{50} = k(a)^{1-n}$

$$\log T_{50} = (1-n) \log a + \log k$$

$$(1-n) = 0$$

$\therefore$ $n = 1$

57 (*c*)

Order	t_{50}	t_{75}	t_{75}/t_{50}
0	$a/2k_0$	$\frac{3a}{4k_0}$	1.5
1	$0.693/k_1$	$2 \times 0.693/k_1$	2
2	$\frac{1}{k_2 a}$	$\frac{3}{k_2 a}$	3

58 (*b*) 0.12 M $\longrightarrow$ 0.06 M 10 h (50% reaction)

0.12 M $\longrightarrow$ 0.03 M 20 h (75% reaction)

Time of 75% reaction = 2 × time of 50% reaction

Which is property of first order reaction.

59 (*d*) $\left(\frac{dx}{dt}\right) = k[G]^a[H]^b$

$$8\left(\frac{dx}{dt}\right) = k[2G]^a[2H]^b$$

$$2\left(\frac{dx}{dt}\right) = k[2G]^a[H]^b$$

$\therefore$ $2 = (2)^a \Rightarrow a = 1 \Rightarrow 8 = (2)^a(2)^b$

$$4 = (2)^b \Rightarrow b = 2$$

Thus, overall order = 1 + 2 = 3.

60 (*a*) $k = 8.0 \times 10^{-15} \text{ cm}^3 \text{ molecule}^{-1} \text{ s}^{-1}$

$$= \frac{8.0 \times 10^{-15}}{1000} \text{ dm}^3 \text{ molecule}^{-1} \text{ s}^{-1}$$

$$= 4.8 \times 10^6 \text{ dm}^3 \text{ mol}^{-1} \text{ s}^{-1}$$

$$\left(\begin{array}{l} 1 \text{ cm} = \frac{1}{10} \text{ dm} \\ 1 \text{ molecule} = \frac{1}{N_0} \text{ mol} = \frac{1}{6.0 \times 10^{23}} \text{ mol} \\ \therefore \frac{1}{\text{molecule}} = \frac{6.0 \times 10^{23}}{\text{mol}} \end{array}\right)$$

61 (*b*)

pH	$[H^+]$
2	10^{-2} M
1	10^{-1} M

$$\left(\frac{dx}{dt}\right) = k(10^{-2})^n$$

$$100\left(\frac{dx}{dt}\right) = k(10^{-1})^n$$

$\therefore$ $100 = (10)^n$

$$(10)^2 = (10)^n$$

$\therefore$ $n = 2$

65 (*d*)

I. $\left(\frac{dx}{dt}\right) = k$ for zeroth order reaction, i.e. C.

II. Half-life is independent of concentration for the first order reaction, i.e. A.

III. Graph between $(a-x)^{-1}$ and time is linear for second order reaction, i.e. B.

66 (*b*) Half-life period is independent of concentration of zinc thus, order *wrt* Zn = 1

pH	$[H^+]$	t_{50}
2	10^{-2} M	10 min
3	10^{-3} M	100 min

$$\frac{(t_{50})_2}{(t_{50})_1} = \left(\frac{[H^+]_1}{[H^+]_2}\right)^{n-1}$$

$$\left(\frac{100}{10}\right) = \left(\frac{10^{-2}}{10^{-3}}\right)^{n-1}$$

$$(10) = (10)^{n-1}$$

$\therefore$ $n - 1 = 1$

$\therefore$ $n = 2$

Thus, rate law = $k[\text{Zn}][H^+]^2$

67 (*a*) $\frac{d[A_2]}{dt} = k[A_2]^x[B]^y$

On doubling the concentration of A_2, rate is doubled hence, $x = 1$.

On doubling the concentration of B, rate is four times, hence, $y = 2$.

68 (b) Rate of reaction is changed in concentration of reactant with respect to time.

$$r = k[A]^x[B]^y$$

$$\frac{\text{Rate of exp.1}}{\text{Rate of exp.2}} = \frac{[0.30]^x[0.30]^y}{[0.30]^x[0.60]^y}$$

$$\frac{0.10}{0.40} = \frac{[0.30]^y}{[0.60]^y}$$

$$\Rightarrow \quad \frac{1}{4} = \left[\frac{1}{2}\right]^y$$

$$\left[\frac{1}{2}\right]^2 = \left[\frac{1}{2}\right]^y$$

$$y = 2$$

$$\frac{\text{Rate of exp.1}}{\text{Rate of exp.3}} = \frac{[0.30]^x[0.30]^y}{[0.60]^x[0.30]^y}$$

$$\frac{0.10}{0.20} = \left[\frac{0.30}{0.60}\right]^x\left[\frac{0.30}{0.30}\right]^y$$

$$\frac{1}{2} = \left[\frac{1}{2}\right]^x [1]^y \frac{1}{2} = \left[\frac{1}{2}\right]^x$$

i.e. $x = 1$

$\therefore$ Rate $= k[A]^x[B]^y$

Rate $= k[A]^1[B]^2$

70 (b) Activation energy of a chemical reaction is related to rate constant of a reaction at two different temperatures, i.e. k_1 and k_2 respectively

$$\ln\left(\frac{k_2}{k_1}\right) = \frac{E_a}{R}\left[\frac{1}{T_1} - \frac{1}{T_2}\right]$$

where, E_a = activation energy
T_2 = higher temperature
T_1 = lower temperature
k_1 = rate constant at temperature T_1
k_2 = rate constant at temperature T_2

This equation is known as Arrhenius equation.

71 (a) Activation energy is the minimum energy required to convert reactant molecules to product molecules. Here, the energy gap between reactants and activated complex is sum of E_1 and E_2.

$\therefore$ Activation energy $= E_1 + E_2$

Product is less stable than reactant as energy of product is greater than the reactant.

72 (a) According to Arrhenius equation, $k = Ae^{-E_a/RT}$

Taking log on both side, $\ln k = \ln(A.e^{-\frac{E_a}{RT}})$

$$\ln k = \ln A - \frac{E_a}{RT}$$

$$\ln k = -\frac{-E_a}{R} \times \frac{1}{T} + \ln A$$

$$y = mx + c$$

This equation can be related to equation of straight line as shown in graph.

From the graph, it is very clear that slope of the plot $= \frac{-E_a}{R}$ and intercept $= \ln A$.

75 (d) $k = Ae^{-E_a/RT} = Ae^{RT/E_a}$

k is least when T is low and E_a is high.

76 (d) $E_a = \frac{2.303\,RT_1T_2}{(T_2 - T_1)} \log \frac{k_2}{k_1}$

If $E_a = 0$, then $\log \frac{k_2}{k_1} = 0$

$$\frac{k_2}{k_1} = \text{antilog}\,[0]$$

$$\therefore \quad \frac{k_2}{k_1} = 1$$

$$\therefore \quad k_1 = k_2$$

Hence, the rate constant is same for both temperatures.

79 (c) In the presence of catalyst, the heat absorbed, or evolved during the reaction remains unchanged as there is no change in stability of reactant and product.

82 (b) Reaction is endothermic, thus option (b) is correct.

83 (c) If $E_a = 10$ kJ mol^{-1} for $A \rightarrow B$

and $\Delta H = 5$ kJ mol^{-1} for $A \rightarrow B$

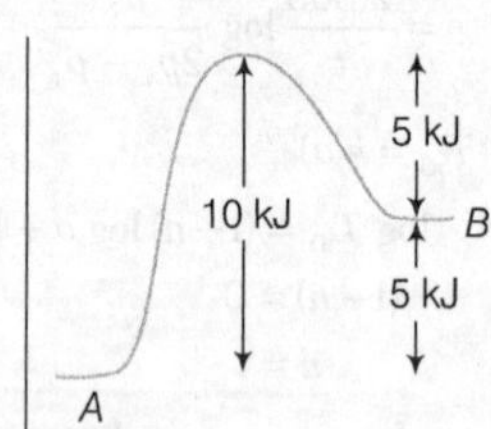

then, $E_a = 5$ kJ mol^{-1} for $B \rightarrow A$

88 (d) According to Arrhenius equation, $k = Ae^{-E_a/RT}$

Here, $k \propto e^{-E_a}$

$$k \propto e^{-\frac{1}{T}}, \propto e^T$$

which indicates that as activation energy decreases rate constant increases and as temperature increases rate of reaction increases.

89 (b) Characteristics of catalyst

(a) It catalyses the forward and backward reaction to the same extent as it decreases energy of activation hence, increases the rate of both the reactions.

(b) Since, reaction quotient is the relation between concentration of reactants and products. Hence, catalyst does not alter Gibbs free energy as it is related to reaction quotient. Thus, Gibbs free energy does not change during the reaction when catalyst is added to it.

$$\Delta G = -RT \ln Q$$

where, Q = reaction quotient

(c) It doesn't alter equilibrium of reaction as equilibrium constant is also concentration dependent term.

(d) It provides an alternate mechanism by reducing activation energy between reactants and products.

90 (*c*) According the postulates of collision theory, there are following necessary conditions for any reaction to be occur

(i) Molecule should collide with sufficient threshold energy.

(ii) Their orientation must be proper.

(iii) The collision must be effective.

92 (*b*) For reaction (*A*),

$$E_a = \frac{2.303R\times 300\times 310}{10}\log 2 \quad ...(i)$$

For reaction (*B*),

$$2E_a = \frac{2.303R\times 300\times T_2}{(T_2-300)}\log 2 \quad ...(ii)$$

$\therefore$ From (i) and (ii),

$$\frac{300\times 310}{10} = \frac{300\times T_2}{2(T_1-300)}$$

$$\left(\frac{T_2-300}{T_2}\right) = \frac{1}{62}$$

$$\therefore \quad T_2 = 304.92\text{ K}$$

Thus, increase = 304.92 − 300 = 4.92 K

96 (*c*) $\log R = \log_{10} A - \frac{E_a}{2.303\,RT}$

$$\text{slope} = -\frac{E_a}{2.303\,R} = -\frac{1}{2.303}$$

$$\therefore \quad E_a = R = 2\text{ cal}$$

97 (*d*) $\Delta H(A\to B) = -10$ kJ (exothermic reaction)

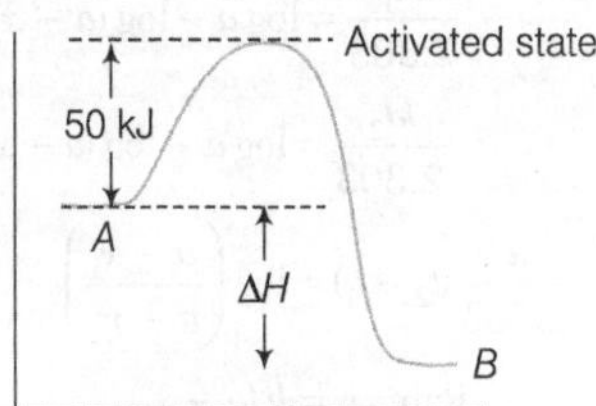

$$\therefore \quad \Delta H(B\to A) = 10\text{ kJ}$$

$$E_a(A\to \text{activated state}) = 50\text{ kJ}$$

$$\therefore \quad E_a\,(B\to\text{activated state}) = 50 + 10 = 60\text{ kJ}$$

98 (*b*) $k = Ae^{-E_a/RT}$

when, $T\to\infty$

$$\therefore \quad k = Ae^0 = A$$

$$\therefore \quad A = k = 1.2\times 10^3\text{ mol}^{-1}\text{ L s}^{-1}$$

100 (*b*) $k_B = Ae^{-E_a/RT}$

$$= 10^{12}\times 4.35\times 10^{-8}$$

$$= 4.35\times 10^4\text{ s}^{-1}$$

Also, equilibrium constant $= \frac{k_A}{k_B} = 10^4$

$$\therefore \quad k_A = k_B\times 10^4 = 4.35\times 10^8\text{ s}^{-1}.$$

$$\therefore \quad k_A + k_B = (4.35\times 10^8 + 4.35\times 10^4)$$

$$\approx 4.35\times 10^8\text{ s}^{-1}$$

(B) Miscellaneous questions

1 (*a*) $T_{50} \propto \left(\frac{1}{a}\right)^{n-1}$

$$T_{50}\propto a \qquad n = 0$$

$$T_{50} = ka$$

$$\log T_{50} = \log a + \log k$$

Thus, a graph between $\log T_{50}$ and $\log a$ is a straight line of slope 45°.

2 (*b*) Graph I indicates zeroth order reaction.

Hence, $-\frac{d[A]}{dt} = k$

4 (*d*) If $k_A = k_C$, then

$$10 = e^{1000/T}$$

$$\therefore \quad \log_e 10 = \frac{1000}{T}$$

$$2.303 = \frac{1000}{T}$$

$$T = \frac{1000}{2.303}\text{ K}$$

5 (*c*) $k = \frac{1}{t}\left[\frac{1}{a-x} - \frac{1}{a}\right]$ for second order reaction

$$\therefore \quad \frac{1}{a-x} = kt + \frac{1}{a}$$

Graph between $(a-x)^{-1}$ and time t is linear.

$$\therefore \quad \text{Slope} = k = \tan\theta = 0.5\text{ L mol}^{-1}\text{ min}^{-1}$$

$$OA = \frac{1}{a} = 2\text{ L mol}^{-1}$$

$$\therefore \quad a = 0.5\text{ mol L}^{-1}$$

$$\text{Rate} = k(a)^2 = 0.5\times 0.5\times 0.5$$

$$= 0.125\text{ mol L}^{-1}\text{ min}^{-1}$$

8 (*d*) $k_2 = Ae^{-E_2/RT}$

$$k_1 = Ae^{-(E_2+10)/RT}$$

$$\therefore \quad \frac{k_2}{k_1} = e^{10/RT}$$

$$\therefore \quad \ln\frac{k_2}{k_1} = \frac{10}{RT} = \frac{10}{8.314\times 10^{-3}\times 300} = 4.00$$

10 (*a*) $\log k\,(\text{min}^{-1}) = 5 - \frac{2000\text{ K}}{T}$

$$\log_{10} k = \log_{10} A - \frac{E_a}{2.303\,RT}$$

I. $\log_{10} A = 5$, $A = 10^5$, true

II. $\frac{E_a}{2.303\,R} = 2000\,E_a = 2000\times 2.303\times 0.002$ kcal

= 9.212 kcal, true

III. Equation represents straight line, hence true.

11 (*a*) $k = \frac{1}{t}\left[\frac{1}{a-x} - \frac{1}{a}\right]$

or, $$k = \frac{1}{t}\left[\frac{1}{[A]} - \frac{1}{[A_0]}\right]$$

$$kt = \frac{1}{[A]} - \frac{1}{[A_0]}$$

$$kt + \frac{1}{[A_0]} = \frac{1}{[A]}$$

$\therefore$ $$\frac{kt[A_0]+1}{[A_0]} = \frac{1}{[A]}$$

$\therefore$ $$[A] = \frac{[A_0]}{kt[A_0]+1}$$

14 (*a*) The chemical reaction in which energy is evolved during the reaction is known as exothermic reaction, i.e. activation energy of product is greater than activation energy of reactants.

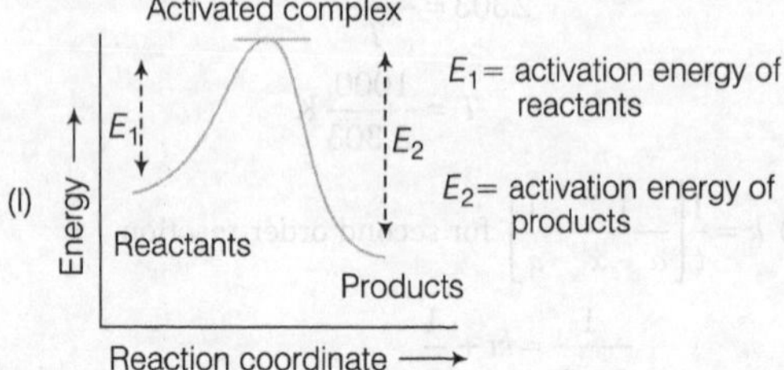

Also, $\Delta H < 0$

Here, only (I) denotes correct picture of exothermic reaction.

18 (*c*) $k = Ae^{-E_a/RT}$ in absence of catalyst

$$k' = Ae^{-(E_a-2)/RT} \text{ in presence of catalyst}$$

$$R = k[A]$$

$$R' = k'[A]$$

$\because$ $$\frac{R'}{R} = \frac{k'}{k} = e^{2/RT}$$

$$= e^{2/0.002\times300}$$

$$= 28.03$$

19 (*a*) When there is no radioactive substance in G.M. counter, it records = 3 dps. Thus, it is the **zero-error** in the counter and in each it is added.

Thus, initial counts = 23 − 3 = 20 dps.

At time *t* (after 10 min) = 13 − 3 = 10 dps

Thus, half-life is **10 min** = ***y***

Hence, after next 10 min, dps = 5

Hence, reading in G.M. Counter = 5 + 3 = 8 dps = *x*

$x = 8$ dps, $y = 10$ min

Hence, option is correct.

20 (*a*) r_0 = rotation at the start, r_t = rotation at time *t*

r_∞ = rotation at the end

then $(r_\infty - r_0) = a$ and $(r_\infty - r_t) = (a-x)$

When 50% inversion has taken place

$$a - x = \frac{a}{2}$$

$\therefore$ $$2(a-x) = a$$

$$2(r_\infty - r_t) = r_\infty - r_0$$

$$2r_\infty - 2r_t = r_\infty - r_0$$

$\therefore$ $$r_0 = (2r_t - r_\infty)$$

22 (*c*) When *B* is formed from *A*, two α-particles (He) are formed.

A taken initially = 1 mol

A decomposed in two half-lifes = 75% = $\frac{3}{4}$ mol

Thus, helium formed = $2 \times \frac{3}{4}$ mol = $\frac{3}{2}$ mol

This helium remains in closed vessel.

Thus, pressure set up is due to helium.

$$p = \frac{n}{V}RT$$

$$= \frac{3}{2} \times \frac{\text{mol} \times 0.0821 \text{ L atm mol}^{-1}\text{ K}^{-1} \times 300 \text{ K}}{1 \text{ L}}$$

$$= 36.945 \text{ atm} \approx 37.0 \text{ atm}$$

23 (*a*) $k = \frac{2.303}{t}\log\left(\frac{a}{a-x}\right)$

$$\frac{kt}{2.303} = \log a - \log(a-x)$$

$$\frac{kt_1}{2.303} = \log a - \log(a-x_1) \text{ at time } t_1$$

$$\frac{kt_2}{2.303} = \log a - \log(a-x_2) \text{ at time } t_2$$

$\therefore$ $$\frac{k}{2.303}(t_2 - t_1) = \log\left(\frac{a-x_1}{a-x_2}\right)$$

$$\text{Rate, } r_1 = k(a-x_1)$$

$$r_2 = k(a-x_2)$$

$\therefore$ $$\frac{(a-x_1)}{(a-x_2)} = \frac{r_1}{r_2} \Rightarrow \frac{0.04}{0.03} = \frac{4}{3}$$

$$\frac{k(20-10)}{2.303} = \log\frac{4}{3}$$

$\therefore$ $$k = \frac{2.303}{10}\log\frac{4}{3} = \frac{2.303}{10}\log 1.33$$

$$= \frac{2.303 \times 0.125}{10} = 0.02878 \text{ min}^{-1}$$

$$T_{50} = \frac{0.693}{0.02878} = 24.1 \text{ min}$$

24 (*b*) $$X \underset{k_2}{\overset{k_1}{\rightleftharpoons}} Y$$

	X	Y
At $t = 0$	$[X]_0$	0
at time t	$[X]$	$[X]_0 - [X]$

Thus, net rate $\frac{-d[X]}{dt} = k_1[X] - k_2([X]_0 - [X])$

$$= k_1[X] - k_2[X]_0 + k_2[X] = (k_1 + k_2)[X] - k_2[X_0]$$

25 (*a*) $A \longrightarrow C_6H_5-Cl + N_2$

	A	N_2
Intial	a	0
After time t	$t(a-x)$	x
At complete reaction $(a-a)$		a

Thus, $a = 50L$

$x = 10L$ at time 10 min

$(a-x) = 40\,L$

$$k = \frac{1}{t}\ln\left(\frac{a}{a-x}\right) = \frac{1}{t_0}\left(\frac{50}{40}\right) = \frac{\ln 1.25}{10}$$

(C) Medical entrances' gallery

1 (*c*) An increase in the concentration of the reactants of a reaction leads to change in collision frequency. If the concentration of one or more reactants is increased, the rate of reaction increases. This happens because higher concentration of a reactant leads to more collisions of that reactant with in a particular time period.

2 (*b*) For a first order reactions

Rate constant $(k) = 4.606 \times 10^{-3}\,s^{-1}$

Initial amount $(a_0) = 2\,g$

Final amount $(a) = 0.2\,g$

So, time required,

$$t = \frac{2.303}{k}\log\frac{a_0}{a} = \frac{2.303}{4.606\times 10^{-3}} \times \log\frac{2}{0.2} = 500\text{ s}$$

3 (*a*) For a zero order reaction, $t = \frac{1}{k}(a_0 - a)$

$$\Rightarrow \quad t_{50} = \frac{1}{k} \times \frac{a_0}{2} \qquad [\because \text{ at } t_{50},\ a = a_0/2]$$

$$\Rightarrow \quad k = \frac{a_0}{2\times t_{50}} = \frac{0.02\text{ M}}{2\times 100\text{ s}}$$

$$= 1\times 10^{-4}\,Ms^{-1} = 1\times 10^{-4}\,\text{mol L}^{-1}\,s^{-1}$$

4 (*b*) From collision theory of chemical reaction, Arrhenius equation we get, $k = A \times e^{-E_a/RT}$

where, k = rate constant,

$e^{-E_a/RT}$ = fraction of molecules with energy greater than E_a [option-(a)].

A = Arrhenius factor or frequency factor $= P \times Z_{AB}$

when, P = steric factor [option-(c)]

Z_{AB} = collision frequency of reactants, A and B

= number of effective binary collisions between A and B in one second in unit volume [option-(b)].

Hence, option (b) is the correct.

5 (*b*) Rate of reaction,

$$= \frac{\text{Decrease in the concentration of reactants or increase in the concentration of products}}{\text{Time taken}}$$

For the given reaction,

$$N_2(g) + 3H_2(g) \rightleftharpoons 2NH_3(g)$$

$$\text{Rate of reaction} = \frac{-d[N_2]}{dt} = -\frac{1}{3}\frac{d[H_2]}{dt} = +\frac{1}{2}\frac{d[NH_3]}{dt}$$

Hence, option (b) is correct.

6 (*d*) For first order reaction ,

$$t = \frac{2.303}{k}\log\frac{a}{a-x} \qquad \text{...(i)}$$

Given: $k = 2.303\times 10^{-3}\,s^{-1}$

$a = 40\,g$

$a - x = 10\,g$

On substituting the given values in Eq. (i) ,

$$t = \frac{2.303}{2.303\times 10^{-3}}\log\frac{40}{10}$$

$$= 10^3 \log 2^2 = 2\times 10^3 \times \log 2$$

$$= 2\times 10^3 \times 0.3010 = 602\text{ s}$$

Alternative method

For first order reaction,

$$t_{1/2} = \frac{0.693}{k}$$

$$t_{1/2}(t_{50\%}) = \frac{0.693}{2.303\times 10^{-3}} = 301\text{ s}$$

Also, $t_{75\%} = 2t_{50\%}$

$\therefore$ $t_{75\%} = 2\times 301 = 602\text{ s}$

7 (*b*) For first order reaction,

$$t = \frac{2.303}{k}\log\frac{a}{a-x}$$

where, a = initial concentration,

$a - x$ = final concentration.

Let the initial concentration $(a) = 100$

After time t, final concentration $(a-x) = 100 - 99 = 1$

We know that,

$$t = \frac{2.303}{k}\log\frac{a}{a-x}$$

On substituting the given values in above Eq. ,

$$t = \frac{2.303}{k}\log\frac{100}{1} = \frac{2.303}{k}\log 10^2$$

$$= \frac{2.303}{k} \times 2\log 10 = \frac{4.606}{k}$$

Thus, option (b) is correct.

8 (*b*) Rate constants at two different temperatures is given by Arrhenius equation as follows

$$\log\frac{k_2}{k_1} = \frac{E_a}{2.303R}\left[\frac{1}{T_1} - \frac{1}{T_2}\right]$$

$$T_2 > T_1$$

Given, $E_a = 0$, $T_1 = 200\text{ K}$

$k_1 = 1.6\times 10^6\,s^{-1}$, $T_2 = 400\text{ K}$

$R = 8.314\,JK^{-1}\,mol^{-1}$

On substituting the given values in above equation,

$$\log\frac{k_2}{1.6\times 10^6} = \frac{0}{2.303\times 8.314}\left[\frac{1}{200} - \frac{1}{400}\right]$$

$$\log\left(\frac{k_2}{1.6\times10^6}\right)=0 \Rightarrow \frac{k_2}{1.6\times10^6}=10^0=1$$

$$k_2 = 1.6\times10^6\ s^{-1} \text{ at } 400K$$

Thus, option (b) is correct.

9 (b) $\log\frac{k_2}{k_1}=\frac{E_a}{2.303R}\left[\frac{1}{T_1}-\frac{1}{T_2}\right]$

where T_1 and T_2 are initial and final temperature and k_1 and k_2 are initial and final rate constants, E_a = activation energy and R = gas constant.

Given, $T_1 = 300$ K,

$T_2 = 400$ K

$R = 8.314\ J\ mol^{-1}\ K^{-1}$

$$\frac{k_2}{k_1}=\frac{2}{1}$$

$$\log\frac{k_2}{k_1}=\frac{E_a}{2.303\times R}\left[\frac{1}{T_1}-\frac{1}{T_2}\right]$$

Thus, $\log\frac{2}{1}=\frac{E_a}{2.303\times8.314}\left[\frac{1}{300}-\frac{1}{400}\right]$

$$E_a = 0.30\times2.303\times8.314\times3\times400$$

$$E_a = 6.88\ kJ\ mol^{-1}$$

10 (b) For zero order reaction, if concentration of product is $[C_0]$ at time t_1 and $\frac{[C_0]}{2}$ at time t_2, then

$$k=\frac{\left[\frac{C_0}{2}\right]-[C_0]}{t_2-t_1}=\frac{[C_0]-\left[\frac{C_0}{2}\right]}{t_{1/2}}$$

$\therefore$ $t_{1/2}=\frac{[C_0]}{2k}$

Thus, $t_{1/2}$ for zero order reaction is directly proportional to the initial concentration of the reactant.

11 (a) Given, $t_{1/2} = 36$ min

$t = 2\text{hrs} = 120$ min, $[A_0] = 10$ g

For first order reaction, $k=\frac{0.693}{t_{1/2}}$

$\therefore$ $k=\frac{0.693}{36}=0.01925$

$$t=\frac{2.303}{k}\log\frac{[A]_0}{[A]}\ g$$

where, $[A]_0$ = initial amount = 10 g

$[A]$ = final amount

$\therefore$ $120=\frac{2.303}{0.01925}\log\frac{10}{[A]}$

$\log 10-\log[A]=1$

$1-\log[A]=1$

$\Rightarrow$ $\log[A]=0$

$[A]=1$g

12 (d) For a first order reaction,

$$t=\frac{2.303}{k}\log\frac{a}{a-x}$$

where, a = initial concentration

$a-x$ = final concentration

Given, rate constant $(k) = 4.3\times10^{-4}s^{-1}$

Let, the initial concentration = a

$\therefore$ Final concentration $(a-x)=a-\frac{2}{3}a=\frac{a}{3}$

For first order reaction,

$$t=\frac{2.303}{k}\log\frac{a}{a-x}$$

$$=\frac{2.303}{4.3\times10^{-4}}\log\frac{a}{a/3}=2.5\times10^3\ s$$

Thus, option (d) is correct.

13 (b) For first order reactions, the rate of reaction is proportional to the first power of the concentration of the reactant .

For, $A\rightarrow B$

Rate $=-\frac{d[A]}{dt}=k[A]$ [where, k = constant]

Half-life $(t_{1/2})=\frac{0.693}{k}$

$\therefore$Rate of first order reaction depends upon reactant concentrations and half life does not depend upon initial concentration of reactant, $[A]_0$.

For second order reactions, the rate of reaction is proportional to the second power of the concentration of the reactant.

For, $2A\rightarrow B$

Rate = $k[A]^2$

Half-life $(t_{1/2})=\frac{1}{k[A]_0}$

$\therefore$Rate of second order reaction depends upon reactant concentration and half-life also does depend on $[A]_0$.

14 (b) For zero order reaction, $t_{1/2}=\frac{[R]_0}{2k}$

where, II $[R]_0$ = Initial concentration of the reactant.

k = Rate constant.

Thus, $t_{1/2}$ for zero order reaction is directly proportional to the initial concentration of the reactant.

$$t_{1/2}\propto[R]_0$$

$\therefore$For zero order reaction, when the concentration of reactant is doubled, the half-life $(t_{1/2})$ will also get doubled.

15 (a) For a first order reaction, the integrated rate law is written as,

$$k=\frac{2.303}{t}\log\frac{a}{a-x}$$

where, k = rate constant

a = initial concentration

$a-x$ = concentration after time 't'

For half-life, $t = t_{1/2}$, $x = \frac{a}{2}$

On substituting the values,

$$k = \frac{2.303}{t_{1/2}} \log \frac{a}{a - \frac{a}{2}} = \frac{2.303}{t_{1/2}} \log 2$$

$$\therefore \quad k = \frac{0.693}{t_{1/2}}$$

Thus, $t_{1/2}$ of a first order reaction does not depend upon the concentration.

16 (*b*) $2NO_2 \underset{k_2}{\overset{k_1}{\rightleftharpoons}} N_2O_4$

$$\text{Rate of reaction} = -\frac{1}{2}\frac{d[NO_2]}{dt} = k_1[NO_2]^2 - k_2[N_2O_4]$$

$\therefore$ Rate of disappearance of NO_2

i.e. $$\frac{-d[NO_2]}{dt} = 2k_1[NO_2]^2 - 2k_2[N_2O_4]$$

17 (*b*) For every 10°C rise, rate constant gets doubled. Hence, for every 20° C rise, rate constant will be 4 times.

Therefore, according to the given statement,

Temperature	280 K	300 K
Rate constant	k_2	$k_1(4k_2)$

Thus, $k_1 = 4k_2 \Rightarrow k_2 = 0.25k_1$

18 (*d*) Given that,

$$\ln k\,(s^{-1}) = 14.34 - \frac{125 \times 10^4}{T}\,\text{K} \quad \text{...(i)}$$

We know that,

$$\ln k(s^{-1}) = \ln A - \frac{E_a}{RT} \quad \text{...(ii)}$$

On comparing equation (i) and (ii), we get

$$\frac{E_a}{R} = 1.25 \times 10^4\ \text{K}$$

$$\therefore \quad E_a = 1.25 \times 10^4\ \text{K} \times R\,\text{cal K}^{-1}\ \text{mol}^{-1}$$

$$= 1.25 \times 10^4\ \text{K} \times 2\,\text{cal K}^{-1}\ \text{mol}^{-1}$$

$$[\because R = 2\ \text{cal K}^{-1}\ \text{mol}^{-1}]$$

$$= 2.50 \times 10^4\,\text{cal mol}^{-1}$$

19 (*d*) Both Assertion and Reason are false.

The reaction $2NO + O_2 \longrightarrow 2NO_2$ and $2CO + O_2 \longrightarrow 2CO_2$ proceeds at the different rates. Both the reactions have different activation energies.

20 (*c*) $k = 0.01\,s^{-1}$

$$\therefore \quad t_{50} = \frac{0.693}{k} = \frac{0.693}{0.01} = 69.3\ \text{s}$$

20 g reduced to 5g, thus 75% reaction.

$$\therefore \quad t_{75} = 2 \times t_{50} = 138.6\ \text{s}$$

21 (*a*) From (ii) for slow step,

$$\left(\frac{dx}{dt}\right) = k[X][Y_2]$$

From (i),

$$\frac{[X]^2}{[X_2]} = k_C$$

$$\therefore \quad [X]^2 = k_C[X_2] \Rightarrow [X] = k_C'[X_2]^{\frac{1}{2}}$$

$$\therefore \quad \left(\frac{dx}{dt}\right) = k_C'\,k[X_2]^{\frac{1}{2}}[Y_2]$$

order $$= \frac{1}{2} + 1 = 1.5$$

22 (*c*) A catalyst is a substance which alters the reaction rate but itself remains unchanged in amount and chemical composition at the end of the reaction. It provides a new reaction path with a lower energy barrier (lowering of activation energy).

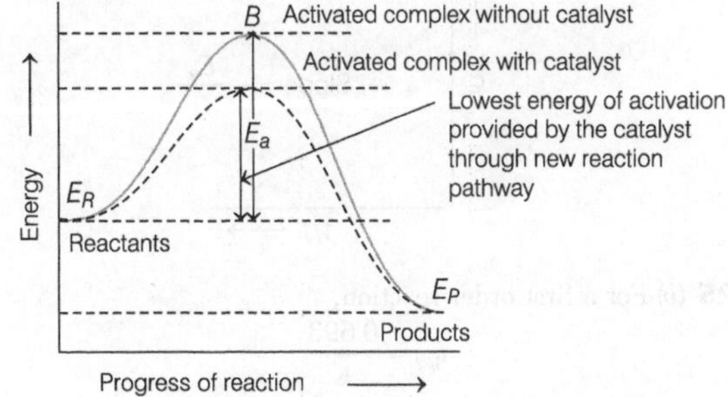

23 (*d*) Given, order of reaction = 1

Rate of reaction at 10 s = 0.04 mol $L^{-1}\,s^{-1}$

Rate of reaction at 20 s = 0.03 mol $L^{-1}\,s^{-1}$

$$\left(\frac{dx}{dt}\right)_1 = k(a - x_1) \text{ at time } t_1$$

$$\left(\frac{dx}{dt}\right)_2 = k(a - x_2) \text{ at time } t_2$$

$$\left(\frac{a - x_1}{a - x_2}\right) = \frac{4}{3}$$

$\therefore$ Half-life period $(t_{1/2}) = ?$

$$k = \frac{2.303}{t_2 - t_1} \log \left(\frac{a - x_1}{a - x_2}\right)$$

$\therefore$ Half-life period $(t_{1/2}) = ?$

$$k = \frac{2.303}{10} \log \frac{0.04}{0.03} = \frac{2.303}{10} \times 0.124$$

$$k = 0.028\,s^{-1}$$

We know that,

$$t_{1/2} = \frac{0.693}{k} = \frac{0.693}{0.0287\ s^{-1}}$$

$$= 24.14\ \text{s} \approx 24.1\ \text{s}$$

24 (*c*) By Arrhenius equation,

$$k = Ae^{-E_a/RT}$$

where, E_a = energy of activation

Applying log on both sides,

In $$k = \ln A - \frac{E_a}{RT} \quad ...(i)$$

or $$\log k = -\frac{E_a}{2.303R} + \log A \quad ...(ii)$$

This equation (ii) is of the form of $y = -mx + C$, i.e. the equation of a straight line.

Thus, if a plot of log k *vs* $\frac{1}{T}$ is a straight line, the validity of the equation is confirmed by slope of the line

$$= -\frac{E_a}{2.303\,R}$$

Thus, by measuring the slope of the line, the value of E_a can be calculated.

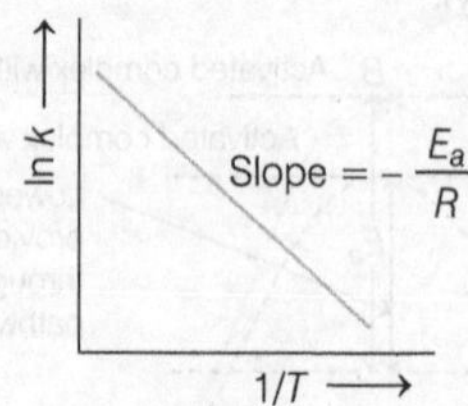

25 (*b*) For a first order reaction,

$$t_{1/2} = \frac{0.693}{k}$$

i.e. half-life period is independent of initial concentration of a reactant.

26 (*d*) For a zero order reaction, unit of rate constant is $ML^{-1}s^{-1}$. Hence, we can easily calculate concentration of *B* after 20 minutes by the following formula

$$x = kt$$

$$x = kt = 0.6 \times 10^{-3} \times 20 \times 60 = 0.72 \text{ M}$$

27 (*a*) For the reaction,

Change in concentration $\underset{0.4}{2N_2O_5} \longrightarrow \underset{0.8}{4NO_2} + O_2$

Rate of production of NO_2

$$= \frac{0.8}{200} = 4 \times 10^{-3} \text{ mol L}^{-1}\text{min}^{-1}$$

28 (*a*) The time taken for change for concentration from 0.6 M to 0.3 M $(t_{1/2}) = 30$ min

Now, the time taken for change the concentration from

$$0.1 \to 0.05 \to 0.025 = (2t_{1/2}) = 60 \text{ min}$$

time of 75% reaction $= 2t/2$ i.e. 60 min.

29 (*a*) For first order reaction,

$$t = \frac{2.303}{k} \log \frac{a}{a-x}$$

Here, let the initial amount of reactant, $a = 100$

After time *t*, dissociated amount of reactant, $x = 90$

Now, putting the values

$$t_{90\%} = \frac{2.303}{k} \log\left(\frac{100}{100-90}\right)$$

$$t_{99\%} = \frac{2.303}{k} \log\left(\frac{100}{100-99}\right)$$

$$\frac{t_{90\%}}{t_{99\%}} = \frac{\frac{2.303}{k}\log\left(\frac{100}{10}\right)}{\frac{2.303}{k}\log\left(\frac{100}{1}\right)} = \frac{1}{2}$$

$$t_{99\%} = 2t_{90\%} \Rightarrow t_{99\%} = 2t$$

30 (*c*) The sum of power of the concentration of the reactants in the rate law expression is called the order of that chemical reaction.

$$X + Y \longrightarrow XY$$

The order of reaction with respect to *X* is

$$2. \quad \frac{-d[X]}{dt} = k[X]^2[Y]$$

The order of reaction $= 2 + 1 = 3$.

31 (*a*) $X \xrightarrow{\text{I}} Y \xrightarrow{\text{II}} Z$, is a complex reaction.

Total order of reaction is 2.

Molecularity of slowest step II is also 2 because for complex reaction, molecularity of the slowest step is same as the order of the overall reaction.

32 (*b*) I. $ClO^- + ClO^- \xrightarrow{k_1} ClO_2^- + Cl^-$ (slow step)

II. $ClO_2^- + ClO^- \xrightarrow{k_2} ClO_3^- + Cl^-$ (last step)

Since, slowest step is the rate determining step, so the rate of reaction $= k_1[ClO^-]^2$.

33 (*c*) For reaction, $2A \longrightarrow 3B$

$$\therefore \quad -\frac{1}{2}\frac{d[A]}{dt} = \frac{1}{3}\frac{d[B]}{dt}$$

Given, $$\frac{d[B]}{dt} = x \text{ mol / L}$$

$$\therefore \quad \frac{d[A]}{dt} = -\frac{2}{3}\frac{d[B]}{dt} = -\frac{2}{3}x$$

Hence, rate of consumption of $A = \left[-\frac{d[A]}{dt} = \frac{2}{3}x\right]$

34 (*a*) Given, $E_f = 90$ kJ, $E_b = ?$, $\Delta H = 50$ kJ

$$\Delta H = E_f - E_b$$

$$50 = 90 - E_b$$

$$E_b = (90 - 50) \text{ kJ}$$

$$E_b = 40 \text{ kJ}$$

35 (*c*) Activation energy of forward reaction = *A*

Activation energy of backward reaction = *B*

$\therefore$ Heat of reaction $= B - A = C (B > A)$

$\therefore$ Reaction is exothermic.

36 (*d*) If a plot is drawn between $1/T$ and in *k*, then according to Arrhenius equation,

$$\ln k = \ln A - \frac{E_a}{RT}$$

Here, slope $= -\dfrac{E_a}{R}$

$\Rightarrow \quad -1\times10^4 = -\dfrac{E_a}{8.314}$

$\therefore \quad E_a = 8.314\times10^4$ J mol^{-1}

or $= 83.14$ kJ mol^{-1}.

37 (c) In reaction M, there is only an intermediate step, so it is a fast reaction. Further, in reaction M energy is much lesser than that of the reactant, as compared to as shown in reaction N. Thus, reaction M is faster and more exothermic than reaction N.

38 (a) Arrhenius equation is $k = Ae^{-E_a/RT}$

On taking log both sides,

We get, $\log k = \log A - \dfrac{E_a}{2.303RT}$

When log k is plotted with $1/T$, the slope of the reaction will be $-E_a / 2.303R$ and intercept given by log A.

39 (a) Rate constant is affected by temperature only.

40 (b) For reaction, $A + B \longrightarrow$ Product

$$\frac{dx}{dt} = k[A]^2[B] \Rightarrow 0.22 = k(1)^2\ (0.25)$$

$$\therefore \quad k = \frac{0.22}{0.25} = 0.88\ \text{M}^{-2}\text{s}^{-1}$$

41 (b) Reaction is of first order

So, $\ln\left(\dfrac{a}{a-x}\right) = kt$

When a = initial concentration

$a = 100$

Let $\Rightarrow (a - x) = 10\%$ $\quad (\because x = 90\%)$

$a - x$ = concentration at instant

$$\ln\left(\frac{100}{10}\right) = k\times3600 \text{ as } (t = 3600\text{ s})$$

$$2.303\log 10 = k\times3600 \qquad (\log 10 = 1)$$

$$k = \frac{2.303}{3600} = 6.4\times10^{-4}\text{s}^{-1}$$

42 (c) For zero order reaction, $t_{50} = \dfrac{[A_0]}{2k}$

$$k = \frac{[A_0]}{2t_{50}}$$

$$\therefore \quad k = \frac{6\times10^{-3}}{2\times60}$$

$$\Rightarrow \quad k = 0.5\times10^{-4}\ \text{Ms}^{-1} \text{ or } 5\times10^{-5}\ \text{mol dm}^{-3}\text{s}^{-1}$$

43 (d) Rate $= k[A]^a[B]^b$

Overall order of reaction $= a + b$.

Inversion of cane sugar is first order reaction (given).

Hence, rate $= k\,[\text{sugar}]^x[\text{H}^+]^y$

and $x + y = 1$

In given option, only (d) value of $x + y = 1$.

44 (b) $400\text{ g} \underset{(A)}{} \longrightarrow \underset{\text{1half-life}}{200\text{ g}} \longrightarrow \underset{2t_{1/2}}{100\text{ g}} \longrightarrow \underset{3t_{1/2}}{50\text{ g}} \longrightarrow \underset{4t_{1/2}}{25\text{ g}}$

$\therefore$ A undergoes 4 half-lifes.

For B, $\underset{(B)}{400\text{ g}} \longrightarrow \underset{\text{Half-life}}{200\text{ g}}$

$\therefore$ B undergoes 1 half-life.

Thus, $t_{1/2}$ of $B = 4t_{1/2}$ of A.

45 (c) Given, rate $= k[\text{H}^+]^n$

Initial pH = 3, so $[\text{H}^+]^n = 1\times10^{-3}$, initial rate $= r_1$

Final pH = 1, so $[\text{H}^+] = 1\times10^{-1}$,

Final rate $= r_2 = 100r_1$

On substituting values, we get

$$r_1 = k[1\times10^{-3}]^n \qquad \text{...(i)}$$

$$r_2 = 100r_1 = k[10^{-1}]^n \qquad \text{...(ii)}$$

On dividing Eq. (i) by Eq. (ii), we get

$$\frac{r_1}{100r_1} = \left[\frac{1\times10^{-3}}{1\times10^{-1}}\right]^n = \frac{1}{100} = \left[\frac{10^{-2}}{1}\right]^n \Rightarrow \left[\frac{1}{100}\right]^1 = \left[\frac{1}{100}\right]^n$$

$\therefore \quad n = 1$

Thus, the reaction is of first order.

46 (a) The reaction is of first order,

If $a = 100$, $x = 60$

$$(a - x) = 40$$

$$\therefore \quad k = \frac{2.303}{t}\log\left(\frac{a}{a-x}\right)$$

$$\text{Thus,} \quad k = \frac{2.303}{60}\log\frac{[100]}{[40]}$$

$$k = \frac{2.303}{60}\times0.398 \qquad \text{...(i)}$$

Later on 50% of the reaction would be completed, thus,

$$k = \frac{0.693}{t_{50}} \qquad \text{...(ii)}$$

$$\therefore \quad t_{50} = \frac{0.693}{k} = \frac{0.693\times60}{2.303\times0.398}$$

$$= 45\text{ min}$$

47 (d) $A(g) \longrightarrow P(g) + Q(g) + R(g)$

	$A(g)$	$P(g)$	$Q(g)$	$R(g)$
At $t = 0$	0.4 atm	0	0	0
At time, t	$(0.4 - x)$	x	x	x

If initial pressure $= p_i$

Moles of gaseous product from 1 mole reactant $= y = 3$

Then, $(a - x) = \dfrac{yp_i - p_t}{(y-1)} = \dfrac{3p_i - p_t}{2}$

Also, $t_{1/2} = 69.3$ s

$$\therefore \quad k = \frac{0.693}{t_{1/2}} = \frac{0.693}{69.3} = 1\times10^{-2}\text{ s}^{-1}$$

$$k = \frac{2.303}{t}\log\frac{a}{a-x}$$

$$1\times10^{-2}=\frac{2.303}{230}\log\frac{2p_i}{(3p_i-p_t)}$$

$$1=\log\frac{2p_i}{3p_i-p_t}$$

$$\therefore\quad \frac{2p_i}{3p_i-p_t}=10$$

$$p_i=15p_i-5p_t$$

$$\therefore\quad 5p_t=14p_i$$

$$p_t=\frac{14\times0.4}{5}=1.12\,\text{atm}$$

48. (a) For a first order reaction,

$$\therefore\quad k=\frac{0.693}{t_{1/2}}$$

$$t_{1/2}=\frac{0.693}{k}$$

Thus, $t_{1/2}$ of a first order reaction does not depend upon the concentration.

49 (a) Given, rate constant for a first order reaction,

$$k=2.303\times10^{-2}\ \text{s}^{-1}$$

Let the initial concentration be a

Then, final concentration $(a-x)=\frac{1}{10}$ of $a=\frac{a}{10}$

Integrated rate law expression,

$$k=\frac{2.303}{t}\log\frac{a}{a-x}$$

or $$t=\frac{2.303}{k}\log\frac{a}{(a-x)}$$

On substituting values, we get

$$t=\frac{2.303}{2.303\times10^{-2}}\log\frac{a}{a/10}$$

or $$t=10^2\log10$$

$$t=100\ \text{s}\qquad(\because \log 10=1)$$

50 (c) Given, initial temperature

$$T=20+273\ \text{K}=293\ \text{K}$$

Final temperature $T_2=35+273=308\,\text{K}$

$$R=8.314\ \text{J mol}^{-1}\text{K}^{-1}$$

Since, rate becomes double on raising temperature

$$r_2=2r_1\Rightarrow\frac{r_2}{r_1}=2$$

As rate constant $k\propto r$

$$\frac{k_2}{k_1}=2$$

From Arrhenius equation, we know that

$$\log\frac{k_2}{k_1}=-\frac{E_a}{2.303\,R}\left(\frac{T_2-T_1}{T_1T_2}\right)$$

$$\log2=-\frac{E_a}{2.303\times8.314}\left[\frac{293-308}{293\times308}\right]$$

$$0.3010=\frac{E_a}{2.303\times8.314}\left[\frac{15}{293\times308}\right]$$

$$E_a=\frac{0.3010\times2.303\times8.314\times293\times308}{15}$$

$$=34673.48\ \text{mol}^{-1}\text{or}\ =34.7\ \text{kJ mol}^{-1}.$$

51 (c) Energy profile diagram for a reaction is as from the figure, it is clear that.

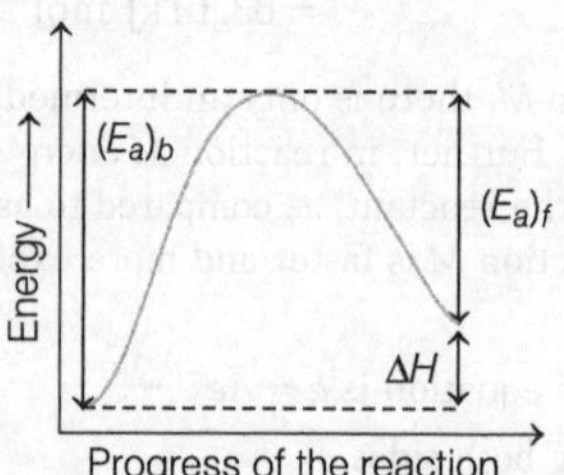

[Here, $(E_a)_b$ = activation energy of backward reaction and $(E_a)_f$ = activation energy of forward reaction]

If $(E_a)_b=(E_a)_f$

then, $\Delta H=0$

52. (d) Time of 75% reaction $=2\times t_{1/2}$

$$\therefore\quad t_{1/2}=1.5\,\text{min}$$

if $x=93.75\%$,

$$\therefore\quad (a-x)=6.25\%=C$$

$$C=C_0\left(\frac{1}{2}\right)^y$$

$$\frac{6.25}{100}=\left(\frac{1}{2}\right)^y$$

$$\left(\frac{1}{16}\right)=\left(\frac{1}{2}\right)^y$$

$$\therefore\quad\left(\frac{1}{2}\right)^4=\left(\frac{1}{2}\right)^y$$

$$\therefore\quad y=4$$

$$y=\frac{\text{total}}{t_{1/2}}$$

$$\therefore\quad \text{total}=60\ \text{min}$$

53 (d) In general, $t_{1/2}\propto\frac{1}{(a)^{n-1}}$

here, n = order of reaction

Given, $$t_{1/2}\propto\frac{1}{a}$$

On comparing $(a)^{n-1}=a\Rightarrow n-1=1$

$$\therefore\quad n=2$$

54 (a) According to Arrhenius, the rate of reaction increases with rise in temperature because on rising the temperature, the number of active reactant molecules increases.

55 (c) For every 10°C rise, rate constant gets doubled.

Hence, for 20°C rise, rate constant will be 4 times.

Thus, according to given statement,

Temperature	280	300
Rate constant	k_2	$k_1(4k_2)$

Thus, $k_1=4k_2$ or $k_2=0.25\,k_1$

56 (*d*) Let the order of reaction with respect to A and B is x and y, respectively

So, the rate law can be given as

$$\text{Rate} = k[A]^x[B]^y \quad ...(i)$$

When the concentration of only B is doubled the rate is doubled.

So, $$\text{Rate}_1 = k[A]^x[2B]^y = 2\text{Rate} \quad ...(ii)$$

If concentration of both the reactants A and B are doubled the rate increases by a factor 8. So,

$$\text{Rate}_2 = k[2A]^x[2B]^y = 8\text{Rate} \quad ...(iii)$$

Divide Eq. (iii) by Eq. (i) to get,

$$\frac{8\text{Rate}}{2\text{Rate}} = \frac{k[2A]^x[2B]^y}{k[A]^x[2B]^y}$$

$$4 = (2)^x = (2)^2 = (2)^x \Rightarrow x = 2$$

Now, divide Eq. (ii) by Eq. (i) to get,

$$\frac{2\text{Rate}}{\text{Rate}} = \frac{k[A]^x[2B]^y}{k[A]^x[B]^y} \Rightarrow (2)^1 = (2)^y \Rightarrow y = 1$$

On substitution of values of x and y in Eq. (i) gives

$$\text{Rate} = k[A]^2[B]$$

57 (*b*) For $10°$ rise in temperature $n = 1$

So, $$\text{rate} = 2^n = 2^1 = 2$$

When temperature is increased from 10°C to 100°C change in temperature $= 100 - 10 = 90°$ C

$$n = 9$$

So, $$\text{rate} = 2^9 = 512 \text{ times}$$

58 (*b, d*) According to Arrhenius equation, activation energy (E_a) and rate constant (k_1 and k_2) of a chemical reaction at two different temperature (T_1 and T_2) are related as,

$$\ln\frac{k_2}{k_1} = \frac{E_a}{R}\left(\frac{1}{T_1} - \frac{1}{T_2}\right) = -\frac{E_a}{R}\left(\frac{1}{T_2} - \frac{1}{T_1}\right)$$

Hence, option (d) is also correct.

59 (*a*) According to Arrhenius equation,

$$k = Ae^{-E_a/RT}$$

Rate constant is dependent upon temperature only.

60 (*c*) $$\log\frac{k_2}{k_1} = \frac{E_a}{2.303R}\left(\frac{T_2 - T_1}{T_1 \times T_2}\right)$$

Here, $$\frac{k_2}{k_1} = 10,$$

$$T_2 = 27°\text{ C} = 300 \text{ K}, \; T_1 = -3°\text{ C} = 270 \text{ K}$$

$$\log 10 = \frac{E_a}{2.303R}\left[\frac{300 - 270}{300 \times 270}\right]$$

$$1 = \frac{E_a}{2.303R} \times \frac{30}{300 \times 270}$$

$$E_a = +2.303R \times 2700 \text{ Jmol}^{-1}$$

$$E_a = 2.303 \times 2.7\, R \text{ kJ mol}^{-1}.$$

61 (*b*) It is true that order of reaction is defined as the sum of powers of concentration present in the rate law.

$$\text{Rate} = k[A]^m[B]^n$$

Order of reaction $= m + n$

62 (*d*) It is a pseudo first order reaction. In this reaction, water takes part in large amount, so the concentration change in water is not measurable.

63 (*c*) If $t_{1/2} = \frac{1}{a^{n-1}}$, (order $= n$)

Given that, $$t_{1/2} \propto \frac{1}{a^3}$$

$$\Rightarrow \quad n - 1 = 3$$

$\therefore$ Order of reaction $= 3 + 1 = 4$

64 (*d*) For second order reaction,

$$\frac{dx}{dt} = k[R]^2$$

$$k = \frac{\text{conc.}}{\text{time} \times (\text{conc.})^2} = \text{conc.}^{-1}\text{time}^{-1}$$

$$k = \text{L mol}^{-1}\text{s}^{-1}$$

65 (*b*) For first order reaction, $t_{1/2}$ is independent of initial concentration

$$t_{1/2} = \frac{0.693}{k}$$

66 (*d*) Let order of reaction with respect to A and B is x and y respectively. Then,

$$\text{rate} = k[A]^x[B]^y$$

$$2 = k[0.2]^x[0.2]^y \quad ...(i)$$

$$4 = k[0.2]^x[0.4]^y \quad ...(ii)$$

$$36 = k[0.6]^x[0.4]^y \quad ...(iii)$$

On comparing Eqs. (i) and (ii), we get

$$\frac{4}{2} = \frac{(0.4)^y}{(0.2)^y} \Rightarrow 2 = 2^y \Rightarrow y = 1.$$

On comparing Eqs. (ii) and (iii), we get

$$\frac{36}{4} = \frac{(0.6)^x}{(0.2)^x}$$

$$9 = 3^2 = 3^x \Rightarrow x = 2$$

$\therefore$ Rate law, $r = k[A]^2[B]$

67 (*a*) Order of reaction can be 0, 1, 2, 3 or even or a fractional number and negative

68 (*d*) Taking slow step,

$$\frac{dx}{dt} = k'[X][B]$$

From step I, $$K_{eq} = \frac{[X]}{[A]^2}$$

$$[X] = K_{eq}[A]^2$$

$$\frac{dx}{dt} = k' \cdot K_{eq}[A]^2[B] = k[A]^2[B]$$

69 (*c*) For a zero order reaction, $\frac{dx}{dt} = k$,

or $$dx = k\,dt \Rightarrow \int dx = \int k\,dt$$

$$x = kt$$

For 100% completion of the reaction, $x = a$

therefore, $a = kt$ or $t = \frac{a}{k}$

70 (*a*) $$p_t = p_c - 2x + x = p_i - x = 300 \text{ Pa}$$

$$p_\infty = p_i - p_c + \frac{p_c}{2} = \frac{p_i}{2} = 200 \text{Pa}$$

Thus, $p_c = 400$ Pa

$x = 100$ Pa

Thus, $p_i - x = 300 - 100 = 200$ Pa

$$k = \frac{2.303}{t} \log \frac{p_i}{p_c - x}$$

$$= \frac{2.303}{10} \log \frac{400}{200} = 0.0693 \text{ min}^{-1}$$

71 (*c*) Order is the sum of the powers to which the concentration terms are raised in the rate equation.

72 (*e*) $\left(\frac{dx}{dt}\right) = k[A]^x[B]^y[C]^z$

Rate changes to m^n times (order = *n*) if concentration is multiplied by *m*.

From (1) and (2), $m = 1.5$ for the reactant (*C*)

$$m^z = 0 = (1.5)^0$$

Thus, $Z = 0$

From (1) and (3), $m = 2$ for the reactant (*B*)

and $m^y = 2 = 2^y \Rightarrow y = 1$

From (1) and (4), $m = \frac{1}{2}$ for the reactant (*A*)

and $$m^x = \left(\frac{1}{4}\right) \Rightarrow \left(\frac{1}{2}\right)^x = \left(\frac{1}{4}\right) = \left(\frac{1}{2}\right)^2$$

$\Rightarrow$ $x = 2$

Thus, order *wrt A, B* and *C* are 2, 1, 0.

73 (*d*) $C = C_0\left(\frac{1}{2}\right)^y$

[where, *C* is the concentration at time *t* and C_0 at $t = 0$]

$$y = \frac{\text{total time}}{T_{50}} = \frac{t}{T_{50}}$$

Case I $$C = \frac{40}{4}$$

$$\frac{C_0}{4} = C_0\left(\frac{1}{2}\right)^y$$

$$y = 2 = \frac{20}{T_{50}}$$

$\therefore$ $T_{50} = 10$ min

Case II $$C = \frac{C_0}{16}$$

$\therefore$ $$\frac{C_0}{16} = C_0\left(\frac{1}{2}\right)^y$$

$$\left(\frac{1}{16}\right) = \left(\frac{1}{2}\right)^y \Rightarrow \left(\frac{1}{2}\right)^4 = \left(\frac{1}{2}\right)^y$$

$$y = 4$$

$$t = T_{50} \times 4 = 40 \text{ min}$$

74 (*b*) $t_{50} \propto \left(\frac{1}{R_0}\right)^{x-1}$

where, x = order of reaction,

R_0 = initial concentration.

If $x = n - 1$, $$t_{50} \propto \left(\frac{1}{R_0}\right)^{n-2}$$

or $$t_{50} \propto (R_0)^{2-n}$$

75 (*a*) $k = \frac{2.303}{45} \log \frac{0.05}{0.035}$

$$k = \frac{2.303}{45} \times 0.1549 = 0.00793$$

$$t_{1/2} = \frac{0.693}{0.00793} = 87.39 \text{ min}$$

76 (*b*) $\log k = \log A - \frac{E_a}{2.303RT}$

$$\log k - \log A = -\frac{E_a}{2.303RT}$$

$$\log\left(\frac{k}{A}\right) = -\frac{2.303RT}{2.303RT} = -1$$

$$\frac{k}{A} = \text{antilog } (-1.0) = 1 \times 10^{-1} = 0.1$$

CHAPTER 05

Surface Chemistry

The branch of science dealing with the study of physical and chemical phenomenon occurring at the interface of two phases (*viz.* solid-liquid, liquid-gas, solid-gas) is called **surface chemistry**. Phenomena like adsorption, catalysis, corrosion and colloidal properties occur at the interface and therefore are the subject of interest under surface chemistry.

Inside

- Adsorption
  - *Thermodynamics of adsorption*
  - *Adsorption isotherms*
- Catalyst and catalysis
  - *Types of catalysis*
  - *Substance affecting catalytic activity*
  - *Important features of solid catalysts*
- Enzyme catalysis or biocatalysis
  - *Mechanism of enzyme catalysis*
  - *Diseases caused by enzyme deficiency*
- Colloidal state : classification and method of preparation
  - *Classification of colloids*
  - *Mechanism of micelle formation Colloids*
  - *Purification of colloidal sols*
- Charge on colloidal particles
- Properties of colloids
  - *Mechanical properties : Brownian movement*
  - *Optical property : Tyndall effect*
  - *Electrical properties*
- Emulsions
  - *Types of emulsions*
  - *Emulsifiers*
- Applications of colloidal sols
  - *Applications of emulsion*

ADSORPTION

Surface of solids (and sometimes liquids) tends to attract and hold the molecules of other substance when come in their contact. This phenomenon of accumulation of a substance at the surface of another substance rather than in its bulk is called **adsorption**. Thus, adsorption is essentially a surface phenomenon.

Examples of adsorption

(*i*) When dilute solution of methylene blue is stirred with finely divided charcoal, it gets adsorbed by the charcoal molecules, which is proved by decrease in the intensity of colour of solution of methylene blue.

(*ii*) Gases like ammonia (NH_3), chlorine (Cl_2), sulphur dioxide (SO_2) etc., get adsorbed over the powdered charcoal. This can be proved by decrease in gas pressure, if these are taken in a closed vessel.

(*iii*) Silica gel is generally used to adsorb moisture from the air or atmosphere.

Cause of adsorption

In case of solids or liquids, the particles present in their bulk are surrounded by same kind of species (atoms, molecules etc.) from all the sides. Thus, all are surrounded by the same environment. But this fact is not true for the particles present at their surface.

These particles are not surrounded by species (atoms, molecules etc.) of same kind from all the sides and hence, they possess some unbalanced or residual attractive forces. These forces are responsible for attracting particles of another substance at the surface of solids or liquids.

Some terms related to adsorption

(*i*) **Adsorbent** The solid/liquid that takes up a gas or vapour or a solute from solution, is called the adsorbent. Some important adsorbents are charcoal (both vegetable and animal), silica gel (obtained by heating a mixture of sodium silicate and 10% hydrochloric acid at $\approx 50°$ C), alumina, metals such as Ni, Cu, Ag, Au, Pt colloids etc.

Note *Colloids, on account of their extremely small dimensions, have larger surface area per unit mass and these are good adsorbents.*

(*ii*) **Adsorbate** The gas or the solute, which is held to the surface of the solid is called the adsorbate.

(*iii*) **Desorption** The removal of the adsorbed substance from a surface is called desorption. When equilibrium is reached,

Rate of adsorption = Rate of desorption

We can say, it is a reverse term of adsorption.

(*iv*) **Capillary condensation** If adsorbent is porous, the adsorbate gets condensed into its pores. This process is called capillary condensation.

(*v*) **Positive and negative adsorption** When concentration of adsorbate is more at the surface of adsorbent than its bulk (or interior phase), the adsorption is called positive adsorption and when concentration of adsorbate is less at the surface of adsorbent than its bulk (or interior phase) the adsorption is called negative adsorption.

(*vi*) **Sorption** When adsorption and absorption take place simultaneously and it is difficult to determine the relative extent of adsorption and absorption, the term sorption is used. This term was introduced by MC Bain.

Difference between adsorption and absorption

The main points of difference between adsorption and absorption are as follows

Table 5.1 Adsorption *versus* absorption

Adsorption	Absorption
It is surface phenomenon.	It is bulk phenomenon.
Adsorbate is accumulated at the surface.	The substance getting absorbed, is uniformly distributed throughout the bulk of the substance.
The rate of this phenomenon is very rapid in the beginning. However, which decreases gradually until equilibrium is reached.	Absorption proceeds at a steady rate.
e.g. adsorption of water vapour by silica gel.	e.g. absorption of a water vapour by anhydrous calcium chloride.

Thermodynamics of adsorption

During adsorption, following thermodynamical changes may observe.

(*i*) Heat is evolved, i.e. adsorption is an exothermic process and for an exothermic process, ΔH is negative.

(*ii*) When adsorbate is adsorbed, freedom of movement of its molecules gets restricted. It results in decrease in entropy of adsorbate or enthalpy of adsorption, i.e. ΔS is negative for adsorption.

(*iii*) From Gibbs Helmholtz equation,

$$\Delta G = \Delta H - T\Delta S$$

If $\Delta H = -\text{ve}$ and $\Delta S = -\text{ve}$

$$\therefore \quad \Delta G = -\text{ve} - [T(-\text{ve})]$$
$$= -\text{ve} + T$$

i.e. ΔG is negative only when ΔH has sufficiently high negative value, and temperature is not very high.

(*iv*) $\Delta G = -\text{ve}$ means, Gibbs free energy decrease

(v) When ΔG becomes zero, a state of adsorption equilibrium is established.

Types of adsorption

Depending on the nature of attractive forces involved between adsorbent and adsorbate, adsorption can be classified in two ways, i.e. Physical adsorption (physisorption) and chemical adsorption (chemisorption). Their differences are given is table 5.2.

Table 5.2 Difference between physical and chemical adsorptions

Physical adsorption	Chemical adsorption
Reversible	Irreversible
Formation of multimolecular layer/monolayer, depends on pressure.	Formation of unimolecular layer.
The extent of adsorption is related to the ease of liquefaction of the gas. A highly liquefiable gas gets adsorbed more.	No correlation exists with ease of liquefaction.
Forces of attraction between adsorbate and adsorbent are weak (van der Waals' forces).	Forces of attraction between adsorbate and adsorbent are strong (as a chemical bond).
Non-specific in nature due to weak forces.	Highly specific due to strong bond.
Low heat of adsorption (20-40 kJ mol^{-1}).	High heat of adsorption (80-240 kJ mol^{-1}).

Physical adsorption	Chemical adsorption
Occurs at low temperature and decrease with increasing temperature. 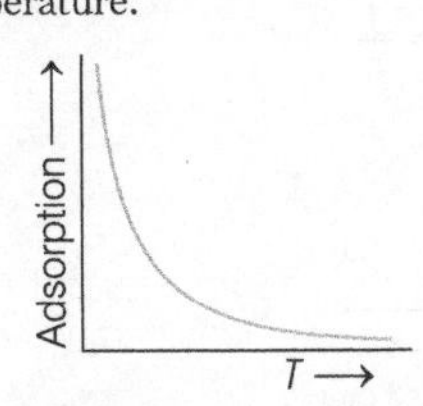	Occurs at higher temperature, extent of adsorption first increases and then decreases with increase in temperature. 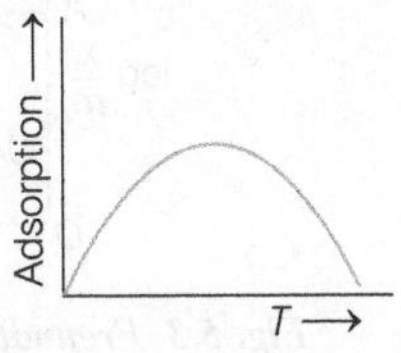
No appreciable activation energy is required.	High activation energy is required.
Favoured by high pressure and decrease in pressure results in desorption.	Favoured by high pressure but there is no desorption with decrease in pressure.

Note *The enthalpy change for the adsorption of one mole of an adsorbate on an adsorbent surface is called the molar enthalpy of adsorption.*

Simultaneous occurrence of physisorption and chemisorption

There are several cases in which physisorption and chemisorption take place simultaneously, e.g. hydrogenation reaction, (Ni as catalyst) in which hydrogen molecules are first adsorbed over nickel surface by weak van der Waals' forces (i.e. physisorption takes place). After that it dissociates to give hydrogen atoms.

These atoms are then bound to the nickel surface by strong forces, (i.e. chemisorption takes place). Similarly, at 83 K. N_2 is physisorbed and at 773 K it is chemisorbed over iron surface.

Factors affecting adsorption over solid

The extent of adsorption of gases over solid adsorbent is affected by the following factors

(*i*) **Nature of adsorbent** Some solid substances adsorb gases more efficiently than the other. It depends on how attractive the solid surface is for gaseous molecules, e.g. charcoal is a better adsorbent than many metals. That's why charcoal is used in gas masks.

(*ii*) **Nature of adsorbate** Because of non-specific nature of physisorption, in general, every gas will get physisorbed over the adsorbent surface to a lesser or greater extent. Generally, readily soluble and easily liquefiable gases such as NH_3, HCl, Cl_2, SO_2 etc., are adsorbed more than the so called permanent gases such as H_2, N_2 and O_2.

The ease with which a gas can be liquefied is dependent on T_C (critical temperature). Higher the value of T_C of a gas, the more easily it will be liquefied and therefore, more readily it will be adsorbed on the solid.

The process of chemisorption is highly specific in nature hence, a gas adsorbed on the solid only if it enters into chemical combination with it.

The process of adsorption of certain gases like H_2 on the surface of finely divided metals such as Ni, Pt, Pd etc., is called **occlusion.**

(*iii*) **Surface area of the solid adsorbent** The surface area per gram of the adsorbent is called specific surface area of the adsorbent. Larger the specific surface area of the absorbent greater is the extent of adsorption. The surface area can be increased by making the surface of adsorbent rough or by breaking it into small pieces. That's why an adsorbent has maximum adsorption power in its porous or powdered form.

Surface used for adsorption can be activated by mechanical rubbing or superheated steam.

(*iv*) **Temperature** Adsorption is an exothermic process (taking place with evolution of heat), therefore in accordance with Le-Chatelier's principle, the magnitude of adsorption should increase with fall in temperature.

This actually happens except in chemisorption where extent of adsorption first increases and then decreases with increase in temperature. The curve showing the effect of temperature on the extent of adsorption at a given pressure is called an adsorption isobar.

(*v*) **Pressure of the adsorbate gas** Since, adsorption of a gas leads to net decrease of pressure, thus, according to Le-Chatelier's principle, the magnitude of adsorption increases with increase in pressure. More detail of effect of pressure on extent of adsorption is given under the heading adsorption isotherms.

Is chemisorption endothermic?

In a few cases, chemisorption may be endothermic. Some of the examples are given below

(*i*) **Adsorption of hydrogen gas over glass** It is an endothermic process, i.e. ΔH is positive. So, to make a process spontaneous (ΔG negative); ΔS of the process is sufficiently high (positive).

(*ii*) **Adsorption of highly hydrate solute over solids** It is also an endothermic process and high positive value of ΔS makes the process spontaneous.

Adsorption isotherms

The variation of extent of adsorption with pressure at a given constant temperature is called adsorption isotherm for that temperature.

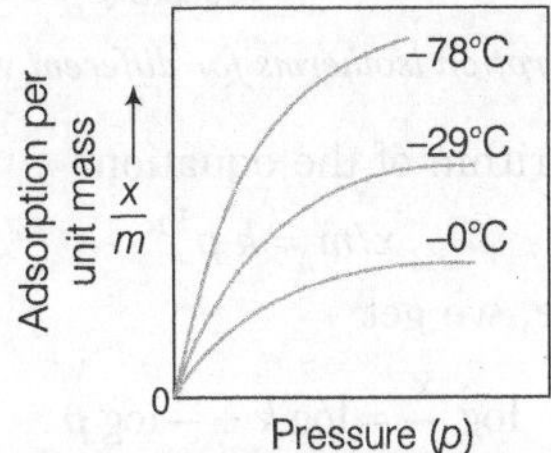

Fig. 5.1 Adsorption isotherms

From the Fig. 5.1, it can be concluded that at low temperature, the extent of adsorption increases very rapidly with pressure. Over a small range of pressure, the extent of adsorption is found to be directly proportional to the pressure. However, at higher pressures (closer to the saturation point of the vapour pressure the adsorption tends to achieve a limiting value. At this point, formation of multilayers starts and a stage of desorption may be observed.

There are several adsorption isotherms, but two are important of them which are as follows

Freundlich adsorption isotherm

Freundlich in 1909, gave the following relation between extent of adsorption and pressure.

$$x/m = kp^{1/n} \quad \text{...(i)}$$

where, $\left(\frac{x}{m}\right)$ is the amount of the gas adsorbed per unit mass of the adsorbent at pressure *p*; *k* and *n* are constants, $n \geq 1$.

All three possible cases of '*n*' are as follows

***Case* I** When $n = 1$,

$$\frac{x}{m} = kp \Rightarrow \frac{x}{m} \propto p$$

This is observed in **lower pressure range.** (Fig. 5.2) It is observed at the initial stage of adsorption over a free surface of adsorbent.

***Case* II** When $n = \infty$ (large enough)

$$x/m = k$$

This is observed at **high pressure** when saturation point is reached (Fig. 5.2).

***Case* III** When $n =$ intermediate,

$$x/m = kp^{1/n}$$

All the above conditions are shown in following figure.

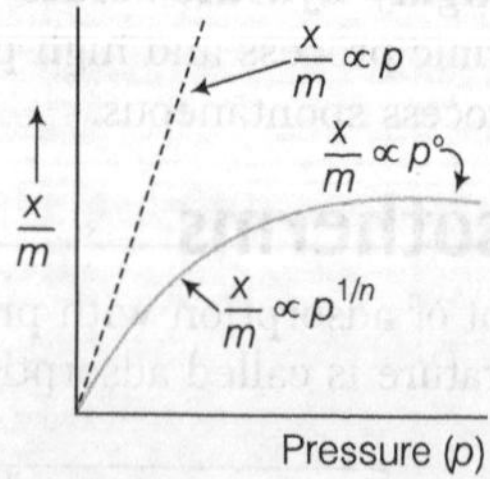

Fig. 5.2 Adsorption isotherms for different values of n

Taking logarithm of the equation

$$x/m = k\,p^{1/n} \quad \text{...(i)}$$

on both side, we get

$$\log\frac{x}{m} = \log k + \frac{1}{n}\log p$$

Graph between log (*x*/*m*) and log *p* is a straight line with slope 1/*n* and intercept log *k* (Fig. 5.3).

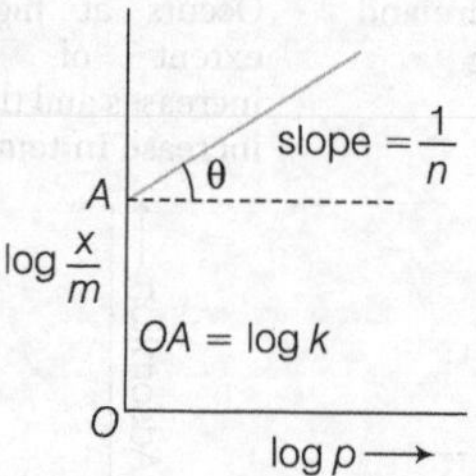

Fig. 5.3 Freundlich adsorption isotherm

The factor $1/n$ can have values between 0 and 1, thus, this equation holds good over a limited range of pressure.

Example 5.1 *Plot of log (x/m) against log p gives a straight line inclined at an angle of* 45°. *When the pressure is* 0.5 atm *and Freundlich parameter k is* 10, *what will be the amount of solute adsorbed per gram of adsorbent* (*log* 5 = 0.6990)?

Sol. According to Freundlich adsorption isotherm

$$\log\left(\frac{x}{m}\right) = \log k + \frac{1}{n}\log p$$

When log (*x*/*m*) is plotted against log *p*, we get straight line of slope (1/*n*) and intercept (log *k*).

$$\frac{1}{n} = \tan 45° = 1$$

$$\Rightarrow \quad \log k = \log 10 = 1$$

$$\frac{x}{m} = k(p)^{1/n} = 10 \times (0.5)^1 = 5$$

where $m = 1$ g, $x = 5$ g.

Langmuir adsorption isotherm

To overcome the limitation of Freundlich adsorption isotherm (i.e. failure at high pressure and no consideration of surface area), Langmuir developed a new isotherm which, is based on following kinetic considerations.

(*i*) There is no interaction between the adjacent absorbed molecule. i.e. gas behaves ideally in vapour phase.

(*ii*) All sites are equivalent on a surface.

(*iii*) The layer of gas adsorbed on the surface of the solid is only one molecule thick and uniformly distributed all over. i.e. unimolecular. Thus, it is the case of chemisorption.

(*iv*) At equilibrium, rate of adsorption = rate of desorption

(*v*) One site can accommodate only one molecule of adsorbate.

Based on above facts, Langmuir adsorption isotherm gives an equation which is represented as,

$$\frac{x}{m} = \frac{ap}{1 + bp} \quad \text{...(ii)}$$

where, *a* and *b* are Langmuir parameters.

Different cases are possible at different pressure. These are discussed below

Case I *If pressure p is very high then bp>>> 1,*

hence $1 + bp \simeq bp$

$\therefore \quad x/m = ap/bp = a/b = \text{constant}$

This indicates saturation point as shown along *BC* in the Fig. 5.4.

Case II *At very low pressure,*

$bp << 1$

$\therefore \quad 1 + bp \simeq 1 \quad \therefore \quad \frac{x}{m} = ap$

$\therefore \quad \frac{x}{m} \propto p$

This is shown along *OA* in the Fig. 5.4.

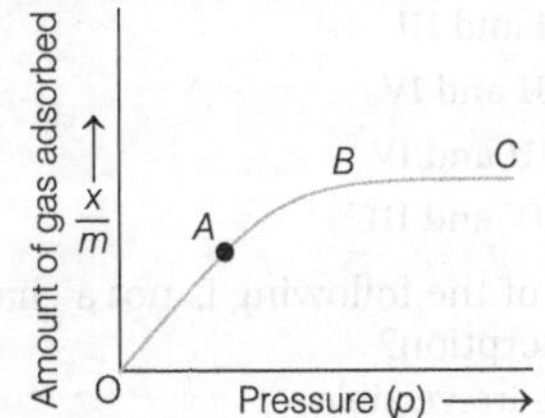

Fig. 5.4 Langmuir adsorption isotherm

Case III *At moderate pressure.*

$$\frac{x}{m} \propto kp^{1/n}$$

($1/n$ lies between 0 and 1).

This is shown along *AB* in the Fig. 5.4.

An another form of eq. (ii) is

$$\frac{p}{x/m} = \frac{1+bp}{a} \quad \text{or} \quad \frac{1}{a} + \frac{b}{a}p$$

So, when $\frac{p}{x/m}$ is plotted against p, a straight line is obtained with slope b/a and intercept $1/a$.

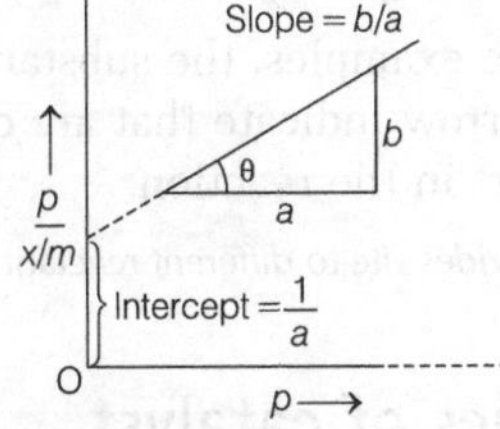

Fig. 5.5 Langmuir isotherm

Note *Freundlich adsorption isotherm explains physisorption, while Langmuir isotherm explains chemisorption.*

Adsorption from solution phase

Solids can adsorb solutes from solution and the extent of adsorption from solution depends upon the concentration of solute in the solution as given by Freundlich isotherm.

$$\frac{x}{m} = kC^{1/n}$$

where, C is the equilibrium concentration of the solute in solution.

Variation of $\left(\frac{x}{m}\right)$ with concentration at constant temperature is shown in Fig. 5.2 [take concentration (C) instead of pressure p].

Note *Adsorption from solution phase decrease with rise in temperature and decrease in surface area.*

Application of adsorption

Some of the important applications of adsorption are as follows

(*i*) Activated charcoal is used in gas masks (to remove poisonous gases) and to generate high vacuum.

(*ii*) Animal charcoal, fuller's earth etc., are used to adsorb colouring impurities during the production of sugar, vinegar etc.

(*iii*) In froth floatation process of concentration of ore, the sulphide are wetted with oil are adsorbed in air bubbles.

(*iv*) Ion exchange resins (used to remove hardness of water) also works on the principle of adsorption.

(*v*) Silica gel and alumina are used to remove moisture (i.e. as drying agents).

(*vi*) Heterogeneous catalysis and chromatographic analysis are based on the principle of adsorption.

(*vii*) All types of chromatographic techniques are based upon selective adsorption of different types of substances on a given adsorbent.

(*viii*) In analytical chemistry.

(*ix*) In medicines etc.

CHECK POINT 5.1

1. Adsorption is the phenomenon in which a substance
   (a) accumulates on the surface of the other substance
   (b) goes into the body of the other substance
   (c) remains close to the other substance
   (d) None of the above
2. Ink is adsorbed on the chalk. In this, ink and chalk are respectively
   (a) adsorbate and adsorbent (b) adsorbent and adsorbate
   (c) Both adsorbent (d) Both adsorbate
3. Sorption is the term used when [NCERT Exemplar]
   (a) adsorption takes place (b) absorption takes place
   (c) Both (a) and (b) (d) desorption takes place
4. The heat of adsorption of physical adsorption is nearly
   (a) 20-40 kJ/mol (b) 40-100 kJ/mol
   (c) 40-200 kJ/mol (d) 40-400 kJ/mol
5. Two gases A and B are exerting pressure p_1 and p_2 in two different container having equal volume. If equal amount of charcoal is added to both container and pressure reduced to $\frac{p_1}{2}$ and $\frac{p_2}{4}$ respect. Which statement is true?
   (a) B is easily liquefiable than A
   (b) Critical temperature of B is more than A
   (c) B adsorbs easily than A on charcoal
   (d) All of the above
6. Which of the following statements is/are true?
   (a) Water vapour is absorbed by anhydrous calcium chloride but adsorbed by silica gel
   (b) NH_3 is absorbed by water but adsorbed by charcoal
   (c) Sugar is decolourised by animal charcoal based on adsorption
   (d) All of the above
7. Physical adsorption is inversely proportional to
   (a) temperature (b) volume
   (c) concentration (d) All of these
8. There are certain properties related to adsorption
   I. reversible
   II. formation of unimolecular layer
   III. low heat of adsorption
   IV. occurs at low temperature and decreases with increasing temperature.

   Which of the above properties are for physical adsorption?
   (a) I, II and III
   (b) I, III and IV
   (c) II, III and IV
   (d) II, IV and III
9. Which of the following is not a characteristic of chemisorption?
   (a) It is irreversible
   (b) It is specific
   (c) It is multilayer phenomenon
   (d) Heat of adsorption is of about – 400 kJ
10. The rate of chemisorption
    (a) decreases with increase of pressure
    (b) increases with increase of pressure
    (c) is independent of pressure
    (d) is independent of temperature

CATALYST AND CATALYSIS

A substance that can influence the rate of a chemical reaction but is not consumed itself and remains unchanged is knowna s **catalyst** and the phenomenon is known as **catalysis.**

Berzelius (1836) introduced the term catalysis and catalyst. Ostwald (1895) redefined a catalyst as, a substance which changes the reaction rate without affecting the overall energetics of the reaction. e.g.

(i) Decomposition of potassium chlorate is catalysed by MnO_2.

$$2KClO_3 \xrightarrow{MnO_2} 2KCl + 3O_2$$

(ii) Similarly, manufacture of chlorine from Deacon's process is catalysed by cupric chloride.

$$4HCl + O_2 \xrightarrow{CuCl_2} 2H_2O + 2Cl_2$$

(iii) Platinum catalysed the formation of water from its constituents.

$$2H_2 + O_2 \xrightarrow{Pt} 2H_2O$$

In the above examples, the substance mentioned above the arrow indicate that are catalyst, that does not take part in the reaction.

Note *Catalyst provides site to different reactant to approach each other and react.*

Characteristics of catalyst

(i) Catalyst increases the rate of reaction by providing an alternate pathway of lower activation energy. Thus, they temporarily involved in a chemical reaction.

(ii) A catalyst does not effect the equilibrium constant, or enthalpy change of a reaction. However, in the presence of catalyst, equilibrium is attained quickly.

(iii) The catalyst remains unchanged in amount as well as in chemical composition at the end of the reaction. It may undergo some physical change.

(*iv*) Only a small quantity of the catalyst is generally needed, e.g. one mole of colloidal platinum in 10^8 L can catalyse decomposition of H_2O_2.

Note *In certain reactions, the rate of the reaction is dependent on the concentration of the catalyst, e.g. rate of inversion of cane sugar is dependent on the concentration of* H^+ *used as catalyst.*

(*v*) The catalyst does not initiate the reaction.

(*vi*) The catalyst is specific in its action. e.g.

- MnO_2 can catalyse decomposition of $KClO_3$ but not of KNO_3 or $KClO_4$.
- Carbon monoxide and hydrogen combine to form CH_4, CH_3OH and HCHO depending upon the nature of the catalyst used.

$$CO + 3H_2 \xrightarrow{Ni} CH_4 + H_2O$$

$$CO + 2H_2 \xrightarrow{ZnO+Cr_2O_3} CH_3OH$$

$$CO + H_2 \xrightarrow{Cu} HCHO$$

(*vii*) The catalyst cannot alter the nature of the products of the reaction. However, there are certain exceptions. The chlorination of toluene in the absence of catalyst and in the presence of sunlight forms benzyl chloride in first step by replacing H atom of the methyl group in side chain.

$$C_6H_5CH_3 + Cl_2 \xrightarrow{Sunlight} C_6H_5CH_2Cl$$

Toluene → Benzyl chloride

But chlorination of toluene in the presence of halogen carrier like Fe or I_2 and in the absence of sunlight forms *o*-and *p*-chlorotoluene by replacing H atom of the benzene nucleus.

$$2\,C_6H_5CH_3 + 2Cl_2 \xrightarrow[\text{dark}]{\text{Iodine}} o\text{-}ClC_6H_4CH_3 + p\text{-}ClC_6H_4CH_3 + 2HCl$$

Toluene → *o*-chlorotoluene + *p*-chlorotoluene

Types of catalysis

Catalysis can be classified into different types depending upon different factors

1. Homogeneous catalysis

When the catalyst and reactants are in the same phase, the catalyst is called homogeneous catalyst and the phenomenon is called homogeneous catalysis.
e.g. Iodide ion is a homogeneous catalyst for the decomposition of aqueous H_2O_2 because both I^- and H_2O_2 are present in the same aqueous solution phase

$$2H_2O_2(aq) \xrightarrow{I^-(aq)} 2H_2O(l) + O_2(g)$$

Other examples have been summarised in Table 5.3

Table 5.3 Homogeneous catalysis

S.No.	Process	Product(s)	Typical catalyst
1.	Oxidation of $SO_2(g)$ to $SO_3(g)$ in lead-chamber process	SO_3	NO (*g*)
2.	Oxidation of CO(*g*) by $O_2(g)$	$CO_2(g)$	NO (*g*)
3.	Inversion of sucrose solution (*aq*)	Glucose and fructose	$H^+(aq)$
4.	Hydrolysis of an ester (*aq*)	Acid and alcohol	$H^+(aq)$
5.	Aldol condensation of acetaldehyde (*aq*)	Crotonic aldehyde	$OH^-(aq)$
6.	Enolisation of ketone (*aq*)	Enolic form	H^+ or $OH^-(aq)$

2. Heterogeneous catalysis

When the catalyst is in a phase different from that of the reactants, the catalyst is called heterogeneous catalyst and the phenomenon is called heterogeneous catalysis. It is also called **surface catalysis** because in most of the cases, the catalyst is in solid state and the reaction starts from the surface of the solid catalyst. e.g. hydrogenation of vegetable oils to give vegetable fats (ghee) in the presence of finely divided nickel catalyst. Here, vegetable oil is present in liquid phase while nickel is present in solid phase.

$$\text{Vegetable oil }(l) + H_2(g) \xrightarrow{Ni(s)} \text{vegetable ghee }(s).$$

This process is also called hardening of oils. Generally, saturated hydrocarbons have high boiling point than their corresponding unsaturated hydrocarbons. Due to more van der Waals' forces.

Other examples of heterogeneous catalysis have been summarised in Table 5.4.

Table 5.4 Heterogeneous catalysis

S.No.	Process	Product(s)	Typical catalyst
1.	Methanol synthesis	$CH_3OH(g)$	ZnO/Cr_2O_3
2.	Oxidation of $SO_2(g)$ (contact process)	$SO_3(g)$	$V_2O_5 + K_2SO_4$ on SiO_2
3.	Ammonia synthesis (Haber's process)	$NH_3(g)$	Fe with Al_2O_3, MgO, CaO, K_2O
4.	Ammonia oxidation (Ostwald's process for manufacturing HNO_3)	$NO(g)$	90% Pt-10% Rh wire gauge
5.	Alkylation of benzene	$C_6H_5—CH_3$	H_3PO_4 on SiO_2
6.	Oxidation of carbon monoxide in automobile exhaust	CO_2	Pt (or Pd)
7.	Manufacture of H_2 (Bosch's process)	$CO_2 + H_2$	$Fe_2O_3 + Cr_2O_3$
8.	Polymerisation of ethene	Polyethene	$TiCl_4$ and AlR_3 (*Ziegler-Natta* catalyst) [R = propyl, ethyl]
9.	Formation of hydrocarbons from CO and H_2 (Fischer-Tropsch process)	Hydrocarbon	Fe, Co

Note ***Raney Ni*** *It is prepared by reacting Ni-Al alloy with NaOH which make it porous and enhance its catalytic properties.*

$$2Al + 2NaOH + 6H_2O \longrightarrow 2\,Na[Al(OH)_4] + 3H_2\uparrow$$

Activated charcoal *Raw charcoal is heated in absence of O_2 strongly at high temperature about 1700-1800°C which make it porous and enhance adsorbing capacity.*

It is also activated chemically by heating at 450-900°C in presence of phosphoric acid.

Adsorption theory of heterogeneous catalysis

Basically, heterogeneous catalysis involves following steps

Step I Adsorption of reactants on the solid surface is thought to involve chemical bonding (chemisorption) of reactants to the highly reactive metal atoms on the surface with accompanying breaking or at least weakening of bonds in the reactants.

Step II Diffusion of reactants along the surface.

Step III Reaction at an active site to form adsorbed product called the activated surface complex.

Step IV Desorption of the product and the catalyst surface is regenerated.

e.g. hydrogenation of ethene at nickel surface at 400 K takes place in following stages (a) to (d).

$$CH_2 = CH_2(g) + H_2(g) \xrightarrow{Ni(s)} CH_3—CH_3(g)$$

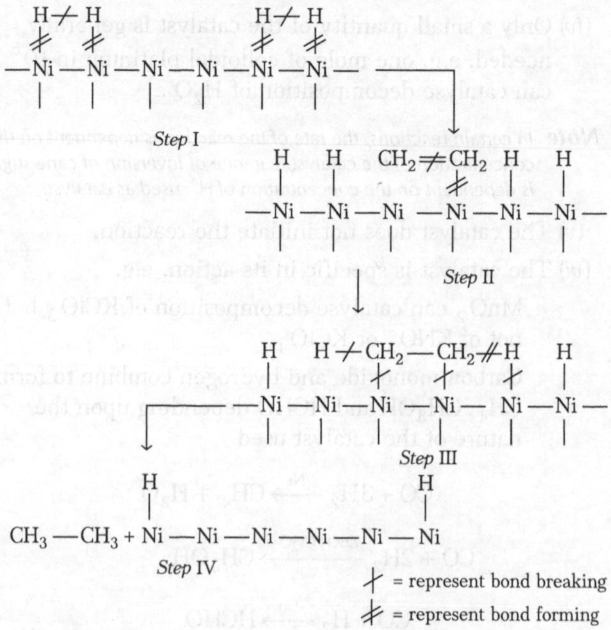

H—H adsorbed on —Ni—Ni—Ni—Ni—Ni—Ni—

Step I

H H $CH_2=CH_2$ H H on —Ni—Ni—Ni—Ni—Ni—Ni—

Step II

H H—CH_2—CH_2—H H on —Ni—Ni—Ni—Ni—Ni—Ni—

Step III

$CH_3—CH_3$ + Ni—Ni—Ni—Ni—Ni—Ni (with H H)

Step IV

⫽ = represent bond breaking

⫽⫽ = represent bond forming

Depending upon the nature of catalyst, whether it increases or decreases the rate of reaction, the catalysis is of following two types

(*i*) **Positive catalysis** When the rate of a reaction is accelerated by the foreign substance, the substance is called *positive catalyst* and the phenomenon is called *positive catalysis,* e.g. in lead chamber process, presence of NO accelerate the oxidation of SO_2 into SO_3.

(*ii*) **Negative catalysis** If rate of a reaction is retarded by the foreign substance, the substance is called *negative catalyst* and the phenomenon is called *negative catalysis,* e.g.

- Presence of phosphoric acid or glycerine or acetanilide retards the decomposition of H_2O_2 (and hence, help in its storage).

$$2H_2O_2(l) \xrightarrow[\text{or } H_3PO_4]{\text{Glycerine or acetanilide}} 2H_2O(l) + O_2(g)$$

- Presence of ethanol (or alcohol) retards the oxidation of chloroform into poisonous phosgene ($COCl_2$).
- Ethyl alcohol slows down the oxidation of sodium sulphite to sodium sulphate by atmospheric oxygen.

Depending upon whether the substance is a reactant or product, the catalysis can be of following two types

(*i*) **Autocatalysis** When one of the products formed, itself acts as a catalyst, then it is called autocatalyst and the phenomenon is called autocatalysis.

e.g. the oxidation of oxalic acid by acidified potassium permanganate, represented by the following equation

$$2MnO_4^- + 5C_2O_4^{2-} + 16H^+ \longrightarrow 2Mn^{2+} + 10CO_2 + 8H_2O$$

This reaction accelerated. It is due to the presence of Mn^{2+} ions which are formed in the reaction.

In case of homogeneous autocatalysis reactions, the rate of reaction is directly proportional to the concentration of catalyst, i.e. as the concentration of catalyst increases, rate increases.

(*ii*) **Induced catalysis** When a reaction accelerates another reaction which may not otherwise occur under the same conditions then the former reaction is called induced catalyst and the phenomenon is called induced catalysis.

e.g. sodium arsenite is not oxidised by air but if air is passed through a mixture of solution of sodium arsenite and sodium sulphite, both of them undergo simultaneous oxidation. Here, oxidation reaction of sodium sulphite thus induces the oxidation of sodium arsenite.

Substance affecting catalytic activity

Some substances increase the activity of catalyst while some other retard it. On this basis the substances which affect catalytic activity are of following two types

Catalytic promoters or activators

A substance, which (though not a catalyst by itself) promotes the activity of a catalyst, is called a promoters or activators. e.g.

(*i*) Molybdenum (Mo) promotes the catalytic activity of iron catalyst in the Haber's process for the manufacture of ammonia.

(*ii*) Copper and tellurium promote the catalytic activity of nickel in the hydrogenation of oils.

Catalytic poisons or inhibitors

A substance which itself is not a poison or inhibitors but in its presence, the activity of the catalyst is reduced, is called catalytic poison or inhibitor. e.g.

(*i*) Presence of H_2S or CO in the manufacture of ammonia by Haber's process retards (or even kills) the catalytic activity of iron.

(*ii*) Presence of arsenic or its compounds retard the activity of Pt (which is used as a catalyst in contact process for the oxidation of SO_2 into SO_3).

(*iii*) $BaSO_4$ or S in quinoline poison palladium catalyst, which in their presence reduce *R*COCl into *R*CHO (aldehyde) but not in RCH_2OH (alcohol), in Rosenmund's reaction.

Important features of solid catalysts

The important features solid catalysts are as follows

1. Activity

It is the ability of catalysts to accelerate chemical reactions. The degree of acceleration can be as high as 10^{10} times in certain reactions.

In general, activity of a catalyst depends upon the strength of chemisorption. For higher catalytic activity, the adsorption should be reasonably strong but not so strong, so that the adsorbed molecules become immobile and have no space for the adsorption of other reactants.

2. Selectivity

It is the ability of catalysts to direct a reaction to yield particular product(s) (excluding other), e.g. CO and H_2 react to form different products in the presence of different catalysts as follows

$$CO(g) + H_2(g) \xrightarrow{Cu} HCHO(g)$$

$$CO(g) + 2H_2(g) \xrightarrow{Cu/ZnO\text{-}Cr_2O_3} CH_3OH(g)$$

$$CO(g) + 3H_2(g) \xrightarrow{Ni} CH_4(g) + H_2O(g)$$

3. Shape selective catalysis

The phenomenon of dependence of catalytic reaction upon the pore structure of the catalyst and size of reactant and product molecules is called shape selective catalysis. In general zeolites are shape selective catalyst and the reaction catalysed by them comes under shape selective catalysis.

4. Zeolites

Zeolites are alumino-silicates of the general formula

$$M_{x/n}[(AlO_2)_x \cdot (SiO_2)_y] \cdot mH_2O$$

where, M may be simple cation, like Na^+, K^+, or Ca^{2+} n is the charge on the simple cation m is the number of molecules of water of crystallisation.

(*i*) The term zeolite has been taken from Greek language which means boiling stone as water trapped in their pores which get boil off when zeolite is heated.

(*ii*) Some well known zeolites are

- Erionite-$Na_2K_2CaMg(AlO_2)_2(SiO_2)_2 \cdot 6H_2O$
- Fauzasite-(natural) $Na_{56}[(Al_2O)_{56}(SiO_2)_{136}] \cdot 250H_2O$
- ZSM-5-$H_x[(AlO_2)_x(SiO_2)_{96-x}] \cdot 16H_2O$

(*iii*) In the structure of zeolite truncated octahedra (eubooctahedra) is present as building block and such a structure is called β-cage or sodalite cage.

(iv) The characteristic feature of zeolites is the openness of the structure, which permits cavities of different sizes (honeycomb like structure). The open structure is provided by silica in which aluminium occupies $x/(x+y)$ fraction of the tetrahedral sites. The negative charge of the alumino-silicate framework is neutralised by the replaceable cations. The void space forms more than 50% of the total volume, which is occupied by water molecules.

(v) The reaction selectivity of zeolites depends upon the size of cavities (cages), pores (apertures) and the distribution of pores in the structure. The pore size in zeolites generally varies from 260 pm to 740 pm.

(vi) There is a new class of highly siliceous zeolites with an optimal pore diameter of 550 pm. ZSM-5 is one such zeolite having the formula $H_x[(AlO_2)_x \cdot (SiO_2)_{96-x}] \cdot 16H_2O$. This converts alcohol to gasoline (petrol) by dehydrating the alcohol and producing a mixture of wide variety of hydrocarbons.

CHECK POINT 5.2

1. Catalyst increases the rate
   (a) by decreasing E_a
   (b) by increasing E_a
   (c) by increasing entropy
   (d) by both (a) and (c)

2. Equilibrium constant change
   (a) if some catalyst is added
   (b) if temperature is changed
   (c) if pressure is changed
   (d) if concentration is changed

3. Following reaction is catalysed by $Br^-(aq)$.
   $$2H_2O_2(aq) \longrightarrow 2H_2O(l) + O_2(g)$$
   This is an example of
   (a) homogeneous catalysis
   (b) heterogeneous catalysis
   (c) autocatalysis
   (d) enzyme catalysis

4. Which of the following is an example for homogeneous catalysis?
   (a) Manufacture of sulphuric acid by contact process
   (b) Manufacture of ammonia by Haber's process
   (c) Hydrolysis of sucrose in the pressure of dilute hydrochloric acid
   (d) Hydrogenation of oil

5. Which of the following catalyst is used during the hydrogenation of oil?
   (a) Fe (b) Ni
   (c) Pt (d) Mo

6. H_2O_2 can be prevented from decomposition in presence of
   (a) glycerine
   (b) acetanilide
   (c) Both (a) and (b)
   (d) None of the above

7. The ability of the catalyst is to direct the reaction to yield particular product is called
   (a) reactivity
   (b) selectivity
   (c) activity
   (d) fugacity

8. Zeolites are microporous aluminosilicates with general formula
   (a) $M[(AlO_2)_x] \cdot mH_2O$
   (b) $M_x[(AlO_2)_x(SiO_2)_y]$
   (c) $M_x[(SiO_2)_y \cdot mH_2O]$
   (d) $M_{x/n}[(AlO_2)_x(SiO_2)_y] \cdot mH_2O$

Enzyme catalysis or biocatalysis

Enzymes are the complex nitrogenous substances produced by living organisms and catalyse the reactions (occurring in their body). That's why these are called biocatalyst and the phenomenon of catalysis of reaction occurring in the body of living organisms by some enzymes is called biocatalysis or enzyme catalysis, e.g. conversion of starch into sugars is catalysed by amylase, and break down of polysaccharide cell walls of bacteria is catalysed by lysozyme (which is found in the tears of eyes).

The multiplication of the HIV virus that causes **AIDS** is dependent on the action of many enzymes. Some other examples of enzyme catalysis have been summarised in Table 5.5.

Table 5.5 Enzymes catalysis

S.No.	Reaction	Catalyst	Source/Place
1.	$\underset{\substack{\text{Sucrose} \\ \text{(sugar cane)}}}{C_{12}H_{22}O_{11}} + H_2O \xrightarrow{\text{(inversion of sugar cane)}} \underset{\text{Glucose}}{C_6H_{12}O_6} + \underset{\text{Fructose}}{C_6H_{12}O_6}$	Invertase	Yeast
2.	$\underset{\text{Maltose}}{C_{12}H_{22}O_{11}} + H_2O \longrightarrow \underset{\text{Glucose}}{2C_6H_{12}O_6}$	Maltase	Yeast
3.	$\underset{\text{Starch}}{2(C_6H_{10}O_5)_n} + nH_2O \longrightarrow \underset{\text{Maltose}}{n(C_{12}H_{22}O_{11})}$	Diastase	Malt
4.	$\underset{\text{Glucose}}{C_6H_{12}O_6} \longrightarrow \underset{\text{Ethanol}}{2C_2H_5OH} + CO_2\uparrow$	Zymase	Yeast
5.	$\underset{\text{Urea}}{NH_2CONH_2} + H_2O \longrightarrow 2NH_3 + CO_2\uparrow$	Urease	Soyabean
6.	$CO_2(aq) + H_2O(l) \rightleftharpoons H^+(aq) + HCO_3^-(aq)$	Carbonic anhydrides (in human body)	(Human body)
7.	N_2 (atmospheric) $\longrightarrow NH_3\uparrow$	Nitrogenase	Bacteria
8.	Alcohol $\longrightarrow$ Acetic acid	Alcohol dehydrogenase (ADH) and other enzymes	Mycoderma aceti
9.	Protein $\longrightarrow$ Peptide	Pepsin	Stomach
10.	Protein $\longrightarrow$ Amino acid	Trypsin	Intestine
11.	Milk $\longrightarrow$ Curd	Lactase	Lacto bacilli

Characteristics of enzyme catalysts

The important characteristics of enzyme catalysts are as follows

(*i*) They have high molecular mass ranging from 15000 to 1000000 g mol^{-1}.

(*ii*) Their aqueous solution is colloidal in nature.

(*iii*) They are highly efficient, e.g. an enzyme molecules may convert one million molecules of reactants into products per minute.

Note *In absence of digestive enzymes, a period of 50 years would be required to digest a single meal.*

(*iv*) Enzymes are highly specific in nature. Their specificity is because of the presence of specific *R* group (active site) over their surface.

(*v*) The enzyme reactions are sensitive to temperature, optimum temperature for maximum enzyme activity varies between 15°C to 25°C.

(*vi*) The activity of enzyme is maximum at optimum pH (which is specific for a particular enzyme). Below and above this pH, enzyme activity greatly reduced.

(*vii*) The activator or promoters for enzyme reactions are called *co-enzyme* and for every enzyme there is a co-enzyme; if co-enzyme is separated from the enzyme, the activity of the enzyme is paralysed.

(*viii*) The addition of electrolyte deactivates enzyme. It is due to coagulation of enzymes.

(*ix*) The activity of enzymes is retarded in the presence of certain substance, called **inhibitors or poisons**. These substances interact with the active functional group or active site of the enzyme and thus, retard the catalytic activity of enzyme.

The inhibitors are of following two types

(*i*) **Competitive inhibitor** Which competes with enzyme to catalyse a reaction. e.g. penicillin. The study of competitive inhibitor is crucial to design some pharmacologically active substance.

(*ii*) **Non-competitive inhibitor** (e.g. H_2S, HCN, CS_2 etc.)

Mechanism of enzyme catalysis

Enzymes are highly specific in their action because of the presence of active sites over their surface like —COOH, —SH, —OH etc. These are the sites at which reactant molecules bind to give an intermediate with the enzymes. The intermediate then decomposed to give free enzyme and products. For any enzyme E having a reaction with substrate S and product P, the reaction is

$$E + S \underset{k_2}{\overset{k_1}{\rightleftharpoons}} ES \xrightarrow{k_3} E + P$$

The enzyme binds the substrate with a rate constant k_1 to produce the complex (ES) which may either reverts to E and S by a backward reaction with rate constant k_2 or decomposes to product P by rate constant k_3 (in which enzyme E is also regenerated).

Rate of formation of the product $= k_3\,[ES]$

ES complex formed, has maximum potential energy in this mechanism.

- In most of the cases, the substrate binds to the active site through intermolecular forces like H-bonding, dipole forces and other weak forces.

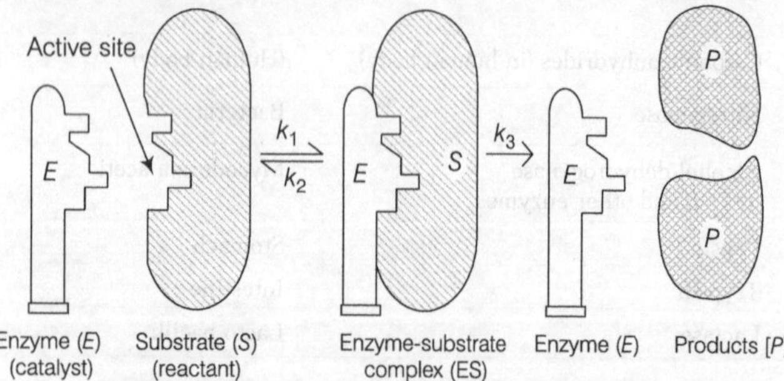

Fig. 5.6 Mechanism of enzyme-catalysed reaction

- Two theories have been proposed to explain the specificity of enzyme and mechanism of enzyme catalysis.
  (*i*) **Lock and key model** According to this theory the substrate (or reactant) fits into the active site of the enzyme just like a key fits into a lock along with its specificity.
  (*ii*) **Induced fit model** According to this theory, the enzyme molecule slightly change the shape when a substrate (reactant) fits into the active site.
  The ability of an enzyme to undergo the correct distortion also determines whether the key will fit or not. In other words, we can say a substrate induces the active site to adopt a perfect fit rather than a rigidly shaped lock and key.

Diseases caused by enzyme deficiency

Generally enzymes are secreted in the body and they catalyse the reactions occurring in the living body. They are required in very small quantity but their deficiency may lead to several diseases, e.g.

(*i*) Deficiency of enzyme phenylalanine hydroxylase leads a congenital disease, called phenyl ketone urea. In this disease, the compounds get accummulated in the body leading to severe brain damage and retardation in children.

This disease can be prevented by using a diet with low phenylalanine content.

(*ii*) Deficiency of tyrosinase causes albinism.

Note *Some enzymes are helpful in treating heart diseases, e.g. streptokinase is used to dissolve blood clot.*

CHECK POINT 5.3

1. Which of the following statements is true?
(a) Enzymes are non-proteinaceous nitrogen containing compounds
(b) Amylase convert starch into amylose
(c) Biocatalysts, catalyse the reaction occurring in living beings
(d) Enzymes are not found in plants

2. Surface of the eye is protected from bacterial infection by the enzyme
(a) carbonic anhydrate (b) urease
(c) lysozyme (d) zymase

3. Which of the following catalyst catalyses the conversion of glucose into ethanol?
(a) Zymase (b) Invertase (c) Maltase (d) Diastase

4. In the human body, enzyme that catalyses the reaction of CO_2 with H_2O is
(a) invertase (b) urease
(c) nitrogenase (d) carbonic anhydrase

5. The true statement(s) about enzymes is/are
(a) they are high molecular weight molecules
(b) they are specific but not efficient
(c) in aqueous solution, they exhibit colloidal nature
(d) Both (a) and (c)

6. Enzyme activity is highest in the temperature range of
(a) 0 - 15° C (b) 15 - 25° C
(c) 25 - 45° C (d) remain same in all

7. Co-enzymes are the substances
(a) which protect the enzyme activity
(b) in absence of which enzyme is paralysed
(c) which is always associated with an enzyme
(d) All of the above

8. In a enzyme catalysed reaction, carbon disulphide acts as
(a) activator (b) competitive inhibitor
(c) non-competitive inhibitor (d) None of these

COLLOIDAL STATE : CLASSIFICATION AND METHODS OF PREPARATION

Initially, Thomas Graham classified the soluble substances (depending upon their tendency to get diffused through parchment paper or animal or vegetable membranes) into following two categories

(*i*) Substances that get diffused through parchment paper are called **crystalloids**. e.g. all organic acids, bases, salt and organic compounds like sugar, urea etc.

(*ii*) Substances that cannot be diffused through parchment paper are called **colloids**. e.g. starch, gelatin, gums, haemoglobin.

But this classification failed because certain substance at some concentration behaves as crystalloids and at some other concentration as colloids. This shows that colloid is not a class of some substances but it is a state and any substance regardless of its nature could be converted into a colloid.

Later on soluble substances are classified into true solution, colloidal solution and suspension depending upon the size of their particle. The main points of difference among these three have been summarised in Table 5.6.

Table 5.6 Properties of true solutions, colloidal solutions and suspensions

S. No.	Properties	True solutions (molecular solutions)	Colloidal solutions (Colloidal disperison)	Suspensions (Coarse dispersion)
1.	Size	Less than 1 nm or 10^{-7} cm	Between 1 nm and 1000 nm or 10^{-7}-10^{-4} cm	Greater than 1000 nm or 10^{-4} cm
2.	Visibility	Invisible even under ultra-microscope	Visible only under ultra-microscope	Visible even by naked eye
3.	Diffusion	Diffuse readily through parchment membrane	Diffuse slowly through parchment membrane	No diffusion
4.	Filtration	Possible through filter paper and parchment membrane	Possible through filter paper but not through parchment membrane	Impossible through filter paper and parchment membrane
5.	Scattering of light	No scattering	Scattering takes place (Tyndall effect)	Do not show Tyndall effect
6.	Settling	Do not settle	Do not settle but settle when centrifuged	Settle under gravity at normal condition
7.	Nature	Homogeneous	Heterogeneous	Heterogeneous
8.	Brownian motion	Do not show	Show	May show
9.	Appearance	Clear	Generally clear or opaque	Opaque

Colloids

Finely divided particles of any substance with diameter lying within 1 to 1000 nm range dispersed in any medium constitute colloidal systems or colloids. These are heterogeneous systems consisting of following two phases

1. Dispersed phase

The substance which is distributed as the colloidal particles is called the dispersed phase. It is also called discontinuous phase or internal phase.

2. Dispersion medium

The second phase in which the colloidal particles (or the dispersed phase) is dispersed (or scattered) is called the dispersion medium. It is also called continuous phase or external phase, e.g. in the smoke (a colloid), dust acts as the dispersed phase which is dispersed in air (dispersion medium).

Classification of colloids

Colloids can be classified into three different classes depending upon different parameters as shown below

1. Depending upon the phases of dispersed phase and dispersion medium

According to this bases, colloids can be classified into following eight classes.

Table 5.7 Different colloidal systems

Dispersed phase	Dispersio n medium	Colloidal system	Examples
Liquid	Gas	Aerosol of liquids	Fogs, clouds, mists, fine insecticide sprays.
Solid	Gas	Aerosol of solids	Smoke, volcanic dust, storm.
Gas	Liquid	Foam or froth	Soap suds, lemonade froth, foam, whipped cream.
Liquid	Liquid	Emulsions	Milk, emulsified oils, medicines, bloods, mayonnaise
Solid	Liquid	Sols	Most paints, starch in water, proteins, gold sol, arsenic, sulphide sol, ink.
Gas	Solid	Solid foam	Pumice stone, styrene rubber
Liquid	Solid	Gels	Cheese, butter, boot polish.
Solid	Solid	Solid sols	Ruby glass, some gem stones.

Note *A colloidal system of gas in gas is liquible as gases are completely miscible and always form homogeneous true solution.*

2. Depending upon the interaction between dispersed phase and dispersion medium

According to this basis, colloids can be classified into following two types

(*i*) **Lyophilic colloids** (*lyo* means solvent and *philic* means *loving*). These colloids have strong interactions with the dispersion medium and are called lyophilic colloids. (hydrophilic when continuous phase is water). e.g. gum, gelatin, starch, rubber etc.

The affinity of the sol particles for the medium is due to hydrogen bonding in water.

(*ii*) **Lyophobic colloids** (*lyo* means solvent and *phobic* means hating). Substances like metals, their sulphides etc. do not pass into colloidal state readily and can be prepared by special method (discussed latter). Such substances called lyophobic colloids (hydrophobic when continuous phase is water).

Table 5.8 Distinction between lyophilic and lyophobic colloids

Property/ characteristic	Lyophilic colloids	Lyophobic colloids
Interaction	These are liquid loving colloids. Term emulsoil is used.	These are liquid hating colloids. Term suspensoid is used.
Charge	Little or no charge	Have characteristic positive or negative charge
Surface tension	Lower than that of the medium	Same as that of the medium
Viscosity	Much higher than that of the medium	Same as that of the medium
Reversibility	Reversible	Irreversible
Stability	They are much stable. On evaporation of the dispersion medium from colloidal sol, residue obtained again pass into colloidal state.	Less stable easily precipitated on adding small amount of suitable electrolyte
Action of electrolyte	Addition of smaller quantity of electrolyte has little effect	Coagulation takes place
Visibility	Particles can't be detected even under ultra-microscope	Particles can be detected under ultra-microscope
Migration on electric field	Particles may migrate in either direction or do not migrate in an electric field	Particles migrate either towards cathode or anode in an electric field.
Hydration	Extensive hydration takes place	No hydration takes place
Preparation method	By mixing with dispersion medium	No specific method

3. Depending upon the type of particles of dispersed phase

According to this basis colloids can be classified into following three types

(*i*) **Multimolecular colloids** The colloids in which an individual particle consists of an aggregate of atoms or small molecules with molecular size less than 1nm are called multimolecular colloids.

In these colloids, the atoms or molecules are held together by van der Waals' forces. These are generally lyophilic in nature. e.g.

- Gold sol contains particles of various sizes having several atoms of gold.
- Sulphur sol consists of particles containing about a thousand of S_8 molecules.

(*ii*) **Macromolecular colloids** The colloids, in which particles of dispersed phase are large enough and have size of colloidal dimension (e.g. polymers) are called macromolecular colloids.

They generally exhibit lyophobic character, e.g.

- Starch, proteins, cellulose, enzyme, gelatin etc. form natural macromolecular colloids.
- Polymers the nylon, polythene, plastics, polystyrene etc., form synthetic macromolecular colloids.

(*iii*) **Associated colloids or micelles** There are certain colloids which at low concentrations, behave as strong electrolytes but at higher concentrations exhibit colloidal properties due to the formation of aggregated particles. These are known as micelles or associated colloids.

The minimum temperature required for micelle formation is called the **Kraft temperature** (T_k) and the minimum concentration required called the **critical micelle concentration** (CMC). CMC increases with increase in concentration of substances.

Depending upon the concentration, two type of micelles are formed

- At concentration higher than CMC, they form extended parallel sheets known as **lamellar micelles** which resemble biological membranes.
- In concentrated solutions, micelles take the form of long cylinders packed in hexagonal arrays and are called **lytotropic mesomorphs**.

Surfactants, i.e. surface active agents like soaps and synthetic detergents belong to this class.

Mechanism of micelle formation

Soap (sodium salt of higher fatty acid) can be represented as $RCOO^-Na^+/K^+$, e.g. $C_{17}H_{35}COONa^+/K^+$.

In aqueous solution, it forms $RCOO^-$ and Na^+ ions. $RCOO^-$ consists of long hydrocarbon chain R (non-polar tail) which is hydrophobic (water-repelling) and the polar group COO^-(polar-ionic 'head') which is hydrophilic (water-loving).

At lower concentration of soap, $RCOO^-$ ions remain on the surface with COO^- group in water and the hydrocarbon chains R staying away from it, and remain at the surface At high concentration (i.e. at or above CMC) these ions are pulled, into the bulk of the solution and thus form 'ionic-micelles' with COO^- part remaining outward and hydrocarbon chain pointing towards the centre (Fig. 5.7).

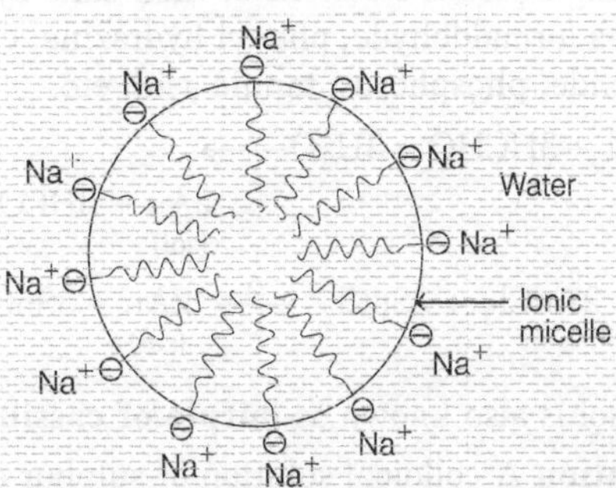

Fig. 5.7 Aggregation of $RCOO^-$ ions to form an ionic micelle

Methods of preparation of colloids

Lyophilic colloids are quite stable and can be easily prepared by just shaking the substance with dispersion medium, e.g. colloidal sols of gelatin, gum, starch etc. are prepared by this method.

Since, lyophobic colloids are generally unstable, so these cannot be prepared by just shaking with dispersion medium some specific methods are required for their preparation. Lyophobic colloidal solution can be prepared by following two types of methods.

Condensation methods

In these methods, particles of smaller size, combine to form aggregates of colloidal size. These methods may be physical or chemical.

Chemical methods

(*i*) **Double decomposition** As_2S_3 colloidal sol is obtained on passing H_2S gas into cold As_2O_3 solution.

$$\underset{\text{(Cold)}}{As_2O_3} + 3H_2S \longrightarrow \underset{\text{(Sol)}}{As_2S_3} + 3H_2O$$

(*ii*) **Hydrolysis** On adding $FeCl_3$ to a large excess of hot distilled water, red sol of hydrated ferric oxide, commonly known as ferric hydroxide sol, is obtained.

$$FeCl_3 + 3H_2O \xrightarrow{\Delta} \underset{\text{(Sol)}}{Fe(OH)_3} + 3HCl$$

Note *Boiling promotes the reaction because HCl formed is removed alongwith water vapours from the system. FeOCl (iron oxy chloride) is formed due to incomplete hydrolysis of $FeCl_3$ and acts as the stabiliser.*

$$FeCl_3 + H_2O \longrightarrow FeOCl + 2HCl$$

(*iii*) **Reduction** A number of metals, such as silver, gold and platinum have been obtained in colloidal state by reducing aqueous solutions of their salts by reagents, such as formaldehyde, phenyl hydrazine, carbon monoxide, phosphorus, tannic acid, etc.

$$2AuCl_3 + 3HCHO + 3H_2O \longrightarrow \underset{(sol)}{2Au} + 3HCOOH + 6HCl$$

$$2AuCl_3 + 3SnCl_2 \xrightarrow{\Delta} \underset{(sol)}{2Au} + 3SnCl_4$$

Gold sol is also known as **purple of cassius.**

(*iv*) **Oxidation** A colloidal solution of sulphur can be obtained by bubbling oxygen through a solution of hydrogen sulphide in water

$$2H_2S + O_2 \longrightarrow 2H_2O + 2S$$

$$2H_2S + SO_2 \longrightarrow 2H_2O + \underset{(Sol)}{3S}$$

Physical methods

Some physical methods involved in the preparation of lyophobic sols are as follows

(*i*) **Exchange of solvents** True solution of a substance can be converted into its colloidal solution by adding such a solvent in which the substance in less soluble, e.g. when water is added to the solution of sulphur in carbon disulphide colloidal sol of sulphur is obtained.

(*ii*) **Excessive cooling** This method is used to prepare colloidal solution of ice. For this, water is cooled in the presence of some organic solvents like ether by using freezing mixture. In the presence of non-polar organic solvents, water molecules aggregates to some extent to give particles of colloidal range.

Dispersion methods

In this method large particles are converted into particles of colloidal range. This can be achieved by following methods

1. Mechanical dispersion

This involves mechanical grinding of substance particles into the size of the colloidal particles using **colloid-mill** which consists of two metal revolving discs at high speed (7000 rpm) in opposite direction. The materials to be converted into colloidal sol is fed in between the two discs in the form of a wet slurry. The particles get broken to colloidal dimensions by the operating shearing force.

Note *Some sols can be prepared by mechanical dispersion in a high intensity ultrasonic generators operating at a frequency of 20 kHz and above. This method is useful in cases where the substance being dispersed is of low mechanical strength such as sulphur, graphite, resins and gypsum Colloidal solution of graphite is also called aqua dag.*

2. Electrical dispersion (Bredig's arc method)

In this method an arc is struck between two electrodes of a metal like platinum, gold, silver and copper in water containing traces of an alkali, when the metal passes into colloidal solution (Fig. 5.8).

The metal first changes into vapours (molecular state) by the heat of the spark and the vapours then condense in water to give aggregate of colloidal range. Here alkali acts as stabiliser.

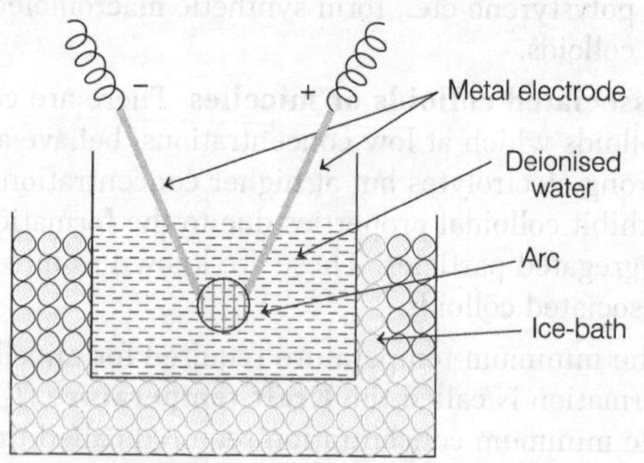

Fig. 5.8 Bredig's arc method

3. Peptisation

The process of conversion of a freshly formed precipitate (such as AgCl, $Fe(OH)_3$, $Al(OH)_3$) into a sol by the addition of a small amount of a suitable electrolyte is called **peptisation**. An electrolyte having an ion in common with the material to be dispersed is required for sol formation and called the **peptising agent**. e.g. colloid of ferric hydroxide (ferric hydroxide sol) is obtained by adding a small amount of ferric chloride to it.

$$\underset{ppt.}{Fe(OH)_3} + xFe^{3+} \longrightarrow \underset{\text{Colloidal sol}}{Fe(OH)_3 \cdot xFe^{3+}}$$

- Sugar, gum gelatin and electrolytes are the important peptising agents.
- These peptisation action is due to preferential adsorption of the common ion of the electrolyte and the precipitate to be dispersed.

Purification of colloidal sols

The colloidal solutions obtained by the above mentioned techniques remain associated with some impurities of electrolytes. These impurities makes the sol unstable. Thus, their removal is necessary. This process is called purification of colloidal sols.

The sols are purified by following techniques

1. Dialysis

The process of separation of an electrolyte (crystalloid) from a colloid by means of diffusion or filtration using a fine membrane is called dialysis and the apparatus used for this purpose is called **hemodialysis.**

In this method, the sol is kept in a bag made up of parchment paper or cellophane and the bag is suspended in continously flowing water. Small molecules and ions can pass through the membrane but the sol is retained.

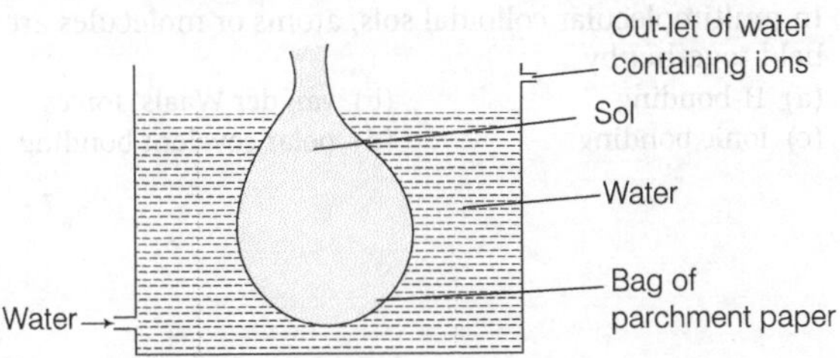

Fig. 5.9 Process of dialysis

The process of dialysis is similar to osmosis except that both solvent molecules and small solute particles can pass through the semi-permeable dialysis membrane. Only large colloidal particles such as cells and large molecules such as proteins cannot pass.

Artificial kidney machines

The most important medical use of dialysis is in artificial kidney machines, where **haemodialysis** is used to cleanse the blood of patients whose kidneys have malfunctioned. Blood is diverted from the body and pumped through a cellophane dialysis tube suspended in a solution formulated to contain many of the same components as blood plasma. These substances (glucose, NaCl, $NaHCO_3$ and KCl) have the same concentrations in the dialysis solutions as they do in blood so that they have no net passage through the cellophane membrane.

2. Electrodialysis

When an electric field is applied during dialysis, we refer it as **electrodialysis.** In this process charged impurities of small size pass through the cellophane bag under the influence of electric field, while pure sol remains inside the cellophane bag.

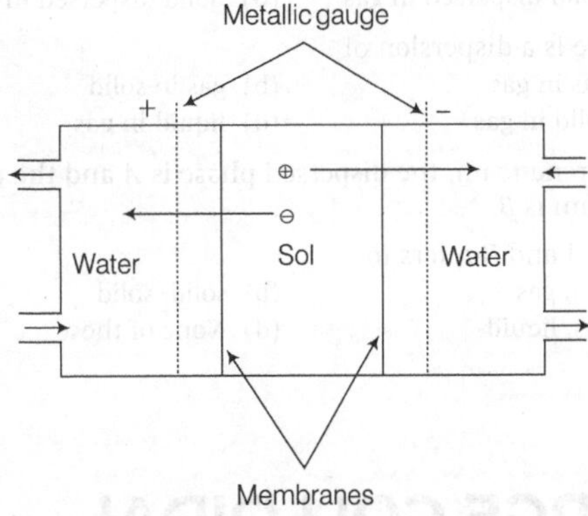

Fig. 5.10 Electrodialysis

3. Ultrafiltration

Colloidal sols can also be purified by carrying out filtration through special type of graded filters called **ultra-filters** and the method is called ultra-filtration.

The ultra-filters are prepared by soaking the filter paper in a **colloidion solution** (4% solution of nitro cellulose in a mixture of alcohol and ether), followed by hardening using formaldehyde and then finally drying.

These ultra-filter papers allow only the electrolytes to pass through under the influence of pressure or suction pump.

4. Ultra centrifugation

In this method, colloidal sol is taken in a tube placed in an ultra-centrifuge. Under the influence of high speed, the colloidal particles settle down at the bottom while impurities remain in solution (centrifugate). Sol is regenerated by mixing the settled colloidal partial with suitable dispersing medium.

CHECK POINT 5.4

1. Consider the following systems.

I. Air II. Sea water
III. Gasoline IV. Wine
V. Pearl VI. Blood

Select true solutions.
(a) I, II, III and IV
(b) II, III, IV and V
(c) I, IV, V and VI
(d) II, IV and VI

2. The diameter of colloidal particle is of the order
(a) 10^{-3} m (b) 10^{-6} m
(c) 10^{-15} m (d) 10^{-7} m

3. If the size of a particle in a solution was found to be 180 nm, then the solution would be
(a) a true solution (b) a colloidal sol
(c) a suspension (d) Any of these

4. Which type(s) of solutions is/are not filtrable through filter paper and parchment membrane both?
(a) True solution and colloidal sol
(b) Colloidal sol and suspension
(c) True solution and suspension
(d) Suspension only

5. Mayonnaise represents
(a) a sol (b) an aerosol
(c) a foam (d) an emulsion

6. Which of the following colloidal systems, fog is an example?
(a) Liquid dispersed in gas (b) Gas dispersed in gas
(c) Solid dispersed in gas (d) Solid dispersed in liquid

7. Smoke is a dispersion of
(a) gas in gas (b) gas in solid
(c) solid in gas (d) liquid in gas

8. In styrene foam, the dispersed phase is *A* and the dispersion medium is *B*.
Here, *A* and *B* refers to
(a) gas, gas (b) solid, solid
(c) gas, liquid (d) None of these

9. Which of the following statements is false for hydrophilic sols?
(a) They do not require electrolytes for stability
(b) Their coagulation is reversible
(c) Their viscosity is of the order of that of water
(d) Their surface tension is usually lower than that of dispersion medium

10. In multimolecular colloidal sols, atoms or molecules are held together by
(a) H-bonding (b) van der Waals' forces
(c) ionic bonding (d) polar covalent bonding

CHARGE COLLOIDAL PARTICLES

Colloidal particle are either positively charged or negatively charged. Examples of positively and negatively charged sols are tabulated below

Table 5.9 Some common sols with type of charge

Positively charged sol	Negatively charged sol
Oxides like TiO_2 sol	Metals like Cu, Au, Ag, sols
Metallic hydroxides like $Cr(OH)_3$, $Al(OH)_3$, $Fe(OH)_3$	Metallic sulphides like As_2S_3, CdS and Sb_2S_3 sols
Basic dye stuffs: methylene blue	Acid dye stuffs : eosin, congo red
Haemoglobin	Sols obtained from starch, charcoal, gum

The charge of colloidal particles is due to preferential adsorption of ions (either positive or negative) on their surface, i.e. if two or more ions are present in the dispersion medium, then that ion will be adsorbed preferentially which is in excess and common to the colloidal particles.

e.g. (*i*) When $AgNO_3$ solution is added to KI solution AgI colloidal sol is obtained.

$$AgNO_3 + KI \longrightarrow \underset{\text{Sol}}{AgI} + KNO_3$$

There are two possibilities

If KI is in excess, (i.e. acts as dispersion medium) the obtained AgI adsorbs iodide ion (I^-) from the dispersion medium and forms a negatively charged sol as shown below

$$AgI + KI \longrightarrow \underset{\text{Fixed part}}{[AgI]I^-} + \underset{\text{Diffused part}}{K^+}$$

If $AgNO_3$ is in excess (i.e. acts as dispersion medium), the obtained AgI adsorb Ag^+ ions from the dispersion medium leading to the formation of a positively charged sol as shown below

$$AgI + AgNO_3 \longrightarrow \underset{\text{Fixed part}}{[AgI]Ag^+} : \underset{\text{Diffused part}}{NO_3^-}$$

(*ii*) When As_2O_3 is saturated with H_2S the As_2S_3 colloidal sol is obtained.

$$As_2O_3 + 3H_2S \longrightarrow \underset{\text{Sol}}{As_2S_3} + 3H_2O$$

As_2S_3 adsorbs S^{2-} ion (common between H_2S and As_2S_3) and thus, a negatively charged sol is formed.

$$As_2S_3 + H_2S \longrightarrow \underset{\text{Fixed part}}{[As_2S_3]S^{2-}} : \underset{\text{Diffused part}}{2H^+}$$

(*iii*) If ferric chloride is added to excess of hot water, a positively charged sol of ferric oxide is obtained due to preferential adsorption of Fe^{3+} ions.

$$FeCl_3 + 3H_2O \rightleftharpoons Fe(OH)_3 + 3HCl$$

$$Fe(OH)_3 + FeCl_3 \longrightarrow Fe_2O_3 \cdot xH_2O : Fe^{3+} + \underset{\text{Diffused Part}}{3Cl^-}$$

However, if ferric chloride is added to excess of sodium hydroxide, a negatively charged sol is obtained due to preferential adsorption of OH^- ions.

$$FeCl_3 + 3NaOH \longrightarrow Fe_2O_3 \cdot xH_2O + 3NaCl$$

$$Fe_2O_3 \cdot xH_2O + NaOH \longrightarrow \underset{\text{Fixed part}}{[Fe_2O_3 \cdot xH_2O]\,OH^-} + \underset{\text{Diffused part}}{Na^+}$$

(*iv*) In acidic medium, SnO_2 forms positively charged colloidal sol due to adsorption of Sn^{4+} formed.

$$SnO_2 + 4H^+ \longrightarrow Sn^{4+} + 2H_2O$$

$$SnO_2 + Sn^{4+} \longrightarrow [SnO_2]\,Sn^{4+}$$

In alkaline medium, SnO_2 forms negatively charged colloidal sol due to adsorption of SnO_3^{2-} formed.

$$SnO_2 + 2OH^- \longrightarrow SnO_3^{2-} + H_2O$$

$$SnO_2 + SnO_3^{2-} \longrightarrow [SnO_2]\,SnO_3^{2-}$$

PROPERTIES OF COLLOIDS

Colloids exhibit properties different from that of the true solutions and hence, it is important to study their properties separately. Some of the important properties of colloids are as follows

Physical properties

The important physical properties of colloids are

(i) **Heterogeneous nature** Colloidal sols are heterogeneous in nature.

(ii) **Filterability** They can pass through (ordinary) filter paper, but not through ultrafilter.

(iii) **Colour** Colloidal particles exhibit different colours depending upon the wavelength of light scattered by the dispersed phase (colloidal particles), which in turn depends upon the size and nature of particles, e.g. colour of finest gold sol is red and as the size of gold particle increase, it appears purple and finally golden.

(iv) **Colligative properties** Due to formation of associated molecules (i.e. reduction in number of particles) observed values of colligative properties like relative decrease in vapour pressure, elevation in boiling point, depression in freezing point, osmotic pressure are smaller than expected for colloids.

Mechanical property : Brownian movement

Brownian movement is motion of colloidal particles in *zig-zag* paths. It can be viewed by using powerful ultra microscope.

This type of erratic motion is due to the constant collision of moving molecules of the dispersion medium with colloidal particles. During collision, molecules of dispersion medium impart momentum to colloidal particles.

An English botanist, **Robert Brown**, observed this phenomenon, in 1827, with pollen grains suspended in water. This motion does not depend upon the nature of colloid but depends upon the size of particles and viscosity of the solution. Smaller the size, lesser is the viscosity and faster is the motion.

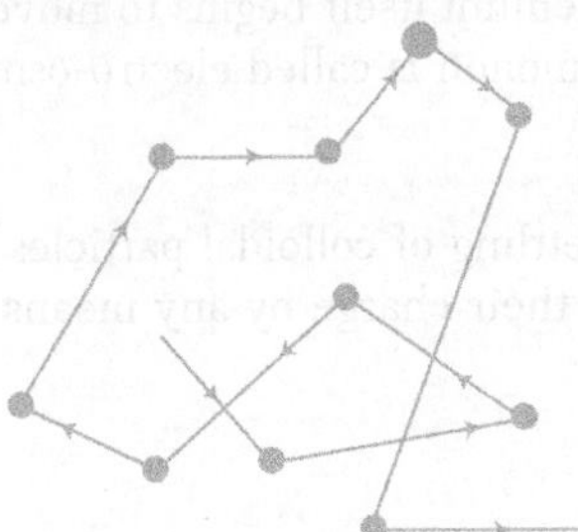

Fig. 5.11 Brownian movement

Brownian movement is not observed in ordinary suspensions or true solution. Because in case of suspension the size of each particle in this case is so large (2000 Å) that the bombardment by molecules of the dispersion medium produces little effect on them. In case of true solution this size is very small [in the range of 1 Å to 10 Å].

Advantages of brownian motion or movement

(i) This offers a visible proof of the random kinetic motion of molecules in a liquid, as postulated by kinetic theory.

(ii) It counters the force of gravity acting on colloidal particles and thus, helps in providing stability to colloidal sol by not allowing them to settle down.

Optical property : Tyndall effect

When light passes through a sol, its path becomes visible because of the scattering of light by sol particles. This phenomenon of the scattering of light by the particles is called **Tyndall effect** and illuminated beam or cone formed by the scattering of light by the sol particles is referred as **Tyndall beam** or **Tyndall cone.**

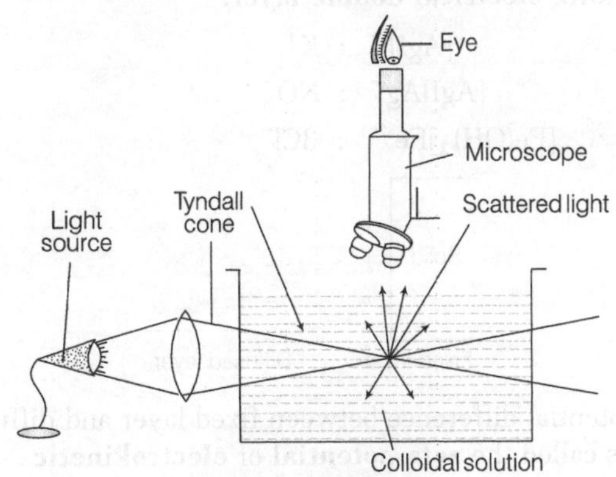

Fig. 5.12 Tyndall effect

(i) This effect was first observed by Faraday and later studied by Tyndall in detail.

(ii) To show this effect, a solution must satisfy following two conditions.

- The diameter of colloidal particles (or dispersed particles) is not much smaller than the wavelength of light used.
- There should be large difference between the refractive indices of dispersed phase and dispersion medium.

(iii) In lyophobic colloids, the difference is appreciable and, therefore, the Tyndall effect is well defined. But in lyophilic sols, the difference is very small and the Tyndall effect is not well defined.

(*iv*) Examples of Tyndall effect are

- blue colour of sky and sea water.
- visibility of tails of comets and twinkling of stars.
- visibility of a beam of light passing through a slit in dark room.
- visibility of projector path and circus light.

Note *In a true solution, there are no particles of sufficiently large diameter to scatter light and hence, no Tyndall effect is observed.*
In 1903, Zsigmondy set up an apparatus, called ultra microscope, using Tyndall effect.

Electrical properties

Presence of charge

Colloidal particles carry an electric charge and the dispersion medium has an equal but opposite equal charge thus, the system as a whole being electrically neutral.

The charged sol attracts the counter ion from the dispersion medium, forming a secondary layer and is called the **fixed layer.** While the second layer is mobile and called the **diffused layer.** This electrical double layer of opposite charge around the sol particles is called **Helmholtz electrical double layer.**

$[AgI]I^- \; : \; K^+$

$[AgI]Ag^+ \; : \; NO_3^-$

$[Fe(OH)_3]Fe^{3+} \; : \; 3Cl^-$

Solid + + + + + : − − − − −

Fixed layer   Diffused layer

The potential difference between fixed layer and diffused layer is called the **zeta potential** or **electrokinetic potential** and is given by

$$Z = \frac{4\pi\eta u}{D}$$

where, η is coefficient of viscosity, D is dielectric constant of the medium and u is velocity of the colloidal particles when an electric field is applied.

The presence of equal and similar charges on colloidal particles is largely responsible in giving stability to the system because the mutual forces of repulsion between similarly charged particles prevent them from coalescing and coagulating when they come closer to one another.

Some of the electrical properties of colloids are as follows

Electrophoresis

Colloidal particles possess some type of electric charge. Depending on the charge they move towards a positive or negative electrode when an electric field is applied to colloidal dispersion. A positively charged particle (i.e. cation) moves towards the cathode while a negatively charged particle (i.e. anion) moves towards anode.

(*i*) The process of movement of colloidal particles towards an electrode under the influence of an applied electric field is called electrophoresis or **cataphoresis.**

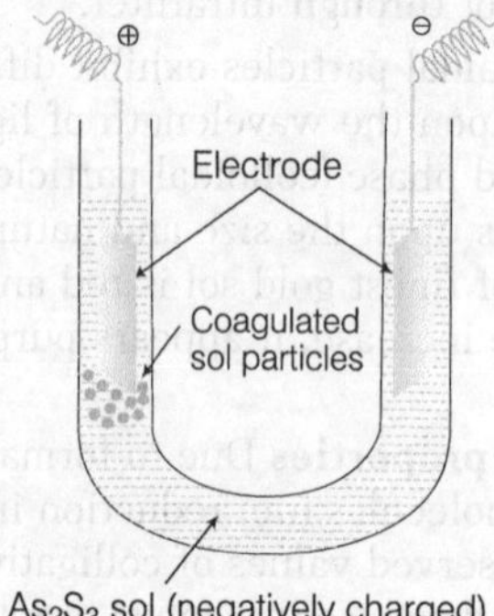

Fig. 5.13 Electrophoresis

(*ii*) The distance travelled by the colloidal particles in one second under a potential gradient of one volt per cm is called **electrophoretic mobility** of colloidal particles.

Sedimentation potential or Dorn effect

- It is reverse of electrophoresis and set up when a particle is forced to move in a resting liquid.
- It was discovered by Dorn so also called Dorn effect.

(*iii*) The direction of migration of ion indicates the type of charge present on colloidal particles.

Electro-osmosis

When electrophoresis of dispersed particles in a colloidal system is prevented by suitable means, it is observed that the dispersion medium itself begins to move in an electric field. This phenomenon is called electro-osmosis.

Coagulation

The process of settling of colloidal particles due to the neutralisation of their charge by any means is called coagulation.

It can be achieved by following means

- By electrophoresis
- By mixing two oppositely charged sol
- By boiling
- By persistent dialysis
- By addition of electrolytes

Out of these the last one is most common and is governed by **Hardy Schulze rule.**

Hardy–Schulze rule

According to this rule "greater the valency of the coagulating ion or flocculating ion (oppositely charged ion to that of colloidal particles) added, the greater is its power to cause coagulation".

e.g.

To coagulate a positively charged sol, the order of coagulating power of negative ion is

$$I^- < SO_4^{2-} < PO_4^{3-} < [Fe(CN)_6]^{4-}$$

Similarly, to coagulate a negatively charged sol, the order of coagulating power of positive ions is

$$Ag^+ < Pb^{2+} < Fe^{3+} < Si^{4+}$$

or $$Na^+ < Mg^{2+} < Al^{3+}$$

Flocculation value

The minimum concentration of an electrolyte required to cause coagulation or flocculation of a sol is called its flocculation value. This is usually expressed in terms of **millimoles per litre**.

In general, smaller the flocculation or coagulation value, higher is the coagulating power and *vice-versa*, i.e.

$$\text{Flocculating value} \propto \frac{1}{\text{coagulating power}}$$

Protective colloids and protection

Lyophilic colloidal sols are much more stable than lyophobic colloidal sols. This is due to the extensive solvation of lyophilic colloidal sols, which forms a protective layer outside it. Thus, prevents it from forming associated colloids. Lyophilic colloidal sols also protect lyophobic colloidal sols from precipitation by the action of electrolytes.

This is due to formation of a protective layer by lyophilic sols outside lyophobic sols. This process is called **protection.**

The lyophilic colloid used is called **protective colloid.** e.g. gelatin (lyophilic) protects gold sol (lyophobic) from coagulation on the addition of sodium chloride solution. The protective colloids play a key role in the stabilisation of non-aqueous dispersion like printing inks.

The protective powers of different colloidal sols are measured in terms of **'gold number'** as discussed further.

Gold number

- It was given by Zsigmondy.
- Gold number is defined as the amount of protective sol in milligrams that prevents the coagulation of 10 mL of a given gold sol on adding 1 mL of 10 per cent solution of sodium chloride. Thus, smaller the gold number of a lyophilic sol, the greater is the protective power.

$$\text{Protective power} \propto \frac{1}{\text{gold number}}$$

Gold number of some lyophilic colloidal sols are given in Table 5.10.

Table 5.10 Gold numbers

Lyophilic sols	Gold number
Gelatin	0.005-0.01
Casein	0.01
Haemoglobin	0.03
Gum arabic	0.15
Sodium oleate	0.4
Gum tragacanth	2
Potato starch	25

Example 5.2 0.025 g *of starch sol is required to prevent coagulation of* 10 mL *gold sol when* 1 mL *of* 10% NaCl *solution is present. What is gold number of starch sol?*

Sol. Starch sol here is used as protective sol. By definition of gold number, it is the amount of protective sol in milligrams that prevents coagulation of 10 mL of gold sol on the addition of 1 mL of 10% NaCl solution.

For such case amount of starch required = 0.025 g = 25 mg

Hence, gold number of starch = 25

Congo Rubin number

It is defined as the amount of protective colloid in milligram which prevents colour change in 100 mL of 0.01% Congo Rubin due to which 0.16 g equivalent of KCl is added.

CHECK POINT 5.5

1. Compared to common colloidal sols, micelles have
(a) higher colligative properties
(b) lower colligative properties
(c) same colligative properties
(d) None of the above

2. Which of the following is/are the kinetic phenomenon?
(a) Brownian motion (b) Tyndall effect
(c) Both (a) and (b) (d) None of these

3. Tyndall effect is not observed in
(a) sugar solution
(b) gold solution
(c) suspension
(d) emulsion

4. By passing a beam of light through a colloidal solution, it is
(a) refracted (b) adsorbed
(c) scattered (d) reflected

5. Intensity of the scattered light depends on the difference between .... of dispersed phase and dispersion medium.
(a) densities (b) refractive indices
(c) viscosity (d) surface tension

6. The electrical charge on a colloidal particle is observed by
(a) ultramicroscope (b) scattering
(c) Brownian movement (d) electrophoresis

7. Electro-osmosis is observed when
(a) dispersion medium begins to move in an electric field
(b) dispersed phase begins to move in an electric field
(c) Both (a) and (b)
(d) None of the above

8. The colloidal sol of As_2S_3 prefers to adsorb
(a) NO_3^- (b) K^+ (c) S^{2-} (d) H^+

9. Potential difference of the electrical double layer formed in a colloidal sol is called
(a) EMF (b) Zeta potential
(c) Brownian potential (d) Nernst potential

10. Which of the following properties is/are related to colloidal sols?
(a) Tyndall effect
(b) Brownian motion
(c) Maxwell distribution of molecules
(d) Both (a) and (b)

EMULSIONS

If a pair of immiscible liquids such as olive oil and water are mixed and shaken vigorously, it will result a dispersion of very fine droplets of one liquid in the other. Such a dispersion is termed as emulsion.

(*i*) One of the components in emulsion is water and the other is an oily substance immiscible in water.

(*ii*) Emulsion droplets are bigger than sol particles ($\approx 10^{-6}$ m) and can be seen under an ordinary microscope or sometimes even with a magnifying glass.

(*iii*) Emulsions resemble lyophobic sols in some of the properties.

Types of emulsions

Depending on the nature of the dispersed phase, the emulsions are classified as

(*i*) **Oil-in-water emulsions** (O/W) This type of emulsion is formed when oil (dispersed phase) is dispersed in water (dispersion medium). Fig 5.14 (I) Milk and vanishing cream are O/W type emulsions. In milk liquid fat globules are dispersed in water.

(*ii*) **Water-in-oil emulsion** (W/O) This type of emulsion is formed when water is dispersed in oil (dispersion medium). These are also called oily emulsion. Fig 5.14 (II)

Cold cream and butter are W/O type emulsions.

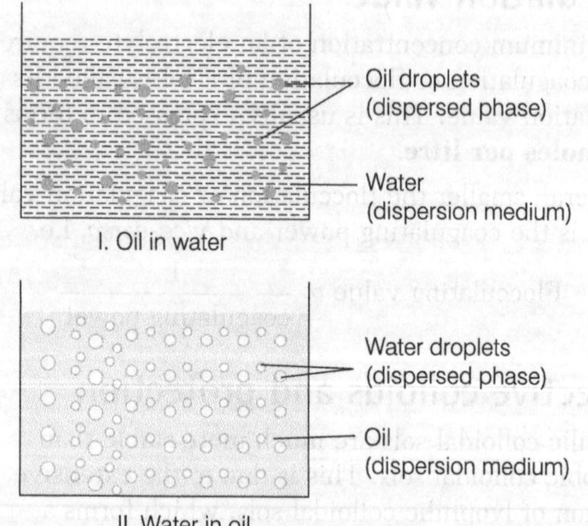

Fig. 5.14 Types of emulsions

Identification of the type of emulsions

The two types may be identified by

(*i*) **Dilution test** An emulsion can be diluted with any amount of the dispersion medium whereas the addition of dispersed phase forms a separate layer.

(*ii*) **Dye test** An oil-soluble dye is shaken with the given emulsion to be identified. If the whole background is coloured, emulsion is (W/O) type. If coloured drops are seen on looking under microscope, the emulsion is of (O/W) type.

Physical tests

(*i*) Surface tension of water is greater than that of the oil hence, formation of drops in (O/W) emulsion is easier.

(*ii*) Oily emulsions are more *viscous* than the aqueous emulsions.

(*iii*) Electrical conductivity of aqueous emulsion is higher than that of oily emulsion and increases if any electrolyte (say KCl) is added due to its dissolution into aqueous dispersion medium.

Characteristics of emulsion

Emulsions show all the properties of sols. Their important characteristics are as follows

(*i*) They can be diluted with liquid forming the dispersion medium in the emulsion.

(*ii*) Their particles size is larger than those of other size. It ranges from 1000 Å to 10000 Å.

(*iii*) They scatter light and thus, exhibit Tyndall effect.

(*iv*) Brownian motion is also observed in emulsions where size of the particle is too near to the limit of 10^{-6} m.

(*v*) Stability of an emulsion is determined by the toughness of the emulsifier film (which helps to prevent the coalescence of droplets of the dispersed phase) and by the electric charge on the emulsoid particles.

(*vi*) These can be coagulated by adding an electrolyte which would destroy the charge or by adding a dehydrating agent or alcohol [for (W/O) emulsion only].

(*vii*) de-emulsification of an emulsion can be done
- by addition of a substance which would destroy the emulsifier.
- by heating, freezing and centrifuging.

Emulsifiers

Emulsions are generally prepared by vigorously agitating a mixture of the relevant oil and water. The emulsions obtained by simple mechanical stirring are unstable and the two components (oil and water) tend to separate out on standing. To obtain a stable emulsion, a suitable stabilising substance, called emulsifier is added.

Substances which can act as emulsifiers are soaps, detergents, long-chain sulphonic acids, lyophilic colloids like gelatin, albumin, casein etc. In milk, casein acts as the emulsifier.

Most emulsifiers are long-chain hydrocarbon derivatives containing a polar end group such as, sulphonic ($—SO_3^{\ominus}$), carboxylic ($—CO_2^{\ominus}$), etc. The polar end group is hydrophilic and thus, water-soluble. The hydrocarbon chain is hydrophobic and thus, oil-soluble. When a suitable emulsifier is added to an oil-water system, the molecules of the emulsifier arrange themselves in such a way that the polar end groups dip in water and the hydrocarbon chains dip into the oil (Fig 5.15). Thus, the emulsifier forms an adsorbed film around the dispersed droplets and prevents their coagulation (Fig.5.16).

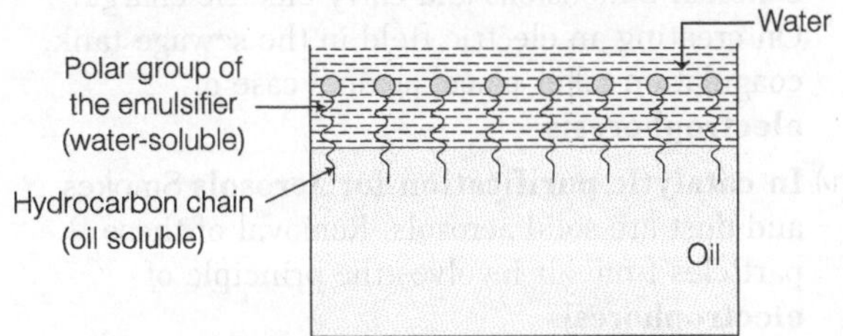

Fig. 5.15 Orientation of an emulsifier at water-oil interface

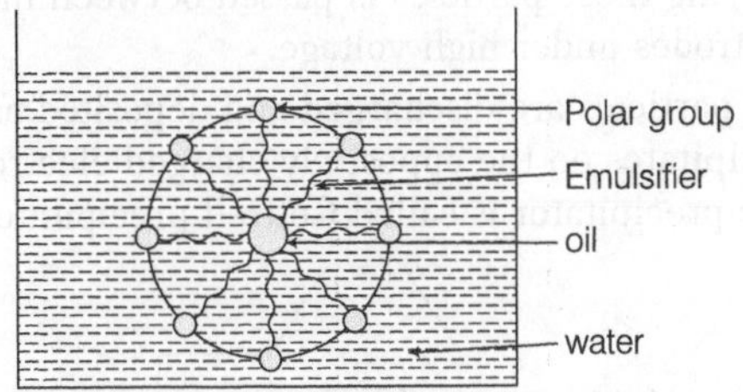

Fig. 5.16 Emulsification of an oil droplet

Soaps and detergents also lower the interfacial surface tension between oil and water, hence, there is a lesser tendency of coagulation so as to reduce surface area.

APPLICATIONS OF COLLOIDAL SOLS

Colloids are important in every aspect of life. Some of their important applications are as follows

(*i*) **In water purification** Water carrying impurities is purified by addition of potash alum. Al^{3+} of the potash alum causes coagulation of negatively charged turbidity (or clay particles).

(*ii*) **In industries** So many industrial goods, e.g. detergents, varnishes, paints, gums, glues, etc., are colloidal in nature. Rubber plating (used for insulation) and chrome tanning of leather are made by the application of **electrophoresis**.
- Rubber is obtained from latex (a colloidal suspension of rubber particles in water) by coagulating it through boiling or adding some electrolyte.

- Animal hides or skins are positively charged colloids thus, mutual coagulation takes place when negatively charged tannin sol is added. This makes skin a hardened leather.

(*iii*) **In artificial rains and artificial smoke screens** Artificial rain can be made by throwing electrified sand or silver halide over clouds from aeroplanes. Similarly, artificial smoke screens can be made by spraying titanium oxide in air.

(*iv*) **In sewage treatment** Sewage water consists of particles of dirt, rubbish, mud, etc., which are of colloidal dimensions and carry electric charge. On creating an electric field in the sewage tank, coagulation takes place, again a case of **electrophoresis.**

(*v*) **In catalytic purification for aerosols** Smokes and dust are solid aerosols. Removal of these particles from air involves the principle of **electrophoresis.**

The air from a furnace or an industrial plant carrying these particles is passed between metal electrodes under high voltage.

The particles are discharged and deposited as precipitates on the oppositely charged electrodes. This precipitator is called Cottrell precipitator.

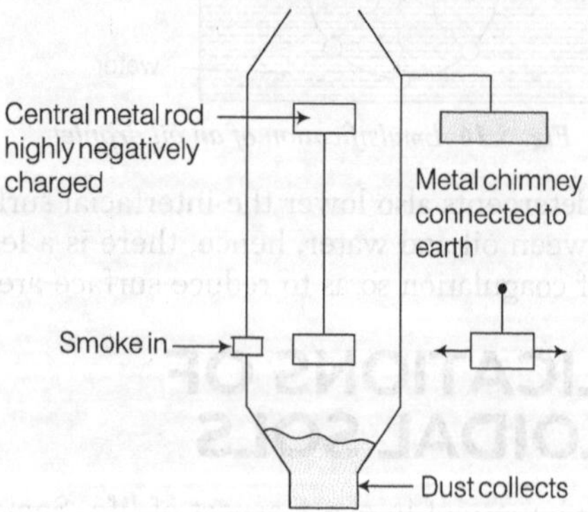

Fig. 5.17 Cottrell precipitator

(*vi*) **In coagulation of blood** Blood is negatively charged sol and thus, bleeding from a cut is stopped by coagulating it using alum or ferric chloride which provides positive Al^{3+} or Fe^{2+} ion for coagulation or clotting of blood.

(*vii*) **Delta formation** The deltas at the river beds are formed by the precipitation of the charged clay particles (negatively charged) carried in suspension in the river water, by the action of salts (Na^+, K^+, Mg^{2+} ion) present in sea-water.

(*viii*) **In food products** Many of our foods are colloidal in nature. Milk is an emulsion of fat dispersed in water. It is stabilised by casein which itself is a lyophilic colloid and being a protein is a nutrient of great value.

(*ix*) **In medicinal and pharmaceutical industries** A number of medicinal and pharmaceutical preparations are colloidal in nature, which are more effective and are easily assimilated, e.g.

- Colloidal gold and calcium are used to increase the vitality of the human system.
- Colloidal antimony is used to treat Kala azar.
- Argyrol and protargyrol are the colloidal solution of silver and are used as eye lotions.
- Colloidal gold is used in intramuscular injection.
- Colloidal sulphur is used as disinfectant.
- Milk of magnesia is used for stomach disorder.

(*x*) **In photography** Photographic plates are prepared by coated an emulsion of light sensitive AgBr in gelatin over glass plates or celluloid films. Here, gelatin acts as protective colloids.

(*xi*) **Other applications** Cleansing action of surfactants (soap and detergents), blood, colour of sky, tail of comets are also the examples of application of colloids.

Applications of emulsion

(*i*) Cleansing action of soap is due to formation of emulsions. Soaps and detergents emulsify the grease along with the adhering dirt and carry them away in the excess water.

(*ii*) For concentration of sulphide ore the finely powdered ore is treated with an oil. Oil forms emulsion with the ore particles. When air is bubbled into the mixture, emulsion containing the particles of the mineral are carried to the surface.

(*iii*) Several oily drugs are prepared in the form of emulsions to facilitate their adsorption in the body. Many of the lotions, creams and ointments are emulsions of oil-in-water or water-in-oil type.

(*iv*) Petroleum products get emulsified in the oil-well, causing problems. However, they can be demulsified into constituent liquids by physical methods such as centrifugation, freezing, filtration, electrostatic precipitation, etc.

CHECK POINT 5.6

1. Milk is an emulsion of fat dispersed in water. It is stabilised by
   (a) casein – a lyophilic colloidal sol
   (b) casein – a lyophobic colloidal sol
   (c) lactose – a lyophilic colloidal sol
   (d) lactose – a lyophobic colloidal sol
2. Which of the following is not the correct matching of emulsions?
   (a) Milk : O/W
   (b) Cold cream : W/O
   (c) Butter : O/W
   (d) Vanishing cream : O/W
3. An emulsion can be diluted with H_2O (dispersion medium), then it is
   (a) O/W type
   (b) W/O type
   (c) Both (a) and (b)
   (d) None of the above
4. Oil soluble dye is mixed with emulsion and emulsion remains colourless. Then it is
   (a) O-in-W (b) W-in-O
   (c) O-in-O (d) W-in-W
5. Oil soluble dye is mixed with water-in-oil emulsion then
   (a) dispersion medium is coloured
   (b) dispersed phase is coloured
   (c) Both (a) and (b)
   (d) None of the above
6. An emulsifier is an agent which
   (a) accelerates the dispersion
   (b) tends to homogenise an emulsion
   (c) destablises an emulsion
   (d) aids the flocculation of an emulsion
7. Select the correct statement(s).
   (a) Emulsifiers stabilise the emulsion
   (b) Soaps, detergents, long-chain sulphonic acids and lyophilic colloids are emulsifiers
   (c) Cleansing action of soap is due to formation of emulsions (Micelle)
   (d) All of the above are correct statements
8. Water carrying impurities is purified by addition of potash alum. Al^{3+} of the potash alum causes
   (a) peptisation of negatively charged turbidity
   (b) coagulation of negatively charged turbidity
   (c) peptisation of positively charged turbidity
   (d) coagulation of positively charged turbidity
9. Rubber plating and chrome tanning of leather are based on
   (a) electro-osmosis (b) Tyndall effect
   (c) electrophoresis (d) Brownian motion
10. Artificial rain is caused by spray of
    (a) electrified sand or salt
    (b) neutral salt or sand
    (c) negatively charged sand or salt
    (d) positively charged sand or salt

Chapter Exercises

(A) Topical exercise

I. Adsorption and types of adsorption

1 Which of the following interface cannot be obtained? **[NCERT Exemplar]**

(a) Liquid-liquid (b) Solid-liquid
(c) Liquid-gas (d) Gas-gas

2 'Silica-garden' is developed by setting of coloured ions in the void of silicates. It is a case of

(a) adsorption (b) absorption
(c) Both (a) and (b) (d) None of these

3 Which of the following is an example of absorption? **[NCERT Exemplar]**

(a) Water on silica gel
(b) Water on calcium chloride
(c) Hydrogen on finely divided nickel
(d) Oxygen on metal surface

4 Which one of the following is most readily adsorbed on activated charcoal?

(a) N_2 (b) H_2
(c) SO_2 (d) O_2

5 Energy of activation of forward and backward reaction are equal in cases (numerical values) where

(a) $\Delta H = 0$
(b) no catalyst present
(c) $\Delta S = 0$
(d) stoichiometry is the mechanism

6 For the adsorption, generally

(a) $\Delta H = +ve, \Delta S = +ve, \Delta G = -ve$
(b) $\Delta H = +ve, \Delta S = -ve, \Delta G = -ve$
(c) $\Delta H = -ve, \Delta S = -ve, \Delta G = -ve$
(d) $\Delta H = -ve, \Delta S = -ve, \Delta G = +ve$

7 Which of the following gases is adsorbed to maximum extent on the given surface?

(a) NH_3 (b) H_2 (c) N_2 (d) O_2

8 The process of adsorption of hydrogen over palladium surface is called

(a) sorption (b) occlusion
(c) desorption (d) None of these

9 Surface area per gram of the adsorbent is called

(a) molar surface area
(b) normal surface area
(c) specific surface area
(d) equivalent surface area

10 Select the correct statement(s).

(a) Residual force-field on a surface is responsible for adsorption.
(b) Rate of adsorption is very rapid in the beginning but decreases until equilibrium is attained.
(c) Desorption and adsorption are two opposite terms.
(d) All of the above are correct statements.

11 Multimolecular layers are not formed in

(a) physisorption (b) chemisorption
(c) Both (a) and (b) (d) None of these

12 Which of the following is the variation of physical adsorption with temperature?

(a)
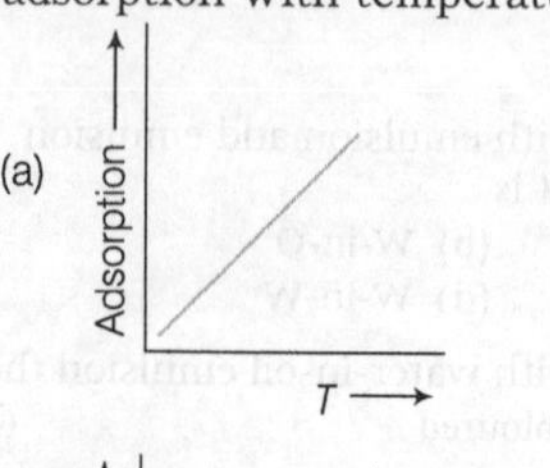

(b)
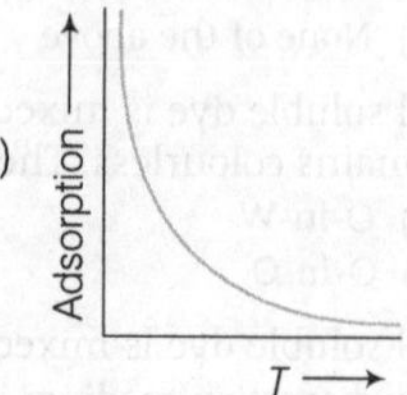

(c)
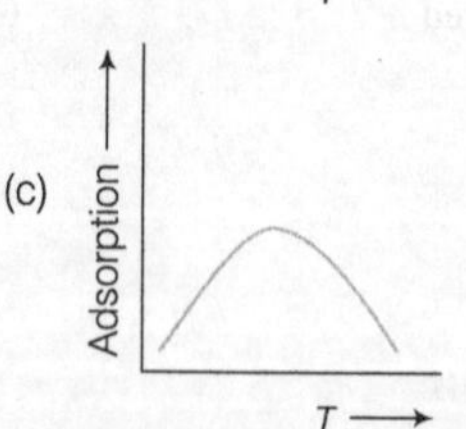

(d)
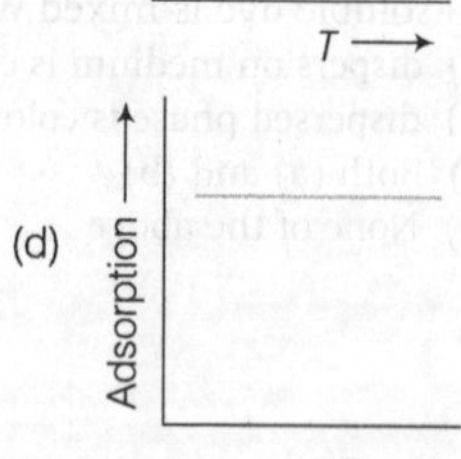

13 **Statement** (I) The adsorption of a vapour on a clean surface is a spontaneous process.

Statement (II) Change in the entropy of the process is highly positive.

(a) Both statements I and II are true
(b) Only statement I is true
(c) Only statement II is true
(d) Neither statement I nor statement II are true

14 Soaking of water by a sponge is an example of

(a) simple adsorption (b) physical adsorption
(c) chemisorption (d) absorption

15 The physical adsorption of gases on the solid surface is due to

(a) covalent bond (b) hydrogen bond
(c) ionic bond (d) van der Waals' forces

16 Extent of physisorption of a gas increases with

(a) increase in temperature **[NCERT Exemplar]**
(b) decrease in temperature
(c) decrease in surface area of adsorbent
(d) decrease in strength of van der Waals' forces

17 There is desorption of physical adsorption when
(a) temperature is increased
(b) temperature is decreased
(c) pressure is increased
(d) concentration is increased

18 Which of the following is not a favourable condition for physical adsorption? **[NCERT Exemplar]**
(a) High pressure
(b) Negative ΔH
(c) Higher critical temperature of adsorbent
(d) High temperature

19 Physical adsorption of a gaseous species may change to chemical adsorption with
(a) decrease in temperature
(b) increase in temperature
(c) increase in surface of adsorbent
(d) decrease in surface area of adsorbent

20 The statement not applicable to chemisorption is that it is
(a) highly specific
(b) independent of temperature
(c) irreversible
(d) slow comparative to physio

21 Adsorption is
(a) an exothermic process hence, increase in temperature decreases adsorption in cases where van der Waals' forces exist between adsorbate and adsorbent
(b) an endothermic process hence, increase in temperature increases adsorption
(c) an exothermic process hence, increase in temperature increases adsorption
(d) None of the above

22 For the adsorption of a gas on a solid, the plot of log (x/m) *versus* log p is linear with slope equal to
(a) k (b) $\log k$
(c) n (d) $1/n$

23 At the equilibrium position, in the process of adsorption
(a) $\Delta H > 0$ (b) $\Delta H = T\Delta S$
(c) $\Delta H > T\Delta S$ (d) $\Delta H < T\Delta S$

24 In Freundlich adsorption isotherm intercept is

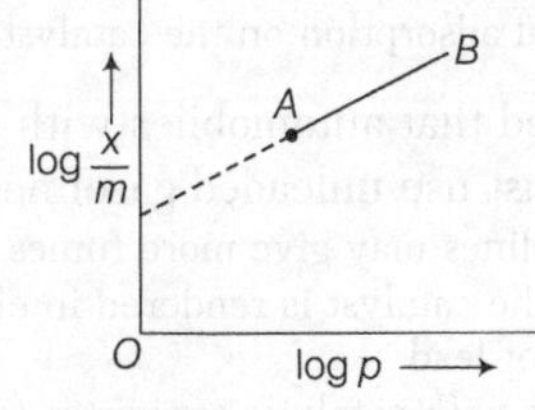

(a) $\log k$ (b) $\log \frac{1}{n}$ (c) $\log n$ (d) $\log \frac{x}{m}$

25 Freundlich adsorption isotherm gives a straight line on plotting against
(a) $\frac{x}{m}$ and $\frac{1}{p}$ (b) $\frac{x}{m}$ and p
(c) $\log \frac{x}{m}$ and p (d) $\log \frac{x}{m}$ and $\log p$

26 Which of the following is correct about adsorption of N_2 over iron?
(a) It is always physically adsorbed
(b) Extent of adsorption over iron first decreases and then increases with rise in temperature
(c) It is always chemically adsorbed
(d) N_2 is never adsorbed over iron

27 When saturation is attained in terms of adsorption, variation of $\left(\frac{x}{m}\right)$ and C (concentration) is given by the portion of the isotherm.

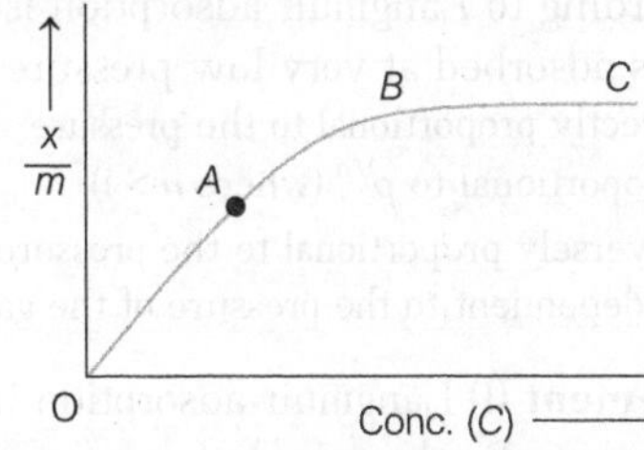

(a) OA (b) OB
(c) AB (d) BC

28 For the adsorption of solution on a solid surface $\frac{x}{m} = kC^{1/n}$. Adsorption isotherm of $\log\left(\frac{x}{m}\right)$ and $\log C$ was found of the type

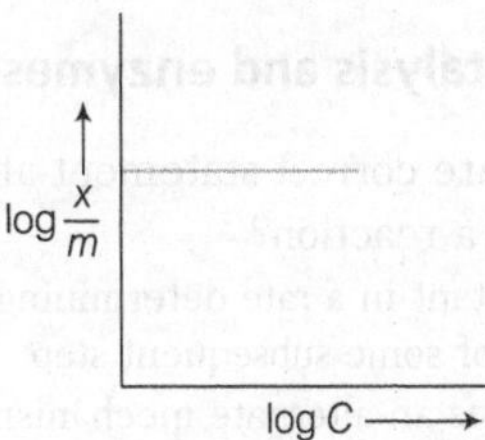

This is when
(a) $C = 0$ (b) $\frac{1}{n} = 0$
(c) $C =$ constant (d) $C = 2\,M$

29 Graph between $\log\left(\frac{x}{m}\right)$ and $\log p$ is a straight line at angle 45° with intercept OA as shown, Hence, $\left(\frac{x}{m}\right)$ at a pressure of 0.2 atm is

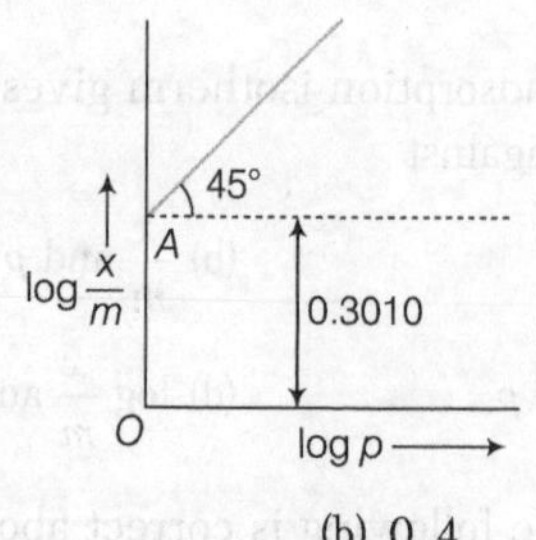

(a) 0.2 (b) 0.4
(c) 0.6 (d) 0.8

30 Statement (I) Heat of adsorption is low (i.e. 20-40 kJ/mol) in case of physisorption.

Statement (II) With rise in temperature, physisorption increases.

(a) Both statements I and II are true
(b) Only statement I is true
(c) Only statement II is true
(d) Neither statement I nor statement II are true

31 According to Langmuir adsorption isotherm amount of gas adsorbed at very low pressure
(a) directly proportional to the pressure
(b) proportional to $p^{1/n}$ (where $n > 1$)
(c) inversely proportional to the pressure
(d) independent to the pressure of the gas

32 Statement (I) Langmuir adsorption involves formation of only a single layer (i.e. unimolecular).

Statement (II) Langmuir adsorption involves van der Waals' force.

(a) Both statements I and II are true
(b) Only statement I is true
(c) Only statement II is true
(d) Neither statement I nor statement II are true

II. Catalyst, catalysis and enzymes catalysis

33 Which is/are correct statement about the role of a catalyst in a reaction?
(a) It is reactant in a rate determining step and then a product of some subsequent step
(b) It provides an alternate mechanism with a lower energy of activation
(c) It increases the rate of chemical reaction but does not itself undergo a permanent change during the course of the reaction
(d) All of the above

34 A chemical reaction is catalysed by a catalyst X. Hence, X
(a) decreases rate constant of the reaction
(b) increases activation energy of the reaction
(c) reduces enthalpy of the reaction
(d) does not affect equilibrium constant of reaction

35 In which of the following reaction, heterogeneous catalysis is involved?

I. $2SO_2(g) + O_2(g) \xrightarrow{NO(g)} 2SO_3(g)$

II. $2SO_2(g) \xrightarrow{Pt(s)} 2SO_3(g)$

III. $N_2(g) + 3H_2(g) + \xrightarrow{Fe(s)} 2NH_3(g)$

IV. $CH_3COOCH_3(l) + H_2O(l) \xrightarrow{HCl(l)} CH_3COOH(aq) + CH_3OH(aq)$

(a) Both II and III (b) II, III and IV
(c) I, II and III (d) Only IV

36 Consider the following reaction.

$$CH_3C{\equiv}CCH_3 \xrightarrow{H_2} \begin{matrix} H & & CH_3 \\ & C{=}C & \\ H_3C & & H \end{matrix}$$

Select suitable catalyst for above reaction.
(a) Li/NH_3 (b) Lindlar (c) Cu (d) Pd/PdO

37 In the atmosphere, there is formation of ozone from oxygen.

$$3O_2 \longrightarrow 2O_3$$

This is catalysed by
(a) NO (b) N_2 (c) CO (d) sunlight

38 Statement (I) In the presence of promoter, the activity of the catalyst is enhanced.

Statement (II) The promoter increases the surface area of the catalyst and thus, enhancing the number of active centres.

(a) Both statements I and II are true
(b) Only statement I is true
(c) Only statement II is true
(d) Neither statement I nor statement II are true

39 $\ldots{-}CH{=}CH{-}\ldots + H_2 \xrightarrow{Ni} \ldots{-}CH_2{-}CH_2{-}\ldots$

Promoters used in the above reaction can be
(a) Cu, Te (b) Cu, Fe (c) Mo, Fe (d) Al_2O_3, Cr_2O_3

40 Catalytic poisons act by
(a) making the products chemically inactive
(b) increasing the rate of the backward reactions
(c) chemical combination with any one of the reactants
(d) preferential adsorption on the catalyst surface

41 It is instructed that automobiles with catalytic converter must use unleaded gasolines because
(a) leaded gasolines may give more fumes
(b) surface of the catalyst is rendered in effective by adsorption of lead
(c) automobiles with catalytic converter can not run on leaded-gasolines
(d) unleaded-gasoline is cheaper

42 The zeolites have shapes selectivity depending on
(a) pore structure (b) atomic structure
(c) molecular structure (d) None of these

43 Which of the following is an example of zeolite?
(a) ZSM-5 (b) $AgNO_3$ (c) $Mg(OH)_3$ (d) $Co(OH)_3$

44 Which of the following small sized elements can replace the silicon and aluminium in the frame work at zeolites?
(a) Boron (b) Magnesium
(c) Phosphorus (d) All of these

45 Which of the following is present at the time of cracking of hydrocarbons?
(a) Copper (b) Zeolite
(c) Nickel (d) Molybdenum

46 Statement (I) Alcohols are dehydrated to a mixture of hydrocarbons in the presence of acidic zeolites.
Statement (II) Zeolites are porous catalysts.
(a) Both statements I and II are true
(b) Only statement I is true
(c) Only statement II is true
(d) Neither statement I nor statement II are true

47 Match the following Column I (catalyst) with Column II (process).

Column I	Column II
A. SiO_2	1. Cracking of hydrocarbon
B. Pt	2. Alkylation of benzene
C. Zeolites	3. Automobile converter

	A	B	C		A	B	C
(a)	3	2	1	(b)	3	1	2
(c)	2	3	1	(d)	2	1	3

48 The efficiency of an enzyme in catalysing a reaction is due to its capacity.
(a) To form a strong enzyme-substrate complex
(b) To decrease the bond energies of substrate molecule
(c) To change the shape of the substrate molecule
(d) To lower the activation energy of the reaction

49 'Lock-key fit' theory is for
(a) homogeneous catalysis (b) heterogeneous catalysis
(c) enzyme action (d) None of these

50 Statment (I) Activity of enzyme is affected by the change in pH.
Statment (II) Change in pH affects the solubility of the enzyme in water.
(a) Both statement I and II are true
(b) Only statement I is true
(c) Only statement II is true
(d) Neither statement I nor statement II are true

III. Colloidal State

51 Match the types of colloidal systems given in Column I with the name given in Columns II. **[NCERT Exemplar]**

Column I	Column II
A. Solid in liquid	1. Foam
B. Liquid in solid	2. Sol
C. Liquid in liquid	3. Gel
D. Gas in liquid	4. Emulsion

Choose the correct match.

	A	B	C	D		A	B	C	D
(a)	2	3	4	1	(b)	3	1	2	4
(c)	4	3	1	2	(d)	3	4	2	1

52 Which is/are correctly matched?
(a) Aerosol — liquid dispersed in a gas
(b) Smoke — solid dispersed in a gas
(c) Emulsion — liquid dispersed in liquid
(d) All of the above are correctly matched

53 Butter is
(a) water dispersed in oil (b) fat dispersed in water
(c) fat dispersed in milk (d) water dispersed in fat

54 Opal (mineral with liquid inclusion) is a
(a) gel (liquid dispersed in solid phase)
(b) solid sol (solid dispersed in solid phase)
(c) sol (solid dispersed in liquid)
(d) foam (gas dispersed in liquid)

55 Consider the following systems.
I. Fog II. Air
III. Blood IV. Paint
V. Aerosol sprays VI. Pearl
True colloids are
(a) I, II and III (b) III, IV and V
(c) I, III and V (d) I, II and V

56 Which substance is not used for preparing lyophilic sols?
(a) Starch (b) Gum
(c) Gelatin (d) Metal sulphide

57 Method by which lyophobic sol can be protected **[NCERT Exemplar]**
(a) by addition of oppositely charged sol
(b) by addition of an electrolyte
(c) by addition of lyophilic sol
(d) by boiling

58 Select the correct statement(s).
(a) A hydrophilic colloid is a colloid in which there is a strong attraction between the dispersed phase and water.
(b) A hydrophobic colloid is a colloid in which there is a lack of attraction between the dispersed phase and water.

(c) Hydrophobic sols are often formed when a solid crystallises rapidly from a chemical reaction or a supersaturated solution.
(d) All of the above

59 When SO_2 gas is bubbled into H_2S gas
(a) lyophilic sol of sulphur is formed
(b) lyophobic sol of sulphur is formed
(c) suspension of water and sulphur is formed
(d) a true solution of sulphur in water is formed

60 The extra stability of lyophilic colloids is due to
(a) charge on the particles
(b) a protective film of the dispersion medium on the particles
(c) the smaller size of the particles
(d) the larger size of the particles

61 Which one of the following is a natural colloid?
(a) Sodium chloride solution (b) Cane sugar solution
(c) Urea solution (d) Blood

62 Match the following Column I with Column II.

Column I (Colloids)	Column II (Classification)
A. Rain cloud	1. Sol
B. Milk of magnesia	2. Aerosol
C. Soap suds	3. Gel
D. Butter	4. Foam

Choose the correct match.

	A	B	C	D
(a)	2	1	4	3
(b)	3	4	1	2
(c)	4	3	2	1
(d)	3	4	2	1

63 At the critical micelle concentration (CMC) the surfactant molecules
(a) decompose (b) dissociate
(c) associate (d) become completely soluble

64 Select the correct statement(s).
(a) The process of dialysis is similar to osmosis, except that both solvent and small solute particles can pass through the semi-permeable dialysis membrane.
(b) Proteins can be purified by dialysis.
(c) Haemodialysis is used to cleanse the blood of patients whose kidneys have malfunctioned.
(d) All of the above

65 When a few typical solutes are separated by a particular selective membrane such as protein particles, blood corpuscles, this process is called
(a) dialysis (b) diffusion
(c) transpiration (d) endosmosis

66 Which of the following metallic sols cannot be prepared by Bredig's Arc method?
(a) Gold (b) Silver
(c) Platinum (d) Sodium

67 Haemodialysis is used to
(a) separate colloidal sol from water
(b) separate charged and uncharged particles
(c) cleanse the blood of patients whose kidneys have malfunctioned
(d) None of the above

68 If a freshly formed ppt. of SnO_2 is peptised by a small amount of NaOH, these colloidal particles may be represented as
(a) $[SnO_2]\,SnO_3^{2-} : 2Na^+$ (b) $[SnO_2]\,Sn^{4+} : O^{2-}$
(c) $[SnO_2]\,Na^+ : OH^-$ (d) $[SnO_2]\,Sn^{4+} : OH^-$

69 Select the incorrect statement(s).
(a) Surface active agents like soaps and synthetic detergents are micelles.
(b) Soaps are emulsifying agents.
(c) $C_{17}H_{35}$ (hydrocarbon part) and $-COO^-$ (carboxylate) part of stearate ion ($C_{17}H_{35}COO^-$) both are hydrophobic.
(d) All of the above are incorrect statements.

70 Which part of the soap ($RCOO^-$) dissolves grease and forms micelle?
(a) *R* part (called tail of the anion)
(b) $-COO^-$ part (called head of the anion)
(c) Both (a) and (b)
(d) None of the above

71 If dilute HCl is added to a precipitate of stannic oxide
(a) $SnCl_2$ and $SnCl_4$ are formed
(b) a stable sol of stannic oxide is formed
(c) $Sn(OH)_2$ is formed
(d) Sn_2O_3 is formed

72 Collodion is obtained
(a) when cellulose nitrate is peptised by organic solvents as ethanol
(b) when high electric current is passed between metal electrodes
(c) colloidal sol is heated
(d) *aq.* NaOH is added to cellulose nitrate

73 Choose the correct statement.
(a) The ultra-filter papers allow only the electrolytes to pass through under the influence of pressure or suction pump.
(b) When an electric field is applied during dialysis, the process is refered to as electrodialysis.
(c) None of the above are correct
(d) Both (a) and (b) are correct

74 Match the method of formation of solution given in Column I with the type of solution given in Column II. **[NCERT Exemplar]**

Column I		Column II
A. Sulphur vapours passed through cold water	1.	Normal electrolyte solution
B. Soap mixed with water above CMC	2.	Molecular colloids
C. White of egg whipped with water	3.	Associated colloid
D. Soap mixed with water below CMC	4.	Macromolecular colloids

Choose the correct match.

	A	B	C	D		A	B	C	D
(a)	2	3	4	1	(b)	3	1	4	2
(c)	4	1	2	3	(d)	2	4	1	3

75 A colloid of gold particles is
(a) hydrophobic in nature
(b) a sol (of solid dispersed in water)
(c) Both (a) and (b)
(d) None of the above

76 Protein solutions are
(a) hydrophobic in nature (b) hydrophilic in nature
(c) Both (a) and (b) (d) None of the above

IV. Charge on colloidal particles and properties of colloids

77 Which of the following is positively charged colloidal particle?
(a) As_2S_3 (b) $Al_2O_3 \cdot xH_2O$ (c) Au
(d) Pt

78 Which of the following processes is not responsible for the process of electric charge on the sol particle?
(a) Electron capture by sol particular **[NCERT Exemplar]**
(b) Adsorption of ionic species from solution
(c) Formation of Helmholtz electrical double layer
(d) Absorption of ionic species from solution

79 Statement (I) Arsenic sulphide sol is coagulated by Fe^{3+} ions.

Statement (II) Fe^{3+} reacts with arsenic sulphide sol to give ferric sulphide.
(a) Both statements I and II are true
(b) Only statement I is true
(c) Only statement II is true
(d) Neither statement I nor statement II are true

80 Fixed parts of a colloidal sol of AgI are respectively $[AgI]Ag^+$ and $[AgI]I^-$ in presence of
(a) KI and $AgNO_3$ (b) KI and KI
(c) AgI and KI (d) $AgNO_3$ and KI

81 Arsenic (III) sulphide forms a sol with a negative charge. Which of the following ionic substances should be most effective in coagulating the sol?
(a) KCl (b) $MgCl_2$ (c) $Al_2(SO_4)_3$ (d) Na_3PO_4

82 Colloidal sulphur particles are negatively charged with $S_2O_3^{2-}$ and other ions on the surface of sulphur. Hence, most effective electrolyte in coagulating colloidal sulphur is
(a) NaCl (b) $MgCl_2$ (c) $AlCl_3$ (d) Na_2SO_4

83 Aluminium hydroxide forms a positively charged sol. Which of the following ionic substances should be most effective in coagulating the sol?
(a) NaCl (b) $Fe_2(SO_4)_3$
(c) $CaCl_2$ (d) K_3PO_4

84 Gold number of haemoglobin is 0.03. Hence, 100 mL of gold sol will require haeomoglobin so that gold is not coagulated by 10 mL of 10% NaCl solution
(a) 0.03 mg (b) 30 mg
(c) 0.30 mg (d) 3 mg

85 $[AgI]I^-$ colloidal sol can be coagulated by the addition of a suitable cation. 1 mol of $[AgI]I^-$ requires mol of $AgNO_3$, $Pb(NO_3)_2$ and $Fe(NO_3)_3$ as
(a) 1, 1, 1 (b) 1, 2, 3
(c) $1, \frac{1}{2}, \frac{1}{3}$ (d) 6, 3, 2

86 A positively charged colloidal sol is formed in which of the following?
(a) 100 mL of 0.1 M $AgNO_3$ + 100 mL of 0.1 M KI
(b) 100 mL of 0.1 M $AgNO_3$ + 50 mL of 0.1 M KI
(c) 50 mL of 0.1 M $AgNO_3$ + 100 mL of 0.1 M KI
(d) 200 mL of 0.1 M $AgNO_3$ + 100 mL of 0.2 M KI

87 100 mL of a colloidal solution is completely precipitated by addition of 5 mL of 1 M NaCl solution. What is the coagulation value of NaCl?
(a) 50 (b) 5
(c) 0.005 (d) None of these

88 The method usually employed for the destruction of a colloidal solution is
(a) dialysis
(b) addition of electrolytes
(c) diffusion through animal membrane
(d) condensation

89 The ability of anion to bring about coagulation of a given colloid depends upon
(a) its charge
(b) the sign of the charge alone
(c) the magnitude of charge
(d) Both magnitude and charge

90 Aluminium hydroxide forms a positively charged sol. Which of the following ionic substances should be most effective in coagulating the sol?
(a) NaCl (b) $CaCl_2$
(c) $Fe_2(SO_4)_3$ (d) K_3PO_4

91 Flocculation value of a coagulating electrolyte is expressed in
(a) millimol L^{-1} (b) mol L^{-1}
(c) mg K^{-1} (d) microgram mL^{-1}

92 Protective sols are
(a) lyophilic (b) lyophobic
(c) Both (a) and (b) (d) None of these

93 Which of the following can act as a protective colloid?
(a) Gelatin (b) Silica gel
(c) Oil-in-water emulsion (d) All of these

94 Gold number is associated with
(a) electrophoresis
(b) protective colloids
(c) Tyndall effect
(d) isotonic solutions

95 Gold number of a lyophilic sol is such property that
(a) the larger its value, the greater is the peptising power
(b) the lower its value, the greater is the peptising power
(c) the lower its value, the greater is the protecting power
(d) the larger its value, the greater is the protecting power

96 What happens when gelatin is added to gold sol?
(a) Gold sol is destabilised and peptisation occurs
(b) Gold sol is stabilised and no coagulation takes place
(c) Gold sol is destabilised and coagulation takes place
(d) Gold sol is stabilised and peptisation takes place

97 Gold number is maximum for which of the following lyophilic sol?
(a) Gelatin (b) Haemoglobin
(c) Sodium oleate (d) Potato starch

98. Statement (I) AgI changes to positively charged colloidal sol in presence of KI.

Statement (II) It is due to adsorption of I^- on AgI.
(a) Both statement I and II are true
(b) Only statement I is true
(c) Only statement II is true
(d) Neither statement I nor statement II are true

99 The process of removing dissolved impurities from a colloidal system, by means of diffusion through suitable membrane under the influence of an electric field is called
(a) electro-osmosis (b) electrodialysis
(c) electrophoresis (d) peptisation

100 The process of passing of a precipitate into colloidal solution, on adding an electrolyte is called
(a) dialysis (b) peptisation
(c) electrophoresis (d) electro-osmosis

101 Which one of the following statements is correct?
(a) Brownian movement is more pronounced for smaller particles than for bigger ones.
(b) Sols of metal sulphides are lyophilic.
(c) Hardy-Schulze law states, the bigger the size of the ion, the greater is its coagulating power.
(d) One would expect charcoal to adsorb chlorine more strongly than hydrogen sulphide.

102 Statement (I) A colloidal gets coagulated by addition of an electrolyte.

Statement (II) The rate of coagulation depends on the magnitude and sign of the charge of the coagulation.
(a) Both statement I and II are true
(b) Only statement I is true
(c) Only statement II is true
(d) Neither statement I nor statement II are true

103 Match the items of Column I and Column II. **[NCERT Exemplar]**

Column I	Column II
A. Dialysis	1. Cleansing action of soap
B. Peptisation	2. Coagulation
C. Emulsification	3. Colloidal sol formation
D. Electrophoresis	4. Purification

Choose the correct match.

	A	B	C	D		A	B	C	D
(a)	2	3	4	1	(b)	3	1	4	2
(c)	4	1	2	3	(d)	4	3	1	2

104 Which of the following will show Tyndall effect? **[NCERT Exemplar]**
(a) Aqueous solution of soap below critical micelle concentration
(b) Aqueous solution of soap above critical micelle concentration
(c) Aqueous solution of sodium chloride
(d) Aqueous solution of sugar

105 Coagulation or de-emulsification can be done by some of the methods given below.
I. By addition of a substance which would destroy the emulsifier.
II. By addition of an electrolyte which would destroy the charge.
III. By heating, freezing and centrifuging.

Select the correct methods.
(a) Both I and II (b) I, II and III
(c) Only II (d) Only III

106 Some of the properties given below are for colloidal sols

I. viscosity same as that of the medium.
II. extensive hydration takes place.
III. Irreversible
IV. particles cannot be detected even under ultra-microscope.

Select the properties which are true for lyophilic colloidal sols.

(a) Both I and II (b) Both III and IV
(c) Both I and III (d) Both II and IV

107 **Statement** (I) Colloidal solution does not show Brownian motion. **[NCERT Exemplar]**

Statement (II) Brownian motion is responsible for stability of sols.

(a) Both statement I and II are true
(b) Only statement I is true
(c) Only statement II is true
(d) Neither statement I nor statement II are true

108 Match the following Column I with Column II.

Column I (Term)		Column II (Related property)	
A.	Coagulation	1.	Scattering of light
B.	Lyophilisation	2.	Washing of precipitate
C.	Peptisation	3.	Purification of colloids
D.	Tyndall effect	4.	Electrolyte

Choose the correct match.

	A	B	C	D		A	B	C	D
(a)	2	4	3	1	(b)	3	1	2	4
(c)	4	1	2	3	(d)	4	3	2	1

V. Emulsions

109 Some of the properties of emulsions are

I. Emulsions scatter light and thus exhibit Tyndall effect.
II. Oily emulsions are more viscous than the aqueous emulsions.
III. Electrical conductance of aqueous emulsions is higher than that of oily emulsion and increases on the addition of electrolyte.

Select the correct properties.

(a) Both I and II (b) Both I and III
(c) Both II and III (d) I, II and III

110 Select the correct statement(s).

(a) A gel is a system in which one component, at least, has a low rigidity and at least one component has a high mobility
(b) A surfactant is an amphipathic substance that means it has both hydrophobic and hydrophilic regions
(c) Both (a) and (b)
(d) None of the above

111 Artificial smoke screens are made by

(a) aluminium oxide (b) lead oxide
(c) sodium oxide (d) titanium oxide

112 Smoke has generally blue tinge. It is due to

(a) scattering (b) coagulation
(c) Brownian motion (d) electro-osmosis

113 Bleeding is stopped by the application of ferric chloride. This is because

(a) the blood starts flowing in opposite direction
(b) the blood reacts and forms a solid, which seals the blood vessel
(c) the blood is coagulated and thus, the blood vessel is sealed
(d) the ferric chloride seals the blood vessel

114 Select the correct statement(s).

(a) Clays may be peptised by alkalis, the OH^- ion being the active agent.
(b) Milk is an emulsion of water in fats.
(c) Both (a) and (b)
(d) None of the above

115 Colloidal solution commonly used in treatment of eye disease is

(a) colloidal sulphur (b) colloidal silver
(c) colloidal gold (d) colloidal antimony

116 Which of the following processes is responsible for the formation delta at a place where rivers meet the sea? **[NCERT Exemplar]**

(a) Emulsification (b) Colloid formation
(c) Coagulation (d) Peptisation

117 At high concentration of soap in water, soap behaves as **[NCERT Exemplar]**

(a) molecular colloid (b) associated colloid
(c) macromolecular colloid (d) lyophilic colloid

118 Which is not the example of coagulation?

(a) Curdling of milk
(b) Purification of water by addition of alum
(c) Rubber plating and chrome tanning
(d) Formation of deltas at the river beds

(B) Miscellaneous questions

1 When a litmus solution is shaken with a piece of charcoal

(a) it is decolourised (b) it turns red to blue
(c) it turns blue to red (d) None of these

2 Pieces of wood burn faster than a log of wood of the same mass because

(a) surface area of log of wood is larger and needs more time to burn
(b) pieces of wood have large surface area
(c) all pieces of wood catch fire at the same time
(d) block of wood has higher density than pieces of the same wood

3 Chromatography is a technique based on
(a) adsorption and then dispersion of solute
(b) absorption of solute
(c) hydration of solute
(d) evaporation of solute

4 The void space in zeolites forms more than 50% of the total volume which is occupied by
(a) silicates (b) aluminates
(c) water molecules (d) Na^+, Mg^{2+} and Ca^{3+} ions

5 Chlorine atoms catalyse the decomposition of ozone in the ozone layer of the earth's atmosphere and thus causing global heating. This Cl comes from
(a) teflon (b) freon
(c) chloroform (d) pyrene

6 During purification of colloidal sol by ultracentrifugation which of the following is observed?
(a) Colloidal particles are settled at the bottom of ultracentrifuge tube
(b) Impurities are settled at the bottom of the ultracentrifuge tube
(c) Impurities are removed through ultrafilters
(d) None of the above

7 Following are the events taking place to explain adsorption theory
I. desorption.
II. diffusion of the reactants along the surface.
III. adsorption of the reactants.
IV. formation of the activated surface complex.
These events are taking place in the following order
(a) I, II, III and IV
(b) III, II, IV and I
(c) III, IV, I and II
(d) IV, III, II and I

8 Which of the following can catalyse decomposition of ozone in ozone layer?
(a) Cl from decomposition of freon
(b) NO formed near ozone layer due to interaction of N_2 and O_2 or during lightning
(c) Both (a) and (b)
(d) Nitrogen radical formed during lightning

9 Colloidal solutions of gold prepared by different methods are of different colours because of
(a) different diameters of colloidal gold particles
(b) variable valency of gold
(c) different concentration of gold particles
(d) impurities produced by different methods

10 Gelatin is often used as an ingredient in the manufacture of ice-cream. The reason for this is
(a) to prevent the formation of a colloid
(b) to stabilise the colloid and prevent crystal growth
(c) to cause the mixture to solidify
(d) to improve the flavour

11 Gold number of some lyophilic sols are
I. Casein : 0.01
II. Haemoglobin : 0.03
III. Gum arabic : 0.15
IV. Sodium oleate : 0.40

Which has maximum protective power?
(a) Only I (b) Only II
(c) Only III (d) Only IV

12 Which is not the correct statement for a catalyst?
(a) It does not alter E_a
(b) The surface of a catalyst adsorbs reactants
(c) Catalyst may form intermediates with the reactants
(d) Action of enzyme catalyst is always specific

13 Which of the following electrolytes will have maximum coagulating power for AgI/Ag^+ sol? **[NCERT Exemplar]**
(a) Na_2S (b) Na_3PO_4
(c) Na_2SO_4 (d) NaCl

14 Amount of gas adsorbed per gram of adsorbent increases with pressure, but after certain limit is reached, adsorption becomes constant. It is where
(a) multilayers are formed
(b) desorption takes place
(c) temperature is increased
(d) absorption also starts

15 Match the reaction (given in Column I) with catalyst employed (given in Column II).

	Column I		Column II
A.	Hydrogenation of oil	1.	Iron
B.	Friedel-Craft's reaction	2.	Anhydrous $AlCl_3$
C.	Oxidation of SO_2 to SO_3 by contact process	3.	Vanadium pentoxide
D.	Haber's process	4.	Nickel

Choose the correct match.

	A	B	C	D
(a)	2	1	4	3
(b)	3	4	1	2
(c)	4	2	3	1
(d)	4	3	2	1

16 Which of the following curves is in accordance with Freundlich adsorption isotherm? **[NCERT Exemplar]**

(a) 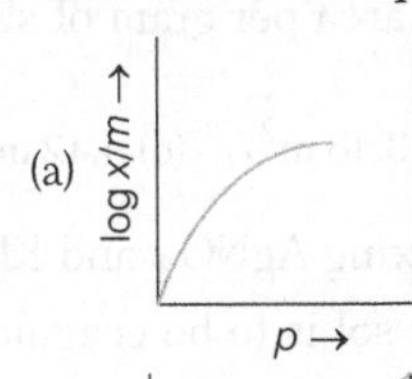

(b) 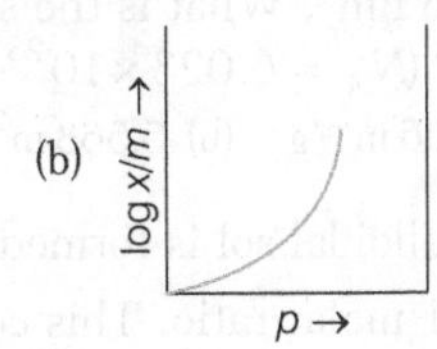

(c) 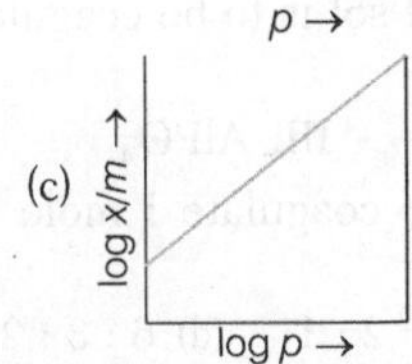

(d) 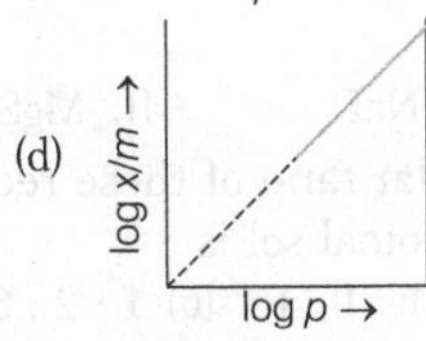

17 In the following isotherm

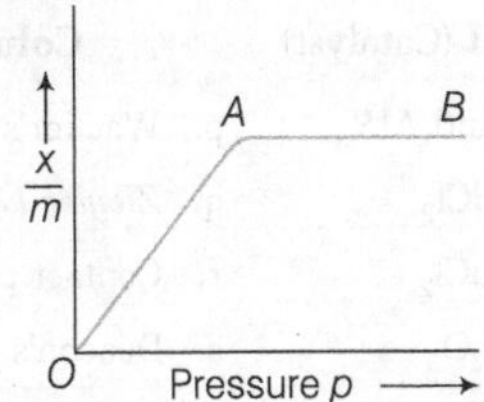

(a) $\frac{x}{m} \propto p^{\circ}$ when point A is reached
(b) desorption may start along AB
(c) $\frac{x}{m} \propto p^{1/n}$ along OA
(d) All of the above

18 Zeta potential or electrokinetic potential is dependent on
(a) viscosity
(b) dielectric constant
(c) velocity of the colloidal particles when an electric field is applied
(d) All of the above

19 In homogeneous catalytic reactions, there are three alternative paths A, B and C (shown in the figure). Which one of them indicates the relative ease with which the reaction can take place?

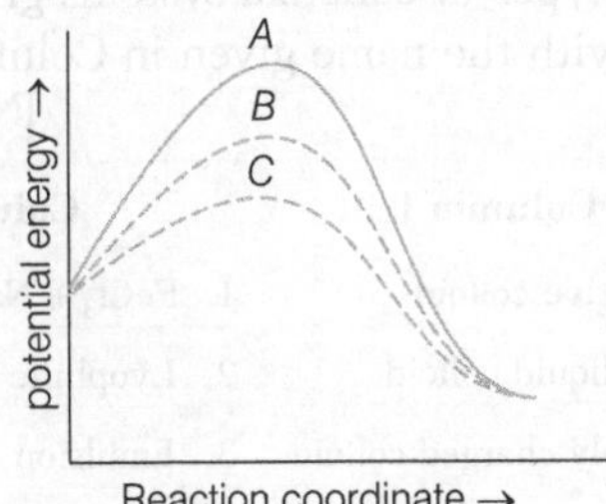

(a) $A > B > C$
(b) $C > B > A$
(c) $B > C > A$
(d) $A = B = C$

20 **Statement I** Colloidal AgI is prepared by adding $AgNO_3$ in slight excess to KI solution.
Statement II When subjected to an electric field, the colloidal particles migrate to the anode.
(a) Both statement I and II are true
(b) Only statement I is true
(c) Only statement II is true
(d) Neither statement I nor statement II are true

21 Match the following Column I (names of the processes) with the Column II (catalysts used).

Column I		Column II
A. Ostwald	1.	Ni
B. Steam hydrocarbon re-forming	2.	NO
C. Fischer-Tropsch	3.	Pt
D. Lead-chamber	4.	Co/Al_2O_3

Choose the correct match.

	A	B	C	D		A	B	C	D
(a)	2	4	1	3	(b)	3	1	4	2
(c)	4	3	2	1	(d)	4	1	2	3

22 When NaCl solution is added to $Fe(OH)_3$ sol
(a) $[Fe(OH)_3]Fe^{3+}$ is formed
(b) $[Fe(OH)_3]\ Cl^-$ is formed
(c) $[Fe(OH)_3]Na^+$ is formed
(d) $Fe(OH)_3$ is coagulated

23 1 mol of $[AgI]\ Ag^+$ sol is coagulated by
(a) 1 mol of KI
(b) 500 mL of 1 M K_2SO_4
(c) Both (a) and (b)
(d) 100 mL of 1M K_2SO_4

24 SnO_2 is taken in basic medium and current is passed then colloidal sol migrates towards
(a) anode (+ve plate) (b) cathode (–ve plate)
(c) Both (a) and (b) (d) None of these

25 The coagulation values of the electrolytes $AlCl_3$ and NaCl for As_2S_3 sol are 0.093 and 52 respectively. How many times has $AlCl_3$ greater coagulating power than NaCl?
(a) 100 (b) 559
(c) 993 (d) 852

26 Which of the following pair is correctly matched?
(a) $[Fe(OH)_3]:Fe^{3+}$
(b) $[As_2S_3]:As^{3+}$
(c) $[SnO_2]:SnO_3^{2-}$ in acidic medium
(d) $[AgI] : I^-$ in excess of $AgNO_3$

27 What are the products X, Y, Z (aliphatic class) formed in the reaction given below?

$$CO + H_2 \xrightarrow{Ni} X$$
$$CO + H_2 \xrightarrow{Cu} Y$$
$$CO + H_2 \xrightarrow{ZnO + Cr_2O_3} Z$$

(a) CH_3OH in all cases (b) CH_3OH, HCHO, CH_4
(c) CH_4, HCHO, CH_3OH (d) CH_4, CH_3OH, HCHO

28 A reddish brown sol (containing Fe^{3+}) is obtained by
(a) the addition of small amount of $FeCl_3$ solution to freshly prepared $Fe(OH)_3$ precipitate
(b) the addition of $Fe(OH)_3$ to freshly prepared $FeCl_3$ solution
(c) the addition of NH_4OH to $FeCl_3$ solution dropwise
(d) the addition of NaOH to $FeCl_3$ solution dropwise

29 **Statement I** The colour of aqueous gold sol is red.
Statement II The colour is due to scattering of light by the sol particles.
(a) Both statement I and II are true
(b) Only statement I is true
(c) Only statement II is true
(d) Neither statement I nor statement II are true

30 **Statement I** Proteins, starch and rubber are lyophilic colloids.
Statement II They have strong interaction with the dispersion medium.
(a) Both statement I and II are true
(b) Only statement I is true
(c) Only statement II is true
(d) Neither statement I nor statement II are true

31 Example of intrinsic colloid is
(a) As_2S_3 sol (b) S sol
(c) egg albumin (d) $Fe(OH)_3$ sol

32 In the following reactions, catalysts used have been shown above the arrow. Which of them are not the suitable catalysts ?

I. $2H_2O_2(aq) \xrightarrow{I^-(aq)} 2H_2O(l) + O_2(g)$

II. $HCl(g) + O_2(g) \xrightarrow{NO(g)} 2H_2O(l) + Cl_2(g)$

III. $2H_2(g) + O_2(g) \xrightarrow{MnO_2} 2H_2O(l)$

(a) Both I and II (b) Both II and III
(c) Both I and III (d) I, II and III

33 50 mL of 1 M oxalic acid (molar mass = 126) is shaken with 0.5 g of wood charcoal. The final concentration of the solution after adsorption is 0.5 M. What is the amount of oxalic acid adsorbed per gram of carbon?
(a) 3.15 (b) 1.575 (c) 6.30 (d) 12.60

34 1.30 cm^3 of N_2 gas at STP is adsorbed per gram of silica gel. The area occupied by nitrogen molecules is 0.16 nm^2. What is the surface area per gram of silica gel? ($N_A = 6.023 \times 10^{23}$)
(a) 1.6 m^2/g (b) 5.568 m^2/g (c) 3.48 m^2/g (d) 4.42 m^2/g

35 A colloidal sol is formed by mixing $AgNO_3$ and KI in 2 : 1 molar ratio. This colloidal sol is to be coagulated by
I. NaI II. $MgSO_4$ III. $AlPO_4$
Molar ratio of these required to coagulate 1 mole colloidal sol is
(a) 1 : 1 : 1 (b) 1 : 2 : 3 (c) 3 : 2 : 1 (d) 6 : 3 : 2

36 Match the catalysts to the correct processes.

	Column I (Catalyst)		Column II (Process)
i.	$TiCl_3$ and AlR_3	p.	Wacker's process
ii.	$PdCl_2$	q.	*Ziegler-Natta* polymerisation
iii.	$CuCl_2$	r.	Contact process
iv.	V_2O_5	s.	Deacon's process

Codes

	i	ii	iii	iv		i	ii	iii	iv
(a)	p	q	r	s	(b)	q	p	s	r
(c)	q	r	s	p	(d)	r	p	q	s

37 The volume of a colloidal particle, V_C as compared to the volume of a solute particle in a true solution V_S, could be
(a) $\frac{V_C}{V_S} \approx 10^3$ (b) $\frac{V_C}{V_S} \approx 10^{-3}$ (c) $\frac{V_C}{V_S} \approx 10^{23}$ (d) $\frac{V_C}{V_S} \approx 1$

38 3 g of activated charcoal was added to 50 mL of 0.06 N acetic acid solution in a flask.
After an hour, it was filtered and strength of the filtrate was found to be 0.042 N. The amount of acetic acid adsorbed per gram of charcoal is
(a) 18 mg (b) 36 mg (c) 42 mg (d) 54 mg

39 Match the types of colloidal systems given in Column I with the name given in Column II. **[NCERT Exemplar]**

Column I	Column II
A. Proctective colloid	1. $FeCl_3$ + NaOH
B. Liquid-liquid colloid	2. Lyophilic colloid
C. Positively charged colloid	3. Emulsion
D. Negatively charged colloid	4. $FeCl_3$ + hot water

Choose the correct match.

	A	B	C	D		A	B	C	D
(a)	2	3	4	1	(b)	3	4	1	2
(c)	4	3	1	2	(d)	4	1	2	3

40 Statement I An ordinary filter paper impregnated with colloidion solution stops the flow of colloidal particle.

Statement II Pore size of the filter paper becomes more than the size of colloidal particle.

(a) Both statement I and II are true
(b) Only statement I is true
(c) Only statement II is true
(d) Neither statement I nor statement II are true

(C) Medical entrances' gallery

1 Measuring zeta potential is useful in determining which property of colloidal solution? **[NEET 2020]**

(a) Solubility
(b) Stability of the colloidal particles
(c) Size of the colloidal particles
(d) Viscosity

2 Identify the correct statements from the following

1. $CO_2(g)$ is used as refrigerant for ice-cream and frozen food.
2. The structure of C_{60} contains twelve six carbon rings and twenty five carbon rings.
3. ZSM-5, a type of zeolite, is used to convert alcohols into gasoline.
4. CO is colourless and odourless gas. **[NEET 2020]**

(a) (1) and (3) only (b) (2) and (3) only
(c) (3) and (4) only (d) (1), (2) and (3) only

3 In which of the sols, the colloidal particles are with negative charge? **[NEET 2020]**

(a) TiO_2 (b) Haemoglobin
(c) Starch (d) Hydrated Al_2O_3

4 The correct option representing a Freundlich adsorption isotherm is **[NEET (Odisha) 2019]**

(a) $\frac{x}{m} = kp^{0.3}$ (b) $\frac{x}{m} = kp^{2.5}$

(c) $\frac{x}{m} = kp^{-0.5}$ (d) $\frac{x}{m} = kp^{-1}$

5 Butter is a colloidal solution of **[UP CPMT 2013; JIPMER 2019]**

(a) solid-solid (b) liquid-solid
(c) solid-liquid (d) gas-solid

6 On which of the following properties does the coagulating power of an ion depend? **[NEET 2018]**

(a) Both magnitude and sign of the charge on the ion
(b) Size of the ion alone
(c) The magnitude of the charge on the ion alone
(d) The sign of charge on the ion alone

7 When an excess and a very dilute aqueous solution of KI is added to very dilute aqueous solution of silver nitrate. The colloidal particles of silver iodide which are associated with the Helmholtz double layer in the form of **[AIIMS 2018]**

(a) $AgI / Ag^+ : I^-$ (b) $AgI / K^+ : NO_3^-$
(c) $AgI / NO_3^- : Ag^+$ (d) $AgI / I^- : K^+$

8. Assertion (A) Micelles are formed by surfactant molecules above the critical micellar concentration (CMC).

Reason (R) The conductivity of a solution having surfactant molecules decreases sharply at the CMC.

(a) Both Assertion and Reason are true and Reason is the correct explanation of the Assertion. **[AIIMS 2018]**
(b) Both Assertion and Reason are true but Reason is not the correct explanation of the Assertion.
(c) Assertion is true but Reason is false.
(d) Both Assertion and Reason are false.

9. Bleeding due to a cut can be stopped by applying ferric chloride solution in the laboratory. This is due to **[JIPMER 2018]**

(a) coagulation of negatively charged blood particles by Fe^{3+} ions
(b) coagulation of positively charged blood particles by Cl^- ions
(c) reaction taking place between ferric ions and the haemoglobin forming a complex
(d) common element, iron, in both $FeCl_3$ and haemoglobin

10. Which one the following statements(s) is/are not correct? **[NEET 2017]**

(a) Catalyst does not initiate any reaction.
(b) The value of equilibrium constant is changed in the presence of a catalyst in the reaction equilibrium.
(c) Enzymes catalyse mainly biochemical reaction.
(d) Coenzymes increase the catalytic activity of enzyme.

11. Which of the following curves is in accordance with Freundlich adsorption isotherm? **[AIIMS 2017]**

(a) log x/m ⟶ vs log p ⟶ (b) log x/m ⟶ vs log p ⟶

(c)

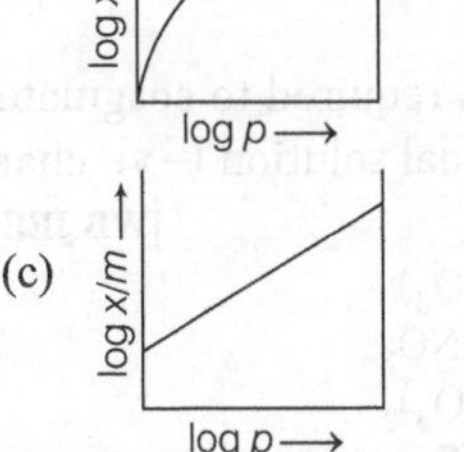

(d)

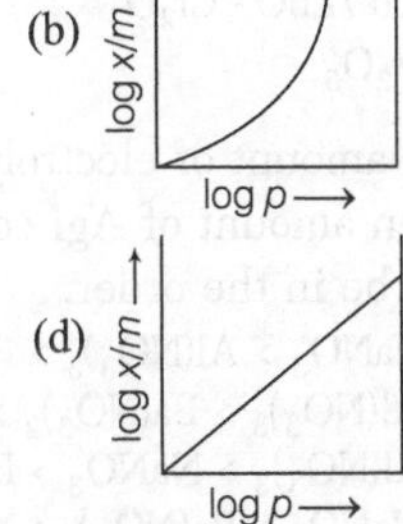

12 Fog is colloidal solution of **[NEET (I) 2016]**

(a) gas in liquid (b) solid in gas
(c) gas in gas (d) liquid in gas

13 The coagulation values in millimoles per litre of the electrolytes used for the coagulation of As_2S_3 are given below **[NEET (II) 2016]**

I. (NaCl) = 52 II. $(BrCl_3) = 0.69$
III. $(MgSO_4) = 0.22$

The correct order of their coagulating power is
(a) I > II > III (b) II > I > III
(c) III > II > I (d) III > I > II

14 Which property of colloids solution is independent of charge on the colloidal particles? **[CBSE AIPMT 2015]**
(a) Coagulation (b) Electrophoresis
(c) Electro-osmosis (d) Tyndall effect

15 Which one of the following is not explained by adsorption? **[Kerala CEE 2014]**
(a) When acetic acid solution is shaken with charcoal, the concentration of the acid decreases
(b) The white precipitate of $Mg(OH)_2$ attains blue colour when precipitated in the presence of magnesion reagent
(c) The air becomes dry in the presence of silica gel
(d) An aqueous solution of NaOH attains pink colour with a drop of phenolphthalein
(e) When animal charcoal is shaken with coloured methylene blue solution, the solution turns colourless.

16 For Freundlich isotherm a graph of $\log x/m$ is plotted against $\log p$. The slope of the line and its Y-axis intercept, respectively corresponds to **[KCET 2014]**
(a) $\frac{1}{n}, k$ (b) $\log \frac{1}{n}, k$ (c) $\frac{1}{n}, \log k$ (d) $\log \frac{1}{n}, \log k$

17 Which of the following statements is incorrect for physical adsorption? **[Guj. CET 2014]**
(a) It is instantaneous
(b) Monomolecular layer forms on the adsorbent
(c) Less activation energy is required for it
(d) Generally it results at low temperature and adsorption decreases with increase in temperature

18 *Zeigler-Natta* catalyst is **[Kerala CEE 2014]**
(a) $ZnCl_2$ (b) $AlR_3 + TiCl_4$
(c) $Cu/ZnO - Cr_2O_3$ (d) Pt
(e) V_2O_5

19 The amount of electrolytes required to coagulate a given amount of AgI colloidal solution (– ve charge) will be in the order. **[WB JEE 2014]**
(a) $NaNO_3 > Al(NO_3)_3 > Ba(NO_3)_2$
(b) $Al(NO_3)_3 > Ba(NO_3)_2 > NaNO_3$
(c) $Al(NO_3)_3 > NaNO_3 > Ba(NO_3)_2$
(d) $NaNO_3 > Ba(NO_3)_2 > Al(NO_3)_3$

20 Protective sols are **[UK PMT 2014]**
(a) lyophilic (b) lyophobic
(c) Both (a) and (b) (d) None of these

21 Gold sol is not **[KCET 2014]**
(a) a macromolecular colloid
(b) a lyophobic colloid
(c) a multimolecular colloid
(d) negatively charged colloid

22. In adsorption of oxalic acid on activated charcoal, the activated charcoal is called **[UP CPMT 2013]**
(a) adsorber (b) adsorbate
(c) adsorbent (d) occlusion

23 The colloid is **[AIIMS 2013]**
(a) urea (b) blood (c) cane sugar (d) NaCl

24 A colloidal solution is subjected to an electric field. The colloidal particles move towards anode. The coagulation of the same solution studied by using NaCl, $BaCl_2$ and $AlCl_3$ solutions. The coagulation power is in the order of **[Manipal 2013]**
(a) $BaCl_2 > NaCl > AlCl_3$ (b) $NaCl > BaCl_2 > AlCl_3$
(c) $AlCl_3 > BaCl_2 > NaCl$ (d) $NaCl > AlCl_3 > BaCl_2$

25 The random motion of colloidal particles in the dispersion medium is known as **[Manipal 2013]**
(a) Tyndall effect (b) Coagulation
(c) Adsorption (d) Brownian movement

26 Amount of protective colloid in milligram is called **[WB JEE 2013]**
(a) gold number (b) congo rubin number
(c) coagulating power (d) None of these

27 Which of the following ions can cause coagulation of proteins? **[KCET 2013]**
(a) Ag^+ (b) Na^+ (c) Mg^{2+} (d) Ca^{2+}

28 In Freundlich adsorption isotherm, the value of $1/n$ is **[AIPMT 2012]**
(a) between 0 and 1 in all cases
(b) between 2 and 4 in all cases
(c) 1 in case of physical adsorption
(d) 1 in case of chemisorption

29 Adsorption is accompanied by the evolution of heat, so according to Le-Chatelier's principle, the amount of substance adsorption should **[AIIMS 2012]**
(a) increase with decrease in T
(b) increase with increase in T
(c) decrease with decrease in T
(d) decrease with increase in T

30 Which among the following statements is false? **[UP CPMT 2012]**
(a) The adsorption may be monolayered or multilayered
(b) Particle size of adsorbent will not affect the amount of adsorption
(c) Increase of pressure increases amount of adsorption
(d) All of the above

31 The gas that is adsorbed to a larger extent on charcoal is **[AMU 2012]**
(a) H_2 (b) CO_2 (c) CO (d) NH_3

32 The protecting power of lyophilic colloidal sol is expressed in terms of **[CBSE AIPMT 2012]**
(a) coagulation value
(b) gold number
(c) critical micelle concentration
(d) oxidation number

33 Cow milk is an example of natural emulsion stabilised by **[AFMC 2012]**
(a) fat (b) water
(c) casein (d) Mg^{2+} ions

34 The simplest way to check whether a system is colloidal is **[UP CPMT 2012]**
(a) electrodialysis
(b) finding out particle size
(c) Tyndall effect
(d) Brownian movement

35 Which of the following is not correct? **[Manipal 2012]**
(a) Milk is a naturally occurring emulsion
(b) Gold sol is a lyophilic sol
(c) Physical adsorption decreases with rise in temperature
(d) Chemical adsorption is unlayered

36 Milk is a kind of **[OJEE 2012]**
(a) sol (b) gel
(c) emulsion (d) mixture

37 All colloidal dispersion have **[KCET 2012]**
(a) low osmotic pressure
(b) no osmotic pressure
(c) high osmotic pressure
(d) very high osmotic pressure

38 If x is the amount of adsorbate and m is the amount of adsorbent, which of the following relations is not related to adsorption process? **[CBSE AIPMT 2011]**
(a) $\frac{x}{m} = p \times T$
(b) $\frac{x}{m} = f(p)$ at constant T
(c) $\frac{x}{m} = f(T)$ at constant p
(d) $p = f(T)$ at constant $\left(\frac{x}{m}\right)$

39 Choose the incorrect statement in respect of physisorption. **[Kerala CEE 2011]**
(a) It is not specific in nature
(b) It arises because of van der Waals' forces
(c) It is reversible in nature
(d) No appreciable activation energy is needed
(e) Enthalpy of adsorption is in the range 80-240 kJ mol^{-1}

40 For the following reaction,
$$C_6H_{12}O_6(aq) + H_2(g) \rightleftharpoons C_6H_{14}O_6(aq)$$
Which one of the following is not affected by the addition of catalyst? **[UP CPMT 2011]**
(a) Rate of forward reaction
(b) Rate of backward reaction
(c) Time required to reach the equilibrium
(d) Spontaneity

41 Match the following columns.

Column I (Catalyst)	Column II (Industrial product)
A. V_2O_5	1. High density polyethylene
B. *Ziegler-Natta*	2. Polyacrylonitrile
C. Peroxide	3. NH_3
D. Finely divided Fe	4. H_2SO_4

Codes **[Kerala CEE 2011]**

	A	B	C	D
(a)	4	1	2	3
(b)	4	3	2	1
(c)	3	1	2	4
(d)	4	2	1	3
(e)	4	1	3	2

42 Which one of the following is correctly matched? **[Kerala CEE 2011]**
(a) Emulsion - smoke (b) Gel - butter
(c) Aerosol - hair cream (d) Sol - whipped cream
(e) Foam - mist

43 Which one of the following does not involve coagulation? **[KCET 2011]**
(a) Clotting of blood by the use of ferric chloride
(b) Formation of delta region
(c) Treatment of drinking water by potash alum
(d) Peptisation

44 Which of the following has maximum coagulation power with ferric hydroxide sol? **[Guj. CET 2011]**
(a) Cryolite (b) $K_2C_2O_4$
(c) $K_3[Fe(CN)_6]$ (d) $K_4[Fe(CN)_6]$

ANSWERS

Check Point 5.1

1. (a)	2. (a)	3. (c)	4. (a)	5. (d)	6. (d)	7. (a)	8. (b)	9. (c)	10. (b)

Check Point 5.2

1. (a)	2. (b)	3. (a)	4. (c)	5. (b)	6. (c)	7. (b)	8. (d)

Check Point 5.3

1. (c)	2. (c)	3. (a)	4. (d)	5. (d)	6. (b)	7. (d)	8. (c)

Check Point 5.4

1. (a)	2. (b)	3. (b)	4. (d)	5. (d)	6. (a)	7. (c)	8. (c)	9. (c)	10. (b)

Check Point 5.5

1. (b)	2. (a)	3. (a)	4. (c)	5. (b)	6. (d)	7. (a)	8. (c)	9. (b)	10. (d)

Check Point 5.6

1. (a)	2. (c)	3. (a)	4. (a)	5. (a)	6. (b)	7. (d)	8. (b)	9. (c)	10. (a)

(A) Topical exercise

1. (d)	2. (a)	3. (b)	4. (b)	5. (a)	6. (c)	7. (a)	8. (b)	9. (c)	10. (d)
11. (b)	12. (b)	13. (a)	14. (d)	15. (d)	16. (b)	17. (a)	18. (d)	19. (b)	20. (b)
21. (a)	22. (d)	23. (b)	24. (a)	25. (d)	26. (b)	27. (d)	28. (b)	29. (b)	30. (b)
31. (a)	32. (b)	33. (d)	34. (d)	35. (a)	36. (a)	37. (a)	38. (a)	39. (a)	40. (d)
41. (b)	42. (a)	43. (a)	44. (d)	45. (b)	46. (a)	47. (c)	48. (d)	49. (c)	50. (a)
51. (a)	52. (d)	53. (a)	54. (a)	55. (b)	56. (d)	57. (c)	58. (d)	59. (a)	60. (d)
61. (d)	62. (a)	63. (c)	64. (d)	65. (a)	66. (d)	67. (c)	68. (a)	69. (c)	70. (a)
71. (b)	72. (a)	73. (d)	74. (a)	75. (c)	76. (c)	77. (b)	78. (d)	79. (a)	80. (d)
81. (c)	82. (c)	83. (d)	84. (c)	85. (c)	86. (b)	87. (a)	88. (b)	89. (d)	90. (d)
91. (a)	92. (a)	93. (a)	94. (b)	95. (b)	96. (b)	97. (d)	98. (c)	99. (b)	100. (b)
101. (a)	102. (a)	103. (d)	104. (b)	105. (b)	106. (d)	107. (c)	108. (d)	109. (d)	110. (c)
111. (d)	112. (a)	113. (c)	114. (a)	115. (b)	116. (c)	117. (b)	118. (c)		

(B) Miscellaneous questions

1. (a)	2. (b)	3. (a)	4. (c)	5. (b)	6. (a)	7. (b)	8. (c)	9. (a)	10. (b)
11. (a)	12. (a)	13. (b)	14. (a)	15. (c)	16. (c)	17. (d)	18. (d)	19. (b)	20. (a)
21. (b)	22. (d)	23. (c)	24. (a)	25. (b)	26. (a)	27. (c)	28. (a)	29. (a)	30. (a)
31. (c)	32. (b)	33. (c)	34. (b)	35. (d)	36. (b)	37. (a)	38. (a)	39. (a)	40. (b)

(C) Medical entrances' gallery

1. (b)	2. (c)	3. (c)	4. (a)	5. (b)	6. (a)	7. (d)	8. (b)	9. (a)	10. (b)
11. (c)	12. (d)	13. (c)	14. (d)	15. (d)	16. (c)	17. (b)	18. (b)	19. (b)	20. (a)
21. (c)	22. (c)	23. (b)	24. (c)	25. (d)	26. (b)	27. (d)	28. (a)	29. (a,d)	30. (b)
31. (d)	32. (b)	33. (c)	34. (c)	35. (b)	36. (c)	37. (a)	38. (a)	39. (e)	40. (d)
41. (a)	42. (b)	43. (d)	44. (d)						

Hints & Explanations

(A) Topical exercise

23 (*b*) $\Delta G = \Delta H - T\Delta S$

At equilibrium, $\Delta G = 0$

$\therefore \quad \Delta H = T\Delta S$

29 (*b*) $\log \frac{x}{m} = \log K + \frac{1}{n} \log p$

From $y = mx + c$, slope $\frac{1}{n} = \tan 45° = 1$ and $\log K = 0.3010$

$K = 2$

$\therefore \quad \log \frac{x}{m} = \log 2 + 1 \log 0.2$

$= \log 2 + \log 0.2 = \log 0.4$

$\therefore \quad \frac{x}{m} = 0.4$

32 (*b*) Langmuir adsorption involves formation of only a single layer. It may be due to van der Waals' forces or due to chemical bonding.

46 (*a*) Both the statements are correct but dehydration is not due to porous nature but due to acidic group present in zeolite.

50 (*a*) Activity of enzyme is affected by change in pH and is maximum at optimum pH.

Further with change in pH, the extent of protonation of the base changes.

68 (*a*) SnO_2 dissolves in NaOH forming Na_2SnO_3.

$$SnO_2 + 2NaOH \longrightarrow Na_2SnO_3 + H_2O$$

SnO_3^{2-} is adsorbed by SnO_2 forming negatively charged colloidal sol $[SnO_2]SnO_3^{2-} : 2Na^+$.

76 (*c*) Protein solution are both type hydrophobic and hydrophilic.

Hydrophobic amino acid - insoluble.

Hydrophilic amino acid - water soluble.

79 (*a*) Fe^{3+} neutralises the charge on the negatively charged particles in the sol and causes its coagulation.

$$2Fe^{3+} + 3[As_2S_3]S^{2-} \longrightarrow Fe_2S_3 + 3As_2S_3$$

81 (*c*) Arsenic sulphide is a negatively charged sol, so $Al_2(SO_4)_3$ with Al^{3+} is most effective to coagulate this sol.

84 (*c*) From the definition of gold number 10 mL of gold sol is prevented from coagulation in the presence of 1 mL of 10% NaCl by 0.03 mg haemoglobin.

$\therefore$ 100 mL of gold sol is prevented from coagulation in 10 mL of 10% NaCl by 0.3 mg haemoglobin.

85 (*c*) Negatively charged colloidal sol is formed when $AgNO_3$ is completely precipitated as AgI and extra KI is adsorbed on AgI.

$$Ag^+ + I^- \longrightarrow AgI$$

$$AgI + I^- \longrightarrow [AgI]I^-$$

Thus, $[Ag^+] < [I^-]$

$$\underset{\substack{3\text{ mol} \\ \text{or } 1\text{ mol}}}{3[AgI]I^-} + \underset{\substack{1\text{mol} \\ \frac{1}{3}\text{ mol}}}{Fe^{3+}} \longrightarrow 3AgI + FeI_3$$

Thus, 1 mol of $[AgI]I^-$ is coagulated by the addition of 1 mol of $AgNO_3$ or 1/2 mol $Pb(NO_3)_2$ or 1/3 mol of $Fe(NO_3)_3$.

86 (*b*) $M_1V_1 > M_2V_2$ while in other option this is not follow

(a) $100 \times 0.1 = 10$

$100 \times 0.1 = 10$

$\therefore 10 = 10$

(b) $100 \times 0.1 = 10$

$50 \times 0.1 = 5$

$\therefore 10 > 5$

(c) $50 \times 0.1 = 5$

$100 \times 0.1 = 10$

$\therefore 5 < 10$

(d) $200 \times 0.1 = 20$

$100 \times 0.2 = 20$

$\therefore 20 = 20$

87 (*a*) Coagulating value is expressed in millimol L^{-1}.

$\therefore \quad$ millimol of NaCl $= 5 \times 1 = 5$

colloidal sol = 100 mL = 0.1 L

$\therefore \quad$ Coagulating value $= \frac{5}{0.1} = 50$

102 (*a*) Coagulation of colloidal sol takes place due to neutralisation of its charge by electrolyte.

116 (*c*) Clay of river is coagulated by the salts present in sea water so delta is formed at the place where river water meets with sea water.

(B) Miscellaneous questions

13 (*b*) For a positively charged sol, anion carrying more negative charge is more effective and thus have maximum coagulating power. Hence, coagulating power is maximum for Na_3PO_4 (with PO_4^{3-} anion).

17 (*d*) From Freundlich adsorption isotherm

$$\frac{x}{m} = Kp^{1/n}$$

At low p, $\quad \frac{x}{p} \propto p$

and at high $\quad p\frac{x}{m} = K \propto p^\circ$

19 (*b*) Order of activation energy is $C < B < A$ and a catalyst increases the rate of reaction by decreasing activation energy.

So, the order of ease of reaction is $A < B < C$.

20 (*a*) $AgI + I^- \longrightarrow \underset{\text{Negatively charged}}{[AgI]I^-}$

25 (*b*) Coagulating power of the electrolyte is inversly proportional to its coagulating value

Thus, $\dfrac{\text{Coagulating power of } AlCl_3}{\text{Coagulating power of NaCl}} = \dfrac{52}{0.093} = 559$

Thus, coagulating power of $AlCl_3$ is 559 times more than that of $NaCl_3$.

29 (*a*) The colour of sol is due to scattering of light. It depends upon the size of particle highest gold sol has red colour. As the size of the particles increases, it becomes purple and finally golden yellow.

33 (*c*) Initial amount of oxalic acid present

in 50 mL = 50×1 millimol

$= 0.050 \text{ mol} = 0.050 \times 126 = 6.3 \text{ g}$

Amount of oxalic acid in 50 mL after adsorption

$= \dfrac{6.3}{2} \text{ g} = 3.15 \text{ g}$

$\therefore$ Amount of oxalic acid adsorbed per gram of charcoal

$\dfrac{3.15}{0.5} = 3.15 \times 2 = 6.3 \text{ g}$

34 (*b*) Number of N_2 molecules adsorbed per gram of silica gel

$= \dfrac{1.30 \times 6 \times 10^{23}}{22400} = 3.48 \times 10^{19}$

Area of cross-section of molecule $= 0.16 \text{ nm}^2 = 1.6 \times 10^{19} \text{ m}^2/\text{g}$

Surface area covered per gram of silica gel

$= 1.6 \times 10^{-19} \times 3.48 \times 10^{19} = 5.568 \text{ m}^2/\text{g}$

35 (*d*) $AgNO_3 + KI \longrightarrow AgI \xrightarrow[\text{(excess)}]{AgNO_3} \underset{\text{Colloidal sol}}{[AgI]Ag^+}$

Sol is coagulated by anion

$\underset{1}{[AgI]Ag^+} + \underset{1}{NaI} \longrightarrow$

$\underset{1}{[AgI]Ag^+} + \dfrac{1}{2} MgSO_4 \longrightarrow$

$[AgI]Ag^+ + \dfrac{1}{3} AlPO_4 \longrightarrow$

Thus ratio $\quad 1 : \dfrac{1}{2} : \dfrac{1}{3}$

$6 : 3 : 2$

37 (*a*) Size of colloidal sol ≈ 10 nm $\quad \therefore V_C \approx (10)^3 \text{ nm}^3$

Size of true solution = 1nm $\quad V_S \approx (1) \text{ nm}^3$

$\therefore \quad \dfrac{V_C}{V_S} = 1000$

38 (*a*) CH_3COOH adsorbed $= (0.06 - 0.042) \times 50 \times 60$ mg

$= 54$ mg

$\therefore$ CH_3COOH adsorbed per gram of charcoal

$= 54/3 = 18$ mg

(C) Medical entrances' gallery

1 (*b*) Stability of a colloidal solution is explained by zeta potential. The particles of colloid carry an electric charge, i.e. acquire positive or negative charge by preferential adsorption of positive or negative ions from dispersion medium. The combination of two layer of positive and negative charges around the colloid is called Helmholtz electrical double layer and the potential difference of this double layer is called zeta potential.

2 (*c*) 3 and 4 are correct statements, whereas 1 and 2 are incorrect.

1. Dry ice or $CO_2(s)$ [not $CO_2(g)$] is used as refrigerant for ice-cream and frozen food.
2. C_{60} contains 20 hexagones (not 12) and 12 pentagones (not 20).
3. ZSM-5, a type of zeolite, is used to convert alcohols directly into gasoline. It is true.
4. CO is a colourless and odourless gas. It is true.

3 (*c*) (a) In TiO_2 sol, TiO_2 particles are positively charged.

(b) In blood, haemoglobin is positively charged.

(c) In starch sol, starch is negatively charged.

(d) In hydrated Al_2O_3, sol $Al_2O_3 \cdot xH_2O$ is positively charged.

Hence, option (c) is the correct.

4 (*a*) According to Freundlich adsorption isotherm,

$$\frac{x}{m} = kp^{1/n}$$

where, x/m = amount of the gas adsorbed per unit

m = mass of adsorbent

p = pressure, k and n = constants

The value of n lies in between 0 to 1.

Thus, $\dfrac{x}{m} = kp^{0.3}$ and option (*a*) is correct.

5 (*b*) In butter, the dispersed phase is liquid and dispersion medium is solid. This type of colloid is called gel. It is liquid-solid emulsion.

6 (*a*) The process of settling of colloidal particles due to the neutralisation of their charge by any means is called coagulation.

Coagulation power of an ion depends on both magnitude and sign of the charge (positive or negative) on the ion. This fact can be explained by Hardy-Schulze rule.

According to this rule "greater the valency of the coagulating ion/flocculating ion (oppositely charged ion) added, the greater is its power to cause coagulation.

To coagulate a positively charged sol, the order of coagulating power of negative ion is

$$I^- < SO_4^{2-} < PO_4^{3-} < [Fe(CN)_6]^{4-}$$

Similarly, to coagulate a negatively charged sol, the order of coagulating power of positive ions is

$$Ag^+ < Pb^{2+} < Fe^{3+} < Si^{4+}$$

7 (*d*) When excess KI solution is added to $AgNO_3$ solution, negatively charged solution results due to adsorption of I^- from dispersion medium forming AgI/I^-. Having acquired a negative charge, this layer attracts counter ions from the medium forming a second layer as $AgI/I^- : K^+$.

8 (*b*) Both assertion and reason are true but reason is not the correct explanation of assertion.

At a certain concentration, surfactant molecule aggregates and forms micelle. This concentration is called critical micelle concentration (CMC). Aggregation of surfactant molecules (ions; $RCOO^-$), i.e. micelle formation causes effective fall in number of free ions to conduct electricity, thus conductivity decreases at CMC.

9 (*a*) Bleeding due to a cut can be stopped by applying $FeCl_3$ or alum solution. This is due to coagulation of negatively charged colloidal blood particles (albuminoid substance) by positively charged Fe^{3+} ions. These substances are used as styptic (which check the flow of blood).

10 (*b*) For a reversible reaction, catalyst accelerates the speed of forward as well as backward reaction to the same extent. Hence, it does not disturb the equilibrium, i.e. does not change the equilibrium constant of the reaction but helps to attain the equilibrium faster.

11 (*c*) Curve (c) is in accordance with Freundlich adsorption isotherm as:

$$\frac{x}{m} = k \cdot p^{1/n}$$

or $$\log \frac{x}{m} = \log k + \frac{1}{n} \log p$$

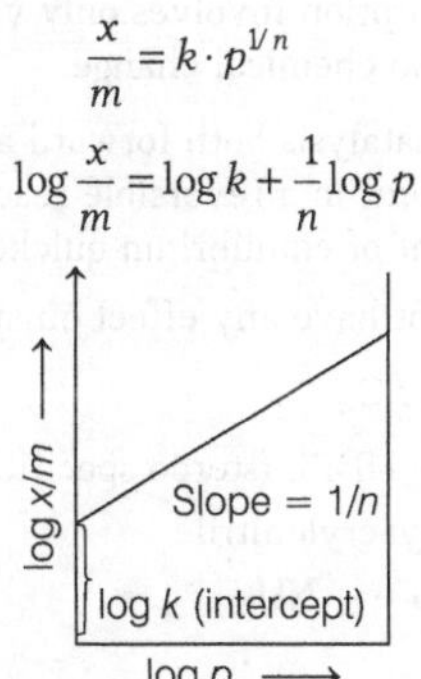

12 (*d*) Fog is a colloidal solution of liquid in a gas. In which liquid is the dispersed phase whereas gas is the dispersion medium.

Example of other options are as follows :

Gas in liquid ; Shaving cream soda water froth

Solid in gas ; Dust in air

Gas in gas ; Atmospheric air

13 (*c*) Lower the coagulating power higher is the coagulation value in mm per litre, i.e. coagulating power is inversely proportional to coagulation values. Thus, correct order of coagulating power is

$$MgSO_4 > BaCl_2 > NaCl \text{ or III} > \text{II} > \text{I}$$

because order of coagulating value is I > II > III.

14 (*d*) Tyndall effect is associated with the scattering of light by the colloidal particles hence, it is independent of charge present on the particles.

15 (*d*) An aqueous solution of NaOH attains pink colour with a drop of phenolphthalien due to its basic nature and presence of OH^- ions.

16 (*c*) According to Freundlich adsorption isotherm,

$$\frac{x}{m} = kp^{1/n}$$

where, $\frac{x}{m}$ is the amount of the gas adsorbed per gram of the adsorbent (x/m) and k and n are the constants depending upon nature of the adsorbate and adsorbent at a particular temperature. The factor $1/n$ has values between 0 and 1 (probable range is 0.1 to 0.5).

On taking log both sides, we get

$$\log \frac{x}{m} = \frac{1}{n} \log p + \log k$$

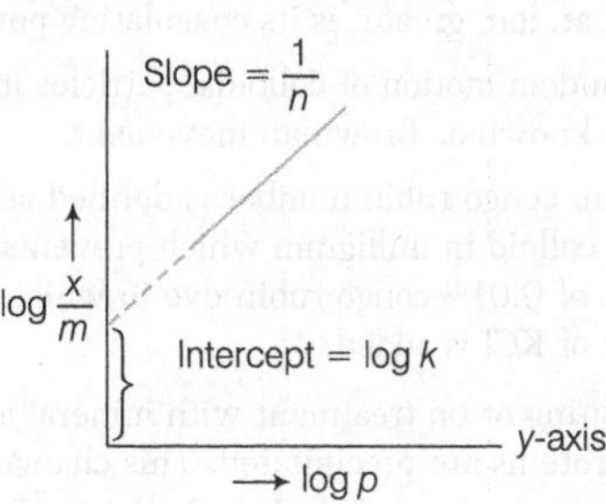

A graph between $\log \frac{x}{m}$ against $\log p$ should therefore give a straight line with slope equal to $1/n$ and ordinate intercept equal to $\log k$.

17 (*b*) Physical adsorption is an instantaneous process in which van der Waals' forces exist between the adsorbate and adsorbent. Thus, activation energy of such adsorption is low and it results in the formation of multimolecular layer. Physisorption occurs at low temperature and decreases with rise in temperature.

18 (*b*) $TiCl_4$ and AlR_3 [R = propyl, ethyl] represents *Ziegler-Natta* catalyst which acts as a catalyst in polymerisation of ethene.

19 (*b*) Since, AgI is a negatively charged sol, so according to Hardy Schulze law, higher the positive charge on the metal atoms, more is the tendency of metal atom to cause coagulation of negatively charged colloid.

The order of positive charge ions is $Na^+ < Ba^2 < Al^{3+}$.

Hence, the order for coagulation of AgI solution by given electrolyte

$$Al(NO_3)_3 > Ba(NO_3)_2 > NaNO_3$$

20 (*a*) Lyophobic colloids a readily coagulated but it is difficult to coagulate lyophilic colloid. When lyophilic colloid is added to lyophobic one. The lyophobic colloid gets enveloped by lyophilic colloid. Therefore, former remains protected against action of electrolytes.

21 (*c*) Gold sol contains particles of various sizes having several atoms of gold. Thus, it is a multimolecular colloid instead of macromolecular colloid.

22 (*c*) **Adsorption** is the phenomenon of attracting and retaining the molecules of a substance on the surface of solid or liquid resulting into higher concentration of molecules on the surface. The substance thus adsorbed on the surface is called **absorbate** and the substance on which it is adsorbed is called **adsorbent.**

So, activated charcoal is adsorbent here because oxalic acid is adsorbed on its surface.

23 (*b*) Blood is an example of colloid. Particles which range between 1 mm-100 mm from colloidal solution. It is a heterogeneous solution.

Urea, cane sugar and NaCl are crystalloids, i. e. form true solution. Crystalloids can easily pass through parchment membranes while colloids do not diffuse or diffuse slowly.

24 (*c*) The correct order for coagulation power is $AlCl_3 > BaCl_2 > NaCl$. The reason for this is that greater the valency of an ion, greater is its coagulating power.

25 (*d*) The random motion of colloidal particles in the dispersion medium is known as Brownian movement.

26 (*b*) The term congo rubin number is defined as the amount of protective colloid in milligram which prevents colour change in 100 mL of 0.01% congo rubin dye to which 0.16 g equivalent of KCl is added.

27 (*d*) On heating or on treatment with mineral acids the colloidal proteins are precipitated. This change is irreversible and is known as coagulation. It is due to Ca^{2+} ions.

28 (*a*) In Freundlich adsorption isotherm

$$x/m = kp^{1/n}$$

So, the constant n is a positive integer and thus, its value is always greater than 1. The factor $1/n$ has values between 0 and 1 (probably 0.1 to 0.5). So, the value of $1/n$ lies between 0 and 1 in all cases.

29 (*a, d*) As the adsorption process is exothermic, the physical adsorption occurs readily at low temperature and **decreases** with increasing temperature (in accordance to Le-chatelier's principle, exothermic process favours backward reaction, thus adsorption decreases with increase in temperature). In other words, adsorption **increases** with **decrease** in temperature.

30 (*b*) The true statement is that the particle size of adsorbent will affect the amount of adsorption.

31 (*d*) It is seen that, higher the critical temperature of the gas, more easily it can be liquefied, i.e. greater are the intermolecular forces of attraction of gaseous molecules on the surface of adsorbent. Hence, greater will be the adsorption.

Gas	Critical temperature
H_2	4.7
CO_2	4.8
CO	9.3
NH_3	181

Hence, NH_3 is adsorbed to a larger extent on charcoal.

32 (*b*) Lyophobic sols unstable, so they are stabilised by adding some lyophilic colloids which protect them from precipitation. Thus, lyophilic colloids are called protecting colloids. Their protecting power is expressed in terms of gold number.

33 (*c*) Cow milk is a natural emulsion stabilised by casein, milk protein.

34 (*c*) The simplest way to check whether a system is colloidal is to pass a beam of light through the colloidal solution, the path becomes visible as a bright streak. This is known as Tyndall effect.

35 (*b*) Gold sol is a lyophobic sol.

36 (*c*) An emulsion is a liquid-liquid colloidal system. If a mixture of two immiscible or partially miscible liquid is shaken, a coarse dispersion of one liquid in the other is obtained which is termed as emulsion, e. g. milk, vanishing cream, butter cream etc.

Milk is an emulsion of liquid for dispersed in water.

37 (*a*) Colloidal particles are bigger aggregate. Thus, the number of particles in the colloidal solution is comparatively smaller as compared to true solution. Hence, the values of colligative properties such osmotic pressure are of small order as compared to values shown by true solution at the same concentration.

38 (*a*) $\frac{x}{m} = p \times T$ is the incorrect relation.

39 (*e*) No doubt, physisorption is an exothermic process but its enthalpy of adsorption is quite low (20-40 kJ mol^{-1}). This is because physisorption involves only van der Waals' forces of attraction and no chemical change.

40 (*d*) A catalyst catalysis both forward and backward reactions to the same extent in a reversible reaction and thus, helps in the achievement of equilibrium quicker.

Thus, it does not have any effect on spontaniety of a reaction.

41 (*a*) V_2O_5 – H_2SO_4

Ziegler-Natta – HDPE (stereo specific)

Peroxide – Polyacrylonitrile

Finely divided Fe – NH_3

42 (*b*) Gel-butter

Butter is an example for gel.

43 (*d*) Peptisation is the process of converting a precipitate into colloidal sol by shaking it with dispersion medium in the presence of a small amount of electrolyte. It is the method of preparation of colloids.

44 (*d*) Ferric hydroxide is a positive solution. Thus, a negative ion cause the precipitation of positively charged sol and *vice-versa.* According to Hardy-Shulze rule, greater is the valency of the coagulating or flocculating ion, greater is its power to bring about coagulation.

Thus, in coagulation of a positive solution ferric hydroxide, the flocculating power is maximum for $[Fe(CN)_6]^{4-}$ ion. Thus $K_4[Fe(CN)_6]$ possesses the maximum coagulating power for $Fe(OH)_3$ solution.

CHAPTER 06

General Principles and Processes of Isolation of Elements

About 118 elements are known till today and out of them mostly are found associated with some other elements, i.e. in the combined state. This is due to their highly reactive nature. Only a few are found in their native state.

The processes through which a metal or element is obtained from its compounds are the topic of interest of this chapter. But before going in detail first we will discuss about their mode of occurrence and some important terminology.

OCCURRENCE OF METALS

Depending upon their reactivity, elements may be found in native (or free) state or in combined state

(*i*) In general less reactive metals like gold (Au), platinum (Pt) other metals of their group (Os, Rh, Ir, Pd), noble gases etc. are found in native states as these are not attacked by oxygen, moisture or carbon dioxide present in atmosphere.

(*ii*) More reactive metals such as Na, Mg, Zn etc. are easily attacked by oxygen, moisture or carbon dioxide present in atmosphere hence, these are found in combined state.

(*iii*) Earth crust is the main source of many elements and relative abundance (by weight) of some of them are shown in the following pie charts.

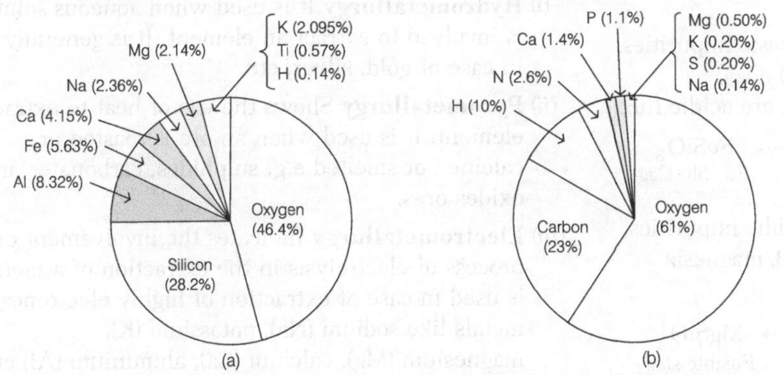

Fig. 6.1 Ten most abundant elements (a) in the earth's crust (b) in the human body

Inside

- Occurrence of metals
- Metallurgy
  - *Crushing and grinding of the ore*
  - *Conversion of concentrated ore into oxide*
- Methods of reduction of oxides or other salts to free metals
- Thermodynamic principles of metallurgy
  - *Ellingham diagram*
- Refining of crude metal
- Extraction and uses of some important metals
  - *Extraction of copper*
  - *Extraction of zinc*
  - *Extraction of aluminium*
  - *Extraction of iron*
  - *Extraction of tin*
  - *Extraction of silver*

(*iv*) From the above charts it is clear that

- oxygen is the most abundant **element.**
- aluminium is the most abundant **metal** in earth crust.
- first five elements comprise almost 92% by weight and first ten elements make up over 99.5% of the earth's crust.
- hydrogen is the 9th most abundant element in the earth's crust. (However, it is most abundant in cosmic region, i.e. in large planets like Jupiter, Saturn etc., due to their greater gravitational pull).

(*v*) In earth's atmosphere nitrogen is most abundant (78%) element. While in ocean hydrogen is most abundant element.

Important terminology related with extraction of elements

Before going in detail of extraction processes, it is very important to become aware of the terminology used in extraction process.

(*i*) **Minerals** Naturally occurring solid substances containing metals either in combined or in native state are called minerals.

(*ii*) **Ores** The minerals from which a metal can be extracted conveniently, economically and in large quantity are called ores. Thus, **"all ores are minerals but all minerals are not ores"**. Types and composition of some important ores/minerals are tabulated in table 6.1.

(*iii*) **Gangue or matrix** The earthly impurities like sand, clay, mica etc. associated with ores are called gangue or matrix.

(*iv*) **Flux** The substance added to convert infusible mass (impurities) into some fusible mass is called flux.

Infusible mass + flux $\longrightarrow$ fusible mass (slag)

Depending upon the nature of impurity, it may be acidic or basic.

- **Acidic flux** It is used to remove basic impurities. e.g., silica (SiO_2), boron trioxide (B_2O_3), phosphorus pentaoxide (P_2O_5) etc., are acidic flux.

e.g. $\underset{\text{Basic impurity (infusible)}}{FeO} + \underset{\text{Acidic flux}}{SiO_2} \longrightarrow \underset{\text{Fusible slag}}{FeSiO_3}$

- **Basic flux** It is used to remove acidic impurities. e.g., lime (CaO), lime stone ($CaCO_3$), magnesia (MgO) etc., are basic flux.

e.g. $\underset{\text{Acidic impurity (infusible)}}{SiO_2} + \underset{\text{Basic flux}}{MgO} \longrightarrow \underset{\text{Fusible slag}}{MgSiO_3}$

(*v*) **Slag** The fusible mass obtained by the reaction of flux and infusible mass is called slag and this process is called **slagging operation.**

Table 6.1 Important ores, their type and composition

Ore type	Ore/mineral and their composition
Uncombined metals	Ag, Au, Bi, Cu, Pd, Pt
Carbonates	$CaCO_3$ (calcite, limestone), $MgCO_3$ (magnesite), $CaCO_3 \cdot MgCO_3$ (dolomite), $PbCO_3$ (cerussite), $ZnCO_3$ (smithsonite or calamine), $FeCO_3$ (siderite), $2CuCO_3 \cdot Cu(OH)_2$ (azurite), $CuCO_3 \cdot Cu(OH)_2$ (malachite)
Halides	CaF_2 (fluorspar), NaCl (rock salt), KCl (sylvite), Na_3AlF_6 (cryolite), AgCl (horn silver or cerargyrite), $MgCl_2 \cdot 6H_2O$ (carnallite)
Oxides	$Al_2O_3 \cdot 2H_2O$ (bauxite), Cu_2O (cuprite), MnO_2 (pyrolusite), SnO_2 (cassiterite or tin stone), Al_2O_3 (corundum), Fe_2O_3 (haematite), TiO_2 (rutile), ZnO (zincite or red zinc), Fe_3O_4 (magnetite or load stone or magnetic oxide of iron)
Phosphates	$Ca_3(PO_4)_2$ (phosphate rock), $Ca_5(PO_4)_3OH$ (hydroxyapatite)
Silicates	Zn_2SiO_4 (willemite), $NaAlSi_3O_8$ (albite), $Mg_3(Si_4O_{10})(OH)_2$ (talc)
Sulphides	Ag_2S (argentite or silver glance), CdS (greenockite), Cu_2S (chalcocite or copper glance), FeS_2 (iron pyrite), HgS (cinnabar), PbS (galena), ZnS (sphalerite or zinc blende or black zinc), $CuFeS_2$ (copper pyrites), $Ag_3.SbS_3$ (pyrargyrite or ruby silver)
Sulphates	$BaSO_4$ (barite), $CaSO_4$ (anhydrite), $MgSO_4 \cdot 7H_2O$ (epsomite), $PbSO_4$ (anglesite).

METALLURGY

The complete scientific and technological process of extraction of a metals in its pure form from its ore is called metallurgy. It depends upon the nature of ore as well as associated impurities.

Depending upon the methods employed for extraction, metallurgical processes may be of following three types

(*i*) **Hydrometallurgy** It is used when aqueous solution is involved to extract an element. It is generally used in case of gold, silver, etc.

(*ii*) **Pyrometallurgy** Shows the use of heat to extract an element. It is used when an ore is roasted or calcined or smelted e.g. sulphides, carbonates and oxides ores.

(*iii*) **Electrometallurgy** Indicates the involvement of process of electrolysis in the extraction of a metal. It is used in case of extraction of highly electropositive metals like sodium (Na), potassium (K), magnesium (Mg), calcium (Ca), aluminium (Al) etc.

Common steps involved in metallurgy

Since, extraction of a metal depends upon the nature of ore and the associated impurities, so it is not possible to draw out a universal scheme for the extraction of all the metals. However, a general schematic representation of steps involved in the extraction process (depending upon the reactivity of metals) is as follows

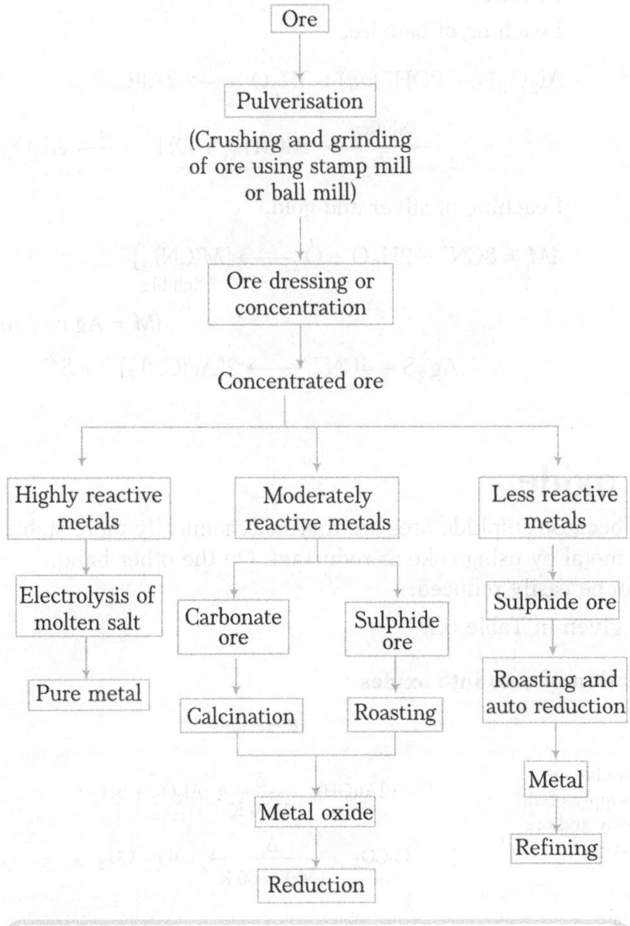

Weathering

The process of selective extraction of igneous rocks by water, CO_2 and humic acid is called weathering of rocks. It removes soluble ions like Na^+, K^+, Ca^{2+}, Cl^-, SO_4^{2-} etc. from the rocks while the insoluble residue contains mainly SiO_2, TiO_2, Fe_2O_3 etc.

Crushing and grinding of the ore

Ores are graded pulverised into powdered form by using crusher and grinder

Concentration of ore or beneficiation

The pretreatment of ores, based on physical properties and without bringing out any major chemical change in the ore is known as benefaction or concentration of the ore or ore dressing.

The various methods adopted for the ore dressing are as follows

1. Gravity separation or hydraulic washing or levigation

- Ores specification is based on the difference in the specific gravities or densities of the ore and gangue and usually employed for oxides and carbonate ores. Cassiterite or tin stone, chromite and haematite are concentrated by this method.
- **Process** Crushed ore is washed in a stream of water. The lighter particles are removed by water stream and heavier particles settle down.

2. Magnetic separation

- Ores specification is based on the difference in magnetic properties of minerals. If either the ore or the gangue is attracted by a magnetic field, it can be concentrated to yield a sample which is rich in metal. Magnetite (Fe_3O_4), haematite (Fe_2O_3), wolframite ($FeWO_4$), chromite ($FeO \cdot Cr_2O_3$) and illmenite ($FeO \cdot TiO_2$) are separated from non-magnetic impurities by this method.
- **Process** The pulverised mineral is passed over a rubber belt which moves on a pulley in a magnetic field. The non-magnetic gangue particles fall off in a vertical position when the belt passes over a pulley but the magnetic ore clings to the belt. When the belt is out of magnetic field, the ore separates. (Fig. 6.2)

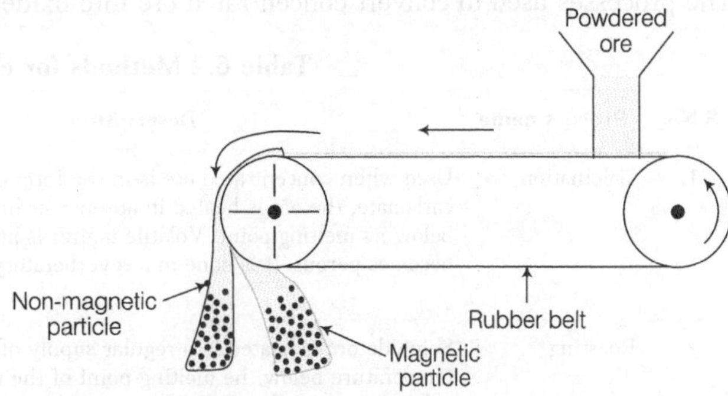

Fig. 6.2 Magnetic separation of ores

3. Froth floatation process

- Ores specification is based on the difference in wettability of different minerals in oil and water. It is used to concentrate sulphide ore.
- **Process** A suspension of powdered ore in water along with collector like pine oil, fatty acid, xanthate etc. (to enhance non-wettability of mineral particles towards water) and froth stabilisers like cresols, aniline etc. The mineral particles become wet by oils and gangue by water.

As a result the froth is formed and carries mineral particles. Being lighter, it comes at the surface.

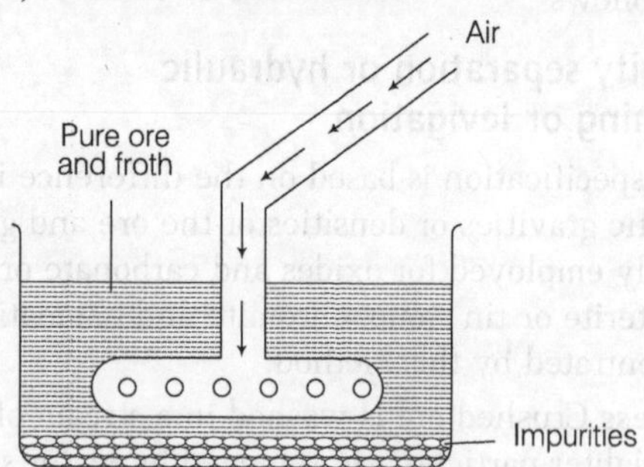

Fig. 6.3 Froth floatation process

Note *Some times depressants are used to prevent certain type of particles from forming the froth with the bubbles. NaCN is used as a depressants in the separation of ZnS and PbS. NaCN acts a depressant for ZnS but does not prevent PbS from the formation of the froth, due to complex formation.*

$$\underset{}{ZnS} + 4NaCN \longrightarrow \underset{\text{Sodium tetracyanozincate (II)}}{Na_2[Zn(CN)_4]} + Na_2S$$

After removing lead PbS, $CuSO_4$ is added to activate the depressed zinc sulphide and air is blown when ZnS floats. This method is known as ***differential floatation.***

4. **Leaching**

- Ores specification is based on solubility in some suitable solvent. Pure Al_2O_3 is obtained from the bauxite ore in the Bayer's process or Serpeck's method by leaching; also employed in concentration of silver and gold ores when leaching is done using NaCN in presence of air.
- **Process**

Leaching of bauxite,

$$Al_2O_3(s) + 2OH^-(aq) + 3H_2O \longrightarrow 2Al(OH)_4^-$$

$$\xrightarrow[CO_2]{\text{Dilution}} Al(OH)_3 + OH^- \xrightarrow{\Delta} Al_2O_3$$

Leaching of silver and gold,

$$4M + 8CN^- + 2H_2O + O_2 \longrightarrow \underset{\text{Soluble}}{[M(CN)_2]^-}$$

(M = Ag or Au)

$$Ag_2S + 4CN^- \longrightarrow 2[Ag(CN)_2]^- + S^{2-}$$

Conversion of concentrated ore into oxide

After concentration, the ores are converted into oxide. This is because sulphide ores are thermodynamically more stable than carbon disulphide (CS_2) and hence, cannot be reduced to metal by using coke as reductant. On the other hand, CO_2 is more volatile and thermodynamically more stable, so can be easily reduced.

The processes used to convert concentrated ore into oxides are given in Table 6.2

Table 6.2 Methods for conversion of ores into oxides

S.No.	Process name	Description	Reactions
1.	Calcination	Used when concentrated ore is in the form of hydroxide or carbonate, the ore is heated in absence or limited supply of air below its melting point. Volatile matter is burnt away and ore becomes porous. It is done in a reverberatory furnace.	$2Al(OH)_3 \xrightarrow[1500\text{ K}]{\Delta} \underset{\text{Alumina}}{Al_2O_3} + 3H_2O$ $\underset{\text{Limestone}}{CaCO_3} \xrightarrow[1300\text{–}1500\text{ K}]{\Delta} CaO + CO_2$
2.	Roasting	Sulphide ore is heated in a regular supply of air at a temperature below the melting point of the metal. It converts sulphide into oxide or sulphate; part of sulphide may also act as a reducing agent. Volatile impurities is burnt away and ore becomes porous. It is carried out in a reverberatory furnace.	$2ZnS + 3O_2 \xrightarrow{\Delta} 2ZnO + 2SO_2$ $2Cu_2S + 3O_2 \xrightarrow{\Delta} 2Cu_2O + 2SO_2$ $2Cu_2O + Cu_2S \xrightarrow{\Delta} 6Cu + SO_2$ **Removal of impurities** $S + O_2 \longrightarrow SO_2 \uparrow$ $4As + 3O_2 \longrightarrow 2As_2O_3 \uparrow$ $P_4 + 5O_2 \longrightarrow 2P_2O_5 \uparrow$

Note *Although, cassiterite (SnO_2) is an oxide ore but it is roasted because of the presence of a large amount of sulphur in it as impurity.*

CHECK POINT 6.1

1. Followed by oxygen which element is the most abundant in human body ?
(a) C (b) H
(c) Ca (d) P

2. Which of the following metals are found in native state Ag, Pt, C, Si, N, O, Mg, Na, Pb?
(a) Ag, Pt, C, N, O (b) Ag, Pt, Mg
(c) Ag, Pt, Pb, Mg (d) Ag, Pt

3. Corundum is .... mineral of Al.
(a) silicate (b) oxide
(c) double salt (d) sulphate

4. Which of the following is not an ore?
(a) Malachite (b) Calamine
(c) Stellite (d) Cerussite

5. Malachite is an ore of
(a) silver (b) mercury
(c) magnesium (d) copper

6. A carbonate ore is
(a) carnallite (b) limonite
(c) siderite (d) horn silver

7. Cassiterite is an ore of
(a) Mn (b) Ni (c) Sb (d) Sn

8. Chemical leaching is useful in the concentration of
(a) copper pyrites (b) bauxite
(c) galena (d) cassiterite

9. Calcination is the process of heating the ore
(a) in inert gas
(b) in the presence of air
(c) in the absence of air
(d) in the presence of CaO and MgO

10. Which of the following processes is related with the removal of sulphur present in the ore by heating it in the air?
(a) Smelting (b) Calcination
(c) Annealing (d) Roasting

Methods of reduction of oxides or other salts to free metals

Thermodynamic aspects show which reductant is better for which metal oxides (or salt) depending upon the nature of metal oxides (or salt) and reductants, the reduction processes are of following types

1. Smelting

The process of reduction of metal oxides (like oxides of less electropositive metals such as zinc, iron, tin, lead etc.) by using coke, coal or CO as reductant at high temperature is called **smelting.**

During smelting process, the roasted or calcined ore is heated with powdered coke or CO in the presence of a flux (acidic or basic depending upon the nature of infusible mass formed).

e.g.

(*i*) $ZnO + C \xrightarrow{\approx 1400^\circ C} CO + Zn$

$ZnO + CO \xrightarrow{>1800^\circ C} CO_2 + Zn$

(*ii*) $SnO_2 + 2C \longrightarrow Sn + 2CO\uparrow$

(*iii*) $Fe_2O_3 + CO \xrightarrow{823K} 2FeO + CO_2\uparrow$

$Fe_2O_3 + 3C \longrightarrow 2Fe + 3CO\uparrow$

(*iv*) $PbO + C \longrightarrow Pb + CO\uparrow$

or $PbO + CO \xrightarrow{\text{High temp.}} Pb + CO_2\uparrow$

This process is usually carried out in blast or reverberatory furnace. Blast furnace is used in the reduction of iron oxide while reverberatory furnace is used in the reduction of the oxide.

2. Reduction by aluminium

(Goldschmidt aluminothermic process)

The process of reduction of oxides of chromium, iron, manganese etc. (Cr_2O_3, Fe_2O_3, Mn_3O_4 etc.) by using aluminium metal as reductant is called **Goldschmidt aluminothermic process.**

$$Cr_2O_3 + 2Al \longrightarrow Al_2O_3 + 2Cr\downarrow$$

The mixture of iron and aluminium in the ratio of 3:1 is called **thermite mixture** and is used for repairing and welding of iron. The ignition is triggered off by burning a mixture of Mg ribbon and BaO_2

$$Fe_2O_3 + 2Al \longrightarrow Al_2O_3 + 2Fe\downarrow$$

Note *In this process, a mixture of BaO_2 and Mg is used as ignition mixture to supply heat.*

3. Self or auto-reduction

The process of reduction of metal oxides on their own without the use of any external reducing agent is called **auto-reduction.** This is used for reduction of sulphide ores of less electropositive metals like Hg, Pb, Cu, Ag, Au, Pt etc., undergo this process, as these are unstable at high temperature.

e.g.

(*i*) $2HgS + 3O_2 \longrightarrow 2HgO + 2SO_2$

$2HgO \xrightarrow{\Delta} 2Hg + O_2$

(*ii*) $2Cu_2S + 3O_2 \longrightarrow 2Cu_2O + 2SO_2$

The Cu_2O formed here reduces the remaining Cu_2S into Cu.

$$2Cu_2O + Cu_2S \longrightarrow 6Cu + SO_2$$

(*iii*) Similarly, $2PbS + 3O_2 \longrightarrow 2PbO + 2SO_2$

$$2PbO + PbS \longrightarrow 3Pb + SO_2$$

4. Electrolytic reduction

Highly electropositive metals like elements of group 1 and 2 are themselves good reductant, so these metals are extracted by electrolysis of their salts (generally halides, oxides etc.) in fused state. In this process the metal is deposited at cathode.

e.g.

(*i*) Sodium metal is obtained by electrolysis of fused NaCl or brine.

$$NaCl \rightleftharpoons Na^+ + Cl^-$$

$$Na^+ + e^- \longrightarrow Na \quad \text{(at cathode)}$$

$$Cl^- \longrightarrow Cl \quad \text{(at anode)}$$

$$Cl + Cl \xrightarrow{-e^-} Cl_2$$

(*ii*) Al is obtained by the electrolysis of fused mixture of Al_2O_3, cryolite and fluorspar.

Note *Fluorine (a non-metal) is obtained by the electrolysis of KHF_2 in HF as it reacts with water.*

5. Reduction by hydrogen

Hydrogen being inflammable can be used as a reductant only in a very few cases.

e.g. $WO_3 + 3H_2 \longrightarrow W + 3H_2O$

$$MoO_3 + 3H_2 \longrightarrow Mo + 3H_2O$$

$$CuO + H_2 \longrightarrow Cu + H_2O$$

$$Co_3O_4 + 4H_2 \longrightarrow 3Co + 4H_2O$$

This method is not widely used because many metals react with H_2 at elevated temperatures forming hydrides. Also there are chances of explosion of H_2 with O_2.

6. Displacement or precipitation method

In this method, a more reactive metal displaces the less reactive metals from their salt solutions and the less reactive metal become free,

e.g.

(*i*) Silver (Ag) and gold (Au) are obtained from their complex cyanides when more reactive zinc (Zn) metal displaces them.

$$2[M(CN)_2]^- + Zn \longrightarrow [Zn(CN)_4]^{2-} + 2M$$

[where, M = Ag, Au]

This type of precipitation process is called **cementation.**

(*ii*) Chlorine is used to displace bromine from sea water.

$$2Br^- + Cl_2 \longrightarrow 2Cl^- + Br_2$$

7. Other methods

Besides the above methods, some other methods can also be used to convert metal oxide or other salts into free metal. A small description of these methods is as follows

(*i*) Magnesium is used to reduce oxides in certain cases where the oxide is too stable,

$$TiO_2 + 2Mg \longrightarrow Ti + 2MgO$$

$$TiO_2 + 4Na \longrightarrow Ti + 2Na_2O$$

(*ii*) More electropositive metals are also used to reduce halides,

$$TiCl_4 + 2Mg \xrightarrow[\text{Kroll process}]{1000\text{-}1150°C} Ti + 2MgCl_2$$

$$TiCl_4 + 4Na \xrightarrow{\text{IMI process}} Ti + 4NaCl$$

(*iii*) Nickel oxide is reduced by using water gas,

$$2NiO + \underbrace{CO + H_2}_{\text{Water gas}} \longrightarrow 2Ni + CO_2 + H_2O$$

(*iv*) Amalgamation method is used in case of noble metals.

(*v*) A few compounds will decompose into their constituent elements simply by heating,

$$2AsH_3 \xrightarrow{\text{Red hot tube}} 2As + 3H_2$$

$$2NaN_3 \xrightarrow{\Delta} 2Na + 3N_2$$

Table 6.3 A short description of reduction methods for producing some common metals

	Metal	Reduction method
Least active	Au, Pt	Found in nature as the free metal.
↓	Cu, Ag, Hg	Roasting of the metal sulphide.
	V, Cr, Mn, Fe	Reduction of the metal oxide with carbon, hydrogen or a more active metal.
	Al	Electrolysis of molten Al_2O_3 in cryolite.
Most active	Li, Na, Mg	Electrolysis of the molten metal chloride.

THERMODYNAMIC PRINCIPLES OF METALLURGY

Thermodynamics plays a key role in deciding which reductant is suitable for a certain metal oxide to reduce it into free metal. It also help in deciding the temperature at which reduction of metal oxide is possible for a particular reductant (reducing agent).

We know that for a reaction to be spontaneous, Gibbs free energy change (ΔG) must be negative and ΔG is given by the expression

$$\Delta G = \Delta H - T\Delta S$$

where,

ΔH = Enthalpy change ΔG = Gibbs free energy change

T = Temperature ΔS = Enthalpy change

ΔG will be negative only under following two situations.

(i) If ΔS = + ve and temperature (T) is high enough, so that $T\Delta S > \Delta H$.

(ii) If reactants and products of two reactions are put together in a system and the net ΔG of the two possible reactions is negative, the overall reaction will be spontaneous.

Ellingham diagram

HJT Ellingham plots Gibbs free energy change (ΔG) values for formation of oxides against temperature and the plots obtained are called **Ellingham diagram** (Fig 6.4)

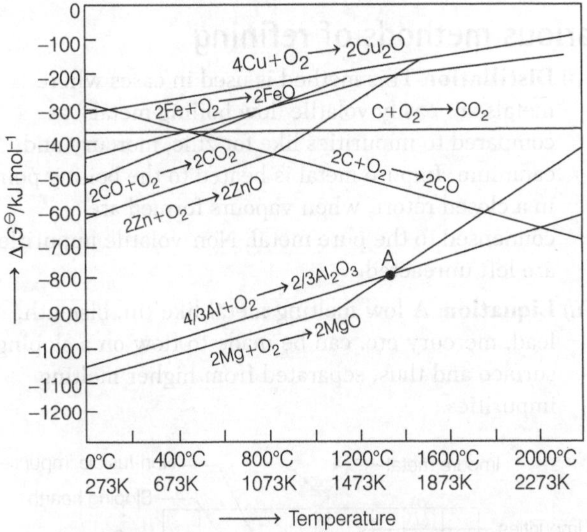

Fig. 6.4 Gibbs energy ($\Delta G°$) vs T-plots (schematic) for the formation of some oxides (Ellingham diagram)

From the figure, it is clear that

(i) Each plot is a straight line except when there is a phase change (solid to liquid or liquid to gas).

(ii) The phase change is shown by an abrupt change in the slope of the line or curve.

(iii) All the plots have positive slope because $\Delta G°$ becomes more positive with rise in temperature, i.e. stability of oxides decreases.

(iv) In each Ellingham plot, there is a point (temperature) below which the oxide is stable and above which it is unstable.

(v) The temperature at which two graphs intersect shows that the metal will reduce the oxide of other metals which lie above it in Ellingham diagram, i.e. the metal oxide of which having more negative value of $\Delta G_f°$ can reduce the oxide having less negative $\Delta G_f°$.

(vi) The decreasing order of negative values of $\Delta G_f°$ of metal oxides is as follows :

Ca > Mg (below 1773 K) > Al > Ti > Cr > C > Fe > Ni > Hg > Ag.

Thus,

- Al can reduce CrO, FeO, NiO but not MgO at a temperature lower than 1773 K.
- Mg can reduce Al_2O_3 below 1623K but above this temperature, MgO gets reduced by Al.

(vii) CO is a good reductant below 1073K but above this temperature, reducing power of coke is good e.g. Fe_2O_3 is reduced by CO below 1073K but above it, it is reduced by coke. Coke reduces ZnO above 1270 K.

Application of Ellingham diagram

Ellingham diagram are useful in deciding the reductant and temperature for reducing a specific metal oxide. e.g. For the reduction of zinc oxide with C, there is a competition between Zn and C for O atoms.

We consider following values at 1100°C for the oxidation of Zn and C

I. $2Zn + O_2 \longrightarrow 2ZnO$; $\Delta G° = -360$ kJ

II. $2C + O_2 \longrightarrow 2CO$; $\Delta G° = -460$ kJ

Reverse the equation (II) and subtract (II) from (I) gives

$2Zn + 2CO \longrightarrow 2ZnO + 2C$; $\Delta G° = +100$ kJ

or   III. $Zn + CO \longrightarrow ZnO + C$; $\Delta G° = +50$ kJ mol^{-1}

Reverse the equation (II) and subtract (II) from (I) gives

IV. $ZnO + C \longrightarrow Zn + CO$; $\Delta G° = -50$ kJ mol^{-1}

For reaction III in which Zn reduces CO, $\Delta G°$ is positive indicating that this reduction is not spontaneous (in forward direction), but for the reaction IV with negative value of $\Delta G°$, in which C reduces ZnO is spontaneous (in forward direction).

From an energy stand point it would certainly be more economical to carry it out at a lower temperature.

From the Ellingham diagram, it is observed that the lines ($2C + O_2 \longrightarrow 2CO$) and ($2Zn + O_2 \longrightarrow 2ZnO$) cross at about 950°C.

At about this temperature Zn and C have equal affinities for O atoms $\Delta G°$ for reaction (IV) is zero. Above this temperature $\Delta G°$ is negative and below this temperature $\Delta G°$ is positive. To make reaction (IV) go essentially to completion, we need a temperature somewhat in excess of 950°C.

Limitations

Although, Ellingham diagram are very helpful in deciding feasibility of a reaction (or in selection of a reductant for the metal oxides) but it fails to explain the following facts.

(*i*) It gives no idea about the kinetics of the reduction reaction (i.e. with what rate the reaction is going).

(*ii*) It assumed an equilibrium between the reactants and products but this situation is not always true because the reactant and or product may be in solid state.

Example 6.1 *The following reaction has negative Gibbs free energy change*

$Cr_2O_3(s) + 2Al(s) \longrightarrow Al_2O_3(s) + 2Cr(s)$; $\Delta G° = -421$ kJ

but it does not occur at room temperature, why?

Sol. In the given reaction, the reactants are in solid state and in this state they are unable to react at room temperature. At high temperature they get melted and hence, reaction proceeds.

Example 6.2 *Given* $\Delta_f G°(CuO) = -129.7 \text{ kJmol}^{-1}$, $\Delta_f G°(H_2O) = -237.2 \text{ kJmol}^{-1}$ *and* $\Delta_f G°(CO) = -137.2 \text{ kJmol}^{-1}$.
Among hydrogen, carbon monoxide and carbon, which one is good reductant for the reduction of CuO? *and why?*

Sol. (i) $CuO + C \longrightarrow Cu + CO$

$$\Delta G° = \Delta_f G°(CO) - \Delta_f G°(CuO)$$
$$= -137.2 + 129.7 = -7.5 \text{ kJmol}^{-1}$$

(ii) $CuO + H_2 \longrightarrow Cu + H_2O$

$$\Delta G° = \Delta_f G°(H_2O) - \Delta_f G°(CuO)$$
$$= 237.2 + 129.7 = -107.5 \text{ kJmol}^{-1}$$

More negative value of $\Delta G°$ for reaction (II) indicates that hydrogen is good reductant as compared to C for the reduction of CuO.

Example 6.3 *Following method of extracting Zn is based on thermodynamics:*

(*A*) $2ZnS + 3O_2 \longrightarrow 2ZnO + 2SO_2$

(*B*) $ZnO + C \longrightarrow Zn + CO$

$\Delta G°_f$ *(standard free energies of formation, in* kJ mol^{-1}*) of*

ZnS = - 205.4; ZnO = - 318.2

SO_2 = -300.4; CO = - 137.3

Calculate free energy changes of the above reactions and comment on the result.

Sol. For reaction (*A*)

$$\Delta G° = 2\Delta G°_f(ZnO) + 2\Delta G°_f(SO_2) - 2\Delta G°_f(ZnS)$$
$$= 2[-318.2 - 300.4 + 205.4]$$
$$= -826.4 \text{ kJ} \quad (\Delta G°_f \text{ (element)} = 0)$$

For reaction (*B*)

$$\Delta G° = \Delta G°_f(CO) - \Delta G°_f(ZnO)$$
$$= -137.3 + 328.2 = +180.9 \text{ kJ}$$

For reaction (*A*), $\Delta G°$ is negative hence, reaction is spontaneous; reaction (*B*) is non-spontaneous $\Delta G°$ being positive for it.

REFINING OF CRUDE METAL

The metals obtained by the extraction process is also associated with some impurities. The process of removal of impurities to get more pure metal is called refining. In other words, refining is the process of purification of metals. Depending upon the nature of metals and associated impurities, different methods are employed to purify metals.

Various methods of refining

(*i*) **Distillation** This method is used in cases where metals are easily volatile (low boiling metals) as compared to impurities like for zinc, mercury and cadmium. Impure metal is heated to the boiling point in a closed retort, when vapours formed are condensed to the pure metal. Non-volatile impurities are left unreacted.

(*ii*) **Liquation** A low melting metal like tin, bismuth, lead, mercury etc. can be made to flow on a sloping surface and thus, separated from higher melting impurities.

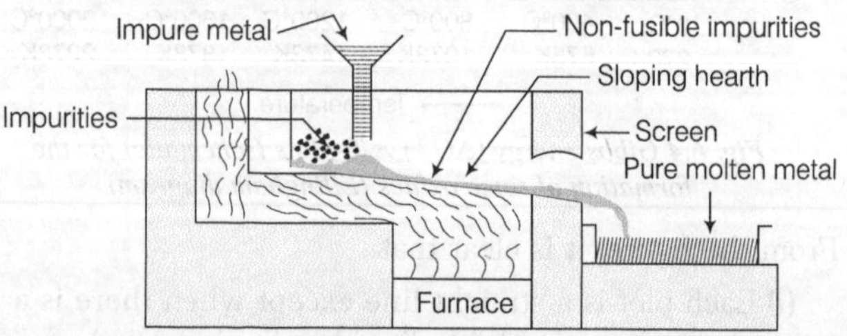

Fig. 6.5 Liquation

(*iii*) **Electrolysis** Used for copper, gold, silver, lead, zinc, aluminium and nickel etc. Impure metal is made anode and the pure metal is made cathode. A solution of metal salt is used as electrolyte. The net result is the transfer of pure metal from anode to cathode.

At anode, $\underset{\text{Impure}}{Cu} \longrightarrow Cu^{2+} + 2e^-$

At cathode, $Cu^{2+} + 2e^- \longrightarrow \underset{\text{Pure}}{Cu(s)}$

$$Cu\ (Impure) \longrightarrow Cu\ (Pure)$$

The insoluble impurities settle down below the anode as anode-mud.

Anode-mud contains Au, Pt, Ag and other precious, metals (noble metals).

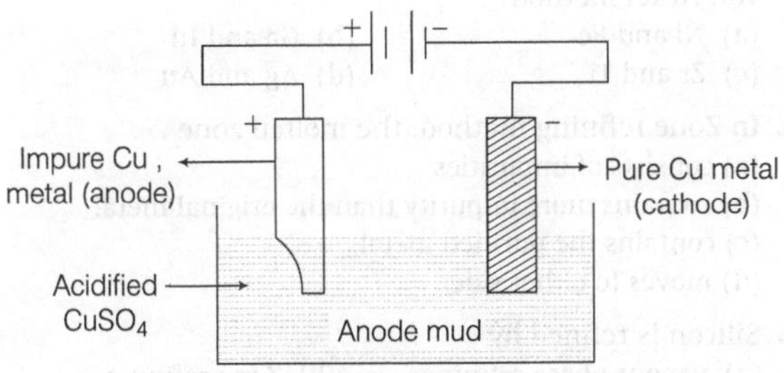

Fig. 6.6 Electrolysis

(*iv*) **Vapour phase refining** Involve conversion of metal to volatile compound and decomposition (at high temperature) of compound to pure metal.

- In Mond's process for the refining of nickel.

$$\underset{\text{Impure}}{Ni} + 4CO \xrightarrow{330\text{-}350\,K} \underset{\substack{\text{Nickel tetracarbonyl}\\ \text{(volatile)}}}{Ni(CO)_4}$$

$$Ni(CO)_4 \xrightarrow{450\text{-}470\,K} \underset{\text{Pure}}{Ni} + 4CO\uparrow$$

- In van Arkel method for zirconium (Zr) or titanium (Ti).

$$\underset{\text{Impure}}{Zr(s)} + 2I_2 \xrightarrow{870\,K} ZrI_4 \xrightarrow{1800\,K} \underset{\text{Pure}}{Zr(s)} + 2I_2(g)$$

$$\underset{\text{Impure}}{Ti(s)} + 2I_2 \xrightarrow{523K} TiI_4 \xrightarrow[\text{on W filament}]{1700\,K} \underset{\text{Pure}}{Ti(s)} + 2I_2(g)$$

(*v*) **Zone refining** This method is based upon the principle of fractional crystallisation, it means this method is based on difference in solubilities of impurities in molten and solid state of metal.

This is very useful for producing semiconductor and other metals of very high purity e.g. silicon, germanium, boron, gallium and indium. It is based on the fact that impurities are more soluble in the melt state than in the pure metal. Impure metal (in the form of rod) is moved through a circular heater. When melting point is reached, pure metal melts. The impurities are concentrated in a particular zone Further, pure metal from molten zone crystallises out in pure form. Process is repeated several times to concentrate impurities at one end which is cut off.

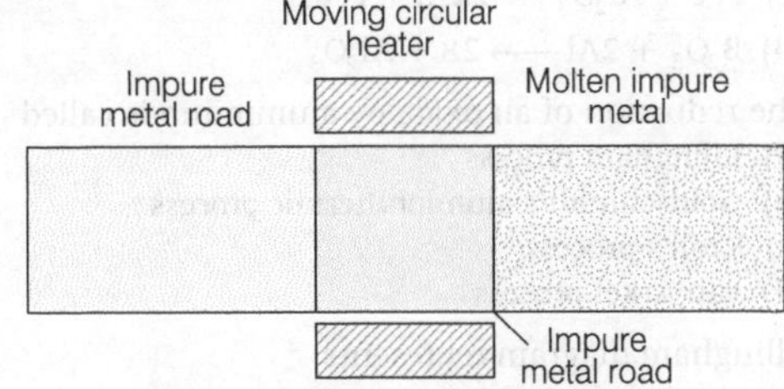

Fig. 6.7 Zone refining of metals

(*vi*) **Poling** Used to purify metal having impurity of its oxide like copper, tin etc. Molten metal is stirred with green logs of wood and the hydrocarbon gases (such as CH_4) released from the wood reduce the metal oxides.

$$3Cu_2O + CH_4 \longrightarrow CO + 6Cu + 2H_2O$$

(*vii*) **Cupellation** Used when metal is associated with volatile impurities e.g. removal of traces of lead from silver, (as volatile PbO).

(*viii*) **Oxidative refining** Here, impurities are oxidised used to convert pig iron and cast iron to wrought iron using Bessemer converter.

(*ix*) **Chromatography** Based on differential adsorption of different components over an adsorbent. Weakly adsorbed component is eluted first and the strongly adsorbed component is eluted afterwards. Its different types are column chromatography, paper chromatography, thin layer chromatography etc. Adsorbent used are alumina, silica etc.

CHECK POINT 6.2

1. The metal which cannot be extracted by smelting process?
 (a) Zn (b) Al (c) Pb (d) Fe
2. Thermite is a mixture of iron oxide and
 (a) aluminium powder (b) zinc powder
 (c) potassium metal (d) sodium metal
3. Which of the following reactions is not involved in thermite process?
 (a) $3Mn_3O_4 + 8Al \longrightarrow 9Mn + 4Al_2O_3$
 (b) $Cr_2O_3 + 2Al \longrightarrow Al_2O_3 + 2Cr$
 (c) $2Fe + Al_2O_3 \longrightarrow 2Al + Fe_2O_3$
 (d) $B_2O_3 + 2Al \longrightarrow 2B + Al_2O_3$
4. The reduction of an oxide by aluminium is called
 (a) Ellingham process
 (b) Goldschmidt's aluminothermic process
 (c) Kroll's process
 (d) van Arkel process
5. Ellingham diagram represents
 (a) change of ΔG with temperature
 (b) change of ΔH with temperature
 (c) change of ΔG with pressure
 (d) change of $(\Delta G - T\Delta S)$ with temperature
6. van Arkel method of purification of metal involves converting the metal to a
 (a) volatile stable compound
 (b) volatile unstable compound
 (c) non-volatile stable compound
 (d) None of the above
7. Which of the following pairs of metals is purified by van Arkel method?
 (a) Ni and Fe (b) Ga and In
 (c) Zr and Ti (d) Ag and Au
8. In Zone refining method, the molten zone
 (a) consists of impurities
 (b) contains more impurity than the original metal
 (c) contains the purified metal
 (d) moves to either side
9. Silicon is refined by
 (a) vapour phase refining (b) Zone refining
 (c) liquation (d) van Arkel method
10. Chromatography is based on the principle of
 (a) Adsorption (b) Absorption
 (c) Concentration (d) Vaporisation

EXTRACTION AND USES OF SOME IMPORTANT METALS

Extraction of copper

Occurrence

- Cuprite : Cu_2O
- Copper glance : Cu_2S
- Malachite : $CuCO_3 \cdot Cu(OH)_2$
- Azurite : $2CuCO_3 \cdot Cu(OH)_2$
- Copper pyrites or chalcopyrites : $CuFeS_2$

Main source is **copper pyrites.**

Pyrometallurgical method (Dry method)

The crushed and sieved sulphide ore is converted into copper metal by following steps

Step I Concentration by **froth floatation process.**

Step II **Roasting** is done in multiple hearth furnace. Volatile impurities are removed as thin oxide (As_2O_3, Sb_2O_3, P_2O_5). It also converts iron sulphide to iron oxide. The copper remains as the sulphide if the temperature is kept below 800°C.

$$2FeS + 3O_2 \xrightarrow{\Delta} 2FeO + 2SO_2 \uparrow$$

Step III **Smelting** of the roasted ore is done in a furnace at 1400°C. In this furnace charge melts and separates into two layers. The bottom layer is **copper-matte,** consisting chiefly of the molten sulphides of copper and iron.

The top layer is a silicate slag formed by the reaction of oxides of Fe, Ca, and Al with SiO_2 (which either is present in the ore or is added) and is removed.

$$FeO + SiO_2 \longrightarrow FeSiO_3$$

Step IV **Bessemerisation** occurs in another furnace where air is blown through the molten copper matte. First, the remaining iron sulphide is converted to oxide (as in step II) followed by slag formation (as in step III). The slag is poured off and air is again blown through the furnace. Now the following reactions occur, yielding a product that is about 98 to 99% Cu.

$$2Cu_2S + 3O_2 \longrightarrow 2Cu_2O + 2SO_2$$

$$2Cu_2O + Cu_2S \xrightarrow{\text{Auto-reduction}} 6Cu + SO_2$$

The product of this step is called **Blister copper** because of frozen bubbles of SO_2. It can be used where high purity is not required.

Step V **Electrorefining**

Anode	:	impure copper
Cathode	:	pure copper
Electrolyte	:	3.5% $CuSO_4$ + 15% H_2SO_4
Voltage	:	0.3 to 0.5 V
Temperature	:	50.55°C

Cu from impure copper anode goes into solution and is deposited at the cathode

$$\underset{\text{Impure}}{Cu} \xrightarrow{\text{Anode}} Cu^{2+} + 2e^-$$

$$Cu^{2+} + 2e^- \xrightarrow{\text{Cathode}} \underset{\text{Pure}}{Cu}$$

Anode-mud contains Au, Pt, Ag and other precious metals (noble metals).

Hydrometallurgical process (Wet process)

Step I Leaching with H_2SO_4 or bacteria converts salts into sulphates

$$CuCO_3 + H_2SO_4 \xrightarrow{\Delta} CuSO_4 + H_2O + CO_2$$

$$Cu(OH)_2 + H_2SO_4 \xrightarrow{\Delta} CuSO_4 + 2H_2O$$

$$CuFeS_2 + 2H_2SO_4 \xrightarrow{\Delta} CuSO_4 + FeSO_4 + 2H_2S$$

Step II Electrochemical displacement; Fe is placed above Cu in electrochemical series hence, if Fe scrap is added to $CuSO_4$ solution, Cu is displaced

$$CuSO_4 + Fe \longrightarrow FeSO_4 + Cu\downarrow$$

Extraction of zinc

Zinc occurs in the earth's crust to the extent of 132 ppm by weight. It is the twenty-fourth most abundant element.

Occurrence

- Zinc blende (Sphalerite) : (ZnFe)S or ZnS
- Smithsonite (Calamine) : $ZnCO_3$
- Zincite : ZnO
- Willemite : Zn_2SiO_4
- Hemimorphite : $Zn_4(OH)_2(Si_2O_7)\cdot H_2O$

Main source is **zinc blende** from which the metal is extracted in the following steps

Step I Concentration by froth floatation process

Step II Roasting

$$2ZnS + 3O_2 \longrightarrow 2ZnO + 2SO_2$$

Step III Reduction of ZnO in smelter

$$ZnO + C \overset{1400°C}{\rightleftharpoons} Zn + CO$$

Above reaction is reversible and thus high temperature is required to shift the equilibrium towards formation of Zn. Thermodynamically,

$$\Delta G° = \Delta H° - T\Delta S°$$

To make $\Delta G°$ negative (for spontaneous reaction)

$$T\Delta S > \Delta H \quad \text{and} \quad T > \frac{\Delta H°}{\Delta S°}$$

If we use standard thermodynamic data, T should exceed 1200°C. Zn and CO_2 both are in gaseous state at this temperature. Reversible nature of the above reaction can be minimised

- by having excess of carbon so that CO_2 formed is converted to CO

$$C + CO_2 \longrightarrow 2CO$$

- by cooling the gaseous products so that equilibrium is not attained immediately.

The impure zinc metal is called **spelter** and is associated with impurities of Cd, Pb and Fe. Zinc obtained in this way is of 99% purity.

Step IV Refinement

(*i*) **By electrolysis** ZnO is dissolved in dil. H_2SO_4 and electrolysed.

Anode : Impure zinc plate

Cathode : Pure zinc plate

Electrolyte : $ZnSO_4$ / H_2SO_4

Zinc from impure zinc plate is oxidised at anode, goes into solution and reduced to pure zinc at cathode.

Anodic reaction,

$$\underset{\text{Impure}}{Zn} \longrightarrow Zn^{2+} + 2e^-$$

Cathodic reaction,

$$Zn^{2+} + 2e^- \longrightarrow \underset{\text{Pure}}{Zn}$$

(*ii*) **By distillation** When it is distilled with a dilute KNO_3, we get pure zinc metal.

Extraction of aluminium

Occurrence

Bauxite	:	$Al_2O_3\cdot 2H_2O$
Kaolin	:	$Al_2O_3\cdot 2SiO_2\cdot 2H_2O$
Cryolite	:	Na_3AlF_6
Feldspar	:	$K_2O\cdot Al_2O_3\cdot 6SiO_2$
Corundum	:	Al_2O_3
Orthoclase	:	$KAlSi_3O_8$
Beryl	:	$Be_3Al_2Si_6O_{18}$

Feldspar or orthoclase on weathering gives rise to porcelain-clay, china-clay or kaolin.

Concentration of bauxite

Depending upon the nature of impurity associated with bauxite ore, the following three methods can be used to concentrate the ore.

1. **Baeyer's process**

This method is used when the bauxite ore is associated with the impurity of Fe_2O_3 and TiO_2.

Step I In this method the powdered ore is treated with a concentrated solution of NaOH resulting in the dissolution of alumina. The other oxides (Fe_2O_3 and TiO_2) being basic remain insoluble in NaOH.

$$\underset{\text{Impure alumina}}{Al_2O_3} + 2OH^- + 3H_2O \longrightarrow \underset{\text{Soluble}}{2[Al(OH)_4]^-}$$

Step II Soluble $[Al(OH)_4]^-$ is diluted and then acidified. $Al(OH)_3$ is precipitated which on ignition at 1250° C gives pure alumina.

$$Al(OH)_4^- + H_3O^+ \longrightarrow Al(OH)_3 \downarrow + 2H_2O$$

$$2Al(OH)_3 \xrightarrow[\Delta]{1250°C} \underset{\text{Pure alumina}}{Al_2O_3} + 3H_2O$$

2. **Serpeck's process**

This process is used when SiO_2 is present as a major impurity in the bauxite ore.

In this process, the bauxite ore is mixed with coke and heated in the presence of nitrogen. The obtained aluminium nitride when digested gives ppt. of $Al(OH)_3$, which on ignition at 1250°C gives pure alumina.

$$\underset{\text{Impure alumina}}{Al_2O_3} + 3C + N_2 \xrightarrow{\Delta} 2AlN + 3CO\uparrow$$

$$AlN + 3H_2O \xrightarrow{\text{Digest}} Al(OH)_3 + NH_3$$

$$2Al(OH)_3 \xrightarrow[100°C]{\Delta} \underset{\text{Pure alumina}}{Al_2O_3} + 3H_2O$$

3. **Halls' process**

This method is used when **iron oxide (Fe_2O_3)** is present as the major impurity. In this method the bauxite ore is fused with Na_2CO_3 in the presence of carbon dioxide. The aluminium hydroxide $Al(OH)_3$ obtained on ignition gives pure alumina.

$$\underset{\text{Impure alumina}}{Al_2O_3 \cdot 2H_2O} + Na_2CO_3 \xrightarrow{\text{Fuse}} 2NaAlO_2 + CO_2\uparrow + 2H_2O$$

$$2NaAlO_2 + 3H_2O + CO_2 \xrightarrow{50°C} 2Al(OH)_3 \downarrow + Na_2CO_3$$

$$2Al(OH)_3 \xrightarrow{\Delta} \underset{\text{Pure alumina}}{Al_2O_3} + 3H_2O$$

Electrolysis of Al_2O_3

(Hall-Heroult process)

Al is usually extracted by this process. Al_2O_3 is melted with cryolite $Na_3[AlF_6]$, CaF_2 and AlF_3 and electrolysed in a graphite lined steel tank, which serves as cathode. The anodes are also made of graphite.

The cell runs continuously and at intervals molten Al (mp 660° C) is drained from the bottom of the cell and more bauxite is added (Fig. 6.8).

Cryolite improves the electrical conductivity of the cell as Al_2O_3 is poor conductor. This also serves as an added impurity and lowers the melting point of the mixture to about 950° C.

By this process Al of 99.6 to 99.8% purity is obtained. Probable net reactions are

Oxidation at anode $\quad C + 2O^{2-} \longrightarrow CO_2 + 4e^-] \times 3$

Reduction at cathode $\quad Al^{3+} + 3e^- \longrightarrow Al] \times 4$

Overall $\quad 3C + 4Al^{3+} + 6O^{2-} \longrightarrow 4Al + 3CO_2$

Reaction can also be written as

Anode (oxidation) $\quad 2O^{2-} \longrightarrow O_2 + 4e^-] \times 3$

Cathode (reduction) $\quad Al^{3+} + 3e^- \longrightarrow Al] \times 4$

Overall $\quad 2Al_2O_3 \longrightarrow 4Al + 3O_2$

The liberated oxygen gas reacts with carbon anodes (at elevated temperature) to form CO.

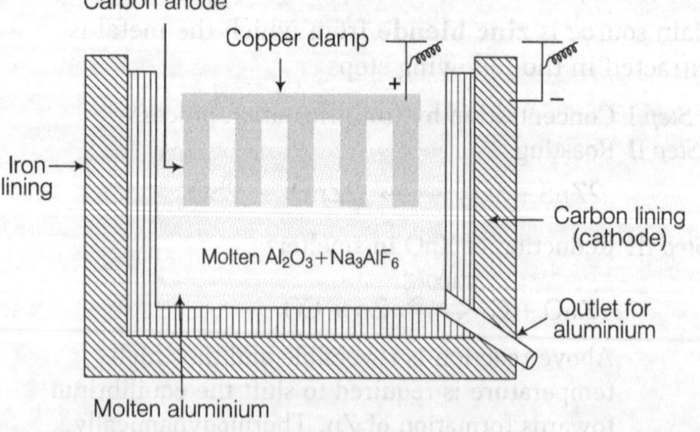

Fig. 6.8 Hall-Heroult process

Electrolytic refining (Hoope's process)

The obtained aluminium metal is not 100% pure. To purify it, carbon electrodes are used as cathode and iron tank linked with carbon bricks as anode.

The electrolyte contains three layers

(*i*) Bottom layer is made by using impure aluminium.

(*ii*) A molten mixture of fluorides of sodium (Na_3AlF_6), Ba and Al and Al_2O_3 is used to make middle layer.

(*iii*) At the top, there is a layer of pure molten aluminium.

When electric current is passed pure aluminium metal is deposited at cathode from the middle layer and an equivalent amount of Al moves from bottom layer to middle layer leaving behind the impurities.

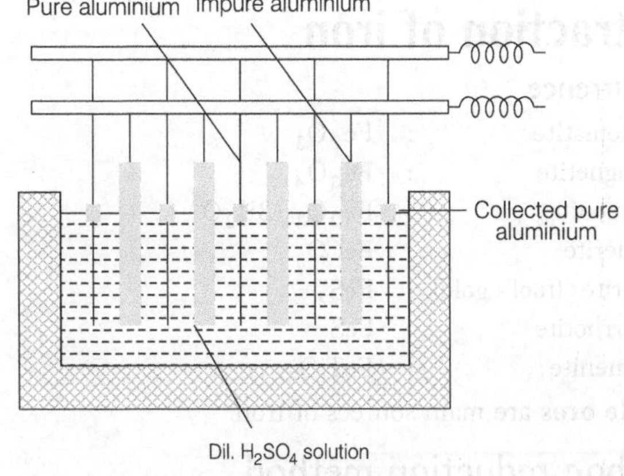

Fig. 6.9 Hoopes process

CHECK POINT 6.3

1. Consider the following steps

$$CuS \xrightarrow{\text{Roast in air}} (A) \xrightarrow{\text{Roast without air}} (B)$$

Which is not the correct statement?

(a) It is self reduction.
(b) It involves disproportionation, $Cu_2S \longrightarrow Cu + CuS$
(c) (*A*) is a mixture of CuO and CuS and (*B*) is a mixture of Cu and SO_2.
(d) All the above are incorrect statements.

2. In the extraction of Cu, the reaction takes place in Bessemer converter is

(a) $2Cu_2O + Cu_2S \longrightarrow 6Cu + SO_2$
(b) $2CuFeS_2 + O_2 \longrightarrow Cu_2S + FeS + SO_2$
(c) $2Cu_2S + 3O_2 \longrightarrow 2Cu_2O + 2SO_3$
(d) $2FeS + 3O_2 \longrightarrow 2FeO + 2SO_2$

3. Pyrometallurgical method used for the extraction of copper from sulphide ore

(a) is a dry method
(b) involves concentration by leaching the sulphide ore with dil. H_2SO_4
(c) involves concentration of the sulphide ore by Froth floatation process
(d) statement given in (b) is incorrect

4. Blister Cu is about

(a) 60% Cu (b) 90% Cu (c) 98% Cu (d) 100% Cu

5. Blister copper is

(a) impure copper
(b) obtained in self reduction process during Bessemerisation
(c) Both (a) and (b) are correct
(d) None of the above is correct

6. Extraction of zinc from zinc blende is achieved by

(a) electrolytic reduction
(b) roasting followed by reduction with carbon
(c) roasting followed by reduction with another metal
(d) roasting followed by self reduction

7. Consider the following flow-chart

$$\boxed{\text{Bauxite}} \xrightarrow[\uparrow N_2]{+C} \boxed{\text{AlN}} \xrightarrow{H_2O} Al(OH)_3 \xrightarrow{\Delta} Al_2O_3$$
(CO is released from the AlN step)

This flow-chart is for

(a) Baeyer's process (b) Serpeck's process
(c) Hall's process (d) Kroll's process

8. In electrolysis of Al_2O_3 by Hall-Heroult process

(a) cryolite $Na_3[AlF_6]$ lowers the mp of Al_2O_3 and increases its electrical conductivity.
(b) Al is obtained at cathode and probably CO_2 at anode
(c) Both (a) and (b) are correct
(d) None of the above is correct

Extraction of iron

Occurrence

- Haematite : Fe_2O_3
- Magnetite : Fe_3O_4
- Limonite : $Fe_2O_3 \cdot 3H_2O$
- Siderite : $FeCO_3$
- Pyrites (fool's gold) : FeS_2
- Pyrrhotite : FeS
- Illmenite : $FeTiO_3$

Oxide ores are main sources of iron.

Carbon reduction method

Step I **Concentration by**

- gravity separation method (which removes gangue)
- magnetic separator (which enriches it by removing non-magnetic impurities)

Step II **Calcination and roasting is done**

- Calcination is done to remove moisture and if extraction is carried out from carbonate ore, it is concentrated to oxide.

$$Fe_2O_3 \cdot 3H_2O \xrightarrow{\Delta} Fe_2O_3 + 3H_2O$$

$$FeCO_3 \xrightarrow{\Delta} FeO + CO_2$$

- Calcined ore is heated in presence of air to oxidise volatile impurities.

$$4P + 5O_2 \longrightarrow 2P_2O_5$$

$$4As + 3O_2 \longrightarrow 2As_2O_3$$

$$S + O_2 \longrightarrow SO_2$$

and ferrous oxide is oxidised to ferric oxide.

$$4FeO + O_2 \longrightarrow 2Fe_2O_3$$

Step III **Reduction or smelting**

It is done in a blast furnace. For this iron ore, coke and limestone are added at the top of the furnace and hot air is introduced through the bottom. The reactions taking place in four zone of blast furnace (alongwith working temperature) are as follows.

(*i*) **Combustion zone** It is the lowest part of furnace where the temperature is about 2170K. Here coke burns to form carbon dioxide. This is an exothermic reaction thus a lot of heat is produced.

$$C(s) + O_2(g) \longrightarrow CO_2(g) ; \Delta H = -393.3 \text{ kJ}$$

This hot CO_2 gas rises upwards and meets with red hot coke.

(*ii*) **Fusion zone** This zone is just above the combustion zone, where the temperature is about 1570 K. In this zone formation of reducing agents principally $CO(g)$ and $H_2(g)$ take place.

$$C + O_2 \longrightarrow 2CO \quad (>1270K)$$

(*iii*) **Slag formation zone** This zone is present at the middle part of the furnance. (where temperatures are about 1070 – 1470K). In this region, limestone decomposes to form CuO and CO_2. CaO acts as flux and combined with silica to form fusible calcium silicate slag.

$$CaCO_3 \longrightarrow CaO + CO_2 \quad (1070-1170K)$$

$$CaO + SiO_2 \longrightarrow CaSiO_3(l) \quad (1470K)$$

$$3CaO + P_2O_5 \longrightarrow Ca_3(PO_4)_2(l) \quad (1470K)$$

(*iv*) **Reduction zone** This is the uppermost part of the furnance. The temperature is in between 870 – 1070 K. Here CO coming upwards comes into contact with iron oxide and reduces it to iron.

$$3Fe_2O_3(s) + CO(g) \xrightarrow{500\text{-}600\,K} 2Fe_3O_4(s) + CO_2(g)$$

$$Fe_3O_4(s) + CO(g) \xrightarrow{773\text{-}873\,K} 3Fe(s) + CO_2(g)$$

$$Fe_2O_3(s) + CO(g) \xrightarrow{773\text{-}873\,K} 2FeO(s) + CO_2(g)$$

$$FeO(s) + CO(g) \xrightarrow{>1070K} Fe(s/l) + CO_2(g)$$

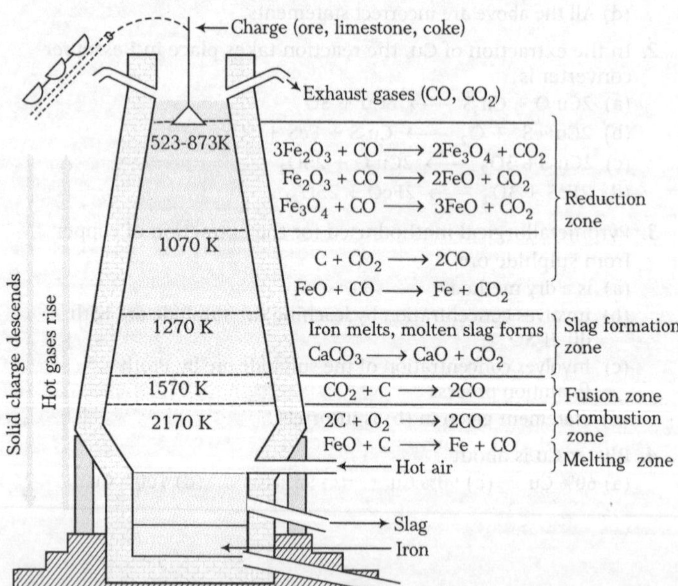

Fig. 6.10 Reduction or smelting

Maximum temperature are attained near the bottom of the furnance where molten iron and slag are drained off through different notes.

Iron obtained from a blast furnace is called **pig iron.** It contains about 95% Fe, 4% C and varying quantities of other impurities.

Conversion of pig iron into cast iron

Cast iron can be obtained by pouring pig iron directly into moulds of the desired shape. Cast iron is very hard and brittle and can be used where it will not be subjected to mechanical or thermal shock.

Conversion of cast iron into wrought iron

Wrought iron is the purest form of iron and contains about 0.15-0.28% impurity. It is malleable and fibrous.

Wrought iron is prepared by **"Puddling process"** in which cast iron is fused in a reverberatory furnace, the hearth of which is lined with haematite, which oxidises carbon into CO. Lime stone is added to remove Si, S and P as slag. The wrought iron is separated from slag by passing through rollers.

Conversion of pig iron into steel

Pig iron is converted into steel by following changes by

(*i*) reducing the carbon content from 3–4% in pig iron to 0–1.5% in steel.

(*ii*) removing the impurities like S, P, Si, Mn etc. by oxidising them to volatile oxide or through slag formation.

(*iii*) by adding alloying elements (such as Cr, Ni, Mn, V, Mo, and W) to give the steel its desired end properties.

Following are the methods used for manufactures of steel.

(a) Bessemer's process through Bessemer's converter

(b) Siemens Martin's open hearth process

(c) Oxygen top blowing process

(d) Electric arc process

(e) High frequency induction process

(f) Basic oxygen process (BOP)

It is latest and more convenient and faster. It gives purest form and surface is free from nitrides.

> **Refractory materials**
>
> The substance that can withstand very high temperature without being melted or soft is called refractory materials. These materials may be acidic, basic or neutral .
>
> - Acidic refractories—silica, sand, quartz etc.
> - Basic refractories—lime, dolomite, magnesite etc.
> - Neutral refractories—graphite, chromite, bone ash etc.

Properties of steel

The properties of steel not only depend on its chemical composition but also on **heat treatment** by which its hardness can be changed. At high temperatures, iron and carbon in steel combine to form iron carbide, Fe_3C called **cementite.**

$$3Fe(s) + C(s) \rightleftharpoons \underset{\text{Cementite}}{Fe_3C(s)}$$

The forward reaction is endothermic, so that the formation of cementite is favoured at high temperatures. When steel containing cementite is cooled slowly, the above equilibrium shifts to the left, and the carbon separates as small particles of graphite, which gives the steel a gray colour.

If the steel is cooled rapidly, equilibrium is not attained and the carbon remains large in the form of cementite, Fe_3C. Steel containing cementite is light in colour and is harder and more brittle than that containing graphite.

Heating the steel to appropriate temperature for a short time and then cooling it rapidly in order to give it the desired mechanical properties is called **tempering.**

In this way, the ratio of carbon present as graphite and as cementite can be varied . Table 6.4 shows the composition, properties and uses of various types of steel.

Table 6.4 Types of steel

Composition	Type-mass%		
	Plain	High strength	Stainless
C	1.35	0.25	0.03-1.2
Mn	1.65	1.65	1.0-10
P	0.04	0.04	0.04-0.06
S	0.05	0.05	0.03
Si	0.06	0.15-0.90	1-3
Ni	–	0.4 -1.0	1.22
Cr	–	0.3 -1.3	4.0-27
Others	Cu (0.2 - 0.6)	Cu (0.01-0.08)	–
Uses	Sheet products, tools	Construction, steel turbines	Kitchen utensils, razor blades

Extraction of tin

Occurrence

Cassiterite (tin stone) : SnO_2

It contains $FeWO_4 \cdot MnWO_4$ (wolframite) as a magnetic impurity. A schematic diagram of extraction process of tin is shown below.

SnO_2 tin ore

↓ Concentration by electromagnetic separation

Concentrated ore

↓ Roasted to remove volatile impurities

Smelting ore mixed with coal, limestone and sand

↓ Poling to reduce SnO_2 to Sn

Electrorefining

Extraction of silver

Occurrence

Argentite (silver glance) : Ag_2S
Horn silver (chlorargyrite) : AgCl
Argentiferrous galena : $PbS \cdot Ag_2S$
Pyrargyrite or ruby silver : $3Ag_2S \cdot Sb_2S_3$
Main source is argentite.

Silver can be extracted by following two processes.

1. Cyanide process

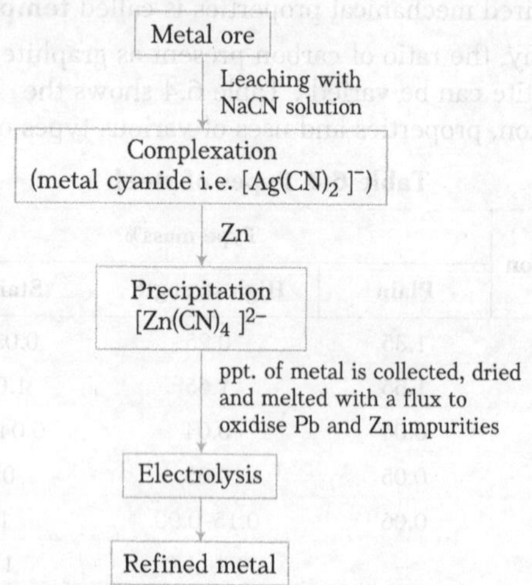

Note *In case of gold, $O_2(g)$ oxidises free metal to Au^+, which complexes with CN^-.*

$$4Au(s) + 8CN^- + O_2(g) + 2H_2O \longrightarrow 4[Au(CN)_2]^- + 4OH^-$$

The pure metal is then displaced from the solution by an active metal.

$$2[Au(CN)_2]^- + Zn \longrightarrow 2Au(s)\downarrow + [Zn(CN)_4]^{2-}$$

2. Lead process

This process depends upon the fact that when ores of silver are smelted down with lead, an alloy of lead and silver is formed. From this alloy lead is removed by oxidation.

- Ores of silver $\xrightarrow[\text{with lead}]{\text{Smelted}}$ (Pb – Ag) alloy $\xrightarrow{\text{Cupellation}}$ silver
- (Pb-Ag) alloy rich in silver : Cupellation process is used
- (Pb-Ag) alloy poor in silver : Parke's or Pattinson's process is used

Desilverisation of lead is done by following methods

(*i*) **By Parke's process** When lead-silver alloy is poor in silver (such as argentiferrous lead or galena), desilverisation of lead is done by Parke's process.

This process takes advantage of two liquid state properties of zinc that are explained below.

- When zinc is added to a melted alloy of silver and lead, zinc takes away silver from the alloy. Zinc itself forms an alloy with silver and not with lead.
- Alloy of zinc and silver melts at a higher temperature than lead.
- The alloy of silver and zinc is lighter than lead and forms two separate layers.
- On distillation of silver and zinc, zinc separates.

(*ii*) **By Pattinson's process** It depends upon the fact that the alloy of lead and silver containing less than 1.8 per cent of silver, melts at a temperature lower than pure lead. If this type of alloy is melted and the melted alloy is allowed to cool, lead first separates and repeating the process, one becomes richer in silver. Further recovery is done by Cupellation process.

Uses

(*i*) **Copper**

- For the preparation of steam pipes
- In making electrical wires
- In making calorimeters
- In making alloys, like brass (with Zn) bronze (with tin) coinage alloy (with Ni), german silver (with Zn and Ni) etc.

(*ii*) **Zinc**

- Making alloys like brass (with Cu), german silver (with Cu 25-30%, Zn 25-30%, Ni 40-50%) etc.
- For galvanisation of iron
- For making dry cell
- For extraction of Ag and Au as reducing agent etc.

(*iii*) **Aluminium**

- Making electric wires
- Making silver paint
- Kitchen utensils
- Extraction of Mo, Cr, etc.
- Making food packing etc.

(*iv*) **Iron**

- Making wrought iron and different varieties of steel. Wrought iron is used for making agricultural implements, anchors, bolts etc.
- Steel is used for making parts of automobiles and aeroplanes, cutting tools, utensils etc.
- Cast iron is used for making gutter pipes, railway sleepers, etc.

CHECK POINT 6.4

1. Purest form of iron is
(a) cast iron (b) wrought iron
(c) pig iron (d) None of these

2. Consider the following statements.
I. Pig iron is malleable and ductile.
II. Pig iron can be melted to produce cast iron of desired shapes.
III. Wrought iron is obtained by heating pig iron with iron oxide in a furnace.

Which of the statements given above is/are correct?
(a) Only I
(b) Both I and II
(c) Both II and III
(d) Only III

3. Cementite is
(a) predominant content of cement
(b) a mineral of calcium
(c) a carbide of iron in steel
(d) None of the above

4. Tempering of steel
(a) is the heating the steel to appropriate temperature and then cooling it rapidly
(b) increases mechanical strength
(c) changes ratio of carbon in cementite
(d) follows all of the above

5. SnO_2 is reduced to metallic Sn on smelting oxide with anthracite, limestone and sand.
In this, function of sand is
(a) it acts as a flux
(b) it removes basic impurities as slag
(c) Both as in (a) and (b)
(d) None of the above

6. In the process of extraction of gold,

$$\text{Roasted gold ore} + CN^- + H_2O \xrightarrow{O_2} [X] + OH^-$$
$$[X] + Zn \longrightarrow [Y] + Au$$

Identify the complexes $[X]$ and $[Y]$.
(a) $X = [Au(CN)_2]^-$, $Y = [Zn(CN)_4]^{2-}$
(b) $X = [Au(CN)_4]^{3-}$, $Y = [Zn(CN)_4]^{2-}$
(c) $X = [Au(CN)_2]^-$, $Y = [Zn(CN)_6]^{4-}$
(d) $X = [Au(CN)_4]^-$, $Y = Zn(CN)_4]^{2-}$

7. Softened lead is desilverised by
(a) Parke's method (b) Pattinson' method
(c) Both (a) and (b) (d) None of these

8. The silver is extracted by Parke's process. The basis of this method is
(a) silver is immiscible in molten Zn
(b) Ag is miscible in NaCN
(c) Ag is more miscible in molten zinc than in molten Pb
(d) Ag is more miscible in molten Pb in comparison to molten zinc

Chapter Exercises

(A) Topical exercise

I. Occurrence of metals and metallurgy

1 Most abundant metal in earth crust is
(a) Al (b) O (c) Fe (d) Si

2 A number of elements are available in earth's crust but most abundant elements are
(a) Al and Fe (b) Al and Cu
(c) Fe and Cu (d) Cu and Ag

3 To dissolve argentite ore which of the following is used?
(a) $Na[Ag(CN)_2]$ (b) NaCN
(c) NaCl (d) HCl

4 Which is the sulphate ore of Mg?
(a) Dolomite (b) Carnallite
(c) Magnesite (d) Epsomite

5 Which of the following represents correct matching of metals with their minerals?

	Mg	Na	Cu	Al
(a)	chloride	chloride	sulphide	silicate
(b)	carbonate	chloride	sulphide	oxide
(c)	carbonate	carbonate	oxide	phosphate
(d)	oxide	chloride	sulphide	oxide

6 Impurities present in the ore react to form a fusible substance known as
(a) flux (b) gangue (c) nugget (d) mineral

7 Flux is used to remove
(a) basic impurities
(b) acidic impurities
(c) all types of impurities
(d) acidic and basic both impurities

8 Which of the following is not a basic flux ?
(a) $CaCO_3$ (b) Lime (c) SiO_2 (d) CaC_2

9 During smelting an additional substance is added which combines with impurities to form a fusible product which is known as
(a) mud (b) slag (c) flux (d) gangue

10 Which step is not involved in hydrometallurgical process?
(a) $Cu_2S + 2Cu_2O \longrightarrow 6Cu + SO_2$
(b) $CuFeS_2 + 2H_2SO_4 \longrightarrow CuSO_4 + FeSO_4 + 2H_2S$
(c) $CuSO_4 + Fe \longrightarrow FeSO_4 + Cu$
(d) $CuCO_3 + H_2SO_4 \longrightarrow CuSO_4 + H_2O + CO_2$

11 The process used for purification of $Al_2O_3 \cdot 2H_2O$ is
(a) magnetic separation (b) froth floatation
(c) leaching (d) liquation

12 Hydrometallurgical process of extraction of metals is based on
(a) complex formation (b) hydrolysis
(c) dehydration (d) dehydrogenation

13 Gravity separation process is used for the concentration of
(a) calamine (b) haematite
(c) chalcopyrites (d) bauxite

14 Froth floatation is a process of
(a) oxidation (b) reduction
(c) refining (d) concentration

15 The role of pine oil and xanthates in froth floatation process is to enhance the non-wettability of
(a) mineral particles in froth
(b) mineral particles in water
(c) gangue particles in froth
(d) gangue particles in water

16 Which one of the following is not a method of concentration of metals?
(a) Gravity separation
(b) Froth floatation process
(c) Electromagnetic separation
(d) Smelting

17 Consider the following statements.

Statement I Gold occurs in native state

Statement II Gold dissolves in *aqua-regia.*

(a) Both statements I and II are true
(b) Only statement I is true
(c) Only statement II is true
(d) Neither statement I nor statement II is true

18 Consider the following statements.

Statement I Hydrometallurgy involves dissolving the ore in a suitable reagent followed by precipitation by a more electropositive metal.

Statement II Copper is extracted by hydrometallurgy.

(a) Both statements I and II are true
(b) Only statement I is true
(c) Only statement II is true
(d) Neither statement I nor statement II is true

19 Consider the following statements.

Statement I Froth floatation process is used to concentrate sulphide ores.

Statement II There is difference in the wettability of different minerals.

(a) Both statements I and II are true
(b) Only statement I is true
(c) Only statement II is true
(d) Neither statement I nor statement II is true

20 Match the method of concentration of the ore given in Column I with the ore given in Column II.

Column I	Column II
A. Magnetic separation	1. Ag_2S
B. Froth floatation	2. $FeCr_2O_4$
C. Gravity separation	3. $Al_2(SiO_3)_3$

Choose the correct match.

	A	B	C		A	B	C
(a)	1	2	3	(b)	2	1	3
(c)	3	1	2	(b)	3	2	1

21 Consider the following statements.

Statement I Sulphide ores are concentrated by Froth floatation method.

Statement II Cresol stabilises the froth in Froth floatation.

(a) Both statements I and II are true
(b) Only statement I is true
(c) Only statement II is true
(d) Neither statement I nor statement II is true

II. Methods of reduction of oxides or salts to free metals

22 Main source of lead is PbS. It is converted to Pb by

(A) $PbS \xrightarrow[\Delta]{Air} PbO + SO_2 \xrightarrow{C} Pb + CO_2$

(B) $PbS \xrightarrow[\Delta]{Air} PbO + SO_2 \xrightarrow{PbS} Pb + SO_2$

Self reduction process is

(a) Only A
(b) Only B
(c) Both (a) and (b)
(d) None of these

23 Oxides of the various metals are converted into metal by heating but not in case of

(a) Ag_2O
(b) CuO
(c) HgO
(d) All of these

24 Self reduction of CuS to Cu can be carried out in

(a) Bessemer converter
(b) Pierce-Smith converter
(c) Both (a) and (b)
(d) None of these

25 Select the incorrect reduction process.

(a) $2[Ag(CN)_2]^- + Zn \longrightarrow [Zn(CN)_4]^{2-} + 2Ag$
(b) $CuO + H_2 \longrightarrow Cu + H_2O$
(c) $ZnO + H_2 \longrightarrow Zn + H_2O$
(d) $MgO + C \longrightarrow Mg + CO$

26 In the extraction of copper from its sulphide ore, the metal is formed by the reduction of Cu_2O with **[NCERT Exemplar]**

(a) FeS
(b) CO
(c) Cu_2S
(d) SO_2

27 Extraction of aluminium from aluminium oxide (Al_2O_3) is best done by

(a) electrolytic reduction of Al_2O_3
(b) reduction of Al_2O_3 with carbon
(c) reduction of Al_2O_3 with sodium
(d) reduction of Al_2O_3 with CO

28 Which of the following ore is concentrated by both, magnetic and gravimetric separation?

(a) Dolomite (b) Tinstone (c) Galena (d) Bauxite

29 Which of the following is not the correct process–mineral matching in metallurgical extraction?

(a) Leaching : Ag
(b) Zone refining : Sn
(c) Liquation : Si
(d) van Arkel : Zr

30 When $FeCr_2O_4$ (chromite) is reduced with C in an electric arc furnace

(a) Cr and Fe_2O_3 are formed
(b) Fe and Cr_2O_3 are formed
(c) Fe and Cr (ferrochrome) are formed
(d) $FeCrO_4$ is formed

31 The element which could be extracted by electrolytic reduction of its oxide dissolved in a high temperature melt is

(a) sodium
(b) magnesium
(c) fluorine
(d) aluminium

32 Carbon cannot be used in the reduction of Al_2O_3 because

(a) it is an expensive proposition
(b) the enthalpy of formation of CO_2 is more than that of Al_2O_3
(c) pure carbon is not easily available
(d) the enthalpy of formation of Al_2O_3 is too high

33 Wolframite ore is separated from tinstone ore by the process of

(a) roasting
(b) electromagnetic separation
(c) smelting
(d) calcination

34 The chemical composition of 'slag' formed during the smelting process in the extraction of copper is
(a) $Cu_2O + FeS$
(b) $FeSiO_3$
(c) $CuFeS_2$
(d) $Cu_2S + FeO$

35 Out of Cu_2S, HgS, Ag_2S and ZnS, roasting will convert the minerals into metal in case of
(a) Cu_2S, ZnS (b) HgS, ZnS
(c) Cu_2S, Ag_2S (d) HgS

36 Consider the following reactions.

$$Ag_2S + NaCN \longrightarrow (A)$$
$$(A) + Zn \longrightarrow (B)$$

(*B*) is a metal. Hence, (*A*) and (*B*) are
(a) $Na_2[Zn(CN)_4]$, Zn
(b) $Na[Ag(CN)_2]$, Ag
(c) $Na_2[Ag(CN)_4]$, Ag
(d) $Na_3[Ag(CN)_4]$, Ag

37 In the leaching of Ag_2S with NaCN, a stream of air is also passe.d. It is because
(a) reversible nature of reaction between Ag_2S and NaCN.
(b) to oxidised Na_2S formed into $Na_2S_2O_3$ and sulphur.
(c) Both (a) and (b)
(d) None of the above

38 In the extraction of chlorine by electrolysis of brine **[NCERT Exemplar]**
(a) oxidation of Cl^- ion to chlorine gas occurs
(b) reduction of Cl^- ion to chlorine gas occurs
(c) for overall reaction $\Delta G^\ominus$ has negative value
(d) a displacement reaction takes place

39 The incorrect statement among the following is
(a) hydrogen is used to reduce ion
(b) zirconium is refined by van Arkel method
(c) the sulphide ore galena is concentrated by froth flotation
(d) in the metallurgy of iron, the flux used is SiO_2

40 Which of the following does not represent correct method?

(a) $TiCl_2 + 2Mg \longrightarrow Ti + 2MgCl_2$	:	Kroll
(b) $Ni(CO)_4 \longrightarrow Ni + 4CO$	:	Mond
(c) $Ag_2CO_3 \longrightarrow 2Ag + CO_2 + \frac{1}{2}O_2$	:	van Arkel
(d) $ZrI_4 \longrightarrow Zr + 2I_2$	:	van Arkel

41 Froth floatation process is used for the concentration of sulphide ore. Which of the following processes is correct?
(a) It is based on the difference in wettability of different minerals.
(b) In this process sodium ethyl xanthate, $C_2H_5OCS_2Na$ is used as collector.
(c) In this process NaCN is used as depressant in the mixture of ZnS and PbS when ZnS forms soluble complex and PbS forms froth.
(d) All of the above.

42 Sodium is made by the electrolysis of molten mixture of about 40% NaCl and 60% $CaCl_2$ because
(a) $CaCl_2$ helps in conduction of electricity
(b) Ca^{2+} can reduce NaCl to Na
(c) Ca^{2+} can displace Na from NaCl
(d) this mixture has a lower melting point than NaCl

43 Four metals and their methods of refinement are given
(i) Ni, Cu, Zr, Ga
(ii) electrolysis, van Arkel process, zone refining, Mond's process

Choose the right method for each and choose the correct option.
(a) Ni : Electrolysis, Cu : van Arkel process, Zr : Zone refining, Ga : Mond's process
(b) Ni : Mond's process, Cu : Electrolysis, Zr : van Arkel process, Ga : Zone refining
(c) Ni : Mond's process, Cu : van Arkel process, Zr : Zone refining, Ga : Electrolysis
(d) Ni : Electrolysis, Cu : Zone refining, Zr : van Arkel process, Ga : Mond's process

44 Consider the following statements.

Statement I Al_2O_3 is converted into Al by reduction with carbon.

Statement II Carbon (graphite) has greater affinity for oxygen than aluminium has.
(a) Both statements I and II are true
(b) Only statement I is true
(c) Only statement II is true
(d) Neither statement I nor statement II is true

III. Thermodynomic principles of metallurgy and refining of crude metal

45 Which of the following statements is correct *wrt* the following graph?

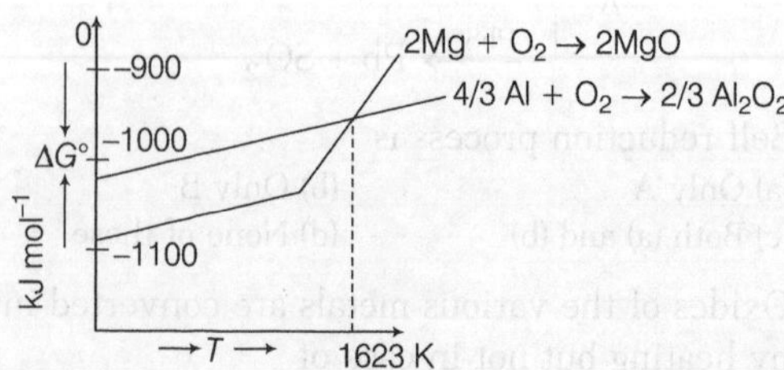

(a) Below 1623 K, Mg reduces Al_2O_3
(b) Above 1623 K, Al reduces MgO
(c) Both (a) and (b) are correct
(d) Both (a) and (b) are wrong

46 Choose the correct option of temperature at which carbon reduces FeO to iron and produces CO wrt the following graph **[NCERT Exemplar]**

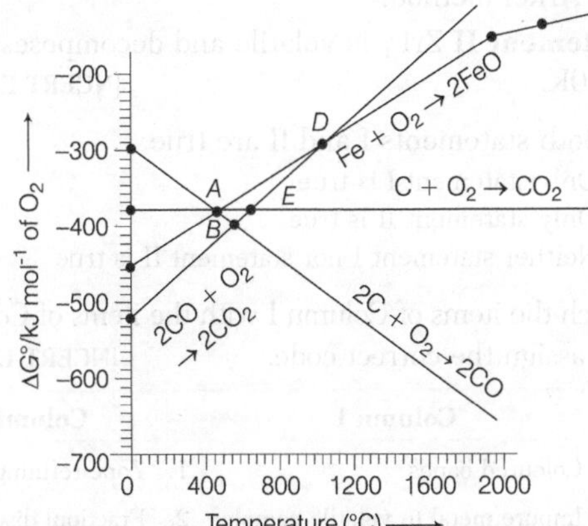

(a) Below temperature at point *A*.
(b) Approximately at the temperature corresponding to point *A*.
(c) Above temperature at point *A* but below temperature at point *D*.
(d) Above temperature at point *A*.

47 Given, $\Delta G_f^\circ(\text{CuO}) = -129.7 \text{ kJ mol}^{-1}$,
$\Delta G_f^\circ(H_2O) = -237.2 \text{ kJ mol}^{-1}$ and
$\Delta G_f^\circ(\text{CO}) = -137.2 \text{ kJ mol}^{-1}$

The better reducing agent for the reduction of CuO is
(a) H_2 (b) CO (c) C (d) Any of these

48 The reaction,
$\text{FeO}(s) + \text{C}(s) \longrightarrow \text{Fe}(s) + \text{CO}(g)$;
$\Delta_r G°$ will take place when
(a) $\Delta_r G°_{(\text{C,CO})} + \Delta rG°_{(\text{FeOFe})} = \Delta_r G°$
(b) $\Delta_r G°_{(\text{C,CO})} + \Delta rG°_{(\text{FeOFe})} > \Delta_r G°$
(c) $\Delta_r G°_{(\text{C,CO})} + \Delta rG°_{(\text{FeOFe})} < \Delta_r G°$
(d) $\Delta_r G°_{(\text{C,CO})} + \Delta rG°_{(\text{FeOFe})} = 0$

49 Sodium, magnesium and aluminium can be obtained from their ores by
(a) electrometallurgy (b) pyrometallurgy
(c) hydrometallurgy (d) smelting

50 Which one of the following is manufactured by the electrolysis of fused sodium chloride?
(a) NaOH (b) NaClO (c) Na (d) $NaClO_3$

51 Si of high purity to be used in semiconductor can be prepared by following methods
(A) $SiO_2 + 2C \longrightarrow Si + 2CO$
(B) $Si + 2Cl_2 \longrightarrow SiCl_4$;
$SiCl_4 + 2Mg \longrightarrow Si + 2MgCl_2$
Better method is
(a) Only A (b) Only B
(c) Both (A) and (B) (d) None of these

52 When the sample of Cu with Zn impurity is to be purified by electrolysis, the appropriate electrodes are

	Cathode	Anode
(a)	Pure Zn	Pure Cu
(b)	Impure sample	Pure Cu
(c)	Impure Zn	Impure sample
(d)	Pure copper	Impure sample

53 The following equation represents a method of purification of nickel by

$$\underset{\text{Impure}}{\text{Ni}} + 2\text{CO} \xrightarrow{320\text{ K}} \text{Ni(CO)}_4 \xrightarrow{420\text{ K}} \underset{\text{Pure}}{\text{Ni}} + 4\text{CO}$$

The name of method is
(a) cupellation (b) Mond's process
(c) van Arkel method (d) zone refining

54 Which of the following pair can be refined using same basic principle?
(a) Cu and Ag (b) Si and Ge
(c) Ti and Ni (d) All of these

55 Which method is not correct given for refining of crude metals?
(a) Distillation : zinc and mercury
(b) Liquation : tin
(c) van Arkel : zirconium
(d) Mond's process : lead

56 In the metallurgy of sodium by electrolysis, excess of calcium chloride is mixed with sodium chloride to
(a) make the latter a good conductor
(b) make the latter soft
(c) generate more energy for the electrolytic cell
(d) assist liquefaction of the latter at a much lower temperature

57 Match Column I with Column II.

Column I	Column II
A. van Arkel method	1. Manufacture of caustic soda
B. Solvay process	2. Purification of titanium
C. Cupellation	3. Manufacture of Na_2CO_3
D. Poling	4. Purification of copper
	5. Refining of silver

Choose the correct match.

	A	B	C	D		A	B	C	D
(a)	2	3	5	4	(b)	3	1	4	5
(c)	4	5	3	1	(b)	5	4	2	3

58 Refining of silver is done by
(a) Liquation
(b) Poling
(c) Cupellation
(d) van Arkel method

59 In the extraction of nickel by Mond's process, the metal is obtained by
(a) electrochemical reduction
(b) thermal decomposition
(c) chemical reduction by aluminium
(d) reduction by carbon

60 Refining of tin cannot be done by
(a) cupellation (b) liquation
(c) poling (d) electro refining

61 Main process for extracting nickel in the pure form uses which one of the following?
(a) Vapour phase refining (b) Zone refining
(c) Electrolysis (d) Solvent extraction

62 Consider the following statements.
Statement I Metals of high impurity are obtained by Zone refining.
Statement II Impurities in the ore are more soluble in the melt than in the pure metal.
(a) Both statements I and II are true
(b) Only statement I is true
(c) Only statement II is true
(d) Neither statement I nor statement II is true

63 Consider the following statements.
Statement I Zone refining method is very useful for producing semiconductors.
Statement II Semiconductors are of high purity. **[NCERT Exemplar]**
(a) Both statements I and II are true
(b) Only statement I is true
(c) Only statement II is true
(d) Neither statement I nor statement II is true

64 Electrolytic refining is used to purify which of the following metals? **[NCERT Exemplar]**
(a) Cu and Zn (b) Ge and Si
(c) Zr and Ti (d) Zn and Hg

65 Formation of $Ni(CO)_4$ and its subsequent decomposition into Ni and CO (recycled) makes the basis of Mond's process

$$Ni + 4CO \xrightarrow{T_1} Ni(CO)_4 \xrightarrow{T_2} Ni + 4CO$$

T_1 and T_2 are
(a) 100° C, 50° C (b) 50° C, 100° C
(c) 50° C, 230° C (d) 230° C, 50° C

66 Zone refining is based on the principle that **[NCERT Exemplar]**
(a) Impurities of low boiling metals can be separated by distillation.
(b) Impurities are more soluble in molten metal than in solid metal.
(c) Different components of a mixture are differently adsorbed on an absorbent.
(d) Vapour of volatic compound can be decomposed in pure metal.

67 Consider the following statements.
Statement I Zirconium can be purified by van Arkel method.
Statement II ZrI_4 is volatile and decomposes at 1800K. **[NCERT Exemplar]**
(a) Both statements I and II are true
(b) Only statement I is true
(c) Only statement II is true
(d) Neither statement I nor statement II is true

68 Match the items of Column I with the items of Column II and assign the correct code. **[NCERT Exemplar]**

Column I	Column II
A. Coloured bands	1. Zone refining
B. Impure metal to volatile complex	2. Fractionl distillation
C. Purification of Ge and Si	3. Mond's process
D. Purification of mercury	4. Chromatography
	5. Liquation

Choose the correct match

	A	B	C	D		A	B	C	D
(a)	5	1	2	3	(b)	4	3	1	2
(c)	3	5	4	1	(b)	2	4	3	5

69. Consider the following statements.
Statement I Nickel can be purified by Mond's process.
Statement II $Ni(CO)_4$ is a volatile compound which decomposes at 460 K to give pure Ni.
(a) Both statements I and II are true **[NCERT Exemplar]**
(b) Only statement I is true
(c) Only statement II is true
(d) Neither statement I nor statement II is true

IV. Extraction and uses of some important metals

70 When copper ore is mixed with silica in a reverberatory furnace, copper matte is produced. The copper matte contains
(a) sulphides of copper (II) and iron (II)
(b) sulphides of copper (II) and iron (III)
(c) sulphides of copper (I) and iron (II)
(d) sulphides of copper (I) and iron (III)

71 A metal used to recover copper from a solution of $CuSO_4$ is
(a) Fe (b) Hg (c) Na (d) Ag

72 Identify the reaction that does not take place during smelting process of copper extraction
(a) $2FeS + 3O_2 \longrightarrow 2FeO + 2SO_2$
(b) $Cu_2O + FeS \longrightarrow Cu_2S + FeO$
(c) $2Cu_2S + 3O_2 \longrightarrow 2Cu_2O + 2SO_2$
(d) $FeO + SiO_2 \longrightarrow FeSiO_3$

73 In the extraction of copper, the reductant used is
(a) C (b) CO (c) Cu_2S (d) Cu_2O

74 Which of the following process is involved in manufacturing of steel?
(a) Bessemer process (b) Open hearth process
(c) BOP (d) Thermite process

75 The slag obtained during the extraction of copper from copper pyrites is composed of
(a) Cu_2S (b) $CuSiO_3$ (c) $FeSiO_3$ (d) SiO_2

76 In the Baeyer's process
(a) Al_2O_3 goes into solution as soluble $Al(OH)_4^-$ while other basic oxides as TiO_2 and Fe_2O_3 remain insoluble.
(b) Al_2O_3 changes to AlN which in turn decomposed by H_2O
(c) Al_2O_3 changes to $Al_2(CO_3)_3$ which changes to $AlCl_3$
(d) None of the above is correct

77 Which of following methods is not used for extraction of Al?
(a) van Arkel (b) Serpeck
(c) Baeyer (d) Hall-Heroult

78 In electrolytic method of obtaining aluminium from purified bauxite, cryolite is added to the charge in order to
(a) minimise the heat loss due to radiation
(b) protect aluminium produced from oxygen
(c) dissolve bauxite and render it conductor of electricity
(d) lower the melting point of bauxite

79 In aluminium extraction by the Baeyer's process, alumina is extracted from bauxite by sodium hydroxide at high temperature and pressure.
$Al_2O_3(s) + 2OH^-(aq) \longrightarrow 2Al_2O_2^-(aq) + H_2O(l)$
Solid impurities such as Fe_2O_3 and SiO_2 are removed and then $Al(OH)_4^-$ is reprecipitated.
$2Al(OH)_4^- \longrightarrow Al_2O_3 \cdot 3H_2O(s) + 2OH^-(aq)$
In the industrial world
(a) carbon dioxide is added to precipitate the alumina
(b) temperature and pressure are dropped and the supersaturated solution seeded
(c) Both (a) and (b) are practised
(d) the water is evaporated

80 In which one of the following minerals, aluminium is not present?
(a) Cryolite (b) Mica (c) Feldspar (d) Fluorspar

81 Electrolytic reduction of alumina to aluminium by Hall-Heroult process is carried out
(a) in the presence of NaCl
(b) in the presence of fluorite
(c) in the presence of cryolite which forms a melt with lower melting temperature
(d) in the presence of cryolite which forms a melt with higher melting temperature

82 Purification of aluminium by electrolytic refining, is known as
(a) Hall's process (b) Baeyer's process
(c) Hoope's process (d) Serpeck's process

83 On igniting Fe_2O_3 at near temperature of about 1400°C the product obtained is
(a) Fe_2O_3 melt (b) FeO
(c) Fe_3O_4 (d) metallic iron

84 In the metallurgy of iron, when $CaCO_3$ is added to blast furnace, calcium ion appears as
(a) CaO (b) metallic Ca
(c) gangue (d) slag

85 Match the items of Column I with items of Column II **[NCERT Exemplar]**

Column I	Column II
A. Pendulum	1. Chrome steel
B. Malachite	2. Nickel steel
C. Calamine	3. Na_3AlF_6
D. Cryolite	4. $CuCO_3 \cdot Cu(OH)_2$
	5. $ZnCO_3$

Choose the correct match.

	A	B	C	D		A	B	C	D
(a)	2	4	5	3	(b)	3	1	4	2
(c)	4	3	1	5	(b)	5	2	4	1

86 Which one of the following elements constitutes a major impurity in pig iron?
(a) Silicon (b) Oxygen
(c) Sulphur (d) Graphite

87 Iron obtained from blast furnace is
(a) wrought iron (b) cast iron
(c) pig iron (d) steel

88 Which of the following is not an iron ore?
(a) Cassiterite (b) Magnetite
(c) Limonite (d) Copper pyrite

89 The temperature of the slag zone in the metallurgy of iron using blast furnace is
(a) 1500-1600°C (b) 400-700°C
(c) 800-1000°C (d) 1200-1500°C

90 Which mineral has been named incorrectly? **[NCERT Exemplar]**
(a) Bauxite : $Al_2O_3 \cdot 2H_2O$
(b) Corundum : Al_2O_3
(c) Cryolite : $3NaF \cdot AlF_3$
(d) Feldspar : $Be_3Al_2Si_6O_{18}$

91 Bessemer converter is used for preparation of
(a) steel (b) wrought iron
(c) pig iron (d) cast iron

92 When alloy of silver and lead is rich in silver,
(a) Cupellation process is used
(b) Parke's method is used
(c) Pattinson's method is used
(d) any of the above methods can be used

93 These are the extraction processes of silver except
(a) as a side product in electrolytic refining of copper
(b) Parke's process in which Zn is used to extract silver by solvent extraction from molten lead
(c) by reaction of silver sulphide with KCN and then reaction of soluble complex with Zn
(d) by heating $Na[Ag(CN)_2]$

94 Lead pencil contains
(a) graphite (b) softened lead
(c) galena (d) anglesite

95 Match the following Column I with Column II.

	Column I		Column II
A.	Carnallite	1.	Zinc
B.	Calamine	2.	Titanium
C.	Illmenite	3.	Magnesium
D.	Chalcopyrite	4.	Copper

Choose the correct match.

	A	B	C	D		A	B	C	D
(a)	2	4	1	3	(b)	3	1	2	4
(c)	4	3	2	1	(b)	4	1	3	2

96 Which of the following does not contain Mg?
(a) Magnetite (b) Magnesite
(c) Asbestos (d) Carnallite

97 Magistral is the burnt pyrites containing
(a) sulphates of iron and copper
(b) sulphates and oxides of iron and copper
(c) oxides of iron and copper
(d) sulphides of silver and lead

98 Match the items of Column I with items of Column II. **[NCERT Exemplar]**

	Column I		Column II
A.	Blistered Cu	1.	Aluminium
B.	Blast furnace	2.	$2Cu_2O + Cu_2S \longrightarrow 6Cu + SO_2$
C.	Reverberatory furnace	3.	Iron
D.	Hall-Heroult process	4.	$FeO + SiO_2 \longrightarrow FeSiO_3$
		5.	$2Cu_2S + 3O_2 \longrightarrow 2Cu_2O + 2SO_2$

Choose the correct match.

	A	B	C	D		A	B	C	D
(a)	2	3	4	1	(b)	3	5	1	2
(c)	4	1	5	3	(b)	5	2	3	4

(B) Miscellaneous questions

1 Self reduction of the sulphide ore takes place during
(a) roasting (b) smelting
(c) calcination (d) cupellation

2 Selection of temperature to carry out a reduction process depends so as to make
(a) ΔG negative (b) ΔG positive
(c) ΔH negative (d) ΔH positive

3 By annealing, steel
(a) becomes soft
(b) becomes liquid
(c) becomes hard and brittle
(d) is covered with a thin film of Fe_3O_4

4 Match the following Column I with Column II.

	Column I Ore/mineral		Column II Formula of ore/mineral
A.	Feldspar	1.	$[Ag_3SbS_3]$
B.	Asbestos	2.	$Al_2O_3 \cdot H_2O$
C.	Pyrargyrite	3.	$MgSO_4 \cdot H_2O$
D.	Diaspore	4.	$KAlSi_3O_8$
		5.	$CaMg_3(SiO_3)_4$

Choose the correct match.

	A	B	C	D		A	B	C	D
(a)	5	3	2	1	(b)	4	5	1	2
(c)	3	4	5	1	(b)	2	1	4	3

5 Poling process
(a) reduces SnO_2 to Sn
(b) oxidises impurities like iron and removes as scum
(c) uses green poles
(d) involves all of the above

6 The chief impurity present in red bauxite is
(a) SiO_2 (b) Fe_2O_3 (c) K_2SO_4 (d) NaF

7 Phosphorus separates in the extraction of iron, as
(a) slag, $Ca_3(PO_4)_2$ (b) volatile, P_2O_5
(c) slag, $Mg_3(PO_4)_2$ (d) Ca_3P_2

8 Which one of the following processes is used in the extraction of magnesium?
(a) Self reduction (b) Thermite reduction
(c) Fused salt electrolysis (d) Aqueous solution electrolysis

9 Thomas slag is
(a) $Ca_3(PO_4)_2 \cdot 2H_2O$ (b) $Ca_3(PO_4)_2 \cdot CaSiO_3$
(c) $MgSiO_3$ (d) $CaSiO_3$

10 Reduction of a metal oxide by excess carbon at high temperature is a method for the commercial preparation of some metals. This method can be successfully applied in the case of
(a) BeO and Al_2O_3 (b) ZnO and Fe_2O_3
(c) CaO and Cr_2O_3 (d) BaO and U_3O_8

11 Which of the following reactions is an example of auto-reduction? **[NCERT Exemplar]**

(a) $Fe_3O_4 + 4CO \longrightarrow 3Fe + 4CO_2$

(b) $Cu_2O + C \longrightarrow 2Cu + CO$

(c) $Cu^{2+}(aq) + Fe(s) \longrightarrow Cu(s) + Fe^{2+}(aq)$

(d) $Cu_2O + \frac{1}{2}Cu_2S \longrightarrow 3Cu + \frac{1}{2}SO_2$

12 Extraction of gold and silver involves leaching the metal with CN^- ion. The metal is recovered by

(a) displacement of metal by some other metal from the complex ion **[NCERT Exemplar]**

(b) roasting of metal complex

(c) calcination followed by roasting

(d) thermal decomposition of metal complex

13 In which of the following isolations no reducing agent is required?

(a) Iron from haematite (b) Aluminium from bauxite

(c) Mercury from cinnabar (d) Zinc from zinc blende

14 Match the Columns I reactions taking place in blast furnace with Column II (temperature-range) of operations.

	Column I (Reactions)		Column II (Temperature range)
A.	$Fe_2O_3 + CO \longrightarrow CO_2 + Fe$	1.	1200°C
B.	$C + CO_2 \longrightarrow CO$	2.	900°C
C.	$C + H_2O \longrightarrow CO + H_2$	3.	600°C
D.	$CaO + SiO_2 \longrightarrow CaSiO_3$	4.	1000°C

Choose the correct match.

	A	B	C	D		A	B	C	D
(a)	2	3	1	4	(b)	2	4	3	1
(c)	3	1	4	2	(b)	4	3	2	1

15 Consider the following statements.

Statement I Reduction of ZnO with carbon is done at 1100°C.

Statement II At this temperature, $\Delta G°$ is negative thus, process is spontaneous.

(a) Both statements I and II are true

(b) Only statement I is true

(c) Only statement II is true

(d) Neither statement I nor statement II is true

16 Brine is electrolysed by using inert electrodes. The reaction at anode is **[NCERT Exemplar]**

(a) $Cl^-(aq) \longrightarrow \frac{1}{2}Cl_2(g) + e^-; E_{cell} = 1.36\text{ V}$

(b) $2H_2O(l) \longrightarrow O_2(g) + 4H^+ + 4e^-; E^\circ_{cell} = 1.23\text{ V}$

(c) $Na^+(aq) + e^- \longrightarrow Na(s); E^\circ_{cell} = 2.71\text{ V}$

(d) $H^+(aq) + e^- \longrightarrow \frac{1}{2}H_2(g); E^\circ_{cell} = 0.00\text{ V}$

17 Carbon can reduce ferric oxide to iron at a temperature above 983 K because

(a) carbon monoxide formed is thermodynamically less stable than ferric oxide.

(b) carbon has a higher affinity towards oxygen than iron.

(c) free energy change for the formation of carbon dioxide is less negative than that for ferric oxide.

(d) iron has a higher affinity towards oxygen than carbon.

18 Match the items of Column I with the items of Column II. **[NCERT Exemplar]**

	Column I		Column II
A.	Siderite	1.	Al_2O_3
B.	Sphalerite	2.	NaCN
C.	Depressant	3.	$FeCO_3$
D.	Corundum	4.	ZnS
		5.	Fe_2O_3

Choose the correct match.

	A	B	C	D
(a)	4	2	1	3
(b)	3	1	4	2
(c)	2	3	4	1
(d)	3	4	2	1

19 In the metallurgy of aluminium **[NCERT Exemplar]**

(a) Al^{3+} is oxidised to Al (s)

(b) graphite anode is oxidised to carbon monoxide and carbon dioxide

(c) oxidation state of oxygen changes in the reaction at anode

(d) oxidation state of oxygen changes in the overall reaction involved in the process

20 Among the following statements, the incorrect one is,

(a) zinc blende and iron pyrites are sulphides

(b) calamine and siderite are carbonates

(c) malachite and chalco pyrite are the copper ore

(d) argentite and cuprite are oxides

21 Which of the following is/are correct reduction processes

I. $Fe_2O_3 + C \longrightarrow Fe$

II. $ZnO + C \longrightarrow Zn$

III. $Ca_3(PO_4)_2 + C \longrightarrow P$

IV. $PbO + C \longrightarrow Pb$

(a) Both II and IV (b) All but III

(c) All but IV (d) All of these

22 Consider the following statements.

Statement I Desilverisation of lead is done by Parke's method.

Statement II When lead-silver alloy is poor in silver, zinc is added to molten ore.

(a) Both statements I and II are true

(b) Only statement I is true

(c) Only statement II is true

(d) Neither statement I nor statement II is true

23 Bessemerisation is carried out for

I. Fe II. Cu
III. Al IV. silver

(a) Both I and II (b) Both II and III
(c) Both III and IV (d) Both I and III

24 An impure metal is allowed to react with carbon monoxide at 50°C and the volatile gas thus, formed is collected and heated further to about 200°C. This process gives the metal of 99.99% purely. What is the metal?

(a) Fe (b) Cr (c) Co (d) Ni

25 Match items of Column I with the items of Column II. **[NCERT Exemplar]**

	Column I		Column II
A.	Cyanides process	1.	Ultrapure Ge
B.	Froth floatation process	2.	Dressing of ZnS
C.	Electrolytic reduction	3.	Extraction of Al
D.	Zone refining	4.	Extraction of Au
		5.	Purification of Ni

Choose the correct match.

	A	B	C	D		A	B	C	D
(a)	4	2	3	1	(b)	3	1	4	2
(c)	2	4	1	3	(b)	1	3	2	4

26 Select the correct statement(s), about differentiation floatation

(a) is used to concentrate mixture of sulphides at different stages.
(b) zinc sulphate is used to depress zinc sulphide and PbS forms froth.
(c) copper sulphate is used to activate depressed lead sulphide and O_2 is blown to make froth.
(d) All of the above are correct statements.

27 Consider the following reaction,

$$(\text{Ag} + \text{Pb})\text{ alloy} \xrightarrow{\text{Melt and add zinc}} (\text{Ag} + \text{Pb} + \text{Zn})$$

$$\text{melt} \xrightarrow{\text{Cool}} \frac{\text{Layer } X}{\text{Layer } Y}$$

Select the correct statement(s) based on above scheme.

(a) Layer X contains zinc and silver.
(b) Layer Y contains lead and silver but amount of silver in this layer is smaller than in the layer X.
(c) X and Y are immiscible layers.
(d) All of the above are correct statements.

28 Which of the following is not the correct statement?

(a) Cassiterite, chromite and pitchblende are concentrated by hydraulic washing (Tabling).
(b) Pure Al_2O_3 is obtained from the bauxite ore by leaching in the Baeyer's process.
(c) Sulphide ore is concentrated by calcination method.
(d) Roasting can convert sulphide into oxide or sulphate and part of sulphide may also act as a reducing agent.

29 The chemical process in the production of steel from haematite ore involves

(a) reduction
(b) oxidation
(c) reduction followed by oxidation
(d) oxidation followed by reduction

30 Select the correct statement(s).

(a) In the decomposition of an oxide, into oxygen and gaseous metal, entropy increases.
(b) Decomposition of an oxide is an endothermic change.
(c) To make $\Delta G°$ negative, temperature should be high enough so that $T\Delta S° > \Delta H°$.
(d) All of the above are correct statements.

31 Consider the following statements.

Roasting is carried out to

I. convert sulphide to oxide and sulphate.
II. remove water of hydration.
III. melt the ore.
IV. remove arsenic and sulphur impurities.

Of these statements

(a) I, II and III are correct
(b) I and IV are correct
(c) I, II and IV are correct
(d) II, III and IV are correct

32 Consider the following metallurgical processes.

I. Heating impure metal with CO and distilling the resulting volatile carbonyl, (boiling point 43° C and finally decomposing at 150° to 200° C to get the pure metal.
II. Heating the sulphide ore in air until a part is converted to oxide and then further heating in the absence of air to let the oxide react with unchanged sulphide.
III. Electrolysing the molten electrolyte containing approximately equal amounts of the metal chloride and $CaCl_2$ to obtain the metal.

The processes used for obtaining sodium, nickel and copper are respectively

(a) I, II and III (b) II, III and I
(c) III, I and II (d) II, I and III

(C) Medical entrances' gallery

1 Identify the correct statement from the following **[NEET 2020]**

(a) Blister copper has blistered appearance due to evolution of CO_2.
(b) Vapour phase refining is carried out for nickel by van Arkel method.
(c) Pig iron can be moulded into a variety of shapes.
(d) Wrought iron is impure iron with 4% carbon.

2 Which of the following has maximum iron content?

(a) Cast iron (b) Wrought iron **[AIIMS 2019]**
(c) Pig iron (d) Stainless steel

3 Considering Ellingham diagram, which of the following metals can be used to reduce alumina? **[NEET 2018]**
(a) Mg (b) Zn (c) Fe (d) Cu

4 In the extraction of copper from its sulphide ore, the metal is formed by the reduction of Cu_2O with
(a) FeS (b) CO (c) Cu_2S (d) SO_2

5 Lead is extracted from galena (PbS) by following ways :

I. $PbS \xrightarrow[-SO_2]{\text{Roasted in air}} PbO \xrightarrow[-CO_2]{\text{Heated with C}} Pb$

II. $PbS \xrightarrow[-SO_2]{\text{Roasted in air}} PbO \xrightarrow{\text{Heated with PbS}} Pb + SO_2$

Auto-reduction process is **[JIPMER 2018]**
(a) Only I (b) Only II
(c) Both (a) and (b) (d) None of these

6 Extraction of gold and silver involves leaching with CN^- ion. Silver is later recovered by **[NEET 2017]**
(a) displacement with Zn (b) liquation
(c) distillation (d) zone refining

7 Metals are usually not found as nitrates in their ores. Out of the following two (I and II) reasons which is/are true for the above observation? **[NEET 2015]**
I. Metal nitrates are highly unstable.
II. Metal nitrates are highly soluble in water.
(a) I and II are true (b) I and II are false
(c) I is false but II is true (d) I is true but II is false

8 Sphalerite is concentrated by **[Kerala CEE 2014]**
(a) gravity separation (b) froth floatation
(c) magnetic separation (d) hydroulic washing
(e) leaching

9 Which one of the following methods is used in the concentration of sulphide ore? **[EAM CET 2014]**
(a) Froth floatation (b) Smelting
(c) Roasting (d) Leaching

10 Which of the following statements is incorrect? **[KCET 2014]**
(a) A furnace lined with haematite is used to convert cast iron to wrought iron.
(b) Collectors enhance the wettability of mineral particles during froth floatation.
(c) In vapour phase refining, metal should form a volatile compound.
(d) Copper from its low grade ores is extracted by hydrometallurgy.

11 Which among the following metals is refined by electrolytic method? **[MHT CET 2014]**
(a) Aluminium (b) Bismuth
(c) Tin (d) Lead

12 The following set of reactions are used in refining of zirconium

$Zr\ (\text{impure}) + 2I_2 \xrightarrow{523K} ZrI_4 \xrightarrow{3800\,K} Zr\ (\text{pure}) + ZI_2.$

This method is known as **[KCET 2014]**
(a) Distillation (b) Liquation
(c) Hall-Heroult method (d) van Arkel method

13 The ore of aluminium is **[UP CPMT 2013]**
(a) carnallite (b) malachite
(c) galena (d) bauxite

14 The formula of malachite ore is **[Manipal 2013]**
(a) Fe_3O_4 (b) Mn_3O_4
(c) NiAs (d) $Cu(OH)_2 \cdot CuCO_3$

15 Azurite is an ore of **[WB JEE 2013]**
(a) gold (b) silver
(c) copper (d) zinc

16 The process of converting hydrated into anhydrous alumina is called **[CBSE AIPMT 2013]**
(a) roasting (b) smelting
(c) dressing (d) calcination

17 NaCN is sometimes added in the froth floatation process as a depressant, when ZnS and PbS mineral are extracted because **[AIIMS 2013]**
(a) ZnS forms soluble complex $Na_2[Zn(CN)_4]$ while PbS forms froth.
(b) $Pb(CN)_2$ is precipitated while no effect on ZnS.
(c) PbS forms soluble complex $Na_2[Pb(CN)_4]$ while ZnS forms froth.
(d) NaCN is never added in froth floatation process.

18 Auto-reduction process is used in the extraction of **[AFMC 2013]**
(a) Cu and Hg (b) Zn and Hg
(c) Cu and Al (d) Fe and Pb

19 Which of the following pairs of metals is purified by van Arkel method? **[CBSE AIPMT 2013]**
(a) Ni and Fe (b) Ga and In
(c) Zr and Ti (d) Ag and Au

20 Which one of the following is a mineral of iron? **[AIPMT 2012]**
(a) Malachite (b) Cassiterite
(c) Pyrolusite (d) Magnetite

21 Identify the ore not containing iron. **[UP CPMT 2012]**
(a) Limonite (b) Siderite
(c) Carnallite (d) Chalcopyrite

22 Which of the following is an oxide ore? **[OJEE 2012]**
(a) Malachite (b) Haematite
(c) Copper glance (d) Zinc blende

23 Which one of the following does not occur as sulphide ore? **[Kerala CEE 2011]**
(a) Zn (b) Cr (c) Ag (d) Fe
(e) Co

24 Which of the following elements is present as the impurity to the maximum extent in the pig iron? **[CBSE AIPMT 2011]**

(a) Phosphorus (b) Manganese
(c) Carbon (d) Silicon

25 Carbon cannot reduce Fe_2O_3 to Fe at a temperature below 983 K because **[KCET 2011]**

(a) free energy change for the formation of CO is more negative than that of Fe_2O_3
(b) CO is thermodynamically more stable than Fe_2O_3
(c) carbon has higher affinity towards oxygen than iron
(d) iron has higher affinity towards oxygen than carbon

26 Which one of the following statements is false? **[KCET 2011]**

(a) During roasting, moisture is removed from the ore
(b) The ore is freed from almost all non-metallic impurities
(c) Calcination of ore is carried out in the absence of any blast of air
(d) The concentrated zinc blende is subjected to calcination during its extraction by pyrometallurgy.

ANSWERS

Check Point 6.1

1. (a)	2. (d)	3. (b)	4. (c)	5. (d)	6. (c)	7. (d)	8. (b)	9. (c)	10. (d)

Check Point 6.2

1. (b)	2. (a)	3. (c)	4. (b)	5. (a)	6. (a)	7. (c)	8. (b)	9. (b)	10. (a)

Check Point 6.3

1. (b)	2. (a)	3. (d)	4. (c)	5. (a)	6. (a)	7. (b)	8. (c)

Check Point 6.4

1. (b)	2. (c)	3. (c)	4. (d)	5. (c)	6. (a)	7. (a)	8. (c)

(A) Topical exercise

1. (a)	2. (a)	3. (b)	4. (d)	5. (b)	6. (b)	7. (d)	8. (c)	9. (b)	10. (a)
11. (c)	12. (a)	13. (b)	14. (d)	15. (b)	16. (d)	17. (a)	18. (a)	19. (a)	20. (b)
21. (a)	22. (b)	23. (b)	24. (c)	25. (c)	26. (c)	27. (a)	28. (b)	29. (b)	30. (c)
31. (d)	32. (d)	33. (b)	34. (b)	35. (d)	36. (b)	37. (c)	38. (a)	39. (d)	40. (c)
41. (a)	42. (b)	43. (b)	44. (d)	45. (c)	46. (d)	47. (a)	48. (b)	49. (a)	50. (c)
51. (b)	52. (d)	53. (b)	54. (d)	55. (d)	56. (d)	57. (a)	58. (c)	59. (b)	60. (a)
61. (a)	62. (a)	63. (a)	64. (a)	65. (c)	66. (b)	67. (a)	68. (b)	69. (a)	70. (a)
71. (a)	72. (a)	73. (c)	74. (c)	75. (c)	76. (a)	77. (a)	78. (d)	79. (c)	80. (d)
81. (c)	82. (c)	83. (d)	84. (d)	85. (a)	86. (d)	87. (c)	88. (a)	89. (c)	90. (d)
91. (a)	92. (a)	93. (d)	94. (a)	95. (b)	96. (a)	97. (b)	98. (a)		

(B) Miscellaneous questions

1. (b)	2. (a)	3. (a)	4. (b)	5. (d)	6. (b)	7. (a)	8. (c)	9. (a)	10. (b)
11. (d)	12. (a)	13. (c)	14. (b)	15. (a)	16. (a)	17. (b)	18. (d)	19. (b)	20. (d)
21. (d)	22. (a)	23. (a)	24. (d)	25. (a)	26. (d)	27. (d)	28. (c)	29. (d)	30. (d)
31. (c)	32. (c)								

(C) Medical entrances' gallery

1. (c)	2. (b)	3. (a)	4. (c)	5. (b)	6. (a)	7. (c)	8. (b)	9. (a)	10. (b)
11. (a)	12. (d)	13. (d)	14. (d)	15. (c)	16. (d)	17. (a)	18. (a)	19. (c)	20. (d)
21. (c)	22. (b)	23. (b)	24. (c)	25. (d)	26. (d)				

Hints & Explanations

Check point 6.1

3 (*b*) Corundum (Al_2O_3) is an oxide mineral of Al.

10 (*d*) Because the ore is heated in excess of air to convert into oxide.

Check point 6.3

1 (*b*) No disproportionation is involved in the extraction of Cu.

7 (*b*) Because zinc blende ZnS is a sulphide ore.

(A) Topical exercise

10 (*a*) Because reactants and product are not in aqueous form.

17 (*a*) Being very less reactive, metals like gold, platinum etc. are found in native (or free) states.

25 (*c*) Because Zn has more affinity towards O as compared to H. Moreover Zn is more reactive than H. So, hydrogen is not able to displace zinc from its oxide.

26 (*c*) $\underset{\text{Reductant}}{2Cu_2O + Cu_2S} \longrightarrow 6Cu + SO_2$ (Reduction: $Cu_2O \to Cu$; Oxidation: $Cu_2S \to SO_2$)

29 (*b*) Zone refining is generally used for semiconductors like Si, Ge, etc.

33 (*b*) Because it is electromagnetic separation.

39 (*d*) In the metallurgy of iron the impurities present (i.e. SiO_2) is acidic so basic flux (SiO_2) is needed.

40 (*c*) van Arkel process is used for Zr or Ti.

44 (*d*) Al can not obtained from Al_2O_3 by C reduction because carbon (graphite or coke) has lesser affinity towards oxygen as compared to aluminium.

52 (*d*) **Anode** $\underset{\text{(Impure sample)}}{Cu} \longrightarrow Cu^{2+} + 2e^-$

Cathode $Cu^{2+} + 2e^- \longrightarrow \underset{\text{(Pure copper)}}{Cu}$

59 (*b*) $\underset{\text{Impure}}{Ni} + 4CO \longrightarrow Ni(CO)_4$

$Ni(CO)_4 \xrightarrow{\Delta} \underset{\text{Pure}}{Ni} + 4CO$ (Mond' s process)

70 (*a*) Matte is CuS + FeS

71 (*a*) Because Fe is more reactive than Cu and does not react with water.

75 (*c*) Iron is an impurity in copper pyrites so slag is

$$FeO + SiO_2 \longrightarrow \underset{\substack{\text{Slag} \\ \text{(Ferrous silicate)}}}{FeSiO_3}$$

77 (*a*) van Arkel method is a refining process. Al is refined by electrolysis.

80 (*d*) Fluorspar is CaF_2 but feldspar is $K_2O \cdot Al_2O_3 \cdot 6SiO_2$.

86 (*d*) Carbon (graphite) is the major impurity (~ 4%) present in pig iron.

88 (*a*) Cassiterite (tin stone) is SnO_2, so it is not an iron ore.

90 (*d*) Feldspar is $K_2O \cdot Al_2O_3 \cdot 6SiO_2$.

96 (*a*) Magnetite is Fe_3O_4, so it is not contain Mg.

(B) Miscellaneous questions

1 (*b*) Because only smelting is a reduction process among the given.

2 (*a*) For a spontaneous process, $\Delta G = -$ ve

8 (*c*) Being highly reactive, Mg is extracted by electrolysis of fused salt.

13 (*c*) Because HgO decomposes by heat.

16 (*a*) Because anode is always the site of oxidation and E°_{cell} of Cl^- is greater than that of water.

19 (*b*) In the metallurgy of aluminium, the graphite anode is oxidised to CO and CO_2 by the released O_2.

20 (*d*) Argentite is Ag_2S (a sulphide ore) and cuprite is Cu_2O (an oxide ore).

(C) Medical entrances' gallery

1 (*c*) (a) Matte on auto-reduction produces blister copper, which has blistered appearance due to evolution of SO_2 (not CO_2).

$$\underset{\text{Matte}}{Cu_2S} + 2Cu_2O \longrightarrow \underset{\text{Blister copper}}{6\,Cu} + SO_2\uparrow$$

(b) Vapour phase refining is carried out for Ni by Mond's process (not by van Arkel method).

$$\underset{\text{Impure}}{Ni} \xrightarrow[80°C]{4CO} Ni(CO)_4 \xrightarrow[-\,4CO]{200°C} \underset{\text{Pure}}{Ni}$$

(c) Pig iron can be moulded into variety of shapes.

(d) Wrought iron the purest form of iron which contains 0.2-0.5% carbon (not 4%).

So, option (c) is correct.

2 (*b*) The percentage of other elements in iron are as follows:

(a) Cast iron = about 2.1%

(b) Wrought iron = about 0.08%

(c) Pig iron = about 4%

(d) Stainless steel = about 1.7%

On the basis of above information wrought iron has maximum iron content.

Hence, option (b) is the correct.

3 (*a*) Ellingham diagrams help us in predicting the feasibility of thermal reduction of an ore. The criterion of feasibility is that at a given temperature, Gibbs energy of the reaction must be negative.

Gibbs energy $\Delta G°$ *vs* T plots (schematic) for formation of some oxides (Ellingham diagram) is given below :

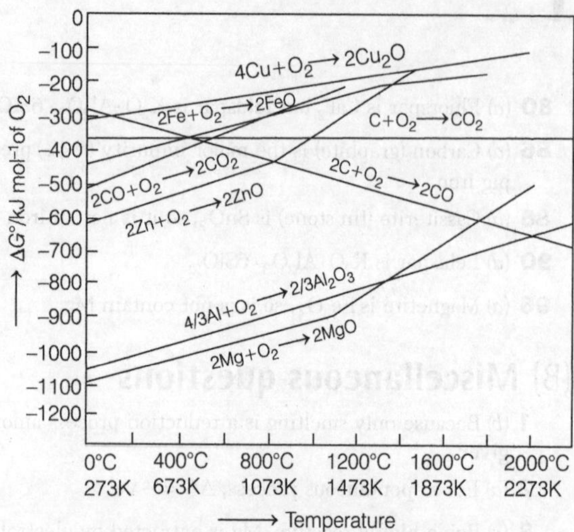

Gibbs energy $\Delta G°$ vs T plots (schematic) for formation of some oxides (Ellingham diagram).

According to Ellingham diagram, the temperature at which two lines intersect shows that the metal will reduce the oxide of other metals which lie above it in Ellingham diagram.

In other words, the metal oxide having more negative value of ΔG_f° can reduce the oxide having less negative ΔG_f°. As, Mg has more $-\Delta G°$ value than alumina, so it will be in lower part of Ellingham diagram. Hence, Mg will be used to reduce alumina.

4 (*c*) In the extraction of copper from its sulphide ore, the metal is formed by the reduction of Cu_2O with Cu_2S. This reaction completes with the process of auto-reduction. Chemical reaction occuring in this reaction is as follows

$$2Cu_2S + 3O_2 \longrightarrow 2Cu_2O + 2SO_2 \uparrow$$

$$Cu_2O + \frac{1}{2}Cu_2S \longrightarrow 3Cu + \frac{1}{2}SO_2$$

The molten copper is allowed to cool. During cooling, dissolved SO_2 comes out in the form of blisters, so it is called blister copper.

5 (*b*) In auto-reduction, no external reducing agent is required. In such reduction reaction, the product formed acts as a reducing agent and reduces the oxide to metal. Thus, only reaction II is called auto-reduction.

$$2\,PbS + 3O_2 \xrightarrow{\text{Roasted in air}} 2PbO + 2SO_2$$

$$2\,PbO + PbS \longrightarrow 3Pb + SO_2$$

(Act as a reducing agent)

6 (*a*) $4Ag(s) + 8CN^- + O_2(g) + 2H_2O \longrightarrow 4[Ag(CN)_2]^- + 4OH^-$

$2[Ag(CN)_2]^- + Zn \longrightarrow [Zn(CN)_4]^{2-} + 2Ag(s)$

7 (*c*) Metals nitrates are generally stable but most of them are soluble in water. That's why metal are usually not found as nitrates.

8 (*b*) Sphalerite or zinc blende (ZnS) is concentrated by froth floatation method.

10 (*b*) Collectors enhance the non-wettability of mineral particles towards water.

13 (*d*) Bauxite ($Al_2O_3 \cdot 2H_2O$.)

15 (*c*) Azurite is $2CuCO_3 \cdot Cu(OH)_2$.

16 (*d*) Calcination is used for hydrated oxide ores or carbonate ore to convert them oxides (anhydrous).

18 (*a*) Oxides of metals less reactive than being less stable undergo auto-reduction.
So, carnallite does not contain iron.

21 (*c*) Carnallite ($KCl \cdot MgCl_2 \cdot 6H_2O$) ; Limonite ($Fe_2O_3 \cdot 3H_2O$) ; Siderite ($FeCO_3$) ; Chalcopyrite ($CuFeS_2$).

24 (*c*) Pig iron contains upto 4% C.

25 (*d*) Iron has higher affinity towards oxygen than carbon. So, it cannot reduce Fe_2O_3 to Fe at a temperature below 983K.

26 (*d*) Zinc blende is a sulphide ore (i.e. ZnS), so it is roasted not calcined.

CHAPTER 07

p-Block Elements 1 (Nitrogen Family)

Group 15 of the periodic table contains five elements namely, nitrogen (N), phosphorus (P), arsenic (As), antimony (Sb) and bismuth (Bi). They are commonly known as **pnictogens** (pnicogen) meaning choking. The first two elements, N and P are non-metals, arsenic and antimony are metalloids, whereas the last element bismuth is a metal. The elements N and P are essential constituents of living system. Heavier members of this group, particularly arsenic are extremely toxic.

Inside

Occurrence

Nitrogen is highly abundant (about 78% by volume and 75% by mass) in air as N_2. However, abundance of nitrogen in rocks and soils is low (19 ppm), present as KNO_3 and $NaNO_3$.

Naturally occurring nitrogen consists of ^{14}N stable isotope, ^{15}N also occurs slightly, relative abundance being $^{14}N/^{15}N \simeq 272:1$.

Phosphorus is highly abundant in crystal rocks (≃ 1120 ppm) available as orthophosphate. It also occurs in plants and animals where it is an essential constituent of life. ^{31}P is the only stable isotope of phosphorus. Among six radioactive isotopes ^{32}P (β-emitter) is most important.

The elements As, Sb, Bi have very low abundances

As < 2 ppm

Sb 0-2 ppm

Bi 0-0.1 ppm

They occur as sulphides (As_2S_3, Sb_2S_3, Bi_2S_3) and are thus chalcophilic (i.e. ore forming) in nature.

GENERAL AND PHYSICAL PROPERTIES

The variations of general and physical properties of group 17 elements are summarised in Table 7.1

Propertry		N	P	As	Sb	Bi
Atomic number		7	15	33	51	83
Atomic mass/u		14.01	30.97	74.92	121.76	208.98
Electronic configuration		[He] $2s^2 2p^3$	[Ne]$3s^2 3p^3$	[Ar]$3d^{10} 4s^2 4p^3$	[Kr]$4d^{10} 5s^2 5p^3$	[Xe]$4f^{14} 5d^{10} 6s^2 6p^3$
Ionisation enthalpy	I	1402	1012	947	834	703
($\Delta_i H$/kJ mol^{-1})	II	2856	1903	1798	1595	1610
	III	4577	2910	2736	2443	2466
Electronegativity		3.0	2.1	2.0	1.9	1.9
Covalent radius/pm[a]		70	110	121	141	150
Ionic radius/pm		171^b	212^b	222^b	76^c	103^c
Melting point/K		63^*	317^d	1089^e	904	544
Boiling point/K		77.2^*	554^d	888^f	1860	1837
Density/g cm^{-3} (298 K)		0.879^g	1.823	5.778^h	6.697	9.808

Note *a - E(+3) single bond (E = element); b - E^{3-}; c - E^{3+}; d-White phosphours; e - Grey α - form; f - sublimation temperature at 38.6 atmosphere pressure; g - At 63 K; h - Grey α- form; *Molecular N_2.*

Electronic configuration

The general electronic configuration of this group is ns^2np^3. Therefore, they are known as *p*-block elements as the last electron enters in *p*-subshell.

Physical appearance

Nitrogen is gaseous in nature and exist as a diatomic molecule (N≡N). Phosphorus exist as tetraatomic P_4 molecule in the gaseous state, whereas in solid state, it exists in different allotropic forms. Arsenic, antimony and bismuth are solids.

Atomic radii

On moving down the group i.e. from N to Bi, atomic radii increases due to addition of extra principal energy shell in each succeeding element.

Ionisation enthalpy

The ionisation values of elements decreases down the group due to gradual increase in atomic size. Nitrogen, being the first member of this group has very high value i.e. 1403 kJ /mol, due to its small atomic size.

Metallic nature

Metallic nature increases down the group. Here, nitrogen and phosphorus are non-metals, arsenic and antimony are metalloids and bismuth is a typical metal.

Electronegativity

The elements of nitrogen family are electronegative in nature. The value of electronegativity decreases on moving down the group from nitrogen to bismuth. Nitrogen is the most electronegative element in group 15.

Melting and boiling points

Elements of group 15 do not show a regular trend in the values of melting and boiling points. Melting point first increases from N to As and then decreases to Sb and Bi (this is because of their tendency to form three covalent bond instead of five covalent bond due to inert pair effect). Boiling point first increases from N to Sb but boiling point of Bi is less than Sb.

The correct order for melting and boiling point is given below

Melting point $N < P < Bi < Sb < As$

Boiling point $N < P < As < Bi < Sb$

Catenation

All the elements of this group show property of catenation (self linkage) but to a much smaller extent than carbon. The decrease in tendency to form chains is mostly due to a decrease in the element-to-element bond energies. The catenation in stable nitrogen compounds is restricted to two or three N atoms because of weak N—N single bond N_2H_4, $N_3^{\ominus}$ etc. Among group 15 elements, phosphorus has maximum tendency to catenate.

Allotropy

All the members of group 15 (except Bi) show the phenomenon of allotropy. Nitrogen exists in two solid and one gaseous allotropic forms. Phosphorus exists in several allotropic forms such as white, red, scarlet, violet and black phosphorus. Arsenic exists in three allotropic forms. i.e. grey, yellow and black. Similarly, antimony (Sb) exists as metallic, α- or yellow and black or β-antimony.

Oxidation state

All the elements of group 15 have five electrons in their valence shell. Thus, oxidation states of these elements are +3 and +5. On moving down the group, stability of +5 oxidation state decreases and that of +3 oxidation state increases due to **inert pair effect,** i.e. reluctance of *s*-electrons to participate in bonding.

Therefore, Bi^{3+} is more stable than Bi^{5+}. As a result, Bi^{5+} salts are very rare and are good oxidising agents

$$Bi^{5+} + 2e^- \longrightarrow Bi^{3+}$$

As^{3+} salts are good reducing agents,

$$As^{3+} \longrightarrow As^{5+} + 2e^-$$

Nitrogen shows a wide range of oxidation states, i.e. from −3 to +5 due to its high electronegativity, small size and strong tendency to form $p\pi$- $p\pi$ bonds as shown below.

Table 7.2 The oxidation state of nitrogen element in some compounds

Compound	Oxidation state of N
NH_3	−3
N_2H_4	−2
NH_2OH	−1
N_2	0
N_2O	+1
NO	+2
NO_2	+4
HNO_3	+5
N_3H	−1/3

Chemical Properties

Reaction with hydrogen

Elements of group 15 react with hydrogen to form hydrides of formula MH_3. All the hydrides are volatile, poisonous and foul smelling gases. These are pyramidal in shape.

The trihydrides of group 15 elements are prepared from binary metal compounds such as Mg_3N_2, Zn_3As_2, Mg_3Sb_2 etc. by the action of water or dilute acids.

e.g. $$Mg_3N_2 + 6H_2O \longrightarrow 3Mg(OH)_2 + 2NH_3$$

$$Mg_3N_2 + 6HCl \longrightarrow 3MgCl_2 + 2NH_3$$

$$Zn_3As_2 + 6HCl \longrightarrow 3ZnCl_2 + 2AsH_3$$

On descending the group from NH_3 to BiH_3 following trends are observed.

(*i*) Bond angle decreases in the order

$$NH_3 > PH_3 > AsH_3 > SbH_3 > BiH_3$$

The bond angles in PH_3, AsH_3, SbH_3, BiH_3 are slightly greater than 90°. This shows that the orbitals are not sp^3-hybridised but are close to *p*-orbitals (mutually perpendicular).

∴ Lone pair is occupied by *s*-orbital and hence they are strongly bound to nucleus.

(*ii*) Basic strength decreases in the order as follows

$$NH_3 > PH_3 > AsH_3 > SbH_3 > BiH_3$$

This is due to the decrease in the availability of the lone pair of electrons.

(*iii*) Thermal stability decreases in the order

$$NH_3 > PH_3 > AsH_3 > SbH_3 > BiH_3$$

due to increase in size of central atom strength of M—H bond decreases.

(*iv*) Reducing character increases in the order

$$NH_3 < PH_3 < AsH_3 < SbH_3 < BiH_3$$

This is due to the decrease in the stability of hydrides and increase in ionisation of hydrides that increases the reducing character.

(*v*) Order of boiling points is as follows

$$PH_3 < AsH_3 < NH_3 < SbH_3 < BiH_3$$

Except for ammonia, the boiling point increases down the group. The abnormally high boiling point of NH_3 is due to the association of molecules through intermolecular hydrogen bonding.

(*vi*) Order of melting point

$$NH_3 > BiH_3 > SbH_3 > AsH_3 > PH_3.$$

NH_3 has the highest melting point this is due to intermolecular H-bonding. The other hydrides of this group do not show this H-bonding hence their melting point are lower than NH_3 and it decreases down the group with increase in the molecular size and their van der Waals' forces of attraction.

Reaction with halogens

Elements of group 15 react with halogens to form trihalides (MX_3) and pentahalides (MX_5). All trihalides except that of nitrogen are stable. In case of nitrogen only NF_3 is stable and inert.

(*i*) Trihalides are sp^3-hybridised with distorted tetrahedral geometry and pyramidal shape. Pentahalides are sp^3d-hybridised with trigonal bipyramidal geometry.

(*ii*) Trihalides are less covalent than pentahalides as $M(+5)$ are more polarising than $M(+3)$.

(*iii*) Trihalides are easily hydrolysed than pentahalides. Ease of hydrolysis decreases down the group. Therefore, NCl_3 is easily hydrolysed but $SbCl_3$ and $BiCl_3$ are partly hydrolysed. NF_3 and PF_3 do not undergo hydrolysis.

(*iv*) Basic character increases down the group as the electronegativity of halogen decreases that allows the availability of the lone pair of electrons as follows

$$NF_3 < NCl_3 < NBr_3 < NI_3$$

(*v*) The acidic strength increases with increase in electronegativity of the central atom.

$$PCl_3 > AsCl_3 > SbCl_3$$

(*vi*) Nitrogen does not form pentahalides due to non-availability of *d*-orbitals. The pentahalide of phosphorus, i.e. PCl_5 is unstable and decomposes to give PCl_3 and Cl_2.

(*vii*) The trihalides of P, As and Sb especially, fluorides and chlorides act as Lewis acids, since they can accept a lone pair of electrons from halide ion.

$$PF_3 + F_2 \longrightarrow PF_5$$

$$SbF_3 + 2F^- \longrightarrow [SbF_5]^{2-}$$

$$SbCl_3 + 2Cl^- \longrightarrow [SbCl_5]^{2-}$$

Note
- *Solid PCl_5 is an ionic compound and exists as $[PCl_4^+][PCl_6]^-$.*
- *PCl_5 (bipyramidal) changes to PCl_4^+ (tetrahedral) and PCl_6^- 1 (octahedral).*
- *It readily reacts with moisture, therefore is kept in stoppered bottles.*

Reaction with oxygen

Elements of group 15 react with oxygen to form several oxides such as R_2O_3 and R_2O_5.

Table 7.3 Various oxides formed by groups 15 elements

S.No.	Oxidation state	N	P	As	Sb	Bi
1.	+1	N_2O	–	–	–	–
2.	+2	NO	–	–	–	–
3.	+3	N_2O_3	P_2O_6	As_2O_3	Sb_2O_3	Bi_2O_3
4.	+4	N_2O_4	P_4O_8	–	–	–
5.	+5	N_2O_5	P_4O_{10}	As_2O_5	Sb_2O_5	Bi_2O_5

(*i*) The oxide in the higher oxidation state of the element is more acidic than that of lower oxidation state. Order for the acidic strength of oxides

$$N_2O < NO < N_2O_3 < N_2O_4 < N_2O_5$$

(*ii*) On moving down the group, acidic character decreases. This is because down the group, metallic character increases. Trioxides of N and P are acidic, trioxides of As and Sb are amphoteric and that of Bi is basic in nature.

(*iii*) All the pentaoxide (except P_2O_5) can readily loose oxygen to form lower oxides, thus they act as strong oxidising agents. The stability of pentaoxides follows the order

$$P_2O_5 > As_2O_5 > Sb_2O_5 > N_2O_5 > Bi_2O_5$$

CHECK POINT 7.1

1. Which of the following elements occur as sulphides in nature?
(a) Nitrogen (b) Arsenic
(c) Antimony (d) Both (b) and (c)

2. Which element of group 15 of the periodic table is a typical metal?
(a) Nitrogen (b) Phosphorus
(c) Antimony (d) Bismuth

3. Which of the following element is least electronegative in group 15 element?
(a) Arsenic (b) Nitrogen
(c) Bismuth (d) Antimony

4. Select the correct statement about order of stability of oxidation states.
(a) $Bi^{3+} > Sb^{3+} > As^{3+}$ (b) $As^{3+} > Sb^{3+} > Bi^{3+}$
(c) $Sb^{3+} > As^{3+} > Bi^{3+}$ (d) $Bi^{5+} > Sb^{5+} > As^{5+}$

5. Among the following, which is a good reducing agent?
(a) As^{3+} (b) As^{5+}
(c) Bi^{3+} (d) Bi^{5+}

6. Boiling point of $NH_3 > PH_3$ due to
(a) small size of nitrogen
(b) large size of phosphorus
(c) presence of intermolecular H-bonding
(d) higher electronegativity of nitrogen than phosphorus

7. Among the following trihalides, which one of them is easily hydrolysed?
(a) PCl_3 (b) NCl_3
(c) $SbCl_3$ (d) $BiCl_3$

8. Among the following oxides of nitrogen, which is the most acidic in nature?
(a) N_2O (b) N_2O_3
(c) N_2O_4 (d) N_2O_5

9. Which of the following oxide is amphoteric in nature?
(a) Bi_2O_3 (b) N_2O_3
(c) P_2O_3 (d) Sb_2O_3

10. Which of the following oxides is most stable?
(a) As_2O_5
(b) N_2O_5
(c) Bi_2O_5
(d) P_2O_5

DINITROGEN (N_2)

Occurrence

Nitrogen exists as a diatomic gas in the elemental state. Therefore, it is called dinitrogen.

Nitrogen occurs in free as well as combined state. In the combined state, it occurs mostly as nitrates such as potassium nitrate, sodium nitrate etc.

It is an important constituent of proteins in plants and animals which contains about 16 per cent of combined nitrogen.

Preparation

(i) **Laboratory method** Nitrogen is prepared by heating aqueous solution of NH_4Cl and $NaNO_2$.

$$NH_4Cl(aq) + NaNO_2(aq) \xrightarrow{\Delta} N_2(aq)\uparrow + 2H_2O(aq) + NaCl(aq)$$

(ii) **Commercial method** Commercially, nitrogen is prepared by fractional distillation of liquid air.

(iii) Pure nitrogen can be obtained by thermal decomposition of sodium azide or barium azide.

$$Ba(N_3)_2 \xrightarrow{\Delta} Ba + 3N_2$$

Physical properties

(i) It is a colourless, odourless and tasteless gas.

(ii) It is a non-toxic gas and has very low solubility in water.

(iii) Liquid nitrogen boils at –195.8°C. On rapid evaporation, the liquid nitrogen changes into an ice-like solid which melts at –209.9°C. At –238.5°C, it changes into another solid form.

(iv) It is slightly lighter than air as its vapour density is 140.

Chemical properties

(i) It is chemically unreactive at ordinary temperature due to very high bond energy of N≡N bond, i.e. 946 kJ mol^{-1} with a bond distance of 109.8 pm.

(ii) N_2 combines with certain compounds on strong heating.

e.g. $$\underset{\text{Calcium carbide}}{CaC_2} + N_2 \xrightarrow{1300\text{ K}} \underset{\text{Calcium cyanamide}}{CaCN_2} + C$$

These compounds are hydrolysed on boiling with water to give ammonia.

$$CaCN_2 + 3H_2O \longrightarrow CaCO_3 + 2NH_3$$

Therefore, calcium cyanamide is used as a fertiliser under the name of **nitrolim** ($CaCN_2 + C$).

(iii) Nitrogen combines with highly electropositive metal to form the corresponding nitrides. These nitrides are ionic in nature in which nitrogen is present as N^{3-}.

$$6Li + N_2 \xrightarrow{\Delta} 2Li_3N$$

N_2 combines with less electropositive elements like Mg, Sr, Ba etc., to form nitrides.

$$3Mg + N_2 \xrightarrow{\Delta} Mg_3N_2$$

The nitrides formed with metals of group 2 are generally ionic while those formed with elements of groups 13 and 14 are covalent in character.

(iv) It combines with H_2 at about 773 K in presence of catalyst to form ammonia (Haber's process).

$$N_2(g) + 3H_2(g) \overset{773\text{K}}{\rightleftharpoons} 2NH_3(g)$$

(v) Nitrogen combines with oxygen to give nitric oxide.

$$N_2 + O_2 \rightleftharpoons 2NO$$

This reaction is highly endothermic and takes place at elevated temperatures (about 3000°C).

Uses

(i) N_2 is mainly used in manufacture of compounds such as NH_3, HNO_3 etc.

(ii) It is also used to create an inert atmosphere.

(iii) Liquid N_2 is used as refrigerant to preserve food items, biological specimen and also used in cryosurgery.

Anomalous behaviour of nitrogen

Nitrogen differs from rest of the members of the group because of its small size, high electronegativity, high ionisation energy, absence of vacant *d*-orbitals and capacity to form *pπ-pπ* multiple bonds.

The properties that are different from other members are as follows

(*i*) Nitrogen is a gas while others are solid.

(*ii*) It exists as diatomic molecule with triple bond while others form single bond.

(*iii*) Nitrogen is chemically inert under ordinary conditions due to high bond dissociation energy, while others are reactive due to presence of single bond.

(*iv*) Nitrogen does not show penta valency because of non-availability of *d*-orbitals. Therefore, NCl_3 exists but NCl_5 doesn't.

(*v*) Because of its high electronegativity, nitrogen is distinctly non-metallic while the other two members (As, Sb) are metalloids and bismuth is distinctly a metal.

(*vi*) Nitrogen exhibits a large number of oxidation states such as −3 to +5. Other elements do not exhibit such a variety of oxidation states.

(*vii*) Hydride of nitrogen (NH_3) is stable. The hydrides of other elements are not stable. Moreover, only hydrides of nitrogen form H-bonding while other members do not.

(*viii*) Nitrogen forms oxides such as N_2O, NO, N_2O_4. NO such oxides are formed by other elements. Nitrogen forms oxides that are monomeric such as N_2O, NO, N_2O_4. P forms 3 dimeric oxides such as ($P_4O_6, P_4O_8, P_4O_{10}$) As and Sb form 2 dimeric oxides only whereas Bi forms monomeric oxide.

COMPOUNDS OF NITROGEN

Nitrogen reacts with oxygen, hydrogen to form various compounds like oxides, hydrides and oxoacids.
Some of the compounds are discussed below

Oxides of nitrogen

Nitrogen combines with oxygen under different conditions to form several oxides. These oxides show wide range of oxidation states from +1 to +5. All of the oxides are covalent in nature.

1. Nitrous oxide (N_2O)

N_2O is a triatomic molecule with 16 electrons in outer shell. Molecule has linear geometry (as for CO_2).

$$:N\equiv\overset{\oplus}{N}-\ddot{\underset{\cdot\cdot}{O}}:^{-} \longleftrightarrow :\overset{-}{\ddot{N}}=\overset{\oplus}{N}=\ddot{O}:$$

$$N\overset{1.12\,Å}{\text{——}}N\overset{1.19\,Å}{\text{——}}O$$

Preparation

(*i*) It is prepared by heating ammonium nitrate upto 240°C.

$$NH_4NO_3 \xrightarrow{\Delta, 250°C} N_2O + 2H_2O$$

(*ii*) It can also be prepared by hydroxyl amine by heating it with HNO_2.

$$NH_2OH + HNO_2 \xrightarrow{aq.} N_2O + 2H_2O$$

Properties

(*i*) It is stable, relativity unreactive, colourless gas and a neutral oxide.

(*ii*) It is soluble in water (1 : 1 by volume) but solution is neutral probably the following equilibrium lies in left-hand side.

$$N_2O + H_2O \rightleftharpoons \underset{\text{Hyponitrous acid}}{H_2N_2O_2}$$

It may be taken as anhydride of $H_2N_2O_2$.

(*iii*) It forms sodium azide on reaction with $NaNH_2$.

$$NaNH_2 + N_2O \longrightarrow NaN_3 + H_2O$$
$$NaNH_2 + H_2O \longrightarrow NaOH + NH_3$$
$$2NaNH_2 + N_2O \longrightarrow NaN_3 + NaOH + NH_3$$

(*iv*) N_2O resembles molecular oxygen (O_2), in that case it supports combustion. It does so because it decomposes to molecular nitrogen and molecular oxygen when heated.

$$2N_2O \longrightarrow 2N_2 + O_2$$

Uses

(*i*) The largest use of N_2O is as a propellant for whipped ice-cream. Because it has no taste and is non-toxic, it meets the strict food and health regulations.

(*ii*) N_2O is also used as an anaesthetic, particularly by dentists. It is also called **laughing gas**, because small amount causes **euphoria**.

Preparation and properties of N_2O

At a glance are shown in the figure given below

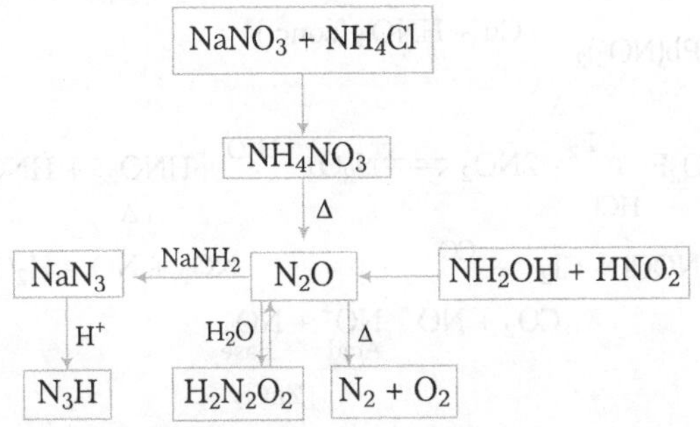

2. Nitric oxide (NO)

It can be represented by the following resonance structures. In solid and liquid state, it is diamagnetic as it dimerises to form N_2O_2.

$$\cdot\ddot{N}=\ddot{O}: \longleftrightarrow \cdot\overset{\oplus}{N}\equiv\ddot{O}:^{-} \equiv N \overset{\cdots}{=} O \quad (1.15\ \text{Å})$$

$$\begin{array}{c} N-O \\ | \quad\ | \\ O-N \end{array} \quad \text{Dimer}$$

$$\underset{\text{(I)}}{O-N-N-O} \qquad \underset{\text{(II)}}{O-N-O-N}$$

Structure (II) is asymmetrical and is formed as a red solid in the presence of HCl.

Preparation

In laboratory, it is prepared by reaction of dil.HNO_3 on copper.

$$3Cu + 8HNO_3 \longrightarrow 2NO + 3Cu(NO_3)_2 + 4H_2O$$

Some other methods of preparation are

(*i*) $N_2 + O_2 \longrightarrow 2NO$

(*ii*) $3Cu + \text{dil.}8HNO_3 \longrightarrow 3Cu(NO_3)_2 + 2NO + 4H_2O$

(*iii*) $2HNO_2 + 2I^- + 2H^+ \longrightarrow 2NO + I_2 + 2H_2O$

(*iv*) NO can be prepared in the laboratory by reacting sodium nitrite ($NaNO_2$) with a reducing agent such as Fe^{2+} in an acidic medium.

$$NO_2^-(aq) + Fe^{2+}(aq) + 2H^+(aq) \longrightarrow NO(g) + Fe^{3+}(aq) + H_2O(l)$$

$$2NO_3^- + 6Hg + 8H^+ \longrightarrow 3Hg_2^{2+} + 2NO + 4H_2O$$

Properties

(*i*) Lighting also contributes to the atmospheric concentration of NO. Exposed to air, nitric oxide quickly forms brown nitrogen dioxide gas.

$$2NO(g) + O_2(g) \longrightarrow \underset{\text{(Brown fumes)}}{2NO_2(g)}$$

Nitrogen dioxide is a major component of smog.

(*ii*) NO is an odd electron molecule alternately we can say it is a free radical. Hence, it is highly reactive.

$$2NO + Cl_2 \longrightarrow 2NOCl$$

$$3NO \xrightarrow[\text{High pressure}]{30-50°C} N_2O + NO_2$$

NO readily forms coordination complexes with transition metal ions. These complexes are called **nitrosyls.** Example, Brown ring

$$[Fe(H_2O)_5NO]^{2+}$$

(*iii*) Bond order is 2.5 (as described by MO theory) and bond length is 1.15Å, which is intermediate between a double bond and a triple bond.

(*iv*) It is paramagnetic in gaseous state due to presence of one unpaired electron (total 11 valence electron).

Preparation and properties of NO

At a glance are shown in the figure given below

N2 + O2 —Lighting→ NO
Dil. HNO3 —Cu or Hg→ NO
HNO2 —I⁻/H⁺→ NO
NO2 —Fe²⁺/H⁺/NO2⁻→ NO
NO —O2→ NO2
NO —Cl2→ NOCl
NO —30-50°C, High pressure→ N2O + NO2
NO —Fe²⁺→ [Fe(H2O)5NO]²⁺ (Brown ring)

3. Nitrogen dioxide (NO_2)

NO_2 is a V-shaped molecule with bond angle 134° and N—O bond length 119 pm. NO_2 is regarded as a resonance hybrid. Its structure can be represented as,

$$O\overset{\cdot}{N}{=}O \ (134°) \longleftrightarrow O{=}\overset{\cdot}{N}O$$

It is an odd electron species and paramagnetic in nature. It polymerises to form a colourless dimer N_2O_4 as shown below.

$$\begin{array}{c} O \quad\quad O \\ \ \ N-N \\ O \quad\quad O \end{array}$$

175 pm
Planar

Preparation

$$Cu + 4H^+ + 2NO_3^- \xrightarrow{\Delta} Cu^{2+} + 2H_2O + 2NO_2$$

$$2Pb(NO_3)_2 \xrightarrow{\Delta} 2PbO + 4NO_2 + O_2$$

On condensation, NO_2 is obtained as a pale liquid and ultimately a colourless solid (N_2O_4 due to dimerisation) is formed.

Properties

(*i*) It is a reddish brown gas. N_2O_4 is a mixed anhydride as indicated by its reaction on reaction with H_2O, HNO_2 (nitrous acid) and HNO_3 (nitric acid) are formed.

$$\underset{\underset{2NO_2}{\uparrow\downarrow}}{N_2O_4} + H_2O \longrightarrow HNO_2 + HNO_3$$

The HNO_2 formed decomposes to give NO and NO_2.

$$2HNO_2 \longrightarrow NO_2 + NO + H_2O$$

Thus, moist NO_2 and N_2O_4 gases are strongly acidic.

(*ii*) Liquid N_2O_4 is useful as a non-aqueous solvent and self ionises as,

$$N_2O_4 \rightleftharpoons \underset{\text{Acid}}{NO^+} + \underset{\text{Base}}{NO_3^-}$$

Thus, $ZnCl_2 + 2N_2O_4 \longrightarrow 2NOCl + Zn(NO_3)_2$

(*iii*) NO_2—N_2O_4 system is a strong oxidising agent.

$$2NO_2 \xrightarrow{150°C} 2NO + O_2$$

$$2NO_2 + F_2 \longrightarrow \underset{\text{Nitryl fluoride}}{2NO_2F}$$

$$2NO_2 + 4HCl \longrightarrow \underset{\text{Nitrosyl chloride}}{2NOCl} + Cl_2 + 2H_2O$$

$$NO_2 + CO \longrightarrow CO_2 + NO$$

$$2NO_2 \rightleftharpoons N_2O_4$$

N_2O_4 solidifies at $-11.2°C$ and is colourless. As the temperature increases, proportion of brown NO_2 increases and the colour of the liquid darkens. At around $140°C$, the mixture is almost entirely NO_2. NO_2 is paramagnetic due to unpaired electron but in dimer formation paramagnetism disappears due to pairing of unpaired electrons.

$$2NO_2 \rightleftharpoons N_2O_4$$

O₂N—NO₂ structure: N—N 175 pm, planar

This reaction occurs in both the gas phase and the liquid phase. (N—N) bond length in N_2O_4 is larger than normal (N—N) covalent bond. It is due to dipole-dipole repulsion.

Preparation and properties of NO_2

At a glance are shown in the figure given below

$Pb(NO_3)_2 \xrightarrow{\Delta} 2NO_2$; Cu + HNO_3 (conc.) $\rightarrow 2NO_2$

$2NO_2 \rightleftharpoons N_2O_4$

$2NO_2 \xrightarrow{F_2} NO_2F$

$2NO_2 \xrightarrow{HCl} NOCl + Cl_2$

$NO_2 \xrightarrow{CO} CO_2 + NO$

$N_2O_4 \xrightarrow{H_2O} HNO_2 + HNO_3$; $HNO_2 \xrightarrow{\Delta} NO_2 + NO + H_2O$

$N_2O_4 \rightarrow \underset{\text{Acid}}{NO^+} + \underset{\text{Base}}{NO_3^-} \xrightarrow{ZnCl_2} Zn(NO_3)_2 + NOCl$

4. Dinitrogen trioxide (N_2O_3)

The N_2O_3 molecule has an O_2N-NO structure in the gas phase and at low temperature, with an extremely long (1.86 Å) N—N bond, which is consistent with its easy dissociation.

(Structure: O–N–O angle 128.6°, N–O 1.21 Å, N—N 1.86 Å, angle 105°, N–O 1.12 Å)

Two interconvertible structures (in presence of light radiations of a particular wavelength) are assigned to N_2O_3. These are

$$O=N^{+}(-O^{-})-N=O \rightleftharpoons O=N-O-N=O$$

Preparation

It is obtained by condensing equimolar amount of NO and NO_2 together or by reacting NO with O_2 in a fixed ratio.

$$NO + NO_2 \longrightarrow N_2O_3$$

$$4NO + O_2 \longrightarrow 2N_2O_3$$

Properties and uses

(*i*) It is an acidic oxide and is the anhydride of HNO_2.

$$N_2O_3 + H_2O \longrightarrow 2HNO_2$$

$$N_2O_3 + 2NaOH \longrightarrow \underset{\text{Sodium nitrite}}{2NaNO_2} + H_2O$$

(ii) It is an intensely blue liquid or a pale blue solid. It exists only at low temperature and readily dissociates to give NO and NO_2.

$$N_2O_3 \rightleftharpoons NO + NO_2$$

(iii) It is used as special purpose fuel.

Preparation and properties of N_2O_3

At a glance are given below

$NO + NO_2 \to N_2O_3$; $NH_4NO_2 \xleftarrow{NH_4OH} N_2O_3 \xrightarrow{H_2O} HNO_2$; $NH_4NO_2 \xrightarrow{\Delta} N_2 + H_2O$; $N_2O_3 \xrightarrow{NaOH} NaNO_2$

5. Dinitrogen pentoxide (N_2O_5)

It exists as a symmetrical molecule having the structure $O_2N—O—NO_2$ in the gaseous state. It may be represented as,

(Structure: $O_2N-O-NO_2$; N–O (bridge) 151 pm, N–O (terminal) 119 pm, N–O–N angle 112°)

Each N-atom is sp^2-hybridised. X-ray diffraction shows that solid N_2O_5 is ionic in nature and exists as nitronium nitrate, i.e. $NO_2^+NO_3^-$. In the solution and gaseous phase, it is covalent in nature.

Preparation

N_2O_5 is the anhydride of HNO_3 and prepared by following reaction,

$$2HNO_3 \xrightarrow[\text{Low temp.}]{P_2O_5} N_2O_5 + H_2O$$

Properties

(i) It is colourless deliquescent solid and highly reactive.

(ii) In the gaseous phase,

$$N_2O_5 \xrightarrow{\Delta} NO_2 + NO + O_2$$

$$N_2O_5 + O_3 \longrightarrow 2NO_3 + O_2$$

(iii) In solid state,

$$N_2O_5 + H_2O \longrightarrow 2HNO_3$$

$$N_2O_5 + Na \longrightarrow NaNO_3 + NO_2$$

$$N_2O_5 + NaCl \longrightarrow NaNO_3 + NO_2Cl$$

(iv) With conc. H_2SO_4, it forms NO_2^+ (nitronium ion) which can attack benzene nucleus forming nitrobenzene.

$$N_2O_5 + 3H_2SO_4 \longrightarrow 2NO_2^+ + 3HSO_4^- + H_3O^+$$

$$C_6H_6 + NO_2^+ \longrightarrow C_6H_5NO_2 + H^+$$

Preparation and properties of N_2O_5

At a glance are shown in the figure given below

$HNO_3 \xrightarrow{P_2O_5} N_2O_5$; N_2O_5 (in solid phase) $\xrightarrow{H_2O} HNO_3$; $\xrightarrow{Na} NaNO_3 + NO_2$; $\xrightarrow{NaCl} NaNO_3 + NO_2Cl$; $(NO_2^+ + NO_3^-)$; N_2O_5 (in gaseous phase) $\xrightarrow{D} NO_2 + NO + O_2$; $\xrightarrow{O_3} NO_3^\bullet + O_2$ (Radical); $N_2O_5 \xrightarrow{H_2SO_4} NO_2^+$; NO_2^+ + benzene $\to$ nitrobenzene ($C_6H_5—NO_2$)

Nitrogen oxides and air pollution

Collectively, nitrogen oxides are often given the symbol NO_x. Two of them, NO and NO_2 are of special importance in environmental chemistry, causing considerable amount of air pollution. They are given off in car exhaust fumes and when fossil fuels are burnt, as well as being made during thunderstorms. In each case NO is made and then NO_2.

$$N_2 + O_2 \longrightarrow 2NO$$

NO_2 is more unpleasant and gives rise to acid rain and also takes part in photochemical reactions forming smog. The photons making up sunlight, can cause the NO_2 molecules to fall apart, releasing oxygen and no radicals.

$$NO_2 \xrightarrow{h\nu} N\dot{O} + \dot{O}$$

Oxygen radicals are extremely reactive and will attack other molecules, especially those in car exhaust fumes. For examples, hydrocarbons that escape combustion can be changed into aldehydes. The oxygen atoms can also react with O_2 to form O_3. NO_2 is also involved in the production of one of the most unpleasant of all chemicals present in photochemical smog. It is known as PAN (peroxyacetyl nitrate).

$$CH_3—C(=O)—O—O—NO_2$$

PAN

Its presence in the atmosphere irritates eyes and lungs and generally makes breathing difficult. In early of 2017, NCR was under the dense effect of **SMOG** due to PAN.

CHECK POINT 7.2

1. Which of the following are the applications of dinitrogen gas?
(a) Preservation of biological materials and food items
(b) Production of inert atmosphere in copper and steel industry
(c) In the preparation of explosives
(d) Etching of metals

2. Nitrogen is relatively inactive element because
(a) its atom has a stable electronic configuration
(b) it has low atomic radius
(c) its electronegativity is fairly high
(d) dissociation energy of its molecule is fairly high

3. Which statement is wrong for NO?
(a) It is anhydride of nitrous acid (b) It forms dimer
(c) It is paramagnetic (d) Both (b) and (c)

4. Reddish brown gas which is a mixed anhydride of HNO_2 and HNO_3
(a) N_2O (b) NO_2 (c) NO (d) N_2O_3

5. Dinitrogen oxide is produced by the
(a) thermal decomposition of NH_4NO_3
(b) disproportionation of N_2O_4
(c) thermal decomposition of NH_4NO_2
(d) interaction of hydroxyl amine and HNO_3

6. On heating HNO_3 with P_4O_{10}, the oxide of nitrogen produced is
(a) NO_2 (b) N_2O_5
(c) N_2O_4 (d) N_2O_3

7. The correct order for the decreasing acidic strength of oxides of nitrogen is
(a) $N_2O_5 > N_2O_3 > N_2O_4 > NO > N_2O$
(b) $NO > N_2O > N_2O_3 > N_2O_4 > N_2O_5$
(c) $N_2O > NO > N_2O_3 > N_2O_4 > N_2O_5$
(d) $N_2O_5 > N_2O_4 > N_2O_3 > NO > N_2O$

8. Consider the following reactions :

I. $2NaNO_2 + 2FeSO_4 + 3H_2SO_4 \longrightarrow Fe_2(SO_4)_3 + 2NaHSO_4 + 2H_2O + 'A'$

II. $2NO + N_2O_4 \xrightarrow{250K} 'B'$

The products A and B are respectively
(a) NO and N_2O_3
(b) NO and NO_2
(c) N_2O_3 and N_2O_5
(d) N_2O_3 and NO_2

Hydrides of nitrogen

Nitrogen forms hydrides, when reacts with hydrogen. Some of the hydrides of nitrogen are given below

1. Ammonia (NH_3)

NH_3 is a covalent molecule in which nitrogen possess sp^3-hybridisation. But due to the presence of one lone pair, it acquires pyramidal structure. The bond angle is 107° due to *lp-bp* repulsion.

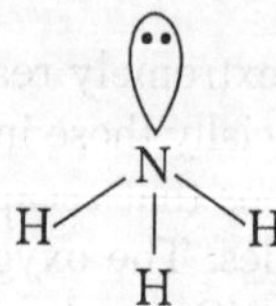

Preparation

(*i*) Ammonia is obtained from ammonium salts, when it is heated with caustic soda or lime.

$$NH_4Cl + NaOH \longrightarrow NH_3 + NaCl + H_2O$$

$$2NH_4Cl + \underset{\text{Slaked lime}}{Ca(OH)_2} \longrightarrow 2NH_3 + CaCl_2 + 2H_2O$$

(*ii*) Commercially, ammonia can be obtained by Haber's process. This process is discovered by a German chemist Fritz Haber. It involves direct combination of nitrogen and hydrogen. It is carried out at 400-500°C and at a pressure of 10^2 - 10^3 atm in presence of iron as catalyst.

$$N_2(g) + 3H_2(g) \xrightleftharpoons[P = 100 - 100 \text{ atm}, Fe_2O_3]{400\text{-}500°C,} 2NH_3(g); \quad \Delta H = -22.4 \text{ kcal}$$

The reaction is reversible and exothermic in nature. Thus, according to Le-Chatelier's principle the optimum conditions for the formation of NH_3 are
High pressure $= 200 \times 10^5$ Pa (about 200 atm).
Temperature = 700 K. Catalyst iron oxide with small amount of K_2O and Al_2O_3 or with molybdenum.

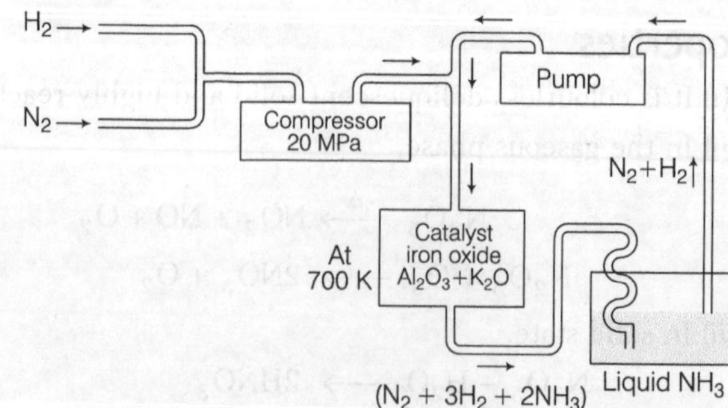

Fig 7.1 Flow chart for the manufacturing of ammonia

(*iii*) $\underset{\text{Calcium cyanamide}}{CaCN_2} + 3H_2O \longrightarrow CaCO_3 + 2NH_3$

Physical properties

(i) Ammonia is a colourless gas with a characteristic pungent odour. It is quite poisonous gas.

(ii) It's boiling point is −33.35°C.

(iii) It is lighter than air.

(iv) It is highly soluble in water. The high solubility is due to presence of hydrogen bonding.

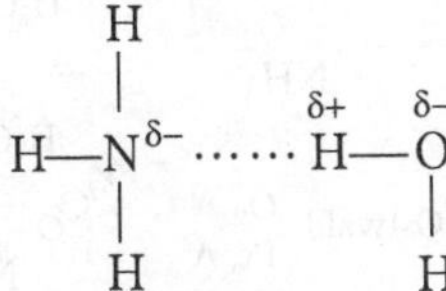

The solubility of ammonia increases with increase of pressure and decreases with increase of temperature.

(v) It can be easily liquefied at room temperature by application of pressure. It freezes at −78°C and has $[NH_3(l)]$ large heat of vaporisation (327 cal/g). It is therefore, used in ice plants.

(vi) Liquid NH_3 resembles water as being highly associated *via* strong hydrogen bonding. It self-ionises like water.

$$H_2O + H_2O \rightleftharpoons H_3O^+ + OH^-$$

$$NH_3 + NH_3 \rightleftharpoons NH_4^+ + NH_2^-$$

or

$$NH_3 \rightleftharpoons H^+ + NH_2^-$$

Chemical properties

The chemical properties of ammonia are as follows

(i) **Stability** It is highly stable. It decomposes into nitrogen and hydrogen at red heat or when electric sparks are passed through it.

$$2NH_3 \xrightleftharpoons{\text{Red heat or electric spark}} N_2 + 3H_2$$

(ii) **Basic nature** Ammonia is a Lewis base because it has tendency to donate an electron pair.

$$H_3N: + H^+ \longrightarrow [H_3N \rightarrow H]^+$$

It forms salts when react with acids,

$$NH_3 + HCl \longrightarrow \underset{\text{(Ammonium chloride) Dense white fumes}}{NH_4Cl}$$

(iii) **As a ligand** Ammonia acts as a ligand because it has much lower dielectric constant than water. It is a better solvent for organic compounds and generally a poorer one for ionic inorganic compounds but complex formation by NH_3 (as ligand) is better than that by H_2O. AgCl is insoluble in H_2O but soluble in NH_3 due to complex formation.

$$AgCl + 2NH_3 \longrightarrow [Ag(NH_3)_2]Cl$$

$$Cu^{2+} + 2NH_3 \longrightarrow [Cu(NH_3)_2]^{2+}$$

$NH_3(aq)$ precipitates salts as hydroxides

$$AlCl_3 + 3NH_4OH \longrightarrow \underset{\text{ppt.}}{Al(OH)_3} + 3NH_4Cl$$

$$FeCl_3 + NH_4OH \longrightarrow \underset{\text{ppt.}}{Fe_2O_3xH_2O\downarrow} + NH_4Cl$$

In some cases, precipitates dissolves due to formation of complex.

$$ZnCl_2 + 2NH_4OH \longrightarrow \underset{\text{ppt.}}{Zn(OH)_2} + 2NH_4Cl$$

$$Zn(OH)_2 + 4NH_4OH \longrightarrow \underset{\text{Soluble}}{[Zn(NH_3)_4]^{2+}} + 6H_2O$$

(iv) **Oxidation** Ammonia reacts with oxygen to produce nitrogen and water molecules.

$$4NH_3 + 3O_2 \longrightarrow 2N_2\uparrow + 6H_2O$$

This reaction is thermodynamically favoured under normal conditions. However, at 750°-900°C in the presence of Pt catalyst NO is formed.

$$4NH_3 + 5O_2 \xrightarrow{\text{Pt, }\Delta} 4NO + 6H_2O$$

Thus, NO can be converted into NO_2 which in turn convert into HNO_3.

$$2NO + O_2 \longrightarrow 2NO_2$$

$$3NO_2 + H_2O \longrightarrow 2HNO_3 + NO$$

Thus, the sequence in industrial utilisation of atmospheric nitrogen is as follows

$$N_2 \xrightarrow[\text{Haber's process}]{H_2} NH_3 \xrightarrow[\text{Ostwald process}]{O_2} NO \xrightarrow{O_2 + H_2O} HNO_3$$

(v) **Reducing nature** NH_3 can reduce oxides into metal. Thus, it exhibits reducing nature.

$$3CuO + 2NH_3 \xrightarrow{\Delta} 3Cu + N_2 + 3H_2O$$

$$3PbO + 2NH_3 \xrightarrow{\Delta} 3Pb + N_2 + 3H_2O$$

When chlorine is passed in excess, an explosive substance nitrogen trichloride is formed.

$$NH_3 + \underset{\text{Excess}}{3Cl_2} \longrightarrow \underset{\text{Nitrogen trichloride}}{NCl_3} + 3HCl$$

When NH_3 is in excess, N_2 and NH_4Cl are formed

$$8NH_3 + 3Cl_2 \longrightarrow N_2 + 6NH_4Cl$$

On warming, ammonia reacts with bleaching powder, forming N_2 and Cl^-.

Thus, ammonia acts as a reducing agent.

$$3Ca(OCl)_2 + 2NH_3 \longrightarrow 3CaCl_2 + N_2\uparrow + 3H_2O$$

NH_3 also reduces hypochlorite to Cl^- and NH_3 itself oxidised to N_2H_4

$$2NH_3 + NaClO \longrightarrow N_2H_4 + NaCl + H_2O$$

(vi) Dilute solutions of alkali metals in liquid NH_3 are dark blue in colour. This is due to the presence of solvated metals ions and solvated electrons.

$$2NH_3 + 2Na \longrightarrow \underset{\text{Sodamide}}{2NaNH_2} + H_2$$

(vii) Aqueous solution of NH_3 is generally referred to as solution of weak base NH_4OH. Undissociated NH_4OH does not exist. The solutions are best described as equilibrium of $NH_3(aq)$ with NH_4OH.

$$NH_3(aq) + H_2O \rightleftharpoons NH_4^+ + OH^-$$

$$K_b = \frac{[NH_4^+][OH^-]}{[NH_3]} = 1.8\times 10^{-5} \text{ mol L}^{-1}$$

(viii) Liquid ammonia functions as a good solvent for many substances and also for many types of reactions. All those substances, which dissolve in liquid NH_3 to give NH_4^+ ions are acids and those which give amides, NH_2^- ions are bases. NH_4Cl, NH_4NO_3 are acids in liquid NH_3 due to existence of NH_4^+ ions.

$$NH_4Cl \rightleftharpoons NH_4^+ + Cl^-$$

$$NH_4NO_3 \rightleftharpoons NH_4^+ + NO_3^-$$

$NaNH_2$ is a base in liquid NH_3 due to existence of NH_2^- ion.

$$NaNH_2 \rightleftharpoons Na^+ + NH_2^-$$

Acid-base neutralisation reaction in liquid NH_3 can be thus represented as,

$$\underset{\text{Acid}}{2NH_4Cl} + \underset{\text{Base}}{PbNH} \xrightarrow{NH_3} PbCl_2 + 3NH_3$$

$$3NH_4Cl + AlN \xrightarrow{NH_3} AlCl_3 + 4NH_3$$

(ix) Acids such as acetic acid (CH_3COOH) is a weak acid in water but behaves as a strong acid in liquid ammonia.

$$\underset{\text{Weak acid}}{CH_3COOH} + H_2O \rightleftharpoons CH_3COO^- + H_3O^+$$

$$\underset{\text{Strong acid}}{CH_3COOH} + NH_3 \rightleftharpoons CH_3COO^- + NH_4^+$$

Some other reactions involving NH_3 are represented below

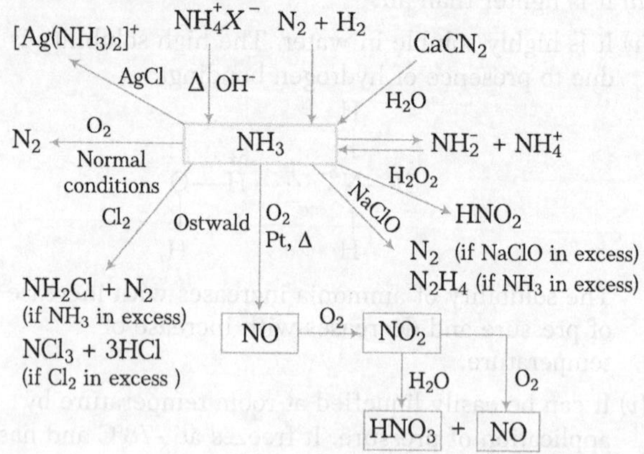

Test

$$\underset{\text{Nessler's reagent}}{2K_2HgI_4} + NH_3 + 3KOH \longrightarrow \underset{\text{Iodide of Millon's base (brown ppt.) / (yellow ppt.)}}{H_2NHgOHgI\downarrow} + 7KI + 2H_2O$$

Uses

(i) Ammonia is used in the form of aqueous solution of NH_4OH in laboratory in qualitative and quantitative analysis.

(ii) Liquid NH_3 is used in refrigeration because of its large heat of evaporation.

(iii) It is used for making artificial silk.

(iv) It is used as a cleaning agent for removing grease.

2. Hydrazine (N_2H_4)

Its structure can be represented as,

$$\begin{array}{c} H \quad\quad \\ \ddot{}\,N - N - H \\ H \quad\quad H \end{array}$$

Each N-atom is ***sp*³-hybridised** and tetrahedrally surrounded by one nitrogen two hydrogen and one lone pair of electrons.

Preparation

Raschig process

It is obtained by controlled oxidation of NH_3 (in excess) by alkaline solution of sodium hypochlorite (NaOCl).

$$\underset{\text{(in excess)}}{2NH_3} + NaOCl \longrightarrow N_2H_4 + NaCl + H_2O$$

Properties

(i) Colourless fuming liquid with faint ammoniacal odour.

(ii) Pure hydrazine and its aqueous solution are kinetically stable inspite of its endothermic nature.

(iii) **Basic nature** It is a weaker base than NH_3 but its behaviour is bifunctional.

Hydrazinium salts of $N_2H_5^+$ are stable in water but those of $N_2H_6^{2+}$ are hydrolysed in aqueous solution. Such salts as $N_2H_6^{2+}$ are less soluble than those of $N_2H_5^+$ and may be crystallised in presence of acid.

(iv) N_2H_4 is oxidised to N_2 and NH_3, thus it is a better reducing agent.

(v) In presence of a powerful reducing agent (as Zn/HCl), it is an oxidising agent.

Uses

As a rocket fuel Hydrazine being an endothermic compound burns rapidly and completely in air with evolution of heat.

$$N_2H_4 + \underset{\text{Air}}{O_2} \longrightarrow N_2 + 2H_2O; \ \Delta H = -620 \text{ kJ mol}^{-1}$$

Thus, N_2H_4 is used as a rocket fuel in guided missile space shuttles and lunar missions.

(i) Methyl derivatives $MeNHNH_2$ and Me_2NNH_2 are also used in combination with it. Oxidants are O_2 (liquid), N_2O_4 or H_2O_2.

(ii) As a reducing agent to prepare copper and silver mirror to prepare platinum metal.

(iii) Isonicotinic hydrazine ($C_5H_4NCONHNH_2$) is an antitubercular drug.

Table 7.4 Similarity between hydrazine and hydrogen peroxide

S. No.	Condition	N_2H_4	H_2O_2
1.	Structures	$H_2N{-}NH_2$	$H{-}O{-}O{-}H$
2.	Association	Forms liquid and solid through hydrogen-bonding	Forms liquid and gas through hydrogen bonding
3.	Hybridisation	N-atom is sp^3-hybridised	O-atom is sp^3-hybridised
4.	Decomposition	$3N_2H_4 \longrightarrow N_2 + 4NH_3$	$2H_2O_2 \longrightarrow 2H_2O + O_2$
5.	Oxidising nature	$N_2H_5^+ + 3H^+ + 2e^- \longrightarrow 2NH_4^+$	$H_2O_2 + 2H^+ + 2e^- \longrightarrow 2H_2O$
6.	Reducing nature	$N_2H_5^+ \longrightarrow N_2 + 5H^+ + 4e^-$	$H_2O_2 \longrightarrow O_2 + 2H^+ + 2e^-$

3. Hydrogen azide (N_3H)

Its structure can be represented as

H–N (1.01Å) ; N–N 1.24Å ; N–N 1.13Å ; angle 120°

$$H{-}\ddot{\underset{\cdot\cdot}{N}}{-}\overset{+}{N}{\equiv}N{:} \longleftrightarrow H{-}\ddot{N}{=}\overset{+}{N}{=}\overset{-}{N}{:}$$

Azide ion is similar to halide and cyanide ion and is known as **pseudo halide.**

Preparation

(i) By oxidation of aqueous hydrazine with nitrous acid,

$$N_2H_4 + HNO_2 \longrightarrow N_3H + 2H_2O$$

Oxidising agent can be H_2O_2, HNO_3, ClO_3^- etc.

(ii) By careful distillation of sodium azide (NaN_3) with dilute H_2SO_4,

$$NaN_3 + H_2SO_4 \longrightarrow N_3H + NaHSO_4$$

Properties

(i) Colourless liquid, boiling point 37°C. The liquid and vapour are deadly poisonous and explode on heating,

$$2N_3H \longrightarrow H_2\uparrow + 3N_2\uparrow$$

(ii) Weak acid but stronger than acetic acid.

(iii) N_3H dissolves Mg forming H_2

$$Mg + 2N_3H \longrightarrow Mg(N_3)_2 + H_2\uparrow$$

(iv) Metals like Cu, Zn, Fe form NH_3 and N_2 with N_3H

$$Cu + 3N_3H \longrightarrow Cu(N_3)_2 + N_2 + NH_3$$

Reducing nature

- $$\underset{+3}{HNO_2} + \underset{-\frac{1}{3}}{N_3H} \longrightarrow \underset{0}{N_2} + \underset{+1}{N_2O} + H_2O$$

Oxidation ($-\frac{1}{3} \to 0$); Reduction ($+3 \to +1$)

N_3H may be estimated by titration with $NaNO_2$ solution using $FeCl_3$ as an indicator N_3H gives red colour with $FeCl_3$.

- $HOCl + 2N_3H \longrightarrow 3N_2 + HCl + H_2O$

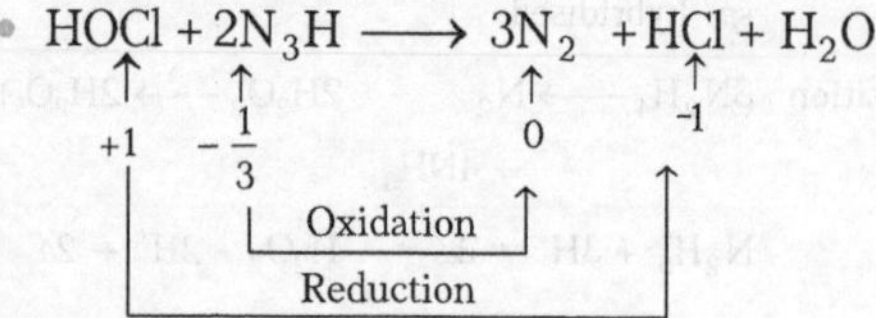

4. Hydroxylamine (NH_2OH)

Its structure may be represented as,

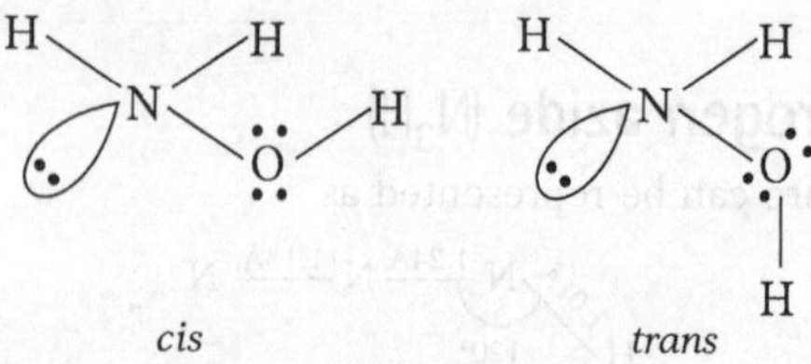

NH_2OH exhibits *cis* and *trans*-isomerism. N-atom is sp^3-hybridised with one lone pair of electrons. The N—O bond distance is 1.47Å corresponding to single bond.

Preparation

It is stable in the form of acid salt as $[NH_3OH]^+ Cl^-$ or $[NH_3OH]^+ HSO_4^-$.

NH_2OH in the form of salt is formed, when ammonium nitrite (NH_4NO_2) reacts with ammonium hydrogen sulphite (NH_4HSO_3) in presence of sulphur dioxide (SO_2).

$$NH_4NO_2 + NH_4HSO_3 + SO_2 + 2H_2O \longrightarrow [NH_3OH]^+ HSO_4^- + (NH_4)_2SO_4$$

Properties

(*i*) Pure NH_2OH is colourless, odourless deliquescent crystals with melting point= 33°C
Boiling point = 58°C at 72 mmHg

(*ii*) Dielectric constant = 78
Highly soluble in water and alcohol but slightly in ether or benzene.

(*iii*) Its aqueous solution is less basic than NH_3 and N_2H_4.

(*iv*) In absence of H_2O, it also behaves as a weak acid.

$$NH_2OH \rightleftharpoons NH_2O^- + H^+$$

$$\underset{\text{Acid}}{NH_2OH} + \underset{\text{Base}}{Ca(OH)_2} \longrightarrow \underset{\text{Salt}}{Ca(OH)(ONH_2)}$$

(v) **NH_2OH as an oxidising agent** NH_2OH can be reduced to NH_3, hence it can be used as an oxidising agent.

Example In acidic medium

$$NH_2OH + 2H^+ + 2e^- \longrightarrow NH_3 + H_2O$$

In basic medium

$$NH_2OH + H_2O + 2e^- \longrightarrow NH_3 + OH^-$$

$$\text{Equivalent mass} = \frac{\text{molar mass}}{2}$$

(a) $SnCl_2$ is oxidised to $SnCl_4$ in acidic medium.

$$SnCl_2 + NH_2OH + 2HCl \longrightarrow SnCl_4 + NH_3 + H_2O$$

(b) SO_2 is oxidised to SO_4^{2-} and $(NH_4)_2SO_4$ is the end product.

$$NH_2OH + 2H^+ + 2e^- \longrightarrow NH_3 + H_2O$$

$$SO_2 + 2H_2O \longrightarrow SO_4^{2-} + 4H^+ + 2e^-$$

$$2NH_3 + 2H^+ + SO_4^{2-} \longrightarrow (NH_4)_2SO_4$$

(c) HI is oxidised to I_2

$$NH_2OH + 2HI \longrightarrow NH_3 + H_2O + I_2$$

(d) $Fe(OH)_2$ is oxidised to $Fe(OH)_3$ in basic medium

$$2Fe(OH)_2 + NH_2OH + H_2O \longrightarrow 2Fe(OH)_3 + NH_3$$

(e) Na_3AsO_3 (sodium argenite) is oxidised to Na_3AsO_4 (sodium argenate) in basic medium.

$$Na_3AsO_3 + NH_2OH \longrightarrow Na_3AsO_4 + NH_3$$

(*vi*) **NH_2OH as a reducing agent** NH_2OH can be oxidised to N_2O, hence it can be used as a reducing agent.

In acidic medium,

$$2NH_2OH \longrightarrow N_2O + H_2O + 4H^+ + 4e^-$$

In basic medium,

$$2NH_2OH + 4OH^- \longrightarrow N_2O + 5H_2O + 4e^-$$

$$\text{Equivalent mass} = \frac{\text{molar mass}}{4}$$

e.g.
CuO(II) is reduced to Cu_2O(I).

$$4CuO + 2NH_2OH \longrightarrow \underset{\text{Red ppt.}}{Cu_2O\downarrow} + N_2O + 3H_2O$$

Fe^{3+}(III) is reduced to Fe^{2+}(II).

$$2NH_2OH + 4FeCl_3 \longrightarrow N_2O + 4FeCl_2 + H_2O + 4HCl$$

Uses

(*i*) As a reducing agent anti-oxidant in photographic developers and reduction of Cu(II) in the dyeing of acrylic fibres.

(*ii*) In the preparation of caprolactam (is a monomer in the manufacture of nylon-6).

Cyclohexanone (=O) $\xrightarrow{NH_2OH}$ oxime (=NOH) $\xrightarrow{\text{Oleum}}$ Caprolactam (C=O, NH)

Oxoacids of nitrogen

Nitrogen forms several oxoacids such as $H_2N_2O_2$, HNO_2, HNO_3 and HNO_4 that is given in the following table

Table 7.5 Oxoacids of nitrogen

Name of oxoacid	Molecular formula	Structure	Oxidation state of N	Basicity	Nature
Hyponitrous acid	$H_2N_2O_2$	HO—N=N—OH exists as *trans*-isomer dipole moment being zero	+1	2(dibasic)	Highly explosive
Nitrous acid	HNO_2	H—N=O (N→O)	+3	1(monobasic)	Unstable weak acid
Nitric acid	HNO_3	H—O—N=O (N→O)	+5	1(monobasic)	Stable, strong acid
Pernitric acid	HNO_4	O=N—O—O—H (N→O)	+5	1(monobasic)	Unstable and explosive

Nitric acid (HNO_3)

It is also known as aqua fortis or spirit of nitre. Commercially, nitric acid has a concentration of 68% in water. When the solution contains more than 68% HNO_3, it is referred as fuming nitric acid.

It is commonly used as a strong oxidising agent.

Structure

Nitrogen present in nitric acid is sp^2-hybridised and has a trigonal planar geometry. Structure of HNO_3 can be represented as,

H—Ö—N(=O)(→O) ⟷ H—Ö—N(→O)(=O)   [N–O 1.22Å, N–OH 1.41Å, sp^2]

Preparation

Nitric acid can be prepared by following methods

(*i*) **Laboratory method** Nitric acid can be prepared in laboratory by heating KNO_3 with conc.H_2SO_4.

$$KNO_3 + \text{conc.}H_2SO_4 \xrightarrow{\Delta} KHSO_4 + HNO_3$$

On condensation, vapours of HNO_3 changes to liquid state.

(*ii*) **Birkeland and Eyde process** This process involves electric sparking of N_2 and O_2 and passing the gas produced into water.

$$N_2 + O_2 \overset{\text{Spark}}{\rightleftharpoons} 2NO, \ \Delta H = 181 \text{ kcal mol}^{-1}$$

$$2NO + O_2 \longrightarrow 2NO_2$$

$$2NO_2 + H_2O \longrightarrow HNO_2 + HNO_3$$

$$3HNO_2 \longrightarrow HNO_3 + H_2O + 2NO$$

(*iii*) **Ostwald's process** This process consist of following two steps

Step I. NH_3 obtained by Haber's process is oxidised to NO by atmospheric O_2 in the presence of platinum-guaze catalyst electrically heated at 750°-900°C.

$$4NH_3 + 5O_2 \xrightarrow[850°C]{Pt/Rh, 5 \text{ atm}} 4NO + 6H_2O$$

Steps II. NO produced above is oxidised by atmospheric O_2 to NO_2, which is cooled to about 50°C and then absorbed in water to give HNO_3.

$$2NO + O_2 \longrightarrow 2NO_2$$

$$2NO + O_2 \rightleftharpoons 2NO_2$$

$$2NO_2 + H_2O \longrightarrow HNO_3 + HNO_2$$

$$2HNO_2 \longrightarrow H_2O + NO_2 + NO$$

$$3NO_2 + H_2O \longrightarrow 2HNO_3 + NO$$

Overall reaction being

$$NH_3 + 2O_2 \longrightarrow HNO_3 + H_2O$$

Acid is 60% by weight and concentrated HNO_3 (98% by weight) is obtained by dehydration with conc. H_2SO_4. Fuming HNO_3 contains dissolved NO_2 in concentrated nitric acid.

Physical properties

Important physical properties of nitric acid are as follows

(*i*) Anhydrous HNO_3 is a colourless fuming and pungent smelling liquid, soluble in water.

(*ii*) It's boiling point is 183°C and freezing point is 42°C.

(*iii*) It undergoes photochemical decomposition (in sunlight) and gradually turns yellow in colour.

$$2HNO_3 \xrightarrow{h\nu} 2NO_2 + H_2O + \frac{1}{2}O_2$$

(*iv*) HNO_3 is highly corrosive in nature and causes burning sensation, when falls on skin.

(*v*) It undergoes auto-ionisation and produces NO_2^+ (nitronium ion) and NO_3^-.

$$2HNO_3 \longrightarrow NO_2^+ + NO_3^- + H_2O$$

Chemical properties

Various chemical properties shown by HNO_3 are given below

(*i*) Nitric acid is a powerful oxidising agent. It gives following reactions

- Pure HNO_3 undergoes dissociation to give N_2O_5 and O_2 at room temperature.

$$2HNO_3 \longrightarrow N_2O_5 + H_2O$$

$$N_2O_5 \rightleftharpoons N_2O_4 + \frac{1}{2}O_2$$

- On heating, nitric acid produces brown fumes of NO_2.

$$4HNO_3 \xrightarrow{\Delta} \underset{\text{Brown fumes}}{4NO_2\uparrow} + O_2\uparrow + 2H_2O$$

- It readily gives nascent oxygen in both concentrated and dilute solutions.

$$2HNO_3 \longrightarrow 2NO_2 + H_2O + [O]$$

- HNO_3 oxidises non-metals to their oxoacids.

  Carbon is oxidised to carbonic acid.

$$C + 4HNO_3 \longrightarrow H_2CO_3 + 4NO_2 + H_2O$$

  Sulphur is oxidised to sulphuric acid.

$$S + 6HNO_3 \longrightarrow H_2SO_4 + 6NO_2 + 2H_2O$$

  Iodine is oxidised to iodic acid.

$$I_2 + 10HNO_3 \longrightarrow 2HIO_3 + 10NO_2 + 4H_2O$$

- Nitric acid oxidises reactive metals to metal nitrates.

  Metals like Zn, Fe, Pb are oxidised to their respective nitrates, when reacts with HNO_3.

  e.g.

$$4Zn + \underset{\text{Dilute}}{10HNO_3} \longrightarrow 4Zn(NO_3)_2 + NH_4NO_3 + 3H_2O$$

$$4Zn + \underset{\text{Cold and dilute}}{10HNO_3} \longrightarrow 4Zn(NO_3)_2 + N_2O + 5H_2O$$

$$Zn + \underset{\text{Cold and conc.}}{4HNO_3} \longrightarrow Zn(NO_3)_2 + 2NO_2 + 2H_2O$$

$$3Zn + \underset{\text{Conc.}}{8HNO_3} \longrightarrow 3Zn(NO_3)_2 + 4H_2O + 2NO$$

  HNO_3 is reduced to NO_2, NO, N_2O, N_2 or NH_3 depending on the nature of metal, temperature and concentration of acid.

- $\underset{\text{Concentrated}}{2NO_3^-} + 4H^+ + 2e^- \longrightarrow 2NO_2 + 2H_2O$

- $\underset{\text{Dilute}}{NO_3^-} + 4H^+ + 3e^- \longrightarrow NO + 2H_2O$

- $\underset{\text{Very dilute}}{NO_3^-} + 10H^+ + 8e^- \longrightarrow NH_4^+ + 3H_2O$

  Thus, oxidising power of HNO_3 depends on the extent of dilution.

$$\underset{+5}{NO_3^-} \longrightarrow \underset{+4}{NO_2} \longrightarrow \underset{+2}{NO} \longrightarrow \underset{-3}{NH_3}$$

- HNO_3 oxidises less reactive metals (than hydrogen) in electrochemical series.

  Metals like Cu, Ag, Hg when reacts with HNO_3 are oxidised to their corresponding nitrates and it itself reduces to NO_2, NO, N_2O, N_2 or NH_3.

  For example, when conc. HNO_3 is taken, NO_2 is formed.

$$2HNO_3 \longrightarrow H_2O + 2NO_2 + [O]$$

$$Cu + [O] \longrightarrow CuO$$

$$CuO + 2HNO_3 \longrightarrow Cu(NO_3)_2 + H_2O$$

$$\overline{Cu + 4HNO_3 \longrightarrow Cu(NO_3)_2 + 2H_2O + 2NO_2\uparrow}$$

  Similarly,

$$Ag + 2HNO_3 \longrightarrow AgNO_3 + H_2O + NO_2\uparrow$$

$$Hg + 4HNO_3 \longrightarrow Hg(NO_3)_2 + 2H_2O + 2NO_2\uparrow$$

  But when dil.HNO_3 is taken in the reaction then NO is formed.

$$3Cu + 8HNO_3 \longrightarrow 3Cu(NO_3)_2 + 4H_2O + 2NO$$

$$3Ag + 4HNO_3 \longrightarrow 3AgNO_3 + 2H_2O + NO$$

$$6Hg + 8HNO_3 \longrightarrow 3Hg_2(NO_3)_2 + 4H_2O + 2NO$$

- Nitric acid oxidises compounds such as CuS, H_2S, sugar cane, toluene and itself reduces to NO (dilute) and NO_2 (conc.) depending on the concentration.

$$3CuS + 8HNO_3 \longrightarrow 2NO + 4H_2O + 3S + 3Cu(NO_3)_2$$

$$2HNO_3 + H_2S \longrightarrow 2NO_2 + S + 2H_2O$$

$$36HNO_3 + \underset{\text{Sugar cane}}{C_{12}H_{22}O_{11}} \longrightarrow 23H_2O + 36NO_2 + 6H_2C_2O_4$$

$$2HNO_3 + \underset{\text{Toluene}}{C_6H_5CH_3} \longrightarrow 2NO + 2H_2O + C_6H_5COOH$$

When $SnCl_2$ reacts with nitric acid in presence of HCl, Sn^{2+} is oxidised to Sn^{4+} and HNO_3 is reduced to NH_2OH.

$$3SnCl_2 + 6HCl + HNO_3 \longrightarrow 2H_2O + 3SnCl_4 + NH_2OH$$

(*ii*) HNO_3 is a **strong monobasic acid** and form nitrates with carbonates, oxides and hydroxides.

$$Na_2CO_3 + 2HNO_3 \longrightarrow 2NaNO_3 + H_2O + CO_2$$
$$CaO + 2HNO_3 \longrightarrow Ca(NO_3)_2 + H_2O$$

(*iii*) When HNO_3 is treated with metals like Cr, Ni, Al, no reaction occurs. This is due to formation of their oxide layer over metal surface which makes metal inert in nature.

(*iv*) Noble metals like Au, Pt, Rn and Ir are not attacked by nitric acid. *Aqua-regia* (mixture of one part of conc. HNO_3 and 3 parts of conc. HCl) is used to dissolve these metals.

This is due to coupling of oxidising power of HNO_3 with Cl^- ion to form complexes with metal ions.

$$\underbrace{HNO_3 + 3HCl}_{\text{Aqua-regia}} \longrightarrow 2H_2O + 2Cl + \underset{\text{Nitrosyl chloride}}{NOCl}$$

$$Au + 3Cl \longrightarrow AuCl_3$$

$$AuCl_3 + HCl \longrightarrow \underset{\text{Chloroauric acid}}{HAuCl_4}$$

$$Pt + 4Cl \longrightarrow PtCl_4$$

$$PtCl_4 + 2HCl \longrightarrow \underset{\text{Chloroplatinic acid}}{H_2PtCl_6}$$

(*v*) **Brown ring test** It is a characteristic test used to detect presence of nitrates. It depends on the ability of Fe^{2+} ions to reduce NO_3^- ion to nitric oxide [NO] which reacts with Fe^{2+} to form brown coloured complex.

$$NO_3^- + 3Fe^{2+} + 4H^+ \longrightarrow NO + 3Fe^{3+} + 2H_2O$$

$$[Fe(H_2O)_6]^{2+} + NO \longrightarrow \underset{\text{Brown ring}}{[Fe(H_2O)_5NO]^{2+}} + H_2O$$

The colour is due to charge transfer. This complex formally contains Fe (+ I) and NO^+, since Fe (II) gains one electron from NO, which changes to NO^+. Its magnetic moment is 3.87 BM confirming the presence of three unpaired electrons. Nitrites give the brown colour before H_2SO_4 is added.

$$NO \longrightarrow NO^+ + e^-$$

$$Fe^{2+} + e^- \longrightarrow Fe^+$$

⥮	1	1	1	1

⥮	⥮	1	1	1

The ring test is unreliable in the presence of bromide, iodide, nitrite, chlorate and chromate, hence must be removed before ring-test of NO_3^- is performed.

Reactions of HNO_3 can be summarised as shown below

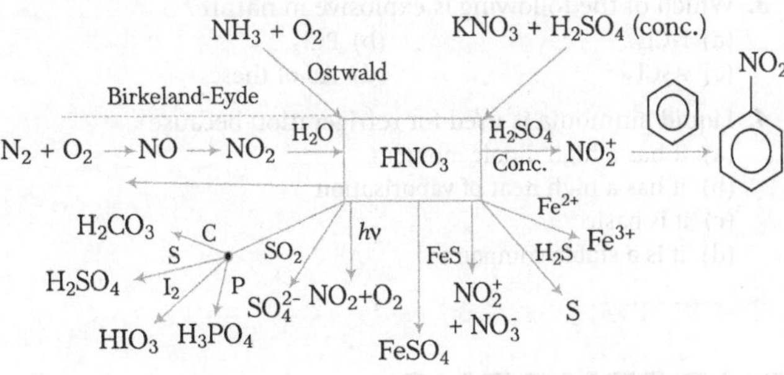

Uses

(*i*) HNO_3 is used in preparation of explosive such as TNT, picric acid, gun cotton and dynamite.

(*ii*) It is used in purification of Ag and Au.

(*iii*) The major use of nitric acid is in the manufacture of ammonium nitrate for fertilizers.

(*iv*) Other major uses are in the pickling of stainless steel, etching of metals and as an oxidiser in rocket fuels.

$NaNO_2$ and $NaNO_3$ as food additives

The brown colour of "old" meat is the result of oxidation of blood and is objectionable to many consumers. Nitrites and nitrates are added to food to retard this oxidation and also to prevent growth of botulism bacteria. NO_3^- ions are reduced to NO_2^- ions, which are then converted to NO. This in turn reacts with the brown oxidised form of the heme in blood. This reaction keeps meat red longer. However, there are doubts that nitrites may combine with amines under the acidic conditions in the stomach to give carcinogenic nitrosoamines.

$$HNO_2 + R_2NH \longrightarrow \underset{\text{Nitrosoamine}}{R_2N{-}N{=}O} + H_2O$$

CHECK POINT 7.3

1. In case of nitrogen, NCl_3 is possible but not NCl_5 while in case of phosphorus, PCl_3 and PCl_5 are possible. It is due to
(a) availability of vacant *d*-orbital in P but not in N
(b) lower electronegativity of P than N
(c) lower tendency of H bond formation in P than N
(d) occurrence of P in solid while N in gaseous state at room temperature

2. A solution of ammonia in water contains
(a) H^+
(b) OH^-
(c) only NH_4^+
(d) OH^-, NH_4^+ and NH_4OH molecules

3. Which of the following is explosive in nature?
(a) NCl_3 (b) PCl_3
(c) $AsCl_3$ (d) All of these

4. Liquid ammonia is used for refrigeration because
(a) it has a high dipole moment
(b) it has a high heat of vaporisation
(c) it is basic
(d) it is a stable compound

5. Auto-ionisation of nitric acid produces
(a) nitronium ion
(b) nitrate ion
(c) Both (a) and (b)
(d) None of the above

6. Which of the following is used for dissolution of gold?
(a) Sulphuric acid
(b) Hydrochloric acid
(c) Nitric acid
(d) *Aqua-regia*

7. Brown ring test for nitrates depends on
(a) reduction of ferrous sulphate to iron
(b) oxidation of nitric oxide to nitrogen dioxide
(c) reduction of nitrate to nitric oxide
(d) oxidising action of sulphuric acid

8. A metal '*M*' reacts with nitrogen gas to afford 'M_3N'. 'M_3N' on heating at high temperature gives back '*M*' and on reaction with water produces gas '*B*'. Gas '*B*' reacts with aqueous solution of $CuSO_4$ to from a deep blue compound. '*M*' and '*B*' respectively are :
(a) Li and N_2 (b) Ba and N_2
(c) Na and NH_3 (d) Al and N_2

PHOSPHORUS

Phosphorus (P) is the second member of group 15 of the periodic table. It is an important constituent in living system and is the sixth most abundant element in the human body.

Our bones are mostly made up of $Ca_3(PO_4)_2$ and tooth enamel is almost pure $Ca_5(PO_4)_3OH$. Phosphate groups are an integral part of the nucleic acids DNA and RNA, the molecules that pass genetic information from one generation to other generation.

Due to larger size of P, it cannot form stable *pπ-pπ* bonds with other phosphorus atoms.

Occurrence

Phosphorus occurs mainly in the form of phosphate minerals in the crust of earth.

Some of these are

(*i*) phosphorite $[Ca_3(PO_4)_2]$
(*ii*) fluorapatite $[3Ca_3(PO_4)_2 \cdot CaF_2]$
(*iii*) chlorapatite $[3Ca_3(PO_4)_2 \cdot CaCl_2]$
(*iv*) hydroxyapatite $[Ca_5(PO_4)_3OH]$

Allotropic forms of phosphorus

Phosphorus exists in three main allotropic forms. These are as follows

1. White phosphorus

It is a toxic, waxy, white solid containing discrete tetrahedral P_4 molecules. It is obtained by rapidly cooling the vapours of phosphorus. When exposed to air it undergoes oxidation, which gradually raises its temperature and ultimately it catches fire, when the temperature exceeds 30°C. It is extremely poisonous. It is insoluble in water but soluble in CS_2 and glows in dark on account of its slow oxidation(phosphorescence). It dissolves in boiling NaOH solution in an inert atmosphere giving PH_3.

$$P_4 + 3NaOH + 3H_2O \longrightarrow PH_3 + \underset{\text{Sodium hypophosphite}}{3NaH_2PO_2}$$

It can be represented as,

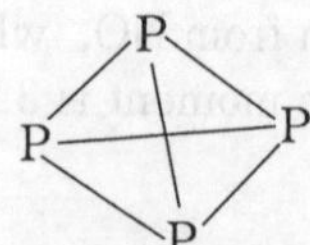

White phosphorus

2. Red phosphorus

It is essentially non-toxic and has a polymeric structure. It is prepared by heating white phosphorus in the absence of air at 250°C with a trace of iodine as catalyst. It is less reactive than white phosphorus. It is stable under ordinary conditions. It does not ignite spontaneously in air. When red phosphorus is heated under high pressure, a series of phases of red phosphorus is formed. It possesses iron grey lustre. It is insoluble in water as well as in CS_2. It does not glow in the dark. Its structure can be represented as,

Red phosphorus

$$\underset{\text{White}}{P_4(s)} \xrightarrow{\Delta} \underset{\text{(Red phosphorus)}}{P_4(s)}$$

3. Black phosphorus

It is an extented layer structure in which each phosphorus is bonded to three neighbours by single covalent bonds. It has two forms α- black phosphorus and β-black phosphorus.

α-black phosphorus is formed when red phosphorus is heated in a sealed tube at 803 K. It can be sublimed in air and has opaque monoclinic or rhombohedral crystals. It does not oxidise in air.

β-black phosphorus is obtained by heating white phosphorus at 473 K under high pressure. It does not burn in air upto 673 K. Thermodynamically, black phosphorus is the most stable form. Its structure can be represented as,

Black phosphorus

Properties of three allotropic forms of phosphorus are described in the table given below

Table 7.6 Comparison of white, red and black phosphorus

Properties	White phosphorus	Red phosphorus	Black phosphorus
Colour	White	Reddish-violet	Black
Smell	Garlic-like	Odourless	Odourless
Specific gravity	1.83–1.85	2.06–2.39	≈ 3.9
Solubility in CS_2	Soluble	Insoluble	Insoluble
mp/°C	44	~ 600	~ 610
Ignition temperature/°C	35	260	~ Stable
Action of air	Phosphorescence	No phosphorescence	No phosphorescence
With Cl_2	Reacts at room temperature	Reacts only when heated	Reacts at melting point
With NaOH	PH_3 is formed	No reaction	No reaction
Physiological effect	Poisonous	Non-poisonous	Non-poisonous

4. Scarlet phosphorus

It resembles the red phosphorus variety in its physical properties and white phosphorus in its chemical properties. It shows slow oxidation in air. It can be prepared by heating phosphorus tribromide with mercury at 240°C.

$$2PBr_3 + 3Hg \xrightarrow{240°C} 3HgBr_2 + \underset{\text{Scarlet phosphorus}}{2P}$$

5. Violet phosphorus

It is crystalline in nature. It can be obtained by heating white phosphorus with a trace of sodium at 230°C under high pressure.

COMPOUNDS OF PHOSPHORUS

Phosphorus can use *d*-orbitals for bonding. Therefore, it can exhibit oxidation state of +5. This can be represented as

P_{15}[Ne] ⇅ | 1 1 1 | (empty 3d) — $3s^2$ $3p^3$ $3d$ — Three electrons unpaired in 3*p*, hence PCl_3 is formed

↓ Excite

1 | 1 1 1 | 1 — $3s^1$ $3p^3$ $3d^1$ — Five electrons unpaired in outershell thus, PCl_5 is formed

Phosphorus reacts with hydrogen, oxygen and halogens to form compounds such as oxides, halides, hydrides etc. Some of them are discussed below;

Phosphine (PH_3)

It is discovered by Gembre in 1783.

Structure

Phosphine is a covalent molecule having pyramidal structure like ammonia. The bond angle H—P—H is 93°. It can be represented as ;

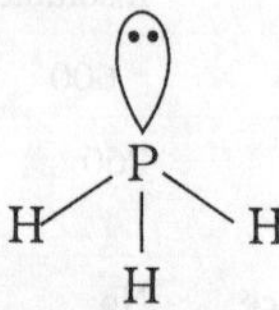

P—H(142 pm) bonds is longer than N—H(101.7pm) because of bigger size of P than N.

Preparation

(*i*) In the laboratory it is prepared by boiling yellow phosphorus with a concentrated solution of sodium hydroxide in an inert atmosphere.

$$P_4 + 3NaOH + 3H_2O \longrightarrow \underset{\text{Sodium hypophosphite}}{3NaH_2PO_2} + PH_3$$

Note *Red phosphorus does not react with NaOH hence reaction can be used to separate white and red phosphorus.*

In this reactions, phosphorus also forms P_2H_4, besides PH_3.

$$P_4 + 4NaOH + 4H_2O \longrightarrow 4NaH_2PO_2 + 2H_2$$

$$3P_4 + 8NaOH + 8H_2O \longrightarrow 8NaH_2PO_2 + 2P_2H_4$$

It is necessary to exclude air from the apparatus because, although pure PH_3 will not burn unless ignited, it is often accompanied with P_2H_4 and H_2 which are spontaneously flammable.

(*ii*) To purify it from the impurities, it is absorbed in HI to form phosphonium iodide (PH_4I) which on treating with KOH gives of phosphine.

$$PH_4I + KOH \longrightarrow KI + H_2O + PH_3$$

(*iii*) It can also be obtained by action of water or acid on calcium phosphide or sodium phosphide.

$$Ca_3P_2 + 6H_2O \longrightarrow 2PH_3 + 3Ca(OH)_2$$

Physical properties

(*i*) It is a colourless, highly reactive and extremely toxic gas, which smells slightly of garlic or rotten fish.

(*ii*) It is poisonous in nature.

(*iii*) It liquefies at −89°C and solidifies at −134°C.

(*iv*) It is heavier than air and sparingly soluble in water.

Chemical properties

Various chemical properties shown by PH_3 are as follows

(*i*) When PH_3 is heated at 440°C in absence of air or when electric spark is passed through PH_3 then it decomposes into red phosphorus and hydrogen.

$$4PH_3 \xrightarrow{440^\circ C} \underset{\text{Red}}{P_4} + 6H_2$$

(*ii*) PH_3 burns in air or oxygen when heated at 150°C,

$$2PH_3 + 4O_2 \longrightarrow P_2O_5 + 3H_2O$$

P_2O_5 dissolves in H_2O forming H_3PO_4

(*iii*) PH_3 is weakly basic and gives phosphonium compounds with acids, e.g.

$$PH_3 + HBr \longrightarrow PH_4Br$$

(*iv*) When PH_3 is absorbed in copper sulphate or mercuric chloride solution the corresponding phosphides are obtained.

$$3CuSO_4 + 2PH_3 \longrightarrow Cu_3P_2 + 3H_2SO_4$$

$$3HgCl_2 + 2PH_3 \longrightarrow Hg_3P_2 + 6HCl$$

(*v*) It is better reducing agent than NH_3.

- Ag^+ is reduced to Ag metal,

$$6Ag^+ + PH_3 + 3H_2O \longrightarrow H_3PO_3 + 6Ag\downarrow + 6H^+$$

- Cu^{2+} is reduced to Cu,

$$4Cu^{2+} + PH_3 + 4H_2O \longrightarrow H_3PO_4 + 4Cu\downarrow + 8H^+$$

(*vi*) In liquid NH_3, PH_3 dissolves to give $NH_4^+PH_2^-$,

$$PH_3 + NH_3 \longrightarrow NH_4^+PH_2^-$$

Some of the chemical reactions are shown below

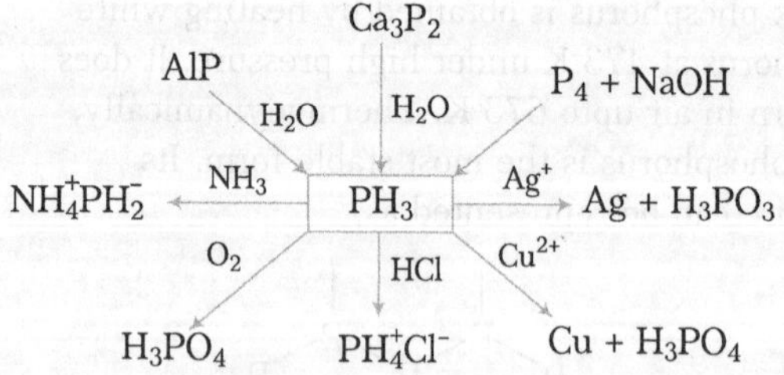

Uses

(*i*) It is used for making smoke screens, metallic phosphides.

(*ii*) It is also used in **Holme signals** due to spontaneous combustion of PH_3.

Halides of phosphorus

Two series of halides of phosphorus are known such as PX_3 and PX_5 (X = halogen). These halides may be gases, volatile liquids or low melting solids.

Some of the halides are given below

Phosphorus trichloride (PCl_3)

It is a colourless, pungent smelling oily liquid (bp 347 K).

Structure

Structure of PCl_3 is similar to ammonia. P-atom is ***sp*³-hybridised**. One of the position in tetrahedral configuration is occupied by lone pair. It has pyramidal shape. Its structure can be represented as

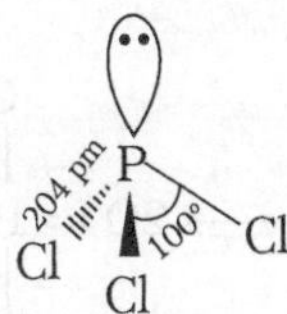

Preparation

(*i*) It is prepared by passing dry Cl_2 over heated white phosphorus,

$$\underset{\text{White phosphorus}}{P_4(s)} + 6Cl_2(g) \longrightarrow 4PCl_3(l)$$

(*ii*) Thionyl chloride when treated with P_4 also produces PCl_3,

$$P_4 + 8SOCl_2 \longrightarrow 4PCl_3 + 4SO_2 + 2S_2Cl_2$$

Chemical properties

(*i*) It acts as a reducing agent,

$$PCl_3 + SO_3 \longrightarrow POCl_3 + SO_2$$

(*ii*) It produces chlorosulphuric acid when reacts with conc. H_2SO_4,

$$PCl_3 + 2H_2SO_4 \longrightarrow HSO_3Cl + SO_2 + HPO_3 + 2HCl$$

(*iii*) It is widely used in organic chemistry to convert carboxylic acid to acid chlorides and alcohols to alkyl halides,

$$3CH_3COOH + PCl_3 \longrightarrow 3CH_3COCl + H_3PO_3$$

$$3C_2H_5OH + PCl_3 \longrightarrow 3C_2H_5Cl + H_3PO_3$$

(*iv*) It fumes in moist air,

$$PCl_3 + 3H_2O \longrightarrow H_3PO_3 + 3HCl$$

Uses

(*i*) It is used in the manufacture of $POCl_3$ (chloroanhydride of phosphoric acid) and in preparation of organic compounds.

(*ii*) These are used during electrodeposition of metal on rubber and for making pesticides.

Phosphorus pentachloride (PCl_5)

It is a yellowish white crystalline solid with pungent odour

Structure

In gaseous and liquid phases, it has a trigonal bipyramidal structure with sp^3d-hybridisation. The three equatorial P—Cl bonds are equivalent while the two axial bonds are longer than equatorial bonds. This is due to the fact that the axial bond pairs suffer more repulsion as compared to equatorial bond pairs.

In the solid state, it exist as ionic solid $[PCl_4]^+[PCl_6]^-$ in which the cation $[PCl_4]^+$ is tetrahedral and the anion $[PCl_6]^-$ is octahedral.

Cl
90°
202 pm
Cl—P 120° Cl
240 pm Cl
Cl

Preparation

(*i*) PCl_5 can be prepared by reaction of white phosphorus or PCl_3 with excess of dry Cl_2.

$$\underset{\text{White phosphorus}}{P_4} + 10Cl_2 \longrightarrow 4PCl_5$$

$$PCl_3(l) + Cl_2(g) \rightleftharpoons PCl_5(s)$$

(*ii*) Sulphuryl chloride, SO_2Cl_2 on reaction with white phosphorus or PCl_3 produces PCl_5.

$$\underset{\text{White phosphorus}}{P_4} + 10SO_2Cl_2 \longrightarrow 4PCl_5 + 10SO_2$$

$$PCl_3 + SO_2Cl_2 \longrightarrow PCl_5 + SO_2$$

Chemical properties

(*i*) It sublimes but decomposes to PCl_3 and Cl_2 on strong heating.

$$PCl_5 \xrightarrow{\Delta} PCl_3 + Cl_2$$

(*ii*) In moist-air, PCl_5 hydrolyses to $POCl_3$ and gets converted to H_3PO_4.

$$PCl_5 + H_2O \longrightarrow POCl_3 + 2HCl$$

$$POCl_3 + 3H_2O \longrightarrow H_3PO_4 + 3HCl$$

With excess of water, it reacts violently.

$$PCl_5 + \underset{\text{Excess}}{4H_2O} \longrightarrow H_3PO_4 + 5HCl$$

(*iii*) PCl_5 reacts with metals to give corresponding chlorides

$$2Ag + PCl_5 \longrightarrow AgCl + PCl_3$$

$$Sn + 2PCl_5 \longrightarrow SnCl_4 + 2PCl_3$$

(*iv*) PCl_5 react with SO_2 to give thionyl chloride.

$$SO_2 + PCl_5 \longrightarrow POCl_3 + \underset{\text{Thionyl chloride}}{SOCl_2}$$

(*v*) PCl_5 reacts with organic compounds containing —OH group converting them into chloro derivative.

$$C_2H_5OH + PCl_5 \longrightarrow C_2H_5Cl + POCl_3 + HCl$$

$$RCOOH + PCl_5 \longrightarrow RCOCl + POCl_3 + HCl$$

Uses

It is used as a chlorinating agent in organic chemistry.

Oxides of phosphorus

Two important oxides of phosphorus are as follows

Phosphorus (III) oxide (P_4O_6)

- It is formed when phosphorus is burnt in limited supply of air.

$$P_4 + 3O_2 \xrightarrow[\text{(Limited)}]{\Delta} P_4O_6$$

- It is a crystalline solid with garlic odour. It dissolves in cold and hot water to form phosphorous acid and phosphoric acid respectively.

$$P_4O_6 + \underset{\text{(Cold)}}{6H_2O} \longrightarrow \underset{\text{Phosphorous acid}}{4H_3PO_3}$$

$$P_4O_6 + \underset{\text{(Hot)}}{6H_2O} \longrightarrow \underset{\text{Phosphoric acid}}{3H_3PO_4} + PH_3$$

Structure

It has 6 P—O—P bonds and 16 lone pairs of electrons (one lone pair on each P atom and two lone pairs of electrons on each O atom). The bond length P—O is 1.638 Å.

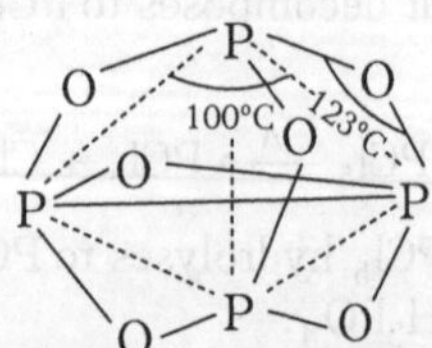

In P_4O_6, oxygen atoms act as a bridge between the corners of the tetrahedron.

Phosphorus (V) oxide (P_4O_{10})

- It is formed when white phosphorus is heated in excess of air.

$$\underset{\text{White phosphorus}}{P_4} + \underset{\text{Excess}}{5O_2} \longrightarrow P_4O_{10}$$

- It is a white solid with snow like appearance and sublimes on heating.
- It readily dissolves in hot and cold water to give metaphosphoric acid and phosphoric acid.

$$P_4O_{10} + \underset{\text{(Cold)}}{2H_2O} \longrightarrow \underset{\text{Metaphosphoric acid}}{4HPO_3}$$

$$P_4O_{10} + \underset{\text{(Cold)}}{6H_2O} \longrightarrow \underset{\text{Phosphoric acid}}{4H_3PO_4}$$

(*iv*) It is a powerful dehydrating agent.

(*v*) P_4O_{10} reacts with alcohols and ethers forming phosphate esters

$$P_4O_{10} + 6EtOH \longrightarrow 2O{=}P(OH)_2(OEt) + 2O{=}P(OH)(OEt)_2$$

$$P_4O_{10} + 6Et_2O \longrightarrow 2O{=}P(OEt)_3$$

Formation of esters confirms that H_3PO_4 has structure

$$O{=}P(OH)_3$$

Structure

It has 16 sigma bonds, 4π-bonds and 24 lone pairs of electrons on oxygen atoms.

In P_4O_{10} each P-atom is also bonded to an extra oxygen atom. Like P_4O_6 it has also **six** POP bonds.

Oxoacids of phosphorus

Phosphorus forms a number of oxoacids, which differs in their structure and oxidation state of phosphorus.

Distinction between various oxoacids is given in the table shown below

Table 7.7 Oxoacids of phosphorus

Name of acid	Formula	Structure	Oxidation state	Basicity
Hypophosphorous acid	H_3PO_2		+1	Monobasic
Phosphoric acid (*o*-phosphorus acid)	H_3PO_3		+3	Diabasic
Phosphoric acid	H_3PO_4		+5	Tribasic
Pyrophosphorus acid	$H_4P_2O_5$		+3	Dibasic
Hypophosphoric acid	$H_4P_2O_6$		+4	Tetrabasic
Pyrophosphoric acid (diphosphoric acid)	$H_4P_2O_7$		+5	Tetrabasic
Cyclo tri-metaphosphoric acid	$(HPO_3)_3$		+5	Tribasic

Preparation

(*i*) Oxoacids can be prepared by following methods

$$P_4 + 5O_2 \longrightarrow P_4O_{10}$$

$$P_4O_{10} + 6H_2O \longrightarrow 4H_3PO_4$$

(*ii*) H_3PO_2 and H_3PO_3 disproportionate to PH_3 and H_3PO_4 on heating.

$$2H_3PO_2 \xrightarrow{\Delta} H_3PO_2 + PH_3$$

$$4H_3PO_3 \xrightarrow{\Delta} 3H_3PO_4 + PH_3$$

(*iii*) $P_4 + 20HNO_3 \longrightarrow 4H_3PO_4 + 20NO_2 + 4H_2O$

H_3PO_3 is formed by hydrolysis of PCl_3.

$$PCl_3 + 3H_2O \longrightarrow H_3PO_3 + 3HCl$$

(*iv*) H_3PO_2 is obtained by alkaline hydrolysis of P_4.

$$P_4 + 3OH^- + 3H_2O \longrightarrow PH_3 + 3H_2PO_2^-$$

Properties

(*i*) Phosphoric acid H_3PO_4, is hydrogen bonded in aqueous solution and because of this the 'concentrated acid' is syrupy and viscous.

(*ii*) In all of these acids, P is coordinated by four other atom and tetrahedrally surrounded wherever possible *pπ-dπ* back bonding gives rise to P=O bonds. The hydrogen atoms in O—H groups are ionisable and are acidic but P—H bonds found in phosphorous acids have reducing and not the acidic properties.

(*iii*) Basicity = number of ionisable H (protonic) = H attached to O-atoms (H encircled are acidic and H squared indicates reducing properties)

Tribasic Dibasic Monobasic

(*iv*) Thus, all P atoms in the oxoacids and oxoanions (conjugate base) are 4-coordinate and contain at least one P—O unit, all P atoms in the oxoacids have at least one P—OH group. In every case, the H atom is ionisable as H^+. Some oxoacids have one or more P—H bonds. This H atom is not ionisable as H^+. Polymerisation can occur by P—O—P bond formation to give both linear and cyclic species. Two P atoms are never joined by more than one P—O—P bridge. When P atom is surrounded by O atoms (as in H_3PO_4), its oxidation number is +5. For each P—OH that is replaced by P—H bond, its oxidation number drops by 2.

Table 7.8 Distinctive properties of oxoacids of phosphorus

S.No.	Properties	H_3PO_2	H_3PO_3	H_3PO_4
1.	Basicity	Monobasic $H_3PO_2 \rightleftharpoons H^+ + H_2PO_2^-$ Hence, only one type of salt (like NaH_2PO_2) is formed. NaH_2PO_2 is normal salt (which does not further react with base	Dibasic $H_3PO_3 \rightleftharpoons H^+ + H_2PO_3^-$ $H_2PO_3^- \rightleftharpoons H^+ + HPO_3^{2-}$ two types of salts, NaH_2PO_3 and Na_2HPO_3 are formed. NaH_2PO_3 is an acid salt but Na_2HPO_3 is normal salt.	Tribasic $H_3PO_4 \rightleftharpoons H^+ + H_2PO_4^-$ $H_2PO_4^- \rightleftharpoons H^+ + HPO_4^{2-}$ $HPO_4^{2-} \rightleftharpoons H^+ + PO_4^{3-}$ Three series of salts, NaH_2PO_4, Na_2HPO_4 and Na_3PO_4 are formed. NaH_2PO_4 and Na_2HPO_4 are acid salts but Na_3PO_4 is normal salt.
2.	Effect of heat	$2H_3PO_2 \xrightarrow{\Delta} H_3PO_4 + PH_3$ Disproportionation	$4H_3PO_3 \xrightarrow[200°C]{\Delta} 3H_3PO_4 + PH_3$ Disproportionation	$H_3PO_4 \xrightarrow[600°C]{\Delta} HPO_3 + H_2O$ Dehydration
3.	Reducing properties	Most reducing agent $AgNO_3 \longrightarrow Ag$ $HgCl_2 \longrightarrow Hg$ $CuSO_4 \longrightarrow Cu_2H_2$ $Cl_2 \longrightarrow HCl$	$AgNO_3 \longrightarrow Ag$ $HgCl_2 \longrightarrow Hg_2Cl_2$ $CuSO_4 \longrightarrow Cu$ $MnO_4^- \longrightarrow Mn^{2+}$	Does not reduce salts into metals.

Table 7.9 Comparison between nitrogen and phosphorus

S. No.	Properties	Nitrogen	Phosphorus
1.	Physical state	N_2 is a gas which can be condensed to a liquid at very low temperature.	P_4 is a solid at room temperature.
2.	Bond energy	N≡N (multiple bonds) are present in N_2 due to $p\pi - p\pi$ N≡N bond energy 941.4 kJ mol^{-1} Due to high value of bond energy (N≡N), it is less reactive and more stable. $N_2 + O_2 \xrightarrow{3300°C} 2NO$	Multiple bonds are absent in P P—P bond energy 79.08 kJ mol^{-1} Due to low value of (P—P) bond energy, P_4 is less stable and more reactive. $P_4 + 5O_2 \longrightarrow P_4O_{10}$ This reaction takes place at room temperature.
3.	Formation of compounds	NCl_3 is formed but NCl_5 is not formed since N cannot expand its octet due to lack of *d*-orbital.	Both PCl_3 and PCl_5 are formed due to expansion of octet.

S. No.	Properties	Nitrogen	Phosphorus
4.	Hydride	NH_3 is more stable than PH_3.	Also PH_4^+ is less stable than NH_4^+.
		NH_3 is highly soluble in water due to H-bonding with water, giving alkaline solution.	PH_3 is sparingly soluble in water and the solution is neutral-the hydrogen bond in latter is very weak.
5.	Oxyacids	Oxyacids of nitrogen are stronger than those of phosphorus. HNO_3 is a strong oxidising agent.	Oxy acids are weaker than those of nitrogen H_3PO_4 is weaker acid than HNO_3. H_3PO_4 is not oxidising agent.

Fertilisers

(i) Nitrogen, phosphorus and potassium are three most essential plant nutrients consumed by them. Fertility of soil is maintained by using these nutrients in the form of fertilisers.

(ii) Fertiliser used should
- be water soluble.
- not disturb alkaline nature (pH).
- be stable.

(iii) Examples NH_2CONH_2 (urea)
NH_4NO_3 (ammonium nitrate)
$(NH_4)_2SO_4$ (ammonium sulphate)
$NH_4H_2PO_4$ (monoammonium hydrogen phosphate)
These are either used on their own or in conjugation with other chemicals, often to make **NPK** fertilisers. These are designed to release nitrogen (N), phosphorus (P) and potassium (K) into the soil.

If the NPK fertiliser is marked 4–8–2, it means it has
4% — N_2
8% — P_2O_5
2% — K_2O

(iv) Phosphate rocks such as fluoroapatite $(Ca_3(PO_4)_2 \cdot CaF_2)$ are very insoluble and thus are not used as fertilisers. **Superphosphate** is made by treating phosphate rock with conc. H_2SO_4.
The acid salt $Ca(H_2PO_4)_2$ is more soluble and over a period of weeks this would dissolve in the soil water. $CaSO_4$ is waste product of this fertiliser :

$$3[Ca_3(PO_4)_2 \cdot CaF_2] + 7H_2SO_4 \longrightarrow \underbrace{3Ca(H_2PO_4)_2 + 7CaSO_4}_{\text{Superphosphate}} + 2HF$$

(v) **'Triple superphosphate'** is made using H_3PO_4 and thus waste product $CaSO_4$ is not formed.

$$[Ca_3(PO_4)_2 \cdot CaF_2] + 6H_3PO_4 \longrightarrow \underset{\text{Triple superphosphate}}{4Ca(H_2PO_4)_2} + 2HF$$

CHECK POINT 7.4

1. Most stable allotropic form of phosphorus is
(a) white (b) red (c) black (d) violet

2. Bond angle in phosphine is
(a) 93° (b) 107° (c) 109°.5 (d) None of these

3. Compound of phosphorus which act as reducing agent.
(a) PCl_3 (b) PCl_5 (c) Both (a) and (b) (d) None of these

4. PCl_3 on hydrolysis gives
(a) HPO_3 (b) H_3PO_2 (c) H_3PO_4 (d) H_3PO_3

5. Solid PCl_5 exists as
(a) PCl_5 (b) PCl_4^+ (c) PCl_6^- (d) PCl_4^+ and PCl_6^-

6. Number of H-atom (s) attached to phosphorus atom in pyrophosphorous acid
(a) 0 (b) 1 (c) 2 (d) 3

7. Basicity of orthophosphoric acid
(a) 2 (b) 3 (c) 4 (d) 5

8. Among the following, which is the tribasic acid?
(a) H_3PO_2 (b) H_3PO_3 (c) H_3PO_4 (d) $H_4P_2O_7$

9. Oxidation state of phosphorus in diphosphoric acid is
(a) +1 (b) +3 (c) +4 (d) +5

10. The number of P – OH bonds and the oxidation state of phosphorus atom in pyrophosphoric acid ($H_4P_2O_7$) respectively are
(a) four and four
(b) five and four
(c) five and five
(d) four and five

Chapter Exercises

(A) Topical Exercise

I. Physical and chemical properties

1 Essential constituents of plants and animal tissues are
(a) N and P (b) N and As
(c) Cu and Mg (d) Ca and Mg

2 Major minerals containing nitrogen are
(a) DNA (b) RNA
(c) KNO_3, $NaNO_3$ (d) All of these

3 Most abundant uncombined element present in atmosphere is
(a) Ca (b) Mg
(c) Al (d) N

4 Which of the following elements does not show allotropy? **(NCERT Exemplar)**
(a) Nitrogen (b) Bismuth
(c) Antimony (d) Arsenic

5 Maximum number of covalent bonds formed by N and P respectively are
(a) 3, 5 (b) 3, 6
(c) 4, 5 (d) 4, 6

6 For the hydrides of nitrogen family going down the group
(a) stability decreases
(b) reducing activity increases
(c) bond angle H*M*H decreases
(d) All of the above statements are correct

7 Which forms *pπ–pπ* multiple bonds with itself and with C and O?
(a) P, As (b) N, As
(c) N, P (d) N

8 Which of the following compounds do not exist?
N_4, P_4, PCl_5, NCl_5, NCl_3, P_2O_5, NO_2, PO_2
(a) N_4, NCl_5, NO_2 (b) N_4, NCl_5, PO_2
(c) PCl_5, NCl_5, PO_2 (d) All of these

9 Which of the following is least basic?
(a) NF_3 (b) NCl_3
(c) NBr_3 (d) NI_3

10 Which of the following is most acidic?
(a) N_2O_5 (b) P_2O_5
(c) As_2O_5 (d) Sb_2O_5

11 Which of the following is arranged in the increasing order of enthalpy of vaporisation?
(a) NH_3, PH_3, AsH_3 (b) AsH_3, PH_3, NH_3
(c) NH_3, AsH_3, PH_3 (d) PH_3, AsH_3, NH_3

12 Both NF_3 and NCl_3 are covalent but they differ in the extent of hydrolysis because
(a) NF_3 is more stable than NCl_3 and hydrolysis product of NF_3, H_2O, does not exist
(b) dipole moment of NF_3 is greater than that of NCl_3
(c) electronegativity of F is greater than that of Cl
(d) Cl can expand its octet by using *d*-orbitals

13 Nitrogen differs from P, As, Sb and Bi in the following properties
(a) it forms diatomic molecule
(b) it is not able to extend its coordination number beyond four
(c) nitrogen does not function as a Lewis acid whereas P, As, Sb and Bi do so
(d) N differs in all the properties given above

14 Consider the following statements.
Statement I $(SiH_3)_3N$ is planar while $(CH_3)_3N$ is pyramidal.
Statement II (*lp-bp*) repulsion in $(CH_3)N$ makes it pyramidal in case of $(SiH_3)_3N$ lone pair on N-atom is transferred to empty *d*-orbital of Si, thus minimising repulsion.
(a) Only statement I is correct
(b) Only statement II is correct
(c) Both the statements are correct
(d) Both the statements are incorrect

II. Nitrogen and its oxides

15 In plants, atmospheric nitrogen is converted into
(a) carbohydrate (b) proteins
(c) amino acids (d) fats

16 Pure N_2 is prepared in the laboratory by heating a mixture of
(a) NH_4Cl and NaOH (b) NH_4OH and NaCl
(c) NH_4Cl and $NaNO_2$ (d) NH_4Cl and $NaNO_3$

17 Which is/are used as an anaesthetic?
(a) N_2O (b) C_2H_4
(c) $CHCl_3$ (d) All of these

18 Following are the neutral oxides except
(a) NO (b) N_2O
(c) CO (d) NO_2

19 Which of the following oxides of nitrogen is solid?
(a) NO_2 (b) N_2O
(c) N_2O_3 (d) N_2O_5

20 Nitrosonium ion (NO^+) is isoelectronic with
(a) carbon dioxide (b) carbon monoxide
(c) nitrogen dioxide (d) nitric oxide

21 $p\pi$-$p\pi$ multiple bonding is present in
(a) oxides of nitrogen (b) oxides of phosphorus
(c) halides of nitrogen (d) halides of phosphorus

22 Which of the following is/are paramagnetic?
$NO_2, NO, N_2O_4, N_2O_2, N_2O_5$
(a) Only NO_2 (b) NO_2, NO
(c) NO, NO_2, N_2O_5 (d) All are paramagnetic

23 Nitric oxide is formed by the reaction of
(a) $C + N_2O$ (b) $Cu + HNO_3$
(c) $Cu + N_2O$ (d) $Cu + NO_2$

24 The wrong statement about N_2O is
(a) it is nitrous oxide
(b) it is least reactive oxide of nitrogen
(c) it is not a linear molecule
(d) it is known as laughing gas

25 N_2O is formed when
(a) moist Fe reacts with NO
(b) Sn^{2+} reacts with conc. HNO_3 in presence of HCl
(c) cold dil. HNO_3 reacts with Cu and Zn
(d) by all the reactions

26. N_2O (laughing gas) finds use in the following except.
(a) as a propellant for whipped ice-cream
(b) as an anaesthetic
(c) for the preparation of N_3H
(d) as fuel for rockets

27 Nitrogen is obtained by the thermal decomposition of
(a) NH_4NO_2 (b) NH_4NO_3
(c) $AgNO_3$ (d) $Pb(NO_3)_3$

28 On heating lead nitrate the product formed is
(a) NO (b) NO_2
(c) N_2O (d) N_2O_5

29 Laughing gas is prepared by heating
(a) $NH_4Cl + NaNO_3$ (b) NH_4Cl
(c) $(NH_4)_2SO_2$ (d) NH_4NO_2

30 Which of the following oxides of nitrogen is paramagnetic?
(a) N_2O (b) N_2O_5
(c) NO_2 (d) N_2O_3

31 A gaseous substance dissolves in water giving a pale blue solution, which decolorises $KMnO_4$ and oxidises KI to I_2. Gaseous substances are
(a) N_2O_5 (b) NH_3
(c) N_2O_3 (d) HNO_3

32 Which blue liquid is obtained on reacting equimolar amounts of two (NO and NO_2) gases at −30°C ?
(a) N_2O (b) N_2O_3
(c) N_2O_4 (d) N_2O_5

33 NO_2 is not obtained when following is heated
(a) $Pb(NO_3)_2$ (b) $AgNO_3$
(c) $LiNO_3$ (d) KNO_3

III. Hydrides of nitrogen

34 In liquid NH_3
(a) NH_4Cl is an acid
(b) $NaNH_2$ is a base
(c) CH_3COOH behaves as strong acid
(d) All of the above facts are true

35 All of the following are bases except
(a) N_2H_4 (b) NH_2OH (c) NH_3 (d) N_3H

36 Which of the following is used as a rocket fuel?
(a) N_2H_4 (b) NH_3
(c) Both (a) and (b) (d) None of these

37 Calcium cyanamide on treatment with steam produces
(a) NH_3 + CaO (b) $NH_3 + CaHCO_3$
(c) $NH_3 + CaCO_3$ (d) $NH_3 + Ca(OH)_2$

38 Hydrazine is not
(a) a reducing agent (b) an oxidising agent
(c) an acid (d) a base

39 Bond angle of 109°28′ is found in
(a) NH_3 (b) H_2O
(c) $\overset{\oplus}{C}H_5$ (d) $\overset{\oplus}{N}H_4$

40 Consider the following statements.
Statement I N_2H_4 is kineticaly stable compound.
Statement II Two N-atoms are bonded by a triple bond in N_2H_4.
(a) Only statement I is correct
(b) Only statement II is correct
(c) Both the statements are correct
(d) Both the statements are incorrect

41 With excess of Cl_2 ammonia forms
(a) ammonium chloride (b) nitrogen trichloride
(c) nitrosyl chloride (d) nitrogen

42 NaOCl reacts with excess of NH_3 to produce
(a) NH_2OH (b) NH_2NH_2
(c) N_2 (d) NO

43 Fe^{2+} reduces NH_2OH to
(a) NH_3 (b) N_3H
(c) N_2H_4 (d) N_2

44 The true statement for N_3^- is
(a) it has non-linear structure
(b) it is called pseudohalogen
(c) the formal oxidation state of nitrogen in this anion is +1
(d) it is isoelectronic with N_2O

45 $NH_4Cl(s)$ is heated in a test tube. Vapours are brought in contact with red litmus paper, which changes to blue and then to red. It is because of
(a) formation of NH_4OH and HCl
(b) formation of NH_3 and HCl
(c) greater diffusion of NH_3 than HCl
(d) greater diffusion of HCl than NH_3

46 Select the incorrect statement about hydrides of group 15 elements.
(a) The central atom in the hydride is sp^2-hybridised
(b) NH_3 readily form NH_4^+ salts with H^+, PH_4^+ salts are formed with H^+ under anhydrous condition
(c) The tetrahedron is distorted due to repulsion between the lone pair of electrons and the bond pairs
(d) The bond energy of the M—H bond decreases from NH_3 to BiH_3 because of increase in the size of the element

47 Consider the following statements.
Statement I N_3H explodes violently on slight heating.
Statement II N_3^- has linear structure and is stable due to resonance.
(a) Only statement I is correct
(b) Only statement II is correct
(c) Both the statements are correct
(d) Both the statements are incorrect

IV. Oxoacids of nitrogen

48 Which of the following metals are soluble in *aqua regia* ?
(a) Pt (b) Au
(c) Ag (d) All of these

49 Acid rain may cause
(a) rusting easier (b) stone cancer in Taj Mahal
(c) non-fertility of soil (d) All of these

50 Which one of the following acid possesses oxidising, reducing and complex forming properties ?
(a) HNO_3 (b) HCl
(c) H_2SO_4 (d) HNO_2

51 $p\pi$-$p\pi$ multiple bonding between nitrogen atoms is present in **[NCERT Exemplar]**
(a) hyponitrous acid (b) nitrous acid
(c) nitric acid (d) in all of these

52 On photochemical decomposition of HNO_3 the products formed are
(a) NO and O_2 (b) NO_2 and O_2
(c) N_2O_4 and NO_2 (d) N_2O_5 and NO_2

53 $4HNO_3 + P_4O_{10} \longrightarrow 4HPO_3 + X$
In the above reaction the product X is
(a) NO_2 (b) N_2O_3
(c) N_2O_4 (d) N_2O_5

54 Nitration of aromatic compounds is done using a mixture of conc. H_2SO_4 and conc. HNO_3. Intermediate formed in the nitration process is
(a) NO_2^+ (nitronium ion) (b) NO_3^- (nitrate ion)
(c) NO^+ (nitrosonium ion) (d) NO_2^- (nitrite ion)

55 $NaNO_2$ and $NaNO_3$ are used as food additives and generally added to meat to
(a) retard oxidation of meat to brown coloured material
(b) prevent growth of botulism bacteria
(c) impart red colour to oxidised meat by NO by reaction of heme in blood
(d) make all of the functions given above effective

56 Nitrites in water may react with amines under acidic conditions in the stomach to
(a) give carcinogenic compounds
(b) give proteins
(c) keep stomach acidity free
(d) produce free amino acids

57 HNO_3 in pure state is colourless but it is often of yellow colour. It is due to
(a) unstable structure of HNO_3, which immediately changes to NO_2^-
(b) photochemical decomposition in presence of sunlight
(c) interaction of atmospheric gases with HNO_3
(d) conversion of HNO_3 into NO_2^+

58 Which of the following statements are true?
(a) Cold and very dilute HNO_3 forms NH_4NO_3 with Zn or Sn
(b) Concentrated HNO_3 forms H_2SnO_3 with Sn
(c) Cold and more concentrated HNO_3 forms NO_2 with Cu
(d) All the above are correct

59 Match the reactions of metals with dilute HNO_3 (in Column I) with the nitrogen compounds, obtained by oxidation/reduction (in Column II).

Column I	Column II
A. Mg + dil. HNO_3	1. NO
B. Zn + dil. HNO_3	2. H_2
C. Sn + dil. HNO_3	3. N_2O
D. Pb + dil. HNO_3	4. NH_4NO_3

	A	B	C	D		A	B	C	D
(a)	2	3	4	1	(b)	1	2	4	3
(c)	3	2	1	4	(d)	1	4	3	2

V. Phosphorus and its compound

60 Among these ores the highest phosphorus content is present in
(a) chlorapatite (b) phosphorite
(c) fluorspar (d) equal in (a), (b) and (c)

61 A person working with phosphorus suffers from a disease in which bones decay. It is known as
(a) arthrities (b) phossy jaw
(c) rickets (d) cancer

62 The salt of oxoacids of phosphorus that is used as washing detergent is
(a) wavellite (b) microcosmic salt
(c) calgon (d) chlorapatite

63 Glacial phosphoric acid is
(a) HPO_3 (b) H_3PO_3
(c) H_3PO_4 (d) $H_4P_2O_7$

64 An example of an "*ortho*" acid is
(a) $HAsO_2$ (b) H_3AsO_4
(c) $(HPO_3)_n$ (d) $H_4As_2O_7$

65 Compound used in safety matches is
(a) P_4S_3 (b) P_4
(c) P_2O_5 (d) PCl_3

66 The structural formula of hypophosphorus acid is

(a) 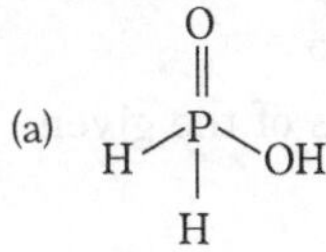

(b) 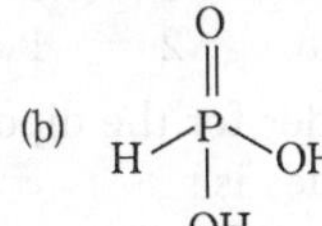

(c) 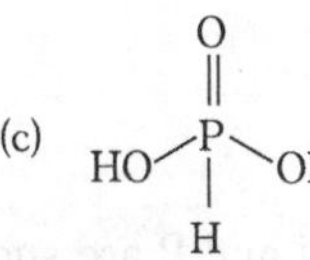

(d) O ‖ HO–P–OH | OH

67 Which of the following phosphorus is the most reactive?
(a) Red phosphorus (b) White phosphorus
(c) Scarlet phosphorus (d) Violet phosphorus

68 What is the role of phosphate ion in a detergent?
(a) It reduces pH of the water
(b) It increases pH of the water
(c) It removes Ca^{2+} and Mg^{2+} ions from water that causes hardness
(d) It increases its solubility in water

69 A substance which gives an yellow precipitate, when boiled with an excess of nitric acid and ammonium molybdate and red precipitate with $AgNO_3$ is
(a) orthophosphate (b) pyrophosphate
(c) metaphosphate (d) hypophosphate

70 With $FeCl_3$ a neutral solution of orthophosphate gives an yellow precipitate insoluble in
(a) CH_3COOH (b) HCl
(c) H_2SO_4 (d) NaOH

71 Phosphorus trichloride PCl_3 undergoes hydrolysis to produce an oxoacid. It has the formula
(a) HPO_3 (b) H_3PO_3
(c) H_3PO_4 (d) H_3PO_2

72 Choose the correct statement.
(a) Superphosphate is $[3Ca(H_2PO_4)_2 + 7CaSO_4]$
(b) Triple superphosphate is $Ca(H_2PO_4)_2 \cdot H_2O$
(c) Both (a) and (b) are correct
(d) None of the above is correct

73 The high reactivity and low volatility of white phosphorus is due to
(a) tetrahedrally arranged P_4 units
(b) bond angle of 60°
(c) weak van der Waals' forces of attraction
(d) increased steric (strain) factor

74 Red and white phosphorus will differ but not in
(a) smell
(b) solubility in $CHCl_3$
(c) exhibiting phosphorescence
(d) reactions with HNO_3

75 Repeated use of which one of the following fertilisers would increase the acidity of the soil?
(a) Urea
(b) Superphosphate of lime
(c) Ammonium sulphate
(d) Potassium nitrate

76 The percentage of *p*-character in the orbitals forming P—P bonds in P_4 is
(a) 25 (b) 33 (c) 50 (d) 75

77 Polyphosphates are used as water softening agents because they
(a) form soluble complexes with anionic species
(b) precipitate anionic species
(c) form soluble complexes with cationic species
(d) precipitate cationic species

78 PI_3 upon hydrolysis gives
(a) dibasic acid and monobasic acid
(b) monobasic acid and tribasic acid
(c) monbasic acid and a salt
(d) dibasic acid and tribasic acid

79 In the preparation of red phosphorus from white phosphorus
(a) MnO_2 is used as a catalyst
(b) the white phosphorus is treated in an electric furnace
(c) a little iodine is used as catalyst
(d) the gas P_4 is released.

80 Compound *A* undergoes hydrolysis to produce a colourless gas with rotten fish smell. The gas gives a vortex ring. The gas is
(a) PH_3 (b) P_2O_3 (c) P_2O_5 (d) P_2S_5

81 One of the acid listed below is formed only from P_2O_3 the rest are formed from P_2O_5. Acid formed from P_2O_3 is
(a) HPO_3 (b) $H_4P_2O_7$
(c) H_3PO_4 (d) H_3PO_3

(B) Miscellaneous questions

1 In P_4 (tetrahedral)
(a) each P is joined to four P
(b) each P is joined to three P
(c) each P is joined to two P
(d) P_4 does not exist

2 The number of P—O—P bonds in cyclic metaphosphoric acid is
(a) zero (b) two (c) three (d) four

3 The element which forms oxides in all oxidation states +1 to +5 is
(a) N (b) P (c) As (d) Sb

4 Pentavalence phosphorus is more stable when compared to that of nitrogen even though they belong to same group. It is due to
(a) reactivity of phosphorus
(b) inert nature of nitrogen
(c) presence of *d*-orbital
(d) large size of phosphorus atom

5 Consider the following statements :
Statement I Each N-atom is sp^3-hybridised in N_2H_4.
Statement II It has gauche conformation.
Choose the correct option.
(a) Only statement I is correct
(b) Only statement II is correct
(c) Both the statements are correct
(d) Both the statements are incorrect

6 Nitric oxide is paramagnetic in
(a) gaseous state (b) liquid state
(c) solid state (d) polymeric state

7 Stability of pentahalides is in order
(a) P > As > Sb > Bi for given *X* (Halogen)
(b) Br > Cl > F for given element
(c) Both (a) and (b)
(d) None of the above

8 $d\pi$-$p\pi$ bonding is shown in
(a) NO_3^-, NO_2^-, N_3^{3-}, CN^- (b) P_2O_3, P_2O_5, PO_4^{3-}
(c) NH_3, PH_3, BiH_3 (d) CO, NO, CO_2, NO_2

9 Paramagnetic species are
(a) NO, NO_2, O_2 (b) N_2O_2, N_2O_4, O_3
(c) NO^+, NO_2^+, O_2^+ (d) O_2, CO, CO_2

10 The dipole moment of NF_3 is less than that of NH_3 because
(a) F is more reactive than H
(b) NH_3 forms associated molecules
(c) the resultant of the individual bond polarities is opposed by the polarity of lone pair in NF_3
(d) None of the above

11 The oxidation state of central atom in the anion of compound NaH_2PO_2 will be ....
(a) +3 (b) +5
(c) +1 (d) −3

12 Match the compounds (in Column I) with their oxidation number of N (in Column II).

	Column I		Column II
A.	N_2H_4	1.	−1
B.	NH_2OH	2.	3
C.	HNO_2	3.	4
D.	NO_2	4.	5
E.	HNO_3	5.	−2

	A	B	C	D	E
(a)	5	1	2	3	4
(b)	1	5	3	4	2
(c)	1	2	5	4	3
(d)	3	2	1	4	5

13 The order for the dipole moments of the given molecules is
(a) $BF_3 > NF_3 > NH_3$
(b) $NF_3 > BF_3 > NH_3$
(c) $NH_3 > NF_3 > BF_3$
(d) $NH_3 > BF_3 > NF_3$

14 The electronegativities of N, C, Si and P are such that
(a) P < Si < C < N (b) Si < P < N < C
(c) Si < P < C < N (d) P < Si < N < C

15 Select incorrect statement about N_2O_4.
(a) It self ionises as NO^+, NO_3^-.
(b) Substance containing NO^+ is said to be acid and that containing NO_3^- is said to be base.
(c) N_2O_4 is paramagnetic.
(d) NO_2 dimerises to N_2O_4 with disappearance in paramagnetism.

16 The true statement for the acids of phosphorus. H_3PO_2, H_3PO_3 and H_3PO_4 is
(a) their acidic nature is : $H_3PO_4 < H_3PO_2 < H_3PO_3$
(b) all of them are reducing in nature
(c) all of them are tribasic acids
(d) the geometry of phosphorus is tetrahedral in all the three cases

17 Holme's signals are based on the following chemicals
(a) $Ca_3P_2 + Al_4C_3$
(b) $Ca_3P_2 + CaC_2$
(c) $NH_4Cl + NaNO_2$
(d) $NH_4Cl + CaNO_3$

18 Consider the following statements.

Statement I On cooling, the brown colour of NO_2 disappears.

Statement II On cooling, NO_2 forms dimer (N_2O_4) resulting in the pairing of the odd electrons in NO_2.

Choose the correct option.
(a) Only statement I is correct
(b) Only statement II is correct
(c) Both the statements are correct
(d) Both the statements are incorrect

19 Which of the following structure is exhibited by solid crystalline PCl_5?
(a) Bi pyramidal moieties
(b) Octahedral and tetrahedral ions
(c) Square pyramidal moieties
(d) Pentagonal moieties

20 The bond angle in AsH_3 is greater than that in
(a) NH_3 (b) H_2O
(c) BCl_3 (d) None of these

21 When PCl_3 reacts with sulphuric acid, sulphuryl chloride is formed as the final product. This shows that sulphuric acid
(a) has two hydroxyl groups in its structure
(b) is a derivative of sulphur dioxide
(c) is a dibasic acid
(d) has greater affinity for water

22 Which of the following acids forms three series of salts?
(a) H_3PO_2 (b) H_3BO_3
(c) H_3PO_4 (d) H_3PO_3

23 Which of the following statements is wrong?
(a) Single N—N bond is stronger than the single P—P bond.
(b) PH_3 can act as a ligand in the formation of coordination compound with transition elements.
(c) NO_2 is paramagnetic in nature.
(d) Covalency of nitrogen in N_2O_5 is four.

24 Strong reducing behaviour of H_3PO_2 is due to
(a) low oxidation state of phosphorus
(b) presence of two — OH group and one P — H bond
(c) presence of one — OH group and two P—H bonds
(d) high electron gain enthalpy of phosphorus

25 Substances burn more readily in N_2O than in air because N_2O
(a) is reactive at high temperature
(b) dissociates to give O_2 that supports combustion
(c) the activation energy is increased on increasing temperature
(d) acts as a catalyst

26 MnO_4^- / H^+ oxidises NO_2^- to
(a) N_2 (b) NH_3
(c) NO_3^- (d) N_2O

27 A colourless salt gives a white ppt. (soluble in ammonium acetate) and a brown coloured pungent gas on reaction with conc. H_2SO_4. Salt is
(a) $Ba(NO_3)_2$ (b) $Pb(NO_3)_2$
(c) $NaNO_3$ (d) NH_4NO_3

28 Consider the reaction,

$$(NH_4)_2HAsO_3 + CuSO_4 \longrightarrow \underbrace{CuHAsO_3}_{\text{Green ppt.}} + (NH_4)_2SO_4$$

The green precipitate is also known as
(a) Pair's green (b) Scheele's green
(c) Verdigris' green (d) Rinmann's green

29 Match the acids (in Column I) with number of acid salts formed by them (in Column II).

Column I	Column II
A. Hypophosphorus acid	1. Two
B. Orthophosphorus acid	2. Three
C. Orthophosphoric acid	3. One
D. Mellitic acid	4. Six

	A	B	C	D		A	B	C	D
(a)	3	1	2	4	(b)	1	3	2	1
(c)	3	1	4	2	(d)	1	2	3	4

30 Which of the following is/are correct statements about P_4O_6 and P_4O_{10}?
(a) Both form oxyacids H_3PO_3 and H_3PO_4 respectively
(b) In P_4O_6 each P is joined to four O and in P_4O_{10} each P is joined to six O atoms
(c) Both (a) and (b)
(d) None of the above

31 1 mole each of H_3PO_2, H_3PO_3 and H_3PO_4 will neutralise *x* moles of NaOH, *y* moles of $Ca(OH)_2$ and *z* moles of $Al(OH)_3$ (assuming all as strong electrolytes). *x, y, z* are in the ratio of
(a) 3 : 1.5 : 1 (b) 1 : 2 : 3 (c) 3 : 2 : 1 (d) 1 : 1 : 1

32 Which of the following are acid salts?
(a) NaH_2PO_2, Na_2HPO_3 (b) Na_2HPO_3, Na_2HPO_4
(c) $NaHCO_3$, Na_2HPO_4 (d) All of these

33 NH_3 cannot be obtained by
(a) heating of NH_4NO_3 or NH_4NO_2
(b) heating of NH_4Cl or $(NH_4)_2CO_3$
(c) heating of NH_4NO_3 with NaOH
(d) reaction of AlN or Mg_3N_2 or $CaCN_2$ with H_2O

34 Consider the following statements.

Statement I Both H_3PO_3 and H_3PO_4 have the same number of hydrogen atoms but H_3PO_4 is a tribasic acid and H_3PO_3 is a dibasic acid.

Statement II 1 mole of H_3PO_3 is neutralised by 2 moles of NaOH while 1 mole of H_3PO_4 is neutralised by 3 moles of NaOH.

(a) Only statement I is correct
(b) Only statement II is correct
(c) Both the statements are correct
(d) Both the statements are incorrect

(C) Medical entrances' gallery

1 Among the compounds shown below which one revealed a linear structure? **[NEET 2020]**

(a) NO_2 (b) HOCl
(c) O_3 (d) N_2O

2 Identify the incorrect statement related to PCl_5 from the following. **[NEET 2019]**

(a) Two axial P— Cl bonds make an angle of 180° with each other.
(b) Axial P— Cl bonds are longer than equatorial P— Cl bonds.
(c) PCl_5 molecule is non-reactive.
(d) Three equatorial P— Cl bonds make an angle of 120° with each other.

3 A compound 'X' upon reaction with H_2O produced a colourless gas 'Y' with rotten fish smell. Gas 'Y' is absorbed in a solution of $CuSO_4$ to give Cu_3P_2 as one of the products. Predict the compound 'X'. **[NEET (Odisha) 2019]**

(a) Ca_3P_2 (b) NH_4Cl
(c) As_2O_3 (d) $Ca_3(PO_4)_2$

4 Which of the following oxoacids of phosphorus has strongest reducing property? **[NEET (Odisha) 2019]**

(a) $H_4P_2O_7$ (b) H_3PO_3
(c) H_3PO_2 (d) H_3PO_4

5 In which of the following, oxidation state of phosphorus is +3 ? **[JIPMER 2019]**

(a) Ortho phosphoric acid
(b) Pyro phosphoric acid
(c) Ortho phosphorous acid
(d) Meta phosphoric acid

6 The correct order of N-compounds in its decreasing order of oxidation states is **[NEET 2018]**

(a) HNO_3, NH_4Cl, NO, N_2
(b) HNO_3, NO, NH_4Cl, N_2
(c) HNO_3, NO, N_2, NH_4Cl
(d) NH_4Cl, N_2, NO, HNO_3

7 In which of the following arrangements, the order is not strictly according to the property written against it? **[AIIMS 2018]**

(a) $CO_2 < SiO_2 < SnO_2 < PbO_2$ (oxidising power)
(b) HF < HCl < HBr < HI (acidic strength)
(c) $NH_3 > PH_3 < AsH_3 < SbH_3$ (basic strength)
(d) B < C < O < N (first ionisation enthalpy)

8 Which of the following products will be obtained, when copper metal is reacted with HNO_3? **[AIIMS 2018]**

(a) NO and N_2O_5 (b) NO and NO_2
(c) NO_2 and N_2O_5 (d) HNO_2 and N_2

9 Which of the following oxyacid does not contain P—O—P bond? **[AIIMS 2018]**

(a) Isohypophosphoric acid (b) Pyrophosphorus acid
(c) Diphosphoric acid (d) Hypophosphoric acid

10 Which of the following chemicals are used to manufacture methyl isocyanate that caused "Bhopal Gas Tragedy"? **[JIPMER]**

(i) Methylamine (ii) Phosgene
(iii) Phosphine (iv) Dimethylamine

(a) (i) and (ii) (b) (iii) and (iv)
(c) (i) and (iii) (d) (ii) and (iv)

11 Which of the following pair behaves as strong and weak base hydride respectively ? **[JIPMER]**

NH_3, PH_3, AsH_3, SbH_3, BiH_3

(a) NH_3, PH_3 (b) NH_3, BiH_3
(c) PH_3, BiH_3 (d) BiH_3, AsH_3

12 The species having bond angles of 120° is **[NEET 2017]**

(a) BCl_3 (b) PH_3
(c) ClF_3 (d) NCl_3

13 The product obtained as a result of a reaction of nitrogen with CaC_2 is **[NEET (I) 2016]**

(a) CaCN (b) $CaCN_3$
(c) Ca_2CN (d) $CaCN_2$

14 Among the following, which one is a wrong statement? **[NEET (II) 2016]**

(a) PH_5 and $BiCl_5$ do not exist
(b) $p\pi$ - $d\pi$ bonds are present in SO_2
(c) SeF_4 and CH_4 have same shape
(d) I_3^+ has bent geometry

15 Which is the correct statement for the given acids? **[NEET (I) 2016]**

(a) Phosphinic acid is a monoprotic acid while phosphonic acid is a diprotic acid
(b) Phosphinic acid is a diprotic acid while phosphonic acid is a monoprotic acid
(c) Phosphinic acid and phosphonic acid are triprotic acids
(d) Phosphinic acid and phosphonic acid are diprotic acids

16 On heating with concentrated NaOH solution in an inert atmosphere of CO_2, white phosphorus gives a gas. Which of the following statement is incorrect about the gas? **[Karnataka CET 2015]**

(a) It is less basic than NH_3
(b) It is more basic than NH_3
(c) It is highly poisonous and has smell like rotten fish
(d) Its solution in water decomposes in the presence of light

17 Which of the following oxides of nitrogen contains N—O—N bond? **[Kerala CEE 2015]**

(a) Dinitrogen oxide (b) Nitrogen monoxide
(c) Dinitrogen pentoxide (d) Dinitrogen trioxide
(e) Dinitrogen tetraoxide

18 The reaction below represents the Haber process for the industrial production of ammonia.

$$N_2(g) + 3H_2(g) \rightleftharpoons 2NH_3(g) \quad \Delta H° = -92 \text{ kJ}$$

The optimum conditions of temperature and pressure are choosen as a compromise between those that favour a high yield of ammonia and those that favour a fast rate of production. Economic considerations are also important. Which statement is correct? **[GGSIPU 2015]**

(a) A higher temperature would ensure a higher yield and a faster rate
(b) A lower pressure would ensure a higher yield at a lower cost
(c) A lower temperature would ensure a higher yield and a faster rate
(d) A higher pressure would ensure a higher yield at a higher cost

19 The species that is not hydrolysed in water is **[KCET 2014]**

(a) P_4O_{10} (b) BaO_2
(c) Mg_3N_2 (d) CaC_2

20 The correct arrangement of following acids of phosphorus in the increasing order of oxidation state of phosphorus is **[Guj CET 2014]**

(a) Hypophosphorus acid (<) orthophophorus acid < pyrophosphoric acid
(b) Hypophosphorus acid (<) orthophosphorus acid = pyrophosphoric acid
(c) Pyrophosphoric acid (<) hypophosphorus acid < orthophosphorus acid
(d) Pyrophosphoric acid (<) orthophosphorus acid < hypophosphorus acid

21 The mixture of concentrated HCl and HNO_3 made in 3 : 1 ratio contains **[AIIMS 2013]**

(a) ClO_2 (b) NOCl
(c) NCl_3 (d) N_2O_4

22 In pyrophosphoric acid the number of hydroxyl groups are **[UP CPMT 2013]**

(a) 2 (b) 3
(c) 4 (d) 5

23 A neutral fertiliser among the following is **[WB JEE 2013]**

(a) CAN
(b) ammonium sulphate
(c) ammonium nitrate
(d) urea

24 Which of the following statements is not valid for oxoacids of phosphorus? **[Manipal 2012]**

(a) Orthophosphoric acid is used in the manufacture of triple super phosphate
(b) Hypophosphorus acid is a diprotic acid
(c) All oxoacids contain tetrahedral four coordinated phosphorus
(d) All oxoacids contain atleast one P=O unit and one P—OH group

25 A hydride of nitrogen, which is acidic is **[AIIMS 2012]**

(a) NH_3 (b) N_3H
(c) N_2H_2 (d) N_2H_4

26 The oxidation state of phosphorus in cyclotrimetaphosphoric acid is **[Kerala CEE 2011]**

(a) + 3 (b) + 5
(c) − 3 (d) + 2
(e) − 2

27 The basicity of pyrophosphorus acid is **[Kerala CEE 2011]**

(a) 2 (b) 4
(c) 1 (d) 5
(e) 3

ANSWERS

Check point 7.1

1. (d)	2. (d)	3. (c)	4. (a)	5. (a)	6. (c)	7. (b)	8. (d)	9. (d)	10. (d)

Check point 7.2

1. (a)	2. (d)	3. (a)	4. (b)	5. (a)	6. (b)	7. (d)	8. (a)

Check point 7.3

1. (a)	2. (d)	3. (a)	4. (b)	5. (c)	6. (d)	7. (c)	8. (a)

Check point 7.4

1. (c)	2. (a)	3. (a)	4. (d)	5. (d)	6. (a)	7. (b)	8. (c)	9. (d)	10. (d)

(A) Topical Exercise

1. (a)	2. (c)	3. (d)	4. (b)	5. (a)	6. (d)	7. (d)	8. (b)	9. (a)	10. (a)
11. (d)	12. (a)	13. (d)	14. (c)	15. (b)	16. (c)	17. (d)	18. (d)	19. (c)	20. (b)
21. (a)	22. (b)	23. (b)	24. (c)	25. (d)	26. (d)	27. (a)	28. (b)	29. (a)	30. (c)
31. (c)	32. (b)	33. (d)	34. (d)	35. (d)	36. (a)	37. (c)	38. (c)	39. (d)	40. (a)
41. (b)	42. (b)	43. (a)	44. (d)	45. (c)	46. (a)	47. (c)	48. (d)	49. (d)	50. (d)
51. (a)	52. (b)	53. (d)	54. (a)	55. (d)	56. (a)	57. (b)	58. (d)	59. (a)	60. (b)
61. (b)	62. (c)	63. (a)	64. (b)	65. (a)	66. (a)	67. (b)	68. (d)	69. (a)	70. (a)
71. (b)	72. (c)	73. (d)	74. (d)	75. (c)	76. (d)	77. (c)	78. (a)	79. (c)	80. (a)
81. (d)									

(B) Miscellaneous questions

1. (b)	2. (c)	3. (a)	4. (c)	5. (c)	6. (a)	7. (c)	8. (b)	9. (a)	10. (c)
11. (c)	12. (a)	13. (c)	14. (c)	15. (c)	16. (d)	17. (b)	18. (c)	19. (b)	20. (d)
21. (a)	22. (c)	23. (a)	24. (c)	25. (b)	26. (c)	27. (b)	28. (b)	29. (a)	30. (c)
31. (d)	32. (c)	33. (a)	34. (c)						

(C) Medical entrances' gallery

1. (d)	2. (c)	3. (a)	4. (c)	5. (c)	6. (c)	7. (c)	8. (b)	9. (d)	10. (a)
11. (b)	12. (a)	13. (d)	14. (c)	15. (a)	16. (b)	17. (c)	18. (d)	19. (a)	20. (a)
21. (b)	22. (c)	23. (a)	24. (b)	25. (b)	26. (b)	27. (a)			

Hints & Explanations

Check point 7.2

1 (*a*) Liquid N_2 is used as refrigerant to preserve biological materials, food items and in cryosurgery.

It also finds use where an inert atmosphere is required (e.g. in iron and steel industry, inert diluent for reactive chemicals).

5 (*a*) Dinitrogen oxide is produced by the thermal decomposition of NH_4NO_3. The reaction is given below

$$NH_4NO_3 \xrightarrow{\Delta} N_2O + 2H_2O$$

6 (*b*) On heating HNO_3 with P_4O_{10}, the oxide of nitrogen produced is N_2O_5.

The reaction is given below

$$2HNO_3 + \frac{1}{2}P_4O_{10} \longrightarrow \underset{\text{Dinitrogen pentaoxide}}{N_2O_5} + \underset{\text{Metaphosphoric acid}}{2HPO_3}$$

7 (*d*) Acidic strength of oxides generally decreases with decrease in oxidation states. Thus, the correct order for decreasing acidic strength of oxides of nitrogen is

$$N_2O_5 > N_2O_4 > N_2O_3 > NO > N_2O$$

8 (*a*) The products *A* and *B* are NO and N_2O_3 respectively. The two reactions complete as follows

I. $2NaNO_2 + 2FeSO_4 + 3H_2SO_4 \longrightarrow Fe_2(SO_4)_3 + 2NaHSO_4 + 2H_2O + \underset{'A'}{2NO}$

II. $2NO + N_2O_4 \xrightarrow{250K} \underset{'B'}{2N_2O_3}$

(A) Topical Exercise

5 (*a*)

N (7): $1s^2$ [↿⇂] $2s^2$ [↿⇂] $2p^3$ [↿ | ↿ | ↿] Three unpaired electrons, thus maximum covalency of N is 3. It doesn't have vacant *d*-orbital

P (15) [Ne]: $3s^2$ [↿⇂] $3p^3$ [↿ | ↿ | ↿] $3d$ [| | | |]

↓ excite

$3s^1$ [↿] $3p^3$ [↿ | ↿ | ↿] $3d^1$ [↿ | | | |] Five unpaired electrons, thus maximum covalency of P is 5. It has vacant *d*-orbital.

7 (*d*) C≡N, N≡N, N=O

P, As have *d*-orbitals, thus they form $d\pi$-$p\pi$, $d\pi$-$d\pi$ multiple bonds.

9 (*a*) Fluorine is the most electronegative element, thus tendency of N-atom to donate lone pair (to behave as Lewis acid) is least.

10 (*a*) Electronegativity of N is highest.

$\therefore$ It is most acidic.

11 (*d*) In NH_3 there is strong intermolecular H-bonding. Thus, it has maximum heat of vaporisation.

In PH_3, there is no H-bonding. Thus, it is in gaseous state and thus it has minimum heat of vaporisation.

Thus, correct order is

$$PH_3 < AsH_3 < NH_3$$

16 (*c*) $NH_4Cl + NaNO_2 \xrightarrow{\Delta} NH_4NO_2 + NaCl$

$NH_4NO_2 \xrightarrow{\Delta} N_2 + 2H_2O$

20 (*b*) Total number of electrons in $NO^+ = 7 + 8 - 1 = 14$

	Electrons
(a) CO_2	22
(b) CO	14
(c) NO_2	23
(d) NO	15

21 (*a*) There are only valence *p*-orbitals in N and O. Thus, there is $p\pi$-$p\pi$ bonding in oxides of nitrogen.

23 (*b*) $Cu + \text{dil. } HNO_3 \longrightarrow Cu(NO_3)_2 + H_2O + NO$

24 (*c*) N≡N—O is a linear molecule.

27 (*a*) $NH_4NO_2 \xrightarrow{\Delta} N_2 + 2H_2O$

28 (*b*) $2Pb(NO_3)_2 \xrightarrow{\Delta} 2PbO + 4NO_2 + O_2$

29 (*a*) $NH_4Cl + NaNO_3 \longrightarrow NH_4NO_3 + NaCl$

$NH_4NO_3 \xrightarrow{\Delta} \underset{\text{Laughing gas}}{N_2O}$

30 (*c*) NO_2 has one unpaired electron on N-atom. Thus, it is paramagnetic in nature.

•N(=O)O

31 (*c*) $N_2O_3 + H_2O \longrightarrow \underset{\text{Pale blue}}{HNO_2}$

HNO_2 is a reducing agent

$$MnO_4^- \xrightarrow{HNO_2} Mn^{2+}$$

HNO_2 is also an oxidising agent.

$$I^- \xrightarrow{HNO_2} I_2$$

32 (*b*) $NO + NO_2 \longrightarrow N_2O_3$ blue-liquid at -30°C

33 (*d*) $2KNO_3 \xrightarrow{\Delta} 2KNO_2 + O_2$

34 (*d*) (a) $\underset{\text{Acid}_1}{NH_4^+} + \underset{\text{Base}_2}{NH_3} \longrightarrow \underset{\text{Base}_1}{NH_3} + \underset{\text{Acid}_2}{NH_4^+}$

(b) $\underset{\text{Base}_1}{NH_2^-} + \underset{\text{Acid}_2}{NH_3} \longrightarrow \underset{\text{Acid}_1}{NH_3} + \underset{\text{Base}_2}{NH_2^-}$

(c) $\underset{\text{Strong acid}}{CH_3COOH} + \underset{\text{Weak base}}{NH_3} \longrightarrow \underset{\text{Weak base}}{CH_3COO^-} + \underset{\text{Strong acid}}{NH_4^+}$

35 (*d*) $\underset{\text{Acid}}{N_3H} \longrightarrow N_3^- + H^+$

37 (*c*) $CaCN_2 + 3H_2O \longrightarrow CaCO_3 + 2NH_3$

38 (*c*) Hydrazine (N_2H_4) does not behave as an acid.

39 (*d*) $\overset{\oplus}{N}H_4$ has tetrahedral structure and N-atom is sp^3-hybridised.

Thus, bond-angle is 109° 28′.

42 (*b*) $NaOCl + NH_3 \longrightarrow NH_2Cl + NaOH$

$NH_2Cl + NH_3 \longrightarrow NH_2—NH_2 + HCl$

44 (*d*) N_3^- has total electrons = 22

Thus, it is isoelectronic with N_2O.

45 (*c*) $NH_4Cl(s) \xrightarrow{\Delta} NH_3(g) + HCl$

NH_3 (basic) changes red litmus blue and HCl (acidic) changes blue litmus red.

Since, diffusion of NH_3 (lighter gas) is more than that of HCl (heavier gas). Thus, this change is observed.

46 (*a*) $:NH_3$

There is lone pair on central atom resulting in sp^3-hybridised MH_3, where M = Group 15 elements.

51 (*a*) Two N-atoms are present in hyponitrous ($H_2N_2O_2$) acid. Thus, multiple bond between two N-atoms is present in $H_2N_2O_2$.

$H—O—N{=}N—O—H$

52 (*b*) $2HNO_3 \xrightarrow{h\nu} H_2O + 2NO_2 + \frac{1}{2}O_2$

53 (*d*) HNO_3 (oxidation number of N = + 5) changes to N_2O_5 by P_4O_{10}.

54 (*a*) $HNO_3 + H_2SO_4 \longrightarrow NO_2^+ + HSO_4^- + H_2O$

$C_6H_6 + NO_2^+ \longrightarrow C_6H_5NO_2$

66 (*a*) Hypophosphorus acid is H_3PO_2.

67 (*b*) Due to strain, white phosphorus is most reactive.

69 (*a*) It is analytical qualitative test of phosphate.

$\underset{\text{(Orthophosphate)}}{H_3PO_4} + 12(NH_4)_2MoO_4 + 21HNO_3 \longrightarrow \underbrace{(NH_4)_3PO_4 \cdot 12MoO_3}_{\text{Yellow ppt.}} + 21NH_4NO_3$

70 (*a*) $FePO_4$ (yellow precipitate) is precipitated in ($CH_3COOH + CH_3COONa$) buffer.

71 (*b*) $PCl_3 + 3H_2O \rightarrow H_3PO_3 + 3HCl$

73 (*d*)

The high reactivity of white phosphorus is due to unusual bonding that produces strain in P_4 molecule.

76 (*d*) Phosphorus atom has sp^3-hybridisation hence, it has 75% *p*-character.

78 (*a*) $PI_3 + 3H_2O \longrightarrow \underset{\substack{\text{Dibasic}\\ \text{acid}}}{H_3PO_3} + \underset{\substack{\text{Monobasic}\\ \text{acid}}}{3HI}$

81 (*d*) Preparation of H_3PO_3 is from P_2O_3 and all other acids i.e. HPO_3, $H_4P_2O_7$, H_3PO_4 formed from P_2O_5. In these preparation neither oxidation nor reduction occur (i.e. these are not a redox reactions). Thus, we can check the products by finding out the oxidation states. Here, we used elimination method to pick the odd one.

$$P_2O_3 + 3H_2O \longrightarrow 2H_3PO_3$$
$$P_2O_5 + H_2O \longrightarrow HPO_3$$

Species	Oxidation state of P
(a) HPO_3	+5
(b) $H_4P_2O_7$	+5
(c) H_3PO_4	+5
(d) H_3PO_3	+3

Oxidation state of P in H_3PO_3 is +3.

(B) Miscellaneous questions

2 (*c*) Cyclic metaphosphoric acid is $(HPO_3)_3$.

3 (*a*)

	Oxidation state
N_2O	+1
NO	+2
N_2O_3	+3
NO_2	+4
N_2O_5	+5

4 (*c*) Due to lack of *d*-orbital in N-atom, NCl_5 is not formed.

6 (*a*) By MO theory there is one unpaired electron, thus it is paramagnetic. But it is in gaseous state.

NO ($15e^-$)

$(\sigma 1s)^2 (\overset{*}{\sigma} 1s)^2 (\sigma 2s)^2 (\overset{*}{\sigma} 2s)^2 (\sigma 2p_x)^2 (\pi 2p_y)^2 (\pi 2p_z)^2 (\overset{*}{\pi} 2p_y)^1$.

In solid state, it forms dimer. Hence, unpaired electron gets paired and paramagnetism is lost.

$$\underset{\text{(Gas)}}{2NO} \rightleftharpoons \underset{\text{(Solid)}}{\begin{matrix} N—O \\ \vdots \quad \vdots \\ O—N \end{matrix}}$$

7 (*c*) $PCl_5 \quad AsCl_5 \quad SbCl_5 \quad BiCl_5$

$P^{5+} < As^{5+} < Sb^{5+} < Bi^{5+}$

Size $\xrightarrow{\text{Maximum in this order}}$

Charge character $\xrightarrow{\text{Same}}$

Polarisation $\xrightarrow{\text{Maximum}}$

$PCl_5 > AsCl_5 > SbCl_5 > BiCl_5$

Covalent $\longleftarrow$

Stability $\xleftarrow{\text{Maximum}}$

$PCl_5 > AsCl_5 > SbCl_5 > BiCl_5$

Size of anion $F^- < Cl^- < Br^-$

Polarisation Maximum

Covalent character $\longrightarrow$

$PF_5 < PCl_5 < PBr_5$

Stability $\longrightarrow$

$PF_5 < PCl_5 < PBr_5$

Observation is based, on Fajan's rule.

8 (*b*) C, N and O lack *d*-orbitals hence there is no $d\pi$-$p\pi$ bonding in compounds of N and O.

Thus, $d\pi$-$p\pi$ is present in P_2O_3, P_2O_5 and PO_4^{3-}.

9 (*a*)

	Electrons	Unpaired electron	
NO	15	1	paramagnetic
NO_2	23	1	paramagnetic
O_2	16	2	paramagnetic

10 (*c*) [Structures of NH_3 and NF_3 with lone pairs and bond dipoles]

$\mu \neq 0$ $\quad\quad$ $\mu = 0.24$ D

11 (*c*) Let oxidation state of P in $[NaHPO_2]^-$ is *x*.

$[NaHPO_2]^-$

$1 + 1 + x - 4 = -1$

$x - 2 = -1 \Rightarrow x = +1$

13 (*c*) Dipole moment of BF_3 is zero while that of NH_3 is maximum.

$$NH_3 > NF_3 > BF_3$$

14 (*c*) Electronegativity increases along a period left to right. Thus,

$$C < N$$
$$Si < P$$

Electronegativity decreases down the group.

Then $\quad C > Si$

$N > P$

Thus, $Si < P < C < N$

15 (*c*) NO_2 is paramagnetic due to one unpaired electron. When NO_2 dimerises to N_2O_4, paramagnetism is lost due to sharing of unpaired electron between two N-atoms. Thus, N_2O_4 is diamagnetic.

$$\underset{\text{Paramagnetic}}{2NO_2} \rightleftharpoons \underset{\text{Diamagnetic}}{N_2O_4}$$

16 (*d*)

(a) Acidic nature $H_3PO_2 < H_3PO_3 < H_3PO_4$. Thus, incorrect.

(b) H_3PO_2, H_3PO_3 are reducing agent but H_3PO_4 is not. Thus, incorrect.

(c) Only H_3PO_4 is tribasic. Thus, incorrect.

(d) Geometry is tetrahedral in all cases. Thus, true.

17 (*b*) The spontaneous combustion of phosphine is technically used is Holme's signals containers containing Ca_3P_2 (calcium phosphide) and CaC_2 (calcium carbide) are pierced and thrown in the sea when the gases evolved burn and serve as a signal.

$$Ca_3P_2 \xrightarrow{H_2O} PH_3$$

$$CaC_2 \xrightarrow{H_2O} C_2H_2$$

$$PH_3 + C_2H_2 \longrightarrow \text{Fire (signal)}$$

19 (*b*) $\underset{\text{Solid}}{2PCl_5} \rightleftharpoons \underset{\text{Tetrahedral}}{PCl_4^{\oplus}} + \underset{\text{Octahedral}}{PCl_6^{\ominus}}$

20 (*d*) $AsH_3 \approx 90°$ due to layer size of As

NH_3 — 107°

H_2O — 104.5°

BCl_3 — 120°

21 (*a*) $SO_2(OH)_2 + PCl_5 \longrightarrow SO_2Cl_2 + 2HCl + POCl_3$

Formation of SO_2Cl_2 indicates that there are two —OH groups in H_2SO_4.

$$H—\ddot{O}—\overset{\overset{\displaystyle :\ddot{O}}{\|}}{\underset{\underset{\displaystyle O}{\|}}{S}}—\ddot{O}—H$$

22 (c) Structure of H_3PO_4 is

$$\begin{array}{c} O \\ \| \\ HO-P-OH \\ | \\ OH \end{array}$$

H_3PO_4 has 3 — OH groups i.e. has three ionisable H-atoms and hence forms three series of salts. These three possible series of salts for H_3PO_4 are as follows.

NaH_2PO_4, Na_2HPO_4 and Na_3PO_4

23 (a) True statement is that single N—N bond is weaker than the single P—P bond.

This is why phosphorus show allotropy but nitrogen does not.

24 (c) Strong reducing behaviour of H_3PO_2 is due to presence of two P—H bonds and one P— OH bond

$$\begin{array}{c} O \\ \| \\ H-P-H \\ | \\ OH \end{array}$$

Hypophosphorus acid (Monobasic)

H shows reducing behaviour.

26 (c) Reaction is,

$$5NO_2^- + 2MnO_4^- + 6H^+ \longrightarrow 6NO_3^- + 2Mn^{2+} + 3H_2O$$

But alternatively we can also check oxidation state of all the option and find out the oxidised product.

Species	Oxidation number of N
NO_2^-	+3
(a) N_2	0
(b) NH_3	−3
(c) NO_3^-	+5
(d) N_2O	+1

Thus, oxidised part is NO_3^-.

27 (b) $Pb(NO_3)_2 + H_2SO_4 \longrightarrow \underset{\text{White ppt.}}{PbSO_4} + 2HNO_3$

Soluble in $CH_3COO\overset{+}{N}H_4$

$$PbSO_4 + 2CH_3COONH_4 \longrightarrow (NH_4)_2SO_4 + (CH_3COO)_2Pb$$

31 (d) H_3PO_2 is monobasic acid,

$$\underset{1\text{mole}}{H_3PO_2} + \underset{1\text{mole}}{NaOH} \longrightarrow NaH_2PO_2 + H_2O$$

$\therefore \quad x = 1$

H_3PO_3 is dibasic acid,

$$\underset{1\text{ mole}}{H_3PO_3} + \underset{1\text{ mole}}{Ca(OH)_2} \longrightarrow CaHPO_3 + 2H_2O$$

$y = 1$

H_3PO_4 is tribasic acid,

$$\underset{1\text{mole}}{H_3PO_4} + \underset{1\text{mole}}{Al(OH)_3} \longrightarrow AlPO_4 + 3H_2O$$

$z = 1$

Thus, $x : y : z :: 1 : 1 : 1$

32 (c) Acid salts are those which have ionisable hydrogen and thus react with alkali.

	Acid salt	Normal salt
H_3PO_2 (monobasic acid)	×	NaH_2PO_2
H_3PO_3 (dibasic acid)	NaH_2PO_3	Na_2HPO_3
H_3PO_4 (tribasic acid)	NaH_2PO_4	Na_3PO_4
	Na_2HPO_4	
H_2CO_3 (dibasic acid)	$NaHCO_3$	Na_2CO_3

33 (a) (a) $NH_4NO_3 \xrightarrow{\Delta} N_2O + 2H_2O$

$NH_4NO_2 \xrightarrow{\Delta} N_2 + 2H_2O$

(b) $NH_4Cl \xrightarrow{\Delta} NH_3 + HCl$

$(NH_4)_2CO_3 \xrightarrow{\Delta} 2NH_3 + CO_2 + H_2O$

(c) $NH_4NO_3 + NaOH \xrightarrow{\Delta} NH_3 + NaNO_3 + H_2O$

(d) $\left.\begin{array}{l} AlN \\ Mg_3N_2 \\ CaCN_2 \end{array}\right] \xrightarrow{H_2O} NH_3$

(C) Medical entrances' gallery

1 (d)

(a) $NO_2 \Rightarrow$ $O \leftarrow \overset{sp^2}{N} = O$ → Bent structure

(b) HOCl ⇒ $H-\overset{sp^3}{O}-Cl$ → Bent structure

(c) $O_3 \Rightarrow$ $O \leftarrow \overset{sp^2}{O} = O$ → Bent structure

(d) $N_2O \Rightarrow :N \equiv \overset{sp}{N} \rightarrow O$ → Linear structure

Hence, option (d) is correct.

2 (c) In gaseous and liquid phases PCl_5 has a trigonal bipyramidal structure with sp^3d-hybridisation.

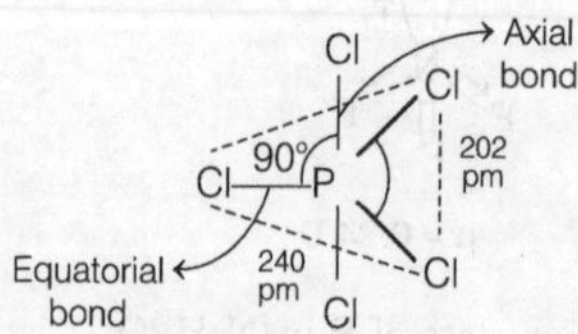

Due to presence of longer and weaker axial bonds PCl_5 is a reactive molecule. Hence, statement in option (c) is incorrect while the remaining options contain correct statements.

3 (*a*) The road map of given problem is

$$\text{'}X\text{'} + H_2O \xrightarrow[\text{Rotten fish smell}]{} \text{'}Y\text{' (colourless)}$$

$$\downarrow CuSO_4$$

$$Cu_3P_2 + \text{Product}$$

As compound 'X' produces a colourless gas 'Y' with rotten fish smell X can be Ca_3P_2 (calcium phosphide). The reaction takes place as follows

$$Ca_3P_2 + H_2O \longrightarrow Ca(OH)_2 + PH_3$$

Colourless gas with rotten fish smell 'Y'

$$CuSO_4 + \underset{\text{'}Y\text{'}}{PH_3} \longrightarrow Cu_3P_2 + H_2SO_4$$

4 (*c*) P—H bonds found in phosphorus acids have reducing properties. Thus, reducing property is directly proportional to number of P—H bonds. The structures of given oxoacids of phosphorus are as follows:

($H_4P_2O_7$) (H_3PO_3)

(H_3PO_2) (H_3PO_4)

H_3PO_2 contains 2P–H bonds which is maximum among the given options. Thus, H_3PO_2 has strongest reducing property.

5 (*c*) The structure and oxidation state of oxoacids of phosphorus are as follows

Orthophosphoric acid (H_3PO_4) (+5)

Pyrophosphoric acid (+5)

Orthophosphorous acid (H_3PO_3) (+3)

Metaphosphoric acid $(HPO_3)_3$ (+5)

Thus, oxidation state of orthophosphorous acids is +3.

6 (*c*) Let the oxidation state of nitrogen in each of the given N-compounds be x.

(i) $HNO_3 : +1 + x + 3(-2) = 0$

$$x = +5$$

$\therefore$ Oxidation state of N in HNO_3 is +5.

(ii) $NO : x + 1(-2) = 0$

$$x = +2$$

$\therefore$ Oxidation state of N in NO is +2.

(iii) $NH_4Cl : x + 4(+1) + 1(-1) = 0$

$$x = -3$$

$\therefore$ Oxidation state of N in NH_4Cl is -3.

(iv) $N_2 : x = 0$

[$\because N_2$ is present in elemental state]

$\therefore$ Oxidation state of N in N_2 is 0.

Thus, the correct decreasing order of oxidation states of given N compounds will be $H\overset{+5}{N}O_3 > \overset{+2}{N}O > \overset{0}{N_2} > \overset{-3}{N}H_4Cl$

7 (*c*) The correct decreasing order of basic strength is

$$NH_3 > PH_3 > AsH_3 > SbH_3$$

NH_3 is the most basic because of its small size, the electron density of electron pair is concentrated over small region. As the size increases, the electron density gets diffused over a large surface area.

Hence, the ability to donate the electron pair (basicity) decreases. The other three statements given are correct.

8 (*b*) The products of the reaction of copper with HNO_3 depend upon the concentration of HNO_3 used.

Copper metal reacts with dilute HNO_3 to form nitrogen (II) oxide (NO).

$$3Cu + 8HNO_3(\text{dil.}) \longrightarrow 3Cu(NO_3)_2 + 2NO + 4H_2O$$

Copper metal reacts with conc. HNO_3 to form nitrogen (IV) oxide or nitrogen dioxide (NO_2).

$$Cu + 4HNO_3 \text{ (conc.)} \longrightarrow Cu(NO_3)_2 + 2NO_2 + 2H_2O$$

9 (*d*) Hypophosphoric acid ($H_4P_2O_6$) does not contain P — O — P bond whereas, isohypophosphoric acid ($H_4P_2O_6$), diphosphorus acid ($H_4P_2O_5$) and diphosphoric acid ($H_4P_2O_7$) all contain P — O — P bonds.

The structures of the following oxyacids of phosphorus is shown below

(a) Isohypophosphoric acid (b) Pyrophosphorus acid

(c) Diphosphoric acid (d) Hypophosphoric acid

10 (*a*) Methyl isocyanate (CH_3NCO) is industrially prepared by the reaction of methylamine and phosgene (carbonyl chloride).

$$\underset{\text{Methyl amine}}{CH_3NH_2} + \underset{\text{Phosgene}}{COCl_2} \xrightarrow{-HCl} [CH_3NH — CO — Cl] \xrightarrow{\Delta, -HCl} \underset{\text{Methyl isocyanate}}{CH_3N{=}C{=}O}$$

11 (*b*) The basic strength of given hydrides decreases in the order as follows

$$NH_3 > PH_3 > AsH_3 > SbH_3 > BiH_3$$

This is due to the decrease in the availability of the lone pair of electrons which depends on the size of central atom that increases from N to Bi. Therefore, NH_3 and BiH_3 acts as a strong and weak base hydride respectively.

12 (*a*) Cl–B(–Cl)–Cl, 120°

Trigonal planar

13 (*d*) When calcium carbide (CaC_2) reacts with nitrogen (N_2) under high temperature it forms calcium cyanamide, which is also called nitrolim.

$$CaC_2 + N_2 \xrightarrow[\text{Temperature}]{\text{High}} \underset{\text{Calcium cyanamide}}{CaCN_2} + C$$

14 (*c*)
- PH_5 does not exist due to very less electronegativity difference between P and H. Hydrogen is slightly more electronegative than phosphorus, thus could not hold significantly the sharing electrons. On the other hand, $BiCl_5$ does not exist due to inert pair effect.
- On moving down the group, +5 oxidation state becomes less stable while +3 oxidation state becomes more stable.

In SO_2, $p\pi - d\pi$ and $p\pi - p\pi$ both types of bonds are present.
- SeF_4 has sp^3d hybridisation whereas CH_4 has sp^3-hybridisation. This, shapes are different.

15 (*a*) Phosphinic acid

H–P(=O)(H)–OH

Phosphonic acid

HO–P(=O)(H)–HO

Due to the presence of one replaceable proton in phosphinic acid, it is monoprotic acid and due to presence of two replaceable protons in phosphonic acid, it is diprotic acid.

16 (*b*)
$$\underset{\text{(White)}}{P_4} + 3NaOH + 3H_2O \xrightarrow{\text{Inert atmosphere}} \underset{\text{Phosphine}}{PH_3} + \underset{\text{Sodium hypophosphite}}{3NaH_2PO_2}$$

Phosphine is a colourless gas with rotten fish smell and is highly poisonous. Its solution in water decomposes in the presence of light.

Phophine is weakly basic. So, statement (II) is incorrect. It is less basic than NH_3.

17 (*c*) Dinitrogen pentoxide contains a N—O—N bond.

O=N(→O)–O–N(=O)→O ⟷ O←N(=O)–O–N(→O)=O (resonance structures)

18 (*d*) Applying Le-Chatelier's principle on the reaction of Haber's process, we get that high pressure would ensure higher yield but to generate and maintain high pressure it will require high cost to be invested on infrastructure and for running the process. The economic consideration for pressure is 200 atm.

19 (*a*) Chemical reaction occurs when P_4O_{10}, Mg_3N_2, CaC_2 and BaO_2 react with water as shown below

$$P_4O_{10} + 6H_2O \longrightarrow 4H_3PO_4$$
$$Mg_3N_2 + 6H_2O \longrightarrow 3Mg(OH)_2 + 2NH_3$$
$$CaC_2 + 2H_2O \longrightarrow Ca(OH)_2 + C_2H_2$$
$$BaO_2 + 2H_2O \longrightarrow Ba(OH)_2 + H_2O_2$$

20 (*a*)

Hypophosphorus acid (H_3PO_2)	Orthophosphorus acid (H_3PO_3)	Pyrophosphoric acid ($H_4P_2O_7$)
H–P(=O)(H)–OH	H–P(=O)(OH)–OH	HO–P(=O)(OH)–O–P(=O)(OH)–OH
$+3 + x + 2(-2) = 0$, $x = +1$	$+3 + x + 3(-2) = 0$, $x = +3$	$+4 + 2x + 7(-2) = 0$, $x = +5$

Thus, correct order of acids of phosphorus in the increasing order of oxidation state of phosphorus is

hypophosphorus acid < orthophosphorus acid < pyrophosphoric acid

21 (*b*) The mixture of one part of conc.HNO_3 and three parts of conc.HCl is known as *aqua-regia*. It contains NOCl.

$$HNO_3 + 3HCl \longrightarrow 2H_2O + NOCl + 2[Cl]$$

The enhanced ability to dissolve metals shown by aqua regia arises from the oxidising power of HNO_3 coupled with the ability of Cl^- to form complexes with the metal ions.

22 (*c*) Draw the structure of pyrophosphoric acid and count —OH groups in it.

HO–P(=O)(OH)–O–P(=O)(OH)–OH

Pyrophosphoric acid (Tetrabasic)

∴ Pyrophosphoric acid has 4—OH groups.

23 (*a*) CAN or calcium ammonium nitrate is a neutral fertiliser. It gives non-acid hydrolysis products and it decreases acidity in the soil.

24 (*b*) Hypophosphorus acid H_3PO_2, has the following structure

$$H-\overset{\overset{\displaystyle O}{\|}}{\underset{\underset{\displaystyle H}{|}}{P}}-OH$$

As it contains only one replaceable H-atom (that is attached with O, not with P directly) so, it is a monoprotic acid.

All other given statements are true.

25 (*b*) Hydrazoic acid (N_3H) is a weak acid. Its weak acidic character is due to the formation of azide ion (N_3^-) after the loss of a proton. Azide ion is more stabilised by resonance as compared to hydrazoic acid itself.

$$[N{\equiv}N^+{-}N^-]^- \longleftrightarrow [N^-{-}\overset{+}{N}{\equiv}N]^-$$

26 (*b*) Phosphorus is + 5 oxidation state. (trimer of HPO_3 formula is $(HPO_3)_3$

(Cyclic structure: three P atoms joined by bridging O atoms in a six-membered P–O ring; each P carries =O and –OH.)

27 (*a*) Pyrophosphorus acid is $H_4P_2O_5$

$$HO-\overset{\overset{\displaystyle H}{|}}{\underset{\underset{\displaystyle O}{\|}}{P}}-O-\overset{\overset{\displaystyle H}{|}}{\underset{\underset{\displaystyle O}{\|}}{P}}-OH$$

Basicity is equal to the number of —OH bonds present in a unit/compound.

CHAPTER 08

p-Block Elements 2 (Oxygen Family)

Group 16 of the periodic table contains five elements namely, oxygen (O), sulphur (S), selenium (Se), tellurium (Te) and polonium (Po). The first four elements are non-metals while the last element, polonium is a **radioactive metal.** These non-metallic elements are commonly referred to as **chalcogens** (meaning ore forming) because many metal ores occur as oxides, sulphides etc. They are also known as members of **oxygen family**.

Occurrence

Oxygen is the most abundant element in the earth's crust, representing slightly less than 50% of it by weight. It is present as elemental oxygen in the atmosphere and is combined with other elements in water and in many minerals. Sulphur is the seventeenth in abundance and in the earth's crust occurs as sulphide (HgS, PbS etc.) in natural gas and oil as disulphide ion (FeS_2 –**"fools gold"**) as sulphate ($CaSO_4$).

Selenium and tellurium are comparatively rare on earth's crust occur in sulphide ores in the form of selenides and telluride.

Inside

GENERAL AND PHYSICAL PROPERTIES

The general and physical properties of group 16 elements are as follows

Electronic configuration

The general electronic configuration of all the elements of this group is ns^2np^4. Thus, they all are p-block elements because last electron enters in p-subshell as shown in Table 8.1.

The electronic configuration of members of this family are given as

Physical state

Oxygen occurs as diatomic molecule O_2 and is therefore, a gas while sulphur occurs as polyatomic molecule S_8 and therefore, is a solid. All other elements also exist as solids.

Atomic radii

Atomic radii increase gradually on moving down the group, i.e. from O to Po. This occurs due to an increase in the number of electronic shells.

Group 16 elements have shorter atomic radii than those of corresponding elements of nitrogen family but larger than those of halogens due to the increase in effective nuclear charge from left to right in a period.

Ionisation energy

I and II ionisation energy of group 16 elements decrease down the group due to the combined effect of the increased atomic size and shielding effect.

The ionisation energy of the first element (oxygen) is very high, i.e. 1313.7 kJ per mole. The values for the succeeding elements decreases systematically from 999.4 to 813.5 kJ per mole on moving down the group. These high values indicate reluctance of these elements to form cations.

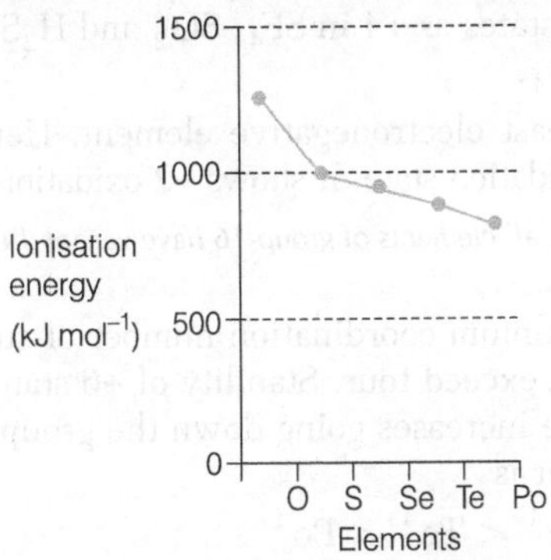

Metallic and non-metallic character

Down the group metallic character increases due to decrease in ionisation enthalpy.

The first four elements are non-metals but the non-metallic character is stronger in oxygen and sulphur. It is weaker in selenium and tellurium. The last element, polonium is distinctly a metal however, it is radioactive and is only short-lived.

Electronegativity

Electronegativity decreases from O to Po as the atomic size increases. Oxygen is the second most electronegative element after fluorine.

Due to high electronegativity, the elements of oxygen family (except Po) form ionic compounds with metals and covalent compounds with non-metals.

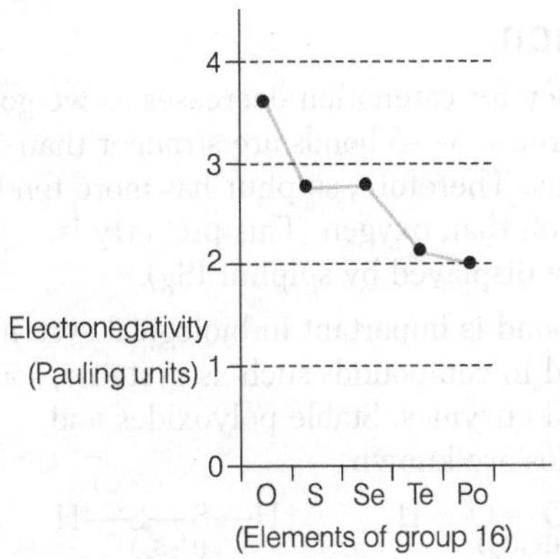

Electron affinity

Electron affinity of group-16 elements decreases as the atomic size increases.

However, electron affinity of O is lesser as compared to S. This is because of the small size and high charge density of O atom.

Nature of bonding

Compounds of oxygen with non-metals are predominantly covalent. S, Se and Te because of low electronegativities show more covalent character.

Melting and boiling point

Melting and boiling points show a regular increase from oxygen to polonium. There is a large difference in melting and boiling points of oxygen and sulphur that can be explained on the basis of their structure.

Oxygen exists mostly as a diatomic molecule held together by weak van der Waals' forces while sulphur exist as polyatomic molecules, e.g. S_8, Se_8, etc. where the atoms are bonded by covalent bonds associated with high dissociation energy.

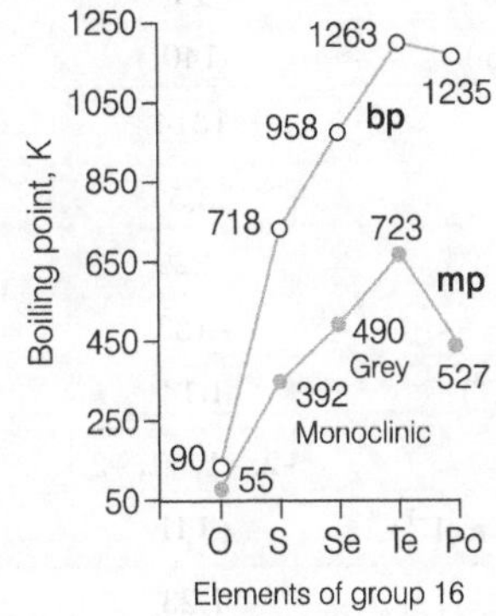

Catenation

The tendency for catenation decreases as we go down the group. S—S bonds are stronger than O—O bonds. Therefore, sulphur has more tendency for catenation than oxygen. This property is prominently displayed by sulphur (S_8).

The S—S bond is important in biological systems and is found in compounds such as cysteine, some proteins and enzymes. Stable polyoxides and polysulphides are known.

e.g. H—O—O—H, (H_2O_2)   H—S—S—H (H_2S_2)

H—S—S—S—H, (H_2S_3)   H—S—S—S—S—H (H_2S_4)

Note *H_2S_2 and H_2S_3 are known as yellow oils of sulphur.*

Allotropy

Like the elements of groups 14 and 15, all the elements of group 16 show allotropy. Thus, oxygen exists in two allotropic forms O_2 and O_3 (ozone). Sulphur occurs in a number of allotropic forms such as rhombic (α) sulphur, monoclinic (β) sulphur, plastic (γ) sulphur, colloidal (δ) sulphur. All these allotropes are non- metallic in character. Selenium exists in two forms, i.e. red form which is non-metallic and grey form which is metallic in nature.

Tellurium also exists in two forms, i.e. non-metallic and metallic (more stable) form. While allotropes of polonium are α and β forms, which are metallic in nature.

Oxidation state

With valence electronic configuration ns^2np^4, the common oxidation states are +4 and +6. The group-16 elements can attain a noble gas configuration either by gaining or by sharing of two electrons.

Since, oxygen has a very high electronegativity, it tends to complete its valence shell preferably by gaining electrons. Even if the octet is completed by sharing of electrons, these electrons for the purpose of determining the oxidation state are counted towards oxygen (excepting F_2O and H_2O_2).

In that case, the oxidation state exhibited by oxygen is −2. Hence, act as powerful oxidising agent. Also, oxygen shows an oxidation state of −1 in H_2O_2, 0 in O_2 and O_3, +1 in O_2F_2 and +2 in OF_2.

The electronegativities of other elements of this group are low. There is much less probability of the formation of di-negative ions in the case of S, Se and Te.

S, Se and Te are much less electronegative than oxygen and due to availability of vacant *d*-orbitals they are commonly found in positive oxidation states as +4 in SF_4, SO_2 and H_2SO_3 and + 6 in SF_6, SO_3, H_2SO_4.

Polonium is the least electronegative element. Hence, does not exhibit negative oxidation state. It shows +2 oxidation state.

Note *Except oxygen, all elements of group-16 have vacant d-orbitals in their valence shells.*

Therefore, the maximum coordination number of sulphur and other elements can exceed four. Stability of +6 state decreases and that of +4 state increases going down the group due to inert pair effect. Its order is

$$Se^{4+} < Te^{4+} < Po^{4+}$$
$$Se^{6+} > Te^{6+} > Po^{6+}$$

Table 8.1 Physical properties of the elements in group 16 (VIA)

Property	O_8 [Oxygen]	S_{16} [Sulphur]	Se_{34} [Selenium]	Te_{52} [Tellurium]	Po_{84} [Polonium]
Electronic configuration	[He] $2s^22p^4$	[Ne] $3s^23p^4$	[Ar] $3d^{10}4s^24p^4$	[Kr] $4d^{10}5s^25p^4$	[Xe]$4f^{14}5d^{10}6s^26p^4$
Covalent radius (pm)	74	104	114	137	168
Ionic radius (M^{2-}, pm)	140	184	198	221	—
$(IE)_1$ /kJ mol^{-1}	1314	999	941	869	813
Electronegativity	3.5	2.5	2.4	2.1	2.0
Melting point (°C)	−229	114*	221^α	452	254
Boiling point (°C)	−183	445	685	1087	962
Density ($10^3 \times$ kg m^{-3})	1.14	2.07	4.79	6.25	9.4
Oxidation states	−2, −1, +1, +2	−2, +2, +4, +6	−2, +2, +4, +6,	−2, +2, +4, +6	+2, +4
Electron affinity (kJ mol^{-1})	−141	−200	−195	−190	−183
$E°$/ V for $X + 2H^+ + 2e^- \rightarrow H_2X$	1.23	0.14	− 0.40	− 0.79	—

* Rhombic sulphur
α Grey Se

CHEMICAL PROPERTIES

The chemical properties shown by group 16 elements are as follows

Reaction with hydrogen

Group 16 elements react with H_2 to form binary hydrides of general formula H_2M (where $M = O, S, Se, Te, Po$).

$$M + H_2 \longrightarrow H_2M$$

Hydrides of H_2M_2 type are also known. Hydride of oxygen is readily available in nature, while hydrides of S, Se and Te are obtained by the action of acids on metal sulphides, selenides and tellurides.

$$FeS + H_2SO_4 \longrightarrow FeSO_4 + H_2S$$
$$K_2Se + H_2SO_4 \longrightarrow K_2SO_4 + H_2Se$$

Characteristics of these hydrides are discussed below

(*i*) **Physical state** H_2O is liquid at room temperature but all other hydrides are colourless and foul smelling toxic gases.

(*ii*) **Volatility** It increases from H_2O to H_2S and then decreases from H_2S to H_2Te. This can be shown by their boiling points given below

H_2O	H_2S	H_2Se	H_2Te	H_2Po
100°C	-60.75°C	-41.5°C	-1.8°C	–

High boiling point of water is due to the association of molecules through intermolecular hydrogen bonding.

(*iii*) **Acidic nature** They all are weak diprotic acids and dissociate to varying degrees to give H^+ ions.

$$H_2M \rightleftharpoons HM^- + H^+$$
$$HM^- \rightleftharpoons M^{2-} + H^+$$

[Where, M = O, S, Se, Te, Po]

The acidic strength increases from H_2O to H_2Se. This can be explained on the basis of size of the central atom. The atomic size increases gradually as we move from O to Te. As a result, the length of the X—H bond also increases. Since, the strength of a bond is inversely proportional to its length, the strength of the X—H bond decreases as we move from O to Te. Consequently, the tendency of the hydrides to dissociate to yield H^+ ions increases from H_2O to H_2Te.

Order of acidic strength $H_2O < H_2S < H_2Se < H_2Te$

(*iv*) **Thermal stability** Hydrides with high molar mass (H_2Te) is less thermally stable than hydride with low molar mass (H_2O). This is due to the increase in size of central atom. As a result, the M—H bond becomes weaker and longer. Consequently, thermal stability decreases.

Order of thermal stability of hydride of this group is as follow

$$H_2O > H_2S > H_2Se > H_2Te$$

(*v*) **Reducing character** All the hydrides, except water are reducing agents. The reducing character increases from H_2S to H_2Te. This is due to decrease in stability of hydride.

Reaction with oxygen

They react with oxygen to form oxides of types MO_2 and MO_3 and in some cases MO.

$$M + O_2 \longrightarrow MO_2$$
$$2M + 3O_2 \longrightarrow 2MO_3$$
$$M + \frac{1}{2}O_2 \longrightarrow MO$$

Table 8.2 Different oxides of 16 group elements

S.No.	Type of oxide	O	S	Se	Te	Po
1.	Monoxide	–	SO	–	TeO	PoO
2.	Dioxide	O_2	SO_2	SeO_2	TeO_2	PoO_2
3.	Trioxide	O_3	SO_3	SeO_3	TeO_3	PoO_3
4.	Heptoxide	–	S_2O_7	–	–	–

Some important characteristics of oxides of group 16th elements are given below

(*i*) O_3 and SO_2 are gases but SeO_2 is solid. On moving from SO_2 to TeO_2, reducing property decreases. As a result, TeO_2 acts as an oxidising agent. MO_3 type oxide are formed by sulphur, selenium and tellurium but tendency to form MO_3 type of oxides decreases down the group due to inert pair effect.

(*ii*) These oxides are acidic in nature, i.e. form acids with water and this nature decreases down the group due to decrease in electronegativity and increase of metallic character.

Hence, order of acidic nature of oxides is

$$SO_3 > SeO_3 > TeO_3$$

Reaction with halogen

They react with halogens to form various types of halides such as MX_6, MX_4 and MX_2. Oxygen forms halides only with fluorine (OF_2 or F_2O) and its compounds with rest of the halogens are called **oxides** (not halides) as its electronegativity is more than any other halogen. Other elements form halides of different oxidation number, i.e. + 1, + 2, + 4 and + 6.

$$M + 3F_2 \longrightarrow MF_6, \quad 2M + Cl_2 \longrightarrow M_2Cl_2$$
$$M_2Cl_2 + Cl_2 \longrightarrow 2MCl_2, \quad M + 2Cl_2 \longrightarrow MCl_4$$

Table 8.3 Different oxidation numbers of other elements and their respectively halides

Halides	X	Oxidation numbers
S_2X_2	F, Cl, Br, I	+1
SX_2	F, Cl	+2
SX_4	F	+4
SX_6	F	+6
Se_2X_2	Cl, Br	+1
SeX_2, TeX_2	Cl, Br	+2
SeX_4, TeX_4	F, Cl, Br	+4
TeX_4	I	+4
SeX_6, TeX_6	F	+6

Some important characteristic of 16th group halides are as follows

(*i*) For same elements the stability of the halides decreases in the order F > Cl > Br > I.

(*ii*) SF_6 is thermodynamically inert, non-toxic gas. Inertness of SF_6 is due to the presence of sterically protected sulphur atom. SF_6 is used as a gaseous insulator in high voltage generators. SF_4 and SeF_4 are good fluorinating agents by which —COOH, >C=O and —P=O groups can be converted into —CF_3, —CF_2 and —PF_2 groups respectively.

Formation of oxoacids

Group 16 elements such as sulphur, selenium and tellurium form a variety of oxoacids which are tabulated below.

Table 8.4 Oxoacids of sulphur, selenium and tellurium

Sulphur	Selenium	Tellurium
H_2SO_3 Sulphurous acid	H_2SeO_3 Selenous acid	H_2TeO_3 Tellurous acid
H_2SO_4 Sulphuric acid	H_2SeO_4 Selenic acid	H_6TeO_6 Telluric acid
H_2SO_5 Peroxomonosulphuric acid	—	—
$H_2S_2O_8$ Peroxodisulphuric acid	—	—
$H_2S_2O_3$ Thiosulphuric acid	—	—
$H_2S_2O_5$ Dithionic acid	—	—
$H_2S_2O_7$ Pyrosulphuric acid	—	—

Description of oxoacides of sulphur will be discussed later in this chapter.

CHECK POINT 8.1

1. Which of the following element possesses highest ionisation energy?
(a) O (b) S
(c) Te (d) Po

2. Among the following, which is not an allotropic form of sulphur?
(a) α-sulphur (b) β-sulphur
(c) μ-sulphur (d) ν-sulphur

3. Oxidation state of oxygen in O_2F_2 is
(a) +1 (b) +2
(c) +4 (d) −2

4. The correct order of boiling point of hydrides of group 16 is
(a) $H_2O > H_2Te > H_2S > H_2Se$
(b) $H_2O > H_2S > H_2Se > H_2Te$
(c) $H_2O > H_2Te > H_2Se > H_2S$
(d) $H_2O > H_2Se > H_2S > H_2O$

5. Which of the following hydrides is least volatile?
(a) H_2S (b) H_2Te
(c) H_2O (d) H_2Se

6. H_2S is more acidic than H_2O because
(a) oxygen is more electronegative than sulphur
(b) atomic number of sulphur is higher than oxygen
(c) H—S bond is weaker as compared to H—O bond
(d) H—O bond is weaker as compared to H—S bond

7. Which of the following hydride does not act as reducing agent?
(a) H_2S (b) H_2Se
(c) H_2O (d) H_2Te

8. Which of the following statements is incorrect?
(a) Oxygen has very high electronegativity
(b) Oxygen has vacant *d*-orbitals
(c) Oxygen exists in two allotropic forms such as O_2 and O_3
(d) Selenium exists in three allotropic forms

9. Among the following elements of group 16 which will form halide with high oxidation state?
(a) Only O (b) Only Se
(c) Only Te (d) Both (b) and (c)

ANOMALOUS BEHAVIOUR OF OXYGEN

Oxygen, the first member of 16th group, differs in some important ways from other group members.
This is due to its

- small size,
- high electronegativity and
- non-availability of *d*-orbitals for bonding.

The main points of difference between oxygen and other members of 16th group are given below

(*i*) Oxygen exists as O_2 (diatomic) gas while all other members are solid.

(*ii*) Oxygen is paramagnetic while others are not (S_2 is also paramagnetic).

(*iii*) Compounds of oxygen are ionic and polar covalent due to its high electronegativity.

(*iv*) Oxygen exhibit – 2 oxidation state while other members show + 6 oxidation states.

(*v*) Oxygen form *pπ-pπ* bonds while sulphur forms *dπ-pπ* bonds.

(*vi*) Only hydride of oxygen is non-poisonous, i.e. H_2O, while other hydrides are poisonous gases.

Dioxygen

Oxygen is the most abundant of all elements of this group. It exists in the free form as dioxygen molecules, O_2 and makes up 20.9% by volume and 23% by weight of the atmosphere. It was discovered by Priestley and Scheele but named by A L Lavoisier. In diatomic oxygen molecule, the two oxygen atoms are joined together through a covalent double bond as shown below

$$:\ddot{O}=\ddot{O}:$$

Methods of preparation

Dioxygen is produced industrially by the fractional distillation of liquid air. In laboratory, it is prepared on a small scale by thermal decomposition of $KClO_3$.

$$2KClO_3 \xrightarrow{150°C,\ MnO_2(\text{catalyst})} 2KCl + 3O_2\uparrow$$

It can also be prepared by the catalytic decomposition of hypochlorites by electrolysis of water with traces of H_2SO_4 or $Ba(OH)_2$ solution by decomposition of oxygen rich compounds ($KMnO_4$, KNO_3, BaO_2 etc.).

The reactions are as follows :

By electrolysis of water with mineral acids,

$$2H_2O(l) \xrightarrow{\text{Electrolysis}} 2H_2(g) + O_2(g)$$

By thermal decomposition of oxides,

$$2HgO \xrightarrow{\Delta} 2Hg + O_2\uparrow$$

$$3MnO_2 \xrightarrow{\Delta} Mn_3O_4 + O_2\uparrow$$

Physical properties

Oxygen is colourless, odourless and tasteless gas. It can be easily liquefied by applying pressure, when cooled below its critical temperature (– 118.7°C). It is slightly soluble in water and is slightly heavier than air.

The O—O bond distance in O_2 is 1.21Å. Liquid oxygen is pale blue in colour and is strongly magnetic. It boils at −183°C and freezes to a bluish white solid at −219°C. It has three stable isotopes :

$$^{16}_{8}O,\ ^{17}_{8}O \text{ and } ^{18}_{8}O$$

It is paramagnetic in nature through it contains even number of electrons.

Chemical properties

(*i*) Oxygen is a supporter of combustion. It does not burn itself but helps in burning.

(*ii*) It is quite stable in nature and its dissociation energy is very high. Therefore, it is not very reactive.

$$O_2 \longrightarrow O + O$$

(*iii*) It brings about oxidation of several organic and inorganic compounds on heating in the presence of a catalyst.

(*iv*) It reacts at high temperature. Once the reaction starts, it proceeds on its own. Hence, exothermic in nature.

(*v*) It is a neutral gas and has no colour change with blue or red litmus.

(*vi*) O_2 readily react with active metals such as Na and Ca to form respective metal (basic) oxides and give base with water.

$$4Na + O_2 \longrightarrow 2Na_2O$$

$$2Ca + O_2 \longrightarrow 2CaO$$

$$CaO + H_2O \longrightarrow Ca(OH)_2$$

$$(\text{K forms superoxide})\ K + O_2 \longrightarrow KO_2$$

Note *Some metallic oxides show dual behaviour of both acid as well as base e.g. Al_2O_3 known as amphoteric oxide.*

(*vii*) O_2 reacts with non-metals to form acidic oxides and give acid with water.

$$2H_2 + O_2 \xrightarrow[\text{Electric discharge}]{1073\text{ K}} 2H_2O$$

$$N_2 + O_2 \xrightarrow{3273\text{ K}} \underset{\text{Nitric oxide}}{2NO}$$

$$S + O_2 \xrightarrow{\text{Heat}} SO_2$$

$$C + O_2 \xrightarrow{\text{Heat}} CO_2$$

$$SO_2 + H_2O \longrightarrow H_2SO_3$$

$$C + O_2 \xrightarrow{\text{Heat}} CO_2$$
$$SO_2 + H_2O \longrightarrow H_2SO_3$$

(*viii*) Dioxygen is an oxidising agent and oxidises many compounds under specific conditions.

$$4HCl + O_2 \xrightarrow{700\text{ K}} 2H_2O + 2Cl_2$$
$$CH_4 + 2O_2 \xrightarrow{CuCl_2} CO_2 + 2H_2O$$
$$CS_2 + 3O_2 \xrightarrow{\text{Heat}} CO_2 + 2SO_2$$

Uses

(*i*) Dioxygen is essential for respiration (for the release of energy in the body) by both animals and plants. It is essential for life.

$$\underset{\text{Glucose}}{C_6H_{12}O_6} + 6O_2 \xrightarrow{\text{Respiration}} 6CO_2 + 6H_2O + \text{Energy}$$

(*ii*) It is used as a fuel in the form of liquid oxygen.

(*iii*) It is used in welding and cutting of metals as oxyhydrogen or oxy acetylene torch.

(*iv*) It is used as an oxidising and bleaching agent.

(*v*) It is used for artificial respiration.

(*vi*) It is used in metallurgical processes to remove the impurities of metals by oxidation.

OZONE (O_3)

It is an unstable gas that absorbs the UV rays. This is particularly important since, there is a layer of O_3 in the upper atmosphere, which absorbs harmful UV radiation from the sun, thus protecting people on the earth.

Structure

The molecule O_3 is sp^3 hybridised and has bent space with an O–O–O bond angle of 116.8° due to resonance both the O—O bonds are equal and lie in between those of O—O single and O=O double bond length. Its canonical forms are

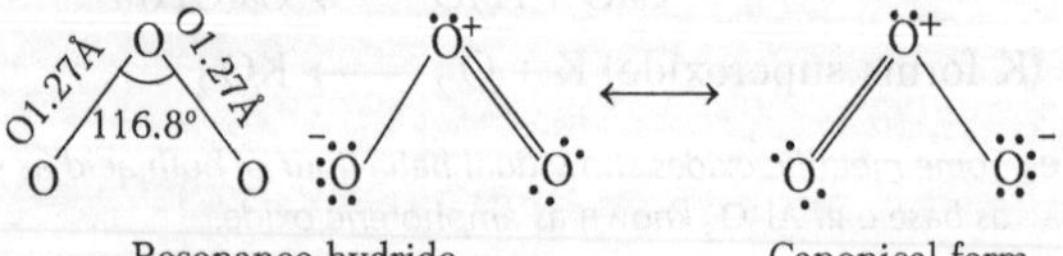

Resonance hydride   Canonical form

Methods of preparation

O_3 is prepared by the action of a silent electric discharge upon dioxygen between two concentric metallised tubes in an apparatus called **an ozoniser.**

$$3O_2 \xrightleftharpoons{\text{Silent electric discharge}} 2O_3$$

It is only a metastable allotrope, always having a tendency to convert back into oxygen.

Some 30 km above the earth's surface oxygen molecules can be split apart by UV light from the sun. Some of the oxygen atoms join with other oxygen molecules to make ozone.

$$O_2(g) + O(g) \longrightarrow O_3(g)$$

Physical properties

O_3 is dark blue **diamagnetic gas** with boiling point – 112°C. The colour is due to intense absorption of red light. It has a characteristic sharp smell.

Chemical properties

(*i*) It is thermodynamically unstable and decomposes to O_2.

$$2O_3 \xrightarrow{\Delta} 3O_2;\ \Delta G = -163\text{ kJ mol}^{-1}$$

(*ii*) It is an extremely powerful oxidising agent.

i.e. $O_3 + 2H^+ + 2e^- \rightleftharpoons O_2 + H_2O$

$$O_3 + H_2O + 2e^- \rightleftharpoons O_2 + 2OH^-$$
$$PbS + 4O_3 \longrightarrow PbSO_4 + 4O_2$$
$$6NO_2 + O_3 \longrightarrow 3N_2O_5$$
$$S + H_2O + O_3 \longrightarrow H_2SO_4$$
$$2KOH + 5O_3 \longrightarrow 2KO_3 + 5O_2 + H_2O$$

(Orange solid have paramagnetic O_3^- ions)

$$O_3 + 2KI + H_2O \longrightarrow I_2 + 2KOH + O_2$$
$$3SnCl_2 + 6HCl + O_3 \longrightarrow 3SnCl_4 + 3H_2O$$
$$2[Fe(CN)_6]^{4-} + H_2O + O_3 \longrightarrow 2[Fe(CN)_6]^{3-} + O_2 + 2OH^-$$

(*iii*) It acts as a bleaching agent as well as a sterilising agent due to the oxidising power. It acts as a bleaching agent for vegetable colouring matter.

Vegetable Colouring matter + $O_3 \longrightarrow$

Oxidise colouring matter to colourless + O_2

(*iv*) It is also added to unsaturated organic compounds at room temperature in presence of inert solution forming **ozonides**, which can be cleaved to give aldehydes and ketones in solution.

$$CH_2{=}CH_2 + O_3 \xrightarrow{\text{Inert solution}} \underset{\text{Ozonides}}{CH_2(O)(O{-}O)CH_2} \xrightarrow{H_2O/Zn} 2HCHO$$

$$CH_3{-}CH{=}CH{-}CH_3 + O_3$$

$$\xrightarrow{\text{Inert solution}} \underset{\text{Ozonides}}{(CH_3)(H)C(O)(O{-}O)C(CH_3)(H)} \xrightarrow{H_2O/Zn} 2CH_3CHO$$

The amount of O_3 in a gas mixture can be determined by passing the gas into a KI solution (at a constant pH 9.2 using borate buffer). The iodine that is liberated is titrated with sodium thiosulphate solution.

$$O_3 + 2I^- + H_2O \longrightarrow I_2 + 2OH^- + O_2$$

$$I_2 + 2S_2O_3^{2-} \longrightarrow 2I^- + S_4O_6^{2-}$$

$$O_3 \equiv I_2 \equiv 2I \equiv 2S_2O_3^{2-}$$

$$\therefore \quad \text{Equivalent weight of } O_3 = \frac{\text{molecular weight}}{2}$$

Preparation and properties of ozone are summarised in the given below

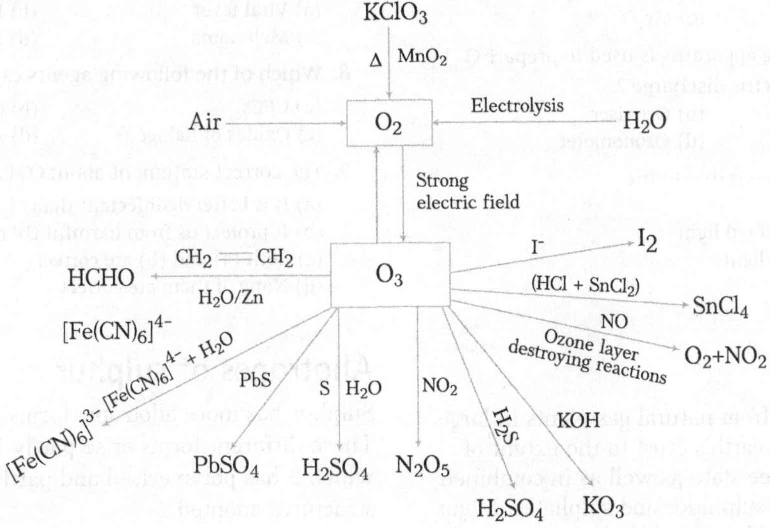

Uses

(*i*) Ozone layer is present in the upper atmosphere, which absorbs harmful UV radiation from the sun, thus protecting people on the earth. Exposure of an excessive amount of UV light to the earth will cause **melanoma** (skin cancer) in humans.

Oxides of nitrogen (from car exhausts) and the halogens can also damage the O_3 layer.

(*ii*) O_3 is also used as a disinfectant to purify drinking water since, it can destroy bacteria and viruses. It is better disinfectant than Cl_2 since, it (O_3) avoids the unpleasant smell and taste of Cl_2 and any excess O_3 soon decomposes to O_2.

(*iii*) It is used in industry for the manufacture of $KMnO_4$, artificial silk and synthetic camphor.

(*iv*) It is used to purify air of crowded areas such as railway stations, cinemas etc.

(*v*) It is used in many organic reactions.

CHECK POINT 8.2

1. Which of the following statements for oxygen is incorrect?
   (a) It is paramagnetic in nature
   (b) It exhibits – 2 oxidation state
   (c) It forms $d\pi$ - $p\pi$ bonding
   (d) It gives non-poisonous hydrides

2. Percentage of O_2 by volume in the atmosphere is
   (a) 18 (b) 19
   (c) 24.15 (d) 20.9

3. The catalyst used in the following reaction is
$$2KClO_3 \longrightarrow 2KCl + 3O_2$$
   (a) Fe (b) ZnS
   (c) MnO_2 (d) Mo_2O_3

4. Which of the following apparatus is used to prepare O_3 by the action of silent electric discharge ?
   (a) Ballastometer (b) Ozoniser
   (c) Richter meter (d) Ozonometer

5. The blue colour of ozone is due to the
   (a) diffraction
   (b) intense absorption of red light
   (c) absorption of visible lights
   (d) spliting of light

6. The structure of ozone is best represented by
   (a) $:\ddot{O}=\overset{+}{\ddot{O}}-\ddot{O}:^{-}$
   (b) 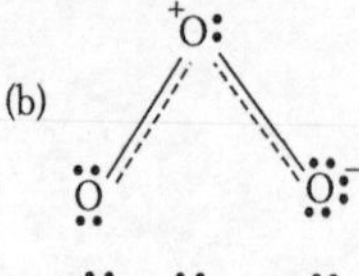
   (c) O O O
   (d) $:\ddot{O}=\ddot{O}\longrightarrow\ddot{O}:$

7. Which of following diseases will take place if O_3 layer of atmosphere get holes?
   (a) Viral fever (b) HIV
   (c) Melanoma (d) Brain cancer

8. Which of the following agents can destroy O_3 layer?
   (a) CFCs (b) Oxides of nitrogen
   (c) Oxides of halogen (d) All of these

9. The correct statement about O_3 is/are
   (a) It is better disinfectant than Cl_2
   (b) It protect us from harmful UV rays
   (c) Both (a) and (b) are correct
   (d) None of them are correct

SULPHUR

Sulphur can be obtained from natural gas plants in large amounts. It occurs in the earth's crust to the extent of 0.05%. It occurs in the free state as well as in combined state. It mainly occurs as sulphides and sulphates. Major deposits of S were formed by anaerobic bacteria, which metabolise $CaSO_4$ to form H_2S and S.

Methods of extraction

Frasch process is used to mine the underground deposits of S. In this method, a mixture of superheated water and steam (at about 160°C and 16 atm) is forced down the outermost of three concentric pipes into an underground bed of sulphur containing rock. The sulphur melts and forms a liquid pool. Compressed air (at 20–25 atm) is pumped down the innermost pipe and forces liquid sulphur up the remaining pipe.

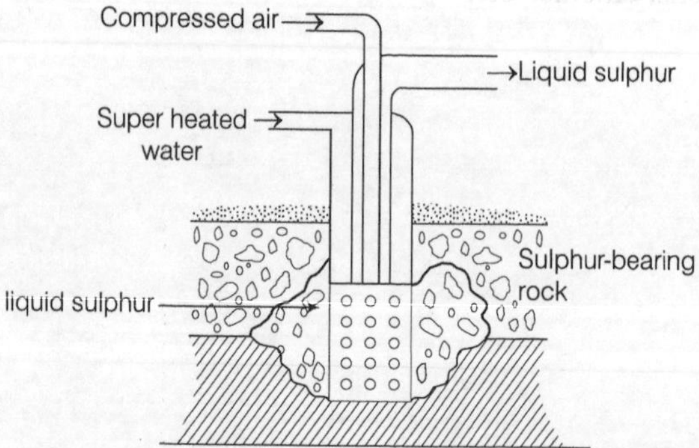

Fig. 8.1 Frasch process for extraction of sulphur

Allotropes of sulphur

Sulphur has more allotropic forms than any other element. These different forms arise partly from the extent to which S has polymerised and partly from the crystal structures adopted.

The properties of these allotropic forms are as follows

(i) **α-rhombic sulphur (octahedral)** It is stable at room temperature. It occurs naturally as large yellow crystals in volcanic area. It is soluble in carbon disulphide. It has sixteen S_8 rings in a unit cell and at 95.5°C converts into monoclinic sulphur.

(ii) **β-monoclinic (prismatic) sulphur** It is stable above 95.5°C like α-sulphur. Its crystal structure also contains cyclic S_8 rings but the mode of packing is different than that in α-sulphur. It has six S_8 rings in its unit cell. It melts at 119°C yielding liquid crystals.

(iii) **γ- monoclinic sulphur** It is obtained by chilling hot concentrated solutions of S in solvents such as CS_2, toluene or EtOH. All three forms contain puckered S_8 rings with a crown conformation and differ only in the overall packing of the rings in the crystal.

(iv) **Liquid sulphur** (S_λ and S_μ) Liquid sulphur (S_λ) comprised of S_8 molecules. This is yellow, transparent, mobile liquid. At 160°C, a remarkable transformation occurs. The S_8 rings open up and join together into long spiral chain molecules, resulting in liquid sulphur (S_μ) which is a dark in colour and very thick and viscous. The chain length and viscosity of the liquid reach a maximum at about 180°C. At higher temperatures the chains break up and viscosity decreases again.

(v) **Plastic or χ-sulphur** Plastic sulphur forms, if liquid sulphur (μ) is poured into cold water. It consists of chain-like molecules and has rubber-like properties when first formed. On standing, it becomes brittle and finally converts to rhombic sulphur.

Above transformations can be written as

$$S_\alpha \xrightarrow{95.5°C} S_\beta \xrightarrow{119°C} S_\lambda \xrightarrow{160°C} S_\mu \xrightarrow{445°C} S_8(g)$$

$$S_8(g) \longrightarrow S_6$$

$$S \xleftarrow{2000°C} S_2 \xleftarrow{1000°C} S_4 \longleftarrow S_6$$

(vi) **δ-or colloidal sulphur** It is obtained by passing H_2S through an oxidising solution such as nitric acid, potassium permanganate etc. or water or by heating sodium thiosulphate with dil. HCl.

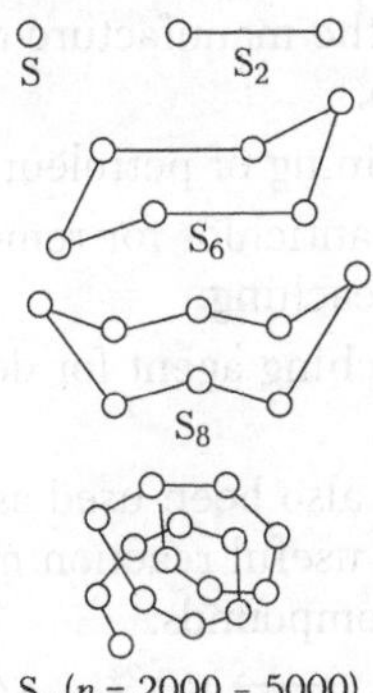

Properties

(i) Sulphur on heating with NaOH solution and produces sodium sulphide and sodium thiosulphate.

$$S_8 + 12NaOH \xrightarrow{\Delta} 4Na_2S + 2Na_2S_2O_3 + 6H_2O$$

Na_2S produced in the reaction again reacts with excess of sulphur (S_8) to give Na_2S_5.

$$2Na_2S + S_8 \longrightarrow 2Na_2S_5$$

(ii) It burns in air with a blue flame forming SO_2 and gives sulphur hexafluoride with F_2 and sulphur monochloride with Cl_2.

(iii) It reacts with metals to give metal sulphides.

$$2Na + S \longrightarrow Na_2S$$

$$Ca + S \longrightarrow CaS$$

$$Zn + S \longrightarrow ZnS$$

(iv) It reduces HNO_3 to NO_2 and also reduces H_2SO_4 to SO_2.

Uses

(i) Sulphur is important for the synthesis of vulcanised rubber.

(ii) It is important in the manufacture of matches, gun powder, explosives and fire works, SO_2, H_2SO_4, CS_2 and dyes, sulpha drugs.

(iii) It is used in ointments for curing skin diseases.

COMPOUNDS OF SULPHUR

Sulphur forms a number of compounds. It readily form oxides and oxoacids.

Oxides

Sulphur forms two main gaseous oxides.

(i) SO_2 (anhydride of H_2SO_3)

(ii) SO_3 (anhydride of H_2SO_4)

Sulphur dioxide (SO_2)

Methods of preparation

(i) SO_2 is produced commercially on a vast scale

- by burning sulphur and H_2S in air,

$$S + O_2 \longrightarrow SO_2$$

$$2H_2S + 3O_2 \longrightarrow 2H_2O + 2SO_2$$

- by roasting metal sulphide ores with air,

$$2ZnS + 3O_2 \longrightarrow 2ZnO + 2SO_2$$

(ii) In the laboratory, it can be prepared by the action of dilute acid on a sulphite ion,

$$Na_2SO_3 + 2HCl \longrightarrow 2NaCl + H_2O + SO_2$$

or by the action of concentrated H_2SO_4 on Cu.

$$Cu + 2H_2SO_4 \longrightarrow CuSO_4 + 2H_2O + SO_2$$

Physical properties

(i) SO_2 is a colourless gas (bp -10°C, mp -75.5°C) which has a choking smell and is very soluble in water.

(ii) SO_2 in solution is almost completely present as $SO_2 \cdot 6H_2O$ and only traces of H_2SO_3 (sulphurous acid) may be present.

(iii) SO_2 levels above 5 ppm are poisonous to man but plants are harmed at appreciably lower levels.

Chemical properties

(i) It is an acidic oxide which turns blue litmus red.

(ii) SO_2 acts as reducing agent as well as oxidising agent.

Some of the reactions are given below

Reducing properties

(*i*) SO_2 is a reducing agent and this property is due to formation of nascent [H].

$$2H_2O + SO_2 \longrightarrow H_2SO_4 + 2[H]$$

or in ionic reaction

$$SO_2 + 2H_2O \longrightarrow SO_4^{2-} + 4H^+ + 2e^-$$

It reduces acidified $Cr_2O_7^{2-}$ to green Cr^{3+}

$$Cr_2O_7^{2-} + 14H^+ + 6e^- \longrightarrow \underset{\text{Green}}{2Cr^{3+}} + 7H_2O$$

- It reduces acidified MnO_4^- to colourless Mn^{2+}

$$MnO_4^- + 8H^+ + 5e^- \longrightarrow \underset{\text{Colourless}}{Mn^{2+}} + 4H_2O$$

- It turns starch iodate paper blue (due to I_2)

$$2IO_3^- + 12H^+ + 10e^- \longrightarrow I_2 + 6H_2O$$

This reaction can also be used to determine SO_2 quantitatively. Liberated I_2 is titrated using hypo ($Na_2S_2O_3$).

- Fe^{3+} is reduced to Fe^{2+},

$$SO_2 + 2Fe^{3+} + 2H_2O \longrightarrow SO_4^{2-} + 2Fe^{2+} + 4H^+$$

(*ii*) SO_2 is oxidised to SO_3 and thus used to manufacture H_2SO_4.

$$2SO_2 + O_2 \rightleftharpoons 2SO_3$$

$$SO_3 + H_2O \longrightarrow H_2SO_4$$

(*iii*) $SO_2 + Cl_2 \xrightarrow{\text{Charcoal}} SO_2Cl_2$

Oxidising properties

(*i*) H_2S is oxidised to colloidal sulphur.

$$2H_2S + SO_2 \longrightarrow 2H_2O + 3S$$

(*ii*) SO_2 is used to make other products.

$$2SO_2 + Na_2CO_3 + H_2O \longrightarrow 2NaHSO_3 + CO_2$$

$$2NaHSO_3 + Na_2CO_3 \longrightarrow 2Na_2SO_3 + H_2O + CO_2$$

$$Na_2SO_3 + S \longrightarrow Na_2S_2O_3 \text{ (hypo)}$$

Structure

SO_2 molecule has a bent structure (given below) where S atom utilises sp^2-hybrid orbitals for sigma bonding.

$$S\ (Z = 16) = [Ne]\ 3s^2 3p^4$$

$$O\ (Z = 8) = [He] 2s^2\ 2p^4$$

S (Ground state) = s^2 [⇅] p^4 [⇅ | ↑ | ↑]

S (Excited) = [⇅] [↑ | ↑ | ↑] [↑ | | | |]

sp^2-hybridisation

Among three sp^2-hybrid orbitals of S-atom, one is occupied by lone pair of electrons. The remaining two half-filled hybrid orbitals overlap with one of the p-orbitals of each of the two O-atoms to form a σ-bond. Sulphur atom has still two half-filled orbitals (p and d) each oxygen atom has also one half filled p-orbital. So, p-orbital of one of the oxygen atoms overlaps with p orbital of the sulphur atom forming $p\pi$-$p\pi$ bond.

sp^2 ; $p\pi$-$p\pi$; $p\pi$-$d\pi$; σsp^2-p

Similarly, p-orbital of the second oxygen atom overlaps with d-orbital of sulphur atom forming *a* $p\pi$-$d\pi$ bond. In this manner, both sulphur and oxygen atoms get their electrons fully paired.

Uses

(*i*) SO_2 is used in the manufacture of H_2SO_4 and paper from wood pulp.

(*ii*) It is used in refining of petroleum and sugar.

(*iii*) It is used as an antichlor for removing chlorine from a fabric after bleaching.

(*iv*) It acts as a bleaching agent for delicate articles like silk and straw.

(*v*) Liquid SO_2 has also been used as a non-aqueous solvent and is a useful reaction medium for organic and inorganic compounds.

$$2SO_2 \rightleftharpoons SO^{2+} + SO_3^{2-}$$

Table 8.5 Structure of sulphur dioxide and sulphur trioxide

Molecular formula	Name	Structure
SO_2	Sulphur dioxide	
SO_3	Sulphur trioxide	

Note *In gaseous phase, SO_3 has planar triangular geometry whereas in solid state it exists as either polymeric chain or cyclic trimer as shown below.*

Oxoacids of sulphur

Sulphurous acid series

	Formula	Oxidation state	Name	Structure
(a)	H_2SO_3	S (+ 4)	sulphurous acid	HO >S=O HO
(b)	$H_2S_2O_5$	S (+ 5) *S (+ 3)	disulphurous acid or pyrosulphurous acid	O O ‖ ‖ H—O—S—S—OH ‖ * O
(c)	$H_2S_2O_4$	S (+ 3)	dithionous acid	O O ‖ ‖ HO—S—S—OH

Sulphuric acid series

	Formula	Oxidation state	Name	Structure
(a)	H_2SO_4	S (+ 6)	sulphuric acid	O ‖ HO—S—OH ‖ O
(b)	$H_2S_2O_3$	S (+ 4) *S(0)	thiosulphuric acid	S * ‖ HO—S—OH ‖ O
(c)	$H_2S_2O_7$	S (+ 6)	pyrosulphuric acid	O O ‖ ‖ HO—S—O—S—OH ‖ ‖ O O

Peroxo acid series

	Formula	Oxidation state	Name	Structure
(a)	H_2SO_5	S (+ 6)	peroxomono-sulphuric acid (Caro's acid)	O ‖ HO—S—O—OH ‖ O
(b)	$H_2S_2O_8$	S (+ 6)	peroxodisulphuric acid (Marshall's acid)	O O ‖ ‖ HO—S—O—O—S—OH ‖ ‖ O O

Thionic acid series

	Formula	Oxidation state	Name	Structure
(a)	$H_2S_2O_6$	S (+5)	Dithionic acid	O O ‖ ‖ HO—S—S—OH ‖ ‖ O O
(b)	$H_2S_nO_6$	S (+5) S* (0)	Polythionic acid	O O ‖ ‖ HO—S—(S*)$_n$—S—OH ‖ ‖ O O

Sulphuric acid or oil of vitriol

It is a very stable oxoacid of sulphur. It is often called **king of chemicals.**

Method of preparation

The following two methods are used for the preparation of H_2SO_4.

(*i*) **Contact process** (a commercial method)

Involves four stages as shown in fig. 8.2.

Stage I $\quad S + O_2 \xrightarrow{\Delta} SO_2$

Stage II $\quad 2SO_2 + O_2 \xrightarrow{V_2O_5} 2SO_3$

In accordance with Le-Chatelier's principle the favourable conditions for maximum yield of SO_3 are

- high pressure of about 2 atomospheres
- low temperature of 676-723 K and
- high concentration of SO_2 and O_2

Stage III $\quad SO_3 + H_2O \longrightarrow H_2SO_4$

This is a violent reaction as it produces a dense fog, which does not easily condense. Therefore, SO_3 is passed into 98% H_2SO_4.

$$H_2SO_4 + SO_3 \longrightarrow \underset{\text{Oleum (pyrosulphuric acid)}}{H_2S_2O_7}$$

Some trisulphuric acid $H_2S_3O_{10}$ is also formed .

Stage IV $H_2S_2O_7 + H_2O \longrightarrow 2H_2SO_4$

$H_2S_3O_{10} + 2H_2O \longrightarrow 3H_2SO_4$

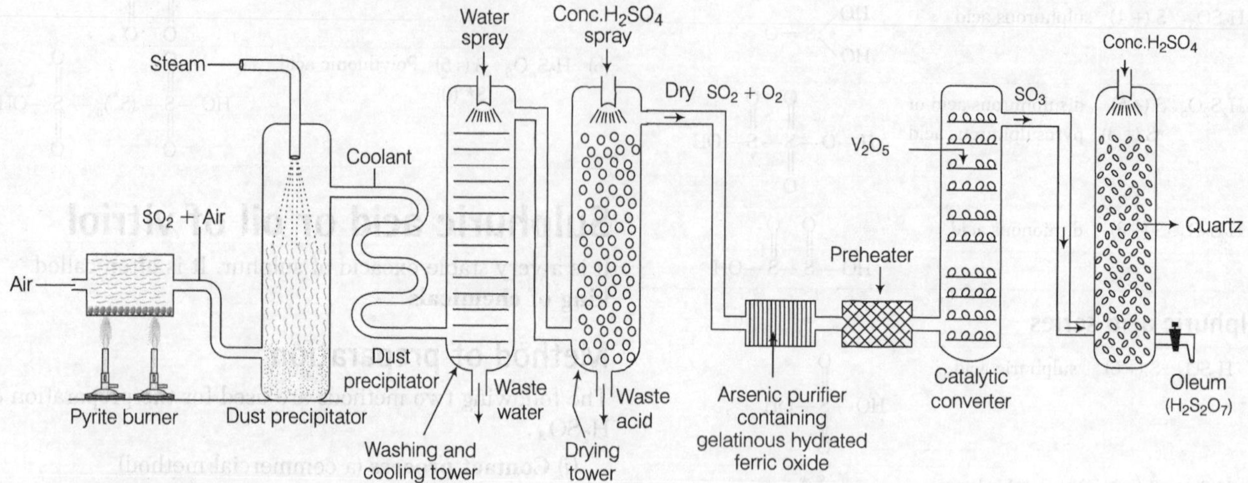

Fig. 8.2 Contact process for extraction of sulphur

(*ii*) Lead chamber process (which is no longer in use)

This process involves three stages

Stage I $2NO\ (\text{catalyst}) + O_2 \longrightarrow 2NO_2$

Stage II $2SO_2 + 2NO_2 \longrightarrow 2SO_3 + 2NO$

$$2SO_2 + O_2 \xrightarrow{NO} 2SO_3$$

Stage III $SO_3 + H_2O \longrightarrow H_2SO_4$

$NO_2 + SO_2 + H_2O \longrightarrow H_2SO_4 + NO\ (\text{recycled})$

$2NO + O_2 \longrightarrow 2NO_2$

Physical properties

(*i*) It is a diprotic acid. It is colourless viscous liquid with melting and boiling points of 283 K and 611 K respectively.

(*ii*) It is highly soluble in water and have high boiling point due to hydrogen bonding.

(*iii*) The concentrated sulphuric acid we use in the laboratory is 98 per cent H_2SO_4 by mass (density 1.84 g/cm^3) which corresponds to a concentration of 18 M.

Structure

$$O{=}S({=}O)(OH)\text{–}O\text{–}H \underset{}{\overset{-H^+}{\rightleftharpoons}} O{=}S({=}O)(O^-)\text{–}O^-$$

S-atom is sp^3-hybridised. SO_4^{2-} ion is tetrahedral. The bond lengths are all equal (1.49 Å) and are all rather short. The bond order of the S—O bonds is approximately 1.5. There are four σ-bonds and two π-bonds between S and O atoms (with two π-bonds delocalised over the S and the four O atoms).

Chemical properties

The various chemical properties shown by H_2SO_4 are as follows

(*i*) **As a strong acid** In aqueous solution, it is regarded as a strong acid ionising in two stages.

$$H_2SO_4 + H_2O \longrightarrow HSO_4^- + H_3O^+$$

$$HSO_4^- + H_2O \rightleftharpoons SO_4^{2-} + H_3O^+$$

Only the first dissociation is complete, the second is partial. When it is dilute, it shows all the usual properties of acids.

- It reacts with bases to form bisulphates and sulphates.

$$H_2SO_4 + NaOH \longrightarrow NaHSO_4 + H_2O$$
$$H_2SO_4 + 2NaOH \longrightarrow Na_2SO_4 + 2H_2O$$

- H_2SO_4 decomposes carbonates and bicarbonates to evolve carbon dioxide.

$$H_2SO_4 + Na_2CO_3 \longrightarrow Na_2SO_4 + H_2O + CO_2\uparrow$$
$$H_2SO_4 + 2NaHCO_3 \longrightarrow Na_2SO_4 + 2H_2O + 2CO_2\uparrow$$

- It reacts with metals like Zn, Fe, Mg, etc. to liberate H_2 gas. [displacement reaction]

$$Zn + H_2SO_4 \longrightarrow ZnSO_4 + H_2\uparrow$$

- Sulphuric acid being a strong acid, decomposes the salts of volatile acids.

$$H_2SO_4 + 2NaCl \longrightarrow Na_2SO_4 + 2HCl$$
$$H_2SO_4 + 2KNO_3 \longrightarrow K_2SO_4 + 2HNO_3$$

(*ii*) **As an oxidising agent** The oxidising strength of sulphuric acid depends on its temperature and concentration. A cold, dilute sulphuric acid solution reacts with metals above hydrogen in the activity series (as shown above with Zn), thereby liberating molecular hydrogen (H_2) in a displacement reaction, Mg reacts in the same way.

$$Mg + 2H_3O^+ \longrightarrow Mg^{2+} + H_2 + 2H_2O$$

This is a typical reaction of an active metal with an acid.

The strength of sulphuric acid as an oxidising agent is greatly enhanced when it is hot and concentrated too. In such a solution, the oxidising agent is actually the sulphate ion rather than the hydrated proton, $H^+ (aq)$.

Thus, $$SO_4^{2-} + 4H^+ + 2e^- \longrightarrow SO_2 + 2H_2O$$

$$Cu + 2H_2SO_4 \longrightarrow CuSO_4 + 2H_2O + SO_2$$
$$S + 2H_2SO_4 \longrightarrow 2H_2O + 3SO_2$$
$$C + 2H_2SO_4 \longrightarrow CO_2 + 2H_2O + 2SO_2$$
$$2HBr + H_2SO_4 \longrightarrow Br_2 + SO_2 + 2H_2O$$
$$K_4[Fe(CN)_6] + 6H_2SO_4 + 6H_2O \longrightarrow 2K_2SO_4 + FeSO_4 + 3(NH_4)_2SO_4 + 6CO$$

Depending on the nature of the reducing agents, the sulphate ion may be further reduced to elemental sulphur or the sulphide ion.

$$8HI + H_2SO_4 \longrightarrow H_2S + 4I_2 + 4H_2O$$

(*iii*) **As a dehydrating agent**

Concentrated H_2SO_4 is also a dehydrating agent

$$C_6H_{12}O_6 \xrightarrow{\text{Conc. } H_2SO_4} 6C + 6H_2O$$

$$CH_3CH_2OH \xrightarrow[160°C]{\text{Conc. } H_2SO_4} CH_2{=}CH_2 + H_2O$$

$$2CH_3CH_2OH \xrightarrow[140°C]{\text{Conc. } H_2SO_4} CH_3CH_2{-}O{-}CH_2CH_3 + H_2O$$

Materials like cloth, wood etc get charred in presence of H_2SO_4 as it removes water from these substances.

(*iv*) A mixture of conc. HNO_3 and conc. H_2SO_4 is used as a nitrating mixture in the nitration of aromatic compounds.

$$HNO_3 + 2H_2SO_4 \longrightarrow \underset{\text{Nitronium ion}}{NO_2^+} + H_3O^+ + 2HSO_4^-$$

Nitronium ion is a sufficiently good electrophile to attack a benzene ring.

$$C_6H_6 + NO_2^+ \longrightarrow \underset{\text{Nitrobenzene}}{C_6H_5{-}NO_2} + H^+$$

(*v*) Conc. H_2SO_4 is also used for sulphonation of organic compounds.

$$C_6H_6 + H_2SO_4 \longrightarrow \underset{\text{Benzene sulphonic acid}}{C_6H_5{-}SO_3H} + H_2O$$

(*vi*) $$HO{-}SO_2{-}OH + 2PCl_5 \longrightarrow \underset{\text{Sulphuryl chloride}}{SO_2Cl_2} + 2POCl_3 + 2HCl$$

(*vii*) When conc. H_2SO_4 is subjected to electrolysis there is formation of peroxy disulphuric acid and H_2O_2 at the anode.

$$H_2SO_4 \rightleftharpoons H^+ + HSO_4^-$$

At anode,

$$2HSO_4^- \longrightarrow H_2S_2O_8 + 2e^-$$

$$H_2S_2O_8 + 2H_2O \longrightarrow 2H_2O + 2H_2SO_4$$

At cathode, $2H^+ + 2e^- \longrightarrow H_2$

Uses

(*i*) It is used as a drying and dehydrating agent.

(*ii*) It is used in the manufacture of paints, pigments, etc.

(*iii*) It is used in fertilizer industry to prepare $(NH_4)_2PO_4$, $(NH_4)_2SO_4$, etc.

(*iv*) It is used in the manufacture of explosives, such as dynamite, TNT etc.

Preparation and properties of H_2SO_4 are summarised in figure given below.

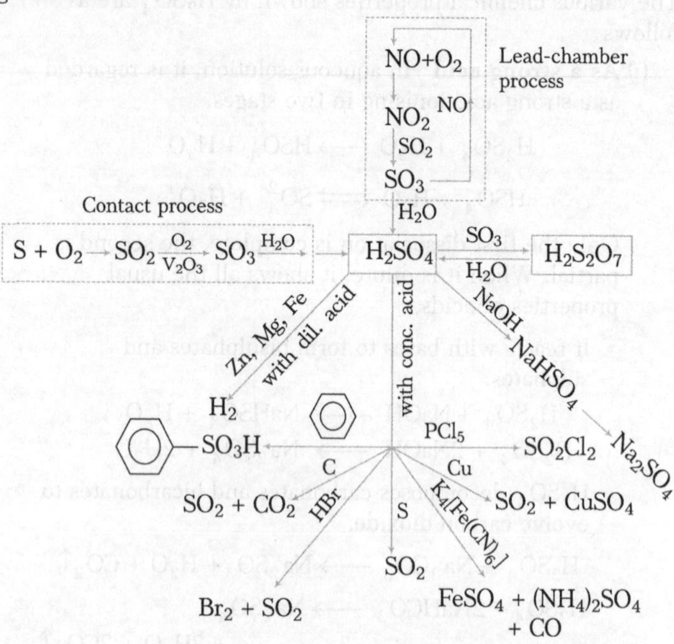

CHECK POINT 8.3

1. In Frasch process a mixture of superheated water and steam is force down. The temperature range for this mixture is

(a) 50° to 100°C (b) 50° to 70°C
(c) 150° to 160°C (d) 500° to 600°C

2. Which of the following forms of sulphur exists as a *zig-zag* chains form?

(a) Rhombic sulphur
(b) Plastic sulphur
(c) Monoclinic
(d) Liquid sulphur

3. An allotrope of sulphur (*X*) is stable at room temperature and has sixteen S_8 rings in a unit cell and at 95.5°C converts into another sulphur (*Y*). *X* and *Y* respectively are

(a) α and β-sulphur
(b) λ and μ-sulphur
(c) γ and δ-sulphur
(d) Liquid and plastic sulphur

4. Which of the following compounds is not produced when sulphur reacts with boiling NaOH?

(a) Na_2S (b) SO_2
(c) $Na_2S_2O_3$ (d) H_2O

5. Which of the following is not a use of sulphur?

(a) Synthesise vulcanised rubber
(b) Manufacture gun powder
(c) Drug for harmful viruses
(d) Ointment for curing skin diseases

6. Which of the following statements is incorrect for SO_2 gas?

(a) It is a colourless gas
(b) It's melting point is –75.5°C
(c) SO_2 level below 2 ppm is poisonous to man
(d) It has a choking smell

7. Among the following, the correct formula of peroxydisulphuric acid is

(a) $H_2S_2O_7$ (b) $H_2S_2O_8$
(c) H_2SO_5 (d) H_2SO_3

8. Choose the correct statement regarding H_2SO_4.

(a) It is a diprotic acid
(b) It shows H-bonding when dissolved in water
(c) It is also called as oil of vitrol
(d) All of the above are correct

9. The number of S=O and S—OH bonds present in peroxodisulphuric acid and pyrosulphuric acid respectively are

(a) (2, 2) and (2, 2) (b) (2, 4) and (2, 4)
(c) (4, 2) and (2, 4) (d) (4, 2) and (4, 2)

10. The correct sequence of decreasing number of π-bonds in the structures of H_2SO_3, H_2SO_4 and $H_2S_2O_7$ is

(a) $H_2SO_3 > H_2SO_4 > H_2S_2O_7$
(b) $H_2SO_4 > H_2S_2O_7 > H_2SO_3$
(c) $H_2S_2O_7 > H_2SO_4 > H_2SO_3$
(d) $H_2S_2O_7 > AH_2SO_3 > H_2SO_4$

Chapter Exercises

(A) Topical exercise

I. General, physical and chemical properties

1 The most abundant element in the earth's crust is
(a) oxygen (b) aluminium
(c) iron (d) sulphur

2 Which of the following general electronics configuration is correct for group 16 elements?
(a) ns^2np^2 (b) ns^2np^3 (c) ns^2np^4 (d) ns^2np^5

3 Tetrafluorides of elements of group-16 have hybridisation and structure respectively are
(a) sp^3 and trigonal pyramidal
(b) sp^3d and tetrahedral
(c) sp^3d and trigonal bipyramidal
(d) sp^3 and tetrahedral

4 The element which is radioactive in nature is
(a) S (b) Te (c) Se (d) Po

5 Which element of chalcogens has maximum tendency to show catenation property?
(a) Oxygen (b) Sulphur
(c) Tellurium (d) Polonium

6 A chalcogen combines directly with hydrogen to form a hydride but with a great difficulty. On burning in air, this chalcogen forms a solid polymeric dioxide and it also has the highest electrical resistance amongst the metals. This chalcogen is
(a) sulphur (b) selenium
(c) polonium (d) tellurium

7 Which of the following elements does not show an oxidation state higher than +2?
(a) Oxygen (b) Sulphur
(c) Selenium (d) Tellurium

8 Water is much less volatile than H_2S because
(a) H_2O has a bond angle of nearly 150°
(b) hydrogen is loosely bonded with the sulphur
(c) S-atom is less electronegative than O-atom
(d) S-atom is more electronegative than O-atom

9 Which of the following physical property is not possessed by group 16 elements?
(a) Multiple bonding (b) Catenation
(c) Allotropy (d) High viscosity

10 Maximum coordination number of oxygen in liquid water is
(a) two (b) three (c) four (d) five

11 There is $p\pi$-$d\pi$ multiple bonding in
(a) CS_2 (b) CO_2
(c) NO_2 (d) NO

12 Consider the following statements.
Statement I Oxygen is the most abundant element on the surface of earth.
Statement II Oxygen is paramagnetic with double bond character.
Choose the correct option.
(a) Only Statement I is correct
(b) Only Statement II is correct
(c) Both the Statements are correct
(d) None of the above Statements are correct

13 The best reducing agent of the following is
(a) H_2S (b) Cl^- (c) SO_4^{2-} (d) O_3

14 Which of the following is not a chalcogen?
(a) Se (b) O (c) S (d) Na

15 Which of the following gas is insoluble in water?
(a) SO_2 (b) H_2
(c) NH_3 (d) CO_2

16 Group 16 element exists in the monoatomic state in metallic lattice. It also exists in two crystalline forms. The metal is
(a) sulphur (b) polonium (c) selenium (d) tellurium

17 In presence of H_2O, SO_2 acts as
(a) an oxidising agent (b) a reducing agent
(c) a hydrolysing agent (d) a redox reagent

18 In case of oxygen family (group 16)
(a) the tendency for catenation decreases markedly as we go down the group
(b) maximum coordination of oxygen is four due to lack of *d*-orbital but that of other elements is six due to presence of *d*-orbital
(c) the tendency to form multiple bonds with C, N and O decreases as going down the group from S to Te
(d) All of the above are correct statements

19 Consider the following statements.
Statement I SF_6 cannot be hydrolysed but SF_4 can be.
Statement II Six F-atoms in SF_6 prevent the attack of H_2O on sulphur atom of SF_6.
Choose the correct option.
(a) Only Statement I is correct
(b) Only Statement II is correct
(c) Both the Statements are correct
(d) None of above Statements are correct

20 Which of the following has maximum pK_a value?
(a) H_2O (b) H_2S (c) H_2Se (d) H_2Te

21 Which of the following bonds has the least bond energy?
(a) Se—Se (b) Te—Te (c) S—S (d) O—O

22 Electron affinity is positive, when
(a) O changes into O^-
(b) O^- changes into O^{2-}
(c) O changes into O^+
(d) electron affinity is always negative

23 Among CH_4, NH_3 and H_2O, acidity of H_2O is maximum because
(a) oxygen contains two lone pairs of electrons
(b) bond angle in H_2O is less than 109°28′ due to VSEPR
(c) oxygen is more electronegative than C or N
(d) H_2O is an associated liquid

II. Ozone (O_3) and anomalous behaviour of oxygen

24 Which of the following properties makes oxygen different from other group members?
(a) Small size
(b) High electronegativity
(c) Non-availability of *d*-orbitals for bonding
(d) All of the above

25 Who discovered the dioxygen molecule?
(a) Mendeleev's (b) Priestley and Scheele
(c) A.L Lavoisier (d) Albert Einstein

26 Which of the following methods is not used in the preparation dioxygen?
(a) Catalytic decomposition of hypochlorite
(b) Electrolysis of water with traces of H_2SO_4
(c) Thermal decomposition of $KClO_3$
(d) None of the above

27 Select the correct statement about O_2 gas.
(a) It is stable in nature
(b) It is neutral in nature
(c) It is used as bleaching agent
(d) All of the above are correct

28 In a chemical reaction, scientist accidently synthesised a dark blue, diamagnetic gas with boiling point −112°C. Name the gas
(a) O_2 (b) SO_2 (c) N_2 (d) O_3

29 The O — O distance in O_3 molecule is
(a) 128 pm (b) 130 pm
(c) 136 pm (d) 125 pm

30 Which of the following gas mixture is used by the divers inside the sea?
(a) O_2 + He (b) O_2 + Xe
(c) O_2 + Ar (d) O_2 + N_2

31 Ozone reacts with all the following reagents except
(a) K_2MnO_4
(b) moist iodine
(c) potassium ferrocyanide
(d) mercury

32 The absorption of UV radiation by O_3
(a) protects the inhabitants of our planet from injurious radiation
(b) maintains an equilibrium between the concentrations of O_2 and O_3
(c) makes both the function effective
(d) makes no function effective

33 Ozone layer is being depleted. This is due to
(a) NO emission from supersonic jets
(b) chlorofluorocarbon used as aerosols
(c) Both (a) and (b)
(d) None of the above

34 Oxidative ozonolysis of 2-butene in presence of Zn or dimethyl sulphide forms
(a) CH_3CHO (b) CH_3COOH
(c) CO_2 (d) $(CH_3)_2CO$

35 Consider the following compounds
1. Sulphur dioxide 2. Hydrogen peroxide 3. Ozone
Among these compounds, those which can act as bleaching agents would include
(a) 1 and 3 (b) 2 and 3
(c) 1 and 2 (d) 1, 2 and 3

36 A considerable part of the harmful UV rays of the sun does not reach the surface of the earth. This is because high above the earth's atmosphere, there is a layer of
(a) O_3 (b) CO_2
(c) SO_2 (d) NO

37 Select the correct statement(s).
(a) O_3 molecule is resonance stabilised
(b) There is 3-centre, 4-electron π-bond system in O_3
(c) Ozone layer is being depleted by fluorocarbons
(d) All of the above are correct statements

38 The correct order of O—O bond length in O_2, H_2O_2 and O_3 is
(a) $H_2O_2 > O_3 > O_2$ (b) $O_2 > H_2O_2 > O_3$
(c) $O_3 > H_2O_2 > O_2$ (d) $O_2 > O_3 > H_2O_2$

39 Consider the following statements.
Statement I O_2 gas is affected in magnetic field.
Statement II It has north and south pole.
Choose the correct option
(a) Only Statement I is correct
(b) Only Statement II is correct
(c) Both the Statements are correct
(d) None of the above Statements are correct

III. Sulphur, its allotropes and its compounds (except oxoacids)

40 What is the proportion of sulphur in the earth's crust?
(a) 0.10% (b) 0.02% (c) 0.05% (d) 0.06%

41 Which of the following processes is used to extract sulphur from underground sulphur bearing rocks?
(a) Frasch process (b) Contact process
(c) Spring process (d) Bosch process

42 Consider the following chain of reactions,

$$S_\alpha \xrightarrow{95.5°C} S_\beta \xrightarrow{119°C} A \xrightarrow{160°C} B$$

A and *B* respectively are
(a) S_λ and S_μ (b) S_μ and S_λ
(c) S_8 and S_6 (d) S_6 and S_8

43 The process of obtaining sulphur by the borehole method is called
(a) the Frasch process (b) the Lablanc process
(c) the Calcaroni process (d) the Mannheium process

44 When molten sulphur is suddenly cooled by pouring into water, it takes the form of
(a) milk of sulphur (b) colloidal sulphur
(c) flower of sulphur (d) plastic sulphur

45 Yellow oils of sulphur is/are
(a) H_2S (b) H_2S_2, H_2S_3
(c) H_2SO_4 (d) CS_2, NH_2CSNH_2

46 When an article is bleached by SO_2 it loses its colour. The colour can be restored by
(a) exposure to air (b) heating
(c) dilution (d) None of these

47 The compound of sulphur used as a solvent in rubber industry is
(a) $SO_2(OH)Cl$ (b) SO_2 (c) SO_3 (d) S_2Cl_2

48 Consider the following statements.
Statement I Both rhombic and monoclinic sulphur exist as S_8 but oxygen exists as O_2.
Statement II Oxygen forms $p\pi$ - $p\pi$ multiple bond due to small size and small bond length but $p\pi$ - $p\pi$ bonding is not possible in sulphur.
Choose the correct option.
(a) Only Statement I is correct
(b) Only Statement II is correct
(c) Both the Statements are correct
(d) None of above Statements are correct

49 In SO_3^{2-}
(a) $d\pi$-$p\pi$ bond between S and O is delocalised.
(b) bonds between S and O are equivalents.
(c) there is sp^3-hybridised sulphur atom.
(d) All of the facts given above are true.

50 Sulphur dioxide levels in the atmosphere can be reduced by using
(a) catalytic converters in industry
(b) static electricity to attract it in factory chimneys
(c) more efficient car engines
(d) low sulphur fuels

51 Which of the following is used in vulcanisation of rubber?
(a) SF_6 (b) CF_4
(c) Cl_2F_2 (d) C_2F_2

52 The S—S—S bond angle in S molecule is
(a) 109.5° (b) 105°
(c) 110° (d) 60°

53 H_2S precipitates aqueous $ZnCl_2$ solution in
(a) acidic medium
(b) neutral medium
(c) Both (a) and (b)
(d) None of the above

54 Consider the following statements.
Statement I Among all the elements of the group 16, sulphur has the largest tendency to catenation.
Statement II Sulphur-sulphur bond energy is very high.
Choose the correct option.
(a) Only Statement I is correct
(b) Only Statement II is correct
(c) Both the Statements are correct
(d) None of above Statements are correct

55 Which of the following can be used as dehydrating agents?
(a) Conc. H_2SO_4 (b) $POCl_3$
(c) P_2O_5 (d) All of these

56 SO_2 behaves as a reducing agent when
(a) passed over hot CuO
(b) mixed with moist H_2S
(c) passed through acidified $KMnO_4$ solution
(d) passed through $FeSO_4$ solution

57 To a piece of charcoal, sulphuric acid is added. Then
(a) there is no reaction
(b) water gas is formed
(c) SO_2 and CO_2 are evolved
(d) CO and SO_2 are evolved

58 Aqueous solutions of hydrogen sulphide and sulphur dioxide when mixed together yield
(a) sulphur trioxide and water
(b) sulphurous acid
(c) sulphur and water
(d) hydrogen preroxide and sulphur

59 Which is thermocromic?
(a) S_4N_4 (b) $B_3N_3H_6$
(c) S_8 (d) None of these

60 Consider the following statements.

Statement I SO_2 turns acidified $K_2Cr_2O_7$ green.

Statement II $K_2Cr_2O_7$ is reduced to Cr^{3+} (green) salt by SO_2.

Choose the correct option.

(a) Only Statement I is correct
(b) Only Statement II is correct
(c) Both the Statements are correct
(d) None of the above Statements are correct

IV. Oxoacids of sulphur

61 In laboratory, SO_2 is prepared easily by

(a) the action of moderately concentrated H_2SO_4 on copper
(b) the action of H_2SO_4 on sodium sulphide
(c) the action of H_2SO_4 on sodium sulphate
(d) None of the above

62 The correct formula of Caro's acid is

(a) H_2SO_3 (b) H_2SO_5 (c) $H_2S_2O_8$ (d) $H_2S_2O_7$

63 Which of the following property is not shown by H_2SO_4?

(a) Reducing agent (b) Dehydrating agent
(c) Sulphonating agent (d) Highly viscous

64 Consider the following reaction,

$$C_6H_{12}O_6 \xrightarrow{A} 6C + 6H_2O$$

Here, A is

(a) conc. H_2SO_4 (b) H_2S
(c) H_2SO_3 (d) H_2

65 S—S bond is present in

(a) α - $(SO_3)_n$ (b) γ - (S_3O_9)
(c) $H_2S_2O_3$ (d) $H_2S_2O_8$

66 Oil of vitriol is

(a) HNO_3 (b) HCl
(c) H_2SO_4 (d) H_3PO_4

67 $S_2O_3^{2-}$ has

(a) S–S linkage (b) S–O linkage
(c) Both (a) and (b) (d) None of these

68 The compound which does not react with $KMnO_4$ is

(a) perdisulphuric acid (b) sulphurous acid
(c) hydrogen sulphide (d) hydrogen peroxide

69 Consider the following statements.

Statement I $S_3O_6^{2-}$ (trithionate) has S—S bond.

Statement II It is a conjugate base of $HS_3O_6^-$.

Choose the correct option.

(a) Only Statement I is correct
(b) Only Statement II is correct
(c) Both the Statements are correct
(d) None of above Statements are correct

70 Out of $H_2S_2O_3$, $H_2S_4O_6$, H_2SO_5 and $H_2S_2O_8$ peroxy acids are

(a) $H_2S_2O_3$, $H_2S_4O_6$ (b) $H_2S_4O_6$, H_2SO_5
(c) H_2SO_5, $H_2S_2O_8$ (d) $H_2S_2O_3$, $H_2S_2O_8$

71 Liver of sulphur is a mixture of

(a) potassium penta sulphide (*A*) and potassium thiosulphate (*B*)
(b) potassium carbonate (*C*) and (*B*) above
(c) potassium disulphide (*D*) and (*B*) above
(d) (*B*) and (*D*) above

72 Which of the following is not a hypervalent ion?

(a) SO_3^{2-} (b) PO_4^{3-} (c) SO_4^{2-} (d) ClO_4^-

73 Which of the following does not have (S—S) bond?

(a) $H_2S_2O_3$ (b) $H_2S_2O_5$
(c) $H_2S_2O_4$ (d) S_3O_9

74 Oxalic acid on heating with conc. H_2SO_4 produce

(a) H_2O and CO_2 (b) CO, CO_2 and H_2O
(c) CO_2 and H_2S (d) CO and CO_2

75 Excess of PCl_5 reacts with conc. H_2SO_4 giving

(a) sulphuryl chloride (b) sulphurous acid
(c) chlorosulphuric acid (d) thionyl chloride

(B) Miscellaneous exercise

1 Bleaching of a fabric cloth is done by using *A* and excess of chlorine is removed by using *B* is

(a) $CaOCl_2$, Na_2SO_3 (b) $Na_2S_2O_3$, $CaOCl_2$
(c) $CaCl_2$, $Na_2S_2O_3$ (d) $CaOCl_2$, $Na_2S_2O_3$

2 $Na_2S_2O_3$ is used in photography for fixing the negative. It removes the AgBr by

(a) complex formation $Na_3[Ag(S_2O_3)_2]$
(b) oxidation of AgBr to Br_2
(c) reduction of AgBr to Ag
(d) formation of double salt

3 Aqueous hyposolution on reaction with aqueous $AgNO_3$ gives

(a) yellow ppt changing to black
(b) white ppt changing to black
(c) orange ppt changing to blue
(d) no ppt

4 Cold solution of barium nitrite on mixing with sulphuric acid produces

(a) $BaSO_4 + HNO_2$ (b) $BaSO_4 + HNO_3$
(c) $BaSO_4 + NO_2$ (d) $BaSO_4 + N_2 + O_2$

5 Which of the following products is formed by the reaction of sulphur dioxide with chlorine in presence of sunlight?

(a) SO_2Cl (b) SO_2Cl_2
(c) $SOCl_2$ (d) SO_3Cl

6 Na_2CO_3 reacts with SO_2 in alkaline medium to produce
(a) Na_2SO_4 (b) Na_2SO_3
(c) $NaHSO_4$ (d) $NaHSO_3$

7 Consider the following statements.
Statement I In aqueous solution, SO_2 reacts with H_2S forming sulphur.
Statement II SO_2 is an effective reducing agent.
Choose the correct option.
(a) Only Statement I is correct
(b) Only Statement II is correct
(c) Both the Statements are correct
(d) None of above Statements are correct

8 Consider the following statements.
Statement I SO_2 is a bleaching agent.
Statement II Nascent oxygen causes bleaching action in SO_2.
Choose the correct option
(a) Only Statement I is correct
(b) only Statement II is correct
(c) Both the Statements are correct
(d) None of above Statements are correct

9 Match the species given in Column I with their corresponding properties given in Column II.

Column I	Column II
A. SO_2	1. Sulphonating agent
B. $H_2S_2O_8$	2. Precipitating agent
C. H_2S	3. Oxidising agent
D. SO_3	4. Reducing agent

Codes

	A	B	C	D		A	B	C	D
(a)	4	3	2	1	(b)	1	2	3	4
(c)	1	3	2	4	(d)	4	3	1	2

10 The number of S—S bond in sulphur trioxide trimer (S_3O_9) is
(a) three (b) two (c) one (d) zero

11 Consider the following compounds.
1. $Na_2S_2O_3$ 2. $Na_2S_4O_6$ 3. $CaOCl_2$
Among these compounds identify those that have different oxidation states for same type of atoms.
(a) Both 1 and 3 (b) Both 2 and 3
(c) Both 1 and 2 (d) 1, 2 and 3

12 Which of the following can convert acidified $Cr_2O_7^{2-}$ to green colour ?
(a) $SO_2/H_2SO_3/H_2SO_4$ (b) $SO_3/H_2SO_3/H_2S$
(c) $SO_3^{2-}/H_2S/Fe^{2+}$ (d) $S_2O_3^{2-}/SO_3/Fe^{3+}$

13 A bottle completely filled with conc. H_2SO_4 is left unstoppered for several days and we observe spontaneous overflow of acid. It is due to
(a) change in temperature
(b) hygroscopic nature of H_2SO_4
(c) absorption of air by H_2SO_4
(d) dehydration of H_2SO_4

14 Consider the following statements.
Statement I Anhydrous sulphuric acid has very high viscosity.
Statement II H-bonding makes it viscous.
Choose the correct option.
(a) Only Statement I is correct
(b) Only Statement II is correct
(c) Both the Statements are correct
(d) None of above Statements are correct

15 Consider the following statements.
Statement I Structure of SO_4^{2-} is identical to $S_2O_3^{2-}$.
Statement II Sulphur atoms in $S_2O_3^{2-}$ are in different oxidation states.
Choose the correct option
(a) Only Statement I is correct
(b) Only Statement II is correct
(c) Both the statements are correct
(d) None of the above Statements are correct

16 $FeS + HCl \longrightarrow A + B(g)$
Gas B is passed into aqueous solution of C to form A
Select the correct statements based on the above facts.
(a) Colour of C changes from yellow to very light green
(b) Gas B turns lead acetate paper black
(c) A form deep blue colour with $K_3[Fe(CN)_6]$
(d) All of the above facts are true

17 Hot conc. H_2SO_4 acts as moderately strong oxidising agent. It oxidises both metals and non-metals. Which of the following element is oxidised by conc. H_2SO_4 into two gaseous products?
(a) Cu (b) S
(c) C (d) Zn

18 In qualitative analysis when H_2S is passed through an aqueous solution of salt acidified with dil. HCl, a black precipitate is obtained. On boiling the precipitate with dil. HNO_3, it forms a solution of blue colour. Addition of excess of aqueous solution of ammonia to this solution gives ....
(a) deep blue precipitate of $Cu(OH)_2$
(b) deep blue solution of $[Cu(NH_3)_4]^{2+}$
(c) deep blue solution of $Cu(NO_3)_2$
(d) deep blue solution of $Cu(OH)_2 \cdot Cu(NO_3)_2$

19 The term 'thio' is used in the names of all of the following compounds except
(a) $Na_2S_2O_3$ (b) $NaCS_3$
(c) NaSCN (d) Na_2SO_3

20 Consider the following statements.
Statement I Anhydrous H_2SO_4 undergoes autoprotolysis.

$$H_2SO_4 + H_2SO_4 \rightleftharpoons H_3SO_4^+ + HSO_4^-$$

Statement II Anhydrous H_2SO_4 has very high electrical conductivity.
Choose the correct option.
(a) Only Statement I is correct
(b) Only Statement II is correct
(c) Both the Statements are correct
(d) None of the above Statements are correct

21 Match the species in given Column I with their corresponding properties given in Column II.

Column I	Column II
A. SO_3	1. sp^2-hybridised atom
B. H_2SO_5	2. Sulphur in different oxidation states
C. $S_2O_3^{2-}$	3. No (S—S) linkage
D. S_3O_9	4. Peroxy linkage

Codes

	A	B	C	D
(a)	1	4	2	3
(b)	1	4	3	2
(c)	4	1	2	3
(d)	4	1	3	2

(C) Medical entrances' gallery

1 Which of the following oxoacid of sulphur has —O—O— linkage? **[NEET 2020]**
(a) H_2SO_4, sulphuric acid
(b) $H_2S_2O_8$, peroxodisulphuric acid
(c) $H_2S_2O_7$, pyrosulphuric acid
(d) H_2SO_3, sulphurous acid

2 Which is the correct order of thermal stability for H_2E (E = O, S, Se, Te and Po)? **[NEET 2019]**
(a) $H_2O < H_2S < H_2Se < H_2Te < H_2Po$
(b) $H_2Po < H_2Te < H_2Se < H_2S < H_2O$
(c) $H_2Se < H_2Te < H_2Po < H_2O < H_2S$
(d) $H_2S < H_2O < H_2Se < H_2Te < H_2Po$

3 Which of the following statement is correct for oleum? **[AIIMS 2019]**
(a) It is prepared by absorption of SO_3 in conc. H_2SO_4
(b) It contains O— O bond
(c) It has six OH groups
(d) None of the above

4 Identify the correct formula of oleum from the following **[NEET (Odisha) 2019]**
(a) $H_2S_2O_7$ (b) H_2SO_3
(c) H_2SO_4 (d) $H_2S_2O_8$

5 Conversion of oxygen into ozone is non-spontaneous at **[CBSE AIPMT 2018]**
(a) low temperature (b) all temperature
(c) high temperature (d) room temperature

6 Among the following, the correct formula of Marshall's acid is **[JIPMER 2018]**
(a) $H_2S_2O_6$ (b) $H_2S_2O_8$
(c) H_2SO_4 (d) H_2SO_5

7 Name the gas that can readily decolourise acidified $KMnO_4$ solution. **[NEET 2017]**
(a) P_2O_5 (b) CO_2 (c) SO_2 (d) NO_2

8 In which pair of ions both the species contain S—S bond?
(a) $S_4O_6^{2-}$, $S_2O_7^{2-}$ (b) $S_2O_7^{2-}$, $S_2O_3^{2-}$
(c) $S_4O_6^{2-}$, $S_2O_3^{2-}$ (d) $S_2O_7^{2-}$, $S_2O_8^{2-}$

9 Which one of the following statement is correct when SO_2 is passed through acidified $K_2Cr_2O_7$ solution? **[NEET (I) 2016]**
(a) The solution is decolourised
(b) SO_2 is reduced
(c) Green $Cr_2(SO_4)_3$ is formed
(d) The solution turns blue

10 Nitrogen dioxide and sulphur dioxide have some properties in common. Which property is shown by one of these compounds, but not by the other?
(a) Forms' acid rain **[AIPMT 2015]**
(b) Is a reducing agent
(c) Is soluble in water
(d) Is used as a food preservative

11 Acidity of diprotic acids in aqueous solutions increases in the order **[CBSE AIPMT 2014]**
(a) $H_2S< H_2Se< H_2Te$
(b) $H_2Se< H_2S< H_2Te$
(c) $H_2Te< H_2S< H_2Se$
(d) $H_2Se< H_2Te< H_2S$

12 Which among the following group 16 elements exists in more than two allotropic states? **[JCECE 2014]**
(a) Polonium (b) Tellurium
(c) Selenium (d) Oxygen

13 Aqueous solution of hydrogen sulphide and sulphur-dioxide when mixed together, yield
(a) sulphur trioxide and water **[KCET 2013]**
(b) hydrogen and sulphurous acid
(c) sulphur and water
(d) hydrogen peroxide and sulphur

14 When conc. H_2SO_4 is heated with P_2O_5, the acid is converted to **[UP CPMT 2012]**
(a) sulphur trioxide
(b) sulphur dioxide
(c) sulphur
(d) a mixture of sulphur dioxide and sulphur trioxide

15 Which is not the correct statement? **[OJEE; DUMET 2011]**
(a) The S_8 ring is not planar
(b) Oxygen is more electronegative than sulphur
(c) SF_4 exists but OF_4 does not exist
(d) SO_3 and SO_3^{2-} both have trigonal planar geometry

16. When nitric acid reacts with nitric oxide a gas is released, which converts H_2S into **[AIIMS 2009]**
(a) SO_4^{2-} (b) S^{2-}
(c) S (d) $S_2O_5^{2-}$

ANSWERS

Check point 8.1

1. (a)	2. (d)	3. (a)	4. (c)	5. (c)	6. (c)	7. (c)	8. (b)	9. (d)

Check point 8.2

1. (c)	2. (d)	3. (c)	4. (b)	5. (b)	6. (c)	7. (c)	8. (d)	9. (c)

Check point 8.3

1. (c)	2. (d)	3. (a)	4. (b)	5. (c)	6. (c)	7. (b)	8. (d)	9. (b)	10. (c)

(A) Topical exercise

1. (a)	2. (c)	3. (c)	4. (d)	5. (b)	6. (d)	7. (a)	8. (c)	9. (d)	10. (c)
11. (a)	12. (c)	13. (a)	14. (d)	15. (b)	16. (b)	17. (b)	18. (d)	19. (c)	20. (a)
21. (b)	22. (b)	23. (c)	24. (d)	25. (b)	26. (d)	27. (d)	28. (d)	29. (a)	30. (a)
31. (b)	32. (c)	33. (c)	34. (a)	35. (c)	36. (a)	37. (d)	38. (a)	39. (a)	40. (c)
41. (a)	42. (a)	43. (a)	44. (d)	45. (b)	46. (a)	47. (d)	48. (c)	49. (d)	50. (d)
51. (a)	52. (b)	53. (b)	54. (c)	55. (d)	56. (c)	57. (c)	58. (c)	59. (a)	60. (c)
61. (a)	62. (b)	63. (a)	64. (a)	65. (c)	66. (c)	67. (c)	68. (a)	69. (c)	70. (c)
71. (d)	72. (a)	73. (d)	74. (b)	75. (a)					

(B) Miscellaneous exercise

1. (d)	2. (a)	3. (b)	4. (a)	5. (b)	6. (b)	7. (c)	8. (a)	9. (a)	10. (d)
11. (d)	12. (c)	13. (b)	14. (c)	15. (c)	16. (d)	17. (c)	18. (b)	19. (d)	20. (c)
21. (a)									

(C) Medical entrances' gallery

1. (b)	2. (b)	3. (a)	4. (a)	5. (c)	6. (b)	7. (c)	8. (c)	9. (c)	10. (d)
11. (a)	12. (c)	13. (c)	14. (a)	15. (d)	16. (c)				

Hints & Explanations

(A) Topical exercise

10 (*c*) Oxygen in H_2O is attached to two H_2O molecules by H-bonding, thus coordination number of oxygen in H_2O is four.

$$\begin{array}{c} \quad H \\ \delta+ \;\; | \\ H—O \\ \vdots \\ \delta- \quad \delta+ \\ H—O\cdots H—O \\ \delta- \\ | \qquad\quad | \\ H \qquad\quad H \end{array}$$

11 (*a*) Sulphur has vacant *d*-orbital in which electrons are excited and take part in π-bonding.

$$S{=}C{=}S$$

13 (*a*) Best reducing agent is the species that is in least oxidation state and can be oxidised to maximum oxidation state.

	Oxidation state
H_2S	$S = -2$
Cl^-	$Cl = -1$
SO_4^{2-}	$S = +6$
O_3	$O = 0$

14 (*d*) Chalcogens are the elements of oxygen family. i.e. O, S, Se, Te, Po

Thus, Na is not the chalcogen.

15 (*b*) $SO_2 + H_2O \longrightarrow H_2SO_3$

$H_2 + H_2O \longrightarrow$ No reaction

$NH_3 + H_2O \longrightarrow NH_4OH$

$CO_2 + H_2O \longrightarrow H_2CO_3$

17 (*b*) $SO_2 + 2H_2O \longrightarrow H_2SO_4 + \underset{\text{(Reducing)}}{2H}$

20 (*a*) Acidic nature of hydride

$H_2O < H_2S < H_2Se < H_2Te$

$K_a \longrightarrow$

$pK_a \longleftarrow$

$(= -\log K_a)$

maximum

23 (*c*) Oxygen is more electronegative than C or N, thus bonding pair is pulled by oxygen making ionisation of H_2O easier and thus making it more acidic than NH_3 and CH_4.

34 (*a*) $CH_3CH{=}CHCH_3 \xrightarrow{O_3/Zn} 2CH_3CHO$

Rule (Break (C=C) bond and place oxygen at the breaking point).

35 (*c*) $SO_2 + H_2O \longrightarrow H_2SO_3 + 2H$

Bleaching action is due to H.

$H_2O_2 \longrightarrow H_2O + [O]$

Bleaching action is due to O.

O_3 is not a bleaching agent.

38 (*a*)

		Bond order
O=O	(double bond)	2
H—O—O—H	(single bond)	1
O=O → O	(between double bond and single bond)	1.5

Larger the bond order, smaller the bond length.
Thus, (O—O) bond length in $H_2O_2 > O_3 > O_2$.

46 (*a*) Bleaching action is due to reduction. Thus, colour is restored by oxidation.

49 (*d*) $\left[O \leftarrow \underset{\underset{O}{\downarrow}}{S} = O \right]^{2-}$ $d\pi$-$p\pi$ bond due to presence of *d*-orbital in sulphur and this is delocalised by resonance. Thus, (a) is true.

$$\left[O = \underset{\underset{O}{\downarrow}}{S} \rightarrow O \right]^{2-}$$

(b) Due to resonance (S—O) bonds are equivalent. Thus, (b) is true.

(c) S-atom is sp^3-hybridised.

Thus, (c) is also true.

53 (*b*) $ZnCl_2 + H_2S \longrightarrow ZnS + 2HCl$

ZnS is soluble in HCl. When H_2S is passed into $ZnCl_2$ solution, both ZnS and HCl are formed and ZnS dissolves in HCl. To precipitate Zn^{2+} as ZnS, aqueous solution of $ZnCl_2$ is made neutral by NH_4OH.

55 (*d*) Conc. H_2SO_4 is used as a dehydrating agent in the synthesis of ether and alkene.

$CH_3CH_2OH \xrightarrow{110°C,\ \text{conc. } H_2SO_4} CH_3CH_2OCH_2CH_3$

$CH_3CH_2OH \xrightarrow{140°C,\ \text{conc. } H_2SO_4} CH_2{=}CH_2$

$POCl_3$, P_2O_5 are also act as dehydrating agents.

$CH_3CONH_2 \xrightarrow{P_2O_5 \text{ or } POCl_3} CH_3CN$

56 (*c*)

(a) $CuO + SO_2 \longrightarrow CuSO_3$
It is neutralisation reaction.

(b) $2H_2S + SO_2 \longrightarrow 2H_2O + 3S$
It is a redox reaction. H_2S has been oxidised and SO_2 has been reduced. In this case SO_2 is used as oxidising agent

(c) $\underset{+7}{2MnO_4^-} + \underset{+2}{5SO_2} + 16H^+ \longrightarrow 2Mn^{2+} + 5SO_4^{2-} + 8H_2O$

SO_2 is a reducing agent.

57 (*c*) H_2SO_4 acts as oxidising agent and give out CO_2 and SO_2.

58 (*c*) $2H_2S + SO_2 \longrightarrow 2H_2O + 3S$

65 (*c*) $\left[O-\overset{\overset{\displaystyle S}{\|}}{\underset{\underset{\displaystyle O}{\|}}{S}}-O \right]^{2-}$

67 (*c*) $\left[O=\overset{\overset{\displaystyle O}{|}}{\underset{\underset{\displaystyle O}{|}}{S}}=S \right]^{2-}$

68 (*a*) $KMnO_4$: Oxidation number of Mn = + 7

$H_2S_2O_8$ (perdisulphuric acid) : Oxidation number of S = + 6

Both are in maximum oxidation state.

Thus, $H_2S_2O_8$ can't be reduced by $KMnO_4$.

Thus, (a).

70 (*c*) $H_2SO_5 + H_2O \longrightarrow H_2SO_4 + H_2O_2$

$H_2S_2O_8 + 2H_2O \longrightarrow 2H_2SO_4 + H_2O_2$

Both have peroxy linkages.

72 (*a*) Compounds in which the central atom has more than eight electrons in the valence shell are known as hypervalent compounds. So, to know which one of the given ions are hypervalent first calculate their valence shell electron (of central atom).

(a) Structure of sulphite ion (SO_3^{2-}) $\bar{O}-S(=O)-\bar{O}$

In this case S has 8 electrons in valence shell, therefore it not hypervalent ion.

(b) Structure of phosphate ion (PO_4^{3-}) $\bar{O}-P(\bar{O})(=O)-\bar{O}$

In this case P has 10 electrons in valence shell, therefore, it is a hypervalent ion.

(c) Structure of sulphate ion (SO_4^{2-}) $\bar{O}-S(=O)_2-\bar{O}$

In this case S, has 12 electrons in valence shell, therefore, it is also a hypervalent ion.

(d) Structure of perchlorate ion (ClO_4^-) $\bar{O}-Cl(=O)_3$

In this case, Cl has 14 electrons in valence shell, therefore it is also a hypervalent ion.

73 (*d*) S_3O_9 is a trimer of SO_3 joined together.

(Cyclic structure: three SO_2 units joined by bridging O atoms in a six-membered S–O ring.)

74 (*b*) $H_2C_2O_4 \xrightarrow{H_2SO_4} H_2O + CO + CO_2$

(B) Miscellaneous exercise

1 (*d*) $\underset{(A)}{CaOCl_2} + H_2O \longrightarrow Ca(OH)_2 + Cl_2$

$Cl_2 + H_2O \longrightarrow 2HCl + [O]$

Bleaching action is due to oxidation by Cl_2.

Excess of Cl_2 is removed by hypo (*B*)

$\underset{(B)}{Na_2S_2O_3} + Cl_2 \longrightarrow 2NaCl + Na_2S_4O_6$

2 (*a*) AgBr is soluble in $Na_2S_2O_3$ due to complex formation $Na_3[Ag(S_2O_3)_2]$.

$AgBr + 2Na_2S_2O_3 \longrightarrow Na_3[Ag(S_2O_3)_2] + NaBr$

3 (*b*) $2AgNO_3 + Na_2S_2O_3 \longrightarrow \underset{\text{White}}{Ag_2S_2O_3} + 2NaNO_3$

$Ag_2S_2O_3 + H_2O \longrightarrow \underset{\text{Black}}{Ag_2O} + H_2SO_4$

4 (*a*) $Ba(NO_2)_2 + H_2SO_4 \longrightarrow BaSO_4 + 2HNO_2$

5 (*b*) $SO_2 + Cl_2 \xrightarrow{\text{Sunlight}} SO_2Cl_2$

6 (*b*) $Na_2CO_3 + H_2O \longrightarrow NaOH + H_2CO_3$

$2NaOH + SO_2 \longrightarrow Na_2SO_3 + H_2O$

10 (*d*) S_3O_9 is a trimer of SO_3 and there is no (S—S) bond. Its structure is shown below

(Cyclic structure: three SO_2 units joined by bridging O atoms.)

11 (*d*) $Na_2S_2O_3$, oxidation number of sulphur = – 2, + 6

$^{\ominus}O-\overset{\overset{\displaystyle O}{\|}}{\underset{\underset{\displaystyle O^{\ominus}}{|}}{S}}=S$ (central S: +6, terminal S: –2)

$Na_2S_4O_6$, oxidation number of sulphur = 0, + 5

$^{\ominus}O-S(=O)_2-S-S-S(=O)_2-O^{\ominus}$

$CaOCl_2$, oxidation number of chlorine (ClO^-, Cl^-) = –1, + 1

12 (*c*) $Cr_2O_7^{2-} + H^+ + e^- \longrightarrow \underset{\text{Green}}{Cr^{3+}}$

Green colour is due to formation of Cr^{3+} by reduction of $Cr_2O_7^{2-}$ using a suitable oxidising agent.

$$SO_2 + 2H_2O \longrightarrow SO_4^{2-} + 4H^+ + 2e^-$$

H_2SO_3 is dissolved SO_2 in H_2O.

H_2SO_4 is not the reducing agent.

Thus, (a) is incorrect.

SO_3 is not the reducing agent.

H_2S can reduce $Cr_2O_7^{2-}$ to Cr^{3+}

$$H_2S + Cr_2O_7^{2-} \longrightarrow S + Cr^{3+}$$

Thus, (b) is incorrect (due to SO_3)

$$SO_3^{2-} + Cr_2O_7^{2-} + H^+ \longrightarrow SO_4^{2-} + Cr^{3+}$$

$$Fe^{2+} + Cr_2O_7^{2-} + H^+ \longrightarrow Fe^{3+} + Cr^{3+}$$

Thus (c) is correct.

13 (*b*) H_2SO_4 absorbs H_2O from atmosphere (hygroscopic), hence overflows.

16 (*d*) $FeS + 2HCl \longrightarrow \underset{(A)}{FeCl_2} + \underset{(B)}{H_2S(g)}$

(a) $\underset{\underset{\text{Yellow}}{(C)}}{2FeCl_3} + H_2S \longrightarrow \underset{\underset{\text{Green}}{(A)}}{2FeCl_2} + 2HCl + S$

(b) $(CH_3COO)_2Pb + \underset{(B)}{H_2S} \longrightarrow CH_3COOH + \underset{\text{Black}}{PbS}$

(c) $FeCl_2 + K_3[Fe(CN)_6] \longrightarrow \underset{\text{Turnbull's blue}}{Fe_3[Fe(CN)_6]_2}$

(a), (b) and (c) are true.

17 (*c*) H_2SO_4 is a moderately strong oxidising agent which oxidises both metals and non-metals as shown below,

$$Cu + 2H_2SO_4\,(\text{conc.}) \longrightarrow CuSO_4 + SO_2 + 2H_2O$$

$$S + 2H_2SO_4\,(\text{conc.}) \longrightarrow 3SO_2 + 2H_2O$$

While carbon on oxidation with H_2SO_4 produces two types of oxides CO_2 and SO_2.

$$C + 2H_2SO_4\,(\text{conc.}) \longrightarrow CO_2 + 2SO_2 + 2H_2O$$

18 (*b*) In qualitative analysis when H_2S is passed through an aqueous solution of salt acidified with dil. HCl a black ppt of CuS is obtained.

$$CuSO_4 + H_2S \xrightarrow{\text{Dil. HCl}} \underset{\text{Black ppt}}{CuS} + H_2SO_4$$

On boiling CuS with dil. HNO_3 it forms a blue coloured solution and the following reactions occur

$$3CuS + 8HNO_3 \longrightarrow 3Cu(NO_3)_2 + 2NO + 3S + 4H_2O$$

$$S + 2HNO_3 \longrightarrow H_2SO_4 + NO$$

$$2Cu^{2+} + SO_4^{2-} + 2NH_3 + 2H_2O \longrightarrow Cu(OH)_2 \cdot CuSO_4 + 2NH_4OH$$

$$Cu(OH)_2 \cdot CuSO_4 + 8NH_3 \longrightarrow \underset{\substack{\text{Tetraamminecopper (II)} \\ \text{(Deep blue solution)}}}{2\,[Cu\,(NH_3)_4]SO_4} + 2OH^- + SO_4^{2-}$$

(C) Medical entrances' gallery

1 (*b*) The structure of given sulphur oxoacids are

- H_2SO_4 (Sulphuric acid) ⇒ HO—S(=O)(=O)—OH
- $H_2S_2O_8$ (Peroxodisulphuric acid) ⇒ HO—S(=O)(=O)—O—O—S(=O)(=O)—OH (Peroxy linkage: O—O)
- $H_2S_2O_7$ (Pyrosulphuric acid) ⇒ HO—S(=O)(=O)—O—S(=O)(=O)—OH
- H_2SO_3 (Sulphurous acid) ⇒ HO—S(=O)—OH

2 (*b*) For group 16 elements, the hydrides with higher molar mass (e.g. H_2Po) are less thermally stable than hydride with lower molar mass (e.g. H_2O). This is due to the increase in size of central atom, which results to the weakening of M—H bond due to increased bond length.

Thus, the correct order of thermal stability for H_2E (E = O, S, Se, Te and Po) is as follows:

$H_2Po < H_2Te < H_2Se < H_2S < H_2O$

3 (*a*) Oleum, i.e. $H_2S_2O_7$ can be prepared by absorption of SO_3 in conc. H_2SO_4 as follows:

$$SO_3 + H_2SO_4 \longrightarrow \underset{\text{Oleum}}{H_2S_2O_7}$$

It's structure is as follows:

HO—S(=O)(=O)—O—S(=O)(=O)—OH

Thus, it has no O— O bond and has 2 —OH groups.

Thus, option (a) is the correct.

4 (*a*) The correct formula of oleum is $H_2S_2O_7$ which is also known as pyrosulphuric acid.

HO—S(=O)(=O)—O—S(=O)(=O)—OH

5 (*c*) Ozone is not stable at high temperature.

So, it decomposes into oxygen at high temperature.

$$2O_3 \longrightarrow 3O_2$$

Thus, the reverse of this reaction is non-spontaneous at high temperature.

6 (*b*) The correct formula of Marshall's acid is $H_2S_2O_8$. It is known as peroxodisulphuric acid having sulphur (S) in +6 oxidation state. The structure of this acid is shown below:

$$HO-\overset{\overset{O}{\|}}{\underset{\underset{O}{\|}}{S}}-O-O-\overset{\overset{O}{\|}}{\underset{\underset{O}{\|}}{S}}-OH$$

$(H_2S_2O_8)$

9 (*c*) When SO_2 is passed through acidified $K_2Cr_2O_7$ solution, green chromium sulphate is formed. In this reaction, oxidation state of Cr changes from + 6 to + 3.

$$\underset{\text{O.S of Cr} = +6}{K_2Cr_2O_7} + H_2SO_4 + 3SO_2 \longrightarrow K_2SO_4 + \underset{\substack{\text{O.S of Cr} = +3 \\ \text{(Green)}}}{Cr_2(SO_4)_3} + H_2O$$

The appearance of green colour is due to the reduction of chromium metal.

10 (*d*) Nitrogen dioxide and sulphur dioxide forms acid rain. 'Acid rain' is the rain water containing sulphuric acid and nitric acid.

(a) $HNO_2 + 2H_2O + O_2 \longrightarrow 4HNO_3$

$2SO_2 + 2H_2O + O_2 \longrightarrow 2H_2SO_4$

(b) NO_2 and SO_2 act as a good reducing agent e.g. SO_2 reduces halogens to halogen acid

$Cl_2 + 2H_2O + SO_2 \longrightarrow 2HCl + H_2SO_4$

(c) NO_2 and SO_2 both are soluble in water.

(d) SO_2 is used in the manufacture of sodium bisulphite $(NaHSO_3)$ which is used as a preservative for jams, jellies and squashes.

11 (*a*) Acidic strength of hydrides increases as the size of central atom increases which weakens the *M*—H bond. Since, the size increases from S to Te, thus acidic strength follows the order

$$H_2S < H_2Se < H_2Te$$

or as we go down the group, acidic strength of hydride increases.

12 (*c*)

Element	Allotropes
(a) Polonium	α-and β-polonium
(b) Tellurium	None
(c) Selenium	Trigonal, rhombohedral, three deep red monoclinic forms (α, β and γ-amorphous red, black vitrous selenium).
(d) Oxygen	Dioxygen and ozone

13 (*c*) On mixing aqueous solution of H_2S and SO_2, colloidal sulphur is obtained.

$$2H_2S + SO_2 \longrightarrow 3S + 2H_2O$$

14 (*a*) $2H_2SO_4 + 2P_2O_5 \longrightarrow \underset{\substack{\text{Sulphur} \\ \text{trioxide}}}{2SO_3} \uparrow + 4HPO_3$

15 (*d*) SO_3 is trigonal planar due to sp^2-hybridised sulphur. SO_3^{2-} is sp^3-hybridised, but pyramidal due to presence of lone pair of electron.

16 (*c*) $2HNO_3 + NO \longrightarrow H_2O + 3NO_2 \uparrow$

$$NO_2 + H_2S \longrightarrow H_2O + S + NO$$

CHAPTER 09

p-Block Elements 3 (Halogen Family)

Group 17 (earlier called VII A) consists six elements, namely fluorine (F), chlorine (Cl), bromine (Br), iodine (I), astatine (At) and the recently discovered element, tennessine (Ts, earlier called ununseptium-Uus). These are collectively called **halogens** due to their property of forming salts (*Greek halo*-means salt). The first four members (F, Cl, Br and I) are non-metals and show similar properties. However, not much is known about the rest two members (At and Ts) except that astatine (At) is radioactive.

Occurrence

Fluorine is found mainly as insoluble fluorides, like cryolite (Na_3AlF_6), fluorspar (CaF_2) and fluoroapatite $[Ca_3(PO_4)_2 \cdot CaF_2]$. Fluorine is also present in bones and teeth of animals, plants, river water in low concentrations. Chlorine is present in minerals like rock salt (NaCl), carnallite ($KCl \cdot MgCl_2 \cdot 6H_2O$) and in sea water as soluble NaCl. Sea water also contains different salts of bromine and iodine. The main sources of iodine are certain sea weeds (like kelp that contains 0.5% iodine) and chile saltpetre (contains about 0.2% sodium iodate).

Inside

- Occurrence
- General and physical properties
- Chemical properties
  - *Reaction with oxygen*
  - *Reaction with water*
  - *Reaction with hydrogen*
  - *Reaction with alkalies*
  - *Reaction with metals*
- Anomalous behaviour of fluorine
- Uses
- Chlorine
  - *Occurrence*
  - *Preparation*
  - *Manufacture*
  - *Properties*
  - *Uses*
- Compounds of halogens
  - *Hydrogen chloride*
  - *Bleaching powder*
  - *Oxoacids of halogens*
  - *General properties of oxoacids*
  - *Interhalogen compounds*

GENERAL AND PHYSICAL PROPERTIES

The variations of general and physical properties of group 17 elements are summarised in Table 9.1

Table 9.1 Physical properties of the halogens

Properties	F	Cl	Br	I
Atomic number	9	17	35	53
Electronic configuration	[He] $2s^2 2p^5$	[Ne] $3s^2 3p^5$	[Ar] $3d^{10} 4s^2 4p^5$	[Kr] $4d^{10} 5s^2 5p^5$
Electronegativity (Pauling)	4.0	3.0	2.8	2.5
$(IE)_1$/kJ mol^{-1}	1681	1256	1143	1009
EA/kJ mol^{-1}	– 333	– 349	– 325	– 296
Hydration energy of X^-/kJ mol^{-1}	– 515	– 381	– 347	– 305
Atomic radius/pm	72	99	114	133
Ionic radius (X^-)/pm	133	184	196	220
Enthalpy of atomisation (kJ mol^{-1})	79.1	122	111	106

Properties	F	Cl	Br	I
$X_2 + 2e^- \rightarrow 2X^-$ $E°$/V	2.87	1.40	1.09	0.62
Bond energy (X–X)/ kJ mol^{-1}	126	210	158	118
Melting point (X_2)/°C	– 219	– 101	– 7	114
Boiling point (X_2)/°C	– 188	– 34	60	185
Density of solid X_2 (g/cc)	1.3	1.9	3.4	4.93
Oxidising power (X_2)	Most	–	–	Least
Common oxidation state	– 1	–1, +1, +3, +5, +7	–1, +1, +3, +5, +7	–1, +1, +3, +5, +7
Physical form (X_2) at room temperature	Pale yellow gas	Yellow green gas	Dark red liquid	Violet-black

Electronic configuration

The general electronic configuration of group 17 elements is ns^2np^5, where n = 2, 3, 4 and so on. Electronic configurations of each element of group 17 are listed in Table 9.1.

Physical state

Halogens exist as diatomic covalent molecules, held together by weak van der Waals' forces of attraction. Strength of these forces increases as the size of halogen increase from F to I. Hence, F_2 and Cl_2 exist as gases at room temperature, Br_2 is a liquid whereas I_2 exists as solid.

Atomic and ionic radii

The atomic or covalent radii of group 17 elements increase down the group as the atomic number increases. The similar trend is found in case of ionic radii as well. The increase in radius is due to the addition of an extra shell of electron as we go down in a group.

As the size of atom decreases going right in a periodic table, halogens are smaller than the elements at their left in respective periods. This happens due to increase in effective nuclear charge while going right in a period.

Ionisation energy

The first ionisation energy decreases down the group as expected.

$$F > Cl > Br > I$$

This is due to increase in atomic size down the group. As the size increases, it is easier to pull an electron out from the outermost shell.

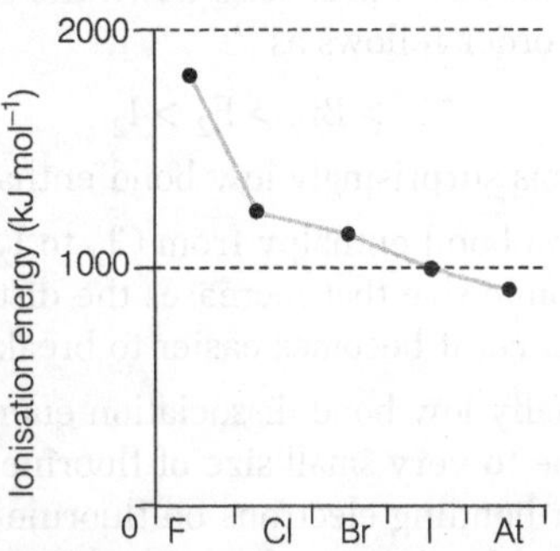

Electron affinity

Halogens have seven electrons in their valence shell, they have a very strong tendency of gaining an electron to acquire a stable noble gas configuration. Therefore, they have very high negative electron affinities. In fact, their electron affinities are the highest in their respective periods. The order follows as

$$Cl > F > Br > I$$

The smaller electron affinity of fluorine than that of chlorine is a consequence of the small size of fluorine and high charge density.

Electronegativity

Electronegativity decreases on moving down in a group, making fluorine the most and iodine the least electronegative among the halogens.

$$F > Cl > Br > I$$

As a result of decrease of electronegativity, non-metallic character decreases from F to I. As we go across the *p*-block elements in a period, the electronegativity increases reaching maximum at the halogen group. Thus, halogens are the most electronegative elements in their respective periods and fluorine is the most electronegative element in the periodic table.

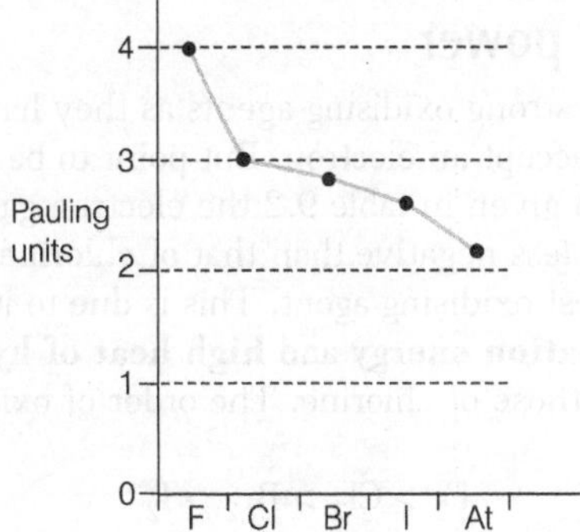

Bond enthalpy

In general, bond enthalpy or bond dissociation energy of halogen molecules (X_2) decreases down the group from Cl_2 to I_2. The order follows as

$$Cl_2 > Br_2 > F_2 > I_2$$

However, F_2 has surprisingly low bond enthalpy.

The decrease in bond enthalpy from Cl_2 to I_2 is due to increase in atomic size that increases the distance between two atoms. Hence, it becomes easier to break them apart.

The exceptionally low bond dissociation energy of fluorine molecule is due to very small size of fluorine atom. This brings the non-bonding electrons on fluorine nearer to each other, resulting in a much greater lone pair-lone pair repulsion, which weakens the covalent bond and lowers its dissociation energy.

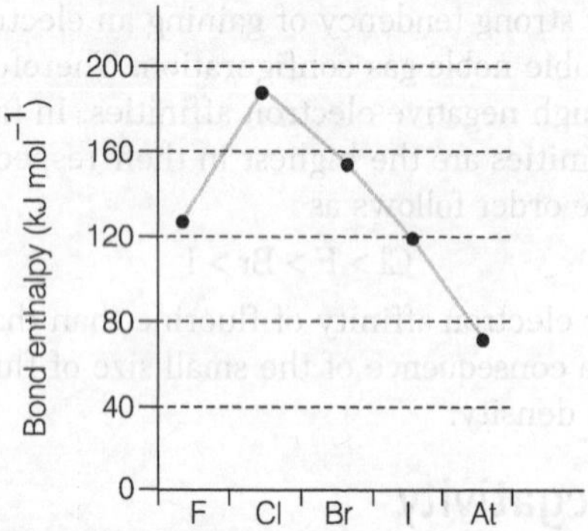

Oxidation state

The most common oxidation state of halogens is −1. However, Cl, Br and I exhibit +1, +3, +5 and +7 oxidation states as well. Fluorine being most electronegative element and devoid of *d*-orbitals does not show oxidation state other than −1 (except in **HOF**, where it shows an oxidation state of +1).

While different other oxidation states shown by the other group members are possible due to presence of *d*-orbitals in them, in which *p*-electrons can be excited to *d*-orbital.

Oxidising power

Halogens are strong oxidising agents as they have strong tendency to accept an electron. But point to be noted that from the data given in table 9.2 the electron gain enthalpy of fluorine is less negative than that of chlorine but still it is the strongest oxidising agent. This is due to its **low bond dissociation energy** and **high heat of hydration** as compared to those of chlorine. The order of oxidising power is

$$F_2 > Cl_2 > Br_2 > I_2$$

F_2 has the most negative $\Delta G°$ value, i.e. −730 kJ mol^{-1}, hence is strongest oxidising agent and can oxidise Cl^-, Br^- and I^- to Cl_2, Br_2 and I_2 respectively. F_2 can also oxidise H_2O to O_2,

$$F_2 + H_2O \longrightarrow 2HF + \frac{1}{2}O_2$$

Similarly, Cl_2 having $\Delta G°$ value −592 kJ mol^{-1} will oxidise Br^- and I^- to Br_2 and I_2 respectively and Br_2 will oxidise I^-. In general, any halogen (X_2) of lower atomic number will oxidise halide ions (X^-) of higher atomic number.

Table 9.2 Free energy ($\Delta G°$) values for $1/2X_2 \longrightarrow X^-$ (hydrated) (All values in kJ mol^{-1})

	1/2 free energy			Electron affinity	Free energy of hydration	$\Delta G°$
	Fusion	Evaporation	Dissociation			
F_2	—	—	126/2	−333	−460	−730
Cl_2	—	—	210/2	−349	−348	−592
Br_2	—	31/2	158/2	−325	−318	−548.5
I_2	15/2	44/2	118/2	−296	−279	−486.5

Solubility

Halogens being non-polar in molecular form are readily soluble in organic solvents like hexane.

However, they're not as much soluble in polar solvents like water. Their solubility in water decreases down the group as

$$F_2 > Cl_2 > Br_2 > I_2$$

Fluorine reacts with water violently. Chlorine and bromine are fairly soluble, while iodine is least soluble.

Melting and boiling points

Melting and boiling points of halogens increase down the group. This trend is in accordance with their increasing atomic size. As the size increases, interparticle attraction increases, which further increase the melting and boiling points.

Colours

All these halogens are coloured. F_2 is yellow, Cl_2 is greenish yellow, Br_2 is reddish brown and I_2 is violet. The colour is because of absorption of energy from visible light by halogen molecules for exicitation of outer electron from the highest occupied π^* molecular orbital to the lowest unoccupied σ^* molecular orbital. As one moves from F_2 to I_2, the difference between the two energy levels decreases and probability of this transition also increases at the same time.

Hence, the absorption maximum gradually shifts from the near UV to the red region of the visible spectrum thus, the intensity of absorption also increases side by side. Hence, we observe the complimentary colour with increasing intensity. Thus, fluorine absorbs violet portion of light and appears yellow while iodine absorbs yellow and green light and appears violet. A general trend is shown below

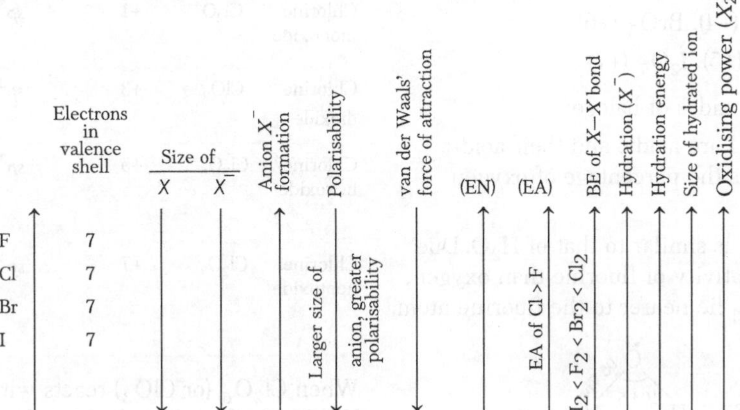

CHECK POINT 9.1

1. Which of the following halogen is a synthetic radioactive element?
(a) Bromine (b) Iodine (c) Astatine (d) Tennessine

2. In the periodic table, on moving down the group 17
(a) ionic radius increases
(b) electronegativity increases
(c) ionisation energy increases
(d) reactivity increases

3. With increase in molecular weight the boiling point of halogens increases, it happens because
(a) bond length increases with increase in size of halogen atom
(b) bond strength increases due to increase in electronegativity
(c) with the increase in size, molecules undergo association leading to higher stability
(d) None of the above

4. The correct order of electron affinity for halogen family is
(a) Cl > F > I > Br (b) F > Cl > Br > I
(c) Cl < F < Br < I (d) Cl > F > Br > I

5. Which of the following statements is/are correct?
(a) All the halogens are gas at room temperature
(b) All the halogens are coloured
(c) All the halogens are radioactive
(d) All of the above statements are true

6. The greater reactivity of F_2 is due to
(a) lower electron-affinity of F
(b) lower bond energy of F—F bond
(c) higher electronegativity of F
(d) gaseous state of F_2

7. Which of the following is the most ionic chloride?
(a) $AlCl_3$ (b) $SnCl_2$
(c) NaCl (d) $SnCl_4$

8. Which of the following is the correct order for solubility in water?
(a) $F_2 > Cl_2 > Br_2 > I_2$ (b) $Cl_2 > F_2 > I_2 > Br_2$
(c) $Cl_2 < F_2 < Br_2 < I_2$ (d) $Cl_2 > F_2 > Br_2 > I_2$

CHEMICAL PROPERTIES

Halogen group is one of the most reactive groups in the periodic table. Having high electron affinity and low bond enthalpy, these can react with most of elements in the periodic table. However, the order of reactivity of the group members decreases down the group as

$$F > Cl > Br > I$$

Having high oxidising power, halogens can oxidise even noble metals (like Ag, Au, Pt) and some noble gases (like Kr, Xe, Rn). Reactivity and chemical properties of halogens are discussed as follows.

Reaction with oxygen

All the halogens are known to form oxides of varying oxidation states but most of them are unstable. Stability of particular type of oxide increases down the group as

$$F < Cl < Br < I$$

(*i*) Fluorine forms two oxides OF_2 and O_2F_2. Among them OF_2 is thermally stable at 298 K. These oxides are essentially oxygen fluorides. This is because of the higher electronegativity of fluorine than oxygen.

(*ii*) Chlorine, bromine and iodine form oxides in which the oxidation states of these halogens range from +1 to +7.

Some of the known oxides of halogens with oxidation state of the respective halogen (atom) are as follows

F : OF_2 (−1), O_2F_2 (−1)

Cl : Cl_2O (+1), ClO_2 (+4), Cl_2O_6 (+6), Cl_2O_7 (+7)

Br : Br_2O (+1), BrO_2 (+4), BrO_3 (+6)

I : I_2O_4 (+4), I_2O_5 (+5), I_2O_7 (+7)

General characteristics of oxides of halogen

(*i*) Oxides of Cl, Br and I are acidic and their acidic character increases as the percentage of oxygen increases in them.

(*ii*) The structure of OF_2 is similar to that of H_2O. Due to greater electronegativity of fluorine than oxygen, the bond pairs in OF_2 lie nearer to the fluorine atom.

O–F 141 pm, F–O–F 103°; O–H 95.7 pm, H–O–H 104.5

But in H_2O, electronegativity of O is greater than H, the bond pair lie nearer to O-atom.

(*iii*) The structure of O_2F_2 is similar to that of H_2O_2 involving sp^3-hybridisation.

H_2O_2: O–O 148 pm; O_2F_2: O–F 122 pm, O–O 158 pm

In this case, the bond length of O—O bond in O_2F_2 is shorter than H_2O_2 because F is much more electronegative than O, therefore it attracts the *lp* of electrons on the O-atom towards itself, which result in decrease in *lp-lp* repulsion between the two O-atom in O—O bond.

O_2F_2 oxidises plutonium to PuF_6 and the reaction is used in removing plutonium as PuF_6 from spent nuclear fuel.

(*iv*) Cl oxides are highly reactive oxidising agent. Their structure and hybridisation are shown in Table 9.3.

(*v*) The oxides of bromine are least stable and are powerful oxidising agents.

(*vi*) Out of three oxides of iodine, I_2O_5 is the most stable oxide. It is the anhydride of iodic acid. It dissolve in water to produce iodic acid. It is also a good oxidising agent and used in estimation of CO from mixture of gases.

Table 9.3 Oxides of chlorine

Name	Formula	Oxidation state of chlorine	Hybridisation of chlorine	Structure
Chlorine monoxide	Cl_2O	+1	sp^3	Cl–O–Cl (bent)
Chlorine dioxide	ClO_2	+3	sp^2	O–Cl–O ⟷ O–Cl–O (resonance)
Chlorine hexoxide	Cl_2O_6	+5	sp^3	O_3Cl—ClO_3
Chlorine heptoxide	Cl_2O_7	+7	sp^3	O_3Cl–O–ClO_3

When Cl_2O_6 (or ClO_3) reacts with KOH, $KClO_3$ and $KClO_4$ are formed indicating that Cl_2O_6 is a mixed anhydride of $HClO_3$ and $HClO_4$

$$2KOH + Cl_2O_6 \longrightarrow KClO_3 + KClO_4 + H_2O$$

In the liquid state and in aqueous solution, there is paramagnetism indicating existence of ClO_3 formed due to dissociation of Cl_2O_6

$$Cl_2O_6 \rightleftharpoons 2ClO_3$$

Paramagnetism is due to unpaired electrons in ClO_3, Cl_2O_6 is diamagnetic in solid state, which is due to paired electrons.

$$O_3Cl—ClO_3 \rightleftharpoons 2\,O_3Cl\bullet$$

Solid has the ionic composition $ClO_2^+ \ ClO_4^-$ confirmed by X-ray study. The liquid may have either of two structures

O_3Cl—ClO_3 or $O_2Cl(\mu\text{-}O)_2ClO_2$

- ClO_2 reacts with KOH forming $KClO_3$ and $KClO_2$.

$$2KOH + 2ClO_2 \longrightarrow KClO_2 + KClO_3 + H_2O$$

This reaction indicates that ClO_2 is a mixed anhydride of $HClO_2$ and $HClO_3$.

Reaction with water

Different halogens react with water in different manners. These reactions represent relative oxidising power of halogens. Fluorine oxidises water to give oxygen gas.

$$F_2 + 2H_2O \longrightarrow 4H^+F^- + O_2$$

Chlorine and bromine form hydrohalic acid and hypohalous acid, upon reactions with water

$$X_2 + H_2O \longrightarrow HX + HOX \text{ (Here, } X = \text{Cl, Br)}$$

Whereas, iodine does not react with water spontaneously.

$$I_2 + H_2O \longrightarrow \text{No reaction}$$

Reaction with hydrogen

Halogens react with hydrogen to form hydrogen halide.

$$X_2 + H_2 \longrightarrow 2HX$$

(Where, X = F, Cl, Br or I)

Fluorine reacts violently with hydrogen while iodine gives very poor yield with hydrogen. The affinity of halogens toward hydrogen decreases in the order

$$F_2 > Cl_2 > Br > I_2$$

General characteristics of hydrogen halides

(*i*) **Physical state** HF is a low boiling liquid and others are colourless, irritating gases at room temperature. HF exists as an associated molecules $[HF]_n$, due to the presence of intermolecular hydrogen bonding. In solid state, $[HF]_n$ has the *zig-zag* structure as shown below.

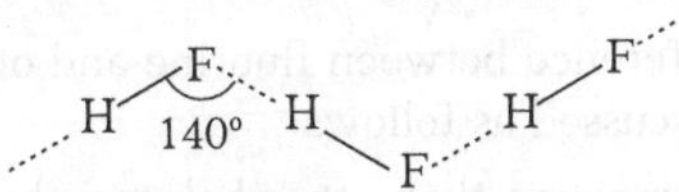

(*ii*) **Boiling point** It increases as we move down in the group due to increase in magnitude of van der Waals' forces of attraction as the size of halogen increase. Boiling point of HF is also normally high due to H-bonding.

(*iii*) **Dipole moment** The order of dipole moment is as follow.

$$\underset{1.86\text{ D}}{HF} > \underset{1.11\text{ D}}{HCl} > \underset{0.79\text{ D}}{HBr} > \underset{0.38\text{ D}}{HI}$$

As the electronegativity of the halogen decreases down the group, dipole moment of halogen halide also decreases in the same order.

(*iv*) **Percentage of ionic character** It increases in the order

$$HI < HBr < HCl < HF$$

As the electronegativity of the halogen decreases down the group, ionic character also decreases in the same direction.

(*v*) **Bond length** It increases in the order

$$\underset{91.7\text{pm}}{HF} < \underset{127.4\text{ pm}}{HCl} < \underset{141.4\text{ pm}}{HBr} < \underset{160.9\text{ pm}}{HI}$$

With the increase in the size of the halogen atom, bond length increases in the same direction.

(*vi*) **Bond strength or bond dissociation energies** It increases in the order

$$HI < HBr < HCl < HF$$

It increases with decrease in the bond length. In simple words, shorter the bond length, higher is the bond strength.

(*vii*) **Thermal stability** It increase in the order

$$HI < HBr < HCl < HF$$

Thermal stability increases with increase in the bond dissociation energy or bond strength.

Note *ΔG_f° of HF, HCl and HBr are large and negative, suggesting that for them $H_2 + X_2 \rightarrow 2HX$ goes to completion.*
For HI (g), ΔG_f° is small and positive suggesting the HI should dissociate

$$2HI(g) \longrightarrow H_2(g) + I_2(g)$$

to some extent. However, because of the high activation energy the dissociation occurs only very slowly in the absence of a catalyst. As a result, HI(g) is quite stable at room temperature. Thus, at room temperature, decomposition of HI(g) is kinetically controlled (rather than thermodynamically controlled).

(*viii*) **Acidic strength** It increases in the order

$$HF < HCl < HBr < HI$$

Higher is the bond dissociation energy, lower is the degree of ionisation. Hence, weaker is the acid. Since, the bond dissociation decreases down the group, therefore ionisation of halogen halide increases in same order.

- Liquid HF has been used as a non-aqueous solvent. It undergoes self ionisation.

$$2HF \rightleftharpoons [H_2F]^+ + F^-$$

Acid-base reactions occur in this solvent system. However, the solvent itself has a very strong tendency to donate protons. Mineral acids like HNO_3, H_2SO_4 and HCl, behave as bases when dissolved in HF, as they accept proton from HF.

$$\underset{\text{Base}}{HCl} + \underset{\text{Acid}}{HF} \rightleftharpoons \underset{\text{Acid}}{H_2Cl^+} + \underset{\text{Base}}{F^-}$$

$$\underset{\text{Base}}{H_2SO_4} + \underset{\text{Acid}}{HF} \rightleftharpoons \underset{\text{Acid}}{H_3SO_4^+} + \underset{\text{Base}}{F^-}$$

$$\underset{\text{Base}}{HNO_3} + \underset{\text{Acid}}{HF} \rightleftharpoons \underset{\text{Acid}}{H_2NO_3^+} + \underset{\text{Base}}{F^-}$$

$HClO_4$ (perchloric acid) however behaves as an acid in liquid HF which acts like a base

$$\underset{\text{Acid}}{HClO_4} + \underset{\text{Base}}{HF} \rightleftharpoons \underset{\text{Acid}}{H_2F^+} + \underset{\text{Base}}{ClO_4^-}$$

(*ix*) **Reducing power** It depends upon the ease with which they decomposes to give H_2 and X_2.

$$2HX \longrightarrow H_2 + X_2$$

Reducing power increases in the order

$$HF < HCl < HBr < HI$$

As the bond dissociation energy decreases down the group, the ease of decomposition increases and hence reducing character increases in same direction.

Reaction with alkalies

Fluorine reacts with alkalies at different conditions. Reaction involved are as follows

$$2F_2 + \underset{\text{Cold dil.}}{2NaOH} \longrightarrow 2NaF + OF_2 + H_2O$$

$$2F_2 + \underset{\text{Hot conc.}}{4NaOH} \longrightarrow 4NaF + O_2 + 2H_2O$$

While reaction involving other halogens (Cl_2, Br_2, I_2) are as follow

$$X_2 + \underset{\text{Cold dil.}}{2NaOH} \longrightarrow NaX + \underset{\text{Hypohalite}}{NaOX} + H_2O$$

$$3X_2 + \underset{\text{Hot conc.}}{6NaOH} \longrightarrow 5NaX + \underset{\text{Halate}}{NaXO_3} + 3H_2O$$

Reaction with metals

Most of the metals react with halogens to form metal halides as

$$2M + nX_2 \longrightarrow 2MX_n$$

(Here M = metal, n = oxidation state of metal, X = F, Cl, Br or I)

The ionic character of halides decreases as the size of halogen ion increases

$$MF_n > MCl_n > MBr_n > MI_n$$

Moreover, the ionic character decreases (and hence the covalent character increases) as the oxidation state of metals combining a halogen increases. Thus, for chlorides of metals with different oxidation state, the ionic character decreases as

$$MCl > MCl_2 > MCl_3 > MCl_4$$

Reaction with some non-metal

(*i*) **With phosphorus** Phosphorus exhibits an oxidation state of +3, when reacts with F_2, Cl_2, Br_2 and I_2.

$$2P + 3X_2 \longrightarrow 2PX_3 \quad \text{(Here, } X = \text{F, Cl, Br, I)}$$

However, phosphorus forms halides with the oxidation state of +5 only with F_2, Cl_2 and Br_2.

$$2P + 5X_2 \longrightarrow 2PX_5 \quad \text{(Here, } X = \text{F, Cl, Br)}$$

(*ii*) **With sulphur** Sulphur forms halides of different oxidation states with different halogens except the iodine. Sulphur does not form a stable compound with iodine. Different reactions involving sulphur and halogens are as follows

$$3F_2 + S \longrightarrow SF_6$$

$$2Cl_2 + S \longrightarrow SCl_4$$

$$X_2 + 2S \longrightarrow S_2X_2 \quad \text{(Here, } X = \text{Cl, Br)}$$

Reaction with a different halogen

A halogen can react with other family members to form stable compounds. These types of compounds are called **interhalogens,** e.g. ClF, ClF_3, BrF_3, BrF_7, IF_7, etc. We will discuss these compounds further in this chapter.

Anomalous behaviour of fluorine

Fluorine differs from other elements of the group because of its exceptionally small atomic and ionic size and low fluorine—fluorine bond dissociation energy. The result of these differences is that fluorine is the most electronegative element in the periodic table and a powerful oxidant.

Some of the difference between fluorine and other halogens are discussed as follows

(*i*) It is more reactive than other halogens because of low F—F bond energy.

(*ii*) Due to its high electronegativity, the bond between fluorine and other elements is very strong, so its compounds are more stable. Some of them are inert, e.g. SF_6.

(*iii*) Fluorine is almost invariably monocoordinate (coordination number = 1).

(*iv*) It always has oxidation state of −1 (exception HOF) and is never more than monocovalent.

(*v*) Fluorine is the strongest oxidising agent and oxidises the elements to their highest oxidation state, e.g. in IF_7, iodine has oxidation state of +7.

(*vi*) Fluorides are more ionic.

(*vii*) Fluorine forms strong hydrogen bonds resulting in the properties of hydrogen fluoride being anomalous (at high bp).

(*viii*) Fluorine forms fumes in moist air and decomposes water to give oxygen, whereas other halogens are sparingly soluble in water and react partly to give hydrohalic and hypohalous acids.

$$2F_2 + 2H_2O \longrightarrow 2HF + O_2$$

$$X_2 + H_2O \rightleftharpoons HOX + HX$$

(*ix*) Fluorine reacts with alkalies to yield the oxide, OF_2

$$2F_2 + 2NaOH \longrightarrow OF_2 + 2NaF + H_2O$$

The other halogens react with alkalies to yield a solution of hypohalite ions, which may disproportionate.

$$Cl_2 + 2NaOH \longrightarrow NaClO + NaCl + H_2O$$

(*x*) The oxides of fluorine are not acidic, while those of other halogens are acidic.

Uses of halogens (Other than chlorine)

The important uses of halogens are given below

Fluorine

(*i*) It is used in the preparation of UF_6, which is used in the gaseous diffusion process for separating isotopes of uranium (U-235 and U -238).

(*ii*) In the preparation of teflon.

Bromine

(*i*) In preparation of dye stuff.

(*ii*) In bromination of organic compounds.

(*iii*) In preparation of AgBr (which used in photographic plate).

Iodine

(*i*) **To prepare tincture of iodine** An antiseptic, (50% by mass) alcohol solution of iodine is called tincture of iodine.

(*ii*) **In preparation of iodised salt** Iodine is necessary for the normal functioning of the thyroid gland. Insufficient iodine in the diet leads to Goitre (enlargement of thyroid gland). 'Iodised salt' contain 0.01% KI or NaI, which is more than sufficient to satisfy 1mg iodine per week required for the formation of thyroxine in the human body.

(*iii*) **In prepare of AgI** AgI is used in cloud seeding (process for inducing rainfall).

Pseudohalogens and pseudohalides

Some molecules like cyanogen, $(CN)_2$, thiocyanogen, $(SCN)_2$ and selenocyanogen, $(SeCN)_2$, have properties similar to those of the halogens. Therefore, these are called **pseudohalogens.**

Pseudohalogens form hydro acids similar to the hydrohalic acids, e.g. H—CN. They also form anions such as cyanide, CN^-, cyanate, OCN^-, thiocyanate, SCN^-, selenocyanate, $SeCN^-$, tellurocyanate, $TeCN^-$ and azide, N_3^-, which resemble the halide ions to some extent.

The best known pseudohalide is the cyanide, CN^-, which resembles Cl^-, Br^- and I^- in the following respects:

- It can be oxidised to form a molecule, $(CN)_2$.
- It forms an acid HCN similar to HCl, HBr, etc.
- It forms insoluble salts with Ag^+, Pb^{2+} and Hg^+.
- Interpseudohalogen compounds ClCN, BrCN and ICN can also be formed.
- Like AgCl, AgCN is insoluble in water but soluble in liquid ammonia.
- It forms a large number of complexes similar to the halide complexes, e.g. $[Cu(CN)_4]^{2-}$ is similar to $[CuCl_4]^{2-}$ and $[Co(CN)_6]^{3-}$ is similar to $[CoCl_6]^{3-}$.

CHLORINE

Chlorine is one of the most important element (or halogen) as it uses in many industrial applications.

Occurrence

Being a reactive element, chlorine is found usually in combined state as metal chlorides. The most common source of chlorine on earth is the common salt (NaCl), which is found in rocks, salty lake water and sea water. Other important minerals composed of chlorine are carnallite ($KCl\cdot MgCl_2\cdot 6H_2O$), sylvine (KCl), chlorapatite $[3Ca_3(PO_4)_2\cdot CaCl_2]$, horn silver (AgCl), etc.

Preparation

(*i*) By heating manganese dioxide with concentrated hydrochloric acid

$$MnO_2 + 4HCl \xrightarrow{\Delta} MnCl_2 + 2H_2O + Cl_2\uparrow$$

(*ii*) By action of concentrated H_2SO_4 on the mixture of MnO_2 and NaCl in presence of heat

$$2NaCl + MnO_2 + 3H_2SO_4 \xrightarrow{\Delta} 2NaHSO_4 + MnSO_4 + 2H_2O + Cl_2\uparrow$$

(*iii*) By action of hydrochloric acid on potassium permanganate

$$2KMnO_4 + 16HCl \longrightarrow 2KCl + 2MnCl_2 + 8H_2O + 5Cl_2\uparrow$$

(*iv*) By action of sulphuric acid on bleaching powder

$$CaOCl_2 + H_2SO_4 \longrightarrow CaSO_4 + H_2O + Cl_2\uparrow$$

(*v*) By action of carbon dioxide on bleaching powder

$$CaOCl_2 + CO_2 \longrightarrow CaCO_3 + Cl_2\uparrow$$

Manufacture

The commercial production of chlorine is done by the following processes

(i) **Electrolysis** This is the most common process to obtain chlorine commercially. In this process chlorine is obtained by the electrolysis of concentrated NaCl solution, where chlorine is liberated at anode.

$$NaCl \xrightarrow{H_2O} Na^+(aq) + Cl^-(aq)$$

At anode, $2Cl^-(aq) \longrightarrow Cl_2(g) + 2e^-$

At cathode, $Na^+(aq) + e^- \longrightarrow Na(s)$

(ii) **Deacon's process** In this process, hydrogen chloride gas is oxidised by the atmospheric oxygen in the presence of $CuCl_2$ (used as catalyst) at 450°C.

$$4HCl(g) + O_2(g) \xrightarrow[450°C]{CuCl_2} 2Cl_2(g) + 2H_2O(l)$$

(iii) **Nitrosyl chloride process** In this process, NaCl is treated with concentrated nitric acid. As a result, chlorine gas along with nitrosyl chloride (NOCl) gas evolves.

$$3NaCl(s) + 4HNO_3(l) \longrightarrow 3NaNO_3(s) + NOCl(g) + Cl_2(g) + 2H_2O(l)$$

Nitrosyl chloride on oxidation with oxygen gives chlorine gas.

$$2NOCl + O_2 \longrightarrow 2NO_2 + Cl_2$$

Properties

(i) Chlorine is a greenish yellow gas with pungent and suffocating smell. At STP it is about 2-5 times heavier than air. Its boiling point is 239 K and is fairly soluble in water.

(ii) Chlorine is almost as electronegative (3.0) as nitrogen. However, Cl does not form hydrogen bond (as N-atoms do) due to its larger atomic size.

(iii) It has high affinity for hydrogen, even when the hydrogen atom is in combined state.

$$H_2 + Cl_2 \longrightarrow 2HCl$$

$$H_2S + Cl_2 \longrightarrow 2HCl + S$$

$$\underset{\text{Turpentine}}{C_{10}H_{16}} + 8Cl_2 \longrightarrow 16HCl + 10C$$

(iv) It forms chlorides with various metals and non-metals.

$$2Na + Cl_2 \longrightarrow 2NaCl$$

$$2Mg + 2Cl_2 \longrightarrow 2MgCl_2$$

$$2Al + 3Cl_2 \longrightarrow 2AlCl_3$$

$$2Fe + 3Cl_2 \longrightarrow 2FeCl_3$$

$$P_4 + 6Cl_2 \longrightarrow 4PCl_3$$

$$S_8 + 4Cl_2 \longrightarrow 4S_2Cl_2$$

(v) Ammonia reacts with chlorine in two different manners as follows

$$NH_3 + \underset{\text{(Excess)}}{3Cl_2} \longrightarrow \underset{\text{(Explosive)}}{NCl_3} + 3HCl$$

$$\underset{\text{(Excess)}}{8NH_3} + 3Cl_2 \longrightarrow 6NH_4Cl + N_2$$

(vi) With sodium hydroxide, chlorine reacts to form sodium hypochlorite (NaOCl) or sodium chlorate ($NaClO_3$) along with NaCl and H_2O.

$$\underset{\text{(Cold and dil.)}}{2NaOH} + Cl_2 \longrightarrow NaCl + NaOCl + H_2O$$

$$\underset{\text{(Hot and conc.)}}{6NaOH} + 3Cl_2 \longrightarrow NaCl + NaClO_3 + 3H_2O$$

(vii) Chlorine reacts with dry slaked lime [$Ca(OH)_2$] to form bleaching powder [Ca(OCl)Cl].

$$2Ca(OH)_2 + 2Cl_2 \longrightarrow Ca(OCl)Cl + CaCl_2 + 2H_2O$$

(viii) Chlorine reacts with hydrocarbons forming substitution product with saturated ones and forming addition product with unsaturated ones.

$$CH_4 + Cl_2 \xrightarrow{\text{UV rays}} CH_3Cl + HCl$$

$$C_2H_4 + Cl_2 \longrightarrow CH_2Cl—CH_2Cl$$

(ix) Chlorine water on standing loses yellow colour due to formation of HCl and HOCl, which in turns provide nascent oxygen [O].

$$Cl_2 + H_2O \longrightarrow HCl + HOCl$$

$$HOCl \longrightarrow HCl + [O]$$

The nascent oxygen, thus formed, acts as oxidising and bleaching agent.

Important examples of oxidising property are

$$\underset{\text{Ferrous sulphate}}{2FeSO_4} + H_2SO_4 + Cl_2 \longrightarrow \underset{\text{Ferric sulphate}}{Fe_2(SO_4)_3} + 2HCl$$

$$\underset{\text{Sodium sulphite}}{Na_2SO_3} + Cl_2 + H_2O \longrightarrow \underset{\text{Sodium sulphate}}{Na_2SO_4} + 2HCl$$

$$SO_2 + 2H_2O + Cl_2 \longrightarrow H_2SO_4 + 2HCl$$

$$I_2 + 6H_2O + 5Cl_2 \longrightarrow 2HIO_3 + 10HCl$$

Uses

(i) It is used in paper and textile industries to bleach wood pulp and cottons.

(ii) It is used in manufacture of dyes, drugs, pesticides (DDT), refrigerants (CCl_4) and other organic compounds.

(iii) As it can react with noble metals, Cl is used in extraction of gold and platinum.

(iv) It is used to sterilise drinking water.

(v) It is used to produce tear gas (CCl_3NO_2), which finds application in controlling violent mobs. It has also been used to produce poisonous gases like phosgene ($COCl_2$) and mustard gas ($ClCH_2CH_2SCH_2CH_2Cl$).

CHECK POINT 9.2

1. Which of the following minerals does not consist of chlorine?
(a) Carnallite
(b) Sylvine
(c) Horn silver
(d) All of these, consist chlorine

2. Which of the following does not react with HCl to produce Cl_2 gas?
(a) MnO_2 (b) $KMnO_4$
(c) $CaOCl_2$ (d) Both (b) and (c)

3. Identify the compound which is not formed by the action of sulphuric acid upon the mixture of NaCl and MnO_2?
(a) $NaHSO_4$ (b) Na_2SO_4
(c) $MnSO_4$ (d) H_2O

4. Choose the suitable reagent(s) that produce(s) Cl_2 gas with bleaching powder.
(a) CO_2 (b) H_2SO_4
(c) NaCl (d) Both (a) and (b)

5. During electrolysis of salt solution the element(s) produced at anode is/are
(a) Cl_2 (b) O_2
(c) Na (d) All of these

6. The catalyst used in Deacon's process of manufacture of chlorine is
(a) $MnCl_2$ (b) MnO_2
(c) $CuCl_2$ (d) $CuSO_4$

7. Which of the following pairs represents the oxidation products of nitrosyl chloride?
(a) $NO + Cl_2$ (b) $NO_2 + Cl_2$
(c) $NO + HCl$ (d) $NO_2 + HCl$

8. Which of the following is/are produced when Cl_2 reacts with excess of ammonia?
(a) NCl_3 (b) HCl
(c) NH_4Cl (d) All of these

9. When Cl_2 is passed through cold dil. NaOH, the products are
(a) NaCl, NaOCl and H_2O (b) NaCl, $NaClO_2$ and H_2O
(c) NaCl, $NaClO_3$ and H_2O (d) NaCl, $NaClO_4$ and H_2O

10. When chlorine is dissolved in water, which of following compounds formed has bleaching property?
(a) $HClO_2$ (b) HOCl
(c) $HClO_3$ (d) ClO_2

COMPOUNDS OF HALOGENS

Hydrogen chloride

This was first prepared by **Glauber** in 1648, when he heated NaCl with conc. H_2SO_4.

Preparation

Laboratory method It is done by the same method as used by Glauber.

$$NaCl + H_2SO_4 \xrightarrow{420\text{ K}} NaHSO_4 + HCl$$
$$NaHSO_4 + NaCl \xrightarrow{823\text{ K}} Na_2SO_4 + HCl$$

HCl gas thus, produced is dried by passing through the conc. H_2SO_4.

Properties

(*i*) HCl is a colourless, pungent smelling gas, which is readily soluble in water and has a boiling point of 190 K.

(*ii*) Although it is a covalent molecule in gaseous state but it dissociates into ions, when dissolved in water.

$$HCl(g) + H_2O(l) \longrightarrow H_3O^+(aq) + Cl^-(aq);\ K_a = 10^7$$

Thus, it acts as strong acid in water.

(*iii*) When reacted with NH_3, it forms fumes of NH_4Cl.

$$NH_3 + HCl \longrightarrow NH_4Cl$$

(*iv*) *Aqua-regia*, formed by mixing 3 parts of HCl with one part of HNO_3, can dissolve even noble metals.

$$Au + 4H^+ + NO_3^- + 4Cl^- \longrightarrow [AuCl_4^-] + NO + 2H_2O$$
$$3Pt + 16H^+ + 4NO_3^- + 18Cl^- \longrightarrow 3[PtCl_6]^{2-} + 4NO + 8H_2O$$

(*v*) It can decompose salts of weaker acids (like carbonates, hydrogen carbonates, sulphites etc.)

$$Na_2CO_3 + 2HCl \longrightarrow 2NaCl + H_2O + CO_2$$
$$NaHCO_3 + HCl \longrightarrow NaCl + H_2O + CO_2$$
$$Na_2SO_3 + 2HCl \longrightarrow 2NaCl + H_2O + SO_2$$

Uses

(*i*) It is used to produce chlorine, ammonium chloride, glucose (from starch) and various other chemical compounds.

(*ii*) It is used as an important laboratory reagent and also in production of various drugs.

(*iii*) It is also employed in extraction of glue from bones and purification of bone black.

Bleaching powder

It is commonly represented as Ca(OCl) Cl or $CaOCl_2$. However, its actual composition is $Ca(OCl)_2 \cdot CaCl_2 \cdot Ca(OH)_2 \cdot 2H_2O$.

Preparation

It is prepared by passing Cl_2 gas over dried slaked lime.

$$3Ca(OH)_2 + 2Cl_2 \longrightarrow Ca(OCl)_2 \cdot CaCl_2 \cdot Ca(OH)_2 \cdot 2H_2O$$

Properties

(*i*) It is pale yellow solid with odour of Cl_2.

(*ii*) In aqueous solution, it gives Cl^-, OCl^- and Ca^{2+} ions.

$$CaOCl_2 \xrightleftharpoons{H_2O} Ca^{2+} + Cl^- + OCl^-$$

(*iii*) It undergoes disproportionation on standing.

$$6CaOCl_2 \longrightarrow Ca(ClO_3)_2 + 5CaCl_2$$

(*iv*) When CO_2 is passed into suspension of bleaching powder in water, HOCl is formed.

$$2CaOCl_2 + H_2O + CO_2 \longrightarrow CaCl_2 + CaCO_3 + 2HOCl$$

(*v*) It is a bleaching agent as well as oxidising agent due to the fact that it forms nasant oxygen in acidic medium/neutral medium.

$$CaOCl_2 + H_2O \longrightarrow Ca(OH)_2 + Cl_2$$
$$Cl_2 + H_2O \longrightarrow 2HCl + [O]$$
$$CaOCl_2 + HCl \longrightarrow CaCl_2 + HOCl$$
$$HOCl \longrightarrow HCl + [O]$$

(*vi*) Important oxidising properties of $CaOCl_2$ include conversions of

- H_2S into S
  $$CaOCl_2 + H_2S \longrightarrow CaCl_2 + H_2O + S$$
- $FeSO_4$ into $Fe_2(SO_4)_3$
  $$CaOCl_2 + 2FeSO_4 + H_2SO_4 \longrightarrow Fe_2(SO_4)_3 + CaCl_2 + H_2O$$
- NH_3 into N_2
  $$3CaOCl_2 + 2NH_3 \longrightarrow 3CaCl_2 + 3H_2O + N_2$$
- KNO_2 into KNO_3
  $$KNO_2 + CaOCl_2 \longrightarrow KNO_3 + CaCl_2$$

Uses

(*i*) It is used as bleaching agent, like chlorine, in textile, paper and related industries.

(*ii*) It is used to disinfect water and sanitise stagnant water including swimming pool.

(*iii*) Due to its oxidising property, it is used in many organic reaction e.g. haloform reaction, which involves synthesis of chloroform. This reaction also serves as a test reaction for many organic compounds.

$$CaOCl_2 + CH_3CH_2OH + H_2O \longrightarrow CHCl_3$$

Oxoacids of halogens

Being highly electronegative and small size, fluorine forms only one oxoacid, HOF, as fluoric (I) acid or hypofluorous acid on reaction with water.

Other member of halogen form four series of oxoacids with formula HO*X* [halic (I) acid or hypohalous acid], HO*XO* [halic (III) acid or halous acid], $HOXO_2$ [halic (V) acid or halic acid] and $HOXO_3$ [halic (VII)] acid or perhalic acid].

Note *Many of these are not known at normal condition.*

Table 9.4 Oxoacids of halogens

Properties	HO*X* hypohalous acid	HXO_2 halous acid	HXO_3 halic acid	HXO_4 perhalic acid	H_5XO_6 para perhalic acid
	HOF	–	–	–	–
	HOCl	$HClO_2$	$HClO_3$	$HClO_4$	–
	HOBr	–	$HBrO_3$	$HBrO_4$	–
	HOI	–	HIO_3	HIO_4	H_5IO_6
Anhydride	X_2O	X_2O_3	X_2O_5	X_2O_7	I_2O_7
Oxidation state of *X*	+I	+III	+V	+VII	+VII
Oxidising power	strong	⟶	⟶	weak	
Acid strength*	weak	⟶	⟶	strong	
Thermal stability	weak	⟶	⟶	good	
Hydrolysis of anion	max	⟵	⟵	mini	
Structure of anion	(OCl^-) linear	(ClO_2^-) angular	(ClO_3^-) trigonal pyramidal	(ClO_4^-) tetrahedral	

**Acid strength of one type oxoacid increases as size of halogen atom increases.*

The important oxyacids of chlorine, their structures, anionic constituents alongwith oxidation state of chlorine, are listed in Table 9.5.

Table 9.5 Oxyacids of chlorine

Name	Formula	Oxidation state of chlorine	Anionic part	Shape and structure
Hypochlorous acid	$HClO$	+1	Hypochlorite, ClO^-	Angular 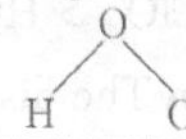
Chlorous acid	$HClO_2$	+3	Chlorite, ClO_2^-	Angular 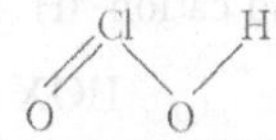
Chloric acid	$HClO_3$	+5	Chlorate, ClO_3^-	Pyramidal 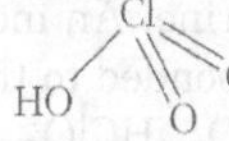
Perchloric acid	$HClO_4$	+7	Perchlorate, ClO_4^-	Tetrahedral 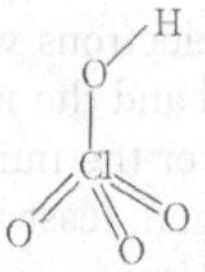

General properties of oxoacids

Thermal stability

It depends on the following factors

(*i*) **Oxidation state and thermal stability** With increase in oxidation state of halogen, thermal stability will also increase and *vice-versa*.

(*ii*) **Electronegativity** With increase in electronegativity of halogen atom, thermal stability will also increase and *vice-versa*.

Case I. **In case of oxoacids of a given halogen atom** i.e. $HClO$, $HClO_2$, $HClO_3$, $HClO_4$

Their thermal stability order is

$$\overset{+1}{HClO} < \overset{+3}{HClO_2} < \overset{+5}{HClO_3} < \overset{+7}{HClO_4}.$$

Reason The electronegativity of the central halogen atom, i.e. Cl-atom in above case remains the same. But the oxidation number of the central atom increases from left to right. With increase in the oxidation state, *X*—O bond becomes more and more covalent and hence, thermal stability of the acids increases in same direction.

Case II. **In case of oxoacids of different halogen having same oxidation state.** Their stability order is

$$\underset{+1}{HOCl} > \underset{+1}{HOBr} > \underset{+1}{HOI}$$

Reason Here, the oxidation state of the halogen atom remains the same, thermal stability depends only upon the electronegativity of the halogen atom. Thermal stability decreases with the decrease in electronegativity of halogen atom.

Oxidising power

Case I. **Oxidising power of oxoacids of a common halogen atom.**

e.g. $HClO > HClO_2 > HClO_3 > HClO_4$.

Reason Here, electronegativity of central atom is same but oxidation state of central atom increases from left to right. As the oxidation state increase, the oxidising power decreases. This is due to increase in the more covalent character of *X*—O bond

Case II. **Oxidising power of oxoacids having same formula and different halogen atom.**

$$HClO_3 > HBrO_3 > HIO_3$$

Reason Oxidising power decreases with the decrease in the electronegativity of the halogen atom. In this case oxidation state of all central atom is same.

Acidity

Case I. **Oxoacids of different halogen atom with same oxidation state.**

$$\overset{+7}{HClO_4} > \overset{+7}{HBrO_4} > \overset{+7}{HIO_4}.$$

Reason As the electronegativity of the halogen decreases, the tendency of the XO_3 group to withdraw electrons of the O — H bonds towards itself decreases.

Case II. **Oxoacids of same halogen in different oxidation state.**

$$\overset{+7}{HClO_4} > \overset{+5}{HClO_3} > \overset{+3}{HClO_2} > \overset{+1}{HClO}$$

Reason The dissociation of an oxoacid involves breaking an O–H bond to produce an anion (OX^-) and cation (H^+ ion)

$$HOX \rightleftharpoons H^+ + OX^-$$

Electronegativity of oxygen is greater than that of chlorine. An increasing number of oxygen atoms are bonded to the chlorine atom is in order HOCl, $HClO_2$, $HClO_3$, $HClO_4$.

The more oxygen atoms that are bonded, the more the electrons will be pulled away from the O–H bond and the more this bond is weakened. Thus, greater the number of oxygen bonds attached to halogen, least is the energy required to break O–H bond.

Note *According to Bronsted-Lowry concept, stronger is the acid, weaker is its conjugate base and vice-versa. The conjugate base of* $HClO$, $HClO_2$, $HClO_3$ *and* $HClO_4$ *are* ClO^-, ClO_2^-, ClO_3^- *and* ClO_4^- *respectively.*

These anions attain stability by dispersal of negative charge. Greater is the number of oxygen atom in anion greater will be the dispersal of negative charge through $p\pi$-$d\pi$ back bonding, more is the stability of anion and hence, stronger is the acid. The order of stability of anion is as follows

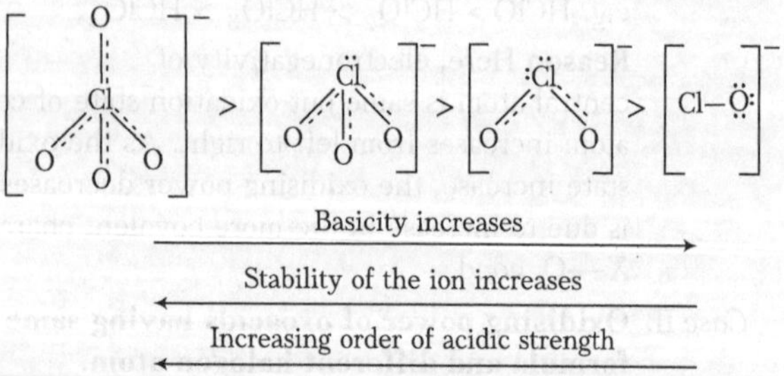

$$[ClO_4]^- > [ClO_3]^- > [ClO_2]^- > [Cl–O]^-$$

Basicity increases →

← Stability of the ion increases

← Increasing order of acidic strength

Since, the stability of anions decreases in the order $ClO_4^- > ClO_3^- > ClO_2^- > ClO^-$, therefore their acid strength also decreases in the same order.

Interhalogen compounds

Interhalogens are the compounds that are formed by two different halogens. Depending upon the ratio in which two halogens combine, they are known to be of four types *viz.* larger halogen always be the central atom. *XY*, XY_3, XY_5 and XY_7 of which *A* is of higher atomic number as shown.

XY	XY_3	XY_5	XY_7
ClF	ClF_3	ClF_5	
BrF	BrF_3	BrF_5	
ICl	ICl_3 (I_2Cl_6)	IF_5	IF_7
IBr			
IBrCl			

Nomenclature

They are named as halogen halides. The halogen with positive oxidation state is named as such while, the halogen with negative oxidation state is named as halide.

e.g. BrCl — Bromine chloride

ICl_3 — Iodine trichloride

Preparation

The type of interhalogen formed, however, depends upon the conditions. Some of the reactions are discussed below

$$\underset{\text{(Equal volume)}}{Cl_2(g) + F_2(g)} \xrightarrow{437\text{ K}} \underset{\text{Chlorine monofluoride}}{2ClF(g)}$$

$$Cl_2(g) + \underset{\text{Excess}}{3F_2(g)} \xrightarrow{573\text{ K}} \underset{\text{Chlorine trifluoride}}{2ClF_3(g)}$$

$$ClF(g) + F_2(g) \xrightarrow{475\text{-}575\text{ K}} ClF_3(g)$$

$$Br_2(l) + \underset{\text{Diluted with water}}{3F_2(g)} \longrightarrow \underset{\text{Bromine trifluoride}}{2BrF_3(l)}$$

$$Br_2(l) + \underset{\text{Excess}}{5F_2} \longrightarrow \underset{\text{Bromine pentafluoride}}{2BrF_5}$$

$$\underset{\text{Equimolar}}{I_2 + Cl_2} \longrightarrow \underset{\text{Iodine monochloride}}{2ICl}$$

$$I_2 + \underset{\text{Excess}}{3Cl_2} \longrightarrow \underset{\text{Iodine trichloride}}{2ICl_3}$$

General and physical properties

(*i*) These are covalent compounds because of small electronegativity difference.

(*ii*) These compounds are of the type XY_n where, *n* is an odd number and *Y* is always lighter halogen and more electronegative, while *X* is larger and less electronegative (when $n > 1$). Because *n* is odd the compounds are diamagnetic, their valence electrons are present as bonding pairs or as shared pairs.

(*iii*) Their physical properties are intermediate between those of constituent halogens, which are summarised in Table 9.6.

Exception Their mp and bp are greater than their constituent halogens because these molecules are more polar in nature.

(*iv*) Interhalogen compounds can never have more than two different halogens in a molecule.

(*v*) Interhalogen ions can be either cations or anions.

Chemical properties

(*i*) **Reactivity** These compounds are generally more reactive than the halogens (except F_2). Their order is as follows

$ClF_3 > BrF_3 > IF_7 > ClF > BrF_3 > IF_5 > BrF > IF_3 > IF$

Reason The *X*—*Y* bond between two dissimilar electronegative elements is weaker than the bond between two similar atoms . This occurs because overlapping of orbitals of two dissimilar atoms is less effective than the overlapping of orbitals of similar atoms.

(*ii*) **Thermal stability** As the size difference or the electronegativity difference between the two halogen atoms decreases, thermal stability also decreases. As the size of central atom increases, stability of the interhalogen compound increases. General order of thermal stability follows as

Thermal stability	IF	> BrF	> ClF	> ICl	> IBr	> BrCl
Electronegativity difference	1.5	1.2	1.0	0.5	0.3	0.2

(*iii*) **Hydrolysis** Interhalogen compounds, when undergo hydrolysis gives halide ion derived from the *smaller halogen* along with a hypohalite (when *XY*) or halite (when XY_3) or halate (when XY_5) or perhalate (when XY_7) anion derived from the larger halogen.

$$ClF + H_2O \longrightarrow HF + HClO$$

(*iv*) Interhalogens compound fluorinate many metal oxides, metal halides and metals.

Table 9.6 Formula, name, physical properties and structure of interhalogen compounds

Type of interhalogen compound	Formula and name	Physical state and colour at ordinary temperatures	Shape	Structure
XY	ClF- chlorine fluoride	Gas, colourless	Linear	*X*—*Y*
	BrF-bromine fluoride	Liquid below 293 K, pale-brown gas above 293 K		
	BrCl-bromine chloride	Liquid, reddish yellow		
	ICl-iodine chloride	Solid, rubby red		
	IBr-iodine bromide	Solid, brown like iodine		
XY_3	ClF_3- chlorine trifluoride	Liquid below 261 K, colourless gas above 261 K	T-shaped	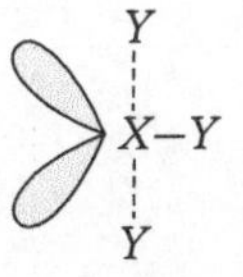
	BrF_3- bromine trifluoride	Liquid, colourless		
	$BrCl_3$- bromine trichoride	Solid, orange		
XY_5	ClF_5- chlorine pentafluoride	Gas, colourless	Square pyramidal	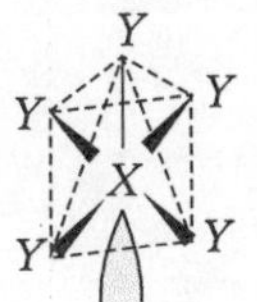
	BrF_5- bromine pentafluoride	Liquid, colourless		
	IF_5- Iodine pentafluoride	Liquid, colourless		
XY_7	IF_7 -iodine heptafluoride	Gas, colourless	Pentagonal bipyramidal	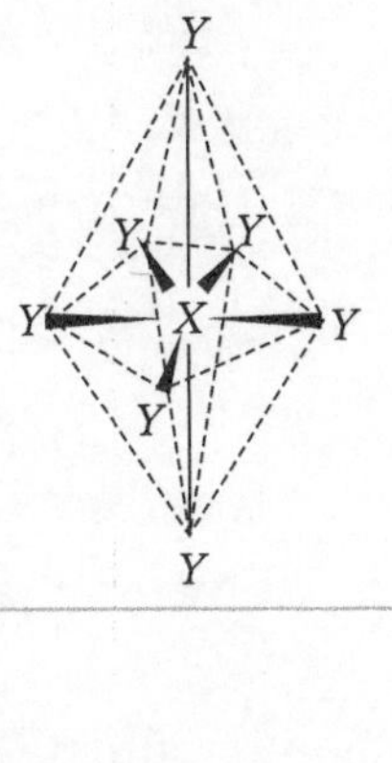

Uses

- These compounds can be used as non-aqueous solvents.
- Interhalogen compounds are used as fluorinating agents.
- ClF_3 and BrF_3 are used in production of UF_6 in enrichment of ^{235}U.

$$U(s) + 3ClF_3(l) \longrightarrow UF_6(g) + 3ClF(g)$$

CHECK POINT 9.3

1. On which of the following compounds, action of sulphuric acid results in formation of hydrochloric acid?
(a) NaCl (b) $Ca(OCl)_2$
(c) NH_4Cl (d) HOCl

2. HCl in gaseous and aqueous states behaves as
(a) ionic and covalent molecules respectively
(b) covalent and ionic molecules respectively
(c) ionic molecules
(d) covalent molecules

3. Bleaching powder contains
(a) CaO and Cl_2 molecules
(b) Ca^{2+} and OCl_2^{2-} ions
(c) Ca^{2+}, Cl^- and OCl^- ions
(d) Ca^{2+}, O^- ions and Cl_2 molecule

4. $CaOCl_2 + H_2O + CO_2 \longrightarrow CaCO_3 + CaCl_2 + HOCl$
This reaction is simply
(a) oxidation
(b) reduction
(c) redox
(d) hydrolysis in presence of CO_2

5. The oxidation state of Cl-atom in $HClO_4$ is
(a) +1 (b) +3
(c) +5 (d) +7

6. Which of the following acids consists chlorite ion?
(a) HClO (b) $HClO_2$
(c) $HClO_3$ (d) $HClO_4$

7. The shape of chloric acid is
(a) angular (b) linear
(c) tetrahedral (d) pyramidal

8. Correct the order of thermal stability is
(a) $HClO > HClO_2 > HClO_3 > HClO_4$
(b) $HClO_4 > HClO_3 > HClO_2 > HClO$
(c) $HClO_3 > HClO_4 > HClO_2 > HClO$
(d) $HClO_2 > HClO_3 > HClO_4 > HClO$

9. The shape of ClF_5 molecule is
(a) tetrahedral (b) trigonal bipyramidal
(c) square pyramidal (d) pentagonal bipyramidal

10. Which of the following interhalogen is not known?
(a) IBr (b) BrF_3
(c) IF_5 (d) ICl_7

Chapter Exercises

(A) Topical exercise

I. General and physical properties

1 The general electronic configuration of the valence shell of a halogen atom is

(a) s^2p^3 (b) s^2p^4
(c) s^2p^5 (d) s^2p^6

2 The halogen, which is liquid at room temperature is

(a) Cl_2 (b) Br_2 (c) I_2 (d) F_2

3 Which of the following atoms has the highest ionisation energy ?

(a) F (b) Cl (c) Br (d) I

4 Which of the following has highest bond energy?

(a) F_2 (b) Cl_2 (c) Br_2 (d) I_2

5 Which is the strongest oxidising agent out of the following?

(a) I_2 (b) Cl_2 (c) Br_2 (d) F_2

6 Consider the following statements.

Statement I Molten I_2 is conducting in nature.

Statement II Bond dissociation energy increases in the order $Br_2 < Cl_2 < I_2$.

(a) Only Statement I is true
(b) Only Statement II is true
(c) Both Statements I and II are true
(d) Both the Statements are false

7 Consider the following statements.

Statement I Fluorine is not super halogen.

Statement II It is most reactive of all the halogen.

(a) Only Statement I is true
(b) Only Statement II is true
(c) Both Statements I and II are true
(d) Both the Statements are false

8 Dipole moment and ionisation constant are maximum in case of ... and ... respectively

(a) HF, HF (b) HF, HI
(c) HI, HF (d) HI, HI

9 Chlorine dioxide

(a) is paramagnetic in nature
(b) has odd electron bond
(c) is stable
(d) has all the above properties true

10 Shape of O_2F_2 is similar to that of

(a) C_2F_2 (b) H_2O_2
(c) H_2F_2 (d) C_2H_2

11 Which of the following has the strongest bond?

(a) Br—F (b) F—F
(c) Cl—F (d) Br—Cl

12 Which of the following is isoelectronic pair? **[NCERT Exemplar]**

(a) ICl_2, ClO_2 (b) BrO_2^-, BrF_2^+
(c) ClO_2, BrF (d) CN^-, O_3

13 Match the species in Column I with the best satisfied corresponding properties in Column II.

Column I	Column II
A. HF	1. Strongest acid
B. HCl	2. Maximum boiling point
C. HBr	3. Oxidised by H_2SO_4
D. HI	4. Least boiling point

Codes

	A	B	C	D		A	B	C	D
(a)	2	4	3	1	(b)	2	4	1	3
(c)	1	3	2	4	(d)	4	3	2	1

14 Consider the following statements.

Statement I HI cannot be prepared by the reaction of KI with concentrated H_2SO_4. **[NCERT Exemplar]**

Statement II HI has lowest H—X bond strength among halogen acids.

(a) Only Statement I is true
(b) Only Statement II is true
(c) Both Statements I and II are true
(d) Both the Statements are false

15 Consider the following statements.

Statement I Fluorine has only –1 oxidation state in compounds.

Statement II Fluorine has maximum electronegativity and electron affinity.

(a) Only Statement I is true
(b) Only Statement II is true
(c) Both Statements I and II are true
(d) Both the Statements are false

16 Which of the following statement(s) is/are correct for all three halogens (X), Cl, Br and I?
(a) They all need to gain one electron to acquire stable configuration.
(b) They all form strong acid of the type HX.
(c) Both (a) and (b) are correct.
(d) They are more electronegative than F.

17 Match the species in Column I with the best satisfied corresponding properties in Column II.

	Column I		Column II
A.	Fluorine (F_2)	1.	Exist in liquid state
B.	Chlorine (Cl_2)	2.	Least (X–X) bond energy
C.	Bromine (Br_2)	3.	Maximum electron gain enthalpy in atomic state
D.	Iodine (I_2)	4.	Best oxidising agent

Codes

	A	B	C	D		A	B	C	D
(a)	4	3	1	2	(b)	4	3	2	1
(c)	3	4	1	2	(d)	3	4	2	1

18 Consider the following statements.

Statement I Cl_2 or Br_2 changes KI into I_2 (violet in CCl_4 layer), Cl_2 changes KBr into Br_2 (orange-yellow in CCl_4 layer). Thus, Cl_2 would change mixture of KI and KBr into violet layer in CCl_4.

Statement II Oxidising power is in order
$$I_2 < Br_2 < Cl_2 < F_2.$$
(a) Only Statement I is true
(b) Only Statement II is true
(c) Both Statements I and II are true
(d) Neither Statement I nor Statement II is true

II. Chemical properties

19 Which of the following is/are interhalogen compounds ?
(a) ICl_3 (b) BrF_5
(c) IF_7 (d) All of these

20 Which of the following will liberate Br_2 from KBr?
(a) I_2 (b) SO_2
(c) HI (d) Cl_2

21 An important product in the ozone depletion by chlorofluorocarbons is
(a) Cl_2 (b) OCl
(c) F_2 (d) O_2F_2

22 Which of the following will displace the halogen from the solution of the halide?
(a) Br_2 added to NaI (b) Br_2 added to NaCl
(c) Cl_2 added to KCl (d) Cl_2 added to NaF

23 The high oxidising power of fluorine is due to
(a) high electron affinity
(b) high heat of dissociation and low heat of hydration
(c) low heat of dissociation and high heat of hydration
(d) high heat of dissociation and high heat of hydration

24 Dissociation of HI into H_2 and I_2 is
(a) kinetically controlled
(b) thermodynamically controlled
(c) kinetically as well as thermodynamically controlled
(d) neither kinetically nor thermodynamically controlled

25 Acid strength of HX (X = F, Cl, Br, I) is dependent on
(a) the electronegativity differences of H and X
(b) the tendency of the hydrated molecule to form H^+
(c) Both (a) and (b)
(d) None of the above

26 Consider the following statements.

Statement I NaCl reacts with concentrated H_2SO_4 to give colourless fumes with pungent smell. But on adding MnO_2 the fumes become greenish yellow. **[NCERT Exemplar]**

Statement II MnO_2 oxidises HCl to chlorine gas, which is greenish yellow.
(a) Only Statement I is true
(b) Only Statement II is true
(c) Both Statements I and II are true
(d) Neither Statement I nor Statement II is true

27 Consider the following statements.

Statement I Hydrofluoric acid forms two series of salts KHF_2, K_2F_2.

Statement II It is a dibasic acid due to hydrogen bonding.
(a) Only Statement I is true
(b) Only Statement II is true
(c) Both Statements I and II are true
(d) Neither Statement I nor Statement II is true

28 Select the incorrect statement.
(a) Cl_2O_6 is diamagnetic in solid state.
(b) ClO_3 is paramagnetic and exists in liquid state.
(c) Cl_2O_6 is the anhydride of $HClO_3$.
(d) ClO_3 forms dimer due to unpaired electron.

29 Which of the following is incorrect statement?
(a) All halogens form oxyacids.
(b) All halogens show –1, +1, +3, +5, +7 oxidation states.
(c) Hydrofluoric acid forms KHF_2 and K_2F_2 and attacks glass.
(d) Oxidising power is in order $F_2 > Cl_2 > Br_2 > I_2$.

30 Which one of the following is the weakest base as per the Bronsted concept?
(a) $[ClO_4]^-$ (b) $[ClO_3]^-$ (c) $[ClO_2]^-$ (d) $[ClO]^-$

31 Which of the following is/are true statement(s)?
(a) Basic nature of X^- is in order $F^- > Cl^- > Br^- > I^-$.
(b) Among all halogen halides, HI is a strongest acid.
(c) The ionic character of M–X bond decreases in the order M–F > M–Cl > M–Br > M–I.
(d) All the above are correct statements.

32 Select the incorrect statement.
(a) ClO_2 and Cl_2O are used as bleaching agents for paper pulp and textiles.
(b) OCl^- (hypohalites) salts are used as detergent.
(c) OCl^- disproportionates in alkaline medium.
(d) BrO_3^- is oxidised to Br_2 by Br^- in acidic medium.

33 Consider the following statements.
Statement I CaF_2 has been given the name fluorspar.
Statement II Solid CaF_2 emits light when heated.
(a) Only Statement I is true
(b) Only Statement II is true
(c) Both Statements I and II are true
(d) Neither Statement I nor Statement II is true

34 Consider the following statements.
Statement I HF is stored in the vessel made of wax.
Statement II HF is a weak acid.
(a) Only Statement I is true
(b) Only Statement II is true
(c) Both Statements I and II are true
(d) Neither Statement I nor Statement II is true

III. Chlorine and Compounds of halogen

35 Which of the following is used to dry HCl gas?
(a) $Ca(OCl)_2$ (b) NaCl
(c) H_2SO_4 (d) $NaHSO_4$

36 When HCl gas reacts with NH_3 gas white fumes are formed. This white fume arises due to formation of
(a) aqueous HCl (b) NH_4OH
(c) water vapour (d) NH_4Cl

37 Which of the following correctly represents the composition of *aqua-regia*?
(a) 1 part HCl, 3 parts H_2SO_4
(b) 1 part HCl, 3 parts HNO_3
(c) 3 parts HCl, 1 part H_2SO_4
(d) 3 parts HCl, 1 part HNO_3

38 Bleaching powder formed when Cl_2 gas is passed over
(a) CaO (b) $Ca(OH)_2$
(c) $CaCO_3$ (d) All of these

39 Which of the following compounds can be use to perform haloform test?
(a) HCl (b) $CaOCl_2$
(c) ClO_2 (d) HClO

40 Which of the following has peroxy linkage?
(a) Perchloric acid
(b) Hypochlorous acid
(c) *para* perchloric acid
(d) None of the above

41 ClO_2 is the anhydride of
(a) HOCl
(b) $HClO_2$
(c) $HClO_3$
(d) $HClO_2$ and $HClO_3$

42 The active constituent of bleaching powder is
(a) $Ca(OCl)_2$ (b) Ca(OCl)Cl
(c) $Ca(ClO_2)_2$ (d) $Ca(ClO_2)Cl$

43 The interhalogen compound that does not exist is
(a) ICl (b) IF_5
(c) BrF_5 (d) $BrCl_7$

44 The tetrahedral arrangement of perchlorate ion is due to
(a) presence of a lone pair of electrons
(b) trigonal bipyramidal shape of the ion
(c) sp^3-hybridisation
(d) sp^2-hybridisation

45 Acidic strength of oxoacids of halogen is in order
(a) HOF > HOCl > HOBr > HOI
(b) HOI > HOBr > HOCl > HOF
(c) HOI > HOCl > HOBr > HOF
(d) HOCl > HOBr > HOI > HOF

46 CH_3CH_2OH on reaction with bleaching powder forms
(a) $CHCl_3$ (b) CCl_3CHO
(c) CH_2ClCHO (d) CCl_4

47 "Chlorine-type" laundry bleaches are in reality aqueous solution of
(a) HOCl (b) $HClO_3$
(c) $NaClO_3$ (d) NaOCl

48 Match the species in Column I with the corresponding properties in Column II.

Column I	Column II
A. Cl_2O_4	1. Anhydride of $HClO_3$ and $HClO_4$
B. Cl_2O_6	2. Paramagnetic
C. ClO_2	3. Anhydride of $HClO_2$ and $HClO_3$
D. Cl_2O_2	4. Unstable and explodes on heating

Codes

	A	B	C	D
(a)	3	1	2	4
(b)	3	1	4	2
(c)	1	3	2	4
(d)	1	3	4	2

(B) Miscalleneous questions

1 The mixture of concentrated HCl and HNO_3 made in 3 : 1 ratio contains

(a) ClO_2 (b) NOCl
(c) NCl_3 (d) N_2O_4

2 Which one of the following oxoacids of chlorine is the least oxidising in nature?

(a) HClO (b) $HClO_2$
(c) $HClO_3$ (d) $HClO_4$

3 Which of the following reaction is not feasible?

(a) $2KI + Br_2 \longrightarrow 2KBr + I_2$
(b) $2KBr + I_2 \longrightarrow 2KI + Br_2$
(c) $2KBr + Cl_2 \longrightarrow 2KCl + Br_2$
(d) $2H_2O + 2F_2 \longrightarrow 4HF + O_2$

4 Affinity for hydrogen decreases in the group from fluorine to iodine. Which of the halogen acids should have highest bond dissociation enthalpy? **[NCERT Exemplar]**

(a) HF (b) HCl (c) HBr (d) HI

5 Bleaching powder is disinfectant for purification of water when water-borne germs are killed. But after some time its disinfectant activity is destroyed. It is due to its disproportionation into

(a) $CaCl_2$ and Cl_2 (b) $CaCl_2$ and $Ca(ClO_3)_2$
(c) CaO and Cl_2 (d) CaO, Cl_2 and $CaCl_2$

6 $HClO_4$, HNO_3 and HCl are all very strong acids in aqueous solution. In glacial acetic acid medium, their acid strength is such that

(a) $HClO_4 > HNO_3 > HCl$
(b) $HNO_3 > HClO_4 > HCl$
(c) $HCl > HClO_4 > HNO_3$
(d) $HCl > HClO_4 > HNO_3$

7 A black compound of manganese reacts with a halogen acid to give greenish yellow gas. When excess of this gas reacts with NH_3, an unstable trihalide is formed. In this process the oxidation state of nitrogen changes from .... **[NCERT Exemplar]**

(a) – 3 to + 3 (b) – 3 to 0
(c) – 3 to + 5 (d) 0 to – 3

8 Which of the following compounds possesses Lewis acid character?

I. BF_3 II. SiF_4
III. PF_5

Select the correct answer using the codes given below.

(a) Only I (b) I, II and III
(c) Both II and III (d) Both I and III

9 Consider the following statements.

Statement I CN^- can be oxidised to $(CN)_2$.

Statement II CN^- is pseudohalide.

(a) Only Statement I is true
(b) Only Statement II is true
(c) Both Statements I and II are true
(d) Both the Statements are false

10 Consider the following statements.

Statement I Aqueous solution of $NaOX$ is alkaline.

Statement II It is a salt of weak acid (HOX) are strong base (NaOH).

(a) Only Statement I is true
(b) Only Statement II is true
(c) Both Statements I and II are true
(d) Both the Statements are false

11 Consider the following boron halides.

I. BF_3 II. BCl_3 III. BBr_3 IV. BI_3

The Lewis acid characters of these halides are such that

(a) I < II< III< IV (b) I< III< II < IV
(c) IV < III< II< I (d) IV < II< III< I

12 Oxidation of hydrogen halide, HX affords a method for the industrial and laboratory preparation of the halogen, X_2 in the free state in respect of all of the following except

(a) fluorine (b) chlorine
(c) bromine (d) iodine

13 Match the items of Columns I and II and mark the correct option. **[NCERT Exemplar]**

Column I	Column II
A. H_2SO_4	1. Highest electron gain enthalpy
B. CCl_3NO_2	2. Chalcogen
C. Cl_2	3. Tear gas
D. Sulphur	4. Storage batteries

Codes

	A	B	C	D		A	B	C	D
(a)	4	3	1	2	(b)	4	3	2	1
(c)	3	4	1	2	(d)	3	4	2	1

14 Consider the following statements.

Statement I Bond dissociation energy of F_2 is smaller than that of Cl_2.

Statement II Bond dissociation energy is in order $F_2 < Cl_2 < Br_2 < I_2$.

(a) Only Statement I is true
(b) Only Statement II is true
(c) Both Statements I and II are true
(d) Both the Statements are false

15 Which of the following arrangements does not truly represent the property indicated against it?

(a) $Br_2 < Cl_2 < F_2$: Oxidising power
(b) $Br < Cl < F$: Electronegativity
(c) $Br < F < Cl$: Electron affinity
(d) $Br_2 < Cl_2 < F_2$: Bond energy

16 Reduction potentials of some ions are given below. Arrange them in decreasing order of oxidising power.

Ion	ClO_4^-	IO_4^-	BrO_4^-
Reduction potential E^-/V	$E^\circ = 1.19$ V	$E^\ominus = 1.65$ V	$E^\ominus = 1.74$ V

[NCERT Exemplar]

(a) $ClO_4^- > IO_4^- > BrO_4^-$
(b) $IO_4^- > BrO_4^- > ClO_4^-$
(c) $BrO_4^- > IO_4^- > ClO_4^-$
(d) $BrO_4^- > ClO_4^- > IO_4^-$

17. Match the species given in Column I with the shape given in Column II and mark the correct option. **[NCERT Exemplar]**

	Column I	Column II
A.	SF_4	1. Tetrahedral
B.	BrF_3	2. Pyramidal
C.	BrO_3^-	3. *Sea-saw* shaped
D.	NH_4^+	4. Bent T-shaped

Codes

	A	B	C	D		A	B	C	D
(a)	3	4	2	1	(b)	3	4	1	2
(c)	4	3	2	1	(d)	4	3	1	2

(C) Medical entrances' gallery

1 The correct structure of tribromooctaoxide is **[NEET 2019]**

(a)
```
  O     O     O
   \\   ||   //
O=Br—Br—Br—O⁻
   //   |    \
  O    O⁻    O⁻
```
(b)
```
  O     O     O⁻
   \\   ||   /
⁻O—Br—Br—Br=O⁻
   /    |    \
 ⁻O    O⁻    O⁻
```
(c)
```
  O     O⁻    O
   \\   ||   //
O=Br—Br—Br—O⁻
   //   |    \\
  O    O⁻    O⁻
```
(d)
```
  O     O     O
   \\   ||   //
O= Br—Br—Br= O
   //   ||   \\
  O     O     O
```

2 Match the following : **[NEET 2019]**

A. Pure nitrogen	(i) Chlorine
B. Haber's process	(ii) Sulphuric acid
C. Contact process	(iii) Ammonia
D. Deacon's process	(iv) Sodium azide or barium azide

Which of the following is the correct option?

	A	B	C	D
(a)	(ii)	(iv)	(i)	(iii)
(b)	(iii)	(iv)	(ii)	(i)
(c)	(iv)	(iii)	(ii)	(i)
(d)	(i)	(ii)	(iii)	(iv)

3 Which of the following statements is not true for halogens? **[NEET 2019]**

(a) Fluorine show positive oxidation states
(b) All are oxidising agents
(c) All form monobasic oxyacids
(d) Chlorine has the highest electron gain enthalpy

4 Correct order of bond dissociation energy is **[AIIMS 2019]**

(a) $Br_2 > Cl_2$ (b) $F_2 > Cl_2$ (c) $I_2 > F_2$ (d) $F_2 > I_2$

5 Yellow colour of chlorine water fades because of **[AIIMS 2019]**

(a) Formation of HCl and HOCl
(b) chlorine gas escapes
(c) $ClO_2 + H_2$
(d) Cl_2O

6 Affinity for hydrogen decreases in the group from fluorine to iodine. Which of the halogen acids should have highest bond dissociation enthalpy? **[AIIMS 2019]**

(a) HF (b) HCl (c) HBr (d) HI

7 $\underset{\text{(Excess)}}{8NH_3} + 3Cl_2 \longrightarrow X$

$NH_3 + \underset{\text{(Excess)}}{3Cl_2} \longrightarrow Y$

What is X and Y in the above reaction? **[AIIMS 2019]**

(a) $X = 6NH_4Cl + N_2, Y = NCl_3 + 3HCl$
(b) $X = NCl_3 + 3HCl, Y = 6NH_4Cl + N_2$
(c) $X = NCl_3 + N_2, Y = 6\,NH_4Cl + 3HCl$
(d) $X = 6NH_4Cl + 3HCl, Y = NCl_3 + N_2$

8 A green yellow gas reacts with an alkali metal hydroxide to form a halate which can be used in fireworks and safety matches. The gas and halate are, respectively **[AIIMS 2019]**

(a) $Br_2, KBrO_3$ (b) $Cl_2, KClO_3$
(c) $I_2, NaIO_3$ (d) $Cl_2, NaClO_3$

9 Match the interhalogen compounds of Column I with the geometry in Column II and assign the correct code. **[NEET 2017]**

	Column I	Column II
p.	XX'	(i) T-shape
q.	XX'_3	(ii) Pentagonal bipyramidal
r.	XX'_5	(iii) Linear
s.	XX'_7	(iv) Square-pyramidal
		(v) Tetrahedral

Code

	p	q	r	s
(a)	(iv)	(iii)	(ii)	(i)
(b)	(iii)	(iv)	(i)	(ii)
(c)	(iii)	(i)	(iv)	(ii)
(d)	(v)	(iv)	(iii)	(ii)

10 $HgCl_2$ and I_2 both when dissolved in water containing I^- ions the pair of species formed is **[NEET 2017]**
(a) Hg_2I_2, I^- (b) HgI_2, I_3^-
(c) HgI_2, I^- (d) HgI_4^{2-}, I_3^-

11 Which one of the following order is correct for the bond dissociation enthalpy of halogen molecules? **[NEET (I) 2016]**
(a) $Cl_2 > Br_2 > F_2 > I_2$ (b) $Br_2 > I_2 > F_2 > Cl_2$
(c) $F_2 > Cl_2 > Br_2 > I_2$ (d) $I_2 > Br_2 > Cl_2 > F_2$

12 Among the following, the correct order of acidity is **[NEET (I) 2016]**
(a) $HClO < HClO_2 < HClO_3 < HClO_4$
(b) $HClO_2 < HClO < HClO_3 < HClO_4$
(c) $HClO_4 < HClO_2 < HClO < HClO_3$
(d) $HClO_3 < HClO_4 < HClO_2 < HClO$

13 Which of the following is the correct Lewis structure for the ionic compound $Ca(ClO_2)_2$? **[GGSIPU 2015]**
(a) $\left[:\ddot{Cl}-\ddot{O}-\ddot{O}:\right] Ca^{2+} \left[:\ddot{O}-\ddot{O}-\ddot{Cl}:\right]$
(b) $:\ddot{Cl}-\ddot{O}-\ddot{O}-Ca-\ddot{O}-\ddot{O}-\ddot{Cl}:$
(c) $\left[O_2\ddot{Cl}\right]^- Ca^{2+} \left[\ddot{Cl}O_2\right]^-$ (each Cl bonded to two O atoms)
(d) $O_2Cl\,Ca-ClO_2$ (each Cl bonded to two O atoms)

14 The aqueous solution of following salt will has the lowest pH **[CET Karnataka 2015]**
(a) $NaClO_3$ (b) NaClO
(c) $NaClO_2$ (d) $NaClO_4$

15 The gaseous product formed when HOCl reacts with H_2O_2 in acidic medium is **[Kerala CEE 2014]**
(a) H_2 (b) Cl_2
(c) O_2 (d) $HClO_2$
(e) $HClO_3$

16 The increasing order of reducing power of halogen acid is **[Manipal 2013]**
(a) HCl < HBr < HF < HI (b) HF < HCl < HBr < HI
(c) HI < HBr < HCl < HF (d) HBr < HCl < HF < HI

17 Which of the following statement is incorrect? **[WB JEE 2013]**
(a) ICl is a good conductor of electricity in fused state.
(b) Cl_2O_7 is anhydride of perchloric acid.
(c) Melting and boiling points of HBr is less than HCl.
(d) F_2 forms only one oxoacid.

18 Sea weeds are an important source of **[WB JEE 2013]**
(a) chlorine (b) bromine
(c) iodine (d) zinc

19 Bleaching powder loses its power on keeping for a long time because **[KCET 2013]**
(a) it changes into calcium hypochlorate
(b) it changes into calcium chloride and calcium hydroxide
(c) it absorbs moisture
(d) it changes into calcium chloride and calcium chlorate

20 When Cl_2 gas reacts with hot and concentrated sodium hydroxide solution, the oxidation number of chlorine changes from **[CBSE AIPMT 2012]**
(a) zero to +1 and zero to −5
(b) zero to −1 and zero to + 5
(c) zero to −1 and zero + 3
(d) zero to + 1 and zero to − 3

21 When a colourless gas is passes through bromine water, only decolourisation takes place. The gas is **[AFMC 2012]**
(a) SO_2 (b) HBr
(c) HCl (d) H_2S

22 Which of the following is not a peroxy acid? **[Manipal 2012]**
(a) Perphosphoric acid (b) Pernitric acid
(c) Perdisulphuric acid (d) Perchloric acid

23 Which of the following is not a pseudohalide? **[OJEE 2012]**
(a) $RCOO^-$ (b) N_3^-
(c) CNO^- (d) CN^-
(e) Cl_2 is liberated (f) Cl_2 and I_2 are liberated

24 When Br_2 is treated with aqueous solutions of NaF, NaCl and NaI separately **[Kerala CEE 2011]**
(a) F_2, Cl_2 and I_2 are liberated
(b) F_2 and Cl_2 are liberated
(c) I_2 is liberated
(d) Cl_2 is liberated
(e) Cl_2 and I_2 are liberated

ANSWERS

Check Point 9.1

1. (d)	2. (a)	3. (a)	4. (d)	5. (b)	6. (b)	7. (c)	8. (a)

Check Point 9.2

1. (d)	2. (c)	3. (b)	4. (d)	5. (a)	6. (c)	7. (b)	8. (c)	9. (a)	10. (b)

Check Point 9.3

1. (a)	2. (b)	3. (c)	4. (d)	5. (d)	6. (b)	7. (d)	8. (b)	9. (c)	10. (d)

(A) Topical exercise

1. (c)	2. (b)	3. (a)	4. (b)	5. (d)	6. (a)	7. (c)	8. (b)	9. (d)	10. (b)
11. (b)	12. (b)	13. (a)	14. (c)	15. (a)	16. (c)	17. (a)	18. (c)	19. (d)	20. (d)
21. (b)	22. (a)	23. (c)	24. (a)	25. (c)	26. (c)	27. (c)	28. (c)	29. (b)	30. (a)
31. (d)	32. (b)	33. (c)	34. (c)	35. (c)	36. (d)	37. (d)	38. (b)	39. (b)	40. (d)
41. (d)	42. (b)	43. (d)	44. (c)	45. (a)	46. (a)	47. (a)	48. (a)		

(B) Miscellaneous questions

1. (b)	2. (d)	3. (b)	4. (a)	5. (b)	6. (a)	7. (a)	8. (a)	9. (c)	10. (c)
11. (a)	12. (a)	13. (a)	14. (a)	15. (d)	16. (c)	17. (a)			

(C) Medical entrances' gallery

1. (d)	2. (c)	3. (a)	4. (d)	5. (a)	6. (a)	7. (a)	8. (b)	9. (c)	10. (d)
11. (a)	12. (a)	13. (c)	14. (d)	15. (c)	16. (b)	17. (c)	18. (c)	19. (d)	20. (b)
21. (a)	22. (d)	23. (a)	24. (c)						

Hints & Explanations

(A) Topical exercise

6 (*a*) I^+ and I_3^- are formed that's why molten I_2 is conducting in nature. Statement II is false the correct order is $Cl_2 > Br_2 > I_2$

8 (*b*) HI is the strongest acid hence has highest ionisation constant. HF has maximum dipole moment.

9 (*d*) ClO_2 has following structure

$$:\ddot{O}: \;{}^{\times\times}\;\overset{\times\times}{\underset{\times}{Cl}}\;{}^{\times\times}\; :\ddot{O}:$$

- Each Cl—O bond is formed by three electrons, thus is an odd electron bond.
- There is one unpaired electron on Cl-atom thus, paramagnetic in nature.
- There are resonance structures due to 3-electron bonds and thus stable.

12 (*b*) Isoelectronic pair has same number of electrons.

	BrO_2^-	BrF_2^+
Total number of electrons	$= 35 + 2 \times 8 + 1 = 52$	$= 35 + 9 \times 2 - 1 = 52$

Hence, (b) is the correct choice, while in another cases this value is not equal.

ICl_2	ClO_2
$53 + 2 \times 17 = 87$	$17 + 16 = 33$

ClO_2	BrF
$17 + 16 = 33$	$35 + 9 = 44$

CN^-	O_3
$= 6 + 7 + 1 = 14$	$= 8 \times 3 = 24$

14 (*c*) Both are correct statements.
HI cannot be prepared by the reaction of KI with concentrated H_2SO_4 because formed HI is not stable and converted into I_2 on reaction with H_2SO_4.

15 (*a*) Cl has maximum electron affinity.

16 (*c*) $:\ddot{X}\cdot \;+\; e^- \longrightarrow [:\ddot{X}:]^-$

Outer-shell has seven electrons — Octet complete in X^-

HCl, HBr and HI are all strong acids.

18 (*c*) Cl_2 oxidises Br^- to Br_2 and Br_2 oxidises I^- to I_2 (violet).

19 (*d*) Interhalogen compound (*XY*) has two halogens *X* and *Y*, such that *X* is more electropositive than *Y*.

20 (*d*) $2KBr + Cl_2 \longrightarrow 2KCl + Br_2$

22 (*a*) Oxidising power of halogen is in order

$$I_2 < Br_2 < Cl_2 < F_2$$

Thus, $Br_2 + 2NaI \longrightarrow I_2 + 2NaBr$ is feasible.

26 (*c*) Both are correct statements.
NaCl reacts with concentrated H_2SO_4 to give colourless fumes with pungent smell. Pungent smell is due to formation of HCl.

$$NaCl + H_2SO_4 \longrightarrow Na_2SO_4 + 2HCl$$

But on adding MnO_2 the fumes become greenish yellow due to formation of chlorine gas.

28 (*c*) Cl_2O_6 (dimer of ClO_3) is mixed anhydride of $HClO_3$ and $HClO_4$.

$$Cl_2O_6 + H_2O \longrightarrow HClO_3 + HClO_4$$

Thus, statement (c) is incorrect.

29 (*b*) Fluorine shows only (−1) oxidation state due to lack of *d*-orbital.

30 (*a*) If acid is strong, conjugate base is weak.

Base	ClO_4^-	ClO_3^-	ClO_2^-	ClO^-
Acid	$HClO_4$	$HClO_3$	$HClO_2$	HClO
Oxidation number of halogen	+7	+5	+3	+1
Acidic strength	Maximum ⟵			Minimum
Basic strength	Minimum			⟶ Maximum

Thus, ClO_4^- is the weakest (Bronsted) base.

32 (*b*) OCl^- is used as bleaching agent.

33 (*c*) Fluorspar name is given after fluorescence (substance that unit light).

41 (*d*) ClO_2 is the mixed anhydride of $HClO_2$ and $HClO_3$ as shown by following reaction.

$$2KOH + 2ClO_2 \longrightarrow KClO_2 + KClO_3 + H_2O$$

42 (*b*) Bleaching powder $CaOCl_2$.

45 (*a*) HOF > HOCl > HOBr > HOI

46 (*a*)

$$CaOCl_2 + H_2O \longrightarrow Ca(OH)_2 + Cl_2$$
$$Cl_2 + H_2O \longrightarrow 2HCl + [O]$$
$$CH_3CH_2OH + [O] \longrightarrow CH_3CHO + H_2O$$
$$CH_3CHO + 3Cl_2 \longrightarrow \underset{\text{Chloral}}{CCl_3CHO} + 3HCl$$
$$2CCl_3CHO + Ca(OH)_2 \longrightarrow (HCOO)_2Ca + \underset{\text{Chloroform}}{CHCl_3}$$

(B) Miscellaneous questions

1 (b) Mixture is called *aqua-regia*.

$$\underset{\text{Conc.}}{HNO_3} + \underset{\text{Conc.}}{3HCl} \longrightarrow NOCl + Cl_2 + 2H_2O$$

2 (d) $HClO_4$ has Cl in maximum oxidation state. Thus, is not further oxidised. i.e. least oxidising in nature.

3 (b) Oxidising power of X_2

$$I_2 < Br_2 < Cl_2 < F_2.$$

Thus, I_2 cannot oxidised KBr as it is least oxidising in nature.

4 (a) HF, HCl, HBr, HI — On moving top to bottom
- Size of halogen atom increases
- H–X bond length increases
- Bond dissociation enthalpy decreases

5 (b) $\underset{+1}{ClO^-}$ (disinfectant property) $\longrightarrow \underset{-1}{Cl^-} + \underset{+5}{ClO_3^-}$

ClO^- disproportionates to Cl^-, ClO_3^-.

6 (a) Greater is the oxidation state of the central atom in oxy acid, greater is the acidic strength.

$$\underset{+7}{HClO_4} > \underset{+5}{HNO_3} > \underset{-1}{HCl}$$

7 (a) Black coloured compound MnO_2 reacts with HCl to produce greenish yellow coloured gas of Cl_2.

$$\underset{\text{(Black)}}{MnO_2} + 4HCl \longrightarrow MnCl_2 + 2H_2O + \underset{\text{(Greenish yellow gas)}}{Cl_2}$$

Cl_2 on further treatment with NH_3 produces NCl_3.

$$\overset{-3}{N}H_3 + 3Cl_2 \longrightarrow \overset{+3}{N}Cl_3 + 3HCl$$

NH_3 (–3) changes to NCl_3 (+3) in the above reaction.

Hence, (a) is the correct choice.

8 (a) Lewis acid has electron deficient central atom.

I. BF_3 (electron deficient B)

II. (SiF_4 octet of Si is complete)

III. (PF_5 expanded octet of P)

Thus, I only show Lewis acid character.

9 (c) Both the statements are correct. Example for statement I is

$$2CuSO_4 + 10KCN \longrightarrow 2K_3[Cu(CN)_4] + 2K_2SO_4 + (CN)_2$$

Statement II is also correct because CN^- resembles to Cl^-, Br^- and I^- in many of its properties. That's why it is known as pseudo halide.

11 (a) BX_3 has electron deficient boron atom and thus behaves as Lewis acid (lone pair acceptor).

As size of halogen increases, Lewis acid character increases.

Thus, $BF_3 < BCl_3 < BBr_3 < BI_3$

Also in BF_3, there is $p\pi$-$p\pi$ backbonding which reduces the acidic property.

12 (a) Due to maximum bond energy of H–F, it is not oxidised easily.

15 (d)

Bond	Energy
Br–Br	158 kJ mol^{-1}
Cl–Cl	210 kJ mol^{-1}
F–F	126 kJ mol^{-1}

Thus, $F_2 < Br_2 < Cl_2$

16 (c) Greater the SRP value of species, higher will be its oxidising power.

Species	E°_{cell}
ClO_4^-	1.19V
IO_4^-	1.65V
BrO_4^-	1.74V

Here, SRP = standard reduction potential.

17 (a)

	Species	Shape	Structure
A.	SF_4	*Sea-saw* shaped	S with one lone pair, four F
B.	BrF_3	Bent T-shaped	central atom with two lone pairs, three F
C.	BrO_3^-	Pyramidal	Br with lone pair, three O
D.	NH_4^+	Tetrahedral	$[NH_4]^+$

(C) Medical entrances' gallary

1 (d) The correct structure of tribromooctaoxide is

$$O{=}\overset{\overset{O}{\|}}{\underset{\underset{O}{\|}}{Br}}{-}\overset{\overset{O}{\|}}{\underset{\underset{O}{\|}}{Br}}{-}\overset{\overset{O}{\|}}{\underset{\underset{O}{\|}}{Br}}{=}O$$

In this compound, Br exhibits variable oxidation states. The oxidation state of Br from left to right are + 6, + 4 and + 6, respectively.

2 (c) The correct match is as follows

(A) **Pure nitrogen** It can be obtained by thermal decomposition of sodium or barium azide.

$$Ba(N_3)_2 \xrightarrow{\Delta} Ba + 3N_2$$

$$2NaN_3 \xrightarrow{\Delta} 2Na + 3N_2$$

(B) **Haber's process** On large scale, ammonia is manufactured by Haber's process.

$$N_2(g) + 3H_2(g) \underset{723\text{–}773\text{ K}}{\overset{\text{Fe, (MO), 200 bar}}{\rightleftharpoons}} 2NH_3(g)$$

(C) **Contact process** Sulphuric acid is manufactured by the contact process.

(D) **Deacon's process** Chlorine is prepared by Deacon's process.

$$4HCl + O_2 \xrightarrow{CuCl_2} 2Cl_2 + 2H_2O$$

Hence, the correct match is

(A) → (iv), (B) → (iii), (C) → (ii), (D) → (i)

3 (*a*)

(a) Fluorine is the most electronegative element and cannot exhibit any positive oxidation state. Other halogens have *d*-orbitals and therefore, can expand their octets and show +1, +3, +5 and +7 oxidation states. Thus, option (a) is incorrect.

Note *Fluorine can form an oxoacid, HOF in which oxidation state of F is +1 But HOF is highly unstable compound.*

(b) All halogens are strong oxidising agents as they have strong tendency to accept an electron. Thus, option (b) is correct.

(c) All halogens form monobasic oxyacids. Thus, option (c) is also correct.

(d) Electron gain enthalpy of halogens become less negative down the group. However, the negative electron gain enthalpy of fluorine is less than chlorine due to small size of fluorine atom. Thus, option (d) is also correct.

4 (*d*) In general, bond enthalpy or bond dissociation energy of halogen molecules (X_2) decreases down the group from Cl_2 to I_2.

$$Cl_2 > Br_2 > F_2 > I_2$$

The decrease in bond enthalpy from Cl_2 to I_2 is due to increase in atomic size that increases the distance between two atoms. Hence, it becomes easier to break them apart. The exceptionally low bond dissociation energy of fluorine molecule is due to very small size of fluorine atom. This brings the non-bonding electrons on fluorine nearer to each other, resulting in a much greater lone pair-lone pair repulsion, which weakens the covalent bond and lowers its dissociation energy.

Hence, option (d) is correct.

5 (*a*) Chlorine water on standing loses yellow colour due to formation of HCl and HOCl, which in turn provides nascent oxygen.

$$Cl_2 + H_2O \longrightarrow HCl + HOCl$$
$$HClO \longrightarrow HCl + \underset{\text{Nascent oxygen}}{[O]}$$

6 (*a*) As the size of the halogen atom increases from F to I, $H-X$ bond length in halogen acids also increases from H–F to HI (HF < HCl < HBr < HI). The increase in HX bond length decreases the strength of HX bond from HF to HI (HF > HCl > HBr > HI).

Due to successive decrease in the strength of HX bonds, bond dissociation enthalpy decreases from HF to HI.

HX :	HF	>	HCl	>	HBr	>	HI
Bond dissociation enthalpy (kJ / mol)	574.0		428.1		362.5		294.6

7 (*a*) $\underset{\text{(Excess)}}{8NH_3} + 3\,Cl_2 \longrightarrow 6NH_4Cl + N_2\uparrow$

$NH_3 + \underset{\text{(Excess)}}{3\,Cl_2} \longrightarrow NCl_3 + 3HCl$

Hence, $X = 6NH_4Cl + N_2\uparrow$

$Y = NCl_3 + 3\,HCl$

8 (*b*) A halate will be formed from halogen and the greenish yellow gas is Cl_2. The halate which is used in fireworks and safety matches is $KClO_3$. The reaction is as follows :

$$3Cl_2 + 6KOH \longrightarrow KClO_3 + 5KCl + 3H_2O$$

10 (*d*) $HgCl_2 + 4KI \longrightarrow K_2HgI_4 + 2KCl$

$$KI + I_2 \longrightarrow KI_3^-$$

Thus, HgI_4^{2-} and I_3^- ions exist.

11 (*a*) As the size increases, bond dissociation enthalpy becomes lower. Also, as the size of atoms get smaller, ion pairs on the two atoms get close enough together to experience repulsion in case of F_2, this repulsion is bigger and bond becomes weaker.

Hence, the correct order is $Cl_2 > Br_2 > F_2 > I_2$.

12 (*a*) As the oxidation state of halogen increases, (i.e. — Cl, in this case) acidity of oxyacid increases.

HClO : Oxidation state of Cl = +1

$HClO_2$: Oxidation state of Cl = +3

$HClO_3$: Oxidation state of Cl = +5

$HClO_4$: Oxidation state of Cl = +7

Therefore, the correct order of acidity would be

$$HClO < HClO_2 < HClO_3 < HClO_4$$

13 (*c*) Number of valence electron

$$= 1\times Ca + 2(1Cl + 2O)$$
$$= 1\times 2 + 2\,(1\times 7 + 2\times 6)$$
$$= 2 + 2\,(7 + 12)$$
$$= 2 + 2\,(19) = 40$$

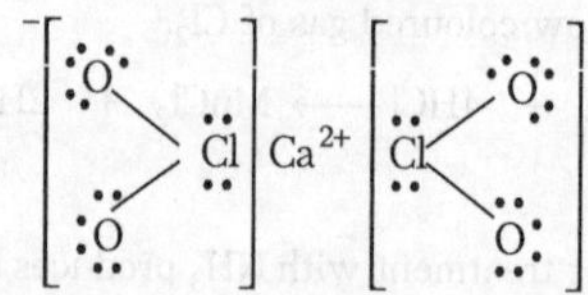

14 (*d*) The salts of strong acids give acidic solution (having pH less than 7).

In the given salts, acidic strength of oxoacids of chlorine increases with increase in oxidation number of chlorine.

Oxidation number (x) of Cl in $HClO_4$ (perchloric acid)

$$1 + x - 8 = 0$$
$$\Rightarrow \quad x = +7$$

Oxidation number of Cl in $HClO_3$ (chloric acid)

$$1 + x - 6 = 0$$
$$\Rightarrow \quad x = +5$$

Oxidation number of Cl in $HClO_2$ (chlorous acid)

$$1 + x - 4 = 0$$
$$\Rightarrow \quad x = +3$$

Oxidation number of Cl in HClO (hypochlorous acid)

$$1 + x - 2 = 0$$
$$\Rightarrow \quad x = +1$$

So, acidic strength decreases in the order.

$$\underset{\text{Most acidic}}{HClO_4} > HClO_3 > HClO_2 > \underset{\text{Less acidic}}{HClO}$$

So, $HClO_4$ is the strongest acid and salt of this acid $NaClO_4$ will has the lowest pH.

15 (*c*) When HOCl is treated with H_2O_2 in acidic medium it gets reduced to Cl^- with the evolution of oxygen gas.

$$HOCl(aq) + H_2O_2(aq) \longrightarrow HCl(aq) + H_2O(l) + O_2\,(g)$$

16 (*b*) The increasing order of reducing power of halogen acid is

$$HF < HCl < HBr < HI$$

The reason is that by increasing the size of halogen atom, the bond strength of H—*X* bond decreases. Thus, the ease of hydrogen donation increases and hence, reducing power increases.

17 (*c*) Melting and boiling points of HCl is the lowest amongst H*X* due to weak intermolecular H—bonding in H— Cl.

18 (*c*) Sea weeds contain iodine.

19 (*d*) Bleaching powder losses its power on keeping for a long time because it changes into calcium chloride and calcium chlorate.

$$ClO^- \longrightarrow Cl^- + ClO_3^-$$

20 (*b*) When chlorine gas reacts with hot and concentrated NaOH solution, it disproportionates into chloride (Cl^-) and chlorate (ClO_3^-) ions.

$$6NaOH + 3Cl_2 \longrightarrow NaClO_3 + 5NaCl + 3H_2O$$

In this process, oxidation number of chlorine changes from 0 to – 1and 0 to +5.

Note *It is a disproportion reaction, the same element undergoes oxidation as well as reduction.*

21 (*a*) Bromine water is an oxidising agent. It oxidises SO_2 to SO_3 and itself gets reduces to colourless HBr.

$$\underset{\text{Orange red}}{Br_2} + H_2O \longrightarrow \underset{\text{Colourless}}{2HBr} + [O]$$

$$SO_2 + [O] \longrightarrow SO_3$$

$$H_2O + SO_3 \longrightarrow H_2SO_4$$

H_2S is also oxidised by bromine water but in this case, decolourisation is also accompanied by the formation of sulphur.

$$H_2S + [O] \longrightarrow H_2O + S$$

22 (*d*) A peroxy acid is one that contains peroxide [—O—O—] linkage. As perchloric acid ($HClO_4$) doesn't contain a peroxide linkage. It is not a peroxy acid, whereas others having a peroxide linkage are peroxy acids.

Structure of $HClO_4$

23 (*a*) Pseudohalides are the combinations of more than one electronegative atoms which have one unit negative charge, e.g. N^-, NO^- N^- etc., whereas *R*COO⁻ is not a pseudohalide.

24 (*c*) Only I_2 is liberated.

Br_2 liberates I_2 from NaI, because iodine lies below bromine in electrochemical series.

Oxidising power : $F_2 > Cl_2 > Br_2 > I_2$

CHAPTER 10

p-Block Elements 4 (Noble Gases)

Group 18 or zero group of periodic table contains six elements collectively known as noble gases. These are helium (He), neon (Ne), argon (Ar), krypton (Kr), xenon (Xe) and radon (Rn). They are also known as **inert gases** as they have stable configuration due to which they are inert in nature at normal condition. However, some of these such as xenon, forms compounds under specific conditions.

Discovery

Due to their inertness, these gases were discovered on the basis of study of their physical properties such as density. W Ramsay was awarded by Nobel prize in chemistry for the discovery of the inert gases in air. Some of the chemists associated with discovery of inert gases are given is table 10.1.

Occurrence

Due to the inert nature of these gases, they always occur in free state. Except **radon,** all the noble gases occur in atmosphere in atomic state. Abundance of these gases in dry air is approximately 1% by volume. **Argon** is the main constituent of dry air. Helium is present in natural gas.

He and Ne are found in minerals of radioactive origin, e.g. pitchblende, monazite, cleveite, etc. and xenon and radon are rarest elements of this group.

Isolation of elements

(*i*) Helium is commercially obtained from natural gas. Natural gas contains hydrocarbons (such as methane), CO_2, H_2S and He as main constituents. The hydrocarbon and other impurities are liquefied at about 120 K and helium is pumped out.

(*ii*) Ar, Ne, Kr and Xe are prepared by fractional distillation of liquid air. It gives O_2, N_2 and mixture of the noble gases. The separation depends upon the boiling points of the noble gases and to enhance the separation, coconut charcoal is used.

Inside

The individual gas is obtained as shown below

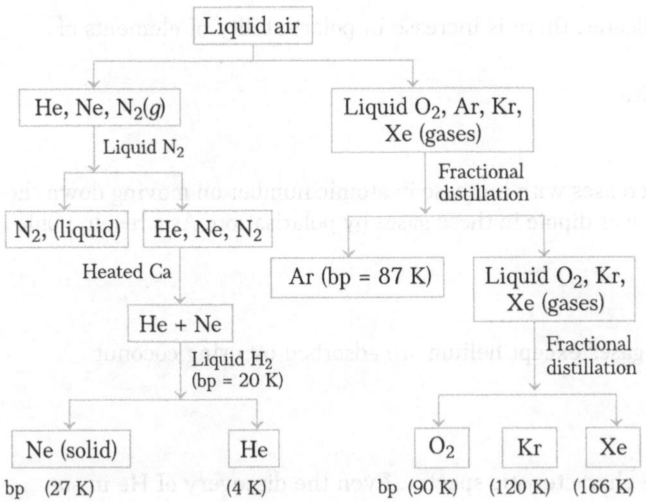

(*iii*) Radon (Rn) is produced by α-decay of an aqueous solution of radium chloride ($RaCl_2$).

$$^{226}_{88}Ra \longrightarrow ^{222}_{86}Rn + ^{4}_{2}He$$
(α-particles)

GENERAL AND PHYSICAL PROPERTIES

These are monoatomic, colourless, odourless and tasteless gases. They are sparingly soluble in water and have low melting and boiling points due to presence of weak dispersion forces. Some of the trends in properties of group 18 of the periodic table are given below in table 10.1.

Electronic configuration

The general electronic configuration of this group is ns^2np^6. (except helium for which it is $1s^2$). Due to presence of completely filled valence shell, these gases are highly stable. Electronic configuration of these elements are given in table 10.1.

Atomic radii

Atomic radii increase with increase in atomic number, i.e. down the group from He to Rn due to addition of extra principal shell in each succeeding element. Atomic radii of these elements correspond to van der Waals' radii.

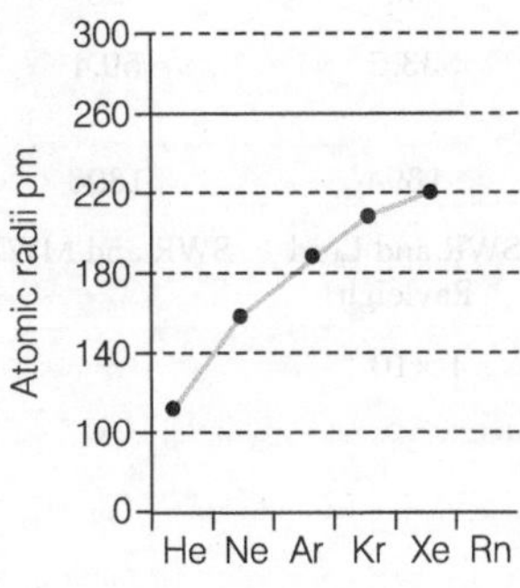

Boiling point

Noble gases are low boiling gases. Boiling point increases on descending the group due to increase in magnitude of van der Waals' radii. Trend of boiling points down the group is shown below.

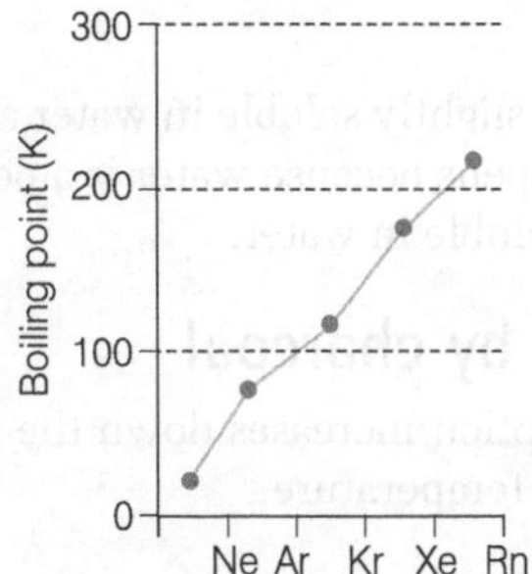

Helium has the lowest boiling point of any element and its behaviour as a liquid is most unusual. It has no triple point that is there is no single combination of temperature and pressure at which the solid, the liquid and the gas all exist. At about 2K, helium undergoes a transition to a form of helium, labelled He_{II}. It will creep up the walls of the container in which it is placed and thus it exists in two phases in liquid helium, one of which has an extremely low viscosity.

Ionisation energy and electron affinity

The noble gases have high values of ionisation energy and electron affinity is zero due to presence of fully filled valence shells with electronic configuration ns^2np^6. They have no tendency to add or lose electron thus, lack chemical reactivity as compared to other elements of periodic table. Down the group the ionisation enthalpies decrease due to increase in atomic radii and shielding effect. Also on moving down the group atomic size increase and electron affinity become less positive.

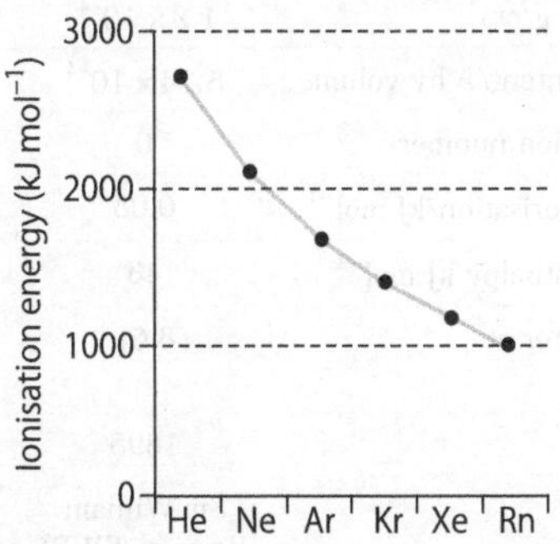

Heat of vaporisation

They have very low values of heat of vaporisation due to presence of weak van der Waals' forces of attraction. Values of heat of vaporisation increases with increase in atomic number, i.e. down the group.

Polarisability

Since, heat of vaporisation increases down the group, this indicates, there is increase in polarisability of elements of larger electronic clouds.

The order of polarisability is as follows He < Ne < Ar < Kr < Xe

Solubility

Noble gases are slightly soluble in water and their solubility increases with increase in atomic number on moving down the group. This happens because water is a polar molecule and induces dipole in these gases by polarisation. As a result, noble gas is slightly soluble in water.

Adsorption by charcoal

Extent of adsorption increases down the group. All the noble gases except helium are adsorbed by using coconut charcoal at low temperature.

Spectra

These gases can be identified by spectra. All of the gases give characteristic spectra. Even the discovery of He in sun was a result of its characteristics spectra.

Liquefaction of gases

It is difficult to liquefy noble gases due to presence of weak van der Waals' forces of attraction. Tendency to be liquefied, increases on moving down the group, i.e. from He to Rn because of increased intermolecular forces.

The attractive forces among the atoms are also so small that they remain liquid at 1 atm pressure even at 0.001K.

Table 10.1 Physical properties of group 18 element

Property	He	Ne	Ar	Kr	Xe	Rn
Atomic number	2	10	18	36	54	86
Electronic configuration	$1s^2$	[He] $2s^2 2p^6$	[Ne] $3s^2 3p^6$	[Ar] $3d^{10} 4s^2 4p^6$	[Kr] $4d^{10} 5s^2 5p^6$	[Xe] $4f^{14} 5d^{10} 6s^2 6p^6$
$(IE)_1$/kJ mol^{-1}	2372	2080	1521	1351	1170	1040
Melting point/°C	—	– 248.6	– 189.4	–157.2	– 111.8	– 71
Boiling point/°C	– 269	– 246	– 186	– 153.6	–108.1	– 62
Atomic radius/pm	120	160	191	200	220	—
Density at STP/ g/cm^3	1.8×10^{-4}	9.0×10^{-4}	1.8×10^{-3}	3.7×10^{-3}	5.9×10^{-3}	9.7×10^{-3}
Atmospheric content/% by volume	5.24×10^{-4}	1.82×10^{-3}	0.934	1.14×10^{-3}	8.7×10^{-6}	
Principal oxidation number	0	0	0	0	+2, +4, +6	0
Enthalpy of vaporisation/kJ mol^{-1}	0.08	1.7	6.5	9.1	12.7	18.1
Electron gain enthalpy kJ mol^{-1}	48	116	96	96	97	68
Solubility in water at 20°C/cm^3 kg^{-1}	8.61	10.5	33.6	59.4	108.1	230
Discovery year	1895	1898	1894	1898	1898	1902
Discovered by	Sir William Ramsay (SWR)	SWR and MW Travers (MWT)	SWR and Lord Rayleigh	SWR and MWT	SWR and MWT	Friedrich Dorn
Abundance in igneous rocks/ppm by weight	3×10^{-3}	7×10^{-5}	4×10^{-2}	—	—	1.7×10^{-10}

CHEMICAL PROPERTIES

Noble gases are chemically inert in nature. Therefore, they do not have tendency to lose or add electrons. This is because of completely filled valence shells, high ionisation enthalpies and positive electron gain enthalpies.

Hence, noble gases do not form compounds.

But in 1962, **Neil Bartlett** prepared a red compound $O_2^+[PtF_6]^-$ by reacting platinum hexafluoride (PtF_6) with dioxygen.

$$O_2(g) + PtF_6(g) \longrightarrow \underset{\text{Dioxygen hexafluoro platinate (V)}}{O_2^+[PtF_6]^-(s)}$$

He found that the ionisation enthalpy of xenon (1170 kJ/mol) is fairly close to molecular oxygen (1175 kJ/mol). Therefore, he was able to form compound of xenon.

$$Xe + PtF_6 \longrightarrow \underset{\text{Xenon hexafluoro platinate}}{Xe^+[PtF_6]^-}$$

As a results, number of compounds of xenon with fluorine and oxygen are formed.

CHECK POINT 10.1

1. Noble gases are group of elements which exhibit very
   (a) high chemical reactivity
   (b) low chemical reactivity
   (c) minimum electronegativity
   (d) much paramagnetic properties
2. Noble gases can be separated by
   (a) passing them through some solutions
   (b) electrolysis of their compounds
   (c) adsorption and desorption on coconut charcoal
   (d) None of the above
3. A radioactive element X decays to give inert gases. X is
   (a) ${}_{92}U^{238}$ (b) ${}_{88}Ra^{226}$
   (c) ${}_{90}Th^{234}$ (d) ${}_{89}Ac$
4. Which of the following inert gas having highest boiling point?
   (a) Xe (b) Ar
   (c) Kr (d) He
5. Gradual addition of electronic shells in the noble gases causes a decrease in their
   (a) ionisation energy (b) atomic radius
   (c) boiling point (d) density
6. Which of the following is highly polarised?
   (a) Kr (b) He (c) Ar (d) Xe
7. Which of the following noble gas is most soluble in water?
   (a) He (b) Ar (c) Ne (d) Xe
8. The correct order of solubility in water for He, Ne, Ar, Kr, Xe is
   (a) He >Ne >Ar >Kr >Xe (b) Ne >Ar >Kr >He >Xe
   (c) Xe >Kr >Ar >Ne >He (d) Ar >Ne >He >Kr >Xe
9. Which of the following exhibits the weakest intermolecular forces?
   (a) He (b) HCl (c) NH_3 (d) H_2O
10. Which of the following inert gas liquefies easily?
    (a) Kr (b) He (c) Ne (d) Ar

COMPOUNDS OF GROUP 18 ELEMENTS

No true compounds of He, Ne or Ar are known but few compounds of Xe are known, which are disscussed below

Compounds of xenon (Xe)

It reacts with electronegative elements such as fluorine and oxygen to form various compounds like xenon fluorides, xenon oxides and xenon oxyfluorides.

Xenon fluorides

Xenon reacts with fluorine under different conditions to form three binary fluorides, i.e. XeF_2, XeF_4 and XeF_6. In these compounds, Xe present in + 2, + 4 and + 6 oxidation state respectively. The products formed depends on ratio of Xe and F.

$$Xe + F_2 \begin{cases} \text{Xenon in excess} \xrightarrow{400^\circ C,\ 1\ atm} XeF_2 \\ 1:5 \text{ mixture} \xrightarrow{600^\circ C,\ 7\ atm} XeF_4 \\ 1:20 \text{ mixture} \xrightarrow{300^\circ C,\ 60\ atm} XeF_6 \end{cases}$$

XeF_2 can also be prepared by irradiating mixture of xenon and fluorine in sunlight from a high pressure mercury lamp. XeF_6 can be prepared by reaction of XeF_4 and O_2F_2 at 143 K.

$$XeF_4 + O_2F_2 \xrightarrow{143K} XeF_6 + O_2$$

Structure

Both VSEPR theory and hybridisation are applied to predict the structure of Xe compounds.

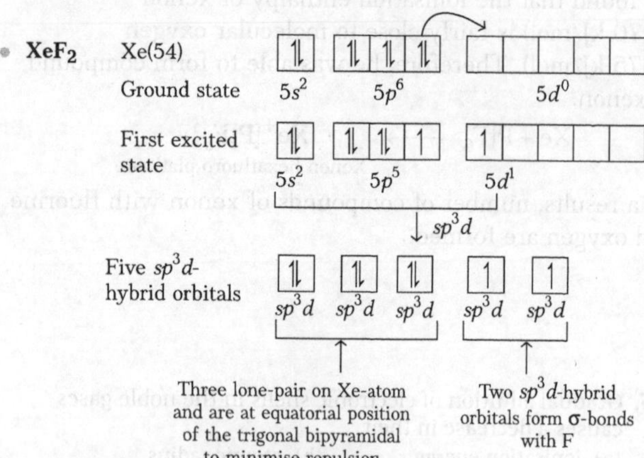

Thus, structure is linear

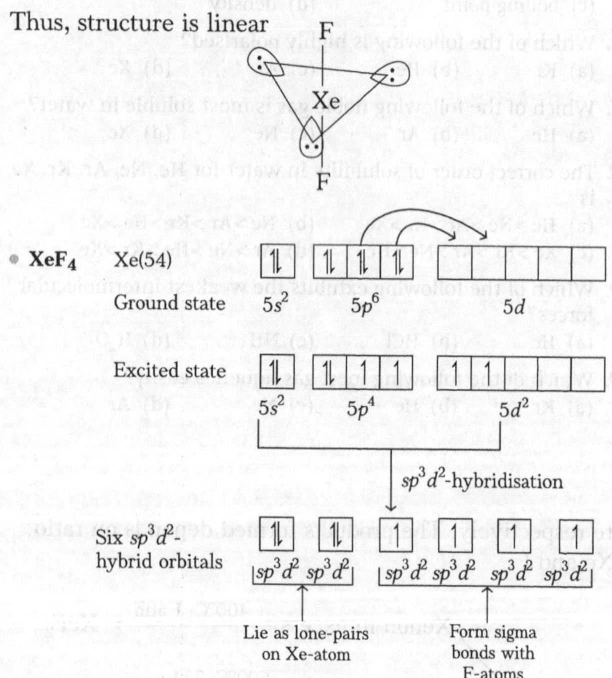

Thus, XeF_4 assumes square planar structure.

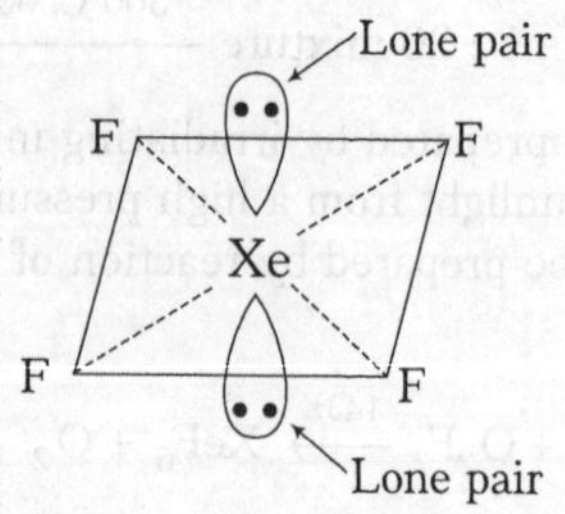

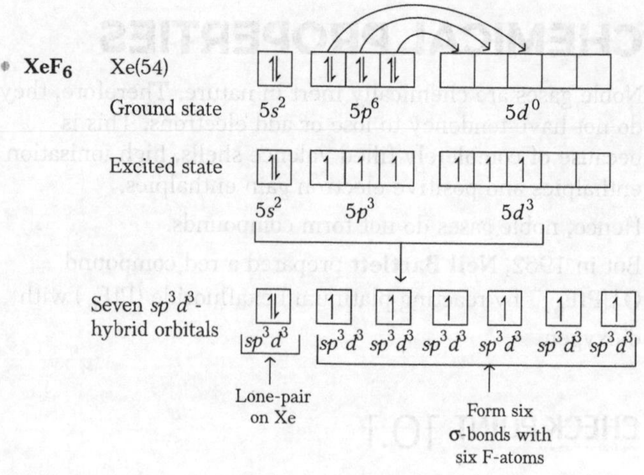

Structure is distorted octahedron

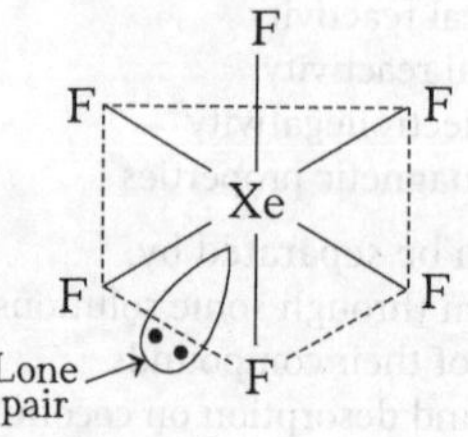

Properties

(*i*) These are colourless (white) volatile solids except XeF_6, which is yellow in colour and present in liquid or in gaseous phases. These fluorides form large crystals on standing in closed vessels. XeF_2 and XeF_4 can be stored conveniently in **glass vacuum** system but XeF_6 cannot. They can be sublimed at room temperature and hence, stored indefinitely in nickel or monel (an alloy of nickel) containers.

(*ii*) All the fluorides are strong oxidising agents. Their oxidising power increases in the order as follow

$$XeF_2 < XeF_4 < XeF_6$$

(*iii*) Hydrogen reduces xenon fluorides to xenon at room temperature.

$$XeF_2 + H_2 \longrightarrow Xe + 2HF$$
$$XeF_4 + 2H_2 \longrightarrow Xe + 4HF$$
$$XeF_6 + 3H_2 \longrightarrow Xe + 6HF$$

(*iv*) They oxidise Cl^- to Cl_2 and I^- to I_2.

$$XeF_2 + 2HCl \longrightarrow 2HF + Xe + Cl_2$$
$$XeF_4 + 4KI \longrightarrow 4KF + Xe + 2I_2$$

(*v*) Xenon fluorides are strong fluorinating agents.

$$2SF_4 + XeF_4 \longrightarrow 2SF_6 + Xe$$
$$Pt + XeF_4 \longrightarrow PtF_4 + Xe$$
$$C_6H_6 + XeF_2 \longrightarrow C_6H_5F + Xe + HF$$

(*vi*) **Reaction with water** Xenon fluorides are easily hydrolysed by water. These differ in reactivity with water. XeF_2 is highly soluble in water but undergoes slow hydrolysis.

$$2XeF_2 + 2H_2O \longrightarrow 2Xe + 4HF + O_2$$

(*vii*) XeF_4 and XeF_6 react violently with water and undergoes disproportion reaction.

$$3XeF_4 + 6H_2O \longrightarrow 2Xe + \underset{\text{Explosive}}{XeO_3} + 12HF + \frac{3}{2}O_2$$

$$XeF_6 + H_2O \longrightarrow XeOF_4 + 2HF$$
$$XeOF_4 + H_2O \longrightarrow XeO_2F_2 + 2HF$$
$$XeO_2F_2 + H_2O \longrightarrow XeO_3 + 2HF$$

$$XeF_6 + 3H_2O \longrightarrow XeO_3 + 6HF$$

$$XeF_6 + 2H_2O \xrightarrow{\text{Atmospheric moisture}} XeO_2F_2 + 4HF$$

(*viii*) Xenon fluorides react with fluoride ion acceptor to form cationic species and with fluorides ion donors to form fluoro anion.

$$XeF_2 + MF_5 \longrightarrow [XeF]^+[MF_6]^-$$
$$(M = As^{3+}, Sb^{5+}, P^{5+}, \text{etc.})$$

$$XeF_6 + MF \longrightarrow M^+[XeF_7]^- \ [M = Na, K, Rb \text{ or } Cs]$$

(*ix*) SiO_2 converts XeF_6 into $XeOF_4$

$$2XeF_6 + SiO_2 \longrightarrow SiF_4 + 2XeOF_4$$
$$2XeOF_4 + SiO_2 \longrightarrow 2XeO_2F_2 + SiF_4$$
$$2XeO_2F_2 + SiO_2 \longrightarrow 2XeO_3 + SiF_4$$

SiO_2 is part of glass or quartz. Hence, XeF_6 cannot be stored in glass or quartz vessel.

Here, XeF_6 behaves as an acid (oxide ion acceptor) according to **Lux–Flood** definition.

(*x*) They are also hydrolysed in alkaline solution rapidly.

$$2XeF_2 + 4OH^- \xrightarrow{\text{Rapid}} 2Xe + 4F^- + 2H_2O + O_2$$

$$XeF_6 + 7OH^- \longrightarrow \underset{\text{Xenate ion}}{HXeO_4^-} + 3H_2O + 6F^-$$

$$2HXeO_4^- + 2OH^- \xrightarrow{\text{Slow}} \underset{\text{Perxenate ion}}{XeO_6^{4-}} + Xe + 2H_2O + O_2$$

Xenon oxides

Xenon forms following two oxides

Xenon trioxide (XeO_3)

It is a colourless solid and highly explosive in nature. It is a powerful oxidising agent.

Structure

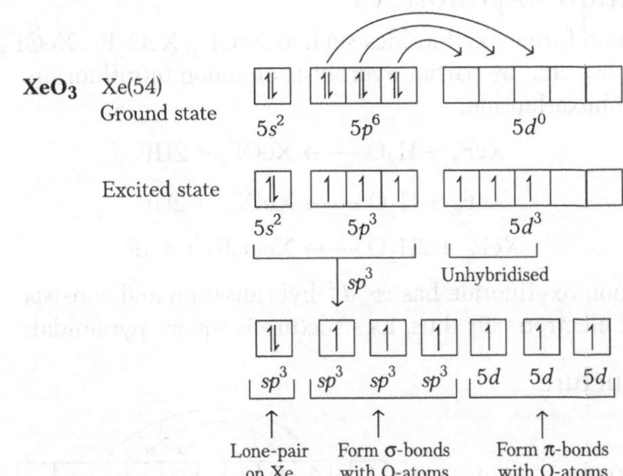

Due to lone pair-bond pair repulsion XeO_3 assumes trigonal pyramidal structure (instead of tetrahedral structure $-sp^3$-hybridisation) and bond angle is 103° . (VSEPR theory)

Lone-pair
Xe
O O O

Xenon tetrafluoride (XeO_4)

It is a unstable as compare to XeO_3 and decomposes to give Xe and O_2.

Structure

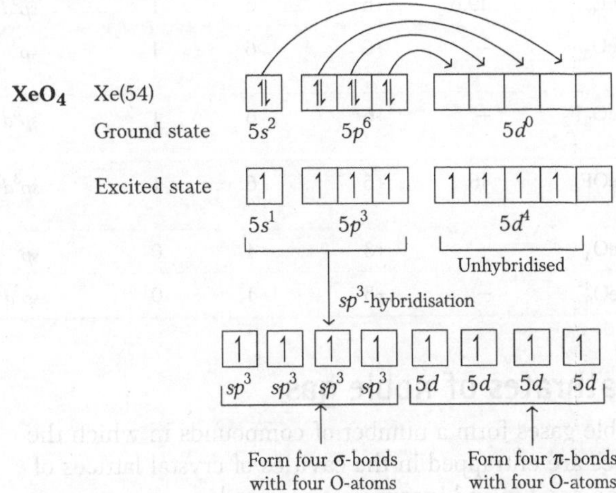

Structure is tetrahedral,

O
Xe
O O O

Xenon oxyfluorides

Xenon forms oxyfluorides such as $XeOF_2$, XeO_2F_2, $XeOF_3$, $XeOF_4$ etc., by partial hydrolysis of xenon tetrafluoride and hexafluoride.

$$XeF_4 + H_2O \longrightarrow XeOF_2 + 2HF$$

$$XeF_6 + H_2O \longrightarrow XeOF_4 + 2HF$$

$$XeF_6 + 2H_2O \longrightarrow XeO_2F_2 + 4HF$$

Xenon oxyfluoride has sp^3d^2-hybridisation and consists one electron pair thus, its structure is square pyramidal.

Structure

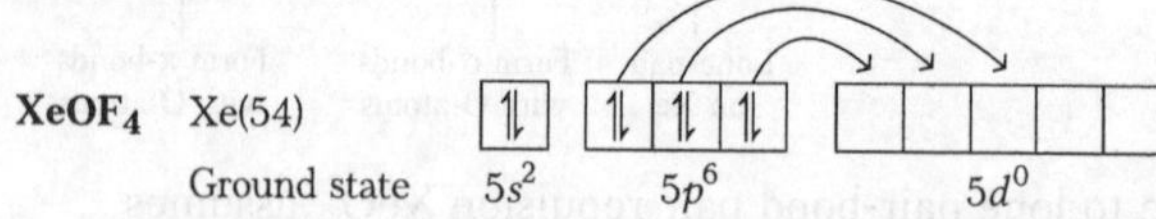

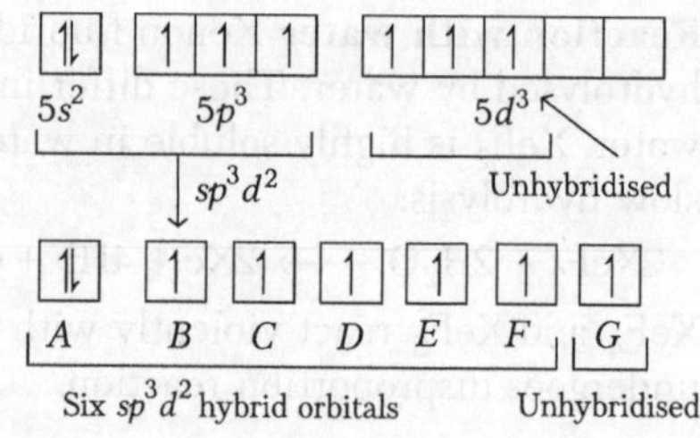

A – lone-pair on Xe
B to E – form four σ-bonds with four F-atoms
F – forms σ-bond with O-atom
G – forms π-bond with O-atom

Structure is square pyramidal,

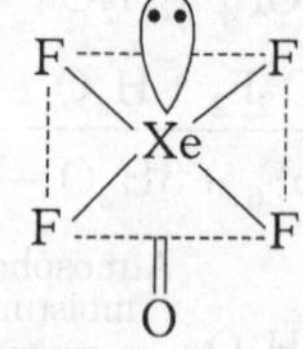

Table 10.2 Structures of some xenon compounds

Formula	(mp)/°C	Oxidation state	Number of bonding pairs	Number of lone pairs	Hybridisation of Xe	Structure	Form
XeF_2	129	+2	2	3	sp^3d	linear with three lone pairs in equatorial position	colourless crystal
$XeF_2 \cdot 2SbF_3$	63	–	–	–	–	–	yellow solid crystals
XeF_4	117	+4	4	2	sp^3d^2	square planar with two positions occupied by lone pairs	colourless crystals
$XeOF_2$	31	+4	4	2	sp^3d	T-shaped	colourless crystals
XeF_6	49.6	+6	6	1	sp^3d^3	distorted octahedron	colourless crystals
XeO_3	–	+6	6	1	sp^3	trigonal pyramidal with three π-bonds and one lone pair	colourless liquid
XeO_2F_2	–	+6	6	1	sp^3d	trigonal pyramidal with two π-bonds and one lone pair	colourless liquid
$XeOF_4$	46	+6	6	1	sp^3d^2	square pyramidal with one π-bond and one lone pair	colourless crystals
XeO_4	–	+8	4	0	sp^3	tetrahedral	colourless gas
XeO_6^{4-}	–	+8	4	0	sp^3d^2	octahedral	ions

Clathrates of noble gas

Noble gases form a number of compounds in which the gases are entrapped in the cavities of crystal lattices of some organic and inorganic compounds.

Such compounds are called **clathrates** or **cage compounds** or **host-guest compounds**, e.g. $Xe \cdot 6H_2O$, quinol clathrate.

The organic or inorganic compound with cavities is called **host** and the gas entrapped in them is called **guest**.

He and Ne do not form such compounds because of small size. Only Ar, Kr and Xe form clathrates.

Uses

(*i*) Clathrates are used in separation of noble gases as an anaesthetic.

(*ii*) These are used in processing and transporting of isotopes of noble gases.

(*iii*) Clathrate of Kr-85 is used for obtaining β-particles and clathrate of Xe-133 is chief source of γ-rays.

Uses of Noble Gas

There are many uses of noble gases. Some of the uses are summarised in the table given below.

Noble gas	Uses	Reasons
Helium	1. Filling of observation balloons and other lighter than air craft	Non-flammable, 93% lifting power as compared to flammable H_2
	2. He/O_2 mixture use for deep-sea breathing, instead of N_2/O_2 mixture	Low solubility in blood, prevents nitrogen narcosis and "bends"
	3. Diluent use for gaseous anaesthetics	Non-flammable, non-reactive
	4. He/O_2 mixtures use for respiratory patients	Low density, flows easily through restricted passages
	5. Heat transfer medium for nuclear reactions	Transfers heat readily, does not become radioactive, chemically inert
	6. (Liquid He) To maintain very low temperature in research (cryogenics)	Extremely low boiling point
Neon	Neon signs	Even at low Ne pressure, moderate electric current causes bright orange-red glow, can be modified by coloured glass or mixing with Ar or Hg vapour.
Argon	1. Inert atmosphere for welding	Chemically inert
	2. Filling incandescent light bulbs	Inert, prevents vaporisation of tungsten and blackening of bulbs
Krypton	Airport runway and approach lights	Gives longer life to incandescent lights than Ar but more expensive
Xenon	Xe and Kr mixture in high intensity, short exposure photographic flash tubes	Both have fast response to electric current
Radon	Radiotherapy of cancerous-tissues	Radioactive

CHECK POINT 10.2

1. What is formed when xenon combines with fluorine in the presence of mercury vapour?
(a) XeF (b) Xe_2F (c) XeF_2 (d) XeF_4

2. In XeF_2, XeF_4, XeF_6 the number of lone pairs on Xe respectively are
(a) 2, 3, 1 (b) 1, 2, 3
(c) 4, 1, 2 (d) 3, 2, 1

3. XeF_2 molecule is
(a) square planar (b) trigonal bipyramidal
(c) trigonal planar (d) linear

4. Which of the following is not obtained by direct reaction of constituent elements?
(a) XeF_2 (b) XeF_4
(c) XeO_3 (d) XeF_6

5. XeF_4 on partial hydrolysis produces
(a) XeF_2 (b) $XeOF_2$
(c) $XeOF_4$ (d) XeO_3

6. In XeO_3, Xe is
(a) sp^3-hybridised (b) sp^3d-hybridised
(c) sp^3d^3-hybridised (d) sp^3d^2-hybridised

7. Geometry of $XeOF_4$ molecule is
(a) square planar
(b) square pyramidal
(c) triangular bipyramidal
(d) octahedron

8. Among the following molecule
(i) XeO_3 (ii) $XeOF_4$
(iii) XeF_6
Those having same number of lone pairs on Xe are
(a) Both (i) and (ii)
(b) Both (i) and (iii)
(c) Both (ii) and (iii)
(d) Both (i), (ii) and (iii)

9. Which of the following is planar?
(a) XeF_2 (b) XeO_3F
(c) XeO_2F_2 (d) XeF_4

10. Which of the following are correctly matched?
(a) $XeOF_2$ – T-shaped and sp^3d-hybridisation
(b) $XeOF_4$ – Square pyramidal and sp^3d^2
(c) XeO_2F_2 – Trigonal bipyramidal and sp^3d
(d) XeO_3F_2 – Tetrahedral and sp^3

Chapter Exercises

(A) Topical exercise

I. Discovery, occurrences and its properties

1 The noble gas was first time discovered by
(a) Cavendish (b) William Ramsay
(c) Rayleigh (d) Frankland

2 Argon was discovered by
(a) Rayleigh (b) Ramsay
(c) Both (a) and (b) (d) Frankland and Lockeyer

3 Which of the following configurations represents a noble gas?
(a) $1s^2, 2s^2 2p^6, 3s^2$ (b) $1s^2, 2s^2 2p^6, 3s^1$
(c) $1s^2, 2s^2 2p^6$ (d) $1s^2, 2s^2 2p^6, 3s^2 3p^6, 4s^2$

4 Which has same electronic configuration as of noble gas?
(a) Ag^{3+} (b) Cu^{3+}
(c) Pb^{4+} (d) Ti^{4+}

5 The forces acting between noble gas atoms are
(a) van der Waals' forces
(b) ion-dipole forces
(c) London-dispersion forces
(d) magnetic force

6 Who was awarded the nobel prize in chemistry for the discovery of inert gas elements in air?
(a) Lord Rayleigh (b) W Ramsay
(c) M Curie (d) Grignard

7 Who was awarded the nobel prize in physics for discovery of argon?
(a) Lord Rayleigh (b) Grignard
(c) Friedel-Crafts (d) Williamson

8 Isolation and separation of inert gases from air depends on
(a) viscosity (b) surface tension
(c) boiling point (d) melting point

9 What is X in the following reaction?

$$^{226}_{88}\text{Ra} \longrightarrow {}^{222}_{86}\text{Rn} + X$$

(a) ^1_1H (b) ^2_1H
(c) ^4_2He (d) ^3_1H

10 Variable oxidation state is shown by
(a) He (b) Ne
(c) Ar (d) Xe

11 Which of the following has lowest boiling point?
(a) Xe (b) Ar (c) He (d) Kr

12 A rare gas that was detected in the sun before its discovery on earth is
(a) He (b) Ne
(c) Ar (d) Kr

13 The inert gas abundantly found in atmosphere is
(a) Kr (b) He
(c) Ar (d) Xe

14 Second most abundant element in the universe (after hydrogen) is
(a) Al (b) Fe
(c) He (d) Ne

15 Radioactive isotopes of Kr and Xe can be stored in
(a) nuclear reactor (b) charcoal cavity
(c) clathrates (d) inert atmosphere of N_2

16 Thermal stability of hydrates of group 18 elements
(a) increases down the group
(b) decreases down the group
(c) remains unchanged
(d) first increases up to Ar and then decreases

17 Helium accumulates in the earth's crust through
(a) α-decay of radioactive element
(b) thermonuclear reactions
(c) β-decay of radioactive element
(d) γ-decay of radioactive element

18 In the preparation of compounds of Xe, Bartlett had taken $O_2^+PtF_6^-$ as a base compound. This is because **[NCERT Exemplar]**
(a) both O_2 and Xe have same size
(b) both O_2 and Xe have same electron gain enthalpy
(c) both O_2 and Xe have almost same ionisation enthalpy
(d) both Xe and O_2 are gases

19 Select the correct statement(s).
(a) There exists van der Waals' force of attraction in noble gases.
(b) Polarisability and interatomic attraction increases on going down the group.
(c) They remain liquid at 1 atm pressure even at 0.001 K.
(d) All of the above are correct statements

II. Compounds of group 18 elements

20 Which among the following first prepared a stable compound of noble gas?
(a) Rutherford (b) Rayleigh
(c) Ramsay (d) Neil Bartlett

21 Reaction between xenon with fluorine in excess molar ratio at 400° C gives
(a) XeF_4
(b) XeF_6
(c) XeF_2
(d) they do not react as Xe is inert

22 XeF_6 on hydrolysis gives
(a) XeO_3 (b) XeO
(c) XeO_2 (d) Xe

23 Which of the following is not known?
(a) XeF_6 (b) XeF_4 (c) KrF_6 (d) Cs_2XeF_6

24 XeF_6 on partial hydrolysis with water produces a compound *X*. The same compound *X* is formed when XeF_6 reacts with silica. The compound *X* is
(a) XeF_2 (b) XeF_4
(c) $XeOF_4$ (d) XeO_3

25 Which of the following is the best fluorinating agent?
(a) XeF_2 (b) XeF_4
(c) XeF_6 (d) XeO_2F_2

26 XeF_6 is converted into XeO_3 by reaction with
(a) moisture (b) SiO_2
(c) Both (a) and (b) (d) H_2SO_4

27 Xenon fluoride acts as fluoride acceptor with
(a) RbF (b) CsF
(c) Both (a) and (b) (d) None of these

28 XeF_6 is conducting in presence of
(a) BF_3 (b) RbF
(c) HF (d) All of these

29 Which of the following is most basic fluoride?
(a) XeF_6
(b) XeF_2
(c) XeF_4
(d) None is basic, all are acids

30 Central atom is sp^3-hybridised in
(a) $Ni(CO)_4$ (b) XeO_3
(c) XeO_4 (d) in all cases

31 Central atom is sp^3d^2-hybridised in
(a) XeF_4 (b) XeO_2F_4
(c) $XeOF_4$ (d) in all cases

32 Lone pair and π-bonds exist in
(a) XeF_2 (b) XeO_3 (c) XeF_6 (d) XeO_4

33 Consider the following statements.
Statement I Fluorides of Xe are thermodynamically stable.
Statement II They are endothermic compounds.
(a) Only statement I is correct
(b) Only statement II is correct
(c) Both the statements are correct
(d) Both the statements are incorrect

34 XeF_2 forms .... with AsF_5.
(a) XeF^+ cation and AsF_4^+ anion
(b) XeF_2^- anion and AsF_4^+ anion
(c) AsF_4^+ cation and AsF_4^- anion
(d) XeF^+ cation and AsF_6^- anion

35 In which of the following group of species central atom is sp^3d-hybridised in
(a) PCl_5, BrF_3, XeF_2 (b) PCl_6^-, PCl_4^+, XeF_2
(c) XeF_2, XeO_3, $XeOF_4$ (d) PCl_5, PCl_3, XeO_3

36 Consider the following statements.
Staement I XeF_6 can be used as an oxidising agent.
Statement II Oxidation state of Xe in XeF_6 is +6.
(a) Only statement I is correct
(b) Only statement II is correct
(c) Both the statements are correct
(d) Both the statements are incorrect

37 Consider the following statements.
Statement I XeF_6 cannot be kept in glass or quartz apparatus.
Statement II Reaction between XeF_6 and SiO_2 of glass form explosive XeO_3.
(a) Only statement I is correct
(b) Only statement II is correct
(c) Both the statements are correct
(d) Both the statements are incorrect

38 Match the compounds (in Column I) with the structures (in Column II).

Column I	Column II
A. XeF_4	1. Distorted octahedral
B. XeF_6	2. Tetrahedral
C. XeO_3	3. Square planar
D. XeO_4	4. Pyramidal

	A	B	C	D
(a)	3	1	4	2
(b)	1	2	3	4
(c)	3	1	2	4
(d)	1	3	4	2

III. Clathrates and uses of noble gases

39 Which of the following statements regarding helium is incorrect?
(a) It is used to produce and sustain powerful superconducting magnets.
(b) It is used as a cryogenic agent for carrying out experiments at low temperatures.
(c) It is used to fill gas balloons instead of hydrogen because it is lighter and non-inflammable.
(d) It is used in gas-cooled nuclear reactors.

40 Deep sea divers used a mixture of which gases to respirate?
(a) Oxygen and argon (b) Oxygen and helium
(c) Oxygen and nitrogen (d) Oxygen and hydrogen

41 Neon lamps are used in botanical gardens and in green houses as they
(a) provide oxygen
(b) provide better light
(c) stimulate growth of the plants
(d) None of the above

42 For advertisement the coloured discharged tubes contain
(a) He (b) Ne
(c) Ar (d) Kr

43 Inert atmosphere for welding of metal is obtained by using
(a) Ar, He (b) N_2, He
(c) O_2, N_2 (d) Ar, N_2

44 Clathrates compounds are not formed by
(a) He,Ne (b) Kr, Xe
(c) He, Kr (d) He, Xe

45 Select the correct statement.
(a) Helium has the lowest melting point and boiling point.
(b) Helium can diffuse through rubber, PVC and even glass.
(c) Ar, Kr and Xe form clathrate.
(d) All of the above are correct statements.

46 When water is allowed to freeze in the presence of Ar, Kr or Xe under pressure then
(a) atoms of noble gas get trapped in the crystal lattice of ice
(b) the clathrates of noble gas (G) and water having composition $8G \cdot 26H_2O$ are formed
(c) the resulting substance is called noble gas hydrates
(d) All of the above statements are true

47 Consider the following properties of the noble gases.
I. They readily form compounds which are colourless.
II. They generally do not form ionic compounds.
III. They have variable oxidation states in their compounds.
IV. Generally do not form covalent compounds.

Select the correct properties and choose the correct option.
(a) I, II and III (b) Both II and III
(c) Both I and III (d) Only I

(B) Miscellaneous questions

1 End-product of the hydrolysis of XeF_6 is
(a) XeF_4O (b) XeF_2O_2
(c) XeO_3 (d) XeO_3^-

2 Which of the following compound attacks glass?
(a) XeF_6 (b) HF
(c) Both (a) and (b) (d) HCl

3 Xenon fluorides act as fluoride donor with
(a) AsF_5 (b) BF_3
(c) HF (d) All of these

4 Basic character of fluorides increases in the order
(a) $XeF_6 < XeF_4 < XeF_2$ (b) $XeF_2 < XeF_4 < XeF_6$
(c) $XeF_2 = XeF_4 < XeF_6$ (d) $XeF_2 > XeF_4 < XeF_6$

5 Maximum distortion in structure takes place in
(a) XeF_2 (b) XeF_4
(c) XeF_6 (d) XeO_3

6 In alkaline solution $HXeO_4^-$ disproportionates to
(a) XeO_6^{4-} and Xe (b) XeO_4^{2-} and XeO_3
(c) XeO_3 and Xe (d) XeO_6^{4-} and XeO_3

7 Consider the following statements.
Statement I He and Ne do not form clathrates.
Statement II They are difficult to polarise due to smaller size.
(a) Only statement I is correct
(b) Only statement II is correct
(c) Both the statements are correct
(d) Both the statements are incorrect

8 Hydrates of helium and neon have not been prepared because of
(a) low polarisability (b) small size
(c) low boiling point (d) All of these

9 Which have distorted geometry based on VSEPR model?
(a) XeF_2, XeF_4, XeO_3 (b) $XeOF_2$, XeO_3, XeF_2
(c) NH_3, SO_2, H_2O (d) All of these

10 The $[XeF]^+$ ions are attached to anion $[AsF_6]^-$
(a) with a linear F-bridge
(b) with an angular F-bridge
(c) with H-bond
(d) with tetrahedral F-bridge

11 In the following reaction,

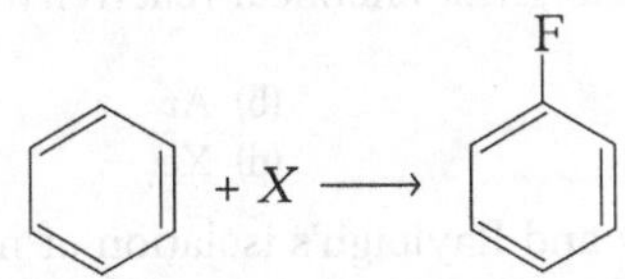

X is

(a) HF (b) F_2 (c) $NaBF_4^-$ (d) XeF_2

12 Consider the following transformation.

I. $XeF_6 + NaF \longrightarrow Na^+[XeF_7]^-$

II. $2PCl_5(s) \longrightarrow [PCl_4]^+ + [PCl_6]^-$

III. $[Al(H_2O)_6]^{3+} + H_2O \longrightarrow [Al(H_2O)_5OH]^{2+} + H_3O^+$

Possible transformations are

(a) I, II and III (b) Both I and III
(c) Both I and II (d) Both II and III

13 Match the molecules/ions (in Column I) with their shapes (in Column II).

Column I		Column II	
A.	ClF_3	1.	Square planar
B.	$[PCl_4]^+$	2.	T-shaped
C.	XeF_4	3.	Tetrahedral
D.	XeO_3	4.	Pyramidal

	A	B	C	D		A	B	C	D
(a)	2	3	1	4	(b)	2	3	4	1
(c)	3	2	4	1	(d)	3	2	1	4

14 Match the items of Columns I and II and mark the correct option. **[NCERT Exemplar]**

Column I	Column II
A. Its partial hydrolysis does not change oxidation state of central atom.	1. He
B. It is used in modern diving apparatus.	2. XeF_6
C. It is used to provide inert atmosphere for filling electrical bulbs.	3. XeF_4
D. Its central atom is in sp^3d^2-hybridisation.	4. Ar

	A	B	C	D		A	B	C	D
(a)	2	1	4	3	(b)	1	2	4	3
(c)	2	1	3	4	(d)	1	2	3	4

15 Consider the following statements.

Statement I XeF_4 may be estimated by titrating the I_2 by reaction with KI.

Statement II Equivalent weight of $Na_2S_2O_3$ can be determined by iodimetric titration.

(a) Only statement I is correct
(b) Only statement II is correct
(c) Both the statements are correct
(d) Both the statements are incorrect

(C) Medical entrances' gallery

1 Match the compounds of Xe in column I with the molecular structure in column II. **[NEET 2020]**

Column I	Column II
A. XeF_2	I. Square planar
B. XeF_4	II. Linear
C. XeO_3	III. Square pyramidal
D. $XeOF_4$	IV. Pyramidal

	A	B	C	D		A	B	C	D
(a)	II	I	III	IV	(b)	II	IV	III	I
(c)	II	III	I	IV	(d)	II	I	IV	III

2. Match the xenon compounds in Column I with its structure in Column II and assign the correct code **[NEET 2019]**

Column I		Column II	
A.	XeF_4	(i)	Pyramidal
B.	XeF_6	(ii)	Square planar
C.	$XeOF_4$	(iii)	Distorted octahedral
D.	XeO_3	(iv)	Square pyramidal

Codes

	A	B	C	D		A	B	C	D
(a)	(ii)	(iii)	(iv)	(i)	(b)	(ii)	(iii)	(i)	(iv)
(c)	(iii)	(iv)	(i)	(ii)	(d)	(i)	(ii)	(iii)	(iv)

3 Which of the following inert gas participate in chemical reaction? **[AIIMS 2019]**

(a) Xe (b) He
(c) Ne (d) None of these

4 In how many of the following number of lone pairs at central atom is zero

XeO_3, XeO_2F_2, XeO_4, XeO_3F_2, XeF_6? **[AIIMS 2019]**

(a) 2 (b) 3
(c) 4 (d) Zero

5 Which of the following rare gas is the abundant in air? **[JIPMER 2019]**

(a) He (b) Ne (c) Ar (d) Kr

6 A gas 'X' is used in filling balloons for meteorological observations. It is also used in gas-cooled nuclear reactors. Here, the gas X is

(a) neon (b) argon **[AIIMS 2018]**
(c) krypton (d) helium

7 Xenon hexafluoride reacts with silica to form a xenon compound X. The oxidation state of xenon in X is **[JIPMER 2018]**

(a) + 2 (b) + 4 (c) + 6 (d) 0

8 Consider the following reactions

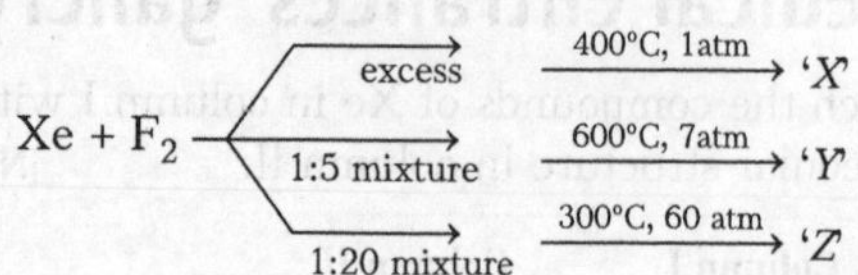

Here, *X*, *Y* and *Z* respectively are **[AIIMS 2017]**

(a) XeF_2, XeF_6, XeF_4 (b) XeF_2, XeF_4, XeF_6
(c) XeF_4, XeF_2, XeF_6 (d) XeF_6, XeF_4, XeF_2

9. Which of the following compound of xenon has pyramidal geometry? **[Karnataka CET 2015]**

(a) $XeOF_4$ (b) XeF_2
(c) XeO_3 (d) XeF_4

10. Among the following inert gas elements, the element that shows highest chemical reactivity is **[EAMCET 2014]**

(a) Ne (b) Ar
(c) He (d) Xe

11. In Ramsay and Rayleigh's isolation of noble gases from air, the nitrogen of the air is finally converted into **[KCET 2011]**

(a) $NaNO_2$ only (b) NO and NO_2
(c) $NaNO_3$ only (d) $NaNO_2$ and $NaNO_3$

12. Number of non-bonding electron pair on Xe in XeF_6, XeF_4 and XeF_2 respectively will be **[Guj. CET 2011]**

(a) 6, 4, 2 (b) 1, 2, 3
(c) 3, 2, 1 (d) 0, 3, 2

ANSWERS

Check Point 10.1

1. (b)	2. (c)	3. (b)	4. (a)	5. (a)	6. (d)	7. (d)	8. (c)	9. (a)	10. (a)

Check Point 10.2

1. (c)	2. (d)	3. (d)	4. (c)	5. (b)	6. (a)	7. (b)	8. (d)	9. (d)	10. (a)

(A) Topical exercise

1. (b)	2. (c)	3. (c)	4. (d)	5. (a)	6. (b)	7. (a)	8. (c)	9. (c)	10. (d)
11. (c)	12. (a)	13. (c)	14. (c)	15. (c)	16. (a)	17. (a)	18. (c)	19. (d)	20. (d)
21. (b)	22. (a)	23. (c)	24. (c)	25. (c)	26. (c)	27. (c)	28. (d)	29. (b)	30. (d)
31. (d)	32. (b)	33. (a)	34. (d)	35. (a)	36. (c)	37. (c)	38. (a)	39. (a)	40. (b)
41. (c)	42. (b)	43. (a)	44. (a)	45. (d)	46. (d)	47. (b)			

(B) Miscellaneous questions

1. (c)	2. (c)	3. (d)	4. (a)	5. (a)	6. (a)	7. (c)	8. (d)	9. (d)	10. (a)
11. (d)	12. (a)	13. (a)	14. (a)	15. (c)					

(C) Medical entrances' gallery

1. (d)	2. (a)	3. (a)	4. (a)	5. (c)	6. (d)	7. (c)	8. (b)	9. (c)	10. (d)
11. (d)	12. (b)								

Hints & Explanations

(A) Topical exercise

9 (*c*) $^{226}_{88}Ra \longrightarrow ^{222}_{86}Rn + ^{b}_{a}X$

$$86 + a = 88, \text{ thus } a = 2$$

$$222 + b = 226, \text{ thus } b = 4$$

Thus, $^{b}_{a}X$ is $^{4}_{2}He$.

10 (*d*) Xe forms compounds (XeF_2, XeF_4, XeF_6) of variable oxidation states.

17 (*a*) $\underset{\text{(Radioactive)}}{^{226}_{88}Ra} \longrightarrow \underset{(\alpha\text{-particles})}{^{4}_{2}He} + ^{222}_{86}Rn$

18 (*c*) Bertlett had taken $O_2^+PtF_6^-$ as a base compound because O_2 and Xe both have almost same ionisation enthalpy. The ionisation enthalpies of noble gases are the highest in their respective periods due to their stable electronic configurations.

26 (*c*) $XeF_6 + H_2O \longrightarrow XeOF_4 \xrightarrow{H_2O} XeO_2F_2 \xrightarrow{H_2O} XeO_3$

SiO_2 is a part of glass.

$XeF_6 + SiO_2 \longrightarrow XeOF_4 \xrightarrow{SiO_2} XeO_2F_2 \xrightarrow{SiO_2} XeO_3$

XeF_6 behaves as an acid. Thus, it does not react with H_2SO_4. (Oxide ion acceptor—Lux flood definition)

27 (*c*) $RbF + XeF_6 \longrightarrow [Rb]^+ [XeF_7]^-$

$CsF + XeF_6 \longrightarrow [Cs]^+ [XeF_7]^-$

28 (*d*) BF_3, RbF, HF form ionic compounds with XeF_6, thus conducting in nature.

29 (*b*) Due to availability of lone pairs on Xe.

XeF_6 – One lone pair $\quad XeF_4$ – Two lone pairs

XeF_2 – Three lone pairs

32 (*b*)

Three π-bonds and one lone pair

35 (*a*) PCl_5, sp^3d-hybridised P, no lone pair

BrF_3, sp^3d-hybridised Br, two lone pairs

XeF_2, sp^3d-hybridised Xe, three lone pairs

47 (*b*) Since, they have stable configuration ($1s^2$ in case of helium and s^2p^6 in case of others), hence ionisation energy is maximum in their respective periods, hence formation of compounds is not easy.

Thus, I is incorrect.

Compounds are covalent. Thus, II is correct and IV is incorrect.

XeF_2, XeF_4, XeF_6, XeO_3 have variable oxidation states of Xe. Thus, III is also correct.

(B) Miscellaneous questions

1 (*c*) $XeF_6 \xrightarrow{H_2O} XeOF_4 \xrightarrow{H_2O} XeO_2F_2 \xrightarrow{H_2O} XeO_3$

2 (*c*) $XeF_6 + \underset{\text{Glass/quartz}}{SiO_2} \longrightarrow SiF_4 + XeOF_4$

$6HF + Na_2SiO_3 \longrightarrow Na_2SiF_6 + 3H_2O$

$HCl + SiO_2 \longrightarrow$ No reaction

3 (*d*) $XeF_2 + AsF_5 \longrightarrow [XeF]^+ [AsF_6]^-$

$XeF_6 + BF_3 \longrightarrow [XeF_5]^+ [BF_4]^-$

$XeF_6 + HF \longrightarrow [XeF_5]^+ [HF_2]^-$

5 (*a*)

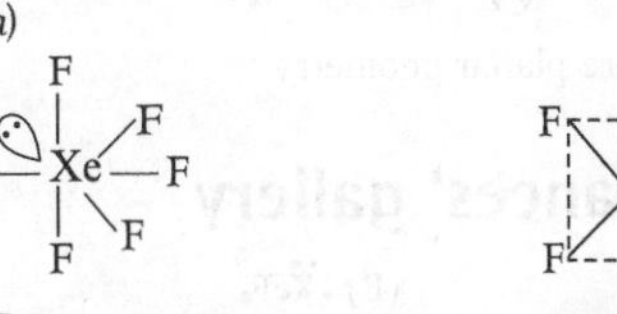

XeF_6
Distorted octahedral
[4 bond pairs, 1 lone pair]

XeF_4
Square planar
[4 bond pairs, 2 lone pairs]

XeF_2
Linear
[2 bond pairs, 3 lone pair]

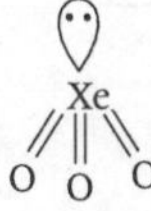

XeO_3
Pyramidal
[3 bond pairs, 1 lone pair]

Due to presence of three lone pairs. XeF_2 is distorted more as compared to other.

6 (*a*) $\underset{+6}{HXeO_4^-} + 2OH^- \longrightarrow \underset{+8}{XeO_6^{4-}} + \underset{0}{Xe} + H_2O + O_2$

Thus, XeO_6^{4-} disproportionates to XeO_6^{4-} (by oxidation) and Xe (by reduction).

9 (*d*) All the compounds given have lone pairs on the central atom thus, structures are distorted.

XeF_2, XeF_4, XeF_6, $XeOF_2$, XeO_3 [For structure see theory part]

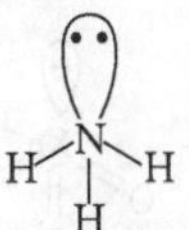

Pyramidal
(instead of tetrahedral)

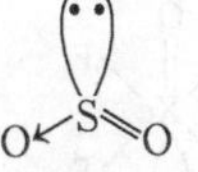

Slightly distorted, bond angle 119°
(instead of 120°)

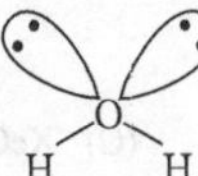

V-shaped
(instead of tetrahedral)

12 (*a*)

I. XeF_6 form complex with NaF.
Thus, I is true.

II. PCl_5 in the solid state exists as $[PCl_4]^+$ and $[PCl_6]^-$.

Thus, II is also true.

III. $[Al(H_2O)_6]^{3+}$ is proton donor.

Thus, III is also true.

14 (*a*)

(A) Partial hydrolysis of XeF_6 does not change oxidation state of central atom.

$$\overset{+6}{XeF_6} + 2H_2O \longrightarrow \overset{+6}{XeO_3} + 6HF$$

(B) He is used in modern diving apparatus.

(C) Ar is used to provide inert atmosphere for filling electrical bulbs.

(D) Central atom (Xe) of XeF_4 is in sp^3d^2-hybridisation.

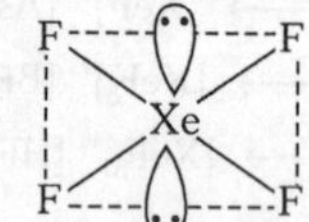

Square planar geometry

(C) Medical entrances' gallery

1 (*d*) (A) $:\ddot{X}eF_2$ (B) $:\ddot{X}eF_4$

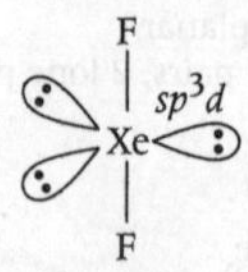

Linear (II) Square planar (I)

(C) $\ddot{X}eO_3$ (D) $\ddot{X}eOF_4$

Pyramidal (IV) Square pyramidal (III)

Hence, option (d) is the correct match.

2 (*a*) The given xenon compounds with its structures are as follows:

(A) XeF_4 — Square planar (B) XeF_6 — Distorted octahedral

(C) $XeOF_4$ — Square pyramidal (D) XeO_3 — Pyramidal

Hence, the correct match is option (c) i.e.,

(A) → (ii), (B) → (iii), (C) → (iv), (D) → (i)

3 (*a*) Neil Bartlett prepared a red compound $O_2^+[PtF_6]^-$ by reacting platinum hexafluoride (PtF_6) with oxygen.

He found that, ionisation energy of Xe(1170 kJ mol^{-1}) is nearly equal to that of molecular oxygen (1175 kJ mol^{-1}) available.

Therefore, He was able to form compound of xenon.

$$Xe + PtF_6 \longrightarrow Xe^+[PtF_6],$$

On the other hand, He and Ne has very high values of ionisation energy. Therefore, only Xe can participate in chemical reactions and form compounds.

Hence, option (a) is correct.

4 (*a*) Structures of XeO_4 and XeO_3F_2 do not have any lone pair at central-atom. i.e.

(XeO_4) (XeO_3F_2)

While in XeO_3, XeO_2F_2 and XeF_6 one lone pair of electrons is present on central atom of each compound.

(XeO_3, one lone pair of electrons) (XeO_2F_2, one lone pair of electrons)

(XeF_6, one lone pair of electrons)

Thus, option (a) is the correct.

5 (*c*) The % by volume of rare gases are

	% by volume	% by weight
He	0.0005	0.000037
Ne	0.0015	0.001
Ar	0.932	1.285
Kr	0.0001	0.00028
Xe	0.0001	0.00004

So, Ar is most abundant in air.

6 (*d*) Helium is used in filling balloons for meteorological observations. It is also used in gas-cooled nuclear reactors. Liquid helium is used as cryogenic agent for carrying out various experiments at low temperature.

7 (*c*) Xenon hexafluoride reacts with silica to form $XeOF_4$ as

$$2XeF_6 + SiO_2 \longrightarrow 2XeOF_4 + SiF_4$$

The oxidation state of xenon in $XeOF_4$ is + 6 as calculated below

Let the oxidation state of Xe is *a*.

$$\therefore \quad a + 1\times(-2) + 4\times(-1) = 0$$

$$a - 2 - 4 = 0 \Rightarrow a = +6$$

8 (*b*) Xenon reacts with fluorine under different conditions to form binary fluorides, i.e., XeF_2, XeF_4 and XeF_6. In these compounds, Xe is present in + 2, + 4 and + 6 oxidation state respectively. The products formed depends on ratio of Xe and F.

$$Xe + F_2 \begin{cases} \xrightarrow[\text{in excess}]{} \xrightarrow{400°C,\ 1atm} XeF_2 \\ \xrightarrow[\text{1:5 mixture}]{} \xrightarrow{600°C,\ 7atm} XeF_4 \\ \xrightarrow[\text{1:20 mixture}]{} \xrightarrow{300°C,\ 60\ atm} XeF_6 \end{cases}$$

9 (*c*)

$XeOF_4$ — Square pyramidal (sp^3d^2)

XeF_2 — Linear (sp^3d^2)

XeO_3 — Pyramidal (sp^3)

XeF_4 — Square planar (sp^3d^2)

10 (*d*) Xe, because of its low ionisation energy forms many compounds. However, no true compounds of He, Ne and Ar are known. Xe forms most of the compounds with F and O, like XeF_2, XeF_4, XeF_6 and XeO_3 etc.

11 (*d*) Nitrogen of air is finally converted into $NaNO_2$ and $NaNO_3$.

12 (*b*)

XeF_6 XeF_4 XeF_2

Hence, number of non-bonding pair on Xe in XeF_6, XeF_4 and XeFe respectively will be 1, 2, 3.

CHAPTER 11

d- and f- Block Elements

The elements positioned between *s*-block (group 1 and 2) and *p*-block (group 13-18) elements are classified as *d*-block and *f*-block elements. The *d*-block elements are situated from groups 3 to 12 and are also called **transition elements** because of their position in the periodic table is between the *s*-block and *p*-block elements.

The *f*-block elements are present in group 3 (periods 6 and 7) only and are also called **inner-transition elements.** The *f*-block elements are further classified as **lanthanides** (lying in 6th period) and **actinides** (lying in 7th period).

d-BLOCK ELEMENTS (TRANSITION ELEMENTS)

The elements which have incompletely filled *d*-orbitals in their ground state or in their most common oxidation state are called ***d*-block** or **transition elements.**

As discussed in the **periodicity (volume I)**, three series of elements are formed by filling the 3*d*, 4*d* and 5*d*-shells of electrons. Together these comprise the *d*-block elements.

The electrons are added to the penultimate shell, expanding it from 8 to 18 electrons. Thus, transition metals typically have incompletely filled *d*-subshells or readily give rise to ions with incompletely filled *d*-subshells.

The group 12 (II B) metals—Zn, Cd and Hg do not have this characteristic electronic configuration. They have completely filled *d*-orbitals and so they really do not belong in this category. However, they are still sometimes called **transition metals.**

General properties of transition elements

Important general properties of the *d*-block elements are discussed as follows

Electronic configuration

The general electronic configuration of *d*-block elements is $(n-1)\,d^{1\text{-}10}\ ns^{1\text{ or }2}$. Based on the energy level in which outermost *d*-orbitals lie, transition elements are classified into different series. These series with examples and their outer shell electronic configurations are listed in table 11.1 Cr and Cu show anomalous configuration, which is shown below

	Expected configuration	Observed configuration
Cr	$[Ar]4s^2 3d^4$	$[Ar]4s^1 3d^5$
Cu	$[Ar]4s^2 3d^9$	$[Ar]4s^1 3d^{10}$

Inside

These anomalous configuration is because of half-filled configuration in Cr and completely filled configuration in Cu which are more stable. Though Cu atom has completely filled *d*-orbital but still considered as a transition element because Cu in its +2 oxidation state shows $4d^9$ configuration.

Table 11.1 Types of transition element

Series name	Group-3	Group-4	Group-5	Group-6	Group-7	Group-8	Group-9	Group-10	Group-11	Group-12
	(VII B)	(IV B)	(V B)	(VI B)	(VII B)		VIII B		(I B)	(II B)
3*d*-series	Sc-21 ($3d^1 4s^2$)	Ti-22 ($3d^2 4s^2$)	V-23 ($3d^3 4s^2$)	Cr-24 ($3d^5 4s^1$)	Mn-25 ($3d^5 4s^2$)	Fe-26 ($3d^6 4s^2$)	Co-27 ($3d^7 4s^2$)	Ni-28 ($3d^8 4s^2$)	Cu-29 ($3d^{10} 4s^1$)	Zn-30 ($3d^{10} 4s^2$)
4*d*-series	Y-39 ($4d^1 5s^2$)	Zr-40 ($4d^2 5s^2$)	Nb-41 ($4d^4 5s^1$)	Mo-42 ($4d^5 5s^1$)	Tc-43 ($4d^6 5s^1$)	Ru-44 ($4d^7 5s^1$)	Rh-45 ($4d^8 5s^1$)	Pd-46 ($4d^{10} 5s^0$)	Ag-47 ($4d^{10} 5s^1$)	Cd-48 ($4d^{10} 5s^2$)
5*d*-series	La-57 ($5d^1 6s^2$)	Hf-72 ($5d^2 6s^2$)	Ta-73 ($5d^3 6s^2$)	W-74 ($5d^4 6s^2$)	Re-75 ($5d^5 6s^2$)	Os-76 ($5d^6 6s^2$)	Ir-77 ($5d^7 6s^2$)	Pt-78 ($5d^9 6s^1$)	Au-79 ($5d^{10} 6s^1$)	Hg-80 ($5d^{10} 6s^2$)
6*d*-series	Ac-89 ($6d^1 7s^2$)	Rf-104 ($6d^2 7s^2$)	Db-105 ($6d^3 7s^2$)	Sg-106 ($6d^4 7s^2$)	Bh-107 ($6d^5 7s^2$)	Hs-108 ($6d^6 7s^2$)	Mt-109 ($6d^7 7s^2$)	Ds-110 ($6d^8 7s^2$)	Rg-111 ($6d^{10} 7s^1$)	Cn-112 ($6d^{10} 7s^2$)

Atomic and ionic sizes

- The atomic and ionic radii of the transition elements decrease from left to right across a row in the transition series, until near the end when the size increases slightly.
- On passing from left to right, extra protons are placed in the nucleus and extra orbital electrons are added. Due to poor shielding effect of the *d*-orbital electrons ($d < p < s$) the attractive force between nucleus and valence electrons increases. Hence size decreases.
- The radii of the elements from chromium to copper, however, are very close to one another. This is due to the fact that successive addition of *d*-electrons screen the outer electrons (4*s*) from the inward pull of the nucleus. This counterbalances the increased nuclear charge due to increase in atomic number. As a result of this, the size of the atom does not alter much in moving from chromium to copper. [see fig 11.1]
- In the end of the series size of elements increases. This is because the increased in the electron-electron repulsion between added electrons in the same orbitals are greater than the attractive forces (due to increased in nuclear charge).
- While going down the group, the atomic radii increase. This means that the atomic radii of second transition series are larger than those of first transition series. However, atomic radii of third transition series elements (5*d*-series) have nearly equal to those of second transition elements (4*d*-series). This is due to a special effect called **lanthanide contraction.** e.g., Zn(160 pm) and HF (159 pm) see fig 11.1. [Detail discussed later]

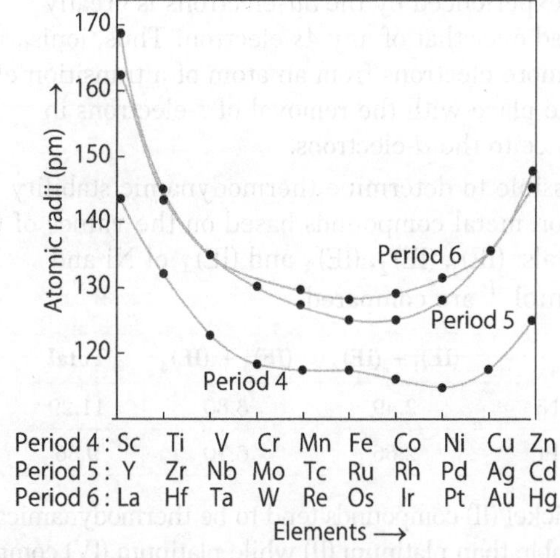

***Fig. 11.1** Comparison of the atomic radii for transition elements of different periods. Note that elements in the same column of the fifth and sixth periods have approximately the same radii.*

The size of transition metal, ions those who having same charges, decrease gradually with the increase in atomic number across a period. It is due to the gradual increase in the effective nuclear charge.

Ionisation enthalpy

The ionisation energies of transition elements usually lie between those of *s*-and *p*-block elements. Their ionisation energy increases in a period with atomic number. This is due to increase in nuclear charge with atomic number. First ionisation energy $(IE)_1$ of 5*d*-elements are higher than those of the 3*d* and 4*d*-elements. This is due to greater effective nuclear charge acting on outer valence electrons because of the weak shielding of the nucleus by $4f$-electrons. The ionisation energies of the $3d$- and $4d$-elements are irregular. (Fig 11.2)

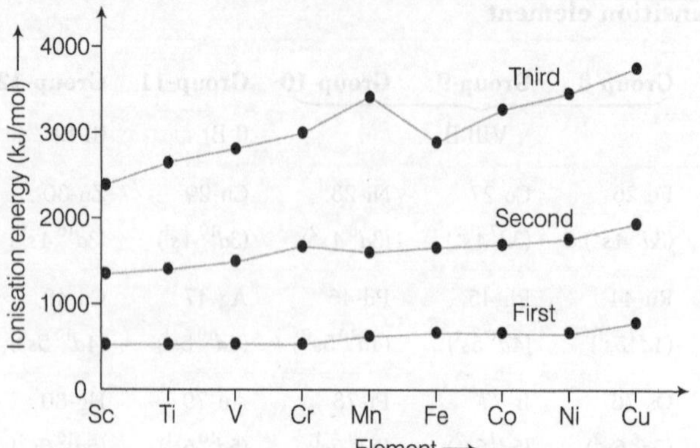

***Fig. 11.2** Variation of the first, second and third ionisation energies for the first transition metals.*

When the electron is ionised from any transition element, say from iron ($Z = 26$) of $3d$-series, the effective nuclear charge experienced by the $3d$-electrons is greatly enhanced over that of any $4s$-electron. Thus, ionisation of two or more electrons from an atom of a transition element will take place with the removal of s-electrons in preference to the d-electrons.

It is possible to determine thermodynamic stability of the transition metal compounds based on the values of (IE) of the metals. $(IE)_1$, $(IE)_2$, $(IE)_3$ and $(IE)_4$ of Ni and Pt in J mol^{-1} are compared.

	$(IE)_1 + (IE)_2$	$(IE)_3 + (IE)_4$	Total
Ni	2.49	8.80	11.29
Pt	2.66	6.70	9.36

Thus, nickel (II) compounds tend to be thermodynamically more stable than platinum (II) while platinum (IV) compounds are relatively more stable than nickel (IV). $K_2\overset{IV}{Pt}Cl_6$ is well known but such type of nickel compound is not formed.

Following factors also decide the stability of the compounds

- the energy of the sublimation,
- the lattice energy,
- the solvation energy.

Electronegativity

Transition elements have fairly low values of electronegativity. It increases from Sc to Cu with a exceptional fall at Mn and Zn. However, this increase in electronegativity is much slower because the additional electron is being added to an inner shell which provides relatively good shielding to the outer electrons from the nucleus. The increasing electronegativity from Sc to Cu means that the elements become slightly less metallic and this is reflected in the increasing positive electrode potentials of their ions M^{2+} and M^{3+}.

Oxidation states

A characteristic property of the d-transition metals is their ability to exhibit several oxidation states differing usually by units of one. This is due to the fact that $(n-1)d$ electrons may get involved along with ns-electrons in bonding, as electrons in $(n-1)d$-orbitals are in an energy state comparable to ns-electrons.

The stability of a given oxidation state depends on the nature of the element with which the transition metal is combined. The highest oxidation states are found in the compounds of transition elements with most electronegative elements like F and O.

The different oxidation state of $3d$-series are given below

Sc	Ti	V	Cr	Mn	Fe	Co	Ni	Cu	Zn
			I					I	
(II)	(II)	II	II	II	II	II	II	II	II
III	III	III	III	(III)	III	III	(III)	(III)	
	IV	IV	IV	IV	(IV)	(IV)	(IV)		
		V	(V)	(V)	(V)	(V)			
			VI	VI	VI				
				VII					

Oxidation states given in brackets are very rare. The lesser number of oxidation states in the beginning of the series can be due to the presence of too few electrons to lose or share. While towards the end of the series it can be described by the presence of too many electrons and thus fewer empty orbitals to share electrons with the ligands.

Within a group, the maximum oxidation state increases with atomic numbers.

e.g. Fe II, III
Ru, Os II, III, IV, VI, VIII

The reduced tendency of higher oxidation states towards the end of the series is due to steady increase in the effective nuclear charge along the series thus pulling the d-orbitals electron into the electron core and not letting them readily available for bonding. Thus, only oxidation of zinc is Zn(II) where no d-orbital electrons are involved.

Note *When an element exists in more than one oxidation states, standard electrode potential data provides a clue to the relative stabilities of different states with respect to oxidation.*

Electrode potential

- When a metal is placed in a solution of its ions, a potential difference is set up between the metal and the solution.
- The value of this potential difference for a particular metal depends upon the nature of metal, the concentration of the metal ions in solution and the temperature.
- More negative the value of the electrode potential for the couple M^{n+}/M, usually more is the reducing power of the element. Similarly, more positive is the value of electrode potential for the couple M^{n+}/M, usually more is the oxidising power of the element. Electrode potential is a measure of the electropositive character and the reactivity of the metals.
- In general along a period, there is a decrease in electropositive character. The reactivity of metals also decreases along a period. All the elements of the first transition series, except copper, have negative values and can react with acids (H^+) producing hydrogen.
- Although, most of the transition metals have negative electrode potential and can react with acid but these are inert or less reactive with acids due to formation of protective layers, e.g. despite a negative standard reduction potential Cr is quite inert chemically because of the formation of chromium (III) oxide, Cr_2O_3 on its surfaces. This fact finds application of Cr to use as a protective and non-corrosive plating on other metals.

Metallic character and melting, boiling points

- Nearly all the transition metals have the simple **hcp, ccp** or **bcc** lattice characteristics of true metals. They have relatively high densities. They are malleable, ductile and have high tensile strength, thermal and electrical conductivity and lustre.
- Melting points of 3*d*-series rise to a maximum and fall as the atomic number increases except in Mn where melting point is minimum.
  The periodic trends in the boiling points are similar to those in the melting points. But the boiling points of respective metals are much higher than their melting points.
- High melting point, boiling point and high heat of atomisation indicate that the atoms in these elements are held together by strong metallic bonds. These bonds are present in the molten states as well. Greater the number of unpaired valence shell *d*-electrons, stronger is the resultant bonding higher is the melting and boiling points.

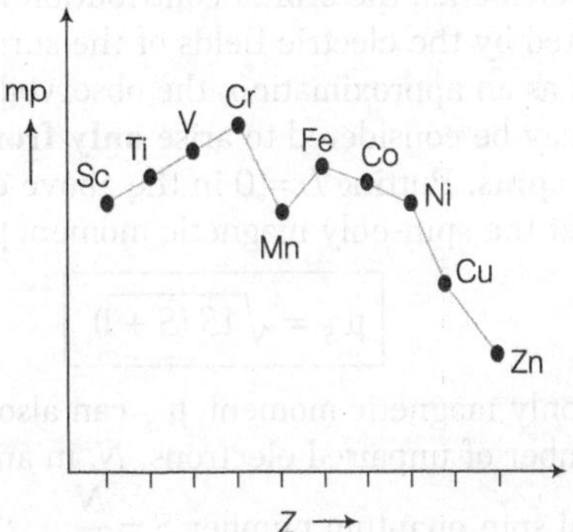

Fig. 11.3 Melting point of 3d-series elements (not according to scale)

Magnetic character

- Magnetic moment (μ) increases as number of unpaired electrons (*n*) increases. As chromium (Cr) has the highest number of unpaired electrons ($N = 6$) in the 3*d*-series thus its magnetic moment is maximum.

	Sc	Ti	V	Cr	Mn	Fe	Co	Ni	Cu	Zn
Z	21	22	23	24	25	26	27	28	29	30
N	1	2	3	6	5	4	3	2	1	0
μ	$\sqrt{3}$	$\sqrt{8}$	$\sqrt{15}$	$\sqrt{48}$	$\sqrt{35}$	$\sqrt{24}$	$\sqrt{15}$	$\sqrt{8}$	$\sqrt{3}$	0

- In some cases (e.g. Mn^{2+}, or Fe^{3+}, in which all the *d*-orbitals are occupied singly by electrons for which $m_l = 2, 1, 0, -1$ and -2, giving $L = 0$) the observed magnetic moment values agree very well with the spin-only values. But generally, experimental values differ from the spin-only values. This is because the orbital motion of the electron also makes some contribution to the moment.
  e.g. In $[Ti(H_2O)_6]^{3+}$ the electronic configuration of Ti is $3d^1$ having one unpaired electron. Its calculated magnetic moment (μ) is 1.73 BM while its experimental magnetic moment is 1.75 BM.
- The magnetic moment is usually expressed in unit called **Bohr magneton** (BM). The general equation for the magnetic moment is given by

$$m_{(S+L)} = \sqrt{4S(S+1) + L(L+1)}$$

In the above expression, S is the sum of the spin quantum numbers and L is the sum of orbital angular momentum of quantum numbers of all the electrons.

- In many compounds including those of the first row transition elements, the orbital contribution is compensated by the electric fields of the surrounding atoms and as an approximation, the observed magnetic moment may be considered to arise **only from unpaired spins.** Putting $L = 0$ in the above expression, you can get the spin-only magnetic moment μ_S.

Thus, $$\boxed{\mu_S = \sqrt{4S(S+1)}}$$

The spin-only magnetic moment, μ_S can also be related to the number of unpaired electrons, N, in any species, as the total spin quantum number $S = \frac{N}{2}$

Hence, $$\mu_S = \sqrt{4S(S+1)}$$

$$= \sqrt{\frac{4N}{2}\left(\frac{N}{2}+1\right)}$$

$$\boxed{\mu_S = \sqrt{N(N+2)}}$$

Above expression gives the values of magnetic moment in Bohr magnetons (BM) which can be converted into SI unit of Ampere square meter (Am^2) by the following relationship

$$\boxed{1\ BM = 9.274 \times 10^{-24}\ Am^2}$$

Colour

- Many ionic and covalent compounds of transition elements (and also inner transition elements) are coloured. In contrast, compounds of the s- and p- block elements are almost always white.
- Colour in d-block element complexes arises due to transition of electron from one d-orbital to another. This transition is called ***d-d* transition*** which we will discuss in next chapter.
- The colour of a transition metal complex is dependent on how big the energy difference is between the two d-levels. This in turn depends on the nature of the ligand, and the type of the complex formed. The colour which we see is the colour of the transmitted wavelength and is the complimentary colour absorbed by the compounds.

e.g. $[Ni(NH_3)_6]^{2+}$ → Blue

$[Ni(H_2O)_6]^{2+}$ → Green

$[Ni(NO_2)_6]^{2-}$ → Brown-red

- In $ZnSO_4$ (Zn^{2+} with $3d^{10}$) and TiO_2 (Ti^{4+} with $3d^0$-) d-d^* transition is not possible hence, they are colourless or white.
- In the series Sc (+III), Ti (+IV), V (+V) Cr (+VI) and Mn (+VII), all have empty d-orbitals, hence d-d^* spectra are impossible and they should be colourless. But VO_2^{2+}(pale yellow), CrO_4^{2-} (deep yellow) and MnO_4^- (dark purple) are coloured in aqueous solutions.

  The **colours arise by charge transfer,** for example, in MnO_4^- an electron is momentarily transferred from O to the metal and thus oxygen changes from O^{2-} to O^- and manganese from Mn(+VII) to Mn(+VI).

Catalytic property

Transition metals can function as catalysts because they can utilise both d and s-electrons for the formation of bonds between reactant molecules and the surface catalyst atoms. This increases the concentration of the reactants at the catalyst surface and weakens the bonds in the reactant molecules. As a result, the activation energy of reaction is lowered.

Compounds of transition metals are able to act as catalysts because of the ease with which the metal can adopt different oxidation states and also because of their ability to form complexes.

Table 11.2 Transition metals and their compounds as catalyst

Catalyst	Function
$TiCl_4$	Used as ***Ziegler-Natta* catalyst** in the production of polythene
$FeSO_4 + H_2O_2$	Used as **Fenton's reagent** in the oxidation of alcohols to aldehydes.
Ni	**Raney nickel** in the reduction processes –manufacture of hexamethylenediamine, –production of H_2 from NH_3, –reducing anthraquinone to anthraquinol in the production of H_2O_2
Ni complexes	**Reppe** synthesis (polymerisation of alkynes to give benzene, cyclooctatetraene).
Cu	Used in the manufacturing of $(CH_3)_2SiCl_2$ being used to make silicones.
Cu/V	Used in the oxidation of cyclohexanol/ cyclohexanone mixture to acetic acid which is used to prepare Nylon-66.
$CuCl_2$	**Deacon's** process of making Cl_2 from HCl.
MnO_2	Used in the decomposition of $KClO_3$ to KCl and O_2.
V_2O_5	Used in contact process of H_2SO_4.

Catalyst	Function
Fe/Mo	Promoted Fe used in the **Haber-Bosch** process of NH_3.
$FeCl_3$	Used in the production of CCl_4 from CS_2 and Cl_2.
$PdCl_2$	**Wacker** process for converting C_2H_4 into CH_3CHO.
H_2/Pd/C	**Lindlar's catalyst** used in reduction of alkene.
Pt	In three stage convertors for cleaning car exhaust fumes.
Pt/Rh	In **Ostwald** process for HNO_3 from NH_3.
ZnO/CuO	In the synthesis of CH_3OH from CO and H_2.
Co/Fe	In the synthesis of hydrocarbons by CO and H_2 in **Fischer-Tropsch** process.
Pt/PtO	Adams catalyst for reduction

Transition metal catalyst as enzyme

Enzymes are catalysts that enhance the rates of specific reactions. Some enzymes require the presence of metal ions as cofactors called as **metalloenzymes** which generally contain transition metal, e.g. Mo in xanthine oxidase (metabolism of purines).

A cobalt atom lies at the centre of the vitamin B_{12} coenzyme. Iron atom is an important component of the haemoglobin of blood and the ferredoxins of photosynthetic process. Fe and Mo are present in nitrogen fixing enzyme.

Complex formation

The transition elements have an unparalleled tendency to form coordination compounds with Lewis bases. These Lewis base are able to donate an electron pair and are called **ligand**. A ligand may be a neutral molecule such as NH_3, or an ion such as Cl^- or CN^-.

$$Co^{3+} + 6NH_3 \longrightarrow [Co(NH_3)_6]^{3+}$$

$$Fe^{2+} + 6CN^- \longrightarrow [Fe(CN)_6]^{4-}$$

$$Cr^{3+} + 6H_2O \longrightarrow [Cr(H_2O)_6]^{3+}$$

This ability to form complexes is in marked contrast to the *s*- and *p*-block elements which form only a few complexes.

Tendency to form complexes by transition metal is due to

- their smaller size,
- higher nuclear charge,
- presence of low energy vacant orbitals to accept lone pair of electrons donated by ligands.

Complexes where the metal is in the +3 oxidation states are generally more stable than those where the metal is in the +2 state. Thus, $[Fe^{III}(CN)_6]^{3-}$ is more stable than $[Fe^{II}(CN)_6]^{4-}$.

Alloy formation

An alloy is a blend of metals prepared by mixing the components. Alloys are homogeneous solid solutions in which the atoms of one metal are distributed randomly among the atoms of the other.

Such alloys are formed by atoms with metallic radii that are within about 15 per cent of each other. Because of similar radii and other characteristics of transition metals, alloys are readily formed by these metals. The alloys so formed are have better property.

The best known are ferrous alloys (steels). Chromium, vanadium, tungsten, molybdenum and manganese are used for the production of variety of steels and stainless steel.

Alloys of transition metals with non-transition metals as brass (copper-zinc) and bronze (copper-tin) are also of considerable industrial importance.

Table 11.3 Some alloys of copper

S.No.	Name	Composition	Uses
1.	Brass	Cu = 66-80% Zn = 20-40%	Domestic utensils, condensers, tubes, costing cartridges.
2.	Bronze	Cu = 75-90% Sn = 10-25%	Statues, coins, utensils, bells, machine parts, propellers, blades.
3.	Aluminium bronze	Cu = 90% Al = 10%	Utensils, coins frames, statues, jewellery bearings and castings.
4.	Phosphorus bronze	Cu = 85% Sn = 13% P = 2%	Suspension wire instrument, gears, aerials, ships, propellers.
5.	Silica bronze	Cu = 97% Sn = 2% Si = 1%	Telephone and telegraph wires, suspensions wires.
6.	German silver or nickel silver	Cu = 50% Zn = 30% Ni = 20%	Utensils, plates, ornaments.
7.	Gun metal	Cu = 88% Sn = 10% Zn = 2%	For making cannons, guns, machine parts.
8.	Monel metal	Cu = 30% Si = 67% Fe + Mn = 3%	Alkali industry resistance wires domestic wires valves, automobile engine parts.
9.	Bell metal	Cu = 77% Sn = 23%	Casting bells
10.	Solder	Sn = 70% Pb = 30%	Joining circuits
11.	Duraluminium	Al = 95% Cu = 4.0% Mg = 0.5% Mn = 0.5%	In utensils

Table 11.4 Some important alloys of steels

S.No.	Name of the alloy (steel)	Metal added	Uses
1.	Manganese steel	1.2-15% Mn	Rock crushing machinery, safes, helmets.
2.	Nickel steel	2.5-5% Ni	Armour plate, cables, propeller shafts, automobile parts, heavy guns.
3.	Invar	3.6% Ni	Pendulum, measuring tapes, instruments
4.	Chrome steel	1.5-2% Cr	Cutting tools, crushing machinery.
5.	Stainless steel	13-14% Cr, 1% Ni	Cutlery, food processing machinery.
6.	Tungsten steel	14-20% W	High speed tools.
7.	Chrome-vanadium steel	1% Cr, 0.15% V	For making spring, shafts, axles.
8.	Silicon steel	3.5 % Si	Electromagnets and transformers
9.	Cobalt steel	35 % cobalt	Permanent magnets
10.	Plantinite	46 % Ni	Electric bulb
11.	Molybdenum steel	6-7 % Mo	High speed tools cutting

Formation of interstitial compounds

Transition metals can trap some small atoms like hydrogen, boron, carbon, nitrogen, etc., in vacant spaces in their crystal lattice forming interstitial compounds. Carbon and nitrogen always occupy octahedral holes; hydrogen is smaller and occupies tetrahedral holes. As only transition metals form such compounds, the *d*-electrons are, therefore, presumably involved in the bonding. The structure of the metal often changes during the formation of such compounds. The composition of these compounds is generally non-stoichiometric.

e.g. $TiH_{1.73}$, $PdH_{0.56}$, $VH_{0.56}$, but may approach regular stoichiometry and a regular structure, e.g. TiC and VN.

The later transition elements of the first series form non-stoichiometric carbides with irregular structures, such as Cr_7C_3, which are more reactive than the interstitial carbides of the early transition elements. These interstitial compounds are of much importance, e.g. carbon steels are interstitial iron-carbon compounds in which the interstitial carbon prevents the iron atoms from sliding over one another, making iron harder, stronger but more brittle.

Nature of oxides/hydroxides

(*i*) While moving along a period from Sc to Zn basic nature decreases and acidic nature increases.

Sc_2O_3	TiO_2	V_2O_5	Cr_2O_3	MnO_2	Fe_2O_3	Co_2O_3	NiO	CuO	ZnO
A	*A*	*A*	*A*	*A*	*B*	*B*	*B*	*B*	*A*

A : *Amphoteric*, *B* : *Basic*

This behaviour differs when oxidation states are different.

With increase in the oxidation state of a given transition metal, covalent character of its compound increases and thus acidic character also increases.

	MnO	Mn_3O_4	Mn_2O_3	MnO_2	MnO_3	Mn_2O_7
Oxidation state of Mn	+2	+8/3	+3	+4	+6	+7
Nature of oxide	Basic	Amphoteric			Weakly acidic	Strongly acidic

Characteristics of oxides and ions of vanadium and chromium have been summarised in Table 11.6.

(*ii*) The basic and amphoteric oxides dissolve in non-oxidising acids and form hexaaquo ions $[M(H_2O)_6]^{n+}$.

(*iii*) The acidic oxides like CrO_3 and Mn_2O_3 dissolve in water to form the oxyacids. These oxides also react with bases to form oxosalts.

$$CrO_3 + H_2O \longrightarrow \underset{\text{(Chromic acid)}}{H_2CrO_4}$$

$$CrO_3 + 2KOH \longrightarrow K_2CrO_4 + H_2O$$

$$2CrO_3 + H_2O \longrightarrow \underset{\text{(Dichromic acid)}}{H_2Cr_2O_7}$$

$$2CrO_3 + 2KOH \longrightarrow \underset{\text{Potassium dichromate}}{K_2Cr_2O_7} + H_2O$$

$$Mn_2O_7 + H_2O \longrightarrow \underset{\text{(Permanganic acid)}}{2HMnO_4}$$

$$Mn_2O_7 + 2KOH \longrightarrow 2KMnO_4 + H_2O$$

(*iv*) As we move along a given transition series, +2 oxidation state of the metals becomes more and more stable and hence the tendency of these metals to lose electrons increases hence, reducing nature of the metal oxide increases. VO and CrO are strong reducing agents.

Table 11.5 Characteristics of oxides and some ions of V and Cr

O.S.	Oxide/ Hydroxide	Behaviour	Ion	Name of ion	Colour of ion
+2	VO	Basic	V^{2+}	Vanadium (II) (vanadous)	Violet
+3	V_2O_3	Basic	V^{3+}	Vanadium (III) (vanadic)	Green
+4	VO_2	Amphoteric	VO^{2+}	Oxovanadium (IV) (vanadyl)	Blue
			$V_4O_9^{2-}$	Hypovanadate (vanadite)	Brown
+5	V_2O_5	Amphoteric	VO_2^{+}	Dioxovanadium (V)	Yellow
			VO_4^{3-}	Orthovanadate (V)	Colourless
+2	CrO $Cr(OH)_2$	Basic	Cr^{2+}	Chromium (II) (chromous)	Light blue
+3	Cr_2O_3 $Cr(OH)_3$	Amphoteric	Cr^{3+} $Cr(OH)_4^-$	Chromium (III) (chromic) Chromite	Violet Green
+6	CrO_3 $CrO_2(OH)_2$ $H_2Cr_2O_7$	Acidic	CrO_4^{2-} $Cr_2O_7^{2-}$	Chromate Dichromate	Yellow Orange

CHECK POINT 11.1

1. The electronic configuration of Cu(II) is $3d^9$ whereas that of Cu(I) is $3d^{10}$. Which of the following is correct?
(a) Cu(II) is more stable
(b) Cu(II) is less stable
(c) Cu(I) and Cu(II) are equally stable
(d) Stability of Cu(I) and Cu(II) depends on nature of copper salts

2. Metallic radii of some transition elements are given below. Which of these elements will have highest density?

Element	Fe	Co	Ni	Cu
Metallic radii/pm	126	125	125	128

(a) Fe (b) Ni
(c) Co (d) Cu

3. Four successive members of the first row transition elements are listed below with the atomic numbers. Which one of them is expected to have the highest third ionisation enthalpy?
(a) Vanadium (Z = 23) (b) Chromium (Z = 24)
(c) Iron (Z = 26) (d) Manganese (Z = 25)

4. Which of the following transition metal ions is not coloured?
(a) Cu^+ (b) V^{3+}
(c) Co^{2+} (d) Ni^{2+}

5. The magnetic nature of elements depends on the presence of unpaired electrons. Identify the configuration of transition element, which shows highest magnetic moment?
(a) $3d^7$ (b) $3d^5$ (c) $3d^8$ (d) $3d^2$

6. Among the given compounds the highest magnetic moment will be shown by
(a) Ni (b) Co (c) Fe (d) Sc

7. The stability of ferric ion is due to
(a) half-filled *f*-orbitals
(b) half-filled *d*-orbitals
(c) completely filled *f*-orbitals
(d) completely filled *d*-orbitals

8. Maximum number of unpaired electrons are present in
(a) Fe^{2+} (b) Mn^{2+} (c) Cu^+ (d) Cr^{2+}

9. The magnetic moment of a transition metal ion is $\sqrt{15}$ BM. Therefore, the number of unpaired electrons present in it is
(a) 4 (b) 1 (c) 2 (d) 3

10. The magnetic moment μ, of transition metals is related to the number of unpaired electron, n as
(a) $\mu = n(n+2)^2$ (b) $\mu = n^2(n+2)$
(c) $\mu = \frac{n}{n+2}$ (d) $\mu = \sqrt{n(n+2)}$

IMPORTANT COMPOUNDS OF d-BLOCK ELEMENTS

Potassium dichromate ($K_2Cr_2O_7$)

Preparation

$K_2Cr_2O_7$ is obtained from chromite ore $[FeCr_2O_4]$ in the following steps

Step I. Chromite ore is fused with molten Na_2CO_3 in the presence of air to give Na_2CrO_4 and Fe_2O_3. Fe_2O_3 is water insoluble while Na_2CrO_4 is water soluble hence, separation can easily be done.

$$4FeCr_2O_4 + 8Na_2CO_3 + 7O_2 \xrightarrow{1100°C} \underset{\text{Water soluble}}{8Na_2CrO_4} + \underset{\text{Water insoluble}}{Fe_2O_3} + 8CO_2\uparrow$$

Step II. Aqueous solution of Na_2CrO_4 is acidified when $Na_2Cr_2O_7$ is formed.

$$2Na_2CrO_4 + \underset{\text{Dil.}}{H_2SO_4} \rightarrow Na_2Cr_2O_7 + Na_2SO_4 + H_2O$$

Step III. Sodium dichromate ($Na_2Cr_2O_7$) is more soluble and less stable than $K_2Cr_2O_7$ hence, $Na_2Cr_2O_7$ changes to $K_2Cr_2O_7$ on reaction with KCl. On cooling concentrated solution of $K_2Cr_2O_7$, orange crystals are obtained.

$$Na_2Cr_2O_7 + 2KCl \longrightarrow \underset{\text{Orange crystals}}{K_2Cr_2O_7\downarrow} + 2NaCl$$

Structure

A dichromate ion is made up of two chromate ions. Each of the chromate groups has a tetrahedral structure.

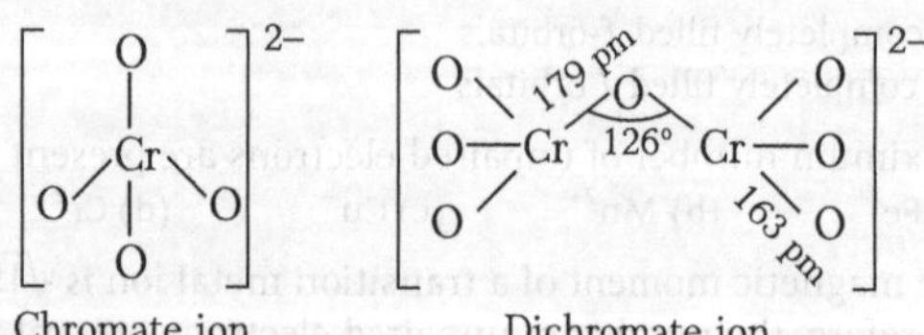

Chromate ion Dichromate ion

Physical properties

(*i*) It is orange-red coloured crystals, having melting point of 699 K. It is moderately soluble in cold water but highly soluble in hot water.

(*ii*) It is generally used as primary standard in the volumetric analysis because it is very stable at room temperature.

Chemical properties

(i) Effect of heat

When potassium dichromate strongly heated, it decomposes to give oxygen.

$$\underset{\text{Orange red}}{4K_2Cr_2O_7(s)} \xrightarrow{\Delta} \underset{\text{Yellow}}{4K_2CrO_4(s)} + 2Cr_2O_3(s) + 3O_2$$

(ii) Action of concentrated sulphuric acid

- In cold conc., red crystals of chromic anhydride (or chromium trioxide) are formed.

$$K_2Cr_2O_7 + \underset{\text{(Cold and conc.)}}{2H_2SO_4} \longrightarrow 2CrO_3 + 2KHSO_4 + H_2O$$

- On heating the mixture, chromium (III) sulphate is formed.

$$\underset{\text{Yellow}}{2K_2Cr_2O_7} + 8H_2SO_4 \xrightarrow{\Delta} 2K_2SO_4 + \underset{\text{Green}}{2Cr_2(SO_4)_3} + 8H_2O + 3O_2$$

- On adding an alkali, the concentration of H^+ ions decreases or we can say pH of the solution increases. In that case yellow colour chromate solution is formed. While, on adding acid, the concentration of H^+ ions increases or we can say pH of the solution decreases. Hence, the reaction proceeds in the backwards direction producing the orange solution containing $Cr_2O_7^{2-}$ ions.

$$\underset{\text{Yellow}}{2CrO_4^{2-}} + \underset{(pH<7)}{2H^+} \longrightarrow \underset{\text{Orange}}{Cr_2O_7^{2-}} + H_2O$$

$$\underset{\text{Orange}}{Cr_2O_7^{2-}} + \underset{(pH>7)}{2OH^-} \longrightarrow \underset{\text{Yellow}}{2CrO_4^{2-}} + H_2O$$

(iii) Chromyl chloride test

- $K_2Cr_2O_7$ when heated with conc. H_2SO_4 and soluble chlorides (like NaCl, KCl), gives deep red vapours of CrO_2Cl_2 (chromyl chloride). CrO_2Cl_2 is an acid chloride and is analogous to SO_2Cl_2 (sulphuryl chloride).

$$K_2Cr_2O_7 + \underset{\text{(conc.)}}{6H_2SO_4} + 4NaCl \longrightarrow 2KHSO_4 + NaHSO_4 + \underset{\substack{\text{Chromyl chloride}\\\text{(Red vapours)}}}{2CrO_2Cl_2} + 3H_2O$$

$$NaCl + \underset{\text{Conc.}}{H_2SO_4} \longrightarrow NaHSO_4 + HCl$$

$$CrO_3 + 2HCl \xrightarrow{\Delta} CrO_2Cl_2 + H_2O$$

- CrO_2Cl_2 (chromyl chloride) when passed into aqueous NaOH solution, yellow colour solution of Na_2CrO_4 is obtained. This on reaction with $(CH_3COO)_2Pb/CH_3COOH$ gives yellow ppt. of $PbCrO_4$.

$$CrO_2Cl_2 + 2NaOH \longrightarrow \underset{\text{Yellow}}{Na_2CrO_4} + 2HCl$$

$$Na_2CrO_4 + (CH_3COO)_2Pb \longrightarrow PbCrO_4\downarrow + 2CH_3COONa$$

This test called '**chromyl-chloride test**' is used to identify Cl^- in inorganic salt-analysis.

- On action with concentrated HCl, Cl_2 gas is evolved and chromic chloride is obtained.

$$K_2Cr_2O_7 + 14HCl \longrightarrow 2KCl + \underset{\text{Chromic chloride}}{2CrCl_3} + 7H_2O + 3Cl_2\uparrow$$

(iv) **Action of dilute sulphuric acid**

- Iodides to iodine

$$K_2Cr_2O_7 + 7H_2SO_4 + 6KI \longrightarrow 4K_2SO_4 + Cr_2(SO_4)_3 + 3I_2 + 7H_2O$$

- Hydrogen sulphide to sulphur

$$K_2Cr_2O_7 + 4H_2SO_4 + 3H_2S \longrightarrow H_2SO_4 + Cr_2(SO_4)_3 + 7H_2O + 3S$$

- Sulphite to sulphate

$$K_2Cr_2O_7 + 4H_2SO_4 + 3Na_2SO_3 \longrightarrow K_2SO_4 + Cr_2(SO_4)_3 + 3Na_2SO_4 + 4H_2O$$

- Ferrous salts to ferric salts

$$K_2Cr_2O_7 + 4FeSO_4 + 7H_2SO_4 \longrightarrow K_2SO_4 + Cr_2(SO_4)_3 + 3Fe_2(SO_4)_3 + 7H_2O$$

- Hydrogen halides to halogens

$$K_2Cr_2O_7 + 4H_2SO_4 + 6HX \longrightarrow K_2SO_4 + Cr_2(SO_4)_3 + 3X_2 + 7H_2O$$

where, X is Cl, Br and I.

- SO_2 is oxidised to SO_4^{2-} and $Cr_2O_7^{2-}$ solution turns green.

$$Cr_2O_7^{2-} + 3SO_2 + 2H^+ \longrightarrow \underset{\text{Green}}{2Cr^{3+}} + 3SO_4^{2-} + H_2O$$

(v) **Inter conversion of K_2CrO_4 and $K_2Cr_2O_7$**

On acidifying (i.e pH < 7) the K_2CrO_4 solution, the yellow colour changes to orange-red due to the formation of dichromate.

$$2K_2CrO_4 + H_2SO_4 \longrightarrow K_2Cr_2O_7 + K_2SO_4 + H_2O$$

or $$\underset{\text{Yellow}}{2CrO_4^{2-}} + 2H^+ \longrightarrow \underset{\text{Orange}}{Cr_2O_7^{2-}} + H_2O$$

Reason for inter conversion In dichromate solution, the orange-red $Cr_2O_7^{2-}$ ions are invariably in equilibrium with yellow CrO_4^{2-} ions.

$$\underset{\text{Orange-red}}{Cr_2O_7^{2-}} + H_2O \rightleftharpoons \underset{\text{Yellow}}{2CrO_4^{2-}} + 2H^+$$

(vi) **Oxidising properties**

$K_2Cr_2O_7$ is a good oxidising agent in acidic medium.

$$K_2Cr_2O_7 + 4H_2SO_4 \longrightarrow K_2SO_4 + Cr_2(SO_4)_3 + 4H_2O + 3[O]$$

or $$\underset{\text{Orange}}{Cr_2O_7^{2-}} + 14H^+ + 6e^- \longrightarrow \underset{\text{Green}}{2Cr^{3+}} + 7H_2O$$

The reaction is spontaneous ($E° > 0$) and $Cr_2O_7^{2-}$ is used as a primary standard in volumetric (titrimetric) analysis.

$$Cr_2O_7^{2-} \equiv 6e^-;$$

$$\therefore \text{ Eq. wt. } (Cr_2O_7^{2-}) = \frac{\text{Mol. wt.}}{\text{Electron transfer}} = \frac{294}{6} = 49$$

Means one sixth of the molecular weight of $K_2Cr_2O_7$ dissolved in 1 litre of water would give one normal (1M) solution (containing 1g-equivalent of $K_2Cr_2O_7$)

(vii) **Action of alkalies**

In alkaline solution (pH > 7), orange colour of $Cr_2O_7^{2-}$ (dichromate ion) changes to yellow colour due to the formation of CrO_4^{2-} (chromate ion).

$$\underset{\text{Oranged red}}{K_2Cr_2O_7} + 2KOH \longrightarrow \underset{\text{Yellow}}{2K_2CrO_4} + H_2O$$

or $$Cr_2O_7^{2-} + 2OH^- \longrightarrow 2Cr_2O_7^{2-} + H_2O$$

(viii) **Reaction with hydrogen peroxide**

When H_2O_2 is added to an acidified solution of a dichromate [or any other Cr(VI) species], a complicated reaction occurs. The products depend on the pH and the concentration of Cr(VI) ion.

$$Cr_2O_7^{2-} + 2H^+ + 4H_2O_2 \longrightarrow \underset{\text{Deep blue-violet coloured}}{2CrO(O_2)_2} + 5H_2O$$

A deep blue-violet coloured peroxo compound $CrO(O_2)_2$, called chromic peroxide is formed. This decomposes rapidly acidic solution into Cr^{3+} ions and oxygen.

$$4CrO_5 + 6H_2SO_4 \longrightarrow 2Cr_2(SO_4)_3 + 6H_2O + 7O_2$$

It reacts with pyridine and form an adduct, i.e. $PyCrO(O_2)_2$.

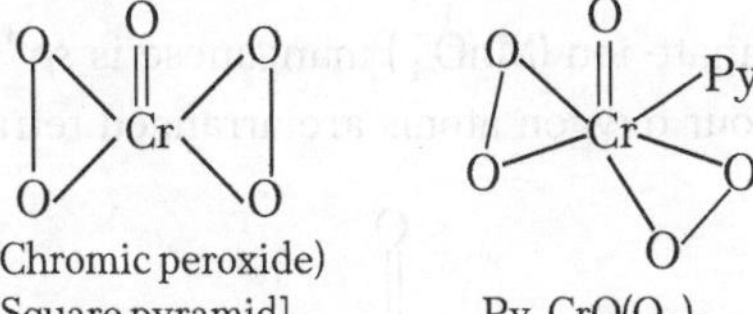

(Chromic peroxide) [Square pyramid] Py. $CrO(O_2)_2$

Uses

(*i*) In volumetric analysis, it is used as standard solution for the estimation of Fe^{2+}, I^- etc.

(*ii*) In organic reactions, it is used as an oxidising agent.

(*iii*) It is used in preparation of chrome alum.

It is deep purple compound used for leather tanning.

$$K_2Cr_2O_7 + H_2SO_4 + 3SO_2 \xrightarrow{\Delta} \underbrace{K_2SO_4 + Cr_2(SO_4)_3}_{\text{Crystallises as}}$$

$$\xrightarrow{+H_2O} \underset{\text{Chrome alum}}{KCr(SO_4)_2 \cdot 12H_2O}$$

(iv) In calico printing and dyeing. It is deep purple compound used for leather tanning.

(v) In photography.

(vi) In hardening of gelatine flim.

Potassium permanganate [$KMnO_4$]

Preparation

$KMnO_4$ is prepared from MnO_2 (pyrolusite) in the following steps.

Step I MnO_2 when is fused with KOH and $KClO_3$ or KNO_3, potassium manganate (K_2MnO_4) is formed which is indicated by its green colour. In this reaction, $KClO_3/KNO_3$ acts as an oxidising agent.

$$\underset{\text{Brown}}{3MnO_2} + 6KOH + KClO_3 \xrightarrow{\Delta} \underset{\text{Green}}{3K_2MnO_4} + KCl + 3H_2O$$

The fused mass is extracted with water and solution is green (MnO_4^{2-}).

Step II • When Cl_2 gas is passed, K_2MnO_4 is oxidised to $KMnO_4$ (purple).

$$\underset{\text{Green}}{2K_2MnO_4} + Cl_2 \longrightarrow 2KCl + \underset{\text{Purple}}{2KMnO_4}$$

• MnO_4^{2-} can also be oxidised to MnO_4^- electrochemically at anode.

At anode $MnO_4^{2-} \longrightarrow MnO_4^- + e^-$

At cathode $2H^+ + 2e^- \longrightarrow H_2$

Structure

In permanganate ion (MnO_4^-) manganese is sp^3-hybridised and hence four oxygen atoms are arranged tetrahedrally.

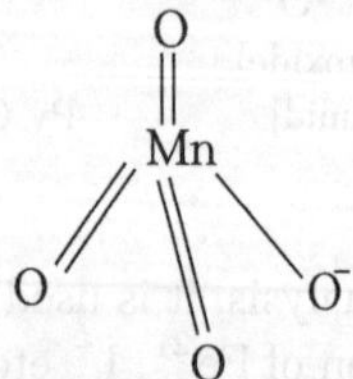

Structure of permanganate ion

Physical properties

(i) It exists as deep purple black prisms with a greenish lustre which becomes dull in air due to reduction.

(ii) It is moderately soluble in water at room temperature and more soluble in hot water.

Chemical properties

(i) Effect of heat

When heated to 513 K, it readily decompose to give potassium manganate, manganese (IV) oxide and oxygen.

$$2KMnO_4 \xrightarrow{513\,K} K_2MnO_4 + MnO_2 + O_2$$

(ii) Effect of conc. H_2SO_4

- With cold conc. H_2SO_4, it gives a green oily compound (Mn_2O_7) which is highly explosive in nature.

$$2KMnO_2 + \underset{\text{Cold conc.}}{2H_2SO_4} \longrightarrow \underset{\text{Green}}{Mn_2O_7} + 2KHSO_4 + H_2O$$

$$2Mn_2O_7 \xrightarrow{\Delta} 4MnO_2 + 3O_2$$

- With warm conc. H_2SO_4

$$4KMnO_4 + \underset{\text{Warm conc.}}{6H_2SO_4} \longrightarrow 2K_2SO_4 + 4MnSO_4 + 6H_2O + 5O_2$$

(iii) Stability in acidic medium

Stability in acidic medium Permanganate solutions are intrinsically unstable in acidic solution and decompose slowly. Decomposition is catalysed by sunlight, so $KMnO_4$ solution should be stored in dark bottles and they must be standardised at the moment of its use in volumertic analysis.

$$4MnO_4^- + 4H^+ \longrightarrow 4MnO_2 + 3O_2 + 2H_2O$$

(iii) Oxidising property

$KMnO_4$ is a good oxidising agent. The actual oxidising action depends upon the use of the $KMnO_4$ in (a) neutral (b) alkaline or (c) acidic medium.

In neutral solution It acts as moderate oxidising agent

$$2KMnO_4 + H_2O \longrightarrow 2KOH + 2MnO_2 + 3O$$

$$MnO_4^- + 2H_2O + 3e^- \longrightarrow MnO_2 + 4OH^-$$

During the course of reaction, alkali is generated. Which make the medium weakly alkaline (even though we started with neutral solutions).

Hence eq. wt. of $KMnO_4$ in neutral or weakly alkaline medium $= \dfrac{\text{mol. wt}}{3} = \dfrac{158}{3} = 52.67$

Some oxidising properties of $KMnO_4$ in the neutral medium.

- It oxidises sodium thiosulphate to sodium sulphate

$$2KMnO_4 + 3Na_2S_2O_3 + H_2O \longrightarrow K_2SO_4 + 3Na_2SO_4 + 2KOH$$

- H_2O_2 is oxidised to O_2

$$2MnO_4^- + 3H_2O_2 \longrightarrow 2\,MnO_2 + 3O_2 + 2OH^- + 2H_2O$$

- $S_2O_3^{2-}$ is oxidised to SO_4^{2-}

$$8\,MnO_4^- + 3S_2O_3^{2-} + H_2O \longrightarrow 8MnO_2 + 6SO_4^{2-} + 2OH^-$$

- Mn^{2+} is oxidised to MnO_2

$$2MnO_4^- + 3Mn^{2+} + H_2O \longrightarrow 5MnO_2 + 4H^+$$

- It oxidises hydrogen sulphide to sulphur

$$2KMnO_4 + 4H_2S \longrightarrow 2MnS + S + K_2SO_4 + 4H_2O$$

In acidic medium It acts as a strong oxidising agent

$$2KMnO_4 + 3H_2SO_4 \longrightarrow K_2SO_4 + 2MnSO_4 + 3H_2O + 5O$$

$$\underset{\text{(Purple)}}{MnO_4^-} + 8H^+ + 5e^- \longrightarrow \underset{\text{(Colourless or extremely light pink)}}{Mn^{2+}} + 4H_2O$$

- The purple colour of MnO_4^- acts as its own indicator (self) in titrations.
- Since, in above reaction, MnO_4^- ion gains 5 electrons or 5 atoms of oxygen are available for two molecules of $KMnO_4$.

Thus, equivalent weight of MnO_4^-

$$= \frac{\text{mol. wt.}}{5} \text{ in acidic medium.}$$

Some oxidising properties of $KMnO_4$ in acidic medium are given below.

(*i*) It oxidises ferrous sulphate to ferric sulphate (i.e. ferrous salt to ferric salt).

$$2KMnO_4 + 8H_2SO_4 + 10FeSO_4 \longrightarrow K_2SO_4 + 2MnSO_4 + 5Fe_2(SO_4)_3 + 8H_2O$$

or $$2MnO_4^- + 16H^+ + 10Fe^{2+} \longrightarrow 2Mn^{2+} + 8H_2O + 10Fe^{3+}$$

(*ii*) It oxidises oxalates or oxalic acid to CO_2

$$2KMnO_4 + 3H_2SO_4 + 5C_2H_2O_4 \longrightarrow K_2SO_4 + 2MnSO_4 + 8H_2O + 10CO_2$$

or $$2MnO_4^- + 16H^+ + 5C_2O_4^{2-} \longrightarrow 2Mn^{2+} + 8H_2O + 10CO_2$$

(*iii*) It oxidises potassium iodide to iodine

$$2KMnO_4 + 3H_2SO_4 + 10KI \longrightarrow K_2SO_4 + 2MnSO_4 + 5H_2O + 5I_2$$

or $$2MnO_4^- + 16H^+ + 10I^- \longrightarrow 2Mn^{2+} + 8H_2O + 5I_2$$

(*iv*) It also oxidises H_2O_2 to H_2O and O_2

$$2KMnO_4 + 3H_2SO_4 + 5H_2O_2 \longrightarrow K_2SO_4 + 2MnSO_4 + 8H_2O + 5O_2$$

or $$2MnO_4^- + 6H^+ + 5H_2O_2 \longrightarrow 2Mn^{2+} + 8H_2O + 5O_2$$

(*v*) It oxidises nitrites (NO_2^-) to nitrates [NO_3^-].

$$2KMnO_4 + 3H_2SO_4 + 5KNO_2 \longrightarrow K_2SO_4 + 2MnSO_4 + 3H_2O + 5KNO_3$$

or $$2MnO_4^- + 6H^+ + 5NO_2^- \longrightarrow 2Mn^{2+} + 3H_2O + 5NO_3^-$$

(*vi*) It oxidises sulphur dioxide to sulphuric acid

$$2KMnO_4 + 5SO_2 + 2H_2O \longrightarrow K_2SO_4 + 2MnSO_4 + 2H_2SO_4$$

or $$2MnO_4^- + 5SO_2 + 2H_2O \longrightarrow 5SO_4^{2-} + 2Mn^{2+} + 4H^+$$

(*vii*) It oxidises hydrogen halide (HX, where X = Cl, Br, I) to dihalogen (X_2)

$$2KMnO_4 + 3H_2SO_4 + 10HX \longrightarrow K_2SO_4 + 2MnSO_4 + 8H_2O + 5X_2$$

or $$2MnO_4^- + 16H^+ + 10X^- \longrightarrow 2Mn^{2+} + 8H_2O + 5X_2$$

(*viii*) It oxidises ethyl alcohol to aldehyde

$$2KMnO_4 + 3H_2SO_4 + 5CH_3CH_2OH \longrightarrow K_2SO_4 + 2MnSO_4 + 5CH_3CHO + 8H_2O$$

In alkaline medium In strongly alkaline medium first potassium manganate is formed which further reduced to MnO_2

$$2KMnO_4 + 2KOH \longrightarrow 2K_2MnO_4 + H_2O + O$$

or $$MnO_4^- + e^- \longrightarrow MnO_4^{2-}$$

$$2K_2MnO_4 + 2H_2O \longrightarrow 2MnO_2 + 4KOH + 2[O]$$

Equivalent weight (in strongly basic medium)

mol. wt. = 158

It oxidises potassium iodide to potassium iodate

$$2KMnO_4 + H_2O + KI \longrightarrow 2MnO_2 + 2KOH + KIO_3$$

Uses

(*i*) Alkaline $KMnO_4$ is called **Baeyer's reagent** and is used to test **unsaturation** in an organic compound, when it decolorises $KMnO_4$.

$$CH_2{=}CH_2 + \underbrace{H_2O + O}_{\text{Alkaline } KMnO_4} \longrightarrow \underset{OH}{CH_2} - \underset{OH}{CH_2}$$

(*ii*) $KMnO_4$ is also widely used as a disinfectant and germicide.

(*iii*) A very dilute solution of $KMnO_4$ is used for washing wounds and gargling of mouth sore.

(*iv*) It is also used for purifying water and treat stinking wells.

(v) Being good oxidising agent it is also used for bleaching of wool, cotton, silk etc.

(vi) It is used in volumetric analysis for the estimation of ferrous salts, oxalates, iodides and hydrogen peroxides.

Note *$KMnO_4$ is not used as primary standard solution. This is due to the fact that it is difficult to obtain it in the pure state and free from traces of MnO_2^-. That's why it is always first standardised with a standard solution of oxalic acid.*

Zinc oxide (ZnO) or zinc white

Preparation

(i) ZnO is formed when ZnS is oxidised with O_2

$$2ZnS + 3O_2 \longrightarrow 2ZnO + 2SO_2$$

(ii) $Zn(OH)_2$ on strongly heating gives ZnO

$$Zn(OH)_2 \xrightarrow{\Delta} ZnO + H_2O$$

(iii) Zinc on burning in air gives ZnO (commercial method)

$$2Zn + O_2 \longrightarrow 2ZnO$$

Properties

(i) ZnO is white when it is cold, due to this property it is used as a pigment in paints. However, it changes colour, when hot, to a pale yellow. This is due to change in the structure of lattice.

(ii) ZnO is soluble both in acid and alkali medium and is thus **amphoteric** in nature.

(iii) It is preferred over white lead as it is not blackened by H_2S. It is also used in medicine and in the preparation of **Rinmann's green** ($ZnCo_2O_4$).

Zinc sulphate or white vitriol ($ZnSO_4 \cdot 7H_2O$)

Preparation

(i) By decomposing $ZnCO_3$ with dil. H_2SO_4.

$$ZnCO_3 + \underset{\text{(dil.)}}{H_2SO_4} \longrightarrow ZnSO_4 + H_2O + CO_2$$

(ii) By heating ZnS (zinc blende) in air at lower temperature and dissolving the product in dil. H_2SO_4.

$$2ZnS + 3.5\,O_2 \xrightarrow{\Delta} ZnO + ZnSO_4 + SO_2$$

$$ZnO + \underset{\text{(dil.)}}{H_2SO_4} \longrightarrow ZnSO_4 + H_2O$$

Properties

(i) Highly soluble in water and solution is acidic in nature due to hydrolysis.

(ii) It slowly effloresces when exposed to air.

(iii) It is isomorphous with epsom salt and used in the manufacture of **lithopone** (which is a mixture of $BaS + ZnSO_4$ and is used as white pigment).

Silver nitrate or Lunar caustic ($AgNO_3$)

Preparation

When Ag is heated with dil. HNO_3, $AgNO_3$ is formed. Crystals separate out on cooling the concentrated solution of $AgNO_3$.

$$3Ag + \underset{\text{(dil.)}}{4HNO_3} \xrightarrow{\Delta} 3AgNO_3 + NO + 2H_2O$$

Physical properties

(i) It is colourless crystalline compound.

(ii) It is soluble in water as well as in alcohol.

(iii) It has a moderate melting point of 212°C.

Chemical properties

(i) When exposed to light, it decomposes hence, stored in a brown coloured bottle.

$$2AgNO_3 \xrightarrow{\Delta,\ \text{Red hot}} 2Ag + 2NO_2 + O_2$$

$$2AgNO_3 \xrightarrow{\Delta,\ T > 212^\circ C} 2AgNO_2 + O_2$$

(ii) It is reduced to metallic Ag by more electropositive metals like Cu, Zn, Mg and also by PH_3.

$$2AgNO_3 + Cu \longrightarrow Cu(NO_3)_2 + 2Ag$$

$$6AgNO_3 + PH_3 + 3H_2O \longrightarrow 6Ag + 6HNO_3 + H_3PO_3$$

(iii) It dissolves in excess of KCN to form an complex compound.

$$AgNO_3 \xrightarrow{KCN} \underset{\text{White ppt.}}{AgCN} \xrightarrow{KCN} \underset{\text{Soluble potassium argentocyanide}}{K[Ag(CN)_2]}$$

(iv) $AgNO_3$ gives white precipitate with $Na_2S_2O_3$ which turns black after sometime.

$$2AgNO_3 + Na_2S_2O_3 \longrightarrow \underset{\text{White ppt.}}{Ag_2S_2O_3} + 2NaNO_3$$

$$Ag_2S_2O_3 + H_2O \longrightarrow \underset{\text{Black}}{Ag_2S} + H_2SO_4$$

(v) It possesses powerful corrosive action on organic tissues, which it turns black especially in presence of light. The blackening is due to finely divided metallic silver, reduced by the organic matter.

(vi) Ammoniacal $AgNO_3$ is called **Tollen's reagent** and is used to identify reducing sugars (including aldehydes)

$$RCHO + 2Ag^+ + 3OH^- \xrightarrow{\Delta} RCOO^- + 2Ag\downarrow + 2H_2O$$

It is called **'silver mirror test'** of aldehydes and reducing sugar (like glucose, fructose).

Ferric chloride ($FeCl_3$)

Preparation

It is prepared by the direct action of chlorine gas on iron metal.

$$2Fe + 3Cl_2 \xrightarrow{\Delta} 2FeCl_3 \text{ (Anhydrous)}$$

Physical properties

(i) Anhydrous $FeCl_3$ is black but its hydrate form i.e., $FeCl_3 \cdot 6H_2O$, is yellowish brown.

(ii) It is highly deliquescent crystalline solid.

(iii) It sublimes at 300°C giving a dimeric gas.

$$2FeCl_3 \xrightarrow{300°C} (FeCl_3)_2 \text{ Gas dimer}$$

(Cl₂Fe(μ-Cl)₂FeCl₂ structure: each Fe bonded to two terminal Cl and two bridging Cl)

(iv) $FeCl_3$ dissolves in both ether and water giving solvated monomeric species.

$$(H_5C_2)_2O \longrightarrow FeCl_3 \text{ (Solvated } FeCl_3\text{)}, \quad [Fe(H_2O)_4Cl_2]Cl \cdot 2H_2O \text{ (Hydrate } FeCl_3\text{)}$$

Its hydrated monomer normally consists of *trans* $[Fe(H_2O)_4Cl_2]Cl \cdot 2H_2O$ but in strong HCl it forms tetrahedral $(FeCl_4)^-$ ions.

Chemical properties

(i) Its aqueous solution is **acidic** due to hydrolysis.

$$FeCl_3 + 3H_2O \rightleftharpoons Fe(OH)_3 + 3HCl$$

$$\text{or } \underset{\text{Acid}}{[Fe(H_2O)_6]^{3+}} + \underset{\text{Base}}{H_2O} \rightleftharpoons \underset{\text{Base}}{[Fe(H_2O)_5(OH)]^{2+}} + \underset{\text{Acid}}{H_3O^+}$$

(ii) $FeCl_3$ is also act as an **oxidising agent.**

$$2FeCl_3 + H_2S \longrightarrow 2FeCl_2 + 2HCl + S$$

$$2FeCl_3 + SnCl_2 \longrightarrow 2FeCl_2 + SnCl_4$$

Yellow colour of aqueous Fe (III) changes to light green aqueous Fe (II).

(iii) Fe^{3+} solution gives blood red colour with SCN^- ions.

$$Fe^{3+} + SCN^- \longrightarrow \underset{\text{(Blood red colouration)}}{[Fe(SCN)]^{2+}}$$

This red colour solution also contains $[Fe(SCN)(H_2O)_5]^{2+}$, $[Fe(SCN)_3]$ and $[Fe(SCN)_4]^-$.

However, when F^- is added, colour fades due to formation of $[FeF_6]^{3-}$.

(iv) Fe^{3+} solution also gives deep blue precipitate of **Prussian blue** with $K_4[Fe(CN)_6]$ (potassium ferrocyanide).

$$4\overset{III}{Fe}Cl_3 + 3K_4[\overset{II}{Fe}(CN)_6] \longrightarrow \underset{\substack{\text{Ferric ferrocyanide} \\ \text{(Prussian blue)}}}{\overset{III}{Fe}_4[\overset{II}{Fe}(CN)_6]_3} + 12KCl$$

$$\text{or } FeCl_3 + K_4[Fe(CN)_6] \longrightarrow \underset{\substack{\text{Potassium} \\ \text{ferrocyanide)} \\ \text{(Prussian blue)}}}{K\overset{III}{Fe}[\overset{II}{Fe}(CN)_6]} + 3KCl$$

Recent X-ray work, IR and other spectroscopic methods have proved that Turnbull's blue is identical to Prussian blue. The intense colour arises from electron transfer between Fe (+II) and Fe (+III).

Fe^{3+} partly oxidises $[\overset{II}{Fe}(CN)_6]^{4-}$ to $[\overset{III}{Fe}(CN)_6]^{3-}$ and itself is reduced to Fe^{2+} and also Fe^{2+} partly reduces $[\overset{III}{Fe}(CN)_6]^{3-}$ to $[\overset{II}{Fe}(CN)_6]^{4-}$ and itself is oxidised to Fe^{3+}.

Thus, $\overset{II}{Fe}[\overset{III}{Fe}(CN)_6]^-$ and $\overset{III}{Fe}[\overset{II}{Fe}(CN)_6]^-$ are identical.

Ferrous sulphate or Green vitriol ($FeSO_4 \cdot 7H_2O$)

Preparation

Industrially, it is prepared by slow oxidation of iron pyrites (FeS_2) in presence of air.

$$2FeS_2 + 2H_2O + 7O_2 \longrightarrow 2FeSO_4 + 2H_2S_4O_4$$

Chemical properties

(i) $2FeSO_4 \xrightarrow{\Delta} Fe_2O_3 + SO_2 + SO_3$

(ii) $2[FeSO_4 \cdot 7H_2O] \xrightarrow{\Delta} Fe_2O_3 + H_2SO_4 + SO_2 + 13H_2O$

(iii) $Fe^{2+} + 2H_2O \longrightarrow Fe(OH)_2 + 2H^+$ (Acidic)

(iv) $FeSO_4 + 2KCN \longrightarrow Fe(CN)_2\downarrow + K_2SO_4$

(v) Fe^{2+} salts or $FeSO_4$ form blue precipitate of Turnbull's blue with $K_3[Fe(CN)_6]$.

Turnbull's blue is used as pigment in ink and paint (also refer $FeCl_3$).

$$Fe^{2+} + [Fe^{III}(CN)_6]^{4-} \longrightarrow \underset{\text{Turnbull's blue}}{Fe^{II}[Fe^{III}(CN)_6]^-}$$

(*vi*) **Atmospheric oxidation** Anhydrous $FeSO_4$ is colourless/white. On exposure to atmosphere, it turns brownish-yellow due to formation of basic ferric sulphate :

$$4FeSO_4 + 2H_2O + O_2 \longrightarrow 4Fe(OH)SO_4$$

$FeSO_4 \cdot 7H_2O$ is isomorphous with $ZnSO_4 \cdot 7H_2O$ (white vitriol) and $MgSO_4 \cdot 7H_2O$ (Epsom salt).

Uses

(*i*) Mixture of $FeSO_4$ and H_2O_2 is used as **Fenton's reagent** for producing hydroxyl radicals and for oxidising alcohols to aldehydes.

(*ii*) Saturated solution of $FeSO_4$ and $(NH_4)_2SO_4$ on concentration and cooling gives crystals of Mohr's salt, $FeSO_4 \cdot (NH_4)_2SO_4 \cdot 6H_2O$ which is widely used as **primary standard** for standardisation of $KMnO_4$ and $K_2Cr_2O_7$ solutions.

(*iii*) **In detection of NO_3^- ion (Brown ring test).** The complex $[Fe(H_2O)_5NO]^{2+}$ is formed in the brown ring test for nitrates when freshly prepared $FeSO_4$ solution is added to aqueous solution of NO_3^- followed by addition of conc. H_2SO_4. The colour is due to charge transfer. Its magnetic moment is approximately 3.873 BM confirming the presence of three unpaired electrons in iron. Thus, iron exists as Fe(+1) and nitrosyl as NO^+.

$$NO \longrightarrow NO^+ + e^-$$

$$Fe^{2+} + e^- \longrightarrow Fe^+$$

$$[Fe(H_2O)_6]^+ + NO^+ \longrightarrow \underset{\text{Brown ring}}{[Fe(H_2O)_5NO]^{2+}} + H_2O$$

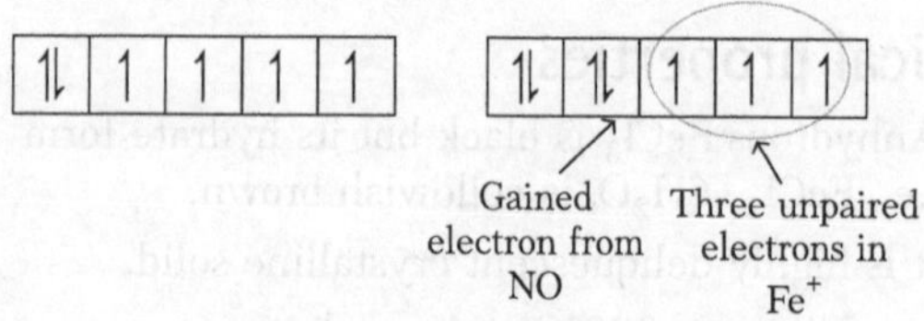

CHECK POINT 11.2

1. $4K_2Cr_2O_7 \xrightarrow{\text{Heat}} 4K_2CrO_4 + 3O_2 + X$;
In the above reaction, X is
(a) CrO_3 (b) Cr_2O_7
(c) Cr_2O_3 (d) CrO_5

2. At pH = 4, CrO_4^{2-} exists as
(a) CrO_4^{2-} (b) CrO_3 (c) CrO_2^{2+} (d) $Cr_2O_7^{2-}$

3. At pH = 12, $Cr_2O_7^{2-}$ changes to
(a) CrO_3 (b) CrO_3^{2+}
(c) CrO_4^{2-} (d) no change

4. $K_2Cr_2O_7$ reacts with NH_4Cl in presence of H_2SO_4. The product formed is
(a) chromyl chlorate with green vapour
(b) chromous chloride with white vapour
(c) chromyl chloride with deep red colour
(d) chromous chloride with blue vapour

5. For reaction, $Cr_2O_7^{2-} \xrightarrow{H^+} Cr^{3+}$
the equivalent weight of $Cr_2O_7^{2-}$ is
(a) mol. wt/6 (b) mol. wt/3
(c) mol.wt/4 (d) mol.wt/1

6. On addition of small amount of $KMnO_4$ to concentrated H_2SO_4, a green oily compound is obtained which is highly expolsive in nature. Identify the compounds from the following.
(a) Mn_2O_7 (b) MnO_2 (c) $MnSO_4$ (d) Mn_2O_3

7. $KMnO_4$ acts as an oxidising agent in acidic medium. The number of moles of $KMnO_4$ that will be needed to react with one mole of sulphide ions in acidic solution is
(a) $\frac{2}{5}$ (b) $\frac{3}{5}$ (c) $\frac{4}{5}$ (d) $\frac{1}{5}$

8. MnO_4^{2-} (1 mol) in neutral aqueous medium disproportionates to
(a) $\frac{2}{3}$ mol of MnO_4^- and $\frac{1}{3}$ mol of MnO_2
(b) $\frac{1}{3}$ mol of MnO_4^- and $\frac{2}{3}$ mol of MnO_2
(c) $\frac{1}{3}$ mol of Mn_2O_7 and $\frac{1}{7}$ mol of MnO_2
(d) $\frac{2}{3}$ mol of Mn_2O_7 and $\frac{1}{3}$ mol of MnO_2

9. When $KMnO_4$ solutions is added to oxalic acid solution, the decolourisation is slow in the beginning but becomes instantaneous after some time because
(a) CO_2 is formed as the product
(b) reaction is exothermic
(c) MnO_4^- catalyses the reaction
(d) Mn^{2+} acts as autocatalyst

10. $KMnO_4$ acts as oxidising agent in
(a) acidic medium only
(b) neutral and acidic medium
(c) neutral and alkaline medium
(d) neutral, acidic and alkaline medium

f-BLOCK ELEMENTS (INNER-TRANSITION ELEMENTS)

The elements in which differentiating electron enters in $(n-2)f$ subshell are called ***f*-block elements**. They are also called **inner-transition elements** because in periodic table they are placed inside the transition elements. These elements are divided into two series

(a) 4*f*-series or lanthanides (b) 5*f*-series or actinides

4*f*-series or lanthanides or first inner transition series

Fourteen elements of the first row (4*f*-series) from lanthanum ($Z = 57$) to lutetium ($Z = 71$) which follow lanthanum ($Z = 57$) in the periodic table are called **lanthanides** or **lanthanones.** These elements are characterised by successive filling of 4*f*-orbitals in their electronic configuration.

General characteristics of lanthanides

1. Electronic configuration

The general electronic configuration of these elements is $[Xe]4f^{1-14}\ 5d^{0-1}\ 6s^2$. Members of this series along with their electronic configurations are summarised in Table 11.7.

Table 11.6 Electronic configuration of lanthanides

Element	Symbol	Atomic number	Outer configuration
Lanthanum	La	57	$[Xe]5d^1, 6s^2$
Cerium	Ce	58	$[Xe]4f^2, 6s^2$
Praseodymium	Pr	59	$[Xe]4f^3, 6s^2$
Neodymium	Nd	60	$[Xe]4f^4, 6s^2$
Promethium	Pm	61	$[Xe]4f^5, 6s^2$
Samarium	Sm	62	$[Xe]4f^6, 6s^2$
Europium	Eu	63	$[Xe]4f^7, 6s^2$
Gadolinium	**Gd**	**64**	$[Xe]4f^7, 5d^1, 6s^2$
Terbium	Tb	65	$[Xe]4f^9, 6s^2$
Dysprosium	Dy	66	$[Xe]4f^{10}, 6s^2$
Holmium	Ho	67	$[Xe]4f^{11}, 6s^2$
Erbium	Er	68	$[Xe]4f^{12}, 6s^2$
Thulium	Tm	69	$[Xe]4f^{13}, 6s^2$
Ytterbium	Yb	70	$[Xe]4f^{14}, 6s^2$
Lutetium	**Lu**	**71**	$[Xe]4f^{14}, 5d^1, 6s^2$

2. Physical state

(*i*) All are silvery white (except Eu and Yb which are pale yellow) metals having tensile strength and are good conductor of heat and electricity.

(*ii*) They have higher density values which increases with increase in atomic number.

(*iii*) They have high melting point with no regularity.

(*iv*) They form alloys easily with other metals (especially with iron).

3. Atomic size, ionic size and lanthanide contraction

There is a regular decrease in the atomic and ionic radii of lanthanide from La to Lu. This decrease in radii is a unique feature and is known as **lanthanide contraction.** It is due to greater effect of the increased nuclear charge than that of the screening effect.

Consequence of lanthanide contraction

(*i*) As a result of lanthanide contraction, the normal increase in size from Sc→ Y→ La disappears after the lanthanide and the pairs of elements such as Zr and Hf, Nb and Ta, Mo and W etc., have nearly same size. The properties of these elements are very similar.

(*ii*) The similarities in properties within these pair make their separation very difficult. However, due to this contraction, their ability to form complexes varies slightly. This slight variation is utilised to separate lanthanides by ion-exchange method.

(*iii*) Due to lanthanide contraction, the elements of 5*d* and 4*d*-transition series resemble each other much more closely than to that of 4*d* and 3*d* series.

(*iv*) Covalent character of hydroxides of lanthanides increases as the size decreases, from La^{3+} to Lu^{3+} but basic strength decreases, i.e. $La(OH)_3$ is most basic whereas $Lu(OH)_3$ is least basic.

(*v*) Tendency to form stable complexes from La^{3+} and Lu^{3+} increases as the size decreases in that order.

4. Ionisation enthalpy

These elements have low ionisation enthalpy and thus, have high electropositive character.

5. Oxidation state

Lanthanides have limited oxidation state because there is a large gap in the energy of 4*f* and 5*d*-subshells. Most stable oxidation state of lanthanides is +3 but some of them also show +2 and +4 oxidation state.

(*i*) In aqueous solution Eu^{2+}, Sm^{2+}, Tm^{2+} and Yb^{2+} ions are present but they revert back to +3 oxidation state. Hence, they are good reducing agents.

(ii) Similarly Pr, Ce, Tb, Dy, in aqueous solution are present in +4 oxidation state but they revert back to +3 oxidation state. Hence, they are good oxidising agents.

7. Colour

Several trivalent (M^{3+}) lanthanide ions are coloured in solid state as well as in aqueous solution. Colour of member of this series arise due to f-f electronic transition, as f-subshells are partially filled.

Generally lanthanide ions with xf electrons have a similar colour to those with $(14-x)f$ electrons, e.g. La^{5+} and Lu^{3+} are colourless, Eu^{3+} and Tb^{3+} are pale pink etc. (See table 11.8)

7. Magnetic properties

Many lanthanide ions are paramagnetic due to the presence of unpaired electrons in the 4f-orbital (except Ce^{4+}, Lu^{3+} and Yb^{3+}). There is difference in calculating the magnetic moment of these elements with respect to transition elements.

Hence, both spin as well as orbital moments give contribution and μ_{eff} is calculated.

$$\mu_{eff} = \sqrt{4s(s+1) + l(l+1)} \text{ B.M}$$

where, l is orbital quantum number and s is the spin quantum number.

also, $\mu_{eff} = \sqrt{N(N+2)}$ B.M.

for N unpaired electrons.

Table 11.7 The colour and magnetic moment of lanthanide (III) ions

Ion	Colour (aqua complex)	Electronic configuration	N (unpaired electrons)	Magnetic moment of N unpaired electrons/ μ BM: Calculated (only total spin) $\sqrt{N(N+2)}$	Observed
La^{3+}	Colourless	$4f^0$	0	00	0
Ce^{3+}	Colourless	$4f^1$	1	1.73	2.3 - 2.5
Pr^{3+}	Green	$4f^2$	2	2.83	3.4 - 3.6
Nd^{3+}	Lilac	$4f^3$	3	3.87	2.5 - 3.6
Pm^{3+}	Yellow	$4f^4$	4	4.90	2.70
Sm^{3+}	Yellow	$4f^5$	5	5.92	1.5 - 1.6
Eu^{3+}	Pale pink	$4f^6$	6	6.93	3.4 - 3.6
Gd^{3+}	Colourless	$4f^7$	7	7.94	7.8 - 8.0
Tb^{3+}	Pale pink	$4f^8$	6	6.93	9.4 - 9.6
Dy^{3+}	Yellow	$4f^9$	5	5.92	10.4 - 10.5
Ho^{3+}	Pale yellow	$4f^{10}$	4	4.90	10.3 - 10.5
Er^{3+}	Rose-red (pink)	$4f^{11}$	3	3.87	9.4 - 9.6
Tm^{3+}	Pale-green	$4f^{12}$	2	2.83	7.1 - 4.4
Yb^{3+}	Colourless	$4f^{13}$	1	1.73	4.4 - 4.9
Lu^{3+}	Colourless	$4f^{14}$	0	0.00	0

8. Complex formation

Lanthanide ions have very less tendency to form complexes. This is because of their low charge density, i.e. charge to size ratio. They form complexes with strong chelating agents like EDTA, β-diketones, oxime etc. but does not form complex with π-bonding ligands are known.

9. Alloy formation

An alloy of lanthanide metal is **mischmetal.** It has lanthanide metal (95%) and iron about 5% with traces of S, C, Ca and Al are used in the production of bullets, shells and lighter flint. Mixed oxides of lanthanides are used as catalysts in petroleum cracking.

10. Chemical reactivity

(i) They are highly electropositive and reactive. All of them burn in air to form oxides Ln_2O_3 and nitrides LnN (cerium gives CeO_2). Powdered cerium is even pyrophoric. Yttrium usually forms a protective oxide coating on it and remains unattacked by air even at 1273 K.

(ii) They combine with H_2 on gentle heating and form hydrides, have a defect lattice and are non-stoichiometric.

(iii) They form trihalides on burning with halogens but Ce and Tb give CeF_4 and TbF_4 respectively. Complex fluorides of Pr(IV), Nd(IV) and Dy(IV) are also formed.

(iv) They react with H_2O form hydroxide and liberate H_2 gas.

$$Ln + H_2O \longrightarrow Ln(OH)_3 + H_2\uparrow$$

(v) The hydroxides are ionic and basic. They are less basic than $Ca(OH)_2$. The basic strength decreases from $Ce(OH)_3$ to $Lu(OH)_3$.

(vi) Lanthanides on reaction with carbon at high temperature (approx. 2500°C) form carbides of the type Ln_2C, Ln_2C_3 and LnC_2.

(vii) Non-availability of d-electrons for π-bonding does not favour the formation of carbonyls by most of the lanthanides. However, an unstable $Nd(CO)_6$ has been prepared.

(viii) The reaction with acids are very vigorous, but do not dissolve in alkalies. They are also stable in HF and H_3PO_4 owing to the formation of a protective film of insoluble salts.

Note *Europium and ytterbium dissolve in liquid ammonia at 195 K forming blue coloured solutions which give spectra characteristic of M^{2+} and ammoniated electrons.*

Uses

They are used in the production of alloys like mischmetal. Their oxides are lessed in glass industry for polishing, for colouring). Their mixed oxides are used in petroleum cracking etc.

CHECK POINT 11.3

1. Lanthanoids are
(a) 14 elements in the sixth period (atomic no. = 90 to 103) that are filling 4f-sublevel
(b) 14 elements in the seventh period (atomic no. = 50 to 11) that are filling 5f-sublevel
(c) 14 elements in the sixth period (atomic no. = 58 to 71) that are filling 4f-sublevel
(d) 14 elements in the seventh period (atomic no. = 40 to 103) that are filling 4f-sublevel

2. Although, zirconium belong to 4d-transition series and hafnium to 5d-transition series even then they show similar physical and chemical properties because ....
(a) Both belong to d-block
(b) Both have same number of electrons
(c) Both have similar atomic radius
(d) Both belong to the same group of the periodic table

3. The atomic size of cerium and promethium is quite close because
(a) they are is same period in periodic table
(b) their electronic configuration is same
(c) f-electrons have poor shielding effect
(d) nuclear charge is higher on cerium than promethium

4. Which of the following trivalent ion has the largest atomic radii in the lanthanide series?
(a) Ce (b) Lu
(c) La (d) Pm

5. Ce^{4+} is stable ion. This is because of
(a) half-filled d-orbitals
(b) all paired electrons in d-orbitals
(c) empty orbitals
(d) fully filled d-orbitals

6. Across the lanthanide series, the basicity of the lanthanide hydroxides
(a) increases
(b) decreases
(c) first increases and then decreases
(d) first decreases and then increases

7. In aqueous solution, Eu^{2+} and Sm^{2+} ions act as
(a) a reducing agent (b) an oxidising agent
(c) an acid (d) All of the above

8. Most basic hydroxide among the following is
(a) $Lu(OH)_3$ (b) $Eu(OH)_3$
(c) $Yb(OH)_3$ (d) $Ce(OH)_3$

9. Which of the following ions is diamagnetic?
(a) Nd^{3+} (b) La^{3+} (c) Tb^{3+} (d) Er^{3+}

10. Mischmetal is
(a) an alloy of lanthanide and copper
(b) an alloy of lanthanide and nickel
(c) an alloy of lanthanide, iron and carbon
(d) an alloy of calcium and copper

5*f*-Series or actinides or second inner-transition series

- In 5*f*-series elements, the differentiating electron occupies 5*f*-orbital, thus these elements are said to belong to 5*f*-series. These are also called ***actinides*** because of their close resemblance in properties with actinium.
- This series includes 14-elements from Th to Lr following actinium in the seventh period of the periodic table. Most of them are radioactive and therefore have the greatest importance in the study of *nuclear chemistry.*
- Elements with atomic numbers greater than 92 (atomic number of uranium) are called the **transuranium elements.** All these elements are man-made, i.e. they are produced in the laboratory through artificial nuclear reactions.
- At present, transuranium elements with atomic number upto 110 have been reported. Recently, on August 16, 2003, IUPAC approved the name for the element with atomic number 110 as darmstadtium (Ds). It is placed in *d*-block. The IUPAC proposed a system for naming elements with $Z > 100$.

General characteristics of actinides

(*i*) **Electronic configuration**

- The general electronic configuration of actinides is $[Rn]5f^{1-14}6d^{0-2}7s^2$. The electrons are formally added to 5*f* but not from, thorium (Z - 90) but from Pa(Z - 91) onward.
- The irregularities in the electronic configuration of the actinides like those in lanthanides is due to the extra stabilities of f^0, f^7 and f^{14} occupancies of the 5*f*-orbitals. Members of this series along with their electronic configuration are tabulated in Table 11.9.
- The **allocation** of electrons to the 5*f* or 6*d*-orbitals in the actinide atoms is more uncertain than the allocation to the 4*f* or 5*d*-orbitals in the lanthanide atoms, because the external shielding of the 5*f*-electrons is less than that of the 4*f*-electrons.
- In simple words 5*f* is less deeply burried than 4*f*, thus, 4*f* electrons can participate in bonding to a far greater extent.

Table 11.8 Electronic configuration of actinides

Name	Symbol	At. No.	Configuration
Thorium	Th	90	$[Rn]5f^0 6d^2 7s^2$
Protactinium	Pa	91	$[Rn]5f^2 6d^1 7s^2$
Uranium	U	92	$[Rn]5f^3 6d^1 7s^2$
Neptunium	Np	93	$[Rn]5f^4 6d^1 7s^2$
Plutonium	Pu	94	$[Rn]5f^6 6d^0 7s^2$
Americium	Am	95	$[Rn]5f^7 6d^0 7s^2$
Curium	**Cm**	**96**	$\mathbf{[Rn]5f^7 6d^1 7s^2}$
Berkelium	Bk	97	$[Rn]5f^9 6d^0 7s^2$
Californium	Cf	98	$[Rn]5f^{10} 6d^0 7s^0$
Einsteinium	Es	99	$[Rn]5f^{11} 6d^0 7s^2$
Fermium	Fm	100	$[Rn]5f^{12} 6d^0 7s^2$
Mendelenium	Md	101	$[Rn]5f^{13} 6d^0 7s^2$
Nobelium	No	102	$[Rn]5f^{14} 6d^0 7s^2$
Lawrencium	**Lr**	**103**	$\mathbf{[Rn]5f^{14} 6d^1 7s^2}$

(*ii*) **Physical state** They are silvery white in appearance and display a variety of structures.
Generally, actinides have high density (except thorium and americium).

(*iii*) **Atomic, ionic size and actinide contraction**

- Like lanthanides the atomic and ionic size of actinides too decreases gradually from Th to Lr. This is called as **actinide contraction.**
- It is due to poor shielding effect of the 5*f*-electrons which result in increase in effective nuclear charge (which causes contraction in size of the electron cloud).

(*iv*) **Oxidation state** The actinoids in general show +3 oxidation state. There is also a greater range of oxidation states due to comparable energies of 5*f*, 6*d* and 7*s* levels.

- The maximum oxidation state increases from Th to Pu and then decreases in succeeding elements (see Table 11.10).
- The actinoids resemble the lanthanoids in having more compounds in +3 state than in +4 state but they tend to hydrolyse.

- U^{3+} in aqueous solution evolves hydrogen on standing; aqueous solutions of Np^{3+} and Pu^{3+} are stable, but are easily oxidised to Np^{4+} and Pu^{4+}. All the other trivalent ions up to No^{3+} are stable.
- U^{4+} and Np^{4+} are stable to water, but are slowly oxidised by air to UO_2^{2+} and NpO_2^{2+}.

Table 11.9 Oxidation states of actinium and actinides

Th	Pa	U	Np	Pu	Am	Cm	Bk	Cf	Es	Fm	Md	No	Lr	Ds
	3	3	3	3	3	3	3	3	3	3	3	3	3	3
4	4	4	4	4	4	4	4							
	5	5	5	5	5									
		6	6	6	6									
			7	7										

(*v*) **Ionisation energy** Ionisation energy of the actinoids are smaller than those of lanthanides. When electrons are being filled into 5*f*-orbitals they will penetrate less into the inner core of the electrons and thus these occupied 5*f*-electrons will be more effectively shielded from the nuclear charge than are the 4*f*-electrons of the corresponding lanthanoids.

(*vi*) **Magnetic properties** The magnetic properties of the actinides are again dependent on unpaired electrons (*N*) and are given by

μ (magnetic moment) $= \sqrt{N(N+2)}$ BM

but observed values are more complex than those of lanthanides. This is due to the fact that 5*f*-electrons of actinides are less effectively shield.

Table 11.10 Magnetic moment of actinides

Element	*Z*	*N* (unpaired electrons)	$\mu = \sqrt{N(N+2)}$ BM (only total spin)
Th	90	2	2.83
Pa	91	3	3.87
U	92	4	4.90
Np	93	5	5.92
Pu	94	6	6.93
Am	95	7	7.94
Cm	96	8	8.94
Bk	97	5	5.92
Cf	98	4	4.90
Es	99	3	3.87
Fm	100	2	2.83
Md	101	1	1.73
No	102	0	0.00
Lr	103	1	1.73

(*vii*) **Complex formation** These elements form complex because they have higher nuclear charge and small size.

(*viii*) **Colour** Actinides ion are generally coloured because of the *f*-*f* electronic transition. But those ion having $5f^0$, $5f^7$ and $5f^{14}$ electronic configuration are colourless.

(*ix*) **Reactivity**

- Actinoids are more electropositive and are reactive. Their reactivity increases with increase in atomic number. They tarnish rapidly in air, forming an oxide coating which is protective in case of Th but less for the other elements.
- The dominant feature of the actinides is their nuclear instability as they have tendency to undergo spontaneous fission and decay becomes more pronounced with the heavier elements. The radioactivity of Th and U is probably responsible for much of the earth's internal heat.
- The metals react with most non-metals especially on heating, but resist alkali attack and are less reactive towards acids than might be expected. Concentrated HCl react rapidly but concentrated HNO_3 makes Th, U and Pu passive, which can be prevented in presence of F^-.
- With boiling water or steam, H_2 is obtained and metal is converted into oxide.
- Metals react rapidly with H_2 forming hydrides which are decomposed by H_2O.

Uses

The most useful actinides are thorium, uranium and plutonium. Thorium is used in the atomic reactors and in the treatment of cancer; uranium is used as nuclear fuel. Plutonium can be used as fuel.

CHECK POINT 11.4

1. Which of the following is not an actinide?
(a) Curium (b) Californium
(c) Uranium (d) Terbium

2. Which of the following belongs to the actinide series?
(a) Ce (b) Cf
(c) Ca (d) Cs

3. There are 14 elements in actinide series. Which of the following elements does not belong to this series?
(a) U (b) Np
(c) Tm (d) Fm

4. The general electronic configuration of actinides is
(a) $[Rn]5f^{0-14}, 6d^{0-2}, 7s^2$
(b) $[Rn]6d^{0-2}, 5f^{0-14}, 7s^2$
(c) $[Rn]5d^{0-2}, 5f^{0-14}, 7s^2$
(d) $[Rn]5f^{0-14}, 5d^{0-2}, 7s^2$

5. Like lanthanide contraction, actinides too show actinide contraction due to
(a) poor-shielding effect of $5f$-orbitals
(b) poor-shielding effect of $4f$-orbitals
(c) poor-shielding effect of $6f$-orbitals
(d) poor-shielding effect of $7f$-orbitals

6. The most common oxidation state of actinide is
(a) +4 (b) +3 (c) +2 (d) +1

7. Actinides have a greater range of oxidation states, this is because
(a) $6f$, $5d$ and $6s$ levels are of comparable energy
(b) $5f$, $6d$ and $7s$ levels are of comparable energy
(c) $7f$, $8d$ and $9s$ levels are of comparable energy
(d) None of the above options are correct

8. In general, the actinides exhibit more number of oxidation states as compared to the lanthanides. This is because
(a) the $5f$ orbital extends farther from the nucleus than the $4f$ orbitals
(b) the $5f$ orbitals are more reactive than $4f$ orbital
(c) the actinids are more reactive than the lanthanides
(d) there is a similarity between $4f$ and $5f$ orbital in their angular part of the wave function

9. The magnetic property of actinides are
(a) more complex than those of lanthanides
(b) less complex than those of lanthanides
(c) similar to lanthanides
(d) cannot be comparable

10. Choose the incorrect statement about actinides.
(a) Most of these ions are colourless
(b) These possess greater tendency for complexation as compared to lanthanides
(c) Both (a) and (b) are incorrect
(d) Both (a) and (b) are correct

Chapter Exercises

(A) Topical exercise

I. General properties of transition elements-I

1 Among the following elements the most abundant element is

(a) Cu (b) Hg (c) Cd (d) Fe

2 Among the given trivalent ion, the ion having largest size is

(a) Ti (b) Zr (c) Hf (d) La

3 Which of the following elements do not show the characteristic properties of *d*-block elements?

(a) Cu, Ag, Au (b) Zn, Hg, Cd
(c) Sc, Ti, V (d) Fe, Co, Ni

4 The lanthanide contraction is responsible for the fact that

(a) Zn and Y have about the same radii
(b) Zr and Nb have similar oxidation state
(c) Zr ahd Hf have about the same radii
(d) Zr and Zn have the same oxidation state

5 The reactivity of transition elements decreases with

(a) the decrease in the atomic number
(b) the increase in the atomic number
(c) low heat of hydration
(d) None of the above

6 First IE of 5*d*-elements are higher than those of 3*d* and 4*d*-elements. This is due to

(a) greater effective nuclear charge acting on outer valence electrons
(b) greater effective nuclear charge which is experienced because of the weak shielding of the nucleus by 4*f*-electrons
(c) Both (a) and (b)
(d) None of the above

7 Select the incorrect statement.

(a) Nickel (II) compounds tend to be more stable than platinum (II)
(b) Platinum (IV) compounds are relatively more stable than nickel (IV) compounds
(c) $K_2Pt^{IV}Cl_6$ exists
(d) $K_2Ni^{IV}Cl_6$ exists

8 Consider the following statements.

Statement I The $E°(M^{2+}/M)$ of copper is positive (0.34 V).

Statement II Copper (s) has high heat of atomisation and low heat of hydration of Cu^{2+}.

(a) Both Statements I and II are correct
(b) Only Statement I is correct
(c) Only Statement II is correct
(d) Both Statements are incorrect

9 Factors which affect the stability of the compounds are

(a) the energy of sublimation (b) the lattice energy
(c) the solvation energy (d) All of these

10 Which of the following statements is not correct? **[NCERT Exemplar]**

(a) Copper liberates hydrogen from acids.
(b) In its higher oxidation states, manganese forms stable compounds with oxygen and fluorine.
(c) Mn^{3+} and Co^{3+} are oxidising agents in aqueous solution.
(d) Ti^{2+} and Cr^{2+} are reducing agents in aqueous solution.

11 The colour imparted by Co(II) compounds to glass is

(a) green (b) deep blue
(c) yellow (d) red

12 Consider the following statements.

Statement I Most of the compounds of the transition elements are coloured in the solid and solution state.

Statement II During *d-d* transition, the *d*-electrons absorb certain radiation from the visible region of the spectrum.

(a) Both statements I and II are true
(b) Only statement I is true
(c) Only statement II is true
(d) Neither statement I nor statement II is true

13 Which one of the following ionic species will impart colour to an aqueous solution?

(a) Ti^4 (b) Cu^+
(c) Zn^{2+} (d) Cr^{3+}

14 Which of the following is colourless?

(a) $[Zn(H_2O)_6]^{2+}$ (b) $[V(H_2O)_6]^{2+}$
(c) $[Mn(H_2O)_6]^{3+}$ (d) $[Fe(H_2O)_6]^{2+}$

15 Which of the following is the coloured ion?

(a) $[Cu(CN)_4]^{3-}$ (b) $[Sc(H_2O)_6]^{3+}$
(c) $[Mn(H_2O)_6]^{3+}$ (d) All of these

16 Which of the following is paramagnetic as well as coloured ion?

(a) Cu^+ (b) Cu^{2+}
(c) Sc^{3+} (d) Ti^{4+}

17 Generally, transition elements form coloured salts due to the presence of unpaired electrons. Which of the following compounds will be coloured in solid state? **[NCERT Exemplar]**

(a) Ag_2SO_4 (b) CuF_2 (c) ZnF_2 (d) Cu_2Cl_2

18 Select the correct statement(s).

(a) Colour of the ion arises due to *d-d** transition
(b) We observe complimentary colour absorbed by the compound
(c) Both (a) and (b) are correct
(d) None of the above is correct

19 Catalyst used in making H_2SO_4 in contact process is

(a) V_2O_5 (b) Fe_2O_3
(c) Cr_2O_3 (d) CrO_3

20 Given below, catalyst and corresponding process/ reaction are matched. The mismatch is

(a) $[RhCl(PPh_3)_2]$: Hydrogenation
(b) $TiCl_4 + Al(C_2H_5)_3$: Polymerisation
(c) V_2O_5 : Haber-Bosch process
(d) Raney Nickel : Hydrogenation

21 In the following reaction,

$$nCH_2{=}CH_2 \longrightarrow -\!\!\!\!\![CH_2{-}CH_2]_n$$

Catalyst is

(a) Adam (b) *Ziegler-Natta*
(c) Ni/Pd (d) Fe/Mo

22 Match the catalysts given in Column I with the processes given in Column II. **[NCERT Exemplar]**

Column I (Catalyst)	**Column II** (Process)
A. Ni in the presence of hydrogen	1. *Ziegler-Natta* catalyst
B. Cu_2Cl_2	2. Contact process
C. V_2O_5	3. Vegetable oil to ghee
D. Finely divided iron	4. Sandmeyer reaction
E. $TiCl_4 + Al(CH_3)_3$	5. Haber's process
	6. Decomposition of $KClO_3$

	A	B	C	D	E
(a)	3	4	2	5	1
(b)	3	4	2	1	5
(c)	3	4	1	5	2
(d)	4	3	2	5	1

23 Consider the following statements.

Statement I $[Fe(CN)_6]^{3-}$ is more stable than $[Fe(CN)_6]^{4-}$.

Statement II Complexes where the iron is in the (III) oxidation state are more stable than those in (II) oxidation state.

(a) Both Statements I and II are correct
(b) Only Statement I is correct
(c) Only Statement II is correct
(d) Both Statements are incorrect

24 Which of the following plays a major role in the formation of complex compound?

(a) Transition metal
(b) Lanthanides and actinides
(c) Representative elements
(d) *p*-block element

25 The metal present in vitamin B_{12} is

(a) cobalt (b) iron
(c) manganese (d) magnesium

26 Which among the following is consumed by humans in the elemental form?

(a) Cu (b) Pb (c) Ag (d) Hg

27 Consider the following statements.

Statement I Nickel (II) compounds tend to be thermodynamically more stable than platinum (II).

Statement II Platinum (IV) compounds are relatively more stable than nickel (IV) compounds.

(a) Both Statements I and II are correct
(b) Only Statement I is correct
(c) Only Statement II is correct
(d) Both Statements are incorrect

28 Stability of Cu^+ and Ag^+ halide complexes are in order

(a) I > Br > Cl > F (b) F > Cl > Br > I
(c) Cl > F > I > Br (d) Br > I > Cl > F

29 Which one of the following elements is the main metallic constituent of haemoglobin?

(a) Al (b) Cu (c) Fe (d) Mn

30 Select the incorrect statement.

(a) Ionisation energies of 5*d*-elements are greater than those of 3*d* and 4*d*-elements.
(b) Cu(I) is diamagnetic while Cu(II) is paramagnetic.
(c) $[Ti(H_2O)_6]^{3+}$ is coloured while $[Sc(H_2O)_6]^{3+}$ is colourless.
(d) Transition elements cannot form complexes.

31 The element present in haemoglobin and chlorophyll respectively are

(a) Fe, Co (b) Fe, Mn (c) Mg, Fe (d) Fe, Mg

32 The incorrect statement for *d*-block element is

(a) it shows magnetic property
(b) it has variable valency
(c) it has tendency of formation of coloured ions
(d) all element it has completly filled *d*-orbitals

33 German silver is an alloy of

(a) Fe, Cr, Ni (b) Ag, Cu, Au
(c) Cu, Zn, Ni (d) Cu, Zn, Sn

34 The composition of duralumin is
(a) Al 94%, Mg 6%
(b) Cu 56%, Zn 24%, Ni 20%
(c) Cu 95, Al 5%
(d) Al 95%, Cu 4%, Mn 0.5%, Mg 0.5%

35 Solder is an alloy of
(a) 70% lead 30% tin (b) 30% lead, 70% tin
(c) 80% lead, 20% tin (d) 90% copper, 10% tin

36 Which of the following pairs of element cannot form an alloy?
(a) Zn, Cu (b) Fe, Hg
(c) Fe, C (d) Hg, Na

37 Which is wrongly matched?
(a) Duralumin – Al + Cu + Mg + Mn
(b) Alnico – Fe + Al + Ni + Cu
(c) German silver – Cu + Zn + Ni
(d) Monel metal – Cu + Zn + Sn

38 Aluminium appears like gold when mixed with
(a) 90% Cu (b) 75% Ni (c) 80% Sn (d) 80% Co

39 Match the alloys (in Column I) with the constituents metal (in Column II).

Column I	Column II
A. Gun metal	1. Pb, Sn
B. German silver	2. Cu, Sn, Zn
C. Brass	3. Cu, Zn
D. Solder	4. Cu, Zn, Ni

	A	B	C	D		A	B	C	D
(a)	2	4	3	1	(b)	2	4	1	3
(c)	4	2	3	1	(d)	4	2	1	3

40 Non-stoichiometry is shown
(a) due to variable valency of transition elements
(b) due to defects in solid structures
(c) due to both (a) and (b)
(d) by only transition metals

41 Which of the following is the set of non-stoichiometric compounds?
(a) NaCl, FeO, $MgCl_2$ (b) FeO, CuS, VSe
(c) Fe_3O_4, NaCl, CuS (d) CuCl, CuS, MgO

42 Interstitial compound is formed by
(a) Fe, Co (b) Co, Ni
(c) Fe, Ni (d) All of these

43 In an atmosphere with industrial smog, Cu corrodes to
(a) basic sulphate $Cu_2(OH)_2SO_4$
(b) basic carbonate $Cu_2(OH)_2CO_3$
(c) Both (a) and (b)
(d) basic oxide, Cu_2O

44 The element which forms ions in dimeric state is
(a) iron (b) mercury (c) cadmium (d) nickel

45 $TiCl_4$ is a
(a) Bronsted-Lowry acid (b) Bronsted-Lowry base
(c) Lewis acid (d) Lewis base

46 Fe_3O_4 is called
(a) haematite (b) magnetite
(c) casserite (d) dolomite

47 Which of the following elements is responsible for oxidation of water to O_2 is biological processes?
(a) Fe (b) Cu
(c) Mn (d) Mo

48 Stainless steel does not rust because
(a) chromium and nickel combine with iron
(b) chromium forms an oxide layer and protects iron from rusting
(c) nickel present in it, does not rust
(d) iron forms a hard chemical compound with chromium present in it

49 Which of the following is most stable ion?
(a) Cu^{2+} (b) Fe^{+}
(c) Fe^{2+} (d) Fe^{3+}

50 Select the correct statement for the lesser number of oxidation states in 3 *d*-series.
(a) In the beginning of the series it can be due to the presence of too few electrons to loose or share.
(b) Towards the end of the series, it can be ascribed to the presence of too many electrons and thus fewer empty orbitals to share electrons with the ligands.
(c) Both (a) and (b) are correct
(d) None of the above is correct

51 Among the following pairs of ions, the lower oxidation state in aqueous solutions is more stable than the other in
(a) Ti^{+}, Ti^{3+} (b) Cu^{+}, Cu^{2+}
(c) Cr^{2+}, Cr^{3+} (d) V^{2+}, VO^{2+}

52 Cu^{2+} has a stronger polarising power than that of Ca^{2+} because
(a) Cu^{2+} ion is smaller than Ca^{2+} ion
(b) Ca^{2+} ion has inert gas configuration whereas Cu^{2+} ion does not
(c) copper shows variable valency, calcium does not
(d) Cu^{2+} ion is smaller than Ca^{2+} ion and the *d*-electrons in Cu^{2+} ion shield the nucleus poorly

II. General properties of transition element-II
(Oxidation and nature of magnetic property)

53 Oxidation state of Fe in Fe_3O_4 is
(a) 3/2 (b) 4/5
(c) 5/4 (d) 8/3

54 Of the following transition metals, the maximum number of oxidation states are exhibited by
(a) chromium ($Z = 24$)
(b) manganese ($Z = 25$)
(c) iron ($Z = 26$)
(d) titanium ($Z = 22$)

55 The pair of compounds having metals in their highest oxidation state is
(a) MnO_2 and CrO_2Cl_2
(b) $[NiCl_4]^{2-}$ and $[CoCl_4]^{2-}$
(c) $[Fe(CN)_6]^{3-}$ and $[Cu(CN)_4]^{2-}$
(d) $[FeCl_4]^{2-}$ and Co_2O_3

56 Consider the following statements.
Statement I Oxidising power of the oxides is in order.

$$VO_2^+ < Cr_2O_7^{2-} < MnO_4^-$$

Statement II Stability of the reduced species is in order

$$V^{3+} < Cr^{3+} < Mn^{2+}$$

(a) Both statements I and II are true
(b) Only statement I is true
(c) Only statement II is true
(d) Neither statement I nor statement II is true

57 Maximum oxidation state of Os is
(a) +6 (b) +7
(c) +8 (d) +5

58 Which of the following metals has more than one oxidation state?
(a) Zn (b) Mn
(c) Sr (d) Ca

59 Which of the following metals have only one oxidation state?
(a) Co (b) Fe
(c) Sc (d) Al

60 Only +2, +3, +4 oxidation states are shown by
(a) Sc, Ti (b) Fe, Ni
(c) Ti, Ni (d) Zn, Ni

61 Maximum oxidation state is shown by
(a) Os (b) Mn
(c) Cr (d) Co

62 Consider the following statements. **[NCERT Exemplar]**
Statement I The highest oxidation state of osmium is +8 .
Statement II Osmium is a 5*d*-block element.
(a) Both Statements I and II are correct
(b) Only Statement I is correct
(c) Only Statement II is correct
(d) Both Statements are incorrect

63 Match the compounds of Column I with oxidation state of metal of Column II.

	Column I		Column II
A.	$[Cr(H_2O)_6]Cl_3$	1.	7
B.	CrO_5	2.	4
C.	MnO_4^-	3.	6
D.	MnO_2	4.	3

	A	B	C	D
(a)	4	3	1	2
(b)	4	3	2	1
(c)	3	4	1	2
(d)	3	4	2	1

64 Match the statements given in Column I with the oxidation states given in Column II. **[NCERT Exemplar]**

	Column I		Column II
A.	Oxidation state of Mn in MnO_2 is	1.	+2
B.	Most stable oxidation state of Mn is	2.	+3
C.	Most stable oxidation state of Mn in oxides is	3.	+4
D.	Characteristic oxidation state of lanthanide is	4.	+5
		5.	+7

	A	B	C	D
(a)	3	4	5	2
(b)	3	4	2	5
(c)	4	3	5	2
(d)	4	3	2	5

65 Match the underlined atoms in Column I with oxidation number in Column II.

	Column I		Column II
A.	$\underline{Mn}O_4^-$	1.	1
B.	$\underline{Cr}O_4^{2-}$	2.	2
C.	$\underline{Fe}[Fe^{II}(CN)_6]^-$	3.	3
D.	$\underline{Zn}O_2^{2-}$	4.	6
E.	$[Ag(CN)_2]^-$	5.	7

	A	B	C	D	E
(a)	5	4	3	2	1
(b)	5	4	2	3	1
(c)	5	3	4	2	1
(d)	4	5	3	1	2

66 The nature of Fe_2O_3 is
(a) acidic (b) basic
(c) amphoteric (d) None of these

67 VO_2 is an amphoteric oxide and in acidic medium it forms
(a) VO^{2+} (b) VO_2^+
(c) V^{3+} (d) VO_2^{2+}

68 Which of the following is amphoteric oxide?
Mn_2O_7, CrO_3, Cr_2O_3, CrO, V_2O_5, V_2O_4 **[NCERT Exemplar]**

(a) V_2O_5, Cr_2O_3 (b) Mn_2O_7, CrO_3
(c) CrO, V_2O_5 (d) V_2O_5, V_2O_4

69 The basic character of the transition metal monoxides follows the order
(Atomic number Ti = 22, V = 23, Cr = 24, Fe = 26)
(a) TiO > FeO > VO > CrO
(b) TiO > VO > CrO > FeO
(c) TiO > CrO > TiO > FeO
(d) CrO > VO > FeO > TiO

70 Maximum magnetic moment is shown by
(a) d^5 (b) d^6
(c) d^7 (d) d^8

71 Which of the following shows maximum magnetic moment among the bivalent ions of the first transition series?
(a) Fe^{2+} (b) Co^{2+}
(c) Ni^{2+} (d) Mn^{2+}

72 Magnetic moment of Cr (Z = 24), Mn^+ (Z = 25) and Fe^{2+} (Z = 26) are x, y, z. They are in order
(a) $x < y < z$ (b) $x = y < z$
(c) $z < x = y$ (d) $x = y = z$

73 Magnetic moment of $[Ti(H_2O)_6]^{3+}$ is
(a) 1.73 BM (b) 2.83 BM (c) 3.87 BM (d) 4.90 BM

74 Spin only magnetic moment of cobalt in the compound $Hg[Co(SCN)_4]$ is
(a) $\sqrt{3}$ (b) $\sqrt{15}$
(c) $\sqrt{24}$ (d) $\sqrt{8}$

75 The magnetic moment is associated with its spin angular momentum and orbital angular momentum. Spin only magnetic moment value of Cr^{3+} ion is
(a) 2.87 BM (b) 3.87 BM **[NCERT Exemplar]**
(c) 3.47 BM (d) 3.57 BM

76 Magnetic moment of Fe is similar to that of
(a) $[Fe(H_2O)_6]^{2+}$ (b) $[Cr(H_2O)_6]^{2+}$
(c) Both (a) and (b) (d) $[Ti(H_2O)_6]^{3+}$

77 There are three unpaired electrons in $[Co(H_2O)_6]^{2+}$ and calculated value of magnetic moment is 3.87 BM which is quite different from the experimental value of 4.40 BM. This is because of
(a) increase in number of unpaired electrons
(b) some contribution of the orbital motion of the electron to the magnetic moment
(c) change in orbital spin of the electron
(d) *d*-*d** transition

III. Important compounds of *d*-block element-I ($K_2Cr_2O_7$, $KMnO_4$)

78 CrO_2Cl_2 is formed while testing
(a) NO_3^- (b) Cl^- (c) Cr^{3+} (d) Fe^{3+}

79 Which of the following mixture is chromium acid?
(a) $K_2Cr_2O_7$ + HCl (b) K_2SO_4 + conc. H_2SO_4
(c) $K_2Cr_2O_7$ + conc. H_2SO_4 (d) H_2SO_4 + HCl

80 In reduction of dichromate by Fe(II), the number of electrons involved per chromium atom is
(a) 2 (b) 3
(c) 4 (d) 1

81 $(NH_4)_2Cr_2O_7$ on heating gives a gas which is also given by
(a) heating NH_4NO_2 (b) heating NH_4NO_3
(c) $Mg_3N_2 + H_2O$ (d) $Na + H_2O_2$

82 The oxide Cr_2O_3 is
(a) acidic (b) amphoteric
(c) basic (d) ozonide

83 $K_2Cr_2O_7$ on heating with aqueous NaOH, gives
(a) CrO_4^{2-} (b) $Cr(OH)_3$
(c) $Cr_2O_7^{2-}$ (d) $Cr(OH)_2$

84 Which of the following compounds is amphoteric?
(a) $Cr(OH)_2$ (b) $Fe(OH)_2$
(c) $Cr(OH)_3$ (d) $Fe(OH)_3$

85 $Cr_2O_7^{2-} \xrightarrow{pH = x} CrO_4^{2-} \xrightarrow{pH = y} Cr_2O_7^{2-}$

pH values x and y can be
(a) 4 and 5 (b) 4 and 8
(c) 8 and 4 (d) 8 and 9

86 $K_2Cr_2O_7 / H^+$ changes to green by
(a) Fe^{2+} (b) SO_2
(c) Both (a) and (b) (d) CO_2

87 Consider the following statements.
Statement I CrO_3 is an acidic oxide.
Statement II CrO_3 liberates CO_2 with Na_2CO_3.
(a) Both Statements I and II are correct
(b) Only Statement I is correct
(c) Only Statement II is correct
(d) Both Statements are incorrect

88 When acidified $K_2Cr_2O_7$ solution is added to Sn^{2+} salt then Sn^{2+} changes to **[NCERT Exemplar]**
(a) Sn (b) Sn^{3+} (c) Sn^{4+} (d) Sn^+

89 Which forms protective and non-corrosive oxide layer?
(a) Cr (b) Ni (c) Zn (d) Cu

90 Consider the following statements.

Statement I Chromium can be used as a protective and non-corrosive plating on other metals.

Statement II Due to atmospheric oxidation, chemically a inert layer of Cr_2O_3 is formed.

(a) Both Statements I and II are correct
(b) Only Statement I is correct
(c) Only Statement II is correct
(d) Both Statements are incorrect

91 In alkaline H_2O_2, $Cr_2O_7^{2-}$ changes to tetraperoxo species ... having oxidation number of Cr as ...

(a) CrO_4^{2-}, 6 (b) CrO_5, 6
(c) CrO_8^{3-}, 5 (d) CrO_8^{3-}, 11

92 In $Cr_2O_7^{2-}$, every Cr is linked to

(a) two O atoms (b) three O atoms
(c) four O atoms (d) five O atoms

93 Cl_2 gas is obtained by various reactions but not by

(a) $KMnO_4 + \text{conc. HCl} \xrightarrow{\Delta}$
(b) $KCl + K_2Cr_2O_7 + \text{conc. } H_2SO_4 \xrightarrow{\Delta}$
(c) $MnO_2 + \text{conc. HCl} \xrightarrow{\Delta}$
(d) $KCl + F_2 \longrightarrow$

94 Amongst the following, identify the species with an atom in + 6 oxidation state

(a) MnO_4^- (b) $Cr(CN)_6^{3-}$
(c) NiF_6^{2-} (d) CrO_2Cl_2

95 In acidic medium, H_2O_2 changes $K_2Cr_2O_7$ to CrO_5 (deep violet solution in ether) having two peroxy linkage. Oxidation number of Cr in CrO_5 is

(a) +10 (b) −10 (c) +4 (d) +6

96 Consider the following statements.

Statement I Green solution of potassium manganate (VI), K_2MnO_4 turns purple and brown solid is precipitated when CO_2 is bubbled into it.

Statement II K_2MnO_4 changes to K_3MnO_8 by CO_2.

(a) Both Statements I and II are correct
(b) Only Statement I is correct
(c) Only Statement II is correct
(d) Both Statements are incorrect

97 In dilute alkaline solution MnO_4^- changes to

(a) MnO_4^{2-} (b) MnO_2
(c) Mn_2O_3 (d) MnO

98 $KMnO_4$ spot can be bleached by

(a) H_2O_2 / H^+ (b) SO_2 / H^+
(c) $C_2O_4^{2-}$ / H^+ (d) All of these

99 Traces of MnO_4^- in conc.H_2SO_4 may change to

(a) Mn_2O_7 (b) MnO_3^+
(c) MnO_2 (d) MnO_4^{2-}

100 Highest oxidation state of manganese in fluoride is +4 (MnF_4) but highest oxidation state in oxides is +7(Mn_2O_7) because

(a) fluorine is more electronegative than oxygen
(b) fluorine does not possess *d*-orbitals
(c) fluorine stabilises lower oxidation state
(d) in covalent compounds, fluorine can form single bond only while oxygen forms double bond

101 MnO_4^{2-} can be converted to MnO_4^-

(a) by oxidation with Cl_2
(b) by electrochemical oxidation at anode
(c) by electrochemical oxidation at cathode
(d) by both (a) and (b) methods

102 Consider the following statements.

Statement I The purple colour of $KMnO_4$ is due to the charge transfer transition.

Statement II The intense colour, in most of the transition metal complexes is due to *d-d* transition.

(a) Both Statements I and II are correct
(b) Only Statement I is correct
(c) Only Statement II is correct
(d) Both Statements are incorrect

103 When MnO_2 is fused with KOH and $KClO_3$, a coloured compound is formed, the product and its colour is

(a) K_2MnO_4, green
(b) $KMnO_4$, purple
(c) Mn_2O_3, brown
(d) Mn_3O_4 black

104 Oxidation state of Mn in MnO_4^- is +7 indicating all electrons paired in Mn but MnO_4^- is coloured. It is due to

(a) charge transfer
(b) presence of unpaired electron in *d*-orbital in oxygen
(c) Both (a) and (b)
(d) None of the above

105 The pair of compounds having metals in their highest oxidation state is

(a) MnO_2, $FeCl_3$ (b) $[MnO_4]^-$, CrO_2Cl_2
(c) $[Fe(CN)_6]^{3-}$, $[Co(CN)_3]$ (d) $[NiCl_4]^{2-}$, $[CoCl_4]^-$

106 Which represents correct comparison of the stability of ions?

(a) $MnO_4^- < Mn^{2+}$ (b) $Cr^{2+} < Cr^{3+}$
(c) $CrO_4^{2-} < Cr^{3+}$ (d) All of these

IV. Important compounds of *d*-block element-II

107 The Mohr's salt is represented by the formula
(a) $FeSO_4(NH_4)_2SO_4 \cdot 6H_2O$
(b) $FeSO_4(NH_3)_2SO_4 \cdot 6H_2O$
(c) $K_2SO_4Al_2(SO_4)_3 \cdot 24H_2O$
(d) $FeSO_2(NH_2)_4SO_4 \cdot 6H_2O$

108 Which of the following is a highly corrosive salt?
(a) $FeCl_2$ (b) $PbCl_2$
(c) Hg_2Cl_2 (d) $HgCl_2$

109 The colour of $FeSO_4 \cdot (NH_4)_2SO_4 \cdot 6H_2O$ is
(a) blue (b) green
(c) white (d) red

110 Finely divided iron combines with CO to give
(a) $Fe(CO)_5$ (b) $Fe_2(CO)_9$
(c) $Fe_2(CO)_{12}$ (d) $Fe(CO)_6$

111 Green vitriol is formed by
(a) $FeS_2 + H_2O + O_2$ (b) $FeS_2 + H_2O + CO_2$
(c) $FeS_2 + CO + CO_2$ (d) $FeS_2 + CO$

112 Thermite is a mixture of
(a) MgO + Al (b) Fe_2O_3 + Al
(c) Zn + $CaCO_3$ (d) Zn + P_2O_5

113 Reason of passivity of iron is
(a) Fe_2O_3 (b) Fe_3O_4
(c) FeO (d) $Fe_2O_3 \cdot 3H_2O$

114 Among K, Ca, Fe and Zn, the element which can form more than one binary compound with chlorine is
(a) Fe (b) Zn
(c) K (d) Ca

115 $FeSO_4$ solution gives brown colour ring in testing nitrates or nitrites. This is
(a) $[Fe(H_2O)_5NO]^{2+}$ (b) $[Fe(H_2O)_5NO_2]^{2+}$
(c) $[Fe(H_2O)_4(NO)_2]^{2+}$ (d) $[Fe(H_2O)_4NO]^{2+}$

116 The compound that gets oxidised even on exposure to air is
(a) Fe_2O_3 (b) $FeCl_2$
(c) $FeCl_3$ (d) $Fe_2(SO_4)_3$

117 Rust contains $Fe_2O_3 \cdot H_2O$. Rust spots can be removed by
(a) $Na_2S_2O_3$ (hypo) (b) SO_2
(c) $H_2C_2O_4$ (oxalic acid) (d) $KMnO_4$

118 Iron is dropped in very dil. HNO_3, it gives
(a) ferric nitrate
(b) ferric nitrate and NO_2
(c) ferrous nitrate and ammonium nitrate
(d) ferrous nitrate and nitric oxide

119 If a person is asked to prepare the blue print of a building plan, he can use
(a) $FeCl_3 + K_4[Fe(CN)_6]$
(b) $FeCl_3 + K_3[Fe(CN)_6]$
(c) $FeCl_2 + K_4[Fe(CN)_6]$
(d) Fe(Ct) + $K_3[Fe(CN)_6]$, Ct is citrate

120 Coagulation of blood takes place by
(a) ferric alum (b) potash alum
(c) Both (a) and (b) (d) ferrous sulphate

121 Anhydrous ferric chloride is prepared by
(a) heating hydrated ferric chloride at a high temperature in a stream of air
(b) heating metallic iron in a stream of dry chlorine gas
(c) reaction of feric oxide with HCl
(d) reaction of metallic iron with HCl

122 Consider the following statements.
Statement I Aqueous solution of ZnO is purely basic in nature.
Statement II ZnO cannot be dissolved in dil. acid.
(a) Both Statements I and II are correct
(b) Only Statement I is correct
(c) Only Statement II is correct
(d) Both Statements are incorrect

123 Which of the following dissolves in hot conc. NaOH solution?
(a) Fe (b) Zn
(c) Cr (d) Ag

124 To protect iron against corrosion, the most durable metal plating on it, is
(a) nickel plating (b) tin plating
(c) copper plating (d) zinc plating

125 Which oxides will not give metal on heating?
(a) HgO (b) ZnO
(c) Ag_2O (c) All of these

126 Which is not blackened by atmospheric H_2S?
(a) $Pb(CH_3COO)_2$ (b) $Zn(CH_3COO)_2$
(c) $Cu(CH_3COO)_2$ (d) $Hg(CH_3COO)_2$

127 Aqueous ZnO can neutralise HCl as well as NaOH solution. Thus, ZnO is an
(a) acidic oxide
(b) basic oxide
(c) amphoteric oxide
(d) amphiprotic oxide

128 A white solid *Y*, on heating gives off a gas which turns lime water milky; the residue is yellow when hot; white when cold. The solid *Y* is probably
(a) $ZnCO_3$ (b) $PbCO_3$
(c) $ZnSO_4$ (d) $Zn(NO_3)_2$

129 ZnO in CH_3COOH gives bluish white ppt. with X. X is
(a) $K_4[Fe(CN)_6]$ (b) $K_3[Fe(CN)_6]$
(c) $Na_2[Fe(CN)_5NO]$ (d) $Fe(CO)_5$

130 Which of the following does not react with AgCl?
(a) NH_4OH (b) $NaNO_3$
(c) $Na_2S_2O_3$ (d) Na_2CO_3

131 Which of the following is used as purgative?
(a) HgS (b) Hg_2Cl_2
(c) $HgCl_2$ (d) $ZnSO_4$

132 Silver ornaments turn black by atmospheric
(a) O_2 (b) N_2
(c) Cl_2 (d) H_2S

133 Out of $AgNO_3$, AgF and $AgClO_4$, water soluble salts are
(a) AgF (b) AgF, $AgNO_3$
(c) AgF, $AgNO_3$, $AgClO_4$ (d) None of these

134 Which of the following exists as white salt in anhydrous state?
(a) CuF_2 (b) $CuSO_4$
(c) Both (a) and (b) (d) None of these

135 A blue solution of copper sulphate becomes darker when treated with excess of ammonia. This is because
(a) ammonia molecules replace water molecules in the solution
(b) ammonia is stronger ligand than water
(c) ammonia forms a stable complex ion $[Cu(NH_3)_4]^{2+}$ with Cu^{2+} ions
(d) All of the above are correct

136 An aqueous solution of $CoCl_2$ on addition of excess of concentrated HCl turn blue due to formation of
(a) $[Co(H_2O)_4Cl_2]$ (b) $[Co(H_2O)_2Cl_4]^{2-}$
(c) $[CoCl_4]^{2-}$ (d) $[Co(H_2O)_2Cl_2]$

137 TiO_2 dissolves in H_2SO_4 forming
(a) $TiO \cdot SO_4$ (b) $TiO_2 \cdot SO_4$
(c) $Ti(SO_4)_2$ (d) $TiSO_4$

138 Which pair of the compounds is expected to show similar colour in aqueous medium?
(a) $FeCl_3$ and $CuCl_2$
(b) $VOCl_2$ and $CuCl_2$
(c) $VOCl_2$ and $FeCl_2$
(d) $FeCl_2$ and $MnCl_2$

139 Which are not blackend by atmospheric H_2S?
(a) TiO_2 (b) ZnO
(c) $ZnSO_4$ + BaS (d) All of these

140 Match the compounds in Column I with their uses in Column II.

Column I	Column II
A. Hg_2Cl_2	1. Metallurgical extraction of Ag
B. ZnO	2. Electrode
C. $ZnSO_4$	3. Luminous paints
D. Zn	4. Lithopone

	A	B	C	D		A	B	C	D
(a)	2	3	4	1	(b)	2	3	1	4
(c)	3	2	4	1	(d)	3	2	1	4

V. *f*-block elements (Inner transition elements)

141 The correct order of ionic radii of Y^{3+}, La^{3+}, Eu^{3+} and Lu^{3+} is
(Atomic number of Y= 39, La= 57, Eu= 63, Lu= 71)
(a) $Lu^{3+} < Eu^{3+} < La^{3+} < Y^{3+}$
(b) $La^{3+} < Eu^{3+} < Lu^{3+} < Y^{3+}$
(c) $Y^{3+} < La^{3+} < Eu^{3+} < Lu^{3+}$
(d) $Y^{3+} < Lu^{3+} < Eu^{3+} < La^{3+}$

142 Consider the following statements.
Statement I Ionic radii of Ta and Nb are same.
Statement II The lanthanide contraction cancels almost exactly the normal size increase on descending a group of transition elements.
(a) Both Statements I and II are correct
(b) Only Statement I is correct
(c) Only Statement II is correct
(d) Both Statements are incorrect

143 The lanthanoide contraction is due to
(a) filling of 4f orbital before 4d
(b) filling of 4d orbital before 4f
(c) filling of 4f orbital before 5d
(d) filling of 5d orbital before 4f

144 Which of the following has/have completely filled 5f-orbitals?
(a) No (b) Lr
(c) Cm (d) Both (a) and (b)

145 An^{3+} (trivalent actinoid ions) have configuration
(a) $5f^0$ to $5f^{14}$ (b) $5f^1$ to $5f^{14}$
(c) $5f^2$ to $5f^{14}$ (d) $5f^0$ to $5f^{13}$

146 Lanthanide for which + 2 and + 3 oxidation states are common is
(a) La (b) Nd (c) Ce (d) Eu

147 Which one of the following elements shows maximum number of different oxidation states in its compounds?
(a) Eu (b) La (c) Gd (d) Am

148 Match the lanthanide ions (Ln^{3+}, in Column I) with their calculated magnetic moments (in Column II).

Column I	Column II
A. Ce^{3+}	1. 7.94 BM
B. Nd^{3+}	2. 4.90 BM
C. Gd^{3+}	3. 1.73 BM
D. Pm^{3+}	4. 3.87 BM

	A	B	C	D		A	B	C	D
(a)	3	4	1	2	(b)	4	3	2	1
(c)	1	2	3	4	(d)	1	4	3	2

149 Which of the following lanthanide is commonly used?

(a) Lanthanium (b) Nobelium
(c) Thorium (d) Cerium

150 Match the compounds/elements given in Column I with uses given in Column II. **[NCERT Exemplar]**

Column I (Compound/element)	Column II (Use)
A. Lanthanide oxide	1. Production of iron alloy
B. Lanthanide	2. Television screen
C. Misch metal	3. Petroleum cracking
D. Magnesium based alloy is constituent of	4. Lanthanide metal + iron
E. Mixed oxides of lanthanide are employed	5. Bullets
	6. X-ray screen

	A	B	C	D	E		A	B	C	D	E
(a)	2	1	4	5	3	(b)	2	1	5	4	3
(c)	2	5	4	3	1	(d)	5	4	3	2	1

151 Match the property given in Column I with the element given in Column II. **[NCERT Exemplar]**

Column I (Property)	Column II (Element)
A. Lanthanide which shows +4 oxidation state.	1. Pm
B. Lanthanide which can shows +2 oxidation state.	2. Ce
C. Radioactive lanthanide.	3. Lu
D. Lanthanide which has $4f^7$ electronic configuration in +3 oxidation state.	4. Eu
E. Lanthanide which has $4f^{14}$ electronic configuration in +3 oxidation state.	5. Gd
	6. Dy

	A	B	C	D	E		A	B	C	D	E
(a)	2	4	1	5	3	(b)	2	4	3	5	1
(c)	4	2	3	1	5	(d)	3	1	5	4	2

152 Consider the following statements.

Statement I Actinide form relatively less stable complexes as compared to lanthanides.

Statement II Actinides can utilise their 5*f*-orbitals alongwith 6*d*-orbitals in bonding but lanthanides do not use their 4*f*-orbital for bonding. **[NCERT Exemplar]**

(a) Both Statements I and II are correct
(b) Only Statement I is correct
(c) Only Statement II is correct
(d) Both Statements are incorrect

(B) Miscellaneous questions

1 Atoms of the transition elements are smaller than those of the *s*-block elements. This is because of

(a) usual contraction in size across a horizontal period
(b) orbital electrons added to the penultimate *d*-shell rather than to the outer shell of the atom
(c) Both (a) and (b)
(d) None of the above

2 The radii of the elements from chromium ($Z = 24$) to copper ($Z = 29$) are very close to one another. This is due to

(a) lanthanide contraction
(b) the fact that successive addition of *d*-electrons screen the outer electrons (4*s*) from the inward pull of the nucleus
(c) increase in radii due to increase in *n* is compensated by decrease in radii due to increase in *Z*
(d) atomic radii do not remain constant but decrease in a normal gradation.

3 Select the correct statement(s).

(a) Fe and Mo atoms are present in nitrogen fixing enzymes
(b) A cobalt atom lies at the centre of the vitamin B_{12} coenzyme
(c) Fe atoms are involved in the ferredoxins of photosynthetic process
(d) All of the above are correct statements

4 CrO_3 liberates CO_2 with Na_2CO_3 solution. Thus, CrO_3 is

(a) an acidic oxide (b) a basic oxide
(c) an amphoteric oxide (d) a non-stoichiometric oxide

5 Alkaline $KMnO_4$ (Baeyer's reagent) can be used to test unsaturation in (*A*).

C_6H_5–CH=CH$_2$

(*A*)

In this case,
(a) unsaturation in side-chain is affected
(b) unsaturation in benzene nucleus is affected
(c) unsaturation in both is affected
(d) Baeyer's reagent cannot be used

6 Silver nitrate solution is kept in brown bottles in the laboratory because
(a) it reacts with ordinary bottles
(b) ordinary bottles catalyse the decomposition
(c) the materials of brown bottles do not react with it
(d) brown bottles cut the passage of light

7 Consider the following statements.

Statement I Cu cannot liberate hydrogen from acids.

Statement II Because it has positive electrode potential. **[NCERT Exemplar]**
(a) Both Statements I and II are correct
(b) Only Statement I is correct
(c) Only Statement II is correct
(d) Both Statements are incorrect

8 On addition of $AgNO_3$ to four different test tubes containing different solutions, one of them gave a white precipitate. It may be
(a) $CHCl_3$ (b) $CaCl_2$
(c) KNO_3 (d) CCl_4

9 Select the correct statement(s).
(a) PH_3 reduces $AgNO_3$ to metallic Ag
(b) Organic tissues turn $AgNO_3$ black by reducing it to Ag
(c) AgCN is soluble in KCN
(d) All of the above are correct statements

10 If $FeCl_3 \cdot 6H_2O$ would exist as a complex, one mole of it in aqueous solution on reaction with excess of $AgNO_3$ forms
(a) 1 mol of AgCl
(b) 2 mol of AgCl
(c) 3 mol of AgCl
(d) no reaction

11 $K_3[Fe(CN)_6]$ is used as an external indicator in the dichromate estimation of Fe^{2+}. The change is observed from
(a) colourless to blue
(b) blue to red
(c) colourless to red
(d) blue to colourless

12 Ti^{2+} is purple while Ti^{4+} is colourless, because
(a) there is zero upaired electron in Ti^{4+}
(b) Ti^{2+} has $3d^2$ configuration
(c) Ti^{4+} has $3d^2$ configuration
(d) Ti^{4+} is a very small cation when compared to Ti^{2+} and hence, does not absorb any radiation

13 On passing Cl_2 gas into alkaline hydrated Fe_2O_3 solution
(a) red purple solution of FeO_4^{2-} is formed
(b) colourless $NaFeO_2$ is formed due to dissolution of Fe_2O_3 into NaOH
(c) $Fe(OH)_3$ is formed
(d) no reaction takes place

14 Isomorphous salts are
(a) green vitriol, blue vitriol and epsom salt
(b) green vitriol, white vitriol and blue vitriol
(c) green vitriol, white vitriol and epsom salt
(d) blue vitriol, white vitriol and epsom salt

15 $[Fe(H_2O)_5NO]^{2+}$ is brown-ring complex in NO_3^-. In this complex,
(a) NO transfers its electron to Fe^{2+} so that there are three electrons unpaired making iron as Fe(I) and nitrosyl as NO^+
(b) magnetic moment of Fe is 3.87 BM
(c) the colour is due to charge transfer
(d) All the above are correct statements

16 Which of the following are sets of diamagnetic species?
(a) $TiCl_4$, O_2, $[Ni(CN)_4]^{2-}$, $[Ni(H_2O)_4]^{2+}$
(b) $TiCl_4$, $[Ni(CN)_4]^{2-}$, CO, $[Ni(H_2O)_4]^{2+}$
(c) $TiCl_4$, $[Ni(CN)_4]^{2-}$, CO
(d) $TiCl_4$, $[Ni(CN)_4]^{2-}$, O_2

17 Select the correct statement(s).
(a) Pu^{4+} disproportionates to Pu^{3+} and PuO_2^{2+} in strongly acidic solution
(b) Maximum oxidation state of Np is +7
(c) UO_2^{2+} is stable
(d) All of the above are correct statements

18 When KCN is added to $CuSO_4$ solution
(a) KCN acts a reducing agent
(b) KCN acts as complexing agent
(c) $K_3[Cu(CN)_4]$ is formed
(d) All of the above are correct

19 Ni^{2+}, in traces, can be tested by using
(a) sodium nitroprusside
(b) dimethyl glyoxime
(c) ammonium sulphocyanide
(d) potassium ferrocyanide

20 Which of the following is not true statement about FeO?
(a) It is non-stoichiometric and is metal deficient
(b) It is basic oxide
(c) Its aqueous solution changes to $Fe(OH)_3$ and then to $Fe_2O_3 \cdot (H_2O)_n$ by atmospheric oxygen
(d) It gives red colour with KCNS

21 Due to lanthanide contraction
(a) Fe, Co, Ni have equal size
(b) Zr and Hf have equal size
(c) all *f*-block ions have equal size
(d) all isoelectronic ions have equal size

22 Paramagnetism is given by the relation $\mu = \sqrt{2s(s+1)}$ magnetons where '*s*' is the total spin. On this basis, the paramagnetism of Cu^+ ion is
(a) 3.88 magnetons (b) 2.83 magnetons
(c) 1.41 magnetons (d) zero

23 The oxygen carrying pigment, oxy-haemocyanin, containing two copper ions is diamagnetic, because
(a) the two copper ions are in +1 oxidation state
(b) one of the copper ions is in +1 oxidation state and the other is in +2 oxidation state
(c) there are strong anti-ferromagnetic interactions between the two copper ions
(d) there are ferromagnetic interactions between the two copper ions

24 If zinc pieces are added in yellow $FeCl_3$ solution, it turns to very light green. It is because
(a) $FeCl_3$ solution is acidic due to hydrolysis and zinc added produces nascent hydrogen causing reduction of $FeCl_3$ to $FeCl_2$
(b) $FeCl_3$ solution is basic and zinc added produces H_2 which reduces $FeCl_3$ to $FeCl_2$
(c) zinc reduces $FeCl_3$ to $FeCl_2$
(d) None of the above is correct

25 Consider the following statements.
Statement I Separation of Zr and Hf is difficult.
Statement II Because Zr and Hf lie in the same group of the periodic table. **[NCERT Exemplar]**
(a) Both Statements I and II are correct
(b) Only Statement I is correct
(c) Only Statement II is correct
(d) Both Statements are incorrect

26 An acidic solution contains Cu^{2+}, Pb^{2+} and Zn^{2+}. If hydrogen sulphide gas is passed through this solution, the precipitate will contain
(a) CuS and ZnS (b) PbS and ZnS
(c) CuS and PbS (d) CuS, PbS and ZnS

27 In $[Fe(CN)_5(NO^+)]^{2-}$, Fe has +3 state. It can be decided by
(a) magnetic moment measurement
(b) colligative property
(c) colour
(d) hybridisation

28 In 3*d*-transition series, if nuclear charge increases, the screening effect
(a) increases (b) decreases
(c) first decreases and then increases
(d) first increases and then decreases

29 Which one of the following characteristics of the transition metals is associated with their catalytic activity?
(a) Colour of hydrated ions
(b) Variable oxidation states
(c) High enthalpy of atomisation
(d) Paramagnetic behaviour

30 Consider the following statements.
Statement I Prussian blue and Turnbull's blue are formed by different reagents, yet they are identical.
Statement II They have formula as $Fe[Fe(CN)_6]^-$.
(a) Both Statements I and II are correct
(b) Only Statement I is correct
(c) Only Statement II is correct
(d) Both Statements are incorrect

31 Sugar in urine-sample can be detected by
I. Fehling's solution II. Benedict's solution
III. Tollen's solution.
Select the correct alternate.
(a) I, II and III (b) Both I and III
(c) Both I and II (d) Both II and III

32 The ability of *d*-block elements to form complexes is due to
(a) small and highly charged ions
(b) vacant low energy orbitals to accept lone pair of electrons from ligands
(c) Both (a) and (b) are correct
(d) None of the above is correct

33 Consider a titration of potassium dichromate solution with acidified Mohr's salt solution using diphenylamine as indicator. The number of moles of Mohr's salt required per mole of dichromate is
(a) 3 (b) 4
(c) 5 (d) 6

34 AgCl and NaCl are colourless. NaBr and NaI are also colourless but AgBr and AgI are coloured. This is because,
(a) Ag^+ polarises Br^- and I^-
(b) Ag^+ has unpaired *d*-orbital
(c) Ag^+ depolarises Br^- and I^-
(d) None of the above

35 Which is not the true statement about $KMnO_4$?
(a) Its solution is unstable in acidic medium
(b) Its small quantity added to conc. H_2SO_4, a green coloured solution containing MnO_3^+ ions is formed
(c) MnO_4^- changes to Mn^{2+} in basic solution
(d) It is self-indicator in Fe^{2+} or $C_2O_4^{2-}$ titration

36 Pyrolusite in MnO_2 is used to prepare $KMnO_4$.

Steps are :

$$MnO_2 \xrightarrow{I} MnO_4^{2-} \xrightarrow{II} MnO_4^-$$

I and II are

(a) fuse with KOH/air, electrolytic oxidation
(b) fuse with KOH/air, electrolytic reduction
(c) fuse with conc HNO_3/air, electrolytic reduction
(d) All of the above are correct

37 Which is not the true statement?

(a) Ions of *d*-block elements are coloured due to *d-d* transition
(b) Ions of *f*-block elements are coloured due to *f-f* transition
(c) $[Sc(H_2O)_6]^{3+}$, $[Ti(H_2O)_6]^{4+}$ are coloured complexes
(d) Cu^+ is colourless ion

38 Consider the following statements.

Statement I Cu^{2+} *(aq)* and Cd^{2+} *(aq)* both form stable complexes with CN^- *(aq)*.

Statement II $[Cu(CN)_4]^{3-}$ is not decomposed by H_2S.

(a) Both Statements I and II are correct
(b) Only Statement I is correct
(c) Only Statement II is correct
(d) Both Statements are incorrect

39 When H_2O_2 is added to an acidified solution of $K_2Cr_2O_7$

(a) solution turns green due to formation of Cr_2O_3
(b) solution turns yellow due to formation of K_2CrO_4
(c) a deep blue-violet coloured compound $CrO(O_2)_2$ is formed
(d) solution gives green ppt. of $Cr(OH)_3$

40 Which is the correct statement?

(a) In less acidic solution, $K_2Cr_2O_7$ and H_2O_2 give violet coloured diamagnetic $[CrO(O_2)(OH)]^-$ ion
(b) In alkaline H_2O_2, K_3CrO_8 (with tetraperoxo species) $[Cr(O_2)_4]^{3-}$ is formed
(c) In ammoniacal solution, $K_2Cr_2O_7$ gives $(NH_3)_3CrO_4$
(d) All of the above are correct statements

41 For CrO_3 following is not true statement.

(a) It is called chromic acid
(b) It is colourless due to $3d^{10}$ configuration
(c) It is bright orange solid and colour arises due to charge transfer
(d) It is toxic and corrosive

42 When KCN comes in contact with blood, one dies immediately, it is due to

(a) CN^- forms stable complex with iron of haemoglobin of blood
(b) CN^- combines with H_2O of blood causing poisoning
(c) Both (a) and (b) are correct
(d) None of the above is correct

43 MnO_4^- is of intense pink colour, though Mn is in (+7) oxidation state. It is due to

(a) oxygen gives colour to it
(b) charge transfer when Mn gives its electron to oxygen
(c) charge transfer when oxygen gives its electron to Mn making it Mn (+VI) hence, coloured
(d) None of the above is correct

44 Which one of the following statements is not correct?

(a) Zinc dissolves in sodium hydroxide solution
(b) Carbon monoxide reduces iron (III) oxide to iron
(c) Mercury (II) iodide dissolves in excess of potassium iodide solution
(d) Tin(IV) chloride is made by dissolving tin solution in concentrated hydrochloric acid

45 Which of the following statements concerning lanthanides elements is false?

(a) Lanthanides are separated from one another by ion-exchange method
(b) Ionic radii of trivalent lanthanides steadily increases with increase in the atomic number
(c) All lanthanides are highly dense metals
(d) More characteristic oxidation state of lanthanides elements is +3

46 The main reason for larger number of oxidation state exhibited by the actinides than that corresponding lanthanides, is

(a) lesser energy difference between 5*f* and 6*d*-orbitals than between 4*f* and 5*d*-orbitals
(b) larger atomic size of actinides than the lanthanides
(c) more energy difference between 5*f* and 6*d*-orbitals than between 4*f* and 5*d*-orbitals
(d) greater reactive nature of the actinides than the lanthanides

47 What would happen, when as solution of potassium chromate is treated with an excess of dilute nitric acid?

(a) Cr^{3+} and $Cr_2O_7^{2-}$ are formed
(b) $Cr_2O_7^{2-}$ and H_2O are formed
(c) CrO_4^{2-} is reduced to +3 state of Cr
(d) None of the above

48 Which of the following factors may be regarded as the main cause of lanthanide contraction?

(a) Greater shielding of 5*d*-electron by 4*f*-electrons
(b) Poorer shielding of 5*d*-electron by 4*f*-electrons
(c) Effective shielding of one 4*f*-electrons by another in the subshell
(d) Poor shielding of one of 4*f*-electrons by another in the subshell

49 In which case(s) there is change in oxidation number?

(a) Aqueous solution of CrO_4^{2-} is acidified

(b) SO_2 gas is passed into $Cr_2O_7^{2-}/H^+$

(c) $Cr_2O_7^{2-}$ solution is made alkaline

(d) CrO_2Cl_2 is dissolved in NaOH

50 Which of the following reactions are disproportionation reactions?

(i) $Cu^+ \longrightarrow Cu^{2+} + Cu$

(ii) $3MnO_4^{2-} + 4H^+ \longrightarrow 2MnO_4^- + MnO_2 + 2H_2O$

(iii) $2KMnO_4 \longrightarrow K_2MnO_4 + MnO_2 + O_2$

(iv) $2MnO_4^- + 3Mn^{2+} + 2H_2O \longrightarrow 5MnO_2 + 4H^+$

(a) Only (i)
(b) (i), (ii) and (iii)
(c) (ii), (iii) and (iv)
(d) Both (i) and (ii)

51 A solution of a metal ion when treated with KI gives a red precipitate which dissolves in excess KI to give a colourless solution. Moreover, the solution of metal ion on treatment with a solution of cobalt(II) thiocyanate gives rise to a deep blue crystalline precipitate. The metal ion is

(a) Pb^{2+} (b) Hg^{2+}
(c) Cu^{2+} (d) Co^{2+}

52 Consider the following statements.

Statement I Copper(I) chloride is white, Cu(I) oxide is red.

Statement II In Cu(I) oxide, crystal structure has excess copper (I) ions making it defective.

(a) Both Statements I and II are correct
(b) Only Statement I is correct
(c) Only Statement II is correct
(d) Both Statements are incorrect

53 Match the alloys (in Column I) with their constituents (in Column II).

Column II		Column II
A. Invar	1.	70% Cu + 30% Zn
B. Brass	2.	66% Ni + 33% Cu
C. Monel	3.	36% Ni
D. Coinage metal	4.	60% Ni + 20% Fe + 20% Cr
E. Nichrome	5.	75% Cu + 25% Ni

	A	B	C	D	E
(a)	3	1	2	5	4
(b)	3	1	2	4	5
(c)	3	2	1	5	4
(d)	3	2	1	4	5

54 Match the properties given in Column I with the metals given in Column II. **[NCERT Exemplar]**

Column I (Property)		Column II (Metal)
A. An element which can show +8 oxidation state	1.	Mn
B. 3*d*-block element that can show upto +7 oxidation state	2.	Cr
C. 3*d*-block element with highest melting point	3.	Os
	4.	Fe

	A	B	C		A	B	C
(a)	3	1	2	(b)	1	2	3
(c)	3	2	1	(d)	2	1	3

(C) Medical entrances' gallery

1 Identify the incorrect statement from the following. **[NEET 2020]**

(a) Zirconium and hafnium have identical radii of 160 pm and 159 pm, respectively as a consequence of lanthanoid contraction.

(b) Lanthanoids reveal only +3 oxidation state.

(c) The lanthanoid ions other than the f^0 type and the f^{14} type are all paramagnetic.

(d) The overall decrease in atomic and ionic radii from lanthanum to lutetium is called lanthanoid contraction.

2 Match the followings aspects with the respective metal.

Aspects	Metal
A. The metal which reveals a maximum number of oxidation states	I. Scandium
B. The metal although placed in 3*d* block is considered not as a transition element.	II. Copper
C. The metal which does not exhibit variable oxidation states	III. Manganese
D. The metal which in +1oxidation state in aqueous solution undergoes disproportionation	IV. Zinc

Select the correct option. **[NEET 2020]**

	A	B	C	D		A	B	C	D
(a)	I	IV	II	III	(b)	III	IV	I	II
(c)	III	I	IV	II	(d)	II	IV	I	III

3 Identify the incorrect statement. **[NEET 2020]**

(a) The transition metals and their compounds are known for their catalytic activity due to their ability to adopt multiple oxidation states and to form complexes.

(b) Interstitial compounds are those that are formed when small atoms like H, C or N are trapped inside the crystal lattices of metals.

(c) The oxidation states of chromium in CrO_4^{2-} and $Cr_2O_7^{2-}$ are not the same.

(d) $Cr^{2+}(d^4)$ is a stronger reducing agent than $Fe^{2+}(d^6)$ in water.

4 The manganate and permanganate ions are tetrahedral, due to **[NEET 2019]**

(a) there is no π-bonding
(b) the π-bonding involves overlap of p-orbitals of oxygen with p-orbitals of manganese
(c) the π-bonding involves overlap of d-orbitals of oxygen with d-orbitals of manganese
(d) the π-bonding involves overlap of p-orbitals of oxygen with d-orbitals of manganese

5 When neutral or faintly alkaline $KMnO_4$ is treated with potassium iodide, iodide ion is converted into 'A'. 'A' is **[NEET (Odisha) 2019]**

(a) I_2 (b) IO_4^- (c) IO_3^- (d) IO^-

6 Which of the following can react with $K_2Cr_2O_7$? **[AIIMS, 2019]**

(a) SO_3^{-2} (b) CO_3^{-2} (c) SO_4^{-2} (d) NO_3^-

7 Which element can have oxidation state from 4 to 6? **[AIIMS 2019]**

(a) Fe (b) Mg (c) Co (d) Cr

8 Which have melting point below 500°C? **[AIIMS 2019]**

(a) Ag, Cu (b) Zn, Cd
(c) Cd, Cu (d) Ag, Zn

9 Match the metal ions given in Column I with the spin magnetic moments of the ions given in Column II and assign the correct code. **[NEET 2018]**

Column I	Column II
1. Co^{3+}	(i) $\sqrt{8}$ BM
2. Cr^{3+}	(ii) $\sqrt{35}$ BM
3. Fe^{3+}	(iii) $\sqrt{3}$ BM
4. Ni^{2+}	(iv) $\sqrt{24}$ BM
	(v) $\sqrt{15}$ BM

	1	2	3	4
(a)	(iv)	(i)	(ii)	(iii)
(b)	(i)	(ii)	(iii)	(iv)
(c)	(iv)	(v)	(ii)	(i)
(d)	(iii)	(v)	(i)	(ii)

10 Which one of the following ions exhibits d-d transition and paramagnetism as well? **[NEET 2018]**

(a) MnO_4^- (b) $Cr_2O_7^{2-}$ (c) CrO_4^{2-} (d) MnO_4^{2-}

11 Statement I NaCl reacts with concentrated H_2SO_4 to give colourless fumes with pungent smell. But on adding MnO_2 the fumes become greenish yellow.

Statement II MnO_2 oxidises HCl to chlorine gas which is greenish yellow. **[AIIMS 2018]**

(a) Both Statements I and II are correct
(b) Only Statement I is correct
(c) Only Statement II is correct
(d) Both Statements are incorrect

12 When hydrogen peroxide is added to acidified solution of dichromate $Cr_2O_7^{2-}$, a deep blue coloured complex of chromic peroxide CrO_5 is formed. Oxidation state of Cr in CrO_5 is **[JIPMER 2018]**

(a) + 5 (b) + 3 (c) + 6 (d) −10

13 The correct order of spin only magnetic moment (in BM) for Mn^{2+}, Cr^{2+} and Ti^{2+} ions is **[AIIMS 2018]**

(a) $Mn^{2+} > Ti^{2+} > Cr^{2+}$ (b) $Ti^{2+} > Cr^{2+} > Mn^{2+}$
(c) $Mn^{2+} > Cr^{2+} > Ti^{2+}$ (d) $Cr^{2+} > Ti^{2+} > Mn^{2+}$

14 Which one of the following species is paramagnetic in nature? **[JIPMER 2015]**

(a) V^{4+} (b) Sc^{3+} (c) V^{5+} (d) Zn^{2+}

15 Among the following species, the one which become paramagnetic on heating at 850 K is **[JIPMER 2018]**

(a) Fe_3O_4 (b) MnO
(c) $ZnFe_2O_4$ (d) Both (a) and (c)

16 The reason for greater range of oxidation states in actinoids is attributed to **[NEET 2017]**

(a) $4f$ and $5d$ levels being close in energies
(b) the radioactive nature of actinoids
(c) actinoid contraction
(d) $5f$, $6d$ and $7s$ levels having comparable energies

17 The electronic configurations of Eu (at. no. 63), Gd (at. no. 64) and Tb (at. no. 65) are **[NEET-I 2016]**

(a) $[Xe]\,4f^6 5d^1 6s^2$, $[Xe]\,4f^7 5d^1 6s^2$ and $[Xe]\,4f^9 6s^2$
(b) $[Xe]\,4f^6 5d^1 6s^2$, $[Xe]\,4f^7 5d^1 6s^2$ and $[Xe]\,4f^8 5d^1 6s^2$
(c) $[Xe]\,4f^7 6s^2$ $[Xe]\,4f^7 5d^1 6s^2$ and $[Xe]\,4f^9 6s^2$
(d) $[Xe]\,4f^7 6d^2$, $[Xe]\,4f^8 6s^2$ and $[Xe]\,4f^8 5d^1 6s^2$

18 Because of lanthanide contraction, which of the following pairs of elements have nearly same atomic radii? (Numbers in the parenthesis are atomic numbers). **[NEET 2015]**

(a) Ti (22) and Zr (40) (b) Zr (40) and Nb (41)
(c) Zr (40) and Hf (72) (d) Zr (40) and Ta (73)

19 Which of the following process does not involve oxidation of iron? **[NEET 2015]**

(a) Rusting of iron sheets
(b) Decolourisation of blue $CuSO_4$ solution by iron
(c) Formation of $Fe(CO)_5$ from Fe
(d) Liberation of H_2 from steam by iron at high temperature

20 Magnetic moment 2.84 BM is given by (at. no. Ni = 28, Ti = 22, Cr = 24 , Co = 27) **[NEET 2015]**

(a) Ni^{2+} (b) Ti^{3+} (c) Cr^{3+} (d) Co^{2+}

21 Which one of the following transition metal ions is colourless in aqueous solutions? **[Kerala CEE 2014]**

(a) Ti^{4+} (b) V^{4+} (c) Mn^{2+} (d) Fe^{3+}
(e) Ni^{2+}

22 The magnetic moment of Ni^{2+} ion (atomic number of Ni = 28) in BM unit is **[Kerala CEE 2014]**
(a) 1.73 (b) 4.81
(c) 5.96 (d) 2.84
(e) 3.86

23 Reason of lanthanoid contraction is **[CBSE AIPMT 2014]**
(a) negligible screening effect of *f*-orbitals
(b) increasing nuclear charge
(c) decreasing nuclear charge
(d) decreasing screening effect

24 Which of the following oxidation states is the most common among the lanthanides? **[UP CPMT 2014]**
(a) 3 (b) 4
(c) 2 (d) 5

25 The correct statement is **[KCET 2014]**
(a) The earlier members of lanthanide series resemble calcium in their chemical properties
(b) The extent of actinide contraction is almost the same as lanthanide contraction
(c) In general, lanthanide and actinide do not show variable oxidation states
(d) Ce^{4+} in aqueous solution is not known

26 Lanthanide for which + II and + III oxidation states are common is **[AIIMS 2013]**
(a) La (b) Nd
(c) Ce (d) Eu

27 The trivalent ion having largest size in lanthanide series is **[Manipal 2013]**
(a) Ti (b) Zr (c) Hf (d) La

28 Which of the following lanthanide ions is diamagnetic? [atomic numbers Ce = 58, Sm = 62, Eu = 63 and Yb = 70] **[NEET 2013]**
(a) Ce^{2+} (b) Sm^{2+} (c) Eu^{2+} (d) Yb^{2+}

29 Which of the following statements about the interstitial compounds is incorrect? **[NEET 2013]**
(a) They retain metallic conductivity
(b) They are chemically reactive
(c) They are much harder than the pure metal
(d) They have higher melting points than the pure metal

30 Which of the following does not give oxygen on heating? **[NEET 2013]**
(a) $KClO_3$ (b) $Zn(ClO_3)_2$
(c) $K_2Cr_2O_7$ (d) $(NH_4)_2Cr_2O_7$

31 $KMnO_4$ can be prepared from K_2MnO_4 as per the reaction,
$$3MnO_4^{2-} + 2H_2O \rightleftharpoons 2MnO_4^- + MnO_2 + 4OH^-$$
The reaction can go completion by removing OH^- ions by adding **[NEET 2013]**
(a) HCl (b) KOH (c) CO_2 (d) SO_2

32 The number of water molecules in Mohr's salts is **[AIIMS 2013]**
(a) 2 (b) 4 (c) 6 (d) 8

33 **Statement I** Solution of Na_2CrO_4 in water is intensely coloured.
Statement II Oxidation state of Cr in Na_2CrO_4 is (+ VI). **[AIIMS 2013]**
(a) Both Statements I and II are correct
(b) Only Statement I is correct
(c) Only Statement II is correct
(d) Both Statements are incorrect

34 The correct one for *d*-orbital is **[UP CPMT 2013]**
(a) $(n-1)d^{1-9}ns^1$
(b) $(n-1)d^{1-10}ns^{1-2}$
(c) $(n-1)d^{1-5}$
(d) $(n-1)d^{1-10}ns^2$

35 Transition elements does not show **[Manipal 2013]**
(a) paramagnetism (b) colour
(c) fixed valency (d) All of these

36 Cuprous ion is colourless while cupric ion is coloured because **[KCET 2013]**
(a) both have half-filled *p* and *d*-orbitals
(b) cuprous ion has incomplete *d*-orbital and cupric ion has a complete *d*-orbital
(c) both have unpaired electrons in the *d*-orbitals
(d) cuprous ion has a complete *d*-orbital and cupric ion has an incomplete *d*-orbital

37 Across the lanthanide series, the basicity of the lanthanide hydroxides **[UP CPMT 2013]**
(a) increases
(b) decreases
(c) first increases and then decreases
(d) first decreases and then increases

38 Which of the following statements is not true? **[CBSE AIPMT 2012]**
(a) On passing H_2S through acidified $K_2Cr_2O_7$ solution, a milky colour is observed
(b) $Na_2Cr_2O_7$ is preferred over $K_2Cr_2O_7$ in volumetric analysis
(c) $K_2Cr_2O_7$ solution in acidic medium is orange
(d) $K_2Cr_2O_7$ solution becomes yellow on increasing the pH beyond 7

39 Which one of the following does not correctly represent the correct order of the property indicated against it? **[CBSE AIPMT 2012]**
(a) Ti < V < Cr < Mn increasing number of oxidation states
(b) $Ti^{3+} < V^{3+} < Cr^{3+} < Mn^{3+}$ increasing magnetic moment
(c) Ti < V < Cr < Mn increasing melting points
(d) Ti < V < Mn < Cr increasing 2nd ionisation enthalpy

40 Four successive members of the first series of the transition metals are listed below. For which one of them, the standard potential ($E^{\circ}_{M^{2+}/M}$) value has a positive sign? **[CBSE AIPMT 2012]**

(a) Co ($Z = 27$) (b) Ni ($Z = 28$)
(c) Cu ($Z = 29$) (d) Fe ($Z = 26$)

41 The oxidation state of chromium in the final product formed by the reaction between KI and acidified potassium dichromate solution is **[CBSE AIPMT 2012]**

(a) + 3 (b) + 2
(c) + 6 (d) + 4

42 When H_2O_2 is shaken with an acidified solution of $K_2Cr_2O_7$ in the presence of ether, the ethereal layer turns blue due to the formation of **[WB JEE 2012]**

(a) Cr_2O_3 (b) CrO_4^{2-}
(c) $Cr_2(SO_4)_3$ (d) CrO_5

43 Which one of the following has a magnetic moment of 1.75 BM? **[KCET 2012]**

(a) V^{3+} (b) Cr^{3+}
(c) Fe^{3+} (d) Ti^{3+}

44 Stainless steel contains **[KCET 2012]**

(a) Fe + Cr + Cu (b) Fe + C + Ni
(c) Fe + Cr + Ni (d) Fe + Ni + Cu

45 For the four successive transition elements (Cr, Mn, Fe and Co), the stability of +2 oxidation state will be there in which of the following order?
[Atomic number Cr = 24, Mn = 25, Fe = 26 and Co = 27 **[CBSE AIPMT 2011]**

(a) Cr> Mn> Co> Fe (b) Mn> Fe> Cr> Co
(c) Fe> Mn> Co> Cr (d) Co> Mn> Fe> Cr

46 Acidified $K_2Cr_2O_7$ solution turns green when Na_2SO_3 is added to it. This is due to the formation of **[CBSE AIPMT 2011]**

(a) $CrSO_4$ (b) $Cr_2(SO_4)_3$
(c) CrO_4^{2-} (d) $Cr_2(SO_3)_3$

47 Which of the following species is/are paramagnetic? Fe^{2+}, Zn^0, Hg^{2+}, Ti^{4+} **[Kerala CEE 2011]**

(a) Only Fe^{2+}
(b) Zn^0 and Ti^{4+} (c) Fe^{2+} and Hg^{2+}
(d) Zn^0 and Hg^{2+} (e) Fe^{2+} and Ti^{4+}

48 The acidic, basic or amphoteric nature of Mn_2O_7, V_2O_5 and CrO, are respectively, **[Kerala CEE 2011]**

(a) acidic, acidic and basic
(b) basic, amphoteric and acidic
(c) acidic, amphoteric and basic
(d) acidic, basic and amphoteric
(e) acidic, basic and basic

49 Ammonia will not form complex with **[Kerala CEE 2011]**

(a) Ag^{2+} (b) Pb^{2+} (c) Cu^{2+} (d) Cd^{2+}
(e) Fe^{2+}

50 The titanium (atomic number 22) compound that does not exist is **[Kerala CEE 2011]**

(a) TiO (b) TiO_2
(c) K_2TiF_6 (d) $TiCl_3$
(e) K_2TiO_4

51 The spin only magnetic moment of Fe^{3+} ion (in BM) is approximately **[KCET 2011]**

(a) 4 (b) 7
(c) 5 (d) 6

52 Which is the correct statement about $Cr_2O_7^{2-}$ structure? **[DUMET 2011]**

(a) It has neither Cr—Cr bonds nor O—O bonds
(b) It has one Cr—Cr bond and six O—O bonds
(c) It has no Cr—Cr bond and has six O—O bonds
(d) It has one Cr—Cr bond and seven Cr—O bonds

53 Copper exhibits only +2 oxidation state in its stable compounds. Why? **[Guj CET 2011]**

(a) Copper is transition metal in +2 state
(b) +2 state compounds of copper are formed by exothermic reactions
(c) Electronic configuration of copper in +2 state is $[Ar]3d^9, 4s^0$
(d) Copper gives coloured compounds in +2 state

54 When dil. H_2SO_4 is added to aqueous solution of potassium chromate, yellow colour of solution turns to orange colour. It indicates **[Guj CET 2011]**

(a) chromate ions are reduced
(b) chromate ions are oxidised
(c) monocentric complex is converted into dicentric complex
(d) oxygen gets removed from chromate ions

55 German silver alloy contains **[Guj CET 2011]**

(a) zinc, silver and copper
(b) nickel, silver and copper
(c) germanium, silver and copper
(d) zinc, nickel and copper

56 Experimental value of magnetic moment of Mn^{2+} complex is 5.96 BM. This indicates **[Guj CET 2011]**

(a) axial and orbital motion of electron in same direction
(b) axial and orbital motion of electron in opposite direction
(c) electron does not exhibit orbital motion, it only exhibits axial motion
(d) electron does not exhibit axial motion, it only exhibits orbital motion

ANSWERS

Check point 11.1

1. (a) 2. (d) 3. (d) 4. (a) 5. (b) 6. (c) 7. (b) 8. (b) 9. (d) 10. (d)

Check point 11.2

1. (c) 2. (d) 3. (c) 4. (c) 5. (a) 6. (a) 7. (a) 8. (a) 9. (d) 10. (d)

Check point 11.3

1. (c) 2. (c) 3. (c) 4. (c) 5. (c) 6. (b) 7. (a) 8. (d) 9. (b) 10. (c)

Check point 11.4

1. (d) 2. (b) 3. (c) 4. (a) 5. (a) 6. (b) 7. (b) 8. (a) 9. (a) 10. (a)

(A) Topical exercise

1. (d) 2. (d) 3. (b) 4. (c) 5. (b) 6. (b) 7. (d) 8. (a) 9. (d) 10. (a)
11. (b) 12. (a) 13. (d) 14. (a) 15. (c) 16. (b) 17. (b) 18. (c) 19. (a) 20. (c)
21. (b) 22. (a) 23. (a) 24. (a) 25. (a) 26. (c) 27. (a) 28. (a) 29. (c) 30. (d)
31. (d) 32. (d) 33. (c) 34. (d) 35. (b) 36. (d) 37. (d) 38. (b) 39. (a) 40. (c)
41. (b) 42. (d) 43. (c) 44. (b) 45. (c) 46. (b) 47. (d) 48. (b) 49. (d) 50. (c)
51. (b) 52. (a) 53. (a) 54. (b) 55. (c) 56. (a) 57. (c) 58. (b) 59. (d) 60. (c)
61. (a) 62. (a) 63. (a) 64. (a) 65. (a) 66. (b) 67. (a) 68. (a) 69. (b) 70. (a)
71. (d) 72. (c) 73. (a) 74. (b) 75. (b) 76. (c) 77. (b) 78. (b) 79. (c) 80. (b)
81. (a) 82. (b) 83. (a) 84. (c) 85. (c) 86. (c) 87. (a) 88. (c) 89. (a) 90. (a)
91. (c) 92. (c) 93. (b) 94. (d) 95. (d) 96. (b) 97. (a) 98. (d) 99. (b) 100. (d)
101. (d) 102. (a) 103. (a) 104. (a) 105. (b) 106. (d) 107. (a) 108. (d) 109. (b) 110. (a)
111. (a) 112. (b) 113. (b) 114. (a) 115. (a) 116. (b) 117. (c) 118. (d) 119. (a) 120. (c)
121. (b) 122. (d) 123. (b) 124. (d) 125. (b) 126. (b) 127. (c) 128. (a) 129. (a) 130. (b)
131. (b) 132. (d) 133. (c) 134. (c) 135. (d) 136. (c) 137. (a) 138. (b) 139. (d) 140. (a)
141. (d) 142. (a) 143. (c) 144. (d) 145. (a) 146. (d) 147. (d) 148. (a) 149. (d) 150. (a)
151. (a) 152. (b)

(B) Miscellaneous questions

1. (c) 2. (b) 3. (d) 4. (a) 5. (a) 6. (d) 7. (a) 8. (b) 9. (d) 10. (a)
11. (d) 12. (a) 13. (a) 14. (c) 15. (d) 16. (c) 17. (a) 18. (d) 19. (b) 20. (d)
21. (b) 22. (d) 23. (a) 24. (a) 25. (a) 26. (c) 27. (a) 28. (a) 29. (b) 30. (a)
31. (a) 32. (c) 33. (d) 34. (a) 35. (c) 36. (a) 37. (c) 38. (c) 39. (c) 40. (d)
41. (b) 42. (a) 43. (c) 44. (d) 45. (b) 46. (a) 47. (b) 48. (d) 49. (b) 50. (d)
51. (b) 52. (a) 53. (a) 54. (a)

(C) Medical entrances' gallery

1. (b) 2. (b) 3. (c) 4. (d) 5. (c) 6. (a) 7. (d) 8. (b) 9. (c) 10. (d)
11. (a) 12. (c) 13. (c) 14. (a) 15. (d) 16. (c) 17. (c) 18. (c) 19. (c) 20. (a)
21. (a) 22. (d) 23. (a) 24. (a) 25. (a) 26. (d) 27. (d) 28. (d) 29. (b) 30. (d)
31. (c) 32. (c) 33. (a) 34. (b) 35. (c) 36. (d) 37. (b) 38. (b) 39. (c) 40. (c)
41. (a) 42. (d) 43. (d) 44. (c) 45. (b) 46. (b) 47. (a) 48. (c) 49. (b) 50. (e)
51. (d) 52. (a) 53. (a) 54. (c) 55. (d) 56. (a)

Hints & Explanations

(A) Topical exercise

3 (*b*) transition metals have incompletely filled *d*-subshell in their ground state or in their most probable oxidation state but Zn, Cd and Hg don't have this characteristic.

7 (*d*) IE of formation of Ni^{2+} < IE of formation of Pt^{2+}. Thus, Ni^{2+} complexes are more stable than that of Pt^{2+} complexes.

(IE) of formation of Pt^{4+} < (IE) of formation of Ni^{4+}. Thus, Pt^{4+} complexes are more stable than that of Ni^{4+} complexes.

Thus (a), (b) and (c) are true.

10 (*a*) Copper lies below hydrogen in the electrochemical series and hence does not liberate H_2 from acids. Therefore, option (a) is not correct.

Other three options (b, c, d) are correct.

13 (*d*) Maximum number of unpaired electrons

Cr^{3+}= | 1 | 1 | 1 | | | d^3

14 (*a*) $[Zn(H_2O)_6]^{2+}$

Oxidation state of Zn = +2 and its electronic configuration is [Ar] $3d^{10}$

All electrons are paired, thus colourless.

15 (*c*) $[Mn(H_2O)_6]^{3+}$

Oxidation state of Mn in given compound = +3

Mn^{3+} has [Ar] $3d^3$ electronic configuration and three unpaired electrons.

Thus, *d*-*d** transition makes it coloured.

Cu^+ in (a) and Sc^{3+} in (b) have paired electrons in *d*-orbitals thus colourless.

16 (*b*) Electronic configuration of Cu^{2+} is [Ar] $3d^9$

Thus, it has one unpaired electron.

Hence, it coloured and paramagnetic in nature.

17 (*b*) Transition elements form coloured salt due to the presence of unpaired electrons. In CuF_2, Cu(II) contain one unpaired electron hence, CuF_2 is coloured in solid state.

43 (*c*) In atmosphere there are SO_2, CO_2 which react with Cu and causing its corrosion.

49 (*d*) Fe^{3+} = $[Ar]3d^5$

52 (*a*) It is due to screening effect based on Fajan's rule smaller the cationic size.

$$(Cu^{2+} < Ca^{2+})$$

Presence of *d*-orbital electrons $3d^{10}$ in Cu^{2+}.

Hence, larger the polarising power.

58 (*b*) Mn has more than one oxidation state (i.e. +2, + 3, + 4, + 5, + 6, + 7).

Zn^{2+} has completely filled *d*-orbital which is a stable configuration. Thus, it only show +2 oxidation state. Sr and Ca show +1and +2 oxidation states.

67 (*a*) $VO_2 + 2H^+ \longrightarrow VO^{2+} + H_2O$

ON of V = + 4

72 (*c*) Magnetic moment, $\mu = \sqrt{N(N+2)}$ BM

where, *N* = number of unpaired electrons

Cr(24) = [Ar] $3d^5 4s^1$ $N = 6$, $\mu = \sqrt{48}$ BM = x

Mn^+ = [Ar] $3d^5 4s^1$ $N = 6$, $\mu = \sqrt{48}$ BM = y

Fe^{2+} = [Ar] $3d^6$ $N = 4$, $\mu = \sqrt{24}$ BM = z

For Fe^{2+} | 1⇂ | 1 | 1 | 1 | 1 |

$N = 4$

Thus, $z < x = y$

73 (*a*) $[Ti(H_2O)_6]^{3+}$

Ti (22) = [Ar] $4s^2$ $3d^2$

Ti^{3+} (19) = [Ar] $4s^0$ $3d^1$

N = unpaired electron = 1

Magnetic moment = $\sqrt{N(N+2)}$ BM number of unpaired electrons

$= \sqrt{3}$ BM

= 1.73 BM

74 (*b*) $[Co(SCN)_4]^{2-}$, cobalt is +2 with three unpaired electrons in 3*d*. Thus, $\mu = \sqrt{15}$ BM.

Note Due to reaction with acid or base oxidation number does not change.

75 (*b*) The magnetic moment is associated with its spin angular momentum and orbital angular momentum.

Spin only magnetic moment value of Cr^{3+} ion is 3.87.

76 (*c*)

Species	Oxidation states	Unpaired electrons	Magnetic moment $\sqrt{N(N+2)}$ BM
$[Fe(H_2O)_6]^{2+}$	Fe = + 2	4	$\sqrt{32}$ BM
$[Cr(H_2O)_6]^{3+}$	Cr = + 3	4	$\sqrt{32}$ BM
Fe	Fe = 0	4	$\sqrt{32}$ BM

80 (*b*) $Cr_2O_7^{2-} + 6e^- + 14H^+ \longrightarrow 2Cr^{3+} + 7H_2O$

electron per Cr atom = 3

81 (*a*) $(NH_4)_2Cr_2O_7 \xrightarrow{\Delta} Cr_2O_3 + N_2\uparrow + 4H_2O$

$NH_4NO_2 \xrightarrow{\Delta} N_2\uparrow + 2H_2O$

82 (*b*) $\underset{\text{Base}}{Cr_2O_3} + \underset{\text{Acid}}{6HCl} \longrightarrow CrCl_3 + 3H_2O$

$\underset{\text{Base}}{2NaOH} + \underset{\text{Acid}}{Cr_2O_3} + 3H_2O \longrightarrow 2Na\,[Cr(OH)_4]$

Thus, Cr_2O_3 is an amphoteric oxide.

85 (*c*) $Cr_2O_7^{2-} + 2OH^- \longrightarrow 2CrO_4^{2-} + H_2O$

$pH > 7\ (x > 7)$

$2CrO_4^{2-} + 2H^+ \longrightarrow Cr_2O_7^{2-} + H_2O$

$pH < 7\ (y < 7)$

86 (*c*) $Cr_2O_7^{2-} + SO_2 + H^+ \longrightarrow \underset{\text{Green}}{Cr^{3+}} + SO_4^{2-}$

$\xrightarrow{Fe^{2+} + H^+} \underset{\text{Green}}{Cr^{3+}} + Fe^{3+}$

SO_2 and Fe^{2+} both reduce $Cr_2O_7^{2-}$ to Cr^{3+} (green).

88 (*c*) When acidified $K_2Cr_2O_7$ solution is added to Sn^{2+} salts then Sn^{2+} changes to Sn^{4+}.

The reaction is given below

$\overset{+6}{Cr_2}O_7^{2-} + 14H^+ + 3Sn^{2+} \longrightarrow 2Cr^{3+} + 3Sn^{4+} + 7H_2O$

(Oxidation: $Sn^{2+} \to Sn^{4+}$; Reduction: $Cr^{+6} \to Cr^{3+}$)

89 (*a*) There is the formation of oxide layer (Cr_2O_3) on chromium.

91 (*c*) Tetra peroxo species $[Cr(O_2)_4]^{3-}$

Let the oxidation state of Cr be *x*.

Then, oxidation number of Cr $= x - 8 = -3$

$x = +5$

93 (*b*) In this case, CrO_2Cl_2 is formed.

95 (*d*) [Structure: Cr with one double-bonded O and two peroxo (O–O) groups]

$Cr(O_2)_2O$

Oxidation number of oxygen in peroxide = – 1

97 (*a*) $MnO_4^- + e^- \longrightarrow MnO_4^{2-}$

99 (*b*) $MnO_4^- + H_2SO_4 \longrightarrow \underbrace{MnO_3^+ + HSO_4^-}_{\text{Green}}$

100 (*d*) Highest oxidation state of manganese in fluoride is +4 (MnF_4) but highest oxidation state in oxides is +7(Mn_2O_7). The reason is that in covalent compounds fluorine can form single bond while oxygen forms double bond.

101 (*d*) $2MnO_4^{2-} + Cl_2 \longrightarrow 2MnO_4^- + 2Cl^-$

$MnO_4^{2-} + e^- \longrightarrow MnO_4^-$

116 (*b*) $Fe^{2+} \longrightarrow Fe^{3+} + e^-$

Others given compound are in oxidised state.

117 (*c*) $Fe_2O_3 \cdot H_2O$ forms soluble complex with oxalic acid.

$Fe_2O_3 + \underset{\text{Oxalic acid}}{H_2C_2O_4} \longrightarrow \underset{\text{Soluble}}{[Fe(C_2O_4)_3]^{3-}}$

119 (*a*) $FeCl_3 + K_4[Fe(CN)_6] \longrightarrow \underset{\text{Blue}}{KFe\,[Fe(CN)_6]} + 3KCl$

123 (*b*) $Zn + 2NaOH \longrightarrow Na_2ZnO_2 + H_2$

125 (*b*) Metals lying above H in electrochemical series are not obtained from their oxides by heating.

$HgO \xrightarrow{\Delta} Hg + O_2$

$Ag_2O \xrightarrow{\Delta} Ag + O_2$

126 (*b*) PbS—black ZnS—white

CuS—black HgS—black

127 (*c*) $\underset{\text{Base}}{ZnO} + \underset{\text{Acid}}{2HCl} \longrightarrow ZnCl_2$

$\underset{\text{Acid}}{ZnO} + \underset{\text{Base}}{2NaOH} \longrightarrow Na_2ZnO_2$

Thus, ZnO is an amphoteric (both acidic and basic) oxide.

128 (*a*) White solid $\xrightarrow{\text{heat}}$ Gas

This gas turns lime water milky. So, it can be CO_2 or SO_2. Residue is yellow when hot. Thus, the residue is ZnO.

$ZnO + CO_2 \longrightarrow \underset{(Y)}{ZnCO_3}$

$ZnO + SO_2 \longrightarrow ZnSO_3$

129 (*a*) $K_4[Fe(CN)_6] + ZnO \longrightarrow \underset{\text{Bluish white ppt.}}{Zn_2[Fe(CN)_6]}$

132 (*d*) $2Ag^+ + H_2S \longrightarrow Ag_2S$ (black)

139 (*d*) TiO_2, ZnO, $ZnSO_4$ + BaS are not affected by H_2S.

(B) Miscellaneous questions

4 (*a*) $CrO_3 + H_2O \rightleftharpoons \underset{\text{Acid}}{H_2CrO_4} \rightleftharpoons 2H^+ + CrO_4^{2-}$

$\underset{\text{Base}}{Na_2CO_3} + \underset{\text{Acid}}{2H^+} \longrightarrow CO_2$

Thus, CrO_3 is an acidic oxide.

5 (*a*) In the given compound, $KMnO_4$ is decolourised by unsaturated side chain.

6 (*d*) $AgNO_3$ is very light-sensitive and is decomposed by light.

8 (*b*) $AgNO_3 + CaCl_2 \longrightarrow \underset{\text{White ppt.}}{AgCl\downarrow} + Ca(NO_3)_2$

$CaCl_2$ is ionic compound giving free Cl^- that gives white precipitate with Ag^+ ion.

$CHCl_3$, CCl_4 are covalent and no Cl^- is formed.

KNO_3 does not react with $AgNO_3$.

9 (*d*) (a) $AgNO_3 + PH_3 \longrightarrow Ag$ True

(b) True

(c) $AgCN + KCN \longrightarrow K\,[Ag(CN)_2]$ True.

10 (*a*) $FeCl_3 \cdot 6H_2O$ exists as $[Fe(H_2O)_4Cl_2]Cl \cdot 2H_2O$

$$\underset{1\text{ mol}}{[Fe(H_2O)_4Cl_2]Cl \cdot 2H_2O} \rightleftharpoons [Fe(H_2O)_4Cl_2]^+ + Cl^-$$

$$Cl^- + Ag^+ \longrightarrow \underset{1\text{ mol}}{AgCl}$$

11 (*d*) Fe^{2+} is oxidised to Fe^{3+}

$$Fe^{2+} + K_3[Fe(CN)_6] \longrightarrow \text{Blue}$$

$$Fe^{3+} + K_3[Fe(CN)_6] \longrightarrow \text{No colour}$$

Thus, option (d) is correct.

12 (*a*) Ti(22) : $[Ar]4s^2 3d^2$

1	1			

$Ti^{2+} = [Ar]3d^2$

$Ti^{4+} = [Ar]3d^0$

Ti^{2+} has two unpaired electrons in $3d$ and thus d-d * transition is possible due to absorption of light in visible region. But this is not possible in Ti^{4+} as it has zero unpair electrons.

13 (*a*) $Fe_2O_3 + 10OH^- + 3Cl_2 \longrightarrow \underset{\text{Red purple}}{2[FeO_4]^{2-}} + 6Cl^- + 5H_2O$

14 (*c*) Green vitriol : $FeSO_4 \cdot 7H_2O$

White vitriol : $ZnSO_4 \cdot 7H_2O$

Epsom salt : $MgSO_4 \cdot 7H_2O$

16 (*c*) $TiCl_4$: Oxidation state of Ti = +4

Thus, no electrons are present in d-orbital $[Ar]3d^0$

Thus, it is diamagnetic in nature.

O_2 : Two electrons unpaired (by MO Theory)

Thus, paramagnetic.

$[Ni(CN)_4]^{2-}$: dsp^2-hybridisation of Ni^{2+}; square planar; all electrons paired

Thus, diamagnetic.

$[Ni(H_2O)_4]^{2+}$: sp^3-hybridisation; two electrons unpaired.

Thus, it is paramagnetic in nature.

CO : Isoelectronic to N_2, all paired electrons thus it is diamagnetic in nature.

18 (*d*) Cu^{2+} is reduced to Cu^+ by CN^-

Thus, (a) is true.

Cu^+ undergoes complex ion with CN^- forming $K_3[Cu(CN)_4]$.

Thus, (b) and (c) are true.

The reactions involve are given as follows

$$CuSO_4 + 2KCN \longrightarrow Cu(CN)_2 + K_2SO_4$$

$$2Cu(CN)_2 \longrightarrow 2CuCN + (CN)_2$$

$$CuCN + 3KCN \longrightarrow K_3Cu(CN)_4$$

19 (*b*) $Ni^{2+} + DMG \longrightarrow \underset{\text{Cherry red ppt.}}{[Ni(DMG)_2]^{2+}}$

20 (*d*) Fe(III) gives red colours with KCNS hence, (d) is incorrect.

22 (*d*) $Cu^+ = [Ar]3d^{10}$

No unpaired electron, thus $s = 0$

Thus, (d)

24 (*a*) $FeCl_3 + 3H_2O \longrightarrow Fe(OH)_3 + HCl$

$$Zn + 2HCl \longrightarrow ZnCl_2 + 2H$$

$$FeCl_3 + 2H \longrightarrow \underset{\text{Light green}}{FeCl_2} + HCl$$

26 (*c*) II group (of cation-analysis) contains Pb^{2+} and Cu^{2+} and are precipitated as sulphides on passing H_2S in acidic medium. ZnS is precipitated in presence of NH_4OH and NH_4Cl.

31 (*a*) Sugar in urine sample is in the form of glucose which reduces.

I. Fehling solution to red ppt.

II. Benedict's solution to red ppt.

III. Tollen's reagent to metallic silver (silver-mirror).

33 (*d*) $6Fe^{2+} + Cr_2O_7^{2-} \longrightarrow 6Fe^{3+} + 2Cr^{3+}$

35 (*c*) MnO_4^- changes to MnO_4^{2-} or MnO_2 in basic medium.

47 (*b*) $CrO_4^{2-} + 2H^+ \longrightarrow Cr_2O_7^{2-} + H_2O$

49 (*b*) $\underset{+6}{Cr_2O_7^{2-}} + \underset{+4}{SO_2} \longrightarrow \underset{+6}{SO_4^{2-}} + \underset{+3}{Cr^{3+}}$

51 (*b*) $Hg^{2+} + KI \longrightarrow HgI_2$ (Red ppt.)

$HgI_2 + KI \longrightarrow K_2HgI_4$ (Soluble)

$HgI^{2+} + Co(SCN)_2 \longrightarrow Hg(SCN)_2$ (Blue crystalline ppt.)

(C) Medical entrances' gallery

1 (*b*) Due to the presence of poor shielding f-orbitals, lanthanide contraction is observed in lanthanoids ($4f$-block elements). So, overall a decrease in atomic and ionic radii is observed from $_{57}La(d^1)$ to $_{58}Ce$ and upto $_{71}Lu$.

For this reason, d^2-group (Group 4) members of 5th period ($_{40}Zr$) and 6th period ($_{72}Hf$) will have identical atomic radii.

Due to the absence of unpaired electrons, f^0 and f^{14} type lanthanoids are diamagnetic.

+ 3 oxidation state is common to lanthanoids in their compounds. However, + 4 and + 2 ions in solution or in solid are also known.

Hence, (b) is the incorrect statement.

2 (*b*) (A) Mn shown six oxidation states in its compounds, *viz*, + 2, + 3, + 4, + 5, + 6 and + 7 ⇒ (III)

(B) Zn is a pseudo transition element like Cd and Hg of group12 (d^{10}-configuration) ⇒ (IV)

(C) Sc does not exhibit variable oxidation states. Only + 3 state is shown by to Sc in its compounds ⇒ (I)

(D) Cu^+ disproportionaton in aqueous solution as,

$$2Cu^+ \longrightarrow Cu^0 + Cu^{2+} \Rightarrow \text{(II)}$$

3 (*c*) Statements in options (a), (b) and (d) are correct.

Statement in option (c) is incorrect, because oxidation states of Cr in CrO_4^{2-} (chromate) and $Cr_2O_7^{2-}$ (dichromate) are same and it is +6.

i.e., $\overset{x}{Cr}O_4^{2-} \Rightarrow x + 4(-2) = 2 \Rightarrow x = +6$

$\overset{y}{Cr_2}O_7^{2-} \Rightarrow 2y + 7(-2) = -2 \Rightarrow y = +6$

4 (*d*) The structures of manganate and permanganate ions are as follows

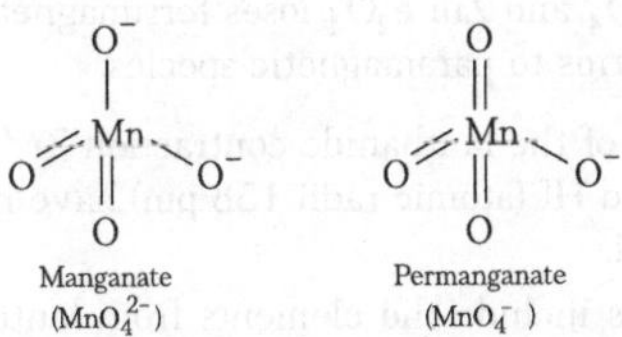

The manganate and permanganate ions are tetrahedral and contain π-bonds of *d*π - *p*π-type, oxygen does not have any *d*-orbital.

Hence, the π-bonding takes place by overlap of *p*-orbitals of oxygen and *d*-orbitals of manganese.

5 (*c*) When neutral or faintly alkaline $KMnO_4$ is treated with potassium iodide, $KMnO_4$ is converted into MnO_2 while iodide (I^-) ion is converted into IO_3^- (iodate). The reaction takes place as follows :

$$2KMnO_4 + H_2O + KI \longrightarrow 2MnO_2 + 2KOH + KIO_3$$

6 (*a*) Among the given options, oxidation state of

(a) sulphur in $SO_3^{-2} = +4$

(b) carbon in $CO_3^{-2} = +4$

(c) sulphur in $SO_4^{-2} = +6$

(d) nitrogen in $NO_3^- = +5$

Since, $K_2Cr_2O_7$ is a strong oxidising agent it can oxidise the species, which is not in its most possible positive oxidising state.

Thus, SO_3^{-2} can react with $K_2Cr_2O_7$ as follows:

$$Cr_2O_7^{2-}(aq) + 3SO_3^{2-}(aq) + 8H^+(aq) \longrightarrow 2Cr^{3+}(aq) + 3SO_4^{2-}(aq) + 4H_2O(l)$$

7 (*d*) Oxidation state of *d*-block elements can vary, i.e. of Fe, Co and Cr as follows:

(a) (c) Oxidation state of Co = (+)2 to (+) 5 and oxidation state of Fe = (+) 2 to (+) 6, but (+) 6 is less stable. Also, (+) 4 and + 5 oxidationi states of Fe are very rare.

(d) Oxidation state of Cr = (+) 2 to (+) 6, where (+) 6 state is more stable.

(b) Oxidation state of alkaline earth metals (i.e. Mg) is fixed and equal to (+)2.

Hence, option (d) is the correct.

8 (*b*) Since, Zn and Cd both have completely filled outermost shells, i.e. have no unpaired electrons. Thus, they cannot form covalent and metallic bonds strongly and show boiling point less than 500°C [Zn (146°C) and Cd (321°C)]. Hence, option (b) is the correct.

9 (*c*) Spin magnetic moment can be calculated as,

$$\mu = \sqrt{n(n+2)} \text{ BM}$$

where, μ = magnetic moment,

BM = Bohr Magneton (unit of μ)

and *n* = number of unpaired electrons in *d*-orbital.

1. The electronic configuration of Co^{3+} is $[Ar]\,3d^6$.
   Here, $n = 4$
   $\mu = \sqrt{4(4+2)} = \sqrt{24}$ BM
2. The electronic configuration of Cr^{3+} is $[Ar]\,3d^3$.
   Here, $n = 3$
   $\mu = \sqrt{3(3+2)} = \sqrt{15}$ BM
3. The electronic configuration of Fe^{3+} is $[Ar]\,3d^5$.
   Here, $n = 5$
   $\mu = \sqrt{5(5+2)} = \sqrt{35}$ BM
4. The electronic configuration of Ni^{2+} is $[Ar]\,3d^8$.
   Here, $n = 2$
   $\mu = \sqrt{2(2+2)} = \sqrt{8}$ BM

So, the correct option is (c).

10 (*d*) In *d-d* transition, an electron in a *d*-orbital of the metal is excited by a photon to another *d*-orbital of higher energy. The complex compound which contains unpaired electrons shows paramagnetism while which contains paired electrons shows diamagnetism.

The complex which contains unpaired electrons exhibit *d-d* transition and paramagnetism.

(a) In MnO_4^-,
the electronic configuration of Mn^{7+} is $[Ar]\,3d^0$.
Number of unpaired electrons = 0
Therefore, it will be diamagnetic and will not show *d-d* transition.

(b) In $Cr_2O_7^{2-}$,
the electronic configuration of Cr^{6+} is $[Ar]\,3d^0$.
Number of unpaired electrons = 0
Therefore, it will be diamagnetic and will not show *d-d* transition.

(c) In CrO_4^{2-},
the electronic configuration of Cr^{6+} is $[Ar]\,3d^0$.
Number of unpaired electrons = 0
Therefore, it is also diamagnetic and will not show *d-d* transition.

(d) In MnO_4^{2-},
the electronic configuration of Mn^{6+} is $[Ar]\,3d^1$.
Number of unpaired electrons = 1

Since, it contains one unpaired electron, so it will exhibit both *d-d* transition and paramagnetism.

11 (*a*) Both Assertion and Reason are correct and Reason is the correct explanation of the Assertion.

NaCl reacts with concentrated H_2SO_4 to give colourless fumes with pungent smell. This pungent smell is due to formation of HCl.

$$NaCl + H_2SO_4 \longrightarrow Na_2SO_4 + \underset{\text{Pungent smell}}{2HCl}$$

But on adding MnO_2, the fumes become greenish yellow due to formation of chlorine gas.

12 (*c*) When H_2O_2 is added to an acidified solution of dichromate $Cr_2O_7^{2-}$, a deep blue complex chromic peroxide, CrO_5 is formed.

$$Cr_2O_7^{2-} + 2H^+ + 4H_2O_2 \longrightarrow 2CrO(O_2)_2 + 5H_2O$$

This deep blue coloured complex has the following structure

In $CrO(O_2)_2$, let the oxidation state of Cr be x.

$\therefore \quad x + (-1)4 + (-2) = 0$

$x - 6 = 0 \Rightarrow x = +6$

13 (*c*) Spin only magnetic moment depends upon the number of unpaired electrons.

Magnetic moment, $\mu = \sqrt{n(n+2)}$

where, n = number of unpaired electrons

The electronic configuration for the given ions is as follows:

(i) For $Mn^{2+}[Z = 25] = 1s^2, 2s^2\,2p^6, 3s^2\,3p^6, 3d^5\,4s^0$

$3d^5$

1	1	1	1	1

$\mu = \sqrt{5(5+2)} = \sqrt{35}$

(ii) For $Cr^{2+}[Z = 24] = 1s^2, 2s^2 2p^6, 3s^2 3p^6, 3d^4, 4s^0$

$3d^4$

1	1	1	1	

$\mu = \sqrt{4(4+2)} = \sqrt{24}$

(iii) For $Ti^{2+}[Z = 22] = 1s^2, 2s^2\,2p^6, 3s^2\,3p^6, 3d^2, 4s^0$

$3d^2$

1	1			

$\mu = \sqrt{2(2+2)} = \sqrt{8}$

$\therefore$ Correct order of spin only magnetic moment is

$Mn^{2+} > Cr^{2+} > Ti^{2+}$.

14 (*a*) The number of unpaired electrons and electronic configuration of the given species is shown below :

(a) V^{4+} $[Z = 23] = [Ar]_{18}\,3d^1$

No. of unpaired electron = 1

(b) Sc^{3+} $[Z = 21] = [Ar]_{18}\,3d^0$

No. of unpaired electron = 0

(c) V^{5+} $[Z = 23] = [Ar]_{18}\,3d^0$

No. of unpaired electron = 0

(d) Zn^{2+} $[Z = 30] = [Ar]_{18}\,3d^{10}$

No. of unpaired electron = 0

$\therefore$ The correct option is (a).

15 (*d*) All magnetically ordered solids, i.e. ferromagnetic; antiferromagnetic; ferrimagnetic changes to paramagnetic state at high temperature because of the randomisation of spins.

Ferrimagnetic species such as Fe_3O_4 (magnetite) and ferrites, i.e. $MgFe_2O_4$ and $ZnFe_2O_4$ loses ferrimagnetism on heating and transforms to paramagnetic species.

18 (*c*) Because of the lanthanide contraction Zr (atomic radii 160 pm) and Hf (atomic radii 158 pm) have nearly same atomic radii.

Lanthanides include the elements from lanthanum La($Z = 57$) to lutetium Lu($Z = 71$). Zirconium Zr(40) belong to the second transition series (4*d*) and Hf (72) belongs to third transition series (5*d*).

Lanthanide contraction is associated with the intervention of the 4*f*-orbitals which are filled before the 5*d*-series of elements starts.

The filling of 4*f*-orbitals before 5*d*-orbitals results in regular decrease in atomic radii which compensates the expected increase in atomic size with increasing atomic number. As a result of this lanthanide contraction, the elements of second and third transition series have almost similar atomic radii.

19 (*c*)

(a) $\overset{0}{Fe} + \underbrace{H_2O + O_2}_{\text{Form air}} \longrightarrow \overset{III}{Fe_2}O_3 \cdot xH_2O$

(b) $\overset{0}{Fe} + CuSO_4 \longrightarrow \overset{II}{Fe}SO_4 + Cu$

(c) $\overset{0}{Fe} + 5CO \longrightarrow \overset{0}{Fe}(CO)_5$

(d) $\overset{0}{Fe} + \underset{\text{Steam}}{H_2O} \longrightarrow \overset{III}{Fe_2}O_3 + H_2$

Formation of $Fe(CO)_5$ from Fe does not involve oxidation of iron because there is no change in oxidation state.

20 (*a*) Magnetic moment, $\mu = \sqrt{n(n+2)}$ BM

where, n = number of unpaired electrons

$\mu = 2.84$(given)

$\therefore \quad 2.84 = \sqrt{n(n+2)}$ BM

$(2.84)^2 = n(n+2)$

$8 = n^2 + 2n$

$n^2 + 2n - 8 = 0$

$(n+4)(n-2) = 0$

$n = 2$

$Ni^{2+} = [Ar]\,3d^8\,4s^0$ (two unpaired electrons)

$Ti^{3+} = [Ar]\,3d^1 4s^0$ (one unpaired electrons)

$Cr^{3+} = [Ar]\,3d^3$, (three unpaired electrons)

$Co^{2+} = [Ar]\,3d^7, 4s^0$, (three unpaired electrons)

So, only Ni^{2+} has 2 unpaired electrons.

21 (*a*) Ti^{4+} $(22 - 4 = 18) = [Ar]3d^0$ (no unpaired electron)

V^{4+} $(23 - 4 = 19) = [Ar]3d^1$ (one unpaired electron)

1				

Mn^{2+} $(25 - 2 = 23) = [Ar]3d^5$ (five unpaired electrons)

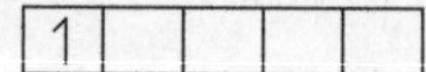

Fe^{3+} $(26 - 3 = 23)$ = same as Mn^{2+} (five unpaired electrons)

Ni^{2+} (28 − 2 = 26) = [Ar]$3d^8$ (two unpaired electrons)

↑↓	↑↓	↑↓	↑	↑

Since, Ti^{4+} does not have any unpaired electron.

So, *d-d* transition is not possible and hence, its aqueous solution is colourless.

22 (*d*) $Ni_{28} = [Ar]_{18} 3d^8, 4s^2$; $Ni^{2+} = [Ar]_{18} 3d^8, 4s^0$

$3d^8$

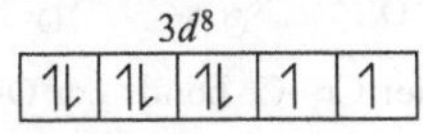

Number of unpaired electrons = 2

Magnetic moment, $\mu = \sqrt{n(n+2)} = \sqrt{2(2+2)}$

$= \sqrt{8} = 2.84$ BM

28 (*d*) Ce^{2+} [58 − 2 = 56] = [Xe] $4f^1 5d^1 6s^0$

↑				

5*d*

It has one unpaired electron, thus it is paramagnetic

Sm^{2+} [62 − 2 = 60] = [Xe] $4f^6 6s^0$

↑	↑	↑	↑	↑	↑	

4*f*

It has 6 unpaired electrons, thus it is also an paramagnetic.

Eu^{2+} [63 − 2 = 61] = [Xe] $4f^7 6s^2$

↑	↑	↑	↑	↑	↑	↑

4*f*

It has 7 unpaired electrons, thus it is also an paramagnetic

Yb^{2+} [70 − 2 = 68] = [Xe] $4f^{14} 6s^2$

↑↓	↑↓	↑↓	↑↓	↑↓	↑↓	↑↓

It has no unpaired electron avilable thus, it is diamagnetic in nature.

29 (*b*) Interstitial compounds are obtained when small atoms like H, B, C, N fit into the lattice of other elements. These compounds retain their metallic conductivity but, these chemically inert.

30 (*d*) All compounds give O_2 on heating except $((NH_4)_2Cr_2O_7$.

$$2KClO_3 \xrightarrow{\Delta} 2KCl + 3O_2 \uparrow$$

$$Zn(ClO_3)_2 \xrightarrow{\Delta} ZnCl_2 + 3O_2 \uparrow$$

$$K_2Cr_2O_7 \longrightarrow 2Cr_2O_2 + K_2O + O_2 \uparrow$$

$$(NH_4)_2Cr_2O_7 \xrightarrow{\Delta} Cr_2O_3 + N_2 \uparrow + 4H_2O \uparrow$$

On heating $(NH_4)_2Cr_2O_7$, the products formed are N_2, Cr_2O_3 and H_2O. Oxygen is not formed in this reaction.

31 (*c*) $3MnO_4^{2-} + 2H_2O \rightleftharpoons 2MnO_4^- + MnO_2 + 4OH^-$

The reaction can go completion by removing OH^- ions by adding CO_2 as it reacts with OH^- ions to form CO_3^{2-} ions.

$$4OH^- + 2CO_2 \longrightarrow 2CO_3^{2-} + 2H_2O$$

32 (*c*) Mohr's salt is a double salt.

$FeSO_4 \cdot (NH_4)_2SO_4 \cdot 6H_2O \longrightarrow$ Ferrous ammonium sulphate

It contains six water molecules of crystallisation, which is present here in hydrate form, i.e. $6H_2O$.

33 (*a*) In Na_2CrO_4, Cr is in +6 oxidation state with d^0 configuration. The CrO_4^{2-} ions in aqueous solution gives yellow colour which is not due to *d-d* transition but due to charge transfer, i.e. momentary transfer of charge from O-atom to metal atom there by changing O^{2-} ion momentarily to O^- ion and reducing the oxidation state of Cr from +6 to + 5.

34 (*b*) The general electronic configuration of *d*-block elements is $(n-1)d^{1\text{-}10}, ns^{1\text{-}2}$.

35 (*c*) Transition elements show variable valencies.

e.g. $Fe^{2+}, Fe^{3+}, Co^{2+}, Co^{3+}$, etc.

36 (*d*) Cuprous ion is colourless while cupric ion is coloured because cuprous ion has a complete *d*-orbital, i.e. $3d^{10}$, $4s^0$ configuration and cupric ion has an incomplete *d*-orbitals due to $3d^9$, $4s^0$ configuration.

38 (*b*) Being hygroscopic or deliquescent sodium dichromate, $Na_2Cr_2O_7$ cannot be used in volumetric analysis. And all, other statements are true.

39 (*c*) The correct order is Ti < V < Cr < Mn. The melting point of transition elements first increase to a maximum extent and then fall as atomic number increases Mn, however, has abnormally low melting point, due to stable configuration and weak intermetallic bonding.

40 (*c*) In electrochemical series, metals with positive standard potential are placed below hydrogen. Out of given transition metals, only Cu is lower than hydrogen in electrochemical series. Therefore, it must have positive sign for standard potential. $[E^\circ_{M^{2+}/M}]$.

41 (*a*) $Cr_2O_7^{2-} + 14H^+ + 6I^- \longrightarrow 2Cr^{3+} + 7H_2O + 3I_2$

Thus, final oxidation state of Cr is +3.

42 (*d*) Acidified $K_2Cr_2O_7$ solution reacts with H_2O_2 to give a deep blue solution due to the formation of peroxo compound, $CrO(O_2)_2$ or CrO_5.

$$Cr_2O_7^{2-} + 2H^+ + 4H_2O_2 \longrightarrow 2CrO_5 + 5H_2O$$

Note *The blue colour fades away gradually due to the decomposition of CrO_5 into Cr^{3+} ions and oxygen.*

43 (*d*) Magnetic moment,

$$\mu = \sqrt{n(n+2)} \text{ BM}$$

where, n = number of unpaired electrons

Ion	Outermost electronic configuration	Number of unpaired electrons	Magnetic moment, μ (BM)
V^{3+}	$3d^2$ [1][1][ ][ ][ ]	2	$\sqrt{2(2+2)} = 2.82$
Cr^{3+}	$3d^3$ [1][1][1][ ][ ]	3	$\sqrt{3(3+2)} = 3.87$
Fe^{3+}	$3d^5$ [1][1][1][1][1]	5	$\sqrt{5(5+2)} = 5.91$
Ti^{3+}	$3d^1$ [1][ ][ ][ ][ ]	1	$\sqrt{1(1+2)} = 1.73$

44 (*c*) Stainless steel contains Fe + Cr + Ni.

45 (*b*) The stability of + 2 oxidation state in different ions can be determined on the basis of electrode potential data. On the basis of electrode potentials, the correct order is

Mn > Fe > Cr > Co. Thus, Mn^{2+} is highly stable ion due to d^5-configuration.

46 (*b*) Acidified $K_2Cr_2O_7$ solution turns green when Na_2SO_3 is added to it. This is due to the formation of $Cr_2(SO_4)_3$.

$$K_2Cr_2O_7 + 3Na_2SO_3 + 4H_2SO_4 \longrightarrow 3Na_2SO_4 + K_2SO_4 + Cr_2(SO_4)_3 + 4H_2O$$

51 (*d*) $Fe^{3+} = [Ar]\, 3d^5$

[1][1][1][1][1] (5 unpaired electrons)

Magnetic moment, $\mu = \sqrt{n(n+2)}$

n = unpaired electrons = 5

$\Rightarrow \sqrt{5(5+2)} = \sqrt{5 \times 7} = 5.92 \simeq 6$

52 (*a*) Structure of $Cr_2O_7^{2-}$

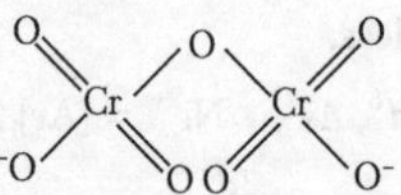

Thus, it has neither Cr—Cr bonds nor O—O bonds

53 (*a*) $Fe^{2+} = [Ar]\, 3d^6\, 4s^0$

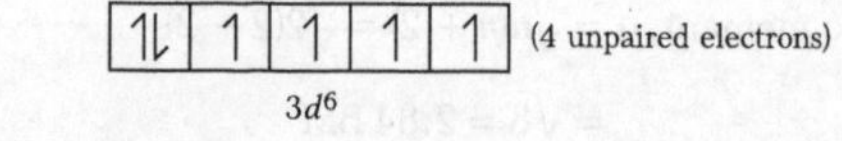

$Zn^0 = [Ar] 3d^{10}\, 4s^2$

[1↓][1↓][1↓][1↓][1↓] (no unpaired electrons)

$Ti^{4+} = [Ar]\, 3d^0\, 4s^0$

[ ][ ][ ][ ][ ] (no unpaired electrons)

$Hg^{2+} = [Xe] 4f^{14}\, 5d^{10}$

[1↓][1↓][1↓][1↓][1↓] (no unpaired electrons)

∴ Only Fe^{2+} is paramagnetic its is because the presence of 4 unpaired electrons

54 (*c*) $K_2Cr_2O_4 + H_2SO_4 \longrightarrow K_2Cr_2O_7 + K_2SO_4 + H_2O$

$K_2Cr_2O_4$: Monocentric compound; $K_2Cr_2O_7$: Dicentric compound

CHAPTER 12

Coordination Compounds

Some metals, specially transition metals, have tendency to form large compounds. These compounds are formed by either combination of two or more simpler compounds of metals or by combination of few groups to a metal ion. These are combinedly known as **addition compounds** and play important roles in natural processes as well as in our daily life.

Addition compounds

When solution having two or more stable compounds, combine in a simple molecular (stoichiometric) proportions, they form a new compound called addition or molecular compounds.

e.g. $$Fe(CN)_2 + 4KCN \longrightarrow Fe(CN)_2 \cdot 4KCN$$
$$K_2SO_4 + Al_2(SO_4)_3 + 24H_2O \longrightarrow K_2SO_4 \cdot Al_2(SO_4)_3 \cdot 24H_2O$$

These addition compounds can be divided into two types

(*i*) **Double salts or lattice compounds** The addition compounds which dissociate and lose their identity in solution are called double salts.

e.g. Mohr salt $[FeSO_4 \cdot (NH_4)_2SO_4 \cdot 6H_2O]$

Alums $[K_2SO_4 \cdot Al_2(SO_4)_3 \cdot 24H_2O]$

(*ii*) **Coordination or complex compounds** The addition compounds which do not dissociate and retain their identity in solution are called complex compounds.

e.g. $K_2[PtCl_6]$, $K_4[Fe(CN)_6]$

Transition metals form large number of complex compounds with anions or neutral molecules. In complex compounds, metal ion acts as Lewis acid while the other groups attached to it act as Lewis base.

Difference between double salts and coordination compounds

The important difference between double salts and coordination compounds are as follows

(*i*) Although, both the double salts and coordination compounds are formed by combination but basic difference in both the compounds is that double salts dissociate into ions when dissolved in water whereas complex ions do not dissociate.

Inside

- **Addition compounds**
- **Important terms related to coordination compounds**
  - *– Ligand*
  - *– Coordination number*
  - *– Oxidation number*
  - *– Effective atomic number (EAN)*
- **Nomenclature of coordination compounds**
- **Isomerism in coordination compounds**
  - *– Structural isomerism*
  - *– Stereoisomerism*
  - *– Optical isomerism*
- **Bonding in coordination compounds**
  - *– Werners theory*
  - *– Valence bond theory*
- **Crystal field theory (CFT)**
  - *– Calculation of CFSE*
  - *– Magnetic nature*
  - *– John-Teller effect*
- **Stability and importance of coordination compounds**
  - *– Importance of coordination compounds*

For instance, Mohr salt, $FeSO_4 \cdot (NH_4)_2SO_4 \cdot 6H_2O$ on dissolving in water dissociates into Fe^{2+}, NH_4^+ ions while complex ion such as $[Fe(CN)_6]^{4-}$ does not dissociate into Fe^{2+} and CN^- ions.

(*ii*) The properties of double salt are same as that of its constituents whereas it is different in coordination compounds.

(*iii*) In double salt, the metal ion show it's normal valency whereas, in complex compounds it satisfies two types of valencies i.e. primary and secondary.

Important terms related to coordination compounds

Complex ion

It is a metal ion with a Lewis base attached to it through coordinate bonds, e.g. $[Fe(CN)_6]^{4-}$.

Different types of complex ions are as follows

(*i*) Complex ion which bears positive charge is called **cationic complex ion,** e.g. $[Co(NH_3)_6]^{3+}$.

(*ii*) Complex ion which bears negative charge is called **anionic complex ion,** e.g. $[Ag(CN)_2]^-$.

(*iii*) Complex ion which has no charge is called **neutral complex ion,** e.g. $[Ni(CO)_4]$.

Central metal ion

The atom (or ion) to which number of groups or ions are bonded in definite arrangement, is called central metal ion.

e.g. in $[Fe(CN)_6]^{3-}$, Fe^{3+} is a central metal ion.

Ligands

The groups of ions or molecules which are bonded to central metal ion are known as ligands. They donate pair of electrons to metal ion. Therefore, are also called Lewis bases.

e.g. in $[Fe(CN)_6]^{3-}$, CN^- ion acts as ligand.

Ligands can be further divided on the basis of number of donor atoms present.

(*i*) **Mono or unidentate ligands** Ligands which have only one donar atom and donates one electron pair to central metal ion are called monodentate ligands.

e.g. NH_3, H_2O, Cl^- etc.

(*ii*) **Bidentate ligands** Ligands which can bind metal ion through two donar atoms are called bidentate ligands.

e.g. oxalate $(C_2O_4^{2-})$, ethane-1, 2-diamine

$$\left(\begin{array}{l} CH—NH_2 \\ | \\ CH_2—NH_2 \end{array}\right) \text{etc.}$$

(*iii*) **Polydentate ligands** Ligands which have more than two donor atoms are called polydentate ligands. e.g. $EDTA^{4-}$ (ethylene diamine tetraacetate ion) is a hexadentate ligand.

$$\begin{array}{l} ^-OOCH_2C \\ ^-OOCH_2C \end{array}\!\!>N—CH_2—CH_2—N<\!\!\begin{array}{l} CH_2COO^- \\ CH_2COO^- \end{array}$$

(*iv*) **Ambidentate ligands** Ligands which have two donor atoms and can bind through any of the atom is called ambidentate ligand.

e.g. NO_2^- (nitrite ion) can bind through nitrogen as well as through oxygen. Other examples are CN^-, SCN^- etc.

(*v*) **Chelating ligands** When a poly dentate ligand binds to a central metal ion and forms a ring, then it is called chelating ligands.

They usually form five or six membered rings called **chelates**.

e.g.

$$\begin{array}{ccc} & O\text{---}H & —O \\ & \uparrow & | \\ CH_3—C{=}N & & N{=}C—CH_3 \\ & Ni & \\ CH_3—C{=}N & & N{=}C—CH_3 \\ & | & \downarrow \\ & O—H & \text{------}O \end{array}$$

Nickel (II) dimethyl glyoximate (chelate)
[*bis*-(dimethylglyoximato) nickel (II)]

$$\begin{array}{l} CH_2NH_2 \qquad\qquad O—C{=}O \\ O{=}C—O—Cu—NH_2CH_2 \end{array}$$

Copper (II) glycinate (chelate)
[*trans*-diglycinato copper (II)]

The number of such donar groups present in a single ligand that bind to central metal atom is called **denticity**.

Coordination sphere The central metal ion and ligands attached to it are enclosed in square brackets called as coordination sphere. It represents single entity in coordination compounds.

e.g. $[Co(NH_3)_6]^{2+}$ is a coordination sphere.

Coordination number

It is the number of ligands to which a central metal ion is directly bonded.

e.g. in $[Ni(NH_3)_4]^{2+}$, here coordination number of Ni is 4.

In case of polydentate ligands, C.N.

= number of ligand × denticity of ligand

e.g. in $[Fe(C_2O_4)_3]^{3-}$, coordination number of Fe is 6.

Oxidation number

It is the charge present on central metal ion if all the ligands bonded to it are removed along with pairs of electrons.

e.g. oxidation number of Cu in $[Cu(NH_3)_4]^{2+}$ is +2, as NH_3 is a neutral ligand, in $[Cu(CN)_4]^{3-}$ oxidation number of Cu is +1 as CN is an anion ligand.

Representation of complex ion

Complex ion can be represented as

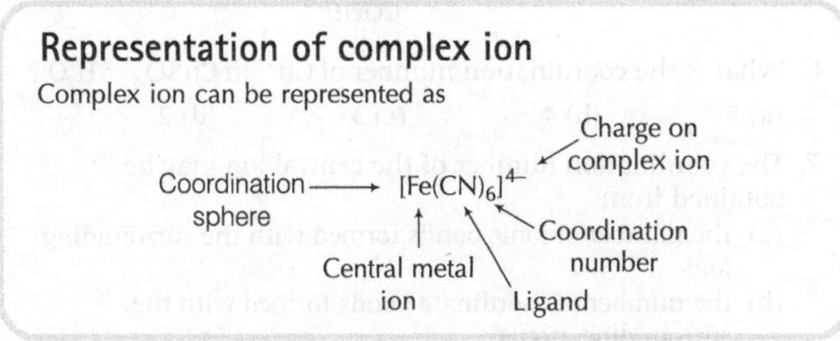

Homoleptic and Heteroleptic complexes

Complexes in which a metal is bound to only one kind of donor atom are known as homoleptic. e.g. $[Co(NH_3)_6]^{3+}$ whereas complexes in which a metal is bound to more than one kind of donor groups are known as heteroleptic. e.g. $[Co(NH_3)_4Cl_2]^+$.

Effective atomic number (EAN)

Each ligand donates an electron pair to the metal ion, thus forming a coordinate bond. Transition metals form coordination compounds readily because they have vacant *d*-orbitals which can accommodate these electron pairs.

Metal ion in the complex tends to attain nearest stable inert gas configuration by gaining electrons from ligands.

Effective atomic number (EAN) of the metal in a complex is given by

$$EAN = Z - (ON) + 2(CN)$$

(where, ON is oxidation number, CN is the coordination number and Z is the atomic number).

EAN can also be given by

$$EAN = Z - (ON) + 2(\text{ligands})$$

assuming that each ligand is unidentate.

EAN Rule Complexes following EAN rule have EAN of central metal atom/ion equal to nearest inert gas configuration. e.g. $[Fe(CN)_6]^{4-}$ does follow EAN rule whereas $[Fe(CN)_6]^{3-}$ doesn't.

EAN of some metals in complexes are given in the table below

Table 12.1 EAN of some complex

Element	Atomic number (Z)	Complex	C.N	EAN
Cr	24	$[Cr(CO)_6]$	6	36
Ni	28	$Ni(CO)_4$	4	36
Fe	26	$[Fe(CN)_6]^{4-}$	6	36
Fe	26	$[Fe(NH_3)_6]^{2+}$	6	36 } [Kr]
Cu	29	$[Cu(CN)_4]^{3-}$	4	36
Co	27	$[Co(en)_3]^{3+}$, (en) is bidentate	6	36
Pd	46	$[Pd(H_2O)_6]^{4+}$	6	54[Xe]
Pt	78	$[PtCl_6]^{2-}$	6	86[Rn]
Fe	26	$[Fe(CN)_6]^{3-}$	6	35
Ag	47	$[Ag(CN)_2]^-$	2	50

CHECK POINT 12.1

1. Which of the following pair contains complex salt and double salt respectively?
(a) $FeSO_4, K_4[Fe(CN)_6]$
(b) $[Cu(NH_3)_4]SO_4, FeSO_4 \cdot 7H_2O$
(c) $[Cu(NH_3)_4]SO_4, K_2SO_4 \cdot Al_2(SO_4)_3 \cdot 24H_2O$
(d) $MgSO_4 \cdot 7H_2O, CuSO_4$

2. In the complex $K_4[Fe(CN)_6]$ identify the coordination sphere and the counter ion.
(a) K^+ coordination sphere, $[Fe(CN)_6]^{4-}$ counter ion
(b) $[Fe(CN)_6]^{4-}$ coordination sphere, K^+ counter ion
(c) $(CN)_6$ coordination sphere; K^+ counter ion
(d) None of the above

3. Which is the pair of ambident ligands?
(a) CN^-, NO_2^- (b) NO_3^-, CNS^-
(c) NO_3^-, NO_2^- (d) $NCS^-, C_2O_4^{2-}$

4. Which of the following represents a hexadentate ligand?
(a) 2, 2 -bipyridyl (b) DMG
(c) Ethylenediamine (d) None of these

5. The correct structure of ethylene diamminetetra acetic acid (EDTA) is
(a) $(HOOCCH_2)_2N{-}CH{=}CH{-}N(CH_2COOH)_2$
(b) $(HOOH)(HOOC)N{-}CH_2{-}CH_2{-}N(COOH)_2$
(c) $(HOOCCH_2)_2N{-}CH_2{-}CH_2{-}N(CH_2COOH)_2$
(d) $(HOOC{-}CH_2)(H)N{-}CH(CH_2COOH){-}CH(CH_2COOH){-}N(H)(CH_2COOH)$

6. What is the coordination number of Cu^{2+} in $CuSO_4 \cdot 5H_2O$?
(a) 5 (b) 4 (c) 3 (d) 2

7. The coordination number of the central ion may be obtained from
(a) the number of ionic bonds formed with the surrounding ions
(b) the number of coordinate bonds formed with the surrounding atoms
(c) the number of ions of opposite charge immediately surrounding the specific ion
(d) None of the above

8. The coordination number and oxidation number of Cr in $K_3[Cr(C_2O_4)_3]$ respectively are
(a) 6 and +3 (b) 4 and +2 (c) 3 and +3 (d) 3 and 0

9. In which of the following compounds, the metal is in the lowest oxidation state?
(a) $[Co(NH_3)_5Br]_2SO_4$ (b) $Fe_3[Fe(CN)_6]_2$
(c) $Mn_2(CO)_{10}$ (d) $K[PtCl_3(C_2H_4)]$

10. Which of the following does not follow EAN rule?
(a) $V(CO)_6$ (b) $Cr(CO)_6$
(c) $Ni(CO)_4$ (d) $W(CO)_6$

NOMENCLATURE OF COORDINATION COMPOUNDS

Naming of coordination compounds are done as per some rules set by the IUPAC. These rules are as follows

(*i*) Coordination compounds contain two parts : cation and anion. Cations are named before anions.
e.g. in $K_4[Fe(CN)_6]$, K^+ is a cation, it is named before the anion $[Fe(CN)_6]^{4-}$.
Hence, IUPAC name of compound is potassium hexacyanoferrate (II).

(*ii*) Names of anionic ligands ending with 'ide' changes to 'o', 'ite' changes to 'o' 'ate' changes to 'o'
Positive ligands end with '**-ium**' and neutral ligands like H_2O, NH_3 are given names such as aqua, ammine respectively.
In latest trend **e** is simply replaced by **o**.

Table 12.2 Name of some common ligands

Neutral ligands		Negative ligands	
Formula	**Name as ligand**	**Formula**	**Name as ligand**
OH_2	aqua		
NH_3	ammine	F^- (fluoride)	fluoro
CO	carbonyl	Cl^-	chloro
NO	nitrosyl	Br^-	bromo
NH_2CH_3	methylamine	I^-	iodo
NC_5H_5 or Py	pyridine	O^{2-} (oxide)	oxo
$N(C_2H_5)_3$	triethylamine	OH^- (hydroxide)	hydroxo
$N(CH_3)_3$	trimethylamine	CN^- (cyanide)	cyano*
$P(C_6H_5)_3$	triphenyl phosphine	NC^-	isocyano *
PH_3	phosphine	S^{2-}	sulphido

Neutral ligands		Negative ligands	
Formula	Name as ligand	Formula	Name as ligand
(positive ligands) NO_2^+	Nitronium	$\underline{H}^-$	hydrido
$NH_2—NH_3^+$	Hydrozinium	SO_4^{2-}	Sulphato
NO^+	Nitrosonium	$C_2O_4^{2-}$	oxalato
		NO_2^- ONO^-	nitro nitrito
		CO_3^{2-}	carbonato

** Ambident, underlined atoms are electron donating.*

(*iii*) If more than one type of ligands are present then they are written in alphabeticall order.

(*iv*) If several ligands of same types are present, prefixes di, tri, tetra etc., are used. When the name of ligand involves a number than prefixes *bis, tris, tetrakis* are used and name of ligands is placed in brackets, e.g. compound $[Co(en)_3]Cl_3$ is named as *tris* (ethylenediamine)cobalt (III) chloride.

Here, *tris* is used to indicate three ligands since **en** (ethylenediamine) also carries **di** in its name.

(*v*) In complex ion, ligands are named first followed by name of central metal ion with its oxidation state in roman numeral.

(*vi*) If complex ion is cation, the name of central metal ion remains the same, e.g.

- $[Fe(NH_3)_6]Cl_3$
  hexaammineiron (III) chloride
- $[CoCl(NH_3)_5]^{2+}$
  pentaamminechlorocobalt (III) ion

(*vii*) If complex ion is anion, the name of central metal ion ends with '-ate'. e.g.

- $[ZnCl_4]^{2-}$ tetrachlorozincate (II) ion
- $[AlH_4]^-$ tetrahydridoaluminate (III) ion
- $K_4[Fe(CN)_6]$ potassium hexacyanoferrate (II)

(*viii*) If the complex contains two or more metal atoms, it is termed polynuclear. The bridging ligands which link the two metal atoms together are indicated by the prefix μ.

e.g. $[(NH_3)_5Co—NH_2—Co(NH_3)_5](NO_3)_5$

bridging ligand will start with μ

bis [pentaammine] μ-amido cobalt (III) nitrate

(*ix*) Ambidentate ligands are named as follows.

M—NO_2 (NO_2 joined to metal *M* through N is **nitro**).

M–ONO (NO_2 joined to metal *M* through O is **nitrito**).

Similarly, the SCN group may bond *M*–SCN (thiocyanato) or *M*–NCS (isothiocyanato).

They may be named thiocyanato-S [to indicate S coordinated to metal *M*]

thiocyanato-N [to indicate N coordinated to metal *M*]

(*x*) The ligands and metal ion inside coordination sphere are written without any space. While oxidation number and the ion outside are separated by single space. e.g. $[Co(NH_3)_5SO_4]Br$ (Pentaamminesulphatocobalt (III) bromide).

Complex ion should be enclosed by square brackets and ligands are placed after metal in the alphabetical order but first negative ligands, then neutral, then positive.

$[Pt(H_2O)_4(NH_3)_2]^{4+}$ → Wrong

Aqua Ammine

$[Pt(NH_3)_2(H_2O)_4]^{4+}$ → Right

$[Cr(CN)_4(NH_3)_2]^-$ → Right

Negative Neutral

e.g.

(a) Metal K_4 (Cation) Ligands $[Fe(CN)_6]$ (Anion)

potassium (Cation) hexacyanoferrate (II) (Anion)

Here, anion is a complex hence, its metal ends with **ate** (ferrate), ligands are named before metal ion of the complex.

(b) $[Co(NH_3)_6]Cl_3$

Cation Anion

(Hexaamminecobalt (III) chloride)

Cation is a complex anion will not change its name (since it is not the ligand)

(c) $[PtCl_2(NH_3)_4]Cl_2$

(Tetraamminedichloroplatinum (IV) chloride)

NH_3 (ammine) and Cl (chloro) are two types of ligands named in alphabetical order.

Naming of coordination compounds having cationic complex ion

In the following examples, complex ion exists as cation

- $[Fe(NH_3)_6]Cl_3$
  Hexaammineiron (III) chloride
- $[CoCl(NH_3)_5]^{2+}$
  Pentaamminechloridocobalt (III) ion
- $[CoSO_4(NH_3)_4]NO_3$
  Tetraamminesulphatocobalt (III) nitrate

Naming of coordination compounds having anionic complex ion

In the following examples, complex ion exists as anion

- $K_2[Cr(CN)_2(O)_2(O_2)NH_3]$
  Potassium amminedicyanodioxoperoxo chromate (VI)
- $[Zn(NCS)_4]^{2-}$
  Tetrathiocyanato-N-zincate (II) ion
  (N indicates that ligand joined to metal ion through N-side)

Naming of coordination compounds having organic ligands

In the following examples, organic ligands have been used

- $[Pt(py)_4][PtCl_4]$
  Tetrapyridineplatinum (II) tetrachloroplatinate (II)
- $[Cr(en)_3]Cl_3$
  d or *l-tris*(ethylenediamine) chromium (III) chloride
- $[Pt(en)_2]^{2+}$
  bis(ethylenediamine) platinum (II) ion
- $[Cr(C_6H_6)_2]$
  bis(benzene) chromium (0)
- $[Fe(C_5H_5)_2]$
  bis(cyclopentadienyl) iron (II)

Naming of coordination compounds having bridging complex groups

In the following examples, bridging groups have been used

- $\left[(en)_2Co\langle{}^{NH}_{OH}\rangle Co(en)_2\right]^{4+}$
  Tetrakis(ethylenediamine)-μ- imido- μ-hydroxodicobalt (III) ion
- $(C_6H_5)_3P$, Cl, Pd, Cl, Pd, Cl, Cl, $P(C_6H_5)_3$ (bridged dimer)
  (.... line used to indicate symmetry)
  trans-bis(chlorotriphenylphosphine-μ-chloropalladium(I))
  or
  Chlorotriphenylphosphinepalladium (II) -μ-dichlorotriphenylphosphinepalladium (II)

ISOMERISM IN COORDINATION COMPOUNDS

Compounds which have same molecular formula but different arrangements of atoms or molecules or structures having different properties are known as **isomers**. This phenomenon is called **isomerism.**

It is classified into two types namely.

(*i*) Structural isomerism (*ii*) Stereoisomerism

Structural isomerism

The isomers which differ in their linkages, distribution of ligands around central metal atom are called structural isomers. These are further divided into following types.

Ionisation isomerism

Isomers which give different ions in solution or in molten state when ionised are called ionisation isomers. There is an interchange of anions between the coordination sphere of the metal ion and the ions outside the sphere.

e.g.

$[Co(NH_3)_5(SO_4)]Br$ Red	$[Co(NH_3)_5Br]SO_4$ Violet
(Pentaamminesulphatocobalt (III) bromide)	(Pentaamminebromocobalt (III) sulphate)
• SO_4^{2-} inside coordination sphere and Br^- outside it	• Br^- inside coordination sphere and SO_4^{2-} outside it.
• gives yellow ppt. with Ag^+ ionisation being as $[Co(NH_3)_5SO_4]^+ + Br^-$	• gives white ppt. with Ba^{2+} ionisation being as $[Co(NH_3)_5Br]^{2+} + SO_4^{2-}$

Other examples are

$[Pt(NH_3)_4SO_4](OH)_2$ and $[Pt(NH_3)_4(OH)_2]SO_4$
$[Co(NH_3)_5NO_2]SO_4$ and $[Co(NH_3)_5SO_4]NO_2$
$[Cr(NH_3)_5SO_4]Br$ and $[Cr(NH_3)_5Br]SO_4$

Linkage isomerism

When the ligand is attached to the central metal ion of a complex in different ways, it gives rise to linkage isomerism. Such ligands are called ambidentate ligands, e.g. nitrite ion has electron pairs available for coordination on both N and O atoms.

$[Co(NH_3)_5ONO]Cl_2$	and	$[Co(NH_3)_5NO_2]Cl_2$
Pentaamminenitritocobalt (III) chloride (red), decomposes in acids		Pentaamminenitrocobalt (III) chloride (yellow), stable in acids

Note *Irradiation with ultraviolet light of the yellow isomer results in interconversion into red isomer. This isomerisation reverses on warming.*

Coordination isomerism

When both the positive and negative ions are complex ions, isomerism may be caused by the interchange of ligands between the anion and cation. This arises only if cationic and anionic metals have same C.N. and charge.

e.g. $[Cr(NH_3)_6]^{3+}$ $[Co(CN)_6]^{3-}$

Hexaamminecobalt (III) hexacyanochromate (III)

$[Co(NH_3)_6]^{3+}$ $[Cr(CN)_6]^{3-}$

Hexaamminechromium (III) hexacyanocobaltate (III)

Both above complex have same charge ; same C.N. = 6

Coordination position isomerism

In polynuclear complexes, on interchange of ligands between the different metal nuclei gives rise to position isomerism.

e.g. $\left[(NH_3)_4Co\langle^{NH_2}_{O_2}\rangle Co(NH_3)_2Cl_2\right]Cl_2$

$\left[Cl(NH_3)_3Co\langle^{NH_2}_{O_2}\rangle Co(NH_3)_3Cl\right]Cl_2$

Hydrate (solvate) isomerism

Hydrate isomers are isomers of a complex that differ in the placement of water molecules in the complex. e.g. $CrCl_3 \cdot 6H_2O$ have following hydrate isomers.

$[Cr(H_2O)_6]Cl_3$ – violet (anhydrous)

$[Cr(H_2O)_5Cl]Cl_2 \cdot H_2O$ – light green (monohydrate)

$[Cr(H_2O)_4Cl_2]Cl \cdot 2H_2O$ – dark green (dihydrate)

This is closely related to ionisation isomerism and principle is the same except that H_2O (a neutral ligand) is present in this case. While in case of ionisation isomerism anionic ligand is present.

Stereoisomerism

Isomers which have same molecular formula but different spatial arrangements are called stereoisomers. This phenomenon is known as stereoisomerism.

It is divided into two types namely.

(*i*) Geometrical isomerism (*ii*) Optical isomerism

Geometrical isomerism

Isomers in which the atoms or ligands occupy different positions around central metal ion are called geometrical isomers. Complexes in which the two same ligands occupy adjacent positions to each other, are called *cis* - isomers, and in which they occupy opposite positions to each other, are called *trans* - isomers.

Coordination compounds having 2 (linear molecule), 3 (trigonal planar) and 4 (tetrahedral geometry) coordination numbers does not exhibit geometrical isomerism.

This is because, it is not possible to have more than one arrangement of ligands in space around central metal ion.

Geometrical isomerism in C.N. 4

For square planar complex MA_4, MA_3B or MAB_3 where, A and B etc., are monodentate ligands, geometrical isomerism is not possible. However, square planar complexes of the type MA_2B_2, MA_2BC, $MABCD$ and $M(AA)_2$, $M(AB)_2$ where, AA and AB represent symmetrical and unsymmetrical chelating agents respectively gives geometrical isomers.

(*i*) $\boldsymbol{MA_2B_2}$ **type** $[Pt(NH_3)_2Cl_2]$ can exist in two geometric isomers.

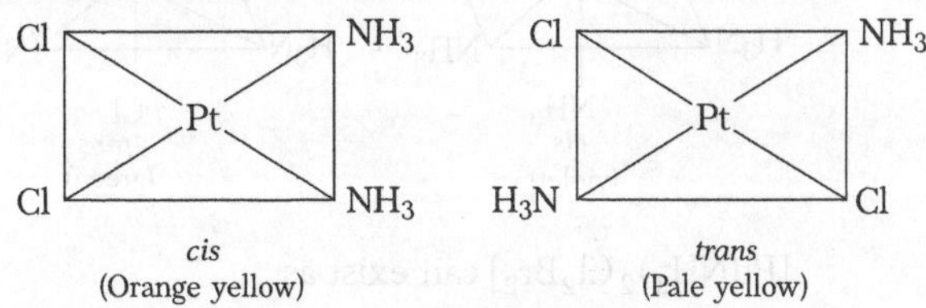

cis (Orange yellow) *trans* (Pale yellow)

(*ii*) $\boldsymbol{MA_2BC}$ **type** $[Pt(NH_3)_2ClBr]$ can exist as shown below.

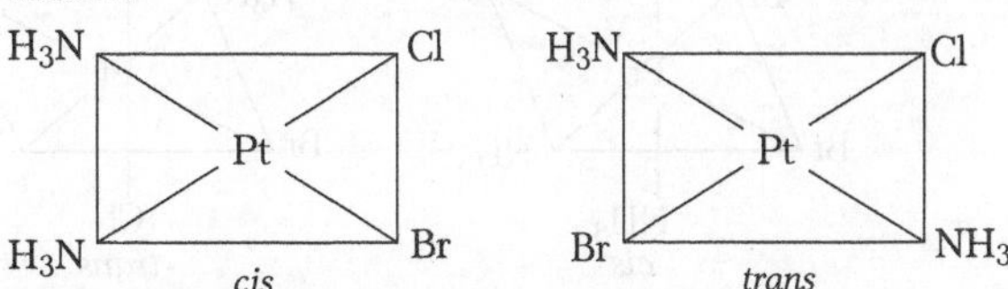

cis *trans*

(*iii*) ***MABCD*** **type** $[Pt(NH_3)(C_5H_5N)(Cl)(Br)]$ can exist in three geometric isomers.

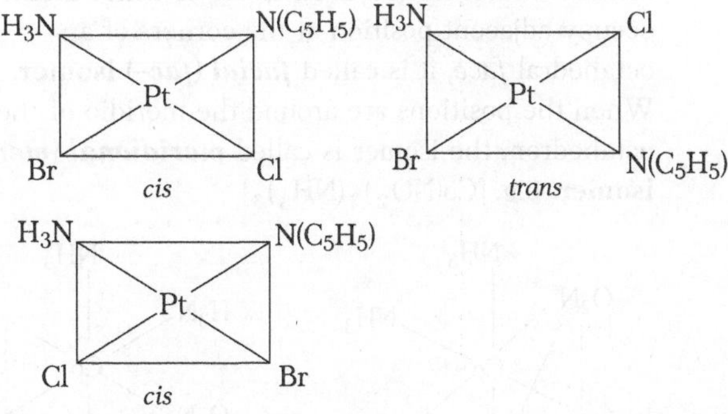

cis *trans* *cis*

(*iv*) $\boldsymbol{M(AB)_2}$ **type** Platinum glycinato complex, $[Pt(gly)_2]$ exhibit geometrical isomerism.

$CH_2—NH_2$ and $NH_2—CH_2$ on one side, $CO—O$ and $O—CO$ on the other, around Pt

cis

$CO—O$ and $NH_2—CH_2$ on one side, $CH_2—NH_2$ and $O—CO$ on the other, around Pt

trans

Geometrical isomerism in C.N. 6

Complexes with formula MA_6 and MA_5B do not show geometrical isomerism, where both A and B are monodentate ligands.

(*i*) Complexes with formula MA_4B_2 or $MA_2B_2C_2$ exhibit geometrical isomerism.

e.g.

- $[Co(NH_3)_4Cl_2]^+$ can exist as

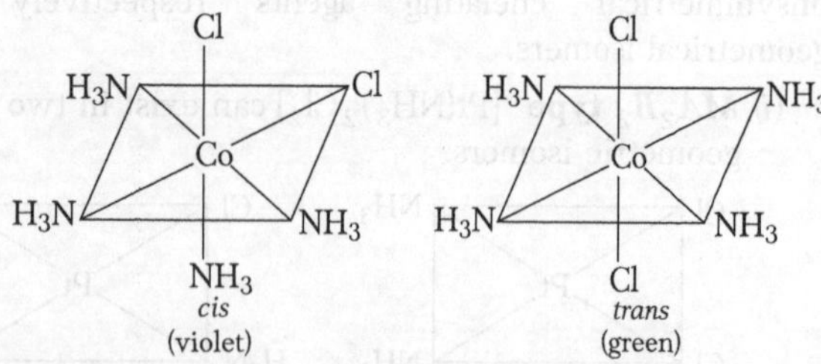

cis (violet) *trans* (green)

- $[Pt(NH_3)_2Cl_2Br_2]$ can exist as

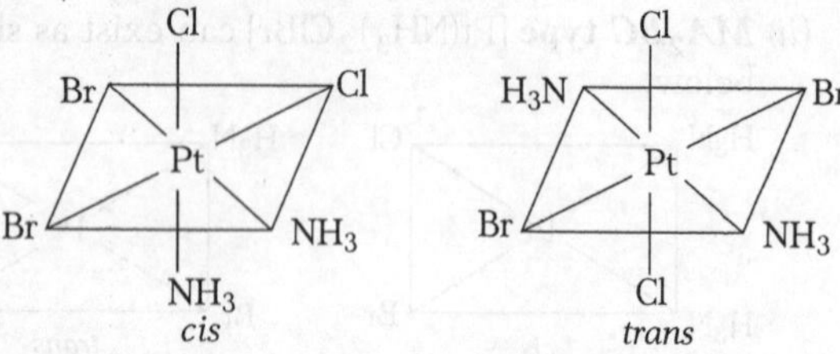

cis *trans*

(*ii*) Complexes with formula MA_3B_3 exhibit *facial* and *meridional* isomerism. If each trio of donor atoms occupy adjacent position at the corners of an octahedral *face*, it is called ***facial (fac–) isomer.*** When the positions are around the meridio of the octahedron, the isomer is called ***meridional (mer–) isomer.*** e.g. $[Co(NO_2)_3(NH_3)_3]$

fac-isomer *mer*-isomer

(*iii*) Complexes with formula *MABCDEF* exist in 15 isomeric forms but practically only three of them have been isolated, e.g. $[Pt(Br)(Cl)(I)(NO_2)(NH_3)(py)]$.

(*iv*) Complexes with formula $M(AA)_2B_2$, $M(AA)_2BC$ also exhibit geometrical isomerism (where, AA is a symmetrical bidentate ligand),

e.g. $[Co(en)_2Cl_2]^+$

cis-isomer *trans*-isomer

(*v*) Complexes with formula $M(AB)_3$ (where, AB is an unsymmetrical bidentate ligand) also exhibit geometrical isomerism, e.g. $[Cr(gly)_3]$.

Optical isomerism

Complexes which are non-superimposable on their mirror images and do not possess the plane of symmetry are called **optical isomers** and the phenomenon is known as optical isomerism. They have same chemical properties but differ in action towards plane polarised light.

The optically active complex which rotates plane polarised light to left is laevorotatory (*l* or –) and which rotates to right is dextrorotatory (*d* or +). These '*d*' and '*l*' isomers of a complex are called **enantiomorphs** or **enantiomers.**

Optical isomerism in tetrahedral complexes

Tetrahedral complexes with unsymmetrical bidentate ligand are optically active, e.g. *bis* (benzoyl acetonato) beryllium (II)

Optical isomerism in square planar complexes

Optical isomerism is rarely shown by square planar complexes because they have all the ligands and metal ion in one plane. Hence, there is plane of symmetry. Complexes that is given below show optical isomerism as it has no plane of symmetry.

Optical isomerism in octahedral complexes

It is common in octahedral complexes involving bidentate ligands. These are as follows

(*i*) Complex with formula MA_2B_2 can exist as *cis* and *trans* isomers. But only *cis*-isomer can have *d*- and *l*-optical isomers. *trans*-form is optically inactive due to presence of plane of symmetry, e.g. $[Co(en)_2Cl_2]$.

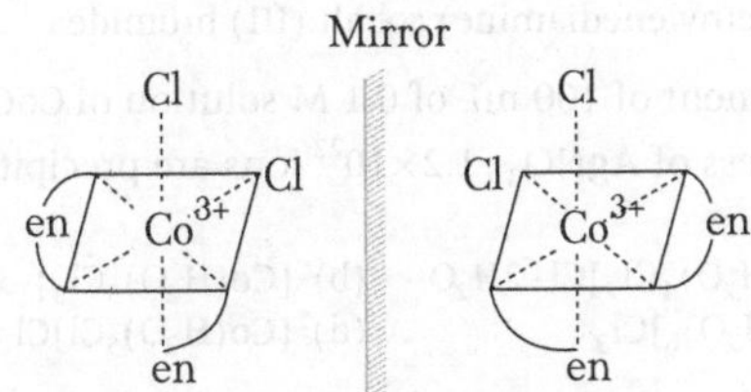

Enantiomorphic pair *d* and *l* of *cis*-dichloro *bis* (ethylenediamine) cobalt (III) ion-violet

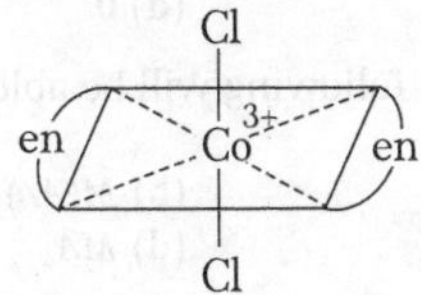

(Symmetrical) *trans*–(no optical isomer possible)–green

(*ii*) Complex with formula $M(AA)_3$ also show optical isomerism, e.g. $[Co(en)_3]^{3+}$.

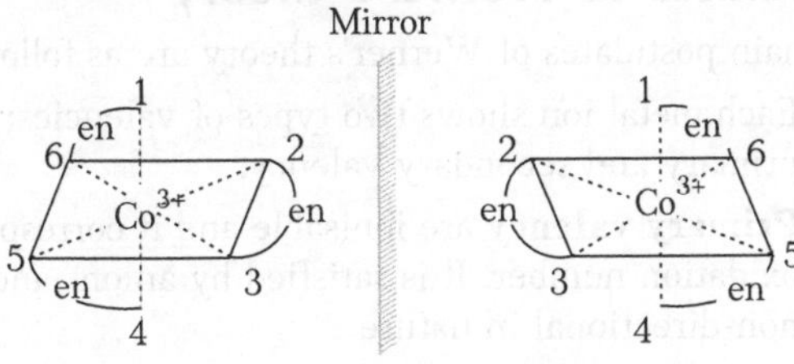

(*iii*) Complexes with formula MA_3B_3 can exist in *facial* and *meridional* forms but both the isomers are optically inactive.

(*iv*) Complexes with formula $MA_2B_2C_2$ are optically active, e.g. $Pt(Cl_2)(NH_3)_2(py)_2]^{2+}$ can exist in five geometrical isomers but only three of them are isolated. These isomers are optically active in their *cis*-form.

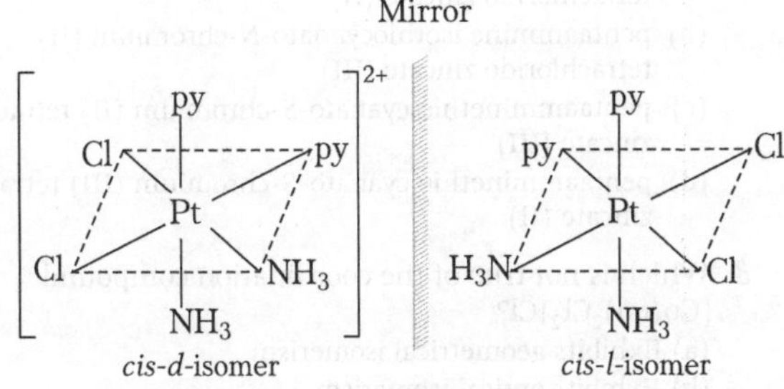

(*v*) Complexes with formula *MABCDEF* exist in 15 geometrical isomers and each isomer is optically active. Thus, there are 30 optical isomers possible. e.g. $[Pt(Br)(Cl)(I)(NO_2)(NH_3)(py)]$

(*vi*) Complexes with formula $M(AA)_3$ (where, '*AA*' is symmetrical bidentate ligand) exist as *d*- and *l*-forms. e.g. $[Co(en)_3]^{3+}$ in its *cis*-form shows optical isomerism.

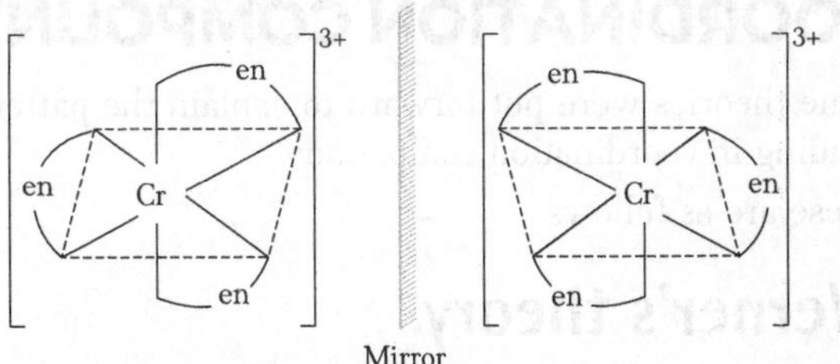

CHECK POINT 12.2

1. Pick the correct name for $[Co(NH_3)_5Cl]Cl_2$
   (a) pentaamminechlorido cobalt(II) chloride
   (b) pentaamminechlorido cobalt(III) chloride
   (c) pentaamminechlorido cobalt(III)
   (d) pentaammine cobalt(III) chloride

2. The IUPAC name of the compound $[Cr(NH_3)_5NCS][ZnCl_4]$
   (a) pentaammine isothiocyanato-N-chromium (III) tetrachlorido zincate (II)
   (b) pentaammine isothiocyanato-N-chromium (II) tetrachlorido zincate (III)
   (c) pentaamminethio cyanato-S-chromium (II) tetrachlorido zincate (III)
   (d) pentaamminethio cyanato-S-chromium (III) tetrachlorido zincate (II)

3. Which is not true of the coordination compound $[Co(en)_2Cl_2]Cl$?
   (a) Exhibits geometrical isomerism
   (b) Exhibits optical isomerism
   (c) Exhibits ionisation isomerism
   (d) It is an octahedral complex

4. Which of the following ligands can exhibit linkage isomerism?
   (a) NO_2^- (b) CN^- (c) SCN^- (d) All of these

5. $[Co(NH_3)_6][Cr(CN)_6]$ and $[Cr(NH_3)_6][Co(CN)_6]$ are
   (a) linkage isomers (b) ionisation isomers
   (c) coordination isomers (d) None of these

6. Which of the following coordination compounds would exhibit optical isomerism?
   (a) Pentaamminenitro cobalt (III) iodide
   (b) Diamminedinitro platinum (II)
   (c) *trans*-dicyano *bis* (ethylenediamine)
   (d) *Tris*-(ethylenediamine) cobalt (III) bromide

7. On treatment of 100 mL of 0.1 M solution of $CoCl_3 \cdot 6H_2O$ with excess of $AgNO_3$; 1.2×10^{22} ions are precipitated. The complex is
   (a) $[Co(H_2O)_4Cl_2]Cl \cdot 2H_2O$ (b) $[Co(H_2O)_3Cl_3] \cdot 3H_2O$
   (c) $[Co(H_2O)_6]Cl_3$ (d) $[Co(H_2O)_5Cl]Cl_2 \cdot H_2O$

8. The number of geometrical isomers of the complex $[Co(NO_2)_2(NH_3)_2]$ is
   (a) 2 (b) 3
   (c) 4 (d) 0

9. Which one of the following will be able to show *cis-trans* isomerism?
   (a) MA_3B (b) $M(AA)_2$
   (c) $MABCD$ (d) MA_4

BONDING IN COORDINATION COMPOUND

Some theories were put forward to explain the pattern of bonding in coordination compounds.

These are as follows

Werner's theory

In 1892, Alfred Werner proposed a theory about the structures of coordination compounds. He prepared and isolated many different compounds from $CoCl_3$ and NH_3 such as

- $CoCl_3 \cdot 6NH_3$ $[Co(NH_3)_6]Cl_3$ orange-yellow
- $CoCl_3 \cdot 5NH_3$ $[Co(NH_3)_5Cl]Cl_2$ violet
- $CoCl_3 \cdot 4NH_3$ $[Co(NH_3)_4Cl_2]Cl$ violet
- $CoCl_3 \cdot 3NH_3$ $[Co(NH_3)_3Cl_3]$ green

Postulates of Werner's theory

The main postulates of Werner's theory are as follows

(*i*) Each metal ion shows two types of valencies; primary and secondary valency.

(*ii*) **Primary valency** are ionisable and it corresponds to oxidation number. It is satisfied by anions and are non-directional in nature.

(*iii*) **Secondary valency** are non-ionisable and corresponds to coordination number. It is satisfied by neutral or anionic ligands and are directional in nature.

(*iv*) Metal ion should satisfy both primary and secondary valencies. Some negative ions may satisfy primary as well as secondary valencies, are called to have dual nature. e.g. in the complex $[Co(NH_3)_6]Cl_3$, primary valency of Co is three and secondary valency is six.

Werner's representation

Consider the case of $CoCl_3 \cdot xNH_3$ where maximum value of x = C.N (coordination number) of Co(III) = 6 and minimum value of x = coordination number C.N. – (O.N) oxidation number = 3

Various structures are summarised below

	Werner's complex	Modern notation	Ionisation	Secondary valency satisfied by	Primary valency satisfied by
(A)	$CoCl_3 \cdot 6NH_3$	$[Co(NH_3)_6]Cl_3$	$[Co(NH_3)_6]^{3+} + 3Cl^-$	Six (NH_3)	Three (Cl^-)
(B)	$CoCl_3 \cdot 5NH_3$	$[Co(NH_3)_5Cl]Cl_2$	$[Co(NH_3)_5Cl]^{2+} + 2Cl^-$	Five (NH_3) and one (Cl^-)	Three (Cl^-) [including one (Cl^-) with **dual nature**]
(C)	$CoCl_3 \cdot 4NH_3$	$[Co(NH_3)_4Cl_2]Cl$	$[Co(NH_3)_4Cl_2]^+ + Cl^-$	Four (NH_3) and two (Cl^-)	Three (Cl^-) [including two (Cl^-) with **dual nature**]
(D)	$CoCl_3 \cdot 3NH_3$	$[Co(NH_3)_3Cl_3]$	$[Co(NH_3)_3Cl_3]$ (single species)	Three (NH_3) and three (Cl^-)	Three (Cl^-) [all with **dual nature**]

Limitations

Werner theory was unable to account for the following points.

(*i*) Definite geometry of coordination compounds.

(*ii*) Presence of magnetic and optical properties of coordination compounds.

To overcome the limitations of Werner's theory, various theories were put forward such as valence bond theory and crystal field theory.

Valence bond theory

This theory was proposed by L. Pauling in 1930's. According to this, transition metal ions have incompletely filled *d*-orbitals. These orbitals contain unpaired electrons which couple as fully as possible before bond formation takes place.

(*i*) Thus, some $(n-1)\,d$-orbitals becomes vacant and therefore, central metal ion provides number of empty orbitals equal to its coordination number for formation of complex.

(*ii*) Thus, atomic orbitals present on central metal ion hybridise to form hybrid orbitals. There is interaction of hybrid orbitals of metal ion and orbitals of ligands.

(*iii*) Ligands donate its lone pairs of electrons, to form bond with metal ion. It describes the geometric structure of complex ion.

(*iv*) If $(n-1)\,d$-orbitals are involved in hybridisation, then *inner orbital complex* are formed while involvement of *nd*-orbitals results in *outer orbital complex.* Octahedral, square planar and tetrahedral complexes are formed as a result of d^2sp^3, dsp^2 and sp^3 hybridisation respectively of the central metal atom or ion. It helps to predict the geometry, hybridisation, magnetic properties of a complex.

Applications of valence bond theory

Some applications of valence bond theory are as follows

In formation of complex with coordination number 4

Coordination compounds which have C.N. = 4 can have tetrahedral or square planar geometry. It depends upon the orbitals available for hybridisation.

1. In tetrahedral complexes

Complexes which involve one *s* and three *p*-orbitals for bonding, i.e. sp^3, are said to be tetrahedral complexes.

e.g. $[NiCl_4]^{2-}$, $[Ni(CO)_4]$, $[CuCl_4]^{2-}$, $[Zn(NH_3)_4]^{2+}$ etc.

- $[NiCl_4]^{2-}$

Electronic configuration of Ni(Z = 28) = [Ar] $3d^8\,4s^2$.

Due to presence of 4 Cl^-, oxidation state of Ni is +2.

Therefore, electronic configuration of Ni^{2+} is [Ar] $3d^8\,4s^0$.

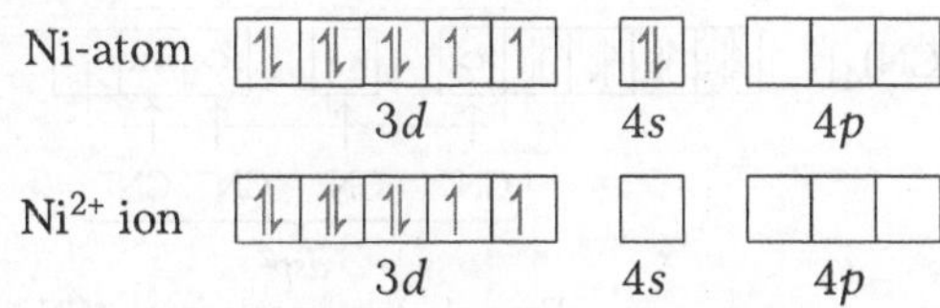

Cl^- is a weak field ligand, thus it cannot pair the unpaired electrons of Ni^{2+}.

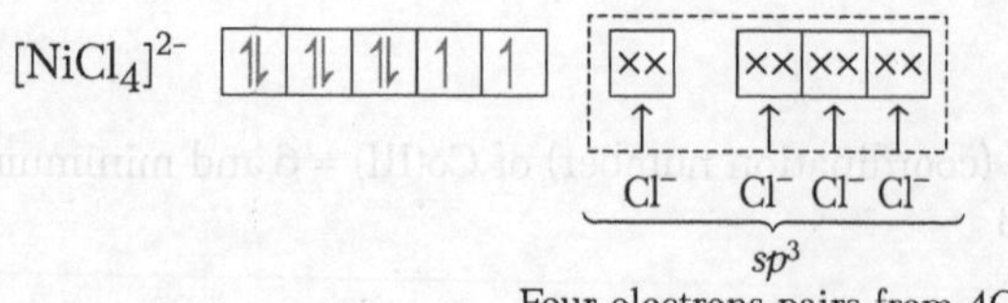

Four electrons pairs from 4Cl⁻

Due to presence of two unpaired electrons, $[NiCl_4]^{2-}$ is paramagnetic in nature.

Magnetic moment of N unpaired electrons

$$= \sqrt{N(N+2)}\text{B.M} = \sqrt{8}\text{ BM}$$

- $[Ni(CO)_4]$

Since, CO is neutral ligand thus, Ni has zero oxidation state.

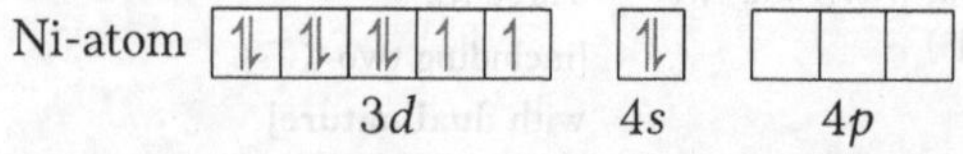

CO is a strong field ligand, it can pair up the unpaired electrons of Ni.

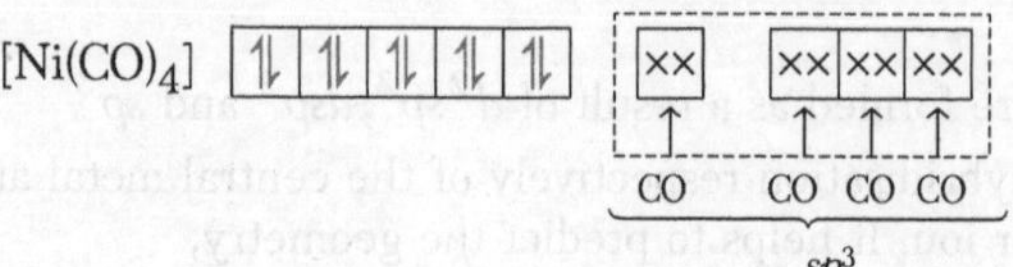

Four electrons pairs from 4CO

There is no unpaired electron, hence, $[Ni(CO)_4]$ is diamagnetic in nature.

2. **In square planar complexes**

Complexes which involves one d, one s and two p-orbitals for bonding, i.e. dsp^2 are said to be square planar complexes.

e.g. $[Ni(CN)_4]^{2-}$, $[Pt(CN)_4]^{2-}$, $[Cu(NH_3)_4]^{2+}$ etc.

- $[Ni(CN)_4]^{2-}]$

Ni is present in +2 oxidation state due to four CN^- ligands.

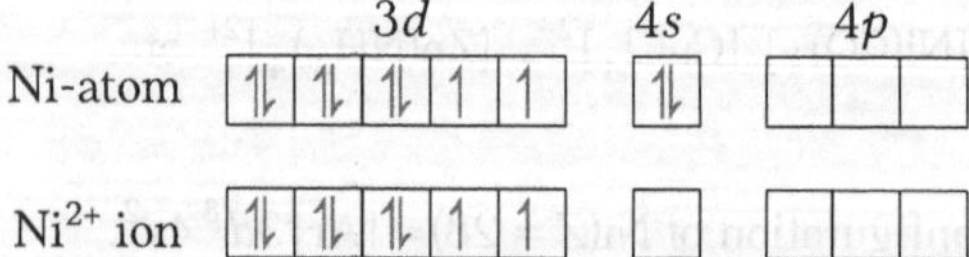

Since, CN^- is a strong field ligand, it pair-up the electrons of Ni^{2+} ion. Thus, one $(n-1)d$ orbital is vacant.

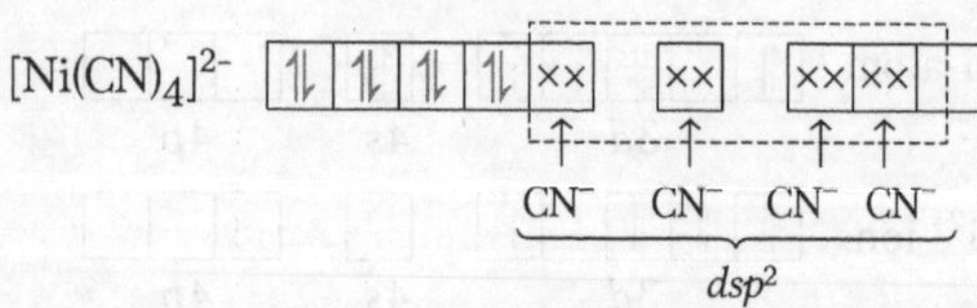

Four electrons pairs from 4CN⁻

All the electrons in $[Ni(CN)_4]^{2-}$ complex are paired. Therefore, it is diamagnetic.

In formation of complex with coordination number 6

Complex which have CN=6, can form inner orbital complex (d^2sp^3-hybridisation) or outer orbital complex (sp^3d^2-hybridisation). These complexes have octahedral geometry.

Some of the examples of octahedral complex are given below

1. Inner orbital complex

These complexes involve two inner d-orbitals, one s and three p-orbitals for bonding. These are formed by strong field ligands hence, are low spin complex.

e.g. $[Co(NH_3)_6]^{3+}$, $[Fe(CN)_6]^{4-}$, $[Cr(NH_3)_6]^{3+}$, $[Fe(CN)_6]^{3-}$ $[Co(ox)_3]^{3-}$ etc.

- $[Co(NH_3)_6]^{3+}$

Since, NH_3 is a neutral ligand, thus, oxidation state of Co is +3.

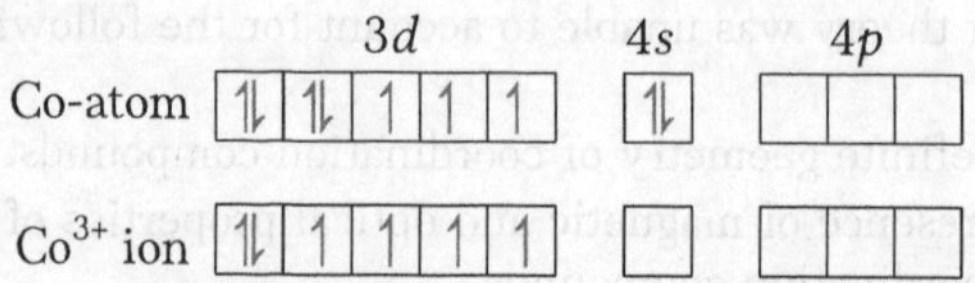

NH_3 is a strong field ligand, therefore, it can pair the electrons of Co^{3+}.

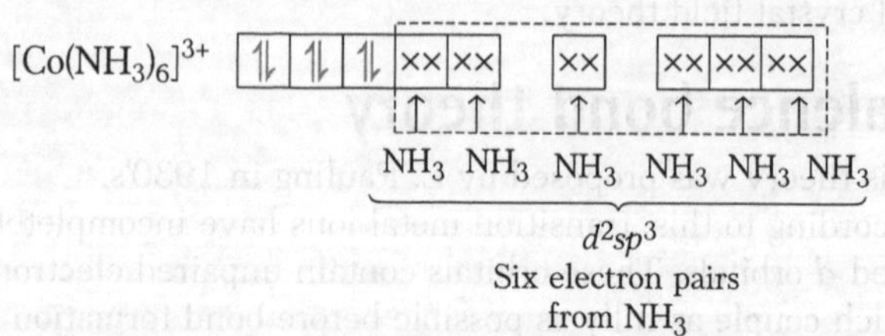

Six electron pairs from NH_3

Hence, hybridisation of $[Co(NH_3)_6]^{3+}$ is d^2sp^3.

2. Outer orbital complex

These complex involves one s, three p and two outer d-orbitals for bonding. These are formed by weak field ligands hence, are high spin complex,

e.g. $[CoF_6]^{3-}$, $[Fe(H_2O)_6]^{3+}$, $[Co(H_2O)_6]^{2+}$ etc.

- $[CoF_6]^{3-}$

Oxidation state of Co is +3 due to presence of 6F⁻ ligands.

	3d	4s	4p	4d
Co-atom	⥮ ⥮ 1 1 1	⥮		
Co^{3+} ion	⥮ 1 1 1 1			

Since, F^- is a weak field ligand, it cannot pair electrons of Co^{3+}.

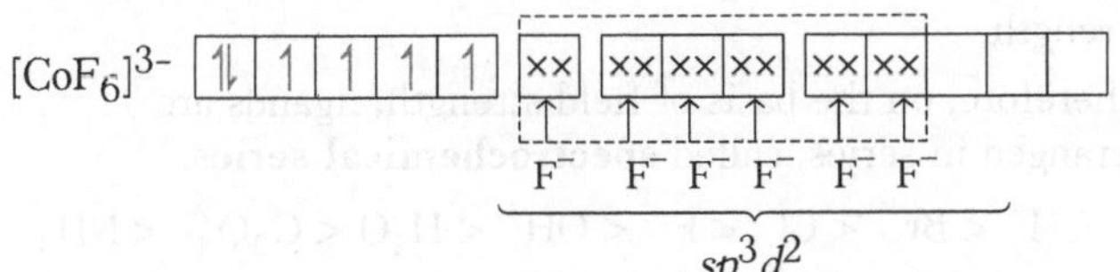

Hence, hybridisation of $[CoF_6]^{3-}$ is sp^3d^2.

Magnetic moment $= \sqrt{4(4+2)}$

$= \sqrt{24}$ BM

Limitations

VBT was unable to account for the following points

(i) It does not provide insight into the origin of characteristic colour of the complex ions.

(ii) It does not distinguish between weak and strong ligands.

(iii) It does not give quantitative interpretation of the thermodynamic or kinetic stabilities of coordination compounds.

(iv) There is no quantitative interpretation of magnetic data.

(v) It does not explain why some complexes are inner orbital complex and some complexes are outer orbital.

(vi) It fails to give a satisfactory explanation for the formation of some complexes like $[Cu(NH_3)_4]^{2+}$

∴ Therefore, the complex is square planar and paramagnetic. Valence bond theory gives no explanation for exciting an electron from 3*d* to 4*p*-orbital which is not energetically favoured.

$[Cu(NH_3)_4]^{2+}$ ion

Expected configuration of $[Cu(NH_3)_4]^{2+}$ ion

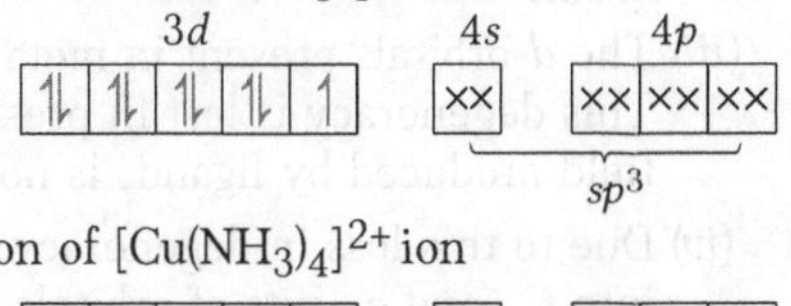

Actual configuration of $[Cu(NH_3)_4]^{2+}$ ion

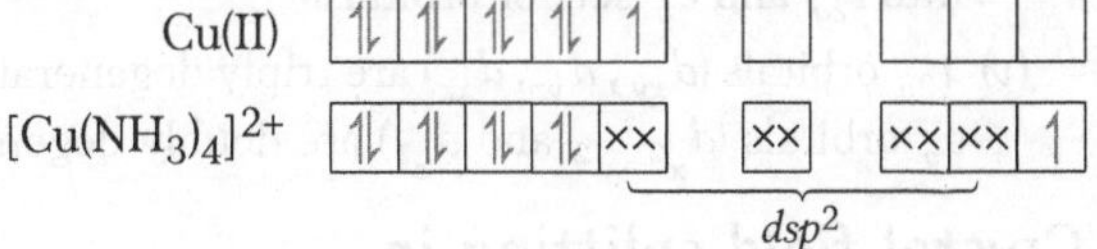

CHECK POINT 12.3

1. Primary and secondary valency of Pt in $[Pt(en)_2Cl_2]Cl_2$ are
(a) 4, 4 (b) 2, 6 (c) 6, 6 (d) 4, 4

2. Which does not give a precipitate with $AgNO_3$ solution?
(a) $[Co(NH_3)_6]Cl_3$ (b) $[Co(NH_3)_5Cl]Cl_2$
(c) $[Co(NH_3)_4Cl_2]Cl$ (d) $[Co(NH_3)_3Cl_3]$

3. Which one of the following has a square planar geometry?
(a) $[CoCl_4]^{2-}$ (b) $[FeCl_4]^{2-}$
(c) $[NiCl_4]^{2-}$ (d) $[PtCl_4]^{2-}$

4. Which of the following statement is incorrect?
(a) $Ni(CO)_4$ – tetrahedral, paramagnetic
(b) $Ni(CN)_4^{2-}$ – square planar, diamagnetic
(c) $Ni(CO)_4$ – tetrahedral, diamagnetic
(d) $[Ni(Cl_4)]^{2-}$ – tetrahedral, paramagnetic

5. The hybridisation of $[Ni(CN)_4]^{2-}$ ion is
(a) dsp^2 (b) sp^2d^2
(c) d^2sp (d) sp^2

6. $[CoF_6]^{3-}$ involves which of the following hybridisation?
(a) d^2sp^3 (b) d^3sp^2 (c) dsp^3 (d) sp^3d^2

7. Which of the following is an inner orbital complex as well as diamagnetic in behaviour?
(a) $[Cr(NH_3)_6]^{3+}$ (b) $[Co(NH_3)_6]^{3+}$
(c) $[Ni(NH_3)_6]^{2+}$ (d) $[Zn(NH_3)_6]^{2+}$

8. The unpaired electron present in $[Ni(CO)_4]$ is
(a) zero (b) 1 (c) 2 (d) 3

9. Out of $Ni(CO)_4$, $[Ni(CN)_4]^{2-}$, $[NiCl_4]^{2-}$, $[Ni(H_2O)_4]^{2+}$. Central metal ion is sp^3-hybridised in
(a) all except $[Ni(CN)_4]^{2-}$ (b) all except $[NiCl_4]^{2-}$
(a) all except $[Ni(CO)_4]$ (b) all of these

10. Which one of the following complex is an outer orbital complex?
(a) $[Fe(CN)_6]^{4-}$ (b) $[Mn(CN)_6]^{4-}$
(c) $[Co(NH_3)_6]^{3+}$ (d) $[Ni(NH_3)_6]^{2+}$

CRYSTAL FIELD THEORY (CFT)

This theory was proposed to overcome the shortcomings of valence bond theory. This theory is based on ionic model. It describes the splitting of degenerate *d*-orbitals in presence of ligands.

Some assumptions of crystal field theory are as follows

(*i*) Ligands are considered as point charges. Anionic ligands as negative point charges.

(*ii*) The interaction between metal ion and ligand is purely electrostatic. Therefore, metal ion and ligands do not mix their orbitals, i.e. no overlap of orbitals occur.

(*iii*) The *d*-orbitals present in metal ion are degenerate. This degeneracy is lost in presence of ligands as the field produced by ligands is not symmetrical.

(*iv*) Due to this loss in degeneracy, *d*-orbitals are splitted into t_{2g} and e_g sets of orbitals.

(*v*) t_{2g} orbitals (d_{xy}, d_{yz}, d_{xz}) are triply degenerate while e_g orbitals ($d_{x^2-y^2}$ and d_{z^2}) are doubly degenerate.

Crystal field splitting in octahedral complex

Complex with C.N. = 6 forms octahedral complex. When a ligand approaches to a metal ion, it distorts the spherically symmetrical cloud of electrons in metal ion and repulsions occur between the electrons of *d*-orbitals of metal ion and electrons of ligands.

The $d_{x^2-y^2}$ and d_{z^2} orbital (i.e e_g orbitals) which point towards the axes along the direction of the ligand will experience more repulsion and will be raised in energy and the d_{xy}, d_{zx}, d_{yz} orbitals (i.e. t_{2g} orbitals) which are directed between the axes will be lowered in energy relative to the average energy in the spherical crystal field. This splitting of degenerate orbitals into two sets is known as crystals field splitting and the difference of energy between the two sets of degenerate orbitals as result of this splitting is known as **crystal field stabilisation energy** (CFSE). It is represented by Δ_O (the subscript *O* is for octahedral.

Hence, the energy of two e_g orbitals is increased by value of $\frac{3}{5}\Delta_O$ and energy of three t_{2g} orbitals is decreased by $\frac{2}{5}\Delta_O$.

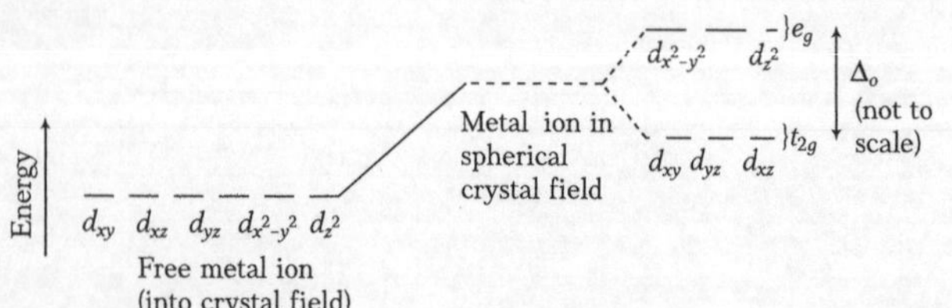

The value of Δ_O depends on the nature of ligand. Some ligands produce strong fields while some have weak field strength.

Therefore, on the basis of field strength, ligands are arranged in series, called **spectrochemical series.**

$$I^- < Br^- < Cl^- < F^- < OH^- < H_2O < C_2O_4^{2-} < NH_3 < en < NO_2^- < CN^-$$

If the metal ion has d^1, d^2 or d^3 configuration, the electron occupies t_{2g} orbitals according to Hund's rule (i.e. each orbital is first singly filled.)

For d^4 configuration, the fourth electron can either occupy t_{2g} or e_g orbital. It depends upon the magnitude of pairing energy, *P*. The two possibilities are

(a) If $\Delta_O > P$, the fourth electron will pair up and occupies the t_{2g} orbital, thus, forms low spin complex.

(b) If $\Delta_O < P$, the fourth electron will not be able to pair-up then it occupy e_g orbital, thus, forms high spin complex.

e.g. in $[Co(NH_3)_6]^{3+}$, NH_3 is a strong field ligand thus, it has ability to pair the electrons, i.e., $\Delta_O > P$. Hence, it will form low spin complex.

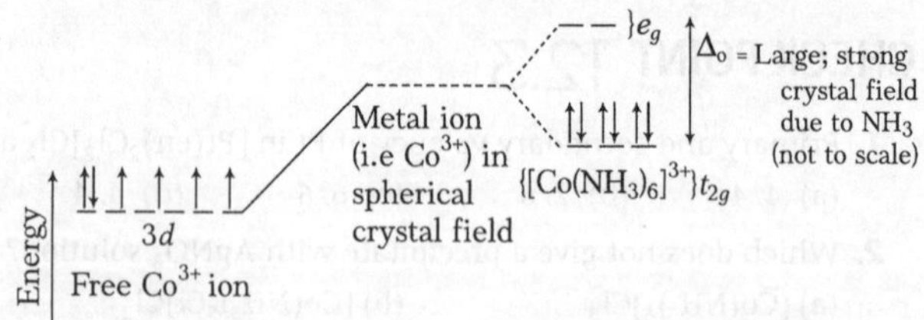

In $[CoF_6]^{3+}$, F^- is a weak field ligand, thus it cannot pair the electrons, i.e. $\Delta_O < P$. Hence, it will form high spin complex.

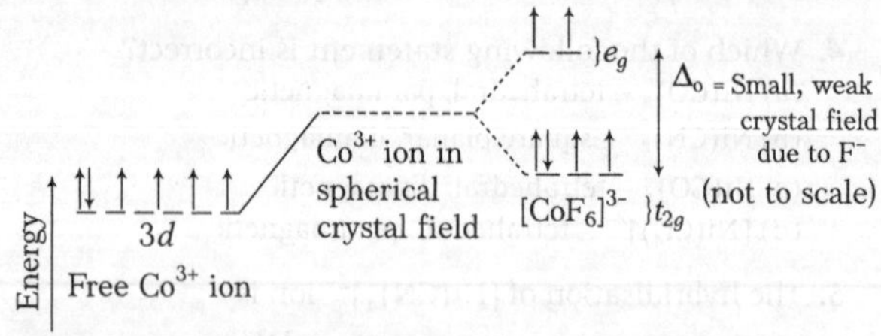

Calculation of CFSE

Electron occupying t_{2g} orbital lowers the energy by $-0.40\,\Delta_O$ and electron occupying e_g orbital increases the energy by $0.60\,\Delta_O$.

Therefore, magnitude of CFSE is given by the relation,

$$CFSE = (-0.40x + 0.60y)\Delta_O + P$$

where, x = number of electrons in t_{2g} orbital

y = number of electrons in e_g orbital

P = pairing energy

e.g. the CFSE for $[Fe(CN)_6]^{3-}$ is $-2.0\,\Delta_O$

$Fe^{3+} = [Ar]3d^5$

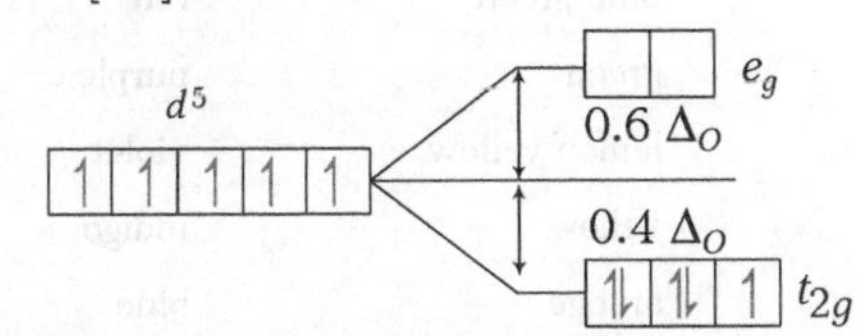

$$CFSE = (-0.40 \times x + 0.60\,y)\,\Delta_O$$

$$= (-0.40 \times 5 + 0.60 \times 0)\,\Delta_O = -2.0\,\Delta_O$$

Crystal field splitting in tetrahedral complex

Complex with C.N. = 4 forms tetrahedral complex.

In tetrahedral complex formation, the *d*-orbital splitting is inverted and smaller as compared to the octahedral field splitting. The magnitude of energy separation in tetrahedral complex (Δ_t) is considerably less than in octahedral. This is because of the following two reasons

- These are only four ligands (instead of six), therefore the ligand field is only $\frac{2}{3}$ to the size. Hence, the ligand field splitting is also $\frac{2}{3}$ the size.
- The directions of the orbitals does not coincide with the direction of the ligand due to which crystal field splitting again reduces to 2/3.

That's why the tetrahedral crystal field splitting Δ_t is $= \frac{2}{3} \times \frac{2}{3}$ of octahedral crystal field splitting (Δ_O)

The relation can be represented as

$$\Delta_t = -\frac{4}{9}\Delta_o$$

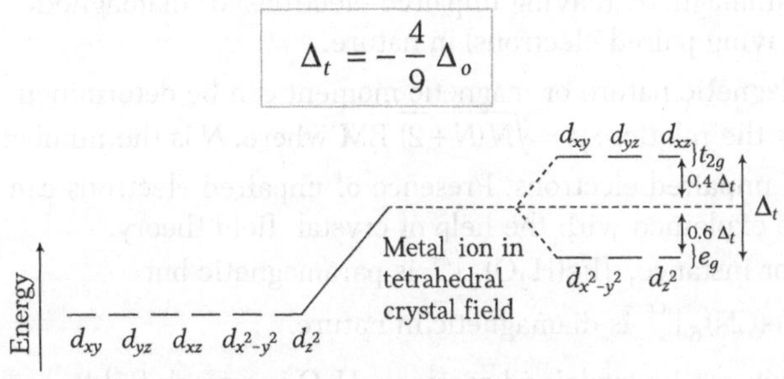

Calculation of CFSE

In this case, low spin configuration is rarely observed.

CFSE of tetrahedral complexes = $[-0.6 \times y + 0.4 \times x]\Delta_t$

Favourable conditions for tetrahedral complexes

- When ligands are large and bulky
- When after splitting in tetrahedral field, t_{2g} and e_g orbitals are either unfilled or half-filled or completely filled, then tetrahedral complexes are favoured.
- When the ligands are weak field and the loss of CFSE is smaller.
- When the metal is in low oxidation state.

Crystal field splitting in square planar complex

Square planar complex can be formed by removing ligands from *z*-axis in octahedral geometry. Therefore, the orbitals which have *z*-component, i.e. d_{z^2}, d_{xz} and d_{yz} are more stable than orbitals which have *x* and *y*-component. As a result, orbitals such as $d_{x^2-y^2}$ and d_{xy} are raised to higher energy levels.

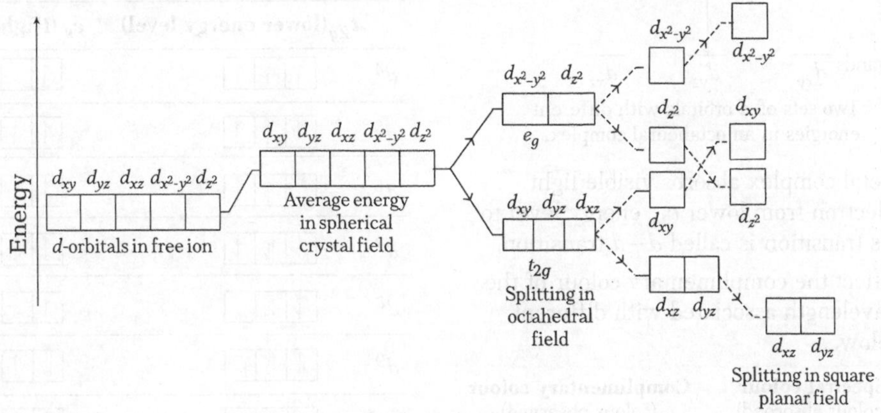

Factors affecting CFSE

Following are the main factors that affect CFSE.

- Nature of ligand
- Charge on the metal ion
- Position of metal ion in transition series

Magnetic nature

Transition metal complexes exhibit magnetic properties. The magnetic nature depends on the number of unpaired electrons present. On this basis, these complexes can be paramagnetic (having unpaired electrons) or diamagnetic (having paired electrons) in nature.

Magnetic nature or magnetic moment can be determined by the relation, $\mu = \sqrt{N(N+2)}$ BM where, N is the number of unpaired electrons. Presence of unpaired electrons can be explained with the help of crystal field theory. For instance, $[Fe(H_2O)_6]^{2+}$ is paramagnetic but $[Fe(CN)_6]^{4+}$ is diamagnetic in nature.

This can be explained as, since, H_2O is a weak field ligand, it does not pair up the electrons and forms high spin complex. While CN^- is a strong field ligand, pairs the electrons and forms low spin complex.

Colour of the complex

A substance appears coloured because it absorbs light that corresponds to wavelength in the visible region and reflects other wavelength. Most transition metal complexes are coloured. In these complexes, *d*-orbitals are not degenerate and split into two sets t_{2g} and e_g which are separated by energies equal to Δ_O.

Energy difference between these two sets is small. Therefore, excitation of electron is quite easy.

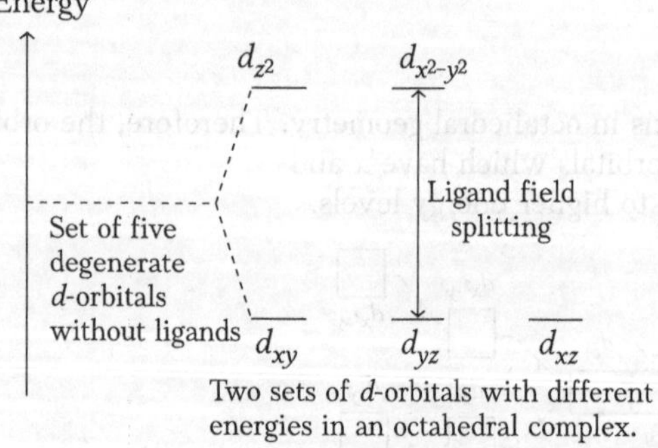

Two sets of *d*-orbitals with different energies in an octahedral complex.

Hence, transition metal complex absorbs visible light which excites the electron from lower t_{2g} energy level to higher e_g level. This transition is called $d-d$ transition. These complexes reflect the complimentary colour of the colour absorbed. Wavelength associated with different colours are given below.

Wavelength absorbed (Å)	Spectral colour (colour absorbed)	Complimentary colour (colour observed)
4100	violet	lemon yellow
4300	indigo	yellow
4800	blue	orange
5000	blue-green	red
5300	green	purple
5600	lemon yellow	violet
5800	yellow	indigo
6100	orange	blue
6800	red	blue-green

In absence of ligand, crystal field splitting does not occur and therefore, the metal complexes appear colourless.

e.g. anhydrous copper sulphate is white while copper sulphate pentanhydrates (i.e. $CuSO_4 \cdot 5H_2O$) is blue coloured solution.

Limitations of CFT

Limitations of CFT are as follows

(*i*) It is assumed that ligands are point charges and anionic ligands should exert the greatest splitting effect. But, anionic ligands are placed at the low end of spectrochemical series.

(*ii*) It could not explain the covalent character of bonding between the ligand and central metal ion. Short comings of crystal field theory was explained by **ligand field theory** (LFT). It is considered as the extention of CFT. LFT is a qualitative approach used for octahedral metal complexes.

John-Teller effect

When there is geometric distortion in unsymmetrical octahedral complexes with high spin, then it is called John-Teller effect. There is no geometrical distortion if high spin complex is symmetrical.

	t_{2g} (lower energy level)	e_g (Higher energy level)	
d^4	1 1 1	1	unsymmetrical
d^5	1 1 1	1 1	symmetrical
d^6	1l 1 1	1 1	unsymmetrical
d^7	1l 1l 1	1 1	unsymmetrical
d^8	1l 1l 1l	1 1	symmetrical
d^9	1l 1l 1l	1l 1	unsymmetrical
d^{10}	1l 1l 1l	1l 1l	symmetrical

Thus, this effect is observed in d^4, d^6, d^7 and d^9 cases. It is not observed in d^5, d^8 and d^{10} cases.

CHECK POINT 12.4

1. Increasing crystal field strength of the different ligands is
(a) $I^- < OH^- < en < CN^-$ (b) $CN^- < en < OH^- < I^-$
(c) $en < CN^- < OH^- < I$ (d) $en < OH^- < CN^- < I^-$

2. Which one of the following complexes is expected to have lowest Δ_o value?
(a) $[Co(NH_3)_6]^{3+}$ (b) $[CoF_6]^{3-}$
(c) $[Rh(NH_3)_6]^{3+}$ (d) $[Ir(NH_3)_6]^{3+}$

3. In spectrochemical series, chloride is above than water, i.e. $Cl > H_2O$, this is due to
(a) good π-acceptor properties of Cl
(b) strong σ-donor and good π-acceptor properties of Cl
(c) good π-donor properties of Cl
(d) larger size of Cl than H_2O

4. The crystal field splitting energy for octahedral (Δ_o) and tetrahedral (Δ_t) complexes is related as
(a) $\Delta_t = -\frac{4}{9}\Delta_o$ (b) $\Delta_t = \frac{1}{2}\Delta_o$ (c) $\Delta_o = -2\Delta_t$ (d) $\Delta_o = -\frac{4}{9}\Delta_t$

5. Among the following ions which has the highest paramagnetism?
(a) $[Cr(H_2O)_6]^{3+}$ (b) $[Fe(H_2O)_6]^{2+}$
(c) $[Cu(H_2O)_6]^{2+}$ (d) $[Zn(H_2O)_6]^{2+}$

6. $[Fe(H_2O)_6]^{2+}$ and $[Fe(CN)_6]^{4-}$ differ in
(a) geometry, magnetic moment
(b) magnetic moment and colour
(c) geometry and hybridisation
(d) None of the above

7. Magnetic moment (in BM) of transition metal complexes can be determined by which of the following relation?
(a) $\mu = \sqrt{N(N+2)}$ (b) $\mu = \sqrt{N(N+1)}$
(c) $\mu = \sqrt{2N}$ (d) $\mu = N^2$

8. The coloured species is
(a) VCl_3 (b) $VOSO_4$
(c) Na_3VO_4 (d) $[V(H_2O)_6]SO_4 \cdot H_2O$

9. Choose the correct statement.
(a) Only few transition metal complexes are coloured
(b) *d*-orbital are degenerated hence, they form complexes
(c) Transition metal complexes reflect the complimentary colour of absorbed colour
(d) Energy difference between t_{2g} and e_g level is very large

10. Complimentary colour of absorbed yellow colour is
(a) violet (b) indigo (c) blue (d) red

STABILITY AND IMPORTANCE OF COORDINATION COMPOUNDS

Stability of coordination complexes

The stability of a complex in a solution is the degree of association between the two species involved in the state of equilibrium.

A coordination complex is formed by stepwise addition of ligands. To determine the stability constant, consider the stepwise formation of the complex ML_4 from the metal ion M^{4+} (Lewis acid) and ligand L^-(Lewis base).

Let k_1, k_2, k_3 and k_4 be stepwise formation constants and $\beta_4 (= k_f)$ be the overall formation constant. Thus,

$$M^{4+} + L^- \rightleftharpoons ML^{3+}, \quad k_1 = \frac{[ML^{3+}]}{[M^{4+}][L^-]}$$

$$ML^{3+} + L^- \rightleftharpoons ML_2^{2+}, \quad k_2 = \frac{[ML_2^{2+}]}{[ML^{3+}][L^-]}$$

$$ML_2^{2+} + L^- \rightleftharpoons ML_3^+, \quad k_3 = \frac{[ML_3^+]}{[ML_2^{2+}][L^-]}$$

$$ML_3^+ + L^- \rightleftharpoons ML_4, \quad k_4 = \frac{[ML_4]}{[ML_3^+][L^-]}$$

$$\therefore \quad M^{4+} + 4L^- \rightleftharpoons ML_4, \quad \beta_4 = \frac{[ML_4]}{[M^{4+}][L^-]^4}$$

$$\therefore \quad \beta_4 = k_1 k_2 k_3 k_4$$

$$\therefore \quad \log \beta_4 = \log k_1 + \log k_2 + \log k_3 + \log k_4$$

- Higher the overall stability constant, higher is the fraction of ML_4 in solution.
- Successive values of stability constants decreases in the order.

$$k_1 > k_2 > k_3 > k_4$$

- Reciprocal of formation constant is known as **instability constant.**

Factors affecting stability

Important factors that affect stability are as follows

(*i*) **Charge on metal ion** Greater the charge on metal ion, greater is the stability.
e.g. $[Fe^{III}(CN)_6]^{3-} > [Fe^{II}(CN)_6]^{4-}$

(*ii*) **Electronegativity of central metal ion** Higher the electronegativity of central metal ion, higher is the stability of complex.
e.g. $Mn^{2+} < Fe^{2+} < Co^{2+} < Ni^{2+} < Cu^{2+} < Zn^{2+}$

(*iii*) **Basic strength of ligand** Greater is the basic strength of ligand, more easily it can donate its lone pairs of electron and hence, greater is the stability of complex.
e.g. amine (NH_3) complexes are more stable than halide complexes because NH_3 is a strong Lewis base.

(*iv*) **Chelate effect** Stability of a complex depends on the formation of chelate rings. Chelating ligands form more stable complex than monodentate ligands. This effect, called chelation effect is maximum for five and six membered rings.

(*v*) **Monocyclic effect** If a multidentate ligand happens to be cyclic and there are no unfavourable steric effects, a further increase in stability occurs. This is termed as **macrocyclic effect.**

Types of stability

Stability of a complex is divided into following two types

- **Thermodynamic stability**
- **Kinetic stability**

Thermodynamic stability can be explained on the basis of following replacement reactions.

$$[M(H_2O)_6]^{2+} + 6NH_3 \rightleftharpoons [M(NH_3)_6]^{2+} + 6H_2O$$

equilibrium constant $= k_1$

$$[M(H_2O)_6]^{2+} + 3en \rightleftharpoons [M(en)_3]^{2+} + 6H_2O$$

equilibrium constant $= k_2$

Note NH_3 *is unidentate ligand and en is bidentate ligand.*

In case of $[M(en)_3]^{2+}$, chelate effect is observed, thus $[M(en)_3]^{2+}$ is more stable than $[M(H_2O)_6]^{2+}$. Hence $k_2 > k_1$ and ΔG_2° is more negative than ΔG_1°

$$(\Delta G = -2.303RT \log k)$$

Since $\Delta G^\circ = \Delta H^\circ - T\Delta S^\circ$

$$\Delta S^\circ = \frac{\Delta H^\circ - \Delta G^\circ}{T}$$

Since, $\Delta G^\circ < 0$, $\Delta S^\circ > 0$. Thus in the formation of $[M(en)_3]^{2+}$ entropy change is more than $[M(NH_3)_6]^{2+}$.

On the basis of kinetic stability, complexes are classified as **labile complexes** and **inert complexes.**

Complex ions in which ligands can be interchanged rapidly are called labile complexes.
e.g. $[Cu(H_2O)_4]^{2+}$

When $NH_3(aq)$ is added to $[Cu(H_2O)_4]^{2+}$, a colour change from pale blue to deep blue is observed. This happens because NH_3 molecules have displaced H_2O molecules as ligands.

$$\underset{\text{Pale blue}}{[Cu(H_2O)_4]^{2+}} + 4NH_3 \longrightarrow \underset{\text{Deep blue}}{[Cu(NH_3)_4]^{2+}} + 4H_2O$$

A complex ion that exchanges ligands slowly are called inert or non-labile complex.

e.g. when freshly prepared $CrCl_3(aq)$ is added to $[Cr(H_2O)_4Cl_2]^+$, it produces a green colour which gradually turns to violet. This colour change results from very slow exchange of H_2O molecules by Cl^- as ligands.

Organometallic compounds

Organometallic compounds are those compounds which contain metal-carbon bond. The first organometallic compound of *d*-block elements was prepared by W.C. Zeise.

Some examples of organometallic compounds are $[Pt(C_2H_4)Cl_3]^-$, $Ni(CO)_4$, $[Fe(C_5H_5)_2]$ (also known as ferrocene).

Metal carbonyls

Many transition metals form metal carbonyl compounds. In these compounds CO act as donor ligand and donates lone pairs of electrons of carbonyl carbon into vacant orbital of metal forming σ-bond.

It also behaves as an acceptor ligands and forms π-bond by back donation of electrons from filled *d*-orbital of metal to vacant antibonding (π^*)-orbital of CO.

This backbonding stabilises the metal-ligands interaction. It is also called **synergic bonding.**

Importance of coordination compounds

Coordination compounds have great importance and widely used in various fields. Some important applications of transition metal complexes are given below

In medicinal chemistry

(*i*) The platinum complex *cis*-$[Pt(NH_3)_2Cl_2]$ commonly known as ***cis*-platin** is used in the treatment of cancer. It has ability to block the uncontrolled division of cancerous cells that results in growth of tumours.

(*ii*) EDTA complex of calcium, $[Ca(EDTA)]^{2-}$ is used to treat a person suffering from lead poisoning.

$$Pb^{2+} + [Ca(EDTA)]^{2-} \longrightarrow [Pb(EDTA)]^{2-} + Ca^{2+}$$

EDTA complex of lead is more stable than $[Ca(EDTA)]^{2-}$ complex. Thus, lead complex is excreted from the body and Ca^{2+} remains as the body nutrient.

In quantitative and qualitative analysis

(*i*) Traces of Ni^{2+} salt can be detected by dimethyl glyoxime when cherry red precipitate of nickel-dimethyl glyoximate (chelate) is formed.

$$Ni^{2+} + 2\ \begin{matrix} HO\!-\!N\!=\!C\!-\!CH_3 \\ | \\ HO\!-\!N\!=\!C\!-\!CH_3 \end{matrix} \xrightarrow{NH_4OH/\ Alcohol}$$

Dimethyl glyoxime

Nickel-dimethyl glyoximate (chelate) (red ppt.)

(*ii*) In qualitative analysis of group I, there is formation of white precipitate of AgCl, Hg_2Cl_2, and $PbCl_2$. $PbCl_2$ is soluble in hot water. AgCl can be removed as soluble $[Ag(NH_3)_2]Cl$ on adding $NH_3(aq)$.

$$AgCl(s) + 2NH_3(aq) \rightleftharpoons \underset{\text{Soluble}}{Ag[(NH_3)_2]Cl}$$

(*iii*) In group II, Cd^{2+} can be precipitated as CdS in presence of Cu^{2+}. By treating the solution of Cu^{2+} and Cd^{2+} with an excess of CN^-, Cu^{2+} forms a stable complex in a redox reaction.

$$2Cu^{2+} + 10CN^- \longrightarrow \underset{\text{Stable}}{2[Cu(CN)_4]^{3-}} + C_2N_2(g)$$

while Cd^{2+} forms less stable complex

$$Cd^{2+} + 4CN^- \rightleftharpoons \underset{\text{Unstable}}{[Cd(CN)_4]^{2-}}$$

When H_2S gas is passed, $[Cd(CN)_4]^{2-}$ is precipitated as CdS without affecting $[Cu(CN)_4]^{3-}$ complex.

In extraction of metals

(*i*) Nickel can be extracted by Mond's process which involves formation of carbonyl complex of nickel.

$$Ni(s) + 4CO(g) \longrightarrow Ni(CO)_4(g)$$

(*ii*) Silver (Ag) and gold (Au) can be extracted by formation of cyanide complex.

$$4Au(s) + 8CN^-(aq) + O_2(g) + 2H_2O(l) \longrightarrow 4OH^-(aq) + 4[Au(CN)_2]^-(aq)$$

$$2[Au(CN)_2]^-(aq) + Zn(s) \longrightarrow [Zn(CN)_4]^{2-}(aq) + 2Au(s)$$

In purification of water

Purification of water can be done through complex formation. It involves chelation with sodium salt of EDTA. The formation constant of $[Ca(EDTA)]^{2-}$ and $[Mg(EDTA)]^{2-}$ are very large.

Hence, on treating water with sodium salt of EDTA, concentration of Ca^{2+} and Mg^{2+} can be reduced to the point that these ions will not precipitate with soap or detergent.

$$Ca^{2+} + (EDTA)^{4-} \rightleftharpoons [Ca(EDTA)]^{2-}$$

$$Mg^{2+} + (EDTA)^{4-} \rightleftharpoons [Mg(EDTA)]^{2-};\ k_f = 4 \times 10^8$$

In photography

The photographic "fixing" process removes unreacted AgBr which forms soluble complex with sodium thiosulphate.

$$AgBr(s) + 2S_2O_3^{2-} \longrightarrow \underset{\text{Soluble}}{[Ag(S_2O_3)_2]^{3-}} + Br^-$$

In biological processes

Many important biological substances are coordination compounds, e.g.

(*i*) Haemoglobin is a protein that carries O_2 in blood. It contains iron (II) ions bound to large porphyrin rings.

(*ii*) Chlorophyll is necessary for photosynthesis in plants. It contains magnesium ions bound to porphyrin rings.

(*iii*) Vitamin B_{12} is a large complex of cobalt.

CHECK POINT 12.5

1. The overall stability constant (β) for formation of ML_4 complex is represented as
(a) $\frac{[ML^{3+}]}{[M^{4+}][L^-]}$ (b) $\frac{[ML_4]}{[M^{4+}][L^-]}$
(c) $\frac{[ML_4]}{[ML_3^+][L^-]}$ (d) $\frac{[ML_4]}{[M^{4+}][L^-]^4}$

2. The closed ring compounds formed by bidentate ligands, on binding to a metal or metal ions, are called
(a) ligands (b) chelates
(c) complexes (d) ambident

3. Which can form chelates?
(a) Ethylenediamine (b) Oxalate
(c) Glycinate (d) All of these

4. Among the following, which factor affects the stability of complex?
(a) Chelate affect
(b) Charge on metal ion
(c) Basic strength of ligand
(d) All of the above

5. Which of the following is the most stable complex?
(a) $[Co(NH_3)_6]^{3+}$ (b) $[Co(NH_3)(en)_2]^{3+}$
(c) $[Co(NH_3)_4(en)]^{3+}$ (d) $[Co(en)_3]^{3+}$

6. Which of the following is an organometallic compound?
(a) Lithium methoxide
(b) Lithium acetate
(c) Lithium dimethylamide
(d) Methyl lithium

7. Which of the following act as a donor as well as acceptor ligand?
(a) H_2O (b) NH_3
(c) CO (d) NO_2

8. Vitamin B_{12} is the coordination comopound of
(a) Mg (b) Fe
(c) Co (d) Zn

9. The complex used as an anticancer agent is
(a) *mer*- $[Co(NH_3)Cl_3]$ (b) *cis* - $[Pt\,Cl_2(NH_3)_2]$
(c) *cis* - $K_2\,[Pt\,Cl_2Br_2]$ (d) $Na_2[COCl_4]$

10. Which of the following statements is incorrect?
(a) Haemoglobin contains an iron complex called *heme*
(b) Chlorophyll in plants contains porphyrin ring with metal Mg^{2+}
(c) The photographic "fixing" process removes unreacted AgBr which forms soluble complex $[Ag(S_2O_3)_2]^{3-}$
(d) Complexes of Ca^{2+} and Al^{3+} is used for the purification of water

Chapter Exercises

(A) Topical exercise

I. Introduction and important terms related to coordination compounds

1 Which of the following is an example of coordination compound?

(a) $CoCl_3 \cdot 6NH_3$ (b) $KCl \cdot MgCl_2 \cdot 6H_2O$
(c) $FeSO_4 \cdot 6H_2O$ (d) $FeSO_4 \cdot (NH_4)_2SO_4 \cdot 6H_2O$

2 $CuSO_4 \cdot 5H_2O$ is like a complex represented as

(a) $[Cu(H_2O)_5] \cdot SO_4$ (b) $[Cu(H_2O)_4]SO_4 \cdot H_2O$
(c) $[Cu(H_2O)_3]SO_4 \cdot 2H_2O$ (d) $[Cu(H_2O)_2]SO_4 \cdot 3H_2O$

3 Which of the following species is not expected to be a ligand? **[NCERT Exemplar]**

(a) NO (b) NH_4^+
(c) $NH_2CH_2CH_2NH_2$ (d) CO

4 Consider the following statements.

Statement I NO_2^- is an ambident ligand.

Statement II NO_2^- is a Lewis base.

(a) Both statements I and II are correct.
(b) Only statement I is correct.
(c) Only statement II is correct.
(d) Both statements I and II are incorrect.

5 A chelating agent has two or more than two donor atoms to bind to a single metal ion. Which of the following is not a chelating agent? **[NCERT Exemplar]**

(a) Thiosulphato (b) Oxalato
(c) Glycinato (d) Ethan-1, 2-diamine

6 Which of the following are chelating ligands?

I. PO_4^{3-} II. (cyclic) $(O—CH_2CH_2)_5$
III. $(CH_3)_3N$ IV. $N(CH_2—CO_2^-)_3$
V. $[NH_3CH_2CH_2NH_3]^{2+}$

(a) Both I and II (b) Both II and IV
(c) II, IV and V (d) I, III and V

7 Consider the following statements.

Statement I A stable complex is formed when ethylenediamine(en) is added to $[Ni(H_2O)_4]^{2+}$.

Statement II (en) being a strong ligand replaces weak ligand H_2O and thus, chelates is formed.

(a) Both statements I and II are correct.
(b) Only statement I is correct.
(c) Only statement II is correct.
(d) Both statements I and II are incorrect

8 The coordination number of a central metal atom in a complex is determined by

(a) the number of ligands around a metal ion bonded by coordinate bonds
(b) the number of ligands around a metal ion bonded by pi-bonds
(c) the number of ligands around a metal ion bonded by sigma and pi-bonds both
(d) the number of only anionic ligands bonded to metal ion

9 The coordination number of Pt in the complex ion $[Pt(en)_2Cl_2]^{2+}$ is

(a) 3 (b) 4 (c) 5 (d) 6

10 In which of the following compounds transition metal has zero oxidation state?

(a) $[Fe(CO)_5]$ (b) $[Ni(CN)_4]^{2-}$
(c) Fe_2O_3 (d) CrO_5

11 In the complex $[Pt(O_2)(en)_2(Br)]^{2+}$, coordination number and oxidation number of platinum are

(a) 4, 3 (b) 4, 5 (c) 4, 6 (d) 6, 4

12 The pair of compounds having metals in their highest oxidation state is

(a) MnO_2, $FeCl_3$
(b) $[MnO_4]^-$, CrO_2Cl_2
(c) $[Fe(CN)_6]^{3-}$, $[Co(CN)_3]$
(d) $[NiCl_4]^{2-}$, $[CoCl_4]^-$

13 Match the compounds given in Column I with the oxidation state of cobalt present in it (given in Column II) and assign the correct code. **[NCERT Exemplar]**

	Column I (Compound)		Column II (Oxidation state of Co)
A.	$[Co(NCS)(NH_3)_5](SO_3)$	1.	+4
B.	$[Co(NH_3)_4Cl_2]SO_4$	2.	0
C.	$Na_4[Co(S_2O_3)_3]$	3.	+2
D.	$[Co_2(CO)_8]$	4.	+3

Codes

	A	B	C	D
(a)	1	2	3	4
(b)	4	1	3	2
(c)	3	4	2	1
(d)	1	2	4	3

14 In the complex $[CoCl_3 \cdot 4NH_3]$
(a) coordination entity is $[Co(NH_3)_4Cl_2]^+$
(b) counter ion is Cl^-
(c) Both (a) and (b) are correct
(d) None of the above is correct

15 $CrCl_3 \cdot 4H_2O$ will exist as
(a) $[Cr(H_2O)_3Cl_3] \cdot H_2O$ (b) $[Cr(H_2O)_4Cl_2]Cl$
(c) Both (a) and (b) (d) None of these

16 In which case EAN of Cr is maximum?
(a) $[Cr(CN)_6]^{3-}$ (b) $[Cr(H_2O)_4(NH_3)_2]^{3+}$
(c) $[Cr(H_2O)_4(C_2O_4)]^+$ (d) Equal in all cases

17 EAN of cobalt is 36 in $[Co(NH_3)_2O_2(en)Cl]$. Thus, O_2 is
(a) dioxide (b) superoxide ion
(c) peroxide ion (d) oxide

18 In $[Mg(EDTA)]^{2-}$, EAN of Mg is
(a) 16 (b) 20
(c) 22 (d) 18

19 The EAN of Zn in $Zn(OH)_4^{2-}$ complex is
(a) 16 (b) 26
(c) 36 (d) 46

20 EAN of a metal carbonyl, $M(CO)_x$ is 36. The atomic number of metal is 26. The value of x is
(a) 4 (b) 5
(c) 6 (d) 7

21 Which of the following ion does not follow EAN rule?
(a) $[Fe(CN)_6]^{4-}$ (b) $[Ag(CN)_2]^-$
(c) $[PtCl_6]^{2-}$ (d) None of these

22 Which has maximum EAN of the underlined atoms? (Cr = 24 , Co = 27 , Fe = 26 , Ni = 28)
(a) $[\underline{Cr}(EDTA)]^-$ (b) $[\underline{Co}(en)_3]^{3+}$
(c) $[\underline{Fe}(C_2O_4)_3]^{3-}$ (d) $[\underline{Ni}(CN)_4]^{2-}$

23 Increasing order of EAN of the metals in $[Ni(CN)_4]^{2-}$, $[Fe(CN)_6]^{3-}$ and $[Cu(CN)_4]^{3-}$ is
(a) $[Ni(CN)_4]^{2-} < [Fe(CN)_6]^{3-} < [Cu(CN)_4]^{3-}$
(b) $[Ni(CN)_4]^{2-} < [Fe(CN)_6]^{3-} = [Cu(CN)_4]^{3-}$
(c) $[Ni(CN)_4]^{2-} < [Cu(CN)_4]^{3-} < [Fe(CN)_6]^{3-}$
(d) $[Cu(CN)_4]^{3-} < [Fe(CN)_6]^{3-} < [Ni(CN)_4]^{2-}$

II. Nomenclature of coordination compounds

24 IUPAC name of $K_4[Fe(CN)_6]$ is
(a) potassium hexacyanoiron (II)
(b) potassium iron (II) hexacyano
(c) tetrapotassium iron (II) hexacyano
(d) potassium hexacyanoferrate (II)

25 $[Fe^{II}(O_2)(CN)_4Cl]^{4-}$ is named as
(a) chlorotetracyanodioxoferrate (II) ion
(b) chlorotetracyanoperoxoferrate (II) ion
(c) chlorotetracyanosuperoxoferrate (II) ion
(d) None of the above is correct

26 Pick the correct name for $[Co(NH_3)_5Cl]Cl_2$
(a) pentaamminechloridocobalt (II) chloride
(b) pentaamminechloridocobalt (III) chloride
(c) pentaamminechloridocobalt (III)
(d) pentaammine cobalt(III) chloride

27 IUPAC name of $[Pt(Br)(NO_2)Cl(NH_3)_3]Cl$ is
(a) triamminechlorobromonitroplatinum (IV) chloride
(b) triamminebromonitrochloroplatinum (IV) chloride
(c) triamminebromochloronitroplatinum (IV) chloride
(d) triamminenitrochlorobromoplatinum (IV) chloride

28 Potassium trisoxalatochromate (III) is
(a) $K_3[Cr(C_2O_4)_3]$ (b) $K[Cr(C_2O_4)_3]$
(c) $K_2[Cr(C_2O_4)_3]$ (d) $K_4[Cr(C_2O_4)_3]$

29 Hexammineplatinum (IV) tetrachloroplatinate (II) is
(a) $[Pt(NH_3)_6][PtCl_4]$ (b) $[Pt(NH_3)_6]_2[PtCl_4]$
(c) $[Pt(NH_3)_6][PtCl_4]_2$ (d) $[Pt(NH_3)_6]_3[PtCl_4]_2$

III. Isomerism in coordination compounds

30 In which case geometrical isomers is possible with M as metal ion?
(a) MX_2Y_2 (b) MX_2Y_4 (c) $MX_2Y_2Z_2$ (d) In all cases

31 Indicate the complex ion which shows geometrical isomerism. **[NCERT Exemplar]**
(a) $[Cr(H_2O)_4Cl_2]^+$ (b) $[Pt(NH_3)_3Cl]$
(c) $[Co(NH_3)_6]^{3+}$ (d) $[Co(CN)_5(NC)]^{3-}$

32 Which of the following does not have optical isomer?
(a) $[Co(NH_3)_3Cl_3]$ (b) $[Co(en)_3]^{3+}$
(c) $[Co(en)_2Cl_2]Cl$ (d) $[Co(en)(NH_3)_2]Cl$

33 Of the following complex ions one exhibits optical isomerism. That one is
(a) *cis*- $[Co(en)_2Cl_2]^-$ (b) $[Co(NH_3)_4Cl_2]^+$
(c) $[Co(NH_3)_2Cl_4]^-$ (d) *trans*-$[Co(en)_2Cl_2]^+$

34 Which one of the following octahedral complexes will not show geometric isomerism? (A and B are monodentate ligands)
(a) $[MA_4B_2]$ (b) $[MA_5B]$ (c) $[MA_2B_4]$ (d) $[MA_3B_3]$

35 Which of the following coordination compounds would exhibit optical isomerism?
(a) Pentaammine nitro cobalt(III) iodide
(b) Diammine dichloroplatinum(II)
(c) *Trans*-dicyano *bis* (ethylenediamine) chromium(III) chloride
(d) *Tris*-(ethylenediamine) cobalt(III) bromide

36 Statement I Complexes of MX_6 and MX_5L type (X and L are unidentate) do not show geometrical isomerism.

Statement II Geometrical isomerism is not shown by complexes of coordination number 6. **[NCERT Exemplar]**

(a) Both statements I and II are correct.
(b) Only statement I is correct.
(c) Only statement II is correct.
(d) Both statements I and II are incorrect

37 $[Pt(Ox)(py)_2(O_2)(H_2O)]$ select correct statement about this complex

(a) Oxidation state of O_2 is -1
(b) EAN of Pt is 86
(c) Complex will show geometrical as well as optical isomerism
(d) Both (b) and (c)

38 What kind of isomerism exists between $[Cr(H_2O)_6]Cl_3$ (violet) and $[Cr(H_2O)_5Cl]Cl_2 \cdot H_2O$ (greyish-green)? **[NCERT Exemplar]**

(a) Linkage isomerism (b) Solvate isomerism
(c) Ionisation isomerism (d) Coordination isomerism

39 Which kind of isomerism is exhibited by octahedral $[Co(NH_3)_4Br_2]Cl$?

(a) Geometrical and ionisation
(b) Geometrical and optical
(c) Optical and ionisation
(d) Geometrical only

40 The compounds $[Co(SO_4)(NH_3)_5]Br$ and $[Co(SO_4)(NH_3)_5]Cl$ represent **[NCERT Exemplar]**

(a) linkage isomerism (b) ionisation isomerism
(c) coordination isomerism (d) no isomerism

41 The number of geometrical isomers for $[Pt(NH_3)_2Cl_2]$ is

(a) 1 (b) 2 (c) 3 (d) 4

42 Due to the presence of ambidentate ligands coordination compounds show isomerism. Palladium complexes of the type $[Pd(C_6H_5)_2(SCN)_2]$ and $[Pd(C_6H_5)_2(NCS)_2]$ are **[NCERT Exemplar]**

(a) linkage isomers (b) coordination isomers
(c) ionisation isomers (d) geometrical isomers

43 Which one of the following complexes will have four different isomers?

(a) $[Co(en)_2Cl_2]Cl$ (b) $[Co(en)(NH_3)_2Cl_2]Cl$
(c) $[Co(PPh_3)_2(NH_3)Cl_2]Cl$ (d) $[Co(en)_3]Cl_3$

44 Consider the following statements.

Statement I Linkage isomerism arises in coordination compounds containing ambidentate ligand.

Statement II Ambidentate ligand has two different donor atoms. **[NCERT Exemplar]**

(a) Both statements I and II are correct.
(b) Only statement I is correct.
(c) Only statement Ii is correct.
(d) Both statements I and II are incorrect.

45 Match the complex species given in Column I with the possible isomerism given in Column II and assign the correct code. **[NCERT Exemplar]**

	Column I (Complex species)		Column II (Isomerism)
A.	$[Co(NH_3)_4Cl_2]^+$	1.	Optical
B.	*cis* - $[Co(en)_2Cl_2]^+$	2.	Ionisation
C.	$[Co(NH_3)_5(NO_2)]Cl_2$	3.	Coordination
D.	$[Co(NH_3)_6][Cr(CN)_6]$	4.	Geometrical

Codes

	A	B	C	D
(a)	1	2	3	4
(b)	4	1	2	3
(c)	4	3	2	1
(d)	2	3	1	4

IV. Werners theory and valence bond theory

46 The complex

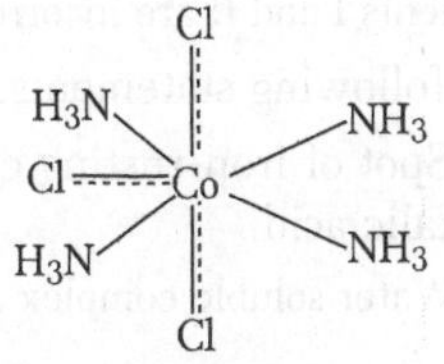

(a) is the molecular complex
(b) has four ions in the aqueous solution
(c) primary valency of cobalt is six
(d) All of the above are correct

47 For the complex $PtCl_4 \cdot 4NH_3$

(a) two Cl^- ligands satisfy primary as well secondary valencies
(b) four NH_3 ligands, satisfy secondary valency
(c) ions are $[Pt(NH_3)_4Cl_2]^{2+}$ and $2Cl^-$
(d) All of the above statements are correct

48 Number of ions present in $K_3[Fe(CN)_6]$ are

(a) 2 (b) 4
(c) 6 (d) 8

49 Which complex gives three chloride ions per formula unit?

(a) $CrCl_3 \cdot 6H_2O$ (b) $CrCl_3 \cdot 5H_2O$
(c) $CrCl_3 \cdot 4H_2O$ (d) All of these

50 Which of the following will exhibit maximum ionic conductivity?

(a) $K_4[Fe(CN)_6]$ (b) $[Co(NH_3)_6]Cl_3$
(c) $[Cu(NH_3)_4]Cl_2$ (d) $[Ni(CO)_4]$

51 When 1 mole of $CrCl_3 \cdot 6H_2O$ is treated with excess of $AgNO_3$, 3 moles of AgCl are obtained. The formula of the complex is **[NCERT Exemplar]**
(a) $[CrCl_3(H_2O)_3] \cdot 3H_2O$
(b) $[CrCl_2(H_2O)_4]Cl \cdot 2H_2O$
(c) $[CrCl(H_2O)_5]Cl_2 \cdot H_2O$
(d) $[Cr(H_2O)_6]Cl_3$

52 One mole of the complex $CoCl_3 \cdot 6H_2O$ on reaction with excess of $AgNO_3$ gives two moles of white precipitate. Thus, complex is
(a) $[Co(H_2O)_6]Cl_3$
(b) $[Co(H_2O)_5Cl]Cl_2 \cdot H_2O$
(c) $[Co(H_2O)_4Cl_2]Cl \cdot 2H_2O$
(d) $[Co(H_2O)_3Cl_3] \cdot 3H_2O$

53 Consider the following statements.
Statement I $CoCl_3 \cdot 4NH_3$ gives two ions when ionised.
Statement II Complex formed by $CoCl_3 \cdot 4NH_3$ is $[Co(NH_3)_4]Cl_3$.
(a) Both statements I and II are correct.
(b) Only statement I is correct.
(c) Only statement II is correct.
(d) Both statements I and II are incorrect

54 Consider the following statements.
Statement I Spot of iron-rusting can be removed by addition of oxalic acid.
Statement II Water soluble complex is formed with oxalic acid.
(a) Both statements I and II are correct.
(b) Only statement I is correct.
(c) Only statement II is correct.
(d) Both statements I and II are incorrect

55 Match the complex (in Column I) with the equivalent conductance (in Column II).

Column I	Column II
A. $[Pt(NH_3)_5Cl]Cl_3$	1. 229
B. $[Pt(NH_3)_4Cl_2]Cl_2$	2. 97
C. $[Pt(NH_3)_3Cl_3]Cl$	3. 404
D. $[Pt(NH_3)_6]Cl_4$	4. 523

Codes

	A	B	C	D
(a)	4	3	2	1
(b)	1	2	3	4
(c)	3	1	2	4
(d)	1	3	4	2

56 Match the complex (in Column I) with the hybridisation of the central ion (in Column II).

Column I	Column II
A. $Ni(CO)_4$	1. sp^3d^2
B. $[Ni(CN)_4]^{2-}$	2. sp^3
C. $[Fe(CN)_6]^{4-}$	3. d^2sp^3
D. $[MnF_6]^{4-}$	4. dsp^2

Codes

	A	B	C	D
(a)	1	2	3	4
(b)	4	2	1	3
(c)	2	4	3	1
(d)	3	4	2	1

57 Consider the following statements.
Statement I The electronic absorption spectrum of the tetrahedral complex, $[CoCl_4]^{2-}$ shows bands which are more intense compared to the absorption bands for the octahedral complex $[Co(H_2O)_6]^{2+}$.
Statement II The octahedral complexes possess a centre of symmetry.
(a) Both statements I and II are correct.
(b) Only statement I is correct.
(c) Only statement II is correct.
(d) Both statements I and II are incorrect

58 Consider the following statements.
Statement I $Ni(CO)_4$ is diamagnetic complex.
Statement II All the electrons in the complex are paired.
(a) Both statements I and II are correct.
(b) Only statement I is correct.
(c) Only statement II is correct.
(d) Both statements I and II are incorrect.

59 Consider the following statements.
Statement I $[Cr(H_2O)_6]Cl_2$ and $[Fe(H_2O)_6]Cl_2$ are reducing in nature.
Statement II Unpaired electrons are present in their *d*-orbitals. **[NCERT Exemplar]**
(a) Both statements I and II are correct.
(b) Only statement I is correct.
(c) Only statement II is correct.
(d) Both statements I and II are incorrect.

60 In square planar complexes, the hybrid orbitals are dsp^2. The atomic orbitals which give rise to dsp^2-hybrid orbitals are
(a) $d_{x^2-y^2}, s, p_x$ and p_y
(b) d_{z^2}, s, p_x and p_y
(c) $d_{x^2-y^2}, s, p_z$ and p_y
(d) d_{z^2}, s, p_z, p_y

61 $[Ni(CN)_4]^{2-}$ and $[Ni(CO)_4]$ have
(a) sp^3- hybridised Ni in both cases
(b) sp^3 and sp^2d-hybridised Ni
(c) dsp^2 and sp^3-hybridised Ni
(d) dsp^2 in both cases

62 Among $[Ni(CO)_4]$, $[Ni(CN)_4]^{2-}$, $[NiCl_4^{2-}]$ species, the hybridisation states at the Ni atom are, respectively (Atomic number of Ni = 28)
(a) sp^3, dsp^2, dsp^2 (b) sp^3, dsp^2, sp^3
(c) sp^3, sp^3, dsp^2 (d) dsp^2, sp^3, sp^3

63 Which of the following has square planar structure?
(a) $[Ni(CN)_4]^{2-}$ (b) $[Ni(CO)_4]$
(c) $[NiCl_4]^{2-}$ (d) All of these

64 Consider the following statements.
Statement I $[Fe(CN)_6]^{3-}$ ion shows magnetic moment corresponding to two unpaired electrons.
Statement II Because it has d^2sp^3 type hybridisation.
[NCERT Exemplar]
(a) Both statements I and II are correct.
(b) Only statement I is correct.
(c) Only statement II is correct.
(d) Both statements I and II are incorrect.

65 Match the complex ions given in Column I with the hybridisation and number of unpaired electrons given in Column II and assign the correct code. **[NCERT Exemplar]**

	Column I (Complex ion)		Column II (Hybridisation, number of unpaired electrons)
A.	$[Cr(H_2O)_6]^{3+}$	1.	dsp^2, 1
B.	$[Co(CN)_4]^{2-}$	2.	sp^3d^2, 5
C.	$[Ni(NH_3)_6]^{2+}$	3.	d^2sp^3, 3
D.	$[MnF_6]^{4-}$	4.	sp^3d^2, 2

Codes

	A	B	C	D
(a)	3	1	4	2
(b)	1	4	3	2
(c)	4	3	1	2
(d)	1	2	3	4

V. Crystal field theory (CFT)

66 The crystal field-splitting for Cr^{3+} ion in octahedral field increases for ligands I^-, H_2O, NH_3, CN^- and the order is
(a) $I^- < H_2O < NH_3 < CN^-$ (b) $CN^- < I^- < H_2O < NH_3$
(c) $CN^- < NH_3 < H_2O < I^-$ (d) $NH_3 < H_2O < I^- < CN^-$

67 What is the CFSE of a free Co (II) ion on forming the tetrahedral chloro complex $[CoCl_4]^{2-}$?
(in the units of Δ_o)
(a) 0.6 (b) 1.2 (c) 1.8 (d) 2.4

68 If Δ_o and Δ_t represent crystal field splitting energies for d-orbitals for octahedral and tetrahedral geometries respectively, therefore d^6 (high spin in both the cases). Which are the CFSE (ignoring the pairing energies) respectively?
(a) $0.6\Delta_o, 0.6\Delta_t$ (b) $0.4\Delta_o, 0.4\Delta_t$
(c) $0.4\Delta_o, 0.6\Delta_t$ (d) $0.6\Delta_o, 0.4\Delta_t$

69 The CFSE for octahedral $[CoCl_6]^{4-}$ is 18000 cm^{-1}. The CFSE for tetrahedral $[CoCl_4]^{2-}$ will be **[NCERT Exemplar]**
(a) 18000 cm^{-1} (b) 16000 cm^{-1}
(c) 8000 cm^{-1} (d) 20000 cm^{-1}

70 Which one of the following complexes is expected to have lowest Δ_o value?
(a) $[Co(NH_3)_6]^{3+}$ (b) $[CoF_6]^{3-}$
(c) $[Rh(NH_3)_5]^{3+}$ (d) $[Ir(NH_3)_6]^{3+}$

71 Match the complex (in Column I) with the geometry (in Column II).

	Column I		Column II
A.	$[Ni(CN)_4]^{2-}$	1.	Tetrahedral
B.	$[ZnCl_4]^{2-}$	2.	Trigonal bipyramidal
C.	$[Fe(CO)_5]$	3.	Square planar

Codes

	A	B	C		A	B	C
(a)	1	2	3	(b)	3	1	2
(c)	2	1	3	(d)	3	2	1

72 Match the complex ion (in Column I) with number of unpaired electrons (in Column II).

	Column I		Column II
A.	$[Fe(H_2O)_6]^{2+}$	1.	0
B.	$[Fe(H_2O)_6]^{3+}$	2.	1
C.	$[Fe(CN)_6]^{4-}$	3.	2
D.	$[Fe(CN)_6]^{3-}$	4.	4
E.	$[Ni(H_2O)_4]^{2+}$	5.	5

Codes

	A	B	C	D	E
(a)	4	5	1	2	3
(b)	5	4	2	1	3
(c)	2	4	5	1	3
(d)	1	3	5	4	2

73 The number of unpaired electrons in the complex ion $[CoF_6]^{3-}$ is (Atomic number of Co= 27)

(a) Four (b) Zero (c) Two (d) Three

74 The number of unpaired electrons expected for the complex ion $[Cr(NH_3)_6]^{3+}$ is

(a) 2 (b) 3 (c) 4 (d) 5

75 Considering H_2O as a weak field ligand, the number of unpaired electrons in $[Mn(H_2O)_6]^{2+}$ will be (Atomic number of Mn= 25)

(a) three (b) five
(c) two (d) four

76 The complex ion which has no '*d*'-electrons in the central metal atom is

(a) $[MnO_4]^-$ (b) $[Co(NH_3)_6]^{3+}$
(c) $[Fe(CN)_6]^{3-}$ (d) $[Cr(H_2O)_6]^{3+}$

77 Arrange the following in order of decreasing number of unpaired electrons

I. $[Fe(H_2O)_6]^{2+}$ II. $[Fe(CN)_6]^{3-}$
III. $[Fe(CN)_6]^{4-}$ IV. $[Fe(H_2O)_6]^{3+}$

(a) IV, I, II and III (b) I, II, III and IV
(c) III, II, I and IV (d) II, III, I and IV

78 The *d*-electron configurations of Cr^{2+}, Mn^{2+}, Fe^{2+} and Ni^{2+} are $3d^4$, $3d^5$, $3d^6$ and $3d^8$ respectively. Which one of the following aqua complexes will exhibit the minimum paramagnetic behaviour?

(a) $[Mn(H_2O)_6]^{2+}$ (b) $[Fe(H_2O)_6]^{2+}$
(c) $[Ni(H_2O)_6]^{2+}$ (d) $[Cr(H_2O)_6]^{2+}$

79 Which one of the cyano complexes would exhibit the lowest value of paramagnetic behaviour? (Atomic number Cr = 24, Mn = 25, Fe = 26, Co = 27)

(a) $[Co(CN)_6]^{3-}$ (b) $[Fe(CN)_6]^{3-}$
(c) $[Mn(CN)_6]^{3-}$ (d) $[Cr(CN)_6]^{3-}$

80 Among the following ions which has the highest paramagnetism?

(a) $[Cr(H_2O)_6]^{3+}$ (b) $[Fe(H_2O)_6]^{2+}$
(c) $[Cu(H_2O)_6]^{2+}$ (d) $[Zn(H_2O)_6]^{2+}$

81 Match the complex ion (in Column I) with spin only magnetic moment (in Column II).

	Column I		Column II
A.	$[Co(NH_3)_6]^{3+}$	1.	1.73 BM
B.	$[Fe(CN)_6]^{3-}$	2.	5.92 BM
C.	$[Mn(H_2O)_6]^{2+}$	3.	2.83 BM
D.	$[Ni(H_2O)_6]^{2+}$	4.	0.0 BM

Codes

	A	B	C	D
(a)	4	1	2	3
(b)	1	2	3	4
(c)	4	3	2	1
(d)	3	2	1	4

82 Which one of the following salts (high spin for paramagnetic samples) will have the strongest attraction to a magnetic field?

(a) $MnSO_4$ (b) $CoSO_4$
(c) $CuSO_4$ (d) $ZnSO_4$

83 Magnetic moment of the complexes is zero in

(a) $[Ni(CN)_4]^{2-}$, $[Ni(CO)_4]$ (b) $[Ni(H_2O)_4]^{2+}$, $[Ni(CN)_4]^{2-}$
(c) $[Ni(H_2O)_4]^{2+}$, $[Ni(CO)_4]$ (d) $[Fe(CN)_6]^{4-}$, $[Fe(CN)_6]^{3-}$

84 What are the spin-only magnetic moments (in BM) for Ni(II) ion in square-planar and octahedral geometry, respectively?

(a) 0 and 2.83 (b) 2.83 and 2.83
(c) 0 and 1.73 (d) 0 and 0

85 Dipole moment will be zero in complexes

I. $[Ni(CN)_4]^{2-}$
II. *cis*-$Pt[(NH_3)_2Cl_2]$
III. *trans*- $[Pt(NH_3)_2Cl_2]$

(a) Both I and II (b) Both I and III
(c) Both II and III (d) I, II and III

86 $[Cr(H_2O)_6]Cl_3$ (Atomic number of Cr= 24) has a magnetic moment of 3.83 BM, the correct distribution of 3*d*-electrons in the chromium of complex is

(a) $3d^1_{x^2-y^2}, 3d^1_{z^2}, 3d^1_{xz}$ (b) $3d^1_{xy}, 3d^1_{x^2-y^2}, 3d^1_{yz}$
(c) $3d^1_{xy}, 3d^1_{zy}, 3d^1_{xz}$ (d) $3d^1_{xy}, 3d^1_{yz}, 3d^1_{z^2}$

87 The pair in which both species have same magnetic moment (spin only value) is

(a) $[Cr(H_2O)_6]^{2+}$, $[CoCl_4]^{2-}$
(b) $[Cr(H_2O)_6]^{2+}$, $[Fe(H_2O)_6]^{2+}$
(c) $[Mn(H_2O)_6]^{2+}$, $[Cr(H_2O)_6]^{2+}$
(d) $[CoCl_4]^{2-}$, $[Fe(H_2O)_6]^{2+}$

88 Identify the complexes which are expected to be coloured.

(a) $Ti(NO_3)_4$ (b) $[Cu(NCCH_3)]^+ BF_4^-$
(c) $[Cr(NH_3)_6]^{3+} 3Cl^-$ (d) $K_3[VF_6]$

89 Amongst TiF_6^{2-}, CoF_6^{3-}, Cu_2Cl_2 and $NiCl_4^{2-}$, which are the colourless species? (Atomic number of Ti = 22, Co = 27, Cu= 29, Ni = 28)

(a) Cu_2Cl_2 and $NiCl_4^{2-}$ (b) TiF_6^{2-} and CoF_6^{3-}
(c) CoF_6^{3-} and $NiCl_4^{2-}$ (d) TiF_6^{2-} and Cu_2Cl_2

90 The colour of the coordination compounds depends on the crystal field splitting. What will be the correct order of absorption of wavelength of light in the visible region, for the complexes, $[Co(NH_3)_6]^{3+}$, $[Co(CN)_6]^{3-}$, $[Co(H_2O)_6]^{3+}$ **[NCERT Exemplar]**

(a) $[Co(CN)_6]^{3-} > [Co(NH_3)_6]^{3+} > [Co(H_2O)_6]^{3+}$
(b) $[Co(NH_3)_6]^{3+} > [Co(H_2O)_6]^{3+} > [Co(CN)_6]^{3-}$
(c) $[Co(H_2O)_6]^{3+} > [Co(NH_3)_6]^{3+} > [Co(CN)_6]^{3-}$
(d) $[Co(CN)_6]^{3-} > [Co(NH_3)_6]^{3+} > [Co(H_2O)_6]^{3+}$

91 Match the complex ions given in Column I with the colours given in Column II and assign the correct code. **[NCERT Exemplar]**

Column I (Complex ion)	Column II (Colour)
A. $[Co(NH_3)_6]^{3+}$	1. Violet
B. $[Ti(H_2O)_6]^{3+}$	2. Green
C. $[Ni(H_2O)_6]^{2+}$	3. Pale blue
D. $[Ni(H_2O)_4(en)]^{2+}(aq)$	4. Yellowish orange

Codes

	A	B	C	D
(a)	1	4	2	3
(b)	3	2	1	4
(c)	1	2	3	4
(d)	4	3	2	1

92 The stabilisation of coordination compounds due to chelation is called the chelate effect. Which of the following is the most stable complex species? **[NCERT Exemplar]**

(a) $[Fe(CO)_5]$ (b) $[Fe(CN)_6]^{3-}$
(c) $[Fe(C_2O_4)_3]^{3-}$ (d) $[Fe(H_2O)_6]^{3+}$

93 Of the following complex ions, the one that probably has the largest overall formation constant, K_f, is

(a) $[Co(NH_3)_6]^{3+}$
(b) $[Co(H_2O)_6]^{3+}$
(c) $[Co(H_2O)_4(NH_3)_2]^{3+}$
(d) $[Co(en)_3]^{3+}$

94 Which of the following complexes formed by Cu^{2+} ions is most stable? **[NCERT Exemplar]**

(a) $Cu^{2+} + 4NH_3 \longrightarrow [Cu(NH_3)_4]^{2+}$, $\log K = 11.6$
(b) $Cu^{2+} + 4CN^- \longrightarrow [Cu(CN)_4]^{2-}$, $\log K = 27.3$
(c) $Cu^{2+} + 2en \longrightarrow [Cu(en)_2]^{2+}$, $\log K = 15.4$
(d) $Cu^{2+} + 4H_2O \longrightarrow [Cu(H_2O)_4]^{2+}$, $\log K = 8.9$

VI. Stability and importance of coordination compounds

95 Among the following, the most stable ion is

(a) $[Fe(OH)_5]^{3-}$ (b) $[FeCl_6]^{3-}$
(c) $[Fe(CN)_6]^{3-}$ (d) $[Fe(H_2O)_6]^{3+}$

96 For the complex ML_2, stepwise formation constants for

$$M + L \rightleftharpoons ML$$
$$ML + L \rightleftharpoons ML_2$$

are 4 and 3. Hence, overall stability constant for $M + 2L \rightleftharpoons ML_2$ is

(a) 12 (b) 7 (c) 1.33 (d) 0.75

97 Among the following, which is not the π-bonded organometallic compound?

(a) $Cr(\eta^6 - C_6H_6)_2$ (b) $(CH_3)_4Sn$
(c) $K[PtCl_3(\eta^2 - C_2H_4)]$ (d) $Fe(\eta^5 - C_5H_5)_2$

98 In metal carbonyl having general formula $M(CO)_x$ where, M = metal, x = 4, the metal is bonded to

(a) oxygen (b) carbon
(c) carbon and oxygen (d) C≡O triple bond

99 Among the following metal carbonyls, the C—O bond order is lowest in

(a) $[Mn(CO)_6]^+$ (b) $[Fe(CO)_5]$
(c) $[Cr(CO)_6]$ (d) $[V(CO)_6]$

100 A person suffering from lead poisoning should be fed with

(a) DMG (b) *cis*-platin
(c) $[Ca(EDTA)]^{2-}$ (d) hypo

101 Extraction of Ag from sulphide ore and removal of unreacted silver from photographic plate involve complexes

(a) $[Ag(S_2O_3)_2]^{3-}$ in both (b) $[Ag(CN)_2]^-$ in both
(c) $[Ag(S_2O_3)_2]^{3-}$, $[Ag(CN)_2]^-$ (d) $[Ag(CN)_2]^-$, $[Ag(S_2O_3)_2]^{3-}$

102 Extraction of metals or other processes is through the complex formation.

I. Cyanide process II. Mond process
III. Photographic fixing process

Complexes formed in these methods are

	I	II	III
(a)	$[Ag(NH_3)_2]Cl$	$Ni(CO)_4$	$[Ag(CN)_2]^-$
(b)	$[Ag(CN)_2]^-$	$Ni(CO)_4$	$[Ag(S_2O_3)_2]^{3-}$
(c)	$[Ag(CN)_2]^-$	$[Ag(S_2O_3)_2]^{3-}$	$Ni(CO)_4$
(d)	$[Cd(CN)_4]^{2-}$	$Ni(CO)_4$	$[Ag(S_2O_3)_2]^{3-}$

103 Consider the following statements.

Statement I All the iron-carbon bond distances in ferrocene are equal.

Statement II The π-electrons in the cyclopentadienyl group of ferrocene are delocalised.
(a) Both statements I and II are correct.
(b) Only statement I is correct.
(c) Only statement Ii is correct.
(d) Both the statements are incorrect.

104 Match the coordination compounds given in Column I with the central metal atoms given in Column II and assign the correct code. **[NCERT Exemplar]**

Column I (Coordination compound)	Column II (Central metal atom)
A. Chlorophyll	1. Rhodium
B. Blood pigment	2. Cobalt
C. Wilkinson catalyst	3. Magnesium
D. Vitamin B_{12}	4. Iron

Codes

	A	B	C	D
(a)	3	4	1	2
(b)	4	3	1	2
(c)	1	2	3	4
(d)	3	2	1	4

105 Consider the following statements.
Statement I Toxic metal ions are removed by the chelating ligands.
Statement II Chelate complexes tend to be more stable. **[NCERT Exemplar]**
(a) Both statements I and II are correct.
(b) Only statement I is correct.
(c) Only statement II is correct.
(d) Both statements I and II are incorrect.

(B) Miscellaneous questions

1 $FeSO_4$ on treatment with excess of KCN gives a product that does not give test of Fe^{2+} the product formed is
(a) $[Fe(CN)_2]SO_4$ (b) $K_3[Fe(CN)_6]$
(c) $K_4[Fe(CN)_6]$ (d) $KFe[Fe(CN)_6]$

2 Potassium diamminedicyanosulphatosuperoxide platinate(IV) is
(a) $K[Pt(CN)_2(O_2)(SO_4)(NH_3)_2]$
(b) $K_4[Pt(NH_3)_2(CN)_2(SO_4)(O_2)]$
(c) $K_4[Pt(NH_3)_2(CN)_2(SO_4)(O_2)]$
(d) $K_3[Pt(NH_3)_2(CN)_2(SO_4)(O_2)]$

3 Complex $[Co(en)_2Cl_2]^+$ can
(a) exist as *cis*- and *trans*-isomer
(b) enantiomorphic pair *d* and *l* are formed if the isomer in which two unidentate ligands (Cl^-) have to be *cis*- *cis* to each other
(c) Both (a) and (b) are correct
(d) None of the above is correct

4 The correct order of the stoichiometries of AgCl formed when $AgNO_3$ in excess is treated with 1 mole each of the complexes $CoCl_3 \cdot 6NH_3$, $CoCl_3 \cdot 5NH_3$, $CoCl_3 \cdot 4NH_3$ respectively is
(a) 2 AgCl, 3 AgCl, 1 AgCl
(b) 1 AgCl, 3 AgCl, 2 AgCl
(c) 3 AgCl, 1 AgCl, 2 AgCl
(d) 3 AgCl, 2 AgCl, 1 AgCl

5 Which of the following complex is Prussian blue?
(a) $KFe^{III}[Fe^{II}(CN)_6]$ (b) $KFe^{III}[Fe^{III}(CN)_6]$
(c) $K_2[Fe^{II}(CN)_6]$ (d) $K_3[Fe(CN)_6]$

6 Which of the following option has same EAN value?
(a) $\overset{*}{Ni}(CO)_4$, $[\overset{*}{Fe}(CN)_6]^{4-}$ (b) $[\overset{*}{Ni}(en)_2]^{2+}$, $[\overset{*}{Fe}(H_2O)_6]^{2+}$
(c) $[\overset{*}{Co}(CN)_6]^{3-}$, $[\overset{*}{Fe}(CN)_6]^{3-}$ (d) $[\overset{*}{Ni}(en)_2]^{2+}$, $[\overset{*}{Sc}(H_2O)_6]^{3+}$

7 Which is used in cancer-chemotherapy?
(a) *cis*-platin (b) Zeise's salt
(c) Both (a) and (b) (d) None of these

8 Isomerisms exhibited by $[Cr(NH_3)_2(H_2O)_2Cl_2]^+$ are
(a) ionisation, optical (b) hydrate, optical
(c) geometrical, optical (d) coordinate, geometrical

9 Which of the following ligands is expected to be bidentate?
(a) Br (b) $C_2O_4^{2-}$
(c) CH_3NH_2 (d) $CH_3C{\equiv}N$

10 Ink contains Fe^{3+} spot of ink can be removed by addition of $C_2O_4^{2-}$. Complex formed is
(a) $[Fe(C_2O_4)_6]^{3-}$ (b) $[Fe(C_2O_4)_3]^{3-}$
(c) $[Fe(C_2O_4)_3]^{2-}$ (d) $[Fe(C_2O_4)_3]^{4-}$

11 Which compound is formed when excess of KCN is added to aqueous solution of copper sulphate?
(a) $Cu(CN)_2$ (b) $K_2[Cu(CN)_4]$
(c) $K[Cu(CN)_2]$ (d) $K_3[Cu(CN)_4]$

12 The hypothetical complex diaquatriamminechloro cobalt(III) chloride can be represented as
(a) $[CoCl(NH_3)_3(H_2O)_2]Cl_2$ (b) $[Co(NH)_3(H_2O)Cl_3]$
(c) $[Co(NH_3)_3(H_2O)_2Cl]$ (d) $[Co(NH_3)_3(H_2O)_3Cl_3]$

13 Which of the following will give a pair of enantiomorphs?
(a) $[Co(NH_3)_4Cl_2]NO_2$
(b) $[Cr(NH_3)_6][Co(CN)_6]$
(c) $[Co(en)_2Cl_2]Cl$
(d) $[Pt(NH_3)_4][PtCl_6]$ en $= NH_2CH_2CH_2NH_2$

14 According to IUPAC nomenclature sodium nitroprusside is named as
(a) sodium pentacyanonitrosyl ferrate(II)
(b) sodium pentacyanonitrosyl ferrate(III)
(c) sodium nitroferricyanide
(d) sodium nitroferrocyanide

15 Which of the following contains monovalent cationic complex?
(a) $CoCl_3 \cdot 6NH_3$ (b) $CoCl_3 \cdot 5NH_3$
(c) $CoCl_3 \cdot 4NH_3$ (d) $CoCl_3 \cdot 3NH_3$

16 Complexes of which ions are not kinetically labile?
(a) Cu^{2+}, Ni^{2+} (b) Cr^{3+}, Co^{3+}
(c) V^{3+}, Ti^{4+} (d) None of these

17 Which pair is not associated with complimentary colours?
(a) Orange – blue (b) Yellow – purple
(c) Green – purple (d) Red – blue green

18 Pick out the complex compound in which the central metal atom obeys EAN rule strictly.
(a) $K_4[Fe(CN)_6]$ (b) $K_3[Fe(CN)_6]$
(c) $[Cr(H_2O)_6]Cl_3$ (d) $[Cu(NH_3)_4]SO_4$

19 In the complex $[Fe(NH)_6]^{3+}$
(a) Fe^{3+} is a Lewis acid
(b) NH_3 is a Lewis base
(c) Both (a) and (b) are correct
(d) Fe^{3+} is a Lewis base

20 Which of the following compounds show optical isomerism?
1. *cis*-$[Co(NH_3)_4Cl_2]^+$ 2. *trans*-$[Co(en)_2Cl_2]^+$
3. *cis*-$[Co(en)_2Cl_2]^+$ 4. $[Co(en)_3]^{3+}$
Select the correct answer using the codes given below.
(a) Both 1 and 2 (b) Both 2 and 3
(c) Both 3 and 4 (d) 1, 3 and 4

21 The total number of possible isomers for the complex compound $[Cu^{II}(NH_3)_4][Pt^{II}Cl_4]$
(a) 3 (b) 6 (c) 5 (d) 4

22 Which of the following has maximum conductance?
(a) $PtCl_4 \cdot 6NH_3$ (b) $PtCl_4 \cdot 5NH_3$
(c) $PtCl_4 \cdot 4NH_3$ (d) Equal

23 The compound which is not coloured is
(a) $K_4[Fe(H_2O)_6]$ (b) $K_3[Fe(CN)_6]$
(c) $Na_2[CdCl_4]$ (d) $Na_2[CuCl_4]$

24 Which one of the following forms with an excess of CN^-, a complex having coordination number two?
(a) Cu^+ (b) Ag^+ (c) Ni^{2+} (d) $Al(OH)_3$

25 An aqueous solution of $CoCl_2$ on addition of excess of conc. HCl turns blue due to formation of
(a) $[Co(H_2O)_4Cl_2]$ (b) $[Co(H_2O)_2Cl_4]^{2-}$
(c) $[CoCl_4]^{2-}$ (d) $[Co(H_2O)_2Cl_2]$

26 Which of the following is obtained when auric chloride reacts with sodium chloride?
(a) Na[AuCl] (b) $Na[AuCl_2]$
(c) $Na[AuCl_3]$ (d) $Na[AuCl_4]$

27 The most soluble of the following compounds in $NH_3(aq)$ is
(a) $Cu(OH)_2$ (b) $BaSO_4$
(c) SiO_2 (d) $MgCO_3$

28 Of the following complex ions, one is a Bronsted-Lowry acid. That one is
(a) $[Cu(NH_3)_4]^{2+}$ (b) $[FeCl_4]^-$
(c) $[Fe(H_2O)_6]^{3+}$ (d) $[Zn(OH)_4]^{2-}$

29 A violet colour compound is formed in detection of 'S' in a compound
(a) $Na_4[Fe(CN)_5NOS]$ (b) $Na_3[Fe(CN)_5NOS]$
(c) $Na_2[Fe(CN)_5NOS]$ (d) $Na_2[Fe(CN)_5NO]$

30 When 0.1 mol $[CoCl_3(NH_3)_5]$ is treated with excess of $AgNO_3$, 0.2 mol of AgCl are obtained. The conductivity of solution will correspond to
(a) 1 : 3 electrolyte (b) 1 : 2 electrolyte
(c) 1 : 1 electrolyte (d) 3 : electrolyte

31 The coordination number and oxidation number of *X* in the following compound $[X(SO_4)(NH_3)_5]Cl]$ will be
(a) 10 and 3 (b) 2 and 6
(c) 6 and 3 (d) 6 and 4

32 Atomic number of Cr and Fe are respectively 24 and 26. Which of the following is paramagnetic with the spin of electron?
(a) $[Cr(CO)_6]$ (b) $[Fe(CO)_5]$
(c) $[Fe(CN)_6]^{4-}$ (d) $[Cr(NH_3)_6]^{3+}$

33 The geometry of $Ni(CO)_4$ and $[Ni(PPh_3)_2]Cl_2$ are
(a) Both square planar
(b) tetrahedral and square planar respectively
(c) Both tetrahedral
(d) square planar and tetrahedral respectively

34 CN^- is a strong field ligand. This is due to the fact that
(a) it carries negative charge
(b) it is a conjugate base of a weak acid
(c) it can accept electrons from metal species
(d) it forms high spin complexes with metal species

35 The formula of dichloro (urea) copper(II) is
(a) $[Cu\{O{=}C(NH_2)_2\}]Cl_2$
(b) $[CuCl_2\{O{=}C(NH_2)_2\}]$
(c) $[Cu\{O{=}C(NH_2)Cl\}]Cl$
(d) $[CuCl_2]\{O{=}C(NH_2)_2H_2\}$

36 Select the correct statement(s) about the complex $[Co(NH_3)_5SO_4]Br$.
(a) Its ionisation isomer is $[Co(NH_3)_5Br]SO_4$
(b) It gives yellow precipitate with $AgNO_3$
(c) Its ionisation isomer gives white precipitate with $BaCl_2$
(d) All of the above are correct statements

37 The intense blue colour of prussian blue salt arises from which one of the following?
(a) *d-d* transition
(b) Inter valence electron transfer
(c) Ligand to metal charge transfer
(d) Metal to ligand charge transfer

38 $[Co(NH_3)_5(NO_2)]^{2+}$ and $[Co(NH_3)_5(ONO)]^{2+}$ will differ in
(a) colour (b) structure
(c) hybridisation (d) magnetic moment

39 In which case racemic mixture is obtained on mixing its mirror images in 1 : 1 molar ratio?
(a) $[Cr(en)_3]^{3+}$ (b) $[Ni(DMG)_2]$
(c) *cis*- $[Cu(Gly)_2]$ (d) In all cases

40 Which of the following is wrong statement?
(a) $Ni(CO)_4$ has oxidation number +4 for Ni
(b) $Ni(CO)_4$ has zero oxidation number for Ni
(c) Ni is metal
(d) CO is gas

41 In nitroprusside ion the iron and NO exist as Fe^{II} and NO^+ rather than Fe^{III} and NO. These forms can be differentiated by
(a) estimating the concentration of iron
(b) measuring the concentration of CN^-
(c) measuring the solid state magnetic moment
(d) thermally decomposing the compound

42 Cu^{2+} and Cd^{2+} are distinguished through formation of complex $[Cu(CN)_4]^{3-}$ and $[Cd(CN)_4]^{2-}$ when H_2S gas is passed
(a) There is yellow precipitate due to CdS
(b) There is black precipitate due to CuS
(c) There is blue precipitate due to CuS
(d) There is precipitation of CuS and CdS together

43 Fe (III) ion develops intense red colour when it reacts with SCN^- ion while Fe (II) ion does not. It is due to
(a) Fe (III) ion forms a charge transfer complex with SCN^-
(b) Fe (III) is reduced to Fe (I) which is deep-red in colour
(c) SCN^- ion is oxidised to CN^- ion that forms red complex with Fe (II) ion
(d) SCN^- ion does not form any complex with Fe (II) ion

44 $[Co(NH_3)_4(NO_2)_2]Cl$ exhibits
(a) linkage isomerism, ionisation isomerism and optical isomerism
(b) linkage isomerism, ionisation isomerism and geometrical isomerism
(c) ionisation isomerism, geometrical isomerism and optical isomerism
(d) linkage isomerism, geometrical isomerism and optical isomerism

45 The complex $[Ru(H_2O)_6]^{2+}$ and $[RuCl_6]^{3-}$ have nearly the same Δ_o values, because
(a) both the complexes contain the same metal atom
(b) both the complexes have a symmetrical octahedral structure
(c) the presence of weaker field ligand in one is compensated by the presence of the metal in a higher oxidation state
(d) the H_2O and Cl^- occupy adjacent positions in the spectrochemical series

46 Following equilibrium,
$[Ni(NH_3)_6]^{2+} + 3NH_2CH_2CH_2NH_2 \rightleftharpoons [Ni(NH_2CH_2CH_2NH_2)_3]^{2+} + 6NH_3$
(a) favours formation of complex with ethylenediamine
(b) has $T\Delta S > \Delta H$ thus, entropy effect is responsible for making ΔG negative
(c) is an example of chelate effect
(d) follows all the given facts

47 A coordination complex compound of cobalt has the molecular formula containing five ammonia molecules, one nitro group and two chlorine atoms for one cobalt atom. One mole of this compound produces three moles ion in an aqueous solution; on reacting this solution with excess of $AgNO_3$, AgCl precipitates. The ionic formula for this complex would be
(a) $[Co(NH_3)_5(NO_2)]Cl_2$
(b) $[Co(NH_3)_5Cl]\ [Cl(NO_2)]$
(c) $[Co(NH_3)_4(NO_2)Cl]\ [(NH_3)Cl]$
(d) $[Co(NH_3)_5]\ [(NO_2)_2Cl_2]$

48 Consider the following complexes.
1. K_2PtCl_6
2. $PtCl_4 \cdot 2NH_3$
3. $PtCl_4 \cdot 3NH_3$
4. $PtCl_4 \cdot 5NH_3$

Their electrical conductances in aqueous solutions are
(a) 256, 0, 97, 404
(b) 404, 0, 97, 256
(c) 256, 97, 0, 404
(d) 404, 97, 256, 0

49 Which of the following reactions are kinetically favourable?
I. $[Cu(H_2O)_4]^{2+} + 4NH_3 \longrightarrow [Cu(NH_3)_4]^{2+} + 4H_2O$
II. $[Cu(H_2O)_4]^{2+} + 4Cl^- \longrightarrow [CuCl_4]^{2-} + 4H_2O$
III. $[Co(H_2O)_6]^{3+} + 6Cl^- \longrightarrow [CoCl_6]^{3-} + 6H_2O$
(a) Both I and III
(b) Both II and III
(c) Both I and II
(d) I, II and III

50 Ammonia forms the complex ion $[Cu(NH_3)_4]^{2+}$ with copper ions in the alkaline solutions but not in acidic solutions. What is the reason for it?
(a) In acidic solutions hydration protects copper ions
(b) In acidic solutions protons coordinate with ammonia molecules forming NH_4^+ ions and NH_3 molecules are not available
(c) In alkaline solutions insoluble $Cu(OH)_2$ is precipitated which is soluble in excess of any NH_4OH.
(d) Copper hydroxide is an amphoteric substance

51 While Ti^{3+}, V^{3+}, Fe^{3+} and Co^{2+} afford a large number of tetrahedral complexes, Cr^{3+} never does this, the reason being
(a) Cr^{3+} forces high crystal field splitting with a varieties of ligands
(b) Crystal field stabilisation energy in octahedral *vis-a-vis* tetrahedral Cr^{3+} system plays the deciding role
(c) The ionic radius of Cr^{3+} is the largest among the other M^{3+} ions mentioned
(d) Electronegativity of Cr^{3+} is the largest among these trivalent 3*d*-metals and so chromium prefers to be associated with as many ligands as its ionic radius permits

52 Coordination number of Cr is six. A complex with $C_2O_4^{2-}$, en and superoxide O_2^- will be in the ratio to make complex $[Cr(C_2O_4)_x(en)_y(O_2)_z]^-$

	x	*y*	*z*		*x*	*y*	*z*
(a)	1	1	1	(b)	1	1	2
(c)	1	2	2	(d)	2	1	1

53 Select the correct statement.
(a) Complex ion $[MoCl_6]^{3-}$ is paramagnetic
(b) Complex ion $[Co(en)_3]^{3+}$ is diamagnetic
(c) Both (a) and (b) are correct
(d) Both $[MOCl_6]^{3-}$ and $[Co(en)3]^{3+}$ are paramagnetic

54 Select the correct statements.
(a) Chelate ligands form more stable complexes than analogous monodentate ligands
(b) Chelates of higher denticity will give more stable complexes than chelates of lower denticity
(c) Macrocyclic ligands of appropriate size form more stable complexes than chelate ligands
(d) All of the above are correct statements

55 Which of the following statements is incorrect?
(a) In $K_3[Fe(CN)_6]$, the ligand has satisfied only the secondary valency of ferric ion
(b) In $K_3[Fe(CN)_6]$, the ligand has satisfied both primary and secondary valencies of ferric ion
(c) In $K_4[Fe(CN)_6]$, the ligand has satisfied both primary and secondary valencies of ferrous ion
(d) In $[Cu(NH_3)_4]SO_4$, the ligand has satisfied only the secondary valency of copper

56 Select the incorrect statement.
(a) $[Ni(en)_3]^{2+}$ is less stable than $[Ni(NH_3)_6]^{2+}$
(b) Increase in stability of the complexes due to presence of multidentate cyclic ligand is called macrocyclic effect
(c) A complex ion that exchanges ligands slowly is said to be non-labile or inert
(d) For a given ion and ligand, greater the charge on the metal ion, greater the stability

57 Which is the incorrect observation?
(a) For the complex, $[CoF_6]^{3-}$, F^- is a weak-field ligand, so that $\Delta_{oct} < P$ (electron-pairing energy) and thus form high spin complex
(b) For the complex, $[Co(NH_3)_6]^{3+}$, NH_3 is a strong-field ligand, so that $\Delta_{oct} > P$ and thus form low spin complex
(c) $\Delta_t = +\frac{4}{9}\Delta_{oct}$
(d) Greater the ionic charge on the central metal ion, the greater the value of Δ

58 Select the correct statement.
(a) Geometrical isomer may differ in dipole moment and visible/UV spectra
(b) Complexes of the type $[Ma_3b_3]$ can also have facial (*fac*) and meridional (*mer*) isomer
(c) No optical isomer exists for the complex *trans*-$[Co(en)_2Cl_2]^+$
(d) All of the above are correct statement.

59 Select the correct statement.
(a) $[Fe(CN)_6]^{4-}$ and $[Fe(H_2O)_6]^{2+}$ can be distinguished by magnetic moment
(b) $[Ni(CN)_4]^{2-}$ and $[Ni(CO)_4]$ both are diamagnetic
(c) $[Fe(CN)_6]^{4-}$ and $[Co(NH_3)_6]^{3+}$ have equal magnetic moment
(d) All the above are correct statements

60 Coordination compounds have great importance of it in biological systems. In this context which of the following statements is incorrect?
(a) Chlorophylls are green pigments in plants and contain calcium
(b) Haemoglobin is the red pigment of blood and contains iron
(c) Cyanocobalamin is vitamin B_{12} and contains cobalt
(d) Carboxypeptidase-A is an enzyme and contains zinc

61 Consider the following statements in respect of $[CoCl_6]^{4-}$ complex ion
I. It is paramagnetic. II. It is low-spin complex.
III. Oxidation number of cobalt is – 4.
IV. Coordination number of cobalt is 6.
Which of the statements given above are correct?
(a) I, II, III and IV (b) Both III and IV
(c) Both I and IV (d) Both I and II

62 Consider the following facts about chelating ligands.

I. They have more than one donor atom.

II. They are non-linear, and they must form bonds with reasonable angles at the metal atom.

III. Ethylenediamine is a bidentate chelating ligand.

Select the correct facts.

(a) Only I
(b) Both I and II
(c) Both II and III
(d) I, II and III

63 The complex $[Fe(H_2O)_5NO]^{2+}$ is formed in the brown ring test, for nitrates when freshly prepared $FeSO_4$ solution is added to aqueous solution of NO_3^- followed by addition of conc. H_2SO_4.

Select the correct statement about this complex.

(a) Colour change is due to charge transfer
(b) It has iron in +1 oxidation state and nitrosyl as NO^+
(c) It has magnetic moment of 3.87 B.M. confirming three unpaired electrons in Fe
(d) All of the above are correct statements

64 Optical isomer

I. are enantiomers

II. are superimposable

III. have the property of chirality

IV. form racemic mixture if present in 1 : 1 ratio

Select the correct alternate.

(a) I, II and IV are correct
(b) I, II and III are correct
(c) I, III and IV are correct
(d) All are incorrect

(C) Medical entrances' gallery

1 Match the coordination number and type of hybridisation with distribution of hybrid orbitals in space based on valence bond theory.

	Coordination number and type of hybridisation		Distribution of hybrid orbitals in space
A.	4, sp^3	I.	trigonal bipyramidal
B.	4, dsp^2	II.	octahedral
C.	5, sp^3d	III.	tetrahedral
D.	6, d^2sp^3	IV.	square planar

Select the correct option. **[NEET 2020]**

	A	B	C	D
(a)	II	III	IV	I
(b)	III	IV	I	II
(c)	IV	I	II	III
(d)	III	I	IV	II

2 Which of the following is the correct order of increasing field strength of ligands to form coordination compounds? **[NEET 2020]**

(a) $SCN^- < F^- < CN^- < C_2O_4^{2-}$
(b) $F^- < SCN^- < C_2O_4^{2-} < CN^-$
(c) $CN^- < C_2O_4^{2-} < SCN^- < F^-$
(d) $SCN^- < F^- < C_2O_4^{2-} < CN^-$

3 Aluminium chloride in acidified aqueous solution forms a complex '*A*', in which hybridisation state of Al is '*B*'. What are '*A*' and '*B*', respectively? **[NEET (Odisha) 2019]**

(a) $[Al(H_2O)_6]^{3+}$, sp^3d^2
(b) $[Al(H_2O)_4]^{3+}$, sp^3
(c) $[Al(H_2O)_4]^{3+}$, dsp^2
(d) $[Al(H_2O)_6]^{3+}$, d^2sp^3

4 The Crystal Field Stabilisation Energy (CFSE) for $[CoCl_6]^{4-}$ is 18000 cm^{-1}. The CFSE for $[CoCl_4]^{2-}$ will be **[NEET (Odisha) 2019]**

(a) 6000 cm^{-1}
(b) 16000 cm^{-1}
(c) 18000 cm^{-1}
(d) 8000 cm^{-1}

5 $[Co(C_2O_4)_3]^{3-}$ is a **[AIIMS 2019]**

(a) low spin complex
(b) paramagnetic
(c) high spin
(d) sp^3d^2-hybridised

6 Correct increasing order for the wavelength of absorption in the visible region for the complexes of Co^{3+} is **[AIIMS 2019]**

(a) $[Co(CN)_6]^{3-}$, $[Co(NH_3)_6]^{3+}$, $[Co(NH_3)_5(H_2O)]^{3+}$, $[Co(NH_3)_5Cl]^{+2}$
(b) $[Co(CN)_6]^{3-}$, $[Co(NH_3)_5(H_2O)]^{3+}$ $[Co(NH_3)_5Cl]^{2+}$, $[Co(NH_3)_6]^{3+}$
(c) $[Co(NH_3)_6]^{3+}$, $[Co(CN)_6]^{3-}$, $[Co(NH_3)_5(H_2O)]^{3+}$, $[Co(NH_3)Cl]^{+2}$
(d) $[Co(NH_3)_5Cl]^{+2}$, $[Co(NH_3)_5(H_2O)]^{3+}$, $[Co(NH_3)_6]^{3+}$, $[Co(CN)_6]^{3-}$

7 What is the correct electronic configuration of the central atom in $K_4[Fe(CN)_6]$ based on crystal field theory? **[NEET 2019]**

(a) $t_{2g}^6 e_g^0$
(b) $e^3 t_2^3$
(c) $e^4 t_2^2$
(d) $t_{2g}^4 e_g^2$

8 The type of isomerism shown by the complex $[CoCl_2(en)_2]$ is **[NEET 2018]**

(a) ionisation isomerism
(b) coordination isomerism
(c) geometrical isomerism
(d) linkage isomerism

9 In the complex, $[Cr(NH_3)_4Cl_2]^+$, the oxidation number of Cr is **[JIPMER 2018]**

(a) +4
(b) +3
(c) +2
(d) +6

10 Which of the following compound has same oxidation state of the central metal atom in the cationic and anionic part? **[AIIMS 2018]**

(a) $[Pt(NH_3)_4][PtCl_6]$ (b) $[Pt(NH_3)_4Cl_2][PtCl_4]$
(c) $[Pt(Py)_4][PtCl_4]$ (d) $K_4[Ni(CN)_6]$

11 The IUPAC name of the complex $[Pt(NH_3)_3Br(NO_2)Cl]Cl$ is **[AIIMS 2018]**

(a) triammine chloridobromidonitro platinum (IV) chloride
(b) triammine bromidochloridonitro platinum (IV) chloride
(c) triammine bromidochloridonitro platinum (II) chloride
(d) triammine chloridobromidonitro platinum (II) chloride

12 Among the following complex ions, the one which shows geometrical isomerism will be **[AIIMS 2018]**

(a) $[Cr(H_2O)_4Cl_2]^+$ (b) $[Pt(NH_3)_3Cl]$
(c) $[Co(NH_3)_6]^{3+}$ (d) $[Co(CN)_5(NC)]^{3-}$

13 **Assertion** (A) Toxic metal ions are removed by the chelating ligands.
Reason (R) Chelate complexes tend to be more stable.

(a) Both A and R are correct and R is the correct explanation of A **[AIIMS 2018]**
(b) Both A and R are correct and R is not the correct explanation of A
(c) A is correct but R is incorrect
(d) Both A and R are incorrect

14 The geometry and magnetic behaviour of the complex $[Ni(CO)_4]$ are **[NEET 2018]**

(a) square planar geometry and paramagnetic
(b) tetrahedral geometry and diamagnetic
(c) square planar geometry and diamagnetic
(d) tetrahedral geometry and paramagnetic

15 The electronic configuration of central metal atom/ion in $[Co(CN)_6]^{3-}$ is **[AIIMS 2018]**

(a) $t_{2g}^5 e_g^0$ (b) $t_{2g}^4 e_g^2$ (c) $t_{2g}^4 e_g^3$ (d) $t_{2g}^6 e_g^0$

16 The spin only magnetic moment of $[MnBr_4]^{2-}$ is 5.9 BM. The geometry of this complex ion is **[AIIMS 2018]**

(a) tetrahedral (b) octahedral
(c) trigonal pyramidal (d) square planar

17 Iron carbonyl, $Fe(CO)_5$ is **[NEET 2018]**

(a) trinuclear (b) mononuclear
(c) tetranuclear (d) dinuclear

18 Correct increasing order for the wavelengths of absorption in the visible region for the complexes of Co^{3+} is **[NEET 2017]**

(a) $[Co(en)_3]^{3+}$, $[Co(NH_3)_6]^{3+}$, $[Co(H_2O)_6]^{3+}$
(b) $[Co(H_2O)_6]^{3+}$, $[Co(en)_3]^{3+}$, $[Co(NH_3)_6]^{3+}$
(c) $[Co(H_2O)_6]^{3+}$, $[Co(NH_3)_6]^{3+}$, $[Co(en)_3]^{3+}$
(d) $[Co(NH_3)_6]^{3+}$, $[Co(en)_3]^{3+}$, $[Co(H_2O)_6]^{3+}$

19 An example of a sigma bonded organometallic compound is **[NEET 2017]**

(a) cobaltocene (b) ruthenocene
(c) Grignard's reagent (d) ferrocene

20 Pick out the correct statement with respect to $[Mn(CN)_6]^{3-}$ **[NEET 2017]**

(a) It is sp^3d^2 hybridised and octahedral
(b) It is sp^3d^2 hybridised and tetrahedral
(c) It is d^2sp^3 hybridised and octahedral
(d) It is dsp^2 hybridised and square planar

21 Which of the following has longest C— O bond length [Free C— O bond length in CO is 1.128 Å.] **[NEET (I) 2016]**

(a) $[Co(CO)_4]^-$ (b) $[Fe(CO)_4]^{2-}$
(c) $[Mn(CO)_6]^+$ (d) $[Ni(CO)]_4$

22 AlF_3 is soluble in HF only in presence of KF. It is due to the formation of **[NEET (II) 2016]**

(a) $K_3[AlF_3H_3]$ (b) $K_3[AlF_6]$
(c) AlH_3 (d) $K[AlF_3H]$

23 Cobalt (III) chloride forms several octahedral complexes with ammonia. Which of the following will not give test for chloride ions silver nitrate at 25°C? **[CBSE AIPMT 2015]**

(a) $CoCl_3 \cdot 3NH_3$
(b) $CoCl_3 \cdot 4NH_3$
(c) $CoCl_3 \cdot 5NH_3$
(d) $CoCl_3 \cdot 6NH_3$

24 Which one of these statement about $[Co(CN)_6]^{3-}$ is correct? **[CBSE AIPMT 2015]**

(a) $[Co(CN)_6]^{3-}$ has no unpaired electrons and will be in a low-spin configuration
(b) $[Co(CN)_6]^{3-}$ has four unpaired electron and will be in a low-spin configuration
(c) $[Co(CN)_6]^{3-}$ has four unpaired electrons will be in a high-spin configuration
(d) $[Co(CN)_6]^{3-}$ has no unpaired electrons and will be in a high-spin configuration

25 Number of possible isomers for the complex $[Co(en)_2Cl_2]Cl$ will be (en = ethylenediamine) **[CBSE AIPMT (Re exam) 2015]**

(a) 2 (b) 1
(c) 3 (d) 4

26 The sum of coordination number and oxidation number of the metal M in the complex $[M(en)_2(C_2O_4)]Cl$ (where, en is ethylenediamine) is **[CBSE AIPMT (Re exam) 2015]**

(a) 9 (b) 6
(c) 7 (d) 8

27 The name of complex ion, $[Fe(CN)_6]^{3-}$ is **[CBSE AIPMT (Re exam) 2015]**

(a) hexacyanoion
(b) hexacyaniotoferrate (III) ion
(c) tricyanoferrate (III) ion
(d) hexacyanoferrate (III) ion

28 Cold ferrous sulphate solution on absorption of NO develops brown colour due to the formation of **[WB JEE 2015]**

(a) paramagnetic $[Fe(H_2O)_5(NO)]SO_4$
(b) diamagnetic $[Fe(H_2O)_5(N_3)]SO_4$
(c) paramagnetic $[Fe(H_2O)_5(NO_3)]SO_4$
(d) diamagnetic $[Fe(H_2O)_4(SO_4)]NO_3$

29 Which of the following will be able to show geometrical isomerism? **[KCET 2015]**

(a) MA_3B - square planar
(b) MA_2B_2 - tetrahedral
(c) $MABCD$ - square planar
(d) $MABCD$ - tetrahedral

30 Which one among the following is a homoleptic complex? **[KCET 2015]**

(a) *tris* (ethan-1,2-diamine) cobalt (III) chloride
(b) Triaminetriaquachromium-(III) chloride
(c) Dichlorodo *bis* (ethan-1, 2-diamine) cobalt (III) chloride
(d) Pentaminecarbonatocobalt (III) chloride

31 The coordination number, oxidation number and the number of *d*-electrons in the metal ion of the complex $[CoCl_2(en)_2]Cl$ respectively are [atomic number of Co= 27] **[Kerala CEE 2015]**

(a) 4, 3 and 6 (b) 6, 2 and 6
(c) 6, 6 and 3 (d) 4, 2 and 6
(e) 6, 3 and 6

32 The pair of compounds that can exist together is **[CBSE AIPMT 2014]**

(a) $FeCl_3$, $SnCl_2$ (b) $HgCl_2$, $SnCl_2$
(c) $FeCl_2$, $SnCl_2$ (d) $FeCl_3$, KI

33 Among the following complexes, the one which show zero crystal field stabilisation energy (CFSE) is **[CBSE AIPMT 2014]**

(a) $[Mn(H_2O)_6]^{3+}$ (b) $[Fe(H_2O)_6]^{3+}$
(c) $[Co(H_2O)_6]^{2+}$ (d) $[Co(H_2O)_6]^{3+}$

34 Which of the following complexes have lowest molar conductance? **[MHT CET 2014]**

(a) $CoCl_3 \cdot 3NH_3$ (b) $CoCl_3 \cdot 4NH_3$
(c) $CoCl_3 \cdot 5NH_3$ (d) $CoCl_3 \cdot 6NH_3$

35 Select the diamagnetic complex ion amongst the following complexes (Atomic number of Fe = 26, Co = 27) **[MHT CET 2014]**

(a) $K_3[Fe(CN)_6]$ (b) $[Co(NH_3)_6]Cl_3$
(c) $K_3[FeF_6]$ (d) $K_3[CoF_6]$

36 The hybridisation of central metal ion in $K_2[Ni(CN)_4]$ and $K_2[NiCl_4]$ are respectively **[Kerala CEE 2014]**

(a) dsp^2, sp^3 (b) sp^3, sp^3
(c) dsp^2, dsp^3 (d) sp^3, sp^3d^2
(e) sp^3d^2, d^2sp^2

37 Which of the following compounds show optical isomerism ? **[Kerala CEE 2014]**

I. *cis* - $[Co(NH_3)_4Cl_2]^+$ II. *trans* - $[Co(en)_2Cl_2]^+$
III. *cis* - $[Co(en)_2Cl_2]^+$ IV. $[Co(en)_3]^{3+}$

Choose the correct answer from the codes given below.

(a) Both I and II (b) Both II and III
(c) Both III and IV (d) I, III and IV
(e) I, II, III and IV

38 How many isomers are possible in $[Co(en)_2Cl_2]$? **[UK PMT 2014]**

(a) 2 (b) 4
(c) 6 (d) 1

39 A correct statement is **[KCET 2014]**

(a) $[Co(NH_3)_6]^{3+}$ is paramagnetic
(b) $[MnBr_4]^{2-}$ is square planar
(c) $[CoBr_2(en)_2]^-$ exhibits linkage isomerism
(d) $[Ni(NH_3)_6]^{2+}$ is an inner orbital complex

40 In which one of the pairs of ion given, there is an ion that forms a coordination compound with both aqueous sodium hydroxide and ammonia and an other ion that forms a coordination compound only with aqueous sodium hydroxide? **[KCET 2014]**

(a) Pb^{2+}, Cu^{2+} (b) Zn^{2+}, Al^{3+}
(c) Cu^{2+}, Zn^{2+} (d) Al^{3+}, Cu^{2+}

41 The IUPAC name of the complex ion formed when gold dissolves in aqua-regia is **[KCET 2014]**

(a) tetrachloridoaurate (III) (b) tetrachloridoaurate (I)
(c) tetrachloridoaurate (II) (d) dicoloridoaurate (III)

42 In $[CoF_6]^{3-}$, Co^{3+} uses outer *d*-orbitals (4*d*) in sp^3d^2-hybridisation. The number of unpaired electrons present in complex ion is **[EAMCET 2014]**

(a) 0 (b) 4
(c) 2 (d) 3

43 Crystal field theory does not explain which of the following property of coordination compounds? **[EAMCET 2014]**

(a) The covalent character of the bond between metal and the ligand
(b) Magnetic property
(c) Colour
(d) Structure of coordination compounds

44 A magnetic moment of 1.73 BM will be shown by one among the following. **[NEET 2013]**

(a) $[Cu(NH_3)_4]^{2+}$ (b) $[Ni(CN)_4]^{2-}$
(c) $TiCl_4$ (d) $[CoCl_6]^{4-}$

45 The complex used as an anticancer agent is **[AIIMS 2013]**

(a) *meso* - $[Co(NH_3)_3Cl_3]$ (b) *cis* - $[PtCl_2(NH_3)_2]$
(c) *cis* - $K_2[PtCl_2Br_2]$ (d) Na_2CoCl_4

46 The coordination number of platinum in $[Pt(NH_3)_4Cl_2]^{2+}$ ion is **[UP CPMT 2013]**

(a) 2 (b) 4
(c) 6 (d) 8

47 Which of the following ligands is not a chelating agent? **[WB JEE 2013]**

(a) EDTA (b) en
(c) Oxalate (d) Pyridine

48 Which of the following salts is colourless? **[WB JEE 2013]**

(a) $CdCl_2$ (b) $CuSO_4 \cdot 5H_2O$
(c) $MnSO_4 \cdot 7H_2O$ (d) $NiSO_4 \cdot 7H_2O$

49 The oxidation number of Co in $[Co(NH_3)(NO_2)_3]$ is **[WB JEE 2013]**

(a) +3 (b) 0
(c) −3 (d) + 6

50 IUPAC name of $K_3[Fe(CN)_6]$ is **[KCET 2013]**

(a) potassium ferricyanide
(b) potassium hexacyanoferrate (I)
(c) potassium hexacyanoferrate (III)
(d) potassium hexacyanoferrate (II)

51 The IUPAC name of $Ni(CO)_4$ is **[Manipal 2013]**

(a) tetracarbonylnickelate (0) (b) tetracarbonylnickel (II)
(c) tetracarbonylnickel (0) (d) tetracarbonylnickelate (II)

52 What is the oxidation number of Fe in $Fe(CO)_5$? **[UP CPMT 2013]**

(a) + 3 (b) Zero
(c) + 2 (d) + 5

53 The charge on the central metal ion in the complex $[Ni(CO)_4]$ is **[Kerala CEE 2013]**

(a) + 2 (b) + 4
(c) 0 (d) + 3
(e) + 1

54 The non-existant metal carbonyl among the following is **[J&K CET 2013]**

(a) $Cr(CO)_6$ (b) $Mn(CO)_5$
(c) $Ni(CO)_4$ (d) $Fe(CO)_5$

55 Which one of the following is an outer orbital complex and exhibits paramagnetic behaviour? **[CBSE AIPMT 2012]**

(a) $[Ni(NH_3)_6]^{2+}$ (b) $[Zn(NH_3)_6]^{2+}$
(c) $[Cr(NH_3)_6]^{3+}$ (d) $[Co(NH_3)_6]^{3+}$

56. Red precipitate is obtained when ethanol solution of dimethylglyoxime is added to ammoniacal Ni(II). Which of the following statements is not correct? **[CBSE AIPMT 2012]**

(a) Red complex has a square planar geometry
(b) Complex has symmetrical H-bonding
(c) Red complex has a tetrahedral geometry
(d) Dimethylglyoxime functions as bidentate ligand

57 Compounds $[PtCl_2(NH_3)_4]Br_2$ and $[PtBr_2(NH_3)_4]Cl_2$, shows the following type of isomerism. **[AFMC, AMU 2012]**

(a) Hydrate isomerism
(b) Ionisation isomerism
(c) Linkage isomerism
(d) Coordination position isomerism

58 The IUPAC name of $K_2[Ni(CN)_4]$ is **[AFMC, KCET 2012]**

(a) potassium tetracyanonickelate (II)
(b) potassium tetracyanatonickelate (III)
(c) potassium tetracyanatonickel (II)
(d) potassium tetracyanonickel (III)

59 Which of the following compounds shows optical isomerism? **[Manipal 2012]**

(a) $[Co(CN)_6]^{3-}$ (b) $[Cr(C_2O_4)_3]^{3+}$
(c) $[ZnCl_4]^{2-}$ (d) $[Cu(NH_3)_4]^{2+}$

60 In the brown ring complex $[Fe(H_2O)_5(NO)]SO_4$, nitric oxide behaves as **[WB JEE 2012]**

(a) NO^+ (b) neutral, no molecule
(c) NO^- (d) NO^{2-}

61 The *d*-electron configuration of Cr^{2+}, Mn^{2+}, Fe^{2+} and Co^{2+} are d^4, d^5, d^6 and d^7 respectively. Which one of the following will exhibit minimum paramagnetic behaviour? (At. no. Cr = 24, Mn = 25, Fe = 26, Co = 27) **[CBSE AIPMT 2011]**

(a) $[Cr(H_2O)_6]^{2+}$ (b) $[Mn(H_2O)_6]^{2+}$
(c) $[Fe(H_2O)_6]^{2+}$ (d) $[Co(H_2O)_6]^{2+}$

62 Which of the following complex ions, is diamagnetic in nature? **[CBSE AIPMT 2011]**

(a) $[CoF_6]^{3-}$ (b) $[NiCl_4]^{2-}$
(c) $[Ni(CN)_4]^{2-}$ (d) $[CuCl_2]^{2-}$

63 The complexes $[Co(NH_3)_6][Cr(CN)_6]$ and $[Cr(NH_3)_6][Co(CN)_6]$ are the examples of which type of isomerism? **[CBSE AIPMT 2011]**

(a) Geometrical isomerism
(b) Linkage isomerism
(c) Ionisation isomerism
(d) Coordination isomerism

64 The complex, $[Pt(py)(NH_3)BrCl]$ will have how many geometrical isomers? **[CBSE AIPMT 2011]**

(a) 2 (b) 3 (c) 4 (d) 0

65 Which can exist both as diastereoisomer and enantiomer? **[DUMET 2011]**
(a) $[Pt(en)_3]^{4+}$ (b) $[Pt(en)_2ClBr]^{2+}$
(c) $[Ru(NH_3)_4Cl_2]^0$ (d) $[PtCl_2Br_2]^0$

66 Number of isomeric forms (constitutional and stereoisomers) for $[Rh(en)_2(NO_2)(SCN)]^+$ are **[DUMET 2011]**
(a) three (b) six
(c) nine (d) twelve

67 Four transition metal octahedral complexes, (the choice given below) low spin electronic configurations arises only for **[DPMT 2011]**
(a) d^1 to d^3 complexes (b) d^4 to d^7 complexes
(c) d^7 to d^9 complexes (d) d^1, d^2 and d^8 complexes

68 Excess of silver nitrate solution is added to 100 mL of 0.01 M pentaaquachlorochromium (III) chloride solution. The mass of silver chloride obtained (in gram) is [Atomic mass of silver is 108] **[KCET 2011]**
(a) 287×10^{-3} (b) 143.5×10^{-3}
(c) 143.5×10^{-2} (d) 287×1^{-2}

69 The IUPAC name of the complex $[Co(NH_3)_4Cl_2]Cl$ is **[KCET 2011]**
(a) dichlorotetraamminecobalt (III) chloride
(b) tetraamminedichlorocobalt (III) chloride
(c) tetraamminedichlorocobalt (II) chloride
(d) tetraamminedichlorocobalt (IV) chloride

70 The ligand $N(CH_2CH_2NH_2)_3$ is **[Kerala CEE 2011]**
(a) tridentate
(b) pentadentate
(c) tetradentate
(d) bidentate
(e) hexadentate

71 Ammonia gas does not evolve from the complex $FeCl_3\cdot 4NH_3$ but it gives white precipitate with aqueous solution of $AgNO_3$. Coordination number of central metal ion in the above complex is six. Give IUPAC name of the complex. **[Guj. CET 2011]**
(a) Ammonium trichloro triammine ferrate (III)
(b) Tetraamine iron (III) chloride
(c) Dichlorotetraammine iron (II) chloride
(d) Dichlorotetraammine iron (III) chloride

ANSWERS

Check point 12.1

1. (c)	2. (b)	3. (a)	4. (d)	5. (c)	6. (b)	7. (b)	8. (a)	9. (c)	10. (a)

Check point 12.2

1. (b)	2. (a)	3. (c)	4. (d)	5. (c)	6. (d)	7. (d)	8. (a)	9. (c)

Check point 12.3

1. (b)	2. (d)	3. (d)	4. (a)	5. (a)	6. (d)	7. (b)	8. (a)	9. (a)	10. (d)

Check point 12.4

1. (a)	2. (b)	3. (b)	4. (a)	5. (b)	6. (b)	7. (a)	8. (c)	9. (c)	10. (a)

Check point 12.5

1. (d)	2. (b)	3. (d)	4. (d)	5. (d)	6. (d)	7. (c)	8. (c)	9. (b)	10. (d)

(A) Topical exercise

1. (a)	2. (b)	3. (b)	4. (a)	5. (a)	6. (b)	7. (a)	8. (a)	9. (d)	10. (a)
11. (d)	12. (b)	13. (b)	14. (c)	15. (c)	16. (d)	17. (c)	18. (c)	19. (c)	20. (b)
21. (b)	22. (b)	23. (a)	24. (d)	25. (c)	26. (b)	27. (c)	28. (a)	29. (c)	30. (d)
31. (a)	32. (a)	33. (a)	34. (b)	35. (d)	36. (b)	37. (d)	38. (b)	39. (a)	40. (d)
41. (b)	42. (a)	43. (b)	44. (a)	45. (b)	46. (a)	47. (d)	48. (b)	49. (a)	50. (a)
51. (d)	52. (b)	53. (b)	54. (a)	55. (c)	56. (c)	57. (a)	58. (a)	59. (a)	60. (a)
61. (c)	62. (b)	63. (a)	64. (c)	65. (a)	66. (a)	67. (b)	68. (c)	69. (c)	70. (b)
71. (b)	72. (a)	73. (a)	74. (b)	75. (b)	76. (a)	77. (a)	78. (a)	79. (a)	80. (b)
81. (a)	82. (a)	83. (a)	84. (a)	85. (b)	86. (c)	87 (b)	88. (c)	89. (d)	90. (c)
91. (d)	92. (c)	93. (d)	94. (b)	95. (c)	96. (a)	97. (b)	98. (b)	99. (c)	100. (c)
101. (d)	102. (b)	103. (a)	104. (a)	105. (a)					

(B) Miscellaneous questions

1. (c)	2. (a)	3. (c)	4. (d)	5. (a)	6. (b)	7. (a)	8. (c)	9. (b)	10. (b)
11. (d)	12. (a)	13. (c)	14. (b)	15. (c)	16. (b)	17. (c)	18. (a)	19. (c)	20. (d)
21. (a)	22. (a)	23. (c)	24. (b)	25. (c)	26. (d)	27. (a)	28. (c)	29. (a)	30. (b)
31. (c)	32. (d)	33. (c)	34. (b)	35. (b)	36. (d)	37. (c)	38. (a)	39. (a)	40. (a)
41. (c)	42. (a)	43. (a)	44. (b)	45. (d)	46. (d)	47. (a)	48. (a)	49. (c)	50. (c)
51. (b)	52. (b)	53. (c)	54. (d)	55. (a)	56. (a)	57. (c)	58. (d)	59. (d)	60. (a)
61. (c)	62. (d)	63. (d)	64. (c)						

(C) Medical entrances' gallery

1. (b)	2. (d)	3. (a)	4. (d)	5. (a)	6. (a)	7. (a)	8. (c)	9. (b)	10. (c)
11. (b)	12. (a)	13. (a)	14. (b)	15. (d)	16. (a)	17. (b)	18. (a)	19. (c)	20. (c)
21. (b)	22. (b)	23. (a)	24. (a)	25. (c)	26. (a)	27. (d)	28. (a)	29. (c)	30. (a)
31. (e)	32. (c)	33. (b)	34. (a)	35. (b)	36. (a)	37. (c)	38. (b)	39. (a)	40. (b)
41. (a)	42. (b)	43. (a)	44. (a)	45. (b)	46. (c)	47. (d)	48. (a)	49. (a)	50. (a,c)
51. (c)	52. (b)	53. (c)	54. (b)	55. (a)	56. (c)	57. (b)	58. (a)	59. (b)	60. (a)
61. (d)	62. (c)	63. (d)	64. (b)	65. (b)	66. (d)	67. (b)	68. (a)	69. (b)	70. (c)
71. (d)									

Hints & Explanations

(A) Topical exercise

1 (*a*) $CoCl_3 \cdot 6NH_3$ exists as $[Co(NH_3)_6]\,Cl_3$ in which six NH_3 ligands are coordinated to cobalt.

2 (*b*) $CuSO_4 \cdot 5H_2O$ is a monohydrate salt with four H_2O ligands coordinated to Cu^{2+}.

$$[Cu(H_2O)_4]\,SO_4 \cdot H_2O$$

H_2O is joined to SO_4^{2-} by H-bonding.

3 (*b*) Ligand must donate a pair of electron or loosely held electron pair to metal and form a *M*–*L* bond.

e.g. $\ddot{N} = O, \ddot{N}H_2CH_2CH_2\ddot{N}H_2, :CO,$

Among the given species NH_4^+ does not have any pair of electron.

Hence, NH_4^+ is not a ligand.

4 (*a*) NO_2^- is an ambident ligand. It has two binding donor atoms and can coordinate either through nitrogen or through oxygen to a central metal atom or ion.

5 (*a*) A chelating ligand has two or more binding donor atoms attached to a single metal ion.

e.g.

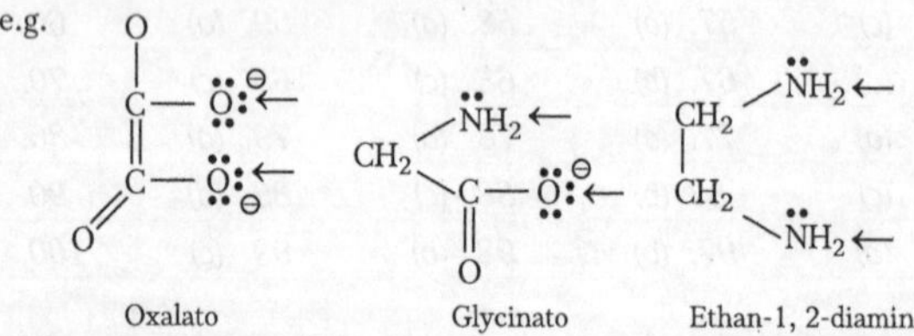

Oxalato Glycinato Ethan-1, 2-diamine

Here, (←) denotes binding site.

Among the given ligands, thiosulphato ($S_2O_3^{2-}$) is not a chelating ligand because geometrically it is not favourable for $S_2O_3^{2-}$ to chelate a metal ion.

11 (*d*) O_2^- (superoxide) — monodentate

en — bidentate

Br — monodentate

Thus, C.N. of Pt is 6

$$[Pt(O_2)\,(en)_2\,(Br)]^{2+}$$

$$x - 1 + 0 - 1 = 2$$

$$\therefore \quad x = +4$$

Thus, O.N. of Pt = + 4.

13 (*b*) Oxidation state of central metal ion can be calculated by considering the oxidation state of whole molecule is equal to charge present on coordination sphere.

A. $[Co(NCS)(NH_3)_5]SO_3$.

Let the oxidation state of Co be *x*.

$$x - 1 + 5 \times 0 = +2$$
$$x = +2 + 1 \Rightarrow +3$$

B. $[Co(NH_3)_4\,Cl_2]SO_4$

Let the oxidation state of Co = *x*

$$\Rightarrow \quad x + 4 \times 0 + 2 \times (-1) = +2$$
$$\Rightarrow \quad x - 2 = +2 \Rightarrow x = +4$$

C. $Na_4[Co(S_2O_3)_3]$

Let the oxidation state of Co = *x*

$$x + 3 \times (-2) = -4$$
$$x - 6 = -4$$
$$x = -4 + 6 \Rightarrow +2$$

D. $[Co(CO)_8]$

Let the oxidation state of Co = *x*

$$x - 8 \times 0 = 0 \Rightarrow x = 0$$

14 (*c*) $[Co(NH_3)_4\,Cl_2]Cl \rightleftharpoons [Co(NH_3)_4\,Cl_2]^+ + Cl^-$.

15 (*c*) C.N. of Cr = 6 and O.N. = 3

Thus, $[Cr(H_2O)_4\,Cl_2]$ Cl and $[Cr(H_2O)_3Cl_3]\cdot H_2O$.

16 (*d*)

Complex	O.N. of Cr	Electron due to ligands	EAN
(a) $[Cr(CN)_6]^{3-}$	+3	12	24 − 3 + 12 = 33
(b) $[Cr(H_2O)_4(NH_3)_2]^{3+}$	+3	12	33
(c) $[Cr(H_2O)_4\,(C_2O_4)]^+$	+3	12	33

17 (*c*) $[(Co\,(NH_3)_2\,(O_2)\,(en)\,Cl]$

$$x \quad 0 \quad y \quad 0 \quad -1 = 0$$
$$x + y = 1$$

Let $x = +2$ then $y = -1$

and EAN = 27 − 2 + 12 = 37

$x = +3$ then $y = -2$

and EAN = 27 − 3 + 12 = 36.

Thus, O_2 is peroxide (O_2^{2-}).

18 (*c*) EDTA is hexadentate

O.N. of Mg = +2

Thus, EAN = 12 − 2 + 12 = 22

21 (*b*) EAN for $[Ag(CN)_2]^-$ = 47 − 1 + 2 × 2 = 50

Hence, it does not follow EAN rule,

22 (*b*)

	O.N.	Electrons donated due to ligands	EAN
(a) $[Cr(EDTA)]^-$	+3	6×2	24 − 3 + 12 = 33
(b) $[Co(en)_3]^{3+}$	+3	6×2	27 − 3 + 12 = 36
(c) $[Fe(C_2O_4)_3]^{3-}$	+3	6×2	26 − 3 + 12 = 35
(d) $[Ni(CN)_4]^{2-}$	+3	6×2	26 − 3 + 12 = 35

23 (*a*)

Complex	*Z*	O.N. of metal	Electron gained due to ligands	EAN
$[Ni(CN)_4]^{2-}$	28	+2	8	28 − 2 + 8 = 34
$[Fe(CN)_6]^{3-}$	26	+3	12	26 − 5 + 12 = 35
$[Cu(CN)_4]^{3-}$	29	+1	8	29 − 1 + 8 = 36

28 (*a*) $[Cr(C_2O_4)_3]$
+3 −6 = −3

Thus, $K_3[Cr(C_2O_4)_3]$

29 (*c*) $[Pt(NH_3)_6]^{4+}[Pt\,Cl_4]^{2-}$

Thus, $[Pt\,(NH_3)_6]\,[PtCl_4]_2$.

31 (*a*) $[Cr(H_2O)_4\,Cl_2]^+$ shows geometrical isomerism because it is a MA_4B_2 type coordination compound which contains two set of equivalent ligands, four H_2O and 2Cl.

Hence, the possible geometrical isomers are

trans-isomer and *cis*-isomer

36 (*b*) Only statement I is correct.

Complexes of MX_6 and MX_5L type (*X* and *L* are unidentate) do not show geometrical isomerism due to presence of plane of symmetry and necessary condition for showing geometrical isomerism is that complex is must of MA_4B_2 type or $[M(AB)_2X_2]$ type with octahedral geometry.

Plane of symmetry

38 (*b*) Solvate isomerism is shown when two compounds having same molecular formula differ by water or solvent molecule and is directly bonded to metal ion or is present as free solvent molecules in the crystal lattice.

When water is present as solvent and show this type of isomerism then it is known as hydrate isomerism.

Coordination compounds $[Cr(H_2O)_6]Cl_3$ and $[Cr(H_2O)_5Cl]H_2O \cdot Cl_2$ are solvate isomers because water is exchanged by chloride ion. This is why both of them show different colour on exposure to sunlight.

39 (*a*) $[Co(NH_3)_4\,BrCl]Br$—ionisation isomer

cis and *trans* are geometrical isomers of the given complex.

40 (*d*) Compounds having same molecular formula but different structural formula are known as isomers. $[Co(SO_4)_2(NH_3)_5]Br$ and $[Co(SO_4)(NH_3)_5]Cl$ do not have same molecular formula. Hence, they are not isomers.

42 (*a*) The ligand(s) which has two different bonding sites are known as ambident ligands, e.g. NCS, NO_2 etc.

Here, NCS has two binding sites at N and S.

Hence, NCS (thiocyanate) can bind to the metal ion in two ways

$$M \leftarrow \text{NCS or } M \leftarrow \text{SNC}$$

Thus, coordination compounds containing NCS as a ligand can show linkage isomerism, i.e. $[Pd(C_6H_5)_2(SCN)_2]$ and $[Pd(C_6H_5)_2(NCS)_2]$ are linkage isomers.

44 (*a*) Both statements I and II are correct. Linkage isomerism arises in coordination compounds containing ambidentate ligands because ambidentate ligand has two different donor atoms.

e.g. SCN, NO_2 etc.

45 (*b*) Isomerism in coordination compound is decided by type of ligands and geometry of coordination and arrangement of ligands.

A. $[Co(NH_3)_4\,Cl_2]^+$ shows geometrical isomerism due to presence of two types of ligand whose $[Co(NH_3)_4\,Cl_2]^+$ arrangement around central metal ion.

cis-isomer *trans*-isomer

B. *cis* − $[Co(en)_2Cl_2]^+$ shows optical isomer due to its non-superimposable mirror image relationship.

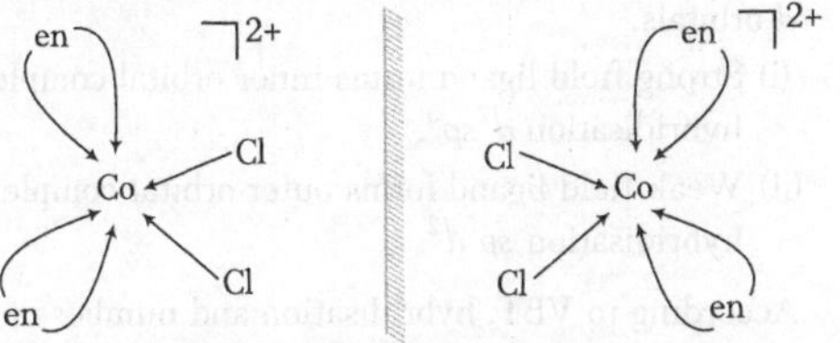

C. $[Co(NH_3)_5(NO_2)]Cl_2$ shows ionisation isomer due to its interchanging ligand from outside the ionisation sphere.

D. $[Co(NH_3)_6][Cr(CN)_6]$ shows coordination isomer due to interchanging of ligand in between two metal ions from one coordination sphere to another coordination sphere.

46 (*a*) $[Co(NH_3)_3Cl_3]$ is molecular complex no ionisation occur. Primary valency is three thus only (a) is correct.

======== Indicates primary and secondary valencies
———— Indicates secondary valencies
-------- Indicates primary valencies

47 (*d*) $[Pt(NH_3)_4Cl_2]Cl_2 \rightleftharpoons [Pt(NH_3)_4Cl_2]^+ + 2Cl^-$

Two Cl^- ligands are also in coordination sphere thus, they satisfy primary as well as secondary valencies. Four NH_3 ligands satisfy secondary valencies.

51 (*d*) 1 mole of $AgNO_3$ precipitates one free chloride ion (Cl^-). Here, 3 moles of AgCl are precipitated by excess of $AgNO_3$. Hence, there must be three free Cl^- ions.

So, the formula of the complex can be $[Cr(H_2O)_6]Cl_3$ and correct choice is (d).

52 (*b*) $\underset{1\ \text{mol}}{CoCl_3 \cdot 6H_2O} + AgNO_3 \longrightarrow 2$ mol AgCl

Thus, two Cl atoms are outside the coordinate sphere.

C.N. of Co = 6

Thus, $[Co(H_2O)_5Cl]Cl_2 \cdot H_2O$.

59 (*a*) Both statements I and II are correct. $[Cr(H_2O)_6]Cl_2$ and $[Fe(H_2O)_6]Cl_2$ are reducing in nature due to formation of more stable complex ion after losing of electron.

64 (*c*) Statement II is correct.

According to VBT, molecular orbit electronic configuration of Fe^{3+} in $[Fe(CN)_6]^{3-}$ is

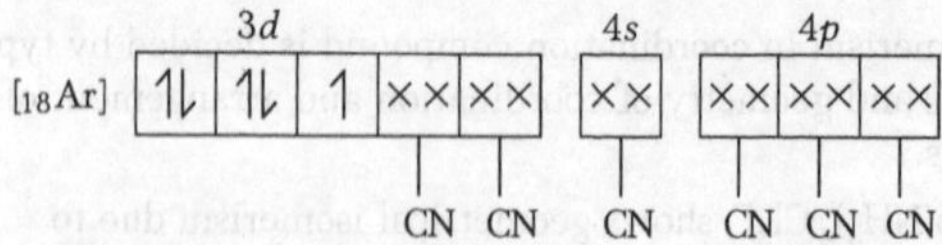

Hybridisation = d^2sp^3, Number of unpaired electrons (N) = 1

Hence, correct statement I is

$[Fe(CN)_6]^{3-}$ ion shows magnetic moment corresponding to one unpaired electron.

i.e. $\mu = \sqrt{N(N+2)} = \sqrt{1(1+2)}$

$= \sqrt{3} = 1.73$ BM

65 (*a*) Formation of inner orbital complex and outer orbital complex determines hybridisation of molecule which in turn depends upon field strength of ligand and number of vacant *d*-orbitals.

(i) Strong field ligand forms inner orbital complex with hybridisation d^2sp^3.

(ii) Weak field ligand forms outer orbital complex with hybridisation sp^3d^2.

According to VBT, hybridisation and number of unpaired electrons of coordination compounds can be calculated as

A. $[Cr(H_2O)_6]^{3+}$

MOEC (Molecular orbital electronic configuration) of Cr^{3+} in $[Cr(H_2O)_6]^{3+}$ is

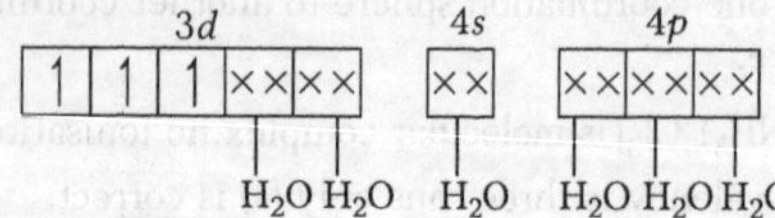

Hybridisation = d^2sp^3

N (number of unpaired electrons) = 3

B. $[Co(CN)_4]^{2-}$

MOEC of Co^{2+} in $[Co(CN)_4]^{2-}$ is

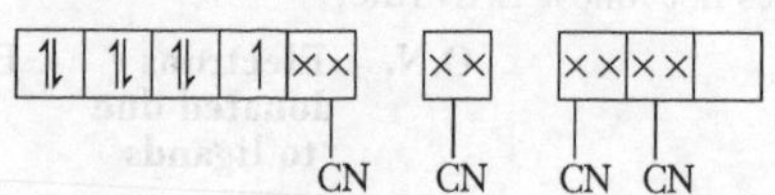

Hybridisation = dsp^2

$N = 1$

C. $[Ni(NH_3)_6]^{2+}$

MOEC of Ni^{2+} in $[Ni(NH_3)_6]^{2+}$ is

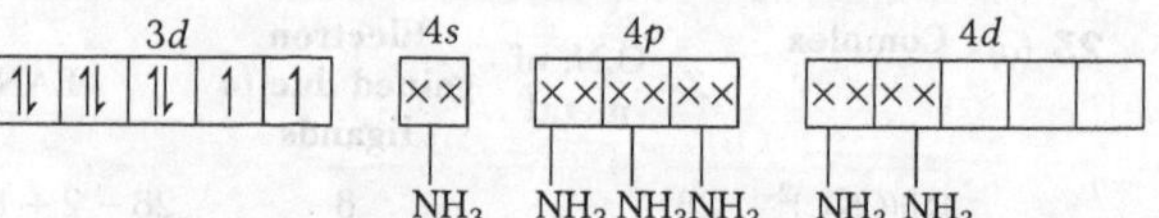

Hybridisation = sp^3d^2

$N = 2$

D. $[MnF_6]^{4-}$

MOEC of Mn^{2+} in $[MnF_6]^{4-}$ is

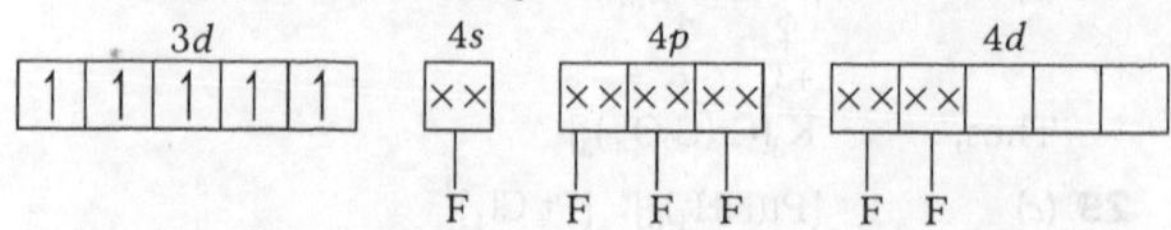

Hybridisation = sp^3d^2

$N = 5$

69 (*c*) CFSE for octahedral and tetrahedral complexes are closely related to each other by formula $\Delta_t = \dfrac{-4}{9}\Delta_o$.

where, Δ_o = CFSE for octahedral complex, Δ_t = CFSE for tetrahedral complex ($\Delta E = h\nu\ \Delta E \propto \nu$)

According to question,

$\Delta_o = 18000\ \text{cm}^{-1}$

$\therefore \Delta_t = \dfrac{4}{9}\Delta_o = \dfrac{4}{9} \times 18000\ \text{cm}^{-1} = 4 \times 2000\ \text{cm}^{-1} = 8000\ \text{cm}^{-1}$

76 (*a*) MnO_4^- (oxidation number) = + 7 thus their is no *d*-electrons in Mn^{7+} ion.

79 (*a*)

(a) $[Co(CN)_6]^{3-}$	d^2sp^3	0	0
(b) $[Fe(CN)_6]^{3-}$	d^2sp^3	1	$\sqrt{3}$ BM
(c) $[Mn(CN)_6]^{3-}$	d^2sp^3	2	$\sqrt{8}$ BM
(d) $[Cr(CN)_6]^{3-}$	d^2sp^3	3	$\sqrt{15}$ BM

Therefore, least paramagnetism is in (a).

82 (*a*) Mn^{2+} has five unpaired electrons.

84 (*a*) In square planar geometry as $[Ni(CN)_4]^{2-}$

Unpaired electron (N) = 0

In octahedral geometry

Unpaired electron (N) = 2

Then μ (magnetic moment) = $\sqrt{2 \times 4}$ BM = $\sqrt{8}$ BM = 2.83 BM

86 (*c*) Magnetic moment (μ) = $\sqrt{N(N+2)}$ BM

$3.83 = \sqrt{N(N+2)}$

$$3.83 \times 3.83 = N^2 + 2N$$
$$14.6689 = N^2 + 2N \Rightarrow N = 3$$

Hence, number of unpaired electrons in d-subshell of penultimate shell of chromium is 3.

$$Cr^{3+} = 1s^2\, 2s^2\, 2p^6\, 3s^2\, 3p^6\, 3d^3$$

In $[Cr(H_2O)_6]Cl_2$ oxidation state of Cr is +3.

Hence, in $3d^3$, the distribution of electrons :

$$3d^1_{xy}\, 3d^1_{yz}\, 3d^1_{zx}\, 3d^0_{x^2-y^2}\, 3d^0_{z^2}$$

87 (*b*) $[Cr(H_2O)_6]^{2+} = Cr^{2+} = 3d^4$; four unpaired electrons.

$[CoCl_4]^{2-} = Co^{2+} = 3d^7 = 3$ unpaired electrons.

$[Fe(H_2O)_6]^{2+} = Fe^{2+} = 3d^6 = 4$ unpaired electrons.

$[Mn(H_2O)_6]^{2+} = Mn^{2+} = 3d^5 = 5$ unpaired electrons.

Thus, $[Cr(H_2O)_6]^{2+}$ and $[Fe(H_2O)_6]^{2+}$ have same magnetic moment as they have same number of unpaired electrons.

90 (*c*) As we know that, strong field ligand split the five degenerate energy levels with more energy separation than weak field ligand, i.e. as strength of ligand increases crystal field splitting energy increases.

Hence, $\Delta E = \dfrac{hc}{\lambda}$

$\Rightarrow \quad \Delta E \propto \dfrac{1}{\lambda} \Rightarrow \lambda \propto \dfrac{1}{\Delta E}$

As energy separation increases, the wavelength decreases. Thus, the correct order for wavelength absorbed is

$$[Co(H_2O)_6]^{3+} > [Co(NH_3)_6]^{3+} > [Co(CN)_6]^{3-}$$

Here, strength of ligand increases, ΔE increases, CFSE increases and λ absored decreases.

Hence, correct choice is (c).

92 (*c*) Chelation (formation of cycle by linkage between metal ion and ligand) stabilises the coordination compound. The ligand which chelates the metal ion are known as chelating ligand. Here, only $[Fe(C_2O_4)_3]^{3-}$ is a coordination compound which contains oxalate ion as a chelating ligand. Hence, it stabilises coordination compound by chelating Fe^{3+} ion.

93 (*d*) Most stable due to chelation.

94 (*b*) Greater the value of log *K*, greater will be stability of complex compound formed.

For reaction, $Cu^{2+} + 4CN^- \longrightarrow [Cu(CN)_4]^{2-}$

$$K = \frac{[(Cu(CN)_4)^{2-}]}{[Cu^{2+}][CN^-]^4} \text{ and } \log K = 27.3$$

For this reaction, log *K* has highest value among the given four reactions. Hence, *K* will also be higher among these four, i.e. stability of the complexes will be highest among these four complexes.

95 (*c*) Stability of complex increases with increase in charge on the central metal ion and with basic strength of ligand.

In $[Fe(CN)_6]^{3-}$, oxidation state of Fe is +3 and basicity of CN^- is higher than OH^-, Cl^- and H_2O.

96 (*a*) $M + L \rightleftharpoons ML \quad k_1 = \dfrac{[ML]}{[M][L]} = 4$

$ML + L \rightleftharpoons ML_2 \quad k_2 = \dfrac{[ML_2]}{[ML][L]} = 3$

$M + 2L \rightleftharpoons ML_2 \quad k_3 = \dfrac{[ML_2]}{[M][L]^2} = k_1k_2 = 12$

102 (*b*) Cyanide process is for Ag and Au

Mond's process is for Ni

Photographic fixing is for AgBr

$$Ag^+ + 2CN^- \longrightarrow [Ag(CN)_2]^-$$
$$Ni + 4CO \longrightarrow [Ni(CO)_4]$$
$$AgBr + 2Na_2S_2O_3 \longrightarrow [Ag(S_2O_3)_2]^{3-}$$

104 (*a*) Central metal ions present in coordination compounds determine the properties of coordination compound and their biological role.

	Column I (Coordination compound)		Column II (Central metal atom)
A.	Chlorophyll	3.	Magnesium
B.	Blood pigment	4.	Iron
C.	Wilkinson catalyst	1.	Rhodium
D.	Vitamin B_{12}	2.	Cobalt

105 (*a*) Both statements I and II are correct.

Toxic metal ions are removed by chelating ligands. When a solution of chelating ligand is added to solution containing toxic metals ligands chelates the metal ions by formation of stable complex.

(B) Miscellaneous questions

1 (*c*) $FeSO_4 + 6KCN \longrightarrow K_4[Fe(CN)_6] + K_2SO_4$

2 (*a*) $K^+\; Pt^{4+}\; (NH_3)\; (CN^-)_2\; (SO_4^{2-})\; O_2^-$

$\downarrow$ 4, +0, −2, −2, −1 → −1

Thus, $K[Pt(NH_3)_2(CN)_2(SO_4)(O_2)]$

10 (*b*) $Fe^{3+} + 3C_2O_4^{2-} \longrightarrow [Fe(C_2O_4)_3]^{3-}$

13 (*c*) A pair of molecules related to each other as an object to its mirror image are known as enantiomorphs or enantiomers. These are not superimposable on its mirror image, e.g. $[Co(en)_2Cl_2]^+$.

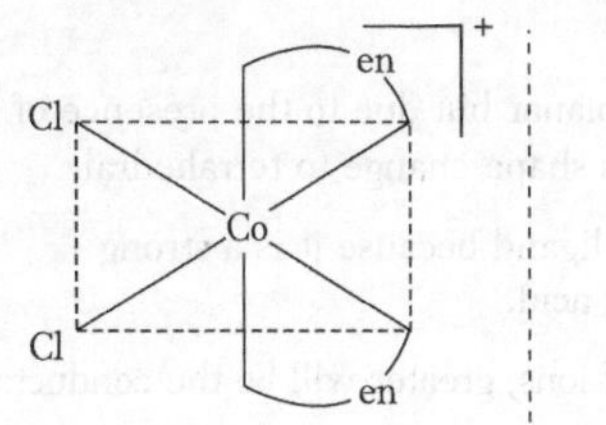

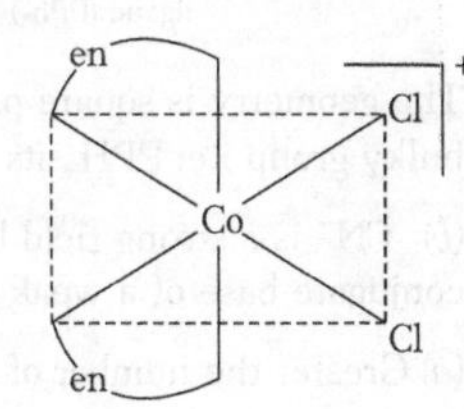

14 (b) IUPAC name of sodium nitroprusside $Na_2[Fe(CN)_5NO]$ is sodium pentacyanonitrosylferrate (III) because in it NO is neutral ligand and oxidation number of Fe is +3.

15 (c) Monovalent cationic complex is one in which oxidation state of metal is one $[Co(NH_3)Cl_2]^+ Cl^-$.

29 (a) $\underset{\text{Sodium nitroprusside}}{Na_2S + Na_2[Fe(NO)(CN)_5]} \longrightarrow \underset{\substack{\text{Sodium thionitroprusside} \\ \text{(violet colour)}}}{Na_4[Fe(CN)_5NOS]}$

30 (b) One mole of $AgNO_3$ precipitates one mole of chloride ion. In the above reaction, when 0.1 mole $CoCl_3(NH_3)_5$ is treated with excess of $AgNO_3$, 0.2 mole of AgCl are obtained thus, there must be two free chloride ions in the solution of electrolyte.

So, molecular formula of complex will be $[Co(NH_3)_5 Cl] Cl_2$ and electrolytic solution must contain $[Co(NH_3)_5Cl]^{2+}$ and two Cl^- as constituent ions.

Thus, it is 1 : 2 electrolyte.

$$[Co(NH_3)_5Cl]Cl_2 \longrightarrow [Co(NH_3)_5Cl]^{2+}(aq) + 2Cl^-(aq)$$

Hence, option (b) is the correct.

32 (d) Atoms, ions or molecules having unpaired electrons are paramagnetic.

In $[Cr(NH_3)_6]^{3+}$ Cr is present as Cr^{3+}.

$$_{24}Cr = 1s^2\ 2s^2\ 2p^6\ 3s^2\ 3p^6\ 3d^5\ 4s^1$$

$$Cr^{3+} = 1s^2\ 2s^2\ 2p^6\ 3s^2\ 3p^6\ 3d^3$$

Number of unpaired electrons = 3, so, it is paramagnetic.

33 (c) $Ni(CO)_4$; Oxidation state of Ni = 0

$Ni = 1s^2\ 2s^2 2p^6\ 3s^2 3p^6\ 4s^2\ 3d^8$

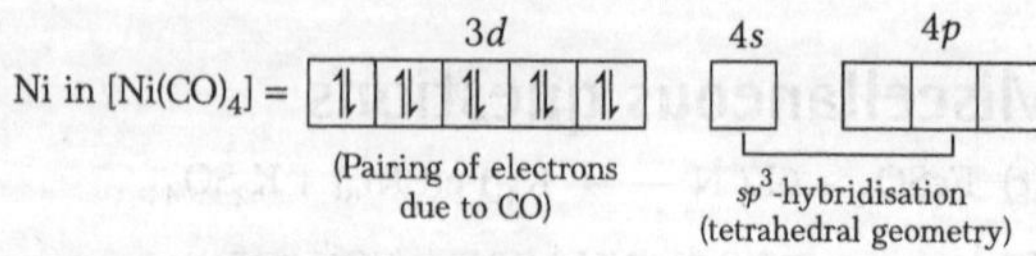

Oxidation state of Ni in $[Ni(PPh_3)_2Cl_2]$ is

$$x + 0 \times 2 + (-2) = 0$$

$$x = +2$$

Electronic configuration of Ni^{2+} in $[Ni(PPH_3)_2]Cl_2$ is

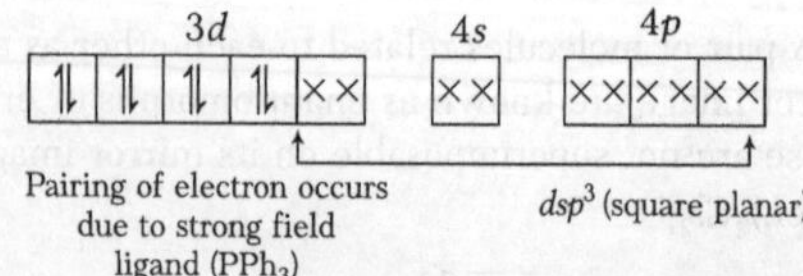

The geometry is square planar but due to the presence of two bulky group i.e. PPH_3, its shape change to tetrahedral.

34 (b) CN^- is a strong field ligand because it is a strong conjugate base of a weak acid.

48 (a) Greater the number of ions, greater will be the conductance.

1. $K_2[PtCl_6] \rightleftharpoons \underbrace{2K^+ + [PtCl_6]^{2-}}_{\text{Three ions}}$

2. $PtCl_4 \cdot 2NH_3 \rightleftharpoons [Pt(NH_3)_2Cl_4] \longrightarrow$ No ion (least conductance)

3. $[Pt(NH_3)_3Cl_3]Cl \rightleftharpoons \underbrace{[Pt(NH_3)_3Cl_3]^+ + Cl^-}_{\text{Two ions}}$

4. $[Pt(NH_3)_5Cl]Cl_3 \rightleftharpoons \underbrace{[Pt(NH_3)_5Cl]^+ + 3Cl^-}_{\text{Four ions}}$

52 (b) $[Cr(C_2O_4)_x\ (en)_y\ (O_2)_z]^{-1}$

Bidentate Monodentate

Thus, $2x + 2y + z = 6$

Also total charge = − 1

Thus, $3 + (-2x) + y(0) - z = -1$

$$3 - 2x - z = -1 \Rightarrow 2x + z = 4$$

Thus, $y = 1, x = 1$ and $z = 2$

Thus, ratio = 1: 1: 2

63 (d) Fe^{2+} [⥮ | ↿ | ↿ | ↿ | ↿]

$$\left.\begin{array}{l} NO \longrightarrow NO^+ + e^- \\ Fe^{2+} + e^- \longrightarrow Fe^+ \end{array}\right] \text{Charge transfer}$$

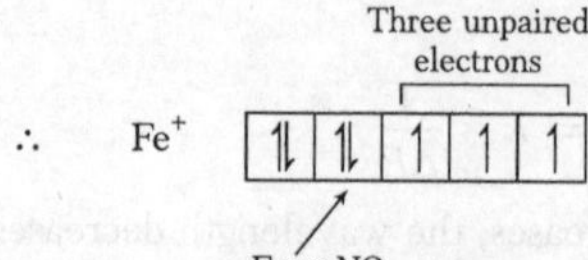

Unpaired electrons (N) = 3

Magnetic moment $= \sqrt{N(N+2)}$

$$= \sqrt{3(3+2)} = \sqrt{15}$$

$$= 3.87 \text{ BM}$$

(a), (b) and (c) are true.

(C) Medical entrances' gallery

1 (b) Correct match is

$$A \rightarrow III, B \rightarrow IV, C\text{–}I, D \rightarrow II$$

(A) CN=4, sp^3 ⟶ or tetrahedral ⇒ (III)

(B) CN=4, dsp^2 ⟶ or square planar ⇒ (IV)

(C) CN=5, sp^3d ⟶ or trigonal bipyramidal ⇒ (I)

(D) CN=6, d^2sp^3 ⟶ or octahedral ⇒ (II)

Hence, option (b) is correct.

2 *(d)* Increasing field strength of ligands to form coordination compounds, we have to follow the spectrochemical series.

$I^- < Br^- < SCN^- < Cl^- < S^{2-} < F^- < OH^- < C_2O_4^{2-} < H_2O < NCS^- < EDTA^{4-} < NH_3 < en < CN^- < CO$

So, the order given in option (d):

$SCN^- < F^- < C_2O_4^{2-} < CN^-$ is correct.

3 *(a)* Aluminium chloride in acidified aqueous solution forms an octahedral complex. Aqueous solution is mostly water and when the compound is dissolved in acidified aqueous solution, the water fills its vacancies and an octahedral complex, '*A*' which is $[Al(H_2O)_6]^{3+}$ is formed.

$Al(13) = 1s^2 2s^2, 2p^6, 3s^2, 3p^1 \Rightarrow Al^{3+} = 1s^2, 2s^2, 2p^6.$

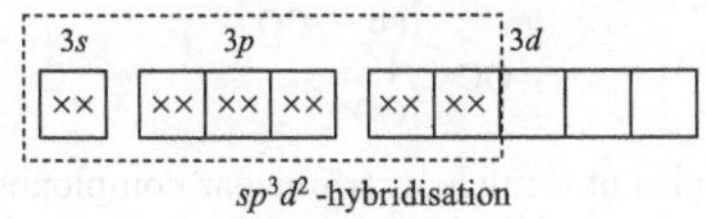

$\therefore$ $[Al(H_2O)_6]^{3+}$ has sp^3d^2 hybridisation.

4 *(d)* $[CoCl_6]^{4-}$ is octahedral while $[CoCl_4]^{2-}$ is a tetrahedral complex.

$$\Delta_{octahedral} = 18000 \text{ cm}^{-1}$$

We know that,

$$\Delta_{tetrahedral} = \frac{4}{9}\Delta_{octahedral}$$

$$= \frac{4}{9} \times 18000 \text{ cm}^{-1} = 8000 \text{ cm}^{-1}$$

Hence, option (d) is correct.

5 *(a)* Oxalate $(C_2O_4^{2-})$ is a bidentate strong field ligand, thus it can use six orbitals of Co^{3+} ion having d^2sp^3-hybridisation.

In $[Co(C_2O_4)_3]^{3-}$, Co^{3+} has all the paired electrons, therefore it is a low spin complex.

Hence, option (a) is correct.

6 *(a)* The wavelength of absorption in visible region for a complex (e.g. Co^{3+}) depends on the nature of ligand, i.e. stronger be the ligand (i.e. higher the value of Δ_o) lower is the wavelength.

$$\text{Wavelength } (\lambda) \propto \frac{1}{\Delta_0}$$

$\therefore$ Strength of ligands in the spectrochemical series are in order: $CN^- > NH_3 > H_2O > Cl^-$.

Therefore, the correct increasing order for wavelength of absorption is, $[Co(CN)_6]^{3-}, [Co(NH_3)_6]^{3+}, [Co(NH_3)_5(H_2O)]^{3+}, [Co(NH_3)_5Cl]^{2+}$

Thus, option (a) is the correct.

7 *(a)* In $K_4[Fe(CN)_6]$,

$Fe^{2+} = [Ar]3d^6 4s^0$.

CN^- is a strong field ligand and as it approaches the metal ion, the electrons must pair up.

The splitting of the *d*-orbitals in two sets orbitals in an octahedral complex, $K_4[Fe(CN)_6]$ may be represented as :

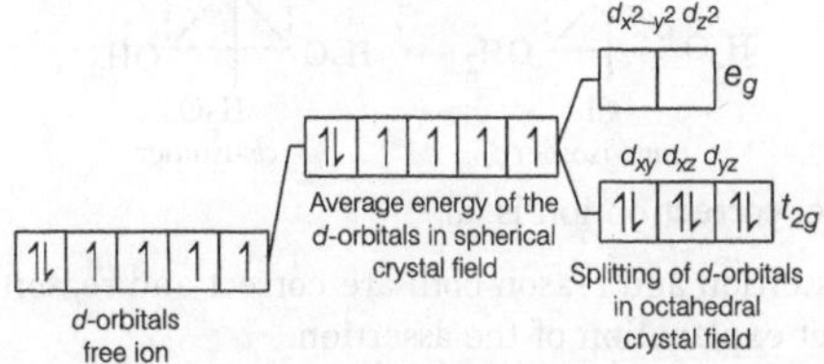

Hence, the electronic configuration of the central atom in $K_4[Fe(CN)_6]$ is $t_{2g}^6 e_g^0$.

8 *(c)* Isomers in which the atoms or ligands occupy different positions around central metal/ion are called geometrical isomers. Complexes having coordination number six of central atom/ion with formula $M(AA)_2B_2$ exhibit geometrical isomerism [where, *AA* is a bidentate ligand].

In $[CoCl_2(en)_2]$, coordination number of Co is 6 with octahedral geometry.

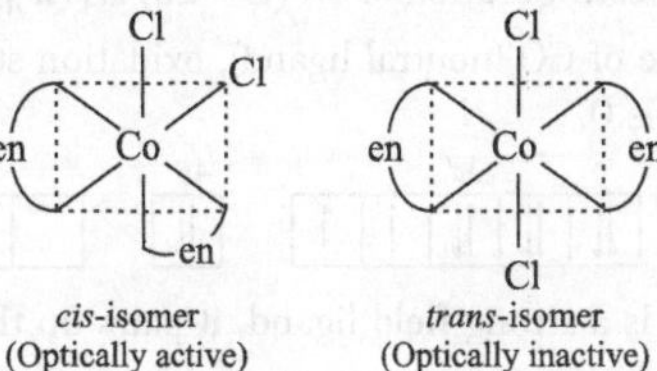

Thus, $[CoCl_2(en)_2]$ show geometrical isomerism.

9 *(b)* Let the oxidation number of Cr be *x* in the complex $[Cr(NH_3)_4Cl_2]^+$

$\therefore$ $x + 4 \times (0) + 2 \times (-1) = +1$

$x - 2 = +1$

$x = +3$

Thus, the oxidation number of Cr is + 3.

10 *(c)* $[Pt(Py)_4][PtCl_4]$

Let the oxidation state of Pt be *x*.

$x + 0 \times 4 + x + (-1) \times 4 = 0$

or, $2x = +4$ or, $x = +2$

Thus, it is the complex in which the central metal atom in cationic and anionic parts have same oxidation state.

11 *(b)* The oxidation state of central atom (Pt) is calculated as

Let the oxidation state of Pt = *x*, then

$(x \times 1) + 3 \times (0) + (1 \times -1) + (1 \times -1) + (1 \times -1) = +1$

$x - 3 = +1$

$x = +4$

Thus, the IUPAC name of the given complex $[Pt(NH)_3Br(NO_2)Cl]$ Cl is triamminebromidochloridonitroplatinum (IV) chloride as the names of the different ligands are arranged alphabetically.

12 (*a*) $[Cr(H_2O)_4Cl_2]^+$ shows geometrical isomerism because it is a MA_4B_2 type coordination compound which contains two set of equivalent ligands, four H_2O and 2Cl.

Hence, the possible geometrical isomers are:

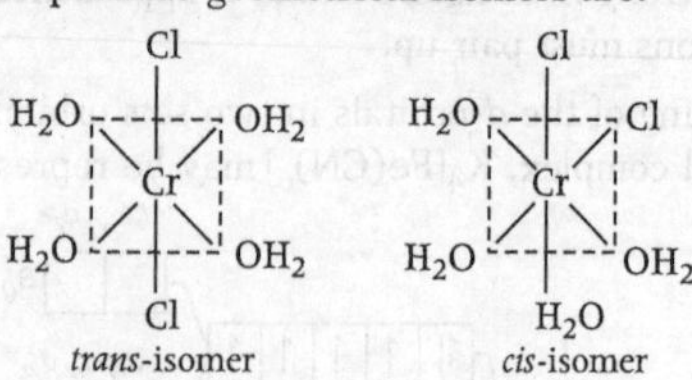

Hence, correct option is (a).

13 (*a*) Assertion and reason both are correct and reason is the correct explanation of the assertion.

Toxic metal ions are removed by chelating ligands. When a solution of chelating ligand is added to solution containing toxic metal, ligands chelate the metal ions by formation of stable complex.

14 (*b*) The complexes having sp^3-hybridisation are tetrahedral while having dsp^2-hybridisation are square planar. The magnetic behaviour of complexes can be paramagnetic or diamagnetic based on the presence or absence of unpaired electrons, respectively.

Electronic configuration of Ni ($Z = 28$) is $[Ar]_{18}\ 3d^8 4s^2$. Due to presence of CO (neutral ligand), oxidation state of Ni in $[Ni(CO)_4]$ is 0.

Ni-atom = 3*d* [⥮ ⥮ ⥮ ↿ ↿] 4*s* [⥮] 4*p* [   ]

Since, CO is a strong field ligand, it pairs up the unpaired electrons of Ni.

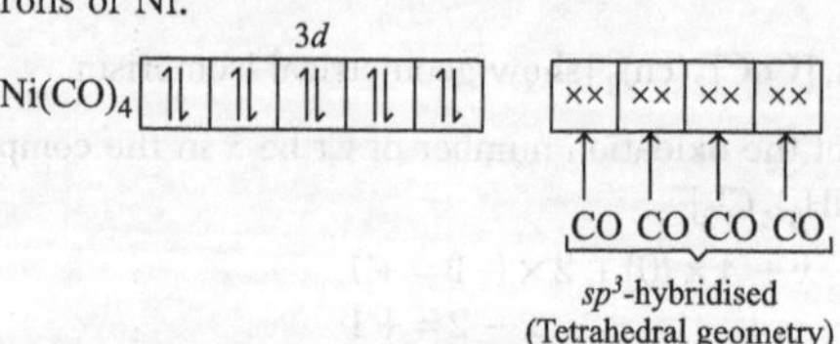

It has tetrahedral geometry as it is sp^3-hybridisation. Since, all the electrons are paired, $[Ni(CO)_4]$ is diamagnetic in nature.

15 (*d*) $[Co(CN)_6]^{3-}$ contains Co^{3+} as central metal ion.

The electronic configuration of Co^{3+} ion = $[Ar]_{18}\ 3d^6$

$$Co^{3+} = 3d^6 \text{ (or } t_{2g}^6\ e_g^0 \text{ shown below)}$$

The splitting in octahedral field is shown below:

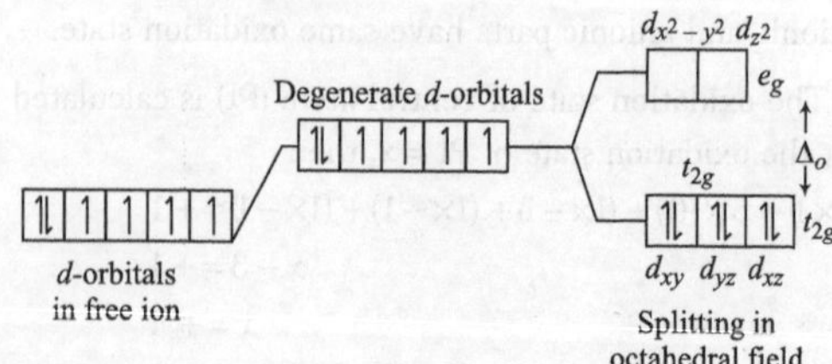

16 (*a*) Since, the coordination number of Mn^{2+} ion in the complex ion is 4, it will have either tetrahedral (sp^3-hybridisation) or square planar (dsp^2-hybridisation) geometry. But the fact that the magnetic moment of the complex ion is 5.9BM suggests that it should have tetrahedral geometry rather than square planar because of the presence of five unpaired electrons in the *d*-orbitals.

The number of unpaired electrons can be determined as follows:

As we know that, magnetic moment $= \sqrt{n(n+2)}$

$$5.9 = \sqrt{n(n+2)}$$

On squaring both sides, the above equation becomes

$$(5.9)^2 = n^2 + 2n \Rightarrow n^2 + 2n - 35 = 0$$

On solving the above equation, we get $n = 5$

17 (*b*) Compounds of transition metal with carbonyls (carbon monoxide) are known as metal carbonyls. These are classified into mononuclear, dinuclear, trinuclear and so on based on the number of central metal atoms/ions present in a complex.

Since, there is only one central metal atom present in iron carbonyl, $Fe(CO)_5$, thus it is mononuclear. The structure of $Fe(CO)_5$ is shown below:

CO, OC, Fe—CO, OC, CO

The examples of dinuclear, trinuclear complexes are $Co_2(CO)_{18}$, $Fe_3(CO)_{12}$ respectively.

18 (*a*) The order of the given ligand in the spectrochemical series is

$H_2O < NH_3 <$ en

Hence, the wavelength of the light observed will be in the order

$[Co(H_2O)_6]^{3+} < [Co(NH_3)_6]^{3+} < [Co(en)_3]^{3+}$

Thus, wavelength absorbed will be in the opposite order i.e., $[Co(en)_3]^{3+}$, $[Co(NH_3)_6]^{3+}$, $[Co(H_2O)_6]^{3+}$.

20 (*c*) $[Mn(CN)_6]^{3-}$

Mn (III) $= [Ar]3d^4$

CN^- being strong field ligand thus, it forces pairing of electrons.

This gives $t_{2g}^4 e_g^0$

∴ Mn (III) = [Ar]

$[Mn(CN)_6]^{3-}$ = [Ar] [⥮ ↿ ↿ | ×× ××] [××] [×× ×× ××]

d^2sp^3

∵ Coordination number of Mn = 6

∴ Structure = Octahedral

21 (*b*) As negative charge on metal carbonyl complex increases, back π-bonding increases and hence, bond order of C — O bond increases while bond length of metal-carbon bond decreases.

Hence, $[Fe(CO)_4]^{2-}$ has longest C — O bond length among the given complexes.

The correct order of bond length of the given complexes is $[Mn(CO)_6]^+ < [Ni(CO)_4 < [Co(CO)_4]^- < [Fe(CO)_4]^{2-}$

22 (*b*) Al^{3+} shows maximum coordination number 6, thus it will form AlF_6^{3-}.

AlF_3 forms $K_3[AlF_6]$ when dissolved in HF in the presence of KF as shown below.

$$AlF_3 + 3KF \xrightarrow{HF} K_3[AlF_6]$$

23 (*a*)

$$[Co(NH_3)_6]Cl_3 \longrightarrow [Co(NH_3)_6]^{3+} + 3Cl^-$$
$$[Co(NH_3)_3Cl_3] \longrightarrow [Co(NH_3)_3Cl_3]$$
$$[Co(NH_3)_4Cl_2]Cl \longrightarrow [Co(NH_3)_4Cl_2]^{1+} + Cl^-$$
$$[Co(NH_3)_5Cl]Cl_2 \longrightarrow [Co(NH_3)_5Cl]^{2+} + 2Cl^-$$

So, $[Co(NH_3)_3Cl_3]$ does not ionise so does not give test for chloride ions.

24 (*a*) $[Co(CN)_6]^{3-}$

$$Co^{3+} = 1s^2, 2s^2 2p^6, 3s^2 3p^6 3d^6$$

CN^- is a strong field ligands and as it approaches the metal ion, the electrons must pair up.

The splitting of the *d*-orbitals into two sets of orbitals in an octahedral complex $[Co(CN)_6]^{3-}$ may be represented as:

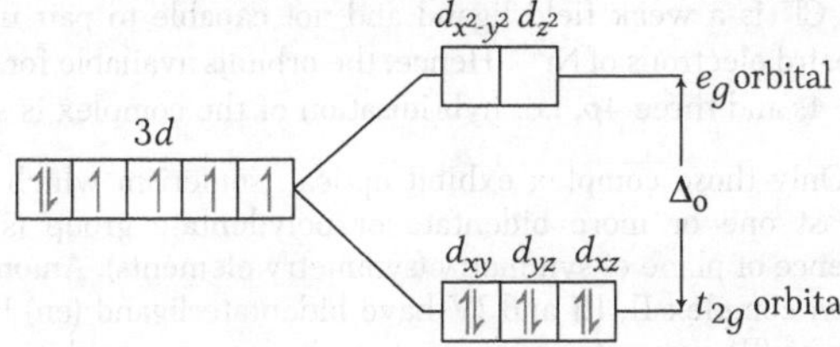

Here, for d^6 ions, three electrons first enter orbitals with parallel spin put the remaining may pair up in t_{2g} orbital giving rise to low spin complex (strong ligand) field.

$\therefore$ $[Co(CN)_6]^{3-}$ has no unpaired electron and will be in a low spin configuration.

25 (*c*) Possible isomers of $[Co(en)_2Cl_2]Cl$ are shown below.

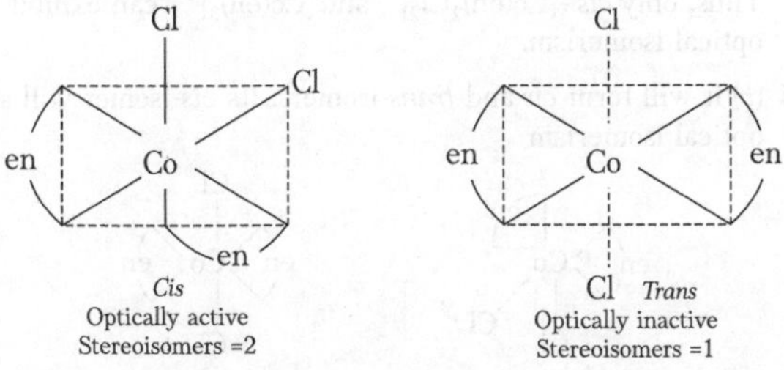

Hence, total number of stereoisomers $= 2 + 1 = 3$

26 (*a*) Given complex compounds is $[M(en)_2C_2O_4]$ Cl.

Let oxidation number of *M* is *x*.

$\therefore$ $x - 2 - 2 = -1$ or $x = +3$

Now, as coordination number is defined as the total number of binding sites attached to the metal.

Hence, in the given complex coordination number is 6.

Thus, the sum of coordination number and oxidation number of the metal *M* in the given complex $= 3 + 6$
$= 9$

27 (*d*) When complex ion is an anion, the name of the metal ends with suffix-ate along with its oxidation number in the complex ion.

$[Fe(CN)_6]^{3-}$ = Hexacyanoferrate (III) ion

29 (*c*) Tetrahedral complexes do not show geometrical isomerism because relative position of unidentate ligands attached to central metal atom are same with respect to each other.

So, options (b) and (d) are incorrect.

Square planar complex MA_3B (option (a)) do not show geometrical isomerism because the possible spatial arrangement are equivalent.

Square planar complex *MABCD* type form three isomers (*cis – trans* isomerism). Two of these are *cis* and one is *trans*.

30 (*a*) Homoleptic complex has only one type of ligand linked to the central metal, e.g. *tris* (ethan-1, 2 - diamine) cobalt (III) chloride.

31 (*e*) Structure of complex $[CoCl_2(en)_2]Cl$ is

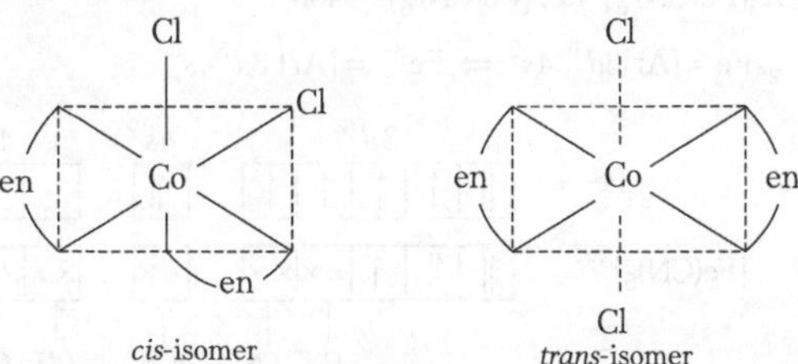

Coordination number = 6,

Let the oxidation state of Co be *x*

$$x + 2 \times (-1) = +1$$
$$x + (-2) = +1$$
$$x = +3$$

Thus, the oxidation number of Co is +3.

Number of *d*-electron in Co^{3+} ion = 6

32 (*c*) The compounds with lower oxidation number and those which cannot be reduced by one another can exist together. Thus, $FeCl_2$ and $SnCl_2$ can exist together as Fe^{2+} cannot be reduced by Sn^{2+}.

33 (*b*) The CFSE for octahedral complex is given by

$$CFSE = [-0.4t_{2g}\, e^- + 0.6e_g\, e^-]$$

(a) For $Mn^{3+}, [3d^4] \rightarrow t_{2g}^3 e_g^1$

$$CFSE = [-0.4 \times 3 + 0.6 \times 1] = -0.6$$

(b) For Fe^{3+}, $[3d^5] \rightarrow t_{2g}^3 e_g^2$

$$CFSE = [(-0.4 \times 3) + (2 \times 0.6)] = 0$$

(c) For $Co^{3+}, [3d^6] \rightarrow t_{2g}^4 e_g^2$

$$CFSE = [(-0.4 \times 4) + (2 \times 0.6)] = -0.4$$

(d) Fo Co^{2+}, $[3d^7] \rightarrow t_{2g}^5 e_g^2$

$$CFSE = [(-0.4 \times 5) + (2 \times 0.6)] = -0.8$$

34 (*a*) Molar conductance is the conducting power of all the ions produced by dissolving 1 g-mole of an electrolyte. Therefore, molar conductance $\propto$ number of ions produced

Werner complex	Modern-notation	Ionisation
(a) $CoCl_3 \cdot 3NH_3$	$[Co(NH_3)_3Cl_3]$	$[Co(NH_3)_3Cl_3]$ (No ionisation)
(b) $CoCl_3 \cdot 4NH_3$	$[Co(NH_3)_4Cl_2]Cl$	$[Co(NH_3)_4Cl_2]^+ + Cl^-$ (2 ions)
(c) $CoCl_3 \cdot 5NH_3$	$[Co(NH_3)_5Cl]Cl_2$	$[Co(NH_3)_5Cl]^{2+} + 2Cl^-$ (3 ions)
(d) $CoCl_3 \cdot 6NH_3$	$[Co(NH_3)_6]Cl_3$	$[Co(NH_3)_6]^{3+} + 3Cl^-$ (4 ions)

From the above table it is clear that, molar conductance of the complexes will be in the following order

$$a < b < c < d$$

35 (*b*) Coordination compound/ion having unpaired electron is paramagnetic in nature while if it has all the electrons paired, it is diamagnetic.

(a) $K_3[Fe(CN)_6]$ i.e. $[Fe(CN)_6]^{3-}$ ion

$$_{26}Fe = [Ar]3d^6\,4s^2 \Rightarrow Fe^{3+} = [Ar]\,3d^5 4s^0$$

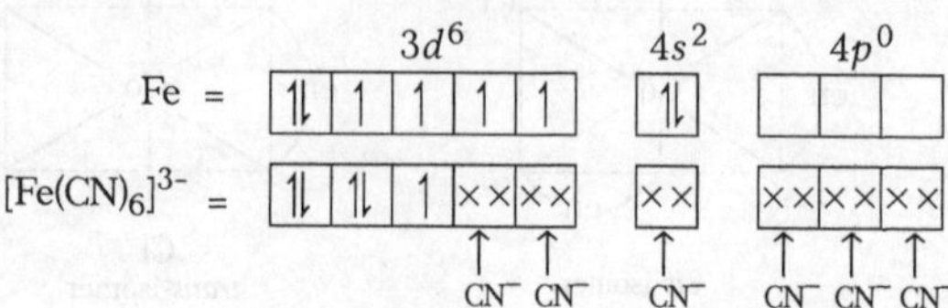

CN^- is a strong field ligand. Thus, pairing of electrons occurs

(b) $[Co(NH_3)_6]Cl_3$

i.e. $[Co(NH_3)_6]^{3+}$ ion

$$_{27}Co = [Ar]3d^7 4s^2 \quad \Rightarrow Co^{3+} = [Ar]\,3d^6\,4s^0$$

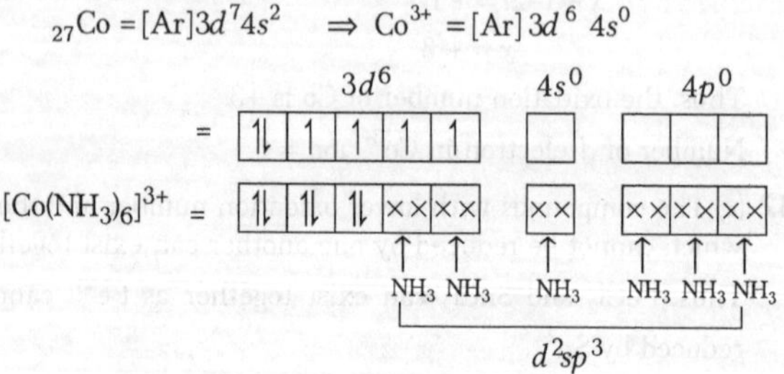

NH_3 is a strong field ligand. Thus, pairing of electrons occurs.

(c) $K_3[FeF_6]$ i.e. FeF_6^{3-} ion

$$_{26}Fe = [Ar]3d^6\,4s^2$$

$$\Rightarrow \quad Fe^{3+} = [Ar]\,3d^5\,4s^0$$

	$3d^5$	$4s^2$	$4p^0$	$4d^0$
=	⥮ ↿ ↿ ↿ ↿ ↿	⥮		

	$3d^5$	$4s^0$		
FeF_6^{3-} =	↿ ↿ ↿ ↿ ↿	×× (F⁻)	×× ×× ×× (F⁻ F⁻ F⁻)	×× ×× (F⁻ F⁻)

sp^3d^2

F^- is a weak field ligand. Thus, pairing of electron will not occur.

(d) $K_3[CoF_6]$ i.e. CoF_6^{3-} ion

$$_{27}Co = [Ar]3d^7,\,4s^2$$

$$\Rightarrow \quad Co^{3+} = [Ar]\,3d^6\,4s^0$$

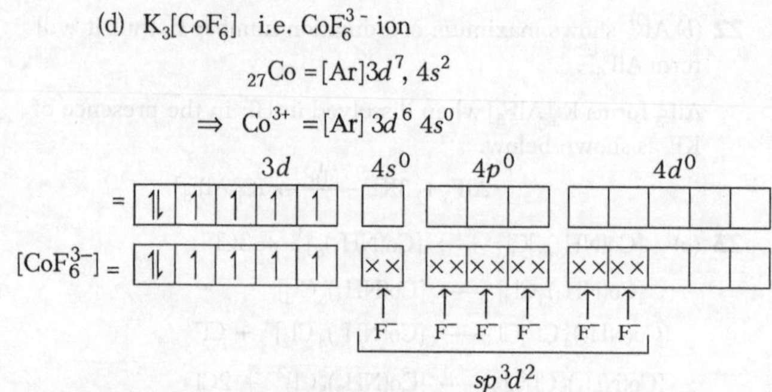

Therefore, from the given complex only $[Co(NH_3)_6]Cl_3$ is diamagnetic in nature.

36 (*a*) In both the complexes Ni is present as Ni^{2+}.

	$3d^8$	$4s^0$	$4p^0$
Ni^{2+} = [Ar]	⥮ ⥮ ⥮ ↿ ↿		

CN^- being strong field ligand pair up the unpaired electrons of Ni^{2+} and hence, $4CN^-$ ions will occupy one 3*d*, one 4*s* and two 4*p*-orbitals, i.e. hybridisation is dsp^2.

But Cl^- is a weak field ligand and not capable to pair up the unpaired electrons of Ni^{2+}. Hence, the orbitals available for it are one 4*s* and three 4*p*, i.e. hybridisation of the complex is sp^3.

37 (*c*) Only those complex exhibit optical isomerism which have atleast one or more bidentate or polydentate group (s) and absence of plane of symmetry (symmetry elements). Among the given complex II, III and IV have bidentate ligand (en) but in case of (II) symmetry is present, so it does not extibit optical isomerism

Cl — en Co en — Cl

Trans- $[Co(en)_2Cl_2]^+$

Thus, only *cis* -$[Co(en)_2Cl_2]^+$ and $[Co(en)_3]^{3+}$ can exhibit optical isomerism.

38 (*b*) It will form *cis* and *trans*-isomers. Its *cis*-isomer will show optical isomerism

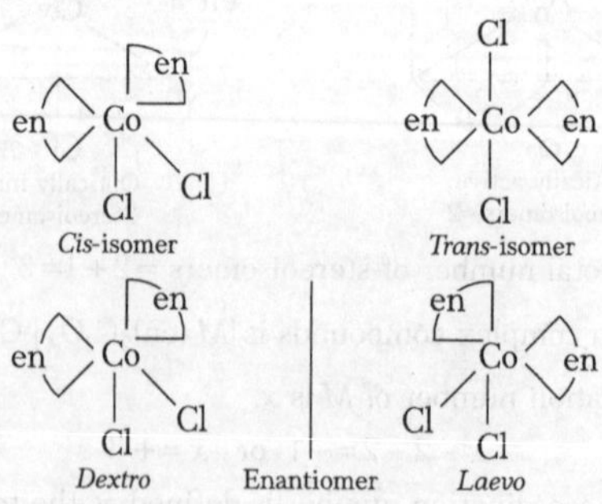

Although *cis*-isomer and *dextro*-isomer are same but these are aligned to different properties, that's why these are counted separately.

39 (*a*) (a) $[Co(NH_3)_6]^{2+}$ in this ion, Co is present as Co^{2+}. NH_3, being strong field ligand, pair up the unpaired electrons.

Electronic configuration of $Co^{2+} = [Ar]\ 3d^7 4s^0 4p^0$

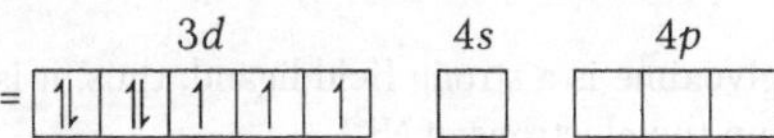

Electronic configuration of Co^{2+} in $[Co(NH_3)_6]^{3+}$ =

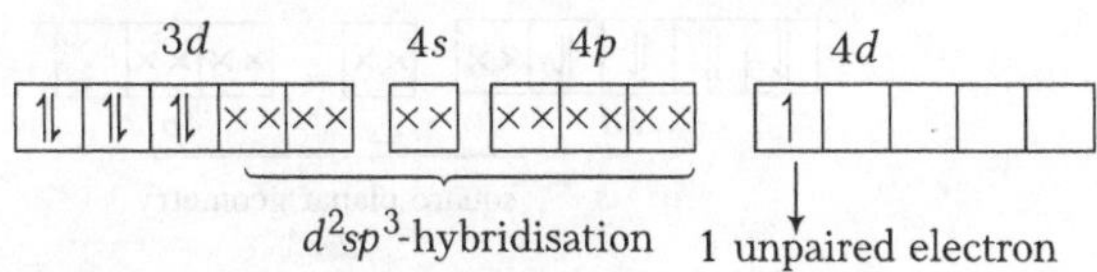

Due to the presence of one unpaired electron, $[Co(NH_3)_6]^+$ is paramagnetic.

(b) $[MnBr_4]^{2-}$ in this ion, Mn is present as Mn^{2+}. Br being weak field ligand, leave unpaired electrons of the metal ion as such.

Electronic configuration of $Mn^{2+} = [Ar]\ 3d^5 4s^0 4p^0$.

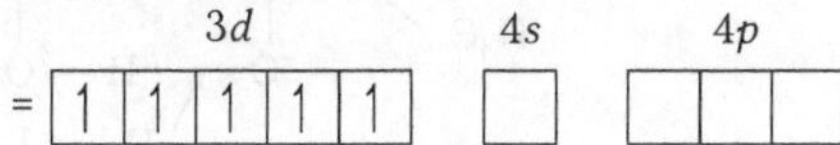

Molecular orbital electronic configuration of Mn^{2+} in $[MnBr_4]^{2-}$ =

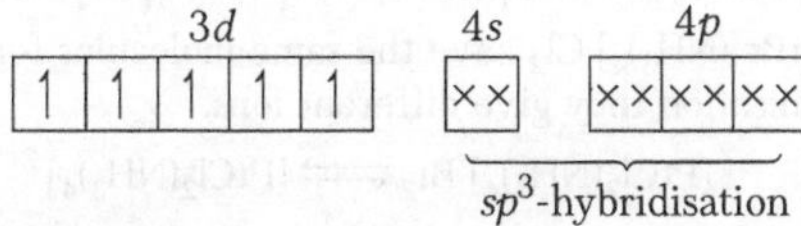

Due to sp^3-hybridisation, shape of $[MnBr_4]^{2-}$ is tetrahedral.

(c) Linkage isomerism occurs when two or more atoms in a monodentate ligand may function as a donor.

Therefore, $[CoBr_2(en)_2]^-$ does not exhibit linkage isomerism.

(d) $[Ni(NH_3)_6]^{2+}$ in this ion, Ni is present as Ni^{2+}.

Electronic configuration of $Ni^{2+} = [Ar]\ 3d^8\ 4s^0$

3d 4s

= ⥮ ⥮ ⥮ 1 1 | ☐

NH_3 is a strong field ligand, but cannot pair up electrons as if it pair up electron then hybridisation comes out to be dsp^3d which is not possible. Hence, here no pairing of e^- takes place.

Molecular orbital electronic configuration of Ni in $[Ni(NH_3)_6]^{2+}$.

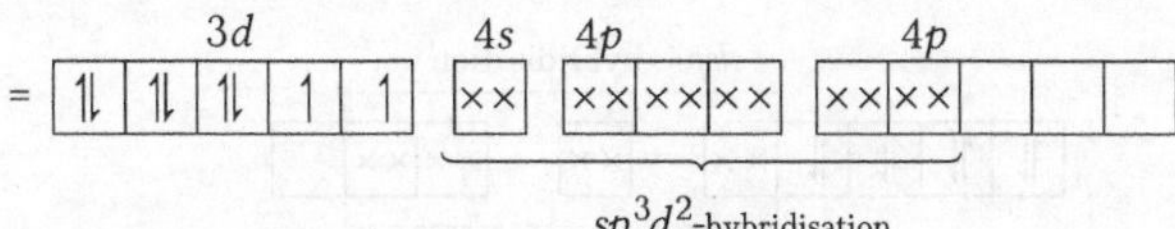

Since, (*n* - 1) *d*-orbital i.e. 3*d*-orbital, does not take part in hybridisation, it is not an inner orbital complex. It is an outer orbital complex.

40 (*b*) Zn^{2+}, Al^{3+}

(*i*) $Zn(OH)_2 + 2NaOH \longrightarrow Na[Zn(OH)_4]^{2-} + 2Na^+$

$\underset{\text{Anhydrous}}{ZnCl_2} + 6NH_3 \longrightarrow \underset{\text{Hexamine zinc chloride}}{[Zn(NH_3)_6\ Cl]}$

(*ii*) $Al(OH)_3 + NaOH \longrightarrow [Al(OH)_4]^-$

$Al^{3+} + 3NH_3 \longrightarrow Al(NH_3)_3$

41 (*a*) *Aqua-regia* is a solution containing conc. HCl and conc. HNO_3 in the ratio 3 : 1.

It dissolves noble metals like Au, Pt etc.

$HNO_3 + 3HCl \longrightarrow 2H_2O + 2Cl + NOCl$

$Au + 3Cl \longrightarrow AuCl_3$

$AuCl_3 + HCl \longrightarrow \underset{\text{Tetra chloridoaurate (III)}}{AuCl_4}$

42 (*b*) $[CoF_6]^{3-}$

Here, Co is present in +3 oxidation state.

Electronic configuration of ${}_{27}Co = [Ar]3d^7\ 4s^2$

$\Rightarrow$ Electronic configuration of $Co^{3+} = [Ar]\ 3d^6\ 4s^0$

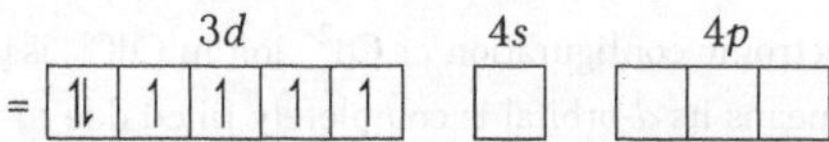

F^- being a weak ligand is unable to pair up the unpaired electrons of Co^{3+} ion thus, occupy 4*s* 4*p* and 4*d* empty orbitals as

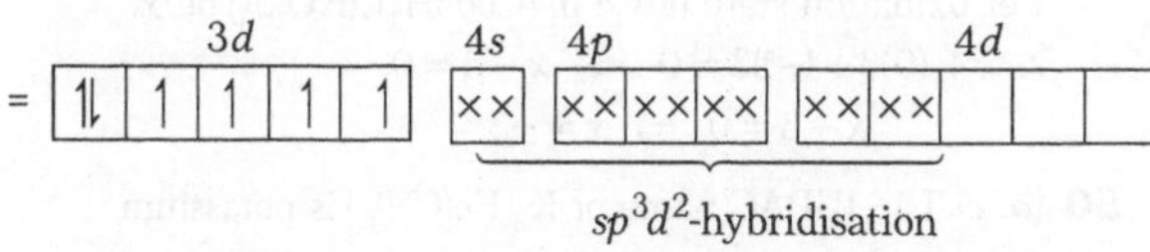

Hence, number of unpaired electrons in complex ion is 4.

43 (*a*) The crystal field theory can explain the spectra, i.e. colour, magnetic property and structure of coordination compounds. However, the position of intensities of spectral bands (or covalent character of covalent bonds) cannot be explained on the basis of this theory.

44 (*a*) Magnetic moment, μ is related with number of unpaired electrons as

$$\mu = \sqrt{N(N+2)}\ \text{BM}$$

$$(1.73)^2 = N(N+2)$$

On solving, $N = 1$

Thus, the complex/compound having one unpaired electron exhibit a magnetic moment of 1.73 BM.

(a) In $[Cu(NH_3)_4]^{2+}$ $Cu^{2+} = [Ar]\ 3d^9$

⥮ ⥮ ⥮ ⥮ 1

(Although in the presence of strong field ligand NH_3, the unpaired electrons gets excited to higher energy level but it still remains unpaired.)

(b) In $[Ni(CN)_4]^{2-}$, $Ni^{2+} = [Ar]\ 3d^8$

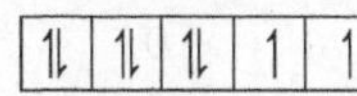

But CN^- being strong field ligand pair up the unpaired electrons and hence, in this complex, number of unpaired electrons = 0

(c) In $[TiCl_4]$, Ti^{2+} = [Ar]; there is no unpaired electron.

(d) In $[CoCl_6]^{4-}$, Co^{2+} = [Ar] $3d^7$

⥮	⥮	1	1	1

It contains three unpaired electrons.

[Thus, $[Cu(NH_3)_4]^{2+}$ is the complex that exhibits a magnetic moment of 1.73 BM.]

45 (*b*) *cis*-platin, i.e. *cis*-$[PtCl_2(NH_3)_2]$ is anti-cancer agent. It destroys cancer cells. So, used as anti-tumour agent.

46 (*c*) Coordination number = number of monodentate ligands.

In $[Pt(NH_3)_4Cl_2]^{2+}$ number of monodentate ligands

$$= 4NH_3 + 2Cl^- = 6$$

47 (*d*) Pyridine is a monodentate ligand. Pyridine coordinates with a metal only from one site, i.e. the lone pair on N. Thus, it is monodentate ligand. While the other three are chelating ligands, i.e. these are attached to a central metal ion from more than one site.

48 (*a*) Electronic configuration of Cd^{2+} ion in $CdCl_2$ is $[Ar]3d^{10}$. That means its *d*-orbital is completely filled due to which *d-d* transition will not occurs.

Thus, $CdCl_2$ is colourless.

49 (*a*) Oxidation state of $NH_3 = 0$ and $NO_2 = -1$

Let oxidation state of Co in $[Co(NH_3)_3(NO_2)_3]$ be *x*.

$$\therefore\ x + (0)3 + (-1)3 = 0 \Rightarrow x - 3 = 0$$

$$x - 3 = 0 \Rightarrow x = +3$$

50 (*a, c*) The IUPAC name of $K_3[Fe(CN)_6]$ is potassium ferricyanide or potassium hexacyanoferrate (III).

51 (*c*) The IUPAC name of $Ni(CO)_4$ is tetracarbonyl nickel (0).

52 (*b*) Let the oxidation state of Fe in $Fe(CO)_5$ be *x*.

$$Fe(CO)_5 \Rightarrow x + 0 \times 5 = 0 \quad (\because \text{Oxidation state of CO} = 0)$$

$$\Rightarrow x = 0$$

In metal carbonyls, oxidation state of metal is always zero.

53 (*c*) Let in $[Ni(CO)_4]$, oxidation state (charge) of Ni is *x*.

$$Ni(CO)_4 \Rightarrow x + (0 \times 4) = 0 \quad (\because \text{Oxidation state of CO} = 0)$$

$$\therefore \quad x = 0$$

In metal carbonyl, oxidation state of metal is always zero.

54 (*b*) By applying the EAN rule, the stability of metal carbonyls can be predicted.

All the given metal carbonyls except $Mn(CO)_5$ follows the EAN rule and thus, it exists as dimer $Mn_2(CO)_{10}$.

55 (*a*)

	Unpaired electron	Hybridisation
(a) $[Ni(NH_3)_6]^{2+}$	2	sp^3d^2
(b) $[Cu(NH_3)_6]^{2+}$	0	sp^3d^2
(c) $[Cr(NH_3)_6]^{3+}$	3	d^2sp^3
(d) $[Co(NH_3)_6]^{3+}$	0	d^2sp^3

Thus $[Ni(NH_3)_6]^{2+}$ is paramagnetic, outer *d*-complex.

56 (*c*) Ni^{2+} = [Ar] $3d^8$

=[Ar]

⥮	⥮	⥮	1	1

3p

Dimethyl glyoxime is a strong field ligand, thus, it is capable in pairing-up the electrons of Ni^{2+}

Molecular orbital electronic configuration of Ni in the complex formed is

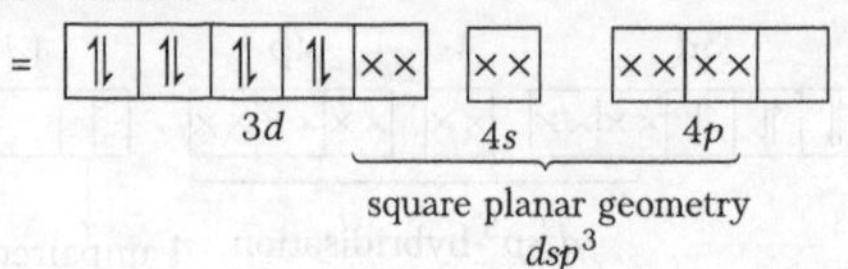

$CH_3-C{=}NOH$ / $CH_3-C{=}NOH$ + Ni^{2+} → [Ni(DMG)$_2$ structure: H$_3$C, C=N, N=C—CH$_3$, Ni^{2+}, O—H...... O$^-$, $^-$O......H—O, H-bond]

DMG acts as a bidentate ligand.

57 (*b*) Since, the complexes $[PtCl_2(NH_3)_4]\,Br_2$ and $[PtBr_2(NH_3)_4]\,Cl_2$ have the same molecular formula but on ionisation they give different ions.

$$[PtCl_2(NH_3)_4]\,Br_2 \rightleftharpoons [PtCl_2(NH_3)_4]^{2+} + 2Br^-$$

$$[PtBr_2(NH_3)_4]\,Cl_2 \rightleftharpoons [PtBr_2(NH_3)_4]^{2+} + 2Cl^-$$

Thus, they exhibit ionisation isomerism.

58 (*a*) The IUPAC name of $K_2[Ni(CN)_4]$ is potassium tetracyanonickelate (II).

59 (*b*) $[Cr(C_2O_4)_3]^{3+}$ shows optical isomerism because its mirror image is not superimposable.

60 (*a*) In given Fe-complex, NO behaves as NO^+.

61 (*d*) $Co^{2+} \longrightarrow 3d^7, 4s^0$

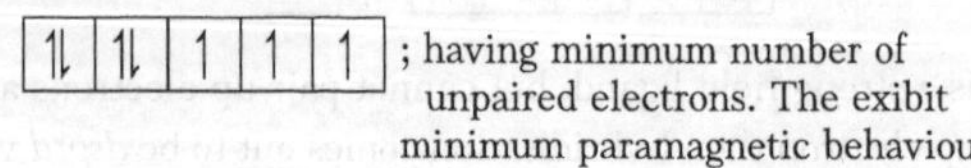

62 (*c*) In $[Ni(CN)_4]^{2-}$, Ni has dsp^2-hybridisation where CN^- is a strong field ligand.

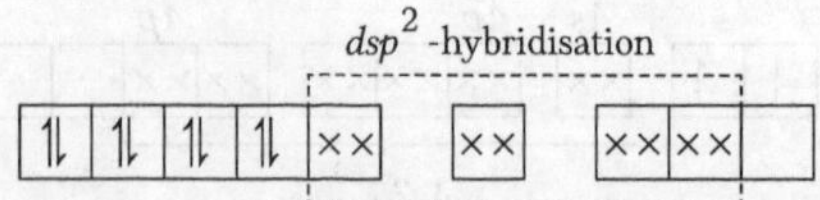

Since, all the electrons are paired, it is diamagnetic.

63 (*d*) When the cation and anion both are complex ions, the coordination compound exhibit coordination isomerism. Thus, the given examples are of coordination isomerism.

64 (*b*) *M(ABCD)* type complex have three geometrical isomers as

Py, NH_3, Cl, Br — Pt ; Py, Br, Cl, NH_3 — Pt ; Py, NH_3, Br, Cl — Pt

65 (*b*)

Cl, Br, en, en — Pt ; Cl, Br, en, en — Pt

Cis

(+) I ; (-) II

Cl, en, en, Br — Pt

Trans

III

- I and II are enantiomers.
- I (or II) and III are diastereomers.

66 (*d*) Ambidentate ligands NO_2 and SCN will give twelve different isomers.

ON=O, SCN, en, en — Rh (I) ; NO_2, SCN, en, en — Rh (II) ; NO_2, NCS, en, en — Rh (III) ; ONO, NCS, en, en — Rh (IV)

I to IV each will have two isomers (*d* and *l*) each.

NO_2, en, en, SCN — Rh ; ONO, en, en, SCN — Rh ; ONO, en, en, NCS — Rh ; NO_2, en, en, NCS — Rh

Total isomeric forms-12.

67 (*b*) The distribution of electrons of d^1, d^2, d^3, d^8, d^9 in t_{2g} and e_g, sets both for strong and weak ligands field, is same.

But for d^4, d^5, d^6 and d^7 there is difference in the arrangement of electrons in weak and strong field ligands. A weak ligand will have greater number of unpaired electrons.

68 (*a*) $[CrCl(H_2O)_5]Cl_2 \longrightarrow [CrCl(H_2O)_5]^{2+} + 2Cl^-$

$$\frac{100 \times 0.01}{1000} = 0.001 \text{ mol} = 0.001 \times 2 \text{ mol}$$

$$\underset{0.002 \text{ mol}}{2Ag^+ + 2Cl^-} \longrightarrow \underset{0.002 \text{ mol}}{2AgCl}$$

0.001 mol of $[Cr \cdot Cl(H_2O)_5]Cl_2$ gives 0.002 mol of AgCl.

$\therefore$ Mass of silver chloride $= 143.5 \times 0.002 = 0.287$ g or 287×10^{-3} g.

69 (*b*) $[Co(NH_3)_4Cl_2]Cl$

Tetraamminedichlorocobalt (III) chloride.

70 (*c*) $N(CH_2CH_2NH_2)_3$ is a tetradentate ligand as it contains four coordinating sites.

71 (*d*) Dichlorotetraamine iron (III) chloride $[Fe(NH_3)_4Cl_2]Cl$.

CHAPTER 13

Haloalkanes

The replacement of hydrogen atom(s) in aliphatic hydrocarbon by halogen atom (s) results in the formation of alkyl halide. These are also known as **haloalkanes.** They contain halogen atom(s) attached to the sp^3-hybridised carbon atom of an alkyl group. Since, halogen atoms are more electronegative than carbon, the carbon-halogen bond of alkyl halide is polarised. Due to this reason, they undergo a number of chemical reactions.

$$-\overset{|}{\underset{|}{C}}^{\delta+}-X^{\delta-}$$

In this chapter, we will study the important methods of preparation, physical and chemical properties and uses of polyhalogen compounds.

Inside

- Classification
- Nomenclature
- Isomerism
- Methods of preparation
  - From alkanes
  - From alkenes
  - From alkynes
  - From silver salts of carboxylic acid
  - From dihalides, Grignard reagent and ether
- Physical properties
- Chemical properties
  - Nucleophilic substitution reactions
  - Elimination reactions
  - Reaction with metals
- Polyhalogen compounds
  - Dichloromethane
  - Trichloromethane
  - Triiodomethane
  - Tetrachloromethane
  - Freons

CLASSIFICATION

On the basis of number of halogen atoms

Haloalkanes or alkyl halides can be divided into following four classes

(*i*) **Monohaloalkanes** [or monohalides] Haloalkanes containing only one halogen atom are called monohaloalkanes. Their general formula is $C_nH_{2n+1}X$. e.g. CH_3Cl [chloromethane], C_2H_5Br [bromoethane]. [X = F, Cl, Br, I]

(*ii*) **Dihaloalkanes** [or dihalides] Haloalkanes containing two halogen atoms are called dihaloalkanes. Their general formula is $C_nH_{2n}X_2$. e.g. CH_2X_2 [methylene dihalides].

Dihaloalkanes are mainly of following two types

- ***Geminal* dihalides** When both the halogen atoms are attached to the same carbon atom, such dihaloalkanes are called *geminal* (or *gem*) dihalides. They are also called alkylidene halides.

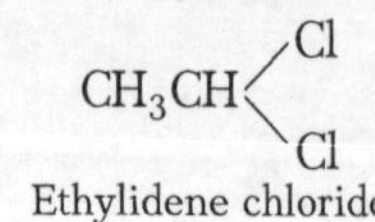

Ethylidene chloride

- ***Vicinal* dihalides** When both the halogen atoms are attached to adjacent carbon atoms [i.e. at *vicinal* position] they are called *vicinal* (or *vic*) dihalides. They are also called alkylene halides.

$$\begin{array}{l} CH_2—Cl \\ | \\ CH_2—Cl \end{array}$$

Ethylene chloride

(*iii*) **Trihaloalkanes** [or trihalides] Haloalkanes containing three halogen atoms are called trihaloalkanes. Their general formula is $C_nH_{2n-1}X_3$ e.g. $CHBr_3$ [bromoform or tribromomethane]. This type of trihalides such as CHX_3 are called haloforms.

(*iv*) **Tetrahaloalkanes** [or tetrahalides] Haloalkanes containing four halogen atoms are called tetrahaloalkanes. Their general formula is $C_nH_{2n-2}X_4$. e.g. CCl_4 (carbon tetrachloride).

On the basis of hydrogen atoms attached to C-atom

Haloalkanes can be classified as primary, secondary and tertiary depending upon the fact that hydrogen is attached to primary, secondary or tertiary C-atom.

$H_3C—CH_2Br$ Primary (1°) $H_3C—CH(CH_3)—Br$ Secondary (2°) $H_3C—C(CH_3)_2—Br$ Tertiary (3°)

On the basis of hybridisation of carbon atom of C—*X* bond

On the basis of hybridisation of carbon atom of C—*X* bond, haloalkanes can be classified into following two classes

(*i*) **Compounds containing C_{sp^3} –*X* bond** In this class of compounds, halogen atom is directly bonded to sp^3-hybridised carbon atom. This class is further divided into the following types of compounds.

- **Alkyl halides** As discussed earlier, in alkyl halides, halogen atom is directly bonded to an alkyl group. If in *R*—*X*, *R* is alicyclic, *R*—*X* is called cycloalkyl halide or halocycloalkane. They may be primary, secondary or tertiary.

Primary (1°) Secondary (2°) Tertiary (3°)

- **Allylic halides** In allylic halides, halogen is attached to sp^3- hybridised carbon atom next to a carbon-carbon double bond. Such C-atom is called an allylic carbon. They may be 1°, 2° or 3°.

$CH_2{=}CHCH_2X$, (1°) $CH_3CH{=}CHCHX$, C_2H_5 (2°) CH_3 (3°)

Allylic C-atom

- **Benzylic halides** In benzylic halides, the halogen atom is attached to a sp^3-hybridised carbon atom, which is linked to an aromatic ring. Such carbon atom is called benzylic carbon atom. They may be 1°, 2° or 3°.

CH_2X (1°) $R—CH—X$ (2°) $C(R)(R)X$ (3°)

Benzylic C-atom

(*ii*) **Compounds containing C_{sp^2} – *X* bond** In this class of compounds, halogen atom is directly bonded to a sp^2-hybridised carbon atom. This class is further divided into the following types of compounds.

- **Vinylic halides** In vinylic halides, halogen is attached to sp^2-hybridised carbon of carbon-carbon double bond [i.e. to a vinylic C-atom].

$CH_2 = CH—X$ Vinylic C-atom; Vinylic C-atom

- **Aryl halides** In aryl halides, the halogen is attached directly to carbon atom of an aromatic ring.

NOMENCLATURE

Commonly, monohaloalkanes are derived by naming the alkyl group followed by the name of halogen.

The prefix *iso, neo, sec, tert* are also used. In IUPAC system, haloalkanes are written by prefixing the word 'halo' to the name of the longest straight chain alkane.

For di-haloalkanes, commonly alkylidene or alkylene dihalides are used. In IUPAC system they are named as dihaloalkanes.

e.g. $CH_3CH_2CH_2Cl$ $\quad$ $CH_3-CH(Cl)-CH_2CH_3$

Common name *n*-propyl chloride, *sec*-butyl chloride
IUPAC name Chloropropane, 2-chlorobutane

e.g. $CH_3-C(CH_3)_2-CH_2Cl$

Common name *neo*-pentyl chloride
IUPAC name 1-chloro-2,2-dimethylpropane

For disubstituted haloalkanes.

CH_3-CHCl_2

Common name Ethylidene dichloride (*gem*)
IUPAC name 1, 1-dichloroethane

CH_2Cl-CH_2Cl

Common name Ethylene dichloride (*vic*)
IUPAC name 1, 2-dichloroethane

ISOMERISM

Haloalkanes exhibit the following three types of isomerism.

1. Chain isomerism

Haloalkanes exhibit chain isomerism due to the difference in the arrangement of carbon atoms in the chain. This type of isomerism is not exhibited by the first three members but is shown only by higher members.

e.g. $\overset{4}{C}H_3-\overset{3}{C}H_2-\overset{2}{C}H_2-\overset{1}{C}H_2Br$
1-bromobutane

$\underset{3}{C}H_3\underset{2}{C}H(CH_3)-\underset{1}{C}H_2Br$
1-bromo-2-methylpropane

2. Position isomerism

In position isomerism, isomers differ in the position of halogen atom.

e.g. $\overset{3}{C}H_3\overset{2}{C}H_2\overset{1}{C}H_2Br$ $\quad$ $\overset{1}{C}H_3-\overset{2}{C}H(Br)-\overset{3}{C}H_3$
1-bromopropane, 2-bromopropane

3. Optical isomerism

This type of isomerism is exhibited by those alkyl halides that contains a chiral carbon atom [i.e. a carbon atom to which four different groups are attached].

e.g. 2-bromobutane contains a chiral carbon and thus, it exhibits two isomers called enantiomers that are non- superimposable mirror images of each other.

CH_3, H, Br, CH_2-CH_3 on C^* | Mirror | CH_3, Br, H, H_3C-H_2C on C^*

Here, C^* indicates chiral C-atom.

Note *Molecules containing a chiral carbon atom are called chiral molecules and this property of molecules is called chirality.*

METHODS OF PREPARATION

Various methods of preparation for haloalkanes are as follows

From alkanes

Alkanes react with halogen in presence of sunlight or high temperature to give a mixture of mono and poly haloalkanes. This reaction proceeds *via* free radical mechanism.

For a given alkane, the reactivity order of halogen is $F_2 > Cl_2 > Br_2 > I_2$ and for a given halogen, the reactivity order of H-atom in alkane is 3°>2°>1° > CH_4.

e.g. $CH_4 \xrightarrow{Cl_2/h\nu} CH_3Cl + CH_2Cl_2 + CHCl_3 + CCl_4$

$\underset{\text{Propane}}{CH_3CH_2CH_3} \xrightarrow{Cl_2/h\nu} \underset{\substack{(45\%)\\\text{Chloropropane}}}{CH_3CH_2CH_2Cl} + \underset{\substack{(55\%)\\\text{2-chloropropane}}}{CH_3CHClCH_3}$

Chlorination of *n*-butane gives *d*-and *l*-optical isomers of *sec*-butyl chloride hence mixture is optically inactive (racemic).

$\underset{n\text{-butane}}{CH_3CH_2CH_2CH_3} \xrightarrow[h\nu]{Cl_2} H-C(CH_3)(CH_2CH_3)-Cl + Cl-C(CH_3)(CH_2CH_3)-H$

and its an enantiomer (racemic mixture).

Note *The relative amounts of these isomeric haloalkanes depend upon the nature of halogen and the number and type of hydrogen atoms (1° , 2° ,3°) which being substituted. For the ease of substitution, various hydrogens follow the sequence,*

3° H-atom > 2° H-atom > 1° H-atom

In this reaction, direct fluorination is not possible because it occurs with breakage of C—C bonds in higher alkanes and reaction is highly exothermic too. This occurs due to low dissociation energy of F_2 molecule and due to strongly exothermic nature of one of termination steps ($R + F \rightarrow RF$).

This step is so exothermic that the energy released breaks a C — C bond resulting in fission of molecule.

Iodine is unique among the halogens. Its reaction with alkanes is endothermic. Also, **iodination of alkanes is a reversible process,** therefore, **iodination** is carried out **in the presence of oxidising agents such as** HIO_3, conc. HNO_3 which oxidises HI produced to form I_2.

$$\underset{\text{Ethane}}{CH_3CH_3} + I_2 \xrightarrow{HIO_3} \underset{\text{Iodoethane}}{CH_3CH_2I} + HI$$

$$5HI + HIO_3 \longrightarrow 3I_2 + 3H_2O$$

From alkenes

Haloalkanes can be prepared from alkenes by the following three ways

1. Addition of hydrogen halides

An alkene is converted to corresponding alkyl halides on reaction with hydrogen halide. Addition to alkene takes place in accordance with **Markownikoff rule.**

$$\underset{\text{Alkene}}{R—\overset{H}{\overset{|}{C}}=\overset{H}{\overset{|}{C}}—R'} + HX \longrightarrow R—\underset{H}{\underset{|}{\overset{H}{\overset{|}{C}}}}—\underset{X}{\underset{|}{\overset{H}{\overset{|}{C}}}}—R'$$

$$\text{e.g.}\quad \underset{\text{Propene}}{CH_3CH{=}CH_2} + HCl \longrightarrow \underset{\text{2-chloropropane}}{CH_3\underset{Cl}{\underset{|}{C}}HCH_3}$$

Mechanism

The reaction follows electrophilic addition mechanism and intermediate is carbocation after H^+ is added.

$$R—\overset{H}{\overset{|}{C}} = \overset{H}{\overset{|}{C}}H + H—X \xrightarrow{\text{Slow step}} \underset{\text{Carbocation}}{R—\underset{+}{\overset{H}{\overset{|}{C}}}—\underset{H}{\underset{|}{\overset{H}{\overset{|}{C}}}}—H} + X:^-$$

$$R—\underset{+}{\overset{H}{\overset{|}{C}}}—\underset{H}{\underset{|}{\overset{H}{\overset{|}{C}}}}—H + X^-: \xrightarrow[\text{step}]{\text{Fast}} R—\underset{X}{\underset{|}{\overset{H}{\overset{|}{C}}}}—\underset{H}{\underset{|}{\overset{H}{\overset{|}{C}}}}—H$$

It should be noted that there is a possibility of rearrangement to attain the greater stability of carbocation.

Stability order of carbocation is being in order as

$$3° > 2° > 1° > {}^+CH_3$$

$$\text{e.g.}\ \underset{\text{3-methaylbutene}}{CH_3\underset{CH_3}{\underset{|}{\overset{CH_3}{\overset{|}{C}}}}—CH{=}CH_2} \xrightarrow{H^+} \underset{2°\ \text{carbocation}}{CH_3\underset{H}{\underset{|}{\overset{CH_3}{\overset{|}{C}}}}\overset{\oplus}{C}HCH_3} \xrightarrow[\text{1, 2 hydride shift}]{\text{Rearrangement}}$$

$$\underset{\substack{3°\ \text{carbocation}\\ \text{(more stable than 2°)}}}{CH_3\underset{\oplus}{\overset{CH_3}{\overset{|}{C}}}CH_2CH_3} \xrightarrow{X^-} \underset{3°\ \text{haloalkane}}{CH_3—\underset{X}{\underset{|}{\overset{CH_3}{\overset{|}{C}}}}—CH_2CH_3}$$

Kharasch effect

Addition to alkene takes place against Markownikoff rule in presence of benzoyl chloride. This is known as Kharasch effect. This is shown by HBr only.

$$\underset{\text{2-methyl propene}}{CH_3—\overset{CH_3}{\overset{|}{C}}=CH_2} \xrightarrow[\text{Peroxide}]{HBr} \underset{\substack{\text{Anti-Markownikoff}\\ \text{addition}}}{CH_3—\overset{CH_3}{\overset{|}{C}}HCH_2Br}$$

2. Addition of halogens

Addition of halogens to alkene produces *vicinal* dihalides. In laboratory, addition of bromine in CCl_4 to an alkene results in discharge of reddish brown colour of bromine. It constitutes an important method for the detection of double bond in a molecule.

The addition results in the synthesis of *vic*-dibromides, which are colourless.

$$\underset{\text{Ethene}}{\overset{H}{\underset{H}{>}}C{=}C\overset{H}{\underset{H}{<}}} + Br_2 \xrightarrow{CCl_4} \underset{vic\text{-dibromide}}{BrCH_2—CH_2Br}$$

3. Addition of Br_2 to alkene

- With *trans*-2-butene, product of Br_2 addition is optically inactive due to formation of symmetric *meso* compound.

$$\underset{(trans\text{-2-butene})}{(H)(H_3C)C{=}C(CH_3)(H)} \xrightarrow{Br_2} \underset{\text{Identical } (meso)}{\left[\begin{array}{c} CH_3 \\ H—\!\!+\!\!—Br \\ H—\!\!+\!\!—Br \\ CH_3 \end{array}\right] + \left[\begin{array}{c} CH_3 \\ Br—\!\!+\!\!—H \\ Br—\!\!+\!\!—H \\ CH_3 \end{array}\right]}$$

- Relative rates of reaction of alkene with Br_2/CH_3OH are in order
  Ethene < 1-butene < *trans*-2-butene
  < 2-methyl-1-butene < 2,3-dimethyl-2-butene
- Anti addition means the Br atoms are on the opposite sides of the plane of the molecule.

- Anti addition of a symmetric reagent to a symmetric *cis*-alkene gives racemic product, while anti addition to a *trans*-alkene gives *meso* product.
- In general, for symmetric alkenes and symmetric reagents (addition of two identical groups)

cis - alkene + *syn* addition ⟶ *meso*

cis - alkene + *anti* addition ⟶ racemic

trans - alkene + *syn* addition ⟶ racemic

trans - alkene + *anti* addition ⟶ *meso*

Allylic substitution of alkenes

Alkenes are heated with Cl_2 or Br_2 at 773 K to give allylic halide. Here, the H-atom of allylic carbon is replaced by a halogen atom. The reaction proceeds *via* free radical mechanism.

e.g. $\underset{\text{Butene}}{CH_3CH_2CH{=}CH_2} + X_2 \xrightarrow{773\text{ K}} CH_3\underset{\substack{|\\X}}{CH}{-}CH{=}CH_2 + HX$

The reactivity of different types of hydrogen follows the order

benzylic ≊ allylic > 3° > 2° > 1° > CH_4 > vinylic ≈ aryl

If in allylic substitution (in diene), there is increase in conjugation, then conjugated product is more stable than the other one due to resonance,

$CH_2{=}CH{-}CH_2{-}CH{=}CH_2$

↓ Br

$CH_2{=}CH{-}\dot{C}H{-}CH{=}CH_2$ (I) — Intermediate (Allylic)

↕

$CH_2{=}CH{-}CH{=}CH{-}\dot{C}H_2$ (II) $\xrightarrow{\dot{B}r}$

More stable than I due to conjugate

$\underset{\text{5 bromo pent-1, 3-diene}}{CH_2{=}CH{-}CH{=}CH{-}CH_2Br}$

Note *NBS (N-bromo succinimide) is used selectively for bromination at allylic and benzylic position.*

(Structure of NBS: succinimide ring with N–Br (NBS))

$CH_3CH{=}CH_2 + NBS \longrightarrow BrCH_2{-}CH{=}CH_2$ (Wohl-Ziegler reaction)

$\underset{\text{Toluene}}{C_6H_5{-}CH_3} + NBS \longrightarrow \underset{\text{Benzyl chloride}}{C_6H_5{-}CH_2Br}$

From alkynes

Alkynes react with Cl_2 or Br_2 dissolved in CCl_4 to give *vicinal* tetrahalide, e.g.

$$\underset{\text{Alkyne}}{CH_3{-}C{\equiv}CH} + 2Br_2 \xrightarrow{CCl_4} \underset{\textit{Vicinal}\text{ tetrahalide}}{CH_3{-}CBr_2{-}CHBr_2}$$

From alcohols

Haloalkanes are best prepared from alcohols, which are easily available.

These reactions are infact nucleophilic substitution reactions and follows S_N1 or S_N2 mechanism depending upon the alcohol used.

Different methods are discussed below

1. By the reaction of halogen acids

Haloalkanes are prepared by passing dry hydrogen chloride gas through a solution of alcohol or by heating a solution of alcohol in concentrated aqueous acid.

$$\underset{\text{Alcohol}}{RCH_2OH} + HX \xrightarrow{ZnCl_2} \underset{\text{Haloalkane}}{RCH_2X} + H_2O$$

The order of reactivity of HX with a given alcohol is in the order

$$HI > HBr > HCl > HF.$$

All 3° and 2° alcohols follow S_N1 mechanism while 1° alcohols follow S_N2 mechanism.

Mechanism

The above reaction occurs in following three steps.

***Step* I Protonation of alcohol**

$$H{-}X \longrightarrow H^+ + X^-$$

$$R{-}CH_2{-}\ddot{O}H \xrightarrow{H^+} R{-}CH_2{-}\overset{+}{\ddot{O}}H_2$$

***Step* II Loss of H_2O molecule** This step leads to the formation of an intermediate carbocation as H_2O is a good leaving group.

$$R{-}CH_2{-}\overset{+}{\ddot{O}}H_2 \xrightarrow[-H_2O]{} \underset{\text{Carbocation}}{R{-}\overset{+}{C}H_2}$$

***Step* III Attack of halide ion**

$$R{-}\overset{+}{C}H_2 \xrightarrow{X^-} \underset{\text{Haloalkane}}{R{-}CH_2X}$$

Note *Intermediate formed in the reaction is carbocation that undergoes rearrangement to attain greater stability of carbocation.*

The reaction of primary and secondary alcohols with HX requires the presence of a catalyst, $ZnCl_2$ which is known as **Grooves' process.** The reaction occurs by simply shaking with tertiary alcohols concentrated HCl at room temperature.

HBr and HI are generated in *situ* as follows

$$\underset{\text{(Conc.)}}{NaBr + H_2SO_4} \longrightarrow NaHSO_4 + HBr$$

$$NaI + H_3PO_4 \longrightarrow NaH_2PO_4 + HI$$

Note *Phosphoric acid (H_3PO_4) is used instead of H_2SO_4 to synthesise HI as HI gets oxidised by H_2SO_4 to I_2.*

The order of reactivity of alcohols with a given haloacid is $3° > 2° > 1°$. This method is not applicable for the preparation of aryl halides because 'C—O' bond in phenols has a partial double bond character and is difficult to break being stronger than a single bond.

Importance of $ZnCl_2$

The Zn^{2+} ion can coordinate to the oxygen atom of the alcohol making it a good leaving group. This leads to carbocation formation which then readily reacts with chloride ion to form alkyl chlorides.

$$\underset{\text{Ethanol}}{CH_3CH_2OH} \xrightarrow{ZnCl_2} CH_3CH_2-\overset{\oplus}{O}(H)-\overset{\ominus}{Z}nCl_2 \xrightarrow{Cl^{\ominus}} \underset{\text{Chloroethane}}{CH_3CH_2Cl}$$

2. By the reaction of phosphorus halides

Haloalkanes can be prepared by treating phosphorus halides with alcohol. PX_3 may be PCl_3, PBr_3 or PI_3 are usually prepared by treating red phosphorus with Cl_2, Br_2 or I_2.

$$ROH \xrightarrow{PCl_5} RCl + POCl_3 + HCl$$

$$ROH \xrightarrow[{[P_4/X_2,\ X = Br,\ I]}]{PX_3} RX + H_3PO_3$$

These reactions follow S_N1 mechanism. The advantage of using phosphorus halides is that, no rearranged product is formed.

This method gives good yield of primary alkyl halides but poor yields of secondary and tertiary alkyl halides because secondary and tertiary alcohols undergo dehydration and forms alkene.

Note *Only chlorides can be prepared by this method as PBr_5 and PI_5 are highly unstable compounds due to steric hinderance.*

3. By the reaction of $SOCl_2$

Haloalkanes can be prepared by using $SOCl_2$ in the presence of either base or pyridine. It is an excellent method because the other two products are escapable gases. Hence, the reaction gives pure alkyl halides.

$$\underset{\text{Alcohol}}{R—OH} + SOCl_2 \xrightarrow{\Delta} \underset{\text{Alkyl halide}}{R—Cl} + SO_2\uparrow + HCl\uparrow$$

This reaction is called **Darzen's reaction.**

Mechanism

Alcohols react with $SOCl_2$ to give alkyl chloride with the retention of configuration as the reaction follows intramolecular nucleophilic substitution mechanism (S_N1).

$$CH_3-\underset{H}{\overset{H}{C}}-\ddot{O}H + Cl-\overset{O}{\overset{\|}{S}}-Cl \longrightarrow CH_3-\underset{H}{\overset{H}{C}}-\overset{+}{\underset{H}{O}}-\underset{Cl}{\overset{O^-}{S}}-Cl$$

$$CH_3-\underset{H}{\overset{H}{C}}-Cl + SO_2\uparrow \xleftarrow{S_N1} CH_3-\underset{H}{\overset{H}{C}}-O-\underset{Cl}{\overset{O^-}{S}} \xleftarrow{-HCl\uparrow}$$

- $SOBr_2$ is less stable and SOI_2 does not exist, thus RBr and RI cannot be prepared by Darzen's reaction.
- Function of PCl_5 or $SOCl_2$,
  (*i*) is to convert the OH group into a good leaving group.
  (*ii*) is to provide a nucleophile, halide ion, to replace the leaving group.

$$H-\underset{CH_2CH_3}{\overset{CH_3}{C}}-OH \xrightarrow{TsCl} H-\underset{CH_2CH_3}{\overset{CH_3}{C}}-OTs \xrightarrow{NaCl} Cl-\underset{CH_2CH_3}{\overset{CH_3}{C}}-H$$

Inversion of configuration

$$H-\underset{CH_2CH_3}{\overset{CH_3}{C}}-OH \xrightarrow[\text{or } SOCl_2]{PCl_5} H-\underset{CH_2CH_3}{\overset{CH_3}{C}}-Cl$$

Retention of configuration

Thus, it is clear that if configuration about chiral carbon is to be retained, use PCl_5 or $SOCl_2$ and to get compound with inverted configuration, first use TsCl and then NaCl.

Note *TsCl structure has $CH_3-C_6H_4-SO_2$ group.*

On reaction with —OH, O — H bond cleaves and configuration of the chiral carbon is retained. But on reaction with NaCl, S_N2 reaction gives inverted product.

From silver salts of carboxylic acid

Silver salt of carboxylic acid is treated with Br_2 or Cl_2 in presence of CCl_4 to give alkyl bromide/chloride. This reaction is mainly used for the preparation of alkyl bromide and is known as **Borodine-Hunsdiecker reaction.**

$$\underset{\text{Silver salt}}{RCO_2Ag} + Br_2 \xrightarrow{CCl_4} \underset{\text{Alkylbromide}}{RBr} + CO_2\uparrow + AgBr$$

This reaction proceed through free radical mechanism.

Initiation $RCOOBr \longrightarrow Br^\bullet + RCOO^\bullet$

Propagation $RCOO^\bullet \longrightarrow R^\bullet + CO_2$

$R^\bullet + Br_2 \longrightarrow RBr + Br^\bullet$ etc.

The yield of alkyl bromide is better than that of alkyl chloride because O—Br bond is weaker than O—Cl bond as well as AgBr is less soluble than AgCl. The yield and ease of formation of RX is 1° > 2° > 3°. With I_2, silver salt of carboxylic acid gives ester as the main product instead of alkyl iodide and it is known as **Birnbaum-Simonini reaction.**

$$2\underset{\text{Silver salt}}{RCO_2Ag} + I_2 \longrightarrow \underset{\text{Ester}}{RCO_2R} + CO_2 + 2AgI$$

Mercury (II) and lead (IV) compounds react similarly with carboxylic acids to give bromoalkanes.

$$2\,\triangleright\!\!-CO_2H + HgO + 2Br_2 \longrightarrow 2\,\triangleright\!\!-Br + 2CO_2 + HgBr_2 + H_2O$$

Cyclopropanoic acid → 2-bromocyclopropane

From dihalides, Grignard reagent and ether

Haloalkanes can also be prepared from dihalides using Zn—Co / HCl, from Grignard reagent using X_2 and from ether using PCl_5.

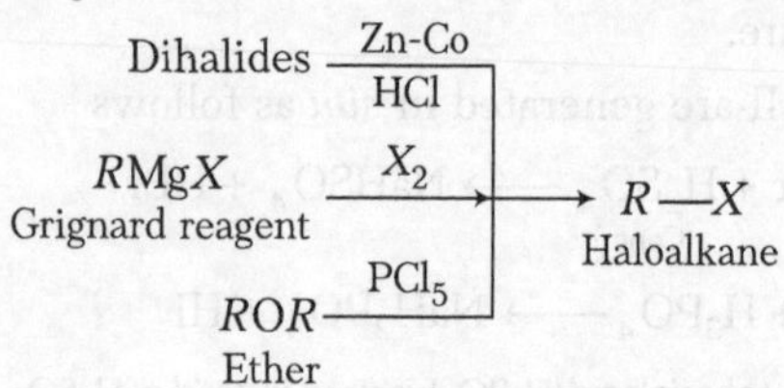

Note *These reactions proceed via nucleophilic substitution reactions.*

By halide exchange method

Alkyl iodides and alkyl fluorides are prepared by the S_N2 mechanism. Alkyl chlorides/bromides reacts with NaI in dry acetone to give alkyl iodide. This reaction is known as **Finkelstein reaction.**

$$\underset{\text{Halo alkane}}{R—X} + NaI \xrightarrow{\text{Dry acetone}} \underset{\text{Alkyl iodide}}{R—I} + NaX\ (X = Cl, Br)$$

Here, Na^+ is a strong metal and I^- is a weak halogen. Thus, Na^+ replaces X^- from $R—X$. NaCl or NaBr thus formed is precipitated in dry acetone. It facilitates the forward reaction according to Le-Chatelier's principle. The synthesis of alkyl fluorides is accomplished by heating an alkyl chloride/bromide in the presence of a metallic fluoride such as AgF, Hg_2F_2, CoF_2 or SbF_3. The reaction is termed as **Swarts reaction.**

$$\underset{\text{Alkyl bromide}}{R—Br} + AgF \xrightarrow[\text{acetone}]{\text{Dry}} \underset{\text{Alkyl flouride}}{R—F} + AgBr$$

In this reaction, F^- act as a strong halogen whereas Ag^+ is a weak metal. Thus, it replaces Br^- from $R—Br$.

CHECK POINT 13.1

1. Only two isomeric monochloro derivatives are possible from
   (a) *n*-pentane (b) 2, 4-dimethylpentane
   (c) benzene (d) 2-methylpropane
2. Which of the following mechanisms is followed by chlorination of methane?
   (a) Electrophilic substitution (b) Nucleophilic substitution
   (c) Free radical mechanism (d) None of these
3. Reactivity of halogens toward alkanes is in order
   (a) $F_2 > Br_2 > Cl_2 > I_2$
   (b) $F_2 > Cl_2 > I_2 > Br_2$
   (c) $Cl_2 > I_2 > Br_2 > F_2$
   (d) $F_2 > Cl_2 > Br_2 > I_2$
4. Addition of HBr gives same product in the presence or absence of peroxide when alkene is
   (a) butene (b) isobutene
   (c) 2,3-dimethylbut-2-ene (d) propene
5. Which of the following reagents helps in the allylic bromination?
   (a) HBr/H_2O_2 (b) HOBr (c) NBS (d) Br_2/CS_2
6. The reaction of primary alcohols with HCl in the presence of anhydrous $ZnCl_2$ is known as
   (a) Swarts reaction (b) Groov's process
   (c) Birnbaum-Simonini reaction (d) Darzen's reaction
7. Which of the following reactants can be used for the preparation of 1-chloropropane?
   (a) Propene and HCl in the presence of peroxide
   (b) Propene and Cl_2 followed by treatment with *aq*.KOH
   (c) Propanol and $SOCl_2$/pyridine
   (d) Any of the above reagent can be used
8. Which of the following reagents when reacts with dihalides give haloalkane?
   (a) Zn-Co and HCl (b) Zn-Hg amalgam
   (c) $FeCl_2$ (d) Cu^{2+}/H_2SO_4

PHYSICAL PROPERTIES

Various physical properties of haloalkanes are as follows

(*i*) **Physical state** Chloride, bromide, fluoride of methane and chloride of ethane are gases at room temperature, rest alkyl halides upto C_{18} are colourless liquids and beyond it are colourless solids.

(*ii*) **Odour** Many volatile halogen compounds have sweet smell.

(*iii*) **Solubility** Haloalkanes have polar nature but are insoluble in polar solvents like water because they have no ability to form H-bonds with water.

However, these are soluble in organic solvents like ether, benzene etc.

(*iv*) **Density** Fluoro and chloro compounds are lighter than water whereas bromo and iodo derivatives are heavier as given below.

$RF < RCl < RBr < RI$ [For the same alkyl group]

The density of alkyl halide decreases as the size of alkyl group increases.

(*v*) **Melting and boiling points** Due to greater polarity as well as higher molecular mass as compared to the parent hydrocarbon, the intermolecular forces of attraction (dipole-dipole and van der Waals') are stronger in the halogen derivative.

As a result, the boiling point of chlorides, bromides and iodides are considerably higher than those of the hydrocarbons of comparable molecular mass.

For the same alkyl group, boiling point increases as the atomic mass of the halogen increases and hence, the magnitude of the van der Waals' forces of attraction increases.

$$RI > RBr > RCl > RF$$

For the same halogen atom, the boiling point increases with increase in the size of the alkyl group.

$$CH_3X < CH_3CH_2X < CH_3CH_2CH_2X < CH_3CH_2CH_2CH_2X$$

(where, X = F, Cl, Br, I)

For isomeric alkyl halides, boiling point decreases with branching due to decrease in surface area of the alkyl halide and hence, the magnitude of van der Waals' forces of attraction decreases.

$$CH_3CH_2CH_2CH_2Cl > CH_3CH_2CH(Cl)CH_3 > (CH_3)_2CHCH_2Cl > (CH_3)_3CCl$$

Boiling point increases with increase in the number of halogen atom.

$$CH_3Cl < CH_2Cl_2 < CHCl_3 < CCl_4$$

(*vi*) **Dipole moment** In an alkyl halide, the halogen atom is bonded to an sp^3-hybrid carbon atom. The halogen is more electronegative than carbon. Thus, C—X bond is polarised with a partial positive charge on carbon and a partial negative charge on the halogen.

$$H_3C^{\delta+} \overset{\mu}{\longrightarrow} X^{\delta-}$$

$$\mu \text{ (dipole moment)} = e \times d$$

(where e is the charge and d is the bond length)

The carbon-halogen bond lengths increase as the halogen atoms become bigger (larger atomic radii) the order is as follows

$$C—F < C—Cl < C—Br < C—I$$

These two effects oppose each other with the larger halogens having longer bonds but weaker electronegativities. The overall result is that the bond dipole moments increase in the order

μ (dipole moment)

C—I	<	C—Br	<	C—F	<	C—Cl
1.29 D		1.48 D		1.51 D		1.56 D
CH_3I	<	CH_3Br	<	CH_3F	<	CH_3Cl
1,64 D		1.79 D		1.82 D		1.94 D
5.40×10^{-30} Cm		6.04×10^{-30} Cm		6.17×10^{-30} Cm		6.24×10^{-30} Cm

CHEMICAL PROPERTIES

The carbon-halogen bond of haloalkane is polar. As a result, the two atoms do not share their bonding electrons equally. This causes the haloalkanes to undergo substitution and elimination reactions.

Nucleophilic substitution reactions

These reactions involve substitution that begins with nucleophilic attack. The atom which departs with its bonding pair of electrons is called leaving group.

$$R—X + Nu^- \longrightarrow R—Nu + X^-$$

(X: Leaving group)

For a nucleophilic substitution reaction to occur, the carbon atom attached to the leaving group must bear a partial positive charge because this is the site where the nucleophile will attack. Most heteroatoms are more electronegative than carbon, so they tend to withdraw electron density from the bond shared with the carbon atom, creating the electrophilic centre.

When a nucleophile is present, it can react with the electrophilic carbon atom and substitution occurs.

$$\ddot{Nu}^- + \overset{\delta+}{C}-\overset{\delta-}{X} \longrightarrow Nu-C + :\ddot{X}:^-$$

The electron-pair on the nucleophile, (Nu^-) is attracted to the electron deficient carbon atom.

Leading to the replacement of the leaving group *X* by Nu. Bromide ion is a good leaving group because it is a weak base.

The halide ions are particularly good leaving groups, thus an alkyl halide is the most common substrate for nucleophilic substitution reactions. The leaving group ability of halide ions follows the order,

$$I^- > Br^- > Cl^- > F^-$$

Iodide ion has the larger size than all the others halides and can accomodate the negative charge easily, thus behaves as the best leaving group.

Types of nucleophilic substitution reaction

There are two types of nucleophilic substitution reactions

S_N2 type (bimolecular nucleophilic substitution)

The term bimolecular means that rate of a reaction depends upon the concentration of two molecules, i.e. on substrate and nucleophile. It follows second order kinetics. These reactions are commonly seen in primary alkyl halides and alcohols. This reaction takes place in one step.

$$\text{Rate} \propto [A]\ [B]$$
$$[A] = RX,\quad [B] = Nu^-$$

Mechanism

These reactions take place through a transition state in which both the reactants are partially bonded to each other.

$$HO^- + H_3\overset{\delta+}{C}-\overset{\delta-}{Br} \longrightarrow \text{(Back side attack of } OH^- \text{ ion)}$$

$$\overset{\delta-}{HO}\text{-----}CH_3\text{-----}\overset{\delta-}{Br} \longrightarrow HO-CH_3 + Br^-$$

Transition state

As the hydroxide ion approaches the molecule from the back side, an effective collision take place. The —OH bond is formed and the —Br bond is broken simultaneously with the liberation of bromide ion.

- **Stereochemistry of S_N2 mechanism** The back side attack by a nucleophile in S_N2 reactions produce a product whose configuration is opposite to that of the reactant. Therefore, inversion of configuration occurs, if there is chiral carbon at alkyl halide. It is also known as Walden inversion.

e.g. optically active, (–)-2-bromooctane was hydrolysed to obtain (+)-2 – octanol.

- **Relative reactivity of alkyl halides toward S_N2 reactions** The relative order of alkyl halides toward S_N2 reaction is

Methyl halides > primary halides > secondary halides > tertiary halides

Factor affecting S_N2 reactions

Various factors affecting S_N2 reactions are as follows

(*i*) **Steric effects** The presence of bulky substituents on or near the C-atom bonded to halogen tends to hinder the approach of the nucleophile to that C-atom and makes the reaction difficult. Amongst all the 1° alkyl halides are reactive towards S_N2 as they are sterically less hindered and has only 3H-atoms on the α-carbon. While tertiary alkyl halides with three bulky alkyl groups are least reactive followed by secondary halides (with two bulky alkyl group) and primary halides (with one bulky alkyl group). Thus, the reactivity order of alkyl halides toward S_N2 reactions is as follows.

$(CH_3)_3C-Br$	$(CH_3)_2CH-Br$	CH_3CH_2-Br	CH_3-Br
(3°)	(2°)	(1°)	Methyl
(< 1)	500	40000	2000000

S_N2 reactivity increases →

(*ii*) **Nucleophilicity** In S_N2 reaction, nucleophile transfers electron pair to carbon of substrate, leading to the formation of transition state. Hence, a stronger nucleophile will react with substrate at a faster rate. e.g. R_3P, $I^\ominus$, $HS^\ominus$, $RS^\ominus$ are very good nucleophile while *R*COOH is a weak nucleophile.

Weak nucleophile fails to promote the S_N2 reaction, therefore reaction with weak nucleophile often proceed by S_N1 mechanism if the substrate is secondary or tertiary.

(*iii*) **Leaving group ability** More the ability of leaving group, faster is the reaction by S_N2 mechanism, e.g. leaving group reactivity for S_N2 reaction given by halides are as follows

F^-	Cl^-	Br^-	I^-
1	200	10000	20000

Relative reactivity → Increases

(*iv*) **Effect of solvent** S_N2 mechanism is favoured by aprotic solvents such as (dimethylsulphoxide) DMSO, (dimethyl formamide) DMF, or (Hexamethylphosphorotriamide) HMPT.

Aprotic solvents dissolve ionic reagents chiefly through their bonding to the cation. They leave the anion free and hence, highly reactive. So, the smallest anion (F^-) is more reactive. Hence, order of reactivity is

$$F^- > Cl^- > Br^- > I^-$$

These solvents form H-bonding with the reactants. This reaction can also proceed through polar protic solvents. More H-bonding takes place with smallest anion (F^- ion) that leads to lesser reactivity. So, the order is

$$I^- > Br^- > Cl^- > F^-$$

S_N1 type (Unimolecular nucleophilic substitution)

The term unimolecular means that rate of nucleophilic substitution reaction depends upon the concentration of the substrate and not on the nucleophile. It follows first order kinetics. This reaction takes place in two steps.

$$\text{Rate} \propto [\text{substrate}] \downarrow RX$$

Mechanism

Step I $R—X \underset{\text{Step}}{\overset{\text{Slowest}}{\rightleftharpoons}} R^+ + X^-$

Step II $R^+ + \underset{\text{(Nucleophile)}}{OH^-} \xrightarrow{\text{Fast}} R—OH$

e.g. $CH_3—C(CH_3)_2—Br$ (Tertiary butyl bromide) $\underset{\text{Slow} \; -Br^-}{\overset{\text{Step I (Ionisation)}}{\rightleftharpoons}}$ $CH_3—\overset{+}{C}(CH_3)_2$ (Planar carbocation)

$\underset{OH^- \text{(Fast)}}{\overset{\text{Step II}}{\rightleftharpoons}}$ $CH_3—C(CH_3)_2—OH$ (3° alcohol)

- **Stereochemistry of S_N1 mechanism** The attack of nucleophile can occur from both the faces with almost equal ease, thereby giving a 50:50 mixture of two enantiomers, i.e. a racemic mixture.

$C(CH_3)(Et)(py)—Br \rightleftharpoons C^+(CH_3)(Et)(py) + Br^-$

$C^+(CH_3)(Et)(py) \xrightarrow{OH^-}$ $C(CH_3)(Et)(py)—OH$ Front attack (Retention) 50% ; $HO—C(CH_3)(Et)(py)$ Rear (back) attack (Inverted product) 50% } Racemic mixture

- **Relative reactivity of haloalkanes towards S_N1 reactions** In S_N1 reactions, carbocations are the intermediates. Greater the stability of carbocation, more easily it is formed and hence, faster will be the rate of reaction. Since, the stability of carbocation decreases in the order.

Allyl ≈ benzyl carbocation > 3° carbocation > 2° carbocation > 1° carbocation

Allyl and benzyl carbocations are more stable due to resonance.

$H_2C{=}CH—\overset{+}{C}H_2 \longleftrightarrow H_2C^+—CH{=}CH_2$

Allyl carbocation

$C_6H_5\overset{+}{C}H_2 \longleftrightarrow \longleftrightarrow \longleftrightarrow \longleftrightarrow$ (resonance structures with CH_2 and + at ortho, para, ortho positions)

Benzyl carbocation

Thus, reactivity of haloalkanes toward S_N1 decreases in the same order.

Allylic halides ≈ benzylic halide > 3° alkyl halide > 2° alkyl halide > 1° alkyl halide.

Factors affecting S_N1 reactions

Various factors affecting S_N1 reactions are as follows

(*i*) **Nucleophilicity** S_N1 reactions do not depend on the concentration of nucleophile. Therefore, these reactions are not affected by strength of nucleophile.

(*ii*) **Leaving group ability** Good leaving groups such as $Cl^\ominus$, $Br^\ominus$, $I^\ominus$ help in proceeding the S_N1 reactions faster.

(*iii*) **Solvent effect** Reaction by S_N1 is favoured by solvents of high ionising power, i.e. by polar protic solvents that help to pull the leaving group out of the molecule.

Table 13.1 Various nucleophilic substitution reactions of alkyl halides ($R—X$)

Reagent	Nucleophile (Nu^-)	Product	Class of product	Reactions
NaOH or KOH	OH^-	$R—OH$	Alcohol	$CH_3CH_2Br + KOH(aq) \xrightarrow{\Delta} CH_3CH_2OH + K^+ Br^-$
H_2O	OH^-	$R—OH$	Alcohol	$CH_3Br + H_2O \longrightarrow CH_3OH + HBr$
$R'\bar{O}Na^+$ (Sodium alkoxide)	$R'O^-$	$R—O—R'$	Ether	$CH_3 - I + N\overset{+}{a}\bar{O}CH_3 \xrightarrow{\Delta} CH_3—OCH_3 + NaI$
NaI	I^-	$R—I$	Alkyl iodide	$CH_3Br + NaI \xrightarrow[\text{Acetone}]{\Delta} CH_3I + NaBr$
NH_3	NH_2^-	$R—NH_2$	Primary amine	$CH_3CH_2Br + H—NH_2 \xrightarrow[C_2H_5OH]{\Delta} CH_3CH_2NH_2 + HBr$
$R'NH_2$	$R'\bar{N}H$	$R'—NH—R'$	Secondary amine	$CH_3CH_2NH—H + BrCH_2CH_3 \xrightarrow[C_2H_5OH]{\Delta} (CH_3CH_2)_2NH + HBr$
$R'NHR''$	$R'\bar{N}R''$	$R'—\underset{R''}{\underset{\mid}{N}}—R'$	Tertiary amine	$(CH_3CH_2)_2NH + Br—CH_2CH_3 \xrightarrow[C_2H_5OH]{\Delta} (CH_3CH_2)_3N + HBr$
KCN (Alcoholic)	CN^-	$R—CN$	Nitrile (Cyanide)	$CH_3—I + KCN\ (alc.) \xrightarrow{\Delta} CH_3CN + KI$
AgCN	$\bar{N}C$	$R—N \overset{\rightarrow}{=} C$	Isonitrile (Isocyanide)	$CH_3—I + AgCN \xrightarrow[\Delta]{C_2H_5OH / H_2O} \underset{\text{Isonitrile}}{R—NC} + AgX$
KNO_2	$O=N—O^-$	$R—O—N=O$	Alkyl nitrite	$CH_3CH_2Br + K^+ —O—N=O \xrightarrow{\Delta} CH_3CH_2—O—N=O + KBr$
$AgNO_2$	NO_2^-	$R'—NO_2$	Nitroalkane	$CH_3CH_2Br + AgO—\ddot{N}=O \xrightarrow[C_2H_5OH / H_2O]{\Delta} CH_3CH_2—N\overset{\nearrow O}{\underset{\searrow O}{}} + AgBr$
$R'—CO\overset{-+}{O}Ag$	$R—COO^-$	$R'—COOR$	Ester	$CH_3—\overset{O}{\overset{\parallel}{C}}—OAg + BrCH_2CH_3 \xrightarrow[C_2H_5OH]{\Delta} CH_3—\overset{O}{\overset{\parallel}{C}}—OCH_2CH_3 + AgBr$
(Sodium hydrosulphide)	HS^-	$R—SH$	Thioalcohol	$CH_3CH_2Br + N\overset{+}{a}\bar{S}H \xrightarrow[\Delta]{C_2H_5OH/H_2O} CH_3CH_2—SH + NaBr$
$R'SNa$	$R'S^-$	$R—S—R'$	Thioether	$CH_3CH_2Br + N\overset{+}{a}\bar{S}CH_2CH_3 \xrightarrow[\Delta]{C_2H_5OH / H_2O} CH_3CH_2S — CH_2CH_3 + NaBr$
$LiAlH_4$	H^- (from $LiAlH_4$)	$R—H$	Alkane	$LiAlH_4 + 4RX \longrightarrow 4R—H + AlX_3 + LiX$
$Na^+—C\equiv C—R'$ Sodium alkynide	$\bar{C}\equiv C—R'$	$R—C\equiv C—R'$	Alkyne	$CH_3CH_2Br + Na^+ — C\equiv CR \longrightarrow CH_3CH_2— C\equiv C—R + NaBr$

Note ***Ambident nucleophiles*** *A cyanide ion can attack as a nucleophile through carbon as well as through nitrogen. Similarly, nitrite can attack as a nucleophile through oxygen and nitrogen. The nucleophiles which can attack a substrate through two possible sites (as in case of cyanide and nitrite) are termed as* ***ambident nucleophiles.***

$$^{\ominus}C\equiv N \longleftrightarrow \ddot{C}=\overset{\ominus}{N};\ O=N-\overset{\ominus}{O} \longleftrightarrow \overset{\ominus}{O}-N=O$$

ELIMINATION REACTIONS

In these reactions, two groups on adjacent atoms or same atoms are lost and a double bond is formed. Therefore, the product is only an alkene or alkyne [multiple bond]. They are just opposite to addition reactions.

This type of reaction is given by the compounds which have a nucleophile as leaving group, i.e. X^-, $\bar{O}H$, $C_2H_5O^-$, $\overset{\oplus}{N}_2$ etc.

$$\text{e.g.}\quad \underset{\text{Ethoxide ion}}{C_2H_5O^-} + H—CH_2—\underset{\substack{|\\CH_3}}{\overset{\substack{Br\\|}}{C}}—CH_3 \longrightarrow \underset{\text{Ethanol}}{C_2H_5OH} + \underset{\text{2-methylpropene}}{H_2C{=}C(CH_3)_2} + Br^-$$

(2-bromo-2-methyl propane)

In haloalkanes, halide is lost from α-carbon and hydrogen is lost from β-carbon. The elimination that involves loss of two groups from adjacent carbon is termed as β-elimination. The two types of elimination reactions are as follows

Bimolecular (E2) elimination

E2 elimination is a concerted one step process.

This elimination abstraction of proton from the β-C atom by base and removal of X^- from the α-C atom take place simultaneously. Due to the formation of transition state, this mechanism is comparable with S_N2 mechanism.

$$\underset{\text{2-chlorobutane}}{CH_3\overset{\substack{Cl\\|}}{\underset{\substack{|\\H}}{C}}—\underset{\substack{|\\H}}{C}HCH_3} + OH^- \longrightarrow \underset{\text{But-2-ene}}{CH_3CH{=}CHCH_3} + H_2O + Cl^-$$

The rate depends on the concentration of alkyl halide as well as on the concentration of base.

$$\text{Rate} \propto [\text{alkyl halide}]\,[\text{base}]$$

Mechanism

$$CH_3—CH—\underset{\substack{|\\H}}{\overset{}{CH_2}} \; (\text{with } L \text{ on CH and } B^\ominus) \xrightarrow{\text{Slow}} \underset{\text{Transition state}}{\left[CH_3—\underset{\substack{\vdots\\L}}{CH}—\underset{\substack{\vdots\\H\cdots B^{\delta-}}}{CH_2}\right]} \xrightarrow{\text{Fast}} B—H + CH_3—CH{=}CH_2 + L^-$$

(Here, L is the leaving group)

Some important points related to E2 reactions are

(*i*) Reactivity depends upon both strength and nature of alkyl halide.

(*ii*) Strong base is required to abstract proton. A more substituted alkyl halide gives more stable alkenes and therefore, more reactive. General order of reactivity is

tertiary > secondary > primary

(*iii*) E2 reaction proceeds at room temperature.

(*iv*) Polar protic solvent retards E2 reaction as it solvates the base.

(*v*) Rate of cleavage of C—X bond is in order

$$\underset{\text{Fastest}}{I} > Br > Cl > \underset{\text{Slowest}}{F}$$

Unimolecular (E1) reaction

Unimolecular (E1) reaction proceed *via* two step. It is named as unimolecular because the rate of E1 reaction depends only on the concentration of alkyl halide.

$$\text{Rate} = k\,[\text{alkyl halide}]$$

In this reaction, firstly, heterolytic dissociation of alkyl halide occurs, which is a slow and rate determining step. Then, abstraction of proton through a base occurs from a carbon adjacent to the positively charged carbon atom. Due to the formation of carbocation this mechanism is comparable with S_N1 mechanism.

Mechanism

Step I Generation of carbocation

$$CH_3—\underset{\substack{|\\CH_3}}{\overset{\substack{CH_3\\|}}{C}}—Br \underset{\text{Slow}}{\overset{\text{Step I}}{\rightleftharpoons}} CH_3—\underset{\substack{|\\CH_3}}{\overset{\substack{CH_3\\|}}{C^{\oplus}}} + \underset{\substack{\text{Leaving}\\\text{group}}}{Br^{\ominus}}$$

This step being slow, is the rate determining step and depends only on the concentration of one reactant.

Step II Abstraction of proton by base

$$CH_3—\underset{\substack{|\\CH_2—H}}{\overset{\substack{CH_3\\|}}{C^{\oplus}}} + OH_2 \underset{\text{Fast}}{\overset{\text{Step II}}{\rightleftharpoons}} CH_3—\underset{\substack{\|\\CH_2}}{\overset{\substack{CH_3\\|}}{C}} + H_3O^{\oplus}$$

Some important points related to E1 are

(*i*) The order of reactivity of alkyl group towards E1 is tertiary > secondary > primary, since the rate-determining step is the formation of a carbocation, which has stability in order : 3° > 2° > 1°.

(*ii*) Highly substituted carbon atom, i.e. tertiary carbon atom favours E1 mechanism.

(*iii*) Use of polar solvents can stabilise the resulting carbocation. It can also stablise a polar transition state involved in the heterolytic bond cleavage.

(iv) In E1 reaction, the intermediate is carbocation and hence, rearrangement occurs before loss of proton.

$$H_3C-\underset{CH_3}{\overset{CH_3}{C}}-\underset{Cl}{CHCH_3} \xrightarrow[+\text{ alc. KOH}]{CH_3OH} H_3C-\underset{CH_3}{\overset{CH_3}{C}}-\overset{\oplus}{C}HCH_3$$

2° carbocation

$$\xrightarrow[\text{1,2-methyl shift}]{\text{Rearrangement}} H_3C-\underset{\oplus}{\overset{CH_3}{C}}-\underset{CH_3}{CH}-CH_3$$

3° carbocation (more stable than 1°)

β-elimination reactions are governed by **Saytzeff 's rule,** which states that dehydrohalogenation occurs in such a way that the major product is more highly substituted product.

$$CH_3CH_2-CH_2-\underset{Br}{CH}-CH_3 \text{ (2-bromopentane)} \xrightarrow[-KBr, -H_2O]{\text{Alc.KOH}} CH_3CH_2CH{=}CH-CH_3 \text{ (Pent-2-ene (major))} + CH_3CH_2CH_2CH{=}CH_2 \text{ (Pent-1-ene (minor))}$$

If substituted alkene formed during Saytzeff elimination is capable of showing *cis-trans* isomerism, *trans*-alkene being more stable is always the major product.

$$CH_3CH_2-\underset{Br}{CH}-CH_3 \text{ (2-bromobutane)} \xrightarrow[-KBr]{KOH(alc)/\Delta} CH_3CH_2CH{=}CH_2 \text{ (19%) Butene}$$

$$\text{(CH}_3\text{)(H)C=C(H)(CH}_3\text{) } trans\text{ (71%) (Major)} + \text{(CH}_3\text{)(H)C=C(CH}_3\text{)(H) } cis\text{ (10%) (Minor)}$$

Elimination reactions of cycloalkyl halides lead exclusively to *cis*-cycloalkenes.

Bromocyclodecene $\xrightarrow[EtOK]{EtOH}$ *cis*-cyclodecene (major) + *trans*-cyclodecene (major)

In presence of bulky base, $(CH_3)_3\overset{-\,+}{COK}$, elimination proceed *via* **Hofmann's rule.** Here, least substituted alkene is obtained because a bulky base suffers steric hinderance in abstracting a proton from more substituted β-carbon.

$$H_3C-\underset{H}{\overset{H}{C}}-\underset{CH_3}{\overset{H}{C}}-\underset{Br}{C}-\underset{H}{\overset{H}{C}}-H + (CH_3)_3CO^{\ominus}$$

(More hindered; Less hindered; Bulky base)

$$\downarrow$$

$$H_3C-C{=}\overset{H}{C}-CH_3 \text{ (with } CH_3 \text{ on C; Minor)} + H_3C-\underset{CH_3}{C}-\overset{H\,H}{C}{=}CH_2 \text{ (Major)}$$

1-bromo-1-methylcyclohexane $\xrightarrow[Et_3N]{\text{Bulky base}}$ 1-methylcyclohexene (Minor (E2)) + methylenecyclohexane (Major (E2))

Substitution *vs* elimination

Elimination reactions are usually accompanied by substitution reactions. When the reagent is a good base, it accept protons to yield elimination products and if it is a good nucleophile, it attacks the carbon to give substitution products.

Various factors that govern the elimination and substitution product are as follows.

- **Substrate** The proportion of elimination increases with increase in the branching of carbon chain because alkenes formed on elimination are stabilised by hyper conjugation.
- **Nature of base** Strong nucleophiles but weak bases promote substitution over elimination whereas strong bases but weak nucleophiles promote elimination reactions over substitution.

Reaction with metals

Alkyl halides undergo various types of reactions. They react with different metals to give various products.

Reaction with magnesium

When a solution of an alkyl halide in dry diethyl ether ($C_2H_5-O-C_2H_5$) is allowed to react with Mg, resulting solution obtained is called **Grignard reagent** (named after Victor Grignard—a Nobelist (1912) for this discovery). Its general formula is $RMgX$ (R can be aryl or alkyl).

e.g. $\underset{\text{Alkyl halide}}{CH_3X} + Mg \xrightarrow{\text{Ether}} \underset{\text{Grignard reagent}}{CH_3MgX}$

Here, X = Cl, Br, I

Reactivity of RI, RBr, RCl with Mg to form Grignard reagent is in order $RI > RBr > RCl$

For a given halide, reactivity order of 'R' is $1° > 2° > 3°$

Note *Zinc also forms organometallic compounds but reactivity is less than Mg. It is due to lower standard reduction electrode potential.*

$$CH_3CH_2CH_2I + Zn \longrightarrow CH_3CH_2CH_2ZnI$$

Wurtz reaction

An ether solution of an alkyl halide is treated with sodium to produce higher alkanes. This reaction involves the formation of new C—C bond.

This reaction is used in general to produce alkanes which have even number of carbon atoms. Methane cannot be synthesised by this method.

e.g. $$\underset{\text{Bromomethane}}{2CH_3—Br} + 2Na \xrightarrow{\text{Ether}} \underset{\text{Ethane}}{CH_3CH_3} + 2NaBr$$

A single product is obtained, when symmetrical alkyl halides are used but if different alkyl halides are used, the product is non separable mixture of alkanes.

$$\underset{\text{Bromoethane}}{CH_3CH_2Br} + 2Na + \underset{\text{Bromomethane}}{CH_3Br} \xrightarrow{\text{Ether}} \underset{\text{Propane}}{CH_3CH_2CH_3}$$

$$\underset{\text{Ethane}}{CH_3—CH_3} + \underset{\text{Butane}}{C_2H_5—C_2H_5} + 2NaBr$$

Other hydrocarbons are also obtained as a result of disproportionation, i.e. intermolecular hydrogenation of free radical, one molecule gaining H at the expense of the other. Although both free radical and ionic mechanism are suggested to explain the reaction but ionic mechanism is preferred as Na does not form free radical in normal practice.

$\dot{C_2}H_5$ intermediate is formed which can

(i) combine to form n-butane (major)

$$CH_3\,\dot{C}H_2 + \dot{C}H_2\,CH_3 \longrightarrow CH_3\,CH_2 — CH_2\,CH_3$$

new (C—C) bond

(ii) undergo disproportionation (intramolecular hydrogenation of free radical)

$$CH_3\dot{C}H_2 + CH_3\dot{C}H_2 \longrightarrow \underset{\substack{\text{Oxidation}\\\text{product}}}{CH_2{=}CH_2} + \underset{\substack{\text{Reduction}\\\text{product}}}{CH_3CH_3}$$

Corey–House synthesis

It is a reaction that involves the reaction of a lithium dialkyl cuprate (R_2CuLi) with an alkyl halide to form a new alkane.

This is a good method to prepare hydrocarbons (alkanes) having odd number of carbon atoms. Methane cannot be synthesised by this method.

$$\underset{\substack{\text{(Alkyl}\\\text{halide)}}}{R—X} \xrightarrow[\substack{\text{Diethyl}\\\text{ether}}]{\text{Li}} RLi \xrightarrow{\text{CuI}} \underset{\text{(Gilman reagent)}}{R_2CuLi}$$

$$\xrightarrow{R'X} R—R' + RCu + LiX$$

(This R' should be methyl, 1° alkyl or 2° cycloalkyl).

$$\underset{\text{2-bromopropane}}{(CH_3)_2CHBr} \xrightarrow[\substack{\text{(ii) Cu}\\\text{step I}}]{\text{(i) Li}} \underset{\substack{\text{Lithium di-}iso\text{-propyl}\\\text{cuprate}}}{[(CH_3)_2CH]\,CuLi}$$

$$\xrightarrow[\text{step II}]{CH_3CH_2CH_2Br} \underset{\text{2-methylpentane}}{(CH_3)_2CHCH_2CH_2CH_3}$$

For the second step to give a good yield of the alkane, the alkyl halide (Say $R'X$) must be either a methyl halide (a primary alkyl halide) or a secondary cycloalkyl halide.

The alkyl groups of the lithium dialkyl cuprate (also called Gilman reagent) may be methyl, 1°, 2° or 3°. Moreover, the two alkyl groups being coupled need not be different.

Reduction of haloalkanes

Haloalkanes can be reduced to alkanes by various reducing agents like

Zn/CH_3COOH, Zn/HCl, $Zn/NaOH$, Zn-Cu couple, $LiAlH_4$, $NaBH_4$ etc.

$$R—X \xrightarrow{\text{Reduction}} R—H$$

e.g. $$2CH_3CH_2\underset{\substack{|\\ Br\\ \text{2-bromobutane}}}{CH}CH_3 + Zn + 2H^+ \longrightarrow \underset{\text{Butane}}{2CH_3CH_2CH_2CH_3} + ZnBr_2$$

CHECK POINT 13.2

1. An optically active halide when allowed to react with $CN^{\ominus}$ gives a racemic mixture, the halide is most likely
   (a) primary halide (b) *sec.* halide
   (c) *tert.* halide (d) None of these

2. When sodium salt of ethanol is treated with ethyl bromide, the product formed is
   (a) methoxyethane (b) ethoxyethanol
   (c) methylethyl keton (d) diethylether

3. Which of the following reactants will yield ethane?
   (a) Methyl bromide and sodium
   (b) Ethyl bromide and magnesium
   (c) Ethanol and H_2SO_4
   (d) Ethyl bromide and KCN

4. Which of the following alkyl halide gives S_N1 reaction at highest role?
   (a) Methyl chloride
   (b) Ethyl chloride
   (c) Isobutyl chloride
   (d) *Tert*-butyl chloride

5. When isopropyl iodide in ethereal solution is warmed with sodium, the product formed is
   (a) *n*-hexane (b) *neo*-hexane
   (c) 2, 3-dimethylbutane (d) All of these

6. The leaving group ability of halide ions for S_N2 reaction is
   (a) $Cl^- > F^- > Br^- > I^-$ (b) $I^- > Br^- > Cl^- > F^-$
   (c) $Cl^- > F^- > I^- > Br^-$ (d) $Br^- > I^- > F^- > Cl^-$

7. Consider the following reaction,

$$(CH_3)_3CBr \xrightarrow[-H_2O,\ -KBr]{\text{Alc. KOH}} (CH_3)_2C{=}CH_2$$

   This is an example of
   (a) nucleophilic substitution (b) electrophilic substitution
   (c) free radical substitution (d) β-elimination

8. Major product obtained when 2-chloro-3-methylbutane is treated with ethanolic KOH is likely to give
   (a) neopentyl alcohol (b) pentene
   (c) 2-methylbut-2-ene (d) no reaction occur

9. Which of the following compounds is not formed when a mixture of methyl bromide and ethyl bromide is treated with sodium metal in the presence of dry ether?
   (a) Methane (b) Ethane
   (c) Propane (d) Butane

10. Reactivity of alkyl halides with Mg to form Grignard reagent is
   (a) RCl > RBr > RI (b) RI > RBr > RCl
   (c) RI > RCl > RBr (d) RBr > RCl > RI

POLYHALOGEN COMPOUNDS

Carbon compounds having more than one halogen atom are usually referred to as polyhalogen compounds. Many of these compounds are useful in industry and agriculture. Some important polyhalogen compounds are described as follows

Dichloromethane (Methylene chloride, CH_2Cl_2)

It is colourless, sweet smelling, volatile liquid having low boiling point of 313 K and specific gravity of 1.37. It is prepared industrially by the direct chlorination of methane in the presence of diffused sunlight.

$$\underset{\text{Methane}}{CH_4} + 2Cl_2 \xrightarrow{h\nu} \underset{\text{Methylene chloride}}{CH_2Cl_2} + 2HCl$$

Uses

It is used in the following ways

(*i*) as a solvent and also as a paint remover.

(*ii*) as a propellant in aerosols.

(*iii*) as a metal cleaning and finishing solvent.

(*iv*) as refrigerant and dewaxing agent.

(*v*) as a process solvent in the manufacture of drugs.

Harmful effects (Physiological effects)

Some of the important harmful effects of methylene chloride are as follows

(*i*) It affects the human central nervous system.

(*ii*) Exposure to low levels of methylene chloride in air can lead to slightly impaired hearing and vision.

(*iii*) Exposure to high levels of methylene chloride can cause dizziness, nausea, tingling and numbness in the fingers and toes.

(*iv*) Direct contact with skin causes intense burning and mild redness of the skin.

(*v*) Direct contact with eyes can burn the cornea.

Trichloromethane (Chloroform, $CHCl_3$)

Trichloromethane is commonly known as chloroform. It is a colourless liquid with sweet odour. It is non-flammable but its vapour burns with green flame. In industry, $CHCl_3$ is produced by heating a mixture of chlorine and either chloromethane or methane.

CH_3CH_2OH on reaction with moist bleaching powder is converted into $CHCl_3$.

$$CaOCl_2 + H_2O \longrightarrow Ca(OH)_2 + Cl_2$$

$$Cl_2 + H_2O \longrightarrow 2HCl + \underset{\text{(Oxidising agent)}}{[O]}$$

$$CH_3CH_2OH + [O] \longrightarrow CH_3CHO + H_2O$$

$$CH_3CHO + 3Cl_2 \longrightarrow \underset{\text{Chloral}}{CCl_3CHO} + 3HCl$$

$$\underset{\text{Chloral}}{2CCl_3CHO} + Ca(OH)_2 \longrightarrow \underset{\text{Chloroform}}{2CHCl_3} + (HCOO)_2Ca$$

It can also be prepared by the partial reduction of carbon tetrachloride.

$$CCl_4 \xrightarrow{Fe/H_2O} CHCl_3$$

Some of the important reactions of chloroform are as follows

(*i*) **Oxidation** Chloroform is slowly oxidised by air in the presence of light to an extremely poisonous gas carbonyl chloride also called phosgene. It is therefore stored in dark coloured bottles. These bottles are completely filled with chloroform upto the brim and are properly stoppered to keep air out.

$$\underset{\text{Chloroform}}{2CHCl_3} + O_2 \xrightarrow{\text{Sunlight}} \underset{\text{Phosgene}}{2COCl_2} + 2HCl$$

1% ethanol is added so, as to convert phosgene gas to diethyl carbonate which is non-volatile and non-toxic.

$$\underset{\text{Phosgene}}{COCl_2} + \underset{\text{Ethanol}}{2C_2H_5OH} \longrightarrow \underset{\text{Diethylcarbonate}}{(C_2H_5)_2CO_3} + 2HCl$$

(*ii*) **Reduction** Partial reduction of chloroform is carried out with zinc and hydrochloric acid that converts chloroform to methylene chloride. But when zinc dust and water is used then is completely reduced to methane.

$$\underset{\text{Methane}}{CH_4} \xleftarrow[-HCl\ 6[H]]{Zn/H_2O} \underset{\text{Chloroform}}{CHCl_3} \xrightarrow[2[H]^+]{Zn/HCl\text{-}C_2H_5OH} \underset{\text{Methylene chloride}}{CH_2Cl_2}$$

(*iii*) **Halogenation** When chlorine is passed through chloroform in the presence of sunlight, it forms carbon tetrachloride.

$$\underset{\text{Chloroform}}{CHCl_3} + Cl_2 \xrightarrow{h\nu} \underset{\text{Tetrachloromethane}}{CCl_4}$$

(*iv*) **Hydrolysis** Chloroform reacts with *aq.* KOH or NaOH and undergoes hydrolysis to form sodium or potassium salts of formic acid.

$$CHCl_3 + 4NaOH \longrightarrow HCOO^-Na^+ + 3NaCl + 2H_2O$$

(*v*) **Nitration** It reacts with HNO_3 to form chloropicrin, which is used as an insecticide.

$$\underset{\text{Chloroform}}{Cl_3C\text{—}H} + HO\text{—}NO_2 \longrightarrow \underset{\text{(Nitro chloroform) Chloropicrin}}{Cl_3C\text{—}NO_2} + H_2O$$

(*vi*) **Reaction with acetone** It condenses with acetone in presence of KOH to form chloretone, a sleep-inducing medicine (hypnotic).

$$\underset{\text{Chloroform}}{CHCl_3} + \underset{\text{Acetone}}{CH_3\text{—}\overset{O}{\overset{\|}{C}}\text{—}CH_3} \xrightarrow{KOH} \underset{\text{Chloretone}}{(CH_3)_2C(OH)\text{—}CCl_3}$$

(*vii*) **Reaction with Ag powder**

$$\underset{\text{Chloroform}}{CHCl_3} + 6Ag + CHCl_3 \longrightarrow \underset{\text{Ethyne}}{CH{\equiv}CH} + 6AgCl$$

(*viii*) **Reimer-Tiemann reaction**

On refluxing alkaline solution of phenol with $CHCl_3$ at 330 K and on subsequent acidification, *o*-hydroxy- benzaldehyde (salicylaldehyde) is obtained. It is called Reimer-Tiemann reaction, *o*-isomer predominates but is separated from *p*-isomer by steam-distillation.

$$\underset{\text{Phenol}}{C_6H_5OH} + \underset{\text{Chloroform}}{CHCl_3} + 3KOH \xrightarrow{\Delta/H^+} \underset{\text{Salicylaldehyde or } o\text{-hydroxybenzaldehyde}}{o\text{-}HOC_6H_4CHO} + 3KCl + 2H_2O$$

If *o*-position is already occupied, aldehyde group goes to *p*-position.

$$\underset{\text{Guaiacol}}{o\text{-}HOC_6H_4OCH_3} \xrightarrow[\text{(ii) } H^+]{\text{(i) } CHCl_3,\ KOH,\ \Delta} \underset{\text{Vanillin}}{4\text{-}HO\text{-}3\text{-}CH_3OC_6H_3CHO}$$

Mechanism

$$CHCl_3 + OH^- \rightleftharpoons :\overset{-}{C}Cl_3 + H_2O$$

$$:\overset{-}{C}Cl_3 \longrightarrow :CCl_2 + Cl^-$$

$$C_6H_5OH \rightleftharpoons C_6H_5O^{\ominus} + H^+$$

$$\text{Phenoxide} + :CCl_2 \longrightarrow \text{(intermediate with } \overset{\ominus}{C}Cl_2, H) \xrightarrow{H_2O} \text{(intermediate with } H, CHCl_2)$$

$$\longrightarrow \text{(phenoxide with } CHCl_2) \xrightarrow{OH^{\ominus}} \text{(phenoxide with } CHO) \xrightarrow{H^+} \text{(phenol with } CHO)$$

If CCl_4 is used instead of $CHCl_3$, product is *o*-hydroxy benzoic acid.

$$\underset{\text{Phenol}}{C_6H_5OH} + \underset{\text{Carbon tetrachloride}}{CCl_4} + 4KOH \xrightarrow{\Delta/H^+} \underset{o\text{-hydroxy benzoic acid}}{C_6H_4(OH)COOH} + 4KCl + 2H_2O$$

(*ix*) **Carbylamine reaction**

On heating $CHCl_3$ with ethanolic KOH and aniline (or 1° amine RNH_2), phenyl isocyanide is formed and is readily detected by its nauseating odour. It is called isocyanide or carbylamine test of 1° amine RNH_2.

$$\underset{\text{Chloroform}}{CHCl_3} + 3KOH + RNH_2 \xrightarrow{\Delta} \underset{\text{Isocyanide}}{RNC} + 3KCl + 3H_2O$$

Mechanism

$$CHCl_3 + OH^- \longrightarrow :\bar{C}Cl_3 + H_2O$$

$$:\bar{C}Cl_3 \longrightarrow :CCl_2 + Cl^-$$

$$RNH_2 + :CCl_2 \longrightarrow R\overset{\oplus}{N}H_2\overset{\ominus}{C}Cl_2$$

$$R\overset{\oplus}{N}H_2\overset{\ominus}{C}Cl_2 \longrightarrow RNH\overset{\ominus}{C}Cl_2 + H^+$$

$$RNH\overset{\ominus}{C}Cl_2 \longrightarrow RN{=}\overset{\ominus}{C}Cl + HCl$$

$$RN{=}\overset{\ominus}{C}Cl \longrightarrow \underset{\text{Isocyanide}}{RNC} + Cl^{\ominus}$$

Uses

It is used in following ways

(*i*) as a solvent for fats, alkaloids, iodine and other substances.

(*ii*) as a laboratory reagent.

(*iii*) In the preparation of chloropicrin, chloretone, etc.

(*iv*) In medicines.

(*v*) In the production of freon refrigerant, *R*-22.

(*vi*) As an anaesthetic but now it has been replaced by other anaesthetics such as ether.

Harmful effects

Harmful effects of chloroform are as follows.

(*i*) Chloroform affects the central nervous system. Inhaling it, 900 ppm (in air), for a short time can causes dizziness, fatigue and headache.

(*ii*) Chronic exposure to chloroform damages liver and kidney due to the formation of phosgene gas inside the body.

(*iii*) Skin develops sore if immersed directly into the chloroform.

Triiodomethane (Iodoform, CHI_3)

It is yellow coloured crystalline solid with characteristic unpleasant odour.

Uses

It is use in following ways.

(*i*) As an antiseptic due to the liberation of free iodine.

$$\underset{\text{Triiodomethane}}{CHI_3} \xrightarrow[\text{heat}]{\text{Air, moisture}} \underset{\text{Iodine}}{I_2}\uparrow + CO_2 + H_2O$$

But due to its unpleasant smell, it has been replaced by other formulations containing iodine.

(*ii*) In the pharmaceutical industry.

Tetrachloromethane (Carbon tetrachloride, CCl_4)

It is colourless oily liquid with sweet smell. Its boiling point is 350 K. It is formed from chloroform and its reaction is as follows

$$CHCl_3 + Cl_2 \xrightarrow{h\nu} CCl_4 + HCl$$

Formation of freon

- $CCl_4 \xrightarrow[(C+FeCl_3)]{HF,\ 570\ K} CCl_2F_2 + CFCl_3$
- $CCl_4 + 2HF \xrightarrow{SbCl_5} CCl_2F_2 + 2HCl$

With steam

CCl_4 is insoluble in H_2O. Its vapour is non-inflammable, hence, it is widely used as industrial solvent. It is also used as a fire-extinguisher under the name of **pyrene.** At about 770 K, CCl_4 in contact with H_2O forms $COCl_2$ (poisonous gas).

$$CCl_4 + H_2O \xrightarrow{770\ K} COCl_2 + 2HCl$$

Hence, after using pyrene to extinguish a fire, the room should be well ventilated.

Alkaline hydrolysis

$$CCl_4 + OH^- \longrightarrow HCOO^- + CO + Cl^- + H_2O$$

Uses

It is used in following ways

(*i*) In large quantities, in the manufacture of refrigerants and propellants for aerosol cans.

(*ii*) As a feed stock in the synthesis of chlorofluoro carbons (freons) and other chemicals.

(*iii*) As a cleaning fluid, as a degreasing agent and as a spot remover.

(*iv*) As fire extinguisher, under the name of pyrene.

(*v*) $CHCl_2CHCl_2$ (westron) and $CCl_2 = CHCl$ (westron sol) are used as solvents.

Harmful effects (Physiological effects)

Its harmful effects are as follows

(*i*) Exposure to CCl_4 causes liver cancer in humans.

(*ii*) The most common effects are dizziness, light headedness, nausea and vomiting which can cause permanent damage to nerve cells. In several cases, these effects can lead to coma, unconsciousness or even death.

(*iii*) Exposure of CCl_4 vapours can make heartbeat irregular or even can stop it. When brought in contact with eyes, it may cause irritation in eyes.

(*iv*) It is also harmful to environment. When it is released in air, it rises in the atmosphere and depletes the ozone layer.

Depletion of ozone layer increases the human exposure to UV rays which causes skin cancer, eye diseases and disorders and possible disruption of the immune system.

Freons

Chlorofluorocarbon compounds (CFC) of methane and ethane are collectively known as freons.
They are extremely stable, unreactive, non-toxic, non-corrosive and easily liquefiable gases.
Freon-12 (dichlorodifluoromethane, CCl_2F_2) is one of the most common freons in industrial use. It is manufactured from tetrachloromethane by Swarts reaction.

$$3CCl_4 + 2SbF_3 \xrightarrow{SbCl_5} 2SbCl_3 + \underset{\text{Freon-12}}{3CCl_2F_2}$$

Uses

It is used in following ways

(*i*) As refrigerants in refrigerators and air conditioners.

(*ii*) As propellants for aerosols and foams to spray out deodorants, cleansers, shaving creams, hair sprays and insecticides.

Harmful effects

Most freon makes its way into the atmosphere where it undergoes photochemical decomposition and initiates radical chain reactions for depletion of the ozone layer. Therefore, the use of freons as propellants and refrigerants has been drastically discouraged.

CHECK POINT 13.3

1. Which of the following statement(s) is correct?
(a) Dichloromethane is a colourless, sweet smelling solid
(b) It has a boiling point of 300 K
(c) Its specific gravity is 1.37
(d) CH_2Cl_2 can be prepared by direct chlorination of methane in the presence of sunlight

2. Which of the following statements is correct?
(a) Methylenechloride is colourless, low boiling liquid
(b) Exposure to low levels of CH_2Cl_2 causes paralysis
(c) Chloroform is stored in light coloured bottles
(d) Iodoform has fruit like pleasent smell

3. Chloroform can be used in medicine as
(a) antipyretic (b) antihistamine
(c) anaesthetic (d) antibiotic

4. Which of the following gases is evolved by the slow oxidation of chloroform?
(a) Hydrogen sulphide (b) Nitrous oxide
(c) Phosgene (d) Chlorine

5. Consider the following reaction,

$$COCl_2 + 2C_2H_5OH \longrightarrow X$$

What is '*X*' in the above reaction?
(a) Acetyl chloride (b) Chloroform
(c) Diethyl carbonate (d) None of these

6. Which one of the following represents freon?
(a) Acetylene tetrachloride (b) Trichloroethylene
(c) Dichlorodifluoromethane (d) Ethylenedichloride

7. Which of the following reagent is used as refrigerant?
(a) CCl_4 (b) CH_2Cl_2 (c) CF_2Cl_2 (d) CF_4

8. Among the following, which type of reaction is given by freons that results in the depletion of ozone layer?
(a) Substitution reaction (b) Elimination reaction
(c) Radical chain reaction (d) Addition reaction

9. Which of the following diseases is caused by the depletion of ozone layer?
(a) It causes skin cancer
(b) Eye diseases and disorders
(c) Disruption of immune system
(d) All of the above

10. Among the following, which freon can be synthesised by Swarts reaction?
(a) Freon-12 (b) Freon-8
(c) Freon-6 (d) Freon-15

Chapter Exercises

(A) Topical exercise

I. Classification, nomenclature and isomerism of haloalkanes

1 Which of the following is an allylic halide?
(a) 3-bromo-2-methylpropene
(b) 3-bromobut-1-ene
(c) 3-bromopent-2-ene
(d) 4-bromobut-1-ene

2 The IUPAC name of tertiary butyl iodide is
(a) 4-iodobutane
(b) 2-iodobutane
(c) 1-iodo-3-methylpropane
(d) 2-iodo-2-methylpropane

3 Which is the correct IUPAC name for

$$CH_3-\underset{\displaystyle C_2H_5}{\underset{|}{CH}}-CH_2-Br?$$

(a) 1-bromo-2-ethylpropane
(b) 1-bromo-2-ethyl-2-methylethane
(c) 1-bromo-2-methylbutane
(d) 2-methyl-1-bromobutane

4 What should be the correct IUPAC name for diethylbromomethane?
(a) 1-bromo-1, 1-diethylmethane
(b) 3-bromopentane
(c) 1-bromo-1-ethylpropane
(d) 1-bromopentane

5 Ethylidene chloride is a/an ....
(a) *vic*-dihalide (b) *gem*-dihalide
(c) allylic halide (d) vinylic halide

6 Which of the following is in accordance to IUPAC system?
(a) $Br-CH_2-CH=CH_2$
1-bromo prop-2-ene

(b) 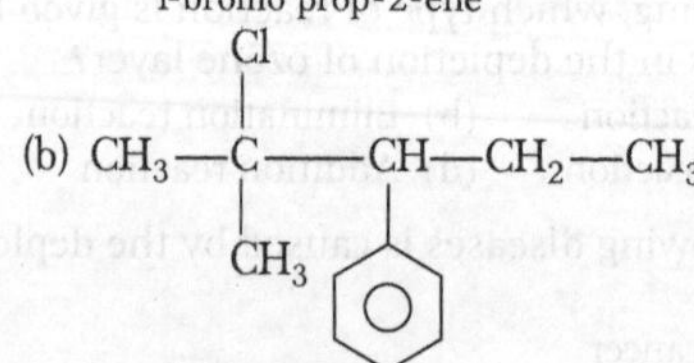
2-chloro-2-methyl-3-phenylpentane

(c) $(CH_3)_3CCl$
1-chloro-1-dimethyl propane

(d) $CH_2=CH-\overset{\displaystyle CH_3}{\overset{|}{CH}}-CH_2Br$
1-bromo-2-methylbutene

7 Which of the following is an example of *vic*-dihalide?
(a) Dichloromethane (b) 1, 2-dichloroethane
(c) Ethylidene chloride (d) Allyl chloride

8 Molecules whose mirror image is non-superimposable over them are known as chiral. Which of the following molecules is chiral in nature?
(a) 2-bromobutane (b) 1-bromobutane
(c) 2-bromopropane (d) 2-bromopropan-2-ol

9 Which of the carbon atoms present in the molecule given below are asymmetric?

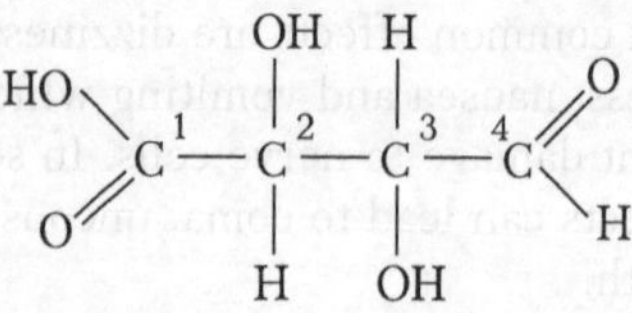

(a) 1, 2, 3 and 4 (b) 2 and 3
(c) 1 and 4 (d) 1, 2 and 3

II. Methods of preparation

10 Which reagent will you use for the following reaction?

$$CH_3CH_2CH_2CH_3 \longrightarrow CH_3CH_2CH_2CH_2Cl + CH_3CH_2CHClCH_3$$

(a) Cl_2 / UV light
(b) $NaCl + H_2SO_4$
(c) Cl_2 gas in dark
(d) Cl_2 gas in the presence of iron in dark

11 Maximum yield of following alkyl bromide is formed by Borodine-Hunsdiecker reaction
(a) CH_3CH_2Br (b) $(CH_3)_2CHBr$
(c) $(CH_3)_3CBr$ (d) C_6H_5Br

12 What is the final product of the following reaction?

$$CH_3CH{=}CH_2 + NBS \xrightarrow{AgCN} \text{(major)}$$

(a) $CH_3CH=CHCN$ (b) $CH_3CH=CHNC$
(c) $CH_2=CHCH_2CN$ (d) $CH_2=CHCH_2NC$

13 Best yield of alkyl halide ($R-X$) is formed in $RCO_2Ag + X_2 \longrightarrow RX + CO_2 + AgX$, when X_2 is
(a) Cl_2 (b) Br_2 (c) I_2 (d) F_2

14 Which of the following is maximum reactive towards HCl?
(a) $CH_2=CHCH_2OH$ (b) $CH_3CH_2CH_2OH$
(c) $CH_3\underset{\displaystyle OH}{\underset{|}{C}H}CH_3$  (d) $CH_2=CHOH$

15 In which of the following alkanes racemic mixture is formed due to halogenation?

(a) $(CH_3)_4C$ (b) $(CH_3)_2CHCH_2CH(CH_3)_2$
(c) $CH_3CH_2CH_2CH_3$ (d) $(CH_3CH_2)_4C$

16 CH_3Br can be prepared by

(a) $CH_3COOAg + Br_2 \xrightarrow{\Delta}$ (b) $CH_4 + NBS \longrightarrow$
(c) Both (a) and (b) (d) None of these

17 Consider the following addition reaction,

$$CH_3\underset{\underset{CH_3}{|}}{C}HCH{=}CH_2 + HBr \longrightarrow$$

Select the incorrect statement about this reaction.

(a) It follows electrophilic addition mechanism
(b) Most stable intermediate is the carbocation, i.e. $CH_3\underset{\underset{CH_3}{|}}{C}H\overset{\oplus}{C}HCH_3$
(c) Final product is $(CH_3)_2\underset{\underset{Br}{|}}{C}CH_2CH_3$
(d) In presence of benzoyl peroxide final product is $(CH_3)_2CHCH_2CH_2Br$

18 The order of reactivity of following alcohols with halogen acids is ....

A. $CH_3CH_2—CH_2—OH$ B. $CH_3CH_2—\underset{\underset{CH_3}{|}}{CH}—OH$

C. $CH_3CH_2—\underset{\underset{CH_3}{|}}{\overset{\overset{CH_3}{|}}{C}}—OH$

(a) A > B > C (b) C > B > A (c) B > A > C (d) A > C > B

19 $CH_2{=}CHCH_2CH{=}CH_2 \xrightarrow{NBS} A$. A is

(a) $CH_2{=}CH\underset{\underset{Br}{|}}{C}HCH{=}CH_2$
(b) $CH_2{=}CHCH{=}CHCH_2Br$
(c) $CH_2{=}CHCH_2CH{=}CHBr$
(d) $CH_2{=}CHCH_2\underset{\underset{Br}{|}}{C}{=}CH_2$

20 What are the chain-propagating steps in the free radical chlorination of methane?

I. $Cl_2 \longrightarrow 2Cl\bullet$
II. $Cl\bullet + CH_4 \longrightarrow CH_3Cl + H\bullet$
III. $Cl\bullet + CH_4 \longrightarrow \bullet CH_3 + HCl$
IV. $H\bullet + Cl_2 \longrightarrow HCl + Cl\bullet$
V. $\bullet CH_3 + Cl_2 \longrightarrow CH_3Cl + Cl\bullet$
IV. $\bullet CH_3 + CH_4 \longrightarrow CH_4 + \bullet CH_3$

(a) Both II and IV (b) Both I and II
(c) Both III and V (d) I, III and V

21 How many monohalogenation products are possible in the following reaction? (Do not consider stereoisomers).

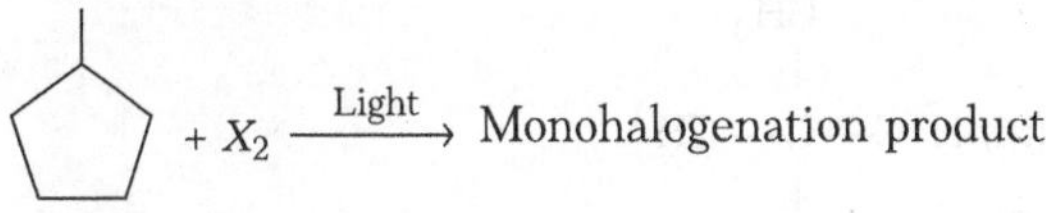

Methylcyclopentane $+ X_2 \xrightarrow{Light}$ Monohalogenation product

(a) 2 (b) 3
(c) 4 (d) 5

22 The reactivity of halogenation of alkanes decreases in order

(a) $F_2 > Cl_2 > Br_2 > I_2$ (b) $I_2 > Br_2 > Cl_2 > F_2$
(c) $Cl_2 > F_2 > I_2 > Br_2$ (d) $F_2 > Br_2 > I_2 > Cl_2$

23 Which hydrogen is abstracted most easily by bromination in the following compound?

$C_6H_5—CH_2—CH{=}CH—CH(CH_3)—H$ (ring hydrogen labelled s; benzylic H labelled r; vinylic H labelled q; terminal C–H labelled p)

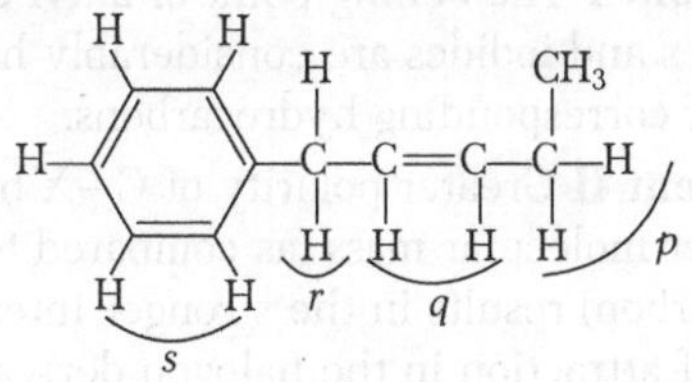

(a) p (b) q
(c) r (d) s

III. Physical and chemical properties

24 Which is the correct increasing order of boiling points of the following compounds?
1-iodobutane, 1-bromobutane, 1-chlorobutane, butane

(a) Butane < 1-chlorobutane < 1-bromobutane < 1-iodobutane
(b) 1-iodobutane < 1-bromobutane < 1-chlorobutane < Butane
(c) Butane < 1-iodobutane < 1-bromobutane < 1-chlorobutane
(d) Butane < 1-chlorobutane < 1-iodobutane < 1-bromobutane

25 Which of the following is halogen exchange reaction?

(a) $RX + NaI \longrightarrow RI + NaX$
(b) $—C{=}C— + HX \longrightarrow —\underset{\underset{H}{|}}{C}—\underset{\underset{X}{|}}{C}—$
(c) $R—OH + HX \xrightarrow{ZnCl_2} R—X + H_2O$
(d) Toluene $+ X_2 \xrightarrow[dark]{Fe}$ o-halotoluene + p-halotoluene

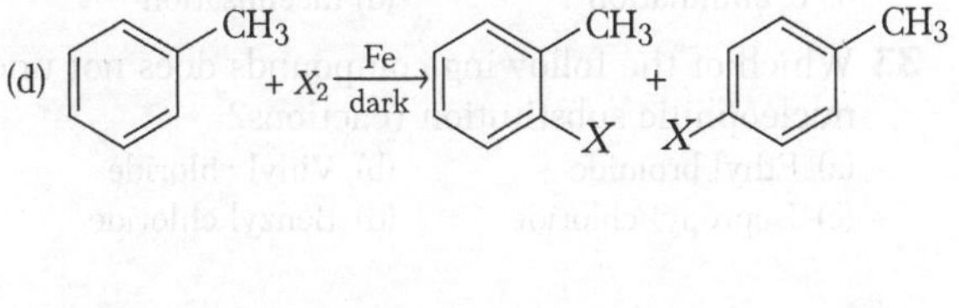

26 Arrange the following compounds in increasing order of their boiling points.

(i) $(CH_3)_2CH—CH_2Br$ (ii) $CH_3CH_2CH_2CH_2Br$

(iii) $H_3C—C(CH_3)(Br)—CH_3$

(a) (ii) < (i) < (iii) (b) (i) < (ii) < (iii)
(c) (iii) < (i) < (ii) (d) (iii) < (ii) < (i)

27 Which of the following cannot form Grignard reagent?
(a) CH_3CH_2Br (b) $CH_2{=}CHBr$
(c) $HC{\equiv}CCH_2Br$ (d) $CH_2{=}CHCH_2Br$

28 Consider the following statements.
Statement I The boiling point of alkyl chlorides, bromides and iodides are considerably higher than those of corresponding hydrocarbons.
Statement II Greater polarity of C—X bond as well as higher molecular mass (as compared to the parent hydrocarbon) results in the stronger intermolecular forces of attraction in the halogen derivatives.
(a) Both statements I and II are correct
(b) Only statement I is correct
(c) Only statement II is correct
(d) Both statements I and II are incorrect

29 Reaction of ethyl bromide with aqueous sodium hydroxide involves
(a) racemisation
(b) S_N1 mechanism
(c) retention of configuration
(d) S_N2 mechanism

30 $RCl + AgCN \longrightarrow A \xrightarrow{\text{Red. P}} B$; A and B are
(a) RCN, RCH_2NH_2 (b) $RNC, RNHCH_3$
(c) $RCN, RNHCH_3$ (d) RNC, RCH_2NH_2

31 CH_3CH_2Cl by Wurtz reaction forms
(a) ethane, ethene and butane
(b) ethene and butane
(c) Only butane
(d) Only ethane

32 A primary alkyl halide would prefer to undergo ..... .
(a) S_N1 reaction (b) S_N2 reaction
(c) α-elimination (d) racemisation

33 Which of the following compounds does not undergo nucleophilic substitution reactions?
(a) Ethyl bromide (b) Vinyl chloride
(c) Isopropyl chloride (d) Benzyl chloride

34 When 3-methyl-2-butanol reacts with $HCl/ZnCl_2$ most stable intermediate formed is
(a) $(CH_3)_2CH\overset{\oplus}{C}HCH_3$ (b) $(CH_3)_2\overset{\oplus}{C}CH_2CH_3$
(c) $(CH_3)_2CHCH_2\overset{\oplus}{C}H_2$ (d) $CH_3CH(\overset{\oplus}{C}H_2)CH_2CH_3$

35 CH_3ONO (methyl nitrite) is formed as major product when CH_3Br reacts with
(a) KNO_2 (b) $AgNO_2$
(c) Both (a) and (b) (d) None of these

36 Consider the following statements.
Statement I Halogenation of vinylic hydrogen is easier than that of allylic hydrogen.
Statement II Allyl radical is stabilised by resonance.
(a) Both statements I and II are correct
(b) Only statement I is correct
(c) Only statement II is correct
(d) Both statements I and II are incorrect

37 Which type of reaction will take place by the following equations?

H—C(Me)(Et)—Br + OH^- $\longrightarrow$ HO—C(Me)(Et)—H

(a) S_N1 (b) E1
(c) E2 (d) S_N2

38 Following reaction is of the type

$$R_2CuLi + R'X \longrightarrow R—R' + LiX + RCu$$

(a) S_N1 (b) E1
(c) E2 (d) S_N2

39 Which reaction is not expected to occur?
(a) $[(CH_3)_2CH]\,CuLi \xrightarrow{C_2H_5—Br}$
(b) $(C_2H_5)_2—CuLi \xrightarrow{(CH_3)_3C—Br}$
(c) Both (a) and (b)
(d) None of the above

40 Following reactions take place in presence of

I. C_6H_5—Cl + $CH_3Cl \longrightarrow$

II. $CH_3CH(Cl)CH_3 + CH_3CH_2\bar{O} \longrightarrow$

(a) $AlCl_3, AlCl_3$ (b) $AlCl_3, CH_3CH_2OH$
(c) $OH^-, AlCl_3$ (d) $CH_3CH_2OH, AlCl_3$

41 Consider the following statements.

Statement I Configuration is retained in the product of S_N1 reaction.

Statement II Intermediate is a carbocation and nucleophile attacks the available side.

(a) Both statements I and II are correct
(b) Only statement I is correct
(c) Only statement II is correct
(d) Both statements I and II are incorrect

42 Which of the following compounds will give racemic mixture on nucleophilic substitution by OH^- ion?

(i) $CH_3—CH(C_2H_5)—Br$

(ii) $CH_3—C(Br)(C_2H_5)—CH_3$

(iii) $CH_3—CH(C_2H_5)—CH_2Br$

(a) Only (i) (b) (i), (ii) and (iii)
(c) Both (ii) and (iii) (d) Both (i) and (iii)

43 $\underset{A}{C_3H_6Cl_2} \xrightarrow[\text{(ii) } H_3O^+ \text{ (iii) } \Delta]{\text{(i) KCN}} CH_3CH(CH_3)COOH$

Hence, A is
(a) 1,1-dichloropropane (b) 1,2-dichloropropane
(c) 2,2-dichloropropane (d) 1,3-dichloropropane

44 Major product of the following S_N1 reaction is

$CH_3CH(Br)CH(CH_3)CH_3 + \bar{O}C_2H_5 \longrightarrow$

(a) $CH_3CH(OC_2H)CH(CH_3)CH_3$ (b) $CH_3CH(CH_3)CH_2CH_2OC_2H_5$
(c) $CH_3CH_2C(CH_3)(OC_2H_5)CH_3$ (d) None is correct

45 Chloromethane on treatment with excess of ammonia yields mainly

(a) N, N-dimethylmethanamine $\left(CH_3—N\begin{matrix}CH_3\\CH_3\end{matrix}\right)$
(b) N-methylmethanamine ($CH_3—NH—CH_3$)
(c) methanamine (CH_3NH_2)
(d) mixture containing all these in equal proportion

46 Consider the following statements.

Statement I Thionyl chloride is preferred for the preparation of halo compounds from alcohol.

Statement II Gaseous by products SO_2 and HCl escape easily from the reaction mixture.

(a) Both statements I and II are correct
(b) Only statement I is correct
(c) Only statement II is correct
(d) Both statements I and II are incorrect

47 (cyclohexane with Br, H on C-1 and CH_3, H on C-4) $+ OH^- \xrightarrow{S_N2} A$, A is

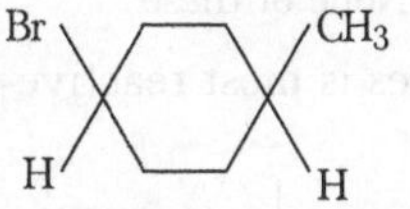

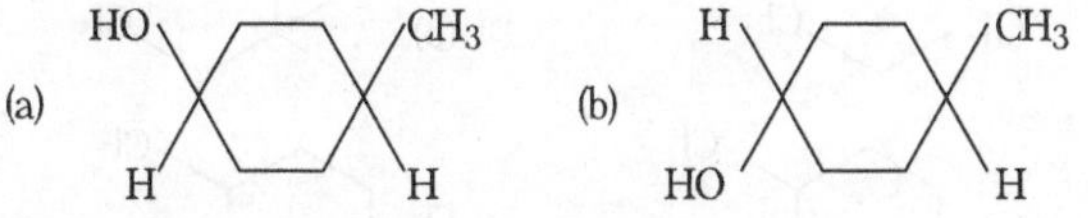

(a) (cyclohexane with HO, H on C-1 and CH_3, H on C-4) (b) (cyclohexane with H, HO on C-1 and CH_3, H on C-4)
(c) Both (a) and (b) (d) None is correct

48 The major product formed in the following reaction is

$CH_3—CH(Cl)—CH_2—CH_2—OH \xrightarrow{aq\ KOH}$

(a) $CH_3CH{=}CHCH_2OH$
(b) $CH_2{=}CH—CH_2—CH_2OH$
(c) $CH_3—CH—CH_2$ (ring: $CH—O—CH_2—CH_2$)
(d) $CH_3—CH(OH)—CH_2—CH_2OH$

49 Which of the following can be debrominated easlily?

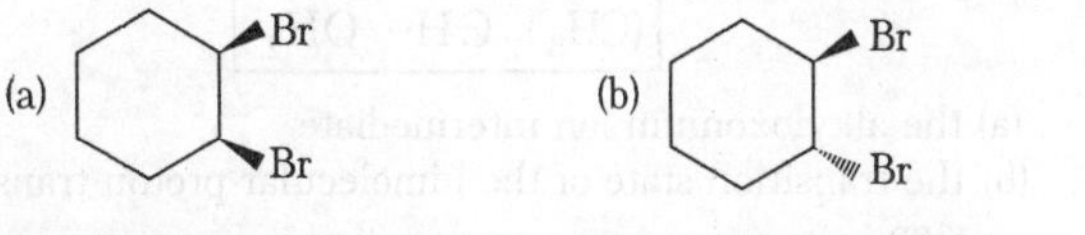

(a) (cyclohexane with two Br on adjacent carbons) (b) (cyclohexane with two Br on adjacent carbons, wedge and hash)
(c) Both (a) and (b) (d) None of these

50 Consider the following statements about *neo*-pentyl bromide.

I. The primary carbon atom that bears the leaving (bromide) group is hindered by the *tert*-butyl group, so backside approach of a nucleophile is prevented, making this a poor S_N2 substrate.

II. It reacts with H_2O by S_N1 reaction forming *neo*-pentyl alcohol.

III. It cannot react by S_N1 reaction because a highly unstable primary carbocation would be formed.

IV. By suitable conditions, it can be made to react by S_N1 reaction, then a rearrangement leads to the formation of a most stable tertiary carbocation.

Select the correct statements.

(a) Both I and III (b) I, III and IV
(c) II, III and IV (d) I, II and III

51 In the following intramolecular S_N reaction nucleophile is

(a) Br (b) N
(c) Both (a) and (b) (d) None of these

52 Which of the following species is most reactive towards S_N2 reaction?

(a) 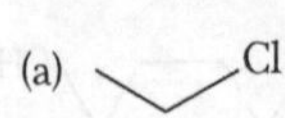 (b) 

(c) 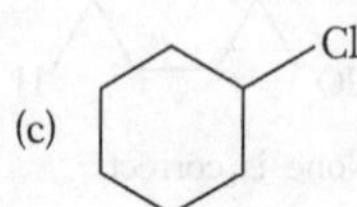 (d) 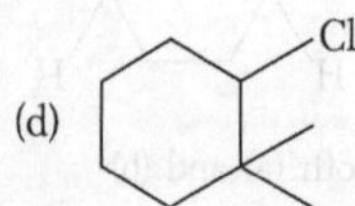

53 What is the decreasing stability order (most stable → least stable) of the following carbocations?

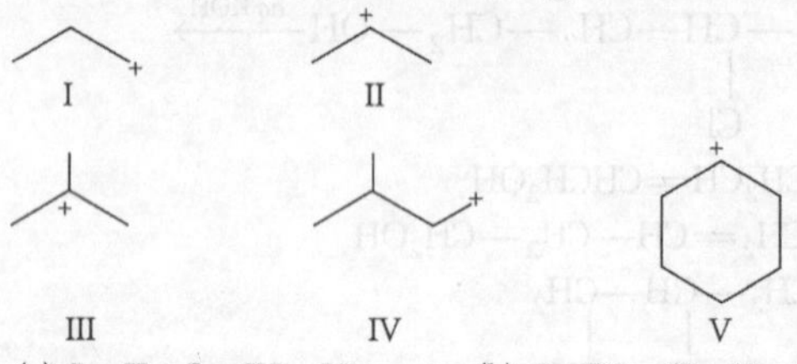

(a) I > II > I > IV > V (b) I ≈ IV > II ≈ V > III
(c) III > II ≈ V > I ≈ IV (d) III > I ≈ IV > II ≈ V

54 The species shown in the box represents ...... of the reaction between isopropyl alcohol and hydrogen bromide.

$$\boxed{(CH_3)_2\overset{\delta+}{C}H\cdots\overset{\delta+}{O}H_2}$$

(a) the alkyloxonium ion intermediate
(b) the transition state of the bimolecular proton transfer step
(c) the transition state of the capture of the carbocation by a nucleophile
(d) the transition state of the unimolecular dissociation step

IV. Elimination reactions and polyhalogen compounds

55 β-elimination is not possible in

(a) $CHCl_3$ (b) $(CH_3)_3CBr$
(c) $(CH_3)_2CHBr$ (d) $(CH_3)_2CHCH_2Br$

56 Following reaction intermediate is used in

$$:\overset{\ominus}{C}Cl_3 \longrightarrow :CCl_2 + Cl^-$$

(a) β-elimination (b) α-elimination
(c) 1,2-elimination (d) E2-elimination

57 When chlorine reacts with ethyl alcohol, it forms

(a) chloroform
(b) chloral
(c) ethyl chloride
(d) carbon tetrachloride

58 Consider the following statements.

Statement I $CHCl_3$ is stored in closed dark coloured bottle completely filled so that air is kept out.

Statement II 1° amines on reaction with ethanolic KOH and $CHCl_3$ give carbylamine.

(a) Both statements I and II are correct
(b) Only statement I is correct
(c) Only statement II is correct
(d) Both statements I and II are incorrect

59 Which cannot undergo E2 reaction?

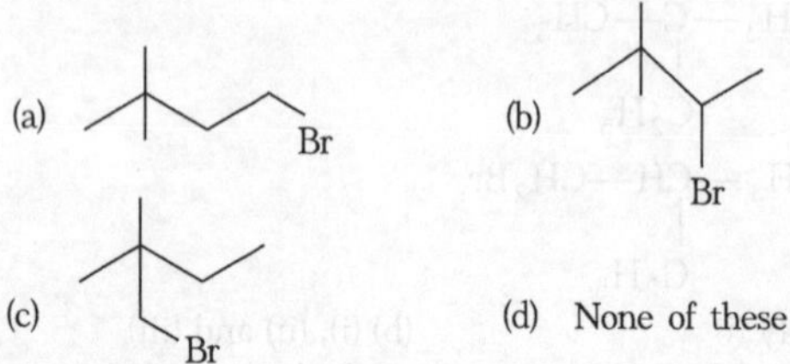

60 Consider the following reaction

$$\underset{\text{(Major)}}{B} \xleftarrow[(CH_3CH_2OH)]{CH_3CH_2\overset{-}{O}\overset{+}{Na}} CH_3\overset{CH}{\underset{CH_3}{\overset{|}{\underset{|}{C}}}}-Br \xrightarrow{CH_3CH_2OH} \underset{\text{(Major)}}{A}$$

A and B are respectively,

(a) $CH_3\overset{CH_3}{\underset{CH_3}{\overset{|}{\underset{|}{C}}}}-OCH_2CH_3$ in both cases

(b) $CH_3\overset{CH_3}{\overset{|}{C}}=CH_2$ in both cases

(c) $CH_3\overset{CH_3}{\underset{CH_3}{\overset{|}{\underset{|}{C}}}}-OCH_2CH_3$ and $CH_3-\overset{CH_3}{\underset{CH_2}{\overset{|}{\underset{\|}{C}}}}$

(d) $CH_3-\overset{CH_3}{\underset{CH_2}{\overset{|}{\underset{\|}{C}}}}$ and $CH_3\overset{CH_3}{\underset{CH_3}{\overset{|}{\underset{|}{C}}}}-OCH_2CH_3$

61 $C_6H_5CH_2—\overset{CH_3}{\underset{Br}{C}}—CH_2—CH_3 \xrightarrow[C_2H_5OH]{C_2H_5ONa}$

Major product (A) is

(a) $C_6H_5CH_2—\overset{CH_3}{\underset{OC_2H_5}{C}}—CH_2CH_3$

(b) $C_6H_5CH{=}\underset{CH_3}{C}—CH_2CH_3$

(c) $C_6H_5CH_2—\underset{CH_3}{C}{=}CHCH_3$

(d) $C_6H_5CH_2—\underset{CH_2CH_3}{C}{=}CH_2$

62 Select the correct statement (s) about following reaction,

$C_2H_5\bar{O} + CH_3CH_2\overset{Br}{C}H\,CH_3 \xrightarrow{C_2H_5OH}$

(a) It is termed as β-elimination reaction
(b) $C_2H_5\bar{O}$ is a Bronsted-base
(c) Products are 1-butene and but-2-ene
(d) All of the above are correct statements

63 Under basic conditions which one suffers elimination the most?

(a) $(CH_3)_3CCl$ (b) $(H_3C)_2CHCH_2Cl$
(c) $CH_3CH_2\underset{Cl}{C}HCH_3$ (d) $CH_3CH_2CH_2CH_2Cl$

64 $CH_3CH_2\underset{Br}{C}HCH_2CH{=}CH_2 \xrightarrow{\text{Alcoholic KOH}} A$

(predominant), A is

(a) $CH_3CH_2CH{=}CHCH{=}CH_2$
(b) $CH_3CH{=}CHCH_2CH{=}CH_2$
(c) $CH_2{=}CHCH_2CH_2CH{=}CH_2$
(d) $CH_3CH_2CH_2CH_2C{\equiv}CH$

65 Consider the following statements.

Statement I After using CCl_4 to extinguish fire, the room should be well ventilated.

Statement II Atmospheric oxygen can convert CCl_4 into poisonous phosgene ($COCl_2$) gas.

(a) Both statements I and II are correct
(b) Only statement I is correct
(c) Only statement II is correct
(d) Both statements I and II are incorrect

66 $CH_3CH_2\underset{Br}{C}HCH_3 \xrightarrow{NaNH_2}$ but-1-ene and but-2-ene

Which are the correct statements?

(a) but-1-ene is Hofmann product
(b) but-2-ene is Saytzeff product
(c) Both (a) and (b)
(d) None of the above

67 $CH_3\overset{\oplus}{\underset{CH_3}{C}}CH_2CH_3$ can undergo elimination to give major product as

(a) $CH_3\underset{CH_3}{C}{=}CHCH_3$ (b) $CH_3\underset{CH_2}{\overset{\|}{C}}CH_2CH_3$
(c) $CH_3\underset{CH_3}{C}HCH{=}CH_2$ (d) None of these

68 $(CH_3)_3C\cdot CH{=}CH_2 \xleftarrow{II} (CH_3)_3C\,\underset{Cl}{C}H\,CH_3 \xrightarrow{I} (CH_3)_2C{=}C(CH_3)_2$

In this reaction

(a) I is E1, II is E2 (b) I is E2, II is E1
(c) Both are E1 (d) Both are E2

69 $CH_3\overset{CH_3}{\underset{CH_3}{C}}Cl + C_2H_5\bar{O}\overset{+}{Na} \longrightarrow$

Major product of this reaction is

(a) $CH_3\overset{CH_3}{\underset{CH_3}{C}}OC_2H_5$ (b) $CH_3—\underset{CH_3}{C}{=}CH_2$
(c) $CH_3\underset{CH_3}{C}HCH_2OC_2H_5$ (d) None is correct

70 For the reaction,

Bromocyclohexane $+ CH_3CH_2O^-Na^+ \longrightarrow$

The major product is formed by

(a) an S_N1 reaction (b) an E1 reaction
(c) an S_N2 reaction (d) an E2 reaction

71 Fischer projection: C_6H_5 (top); CH_3—C—H; Br—C—H; C_6H_5 (bottom) $\xrightarrow[\text{E2}]{\text{Alcoholic, KOH}}$ A.

A is

(a) $H_3C(C_6H_5)C{=}C(H)(C_6H_5)$ [H_3C and H on the same side]

(b) $H_3C(C_6H_5)C{=}C(H_5C_6)(H)$ [H_3C and H_5C_6 on the same side]

(c) $H(CH_3)C{=}C(H_5C_6)(C_6H_5)$

(d) None is correct

72 Select the incorrect statement.

(a) Reactivity of alkyl halides towards E1 or E2 reaction is $3° > 2° > 1°$ alkyl halide

(b) E1 reaction is stereospecific

(c) Dehydrohalogenation is *trans*-elimination

(d) E1 and S_N1 reactions are competitive

73 Which gives stereoselective elimination product?

(a) CH_2Br—CH_2Br

(b) $CH_3CH_2CHCH_2$ (Br on C-3 and C-4: Br Br)

(c) CH_3—CH(Br)—CH(Br)—CH_3 (H—C—Br, H—C—Br)

(d) CH_3CHBr_2

74 Match the items of Column I with the items of Column II.

Column I	Column II
A. Most reactive in E2 reaction	1. (bromocyclohexene)—Br
B. Equally reactive in E1	2. (bromocyclohexadiene)—Br
C. Produces same major product in E2	3. (bromocyclohexene)—Br
D. Loses chirality in both E1 and E2	4. (bromocyclohexene)—Br

Codes

	A	B	C	D
(a)	2	1,2	1,3,4	2,3,4
(b)	1	2,3	2,3,4	1,3
(c)	1,2	1,4	2,3	1,2,3
(d)	1,4	2,3	1,4	1,2

75 Given compound 'A' will give ... by E1 reaction.

(A): cyclohexene bearing CH_2Br

(a) cyclohexene with exocyclic $=CH_2$

(b) cyclohexene with CH_3

(c) cyclohexene with exocyclic $=CH_2$

(d) cyclohexane with exocyclic $=CH_2$

76 (cyclohexyl)cyclohexene $\xrightarrow{Br_2/h\nu}$ A (major) is

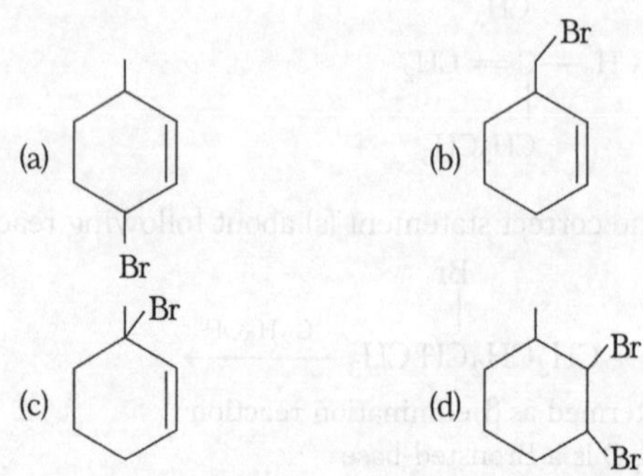

77 An alkane with a molecular formula C_6H_{14} reacts with chlorine in the presence of light and heat to give four constitutionally isomeric monochlorides of molecular formula $C_6H_{13}Cl$. What is the most reasonable structure for the starting alkane?

(a) $CH_3CH_2CH_2CH_2CH_2CH_3$ (b) $(CH_3)_2CHCH_2CH_2CH_3$

(c) $CH_3CH(CH_2CH_3)_2$ (d) $(CH_3)_3CCH_2CH_3$

(B) Miscellaneous questions

1 Carbon-halogen bond lengths increases in the order

(a) C—F < C—Cl < C—Br < C—I

(b) C—I < C—Br < C—Cl < C—F

(c) C—I < C—Br < C—F < C—Cl

(d) C—I < C—F < C—Br < C—Cl

2 Bond dipole moments increase in the order

(a) C—F < C—Cl < C—Br < C—I

(b) C—F < C—Cl < C—I < C—Br

(c) C—I < C—Br < C—F < C—Cl

(d) C—I < C—Br < C—Cl < C—F

3 C—Cl bond is stronger than C—I bond because

(a) C—Cl bond is more ionic than C—I

(b) C—Cl bond has polar covalent bond

(c) C—Cl bond is more covalent than C—I

(d) C—Cl bond length is longer than C—I

4 Alkyl halides can be converted into Grignard reagents by
(a) boiling them with Mg ribbon in alcoholic solution
(b) refluxing them with magnesium powder in dry ether
(c) refluxing them with $MgCl_2$ solution
(d) warming them with $MgCl_2$

5 The major product obtained in the following reaction is

Br, H, C_6H_5, C_6H_5 $\xrightarrow[\Delta]{^tBuOK}$

(a) $(\pm)C_6H_5CH(O^tBu)CH_2C_6H_5$
(b) $C_6H_5CH{=}CHC_6H_5$
(c) $(+)C_6H_5CH(O^tBu)CH_2C_6H_5$
(d) $(-)C_6H_5CH(O^tBu)CH_2C_6H_5$

6 When *neo*-pentyl bromide is subjected to Wurtz reaction, the product formed is
(a) 2, 2, 4, 4-tetramethylhexane
(b) 2, 2, 4, 4-tetramethylpentane
(c) 2, 2, 5, 5-tetramethylhexane
(d) 2, 2, 3, 3-tetramethylhexane

7 Ethyl bromide and *n*-propyl bromide is allowed to react with sodium, in ether, they produce
(a) single alkane
(b) mixture of two alkanes
(c) mixture of three alkanes
(d) mixture of four alkanes

8 Consider the following statements.
Statement I Primary allylic halides show higher reactivity in S_N1 reactions than other primary alkyl halides.
Statement II Intermediate allylic carbocation is stabilised by resonance.
(a) Both statements I and II are correct
(b) Only statement I is correct
(c) Only statement II is correct
(d) Both statements I and II are incorrect

9 Which of the following is least able to serve as a nucleophile in a chemical reaction?
(a) Br^- (b) OH^-
(c) NH_3 (d) CH_3^+

10 Rank the following species in order of decreasing nucleophilicity in a polar protic solvent (most → least nucleophilic)

$\underset{\text{I}}{CH_3CH_2CH_2O^-}$ $\underset{\text{II}}{CH_3CH_2CH_2S^-}$ $\underset{\text{III}}{CH_3CH_2\overset{O}{\overset{\|}{C}}O^-}$

(a) III > I > II (b) II > III > I
(c) I > III > II (d) II > I > III

11 $CH_3—CH_2—Br$ on treatment with $LiAlH_4$ gives ethane gas while $(CH_3)_3C—Br$ on same treatment gives H_2 gas because
(a) the former is S_N2 and later is E2 reaction
(b) the former is E2 and later is S_N2 reaction
(c) the former is S_N1 and later is E2 reaction
(d) the former is E2 and later is S_N1 reaction

12 Consider the following statements.
Statement I *neo*-pentyl halides react slowly in substitution process.
Statement II *neo*-pentyl halides being primary in nature react *via* S_N2 mechanism.
(a) Both statements I and II are correct
(b) Only statement I is correct
(c) Only statement II is correct
(d) Both statements I and II are incorrect

13 Which is the best leaving group (weakest conjugate base)?
(a) $F_3C-C_6H_4-SO_3^{\ominus}$ (b) $H_3C-C_6H_4-SO_3^{\ominus}$
(c) $\overset{\ominus}{C}H_3$ (d) $OH^{\ominus}$

14 To replace circled hydrogen by deuterium (D), reagents used are

$CH_2{=}CH{-}CH_2{-}(H) \xrightarrow{\text{I}} \xrightarrow{\text{II}} \xrightarrow{\text{III}} CH_2{=}CH{\cdot}CH_2D$

	I	II	III
(a)	Br_2 / CCl_4	Na	D_2O
(b)	NBS	Mg/ether	D_2O
(c)	Br_2	Mg/ether	D_2O
(d)	HBr	Mg/ether	D_2O

15 End product of the following sequence of reactions is

Cyclohexyl–Br $\xrightarrow{NH_3} A \xrightarrow{O_3/H_2O} B \xrightarrow[\Delta]{BaO} C$

(a) cyclohexanone (=O) (b) cyclohexylidene =OH
(c) cyclopentanone (=O) (d) HOOC COOH

16 Consider the following reactions.
Statement I 2-bromo-3-methylbutane changes to 2-ethoxy-2-methylbutane on reaction with CH_3CH_2OH.
Statement II 2° carbocation undergoes rearrangement that results in the formation of more stable 3° carbocation.

(a) Both statements I and II are correct
(b) Only statement I is correct
(c) Only statement II is correct
(d) Both statements I and II are incorrect

17 Which one of the following compounds is most rapidly hydrolysed by S_N1 mechanism?

(a) $CH_3CH{=}CHCl$ (b) $ClCH_2CH{=}CH_2$
(c) C_6H_5–CH_2Cl (d) $(C_6H_5)_3CCl$

18 The bimolecular substitution reaction,
$CH_3Br + OH^- \longrightarrow CH_3OH + Br^-$ is represented by the kinetic equation

(a) Rate $= k[CH_3Br]^2$
(b) Rate $= k[CH_3Br]\,[OH^-]$
(c) Rate $= k[CH_3Br] + k[OH^-]$
(d) Rate $= \dfrac{k}{[CH_3Br][OH^-]}$

19 From each of the following pairs select the compound that will react faster with sodium iodide in acetone.

2-chloropropane or 2-bromopropane
I II
1-bromobutane or 2-bromobutane
III IV

(a) Both I and III (b) Both I and IV
(c) Both II and III (d) Both II and IV

20 Which compound undergoes nucleophilic substitution with NaCN at the fastest rate?

(a) Br (b) Br
(c) Br (d) Br

21 Consider the following two paths for preparing compound that are (*A*) given below

A. 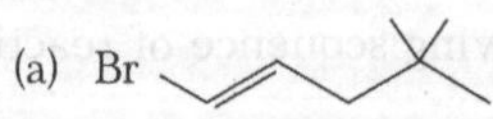

I. 

II. 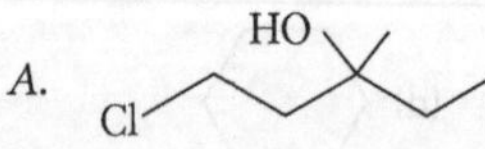

Which path is suitable?

(a) Only I (b) Only II
(c) Both (a) and (b) (d) One of these

22 Identify *Z*, in the following series.

$$C_2H_5I \xrightarrow{\text{Alc. KOH}} X \xrightarrow{Br_2} Y \xrightarrow{KCN} Z$$

(a) CH_3CH_2CN (b) $NCCH_2$—CH_2CN
(c) $BrCH_2$—CH_2CN (d) $BrCH{=}CHCN$

23 What would be the product of the reaction?

Br $\xrightarrow[\text{Ether}]{\text{Li}}$ $\xrightarrow{\text{CuLi}}$ I

(a) 2, 6-dimethyl heptane (b) 2, 2-dimethyl propane
(c) 2, 2-dimethyl octane (d) 2, 2, 5-trimethyl hexane

24 The increasing order of hydrolysis of the following compounds is

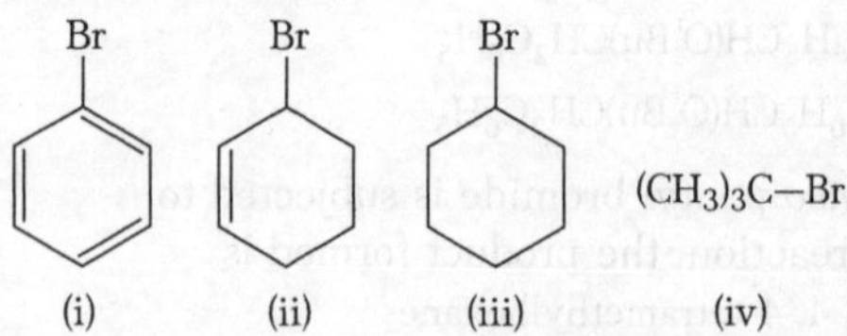

(a) (i)< (iii)< (ii)< (iv) (b) (i)< (iv)< (iii)< (ii)
(c) (iv)< (ii)< (iii)< (i) (d) (i)< (iii)< (iv)< (ii)

25 Match the species in Column I with their commercial names in Column II.

	Column I		Column II
A.	CCl_3NO_2	1.	Freon-12
B.	CCl_4	2.	Chloretone
C.	$(CH_3)_2C(OH)CCl_3$	3.	Pyrene
D.	CCl_2F_2	4.	Chloropicrin

Codes

	A	B	C	D		A	B	C	D
(a)	4	3	2	1	(b)	1	2	3	4
(c)	4	1	2	3	(d)	1	3	4	2

26 Consider the following reaction,

$$\xleftarrow[S_N1]{C_2H_5OH} CH_3\overset{\overset{\large CH_3}{|}}{\underset{\underset{\large CH_3}{|}}{C}}CH_2Br \xrightarrow[S_N2]{C_2H_5O^-}$$

Above reaction shows S_N2 reaction with $C_2H_5O^-$ and S_N1 reaction with C_2H_5OH as given above. Select incorrect statement.

(a) Product obtained by S_N2 reaction has inverted configuration
(b) Product obtained due to S_N1 reaction is a tertiary ether
(c) Both the statements are incorrect
(d) Both the statements are correct

27 Select the correct alternate.

$$CH_3CH_2\text{—}\overset{H}{\underset{CH_3}{C}}\text{—}Cl \xrightarrow[S_N2]{CH_3S^{\ominus}Na^{\oplus}}$$

(a) $CH_3CH_2\text{—}\overset{H}{\underset{CH_3}{C}}\text{—}SNa$ (b) $CH_3S\text{—}\overset{H}{\underset{CH_3}{C}}\text{—}CH_2CH_3$

(c) Both (a) and (b) (d) None of these

28 Select the reagent that will yield the greater amount of substitution on reaction with 1-bromobutane.

(a) CH_3CH_2OK in dimethyl sulphoxide (DMSO)

(b) $(CH_3)_3COK$ in dimethyl sulphoxide (DMSO)

(c) Both (a) and (b) will give comparable amounts of substitution

(d) Neither (a) nor (b) will give any appreciable amount of substitution

29 Which of the following structures is enantiomeric with the molecule (*A*) given below?

$$H_5C_2\text{—}\overset{H}{C}(CH_3)\text{—}Br$$

(*A*)

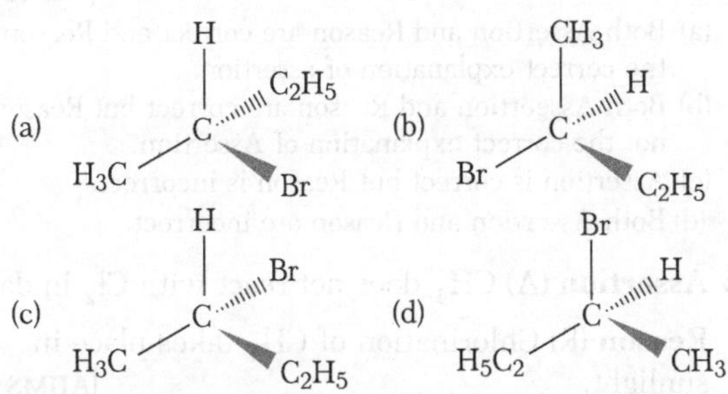

30 Which of the following phrases are not correctly associated with an S_N1 reaction?

I. Rearrangement is possible.

II. Rate is affected by solvent polarity.

III. The strength of the nucleophile is important in determining rate.

IV. The reactivity series is tertiary > secondary > primary.

V. Proceeds with complete inversion of configuration.

(a) Both III and V (b) Only V

(c) II, III and V (d) Only III

31 Which of the following statements pertaining to an S_N2 reaction are true?

I. The rate of reaction is independent of the concentration of the nucleophile.

II. The nucleophile attacks carbon on the side of the molecule opposite the group being displaced.

III. The reaction proceeds with simultaneous bond formation and bond rupture.

IV. Partial racemisation of an optically active substrate results.

(a) Both I and IV (b) I, III and IV

(c) Both II and III (d) All of these

32 Rank the following in order of decreasing rate of solvolysis with aqueous ethanol (fastest → slowest)

$H_2C{=}C(CH_3)\text{—}Br$ (I); 1-bromo-1-methylcyclohexane (II); $CH_3CH(Br)CH_2CH(CH_3)_2$ (III)

(a) II > I > III (b) I > II > III

(c) II > III > I (d) I > III > II

33 1-bromo-2-methylcyclohexane + $CH_3\overset{-}{O}\overset{+}{Na} \xrightarrow{CH_3OH}$ *A*. *A* is

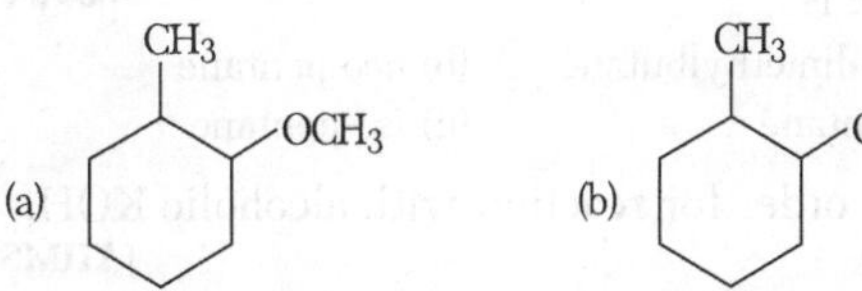

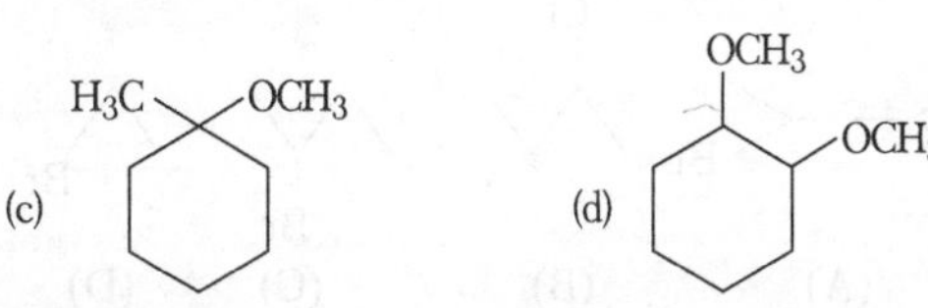

34 Considere the following statements.

Statement I S_N2 reaction on *A* can take place with inversion.

$$H\text{—}\overset{CH_3}{\underset{CH_2CH_3}{C}}\text{—}Br + OH^- \longrightarrow HO\text{—}\overset{CH_3}{\underset{CH_2CH_3}{C}}\text{—}H$$

(*A*)

Statement II Rate of the reaction is dependent on the concentration of (*A*) and OH^-.

Hecne, $\left(\frac{dx}{dt}\right) = k[A][OH^-]$

(a) Both statements I and II are correct

(b) Only statement I is correct

(c) Only statement II is correct

(d) Both statements I and II are incorrect

(C) Medical entrances' gallery

1 Which of the following will not undergo S_N1 reaction with $\overline{O}H$? **[NEET 2020]**

(a) $CH_2 = CH\text{—}CH_2Cl$ (b) $(CH_3)_3CCl$

(c) $C_6H_5CH_2CH_2Cl$ (d) $C_6H_5CH_2Cl$

2 Which of the following is a free radical substitution reaction? **[NEET 2020]**
(a) Benzene with Br_2 / $AlCl_3$
(b) Acetylene with HBr
(c) Methane with Br_2/hv
(d) Propene with $HBr/(C_6H_5COO)_2$

3 Elimination reaction of 2-bromo-pentane to form pent-2-ene is **[NEET 2020]**
1. β-elimination reaction.
2. Follows Zaitsev rule.
3. Dehydrohalogenation reaction.
4. Dehydration reaction.

(a) (1), (3), (4) (b) (2), (3), (4)
(c) (1), (2), (4) (d) (1), (2), (3)

4 The alkane that gives only one mono-chloro product on chlorination with Cl_2 in presence of diffused sunlight is **[NEET 2020]**
(a) 2, 2-dimethylbutane (b) *neo*-pentane
(c) *n*-pentane (d) *iso*-pentane

5 Correct order for reaction with alcoholic KOH. **[AIIMS 2019]**

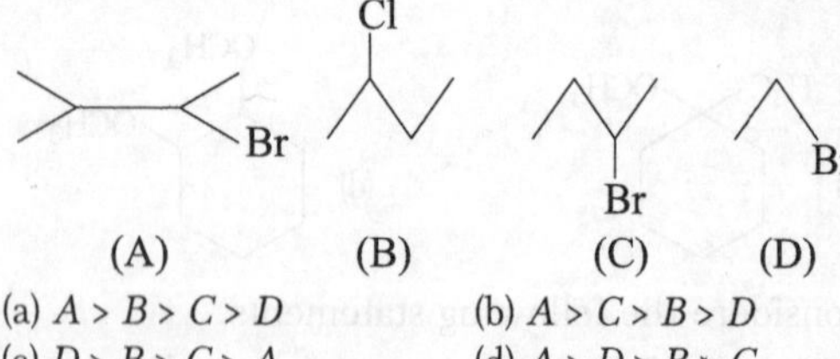

(a) $A > B > C > D$ (b) $A > C > B > D$
(c) $D > B > C > A$ (d) $A > D > B > C$

6 If □ react with Cl_2 in presence of light and then react with Na metal in dry ether, final product is **[JIPMER 2019]**

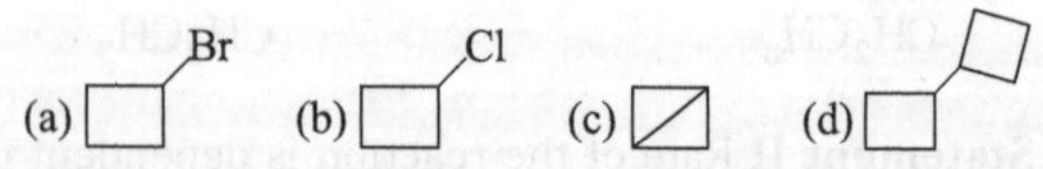

7 Which of the following is not correct? **[JIPMER 2019]**
(a) $PhCH_2Br > PhCHBrCH_3 > PhCBr(CH_3)_2 (S_N1)$
(b) $R—I > R—Br > R—Cl (S_N2)$
(c) $CH_2{=}CH—Cl < CH_2{=}CH—CH_2—Cl < PhCH_2—Cl (S_N1)$
(d) $R—Cl < R—Br < R—I (S_N1)$

8 Which cannot behave as a nucleophile for S_N2 reaction? **[JIPMER 2019]**
(a) H_2O (b) CN^-
(c) NH_2^- (d) I^-

9 $CH_3—CH_2—CH_2—Br \xrightarrow{\text{Alc. KOH}}$ Final product is **[JIPMER 2019]**
(a) propene (b) propanol
(c) cyclopropane (d) propane-1, 2-diol

10 Diethylcarbonate is obtained by the reaction of phosgene with **[MP PMT 2019]**
(a) ethanal (b) methanol
(c) ethanol (d) methanal

11 Hydrocarbon (*A*) reacts with bromine by substitution to form an alkyl bromide which by Wurtz reaction is converted to gaseous hydrocarbon containing less than four carbon atoms. *A* is **[NEET 2018]**
(a) $CH_3—CH_3$ (b) $CH_2=CH_2$
(c) $CH\equiv CH$ (d) CH_4

12 A bromoalkane '*X*' reacts with magnesium in dry ether to form compound '*Y*'. The reaction of '*Y*' with methanal followed by hydrolysis yield an alcohol having molecular formula, $C_4H_{10}O$. The compound '*X*' is **[AIIMS 2018]**
(a) bromomethane (b) bromoethane
(c) 1-bromopropane (d) 2-bromopropane

13 **Assertion** (A) The boiling points of alkyl halides decreases in the order *R*I> *R*Br> *R*Cl> *R*F.
Reason (R) The boiling points of alkyl chlorides, bromides and iodides are considerably higher than that of the hydrocarbon of comparable molecular mass. **[AIIMS 2018]**
(a) Both Assertion and Reason are correct and Reason is the correct explanation of assertion.
(b) Both Asseertion and Reason are correct but Reason is not the correct explanation of Assertion.
(c) Assertion is correct but Reason is incorrect.
(d) Both Assertion and Reason are incorrect.

14 **Assertion** (A) CH_4 does not react with Cl_2 in dark.
Reason (R) Chlorination of CH_4 takes place in sunlight. **[AIIMS 2018]**
(a) Both Assertion and Reason are correct and Reason is the correct explanation of Assertion.
(b) Both Assertion and Reason are correct and Reason is not the correct explanation of Assertion.
(c) Assertion is correct but Reason is incorrect.
(d) Both Assertion and Reason are incorrect.

15 Neopentyl chloride reacts with benzene in presence of a Lewis acid ($AlCl_3$) to form **[JIPMER 2018]**
(a) $C_6H_5—CH_2C(CH_3)_3$ (b) $C_6H_5—C(CH_3)_2CH_2CH_3$
(c) $C_6H_5—C(CH_3)_3$ (d) $C_6H_5—CH_2CH(CH_3)_2$

16 Identify the major product that is obtained on treatment of $CH_3CH_2CH(F)CH_3$ with CH_3O^- / CH_3OH is **[JIPMER 2018]**
(a) $CH_3CH_2CH(OCH_3)CH_3$ (b) $CH_3CH{=}CHCH_3$
(c) $CH_3CH_2CH{=}CH_2$ (d) $CH_3CH_2CH_2CH_2OCH_3$

17 $CH_3—Br \xrightarrow{Mg/ether} \xrightarrow[(ii)\ H_2O]{(i)\ CO_2} \xrightarrow[\Delta]{CH_3OH/H^+} P$

What is the product '*P*' in the above reaction? **[JIPMER 2018]**

(a) Methylethanoate (b) Ethanoic acid
(c) Propanal (d) Methylpropanoate

18 Among the following, the suitable reagent for Wurtz reaction is **[JIPMER 2018]**

(a) Na/alcohol (b) Na/ether
(c) Zn/ether (d) Zn/alcohol

19 Consider the reaction,

$CH_3CH_2CH_2Br + NaCN \longrightarrow CH_3CH_2CH_2CN + NaBr$

This reaction will be the fastest in **[NEET-II 2016]**

(a) ethanol
(b) methanol
(c) N, N′-dimethylformamide (DMF)
(d) water

20 The reaction,

$$CH_3—\underset{CH_3}{\overset{CH_3}{\underset{|}{\overset{|}{C}}}}—ONa + CH_3CH_2Cl \xrightarrow[-NaCl]{} CH_3—\underset{CH_3}{\overset{CH_3}{\underset{|}{\overset{|}{C}}}}—O—CH_2CH_3$$

is called **[CBSE AIPMT 2016]**

(a) Williamson synthesis
(b) Williamson continuous etherification process
(c) Etard reaction
(d) Gatterman-Koch reaction

21 Which one of the following organohalogen compounds when heated with alcoholic potassium hydroxide does not undergo dehydrohalogenation reaction? **[Kerala CEE 2015]**

(a) Secondary butyl chloride (b) *Iso*-propyl chloride
(c) *Neo*-pentyl chloride (d) *Iso*-butyl chloride
(e) Tertiary butyl chloride

22 The products expected to be formed in the wurtz reaction of a mixture of *neo*-pentyl bromide and *iso*-butyl bromide are **[Kerala CEE 2015]**

(i) 2, 2, 4-trimethylpentane
(ii) 2, 2, 5, 5-tetramethylhexane
(iii) 2, 2, 4, 4-tetramethylhexane
(iv) 2, 5-dimethylhexane
(v) 2, 2, 5-trimethylhexane

(a) (ii), (iii) and (v) (b) (ii), (iv) and (v)
(c) (i), (iv) and (v) (d) (i), (iii) and (v)
(e) (i), (ii) and (iv)

23 ⬡—Br + Mg $\xrightarrow{\text{Dry ether}} A \xrightarrow[H^+]{H_2O} B$

[Karnataka CET 2015]

The product *B* is

(a) ⬡—OH (b) ⬡—MgBr
(c) ⬡ (d) ⬡—CH_2OH

24 The arrangement of following compounds. **[Karnataka CET 2015]**

I. Bromomethane
II. Bromoform
III. Chloromethane
IV. Dibromomethane

In the increasing order of their boiling points

(a) III< I< IV < II (b) IV < III< I< IV
(c) II< III< I< IV (d) I< II< III< IV

25 Identify the compound *D* in the following series of reactions, **[MHT CET 2014]**

$$CH_3—\overset{CH_3}{\overset{|}{C}H}—CH_2—CH_2—Br \xrightarrow[\Delta]{\text{Alc. KOH}} A$$

$$\xrightarrow[\text{(ii) } H_2O,\ \Delta]{\text{(i) Conc. } H_2SO_4} \underset{\text{(Major product)}}{B} + \underset{\text{(Minor product)}}{C}$$

$$B \xrightarrow[-H_2O]{HI,\ \Delta} \underset{\text{(Major product)}}{D} + \underset{\text{(Minor product)}}{E}$$

(a) $CH_3—\overset{CH_3}{\overset{|}{C}H}—\underset{I}{\underset{|}{C}H}—CH_3$

(b) $CH_3—\underset{I}{\underset{|}{\overset{CH_3}{\overset{|}{C}}}}—CH_2—CH_3$

(c) $CH_3—\overset{CH_3}{\overset{|}{C}H}—CH_2—CH_2—I$

(d) $CH_3—CH_2—\overset{CH_3}{\overset{|}{C}H}—CH_2—I$

26 An incorrect statement with respect to S_N1 and S_N2 mechanisms for alkyl halide is **[KCET 2014]**

(a) a strong nucleophile in an aprotic solvent increases the rate or favours S_N 2 reaction
(b) competing reaction for S_N2 reaction is rearrangement
(c) S_N1 reactions can be catalysed by some Lewis acids
(d) a weak nucleophile and aprotic solvent increases the rate of S_N1 reaction

27 What is the name of the following reaction? **[EAMCET 2014]**

$$CH_3CH_2CH_2Br \xrightarrow[\text{Dry acetone}]{\text{NaI}} CH_3CH_2CH_2I$$

(a) Sandmeyer reaction
(b) Gattermann reaction
(c) Finkelstein reaction
(d) Swarts reaction

28 Which of the following alkyl halide is used as a methylating agent? **[KCET 2013]**

(a) C_2H_5Br (b) C_6H_5Cl
(c) CH_3Cl (d) C_2H_5Cl

29 In the following sequence of reaction,

$$CH_3—Br \xrightarrow{\text{KCN}} A \xrightarrow{H_3O^+} B \xrightarrow[\text{Ether}]{\text{LiAlH}_4} C,$$

C is **[CBSE AIPMT 2012]**

(a) acetone (b) methane
(c) acetaldehyde (d) ethyl alcohol

30 In alkaline hydrolysis of a tertiary halide by aqueous solution of alkali if concentration of alkali is doubled, then the reaction **[AIIMS 2012]**

(a) will be doubled (b) will be halved
(c) will remain constant (d) None of the above

31 When alkyl halide is heated with dry Ag_2O, it produces **[Manipal 2012]**

(a) ester (b) ether
(c) ketone (d) alcohol

32 Alkyl halide is converted into alcohol. This is an example of reaction **[OJEE 2012]**

(a) addition (b) displacement
(c) substitution (d) complex formation

33 An alkyl halide with molecular formula $C_6H_{13}Br$ on dehydrohalogenation gave two isomeric alkenes X and Y with molecular formula C_6H_{12}. On reductive ozonolysis, X and Y gave four compounds CH_3COCH_3, CH_3CHO, CH_3CH_2CHO and $(CH_3)_2CHCHO$. The alkyl halide is **[Kerala CEE 2012]**

(a) 2-bromohexane
(b) 2,2-dimethyl-1-bromobutane
(c) 4-bromo-2-methylpentane
(d) 2-bromo-2,3-dimethylbutane
(e) 3-bromo-2-methylpentane

34 Given reaction **[DUMET 2011]**

$$\text{(cyclohexyl)}—Br \xrightarrow[\text{Mg}]{\text{Ether}} X \xrightarrow[H_2O]{\text{HCl}} Y \text{ (Main product)}$$

'Y' in the reaction is

(a) hexane
(b) cyclohexane
(c) cyclohexylcyclohexane
(d) cyclohexyl ether

35 Which one of the following is not true for the hydrolysis of *t*-butyl bromide with aqueous NaOH? **[KCET 2011]**

(a) Reaction occurs through the S_N1 mechanism
(b) The intermediate formed is a carbocation
(c) Rate of the reaction doubles when the concentration of alkali is doubled
(d) Rate of the reaction doubles when the concentration of *t*-butyl bromide is doubled

36 Following is the substitution reaction in which —CN replaces —Cl

$$R—Cl + \underset{\text{(Alcoholic)}}{KCN} \xrightarrow{\Delta} R—CN + KCl$$

To obtain propanenitrile, R—Cl should be **[KCET 2011]**

(a) chloroethane
(b) 1-chloropropane
(c) chloromethane
(d) 2-chloropropane

37 When ethyl iodide is heated with dry silver oxide, it forms **[BVP 2011]**

(a) ethyl alcohol (b) diethyl ether
(c) silver ethoxide (d) ethyl methyl ether

38 Which of the following alkyl halides is hydrolysed by S_N2 mechanism? **[BVP 2011]**

(a) $C_6H_5CH_2Br$ (b) CH_3Br
(c) $CH_2{=}CHCH_2Br$ (d) $(CH_3)_3CBr$

39 $CH_3CH_2CH_2Br \xrightarrow[\text{alcohol}]{\text{KOH}} CH_3CH{=}CH_2$

The above reaction is an example of .... reaction. **[OJEE 2011]**

(a) substitution (b) elimination
(c) addition (d) rearrangement

ANSWERS

Check point 13.1

1. (d)	2. (c)	3. (d)	4. (c)	5. (c)	6. (b)	7. (c)	8. (a)

Check point 13.2

1. (b)	2. (d)	3. (a)	4. (d)	5. (c)	6. (b)	7. (d)	8. (c)	9. (a)	10. (b)

Check point 13.3

1. (d)	2. (a)	3. (c)	4. (c)	5. (c)	6. (c)	7. (c)	8. (c)	9. (d)	10. (a)

(A) Topical exercise

1. (b)	2. (d)	3. (c)	4. (b)	5. (b)	6. (b)	7. (b)	8. (a)	9. (b)	10. (a)
11. (a)	12. (d)	13. (b)	14. (a)	15. (c)	16. (a)	17. (b)	18. (b)	19. (b)	20. (c)
21. (c)	22. (a)	23. (c)	24. (a)	25. (a)	26. (c)	27. (c)	28. (a)	29. (d)	30. (b)
31. (a)	32. (b)	33. (b)	34. (b)	35. (a)	36. (c)	37. (d)	38. (d)	39. (a)	40. (b)
41. (c)	42. (a)	43. (c)	44. (c)	45. (c)	46. (a)	47. (b)	48. (d)	49. (a)	50. (b)
51. (c)	52. (a)	53. (c)	54. (d)	55. (a)	56. (b)	57. (b)	58. (a)	59. (c)	60. (c)
61. (c)	62. (d)	63. (a)	64. (a)	65. (a)	66. (a)	67. (a)	68. (a)	69. (b)	70. (d)
71. (b)	72. (b)	73. (c)	74. (a)	75. (c)	76. (c)	77. (c)			

(B) Miscellaneous questions

1. (a)	2. (c)	3. (a)	4. (b)	5. (b)	6. (c)	7. (c)	8. (a)	9. (d)	10. (d)
11. (a)	12. (a)	13. (a)	14. (b)	15. (c)	16. (a)	17. (d)	18. (b)	19. (d)	20. (c)
21. (a)	22. (b)	23. (c)	24. (d)	25. (a)	26. (a)	27. (b)	28. (a)	29. (a)	30. (a)
31. (c)	32. (c)	33. (c)	34. (a)						

(C) Medical entrances' gallery

1. (c)	2. (c)	3. (d)	4. (b)	5. (b)	6. (d)	7. (a)	8. (a)	9. (a)	10. (c)
11. (d)	12. (c)	13. (b)	14. (a)	15. (b)	16. (b)	17. (a)	18. (b)	19. (c)	20. (a)
21. (c)	22. (b)	23. (c)	24. (a)	25. (b)	26. (b)	27. (c)	28. (c)	29. (d)	30. (c)
31. (b)	32. (c)	33. (e)	34. (b)	35. (c)	36. (a)	37. (b)	38. (b)	39. (b)	

Hints & Explanations

(A) Topical exercise

1 (*b*) 3-bromobut-1-ene is an allylic halide.

$$CH_2=CH-\underset{\underset{Br}{|}}{CH}-CH_3$$

2 (*d*) $CH_3-\overset{\overset{^1CH_3}{|}}{\underset{\underset{^3CH_3}{|}}{^2C}}-I$

2-iodo-2-methylpropane

3 (*c*) The correct IUPAC name of the given compound is

$$CH_3-\overset{2}{C}H-\overset{1}{C}H_2-Br, \quad \text{with } {}^3CH_2-\underset{4}{CH_3} \text{ on C-2}$$

1-bromo-2-methylbutane

4 (*b*) Structure of the diethylbromomethane is given below

$$\underset{5}{H_3C}-\underset{4}{H_2C}-\underset{3}{\overset{\overset{Br}{|}}{HC}}-\underset{2}{CH_2}-\underset{1}{CH_3}$$

So, the IUPAC name is 3-bromopentane.

5 (*b*) In *vic*-dihalides, halogen atoms are present on the adjacent carbon atoms.

In *gem*-dihalides, halogen atoms are present on the same carbon atom. They are known as alkylidene halides.

In allylic halides, halogen atom is bonded to sp^3 hybridised carbon atom next to carbon-carbon double bond.

In vinylic halides, halogen atom is bonded to sp^2-hybridised carbon atom of a carbon-carbon double bond.

In ethylidene chloride ($H_3C-CHCl_2$) both halogen atoms are present on same carbon atom, so it is *gem*-dihalide.

6 (*b*) The correct name of the given compounds are as follows

$$Br-\overset{3}{C}H_2-\overset{2}{C}H=\overset{1}{C}H_2$$

3-bromoprop-1-ene

$$CH_3-\overset{\overset{^1CH_3}{|}}{\underset{\underset{^3CH_3}{|}}{^2C}}-Cl$$

2-chloro-2-methyl propane

$$\overset{1}{C}H_2=\overset{2}{C}H-\overset{3}{\underset{\underset{CH_3}{|}}{CH}}-\overset{4}{C}H_2Br$$

4-bromo-3-methyl butane

7 (*b*) *vic*-dihalides are those halides in which two halogen atoms are present on the two adjacent carbon atoms.

The structure of the given compound.

Dichloromethane 1, 2-dichloroethane

$$H-\overset{\overset{H}{|}}{\underset{\underset{H}{|}}{C}}-\overset{\overset{Cl}{|}}{\underset{\underset{Cl}{|}}{C}}-H \qquad \overset{\overset{Cl}{|}}{CH_2}-CH=CH_2$$

Ethylidene chloride Allyl chloride

In 1, 2-dichloroethane, the two chlorine atoms are attached to two adjacent carbon atoms.

Hence, option (b) is correct.

8 (*a*) Chiral/asymmetric carbon is that carbon in which carbon has formed four bonds with four different groups. Let's see the structural formula of the given compounds.

$$\overset{1}{H_3C}-\overset{*2}{\underset{\underset{Br}{|}}{CH}}-\overset{3}{C}H_2-\overset{4}{C}H_3 \qquad \overset{4}{C}H_3-\overset{2}{C}H_2-\overset{2}{C}H_2-\overset{1}{C}H_2Br$$

2-bromobutane 1-bromobutane

$$\overset{1}{C}H_3-\overset{2}{\underset{\underset{Br}{|}}{CH}}-\overset{3}{C}H_3 \qquad \overset{3}{C}H_3-\overset{\overset{Br}{|}}{\underset{\underset{OH}{|}}{\overset{2}{C}}}-\overset{1}{C}H_3$$

2 - bromopropane 2 - bromopropan-2- ol

With the help of these structural formula it is very clear that 2-bromobutane in which asterisk mark carbon atom is bonded to four different atoms or groups. So, this molecule is chiral in nature. Other molecules do not contain four different atoms or groups.

9 (*b*) Carbon has four valencies. If a carbon atom satisfies all of its four valencies with four different groups then it is termed as asymmetric/chiral carbon.

In the given compound, 2nd and 3rd carbon atom are bonded to four different groups, so these are asymmetric.

10 (*a*) The given reaction is a substitution reaction. It involves the replacement of 1° and 2° hydrogen of alkanes by chlorine. It occurs in presence of ultraviolet light or at high temperature.

The chlorination does not occur at room temperature in absence of light. In this reaction, light is absorbed by the chlorine molecule and activated chlorine initiates the reaction as follows

Step 1 $Cl-Cl \xrightarrow[\text{light}]{UV} 2\dot{C}l$

$\dot{C}l + CH_3-CH_2-CH_2-CH_3 \longrightarrow CH_3CH_2CH_2-\dot{C}H_2 + HCl$

Step 2 $CH_3-CH_2-CH_2-\dot{C}H_2 + Cl_2 \longrightarrow CH_3-CH_2-CH_2-CH_2Cl + \dot{C}l$

Step 3 $CH_3-CH_2-CH_2-\dot{C}H_2 + \dot{C}l \longrightarrow CH_3CH_2CH_2CH_2Cl$

Similarly second product is formed.

So, option (a) is the correct.

12 (*d*) $\underset{\text{Propene}}{CH_2{=}CHCH_3} \xrightarrow[\text{Allylic position is affected}]{NBS} CH_2{=}CHCH_2Br \xrightarrow{AgCN} CH_2{=}CHCH_2NC$

13 (*b*) In the given reaction, only best yield of bromoalkanes are obtained when X_2 is Br_2.

14 (*a*) $CH_2{=}CH\overset{\oplus}{C}H_2$ (intermediate) is resonance stabilised.

15 (*c*) $\underset{\text{Butane}}{CH_3CH_2CH_2CH_3} \xrightarrow{Cl_2} CH_3\overset{*}{C}H(Cl)CH_2CH_3$ (* is chiral carbon)

Thus, racemic mixture.

17 (*b*) $\underset{\text{3-methylbutene}}{CH_3CH(CH_3)CH{=}CH_2} \xrightarrow{H^+} \underset{2^\circ\text{carbocation}}{CH_3CH(CH_3)\overset{\oplus}{C}H-CH_3} \xrightarrow{1,2\text{-hydride shift}} \underset{3^\circ\text{carbocation (most stable)}}{CH_3-\overset{\oplus}{C}(CH_3)-CH_2CH_3}$

18 (*b*) Reaction between alcohols and halogen acid follows S_N1 mechanism. In S_N1 mechanism, carbocations are formed as intermediates.

Let us consider the formation of carbocations with the given three alcohols.

$CH_3-CH_2-CH_2-OH \longrightarrow CH_3-CH_2-\overset{+}{C}H_2 + OH^-$

In this case, 1° carbocation is formed. It is least stable. So, here S_N2 mechanism is followed. In this S_N2 mechanism a transition state is observed in α-carbon which is linked with two nucleophiles.

$CH_3-CH_2-CH(CH_3)-OH \longrightarrow CH_3-CH_2-\overset{+}{C}H(CH_3) + OH^-$

2° carbocation (more stable than 1° carbocation)

$CH_3-CH_2-C(CH_3)_2-OH \longrightarrow H_3C-CH_2-C^+(CH_3)_2 + OH^-$

3° carbocation (most stable)

The reaction proceeds with stable carbocation. Higher the stability of carbocation, higher will be the possibilities of attack of X^- ion to the carbocation.

As the tertiary carbocation is most stable, so the possibilities of attack of X^- ion are more prominent in case of tertiary carbocations. Thus, attack of X^- ion to carbocation is proceeded with tertiary carbocation as follows

$\underset{3^\circ\text{ carbocation}}{H_3C-CH_2-C^{\oplus}(CH_3)_2} + X^{\ominus} \longrightarrow H_3C-CH_2-C(CH_3)_2-X$

So, the correct option is (b).

Note Higher the stability of intermediate, higher will be the reactivity of compound and higher will be the yield of the desired product.

Basic nature of alcohol is in order 1° < 2° < 3°

Thus reactivity with HCl (acid) is 1° < 2° < 3°

19 (*b*) Conjugation increases stability of radical intermidiate.

$CH_2{=}CHCH\dot{C}H{=}CH_2 \longleftrightarrow CH_2{=}CHCH{=}CH\dot{C}H_2$

21 (*c*) Methylcyclopentane $\xrightarrow[\text{UV/Light}]{Cl_2}$ (1-chloro-1-methylcyclopentane) + (chloromethyl)cyclopentane + 1-chloro-2-methylcyclopentane + 1-chloro-3-methylcyclopentane

23 (*c*) Since, this H is allylic as well as benzylic.

24 (*a*) Higher the surface area, higher will be the intermolecular forces of attraction and thus, boiling point too. Boiling point increases with increase in molecular mass of halogen atom for the similar type of alkyl halide. Butane has no halogen atom and rest of all three compounds are haloderivatives of butane.

Atomic mass of iodine is highest, so the boiling point of 1-iodobutane is maximum among all the given compounds and hence, option (a) is correct.

25 (*a*) Halogen exchange reactions are those reactions in which one halide replaces another. In option (a) halogen (–*X*) is replaced by iodine. This reaction is named as Finkelstein reaction.

In option (b), there is the addition of hydrogen halide on alkene.

In option (c), halogen replaces alcoholic group.

While in option (d), halogen replaces the hydrogen of benzene ring.

26 (*c*) Boiling point of a compound depends upon the surface area. Higher the surface area, higher will be the boiling point of a compound. Surface area decreases with increase in branching. If the compound has branching then its boiling point will be minimum.

Thus, the increasing order of their boiling points is

$$\underset{\substack{\text{(Maximum branching;}\\ \text{lowest b.p.346 K)}}}{H_3C-\overset{\overset{CH_3}{|}}{\underset{\underset{Br}{|}}{C}}-CH_3} < \underset{\text{(Branching)}}{(H_3C)_2CH-CH_2Br} < \underset{\substack{\text{(Straight chain;}\\ \text{higher bp) 375 K}}}{CH_3CH_2CH_2CH_2Br}$$

27 (*c*) $H-C{\equiv}CCH_2Br \xrightarrow{Mg}$ No reaction

Thus, H is acidic, thus (C—Br) does not react with Mg.

31 (*a*) Intermediate is $CH_3\dot{C}H_2$ (free radical). It can

(i) combine to itself form $CH_3CH_2CH_2CH_3$ (butane).

(ii) undergo disproportionation forming CH_3CH_3 (by reduction) and $CH_2=CH_2$ (by oxidation).

32 (*b*) A primary alkyl halide would prefer to undergo S_N2 reaction.

(a) S_N1 reactions occur only if the intermediate carbocation is stable, i.e. 3° carbocation.

(b) S_N2 reactions occur if there is less steric hinderance on to the α-carbon of alkyl halide. In case of primary alkyl halides, carbocation is highly unstable and steric hinderance is very less. So, primary alkyl halide would prefer to undergo S_N2 reaction.

(c) In α-elimination, proton and the leaving group are present on same atom.

(d) Racemisation is the process of conversion of enantiomer into a racemic mixture.

34 (*b*) $\underset{\text{3-methyl-2-butanol}}{CH_3CH(CH_2OH)CHCH_3} \xrightarrow[-H_2O]{HCl/ZnCl_2} \underset{\text{2° carbocation}}{CH_3CH(CH_3)\overset{\oplus}{C}HCH_3}$

$\downarrow$ Rearrangement

$$\underset{\substack{\text{3° carbocation}\\ \text{(more stable than 2° carbocation)}}}{CH_3-\overset{\oplus}{C}(CH_3)-CH_2CH_3}$$

37 (*d*) Inversion takes place by S_N2 reaction.

39 (*a*) It is Corey House synthesis of alkane. Gilmann reagent $[(CH_3)_2CH]$ CuLi should react with *RX* such that *RX* is Me*X* or 1° or 2° cyclo alkyl halide for good yield.

42 (*a*) A mixture containing two enantiomers in equimolar amount have zero optical rotation, as the rotation due to one isomer is cancelled by the rotation due to other isomer. Such a mixture is known as racemic mixture. All those compounds which follow S_N1 mechanism during nucleophilic substitution reaction form racemic mixture.

Order of reactivity of alkyl halides towards S_N1 reaction follows as.

Tertiary halide > secondary halide > primary halide > CH_3X.

For S_N1 reaction.

Thus, $CH_3-\overset{*}{C}H(C_2H_5)-Br$ contains a chiral carbon and gives a racemic product.

Hence, option (a) is correct.

49 (*a*) In (a) both Br atoms are in the same plane.

50 (*b*) 1° alkyl halide cannot react by S_N1 reaction.

52 (*a*) S_N2 occurs, when alkyl halide is least sterically hindered i.e. 1°.

59 (*c*) Option (a) reaction undergo E2

(skeletal structure with Br) $\xrightarrow{E2}$ (alkene)

- Option (b) undergo E2 reaction

(skeletal structure with Br) $\xrightarrow{E2}$ (alkene)

- E2 elimination allows the formation of an alkene without the formation of an intermediate i.e., unstable carbon cation. Instead this reaction involves one step mechanism in which β-hydrogen (attached to β-carbon) and halogen is removed to form a double bond. So, a β-hydrogen should be present for E2 elimination reaction the given compounds in option (c) β-hydrogen is absent. So, E2 reaction do not occur. Option (c) does not undergo E2.

(skeletal structure with Br) $\xrightarrow{E2}$ ✕

61 (*c*) Strong nucleophile, on reacting with 3°-alkyl halide in presence of polar protic solvent.

Undergo elimination reaction i.e. E1 elimination reaction in this case. The product formed is according to Saytzeff rule.

63 (*a*) Tertiary alkyl halide produces most stable 3° carbocation as an intermediate.

$$CH_3-\overset{\overset{CH_3}{|}}{\underset{\underset{CH_3}{|}}{C^+}} \longrightarrow CH_3-\underset{\underset{CH_3}{|}}{C}=CH_2$$

64 (*a*) Conjugation increases the stability of intermediate formed.

66 (*a*) Due to steric hinderance, less substituted alkene is the major product (Hofmann elimination).

68 (*a*) $CH_3-\overset{\overset{CH_3}{|}}{\underset{\underset{CH_3}{|}}{C}}-\underset{\underset{Cl}{|}}{CH}-CH_3 \xrightarrow[-HCl]{E2} CH_3-\overset{\overset{CH_3}{|}}{\underset{\underset{CH_3}{|}}{C}}-CH=CH_2$

$\downarrow$ E1

$$\underset{\text{2° carbocation}}{CH_3-\overset{\overset{CH_3}{|}}{\underset{\underset{CH_3}{|}}{C}}-\overset{\oplus}{C}H-CH_3} \xrightarrow{\text{Rearrangement}} CH_3-\overset{\overset{CH_3CH_3}{|}}{\underset{\oplus}{C}}-CH-CH_3$$

$\downarrow -H^+$

$(CH_3)_2C=C(CH_3)_2$

71 (*b*) Addition of HBr and elimination of HBr

i.e. $cis \xrightarrow{-HBr} trans$ alkene

$trans \xrightarrow{+HBr} cis$ alkene

Reactant is *cis*, thus elimination product is *trans*.

75 (*c*) (A) $\xrightarrow{E1}$ 1° carbocation ⟶ 2° carbocation $\xrightarrow{-H^{\oplus}}$ Stable due to conjugation

(B) Miscellaneous questions

3 (*a*) μ_p for C—Cl bond = 1.56 D

μ_p for C—I bond = 1.29 D

Due to the greater difference in the electronegativities of C and Cl as compared to that of C and I

C—Cl bond is more ionic than C—I bond. Among halides C—*X* bond length increases as the size of halogen increases. Therefore, C—Cl bond is stronger than C—I bond.

6 (*c*) $CH_3-C(CH_3)_2-CH_2Br + 2Na + Br-CH_2-C(CH_3)_2-CH_3$

neo–pentyl bromide   *nco*–pentyl bromide

$\xrightarrow[-2NaBr]{\text{Dry ether}} CH_3-C(CH_3)_2-CH_2-CH_2-C(CH_3)_2-CH_3$

2, 2, 5, 5-tetramethylhexane

7 (*c*) $2CH_3CH_2Br \xrightarrow{Na/ether} CH_3CH_2CH_2CH_3$

Ethyl bromide   Butane

$2CH_3CH_2CH_2Br \xrightarrow{Na/ether} CH_3CH_2CH_2CH_2CH_2CH_3$

n-propyl bromide   Hexane

$CH_3CH_2Br + CH_3CH_2CH_2Br \xrightarrow{Na/ether} CH_3CH_2CH_2CH_2CH_3$

Pentane

11 (*a*) CH_3-CH_2-Br ; $H-C(H)(H)-C(CH_3)(CH_3)-Br$

H^- H^-

13 (*a*) If acid is strongest, its conjugate base is the weakest and is the best leaving group. Among the given options strongest acid is

$F_3C-C_6H_4-SO_3H$

Thus, $F_3C-C_6H_4-SO_3^{\ominus}$ is the best leaving group.

14 (*b*) $CH_2{=}CHCH_2-(H) \xrightarrow[I]{NBS} CH_2{=}CHCH_2Br$

$\xrightarrow{\text{II Mg/ether}} CH_2{=}CHCH_2MgBr \xrightarrow[III]{D_2O} CH_2{=}CHCH_2D$

15 (*c*) Cyclohexyl–Br $\xrightarrow[\text{Elimination}]{NH_3}$ (A) $\xrightarrow{O_3/H_2O}$ (B) $\xrightarrow[-BaCO_3]{BaO}$ Barium salt of a dibasic acid ⟶ (C) cyclopentanone

(A) This bond is cleaved. H_2O_2 formed causes oxidation of —CHO to —COOH

20 (*c*) CN^- a stronge base undergoes S_N2 reaction is in (c).

21 (*a*) If II path is used alcohol formed may decompose EtMgBr.

(Conjugate system) $\xrightarrow{\overset{+\ -}{HCl}}$ Tautomerisation

Cl–CH₂CH₂CH₂–C(OH)(CH₃)(C₂H₅) $\xleftarrow[(ii)\ H_3O^+]{(i)\ EtMgBr}$ Cl–CH₂CH₂CH₂–CO–CH₃

22 (*b*) $C_2H_5I \xrightarrow[-HI]{\text{Alc. KOH}} CH_2{=}CH_2 \xrightarrow{Br_2} BrCH_2-CH_2Br$

(*X*)   (*Y*)

$\xrightarrow{KCN} NCCH_2-CH_2CN$

(*Z*)

23 (*c*) $((CH_3)_3C-CH_2)_2CuLi$ + I (Correy House)

24 (*d*) Cyclic and tertiary halides undergo hydrolysis by S_N1 mechanism and involve the formation of carbocation intermediate. Greater the stability of carbocation intermediate, higher is the ease of halides to undergo hydrolysis. The order of stability of carbocations is as

cyclohexyl cation $< (CH_3)_3C^{\oplus} <$ cyclohexenyl (allylic) cation

However, C—Br bond in bromobenzene acquires some double bond character, hence it has least tendency to undergo hydrolysis. Therefore, the increasing order of hydrolysis of the given compounds is

Br Br Br
(i) < (iii) < $(CH_3)_3C-Br$ < (ii)
(i) (iii) (iv) (ii)

26 (*a*) There is no chiral carbon, hence inversion does not take place.

27 (*b*) Due to S_N2 reaction, $Nu^\ominus$ attacks from backside giving inverted product.

29 (*a*) The stereoisomers related to each other as non-superimposable mirror images are called enantiomers. Enantiomers possess identical physical properties.

They only differ with respect to the rotation of plane polarised light. If one of the enantiomer is dextrorotatory, the other will be laevorotatory. Here, the enantiomer of molecule (*A*) is shown below

Mirror

H, CH_3, H_5C_2, Br | H, C_2H_5, H_3C, Br

Hence, option (a) is correct.

33 (*c*) CH_3, Br $\xrightarrow[CH_3OH]{CH_3\overset{-}{O}\overset{+}{Na}}$ CH_3, ⊕ (2° carbocation) $\xrightarrow{\text{Rearrangement}}$ CH_3, ⊕ (3° carbocation) $\xrightarrow{CH_3\overset{-}{O}\overset{+}{Na}}$ H_3CO, CH_3 (*A*)

In polor protic solvent S_N1 reaction is preferred.

(C) Medical entrances' gallery

1 (*c*) An S_N1 reaction proceeds through formation of a stable carbocation as an intermediate. Here,

(a) $CH_2{=}CH-CH_2Cl \xrightarrow{-Cl^-} CH_2{=}CH-\overset{\oplus}{C}H_2$
Allyl-1°-carbocation (stable)

(b) $(CH_3)_3CCl \xrightarrow{-Cl^-} CH_3-\overset{\oplus}{C}(CH_3)-CH_3$ (9 hyper conjugations)
tert-butyl carbocation (stable)

(c) $C_6H_5-CH_2CH_2Cl \xrightarrow{-Cl^-} C_6H_5-\overset{\oplus}{C}H_2CH_2$ (−*I*)
1°-carbocation (less stable due to −*I* effect of the phenyl group)

(d) $C_6H_5-CH_2Cl \xrightarrow{-Cl^-} C_6H_5-\overset{\oplus}{C}H_2$
Benzyl-1°-carbocation (stable)

So, option (c) will not undergo S_N1 reaction with OH^-.

2 (*c*) (a) $\underset{\text{Benzene}}{C_6H_5-H} \xrightarrow[-HBr]{Br_2/AlCl_3} \underset{\text{Bromobenzene}}{C_6H_5-Br}$

It is an electrophilic aromatic substitution (ArS_E2) reaction.

(b) $\underset{\text{Acetylene}}{HC\equiv CH} \xrightarrow{HBr} CH_2{=}CH-Br \xrightarrow{HBr} \underset{\text{Dibromo ethane}}{CH_3-CHBr_2}$

It is an example of electrophilic addition reaction.

(c) $\underset{\text{Methane}}{CH_4} \xrightarrow[h\nu]{Br_2} \underset{\text{Bromomethane}}{CH_3Br} + HBr$

It is a free radical substitution reaction, which have free radical intermediates like $Br^\bullet$, $CH_3^\bullet$, $H^\bullet$ etc.

(d) $\underset{\text{Propene}}{CH_3-CH{=}CH_2} \xrightarrow[\text{(Organic peroxide)}]{HBr/(C_6H_5COO)_2} \underset{\text{Bromopropane}}{CH_3CH_2CH_2Br}$ (Major)

It is a free radical addition reaction.
Hence, option (c) is correct.

4 (*b*) Alkanes which contain all equivalent hydrogen atoms form only one monochloro derivative on halogenation.

neo-pentane, $H_3C-C(CH_3)_2-CH_3$ contains all equivalent hydrogen atoms. So, it will give only one monochloro derivative on halogenation.

5 (*b*) When alcoholic KOH reacts with haloalkanes, the reaction (rate) depends on the following factors:

(i) Size of halogen

(ii) Degree of C-atom to which the halogen atom is bonded, i.e. more be the degree of haloalkanes more is its reactivity on the above basis.

As bromine atom is large in size, thus

(Br) and (Br, (2°)) are more reactive than (Cl, (2°)) but due to lesser degree of (Br, (1°)) it is less reactive than (Cl, (2°))

Hence, correct order is $A > C > B > D$.

Thus, option (b) is correct.

6 (*d*) When cyclobutane reacts with Cl_2 in presence of light, substitution reaction takes place.

□ $+ Cl_2 \xrightarrow{UV}$ □–Cl $+ HCl$

The reaction of product formed with Na metal in dry ether takes place as

$$2\,\square\!-\!Cl + 2Na \xrightarrow{\text{dry ether}} \square\!-\!\square$$

7 *(a)* In S_N1 reactions, carbocations are the intermediates. Greater the stability of carbocation, more easily it is formed and hence faster will be the rate of reaction.

The stability of carbocation decreases in the order:

3° carbocation > 2° carbocation > 1° carbocation

$\therefore \quad Ph\overset{+}{C}(CH_3)_2 > Ph\overset{+}{C}H\,CH_3 > Ph\overset{+}{C}H_2$ (3°, 2°, 1°)

Thus, option (a) is incorrect, while trends in other options are correct.

8 *(a)* In S_N2 reaction, nucleophile transfer electron pair to carbon of substrate, leading to the formation of transition state.

$$\ddot{N}u^- + H_2HC{-}X \longrightarrow \left[\overset{\delta-}{Nu}\cdots C(H)(H)(H)\cdots\overset{\delta+}{X}\right] \longrightarrow Nu{-}CH_3 + X^-$$

H_2O is a neutral nucleophile, while CN^-, NH_2^- and I^- are negatively charged nucleophiles.

Thus, H_2O cannot behave as a nucleophile for S_N2 reaction.

9 *(a)* When haloalkanes containing β-hydrogen atom is heated with alc. KOH solution, then alkene is formed.

$$CH_3CH_2CH_2Br \xrightarrow{\text{alc. KOH}} \underset{\text{Propene}}{CH_3CH=CH_2}$$

In this reaction, hydrogen is eliminated from β-carbon and the halogen is lost from α-carbon atom of 1-bromopropane. As a result, an alkene is formed as a product. Due to the involvement of elimination of β-hydrogen atom, the process is often called β-elimination.

10 *(c)* Diethylcarbonate is obtained by the reaction of phosgene with ethanol.

$$COCl_2 + 2C_2H_5OH \longrightarrow O{=}C(OC_2H_5)_2 + 2HCl$$

∴ The correct option is (c).

11 *(d)* The given reaction takes place as follows :

$$\underset{(A)}{CH_4} \xrightarrow[\text{Step 1}]{Br_2\,/\,h\nu} CH_3Br \xrightarrow[\text{Wurtz reaction (Step 2)}]{\text{Na/dry ether}} CH_3{-}CH_3$$

***Step* I** Alkyl halide is formed by free radical halogenation of alkane in the presence of UV-light.

***Step* II** The formed alkyl halide reacts with sodium in the presence of dry ether to form alkane containing double number of carbon atoms present in alkyl halide.

This reaction is known as Wurtz reaction.

From the above reaction, it is concluded that option (d) is correct as in all other cases the hydrocarbon formed in step 2 will have four carbon atoms.

12 *(c)* $\underset{\substack{\text{1-bromopropane}\\(X)}}{CH_3CH_2CH_2Br} + Mg \xrightarrow{\text{Dry ether}} \underset{(Y)}{CH_3CH_2CH_2MgBr}$

$$\xrightarrow{HCHO} CH_3CH_2CH_2CH_2{-}O{-}MgBr$$

$$\xrightarrow{H_2O} \underset{\text{1-butanol}}{CH_3CH_2CH_2CH_2OH} + Mg(Br)(OH)$$

13 *(b)* Assertion and reason both are correct statements but reason is not the correct explanation of assertion. For the same hydrocarbon part, boiling point depends upon the atomic mass of halogen atom. Higher the mass of the halogen atom, higher will be the boiling point. So, boiling point decreases with decrease in atomic mass of halogen atom. Thus, the decreasing order of boiling point is

$$RI > RBr > RCl > RF$$

14 *(a)* Both Assertion and Reason are correct and Reason is the correct explanation of the Assertion. Chlorination of CH_4 does not take place in dark as this reaction follows free radical mechanism and free radicals are generated only in the presence of sunlight.

15 *(b)*

$$\underset{\text{Neopentyl chloride}}{CH_3{-}C(CH_3)_2{-}CH_2Cl} \xrightarrow{\text{Anhy. }AlCl_3} \underset{\text{1° carbocation}}{CH_3{-}C(CH_3)_2{-}\overset{\oplus}{C}H_2}$$

$$\xrightarrow{\text{Rearrangement of carbocation}} \underset{\text{3° carbocation (most stable)}}{CH_3{-}\overset{\oplus}{C}(CH_3){-}CH_2{-}CH_3}$$

$$\xrightarrow{C_6H_5{-}H} C_6H_5{-}C(CH_3)_2CH_2CH_3$$

16 *(b)* According to Saytzeff's rule, dehydrohalogenation occurs in such a way that the major product formed is highly substituted.

$$CH_3CH_2CH(F)CH_3 \xrightarrow[CH_3OH]{CH_3O^-} \underset{\text{(Major)}}{CH_3CH=CHCH_3} + \underset{\text{(Minor)}}{CH_3CH_2CH=CH_2}$$

17 *(a)* The reaction is as follows:

$$\underset{\text{Methyl bromide}}{CH_3{-}Br} \xrightarrow{\text{Mg / ether}} CH_3{-}MgBr \xrightarrow[\text{(ii) }H_2O]{\text{(i) }CO_2} \underset{\text{Ethanoic acid}}{CH_3{-}\overset{O}{\overset{\|}{C}}{-}OH}$$

$$\xrightarrow[\Delta]{CH_3OH/H^+} \underset{\text{Methyl ethanoate}}{CH_3{-}\overset{O}{\overset{\|}{C}}{-}OCH_3}$$

Hence, the product '*P*' is methylethanoate.

18 *(b)* The suitable reagent for Wurtz reaction is Na/ether. An ethereal solution of an alkyl halide is treated with sodium to produce higher alkanes. This reaction involves the formation of new C—C bond. e.g.

$$\underset{\text{Ethylbromide}}{2CH_3CH_2Br} + 2Na \xrightarrow{\text{Ether}} \underset{\text{Butane}}{CH_3CH_2CH_2CH_3}$$

19 (*c*) The given reaction follows S_N2 mechanism and these reactions are favoured in polar aprotic medium like DMSO, DMF... etc.

$$CH_3CH_2CH_2Br + NaCN \xrightarrow{DMF} CH_3CH_2CH_2CN + NaBr$$

So, the correct option is (c).

20 (*a*) The reaction of alkyl halide with sodium alkoxide or sodium phenoxide to form ethers is called Williamson synthesis. Here in this reaction, alkyl halide should be primary and alkoxide should be bulkier as shown below,

$$\underset{\text{Sodium alkoxide}}{R—X + R'—ONa} \longrightarrow R—O—R' + NaX$$

$$CH_3—C(CH_3)(CH_2)—ONa + CH_3CH_2Cl \xrightarrow{-NaCl} CH_3—C(CH_3)_2—O—CH_2—CH_3$$

21 (*c*) *Neo*-pentyl chloride does not readily undergo dehydrohalogenation.

$$CH_3—C(CH_3)=CH_2(CH_3)$$

is not possible.

22 (*b*) Wurtz reaction may be proceed by the following ways

(i) $CH_3—C(CH_3)_2—CH_2—Br + 2Na + Br—CH_2—C(CH_3)_2—CH_3$
(*Neo*-pentyl bromide) (*Neo*-pentyl bromide)

$$\xrightarrow{-2NaBr} \underset{\text{2, 2, 5, 5-tetra methylhexane}}{CH_3—C(CH_3)_2—CH_2—CH_2—C(CH_3)_2—CH_3}$$

(ii) $CH_3—CH(CH_3)—CH_2—Br + 2Na + Br—CH_2—CH(CH_3)—CH_2$
(*Iso*-butyl bromide) (*Iso*-butyl bromide)

$$\xrightarrow{-2NaBr} \underset{\text{2,5-dimethylhexane}}{CH_3—CH(CH_3)—CH_2—CH_2—CH(CH_3)—CH_3}$$

(iii) $CH_3—C(CH_3)_2—CH_2—Br + 2Na + Br—CH_2—CH(CH_3)—CH_3$
(*Neo*-pentyl bromide) (*Iso* - butyl bromide)

$$\xrightarrow{-2\,NaBr} \underset{\text{2, 2, 5-trimethyhexane}}{CH_3—C(CH_3)_2—CH_2—CH_2—CH(CH_3)—CH_3}$$

23 (*c*) I. $C_6H_{11}—Br + Mg \xrightarrow{\text{Dry ether}} \underset{(A)}{C_6H_{11}—MgBr}$

(Reaction or formation of Grignard reagent)

II. $\underset{(A)}{C_6H_{11}—MgBr} \xrightarrow[H^+]{H_2O} \underset{(B)}{C_6H_{12}} + MgBr\,(OH)$

24 (*a*) Boiling point increases with molecular mass

$$\underset{\text{Bromoform}}{CHBr_3} > \underset{\text{Dibromomethane}}{CH_2Br_2} > \underset{\text{Bromomethane}}{CH_3Br} > \underset{\text{Chloromethane}}{CH_3Cl}$$

Thus, the correct order is III < I < IV < II

25 (*b*) $CH_3—CH(CH_3)—CH_2—CH_2—Br \xrightarrow[\Delta]{\text{Alc. KOH}} \underset{(A)}{CH_3—CH(CH_3)—CH=CH_2} + H_2O$

$$\underset{(A)}{CH_3—CH(CH_3)—CH=CH_2} \xrightarrow{\text{Conc. }H_2SO_4}$$

$$CH_3—CH(CH_3)—CH(OH)—CH_3 \xrightarrow{H_2O,\Delta} \underset{\text{(Major)}}{CH_3—C(CH_3)=CHCH_3}\ (B)$$

$$+ \underset{\text{Minor } (C)}{CH_3—CH(CH_3)—CH=CH_2}$$

$$CH_3—C(CH_3)=CHCH_3 \xrightarrow{HI} \underset{(D)\text{ Major}}{CH_3—C(CH_3)(I)—CH_2CH_3}$$

$$\underset{(C)}{CH_3—CH(CH_3)—CH=CH_2} \xrightarrow{HI} \underset{(E)\text{ (Minor)}}{CH_3—C(CH_3)—CH(I)—CH_3}$$

26 (*b*) This problem includes concept of S_N1 and S_N2 reaction and factor affecting S_N1 and S_N2 reaction.

During S_N2 reaction of alkyl halide, inversion of configuration takes place instead of rearrangement.

While S_N1 reaction occurs through the formation of carbocation intermediate which is favoured by the presence of Lewis acid and aprotic solvent.

Rearrangement is favoured in S_N1 reaction.

27 (*c*) Alkyl bromide reacts with sodium iodide (in acetone or methanol) to form alkyl iodides. This is called Finkelstein reaction.

$$CH_3CH_2CH_2Br \xrightarrow[\text{Dry acetone}]{NaI} CH_3CH_2CH_2I + NaBr$$

The high reactivity of alkyl halide is due to highly polarised covalent C—*X* bond.

In nucleophilic substitution $C^{\delta+}$ site is susceptible to attack by nucleophiles.

28 (c) CH_3Cl is used as methylating agent. The methylating agent is one which is used for the introduction of methyl groups.

29 (d)
$$\underset{\text{Methyl bromide}}{CH_3Br} \xrightarrow[-KBr]{KCN} \underset{\underset{(A)}{\text{Methyl cyanide}}}{CH_3CN} \xrightarrow{H_3O^+} \underset{\underset{(B)}{\text{Ethanoic acid}}}{CH_3COOH} \xrightarrow[\text{ether}]{LiAlH_4} \underset{\underset{(C)}{\text{Ethyl alcohol}}}{CH_3CH_2OH}$$

30 (c) In alkaline hydrolysis of a tertiary halide by aqueous alkali, if concentration of alkali is doubled, then the reaction will remain constant because tertiary-alkyl halides with aqueous alkali gives S_N1 reaction and rate of S_N1 reaction is not based upon concentration of nucleophile, (i.e. alkali).

31 (b) Alkyl halide on reaction with dry silver oxide (Ag_2O) gives ether.

$$\underset{\text{Alkyl halide}}{2RX} + Ag_2O \longrightarrow \underset{\text{Ether}}{R—O—R} + 2AgX$$

32 (c) $RCl + KOH \longrightarrow ROH + KCl$

Thus, above reaction is an example of substitution reaction. In this reaction —OH group substitutes —Cl group.

33 (e)
$$\underset{\text{3-bromo-2-methylpentane}}{CH_3—\underset{\underset{CH_3}{|}}{CH}—\underset{\underset{Br}{|}}{CH}—CH_2—CH_3} \xrightarrow{\text{Alc. KOH}}$$

$$\underset{(X)}{CH_3—\underset{\underset{CH_3}{|}}{C}=CH—CH_2CH_3} + \underset{(Y)}{CH_3—\underset{\underset{CH_3}{|}}{CH}—CH=CHCH_3}$$

$$\underset{(X)}{CH_3—\underset{\underset{CH_3}{|}}{C}=CH—CH_2CH_3} \xrightarrow[Zn/H_2O_2]{O_3} CH_3—\underset{\underset{CH_3}{|}}{C}=O + CH_3CH_2CHO$$

$$\underset{(Y)}{CH_3—\underset{\underset{CH_3}{|}}{CH}—CH=CHCH_3} \xrightarrow[Zn/H_2O_2]{O_3} CH_3—\underset{\underset{CH_3}{|}}{CH}—CHO + CH_3CHO$$

34 (b)
$$C_6H_{11}—Br + Mg \xrightarrow{\text{Ether}} \underset{(X)}{C_6H_{11}—MgBr} \xrightarrow[H_2O]{HCl} \underset{\underset{\underset{\text{(Main product)}}{(Y)}}{\underset{(C_6H_{12})}{\text{Cyclohexane}}}}{C_6H_{12}} + Mg$$

35 (c) It is not true because hydrolysis of *t*-butyl bromide takes place through S_N1 mechanism, i.e. rate depends only upon the concentration of alkyl halide.

36 (a)
$$\underset{\text{Chloroethane}}{CH_3CH_2Cl} + \underset{\text{Alcoholic}}{KCN} \xrightarrow{\Delta} \underset{\text{Propanenitrile}}{CH_3CH_2CN} + KCl$$

37 (b)
$$2CH_3CH_2I + Ag_2O \xrightarrow{\Delta} \underset{\text{Diethyl ether}}{(CH_3CH_2)_2O} + 2AgI$$

38 (b) S_N2 mechanism involves the back side attack and formation of a transition state, thus less hindered alkyl halide readily undergoes S_N2 mechanism. Among the given, CH_3Br (methyl bromide) is less hindered, thus hydrolysed by S_N2 mechanism.

39 (b) The given reaction is an example of elimination reaction, in which one molecule of HBr is eliminated from alkyl halide to give alkene in the presence of alcoholic KOH.

CHAPTER 14

Haloarenes

Haloarenes are the aromatic compounds in which halogen replaces hydrogen atom(s) of benzene ring. These are also known as **aryl halide.** The carbon atom to which halogen is attached is sp^2-hybridised.

(where, X = Cl, Br, I, F)

These compounds find wide applications in industries as well as in day-to-day life. These are used as solvents, as antibiotic etc. For instance, chloroquine, a synthetic haloarene, is used for the treatment of malaria.

In this chapter, we will study the important methods of preparations, physical and chemical properties of haloarenes.

CLASSIFICATION

Haloarenes can be classified on the basis of number of halogen atoms. In other words, haloarenes may be classified as mono, di or polyhalogen depending on whether they contain one, two or more halogen atoms in their structure.

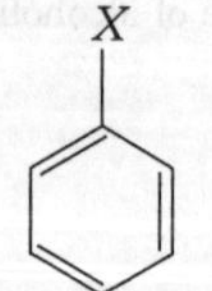
Monohaloarene

Dihaloarene

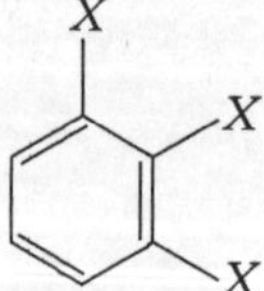
Trihaloarene

NOMENCLATURE

(*i*) Haloarenes are the common as well as IUPAC names of aryl halides. For dihalogen derivatives the prefixes, *o*-, *m*-, *p*- are used in common system but in IUPAC system, the numerals 1, 2; 1,3 and 1,4 are used.

- For mono-haloarenes,

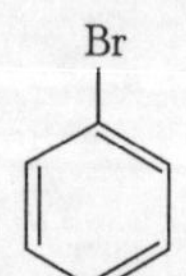
Bromobenzene

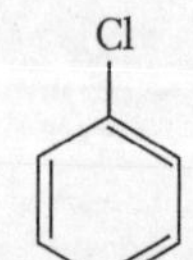
Chlorobenzene

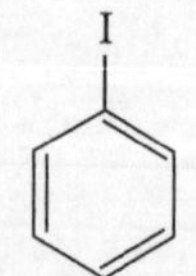
Iodobenzene

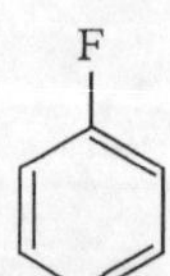
Fluorobenzene

Inside

- For di-haloarenes,

IUPAC name	1, 2-dibromobenzene	1, 3-dibromobenzene
Common name	*o*-dibromobenzene	*m*-dibromobenzene

1,4-dibromobenzene
p-dibromobenzene

(*ii*) Haloarenes are named alphabetically when two different halogen atoms are attached to a benzene ring.

IUPAC name	1-chloro-3-iodobenzene	1-bromo-3-chlorobenzene
Common name	*m*-chloroiodobenzene	*m*-bromochlorobenzene

Benzylic halides

These are the organohalogen compounds in which halogen atom is bonded to sp^3-hybridised C-atom next to an aromatic ring. These are also called **alkyl halides.**

e.g.

(Benzyl chloride) 1-chloro-1-phenyl methane

(Benzal chloride or benzyl dichloride) 1, 1-dichloro-1-phenyl methane

(Benzotrichloride) 1,1,1-trichloro-1-phenyl methane

These are considered as derivative of alkyl halides and while naming, the alkyl chain is considered as main chain as per the IUPAC rules.

e.g. $CH_3—CH(C_6H_5)—CH_2—Br$

1-bromo-2-phenylpropane

METHOD OF PREPARATION OF HALOARENES

Halogenation

Haloarenes can be prepared by electrophilic substitution of arenes with halogens in the presence of Lewis acid catalysts like iron or iron (III) chloride.

$$\text{Benzene} + X_2 \xrightarrow[\text{or anhyd. } AlCl_3]{Fe} \text{Haloarene} + HCl$$

$$\text{Toluene} + X_2 \xrightarrow[\text{Dark}]{Fe} o\text{-halotoluene} + p\text{-halotoluene}$$

Low temperature and the presence of Lewis acid that acts as a halogen carrier to generate an electrophile (halonium ion) favours electrophilic substitution. The chlorides or bromides of Al, Fe, Sb may be used as Lewis acids.

Mechanism

Various steps involve in mechanism are as follows

Step I Generation of electrophile

$$\underset{\text{Lewis acid}}{FeCl_3} + \underset{\text{Lewis base}}{Cl-Cl} \rightleftharpoons \underset{\text{Nucleophile}}{FeCl_4^{\ominus}} + \underset{\text{Electrophile}}{Cl^{\oplus}}$$

Step II Formation of σ-complex

$$C_6H_6 + Cl^{\oplus} \xrightarrow[\text{Rate determining step}]{\text{Slow}} \sigma\text{-complex}$$

σ-complex
(stabilised by resonance)

Resonance hybrid

Step III Loss of a proton from σ-complex

$$FeCl_4^- + H^+ \longrightarrow FeCl_3 + HCl$$

If excess of halogen is used, then dihalogen derivatives are produced.

Chlorobenzene + Cl_2(excess) $\xrightarrow{\text{Anhyd. } AlCl_3}$ *o*-dichloro benzene (minor) + *p*-dichloro benzene (major)

This method is very fast in presence of strongly activating group, i.e. $-OCH_3$ group. The anisole quickly brominates in water without any catalyst.

Anisole $\xrightarrow[H_2O]{Br_2}$ (2,4,6-tribromoanisole) + HBr

Direct **iodination** is not possible since, iodine is least reactive and HI formed makes reaction reversible.
In presence of oxidising agents like HNO_3 or HIO_3, iodination of benzene is possible and HI formed is converted to I_2.

$$\frac{1}{2}I_2 + HNO_3 + C_6H_6 \xrightarrow{[O]} C_6H_5I + H_2O + NO_2$$

$$5HI + HIO_3 \longrightarrow 3I_2 + 3H_2O$$

Iodination probably involves an electrophilic aromatic substitution with iodonium ion (I^+) acting as the electrophile. I^+ is formed from iodine by oxidation with HNO_3

$$H^+ + HNO_3 + \frac{1}{2}I_2 \longrightarrow I^+ + NO_2 + H_2O$$

Fluorination is not possible as reaction is highly explosive.

From benzene diazonium salts

Haloarenes are commonly prepared by replace the nitrogen of a diazonium salt. This method for the preparation of haloarene is more important than direct halogenation because fluorides and iodides can rarely be prepared by direct halogenation and can easily be prepared from diazonium salts.

Some reactions are as follows

Sandmeyer's reaction

When primary aromatic amine dissolved or suspended in cold aqueous mineral acids followed by treatment with sodium nitrite then diazonium salt is formed. This benzene diazonium salt is treated with cuprous chloride or cuprous bromide results in the replacement of the diazonium group by —Cl or —Br. This reaction is called **Sandmeyer's reaction**.

Aniline $\xrightarrow[273\text{-}278\text{ K}]{NaNO_2 + HX}$ Benzene diazonium halide ($C_6H_5\overset{+}{N_2}\overset{-}{X}$) $\xrightarrow{Cu_2X_2}$ Haloarene (X = Cl, Br) + N_2

Gattermann reaction

It is a modification of Sandmeyer reaction. In this reaction, Cu powder and HCl or HBr is used for preparing chlorobenzene and bromobenzene. The yield is only 40%.

Bromobenzene + $N_2\uparrow$ $\xleftarrow[\Delta]{Cu/HBr}$ Benzene diazonium chloride ($C_6H_5N_2^+Cl^-$) $\xrightarrow[\Delta]{Cu/HCl}$ Chlorobenzene + $N_2\uparrow$

Replacement of the diazonium group by iodine does not require the presence of cuprous halide and is done simply by shaking the diazonium salt with potassium iodide.

$$C_6H_5N_2^+Cl^- \xrightarrow{KI} C_6H_5I + N_2\uparrow$$

Balz-Schiemann reaction

It involves the conversion of aryl amines to aryl fluorides *via* diazotisation and subsequent thermal decomposition of the derived tetrafluoroborates.

Benzene diazonium ($C_6H_5N_2^+X^-$) $\xrightarrow[\text{(Sandmeyer)}]{HBF_4}$ $C_6H_5N_2^+BF_4^-$ $\xrightarrow{\Delta}$ Fluorobenzene + $N_2\uparrow$ + BF_3

Other methods

Borodine-Hunsdiecker reaction

Silver salts of aromatic acids on treatment with Br_2 in CCl_4 yields aryl bromides.

$$C_6H_5-\overset{-}{COO}\overset{+}{Ag} + Br_2 \xrightarrow[CCl_4]{\Delta} C_6H_5-Br + AgBr + CO_2$$

Decarboxylation of halogenated acids

When salts of *o*-, *m*- and *p*- halobenzoic acid react with NaOH in presence of $CaCO_3$ haloarenes are formed.

$$X-C_6H_4-\overset{-}{COO}\overset{+}{Na} + NaOH \xrightarrow[\Delta]{CaCO_3} C_6H_5X + Na_2CO_3 + CO_2\uparrow$$

From phenol

Unlike aliphatic alcohols, aromatic hydroxy compound, i.e. phenols do not undergo simple substitution of —OH by —Cl on reaction with HCl or $SOCl_2$ or PCl_3 but with PCl_5 or PBr_5, a poor yield of chlorobenzene and halobenzene is obtained.

This is because $POCl_3$ or $POBr_3$ obtained further reacts with phenol to form a main product, triphenyl phosphate.

$$\underset{\text{Phenol}}{C_6H_5OH} + PCl_5 \xrightarrow{\text{High temp.}} \underset{\text{Less}}{C_6H_5Cl} + POCl_3 + HCl$$

$$POCl_3 + 3\underset{\text{or } (C_6H_5-OH)}{C_6H_5OH} \longrightarrow \underset{\text{Triphenyl phosphate (Major)}}{(C_6H_5)_3PO_4} + 3HBr$$

Raschig process

This is a commercial method of manufacturing chlorobenzene.

$$\underset{\text{Benzene}}{C_6H_6} + HCl + \frac{1}{2}O_2 \xrightarrow[\text{Heat}]{CuCl} \underset{\text{Chloro benzene}}{C_6H_5-Cl} + H_2O$$

CHECK POINT 14.1

1. Which of the following represents aryl alkyl halide?
   (a) *o*-chlorotoluene (b) 1-chloro-2-phenyl ethane
   (c) *o*-bromochlorobenzene (d) None of these

2. Consider the following compound

   (benzene ring with Br at positions 1 and 3)

   The IUPAC nomenclature is
   (a) 1, 3-dibromobenzene (b) 1, 6-dibromobenzene
   (c) *o*-dibromobenzene (d) None of these

3. In the chlorination of benzene in the presence of ferric chloride, function of $FeCl_3$ is
   (a) to provide halogen (b) to form electrophile Cl^+
   (c) resonance stabilisation (d) as catalyst

4. Consider the following reaction,
   $$C_6H_6 + HNO_3 + I_2 \xrightarrow{\Delta}$$
   (a) nitrobenzene (b) iodobenzene
   (c) benzoic acid (d) *p*-iodonitrobenzene

5. The reaction involving the treatment of benzene diazonium chloride with copper powder and HCl is treated as
   (a) Sandmeyer's reaction
   (b) Gattermann's reaction
   (c) Ullmann's reaction
   (d) Kolbe's reaction

6. Chlorobenzene is prepared commercially by
   (a) Dow's process (b) Deacon's process
   (c) Raschig process (d) Etard's process

7. Which of the following is not an example of Sandmeyer's reaction?
   (a) $C_6H_5\overset{+}{N}_2\overset{-}{Cl} \xrightarrow{CuCl} C_6H_5Cl$
   (b) $C_6H_5\overset{+}{N}_2\overset{-}{Cl} \xrightarrow{CuBr} C_6H_5Br$
   (c) $C_6H_5\overset{+}{N}_2\overset{-}{Cl} \xrightarrow{CuCN/KCN} C_6H_5CN$
   (d) $C_6H_5\overset{+}{N}_2\overset{-}{Cl} \xrightarrow{KI/Warm} C_6H_5I$

8. In order to prepare fluorobenzene from benzene chloride which of the following reagent is used?
   (a) Fluorine (b) HF
   (c) Hydrofluorosilic acid (d) Fluoroboric acid

9. The yield of chlorobenzene obtained by reaction of phenols with PCl_5 is less due to formation of
   (a) *o*-chlorophenol
   (b) *p*-chlorophenol
   (c) phosphorus oxychloride
   (d) triphenyl phosphate

10. Chlorobenzene can be prepared by reacting aniline with
    (a) hydrochloric acid
    (b) cuprous chloride
    (c) chlorine in presence of anhydrous aluminium chloride
    (d) ice cold nitrous acid followed by treatment with cuprous chloride and HCl.

PHYSICAL PROPERTIES

1. Physical state

Haloarenes are crystalline solids, less polar compounds, insoluble in water (due to incapability of forming H-bonds) but soluble in organic solvents like ethanol and ether.

2. Boiling point

For the same aryl group, the boiling point increases as the size of the halogen atom increases. Similarly, for the halogen atom, boiling point increases as the size of aryl group increases.

$C_6H_5I > C_6H_5Br > C_6H_5Cl > C_6H_5F$

Boiling point 188°C 156°C 132°C 85°C

The boiling point of isomeric dichlorobenzenes are nearly same.

3. Melting point

When aryl group is same, melting point increases as the size of halogen atom increases. When halogen atom is same, melting point increases as the size of aryl group increases.

For dihalobenzene, melting point trend is different. In general, melting point of the *p*-isomer is always higher than that of *o*- and *m*-isomers because *p*-isomer is more symmetrical and hence, its molecules pack closely in the crystal lattice due to which intermolecular forces of attraction are stronger in *p*-isomer.

4. Dipole moment

The dipole moments of haloarenes increases in the following order

fluorobenzene < chlorobenzene < bromobenzene < iodobenzene.

The dipole moment of fluorobenzene is lowest due to the small size of fluorine. Amongst isomeric dihalobenzenes, the *p*-isomer being symmetrical has zero dipole moment while *o*-isomer has greater dipole moment than *m*-isomer.

$$\mu = 2\mu_{C-X} \cos\left(\frac{\theta}{2}\right)$$

where, μ_{C-X} is vector moment and θ is the angle,

$$\mu \propto \cos\left(\frac{\theta}{2}\right)$$

Isomer	θ
o-	60°
m-	120°
p-	180°

5. Stability

Stability of haloarenes having the same aryl group decreases as follows

aryl fluorides > aryl chlorides > aryl bromides > aryl iodides

This is because the strength of C—*X* bond decreases in the order C—F > C—Cl > C—Br > C—I.

CHEMICAL PROPERTIES

Haloarenes give various types of chemical reactions. This is due to the presence of C—*X* bond in aryl halides that have double bond character due to conjugation between lone pair of halogen and benzene ring.

Nucleophilic substitution reactions

Aryl halides are less reactive than alkyl halides toward nucleophilic substitution reactions due to following reasons

1. Resonance effect

The C—*X* bond of aryl halide is more stronger due to double bond character. The lone pairs of halogen atom are delocalised on the benzene ring.

$$C_6H_5X \xrightarrow{Nu^{\ominus}} \text{No reaction}$$

2. Hybridisation state of C-atom in C—*X* bond

In haloarenes, the C-atom attached to halogen is sp^2-hybridised whereas in alkyl halides, the C-atom attached to halogen is sp^3-hybridised.

sp^2-hybridised carbon | sp^3-hybridised carbon

C_6H_5-X (*X* more electronegative than carbon) $X-CH_3$

R— *X*	Carbon atom of (C—*X*)	*s*-character	C—Cl bond length
Aryl halide	sp^2	33.3%	169 pm
Alkyl halide	sp^3	25.0%	177 pm

Greater the *s*-character, greater is the electronegativity of carbon and thus, bonding pair of (C—*X*) is held tightly and bond length is shortened. Since, it is difficult to break a shorter bond than a longer bond, therefore, haloarenes are less reactive than haloalkanes toward S_N reaction.

3. Electronic repulsions

Arenes are electron-rich and when strong nucleophiles approach aryl halides, repulsion takes place making S_N reactions impossible.

4. Polarity of C—X bond

The C—*X* bond of haloarenes is less polar than that in haloalkanes because of more electronegative nature of sp^2-hybridised C-atom of C—*X* bond. Thus, C—*X* bond is tightly bonded and it has a less tendency to release electrons to the halogen than sp^3-hybrid carbon in alkyl halides. Consequently, the halogen atom present in aryl halides cannot be displaced by nucleophiles.

Because of the above discussed factors, haloarenes do not give the normal reactions of X^-. They undergo nucleophilic substitution only with extreme difficulty.

5. Presence of electron withdrawing group (EWG)

The presence of an electron withdrawing group at *ortho* and *para*-positions increases the reactivity of haloarenes. This effect is more pronounced when $—NO_2$ group is introduced at *o*- and *p*- position. However, no effect on reactivity of haloarenes is observed by the presence of electron withdrawing group at *meta*-position. The reason can be easily understood by the following mechanism.

(*i*) **Attack at *p*-position**

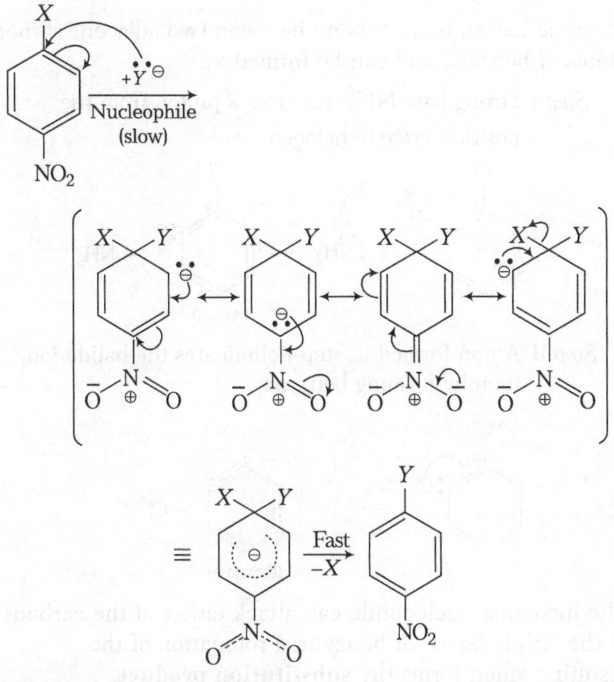

$\xrightarrow[\text{(slow)}]{+\ddot{Y}^{\ominus}\ \text{Nucleophile}}$

$\xrightarrow[-X]{\text{Fast}}$

(*ii*) **Attack at *o*-position**

$\xrightarrow{\text{Fast}}$

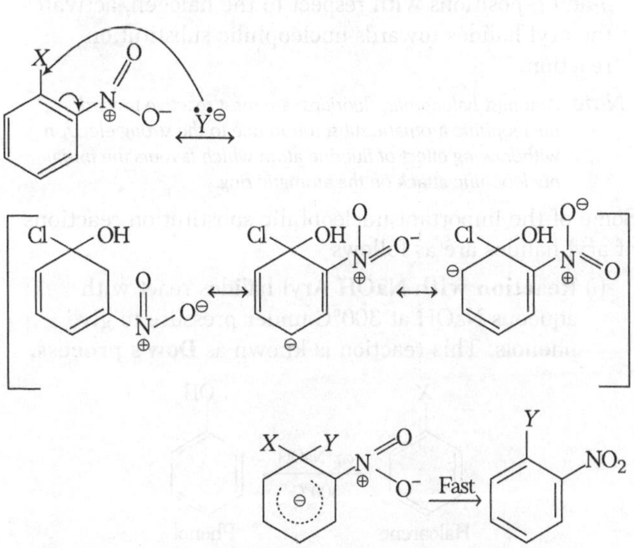

In both *p*-and *o*-positions, one of the structure carries a negative charge on the carbon atom bearing the NO_2 group. Thus, carbanions are stabilised by the $—NO_2$ group as well as π-electron of the benzene ring. Hence, carbanion formed is highly stable.

(*iii*) **Attack at *m*-position**

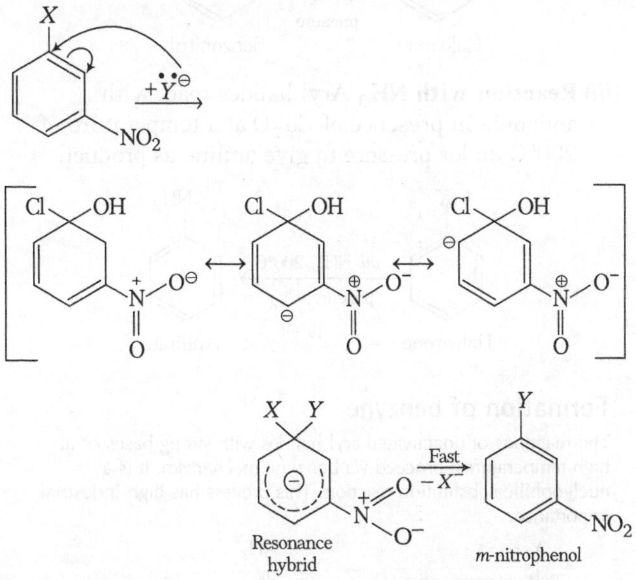

$\xrightarrow[-X]{\text{Fast}}$

Resonance hybrid

m-nitrophenol

In this case, none of the structures carries a negative charge on the carbon atom bearing NO_2 group, carbanion is not stabilised by nitro group but by the π-electrons of the benzene ring. Hence, the carbanion formed in this case is not highly stable as that in *o*- and *p*-cases.

Hence, the presence of electron withdrawing groups *o*-and *p*-positions with respect to the halogen, activates the aryl halides towards nucleophilic substitution reaction.

Note *Amongst haloarenes, fluorides are most reactive towards nucleophilic aromatic substitution due to the strong electron withdrawing effect of fluorine atom which favours the initial nucleophilic attack on the aromatic ring.*

Some of the important nucleophilic substitution reactions of aryl halides are as follows

(*i*) **Reaction with NaOH** Aryl halides react with aqueous NaOH at 300°C under pressure to give phenols. This reaction is known as **Dow's process.**

Haloarene (X) $\xrightarrow[300°C]{aq.\ NaOH}$ Phenol (OH)

(*ii*) **Reaction with CuCN** Aryl halides on reaction with cuprous cyanide at 470 K, yield benzonitrile (or cyanobenzene).

Haloarene (X) $\xrightarrow[470\ K,\ pressure]{CuCN}$ Benzonitrile (CN)

(*iii*) **Reaction with NH_3** Aryl halides react with ammonia in presence of Cu_2O at a temperature of 200°C under pressure to give aniline as product.

Haloarene (X) $\xrightarrow[pressure,\ Cu_2O]{aq.\ NH_3,\ 200°C}$ Aniline (NH_2)

Formation of benzyne

The reactions of unactivated aryl halides with strong bases or at high temperatures, proceed *via* benzyne mechanism. It is a nucleophilic substitution reaction. This process has high industrial importance.

X (*) $+ \overset{\ominus}{N}H_2 \xrightarrow{NH_3\ (liquid)}$ NH_2 (*) + (*) NH_2

(Approximately in equal amounts)

Since, benzyne formation involves an extra π-bond between two adjacent carbon atoms of benzene. Therefore, practically it is not possible. Hence, benzyne cannot be isolated.

Elimination-addition through benzyne intermediate

An aryl halide can undergo a nucleophilic substitution reaction in the presence of a very strong base such as $NH_2^{\ominus}$. There are two surprising features about this reaction the aryl halide does not have an electron withdrawing group and the incoming substituent does not always end up on the carbon vacated by the leaving group.

When chlorobenzene that has the carbon to which chlorine is attached isotopically labelled with ^{14}C is treated with amide ion in liquid ammonia, aniline is obtained as a product.

Half of the product has the amino group attached to the isotopically labelled carbon (*) as expected, but the other half has the amino group attached to the carbon adjacent to the labelled carbon.

Cl (*) $+ \overset{\ominus}{N}H_2 \xrightarrow{NH_3\ (liquid)}$ NH_2 (*) + (*) NH_2

(Approximately in equal amounts)

Above reaction involves formation of a **benzyne intermediate** which has two equivalent carbon atoms to which amino group can be attached.

Benzyne has an extra π-bond between two adjacent carbon atoms of benzene and can be formed as,

Step I Strong base $NH_2^{\ominus}$ removes a proton from the position *ortho* to halogen

Cl (*), H $+ :\overset{\ominus}{N}H_2 \longrightarrow$ Cl (*), $:^{\ominus}$ $+ NH_3$

(Anion)

Step II Anion formed in step I eliminates the halide ion, thereby forming benzyne

Cl (*), $:^{\ominus}$ $\longrightarrow$ (*) $+ Cl^-$

(Benzyne)

The incoming nucleophile can attack either of the carbons of the "triple bond" of benzyne. Protonation of the resulting anion forms the **substitution product.**

The overall reaction is an elimination-addition reaction; benzyne is formed in an elimination reaction and immediately undergoes an addition reaction.

Attack at C* (Labelled)

Attack at C (Unlabelled)

Benzyne

Substitution at the carbon (*) that was attached to the leaving group is called direct substitution. Substitution at the adjacent carbon is called cine substitution.

o-bromotoluene + $NaNH_2$ $\xrightarrow{\text{Liquid } NH_3}$ o-toluidine (by direct substitution) + m-toluidine (by cine substitution)

Electrophilic substitution reactions

The aromatic ring to which halogen is attached can undergo the typical electrophilic aromatic substitution reactions such as nitration, sulphonation, halogenation, Friedel-Craft alkylation. Like any other substituent, halogen also affects the reactivity and orientation in these reactions.

Here, halogen is unusual in being deactivating (due to –*I*-effect) yet *ortho* and *para*-directing (electron density is high). Due to this reaction, haloarenes are less reactive than benzene towards electrophilic substitution reactions.

General reaction

Haloarene $+ E^{\oplus} \rightarrow$ … $\rightarrow$ o-form + p-form

Mechanism

The mechanism of electrophilic substitution reaction has been discussed below

***Step* I Generation of an electrophile**

First of all an electrophile is generated with the help of catalyst.

$$E - Nu \longrightarrow E^{+} + :Nu^{-}$$

***Step* II Formation of carbocation intermediate**

The electrophile E^{+} approaches the π-electron cloud of the aromatic ring and forms a bond with carbon, creating a positive charge on the ring.

This results in the formation of a sigma complex.

***Step* III Removal of proton**

Some important electrophilic substitution reactions of haloarenes are as follows

Chlorobenzene

Conc. HNO_3 / conc. H_2SO_4 (Nitration) → o- chloro nitrobenzene + p- chloro nitrobenzene

Cl_2, Fe (Halogenation) → o-dichlorobenzene + p-dichlorobenzene

H_2SO_4 (Sulphonation) → o-chlorosulphonic acid + p-chlorosulphonic acid

CH_3Cl in presence of $AlCl_3$ (Friedel-Craft alkylation) → o-chlorotoluene + p- chlorotoluene

$CH_3COCl + AlCl_3$ (Friedel-Craft alkylation) → o-chloro acetophenone + p-chloro acetophenone

Reaction with metals

Various types of reactions of haloarenes with metals are discussed below

1. Reaction with Mg

Bromo and iodoarenes can be prepared from Grignard reagents when their ethereal solution is treated with magnesium.

Chloroarenes form Grignard reagents only if the reaction is carried out in dry tetrahydrofuran (THF) as a solvent.

$$C_6H_5-Br + Mg \xrightarrow[\Delta]{\text{Dry ether}} C_6H_5-MgBr$$

$$C_6H_5-Cl + Mg \xrightarrow[\Delta]{\text{Dry THF}} C_6H_5-MgCl$$

2. Ullmann reaction

Reaction of iodoarene with copper powder in a sealed tube to form diaryl is called Ullmann diaryl synthesis.

$$2\,C_6H_5-I + 2Cu \xrightarrow{\Delta} C_6H_5-C_6H_5$$

Biphenyl

$$2\,C_6H_5-CH_2Cl + Cu \xrightarrow{\text{Heat}} C_6H_5-CH_2-CH_2-C_6H_5 + CuCl_2$$

Bibenzyl

Aryl chlorides and bromides also react with Cu when some electronegative substituent (such as $-NO_2$) is attached which has ability to activate the halogen present.

$$2\,(o\text{-}O_2N-C_6H_4-Cl) \xrightarrow[\text{Dimethyl formamide}]{Cu/\Delta} O_2N-C_6H_4-C_6H_4-NO_2$$

3. Wurtz Fittig reaction

Reaction of aryl halide and alkyl halide with sodium metal in the presence of dry ether is called **Wurtz Fittig reaction.**

$$C_6H_5Cl + CH_3Cl + 2Na \xrightarrow{\text{Ether}} C_6H_5CH_3$$

Aryl halide; Alkyl halide; Toluene

$$C_6H_5CH_2Cl + C_6H_5Cl + 2Na \xrightarrow{\text{Ether}} C_6H_5CH_2C_6H_5$$

Alkyl halide (side chain); Aryl halide; Diphenyl methane

4. Fittig reaction

Reaction of aryl halides with sodium metal in the presence of dry ether to form biphenyl is called Fittig reaction.

$$2C_6H_5Cl + 2Na \xrightarrow{\text{Ether}} C_6H_5-C_6H_5$$

Chlorobenzene; Biphenyl

DDT (*p,p'*-dichloro diphenyl trichloroethane)

DDT, the first chlorinated organic insecticide, was originally prepared in 1873. **Paul Muller** was awarded the Nobel prize in medicine and physiology in 1948 for this discovery.

$$CCl_3-CH(C_6H_4Cl)_2$$

(DDT)

It is formed, when chlorobenzene reacts with CCl_3CHO (chloral) in presence of conc. H_2PO_4.

It is a white powder, insoluble in water but soluble in oil.

Uses

(*i*) DDT is a cheap but powerful insecticide.

(*ii*) It is popularly very effective against anopheles mosquitoes (causes malaria) and lices which carry typhus.

Harmful effects

DDT is non-biodegradable and extremely stable compound which acts as water pollutant and kills aquatic animals thus imbalancing the water ecosystem. When higher animals (including humans) feed on dead fishes killed by DDT, it enters in their food chains.

DDT gets stored in fat tissues of animal and increases in amount over the time. Researches have shown that such deposition in long run may cause cancer and other harmful diseases. Hence, DDT has been banned in many countries

CHECK POINT 14.2

1. Which among the following possess highest boiling point?

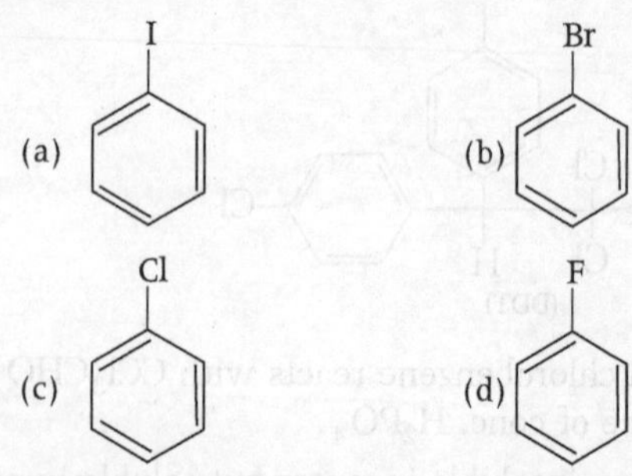

2. The correct trend for the dipole moment of haloarenes is
(a) fluorobenzene < chlorobenzene < bromobenzene
(b) chlorobenzene < fluorobenzene < bromobenzene
(c) bromobenzene < fluorobenzene < chlorobenzene
(d) chlorobenzene < bromobenzene < fluorobenzene

3. Which of the following statements is correct?
(a) Haloarenes are liquids at room temperature
(b) Haloarenes are insoluble in water but soluble in alcohols
(c) The boiling point of isomeric dichlorobenzenes are same
(d) Melting point of haloarenes increases as the size of the halogen atom increases

4. Consider the following reaction,

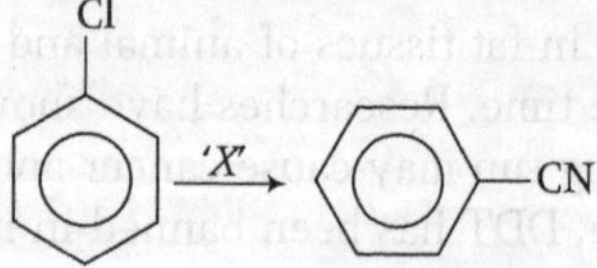

Here, 'X' is
(a) CuCN (b) KCN
(c) AgCN (d) CuCN, 470 K, pressure

5. Consider the following reaction,

Chlorobenzene $\xrightarrow[AlCl_3]{CH_3Cl}$ Product

What is/are the major product in the reaction?
(a) *o*-chlorotoluene (b) *p*-chlorotoluene
(c) *m*-chlorotoluene (d) Both (a) and (b)

6. Iodobenzene can be converted into diphenyl by
(a) Wurtz reaction
(b) Wurtz Fittig reaction
(c) Ullmann reaction
(d) Frankland reaction

7. Find the major product in the following reaction

$C_6H_5CCl_3 \xrightarrow[FeCl_3]{Cl_2}$

(a) o-chloro $C_6H_4CCl_3$ (b) m-chloro $C_6H_4CCl_3$
(c) p-chloro $C_6H_4CCl_3$ (d) benzene

8. Consider the following reaction,

$C_6H_5Cl \xrightarrow{2Na + ether} A$

In this reaction, *A* is
(a) phenol (b) sodium phenoxide
(c) benzol (d) biphenyl

9. Which of the following will give yellow precipitate on shaking with an aqueous solution of NaOH followed by acidification with dil. HNO_3 and addition of $AgNO_3$ solution?

(a) p-Br–C_6H_4–Cl (b) C_6H_5–CH_2—Cl
(c) p-Br–C_6H_4–CH_2–I (d) o-Cl–C_6H_4–CH_3

10. Which of the following statements is incorrect?
(a) DDT is non-biodegradable
(b) DDT get stored in fat tissues of the animal
(c) DDT acts as water pollutant
(d) DDT is not effective against anopheles mosquitoes

Chapter Exercises

(A) Topical exercise

I. Classification and nomenclature of haloarene

1 Compare bond lengths (x) and (y) in

(Benzene ring with —Cl attached by bond x and —CH_2—Cl with bond y)

(a) $x = y$ (b) $x < y$
(c) $x > y$ (d) nothing can be predicted

2 Consider the following statements.

Statement I (C—Cl) bond length in C_6H_5Cl is shorter than that in CH_3—Cl.

Statement II Both C-atoms are sp^3-hybridised.

(a) Both statements I and II are correct.
(b) Only statement I is correct.
(c) Only statement II is correct.
(d) Both statements I and II are incorrect.

3 Match the structures of compound given in Column I with the classes of compound given in Column II.

	Column I	Column II
A.	CH_3—CH(X)—CH_3	1. Aryl halide
B.	CH_2=CH—CH_2—X	2. Alkyl halide
C.	C_6H_5X (benzene ring with X)	3. Vinyl halide
D.	CH_2= CH—X	4. Allyl halide

Codes

	A	B	C	D		A	B	C	D
(a)	2	4	1	3	(b)	1	4	2	3
(c)	3	4	2	1	(d)	1	2	3	4

4 Benzylidene chloride is

(a) $C_6H_5CH_2Cl$ (b) $C_6H_5CHCl_2$
(c) $C_6H_4ClCH_2Cl$ (d) $C_6H_5CCl_3$

5 Number of π-bonds present in BHC (benzene hexa chloride) are

(a) 6 (b) Zero
(c) 3 (d) 12

6 The common and IUPAC names of the compound

(Benzene ring with Br at positions 1, 3 and 5)

respectively are

(a) sym-tribromobenzene and 1,2,5 tribromobenzene
(b) sym-tribromobenzene and 1,3,5-tribromobenzene
(c) *o, m, p*-tribromobenzene and 1,3,5-tribromobenzene
(d) None of the above

II. Preparation of haloarene

7 Following reaction is called

$$C_6H_5\text{—}\overset{-}{COO}\overset{+}{Ag} + Br_2 \xrightarrow[CCl_4]{\Delta} C_6H_5\text{—}Br + AgBr + CO_2$$

(a) Hunsdiecker reaction
(b) Raschig process
(c) Wurtz reaction
(d) Fittig reaction

8 Which of the following cannot be prepared by direct halogenation of benzene in presence of Lewis acid?

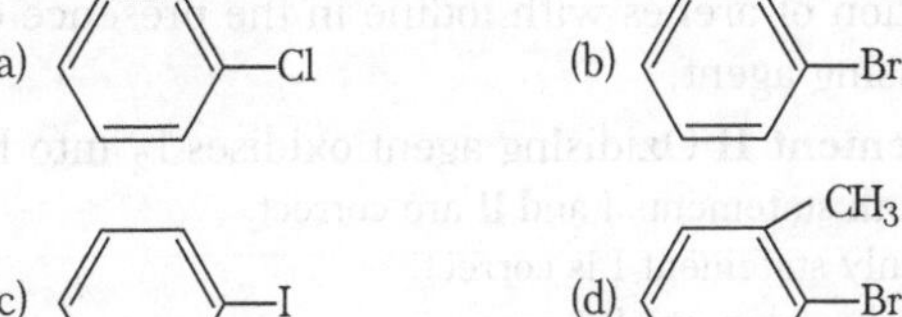

(a) C_6H_5—Cl (b) C_6H_5—Br
(c) C_6H_5—I (d) CH_3, Br substituted benzene

9 σ-complex formed in the chlorination of benzene is

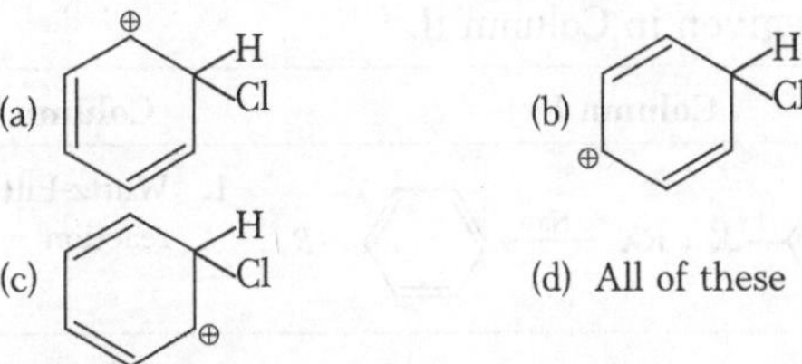

(a) (b)
(c) (d) All of these

10 Chlorobenzene is formed by reaction of chlorine with benzene in the presence of $AlCl_3$. Which of the following species attacks the benzene ring in this reaction? **[NCERT Exemplar]**

(a) Cl^-
(b) Cl^+
(c) $AlCl_3$
(d) $[AlCl_4]^-$

11 The reaction,

$C_6H_5\overset{\oplus}{N_2}\overset{\ominus}{Cl} + CuCl \xrightarrow[HCl]{\Delta} C_6H_5Cl + N_2 + CuCl$

is known as

(a) Wurtz (b) Sandmeyer
(c) Gattermann (d) Friedel-Craft

12 Identify the compound Y in the following reaction. **[NCERT Exemplar]**

$C_6H_5NH_2 \xrightarrow[273\text{-}278\ K]{NaNO_2 + HCl} C_6H_5\overset{+}{N_2}\bar{Cl} \xrightarrow{Cu_2Cl_2} Y + N_2$

(a) chlorobenzene (b) benzene
(c) 1,3-dichlorobenzene (d) 1,4-dichlorobenzene

13 Consider the following statements.

Statement I Aryl iodides can be prepared by reaction of arenes with iodine in the presence of an oxidising agent.

Statement II Oxidising agent oxidises I_2 into HI.

(a) Both statements I and II are correct.
(b) Only statement I is correct.
(c) Only statement II is correct.
(d) Both statements I and II are incorrect.

14 Match the reactions given in Column I with the names given in Column II.

	Column I	Column II
A.	$C_6H_5-X + RX \xrightarrow{Na} C_6H_5-R$	1. Wurtz-Fittig reaction
B.	$C_6H_5\overset{+}{N_2}\bar{X} \xrightarrow{Cu_2X_2} C_6H_5X + N_2$	2. Finkelstein reaction
C.	$C_2H_5Cl + NaI \xrightarrow{\text{Dry acetone}} C_2H_5I + NaCl$	3. Sandmeyer reaction

Codes

	A	B	C		A	B	C
(a)	1	3	2	(b)	1	2	3
(c)	2	3	1	(d)	3	2	1

15 Toluene reacts with a halogen in the presence of iron (III) chloride giving *ortho* and *para* halo compounds. The reaction is **[NCERT Exemplar]**

(a) electrophilic elimination reaction
(b) electrophilic substitution reaction
(c) free radical addition reaction
(d) nucleophilic substitution reaction

16 Toluene is nitrated and the resulting product is reduced with tin and hydrochloric acid. The product so obtained is diazotised and then heated with cuprous bromide. The reaction mixture so formed contains

(a) mixture of *o*-and-*p*-bromotoluenes
(b) mixture of *o*-and-*p*-dibromobenzene
(c) mixture of *o*-and-*p*-bromoaniline
(d) mixture of *o*-and-*m*-bromotoluenes

III. Physical properties

17 Which of the following possesses highest melting point?

(a) *o*- dichlorobenzene
(b) *m*-dichlorobenzene
(c) *p*-dichlorobenzene
(d) All of the above have same melting point

18 Which has maximum dipole moment?

(a) 1,2-dichlorobenzene (b) 1,3-dichlorobenzene
(c) 1,4-dichlorobenzene (d) benzene

19 Highest dipole moment is of the

(a) 1,2-dichlorobenzene (b) 1,3-dibromobenzene
(c) 1,3-dichlorobenzene (d) 1-bromo-2-chlorobenzene

20 Arrange the following compounds in the increasing order of their densities. **[NCERT Exemplar]**

(i) 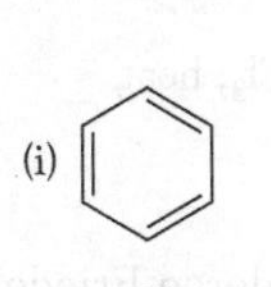

(ii) Cl

(iii) Cl, Cl

(iv) Br, Cl

(a) (i) < (ii) < (iii) < (iv)
(b) (i) < (iii) < (iv) < (ii)
(c) (iv) < (iii) < (ii) < (i)
(d) (ii) < (iv) < (iii) < (i)

21 Consider the following statements.

Statement I *p*-dichlorobenzene is less soluble in organic solvents than the corresponding *o*-isomer.

Statement II *o*-dichlorobenzene is polar while *p*-dichlorobenzene is non-polar.

(a) Both statements I and II are correct.
(b) Only statement I is correct.
(c) Only statement II is correct.
(d) Both statements I and II are incorrect.

IV. Chemical properties and DDT

22 Which Cl is most easily replaced under S_N reaction?

Cl(1), Cl(2), Cl(4)

(a) Cl(1)
(b) Cl(2)
(c) Cl(4)
(d) Cl(2) and Cl(4)

23 S_N reaction is at the fastest rate in

(a) CH_3, H_3C, Cl, CH_3
(b) NO_2, O_2N, Cl, NO_2
(c) NO_2, O_2N, Cl
(d) NO_2, Cl, NO_2

24 Arrange the following in increasing order of reactivity in aromatic nucleophilic substitution?

I. 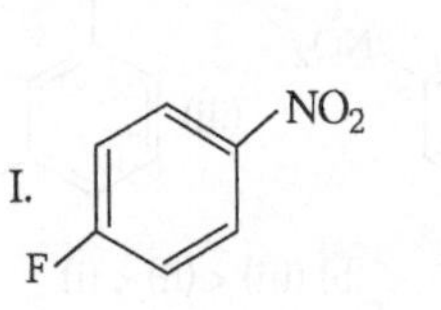

II. Cl, NO_2, NO_2

III. 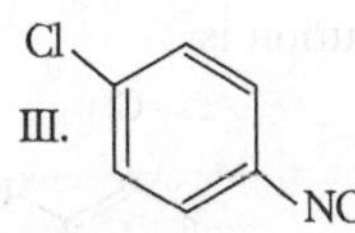

IV. 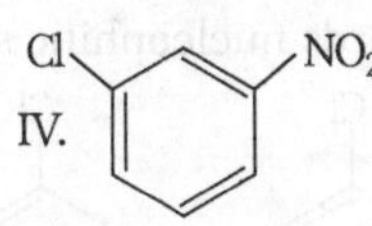

(a) I < IV < III < II
(b) IV < III < I < II
(c) I < IV < II < III
(d) IV < I < II < III

25 Consider the following statements.

Statement I Reactivity in chlorobenzene is controlled by stronger inductive effect and orientation is controlled by resonance effect.

Statement II The resonance effect tends to oppose the inductive effect for the attack at *o*-and *p*-positions.

(a) Both statements I and II are correct.
(b) Only statement I is correct.
(c) Only statement II is correct.
(d) Both statements I and II are incorrect.

26 The correct increasing order of rate of reaction towards nucleophilic substitution is

(i) 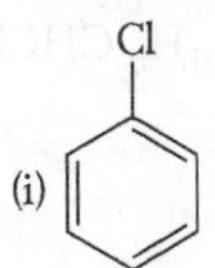

(ii) Cl, CH_3

(iii) 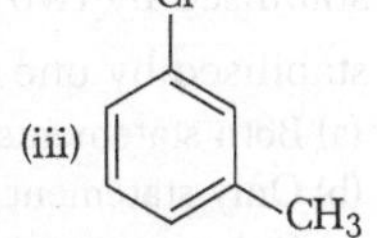

(a) (i) < (ii) < (iii)
(b) (i) < (iii) < (ii)
(c) (iii) < (ii) < (i)
(d) (ii) < (iii) < (i)

27 The correct increasing order of rate of reaction towards nucleophilic substitution is

(i) 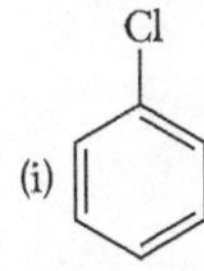

(ii) 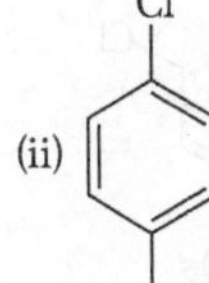

(iii) 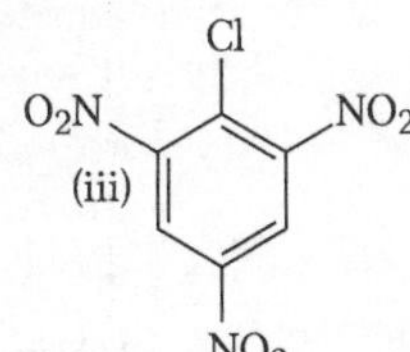

(a) (iii) < (ii) < (i)
(b) (ii) < (iii) < (i)
(c) (i) < (iii) < (ii)
(d) (i) < (ii) < (iii)

28 The correct increasing order of rate of reaction towards nucleophilic substitution is,

(i) Cl (ii) Cl, NO_2 (iii) Cl, NO_2

(a) (i) < (ii) < (iii) (b) (iii) < (ii) < (i)
(c) (i) < (iii) < (ii) (d) (iii) < (i) < (ii)

29 The correct increasing order of rate of reaction towards nucleophilic substitution is,

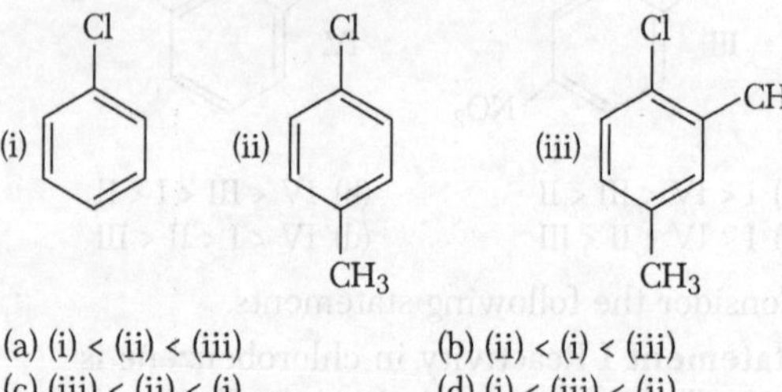

(a) (i) < (ii) < (iii) (b) (ii) < (i) < (iii)
(c) (iii) < (ii) < (i) (d) (i) < (iii) < (ii)

30 Consider the following statements.

Statement I $C_6H_5\underset{\displaystyle Br}{\underset{|}{C}}HC_6H_5$ undergoes S_N1 reaction at a faster rate than $C_6H_5\underset{\displaystyle Br}{\underset{|}{C}}HCH_3$.

Statement II Intermediate $C_6H_5\overset{\oplus}{C}HC_6H_5$ is stabilised by two phenyl rings while $C_6H_5\overset{\oplus}{C}HCH_3$ is stabilised by one phenyl ring.

(a) Both statements I and II are correct.
(b) Only statement I is correct.
(c) Only statement II is correct.
(d) Both statements I and II are incorrect.

31 Which one of the following is the best synthesis of 2-chloro-4-nitrobenzoic acid?

CO_2H, Cl, NO_2

2-chloro-4-nitrobenzoic acid

(a) 1. Heat benzoic acid with HNO_3, H_2SO_4
2. Cl_2, $FeCl_3$, heat

(b) 1. Treat toluene with HNO_3, H_2SO_4
2. $K_2Cr_2O_7$, H_2O, H_2SO_4, heat
3. Cl_2, $FeCl_3$, heat

(c) 1. Treat toluene with HNO_3, H_2SO_4
2. Cl_2, $FeCl_3$, heat
3. $K_2Cr_2O_7$, H_2O, H_2SO_4, heat

(d) 1. Treat nitrobenzene with Cl_2, $FeCl_3$, heat
2. CH_3Cl, $AlCl_3$
3. $K_2Cr_2O_7$, H_2O, H_2SO_4, heat

32 Which of the following will not undergo Friedel-Crafts reaction with chlorobenzene in presence of $FeCl_3$?

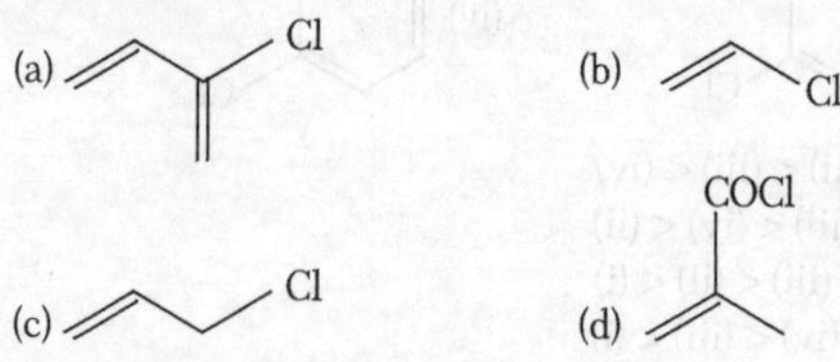

33 Consider the following statements.

Statement I Chlorine in chlorobenzene is deactivating group.

Statement II The inductive effect is stronger than resonance effect and causes net electron withdrawal and thus net activation.

(a) Both statements I and II are correct.
(b) Only statement I is correct.
(c) Only statement II is correct.
(d) Both statements I and II are incorrect.

34 Which one of the following compounds will be most readily hydrolysed in aqueous alkali?

(a) Cl, NO_2 (b) Cl, NO_2

(c) Cl, O_2N, NO_2 (d) Cl, NO_2, NO_2

35 Consider the following statements.

Statement I Presence of a nitro group at *ortho* or *para*-position increases the reactivity of haloarenes towards nucleophilic substitution.

Statement II Nitro group, being an electron withdrawing group decreases the electron density over the benzene ring.

(a) Both statements I and II are correct.
(b) Only statement I is correct.
(c) Only statement II is correct.
(d) Both statements I and II are incorrect.

36 Out of

I: $C_6H_5CH_2Cl$ II: p-$CH_3C_6H_4Cl$ III: p-$ClC_6H_4NO_2$ IV: $CH_2{=}CH{-}Cl$

which undergo S_N reactions easily?

(a) All except III
(b) All except II and IV
(c) All except I and III
(d) All except IV

37 Haloarenes are less reactive than haloalkanes due to the fact that

(a) greater stability of (C—*X*) bond in haloarenes due to delocalisation of electron by resonance
(b) greater electronegativity of C(sp^2) of C—*X* in aryl halides than that of C(sp^3) of C—*X* in alkyl halide, which pulls bonding pair and thus, bond length is shortened
(c) Both (a) and (b)
(d) greater stability of the arene ring

38 What is the end product of the following reaction?

$C_6H_5{-}CH_3 + \underset{\text{Excess}}{Cl_2} \xrightarrow{\text{Boiling}}$

(a) 2,3-dichlorotoluene ($C_6H_3Cl_2{-}CH_3$)
(b) $C_6Cl_5{-}CH_2Cl$ (Cl at all five ring positions)
(c) 2,4-dichlorotoluene (Cl—C_6H_3(Cl)—CH_3)
(d) $C_6H_5{-}CCl_3$

39 Consider the following statements.

Statement I Following S_N reaction is fast reaction

$C_6H_5{-}Cl + NaOH \longrightarrow C_6H_5{-}OH + NaCl$

Statement II C-atom is sp^2-hybridised.

(a) Both statements I and II are correct.
(b) Only statement I is correct.
(c) Only statement II is correct.
(d) Both statements I and II are incorrect.

40 Which of the following compounds gives a single benzyne intermediate on reaction with sodium amide?

I: o-chloroethylbenzene (CH_2CH_3, Cl ortho) II: m-chloroethylbenzene (CH_2CH_3, Cl meta) III: p-chloroethylbenzene (CH_2CH_3, Cl para)

(a) Only I
(b) Both I and III
(c) Only III
(d) Both I and II

41 Product formed in the following reaction is/are

o-bromotoluene (CH_3, Br) $+ NaNH_2 \xrightarrow{\text{liquid } NH_3}$

(a) o-toluidine (CH_3, NH_2 ortho) by direct substitution
(b) m-toluidine (CH_3, NH_2 meta) by Cine substitution
(c) Both of (a) and (b)
(d) None of the above

42 The reaction,

2-bromo-6-nitrotoluene (O_2N, CH_3, Br) $\xrightarrow[NH_3]{KNH_2}$ (O_2N, CH_3, NH_2) + (O_2N, CH_3, NH_2 meta to CH_3)

most likely involves which of the following aromatic substitution mechanisms?

(a) Addition-elimination
(b) Electrophilic substitution
(c) Elimination-addition
(d) Both (a) and (c)

43 Consider the following statements.

Statement I Nitration of chlorobenzene leads to the formation of *m*-nitrobenzene.

Statement II $-NO_2$ group is a *m*-directing group.

(a) Both statements I and II are correct.
(b) Only statement I is correct.
(c) Only statement II is correct.
(d) Both statements I and II are incorrect.

44 In the S_E of haloarene, incorrect intermediate is

(a) (b)

(c) (d)

45 Friedel-Crafts reaction of bromobenzene with methyl iodide gives

(a) *o*-bromotoluene
(b) *p*-bromotoluene
(c) *o*-and *p*-bromotoluene
(d) *m*-bromotoluene

46 The reaction,

H_3C—biphenyl—Cl $\xrightarrow{HNO_3/H_2SO_4}$

gives the major product as

(a) H_3C— ... —Cl (NO_2)
(b) H_3C— ... —Cl (NO_2)
(c) H_3C— ... —Cl (NO_2)
(d) H_3C— ... —Cl (NO_2)

47 Following reaction is called

2 C_6H_5—I $\xrightarrow{Cu}$ C_6H_5—C_6H_5

(a) Fittig
(b) Ullmann
(c) Wurtz
(d) Wurtz Fittig

48 Chlorobenzene can be converted into toluene by

(a) Wurtz reaction
(b) Fittig reaction
(c) Wurtz-Fittig reaction
(d) All of the above

49 Select the correct statement(s) about DDT.

(a) It is synthesised from chlorobenzene and chloral
(b) It is not metabolised very rapidly by animals, instead it is deposited and stored in fatty tissues
(c) It is said to have high toxicity towards fish
(d) All of the above

50 Trichloroacetaldehyde, CCl_3CHO reacts with chlorobenzene in presence of sulphuric acid and produces.

(a) Cl—C_6H_4—C(Cl)(CH_2Cl)—C_6H_4—Cl
(b) Cl—C_6H_4—C(H)(C_6H_4Cl)—C_6H_4—Cl
(c) Cl—C_6H_4—C(OH)(Cl)—C_6H_4—Cl
(d) Cl—C_6H_4—CH(CCl_3)—C_6H_4—Cl

(B) Miscellaneous questions

1 What is the end product of the following reaction?

Cl—C_6H_4—Br + Mg $\longrightarrow$ $\xrightarrow{D_2O}$

(a) D—C_6H_4—D (b) D—C_6H_4—Br
(c) Cl—C_6H_4—D (d)

2 C_6H_5—CH_3 $\xrightarrow{NO_2^+}$ X, C_6H_5—CF_3 $\xrightarrow{NO_2^+}$ Y

X and Y are

(a) X: $O_2N-C_6H_4-CH_3$ (para), Y: $O_2N-C_6H_4-CF_3$ (para)

(b) X: $C_6H_4(CH_3)(NO_2)$ (ortho), Y: $C_6H_4(CF_3)(NO_2)$ (ortho)

(c) X: $O_2N-C_6H_4-CH_3$ (para), Y: $C_6H_4(CF_3)(NO_2)$ (meta)

(d) X: $C_6H_4(CH_3)(NO_2)$ (ortho), Y: $O_2N-C_6H_4-CF_3$ (para)

3 Which sequence of steps describes the best synthesis of the compound shown?

$C_6H_5-CH_2-C_6H_4Br$ (meta-Br)

(a) Benzene $\xrightarrow[AlCl_3]{C_6H_5CH_2Cl}$ $\xrightarrow[FeBr_3]{Br_2}$

(b) Benzene $\xrightarrow[FeBr_3]{Br_2}$ $\xrightarrow[AlCl_3]{C_6H_5CH_2Cl}$

(c) Benzene $\xrightarrow[AlCl_3]{C_6H_5COCl}$ $\xrightarrow[FeBr_3]{}$ $\xrightarrow[HCl]{Zn(Hg)}$

(d) Benzene $\xrightarrow[FeBr_3]{Br_2}$ $\xrightarrow[AlCl_3]{C_6H_5COCl}$ $\xrightarrow[HCl]{Zn(Hg)}$

4 Match the species in Column I with corresponding properties in Column II and choose the correct option.

Column I		Column II
A. C_6H_5Cl	1.	S_N1 reaction
B. $C_6H_5CH_2CH_2Cl$	2.	S_N2 reaction
C. $C_6H_5-CH(Cl)-C_6H_5$	3.	S_E reaction
D. CH_3Cl	4.	Wurtz reaction

Codes

	A	B	C	D
(a)	4	2	3	1
(b)	3	2	1	4
(c)	2	1	4	3
(d)	3	4	1	2

5 Consider the following statements.

Statement I In monohaloarenes, further electrophilic substitution occurs at *ortho*-and *para*-position.

Statement II Halogen atom is a ring deactivator.

(a) Both statements I and II are correct.
(b) Only statement I is correct.
(c) Only statement II is correct.
(d) Both statements I and II are incorrect.

6 The reaction,

2-chloro-1,3-dinitrobenzene $\xrightarrow{CH_3O^-}$ 2-methoxy-1,3-dinitrobenzene

most likely occurs by which of the following mechanisms?

(a) Addition-elimination
(b) Elimination-addition
(c) S_N reaction
(d) Neither of these

7 Which one of the following undergoes nucleophilic aromatic substitution at the fastest rate?

(a) $Cl-C_6H_4-NO_2$ (para)
(b) $Cl-C_6H_4-Cl$ (para)
(c) $Cl-C_6H_4-CH_3$ (para)
(d) $Cl-C_6H_4-OCH_3$ (para)

8 Identify the principal organic product of the following reaction,

$C_6H_3(F)(Br)(NO_2)$ (F, Br ortho to each other; O_2N para to F) + $NaSCH_3$ ⟶

(a) $C_6H_3(F)(Br)(SCH_3)$
(b) $C_6H_2(CH_3S)(F)(Br)(O_2N)$
(c) $C_6H_3(SCH_3)(F)(Br)$ with O_2N
(d) $C_6H_3(SCH_3)(Br)(O_2N)$

9 Match the reaction given in Column I with the types of reactions given in Column II.

	Column I	Column II
A.	Chlorobenzene $\xrightarrow{Fe/Cl_2}$ 1,2-dichlorobenzene + 1,4-dichlorobenzene	1. Nucleophilic aromatic substitution
B.	$CH_3—CH=CH_2 + HBr \longrightarrow CH_3—CH(Br)—CH_3$	2. Electrophilic aromatic substitution
C.	C_6H_5–CH(I)–CH_3 $\xrightarrow{OH^-}$ C_6H_5–CH(OH)–CH_3	3. Saytzeff elimination
D.	4-nitrochlorobenzene + NaOH ⟶ 4-nitrophenol	4. Electrophilic addition
E.	$CH_3CH_2CH(Br)CH_3 \xrightarrow[KOH]{Alkaline} CH_3CH=CH\,CH_3$	5. Nucleophilic substitution (S_N1)

Codes

	A	B	C	D	E
(a)	5	3	1	4	2
(b)	2	4	5	1	3
(c)	2	5	3	1	4
(d)	1	2	3	4	5

10 Which of the following substituted benzene would furnish three isomers when one more substituent is introduced (directly or indirectly)?

I. Chlorobenzene II. 1,2-dichlorobenzene III. 1,3,5-trichlorobenzene IV. 1,4-dichlorobenzene

(a) I, II and III (b) Only I
(c) Both II and IV (d) Both I and III

11 Replacement of Cl of chlorobenzene to give phenol requires drastic conditions but S_N reaction (replacing —Cl) is favoured in 2,4-dinitrochlorobenzene. This is because

(a) NO_2 makes the ring electron-rich at *ortho* and *para*-position
(b) NO_2 withdraws electrons from *meta*-positions
(c) NO_2 donates electron at *meta*-position
(d) NO_2 withdraws electron from *ortho*/*para*-positions

12 Consider the following statements.

I. C-atom in C_6H_5Cl is sp^2-hybridised.
II. C-atom in CH_3Cl is sp^3-hybridised.
III. C_6H_5Cl undergoes S_N reactions faster than that of CH_3Cl.

Select the correct statements.

(a) I, II and III (b) Both I and III
(c) Both II and III (d) Both I and II

13 When 2, 4-dinitrochlorobenzene reacts with methylamine, the major product is

(a) Cl, NO_2, CH_3NH, NO_2 substituted benzene
(b) Cl, NO_2, $NHCH_3$ substituted benzene
(c) $NHCH_3$, NO_2, NO_2 substituted benzene
(d) NH_2, NO_2, H_3C, NO_2 substituted benzene

14 Product formed in the following reaction is/are

2-bromotoluene (CH_3, Br) + $NaNH_2$ $\xrightarrow{\text{Liquid } NH_3}$

(a) 2-methylaniline (CH_3, NH_2) by direct substitution
(b) 3-methylaniline (CH_3, NH_2) by Cine substitution
(c) Both of (a) and (b)
(d) 4-methylaniline (CH_3, NH_2) by tele substitution

15 Consider the following statements.

Statement I It is difficult to replace chlorine by —OH in chlorobenzene in comparison to that in chloroethane.

Statement II Chlorine carbon (C—C) bond in chlorobenzene has a partial double bond character due to resonance.

(a) Both statements I and II are correct.
(b) Only statement I is correct.
(c) Only statement II is correct.
(d) Both statements I and II are incorrect.

16 $X \xrightarrow[\text{Boiling}]{Cl_2}$ Benzotrichloride $\xrightarrow{H_3O^+} Y$

X and Y are

(a) benzene, benzaldehyde (b) toluene, benzaldehyde
(c) toluene, benzoic acid (d) benzene, benzoic acid

17 Benzene can be converted to *p*-bromonitro benzene by path I or II

p-bromonitrobenzene $\xleftarrow[]{NO_2^+}$ bromobenzene $\xleftarrow[\text{path II}]{Fe/Br_2}$ Benzene $\xrightarrow[\text{path I}]{NO_2^+}$ nitrobenzene $\xrightarrow{Fe/Br_2}$ p-bromonitrobenzene

Which path is adopted for the product ?

(a) Only I (b) Only II
(c) Either of I and II (d) None of these

18 Rank the following in order of decreasing rate of reaction with ethoxide ion ($CH_3CH_2O^-$) in a nucleophilic aromatic substitution reaction

I: o-bromonitrobenzene; II: m-bromonitrobenzene; III: 1-bromo-2,4-dinitrobenzene; IV: 1-bromo-3,4-dinitrobenzene

(a) III > IV > I > II (b) III > IV > II > I
(c) II > I > IV > III (d) IV > III > II > I

19 Consider the following statements.

Statement I Chlorine in chlorobenzene withdraws electrons through inductive effect and releases electrons through resonance.

Statement II Carbocation is destabilised by inductive effect.

(a) Both statements I and II are correct.
(b) Only statement I is correct.
(c) Only statement II is correct.
(d) Both statements I and II are incorrect.

20 (benzene ring with —Cl and ortho X)—Cl, C—Cl bond weakens when X is

(a) $—NO_2$ (b) $—CH_3$
(c) —Cl (d) $—NH_2$

21 The reaction of toluene with chlorine in the presence of iron and in the absence of light yields .... **[NCERT Exemplar]**

(a) $C_6H_5CH_2Cl$ (benzyl chloride) (b) o-chlorotoluene (CH_3, Cl)
(c) $H_3C—C_6H_4—Cl$ (p-chlorotoluene) (d) Mixture of (b) and (c)

22 In the following reaction, A is

p-terphenyl $\xrightarrow{Br_2/Fe}$ 'A'

(a) terphenyl with Br on the central ring
(b) terphenyl with Br at para position of terminal ring
(c) Both (a) and (b)
(d) terphenyl with Br on the central ring and Br at para position of terminal ring

23 $\underset{(A)}{C_6H_4Cl_2} \xrightarrow{Cl_2/Fe} \underset{(B)}{C_6H_3Cl_3}$

Only one type of B is formed, when A is

(a) *o*-dichlorobenzene
(b) *m*-dichlorobenzene
(c) *p*-dichlorobenzene
(d) *o*-, *p*-dichlorobenzene

24 Match the species in Column I with corresponding reactants in Column II and choose the correct option.

Column I	Column II
A. Gattermann reaction	1. $C_6H_5COOAg \cdot Br_2$ KCl_4 / heat
B Hunsdiecker reaction	2. $C_6H_5N_2^+Cl^-$, HBF_4, heat
C. Balz-Schiemann reaction	3. $C_6H_5N_2^+Cl^-$, CuCl / HCl, heat
D. Sandmeyer reaction	4. $C_6H_5N_2^+Cl^-$, Cu / HCl, heat

Codes

	A	B	C	D
(a)	1	2	3	4
(b)	4	1	2	3
(c)	4	3	2	1
(d)	1	2	4	3

(C) Medical entrances' gallery

1 Identify compound X in the following sequence of reactions. **[NEET 2020]**

$C_6H_5CH_3 \xrightarrow{Cl_2/h\nu} X \xrightarrow[373\ K]{H_2O} C_6H_5CHO$

(a) $C_6H_5CH_2Cl$ (b) $C_6H_5CHCl_2$

(c) $C_6H_5CCl_3$ (d) C_6H_5Cl

2 Among the following, the reaction that proceeds through an electrophilic substitution is **[NEET 2019]**

(a) $C_6H_6 + Cl_2 \xrightarrow{AlCl_3} C_6H_5Cl + HCl$

(b) $C_6H_6 + Cl_2 \xrightarrow{\text{UV light}} C_6H_6Cl_6$

(c) $C_6H_5CH_2OH + HCl \xrightarrow{\text{Heat}} C_6H_5CH_2Cl + H_2O$

(d) $C_6H_5\overset{+}{N_2}Cl^- \xrightarrow{Cu_2Cl_2} C_6H_5Cl + N_2$

3 The hydrolysis reaction that takes place at the slowest rate, among the following is **[NEET (Odisha 2018)]**

(a) $m\text{-}CH_3C_6H_4Cl \xrightarrow{aq.NaOH} m\text{-}CH_3C_6H_4\overset{\ominus}{O}\overset{\oplus}{N}a$

(b) $H_3C—CH_2—Cl \xrightarrow{aq.\ NaOH} H_3C—CH_2—OH$

(c) $H_2C{=}CH—CH_2Cl \xrightarrow{aq.\ NaOH} H_2C{=}CH—CH_2OH$

(d) $C_6H_5CH_2Cl \xrightarrow{aq.NaOH} C_6H_5CH_2OH$

2018

4 Which of the following represents Wurtz-Fittig reaction? **[AIIMS 2018]**

(a) $C_6H_5I + 2Na + CH_3I \longrightarrow C_6H_5CH_3 + 2NaI$

(b) $2C_6H_5I + 2Na \longrightarrow C_6H_5C_6H_5 + 2NaI$

(c) $2CH_3CH_2I + 2Na \longrightarrow CH_3CH_2CH_2CH_3 + 2NaI$

(d) $CH_3Br + AgF \longrightarrow CH_3F + AgBr$

5 The compound C_7H_8 undergoes the following reactions

$$C_7H_8 \xrightarrow{3Cl_2/\Delta} A \xrightarrow{Br_2\ /\ Fe} B \xrightarrow{Zn\ /\ HCl} C$$

The product 'C' is **[NEET 2018]**

(a) 3-bromo-2, 4, 6-trichlorotoluene

(b) *o*-bromotoluene

(c) *m*-bromotoluene

(d) *p*-bromotoluene

6 Identify A and predict the type of reaction, **[NEET 2017]**

$o\text{-}BrC_6H_4OCH_3 \xrightarrow{NaNH_2} A$

(a) $m\text{-}H_2NC_6H_4OCH_3$ and substitution reaction

(b) $o\text{-}H_2NC_6H_4OCH_3$ and elimination addition reaction

(c) $o\text{-}BrC_6H_4OCH_3$ and cine substitution reaction

(d) $C_6H_5OCH_3$ and cine substitution reaction

7 Assertion (A) Presence of nitro group facilitates nucleophilic substitution reactions in aryl halides.

Reason (R) The intermediate carbocation is stable due to presence of nitro group. **[AIIMS 2017]**

(a) Both Assertion and Reason are true and Reason is the correct explanation of Assertion.

(b) Both Assertion and Reason are true but Reason is not the correct explanation of Assertion.

(c) Assertion is true but Reason is false.

(d) Both Assertion and Reason are false.

8 Assertion (A) Aryl halides undergo nucleophilic substitution with ease. **[AIIMS 2017]**

Reason (R) Hybridisation of C-atom attached to halide is sp^3-hybrid.

(a) Both Assertion and Reason are true and Reason is the correct explanation of Assertion.
(b) Both Assertion and Reason are true but Reason is not the correct explanation of Assertion.
(c) Assertion is true but Reason is false.
(d) Both Assertion and Reason are false.

9 Identify *A* and predict the type of reaction **[NEET 2017]**

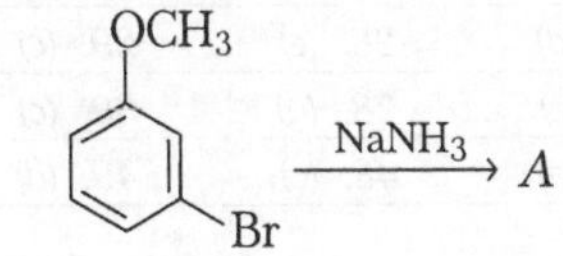

m-bromoanisole (OCH_3, Br) $\xrightarrow{NaNH_3}$ *A*

(a) (OCH_3 benzene) and cine substitution reaction
(b) 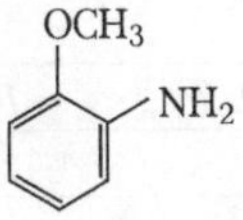 (OCH_3, NH_2 ortho) and elimination addition
(c) 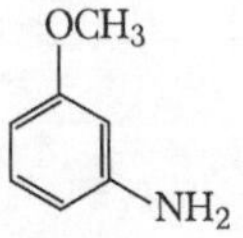 (OCH_3, NH_2 meta) and direct substitution addition
(d) 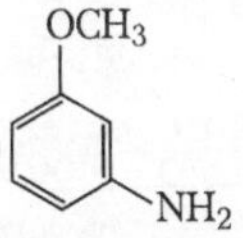 (OCH_3, NH_2 meta) and cine substitution addition

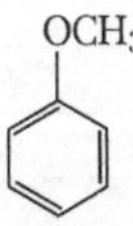

10 Which of the following can be used as the halide component for Friedel-Crafts reactions? **[NEET (II) 2016]**

(a) Chlorobenzene (b) Bromobenzene
(c) Chloroethene (d) Isopropyl chloride

11 The major product of the reaction of **[Manipal 2014]**

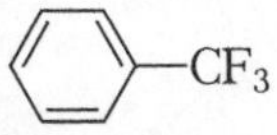

$C_6H_5-CF_3$

with electrophile, E^+, is

(a) 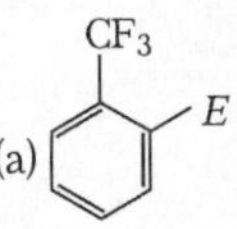 (CF_3, *E* ortho) (b) 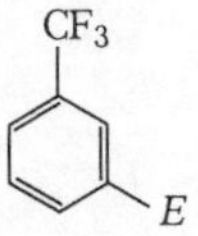 (CF_3, *E* meta) (c) 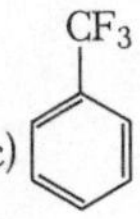 (CF_3 benzene) (d) 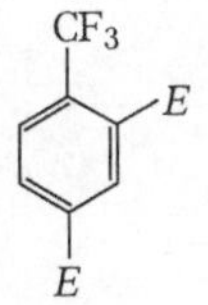 (CF_3, *E* ortho and *E* para)

12 Which of the following organic halogen compounds undergoes hydrolysis with aqueous NaOH predominantly by S_N1 mechanism? **[Kerala CEE 2014]**

(a) Ethyl iodide
(b) Methyl chloride
(c) *Iso*-propyl chloride
(d) Chlorobenzene
(e) Benzyl chloride

13 The compound that reacts the fastest with sodium methoxide is **[KCET 2014]**

(a) 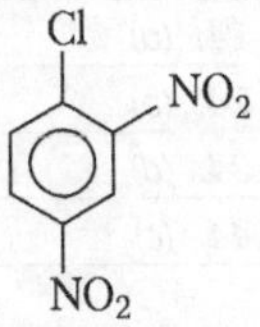 (Cl; NO_2 ortho; NO_2 para) (b) 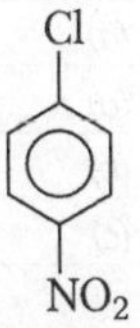 (Cl; NO_2 para)
(c)  (Cl benzene) (d) 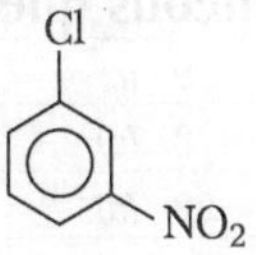 (Cl; NO_2 meta)

14 Statement I Alkyl benzene is not prepared by Friedel-Crafts alkylation of benzene.

Statement II Aryl halides are less reactive than alkyl halide in S_N reaction. **[AIIMS 2013]**

(a) Both A and R are correct and R is the correct explanation of A.
(b) Both A and R are the correct and R is not the correct explanation of A.
(c) A is correct but R is incorrect.
(d) Both A and R are incorrect.

15 Which one of the following is not formed when a mixture of methyl bromide and bromobenzene is heated with sodium metal in the presence of dry ether? **[KCET 2013]**

(a) Diphenyl
(b) Propene
(c) Toluene
(d) Ethane

16 Chlorobenzene on treatment with sodium in dry ether gives diphenyl. The name of the reaction is **[Kerala CEE 2011]**

(a) Fittig reaction
(b) Wurtz-Fittig reaction
(c) Sandmeyer reaction
(d) Gattermann reaction
(e) Wurtz reaction

ANSWERS

Check point 14.1

1. (b)	2. (a)	3. (b)	4. (b)	5. (b)	6. (c)	7. (d)	8. (d)	9. (d)	10. (d)

Check point 14.2

1. (a)	2. (a)	3. (c)	4. (d)	5. (b)	6. (c)	7. (b)	8. (d)	9. (c)	10. (d)

(A) Topical exercise

1. (b)	2. (b)	3. (a)	4. (b)	5. (b)	6. (b)	7. (a)	8. (c)	9. (d)	10. (b)
11. (b)	12. (a)	13. (b)	14. (a)	15. (b)	16. (a)	17. (c)	18. (a)	19. (a)	20. (c)
21. (a)	22. (a)	23. (b)	24. (b)	25. (a)	26. (d)	27. (d)	28. (c)	29. (c)	30. (a)
31. (c)	32. (b)	33. (a)	34. (d)	35. (a)	36. (b)	37. (c)	38. (d)	39. (c)	40. (b)
41. (c)	42. (c)	43. (c)	44. (c)	45. (c)	46. (c)	47. (d)	48. (c)	49. (d)	50. (d)

(B) Miscellaneous questions

1. (c)	2. (c)	3. (c)	4. (b)	5. (a)	6. (c)	7. (a)	8. (d)	9. (b)	10. (d)
11. (d)	12. (d)	13. (c)	14. (c)	15. (a)	16. (c)	17. (b)	18. (a)	19. (b)	20. (a)
21. (d)	22. (a)	23. (c)	24. (b)						

(C) Medical entrances' gallery

1. (b)	2. (a)	3. (a)	4. (a)	5. (c)	6. (a)	7. (a)	8. (d)	9. (c)	10. (d)
11. (b)	12. (e)	13. (a)	14. (a)	15. (b)	16. (a)				

Hints & Explanations

(A) Topical exercise

1 (*b*) This has double bond character

(Structure: benzene ring with $-Cl$ bond labelled x, and $CH_2 \overset{y}{—} Cl$ group)

Hence, bond length of Y is greater than X.

3 (*a*) A. In alkyl halide, halogen atom is bonded to sp^3-hybridised carbon atom, which may be further bonded to one, two or three alkyl group, i.e. $CH_3—CH(X)—CH_3$.

B. Allyl halides are the compounds in which the halogen atom is bonded to sp^3-hybridised carbon atom next to carbon-carbon double bond. i.e. $CH_2=CH—CH_2—X$

C. Aryl halides are the compounds in which the halogen atom is bonded to sp^2-hybridised carbon atom of an aromatic ring, i.e. C_6H_5X.

D. Vinyl halides are the compounds in which the halogen atom is bonded to an sp^2-hybridised carbon atom of a carbon-carbon double bond, i.e. $CH_2=CH—X$.

8 (*c*) $C_6H_5—H + I_2 \rightleftharpoons C_6H_5—I + HI$

Due to the formation of HI in the above reaction, aryl iodide are reduced back to the aromatic hydrocarbon by HI as it is a good reducing agent.

9 (*d*) Benzene $\xrightarrow{Cl^{\oplus}}$ (a) $\longleftrightarrow$ (b) $\longleftrightarrow$ (c)

(Resonance structures of arenium ion with H and Cl on the same carbon, positive charge at different positions: (a), (b), (c))

10 (*b*) In this reaction, $AlCl_3$ acts as a catalyst which activates the chlorine molecule to show heterolytic cleavage. $AlCl_3$ is an electron deficient molecule and form $AlCl_4^-$ and Cl^+ when reacts with Cl_2.

This Cl^+ electrophile attacks on electron rich benzene ring.

$$AlCl_3 + Cl_2 \longrightarrow [AlCl_4]^- + Cl^+$$

$$C_6H_6 + Cl^+ \longrightarrow C_6H_5Cl$$

Chlorobenzene

12 (*a*) When a primary aromatic amine, dissolved or suspended in cold aqueous mineral acid and treated with sodium nitrite, a diazonium salt is formed.

When this freshly prepared diazonium salt is mixed with cuprous chloride, then diazonium group is replaced by —Cl.

Then chlorobenzene is formed which is Y in this reaction.

$$C_6H_5NH_2 \xrightarrow[273\text{-}278\ K]{NaNO_2+HCl} C_6H_5\overset{+}{N}_2Cl \xrightarrow{Cu_2Cl_2} \underset{(Y)}{C_6H_5Cl} + N_2$$

13 (*b*) Statement I is correct but statement II is incorrect statement.

Corrected statement II Oxidising agent oxidises HI into I_2 to prevent the possibility of backward reaction.

14 (*a*) A. A mixture of an alkyl halide and aryl halide when treated with sodium in dry ether and this is called Wurtz-Fittig reaction.

B. Diazonium salt when treated with cuprous chloride or cuprous bromide gives chlorobenzene or bromobenzene.

The reaction is known as Sandmeyer's reaction.

C. Alkyl iodides are prepared by the reaction of alkyl chlorides with sodium iodide in dry acetone. The reaction is known as Finkelstein reaction.

15 (*b*) Toluene reacts with a halogen in the presence of iron (III) chloride giving *ortho* and *para* halo compounds.The reaction is electrophilic substitution reaction.

It has the following mechanism

$$Cl—Cl \xrightarrow{FeCl_3} FeCl_4^- + Cl^+$$

Toluene $+ Cl^+ \longrightarrow$ (arenium ion with H and Cl at *ortho* position) or (arenium ion with H and Cl at *para* position)

(arenium ion, CH_3, H, Cl) $+ [FeCl_4]^- \longrightarrow$ (*o*-chlorotoluene) $+ HCl + FeCl_3$

In this mechanism, electrophile Cl^+ attacks to electron rich benzene ring and replaces hydrogen. So, the reaction is electrophilic substitution reaction.

17 (*c*) Due to symmetry, *para-* isomers fits in crystal lattice better as compared to *ortho* and *meta*-isomers.

18 (*a*) $\mu = 2\mu_{C—Cl} \cos\left(\frac{\theta}{2}\right)$

19 (*a*) Due to large size of Br atoms, repulsion increases, hence θ increases resulting in decrease in dipole moment.
Also, dipole moment of *p*- < *m*- < *o*-.

22 (*a*) Due to (EWG) Cl-atoms at 2 and 4 positions.

24 (*b*) II is most reactive due to two $—NO_2$ at *o* and *p*-positions. IV is least reactive due to each of resonance activation by nitro group present at *meta*-position. I is more reactive than III due to less steric hindrance and greater electrophilicity at α-carbon of *p*-fluoronitrobenzene.

26 (*d*) Presence of electron releasing group at *ortho* or *para*-position decreases the rate of nucleophilic substitution reaction. In compound (iii), electron releasing group is present at *meta*-position *wrt* chlorine, so the impact is less but in compound (ii) it is present at *ortho*-position.
Thus, the rate of reaction towards nucleophilic substitution is least in compound (ii) and highest in compound (i) as there is no electron releasing group in this compound.

27 (*d*) Presence of electron withdrawing group at *ortho* and *para*-position facilitate the nucleophilic substitution reaction and hence, enhances rate of reaction.
Compound (iii) has three electron withdrawing groups at *ortho* and *para*-positions *w.r.t.* chlorine while compound (ii) has only one electron withdrawing group and there is no electron withdrawing group in compound (i). So, the increasing order of rate of reaction towards nucleophilic substitution is
(i) < (ii) < (iii).

28 (*c*) The bond formed between C of benzene ring and halogen is highly stable because due to resonance it acquires partial double bond character.

Therefore, it is difficult to cleave C—*X* bond and hence, the rate of reaction towards nucleophilic substitution is slow. This substitution is facilitated by the presence of electron withdrawing group at *ortho* and *para* position because electron density is high at these positions.
Compound (ii) and (iii) both has one electron withdrawing group but in compound (ii) electron withdrawing ($—NO_2$) group is present at *ortho*-position, so rate of reaction in compound (ii) is more than that of (iii) while (i) has no electron withdrawing group.
Hence, the correct option is (c).

29 (*c*) Presence of electron releasing group at *ortho* and *para*-position *w.r.t.* to chlorine decreases the rate of nucleophilic substitution reaction. Compound (iii) has two electron releasing groups and compound (ii) has one electron releasing group *w.r.t.* chlorine while compound (i) has no electron releasing group.

So, the rate of nucleophilic substitution reaction is highest in compound (i) and order is (iii) < (ii) < (i).

31 (*c*) Check all methods, most suitable is (c).

Toluene (CH_3) $\xrightarrow[(1)]{NO_2^+}$ p-nitrotoluene (CH_3, NO_2) $\xrightarrow[\Delta]{(2)\ Cl_2/FeCl_3}$ (CH_3, Cl, NO_2) $\xrightarrow[\Delta]{(3)\ K_2Cr_2O_7/H^+}$ (COOH, Cl, NO_2)

34 (*d*) (C—Cl) bond is weakened due to presence of $—NO_2$ groups in *o*- and *p*- positions.

35 (*a*) Both are correct statements. Presence of nitro group at *ortho* or *para*-position increases the reactivity of haloarenes towards nucleophilic substitution because — NO_2 group, being an electron withdrawing group decreases the electron density over the benzene ring.

36 (*b*)

I. $C_6H_5—CH_2—Cl$ — Yes (Benzylic (weak))

II. $H_3C—C_6H_4—Cl$ — No (Aryl (strong))

III. $O_2N—C_6H_4—Cl$ (Aryl) — Yes, as $—NO_2$ is EWG making (C—Cl) bond weak

IV. $CH_2═CH—Cl$ (Vinylic) — No

Thus, I and III undergo S_N reaction.

42 (*c*)

(*A*) (ring with CH_3, O_2N, Br) $\longrightarrow$ Elimination product (benzyne with CH_3, O_2N)

Elimination product + $NH_2^{\ominus}$ $\longrightarrow$ ring with CH_3, O_2N, NH_2 (adjacent)

Addition product (By direct substitution from *A*)

Elimination product + $NH_2^{\ominus}$ $\longrightarrow$ ring with CH_3, O_2N, NH_2

Addition product (By Cine substitution from *A*)

43 (*c*) Statement I is incorrect but Statement II is correct.

Correct statement II Chlorination of nitrobenzene leads to the formation of *m*-nitrochlorobenzene because — NO_2 group deactivates the ring because it is *meta*-directing.

44 (*c*) Due to *meta*-attack

46 (*c*) H_3C–(biphenyl–phenyl)–Cl

Activating (both *o*-, *p*-directing) Deactivating

$\xrightarrow{NO_2^{\oplus}}$ H_3C–(ring with O_2N)–(ring)–Cl

(a), (b) and (d) are not formed at all due to the activating nature of CH_3.

48 (*c*) $C_6H_5Cl + CH_3Cl \xrightarrow{\text{Na/ether}} C_6H_5CH_3$

(Wurtz-Fittig)

(B) Miscellaneous questions

1 (*c*) (C—Br) is more reactive than (C—Cl).

$Cl-C_6H_4-Br + Mg \longrightarrow Cl-C_6H_4-Mg-Br \xrightarrow{D_2O} Cl-C_6H_4-D$

2 (*c*) —CH_3 is *o*-, *p*- directing.

—CF_3 is *m*-directing.

3 (*c*) (a) $C_6H_6 \xrightarrow[AlCl_3]{C_6H_5CH_2Cl} C_6H_5-CH_2-C_6H_5$ (*o*-, *p*-directing) $\xrightarrow{Br_2/FeBr_3} C_6H_5-CH_2-C_6H_4-Br + o-$

(b) C_6H_5Br (*o*-, *p*-directing) $\xrightarrow{C_6H_5CH_2Cl}$ $Br-C_6H_4-CH_2-C_6H_5 + o-$

(c) $C_6H_6 \xrightarrow{C_6H_5COCl} C_6H_5-CO-C_6H_5$ (*m*-directing) $\xrightarrow{Br_2/FeBr_3}$

$C_6H_5-CO-C_6H_4Br$ (*m*-Br) $\xrightarrow[\text{HCl reduction (Clemmensen)}]{Zn(Hg)} C_6H_5-CH_2-C_6H_4Br$ (*m*-Br)

5 (*a*) Both statements are correct.

In monohaloarenes, halogen atom increases the electron density at *ortho* and *para*-position. So, further electrophilic substitution occurs ar *ortho* and *para*-positions.

8 (*d*) Due to *para*-effect of —NO_2 group, S_N reaction replaces F-atom.

9 (*b*) A. In this reaction, an electrophile Cl^+ attacks on the benzene ring and substitution takes place.

B. In this reaction, addition of HBr takes place on to the doubly bonded carbons of propene in accordance with Markownikoff's rule and electrophile addition takes place.

C. In this reaction, the reactant is secondary halide. Here, halogen atom is substituted by hydroxy ion. As it is secondary halide so it follows S_N1 mechanism.

D. In this reaction, halogen atom is directly bonded to aromatic ring. So, it is nucleophilic aromatic ring. So, it is nucleophilic aromatic substitution as OH^- group has substituted halogen of given compound.

E. It is an elimination reaction. It follows Saytzeff elimination rule.

10 (*d*) I. Chlorobenzene ⟶ *o*- + *m*- + *p*-dichlorobenzene

Three only

II. 1,2-Dichlorobenzene ⟶ 1, 2, 3 + 1, 2, 4

Two only

III. 1,3-Dichlorobenzene ⟶ 1, 2, 4 + 1, 2, 3 + 1, 3, 5

IV. 1,4-Dichlorobenzene ⟶ 1, 2, 4

One only

12 (*d*) C_6H_5–Cl: Strong bond (sp^2) CH_3–Cl: Weak bond (sp^3)

Thus, I and II are true.

13 (*c*) Statement I is incorrect but Statement II is correct statement.

Chlorination of nitrobenzene leads to the formation of *m*-nitrochlorobenzene because — NO_2 group deactivates the ring because it is *meta*-directing.

15 (*a*) Both are correct statements. It is difficult to replace chlorine by —OH in chlorobenzene in comparison to that in chloroethane because C—Cl bond in chlorobenzene has a partial bond character due to resonance.

16 (*c*) Toluene (X) $+ Cl_2 \xrightarrow{\text{Boiling}}$ $C_6H_5CCl_3 \xrightarrow{H_3O^+}$ C_6H_5COOH (Y)

17 (*b*) Benzene $\xrightarrow{NO_2^+}$ nitrobenzene $\xrightarrow{Fe/Br_2}$ *m*-bromonitro benzene

$-NO_2$ is *m*-directing *m*-bromonitro benzene

Benzene $\xrightarrow{Fe/Br_2}$ bromobenzene $\xrightarrow{NO_2^+}$ *p*-bromonitro benzene

—Br is *o*-, *p*-directing *p*-bromonitro benzene

18 (*a*) $-NO_2$ groups in *o*-, *p*- positions withdraws electrons due to a greater extent, thus S_N reaction in (III) is maximum.

20 (*a*) $-NO_2$ is (EWG), thus S_N reaction is favoured.

21 (*d*) The reaction of toluene with chlorine in the presence of iron is carried out in absence of light, so the substitution occurs in the benzene ring. The $-CH_3$ group of toluene is *o* and *p*-directing then product is the mixture of (b) and (c), i.e. *o*-chlorotoluene and *p*-chlorotoluene.

Toluene $+ Cl_2 \xrightarrow{Fe}$ *o*-chlorotoluene + *p*-chlorotoluene

23 (*c*) *p*-dichlorobenzene $\xrightarrow{Fe/Cl_2}$ 1,2,4-trichlorobenzene, one type $C_6H_3Cl_3$

(C) Medical entrances' gallery

1 (*b*) An 1, 1-*gem* dihalide on hot hydrolysis (H_2O / 373 K) can produce an aldehyde.

Toluene $\xrightarrow[h\nu]{2Cl_2}$ $C_6H_5CHCl_2$ (X) $\xrightarrow[373\text{ K}]{H_2O}$ Benzaldehyde

2 (*a*) Halogenation of benzene is an example of electrophilic substitution reaction.

$C_6H_6 + Cl_2 \xrightarrow{AlCl_3} C_6H_5Cl + HCl$ (Chlorobenzene)

Reaction given in option (b) is an example of addition reaction, reaction in option (c) is an example of nucleophilic substitution reaction and option (d) is an example of substitution reaction.

3 (*a*) In *o*-chlorotoluene Cl atom is directly attached to C atom of benzene ring.
Thus, there is a partial double bond character between sp^2-hybridised C atom on the aromatic ring and Cl. It is most difficult to break this bond and hence, it undergoes hydrolysis reaction with slowest rate.

4 (*a*) The Wurtz-Fittig reaction is the reaction of an aryl halide with alkyl halide and sodium metal to give substituted aromatic compound.

Thus, Wurtz-Fittig reaction is

$$C_6H_5I + 2Na + CH_3I \longrightarrow C_6H_5CH_3 + 2NaI$$

5 (*c*) Given, $C_7H_8 \xrightarrow{3Cl_2/\Delta} A \xrightarrow{Br_2/Fe} B \xrightarrow{Zn/HCl} C$

The reaction in the above road map can be explained by the following steps.

***Step* I** Toluene (*A*) undergoes side chain halogenation with excess of chlorine to give benzotrichloride (*A*).

C_7H_8 (Toluene) $\xrightarrow[\Delta]{3Cl_2}$ $C_6H_5CCl_3$ (*A*) (Benzotrichloride)

***Step* II** In compound (*A*), the substituent CCl_3 is an electron withdrawing group, so the electrophile will attack at *m*-position. Thus, benzotrichloride reacts with bromine in presence of Fe-catalyst to give *m*-bromobenzotrichloride (*B*).

CCl_3 (*A*) $\xrightarrow{Br_2/Fe}$ CCl_3, Br (*B*)
m-bromobenzotrichloride

***Step* III** *m*-bromobenzotrichloride undergoes reduction with Zn in presence of HCl to give *m*-bromotoluene (*C*).

CCl_3, Br (*B*) *m*-bromobenzotrichloride $\xrightarrow{Zn/HCl}$ CH_3, Br (*C*) *m*-bromotoluene

6 (*a*) OCH_3, Br $\xrightarrow{NaNH_2}$ OCH_3, NH_2 (Direct substitution) + OCH_3, NH_2 (Cine substitution)

If nucleophile occupies same position of the leaving group, product is called direct substitution product. If nucleophile occupies adjacent position of the leaving group, product is called cine substitution product. Intermediate is benzyne.

OCH_3, H, Br $+ \overset{\ominus}{N}H_2 \longrightarrow$ OCH_3 (Benzyne) $+ Br^{\ominus} + \overset{\ominus}{N}H_2$

OCH_3 $+ \overset{\ominus}{N}H_2 \longrightarrow$ OCH_3, NH_2 $+ H—NH_2$ $\longrightarrow$ OCH_3, NH_2

Attack of nucleophile at the original position (from where Br^- leaves)

Direct substitution product

OCH_3 $+ \overset{\ominus}{N}H_2 \longrightarrow$ OCH_3, NH_2, $H—NH_2$ $\longrightarrow$ OCH_3, NH_2

Attack of nucleophile at the adjacent carbon

Cine substitution product

7 (*a*) Nitro group is an electron withdrawing (– *I*) group, it stabilises the carbocation because it disperses the negative charge on the carbon by attracting electrons. Hence, both (A) and (R) are correct and (R) is the correct explanation of (A).

8 (*d*) Both (A) and (R) are false as, aryl halides undergo nucleophilic substitution with great difficulty due to partial double bond character between C-atom of ring and halide atom. Also, hybridisation of C-atom attached to halogen is sp^2-hybridisation.

9 (*c*) OCH_3, Br $\xrightarrow{NaNH_2}$ OCH_3, NH_2 (Direct substitution) + OCH_3, NH_2 (cine substitution)

10 (*d*) **Key Idea** In chlorobenzene, bromobenzene and chloroethene, lone pair of halogen is delocalised with π-bonds so C—*X* attains double bond character. Thus, these are not suitable are as halide component for Friedel-Craft reaction.

$$C_6H_6 + CH_3—\underset{CH_3}{CH}—Cl \xrightarrow{Anhyd.AlCl_3} C_6H_5—\underset{H}{\overset{CH_3}{C}}—CH_3$$

Other halides, i.e. chloro and bromobenzene alongwith chloroethene have carbon halogen bond as,

$$>C\overset{...}{=}X$$

11 (*b*) CX_3 group attached to benzene ring are *meta*-directions for electrophiles.

CF_3, *E* (meta)

12 (*e*) Nucleophilic substitution of haloalkane by hydroxide ion is done by two different mechanisms —S_N1 and S_N2 . S_N2 mechanism is most likely to be followed with primary haloalkane.

Tertiary haloalkane tends to react by S_N1 mechanism, involving a carbocation. Secondary haloalkanes react *via* both mechanisms.

But benzyne chloride gives benzyl carbocation $C_6H_5\overset{\oplus}{C}H_2$, when is resonance stabilised hence S_N1 perdominantly.

13 (*a*) The compound *X*—Cl (where *X* aryl, alkyl or their derivatives, on reaction with sodium methoxide produces *X*—OMe and NaCl in accordance with following chemical equation.

$$X—Cl + Me\overset{\ominus}{O}\overset{\oplus}{Na} \longrightarrow X—OMeE + NaCl$$

The above reaction is a type of nucleophilic substitution reaction in which nucleophile [here, OMe] replaces halogen atom of $X - Cl$.

Reactivity of such organic halides in nucleophilic substitution reaction depends upon polarity of C—Cl bond. Higher the polarity of C—Cl bond easier will be the nucleophilic substitution reaction, hence, faster will be the reaction. Here, C—Cl bond of aryl halide has partially double bond character due to conjugation of lone pair of Cl with π-electrons of benzene ring.

Due to partial double bond character breaking of C—Cl bond of aryl halide become impossible. Among the given four substrates, the compound having most polar C—Cl bond will undergo faster nucleophilic substitution reaction. Greater the number of electron withdrawing groups higher will be the polarity of C—Cl bond.

(a) has two electron withdrawing groups hence, (a) has the most polar C—Cl bond and (a) will undergo fastest reaction.

14 (*a*) Alkyl benzene can be prepared by Friedel-Crafts alkylation of benzene

$$C_6H_6 + CH_3Cl \xrightarrow{AlCl_3} C_6H_5CH_3$$

In S_N reaction aryl halides are less reactive than alkyl halides.

15 (*b*) This is an example of Wurtz-Fittig reaction

(i) $C_6H_5Br + 2Na + BrCH_3 \xrightarrow{\text{Ether}} C_6H_5CH_3 + 2NaBr$

Toluene

(ii) $C_6H_5Br + 2Na + C_6H_5Br \xrightarrow{\text{Ether}} C_6H_5\text{—}C_6H_5 + 2NaBr$

Diphenyl

(iii) $CH_3Br + 2Na + BrCH_3 \longrightarrow \underset{\text{Ethane}}{CH_3CH_3} + 2NaBr$

Propene cannot be prepared from this reaction.

16 (*a*) Reaction of aryl halides with sodium in ether to form diaryl is known as Fittig reaction.

CHAPTER 15

Alcohols

Alcohols are hydroxy organic compounds of the alkanes with general formula $C_nH_{2n+2}O$ or $C_nH_{2n+1}OH$ (for monohydric alcohol derived from alkanes) or ROH, where, R is **alkyl or substituted alkyl group.** The name of an alcohol is derived by replacing 'e' of the parent alkane by adding suffix 'ol'.

$$\text{Alkane} \xrightarrow[-e]{+ol} \text{Alkanol}$$

In this chapter, we will study the chemistry of this functional group, its physical and chemical properties in detail.

CLASSIFICATION

On the basis of number of —OH groups present in the molecule, these are classified as follows

(*i*) **Monohydric alcohols** Alcohols that have one —OH group are said to be monohydric alcohols, e.g. $\underset{\text{Methanol}}{CH_3OH}$, $\underset{\text{Ethanol}}{C_2H_5OH}$

(*ii*) **Dihydric alcohols** Alcohols that have two —OH groups are said to be dihydric alcohols, e.g.

$$\begin{array}{c} CH_2OH \\ | \\ CH_2OH \\ \text{Ethylene glycol} \end{array}$$

(*iii*) **Trihydric alcohols** Alcohols that have three —OH groups are said to be trihydric alcohols, e.g.

$$\begin{array}{c} CH_2OH \\ | \\ CHOH \\ | \\ CH_2OH \\ \text{Glycerol} \end{array}$$

(*iv*) **Polyhydric alcohols** Alcohols that have more than three —OH groups are said to be polyhydric alcohols.

$$HO-CH_2-CH(OH)-CH(OH)-CH_2-OH$$

Inside

- Classification
- Structure
- Nomenclature
- Isomerism
- Methods of preparation
- Physical properties
- Chemical properties
  - *Reaction which involve O—H bond cleavage*
  - *Reaction which involve C—OH bond cleavage*
- Distinction between 1°, 2° and 3° alcohols
  - *Action of heated copper*
  - *Victor Meyer's test*
  - *Lucas test*
  - *Haloform test*
- Some commercially important alcohols
  - *Methanol*
  - *Ethanol*
  - *Dihydric alcohol (glycol)*
  - *Trihydric alcohol (glycerol)*

On the basis of position of —OH group in the carbon chain, alcohols are classified as follows

(*i*) **Primary alcohols** In primary (1°) alcohol, the carbon, which carries the —OH group is attached with only one alkyl group, i.e. they contain $—CH_2OH$ group.

e.g. $CH_3—CH_2—OH$ (Ethanol), $CH_3—CH(CH_3)—CH_2—OH$ (2-methylpropan-1-ol)

(*ii*) **Secondary alcohols** In secondary (2°) alcohol, the carbon carrying —OH group is joined directly to two alkyl groups, which may or may not be same. In other words, they have >CHOH group.

e.g. $CH_3—CH(OH)—CH_3$ (Propan-2-ol), $CH_3—CH(OH)—CH_2—CH_3$ (Butan-2-ol)

(*iii*) **Tertiary alcohols** In tertiary (3°) alcohols, the carbon atom holding the —OH group is attached directly to three alkyl groups, which may or may not be same. In other words, they have >C—OH group.

e.g. $CH_3—C(OH)(CH_3)—CH_3$ (2-methylpropan-2-ol), $CH_3—CH_2—C(OH)(CH_3)—CH_3$ (2-methylbutan-2-ol)

Alcohol is of $(3-n)°$, where n is the number of H at carbon that carries OH group. (CH_3OH is always 1°).

STRUCTURE

The structure of an alcohol resembles the structure of water with an alkyl group replacing one of the hydrogen atoms of water. Both have sp^3-hybridised oxygen atoms but the C–O–H bond angle in methanol (108.9°) is considerably larger than the H–O–H bond angle in water (104.5°) because the methyl group is much larger than a hydrogen atom.

The bulky methyl group counter acts the bond angle compression caused by non-bonding (lone) electron pairs of oxygen. The O–H bond lengths are about the same in water and methanol (0.96 Å) but the C–O bond is considerably longer (1.4 Å), reflecting the larger covalent radius of carbon compared to hydrogen.

Water: H–O–H, 104.5°, 0.96 Å   Alcohol: H_3C–O–H, 1.4 Å, 0.96 Å, 108.9°

NOMENCLATURE

Alcohols are named by two principal systems, i.e. common and IUPAC system. For the simpler alcohols, the common names are most often used. A common name consists the name of the alkyl group followed by the word 'alcohol'.

e.g. CH_3CH_2OH Ethyl alcohol (1°); $CH_3CH(OH)CH_3$ Isopropyl alcohol (2°); $(CH_3)_2CHCH_2OH$ i.e. $CH_3CH(CH_3)CH_2OH$ Isobutyl alcohol (1°); $CH_3—C(CH_3)(OH)—CH_3$ *Tert*-butylalcohol (3°)

IUPAC system is the most versatile system. The rules for nomenclature of alcohol are as follows

(*i*) Select the longest continuous carbon chain that contains the —OH group as the parent chain.

(*ii*) Each name is derived by replacing the terminal '–e' of the corresponding alkane name by '–ol'.

(*iii*) Indicate the position of —OH group by a number in the parent chain, generally using the lowest possible number for this purpose.

(*iv*) Indicate the position of other groups by the number in the parent chain.

(*v*) For naming of polyhydric alcohol 'e' is retained.

CH_3OH Methanol; $CH_3CH_2CH_2OH$ Propanol; $CH_3CH(OH)CH_2Br$ 1-bromopropan-2-ol; $CH_2(OH)—CH_2(OH)$ Ethane-1, 2-diol; $CH_3CH(OH)—CH(CH_3)—CH_3$ 3-methylbutan-2-ol; $CH_3CH_2CH(CH_3)CH_2OH$ 2-methylbutan-1-ol

ISOMERISM

Alcohols exhibit following types of isomerism

(*i*) **Chain isomerism** Alcohols with four or more carbon atoms exhibit this type of isomerism in which compounds have similar molecular formula but the carbon skeleton is different.

$$CH_3-CH_2-CH_2-CH_2OH \quad CH_3-\underset{}{\overset{CH_3}{\overset{|}{C}H}}-CH_2OH$$

Butan-1-ol   2-methylbutan-1-ol

(*ii*) **Position isomerism** Alcohols with three or more carbon atoms can exhibit position isomerism. In this type of isomerism, the position of the functional group, i.e. — OH group varies.

$$CH_3CH_2CH_2OH \quad CH_3\underset{\underset{OH}{|}}{C}H-CH_3$$

Propan-1-ol   Propan-2-ol

(*iii*) **Functional isomerism** Alcohols with two or more carbon atoms can exhibit functional isomerism with ethers. Thus, ethers and alcohols have same molecular formula but have different functional groups, hence they are called functional isomers.

$$CH_3CH_2CH_2CH_2OH \quad CH_3CH_2OCH_2CH_3$$

Butan-1-ol   Ethoxyethane

(*iv*) **Optical isomerism** Alcohols containing chiral centres show optical isomerism. These optical isomers can rotate the plane of polarised light in different directions.

$$CH_3-\underset{\underset{OH}{|}}{\overset{*}{C}H}-CH_2-CH_3$$

Butan-2-ol

$$CH_3-CH_2-CH_2-\underset{\underset{OH}{|}}{\overset{*}{C}H}-CH_3$$

Pentan-2-ol

$$CH_3-\underset{\underset{CH_3}{|}}{\overset{*}{C}H}-\underset{\underset{OH}{|}}{\overset{*}{C}H}-CH_3$$

3-methylbutan 2-ol

METHODS OF PREPARATION

Alcohols can be prepared by following methods

Hydrolysis of alkyl halides

The hydrolysis of haloalkanes with aqueous sodium or potassium hydroxide or carbonate or moist silver oxide results in the formation of alcohols.

$$\underset{\text{Alkyl halide}}{R-X} + OH^- \longrightarrow \underset{\text{Alcohol}}{R-OH} + \underset{\text{Leaving group}}{X^- \text{ (or HX)}}$$

$$\underset{\text{Bromopropane}}{CH_3CH_2CH_2Br} \xrightarrow{aq.\ NaOH} \underset{\text{Propanol}}{CH_3CH_2CH_2OH}$$

$$\underset{\text{2-bromo-2-methylpropane}}{CH_3\overset{\overset{CH_3}{|}}{\underset{\underset{CH_3}{|}}{C}}Br} \xrightarrow[\text{Strong base}]{aq.\ NaOH} \underset{\substack{\text{2-methyl-1-ene} \\ \text{(Major)}}}{CH_3\overset{\overset{CH_2}{\|}}{C}CH_3} + \underset{\substack{tert\text{-butyl alcohol} \\ \text{(Minor)}}}{CH_3\overset{\overset{CH_3}{|}}{\underset{\underset{CH_3}{|}}{C}}OH}$$

$$\underset{\text{2-chloromethyl propane}}{(CH_3)_3C-Cl} \xrightarrow[\text{Mild base}]{AgOH} (CH_3)_3C-OH + AgCl$$

Reactivity order of alkyl halides

$$R-I > R-Br > R-Cl > R-F$$

Intermediate is carbocation and the order of stability of carbocation is $1° < 2° < 3°$. In 3° carbocation case elimination is preferred to substitution.

In general, hydrolysis is not preferred for preparation of alcohols from haloalkanes in the presence of strong bases due to the following reasons.

(*i*) The use of strong base (such as NaOH, KOH) may cause elimination as a side reaction and alkenes are the major products along with alcohols as side product.

(*ii*) The use of moist silver oxide (Ag_2O/H_2O) and Na_2CO_3 minimises the possibility of elimination reactions, these being mild bases.

This reaction proceed through nucleophilic substitution reaction that was already discussed in chapter (Haloalkanes).

Hydration of alkenes

An alkene may react with water in acidic medium to form an alcohol, this process is called **hydration.** Hydration of an alkene is the reverse of the dehydration of alcohols. In this process, addition takes place according to Markownikoff's rule in case of unsymmetrical alkene.

$$\underset{\text{Alkene}}{>C=C<} + H_2O \overset{H^+}{\rightleftharpoons} \underset{\text{Alcohol}}{-\overset{\overset{H}{|}}{\underset{|}{C}}-\overset{\overset{OH}{|}}{\underset{|}{C}}-}$$

e.g. $$\underset{\text{Alkene}}{CH_2{=}CH_2} + H_2O \xrightarrow{\text{Dil.}H_2SO_4} \underset{\text{Ethanol}}{\overset{1^\circ}{CH_3CH_2OH}}$$

$$\underset{\text{Propene}}{CH_3CH{=}CH_2} + H_2O \xrightarrow{\text{Dil.}H_2SO_4} CH_3CH(OH)CH_3 \ (2^\circ, \text{Propan-2-ol})$$

$$\underset{\text{2-methylpropene}}{(CH_3)_2C{=}CH_2} + H_2O \xrightarrow{\text{Dil.}H_2SO_4} (CH_3)_3COH \ (3^\circ, \text{2-methylpropan-2-ol})$$

Hydration equilibrium shifts to the formation of an alcohol by adding excess of water.

Mechanism

Step I. Protonation of alkene to form carbocation by electrophilic attack of H_3O^+.

$$H_2O + H^+ \longrightarrow H_3O^+$$

$$>C{=}C< + H{-}\overset{+}{O}(H){-}H \rightleftharpoons {-}\overset{H}{C}{-}\overset{+}{C}< + H_2\ddot{O}$$

Step II. Nucleophilic attack of water on carbocation.

$$-\overset{H}{C}-\overset{+}{C}< + H_2\ddot{O} \rightleftharpoons -\overset{H}{C}-\overset{H}{C}-\overset{+}{O}-H$$

Note *Free carbocation may rearrange itself.*

Step III. Deprotonation to form an alcohol.

$$-\overset{H}{C}-\overset{H}{C}-\overset{+}{O}-H + H_2\ddot{O} \rightarrow -\overset{H}{C}-\overset{\ddot{O}H}{C}- + H_3\overset{+}{O}$$

Many alkenes do not easily undergo hydration in dilute aqueous acid because they are nearly insoluble in water. In such case concentrated sulphuric acid is used, in which alkenes are soluble forming alkyl hydrogen sulphate which on boiling with water changes to alcohol.

$$\underset{\text{Ethene}}{CH_2{=}CH_2} + \underset{\text{Conc. sulphuric acid}}{HOSO_3H} \longrightarrow \underset{\text{Ethyl hydrogen sulphate}}{CH_3CH_2{-}OSO_3H} \xrightarrow{\text{Boiling } H_2O} \underset{\text{Ethanol}}{CH_3CH_2OH} + H_2SO_4$$

Hydroboration-oxidation reaction

Alkenes react with diborane, $(BH_3)_2$ in presence THF or diglyme to form alkyl borane, (by hydroboration) which on oxidation gives alcohol.

$$\underset{\text{Propene}}{CH_3CH{=}CH_2} \xrightarrow{(BH_3)_2 \text{ in THF or diglyme}} \underset{\text{Trialkyl borane}}{(CH_3CH_2CH_2)_3B}$$

$$\xrightarrow{H_2O_2,\ OH^-} \underset{\text{Propanol}}{CH_3CH_2CH_2OH} + \underset{\text{Boric acid}}{B(OH)_3}$$

The addition of boron and hydrogen to double bond is stereospecific that proceed *via* ***syn*-addition.**

$$\underset{\text{2-methyl-but-2-ene}}{(H_3C)_2C{=}C(CH_3)H} \xrightarrow{BH_3\cdot THF} \underset{(A)}{H_3C{-}\overset{CH_3}{\underset{H}{C}}{-}\overset{H}{\underset{BH_2}{C}}{-}CH_3}$$

(*A*) on oxidation with H_2O_2/OH^- changes to alcohol (*B*).

$$\underset{(A)}{H_3C{-}\overset{CH_3}{\underset{H}{C}}{-}\overset{H}{\underset{BH_2}{C}}{-}CH_3} \xrightarrow[H_2O]{H_2O_2,\ OH^-} \underset{\text{3-methyl-but-2-ol}}{\underset{(B)}{H_3C{-}\overset{CH_3}{\underset{H}{C}}{-}\overset{H}{\underset{OH}{C}}{-}CH_3}}$$

This mechanism proceed *via* transition state hence, no rearrangement takes place as seen in the hydration of alkenes.

$$\underset{\text{3, 3-dimethylbutene}}{(CH_3)_3C\,CH{=}CH_2} \xrightarrow[\text{(Hydroboration oxidation reaction)}]{BH_3\cdot THF \quad H_2O_2/OH^-} \underset{\text{3, 3-dimethylbutanol}}{(CH_3)_3C\,CH_2CH_2OH}$$

$$\underset{\text{3, 3-dimethylbutene}}{(CH_3)_3C\,CH{=}CH_2} \xrightarrow[\text{Hydration}]{H_2O,\ H^+} \underset{\text{(Major)}}{\underset{\text{2, 3-dimethylbutanol}}{(CH_3)_2C(OH){-}CH(CH_3)CH_3}}$$

Intermediate is 2° carbocation $CH_3{-}\overset{CH_3}{\underset{CH_3}{C}}{-}\overset{\oplus}{C}H{-}CH_3$

Which changes to more stable 3° carbocation by 1, 2 methyl shift.

The addition of borane to the double bond takes place in such a manner that the boron atom gets attached to the sp^2 carbon carrying greater number of hydrogen atoms. The alcohol so formed looks as if it has been formed by the addition of water to the alkene in a way opposite to the Markownikoff's rule.

Oxymercuration–demercuration reaction

Mercury (II) acetate reacts with alkene in presence of water to give β-hydroxy alkyl mercury (II) acetate. This step is called oxymercuration and is carried out in a mixture of H_2O and THF as solvent, which is treated directly with $NaBH_4$ to form alcohol and this step is called demercuration.

$$\underset{\text{Alkene}}{>C=C<} + \underset{\text{Mercury (II) acetate}}{Hg(CH_3CO_2)_2} + H_2O \xrightarrow[\text{Oxymercuration}]{THF \cdot H_2O} \underset{\beta\text{-hydroxy alkyl mercury (II) acetate}}{-\overset{OH}{\overset{|}{C}}-\overset{|}{C}-HgO\overset{O}{\overset{\|}{C}}CH_3} + \underset{\text{Ethanoic acid}}{CH_3CO_2H}$$

$$-\overset{OH}{\overset{|}{C}}-\overset{|}{C}-HgO\overset{O}{\overset{\|}{C}}CH_3 \xrightarrow[OH^-]{NaBH_4} \underset{\text{Alcohol}}{-\overset{OH}{\overset{|}{C}}-\overset{|}{C}-H} + Hg + CH_3CO_2^-$$

The complete reaction is called oxymercuration-demercuration. It gives product in accordance with Markownikoff's rule without any rearrangement (without formation of carbocation).

Thus,

$$\underset{\substack{\text{3-methylbutan-2-ol} \\ (B)}}{CH_3\underset{CH_3}{\underset{|}{C}H}-\underset{OH}{\underset{|}{C}H}CH_3} \xleftarrow[\text{(ii) } NaBH_4/OH^-]{\text{(i) } Hg(OAc)_2 \cdot THF \cdot H_2O} \underset{\substack{\text{3-methyl-butene} \\ (X)}}{CH_3\underset{CH_3}{\underset{|}{C}H}CH=CH_2} \xrightarrow{H_3O^+} \underset{\substack{\text{2-methylbutan-2-ol} \\ (C)}}{CH_3\overset{OH}{\overset{|}{\underset{CH_3}{\underset{|}{C}}}}CH_2CH_3}$$

(the last arrow drawn leftwards from X to C)

$$\underset{\substack{\text{3-methyl-butene} \\ (X)}}{CH_3\underset{CH_3}{\underset{|}{C}H}CH=CH_2} \xrightarrow[\text{(ii) } H_2O_2/OH^-]{\text{(i) } BH_3 \cdot THF} \underset{\substack{\text{3-methyl butanol} \\ (A)}}{CH_3\underset{CH_3}{\underset{|}{C}H}CH_2CH_2OH}$$

(*A*) is formed by hydroboration-oxidation of *X* and is anti-Markownikoff's addition. (*B*) is formed by oxymercuration- demercuration of *X* and is Markownikoff's product without any rearrangement. (*C*) is formed as major product by hydration of *X* by rearrangement of the carbocation (intermediate).

Reduction of carbonyl compounds (Including acid derivatives)

(*i*) Carbonyl compounds (aldehydes and ketones) can be reduced to alcohols, either by catalytic hydrogenation or by use of chemical reducing agents like lithium aluminium hydride ($LiAlH_4$) and sodium borohydride ($NaBH_4$). These hydride reagents add a hydride ion (H^-), reducing carbonyl group to an alkoxide ion which on protonation gives alcohol. Thus, reaction involves addition of two hydrogen across a $>C=O$ bond.

$$\underset{\text{Hydride ion}}{H:^{\ominus}} + \underset{\text{Ketone}}{\overset{R}{\underset{R}{>}}C=\ddot{O}:} \longrightarrow H-\overset{R}{\overset{|}{\underset{R}{\underset{|}{C}}}}-\ddot{O}:^{\ominus} \xrightarrow{H_3O^+} H-\overset{R}{\overset{|}{\underset{R}{\underset{|}{C}}}}-\ddot{O}H$$

(*ii*) Aldehydes are reduced to 1° alcohol and ketones to 2° alcohol whereas 3° alcohols cannot be prepared by reduction of aldehydes and ketones.

$$\underset{\text{Aldehyde}}{RCHO} \xrightarrow[H_3O^+]{LiAlH_4} \underset{1°}{RCH_2OH}$$

$$\underset{\text{Ketone}}{\overset{R}{\underset{R}{>}}CO} \xrightarrow[H_3O^+]{LiAlH_4} \overset{R}{\underset{R}{>}}\underset{2°}{CHOH}$$

(*iii*) Acid, acid halide, ester and anhydride are reduced to 1° alcohol (but alcohol part of ester retains its character).

$$\underset{\text{Ethanoic acid}}{CH_3\overset{O}{\overset{\|}{C}}OH} \xrightarrow[H_3O^+]{LiAlH_4} \underset{\substack{1° \\ \text{Ethanol}}}{CH_3CH_2OH}$$

$$\underset{\text{Acetyl chloride}}{CH_3\overset{O}{\overset{\|}{C}}Cl} \xrightarrow[H_3O^+]{LiAlH_4} \underset{\substack{1° \\ \text{Ethanol}}}{CH_3CH_2OH}$$

$$\underset{\text{Acetic anhydride}}{(CH_3\overset{O}{\overset{\|}{C}})_2O} \xrightarrow[H_3O^+]{LiAlH_4} \underset{\substack{1^\circ \\ \text{Ethanol}}}{CH_3CH_2OH}$$

$$\underset{\substack{\text{Acid part} \quad \text{Alcohol part} \\ \text{Methyl acetate}}}{CH_3-\overset{O}{\overset{\|}{C}}-OCH_3} \xrightarrow[H_3O^+]{LiAlH_4} \underset{\substack{1^\circ \\ \text{Ethanol}}}{CH_3CH_2OH} + \underset{\substack{1^\circ \\ \text{Methanol}}}{CH_3OH}$$

Alcohol part retains its character

$$\underset{\text{Acid part}}{CH_3CH_2\overset{O}{\overset{\|}{C}}-} \underset{\text{Alcohol part}}{O\overset{H}{C}(CH_3)_2} \xrightarrow[H_3O^+]{LiAlH_4} \underset{\substack{1^\circ \\ \text{Propanol}}}{CH_3CH_2CH_2OH} + \underset{\substack{2^\circ \\ \text{1-methyl-ethanol}}}{(H_3C)_2CHOH}$$

Esters can be reduced to alcohols by Na/C_2H_5OH or by hydrogenolysis (cleavage by hydrogen) in presence of copper chromite ($CuO \cdot CuCr_2O_4$) under high temperature and pressure (350 atm).

Note *$LiAlH_4$ is used for preparing special chemical only as it is an expensive reagent. Commercially, acids are reduced to alcohols as*

$$\underset{\text{Carboxylic acid}}{R-COOH} \xrightarrow[-H_2O]{R'OH/H^+} \underset{\text{Ester}}{RCOOR'} \xrightarrow[\text{or } Na/C_2H_5OH]{H_2/CuO\text{-}Cr_2O_4} \underset{\text{Alcohol}}{RCH_2OH} + R'OH$$

3° alcohol can be produced by reduction of ester by $LiAlH_4$.

$NaBH_4$ *vs* $LiAlH_4$ as reducing agents

(*i*) $LiAlH_4$ can reduce esters, acids, ketones, aldehydes, alkyl halides, epoxides and amides while $NaBH_4$ can reduce ketones and aldehydes only.

$$\underset{\text{1, 2-epoxyethane}}{CH_2-CH_2 \text{ (–O– bridge)}} \xrightarrow[H_3O^+]{LiAlH_4} \underset{\text{ethanol}}{CH_3CH_2OH}$$

$$\underset{\text{1, 2-epoxypropane}}{CH_3\,CH-CH_2 \text{ (–O– bridge)}} \xrightarrow[H_3O^+]{LiAlH_4} \underset{\text{Propan-2-ol}}{CH_3-\underset{OH}{\underset{|}{CH}}-CH_3}$$

(*ii*) $NaBH_4$ is selective and does not reduce carbon-carbon double bond. Thus, unsaturated alcohols can be prepared.

$$\underset{\text{Prop-2-en-1-al-3-cyclohexane}}{C_6H_{11}-CH=CHCHO} \xrightarrow{NaBH_4} \underset{\text{Prop-2-en-1-ol-3-cyclohexane}}{C_6H_{11}-CH=CHCH_2OH}$$

(*iii*) $NaBH_4$ is also selective in reducing more reactive carbonyl group. Keto part of ester is not reduced.

$$\underset{\text{This is reduced}}{O=}C_6H_{10}-\underset{\text{This is not affected}}{CH_2\overset{O}{\overset{\|}{C}}OCH_3} \xrightarrow{NaBH_4} HO-C_6H_{10}-\underset{\text{Ester}}{CH_2\overset{O}{\overset{\|}{C}}OCH_3}$$

$$\xrightarrow[\text{Reduces both}]{LAH} HO-C_6H_{10}-\underset{\text{Alcohol}}{CH_2CH_2OH}$$

Reducing nature of different reagents is summarised in Table 15.1.

Table 15.1 Reducing nature of different reagents

Group	Product	LAH/ Ether	$NaBH_4$/ EtOH	B_2H_6/ THF	H_2/ Metal
$-CHO$	$-CH_2OH$	Yes	Yes	Yes	Yes
$>CO$	$>CHOH$	Yes	Yes	Yes	Yes
$-COOH$	$-CH_2OH$	Yes	No	Yes	Yes (with Ru–C)
$-COCl$	$-CH_2OH$	Yes	Yes	No	Yes
$(RCO)_2O$	RCH_2OH	Yes	No	Yes	Yes
$-CO_2R$	$-CH_2OH$	Yes	No	Yes	Yes
$>C=C<$	$>CH-CH<$	No	No	Yes	Yes

Note *Reducing power of various reagents are increase in the order as*

$$NaBH_4 < LiAlH_4 < Mg(BH_4)_2 < Al(BH_4)_3$$

Diisobutyl aluminium hydride (DIBAL-H) redues ester into aldehydes

From organometallic compounds

Organometallic compounds contain covalent bond between carbon atoms and metal atoms. Organometallic compounds of Li and Mg react with carbonyl compounds forming alcohol.

i.e. $R-Mg-X$ (R, alkyl or aryl, X = Cl, Br, I) is called **Grignard reagent** whereas R — Li is known as organolithium reagents.

Note *The actual species in solution usually contains two, three or four of these units associated together with several molecules of the ether solvent (to solvate and stabilise it after its formation).*

Reaction with Grignard reagent

Grignard reagent reacts with various functional group to give the corresponding alcohol.

These reactions are as follows

(*i*) **Aldehyde and ketones on reaction with *R*Mg*X*** followed by subsequent hydrolysis in acidic medium gives alcohol.

$$HCHO \xrightarrow[\text{(ii) } H_2O/H^+]{\text{(i) } CH_3MgX} CH_3CH_2OH + Mg(OH)X$$
Methanal → Ethanol (1°)

$$CH_3CHO \xrightarrow[\text{(ii) } H_2O/H^+]{\text{(i) } CH_3MgX} CH_3CH(OH)CH_3 + Mg(OH)X$$
Ethanal → Propan-2-ol (2°)

$$CH_3-CO-CH_3 \xrightarrow[\text{(ii) } H_2O/H^+]{\text{(i) } CH_3MgX} (CH_3)_3C-OH + Mg(OH)X$$
Propan-2-one → *Tert*-Butyl alcohol (3°)

The general reaction of carbonyl compound with Grignard reagent can be shown as

$$R_1R_2C{=}O + R_3MgX \xrightarrow{H_3O^+} R_1-C(R_2)(R_3)-OH$$

R_1, R_2 can be H, alkyl, aryl and R_3 can be alkyl, aryl. The carbonyl group is polarised, with a partial positive charge on carbon and a partial negative charge on oxygen. The positively charged carbon is electrophilic; attack by a nucleophile places a negative charge on the electronegative oxygen atom. The product of this nucleophilic attack is an alkoxide ion, a strong base. In all cases ether is the solvent.

$$R:^{\ominus} + \overset{\delta+}{C}{=}\overset{\delta-}{\ddot{O}}: \longrightarrow R-C-\ddot{O}:^{\ominus}$$
Nucleophile — Polarised carbonyl group — Alkoxide ion

Protonation of the alkoxide ion gives alcohol in presence of protic solvent.

$$R-C-\ddot{O}:^{\ominus} + H-\ddot{O}-H \longrightarrow R-C-\ddot{O}H + H\ddot{O}:^{\ominus}$$

(*ii*) **Reaction of Grignard reagent with ethylene oxide** Grignard reagents also react with ethylene oxide (oxirane) to give 1° alcohols.

$$CH_2-CH_2(-O-) + \overset{-}{R}\overset{+}{Mg}X \xrightarrow[\text{Ether}]{\text{Dry}} RCH_2CH_2\overset{-}{O}\overset{+}{Mg}X \xrightarrow{H_2O/H^+} RCH_2CH_2OH$$
1, 2-epoxyethane → (1° alcohol)

Reaction with organolithium reagent

Organolithium reagents react with carbonyl groups in the same way that Grignard reagents do but more readily.

$$RLi + {>}C{=}O \longrightarrow R-C-\overset{-}{O}\overset{+}{Li} \xrightarrow{H_3O^+} R-C-OH$$
Organolithium

e.g. $CH_2{=}CHLi + C_6H_5-CH{=}O \xrightarrow[H_3O^+]{Et_2O} C_6H_5-CH(OH)CH{=}CH_2$

Vinyllithium — Benzaldehyde — 3-phenyl-propene-2-ol

From aliphatic primary amines

Aliphatic primary amines react with nitrous acid to give primary alcohol.

$$\underset{\text{Sodium nitrate}}{NaNO_2} + \underset{\text{Hydrochloric acid}}{HCl} \xrightarrow{0-3°C} \underset{\text{Nitrous acid}}{HNO_2} + NaCl$$

$$\underset{\text{Amine}}{R-NH_2} + \underset{\text{Nitrous acid}}{HNO_2} \xrightarrow{0-3°C} \underset{\text{Alcohol (1°)}}{R-OH} + N_2\uparrow + H_2O$$

But this is not a good method for the preparation of 1° alcohols because a mixture of alkenes, alkyl halides aldehydes are also produced along with alcohols, e.g. methyl amine gives ether and methyl nitrite.

Oxo process

Oxo process is the principal method to get the simple alcohols commercially. This is also called **carbonylation** or **hydroformylation** reaction. A mixture of alkene, carbon monoxide and hydrogen reacts under pressure and elevated temperature in the presence of catalyst, forms aldehyde. Catalyst used in the reaction is cobalt carbonyl hydride $[CoH(CO)_4]$.

Product is a mixture of isomeric straight chain (major) and branched-chain (minor) aldehydes which are separated by fractional distillation. These aldehydes are reduced catalytically to the corresponding alcohols.

$$\underset{\text{Propene}}{2CH_3CH{=}CH_2} + 2CO + 2H_2 \xrightarrow[\text{and temp.}]{\text{High pressure}} \underset{\text{(Major) Butanal}}{CH_3CH_2CH_2CHO} + \underset{\text{(Minor) 2-methylpropanal}}{CH_3CH(CH_3)CHO}$$

$$CH_3CH_2CH_2CHO \xrightarrow[H_2]{Cu-Zn,\,[CaH(CO)_4]} \underset{\text{Butanol}}{CH_3(CH_2)_3(OH)}$$

$$CH_3CH(CH_3)CHO \xrightarrow[H_2]{Cu-Zn} \underset{\text{2-methyl-propanol}}{CH_3CH(CH_3)CH_2OH}$$

Note *Final alcohol product contains one carbon more than the initial substrate.*

CHECK POINT 15.1

1. The C—O—H bond angle in methanol is larger than H—O—H bond angle in water due to
(a) the bulky methyl group counteracts the bond angle compression caused by non-bonding electron pairs of oxygen
(b) due to the lone pair bond pair repulsion
(c) due to the sp^3-hybridised oxygen atom
(d) None of the above

2. 'A' changes to .... in hydroboration-oxidation reaction

$CH_3CH{=}C(CH_3)CH_3$ (A)

(a) $CH_3CH(OH)CH(CH_3)CH_3$ (b) $CH_3CH_2C(OH)(CH_3)CH_3$
(c) $CH_3CH_2CH(CH_3)CH_2OH$ (d) None of these

3. Consider the following reactions,

$CH_3C(CH_3)(OH)-CH(CH_3)CH_3 \xleftarrow{\text{II}} CH_3C(CH_3)_2CH{=}CH_2 \xrightarrow{\text{I}} CH_3-C(CH_3)_2-CH(OH)CH_3$

I and II are respectively
(a) hydroboration-oxidation, hydration
(b) hydroboration-oxidation, oxymercuration-demercuration
(c) oxymercuration-demercuration, hydration
(d) hydration, oxymercuration-demercuration

4. $CH_3-C(CH_3)_2-CH{=}CH_2 \longrightarrow CH_3-C(CH_3)(CH_3)-CH(OH)-CH_3$

This change can be done by
(a) acid catalysed hydration
(b) oxymercuration-demercuration
(c) hydroboration-oxidation
(d) any method mentioned above

5. Alcohol containing least number of carbon which can be prepared using Grignard reagent is
(a) $(CH_3)_2CHOH$
(b) $(CH_3)_3COH$
(c) CH_3OH
(d) CH_3CH_2OH

6. C_2H_4O (A) reacts with CH_3MgBr followed by decomposition with $H_3O^{\oplus}$ to give 2° alcohol. Thus, A is
(a) $CH_2{=}CHOH$
(b) CH_3CHO
(c) CH_2-CH_2 bridged by O (epoxide)
(d) None of the above

7. The alcohol manufactured from water gas followed by reduction is
(a) butanol
(b) ethanol
(c) methanol
(d) isobutanol

8. 2-methyl-2-propanol cannot be synthesised by
(a) $CH_3COCH_3 \xrightarrow{CH_3MgBr} \xrightarrow{H_3O^+}$
(b) $CH_3C(CH_3){=}CH_2 \xrightarrow{H_3O^+}$
(c) $CH_3C(CH_3){=}CH_2 \xrightarrow{BH_3\cdot THF} \xrightarrow{H_2O_2,\, OH^-}$
(d) $CH_3C(CH_3){=}CH_2 \xrightarrow[THF]{(CH_3CO_2)_2Hg/H_2O} \xrightarrow[OH^-]{NaBH_4}$

PHYSICAL PROPERTIES

Various physical properties shown by alcohols are as follows

Physical state

At room temperature, lower alcohols are colourless liquids with characteristic smell and burning taste. The higher members are colourless, odourless waxy liquids.

Solubility

The lower alcohols are completely soluble in water but as the number of carbon atom increases, solubility decreases. This solubility in water is due to intermolecular H-bonding between the water and alcohol molecules due to their polar character.

$$\overset{\delta^-}{O}(R)-\overset{\delta^+}{H}\cdots\overset{\delta^-}{O}(H)-\overset{\delta^+}{H}\cdots\overset{\delta^-}{O}(R)-\overset{\delta^+}{H}$$

Here, the hydroxyl group is **hydrophilic** (meaning water loving) because of its affinity for water and other polar substances. The alkyl group (of alcohol) is called **hydrophobic** (meaning water repelling) and makes the alcohol less hydrophilic.

As the molecular mass increases, the hydrocarbon part (alkyl group) becomes larger which resists the formation of hydrogen bonds with water molecule and hence the solubility decreases which means that the solubility of alcohols in water decreases with lengthening of carbon chain.

$$\underset{n\text{-butyl alcohol}}{CH_3CH_2CH_2CH_2OH} > \underset{n\text{-pentyl alcohol}}{CH_3CH_2CH_2CH_2CH_2OH} > \underset{n\text{-hexyl alcohol}}{CH_3CH_2CH_2CH_2CH_2CH_2OH}$$

Solubility increases with branching amongst isomeric alcohols because as the branching increases, the surface area of the non-polar hydrocarbon part decreases and the solubility increases.

$$\underset{n\text{-butyl alcohol}}{CH_3CH_2CH_2CH_2OH} < \underset{Iso\text{-butyl alcohol}}{CH_3CH(CH_3)CH_2OH} < \underset{Sec\text{-butyl alcohol}}{CH_3CH(OH)CH_2CH_3} < \underset{Tert\text{-butyl alcohol}}{CH_3C(CH_3)(OH)—CH_3}$$

Boiling point

Boiling point of alcohols are higher than that of hydrocarbons and ethers, this is because of the presence of intermolecular hydrogen bonding in alcohols. Dipole-dipole interactions also contribute to the relatively high boiling point of alcohols.

The polarised C—O and H—O bonds and the non-bonding electrons add to produce a dipole moment of 1.69 D in ethanol compared with a dipole moment of 0.08 D in propane.

S.No.	Compound	Molecular weight	Boiling point (K)	Dipole moment (μ)
1.	Propane	44	231	0.08 D
2.	Methoxymethane	46	248	1.30 D
3.	Ethanol	46	351	1.69 D

From the above data, we observe that ethanol and propane have similar molecular weights, yet their boiling points are differ by about 120°. Such a large difference in boiling point indicates that ethanol molecules are attracted to each other much more strongly than propane molecules. In liquid ethanol, the positive and negative ends of the dipoles align to produce attractive interactions.

We can compare the effects of hydrogen bonding and dipole-dipole interactions by comparing ethanol with methoxymethane. Like ethanol, methoxymethane has large dipole moment (1.30 D) but it (methoxymethane) cannot form hydrogen bonding due to absence of —OH hydrogens and only dipole-dipole interaction occurs.

The boiling point of methoxymethane is 248 K about 17° higher than that of propane but still 103° lower than that of ethanol. Hydrogen-bonds are clearly much stronger intermolecular attractions than dipole-dipole attractions.

C_2H_5—O—H, $\mu = 1.69$ D, Ethanol; H_3C—O—CH_3, $\mu = 1.30$ D, Methoxymethane; H_3C—CH_2—CH_3, $\mu = 0.08$ D, Propane

R—O$^{\delta-}$—H$^{\delta+}$ ······ O$^{\delta-}$(H)(R): Hydrogen-bond in alcohol

R—O—R ······ O(R)(R): Dipole-dipole interaction in ether

Acidity of alcohols

Alcohols are acidic due to polar nature of O—H bond. An electron donating group increases the electron density on oxygen, thus, decrease the polarity of —OH bond. Therefore, acidic nature decreases.

The order of acidic strength of alcohols are as follows.

$$\underset{\text{Methanol}}{H—CH_2—OH} > \underset{1^\circ \text{ alcohol}}{R\rightarrow CH_2—OH} > \underset{2^\circ \text{ alcohol}}{R\rightarrow CH(R)—OH} > \underset{3^\circ \text{ alcohol}}{R\rightarrow C(R)(R)—OH}$$

CHEMICAL PROPERTIES

The chemical properties of an alcohol, *R*OH are determined by its functional group, —OH the hydroxyl group. Reactions of an alcohol can involve the breaking of either of two bonds. i.e. the C—OH bond (with removal of the —OH group) and the O—H bond (with removal of —H). Either kind of reaction can involve substitution, in which a group replaces the —OH or —H or elimination in which a double bond is formed.

Various chemical reaction of alcohols are as follows

Reaction which involve O—H bond cleavage

Reaction with metals

Alcohols react with metal to produce alkoxide and hydrogen gas. This reaction shows the acidic nature of alcohols.

$$\underset{\substack{\text{Alcohol}\\ \text{(weak acid)}}}{R—OH} + Na \longrightarrow \underset{\substack{\text{Alkoxide}\\ \text{(stronger base)}}}{R\overset{-}{O}\overset{+}{Na}} + \frac{1}{2}H_2\uparrow$$

Reactivity of alcohols with metal is in the following order

$$CH_3OH > 1° > 2° > 3°$$

This reaction shows that alcohols act as Bronsted acids i.e. they can donate a proton to a strong base ($\ddot{B}^-$)

$$\underset{\text{Base}}{\bar{B}:} + \underset{\text{Acid}}{H—\ddot{O}R} \longrightarrow \underset{\text{Conjugate acid}}{B—H} + \underset{\text{Conjugate base}}{R—\ddot{O}:^-}$$

The alkoxide formed due to the reaction of alcohol with metal can be utilised for the formation of ether. In this reaction, O—H bond breaks and ethyl replaces H. Thus, configuration about chiral carbon (*) is retained.

$$H—\overset{Et}{\underset{Me}{C^*}}—O—H \xrightarrow{H} H—\overset{Et}{\underset{Me}{C^*}}—O^-K^+ \xrightarrow{C_2H_5Br} \underset{(A)}{H—\overset{Et}{\underset{Me}{C^*}}—OC_2H_5}$$

Chiral carbon | This bond breaks | Chiral carbon

Product (*A*) has same configuration as the reactant.

Esterification reaction

Alcohols react with carboxylic acids in the presence of a few drops of conc. H_2SO_4 or HCl gas as a catalyst to form ester. Complete reaction is as follows

$$\underset{\text{Ethanol}}{CH_2CH_2OH} + \underset{\text{Ethanoic acid}}{CH_3\overset{O}{\overset{\|}{C}}OH} \overset{H^+}{\rightleftharpoons} \underset{\text{Ethyl acetate}}{CH_3\overset{O}{\overset{\|}{C}}OC_2H_5} + H_2O$$

Mechanism

Step I. $CH_3—\overset{O}{\overset{\|}{C}}—OH \rightleftharpoons CH_3—\overset{O^-}{\overset{|}{\underset{\oplus}{C}}}—OH$

Step II. $CH_3—\overset{O^-}{\overset{|}{\underset{\oplus}{C}}}—OH + C_2H_5\ddot{O}H \longrightarrow CH_3—\overset{O^-}{\overset{|}{C}}—OH$ with $\underset{\oplus}{H—OC_2H_5}$ attached to C

Step III. $CH_3—\overset{O^-}{\overset{|}{C}}—OH$ (with $C_2H_5—\underset{\oplus}{O}—H$ attached) $\xrightarrow[-H_2O]{} CH_3—\overset{O}{\overset{\|}{C}}—OC_2H_5 + H_2O$

The equilibrium can be shifted towards ester formation by removing water as soon as it is formed. It can be done by adding benzene as a co-solvent and distilling the azeotropic mixture of benzene and water. (The benzene-water azeotrope contains 9% water and boils at 342 K).

The presence of bulky groups near the site of reaction, whether in the alcohol or in the acid, slows down esterification (and reverse hydrolysis) process.

Thus, reactivity of alcohols for given acid is in order

$$CH_3OH > 1° > 2° > 3°$$

and those of acids for given alcohol is in order

$$HCOOH > CH_3COOH > RCH_2COOH > R_2CHCOOH > R_3CCOOH$$

If acid is converted into acid halide and then esterified, total process is then irreversible, hence preferred.

$$\underset{\text{Ethanoic acid}}{CH_3\overset{O}{\overset{\|}{C}}OH} \xrightarrow[PCl_5]{I} \underset{\text{Acetyl chloride}}{CH_3\overset{O}{\overset{\|}{C}}Cl} \xrightarrow[C_2H_5OH]{II} \underset{\text{Ethyl ethanoate}}{CH_3\overset{O}{\overset{\|}{C}}OC_2H_5}$$

Stage II is called acetylation of alcohol and can also be done using acetic anhydride, $(CH_3CO)_2O$.

Note *It is the —OH group of acid and not that of alcohol that cleaves. Thus, isotopic ^{18}O of alcohol in the given reaction remains in ester.*

$$R—\overset{O}{\overset{\|}{C}}—OH + H—\overset{18}{O}R' \xrightarrow{H^+} R—\overset{O}{\overset{\|}{C}}—\overset{18}{O}R' + H_2O$$

Oxidation

The oxidation of an alcohol involves the formation of $>C=O$ group with the cleavage of an O—H and C—H bonds.

$$H—\overset{|}{\underset{|}{C}}—O—H$$

Bond break

The kind of product is formed depends upon how many of these α-hydrogens the alcohol contains, that is upon whether the alcohol is primary, secondary or tertiary.

(*i*) **Primary alcohols** are easily oxidised to aldehyde using oxidising agents like $K_2Cr_2O_7$ or acidified $KMnO_4$.

$$\underset{\text{Ethanol}}{CH_3CH_2OH} \xrightarrow[-H_2O]{[O]} \underset{\text{Ethanol}}{CH_3CHO} \xrightarrow{[O]} \underset{\text{Ethanoic acid}}{CH_3COOH}$$

If reaction is to be stopped at aldehydic stage then it must be distilled from the reaction as it is formed.

(*ii*) **Secondary alcohols** are oxidised to ketone by losing its only α-hydrogen. Ketones are resistant to oxidation and can only be carried out with stronger oxidising agents like conc.HNO_3 to give acids with less number of carbon atoms. In practical use, oxidation of C—C bond is quite difficult.

Therefore, ketone changes to the enolic form which then oxidised to acids as shown below.

$$\underset{\text{2-methylethanol}}{CH_3-\overset{\overset{CH_3}{|}}{C}H-OH} \xrightarrow[-H_2O]{[O]} \underset{\text{Propan-2-one}}{CH_3-\overset{\overset{CH_3}{|}}{C}=O} \xrightleftharpoons{\text{Enolisation}}$$

$$\underset{\text{Propenol}}{CH_3-\overset{\overset{CH_2}{\|}}{C}-OH} \xrightarrow[{[O]}]{-CO_2,\ -H_2O} \underset{\text{Ethanoic acid}}{CH_3\overset{\overset{O}{\|}}{C}OH}$$

(*iii*) **Tertiary alcohols** can be oxidised under vigorous conditions only. Using conc.HNO_3, they get dehydrated to alkenes which further oxidised to give acids with less number of C-atoms.

$$\underset{\textit{Tert}\text{-butylalcohol}}{CH_3-\overset{\overset{CH_3}{|}}{\underset{\underset{CH_3}{|}}{C}}-OH} \xrightarrow[-H_2O]{\text{Conc. } HNO_3} \underset{\text{2-methyl-propene}}{CH_3-\overset{\overset{CH_3}{|}}{\underset{\underset{CH_2}{\|}}{C}}} \xrightarrow{[O]}$$

$$\underset{\text{Acetone}}{CH_3-\overset{\overset{CH_3}{|}}{\underset{\underset{O}{\|}}{C}}} + CO_2 + H_2O$$

$$\underset{\text{Propan-2-one}}{CH_3\overset{\overset{O}{\|}}{C}CH_3} \xrightleftharpoons{\text{Enolisation}} \underset{\text{Enol}}{CH_3-\overset{\overset{OH}{|}}{C}=CH_2} \xrightarrow{[O]}$$

$$\underset{\text{Acetic acid}}{CH_3-\overset{\overset{OH}{|}}{C}=O} + CO_2 + H_2O$$

Different reagents used for oxidation are given in Table15.2.

Table 15.2 Different oxidising agents

S. No.	Oxidising agents	Conversion
1.	Cu at 570 K	1° alcohol to aldehyde; 2° alcohol to ketone, 3° alcohol to alkene
2.	MnO_4^-/H^+ or $Cr_2O_7^{2-}/H^+$	1° alcohol to carboxylic acid 2° alcohol to ketone 2° alcohol is also oxidised to ketone by CrO_3 in acetic acid
3.	PCC (pyridinium chloro chromate : $C_5H_5\overset{\oplus}{N}HCrO_3^{\ominus}$) PCC is soluble in non-polar solvents such as CH_2Cl_2	1° alcohol to aldehyde and 2° alcohol to ketone
4.	PDC (pyridinium dichromate $(C_5H_5NH)_2^{2+}\ Cr_2O_7^{2-}$	1° alcohol to aldehyde 2° alcohol to ketone
5.	H_2CrO_4 (chromic acid)	1° alcohol to carboxylic acid
6.	$CrO_3 \cdot H_2SO_4$/acetone	2° alcohol to ketone
7.	Jones reagent (chromic acid in aqueous acetone solution)	1° alcohol to carboxylic acid and 2° alcohol to ketone without affecting (C=C) bond
8.	MnO_2	1° allylic and benzylic alcohols to aldehyde and 2° allylic and benzylic alcohols to ketone
9.	N_2O_4 in $CHCl_3$ (Field's method)	1° and 2° benzyl alcohol to carbonyl compounds
10.	Collin's reagent (CrO_3 in pyridine) neutral $KMnO_4$, 71% HNO_3	1° alcohol to carboxylic acid and 2° alcohol to ketone

Reactions which involve C—OH bond cleavage

Reaction with hydrogen halide

Alcohols react readily with hydrogen halides to yield alkyl halides and water.

$$\underset{\text{Alcohol}}{R-OH} + HX \xrightarrow{ZnCl_2} \underset{\text{Alkyl halide}}{R-X} + H_2O$$

e.g.
$$\underset{2^\circ\text{alcohol}}{CH_3CH_2\overset{\overset{CH_3}{|}}{C}HOH} \xrightarrow[\text{heat}]{HCl + ZnCl_2} \underset{2^\circ\text{alkyl halide}}{CH_3CH_2\overset{\overset{CH_3}{|}}{C}HCl}$$

Here, mixture of conc. HCl and anhyd. $ZnCl_2$ is called Lucas reagent. Reactivity order for HX

$$HI > HBr > HCl$$

Reactivity order for ROH

$$\text{Allyl} \approx \text{benzyl} > 3^\circ > 2^\circ > 1^\circ < CH_3$$

Mechanism

On the basis of various evidences, it was concluded that all alcohols except methanol and mostly 1° alcohols, follow S_N1 mechanism.

$$ROH + H^+X^- \rightleftharpoons ROH_2^+ + X^-$$

$$ROH_2^+ \rightleftharpoons R^+ + H_2O \text{ (rate determining step)}$$

$$R^+ + X^- \longrightarrow RX$$

In 1° alcohol, rearrangement does not take place because they follow S_N2 mechanism. Protonated 1° alcohols do not form carbocations.

$$\underset{\substack{\text{(Protonated}\\ 1^\circ\text{ alcohol)}}}{ROH_2^+} + X^- \longrightarrow \underset{\text{Intermediate}}{\overset{\delta-}{X}\ldots R\ldots\overset{\delta+}{OH_2}} \longrightarrow \underset{\substack{\text{Alkyl}\\ \text{halide}}}{R-X} + H_2O$$

Reaction with phosphorus halides (PX_3 or PX_5) and thionyl chloride

Alcohols react with phosphorus halide (PX_3 or PX_5) or thionyl chloride ($SOCl_2$) in the presence of pyridine to give alkyl halide i.e., in this reaction hydroxyl group of an alcohol is replaced by halogen. **Thionyl chloride is preferred because the other two by products formed are escapable gases.** Hence, the reaction gives pure alkyl halides.

$$3ROH + PX_3 \longrightarrow 3RX + H_3PO_3 \quad (X = Br, I)$$

$$ROH + PX_5 \longrightarrow RX + POX_3 + HX$$

$$ROH + SOCl_2 \xrightarrow{\text{Dioxane}} RCl + SO_2\uparrow + HCl\uparrow$$

Reaction with PBr_3, P + I_2 and $SOCl_2$ provides a path of preparing 1° alkyl halides which are otherwise not possible with Lucas reagent.

$$\underset{\text{1-bromopropane}}{CH_3CH_2CH_2Br} \xleftarrow{PBr_3} \underset{\text{Propanol}}{CH_3CH_2CH_2OH} \xrightarrow{SOCl_2} \underset{\text{1-chloro propane}}{CH_3CH_2CH_2Cl}$$

$$\downarrow P+I_2$$

$$\underset{\text{1-iodopropane}}{CH_3CH_2CH_2I}$$

No reaction $\xleftarrow[ZnCl_2]{HCl}$ 2-cyclohexane ethanol (1°) $\xrightarrow{SOCl_2}$ 2-cyclohexane chloroethane

1-chloro-1-methyl cyclohexane $\xleftarrow[ZnCl_2]{HCl}$ 1-methyl-1-hydroxy cyclohexane (3°) $\xrightarrow{SOCl_2}$ 1-chloro-1-methyl cyclohexane

Note *Reaction of alcohol (with aryl group) with $SOCl_2$ follows S_N1 (substitution nucleophilic internal) reaction. Its main characteristics retention of stereochemical configuration.*

Table 15.3 Summary of substitution reactions of alcohols leading to alkyl halides

Alcohol	Reagent	Product	Nature of reaction
RCH_2OH 1°	HCl + $ZnCl_2$,	RCH_2X	S_N2
	or HBr or HI, PBr_3	RCH_2Br	S_N2
R_2CHOH 2°	HCl + $ZnCl_2$ or HBr, or HI, PBr_3	R_2CHX	S_N1
		R_2CHBr	S_N2
$R_3C—OH$ 3°	HCl, HBr or HI	$R_3C—X$	S_N1

Substitution reactions of alcohols using $SOCl_2$ and PBr_3 proceed by an S_N2 mechanism. Therefore, there is inversion of configuration at a stereogenic center.

(R) alcohol $\xrightarrow[\text{pyridine}]{SOCl_2}$ (S) chloride — Reaction using $SOCl_2$ proceed by an S_N2 mechanism = **inversion of configuration**

(R) alcohol $\xrightarrow{PBr_3}$ (S) bromide — Reaction using PBr_3 proceed by an S_N2 mechanism = **inversion of configuration**

With TsCl (*p*-toluene sulphonylchloride), TsO^- is formed which is a better leaving group.

H OH $\xrightarrow{TsCl}$ H OTs $\xrightarrow{Cl^-}$ H Cl (Inversion)

$$\left[CH_3—C_6H_4—SO_2O^-\right]$$

TsO^- is a better leaving group, thus in S_N2 reaction, Cl^- (nucleophilic) attacks such that (C — O) bond is cleaved.

Dehydration

Alcohols undergo dehydration to form alkenes on treating with a protic acid e.g., concentrated H_2SO_4 or H_3PO_4 or catalysts such as anhydrous calcium chloride or alumina.

$$\underset{\text{Alcohol}}{—\overset{|}{\underset{H}{C}}—\overset{|}{\underset{OH}{C}}—} \xrightarrow[\text{Heat}]{H^+} \underset{\text{Alkene}}{>C=C<} + H_2O$$

Relative ease of dehydration of alcohols follows the following order

Tertiary > Secondary > Primary

Mechanism

The mechanism follows the following three steps

$$—\underset{|}{\overset{H\ |}{C}}—\underset{|}{\overset{\ddot{O}H\ |}{C}}— \underset{\text{Step I}}{\overset{H^+}{\rightleftharpoons}} —\underset{|}{\overset{H\ |}{C}}—\underset{|}{\overset{\overset{+}{\ddot{O}}H_2\ |}{C}}— \underset{\text{Step II}}{\rightleftharpoons} —\underset{|}{\overset{H\ |}{C}}—\overset{+}{C}< \underset{\text{Step III}}{\rightleftharpoons} >C=C<$$

Step I. Formation of protonated alcohol.

Step II. Formation of carbocation. It is slowest step and hence, the rate determining step of the reaction.

Step III. Formation of ethene by elimination of a proton. In practice, alkene boils at lower temperature than the alcohol.

Hence, separation of the alkene and alcohol mixture is affected by distillation to drive the equilibrium towards alkene formation.

- Polar protic solvent favours equilibrium in forward direction. In all cases intermediate is carbocation and an E1 elimination can occur. There can be rearrangement of carbocation to give more stable carbocation.

$$\underset{\text{Butan-2-ol}}{CH_3CH_2\underset{|}{CH}CH_3} \underset{OH}{} \xrightleftharpoons{H^{\oplus}} CH_3CH_2\underset{\oplus OH_2}{CHCH_3} \xrightleftharpoons[-H_2O]{slow}$$

$$\underset{2^{\circ}\text{carbocation}}{CH_3CH_2\underset{\oplus}{C}HCH_3} \xrightarrow[-H^{\oplus}]{H_2O} \underset{\text{But-2-ene}}{CH_3CH{=}CHCH_3}$$

- In case of 1° alcohol, an unstable primary carbocation would be formed in the E1 dehydration hence, acid-catalysed elimination through a carbocation is difficult. Instead, a proton is lost from an adjacent carbon at the same time as water is lost. This allows the formation of an alkene without the intermediate formation of an unstable carbocation.

Thus, 1° alcohols shows E2 elimination reaction, as shown below

$$\underset{\text{Ethanol}}{H{-}C(H)(H){-}C(H)(H){-}\ddot{O}H} + H^+ \xrightarrow{E2} \underset{\text{Transition state}}{[H^{\delta+}\cdots C\cdots C\cdots O^{\delta-}{-}H]} \longrightarrow \underset{\text{Ethene}}{H^+ + CH_2{=}CH_2 + H_2O}$$

Here, BB = Bond breaking BF = Bond forming

- Alcohols leading to conjugated alkenes are dehydrated to a greater extent than those of alcohols leading to non-conjugated alkenes. Thus, dehydration is in order:

$$CH_2{=}CH\underset{OH}{C}HCH_3 > CH_3CH_2\underset{OH}{C}HCH_3$$

cyclohexadienol (OH) > cyclohexenol (OH) > cyclohexanol (OH)

CHECK POINT 15.2

1. Butanol has higher boiling point than diethyl ether due to
   (a) strong covalent bonding
   (b) intermolecular hydrogen bonding
   (c) electrovalent bonding
   (d) intramolecular hydrogen bonding
2. Boiling points of alcohols are generally high. This is due to
   (a) intermolecular hydrogen bonding attractions
   (b) dipole-dipole attractions
   (c) Both (a) and (b)
   (d) None of the above
3. Which has larger value of dipole moment?
   (a) Methanol (b) Phenol
   (c) Equal in both (d) No dipole moment in both
4. Thiols are alcohol analogs in which the oxygen has been replaced by sulphur (e.g., CH_3SH). Given the fact that the S—H bond is less polar than the O—H bond, which of the following statements comparing thiols and alcohols is correct?
   (a) Hydrogen bonding forces are weaker in thiols
   (b) Hydrogen bonding forces are stronger in thiols
   (c) Hydrogen bonding forces would be the same
   (d) No comparison can be made without additional information
5. Which of the following reaction is not reversible?
   (a) $RCOOH + R'OH \xrightarrow{H^+} RCOOR' + H_2O$
   (b) $CH_3CH_2CH_2OH \xrightarrow{H^+} CH_3CH{=}CH_2 + H_2O$
   (c) $PCl_5 \longrightarrow PCl_3 + Cl_2\uparrow$
   (d) $CH_3OH + O_2 \longrightarrow CO_2\uparrow + H_2O\uparrow$
6. Ester formation of CH_3COOH takes place to maximum extent when alcohol is
   (a) CH_3OH (b) CH_3CH_2OH
   (c) $(CH_3)_2CHOH$ (d) $(CH_3)_3COH$
7. The final product of the oxidation of ethyl alcohol is
   (a) acetic acid (b) ethane
   (c) acetaldehyde (d) acetone
8. Isopropyl alcohol on oxidation produces
   (a) ether (b) acetaldehyde
   (c) acetone (d) ethylene
9. When ethyl alcohol (C_2H_5OH) reacts with thionyl chloride in the presence of pyridine, the products obtained are
   (a) $CH_3CH_2Cl + HCl$
   (b) $CH_3CH_2Cl + Cl_2 + SO_2$
   (c) $CH_3CH_2Cl + H_2O + SO_2$
   (d) $C_2H_5Cl + HCl + SO_2$
10. In the following dehydration of alcohol (*A*) leading to the formation of styrene (*B*)

$$\underset{(A)}{C_6H_5{-}\underset{OH}{C}HCH_3} \xrightarrow[-H_2O]{H^+} \underset{(B)}{C_6H_5{-}CH{=}CH_2}$$

   (a) intermediate is carbocation
   (b) carbocation is resonance stabilised
   (c) it takes place rapidly
   (d) all the above are correct statements

DISTINCTION BETWEEN 1°, 2° AND 3° ALCOHOLS

Distinction among 1°, 2° and 3° alcohols can be made by following methods

Action of heated copper

When vapours of alcohols are passed over heated copper at 573K, primary and secondary alcohols undergo dehydrogenation to give aldehydes and ketones respectively whereas tertiary alcohols undergo dehydration to give alkenes.

$$\underset{\substack{\text{Ethanol}\\(1^\circ\text{ alcohol})}}{CH_3CH_2OH} \xrightarrow{Cu/573\text{ K}} \underset{\text{Ethanal}}{CH_3CHO} + H_2\uparrow$$

$$\underset{\text{Propan-2-ol}(2^\circ)}{(CH_3)_2C(H)OH} \xrightarrow{Cu/573\text{K}} \underset{\text{Propanone}}{(CH_3)_2C{=}O} + H_2\uparrow$$

$$\underset{\text{2-methyl propan-2-ol }(3^\circ)}{CH_3{-}C(CH_3)_2{-}OH} \xrightarrow{Cu/573\text{K}} \underset{\text{2-methylpropene}}{CH_3{-}C(CH_3){=}CH_2} + H_2O$$

Victor Meyer's test

This test consists of the following steps

Setp I. The given alcohol is first converted into its alkyl iodide by treating it with phosphorus and iodine (P/I_2).

Setp II. The alkyl iodide is then treated with silver nitrite $(AgNO_2)$ to convert it into its corresponding nitro alkane.

Setp III. The nitroalkane is then treated with nitrous acid (HNO_2) i.e. $NaNO_2 + HCl$.

Setp IV. The resulting solution is finally made alkaline with *aq.*NaOH or KOH.

Observation

If blood red colouration appears, it indicates a primary alcohol. If blue colouration appears, it indicates a secondary alcohol and if the solution remains colourless, it indicates a tertiary alcohol.

Reagent	Primary (1°) alcohol	Secondary (2°) alcohol	Tertiary (3°) alcohol
	CH_3CH_2OH	$(CH_3)_2CHOH$	$(CH_3)_3COH$
$\downarrow P/I_2$	CH_3CH_2I	$(CH_3)_2CH{-}I$	$(CH_3)_3C{-}I$
$\downarrow AgNO_2$	$CH_3CH_2NO_2$	$(CH_3)_2CHNO_2$	$(CH_3)_3C{-}NO_2$
$\downarrow HNO_2$	$R{-}C(=N{-}OH){-}NO_2$ (Nitrolic acid)	$(CH_3)_2C(NO_2){-}N{=}O$	(No reaction) No colour
$\downarrow NaOH$	Blood red colouration	Blue colouration	

Lucas test

This test is based upon the difference in reactivity of primary, secondary and tertiary alcohols with hydrochloric acid. This test consists of treating an alcohol with Lucas reagent, an equimolar mixture of conc.HCl and anhydrous $ZnCl_2$ at room temperature, turbidity due to the formation of insoluble alkyl chlorides is observed.

Reaction of *R*OH with HCl indicates basic nature of alcohol. Basic nature of alcohol 1° < 2° < 3°

Thus, reactivity with $HCl/ZnCl_2$: 1° < 2° < 3°

Observation

(*i*) When turbidity appears immediately, the alcohol is tertiary.

(*ii*) When turbidity appears in about five minutes, the alcohol is secondary.

(*iii*) When reaction does not take place with Lucas reagent, then it is primary alcohol.

Reactions involved are as follows

$$\underset{3^\circ}{R_3COH} \xrightarrow[\text{anhyd. }ZnCl_2]{\text{Conc.HCl }+} \underset{\substack{\text{Turbidity appears}\\\text{immediately}}}{R_3CCl} + H_2O$$

$$\underset{2^\circ}{R_2CHOH} \xrightarrow[\text{anhyd. }ZnCl_2]{\text{Conc.HCl }+} \underset{\substack{\text{Turbidity appears}\\\text{in 5 min}}}{R_2CHCl}$$

$$\underset{1^\circ}{RCH_2OH} \xrightarrow[\text{anhyd. }ZnCl_2]{\text{Conc.HCl }+} \text{No reaction}$$

Haloform test

Alcohols (also aldehydes and ketones) on heating with X_2 (Cl_2, Br_2 and I_2) and OH^- form CHX_3 (haloform). This reaction can be used to identify the compound (alcohol or carbonyl compound) and is called **haloform test.**

This reaction is given by the following molecules

(*i*) CH_3CH_2OH

(*ii*) $CH_3CH(OH)R$ (2° alcohol with —OH at C_2)

(*iii*) $CH_3C(=O)R$ (keto group at C_2 or compound with acetyl CH_3CO- group)

If we CH_3CHO (Thus, *R* can be H, alkyl or aryl) Since, reaction takes place with alkali solution as one of the reagents hence, alkyl halide like CH_3CH_2Cl and $CH_3CH(Cl)R$ will also give this test.

$$CH_3-CH(X)-R \xrightarrow{\text{Alkali}} CH_3-CH(OH)-R \xrightarrow{[O]} CH_3-C(=O)-R \xrightarrow{X_2} CX_3-C(=O)-R$$

$$CX_3-C(=O)-R \xrightarrow{H_2O/H^+} \underset{\text{(Haloform)}}{CHX_3\downarrow} + R-C(=O)-OH \xrightarrow{\text{Alkali}} R-C(=O)-\overset{-}{O}\overset{+}{Na}$$

e.g.

- $$\underset{\text{Ethanol}}{CH_3CH_2OH} + 4I_2 + 6NaOH \xrightarrow{\Delta} \underset{\text{Iodoform}}{CHI_3\downarrow} + HCO\overset{-}{O}\overset{+}{Na} + 5H_2O + 5NaI$$
- $$\underset{\text{2° alchohol}}{CH_3CH(OH)R} + 4I_2 + 6NaOH \xrightarrow{\Delta} CHI_3\downarrow + RCO\overset{-}{O}\overset{+}{Na} + 5H_2O + 5NaI$$
- $$R-C(=O)-CH_3 + 3I_2 + 4NaOH \xrightarrow{\Delta} CHI_3\downarrow + RCO\overset{-}{O}\overset{+}{Na} + 3H_2O + 3NaI$$

- $$R\overset{*}{C}H(OH)-CH_3 \xrightarrow[\Delta]{X_2,\ OH^-} RCOO^- + CHX_3$$
- $$H\overset{*}{C}H(OH)-CH_3 \xrightarrow[\Delta]{X_2,\ OH^-} HCOO^- + CHX_3$$
- $$R\overset{*}{C}(=O)-CH_3 \xrightarrow[\Delta]{X_2,\ OH^-} RCOO^- + CHX_3$$
- $$H\overset{*}{C}(=O)-CH_3 \xrightarrow[\Delta]{X_2,\ OH^-} HCOO^- + CHX_3$$

CH_3 changes into CHX_3 and part $\overset{*}{C}$ oxidises to $-COO^-$.

This fact can be used to determine nature of alcohol or carbonyl compound.

SOME COMMERCIALLY IMPORTANT ALCOHOLS

Important properties and uses of some commertially important alcohols are as follows

Methanol

Methanol is a colourless liquid and boils at 337K. It is also known as 'wood spirit' as it was produced by destructive distillation of wood. It was also produced by catalytic hydrogenation of carbon monoxide at the high temperature and in the presence of ZnO-Cr_2O_3 catalyst.

It is highly poisonous in nature. Ingestion of even small quantities of methanol is hazardous and large quantities cause even death. It is used as a solvent in paints, varnishes and for manufacturing formaldehyde dyes, medicines and perfumes.

Ethanol

Ethanol is a colourless liquid with boiling point 351K. It is used in alcoholic beverages, produced by fermentation of sugars by yeast. It is used as a solvent in paint industry. It is also a good industrial solvent, hence it is made unfit for drinkable purposes, by addition of a denaturant like CH_3OH which is poisonous.

It gives carbon dioxide and water on burning. Thus, it is used as a fuel. It is safe and can be used to dissolve many organic compounds, which are insoluble in water.

Ethanol is used for medical purposes as a Central Nervous System (CNS) depressant.

Rectified and denatured spirit

- **Rectified spirit** is ≈ 95 per cent *w/w* ethanol. It is also called **95% alcohol.** It is a mixture of 95% alcohol and 5% water (bp 351.15 K). A liquid mixture that has the peculiar property of giving a vapour of the same composition is called an **azeotrope** or a **constant-boiling mixture.** Thus, rectified spirit contains two components and is a **binary azeotrope.**
- **Denatured spirit** (called methylated spirit) is 95% rectified spirit and 5% methanol. **Mineralised methylated** spirit is 90% rectified spirit, 9% methanol and 1% petroleum oil and a purple dye (for distinction). 100% ethyl alcohol is called **absolute alcohol** and is obtained from rectified spirit.

Dihydric alcohol (glycol)

The compounds which contain two —OH groups attached to two successive C—atoms are called **glycols.**

Ethylene glycol

Ethylene glycol contain two — OH groups. It is the first member and simplest dihydric alcohol. It is colourless, viscous and toxic liquid with boiling point 197°C and melting point −13° C.

Preparation

It can be prepared by the following reactions

(*i*) **Oxidation of ethylene** (hydroxylation of double bond)

$$\underset{\text{Ethene}}{CH_2{=}CH_2} + KMnO_4 + KOH \longrightarrow \underset{\text{Ethane-1, 2-diol or (ethylene glycol)}}{\begin{array}{c} CH_2OH \\ | \\ CH_2OH \end{array}}$$

(*ii*) **Hydrolysis of 1, 2-dibromoethane**

$$\underset{\text{1, 2-dibromo ethane}}{\begin{array}{l} CH_2{-}Br \\ | \\ CH_2{-}Br \end{array}} + Na_2CO_3 + H_2O \longrightarrow \underset{\text{Ethane-1, 2-diol or (ethylene glycol)}}{\begin{array}{l} CH_2{-}OH \\ | \\ CH_2{-}OH \end{array}} + 2NaBr + CO_2\uparrow$$

(*iii*) **Hydrolysis of ethylene halohydrin**

$$\underset{\text{Ethylene halohydrin}}{\begin{array}{l} CH_2{-}OH \\ | \\ CH_2{-}X \end{array}} + NaHCO_3 \longrightarrow \underset{\text{Ethane-1, 2-diol or (ethylene glycol)}}{\begin{array}{l} CH_2OH \\ | \\ CH_2OH \end{array}} + NaX + CO_2\uparrow + H_2O$$

(*iv*) **Hydrolysis of ethylene oxide**

$$\underset{\text{1, 2-epoxyethane}}{\begin{array}{l} CH_2 \\ | \quad \rangle O \\ CH_2 \end{array}} \xrightarrow[\text{Nucleophilic addition (Under pressure)}]{H_2O/H^+,\ 200°C} \underset{\text{Ethan-1, 2-diol}}{\begin{array}{l} CH_2OH \\ | \\ CH_2OH \end{array}}$$

Properties

Ethylene glycol gives following chemical reactions

(*i*) Replacement of hydrogen from the second hydroxyl group is difficult. Once the alkoxide is formed by the replacement of one hydrogen, it behaves like an electron-releasing group and thus O—H bond becomes more stronger and thereby decreasing reactivity towards sodium.

$$\underset{\text{(Ethane-1, 2-diol) Ethylene glycol}}{\begin{array}{l} CH_2OH \\ | \\ CH_2OH \end{array}} \xrightarrow{Na} \underset{\text{Bond-energy increases in electron density over oxygen}}{\begin{array}{l} CH_2\overset{\ominus\oplus}{ONa} \\ | \quad \rightarrow +I\text{- effect} \\ CH_2{-}O{-}H \end{array}}$$

(*ii*) Reaction of ethylene glycol with Na at various temperature.

$$\underset{\text{Ethylene glycol disodium salt}}{\begin{array}{l} CH_2\overset{-\,+}{ONa} \\ | \\ CH_2\overset{-\,+}{ONa} \end{array}} \xleftarrow{Na,\ 430\ K} \underset{\text{(Ethane-1, 2-diol) Ethylene glycol}}{\begin{array}{l} CH_2OH \\ | \\ CH_2OH \end{array}} \xrightarrow{Na,\ 320\ K} \begin{array}{l} CH_2\overset{-\,+}{ONa} \\ | \\ CH_2OH \end{array}$$

(*iii*) Reaction of ethylene glycol with HCl at various temperature and with phosphorus halides.

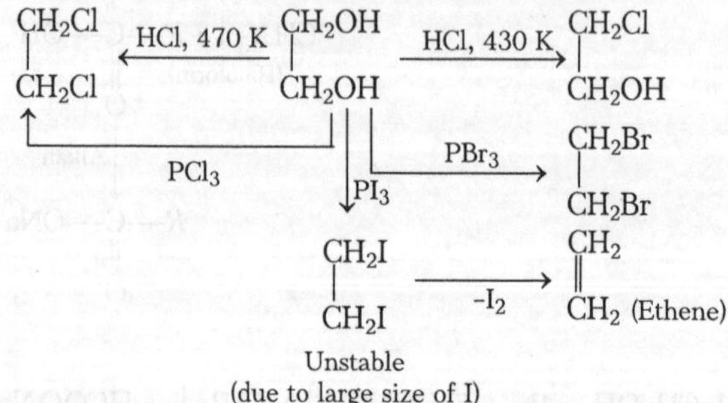

(*iv*) Reaction with acetic acid or acid derivatives.

$$\underset{\text{Ethylene glycol}}{\begin{array}{l} CH_2OH \\ | \\ CH_2OH \end{array}} + \underset{\text{Acetic acid}}{2\,CH_3CO_2H} \xrightarrow{H_2SO_4} \underset{\text{Glycol diacetate}}{\begin{array}{l} CH_2O\overset{O}{\overset{\|}{C}}CH_3 \\ | \\ CH_2O\underset{O}{\underset{\|}{C}}CH_3 \end{array}}$$

(also via CH_3COCl or $(CH_3CO)_2O$)

Two moles of CH_3COCl or $(CH_3CO_2)O$ react per mole of glycol indicating presence of two hydroxyl groups in glycol.

(*v*) On distilling glycol with fuming sulphuric acid 1,4-dioxane is formed.

$$2\begin{array}{c}CH_2OH\\|\\CH_2OH\end{array} \xrightarrow[\text{(oleum)}]{\text{Fuming } H_2SO_4} \text{1,4-dioxane} + 2H_2O$$

Ethylene glycol → 1,4- dioxane

(*vi*) On heating with dehydrating agent (like $ZnCl_2$) glycol is converted into acetaldehyde.

$$\begin{array}{c}CH_2OH\\|\\CH_2OH\end{array} \xrightarrow{ZnCl_2, \Delta} CH_3CHO + H_2O$$

Ethylene glycol → Acetaldehyde

(*vii*) Dehydrating agent like phosphoric acid converts glycol into polyethylene glycols.

$$\begin{array}{c}CH_2OH\\|\\CH_2OH\end{array} + \begin{array}{c}CH_2OH\\|\\HOCH_2\end{array} \xrightarrow[\Delta]{H_3PO_4} \begin{array}{c}OHCH_2 \quad CH_2OH\\| \qquad |\\CH_2OCH_2\end{array} + H_2O$$

Ethylene glycol → Diethylene glycol

Depending on the temperature and amount of dehydrating agent H_3PO_4, polyethylene glycols (upto decaethylene glycol) can be obtained. These polyethylene glycols contain both the alcohol and ether functional groups. They are water soluble (due to alcoholic group) and are very good solvents (ethereal function).

(*viii*) Glycol condenses with aldehydes or ketones in the presence of *p*-toluene sulphonic acid to form cyclic acetals (1,3-dioxolane). At one time it was customary to designate the products of addition of alcohols to ketones as **ketals.** The term has been dropped from the IUPAC system of nomenclature and the term **acetals** is now applied to the adducts of both aldehydes and ketones.

$$\begin{array}{c}CH_2OH\\|\\CH_2OH\end{array} + O{=}C\begin{array}{l}R\\H\end{array} \xrightarrow{p\text{-toluene sulphonic acid}} \begin{array}{c}CH_2{-}O\\|\\CH_2{-}O\end{array}C\begin{array}{l}R\\H\end{array}$$

Ethylene glycol + Aldehyde → Cyclic acetals

$$\begin{array}{c}CH_2OH\\|\\CH_2OH\end{array} + O{=}C\begin{array}{l}R\\R\end{array} \xrightarrow{p\text{-toluene sulphonic acid}} \begin{array}{c}CH_2{-}O\\|\\CH_2{-}O\end{array}C\begin{array}{l}R\\R\end{array}$$

Ethylene glycol + Ketone → Dioxolane

Dioxolane formation provides a path of protecting a carbonyl group in reactions studied in basic medium in which acetals are not affected.

The carbonyl compound may be regenerated by the addition of periodic acid to aqueous solution of the dioxolane or by acidic hydrolysis.

$$\begin{array}{l}R\\R\end{array}C{=}O + \begin{array}{c}CH_2OH\\|\\CH_2OH\end{array} \xleftarrow{H_3O^+} \begin{array}{l}R\\R\end{array}C\begin{array}{c}O{-}CH_2\\|\\O{-}CH_2\end{array}$$

Ethylene glycol ← Cyclic ketal (Dioxolane)

$$\xrightarrow{HIO_4} RCOR + 2HCHO$$

Ketone, Methanal

Aldehyde is more reactive than ketone in dioxolane formation.

(*ix*) Upon treatment with periodic acid, HIO_4, compounds containing two or more —OH or ═O groups attached to the adjacent carbon atoms undergo oxidation with cleavage of carbon-carbon bonds. Products are the same as obtained by ozonolysis of alkenes.

$$\begin{array}{c}CH_2OH\\|\\CH_2OH\end{array} + HIO_4 \longrightarrow HCHO + HCHO + HIO_3 + H_2O$$

$$\begin{array}{c}R'CHOH\\|\\RCHOH\end{array} + HIO_4 \longrightarrow RCHO + R'CHO + HIO_3 + H_2O$$

$$\begin{array}{c}R'\,CH{-}C\,R\\|\quad\;\|\\OH\;\;O\end{array} + HIO_4 \longrightarrow R'CHO + RCOOH + HIO_3$$

Cyclic diols give dicarbonyl compounds. The reactions are faster when the hydroxyl groups are *cis* than when they are *trans* but both stereoisomers are oxidised by HIO_4.

$$\text{Cyclopentane 1,2-diol} \xrightarrow{HIO_4} \overset{O}{\overset{\|}{H C}}CH_2CH_2CH_2\overset{O}{\overset{\|}{CH}}$$

Cyclopentane 1, 2-diol (either stereoisomer) → Pentane-1, 5-dial

Oxidation is particularly useful in determination of structure. $KMnO_4$ in acidic, neutral or alkaline medium, acid dichromate or CrO_3 in acetic acid or lead tetra-acetate has the same role.

(x) Depending on the type of oxidant used, glycol gives variety of products isolated in major as well as minor quantity.

$$\underset{\text{Ethylene glycol}}{\begin{array}{c}CH_2OH\\|\\CH_2OH\end{array}} \xrightarrow{O} \underset{\text{Glycol aldehyde}}{\begin{array}{c}CHO\\|\\CH_2OH\end{array}} \xrightarrow{O} \underset{\text{Glycolic acid}}{\begin{array}{c}COOH\\|\\CH_2OH\end{array}} \xrightarrow{O} \underset{\text{Glyoxalic acid}}{\begin{array}{c}COOH\\|\\CHO\end{array}}$$

$$\xrightarrow{O} \underset{\text{Glyoxal}}{\begin{array}{c}CHO\\|\\CHO\end{array}} \xrightarrow{O} \underset{\text{Glyoxalic acid}}{\begin{array}{c}COOH\\|\\CHO\end{array}} \xrightarrow{O} \underset{\text{Oxalic acid}}{\begin{array}{c}COOH\\|\\COOH\end{array}} \xleftarrow{O}$$

Uses

Glycol is used to prepare polyesters. The most common polyester is **dacron.** This fibre is used to make fabric and tyre cord and mylar film is used to make magnetic recording tape. It is used as an antifreeze for automobile radiators.

Dogs are often poisoned by sweet-tasting ethylene glycol which is metabolished to oxalic acid causing kidneys failure which leads to their death.

Trihydric alcohol (Glycerol)

Compounds containing three —OH groups at successive C-atoms are called triols.

Glycerol

Glycerol is the simplest trihydric alcohol. It contains three OH groups with boiling point 563K. This high boiling point is due to increase in hydrogen bonding as a result of an extra —OH group.

$$\underset{\text{Glycerine (glycerol)}}{\begin{array}{l}CH_2—OH\\|\\CH—OH\\|\\CH_2—OH\end{array}}$$

It is colourless, odourless, thick, syrupy liquid having a sweet taste. It is hygroscopic, soluble in water and insoluble in ether. It decomposes much below its boiling point hence, evaporation is carried under vacuum. It is non-toxic and used as antifreeze in auto-mobile radiators.

Preparation

(i) From fats and oil

$$\underset{\text{Fat or oil}}{\begin{array}{l}CH_2OOCR\\|\\CHOOCR\\|\\CH_2OOCR\end{array}} \xrightarrow{aq.NaOH} \underset{\text{Glycerol}}{\begin{array}{l}CH_2OH\\|\\CHOH\\|\\CH_2OH\end{array}} + \underset{\text{Soap}}{3RCOO^-Na^+}$$

This reaction is known as saponification reaction.

(ii) From propene

$$\underset{\text{Propene}}{\begin{array}{c}CH_3\\|\\CH\\\|\\CH_2\end{array}} \xrightarrow[600°C]{Cl_2} \underset{\text{3-chloropropene}}{\begin{array}{c}CH_2Cl\\|\\CH\\\|\\CH_2\end{array}} \xrightarrow[-NaCl]{Dil.NaOH} \underset{\text{Propen-3-ol}}{\begin{array}{c}CH_2OH\\|\\CH\\\|\\CH_2\end{array}} \xrightarrow{HOCl}$$

$$\underset{\text{3-chloro 1, 2-propan-diol}}{\begin{array}{c}CH_2OH\\|\\CHOH\\|\\CH_2Cl\end{array}} \xrightarrow[-NaCl]{Dil.NaOH} \underset{\text{Glycerol}}{\begin{array}{c}CH_2OH\\|\\CHOH\\|\\CH_2OH\end{array}}$$

Properties

Glycerol gives following chemical reactions.

(i) **Reaction with HNO_3** It reacts with HNO_3 to form trinitroglycerine, an explosive.

$$\underset{\text{Glycerol}}{\begin{array}{l}CH_2O\boxed{H + HO}NO_2\\|\\CHO\boxed{H + HO}NO_2\\|\\CH_2O\boxed{H + HO}NO_2\end{array}} \xrightarrow[-3H_2O]{} \underset{\text{Trinitroglycerine (TNG)}}{\begin{array}{l}CH_2ONO_2\\|\\CHONO_2\\|\\CH_2ONO_2\end{array}}$$

(ii) **Dehydration** Glycerol undergoes dehydration reaction with $KHSO_4$ to yield acrolein.

$$\underset{\text{Glycerol}}{\begin{array}{c}H\\|\\H—C—OH\\|\\HO—C—H\\|\\H—C—OH\\|\\H\end{array}} \xrightarrow[-2H_2O]{KHSO_4,\Delta} \underset{\text{Prop-1, 2-dien-ol}}{\begin{array}{c}H—C—OH\\\|\\C\\\|\\CH_2\end{array}} \xrightleftharpoons{\text{Tautomerise}} \underset{\text{Acrolein}}{\begin{array}{c}H—C=O\\|\\CH\\\|\\CH_2\end{array}}$$

(iii) **Oxidation** Glycerol undergoes oxidation to give number of products.

Different reagents give different oxidation products. Some reagents are as follows

Reagents	Oxidation product
Dil. HNO_3	Glyceric acid + tartonic acid
Conc. HNO_3	Glyceric acid
Bismuth nitrate	Mesoxalic acid
Periodic acid	Formaldehyde + formic acid
Br_2 water or Fenton's reagent($FeSO_4$ + H_2O_2)	Glyceraldehyde + dihydroxyacetone

(*iv*) **With oxalic acid**

- At 383K, oxalic acid is converted to formic acid in presence of glycerol.

$$\begin{array}{l}\text{COOH}\\ |\\ \text{COOH}\end{array} \xrightarrow{\text{383K, glycerol}} CO_2 + \underset{\text{Formic acid}}{HCOOH}$$
Oxalic acid

Glycerol actually takes part in a sequence of reaction but is regenerated at the end.

- At 833K, alkyl alcohol is formed.

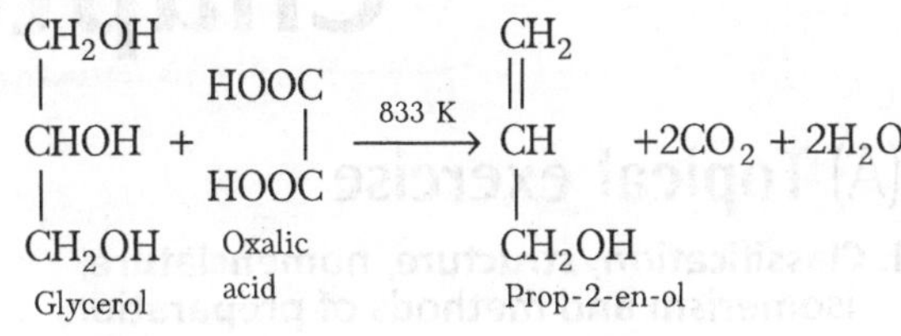

CHECK POINT 15.3

1. The Lucas reagent does not react with
(a) $CH_3(CH_2)_2OH$ (b) $(CH_3)_3COH$
(c) $(CH_3)_2CHOH$ (d) CH_3CH_2OH

2. $C_4H_{10}O$ gives white precipitate immediately with conc. HCl in presence of anhydrous $ZnCl_2$. Alcohol can be

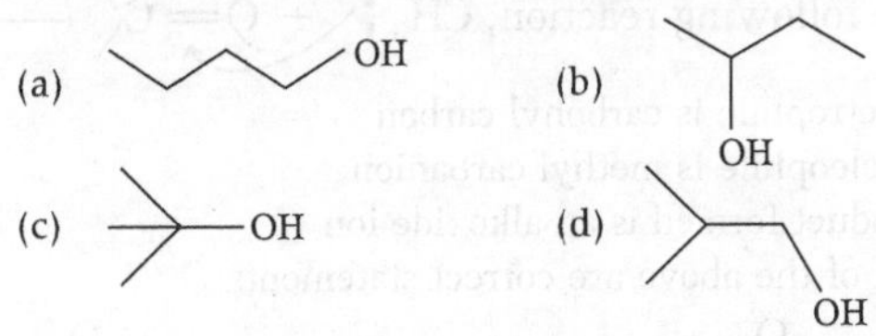

3. Alcohol that gives blue colouration with Victor Meyer's test is
(a) CH_3OH (b) C_2H_5OH
(c) $(CH_3)_2CHOH$ (d) $(CH_3)_3COH$

4. Which one of the following gives iodoform test?
(a) Formaldehyde (b) Benzaldehyde
(c) Ethanol (d) Benzyl alcohol

5. C_2H_5OH and CH_3OH can be differentiated by
(a) I_2 + NaOH (b) I_2 + HCl
(c) H_2O (d) NH_3

6. Select the incorrect alternative.
(a) methanol is very toxic to humans, leading to blindness and even death
(b) 98% methanol —H_2O mixture is called rectified spirit
(c) methanol is used as a premium fuel to power racing car engines
(d) ethanol is mixed with petroleum products to produce non-lead gasoline for use in internal combustion engine

7. Use of methanol for drink may cause blindness and even death. It is due to
(a) its metabolisation to formic acid which causes kidney failure
(b) its reduction to CH_4
(c) its oxidation to CO_2
(d) its oxidation to CO

8. Compound used in manufacture of dacron
(a) ethylene glycol (b) methanol
(c) ethanol (d) glycerol

9. Dogs are often poisoned by sweet-tasting ethylene glycol. Poisoning is due to
(a) its oxidation to glycol aldehyde
(b) its metabolisation to oxalic acid which causes kidney failure leading to death
(c) Both (a) and (b)
(d) None of the above

10. Which of the following compounds can synthesise glycerol?
(a) Fats and oils
(b) Propene
(c) Both (a) and (b)
(d) None of the above

Chapter Exercises

(A) Topical exercise

I. Classification, structure, nomenclature, isomerism and methods of preparation

1 The IUPAC nomenclature for the compound (*A*) is

$$\underset{(A)}{CH_3-CH=\underset{\underset{CH_2-OH}{|}}{C}-CH_2CH_2\,CH_2Cl}$$

(a) 1-chloro-4-(hydroxy methyl) hex-4-ene
(b) 2-(chloropropyl) but-2-en-1-ol
(c) 2-(3-chloropropyl) but-2-en-1-ol
(d) (3-propyl chloro) but-2-en-1-ol

2 Total number of structural isomers shown by $C_4H_{10}O$ are
(a) 3 (b) 7 (c) 4 (d) 6

3 Which of the following statements is correct for hydration of alkene?
(a) It follows S_N1 mechanism
(b) It is a irreversible process
(c) It takes place through carbocation mechanism
(d) It follows all of the above

4 In the hydration of an alkene carbocation is formed from
(a) carbanion (b) oxonium ion
(c) hydroxide ion (d) hydride ion

5. How many alcohols with molecular formula $C_4H_{10}O$ are chiral in nature?
(a) 1 (b) 2 (c) 3 (d) 4

6 The process of converting alkyl halides into alcohols involve ....
(a) addition reaction
(b) substitution reaction
(c) dehydrohalogenation reaction
(d) rearrangement reaction

7 In the following reaction,

$$H-\underset{Et}{\overset{Me}{\underset{|}{\overset{|}{C}}}}-Br \xrightarrow[S_N1]{OH^-}$$

product formed
(a) retains configuration (b) inverts configuration
(c) is a racemic mixture (d) is a *meso* compound

8 Consider the following methods
I. Hydration of olefins
II. Hydrolysis of cyanides
III. Reduction of carbonyl compounds

Which of these can be used to prepare alcohols?
(a) I, II and III (b) Both I and II
(c) Both II and III (d) Both I and III

9 Formation of alcohol by oxymercuration-demercuration of alkene
(a) involves carbocations and rearrangement
(b) involves carbanions and rearrangement
(c) is stereospecific
(d) does not involve rearrangement

10 Which of the following alkyl halide is hydrolysed to alcohol to a maximum extent?
(a) CH_3F (b) CH_3Cl (c) CH_3I (d) CH_3Br

11 In the following reaction, $CH_3\overset{\ominus}{:} + \overset{\delta-}{O}=\overset{\delta+}{C}\langle \longrightarrow$
(a) electrophile is carbonyl carbon
(b) nucleophile is methyl carbanion
(c) product formed is an alkoxide ion
(d) All of the above are correct statements

12 $CH_3\overset{O}{\overset{\|}{C}}CH_2\overset{O}{\overset{\|}{C}}OCH_3$ is reduced to $CH_3\,\underset{OH}{\underset{|}{C}H}CH_2\,\overset{O}{\overset{\|}{C}}O\,CH_3$ by
(a) $LiAlH_4$ (b) $NaBH_4$ (c) Both (a) and (b) (d) None of these

13 Select the incorrect statement about hydroboration-oxidation.
(a) addition is against Markownikoff rule
(b) intermediate is a carbocation
(c) it does not involve rearrangement
(d) addition of boron and hydrogen to double bond is a *syn*-addition

14 Which of the following statement is correct for oxo process?
(a) It is simply carbonylation or hydroformylation reaction
(b) In this cobalt carbonyl hydride $[CoH(CO)_4]$ used as catalyst
(c) Both statements are correct
(d) Both statements are incorrect

15 $CH_3CH=CH_2 \xrightarrow[125°,\ \text{pressure}]{CO,\ H_2,\ [CoH(CO)_4]} \xrightarrow[\text{Cu-Zn}]{H_2}$ Product

This represents oxo method of alcohol synthesis. Product can be
(a) $CH_3CH_2CH_2CH_2OH$ (b) $CH_3\underset{CH_3}{\underset{|}{C}H}CH_2OH$
(c) Both (a) and (b) are true (d) None is true

16 $(CH_3)_3CCH{=}CH_2 \xrightarrow{X} (CH_3)_3CCH(OH)CH_3$

X can be

(a) $BH_3 \cdot THF / H_2O_2 \cdot OH^-$ (b) H_3O^+
(c) $Hg(OAc)_2$ / $NaBH_4$, NaOH (d) None of these

17 Which of the following would yield a secondary alcohol after the indicated reaction followed by hydrolysis if necessary?

(a) $LiAlH_4$ + a ketone
(b) CH_3CH_2MgBr + an aldehyde
(c) but-2-ene +aqueous H_2SO_4
(d) All of the above

18 Which of the following alcohols gives the best yield of dialkyl ether on being heated with a trace of sulphuric acid?

(a) Pentanol (b) Cyclopentanol
(c) Pentan-2-ol (d) 2-methyl butan-2-ol

19 Consider the following reaction,

(2-methylcyclobut-2-en-1-one) $\xrightarrow{KMnO_4/OH^-} \xrightarrow{Cu/\Delta} A.$

A formed is

(a) HO–cyclobutane with CH₃, OH and OH substituents
(b) cyclobutanone with CH₃, OH and =O substituents
(c) cyclobutane-1,3-dione with CH_2OH substituent
(d) cyclobutane-1,3-dione with CHO substituent

20 Cl–◇–Br $\xrightarrow{Mg/ether} \xrightarrow[H_3O^+]{HCHO} A.$

A is

(a) Br–◇–CH_2OH
(b) Cl–◇–CH_2OH
(c) HOH_2C–◇–CH_2OH
(d) OHC–◇–CHO

21 (lactone ring with COOH) $\xrightarrow{A}$ (lactone ring with CH_2OH) *A* is

(a) B_2H_6/H_2O (b) $LiAlH_4$
(c) CH_3OH / Na (d) P/HI

22 $(C_2H_5O)_2CO \xrightarrow[H_3O^+]{CH_3MgBr\ (excess)} A.$

A (alcohol) can also be obtained by

(a) $CH_3CH_2CHO \xrightarrow[H_3O^+]{CH_3MgBr\ (2\ mol)}$

(b) $CH_3COC_2H_5$ (ester, C=O) $\xrightarrow[H_3O^+]{CH_3MgBr\ (2\ mol)}$

(c) CH_3CCH_3 (C=O) $\xrightarrow[H_3O^+]{CH_3MgBr\ (1\ mol)}$

(d) as in (b) and (c)

23 Which of the following reagents would carry out the following transformation? ($D{=}^2H$, the mass-2 isotope of hydrogen)

$C_6H_5COCH_3 \xrightarrow{?} C_6H_5C(OH)(D)CH_3$

(a) $NaBD_4$ in CH_3OH
(b) $NaBD_4$ in CH_3OD
(c) $LiAlH_4$ then D_2O
(d) $LiAlD_4$ then D_2O

24 Consider the following statements.

Statement I Oxymercuration-demercuration of alkene gives alcohol without rearrangement.

Statement II There can be 1,2-hydride or 1,2-alkyl shift to make more stable carbocation.

(a) Both statements I and II are correct
(b) Only statement I is correct
(c) Only statement II is correct
(d) Neither statement I nor statement II is correct

25 CH_3MgBr + (cyclohexanone) $\xrightarrow{H_3O^+} A \xrightarrow{HBr} B \xrightarrow{Mg/ether} C \xrightarrow[H_3O^+]{HCHO} D,$

Major product (*D*) is

(a) cyclopentane with H_3C and CH_2OH on the same carbon
(b) cyclopentane with CH_2OH and CH_3 on adjacent carbons
(c) cyclopentane with OH, CH_3 and CH_3 substituents
(d) cyclohexane with CH_2OH

26 1-Methylcyclohexene $\xrightarrow[\text{Alkaline KMnO}_4]{\text{Cold}} A \xrightarrow[\text{AcOH}]{\text{CrO}_3} B$

A and B respectively are

(a) 1-methylcyclohexane-1,2-diol (CH_3, OH / OH), 2-hydroxy-2-methylcyclohexanone (CH_3, OH / =O)

(b) 1-methylcyclohexane-1,2-diol (CH_3, OH / OH), 2-methylcyclohexane-1,3-dione (CH_3 / =O, =O)

(c) 1-methylcyclohexane-1,2-diol (CH_3, OH / OH), 2-methylcyclohexanol (CH_3 / OH)

(d) no formation of A and B takes place

27 What is the end-product of the following?

F–(cyclobutane)–Br $\xrightarrow{\text{Mg/ether}} \xrightarrow{\text{HCHO}} \xrightarrow{H_3O^{\oplus}}$

(a) HOH_2C–(cyclobutane)–CH_2OH (b) F–(cyclobutane)–CH_2OH

(c) Br–(cyclobutane)–CH_2OH (d) bicyclobutane

28 In the following reaction,

H–C(Et)(C_6H_5)–I $\xrightarrow[S_N2]{OH^-}$

product formed

(a) retains configuration (b) inverts configuration

(c) is a racemic mixture (d) is a *meso* compound

29 In the following reaction, (X) is

(keto-aldehyde: O, CHO) $\xrightarrow{\text{NaB(OAc)}_3\text{H}} X$

(a) (diol: OH, OH) (b) (ketone with OH: O, OH)

(c) (OH, CHO) (d) (cyclic ether: O)

30 $Y \xleftarrow{H_3O^+} H_2C=$(cyclohexane)$=CH_2 \xrightarrow[\text{(ii) } H_2O_2/\overline{O}H]{\text{(i) } B_2H_6 \text{ (excess)}} X.$

X and Y respectively are

(a) HOH_2C–(cyclohexane)–CH_2OH, H_3C–(cyclohexane, HO, OH)–CH_3
(I) (II)

(b) Both II and I

(c) I in both case

(d) II in both case

31 Consider the following statements.

Statement I CH_3OH is more acidic than *t*-butyl alcohol.

Statement II Three methyl electron-attracting groups in *t*-butyl aclohol decreases electron-density.

(a) Both statements I and II are correct

(b) Only statement I is correct

(c) Only statement II is correct

(d) Neither statement I nor statement II is correct

32 (δ-valerolactone, O, =O) $\xrightarrow[H_3O^+]{\text{CH}_3\text{MgBr (excess)}} X$

X is

(a) $(H_3C)_2C(OH)$–$CH_2CH_2CH_2CH_2$–OH (b) HO–$CH(H_3C)$–$CH_2CH_2CH_2$–OH

(c) (tetrahydropyran with OH, CH_3) (d) (tetrahydropyran)

33 (benzene)–CH=CHCH$_3$ $\xrightarrow{H_3O^+}$ A (Major)

(a) (benzene)–CH_2CHCH_3 (OH)

(b) (benzene)–$CHCH_2CH_3$ (OH)

(c) (benzene)–$CHCHCH_3$ (OH OH)

(d) (benzene)–$CH_2CH_2CH_2OH$

II. Physical properties

34 Solubility of alcohol in water is due to
(a) hydrophobic $-R$ group (b) hydrophilic $-$OH group
(c) hydrophilic $-R$ group (d) hydrophobic $-$OH group

35 Ethyl alcohol is soluble in water due to
(a) acidic nature
(b) basic nature
(c) formation of hydrogen bond
(d) dissociation in water

36 Dipole moment of $CH_3CH_2CH_3$ (I), CH_3CH_2OH (II) and CH_3CH_2F (III) is in order
(a) I < II < III (b) I > II > III
(c) I > II > III (d) III < I < II

37 Ethanethiol (CH_3CH_2SH) is a gas at room temperature but ethanol is a liquid. The reason for this is
(a) the C—S—H bonds in ethanethiol are linear
(b) the C—O—H bonds in ethanol are linear
(c) ethanethiol has a higher boiling point
(d) It does not form intermolecular H — bond

38 Rank the following substances in order of increasing boiling point (lowest → highest)

$\underset{1}{CH_3CH_2CH_2CH_2OH}$, $\underset{2}{(CH_3)_2CHOCH_3}$

$\underset{3}{(CH_3)_3COH}$ $\underset{4}{(CH_3)_4C}$

(a) 1 < 3 < 2 < 4 (b) 2 < 4 < 3 < 1
(c) 4 < 2 < 3 < 1 (d) 2 < 3 < 1 < 4

39 Increasing order of dipole-moments of (I) ethanol, (II) methoxymethane and (III) propane is
(a) I < II < III (b) III < II < I
(c) III < II < I (d) III < II < I

40 The correct order of boiling point of primary (1°), secondary (2°) and tertiary (3°) alcohols is of similar molecular mass.
(a) 1° > 2° > 3° (b) 2° > 1° > 3°
(c) 2° > 3° > 1° (d) 3° > 2° > 1°

41 Arrange the following compounds in increasing order of boiling point.
Propan-1-ol, butan-1-ol, butan-2-ol, pentan-1-ol
(a) Propan-1-ol, butan-2-ol, butan-1-ol, pentan-1-ol
(b) Propan-1-ol, butan-1-ol, butan-2-ol, pentan-1-ol
(c) Propan-1-ol, butan-2-ol, pentan-1-ol, butan-1-ol
(d) Propan-1-ol, pentan-1-ol, butan-2-ol, butan-1-ol

42 Which has maximum pK_a value?

(a)  OH (b) OH

(c) OH (d) OH

43 Consider the following statements.
Statement I CF_3OH is more acidic than CH_3OH.
Statement II Strong electron withdrawing effect of the three fluorine atoms in CF_3 group delocalise the negative charge on oxygen.
(a) Both statements I and II are correct
(b) Only statement I is correct
(c) Only statement II is correct
(d) Neither statement I nor statement II is correct

44 Consider the following statements.
Statement I Alcohols are soluble in water.
Statement II Intermolecular hydrogen bonding increases boiling point of alcohols.
(a) Both statements I and II are correct
(b) Only statement I is correct
(c) Only statement II is correct
(d) Neither statement I nor statement II is correct

III. Chemical properties

45 When sodium is added to ethanol
(a) no action occurs
(b) H_2O is formed
(c) $Na^+O^-C_2H_5$ and O_2 are formed
(d) $Na^+O^-C_2H_5$ and H_2 are formed

46 What is the correct order of reactivity of alcohols in the following reaction?

$$R—OH + HCl \xrightarrow{ZnCl_2} R—Cl + H_2O$$

(a) 1° > 2° > 3° (b) 1° < 2° > 3°
(c) 3° > 2° > 1° (d) 3° > 1° > 2°

47 Which of the following is hydrated to a maximum extent?

(a) $CH_3—\overset{CH_3}{\underset{CH_3}{C}}—CH{=}CH_2$ (b) $CH_3—CH{=}CH_2$

(c) $CH_2{=}CH_2$ (d) $CH{\equiv}CH$

48 Which of the following compounds give most stable carbocation (intermediate) on dehydration?

(a) $CH_3\overset{CH_3}{\underset{CH_3}{C}}—OH$

(b) $CH_3\underset{CH_3}{CH}CH_2CH_2OH$

(c) $CH_3\underset{OH}{CH}CH_2CH_3$

(d) $CH_3CH_2CH_2CH_2OH$

49 Which one of the following reactions is not true when heated with copper at 300° C?

(a) secondary alcohol $\longrightarrow$ ketone
(b) tertiary alcohol $\longrightarrow$ olefin
(c) phenol $\longrightarrow$ benzyl alcohol
(d) primary alcohol $\longrightarrow$ aldehyde

50 In the following reaction most stable intermediate formed is

$$CH_3CH(CH_3)CH{=}CH_2 + H_2O \xrightarrow{H^+}$$

(a) $CH_3CH(CH_3)CH_2\overset{\oplus}{C}H_2$
(b) $CH_3CH(CH_3)\overset{\oplus}{C}HCH_3$
(c) $CH_3\overset{\oplus}{C}(CH_3)CH_2CH_3$
(d) $CH_3CH(CH_3)CH_2\overset{\ominus}{C}H_2$

51 $CH_3C(CH_3)_2—CH(OH)CH_3 \xrightarrow{HBr} A$ (predominant). A is

(a) $(CH_3)_3CCH(Br)CH_3$
(b) $(CH_3)_2C(Br)CH(CH_3)_2$
(c) Both (a) and (b)
(d) None is correct

52 $CH_3C(CH_3)_2—CH(OH)CH_3 \xrightarrow{\text{Conc. } H_2SO_4} A$ (predominant).

'A' is

(a) $(CH_3)_3CCH{=}CH_2$
(b) $(CH_3)_2C{=}C(CH_3)_2$
(c) $CH_2{=}C(CH_3)CH_2CH_2CH_3$
(d) None is correct

53 $C_2H_5OH \xrightarrow[443\text{ K}]{\text{Conc. } H_2SO_4} Z$

In the above reaction Z is identified as

(a) $C_2H_5—O—C_2H_5$
(b) C_2H_5CHO
(c) $CH_2{=}CH_2$
(d) $C_2H_5OSO_4$

54 Alcohols given below follow which of the following mechanism with conc. HCl/anhydrous $ZnCl_2$

CH_3OH (I), CH_3CH_2OH (II), $CH_3CH(OH)CH_3$ (III)

(a) S_N1, S_N2, S_N2
(b) S_N2, S_N2, S_N1
(c) S_N2, S_N1, S_N1
(d) S_N1, S_N1, S_N2

55 In the following reaction which bonds are cleaved?

$$\underset{\text{I}}{CH_3CH_2—\overset{O}{\overset{\|}{C}}\overset{q}{—}O\overset{p}{—}H} + \underset{\text{II}}{CH_3CH_2\overset{s}{—}O\overset{r}{—}H}$$

(a) bond p of I and s of II
(b) bond q of I and r of II
(c) bond p of I and r of II
(d) bond q of I and s of II

56 CH_3CH_2OH can be converted into CH_3CHO by ... .

(a) catalytic hydrogenation
(b) treatment with $LiAlH_4$
(c) treatment with pyridinium chlorochromate
(d) treatment with $KMnO_4$

57 Hydrolysis of alkyl halides has been compared. Which is incorrect comparison?

(a) $CH_2{=}CHBr > CH_2{=}CHCH_2Br$
(b) $C_6H_5CH_2Br > C_6H_5Br$
(c) $CH_3I > CH_3Cl$
(d) $(CH_3)_2CHBr > CH_3CH_2CH_2Br$

58 Match the species in Column I with the corresponding method of synthesis in Column II.

Column I	Column II
A. $(CH_3)_3CCH(OH)CH_3$	1. $(CH_3)_3CCH{=}CH_2 + H_3O^+$
B. $(CH_3)_2C(OH)CH(CH_3)_2$	2. $(CH_3)_3CCH{=}CH_2 + Hg(OAc)_2 \cdot$ THF$\cdot$ H_2O / $NaBH_4$, OH^-
C. $(CH_3)_3CCH_2CH_2OH$	3. $(CH_3)_3CCH{=}CH_2 + BH_3\cdot$THF / H_2O_2, OH^-
D. $(CH_3)_3CCH(OCH_3)CH_3$	4. $(CH_3)_3CCH{=}CH_2 + Hg(OAc)_2\cdot$ THF $\cdot$ CH_3OH / $NaBH_4$, OH^-

Choose the correct match.

	A	B	C	D
(a)	4	1	2	3
(b)	2	1	3	4
(c)	3	4	1	2
(d)	4	3	2	1

59 Dehydration of alcohols,

(I) cyclohexanol; (II) cyclohex-1-en-1-ol; (III) phenol; (IV) cyclohex-3-en-1-ol

(I) —OH   (II) —OH   (III) —OH   (IV) —OH

will be in order

(a) I < II < III < IV (b) I > II > III > IV
(c) III < II < I < IV (d) II < III < IV < I

60 What is the product from the following esterification?

$$C_6H_5CH_2\overset{O}{\overset{\|}{C}}OH + CH_3CH_2\overset{18}{O}H \xrightarrow[\text{Heat}]{H^+} ?$$

(a) $C_6H_5CH_2\overset{O}{\overset{\|}{C}}OCH_2CH_3$

(b) $C_6H_5CH_2\overset{O}{\overset{\|}{C}}-\overset{18}{O}CH_2CH_3$

(c) $C_6H_5CH_2\overset{^{18}O}{\overset{\|}{C}}-\overset{18}{O}CH_2CH_3$

(d) $CH_3CH_2\overset{^{18}O}{\overset{\|}{C}}OOCH_2C_6H_5$

61 Which of the esters shown, after reduction with $LiAlH_4$ and aqueous workup will yield two molecules of only a single alcohol?

(a) $CH_3CH_2CO_2CH_2CH_3$ (b) $C_6H_5COCH_2C_6H_5$
(c) $C_6H_5CO_2CH_2C_6H_5$ (d) None of these

62 For the following reaction, select the statement that best describes the situation

$$RCH_2OH + PCC\ [C_5H_5NH^+ ClCrO_3^-] \longrightarrow$$

(a) the alcohol is oxidised to an acid and the Cr(VI) is reduced
(b) the alcohol is oxidised to an aldehyde and the Cr(VI) is reduced
(c) the alcohol is reduced to an aldehyde and the Cr(III) is oxidised
(d) the alcohol is oxidised to a ketone and the Cr(VI) is reduced

63 When 3,3-dimethyl butan-2-ol is heated with H_2SO_4, the major product obtained is

(a) 3,3-dimethyl but-1-ene
(b) 3,3-dimethyl but-2-ene
(c) 2,3-dimethyl but-2-ene
(d) *cis* and *trans*-isomers of 2,3-dimethyl but-1-ene

64 Which of the following is hydrolysed at the fastest rate?

(a) Cl (b) Cl
(c) Cl (d) Cl

65 Dehydration of alcohols takes place more rapidly with $POCl_3$ than with H_2SO_4. Select incorrect statement(s) about the following dehydration reaction.

OH $\xrightarrow[\text{pyridine}]{POCl_3}$

(a) It does not involve carbocation
(b) It involves R—$OPOCl_2$ with —$OPOCl_2$ as a better leaving group
(c) It involves E2 mechanism as pyridine base abstracts proton from the adjacent carbon at the same time at which —$OPOCl_2$ is leaving
(d) It is E1 reaction without formation of carbocation

66 Which is incorrect order of dehydration of alcohol?

(a) —OH > —OH
(b) —OH > —CHCH₃(OH)
(c) —OH > CH_3CH_2OH
(d) $CH_3CH_2\underset{OH}{\underset{|}{CH}}CH_3 > CH_3CH_2CH_2CH_2OH$

67 Carbonyl compounds are reduced to alcohols by hydride reagents. This reduction involves

(a) addition of two hydrogens across a $>C=O$ bond
(b) reduction of carbonyl group to an alkoxide which is protonated to alcohol
(c) Both (a) and (b)
(d) None of the above

68 Major product of the following reactions is

OH, OH $\xrightarrow[\text{(ii) } CH_3I \text{ (1 eq.)}]{\text{(i) } K_2CO_3}$

(a) 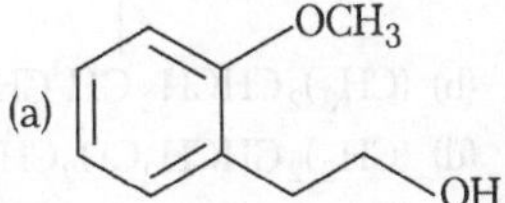

(b) O

(c) 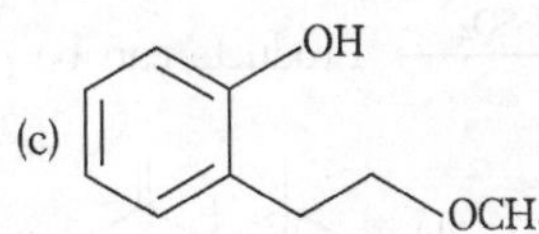

(d) 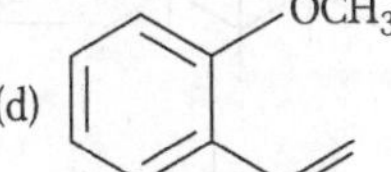

69 Consider the following reactions,

$$(CH_3)_2CHOH \xrightarrow[\text{I}]{PBr_3} (CH_3)_2CHBr$$

$$\text{II: } HBr + ZnCl_2$$

These reactions are of the type

	I	II		I	II
(a)	S_N1	S_N2	(b)	S_N2	S_N1
(c)	S_N2	S_N2	(d)	S_N1	S_N1

70 $CH_2{=}CHCH(OH)CH_2CH_2OH \xrightarrow{MnO_2} A$. A is

(a) $CH_2{=}CHC(=O)CH_2CH_2OH$

(b) $CH_2{=}CHCH(OH)CH_2CHO$

(c) $CH_2{=}CHC(=O)CH_2CHO$

(d) $CH_2{=}CHC(=O)CH_2C(=O)OH$

Select the schemes A, B, C out of

I. acid catalysed hydration II. HBO

III. oxymercuration-demercuration

(a) I in all cases

(b) I, II and III

(c) II, III and I

(d) III, I and II

71 What is the major organic product of the following sequence of reactions?

$$(CH_3)_2CHCH_2OH \xrightarrow{PBr_3} \xrightarrow{Mg} \xrightarrow{H_2C\overset{O}{—}CH_2} \xrightarrow{H_3O^+}$$

(a) $(CH_3)_2CHCH(OH)CH_2CH_3$ (b) $(CH_3)_2CHCH_2CH(OH)CH_3$

(c) $(CH_3)_2CHCH_2CH_2OH$ (d) $(CH_3)_2CHCH_2CH_2CH_2OH$

72 [skeletal structure with OH] $\xrightarrow{\text{Conc. } H_2SO_4}$ Products can be

(a) [skeletal structure] (b) [skeletal structure]

(c) Both are correct (d) None is correct

73 Following reaction represents

$$-\overset{|}{\underset{|}{C}}-\overset{\oplus OH_2}{\overset{|}{\underset{|}{C}}}- + H_2O \longrightarrow -\overset{|}{\underset{|}{C}}-\overset{OH}{\overset{|}{\underset{|}{C}}}- + H_3O^{\oplus}$$

(a) Lowry-Bronsted protonic acid-base reaction

(b) Lewis acid-base reaction

(c) Both (a) and (b)

(d) None of the above

74 Consider the following reactions,

I. $CH_3CH_2CH(OH)CH_3 \xrightarrow{H^+} A\text{(Major)}$

II. $CH_3C(CH_3)_2CH(OH)CH_3 \xrightarrow{H^+} B\text{ (Major)}$

A and B (both alkenes) are respectively

(a) $CH_3CH{=}CHCH_3$, $CH_3C(CH_3)_2CH{=}CH_2$

(b) $CH_3CH_2CH{=}CH_2$, $CH_3C(CH_3)_2CH{=}CH_2$

(c) $CH_3CH{=}CHCH_3$, $CH_3C(CH_3){=}C(CH_3)CH_3$

(d) $CH_3CH_2CH{=}CH_2$, $CH_3C(CH_3){=}C(CH_3)CH_3$

75 Consider oxidation of following alcohols

I. $CH_2{=}CHCH_2OH \xrightarrow{A} CH_2{=}CHCHO$

II. $(CH_3)_3COH \xrightarrow{B} CH_3C(CH_3){=}CH_2$

III. $CH_3CH_2OH \xrightarrow{C} CH_3CO_2H$

IV. $CH_3CH{=}CHCH_3 \xrightarrow{D} CH_3CO_2H$

A, B, C and D are oxidising agents which are respectively

	A	B	C	D
(a)	MnO_2	Cu / Δ	H_2CrO_4	$KMnO_4$ / Δ
(b)	Cu / Δ	MnO_2	H_2CrO_4	$KMnO_4$ / Δ
(c)	MnO_2	Cu / Δ	$KMnO_4$ / Δ	H_2CrO_4
(d)	MnO_2	H_2CrO_4	Cu / Δ	$KMnO_4$

76 2-(1-hydroxyethyl)cyclohexanone $\xrightarrow[\Delta]{I_2/NaOH}$ $\xrightarrow{H^+}$ $\xrightarrow{\Delta}$ A

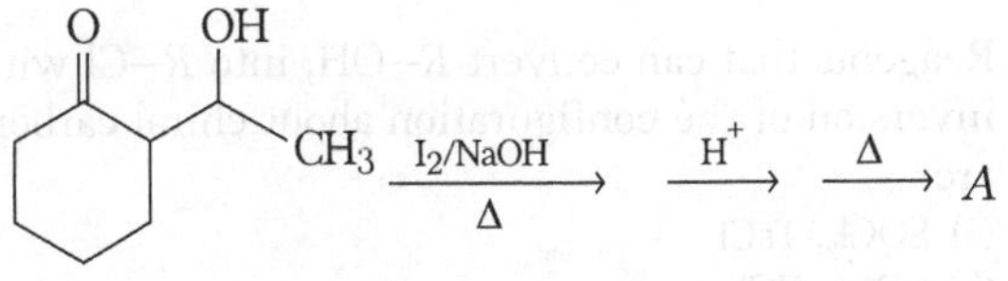

A is

(a) 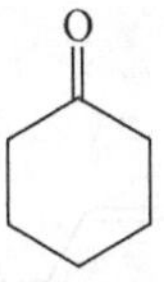 cyclohexanone

(b) 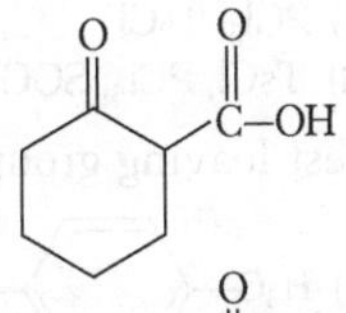 2-oxocyclohexane-1-carboxylic acid (C–OH with C=O on a cyclohexanone ring)

(c) cyclohexanecarboxylic acid (C–OH with C=O)

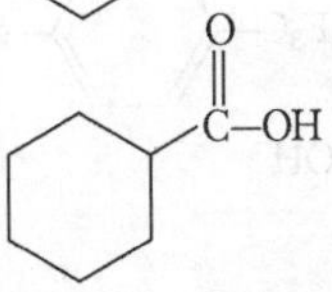

(d) 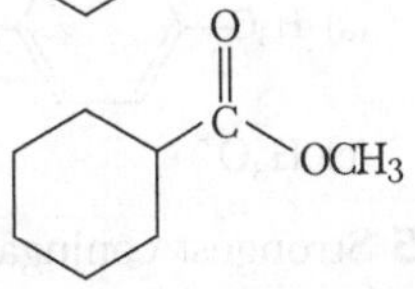 methyl cyclohexanecarboxylate (C–OCH_3 with C=O)

77 Match the reagent in Column I that gives the desired product in Column II from

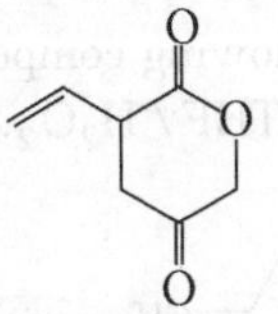

[Structure: a tetrahydropyran-2,4-dione ring carrying a vinyl group and a C=O side chain]

Column I	Column II
A. $LiAlH_4$	1. 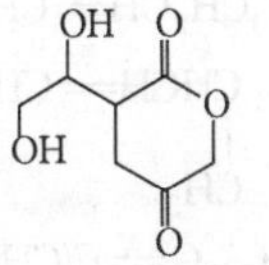 [Structure: OH, O, OH; lactone ring with C=O]
B. $NaBH_4$ /EtOH	2. 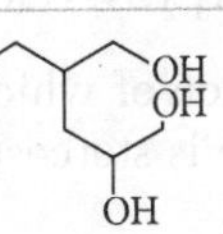 [Structure: OH, OH, OH]
C. H_2 /Pt	3. 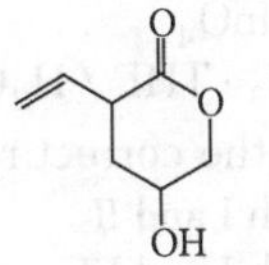 [Structure: O, O, OH]
D. $KMnO_4/OH^-$	4. 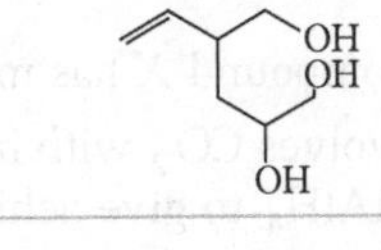 [Structure: vinyl group, OH, OH, OH]

Choose the correct match.

	A	B	C	D
(a)	2	1	4	3
(b)	3	4	1	2
(c)	4	1	2	3
(d)	4	3	2	1

78 An organic compound B is formed by the reaction of ethyl magnesium iodide (CH_3CH_2MgI) with a substance A, followed by treatment with dilute aqueous acid. Compound B does not react with PCC or PDC in dichloromethane. Which of the following is a possible candidate for A?

(a) $CH_3\overset{O}{\overset{\|}{C}}H$ (b) $H_2C{=}O$

(c) 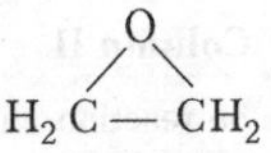 $H_2C{-}CH_2$ (epoxide, bridged by O) (d) $CH_3CH_2\overset{O}{\overset{\|}{C}}CH_3$

IV. Distinction between 1°, 2° and 3° alcohols and some commercially important alcohols

79 Consider the following tests.

I. Iodoform test

II. Victor Meyer test

III. Lucas test

CH_3OH and CH_3CH_2OH can be distinguished by

(a) Both I and II (b) Both I and III

(c) Only I (d) Only III

80 2,2-dimethyloxirane (H_3C, H_3C, CH_2, O) $\xrightarrow{LiAlH_4}$ A, A is

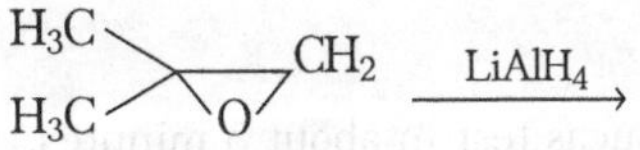

(a) $CH_3CH(CH_3)CH_2OH$ (b) $CH_3CH_2CH_2CH_2OH$

(c) $CH_3C(CH_3)_2{-}OH$ (d) no reaction

81 Ester $A(C_4H_8O_2) + CH_3MgBr$ (2 parts) $\xrightarrow{H_3O^+}$ $C_4H_{10}O$ (Alcohol B)

Alcohol B reacts fastest with Lucas reagent. Hence, A and B are

(a) $CH_3\overset{O}{\overset{\|}{C}}OC_2H_5$, $(CH_3)_3COH$

(b) $H\overset{O}{\overset{\|}{C}}OC_3H_7$, $(CH_3)_2CHOH$

(c) $CH_3\overset{O}{\overset{\|}{C}}OC_2H_5$, $(CH_3)_2CHOH$

(d) $H\overset{O}{\overset{\|}{C}}OC_3H_7$, $(CH_3)_3COH$

82 Dogs are often poisoned by sweet-tasting ethylene glycol. Poisoning is due to

(a) its oxidation to glycol aldehyde

(b) its metabolisation to oxalic acid which causes kidney failure leading to death

(c) Both (a) and (b)

(d) None of the above

83 Which of the following compounds can synthesise glycerol?

(a) Fats and oils (b) Propene

(c) Both (a) and (b) (d) None of these

84 Match the species given in Column I with the corresponding properties given in Column II.

Column I	Column II
A. CH_3OH	1. Gives ketone on reaction with Cu
B. CH_3CH_2OH	2. Gives white ppt. with Lucas reagent immediately
C. $(CH_3)_2CHOH$	3. Cannot be synthesised using Grignard's reagent
D. $(CH_3)_3COH$	4. Gives iodoform test

Choose the correct match.

	A	B	C	D
(a)	3	4	1	2
(b)	2	3	4	1
(c)	4	1	2	3
(d)	3	1	4	2

85 An alcohol gave Lucas test in about 5 minutes. When the alcohol was treated with hot concentrated H_2SO_4 it gave an alkene of molecular formula C_4H_8 which on ozonolysis gave C_2H_4O.

The structure of alcohol is

(a) $CH_3CHOHCH_2CH_3$
(b) $CH_3CH_2CH_2CH_2OH$
(c) $CH_3CHOHCH_2CH_2CH_3$
(d) $(CH_3)_3C-OH$

86 Consider the following statements.

Statement I CH_3OH and CH_3CH_2OH can be distinguished by haloform test.

Statement II Haloform test is given by 2° alcohol.

(a) Both statements I and II are correct
(b) Only statement I is correct
(c) Only statement II is correct
(d) Neither statement I nor statement II is correct

(B) Miscellaneous questions

1 Alkene is also formed as the major product in the hydrolysis of the following with NaOH

(a) CH_3CH_2Br (b) $(CH_3)_2CHBr$
(c) $(CH_3)_3CCH_2Br$ (d) $(CH_3)_3CBr$

2 Boiling point of CH_3CH_2OH (351 K) is much higher than that of isomeric ether. $(CH_3)_2O$ (248 K). This indicates

(a) hydrogen bonds are much stronger intermolecular attractions than dipole-dipole attractions
(b) dipole-dipole attractions are much stronger than hydrogen bonds
(c) $(CH_3)_2O$ has two hydrophobic groups while CH_3CH_2OH has one
(d) $(CH_3)_2O$ has two hydrophilic groups while CH_3CH_2OH has one

3 Reagents that can convert *R*–OH, into *R*–Cl with inversion of the configuration about chiral carbon are

(a) $SOCl_2$, TsCl
(b) $SOCl_2$, PCl_5
(c) PCl_5, TsCl
(d) TsCl, PCl_5, $SOCl_2$

4 Best leaving group is

(a) $H_3C-C_6H_4-SO_3^-$ (b) $F_3C-C_6H_4-SO_3^-$
(c) H_3O^+ (d) OH^-

5 Strongest conjugate base is

(a) $(CH_3)_3CO^-$ (b) $(CH_3)_2CHO^-$
(c) $CH_3CH_2O^-$ (d) CH_3O^-

6 Which of the following compounds on hydration by H_3O^+ and $BH_3 \cdot THF / H_2O_2 . NaOH$ gives same product?

(a) $H_2C=C_6H_8=CH_2$ (cyclohexane ring bearing two exocyclic =CH₂ groups at 1,4 positions)
(b) $CH_3CH_2CH=CH_2$
(c) $CH_3CH(CH_3)CH=CH_2$
(d) $(CH_3)_2C=C(CH_3)_2$

7 Addition of which of the following reagents to alkene is stereospecific.

I. Br_2
II. $KMnO_4$
III. $BH_3 \cdot THF / H_2O_2 \cdot NaOH$

Select the correct reagents.

(a) Both I and II (b) Both II and III
(c) Both I and III (d) I, II and III

8 Compound *X* has molecular formula $C_4H_8O_3$. It evolves CO_2 with *aq*. $NaHCO_3$. *X* reacts with $LiAlH_4$ to give achiral product.

The structure of '*X*' is

(a) $CH_3CH_2CH(OH)COOH$
(b) $CH_3CH(OCH_3)COOH$
(c) $CH_3CH(OH)CH_2COOH$
(d) $CH_3CH(CH_2OH)COOH$

9 $B \xleftarrow{NaBH_4}$ (cyclohexyl)$-CH{=}CHCHO \xrightarrow{H_2/Pt} A$

A and B are respectively

(a) (cyclohexyl)$-CH_2CH_2CHO$, (cyclohexyl)$-CH{=}CHCH_2OH$

(b) (cyclohexyl)$-CH_2CH_2CH_2OH$, (cyclohexyl)$-CH{=}CHCH_2OH$

(c) (cyclohexyl)$-CH{=}CHCH_2OH$ in both cases

(d) (cyclohexyl)$-CH_2CH_2CH_2OH$ in both cases

10 (cyclohexene ring with two OH groups) $\xrightarrow{MnO_2} A$, A is

(a) ring with OH and O

(b) ring with =O and OH

(c) ring with =O and O

(d) ring with =O and OH

11 The correct sequence of reagents for the following conversion will be

(4-formylcyclohexanone, O / CHO) $\longrightarrow$ (HO, CH_3 / $H_3C-C(OH)-CH_3$)

(a) $[Ag(NH_3)_2]^+OH^-$, H^+ / CH_3OH, CH_3MgBr

(b) CH_3MgBr, H^+ / CH_3OH, $[Ag(NH_3)_2]^+OH^-$

(c) CH_3MgBr, $[Ag(NH_3)_2]^+OH^-$, H^+ / CH_3OH

(d) $[Ag(NH_3)_2]^+OH^-$, CH_3MgBr, H^+ / CH_3OH

12 (cyclohexane)$=CH_2$ changes to (cyclohexane with CH_3 and OH) by

(a) $H_3O^{\oplus}$ (b) $BH_3 \cdot THF / H_2O_2$, NaOH

(c) $LiAlH_4$ (d) $NaBH_4$

13 C_2H_4O (A) reacts with CH_3MgBr followed by decomposition with $H_3O^{\oplus}$ to give $1°$ alcohol. Thus, A is

(a) $CH_2{=}CHOH$ (b) CH_3CHO

(c) CH_2-CH_2 (with bridging O) (d) None of these

14 Consider the following reactions

I. $CH_3\overset{O}{\overset{\|}{C}}OH \xrightarrow{PCl_5} \xrightarrow{CH_3OH} CH_3\overset{O}{\overset{\|}{C}}OCH_3$

II. ? (H_3O^+/CH_3OH)

Which path is not reversible?

(a) Only I

(b) Only II

(c) Both I and II

(d) None of the above

15 (diol with OH at C_2 and OH at C_5) In this diol

(a) OH at C_2 is more nucleophilic than that of at C_5

(b) OH at C_2 is more acidic than at C_5

(c) Both behave as a base

(d) Both behave as an acid

16 Rate of hydration of

(cyclopropyl)$-CH{=}CH_2$, (cyclopropyl)$-CH{=}CHCH_3$, (cyclopropyl)$-C(CH_3){=}CH_2$

I II III

will be in order

(a) I < II < III (b) I < III < II

(c) II < I < III (d) III < II < I

17 Major product of the following reaction is

(benzene ring with HO_2C, CH_2OH, $C(=O)NH_2$, OH) $\xrightarrow{HCl(g)/CCl_4}$

(a) (lactone with HO_2C and OH) (b) (HO_2C, CH_2OH, $C(=O)NH_2$, Cl)

(c) (lactone with $C(=O)NH_2$ and OH) (d) (lactone with HO_2C and Cl)

18 Consider the following statements.

Statement I Addition reaction of water to but-1-ene in acidic medium yields butan-2-ol as major product.

Statement II Addition of water in acidic medium proceeds through the formation of primary carbocation.

(a) Both statements I and II are correct
(b) Only statement I is correct
(c) Only statement II is correct
(d) Both statements I and II are incorrect

19 Rank the transition states that occur during the following reaction steps in order of increasing stability (least → most stable)

1. $H_3C—\overset{+}{O}H_2 \longrightarrow CH_3^+ + H_2O$
2. $(CH_3)_3C—\overset{+}{O}H_2 \longrightarrow (CH_3)_3C^+ + H_2O$
3. $(CH_3)_2CH—\overset{+}{O}H_2 \longrightarrow (CH_3)_2\overset{+}{C}H + H_2O$

(a) 1 < 2 < 3 (b) 2 < 3 < 1
(c) 1 < 3 < 2 (d) 2 < 1 < 3

20 What is the major product of the following reaction?

3-oxocyclohexane-1-carboxylic acid (cyclohexanone ring with CO_2H) $\xrightarrow[CH_3OH]{NaBH_4}$

(a) cyclohexane ring with OH and CH_2OH (b) cyclohexanone ring with CH_2OH
(c) cyclohexane ring with OH and CO_2H (d) cyclohexane ring with OH and CO_2CH_3

21 $C_4H_8O(A)$ changes to 3° alcohol by reduction with $LiAlH_4$. Thus, A is

(a) $CH_3\overset{O}{\overset{\|}{C}}CH_2CH_3$ (b) $CH_3CH_2CH_2CHO$
(c) $(CH_3)_2C—CH_2$ (epoxide, bridged by O) (d) $CH_3CH—CHCH_3$ (epoxide, bridged by O)

22 In the reaction given below, the product C is

$$CaC_2 \xrightarrow{H_2O} A \xrightarrow[HgSO_4]{Dil.\ H_2SO_4} B \xrightarrow{H_2/Ni} C$$

(a) C_2H_5OH
(b) CH_3CHO
(c) CH_3OH
(d) CH_3COOH

23 Consider following compound A (below)

(cyclobut-3-ene-1,2-dione ring with two OH groups; 3,4-dihydroxycyclobut-3-ene-1,2-dione)

Select the correct statement(s)
(a) it is more acidic than CH_3OH
(b) it is more acidic than $CH_3\overset{O}{\overset{\|}{C}}OH$
(c) Both (a) and (b)
(d) None of the above

24 (cyclohexyl)–OH $\xleftarrow{B}$ cyclohex-2-en-1-one (=O) $\xrightarrow{A}$ cyclohex-2-en-1-ol (–OH)

A and B are respectively
(a) H_2 / Pt, $LiAlH_4$ / H_2O
(b) H_2 / Pt, H_2 / Pt
(c) $LiAlH_4$ / H_2O, $LiAlH_4$ / H_2O
(d) $LiAlH_4$ / H_2O, H_2 / Pt

25 In which case chiral carbon (*) retains its configuration?

$H—\overset{C_6H_5}{\underset{CH_3}{C}}—OH$ — (I) $SOCl_2$ →; (II) TsCl → NaCl →

(a) In path I (b) in path II
(c) in both of these (d) in none of these

26 In the given reaction

$$B \xleftarrow[(ii)\ H_3O^+]{(i)\ CH_3MgBr\ (two\ mol)} A \xrightarrow{H_3O^+} CH_3\underset{CH_3}{\underset{|}{C}}HCOOH;$$

compound B is
(a) $(CH_3)_2CHCOCH_3$ (b) $(CH_3)_2CH\underset{OH}{\underset{|}{C}}(CH_3)_2$
(c) $(CH_3)_2CHCOCH(CH_3)_2$ (d) $(CH_3)_2CH\underset{OH}{\underset{|}{C}}HCH(CH_3)_2$

27 In which of the following case configuration about chiral C (*) is retained?

(a) $H—\overset{Me}{\underset{Et}{C^*}}—OH \xrightarrow{Na} \xrightarrow{CH_3Br}$
(b) $H—\overset{Me}{\underset{Et}{C^*}}—OH \xrightarrow{TsCl} \xrightarrow{CH_3ONa}$
(c) $H—\overset{Me}{\underset{Et}{C^*}}—OH \xrightarrow{PCl_5} \xrightarrow{CH_3ONa}$
(d) in none of the above case

28 $C \xleftarrow[CH_3OH]{NaBH_4}$ (A) $\xrightarrow{LiAlH_4} B$

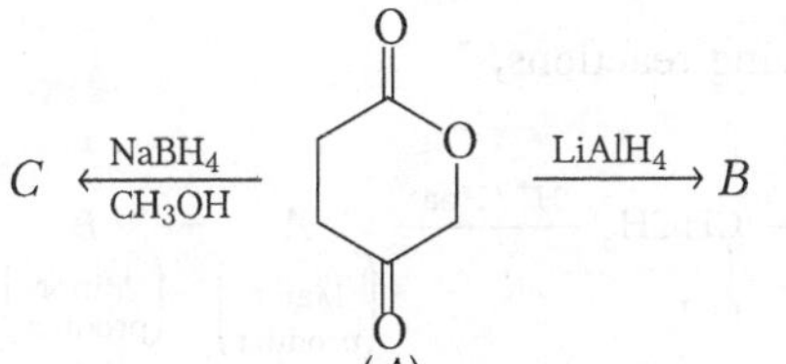

(A)

B and *C* are respectively

(a) I (OH, —OH, OH) II (O, O, OH)

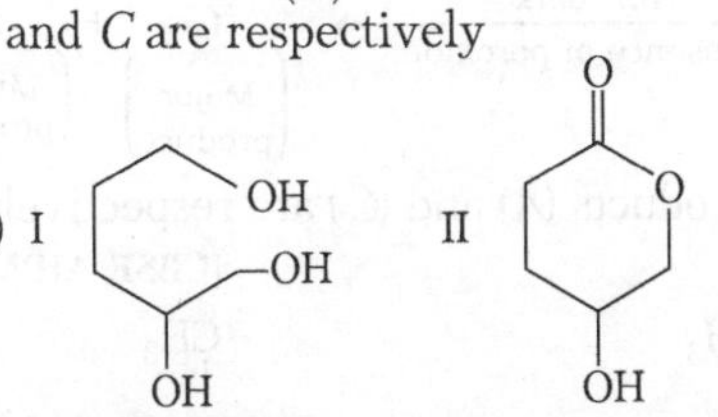

(b) both II and I

(c) I in both cases

(d) II in both cases

(C) Medical entrances' gallery

1 Reaction between acetone and methylmagnesium chloride followed by hydrolysis will give **[NEET 2020]**

(a) secondary butyl alcohol (b) *tert* butyl alcohol
(c) isobutyl alcohol (d) isopropyl alcohol

2 $CH_3CH_2CH{=}CH_2 \xrightarrow[H_2O,\, H_2O_2,\, OH^-]{B_2H_6} Z$

What is *Z*? **[NEET 2020]**

(a) $CH_3CH_2CH_2CH_2OH$ (b) $CH_3CH_2CH(OH)CH_3$
(c) $CH_3CH_2CH_2CHO$ (d) $CH_3CH_2CH_2CH_3$

3 When vapours of a secondary alcohol is passed over heated copper at 573 K, the product formed is **[NEET (Odisha) 2019]**

(a) a carboxylic (b) an aldehyde
(c) a ketone (d) an alkene

4 $CH_3—CH_2—CH{=}CH_2 \xrightarrow{'X'} CH_3CH_2—CH_2—CH_2OH$

What is suitable reagent '*X*' is? **[JIPMER 2019]**

(a) $H_2 \mid Pt$
(b) $B_2H_6 \mid THF,\ H_2O_2 \mid OH^-$
(c) $Br_2 \mid HCl$
(d) $HBr \mid H_2O_2$

5 Assertion (A) CH_3CH_2OH can be distinguished by haloform test.

Reason (R) Haloform test is given by 2° alcohol.

(a) Both Assertion and Reason are true and Reason is the correct explanation of Assertion
(b) Both Assertion and Reason are true but Reason is not the correct explanation of Assertion
(c) Assertion is true but Reason is false
(d) Both Assertion and Reason are false **[AIIMS 2018]**

6 Dehydration of which one of the following alcohols produces an alkene exhibiting *cis-trans* isomerism? **[Kerala CEE 2015]**

(a) *Iso*-propyl alcohol (b) Tertiary butyl alcohol
(c) *n*-butyl alcohol (d) *Neo*-pentyl alcohol
(e) 3-pentanol

7 The hydrolysis of optically active 2-bromobutane results in the formation of **[KCET 2015]**

(a) (+) butan-2-ol (b) (−) butan-2-ol
(c) (±) butan-1-ol (d) (±) butan-2-ol

8 What reagent(s) can be used to convert 2-methyl pentan-1-ol into 2-methylpentanal? **[Manipal 2014]**

(a) $Na_2Cr_2O_7$ (b) $LiAlH_4$
(c) CrO_3-pyridine (d) $KMnO_4$

9 0.44 g of a monohydric alcohol when added to methyl magnesium iodide in ether liberates at STP, 112 cm^3 of methane. With PCC the same alcohol forms a carbonyl compound that answers silver mirror test. The monohydric alcohol is **[KCET 2014]**

(a) $H_3C—CH(OH)—CH_2—CH_3$
(b) $(CH_3)_3C—CH_2OH$
(c) $H_3C—CH(OH)—CH_2—CH_2—CH_3$
(d) $(CH_3)_2CH—CH_2OH$

10 Conversion of ethyl alcohol into acetaldehyde is an example of **[Manipal 2013]**

(a) hydrolysis (b) oxidation
(c) reduction (d) molecular rearrangement

11 Which of the following combinations can be used to synthesise ethanol? **[KCET 2013]**

(a) CH_3MgI and CH_3COCH_3
(b) CH_3MgI and C_2H_5OH
(c) CH_3MgI and $CH_3COOC_2H_5$
(d) CH_3MgI and HCHO

12 Glycerol on oxidation with bismuth nitrate produce **[UP CPMT 2013]**

(a) glyceric acid (b) glyoxalic acid
(c) oxalic acid (d) *meso*-oxalic acid

13 Assertion (A) Glycerol is purified by distillation under reduced pressure.

Reason (R) Glycerol is a trihydric alcohol. **[AIIMS 2013]**

(a) Both A and R are correct and R is correct explanation of A
(b) Both A and R are correct but R is not the correct explanation of A
(c) A is correct but R is incorrect
(d) Both A and R are incorrect

14 Which of the following compounds can be used as an antifreeze in an automobile radiator? **[CBSE AIPMT 2012]**

(a) Methyl alcohol (b) Glycol
(c) Nitrophenol (d) Ethyl alcohol

15 The most suitable reagent for the conversion of $RCH_2OH \longrightarrow RCHO$ is **[AFMC 2012]**

(a) $KMnO_4$ (b) $K_2Cr_2O_7$
(c) PCC (d) CrO_3

16 HCHO was treated with reagent *X*. The product formed upon hydrolysis in the presence of an acid gives C_2H_5OH. The reagent *X* is **[KCET 2012]**

(a) alcoholic KOH (b) alcoholic KCN
(c) CH_3MgI (d) aqueous KOH

17 The alcohol that produces turbidity immediately with Lucas reagent at room temperature is **[AIIMS 2012, KCET 2011]**

(a) butan-1-ol (b) butan-2-ol
(c) 2-methyl propan-2-ol (d) 2-methyl propan-1-ol

18 The best reagent to convert pent-3-en-2-ol into pent-3-en-2-one is **[AIIMS 2012]**

(a) pyridinium chlorochromate
(b) chromic anhydride in glacial acetic acid
(c) acidic dichromate
(d) acidic permanganate

19 Decreasing order of reactivity of hydrogen halide acid in the conversion of $ROH \longrightarrow RX$ is **[DUMET 2011]**

(a) HCl > HBr > HI > HF (b) HI > HBr > HCl > HF
(c) HF > HCl > HBr > HI (d) HF > HBr > HI > HCl

20 Which of the following alcohols gives the best yield of dialkyl ether on being heated with a trace of sulphuric acid? **[Kerala CEE 2011]**

(a) 2-pentanol (b) cyclopentanol
(c) 2-methyl-2-butanol (d) 2-propanol
(e) 1-pentanol

21 In the following reactions,

I. $CH_3-\underset{\displaystyle CH_3}{\overset{}{CH}}-\underset{\displaystyle OH}{CHCH_3}$ with CH_3 on the first CH and OH on the second: $CH_3-CH(CH_3)-CH(OH)CH_3 \xrightarrow{H^+/Heat} \underset{\text{(Major product)}}{A} + \underset{\text{(Minor product)}}{B}$

II. $A \xrightarrow[\text{in absence of peroxide}]{\text{HBr, dark}} \underset{\text{(Major product)}}{C} + \underset{\text{(Minor product)}}{D}$

The major products (*A*) and (*C*) are respectively **[CBSE AIPMT 2011]**

(a) $CH_2=C(CH_3)-CH_2CH_3$ and $CH_3-C(CH_3)(Br)-CH_2CH_3$

(b) $CH_2=C(CH_3)-CH_2CH_3$ and $CH_2(Br)-CH(CH_3)-CH_2CH_3$

(c) $CH_3-C(CH_3)=CH-CH_3$ and $CH_3-C(CH_3)(Br)-CH_2CH_3$

(d) $CH_3=C(CH_3)-CHCH_3$ and $CH_3-CH(CH_3)-CH(Br)-CH_3$

22 The compound that undergoes dehydration very easily is **[Kerala CEE 2011]**

(a) 2-methylpropan-2-ol
(b) ethyl alcohol
(c) 3-methyl-2-butanol
(d) propyl alcohol
(e) methanol

ANSWERS

Check point 15.1

1. (a)	2. (a)	3. (c)	4. (b)	5. (d)	6. (b)	7. (c)	8. (c)

Check point 15.2

1. (b)	2. (c)	3. (a)	4. (a)	5. (d)	6. (a)	7. (a)	8. (c)	9. (d)	10. (d)

Check point 15.3

1. (d)	2. (c)	3. (c)	4. (c)	5. (a)	6. (b)	7. (a)	8. (a)	9. (b)	10. (c)

(A) Topical exercise

1. (c)	2. (b)	3. (c)	4. (b)	5. (a)	6. (b)	7. (c)	8. (d)	9. (d)	10. (c)
11. (d)	12. (b)	13. (b)	14. (c)	15. (c)	16. (c)	17. (d)	18. (d)	19. (b)	20. (b)
21. (a)	22. (d)	23. (b)	24. (b)	25. (d)	26. (a)	27. (b)	28. (b)	29. (b)	30. (a)
31. (b)	32. (a)	33. (b)	34. (b)	35. (c)	36. (a)	37. (d)	38. (c)	39. (c)	40. (a)
41. (a)	42. (d)	43. (a)	44. (a)	45. (d)	46. (c)	47. (a)	48. (a)	49. (c)	50. (c)
51. (b)	52. (b)	53. (c)	54. (b)	55. (b)	56. (c)	57. (a)	58. (b)	59. (c)	60. (b)
61. (c)	62. (b)	63. (c)	64. (a)	65. (b)	66. (b)	67. (c)	68. (a)	69. (b)	70. (a)
71. (d)	72. (c)	73. (a)	74. (c)	75. (a)	76. (a)	77. (d)	78. (d)	79. (c)	80. (c)
81. (a)	82. (b)	83. (c)	84. (a)	85. (a)	86. (b)				

(B) Miscellaneous questions

1. (d)	2. (a)	3. (d)	4. (b)	5. (a)	6. (d)	7. (d)	8. (d)	9. (b)	10. (b)
11. (a)	12. (a)	13. (c)	14. (a)	15. (a)	16. (a)	17. (a)	18. (b)	19. (c)	20. (c)
21. (c)	22. (a)	23. (c)	24. (d)	25. (a)	26. (b)	27. (a)	28. (a)		

(C) Medical entrances' gallery

1. (b)	2. (a)	3. (c)	4. (b)	5. (b)	6. (e)	7. (d)	8. (c)	9. (b)	10. (b)
11. (d)	12. (d)	13. (b)	14. (b)	15. (c)	16. (c)	17. (c)	18. (b)	19. (b)	20. (e)
21. (c)	22. (a)								

Hints & Explanations

• Check point 15.1

4 (*b*) Oxymercurations-demercuration hydrates alkene without rearrangement.

6 (*b*) $CH_3CHO \xrightarrow[\text{(ii) } H_3O^+]{\text{(i) } CH_3MgBr} CH_3\underset{\underset{OH}{|}}{C}HCH_3$ (2° Alcohol)

• Check point 15.3

2 (*c*) 3° alcohol is most basic reacts with conc.HCl forming white precipitate.

6 (*b*) Rectified spirit contains 98% ethanol-water (not methanol) mixture.

(A) Topical exercise

5 (*a*) The three isomers of butanol are possible. Structural formula of these isomers are given below.

(i) $CH_3CH_2—CH_2—CH_2OH$
Butan-1-ol

No carbon is chiral in this compound as none of the four carbon is bonded to four different substituents.

(ii) $CH_3—CH_2—\overset{*}{C}H(OH)—CH_3$
Butan-2-ol

In this compound, asterisk marked carbon is chiral carbon as all four substituents attached to it are different.

(iii) $H_3C—C(CH_3)(OH)—CH_3$
2-methylpropan-2-ol

Here, again carbon is not chiral in nature.

(iv) $CH_3—CH(CH_3)—CH_2OH$, in this case also, no chiral carbon is present. So, only one alcohol is chiral in nature and the correct option is (a).

6 (*b*) An addition reaction is a reaction where two or more molecules combine to form a larger one.
These reactions occur to change the unsaturated compound to saturated compound. In dehydrohalogenation reaction, alkyl halides give alkenes.
Rearrangement gives the structural isomers of the reactant while in substitution reaction one of the groups or atoms is replaced by other group. Therefore, the process of converting alkyl halides into alcohol involves substitution reaction.

$$\underset{\text{Alkyl halide}}{R—X} \xrightarrow{OH^-} \underset{\text{Alcohol}}{R—OH}$$

Reaction of alkyl halides

Primary alkyl halides - S_N2

Secondary alkyl halides - S_N1

Tertiary alkyl halides - S_N1

12 (*b*) $LiAlH_4$ reduces ketonic as well as ester group, thus given conversion is done by reagent (b) only.

16 (*c*) Here X is $Hg(OAc)_2$ /$NaBH_4$, NaOH

$Hg(CH_3COO)_2$ / $NaBH_4$, OH^- - gives alcohol by Markownikoff's rule without rearrangement-no carbocation intermediate.

$BH_3 \cdot HF/H_2O_2/OH^-$ gives alcohol by anti-Markownikoff's rule

H_3O^+ (hydration) - gives alcohol by Markownikoff's rule and carbocation intermediate can undergo rearrangement

17 (*d*) (a) $\underset{\text{Ketone}}{R\overset{O}{\overset{\|}{C}}R} \xrightarrow{LiAlH_4} \underset{\text{2° alcohol}}{R\underset{OH}{\underset{|}{C}}HR}$

(b) $CH_3CH_2MgBr + \underset{\text{Aldehyde}}{RCHO} \xrightarrow{H_3O^+} CH_3CH_2\underset{OH}{\underset{|}{C}}HR$ (2° alcohol)

(c) $\underset{\text{Alkene}}{CH_3CH_2CH{=}CH_2} \xrightarrow{H_3O^+} CH_3CH_2\underset{OH}{\underset{|}{C}}HCH_3$ (2° alcohol)

19 (*b*) 3-methylcyclobut-2-en-1-one (O=, CH₃) $\xrightarrow{KMnO_4/OH^-}$ cyclobutanone with C(CH₃)(OH) (3° alcohol) and CH(OH) (2° alcohol) $\xrightarrow{Cu/\Delta}$ cyclobutane-1,3-dione with CH₃ and OH (*A*)

Cu oxidises 2° alcohol to ketone.

20 (*b*) Cl—◇—Br $\xrightarrow{Mg/ether}$ Cl—◇—Mg—Br $\xrightarrow{HCHO/H_3O^+}$ Cl—◇—CH_2OH (*A*)

Bond energy of (C—Br) < (C—Cl) hence Mg attack here

22 (*d*) $(C_2H_5O)_2C{=}O \xrightarrow{CH_3MgBr} CH_3COOC_2H_5 \xrightarrow[H_3O^+]{CH_3MgBr\ (2nd\ mol)} CH_3COCH_3 \xrightarrow[H_3O^+]{CH_3MgBr\ (3rd\ mol)} (CH_3)_3COH$ (3° alcohol)

23 (*b*) $NaBD_4$ in CH_3OD helps in reductions of ketone to alcohol.

25 (*d*) Cyclopentanone $\xrightarrow[H_3O^+]{CH_3MgBr}$ 1-methylcyclopentanol (*A*) $\xrightarrow{HBr}$ protonated alcohol ($-OH_2^+$) $\xrightarrow{-H_2O}$ 3° carbocation $\xrightarrow{\text{1, 2-methyl shift, ring expansion}}$ cyclohexyl carbocation (*B*) $\xrightarrow{Br^\ominus}$ bromocyclohexane $\xrightarrow{Mg/ether}$ cyclohexylmagnesium bromide (*C*) $\xrightarrow[(ii)\ H_3O^+]{(i)\ HCHO}$ cyclohexylmethanol, CH_2OH (*D*) Predominant

26 (*a*) 1-methylcyclohexene $\xrightarrow[MnO_4^-/OH^-]{Cold}$ 1-methylcyclohexane-1,2-diol (*A*) $\xrightarrow[AcOH]{CrO_3}$ 2-hydroxy-2-methylcyclohexanone (*B*)

Only 2° alcohol is oxidised

27 (*b*) $F-\diamond-Br \xrightarrow[ether]{Mg} F-\diamond-Mg-Br \xrightarrow{HCHO/H_3O^+} F-\diamond-CH_2OH$

This bond has lower B.E.

28 (*b*) S_N2 reaction takes place by nucleophilic attack on the backside resulting in the formation of inverted product.

29 (*b*) Only –CHO group is reduced because $NaB(OAc)_3H$ act as reagent for selective reduction.

30 (*a*) By acid hydration, addition of H_2O is by Markownikoff's rule and by hydroboration oxidation addition of H_2O is by anti-Markownikoff's rule.

40 (*a*) Larger the acidic character, larger the H-bonding and thus greater the boiling point. So, the correct order for boiling point is 1° > 2° > 3°.

41 (*a*) Boiling point increases with increase in the number of carbon atoms because of increase in molecular mass. So, the boiling point of pentan-1-ol is more than that of all other given boiling points than 2°alcohols because of higher surface area in 1°alcohols. Therefore, boiling point increases in the order.

Propan-1-ol < butan-2-ol < butan-1-ol < pentan-1-ol

Here, the correct option is (a).

42 (*d*) $pK_a = -\log K_a$

If K_a is high, pK_a is low.

1° alcohol (butan-1-ol, OH) is the strongest acid thus has maximum K_a value and minimum pK_a value

3° alcohol (2-methylbutan-2-ol, OH) is the weakest acid and thus has minimum K_a value and maximum pK_a value.

46 (*c*) The given reaction is nucleophilic substitution reaction in which —OH group is replaced by —Cl. Tertiary alcohols, when react with HCl in presence of $ZnCl_2$, form tertiary carbocation.

This 3° carbocation intermediate is more stable than 2° carbocation as well as 1°carbocation. Higher the stability of intermediate, higher will be the reactivity of reactant molecule.

So, the order of reactivity of alcohols in the given reaction is 3° > 2° > 1° and correct option is (c).

Or

Basic nature of alcohols 1° < 2° < 3°

Thus, reactivity with acid (HCl) is 1° < 2° < 3° alcohols.

47 (*a*) The compound whose intermediate is most stable carbocation, that hydrated to maximum extent.

$CH_3-C(CH_3)_2-CH{=}CH_2$ (3, 3-dimethyl but-1-ene) $\xrightarrow{H_3O^+} CH_3-C(CH_3)_2-\overset{\oplus}{C}H-CH_3$ (2° carbocation) $\xrightarrow{\text{Rearrangement}} CH_3-\overset{\oplus}{C}(CH_3)-CH(CH_3)-CH_3$ (3° carbocation)

50 (*c*) $CH_3CH(CH_3)CH{=}CH_2$ (3-methyl but-1-ene) $\xrightarrow{H_3O^+} CH_3CH(CH_3)\overset{\oplus}{C}H-CH_3$ (2° carbocation) $\xrightarrow{\text{Rearrangement}} CH_3-\overset{\oplus}{C}(CH_3)-CH_2-CH_3$ (3° carbocation) (most stable)

51 (b) $H_3C\overset{CH_3}{\underset{CH_3}{|}}C—\underset{OH}{CH}CH_3 \xrightarrow{H^+} H_3C\overset{CH_3}{\underset{CH_3}{|}}C—\underset{\oplus OH_2}{CH}CH_3 \xrightarrow{-H_2O} H_3C\overset{CH_3}{\underset{CH_3}{|}}C—\overset{\oplus}{C}HCH_3$ (2° carbocation) $\xrightarrow{\text{rearrangement}} H_3C\overset{CH_3\ CH_3}{\overset{\oplus}{C}}—CHCH_3$ (3° carbocation) $\xrightarrow{Br^\ominus} CH_3—\overset{CH_3\ CH_3}{\underset{Br}{C}}—CHCH_3$

(A)
Predominant

52 (b) $CH_3\overset{CH_3}{\underset{CH_3}{C}}—\underset{OH}{CH}CH_3 \xrightarrow{H^+} CH_3\overset{CH_3}{\underset{CH_3}{C}}—\underset{\oplus OH_2}{CH}CH_3$

3, 3-dimethyl butan-2-ol

$\xrightarrow{-H_2O} CH_3\overset{CH_3}{\underset{CH_3}{C}}—\overset{\oplus}{C}HCH_3$ (2° carbocation) $\xrightarrow[\text{Rearrangement}]{(1,2\text{ methyl shift})} CH_3\overset{CH_3\ CH_3}{\overset{\oplus}{C}}—CHCH_3$

3° carbocation (more stable than 2° carbocation)

$\xrightarrow{-H^\oplus} CH_3\overset{CH_3}{C}=\overset{CH_3}{C}CH_3$

(A)
2, 3-dimethylbut-2-ene

54 (b) 1° alcohol follow S_N2

2°, 3° alcohol follow S_N1

CH_3OH (1° alcohol) CH_3CH_2OH (1°) alcohol $CH_3\underset{OH}{CH}CH_3$ 2° alcohol

55 (b) $CH_3CH_2—\overset{O}{\overset{\|}{C}}\overset{q}{\vdots}O—H$ (I) $+ CH_3CH_2—\overset{18}{O}\overset{r}{\vdots}H$ (II)

$\rightleftharpoons CH_3CH_2\overset{O}{\overset{\|}{C}}—\overset{18}{O}—CH_2CH_3 + H_2O$

If oxygen of CH_3CH_2OH is labelled with ^{18}O, then it goes with ester formed, indicating cleavage of bond q in I and r in II.

56 (c) Ethanal (CH_3CHO) is an oxidised product of ethanol.

Pyridinium chlorochromate ($C_5H_5\overset{+}{N}HCl\overset{-}{Cr}O_3$) oxidises primary alcohols to aldehydes. Strong oxidising agents such as $KMnO_4$ are used for getting carboxylic acid from alcohols.

The oxidation process can be stopped at the aldehyde stage if Cr(VI) reagents such as Collin's reagent ($CrO_3 \cdot 2C_5H_5N$), Corey's reagent or pyridinium chlorochromate and pyridinium dichromate $[(C_5H_5NH)_2^{2+} Cr_2O_7^{2-}]$ in anhydrous medium are used as the oxidising agent. So, the correct option is (c).

$$\underset{\text{Ethanol}}{CH_3CH_2OH} \xrightarrow{PCC} \underset{\text{Ethanal}}{CH_3CHO}$$

57 (a) $CH_2=CHBr$ has stronger C—Br bond thus, hydrolysis is slower.

59 (c) Greater the conjugation, greater the stability and thus, greater is the dehydration. Thus correct order for dehydration of alcohols is III < II < I < IV

60 (b) ^{18}O of alcohol is retained with ester. RO^- of alcohol is attached to ester.

61 (c) $\underset{\text{This is hydrolysed}}{C_6H_5\overset{O}{\overset{\|}{C}}—OCH_2C_6H_5} \xrightarrow{LiAlH_4} C_6H_5CH_2OH + \underset{\text{This is reduced}}{C_6H_5CH_2OH}$

62 (b) $RCH_2OH + Cr(VI) \xrightarrow[\text{(In PCC)}]{} RCHO + Cr(III)$

64 (a) The correct order of hydrolysis is

(Benzylic) most reactive > Vinylic > Allylic > Phenylic

65 (b) Mechanism for dehydration of alcohol using $POCl_3$ with pyridine is as follows.

Proton transfer

Good leaving group

$-Cl^-$

E2 elimination

Alkene

66 (*b*) Cyclohexanol (—OH) ⟶ Cyclohexene + H_2O

C_6H_5—$\underset{|}{C}HCH_3$(OH) ⟶ C_6H_5—$\overset{\oplus}{C}HCH_3$ ⟶ C_6H_5—$CH{=}CH_2$

This dehydration takes place through the formation of resonance stabilised carbocation. Thus, the correct order for dehydration is

$C_6H_5CH(OH)CH_3$ > Cyclohexanol (—OH)

Thus, option (b) is incorrect.

68 (*a*) (acidic H) Phenolic OH ... This is not decomposed (side-chain CH_2CH_2OH)

2-(2-hydroxyethyl)phenol $\xrightarrow{K_2CO_3}$ phenoxide ($\overset{-}{O}\overset{+}{K}$) with CH_2CH_2OH $\xrightarrow{CH_3, \Delta}$ OCH_3 with CH_2CH_2OH

Major product

70 (*a*) There is oxidation of alcohol at allylic or benzylic position by MnO_2. So, product *A* is

$$CH_2 = CH—\overset{O}{\overset{\|}{C}}—CH_2CH_2OH$$

72 (*c*) Alcohol $\xrightarrow{Conc \cdot H_2SO_4}$ protonated alcohol ($\overset{\oplus}{O}H_2$) $\xrightarrow{-H_2O}$ 3° carbocation (*X*)

$H^{\oplus}$ can be lost from this carbon → (option *a*)

3° carbocation (*X*) $\xrightarrow{\text{Rearrangement}}$ 3° carbocation (*Y*); $H^{\oplus}$ can be lost from this carbon also $\xrightarrow{-H^+}$ (option *b*)

Both (a) and (b) options are correct.

74 (*c*) I $CH_3CH_2\underset{OH}{\underset{|}{C}H}CH_3 \xrightarrow{-H_2O} CH_3CH{=}CHCH_3$

But-2-ene
A (Major)

II $CH_3—\overset{CH_3}{\overset{|}{\underset{CH_3}{\underset{|}{C}}}}—\underset{OH}{\underset{|}{C}H}\ CH_3 \xrightarrow[-H_2O]{H^+} CH_3—\overset{CH_3}{\overset{|}{\underset{CH_3}{\underset{|}{C}}}}—\overset{\oplus}{C}H—CH_3$

2° carbocation

↓ 1, 2-methyl shift

$CH_3—\overset{CH_3}{\overset{|}{\underset{\oplus}{C}}}—\underset{CH_3}{\underset{|}{CH}}—CH_3$ (3° carbocation) ⟶ $CH_3\overset{CH_3}{\overset{|}{C}}{=}\overset{CH_3}{\overset{|}{C}}CH_3$

2, 3-dimethyl but-2-ene
B(Major)

78 (*d*) *B* is a tertiary alcohol based on given properties

(a) $CH_3CH_2MgI + CH_3CHO \xrightarrow{H_3O^+} CH_3\underset{OH}{\underset{|}{C}H}CH_2CH_3$ (2° alcohol)

(b) $HCHO + CH_3CH_2MgI \xrightarrow{H_3O^+} CH_3CH_2CH_2OH$ (1° alcohol)

(c) $\underset{\text{(epoxide, O)}}{CH_2—CH_2} + CH_3CH_2MgI \xrightarrow{H_3O^+} CH_3CH_2CH_2CH_2OH$ (1° alcohol)

(d) $CH_3CH_2\overset{O}{\overset{\|}{C}}CH_3 + CH_3CH_2MgI \xrightarrow{H_3O^+} CH_3CH_2\overset{OH}{\overset{|}{\underset{CH_3}{\underset{|}{C}}}}HC\ H_2CH_3$

(*A*) — 3° alcohol (*B*)

79 (*c*) Both are 1° alcohols can respond to tests II and III but iodoform test is given by CH_3CH_2OH

80 (*c*) $(H_3C)_2C—CH_2$ (epoxide, O) — H^- attacks here; This bond is cleaved — $\xrightarrow{LiAlH_4} H_3C—\overset{CH_3}{\overset{|}{\underset{OH}{\underset{|}{C}}}}—CH_3$

(*A*)
2-methyl propan-2-ol

81 (*a*) Alcohol *B* is 3° (reacts fastest with Lucas reagent).

$$CH_3\overset{O}{\overset{\|}{C}}OC_2H_5 \xrightarrow{CH_3MgBr} CH_3\overset{O}{\overset{\|}{C}}CH_3 \xrightarrow[H_3O^+]{CH_3MgBr} \underset{(B)}{(CH_3)_3COH}$$

85 (*a*) Alcohol is 2° as it gives Lucas test in about 5 minutes.

$$\text{Alcohol} \xrightarrow{-H_2O} \underset{\text{Alkene}}{C_4H_8} \xrightarrow{O_3/Zn} CH_3CHO$$

Go reversely.
Thus, alkene is

$$CH_3CH = CHCH_3 \xrightarrow{H_2O/\text{dil. } H_2SO_4} CH_3\underset{OH}{\underset{|}{C}H}CH_2CH_3$$

and alcohol is (a).

(B) Miscellaneous questions

1 (*d*) In case of tertiary halides, elimination is preferred over substitution reaction that results in the formation of alkene by eliminating HBr.

2 (*a*) There is no H-bonding in ether. It show dipole-dipole attractions.

4 (*b*) If acid is strong, its conjugate base is weak and is a better leaving group.

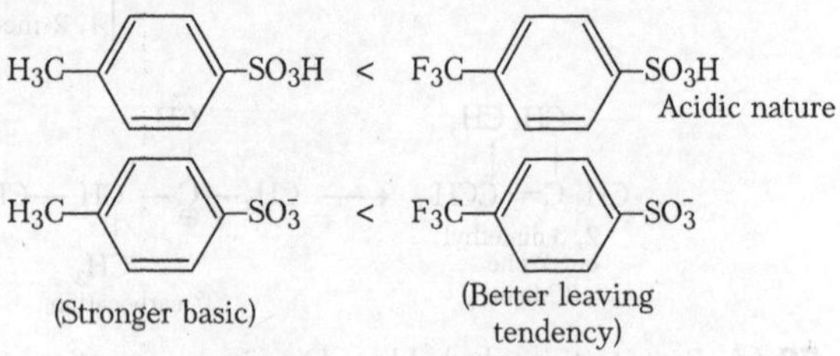

$H_3C-C_6H_4-SO_3H$ < $F_3C-C_6H_4-SO_3H$ Acidic nature

$H_3C-C_6H_4-SO_3^-$ < $F_3C-C_6H_4-SO_3^-$

(Stronger basic) (Better leaving tendency)

5 (*a*) CH_3OH CH_3CH_2OH $(CH_3)_2CHOH$ $(CH_3)_3COH$

Maximum acidic ← Minimum acidic

$CH_3O^\ominus$ $CH_3CH_2O^\ominus$ $(CH_3)_2CHO^\ominus$ $(CH_3)_3CO^\ominus$

Minimum basic → Maximum basic

If acid is weak, its conjugate base is strong.

7 (*d*) A reaction is stereospecific when each of the two specific stereoisomers of a starting material yields a particular or stereoisomer of a product.

8 (*d*) *X* evolves CO_2 with *aq* $NaHCO_3$ indicating –COOH

$X \xrightarrow{LiAlH_4}$ achiral

(a) $\xrightarrow{LiAlH_4}$ $CH_3CH_2CH(OH)CH_2OH$ (chiral). Thus, false

(b) $\xrightarrow{LiAlH_4}$ $CH_3CH(OCH_3)CH_2OH$ (chiral). Thus, false

(c) $\xrightarrow{LiAlH_4}$ $CH_3CH(OH)CH_2CH_2OH$ (chiral). Thus, false

(d) $\xrightarrow{LiAlH_4}$ $CH_3CH(CH_2OH)CH_2OH$ (achiral). Thus, true

11 (*a*) 4-formylcyclohexanone $\xrightarrow{[Ag(NH_3)_2]^+ OH^-}$ $\xrightarrow{CH_3OH}$ 4-(COOCH$_3$)cyclohexanone $\xrightarrow[H_3O^+]{CH_3MgBr}$ 1,4-dimethylcyclohexane-1,4-diol (HO, CH_3 / H_3C, OH, CH_3)

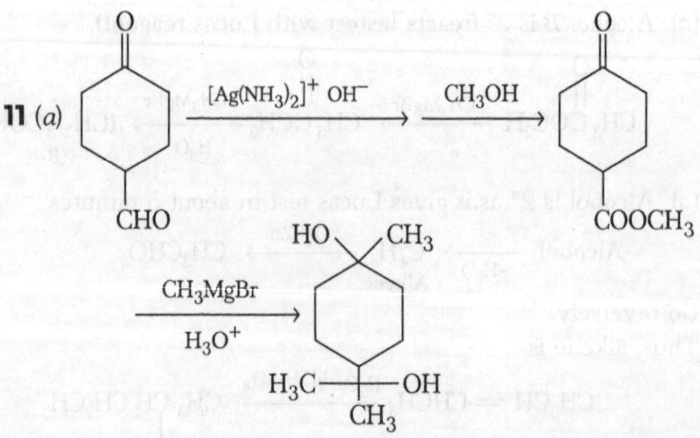

12 (*a*) X $\xleftarrow[\text{or } NaBH_4]{LiAlH_4}$ methylenecyclohexane ($=CH_2$) $\xrightarrow[\text{Hydrolysis}]{H_3O^+}$ 1-methylcyclohexanol (CH_3, OH)

methylenecyclohexane $\xrightarrow{BH_3.THF/H_2O_2, NaOH}$ cyclohexyl–CH_2OH

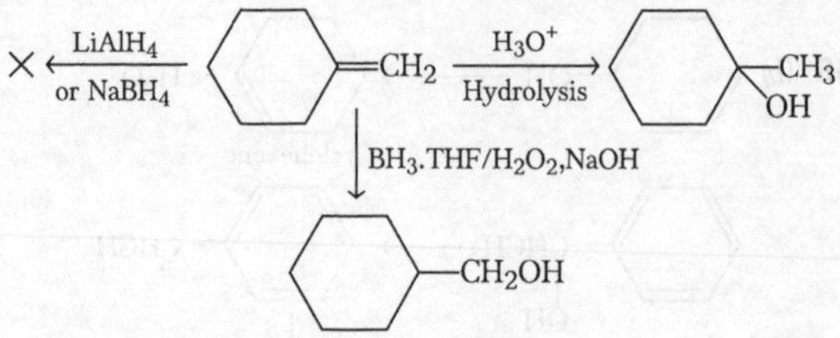

15 (*a*) Basic nature of alcohols is in order 3° > 2° > 1°. Which is the order for nucleophilicity.
Thus, alcohol at $C_2(3°)$ is more nucleophilic than at $C_5(2°)$.

16 (*a*) Greater the stability of the intermediate carbocation, greater the hydration. Substituted alkenes are more hydrated than less substituted. Hence, the correct rate of hydrationis I < II < III.

17 (*a*) Benzamide (HO_2C, CH_2OH, $CONH_2$, OH) $\xrightarrow[\text{Hydrolysis of amido group}]{HCl}$ acid (HO_2C, CH_2OH, COOH, OH) $\xrightarrow{\text{(Intermolecular esterification of alcohol and acid)}}$ lactone (HO_2C, OH)

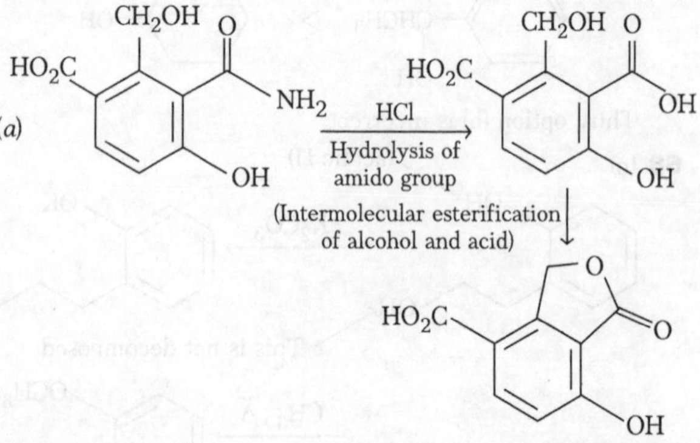

18 (*b*) Statement I is correct and Statement II is incorrect statement.
Correct statement II Addition of water in acidic medium proceeds through the formation of secondary carbocation.

19 (*c*) Greater the stability, smaller the energy.
Thus, the order for stable transition states of the given reaction is,

Reaction 1 < Reaction 2 < Reaction 3.

21 (*c*) $(CH_3)_2C-CH_2$ (epoxide, O) (*A*) $\xrightarrow{LiAlH_4}$ $CH_3-C(CH_3)(OH)-CH_3$ (3° alcohol)

22 (*a*) $CaC_2 \xrightarrow{H_2O} CH\equiv CH \xrightarrow[HgSO_4]{Dil.\ H_2SO_4} CH_3CHO \xrightarrow{H_2/Ni} CH_3CH_2OH$ Ethanol

23 (*c*) Squaric acid (*A*) (O, O, OH, OH) ⇌ dianion (O, O, O^-, O^-) Resonance stabilised

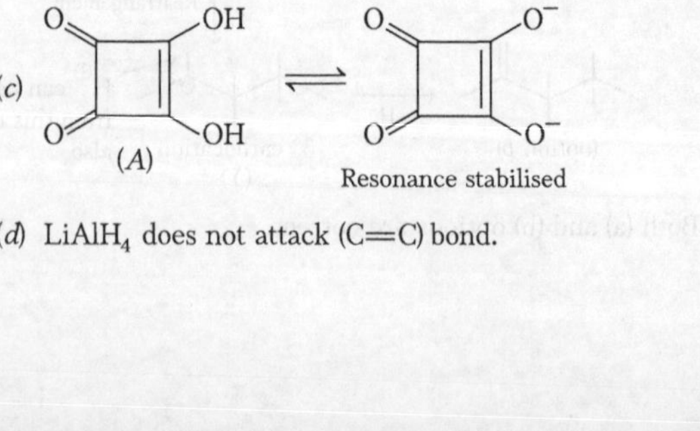

24 (*d*) $LiAlH_4$ does not attack (C=C) bond.

26 (*b*) $A \xrightarrow{H_3O^+} CH_3CH(CH_3)COOH$

Thus, *A* is an ester $CH_3CH(CH_3)COOR$ and *B* is an alcohol

$$A \xrightarrow{CH_3MgBr} CH_3CH(CH_3)COCH_3 \xrightarrow[H_3O^{\oplus}]{CH_3MgBr} CH_3CH(CH_3)C(CH_3)(OH)CH_3 \; (B)$$

27 (*a*) $H-C^*(Me)(Et)-OH \xrightarrow{Na} H-C^*(Me)(Et)-\bar{O}\,|\,\overset{+}{Na}$ (This bond is affected) $\xrightarrow{CH_3Br} H-C^*(Me)(Et)-O-CH_3$

28 (*a*) $NaBH_4$ cannot reduce ester functional group.

(C) Medical entrances' gallery

1 (*b*) A ketone on reaction with Grignard reagent followed by hydrolysis produces a tertiary or *tert.* or 3° alcohol.

$$\underset{\text{Acetone}}{(CH_3)_2\overset{\delta+}{C}=\overset{\delta-}{O}} \xrightarrow{CH_3MgCl} (CH_3)_2C(CH_3)-O-MgCl \xrightarrow[H_2O]{-Mg(OH)Cl} \underset{\text{Tertiary butyl alcohol}}{CH_3-C(CH_3)_2-OH}$$

2 (*a*)

$$CH_3CH_2-CH=CH_2 \xrightarrow[\text{(ii) } H_2O,\, H_2O_2,\, OH^-]{\text{(i) } B_2H_6} \underset{\text{Butan-1-ol (Z)}}{CH_3CH_2-CH(H)-CH_2(OH)}$$

(δ^+ H and δ^- OH added; OH to least substituted sp^2 carbon; Anti-Markownikoff's addition of H_2O (overall))

It is hydroboration-oxidation (HBO) reaction of an alkene, which undergoes hydration to give an alcohol.
Here, anti-Markownikoff's addition of H_2O takes place.

3 (*c*) When vapours of alcohols are passed over heated copper at 573 K, primary and secondary alcohols undergo dehydrogenation to give aldehydes and ketones, respectively. While tertiary alcohols undergo dehydration to give alkenes.

$$\underset{\substack{\text{Propan-2-ol}\\(2^\circ\text{ alcohol})}}{(H_3C)_2CH(OH)} \xrightarrow{Cu/573\text{ K}} \underset{\substack{\text{Propanone}\\\text{(Ketone)}}}{(H_3C)_2C=O} + H_2$$

4 (*b*) The suitable reagent *X* for the conversion is B_2H_6 | THF, H_2O_2 | OH^-.

$$\underset{\text{Butene}}{CH_3CH_2CH=CH_2} \xrightarrow[\text{(ii) } H_2O_2/OH^-]{\text{(i) } B_2H_6/THF} \underset{\text{Butanol}}{CH_3CH_2CH_2CH_2OH}$$

Alkenes react with diborane (B_2H_6) in presence of THF or diglyme to form alkyl borane (by hydroboration) which on oxidation gives alcohol. The reaction is known as hydroboration-oxidation reaction.

5 (*b*) Haloform test is used to identify the compound (alcohol or carbonyl compound). In this reaction, alcohols (or aldehydes and ketones) on heating with X_2 (Cl_2, Br_2 and I_2) and OH^- form CHX_3 (haloform). The alcohols containing the CH_3CHOH-group, linked to carbon or hydrogen atom can be distinguished from other alcohols by this test. Thus, ethanol gives positive haloform test.

Therefore, both Assertion and Reason is correct but Reason is not the correct explanation of the Assertion.

6 (*e*)

$$\underset{\text{3-pentanol}}{CH_3-CH_2-CH(OH)-CH_2-CH_3} \xrightarrow[-H_2O]{\text{Conc.}H_2SO_4} \underset{\text{pent-2-ene}}{CH_3-CH=CH-CH_2-CH_3}$$

This alkene exhibit following *cis/trans*- isomers.

Cis-isomers: $H_3C(H)C=C(H)CH_2-CH_3$ (both H on same side); *Trans*-isomers: $H_3C(H)C=C(H)CH_2-CH_3$ (H on opposite sides)

7 (*d*) $H-C(CH_3)(CH_2CH_3)-Br \longrightarrow H-\overset{\oplus}{C}(CH_3)(CH_2CH_3)$ (carbocation)

2° alkyl halide undergoes S_N1 reaction in polar solvent (as H_2O) and intermediate is 2° carbocation (sp^2-hybridised). Attack of nucleophile can be from either side.

$$\underset{(-)\text{ inversion}}{HO-C(CH_3)(CH_2CH_3)-H} \xleftarrow{OH^-} H\overset{\oplus}{C}(CH_3)(CH_2CH_3) \xrightarrow{OH^-} \underset{(+)\text{ retention}}{H-C(CH_3)(CH_2CH_3)-HO}$$

(+) and (–) isomers are formed in 1 : 1 ratio, thus mixture is optically inactive.

So, 2-bromobutane gives racemic mixture of (±) butan-2-ol.

8 (*c*) Reagent used to convert alcohol to aldehyde is CrO_3-pyridine

$$\underset{\text{2-methylpentan-1-ol}}{CH_3CH_2-CH_2-CH(CH_3)-CH_2OH} \xrightarrow[\text{Pyridine}]{CrO_3/HCl} \underset{\text{2-methylpentanal}}{CH_3CH_2CH_2-CH(CH_3)-CHO}$$

9 (*b*) This problem include conceptual mixing of mole concept, Tollen's reaction of carbonyl compound and function of Grignard reagent. Follow the steps given below to solve this question.

Calculate the molecular mass of alcohol using the mole concept. Now use the concept of Tollen's test to identify the correct substrate as among aldehyde and ketone. Only aldehyde undergo Tollen's test. Aldehyde can be prepared by the reaction of primary alcohol with PCC.

$$\underset{\substack{0.44\text{ g}\\ \text{At STP}}}{ROH} + CH_3MgI \xrightarrow{\text{Ether}} \underset{\substack{112\text{ cm}^3\\ 22400\text{ cm}^3}}{CH_4} + Mg\langle^{I}_{OR}$$

$$\therefore \text{ Molecular mass of alcohol} = \frac{0.44 \times 22400}{112} = 88$$

Since, alcohol oxidised with PCC (pyridinium chlorochromate or Corey's reagent $[C_5H_5NH^+CrO_3Cl^-]$) to form a carbonyl compound (aldehyde), which gives positive silver mirror test hence, carbonyl compound must be an aldehyde. Therefore, it must be a primary alcohol.

$$\underset{m=88}{(CH_3)_3C{-}CH_2OH} \xrightarrow{PCC} (CH_3)_3C{-}CHO$$

$$\underset{m=74}{(CH_3)_2CH{-}CH_2OH} \xrightarrow{PCC} (CH_3)_2CH{-}CHO$$

In the given options, (b) and (c) are primary alcohols having molecular masses 88 and 74, respectively.
Hence, only (b) will be the correct choice.

10 (*b*) Ethyl alcohol is oxidised to produce acetaldehyde

$$\underset{\text{Ethyl alcohol}}{CH_3CH_2OH} \xrightarrow{[O]} \underset{\text{Acetaldehyde}}{CH_3CHO} + H_2O$$

Thus, the conversion of ethyl alcohol into acetaldehyde is an example of oxidation.

11 (*d*) $\underset{\text{Methanal}}{HCHO} + \underset{\text{Grignard reagent}}{CH_3MgI} \xrightarrow{H_3O^+} \underset{\text{Ethanol}}{CH_3CH_2OH}$

12 (*d*) Glycerol gives different products on oxidation with different oxidising agents. On oxidation with bismuth nitrate glycerol gives *meso*-oxalic acid.

$$\underset{\text{Glycerol}}{\begin{matrix}CH_2OH\\ |\\ CHOH\\ |\\ CH_2OH\end{matrix}} \xrightarrow[\text{nitrate}]{\text{Bismuth}} \underset{\textit{Meso-}\text{oxalic acid}}{\begin{matrix}COOH\\ |\\ CO\\ |\\ COOH\end{matrix}}$$

13 (*b*) Glycerol is trihydric alcohol. It is purified by its distillation under reduced pressure because it decomposes below its boiling point to form a foul smelling compound acrolein.

$$\underset{\text{Glycerol}}{\begin{matrix}CH_2OH\\ |\\ CHOH\\ |\\ CH_2OH\end{matrix}} \xrightarrow[-2H_2O]{\Delta} \underset{\substack{\text{Acrolein}\\ \text{(Foul smell)}}}{\begin{matrix}CH_2\\ \|\\ CH\\ |\\ CHO\end{matrix}}$$

14 (*b*) Glycol is widely used as an antifreeze in automobile radiators.

15 (*c*) Pyridinium chlorochromate (PCC) prevents further oxidation of aldehydes to carboxylic acid.

$$\underset{\text{Alcohol}}{RCH_2OH} \xrightarrow[\substack{CH_2Cl_2\\ \text{(Solvent)}}]{PCC} \underset{\text{Aldehyde}}{RCHO} + H_2O$$

16 (*c*) $\underset{\text{Formaldehyde}}{\begin{matrix}H\\ H\end{matrix}\rangle C{=}O} + CH_3MgI \longrightarrow \underset{(X)}{\begin{matrix}H\\ H\end{matrix}\rangle C\langle\begin{matrix}CH_3\\ OMgI\end{matrix}} \xrightarrow{HOH} \underset{\text{Ethanol}}{CH_3CH_2OH} + Mg(OH)I$

17 (*c*) Tertiary alcohol gives turbidity immediately with Lucas reagent. 2-hydroxy-2-methylpropane (or 2-methylpropan-2-ol) is a tertiary alcohol.

$$\underset{(3°\text{ alcohol})}{H_3C{-}\overset{\overset{CH_3}{|}}{\underset{\underset{OH}{|}}{C}}{-}CH_3}$$

18 (*b*) $CH_3\underset{\underset{OH}{|}}{CH}CH{=}CHCH_3 \longrightarrow CH_3{-}\underset{\underset{O}{\|}}{C}{-}CH{=}CHCH_3$

Only suitable reagent is chromic anhydride in glacial acetic acid. Other reagent will also affect $(C{=}C)$ bond.

19 (*b*) $HI > HBr > HCl > HF$.
This is due to the fact that I^- ion is a better nucleophile than Br^- ion, which in turn is better than Cl^- ion.

20 (*e*) 1° alcohols readily form ethers when heated with conc. H_2SO_4. 2° and 3° alcohols on reaction with sulphuric acid give respective alkene.

21 (*c*) $CH_3{-}\overset{\overset{CH_3}{|}}{CH}{-}\underset{\underset{OH}{|}}{CH}{-}CH_3 \xrightarrow[\text{Heat}]{H^{\oplus}} \underset{\substack{(A)\\ \text{Major}}}{CH_3{-}\overset{\overset{CH_3}{|}}{C}{=}CH{-}CH_3} + \underset{\substack{(B)\\ \text{Minor}}}{CH_3{-}\overset{\overset{CH_3}{|}}{CH}{-}CH{=}CH_2}$

$$\underset{\substack{\text{Major}\\ (A)}}{H_3C{-}\overset{\overset{CH_3}{|}}{C}{=}CH{-}CH_3} \xrightarrow{\text{HBr (dark)}} \underset{(C)\text{ Major}}{CH_3{-}\overset{\overset{CH_3}{|}}{\underset{\underset{Br}{|}}{C}}{-}CH_2{-}CH_3} + \underset{(D)\text{ Minor}}{CH_3{-}\overset{\overset{CH_3}{|}}{\underset{\underset{H}{|}}{C}}{-}\underset{\underset{Br}{|}}{CH}{-}CH_3}$$

22 (*a*) 2-methylpropan-2-ol undergo dehydration very easily as it is a 3° alcohol. The relative ease of dehydration of alcohols follows the following order :

tertiary alcohol > secondary alcohol > primary alcohol

CHAPTER 16

Phenols

Organic compounds which contain a hydroxyl group (— OH group) attached directly to an aromatic ring are known as **phenols.** They are also called **carbolic acid** with chemical formula C_6H_5OH. Their general formula is ArOH, where —Ar is phenyl (or aromatic ring) group.

CLASSIFICATION

Phenols may be classified on the basis of number of —OH groups present. These are as follows

1. **Monohydric phenols**

Phenol

2-methylphenol
o-cresol

2-hydroxybenzoic acid
Salicylic acid

2. **Dihydric phenols**

1,2-dihydroxy benzene (catechol)

1, 3-dihydroxy benzene (resorcinol)

1, 4-dihydroxybenzene (quinol)

3. **Trihydric phenols**

1, 2, 3-trihydroxy benzene (pyrogallol)

1, 2, 4-trihydroxy benzene (hydroxyquinol)

1, 3, 5-trihydroxy benzene (phloroglucinol)

Inside

- **Classification**
- **Structure**
- **Nomenclature**
- **Isomerism**
- **Sources of phenols**
  - *Middle oil fraction of coal tar*
- **Methods of preparation**
  - *From benzene sulphonic acids*
  - *From aromatic amines*
  - *From Grignard reagent*
  - *From salicylic acid*
  - *Dow's process*
  - *From Raschig's process*
  - *Cumene*
  - *Oxidation of benzene*
- **Physical properties**
- **Chemical properties**
  - *Acidity of phenols*
  - *Reactions involving —OH group*
  - *Electrophilic substitution reactions*
  - *Uses*

STRUCTURE

In phenols, the oxygen of —OH group is attached to sp^2-hybridised carbon of benzene ring.

The carbon oxygen bond length is slightly less than that of aliphatic alcohol due to presence of partial double bond character between carbon and oxygen.

109° 136 pm

NOMENCLATURE

Phenol is an accepted IUPAC as well as common name. In these substituted phenols the positions of the substituents are indicated by the prefix 1,2 (*ortho*), 1,3 (*meta*) and 1,4 (*para*), e.g.

IUPAC name	2-methylphenol	3-methylphenol	2-bromophenol	4-nitrophenol
Common name	*o*-cresol	*m*-cresol	*o*-bromophenol	*p*-nitrophenol

IUPAC name	2-hydroxybenzoic acid or salicylic acid	Methyl-3-hydroxybenzoate	Benzene-1, 4-diol	Picric acid
Common name	*o*-hydroxybenzoic acid	Methyl *m*-hydroxybenzoate	Hydroquinone	2, 4, 6-trinitrophenol

ISOMERISM

Substituted derivatives of phenols show positional isomerism. In this type of isomerism, the difference lies in the position of substituent group or atom in the same chain, e.g.

1. Positional isomers of cresols

o-cresol *m*-cresol *p*-cresol

Other isomers are functional isomerism

Benzyl alcohol Anisole

2. Positional isomers of xylenols

2, 3-xylenol

2, 4-xylenol

2, 5-xylenol

2, 6-xylenol

3, 4-xylenol

3, 5-xylenol

SOURCES OF PHENOL

Middle oil fraction of coal tar

A number of monohydric phenols occur in coal tar. The middle oil fraction of coal tar consists of phenol in addition to other compounds. The steps involves in separation of phenolic compounds are as follows

Step I Middle oil is washed with H_2SO_4 (to dissolve basic impurities like pyridine).

Step II This fraction is cooled to crystallise out naphthalene (a low melting solid).

Step III Filterate of step II is treated with alkali (to dissolve phenols). Alkaline layer is separated and carbon dioxide gas is bubbled through it to decompose sodium phenoxides into free phenols.

$$C_6H_5OH + NaOH \longrightarrow C_6H_5\overset{-}{O}\overset{+}{Na} + H_2O$$

Phenol

$$C_6H_5ONa \xrightarrow{CO_2,\ H_2O} C_6H_5OH + Na_2CO_3\downarrow$$

Step IV The phenolic mixture is subjected to fractional distillation to isolate individual phenols.

$$\text{Crude phenols} \xrightarrow[\text{distillation}]{\text{Fractional}} \begin{cases} \xrightarrow{450\text{ K}} \text{Phenol} \\ \xrightarrow{463\text{-}476\text{ K}} o,m,p\text{-cresols} \\ \xrightarrow{484\text{-}508\text{ K}} \text{Xylols(hydroxy xylenes)} \end{cases}$$

Methods of preparation

Phenols can be prepared by following methods

From benzene sulphonic acids

Sodium salt of benzene sulphonic acid is fused with NaOH to form sodium phenoxide. This on treatment with dilute acid yields phenol.

This is a nucleophilic aromatic substitution reaction, where the sulphonate group is replaced by hydroxide ion.

$$\underset{\text{Benzene}}{C_6H_6} \xrightarrow{\text{Oleum}} \underset{\text{Benzene sulphonic acid}}{C_6H_5SO_3H} \xrightarrow{\text{NaOH}} \underset{\text{Sodium benzene sulphonate}}{C_6H_5\overset{-}{SO_3}\overset{+}{Na}}$$

$$\xrightarrow{\text{Fuse NaOH}} \underset{\text{Sodium phenoxide}}{C_6H_5\overset{-}{O}\overset{+}{Na}} \xrightarrow{H^+/H_2O} \underset{\text{Phenol}}{C_6H_5OH}$$

From aromatic amines

Aniline, an aromatic amine on treatment with nitrous acid produces diazonium salt. The diazonium salt, on heating in aqueous acidic medium, liberates nitrogen and forms phenol.

Step 1 Formation of diazonium salt

$$\underset{\text{Aniline}}{C_6H_5NH_2} \xrightarrow{HNO_2} \underset{\text{Benzene diazonium chloride}}{C_6H_5N_2^+Cl^-}$$

Step 2 Hydrolysis of benzene diazonium chloride

$$\underset{\text{Benzene diazonium chloride}}{C_6H_5N_2^+Cl^-} \xrightarrow[\text{Warm}]{H_2O} \underset{\text{Phenol}}{C_6H_5OH} + N_2\uparrow$$

From Grignard reagent

The Grignard reagent formed by the reaction of halobenzene and Mg reacts with oxygen followed by subsequent hydrolysis by mineral acid, produces phenols.

$$C_6H_5Br \text{ (Bromobenzene)} + Mg \xrightarrow{\text{Dry ether}} C_6H_5\overset{+}{Mg}\overset{-}{Br} \xrightarrow{1/2\ O_2} C_6H_5\overset{-}{O}\overset{+}{Mg}Br \xrightarrow[H^+]{H_2O} C_6H_5OH \text{ (Phenol)} + Mg(OH)Br$$

From salicylic acid

Phenol is obtained by decarboxylation of sodium salt of salicylic acid with soda-lime followed by acidification with dil.HCl.

$$\text{Salicylic acid } (o\text{-}HOC_6H_4\overset{-}{C}OO\overset{+}{Na}) \xrightarrow[\Delta]{\text{Soda lime NaOH/CaO}} C_6H_5\overset{-}{O}\overset{+}{Na} \xrightarrow{H^+} C_6H_5OH \text{ (Phenol)}$$

Dow's process

In this process, alkaline hydrolysis of chlorobenzene produces sodium phenoxide which on treatment with acid yields phenol. The reaction proceed through nucleophilic aromatic substitution.

$$C_6H_5Cl \text{ (Chlorobenzene)} + 2NaOH \xrightarrow[\text{High pressure, } -NaCl,\ -H_2O]{350°C/300\ atm} C_6H_5\overset{-}{O}\overset{+}{Na} \text{ (Sodium phenoxide)} \xrightarrow{\text{Dil. HCl}} C_6H_5OH \text{ (Phenol)} + NaCl$$

Note • *The reaction is carried out at high temperature and pressure because of stable C — X bond due to resonance.*

• *The presence of electron withdrawing group in chlorobenzene decrease the rate of substitution reaction due to increase in partial double bond character at C — Cl.*

From Raschig's process

In this process, chlorobenzene is hydrolysed at a temperature of 425°C to form phenol. Chlorobenzene can be obtained by treatment of benzene with HCl in air at 250°C in presence of cupric chloride or ferric chloride as catalyst.

$$C_6H_6 \text{ (Benzene)} + HCl + \frac{1}{2}O_2 \xrightarrow[250°C]{CuCl_2/FeCl_3} C_6H_5Cl \text{ (Chlorobenzene)} \xrightarrow[425°C]{\text{Steam}} C_6H_5OH \text{ (Phenol)} + HCl$$

HCl formed in the reaction may be used again to convert more of benzene into chlorobenzene (Cumene).

Oxidation of isopropyl benzene (Cumene)

This is an industrial process for the preparation of phenol. In this method, benzene on reaction with propene in presence of phosphoric acid (or reaction with alkyl chloride in presence of $AlCl_3$) undergoes Friedal-Crafts alkylation to form isopropyl benzene.

$$C_6H_6 \text{ (Benzene)} + CH_3CH_2CH_2Cl \text{ (Chloropropane)} \xrightarrow{AlCl_3} C_6H_5CH(CH_3)_2 \text{ (Isopropyl benzene (Cumene))}$$

$$C_6H_6 \text{ (Benzene)} + CH_3CH{=}CH_2 \text{ (Propene)} \xrightarrow{H_3PO_4} \text{Cumene} \xrightarrow{O_2 \text{ air}} C_6H_5C(CH_3)_2{-}O{-}OH \text{ (Cumene hydroperoxide)}$$

$$\text{Cumene hydroperoxide} \xrightarrow[150°C]{H_2O,\ H^+} CH_3\overset{O}{\overset{\|}{C}}CH_3 \text{ (Acetone)} + C_6H_5OH \text{ (Phenol)}$$

The oxidation of cumene results in the formation of cumene hydroperoxide, which on hydrolysis in acidic medium produces phenol and acetone.The reaction is carried out at 150°C in presence of catalyst.

The reaction produces acetone as a byproduct which is also an important reagent.

Phenol and acetone can be further separated by fractional distillation.

Oxidation of benzene

It is the latest method for the manufacturing of phenol. In this method, the mixture of benzene is passed over vanadium pentaoxide at 315°C and benzene is directly oxidised to phenol.

$$2C_6H_6 \text{ (Benzene)} + O_2 \xrightarrow[315°C]{V_2O_5} 2C_6H_5OH \text{ (Phenol)}$$

CHECK POINT 16.1

1. Carbolic acid is
(a) phenol (b) salol
(c) Phenyl benzoate (d) phenyl acetate

2. Which of the following is a trihydric phenol?
(a) Catechol (b) Phloroglucinol
(c) *p*-cresol (d) Resorcinol

3. Cresols are
(a) trihydric phenol (b) trihydric alcohols
(c) hydroxy alcohols (d) hydroxy toluenes

4. Among the following, which contains phenolic group
(a) picric acid (b) phenylacetic acid
(c) phthalic acid (d) terephthalic acid

5. Picric acid is
(a) nitrophenol
(b) 2, 4-dinitrophenol
(c) 2, 4, 6-trinitrophenol
(d) 1, 3, 5-trinitrophenol

6. Benzene diazonium chloride when treated with dil. H_2SO_4 produces.
(a) Benzene (b) Benzoic acid
(c) Toluene (d) Phenol

7. Which among the following, does not form phenoxide ion?
(a) $C_6H_5SO_3Na$ (b) C_6H_5Cl
(c) $C_6H_5N_2Cl$ (d) C_6H_5COOH

8. Upon acidic hydrolysis, sodium benzene sulphonate when treated with NaOH gives
(a) benzene (b) phenol
(c) benzoic acid (d) benzaldehyde

9. Cumene is the compound used for commercial preparation of phenol. Chemically, cumene is
(a) iso-propyl benzene (b) ethyl benzene
(c) *n*-propyl benzene (d) None of these

10. Decarboxylation of sodium salicylate with soda lime forms
(a) salicylic acid (b) phenol
(c) benzene (d) None of these

PHYSICAL PROPERTIES

Various physical properties shown by phenol are as follows

(*i*) It is a colourless, crystalline, deliquescent solid. It attains pink colour on exposure to air and light.

(*ii*) It has a characteristic carbolic odour and a strong corrosive action on skin.

(*iii*) It is sparingly soluble in water but readily soluble in organic solvents such as alcohol, benzene and ether.

(*iv*) They are capable of forming intermolecular H-bonding. Thus, they have high boiling point.

(intermolecular H-bonding among phenol molecules)

(crossed intermolecular H-bonding between water and phenol molecules).

Due to intramolecular hydrogen bonding, derivatives of phenol, i.e. *o*-nitrophenol has lower solubility in water and lower boiling point. Due to this, association with different molecules decreases hence, comparatively volatile in nature. On the other hand, *m* and *p*- isomer have intermolecular hydrogen bonding. Thus, they have an appreciable value of boiling point and are soluble in water. This may be represented as follows

Intermolecular hydrogen bonding in *p*-nitrophenol molecules

Intramolecular hydrogen bonding in *o*-nitrophenol (chelation)

Table 16.1 Comparison of properties of nitrophenols

Isomer	Boiling point at 760 mm	Solubility g/100g H_2O	Reaction with steam
o-nitrophenol	373 K	0.2	volatile in steam
p-nitrophenol	decomposes	1.69	non-volatile in steam
m-nitrophenol	467 K	1.35	non-volatile in steam

CHEMICAL PROPERTIES

Various chemical properties of phenols are as follows

Acidity of phenols

Phenols are converted into their salts on reaction with aqueous hydroxides. These salts are converted into free phenols by aqueous mineral acids or carboxylic acids.

Phenol (OH) + OH^- ⟶ Phenoxide ion (O^-) + H_2O

Stronger acid — Weaker acid

Phenoxide ion (O^-) + H_2CO_3 ⟶ Phenol (OH) + HCO_3^-

Stronger acid — Weaker acid

Phenols react with active metal to give hydrogen and also turn blue litmus red.

The acidic nature of phenols can be explained by the following resonating structures

As a result of resonance, the oxygen atom acquires a partial positive charge and partial double bond character with sp^2-hybridised carbon of the ring. It weakens the O—H bond and thus, the release of proton is facilitated.

Phenol ⇌ Phenoxide ion + H^+

The phenoxide ion so formed is stabilised through resonance as shown below

Here, phenoxide ion is better resonance stabilised than phenol. This extra stabilisation of phenoxide ion shifts the equilibrium in the forward direction. Thus, phenol is considerably stronger acid than alcohols but weaker acid than carboxylic acid.

This is because the alkoxide ion formed is not stabilised through resonance. The negative charge is localised on oxygen atom only.

e.g.

Cyclohexanoate ion

Effect of substituents on acidity of phenols

Acidity of phenols is attributed to the resonance stabilisation of phenoxide ion relative to phenol. Therefore, if a substituent stabilises the phenoxide ion by dispersal of negative charge, it increases the acidity of phenols. Electron withdrawing groups (EWG) such as NO_2, —CN, —CHO, —*X* (halogens) increases the acidity of phenols, whereas electron donating groups (EDG) such as —NH_2, —O*R* (alkoxy) —*R* (alkyl) decreases the acidity of phenols.

G releases electrons, intensifies the negative charge, destabilises ion, decreases acidity.

$G = -CH_3, -C_2H_5$

G withdraws electrons, disperses negative charge, stabilises ion, increases acidity.

$G = -NO_2, -CN, X, -\overset{\oplus}{N}R_3, -CHO, -COX, -COOR$

Comparison of acidic strength

(*i*) **In nitrophenols** The acidic strength of nitrophenols is in the order as follows

p-nitrophenol > *o*-nitrophenol > *m*-nitrophenol

A nitro group at *o*-and *p*-positions, withdraw electrons of O—H bond by –*R* effect, which increases the acidity of *o*-and *p*-nitrophenol.

p-nitrophenol

o-nitrophenol

o-nitrophenol is less acidic than *p*-nitrophenol due to presence of intramolecular H-bonding which makes removal of proton difficult. Lower acidity in case of *meta*-isomer is explained by the fact that *m*-nitro phenoxide ion is stabilised by inductive effect only and no resonance effect is observed in NO_2.

More the number of nitro groups attached to *o*-and *p*-positions, more is the acidity of nitrophenols.

(*ii*) **In halophenols** All halophenols are more acidic than phenols due to –*I* effect of halogen. –*I* effect decreases with increase in distance. Therefore, order of acidity in halophenols is as follows

o-halophenol > *m*-halophenol > *p*-halophenol

Among different *o*-halophenols, order of acidity is as follows

o-chlorophenol > *o*-bromophenol > *o*-iodophenol > *o*-fluorophenol.

Note *o-fluorophenol is the weakest acid due to strong intramolecular H-bonding.*

(*iii*) **In isomeric cresols** Acidic strength of cresols decreases in the order

Phenol > *m*-cresol > *p*-cresol > *o*-cresol

m-cresol is stronger acid than *o*-and *p*-cresols because electron donating groups such as $—CH_3(R)$, destabilises the phenoxide ion and decreases the acidity. Due to stronger +*I*-effect of $—CH_3$ group at *o*-and *p*-position, *m*-cresol is stronger acid.

A comparison of K_a and pK_a values is given in Table 16.2

Table 16.2 Acidity constants of some phenols

Phenolic compound	Acidity constants (K_a)	pK_a
phenol	1.1×10^{-10}	9.96
o-cresol	0.63×10^{-10}	10.20
m-cresol	0.98×10^{-10}	10.01
p-cresol	0.67×10^{-10}	10.17
o-chlorophenol	77×10^{-10}	8.11
m-chlorophenol	16×10^{-10}	8.80
p-chlorophenol	6.3×10^{-10}	9.20
o-nitrophenol	600×10^{-10}	7.22
m-nitrophenol	50×10^{-10}	8.30
p-nitrophenol	690×10^{-10}	7.16
o-aminophenol	2×10^{-10}	9.70
m-aminophenol	69×10^{-10}	8.16
2,4-dinitrophenol	1×10^{-4}	4.00
2,4,6-trinitrophenol (picric acid)	19.50×10^{-2}	0.71
catechol	1×10^{-10}	10.00
resorcinol	3×10^{-10}	9.52
quinol	2×10^{-10}	9.70
o-methoxyphenol	1.05×10^{-10}	9.98
m-methoxyphenol	2.24×10^{-10}	9.65
p-methoxyphenol	0.62×10^{-10}	10.21

Note *m-methoxyphenol is a stronger acid than m-aminophenol. This is because, $—OCH_3$ and $—NH_2$ groups show –I and +R-effect. But, +R-effect is not observed at m-position. Further, –I-effect of $—OCH_3$ is more than that of $—NH_2$. Therefore, m-methoxy phenol > m-aminophenol > phenol.*

Reactions involving —OH group

Reaction with $FeCl_3$

Neutral ferric chloride solution, when treated with phenol gives violet colouration due to formation of coloured complex of iron. It is a characteristic reaction to the existence of keto-enol tautomerism in phenols predominantly in enolic form.

This reaction is used as a qualitative detection of phenolic functional group.

Enol $\rightleftharpoons$ Keto

$$6\,C_6H_5OH + FeCl_3 \longrightarrow 3H^+ + [Fe(OC_6H_5)_6]^{3-} + 3HCl$$

Enol — Violet colour

Williamson's synthesis (Ether formation)

Phenol reacts with alkyl halides in alkaline solution to form phenyl ethers. This reaction is known as Williamson's synthesis. Halogen of alkyl halide is replaced by phenoxide ion, which acts as a nucleophile.
e.g.

$$C_6H_5OH \xrightarrow{KOH} C_6H_5\bar{O}K^+ \xrightarrow{C_2H_5-I} C_6H_5O-C_2H_5 + KI$$

Phenol — Potassium phenoxide — Ethoxybenzene (phenetol)

Phenols react with allyl bromide in presence of aqueous solution of base to yield aryl allyl ethers.

$$C_6H_5\bar{O}Na^+ + Br-CH_2-CH=CH_2 \xrightarrow{aq.\ KOH} C_6H_5OCH_2CH=CH_2$$

When these ethers are heated at a temperature of 200°C, allyl group migrates from ethereal oxygen to *o*-position. This rearrangement is known as **Claisen rearrangement.**

$$C_6H_5OCH_2-CH=CH_2 \xrightarrow[\text{Diphenyl ether}]{200°C} o\text{-}HOC_6H_4CH_2-CH=CH_2$$

Allyl phenyl ether — *o*-allyl phenol

Note *When both ortho-positions are blocked, migration occurs at para-position.*

$$2,6\text{-}R_2C_6H_3OCH_2-CH=CH_2 \xrightarrow{200°C} 2,6\text{-}R_2\text{-}4\text{-}(CH_2-CH=CH_2)C_6H_2OH$$

In this Claisen rearrangement, carbon which is attached to benzene nucleus is **vinylic** and not **allylic.**

$$C_6H_5OCH_2CH=\overset{14}{C}H_2 \xrightarrow{200°C} o\text{-}HOC_6H_4\overset{14}{C}H_2CH=CH_2$$

Ester formation

Phenol reacts with acid chlorides or anhydrides in alkaline solution to form phenylesters. This reaction is known as **acylation**.

$$C_6H_5OH + CH_3COCl \xrightarrow{NaOH} C_6H_5OCOCH_3 + NaCl + H_2O$$

Phenol — Acid chloride — Methyl benzoate

When phenol reacts with benzoyl chloride, the reaction is known as **benzoylation** or **Schotten-Baumann reaction.**

$$C_6H_5OH + Cl-CO-C_6H_5 \xrightarrow{NaOH} C_6H_5-O-CO-C_6H_5 + NaCl + H_2O$$

Phenol, Benzoyl chloride — Phenyl benzoate

The esters formed, when heated with Lewis acid such as $AlCl_3$, yields *o*-and *p*-acyl phenols.
This reaction is called **Fries arrangement.**

$$C_6H_5O-CO-R \xrightarrow{AlCl_3} o\text{-}HOC_6H_4-CO-R + p\text{-}HOC_6H_4COR$$

o-acyl phenols — *p*-acyl phenols

In this reaction, the acyl group migrates from the phenolic oxygen to an *ortho* or *para* position of ring to produce ketone. RCO^+, acts as an electrophile which is attacked by the ring.

Note *Generally, low temperatures favour the formation of p-isomer whereas high temperature favours the o-isomer.*

Reaction with PCl_5

Phenol reacts with phosphorus pentachloride to produce chlorobenzene.

$$C_6H_5OH + PCl_5 \longrightarrow C_6H_5Cl + POCl_3 + HCl$$

Phenol Chlorobenzene

In this reaction, $POCl_3$ formed, reacts with phenol to form $(C_6H_5)_3PO_4$. As a result, yield of chlorobenzene is poor. Hence, this is not a good method for the synthesis of haloarene.

$$C_6H_5OH + POCl_3 \longrightarrow [C_6H_5]_3PO_4 + 3HCl$$

Phenol or $(C_6H_5)_3PO_4$

Reaction with chromyl chloride

Phenol undergoes oxidation reaction, when reacts with chromyl chloride and yields *p*-benzoquinone (or quinone).

$$\text{Phenol} \xrightarrow[\text{Oxidation}]{CrO_2Cl_2} \text{p-benzoquinone or quinone}$$

Quinone reacts with phenol to give pink colour solution of phenoquinone.

$$\text{p-benzoquinone} + C_6H_5OH + H_2O \longrightarrow C_6H_5OH^{\delta+} \ldots\ O^{\delta-}{=}C_6H_4{=}O$$

Phenoquinone (pink colour solution)

Reaction with zinc dust

Benzene can be obtained by distillation of phenol with zinc dust.

$$C_6H_5OH + Zn \xrightarrow{\Delta} C_6H_6 + ZnO$$

Phenol Benzene

Electrophilic substitution reactions

Phenols undergo electrophilic substitution reactions due to presence of –OH group (an activating group) which directs incoming electrophile to *ortho* and *para*-positions.

It is observed that, due to resonance, electron density is increased on *o*-and *p*-positions as compared to *m*-position. Therefore, electrophile attacks on *o*-and *p*-positions.

$\ddot{O}H \longleftrightarrow \overset{\oplus}{O}H \longleftrightarrow \overset{\oplus}{O}H \longleftrightarrow \overset{\oplus}{O}H$

δ^+ OH, δ^-, δ^-, δ^-

E⁺ **(electrophile) can attack** at *o*-and *p*-sites, hence — OH group is *o*-and *p*-directing and activating group.

Some of electrophilic substitution reactions of phenols are as follows

Halogenation

Phenol reacts with bromine in presence of carbon disulphide (i.e. CS_2) or $CHCl_3$ at low temperature to form a mixture of *ortho* and *para*-bromophenols.

$$C_6H_5OH + Br_2 \xrightarrow[273\ K]{CS_2 \text{ or } CHCl_3} o\text{-}BrC_6H_4OH + p\text{-}BrC_6H_4OH$$

Phenol; *o*-bromophenol (minor products); *p*-bromophenol (major product)

If bromine water is taken in excess, phenol forms a white precipitate of 2, 4, 6-tribromophenol.

$$C_6H_5OH + \underset{(\text{excess})}{3Br_2} \xrightarrow[\text{Solvent}]{\text{Polar protic}} 2,4,6\text{-}Br_3C_6H_2OH + HBr$$

Phenol; 2,4,6-tribromophenol (white ppt.)

In presence of NaOH, phenol is in equilibrium with phenoxide ion, which is more reactive towards electrophilic substitution.

Phenol $\xrightarrow[H_2O]{NaOH}$ Phenoxide ion + Br–Br ⟶ σ-complex (neutral) $\xrightarrow{-H_2O}$ $\xrightarrow{Br_2}$

Sulphonation

The replacement of a hydrogen atom by a sulphonic acid group in a ring is called sulphonation. Phenol reacts with concentrated H_2SO_4 or oleum to form a mixture of *o*-and *p*-hydroxy benzene sulphonic acids.

Phenol + SO_3H^+ ⟶ *o*-hydroxy sulphonic acid + *p*-hydroxy sulphonic acid

At low temperature, *ortho*-isomer is the major product and at high temperature (373 K), *para*-isomer is favoured.

Nitration

Phenol reacts with dilute nitric acid (since —OH group is activating group) to form a mixture of *o*-and *p*-nitrophenols. Nitronium ion (NO_2^+) acts as an electrophile.

Phenol + NO_2^+ $\xrightarrow[288\ K]{CHCl_3}$ *o*-nitrophenol (Major) + *p*-nitrophenol (Minor)

To obtain a better yield of *o*-and *p*-nitrophenol, reaction involves initial nitrosation of phenol with N_2O_3 to form nitrosophenol. It can be oxidised to nitrophenol by HNO_3 as shown below.

Nitrosophenol $\xrightarrow{HNO_3}$ *o*-nitrophenol

When phenol is treated with conc. HNO_3 and conc. H_2SO_4, yields 2, 4, 6-trinitrophenol (picric acid). Although the rate of reaction is very fast but yield of product is very low. It is because, HNO_3 is a strong oxidising agent and may lead to cleavage of benzene ring.

Phenol $\xrightarrow[\text{(Conc. } H_2SO_4 + \text{conc. } HNO_3)]{\text{Conc.} HNO_3 \text{ or}}$ Picric acid

To obtain the better yield of picric acid firstly the ring is deactivated by sulphonation and then nitrated. Presence of SO_3H group prevents oxidation of phenols.

Phenol + H_2SO_4 ⟶ $\xrightarrow[HNO_3]{Conc.}$ 2, 4, 6-trinitrophenol (picric acid)

Friedel-Crafts' alkylation

The Friedel-Crafts' reaction does not occur in case of highly activating rings like phenols. In case of phenol, the $AlCl_3$ forms a complex with oxygen. Thus, direct Friedel-Crafts' reaction generally does not give good results.

So, this reaction is used with modification where phenyl group is protected by methylation before carrying out the reaction.

Phenol + $AlCl_3$ ⟶ $\overset{+}{HO}$: ⟶ $\overset{-}{Al}Cl_3$

Phenol $\xrightarrow[\text{Protection}]{(CH_3)_2SO_4}$ Anisole $\xrightarrow[AlCl_3]{CH_3Cl}$ o-methyl anisole + p-methyl anisole $\xrightarrow[\text{Rearrangement}]{AlCl_3/C_2H_5SH}$ o-methyl phenol + p-methyl phenol

Reimer-Tiemann reaction

Phenol reacts with chloroform in presence of alkaline solution at 330 K followed by acid hydrolysis to give salicylaldehyde as the major product. This reaction is known as Reimer-Tiemann reaction.

Phenol + $CHCl_3$ $\xrightarrow[\text{(ii) } H^+]{\text{(i) } aq. \text{ NaOH, 330 K}}$ Salicylaldehyde (steam volatile) *o*-isomer + *p*-isomer

For mechanism refer Chapter 13 'Haloalkanes' in polyhalogen compound topic.

o-isomer is stabilised by intramolecular H-bonding.

If carbon tetrachloride (CCl_4) is used instead of $CHCl_3$, salicylic acid is formed as a major product.

Phenol $\xrightarrow[330 \text{ K}]{CCl_4, \text{ NaOH}}$ (Disodium salicylate) $\xrightarrow[-2NaCl]{\text{Dil. HCl}}$ (Salicylic acid)

Note *Small amount of p-hydroxybenzoic acid is also formed.*

Kolbe-Schmidt reaction

Sodium phenoxide (formed by action of base on phenol) reacts with carbon dioxide at 120-140°C, high pressure followed by acid hydrolysis to produce salicylic acid. This reaction is known as **Kolbe-Schmidt reaction.**

This is an example of electrophilic substitution reaction where CO_2 acts as an electrophile. It is used for the conversion of phenol into hydroxy carboxylic acid.

Sodium phenoxide + CO_2 $\xrightarrow{400 \text{ K, 4-7 atm}}$ Sodium phenyl carbonate $\xrightarrow{\text{Rearrangement}}$ Sodium salicylate $\xrightarrow{H^+}$ Salicylic acid

Mechanism

The electrophile gets added to phenoxide ion at *ortho*-position. In phenoxide ion, the aromatic ring is strongly activated and easily undergoes electrophilic substitution even with a weak electrophile like CO_2.

Phenoxide ion + $O{=}C{=}O$ ⟶ intermediate + OH^- $\xrightarrow{-H^+}$ COO^- $\xrightarrow{+H^+}$ Salicylic acid

Note *If the reaction is carried out with potassium phenoxide, p-isomer is formed as major product.*

Salicylic acid is used to prepare many compounds which have industrial application.

Salicylic acid

Conc. HNO_3, $-CO_2$ → (Picric acid)

C_6H_5OH, Δ, $POCl_3$ → $CO—C_6H_5 + H_2O$ Salol (phenylsalicylate)

CH_3OH, H_2SO_4/reflux → $COOCH_3 + H_2O$ Methyl salicylate (oil of winter green)

$(CH_3CO)_2O$, $CH_3COO^-Na^+$ → Aspirin (2-acetoxybenzoic acid)

Gattermann synthesis

Phenol reacts with hydrogen cyanide and hydrochloric acid in presence of anhydrous $AlCl_3$ to produce hydroxy benzaldehyde. This reaction is also known as **formylation reaction.**

Phenol + HCN + HCl —Anhyd. $AlCl_3$→ Salicylaldehyde or Hydrobenzaldehyde

Mechanism

This reaction involves attack of formaldinium ion, (formed by HCN and HCl) with the help of $AlCl_3$ to give aldimine. Aldimine on hydrolysis gives hydroxy benzaldehyde. Its mechanism is as follows

$$H—C\equiv N \xrightleftharpoons{H^+} H—\overset{+}{C}=N—H$$

Formaldinium ion

Aldimine —H_3O^+→ (CHO product)

Note *—CHO group prefers para-position but attacks on ortho-position if para-position is blocked.*

Coupling reaction

(i) Coupling with diazonium salts

Phenol reacts with aryl diazonium salts in presence of alkali at low temperature to form azo dyes.

Benzene diazonium salt ($N_2^+Cl^-$) + Phenol —(i) NaOH, (ii) H^+, 273 K→ *p*-hydroxyazobenzene

In this reaction, diazonium salts act as an electrophile and phenol as a nucleophile. Since, phenol is a weak nucleophile, so reaction is carried out in alkaline medium to produce phenoxide ion (strongly nucleophilic).

$-NaCl$ → *p*-hydroxyazobenzene

Note *The reaction should be carried out at optimum pH (i.e. between 9 and 10). Very low pH would suppress the dissociation of phenol to phenoxide ion.*

(ii) Coupling with phthalic anhydride

Phenol reacts with phthalic anhydride in presence of concentrated sulphuric acid to yield phenolphthalein.

Phthalic acid $\xrightarrow{\Delta}$ Phthalic anhydride

Phthalic anhydride + 2 Phenol $\xrightarrow[\Delta]{\text{Conc. } H_2SO_4}$ Phenolphthalein

Phenolphthalein is an acid-base indicator, which gives pink colour in alkaline solution and remains colourless in acidic medium. This is known as **phthalein test** of dibasic acid.

Condensation reaction

Phenol undergoes condensation reaction with formaldehyde (HCHO) in presence of acid to give bakelite as a product. This reaction is carried out in presence of ammonia as catalyst. It also known as **Lederer-Manasse reaction.**

Phenol + HCHO $\xrightarrow[\text{(}NH_3\text{ (catalyst)}]{H^+}$ o-hydroxy benzyl alcohol + p-hydroxy benzyl alcohol $\xrightarrow[\text{(}NH_3\text{ (as catalyst)}]{HCHO/H^+}$ Bakelite

Libermann reaction

This reaction is used to test the presence of phenol. In this reaction, phenol reacts with $NaNO_2$ and concentrated H_2SO_4, to give deep red colour which changes to green or blue on dilution.

Phenol $\xrightarrow{HONO}$ p-nitrosophenol $\rightleftharpoons$ Quinoxim

Quinoxim + phenol $\xrightarrow[H_2O]{H_2SO_4}$ Indophenol (red) $\xrightarrow{H_2O \mid NaOH}$ Sodium salt of indophenol (blue)

This reaction is also known as **Libermann-nitroso test.**

Uses of phenol

(*i*) It is a versatile precursor to large number of drugs such as aspirin, salol, phenacetin etc.

(*ii*) It is an active ingredient in some oral analgesics such as chloroseptic sprays.

(*iii*) Phenol is an important component of industrial paint strippers used in aviation industry for removal of epoxy, polyurethane and other chemical resistant coating.

(*iv*) Phenol derivatives are used in the preparation of cosmetics such as sunscreens, hair colourings and skin lightening.

(*v*) Concentrated phenol liquids are used for permanent treatment of ingrown toe and finger nails.

(*vi*) It is used as an antiseptic in soaps, lotions and ointments. "Dettol" is a phenol derivative (2, 4,-dichloro –3, 5-dimethylphenol).

(*vii*) It is used in manufacture of azodyes, phenolphthalein etc.

(*ix*) It is used in preparation of picric acid (used as an explosive) and for dying silk and wool.

CHECK POINT 16.2

1. Which of the following is the correct order of acidity?

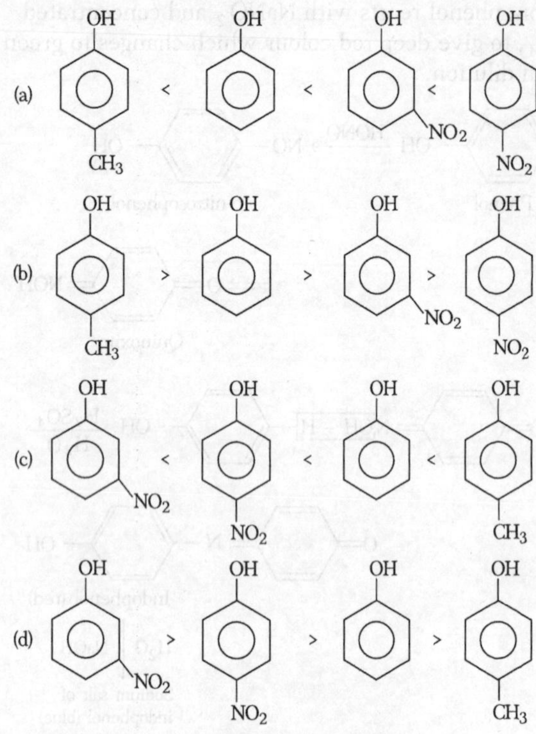

2. Which of the following statement is correct?

(a) C_6H_5OH does not reacts with $NaHCO_3$
(b) C_6H_5OH gives oxime with NH_2OH and HCl
(c) C_6H_5OH is more acidic than C_2H_5OH
(d) C_6H_5OH is less acidic than C_2H_5OH

3. To distinguish phenol from aliphatic alcohol, chemical compound used is

(a) Ferric chloride (b) Fehling's solution
(c) Tollen's reagent (d) Schiff's base

4. Phenol treated with 1 eq. bromine in CS_2 at low temperature gives

(a) 2, 4, 6-tribromophenol
(b) *m*-bromophenol
(c) *p*-bromophenol
(d) *o*-and *p*-bromophenol

5. Reaction of phenol with excess bromine produces

(a) 2, 4, 6-tribromophenol (b) *m*-bromophenol
(c) *p*-bromophenol (d) *o*-and *p*-bromophenol

6. Phenol on reaction with conc. HNO_3 gives

(a) benzoic acid (b) picric acid
(c) salicylic acid (d) salicylaldehyde

7. Phenol on reaction with dil. HNO_3 gives

(a) *o*-and *m*-nitrophenols
(b) *o*-and *p*-nitrophenol
(c) *m*-and *p*-nitrophenols
(d) picric acid

8. Reagents used in Reimer-Tiemann reaction

(a) $CHCl_3/NaOH$ (b) $C_6H_5Cl/NaOH$
(c) $CH_2Cl_2/NaOH$ (d) $CH_3Cl/NaOH$

9. Reaction of phenol with CCl_4 in presence of NaOH produces

(a) benzoic acid (b) salicylaldehyde
(c) benzaldehyde (d) salicylic acid

10. Intermediate formed in the reaction of phenol with chloroform in presence of alkali is

(a) trichlorocarbene (b) chlorine atoms
(c) dichlorocarbene (d) chlorine molecules

Chapter Exercises

(A) Topical exercise

I. Classification, structure, nomenclature and isomerism

1 Which of the following is not classified as phenol?

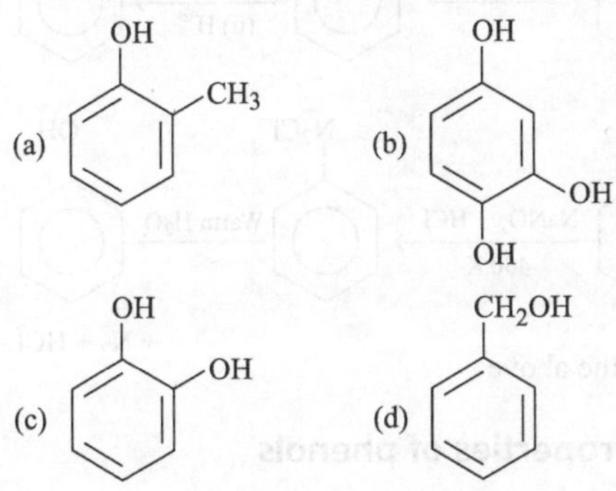

2 Which of the following statements is correct regarding phenol?

(a) The electron density increases on oxygen due to higher electronegativity of sp^2-hybridised carbon of phenol.

(b) Increase in the polarity of O—H bond results in decrease in ionisation of phenols than that of phenols.

(c) In alkoxide ion, the negative charge is localised on oxygen, while in phenoxide ion, the charge is delocalised

(d) In substituted phenols, the presence of electron withdrawing groups decreases the acidic strength of phenol.

3 (Phenyl)$\overset{x}{—}$O—H $H_3C\overset{y}{—}$O—H

Which is true about C—O bond lengths?

(a) $x = y$
(b) $x > y$
(c) $x < y$
(d) is not predictable

4 Consider the following statements.

Statement I C—O bond length in phenol is shorter than that in methanol.

Statement II Phenoxide ion is resonance stabilised.

(a) Both statements I and II are correct.
(b) Only statement I is correct.
(c) Only statement II is correct.
(d) Both statements I and II are incorrect.

5 IUPAC name of *m*-cresol is

(a) 3-methylphenol (b) 3-chlorophenol
(c) 3-methoxyphenol (d) benzene-1,3-diol

6 Match the structures of the compounds given in Column I with the name of the compounds given in Column II.

Column I	Column II
A. (benzene ring with CH_3 and OH)	1. Hydroquinone
B. (benzene ring with OH and OH, meta)	2. o-cresol
C. (benzene ring with OH and OH, para)	3. Resorcinol
D. (benzene ring with OCH_3)	4. Anisole

Codes

	A	B	C	D
(a)	2	3	1	4
(b)	3	2	1	4
(c)	4	1	2	3
(d)	3	2	1	4

II. Sources and methods of preparation of phenols

7 Main source of phenolic compounds is

(c) middle oil (d) heavy oil

8 In addition to phenol, one important solvent acetone is the side product when source is

(a) cumene (b) benzene
(c) cetane (d) octane

9 Which of the following can be used to convert into phenol in one or more steps?

(a) $C_6H_5\overset{+}{N_2}\overset{-}{Cl}$ (b) $C_6H_5SO_3Na$

(c) C_6H_5Cl (d) All of these

10 Phenolic compound is formed in

(a) $C_6H_5\overset{+}{N_2}\overset{-}{Cl} \xrightarrow{H_3O^+}$ (b) $C_6H_5\overset{+}{N_2}\overset{-}{Cl} \xrightarrow{H_3PO_2}$

(c) salicylic acid ($C_6H_4(OH)COOH$) $\xrightarrow{Zn/\Delta}$ (d) $C_6H_5SO_3H + H_2O \xrightarrow[HCl,\ p]{470\ K}$

11 Which of the following hydrocarbon is used for the world wide production of phenol?

(a) *Iso*-butylbenzene (b) *Iso*-propylbenzene
(c) *Iso*-pentylbenzene (d) None of these

12 Identify the major products *P*, *Q* and *R* in the following sequence of reactions :

$$C_6H_6 + CH_3CH_2CH_2Cl \xrightarrow{\text{Anhy. } AlCl_3} P \xrightarrow[\text{(ii) } H_3O^+/\Delta]{\text{(i) } O_2} Q + R$$

	P	*Q*	*R*
(a)	$C_6H_5CH(CH_3)_2$	C_6H_5OH	$CH_3CH(OH)CH_3$
(b)	$C_6H_5CH_2CH_2CH_3$	C_6H_5CHO	C_6H_5COOH
(c)	$C_6H_5CH_2CH_2CH_3$	C_6H_5CHO	$CH_3CH_2—OH$
(d)	$C_6H_5CH(CH_3)_2$	C_6H_5OH	$CH_3—CO—CH_3$

13 Identify which is/are the correct method(s) to synthesise phenol?

(a) $C_6H_5Cl + NaOH \xrightarrow[300\ atm]{623\ K} C_6H_5O^-Na^+ \xrightarrow{HCl} C_6H_5OH$

(b) $C_6H_6 \xrightarrow{Oleum} C_6H_5SO_3H \xrightarrow[\text{(ii) } H^{\oplus}]{\text{(i) } NaOH} C_6H_5OH$

(c) $C_6H_5NH_2 \xrightarrow[400\ K]{NaNO_2 + HCl} C_6H_5\overset{+}{N_2}Cl^- \xrightarrow{\text{Warm } H_2O} C_6H_5OH + N_2 + HCl$

(d) All of the above

III. Physical properties of phenols

14 Which of the following has highest dipole moment?

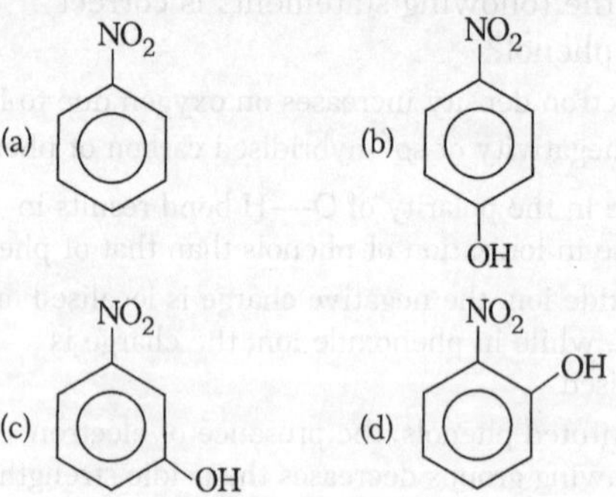

15 *ortho* and *para*-nitrophenols are separated by which of the following method?

(a) Sublimation (b) Crystallisation
(c) Steam-distillation (d) Filteration

16 Consider the following statements.

Statement I *o*-nitrophenol is less soluble in water than the *m* and *p*-isomers.

Statement II *m*- and *p*-nitrophenols exist as associated molecules.

(a) Both statements I and II are correct.
(b) Only statement I is correct.
(c) Only statement II is correct.
(d) Both statements I and II are incorrect.

17 Which of the following has lowest boiling point?

(a) *p*-nitrophenol (b) *m*-nitrophenol
(c) *o*-nitrophenol (d) Phenol

IV. Chemical properties of phenols

18 Which of the following is most acidic?
(a) Benzyl alcohol
(b) Cyclohexanol
(c) Phenol
(d) *m*-chlorophenol

19 Which of the following statements is correct?
(a) Phenol is more acidic than carbonic acid
(b) Phenol is less acidic than ethanol
(c) Phenol is more acidic than ethanol
(d) Phenol is more acidic than acetic acid

20 Which of the following is most acidic?

(a) benzene ring with OH and CHO (ortho)
(b) benzene ring with OH and CHO (meta)
(c) HO–(benzene ring)–CHO
(d) benzene ring with CH_2OH

21 Consider the following statements.

Statement I *p*-nitrophenol is more acidic than phenol.

Statement II Nitro group helps in the stabilisation of the phenoxide ion by dispersal of negative charge due to resonance.

(a) Both statements I and II are correct.
(b) Only statement I is correct.
(c) Only statement II is correct.
(d) Both statements I and II are incorrect.

22 Mark the correct order of decreasing acid strength of the following compounds.

(I) phenol (OH); (II) OH with NO_2 (para); (III) OH with OCH_3 (meta); (IV) OH with NO_2 (meta); (V) OH with OCH_3 (para)

(a) V > IV > II > I > III
(b) II > IV > I > III > V
(c) IV > V > III > II > I
(d) V > IV > III > II > I

23 Rank the following compounds in order of increasing acidity (weakest acid first).

I: Cl, Cl, Cl substituted ring –OH; II: H_3C, CH_3, CH_3 substituted ring –OH; III: O_2N, NO_2, NO_2 substituted ring –OH

(a) II < III < I
(b) III < I < II
(c) III < II < I
(d) II < I < III

24. Which of the following phenols has the largest pK_a value (i.e. least acidic)?

(a) Cl–(ring)–OH
(b) H_3C–(ring)–OH
(c) O_2N–(ring)–OH
(d) N≡C–(ring)–OH

25 Which has maximum pK_a value in phenols?

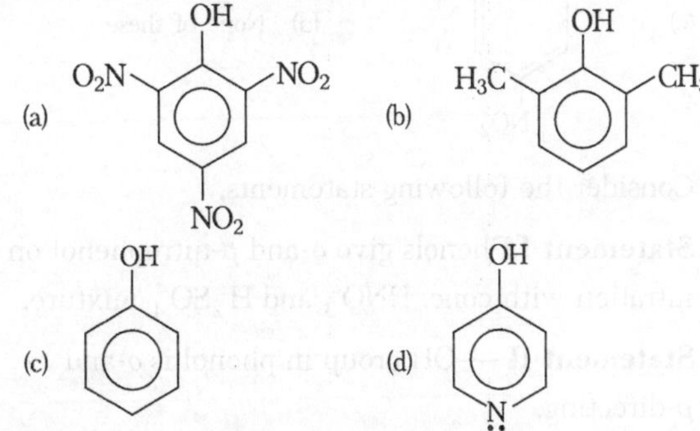

26 Which of the following compounds will react with sodium hydroxide solution in water?
(a) C_6H_5OH (b) $C_6H_5CH_2OH$
(c) $(CH_3)_3COH$ (d) C_2H_5OH

27 Phenol reacts with PCl_5 to give mainly
(a) *p*-chlorophenol only (b) benzene
(c) *o*- and *p*-chlorophenol (d) triphenylphosphate

28 $\text{Phenol} \xrightarrow{\text{Zn dust}} X \xrightarrow[\text{Anhy. AlCl}_3]{CH_3Cl} Y \xrightarrow[KMnO_4]{\text{alcoholic}} Z$

Z is
(a) toluene (b) benzaldehyde
(c) benzoic acid (d) benzene

29 Consider the following statements.

Statement I Phenol forms 2, 4, 6-tribromosphenol on treatment with Br_2 in carbon disuplhide at 273 K.

Statement II Bromine polarises in carbon disulphide.
(a) Both statements I and II are correct.
(b) Only statement I is correct.
(c) Only statement II is correct.
(d) Both statements I and II are incorrect.

30 Phenol reacts with excess of bromine water to produce
(a) *o*-bromophenol
(b) *p*-bromophenol
(c) 2,4,6-tribromophenol
(d) mixture of *o*- and *p*-bromophenol

31 Which of the following is obtained on treating phenol with conc.HNO_3/H_2SO_4 mixture?
(a) OH, NO_2 (b) OH, NO_2
(c) OH, O_2N, NO_2, NO_2 (d) None of these

32 Consider the following statements.

Statement I Phenols give *o*-and *p*-nitrophenol on nitration with conc. HNO_3 and H_2SO_4 mixture.

Statement II —OH group in phenol is *o*-and *p*-directing.
(a) Both statements I and II are correct.
(b) Only statement I is correct.
(c) Only statement II is correct.
(d) Both statements I and II are incorrect.

33 When phenol is treated with $CHCl_3$ and NaOH followed by acidification, salicylaldehyde is obtained. Which of the following species is involved in the above mentioned reaction as intermediates?
(a) O, H, CCl_2 (b) OH, $CHCl_2$
(c) O, H, CCl_2, OH (d) O^-, CH, Cl, Cl

34 A (CH_3, OH) $\xrightarrow{CHCl_3/NaOH} B \xrightarrow[H_3O^+]{HCN} C$, C is
(a) CH_3, CHO, OH (b) CH_3, CH_3, CN
(c) CH_3, CHCOOH, OH, OH (d) CH_3, CHCOOH, OH, OH

35 The product obtained from the following reaction is
OH $\xrightarrow[\text{(ii) } CO_2 \text{ (iii) hydrolysis/}H^+]{\text{(i) NaOH}}$
(a) benzene (b) toluene
(c) salicylic acid (d) benzoic acid

36 When phenol reacts with chloroform and with an alkali like NaOH, the compound formed is salicylaldehyde. If we use pyrene instead of chloroform, the product obtained is
(a) salicylic acid
(b) cyclohexanol
(c) phenolphthalein
(d) salicylaldehyde

37 $X \xrightarrow[\text{Heat}]{NaOH} Y \xrightarrow{CHCl_3 + KOH} Z \xrightarrow{[O]}$ (OH, COOH)

In the above reaction X is
(a) phenol (b) benzoic acid
(c) salicylaldehyde (d) carbonic acid

38 Which of the following phenols will give lower yield during Kolbe's-Schmidt reaction?

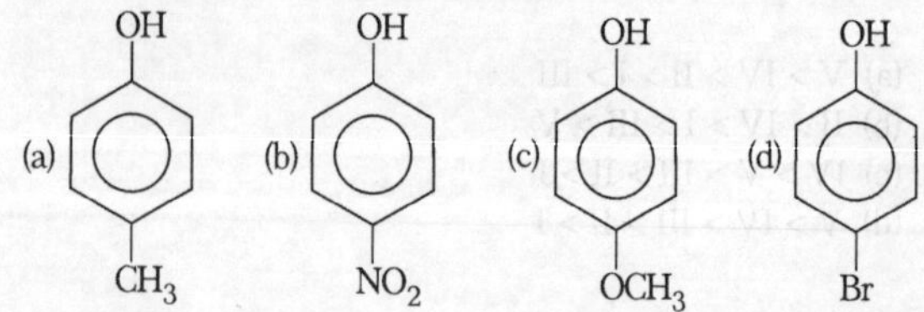

39 Which of the following compound does not form azo dye?

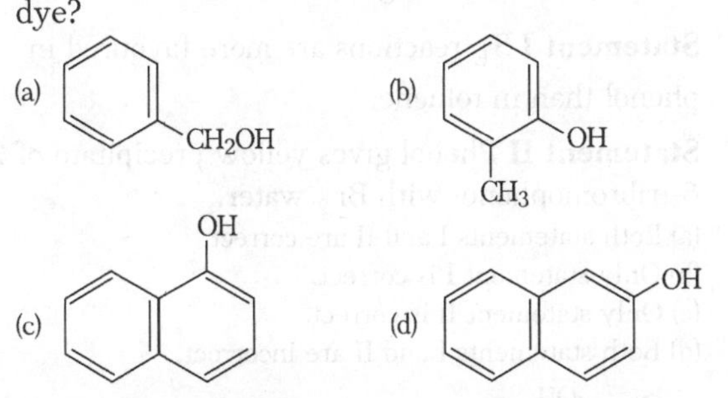

40 Phenol $\xrightarrow{NaNO_2/H_2SO_4} B \xrightarrow{H_2O} C \xrightarrow{NaOH} D.$

Name of the reaction is

(a) Liebermann's reaction (b) Phthalein fusion test
(c) Reimer-Tiemann reaction (d) Schotten-Baumann reaction

41 Phenyl salicylate $\xrightarrow{H_3O^+}$ products are A and B

Product can be distinguished by

(a) Br_2 water (b) $FeCl_3$ solution
(c) $NaHCO_3$ (d) All of these

42 Match the reacting species with phenol (in Column I) with the name of reaction (in Column II) and choose the correct option.

Column I	Column II
A. Phenol + $CHCl_3$ + KOH $\longrightarrow$	1. Friedel-Craft's reaction
B. Phenol + NaOH + $CO_2 \longrightarrow$	2. Phthalein test
C. Phenol + phthalic anhydride $\xrightarrow{\text{Conc. } H_2SO_4}$	3. Kolbe's reaction
D. Phenol + NaOH $\xrightarrow{CH_3Cl/AlCl_3}$	4. Reimer-Tiemann reaction

Codes

	A	B	C	D		A	B	C	D
(a)	1	2	3	4	(b)	2	1	3	4
(c)	4	3	2	1	(d)	3	4	1	2

43 Consider the following statements.

Statement I Phenol is more acid than carboxylic acid.

Statement II Phenol is also called carbolic acid.

(a) Both statements I and II are correct.
(b) Only statement I is correct.
(c) Only statement II is correct.
(d) Both statements I and II are incorrect.

44 CH_3CH_2OH and C_6H_5OH can be distinguished by

(a) neutral $FeCl_3$ solution (b) iodoform test
(c) ceric ammonium nitrate (d) All of these

45 Phenol and benzoic acid are distinguished by

(a) NaOH (b) $NaHCO_3$
(c) Na_2CO_3 (d) H_2SO_4

46 Consider the following statements.

Statement I Phenol gives azo dye with diazonium chloride in basic medium.

Statement II Phenol gives CO_2 with $NaHCO_3$.

(a) Both statements I and II are correct.
(b) Only statement I is correct.
(c) Only statement II is correct.
(d) Both statements I and II are incorrect.

(B) Miscellaneous questions

1 The ionisation constant of phenol is higher than that of ethanol because

(a) phenoxide ion is bulkier than ethoxide ion
(b) phenoxide ion is stronger base than ethoxide ion
(c) phenoxide ion is stabilised through delocalisation
(d) phenoxide ion is less stable than ethoxide ion

2 Phenol is heated with phthalic anhydride in the presence of conc.H_2SO_4. The product gives pink colour with alkali. The product is

(a) phenolphthalein (b) bakelite
(c) salicylic acid (d) fluorescein

3 Phenol and NH_3 reacts in presence of $ZnCl_2$ at 300°C to produce

(a) tertiary amine (b) secondary amine
(c) primary amine (d) All of these

4 Consider the following statements.

Statement I Phenol is more acidic than ethanol.

Statement II Phenoxide ion is stablisied by resonance.

(a) Both statements I and II are correct.
(b) Only statement I is correct.
(c) Only statement II is correct.
(d) Both statements I and II are incorrect.

5 The conversion of *m*-nitrophenol to resorcinol involves

(a) hydrolysis, diazotisation and reduction
(b) diazotisation, reduction and hydrolysis
(c) hydrolysis, reduction and diazotisation
(d) reduction, diazotisation and hydrolysis

6 Which of the following will undergo Friedel-Craft's alkylation reaction?

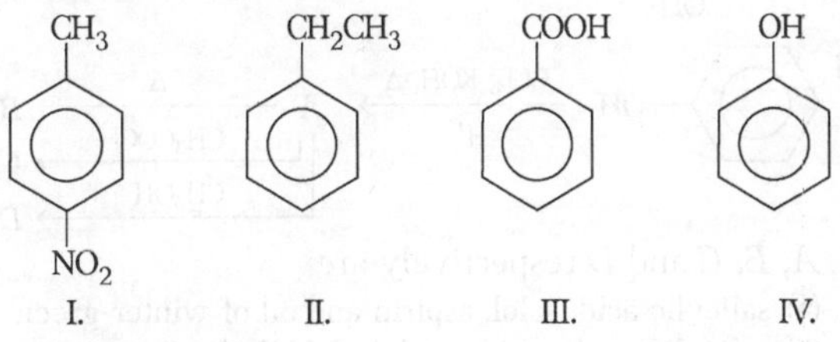

(a) I, II and IV (b) Both I and III
(c) Both II and IV (d) Both I and II

7 OH, CH_3 (o-cresol) and $CH(OH)CH_3$ (1-phenylethanol) can be distinguished by

(a) neutral $FeCl_3$ solution (b) iodoform test
(c) Both (a) and (b) (d) None of these

8 Consider the following statements.

Statement I Benzoic acid is more acidic than phenol.

Statement II Phenol gives violet colour with neutral $FeCl_3$.

(a) Both statements I and II are correct.
(b) Only statement I is correct.
(c) Only statement II is correct.
(d) Both statements I and II are incorrect.

9 Mark the correct increasing order of reactivity of the following compounds with HBr/HCl.

CH_2OH (I), CH_2OH with NO_2 (II), CH_2OH with Cl (III)

(a) I < II < III (b) II < I < III (c) II < III < I (d) III < II < I

10 Consider the following reaction,

OCH_3 $\xrightarrow[\text{Sodamide}]{aq.\ KOH}$ product. Here, product is

(a) OH and $CH_3\overset{-}{O}\overset{+}{K}$ (mixture) (b) OCH_3, OH
(c) OCH_3, OH (d) OCH_3, OH

11 OH $\xrightarrow[H^+]{CCl_4,\ KOH,\ \Delta}$ A $\xrightarrow{\Delta}$ B; A $\xrightarrow{CH_3COCl}$ C; A $\xrightarrow{CH_3OH}$ D

A, B, C and D respectively are

(a) salicylic acid, salol, aspirin and oil of winter-green
(b) salicylic acid, aspirin, salol and oil of winter-green
(c) salicylic acid, oil of winter-green, salol and aspirin
(d) None of the above is correct

12 Consider the following statements.

Statement I S_E reactions are more favoured in phenol than in toluene.

Statement II Phenol gives yellow precipitate of 2, 4, 6-tribromophenol with Br_2 water.

(a) Both statements I and II are correct.
(b) Only statement I is correct.
(c) Only statement II is correct.
(d) Both statements I and II are incorrect.

13 OH $\xrightarrow{K_2S_2O_8 + KOH}$?

The major product is

(a) *p*-benzoquinone
(b) *p*-benzenediol
(c) benzene sulphonic acid
(d) diphenyl ether

14 What is the major product in the following series of reaction?

OH $\xrightarrow[H_2SO_4,\ 100°C]{SO_3}$ $\xrightarrow[FeCl_3]{Cl_2}$ $\xrightarrow[100°C]{\text{Dil. } H_2SO_4}$

(a) HO_3S, OH (b) Cl, OH
(c) Cl, OH (b) Cl, SO_3H

15 A $\xrightarrow[\text{one eq. of } CH_3I]{OH^-}$ $C_9H_{10}O_3$ $\xrightarrow{KMnO_4}$ $C_9H_{10}O_4$ $\xrightarrow{\text{Conc. HI}}$ 3, 4-dihydroxy benzoic acid D

(It gives intense colour with $FeCl_3$ and positive Tollen's test)

Here, A is

(a) OH, OH, H_3C, CH_3 (b) OH, OMe, CHO
(c) OH, OH, OHC, CH_3 (d) OCH_3, OCH_3, CHO

16 In the following reaction,

$$HO-C_6H_4-SO_3H \xrightarrow[H_2O]{Br_2} X$$

X is identified as

(a) 2,4,6-tribromophenol
(b) 2-bromo-4-hydroxybenzenesulphonic acid
(c) 3,5-dibromo-4-hydroxybenzene sulphonic acid
(d) 2-bromophenol

17 Maximum ionisation constant is of

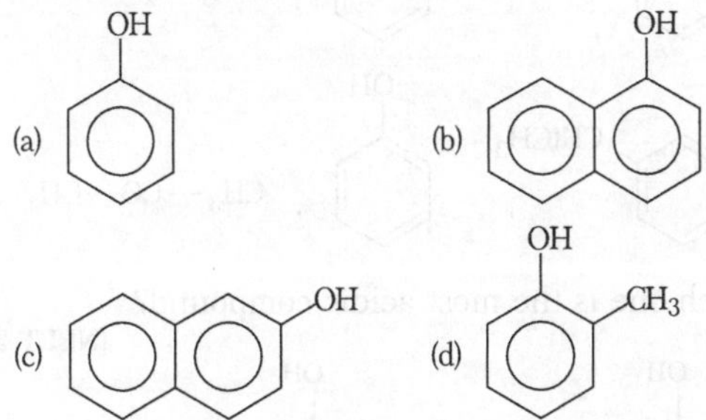

18 Reaction of phenol with CCl_4 and NaOH followed by hydrolysis is likely to give

(a) OH, –Cl (ortho)
(b) OH, –O–H, –C(=O)–H
(c) OH, –CHO (ortho)
(d) OH, –C(=O)–OH (ortho)

19 In the following reaction CO_2 is due to acidic H of

$$C_6H_4(COOH)-OH + NaHCO_3 \longrightarrow CO_2$$

(a) —COOH group (b) — OH group
(c) —HCO_3^- group (d) all the groups

20 Consider the following statements.

Statement I Ethanol is a weaker acid than phenol.

Statement II Sodium ethoxide may be prepared by the reaction of ethanol with aqueous NaOH.

(a) Both statements I and II are correct.
(b) Only statement I is correct.
(c) Only statement II is correct.
(d) Both statements I and II are incorrect.

21 Match the mixtures (in Column I) with the suitable reagents (in Column II) for distinction.

	Column I		Column II
A.	$C_6H_4(CH_3)$–OH, C_6H_5–CH_2OH	1.	Volatile nature/ boiling points
B.	C_6H_5–CH_2CH_2OH, C_6H_5–CH(OH)CH_3	2.	Neutral $FeCl_3$
C.	o-C_6H_4(OH)–NO_2, O_2N–C_6H_4–OH	3.	Iodoform test
D.	C_6H_5–CH_2OH, C_6H_5–OCH_3	4.	Oxidation by $KMnO_4$

	A	B	C	D		A	B	C	D
(a)	4	3	2	1	(b)	2	3	1	4
(c)	2	3	1	4	(d)	3	2	4	1

(C) Medical entrances' gallery

1 Order of acidic nature

(A) p-nitrophenol; (B) o-nitrophenol; (C) p-cresol; (D) phenol **[AIIMS 2019]**

(a) $A > C > D > B$ (b) $A > B > D > C$
(c) $A > B > C > D$ (d) $D > C > B > A$

2 The compound that is most difficult to protonate is **[NEET 2019]**

(a) $H_3C–O–H$ (b) $H_3C–O–CH_3$
(c) Ph–O–H (d) H–O–H

3 The structure of intermediate A in the following reaction is

$$C_6H_5CH(CH_3)_2 \xrightarrow{O_2} A \xrightarrow[H_2O]{H^+} C_6H_5OH + H_3C–CO–CH_3$$

[NEET 2019]

(a) $H_3C–C(CH_3)(C_6H_5)–O–O–H$
(b) $C_6H_5–O–O–CH(CH_3)_2$

(c) C_6H_5—CH(CH_3)—CH_2—O—O—H (d) C_6H_5—O—CH(CH_3)—CH_3

4 In the reaction,

$$C_6H_5OH + CHCl_3 + NaOH \longrightarrow o\text{-}(O^-Na^+)C_6H_4CHO$$

the electrophile involved is **[NEET 2018]**

(a) dichloromethyl anion ($\bar{C}HCl_2$)

(b) formyl cation ($\overset{+}{C}HO$)

(c) dichloromethyl cation ($\overset{+}{C}HCl_2$)

(d) dichlorocarbene ($:CCl_2$)

5 When phenol is treated with $CHCl_3$ and NaOH followed by acidification, salicylaldehyde is obtained. Which of the following species is involved in the above mentioned reaction as intermediates? **[JIPMER 2018]**

(a) cyclohexadienone with C(H)($\overset{\ominus}{C}Cl_2$) at ortho position (b) o-($CHCl_2$)phenol

(c) cyclohexanone with C(H)(CCl_2) and OH at ortho position (d) o-quinone methide with =C(H)Cl, OH

6 Compound A, $C_8H_{10}O$, is found to react with NaOI (produced by reacting Y with NaOH) and yields a yellow precipitate with characteristic smell.

A and Y are respectively, **[NEET 2018]**

(a) C_6H_5—CH(OH)—CH_3 and I_2

(b) C_6H_5—CH_2—CH_2—OH and I_2

(c) H_3C—C_6H_4—CH_2—OH and I_2

(d) CH_3—C_6H_3(CH_3)—OH and I_2

7 Identify the major products P, Q and R in the following sequence of reactions : **[NEET 2018]**

$$C_6H_6 + CH_3CH_2CH_2Cl \xrightarrow[AlCl_3]{Anhy.} P \xrightarrow[(ii)\ H_3O^+/\Delta]{(i)\ O_2} Q + R$$

	P	Q	R
(a)	$C_6H_5CH(CH_3)_2$,	C_6H_5OH,	$CH_3CH(OH)CH_3$
(b)	$C_6H_5CH_2CH_2CH_3$,	C_6H_5CHO,	C_6H_5COOH
(c)	$C_6H_5CH_2CH_2CH_3$,	C_6H_5CHO,	CH_3CH_2—OH
(d)	$C_6H_5CH(CH_3)_2$,	C_6H_5OH,	CH_3—CO—CH_3

8 Which one is the most acidic compound? **[NEET 2017]**

(a) p-cresol (4-$CH_3C_6H_4OH$) (c) p-nitrophenol (4-$NO_2C_6H_4OH$)

(b) phenol (C_6H_5OH) (d) 2,4,6-trinitrophenol (O_2N, NO_2, NO_2)

9 Phenol $\xrightarrow[(ii)\ CO_2/140°C]{(i)\ NaOH} A \xrightarrow{H^+/H_2O} B \xrightarrow{Ac_2O} C$

In the above reaction, end product 'C' is **[JIPMER 2017]**

(a) salicylaldehyde (b) salicylic acid

(c) phenyl acetate (d) aspirin

10 Arrange the following compounds in the increasing order of their acidic strength. **[Karnataka CET 2015]**

I. *m*-nitrophenol II. *m*-cresol

III. phenol IV. *m*-chlorophenol

(a) III < II < I < IV

(b) II < IV < III < I

(c) II < III < IV < I

(d) II < III < I < IV

11 Among the following sets of reactants which one produces anisole? **[CBSE AIPMT 2014]**

(a) CH_3CHO , $RMgX$

(b) C_6H_5OH, NaOH, CH_3I

(c) C_6H_5OH, neutral $FeCl_3$

(d) C_6H_5—CH_3, CH_3COCl , $AlCl_3$

12 When phenol is treated with D_2SO_4/D_2O, some of the hydrogens get exchanged. The final product in this exchange reaction is **[Manipal 2014]**

(a) OD; ring substituents D, D (ortho), H, H (meta), D (para)
(b) OD; ring substituents D, H (ortho), H, H (meta), D (para)
(c) OD; ring substituents D, H (ortho), H, D (meta), H (para)
(d) OD; ring substituents H, H (ortho), D, D (meta), H (para)

13 The main product obtained from phenol with PCl_5 is **[Manipal 2014]**

(a) BHC (b) hexachlorobenzene
(c) chlorobenzene (d) triphenyl phosphate

14 Butylated hydroxy toluene is used in **[MHT CET 2014]**

(a) preventing oxidative rancidity of fats
(b) preserving food grains
(c) killing bacteria living tissues
(d) reducing stress and anxiety

15 Phenol $\xrightarrow{C_6H_5COCl/base} X \xrightarrow{Nitration} Y$ (major product). Y is **[RPMT 2014]**

(a) C_6H_5—COO—C_6H_4—NO_2
(b) OH—C_6H_4—COO—C_6H_4—NO_2
(c) O_2N—C_6H_4—COO—C_6H_5
(d) O_2N—C_6H_4—COO—C_6H_4—NO_2

16 Phenol is treated with excess bromine water and shaken well. The white precipitate formed during the process is **[KCET 2014]**

(a) *m*-bromophenol
(b) 2, 4, 6- tribromophenol
(c) 2, 4- dibromophenol
(d) a mixture of *o*- and *p*- bromophenol

ANSWERS

Check Point 16.1

1. (a)	2. (b)	3. (d)	4. (a)	5. (c)	6. (d)	7. (d)	8. (b)	9. (a)	10. (b)

Check Point 16.2

1. (d)	2. (c)	3. (a)	4. (d)	5. (a)	6. (b)	7. (b)	8. (a)	9. (d)	10. (c)

(A) Topical exercise

1. (d)	2. (c)	3. (c)	4. (a)	5. (a)	6. (a)	7. (c)	8. (a)	9. (d)	10. (a)
11. (b)	12. (d)	13. (d)	14. (d)	15. (c)	16. (c)	17. (c)	18. (d)	19. (c)	20. (c)
21. (a)	22. (b)	23. (d)	24. (b)	25. (b)	26. (a)	27. (d)	28. (c)	29. (c)	30. (c)
31. (c)	32. (a)	33. (d)	34. (c)	35. (c)	36. (a)	37. (a)	38. (b)	39. (a)	40. (a)
41. (c)	42. (c)	43. (c)	44. (d)	45. (b)	46. (b)				

(B) Miscellaneous questions

1. (c)	2. (a)	3. (c)	4. (a)	5. (d)	6. (c)	7. (c)	8. (b)	9. (c)	10. (b)
11. (a)	12. (a)	13. (b)	14. (c)	15. (b)	16. (c)	17. (b)	18. (d)	19. (a)	20. (b)
21. (b)									

(C) Medical entrances' gallery

1. (b)	2. (c)	3. (a)	4. (d)	5. (a)	6. (a)	7. (d)	8. (d)	9. (d)	10. (c)
11. (b)	12. (b)	13. (d)	14. (a)	15. (a)	16. (b)				

Hints & Explanations

(A) Topical exercise

1 (*d*) $C_6H_5CH_2OH$ is not classified as phenols because in this molecule –OH group is directly attached to alkyl group not with aryl group.

Options (a), (b) and (c) are phenols.

Monohydric Dihydric Trihydric

2 (*c*) Statement (c) is correct, regarding phenol whereas other statements (a, b and d) are incorrect.

Corrected statements are as follows :

(a) The electron density decreases on oxygen due to the higher electronegativity of sp^2 hybridised carbon of phenol.

(b) Increase in the polarity of O—H bond results in an increase in ionisation of phenols than that of alcohols.

(d) In substituted phenols, the presence of electron withdrawing groups enhances the acidic strength of phenol.

3 (*c*) There is also double bond character in (C—O) bond of phenol due to resonance. There is no resonance in alcohol.

5 (*a*)

3-methylphenol

9 (*d*)

(a) $C_6H_5\overset{+}{N_2}\overset{-}{Cl} \xrightarrow{H_3O^+} C_6H_5OH$

(b) $C_6H_5\overset{-}{SO_3}\overset{+}{Na} \xrightarrow[\Delta]{NaOH} C_6H_5OH$

(c) $C_6H_5Cl \xrightarrow[\Delta]{NaOH} C_6H_5OH$

10 (*a*)

(a) $C_6H_5\overset{+}{N_2}\overset{-}{Cl} \xrightarrow{H_3O^+} C_6H_5OH$

(b) $C_6H_5\overset{+}{N_2}\overset{-}{Cl} \xrightarrow{H_3PO_2} C_6H_6$

(c) o-HOC$_6H_4$COOH $\xrightarrow{Zn/\Delta}$ C_6H_5COOH

(d) $C_6H_5SO_3H \xrightarrow[470°K, P/HCl]{H_2O} C_6H_6$

12 (*d*) The major products *P*, *Q* and *R* in the given reaction are as follows

$$C_6H_6 + CH_3CH_2CH_2Cl \xrightarrow{\text{Anhy. } AlCl_3} C_6H_5CH(CH_3)_2 \xrightarrow[\text{(ii) } H_3O^+/\Delta]{\text{(i) } O_2} C_6H_5OH + CH_3-\overset{O}{\overset{\|}{C}}-CH_3$$

Benzene Cumene (*P*) Phenol (*Q*) Acetone (*R*)

14 (*d*) *o*-isomer has maximum dipole moment.

15 (*c*) *Ortho*-isomer is volatile due to absence of intermolecular H-bonding. Hence, it is steam volatile.

17 (*c*) In *o*-nitrophenol due to the presence of intramolecular H-bonding association of molecule does not occurs which makes boiling point lower.

18 (*d*) Alpha (α) carbon of benzyl alcohol and cyclohexanol is sp^3-hybridised while in phenol and *m*-chlorophenol, it is sp^2-hybridised. In *m*-chlorophenol electron withdrawing group (— Cl) is present at *meta*-position.

Presence of electron withdrawing group increases the acidic strength. So, *m*-chlorophenol is most acidic among all the given compounds. The correct option is (d).

20 (*c*) The phenoxide ion left from *o*-hydroxybenzaldehyde and *p*-hydroxybenzaldehyde are stabilised by-*I* and *R*-effect of —CHO group but due to chelation in *o*-isomer, it is difficult to remove the H-atom. Hence, *p*-hydroxybenzaldehyde is the strongest acid.

22 (*b*) $—OCH_3$ shows *I*+effect, thus decreases acidic strength.

23 (*d*) EWG : $NO_2 > Cl > CH_3$. Thus, acid strength : II < I < III

25 (*b*) Maximum pK_a value means least acidic (maximum basic).

$—NO_2$: (EWG) increases acidic strength.

$—CH_3$: (ERG) decreases acidic strength.

In (d) lone-pair on N-atom in ring makes it basic (least acidic).

26 (*a*) Phenol is more acidic in nature because by the loss of one proton, it gives phenoxide ion. This phenoxide ion is resonance stabilised. As phenoxide ion is a stable intermediate so, the tendency to give proton is more in phenol than the others. Phenols being more acidic than alcohol dissolves in NaOH.

$$C_6H_5OH + NaOH \longrightarrow C_6H_5\overset{-}{O}\overset{+}{Na} + H_2O$$

Therefore, option (a) is the correct.

34 (*c*) p-Cresol (CH₃ and OH para) $\xrightarrow[\text{Reimer-Tiemann (OH is activating group)}]{CHCl_3/NaOH}$ 2-hydroxy-5-methylbenzaldehyde (CHO ortho to OH) $\xrightarrow{HCN}$ CH(OH)CN derivative $\xrightarrow{H_3O^+}$ CH(OH)COOH derivative (Chiral)

35 (*c*) The given reaction is Kolbe's reaction.

Phenol $\xrightarrow{NaOH}$ Sodium phenolate $\xrightarrow[\text{(iii) Hydrolysis/}H^+]{\text{(ii) }CO_2}$ 2-hydroxybenzoic acid (Salicyclic acid)

36 (*a*) Pyrene is CCl_4, thus by Reimer-Tiemann reaction, –COOH group is introduced into phenol.

38 (*b*) $—NO_2$ deactives benzene ring for electrophilic reaction.

39 (*a*) Phenolic compounds form azo dye.

41 (*c*) $o\text{-}HOC_6H_4COOC_6H_5 + H_3O^+ \longrightarrow \underset{A}{o\text{-}HOC_6H_4COOH} + \underset{B}{C_6H_5OH}$

Compound, *B* does not give CO_2 with $NaHCO_3$.

44. (*d*)

Test	CH_3CH_2OH	C_6H_5OH
(a)	No colour	Violet colour
(b)	Yellow ppt.	No ppt.
(c)	Red colour	No colour

45. (*b*) $C_6H_5—OH + NaHCO_3 \longrightarrow$ No reaction

$C_6H_5—COOH + NaHCO_3 \longrightarrow CO_2$ (brisk effervescence)

(B) Miscellaneous exercises

1 (*c*) $C_6H_5O^- \longleftrightarrow$ (ortho carbanion form) $\longleftrightarrow$ (para carbanion form) ....

Resonance stabilisation in phenoxide ion
No resonance in ethoxide ion.

5 (*d*) *m*-nitrophenol $\xrightarrow[\text{[H]}]{Sn/HCl}$ m-aminophenol (NH_2) $\xrightarrow[0°C]{NaNO_2/HCl}$ m-hydroxybenzenediazonium chloride ($\overset{+}{N_2}\overset{-}{Cl}$) $\xrightarrow[\Delta]{H^+/H_2O}$ Resorcinol

6 (*c*) $C_6H_5CH_2CH_3$ and C_6H_5OH will undergo Friedel-Craft's alkylation on *ortho* or *para*-position because of more electron density.

7 (*c*) o-cresol (OH, CH_3): (a) colour, (b) no ppt.; $C_6H_5CH(OH)CH_3$: no colour, yellow ppt.

9. (*c*) Reaction of the given compounds with HBr/HCl is a nucleophilic substitution reaction. It follows S_N1 mechanism. S_N1 mechanism depends upon the stability of carbocation. Presence of electron withdrawing group decreases the stability of carbocation. In compound (II) and (III), EWG is present at *para*-position.
Since, $—NO_2$ group is a stronger EWG than —Cl.
So, $NO_2—C_6H_5—\overset{+}{C}H_2$ carbocation will be less stable than $Cl—C_6H_5—\overset{+}{C}H_2$ carbocation.
Thus, the order of stability of carbocation is

$$O_2N—C_6H_4—\overset{+}{C}H_2 < Cl—C_6H_4—\overset{+}{C}H_2 < C_6H_5—\overset{+}{C}H_2$$

Therefore, compound (II) is least reactive and correct option is (c).

10 (*b*) Anisole derivative (OCH_3) with δ positions: S_E attack; S_N attack (OH^- comes at this position) $\xrightarrow{KOH}$ m-methoxyphenol (OCH_3, OH)

14 (c) Sulphonation blocks p-position and chlorination gives o-isomer w.r.t. —OH group. In last stage desulphonation gives final product.

15 (b) 4-hydroxy-3-methoxybenzaldehyde (OH, OCH_3, CHO) $\xrightarrow[CH_3I]{OH^-}$ 3,4-dimethoxybenzaldehyde (OCH_3, OCH_3, CHO) $\xrightarrow{KMnO_4}$ 3,4-dimethoxybenzoic acid (OCH_3, OCH_3, COOH) $\xrightarrow{Conc.\ HI}$ 3,4-dihydroxybenzoic acid (OH, OH, COOH)

As 'A' gives intense purple colour with $FeCl_3$ this means A has phenolic group. It also gives Tollen's test, thus —CHO group is also present.

16 (c) $-SO_3H$ group is ring deactivating and —OH group is ring activating. Thus, the compound X formed is

2,6-dibromo-4-hydroxybenzenesulphonic acid (OH, Br, Br, SO_3H)

17 (b) Greater stability of ion is due to resonance.

19 (a) –COOH decomposes $NaHCO_3$.

$$\text{Salicylic acid (OH, COOH)} + NaHCO_3 \longrightarrow \text{(OH, } COO^-Na^+) + CO_2 + H_2O$$

(C) Medical entrances' gallery

1 (b) Acidic nature of aromatic compounds depends on the ability to donate H^+ ions. Presence of electron withdrawing group (i.e. $—NO_2$) increases the acidic nature.

$—NO_2$ at *ortho*-position [i.e. structure (B)] has ability to form hydrogen bond with —OH group. Thus, structure (B) is less acidic than structure (A). On the other hand, electron releasing group bonded with the ring (i.e. $—CH_3$) decreases the acidic nature, thus structure (C) is less acidic than phenol.

Hence, correct order of acidic nature is

p-nitrophenol (A) > o-nitrophenol (B) > phenol (D) > p-cresol (C)

and option (b) is the correct.

2 (c) In CH_3OH and CH_3OCH_3, $—CH_3$ group shows + I-effect due to which oxygen acquires partial negative charge in both the compounds.

$H_3C \rightarrow \overset{\delta-}{O} — H$ (+I-effect) $CH_3 \rightarrow \overset{\delta\delta-}{O} \leftarrow CH_3$ (+I-effect) (+I-effect)

On the other hand in Ph—OH, —Ph shows –I effect due to which oxygen acquire partial positive charge.

$Ph \leftarrow \overset{\delta+}{O} — H$

Alongwith, the lone pair of oxygen also shows conjugation with benzene ring which increases the positive charge on oxygen as

(resonance structure: $\overset{\delta+}{O}—H$ with δ– at ortho and para positions)

Thus, further decreasing the process of protonation. Therefore, incoming proton will not be able to attack easily on partial positive 'O' atom. Hence, phenol is most difficult to protonate. Thus, option (c) is correct.

3 (a) The given reaction is of cumene process for phenol production and intermediate (A) is cumene hydroperoxide. In the process, cumene (isopropylbenzene) is oxidised in the presence of air to cumene hydroperoxide which is then converted to phenol and acetone by treating with dilute acid. Acetone, a by-product of this reaction is also obtained in large quantities by this method. The reaction takes place as follows:

$$\underset{\text{Cumene}}{C_6H_5CH(CH_3)_2} \xrightarrow{O_2} \underset{\substack{\text{Cumene}\\\text{hydroperoxide}\\(A)}}{C_6H_5C(CH_3)_2OOH} \xrightarrow[H_2O]{H^+} \underset{\text{Phenol}}{C_6H_5OH} + \underset{\text{Acetone}}{CH_3COCH_3}$$

4 (d)

$$C_6H_5OH + CHCl_3 + NaOH \longrightarrow \text{(}O^{\ominus}Na^+\text{, CHO ortho)}$$

The above given reaction is known as Riemer-Tiemann reaction. In this reaction, electrophile involved is dichlorocarbene ($:CCl_2$) which is formed in the I[st] step of mechanism.

Mechanism

***Step* I** Generation of electrophile

$$CHCl_3 + OH^- \rightleftharpoons {}^-CCl_3 + H_2O$$

$${}^-CCl_3 \longrightarrow \underset{\substack{\text{(Electrophile)}\\\text{Dichlorocarbene}}}{:CCl_2} + Cl^-$$

***Step* II** Reaction of electrophile with phenoxide

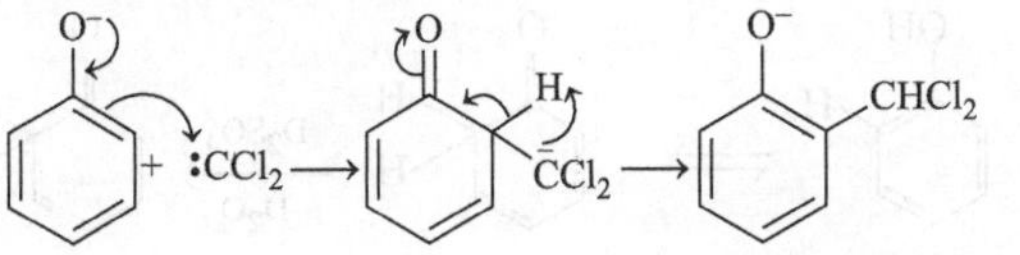

o-dichloromethyl phenoxide

***Step* III** Hydrolysis

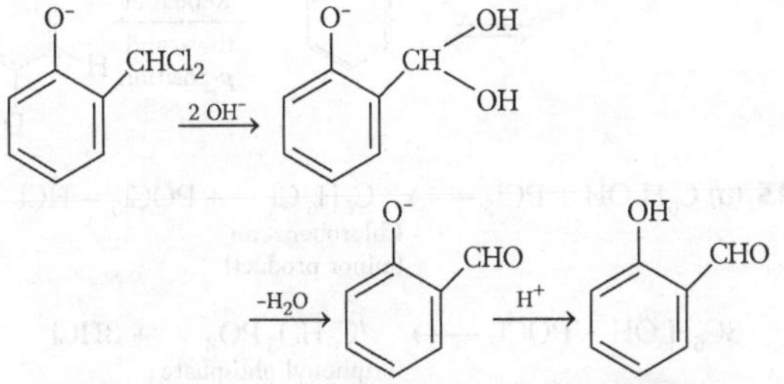

$\xrightarrow{2\,OH^-}$

$\xrightarrow{-H_2O}$ $\xrightarrow{H^+}$

Salicylaldehyde

5 (*a*) On refluxing alkaline solution of phenol with $CHCl_3$ at 330 K and on subsequent acidification, o-hydroxy benzaldehyde (salicylaldehyde) is obtained. This reaction is known as Riemer-Tiemann reaction.

Mechanism

$$CHCl_3 + OH^- \rightleftharpoons CCl_3^- + H_2O$$

$$CCl_3^- \longrightarrow :CCl_2 + Cl^-$$

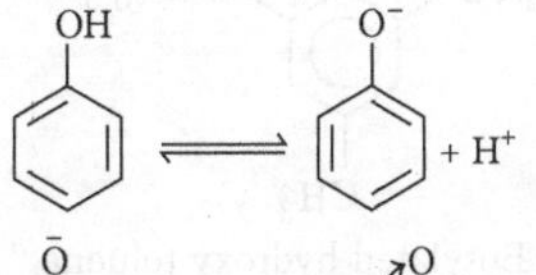

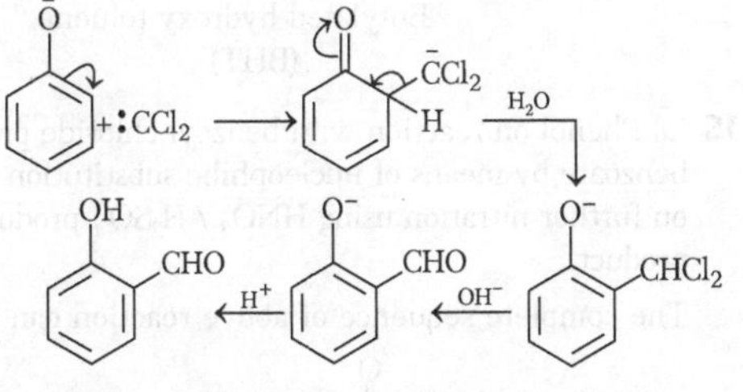

$\therefore$ The correct option is (a).

6 (*a*) Iodoform reaction with sodium hypoiodite (NaOI) is used for the detection of CH_3CO— group. Also compounds containing $CH_3CH(OH)$— group shows positive iodoform test as it produces CH_3CO— group on oxidation. Since, among the given compounds, $CH_3CH(OH)$— group is given only in the substrate of option (a) hence, it is correct.

The reaction of compound *A* with NaOI is given as follows :

$$2NaOH + I_2 \longrightarrow NaOI + NaI + H_2O$$

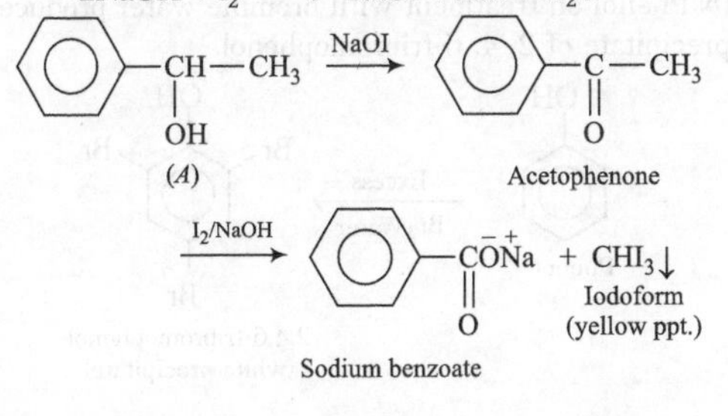

$$C_6H_5-CH(OH)-CH_3 \ (A) \xrightarrow{NaOI} C_6H_5-CO-CH_3 \ \text{(Acetophenone)}$$

$$\xrightarrow{I_2/NaOH} C_6H_5-\overset{-\ +}{COONa} \ \text{(Sodium benzoate)} + CHI_3\downarrow \ \text{(Iodoform, yellow ppt.)}$$

7 (*d*) The given reaction takes place as follows :

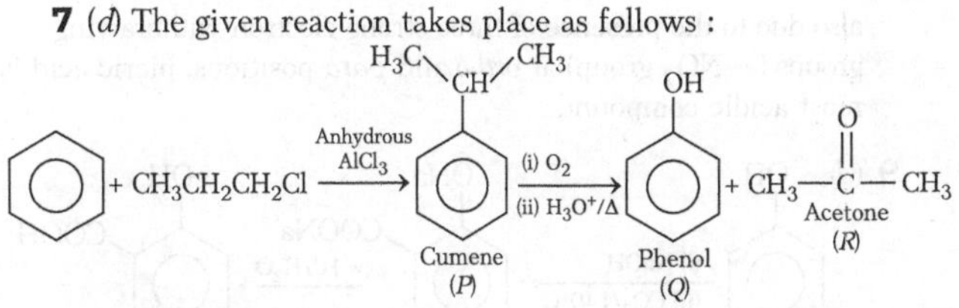

$$C_6H_6 + CH_3CH_2CH_2Cl \xrightarrow{\text{Anhydrous } AlCl_3} \underset{\text{Cumene }(P)}{C_6H_5CH(CH_3)_2} \xrightarrow[\text{(ii) } H_3O^+/\Delta]{\text{(i) } O_2} \underset{\text{Phenol }(Q)}{C_6H_5OH} + \underset{\text{Acetone }(R)}{CH_3-CO-CH_3}$$

Mechanism

***Step* I** Formation of carbocation,

$$CH_3CH_2CH_2-Cl + \underset{\text{(Anhy.)}}{AlCl_3} \longrightarrow CH_3CH_2\overset{+}{C}H_2 + \overset{-}{AlCl_4}$$

$$\underset{\substack{\text{Carbocation}\\(1°)}}{CH_3CH_2\overset{+}{C}H_2} \xrightarrow{1,\,2\,H^-\text{ shift}} \underset{\substack{\text{Carbocation}\\(2°)}}{CH_3\overset{+}{C}H-CH_3}$$

***Step* II** Electrophilic substitution reaction,

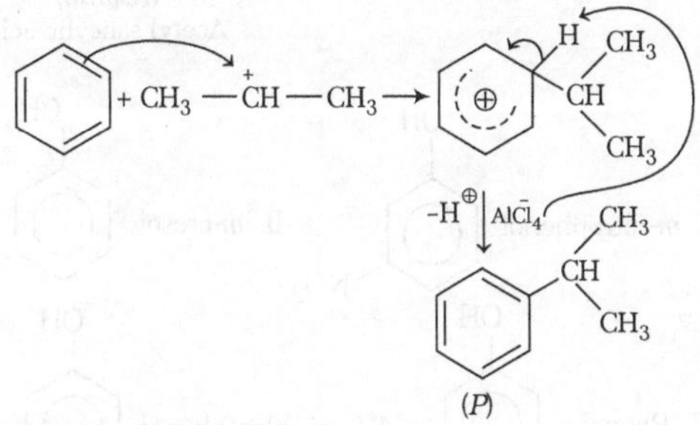

$$C_6H_6 + CH_3-\overset{+}{C}H-CH_3 \longrightarrow \text{(arenium ion)} \xrightarrow{-H^{\oplus}\ (AlCl_4^-)} C_6H_5CH(CH_3)_2 \ (P)$$

***Step* III** Formation of peroxide,

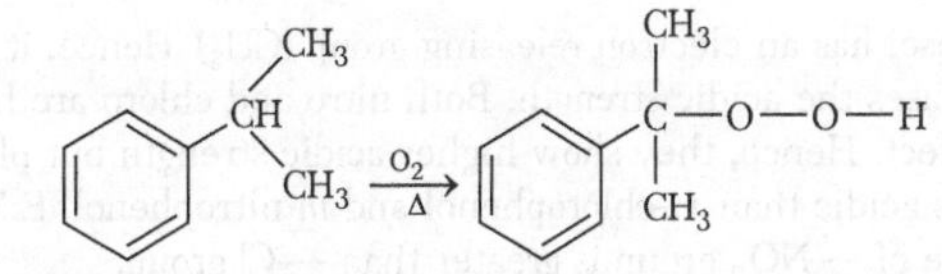

$$C_6H_5CH(CH_3)_2 \xrightarrow[\Delta]{O_2} C_6H_5C(CH_3)_2-O-O-H$$

***Step* IV** Hydrolysis of oxidised product formed in step III,

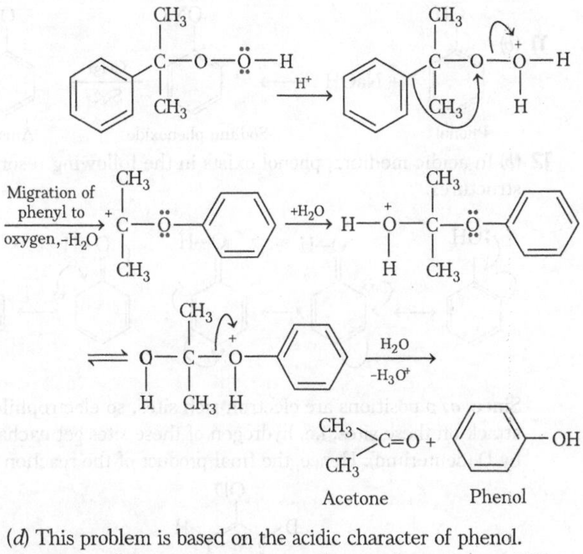

$$C_6H_5C(CH_3)_2-O-\ddot{O}-H \xrightarrow{H^+} C_6H_5C(CH_3)_2-O-\overset{+}{O}H_2$$

$$\xrightarrow[\text{oxygen},\,-H_2O]{\text{Migration of phenyl to}} (CH_3)_2\overset{+}{C}-\ddot{O}-C_6H_5 \xrightarrow{+H_2O} H-\overset{+}{O}(H)-C(CH_3)_2-\ddot{O}-C_6H_5$$

$$\rightleftharpoons H-O-C(CH_3)_2-\overset{+}{O}(H)-C_6H_5 \xrightarrow[-H_3O^+]{H_2O} \underset{\text{Acetone}}{(CH_3)_2C=O} + \underset{\text{Phenol}}{C_6H_5-OH}$$

8 (*d*) This problem is based on the acidic character of phenol. Electron withdrawing group at o and p-position w.r.t. — OH group of phenol, increases the acidic strength.

Picric acid (2,4,6-trinitrophenol) is most acidic compounds than other given compounds because its pK_a value is close to zero and also due to the presence of three strong electron withdrawing groups (—NO_2 group) at *ortho* and *para*-positions, picric acid is most acidic compound.

9 (*d*) Phenol $\xrightarrow[\text{(ii) } CO_2/140°C]{\text{(i) NaOH}}$ (A) (sodium salt, $\overset{-}{O}\overset{+}{Na}$, $\overset{-}{COO}\overset{+}{Na}$) $\xrightarrow{H^+/H_2O}$ (B) (OH, COOH) $\xrightarrow{Ac_2O}$ (C) ($OCOCH_3$, COOH)

(C) (Aspirin) Acetyl salicylic acid

10 (*c*) I. *m*-nitrophenol (OH, NO_2) II. *m*-cresol (CH_3, OH)

III. Phenol (OH) IV. *m*-chlorophenol (OH, Cl)

m-cresol has an electron releasing group (CH_3). Hence, it decreases the acidic strength. Both nitro and chloro are having *–I*-effect. Hence, they show higher acidic strength but phenol is less acidic than *m*-chlorophenol and *m*-nitrophenol, E.W. nature of —NO_2 group is greater than —Cl group.

So, increasing order of acidic strength is II < III < IV < I.

11 (*b*) Phenol (OH) + NaOH ⟶ Sodium phenoxide ($\overset{-}{O}\overset{+}{Na}$) $\xrightarrow[S_N2]{CH_3I}$ Anisole (OCH_3)

12 (*b*) In acidic medium, phenol exists in the following resonating structures.

$\ddot{O}H$ ⟷ $\overset{+}{O}$–H ⟷ $\overset{+}{O}$–H ; $\overset{+}{O}$–H ⟷ $\ddot{O}H$

Since, *o*/*p* positions are electron rich sites, so electrophile will attack on these sites, i.e. hydrogen of these sites get exchanged by D (deuterium). Hence, the final product of the reaction is

(OD, D, H, H, H, D)

Product formed by following reaction

(OH, H) ⇌ (O, H, H) $\xrightarrow[D_2O]{D_2SO_4}$ (O, D, D)

⇌ (OD, D) $\xrightarrow{\text{Repeat at the } o \text{ and } p\text{-position}}$ (OD, D, H, H, H, D)

13 (*d*) $C_6H_5OH + PCl_5 \longrightarrow C_6H_5Cl + POCl_3 + HCl$

Chlorobenzene (minor product)

$3C_6H_5OH + POCl_3 \longrightarrow (C_6H_5)_3PO_4 + 3HCl$

Triphenyl phosphate

The yield of chlorobenzene in the above reaction is very poor. The main product of the reaction is triphenyl phosphate $(C_6H_5)_3PO_4$.

14 (*a*) Butylated hydroxy toluene (BHT) is used as an antioxidant. It retards the action of oxygen on the food material and thereby helps in preventing oxidative rancidity of fats.

(OH, $(CH_3)_3C$, $C(CH_3)_3$, CH_3)

Butylated hydroxy toluene (BHT)

15 (*a*) Phenol on reaction with benzoyl chloride produces phenyl benzoate by means of nucleophilic substitution reaction, which on further nitration using HNO_3 / H_2SO_4 produces the desired product.

The complete sequence of above reaction can be shown as

C_6H_5–OH + Cl–C(=O)–C_6H_5 $\xrightarrow{NaOH}$ C_6H_5–O–C(=O)–C_6H_5 Phenyl benzoate (*X*)

$\xrightarrow{HNO_3/H_2SO_4}$ C_6H_5–O–C(=O)–C_6H_4–NO_2 (*Y*)

16 (*b*) Phenol on treatment with bromine water produces a white precipitate of 2, 4, 6-tribromophenol.

Phenol (OH) $\xrightarrow[\text{Br}_2\text{ Water}]{\text{Excess}}$ (OH, Br, Br, Br)

2,4,6-tribromophenol (white precipitate)

CHAPTER 17

Ethers

Ethers are the organic compounds having two alkyl (or aryl) groups bonded to an oxygen atom with a general molecular structure of $R—O—R'$, [here R and R' may be same or different].

CLASSIFICATION

On the basis of type of attached groups, these are classified as follows

(*i*) **Aliphatic ethers** Both attached groups (R and R') are alkyl groups. They have the general formula as $C_nH_{2n+2}O$.

e.g. $CH_3CH_2OCH_2CH_3$ (Diethyl ether) ; $CH_3—O—C(CH_3)_2—CH_3$ (Methyl *tert*-butyl ether)

(*ii*) **Aromatic ethers** Either one or both the attached groups (R and R') is/are aryl groups.

e.g. $C_6H_5—O—C_6H_5$ (Diphenyl ether) ; $C_6H_5—O—C_2H_5$ (Ethyl phenyl ether)

Aromatic ethers are further classified as follows

- **Phenolic ethers** One of the group is aryl and other is alkyl.

  e.g. $CH_3—O—C_6H_5$
  Phenolic ether

- **Diaryl ethers** Both the groups are aryl.

  e.g. $C_6H_5—O—C_6H_5$
  Diphenyl ether

On the basis of symmetry, these are classified as follows

(*i*) **Simple or symmetrical ethers** Here, both alkyl or aryl groups bonded to O-atom are same.

e.g. $CH_3—O—CH_3$ (Dimethyl ether) ; $C_6H_5—O—C_6H_5$ (Diphenyl ether)

Inside

(*ii*) **Mixed or unsymmetrical ethers** Here, both alkyl or aryl groups bonded to O-atom are different.

e.g. $CH_3—O—CH_2CH_3$; $C_6H_5—O—CH_2C_6H_5$

Ethylmethyl ether; Benzylphenyl ether

STRUCTURE

Ethers are bent molecules with bond angle $\angle ROR'$ is 110° and O-atom is sp^3-hybridised. In ethers, the repulsion between the bulkier alkyl groups almost counter balance the effect of lone pair-lone pair interactions and tetrahedral angle (109.5°) is almost retained.

R—Ö—R′ (110°)

NOMENCLATURE

Common system

The common names of ethers are derived by naming the two alkyl or aryl groups linked to the oxygen atom as separate words in alphabetical order and adding the word ether in the end. In case of symmetrical ether, prefix 'di' is used before the name of alkyl or aryl group.

IUPAC system

(*i*) **Nomenclature of common ethers** In IUPAC system, ethers are called as alkoxyalkanes. Here, ethereal oxygen is taken alongwith smaller alkyl group, which is regarded as substituent on large alkyl group as parent hydrocarbon.

e.g. $CH_3—O—\overset{2}{C}H(\overset{1}{C}H_3)\overset{3}{C}H_3$; $CH_3—O—\overset{2}{C}(\overset{1}{C}H_3)(\overset{3}{C}H_3)—CH_3$

Common name Methyl *iso*-propyl ether; *tert*-butyl methyl ether
IUPAC name 2-methoxypropane; 2-methoxy-2-methylpropane

e.g. $C_6H_5—O—C_2H_5$

Common name Ethyl phenyl ether (phenetole)
IUPAC name Ethoxybenzene

IUPAC names of a few typical ethers are as follows

$CH_3O—CH(OCH_3)—\overset{2}{C}H_3$
1, 1-dimethoxyethane

$\overset{1}{C}H_3—\overset{2}{C}H(OH)—\overset{3}{C}H_2\overset{4}{C}H_2OCH_3$
4-methoxybutan-2-ol

$\overset{4}{C}H_3—\overset{3}{C}H(C_6H_5)—\overset{2}{C}H(I)—\overset{1}{C}H_2OC_2H_5$
1-ethoxy-2-iodo-3-phenylbutane

$C_3H_5—O—\overset{2}{C}H(\overset{1}{C}H_3)—\overset{3}{C}H_2—\overset{4}{C}H_3$;
2-cyclopropoxybutane

(cyclohexane ring: C-1 bearing H_3C and CH_3, C-2 bearing OC_2H_5)
2-ethoxy-1,1-dimethylcyclohexane

(*ii*) **Nomenclature of cyclic ethers** According to IUPAC system, cyclic ethers (heterocyclic compounds) containing three and four membered rings are known as oxiranes and oxetanes respectively. Three membered rings are also called epoxyalkanes. However, oxirane without any substituent can be named as **ethylene oxide**. The numbering starts from oxygen atom.

$H_2C—CH_2$ bridged by O
Oxirane (or ethyleneoxide)

$H_3C—\overset{2}{C}(CH_3)—\overset{2'}{C}H_2$ bridged by $\overset{1}{O}$
2, 2-dimethyloxirane

$H_5C_2—\overset{2}{H}C—\overset{2'}{C}H_2$ bridged by O
2-ethyloxirane

$C_2H_5—\overset{2}{C}H—\overset{2'}{C}H—CH_3$ bridged by $\overset{1}{O}$
2-ethyl-2′ methyloxirane

$\overset{3}{C}H_2—\overset{2}{C}H—\overset{1}{C}H_2Cl$ (C-3 and C-2 bridged by O)
1-chloro-2,3-epoxypropane

$CH_2—O$ / $CH_2—CH_2$ (four-membered ring)
Oxetane

e.g. Few higher cyclic ethers (five and six membered rings) are as follows

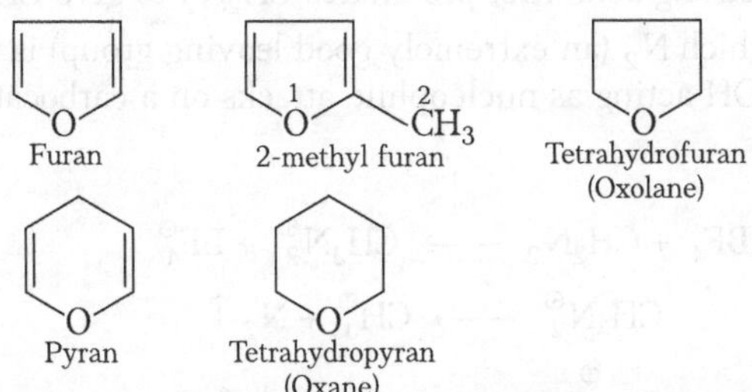

(*iii*) **Nomenclature of crown ethers** These are the heterocyclic polyethers with atleast four O-atoms and are called **crown ethers**. These are named as due to their crown like shape. While writing the names, the word 'crown' is preceded by the total number of atoms in the ring and is followed by the total number of oxygen atoms.

12-crown-4-ether (8 carbon, 4 oxygen) ; 18-crown-6-ether (12 carbon, 6 oxygen)

ISOMERISM

Following types of isomerisms are shown by ethers

(*i*) **Functional isomerism** Both —O— and —OH are the possible functional groups with general molecular formula $C_nH_{2n+2}O$.

e.g. $CH_3—O—CH_3$ (Dimethylether) and CH_3CH_2OH (Ethanol (or ethyl alcohol))

Functional isomers of oxiranes are aldehydes and ketones. e.g.

$CH_3—CH—CH_2$ (epoxide, O bridging), CH_3CH_2CHO, $CH_3—CO—CH_3$

(*ii*) **Metamerism** Ethers having the same molecular formula but different alkyl groups on either side of the divalent atom like oxygen. These are called metamers and the phenomenon is called metamerism. Two pairs of metamers are possible with molecular formula $C_4H_{10}O$.

(a) $CH_3—O—CH_2CH_2CH_3$ (Methyl propylether) and $CH_3CH_2—O—CH_2CH_3$ (Diethylether)

(b) $CH_3—O—\overset{2}{C}H(\overset{3}{C}H_3)—\overset{1}{C}H_3$ (2-methoxypropane) and $CH_3CH_2—O—CH_2CH_3$ (Diethylether)

Note *Minimum 4-C should be present in molecular formula to show metamerism.*

METHODS OF PREPARATION

Ethers are generally prepared by the following methods

Preparation from alcohols

Alcohols are the main source for the preparation of ethers. They react differently in different reaction conditions. Some of the chemical reactions are as follows

By dehydration of alcohols

Symmetrical (or simple) ethers are prepared by heating alcohols at 413 K with sulphuric acid.

$$\underset{\text{Alcohol}}{2ROH} \xrightarrow[\Delta,\ 413K]{H_2SO_4} \underset{\text{Ether}}{R—O—R} + H_2O$$

e.g.
$$\underset{\text{Propanol}}{2CH_3CH_2CH_2OH} \xrightarrow[\Delta,\ 413K]{H_2SO_4} \underset{\text{Dipropyl ether}}{CH_3CH_2CH_2—O—CH_2CH_2CH_3} + H_2O$$

Temperature plays an important role in deciding the product.

$$\underset{\text{Ethanol}}{2CH_3CH_2OH} \xrightarrow[413K]{\Delta,\ H_2SO_4} \underset{\text{Diethylether}}{CH_3CH_2OCH_2CH_3} + H_2O$$

$$\underset{\text{Ethanol}}{2CH_3CH_2OH} \xrightarrow[\Delta,\ 443K]{H_2SO_4} \underset{\text{Ethene}}{2CH_2{=}CH_2} + 2H_2O$$

Thus, by controlling the temperature, nature of the product can be changed.

Mechanism

This reaction is an example of nucleophilic substitution (S_N) reaction in which protonated alcohol is the substrate and second molecule of alcohol acts as a nucleophile. The reaction can follow S_N1, if alcohol is 2° or 3° and S_N2 if alcohol is 1°.

Formation of protonated alcohol

$$R—\ddot{\underset{\cdot\cdot}{O}}—H + H^+ \text{ (from acid)} \rightleftharpoons \underset{\text{Protonated alcohol}}{R—\overset{+}{O}(H)—H}$$

(*i*) **S_N1 mechanism** Reaction is S_N1 if protonated alcohol loses water before the attack of second molecule of alcohol.

$$R-\overset{+}{O}(H)-H \xrightarrow[\text{Slow}]{-H_2O} R^+ \text{ (Carbocation)} \xrightarrow[\text{(Fast)}]{R-\ddot{O}-H} R-\overset{+}{O}(H)-R \xrightarrow{-H^+} R-\ddot{O}-R$$

(Ether)
(S_N1 mechanism)

However, S_N1 mechanism is not followed as dehydration of 2° and 3° alcohols result in elimination. Hence, alkenes are formed instead of ethers.

(*ii*) **S_N2 mechanism** In this case one molecule of alcohol displaces water from another molecule of protonated alcohol (1°).

$$R-\ddot{O}(H): + R-\overset{+}{O}(H)-H \xrightarrow{\text{Slow}} \left[R-\overset{\delta+}{O}(H)\cdots R\cdots\overset{\delta+}{O}H_2\right] \xrightarrow{-H_2O} R-\overset{+}{O}(H)-R \xrightarrow[\text{Fast}]{-H^+} R-\ddot{O}-R$$

Alcohol; (Protonated alcohol); (Transition state); (Ether); (Ether)
(S_N2 mechanism)

Thus, the order of dehydration of alcohols leading to the formation of ethers is

primary > secondary > tertiary.

Note *Secondary alcohols upon dehydration give alkenes as major product while tertiary alcohols upon dehydration give alkenes only due to formation of stable tertiary carbocations.*

Limitation

Dehydration is generally limited to the preparation of symmetrical ethers because a combination of two different alcohols usually yield a mixture of three ethers.

$$ROH + R'OH \longrightarrow ROR + ROR' + R'OR'$$

Note ***Catalytic dehydration*** *of lower alcohols can also be done by passing the vapours of the alcohol over alumina at 510–530 K.*

$$\underset{\text{Ethanol}}{2C_2H_5OH} \xrightarrow[510\text{-}530\text{ K}]{Al_2O_3} \underset{\text{Diethylether}}{C_2H_5OC_2H_5} + H_2O$$

Reaction of diazomethane with alcohols

In the presence of HBF_4 (as catalyst), alcohols react with diazomethane (CH_2N_2) to form methyl ethers in excellent yield.

$$\underset{\text{Alcohol}}{ROH} + \underset{\text{(In ethyl ether)}}{CH_2N_2} \xrightarrow{HBF_4} ROCH_3 + \underset{\text{(Leaving group)}}{N_2}$$

Mechanism

HBF_4 (strong acid) first protonates CH_2N_2 to give $CH_3N_2^+$ from which N_2 (an extremely good leaving group) is lost then *R*OH acting as nucleophile attacks on a carbocation, $\overset{+}{C}H_3$.

$$HBF_4 + CH_2N_2 \longrightarrow CH_3N_2^{\oplus} + BF_4^{\ominus}$$

$$CH_3N_2^{\oplus} \longrightarrow CH_3^{\oplus} + N_2\uparrow$$

$$ROH + \overset{\oplus}{C}H_3 \longrightarrow ROCH_3 + H^{\oplus}$$

It is S_N2 reaction.

Note *Reaction is favoured by polar protic solvent.*

By the addition of alcohols on alkenes

Alcohols when react with alkenes in presence of acid produce an ether.

$$\underset{\text{Propene}}{CH_3-CH=CH_2} \xrightarrow[\substack{H_2SO_4\\ \text{(Markownikoff's addition)}}]{C_2H_5OH} \underset{\text{2-ethoxypropane}}{CH_3-CH(OC_2H_5)-CH_3}$$

$$\underset{\text{2-methylpropene}}{(H_3C)_2C=CH_2} \xrightarrow[\substack{H_2SO_4\\ \text{Markownikoff's addition}}]{CH_3\ddot{O}H} \underset{\text{2-methoxy-2-methylpropane}}{CH_3-C(CH_3)_2-OCH_3}$$

Mechanism

In this reaction, intermediate is carbocation. The mechanism can be illustrated with the help of propene.

$$H^+ + \underset{\text{Propene}}{CH_3-CH=CH_2} \longrightarrow \underset{\text{Carbocation}}{CH_3-\overset{+}{C}H-CH_3} \xrightarrow{C_2H_5-\ddot{O}-H} CH_3-CH(\overset{\oplus}{O}(H)-C_2H_5)-CH_3 \xrightarrow{-H^+} CH_3-CH(O-C_2H_5)-CH_3$$

Ethyl *iso*-propyl ether

Limitations

The main limitation of this method is that initially formed carbocation rearranges to the more stable carbocation which then reacts with alcohol to form rearranged ethers.

$$CH_3-\overset{CH_3}{\underset{OC_2H_5}{C}}-\overset{CH_3}{CH}-CH_3 \xleftarrow[\text{2, 3-methyl shift}]{H_2SO_4}$$

2-ethoxy-2, 3-dimethylbutane (actual product)

$$CH_3-\overset{CH_3}{\underset{CH_3}{C}}-CH=CH_2 + \underset{\text{Ethanol}}{C_2H_5OH}$$

3, 3-dimethyl but-1-ene

$$\downarrow H_2SO_4$$

$$CH_3-\overset{CH_3}{\underset{CH_3}{C}}-\underset{OC_2H_5}{CH}-CH_3$$

3-ethoxy-2, 2-dimethyl butane (expected product)

From alkyl halides

Ethers can also be synthesised from alkyl halides. These reactions are as follows

Williamson ether synthesis

This is one of the best method for the preparation of symmetrical and unsymmetrical ethers. In this method, sodium or potassium alkoxide is heated with an alkyl halide to form an ether.

$$\underset{\text{Ethyl halide}}{C_2H_5-X} + \underset{\text{Sodium ethanoate}}{C_2H_5\overset{-+}{ON}a} \longrightarrow \underset{\text{Diethyl ether}}{C_2H_5OC_2H_5} + NaX$$

$[X = Br, Cl]$

Mechanism

It involves nucleophilic attack of alkoxide ion on alkyl halide according to S_N2 mechanism.

$$\underset{\substack{\text{Nucleophile}\\\text{(alkoxide ion)}}}{H_3C-O^-} + \underset{\substack{\text{Substrate}\\\text{(alkyl halide)}}}{CH_3-X} \xrightarrow{\text{Slow}} \underset{\text{Intermediate}}{[CH_3-\overset{\delta-}{O}\cdots\cdots\overset{\delta+}{CH_3}\cdots\cdots\overset{\delta-}{X}]}$$

$$\downarrow \text{Fast}$$

$$\underset{\text{Ether}}{CH_3-O-CH_3} + X^-$$

Limitations

As this reaction follows S_N2 mechanism and primary alkyl halides are most reactive in S_N2 reactions, therefore, best yields of unsymmetrical ethers are obtained when **alkyl halides** are **primary** and the **alkoxide** may be either **primary**, **secondary** or **tertiary**.

e.g. *tert*-butyl ethyl ether is prepared by treating ethyl bromide with sodium *tert*-butoxide.

$$\underset{\substack{\text{Sodium}\\ tert\text{-butoxide}}}{CH_3-\overset{CH_3}{\underset{CH_3}{C}}-O^-Na^+} + \underset{\text{Bromoethane}}{CH_3\overset{\delta+}{CH_2}-\overset{\delta-}{Br}} \xrightarrow[\text{Substitution}]{\Delta}$$

$$\underset{tert\text{-butylethyl ether}}{CH_3-\overset{CH_3}{\underset{CH_3}{C}}-O-CH_2CH_3} + NaBr$$

This *tert*-butyl ethyl ether cannot be prepared by treating sodium ethoxide with *tert*-butyl chloride or bromide as under these conditions, $C_2H_5O^-$ (ethoxide ion) will cause the elimination of alkyl halide to form an alkene as the main product.

$$\underset{\substack{tert\text{-butyl}\\\text{bromide}}}{CH_3-\overset{CH_2-H}{\underset{CH_3}{C}}-Br} + \underset{\text{Sodium ethoxide}}{C_2H_5O^-Na^+} \longrightarrow \underset{\substack{iso\text{-butylene}\\\text{2-methylpropene}}}{CH_3-\overset{CH_2}{\overset{\|}{\underset{CH_3}{C}}}} + \underset{\text{Ethanol}}{C_2H_5OH} + NaBr$$

Note *Sodium ethoxide behave as a Nu^- as well as strong base.*

Methyl or ethyl sulphate can be used instead of halides.

The tendency of alkyl halides to undergo elimination is $3° > 2° > 1°$.

Reaction of alkyl halides with dry silver oxide

When alkyl halide is heated with silver oxide, symmetrical ether is formed.

$$2CH_3Br + Ag_2O \xrightarrow{\Delta} CH_3OCH_3 + 2AgBr$$

It shows S_N type of reaction and ether obtained is in good yield.

From Grignard reagents

Higher ethers can be prepared by treating α-haloethers with suitable Grignard reagent.

$$\underset{\substack{\text{Methoxy chloro}\\\text{methane}}}{CH_3-O-CH_2Cl} + \underset{\substack{\text{Methyl magnesium}\\\text{iodide}}}{CH_3MgI} \xrightarrow{\text{Dry ether}} \underset{\text{Ethylmethyl ether}}{CH_3-O-CH_2CH_3} + \underset{\substack{\text{Chloro}\\\text{magnesium}\\\text{iodide}}}{Mg\langle{}^{Cl}_{I}}$$

Alkoxymercuration-demercuration of alkenes

In this method, alcohol is added to alkene to form an ether, as per Markownikoff's rule. It proceeds through cyclic transition state formation involving mercury metal. If carbocation ion is intermediate than there is always a possibility of rearrangement. The addition is exactly localised here.

$$\underset{\text{Alkene}}{>C=C<} \xrightarrow[ROH]{Hg(OAc)_2} -\underset{AcOHg}{\overset{|}{C}}-\underset{:\ddot{O}-R}{\overset{|}{C}}- \xrightarrow{NaBH_4,\ OH^-} \underset{\text{Ether}}{-\underset{H}{\overset{|}{C}}-\underset{OR}{\overset{|}{C}}-}$$

This method is similar to oxymercuration-demercuration process for the formation of alcohol by the addition of H_2O to alkene.

e.g. $$\underset{\text{Ether}}{CH_3\overset{OCH_3}{\overset{|}{C}H}\underset{CH_3}{\underset{|}{C}H}CH_3} \xleftarrow[NaBH_4,\ OH^-]{Hg(OAc)_2,\ CH_3OH} \underset{\text{3-methylbut-1-ene}}{CH_3\underset{CH_3}{\underset{|}{C}H}CH=CH_2} \xrightarrow[NaBH_4,\ OH^-]{Hg(OAc)_2,\ H_2O\cdot THF} \underset{\text{Alcohol (3-methylbutan-2-ol)}}{CH_3\underset{CH_3}{\underset{|}{C}H}\overset{OH}{\overset{|}{C}H}CH_3}$$

CHECK POINT 17.1

1. Identify the unsymmetrical ether.
(a) $CH_3—O—CH_3$ (b) $C_2H_5—O—CH_3$
(c) $C_2H_5—O—C_2H_5$ (d) All of these

2. The IUPAC name for the compound having following structure $CH_3—O—CH_2—CH_3$ will be
(a) methoxy ethane (b) ethoxy methane
(c) methoxy propane (d) ethyl methyl ether

3. The structure that represents a cyclic ether, named as 2,2-dimethyloxirane, is

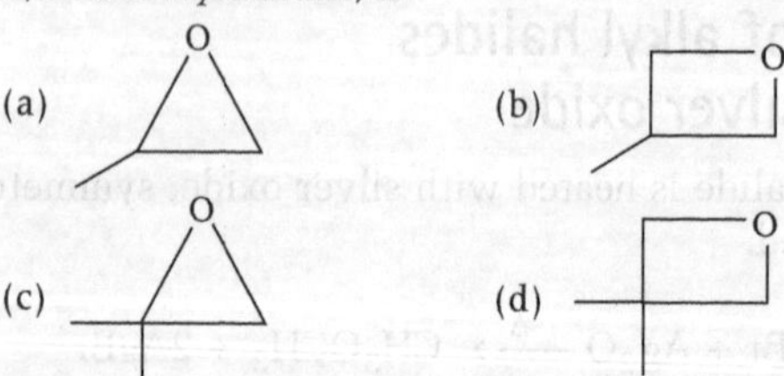

4. The reagent required during the reaction of alcohols with diazomethane is
(a) $NaBH_4$ (b) HBF_4 (c) H_2SO_4 (d) $C_2H_5O^-Na^+$

5. Acid catalysed addition of ethyl alcohol to propene leads to the formation of
(a) 1-methoxy propane (b) 2-methoxy propane
(c) 1-ethoxy propane (d) 2-ethoxy propane

6. Which of the following processes essentially follows the Markownikoff's rule?
(a) Reaction of alkyl halides with dry silver oxide
(b) Alkoxy mercuration demercuration of alkenes
(c) Reaction of α-haloethers with Grignard reagent
(d) All of the above

7. In the following reaction the product formed *A* is

Br–(CH₂ chain)–OH $\xrightarrow[(ii)\ \Delta]{(i)\ Na}$

(a) Br–(chain)–ONa
(b) Br–(chain)–O–(chain)–Br
(c) (cyclic ether, O in ring)
(d) None of the above

8. End product (major) of the following reaction is

$$CH_3\underset{CH_3}{\underset{|}{C}H}CH=CH_2 \xrightarrow[H_2O_2,OH^-]{BH_3\cdot THF} \xrightarrow[HBF_4]{CH_2N_2}$$

(a) $CH_3\underset{CH_3}{\underset{|}{C}H}CH_2CH_2OCH_3$
(b) $CH_3—\underset{CH_3}{\underset{|}{C}H}—\underset{OCH_3}{\underset{|}{C}H}—CH_3$
(c) $CH_3—\overset{CH_3}{\overset{|}{\underset{OCH_3}{\underset{|}{C}}}}—CH_2CH_3$
(d) All of the above

PHYSICAL AND CHEMICAL PROPERTIES

Physical properties

Various physical properties of ether are as follows

(*i*) **Physical state** Dimethyl ether and ethyl methyl ether are both gases. All other ethers are colourless liquids with characteristic odour called ethereal smell.

(*ii*) **Dipole moment** Although ethers lack the polar hydroxyl group of alcohols, they are still significantly polar compounds. The dipole moment of ether is the sum of two polar C–O bonds, with a substantial contribution from the two lone pair of electrons.

(*iii*) **Boiling point** The small polarity of ethers however, does not affect their boiling points appreciably which are much the same as that of alkanes of comparable molecular weights.

Ethers have lower boiling points than the isomeric alcohols since ethers cannot form hydrogen bonds due to absence of –OH groups. However, ethers form hydrogen bonds with the other compounds that have O–H or N–H groups.

(*iv*) **Solubility** Ethers, upto three carbon atoms, are soluble in water due to the formation of intermolecular H-bonds between water and ethereal oxygen.

$$R—\overset{\delta-}{O}.....\overset{\delta+}{H}—\overset{\delta-}{O}—H \quad (R \text{ on O; H-bond})$$

As the molecular mass increases, the solubility of ethers in water decreases due to a corresponding increase in the hydrocarbon portion of the molecule, which is hydrophobic.

Note *Solubility of n-butyl alcohol and diethyl ether is 8g per 100g of water. Actually, the C—O bond has little polarity and the extent of hydrogen bonding with H_2O molecules is very small. However, in dimethyl ether CH_3 groups due to small size mask the polar nature of the C—O bond to a very little extent and is completely miscible with water.*

Table 17.1 Comparison of boiling points and dipole moment

S. No.	Compound	Formula	Molecular weight	Boiling point/K	Dipole moment (*D*)
1.	Water	H_2O	18	373	1.9
2.	Ethanol	CH_3CH_2OH	46	351	1.7
3.	Dimethyl ether	$CH_3—O—CH_3$	46	248	1.3
4.	Propane	$CH_3CH_2CH_3$	44	231	0.1
5.	*n*-butanol	$CH_3(CH_2)_3OH$	74	391	1.7
6.	Tetrahydro-furan	(ring structure with O)	72	339	1.6
7.	Diethylether	$C_2H_5—O—C_2H_5$	74	348	1.2
8.	Pentane	$CH_3(CH_2)_3CH_3$	72	309	0.1

Chemical properties

Ethers, unlike alcohols, phenols and other oxygen containing organic compounds are very less reactive. This is because of the presence of divalent oxygen being atom strongly linked to carbon atoms ($C—\ddot{O}—C$) thus, lacking an active site.

The reagents like ammonia, alkalies, dilute acids and metallic sodium have no action upon ethers. They are not readily oxidised or reduced.

However, under drastic conditions, ethers react mainly because of lone pair of electrons present on the oxygen atom or due to cleavage of the C—O bonds. Some of the chemical reaction of ethers are as follows

Reactions of ethereal oxygen

Due to the presence of lone pairs of electrons on oxygen atom, ethers behave as Lewis bases and gives the following reactions.

1. Action of concentrated acids

(formation of oxonium salts)

Cold concentrated inorganic acids like HCl, H_2SO_4 has no effect on ethers except that it dissolves them forming stable oxonium salts.

$$\underset{\text{Lewis base}}{C_2H_5\ddot{O}C_2H_5} + \underset{\text{(Conc.)}}{H_2SO_4} \xrightarrow{\text{Ice cold}} \underset{\substack{\text{Oxonium salt}\\\text{(diethyl oxonium}\\\text{hydrogen sulphate)}}}{\left[C_2H_5\overset{+}{\underset{H}{O}}C_2H_5\right]HSO_4^-}$$

On heating this solution, alcohol and alkyl hydrogen sulphate is formed.

$$\underset{\text{Diethylether}}{C_2H_5OC_2H_5} + H_2SO_4 \xrightarrow{\Delta} \underset{\text{Ethanol}}{C_2H_5OH} + \underset{\text{Ethyl hydrogen sulphate}}{C_2H_5HSO_4}$$

Similary, with HCl, dialkyl oxonium chloride is formed.

$$\underset{\text{Diethylether}}{(CH_3CH_2)_2\ddot{O}:} + HCl\text{ (conc.)} \longrightarrow \underset{\text{Diethyloxonium chloride}}{\left[(CH_3CH_2)_2\ddot{O}^+—H\right]Cl^-}$$

Note *Secondary and tertiary ethers when heated with conc. H_2SO_4 form alkene.*

$$\underset{\text{2-methoxy-2-methylpropane}}{CH_3—C(CH_3)_2—OCH_3} \xrightarrow[\Delta]{\text{Conc.}H_2SO_4} \underset{\text{2-methylpropene}}{CH_3—C(CH_3)=CH_2} + \underset{\text{Methanol}}{CH_3OH}$$

2. Formation of coordination complexes

Being Lewis bases, ethers form coordination complexes with Lewis acids like BF_3, Grignard reagents, $AlCl_3$, etc.

$$\underset{\text{Diethylether}}{(CH_3CH_2)_2\ddot{O}:} + BF_3 \longrightarrow \underset{\text{Boron trifluoride etherate}}{(CH_3CH_2)_2\ddot{O} \rightarrow BF_3}$$

$$2(CH_3CH_2)_2\ddot{O}: + Mg\langle^R_X \longrightarrow \underset{\text{Grignard reagent etherate}}{(CH_3CH_2)_2\ddot{O} \rightarrow Mg(R)(X) \leftarrow \ddot{O}(CH_2CH_3)_2}$$

Hence, ether acts as a solvent for Grignard reagents, as it is able to solvate and use to dissolve the reagent by acting a base towards the acidic medium.

3. Oxidation with acidified $K_2Cr_2O_7$

Ethers having hydrogen atom at α-carbon, can be oxidised by acidified potassium dichromate to form carbonyl compounds.

e.g. $$\underset{\text{2-ethoxypropane}}{CH_3—\overset{\alpha}{C}H_2—O—\overset{\alpha'}{C}H(CH_3)_2} \xrightarrow[\Delta]{K_2Cr_2O_7/H^+} \underset{\text{Acetic acid}}{CH_3—COOH} + \underset{\text{Acetone}}{CH_3—\overset{O}{\overset{\|}{C}}—CH_3}$$

$$\underset{\text{1-ethoxypropane}}{CH_3—\overset{\alpha}{C}H_2—O—\overset{\alpha'}{C}H_2—CH_2—CH_3} \xrightarrow[\Delta]{K_2Cr_2O_7/H^+} \underset{\text{Acetic acid}}{CH_3COOH} + \underset{\text{Propanoic acid}}{C_2H_5COOH}$$

Reactions involving cleavage of C—O bond

Ethers give a number of reactions involving cleavage of C—O bond. These are as follows

1. Action of halogen acids

Ethers are readily cleaved by halogen acids at 373K to form an alcohol and an alkyl halide.

$$\underset{\text{Ether}}{R—O—R'} + \underset{\text{1 eq.}}{HX} \xrightarrow{373K} \underset{\text{Alcohol}}{R—OH} + \underset{\text{Alkyl halide}}{R'—X}\ ;\ [X = \text{Br or I}]$$

e.g. $$\underset{\text{Dimethyl ether}}{CH_3—O—CH_3} + \underset{\text{1 eq.}}{HI} \xrightarrow{373K} \underset{\text{Methanol}}{CH_3OH} + \underset{\text{Methyl iodide}}{CH_3I}$$

- If excess of acid is used then only alkyl halide is formed because the alcohol formed reacts further with halogen acid to form the corresponding alkyl halide.

$$\underset{\text{Simple ether}}{R—O—R} + \underset{\text{Excess}}{2HI} \xrightarrow{373K} \underset{\text{Alkyliodide}}{2R—I} + H_2O$$

e.g. $$\underset{\text{Diethylether}}{CH_3CH_2—O—CH_2CH_3} + \underset{\text{Excess}}{2HI} \xrightarrow{373K} \underset{\text{Ethyl iodide}}{2\,CH_3CH_2I} + H_2O$$

- In case of unsymmetrical ethers, halogen goes preferentially with smaller alkyl group or more stable carbocation.

e.g. $$\underset{\text{Methoxy ethane}}{CH_3—O—C_2H_5} + HI \longrightarrow \underset{\text{Methyliodide}}{CH_3I} + \underset{\text{Ethanol}}{C_2H_5OH}$$

$$\underset{\text{Methoxy benzene}}{C_6H_5—O—CH_3} + HI \longrightarrow \underset{\text{Phenol}}{C_6H_5—OH} + \underset{\text{Methyl iodide}}{CH_3I}$$

Mechanism

Step I Formation of oxonium ions.

$$CH_3—\ddot{O}—C_2H_5 + \overset{+}{H}—\overset{\ominus}{I} \rightleftharpoons \underset{\text{Oxonium ion}}{CH_3—\overset{H}{\overset{|}{\overset{+}{\ddot{O}}}}—C_2H_5} + I^-$$

The reaction takes place with HBr or HI because Br^- or I^- are good nucleophiles. Hence, the order of reactivity of HX is HI > HBr > HCl.

Step II Attack by iodide ion [or a nucleophile] on oxonium ion. Cleavage involves nucleophilic attack by halide ion on the protonated ether with displacement of the weakly basic alcohol molecule.

$$I^- + CH_3—\overset{H}{\overset{|}{\overset{+}{O}}}—C_2H_5 \xrightarrow[\text{Slow}]{S_N2} \underset{\text{Transition state}}{[\overset{\delta-}{I} \text{---} CH_3 \text{---} \overset{H}{\overset{|}{\overset{\delta+}{O}}}—C_2H_5]}$$

$$\longrightarrow \underset{\substack{\text{Ethanol (weak base)}\\ \text{good leaving group}}}{C_2H_5OH} + \underset{\text{Methyl iodide}}{CH_3I}$$

When excess of HI is used, ethanol formed reacts with another molecule of HI to form ethyl iodide.

$$C_2H_5—\ddot{O}—H + \overset{\delta+}{H}—\overset{\delta-}{I} \rightleftharpoons C_2H_5—\overset{H}{\underset{..}{\overset{|+}{O}}}—H + I^-$$

$$I^- + CH_3CH_2—\overset{H}{\overset{|+}{O}}—H \longrightarrow \underset{\text{Ethyl iodide}}{C_2H_5I} + H_2O$$

In the following reaction,

$$H_3C—\underset{CH_3}{\overset{CH_3}{\underset{|}{\overset{|}{C}}}}—O—CH_3 + HI \longrightarrow (CH_3)_3I + CH_3OH$$

S_N1 mechanism is followed as $(CH_3)_3C^{\oplus}$ is most stable carbocation and I^- is attached to it.

2. Action with dilute sulphuric acid (hydrolysis)

On heating with dil. H_2SO_4 under pressure, ethers are hydrolysed to alcohols.

$$\underset{\text{Ether}}{R—O—R} + H_2O \xrightarrow[\text{Under pressure}]{\text{Dil.}H_2SO_4/\Delta} \underset{\text{Alcohol}}{2R—OH}$$

e.g. $\underset{\text{Diethylether}}{CH_3CH_2—O—CH_2CH_3} \xrightarrow{\text{Dil.}H_2SO_4/\Delta} \underset{\text{Ethanol}}{2CH_3CH_2OH}$

Thus, dehydration of alcohols and hydrolysis of ethers are reversible.

$$\underset{\text{Alcohol}}{2ROH} \overset{\Delta}{\rightleftharpoons} \underset{\text{Ether}}{R_2O} + H_2O$$

Here R is 1° alkyl group.

3. Action of heat (dehydration)

At 673K, ethers in presence of Al_2O_3 undergo dehydration to form alkenes.

e.g., $\underset{\text{Dimethylether}}{CH_3OCH_3} \xrightarrow[673K]{Al_2O_3} \underset{\text{Ethene}}{CH_2{=}CH_2} + H_2O$

4. Action of phosphorus pentachloride (PCl_5)

With hot phosphorus pentachloride, ethers undergo cleavage at both the oxygen bond forming alkyl chlorides.

$$Cl—PCl_3—Cl + C_2H_5OC_2H_5 \longrightarrow \underset{\text{Ethylchloride}}{2C_2H_5Cl} + POCl_3$$

For unsymmetrical ethers, following reaction takes place

$$R–O–R' \xrightarrow{PCl_5} RCl + R'Cl + POCl_3$$

Note *By looking at the products, i.e. alkyl chloride, one can predict the structure of ether. From the products, RCl and R′Cl, remove Cl and join the two by (—O—). Thus*

$\underset{(A)}{C_5H_{12}O} \xrightarrow{PCl_5} (CH_3)_2CHCl + CH_3CH_2Cl + POCl_3$, *then (A) is* $(CH_3)_2CH—O—CH_2CH_3$.

5. Action with acid chlorides

Acid chlorides when heated in the presence of anhydrous $ZnCl_2$ or $AlCl_3$, forms alkyl halide and ester.

$$\underset{\text{Diethyl ether}}{C_2H_5—O—C_2H_5} + \underset{\text{Acetyl chloride}}{CH_3COCl} \xrightarrow[\Delta]{\text{Anhy.}ZnCl_2} \underset{\text{Ethyl chloride}}{C_2H_5Cl} + \underset{\text{Ethyl acetate}}{CH_3COOC_2H_5}$$

6. Action with acid anhydride

With acid anhydride, ethers form only esters.

$$\underset{\text{Diethylether}}{C_2H_5—O—C_2H_5} + \underset{\text{Acid anhydride}}{(CH_3CO)_2O} \xrightarrow[\Delta]{\text{Anhy. }ZnCl_2} \underset{\text{Ester}}{2CH_3COOC_2H_5}$$

Reactions involving alkyl group

Ethers give a number of reaction involving alkyl group. These are as follows

Autoxidation of ethers

When ethers are exposed to atmospheric oxygen, they slowly oxidise to hydroperoxides and dialkylperoxides, both of which are explosives.

Such a spontaneous oxidation by atmospheric oxygen is called auto-oxidation. It is a free radical reaction and oxidation occurs at the C-atom next to ethereal oxygen.

$$R—O—CH_2—R \xrightarrow[\text{slow}]{\text{Excess }O_2} \underset{\text{Hydroperoxide}}{R—O—\underset{OOH}{\underset{|}{CH}}—R} + \underset{\text{Dialkyl peroxide}}{R—O—O—CH_2—R}$$

$$\underset{\text{2-}iso\text{-propoxy propane (ether)}}{(H_3C)_2CH—O—CH(CH_3)_2} \xrightarrow{\text{Atmospheric }O_2} \underset{\text{Hydroperoxide}}{(H_3C)_2CH—O—\overset{OOH}{\overset{|}{C}}(CH_3)_2} + \underset{\text{Diisopropyl peroxide}}{(H_3C)_2CH—O—O—CH(CH_3)_2}$$

Note *These peroxides are highly dangerous compounds as they decompose at low temperature.*

Halogenation

Ethers when react with halogens, give substitution products. Halogen substitution occurs at α-H as this 'H' is acidic in nature. Extent of halogenation depends on the reaction conditions.

(*i*) If reaction occurs in dark, only mono substituted product is formed.

$$\underset{\text{Diethyl ether}}{CH_3CH_2OCH_2CH_3} \xrightarrow[\text{dark}]{Cl_2} \underset{\alpha\text{-chlorodiethyl ether}}{CH_3CH(Cl)OCH_2CH_3}$$

$$\xrightarrow[\text{dark}]{Cl_2} \underset{\alpha,\alpha'\text{-dichlorodiethyl ether}}{CH_3CH(Cl)OCH(Cl)CH_3}$$

(*ii*) If reaction occurs in presence of light and chlorine is present in excess, then substitution of all hydrogen will takes place.

$$(CH_3CH_2)_2O \xrightarrow[\text{light}]{10\,Cl_2\text{ (excess)}} \underset{\text{Pentachlorodiethyl ether}}{(C_2Cl_5)_2O} + 10HCl$$

Electrophilic substitution reactions of aryl ethers

Due to +*R*-effect of alkoxy group (—O*R*) electron density in the benzene ring increases at two *ortho* and one *para* position as compared to *m*-position. Thus, the benzene ring is activated towards electrophilic substitution reactions at *o* and *p*-positions mainly.

:ÖR ⟷ +ÖR ⟷ +ÖR ⟷ +ÖR ⟷ :ÖR ≡ δ+ ÖR with δ– at ortho and para positions

Anisole (methyl phenyl ether):

- Br_2/CH_3COOH or CS_2 (Bromination) → *o*-bromoanisole (10% minor) + *p*-bromoanisole (90% major)
- Conc. HNO_3 + conc. H_2SO_4 (Nitration) → *o*-nitroanisole (minor) + *p*-nitroanisole (major)
- CH_3Cl, Anhy. $AlCl_3$ (Friedel Craft alkylation) → *o*-methylanisole (minor) + *p*-methylanisole (major)
- CH_3COCl, Anhy. $AlCl_3$ (Friedel Craft acylation) → 2-methoxy acetophenone (minor) + 4-methoxy acetophenone (major)

Note *o-products are minor due to steric hinderance experienced by incoming group at o-positions.*

Cleavage of epoxides (oxiranes)

In the formation of oxiranes, bond angle is distorted to 61.5° (from normal value of 110°). Thus, oxiranes (or ethyleneoxide) like cyclopropane are highly strained and therefore, ring can be cleaved under acidic and basic conditions to form 1, 2-diols that are *trans* in nature.

H_2C—CH_2 (epoxide ring): (C—C—O) = 59.2°; (C—O—C) = 61.5°

1. Acid-catalysed cleavage

Like other ethers, an epoxide is converted by acid into the protonated epoxide, which can undergo attack by any nucleophilic reagent.

$$-\overset{|}{C}-\overset{|}{C}-\ (\text{Epoxide, bridged by O}) \underset{}{\overset{H^+}{\rightleftharpoons}} -\overset{|}{C}-\overset{|}{C}-\ (\text{bridged by } \overset{\oplus}{O}-H)$$

Epoxide

Protonated epoxide

Effect of different reagents on the protonated epoxide are as follows

(i) Attack of H_2O gives *trans* 1, 2-diol as product with two (–OH) groups on adjacent carbon atoms.

$$-\overset{|}{C}-\overset{|}{C}-\ (\overset{\oplus}{O}H) + H_2\ddot{O} \xrightarrow{-H^+} >C(OH)-C(OH)<$$

trans 1, 2-diol

(ii) Attack of alcohol gives a compound containing ether and alcohol.

$$-\overset{|}{C}-\overset{|}{C}-\ (\overset{\oplus}{O}H) + R-\ddot{O}(H): \longrightarrow -\underset{OR}{\overset{|}{C}}-\underset{OH}{\overset{|}{C}}- + H^+$$

Alkoxy alcohol

(iii) Attack of halogen acid forms halohydrin.

$$X^- \ -\overset{|}{C}-\overset{|}{C}-\ (\overset{\oplus}{O}H) \longrightarrow -\underset{X}{\overset{|}{C}}-\underset{OH}{\overset{|}{C}}- + H^+$$

Halohydrin

Note *In acid catalysed cleavage on unsymmetrical epoxide, nucleophile attacks at more substituted C-atom.*

$$\text{e.g. } CH_3\ddot{O}H + CH_3-\overset{CH_3}{\overset{|}{C}}-CH_2\ (\text{epoxide O}) \xrightarrow{H^+} CH_3-\underset{OCH_3}{\overset{CH_3}{\overset{|}{C}}}-CH_2-OH$$

2. Base-catalysed cleavage

Unlike ordinary ethers, epoxides can be cleaved under alkaline conditions. It is the epoxide itself that first undergoes nucleophilic attack.

$$-\overset{|}{C}-\overset{|}{C}-\ (\text{O bridge}) + :Z \longrightarrow -\overset{Z}{\overset{|}{C}}-\underset{O^-}{\overset{|}{C}}- \xrightarrow[-Z:]{HZ} -\overset{Z}{\overset{|}{C}}-\underset{OH}{\overset{|}{C}}$$

Different examples of base catalysed reactions are as follows

(i) $$\underset{\text{Sodium ethoxide}}{C_2H_5O^-\overset{+}{Na}} + CH_2-CH_2\ (\text{O bridge}) \longrightarrow \underset{O^-}{CH_2}-\underset{OC_2H_5}{CH_2} \xrightarrow{H^+} \underset{OH}{CH_2}-\underset{OC_2H_5}{CH_2}$$

(ii) $$\underset{\text{Sodium phenoxide}}{C_6H_5-O^-\overset{+}{Na}} + H_2C-CH_2\ (\text{O bridge}) \longrightarrow \underset{O^-}{CH_2}-\underset{O-C_6H_5}{CH_2} \xrightarrow{H^+} \underset{OH}{CH_2}-\underset{O-C_6H_5}{CH_2}$$

(iii) $$NH_3 + CH_2-CH_2\ (\text{O bridge}) \longrightarrow \underset{OH}{CH_2}-\underset{NH_2}{CH_2}$$

In base catalysed cleavage of unsymmetrical epoxide, nucleophile attacks at less substituted C-atom. This is due to the fact that it follows S_N2 reaction mechanism (which favours 1° carbon).

$$\text{e.g. } \underset{\text{Ethoxide}}{CH_3CH_2O^-} + \underset{1^\circ\text{ carbon}}{CH_2}-CH-CH_3\ (\text{O bridge}) \xrightarrow{CH_3CH_2OH} CH_3CH_2-O-CH_2-\underset{OH}{CH}-CH_3$$

Cleavage by Grignard reagents

Epoxide reacts with Grignard reagent followed by hydrolysis, give 1° alcohol containing two more carbons.

$$\overset{\delta-}{R}-\overset{\delta+}{MgX} + H_2C-CH_2\ (\text{O bridge}) \longrightarrow \underset{R}{CH_2}-CH_2-\overset{-}{O}-\overset{+}{Mg}X \xrightarrow{H_3O^+} \underset{R}{CH_2}-\overset{OH}{CH_2}$$

$$C_6H_5-MgBr + H_2C-CH_2\ (\text{O bridge}) \xrightarrow[(ii)\ H_3O^+]{(i)\ \text{Ether}} \underset{\text{2-phenyl ethanol}}{C_6H_5-CH_2CH_2OH}$$

Substituted epoxide can be used in this reaction, with the carbanion attacking the less hindered epoxide carbon atom.

$H_2C—CHCH_3$ (epoxide, O bridging) + cyclohexyl–MgBr

Attack of carbanion on this carbon

$\xrightarrow[\text{(ii) } H_3O^+]{\text{(i) Ether}}$ cyclohexyl–CH_2CHCH_3 (OH)

Grignard reagent behaves as Nu^- due to large polarity difference of C-metal bond.

Deoxygenation of oxiranes

Oxiranes can be deoxygenated (reduced) to alkenes by certain trivalent phosphorus compounds which are highly nucleophilic and possess a high affinity for oxygen.

$$CH_3\underset{H}{C}—\underset{H}{C}\,CH_3\ (\text{O bridging}) + Ph_3P \longrightarrow \underset{\text{But-2-ene}}{CH_3CH{=}CHCH_3} + Ph_3PO$$

[Ph is C_6H_5 (phenyl)]

Deoxygenation involves inversion of configuration *via* the following pathway

$$cis\text{-alkene} \xrightarrow{m\text{-CPBA}} \text{oxiranes} \xrightarrow{Ph_3P} trans\text{-alkene}$$

CHECK POINT 17.2

1. The $R—O—R'$ angle in an ether is
(a) lower than tetrahedral angle
(b) almost equal to tetrahedral angle
(c) much greater than tetrahedral angle
(d) not predicted yet

2. Which of the following statement(s) about the boiling point of dimethyl ether is/are correct?
(a) It is less than that of water
(b) It is less than that of ethanol
(c) It is less than that of ethanol but higher than that of water
(d) both the statements (a) and (b) are correct

3. Which of the following is not produced as an oxidation product of ether when it reacts with acidified $K_2Cr_2O_7$?
(a) CO_2 (b) Alcohol
(c) Ketone (d) Carboxylic acid

4. Predict the product of the following reaction

$$CH_3CH_2OCH_2CH_3 \xrightarrow[\Delta]{\text{Dil. } H_2SO_4} ?$$

(a) CH_3OH (b) C_2H_5OH
(c) C_3H_7OH (d) C_4H_9OH

5. Which of the following is not a product of the following reaction?

$$CH_3—O—C_2H_5 \xrightarrow{PCl_5}$$

(a) CH_3Cl (b) C_2H_5Cl
(c) C_3H_7Cl (d) $POCl_3$

6. Reaction of acetyl chloride with diethyl ether completes in the presence of
(a) anhydrous $ZnCl_2$ (b) Al_2O_3
(c) dilute H_2SO_4 (d) PCl_5

7. Hydroperoxides and dialkyl peroxides are produced when ethers undergo
(a) combustion (b) reduction
(c) autoxidation (d) disproportionation

8. Predict the major product of the following reaction

$C_6H_5OCH_3 \xrightarrow{Br_2/CH_3COOH \text{ or } CS_2} ?$

(a) o-bromoanisole (OCH_3, Br) (b) m-bromoanisole (OCH_3, Br)
(c) p-bromoanisole (OCH_3, Br) (d) Both (a) and (c) produce in equal proportion

9. The reagent required to convert anisole to 4-methoxyacetophenone is
(a) CH_3Cl / H_2SO_4 (b) CH_3COCl / H_2SO_4
(c) $CH_3Cl / AlCl_3$ (d) $CH_3COCl / AlCl_3$

10. What will be the product(s), if NH_3 reacts with epoxide?

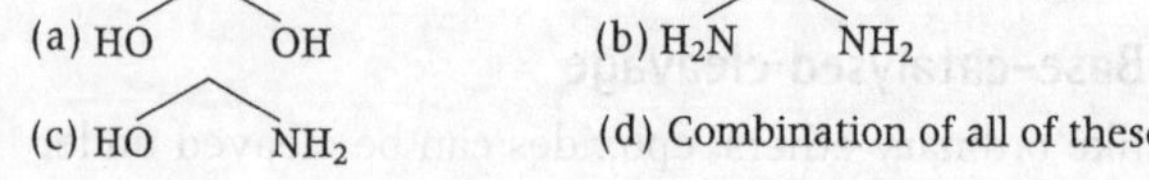

(a) HO–CH₂CH₂–OH (b) H_2N–CH₂CH₂–NH_2
(c) HO–CH₂CH₂–NH_2 (d) Combination of all of these

Chapter Exercises

(A) Topical exercise

I. Classification, structure, nomenclature, isomerism and preparation of ethers

1 IUPAC name of the compound $CH_3-CH(CH_3)-OCH_3$ is [NCERT Exemplar]

(a) 1-methoxy-1-methylethane
(b) 2-methoxy-2-methylethane
(c) 2-methoxypropane
(d) *iso*-propylmethyl ether

2 In the reaction,

$2A$ + dry silver oxide $\xrightarrow{\Delta}$ ether + 2 AgX, A is a/an

(a) primary alcohol (b) acid
(c) alkyl halide (d) alcohol

3 Which is the leaving group in the following reaction?

$$CH_3OH + CH_2N_2 \xrightarrow{HBF_4}$$

(a) H_2O (b) N_2 (c) $\overset{\oplus}{C}H_3$ (d) $H^{\oplus}$

4 Consider the following reaction

$$C_6H_5OH + C_2H_5I \longrightarrow \text{'}P\text{'}$$

Here 'P' is

(a) $C_6H_5OC_2H_5$ (b) $C_2H_5OC_2H_5$
(c) $C_6H_5OC_6H_5$ (d) C_6H_5I

5 Which is the nucleophile in the following reaction?

$$CH_3OH + CH_2N_2 \longrightarrow CH_3OCH_3 + N_2$$

(a) CH_3OCH_3 (b) N_2 (c) CH_3OH (d) CH_2N_2

6 Following step of alkoxy mercuration-demercuration is

$$CH_2{=}CH_2 \xrightarrow[\text{(ii) NaBH}_4,\ OH^-]{\text{(i) Hg(OAc)}_2,\ CH_3OH} CH_3CH_2OCH_3$$

(a) oxidation reaction
(b) reduction reaction
(c) disproportionation reaction
(d) solvation reaction

7 Indicate the reagent in the following conversion

$$C_6H_{11}-CH{=}CH_2 \longrightarrow C_6H_{11}-CH(OCH_3)CH_3$$

(a) $H_3O^{\oplus}$ / CH_3OH
(b) B_2H_6/H_2O_2, OH^- / CH_3OH
(c) $Hg(OAc)_2$, CH_3OH / $NaBH_4$, OH^-
(d) All of the above

8 When ethyl hydrogen sulphate is heated with excess of ethyl alcohol at 410 K, the product obtained is

(a) ethane
(b) ethylene
(c) diethylether
(d) diethyl sulphate

9 Select the incorrect statement about alkoxy mercuration-demercuration of alkenes.

(a) It involves carbocation as the intermediate
(b) It involves no rearrangement
(c) It involves addition of water according to Markownikoff's rule
(d) Identical regiochemistry is seen in product

10 Which of the following alkyl halides is most reactive in Williamson synthesis?

(a) $(CH_3)_3C-Br$ (b) $Br-CH_2-CH=CH_2$
(c) $CH_3CH_2CH_2Cl$ (d) $CH_3CH_2CH_2Br$

11 Addition of methanol to 2-methylpropene in the presence of conc. H_2SO_4 gives

(a) *tert*-butyl alcohol
(b) *tert*-butyl methyl ether
(c) di-*tert*-butyl ether
(d) dimethyl ether

12 Consider the following statements.

Statement I Starting with $(CH_3)_3CBr$ and $CH_3\bar{O}\overset{+}{Na}$, $(CH_3)_3C-O-CH_3$ is prepared.

Statement II Tertiary alkyl bromide loses HBr by CH_3O^- forming alkene.

(a) Both statements I and II are correct.
(b) Only statement I is correct.
(c) Only statement II is correct.
(d) Both statements I and II are incorrect.

13 Phenol on treatment with diethyl sulphate in presence of NaOH gives

(a) phenetole
(b) diphenyl ether
(c) anisole
(d) sodium benzene sulphonate

14 The end product 'C' in the following series of reaction is

$$C_2H_5Br \xrightarrow[\text{NaOH}]{\text{Aqueous}} A \xrightarrow{\text{Na}} B \xrightarrow{CH_3I} C$$

(a) propane (b) dimethyl ether
(c) ethoxy ethane (d) ethylmethyl ether

15 Which of the following cannot be prepared by using Williamson's synthesis?
(a) Methoxybenzene (b) Benzyl *p*-nitrophenylether
(c) Methyl *tert*-butyl ether (d) Ditertiary butyl ether

16 Consider the following statements.
Statement I IUPAC name of the compound

$$CH_3—\underset{\underset{CH_3}{|}}{CH}—O—CH_2—CH_2—CH_3 \text{ is}$$

2-ethoxy-2-methylethane.
Statement II In IUPAC nomenclature, ether is regarded as hydrocarbon derivative in which a hydrogen atom is replaced by —OR or —OAr group [where, R = alkyl group and Ar = aryl group].
(a) Both statements I and II are correct.
(b) Only statement I is correct.
(c) Only statement II is correct.
(d) Both statements I and II are incorrect.

17 Methyl tertiary butyl ether (MTBE) is an important gasoline additive for improving octane number, and can be prepared by following methods.

I. $(CH_3)_3C\overset{-}{O}\overset{+}{Na} + CH_3I \longrightarrow$
II. $(CH_3)_2C{=}CH_2 + CH_3OH \xrightarrow{H^+} (CH_3)_3OCOCH_3$
III. $(CH_3)_3C—Br + CH_3ONa \longrightarrow$

Best method is
(a) Only I (b) Only II (c) Only III
(d) None can be used

18 Consider the following reactions.

I. $C_6H_5—\overset{-}{O}\overset{+}{Na} + CH_3I \longrightarrow$ (benzene ring)
II. $C_6H_5—Br + CH_3\overset{-}{O}\overset{+}{Na} \longrightarrow$ (benzene ring)
III. $(CH_3)_3C—Cl + CH_3\overset{-}{O}\overset{+}{Na} \longrightarrow$
IV. $(CH_3)_3C—\overset{-}{O}\overset{+}{Na} + CH_3Cl \longrightarrow$

Which of the above methods cannot be used in the synthesis of ether?
(a) Both I and IV (b) Both I and III
(c) Both II and IV (d) Both II and III

II. Physical and Chemical properties

19 In H_2O, there is bond-angle $\widehat{HOH}$ of 104.5° due to repulsion, but in ether bond-angle $\widehat{ROR}$ is 110°. It is due to
(a) H is polar while R is non-polar
(b) H is non-polar while R is polar
(c) repulsion between the bulkier alkyl groups that almost counterbalances the effect of lone pair-lone pair interactions
(d) None of the explanations given above is correct

20 Select the correct statement(s).
(a) Ethers are weakly polar compounds
(b) Ethers have lower boiling points than the isomeric alcohols
(c) Dipole moment of ether is the sum of the two polar C—O bonds with a substantial contribution from the two lone pairs of electrons
(d) All of the above statements are correct

21 Ether is a good solvent for Grignard reagent. Which property makes it a good solvent?
(a) It acts as a base towards acidic magnesium
(b) It makes H-bonding with Grignard reagent
(c) It acts as an acid towards basic magnesium
(d) It is an electron rich species

22 Increasing order of dipole moment of (I) ethanol (II) propane and (III) methoxy-methane is
(a) I < II < III
(b) II < III < I
(c) II < I < III
(d) I < II < III

23 Acetic anhydride reacts with diethyl ether in the presence of anhydrous $AlCl_3$ to give
(a) CH_3COOH
(b) $CH_3CH_2COOCH_3$
(c) $CH_3COOCH_2CH_3$
(d) CH_3CH_2OH

24 An ether is more volatile than an alcohol having the same molecular formula. This is due to
(a) dipolar character of ethers
(b) alcohols having resonance structure
(c) intermolecular hydrogen bonding in ethers
(d) intermolecular hydrogen bonding in alcohols

25 What are the products of the reaction given below?

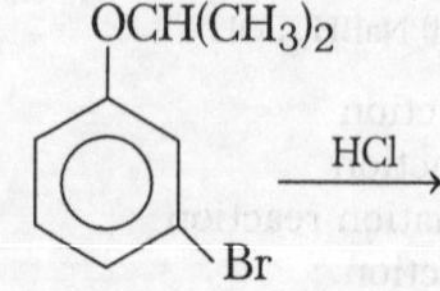

(a) $(CH_3)_2CHCl$ and *m*-bromophenol
(b) $(CH_3)_2CHCl$ and *m*-bromochlorobenzene
(c) $(CH_3)_2CHOH$ and *m*-bromophenol
(d) $(CH_3)_2CHOH$ and *m*-bromochlorobenzene

26 HI cleaves ether (*A*) forming

A : C_6H_5—CH_2—O—CH_3

(a) C_6H_5—CH_2OH, CH_3I (b) C_6H_5—CH_2I, CH_3OH

(c) C_6H_5—CH_2I, CH_3I (d) C_6H_5—CH_2OH, CH_3OH

27 Which one of the following ethers is the most unreactive to cleavage with HBr?

(a) $C_6H_5OCH_2C_6H_5$
(b) $H_2C{=}CHCH_2OCH_2CH{=}CH_2$
(c) $C_6H_5OC_6H_5$
(d) $(CH_3)_3COC(CH_3)_3$

28 Oxiranes are cleaved under acidic and basic conditions because of

(a) decrease in bond-angle to 61.5° from normal tetrahedral angle of sp^3-hybridisation, i.e. 109.5°.
(b) strong, H^+ and OH^-
(c) formation of oxonium ion
(d) formation of carbocation

29 Match the starting materials given in Column-I with the products formed by these (Column II) in the reaction with HI and choose the correct answer using the code given below : **[NCERT Exemplar]**

Column I	Column II
A. CH_3—O—CH_3	1. C_6H_5OH (phenol) + CH_3I
B. $(CH_3)_2CH$—OCH_3	2. CH_3—$C(CH_3)_2$—I + CH_3OH
C. CH_3—$C(CH_3)_2$—OCH_3	3. $(CH_3)_2CH$—OH + CH_3I
D. $C_6H_5OCH_3$ (anisole)	4. CH_3OH + CH_3I

Codes

	A	B	C	D		A	B	C	D
(a)	4	3	2	1	(b)	3	4	1	2
(c)	3	4	2	1	(d)	4	3	1	2

30 Consider the following statements.

Statement I Boiling point of ether is smaller than that of isomeric alcohol.

Statement II Solubility of ether in water is due to H-bonding.

(a) Both statements I and II are correct
(b) Only statement I is correct
(c) Only statement II is correct
(d) Neither statement I nor statement II is correct

31 What is the product of the following reaction?

Epoxide (propylene oxide, with H and CH_3 on the chiral carbon) + $\overset{\ominus}{O}H$

(a) A pure enantiomer of propan-1,2 diol
(b) Racemic mixture of propan-1,2-diol
(c) Unequal amount of enantiomers of propane-1, 2-diol
(d) Formation of propane-1,3-diol

32 Consider the following statements.

Statement I In the cleavage of ether bond (—O—) by HI, the halide formed is a tertiary halide when one of the alkyl groups is a tertiary group.

Statement II Stability of 3° carbocation is maximum.

(a) Both statements I and II are correct.
(b) Only statement I is correct.
(c) Only statement II is correct.
(d) Both statements I and II are incorrect.

(B) Miscellaneous questions

1 Which of the following is the correct statement about the following reaction?

$CH_3OH + CH_3C(H)$—CH_2 (protonated epoxide, $\overset{\oplus}{O}$—H; bonds labelled γ, α, β) $\longrightarrow$

(a) α-bond is cleaved when product formed is $CH_3CH(OH)$—CH_2OCH_3
(b) β-bond is cleaved when product formed is $CH_3CH(OCH_3)$—CH_2OH
(c) α-bond is cleaved when product formed is $CH_3CH(OCH_3)$—CH_2OH
(d) β-bond is cleaved when product formed is $CH_3CH(OH)$—CH_2OCH_3

2 Major addition product of the following reaction is

(cyclohexylidene)$=CH_2 + CH_3OH \xrightarrow{H_2SO_4}$

(a) cyclohexyl$-CH_2OCH_3$ (b) cyclohexane with $-CH_3$ and OCH_3 on same carbon

(c) cyclohexane with $-CH_3$ and adjacent OCH_3 (d) cyclohexane with $-CH_3$ and OCH_3

3 Consider the following statements.

Statement I Cyclic ether can be prepared by intramolecular Williamson synthesis.

Statement II There is nucleophilic attack of alkoxide ion on carbon containing halide. Thus, halide is displaced.

(a) Both statements I and II are correct.
(b) Only statement I is correct.
(c) Only statement II is correct.
(d) Both statements I and II are incorrect.

4 Most stable intermediate in the following reaction is

cyclohexyl$-CH{=}CH_2 \xrightarrow[H_2SO_4]{CH_3CH_2OH}$

(a) cyclohexyl$-\overset{\oplus}{C}H-CH_3$ (b) cyclohexyl$-CH_2\overset{\oplus}{C}H_2$

(c) $\oplus$(cyclohexyl)$-CH_2CH_3$ (d) $\oplus$cyclohexyl$-CH_2CH_3$

5 Select the incorrect statement(s) about the following reaction of ether synthesis.

$$RX + R'ONa \longrightarrow ROR' + NaX$$

(a) It follows S_N2 mechanism
(b) Alkyl halide (RX) should be 2° or 3° while alkoxide (RO^-Na^+) should be 1°
(c) Alkyl halide should be 1° while alkoxide should be 2° or 3°.
(d) 2° and 3° alkyl halides may undergo E2 elimination in the presence of a strong base to form alkenes.

6 Consider the following statements.

Statements I *t*-butyl methyl ether is not prepared by the reaction of *t*-butyl bormide with sodium ethoxide.

Statement II Sodium methoxide is a strong nucleophile.

(a) Both statements I and II are correct.
(b) Only statement I is correct.
(c) Only statement II is correct.
(d) Both statements I and II are incorrect.

7 Select the correct statement(s) about following reaction.

$$2ROH \xrightarrow{H^+} ROR + H_2O$$

(a) It is an example of S_N reaction in which protonated alcohol is the substrate and second molecule of alcohol is the nucleophile.
(b) This reaction can be S_N1 if protonated alcohol loses water before attack of second molecule.
(c) This reaction can be S_N2 if alcohol is 1°.
(d) All of the above statements are correct

8 In the following reaction,

$$-\underset{|}{\overset{\overset{\ddot{O}-H}{|}}{C}}-\underset{\underset{Cl}{|}}{\overset{|}{C}}- + \ominus{:}\ddot{O}-H \xrightarrow{I} A \xrightarrow{II} -\overset{\ddot{O}}{C}\!-\!C-\text{ (epoxide)}$$

the intermediate A is

(a) $-\overset{\overset{\ddot{O}-H}{|}}{C}-\underset{\underset{OH}{|}}{C}-$ (b) $-\overset{\overset{\ddot{O}-H}{|}}{C}-\underset{\underset{:\ddot{O}:^{\ominus}}{|}}{C}-$

(c) $-\overset{\overset{:\ddot{O}:^{\ominus}}{|}}{C}-\underset{\underset{Cl}{|}}{C}-$ (d) $-\underset{\oplus}{\overset{\overset{:\ddot{O}:^{\ominus}}{|}}{C}}-\overset{|}{C}-$

9 Match the species in Column I with the corresponding boiling point in Column II and choose the correct answer using the codes given below.

Column I	Column II
A. CH_3CH_2OH	1. 373 K
B. CH_3-O-CH_3	2. 348 K
C. H_2O	3. 351 K
D. $CH_3CH_2-O-CH_2CH_3$	4. 248 K

Codes

	A	B	C	D		A	B	C	D
(a)	3	4	1	2	(b)	3	4	2	1
(c)	4	3	1	2	(d)	4	3	2	1

10 Consider the following reactions.

(i) $(CH_3)_2CH-CH_2Br \xrightarrow{C_2H_5OH} (CH_3)_2CH-CH_2OC_2H_5 + HBr$

(ii) $(CH_3)_2CH-CH_2Br \xrightarrow{C_2H_5O^-} (CH_3)_2CH-CH_2OC_2H_5 + Br^-$

The mechanisms of reactions (i) and (ii) are respectively

(a) S_N1 and S_N2 (b) S_N1 and S_N1
(c) S_N2 and S_N2 (d) S_N2 and S_N1

11 In the following reaction, final product is

$$ClCH_2\overset{14}{C}H-CH_2 \text{ (epoxide, }-O-) \xrightarrow{\overset{+}{Na}\overset{-}{O}C_2H_5}$$

(a) $ClCH_2\overset{14}{C}H(OH)CH_2OC_2H_5$ (b) $ClCH_2\overset{14}{C}H(OC_2H_5)CH_2ONa$

(c) $\overset{14}{C}H_2-CHCH_2OC_2H_5$ (epoxide, O bridging) (d) $CH_2-\overset{14}{C}HCH_2OC_2H_5$ (epoxide, O bridging)

(C) Medical entrances' gallery

1 The major products C and D formed in the following reactions respectively are

$$H_3C-CH_2-CH_2-OC(CH_3)_3 \xrightarrow[\Delta]{\text{Excess HI}} C+D$$

(a) $H_3C-CH_2-CH_2-I$ and $I-C(CH_3)_3$ **[NEET 2019]**
(b) $H_3C-CH_2-CH_2-OH$ and $I-C(CH_3)_3$
(c) $H_3C-CH_2-CH_2-I$ and $HO-C(CH_3)_3$
(d) $H_3C-CH_2-CH_2-OH$ and $HO-C(CH_3)_3$

2 Assertion (A) Phenol reacts with CH_3I in presence of NaOH to form methoxybenzene. **[NEET 2019]**

Reason (R) Phenoxide is better nucleophile than phenol.

(a) Both Assertion and Reason are true and Reason is the correct explanation of Assertion
(b) Both Assertion and Reason are true, but Reason is not the correct explanation of Assertion
(c) Assertion is true, but Reason is false
(d) Both Assertion and Reason are false

3 Assertion (A) *tert*-butyl methyl ether react with HBr to form *tert* butyl bromide, $(CH_3)_3$ C—Br and CH_3—OH, methanal. **[NEET 2019]**

Reason (R) It follows S_N1 mechanism.

(a) Both Assertion and Reason are true and Reason is the correct explanation of Assertion
(b) Both Assertion and Reason are true, but Reason is not the correct explanation of Assertion
(c) Assertion is true, but Reason is false
(d) Both Assertion and Reason are false

4 The compound A on treatment with Na gives B, and with PCl_5 gives C. B and C react together to give diethyl ether. A, B and C are in the order **[NEET 2018]**

(a) C_2H_5Cl, C_2H_6, C_2H_5OH
(b) C_2H_5OH, C_2H_5Cl, C_2H_5ONa
(c) C_2H_5OH, C_2H_6, C_2H_5Cl
(d) C_2H_5OH, C_2H_5ONa, C_2H_5Cl

5 Assertion (A) IUPAC name of the compound

$$CH_3-\underset{\underset{CH_3}{|}}{CH}-O-CH_2-CH_2-CH_3 \text{ is}$$

1-(2-propoxy) propane. **[NEET 2018]**

Reason (R) In IUPAC nomenclature, ether is regarded as hydrocarbon derivative in which a hydrogen atom is replaced by—OR or —OAr group [where, R=alkyl group and Ar = aryl group]. **[AIIMS 2018]**

(a) Both Assertion and Reason are correct and Reason is the correct explanation of Assertion
(b) Both Assertion and Reason are correct but Reason is not correct Explanation of Assertion
(c) Assertion is correct but Reason is incorrect
(d) Both Assertion and Reason are incorrect

6 The heating of phenyl-methyl ethers with HI produces **[NEET 2017]**

(a) benzene
(b) ethyl chlorides
(c) iodobenzene
(d) phenol

7 The reaction, **[AIPMT 2015]**

$$CH_3-\underset{\underset{CH_3}{|}}{\overset{\overset{CH_3}{|}}{C}}-ONa + CH_3CH_2Cl \xrightarrow[-NaCl]{} CH_3-\underset{\underset{CH_3}{|}}{\overset{\overset{CH_3}{|}}{C}}-O-CH_2-CH_3$$

is called

(a) Williamson synthesis
(b) Williamson continuous etherification process
(c) Etard reaction
(d) Gattermann-Koch reaction

8 Which of the following is added to ether to test its purity? **[JCECE 2015]**

(a) C_2H_5ONa (b) KI
(c) KCl (d) $(CH_3CO)_2O$ [drawn as CH_3CO and CH_3CO joined to O]

9 Select the ether among following that yields methanol as one of the products on reaction with cold hydroiodic acid **[MHT CET 2014]**

(a) 1-methoxybutane
(b) 1-methoxy-2-methylpropane
(c) 2-methoxy-2-methylpropane
(d) methoxybenzene

10 Williamson's synthesis of preparing dimethyl ether is a/an **[Kerala CEE 2014]**
(a) electrophilic substitution (b) S_N1 reaction
(c) electrophilic addition (d) S_N2 reaction
(e) free radical substitution

11 When $CH_2=CH-O-CH_2-CH_3$ reacts with one mole of HI, one of the products formed is **[KCET 2014]**
(a) ethane (b) ethanol (c) iodoethene (d) ethanal

12 Among the following ethers, which one will produce methyl alcohol on treatment with hot concentrated HI? **[NEET 2013]**
(a) $CH_3-CH_2-CH_2-CH_2-O-CH_3$
(b) $CH_3-CH_2-CH(CH_3)-O-CH_3$
(c) $CH_3-C(CH_3)_2-O-CH_3$
(d) $CH_3-CH(CH_3)-CH_2-O-CH_3$

13 One mole of an organic compound A with the formula C_3H_8O reacts completely with two moles of HI to form X and Y. When Y is boiled with aqueous alkali it forms Z. Z answers the iodoform test. The compound A is **[KCET 2010]**
(a) propan-2-ol (b) propan-1-ol
(c) ethoxyethane (d) methoxyethane

14 Which of the following is an approximate set of reactants for the preparation of 1-methoxy-4-nitrobenzene?
A. p-$O_2N-C_6H_4-\overset{-}{O}\overset{+}{Na}$ + CH_3Br B. p-$O_2N-C_6H_4-Br$ + CH_3ONa **[AMU 2010]**
(a) Only A (b) Only B
(c) Both A and B (d) None of these

15 A simple method to remove peroxides from ethers is to treat them with an aqueous solution of **[JCECE 2010]**
(a) KI (b) KCNS (c) $Na_2S_2O_3$ (d) Br_2

ANSWERS

Check point 17.1

1. (b)	2. (a)	3. (c)	4. (b)	5. (d)	6. (b)	7. (c)	8. (a)

Check point 17.2

1. (b)	2. (d)	3. (a)	4. (b)	5. (c)	6. (a)	7. (c)	8. (c)	9. (d)	10. (c)

(A) Topical exercise

1. (c)	2. (c)	3. (b)	4. (a)	5. (c)	6. (d)	7. (c)	8. (c)	9. (a)	10. (b)
11. (b)	12. (c)	13. (a)	14. (d)	15. (d)	16. (c)	17. (a)	18. (d)	19. (c)	20. (d)
21. (a)	22. (b)	23. (c)	24. (d)	25. (a)	26. (b)	27. (c)	28. (a)	29. (a)	30. (b)
31. (a)	32. (a)								

(B) Miscellaneous questions

1. (d)	2. (b)	3. (a)	4. (c)	5. (b)	6. (a)	7. (d)	8. (c)	9. (a)	10. (c)
11. (d)									

(C) Medical entrances' gallery

1. (a)	2. (a)	3. (a)	4. (d)	5. (a)	6. (d)	7. (a)	8. (b)	9. (c)	10. (d)
11. (d)	12. (c)	13. (d)	14. (c)	15. (a)					

Hints & Explanations

(A) Topical exercise

1 (c) $\underset{3}{H_3C} - \underset{2}{CH}(\underset{1}{CH_3}) - OCH_3$

IUPAC name of the above compound is 2-methoxypropane and correct option is (c).

2 (c) An alkyl halide on heating with dry silver oxide gives ether.

$$\underset{(A)}{2R-X} + Ag_2O \xrightarrow[\text{dry}]{\Delta} \underset{\text{Ether}}{R-O-R} + 2AgX$$

3 (b) N_2 escapes easily.

4 (a) The compound '*P*' is $C_6H_5-OC_2H_5$

7 (c) Cyclohexyl$-CH{=}CH_2$ $\xrightarrow{(a)}$ 1-ethyl-1-methoxycyclohexane (CH_2CH_3, OCH_3 on the same ring carbon)

Since, intermediate is 3° carbocation by 1,2-hydride shift.

$\xrightarrow{(b)}$ Cyclohexyl$-CH_2CH_2OCH_3$

Anti-Markownikoff's addition

$\xrightarrow{(c)}$ Cyclohexyl$-CH(OCH_3)-CH_3$

Addition by Markownikoff's rule without arrangement

10 (b) Allylic bromide is a 1° alkyl halide and forms more stable resonance stabilised allyl carbocation.

11 (b) $\underset{\text{2-methyl propene}}{CH_3-C(CH_3){=}CH_2} \xrightarrow{H^+}$

$$CH_3-\overset{+}{C}(CH_3)-CH_2H \xrightarrow[-H^+]{H-O-CH_3} \underset{\text{tert-butylmethyl ether}}{CH_3-C(OCH_3)(CH_3)-CH_3}$$

15 (d) $(CH_3)_3C-Br + \underset{\text{Strong nucleophile}}{(CH_3)_3C-\overset{-}{O}\overset{+}{Na}} \longrightarrow \underset{\text{(Major)}}{CH_3-C(CH_3){=}CH_2}$

18 (d) I is correct.

II is incorrect since C–Br bond is most stable.

III and IV are incorrect since elimination takes place.

21 (a) $\underset{\text{Base}}{R_2O:} + \underset{\text{Acid}}{Mg(R)X} \longrightarrow R_2O \longrightarrow Mg(R)(X) \longleftarrow OR_2$

23 (c) $(CH_3CO)_2O + C_2H_5-O-C_2H_5 \xrightarrow{\text{Anhy. } AlCl_3} \underset{\text{Ethyl acetate}}{2CH_3COOC_2H_5}$

24 (d) Alcohols have stronger intermolecular H-bonding and ethers have weaker intermolecular H-bonding therefore ethers are more volatile than alcohols.

25 (a) 3-bromophenyl $OCH(CH_3)_2$ $\xrightarrow{HCl}$ $\underset{\text{Stable}}{\text{3-bromophenoxide } (O^{\ominus})} + \underset{\text{Stable}}{(CH_3)_2\overset{\oplus}{C}H}$

26 (b) $C_6H_5-CH_2 \mid O-CH_3 \xrightarrow{H^+} \underset{\text{Resonance stabilised carbocation}}{C_6H_5\overset{\oplus}{C}H_2} + CH_3O^{\ominus}$

Thus, products are $C_6H_5-CH_2I + CH_3OH$

27 (c) Due to the delocalisation of one pair of electrons of O-atom with phenyl ring in both side makes it very stable.

31 (a) $OH^- +$ epoxide (H, CH_3) $\longrightarrow$ $HOCH_2-C(OH)(H)(CH_3)$

Only isomer

(B) Miscellaneous questions

1 (d) Attacks least substituted site.

2 (b) Methylenecyclohexane ($=CH_2$) $\longrightarrow$ $\overset{\oplus}{C}$(ring)$-\overset{\ominus}{C}H_2$ $\xrightarrow{H^+}$ $\underset{\text{3° carbocation}}{\overset{\oplus}{C}(\text{ring})-CH_3}$ $\xrightarrow{CH_3O^{\ominus}}$ 1-methoxy-1-methylcyclohexane (CH_3, OCH_3)

4 (*c*) Cyclohexyl—$CH{=}CH_2 \xrightarrow{H^+}$ Cyclohexyl—$\overset{\oplus}{C}H—CH_3$

2° carbocation

↓ 1,2-hydride shift

Cyclohexyl cation ($\oplus$)—$CH_2—CH_3$

3° carbocation (most stable)

5 (*b*) If *RX* is 2° or 3°, *R*ONa will cause elimination. Thus (b) is incorrect statement.

11 (*d*) $ClCH_2\overset{14}{C}H—CH_2$ (epoxide, O) $+ CH_3CH_2O^{\ominus}$ attack at least substituted site

$Cl—CH_2—\overset{14}{C}H—CH_2OCH_2CH_3$ (with $—O^{\ominus}$) $\xrightarrow[\text{Williamson's reaction}]{\text{Intramolecular}}$ $CH_2—\overset{14}{C}HCH_2\,OCH_2CH_3$ (epoxide, O)

(C) Medical entrances' gallery

1 (*a*) Ethers are readily cleaved by heating in presence of halogen acids to form alcohol and an alkyl halide. In case of unsymmetrical ethers, halogen goes preferentially with smaller alkyl group or more stable carbocation.

If excess of acid is used then only alkyl halide is formed because alcohol formed reacts further with halogen acid to form corresponding alkyl halide.

$$CH_3CH_2CH_2OC(CH_3)_3 \xrightarrow[\Delta]{HI} CH_3CH_2CH_2OH + \underset{'D'}{(CH_3)_3CI}$$

$$\downarrow \Delta \text{ Excess HI}$$

$$\underset{'C'}{CH_3CH_2CH_2I}$$

2 (*a*) When phenol reacts with CH_3I in presence of NaOH, it gives methoxybenzene due to formation of phenoxide ion, which is a better nucleophile than phenol.

The reaction takes place as follows:

$$\underset{\text{Phenol}}{C_6H_5OH} + NaOH \longrightarrow \underset{\text{Sodium phenoxide}}{C_6H_5\overset{-}{O}\overset{+}{Na}} \xrightarrow[S_N2]{CH_3I} \underset{\text{Anisole}}{C_6H_5OCH_3}$$

Thus, given Assertion and Reason, both are correct and Reason is the correct explanation of the Assertion.

Hence, option (a) is the correct.

3 (*a*) Ethers are readily cleaved by halogen acids at 373 K to form an alcohol and an alkyl halide.

$$\underset{\textit{tert}\text{-butyl methylether}}{(CH_3)_3COCH_3} + HBr \xrightarrow{373\,K} \underset{\textit{tert}\text{-butyl bromide}}{(CH_3)_3CBr} + \underset{\text{Methanol}}{CH_3OH}$$

S_N1 mechanism is followed as $(CH_3)_3C^+$ is most stable carbocation and I^- is attached to it.

4 (*d*) According to given question and options, (*A*) must be C_2H_5OH, as it reacts with Na to give C_2H_5ONa. The reaction sequence is as follows :

(i) $\underset{\text{Ethanol }(A)}{C_2H_5OH} \xrightarrow{Na} \underset{\text{Sodium ethoxide }(B)}{C_2H_5\overset{-}{O}\overset{+}{Na}} \xrightarrow{PCl_5} \underset{\text{Ethyl chloride }(C)}{C_2H_5Cl}$

(ii) $C_2H_5\overset{-}{O}\overset{+}{Na} + C_2H_5Cl \xrightarrow{S_N2} \underset{\text{Diethyl ether}}{C_2H_5—O—C_2H_5} + NaCl$

The above reaction is known as Williamson's ether synthesis. It involves nucleophilic attack of alkoxide ion on alkyl halide according to S_N2 mechanism.

$$\underset{\substack{\text{Nucleophile}\\ \text{(Alkoxide ion)}}}{C_2H_5O^-} + \underset{\substack{\text{Substrate}\\ \text{(Alkyl halide)}}}{CH_3—CH_2Cl} \xrightarrow{\text{Slow}} C_2H_5—O\ldots\overset{+}{C}H_2\ldots Cl^- \,(\text{with } CH_3)$$

$$\xrightarrow{\text{Fast}} C_2H_5—O—C_2H_5 + Cl^-$$

5 (*a*) IUPAC name of the given compound is correct, i.e. 1-(2-propoxy) propane. These compounds are regarded as hydrocarbon derivative in which a hydrogen atom is replaced by —*O*R or —*O*Ar group [where, *R* = alkyl group and Ar = aryl group]. Therefore, both Assertion and Reason is correct and Reason is the correct explanation of the Assertion.

6 (*d*) $C_6H_5—O—CH_3 + HI \longrightarrow C_6H_5—OH + CH_3I$

7 (*a*) The reaction of alkyl halides with sodium alkoxide or sodium phenoxide to form ethers is called **Williamson synthesis**. Here, in this reaction alkyl halide should be primary and alkoxide, should be bulkier as shown below,

$$\underset{\text{Alkyl halide}}{R—X} + \underset{\text{Sodium alkoxide}}{R'—\overset{-}{O}\overset{+}{Na}} \longrightarrow \underset{\text{Ether}}{R—O—R'} + NaX$$

e.g., $CH_3—C(CH_3)_2—\overset{-}{O}\overset{+}{Na} + CH_3CH_2Cl \xrightarrow[-NaCl]{} CH_3—C(CH_3)_2—OCH_2CH_3$

8 (*b*) KI is added to ether to test its purity because ether generally has peroxide as impurity. On adding KI, ether peroxide, oxidises KI into I_2 and itself gets reduced to ether.

$$2I^- \longrightarrow I_2 + 2e^-$$

$$\text{Ether peroxide} + 2e^- \longrightarrow \text{Ether} + O_2$$

9 (*c*) According to Zeisel method, when an ether reacts with cold HI, alkyl iodide and alcohol are formed. In case of unsymmetrical ether, the alkyl halide is always formed from the smaller alkyl group provided no tertiary (3°) alkyl group is present and if any 3° alkyl group is present, the halogen gets attached with it.

In case of alkyl aryl ethers, the products are always phenol and an alkyl halide.

(a) $CH_3—O—C_4H_9 + HI \xrightarrow{\text{Cold}} \underset{\text{Methyl iodide}}{CH_3I} + \underset{\text{Butanol}}{C_4H_9OH}$
1-methoxybutane

(b) $\underset{\text{1-methoxy-2-methyl propane}}{CH_3—O—CH_2—CH(CH_3)—CH_3} + HI \xrightarrow{\text{Cold}} \underset{\text{Methyl iodide}}{CH_3I} + \underset{\text{2-methyl propanol}}{CH_3—CH(CH_3)—CH_2—OH}$

(c) $\underset{\text{2-methoxy-2-methyl propane}}{CH_3—C(OCH_3)(CH_3)—CH_3} + HI \xrightarrow{\text{Cold}} \underset{\text{Methanol}}{CH_3OH} + \underset{\text{tert-butyliodide}}{CH_3—C(CH_3)_2—I}$

Intermediate is 3° carbocation $CH_3—C^+(CH_3)_2$

(d) Methoxybenzene ($C_6H_5OCH_3$) + HI $\xrightarrow{\text{Cold}}$ Phenol (C_6H_5OH) + CH_3I (Methyliodide)

Intermediate is $C_6H_5O^{\ominus}$

10 (*d*) $R—X \xrightarrow{R\overset{-}{O}\overset{+}{Na}} R—O—R$

This reaction represents a Williamson synthesis. It involves nucleophilic substitution of the halide ion from the alkyl halide by the alkoxide ion *via* S_N2 mechanism.

11 (*d*) Allyl ethyl ether on reaction with HI produces ethanal and ethyl iodide by means of nucleophilic substitution reaction as shown below

$CH_2=CH—O—CH_2CH_3 \xrightarrow{H^+I^-} \underset{\text{Ethyl iodide}}{CH_3CH_2I} + \underset{\text{Ethanol}}{CH_3CHO}$

Intermediate is $CH_2=CHOH$. Which tautomerises to CH_3CHO (ethanal).

12 (*c*) The ether which form more stable carbocation gives CH_3OH as one of the products with hot concentrated HI. The order of stability of carbocation is 3° > 2° > 1°.

Thus, $CH_3—C(CH_3)_2—OCH_3$ gives CH_3OH as one of the product.

The reaction proceed as

$CH_3—C(CH_3)_2—O—CH_3 + H^+ \longrightarrow CH_3—C(CH_3)_2—\overset{+}{O}(H)—CH_3$

$\longrightarrow \underset{\text{3° carbocation}}{CH_3—C^+(CH_3)_2} + CH_3OH \xrightarrow{+I^-} \underset{\text{tert-butyliodide}}{CH_3—C(CH_3)_2—I} + \underset{\text{Methanol}}{CH_3OH}$

13 (*d*) Molecular formula $C_3H_8O(C_nH_{2n+2}O)$ suggests that the organic compound is either alcohol or ether. Since, the compound on reaction with HI gives two different compounds, it must be an unsymmetrical ether and its formula must be $CH_3OC_2H_5$ (methoxyethane).

The reactions are as follows

$\underset{\text{Methoxyethane}}{CH_3OC_2H_5} + 2HI \longrightarrow \underset{X}{CH_3I} + \underset{Y}{C_2H_5I} + H_2O$

$\underset{Y}{C_2H_5I} + \underset{\text{Aqueous}}{2NaOH} \longrightarrow \underset{Z}{C_2H_5OH} + NaI$

$\underset{(Z)}{C_2H_5OH} + \underset{\text{Aqueous}}{6NaOH + 3I_2} \longrightarrow \underset{\text{Iodoform}}{CHI_3} + HCO\overset{-}{O}\overset{+}{N}a + H_2O + 5NaI$

14 (*c*) When the alkyl halide is primary, alkoxide (or aryloxide) attacks on it and ether is obtained as a major product.

$p\text{-}NO_2C_6H_4\overset{-}{O}\overset{+}{Na} + CH_3Br \longrightarrow \underset{\text{1-methoxy-4-nitrobenzene}}{p\text{-}NO_2C_6H_4OCH_3}$

Aryl halides generally do not react with sodium alkoxide but presence of strong electron withdrawing group like $—NO_2$ at *ortho/para* position facilitate the reaction.

Thus, reactants of Set B also gives 1-methoxy-4-nitrobenzen.

$p\text{-}NO_2C_6H_4Br + CH_3\overset{-}{O}\overset{+}{Na} \longrightarrow p\text{-}NO_2C_6H_4OCH_3$

15 (*a*) Ether peroxide oxidises KI into I_2 and itself gets reduced to ether. Therefore, KI is added to remove peroxides from ethers.

$2I^- \longrightarrow I_2 + 2e^-$

Ether peroxide + $2e^- \longrightarrow$ Ether + O_2

CHAPTER 18

Aldehydes and Ketones

Class of organic compounds which contains $-\overset{O}{\overset{\|}{C}}-$ functional group are called **carbonyl compounds.** These compounds are classified into aldehydes and ketones.

In aldehydes, carbonyl carbon is attached to either two hydrogen atoms or one hydrogen atom and one alkyl (or aryl) group. In ketones, carbonyl carbon is attached to two alkyl (or aryl) groups.

$$\underset{\text{Aldehyde}}{Ar/R-\overset{O}{\overset{\|}{C}}-H} \qquad \underset{\text{Ketone}}{Ar/R-\overset{O}{\overset{\|}{C}}-R/Ar}$$

NOMENCLATURE

In IUPAC system, naming of aldehyde is done by replacing the terminal 'e' of the alkanes with suffix 'al' while ketones with suffix 'one'. Position of carbonyl carbon in aldehydes is always C-1 while in ketones, it gets lowest possible number.

e.g. $\underset{\text{Methanal}}{H-\overset{O}{\overset{\|}{C}}-H}$; $\underset{\text{Ethanal}}{CH_3-\overset{O}{\overset{\|}{C}}-H}$

$\underset{\text{Propan-2-one}}{CH_3-\overset{O}{\overset{\|}{C}}-CH_3}$; $\underset{\text{2 - methylpropanal}}{CH_3-\underset{CH_3}{\underset{|}{CH}}-CHO}$; $\underset{\text{Pentan-2-one}}{CH_3CH_2CH_2\overset{O}{\overset{\|}{C}}CH_3}$

ISOMERISM

Carbonyl compounds show following types of isomerism

(*i*) **Chain isomerism** Aldehydes those have at least four carbon atoms and ketones having at least five carbon atoms, show chain isomerism.

Inside

These isomers have same molecular formula but different structural formula.

e.g. C_4H_8O

$CH_3CH_2CH_2CHO$, Butan - 1 - al; $CH_3—CH(CH_3)—CHO$, 2 - methylpropanal

(*ii*) **Metamerism** In this isomerism, the difference lies in nature of alkyl groups attached to a polyvalent functional group. These isomers are called metamers.

$CH_3CH_2—C(=O)—CH_2CH_3$; $CH_3C(=O)—CH_2CH_2CH_3$

Pentan - 3 - one; Pentan - 2 - one

(*iii*) **Functional isomerism** There isomers have same molecular formula but different functional groups. It shows by both ketones and aldehyde.

e.g. $CH_3—C(=O)—CH_3$; $CH_3—CH_2—C(=O)—H$

Propan-2-one; Propanal

Note *If two molecules contain same functional group (polyvalent) never write chain or position isomerism it will always be metamerism.*

STRUCTURE

Carbonyl carbon atom in aldehydes and ketones are sp^2-hybridised. It is attached to three atoms which lie in same plane while the π-electron cloud lies above and below the plane. The bond angle formed between the bond pairs is about 120°.

120° 120° 120°

The double bond between carbon and oxygen is similar to alkene double bond but it is shorter and stronger than double bond of alkene (C=C).

The carbon-oxygen double bond is polar in nature because oxygen is more electronegative than carbon. Thus, π-electron cloud is displaced towards oxygen. Due to this, carbon bears partial positive charge while oxygen bears partially negative charge.

$$\overset{\delta+}{C}=\overset{\delta-}{O} \longrightarrow \overset{+}{C}—\overset{-}{O}$$

METHODS OF PREPARATION

Following methods are used for the preparation of both aldehydes and ketones

Oxidation of alcohols

Alcohols undergo controlled oxidation to give aldehydes and ketones. Reaction is carried out in presence of oxidising agents such as acidified $K_2Cr_2O_7$, alkaline $KMnO_4$, chromic anhydride and H_2SO_4.

Primary alcohols give aldehydes and secondary alcohols give ketones.

$$\underset{1° \text{ alcohol}}{R—CH_2—OH} \xrightarrow[H_2SO_4]{K_2Cr_2O_7} \underset{\text{Aldehyde}}{R—C(=O)—H} + H_2O$$

$$\underset{2° \text{ alcohol}}{R_2CH—OH} \xrightarrow[H_2SO_4]{K_2Cr_2O_7} \underset{\text{Ketone}}{R_2C=O} + H_2O$$

Carbonyl compounds formed in the above reaction is easily oxidised to carboxylic acids. So, to prevent further oxidation, carbonyl compound should be removed from reaction mixture.

Pyridinium dichromate (PDC) or pyridinium chlorochromate (PCC) in anhydrous media such as CH_2Cl_2 oxidises primary alcohols to aldehydes and secondary alcohols to ketones.

PCC : $[C_5H_5N^{\oplus}H]\ CrO_3Cl^{\ominus}$

$$\underset{1° \text{ alcohol}}{RCH_2OH} \xrightarrow[CH_2Cl_2]{\text{PDC or PCC}} \underset{\text{Aldehyde}}{RC(=O)H} \not\longrightarrow \text{No further reaction}$$

$$\underset{2° \text{ alcohol}}{RCH(OH)R'} \xrightarrow[CH_2Cl_2]{PCC} \underset{\text{Ketone}}{RC(=O)R'} + H_2O \xrightarrow{X} \text{No further reaction}$$

Some other examples of reagents that control the oxidation reaction at aldehyde stage are as follows

(*i*) **Sarett reagent** is a 1: 2 complex of chromium trioxide in pyridine as solvent (i.e. $CrO_3 \cdot 2C_5H_5N$).

(ii) **Collin's reagent** is modification of Sarett reagent. In this, CH_2Cl_2 is taken as a solvent.

Another reagent, aluminium *t*-butoxide convert secondary alcohols to ketones as follows

$$\underset{\text{2° alcohol}}{\overset{R}{\underset{R'}{>}}CHOH} \xrightarrow[\text{(ii) 2·Acetone}]{\text{(i) [(CH}_3)_3\text{CO]}_3\text{Al, reflex}} \underset{\text{Ketone}}{\overset{R}{\underset{R'}{>}}C=O}$$

This reagent also oxidises unsaturated secondary alcohols to ketone without affecting double bond.

$$\underset{\text{2° alcohol}}{\text{cyclohexenol (OH)}} \xrightarrow[\text{Unsaturated}]{[(CH_3)_3CO]_3Al} \underset{\text{Ketone}}{\text{cyclohexenone}}$$

Catalytic dehydrogenation

Primary and secondary alcohols react with copper at a temperature of 520-570K, to produce aldehydes and ketones respectively.

$$\underset{\text{1° alcohol}}{RCH_2OH} \xrightarrow[520-570K]{Cu,\Delta} \underset{\text{Aldehyde}}{RCHO} + H_2$$

$$\underset{\text{2° alcohol}}{\overset{R}{\underset{R}{>}}CH-OH} \xrightarrow[520K]{Cu,\Delta} \underset{\text{Ketone}}{\overset{R}{\underset{R}{>}}C=O} + H_2$$

On the other hand, 3° alcohols undergo dehydration to give alkenes as product.

$$\underset{\text{3° alcohol}}{H_3C-\overset{CH_3}{\underset{CH_3}{C}}-OH} \xrightarrow[520\ K]{Cu,\ \Delta} \underset{\text{Alkene}}{CH_3-\overset{CH_2}{\overset{\|}{\underset{CH_3}{C}}}} + H_2O$$

From unsaturated hydrocarbons

(i) **Ozonolysis** Alkenes react with ozone followed by subsequent hydrolysis to form aldehydes and ketones. In this reaction, ozonide is formed as an intermediate.

$$\underset{\text{2 methylbut-2-ene}}{(CH_3)_2C=CH-CH_3} \xrightarrow{O_3} \underset{\text{Ozonide}}{(CH_3)_2C(-O-O-)(-O-)CH-CH_3} \xrightarrow{Zn/H_2O} \underset{\text{Acetone}}{(CH_3)_2C=O} + \underset{\text{Ethanal}}{CH_3-CHO}$$

Easiest method to derive products by ozonolysis of alkenes is to

(a) break (C=C) bond

(b) place O-atoms at breaking point

$$\underset{\text{2, 3-dimethylbut-2-ene}}{(CH_3)_2C=C(CH_3)_2} \xrightarrow{O_3/H_2O/OH^-} \underset{\text{Product before O is placed}}{(CH_3)_2C + (CH_3)_2C} \xrightarrow{\text{Place O}} \underset{\text{Acetone}}{(CH_3)_2CO + (CH_3)_2CO}$$

From products alkene is identified. Place products of ozonolysis in such away that O-atoms are face to face. Replace O-atoms by (=) bond.

$$\underset{\text{Pentene}}{C_5H_{10}} \xrightarrow{O_3/H_2O} \underset{\text{Propan-2-one}}{(CH_3)_2CO} + \underset{\text{Ethanal}}{H-\overset{O}{\overset{\|}{C}}-CH_3}$$

$$\underset{\text{Pentene}}{C_5H_{10}} \longrightarrow \underset{\text{2-methylbut-2-ene}}{(CH_3)_2C=CHCH_3}$$

(ii) **Wacker's process** When alkenes are treated with aqueous $PdCl_2$ in presence of CO, catalyst $CuCl_2$, air or O_2, it forms carbonyl compounds. e.g.,

$$\underset{\text{Ethene}}{CH_2=CH_2} + \frac{1}{2}O_2 \xrightarrow[H_2O]{PdCl_2/CuCl_2} \underset{\text{Ethanal}}{CH_3\overset{O}{\overset{\|}{C}}H}$$

(iii) **By hydration of alkynes** Alkynes in presence of dilute H_2SO_4 and $HgSO_4$ undergo hydration to give aldehydes or ketones. Enol is formed as an intermediate which tautomerises to its keto form.

$$R-C\equiv C-H \xrightarrow[H_2O]{Hg^{2+},\ dil.H_2SO_4} \quad (R = \text{alkyl or aryl})$$

$$\underset{\text{Enol (not isolated)}}{\overset{R}{\underset{HO}{>}}C=C\overset{H}{\underset{H}{<}}} \quad H-\overset{+}{O}(H)-H \xrightarrow{\text{Tautomerises}} \underset{\text{Methyl ketone}}{R-\overset{O}{\overset{\|}{C}}-\overset{H}{\underset{H}{C}}-H}$$

$$CH \equiv CH \xrightarrow[HgSO_4]{H_2O/H^+} CH_3CHO$$
Ethyne → Ethanal

$$CH_3C \equiv CH \xrightarrow[HgSO_4]{H_2O/H^+} CH_3COCH_3$$
Propyne → Propanone

Hydrolysis of *gem*-dihalides

Gem-dihalides undergo hydrolysis in aqueous alkaline solution to give aldehydes and ketones.

$$CH_3CH_2CHCl_2 + 2NaOH \longrightarrow$$
1, 1-dichloropropane

$$CH_3CH_2CH(OH)_2 \xrightarrow[-H_2O]{} CH_3CH_2CHO$$
gem-diol → Propanal

$$CH_3CCl_2CH_3 + 2NaOH \longrightarrow CH_3-C(OH)_2-CH_3 \xrightarrow[-H_2O]{} CH_3\overset{O}{\overset{\|}{C}}CH_3$$
2, 2-dichloropropane → Acetone

Note *vic-dihalides* $CH_2Cl-CH_2Cl + 2NaOH \longrightarrow CH_2OH-CH_2OH$
Glycol

Glycol on oxidation gives two molecular of formaldehyde.

Hydroboration-oxidation

Hydroboration of an alkyne followed by oxidation yields aldehyde or ketone. Symmetrical non-terminal alkynes give a single ketone while unsymmetrical non-terminal alkynes give a mixture of both possible ketones. Terminal alkynes always yield an aldehyde.

$$CH_3C \equiv CCH_3 \xrightarrow{BH_3 \cdot THF} (CH_3CH = C(CH_3)-)_3B$$
But-2-yne
Non-terminal

$$\xrightarrow{H_2O_2,\ OH^-} CH_3CH = C(OH)CH_3 \underset{}{\overset{\text{Tautomerises}}{\rightleftharpoons}} CH_3CH_2\overset{O}{\overset{\|}{C}}CH_3$$
But-2-en-2-ol (Enol) → Butan-2-one

$$CH_3C \equiv CH \xrightarrow{BH_3 \cdot THF} (CH_3CH = C(H)-)_3B \xrightarrow[OH^-]{H_2O_2}$$
Propyne terminal

$$CH_3CH = CHOH \xrightarrow{\text{Tautomerises}} CH_3CH_2CHO$$
Prop-1-en-1-ol (Enol) → Propanal

By calcium salts of carboxylic acid

Carbonyl compounds are formed when mixture of calcium salts of carboxylic acid is heated.

$$(R'COO)_2Ca + (R''COO)_2Ca \xrightarrow{\Delta} 2R'-\overset{O}{\overset{\|}{C}}-R'' + 2CaCO_3$$

(*i*) When calcium salts of formic acid and any other monobasic acid is heated, mixture of carbonyl compounds are formed.

$$(RCOO)_2Ca + (HCOO)_2Ca \xrightarrow{\Delta}$$

$$2RCHO + 2CaCO_3 + RCOR + HCHO$$
Aldehyde, Ketone, Formaldehyde

(*ii*) When calcium salts of monobasic acid (other than formic acid) is heated, ketones are formed.

$$(RCOO)_2Ca \xrightarrow{\Delta} R_2CO + CaCO_3$$
Ketone

$$(HCOO)_2Ca \xrightarrow{\Delta} HCHO + CaCO_3$$
Salt of formic acid → Formaldehyde

(*iii*) When calcium salts of dibasic acid are heated, cyclic ketones are formed.

$$\left(\begin{array}{l} CH_2-\overset{O}{\overset{\|}{C}}-O^- \\ | \\ CH_2-\underset{O}{\underset{\|}{C}}-O^- \end{array}\right) Ca^{2+} \xrightarrow{\Delta} \begin{array}{l} CH_2 \\ | \\ CH_2 \end{array}\!\!>C=O$$
Calcium succinate → Cyclopropanone

$$\left(\begin{array}{l} CH_2-CH_2-\overset{O}{\overset{\|}{C}}-O^- \\ | \\ CH_2-CH_2-\underset{O}{\underset{\|}{C}}-O^- \end{array}\right) Ca^{2+} \xrightarrow{\Delta}$$
Calcium adipate

$$\begin{array}{l} CH_2-CH_2 \\ | \\ CH_2-CH_2 \end{array}\!\!>C=O$$
Cyclopentanone

Catalytic decomposition of carboxylic acids

Carbonyl compounds are also obtained by passing vapours of acid or mixture of acids over heated MnO as catalyst at 570 K.

$$\underset{\text{Carboxylic acid}}{2RCOOH} + MnO \xrightarrow[-H_2O]{\Delta} (RCOO)_2Mn \xrightarrow{\Delta} \underset{\text{Ketone}}{RCOR} + MnCO_3$$

$$MnCO_3 \xrightarrow{\Delta} MnO + CO_2$$

It is assumed, manganese salt of an acid is formed first which decomposes to form carbonyl compounds.

$$\underset{\text{Acetic acid}}{CH_3COOH} + \underset{\text{Formic acid}}{HCOOH} \xrightarrow[570\text{ K}]{MnO}$$

$$\underset{\substack{\text{Ethanal}\\\text{(Major)}}}{CH_3CHO} + \underset{\substack{\text{Methanal}\\\text{(Minor)}}}{HCHO} + \underset{\substack{\text{Propanone}\\\text{(Minor)}}}{CH_3COCH_3} + CO_2 + H_2O$$

$$\underset{\text{Acetic acid}}{2CH_3COOH} \xrightarrow[570\text{ K}]{MnO} \underset{\text{Acetone}}{CH_3COCH_3} + CO_2 + H_2O$$

From Grignard reagent

One equivalent of Grignard reagent reacts with nitriles and esters to produce carbonyl compounds.

(*i*) **Reaction with nitriles** Hydrogen cyanide is used to prepare aldehydes while all other nitriles are used to prepare ketones.

$$\underset{\text{Hydrogen cyanide}}{H—C\equiv N} \xrightarrow[\text{(ii) } H_3O^+]{\text{(i) } \triangleright\!—MgBr} \underset{\text{Aldehyde}}{\triangleright\!—CHO}$$

$$\underset{\text{Methyl cyanide}}{CH_3CN} \xrightarrow[\text{(ii) } H_3O^+]{\text{(i) } C_6H_{11}—MgBr} \underset{\text{Ketone}}{C_6H_{11}—\overset{O}{\overset{\|}{C}}CH_3}$$

Mechanism

$$\underset{\text{Grignard reagent}}{R'—Mg—X} + \underset{\text{Alkyl nitrile}}{R—C\equiv N:} \longrightarrow \begin{matrix}R'\\R\end{matrix}\!\!>\!C=N\!:\!\!—MgX$$

$$\xrightarrow{H^+} \underset{\text{Imine}}{\begin{matrix}R'\\R\end{matrix}\!\!>\!C=\overset{H}{\overset{|}{N}}:} \xrightarrow{H_3O^{\oplus}} \underset{\text{Ketone}}{\begin{matrix}R'\\R\end{matrix}\!\!>\!C=\ddot{O}:} + NH_4^{\oplus}$$

(*ii*) **Reaction with esters** Formic ester is used to prepare aldehyde while all other esters give ketones on treatment with one equivalent of Grignard reagent.

$$\underset{\text{Ethyl formate}}{H\overset{O}{\overset{\|}{C}}OC_2H_5} + \overset{\delta-\ \delta+}{RMgX} \longrightarrow HC(O^{\ominus}\overset{+}{Mg}X)(R)—OC_2H_5 \xrightarrow{H_3O^{\oplus}} \underset{\text{Aldehyde}}{R\overset{O}{\overset{\|}{C}}H} + C_2H_5O^{\ominus}$$

$$\underset{\text{Ethyl acetate}}{CH_3\overset{O}{\overset{\|}{C}}OC_2H_5} + \overset{\delta-\ \delta+}{RMgX} \longrightarrow CH_3C(O^{\ominus}\overset{+}{Mg}X)(R)—OC_2H_5 \xrightarrow{H_3O^{\oplus}} \underset{\text{Ketone}}{CH_3\overset{O}{\overset{\|}{C}}R} + C_2H_5O^{\ominus}$$

Preparation methods for aldehydes only

Reduction of acid chlorides and esters

Acid chlorides on reaction with H_2 in presence of Pd and $BaSO_4$ poisoned by traces of sulphur or quinoline (i.e. Lindlar's catalyst) produces aldehyde. This reaction is known as **Rosenmund reduction.**

e.g. $$\underset{\text{Acid chloride}}{R—\overset{O}{\overset{\|}{C}}—Cl} + H_2 \xrightarrow{Pd/BaSO_4,\text{ xylene}} \underset{\text{Aldehyde}}{R—\overset{O}{\overset{\|}{C}}—H} + HCl$$

Acid chlorides can also be reduced into aldehydes by using lithium tri-*t*-butoxy aluminium hydride $[LiAlH(O\text{-}t\text{-}C_4H_9)_3]$ or *tri-n*-butyl tinhydride $[Sn(C_4H_9)_3H]$.

$$\underset{\text{Acid chloride}}{R—\overset{O}{\overset{\|}{C}}—Cl} \xrightarrow[\text{(ii) } H_2O]{\text{(i) } LiAlH[OC(CH_3)_3]_3,\text{ diethyl ether, 195 K}} \underset{\text{Aldehyde}}{R—\overset{O}{\overset{\|}{C}}H}$$

In this reaction, reduction occurs by transfer of hydride ion from aluminium atom to carbonyl carbon of acyl chloride.

Esters can be reduced to aldehydes selectively using sodium aluminium hydride ($NaAlH_4$) or di-isobutyl aluminium hydride (DIBAL-H).

$$\underset{\text{Ester}}{RCOR'} \xrightarrow{NaAlH_4} \underset{\text{Aldehyde}}{RCHO} + \underset{\text{Alcohol}}{R'OH}$$

$$\underset{\text{Methyl propanoate}}{CH_3CH_2COCH_3} \xrightarrow[H_2O]{\text{DIBAL-H}} \underset{\text{Propanal}}{CH_3CH_2CHO} + \underset{\text{Methanol}}{CH_3OH}$$

Reduction of nitriles

(Stephen method)

Alkyl cyanides react with $SnCl_2$/HCl followed by hydrolysis to give corresponding aldehydes. This method is known as **Stephen reduction**.

$$\underset{\text{Alkyl cyanide}}{R—C\equiv N} \xrightarrow[SnCl_2/HCl]{\text{Reduction}} \begin{matrix} H \\ R \end{matrix}\!\!>C=N—H \xrightarrow{H_3O^+} \underset{\text{Aldehyde}}{R—\overset{O}{\overset{\|}{C}}—H}$$

DIBAL-H is also used to reduce nitriles into aldehydes.

$$\underset{\text{Alkyl cyanide}}{R—C\equiv N} \xrightarrow[230\text{ K}]{i\text{-Bu}_2AlH} \begin{matrix} H \\ R \end{matrix}\!\!>C=N—Al(i\text{-Bu})_2 \xrightarrow{H_3O^+} \underset{\text{Aldehyde}}{R—\overset{O}{\overset{\|}{C}}—H}$$

Oxo process

Alkenes react with water gas ($CO + H_2$) at a temperature of 370 K under pressure in presence of catalyst cobalt carbonyl hydride $[CoH(CO)_4]$ to give aldehyde.

This method is also known as **carbonylation** or **hydroformylation**.

$$>C=C< + CO + H_2 \xrightarrow[\text{[CoH(CO)}_4]]{370\text{ K, pressure}} —\overset{|}{\underset{|}{C}}—\overset{|}{\underset{|}{C}}—CHO$$

In case of unsymmetrical alkenes, mixture of isomeric straight chain and branched chain aldehydes are obtained as products.

$$\underset{\text{Propene}}{2MeCH=CH_2} + 2CO + 2H_2 \longrightarrow \underset{\text{Butanal}}{MeCH_2CH_2CHO} + \underset{\text{2-methylpropanal}}{(Me)_2CHCHO}$$

Preparation methods for ketones only

By cadmium and lithium salts

Cadmium and lithium salts are used to produce carbonyl compounds.

$$2RMgX + CdCl_2 \longrightarrow R—Cd—R + MgCl_2 + MgX_2$$

$$R—Cd—R + R'COCl \longrightarrow RCOR' + RCdCl$$

$$\underset{\text{Lithium acetate}}{C_6H_{11}—COO^- Li^+} \xrightarrow[\text{(ii) }H_3O^+]{\text{(i) }CH_3Li} C_6H_{11}—\overset{OH}{\underset{CH_3}{\overset{|}{\underset{|}{C}}}}—OH \xrightarrow{-H_2O} \underset{\text{Ketone}}{C_6H_{11}—\overset{O}{\overset{\|}{C}}CH_3}$$

$$\underset{\text{Acid chloride}}{C_6H_{11}—\overset{O}{\overset{\|}{C}}—Cl} + \underset{\substack{\text{Lithium}\\\text{dimethyl cuprate}\\\text{(Gilman reagent)}}}{(CH_3)_2CuLi} \xrightarrow[195\text{K}]{\text{Ether}} \underset{\text{Ketone}}{C_6H_{11}—\overset{O}{\overset{\|}{C}}CH_3}$$

Lithium dialkyl cuprate, a weak organometallic reagent is used to stop the reaction at ketone stage. It reacts faster with acid chlorides or esters.

CHECK POINT 18.1

1. High polarity of carbonyl group is explained by
(a) $>C=O$ (b) $>\overset{+}{C}-\overset{-}{O}$
(c) Both (a) and (b) (d) None of these

2. Write the correct IUPAC name for the following compound?
$$CH_3-CH_2-\overset{O}{\overset{\|}{C}}-CH_2CHO$$
(a) 3-oxopentanal (b) 3-oxopentanol
(c) 3-onepentanal (d) 3-onepentanol

3. Reagent which is used to stop the reaction at aldehyde stage
(a) Sarett reagent
(b) Collin's reagent
(c) pyridinium dichromate
(d) All of the above

4. Reagent used in oppenauer oxidation
(a) pyridinium chloro chromate
(b) sarett reagent is
(c) collin's reagent
(d) aluminium *t*-butoxide

5. Product of the following reaction
$$(CH_3)_3COH \xrightarrow[520K]{Cu,\ \Delta} 'X'$$
(a) $(CH_3)_3CHO$ (b) $(CH_3)_2C=CH_2$
(c) $(CH_3)_2C=O$ (d) None of these

6. The starting material in the synthesis of propanone and ethanal by ozonolysis is
(a) 2-methylbut-2-ene (b) 3-methylbut-1-ene
(c) but-2-ene (d) buta-1, 3-diene

7. Catalyst used in carbonylation reaction is
(a) $PdCl_2$ (b) Pd/C
(c) $Rh(PPh)_3$ (d) $[CoH(CO)_4]$

8. Which of the following name reaction is represented by the given equation?
$$CH_2=CH_2+\frac{1}{2}O_2 \xrightarrow[CuCl_2,\ H_2O]{PdCl_2} CH_3CHO$$
(a) Rosenmund reduction
(b) Oppenauer oxidation
(c) Wacker's process
(d) Ethinylation reaction

9. Ethyne + $H_2O \xrightarrow[H_2SO_4]{HgSO_4}$ '*A*'
Product '*A*' formed in the reaction is
(a) ethanoic acid
(b) ethanoyl chloride
(c) acetaldehyde
(d) benzaldehyde

10. What is the product formed during the distillation of calcium formate?
(a) Formaldehyde (b) Ethanoic acid
(c) Formic acid (d) Ethanal

PHYSICAL AND CHEMICAL PROPERTIES

Physical properties

(i) **Physical state** Methanal is a pungent smelling gas while ethanal is a volatile liquid. Rest of the aldehydes and ketones containing upto eleven carbon atoms are colourless liquids and higher members are solids.

(ii) **Solubility** Aldehydes and ketones are soluble in water. Carbonyl group readily forms H-bond with water.

$$\underset{R'}{\overset{R}{>}}\overset{\delta+}{C}=\overset{\delta-}{O}\ldots\overset{\delta+}{H}-\overset{\delta-}{O}-\overset{\delta+}{H}\ldots\overset{\delta-}{O}=\overset{\delta+}{C}\underset{R'}{\overset{R}{<}}$$

Solubility decreases with increase in length of alkyl chain. Therefore, carbonyl compounds upto four carbon atoms are soluble and higher members are insoluble.

Higher carbonyl compounds are soluble in organic solvents like ether, benzene, $CHCl_3$ etc.

(iii) **Boiling point** Boiling points of aldehydes and ketones are lower than alcohols due to absence of intermolecular hydrogen bonding but higher than non-polar compounds such as hydrocarbons due to dipole-dipole interaction. Ketones have slightly higher boiling point than aldehydes because of presence of two electron-donating groups around carbonyl carbon.

(iv) **Dipole moment** Carbonyl compounds have high dipole moment due to polar nature of carbonyl group. *ortho*-benzoquinones are less stable than *para*-benzoquinone because in *o*-isomer, the C=O dipoles are nearly aligned, thus, have a repulsive destabilising interaction, whereas, in *p*-isomer, dipoles are farther apart.

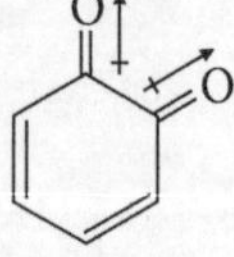
o-benzoquinone
(bond dipoles nearly aligned)

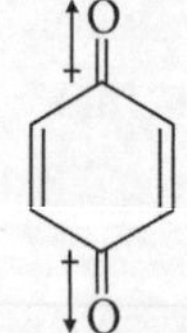
p-benzoquinone
(bond dipoles farther apart)

Chemical properties

Aldehydes and ketones undergo nucleophilic addition reaction due to polar nature of carbonyl carbon.

$$\rangle C{=}O + Nu^- \xrightarrow{\text{Slow}} \rangle C\langle^{O^-}_{Nu} \xrightarrow[\text{Fast}]{H^+} \rangle C\langle^{OH}_{Nu}$$

Also, protonation of carbonyl group increases electrophilic nature.

Aldehydes are more reactive than ketones due to following reasons

(*i*) Alkyl groups of ketone show *+I* effect, which decreases positive charge on the carbonyl carbon. Therefore, tendency of nucleophilic attack decreases.

(*ii*) More the number of alkyl (or aryl) groups, more will be the steric hinderance and due to this, tendency of nucleophilic attack decreases.

Thus, sequence of reactivity towards nucleophilic attack

$$H{-}CHO > CH_3{-}CHO > CH_3{-}\overset{O}{\overset{\|}{C}}{-}CH_3$$

Electron withdrawing substituents ($-CF_3$, $-CCl_3$) cause the carbonyl carbon to be more positive and thus the addition reaction is made more favourable.

Various nucleophilic addition reactions given by carbonyl compounds is discussed below

Addition of hydrogen cyanide

Aldehydes and ketones react with hydrogen cyanide to form cyanohydrins.

e.g.

$$\underset{\text{Acetone}}{(CH_3)_2C{=}O} + \underset{\text{Hydrogen cyanide}}{HCN} \underset{\text{Basic medium}}{\rightleftharpoons} \underset{\text{Cyanohydrin}}{(CH_3)_2C(OH)CN}$$

HCN is a toxic gas. So, it is obtained from reaction of NaCN and HCl.

Mechanism

$$\underset{\text{Ketone}}{R{-}\overset{O}{\overset{\|}{C}}{-}R} + \overset{\ominus}{C}{\equiv}N \longrightarrow R{-}\overset{O^{\ominus}}{\overset{|}{\underset{R}{\underset{|}{C}}}}{-}C{\equiv}N \xrightarrow{H^+} \underset{\text{Cyanohydrin}}{R{-}\overset{OH}{\overset{|}{\underset{R}{\underset{|}{C}}}}{-}CN}$$

It is a nucleophilic addition reaction and the addition of cyanide ion is a rate determining step.

In basic medium, cyano group is eliminated because oxygen atom have partial negative charge.

Thus, cyanohydrins are converted back into carbonyl compounds. While in acidic medium, cyanide group is hydrolysed to carboxylic group to give α-hydroxy carboxylic acid.

$$\underset{\text{Cyanohydrin}}{R_2C(OH)CN} \xrightarrow{H_3O^+} \underset{\alpha\text{-hydroxy carboxylic acid}}{R_2C(OH)COOH}$$

All aldehydes form cyanohydrin but in ketones, only acetone, butanone, pentan-3-one and pinacolone form cyanohydrin. The failure with bulky ketones is largely due to steric effects. Due to nucleophilic attack, hybridisation of carbon atom changes from sp^2 to sp^3 and thus, angle between alkyl groups decreases from 120° to 109.5° and thus, steric hinderance is increased.

Addition of sodium bisulphite

Carbonyl compounds react with sodium bisulphite to give white crystalline addition product, i.e. hydroxy sulphonic acid salt.

$$\underset{\text{Ketone}}{\rangle C{=}O} + \underset{\text{Sodium bisulphite}}{NaHSO_3} \longrightarrow \underset{\text{Bisulphite compound}}{\rangle C\langle^{OH}_{SO_3^-Na^+}}$$

The bisulphite compounds are insoluble in water. These compounds are converted back into carbonyl compounds on heating with acid or sodium carbonate solution. This reaction is used to distinguish carbonyl compounds with others which does not form bisulphite compounds.

$$\rangle C\langle^{OH}_{SO_3^-Na^+} \xrightarrow{H^+} SO_2 + H_2O + \rangle C{=}O + Na^+$$

$$\rangle C\langle^{OH}_{SO_3^-Na^+} \xrightarrow{OH^-} SO_3^{2-} + H_2O + \rangle C{=}O + Na^+$$

Acetophenone does not form bisulphite complex probably due to steric hindrance.

Only methyl ketones undergo addition reaction with $NaHSO_3$. Higher ketones (as acetophenone) do not react due to steric hindrance.

Addition of Grignard reagent

Carbonyl compounds react with Grignard reagent to give alcohols. Aldehydes form secondary alcohols [except formaldehyde which forms primary alcohol] and ketones form tertiary alcohols.

$$R{-}\overset{O}{\overset{\|}{C}}{-}H + R'{-}MgBr \xrightarrow[H_2O]{H^+} R{-}\underset{H}{\overset{OH}{C}}{-}R'$$

Aldehyde — Secondary alcohol

$$R{-}\overset{O}{\overset{\|}{C}}{-}R + R'{-}MgBr \xrightarrow[H_2O]{H^+} R{-}\underset{R}{\overset{OH}{C}}{-}R'$$

Ketone — Tertiary alcohol

Reaction of ketones with Grignard reagent gives 3° alcohol which is prone to E_1 elimination. Therefore, weak acid such as aqueous NH_4Cl is used for hydrolysis.

Addition of water

Aldehydes or ketones react with water to form hydrate. It is also called ***gem*-diols**. The reaction occurs through nucleophilic addition reaction and is catalysed by an acid. Here, water acts as a weak nucleophile.

$$R(H)C{=}O \text{ (Carbonyl compound)} + H_2O \overset{H^+}{\rightleftharpoons} R(R)HC(OH)_2 \text{ (Gem-diol)}$$

If H_2O is enriched with $\overset{18}{O}$, then product contains $\overset{18}{O}$ in carbonyl oxygen.

$$>C{=}O + H_2\overset{18}{O} \overset{H^+}{\rightleftharpoons} >C\overset{x}{-}OH,\ \overset{y}{-}\overset{18}{O}H$$

BE of $(C\overset{x}{-}O)$ bond (without $\overset{18}{O}$) is smaller than that of $(\overset{18}{C}\overset{y}{-}O)$ bond (enriched with $\overset{18}{O}$). Thus, in elimination H_2O is lost and $\overset{18}{O}$ goes with carbonyl carbon.

$$>C(O{-}H)(\overset{18}{O}{-}H) \longrightarrow >C{=}\overset{18}{O} + H_2O$$

Ninhydrin also exists as a hydrate

To minimise dipole-dipole repulsion between adjacent carbon atoms, central $(C{=}O)$ is hydrated.

(*i*) *Gem*-diols are unstable compound and eliminates water to form original compound.

(*ii*) CCl_3CHO (chloral) forms stable hydrate due to $-I$-effect (negative inductive effect) of chlorine and due to the formation of intramolecular hydrogen bonds.

$$Cl_3C{-}CH(OH)_2 \quad (Cl^{\delta-}\cdots H^{\delta+}{-}O)$$

(*iii*) Aldehydes form stable hydrates than ketones. With increase in number of alkyl groups, stability of hydrate decreases. Therefore, order of reactivity of carbonyl compounds toward hydrate formation is as follows.

$$HCHO > RCHO > RCOR$$

Addition of alcohols

Aldehydes react with alcohols to give addition product. When aldehydes react with one equivalent of an alcohol, '**hemiacetals**' are formed as product. The product reacts with second equivalent of alcohol to give 'acetals'. The reaction is catalysed by an acid as alcohol is a poor nucleophile.

$$\underset{\text{Ethanal}}{CH_3{-}\overset{O}{\overset{\|}{C}}{-}H} + CH_3OH \overset{H^+}{\rightleftharpoons} \underset{\text{Hemiacetal}}{CH_3{-}\underset{OCH_3}{\overset{OH}{C}}{-}H} \underset{H^+}{\overset{CH_3OH}{\rightleftharpoons}} \underset{\text{Acetal}}{CH_3{-}\underset{OCH_3}{\overset{OCH_3}{C}}{-}H} + H_2O$$

If ketone is used instead of aldehydes, hemiketal and ketals are formed as products.

$$\underset{\text{Acetone}}{CH_3{-}\overset{O}{\overset{\|}{C}}{-}CH_3} + CH_3OH \overset{H^+}{\rightleftharpoons} \underset{\text{Hemiketal}}{CH_3{-}\underset{OCH_3}{\overset{OH}{C}}{-}CH_3} \underset{H^+}{\overset{CH_3OH}{\rightleftharpoons}} \underset{\text{Ketal}}{CH_3{-}\underset{OCH_3}{\overset{OCH_3}{C}}{-}CH_3} + H_2O$$

This can be used to protect the carbonyl groups in a reaction, since acetal (or ketal) formation is a reversible reaction and can be converted back into carbonyl compounds in aqueous acidic medium.

$$\underset{\text{Ketal}}{CH_3{-}\underset{OCH_3}{\overset{OCH_3}{C}}{-}CH_3} + \underset{\text{(Excess)}}{H_2O} \overset{H^+}{\rightleftharpoons} \underset{\text{Acetone}}{CH_3{-}\overset{O}{\overset{\|}{C}}{-}CH_3} + \underset{\text{Alcohol}}{2CH_3OH}$$

Carbonyl compounds react with 1, 2-diols to form cyclic ketals (or acetals).

e.g. a ketone reacts with ethylene glycol (i.e. ethane-1, 2-diol) to form 5 membered cyclic ketal, i.e. glycol ketal.

$$R_2C{=}O + \underset{\text{Ethylene glycol}}{CH_2OH{-}CH_2OH} \overset{\text{HCl gas}}{\rightleftharpoons} \underset{\text{Glycol ketal}}{R_2C(O{-}CH_2)(O{-}CH_2)}$$

Ketone

Acetal is stable in the presence of alkali but are converted into aldehyde by acid. Thus, acetal formation may be used to protect the aldehyde against alkaline oxidising agents. In acid solution, aldehyde group can be protected by mercaptal formation.

$$\underset{\text{Ethanal}}{CH_3CHO} + 2C_2H_5SH \longrightarrow \underset{\text{Mercaptal}}{CH_3CH(SC_2H_5)_2}$$

Addition of alkynes

Carbonyl compounds react with terminal alkynes in presence of alkoxide ion to form alkynol (and alkyne diol with ethyne). This reaction is known as **ethinylation reaction.**

$$\underset{\text{Alkyne}}{RC{\equiv}CH} + CH_3O^{\ominus} \longrightarrow CH_3OH + RC{\equiv}C^{\ominus}$$

$$\underset{\text{Terminal alkyne}}{RC{\equiv}C^{\ominus}} + H{-}\underset{H}{C}{=}O \rightarrow H{-}\underset{C\equiv CR}{\overset{H}{C}}{-}O^{\ominus} \xrightarrow{H{-}OCH_3} \underset{\text{Alkynol}}{H{-}\underset{C\equiv CR}{\overset{H}{C}}{-}OH}$$

$$\underset{\text{Ethyne}}{HC{\equiv}CH} + 2H{-}\overset{H}{C}{=}O \xrightarrow{CH_3O^{\ominus}} \underset{\text{But-2-yne-1,4-diol}}{HOH_2CC{\equiv}CCH_2OH}$$

Nucleophilic addition-elimination reactions

When carbonyl compounds react with ammonia and its derivatives (primary amine), it undergoes nucleophilic addition-elimination reactions. This result in formation of compounds containing carbon-nitrogen double bond called **imines.**

These are formed due to elimination of water and also called **Schiff bases.**

$$\underset{\text{Ketone}}{>C{=}O} + NH_2G \xrightarrow{H^+} >C(OH)(NHG) \longrightarrow \underset{\text{Imine}}{>C{=}NG}$$

General mechanism of the reaction is shown below

$$>C{=}O + \ddot{N}H_2G \rightleftharpoons >C(O^-)(\overset{+}{N}H_2G) \overset{H^+}{\rightleftharpoons} >C(OH)(\overset{+}{N}H_2G) \overset{-H^+}{\rightleftharpoons} >C(OH)(NHG) \overset{+H^+}{\rightleftharpoons} >C(\overset{+}{O}H_2)(\ddot{N}HG) \rightleftharpoons >C{=}\overset{+}{N}HG \overset{-H^+}{\rightleftharpoons} \underset{\text{Imine}}{>C{=}\ddot{N}G}$$

Table 18.1 Some other nucleophilic addition elimination reactions

Group G	Reagent name	Carbonyl derivative
—H	NH_2—H Ammonia	$>C{=}NH$, $CH_3CH{=}NH$ Imine Aldimine $(H_3C)_2C{=}NH$ Ketimine
—OH	NH_2 OH Hydroxylamine	$>C{=}NOH$ Oxime
—R	NH_2 —R Aliphatic amine	$>C{=}NR$ Azomethine (Schiff base)
—NH_2	NH_2 NH_2 Hydrazine	$>C{=}N\cdot NH_2$ Hydrazone
—NHC_6H_5	NH_2 NHC_6H_5 Phenyl hydrazine	$>C{=}NNHC_6H_5$ Phenyl hydrazone
—$NHC_6H_3(NO_2)_2$	NH_2 $NHC_6H_3(NO_2)_2$ 2,4-dinitrophenyl hydrazine (2, 4-DNP)	$>C{=}NNHC_6H_3(NO_2)_2$ 2,4-dinitro phenylhydrazone
—$NHCONH_2$	NH_2 $NHCONH_2$ Semicarbazide	$>C{=}NNHCONH_2$ Semicarbazone

Note *2, 4-DNP derivatives are coloured solids. These are used for characterisation of carbonyl compounds.*

When HCHO reacts with ammonia, it forms hexamethylene tetramine, i.e. urotropine. It is used as an urinary antiseptic.

$$6HCHO + 4NH_3 \longrightarrow (CH_2)_6N_4 + 6H_2O$$

or

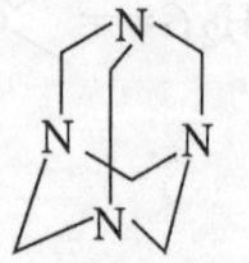

Urotropine

Addition of secondary amines

Carbonyl compounds react with secondary amines to produce **enamines**. It is an α, β-unsaturated tertiary amine i.e. an amino group is bonded to one of the carbons of a carbon-carbon double bond.

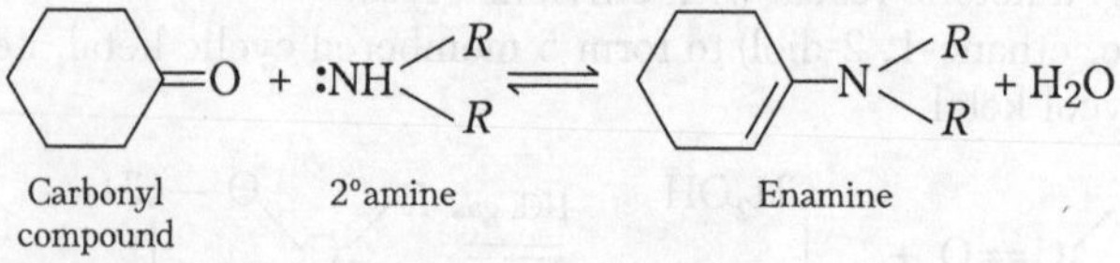

Carbonyl compound + 2°amine ⇌ Enamine + H_2O

In aqueous acidic solution, an enamine is hydrolysed back to carbonyl compound and 2° amine.

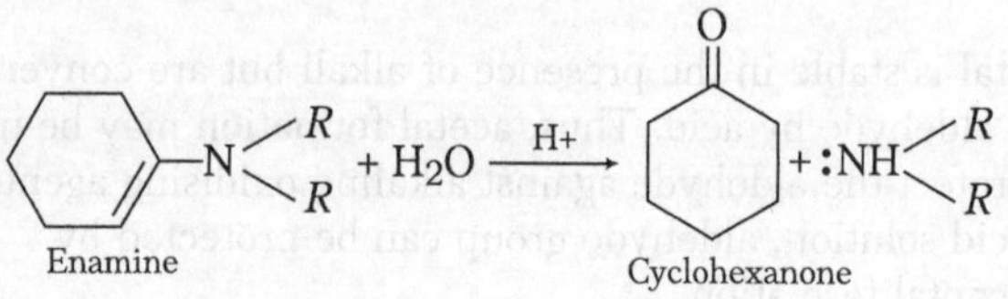

Enamine + H_2O $\xrightarrow{H+}$ Cyclohexanone + $:NHR_2$

CHECK POINT 18.2

1. Carbonyl compounds are soluble in water due to
(a) van der Waals' forces (b) dipole-dipole interaction
(c) hydrogen bonding (d) ion-dipole bonding

2. Ketones have higher boiling points than aldehydes because of
(a) high dipole moment
(b) intermolecular H-bonding
(c) presence of dipole-dipole interaction
(d) presence of two electron-donating groups

3. Order of reactivity towards nucleophilic addition
(a) $CH_3CHO > CH_3COCH_3 > HCHO$
(b) $CH_3COCH_3 > HCHO > CH_3CHO$
(c) $HCHO > CH_3CHO > CH_3COCH_3$
(d) $CH_3CHO > HCHO > CH_3COCH_3$

4. Consider the following reaction,

$$(CH_3)_2C{=}O + HCN \rightleftharpoons 'A'$$

Product of the given reaction
(a) 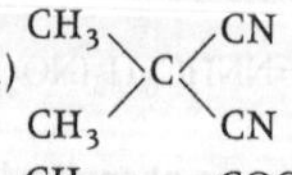 $(CH_3)_2C(CN)_2$ (b) 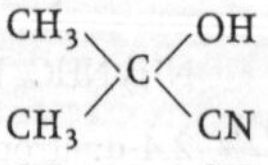 $(CH_3)_2C(OH)(CN)$
(c) 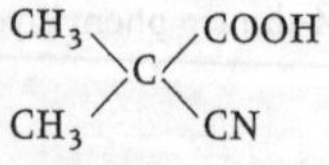 $(CH_3)_2C(COOH)(CN)$ (d) 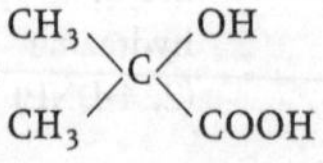 $(CH_3)_2C(OH)(COOH)$

5. Pentan-3-one on reaction with sodium bisulphite gives
(a) no product (b) hydroxy sulphonic acid
(c) sulphonic acid (d) bisulphite compound

6. Acetone reacts with one equivalent of methanol to give
(a) hemiacetal (b) acetal
(c) hemiketal (d) ketal

7. Addition of alcohols to carbonyl compounds is carried out in acidic medium because
(a) reaction is slow in alkaline medium
(b) reaction does not stop at carbonyl stage
(c) yield is poor
(d) alcohol is a poor nucleophile

8. Propanone on reaction with ethylene glycol yields
(a) glycol ketal (b) glycol acetal
(c) glycol hemiketal (d) glycol hemiacetal

9. Aqueous NH_4Cl is used for hydrolysis in reaction of ketones with Grignard reagent as
(a) reaction is slow
(b) alcohol is not formed as a product
(c) tertiary alcohol is formed
(d) 3° alcohol is prone to E1 elimination

10. Chloral hydrate is stable due to
(a) formation of *gem*-diol (b) intramolecular H-bonds
(c) Due to *I*-effect (d) Both (b) and (c)

Oxidation reactions

With strong oxidising agents

Aldehydes undergo oxidation reactions with potassium permanganate ($KMnO_4$) to form carboxylic acids.

$$\underset{\text{Aldehyde}}{R-\overset{\overset{\displaystyle O}{\|}}{C}-H} \xrightarrow[H_3O^+]{KMnO_4/NaOH} \underset{\text{Carboxylic acid}}{R-\overset{\overset{\displaystyle O}{\|}}{C}-OH}$$

Ketones do not undergo oxidation reactions readily. It involves breaking of carbon-carbon bond.

$$\underset{\text{Acetone}}{CH_3-\overset{\overset{\displaystyle O}{\|}}{C}-CH_3} \xrightarrow{[O]} \underset{\text{Methanoic acid}}{H-\overset{\overset{\displaystyle O}{\|}}{C}-OH} + \underset{\text{Ethanoic acid}}{CH_3-\overset{\overset{\displaystyle O}{\|}}{C}-OH}$$

Acetone is not affected by $Cr_2O_7^{2-}/H^+$ but oxidised by $KMnO_4$.

$$\text{(no reaction)} \times \xleftarrow{Cr_2O_7^{2-}} \underset{\text{Acetone}}{CH_3\overset{\overset{\displaystyle O}{\|}}{C}CH_3} \xrightarrow{KMnO_4} \underset{\text{Ethanoic acid}}{CH_3\overset{\overset{\displaystyle O}{\|}}{C}OH}$$

With mild oxidising agents

(i) **Tollen's reagent** Aldehydes reduce Tollen's reagent to Ag and appears in the form of silver mirror. This reaction is used to distinguish aldehydes from ketones and called **silver-mirror test**. It is given by all aldehydes and reducing sugars.

Tollen's reagent is an ammoniacal solution of silver nitrate in the form of $[Ag(NH_3)_2]^+ OH^-$.

$$\underset{\text{Aldehyde}}{RCHO} + \underset{\substack{\text{Tollen's reagent}\\\text{(Colourless solution)}}}{2[Ag(NH_3)_2]^+} + 3OH^- \xrightarrow{\Delta} RCOO^- + \underset{\substack{\text{Silver}\\\text{mirror}}}{2Ag\downarrow} + 4NH_3 + 2H_2O$$

Ketones do not reduce Tollen's reagent. (except α-hydroxy ketones such as fructose.)

(ii) **Fehling's solution** Aldehydes reduce Fehling's solution (alkaline solution of cupric ion complexed with tartarate ion). In this reaction, Cu^{2+} is reduced to Cu^+ (red precipitate).

$$\underset{\text{Aldehyde}}{RCHO} + \underset{\text{Fehling solution}}{2Cu^{2+}} + 3OH^- \xrightarrow{\Delta} RCOO^- + \underset{\text{Red ppt.}}{2Cu^+\downarrow} + 2H_2O$$

Ketones and aromatic aldehydes do not give this reaction.

(iii) **Benedict's solution** Aldehydes reduce benedict's solution (alkaline solution of cupric citrate) to Cu^+. Red precipitate of Cu_2O is formed.

$$\underset{\text{Aldehyde}}{RCHO} + 2Cu^{2+} + 5OH^- \longrightarrow RCOO^- + \underset{\text{Red ppt.}}{Cu_2O\downarrow} + 3H_2O$$

Oxidation with selenium oxide

Aldehydes and ketones with a methyl or methylene group adjacent to carbonyl group are oxidised by SeO_2.

e.g. $$\underset{\text{Ethanal}}{\overset{\alpha}{C}H_3CHO} + SeO_2 \longrightarrow \underset{\substack{|\\CHO}}{CHO} + H_2O + Se$$

$$\underset{\text{Acetone}}{CH_3COCH_3} + SeO_2 \longrightarrow \underset{\text{Methyl glyoxal}}{CH_3\overset{\overset{\displaystyle O}{\|}}{C}CHO} + H_2O + Se$$

Baeyer-villiger oxidation

In this method, ketones are oxidised by peroxy benzoic acid ($C_6H_5CO_3H$) or Caro's acid (H_2SO_5) to esters. Aldehydes are oxidised to acids.

$$\underset{\text{Ketone}}{RCOR'} + [O] \xrightarrow{H_2SO_5} \underset{\text{Ester}}{RCOOR'}$$

$$C_6H_5\overset{\overset{\displaystyle O}{\|}}{C}CH_3 \xrightarrow{C_6H_5CO_3H} C_6H_5-O-\overset{\overset{\displaystyle O}{\|}}{C}-CH_3 + C_6H_5-\overset{\overset{\displaystyle O}{\|}}{C}-O-CH_3$$

$$\underset{\text{Cyclopentanone}}{\text{cyclopentanone}} + [O] \xrightarrow{C_6H_5CO_3H} \text{(cyclic ester, six-membered lactone)}$$

Cyclopentanone

Note *Cyclic esters are reduced by* DIBAL-H*, to aldehydes and alcohol.*

$$\underset{\text{Ethanal}}{CH_3\overset{\overset{\displaystyle O}{\|}}{C}H} \xrightarrow{C_6H_5CO_3H} \underset{\text{Ethanoic acid}}{CH_3\overset{\overset{\displaystyle O}{\|}}{C}OH}$$

Oxygen is inserted between carbonyl carbon and the larger of two alkyl groups attached to it.

$$CH_3-\overset{\overset{\displaystyle O}{\|}}{C}-\boxed{CH_2CH_3} \xrightarrow{C_6H_5CO_3H} CH_3-\overset{\overset{\displaystyle O}{\|}}{C}-OCH_2CH_3$$

Larger alkyl group

Oxygen of peroxy acid behaves as a nucleophile which adds to carbonyl carbon of a ketone, followed by rearrangement which involves migration of an alkyl group of ketone to form ester.

Mechanism

$$R-\overset{O}{\overset{\|}{C}}-R' + CH_3-\overset{O}{\overset{\|}{C}}-O-\ddot{O}H \rightleftharpoons R-\overset{O^\ominus}{\overset{|}{C}}-R' \xrightarrow{-H^+} R-\overset{O^\ominus}{\overset{|}{C}}-R'$$

$$H-\overset{\oplus}{O}-OCOCH_3 \qquad O-OCOCH_3$$

$$\xrightarrow{+H^+} R'-\overset{O}{\overset{\|}{C}}-OR + CH_3COOH$$

Migratory aptitude of different groups is in the following order

$CH_3 <$ 1° alkyl < 2° alkyl ~ phenyl < cyclohexyl < 3° alkyl < H

Haloform reaction

Acetaldehydes and methyl ketones are oxidised by sodium hypohalite solution to give haloform. This is called **haloform reaction**. e.g.

$$\underset{\text{Acetone}}{R-\overset{O}{\overset{\|}{C}}-CH_3} \xrightarrow[ClO^-]{Cl_2,\ OH^-} R-\overset{O}{\overset{\|}{C}}-O^- + \underset{\text{Chloroform}}{CHCl_3}$$

Hypohalites are XO^-,
where, $X = Cl, Br, I$.

This reaction is used to distinguish between higher ketones and methyl ketones. This test is also given by CH_3CH_2OH, 2° alcohol (at C_2) as $CH_3-\underset{OH}{\underset{|}{CH}}-R$ because

NaOI first oxidises them to $CH_3-\overset{O}{\overset{\|}{C}}-H$ and $R-\overset{O}{\overset{\|}{C}}-CH_3$ respectively.

Reduction reactions

Carbonyl compounds are reduced to 1° or 2° alcohols by the use of following type of reducing agents.

1. Reduction with lithium aluminium hydride ($LiAlH_4$)

Hydride ion (H^-) present in $LiAlH_4$ is quite basic. Thus, it reacts vigorously with water, hence, should be used in dry solvents such as dry ether, THF etc.

$$\underset{\text{Aldehyde}}{CH_3CHO} \xrightarrow[H_3O^+]{LiAlH_4/\text{ether}} \underset{\text{Ethanol or Alcohol}}{CH_3CH_2OH}$$

$$\underset{\text{But-2-en-1-ol}}{CH_3CH=CHCHO} \xrightarrow[H_3O^+]{LiAlH_4/\text{ether}} \underset{\text{But-2-en-1-ol}}{CH_3CH=CHCH_2OH}$$

Aldehydes are reduced to 1° alcohols and ketones are reduced to 2° alcohols.

$$\text{Cyclopentanone} \xrightarrow[H^+]{LiAlH_4} \text{Cyclopentanol}$$

$LiAlH_4$ also reduces esters and acid chlorides to alcohols.

2. Reaction with $NaBH_4$

$NaBH_4$ is used for selective reduction. It reduces carbonyl group but doesn't reduce ester, acid chloride and double bonds. Therefore, it is regioselective in nature.

$$\underset{\text{Cyclohexanol}}{\text{C}_6\text{H}_{11}\text{–OH}} \xleftarrow{H_2,\ Ni} \text{cyclohex-3-en-1-one} \xrightarrow[\text{Ethanol}]{NaBH_4} \text{cyclohex-3-en-1-ol}$$

(HO-substituted ring with CH_2OH and $COOC_2H_5$) $\xleftarrow[\text{Ethanol}]{NaBH_4}$ (O= substituted ring with CHO and $COOC_2H_5$) $\xrightarrow{LiAlH_4}$ (HO-substituted ring with CH_2OH and CH_2OH) $+ C_2H_5OH$

Note *$LiAlH_4$ is better reducing agent than $NaBH_4$ but isolated double bonds cannot be reduced by $LiAlH_4$ as well as $NaBH_4$.*

3. Reduction to pinacol

Carbonyl compounds are reduced to **pinacols** (*cis*-1, 2-diol) in presence of Mg/benzene followed by hydrolysis.

$$\underset{\text{Acetone}}{2CH_3COCH_3} \xrightarrow{\text{Mg/benzene}} CH_3\overset{CH_3}{\overset{|}{C}}-\overset{CH_3}{\overset{|}{C}}CH_3 \text{ (O–Mg–O bridged)} \xrightarrow{H_2O/H^+} \underset{\text{Pinacol}}{CH_3\underset{OH}{\underset{|}{\overset{CH_3}{\overset{|}{C}}}}-\underset{OH}{\underset{|}{\overset{CH_3}{\overset{|}{C}}}}CH_3}$$

Note *In presence of conc. H_2SO_4, pinacol changes to pinacolone (A). It is called Pinacol-Pinacolone rearrangement.*

$$CH_3-C(CH_3)_2-CO-CH_3 \ (A)$$

4. **Meerwein-Ponndorf-Verley reduction**

Ketones, when treated with aluminium isopropoxide in isopropyl alcohol, are reduced to 2° alcohols.

$$\underset{\text{Ketone}}{R_2C{=}O} \xrightarrow[(CH_3)_2CHOH]{[(CH_3)_2CHO]_3Al} \underset{\text{2°alcohol}}{R_2CH-OH} + \underset{\text{Acetone}}{CH_3-CO-CH_3}$$

5. **Clemmensen reduction**

Carbonyl compounds are reduced to hydrocarbons in presence of amalgamated zinc (Zn—Hg) and conc. HCl.

e.g

$$C_6H_{11}-COCH_3 \xrightarrow{\text{Zn(Hg) + conc. HCl}} C_6H_{11}-CH_2CH_3$$

6. **Wolff-Kishner reduction**

Carbonyl compounds are treated with hydrazine (NH_2—NH_2) followed by reaction with strong base like KOH or alkaline glycol to give hydrocarbons.

$$\text{e.g;} \ \underset{\text{Ketone}}{>C{=}O} \xrightarrow{NH_2-NH_2} >C{=}N-NH_2 \xrightarrow{KOH} \underset{\text{Hydrocarbon}}{>CH_2}$$

$$\underset{\text{Cyclohexanone}}{C_6H_{10}{=}O} + NH_2-NH_2 \longrightarrow \underset{\text{Hydrazine}}{C_6H_{10}{=}N-NH_2} \xrightarrow[-N_2]{\text{Glycol, OH}^-} \underset{\text{Cyclohexane}}{C_6H_{12}}$$

Note *Clemmensen reduction is used for selective reduction of ketones which contain phenolic or carboxylic groups as are groups sensitive to base. While Wolff-Kishner reduction is preferred for compounds sensitive to acid.*

$$\underset{\text{3-benzoylpropanoic acid}}{Ph-CO-CH_2CH_2-CO_2H} \xrightarrow[\text{+ conc.HCl}]{\text{Zn(Hg)}} \underset{\text{4-phenylbutanoic acid}}{Ph-CH_2CH_2CH_2-CO_2H}$$

Reactions due to α-hydrogen (Condensation reactions)

Carbonyl compounds that have atleast one acidic hydrogen at α-carbon, undergo condensation reactions.

1. **Aldol condensation**

Two molecules of an aldehyde (or ketone) may combine in presence of dilute base to form β-hydroxy aldehyde (or β-hydroxy ketone).

Presence of α-hydrogen in atleast one type of molecules is necessity of reaction. This is known as **aldol reaction**.

$$\underset{\text{Aldheyde}}{CH_3-CH{=}O} + \underset{\alpha\text{-H}}{H}-\underset{\alpha\text{-carbon}}{CH_2}-CHO \xrightarrow{OH^-} \underset{\text{Aldol}}{CH_3-CH(OH)-CH_2-CHO}$$

On heating, aldol loses water and form α, β-unsaturated compounds. This is called aldol condensation.

$$\text{e.g.,} \ CH_3-\overset{\beta}{CH}(OH)-\overset{\alpha}{CH_2}-CHO \xrightarrow{\Delta} \underset{\substack{\text{Crotonaldehyde}\\ (\alpha,\beta\text{-unsaturated aldehyde})}}{CH_3-\overset{\beta}{CH}{=}\overset{\alpha}{CH}-CHO} + H_2O$$

Mechanism

$$\textit{Step I} \quad \overset{\alpha}{CH_2}(H)-CHO + OH^- \longrightarrow \underset{\text{Carbanion}}{\overset{\ominus}{C}H_2-CHO}$$

$$\textit{Step II} \quad CH_3-CHO + \overset{\ominus}{C}H_2-CHO \longrightarrow CH_3-CH(O^-)-CH_2-CHO$$

$$\textit{Step III} \quad CH_3-CH(O^-)-CH_2-CHO \xrightarrow[H_2O]{H^+} CH_3-CH(OH)-CH_2-CHO$$

$$\textit{Step IV} \quad CH_3-CH(OH)-CH_2-CHO \xrightarrow{\Delta} \underset{(\alpha,\ \beta\text{-unsaturated compound})}{CH_3-CH{=}CH-CHO}$$

2. **Crossed-aldol condensation**

If different carbonyl compounds having acidic hydrogen at α-position are condensed to give α, β-unsaturated compounds. Then, reaction is called Crossed-aldol condensation.

$$\underset{\text{Acetaldehyde}}{CH_3-\overset{H}{\overset{|}{C}}=O} + \underset{\text{Acetone}}{HCH_2-\overset{O}{\overset{||}{C}}-CH_3} \xrightarrow{OH^-} CH_3-\underset{OH}{\underset{|}{\overset{H}{\overset{|}{C}}}}-CH_2-\overset{O}{\overset{||}{C}}-CH_3$$

There are number of various products formed by cross-aldol condensation (as both the reactants have α-hydrogen).

$$CH_3CHO + CH_3CH_2CHO \xrightarrow[\Delta]{NaOH}$$

$$\underset{\substack{\text{But-2-enal} \\ \text{(From two molecules of ethanal)}}}{CH_3-CH=CH-CHO} + \underset{\substack{\text{2-methylpent-2-enal} \\ \text{(From two molecules of propanal)}}}{CH_3CH_2CH=\underset{CH_3}{\underset{|}{C}}CHO}$$

Simple or self aldol products

$$+ \underset{\text{2-methylbut-2-enal}}{CH_3-CH=\underset{CH_3}{\underset{|}{C}}-CHO} + \underset{\text{Pent-2-enal}}{CH_3CH_2CH=CHCHO}$$

(From one molecule of ethanal and one molecule of proponal)

Crossed aldol products

3. **Intramolecular aldol condensation**

If two or more carbonyl groups are present in the molecule, there can be intramolecular aldol reaction thus formation of rings is preferred as

six membered > five membered > four membered

e.g.,

$$\text{Nona-2, 8-dione} \xrightarrow[\Delta]{H^+} \text{(1-acetyl-2-methylcyclohexene)}$$

Cannizzaro's reaction

Aldehydes which do not have hydrogen at α-carbon, undergo Cannizzaro's reaction in presence of 50% aqueous or ethanolic alkali solution.

In this reaction, one molecule is reduced to alcohol and other is oxidised to salt of corresponding acid.

e.g. $$2HCHO + \underset{(50\%)}{NaOH} \longrightarrow CH_3OH + HCO\overset{-}{O}\overset{+}{Na}$$

(Reduction: HCHO → CH_3OH; Oxidation: HCHO → $HCOONa$)

$$2(CH_3)_3CCHO + NaOH \xrightarrow{50\%} (CH_3)_3CCH_2OH + (CH_3)_3CCOO\overset{-}{N}\overset{+}{a}$$

Mechanism

Step I $R-\underset{H}{\underset{|}{C}}=\ddot{O} \rightleftharpoons R-\underset{H}{\underset{|}{\overset{\oplus}{C}}}-\overset{\ominus}{O}$

This alkyl/aryl has no H at α-carbon wrt —CHO

Step II $R-\underset{H}{\underset{|}{\overset{\oplus}{C}}}-\overset{\ominus}{O} + OH^- \longrightarrow R-\underset{H}{\underset{|}{\overset{OH}{\overset{|}{C}}}}-\overset{\ominus}{O}$

(First molecule)

Step III $R-\underset{H}{\underset{|}{\overset{OH}{\overset{|}{C}}}}-\overset{\ominus}{O} \longrightarrow R-\overset{OH}{\overset{|}{C}}=O + \underset{\text{Hydride}}{H^-}$

Step IV $R-\overset{OH}{\overset{|}{C}}=O \underset{}{\overset{OH^-}{\rightleftharpoons}} R-\overset{O^-}{\overset{|}{C}}=O + \underset{\text{Proton}}{H^+}$

Step V $R-\overset{H}{\overset{|}{C}}=\ddot{O} \rightleftharpoons R-\underset{\oplus}{\underset{|}{\overset{H}{\overset{|}{C}}}}-\overset{\ominus}{O}$

(Second molecule)

Step VI $R-\underset{\oplus}{\underset{|}{\overset{H}{\overset{|}{C}}}}-O^{\ominus} + H^+ + H^- \longrightarrow RCH_2OH$

$$R-\underset{H}{\underset{|}{\overset{OH}{\overset{|}{C}}}}-\overset{\ominus}{O} + R-\overset{OH}{\overset{|}{C}}=\ddot{O} \xrightarrow{Slow} R-\overset{OH}{\overset{|}{C}}=O + R-\underset{H}{\underset{|}{\overset{OH}{\overset{|}{C}}}}-\overset{\ominus}{O} \longrightarrow RCOO^- + RCH_2OH$$

- Cannizzaro's reaction is a proton-hydride transfer reaction. Since, same molecule is oxidised and reduced, it is a disproportionation reaction.

When the reaction is carried out in D_2O instead of in H_2O, it is found that there is not formation of new C—D bonds. This indicates that the hydrogen must come from the aldehyde and not from the solvent.

If one of the aldehydes is formaldehyde (i.e. HCHO) or better reducing aldehyde in a mixture of aldehydes with no H at α-carbon, then reaction is called

Crossed-Cannizzaro's reaction. In such reactions, it is formaldehyde which is always oxidised.

$$\underset{\text{Formaldehyde}}{HCHO} + \underset{\text{Benzaldehyde}}{C_6H_5CHO} \xrightarrow{OH^-} HCOO^- + C_6H_5CH_2OH$$

In this case HCHO is a better reducing agent that C_6H_5CHO thus HCHO is oxidised and C_6H_5CHO is reduced.

Intramolecular Cannizzaro's reaction

Reaction in which half part is oxidised and other half part is reduced (of the same molecule). These are also called **internal Cannizzaro's reaction**. Generally, α-keto aldehydes or dialdehydes undergoes this types of reaction.

$$\begin{matrix} CHO \\ | \\ CHO \end{matrix} \xrightarrow{\text{Conc.NaOH}} \begin{matrix} CH_2OH \\ | \\ COO^-Na^+ \end{matrix}$$

(Reduction; Oxidation)

Reformatsky reaction

Aldehydes or ketones are treated with α-bromoester in presence of zinc to form β-hydroxy ester followed by dehydration to form α, β-unsaturated ester.

$$\underset{\text{Aldehyde}}{C_2H_5\overset{O}{\overset{\|}{C}}H} + \underset{\alpha\text{-bromo ester}}{Br\overset{CH_3}{\overset{|}{C}H}-\overset{O}{\overset{\|}{C}}OEt} \xrightarrow[\text{(ii) } H_3O^+]{\text{(i) Zn/benzene}}$$

$$C_2H_5\overset{OH}{\overset{|}{C}H}-\overset{CH_3}{\overset{|}{C}H}-\overset{O}{\overset{\|}{C}}OEt \xrightarrow[-H_2O]{\Delta} \underset{\alpha,\beta\text{-unsaturated ester}}{C_2H_5C=\overset{CH_3}{\underset{H}{C}}-COOH}$$

Organo zinc reagent is used because it is less reactive than Grignard reagent and prevents second nucleophilic addition to the ester group.

Mechanism

$$\underset{\text{Carbonyl compound}}{>C=O} + \underset{\alpha\text{-bromo ester}}{Br-\underset{H}{\overset{|}{C}}-\overset{O}{\overset{\|}{C}}OR} \xrightarrow[\text{Benzene}]{Zn} -\overset{BrZnO}{\overset{|}{C}}-\underset{H}{\overset{|}{C}}-\overset{O}{\overset{\|}{C}}OR$$

$$\xrightarrow{H_3O^+} \underset{\beta\text{-hydroxy ester}}{-\overset{HO}{\overset{|\beta}{C}}-\underset{H}{\overset{|\alpha}{C}}-\overset{O}{\overset{\|}{C}}OR} \xrightarrow[-H_2O]{\Delta} \underset{\alpha,\beta\text{-unsaturated ester}}{>C=C-\overset{O}{\overset{\|}{C}}OR}$$

Tischenko reaction

All aldehydes (with or without α-hydrogen) in presence of aluminium ethoxide $[Al(OC_2H_5)_3]$ can be simultaneously oxidised (to acid) and reduced (to alcohol) to form ester. This reaction is known as Tischenko reaction. It is similar to Cannizzaro's reaction.

$$\underset{\text{Aldehyde}}{RCHO + RCHO} \xrightarrow{Al(OC_2H_5)_3} RCOOH + RCH_2OH \longrightarrow \underset{\text{Ester}}{RCOOCH_2R}$$

Wittig reaction

The reaction of triphenyl phosphonium (phosphorus ylide) with carbonyl compounds yield alkenes. This reaction is known as Wittig reaction.

An ylide is a compound which has opposite charges on adjacent atoms with complete octet.

$$\underset{\text{Carbonyl compound}}{R_2C=O} + \underset{\text{Phosphorus ylide}}{Ph_3\overset{\oplus}{P}-\overset{\ominus}{\ddot{C}}H_2} \longrightarrow \underset{\text{Alkene}}{R_2C=CH_2} + Ph_3P=O$$

Mechanism

$$R_2C=O + :\overset{-}{C}H_2-\overset{+}{P}(C_6H_5)_3 \longrightarrow \underset{\text{Betaine (Dipolar intermediate)}}{R_2\overset{O^-\cdots\overset{+}{P}(C_6H_5)_3}{C\cdots CH_2}} \longrightarrow R_2C=CH_2 + O=P-(C_6H_5)_3$$

Note *If geometrical isomerism is possible, mixture of cis-and trans-isomers are formed.*

Polymerisation

Aldehyde undergo polymerisation reactions to give polymers.

e.g. $$\underset{\text{Aldehyde}}{nHCHO} \underset{\Delta}{\overset{\text{Evaporate to dryness}}{\rightleftharpoons}} \underset{\substack{para\text{-formaldehyde} \\ \text{(a white crystalline solid)} \\ (n = 6 \text{ to } 50)}}{(CH_2O)_n \cdot H_2O}$$

$$nHCHO \underset{\Delta}{\overset{\text{Conc. } H_2SO_4}{\rightleftharpoons}} \underset{\substack{\text{Polyoxymethylene}\\ \text{(white solid)}\\ (n > 100)}}{(CH_2O)_n\ Z}$$

$$3HCHO(g) \underset{\Delta}{\overset{\text{Room temp.}}{\rightleftharpoons}} \underset{\substack{\text{Trioxymethylene}\\ (\textit{meta}\text{-formaldehyde, trioxane})}}{(CH_2O)_3}$$

Trioxane exists as a cyclic compound without aldehyde group.

CH_2, O, O, H_2C, CH_2, O (ring)

Trioxane

The synthesis of *para* aldehyde and *meta* aldehyde can also take place from acetaldehyde.

$$3CH_3CHO \underset{\text{Dil. } H_2SO_4\text{, distill.}}{\overset{\text{Conc. } H_2SO_4}{\rightleftharpoons}} \text{(cyclic trimer: } CH_3\text{—}CH\text{, }CH_3\text{—}CH\text{, }CH\text{—}CH_3 \text{ joined by three O)}$$

Para aldehyde

$$4CH_3CHO \underset{\text{dil. } H_2SO_4\text{, distil.}}{\overset{\text{Conc. } H_2SO_4\text{, 273 K}}{\rightleftharpoons}} \begin{array}{c} CH_3\text{—}CH\text{—}O\text{—}CH\text{—}CH_3 \\ | \qquad\qquad | \\ O \qquad\qquad O \\ | \qquad\qquad | \\ CH_3\text{—}CH\text{—}O\text{—}CH\text{—}CH_3 \end{array}$$

meta-aldehyde

Both *para* aldehyde and *meta* aldehyde give ethanal on distillation with dilute H_2SO_4.

- Formaldehyde on reaction with barium hydroxide gives hexose. Formation of hexose is probably through a chain of aldol condensation.

$$6HCHO \xrightarrow{Ba(OH)_2} \begin{array}{c} CHO \\ | \\ (CHOH)_4 \\ | \\ CH_2OH \end{array}$$

Hexose

- Acetone on heating with conc. H_2SO_4 forms mesitylene (1, 3, 5- trimethylene benzene).

$$2CH_3COCH_3 \xrightarrow[\Delta]{\text{Conc. } H_2SO_4} \text{Mesitylene (benzene ring with } CH_3\text{, } H_3C\text{, } CH_3\text{)}$$

Mesitylene

CHECK POINT 18.3

1. When acetaldehyde is treated with Fehling's solution, it gives precipitate of
(a) Cu (b) CuO (c) Cu_2O (d) Cu, Cu_2O and CuO

2. Aldehyde reduces Benedict's solution to
(a) $RCOO^-$ (b) Cu^+ (c) Cu^{2+} (d) Cu

3. Acetaldehyde reacts with selenium oxide to give
(a) ethanoic acid (b) glycolic acid
(c) glyoxal (d) ethylene glycol

4. Consider the following reaction (cyclohexanone) $\xrightarrow[\Delta]{C_6H_5CO_3H}$
Compound '*X*' is

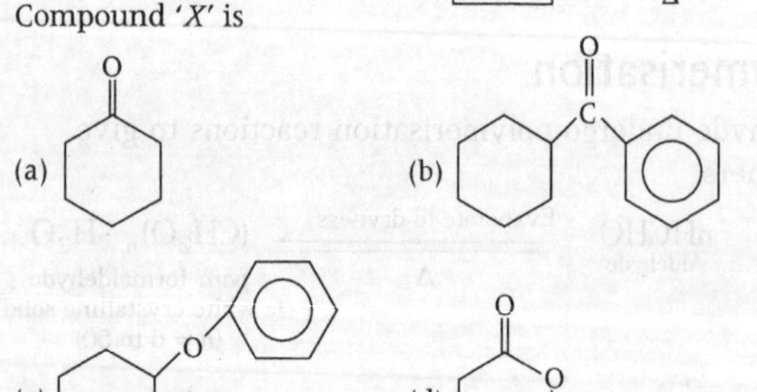

5. A compound that gives a positive iodoform test is
(a) pentanol
(b) pentan-3-one
(c) pentan-2-one
(d) pentanal

6. The compound that will not give iodoform on treatment with alkali and iodine is
(a) acetone (b) ethanol
(c) diethyl ketone (d) isopropyl alcohol

7. Following reaction is carried out in presence of which of the following reagent ?
(a) $NaBH_4$ (b) Ni, H_2
(c) $LiAlH_4$ (d) Aluminium *t*-butoxide

8. Propanone on reaction with magnesium in benzene give
(a) pinacolone (b) pinacols
(c) alcohols (d) *gem*-diols

9. To carry out Clemmensen reduction, reagent used is
(a) $NH_2—NH_2$ (b) $NaBH_4$
(c) Zn(Hg) / HCl (d) H_2 / Ni

10. Which of the following will not undergo aldol condensation ?
(a) Ethanal (b) Hexanal
(c) Phenyl acetaldehyde (d) Chloral

AROMATIC CARBONYL COMPOUNDS

Compounds in which aldehyde or ketone group are directly attached to an aromatic ring, are called aromatic carbonyl compounds.

e.g.

Benzaldehyde (CHO) Acetophenone ($COCH_3$) Benzophenone

Preparation of aromatic carbonyl compounds

Oxidation of toluene

Strong oxidising agents oxidise toluene and its derivatives to benzoic acid. However, to stop the reaction at aldehyde stage following methods are used for this purpose.

(*i*) Toluene undergoes oxidation reaction with CrO_2Cl_2 (mild oxidising agent) in presence of CS_2 to give benzaldehyde. This reaction is known as **Etard's reaction.**

$$\text{Toulene } (C_6H_5CH_3) \xrightarrow[CS_2]{CrO_2Cl_2} \underset{\text{Chromium complex}}{C_6H_5CH(OCrCl_2OH)_2} \xrightarrow{H_2O/H^+} \underset{\text{Benzaldehyde}}{C_6H_5CHO}$$

(ii) Toluene also reacts with chromic oxide in acetic anhydride to give benzaldehyde.

$$\underset{\text{Toluene}}{C_6H_5CH_3} \xrightarrow[\text{Acetic anhydride, 273-283 K}]{CrO_3} C_6H_5CH(OCOCH_3)_2 \xrightarrow{H_2O/H^+} \underset{\text{Benzaldehyde}}{C_6H_5CHO}$$

Chlorination of toluene in side chain followed by hydrolysis

Side chain chlorination of toluene gives benzal chloride, which on hydrolysis gives benzaldehyde. This is the commercial method for the manufacture of benzaldehyde.

$$\underset{\text{Toluene}}{C_6H_5CH_3} \xrightarrow{Cl_2, \Delta} \underset{\text{Benzal chloride}}{C_6H_5CHCl_2} \xrightarrow[373\,K]{H_2O} \underset{\text{Benzaldehyde}}{C_6H_5CHO}$$

Reimer-Tiemann reaction

Phenol reacts with chloroform ($CHCl_3$) in aqueous alkali solution to give *o*-and *p*-hydroxy benzaldehyde.

$$\underset{\text{Phenol}}{C_6H_5OH} + CHCl_3 + 3KOH(aq) \xrightarrow{\Delta} \underset{\substack{\text{(Minor)}\\ p\text{-hydroxy benzaldehyde}}}{p\text{-}HOC_6H_4CHO} + \underset{\substack{\text{(Major)}\\ o\text{-hydroxy benzaldehyde}}}{o\text{-}HOC_6H_4CHO} + 3KCl + 2H_2O$$

o-and *p*-isomers are separated by steam distillation.

Gattermann-Koch reaction

Benzene and its derivative are converted into substituted benzaldehyde, when treated with CO and HCl in presence of anhydrous $AlCl_3$.

$$\underset{\text{Benzene}}{C_6H_6} + CO + HCl \xrightarrow{\text{Anhyd.}AlCl_3/CuCl} \underset{\text{Benzaldehyde}}{C_6H_5CHO} + HCl$$

H ← This goes with Cl; HCl ↑ This goes with (–CHO)

HCl (reactant) and HCl (product) are different. H of HCl (reactant) and CO appears in the form of –CHO in the benzene nucleus. It is confirmed by the fact that using isotopic DCl instead of HCl, would form –CDO and not –CHO.

This reaction succeeds only with benzene and activated benzene derivative.

$$\underset{\text{Benzene}}{C_6H_6} + CO + DCl \xrightarrow{\text{Anhyd.}AlCl_3/CuCl} C_6H_5CDO + HCl$$

H ← this goes with Cl; DCl ← (this goes with –CDO)

Gattermann–aldehyde synthesis

Benzene reacts with HCN and HCl in presence of anhydrous $AlCl_3$ to give benzaldehyde. In this reaction, benzaldimine is formed as an intermediate.

$$C_6H_6 + HCN + HCl \xrightarrow{\text{Anhyd. } AlCl_3} C_6H_5CH{=}NH \xrightarrow{H_2O} C_6H_5CHO + NH_4Cl$$

Benzene, Benzaldimine, Benzaldehyde

Friedel-Crafts acylation

When aromatic hydrocarbon reacts with acid halide in presence of anhydrous $AlCl_3$ as Lewis acid, then aromatic ketones are formed.

$$C_6H_5COC_6H_5 \xleftarrow[\text{Anhyd.}AlCl_3]{C_6H_5COCl} C_6H_6 \xrightarrow[\text{Anhyd.}AlCl_3]{CH_3COCl} C_6H_5COCH_3$$

Benzophenone, Benzene, Acetophenone

Benzene $\xrightarrow{(CH_3CO)_2O,\ AlCl_3}$ Acetophenone

This is the widely used method for the preparation of aromatic ketones.

By Grignard reagent

Phenyl magnesium bromide (Grignard reagent) reacts with ester to produce carbonyl compounds.

$$C_6H_5COCH_3 \xleftarrow{H_3CCOC_2H_5 \ (\text{C=O})} C_6H_5MgBr \xrightarrow{HCOC_2H_5 \ (\text{C=O})} C_6H_5CHO$$

Acetophenone, Benzaldehyde

By diazonium salts

When aromatic diazonium salt react with formaldoxime, it results in the formation of benzaldoxime which on hydrolysis gives aromatic aldehyde.

$$C_6H_5\overset{+}{N}{\equiv}NCl^- + CH_2{=}NOH \longrightarrow C_6H_5CH{=}NOH \xrightarrow{H_2O} C_6H_5CHO$$

Diazonium salt, Formaldoxime, Benzaldoxime, Benzaldehyde

By oxidation of benzyl chloride

Oxidation of benzyl chloride in presence of $Pb(NO_3)_2$ yields benzaldehyde.

$$2C_6H_5CH_2Cl \xrightarrow{Pb(NO_3)_2} C_6H_5CH_2{-}O{-}\overset{\oplus}{N}(=O){-}O^{\ominus} \longrightarrow 2C_6H_5CHO + HNO_2$$

Benzyl chloride, Benzaldehyde

By hydrolysis of *gem*-dihalides

Aromatic *gem*-dihalides on hydrolysis yield benzaldehyde.

$$C_6H_5CHCl_2 + 2NaOH \longrightarrow C_6H_5CHO$$

$$C_6H_5C(Cl_2)CH_3 + 2NaOH \longrightarrow C_6H_5COCH_3$$

Gem-dihalides, Acetophenone

PHYSICAL PROPERTIES

Benzaldehyde is a colourless liquid with bitter almond like odour. It is soluble in organic solvents like ethanol, diethyl ether etc. It is steam volatile. It's boiling point is 179°C. While $C_6H_5COCH_3$ is a colourless liquid, sparingly soluble in water but dissolves readily in ethanol and diethyl ether. Its boiling point is 202°C.

CHEMICAL PROPERTIES

Aromatic aldehydes and ketones are less reactive than aliphatic carbonyl compounds. Due to $-R$ effect of carbonyl group, there is delocalisation of π-electrons and thus, conjugation of carbonyl carbon with aryl ring reduces the electrophilic reactivity of carbonyl carbon.

(Resonance structures of benzaldehyde: $C_6H_5CH{=}O \longleftrightarrow$ $^-O{-}CH{=}$ ring with + charge at ortho $\longleftrightarrow$ $^-O{-}CH{=}$ ring with + charge at para $\longleftrightarrow$ ring + $\equiv$ delocalised ring (+))

Aromatic carbonyl compounds show chemical properties similar to aliphatic carbonyl compounds.

Some common reactions are given below

1. **Nucleophilic addition reactions**

$$C_6H_5CHO + NaHSO_3 \longrightarrow C_6H_5CH(OH)SO_3^-Na^+$$

Sodium bisulphite

2. **Nucleophilic addition-elimination reactions**

$$C_6H_5CHO + H_2N—NH_2 + OHC-C_6H_5 \longrightarrow C_6H_5CH{=}N—N{=}HC-C_6H_5$$

Benzaldehyde, Hydrazine, Benzyledimeazine

3. **Cannizzaro's reaction**

$$2C_6H_5CHO + NaOH \xrightarrow{50\%} C_6H_5CH_2OH + C_6H_5COO^-Na^+$$

Benzaldehyde, Benzyl alcohol, Sodium benzoate

4. **Tischenko reaction**

$$2C_6H_5CHO \xrightarrow{Al(OC_2H_5)_3} C_6H_5COOCH_2C_6H_5$$

5. **Reformatsky reaction**

$$C_6H_5CHO + Br—CH(CH_3)—C(=O)—OEt \xrightarrow[H_3O^+]{Zn/benzene} C_6H_5—CH(OH)—CH(CH_3)—C(=O)—OEt$$

Some of the specific reactions of aromatic carbonyl compounds are as follows.

6. **Perkin condensation**

Aromatic aldehydes react with acid anhydride of an aliphatic acid in presence of sodium salt of same acid to form α, β-unsaturated acid.

$$C_6H_5CHO + CH_3C(=O)—O—C(=O)CH_3 \xrightarrow{CH_3COO^-Na^+} C_6H_5\overset{\beta}{C}H{=}\overset{\alpha}{C}HCOOH + CH_3COOH$$

Acetic anhydride, Cinnamic acid

Mechanism

$$CH_3COO^- + HCH_2C(=O)—O—C(=O)CH_3 \rightleftharpoons CH_3COOH + \overset{\ominus}{C}H_2C(=O)—O—C(=O)CH_3$$

$$C_6H_5—CH{=}O + \overset{\ominus}{C}H_2C(=O)—O—C(=O)CH_3 \longrightarrow C_6H_5—CH(O^{\ominus})—CH_2C(=O)—O—C(=O)CH_3$$

$$C_6H_5—CH(O^{\ominus})—CH_2—C(=O)—O—C(=O)—CH_3 \xrightarrow{H—OAc} C_6H_5—CH(OH)—CH_2—C(=O)—O—C(=O)—CH_3$$

$$\downarrow$$

$$C_6H_5—CH{=}CH—C(=O)—O—C(O^{\ominus}H)—CH_3$$

$$\downarrow -\bar{O}Ac$$

$$C_6H_5—CH{=}CH—C(=O)—OH$$

Cinnamic acid

On heating cinnamic acid with soda lime styrene is produced.

$$C_6H_5—CH{=}CH—C(=O)—OH \xrightarrow[\Delta]{CaO/NaOH} C_6H_5—CH{=}CH_2 + CO_2\uparrow$$

Cinnamic acid, Styrene

If cyclic anhydride is used β, γ-unsaturated acid is formed.

$$C_6H_5CHO + (CH_2CO)_2O \xrightarrow{CH_3COO^-Na^+} C_6H_5\overset{\gamma}{C}H{=}\overset{\beta}{C}HCH_2COOH$$

Benzaldehyde

Benzoin condensation

When benzaldehyde is heated with aqueous ethanolic NaCN or KCN, it dimerises to form α-hydroxy ketone called **benzoin** and reaction is called benzoin condensation. It involves self condensation of an aromatic aldehyde in the presence of CN^- as catalyst.

$$C_6H_5CHO \text{ (Benzaldehyde)} + H\text{—}CO\text{—}C_6H_5 \xrightarrow[C_2H_5OH]{CN^-,\ H_2O} C_6H_5\text{—}CH(OH)\text{—}CO\text{—}C_6H_5 \text{ (Benzoin)}$$

It resembles aldol condensation of aliphatic aldehyde in which α-H condenses, but —CHO group is free.

$$CH_3CHO + H\text{—}\overset{\alpha}{C}H_2CHO \xrightarrow{NaOH} CH_3CH(OH)CH_2CHO$$

Knoevenagel condensation

Aromatic aldehydes or ketones that has no α-hydrogen condense with ester having active methylene group such as diethyl malonate, acetoacetic ester etc. in presence of base (pyridine, piperidine) to form α, β-unsaturated acids. The reaction proceed as follows

$$\ddot{B}^{\ominus} + H\text{—}CH(X)(Y) \longrightarrow :BH + \overset{\ominus}{C}H(X)(Y) \text{ (Carbanion)}$$

This carbanion attacks on C-atom of >C=O group followed by dehydration.

$$C_6H_5CH{=}O + CH_2(COOC_2H_5)_2 \text{ (Diethyl malonate)} \xrightarrow{\text{Piperidine (base)}}$$

Active methylene group

$$C_6H_5\text{—}CH{=}C(COOC_2H_5)_2 \xrightarrow{H_3O^+} C_6H_5\text{—}CH{=}C(COOH)_2 \xrightarrow{\Delta} C_6H_5\text{—}CH{=}CHCOOH$$

Claisen-Schmidt condensation

It is a crossed-aldol condensation in which aromatic aldehyde, i.e., benzaldehyde reacts with an aldehyde or ketone or an ester having α-hydrogen in presence of dilute alkali to give α, β-unsaturated carbonyl compounds.

e.g.

$$C_6H_5\text{—}CHO \text{ (Benzaldehyde)} + H_2CHCHO \text{ (Ethanal)} \xrightarrow[NaOH]{OH^-} C_6H_5\text{—}CH{=}CHCHO \text{ (Cinnamic aldehyde)}$$

$$C_6H_5\text{—}CHO + H_2CHCOR \xrightarrow{\text{Dil. NaOH}} C_6H_5\text{—}CH{=}CHCOR \text{ (Benzylidene ketone)}$$

Electrophilic substitution reactions

Aromatic aldehydes undergo electrophilic substitution reactions at *m*-position, due to presence of —CHO group. —CHO group is deactivating group and directs incoming electrophile at *m*-position.

This can be represented by resonance.

[Resonance structures: H—C=O attached to benzene ⟷ H—C—O⁻ with ⊕ at ortho ⟷ ⊕ at para ⟷ ⊕ at ortho ⟷ resonance hybrid H—C—O δ⊖ with δ+ at ortho and para positions]

E^+ (electrophile) will attack *m*-sites

Some of the electrophilic substitution reactions are as follows

(i) Halogenation

$$C_6H_5CHO \xrightarrow{Br_2/Fe} m\text{-}BrC_6H_4CHO$$

m-bromobenzaldehyde

(ii) **Nitration**

$$C_6H_5CHO \xrightarrow[273\text{-}283\ K]{\text{Conc. } HNO_3 + \text{conc. } H_2SO_4} m\text{-}O_2NC_6H_4CHO$$

m-nitrobenzaldehyde

(iii) **Sulphonation**

$$C_6H_5CHO \xrightarrow[SO_3]{H_2SO_4} m\text{-}HO_3SC_6H_4CHO$$

m-benzaldehyde sulphonic acid

Distinguish tests and uses

(i) Aldehyde and ketone

S. No.	Test	*R* CHO	*R* CO*R*
1.	Schiff's reagent	Magenta colour is restored by *R* CHO	No reaction
2.	Tollen's reagent	Reduced by *R* CHO	Not reduced
3.	Fehling's solution	Reduced by *R* CHO (except C_6H_5CHO)	Not reduced; α-hydroxy ketones reduce Tollen's reagent and Fehling's solution ($-\overset{\alpha}{C}H(OH)-CO-$)
4.	$K_2Cr_2O_7 + H_2SO_4$	Oxidised to *R* COOH	Alkyl group adjacent to keto group and with smaller number of H-atoms is oxidised (denoted by *) $CH_3CO\overset{*}{C}H_2CH_3 \xrightarrow{[O]} 2CH_3COOH$
5.	Iodoform test (Haloform test)	Only CH_3CHO gives this test	$R-\overset{O}{\overset{\Vert}{C}}-CH_3$ (all ketones containing acetyl group CH_3CO) gives this test
6.	Peroxy acid (Caro's acid, peroxy benzoic acid) Baeyer-Villiger oxidation	Acid is formed $RCHO \xrightarrow{[O]} RCOOH$	Ester is formed $R-\overset{O}{\overset{\Vert}{C}}-R' \xrightarrow{O} R-\overset{O}{\overset{\Vert}{C}}-O-R'$
7.	Acetal with alcohol	Formed	Not formed

(ii) Formaldehyde and acetaldehyde

S. No.	Reaction/Test	HCHO	CH_3CHO
1.	Cannizzaro	Yes (with no α-H)	No
2.	Aldol	No	Yes (with at least one α-H)
3.	Haloform	No	Yes
4.	With conc. NaOH solution	Cannizzaro reaction	resinous product
5.	With NH_3	Urotropine	$CH_3CH=NH$ (aldimine) that undergoes trimer formation (cyclic trimer: CH_3CH, $CH(CH_3)$, $CH(CH_3)$ joined by three NH groups)

(iii) Acetaldehyde and acetone

S. No.	Reaction/Test	CH_3CHO	CH_3COCH_3
1.	Tollen's reagent	Reduced	No effect
2.	Fehling's solution	Reduced	No effect
3.	With $CHCl_3$	No reaction	Chloretone is formed $(H_3C)_2C(OH)CCl_3$
4.	With alcohol	Acetal is formed	No reaction

(iv) Benzaldehyde and acetaldehyde

S. No.	Reaction/Test	C_6H_5CHO	CH_3CHO
1.	Fehling solution	is not reduced	is reduced (giving red ppt.)
2.	Cannizzaro	yes (with no α–H)	No
3.	Aldol	No	Yes
4.	Iodoform test	No	Yellow ppt.

Uses

(*i*) Aldehydes and ketones are used in industries as solvents, reagents for synthesising the other products.

(*ii*) Propanone (acetone) is used in nail paint remover. It is used as a solvent in industries and laboratories.

(*iii*) Many aldehydes and ketones are known for their odors and flavours. e.g. vanillin, camphor etc.

(*iv*) Formaldehyde is used as a disinfectant. 40% of aqueous HCHO (known as formalin) is used in preserving biological specimens.

(*v*) Formaldehyde is used in manufacture of polymers like bakelite, when treated with phenol.

(*vi*) Ethanal is used for silvering of mirrors.

(*vii*) Benzaldehyde is used in perfume industry as a flavouring agent.

CHECK POINT 18.4

1. Toluene on reaction with chromyl chloride in presence of carbon disulphide yields

(a) benzophenone (b) benzaldehyde
(c) benzyl alcohol (d) salicylic acid

2. The product (*X*) of the reaction

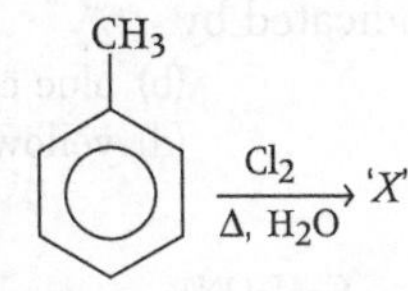

(a) 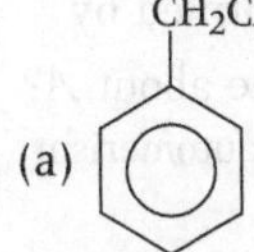 (b) 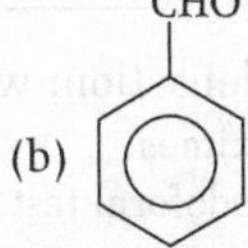

(c) 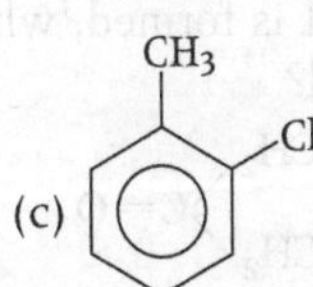 (d) 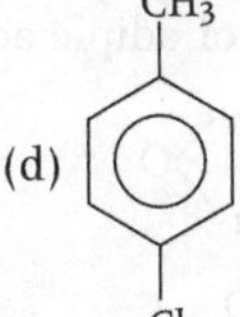

3. Phenol reacts with chloroform in alkaline solution to give benzaldehyde. This reaction is called

(a) Perkin reaction
(b) Knoevenagel reaction
(c) Reimer-Teimann reaction
(d) Gattermann-Koch reaction

4. Reagent used in Gattermann-aldehyde synthesis

(a) CO (b) HCl
(c) HCN (d) Both (b) and (c)

5. Aromatic aldehydes on heating with ethanolic solution of KCN undergoes condensation reaction. This reaction is called

(a) Benzoin condensation
(b) Perkin condensation
(c) Claisen-Schmidt reaction
(d) Cannizzaro's reaction

6. Aromatic aldehydes condense with compound '*X*' to form α, β-unsaturated acids. Compound '*X*' is

(a) malonic acid (b) acetoacetic acid
(c) diethyl malonate (d) succinic acid

7. Claisen-Schmidt reaction is shown by

(a) benzaldehyde and acetoacetic acid
(b) benzaldehyde and ketone
(c) benzaldehyde and KCN
(d) benzaldehyde and acetic anhydride

8. Acetone and acetaldehyde can be distinguished by

(a) Haloform test (b) Cannizzaro's reaction
(c) Tollen's reagent (d) All of these

9. Carbonyl compound used in nail paints remover

(a) propanone (b) formaldehyde
(c) ethanol (d) benzaldehyde

10. Which of the following is used a flavouring agent?

(a) Ethanol (b) Acetone
(c) Formaldehyde (d) Benzaldehyde

Chapter Exercises

(A) Topical exercise

I. Nomenclature, isomerism, structure and methods of preparation

1 The IUPAC name of the compound is

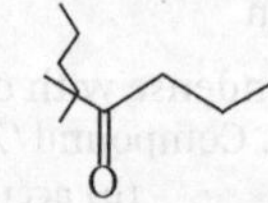

(a) 3, 3-dimethylheptan-4-one
(b) 5, 5-dimethylheptan-4-one
(c) 5-ethyl-5-methylhexan-4-one
(d) 2-ethyl-2-methylhexan-3-one

2 Which of the following compounds is commonly known as phthalaldehyde?

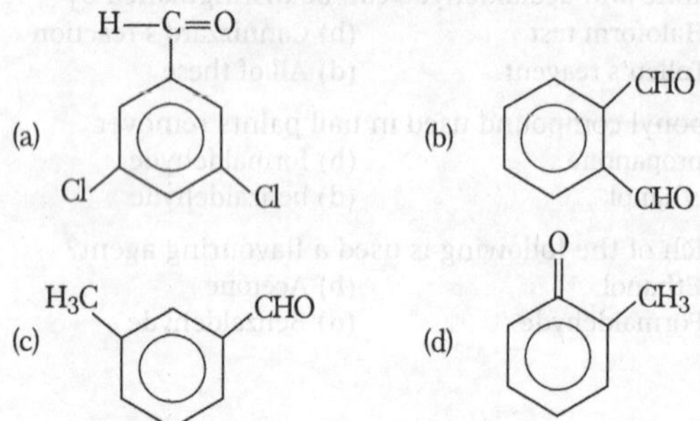

3 Which of the following is a chiral compound?
(a) 2-methylbutanal
(b) Butanal
(c) Butanone
(d) Both (a) and (b)

4 In which of the following compound the carbon bonded to oxygen is sp^2-hybridised?
(a) $H_2C=O$ (b) $H_2C=C=O$
(c) $H_3C-\ddot{\underset{\cdot\cdot}{O}}-CH_3$ (d) $H_3C-N=O$

5 Catalyst used in oxo-process is
(a) $Fe(CO)_5$ (b) $Co(CO)_5$
(c) $CrH(CO)_4$ (d) $[CoH(CO)_4]$

6 In which of the following case, $>C=C<$ (alkene) and $>C=O$ (carbonyl group) do not differ?
(a) Hybridisation of carbon
(b) Dipole moment
(c) Energy of double bond
(d) Bond length of double bond

7 $Cr_2O_7^{2-}/H^+$ is an oxidising agent for alcohol. Stage of completion is indicated by
(a) green colour (b) blue colour
(c) orange colour (d) yellow colour

8 $2CH_3\overset{O}{\overset{\|}{C}}OC_2H_5 \xrightarrow{C_2H_5ONa} A$, A is formed by Claisen condensation; which is/are true about A?
(a) A forms oximes (b) A shows tautomerism
(c) A shows iodoform test (d) All true

9 Which one of the following product is formed, when calcium salt of adipic acid is heated?
(a) $\begin{matrix} CH_2-CH_2 \\ | \\ CH_2-CH_2 \end{matrix}\!\!>O$ (b) $\begin{matrix} CH_2-CH_2 \\ | \\ CH_2-CH_2 \end{matrix}\!\!>C=O$
(c) $\begin{matrix} CH_2CH_2CO \\ | \\ CH_2CH_2CO \end{matrix}\!\!>C=O$ (d) $\begin{matrix} CH_2CH_2COOH \\ | \\ CH_2CH_2COOH \end{matrix}$

10 Ketones $[R-\overset{}{\underset{\underset{O}{\|}}{C}}-R_1]$, where $R = R_1$ = alkyl group, can be obtained in one step by
(a) hydrolysis of esters
(b) oxidation of primary alcohol
(c) oxidation of secondary alcohol
(d) reaction of acid halide with alcohols

11 Pure propanone can be prepared by ozonolysis of
(a) 1,2-dimethyl-2-butene (b) 1-pentene
(c) 2-methyl-2-butene (d) 2,3-dimethyl-2-butene

12 Addition of water to alkynes occurs in acidic medium and in the presence of Hg^{2+} ions as a catalyst. Which of the following products will be formed on addition of water to but-1-yne under these conditions?
(a) $CH_3-CH_2-CH_2-\overset{O}{\overset{\|}{C}}-H$
(b) $CH_3-CH_2-\overset{O}{\overset{\|}{C}}-CH_3$
(c) $CH_3-CH_2-\overset{O}{\overset{\|}{C}}-OH + CO_2$
(d) $CH_3-\overset{O}{\overset{\|}{C}}-OH + H-\overset{O}{\overset{\|}{C}}-H$

13 Which of the following compounds will give butanone on oxidation with alkaline $KMnO_4$ solution?
(a) Butan-1-ol (b) Butan-2-ol
(c) Both (a) and (b) (d) None of these

14 In Clemmensen reduction, carbonyl compound is treated with ....
(a) zinc amalgam + HCl
(b) sodium amalgam + HCl
(c) zinc amalgam + nitric acid
(d) sodium amalgam + HNO_3

15 Select the incorrect statement.
(a) PCC oxidises 1° alcohol to aldehyde and 2° alcohol to ketone
(b) Cu dehydrates 3° alcohol to alkene
(c) MnO_2 can oxidise allyl alcohol ($CH_2{=}CHCH_2OH$) to acrolein ($CH_2{=}CHCHO$)
(d) Oppenauer oxidation method is used to oxidise 1° alcohol to aldehyde

16 Which of the following nitriles will form aldehyde on reaction with Grignard reagent?
(a) HCN
(b) CH_3CN
(c) $(CH_3)_2CHCN$
(d) $CH_2{=}CHCN$

17 An acid chloride on reaction with $H_2/Pd\text{-}BaSO_4$ changes to $(CH_3)_2CHCHO$. This acid chloride on reaction with CH_3MgBr/H_3O^+ gives
(a) $(CH_3)_3COH$
(b) $(CH_3)_2CHOH$
(c) $(CH_3)_2CHCOCH_3$
(d) $(CH_3)_2CHCHO$

18 $NaAlH_4$ reduces an ester into HCHO and $(CH_3)_2CHOH$. This ester changes to .... on reaction with CH_3MgBr / H_3O^+.
(a) $(CH_3)_2CHCOCH_3$ (b) CH_3CHO
(c) CH_3COCH_3 (d) CH_3CH_2CHO

19 Grignard reagent can convert the following into carbonyl compound except in
(a) $CH_3\overset{O}{\overset{\|}{C}}Cl$ (b) $H\overset{O}{\overset{\|}{C}}OCH_3$
(c) $CH_3C{\equiv}N$ (d) CH_3CH_2OH

20 By Oxo-process $(CH_3)_2C{=}CH_2 + CO + H_2 \longrightarrow A$, A formed is
(a) $(CH_3)_2CHCH_2CHO$ (b) $(CH_3)_2CH\overset{O}{\overset{\|}{C}}CH_3$
(c) $(CH_3)_2CHCH_2CH_3$ (d) $(CH_3)_2CH\underset{CH_3}{\underset{|}{C}H}CH_3$

21 Which does not produce aldehyde in the following reactions?
(a) $RCN \xrightarrow{\text{Reduction}/H_3O^+}$
(b) $RCN \xrightarrow{SnCl_2/H_3O^+}$
(c) $RCN \xrightarrow{Sn/HCl}$
(d) $HCN \xrightarrow[H_3O^+]{CH_3MgBr}$

22 Which of the following is the incorrect reaction?
(a) $CH_3CH{=}CH_2 \xrightarrow{\text{Oxo method}} CH_3CH_2CH_2CHO$
(b) $RCN \xrightarrow{\text{DIBAlH}} RCHO$
(c) $(HCOO)_2Ca \xrightarrow{\Delta} HCHO$
(d) $CH_3\overset{O}{\overset{\|}{C}}Cl \xrightarrow{(CH_3)_2CuLi} CH_3CH_2\overset{O}{\overset{\|}{C}}CH_3$

23 Following synthesis
$$CH_3COCl + R_2Cd \longrightarrow RCOCH_3 + RCdCl$$
is effective, when R is of the type
(a) $(CH_3)_2CH—$
(b) $(CH_3)_3CCH_2—$
(c) $(C_6H_5)_2CH—$
(d) All of these

24 Identify A in the following sequence of reactions
$$C_3H_5N(A) \xrightarrow{H_3O^+} \xrightarrow{PCl_5} \xrightarrow{(CH_3)_2CuLi} CH_3CH_2\overset{O}{\overset{\|}{C}}CH_3$$
(a) CH_3CH_2CN (b) CH_3CH_2NC
(c) $CH_2{=}CHCH{=}NH$ (d) None of the above

25 CH_3CHO can be prepared by
(a) oxidation of CH_3CH_2OH using $Cr_2O_7^{2-}$ / H^+
(b) oxidation of $CH_2{=}CH_2$ using $PdCl_2 \cdot CuCl_2/H_2O$
(c) reaction of CH_3CN with $Sn/HCl/H_3O^+$
(d) All of the above

26 In the following sequence,
$$CH_3CN \xrightarrow[\text{(ii) } H_3O^+]{\text{(i) } CH_3MgBr \text{ (2 mol)}} A \xrightarrow{\text{Conc. } H_2SO_4} B;$$
A and B are respectively
(a) CH_3COCH_3 and 1,3,5-trimethylbenzene (benzene ring bearing CH_3, H_3C, CH_3)
(b) $(CH_3)_3COH$ and $(CH_3)_2C{=}CH_2$
(c) $(CH_3)_2CHOH$ and $CH_3CH{=}CH_2$
(d) None of the above is correct

27 Consider the following conversion

A. $CH_3CH_2CH{=}CHCH_2OH \longrightarrow CH_3CH_2CH{=}CHCHO$

B. [cyclohexanol with methyl and isopropyl substituents, OH] $\longrightarrow$ [cyclohexanone with methyl and isopropyl substituents, O]

Suggest suitable oxidising reagents for the above conversions

(a) MnO_2 in *A* and CrO_3 (in glacial acetic acid) in *B*
(b) CrO_3 in *A* and MnO_2 in *B*
(c) Both are correct
(d) None of the above is correct

28 [2-oxocyclohexane with $-COCH_3$ and CH_3] $\xrightarrow[\text{2. } H^+ \quad \text{3. } \Delta]{\text{1. KBrO, } \Delta}$ *A*, *A* is

(a) [2-oxocyclohexane, —COOH, CH_3] (b) [cyclohexanone, —CH_3]
(c) [cyclohexane, —COOH, OH] (d) [cyclohexane, —COOH, CH_3, OH]

29 An olefin on ozonolysis yields a mixture of acetone and methyl ethyl ketone. The possible structure of the olefin is

(a) $CH_3\overset{CH_3}{C}{=}\overset{CH_3}{C}CH_3$ (b) $C_2H_5\overset{C_2H_5}{C}{=}CH_2$
(c) $CH_3\overset{CH_3}{C}{=}\overset{CH_3}{C}CH_2CH_3$ (d) $CH_3CH{=}CHCH_3$

30 Which of the compounds shown is (are) correctly named as pentane derivatives, either as pentanals or pentanones?

[Structures 1, 2, 3]

(a) Only 1 (b) Only 2
(c) Only 3 (d) Both 1 and 3

31 Consider the following statements.

Statement I Formaldehyde is a planar molecule.

Statement II It contains sp^2-hybridised carbon atom.

(a) Both statements I and II are correct
(b) Only statement I is correct
(c) Only statement II is correct
(d) Both statements I and II are incorrect

32 Match the common names given in Column I with the IUPAC names given in Column II.

Column I (Common names)	Column II (IUPAC names)
A. Cinnamaldehyde	1. Pentanal
B. Acetophenone	2. Prop-2-en-al
C. Valeraldehyde	3. 4-methylpent-3-en-2-one
D. Acrolein	4. 3-phenylprop-2- en-al
E. Mesityl oxide	5. 1-phenylethanone

Codes

	A	B	C	D	E
(a)	4	5	1	2	3
(b)	3	4	5	1	2
(c)	2	3	4	1	5
(d)	5	4	1	3	2

II. Physical and chemical properties : Addition reaction

33 In the following compound, electrophilic site is

$$\underset{\gamma}{H_3C}\!\!\!\!\overset{H_3C}{\ }\!\!>\underset{\beta}{C}{=}\underset{\alpha}{\ddot{O}}:$$

(a) α-oxygen (b) β-carbon
(c) γ-carbon (d) None of these

34 In the above compound [in question (33)], nucleophilic site is

(a) α-oxygen (b) β-carbon
(c) γ-carbon (d) All of these

35 Select the incorrect statement(s).

(a) Protonation increases electrophilic nature of the carbonyl group
(b) Aldehydes are more reactive than ketones
(c) Carbonyl compounds cannot form hydrogen-bonding with H_2O
(d) Chloral hydrate formation is possible due to hydrogen bonding

36 Boiling points of carbonyl compounds are higher than those of alkanes due to

(a) hydrogen bonding (b) dipole-dipole interactions
(c) van der Waals' forces (d) All of the above

37 Reactivity of HCHO (I), CH_3CHO (II) and C_6H_5CHO(III) is in order

(a) I > III > II (b) III > II > I
(c) I = II > III (d) I > II > III

38 Nucleophilic attack on carbonyl carbon changes its hybridisation from

(a) sp to sp^2 (b) sp^2 to sp^3
(c) sp^3 to sp^2 (d) sp to sp^3

39 Grignard reagent forms 3° alcohol with
(a) benzophenone
(b) acetophenone
(c) acetone
(d) All of the above

40 Maximum hydration takes place

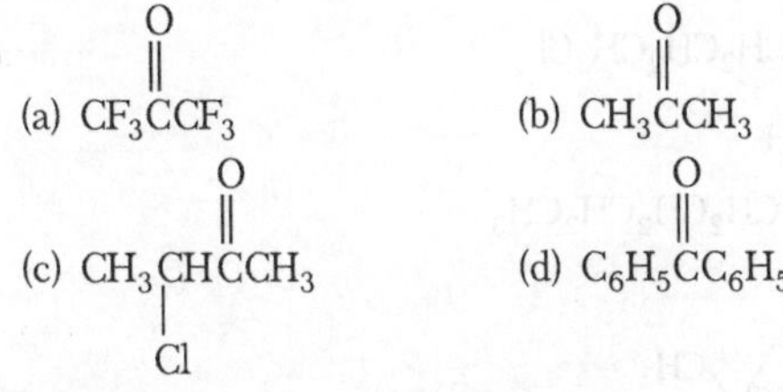

41 Which of the following is correct regarding an acetal?
(a) It is the condensation product from 2 molecules of aldehyde
(b) It is the condensation product from 2 molecules of ketone
(c) It is the condensation product from 2 molecules of alcohol
(d) It is a diether

42 Aldehyde group can be protected
(a) by acetal formation against attack by alkaline oxidising agents
(b) by mercaptal formation against attack by acidic oxidising agents
(c) Both statements are correct
(d) None of the above is correct

43 Mixture of CH_3CH_2OH and CH_3CHO can be separated by using
(a) $NaHSO_3$ (b) NH_2OH
(c) HCN (d) $NH_2—NH_2$

44 In the reaction,
$CH_3CHO + HCN \longrightarrow CH_3CH(OH)CN$, a chiral centre is produced. This product is
(a) *meso* compound (b) laevorotatory
(c) dextrorotatory (d) racemic mixture

45 Acetone and methyl magnesium bromide react to produce a product which on hydrolysis gives
(a) *n*-butyl alcohol (b) secondary butyl alcohol
(c) tertiary butyl alcohol (d) None of these

46 When acetaldehyde is heated with Fehling's solution, it gives a precipitate of
(a) Cu_2O (b) $Cu + Cu_2O + CuO$
(c) Cu (d) CuO

47 Schiff's reagent is
(a) rosaniline solution in water decolourised by SO_2
(b) resorcinol solution in dilute HCl
(c) ammoniacal silver nitrate
(d) alkaline $CuSO_4$ stabilised by Rochelle salt

48 The compound shown is best classified as a(an)

$$(CH_3)_3CCH_2CH{=}NCH_3$$

(a) carbinolamine (b) enamine
(c) hydrazone (d) imine

49 When a nucleophile encounters a ketone, the site of attack is
(a) the carbon atom of the carbonyl
(b) the oxygen atom of the carbonyl
(c) Both the carbon and oxygen atoms, with equal probability
(d) No attack occurs-ketones do not react with nucleophiles

50 Rank the following in order of increasing value of the equilibrium constant for hydration, K_{hyd} (smallest value first)

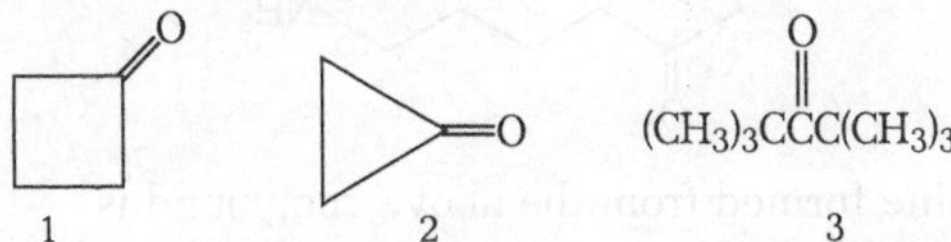

Choose the correct alternate
(a) $1 < 2 < 3$ (b) $3 < 1 < 2$ (c) $2 < 1 < 3$ (d) $2 < 3 < 1$

51 The structure

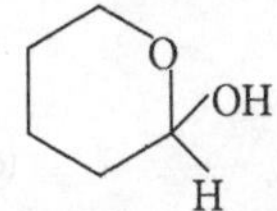

would be best classified as a(an)
(a) acetal (b) hemiacetal (c) hydrate (d) cyanohydrin

52 Which of the following compounds is most reactive towards nucleophilic addition reactions?

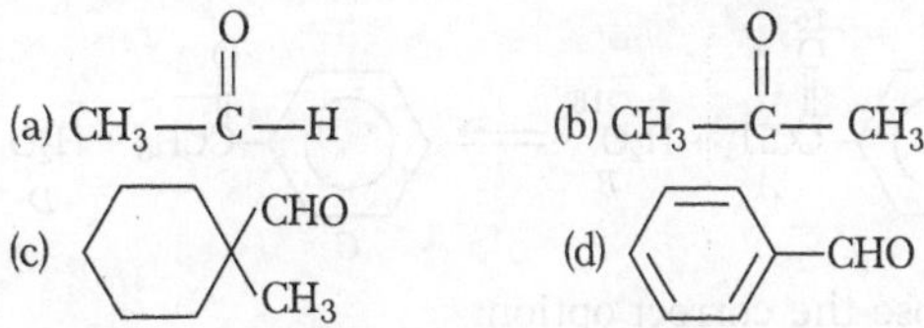

53 Increasing order of boiling points of the following

$$\underset{\text{I}}{CH_3CHO}, \quad \underset{\text{II}}{CH_3CH_2OH}, \quad \underset{\text{III}}{CH_3OCH_3}, \quad \underset{\text{IV}}{CH_3CH_2CH_3}$$

(a) I < II < III < IV (b) IV < III < II < I
(c) IV < III < I < II (d) II < I < III < IV

54 Which is maximum reactive towards nucleophilic attack of H_2O forming hydrate?

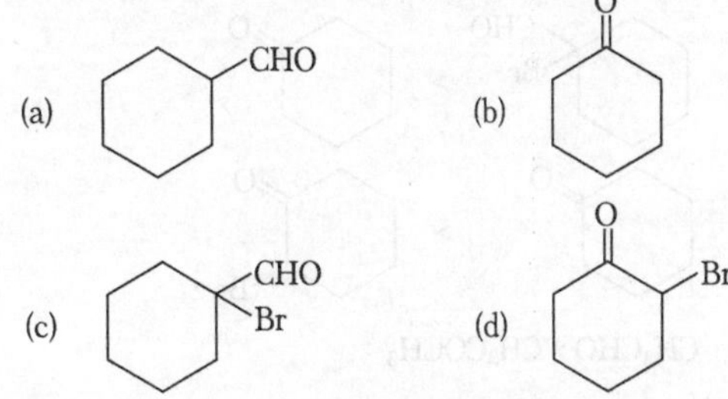

55 Imine (*A*)

A : (cyclohexylidene)$=\ddot{N}-CH_3$

can be prepared by using

(a) (methylenecyclohexane, $=CH_2$), CH_3CHO (b) (cyclohexanone, $=O$), CH_3NH_2

(c) (cyclohexanone, $=O$), CH_3NHCH_3 (d) None of these

56 Consider the following compound.

$CH_3CO(CH_2)_4NH_2$

Imine formed from the above compound is

(a) N, CH_3 (b) N, CH_3

(c) NH, CH_3 (d) H N, CH_3

57 Trace the position of enriched oxygen (^{18}O) in following reaction.

$$\underset{A}{C_6H_5\overset{\overset{16}{O}}{\overset{\|}{C}}CH_3} + \underset{B}{H_2\overset{18}{O}} \rightleftharpoons \underset{C}{C_6H_5\overset{O}{\overset{\|}{C}}CH_3} + \underset{D}{H_2O}$$

Choose the correct option.

(a) in *C* (b) in *D*

(c) in both (a) and (b) (d) in none of these

58 Which of the following is not the correct comparison for the value of equilibrium constant for hydrate formation?

(a) (cyclohexyl–CHO) > (cyclohexanone, =O)

(b) (cyclohexyl with CHO and Br) > (cyclohexanone, =O)

(c) (cyclohexanone, =O) > (2-bromocyclohexanone, =O, Br)

(d) $CH_3CHO > CH_3COCH_3$

59 Consider the following reaction

$$CH_3\underset{\underset{O}{\|}}{C}CH_2CH_2CH_2Cl \xrightarrow[\text{(ii) } H_3O^+]{\text{(i) } CH_3MgBr} A.\ \text{Here is}$$

(a) $CH_3\overset{\overset{CH_3}{|}}{\underset{\underset{OH}{|}}{C}}CH_2CH_2CH_2Cl$

(b) $CH_3\underset{\underset{O}{\|}}{C}CH_2CH_2CH_2CH_3$

(c) (cyclic ether, O, CH_3, CH_3)

(d) (H_3C, CH_3, H_3C, O)

60 Consider the following reaction

$$CH_3\overset{\overset{O}{\|}}{C}CH_2CH_2\overset{\overset{O}{\|}}{C}OCH_2CH_3 \xrightarrow[\text{(ii) } H_3O^+]{\text{(i) } CH_3MgBr \text{ (one mole)}} A,$$

A formed in this reaction is

(a) $CH_3\overset{\overset{OH}{|}}{\underset{\underset{CH_3}{|}}{C}}CH_2CH_2\overset{\overset{O}{\|}}{C}OCH_2CH_3$

(b) $CH_3\overset{\overset{O}{\|}}{C}CH_2CH_2\overset{\overset{O}{\|}}{C}CH_3$

(c) (H_3C, H_3C, O, =O)

(d) $CH_3\overset{\overset{CH_3}{|}}{\underset{\underset{OH}{|}}{C}}CH_2CH_2\overset{\overset{CH_3}{|}}{\underset{\underset{OH}{|}}{C}}CH_3$

61 Consider the following statements.

Statement I Protonation of the carbonyl group increases its electrophilic nature.

Statement II This is an example of electrophilic addition on nucleophilic oxygen.

(a) Both statements I and II are correct

(b) Only statement I is correct

(c) Only statement II is correct

(d) Both statements I and II are incorrect

62 Greatest amount of hydration is in

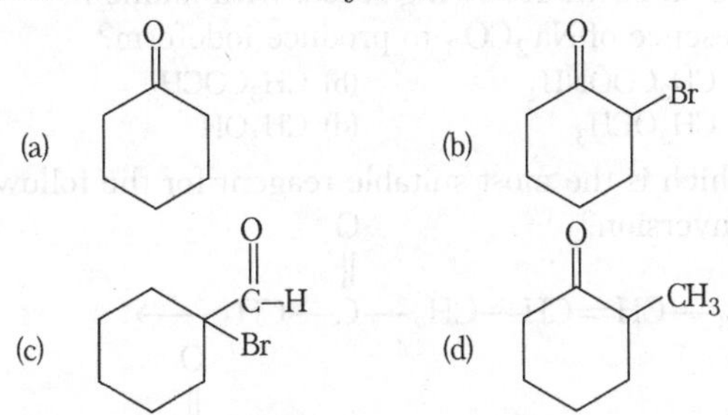

63 Consider the following reaction

$$CH_3MgBr + CH_2{=}CH—\overset{\overset{\displaystyle O}{\|}}{C}—H \xrightarrow{H_3O^+} \text{Product}$$

The product formed in above reaction is

(a) $CH_2{=}CH—\underset{\underset{\displaystyle CH_3}{|}}{\overset{\overset{\displaystyle OH}{|}}{C}}—H$ (b) $CH_2CH{=}\ CH—\underset{\underset{\displaystyle OH}{|}}{}CH_3$

(c) $CH_3CH_2CH_2CHO$ (d) None is correct

64 In the following reaction

$$\underset{\diagdown O \diagup}{CH_2—CHCH_3} + PCl_5 \longrightarrow A,$$

$CH_3CH_2CHO + PCl_5 \longrightarrow B$, *A* and *B* are

(a) *vic* halides (b) *gem* halides

(c) *A gem* halide, *B vic* halide (d) *A vic* halide, *B gem* halide

65 Product of the following reaction

$$\text{(cyclohexanone)}{=}O + (C_6H_5)_3P{=}CH_2 \xrightarrow{DMSO}$$

is

(a) (cyclohexylidene)$=CH_2$ (b) (cyclohexylidene)$=C{=}P(C_6H_5)_3$

(c) Both (a) and (b) (d) None of these

66 Which of the following pairs of reactants is most effective in forming an enamine?

(a) $CH_3CH_2\overset{\overset{\displaystyle O}{\|}}{C}H + [(CH_3)_2CH]_2NH$

(b) $(CH_3)_3C\overset{\overset{\displaystyle O}{\|}}{C}H + (CH_3)_2NH$

(c) (cyclohexanone) $+ (CH_3)_3CNH_2$

(d) None of the above forms an enamine

67 When HCHO reacts with a mixture of NH_4Cl and NaCN, then end product is

(a) $CH_2{=}\overset{\oplus}{N}H_2$ (b) NH_2CH_2CN

(c) $CH_2{=}N—CH_2CN$ (d) $HCH\begin{smallmatrix}\diagup OH \\ \diagdown CN\end{smallmatrix}$

III. Oxidation reaction

68 Silver-mirror test is not positive with

(a) benzaldehyde (b) acetaldehyde

(c) acetone (d) fructose

69 HCHO reduces Fehling's solution giving red precipitate of

(a) Ag^{2+} (b) Cu^{2+} (c) $HCOO^-$ (d) Cu^+

70 For the following transformation

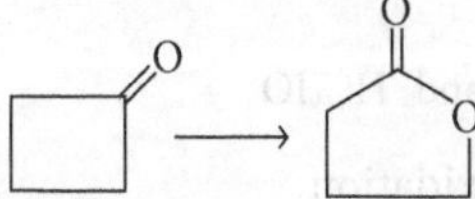

reagent used is

(a) SeO_2 (b) $C_6H_5CO_3H$ (c) ClO^- (d) Se

71 Identify correct product of the following oxidation reaction

$$CH_3CH_2COCH_3 \xrightarrow{SeO_2}$$

(a) CH_3CH_2COCHO (b) $CH_3CO \cdot COCH_3$

(c) CH_3COCH_2CHO (d) CH_3CH_2COOH

72 Nucleophile and Lewis acid in $LiAlH_4$ are respectively

(a) H^-, AlH_4^- (b) Li^+, AlH_4^-

(c) H^-, Li^+ (d) H^+, Al^{3+}

73 Reagents used in Clemmensen reduction and Wolff-Kishner reduction are respectively

(a) Zn(Hg) - conc. HCl; N_2H_4 - glycol, OH^-

(b) $LiAlH_4$, $NaBH_4$

(c) $NaBH_4$; $LiAlH_4$

(d) Pd - C / H_2; Li / NH_3

74 The reaction of aldehydes and ketones with $LiAlH_4$ and $NaBH_4$ is

(a) nucleophilic addition (b) nucleophilic substitution

(c) elimination (d) electrophilic substitution

75 Consider the following reaction

$$CH_3CHO + LiAlH_4 \longrightarrow CH_3CH_2OH$$

Nucleophile added in this reaction is

(a) AlH_4^- (b) Li^+ (c) H^+ (d) H^-

76 In the Wacker's process, which one of the following used as a co-catalyst in

(a) $PdCl_2$ (b) $CuCl_2$ (c) MnO (d) $CoH(CO)_4$

77 Consider the following statements.

Statement I Aldehydes and ketones, both react with Tollen's reagent to form silver mirror.

Statement II Both, aldehydes and ketones contain a carbonyl group.

(a) Both statements I and II are correct
(b) Only statement I is correct
(c) Only statement II is correct
(d) Neither statement I nor statement II is correct

78 Fehling's solution can make distinction between

(a) CH_3CHO and C_6H_5CHO
(b) CH_3CHO and $CH_3\overset{O}{\overset{\|}{C}}CH_2OH$
(c) $CH_2(OH)CH(OH)\overset{O}{\overset{\|}{C}}CH_3$ and HCHO
(d) CH_3CHO and HCHO

79 Oppenauer oxidation

(a) is used to prepare ketone
(b) uses aluminium *t*-butoxide as catalyst
(c) Both are correct
(d) None of the above is correct

80 At room temperature, formaldehyde changes to

(a) trioxane which does not reduce Tollen's reagent
(b) paraldehyde
(c) hexose
(d) None of the above is correct

81 Cyanohydrin of which of the following forms lactic acid?

(a) HCHO (b) CH_3CHO
(c) CH_3CH_2CHO (d) CH_3COCH_3

82 Which pair can be distinguished by using Tollen's reagent?

(a) Ethanol and propanol
(b) Ethanoic acid and propanone
(c) Ethanoic acid and propanal
(d) Methanol and methanoic acid

83 Consider the following statements.

Statement I Compounds containing —CHO group are easily oxidised to corresponding carboxylic acids.

Statement II Carboxylic acids can be reduced to alcohols by treatment with $LiAlH_4$.

(a) Both statements I and II are correct
(b) Only statement I is correct
(c) Only statement II is correct
(d) Both statements I and II are incorrect

84 Acetaldehyde does not respond to

(a) Tollen's test (b) Benedict's test
(c) Lucas test (d) Iodoform test

85 Which of the following reacts with iodine in presence of Na_2CO_3 to produce iodoform?

(a) CH_3COONH_4 (b) CH_3COCH_3
(c) CH_3OCH_3 (d) CH_3OH

86 Which is the most suitable reagent for the following conversion?

$$CH_3—CH{=}CH—CH_2—\overset{O}{\overset{\|}{C}}—CH_3 \longrightarrow CH_3—CH{=}CH—CH_2—\overset{O}{\overset{\|}{C}}—OH$$

(a) Tollen's reagent (b) Benzoyl peroxide
(c) I_2 and NaOH solution (d) Sn and NaOH solution

87 In the following reduction, reducing agent used is

$$CH_3\overset{O}{\overset{\|}{C}}OCH(CH_3)_2 \longrightarrow CH_3CH_2OH + (CH_3)_2CHOH$$

(a) DIBAH (b) $LiAlH_4$
(c) $NaAlH_4$ (d) H_2/Pd-$BaSO_4$

88 $C_3H_6Cl_2$ on reaction with NaOH forms C_3H_6O which gives yellow precipitate on heating with NaOH and I_2. Thus, $C_3H_6Cl_2$ is

(a) 1,1-dichloropropane
(b) 1,2-dichloropropane
(c) 2,2-dichloropropane
(d) 1,3-dichloropropane

89 Match the mixture given in Column I with the reagents used to distinguish them given in Column II.

Column I	Column II
A. HCHO + CH_3CHO	1. Tollen's reagent
B. $CH_3\overset{O}{\overset{\|}{C}}CH_3 + (CH_3)_3COH$	2. Lucas test
C. $C_6H_5CHO + C_6H_5\overset{O}{\overset{\|}{C}}CH_3$	3. Iodoform test
D. $CH_3CH_2OH + CH_3CHO$	4. Cannizzaro reaction

Codes

	A	B	C	D		A	B	C	D
(a)	4	2	3	1	(b)	1	2	3	4
(c)	4	3	2	1	(d)	2	4	3	1

90 In the following reaction,

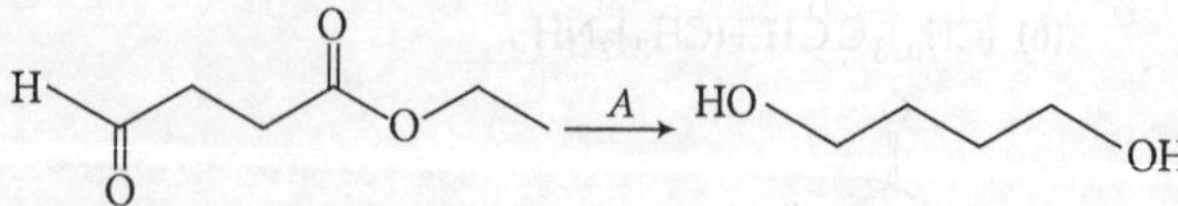

reagent *A* is

(a) $NaBH_4$ (b) $LiAlH_4$
(c) H_2 / Ni (d) Mg/benzene/H_2O

91 Which of the following alcohols cannot be obtained by a hydride reduction of an aldehyde or a ketone?

(a) 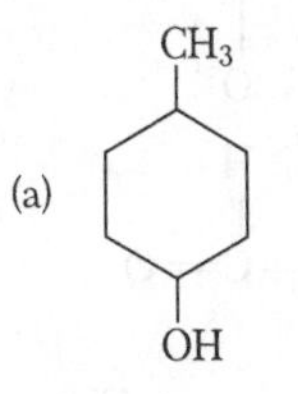

(b) CH_3, OH (cyclohexane ring)

(c) 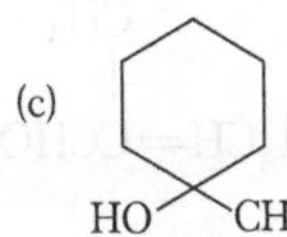

(d) All of these

92 Consider a flow-chart of a reaction equations

$$\underset{A}{C_6H_{10}O_3} \text{ (keto ester)} \xrightarrow[\Delta]{NaOH + I_2} \text{yellow ppt.} + B$$

$$B \xrightarrow{H^+} C \xrightarrow[-CO_2]{\Delta} CH_3COOH.$$

Identify A from the following options

(a) $CH_3\overset{O}{\overset{\|}{C}}CH_2\overset{O}{\overset{\|}{C}}OC_2H_5$ (b) $CH_3CH_2\overset{O}{\overset{\|}{C}}CH_2\overset{O}{\overset{\|}{C}}OCH_3$

(c) Both are correct (d) None is correct

93 In the following reaction

$$C_2H_5O—\overset{O}{\overset{\|}{C}}—OC_2H_5 \xrightarrow{2CH_3MgBr} A.$$

Product A formed can

(a) give iodoform test

(b) further react with CH_3MgBr / H_3O^+ to give *t*-butyl alcohol

(c) be obtained by the ozonolysis of 2,3-dimethyl-2-butene

(d) give all the above reactions

94 Match the reactions given in Column I with the suitable reagents given in Column II.

Column I (Reactions)	Column II (Reagents)
A. Benzophenone → Diphenylmethane	1. $LiAlH_4$
B. Benzaldehyde → 1-phenylethanol	2. DIBAL-H
C. Cyclohexanone → Cyclohexanol	3. Zn(Hg)/Conc. HCl
D. Phenyl benzoate → Benzaldehyde	4. CH_3MgBr

Codes

	A	B	C	D		A	B	C	D
(a)	3	4	1	2	(b)	4	3	1	2
(c)	3	4	2	1	(d)	4	3	2	1

IV. Chemical properties due to α-hydrogen

95 Crossed Cannizzaro reaction can be given by which of the following combination?

(a) CH_3CHO, HCHO (b) C_6H_5CHO, CH_3CHO

(c) C_6H_5CHO, HCHO (d) All of these

96 Cannizzaro reaction is not given by

(a) $(CH_3)_3CCHO$ (b) $(CH_3)_2CHCHO$

(c) C_6H_5CHO (d) $(CH_2OH)_3C \cdot CHO$

97 Which of the following statements is not true about Cannizzaro reaction?

(a) It is a disproportionation reaction

(b) It is a proton-hydride transfer reaction

(c) It is given by all the carbonyl compounds

(d) It takes place with 50% aqueous or ethanolic solution

98 Aldol condensation reaction is not given by

(a) $C_6H_5CH_2CHO$ (b) C_6H_5CHO

(c) CH_3CHO (d) CH_3COCH_3

99 Resonance form of enolate ion (A) of acetone

$$\overset{\ominus}{C}H_2—\overset{O}{\overset{\|}{C}}—CH_3 \quad (A) \text{ is}$$

(a) $CH_2{=}\overset{O^{\ominus}}{\overset{|}{C}}—CH_3$ (b) $\overset{\ominus}{C}H_2—\overset{OH}{\overset{|}{C}}{=}CH_2$

(c) Both (a) and (b) (d) None of these

100 $2\,\underset{CHO}{\overset{COOH}{|}} \xrightarrow{NaOH} \underset{CH_2OH}{\overset{COOH}{|}} + \underset{COONa}{\overset{COOH}{|}}$; it is

(a) Crossed Cannizzaro reaction

(b) Intermolecular Cannizzaro reaction

(c) Intramolecular Cannizzaro reaction

(d) None of the above

101 (A), β-hydroxy carbonyl compound is obtained by the action of NaOH on

(a) $R_3C \cdot CHO$ (b) C_6H_5CHO (c) CH_3CHO (d) HCHO

102 Consider the following statements.

Statement I The α-hydrogen atom in carbonyl compounds is less acidic.

Statement II The anion formed after the loss of α-hydrogen atom is resonance stabilised.

(a) Both statements I and II are correct

(b) Only statement I is correct

(c) Only statement II is correct

(d) Both statements I and II are incorrect

103 Aldol condensation will not take place in

(a) HCHO (b) CH_3CH_2CHO

(c) CH_3CHO (d) CH_3COCH_3

104 Monomer of trioxane is

(a) formaldehyde (b) isoprene

(c) acetone (d) acetaldehyde

105 Only an aldehyde having ... can undergoes the aldol condensation.
(a) at least one α H-atom (b) at least one β H-atom
(c) no α H-atom (d) an aromatic ring

106 Which of the following species is an ylide?
(a) $(C_6H_5)_3\overset{+}{P}CH_2CH_3Br^-$
(b) $(C_6H_5)_3\overset{+}{P}\overset{..}{\underset{-}{C}}HCH_3$
(c) $(C_6H_5)_3P—CHCH_3$ with O—CH_2 bridging P and C
(d) None of the above

107 Cannizzaro's reaction is not given by ....
(a)  (b) benzene ring–CHO
(c) HCHO (d) CH_3CHO

108 In reaction $\underset{(A)}{C_5H_{10}Cl_2} + NaOH \longrightarrow \underset{(B)}{C_5H_{10}O}$
B is an aldehyde with no α-hydrogen. Thus, A is
(a) $(CH_3)_3CCHCl_2$ (b) $CH_3CH_2\underset{CH_3}{\underset{|}{C}H}CHCl_2$
(c) $CH_3\underset{CH_3}{\underset{|}{C}H}CH_2CHCl_2$ (d) $CH_3\overset{Cl}{\overset{|}{\underset{Cl}{\underset{|}{C}}}}CH_2CH_2CH_3$

109 Identify A and B in the following reaction sequence
$$CH_3CH_2\overset{O}{\overset{||}{C}}CH_3 \xrightarrow{C_6H_5CO_3H} \xrightarrow{H_3O^+} A + B$$
Choose the correct code.

	A	B
(a)	$CH_3CH_2CO_2H$,	CH_3OH
(b)	CH_3CO_2H,	CH_3CH_2OH
(c)	CH_3CO_2H	CH_3OH
(d)	CH_3CO_2H,	CH_3CO_2H

110 By Cannizzaro reaction A changes to B and C as given
$$A \xrightarrow{50\% \text{ NaOH}} \underset{B}{\begin{matrix} COONa \\ | \\ COONa \end{matrix}} + \underset{C}{\begin{matrix} CH_2OH \\ | \\ COONa \end{matrix}}$$
Thus, A is
(a) $\begin{matrix} CHO \\ | \\ COOH \end{matrix}$ (b) $\begin{matrix} CHO \\ | \\ CHO \end{matrix}$
(c) $\begin{matrix} CH_2OH \\ | \\ CHO \end{matrix}$ (d) $\begin{matrix} CH_2OH \\ | \\ CH_2OH \end{matrix}$

111 Which of the following is the best hydride donor?
(a) $C_6H_5—\overset{H}{\overset{|}{\underset{OH}{\underset{|}{C}}}}—O^{\ominus}$ (b) $C_6H_5—\overset{H}{\overset{|}{\underset{O^{\ominus}}{\underset{|}{C}}}}—O^{\ominus}$
(c) $C_6H_5—\overset{H}{\overset{|}{\underset{H}{\underset{|}{C}}}}—Cl$ (d) $C_6H_5—\overset{H}{\overset{|}{C}}=O$

112 $C_3H_6O \xrightarrow{OH^-} \xrightarrow[\Delta]{NaHSO_4} CH_3CH_2CH=\overset{CH_3}{\overset{|}{C}}CHO$
Thus, C_3H_6O is
(a) CH_3COCH_3 (b) CH_3CH_2CHO
(c) $CH_2=CHCH_2OH$ (d) $CH_3\underset{OH}{\underset{|}{C}}=CH_2$

113 Consider the following statements.
Statement I Cannizzaro reaction converts HCHO into HCOONa and CH_3OH.
Statement II It is a proton-hydride transfer reaction.
(a) Both statements I and II are correct
(b) Only statement I is correct
(c) Only statement II is correct
(d) Both statements I and II are incorrect

114 What is the final product of the following Crossed Aldol condensation?
$$CH_3\overset{O}{\overset{||}{C}}H + CH_3CH_2\overset{O}{\overset{||}{C}}CH_3 \xrightarrow{OH^-} \xrightarrow{\Delta}$$
(a) $CH_3CH=CHCH_2\overset{O}{\overset{||}{C}}CH_3$
(b) $CH_3CH_2\underset{CH_3}{\underset{|}{C}}=CHCH_3$
(c) $CH_3CH=\underset{CH_3}{\underset{|}{C}}\overset{O}{\overset{||}{C}}CH_3$
(d) $CH_3CH=CHCH=CHCH_3$

115 Consider the following two reactions I and II
$$CH_3CHO + 3HCHO \xrightarrow[(I)]{OH^-} (CH_2OH)_3CCHO$$
$$\downarrow OH^- \text{ (II)}$$
$$(CH_2OH)_4C + (CH_2OH)_3CCOO$$
Reactions at stages I and II are respectively
(a) Cannizzaro and Aldol (b) Aldol and Aldol
(c) Cannizzaro and Cannizzaro (d) Aldol and Cannizzaro

116 Among the following, enolisation is maximum in case of

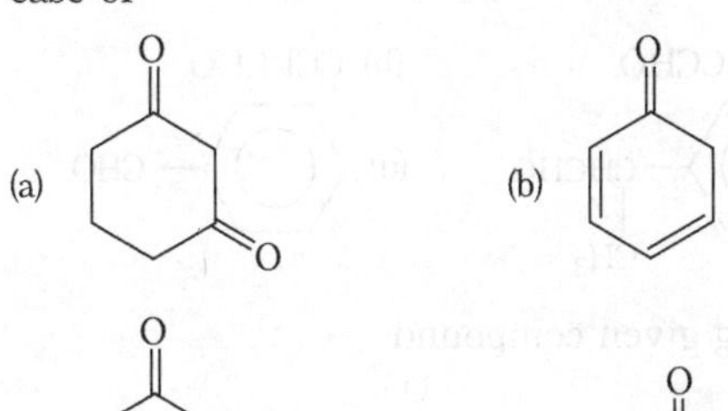

(a) (b)

(c) (d) $C_6H_5\overset{O}{\overset{\|}{C}}C_6H_5$

117 Consider the following reaction

$$CH_3CH_2CHO \xrightarrow[\text{(Aldol)}]{NaOH,\ \Delta} A.\ A \text{ is}$$

(a) $CH_3CH_2CH{=}CHCH_2CHO$ (b) $CH_3CH_2CH{=}\overset{CH_3}{\overset{|}{C}}CHO$

(c) $CH_3CH_2\underset{OH}{\underset{|}{CH}}CH_2CH_2CHO$ (d) $CH_3CH_2\underset{OH}{\underset{|}{CH}}\overset{CH_3}{\overset{|}{CH}}CHO$

118 In reaction,

$$CH_3-\underset{O}{\underset{\|}{C}}-CH_2-CH_2-CH_2-\underset{O}{\underset{\|}{C}}-CH_3 \xrightarrow{OH^-,\Delta}$$

A (formed by Aldol condensation) is

(a) $CH_3\overset{O}{\overset{\|}{C}}C-CH_2$ / CH_3C-CH_2 (b) $CH_3\overset{O}{\overset{\|}{C}}CH-CH_2$ / $CH_3C{=}CH$

(c) $CH_3\overset{O}{\overset{\|}{C}}CH-CH$ / CH_3CH-CH (d) None is correct

119 The compound that can be formed by Aldol condensation of acetaldehyde

(a) $CH_3\underset{O}{\underset{\|}{C}}CH_2COOH$

(b) $CH_3\underset{O}{\underset{\|}{C}}CH_2\underset{O}{\underset{\|}{C}}H$

(c) $CH_3\underset{OH}{\underset{|}{CH}}CH_2\underset{O}{\underset{\|}{C}}H$

(d) $CH_3CH_2\underset{OH}{\underset{|}{CH}}-\underset{O}{\underset{\|}{C}}H$

120 In the presence of an acid or base compound (*A*)

A

changes to

(a) (b)

(c) HO, OH, OH (d)

121 Benzophenone and benzaldehyde can be distinguished by
(a) Tollen's solution
(b) Fehling's solution
(c) Iodoform test
(d) 2,4 DNP solution

122 Acetaldehyde and benzaldehyde can be distinguished by
(a) Tollen's reagent
(b) Fehling's solution
(c) Iodoform test
(d) Both (b) and (c)

123 Benzoin is formed by
(a) benzophenone (b) benzaldehyde
(c) acetophenone (d) quinone

124 Which of the following does not give iodoform test?
(a) CH_3CHO (b) $C_6H_5COCH_3$
(c) CH_3COCH_3 (d) C_6H_5CHO

125 $\text{(cyclohexenyl)}-OCH_3 \xrightarrow{H_3O^{\oplus}} A,\quad A$ is

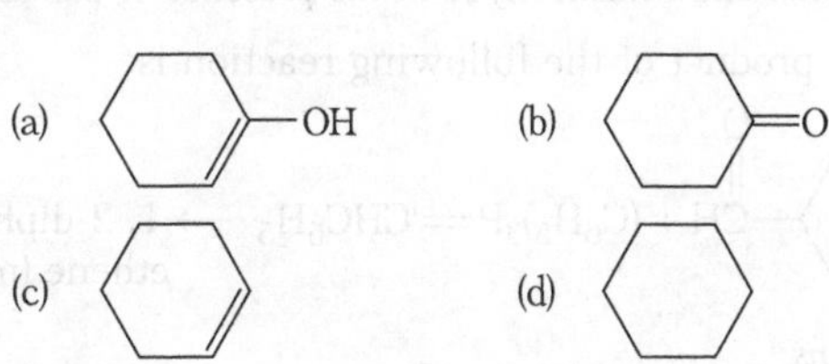

(a) (b)

(c) (d)

126 The reaction $C_6H_4CH{=}CHCHO$ with $NaBH_4$ yields
(a) $C_6H_5CH_2CH_2CH_2OH$
(b) $C_6H_5CH{=}CHCH_2OH$
(c) $C_6H_5CH_2CH_2CHO$
(d) $C_6H_5CH_2CHOHCH_3$

127 Acetone and acetophenone can be distinguished by
(a) Tollen's test
(b) DNP test
(c) Schiff's reagent
(d) Saturated solution of $NaHSO_3$

128 Which of the following species is the conjugate acid of the hemiacetal formed by reaction of benzaldehyde with methanol containing a trace of acid?

(a) $C_6H_5HC\begin{smallmatrix}OH\\OCH_3\end{smallmatrix}$ (b) $C_6H_5CH{=}\overset{+}{O}CH_3$

(c) $C_6H_5HC\begin{smallmatrix}OH\\\overset{+}{O}CH_3\\H\end{smallmatrix}$ (d) $C_6H_5HC\begin{smallmatrix}OH\\\overset{+}{O}H_2\end{smallmatrix}$

129 A single reagent that can distinguish HCHO and C_6H_5CHO is

(a) Tollen's reagent (b) I_2 + NaOH
(c) 50% NaOH (d) Fehling's reagent

130 Benzaldehyde is different from acetaldehyde in reactions towards

(a) $NaHSO_3$ solution (b) Fehling's solution
(c) Schiff's base (d) Tollen's reagent

131 Perkin reaction can be used to prepare

(a) salicylic acid (b) cinnamic acid
(c) picric acid (d) All of these

132 Which of the following can be identified by smell of bitter almonds?

(a) Nitrobenzene (b) Benzaldehyde
(c) Both (a) and (b) (d) None of these

133 Compound $Ph—O—\overset{O}{\overset{||}{C}}—Ph$ can be prepared by the reaction of ....

(a) phenol and benzoic acid in the presence of NaOH
(b) phenol and benzoyl chloride in the presence of pyridine
(c) phenol and benzoyl chloride in the presence of $ZnCl_2$
(d) phenol and benzaldehyde in the presence of palladium

134 Major product of the following reaction is

$C_6H_5—\overset{O}{\overset{||}{C}}H + (C_6H_5)_3P{=}CHC_6H_5 \longrightarrow$ 1, 2-diphenyl ethene (major)

(a) *trans*
(b) *cis*
(c) Both equally
(d) Product formed is in valid

135 Consider the following statements.

Statement I Aromatic aldehydes and formaldehyde undergo Cannizzaro reaction.

Statement II Aromatic aldehydes are almost as reactive as formaldehyde.

(a) Both statements I and II are correct
(b) Only statement I is correct
(c) Only statement II is correct
(d) Both statements I and II are incorrect

136 Which of the following gives Aldol condensation reaction?

(a) $(CH_3)_3CCHO$ (b) CCl_3CHO
(c) C_6H_5—CHCHO with CH_3 on CH (d) C_6H_5—CHO

137 Following given compound (indane-1,2,3-trione; positions 1, 2, 3)

is hydrated maximum at the position

(a) 1 (b) 2 (c) 3 (d) equal

138 What reagent and/or reaction conditions would you choose to bring about the following conversion?

2-phenyl-1,3-dioxolane $\longrightarrow$ C_6H_5—CH=O + $HOCH_2CH_2OH$

(a) $LiAlH_4$, H_2O (b) H_2O, NaOH, heat
(c) H_2O, H_2SO_4, heat (d) PCC, CH_2Cl_2

139 Which of the following does not undergo benzoin condensation?

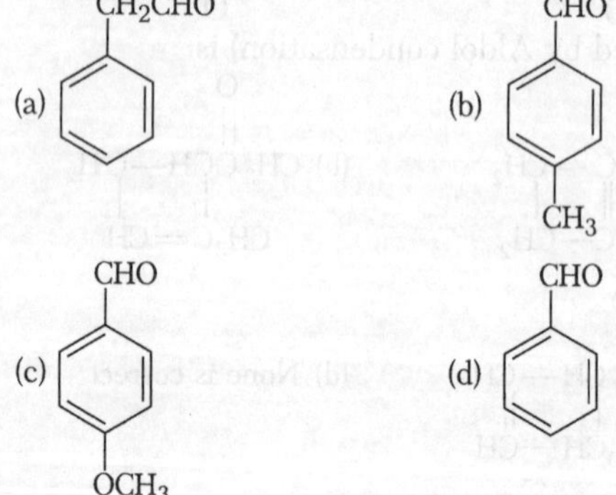

140 In the following reaction

5,8-dihydroxy-1,4-naphthoquinone derivative (• is ^{13}C) $\underset{}{\overset{OH^- \text{ or } H^+}{\rightleftharpoons}}$ A

A is

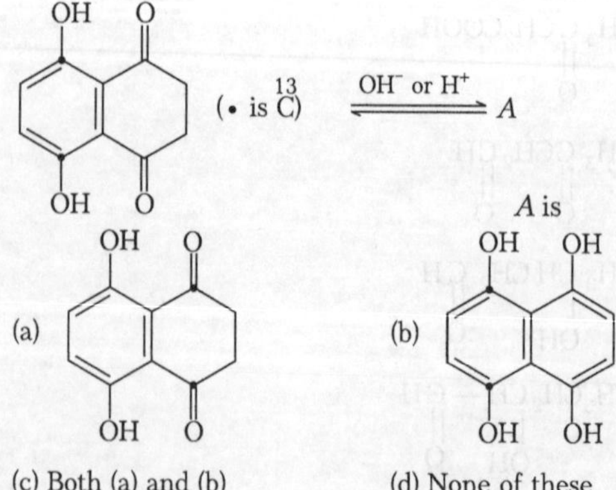

(c) Both (a) and (b) (d) None of these

141 When the following

(Ph–CO–Ph)

is brominated, product formed is

(a) 2-Br on one ring (Ph–CO–C$_6$H$_4$Br, ortho)

(b) 3-Br (Ph–CO–C$_6$H$_4$Br, meta)

(c) 4-Br (Ph–CO–C$_6$H$_4$–Br, para)

(d) Both (a) and (b)

142 Match the reactions given in Column I with their names given in Column II.

	Column I		Column II
A.	$C_6H_5CHO + CH_3CHO \xrightarrow{OH^-}$ $C_6H_5CH{=}CHCHO$	1.	Perkin
B.	$C_6H_5CHO + (CH_3CO)_2O \xrightarrow{CH_3COO^-}$ $C_6H_5CH{=}CHCOOH$	2.	Knoevenagel
C.	$C_6H_5CHO + HCHO \xrightarrow{OH^-}$ $C_6H_5CH_2OH + HCOO^-$	3.	Claisen Schmidt
D.	$C_6H_5CHO + CH_2(COOC_2H_5)_2 \xrightarrow{Base}$ $C_6H_5CH{=}CHCOOH$	4.	Crossed Cannizzaro

Codes

	A	B	C	D
(a)	3	1	4	2
(b)	3	1	2	4
(c)	1	3	2	4
(d)	1	3	4	2

143 When *m*-chloro benzaldehyde is treated with 50% KOH solution, the product(s) obtained is(are)

(a) m-ClC$_6$H$_4$–CH(OH)–CH(OH)–C$_6$H$_4$Cl-m

(b) m-HOC$_6$H$_4$–CH(OH)–CH(OH)–C$_6$H$_4$OH-m

(c) m-HOC$_6$H$_4$–COO$^-$, m-HOC$_6$H$_4$–CH$_2$OH

(d) m-ClC$_6$H$_4$–COO$^-$, m-ClC$_6$H$_4$–CH$_2$OH

144 Identify A in the following reaction,

$$C_8H_8O\,A \xrightarrow{NaOH/I_2} B \xrightarrow[\Delta]{\text{Sodalime}} \text{Benzene}$$

and choose the correct option.

(a) o-CH$_3$C$_6$H$_4$–CHO (b) C$_6$H$_5$–COCH$_3$

(c) o-CH$_3$C$_6$H$_4$–CH$_2$OH (d) All of these

145 End product in the following reaction is

$$C_6H_5\text{–C}{\equiv}\text{N} \xrightarrow{SnCl_2/HCl} X \xrightarrow{\text{Steam distillation}} Y$$

(a) benzamide (b) benzophenone

(c) benzoic acid (d) benzaldehyde

146 Consider the following reaction

$$A \xrightarrow{H_3O^+} \text{glycol} + B$$

(A: cyclohexenone ethylene ketal, O= ring spiro-dioxolane)

in which B can

I. undergoes tautomerism

II. form oximes

III. gives colour with neutral $FeCl_3$

Select the correct statement (s) and choose the correct option.

(a) All are correct (b) I, II and III are correct

(c) Both I and II are correct (d) Both II and III are correct

147 Most electrophilic site is present in

(a) $C_6H_5-\overset{O}{\overset{\|}{C}}-H$ (b) $C_6H_5-\overset{O}{\overset{\|}{C}}-CH_3$

(c) $C_6H_5-\overset{O}{\overset{\|}{C}}-C_6H_5$ (d) $CH_3-\overset{O}{\overset{\|}{C}}-CH_3$

148 Consider the following conversion

$CH_3-\overset{O}{\overset{\|}{C}}-C_6H_3(OCH_3)-CHO \longrightarrow CH_3-\overset{O}{\overset{\|}{C}}-C_6H_3(OCH_3)-\overset{OH}{\overset{|}{C}}HCH_3$

Above conversion can be carried by reactions with

(a) CH_3MgBr, H_3O^+ (b) $\underset{CH_2OH}{\overset{CH_2OH}{|}}$, CH_3MgBr, H_3O^+

(c) Both (a) and (b) (d) None of these

149 In the following reaction,

$C_6H_6 + CO + \underset{x}{\underline{HCl}} \longrightarrow C_6H_5\underset{y}{CHO} + \underset{z}{\underline{HCl}}$

H(x) in the reactant goes as

(a) H(y) in C_6H_5CHO

(b) H(z) in HCl

(c) in C_6H_5CHO and HCl both

(d) None of thes above

150 Consider the following reaction

$C_6H_5-CHO + C_6H_5-NH_2 \longrightarrow X$

Product (X) is called

(a) $C_6H_5-\underset{OH}{\underset{|}{C}H}-NH-C_6H_5$ Schiff's base

(b) $C_6H_5-\underset{OH}{\underset{|}{C}H}-C_6H_4-NH_2$ Pyridine base

(c) $OHC-C_6H_4-C_6H_4-NH_2$ Pyridine base

(d) $C_6H_5-CH{=}N-C_6H_5$ Schiff's base

(B) Miscellaneous questions

1 In the following reaction,

$\underset{A}{>C{=}\ddot{O}:} + Nu^{\ominus} \longrightarrow \underset{Nu}{\underset{|}{C}}-\ddot{O}:^{\ominus}$

rate is maximum, when A is

(a) HCHO (b) CH_3CHO

(c) $(CH_3)_2CO$ (d) C_6H_5CHO

2 Addition reaction is made favourable in the presence of

(a) $-CF_3$

(b) $-CCl_3$

(c) $-NO_2$

(d) All of the above

3 Select the incorrect statement.

(a) Carbonyl compounds are attacked by nucleophiles.

(b) An aldehyde has a greater partial positive charge on its carbonyl carbon.

(c) Protonation of a carbonyl compound decreases electrophilic nature and thus, nucleophilic attack is retarded.

(d) Hydration of CCl_3CHO is more than that of CH_3CHO.

4 Propanal reacts with amino methane forming

(a) *anti*-imine (b) *syn*-imine

(c) Both (a) and (b) (d) None of these

5 Which of the following compounds can be regenerated to carbonyl compounds?

(a) Oximes, hydrazones

(b) Oximes, bisulphite complex

(c) Bisulphite complex, hydrazones

(d) All of the above

6 Urotropine is formed from

(a) urea and formaldehyde

(b) urea and ammonia

(c) formaldehyde and ammonia

(d) urea and acetaldehyde

7 CH_3CN reacts with two moles of CH_3MgBr followed by hydrolysis in each stage. Final product is

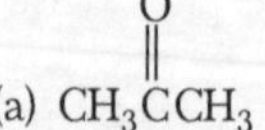

(a) $CH_3\overset{O}{\overset{\|}{C}}CH_3$ (b) $CH_3\overset{O}{\overset{\|}{C}}OH$

(c) $(CH_3)_3COH$ (d) $(CH_3)_2CHOH$

8 Aldehyde group can be protected against alkaline oxidising agents by

(a) acetal formation

(b) mercaptal formation

(c) cyclic acetal formation

(d) All of the above

9 Hydration formation of the following is in order

CCl_3CHO, CF_3CHO, $CH_3\overset{O}{\overset{\|}{C}}CH_3$, $CCl_3\overset{O}{\overset{\|}{C}}CH_3$
I II III IV

(a) I < II < III < IV (b) III < IV < I < II
(c) III < IV < II < I (d) IV < III < I < II

10 Which of the following will reduce Tollen's reagent as well as Fehling's reagent?

(a) o-methylbenzaldehyde ($C_6H_4(CH_3)$–CHO) (b) C_6H_5–CH_2CHO
(c) C_6H_5–$\overset{O}{\overset{\|}{C}}CH_3$ (d) C_6H_5–$\overset{OH}{\overset{|}{C}}=CH_2$

11 In the following reaction,

$$\left[\begin{matrix}H_3C\\H_3C\end{matrix}\!\!>CHO\right]_3 Al + 3\begin{matrix}CH_3CH_2\\CH_3CH_2\end{matrix}\!\!>CHOH \longrightarrow 3\begin{matrix}H_3C\\H_3C\end{matrix}\!\!>CO + \left[\begin{matrix}CH_3CH_2\\CH_3CH_2\end{matrix}\!\!>CHO\right]_3 Al$$
(A) (B)

(a) there is oxidation of *A*
(b) there is oxidation of *B*
(c) there is reduction of *B*
(d) there is disproportionation of *A*

12 Consider the following reagents.
I. $LiAlH_4$
II. $H_2/Pd-BaSO_4$
III. DIBAH
IV. $LiAlH(O-tC_4H_9)_3$

Which of the above reagents can be used to reduce CH_3COCl into CH_3CHO?
(a) Both I and II (b) Both II and III
(c) Both I and IV (d) Only II

13 Consider the following reactions,

I. $CH_3\overset{O}{\overset{\|}{C}}Cl \xrightarrow{CH_3MgBr} \xrightarrow{H_3O^+} CH_3\overset{O}{\overset{\|}{C}}CH_3$

II. $CH_3\overset{O}{\overset{\|}{C}}Cl \xrightarrow{(CH_3)_2CuLi} CH_3\overset{O}{\overset{\|}{C}}CH_3$

Pure yield of CH_3COCH_3 is obtained in
(a) Only I
(b) Only II
(c) Both (a) and (b)
(d) None of the above

14 In the following reaction,

$$(RCOO)_2Ca + (R'COO)_2Ca \longrightarrow CH_3\underset{CH_3}{\underset{|}{CH}}\overset{O}{\overset{\|}{C}}CH_3$$
(A) (B)

A and *B* are respectively

(a) $CH_3\underset{CH_3}{\underset{|}{CH}}COOH$ and $CH_3\overset{O}{\overset{\|}{C}}CH_2COOH$
(b) $\left(CH_3\underset{CH_3}{\underset{|}{CH}}COO\right)_2Ca$ and $\left(CH_3\overset{O}{\overset{\|}{C}}CH_2COO\right)_2Ca$
(c) $\left(CH_3\underset{CH_3}{\underset{|}{CH}}COO\right)_2Ca$ and $(CH_3COO)_2Ca$
(d) $(CH_3COO)_2Ca$ and $\left(CH_3\underset{CH_3}{\underset{|}{CH}}\overset{O}{\overset{\|}{C}}CH_2COO\right)_2Ca$

15 Compound $A(C_5H_{10}O)$ gives yellow precipitate on heating with $Ca(OH)_2$ and I_2, calcium salt thus, formed forms compound *B* on heating

$$B: (CH_3)_2CH\overset{O}{\overset{\|}{C}}CH(CH_3)_2$$

Thus, *A* is

(a) $(CH_3)_2CH\overset{O}{\overset{\|}{C}}CH_3$ (b) $(CH_3)_2CHCH_2CHO$
(c) $CH_3CH_2\overset{O}{\overset{\|}{C}}CH_2CH_3$ (d) $CH_3\overset{O}{\overset{\|}{C}}CH_2CH_2CH_3$

16 What is *A* in the following reaction?

A $\xrightarrow[\text{(ii) } H_2O/Zn]{\text{(i) } O_3}$ $CH_3CO(CH_2)_4COCH_3$ (octane-2,7-dione, skeletal structure)

Choose correct option according

(a) 3,4-dimethylcyclohexene (CH_3, CH_3) (b) 1,2-dimethylcyclohexene (CH_3, CH_3)
(c) tetramethylcyclobutene (H_3C, CH_3, H_3C, CH_3) (d) 1,6-dimethylcyclohexene (CH_3, CH_3)

17 Consider the following statements.

Statement I Hydroboration-oxidation converts terminal alkyne (R—C≡CH) into ketones.

Statement II This is addition of H_2O as per anti-Markownikoff's addition.

(a) Both statements I and II are correct
(b) Only statement I is correct
(c) Only statement II is correct
(d) Neither statement I nor statement II is correct

18 In the following reaction,

m-nitrobenzaldehyde (CHO, NO_2) $\xrightarrow{A}$ m-nitrobenzyl alcohol (CH_2OH, NO_2); $\xrightarrow{B}$ m-aminobenzyl alcohol (CH_2OH, NH_2); $\xrightarrow{C}$ m-nitrotoluene (CH_3, NO_2)

A, *B* and *C* are respectively

(a) $NaBH_4/C_2H_5OH$; $LiAlH_4$/ether; N_2H_4/glycol, OH^-
(b) $LiAlH_4$/ ether; $NaBH_4/C_2H_5OH$; N_2H_4/glycol, OH^-
(c) N_2H_4/glycol, OH^- in all cases
(d) $LiAlH_4/C_2H_5OH$ in all cases

19 Hydrolysis product of *A* (given below)

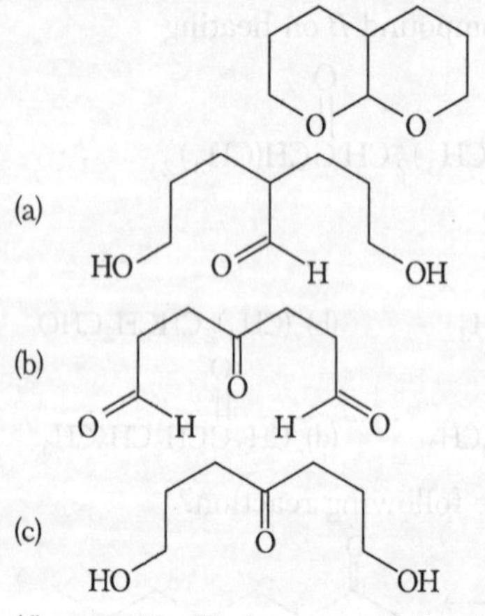

(d) ring is stable hence, no hydrolysis

20 End product in the following sequence of reaction

cyclohexene $\xrightarrow{O_3/H_2O}$ $\xrightarrow{Ca(OH)_2}$ $\xrightarrow{\Delta}$ $\xrightarrow[OH^-/glycol]{NH_2-NH_2}$

is

(a) cyclobutane (b) cyclopentane

(c) cyclohexanone (d) cyclohexane

21 Consider the following reaction with some hint given in bracket.

$$2D-\overset{D}{\overset{|}{C}}=O + OH^- \xrightarrow{\text{Cannizzaro}} X \text{ and } Y$$

(*Y* is alcohol, D is deuterium)

X and *Y* will have structure

(a) $D-\overset{O}{\overset{\|}{C}}-O^-$, $D-\underset{D}{\underset{|}{\overset{D}{\overset{|}{C}}}}-OH$

(b) $D-\overset{O}{\overset{\|}{C}}-O^-$, $D-\underset{H}{\underset{|}{\overset{D}{\overset{|}{C}}}}-OH$

(c) $H-\overset{O}{\overset{\|}{C}}-O^-$, $D-\underset{H}{\underset{|}{\overset{D}{\overset{|}{C}}}}-OH$

(d) None of the above correct

22 Consider the following reaction sequence

$$\text{Aldehyde } A(C_4H_8O) \xrightarrow[\text{Tischenko}]{\text{Al-}t\text{-butoxide}} B(\text{ester}) \xrightarrow{H_3O^+} (H_3C)_2CHCOOH + (H_3C)_2CHCH_2OH$$

Here *A* and *B* are

(a) $(H_3C)_2CHCHO$, $(H_3C)_2CHCOOCH_2CH(CH_3)_2$

(b) $(H_3C)_2CHCHO$, $CH_3CHCOOCH(CH_3)_2$

(c) Both (a) and (b)
(d) None of the above is correct

23 $CH_3CH=CHCHO \xrightarrow[\text{Aldol}]{OH^-} \xrightarrow{\Delta} A$, *A* is

(a) $CH_3(CH=CH)_3CHO$
(b) $CH_3CH_2CH_2(CH=CH)_2CHO$
(c) $CH_3(CH_2CH_2)_3CH=CH-CHO$
(d) None of the above is correct

24 End product of the following sequence of reactions

cyclohexene $\xrightarrow{NBS}$ $\xrightarrow{Mg/ether}$ $\xrightarrow[(ii)\ H_3O^+]{(i)\ CH_3CN}$

is

(a) [cyclohexene with $COCH_3$] (b) [cyclohexene with $COCH_3$]

(c) [cyclohexene with CN] (d) [cyclohexene with COOH]

25 Consider the following statements.

Statement I The C=O bond (0.122 nm) is shorter than the C–O bond (0.141 nm).

Statement II The C=O bond is present in carbonyl compounds and carboxylic acid while C–O bond is in alcohols.

(a) Both statements I and II are correct
(b) Only statement I is correct
(c) Only statement II is correct
(d) Neither statement I nor statement II is correct

26 In the reaction

$$\text{2-ethynylcyclohexanone} \xrightarrow{HgSO_4/H_2SO_4} A,$$

A is

(a) 2-($C(=O)CH_3$)cyclohexanone (b) 2-(CH_2CHO)cyclohexanone

(c) 2-($COOH$)cyclohexanone (d) 2-(CH_2CHO)cyclohexanol (OH)

27 What are *A*, *B* and *C* in the following?

$$B \xleftarrow{\text{Raney Ni, } H_2} \text{[octahydronaphthalenone]} \xrightarrow{\text{Pt, 1 mol } H_2} A$$

$$\downarrow NaBH_4, CH_3OH$$

$$C$$

Choose the correct option.

(a) [decalone] in all cases

(b) [octahydronaphthalenol, OH] in all cases

(c) [decalol, OH] in all cases

(d) *A* : [decalone] , *B* : [decalol, OH] ,

C : [decalol, OH]

28 Which products are formed when the compound [benzene]–CHO is treated with concentrated aqueous KOH solution?

(a) $\overset{+}{K}\overset{-}{O}$–[C$_6H_4$]–CHO

(b) [C$_6H_5$]–C(=O)–$\overset{-}{O}\overset{+}{K}$ + [C$_6H_5$]–CH_2OH

(c) $\overset{+}{K}\overset{-}{O}$–[C$_6H_4$]–C(=O)–$\overset{-}{O}\overset{+}{K}$ + $\overset{+}{K}\overset{-}{O}$–[C$_6H_4$]–$\overset{-}{O}\overset{+}{K}$

(d) [C$_6H_5$]–C(=O)–$\overset{-}{O}\overset{+}{K}$ + [C$_6H_5$]–$\overset{-}{O}\overset{+}{K}$

29 Consider the following reaction

$$CH_3-C\equiv CH \xrightarrow[1\%\ HgSO_4]{40\%\ H_2SO_4} A \xrightarrow{\text{Isomerisation}} CH_3-\underset{\underset{O}{\|}}{C}-CH_3$$

Structure of *A* and type of isomerism in the above reaction are respectively

(a) prop-1-en-2-ol, metamerism
(b) prop-1-en-1-ol, tautomerism
(c) prop-2-en-2-ol, geometrical isomerism
(d) prop-1-en-2-ol, tautomerism

30 Consider the following reaction

$$CH_3CHO \xrightarrow[\text{(ii) } H_2O]{\text{(i) } CH_3MgBr} (A) \xrightarrow{H_2SO_4,\ \Delta} (B) \xrightarrow{\text{Hydroboration oxidation}} (C)$$

Compounds *A* and *C* in the above reaction are ....

(a) identical
(b) positional isomers
(c) functional isomers
(d) optical isomers

31 In the reaction

$$CH_3CN + 2H \xrightarrow[SnCl_2]{HCl} X \xrightarrow{\text{Boiling } H_2O} Y,$$

the term Y is

(a) acetone (b) ethanamine
(c) acetaldehyde (d) dimethyl amine

32 Consider the following statements.

Statement I Hydration of acetone with isotopic ^{18}O enriched water forms isotopic ^{18}O enriched acetone.

Statement II There is S_N reaction replacing ^{16}O (of acetone) by ^{16}O (of water).

(a) Both statements I and II are correct
(b) Only statement I is correct
(c) Only statement II is correct
(d) Neither statement I nor statement II is correct

33 In the reaction sequence

$$CH_3CH_2COOH \xrightarrow{SOCl_2} A$$

$$A \xrightarrow{LiAlH(O\text{-}tBu)_3} B \xrightarrow{HOCH_2CH_2OH} C.$$ 'C' is most likely to be

(a) [structure: H_3C-substituted 1,4-dioxene ring]
(b) CH_3CH_2CH with O—CH_2 / O—CH_2 (1,3-dioxolane)
(c) CH_3CH_2CHO
(d) $CH_3CH_2CH_2$ — O — $CHCH_2$ (epoxide)

34 Consider the following conversion

[cyclohexyl–$\overset{O}{\overset{\|}{C}}CH_3$] $\xrightarrow{?}$ [cyclohexyl–$\underset{OH}{\underset{|}{C}}$ (CH_3) $HCHCH_2CH_3$]

Which of the following sets of reagents, would successfully accomplish the conversion shown?

(a) $CH_3CH_2CH_2MgBr$; H_3O^+; PCC, CH_2Cl_2
(b) $CH_3CH_2CH_2MgBr$; H_3O^+; H_2SO_4, heat; PCC, CH_2Cl_2
(c) $(C_6H_5)_3\overset{+}{P}-\overset{-}{\ddot{C}}HCH_2CH_3$; B_2H_6; H_2O_2, HO^-
(d) $(C_6H_5)_3\overset{+}{P}-\overset{-}{\ddot{C}}HCH_2CH_3$; H_2SO_4, H_2O

35 The amino ketone shown undergoes a spontaneous cyclisation on standing.

[benzene ring with $\overset{O}{\overset{\|}{C}}CH_3$ and $CH_2\underset{CH_3}{\underset{|}{C}}HNH_2$ ortho]

What is the product of this intramolecular reaction?

(a) [structure: O, CH_3] (b) [structure: CH_3, NH_2]
(c) [structure: CH_3, NH_2] (d) [structure: CH_3, N, CH_3]

36 Which of the two routes shown here would actually lead to the product?

Route Y: (i) HCl (ii) EtMgBr; Route X: (i) EtMgBr (ii) HCl; starting material [methyl vinyl ketone, O]; products HO, Cl

(a) Route X (b) Route Y
(c) Both (a) and (b) (d) None of these

37 What is the end product of the following reaction?

[$(CH_3)_3C$–CHO] + [HO–$C(CH_3)_2$–CO_2H] $\longrightarrow$

(a) [structure: OH, O, =O] (b) [structure: O, O, =O]
(c) [structure: O, O–O] (d) No reaction takes place

38 Consider the following transformation,

[R–CH_2–$\overset{O}{\overset{\|}{C}}$–$CH_3$] $\xrightarrow[CH_3ONa]{CH_3NO_2} A \xrightarrow[\text{Pyridine}]{Ac_2O} B$

Here, A and B are respectively

	A	B
(a)	R, OH, NO_2	R, NO_2
(b)	R, OH, NO_2	R, O, NO_2
(c)	R, NO_2	R, NO_2
(d)	Only A R, NO_2 is formed	

39 Product P in the following reaction

$$\text{3-}CH_3O\text{-}C_6H_4\text{-}C{\equiv}N + \underset{1\ mol}{CH_3MgBr} \xrightarrow{H_3O^+} P$$

is

(a) 3-CH_3O-C_6H_4-CH(OH)-CH_3 (b) 3-CH_3O-C_6H_4-CO-CH_3

(c) 3-CH_3O-C_6H_4-CHO (d) 3-CH_3O-C_6H_4-COOH

40 Consider the following statements.

Statement I Carbonyl compounds are more soluble in water than the corresponding alkanes.

Statement II H-bonding between carbonyl oxygen and water renders carbonyl compounds more water-soluble than hydrocarbons.

(a) Both statements I and II are correct
(b) Only statement I is correct
(c) Only statement II is correct
(d) Neither statement I nor statement II is correct

41 Match the species given in Column I with the corresponding reaction shown given in Column II.

Column I	Column II
A. HCHO	1. Benzoin condensation
B. CH_3CHO	2. Iodoform test
C. CH_3COCH_3	3. Cannizzaro reaction
D. C_6H_5CHO	4. Mesitylene with conc. H_2SO_4

Codes

	A	B	C	D		A	B	C	D
(a)	3	2	4	1	(b)	3	2	1	4
(c)	2	3	4	1	(d)	2	3	1	4

42 In the following,

$$CH_3CHO \xrightarrow[I]{CN^-} CH_3-\underset{H}{\overset{CN}{C}}-O^{\ominus} \xrightarrow[II]{H^+} CH_3-\underset{H}{\overset{CN}{C}}-OH$$

slow step is

(a) Only I (b) Only II
(c) Both (a) and (b) (d) None of these

43 The major product obtained in the following reaction is

(bicyclic lactone with COOH) $\xrightarrow{\text{DIBAL-H}}$

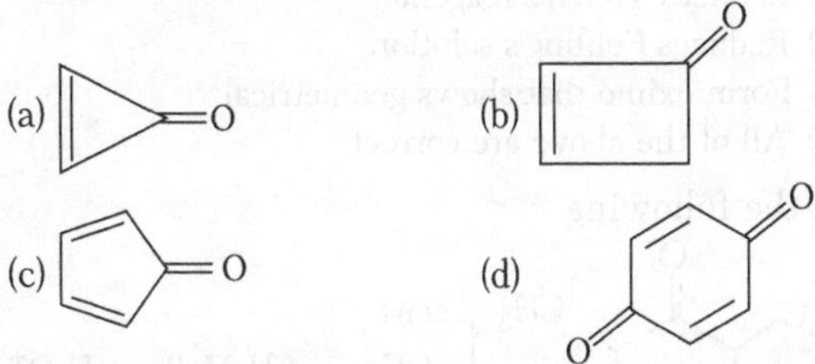

44 Which of the following compounds will show highest dipole moment?

(a) cyclopropenone (b) cyclobutenone (c) cyclopentadienone (d) p-benzoquinone

45 The correct sequence of reagents for the following conversion will be

4-formylcyclohexanone $\longrightarrow$ 1-methyl-4-(1-hydroxy-1-methylethyl)cyclohexan-1-ol

(a) $[Ag(NH_3)_2]^+OH^-$, H^+ / CH_3OH, CH_3MgBr
(b) CH_3MgBr, H^+ / CH_3OH, $[Ag(NH_3)_2]^+OH^-$
(c) CH_3MgBr, $[Ag(NH_3)_2]^+OH^-$, H^+ / CH_3OH
(d) $[Ag(NH_3)_2]^+OH^-$, CH_3MgBr, H^+ / CH_3OH

46 Enol form of (A) is

(A)

(a) 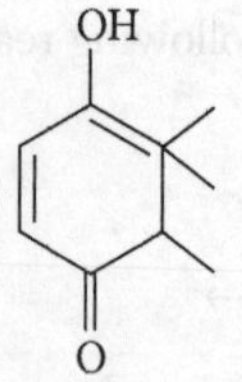 [structure] (b) 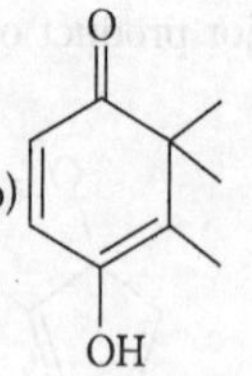 [structure]

(c) 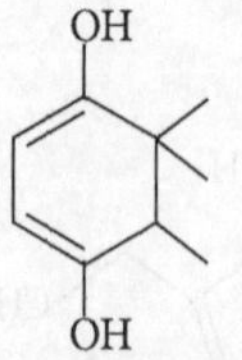 [structure] (d) None of these

47 In the following hydrolysis reaction of (*A*)

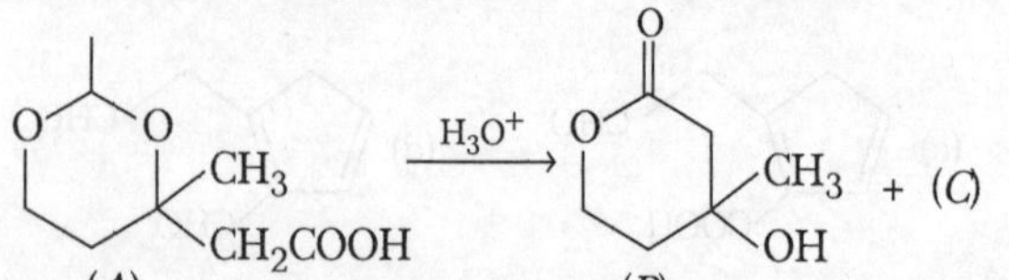

$$(A) \xrightarrow{H_3O^+} (B) + (C)$$

Product *C*

(a) Reduces Tollen's reagent
(b) Reduces Fehling's solution
(c) Form oxime that shows geometrical
(d) All of the above are correct

48 In the following

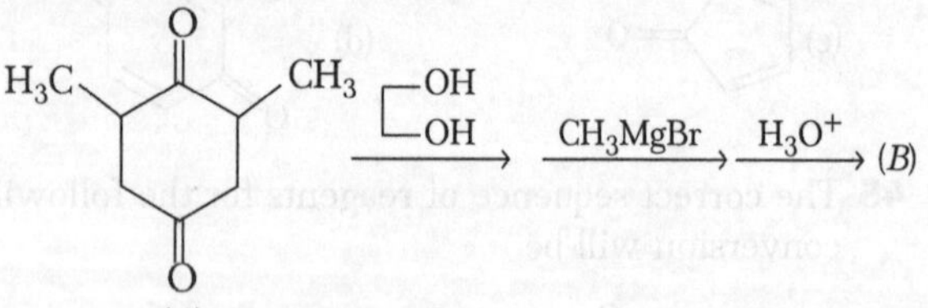

$$(A) \xrightarrow{\text{ethylene glycol}} \xrightarrow{CH_3MgBr} \xrightarrow{H_3O^+} (B)$$

(*B*) is identified as

(a) 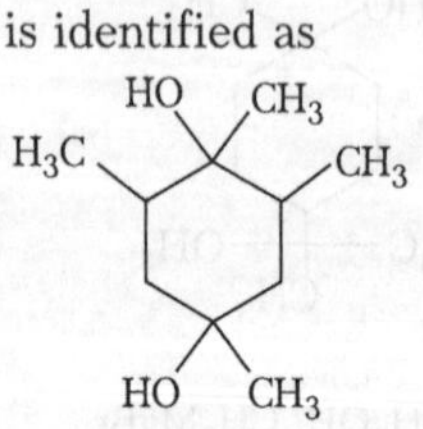 [structure] (b) 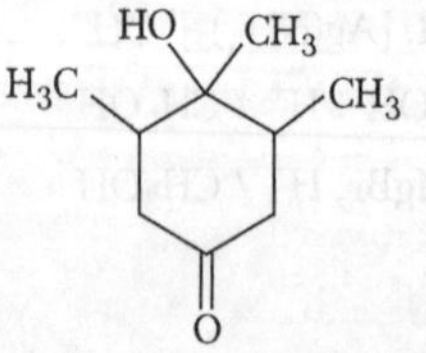 [structure] (c) 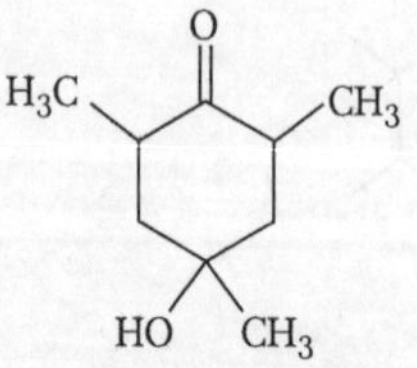 [structure]

(d) None is correct

49 Match the reaction given in column I with the name of reaction given in column II.

Column I	Column II
A. $2CH_3CHO \xrightarrow[\Delta]{\overline{O}H} CH_3CH{=}CHCHO$	1. Gattermann-Koch
B. $2HCHO \xrightarrow{\overline{O}H} HCOO^- + CH_3OH$	2. Benzoin
C. $C_6H_5CHO \xrightarrow[KCN]{\text{Alcoholic}} C_6H_5{-}CH(OH)COC_6H_5$	3. Cannizzaro
D. $C_6H_5 + CO + HCl \xrightarrow{AlCl_3} C_6H_5CHO$	4. Aldol

Codes

	A	B	C	D
(a)	4	3	2	1
(b)	4	2	3	1
(c)	4	1	2	3
(d)	2	3	4	1

(C) Medical entrances' gallery

1 Reaction between benzaldehyde and acetophenone in presence of dilute NaOH is known as **[NEET 2020]**

(a) Cannizzaro's reaction (b) Cross Cannizzaro's reaction
(c) Cross aldol condensation (d) Aldol condensation

2 An alkene on ozonolysis gives methanal as one of the product. Its structure is **[NEET 2020]**

(a) 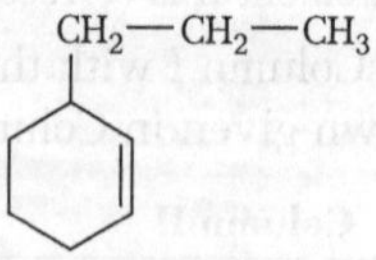 $CH_2{-}CH_2{-}CH_3$ [structure] (b) 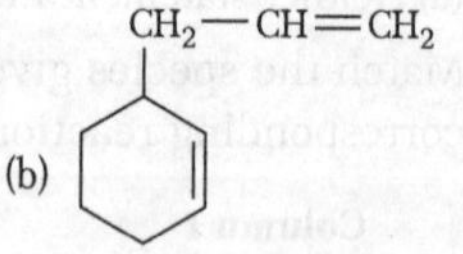 $CH_2{-}CH{=}CH_2$ [structure]
(c) 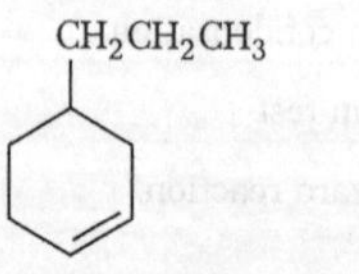 $CH_2CH_2CH_3$ [structure] (d) 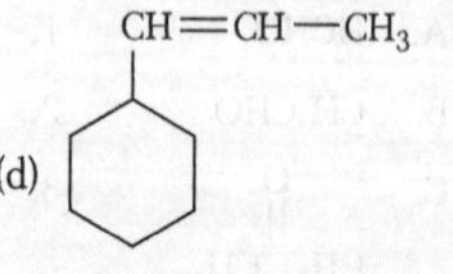 $CH{=}CH{-}CH_3$ [structure]

3 Identify compound (*A*) in the following reaction.

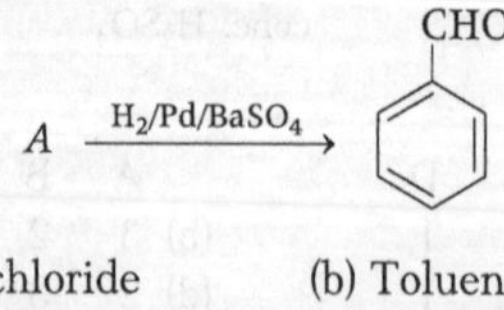

$$A \xrightarrow{H_2/Pd/BaSO_4} C_6H_5CHO$$

[NEET 2020]

(a) Benzoyl chloride (b) Toluene
(c) Acetophenone (d) Benzoic acid

4 $H{-}\overset{O}{\overset{\|}{C}}{-}H + CH_3{-}CH{=}O \xrightarrow{\text{Conc.NaOH}}$ Find out the products of reaction **[AIIMS 2019]**

(a) CH_3CO_2Na and CH_3OH
(b) $CH_3CH_2OH + CH_3OH$
(c) CH_3CH_2OH and HCO_2Na
(d) $CH_3CO_2Na + HCO_2Na$

5 Suitable reagent for conversion of following reaction

$CH_2 = CH—CHO \xrightarrow{?} CH_2 = CH — CH_2 — OH$?

(a) $NaBH_4$ (b) Ni / H_2 **[JIPMER 2019]**
(c) Zn / Hg / HCl (d) Red P + HI

6 Write IUPAC name of the following **[JIPMER 2019]**

(Structure: [1]CHO on C2; chain C2–C3(Br)–C4–C5–C6 with methyl on C2)

(a) 2-bromo-2-ethylhexanal
(b) 3-bromo-2-methylhexanal
(c) 2-methyl-3-bromohexanal
(d) 3-bromo-2-formylhexane

7 The correct IUPAC name of the given compound is **[AIIMS 2018]**

(Structure: cyclohexenone ring with OH and C=O)

(a) 7-hydroxycyclohex -5-en-1-one
(b) 3-hydroxycyclohex-5-en-1-one
(c) 8-hydroxycyclohex -3-en-1-one
(d) 5-hydroxycyclohex -3-en-1-one

8 Which of the following compound does not undergo aldol condensation? **[AIIMS 2018]**

(a) $CH_3 - C(CH_3)_2 - CHO$ (b) $CH_3 - CHO$

(c) $CH_3 - \overset{O}{\overset{||}{C}} - CH_3$ (d) CH_3CH_2CHO

9 Which of the following is the correct sequence of step involved in the mechanism of Cannizzaro reaction? **[JIPMER 2018]**

(a) Nucleophilic attack, transfer of H^- and transfer of H^+
(b) Transfer of H^-, transfer of H^+ and nucleophilic attack
(c) Transfer of H^+, nucleophilic attack and transfer of H^-
(d) Electrophilic attack by OH^-, transfer of H^+ and transfer of H^-

10 In the reaction, **[JIPMER 2018]**

$C_6H_5CH{=}O \xrightarrow[CH_3COONa]{CH_3COOCOCH_3} A,$

The major product formed is
(a) $C_6H_5— CH_2— CHO$
(b) $C_6H_5 — COOH$
(c) $C_6H_5 — CH = CH — COOH$
(d) $C_6H_5 — CH = CH — COCH_3$

11 Cyclohexanone $\xrightarrow[\text{(aldol)}]{NaOH, \Delta}$ A, A can be **[NEET 2017]**

(a) (b) (c) (d) (structures)

12 Predict the correct intermediate and product in the following reaction **[NEET 2017]**

$H_3C—C \equiv CH \xrightarrow[HgSO_4]{H_2O, H_2SO_4}$ intermediate (A) $\longrightarrow$ product (B)

(a) $A : H_3C—\underset{OH}{C}=CH_2$ $B : H_3C—\underset{O}{\underset{||}{C}}—CH_3$

(b) $A : H_3C—\underset{SO_4}{C}=CH_2$ $B : H_3C—\underset{O}{\underset{||}{C}}—CH_3$

(c) $A : H_3C—\underset{OH}{C}=CH_2$ $B : H_3C—\underset{SO_4}{C}=CH_2$

(d) $A : H_3C—\underset{O}{\underset{||}{C}}—CH_3$ $B : H_3C—C \equiv CH$

13 Consider the reactions **[NEET 2017]**

$X (C_2H_6O) \xrightarrow[573 K]{Cu} A \xrightarrow[^-OH, \Delta]{[Ag(NH_3)_2]^+}$ Silver mirror observation

$A \xrightarrow{^-OH, \Delta} Y$

$A \xrightarrow{NH_2—NH—\overset{O}{\overset{||}{C}}—NH_2} Z$

Identify A, X, Y and Z
(a) A-ethanol, X-acetaldehyde, Y-butanone, Z-hydrazone
(b) A-methoxymethane, X-ethanoic acid, Y-acetate ion, Z-hydrazine
(c) A-methoxymethane, X-ethanol, Y-ethanoic acid, Z-semicarbazide
(d) A-ethanal, X-ethanol, Y-but-2-enol, Z-semicarbazone

14 The correct statement regarding a carbonyl compound with a hydrogen atom on its α-carbon, is **[NEET (I) 2016]**

(a) A carbonyl compound with a hydrogen atom on its α-carbon rapidly equilibrates with its corresponding enol and this process is known as aldehyde-ketone equilibration
(b) A carbonyl compound with a hydrogen atom on its α-carbon rapidly equilibrates with its corresponding enol and this process is known as carbonylation
(c) A carbonyl compound with a hydrogen atom on its α-carbon rapidly equilibrates with its corresponding enol and this process is known as keto-enol tautomerism
(d) A carbonyl compound with a hydrogen atom on its α-carbon never equilibrates with its corresponding enol

15 The correct structure of the product *A* formed in the reaction **[NEET(II) 2016]**

$\xrightarrow[\text{Pd/carbon, ethanol}]{H_2\text{ (gas, 1 atmosphere)}} A$ is

(a) OH (b) O (c) OH (d) OH ⊕

16 Which one of the following undergoes iodoform test? **[Kerala CEE 2015]**

(a) Propanal (b) Ethanal
(c) Benzophenone (d) Benzaldehyde
(e) Phenol

17 The formation of cyanohydrins from a ketone is an example of **[KCET 2015]**

(a) nucleophilic substitution (b) nucleophilic addition
(c) electrophilic addition (d) electrophilic substitution

18 A single compound of the structure is **[CBSE AIPMT 2015]**

$OHC—CH_2—C(CH_3)H—CH_2—C(CH_3)=O$

obtainable from ozonolysis of which of the following cyclic compounds?

(a) H_3C, CH_3 (b) H_3C, H_3C
(c) H_3C CH_3 (d) CH_3, CH_3

19 Which one is most reactive towards nucleophilic addition reaction? **[CBSE AIPMT 2014]**

(a) CH_2CHO (b) CHO, CH_3
(c) CHO, OCH_3 (d) CHO

20 In which of the following reactions new carbon-carbon bond is not formed? **[Kerala CEE 2014]**

(a) Cannizzaro reaction (b) Wurtz reaction
(c) Aldol condensation (d) Friedel-Craft reaction
(e) Kolbe's reaction

21 In the following reaction, the product *E* is

$\begin{matrix}CHO\\|\\CHO\end{matrix} \xrightarrow[\text{(ii) } H^+]{\text{(i) NaOH}} E$ **[WB JEE 2014]**

(a) $\begin{matrix}CH_2OH\\|\\CHO\end{matrix}$ (b) $\begin{matrix}CHO\\|\\CO_2H\end{matrix}$

(c) $\begin{matrix}CH_2OH\\|\\CO_2H\end{matrix}$ (d) $\begin{matrix}CO_2H\\|\\CO_2H\end{matrix}$

22 The intermediate *J* in the following Wittig reaction is

$\overset{+}{P}Ph_3Br^- \xrightarrow[\text{(ii) } CH_2=O]{\text{(i) } n\text{-BuLi}} [J] \longrightarrow$

[WB JEE 2014]

(a) PPh_3, O (b) PPh_3, O
(c) O, PPh_3 (d) O^-, $\overset{+}{P}Ph_3$

23 Which of the following reactions cannot be used to increase the number of carbon atoms in a organic compound? **[UK PMT 2014]**

(a) Grignard reaction
(b) Cannizzaro's reaction
(c) Aldol condensation
(d) All of the above

24 $CH_3COC_2H_5 \xrightarrow{C_6H_5CO_3H} A \xleftarrow[\text{reaction}]{\text{Tischenko}} B$

Identify A and B in this reaction. **[UK PMT 2014]**

(a) $CH_3COOC_2H_5$, CH_3CH_2OH
(b) $CH_3COOC_2H_5$, CH_3CHO
(c) CH_3CHO, $CH_3COOC_2H_5$
(d) None of the above

25 $X \xrightarrow[\text{(reductive)}]{\text{Ozonolysis}} Y + Z$

Y can be obtained by Etard's reaction, Z undergoes disproportionation reaction with concentrated alkali. X could be **[KCET 2014]**

(a) 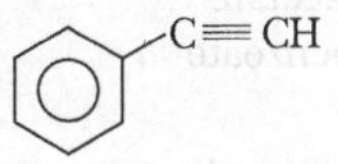 C_6H_5—C≡CH (b) C_6H_5—CH=CH$_2$
(c) C_6H_5—CH=CH—CH$_3$ (d) 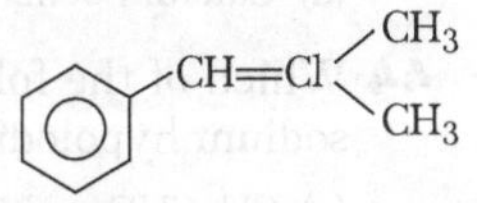 C_6H_5—CH=C(CH_3)$_2$

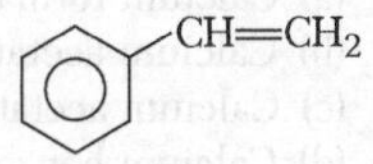

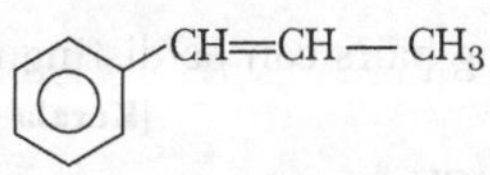

26 $P \xrightarrow[\text{(ii) } H_3O^+]{\text{(i) } CH_3MgBr} R \xrightarrow[\text{(ii) } \Delta]{\text{(i) DilNaOH}}$ 4-methylpent-3-en-2-one **[KCET 2014]**

(a) propanone (b) ethanamine
(c) ethane nitrile (d) ethanal

27 Which one of the following compounds does not give a primary alcohol on reduction? **[Guj CET 2014]**

(a) Propanal
(b) Propanoic acid
(c) Methylpropanoate
(d) Propan-2-one

28 Reaction by which benzaldehyde cannot be prepared is **[NEET 2013]**

(a) 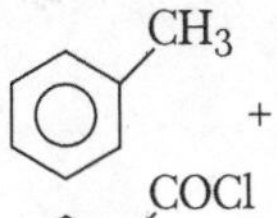 $C_6H_5CH_3$ + CrO_2Cl_2 in CS_2 followed by H_3O^+
(b) 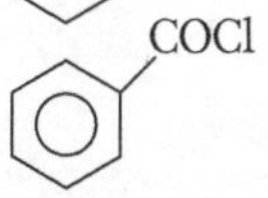 C_6H_5COCl + H_2 in the presence of Pd-$BaSO_4$
(c)  C_6H_6 + CO + HCl in the presence of anhyd. $AlCl_3$
(d) 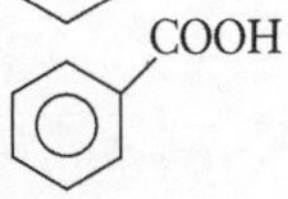 C_6H_5COOH + Zn/Hg and conc. HCl

29 Aldehydes and ketones can be distinguished by **[AIIMS 2013]**

(a) ammonia
(b) H_2SO_4
(c) alkaline $KMnO_4$
(d) Fehling solution

30 Which of the following organic compounds answers both iodoform test and Fehling's test? **[KCET 2013]**

(a) Methanal (b) Ethanol
(c) Propane (d) Ethanal

31 Predict the products in the given reaction,

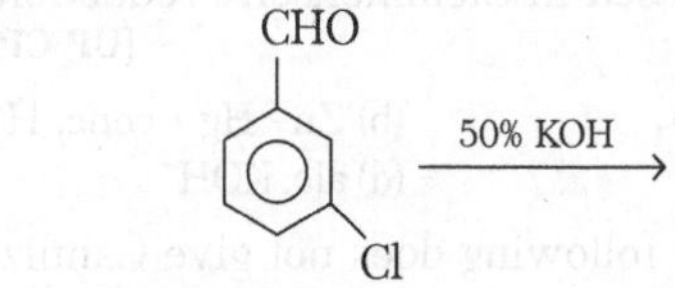

3-chlorobenzaldehyde (CHO, Cl) $\xrightarrow{50\% \text{ KOH}}$ **[CBSE AIPMT 2012]**

(a) 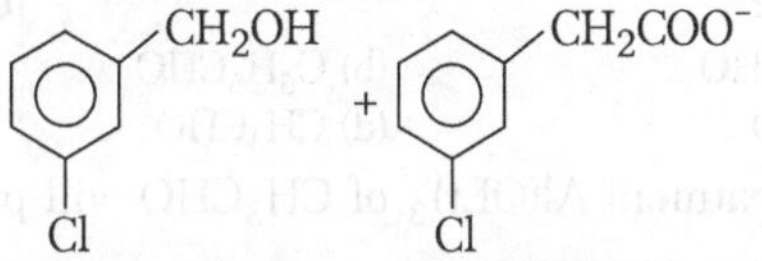 CH_2OH (Cl) + CH_2COO^- (Cl)
(b) 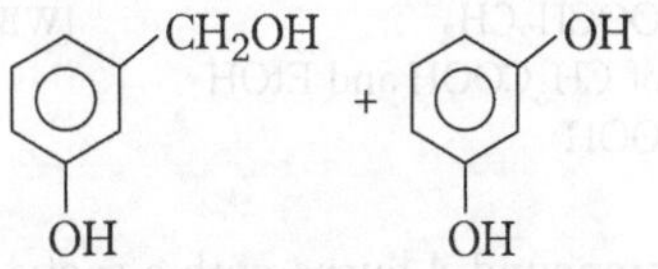 CH_2OH (OH) + OH (OH)
(c) 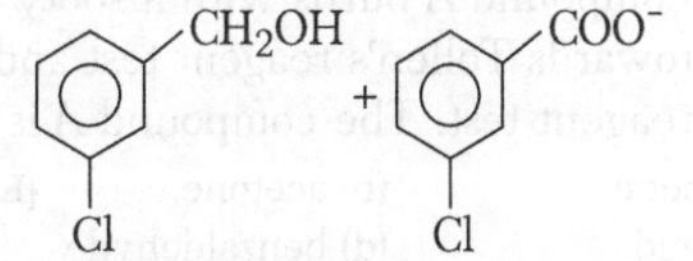 CH_2OH (Cl) + COO^- (Cl)
(d) 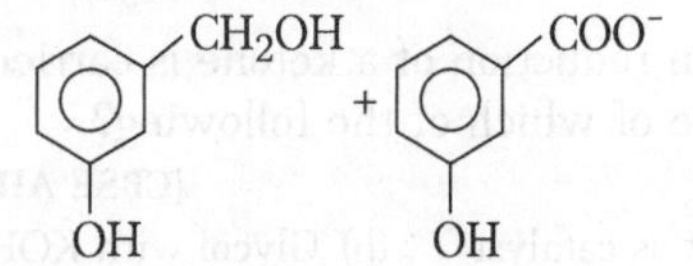 CH_2OH (OH) + COO^- (OH)

32 Consider the following reaction,

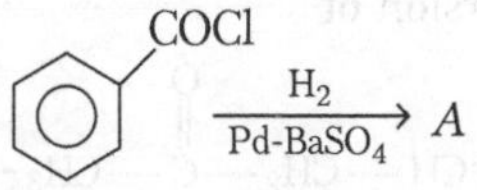

$C_6H_5COCl \xrightarrow[\text{Pd-BaSO}_4]{H_2} A$

The product A is **[CBSE AIPMT/AMU 2012]**

(a) C_6H_5CHO (b) C_6H_5OH
(c) $C_6H_5COCH_3$ (d) C_6H_5Cl

33 Which of the following compounds undergoes both Cannizzaro and Aldol reaction? **[AFMC 2012]**

(a) $(CH_3)_2CH \cdot CHO$ (b) HCHO
(c) C_6H_5CHO (d) CH_3CHO

34 The product formed when acetone reacts nitro methane in the presence of a base is **[AFMC 2012]**

(a) $CH_3—C(CH_3)_2—CH_2NO_2$
(b) $CH_3—C(CH_3)(OH)—CH(OH)—NO_2$
(c) $(H_3C)_2C=CHNO_2$
(d) $CH_3COCH=CHNO_2$

35 The reagent used in Clemmensen's reduction is **[UP CPMT 2012]**

(a) conc. H_2SO_4 (b) Zn - Hg / conc. HCl
(c) *aq.* KOH (d) alc. KOH

36 Which of the following does not give Cannizzaro's reaction? **[OJEE 2012]**

(a) CCl_3CHO (b) C_6H_5CHO
(c) HCHO (d) CH_3CHO

37 Upon treatment $Al(OEt)_3$ of CH_3CHO will produce. **[WB JEE 2012]**

(a) Only $CH_3COOCH_2CH_3$
(b) A mixture of CH_3COOH and EtOH
(c) Only CH_3COOH
(d) Only EtOH

38 An organic compound *A* burns with a sooty flame. It is negative towards Tollen's reagent test and positive for Brady's reagent test. The compound *A* is **[KCET 2012]**

(a) acetophenone (b) acetone
(c) salicylic acid (d) benzaldehyde

39 Clemmensen reduction of a ketone is carried out in the presence of which of the following? **[CBSE AIPMT 2011]**

(a) H_2 and Pt as catalyst (b) Glycol with KOH
(c) Zn–Hg with HCl (d) $LiAlH_4$

40 In the conversion of

$$C_6H_5-CH=CH-CH_2-\overset{O}{\overset{\|}{C}}-CH_3 \xrightarrow{X} C_6H_5-CH=CH-CH_2-\underset{CH_3}{\underset{|}{\overset{OH}{\overset{|}{CH}}}}$$

'*X*' is **[DUMET 2011]**

(a) H_2/Pt (b) Zn—Hg/HCl
(c) Li/NH_3 (d) $NaBH_4$

41 In Tollen's test, aldehydes **[DUMET 2011]**

(a) are oxidised
(b) are reduced to alcohol
(c) neither reduced nor oxidised
(d) precipitate Ag^+ as AgCl

42 The reagent with which both acetaldehyde and acetone react is **[KCET 2011]**

(a) Fehling's solution (b) I_2/NaOH
(c) Tollen's reagent (d) Carbonic acid

43 Which of the following gives an aldehyde on dry distillation? **[KCET 2011]**

(a) Calcium formate + calcium acetate
(b) Calcium acetate + calcium benzoate
(c) Calcium acetate
(d) Calcium benzoate

44 Which of the following pairs can be distinguished by sodium hypoiodite? **[Kerala CEE 2011]**

(a) CH_3CHO and CH_3COCH_3
(b) CH_3CH_2CHO and CH_3COCH_3
(c) CH_3CH_2OH and $CH_3CH_2CHOHCH_3$
(d) CH_3OH and CH_3CH_2CHO
(e) CH_3OH and $CH_3CH_2CH_2OH$

45 The compound that neither forms semicarbazone nor oxime is **[Kerala CEE 2011]**

(a) HCHO (b) CH_3COCH_2Cl
(c) CH_3CHO (d) $CH_3CONHCH_3$
(e) $(CH_3)_2CHCHO$

46 Which of the following reactions convert acetone into hydrocarbon having same number of carbon atoms? **[Guj. CET 2011]**

(a) Wolff-Kishner reaction
(b) Hofmann reaction
(c) Grignard reaction
(d) Reduction with

ANSWERS

Check point 18.1

1. (c)	2. (a)	3. (d)	4. (d)	5. (b)	6. (a)	7. (d)	8. (c)	9. (c)	10. (a)

Check point 18.2

1. (c)	2. (d)	3. (c)	4. (b)	5. (a)	6. (c)	7. (d)	8. (a)	9. (d)	10. (d)

Check point 18.3

1. (c)	2. (b)	3. (c)	4. (d)	5. (c)	6. (c)	7. (a)	8. (b)	9. (c)	10. (d)

Check point 18.4

1. (b)	2. (b)	3. (c)	4. (d)	5. (a)	6. (c)	7. (b)	8. (c)	9. (a)	10. (d)

(A) Topical exercise

1. (a)	2. (b)	3. (a)	4. (a)	5. (d)	6. (a)	7. (a)	8. (d)	9. (b)	10. (c)
11. (d)	12. (b)	13. (b)	14. (a)	15. (d)	16. (a)	17. (c)	18. (b)	19. (d)	20. (a)
21. (c)	22. (d)	23. (b)	24. (a)	25. (d)	26. (b)	27. (a)	28. (b)	29. (c)	30. (c)
31. (a)	32. (a)	33. (b)	34. (a)	35. (c)	36. (b)	37. (d)	38. (b)	39. (d)	40. (a)
41. (d)	42. (c)	43. (a)	44. (d)	45. (c)	46. (a)	47. (a)	48. (d)	49. (a)	50. (b)
51. (b)	52. (a)	53. (c)	54. (c)	55. (b)	56. (a)	57. (a)	58. (c)	59. (c)	60. (c)
61. (a)	62. (c)	63. (a)	64. (d)	65. (a)	66. (a)	67. (c)	68. (c)	69. (d)	70. (b)
71. (a)	72. (c)	73. (a)	74. (a)	75. (d)	76. (b)	77. (c)	78. (a)	79. (c)	80. (a)
81. (b)	82. (c)	83. (a)	84. (c)	85. (b)	86. (c)	87. (b)	88. (c)	89. (a)	90. (b)
91. (c)	92. (a)	93. (d)	94. (a)	95. (c)	96. (b)	97. (c)	98. (b)	99. (a)	100. (b)
101. (c)	102. (c)	103. (a)	104. (a)	105. (a)	106. (b)	107. (d)	108. (a)	109. (b)	110. (a)
111. (b)	112. (b)	113. (a)	114. (c)	115. (d)	116. (b)	117. (b)	118. (a)	119. (c)	120. (b)
121. (a)	122. (d)	123. (b)	124. (d)	125. (b)	126. (b)	127. (d)	128. (c)	129. (d)	130. (b)
131. (b)	132. (c)	133. (b)	134. (a)	135. (b)	136. (c)	137. (b)	138. (b)	139. (a)	140. (c)
141. (b)	142. (a)	143. (d)	144. (b)	145. (d)	146. (a)	147. (c)	148. (b)	149. (a)	150. (d)

(B) Miscellaneous questions

1. (a)	2. (d)	3. (c)	4. (c)	5. (d)	6. (c)	7. (c)	8. (d)	9. (b)	10. (b)
11. (b)	12. (d)	13. (b)	14. (c)	15. (a)	16. (b)	17. (c)	18. (a)	19. (a)	20. (b)
21. (a)	22. (a)	23. (a)	24. (b)	25. (a)	26. (b)	27. (d)	28. (b)	29. (d)	30. (b)
31. (c)	32. (b)	33. (b)	34. (c)	35. (d)	36. (b)	37. (b)	38. (a)	39. (b)	40. (a)
41. (a)	42. (a)	43. (a)	44. (b)	45. (a)	46. (b)	47. (d)	48. (b)	49. (a)	

(C) Medical entrances' gallery

1. (c)	2. (b)	3. (a)	4. (c)	5. (a)	6. (b)	7. (d)	8. (a)	9. (a)	10. (c)
11. (a)	12. (a)	13. (d)	14. (c)	15. (b)	16. (b)	17. (b)	18. (a)	19. (d)	20. (a)
21. (c)	22. (a)	23. (b)	24. (b)	25. (b)	26. (c)	27. (d)	28. (d)	29. (d)	30. (d)
31. (c)	32. (a)	33. (a)	34. (c)	35. (b)	36. (d)	37. (a)	38. (a)	39. (c)	40. (d)
41. (a)	42. (b)	43. (a)	44. (b)	45. (d)	46. (a)				

Hints & Explanations

(A) Topical exercise

6 (*a*) $-\overset{|}{C}=O$ (sp^2) and $-\overset{|}{C}=\overset{|}{C}-$ (sp^2)

They do not differ in hybridisation of carbon.

7 (*a*) $CH_3CH_2OH \xrightarrow{Cr_2O_7^{2-}/H^+} CH_3CHO + Cr^{3+}$
Ethanol; Green

8 (*d*) A is $CH_3\overset{O}{\overset{\|}{C}}\,CH_2\overset{O}{\overset{\|}{C}}\,OC_2H_5$

enol due to H at α-carbon

Oximes due to carbonyl group, iodoform test due to CH_3CO.

9 (*b*) $\begin{array}{l} CH_2CH_2COO \\ | \\ CH_2CH_2COO \end{array}\!\!\Big\rangle Ca \xrightarrow{\Delta} \begin{array}{l} CH_2CH_2 \\ | \\ CH_2CH_2 \end{array}\!\!\Big\rangle C=O$

Calcium adipate

17 (*c*) $R\overset{O}{\overset{\|}{C}}\,Cl \xrightarrow{H_2/Pd/BaSO_4} RCHO$

Thus, $RCOCl$ is,

$(CH_3)_2CH\overset{O}{\overset{\|}{C}}-Cl \xrightarrow{H_2/Pd/BaSO_4} (CH_3)_2CHCHO$

$(CH_3)_2CH\overset{O}{\overset{\|}{C}}-Cl \xrightarrow[H_3O^+]{CH_3MgBr} (CH_3)_2CH\overset{O}{\overset{\|}{C}}-CH_3$

18 (*b*) Ester is

$H-\overset{O}{\overset{\|}{C}}-O-CH(CH_3)_2 \xrightarrow{CH_3MgBr/H_3O^+} CH_3CHO$

19 (*d*) $CH_3CH_2OH + RMgBr \longrightarrow RH$ (alkane)

21 (*c*) $RCN \xrightarrow{Sn/HCl} RCH{=}NH \longrightarrow RCH_2NH_2$
Alkyl nitrile; Imine; Amine

22 (*d*) Product in reaction (d) is $CH_3\overset{O}{\overset{\|}{C}}CH_3$.

24 (*a*) $CH_3CH_2CN \xrightarrow{H_3O^+} CH_3CH_2\overset{O}{\overset{\|}{C}}OH \xrightarrow{PCl_5} CH_3CH_2\overset{O}{\overset{\|}{C}}-Cl$
Ethyl nitrile; Propanoic acid

$\downarrow (CH_3)_2CuLi$

$CH_3CH_2\overset{O}{\overset{\|}{C}}-CH_3$

26 (*b*) $CH_3C\equiv N \xrightarrow[(ii)\ H_3O^+]{(i)\ CH_3MgBr} CH_3\overset{O}{\overset{\|}{C}}CH_3$

$\downarrow CH_3MgBr / H_3O^+$

$\underset{(B)}{CH_3-\overset{CH_2}{\overset{\|}{C}}-CH_3} \xleftarrow[-H_2O]{Conc.\ H_2SO_4} \underset{(A)}{CH_3-\overset{CH_3}{\overset{|}{C}}(OH)-CH_3}$

28 (*b*) 2-acetyl-2-methylcyclohexanone ($COCH_3$, CH_3) $\xrightarrow[(ii)\ H^+]{(i)\ KBrO,\ Haloform\ test}$ 2-oxo-1-methylcyclohexanecarboxylic acid ($COOH$, CH_3) $\xrightarrow{\Delta}$ 2-methylcyclohexanone (CH_3) (*A*)

If keto group is at β-position w.r.t. COOH, CO_2 is lost due to heating.

29 (*c*) Olefin $\xrightarrow{O_3/Zn}$ $\begin{array}{l} H_3C \\ H_3C \end{array}\!\!\Big\rangle C=O + O=C\Big\langle\!\!\begin{array}{l} CH_3 \\ CH_2CH_3 \end{array}$
Acetone; Methyl ethyl ketone

$\downarrow$ On joining

$\begin{array}{l} H_3C \\ H_3C \end{array}\!\!\Big\rangle C=C\Big\langle\!\!\begin{array}{l} CH_3 \\ CH_2CH_3 \end{array}$
Olefin

33 (*b*) $\rangle C=O \rightleftharpoons \rangle \overset{\oplus}{C}-\overset{\ominus}{O}$ Nucleophilic (α at O); Electrophilic (β at C; γ)

38 (*b*) $\rangle C=O \longrightarrow \rangle\overset{\oplus}{C}-\overset{\ominus}{O} \xrightarrow{Nu^{\ominus}} \rangle\underset{Nu}{\overset{|}{C}}-O^{\ominus}$
(sp^2 at carbonyl carbon; sp^3 in product)

39 (*d*) (a) Benzophenone ($C_6H_5\overset{O}{\overset{\|}{C}}C_6H_5$) $\xrightarrow[H_3O^+]{RMgX}$ 3° alcohol

(b) Acetophenone ($C_6H_5\overset{O}{\overset{\|}{C}}-CH_3$) $\xrightarrow[H_3O^+]{RMgX}$ 3° alcohol

(c) Acetone ($CH_3-\overset{O}{\overset{\|}{C}}-CH_3$) $\xrightarrow[H_3O^+]{RMgX}$ 3° alcohol

40 (*a*) $F_3C \leftarrow \overset{O}{\overset{\|}{C}} \rightarrow CF_3$

CF_3 is electron withdrawing group making carbonyl carbon more electrophilic for nucleophilic addition.

43 (*a*) CH_3CHO forms bisulphite (crystalline) complex.

$$\underset{\text{Ethanal}}{CH_3CHO} + \underset{\text{Ethanol}}{CH_3CH_2OH} \xrightarrow[\text{(ii) Filter}]{\text{(i) NaHSO}_3} \underset{\text{Filtrate}}{CH_3CH_2OH}$$

$$\text{Insoluble complex} \xrightarrow{H_3O^+} CH_3CHO$$

44 (*d*) $CH_3CHO + HCN \longrightarrow H-\overset{CH_3}{\underset{CN}{C}}-OH$

(*d*) and (*l*) are formed in equal parts thus, racemic mixture.

50 (*b*) Hydration minimises strain in (2) to maximum extent and then in (1).

52 (*a*) Reactivity of carbonyl compounds can be decided by two factors :

(i) **Steric factor** Lesser the steric hindrance, greater will be its reactivity.

(ii) **Electronic factor** Greater the number of alkyl groups lesser will be its electrophilicity.

Hence, CH_3CHO is most reactive towards nucleophilic addition reaction.

54 (*c*) –Br (EWG) increases positive charge on carbonyl carbon for nucleophilic attack.

59 (*c*) $CH_3\underset{\underset{O}{\|}}{C}CH_2CH_2CH_2Cl \xrightarrow[\text{(ii) } H_3O^+]{\text{(i) } CH_3MgBr} CH_3-\overset{CH_3}{\underset{OH}{C}}-CH_2CH_2CH_2Cl$

Intramolecular cyclisation: $CH_3-\overset{CH_3}{C}(-O-CH_2-CH_2-CH_2-)$ (ring) $\equiv$ 2,2-dimethyltetrahydrofuran

60 (*c*) Keto group is more reactive than ester group for the addition of Grignard reagent.

$$CH_3\overset{O}{\overset{\|}{C}}CH_2CH_2\overset{O}{\overset{\|}{C}}OCH_2CH_3 \xrightarrow[\text{(ii) } H_3O^{\oplus}]{\text{(i) } CH_3MgBr \text{ (one mole)}}$$

$$CH_3-\overset{CH_3}{\underset{OH}{C}}-CH_2CH_2-\overset{}{C}(=O)-OC_2H_5 \xrightarrow[-C_2H_5OH]{\text{Intramolecular cyclisation}} CH_3-\overset{CH_3}{C}(-O-C(=O)-CH_2-CH_2-) \text{ (lactone)}$$

63 (*a*)

$$CH_2=CH-\overset{O}{\overset{\|}{C}}-H \xrightarrow{CH_3\overset{+}{Mg}Br} CH_2=CH-\overset{\overset{-\;+}{OMgBr}}{\underset{CH_3}{C}}-H \xrightarrow{H_3O^+} CH_2=CH-\overset{OH}{\underset{CH_3}{C}}-H$$

64 (*d*) CH_2-CHCH_3 (epoxide, bridged by O) $\xrightarrow{PCl_5} \underset{Cl}{CH_2}-\underset{Cl}{CH}\,CH_3$

[Two halogen at adjacent position (*vic*)]
(*A*)

$$\underset{\text{Propanal}}{CH_3CH_2CHO} \xrightarrow{PCl_5} CH_3CH_2CH\begin{matrix}Cl\\Cl\end{matrix}$$

(Two halogen at same carbon (*gem*))
(*B*)

67 (*c*) $NH_4Cl + NaCN \rightleftharpoons NH_4CN \rightleftharpoons HCN + NH_3$

$$CH_2O + H_2NH \longrightarrow CH_2{=}NH \xrightarrow{H^+CN^-} NCCH_2-NH_2 \xrightarrow[-H_2O]{CH_2O} CH_2{=}N-CH_2-CN$$

70 (*b*) $C_6H_5CO_3H$ (peroxybenzoic acid) oxidises ketone into ester (Baeyer-Villiger oxidation).

75 (*d*) $LiAlH_4 \rightleftharpoons Li^{\oplus} + (AlH_4^{\ominus}) \rightleftharpoons AlH_3 + \underset{\text{Nucleophile}}{H^{\ominus}}$

81 (*b*) Cynohydrin of ethanal forms lactic acid.

$$\underset{\text{Ethanal}}{CH_3CHO} \xrightarrow{HCN} CH_3\underset{OH}{CH}CN \xrightarrow{H_3O^+} \underset{\text{Lactic acid}}{CH_3\overset{OH}{CH}COOH}$$

82 (*c*) (a) Both –No (b) Both –No
(c) Ethanoic acid –No, propanal –Yes (d) Both-No

84 (*c*) Lucas test is for alcohol.

85 (*b*) (a property of $(-\overset{O}{\overset{\|}{C}}CH_3)$ group).

86 (*c*) Iodoform test is used to find the presence of $-COCH_3$ group which is converted into $-COOH$ group.

The reaction is shown as,

$$\underset{\text{Hex-4-en-2-one}}{CH_3-CH=CH-CH_2-\overset{O}{\overset{\|}{C}}-CH_3} \xrightarrow[\text{(Iodoform reaction)}]{I_2/NaOH \text{ solution}} \underset{\text{Pent-3-en-1-oic acid}}{CH_3-CH=CH-CH_2-\overset{O}{\overset{\|}{C}}-OH}$$

87 (*b*) $RCOOR' \xrightarrow{LiAlH_4} R'OH + RCH_2OH$

88 (*c*) C_3H_6O (carbonyl compound) gives iodoform test (yellow ppt. with $I_2/NaOH$), thus it has $-\overset{O}{\overset{\|}{C}}CH_3$.

$-\overset{O}{\overset{\|}{C}}-$ group is due to hydrolysis of $-\underset{Cl}{\underset{|}{\overset{Cl}{\overset{|}{C}}}}-$.

Then, $C_3H_6Cl_2$ is

$$\underset{\text{2,2-dichloropropane}}{CH_3-\underset{Cl}{\underset{|}{\overset{Cl}{\overset{|}{C}}}}-CH_3} \xrightarrow[-H_2O]{2NaOH} \underset{\text{Acetone}}{CH_3-\overset{O}{\overset{\|}{C}}-CH_3}$$

92 (*a*) *A* gives iodoform test, thus *A* has a group $-\overset{O}{\overset{\|}{C}}CH_3$.

C eliminates CO_2 on heating hence, *C* has two $-COOH$ groups at same (*gem*) position.

Only (a) has $\left(-\overset{O}{\overset{\|}{C}}CH_3\right)$ group.

$$\underset{(A)}{CH_3\overset{O}{\overset{\|}{C}}CH_2\overset{O}{\overset{\|}{C}}OC_2H_5} \xrightarrow[\Delta]{NaOH/I_2} \underset{\text{Yellow ppt.}}{CHI_3\downarrow} + \underset{(B)}{Na\overset{+}{}\overset{-}{O}\overset{O}{\overset{\|}{C}}CH_2\overset{O}{\overset{\|}{C}}\overset{-}{O}\overset{+}{Na}}$$

$$\xrightarrow{H^+} \underset{(C)}{HO\overset{O}{\overset{\|}{C}}CH_2\overset{O}{\overset{\|}{C}}OH} \xrightarrow[-CO_2]{\Delta} CH_3\overset{O}{\overset{\|}{C}}OH$$

93 (*d*) $\underset{\text{Diethyl carbonate}}{C_2H_5-O-\overset{O}{\overset{\|}{C}}-O-C_2H_5} \xrightarrow[\text{(ii) } H_3O^+]{\text{(i) } CH_3MgBr \text{ 2moles}} \underset{(A)}{CH_3-\overset{O}{\overset{\|}{C}}-CH_3} \xrightarrow{CH_3MgBr/H_3O^+} \underset{3° \text{ alcohol}}{(CH_3)_3COH}$

$$\underset{\text{2, 3-dimethyl 2-butyne}}{CH_3-\overset{CH_3}{\overset{|}{C}}=\overset{CH_3}{\overset{|}{C}}-CH_3} \xrightarrow{O_3/Zn/H_2O} \underset{(A)}{CH_3-\overset{O}{\overset{\|}{C}}-CH_3}$$

Thus, (a), (b) and (c) are true.

95 (*c*) There should not be H at α-C (or α-C absent) for Cannizzaro reaction.

96 (*b*) $(CH_3)_2CHCHO$

↑ α-C has H thus no Cannizzaro reaction.

98 (*b*) C_6H_5CHO has no H at α-carbon thus no Aldol condensation.

99 (*a*) In resonance form, position of atoms remains unchanged. only electrons shift.

$$\overset{\ominus}{C}H_2-\overset{O}{\overset{\|}{C}}-CH_3 \longleftrightarrow CH_2=\overset{O^{\ominus}}{\overset{|}{C}}-CH_3$$

107 (*d*) Necessary condition for Cannizzaro reaction is absence of α-hydrogen atom. So, CH_3CHO will not give Cannizzaro reaction while other three compounds have no α-hydrogen. Hence, they will give Cannizzaro reaction.

108 (*a*) $\underset{(A)}{(CH_3)_3CCHCl_2} + NaOH \longrightarrow (CH_3)_3CCH(OH)_2 \xrightarrow{-H_2O} (CH_3)_3CCHO(B)$

↑ α- C with no H

109 (*b*) $CH_3CH_2\overset{O}{\overset{\|}{C}}CH_3 \xrightarrow{C_6H_5CO_3H} CH_3CH_2-O-\overset{O}{\overset{\|}{C}}CH_3 \xrightarrow{H_3O^+} \underset{(B)}{CH_3CH_2OH} + \underset{(A)}{CH_3COOH}$

(Oxygen is inserted between carbonyl carbon and larger alkyl group)

110 (*a*) (a) Carbonyl carbon without α-H

$$\begin{matrix}CHO\\|\\COOH\end{matrix} \xrightarrow{NaOH} \underset{\substack{\text{By reduction}\\(C)}}{\begin{matrix}CH_2OH\\|\\COO\overset{-}{}\overset{+}{Na}\end{matrix}} + \underset{\substack{\text{By oxidation}\\(B)}}{\begin{matrix}CO\overset{-}{O}\overset{+}{Na}\\|\\CO\overset{-}{O}\overset{+}{Na}\end{matrix}}$$

(b) $\begin{matrix}CHO\\|\\CHO\end{matrix} \xrightarrow{NaOH} \underset{\text{Single product}}{\begin{matrix}CH_2OH\\|\\CO\overset{-}{O}\overset{+}{Na}\end{matrix}}$

(c) $\begin{matrix}CH_2OH\\|\\CHO\end{matrix} \xrightarrow{NaOH} \underset{\text{By reduction}}{\begin{matrix}CH_2OH\\|\\CH_2OH\end{matrix}} + \underset{\text{By oxidation}}{\begin{matrix}CH_2OH\\|\\CO\overset{-}{O}\overset{+}{Na}\end{matrix}}$

111 (*b*) $R-\underset{O^{\ominus}}{\underset{|}{\overset{H}{\overset{|}{C}}}}-O^{\ominus} \longrightarrow R-\underset{O^{\ominus}}{\underset{|}{C}}=O + H^-$

112 (*b*) C_3H_6O changes to α-β unsaturated aldehyde thus C_3H_6O has $-CHO$ group.

$$\underset{\text{Propanal}}{CH_3CH_2CHO} + H-\underset{CH_3}{\underset{|}{\overset{\alpha}{C}}}HCHO \xrightarrow{OH^-} CH_3CH_2\underset{OH}{\underset{|}{C}}H\overset{CH_3}{\overset{|}{C}}HCHO \xrightarrow[-H_2O]{NaHSO_4} CH_3CH_2CH=\overset{CH_3}{\overset{|}{C}}CHO$$

114 (c)

$CH_3—CHO + CH_3CH_2\overset{O}{\overset{\|}{C}}CH_3$ (More substituted and stable thermodynamic enolate formed.)

$-H_2O$, OH^-

$CH_3—\overset{O^\ominus}{\overset{|}{C^\oplus}}—H$ + $CH_3\bar{C}HCCH_3$ (C=O)

$CH_3CH(CO)CH_3$ with $CH_3—CH(OH)$ $\xleftarrow{H_2O}$ $CH_3CH(CO)CH_3$ with $CH_3—CH(O^-)$

$\xrightarrow{-H_2O}$ $CH_3CH=\underset{CH_3}{\underset{|}{C}}—\overset{O}{\overset{\|}{C}}CH_3$

115 (d) $HCHO + CH_3CHO \xrightarrow[\text{Aldol}]{OH^-} HOH_2C—\underset{CH_2OH}{\overset{CH_2OH}{C}}—CHO$ (α— C without H)

$\xrightarrow[OH^-]{\text{Cannizzaro}}$ $HOH_2C—\underset{CH_2OH}{\overset{CH_2OH}{C}}—COO^- + HOH_2C—\underset{CH_2OH}{\overset{CH_2OH}{C}}—CH_2OH$

116 (b) Cyclohexadienone $\overset{\text{Tautomerism}}{\rightleftharpoons}$ Phenol (OH)

More stable due to conjugation (aromatisation) thus, maximum enolisation.

117 (b) $\underset{\text{Propanal}}{CH_3CH_2CHO} \rightleftharpoons CH_3CH_2\overset{\oplus}{C}—O^\ominus$ with H (A)

$CH_3CH_2CHO \xrightarrow{OH^-} CH_3\bar{C}HCHO$

CH_3CHCHO + $CH_3—CH_2—\underset{H}{\overset{|}{C}}—O^-$ $\xrightarrow{H_2O}$ $CH_3CH_2—\underset{H}{\overset{CH_3CHCHO}{C}}—OH$

$\downarrow -H_2O$

$\underset{(A)}{CH_3CCHO \atop \| \atop CH_3CH_2CH}$

118 (a) $CH_3\underset{O}{\underset{\|}{C}}—CH_2—CH_2—CH_2—\underset{O}{\underset{\|}{C}}—CH_3 \xrightarrow[\text{condensation}]{\text{Intramolecular}}$

H at α-carbon is condensed with this carbonyl group

$CH_3\overset{\ominus}{C}CH—CH_2CH_2\overset{\oplus}{C}—CH_3 \longrightarrow$ $CH_3\overset{O}{\overset{\|}{C}}CH—CH_2$ / $HO—\underset{CH_3}{C}—CH_2$

$\xrightarrow{-H_2O}$ $CH_3\overset{O}{\overset{\|}{C}}\,C—CH_2$ / $\|$ / $C—CH_2$ / CH_3

119 (c) $CH_3CHO \xrightarrow{HO^-} \bar{C}H_2CHO$

$CH_3—\overset{H}{C}=O + \overset{\ominus}{C}H_2CHO \longrightarrow CH_3—\underset{H}{\overset{CH_2CHO}{C}}—O^\ominus$

$\downarrow H^+$

$CH_3—\underset{H}{\overset{CH_2CHO}{C}}—OH$ (Aldol)

120 (b) 3-methylcyclopent-3-enone (C=O) $\xrightarrow{H^-}$ enolate ⟶ enol (OH) $\underset{}{\overset{H^+}{\rightleftharpoons}}$ 3-methylcyclopentenone

121 (a)

	Benzophenone	Benzaldehyde
(a)	No	Yes
(b)	No	No
(c)	No	No
(d)	Yes	Yes

122 (d)

	Acetaldehyde	Benzaldehyde
(a)	Yes	Yes
(b)	Yes	No
(c)	Yes	No

125 (*b*) Cyclohexenyl—$OCH_3 \xrightarrow{H_3O^+}$ Cyclohexenyl—OH $\rightleftharpoons$ (Tautomerises) Cyclohexanone (=O)

Cyclohexane

128 (*c*) $C_6H_5\text{—}\overset{H}{\overset{|}{C}}\text{=O} \xrightarrow{CH_3OH} C_6H_5\text{—}\overset{H}{\overset{|}{C}}\langle^{OH}_{OCH_3}$

Benzaldehyde

$\downarrow H^{\oplus}$

$C_6H_5CH\langle^{OH}_{\overset{\oplus}{O}CH_3\text{—}H}$

129 (*d*)

	HCHO	C_6H_5CHO
(a)	Yes	Yes
(b)	No	No
(c)	Yes	Yes
(d)	Yes	No

133 (*b*) Compound Ph— COO —Ph can be prepared by the reduction of

Phenol + Benzoyl chloride $\xrightarrow{\text{Pyridine}}$ $C_6H_5\text{—O—}\overset{O}{\overset{\|}{C}}\text{—}C_6H_5$

This is an example of **Schotten-Baumann reaction.**

134 (*a*) $C_6H_5\text{(H)C=C(H)}C_6H_5$

trans-product

136 (*c*) Due to presence of H at α-carbon.

139 (*a*) Aliphatic aldehydes do not give benzoin condensation.

140 (*c*) Due to tautomerisation.

143 (*d*) 2 m-Cl-C_6H_4-CHO $\xrightarrow{OH^-}$ m-Cl-C_6H_4-COO^- + m-Cl-C_6H_4-CH_2OH

by Cannizzaro reaction.

144 (*b*)

Benzene $\xleftarrow[\Delta]{\text{Sodalime}}$ $C_6H_5\text{—}\overset{-}{CO}\overset{+}{ONa}$ (B) $\xleftarrow[\text{Iodoform test}]{NaOH + I_2}$ $C_6H_5\text{—}\overset{O}{\overset{\|}{C}}CH_3$

Acetophenone

(A) C_8H_8O

145 (*d*) Product is benzaldehyde (by Stephen's method).

146 (*a*) O=cyclohexadiene spiro-ketal (1,3-dioxolane) (A) $\xrightarrow{H_3O^+}$ $\underset{CH_2OH}{\overset{CH_2OH}{|}}$ + H=cyclohexadienone=O (B) $\rightleftharpoons$ HO—C_6H_4—OH

I : B tautomerises : yes

II : B forms oximes : yes

III : B gives colour wih $FeCl_3$ in enolic form : yes

148 (*b*) —CHO group is not protected by glycol due to steric effect.

$\text{—}\overset{O}{\overset{\|}{C}}CH_3$ group is protected by glycol.

$CH_3\overset{O}{\overset{\|}{C}}\text{—}C_6H_3(OCH_3)\text{—CHO} \xrightarrow{CH_2OH\text{—}CH_2OH}$ $\underset{CH_2\text{—O}}{\overset{CH_2\text{—O}}{|}}\rangle C(CH_3)\text{—}C_6H_3(OCH_3)\text{—CHO}$

$\downarrow CH_3MgBr,\ H_3O^+$ (left) : $H_3C\text{—}\overset{OH}{\overset{|}{C}}(CH_3)\text{—}C_6H_4\text{—}\overset{OH}{\overset{|}{C}}HCH_3$

$\downarrow CH_3MgBr,\ H_3O^+$ (right) : $CH_3\overset{O}{\overset{\|}{C}}\text{—}C_6H_3(OCH_3)\text{—}\overset{CH_3}{\overset{|}{C}}H\text{—OH}$

(B) Miscellaneous questions

3 (*c*) $>C=O \rightleftharpoons >\overset{\oplus}{C}\text{—}\overset{\ominus}{O} \xrightarrow{H^+} >\overset{\oplus}{C}\text{—OH}$

Due to protonation, electrophilic nature of carbonyl carbon is increased, hence nucleophilic attack is increased. Thus, statement (c) is incorrect.

4 (*c*) $CH_3\text{—}CH_2\text{—}\underset{H}{\underset{|}{C}}\text{=O} + H_2NCH_3 \longrightarrow CH_3\text{—}CH_2\text{—}\underset{H}{\underset{|}{C}}H\text{=}\underset{CH_3}{\underset{|}{N}}$

Propanal — *syn* and *anti*

7 (*c*) $CH_3CN \xrightarrow[H_3O^+]{CH_3MgBr} CH_3\overset{O}{\overset{\|}{C}}CH_3 \xrightarrow[H_3O^+]{CH_3MgBr} (CH_3)_3COH$

9 (*b*) Hydration of ($RCHO$) is larger than that of $R—\overset{O}{\overset{||}{C}}CH_3$. (EWG) increases hydration.
Thus, II > I and IV > III.
So, the correct order is III < IV < I < II

10 (*b*)

	Fehling's reagent	Tollen's reagent
(a)	No	Yes (aromatic)
(b)	Yes	Yes (aliphatic)
(c)	No	No
(d)	No	No

14 (*c*) $(RCOO)_2Ca + (R'COO)_2Ca \longrightarrow R'COR$
from the given product

$\underbrace{(CH_3)_2CH}_{R'}\overset{O}{\overset{||}{C}}—\underbrace{CH_3}_{R}$, R is CH_3 and R' is $(CH_3)_2CH$.

15 (*a*) $A\,(C_5H_{10}O)$ gives iodoform test thus A is

$$R\overset{O}{\overset{||}{C}}CH_3 \xrightarrow[\Delta]{Ca(OH)_2/I_2} (R\overset{O}{\overset{||}{C}}O)_2Ca + CHI_3\downarrow$$

$$\xrightarrow{\Delta} R\overset{O}{\overset{||}{C}}R \text{ and given that } (B) \text{ is } (CH_3)_2CH\overset{O}{\overset{||}{C}}CH(CH_3)_2 \; (B)$$

Thus, R is $CH_3)_2CH$ on comparison.

Thus, A is $(CH_3)_2CH\overset{O}{\overset{||}{C}}CH_3$.

16 (*b*) Ozonolysis product is symmetrical and single product with same number of C-atoms. So, reactant 'A' is

[1,2-dimethylcyclohexene structure: CH_3, CH_3]

19 (*a*) [bicyclic diether] $\xrightarrow{H_3O^+}$ [cyclohexane with HO, OH, OH, HO] $\xrightarrow{-H_2O}$ [HO–chain–CH(CHO)–chain–OH]

20 (*b*) [cyclohexene] $\xrightarrow{O_3/H_2O}$ [adipic acid: OH, O, OH, O] $\xrightarrow[\Delta]{Ca(OH)_2}$

Due to ozonolysis, H_2O_2 is also formed which oxidise —CHO to —COOH.

[cyclopentanone] $=O \xrightarrow[\substack{OH^-/glycol \\ \text{(Wolff-Kishner reduction)}}]{NH_2—NH_2}$ [cyclopentane]

21 (*a*) $D—\overset{D}{\overset{|}{C}}=O \longrightarrow D—\overset{D}{\overset{|}{\underset{\oplus}{C}}}—O^\ominus$

$\downarrow OH^-$

$D—\underset{OH}{\underset{|}{C}}=O + D^\ominus \longleftarrow D—\overset{D}{\overset{|}{\underset{OH}{\underset{|}{C}}}}—O^\ominus \rightleftharpoons D—\underset{O^\ominus}{\underset{|}{C}}=O + H^\oplus$ (X)

$D—\overset{D}{\overset{|}{C}}=O \longrightarrow D—\overset{D}{\overset{|}{\underset{\oplus}{C}}}—O^\ominus \xrightarrow[\text{and } D^\ominus]{\text{Addition of } H^\oplus} D—\overset{D}{\overset{|}{\underset{D}{\underset{|}{C}}}}—OH$

Second product (Y)

22 (*a*) By Tischenko reaction (A) $RCHO$ is converted into (B)

$$\underset{\text{Ester}}{R\overset{O}{\overset{||}{C}}OCH_2R} \xrightarrow{H_3O^+} R\overset{O}{\overset{||}{C}}OH + RCH_2OH$$

Thus, R is $(H_3C)_2CH$

Thus, A is $(H_3C)_2CHCHO$ and B is

$(H_3C)_2CH\overset{O}{\overset{||}{C}}OCH_2CH(CH_3)_2$.

23 (*a*) Vinyl α-H is not involved due to maximum (C—H) bond energy.

$$CH_3CH=CH—\overset{H}{\overset{|}{C}}=O + H—CH_2CH=CHCHO$$

$\downarrow$

$$CH_3CH=CH—\overset{H}{\overset{|}{\underset{OH}{\underset{|}{C}}}}—\underset{H}{\underset{|}{CH}}—CH=CHCHO$$

$\downarrow -H_2O$

$$\underset{(A)}{CH_3—CH=CH—CH=CH—CH=CH—CHO}$$

24 (*b*) [cyclohexene] $\xrightarrow[\substack{\text{Allylic} \\ \text{substitution}}]{NBS}$ [3-bromocyclohexene]—Br $\xrightarrow{Mg/ether}$

[cyclohexenyl]—MgBr $\xrightarrow[(ii)\ H_3O^+]{(i)\ CH_3CN}$ [cyclohexenyl]—$\overset{O}{\overset{||}{C}}CH_3$

26 (*b*) (C=O) is electron withdrawing group hence π-electrons are migrated as given

$$\text{(cyclohexanone)}-C\equiv CH \rightarrow \text{(cyclohexanone)}-\overset{\ominus}{C}=\overset{\oplus}{C}H \xrightarrow{H_2O} \text{(cyclohexanone)}-CH=CHOH \xrightarrow{\text{Tautomerises}} \text{(cyclohexanone)}-CH_2CHO$$

28 (*b*) Benzaldehyde is an aromatic aldehyde having no α-hydrogen. So, on reaction with aqueous KOH solution it undergoes Cannizzaro reaction to produce benzyl alcohol and potassium benzoate.

$$\underset{\text{Benzaldehyde}}{C_6H_5CHO} \xrightarrow{KOH\,(aq)} \underset{\text{Benzyl alcohol}}{C_6H_5CH_2OH} + \underset{\text{Potassium benzoate}}{C_6H_5\overset{-}{CO}\overset{+}{OK}}$$

29 (*d*) Chemical reaction can be shown as

$$\underset{\text{Propyne}}{CH_3-C\equiv CH} \xrightarrow[1\%\,HgSO_4]{40\%\,H_2SO_4} \underset{\text{Prop-1-en-2-ol}}{CH_3-\overset{OH}{C}=\overset{H}{C}H} \xrightarrow{\text{Keto- enol tautomerism}} \underset{\text{Acetone}}{CH_3-\overset{O}{\overset{\|}{C}}-CH_3}$$

A is prop-1-en-2-ol, which undergoes tautomerism to form acetone.

30 (*b*) Chemical reaction can be shown as

$$\underset{\text{Ethanal}}{CH_3CHO} \xrightarrow[(ii)\,H_2O]{(i)\,CH_3MgBr} \underset{\text{Propan-2-ol}}{\underset{(A)}{CH_3-\underset{CH_3}{CH}-OH}} \xrightarrow[\text{Dehydration}]{H_2SO_4,\,\Delta}$$

$$\underset{\text{Propene}}{\underset{(B)}{CH_3-\underset{H}{C}=CH_2}} \xrightarrow[\text{(Hydroboration oxidation)}]{(i)\,BH_3 \quad (ii)\,H_2O_2/OH^{\ominus}} \underset{\text{Propan-1-ol}}{\underset{(C)}{CH_3-CH_2-CH_2OH}}$$

Thus, $CH_3\underset{CH_3}{CH}-OH$ and $CH_3-CH_2-CH_2OH$ are positional isomers. Hence, option (b) is correct.

31 (*c*) The product *Y* is acetaldehyde (By Stephen method).

33 (*b*)

$$\underset{\text{Propemoic acid}}{CH_3CH_2COOH} \xrightarrow{SOCl_2} \underset{(A)}{CH_3CH_2\overset{O}{\overset{\|}{C}}Cl} \xrightarrow{LiAlH\,(O\text{-}t\,Bu)_3} \underset{(B)}{CH_3CH_2CHO} \xrightarrow{HOCH_2CH_2OH} \underset{(C)}{CH_3CH_2CH\langle{}^{O-CH_2}_{O-CH_2}}$$

34 (*c*)

$$\text{(cyclohexyl)}-\underset{CH_3}{C}=O \xrightarrow{(C_6H_5)_3P^+\overset{-}{C}HCH_2CH_3} \text{(cyclohexyl)}-\underset{CH_3}{C}=CHCH_2CH_3 \xrightarrow{B_2H_6,\,OH^-} \text{(cyclohexyl)}-\underset{CH_3}{CH}-\underset{OH}{CH}CH_2CH_3$$

35 (*d*)

$$\text{(o-}C_6H_4)(COCH_3)(CH_2CH(CH_3)NH_2) \xrightarrow[-H_2O]{\Delta} \text{cyclic imine: } C(CH_3)=N-CH(CH_3)-CH_2 \text{ fused to benzene}$$

36 (*b*) By route (*X*), there is addition at the conjugate system.

$$CH_2=CH-CO-CH_3 \xrightarrow{Et(MgBr)} CH_2=CH-C(OMgBr)(CH_3)(C_2H_5) \xrightarrow{HCl} CH_2=CH-C(OH)(CH_3)(C_2H_5) + MgClBr$$

Thus, (a) is incorrect.

By route *Y*

$$CH_2=CH-CO-CH_3 \rightleftharpoons \overset{\oplus}{C}H_2-CH=C(\overset{\ominus}{O})-CH_3 \xrightarrow{\overset{\oplus\ominus}{HCl}} Cl-CH_2CH_2COCH_3 \xrightarrow[(ii)\,H_3O^{\oplus}]{(i)\,EtMgBr} Cl-CH_2CH_2C(OH)(CH_3)(C_2H_5)$$

37 (*b*)

$$(CH_3)_3C-CHO + HO-C(CH_3)_2-COOH \longrightarrow \text{hemiacetal} \xrightarrow{-H_2O} \text{2-tert-butyl-5,5-dimethyl-1,3-dioxolan-4-one}$$

38 (*a*)

$$R-CH_2-CHO + HCH_2NO_2 \xrightarrow{CH_3\overset{-}{O}\overset{+}{Na}} \underset{(A)}{R-CH_2-\underset{OH}{CH}-CH_2NO_2} \xrightarrow{-H_2O} \underset{(B)}{R-CH=CH-CH_2NO_2}$$

39 (*b*)

$$\text{m-}CH_3OC_6H_4C\equiv N \xrightarrow[1\,mol]{CH_3MgBr} \text{m-}CH_3OC_6H_4\overset{CH_3}{C}=N-MgBr \xrightarrow{H_3O^+} \underset{(P)}{\text{m-}CH_3OC_6H_4COCH_3}$$

43 (*a*) Ester linkage is cleaved and reduced to —CHO and —OH no effect on —COOH group.

44 (*b*) All have symmetry, except (b). So, this compound show highest dipole moment.

45 (*a*) —CHO is converted to —$COOCH_3$ hence with CH_3MgBr converted to 3° alcohol.

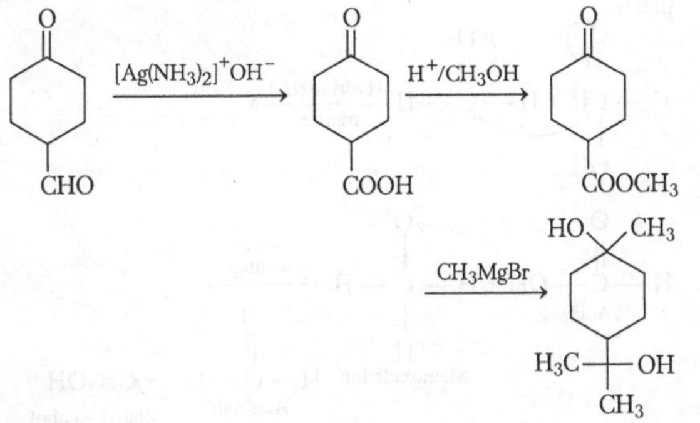

$$\text{(cyclohexanone-4-CHO)} \xrightarrow{[Ag(NH_3)_2]^+OH^-} \text{(cyclohexanone-4-COOH)} \xrightarrow{H^+/CH_3OH} \text{(cyclohexanone-4-COOCH}_3) \xrightarrow{CH_3MgBr} \text{(1-methylcyclohexan-1-ol with } C(CH_3)_2OH)$$

46 (*b*) This H is reactive

Vinylic H difficult to enolise keto group

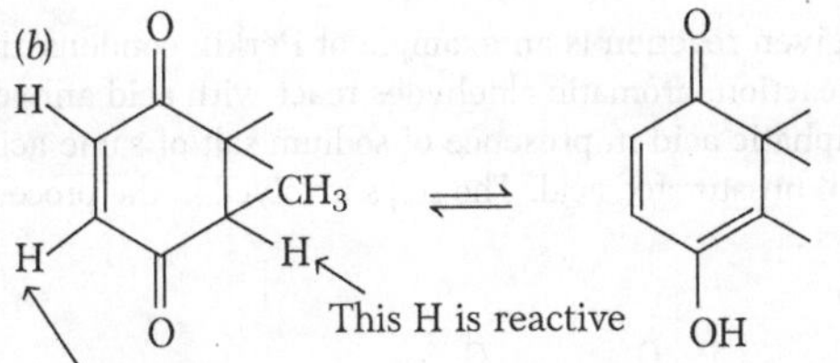

47 (*d*) *C* is CH_3CHO

(a) true (b) true

(c) $CH_3CHO + H_2NOH \longrightarrow CH_3CH{=}NOH$

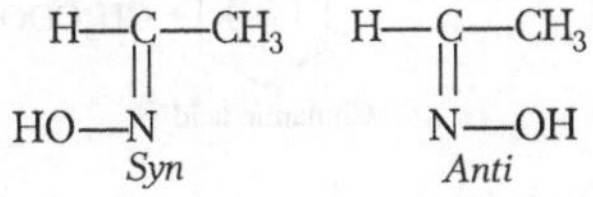

Syn *Anti*

48 (*b*)

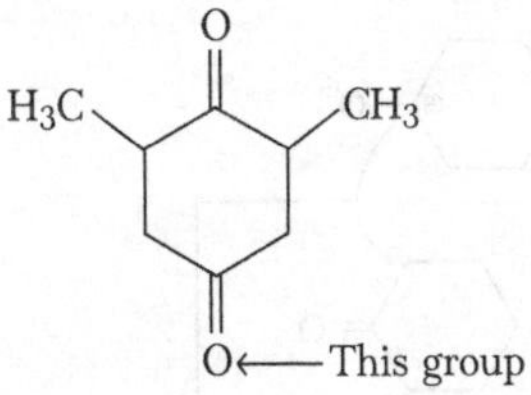

This group is blocked by glycol

So, the product (*B*) is

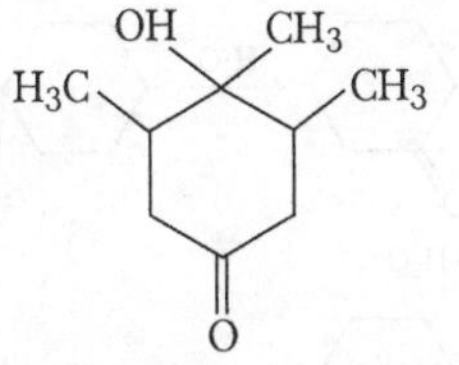

(C) Medical entrances' gallery

1 (*c*) Dilute NaOH is the reagent for aldol condensation. Dilute NaOH process enolate ion from acetophenone which attacks benzaldehyde to give aldol.

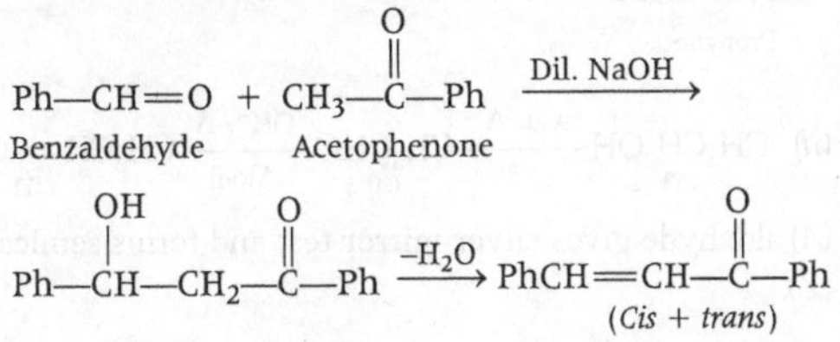

$$\underset{\text{Benzaldehyde}}{Ph{-}CH{=}O} + \underset{\text{Acetophenone}}{CH_3{-}\overset{O}{\overset{\|}{C}}{-}Ph} \xrightarrow{\text{Dil. NaOH}}$$

$$Ph{-}\overset{OH}{\overset{|}{C}H}{-}CH_2{-}\overset{O}{\overset{\|}{C}}{-}Ph \xrightarrow{-H_2O} \underset{(Cis + trans)}{PhCH{=}CH{-}\overset{O}{\overset{\|}{C}}{-}Ph}$$

It is cross aldol condensation or Claisen-Schmidt reaction or Claisen reaction.

Mechanism

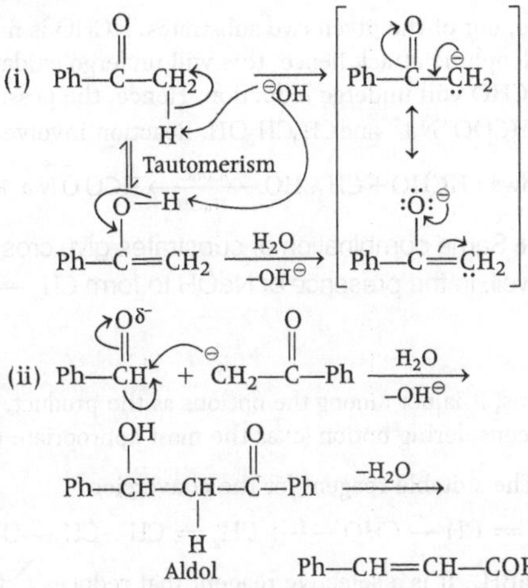

(i) $Ph{-}\overset{O}{\overset{\|}{C}}{-}CH_2{-}H \xrightarrow{^{\ominus}OH} [Ph{-}\overset{O}{\overset{\|}{C}}{-}\overset{\ominus}{C}H_2 \leftrightarrow Ph{-}\overset{:\ddot{O}:^{\ominus}}{\overset{|}{C}}{=}CH_2]$ Enolate ion

Tautomerism: $Ph{-}\overset{O{-}H}{\overset{|}{C}}{=}CH_2 \underset{-OH^{\ominus}}{\overset{H_2O}{\rightleftharpoons}}$ enolate

(ii) $Ph{-}\overset{O^{\delta-}}{\overset{\|}{C}H} + \overset{\ominus}{C}H_2{-}\overset{O}{\overset{\|}{C}}{-}Ph \underset{-OH^{\ominus}}{\xrightarrow{H_2O}}$

$$\underset{\text{Aldol}}{Ph{-}\overset{OH}{\overset{|}{C}H}{-}\underset{H}{\underset{|}{C}H}{-}\overset{O}{\overset{\|}{C}}{-}Ph} \xrightarrow{-H_2O} Ph{-}CH{=}CH{-}COPh$$

2 (*b*) Let us identify the ozonolysis [(i) O_2/THF, (ii) Zn/H_2O] products from each of the following :

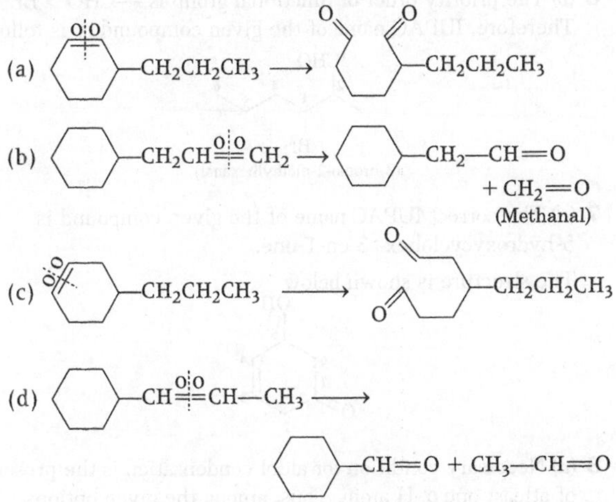

(a) (cyclohexene)—$CH_2CH_2CH_3$ ⟶ $OHC{-}CH_2CH_2CH_2{-}CH(CH_2CH_2CH_3){-}CHO$ type dicarbonyl —$CH_2CH_2CH_3$

(b) (cyclohexyl)—$CH_2CH{=}CH_2$ ⟶ (cyclohexyl)—$CH_2{-}CH{=}O$ + $CH_2{=}O$ (Methanal)

(c) (cyclohexene)—$CH_2CH_2CH_3$ ⟶ dicarbonyl—$CH_2CH_2CH_3$

(d) (cyclohexyl)—$CH{=}CH{-}CH_3$ ⟶ (cyclohexyl)—$CH{=}O$ + $CH_3{-}CH{=}O$

We get, methanal from the hydrocarbon in option (b) only.

3 (*a*)

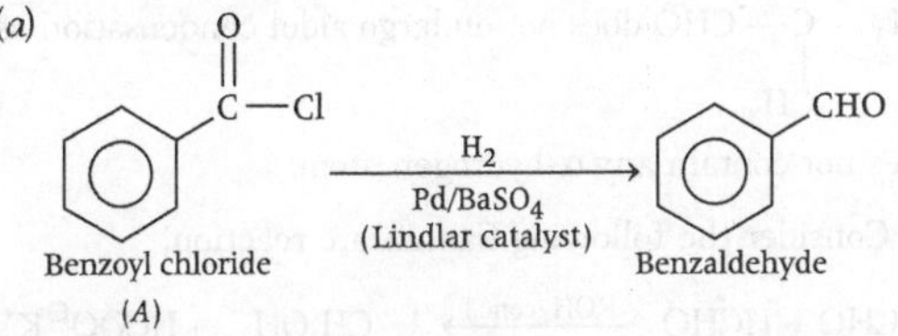

$$\underset{\substack{\text{Benzoyl chloride}\\(A)}}{C_6H_5{-}\overset{O}{\overset{\|}{C}}{-}Cl} \xrightarrow[\text{(Lindlar catalyst)}]{H_2 \; Pd/BaSO_4} \underset{\text{Benzaldehyde}}{C_6H_5CHO}$$

It is Rosenmund reaction, in which an acid chloride gets converted into an aldehyde.

4 (*c*) The given reaction (as per the options) is cross Cannizzaro reaction. A simple Cannizzaro reaction is shown by those aldehydes in which α-hydrogen is absent. These aldehydes in the presence of a strong base like NaOH shows disproportionation, i.e. simultaneous oxidation and reduction of the same molecule. Thus, the product includes an acid alongwith an alcohol.

In cross Cannizzaro reaction however, two different molecules can participate, out of which the one more prone to nucleophilic attack will show oxidation, while the other show reduction.

Here, out of the given two substrates, HCHO is more prone to nucleophilic attack hence, this will undergo oxidation while CH_3CHO will undergo reduction. Hence, the possible products are $HCOO^-Na^+$ and CH_3CH_2OH. Reaction involved is as follows : $HCHO + CH_3CHO \xrightarrow[NaOH]{Conc.} HCO\overset{-}{O}\overset{+}{N}a + CH_3CH_2OH$

Note Same combination of substrates give crossed aldol as well, in the presence of NaOH to form $CH_3—CH(CH_2CHO)—OH$.

But as, it is not among the options as the product, hence we are considering option (c) as the most appropriate option.

5 (*a*) The suitable reagent for the conversion

$CH_2{=}CH—CHO \xrightarrow{?} CH_2{=}CH—CH_2—OH$

is $NaBH_4$. It is a selective reagent that reduces $>C{=}O$ group to —OH and does not reduce carbon-carbon double bond. Thus, unsaturated alcohols can be prepared.

6 (*b*) The priority order of functional group is —CHO > Br Therefore, IUPAC name of the given compound is as follows

(3-bromo-2-methylhexanal)

7 (*d*) The correct IUPAC name of the given compound is 5-hydroxycyclohex -3-en-1-one.

The structure is shown below

8 (*a*) Necessary condition for aldol condensation is the presence of atleast one α-H atom. Thus, among the given options,

$CH_3—C(CH_3)_2—CHO$ does not undergo aldol condensation as it does not contain any α-hydrogen atom.

9 (*a*) Consider the following Cannizzaro reaction,

$$HCHO + HCHO \xrightarrow{KOH\ (conc.)} \underset{\text{Methyl alcohol}}{CH_3OH} + \underset{\text{Potassium formate}}{HCOO^{\ominus}K^+}$$

The mechanism of Cannizzaro reaction is as

***Step* I** Attack of nucleophile ($\overline{O}H$) to the carbonyl carbon.

$$H—\overset{O}{\overset{\|}{C}}—H + \overset{-}{O}H \underset{}{\overset{Fast}{\rightleftharpoons}} \underset{\text{(Anion I)}}{H—\overset{O^{\ominus}}{\overset{|}{\underset{OH}{\underset{|}{C}}}}—H}$$

***Step* II** The transfer of hydride ion from anion (I) to second molecule of aldehyde and finally rapid transfer of proton takes place.

$$H—\overset{O^{\ominus}}{\overset{|}{\underset{OH}{\underset{|}{C}}}}H + H—\overset{O}{\overset{\|}{C}}—H \xrightarrow[\text{transfer}]{\text{Hydride }(H^-)}$$

$$\underset{\text{(Acid)}}{H—\overset{O}{\overset{\|}{C}}—OH} + \underset{\text{Methoxide ion}}{H—\overset{O^-}{\overset{|}{\underset{H}{\underset{|}{C}}}}—H} \xrightarrow{\text{Proton exchange}} \underset{\text{Acid salt}}{H—\overset{O}{\overset{\|}{C}}—O^-} + \underset{\text{Methyl alcohol}}{CH_3OH}$$

10 (*c*) The given reaction is an example of Perkin condensation. In this reaction, aromatic aldehydes react with acid anhydride of an aliphatic acid in presence of sodium salt of same acid to form α, β-unsaturated acid. The steps involved in the process are as follows :

$$\underset{\text{Benzaldehyde}}{C_6H_5CHO} + \underset{\text{Acetic anhydride}}{CH_3—\overset{O}{\overset{\|}{C}}—O—\overset{O}{\overset{\|}{C}}—CH_3} \xrightarrow{CH_3COO^-Na^+} \underset{\text{Cinnamic acid}}{C_6H_5\overset{\beta}{C}H{=}\overset{\alpha}{C}HCOOH} + CH_3COOH$$

Thus, the correct option is (c).

11 (*a*)

α-H is abstracted by $\overline{O}H$

$\xrightarrow[-H_2O]{OH^-}$; H_2O ; $-H_2O$

12 (*a*)

$$\underset{\text{Propyne}}{CH_3C{\equiv}CH} \xrightarrow{H_2O} CH_3\overset{OH}{\overset{|}{C}}{=}CH_2 \xrightleftharpoons{\text{Tautomerises}} \underset{\text{Acetone}}{CH_3\overset{O}{\overset{\|}{C}}CH_3}$$

13 (*d*) $\underset{(X)}{CH_3CH_2OH} \xrightarrow{Cu,\ \Delta} \underset{(A)}{CH_3CHO} \xrightarrow[\text{Aldol}]{OH^-,\ \Delta} \underset{(Y)}{CH_3CH{=}CHCHO}$

(*A*) aldehyde gives silver mirror test and forms semicarbazone (*Z*).

14 (*c*) In keto-enol tautomerism, a carbonyl compound with a hydrogen atom on its α-carbon rapidly equilibrates with its corresponding enol.

$$\underset{\text{Ketone (containing α-hydrogen)}}{R—\overset{O}{\overset{||}{C}}—CH_2R'} \rightleftharpoons \underset{\text{Enol}}{R—\overset{OH}{\overset{|}{C}}=CHR'}$$

15 (*b*) In the presence of Pd-catalyst, selective reduction of α,β-unsaturated carbonyl is observed as hydrogenation takes place of carbon-carbon double bond only.

$$\text{Cyclohexenone} \xrightarrow[\text{Pd/carbon, ethanol}]{H_2\text{(gas 1, atmosphere)}} \underset{A}{\text{Cyclohexanone}}$$

16 (*b*) Those compound which contain $—\overset{O}{\overset{||}{C}}—H$ group gives iodoform test.

So, $CH_3—\overset{O}{\overset{||}{C}}—H$ (ethanal) gives iodoform test.

17 (*b*) $R—\overset{O}{\overset{||}{C}}—R' + CN^- \longrightarrow R—\overset{O^\ominus}{\overset{|}{\underset{CN}{\underset{|}{C}}}}—R' \xrightarrow{H^+} R—\overset{OH}{\overset{|}{\underset{CN}{\underset{|}{C}}}}—R'$

Since, initial attack is by CN^- (nucleophile)

So, it is an example of nucleophilic addition reaction.

18 (*a*) 1-methyl-4-methylcyclopentene (CH_3 at C-1, CH_3 ring substituent) $\xrightarrow{O_3,\ Zn,\ H_2O}$ $OHC—CH_2—\overset{CH_3}{\overset{|}{\underset{H}{\underset{|}{C}}}}—CH_2—\overset{CH_3}{\overset{|}{C}}=O$

$$\text{(3,4-dimethylcyclopentene)} \xrightarrow[Zn,\ H_2O]{O_3} H—\overset{O}{\overset{||}{\underset{1}{C}}}—\overset{CH_3}{\overset{|}{CH}}—\underset{CH_3}{\underset{|}{CH}}—CH_2—\underset{2}{CHO}$$

$$\text{(3,3-dimethylcyclopentene)} \xrightarrow[Zn,\ H_2O]{O_3} \underset{1}{OHC}—\overset{CH_3}{\overset{|}{\underset{CH_3}{\underset{|}{C}}}}—CH_2—CH_2—\underset{2}{CHO}$$

$$\text{(1,2-dimethylcyclopentene)} \xrightarrow[Zn,\ H_2O]{O_3} H_3C—\overset{O}{\overset{||}{\underset{1}{C}}}—CH_2—CH_2—CH_2—\overset{O}{\overset{||}{\underset{2}{C}}}—CH_3$$

Cleave (C=C) bond and place O at cleaved C-atoms.

19 (*d*) Reactivity of carbonyl compounds towards nucleophilic addition reactions depends on the presence of substituted group. Electron withdrawing (*–I, –M*) groups increase reactivity towards nucleophilic addition reactions.

∴ The correct option is (d).

20 (*a*)

(a) When aldehydes having no α-H atom, reacts with concentrated alkali, its one molecule is reduced to respective alcohol and other molecule is oxidised to sodium salt of respective acid. This reaction is called Cannizzaro reaction,

e.g. $HCHO \xrightarrow{\text{Conc. NaOH}} CH_3OH + HCO\overset{-}{O}\overset{+}{N}a$

Here, no new C—C bond is formed.

(b) When alkyl halide is treated with sodium metal in the presence of dry ether, it gives a hydrocarbon having double the number of carbon atoms as present in parent alkyl halide. This reaction is called Wurtz reaction,

e.g. $2CH_3X + Na \xrightarrow{\text{Dry ether}} CH_3—CH_3 + 2NaX$

i. e. it involves formation of a new C—C bond.

(c) Aldehydes having atleast one α-H atom, undergoes Aldol condensation when treated with dilute alkali solution

e.g. $2CH_3CHO + \underset{\text{Dilute}}{NaOH} \longrightarrow \underset{\text{Aldol}}{CH_3CH(OH)CH_2CHO}$

Here, also a new C—C bond is formed.

(d) When benzene is treated with alkyl halide or acyl halide in the presence of anhydrous $AlCl_3$, an alkyl group or acyl group is introduced in the benzene ring. This reaction is called Friedel-Craft reaction, e.g.

$$\underset{\text{Benzene}}{C_6H_6} + CH_3X \xrightarrow[-HX]{\text{Anhyd. }AlCl_3} \underset{\text{Toluene}}{C_6H_5CH_3}$$

This reaction also involves formation of a new C—C bond.

(e) When phenol is treated with carbon dioxide at a temperature of 140-200°C, the $—COO^-$ group is introduced in the benzene nucleus at *ortho* position (mainly) resulting in the formation of salicylic acid.

$$\text{Phenol} + CO_2 \xrightarrow{140\text{-}200°C} \underset{\text{Major}}{\text{o-HOC}_6H_4COOH} + \underset{\text{Minor}}{\text{p-HOC}_6H_4COOH}$$

Here, also a new C—C bond is formed.

21 (*c*) Aldehydes having no α-H atom, undergo Cannizzaro reaction, if treated with concentrated alkali (NaOH), i.e. one mole of the aldehyde is reduced to alcohol and other mole is oxidised to salt of acid. Thus, the given aldehyde (glyoxal) undergoes intramolecular Cannizzaro reaction, i.e. its half part is oxidised and half is reduced.

$$\underset{CHO}{\overset{CHO}{|}} \xrightarrow{\text{(i) NaOH}} \underset{\overset{-\ +}{COONa}}{\overset{CH_2OH}{|}} \xrightarrow{\text{(ii)}H^+} \underset{CO_2H}{\overset{CH_2OH}{|}}$$

22 (*a*) The Wittig reaction allows the preparation of an alkene by the reaction of an aldehyde or ketone with the ylide generate from a phosphonium.

[Cyclopentyl $\overset{+}{P}Ph_3$ with H, Bu⁻ → cyclopentylidene ylide $\overset{-}{C}$–$\overset{+}{P}Ph_3$ + $CH_2=O$ → oxaphosphetane (PPh₃–O spiro ring) → methylenecyclopentane $+ O=PPh_3$]

Hence, the intermediate *J* is

[spiro oxaphosphetane with PPh_3 and O]

23 (*b*) **Grignard reaction** Reaction of ethanal with Grignard reagent.

$$\underset{H}{\overset{H_3C}{>}}C=O + CH_3MgBr \longrightarrow H_3C-\underset{\overset{-}{O}\overset{+}{M}gBr}{CH}-CH_3 \xrightarrow{H_2O} H_3C-\underset{OH}{CH}-CH_3$$

Here, number of carbon atoms in the product is increased.

Cannizzaro's reaction.

$$HCHO + HCHO \xrightarrow{OH^-} CH_2OH + HCO\overset{\ominus}{O}$$

In this reaction, number of carbon atoms in product molecules are same as that in reactant. So, it cannot be used to increase the number of carbon atoms in an organic compound.

Aldol condensation.

$$H_3C-CHO + CH_3CHO \xrightarrow{OH^-} CH_3-\underset{OH}{CH}-CH_2-CHO \xrightarrow[-H_2O]{} H_3C-CH=CH-CHO$$

Again, number of carbon atoms in the products are increased. So, it is also not the correct option.

24 (*b*) Tischenko reaction is a disproportionation reaction that allows the preparation of esters from two equivalents of an aldehyde. In actual, aldehydes with or without α-hydrogens can be made to undergo Cannizzaro reaction on treatment with aluminium alkoxide. However, under these conditions, the alcohol and acid produced combine together to form esters. Thus, Tischenko reaction is a modified form of Cannizzaro reaction.

$$\underset{B}{H_3C-CHO} + H_3C-CHO \xrightarrow{Al(OR)_3} \underset{Acid}{CH_3COOH} + \underset{Alcohol}{CH_3CH_2OH} \longrightarrow \underset{Ester\ A}{CH_3-\overset{O}{\overset{\|}{C}}-OCH_2CH_3} + H_2O$$

So, here the compound *A* is $CH_3COOC_2H_5$ and compound *B* is CH_3CHO.

$$CH_3COC_2H_5 \xrightarrow[(Peracid)]{C_6H_5CO_3H} \underset{(A)}{CH_3-\overset{O}{\overset{\|}{C}}-O-C_2H_5} \xleftarrow{Al(OR)_3} \underset{(B)}{CH_3CHO}$$

25 (*b*) The substrate containing C=C double bond on ozonolysis using O_3 followed by reductive ozonide using Zn / H_2O_2 produces two carbonyl compounds.

Carbonyl compound obtained after reductive ozonolysis of (b), (c) and (d) are as follows

(b) $C_6H_5-CH=CH_2 \xrightarrow[Zn/H_2O_2]{O_3} \underset{Benzaldehyde}{C_6H_5-\overset{H}{C}=O} + \underset{Formaldehyde}{O=C\begin{smallmatrix}H\\H\end{smallmatrix}}$

(c) $C_6H_5-CH=CH-CH_3 \xrightarrow[Zn/H_2O_2]{O_3} C_6H_5-\overset{H}{C}=O + \underset{Acetaldehyde}{CH_3CHO}$

(d) $C_6H_5-CH=C(CH_3)_2 \xrightarrow[Zn/H_2O_2]{O_3} C_6H_5-CHO + \underset{Acetone}{CH_3-\overset{O}{\overset{\|}{C}}-CH_3}$

In each case, benzaldehyde is produced as one of the product but on the other hand formaldehyde, acetaldehyde and acetone are produced as other products. Thus, in order to select the correct choice, the concept of Cannizzaro reaction (disproportionation reaction) is applied. This reaction is a characteristic reaction of carbonyl compounds having absence of α-hydrogen atom.

Hence, correct substrate is $C_6H_5-CH=CH_2$ as both their products are devoid of α-H-atom which undergo Cannizzaro reaction easily. Chemical reaction occurring in the compound can be sequentially written as

$$\underset{X}{C_6H_5-CH=CH_2} \xrightarrow[H_2O_2/Zn]{O_3} \underset{Y}{C_6H_5-\overset{H}{C}=O} + \underset{Z}{O=C\begin{smallmatrix}H\\H\end{smallmatrix}}$$

In Etard reaction, toluene is oxidised to benzaldehyde with a solution of chronyl chloride in CS_2 or CCl_4 as

$$\underset{X}{C_6H_5CH_3} \xrightarrow[H_3O^+]{CrO_2Cl_2/CS_2} \underset{Y}{C_6H_5CHO}$$

Cannizzaro reaction shown by HCHO (*Z*) can be represented as

$$2\ \underset{H}{\overset{H}{>}}C=O \xrightarrow{Conc.\ NaOH} \underset{Alcohol}{CH_2OH} + H\,COO^-Na^+$$

26 (*c*) As the product formed in a given series is alkenone, i.e. Aldol condensation product. This suggests that *R* would be a ketone which undergoes Aldol condensation. The end product contains total six C-atoms, thus, *R* would be a ketone of three carbon atoms i.e. acetone (propanone).

This propanone is produced from ethane nitrile through Grignard reagent. Thus, ethane nitrile (*P*) on reaction with CH_3MgBr followed by hydrolysis produced acetone *R* which on reaction with dil.NaOH undergoes Aldol condensation to produce 4-methylpent-3-ene-2-one as shown below.

$$\underset{\text{Ethane nitrile}}{CH_3-C{\equiv}N} \xrightarrow{CH_3MgBr} (H_3C)_2C{=}NMgBr$$

$$\xrightarrow{H_3O^+} [(H_3C)_2C{=}NH + MgBr(OH)]$$

$$\xrightarrow{H_3O^+} (H_3C)_2C{=}O + NH_3$$

$$(H_3C)_2C{=}O + CH_3-\overset{O}{\overset{\|}{C}}-CH_3 \xrightarrow[\text{Aldol condensation}]{\text{Dil. NaOH}}$$

$$CH_3-\underset{CH_3}{\overset{OH}{C}}-CH_2-\overset{O}{\overset{\|}{C}}-CH_3 \xrightarrow[H^+,\ -H_2O]{\Delta} \underset{\text{4-methylpent-3-en-2-one}}{CH_3-\underset{CH_3}{C}{=}CH-\overset{O}{\overset{\|}{C}}-CH_3}$$

27 (*d*) Ketones always produce secondary or tertiary alcohols on reduction. Thus, propan-2-one does not give a primary alcohol on reduction whereas all the other remaining compounds produce primary alcohols on reduction.

(a) $\underset{\text{Propanal}}{CH_3CH_2CHO} \xrightarrow[2[H]]{LiAlH_4} \underset{\text{Propanol (1° alcohol)}}{CH_3CH_2CH_2OH}$

(b) $\underset{\text{(Propanoic acid)}}{CH_3CH_2COOH} \xrightarrow[4[H]]{LiAlH_4} \underset{\text{Propanol (1° alcohol)}}{CH_3CH_2CH_2OH}$

(c) $\underset{\text{Methyl propanoate}}{CH_3CH_2COOCH_3} \xrightarrow{LiAlH_4} \underset{\text{(Propanol) 1° alcohol}}{CH_3CH_2CH_2OH} + \underset{\text{(Methanol 1° alcohol)}}{CH_3OH}$

(d) $\underset{\text{Propan-2-one}}{CH_3-\underset{O}{\underset{\|}{C}}-CH_3} \xrightarrow{LiAlH_4} \underset{\text{Propan-2-ol (2° alcohol)}}{CH_3\underset{OH}{CH}CH_3}$

28 (*d*) $\underset{\text{Toluene}}{C_6H_5CH_3} \xrightarrow[CS_2]{2CrO_2Cl_2} \underset{\text{Brown (addition product)}}{C_6H_5CH_3\cdot 2CrO_2Cl_2} \xrightarrow{H_2O} \underset{\text{Benzaldehyde}}{C_6H_5CHO}$

This reaction is known as **Etard reaction.**

(b) $C_6H_5COCl \xrightarrow[\text{(Boiling xylene)}]{Pd/BaSO_4/H_2} \underset{\text{Benzaldehyde}}{C_6H_5CHO} + HCl$

This reaction is known as **Rosenmund reduction.**

(c) $C_6H_6 + CO + HCl \xrightarrow{\text{Anhy. } AlCl_3} C_6H_5CHO + HCl$

This reaction is known as **Gattermann-Koch** aldehyde synthesis.

(d) $C_6H_5COOH + Zn/Hg \xrightarrow{\text{Conc. HCl}} \text{No reaction}$

Thus, from the reagents given in option (d) benzaldehyde cannot be prepared.

29 (*d*) Aldehydes and ketones can be distinguished by Fehling's reagent [alkaline solution of $CuSO_4$ containing Rochelle salt (sodium potassium tartarate), mild oxidising agent].

$$\text{Aldehydes} \xrightarrow{\text{Fehling reagent}} \text{Red-brown ppt.}$$

$$\text{Ketones} \xrightarrow{\text{Fehling reagent}} \text{No reaction}$$

Besides this, Tollen's reagent and Benedict's reagent can also be used as they give positive test with aldehydes but not with ketones.

30 (*d*) Iodoform test is given by those compounds which contains $CH_3-\underset{O}{\underset{\|}{C}}-$ group or $CH_3-\underset{OH}{CH}$ group while Fehling's test is given by only aldehydic group. Thus, iodoform test and Fehling's test are performed by only ethanal.

$$\underset{\text{Ethanal}}{CH_3CHO} + \underset{\text{Fehling's solution}}{2Cu^{2+}} + 5OH^- \longrightarrow \underset{\text{Acetate ion}}{CH_3COO^-} + 3H_2O + \underset{\text{Cuprous oxide (red ppt.)}}{Cu_2O}$$

31 (*c*) When benzaldehyde is treated with 50% alkali, it undergoes oxidation to give an acid salt as well as reduction to give an alcohol. This reaction is called Cannizzaro's reaction.

$$m\text{-}ClC_6H_4CHO \xrightarrow{50\%KOH} \underset{m\text{-chlorobenzyl alcohol}}{m\text{-}ClC_6H_4CH_2OH} + \underset{m\text{-chlorobenzoate}}{m\text{-}ClC_6H_4COO^-K^+}$$

32 (*a*) Acid chloride reacts with H_2 in the presence of $Pd/BaSO_4$ to yield aldehyde. This reacts is called **Rosenmund's reduction.**

$$\underset{\text{Benzoyl chloride}}{C_6H_5COCl} \xrightarrow[Pd\text{-}BaSO_4]{H_2} \underset{\text{Benzaldehyde}}{C_6H_5CHO} + HCl$$

33 (*a*) 2- methyl propanol contains one α-H-atom. Hence, it undergoes Aldol condensation. It also undergoes Cannizzaro's reaction.

$$CH_3-\underset{CH_3}{\overset{H}{C}}-CHO + H-\underset{CH_3}{\overset{CH_3}{C}}-CHO \xrightarrow{\text{Dil. NaOH}} CH_3-\underset{CH_3}{CH}-\overset{OH}{CH}-\underset{CH_3}{\overset{CH_3}{C}}-CHO$$

(Aldol)

Product formed has no H at α—C thus it also gives Cannizzaro reaction.

34 (*c*) CH_3COCH_3 (Acetone) + $CH_2=\overset{+}{N}(O^-)O^-$ ⟶ $(CH_3)_2C(O^-)-CH_2-NO_2$

$$(CH_3)_2C(OH)CH_2NO_2 \xrightarrow[\substack{\text{Pyridine} \\ (-H_2O)}]{Ac_2O} CH_3-\underset{CH_3}{C}=CH-NO_2$$

Hence, the product formed when a acetone reacts with nitromethane in the presence of a base is

$$\begin{matrix} H_3C \\ H_3C \end{matrix}\!\!>C=CHNO_2.$$

35 (*b*) Clemmensen's reduction is the reduction of carbonyl group into alkane.

$$\underset{\text{Carbonyl compound}}{>C=O} + 4H \xrightarrow{\text{Zn-Hg/conc. HCl}} \underset{\text{Alkane}}{>CH_2} + H_2O$$

36 (*d*) In Cannizzaro's reaction, aldehydes which have no α-hydrogen atom undergo disproportionation in the presence of strong base to form equal amounts of corresponding alcohol and salt of carboxylic acid. Acetaldehyde due to the presence of 3α-H-atoms does not perform this reaction.

Cannizzaro reaction

$$\underset{(a)}{2CCl_3CHO} + NaOH \rightarrow CCl_3CO\overset{-}{O}\overset{+}{Na} + CCl_3CH_2OH$$

$$\underset{(b)}{2C_6H_5CHO} + NaOH \rightarrow C_6H_5CO\overset{-}{O}\overset{+}{Na} + C_6H_5CH_2OH$$

$$\underset{(c)}{2HCHO} + NaOH \rightarrow HCO\overset{-}{O}\overset{+}{Na} + CH_3OH$$

$$\underset{(d)}{2CH_3CHO + NaOH} \xrightarrow[\substack{\text{condensation} \\ \text{(due to } \alpha\text{-H-atom)}}]{\text{Aldol}} \underset{\text{Aldol}}{CH_3-\overset{OH}{CH}-CH_2-CHO}$$

37 (*a*) $\underset{\text{Acetaldehyde}}{2CH_3CHO} \xrightarrow{Al(OC_2H_5)_3} \underset{\text{Ethyl acetate}}{CH_3COOCH_2CH_3}$ (Tischenko)

38 (*a*) (a) $C_6H_5COCH_3$ Acetophenone (b) CH_3COCH_3 Acetone

(c) $C_6H_4(OH)COOH$ Salicylic acid (c) C_6H_5CHO Benzaldehyde

Since, the given compound *A* burns with sooty flame, thus, it must be aromatic compound. Again when it reacts with Brady's reagent (2, 4-dinitrophenyl hydrazine solution), it yields orange precipitate but gives no precipitate or silver mirror with Tollen's reagent. Therefore, *A* must be an aromatic ketone. Hence, *A* is acetophenone.

39 (*c*) $>C=O \xrightarrow[HCl]{Zn-Hg} CH_2$; Clemmensen reduction.

40 (*d*) $NaBH_4$ does not reduce olefin (alkene) double bond.

41 (*a*) Aldehydes are oxidised in Tollen's test.

42 (*b*) CH_3CHO and CH_3COCH_3 give iodoform test with $NaOH/I_2$ as they contain $CH_3-\overset{O}{\overset{\|}{C}}-$ group (methyl keto group).

43 (*a*) Calcium formate and calcium acetate give aldehyde on dry distillation. (CH_3CHO)

44 (*b*) CH_3CH_2CHO and CH_3COCH_3 can be distinguished by sodium hypoiodite.

CH_3COCH_3 gives yellow precipitate of iodoform when heated with NaOI. CH_3CH_2CHO does not give iodoform.

45 (*d*) $CH_3CONHCH_3$ is an amide. Aldehydes and ketones give semicarbazone and oxime.

46 (*a*) $$\underset{\text{Acetone}}{(H_3C)_2C=O} + NH_2NH_2 \xrightarrow{-H_2O} (H_3C)_2C=N\cdot NH_2$$

$$\xrightarrow[\Delta]{\text{KOH/ethylene glycol}} \underset{\text{Propane}}{CH_3CH_2CH_3} + N_2$$

The reaction is known as Wolff-Kishner reduction.

CHAPTER 19

Carboxylic Acids

The carbonyl compounds in which carbonyl group is bonded to oxygen are known as carboxylic acids. These are named as carboxyl group because a hydroxy group is attached to carboxyl group. The general formula of aliphatic carboxylic acids is C_nH_{2n+1} COOH. (aliphatic class)

Classification

Depending upon the number of carboxylic groups present in molecules, these are classified as monocarboxylic acid, dicarboxylic acid or tricarboxylic acids.

Monocarboxylic acids

	HCOOH ;	CH_3COOH ;	C_6H_5COOH
	Formic acid	Acetic acid	Benzoic acid
IUPAC name	Methanoic acid	Ethanoic acid	

Dicarboxylic acid

$$\begin{array}{l}COOH\\|\\COOH\end{array}\ ;\quad CH_2\begin{array}{l}COOH\\COOH\end{array}\ ;\quad C_6H_4(COOH)_2$$

Oxalic acid; Malonic acid; Phthalic acid

Nomenclature

In IUPAC system, carboxylic acids are named by replacing suffix 'e' of the alkanes with 'oic acid'.

Select longest carbon chain containing carboxyl group as parent chain. The position of a substituent is designated by a number. The carboxyl group is always the C-1carbon. If more than one carboxyl groups are present, prefixes di, tri etc., are used.

Inside

Table 19.1 Some examples of carboxylic acid

Molecular formulae	Common name	IUPAC name
HCOOH	Formic acid	Methanoic acid
CH_3COOH	Acetic acid	Ethanoic acid
CH_3CH_2COOH	Propionic acid	Propanoic acid
$CH_3CH_2CH_2COOH$	Butyric acid	Butanoic acid
$CH_3CH_2CH_2CH_2COOH$	Valeric acid	Pentanoic acid
$(CH_3)_2CHCH_2COOH$	Isovaleric acid	3-methylbutanoic acid
$(CH_3)_3CCOOH$	Pivalic acid	2,2-dimethylpropanoic acid
$CH_3(CH_2)_4COOH$	Caproic acid	Hexanoic acid
$CH_3(CH_2)_8COOH$	Capric acid	Decanoic acid
$CH_3(CH_2)_{10}COOH$	Lauric acid	Dodecanoic acid
$CH_3(CH_2)_{14}COOH$	Palmitic acid	Hexadecanoic acid
$CH_3(CH_2)_{16}COOH$	Stearic acid	Octadecanoic acid
$CH_2{=}CHCOOH$	Acrylic acid	prop-2-enoic acid
$CH_3CH{=}CHCOOH$	Crotonic acid	but-2-enoic acid
COOH–COOH	Oxalic acid	Ethanedioic acid
$H_2C(COOH)_2$	Malonic acid	Propanedioic acid
CH_2COOH–CH_2COOH	Succinic acid	Butanedioic acid
CHCOOH=CHCOOH	Maleic acid	but-2-enedioic acid
$CH_2(CH_2COOH)_2$	Glutaric acid	pentanedioic acid
CH_2CH_2COOH–CH_2CH_2COOH	Adipic acid	hexanedioic acid
CH_2COOH–$CHCH_2CH_2COOH$–CH_2COOH	—	3-(carboxymethyl) hexanedioic acid
CH_2COOH–CHCOOH–CH_2COOH	—	Propane-1,2,3-tri carboxylic acid

Isomerism

Carboxylic acids show functional isomers. The compounds have same molecular formula but different structures due to difference in functional groups attached.

e.g.

(i) Hydroxy carbonyl compound (aldehyde or ketone)

(ii) Hydroxy oxirane

(iii) Esters

e.g. functional isomers of acetic acid are

CH_2OH–CHO (Glycol aldehyde) ; $HCOOCH_3$ (Methyl formate) ; HO—CH(–O–)CH_2 (2-hydroxy oxirane)

Functional isomers of phenyl acetic acid are

I: $C_6H_5COOCH_3$; II: COOH and CH_3 (meta) on benzene ring; III: COOH and CH_3 (ortho) on benzene ring; IV: COOH and CH_3 (para) on benzene ring

I II III IV

Structure

In carboxylic acids, the bonds to the carboxyl carbon lie in one plane. The entire molecule is planar with sp^2-hybridised carbon atom and trigonal planar geometry.

e.g. structure of formic acid is shown below.

Bond angles: H–C=O 124°, O=C–O 125°, C–O–H 106°, H–C–O 111°

Bond lengths: C=O 1.23Å, C–O 1.32Å, C–H 1.10Å, O–H 0.97Å

MONOCARBOXYLIC ACIDS

Methods of preparation

Following methods are used for the preparation of monocarboxylic acids

From alcohols

(i) Primary alcohols are easily oxidised to carboxylic acid with oxidants such as $KMnO_4$ (in neutral, acidic or alkaline medium), CrO_3 and $K_2Cr_2O_7$ (in acidic medium)

e.g.

$$CH_3CH_2OH \xrightarrow{Cr_2O_7^{2-}/H_3O^+} CH_3\overset{O}{\overset{\|}{C}}H \xrightarrow{Cr_2O_7^{2-}/H_3O^+} CH_3\overset{O}{\overset{\|}{C}}OH$$

Secondary alcohols are oxidised to ketones. Tertiary alcohols are not oxidised due to lack of hydrogen at α-carbon.

$$H_3C-\underset{CH_3}{\overset{CH_3}{\underset{|}{\overset{|}{C}}}}-OH$$

α-carbon has no hydrogen

However, strong oxidising agents oxidise secondary and tertiary alcohols to acids of fewer number of carbon atoms with cleavage of C — C bonds.

(*ii*) Many 1° alcohols, on oxidation also form esters due to formation of hemiacetal which is oxidised to give ester.

e.g.

$$\underset{\text{Ethanol}}{CH_3CH_2OH} \xrightarrow{[O]} \underset{\text{Ethanal}}{CH_3CHO} \xrightarrow{RCH_2OH} \underset{\text{Hemiacetal}}{CH_3\underset{OH}{\underset{|}{CH}}-OCH_2R} \xrightarrow{[O]} \underset{\text{Ester}}{CH_3\overset{O}{\overset{||}{C}}OCH_2R}$$

From aldehyde and ketones

Aldehydes can be oxidised to carboxylic acids even with mild oxidants while ketones are oxidised with difficulty. They are oxidised by strong oxidising agents (like $Cr_2O_7^{2-}$) and keto group remains with smaller alkyl group. Common oxidising agents are acidified $K_2Cr_2O_7$, alkaline $KMnO_4$ and Tollen's reagent (for aldehydes only).

$$\underset{\text{Aldehyde}}{RCHO} \xrightarrow[\text{(ii) } H^+]{\text{(i) } Ag_2O/NH_4OH \text{ (Tollen's reagent)}} \underset{\text{Carboxylic acid}}{RCOOH}$$

e.g. $\underset{\text{Butan-2-one}}{CH_3COCH_2CH_3} \xrightarrow{[O]} \underset{\text{Acetic acid}}{2CH_3COOH}$

e.g.

$$\underset{\text{Pentan-2-one}}{CH_3\overset{*}{C}OCH_2CH_2CH_3} \xrightarrow[\Delta]{[O]} \underset{\text{Acetic acid}}{CH_3\overset{*}{C}OOH} + \underset{\text{Propanoic acid}}{CH_3CH_2COOH}$$

$$R\overset{O}{\overset{||}{C}}CH_3 + 3I_2 + 4NaOH \xrightarrow{\Delta} R\overset{O}{\overset{||}{C}}\overset{-}{O}\overset{+}{Na} + 3NaI + \underset{\text{Iodoform}}{CHI_3} + 3H_2O$$

From Grignard reagent

Grignard reagent ($RMgX$) reacts with solid carbon dioxide (i.e. dry ice) in presence of dry ether, followed by acidic hydrolysis yields carboxylic acids.

$$\underset{\text{Carbon dioxide}}{O=C=O} \xrightarrow[\text{dry ether}]{RMgX} O=\underset{R}{\underset{|}{C}}O\overset{-+}{MgX} \xrightarrow{H_3O^+} O=\underset{R}{\underset{|}{C}}OH + Mg^{2+} + X^-$$

Process of reacting Grignard reagent with CO_2 is also known as **carbonation of organometallic reagents.**

Note *This method gives carboxylic acids with one carbon more than starting compound. Thus, HCOOH cannot be prepared by this method.*

From hydrolysis

(i) Nitriles and amides

Nitriles are hydrolysed to amides and then to carboxylic acids in the presence of H^+ or OH^- as catalyst. If conditions used are mild, then reaction can be stopped even at amide.

e.g. $\underset{\text{Methyl cyanide}}{CH_3CN} + 2H_2O \xrightarrow[\Delta]{H^+ \text{ or } OH^-} \underset{\text{Ethanamide}}{CH_3CONH_2} \xrightarrow[\Delta]{H^+ \text{ or } OH^-} \underset{\text{Acetic acid}}{CH_3COOH} + NH_4^+$

Mechanism of acidic hydrolysis of cyanides

$$R-C\equiv\ddot{N} \overset{H^\oplus}{\rightleftharpoons} R-C\equiv\overset{\oplus}{N}H \overset{H_2\ddot{O}:}{\rightleftharpoons} \underset{R\quad\;\; NH}{\overset{H\overset{\oplus}{O}H}{C}} \overset{-H^\oplus}{\rightleftharpoons} \underset{R\quad\;\; NH}{\overset{OH}{C}}$$

$$\xrightleftharpoons{\text{Tautomerises}} \underset{\text{Intermediate}}{R-\overset{O}{\overset{||}{C}}-NH_2} \xrightarrow[\Delta]{H^+, H_2O} R-\overset{O}{\overset{||}{C}}-OH + NH_4^+$$

Different products are obtained in different media.

In alkaline medium, $RCOO^-$ and NH_3 are the products

$$R\overset{O}{\overset{||}{C}}OH + OH^- \longrightarrow R\overset{O}{\overset{||}{C}}O^- + H_2O$$

In acidic medium, RCOOH and NH_4^+ are the products

$$NH_3 + H^+ \longrightarrow NH_4^+$$

(ii) Esters

Hydrolysis of esters can occur *via* both acidic or basic hydrolysis. Alkaline hydrolysis of esters is preferred over acidic hydrolysis, because it is irreversible as base is a strong nucleophile which favours hydrolysis and carboxylate ion cannot attack $R'OH$ (nucleophile).

Reaction in acidic medium is given below

$$R\overset{O}{\overset{||}{C}}OR' + H_2O \xrightarrow{H^+} R\overset{O}{\overset{||}{C}}OH + R'OH$$

Reaction in acidic medium is given below

$$RCOR' \xrightarrow{OH^-} R\overset{O^-}{\underset{OH}{C}}OR' \xrightarrow[-R'O^-]{} \left[RC\left\langle\begin{matrix}O\\ -\\ O\end{matrix}\right.\right] + R'OH$$

(iii) **Acid chlorides**

Acid chlorides are hydrolysed with aqueous base to form carboxylate ion which form carboxylic acid on acidification.

$$\underset{\text{Acid chloride}}{R-\overset{O}{\overset{\|}{C}}-Cl} + H_2O \xrightarrow{H^+} \underset{\text{Carboxylic acid}}{R-\overset{O}{\overset{\|}{C}}-OH} + HCl$$

(iv) **Anhydrides**

Acid anhydrides are hydrolysed with *aq.* base to form carboxylate ions which form carboxylic acid on acidification.

$$\underset{\text{Acid anhydride}}{(RCO)O(COR')} + H_2O \xrightarrow{H^+/OH^-} \underset{\text{Acid}}{RCOOH + R'COOH}$$

Note *Relative reactivity order of acyl compounds toward hydrolysis to form carboxylic acid or carboxylate ion*

$$R-\overset{O}{\overset{\|}{C}}-Cl > (RCO)_2O > R-\overset{O}{\overset{\|}{C}}-OR' > R-\overset{O}{\overset{\|}{C}}-NH_2$$

When acid is strong then its conjugate base is weak and a weak base act as a better leaving group.

$$R-\overset{O}{\overset{\|}{C}}-X + H_2O \rightleftharpoons R-\overset{O}{\overset{\|}{C}}-OH + \underset{\text{Leaving group}}{X^-}$$

Examples of some acids and their conjugate bases with their respective acidic and basic order are given below.

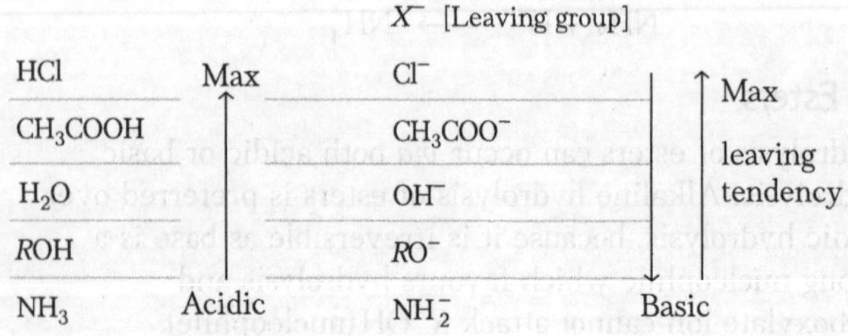

		X^- [Leaving group]	
HCl	Max ↑	Cl^-	↑ Max
CH_3COOH		CH_3COO^-	leaving
H_2O		OH^-	tendency
ROH		RO^-	
NH_3	Acidic	NH_2^-	↓ Basic

(v) **Trihalogen compounds**

Hydrolysis of trihalogen compounds produce carboxylic acids in presence of alkali.

$$R-CX_3 + 3NaOH \rightarrow \left[R-C(OH)_3\right] + 3Na^+X^- \xrightarrow{-H_2O} \underset{\text{Sodium salts of carboxylic acid}}{R-\overset{O}{\overset{\|}{C}}-\overset{-}{O}\overset{+}{Na}}$$

For example, formic acid is obtained by reaction of CHX_3 and alkali.

$$HCX_3 + 3NaOH \xrightarrow{\Delta} \left[H-C(OH)_3\right] \xrightarrow{-H_2O} \underset{\text{Sodium salts of formic acid}}{H-\overset{O}{\overset{\|}{C}}-\overset{-}{O}\overset{+}{Na}} + 3\overset{+}{Na}X^-$$

(vi) **Ozonolysis of unsaturated compounds**

Unsaturated compounds such as alkenes and alkynes undergo ozonolysis to give carboxylic acids.

e.g. $$\underset{\text{Alkene}}{CH_2{=}CH_2} \xrightarrow{O_3} \underset{\text{Ozonide}}{CH_2(O_3)CH_2} \xrightarrow[\text{or } Ag_2O]{H_2O_2} \underset{\text{Formic acid}}{2HCOOH}$$

e.g. $$\underset{\text{Alkyne}}{CH{\equiv}CH} \xrightarrow{O_3} \underset{\text{Ozonide}}{CH(O_3)CH} \xrightarrow{H_2O_2} \underset{\text{Formic acid}}{2HCOOH}$$

(vii) **By oxidative cleavage**

Alkenes, alkynes and cycloalkenes undergo oxidative cleavage in presence of hot basic $KMnO_4$. In this oxidative cleavage, each carbon of the multiple bond becomes the carbon atom of the carbonyl group and a hydrogen substituent on the carbon atom having double bond is replaced by a hydroxyl group.

$$\underset{\text{Alkene}}{RR'C{=}CHR''} \xrightarrow[\text{(ii) } H_3O^+]{\text{(i) } KMnO_4,\ OH^-,\ \Delta} \underset{\text{Ketone}}{RR'C{=}O} + \underset{\text{Carboxylic acid}}{O{=}C(R'')OH}$$

The intermediate may be a diol which is oxidised further with the cleavage of carbon–carbon bond.

The terminal CH_2 group of alkene is completely oxidised to CO_2 and H_2O.

e.g. $\underset{\text{But-1-ene}}{CH_3CH_2CH=CH_2} \xrightarrow[\text{(ii) } H_3O^+]{\text{(i) } KMnO_4, OH^-, \Delta} \underset{\text{Propanoic acid}}{CH_3CH_2COOH} + CO_2 + H_2O$

(viii) Carboxylation of alkenes (Koch reaction)

Mixture of alkene, carbon monoxide and steam under pressure of 500-1000 atm and at a temperature 570-670 K, is heated in presence of phosphoric acid (H_3PO_4), yields carboxylic acid.

$$\underset{\text{Ethene}}{CH_2=CH_2} + CO + H_2O \xrightarrow[\text{570-670 K; 500-1000 atm}]{H_3PO_4} \underset{\text{Propanoic acid}}{CH_3CH_2COOH}$$

This reaction involves electrophilic addition of H^+ ion on alkene to form stable carbocation followed by attack of nucleophile (i.e. CO).

(ix) Carboxylation of sodium alkoxide

Sodium alkoxide is heated with carbon monoxide under pressure to give carboxylic acids.

$$\underset{\text{Sodium alkoxide}}{RO^-Na^+} + CO \xrightarrow[\text{Under pressure}]{\Delta} \underset{\text{Sodium carboxylate}}{RCO\overset{-}{O}\overset{+}{Na}} \xrightarrow{H^+} \underset{\text{Carboxylic acid}}{RCOOH}$$

e.g. $CH_3O^-Na^+ + CO \xrightarrow{H^+} \underset{\text{Acetic acid}}{CH_3COOH}$

Note *Methanol reacts with CO in presence of $Co(CO)_8$ at 480 K to give ethanoic acid.*

$$\underset{\text{Methanol}}{CH_3OH} + CO \xrightarrow[Co(CO)_8]{\text{480 K, pressure}} \underset{\text{Ethanoic acid}}{CH_3COOH}$$

Preparation of important carboxylic acids

(a) HCOOH can be prepared by

$$CO + NaOH \xrightarrow{\text{473 K, 10 atm}} HCO\overset{-}{O}\overset{+}{Na} \xrightarrow{H^+} \underset{\text{Formic acid}}{HCOOH}$$

$$\begin{matrix} COOH \\ | \\ COOH \end{matrix} \xrightarrow{\text{Glycerol, 383 K}} HCOOH + CO_2$$

(b) CH_3COOH is a by-product of alcohol industry.

$$\text{Sugar and starch} \xrightarrow{\text{Fermentation}} CH_3CH_2OH \text{ (Ethanol)} \xrightarrow[\text{Acetobacter}]{\text{Enzyme, } O_2} \underset{\text{(Vinegar)}}{CH_3COOH}$$

$$CH_3OH + CO \xrightarrow[\text{Heat, pressure}]{\text{Catalyst}} CH_3COOH$$

(c) $\underset{\text{Ethyne}}{CH\equiv CH} \xrightarrow{H_2O, HgSO_4, H_2SO_4} \underset{\text{Ethanal}}{CH_3CHO} \xrightarrow[\text{Cobalt acetate, 353 K}]{O_2} \underset{\text{Ethanoic acid}}{CH_3COOH}$

CHECK POINT 19.1

1. General formula for aliphatic carboxylic acid is
(a) $C_nH_{2n+1}COOH$ (b) $C_nH_{2n}COOH$
(c) $C_nH_{2n-1}COOH$ (d) $C_{2n}H_{2n+1}COOH$

2. Common name of carboxylic acid, which is found in butter is
(a) caproic acid (b) valeric acid
(c) stearic acid (d) butyric acid

3. Aldehydes are oxidised to carboxylic acid by
(a) Schiff's base (b) Tollen's reagent
(c) alkaline $KMnO_4$ (d) $K_2Cr_2O_7$

4. Which of the following acids cannot be prepared by Grignard reagent?
(a) CH_3COOH (b) C_6H_5COOH
(c) $C_6H_{11}COOH$ (d) HCOOH

5. Products obtained in acidic hydrolysis of cyanide are
(a) $RCOOH$ and NH_4^+ (b) $RCOOH$ and NH_3
(c) $RCOO^-$ and NH_4^+ (d) $RCOO^-$ and NH_3

6. Select the incorrect statement.
(a) Hydrolysis of ester in alkaline medium is irreversible
(b) Hydrolysis of ester in acidic medium is irreversible
(c) Hydrolysis of ester in acidic medium is reversible
(d) None of the above

7. Order of reactivity of acyl compounds towards hydrolysis to form carboxylic acid is
(a) $RCOCl > RCOOR' > RCOCOR > RCONH_2$
(b) $RCONH_2 > RCOCOR > RCOOR' > RCOCl$
(c) $RCOCl > RCOCOR > RCOOR' > RCONH_2$
(d) $RCOOR' > RCOCl > RCONH_2 > RCOCOR$

8 Intermediate formed in ozonolysis of ethyne is

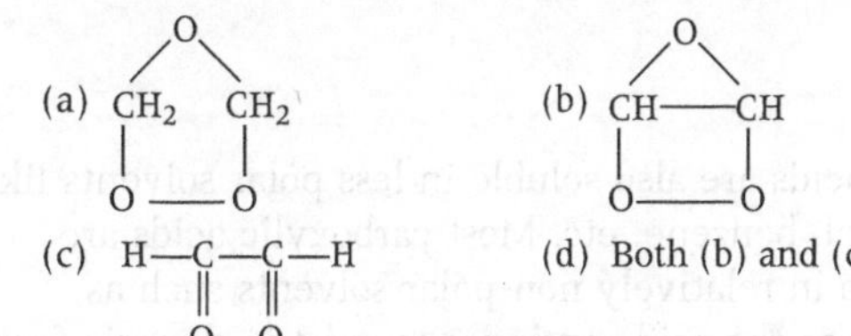

9 Consider the following reaction,

$$CH_3CH_2CH{=}CH_2 \xrightarrow[H_3O^+]{KMnO_4,\ OH^-} A$$

End products formed in above reaction are

(a) CH_3CH_2CHO and HCHO
(b) CH_3CH_2COOH
(c) CH_3CH_2COOH and CO_2
(d) CH_3CH_2COOH, HCHO, CO_2

10. Name of the reaction,

$$CH_2{=}CH_2 + CO + H_2O \xrightarrow[570\text{-}670\ K;\ 500\text{-}1000\ atm]{H_3PO_4} CH_3CH_2COOH$$

is

(a) Arndt-Eistert-reaction
(b) carbonylation
(c) oxidation
(d) Koch reaction

PHYSICAL AND CHEMICAL PROPERTIES

Physical properties

Various physical properties of monocarboxylic acids are as follows.

Physical state

(*i*) Carboxylic acids are polar in nature due to presence of carbonyl ($-\overset{O}{\overset{\|}{C}}-$) and hydroxy (—OH) groups. They have tendency to undergo H-bonding with each other and also with other polar molecules.

$$R-\overset{\overset{\delta-}{O}}{\underset{\delta+}{\overset{\|}{C}}}-\overset{\delta-}{O}-\overset{\delta+}{H}$$

First four members are miscible with water, five carbon acid is partially soluble and higher ones are insoluble in water.

(*ii*) They generally exist in dimeric forms, where hydroxy group of one carboxylic acid is hydrogen bonded to carbonyl oxygen of other carboxylic acid.

(*iii*) They are held together by intermolecular hydrogen bonding.

$$CH_3-\overset{\delta+}{C}\begin{matrix}\overset{\delta-}{=O}\cdots\cdots\overset{\delta+}{H}-\overset{\delta-}{O}\\ -\underset{\delta-}{O}-\underset{\delta+}{H}\cdots\cdots\underset{\delta-}{O}=\end{matrix}\overset{\delta+}{C}-CH_3$$

Therefore, they have high melting and boiling points.

Solubility

Carboxylic acids are also soluble in less polar solvents like ether, alcohol, benzene, etc. Most carboxylic acids are quite soluble in relatively non-polar solvents such as $CHCl_3$ because the acid continues to exist in dimeric form in the non-polar solvent.

Chemical properties

Acidic nature and some chemical reaction of carboxylic acids are as follows.

Acidic nature of carboxylic acids

Carboxylic acids are the most acidic among the organic compounds. These acids ionise in aqueous solution giving carboxylate ion.

$$R-\overset{O}{\overset{\|}{C}}-OH + H_2O \rightleftharpoons \underset{\text{Carboxylate ion Conjugate base}}{R-\overset{O}{\overset{\|}{C}}-O^-} + \underset{\text{Hydronium ion}}{H_3O^+} \quad \ldots(i)$$

It can be expressed in terms of dissociation constant.

$$K_a = \frac{[RCOO^-][H_3O^+]}{[RCOOH]}$$

Greater the value of K_a, stronger is the acid. Therefore, acetic acid is less acidic than sulphuric acid due to lesser value of degree of dissociation. Acidic nature of carboxylic acids can be understood by resonance.

The conjugate base of carboxylic acid, a carboxylate ion is stabilised by two equivalent resonance structures in which the negative charge is at the more electronegative oxygen atom. Thus, this ion is resonance stabilised, resonance hybrid structure being I(C).

$$\underset{\text{I }(A)}{R-C\begin{matrix}O^{\ominus}\\ {=}O\end{matrix}} \longleftrightarrow \underset{\text{I }(B)}{R-C\begin{matrix}{=}O\\ O^{\ominus}\end{matrix}} \equiv \underset{\text{I }(C)}{\left[R-C\begin{matrix}O\\ O\end{matrix}\right]^{\ominus}}$$

Hence, in the above given reaction (i) equilibrium shifts toward the side where stable ion (I) is formed, i.e. R.H.S, hence, carboxylic acids are acidic in nature.

Carboxylic acids are more acidic than alcohols. Both, acid and alcohol involves cleavage of O—H bond, but dissociation of acid gives carboxylate ion in which negative charge is distributed equally over two oxygen atoms whereas, in alkoxide ion, it is concentrated on one oxygen atom. Therefore, delocalisation of charge makes carboxylate ion more stable than alkoxide ion.

Effect of substituents on acidity

Acidity of carboxylic acid depends on the formation of carboxylate ion. Any factor which tends to stabilise the carboxylate ion will increase the acidic strength of carboxylic acid. At the same time, any factor which tends to destabilise the ion, will decrease the acidic strength of carboxylic acid.

(i) Effect of electron withdrawing substituents

An electron withdrawing group (EWG) stabilises the carboxylate ion by dispersal of negative charge. Thus, increases the acidic character.

$$G \longleftarrow C(O)(O)^{\ominus}$$

Some trends of compounds containing —COOH group are as follows

(a) Acidic strength increases with increase in number of electron withdrawing groups.

e.g. $CCl_3COOH > CHCl_2COOH > CH_2ClCOOH > CH_3COOH$

(b) Position of electron withdrawing groups also affect the acidity of carboxylic acids. As the distance between electron withdrawing group and —COOH group increases, acidic nature decreases.

e.g. $CH_3CH(Cl)COOH > CH_2(Cl)CH_2COOH > CH_3CH_2COOH$

(c) $-I$-effect of halogens follows the order as

$$F > Cl > Br > I$$

Therefore, sequence for the acidity of halocarboxylic acids is

$$F—CH_2COOH > Cl—CH_2COOH > Br—CH_2COOH > I—CH_2COOH$$

(ii) Effect of electron releasing substituents

An electron donating group (EDG) destabilises the carboxylate ion and thus, decreases the acidic character.

$$G \longrightarrow C(O)(O)^{\ominus}$$

Acidic strength decreases with increase in number of electron donating groups.

$$HCOOH > CH_3COOH > CH_3CH_2COOH > (CH_3)_2CHCOOH$$

Effect of hybridisation on acidity

(a) Acidic character increases when acetylinic, vinyl or phenyl groups are directly attached to the carboxylic group. This is because, sp^2-hybridised carbon atom is directly attached to —COOH group.

e.g.

$$CH_2=CH—C(=O)OH \longleftrightarrow \overset{\oplus}{C}H_2—CH=C(O^{\ominus})OH$$

Therefore, with increase in % of s-character of α-carbon, acidity of carboxylic acid increases.

e.g.

$$HC\equiv\underset{sp}{\overset{\alpha}{C}}—COOH > CH_2=\underset{sp^2}{\overset{\alpha}{C}H}—COOH > \underset{sp^3}{\overset{\alpha}{C}H_3}COOH$$

(b) Peroxy acids are weaker acids than carboxylic acids because of absence of resonance.

$$CH_3—C(=O)—O—OH \longrightarrow CH_3—C(=O)—O—O^{\ominus}$$

No resonance occurs

Reactions involving acidic hydrogen

(i) Reaction with bases

Carboxylic acids react with bases to form ionic salts which are soluble in water. This reaction shows their acidic character.

$$R—C(=O)—OH + NaHCO_3 \longrightarrow RC(=O)—\bar{O}\overset{+}{Na} + CO_2 + H_2O$$

$$R—C(=O)—OH + NaOH \longrightarrow R—C(=O)—\bar{O}\overset{+}{Na} + H_2O$$

Note *The reaction of carboxylic acids with $NaHCO_3$ is used as a qualitative test for detection of carboxylic functional group in an organic compound.*

(ii) **Reaction with metals**

Carboxylic acids also react with metals like Na or Zn to produce metallic salts with liberation of hydrogen.

$$2R-\overset{O}{\overset{\|}{C}}-OH + 2Na \longrightarrow 2R\overset{O}{\overset{\|}{C}}-\overset{-}{O}\overset{+}{Na} + H_2\uparrow$$

$$2R-\overset{O}{\overset{\|}{C}}-OH + Zn \longrightarrow (RCOO)_2Zn + H_2\uparrow$$

Reactions involving replacement of —OH group

(i) **Esterification**

Carboxylic acids react with alcohols to form esters in presence of conc. H_2SO_4.

$$R-\overset{O}{\overset{\|}{C}}-\boxed{OH + H}-OR' \underset{H_2SO_4}{\overset{\text{Conc.}}{\rightleftharpoons}} \underset{\text{Ester}}{R-\overset{O}{\overset{\|}{C}}-OR'} + H_2O$$

The reaction proceed *via* addition-elimination mechanism that has been discussed in chapter alcohol. *OR'* of alcohol goes with ester.

The order of reactivity of carboxylic acid is follows as

$HCOOH > CH_3COOH > (CH_3)_2CCOOH > (CH_3)_3CCOOH$

The order of reactivity of alcohol follows the sequence as

$$\underset{(1^\circ)}{CH_3OH} > \underset{(2^\circ)}{CH_3-\underset{CH_3}{\underset{|}{CH}}-OH} > \underset{(3^\circ)}{CH_3-\overset{CH_3}{\overset{|}{\underset{CH_3}{\underset{|}{C}}}}-OH}$$

The intramolecular esterification is favourable with γ or δ-hydroxy carboxylic acid spontaneously to give five-membered ring (called γ-lactones) or six-membered ring (called δ-lactones).

$$\underset{\substack{\text{4-hydroxybutanoic acid}\\ (\gamma\text{-hydroxybutyric acid})}}{HO\overset{\gamma}{C}H_2\overset{\beta}{C}H_2\overset{\alpha}{C}H_2\overset{O}{\overset{\|}{C}}OH} \xrightarrow{H_2O/H^+} \underset{\substack{\text{4-butanolide}\\ (\gamma\text{-lactone})}}{\text{(five-membered lactone ring)}}$$

$$\underset{\substack{\text{5-hydroxybutanoic acid}\\ (\delta\text{-hydroxyvaleric acid})}}{HO\overset{\delta}{C}H_2\overset{\gamma}{C}H_2\overset{\beta}{C}H_2\overset{\alpha}{C}H_2\overset{O}{\overset{\|}{C}}OH} \xrightarrow{H_2O/H^+} \underset{\substack{\text{5-pentanolide}\\ (\delta\text{-lactone})}}{\text{(six-membered lactone ring)}}$$

Note *Hydroxy acids, i.e. compounds that contain both hydroxyl and a carboxylic acid functional group within the same molecule, which on intramolecular esterification form cyclic esters called lactones.*

(ii) **Reaction with phosphorous halides**

Carboxylic acid reacts with PCl_3 and PCl_5 to produce acyl halide.

$$3RCOOH + PCl_3 \longrightarrow \underset{\text{Acyl halide}}{3RCOCl} + H_3PO_3$$

$$RCOOH + PCl_5 \longrightarrow RCOCl + POCl_3 + HCl\uparrow$$

(iii) **Reaction with thionyl chloride**

Carboxylic acid reacts with $SOCl_2$ to produce acyl halide.

$$RCOOH + SOCl_2 \longrightarrow RCOCl + SO_2\uparrow + HCl\uparrow$$

$SOCl_2$ is preferred over phosphorus halides as it produces gaseous SO_2 and HCl as byproducts which escape the reaction mixture and makes the purification of product easier.

(iv) **Reaction with ammonia**

The reaction of carboxylic acids with ammonia forms ammonium salt of carboxylic acid, which on heating undergoes elimination to produce amide. The nitrogen of ammonia and its derivatives are better nucleophiles. So, it captures the proton from — OH group of carboxylic acid to form ammonium carboxylate salt.

$$\textbf{Step I } R-\overset{O}{\overset{\|}{C}}-OH + NH_3 \longrightarrow R-\overset{O}{\overset{\|}{C}}-\bar{O}NH_4^+$$

$$\textbf{Step II } R-\overset{O}{\overset{\|}{C}}-\bar{O}NH_4^+ \xrightarrow{\Delta} \underset{\text{Amide}}{R-\overset{O}{\overset{\|}{C}}-NH_2} + H_2O$$

(v) **Formation of anhydride**

Carboxylic acids undergo dehydration, on reaction with phosphorus pentoxide to produce acid anhydride.

$$\underset{\text{Anhydride}}{R-\overset{O}{\overset{\|}{C}}-OH} + HO-\overset{O}{\overset{\|}{C}}-R \underset{-H_2O}{\xrightarrow{P_2O_5,\Delta}} \underset{\text{Acid anhydride}}{R-\overset{O}{\overset{\|}{C}}-O-\overset{O}{\overset{\|}{C}}-R}$$

Other reagents that can be used for dehydration are $Na(NH_4)HPO_4/BPO_4$ at a temperature of 875-895 K.

(vi) **Reaction with diazomethane**

Carboxylic acids, on reaction with diazomethane give high yields of methyl esters at room temperature.

e.g. $$\underset{\text{Propanoic acid}}{CH_3CH_2COOH} + \underset{\text{Diazomethane}}{CH_2N_2} \longrightarrow \underset{\text{Methyl propionate}}{CH_3CH_2COOCH_3} + N_2\uparrow$$

Reactions involving —COOH group

(i) **Decarboxylation**

On heating sodium or potassium salts of carboxylic acid with sodalime (NaOH + CaO) decarboxylation occurs which gives alkane as a final product. The reaction is used to degrade the carbon chain as the product contains one carbon less than parent acid.

$$\underset{\text{Sodium salt of carboxylic acid}}{R\text{—}\overset{\overset{O}{\|}}{C}\text{—}\overset{-}{O}\overset{+}{Na}} + NaOH \xrightarrow[\Delta]{CaO} \underset{\text{Alkane}}{R\text{—}H} + Na_2CO_3$$

If there is a keto group at β-position or (—C = C—) bond between β and γ-position then decarboxylation occurs by simple heating.

$$\text{2-oxocyclohexane ring with } \alpha\text{-carbon bearing COOH and } CH_3 \xrightarrow{\Delta} \text{2-methylcyclohexanone} + CO_2\uparrow$$

$$\overset{\gamma}{C}H_2{=}\overset{\beta}{C}H\overset{\alpha}{C}H_2COOH \xrightarrow{\Delta} CH_2{=}CH—CH_3 + CO_2\uparrow$$

Decarboxylation can also occur by the electrolysis of aqueous solution of sodium or potassium salts of carboxylic acids to give alkanes. This process is known as Kolbe's electrolysis.

$$2CH_3COO^-K^+ \longrightarrow 2CH_3COO^- + 2K^+$$

At anode,

$$2CH_3COO^- \xrightarrow[-2e^-]{} \underset{\text{Unstable}}{2CH_3CO\dot{O}} \longrightarrow CH_3\text{—}CH_3 + CO_2$$

At cathode, $2K^+ \xrightarrow{+2e^-} 2K \xrightarrow{+2H_2O} 2KOH + H_2$

(ii) Schmidt reaction

Carboxylic acid reacts with hydrazoic acid (N_3H) to produce amines. This reaction is known as Schmidt reaction.

$$\underset{\text{Carboxylic acid}}{R\text{—}COOH} + N_3H \xrightarrow{H_2SO_4,\Delta} \underset{\text{Amine}}{RNH_2} + CO_2\uparrow + N_2\uparrow$$

(iii) Reduction

Carboxylic acids undergo reduction to form alkanes, alcohols and carboxyl compounds with different reagents.

$$\underset{\text{Carboxylic acid}}{R\text{—}COOH} \xrightarrow[\Delta]{\text{Red P/HI}} \underset{\text{Alkane}}{R\text{—}CH_3} + 2H_2O + 3I_2$$

$$\underset{\text{Carboxylic acid}}{R\text{—}COOH} \xrightarrow[\text{Ether, } H_3O^+]{LiAlH_4} \underset{\text{Alcohol}}{R\text{—}CH_2OH}$$

$$\underset{\text{Carboxylic acid}}{R\text{—}COOH} \xrightarrow[\Delta]{\text{(i) } RCOOH \text{ (ii) MnO}} \underset{\text{Aldehyde}}{R\text{—}CHO} + CO_2 + 2H_2O$$

(iv) Oxidation

Carboxylic acids undergo oxidation reactions on prolonged heating with powerful oxidising agent, at α - carbon.

$$RCH_2COOH \xrightarrow{SeO_2} \underset{\alpha\text{-keto acid}}{R\overset{\overset{O}{\|}}{C}COOH} + H_2O + Se$$

(v) Hell-Volhard Zelinsky (HVZ) reaction

The reaction in which α-hydrogen of the carboxylic acid is replaced by halogen atom on reaction with red phosphorus in presence of Cl_2 or Br_2 is known as Hell-Volhard Zelinsky (HVZ) reaction.

$$R\underset{\underset{H}{|}}{\overset{\alpha}{C}H}COOH \xrightarrow[HBr]{Br_2,\text{ red P}} R\underset{\underset{Br}{|}}{C}HCOOH \xrightarrow[HBr]{Br_2,\text{ red P}} R\underset{\underset{Br}{|}}{\overset{\overset{Br}{|}}{C}}COOH$$

If no α-hydrogen is present in carboxylic acid, then no reaction occurs.

e.g. $$CH_3\underset{\underset{CH_3}{|}}{\overset{\overset{CH_3}{|}}{C}}COOH + Br_2 \xrightarrow{\text{Red P}} \text{No reaction}$$

α-halogenated acids are good starting material for the preparation of substituted carboxylic acids.

$$\underset{\text{2-bromo propanoic acid}}{CH_3\underset{\underset{Br}{|}}{C}HCOOH} \xrightarrow{\text{(i) } Aq.\text{ KOH (ii) } H^+} \underset{\text{Lactic acid (}\alpha\text{-hydroxy)}}{CH_3\underset{\underset{OH}{|}}{C}HCOOH}$$

$$\underset{\text{2-bromo propanoic acid}}{CH_3\underset{\underset{Br}{|}}{C}HCOOH} \xrightarrow[NH_3]{\text{Large excess of}} \underset{\text{Alanine (}\alpha\text{-amino)}}{CH_3\underset{\underset{NH_2}{|}}{C}HCOOH}$$

$$\underset{\text{2-bromo propanoic acid}}{CH_3\underset{\underset{Br}{|}}{C}HCOOH} \xrightarrow{NaCN} \underset{\text{2-cyano propanoic acid}}{CH_3\underset{\underset{CN}{|}}{C}HCOOH} \xrightarrow{H_3O^+} \underset{\text{Methyl malonic acid}}{CH_3\underset{\underset{COOH}{|}}{C}HCOOH}$$

CHECK POINT 19.2

1. H_2SO_4 is a stronger acid than CH_3COOH
   (a) it has carboxyl group
   (b) it decomposes with increase in temperature
   (c) it has high degree of ionisation
   (d) None of the above

2. Carboxylic acids are stronger acids due to
   (a) formation of carboxylate ion
   (b) high value of degree of ionisation
   (c) presence of H-bonding
   (d) formation of carboxylate ion stabilised by resonance

3. Correct order of acidity is
   (a) $(Cl)_3C—COOH > Cl—CH_2COOH > (Cl)_2CHCOOH > CH_3COOH$
   (b) $(Cl)_3C—COOH > (Cl)_2CHCOOH > Cl—CH_2COOH > CH_3COOH$
   (c) $CH_3COOH > Cl—CH_2COOH > (Cl)_2CHCOOH > (Cl)_3C—COOH$
   (d) $(Cl)_2CHCOOH > Cl—CH_2COOH > CH_3COOH > (Cl)_3C—COOH$

4. 2-bromopropionic acid reacts with ammonia in excess to give
   (a) lactic acid
   (b) malonic acid
   (c) methyl malonic acid
   (d) alanine

5. Reaction of carboxylic acid in presence of red phosphorus and bromine is known as
   (a) HVZ reaction
   (b) Koch reaction
   (c) Decarboxylation
   (d) Arndt-Eistert reaction

6. Product of the following reaction,

$$HOCH_2CH_2CH_2\overset{\overset{O}{\|}}{C}—OH \xrightarrow{H_2O/H^+} ?$$

   is
   (a) CH_3CH_2OH and CH_3CH_2COOH
   (b) $\underset{OH}{\underset{|}{CH_2}}—\underset{OH}{\underset{|}{CH_2}}$
   (c) (five-membered lactone ring: γ-butyrolactone)
   (d) (four-membered lactone ring)

7. Consider the following reaction,

$$R—COOH + PCl_5 \longrightarrow RCOCl + 'X' + HCl$$

   compound '*X*' in the reaction is
   (a) H_3PO_3 (b) $POCl_3$
   (c) H_3PO_4 (d) Cl_2

8. Consider the conversion given below.

$$RCOOH + N_3H \xrightarrow{H_2SO_4, \Delta} RNH_2 + CO_2\uparrow + N_2\uparrow$$

   This conversion is known as
   (a) Schmidt reaction
   (b) Hunsdiecker reaction
   (c) Arndt-Eistert reaction
   (d) Koch reaction

9. Product of the reaction, $(CH_3)_3C—COOH \xrightarrow[Br_2]{\text{Red P}} A$
   (a) $CH_3—\underset{Br}{\underset{|}{CH}}—COOH$
   (b) $CH_3—\overset{Br}{\overset{|}{\underset{Br}{\underset{|}{CH}}}}—COOH$
   (c) $(CH_3)_3C—CH_2OH$
   (d) No product is formed

10. Consider the following reaction

$$CH_3CH_2COOH \xrightarrow{'A'} CH_3\overset{\overset{O}{\|}}{C}COOH + H_2O$$

   Here '*A*' is
   (a) SeO_2 (b) red P / Br_2
   (c) NaOH/CaO (d) AgOH

DICARBOXYLIC ACIDS

Carboxylic acids which have two carboxyl groups, are called dicarboxylic acids. The general formula of saturated dicarboxylic acids is $C_nH_{2n}(COOH)_2$.

The aliphatic dicarboxylic acids are named in IUPAC as *alkanedioic acids.* Most of the dicarboxylic acids are better known by their common names.

$$\begin{array}{l}COOH\\|\\COOH\end{array}, \quad HOOCCH_2COOH, \quad HOOCCH_2—CH_2COOH$$

IUPAC name :	Ethanedioic acid	Propanedioic acid	Butanedioic acid
Common name :	Oxalic acid	Malonic acid	Succinic acid

The unsaturated dicarboxylic acids exhibit geometrical isomerism and named as *E*- or *Z*-alkenedioic acids.

$$\underset{\substack{cis\text{-but-2-enedioic acid}\\ \text{(Maleic acid)}}}{\overset{H\qquad\qquad H}{\underset{HOOC\qquad COOH}{C=C}}} \quad ; \quad \underset{\substack{trans\text{-but-2-enedioic acid}\\ \text{(Fumaric acid)}}}{\overset{H\qquad\qquad COOH}{\underset{HOOC\qquad H}{C=C}}}$$

Methods of preparation

Following methods are used for the preparation of dicarboxylic acids.

Laboratory method

Oxalic acid can be prepared by oxidation of cane sugar in HNO_3 in the presence of V_2O_5.

$$\underset{\text{Cane sugar}}{C_{12}H_{22}O_{11}} + 18[O] \xrightarrow[V_2O_5]{HNO_3} 6 \underset{\text{Oxalic acid}}{\begin{array}{l}COOH\\|\\COOH\end{array}} + 5H_2O$$

Hydrolysis of cyanogen

Cyanogen undergoes hydrolysis in presence of HCl, i.e. acidic hydrolysis.

$$\underset{\text{Cyanogen}}{\begin{array}{l}CN\\|\\CN\end{array}} + 4H_2O \longrightarrow \underset{\text{Oxalic acid}}{\begin{array}{l}COOH\\|\\COOH\end{array}} + 2NH_3$$

$$NH_3 + HCl \longrightarrow NH_4Cl$$

From sodium salt of formic acid

On heating $HCOO^-Na^+$ at 400°C under pressure produces sodium oxalate from which oxalic acid is recovered by treatment with calcium hydroxide followed by conc. H_2SO_4.

$$\underset{\text{Sodium formate}}{2HCOO^-Na^+} \xrightarrow[\text{Under pressure}]{400°C} \underset{\text{Sodium oxalate}}{\begin{array}{l}COO^-Na^+\\|\\COO^-Na^+\end{array}} + H_2\uparrow$$

$$\begin{array}{l}COO^-Na^+\\|\\COO^-Na^+\end{array} + Ca(OH)_2 \longrightarrow \begin{array}{l}COO^-\\|\\COO^-\end{array}Ca^{2+} \xrightarrow[H_2O]{H^+} \underset{\text{Oxalic acid}}{\begin{array}{l}COOH\\|\\COOH\end{array}}$$

From carbon dioxide and sodium or potassium

Na or K are heated in presence of CO_2 at 633 K to produce sodium salt of oxalic acid from which oxalic acid is recovered as mention in the above method.

$$2Na + 2CO_2 \xrightarrow{633K} \begin{array}{l}COO^-Na^+\\|\\COO^-Na^+\end{array}$$

Physical properties

Various physical properties of dicarboxylic acids are as follows

(i) Physical state

Dicarboxylic acids are crystalline solids. Lower acids are readily soluble in water and solubility decreases with increase in number of carbon atoms. Dicarboxylic acids such as oxalic acid exist as dihydrate, i.e. $(COOH_2)\cdot 2H_2O$.

It is an active poison and depresses the central nervous system which causes malfunctioning of kidneys.

(ii) Melting point

The melting points of dicarboxylic acids are relatively high since there is stronger hydrogen bonding due to presence of two carboxylic groups.

e.g.

$$\begin{array}{ccc}
 & \overset{\delta^-}{O} & & \overset{\delta^-}{O}—\overset{\delta^+}{H}--- \\
 & \| & & | \\
CH_2— & C—\overset{\delta^-}{O}—\overset{\delta^+}{H}---\overset{\delta^-}{O}=C & —CH_2 \\
| & & & | \\
CH_2— & C—\overset{\delta^-}{O}—\overset{\delta^+}{H}---\overset{\delta^-}{O}=C & —CH_2 \\
 & \| & & | \\
 & \underset{\delta^-}{O}--- & & \underset{\delta^-}{O}—\underset{\delta^+}{H}---
\end{array}$$

(iii) Acidic strength

Acidity of dicarboxylic acid decreases with increase in number of carbon atoms in between the two carboxyl groups.

$$\begin{array}{l}COOH\\|\\COOH\end{array} > H_2C\begin{array}{l}COOH\\ \\COOH\end{array} > \begin{array}{l}CH_2—COOH\\|\\CH_2—COOH\end{array}$$

Chemical properties

Dicarboxylic acids have similar chemical properties as monocarboxylic acids. A dicarboxylic acid react with two moles of reagent to give corresponding disubstituted derivatives. Some important chemical reactions of dicarboxylic acids are as follows

(i) Reaction with bases

Oxalic acid reacts with alkalies to form sodium salts of oxalic acid.

$$\begin{matrix}COOH\\|\\COOH\end{matrix} \xrightarrow[-H_2O]{NaOH} \begin{matrix}COOH\\|\\COO^-Na^+\end{matrix} \xrightarrow[-H_2O]{NaOH} \begin{matrix}COO^-Na^+\\|\\COO^-Na^+\end{matrix}$$

(ii) Reaction with alcohols

Oxalic acid when treated with alcohols, form esters.

$$\begin{matrix}COOH\\|\\COOH\end{matrix} \xrightarrow[-H_2O]{C_2H_5OH} \begin{matrix}COOH\\|\\COOC_2H_5\end{matrix} \xrightarrow[-H_2O]{C_2H_5OH} \underset{\text{Ethyl oxalate}}{\begin{matrix}COOC_2H_5\\|\\COOC_2H_5\end{matrix}}$$

(iii) Reaction with ammonia

Oxalic acid reacts with NH_3 to form oxamide.

$$\begin{matrix}COOH\\|\\COOH\end{matrix} \xrightarrow[-H_2O]{NH_3} \underset{\text{Oxamic acid}}{\begin{matrix}CONH_2\\|\\COOH\end{matrix}} \xrightarrow[-H_2O]{NH_3} \underset{\text{Oxamide}}{\begin{matrix}CONH_2\\|\\CONH_2\end{matrix}}$$

(iv) Oxidation reaction

Oxalic acid undergoes oxidation reaction with acidified $KMnO_4$ or $K_2Cr_2O_7$ to produce carbon dioxide.

$$5\begin{matrix}COOH\\|\\COOH\end{matrix} + 2KMnO_4 + 3H_2SO_4 \longrightarrow K_2SO_4 + 2MnSO_4 + 10CO_2\uparrow + 8H_2O$$

(v) Reduction reaction

Oxalic acid is reduced to glycolic in presence of Zn and H_2SO_4.

$$\begin{matrix}COOH\\|\\COOH\end{matrix} + 4H \xrightarrow[H_2SO_4]{Zn} \underset{\text{Glycolic acid}}{\begin{matrix}CH_2OH\\|\\COOH\end{matrix}} + H_2O$$

Oxalic acid on electrolytic reduction gives glycolic acid and glyoxalic acid both.

$$2\begin{matrix}COOH\\|\\COOH\end{matrix} + 6H \xrightarrow[\text{Reduction}]{\text{Electrolytic}} \begin{matrix}CH_2OH\\|\\COOH\end{matrix} + \underset{\text{Glyoxalic acid}}{\begin{matrix}CHO\\|\\COOH\end{matrix}} + H_2O$$

(vi) Dehydration

In presence of conc. H_2SO_4, it undergoes dehydration reaction to give water, carbon monoxide and carbon dioxide.

$$\begin{matrix}COOH\\|\\COOH\end{matrix} \xrightarrow[90°C]{\text{Conc. } H_2SO_4} CO_2\uparrow + H_2O + CO\uparrow$$

(vii) Reaction with glycol

Oxalic acid reacts with glycol to form ethylene oxalate.

$$\underset{\text{Oxalic acid}}{\begin{matrix}COOH\\|\\COOH\end{matrix}} + \underset{\text{Glycol}}{\begin{matrix}CH_2OH\\|\\CH_2OH\end{matrix}} \longrightarrow \underset{\text{Ethylene oxalate}}{\begin{matrix}O{=}C-O-CH_2\\|\quad\quad\quad|\\O{=}C-O-CH_2\end{matrix}} + 2H_2O$$

(viii) Reaction with glycerol

Oxalic acids reacts with glycerol at different temperature in following two ways.

(a) When oxalic acid reacts with glycerol at 383 K, monoformate is formed which on hydrolysis gives formic acid and glycerol. Thus, provides a path of converting oxalic acid into formic acid.

$$\underset{\text{Glycerol}}{\begin{matrix}CH_2OH\\|\\CHOH\\|\\CH_2OH\end{matrix}} + \underset{\text{Oxalic acid}}{\begin{matrix}HO-C{=}O\\|\\HO-C{=}O\end{matrix}} \xrightarrow[-H_2O]{383\text{ K}} \underset{\text{Glycerol acid oxalate}}{\begin{matrix}CH_2-O-C{=}O\\|\quad\quad\quad|\\CHOH\quad C{=}O\\|\quad\quad\quad|\\CH_2OH\quad OH\end{matrix}}$$

$$\xrightarrow{-CO_2} \underset{\text{Glycerol monoformate}}{\begin{matrix}CH_2-O-\overset{O}{\overset{\|}{C}}-H\\|\\CHOH\\|\\CH_2OH\end{matrix}} \xrightarrow{H_2O} HCOOH$$

Or the above reaction can directly be written as

$$\underset{\text{Oxalic acid}}{\begin{matrix}COOH\\|\\COOH\end{matrix}} \xrightarrow{\text{Glycerol, 383 K}} \underset{\text{Formic acid}}{HCOOH} + CO_2\uparrow$$

(b) When glycerol si heated with oxalic acid or formic acid at 533 K, it is converted into allyl alcohol.

$$\underset{\text{Glycerol}}{\begin{matrix}CH_2OH\\|\\CHOH\\|\\CH_2OH\end{matrix}} + \underset{\text{Oxalic acid}}{\begin{matrix}COOH\\|\\COOH\end{matrix}} \xrightarrow{533\text{ K}} \underset{\text{Dioxalin}}{\begin{matrix}CH_2OCO\\|\quad\quad|\\CHO\ CO\\|\\CH_2OH\end{matrix}} \xrightarrow{-2CO_2} \underset{\text{Allyl alcohol}}{\begin{matrix}CH_2\\\|\\CH\\|\\CH_2OH\end{matrix}}$$

AROMATIC MONOCARBOXYLIC ACIDS

Organic compounds in which carboxyl group (—COOH) is directly attached to an aromatic ring are known as aromatic carboxylic acids.

In IUPAC nomenclature, these acids are termed as benzene carboxylic acid derivatives. However, the common names of aromatic carboxylic acids are more familiar and are also accepted by IUPAC.

e.g.

Benzene carboxylic acid or Benzoic acid; *o*-toluic acid or 2-methyl benzoic acid; Salicylic acid or 2-hydroxybenzoic acid

Methods of preparation

Monocarboxylic aromatic amines are generally prepared by following methods.

(i) Oxidation of alkyl benzene

Alkyl benzene on reaction with alkaline $KMnO_4$ gives aromatic acid. In this reaction, benzene ring does not undergo oxidation but alkyl group gets oxidised and produces benzoic acid (whatever the length of side chain, oxidation product is benzoic acid).

Toluene $\xrightarrow{\text{Alkaline } KMnO_4}$ Benzoic acid $\xleftarrow{\text{Alkaline } KMnO_4}$ Ethyl benzene

Ethyl benzene $\xrightarrow{KMnO_4,\text{ hot}}$ Acetophenone $\xrightarrow{KMnO_4,\text{ hot}}$ Benzoic acid

Alkyl benzene can also be oxidised using CrO_3 and 40% H_2SO_4 at 373K.

Propyl benzene ($C_6H_5CH_2CH_2CH_3$) $\xrightarrow[373\text{ K}]{K_2Cr_2O_7/H_2SO_4 \text{ or } CrO_3,\ 40\%\ H_2SO_4}$ C_6H_5COOH

(ii) Oxidation of primary alcohols

Primary aromatic alcohols are oxidised by strong oxidising agents such as $K_2Cr_2O_7$ or $KMnO_4$ to give carboxylic acid.

$C_6H_5CH_2OH \xrightarrow[\Delta]{K_2Cr_2O_7} C_6H_5CHO$ (Benzaldehyde) $\xrightarrow[\Delta]{K_2Cr_2O_7} C_6H_5COOH$ (Benzoic acid)

(iii) Hydrolysis of acid derivatives and nitriles

Aromatic acid derivatives on hydrolysis gives benzoic acid. Similarly, benzonitrile on hydrolysis also gives benzoic acid.

Benzoyl chloride (C_6H_5COCl) $\xrightarrow[\text{hydrolysis}]{H_2O}$ Benzoic acid (C_6H_5COOH) + HCl

Benzonitrile (C_6H_5CN) $\xrightarrow[\text{hydrolysis}]{H_2O/H^+}$ Benzoic acid (C_6H_5COOH) + NH_4^+

(iv) From coal-tar or petroleum

Coal-tar or petroleum $\xrightarrow{\text{(Catalytic refining)}}$ Toluene; *o*-xylene or Naphthalene

Toluene $\xrightarrow{Cl_2,\ \Delta}$ Benzotrichloride ($C_6H_5CCl_3$) $\xrightarrow{\text{(i) } OH^-,\ \text{(ii) } H_3O^+}$ Benzoic acid

Toluene $\xrightarrow{\text{Air, 340 K, Co-Mn acetate}}$ Benzoic acid

o-xylene or Naphthalene $\xrightarrow{V_2O_5,\ O_2,\ \Delta,\ H_2O}$ Phthalic acid $\xrightarrow[CO_2]{\text{Catalyst, } \Delta}$ Benzoic acid

Physical properties

The aromatic carboxylic acids exhibit polarity due to presence of polar carbonyl and hydroxyl group. In general, carboxylic acids exist in the *dimer form* due to intermolecular hydrogen bonding. The aromatic carboxylic acids are generally crystalline solids with high melting point.

However, the substituted carboxylic acids may form intramolecular hydrogen bonding as in case of 2-hydroxy benzoic acid (salicylic acid).

Intermolecular hydrogen bonding in benzoic acid

Intramolecular hydrogen bonding

Salicylic acid (or 2-hydroxybenzoic acid)

Acidity of aromatic acids

In aromatic carboxylic acids, carbonyl carbon is sp^2-hybridised and stabilises the charge in carboxylate ion through resonance. Thus, increases the acidity. Resonance in aromatic carboxylic ion is shown below.

Effect of substituents on acidity

- Electron withdrawing groups stabilise the anion by dispersal of negative charge thus, increases the acidity.
- Electron donating groups destabilise the anion by increasing electron density thus, decreases the acidity.

G withdraws electrons, stabilises anion, strengthen acid

G releases electrons, destabilises anion, weakens acid

Note *Both the acid strengthening and weakening effect are more pronounced at the para-positions than at the meta-positions.*

Ortho-effect

o-substituted benzoic acids are stronger acids than benzoic acid. This is known as ***ortho*-effect**. This is due to steric and electronic factors. It does not matter whether the substituent is electron releasing or electron withdrawing.

o-toluic acid > benzoic acid > m-toluic acid > p-toluic acid

Various trends due to ortho-effect are as follows

(*i*) Acidity order for isomers of nitro benzoic acids.

o-nitrobenzoic acid > p-nitrobenzoic acid > m-nitrobenzoic acid > benzoic acid

(*ii*) Acidic strength for isomers of methoxy benzoic acids.

o-methoxybenzoic acid ≅ m-methoxybenzoic acid > benzoic acid > p-methoxybenzoic acid

(*iii*) Acidic strength of aminobenzoic acids.

COOH (benzoic acid) > COOH, NH_2 (*m*-) > COOH, NH_2 (*p*-) > COOH, NH_2 (*o*-)

Explanation *o*-aminobenzoic acid is less acidic as compared to carboxylic acid due to formation of Zwitter ion *via* H-bonding.

δ- δ+ $O^{\ominus}$ $\overset{\oplus}{N}$—H

o-aminobenzoic acid ⟷ Zwitter ion

A comparison of K_a values of aliphatic and aromatic acids has been given in Table 19.2 and 19.3.

Table 19.2 Ionisation constants (K_a) of aliphatic carboxylic acids

Aliphatic Acids	Aliphatic formula	K_a
Formic acid	HCOOH	17.7×10^{-5}
Acetic acid	CH_3COOH	1.75×10^{-5}
Propionic acid	CH_3CH_2COOH	1.40×10^{-5}
n-butyric acid	$CH_3CH_2CH_2COOH$	1.52×10^{-5}
Benzoic acid	C_6H_5COOH	6.30×10^{-5}
Acrylic acid	$CH_2=CHCOOH$	5.5×10^{-5}
Chloroacetic acid	$ClCH_2COOH$	1.36×10^{-3}
Dichloroacetic acid	$Cl_2CHCOOH$	5.53×10^{-2}
Trichloroacetic acid	Cl_3CCOOH	2.32×10^{-1}
Fluoroacetic acid	FCH_2COOH	2.60×10^{-3}
Bromoacetic acid	$BrCH_2COOH$	1.25×10^{-3}
Iodoacetic acid	ICH_2COOH	6.70×10^{-4}
α-chlorobutyric acid	$CH_3CH_2\overset{\alpha}{C}HCOOH$ (Cl on α-C)	1.39×10^{-3}
β-chlorobutyric acid	$CH_3\overset{\beta}{C}HCH_2COOH$ (Cl on β-C)	8.9×10^{-5}
γ-chlorobutyric acid	$\overset{\gamma}{C}H_2CH_2CH_2COOH$ (Cl on γ-C)	2.96×10^{-5}
Pyruvic acid	$CH_3\overset{O}{\overset{\|}{C}}COOH$	3.16×10^{-5}
Oxalic acid	COOH–COOH	5.4×10^{-2} (K_{a_1})
Malonic acid	$CH_2(COOH)_2$	1.4×10^{-3} (K_{a_1})
Succinic acid	$(CH_2)_2(COOH)_2$	6.4×10^{-5} (K_{a_1})

Table 19.3 Ionisation constant of aromatic carboxylic acids

Aromatic acids	*o*-	*m*-	*p*-
Toluic acid	12.4×10^{-5}	5.4×10^{-5}	4.2×10^{-5}
Nitrobenzoic acid	670×10^{-5}	32×10^{-5}	36×10^{-5}
Hydroxybenzoic acid	105×10^{-5}	8.3×10^{-5}	2.6×10^{-5}
Chlorobenzoic acid	120×10^{-5}	15.1×10^{-5}	10.3×10^{-5}
Aminobenzoic acid	1.16×10^{-5}	1.9×10^{-5}	1.4×10^{-5}
Methoxybenzoic acid	8.2×10^{-5}	8.2×10^{-5}	3.3×10^{-5}
Benzoic acid		6.3×10^{-5}	

Diacids have two dissociation constants : K_{a_1} is for the first dissociation, and K_{a_2} is for the second dissociation, to give a dianion. The second carboxyl group is always less acidic than the first ($K_{a_2} << K_{a_1}$) because extra energy is required to create a second negative charge close to another mutually repulsive negative charge.

Chemical properties

Aromatic carboxylic acids show chemical reactions similar to aliphatic carboxylic acids, like reaction with bases, phosphorus halide, thionyl chloride etc. They also give electrophilic substitution reactions. These are as follows

Electrophilic substitution reactions

Aromatic carboxylic acids undergo electrophilic substitution reaction at *m*-position due to presence of electron withdrawing group, i.e. —COOH group.

This can be represented by resonating structures.

Due to delocalisation of π-electrons, electron density is decreased at o-and p-positions. Thus, electrophile attacks at m-position.

Electrophile can attack only at m-sites having decreased reactivity.

If electron donating group (such as OH in salicylic acid) is present, substitution occurs at o-and p-positions. This results in elimination of —COOH group.

Salicylic acid $\xrightarrow[-CO_2]{Br_2 \text{ water}}$ 2,4,6-tribromophenol

Salicylic acid $\xrightarrow[-CO_2]{\text{Fuming } HNO_3}$ 2,4,6-trinitrophenol (picric acid)

Some of the electrophilic substitution reactions are shown below

(i) Nitration

Benzoic acid $\xrightarrow[\text{Conc.}H_2SO_4]{\text{Conc.}HNO_3}$ m-nitrobenzoic acid + H_2O

(ii) Sulphonation

Benzoic acid $\xrightarrow[\Delta]{\text{Conc.}H_2SO_4}$ m-sulphobenzoic acid + H_2O

(iii) Halogenation

Benzoic acid $\xrightarrow[\Delta]{Br_2,FeBr_3}$ m-bromobenzoic acid + HBr

Note *Phthalic acid on heating form an anhydride and on reaction with ammonia it gives phthalimide..*

Tests for distinction

To distinguish between carboxylic acid and other organic compounds such as alcohol and with other carboxylic acids, some characteristic tests are given below

(i) Distinction between alcohol (CH_3CH_2OH) and carboxylic acid (CH_3COOH)

S.No.	Test	CH_3COOH	CH_3CH_2OH
1.	Iodoform test	no	yes
2.	Neutral $FeCl_3$	blood-red colour	no reaction

(ii) Distinction between CH_3COOH and HCOOH

S.No.	Test	CH_3COOH	HCOOH
1.	Neutral $FeCl_3$	blood-red colour	no reaction
2.	Tollen's reagent	no effect	reduced to metallic Ag.
3.	Fehling's solution	no effect	reduced to Cu^+ (Cu_2O, red ppt).
4.	Heat	no effect	gives CO (burns with blue flame) and H_2O.

(iii) Distinction between CH_3COOH (aliphatic) and C_6H_5COOH (aromatic)

S.No.	Test	CH_3COOH	C_6H_5COOH
1.	Neutral $FeCl_3$	blood-red colour	buff coloured ppt.
2.	Sodalime, Δ	CH_4	C_6H_6

(iv) Distinction between HCOOH and C_6H_5COOH

S.No.	Test	HCOOH	C_6H_5COOH
1.	Neutral $FeCl_3$	no colour	buff coloured ppt
2.	Tollen's reagent	is reduced to metallic Ag	no reaction
3.	Fehling's solution	is reduced to red ppt.	no reaction

Uses

(i) Higher fatty acids are used in manufacture of soaps.

(ii) Organic acids are used for preparation of many drugs such as aspirin, phenacetin etc.

(iii) Acetic acid is used as coagulant in manufacture of rubber. It is also used in dye stuffs, perfumes, rayon etc.

(iv) Tannic acid is yellow or light brown powder, is highly soluble in water. It is used for staining wood and dyeing cotton.

(v) Salicylic acid is used in acne creams to reduce acne. It is also used as food preservatives.

(vi) Citric acid is a natural preservative. It is used in foods and soft drinks.

(vii) Oxalic acid is used in bleaching and cleansing solutions.

(viii) Acetic acid is used in foods such as 2-3% solution of CH_3COOH in water known as vinegar.

CONVERSIONS

(i) **CH_3COOH into HCOOH and *vice-versa***

$$CH_3COOH \xrightarrow[\text{Schmidt}]{N_3H} CH_3NH_2 \xrightarrow{HNO_2} CH_3OH$$

$$\xrightarrow{\text{Oxidation}} HCOOH \xrightarrow{LiAlH_4} CH_3OH$$

$$\xrightarrow[KCN]{PCl_3} CH_3CN \xrightarrow{H_3O^+} CH_3COOH$$

(ii) **CH_3COOH into $CH_3CH(OH)COOH$** (lactic acid)

$$CH_3COOH \longrightarrow CH_3COCl \xrightarrow[BaSO_4]{H_2/Pd} CH_3CHO$$

$$\xrightarrow{HCN} CH_3CH(OH)CN \xrightarrow{H_3O^+} CH_3CH(OH)COOH$$

(iii) Butadiene into CH_2CH_2COOH / CH_2CH_2COOH (adipic acid)

$$CH_2{=}CH{-}CH{=}CH_2 \xrightarrow[\text{1,4-addition}]{Br_2} \text{1,4-dibromocyclohexene} \xrightarrow{Pt/H_2} \text{1,4-dibromocyclohexane} \xrightarrow[H_3O^+]{KCN} \text{Adipic acid}$$

(iv) **CH≡CH into** CH_3COCH_3 (through calcium salt)

$$CH{\equiv}CH \xrightarrow{HgSO_4/H_2SO_4} CH_3CHO \xrightarrow{\text{oxidation}}$$

$$CH_3COOH \xrightarrow{Ca(OH)_2} (CH_3COO)_2Ca \xrightarrow{\Delta} CH_3COCH_3$$

(v) **Glyoxal into tartaric acid**

$$\underset{\text{Glyoxal}}{CHO{-}CHO} \xrightarrow{2HCN} CH(OH)CN{-}CH(OH)CN \xrightarrow{H_3O^+} \underset{\text{Tartaric acid}}{CH(OH)COOH{-}CH(OH)COOH}$$

(vi) **2-phenylethanol into 3-phenylpropanoic acid**

$$C_6H_5{-}CH_2CH_2OH \xrightarrow{PCl_5} C_6H_5{-}CH_2CH_2Cl \xrightarrow{KCN}$$

$$C_6H_5{-}CH_2CH_2CN \xrightarrow{H_3O^+} C_6H_5{-}CH_2CH_2COOH$$

(vii) **Phenol into phenyl acetic acid**

$$C_6H_5{-}OH \xrightarrow{Zn,\ \Delta} C_6H_6 \xrightarrow{CH_3Cl/AlCl_3} C_6H_5{-}CH_3 \xrightarrow{Cl_2,\ \Delta} C_6H_5{-}CH_2Cl \xrightarrow{KCN} C_6H_5{-}CH_2CN \xrightarrow{H_3O^+} C_6H_5{-}CH_2COOH$$

- **Bromobenzene into benzoic acid**

$$C_6H_5{-}Br \xrightarrow{Mg/ether} C_6H_5{-}Mg{-}Br \xrightarrow[H_3O^+]{CO_2} C_6H_5{-}COOH$$

- **Phenyl acetic acid into benzoic acid**

$$C_6H_5{-}CH_2COOH \xrightarrow{LiAlH_4} C_6H_5{-}CH_2CH_2OH \xrightarrow{MnO_4^-/H^+} C_6H_5{-}COOH$$

CHECK POINT 19.3

1. Dicarboxylic acids have high melting points than monocarboxylic acid because
(a) of the presence of intermolecular H-bonding
(b) of the presence of two carboxylic groups
(c) of the presence of intramolecular H-bonding
(d) Both (a) and (b)

2. Oxalic acid + $C_2H_5OH \longrightarrow A$,
product of the reaction is
(a) $COOC_2H_5$ | $COOC_2H_5$
(b) $COOH$ | $COOC_2H_5$
(c) Both (a) and (b)
(d) None of these

3. Compound '*X*' in the following reaction

$$2\begin{array}{c}COOH\\|\\COOH\end{array} + 6\,H \xrightarrow[\text{reduction}]{\text{Electrolytic}} X$$

is
(a) CH_2OH | $COOH$
(b) CHO | $COOH$
(c) CH_2OH | CH_2OH
(d) Both (a) and (b)

4. Compound '*X*' in the following reaction

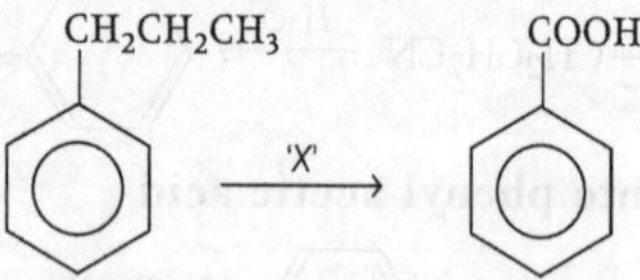

is
(a) Alkaline $KMnO_4$
(b) $K_2Cr_2O_7/H_2SO_4$
(c) CrO_3/40% H_2SO_4
(d) All of these

5. Product '*A*' in the following reaction

(Toluene, CH_3) $\xrightarrow[\Delta]{Cl_2} \xrightarrow[H_3O^+]{OH^-} A$

is

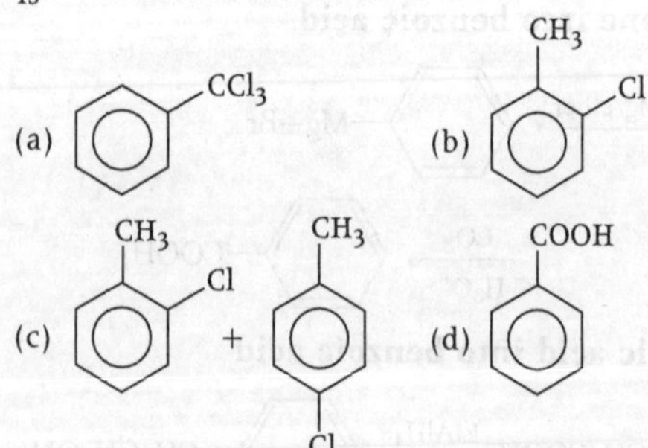

6. The correct acidity order for the following isomers of nitro benzoic acid is

(a) COOH (o-NO_2) > COOH (p-NO_2) > COOH (m-NO_2) > COOH

(b)

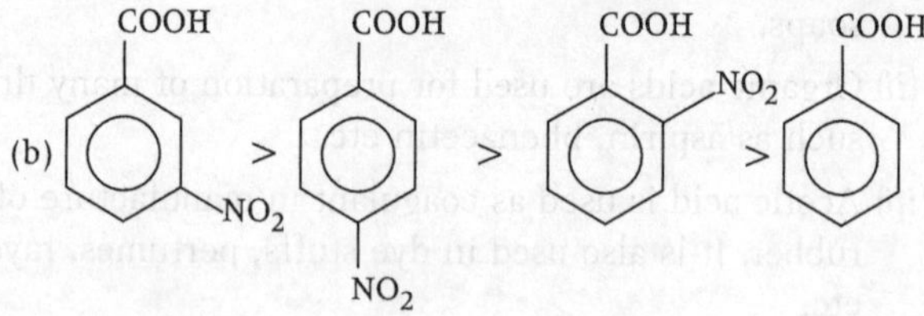

(c)

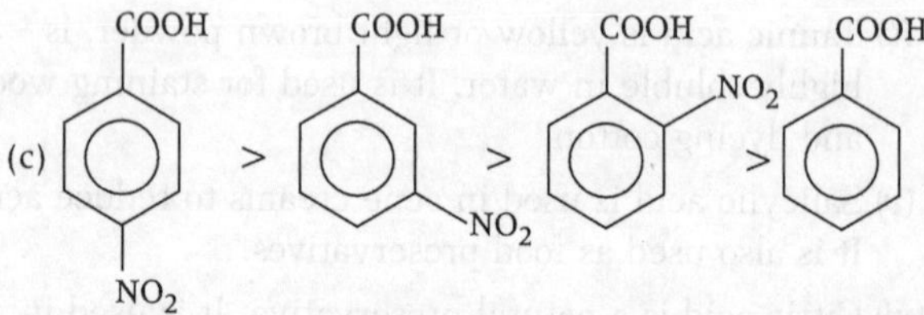

(d) COOH > COOH > COOH (m-NO_2, NO_2) > COOH (p-NO_2)

7. Product formed when salicylic acid reacts with conc.HNO_3
(a) *p*-nitrophenol
(b) *o*-nitrophenol
(c) 2, 4, 6-trinitrophenol
(d) 2, 4-trinitrophenol

8. To distinguish between acetic acid and benzoic acid, reagent used is
(a) Tollen's reagent
(b) Fehling's solution
(c) Neutral $FeCl_3$
(d) Iodoform test

9. Which of the following acid is used in bleaching and cleansing solutions?
(a) Citric acid
(b) Oxalic acid
(c) Salicylic acid
(d) Tannic acid

Chapter Exercises

(A) Topical exercise

I. Classification nomenclature, isomerism and structure

1 Acid present in red ant is
(a) CH_3COOH (b) $HCOOH$ (c) HCN (d) H_3PO_4

2 IUPAC name of valeric acid is
(a) pentanoic acid (b) butanoic acid
(c) hexanoic acid (d) benzoic acid

3 The acid which does not contain —COOH group is
(a) ethanoic acid (b) picric acid
(c) lactic acid (d) palmitic acid

4 Which of the following acids has —COOH group?
(a) Carbolic acid (b) Picric acid
(c) Anthranilic acid (d) Sulphanilic acid

5 Which of the following acid is a part of fat?
(a) Crotonic acid (b) Acrylic acid
(c) Stearic acid (d) Oleic acid

6 Which of the following acid is a part of oil?
(a) Pivalic acid (b) Valeric acid
(c) Stearic acid (d) Oleic acid

7 Which of the following is a correct IUPAC name for the compound shown?

$$(CH_3CH_2)_2\overset{\overset{CO_2H}{|}}{C}CH_2CH(CH_2CH_3)_2$$

(a) 1,1,3-triethylhexanoic acid
(b) 2,2,4-triethylhexanoic acid
(c) 3,5-diethyl-3-heptylcarboxylic acid
(d) 3,5,5-triethyl-6-hexanoic acid

8 Match the acids given in Column I with their correct IUPAC names given in Column II.

	Column I (Acids)	Column II (IUPAC names)
A.	Phthalic acid	1. Hexane- 1, 6-dioic acid
B.	Oxalic acid	2. Benzene-1, 2-dicarboxylic acid
C.	Succinic acid	3. Pentane-1, 5-dioic acid
D.	Adipic acid	4. Butane-1, 4-dioic acid
E.	Glutaric acid	5. Ethane-1, 2-dioic acid

Codes

	A	B	C	D	E
(a)	2	5	4	3	1
(b)	1	3	4	5	2
(c)	5	2	1	3	4
(d)	2	5	4	1	3

9 $CH_3\overset{\overset{O}{\|}}{C}O^-$ ion is stabilised by
(a) resonance effect of the carbonyl group
(b) inductive effect of the methyl group
(c) Both of the above are correct
(d) None of the above is correct

10 Select the correct statement about a carboxylic acid

$$R—\overset{\overset{O}{\|}}{C}—O—H$$

(a) C=O bond length is equal to C—O bond length due to resonance
(b) C=O bond length is larger than C—O bond length
(c) C=O bond length is shorter than C—O bond length
(d) C—O, C=O and C—H bond lengths are equal

11 Consider the following statements.
Statement I Long-chain carboxylic acids are often called fatty acids.
Statement II Many of them were first obtained by hydrolysis of fats.
(a) Both statements I and II are correct
(b) Only statement I is correct
(c) Only statement II is correct
(d) Both statements I and II are incorrect

II. Methods of preparation

12 Which of the following cannot be used for the synthesis of a carboxylic acid?
(a) (cyclopropyl)—MgBr (b) $CH_2=CHCH_2MgBr$
(c) $CH_2\overset{\overset{}{}}{C}CH_2MgBr$ (with C=O) (d) (phenyl)—MgBr

13 Consider the following reaction

$$CH\equiv CH \xrightarrow[(ii) H_2O_2]{(i) O_3} 'X'$$

Here, X is
(a) CH_3COOH (b) CH_3CHO
(c) CH_3COCH_3 (d) HCOOH

14 Alkaline hydrolysis of CH_3CN forms
(a) CH_3COOH
(b) CH_3COO^-
(c) $HCOO^-$
(d) HCOOH

15 Which of the following alcohols is oxidised by alkaline $KMnO_4$ into carboxylic acid?
(a) $CH_2=CHCH_2OH$ (b) $HC\equiv CCH_2OH$
(c) CH_3CH_2OH (d) 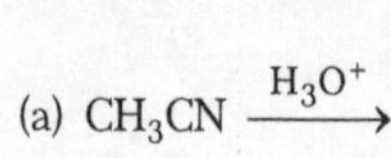$-CH_2OH$

16 Alkaline hydrolysis of ester is irreversible because
(a) carboxylate ion (a nucleophile) cannot attack alcohol (a nucleophile)
(b) base, being strong electrophile favours hydrolysis
(c) Both (a) and (b)
(d) None of the above

17 Which of the following cannot be used for the synthesis of a carboxylic acid?
(a) $CH_3CN \xrightarrow{H_3O^+}$ (b) $CH_3\overset{O}{\overset{\|}{C}}Cl \xrightarrow{H_3O^+}$
(c) $CH_3\overset{O}{\overset{\|}{C}}OC_2H_5 \xrightarrow{OH^-}$ (d) $CH_3\overset{O}{\overset{\|}{C}}O\overset{O}{\overset{\|}{C}}CH_3 \xrightarrow{H_3O^+}$

18 Indicate correct synthesis of carboxylic acid from organometallic compound.
(a) $CH_3MgBr \xrightarrow{CO_2} \xrightarrow{H_3O^+}$
(b) $CH_3MgBr \xrightarrow[\text{(ii) } CO_2]{\text{(i) } H_3O^+}$
(c) $CH_3\overset{O}{\overset{\|}{C}}CH_2CH_2MgBr \xrightarrow[\text{(ii) } H_3O^+]{\text{(i) } CO_2}$
(d) All of the above

19 $CH_2=CHCOOH$ can be obtained in which of the following synthesis?
(a) $CH_2=CHCN \xrightarrow{H_3O^+}$
(b) $CH_2=CHCN \xrightarrow{OH^-}$
(c) Both (a) and (b)
(d) None of the above

20 Compound '*A*' reacts with PCl_5 to produce '*B*' which on treatment with KCN followed by hydrolysis produce propionic acid. Compound '*A*' is
(a) C_2H_5OH (b) C_2H_6
(c) C_3H_8 (d) C_2H_5Cl

21 End product of the following reaction,
$$\underset{Br}{\underset{|}{CH_2}}-CH_2-\underset{Br}{\underset{|}{CHCH_3}} \xrightarrow[H_3O^+]{KCN}$$
is
(a) $CH_2CH_2\underset{CO_2H}{\underset{|}{CH}}\underset{Br}{\underset{|}{CO_2H}}$ (b) $\underset{Br}{\underset{|}{CH_2}}CH_2\underset{CH_3}{\underset{|}{CHCO_2H}}$
(c) $\underset{CO_2H}{\underset{|}{CH_2}}CH=CHCH_3$ (d) $CH_2=CHCH=CH_2$

22 Consider the following reactions,
I. $RCN \xrightarrow{H_3O^+}$ II. $RCN \xrightarrow{OH^-}$
III. $RCN \xrightarrow{DIBAL-H}$ IV. $RCN \xrightarrow{SnCl_2/HCl}$
V. $RCN \xrightarrow{LiAlH_4}$
In which of the above cases RCO_2H is not the product?
(a) All except III (b) All except III, V
(c) All except I (d) All except I, II

23 Isobutyric acid can be synthesised by two methods
I. $CH_3\underset{CH_3}{\underset{|}{CH}}Br \xrightarrow{CN^-} \xrightarrow{H_3O^+}$
II. $CH_3CH=CH_2 + CO + H_2O \xrightarrow[600\text{ K}]{H_3PO_4}$
Which of the above methods can be adopted?
(a) Only I (b) Only II
(c) Both (a) and (b) (d) None of these

24 In which of the following case, the number of carbon atoms is retained?
(a) $CH_3CH=CHCH_3 \xrightarrow{O_3/H_2O/Zn}$
(b) $CH_3C\equiv CCH_3 \xrightarrow[\Delta]{KMnO_4, OH^-}$
(c) (cyclohexene) $\xrightarrow{MnO_4^-, OH^-, \Delta}$
(d) $CH_3CH_2CH=CH_2 \xrightarrow{O_3/H_2O}$

25 Consider the following change
$$CH_3\overset{O}{\overset{\|}{C}}CH_2CH_3 \longrightarrow CH_3\overset{O}{\overset{\|}{C}}OH \text{ and } CH_3CH_2\overset{O}{\overset{\|}{C}}OH$$
Suitable oxidising agent for the above change is ...... and predominant product is ......
(a) MnO_4^-/OH^-, $CH_3CH_2\overset{O}{\overset{\|}{C}}OH$ (b) $Cr_2O_7^{2-}/H^+$, $CH_3\overset{O}{\overset{\|}{C}}OH$
(c) MnO_4^-/OH^-, $CH_3\overset{O}{\overset{\|}{C}}OH$ (d) $Cr_2O_7^{2-}/H^+$, $CH_3CH_2\overset{O}{\overset{\|}{C}}OH$

26 Compound $A(C_8H_{10})$ by ozonolysis forms compound $B(C_4H_6O_2)$. Compound A and B can be
(a) $HC\equiv C(CH_2)_4C\equiv CH$, $CH_2=CHCH_2CO_2H$
(b) $HC\equiv C(CH_2)_4C\equiv CH$, $CH_3CH=CHCO_2H$
(c) ▷—C≡C—◁, ▷—CO_2H
(d) $CH_2=CHCH_2C\equiv CCH_2CH=CH_2$, $CH_2=CHCH_2CO_2H$

27 Consider the following reaction,

$$CH_3CH_2\overset{O}{\overset{\|}{C}}-\overset{18}{O}-CH_2CH_3 \xrightarrow{H_3O^+} CH_3CH_2\overset{O}{\overset{\|}{C}}OH + CH_3CH_2OH$$

Acyl C—O bond Alkyl (—O— bond)

Select the correct statement.
(a) Alkyl C—O bond breaks and O^{18} is contained in acid
(b) Alkyl C—O bond breaks and O^{18} is contained in alcohol
(c) Acyl C—O bond breaks and O^{18} is contained in alcohol
(d) Acyl C—O bond breaks and O^{18} is contained in acid

28 $CH_3\overset{O}{\overset{\|}{C}}CH_3$ can be converted to $CH_3\overset{O}{\overset{\|}{C}}OH$ by following methods

I. $CH_3\overset{O}{\overset{\|}{C}}CH_3 \xrightarrow[\Delta, H^+]{I_2/NaOH}$

II. $CH_3\overset{O}{\overset{\|}{C}}CH_3 \xrightarrow{Cr_2O_7^{2-}/H^+}$

III. $CH_3\overset{O}{\overset{\|}{C}}CH_3 \xrightarrow{Ag(NH_3)_2^+}$

Which methods can be effective?
(a) Both I and III (b) Both II and III
(c) Both I and II (d) I, II and III

29 Which of the following methods is adopted for the synthesis of $(CH_3)_3C-CO_2H$?
(a) $(CH_3)_3C-Br \xrightarrow{CN^-} \xrightarrow{H_3O^+}$
(b) $(CH_3)_3C-Br \xrightarrow{Mg/ether} \xrightarrow{CO_2} \xrightarrow{H_3O^+}$
(c) Both (a) and (b)
(d) None of the above

30 In which of the following cases yield of the carboxylic acid is maximum?
(a) $CH_3CH_2CH_2CH_2Br \xrightarrow{KCN,\ DM\ SO} \xrightarrow{H_3O^+}$
(b) $CH_3\underset{Br}{\underset{|}{C}H}CH_2CH_3 \xrightarrow{KCN,\ DM\ SO} \xrightarrow{H_3O^+}$
(c) $CH_3-\overset{CH_3}{\overset{|}{\underset{CH_3}{\underset{|}{C}}}}-Br \xrightarrow{KCN,\ DMSO} \xrightarrow{H_3O^+}$
(d) Equivalent yield

III. Physical properties

31 Consider the following statements.
Statement I Carboxylic acids in their pure state exist as dimer.
Statement II Osmotic pressure of dimer is increased.
(a) Both statements I and II are correct
(b) Only statement I is correct
(c) Only statement II is correct
(d) Both statements I and II are incorrect

32 Boiling points of CH_3COOH (mol. wt. = 60) is higher than that of $CH_3CH_2CH_2OH$ (mol. wt. = 60). It is due to
(a) intermolecular H-bonding between two carboxylic acid molecules at two points which is at one point only in alcohol
(b) strong van der Waals' forces
(c) intramolecular H-bonding
(d) None of the above

33 Select the correct statement(s).
(a) Carboxylic acids are both hydrogen-bonding donors and hydrogen-bonding acceptors
(b) Carboxylic acids in pure state exist as dimers
(c) Carboxylic acids of low molecular weight are infinitely soluble in water because of extensive H-bonding with water molecules
(d) All of the above statements are correct

34 As the molecular weight of a carboxylic acid increases, solubility in water decreases. It is due to
(a) hydrophobicity of the hydrocarbon portion overweighs the hydrophilicity of the carboxyl group
(b) hydrophobicity of the carboxyl group overweighs the hydrophilicity of the hydrocarbon portion
(c) ionisation of the carboxylic acid is increased
(d) carboxylic acid exists as dimer

35 Increasing order of the boiling points of CH_3COOH (I), $CH_3CH_2CH_2OH$ (II) and CH_3CH_2CHO (III) is in the order of
(a) I < II < III (b) III < II < I (c) II < I < III (d) III < I = II

36 Consider the following statements.
Statement I Carboxylic acids of low molecular weight are infinitely soluble in water molecules.
Statement II Carboxylic acids are both hydrogen-bonding donors and hydrogen-bonding acceptors.
(a) Both statements I and II are correct
(b) Only statement I is correct
(c) Only statement II is correct
(d) Both statements I and II are incorrect

37 The carboxylic acid which has maximum solubility in water is

(a) phthalic acid (b) succinic acid
(c) malonic acid (d) salicylic acid

IV. Acidic nature of carboxylic acid

38 Increasing pK_a values of *o*-, *m*- and *p*-nitrobenzoic acid is

(a) *o*- < *p*- < *m*- (b) *p*- < *m*- < *o*-
(c) *o*- < *m*- < *p*- (d) *m*- < *p*- < *o*-

39 Among the following acids least pK_a value is of

(a) oxalic acid (b) pyruvic acid
(c) malonic acid (d) succinic acid

40 Which of the following has the maximum value of equilibrium constant?

(a) $CH_3COOH + H_2O \rightleftharpoons CH_3COO^- + H_3O^+$
(b) $HCOOH + H_2O \rightleftharpoons HCOO^- + H_3O^+$
(c) $C_6H_5COOH + H_2O \rightleftharpoons C_6H_5COO^- + H_3O^+$
(d) $CF_3COOH + H_2O \rightleftharpoons CF_3COO^- + H_3O^+$

41 Which of the following has maximum pK_a value?

(a) Picric acid (b) Salicyclic acid
(c) Anthranilic acid (d) Carbolic acid

42 Which of the following is the strongest acid?

(a) $CH_3\overset{O}{\overset{\|}{C}}OH$ (b) $CH_3\overset{O}{\overset{\|}{C}}OOH$
(c) $CH_3\underset{CH_3}{\underset{|}{C}}HCH_2\overset{O}{\overset{\|}{C}}OH$ (d) $CH_3CH_2\overset{O}{\overset{\|}{C}}OH$

43 Strongest acid out of CH_3CO_2H, $CH_2{=}CHCO_2H$ and

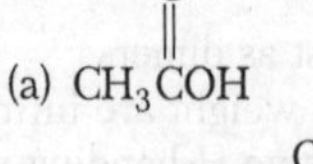
$-CO_2H$ is

(a) CH_3CO_2H (b) $CH_2{=}CHCO_2H$
(c) $C_6H_5-CO_2H$ (d) equal K_a

44 Consider the following statements.

Statement I The characteristic property of carboxylic acids is that they are acidic. They readily give up a proton from the carboxyl group.

Statement II Carboxylate ions formed are stabilised by resonance.

(a) Both statements I and II are correct
(b) Only statement I is correct
(c) Only statement II is correct
(d) Both statements I and II are incorrect

45 Acids with their pK_a values have been given

$\underset{NO_2}{\underset{|}{CH_2}}COOH$; $\underset{CN}{\underset{|}{CH_2}}COOH$; $\underset{Cl}{\underset{|}{CH_2}}COOH$
1.68 2.46 2.86

$\underset{OH}{\underset{|}{CH_2}}COOH$; $\underset{H}{\underset{|}{CH_2}}COOH$
3.83 4.74

Which group has maximum electron withdrawing (acid strength increasing) tendency?

(a) NO_2 (b) H
(c) CN (d) OH

46 Consider the following acids.

I. $CH_3CH_2\overset{O}{\overset{\|}{C}}OH$ II. $CH_2{=}CH\overset{O}{\overset{\|}{C}}OH$

III. $CH{\equiv}C\overset{O}{\overset{\|}{C}}\ OH$

Arrange the above acids in increasing pK_a values.

(a) I < II < III (b) III < II < I
(c) I < III < II (d) III < I < II

47 Increasing values of dissociation constant of K_a of

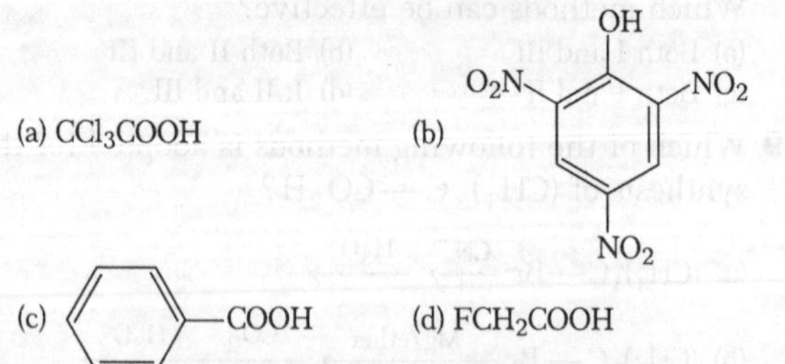

(a) I < II < III (b) III < II < I
(c) I < III < II (d) II < III < I

48 Strongest acid among the following is

(a) CCl_3COOH (b) 2,4,6-trinitrophenol (OH with O_2N, NO_2 and NO_2 on the ring)
(c) C_6H_5-COOH (d) FCH_2COOH

49 For the following acids

$\underset{I}{(CH_3)_3CCH_2CO_2H}$, $\underset{II}{(CH_3)_3\overset{\oplus}{N}CH_2CO_2H}$, $\underset{III}{CH_3COOH}$

pK_a value will be in order

(a) I > II > III
(b) I < II < III
(c) I > III > II
(d) II > III > I

50 Consider the following statements.

Statement I In aliphatic carboxylic acids, presence of strong electron withdrawing groups on the hydrocarbon chain increases the acidity.

Statement II Chloroacetic acid is stronger than acetic acid.

(a) Both statements I and II are correct
(b) Only statement I is correct
(c) Only statement II is correct
(d) Both statements I and II are incorrect

51 Consider the following statements.

Statement I Salicyclic acid (*o*-hydroxy benzoic acid) is stronger than its *m*- and *p*-isomers.

Statement II Carboxylate ion is stabilised by intramolecular H-bonding.

(a) Both statements I and II are true
(b) Only statement I is true
(c) Only statement II is true
(d) Neither statement I nor statement II is true.

52 Peroxyacetic acid (CH_3CO_3H) is a weaker acid than acetic acid (CH_3CO_2H) since

(a) negative charge in $CH_3\overset{O}{\overset{\|}{C}}O\ O^-$ cannot be delocalised into the carbonyl group
(b) CH_3 group in CH_3CO_3H shows *I*-effect
(c) Both of the above are correct
(d) None of the above is correct

53 Compared to phenol (pK_a = 10.02) acetic acid (pK_a = 4.76) is more acidic. This is due to

(a) stabilisation of negative charge on carboxylate ion by induction
(b) resonance effect
(c) Both of the above
(d) None of the above

54 pK_a value is maximum for

(a) $CH_3C{\equiv}CH$ (b) 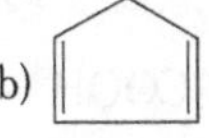
(c) $CH_2{=}CH_2$ (d) CH_3CH_3

55 Which of the following order of acid strength is correct?

(a) *R*COOH > HOH > CH≡CH > *R*OH
(b) *R*COOH > CH≡CH > HOH > *R*OH
(c) *R*COOH > *R*OH > HOH > CH≡CH
(d) *R*COOH > HOH > *R*OH > CH≡CH

56 Rank the following substances in order of decreasing acid strength (strongest → weakest)

$\underset{1}{CH_3CH_2CH_2CO_2H}$ $\underset{2}{CH_3CH{=}CHCO_2H}$

$\underset{3}{CH_3CH_2CH_2CH_2OH}$ $\underset{4}{CH_3C{\equiv}CCO_2H}$

Choose the correct option.

(a) 4 > 2 > 1 > 3 (b) 1 > 2 > 4 > 3
(c) 3 > 1 > 2 > 4 (d) 2 > 4 > 1 > 3

57 Consider the following statements.

Statement I Benzoic acid is stronger than acetic acid.

Statement II Phenyl group is more electron withdrawing in nature due to presence of more electronegative sp^2-hybridised carbon.

(a) Both statements I and II are correct
(b) Only statement I is correct
(c) Only statement II is correct
(d) Both statements I and II are incorrect

V. Chemical properties of monocarboxylic acid

58 When acetic acid reacts with P_2O_5, product formed is

(a) ethyl acetate (b) aceto-acetic ester
(c) acetic anhydride (d) no reaction

59 Decarboxylation will not take place by heating of

(a) methylmalonic acid (b) succinic acid
(c) acetoacetic acid (d) In all cases

60 Carboxylic acids react with diazomethane to yield

(a) alcohols (b) amides
(c) amines (d) esters

61 Which of the following calcium salt will give cyclopentanone on heating?

(a) Calcium succinate (b) Calcium adipate
(c) Calcium glucarate (d) Calcium oxalate

62 Which of the following does not undergo decarboxylation on reaction with soda lime?

(a) Benzoic acid (b) Cinnamic acid
(c) Toluic acid (d) Carbolic acid

63 Acid hydrolysis of an ester is reversible because

(a) alcohol and acid react together to form ester
(b) protonation of ester makes carbonyl carbon more reactive for nucleophilic attack
(c) proton being strong nucleophile, favours hydrolysis
(d) H_3O^+ is a nucleophile as well as electrophile

64 Which of the following does not form Zwitter ion?

(a) Anthranilic acid
(b) Picric acid
(c) Sulphanilic acid
(d) Glycine

65 Which of the following acid is weaker than benzoic acid?

(a) Carbolic acid
(b) Picric acid
(c) Formic acid
(d) *o*-nitrobenzoic acid

66 In which of the following case Hell-Volhard-Zelinsky reaction can take place?

(a) C_6H_5–COOH (benzene ring bearing –COOH)
(b) C_6H_5–$\underset{CH_3}{\overset{CH_3}{C}}$–COOH
(c) C_6H_5–CH_2COOH
(d) $(CH_3)_3C$–COOH

67 RCH_2OH is oxidised to $R\overset{\overset{O}{\|}}{C}OH$ by $Cr_2O_7^{2-}/H_3O^+$. Intermediate formed and final state of $Cr_2O_7^{2-}$ respectively are

(a) $R\overset{\overset{O}{\|}}{C}H$, Cr^{3+} (b) $R\overset{\overset{O}{\|}}{C}H$, CrO_3
(c) $R\overset{\overset{O}{\|}}{C}OH$, Cr^{3+} (d) $R\overset{\overset{O}{\|}}{C}OH$, CrO_3

68 Which of the following gives unsaturated acid on heating?

(a) $CH_3\underset{OH}{\underset{|}{C}}HCO_2H$ (b) $\underset{OH}{\underset{|}{C}}H_2CH_2CO_2H$
(c) $\underset{OH}{\underset{|}{C}}H_2CH_2CH_2CO_2H$ (d) All of these

69 In which of the following case benzene is formed?
(a) Sodium benzoate is heated with sodalime
(b) Calcium benzoate is heated
(c) Mixture of calcium benzoate and calcium acetate is heated
(d) In all the above cases

70 Hell-Volhard-Zelinsky reaction is used to prepare (*A*) (from RCH_2CH_2COOH) and actually takes place through formation of (*B*) which is hydrolysed to get (*A*). (*A*) and (*B*) respectively are

(a) $RCH_2\underset{Br}{\underset{|}{C}}HCOOH$, $RCH_2\underset{Br}{\underset{|}{C}}HCOBr$
(b) $R\underset{Br}{\underset{|}{C}}HCH_2COOH$, $R\underset{Br}{\underset{|}{C}}HCOBr$
(c) $R\underset{Br}{\underset{|}{C}}HCH_2COOH$, $RCH{=}CHCOOH$
(d) $RCH\underset{Br}{\underset{|}{C}}HCOOH$, $RCH{=}CHCOOH$

71 Consider the following acids
$\underset{I}{HCOOH}$, $\underset{II}{CH_3COOH}$, $\underset{III}{CH_3CH_2COOH}$
Ease of esterification of above acids with CH_3OH is
(a) III < II < I (b) I < II < III (c) II < I < III (d) equal

72 Match the acids given in Column I with the properties shown in Column II.

	Column I		Column II
A.	C_6H_5–COOH (benzoic acid)	1.	Gives HVZ reaction
B.	2,4,6-trinitrobenzoic acid (O_2N, NO_2, NO_2 on ring; –COOH)	2.	Weakest acid among the given acids
C.	CH_3COOH	3.	Strongest acid
D.	$(CH_3)_3C$—COOH	4.	Forms dimer

Codes

	A	B	C	D		A	B	C	D
(a)	2	3	4	1	(b)	4	3	1	2
(c)	1	2	3	4	(d)	3	4	1	2

73 Consider the following carboxylic acids.

I. $CH_2{=}CHCH_2\overset{\overset{O}{\|}}{C}OH$ II. $CH_3\overset{\overset{O}{\|}}{C}CH_2\overset{\overset{O}{\|}}{C}OH$
III. C_6H_5–$\overset{\overset{O}{\|}}{C}OH$ IV. (cyclohexadienyl ring)–$CH_2\overset{\overset{O}{\|}}{C}OH$

Indicate the compounds in which decarboxylation takes place on heating.
(a) Both III and IV (b) Both I and III
(c) Both II and IV (d) Both I and II

74 In the following sequence of reactions,

$$RCH_2CH_2COOH \xrightarrow[\text{(ii) } H_2O]{\text{(i) P, } Br_2} A \xrightarrow[\text{(ii) } H^+]{\text{(i) } NaOC_2H_5} B$$

A and *B* respectively are
(a) $R\underset{Br}{\underset{|}{C}}HCH_2COOH$, $RCH{=}CHCOOH$
(b) $R\underset{Br}{\underset{|}{C}}HCH_2COOH$, $R\underset{OC_2H_5}{\underset{|}{C}}HCH_2COOH$
(c) $RCH_2\underset{Br}{\underset{|}{C}}HCOOH$, $RCH{=}CHCOOH$
(d) $RCH_2\underset{Br}{\underset{|}{C}}HCOOH$, $RCH_2\underset{OC_2H_5}{\underset{|}{C}}HCOOH$

75 Which of the following reaction is not given by following carboxylic acids?

(a) $CH_3CH{=}CHCH_2COOH \xrightarrow{LiAlH_4} CH_3CH{=}CHCH_2CH_2OH$

(b) $CH_3CH_2CO_2H \xrightarrow{\Delta} CH_3CH_3 + CO_2$

(c) $\underset{\underset{\underset{O}{\|}}{CCO_2H}}{\overset{\overset{CH_2CO_2H}{|}}{CHCO_2H}} \xrightarrow{\Delta} \overset{CH_2CO_2H}{\underset{CH_2\,\underset{\|}{\underset{O}{C}}CO_2H}{|}}$

(d) $\underset{\underset{Br}{|}}{CH_3CHCO_2H} \xrightarrow{NaOC_2H_5} CH_2{=}CHCO_2H$

76 Match the name-reactions given in Column I that convert acetic acid or its derivative into given compounds in Column II.

	Column I		Column II
A.	Schmidt	1.	$CH_2(Br)COOH$
B.	HVZ	2.	CH_3NH_2
C.	Rosenmund	3.	CH_3COOCH_3
D.	Fischer	4.	CH_3CHO (through CH_3COCl)

Codes

	A	B	C	D
(a)	1	3	4	2
(b)	2	1	4	3
(c)	2	4	3	1
(d)	4	2	1	3

77 Consider the following acids

$\underset{I}{CH_3CH_2OH}, \quad \underset{II}{(CH_3)_2CHOH}, \quad \underset{III}{(CH_3)_3COH}$

Ease of esterification of above alcohols with HCOOH is

(a) I < II < III (b) III < II < I
(c) II < I < III (d) equal

78 Consider the following reaction,

$$CH_3COOH + HCOOH \xrightarrow[570\text{ K}]{MnO}$$

Main product of this reaction is

(a) CH_3CHO (b) CH_3COCH_3
(c) HCHO (d) $(CH_3CO)_2O$

79 Identify final product in the following

$$\underset{\underset{OH}{|}}{CH_3CHCOOH} \xrightarrow{\Delta}$$

Choose the correct option

(a) [cyclic diester (lactide): six-membered ring with two O and two C=O, substituted with CH_3 and H_3C] (b) $CH_2{=}CHCO_2H$

(c) $CH_3\overset{\overset{O}{\|}}{C}CO_2H$ (d) CH_3CH_2OH

80 CH_3Br can be prepared by

(a) $CH_3COOAg + Br_2 \xrightarrow{\Delta}$

(b) $CH_3COOH \xrightarrow{P/Br_2} \xrightarrow{\text{Sodalime, }\Delta}$

(c) Both (a) and (b)

(d) None of the above

81 Consider the following statements.

Statement I β-keto carboxylic acids lose CO_2 upon heating at about 370 K.

Statement II An enol is first formed by loss of CO_2, but it quickly tautomerises to the more stable ketone.

(a) Both statements I and II are correct
(b) Only statement I is correct
(c) Only statement II is correct
(d) Both statements I and II are incorrect

82 Consider the following compounds.

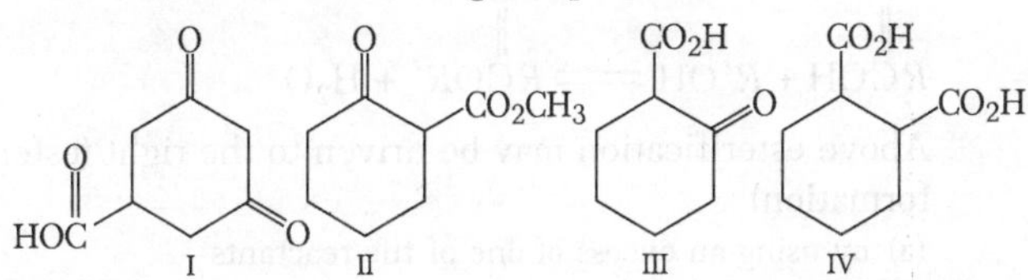

Which of the above compounds will undergo decarboxylation on heating?

(a) Both II and III (b) Both III and IV
(c) Only III (d) Both I and IV

83 Which of the following is least likely to form a lactone?

(a) $\overset{\overset{OH}{|}}{CH_3CHCH_2CH_2CO_2H}$ (b) [cyclohexane with OH and CH_2CO_2H on adjacent carbons]

(c) [cyclohexane with OH and CO_2H in 1,3-positions, cis] (d) [cyclohexane with OH and CO_2H in 1,3-positions, trans]

84 In reaction, $CH_3COOH \xrightarrow{Br_2/P} X \xrightarrow[\text{(ii) } H_3O^+]{\text{(i) KCN}} Y$, Y is identified as

(a) succinic acid (b) malonic acid
(c) glycolic acid (d) α-hydroxy propionic acid

85 CH≡CH on reaction with CH_3COOH in presence of Hg^{2+} gives X which is hydrolysed to give Y. X and Y respectively are

	X	Y
(a)	$CH_3CH(OOCCH_3)_2$	CH_3COCH_3
(b)	$CH_3CH(OOCCH_3)_2$	CH_3CHO
(c)	CH_3CHO	CH_3CHO
(d)	$CH_3CH(OOCCH_3)_2$	CH_3CH_2CHO

86 In the following reactions,

$C_6H_5COOH + NaH\overset{*}{C}O_3 \longrightarrow C_6H_5COONa + H_2O + CO_2$

$\overset{*}{C}$ (isotopes) is present in

(a) $\overset{*}{C}O_2$ (b) $C_6H_5\overset{*}{C}OONa$
(c) Both (a) and (b) (d) None of these

87 Maximum dehydration to form α, β-unsaturated acid takes place in

(a) lactic acid (b) glycolic acid
(c) 3-hydroxypropanoic acid (d) alanine

88 Consider the following reaction.

$$RCOOH + R'OH \rightleftharpoons RCOOR' + H_2O$$

Above esterification may be driven to the right (ester formation)

(a) by using an excess of one of the reactants
(b) by removal of H_2O formed
(c) by both of the above
(d) by none of the above

VI. Dicarboxylic acids

89 Consider the following reaction,

2-oxocyclohexane bearing COOH and CH_2COOH on C-1 $\xrightarrow{\Delta}$ A, here A is

(a) cyclic anhydride (CO–O–CH_2CO) on 2-oxocyclohexane (b) 2-oxocyclohexane with COOH and CH_3 on C-1
(c) 2-oxocyclohexane with $-CH_2COOH$ (d) None of these

90 Consider the following reaction,

$CH_2(CH_2COOH)$–COOH (succinic acid) $+ Ac_2O \xrightarrow{\Delta}$ B, the product B is

(a) $CH_3CH_2CH_2CH_3$ (b) CH_3COOH
(c) cyclic anhydride (d) Both (b) and (c)

91 Consider the following statements.

Statement I *p*-methoxy benzoic acid is stronger than *m*-methoxy benzoic acid.

Statement II $-OCH_3$ displays an electron withdrawing acid strengthening inductive effect from *meta*-position and reverse effect from *para*-position.

(a) Both statements I and II are correct
(b) Only statement I is correct
(c) Only statement II is correct
(d) Both statements I and II are incorrect

92 The reagents used in converting the reaction

$H_2C=CH_2 \longrightarrow HOOC-CH_2-CH_2-COOH$

are

(a) Br_2, KOH (b) HBr, KOH
(c) Br_2, KCN, H_3O^+ (d) Br_2, K_2CO_3

93 Consider the following reactions,

(I) $CH_2(COOH)_2 \xrightarrow[\Delta]{P_2O_5} A$

(II) $HOOC-CH_2-CH_2-COOH + NH_3 \xrightarrow{\Delta} B$

The products (A) and (B) of reactions (I) and (II) are respectively

(a) carbon dioxide, saccharic acid
(b) carbon monoxide, succinimide
(c) carbon suboxide, succinimide
(d) None of the above

VII. Aromatic carboxylic acids

94 Consider the following statements.

Statement I Nitration of salicyclic acid gives picric acid by elimination of $—CO_2H$ group.

Statement II —OH group is strongly activating group hence, S_E reaction takes place is *o*- and *p*-positions.

(a) Both statements I and II are correct
(b) Only statement I is correct
(c) Only statement II is correct
(d) Both statements I and II are incorrect

95 Salicylic acid is subjected to bromination using Br_2 water, product formed is

(a) OH, COOH, Br (b) OH, Br, COOH, Br

(c) Br, Br, COOH, Br (d) OH, Br, Br, Br

96 *o*-toluic acid on reaction with Br_2 /Fe gives

(a) CH_2Br, COOH (b) CH_3, COOH, Br

(c) CH_3, COOH, Br (d) CH_3, COOH, Br

97 Reaction of phenol and CCl_4 and NaOH followed by hydrolysis gives

(a) OH, Cl (b) OH, OH, COOH

(c) OH, COOH (d) OH, CHO

98 Reactivity of (I) C_6H_5COCl, (II) $C_6H_5CONH_2$ and (III) $C_6H_5COOCH_3$ towards hydrolysis is in order

(a) I < II < III
(b) III < II < I
(c) II < III < I
(d) I < III < II

99 Aniline (*A*) can be converted into benzoic acid by which of the following method?

(a) $A \xrightarrow[273\text{ K}]{HNO_2} \xrightarrow{CO_2} \xrightarrow{H_3O^+}$

(b) $A \xrightarrow[273\text{ K}]{HNO_2} \xrightarrow{H_3PO_2} \xrightarrow{CO_2}$

(c) $A \xrightarrow[273\text{ K}]{NaNO_2+HCl} \xrightarrow{CuCN/HCN} \xrightarrow{H_3O^+}$

(d) $A \xrightarrow[273\text{ K}]{HNO_2} \xrightarrow{Mg/ether} \xrightarrow{H_3O^+}$

100 Consider the following sequence of reactions

CH_2Cl, Cl $\xrightarrow{KCN} \xrightarrow{H_3O^+} A$

The end product formed in above reaction is

(a) CH_2COOH, COOH (b) CH_2CN, CN

(c) CH_2COOH, Cl (d) CH_2CN, Cl

101 In the following sequence of reactions,

CH_2Br, Br $\xrightarrow[\text{(iii) } H_3O^+]{\text{(i) Mg/ether, (ii) } CO_2} A$

Identify the end product *A*.

(a) CH_2Br, HOOC (b) CH_2COOH, HOOC

(c) CH_2COOH, Br (d) CH_3, HOOC

102 Consider the following reaction sequence.

$$C_6H_5CH_3 + Cl_2 \xrightarrow{\text{(boiling)}} A \xrightarrow{NaOH} B$$

Here, *A* and *B* respectively are

	(A)	(B)
(a)	$C_6H_5CCl_3$	C_6H_5COOH
(b)	$C_6H_5CCl_3$	$C_6H_5\overset{-}{CO}\overset{+}{O}Na$
(c)	$C_6H_5CHCl_2$	C_6H_5CHO
(d)	$C_6H_5CHCl_2$	$C_6H_5\overset{-}{CO}\overset{+}{O}Na$

103 Phenol and benzoic acid are distinguished by
(a) NaOH (b) $NaHCO_3$ (c) Na_2CO_3 (d) H_2SO_4

104 Phenol and benzoic acid are distinguished by
(a) neutral $FeCl_3$ solution (b) smell
(c) $NaHCO_3$ (d) All of these

105 Match the pair of compounds in given Column I with the reagents given in the Column II used to distinguish them.

Column I	Column II
A. Phenol, benzoic acid	1. Iodine/sodium hydroxide
B. Acetic acid, formic acid	2. Fehling's solution
C. Acetic acid, acetaldehyde	3. Tollen's reagent
D. Benzaldehyde, formic acid	4. Neutral ferric chloride

Codes

	A	B	C	D
(a)	3	4	2	1
(b)	1	2	3	4
(c)	4	2,3	1	2
(d)	4	3	2	1

(B) Miscellaneous questions

1 Fehling's solution can make distinction between following pairs
I. CH_3COOH, HCOOH
II. CH_3COOH, HCHO
III. C_6H_5CHO, HCHO
IV. CH_3CHO, HCHO

Select the correct pair(s) and choose the correct option.
(a) I, II and III (b) II, III and IV
(c) Both II and III (d) Only II

2 Consider the following compounds.
I. $CH_3C(OH){=}CHCOOC_2H_5$
II. $C_6H_5CH_2NH_2$
III. C_6H_5COOH
IV. C_6H_5OH

Which of the above compounds will undergo methylation by using diazomethane?
(a) I, II and IV (b) Both II and III
(c) Both II and IV (d) I, III and IV

3 An optically active compound '*X*' has molecular formula $C_4H_8O_3$. It evolves CO_2 with *aq.* $NaHCO_3$. '*X*' reacts with $LiAlH_4$ to give an achiral compound. '*X*' is
(a) $CH_3CH_2CH(OH)COOH$ (b) $CH_3CH(OMe)COOH$
(c) $CH_3CH(CH_2OH)COOH$ (d) $CH_3CH(OH)CH_2COOH$

4 In the following reaction sequence,

2-ethynylcyclohexanone $\xrightarrow[H_2O_2,\ OH^-]{BH_3.THF} \xrightarrow[H^+]{I_2/NaOH} \xrightarrow{\Delta} X,$

end product *X* is
(a) 2-oxocyclohexane-1-carboxylic acid (ring ketone with COOH)
(b) 2-acetylcyclohexanone (ring ketone with $COCH_3$)
(c) cyclohexanone with CH_2CHO
(d) cyclohexanone

5 Consider the following compounds.

I: cyclobutenedione with two OH (squaric acid)
II: o-hydroxybenzoic acid (OH, COOH on benzene)
III: $CH_3\overset{O}{\overset{\|}{C}}COOH$

Which of the above compounds reacts with $NaHCO_3$ giving CO_2?
(a) I, II and III (b) Both I and III
(c) Both II and III (d) Both I and II

6 Consider a compound

(A) [cyclobutane with two OCH_3 groups on one carbon and two COC_2H_5 (ester, C=O) groups on the opposite carbon]

End product due to hydrolysis and subsequent heating of (A) is

(a) O=⟨cyclobutane⟩ (b) O=⟨cyclobutane⟩–COH (C=O)
(c) $(CH_3O)_2$⟨cyclobutane⟩–COH (C=O) (d) O=⟨cyclobutane⟩(COH)₂ (C=O)

7 Consider the following reactions.

I. $CH_3\overset{O}{\overset{\|}{C}}H \xrightarrow[H^+]{[Ag(NH_3)_2]^+,\ OH^-}$ II. $CH_3CCl_3 \xrightarrow{OH^-}$

III. $CH_3\overset{O}{\overset{\|}{C}}OC_2H_5 \xrightarrow{OH^-}$

Acid is produced in which of the reaction?

(a) I, II and III (b) Both I and III
(c) Both I and II (d) Only I

8 In reaction,

$F_3CCH_2Br \xrightarrow{Mg/ether} \xrightarrow[1\ mol]{^{14}CO_2} \xrightarrow{H_3O^+} X$, (X)

is

(a) $F_2\underset{^{14}CO_2H}{\underset{|}{C}}CH_2Br$ (b) $F_3CCH_2\overset{14}{C}O_2H$
(c) $F_2\underset{CO_2H}{\underset{|}{C}}CH_2\overset{14}{C}O_2H$ (d) No reaction

9 Consider the following reaction sequence

$CH_3C{\equiv}CCO_2H \xrightarrow{H_2,\ \text{Lindlar catalyst}} X$

$CH_3C \equiv CCO_2H \xrightarrow{Li/NH_3} Y$

Select the correct statement about X and Y.

(a) X and Y both have C=C bond but differ in stereoisomerism
(b) X is (Z) but-2-en-1-oic acid and Y is (E) but-2-en-1-oic acid
(c) Dipole moment of X is larger than that of Y
(d) All of the above are correct statements

10 In the following sequence of reactions,`

$CH_3\overset{O}{\overset{\|}{C}}OH \xrightarrow{SOCl_2} A \xrightarrow{C_6H_6,\ AlCl_3} B \xrightarrow{HCN} C \xrightarrow{H_3O^+} D$

Select the correct statements about D.

(a) D has chiral centre (b) D is α-hydroxy acid
(c) Both (a) and (b) (d) None of these

11 Consider the following pairs and reagents.

Pairs	Reagents
I. $CH_3COOH/HCOOH$	P. Tollen's
II. CH_3CHO/CH_3COCH_3	Q. Fehling's
III. Glucose/Sucrose	R. $NaOH/I_2$
IV. Glucose/Fructose	

Which of the above is the correct matching for distinction?

(a) I–P, Q; II–P, Q, R; III–P, Q; IV–P, Q
(b) I–P, Q; II–P, Q; III–P, Q; IV–P
(c) I–P, Q, R; II–P, Q; III–P, Q, R; IV–P
(d) I–P, Q; II–P, Q; III–P; IV–Q

12 Compound (A) given below is to be converted into compound (B),

(A) [5-phenyl-γ-butyrolactone] ⟶ (B) [$C_6H_5CH_2CH_2CH_2CO_2H$]

Suitable reagent is

(a) $LiAlH_4$ (b) $NaBH_4/CH_3OH$
(c) Na/C_2H_5OH (d) H_2, Pd / C

13 If compound A (in Q.12) is to be converted into compound (C)

(C) [$C_6H_5CH(OH)CH_2CH_2CO_2H$]

Then, suitable reagent used is

(a) $NaOH/H_3O^+$
(b) Na/C_2H_5OH
(c) $NaBH_4/CH_3OH$
(d) $LiAlH_4$

14 Identify the product Z in the series given below

$$CH_3CN \xrightarrow{Na/C_2H_5OH} X \xrightarrow{HNO_2} Y \xrightarrow[\text{reagent}]{\text{Baeyer's}} Z$$

Select the correct alternative for Z.

(a) CH_3CHO (b) CH_3CONH_2
(c) CH_3COOH (d) CH_3CH_2NHOH

15 Consider the following reactions,

I. $CH_3NC \xrightarrow{H_3O^+}$

II. $CHCl_3 \xrightarrow{NaOH / H_3O^+}$

III. $CCl_4 \xrightarrow{H_3O^+}$

HCOOH is obtained in

(a) I, II and III (b) Both II and III
(c) Both I and II (d) Both I and III

16 Match the acids in Column I with the properties shown in Column II.

Column I	Column II
A. C_6H_5COOH	1. Red colour, with neutral $FeCl_3$
B. CH_3COOH	2. Reduces Tollen's reagent
C. HCOOH	3. Used as food preservative
D. $C_6H_5CH_2COOH$	4. Decarboxylates to toluene

Codes

	A	B	C	D
(a)	3	1,3	2	4
(b)	2	3	1,4	2
(c)	2	1	3	4
(d)	4	3	2	1

17 Consider the following reactions,

I. $CH_3MgBr \xrightarrow[H_3O^+]{}$ 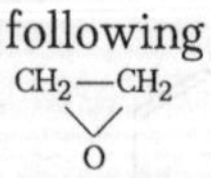 $\xrightarrow[\Delta]{MnO_4^-/H^+}$

II. $CH_3CH_2Br \xrightarrow{KCN} \xrightarrow{H_3O^+}$

III. $CH_3CH_2\overset{O}{\overset{\|}{C}}OC_2H_5 \xrightarrow[H_3O^+]{CH_3MgBr\ (2\ mol)}$

Propanoic acid is the final product in

(a) I, II and III
(b) Both I and II
(c) Both I and III
(d) Both II and III

18 $CH_3CH_2COOH \xrightarrow[\Delta]{NaN_3 + \text{conc. } H_2SO_4} A$, by reaction R_1

$CH_3CH_2COOH \xrightarrow[\Delta]{Br_2/P} B$, by reaction R_2

Which is correct alternate?

	A	B	R_1	R_2
(a)	$CH_3CH_2NH_2$	$CH_3\underset{Br}{CH}COOH$	Schmidt	HVZ
(b)	$CH_3CH_2CONH_2$	CH_3CH_2COBr	HVZ	Schmidt
(c)	$CH_3CH_2NH_2$	CH_3CH_2COBr	HVZ	Schmidt
(d)	None of the above is correct			

19 End product of the conversion

$$CH_3\overset{O}{\overset{\|}{C}}CH_2CH_2CH_2CO_2H \xrightarrow[\text{(ii) } H_2O,\ H^+]{\text{(i) } NaBH_4}$$

is

(a) 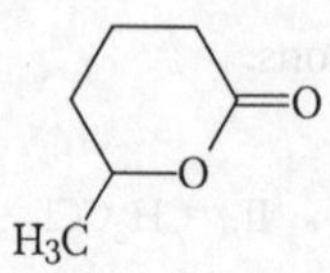

(b) 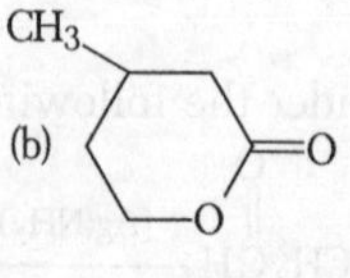

(c) 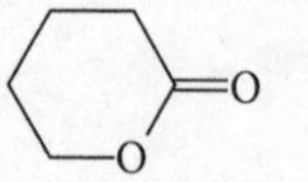

(b) $CH_3\underset{OH}{CH}CH_2CH_2CH_2CO_2H$

20 End products of the following sequence of reactions

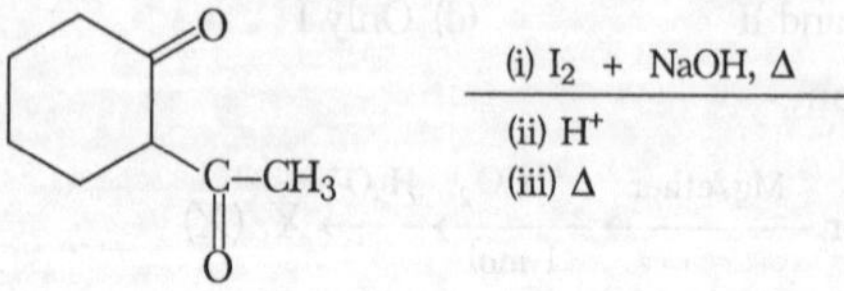

are

(a) yellow ppt. of CHI_3, 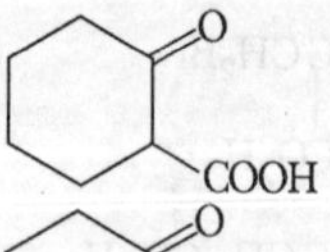

(b) yellow ppt. of CHI_3, 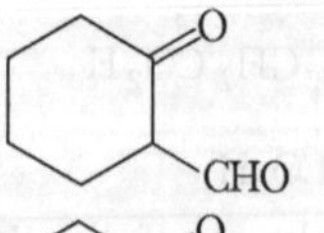

(c) yellow ppt. of CHI_3, 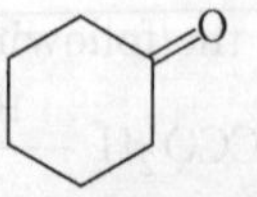

(d) yellow ppt. of CHI_3, 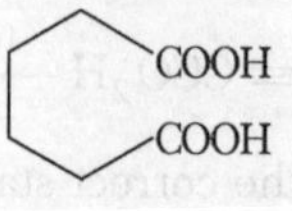

21 For reaction,

$$CH_3—\overset{O}{\overset{\|}{C}}—OH + CH_3CH_2\bullet H \xrightarrow{H^+} \text{ester } (A) + \text{water } (B),$$

A and B are respectively (• is isotopic oxygen)

(a) $CH_3-\overset{O}{\overset{\|}{C}}-{\bullet}C_2H_5, H_2O$

(b) $CH_3-\overset{O}{\overset{\|}{C}}-OC_2H_5, H_2{\bullet}$

(c) $CH_3-\overset{O}{\overset{\|}{C}}-{\bullet}H + CH_3CH_2{\bullet}H$

(d) None of the above is correct

22 Consider following two reactions,

$$CH_3CH_2ONa \xrightarrow[\text{(ii) } H^+]{\text{(i) CO, } \Delta \text{ pressure}} A$$

$$CH_2=CH_2 + CO + H_2O \xrightarrow[670\text{ K}]{H_3PO_4} B$$

A and *B* are respectively,

(a) CH_3CH_2COOH in both cases
(b) CH_3CH_2CHO in both cases
(c) CH_3CH_2COOH, CH_3CHO
(d) CH_3CHO, CH_3COOH

23 Consider the following reaction,

$$CF_3\overset{O}{\overset{\|}{C}}OH + CH_3\underset{OH}{\underset{|}{C}}HCH_3 \xrightarrow{H_2SO_4} CF_3\overset{O}{\overset{\|}{C}}O\underset{CH_3}{\underset{|}{C}}HCH_3 + H_2O$$

Which of the following is not an intermediate in the generally accepted mechanism for the reaction shown?

(a) $CF_3\overset{+\ddot{O}H}{\overset{\|}{C}}-\ddot{O}H$ (b) $CF_3\overset{:\ddot{O}H}{\overset{|}{C}}-\overset{+}{\ddot{O}}CH(CH_3)_2$ with $:\ddot{O}HH$ below

(c) $CF_3\overset{:\ddot{O}H}{\overset{|}{C}}-\ddot{O}CH(CH_3)_2$ with $:\ddot{O}H$ below (d) $CF_3\overset{:\ddot{O}:^-}{\overset{|}{C}}-\ddot{O}CH(CH_3)_2$ with $:\ddot{O}H$ below

24 Identify compound *C* in the following sequence,

$$(CH_3)_2CHCH_2C\equiv N \xrightarrow[\text{Heat}]{H_2O, \text{ HCl}} \text{compound } A$$

$$\xrightarrow[\text{(ii) } H_2O]{\text{(i) } LiAlH_4} \text{compound } B \xrightarrow[CH_2Cl_2]{PCC} \text{compound } C$$

and select the correct alternative that are given below

(a) $(CH_3)_2CH\overset{O}{\overset{\|}{C}}CH_3$ (b) $(CH_3)_2CHCH=C(OH)_2$

(c) $(CH_3)_2CHCH_2\overset{O}{\overset{\|}{C}}H$ (d) $(CH_3)_2CHCH_2\overset{O}{\overset{\|}{C}}OH$

25 Consider the following reaction sequence,

$$\text{Toluene } (C_6H_5CH_3) \xrightarrow[\text{Light}]{Br_2} A \xrightarrow[\text{(ii) } H_3O^+, \text{ heat}]{\text{(i) KCN}} B$$

The final product (*B*) of this sequence is

(a) 2-methylbenzoic acid (CH_3, CO_2H ortho) (b) 3-methylbenzoic acid (CH_3, CO_2H meta)

(c) $C_6H_5CO_2H$ (d) $C_6H_5CH_2CO_2H$

26 Which of the following compounds undergoes decarboxylation (loses carbon dioxide) most readily on being heated?

(a) $C_6H_5COCH_2CH_2COOH$

(b) $HOOC{-}C_6H_4{-}CH_2CH_2COOH$ (meta)

(c) $C_6H_5CH(COOH)CH_2COCH_2CH_3$

(d) $C_6H_5CH(COOH)COCH_2CH_2CH_3$

27 Which of the compounds yields a δ-lactone after being reduced with sodium borohydride?

(a) $C_6H_5COCH_2CH_2COOH$

(b) $C_6H_5CH(COOH)CH_2CH_2COCH_3$

(c) Both (a) and (b) (d) None of these

28 Consider the following reaction sequence,

$$CH_3CH_2CH_2Br \xrightarrow{NaCN} X \xrightarrow[heat]{H_3O^+} Y \xrightarrow[H^+]{CH_3CH_2OH} Z$$

The end product Z formed in above reaction is

(a) $CH_3CH{=}CH\overset{O}{\overset{\|}{C}}OH$

(b) $CH_3CH_2CH_2CH(OCH_2CH_3)_2$

(c) $CH_3CH_2\underset{OCH_2CH_3}{\underset{|}{CH}}C{\equiv}N$

(d) $CH_3CH_2CH_2\overset{O}{\overset{\|}{C}}OCH_2CH_3$

29 Consider the following facts.

I. In both the alcohol and carboxylic acid series, the smaller anion can be better solvated, leading to a larger acidity of the corresponding conjugate acid.

II. H–COOH > CH_3COOH > $(CH_3)_3C$COOH ; (increasing order of acidic strength)

III. $R{-}C{\equiv}C{-}H$ > $R_2C{=}CRH$ > $R_3C{-}H$; (increasing order of acidic strength)

Select the correct facts and mark the correct option.

(a) I, II and III (b) Both I and II

(c) Both II and III (d) Both I and III

30 In the following multisteps synthesis,

$$CH_3CH_2\overset{O}{\overset{\|}{\underset{*}{C}}}CH_2\overset{O}{\overset{\|}{C}}OH \xrightarrow{\Delta} A \xrightarrow[\Delta]{I_2/NaOH} B+C \quad (\overset{*}{C} \text{ is C-14})$$

$$B \xrightarrow[\Delta]{CaO/NaOH} D$$

Products B, C and D are respectively

(a) $CH_3CH_2\overset{O}{\overset{\|}{\underset{*}{C}}}CH_3$, $CH_3CH_2\overset{O}{\overset{\|}{C}}ONa$, $\underset{*}{C}HI_3$

(b) $CH_3CH_2\overset{O}{\overset{\|}{\underset{*}{C}}}ONa$, CHI_3, $CH_3\underset{*}{C}H_3$

(c) $CH_3CH_2\overset{O}{\overset{\|}{C}}ONa$, CHI_3, CH_3CH_3

(d) $CH_3CH_2\overset{O}{\overset{\|}{C}}ONa$, $\underset{*}{C}HI_3$, CH_3CH_3

31 Consider the following reaction,

R–CH(Br)–C(Br)(CO_2H)–CH_2–CO_2H $\xrightarrow[\text{(ii) } H_2SO_4]{\text{(i) NaOH}}$ A,

Here, 'A' is

(a) R–CH(OH)–C(OH)(CO_2H)–CH_2–CO_2H

(b) R–CH(OH)–(lactone ring, C=O, O) with CO_2H

(c) six-membered ring with HOOC, C=O, O and R

(d) five-membered ring with HO_2C, C=O, O and R

32 End-product of the following reaction,

(cyclohexane with CH_2COOH side chain and exocyclic =CH_2) $\xrightarrow[\Delta]{BH_3.THF/H_2O_2,OH}$

is

(a) (cyclohexane with $CH_2CH_2CH_2OH$ chain and OH)

(b) (cyclohexane with CH_2CH_2COOH chain and OH)

(c) (bicyclic lactone)

(d) (bicyclic lactone)

33 Which of the following compounds loses –COOH group on heating ?

(a) (benzene ring with OH and COOH)

(b) $C_6H_5CH_2\overset{O}{\overset{\|}{C}}{-}CH_2COOH$

(c) $C_6H_5CH{=}CHCH_2COOH$

(d) All of these

34 In the following reaction,

(naphthalene) $\xrightarrow[\text{(ii) } NH_3, \Delta]{\text{(i) Air, } V_2O_5, \Delta}$ A,

End product A is

(a) (benzene ring with CO–NH–CO, phthalimide)

(b) (benzene ring with COOH and NH_2)

(c) (benzene ring with COOH and $CONH_2$)

(d) (benzene ring with C=O and NH)

35 In the following reaction,

(o-bromotoluene: CH_3, Br on benzene ring) $\xrightarrow[\text{(iii) } KMnO_4/OH^- \text{ (iv) } H^+]{\text{(i) Mg/ether (ii) } H_3O^+}$ A, end product A is

(a) (benzene ring with COOH and OH, ortho)
(b) (benzene ring with OH and CH_3, ortho)
(c) (benzene ring with COOH)
(d) (benzene ring with OH)

36 An ester (A) on acidic/hydrolysis gives (B) and (C). (B) gives violet colour and (C) gives blood red colour with neutral $FeCl_3$ solution (A), (B) and (C) are respectively

	A	B	C
(a)	C_6H_5—$COOC_6H_5$;	C_6H_5—COOH;	C_6H_5—OH
(b)	C_6H_5—O—C(=O)—CH_3;	C_6H_5—OH;	CH_3COOH
(c)	C_6H_5—O—C(=O)—CH_3;	C_6H_6;	CH_3COOH
(d)	C_6H_5—C(=O)CH_3;	CH_3OH;	C_6H_5—C(=O)OH

37 Which of the following compounds will give significant amount of *meta*-product during mono nitration reaction?

(a) (benzene ring with OH)
(b) (benzene ring with CH_3)
(c) (benzene ring with NH_2)
(d) (benzene ring with $NHCOCH_3$)

(C) Medical entrances' gallery

1 Which of the following acid will form an (i) anhydride on heating and (ii) acid imide on strong heating with ammonia? **[NEET 2020]**

(a) (benzene ring with two COOH, ortho)
(b) (benzene ring with two COOH, meta)
(c) (benzene ring with two COOH, para)
(d) (benzene ring with OH and COOH, meta)

2 The reaction that does not give benzoic acid as the major product is **[NEET (Odisha) 2019]**

(a) $C_6H_5CH_2OH \xrightarrow{K_2Cr_2O_7}$

(b) $C_6H_5COCH_3 \xrightarrow[\text{(ii) } H_3O^+]{\text{(i) NaOCl}}$

(c) $C_6H_5CH_2OH \xrightarrow[\text{(Pyridinium chloro chromate)}]{\text{PCC}}$

(d) $C_6H_5CH_2OH \xrightarrow{KMnO_4/H^+}$

3 Consider the following acids :

(i) F_3C—COOH, (ii) CH_3COOH,
(iii) C_6H_5COOH, (iv) CH_3CH_2COOH

Correct order of pK_a value is **[AIIMS 2019]**

(a) (i) > (iii) > (ii) > (iv)
(b) (iv) > (ii) > (iii) > (i)
(c) (iv) > (iii) > (ii) > (i)
(d) (i) > (ii) > (iv) > (iii)

4 Carboxylic acids have higher boiling points than aldehydes, ketones and even alcohols of comparable molecular mass. It is due to their **[NEET 2018]**

(a) more extensive association of carboxylic acid *via* van der Waals' force of attraction
(b) formation of carboxylate ion
(c) formation of intramolecular H-bonding
(d) formation of intermolecular H-bonding

5 Which of the following reagents is used to distinguish phenol and benzoic acid? **[AIIMS 2017]**

(a) *aq.* NaOH
(b) Tollen's reagent
(c) Molisch reagent
(d) Neutral $FeCl_3$

6 The product of acid hydrolysis of (P) and (Q) can be distinguished by **[JIPMER 2017]**

$P \Rightarrow H_2C{=}C\begin{smallmatrix}O{-}COCH_3\\ CH_3\end{smallmatrix}$

$Q \Rightarrow CH_3{-}CH{=}CH{-}OCOCH_3$

(a) Lucas reagent
(b) 2, 4-DNP
(c) Fehling's solution
(d) $NaHSO_3$

7 The correct order of strengths of the carboxylic acids

I. (cyclohexane-COOH) II. (tetrahydropyran-3-COOH) III. (tetrahydropyran-2-COOH, O adjacent)

is **[NEET (I) 2016]**

(a) I > II > III (b) II > III > I
(c) III > II > I (d) II > I > III

8 Glutaric acid is **[Kerala CEE 2016]**

(a) butane-1, 4-dioic acid
(b) propane-1, 3-dioic acid
(c) pentane-1, 5-dioic acid
(d) hexane-1, 6-dioic acid
(e) ethanedioic acid

9 Write the IUPAC name of the following compounds.

(benzene ring with OH, H_2N, COOH) **[MHT CET 2014]**

(a) 2-amino-4-hydroxy benzoic acid
(b) 6-amino-4-hydroxy benzoic acid
(c) 3-amino-4-carboxyphenol
(d) 2-carboxy-4-hydroxyaniline

10 (cyclohexene) $\xrightarrow{HBr} A \xrightarrow{KCN} B \xrightarrow{H_3O^+} C \xrightarrow[(ii)\ H_2O]{(i)\ Br_2\ /\ red\ P} D$

Identify the compounds D in the above mentioned series of reactions. **[MHT CET 2014]**

(a) cyclohexyl$-\overset{O}{\overset{\|}{C}}-Br$ (b) cyclohexyl(Br)$-\overset{O}{\overset{\|}{C}}-OH$

(c) $Br-$cyclohexyl$-\overset{O}{\overset{\|}{C}}-OH$ (d) (Br)cyclohexyl$-\overset{O}{\overset{\|}{C}}-OH$

11 $C_6H_5COOH \xrightarrow[(ii)\ \Delta]{(i)\ NH_3} P \xrightarrow{NaOBr} Q \xrightarrow[(ii)\ Heat\ to\ 460\ K]{(i)\ Conc.\ H_2SO_4} R$

R is **[KCET 2014]**

(a) o-bromo sulphanilic acid (b) Sulphanilamide
(c) Sulphanilic acid (d) p-bromo sulphanilamide

12 Which of the following is the correct order of acidic strength? **[Guj. CET 2014]**

(a) $Cl_3CCOOH > Cl_2CHCOOH > ClCH_2COOH > CH_3COOH$
(b) $CH_3COOH > ClCH_2COOH > Cl_2CHCOOH > Cl_3CCOOH$
(c) $CH_3COOH > Cl_3 \cdot C \cdot COOH > Cl_2CH \cdot COOH > ClCH_2COOH$
(d) $CH_3COOH > ClCH_2COOH > Cl_2CHCOOH > Cl_3CCOOH$

13 The carboxylic acid which reduces Tollen's reagent is **[UP CPMT 2013]**

(a) $HCOOH$ (b) CH_3COOH
(c) CH_3CH_2COOH (d) $CH_3CH_2CH_2COOH$

14 Glacial acetic acid is obtained by **[WB JEE 2013]**

(a) distilling vinegar
(b) crystallising, separating and melting acetic acid
(c) treating vinegar with dehydrating agents
(d) chemically separating acetic acid

15 When salicylic acid is distilled with Zn-dust, the product obtained is **[KCET 2013]**

(a) zinc salicylate
(b) salicyl aldehyde
(c) phenol
(d) benzoic acid

16 In a set of reactions m-bromobenzoic acid give a product D. Identify the product D. **[CBSE AIPMT 2011]**

A (m-bromobenzoic acid: COOH, Br) $\xrightarrow{SOCl_2} B \xrightarrow{NH_3} C \xrightarrow[Br_2]{NaOH} D$

(a) m-$BrC_6H_4CONH_2$ (CONH$_2$, Br) (b) m-$BrC_6H_4SO_2NH_2$ (SO$_2$NH$_2$, Br)
(c) m-$H_2NC_6H_4COOH$ (COOH, NH$_2$) (d) m-$BrC_6H_4NH_2$ (NH$_2$, Br)

17 Which is correct statement? **[DUMET 2011]**

(a) *o*-nitrobenzoic acid is stronger than 3, 5- dinitrobenzoic acid in H_2O

(b) Branched carboxylic acids are more acidic than unbranched acids

(c) (o-methoxybenzoic acid: benzene ring with OMe and COOH) is stronger acid than (o-hydroxybenzoic acid: benzene ring with OH and COOH)

(d) Butanoic acid is stronger acid than succinic acid

18 Lowest pK_a is associated with **[DUMET 2011]**

(a) benzoic acid (benzene ring–COOH)

(b) H_3C–C_6H_4–COOH (para)

(c) Br–C_6H_4–COOH (para)

(d) H_3CO–C_6H_4–COOH (para)

19 In the conversion,

cyclohexane–COOH ⟶ cyclohexane–NH_2

The sequence of the reagent used are **[DUMET 2011]**

(a) (i) $SOCl_2$ (ii) N_3^- (iii) H_2O, heat

(b) (i) $SOCl_2$ (ii) NH_3

(c) (i) $SOCl_2$ (ii) NH_3 (iii) Heat

(d) (i) $SOCl_2$ (ii) KCN (iii) $LiAlH_4$

20 Give IUPAC name of the product, when acetamide is heated with anhydrous phosphorus pentoxide. **[Guj. CET 2011]**

(a) ethyl amine

(b) propane nitrile

(c) cyano methane

(d) ethane nitrile

21 In a set of reactions, ethyl benzene yielded a product *D*.

$$C_6H_5\text{–}CH_2CH_3 \xrightarrow[KOH]{KMnO_4} B \xrightarrow[FeCl_3]{Br_2} C \xrightarrow[H^+]{C_2H_5OH} D$$

'*D*' would be **[CBSE AIPMT 2010]**

(a) benzene ring with $COOC_2H_5$ and Br (meta)

(b) C_6H_5–CH_2–CH(Br)–$COOC_2H_5$

(c) benzene ring with Br, Br and $CH_2COOC_2H_5$

(d) benzene ring with COOH and OCH_2CH_3 (meta)

ANSWERS

Check point 19.1

1. (a)	2. (d)	3. (b)	4. (d)	5. (a)	6. (b)	7. (c)	8. (b)	9. (c)	10. (d)

Check point 19.2

1. (c)	2. (d)	3. (b)	4. (d)	5. (a)	6. (c)	7. (b)	8. (a)	9. (d)	10. (a)

Check point 19.3

1. (d)	2. (a)	3. (d)	4. (d)	5. (d)	6. (a)	7. (c)	8. (c)	9. (b)

(A) Topical exercise

1. (b)	2. (a)	3. (b)	4. (c)	5. (c)	6. (d)	7. (b)	8. (d)	9. (c)	10. (c)
11. (a)	12. (c)	13. (d)	14. (b)	15. (c)	16. (a)	17. (c)	18. (a)	19. (a)	20. (a)
21. (c)	22. (c)	23. (b)	24. (c)	25. (b)	26. (c)	27. (c)	28. (c)	29. (b)	30. (a)
31. (b)	32. (a)	33. (d)	34. (a)	35. (b)	36. (a)	37. (c)	38. (a)	39. (a)	40. (d)
41. (d)	42. (a)	43. (c)	44. (a)	45. (a)	46. (b)	47. (c)	48. (b)	49. (a)	50. (a)
51. (a)	52. (a)	53. (c)	54. (d)	55. (d)	56. (a)	57. (a)	58. (c)	59. (b)	60. (d)
61. (b)	62. (d)	63. (b)	64. (b)	65. (a)	66. (c)	67. (a)	68. (b)	69. (a)	70. (a)
71. (a)	72. (b)	73. (d)	74. (c)	75. (b)	76. (b)	77. (b)	78. (a)	79. (a)	80. (c)
81. (a)	82. (a)	83. (d)	84. (b)	85. (b)	86. (a)	87. (c)	88. (c)	89. (c)	90. (c)
91. (c)	92. (c)	93. (c)	94. (a)	95. (d)	96. (c)	97. (c)	98. (c)	99. (c)	100. (c)
101. (b)	102. (b)	103. (b)	104. (d)	105. (c)					

(B) Miscellaneous questions

1. (a)	2. (d)	3. (c)	4. (d)	5. (a)	6. (b)	7. (d)	8. (b)	9. (d)	10. (c)
11. (b)	12. (c)	13. (a)	14. (c)	15. (c)	16. (a)	17. (b)	18. (a)	19. (a)	20. (c)
21. (a)	22. (a)	23. (d)	24. (c)	25. (d)	26. (d)	27. (b)	28. (d)	29. (a)	30. (d)
31. (d)	32. (d)	33. (d)	34. (a)	35. (c)	36. (b)	37. (c)			

(C) Medical entrances' gallery

1. (a)	2. (c)	3. (b)	4. (d)	5. (d)	6. (c)	7. (b)	8. (c)	9. (a)	10. (b)
11. (c)	12. (a)	13. (a)	14. (b)	15. (d)	16. (d)	17. (a)	18. (c)	19. (a)	20. (d)
21. (a)									

Hints & Explanations

(A) Topical exercise

2 (*a*) Valeric acid is $CH_3CH_2CH_2CH_2\overset{O}{\overset{\|}{C}}OH$. It IUPAC name is pentanoic acid.

3 (*b*) Picric acid

OH, O_2N, NO_2, NO_2

4 (*c*)

(a) OH — Carbolic acid

(b) OH, N_2O, NO_2, NO_2 — Picric acid

(c) NH_2, COOH — Anthranilic acid

(d) SO_3H — Sulphanilic acid

5 (*c*) Only saturated fatty acid is a part of fat and (a), (b) and (d) are unsaturated acids.

$CH_3CH_2—C—CH_2—CH—CH$

12 (*c*) Grignard reagent and carbonyl group both are present in the same compound. Thus, it is impossible to prepare carboxylic acid from this compound.

15 (*c*) Alkaline $KMnO_4$ also causes hydroxylation of $C=C$ and $C\equiv C$ bond.

17 (*c*) It gives carboxylate ion ($RCOO^-$).

18 (*a*) (a) $CH_3MgBr \xrightarrow{CO_2/H_3O^+} CH_3COOH$

(b) $CH_3MgBr \xrightarrow{H_3O^+} CH_4 \xrightarrow{CO_2}$ no reaction

Grignard reagent is decomposed by H_3O^+ in first step. Thus, incorrect.

(c) Carboxyl group and Grignard reagent both are present in same molecule. Thus, incorrect.

19 (*a*) $CH_2{=}CHCN \xrightarrow{H_3O^+} CH_2{=}CHCOOH$

$\xrightarrow{OH^-} CH_2{=}CHCOO^-$ salt is formed in (b).

20 (*a*) Only possible compound that can react with PCl_5 is

$$\underset{(A)}{C_2H_5OH} \xrightarrow{PCl_5} \underset{(B)}{C_2H_5Cl} \xrightarrow{KCN} C_2H_5CN \xrightarrow{H_3O^+} \underset{\text{Propanoic acid or Propionic acid}}{C_2H_5COOH}$$

21 (*c*) KCN causes elimination of HBr from 2° alkyl halide and 1° alkyl halide changes to $—CO_2H$.

22 (*c*) I : $RCN \xrightarrow{H_3O^+} RCOOH$

II : $RCN \xrightarrow{OH^-} RCOO^-$

III : $RCN \xrightarrow{DIBAH} RCHO$

IV : $RCN \xrightarrow{SnCl_2/HCl} RCHO$

V : $RCN \xrightarrow{LiAlH_4} RCH_2NH_2$

Thus, I gives RCO_2H.

23 (*b*) In II, alkene is also formed alkyl halide being 2°.

24 (*c*) Cyclohexene $\xrightarrow[\Delta]{MnO_4^-, OH^-}$ COOH, COOH

25 (*b*) $CH_3—\overset{O}{\overset{\|}{C}}—CH_2—CH_3 \xrightarrow[\Delta]{Cr_2O_7^{2-}/H^+} CH_3\overset{O}{\overset{\|}{C}}OH$

This bond breaks

26 (*c*) Compound *B* has half of the C-atoms as in compound *A*, thus *A* is symmetrical.

$$\underset{\underset{(A)}{C_8H_{10}}}{\triangleright—C\equiv C—\triangleleft} \xrightarrow{O_3/H_2O} 2\ \underset{\underset{(B)}{C_4H_6O_2}}{\triangleright—COOH}$$

27 (*c*) $CH_3CH_2\overset{O}{\overset{\|}{C}}—\overset{18}{O}—CH_2CH_3 \xrightarrow{H_2O} CH_3CH_2—\overset{O^{\ominus}}{\underset{OH}{C}}—\overset{18}{O}CH_2CH_3$

Acyl (C—O bond) Alkyl (—O—bond)

$CH_3CH_2\overset{O}{\overset{\|}{C}}—OH + CH_3CH_2\overset{18}{O}H \xleftarrow{H^+} CH_3CH_2\overset{18}{O}^{\ominus}$

28 (c) I. $CH_3-\overset{O}{\overset{\|}{C}}-CH_3 + I_2 + NaOH \xrightarrow{\Delta} CH_3\overset{-}{CO}\overset{+}{ONa} \xrightarrow{H^+} CH_3COOH$

II. $\xrightarrow{Cr_2O_7^{2-}/H^+} CH_3COOH$

III. $\xrightarrow{[Ag(NH_3)_2]^+}$ no reaction

Thus I and II.

29 (b) $(CH_3)_3C-Br + CN^- \xrightarrow{H_3O^+} CH_3-\underset{CH_3}{\underset{|}{C}}=CH_2$

Because in (a), CN^- is a strong nucleophile. Alkyl halide is 3° and thus, loss of HBr takes place and alkene is formed.

30 (a) When alkyl halide is 2° or 3°, dehalogenation takes place with strong nucleophile as CN^- giving alkene as major product when alkyl halide is 1° (that is (a)) yield of carboxylic acid is maximum.

32 (a) $CH_3-C\langle^{O\cdots H-O}_{O-H\cdots O}\rangle C-CH_3$ (with $\delta-$, $\delta+$ charges)

Dimer by intermolecular H-bonding

34 (a) $R-\overset{O}{\overset{\|}{C}}-OH$ — Hydrophobic (R), Hydrophilic (COOH)

38 (a) Order of EWG-tendency of $-NO_2$ group at m, p, o is

$m- < p- < o$

Since, $pK_a = -\log K_a$

Thus, pK_a : $o- < p- < m-$

39 (a) Strongest acid is oxalic acid thus, least pK_a value.

40 (d) The maximum value of equilibrium constant is for option (d). Thus, it has least pK_a value.

43 (c) $C_6H_5-\overset{O}{\overset{\|}{C}}O^{\ominus}$ is stabilised due to resonance.

45 (a) $\underset{NO_2}{\underset{|}{CH_2}}COOH$ has minimum pK_a.

Thus, most acidic. Thus $-NO_2$.

46 (b) With the increase in the percentage of *s*-character of α-carbon acidity of carboxylic acid increase. This is because with increase in the *s*-character E.N. of α-carbon atom also increase.

(i.e. $sp^3 < sp^2 < sp$. Thus more easily carboxylic H is releases.

Thus, increasing order of acidic strength of given compounds is

$CH_3CH_2\overset{O}{\overset{\|}{C}}OH < CH_2=CH\overset{O}{\overset{\|}{C}}OH < CH\equiv C\overset{O}{\overset{\|}{C}}OH$

But increasing order of pK_a value is reverse of above order i.e. greater is the acidic strength lower is the pK_a value.

$CH\equiv C\overset{O}{\overset{\|}{C}}OH < CH_2=CH-CH-\overset{O}{\overset{\|}{C}}OH < CH_3CH_2\overset{O}{\overset{\|}{C}}OH$

$\because pK_a = -[\log K_a]$

47 (c) Electron withdrawing tendency of substituted group is as follows

I < II < III

$H < Cl < NO_2$

Thus, oder of dissociation constant K_a is I < III < II.

49 (a) Greater the acidic strength, smaller the value of pK_a.

54 (d) Among the given options least acidic option is CH_3CH_3 hence, has maximum pK_a value.

59 (b) Decarboxylation takes place if there are two —COOH groups attached to same carbon (*gem* position).

Or there is keto group at β-position or (C = C) at β, γ position w.r.t. COOH group.

60 (d) $R\overset{O}{\overset{\|}{C}}OH \xrightarrow{CH_2N_2} R\overset{O}{\overset{\|}{C}}OCH_3 + N_2\uparrow$

Ester

61 (b) $\begin{matrix}CH_2CH_2COO \\ | \\ CH_2CH_2COO\end{matrix}\rangle Ca \longrightarrow CaCO_3 +$ Cyclopentanone

Calcium adipate

63 (b) $R-\overset{O}{\overset{\|}{C}}-O-R' \longrightarrow R-\overset{O^{\ominus}}{\underset{\oplus}{C}}-O-R \xrightarrow{H^+} R-\overset{OH}{\underset{\oplus}{C}}-O-R'$

Active site for further nucleophilic attack.

64 (b) Among the given compound picric acid is the only non-carboxylic acid compound. Thus, no Zwitter ion formation. Zwitter ion is a dipolar ion in which both acidic and basic group are in same molecule.

Picric (2,4,6-trinitrophenol: OH, O_2N, NO_2, NO_2)

65 (a) C_6H_5OH Phenol is also called as carbolic acid.

66 (c) HVZ-reaction is a reaction in which H atom at α- C w.r.t. —COOH group is replaced by halogen atom. This reaction takes place in those carboxylic acid compounds in which at least one H atom present at α- C w.r.t —COOH group. Out of given acids, only (c) has H atom at α-C.

$$C_6H_5CH_2COOH \ (\alpha) \xrightarrow{P/Br_2} C_6H_5CH(Br)COOH$$

67 (*a*) RCH_2OH is oxidised to $R\overset{O}{\overset{||}{C}}OH$ through intermediate $R\overset{O}{\overset{||}{C}}H$ and $Cr_2O_7^{2-}$ is reduced to Cr^{3+} (green).

$$RCH_2OH \longrightarrow RCHO \longrightarrow RCOOH$$
$$Cr_2O_7^{2-} \longrightarrow Cr^{3+}$$

68 (*b*) (a) α-hydroxy acid forms lactide by intermolecular loss of H_2O.

(b) β-hydroxy acid forms unsaturated acids by intramolecular loss of H_2O.

(c) γ-hydroxy acid forms γ-lactone.

69 (*a*) (a) $C_6H_5COO^-Na^+ \xrightarrow{\text{Sodalime, } \Delta} C_6H_6$

Sodium benzoate → Benzene

(b) $(C_6H_5COO)_2Ca \xrightarrow{\Delta} C_6H_5COC_6H_5 + CaCO_3$
Calcium benzoate

(c) $(C_6H_5COO)_2Ca + (CH_3COO)_2Ca \xrightarrow{\Delta} 2C_6H_5COCH_3 + CaCO_3$
Calcium benzoate Calcium acetate

73 (d) Decarboxylation takes place in cases in which their is keto group at β-position or a (C═C) at (β, γ)-position or two —COOH groups at *gem* (same) position.

I : $\underset{\gamma}{CH_2}{=}\underset{\beta}{CH}{-}CH_2\overset{O}{\overset{||}{C}}OH$, Yes, as (C═C) in β, γ-position

II : $CH_3\underset{\beta}{\overset{O}{\overset{||}{C}}}\underset{\alpha}{CH_2}\overset{O}{\overset{||}{C}}OH$ Yes, keto group at β-position

III : (cyclohexene ring with —COOH; positions α, β, γ, δ) No, as (C═C) in γ, δ-position

IV : (cyclohexene ring with —CH_2COOH; positions α, β, γ, δ) No, as in III.

Thus, only in I and II decarboxylation takes place.

74 (*c*) $RCH_2CH_2COOH \xrightarrow[\text{HVZ-reaction}]{\text{(i)}P/Br_2,\ \text{(ii)}H_2O} \underset{(A)}{RCH_2CH(Br)COOH} \xrightarrow[\text{(ii)}H^+]{\text{(i)}NaOC_2H_5} \underset{(B)}{RCH{=}CHCOOH}$

($\bar{O}C_2H_5$ is a strong nucleophile)

78 (*a*) $CH_3COOH \xrightarrow{MnO} (CH_3COO)_2Mn$
$HCOOH \xrightarrow{MnO} (HCOO)_2Mn$
$(CH_3COO)_2Mn + (HCOO)_2Mn \xrightarrow{\Delta} CH_3CHO$

79 (*a*) Reactant is α-hydroxy acid which forms cyclic lactide.

83 (*d*) Being in *trans*-position.

84 (*b*) $CH_3COOH \xrightarrow{Br_2/P} \underset{X}{CH_2(Br)COOH} \xrightarrow[\text{(ii)}H_3O^+]{\text{(i)}KCN} \underset{Y}{CH_2(COOH)COOH}$

85 (*b*) $CH{\equiv}CH + 2CH_3COOH \xrightarrow{Hg^{2+}} \underset{(X)}{CH_3CH(OOCCH_3)_2} \xrightarrow{H_3O^+} CH_3CH(OH)_2 \longrightarrow \underset{(Y)}{CH_3CHO}$

86 (*a*) H of benzoic acid decomposes $H\overset{*}{C}O_3^-$ forming $\overset{*}{C}O_2$.

87 (*c*) Lactic acid $\xrightarrow{\Delta}$ lactide

3-hydroxypropanoic acid $\xrightarrow{\Delta}$ α, β-unsaturated acid

89 (*c*) —COOH group is lost on heating when keto group is at β-position w.r.t. COOH.

95 (*d*) —OH group is strongly activating group, hence, S_E reaction takes place readily at *o*- and *p*-position and this causes elimination of —COOH group.

Salicylic acid (2-hydroxybenzoic acid) $\xrightarrow{3Br_2 \text{ water}}$ 2,4,6-tribromophenol

96 (*c*) 2-methylbenzoic acid $+ Br_2 \xrightarrow{Fe}$ 2-methyl-5-bromobenzoic acid (Br *para* to CH_3)

—CH_3 group is activating *o*-, *p*-directing.

—COOH group is deactivating *m*-directing.

Thus, Br goes into *p*- position w.r.t. —CH_3 group.

97 (*c*) Given reaction is Reimer Tiemann's reaction compound formed is option (c).

98 (*c*) Leaving tendency of conjugate bases is in order.

$$—NH_2 < —OCH_3 < —Cl$$

Therefore, there reactivity order is II < III < I.

99 (*c*) Aniline (*A*) $\xrightarrow[273\text{ K}]{NaNO_2 + HCl}$ $C_6H_5N_2^{\oplus}Cl^-$ $\xrightarrow[\text{HCN}]{CuCN}$ (Sandmeyer reaction) C_6H_5CN $\xrightarrow{H_3O^+}$ C_6H_5COOH (Benzoic acid)

100 (*c*) Cl attached to benzene ring does not undergo S_N reaction, thus, benzylic halide reacts with CN^-.

o-$ClC_6H_4CH_2Cl$ (This is replaced: $-CH_2Cl$) $\xrightarrow{CN^-}$ o-$ClC_6H_4CH_2CN$ $\xrightarrow{H_3O^+}$ o-$ClC_6H_4CH_2COOH$

101 (*b*) m-$BrC_6H_4CH_2Br$ $\xrightarrow{\text{Mg/ether}}$ m-$BrMgC_6H_4CH_2—Mg—Br$ $\xrightarrow{CO_2/H_3O^+}$ m-$HOOCC_6H_4CH_2COOH$ (*A*)

102 (*b*) $C_6H_5CH_3 + Cl_2 \xrightarrow{\text{Boiling}} C_6H_5CCl_3$ (*A*) $\xrightarrow{NaOH} C_6H_5COOH \xrightarrow{NaOH} C_6H_5COO^-Na^+$ (*B*)

103 (*b*) $C_6H_5OH \xrightarrow{NaHCO_3}$ No reaction

$C_6H_5COOH \xrightarrow{NaHCO_3} CO_2$ (brisk effervescence)

104 (*d*)

	Phenol	**Benzoic acid**
	(a) Violet colour	Buff-coloured ppt
	(b) Carbolic smell	No smell
	(c) No reaction	Brisk effervescence

(B) Miscellaneous questions

1. (*a*) Fehling's solution is reduced by

CH_3COOH	No
$HCOOH$	Yes
$HCHO$	Yes
C_6H_5CHO	No
CH_3CHO	Yes

Thus, pair I can be distinguished.
Pair II can be distinguished.
Pair III can be distinguished.
Pair IV cannot be distinguished.

3 (*c*) $X \xrightarrow{NaHCO_3} CO_2$ thus compound X has $-COOH$ group.

$H—C(CH_2OH)(CH_3)—COOH$ (Chiral (*X*); COOH: This is reduced to $—CH_2OH$) $\xrightarrow{LiAlH_4}$ $H—C(CH_3)(CH_2OH)—CH_2OH$ (Achiral)

4 (*d*) 2-ethynylcyclohexanone $\xrightarrow[H_2O_2/OH^-]{BH_3/THF}$ 2-acetylcyclohexanone ($—CCH_3$ with C=O) $\xrightarrow[\text{(ii) } H^+]{\text{(i) } I_2/NaOH}$ 2-oxocyclohexanecarboxylic acid ($—COOH$) $\xrightarrow{\Delta}$ cyclohexanone (*X*)

(C = O) group is electron-withdrawing hence, π-electron migration is from terminal carbon to internal carbon as shown. Also when keto group is at β-position w.r.t. —COOH group, heating eliminates CO_2.

5 (*a*) II, III have —COOH group hence, true. I has keto groups which make —OH group acidic. Thus (a) i.e. I, II and III is correct option.

6 (*b*) H_3CO, H_3CO (cyclobutane) $C(COOC_2H_5)_2$ (*A*) $\xrightarrow{H_3O^+}$ HO, HO (cyclobutane) $C(COOH)_2$

$\longrightarrow$ O= (cyclobutane, β, γ) $C_\alpha(COOH)_2$ $\xrightarrow[-CO_2]{\Delta}$ O= (cyclobutane) —COOH

(When two —COOH groups are at *gem*-position, heating eliminates CO_2).

7 (*d*) I: $CH_3\overset{O}{\overset{\|}{C}}H \xrightarrow[[H^+]]{[Ag(NO_3)_2]^+ OH^-} CH_3\overset{O}{\overset{\|}{C}}OH$

II: $CH_3\overset{O}{\overset{\|}{C}}CCl_3 \xrightarrow{OH^-} CH_3\overset{O}{\overset{\|}{C}}O^\ominus$

III: $CH_3\overset{O}{\overset{\|}{C}}OC_2H_5 \xrightarrow{OH^-} 2CH_3\overset{O}{\overset{\|}{C}}O^\ominus$

8 (*b*) $F_3C—CH_2Br \xrightarrow{Mg/ether} F_3C—CH_2MgBr \xrightarrow[(ii)\ H_3O^\oplus]{(i)\ \overset{14}{C}O_2} F_3C—CH_2\overset{14}{C}OOH$

(C–F bond energy is maximum, does not react with Mg).

9 (*d*) $H_3C(H)C=C(H)COOH$ (*X*) *cis*-(*Z*) $H_3C(H)C=C(COOH)H$ (*Y*) *trans*-(*E*)

10 (*c*) $CH_3\overset{O}{\overset{\|}{C}}OH \xrightarrow{SOCl_2} CH_3\overset{O}{\overset{\|}{C}}Cl$ (*A*) $\xrightarrow{C_6H_6,\ AlCl_3} C_6H_5\overset{O}{\overset{\|}{C}}CH_3$ (*B*)

$\xrightarrow{HCN} C_6H_5C(OH)(CN)CH_3 \xrightarrow{H_3O^+} C_6H_5\overset{\alpha}{C}(OH)(COOH)CH_3$ (*D*) ← Chiral

14 (*c*) $CH_3CN \xrightarrow{Na/C_2H_5OH} CH_3CH_2NH_2$ (*X*)

$\xrightarrow{HNO_2} CH_3CH_2OH$ (*Y*) $\xrightarrow[\text{Baeyer's reagent}]{MnO_4^-/OH^-} CH_3COOH$ (*Z*)

15. (*c*) I: $CH_3NC \xrightarrow{H_3O^+} CH_3NH_2 + HCOOH$

II: $CHCl_3 \xrightarrow{NaOH} H\overset{O}{\overset{\|}{C}}ONa \xrightarrow{H_3O^+} HCOOH$

III: $CCl_4 \xrightarrow{H_3O^+}$ no hydrolysis

17. (*b*) I: $CH_3MgBr \xrightarrow[H_3O^+]{CH_2—CH_2\ (\text{epoxide})} CH_3CH_2CH_2OH \xrightarrow[\Delta]{MnO_4^-/H^+}$

II: $CH_3CH_2Br \xrightarrow{KCN} \xrightarrow{H_3O^\oplus} CH_3CH_2COOH$ (Propanoic acid)

III: $CH_3CH_2\overset{O}{\overset{\|}{C}}OC_2H_5 \xrightarrow[H_3O^+]{CH_3MgBr}$ (2 moles) $CH_3CH_2C(OH)(CH_3)CH_3$ (3° alcohol)

19 (*a*) $CH_3\overset{O}{\overset{\|}{C}}CH_2CH_2CH_2\overset{O}{\overset{\|}{C}}OH \xrightarrow[(ii)\ H_3O^+]{(i)\ NaBH_4}$

$CH_3\overset{OH}{\overset{|}{C}}HCH_2CH_2CH_2—C(=O)—OH \xrightarrow[\text{esterification}]{\text{Intramolecular}}$ (six-membered lactone with CH_3)

20 (*c*) 2-acetylcyclohexanone ($\overset{O}{\overset{\|}{C}}CH_3$) $\xrightarrow[(ii)\ H^+]{(i)\ I_2/NaOH,\ \Delta}$ 2-oxocyclohexanecarboxylic acid (β, α, $\overset{O}{\overset{\|}{C}}OH$) $+ CH_3I$ $\xrightarrow{\Delta}$ cyclohexanone

When keto group is at β-position w.r.t. —COOH, heating eliminates, CO_2.

21 (*a*) $CH_3CH_2O^S$ (alkoxy) group replaces OH group of

$CH_3—\overset{O}{\overset{\|}{C}}—OH$.

24 (*c*) $RC\equiv N \xrightarrow[\Delta]{HCl/H_2O} R\overset{O}{\overset{\|}{C}}OH \xrightarrow[(ii)\ H_2O]{(i)\ LiAlH_4} RCH_2OH$ (*B*)

$\xrightarrow[CH_2Cl_2]{PCC} RCHO$ (*C*)

(where, $R = (CH_3)_2CHCH_2—$)

25 (*d*) (Br_2/light) changes benzyl hydrogen to —Br.

30 (*d*) $CH_3CH_2\overset{O}{\overset{\|}{C}}_*CH_2\overset{O}{\overset{\|}{C}}OH \xrightarrow{\Delta} CH_3CH_2\overset{O}{\overset{\|}{C}}_*CH_3$ (*A*)

$\downarrow I_2/NaOH\ \Delta$

$CO_2\uparrow + CH_3CH_3$ (*D*) $\xleftarrow[\Delta]{CaO/NaOH} CH_3CH_2\overset{O}{\overset{\|}{C}}O^-Na^+$ (*B*) $+ \overset{*}{C}HI_3$ (*C*)

31 (c) R–CH(Br)–C(Br)(CO_2H)–CH_2CO_2H $\xrightarrow{NaOH}$ R–CH(OH)–C(CO_2H)=CH–CO_2H (X)

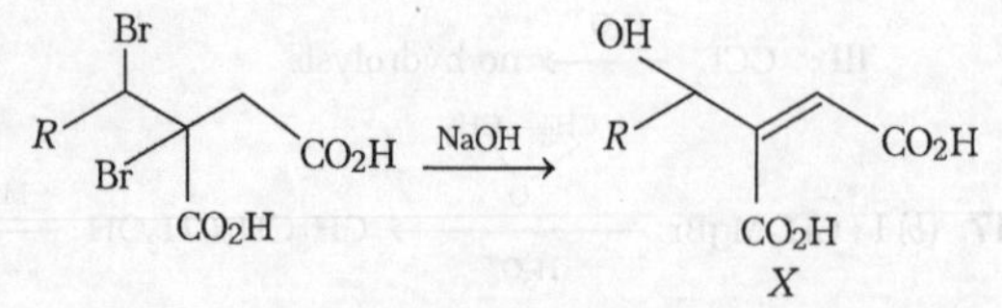

3° alkyl halide undergoes dehydrohalogenation with NaOH and 2° alkyl halide undergoes S_N1 reaction.

$$X \xrightarrow[\text{esterification}]{H_2SO_4} (A)$$

'A' is formed by intermolecular esterification.

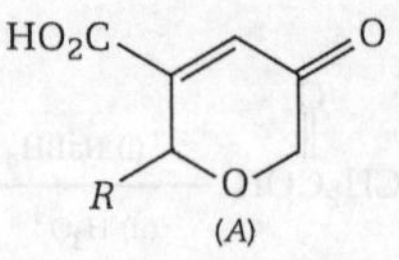

(A): ring with HO_2C, O, R

32. (d) $\xrightarrow{BH_3THF/H_2O_2, OH}$ (OH, OH, O) $\xrightarrow{\Delta}$ (lactone)

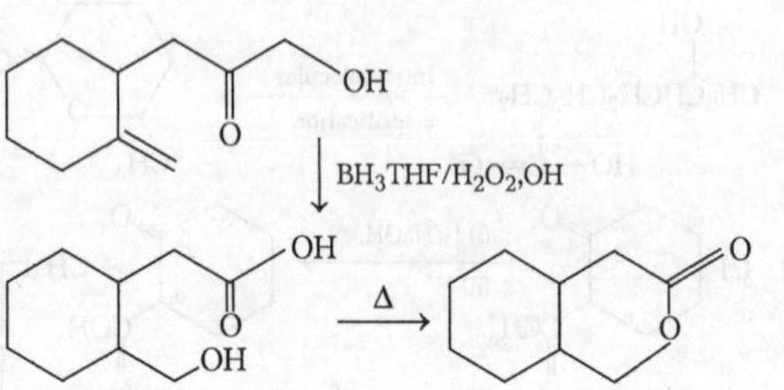

33 (d) (a) o-hydroxybenzoic acid (OH, COOH) $\xrightarrow[-CO_2]{\Delta}$ phenol (OH)

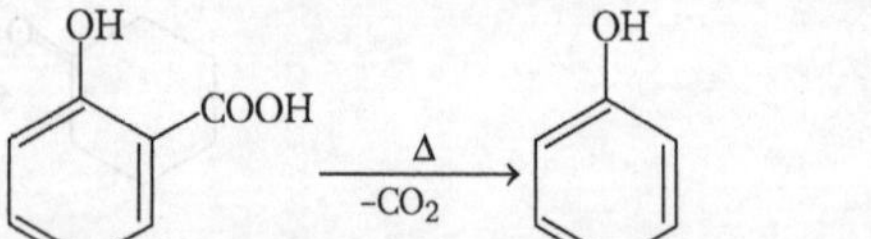

(b) $C_6H_5CH_2\overset{O}{\overset{\|}{C}}CH_2COOH$ (β α) $\xrightarrow{\Delta}$ $C_6H_5CH_2COCH_3$

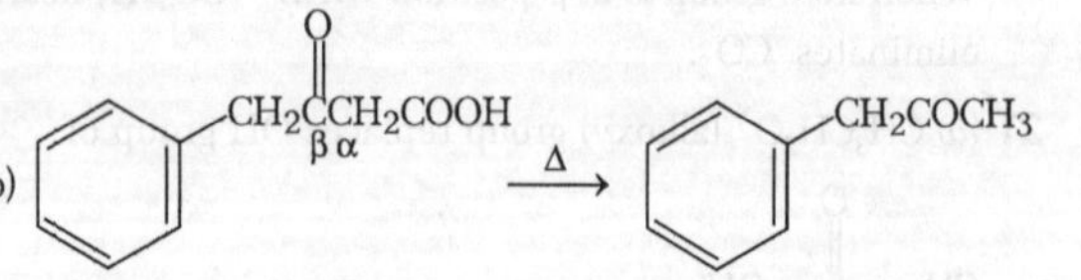

There is loss of $-COOH$ on heating because there is a keto group at β-position.

(c) $C_6H_5\overset{\gamma}{C}H{=}\overset{\beta}{C}HCH_2COOH \xrightarrow{\Delta} C_6H_5CH{=}CHCH_3$

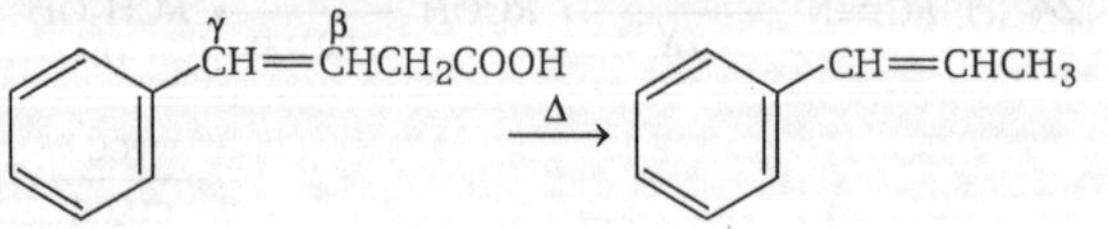

There is a loss $-COOH$ group when there is (C=C) at β-γ position.

35 (c) o-bromotoluene (CH_3, Br) $\xrightarrow{Mg/ether}$ (CH_3, MgBr) $\xrightarrow{H_3O^+}$ toluene (CH_3) $\xrightarrow[\Delta]{KMnO_4/OH^-}$ $\xrightarrow{H^+}$ benzoic acid (COOH)

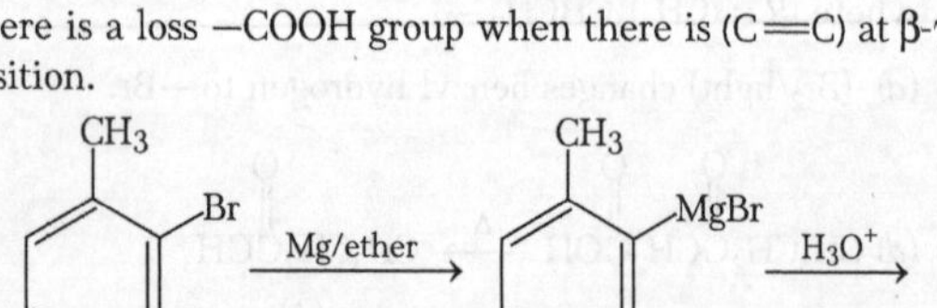

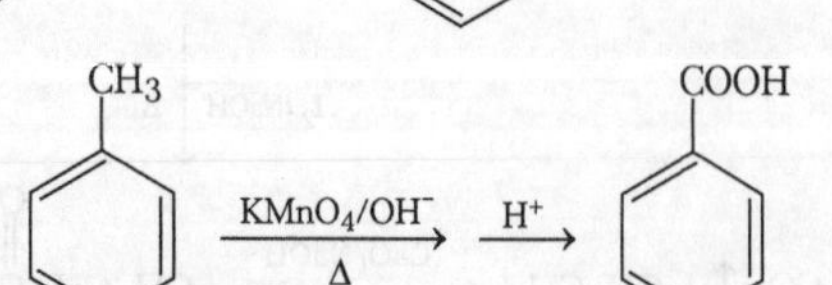

36 (b) (b) Gives violet colour with neutral $FeCl_3$, thus it is phenol.

(c) Gives blood red colour with neutral $FeCl_3$, thus it is CH_3COOH

$$CH_3COOH + C_6H_5OH \rightleftharpoons CH_3\overset{O}{\overset{\|}{C}}{-}O{-}C_6H_5$$

37 (c) In aciddic medium $-\overset{\oplus}{N}H_3$ is *m*-directing.

(C) Medical entrances' gallery

1 (a) An α, β-dicarboxylic acid with same-side (*syn*) orientation of $-COOH$ group is able to form anhydride (cyclic) and imide (cyclic).

Among isomeric benzene dicarboxylic acids, only benzene-1,2-dicarboxylic acid (phthalic acid) will respond to the below reactions.

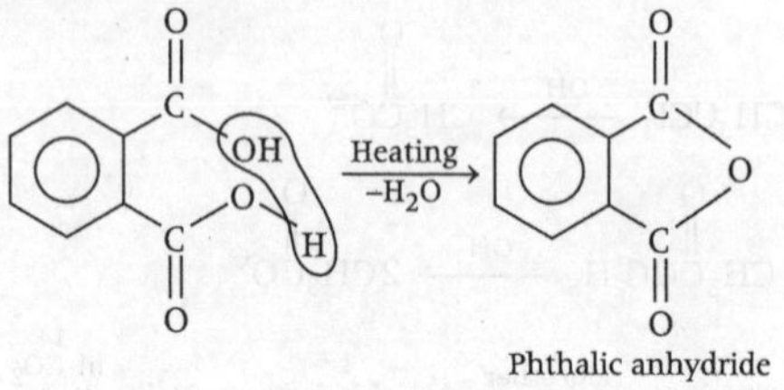

Phthalic acid $\xrightarrow[-H_2O]{\text{Heating}}$ Phthalic anhydride

Phthalic acid (COOH, COOH) $\xrightarrow[NH_3, -H_2O]{\text{Strong heating}}$ (O H_2HN) $\xrightarrow{-H_2O}$ Phthalimide (NH)

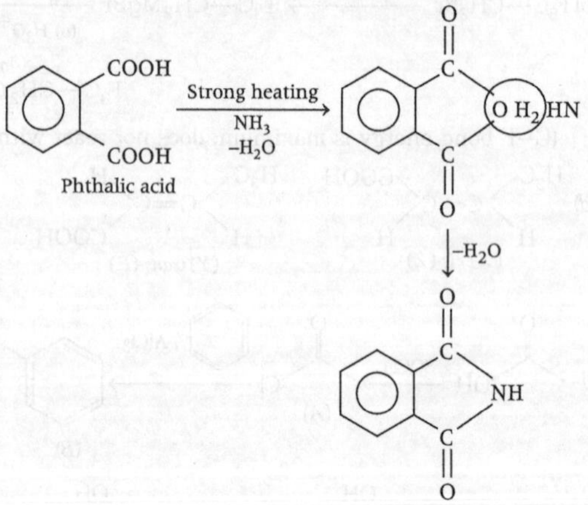

2 (c) Primary aromatic alcohols on reaction with pyridinium chlorochromate (PCC, a mild oxidising agent) forms aromatic aldehydes.

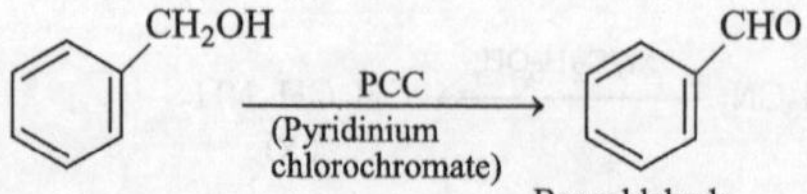

$C_6H_5CH_2OH \xrightarrow[\text{(Pyridinium chlorochromate)}]{PCC} C_6H_5CHO$ (Benzaldehyde)

In the remaining options, benzoic acid is formed as follows:

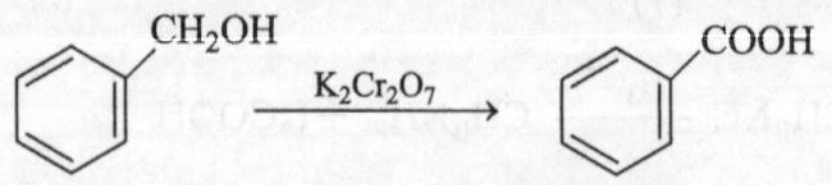

$C_6H_5CH_2OH \xrightarrow{K_2Cr_2O_7} C_6H_5COOH$ (Benzoic acid)

Benzoic acid: $C_6H_5COCH_3 \xrightarrow[(ii)\ H_3O^+]{(i)\ NaOCl} C_6H_5COOH + CHCl_3$

Benzoic acid: $C_6H_5CH_2OH \xrightarrow{KMnO_4/H^+} C_6H_5COOH$

Thus, option (c) is correct.

3 (*b*) Higher the pK_a value or smaller is the K_a value, weaker is the acid.

$\therefore \quad pK_a = -\log[K_a]$

In structure (i), fluorine (F) shows (–*I*) effect, thus is most acidic.

In structure (ii), CH_3COOH, —CH_3 shows (+*I*) effect, and also hyperconjugation, so is weaker acid than (i).

In structure (iii), i.e. C_6H_5—COOH, the —COOH group is in resonance with the benzene ring. Thus, its carbocation is stable, therefore it is more acidic than (ii) and (iv) but is less acidic than (i).

In structure (iv), larger alkyl group bonded with —COOH group, which shows stronger (+*I*) effect, thus is the weakest acid.

Therefore, order of acidic nature is

(i) > (iii) > (ii) > (iv).

Hence, correct order of pK_a values

(iv) > (ii) > (iii) > (i)

Thus, option (b) is correct.

4 (*d*) Carboxylic acids have higher boiling points than aldehydes, ketones and even alcohols of comparable molecular mass, because of the extent of intermolecular-hydrogen bonding with water, due to which they exist as associated molecules.

[Structure: RCOOH hydrogen bonded with water molecules, showing δ+ and δ– charges]

The hydrogen bonds are not completely broken in the vapour state. In fact, most carboxylic acids exist as dimer in the vapour state or in aprotic solvents.

[Structure: carboxylic acid dimer R—C(=O)—O—H ··· linked by H-bonds to C—R; labels δ–, δ+, H-bonds]

5 (*d*) Neutral $FeCl_3$ is used to distinguish phenol and benzoic acid. It gives violet colour with phenol and buff-coloured precipitates with benzoic acid.

6 (*c*)

$P \Rightarrow H_2C{=}C(OCOCH_3)(CH_3) \xrightarrow{H_2O/H^+} H_2C{=}C(OH)(CH_3)$ (Enol form) $+ CH_3COOH$

Tautomerism ⇌

$CH_3—C(=O)—CH_3$ (Acetone)

$Q \Rightarrow CH_3CH{=}CH—OCOCH_3 \xrightarrow{H_2O/H^+} CH_3CH{=}CH—OH$ (Enol-form) $+ CH_3COOH$

Tautomerism ⇌

$CH_3—CH_2—C(=O)—H$ (Aldehyde)

∵ Fehling solution cannot be reduced with acetone.

Thus, it can be used to differentiate ketone (acetone) and aldehyde.

7 (*b*) Order of strength of the given carboxylic acids can be determined by the concept of –*I*-effect. The oxygen atom present in the ring shows –*I*-effect. As the distance between oxygen and —COOH group increases, –*I*-effect of oxygen decreases. Thus, corresponding carboxylic acid will show less acidic nature. The correct order of strengths of the carboxylic acids is

II. (tetrahydropyran-3-carboxylic acid) > III. (tetrahydropyran-4-carboxylic acid) > I. (cyclohexanecarboxylic acid)

II. More *I*-effect (Most acidic); III. Less *I*-effect (Moderately acidic); I. No *I*-effect (Least acidic)

8 (*c*) Glutaric acid is pentane-1,5-dioic acid

$\overset{1}{HOOC}—\overset{2}{CH_2}—\overset{3}{CH_2}—\overset{4}{CH_2}—\overset{5}{COOH}$

9 (*a*) Priority order of functional group is

—COOH > —OH > —NH_2

∴ IUPAC name of the compound is 2-amino-4-hydroxy benzoic acid.

10 (*b*) The complete series of reaction can be represented.

Cyclohexene $+ HBr \longrightarrow$ Cyclohexyl—Br (*A*) $\xrightarrow{KCN}$ Cyclohexyl—CN (*B*) $\xrightarrow{H_3O^+}$ Cyclohexyl—COOH (*C*) $\xrightarrow[\text{HVZ reaction}]{Br_2/\text{Red P}}$ 1-bromocyclohexane-1-COOH (*D*)

11 (*c*) The complete series of given reactions can be represented as

C_6H_5COOH or (Benzoic acid, COOH) $\xrightarrow{NH_3}$ Ammonium benzoate ($COO\overset{-}{}\overset{+}{N}H_4$) $\xrightarrow{\Delta}$ Benzamide ($CONH_2$) (*P*)

$\xrightarrow{NOBr}$ Aniline (NH_2) (*Q*) $\xrightarrow{\text{Conc. } H_2SO_4}$ Anilinium hydrogen sulphate ($\overset{+}{N}H_3HSO_4^-$) $\xrightarrow[-H_2O]{460\ K}$ Sulphanic acid ($NHSO_3H$) $\xrightarrow[-H_2O]{\text{Rearrangement}}$ Sulphanilic acid (NH_2, SO_3H) (*R*)

12 (*a*) Presence of an electron withdrawing group, (–*I*-effect showing group) like-Cl facilitates the removal of proton from carboxylic acid by dispersing the negative charge of the carboxylate ion formed.

Thus, increases the acidity of the carboxylic acids. Greater the number of –*I*-effect showing groups, more will be the acidity of carboxylic acid. Thus, the order of the acidic strength is

$$Cl_3C\cdot COOH > Cl_2CHCOH > ClCH_2COOH > CH_3COOH$$

13 (*a*) Tollen's reagent is alkaline solution of ammoniacal silver nitrate $AgNO_3 + NH_4OH$. It is reduced to silver by compounds having $-\overset{O}{\overset{\|}{C}}-H$ group.

Since, $H-\overset{O}{\overset{\|}{C}}-OH$ (Formic acid) has $-\overset{O}{\overset{\|}{C}}-OH$ group.

So, it reduces Tollen's reagent.

$$[2Ag(NH_3)_2]^+ + H-\overset{O}{\overset{\|}{C}}-OH \longrightarrow 2Ag\downarrow + CO_2 + 2NH_4^+$$

14 (*b*) Below 16.6°C, pure acetic acid is converted into ice-like solid hence it is called 'glacial acetic acid'. So, it is obtained by crystallisation, separating and melting of acetic acid.

15 (*d*) When salicylic acid is treated with zinc dust, benzoic acid is obtained as a product.

Salicylic acid (OH, COOH) $\xrightarrow{\text{Zn dust}}$ Benzoic acid (COOH) + ZnO

16 (*d*) (*A*) (COOH, Br) $\xrightarrow{SOCl_2}$ (*B*) (COCl, Br) $\xrightarrow[-HCl]{NH_3}$ (*C*) ($CONH_2$, Br) $\xrightarrow[Br_2]{NaOH}$ (*D*) (NH_2, Br)

17 (*a*) Due to *ortho* effect *o*-nitrobenzoic acid is stronger than 3,5-dinitrobenzoic acid in H_2O.

18 (*c*) All halogenated acids are stronger acid than benzoic acid due to –*I*-effect of halogen.

19 (*a*) $C_6H_{11}-\overset{O}{\overset{\|}{C}}-OH + SOCl_2 \longrightarrow C_6H_{11}-\overset{O}{\overset{\|}{C}}-Cl \xrightarrow{N_3^-} C_6H_{11}-\overset{O}{\overset{\|}{C}}-N=\overset{\oplus}{N}=\overset{\ominus}{N}$

$\longleftrightarrow C_6H_{11}-\overset{O}{\overset{\|}{C}}-N-\overset{+}{N}\equiv N \xrightarrow{-N_2} C_6H_{11}-\overset{O}{\overset{\|}{C}}-\ddot{N}$ $\xrightarrow{\text{Intramolecular shift}}$ $C_6H_{11}-N=C=O \xrightarrow[\Delta]{H_2O} C_6H_{11}-NH_2 + CO_2$

20 (*d*) Acetamide, CH_3CONH_2 on heating with anhydrous P_2O_5 gives ethane nitrile (CH_3CN).

21 (*a*) (*A*) (CH_2CH_3) $\xrightarrow[KOH]{KMnO_4}$ (*B*) (COOH) $\xrightarrow[FeCl_3]{Br_2}$ (*C*) (COOH, Br) $\xrightarrow[H^+]{C_2H_5OH}$ (*D*) ($COOC_2H_5$, Br)

CHAPTER 20

Amines

Amines constitute an important class of organic compounds. These are derived by replacing one or more hydrogen atom(s) of ammonia molecule by alkyl or aryl group(s). That's why, these are also known as **derivatives of ammonia**. Amines are the most important organic compounds as they occur among proteins, vitamins, alkaloids and hormones.

Inside

- **Aliphatic amines : classification, nomenclature and isomerism**
- **Preparation of amines**
  - *Methods for preparation of mixture of amines (1°, 2°, 3°)*
  - *Methods for preparation of primary amines only*
  - *Methods for preparation of secondary amines only*
  - *Methods for preparation of tertiary amines only*
- **Physical and chemical properties of alkyl amines**
  - *Physical properties*
  - *Chemical properties*
- **Aromatic amines : nomenclature, isomerism and methods of preparation**
- **Physical and chemical properties of aromatic amines**
  - *Distinction and separation of 1°, 2° and 3° amines*

ALIPHATIC AMINES

Classification

Aliphatic amines are called **alkyl derivatives of ammonia**. These are further classified as primary (RNH_2 or 1°), secondary (R_2NH or 2°) and tertiary (R_3N or 3°) depending upon the number of hydrogen atoms replaced by alkyl groups in ammonia molecule. If the number of H-atoms attached to N-atom is n, then degree of amine is $(3-n)°$. Degree and functional group of various amines are shown below

- $R—\overset{\text{H}}{\overset{|}{\text{N}}}—H$; $n = 2$, 1° (primary), $—NH_2$ (amino-group)
- R_2>NH ; $n = 1$, 2° (secondary), $—\underset{|}{\text{NH}}$ (imino-group)
- $R—\underset{R}{\overset{R}{\text{N}}}$; $n = 0$, 3° (tertiary), $—\underset{|}{\overset{|}{\text{N}}}$ (tertiary nitrogen atom)

Structure

The nitrogen atom in amine is sp^3-hybridised. The three alkyl groups are at the corners of a tetrahedron and their shape is trigonal pyramidal. The bond angle becomes lesser than 109.5° due to the bond pair-lone pair repulsion (e.g. bond angle for trimethylamine is 108°).

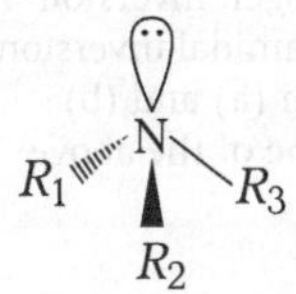

Nomenclature

Commonly aliphatic amines are named as *alkylamines.* But in systematic IUPAC nomenclature, they are named by adding the suffix "amine" to the name of the chain to which NH_2 group is attached with replacement of the final '–e'

e.g.

	$CH_3CH_2NH_2$	$CH_3CH_2CH_2N(CH_2CH_3)—CH_3$
Common name	Ethylamine	Ethylmethylpropylamine
IUPAC name	Ethanamine	N-ethyl-N-methyl-1-propanamine

$NH_2—CH_2—CH_2—NH_2$

Common name Ethylenediamine

IUPAC name Ethane-1,2-diamine

In the IUPAC system, the substituent $—NH_2$ is called the **amino group.**

e.g.

$H_2NCH_2CH_2OH$	$H_2NCH_2CH_2C(=O)—OH$
2-aminoethanol	3-aminopropanoic acid

Isomerism

Amines show chain, position and functional isomerism as well they show metamerism too. Metamerism is present in those molecules that have same molecular formula but differ with respect to the alkyl groups on either sides of the nitrogen atom.

e.g. $C_2H_5NHC_2H_5$ N-ethylethanamine $CH_3—NHCH(CH_3)_2$ N-methylpropan-2-amine

Nitrogen inversion

When alkyl groups of a tertiary amine are different, then amine will be chiral. There should be two enantiomeric forms of the tertiary amine. These enantiomers are impossible to resolute as they interconvert rapidly. The barrier of interconversion is 25 kJ mol^{-1} only. This interconversion is called as **pyramidal or nitrogen inversion.**

$R_1R_2R_3N$: ⇌ :$NR_1R_2R_3$ (cannot be resolved due to rapid interconversion)

However, ammonium salts cannot undergo inversion because they do not have an unshared pair of electron. Therefore, these quaternary NH_4^+ salts with four different groups are chiral and can be resolved into separate enantiomers.

$R_1R_2R_3N^+—P_4$ (can be resolved, chiral)

CHECK POINT 20.1

1. What is the degree of amine in R_2NH?
   (a) 2 (b) 1
   (c) 3 (d) 0

2. Consider the following structure of a molecule. The bond angle for the given molecule is
   $N(CH_3)_3$ (with lone pair on N)
   (a) 108° (b) 125°
   (c) 180° (d) 120°

3. The correct structure for 3-aminopropanoic acid is
   (a) $H_2N(CH_2)_3OH$ (b) $H_2N(CH_2)_3COOH$
   (c) $CH_3CH_2NHCOOH$ (d) $H_2NCH_2CH_2C(=O)—OH$

4. The total isomers exhibited by C_4H_{11} N is
   (a) 4 (b) 8
   (c) 6 (d) 7

5. Which of the following type of isomerism is shown by aliphatic amines?
   (a) Chain isomerism
   (b) Metamerism
   (c) Position isomerism
   (d) Geometrical isomerism

6. Which of the following statement(s) is incorrect for nitrogen inversion?
   (a) It appears in only tertiary amines where all alkyl group are different.
   (b) Amine should be chiral.
   (c) It occurs in those enantiomers that can be able to resolved.
   (d) It occurs in those amines where barrier of interconversion is 25 kJ mol^{-1} only.

7. Ammonium salts cannot undergo inversion because
   (a) they do not have an unshared pair of electron
   (b) they are chiral
   (c) they are resolvable
   (d) None of the above

8. Consider the following reaction,

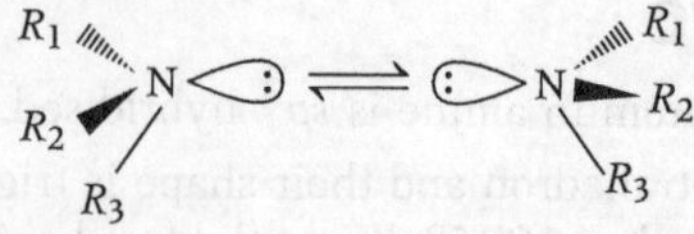

   The correct term used for this reaction is
   (a) nitrogen inversion
   (b) pyramidal inversion
   (c) Both (a) and (b)
   (d) None of the above

PREPARATION OF AMINES

Methods for preparation of mixture of amines (1°, 2°, 3°)

Amine are generally prepared by the following methods

Ammonolysis of alkyl halide

Alkyl halide on reaction with an ethanolic solution of ammonia undergoes nucleophilic substitution reaction by S_N2 mechanism in which the halogen atom is replaced by a amino ($—NH_2$) group. This process of cleavage of the C—X bond by ammonia molecule is known as **ammonolysis.**

The reaction is carried out in a sealed tube at 373 K. The primary amine thus obtained behaves as a nucleophile and can further react with alkyl halide to form secondary and tertiary amines and eventually quaternary ammonium salt.

$$R—Cl + NH_3 \xrightarrow{C_2H_5OH} R—\ddot{N}H_2$$

$$\underset{(1°)}{R\ddot{N}H_2} \xrightarrow[-HX]{RX} \underset{(2°)}{R_2\ddot{N}H} \xrightarrow[-HX]{RX} \underset{(3°)}{R_3\ddot{N}} \xrightarrow{RX} \underset{\text{Quaternary ammonium salt}}{R_4\overset{+}{N}X^-}$$

This reaction involves nucleophilic attack of NH_3 molecule (through lone pair) on alkyl halide by S_N2 mechanism. Amine salt is formed which react with ammonia molecule to give primary amine and ammonium halide. This reaction continues till the quaternary ammonium salt is formed.

Note *Ammonolysis gives the highest yield with primary halides (where substitution predominates) and is of no use with tertiary alkyl halides (where elimination predominates).*

$$\text{CH}_3\text{CH}_2\text{CH}_2\text{CH}_2\text{Br} \xrightarrow{NH_3} \text{CH}_3\text{CH}_2\text{CH}_2\text{CH}_2\text{NH}_2 \text{ (substitution)}$$

$$(CH_3)_3C\text{—Br} \xrightarrow{NH_3} (CH_3)_2C{=}CH_2 \text{ (elimination)}$$

$$\text{Cyclohexyl—Br} \xrightarrow{NH_3} \text{Cyclohexene} \text{ (elimination)}$$

Disadvantages

(*i*) Synthesis of amines by ammonolysis results in the formation of more than one class of amines. It is very difficult to separate them.

(*ii*) This method does not allow the preparation of aniline and other aromatic amines as haloarenes are very little reactive as compared to alkyl halides.

Ammonolysis of alcohols

When alcohol reacts with ammonia over heated alumina (Al_2O_3) or thoria (ThO_2) at 620 K, a mixture of amines is formed.

e.g.

$$CH_3OH + NH_3 \xrightarrow[570\text{-}670\,K]{ThO_2} \underset{1°}{CH_3NH_2}$$

$$CH_3NH_2 + CH_3OH \xrightarrow[570\text{-}670\,K]{ThO_2} \underset{2°}{(CH_3)_2NH}$$

$$(CH_3)_2NH + CH_3OH \xrightarrow[570\text{-}670\,K]{ThO_2} \underset{3°}{(CH_3)_3N}$$

In the above methods, mixture of 1°, 2° and 3° amines is obtained and thus, separation is required. However, by controlling the quantities of reactants, desired products can be obtained.

By reductive amination of aldehydes and ketones

Aldehydes (RCHO) and ketones (R_2CO) are converted into primary amines when they are treated with ammonia in the presence of a catalyst (H_2/Ni), this process is called as **reductive amination.** This reaction probably goes through the intermediate formation of an imine.

$$\underset{\text{Aldehyde}}{RCHO} + NH_3 \xrightarrow[\substack{-H_2O \\ \text{1st step}}]{\Delta} \underset{\text{Imine}}{RCH{=}NH} \xrightarrow[\text{2nd step}]{H_2,\ \text{Raney Ni}} \underset{\text{1° amine}}{RCH_2NH_2}$$

$$\underset{\text{Ketone}}{R—\overset{R}{\overset{|}{C}}{=}O} + NH_3 \xrightarrow[-H_2]{\Delta} \underset{\text{Imine}}{R—\overset{R}{\overset{|}{C}}{=}NH} \xrightarrow[\text{2nd step}]{H_2,\ \text{Raney Ni}} \underset{\text{1° amine}}{R—\overset{R}{\overset{|}{C}}HNH_2}$$

In this reaction, $NaBH_3CN$ (sodium cyanoborohydride) can also be used as a catalyst instead of H_2, raney Ni. Because of the presence of an electron withdrawing cyano group on boron in $NaBH_3CN$. This reagent is less reactive than $NaBH_4$ and does not reduce aldehydes and ketones. Indeed it is the protonated imine that undergoes reduction in these reductive amination reactions.

(*i*) 1° amines can react with aldehydes and ketones to form 2° amine by **reductive amination.** Similarly, 3° amine is formed by reacting 2° amine with aldehyde or ketone.

$$\underset{\text{Methanal}}{CH_3CHO} + RNH_2 + H_2 \xrightarrow{Ni} \underset{\text{2° amine}}{CH_3CH_2NHR}$$

$$\underset{\text{Propan-2-one}}{(H_3C)_2C{=}O} + R_2NH + H_2 \xrightarrow{Ni} \underset{\text{3° Amine}}{(H_3C)_2CHNR_2}$$

(*ii*) A cyclic secondary amine can also be prepared by reductive amination using a compound that contains a ketone and primary amine groups separated by an appropriate number of carbon atoms.

Methods for preparation of primary amines only

Primary amines are generally prepared by a following methods

Reduction of nitro alkanes

Nitro alkanes react with hydrogen in the presence of Raney Ni or finely divided Pt or Pd to give primary amines. This type of reduction is called **catalytic reduction**.

$$R\text{—}NO_2 + 3H_2 \xrightarrow[\text{Ethanol}]{\text{Raney Ni/Pt}} R\text{—}NH_2 + 2H_2O$$

Reduction of nitroalkanes can also occur with active metals like Fe, Sn, Zn in presence of conc. HCl.

$$CH_3CH_2NO_2 + 6[H] \xrightarrow{\text{Sn/conc.HCl}} CH_3CH_2NH_2 + 2H_2O$$

This reaction is preferred in aromatic amines.

Reduction of nitriles

Nitriles can be easily reduced to primary amines by using the following reagents

- H_2 in presence of Raney Ni
- $LiAlH_4$ dissolved in anhydrous ether
- Na and alcohol (reduction of nitriles to primary amines by using Na and alcohol is called **Mendius reaction**).

$$CH_3C{\equiv}N \xrightarrow[\substack{\text{or } LiAlH_4/\text{ether} \\ \text{or Na}/C_2H_5OH}]{H_2/\text{Raney Ni}} CH_3CH_2NH_2$$

This method is an excellent method for the synthesis of primary amines as it increases one carbon more than the parent alkyl halide. This is a step-up synthesis.

Reduction of oximes

Oximes produced by aldehyde and ketones and hydroxylamine can be easily converted into primary amines by using Na/C_2H_5OH or $LiAlH_4$/ether.

$$\underset{\text{Oximes}}{R_1\text{—}\overset{\overset{\displaystyle NOH}{\|}}{C}\text{—}R_2} \xrightarrow[\text{or Na}/C_2H_5OH]{LiAlH_4/\text{ether}} \underset{1^\circ\text{ amines}}{R_1\text{—}\overset{\overset{\displaystyle NH_2}{|}}{CH}\text{—}R_2}$$

$$\text{Cyclohexanone oxime }(C_6H_{10}{=}NOH) \xrightarrow[\text{or Na}/C_2H_5OH]{LiAlH_4/\text{ether}} C_6H_{11}\text{—}NH_2$$

Reduction of primary acid amides

Primary acid amides can easily be reduced to primary amines by using Na/C_2H_5OH or $LiAlH_4$ in presence of dry ether.

$$\underset{\text{Acetamide}}{CH_3\text{—}\overset{\overset{\displaystyle O}{\|}}{C}\text{—}NH_2} \xrightarrow[\text{Na}/C_2H_5OH]{LiAlH_4/\text{dry ether}} \underset{\text{Ethanamine}}{CH_3CH_2NH_2} + H_2O$$

Reduction of azides

Alkyl halide reacts with sodium azide to give alkyl azide which on reduction gives primary amine.

This reaction proceed *via* S_N2 mechanism.

$$\underset{\substack{1^\circ\text{ or }2^\circ \\ \text{alkyl halide}}}{R\text{—}X} + \underset{\substack{\text{Sodium} \\ \text{azide}}}{NaN_3} \xrightarrow[-NaX]{S_N2} \underset{\text{Alkyl azide}}{RN{=}\overset{\oplus}{N}{=}\overset{\ominus}{N}} \xrightarrow[H_2O]{NaBH_4} \underset{1^\circ\text{ amine}}{R\text{—}NH_2} + N_2$$

Instead of $NaBH_4$ / H_2O, $LiAlH_4$ or Na/alcohol or H_2/Pd-C (catalytic hydrogenation) can also be used.

This method is so much useful to prepare primary amines is the laboratory from an alkyl halide.

By hydrolysis of alkyl isocyanides and isocyanates

Both alkyl isocyanides and alkyl isocyanates on hydrolysis give primary amines.

$$\underset{\text{Ethyl isocyanide}}{CH_3CH_2NC} + 2H_2O \xrightarrow{HCl} \underset{\text{Ethanamine}}{CH_3CH_2NH_2} + HCOOH$$

$$\underset{\text{Ethyl isocyanate}}{CH_3CH_2NCO} + 2KOH \xrightarrow{H_2O,\,\Delta} \underset{\substack{\text{Ethanamine} \\ 1^\circ\text{ amine}}}{CH_3CH_2NH_2} + K_2CO_3$$

By decarboxylation of α-amino acid

α-amino acids on heating with $Ba(OH)_2$ decarboxylates to give primary amine.

$$\underset{\alpha\text{-amino acid}}{R\text{—}\underset{\underset{\displaystyle NH_2}{|}}{CH}\text{—}COOH} \xrightarrow[\text{Heat},(-CO_2)]{Ba(OH)_2} \underset{(1^\circ\text{ amine})}{RCH_2NH_2}$$

Hofmann bromamide degradation of amides

Primary acid amides react with an aqueous ethanolic solution of KOH and Br_2 to give primary amines. This method yields amine that contains one carbon atom less than the reactant (amide).

$$\underset{\text{Amide}}{R\text{—}\overset{\overset{\displaystyle O}{\|}}{C}\text{—}NH_2} + Br_2 + \text{aq. }4KOH \xrightarrow{\Delta} \underset{1^\circ\text{ amine}}{RNH_2} + 2KBr + K_2CO_3 + 2H_2O$$

Mechanism

$$R-\overset{O}{\overset{\|}{C}}-NH_2 \xrightarrow[-HBr]{Br_2+KOH} R-\overset{O}{\overset{\|}{C}}-\underset{H}{\underset{|}{\ddot{N}}}-Br + OH^-$$

Amide

$$\longrightarrow R\overset{O}{\overset{\|}{C}}-\ddot{N}^{\ominus}Br + H_2O \longrightarrow R-\overset{O}{\overset{\|}{C}}-\ddot{N}+Br^-$$

Intramolecular rearrangement

$$R-\ddot{N}H_2 \xleftarrow[(ii)\ H_2O]{(i)\ 2KOH} R-\ddot{N}=C=O$$

Amine — Alkyl isocyanate

Gabriels's phthalimide synthesis

This method is used for the preparation of aliphatic as well as aromatic primary amines (having electron withdrawing group at *o*-and *p*-positions). Phthalic anhydride is converted into phthalimide thus, blocking two H-atoms of NH_3 for further reaction with *RX*.

Then this, phthalimide on treatment with *aq.* KOH forms potassium salt of phthalimide which on heating with suitable alkyl halide forms N-substituted phthalimides. These on alkaline or acid hydrolysis produce the corresponding 1° amine and phthalic acid as by product.

Phthalic anhydride $+ H_2NH \longrightarrow$ Phthalimide $\xrightarrow{KOH\ (alc.)}$ Potassium phthalimide (NK^+)

$\xrightarrow{RX,\ \Delta}$ N-alkyl phthalimide $\xrightarrow[pressure]{H^+/H_2O,\ \Delta}$ Phthalic acid $+ RNH_2$ (1° amine)

N-alkyl phthalimide (no active H is available at N for further reaction with *R–X*) — Phthalic acid

In this reaction phthalic acid is reused by which anhydride is prepared and thus, is carrier of the reaction. If hydrolysis is difficult, hydrazine can be used to give the required amine.

N-alkylphthalimide $\xrightarrow[reflux]{NH_2NH_2,\ ethanol}$ (C–NHNH₂ / C–NHR intermediate) $\longrightarrow$ (cyclic hydrazide with NH, NH) $+ RNH_2$ (1° amine)

Note Tert-butylamine $(CH_3)_3CNH_2$ is a primary amine, it cannot be prepared by this method, because it would require S_N2 reaction on a tertiary alkyl halide in first step. Elimination takes place as shown

$$(CH_3)_3CBr + \text{(potassium phthalimide, } \bar{N}\overset{+}{K}) \longrightarrow (CH_3)_2C{=}CH_2 + \text{(phthalimide, NH)} + KBr$$

(Elimination)

Limitations

1° aromatic amine that does not contain electron withdrawing group at *o*-and *p*-position (e.g. aniline, toluidines etc.) cannot be synthesised by this method. This is because aryl halides do not undergo nucleophilic substitution reaction with anion formed by phthalimide.

Curtius degradation method

This method uses acid chloride to prepare 1° amine through formation of acyl azide.

$$R-\overset{O}{\overset{\|}{C}}-Cl \xrightarrow{NaN_3} R-\overset{O}{\overset{\|}{C}}-N_3 \xrightarrow{-N_2}$$

Acyl chloride — Acyl azide

$$R-N=C=O \xrightarrow[\Delta]{2NaOH} RNH_2 + Na_2CO_3$$

Alkyl isocyanate

In this reaction intramolecular alkyl shift is very similar to that of Hofmann bromamide degradation.

Schmidt reaction

This method converts *R*COOH into RNH_2 when warmed with H_3N in the presence of conc. H_2SO_4. Instead of N_3H and conc. H_2SO_4 a mixture of NaN_3 and conc. H_2SO_4 can also be used.

$$CH_3-\overset{O}{\overset{\|}{C}}-OH \xrightarrow[\Delta]{N_3H\ +\ conc.\ H_2SO_4} CH_3NH_2 + N_2\uparrow + CO_2\uparrow$$

Ketones also undergo Schmidt reaction, first it gives N-substituted amide which yields 1° amine on hydrolysis.

$$RCOR + NH_3 \xrightarrow{H_2SO_4} RCONHR + N_2$$

$$RCONHR \xrightarrow{H_2O} RNH_2 + RCOOH$$

Ritter reaction

This method is used to prepare 1° amine by treating a tertiary alkyl alcohol with conc. H_2SO_4 and HCN.

$$CH_3-\underset{CH_3}{\underset{|}{\overset{CH_3}{\overset{|}{C}}}}-OH + HCN + Conc.H_2SO_4 \longrightarrow CH_3-\underset{CH_3}{\underset{|}{\overset{CH_3}{\overset{|}{C}}}}-NH_2$$

3° alcohol — 1° amine

Methods for preparation of secondary amines only

Following methods are used for the preparation of secondary amines

Reduction

(i) **Secondary acid amide** Secondary acid amide or N-alkyl acidamide can be reduced to secondary amine by using $LiAlH_4$ in dry ether.

$$\underset{\text{N-alkyl acid amide}}{R\overset{O}{\overset{\|}{C}}NHR'} \xrightarrow{LiAlH_4/\text{ether}} \underset{2^\circ \text{ amine}}{RCH_2NHR'}$$

(ii) **Isocyanides** Isocyanides can be reduced to secondary amines by using H_2/Ni or Na/C_2H_5OH (as they can easily release nascent hydrogen).

$$\underset{\text{Alkyl isocyanide}}{RNC} + 4H \text{ (nascent)} \longrightarrow \underset{2^\circ \text{ amine}}{RNHCH_3}$$

Mannich reaction

Ketone react with formaldehyde and primary amine in acidic medium to give Mannich base. This reaction is called **Mannich reaction.** This base upon reduction with $LiAlH_4$ yields secondary amine.

$$\underset{\text{Acetone}}{CH_3COCH_3} + \underset{\text{Formaldehyde}}{CH_2{=}O} + \underset{\text{Primary amine}}{RNH_2} \xrightarrow[-H_2O]{H^+}$$

$$\underset{\substack{\text{Mannich base or}\\ \beta\text{-amino carbonyl}\\ \text{compound}}}{CH_3COCH_2CH_2NHR} \xrightarrow[\text{Reduction}]{LiAlH_4} \underset{\text{Secondary amine}}{CH_3CH_2CH_2CH_2NHR}$$

Reductive amination of carbonyl compounds

Reductive amination of aldehydes or ketones (using primary amine followed by reduction) gives secondary amine.

$$\underset{\text{Propan-2-one}}{CH_3{-}\overset{O}{\overset{\|}{C}}{-}CH_3} + \underset{1^\circ \text{ amine}}{RNH_2} \xrightarrow{-H_2O} \underset{\text{Imine}}{CH_3{-}\overset{NR}{\overset{\|}{C}}{-}CH_3}$$

$$\xrightarrow[C_2H_5OH, \text{ pressure}]{H_2, Pt} \underset{2^\circ \text{ amine}}{CH_3{-}\overset{NHR}{\overset{|}{C}H}{-}CH_3}$$

Reduction of the C≡N bond is very similar to reduction of C=O bond. $NaBH_4$ and $NaBH_3CN$ (sodium cyanoborohydride) can be used in reductive amination. Reductive amination with $NaBH_3CN$ is called **Borch reaction.**

Methods for preparation of tertiary amines only

Following methods are used for the preparation of tertiary amines

Reaction of alkyl halide and ammonia

In this method, alcoholic NH_3 solution is heated with excess of alkyl halide resulting to tri alkyl ammonium halide which on treatment with NaOH gives tertiary amine.

$$\underset{\text{Excess}}{3RX} + (alc.)\,NH_3 \longrightarrow R_3\underset{H}{\underset{|}{N^+}}X^- \xrightarrow{NaOH} \underset{3^\circ \text{ amine}}{R_3N} + NaX + H_2O$$

Reduction of N,N-disubstituted amides

$$\underset{\text{N,N-dialkyl acetamide}}{R{-}\overset{O}{\overset{\|}{C}}{-}N\begin{matrix}R'\\R'\end{matrix}} \xrightarrow[\text{ether}]{LiAlH_4} \underset{3^\circ \text{ amine}}{RCH_2N\begin{matrix}R'\\R'\end{matrix}}$$

Hofmann elimination

In this process first ammonolysis of alkyl halide occurs by which quaternary ammonium salts are formed as a final product. When this salt is treated with silver oxide, silver halide is precipitates out.

$$R_4N^+X^- + Ag_2O + H_2O \longrightarrow \underset{\substack{\text{Quaternary}\\ \text{ammonium hydroxide}\\ \text{(basic)}}}{R_4N^+OH^-} + AgX\downarrow$$

Quaternary ammonium hydroxide on heating strongly decomposes to yield water, a tertiary amine and alkene.

$$\left(CH_3{-}\overset{CH_3}{\underset{CH_3}{\overset{|}{\underset{|}{N^+}}}}{-}CH_2CH_2CH_3\right)OH^- \xrightarrow[400\,K]{\Delta} CH_3{-}\overset{CH_3}{\underset{CH_3}{\overset{|}{\underset{|}{N}}}} + CH_2{=}CHCH_3 + H_2O$$

This method proceeds *via* E2 elimination in which the hydroxide ion functions as the base. The formation of quaternary ammonium salts, followed by an elimination is used in the determination of structures of N-containing compounds (1°, 2°, 3° amines).

CHECK POINT 20.2

1. Which of the following mechanism is followed by ammonolysis of alkyl halide?

(a) S_N1 mechanism (b) S_N2 mechanism
(c) E1 mechanism (d) E2 mechanism

2. Which of the following, halide will give highest yield in ammonolysis?

(a) Primary halides (b) Secondary halides
(c) Tertiary halides (d) Quaternary salts of amines

3. Which of the following reagents is used in the ammonolysis of alcohols?

(a) Alumina (b) Thoria
(c) Both (a) and (b) (d) None of these

4. Consider the following reaction,

$$CH_3C\equiv N \xrightarrow{Na/C_2H_5OH} CH_3CH_2NH_2$$

Which of the following is the correct name for the above reaction?

(a) Carbylamine reaction (b) Mendius reaction
(c) Hofmann's reaction (d) Gabriel's reaction

5. RNO_2 is reduced with Sn/HCl.
The product formed is

(a) RNH_2 (b) $RNHOH$ (c) RNH_3^+ (d) R_2NH

6. Consider the following reaction,

(cyclohexanone) $=O + H_2NOH \longrightarrow$ 'A' $\xrightarrow{LiAlH_4}$ (cyclohexyl)$-NH_2$

Here, 'A' is

(a) (cyclohexane with) OH, NH_2OH (b) (cyclohexane)$=NOH$
(c) (cyclohexane)$-NH_2OH$ (d) (cyclohexane)$-NHOH$

7. When acetamide is treated with Br_2 and caustic soda, the product formed is

(a) N-bromoamide (b) bromo acetic acid
(c) methanamine (d) ethane

8. Gabriel phthalimide synthesis is used in the preparation of

(a) 1° amines (b) 2° amines
(c) 3° amines (d) mixture of all amines

9. Which of the following method is used to convert $RCOOH$ into RNH_2?

(a) Curtius degradation (b) Ritter reaction
(c) Schmidt reaction (d) Gabriel's phthalimide

10. Which of the following alkene is formed when

$$\left[CH_3-\underset{CH_3}{\overset{CH_3}{\underset{|}{\overset{|}{N^+}}}}-CH_2CH_2CH_3\right]OH^-$$ is heated?

(a) $CH_2{=}C{=}CH_2$ (b) $CH_2{=}CH_2$
(c) $CH_2{=}CHCH_3$ (d) None of these

PHYSICAL AND CHEMICAL PROPERTIES OF ALKYL AMINES

Physical properties

Some important physical properties of amines are discussed below

(i) **Physical state** Lower amines are gases or low-boiling liquids, with a characteristic ammonia like smell. Higher amines are liquids and have fishy odour.

(ii) **Boiling point** Amines are more polar than alkanes but less polar than alcohols. For similarly constituted compounds alkyl amines have boiling points and dipole moments which are higher than those of alkanes but lower than those of alcohols.

e.g.

Compound	Boiling point	Dipole moment
$CH_3CH_2CH_3$	231 K	0 D
$CH_3CH_2NH_2$	290 K	1.2 D
CH_3CH_2OH	351 K	1.7 D

- Among isomeric amines, primary amines have the highest boiling points and tertiary amines have the lowest.

$$\underset{\substack{(1^\circ\ \text{amine})\\ \text{bp } 323\ \text{K}}}{CH_3CH_2CH_2NH_2} > \underset{\substack{(2^\circ\ \text{amine})\\ \text{bp } 307\ \text{K}}}{CH_3CH_2NHCH_3} > \underset{\substack{(3^\circ\ \text{amine})\\ \text{bp } 276\ \text{K}}}{(CH_3)_3N}$$

- The fact that primary and secondary amines can both donate and accept hydrogen bonds also accounts for the fact that they have higher boiling points than ethers.

$(C_2H_5)_2NH$ diethylamine (2° amine) : bp 329 K
$CH_3(CH_2)_3NH_2$ butylamine (1° amine) : bp 351 K
$(C_2H_5)_2O$ diethylether : bp 310.5 K

- Boiling point of amines get increased by intermolecular hydrogen bonding [except tertiary (R_3N) amines, which lacks active H].

$$\begin{array}{ccccccc} & H--- & & CH_3 & & H\cdots\cdots & \\ & | & & | & & |\delta^+ & \\ CH_3- & \underset{\delta^-}{N} & -\underset{\delta^+}{H}--- & N & -\underset{\delta^+}{H}--- & \underset{\delta}{N} & -CH_3 \\ & & & |\delta & & |\delta & \\ & & & ---H & & H--- & \end{array}$$

As primary amines have two, secondary amines have one while tertiary amines have no hydrogen linked to nitrogen, thus, among isomeric amines, primary amines have the highest while tertiary amines have the lowest boiling point.

(*iii*) **Solubility** All three types of amines (1°, 2° and 3°) having low molecular weight (gases) are highly soluble in water and other organic solvents like alcohols due to intermolecular H-bonding. But solubility decreases with increase in number of carbon atoms (molecular weights).

H-bonding

Chemical properties

Various chemical reactions of amines are due to the difference in electronegativity of nitrogen and hydrogen. They also have unshared pair of electrons over the nitrogen atom that makes amine reactive and behave like a nucleophile.

Some of the chemical reactions of amines are described below

Basic character

Amines react with mineral acid (or in water) to form ammonium salts which shows that they are basic in nature. They behave as Lewis base due to presence of unshared pair of electrons on nitrogen atom.

$$R\text{NH}_2 + \text{H}_2\text{O} \rightleftharpoons R\text{NH}_3^{\oplus} + \text{OH}^-$$

$$K_b = \frac{[R\text{—}\overset{+}{\text{N}}\text{H}_3][\bar{\text{O}}\text{H}]}{[R\text{—}\ddot{\text{N}}\text{H}_2]}$$

Here, K_b is called dissociation constant for base. Greater the value of K_b, stronger will be the base. Basic character of amines, can also be expressed in terms of pK_b values.

$$pK_b = -\log K_b$$

Greater the K_b value, stronger is the base. Smaller the pK_b values, stronger is the base.

Alkylamine *versus* ammonia

Basicity of amines is related to their structure. Basic character of amine depends upon the ease of formation of the cation by accepting a proton from the acid.

The more stable the cation is relative to amine, more basic is amine.

$$R\text{—}\ddot{\text{N}}\text{H}_2 + \text{H}^+ \rightleftharpoons R\text{—N}^+\text{H}_3$$

The order of basicity of alkyl amines and ammonia in gaseous and aqueous phase are shown below follows different order, that the discuss.

***Case* I. Basicity in gaseous phase**

Here, the electron releasing nature of alkyl group (*R*) possess the electrons toward nitrogen and makes the unshared pair more available for sharing with the proton of the acid.

Also, the substituted ammonium ion formed from the amine gets stabilised due to the dispersal of positive charge by +*I*-effect of '*R*' groups. Hence, alkyl amines are stronger bases than ammonia.

We can conclude that the basic nature of aliphatic amines should increase with increase in the number of alkyl groups.

Therefore order of basicity in gaseous phase is as follows

3° amine > 2° amine > 1° amine > NH_3

But the trend is not the same in aqueous state.

***Case* II. Basicity in aqueous phase**

The substituted ammonium cation is stabilised by the following factors.

- Electron releasing effect of the alkyl group (+*I*).
- Solvation effect (greater the size of the ion lesser will be the solvation and the less stabilised is the ion).
- Steric hindrance.

Order of stability of solvated ions

1° > 2° > 3°

Secondly, when alkyl group is small (such as —CH_3 group) there is no steric hinderance to H—bonding but when the alkyl group is bigger then there will be steric hinderance to H-bonding. Therefore, the order of basic strength is different when different alkyl groups (—CH_3 or —C_2H_5) are attached.

The order are given below

- In case of methyl substituted amines.

$$\underset{(2°)}{(CH_3)_2NH} > \underset{(1°)}{CH_3NH_2} > \underset{(3°)}{(CH_3)_3N} > NH_3$$

- In case of ethyl substituted amines.

$$\underset{(2°)}{(C_2H_5)_2NH} > \underset{(3°)}{(C_2H_5)_3N} > \underset{(1°)}{C_2H_5NH_2} > NH_3$$

Summarise basicity constants of aliphatic amines.

Table 20.1 Basicity constants and pK_b of bases

Base	K_b	pK_b
NH_3	0.2×10^{-4}	4.70
CH_3NH_2	4.2×10^{-4}	3.38
$(CH_3)_2NH$	6.0×10^{-4}	3.20
$(CH_3)_3N$	0.6×10^{-4}	4.22
$CH_3CH_2NH_2$	4.7×10^{-4}	3.33
$(CH_3CH_2)_2NH$	9.5×10^{-4}	4.02
$(CH_3CH_2)_3N$	5.5×10^{-4}	3.26
$(CH_3CH_2)_2NCH_2CN$	3.55×10^{-10}	9.45

Alkylation

Aliphatic amines undergo alkylation on reaction with alkyl halides.

$$CH_3NH_2 \xrightarrow{CH_3X} (CH_3)_2NH \xrightarrow{CH_3X} (CH_3)_3N \xrightarrow{CH_3X} [(CH_3)_4]N^+X^-$$

Acylation

Aliphatic amines react with acid chlorides or anhydrides or esters by nucleophilic substitution reaction. This reaction is known as **acylation** to give respective amides. This reaction is carried out in the presence of a base, stronger than the amine like pyridine, which removes HCl so formed and shift the equilibrium to the right hand side.

e.g. $$\underset{\text{ethanamine}}{C_2H_5NH_2} + \underset{\text{ethanoyl chloride}}{CH_3COCl} \longrightarrow \underset{\text{N-methyl ethanamide}}{C_2H_5NH\overset{O}{\overset{\|}{C}}CH_3} + HCl$$

$$C_2H_5NH_2 + \underset{\text{Ethanoic anhydride}}{CH_3\underset{\underset{O}{\|}}{C}-O-\underset{\underset{O}{\|}}{C}CH_3} \longrightarrow \underset{\text{N-methyl ethanamide}}{C_2H_5-\underset{\underset{H}{|}}{\ddot{N}}-\underset{\underset{O}{\|}}{C}\,CH_3} + CH_3COOH$$

Benzoylation

Aliphatic amines react with benzoyl chloride to form benzamide.

e.g. $$\underset{\text{Methanamine}}{CH_3NH_2} + \underset{\text{Benzoyl chloride}}{C_6H_5COCl} \longrightarrow \underset{\text{N-methyl benzamide}}{CH_3NHCOC_6H_5} + HCl$$

Carbylamine reaction

Aliphatic amines on heating with chloroform and alcoholic KOH forms isocyanides or carbylamines which are foul smelling substances. Secondary and tertiary amines do not show this reaction.

e.g. $$\underset{1°\text{ amine}}{CH_3NH_2} + CHCl_3 + \underset{\text{Alcoholic}}{4KOH} \xrightarrow{\Delta} \underset{\text{Isocyanide (carbylamines)}}{CH_3NC} + 3KCl + 3H_2O$$

This reaction is used to test 1°amines (both aromatic and aliphatic).

Reaction with nitrous acid

Three classes of amines, i.e. 1°, 2° and 3° amines react differently with nitrous acid.

(i) Primary aliphatic amines

It reacts with nitrous acid to form aliphatic diazonium salt which is unstable, thus, liberate nitrogen gas.

$$\underset{1°\text{ amine}}{R-NH_2} + HONO \longrightarrow \underset{\text{Alcohol}}{ROH} + N_2\uparrow + H_2O$$

RNH_2 is converted into ROH by HNO_2. Intermediate is a carbocation which may undergo rearrangement to give various products.

$$\underset{\text{Propamamine}}{CH_3CH_2CH_2NH_2} \xrightarrow{HNO_2} CH_3CH_2CH_2N_2^{\oplus}$$

$$\downarrow$$

$$\underset{\substack{2°\text{carbocation}\\\text{(More stable)}}}{CH_3\overset{\oplus}{C}HCH_3} \xleftarrow{\text{1, 2-hydride shift}} \underset{1°\text{carbocation}}{CH_3CH_2CH_2^{\oplus}}$$

2° carbocation $\xrightarrow{-H^+}$ $\underset{\text{Propene}}{CH_3-CH=CH_2}$; 2° carbocation $\xrightarrow{H_2O}$ $\underset{\text{2°alcohol (Major)}}{CH_3\underset{\underset{OH}{|}}{C}HCH_3}$; 1° carbocation $\xrightarrow{H_2O}$ $\underset{\text{1°alcohol (Minor)}}{CH_3CH_2CH_2OH}$

(ii) Secondary aliphatic amines

It reacts with nitrous acid to form N-nitrosoamine which is yellow oily insoluble liquid.

$$R-\underset{\underset{H}{|}}{N}-R + HONO \xrightarrow{-H_2O} \underset{\substack{\text{N-nitrosoamine}\\\text{(yellow oily layer)}}}{R-\overset{\overset{R}{|}}{N}-N=O}$$

Note *N-nitrosoamines on warming with a crystal of phenol and a few drops of conc. H_2SO_4 form a green solution which when made alkaline with aq. NaOH turns deep blue and then red on dilution. This reaction is called **Liebermann's nitroso reaction** and used as a test for secondary amines.*

(iii) Tertiary aliphatic amines

It reacts with nitrous acid to form trialkyl ammonium nitrite, that is soluble in water.

$$R_3N + HONO \longrightarrow \underset{\text{Trialkyl ammonium nitrite (solution in } H_2O)}{R_3\overset{+}{N}HNO_2^-}$$

Reaction with benzenesulphonyl chloride (Hinsberg reagent)

Benzene sulphonyl chloride reacts with primary and secondary amines to form sulphonamide. This reaction is used to distinguish 1°, 2° and 3° amines.

(*i*) Primary amines react with benzene sulphonyl chloride to yield N-alkyl benzene sulphonamide.

$$\underset{\text{Benzene sulphonyl chloride}}{C_6H_5SO_2Cl} + \underset{\text{ethanamine}}{H-NH-C_2H_5} \longrightarrow C_6H_5SO_2-NH-C_2H_5 + HCl$$

N-ethyl benzene sulphonamide (Soluble in alkali as 'H' atom attached to nitrogen is strongly acidic)

(*ii*) Secondary amines react with $C_6H_5SO_2Cl$ to yield N, N-dialkyl benzene sulphonamide.

$$\underset{\text{Benzen sulphonyl chloride}}{C_6H_5SO_2Cl} + H-N(C_2H_5)-C_2H_5 \longrightarrow C_6H_5SO_2-N(C_2H_5)-C_2H_5 + HCl$$

N, N-diethylbenzene sulphonamide (not acidic as there is no H-atom attached to N, thus it is insoluble in alkali, but soluble in ether).

(*iii*) Tertiary amines do not react with $C_6H_5SO_2Cl$.

Reaction with carbonyl compounds

Primary aliphatic amines react with carbonyl compounds to form **Schiff's base**. Whereas secondary amines react with carbonyl compounds to form enamine.

$$\underset{1^\circ \text{ amine}}{R-NH_2} + O=C\langle \longrightarrow \underset{\text{Schiff's base}}{R-N=C\langle} + H_2O$$

$$RCH_2-\overset{O}{\overset{\|}{C}}-R' + \underset{2^\circ\text{amine}}{R_2NH} \longrightarrow \underset{\text{Carbinolamine (Unstable)}}{R-CH_2-\overset{OH}{\overset{|}{\underset{NR_2}{\underset{|}{C}}}}-R'}$$

$$\xrightarrow{-H_2O} \underset{\text{Enamines}}{RCH=C\langle^{R'}_{NR_2}}$$

Tertiary amines do not react with carbonyl compounds. Hence, this reaction can be used to distinguish 3° amine among 1° and 2° amines.

Oxidation reaction

Aliphatic amines readily undergo oxidation reaction with the following reagents

(*i*) With $KMnO_4$ primary amine forms aldehyde or ketone.

$$\underset{1^\circ \text{ amine}}{CH_3CH_2NH_2} \xrightarrow{[O]} CH_3CH{=}NH$$

$$\xrightarrow{H^+} CH_3CHO + NH_3\uparrow$$

Secondary amine forms tetra-alkyl hydrazine

$$2R_2NH \xrightarrow{KMnO_4} \begin{matrix}R\\R\end{matrix}\rangle N-N\langle\begin{matrix}R\\R\end{matrix} + H_2O$$

Tertiary amines are not oxidised by $KMnO_4$.

(*ii*) With Caro's acid (H_2SO_5) primary amine gives different products depending upon the nature of alkyl groups.

e.g. $$\underset{1^\circ \text{ amine}}{CH_3CH_2NH_2} \xrightarrow{[O]} \underset{\text{N-ethylhydroxyl amine}}{CH_3CH_2NHOH} \longrightarrow$$

$$\underset{\text{Aldoxime}}{CH_3CH{=}NOH} \longrightarrow \underset{\text{Hydroxamine acid}}{CH_3C\langle^{OH}_{N-OH}}$$

Secondary amine forms dialkyl hydroxyl amine.

$$\underset{2^\circ\text{amine}}{(CH_3)_2NH} \xrightarrow{H_2SO_5} \begin{matrix}CH_3\\CH_3\end{matrix}\rangle N-OH$$

Tertiary amine gives tertiary amine oxide.

$$\underset{3^\circ\text{amine}}{(CH_3)_3N} + [O] \xrightarrow{\text{Caro's acid}} (CH_3)_3NO$$

CHECK POINT 20.3

1. Amines are polar compounds because of the
(a) difference in electronegativity between N, C and H
(b) hydrogen bonding
(c) $—NH_2$ group
(d) alkyl group

2. Select the correct statement(s).
(a) Primary and secondary amines are hydrogen-bonding acceptors.
(b) All the three classes of amines are hydrogen-bonding donors.
(c) Tertiary amines cannot hydrogen-bond with other amine molecules because of the absence of a hydrogen bonded to nitrogen.
(d) All of the above are correct statements.

3. Consider the following hydrogen-bonded amines I and II

$$\underset{\text{I}}{H_2\overset{\delta^-}{\ddot{O}}\text{:}---H\overset{\delta^+}{N}R_2} \qquad \underset{\text{II}}{R_3\overset{\delta^-}{N}\text{:}---\overset{\delta^+}{H}—\ddot{O}—H}$$

Select the correct statement(s) about above
(a) In I, water is electron-donor while amine is electron acceptor.
(b) In II, amine is electron-donor while water is electron acceptor.
(c) Both are correct statements.
(d) None of the above statement is correct.

4. Which of the following is the general order of basicity for water and amines?
(a) alkanamines < arylamines < water
(b) alkanamines > arylamines > water
(c) arylamines > alkanamines > water
(d) arylamines > water > alkanamines

5. Which has the highest value of pK_b?

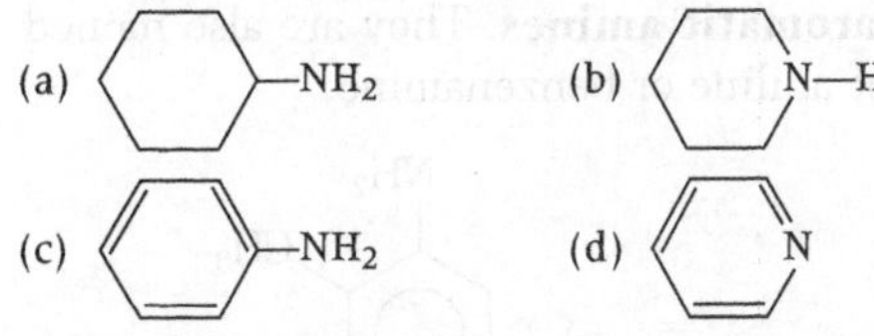

6. Which of the following chemical reaction produces isocyanide as a product?
(a) Hofmann reaction
(b) Gabriel phthalimide reaction
(c) Carbylamine reaction
(d) None of these

7. Which of the following amines will evolve nitrogen gas with nitrous acid?
(a) Primary aliphatic amine
(b) Secondary aliphatic amine
(c) Tertiary aliphatic amine
(d) Aromatic amine

8. Consider the following sequence of reaction.

$$\underset{A}{CH_3CONH_2} \xrightarrow{\text{I}} \underset{B}{CH_3NH_2} \xrightarrow{\text{II}} \underset{C}{CH_3NC}$$

Select the correct statement(s).
(a) Reaction I is called Hofmann bromamide reaction.
(b) Reaction II is called carbylamine reaction.
(c) A can be converted into isomeric C using. P_2O_5 and C can be converted into B by hydrolysis.
(d) All of the above are correct statements.

9. Aliphatic amines react with carbonyl compounds to form
(a) imine
(b) Schiff's base
(c) Enamine
(d) None of these

10. Oxidation of 2-aminopropane with $KMnO_4$ forms a compound that can
(a) give iodoform test
(b) be reduced to secondary alcohol
(c) form oximes
(d) give all the above properties

AROMATIC AMINES

Amines in which 'N' atom is attached to the benzene ring are called **aromatic amines**. They are also named as derivatives of aniline or benzenamine.

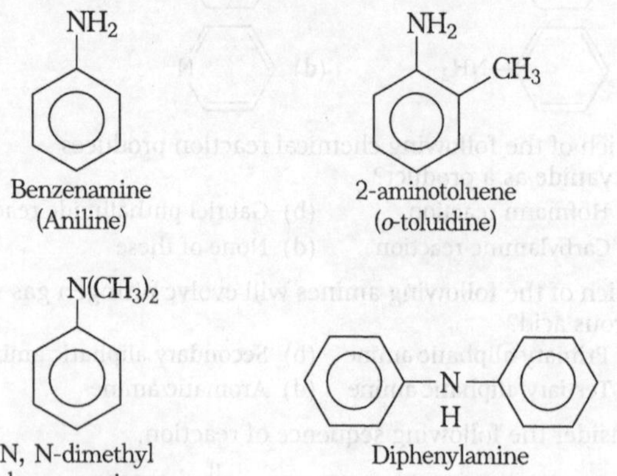

Nomenclature and isomerism

In common system of nomenclature, they are named as derivatives of aniline while in IUPAC system, the names are obtained by replacing the terminal 'e' from the name of the corresponding arene by the suffix 'amine'.

NH_2 $NHCH_3$ CH_3-N-CH_3

Common	Aniline	N-methylaniline	N,N-dimethylaniline
IUPAC	Benzenamine	N-methylbenzenamine	N,N-dimethylbenzenamine

Aromatic amines show position isomerism due to the presence of group at *o*, *m* and *p*-position on benzene ring.

2-methylaniline *o*-toluidines   3-methylaniline *m*-toluidines   4-methylaniline *p*-toluidines

Methods of Preparation

Reduction of nitro compounds

Nitro arenes can be reduced to amines by passing hydrogen gas in the presence of finally divided Ni, Pd or Pt or Sn and also by reduction with metals in acidic medium or by using reducing agents such as $LiAlH_4$. e.g.

$$\text{Nitrobenzene } (C_6H_5NO_2) \xrightarrow[\text{or } LiAlH_4]{H_2/Ni\ 250^\circ C} \text{Aniline } (C_6H_5NH_2)$$

Other products are formed depending on experimental condition that are symmarised below

	Medium	Reagent	Product
1.	Acid	Sn/HCl	I : aniline
2.	Neutral	Zn/NH_4Cl *aq.*	II : N-phenylhydroxylaime
3.	Alkaline	$Na_3AsO_3/NaOH$ $Zn/NaOH, CH_3OH$ $Zn/NaOH, C_2H_5OH$ (cold conc. HCl)	III : azoxybenzene IV : azobenzene V : hydroazobenzene VII : V changes to VII (benzidine) by intramolecular rearrangement
4.	Metallic hydride	$LiAlH_4$	I : aniline
5.	Catalytic	H_2/Pt or Raney Ni	I : aniline
6.	Electrolytic	Dil. H_2SO_4	II changes to VI (*p*-aminophenol) by intramolecular rearrangement

Structure of products I to VII

I: $C_6H_5NH_2$ II: C_6H_5NHOH III: $C_6H_5-\overset{O}{\overset{\uparrow}{N}}=N-C_6H_5$ IV: $C_6H_5-N=N-C_6H_5$

V: $C_6H_5-NH-NH-C_6H_5$ VI: $H_2N-C_6H_4-OH$ (*p*)   VII: $H_2N-C_6H_4-C_6H_4-NH_2$

Note $SnCl_2/HCl$ *converts III, IV and V into I.*

Ammonium sulphide $(NH_4)_2S$ and $SnCl_2/HCl$ can be used for the **selective reduction.**

It is used in those cases where only one $-NO_2$ group is needed to be reduced out of two or three $-NO_2$ groups.

$$\underset{p,m\text{-dinitro benzene}}{C_6H_4(NO_2)_2} + 3NH_4HS \longrightarrow \underset{m\text{-nitroaniline}}{H_2N-C_6H_4-NO_2} + 3NH_3 + 3S + 2H_2O$$

$$\text{m-nitrobenzaldehyde (CHO, } NO_2) \xrightarrow[\text{or } SnCl_2/HCl]{\text{Alkaline } Fe^{2+}} \text{m-aminebenzaldehyde (CHO, } NH_2)$$

m-nitrobenzaldehyde *m*-aminebenzaldehyde

$$\text{o-nitro-p-toluidine } (CH_3, NO_2, NH_2) \xleftarrow[\text{(reduces } NO_2 \text{ para to} -CH_3)]{NH_4HS} \text{2,4,-dinitrotoluene } (CH_3, NO_2, NO_2)$$

o-nitro-*p*-toluidine 2,4,-dinitrotoluene

$$\xrightarrow[\text{(reduces } NO_2 \text{ ortho to } -CH_3)]{SnCl_2/HCl} \text{p-nitro-o-toluidine } (CH_3, NH_2, NO_2)$$

p-nitro-*o*-toluidine

Ammonolysis of aryl chlorides

Aryl chlorides react with ammonia to form aromatic amines.

e.g.

$$C_6H_5Cl + 2NH_3 \xrightarrow[\text{Under pressure}]{CuCl_2/210°C} C_6H_5NH_2$$

Chlorobenzene Aniline

This method can be carried out conveniently under ordinary conditions if a strong electron withdrawing group (e.g. NO_2) is present at *ortho* or *para*-position to —Cl.

Hofmann's bromamide degradation reaction

Aromatic amides react with an aqueous solution of KOH and Br_2 to give aromatic amine.

$$Ar—\overset{O}{\overset{\|}{C}}—NH_2 \xrightarrow{Br_2 + KOH} Ar—NH_2 + CO_2\uparrow$$

This reaction involves intramolecular phenyl migration, the rate of which is increased by the presence of electron releasing substituents.

Reduction of azo compounds

Azo compounds can undergo hydrogenation in presence of catalyst such as Ni to give aromatic amine.

$$Ar—N{=}N—Ar \xrightarrow[Ni, \Delta]{H_2} Ar—\underset{H}{\underset{|}{N}}—\underset{H}{\underset{|}{N}}—Ar \xrightarrow[Ni, \Delta]{H_2} 2ArNH_2$$

Schmidt reaction

This method converts ArCOOH to $ArNH_2$ by using N_3H and conc. H_2SO_4.

$$C_6H_5COOH + N_3H \xrightarrow{Conc.\ H_2SO_4} C_6H_5NH_2 + CO_2\uparrow + N_2\uparrow$$

Benzoic acid Aniline

PHYSICAL AND CHEMICAL PROPERTIES

Physical properties

(*i*) They are colourless liquids or solids, but due to air oxidation, they turn brown.

$$C_6H_5NH_2 \xrightarrow{Air} \text{Benzoquinone}$$

Colourless Benzoquinone

(*ii*) They are slightly soluble or insoluble in water (polar solvent) having larger hydrocarbon part makes their ability low to form H-bonds with water. But highly soluble in benzene and other organic solvents.

(*iii*) They have high boiling point due to intermolecular H-bonding. N,N-dimethyl aniline has lower boiling point due to lack of H-bonding but molecular weight is also important affecting boiling point.

Para-substituted aniline, being the most symmetric, has the highest melting point.

Thus, *p*-toluidine is solid at room temperature while ***o*-and *m*-toluidines are liquids.**

(no intermolecular H-bonding possible)

(intermolecular H-bonding in aniline)

Chemical properties

Basic character and it important reactions of aromatic amines are describe below

Basic character of aromatic amines

Factors that increase the ability of nitrogen atom in an amine to share its electron pair, increase the basicity of amine and *vice-versa*. Generally, aromatic amines are less basic than ammonia and aliphatic amines. It is because; in arylamines, the —NH_2 group is attached directly to the benzene ring. The unshared electron pair on nitrogen atom is in conjugation with benzene ring and thus making it less available for protonation.

Also anilinium ion is less resonance stabilised than aniline.

(resonance stabilised aniline decreases basic character compared to aliphatic amines)

no other resonating structure possible

(less resonance stabilised than aniline)

Relative basic strength of aromatic amines

- Diphenyl amine is less basic than aniline since lone-pair on N-atom is now used by two benzene rings.
- However, pyridine is more basic than aniline. Pyridine does not use lone-pair of N-atom in delocalisation.

Pyridine

- Pyrrole uses lone-pair on N-atom in delocalisation, hence, less electrons are available for protonation. Hence, it is less basic than pyridine.

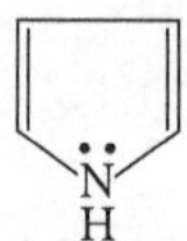

Effect of substituent group on basicity of aromatic amines

- Electron-releasing groups like (–NH_2, –OCH_3, –CH_3) increase electron density at N-atom hence, such substituents increase basic nature of aromatic amines in benzene nucleus.
- Electron-withdrawing groups like (–NO_2, –SO_3H, –COOH, –X) decrease electron density at N-atom thus, decreasing basic nature of aromatic amines.

  When an alkyl group (electron-releasing) is present at **N-atom** of aniline, the basicity of the aniline is increased markedly.
- A carbonyl group attached adjacent to nitrogen stabilises its negative charge. The strongest base is the anion that has no carbonyl groups on nitrogen; the weakest base is the phthalimide anion which has two carbonyl groups.

Strongest base > > Weakest base

Table 20.2 Basicity constants and pK_b of aromatic amines

Base	K_b			pK_b		
C_6H_5—NH_2		4.0×10^{-4}			3.40	
Aniline		4.2×10^{-10}			9.38	
N-methylaniline		7.08×10^{-10}			9.15	
N,N-dimethylaniline		11.50×10^{-10}			8.94	
N-ethylaniline		12.88×10^{-10}			8.89	
N,N-diethylaniline		363×10^{-10}			7.44	
Diphenylamine		0.0008×10^{-10}			13.10	
N-acetylaniline		0.00025×10^{-10}			13.60	
Benzylamine		0.2×10^{-4}			4.70	
	o	***m***	***p***	***o***	***m***	***p***
Toluidine	2.45×10^{-10}	4.9×10^{-10}	12.0×10^{-10}	9.61	9.31	8.92
Nitroaniline	5.12×10^{-15}	3.16×10^{-12}	1.04×10^{-13}	14.29	11.50	12.98
Chloroaniline	4.36×10^{-12}	2.19×10^{-11}	1.50×10^{-10}	11.36	10.66	9.82
Methoxyaniline	3.09×10^{-10}	1.58×10^{-10}	15×10^{-10}	9.51	9.80	8.82
Phenylenediamine	2.95×10^{-10}	7.59×10^{-10}	1.1×10^{-8}	9.53	9.12	7.96

Alkylation

Aromatic amines react with alkyl halide to give respective secondary and tertiary amines. e.g.

$$C_6H_5{-}NH_2 + CH_3I \xrightarrow{OH^-} \underset{(2^\circ)}{C_6H_5{-}NHCH_3} \xrightarrow[CH_3I]{OH^-}$$

$$\underset{(3^\circ)}{C_6H_5{-}N(CH_3){-}CH_3} \xrightarrow{CH_3I} \underset{\text{Quaternary ammonium salts}}{\left[C_6H_5{-}\overset{\oplus}{N}(CH_3)_2{-}CH_3\right] I^{\ominus}}$$

The process of conversion of 1° amines into quaternary ammonium salt is called Hofmann exhaustive alkylation. Quaternary ammonium salt is not affected by hydroxide ion because ammonium ion has no proton to give.

Acylation

Aromatic amine reacts with acetyl chloride in presence of pyridine to form N-phenyl acetamide

e.g.
$$\underset{\text{Aniline}}{C_6H_5{-}NH_2} + \underset{\text{Acetyl chloride}}{CH_3COCl} \longrightarrow \underset{\text{N-phenylacetamide (acetanilide)}}{C_6H_5{-}NHCOCH_3}$$

Note *Acetyl group deactivates —NH_2 group, hence, whenever aniline is subjected to nitration, acetylation is necessary to protect the ring from being oxidised.*

Benzoylation

Aromatic amine reacts with benzoylchloride in presence of *aq.* NaOH to give N-phenyl benzamide.

e.g;
$$C_6H_5{-}NH_2 + \underset{\text{Acetyl chloride}}{C_6H_5COCl} \xrightarrow{aq.\ NaOH} \underset{\text{Benzanilide or N-phenyl benzamide}}{C_6H_5{-}NH{-}COC_6H_5}$$

Sulphonylation

Aromatic amine reacts with benzene sulphonyl chloride to give N-phenyl benzene sulphonamide.

e.g.
$$C_6H_5{-}NH_2 + \underset{\text{Benzene sulphonyl chloride (Hinsberg reagent)}}{C_6H_5{-}SO_2Cl} \longrightarrow \underset{\text{N-phenyl benzene sulphonamide (soluble in alkali)}}{C_6H_5{-}NHSO_2{-}C_6H_5}$$

2° amine also reacts with this reagent but the product formed is insoluble in alkali. This provides a method (called Hinsberg method) for the separation of mixture of 1°, 2° and 3° amines.

Carbylamine reaction

Only 1° amines give this reaction when unpleasant smell of isocyanide is obtained on heating a mixture of alcoholic KOH, $CHCl_3$ and 1° amine.

e.g. $C_6H_5-NH_2 + CHCl_3 + 3KOH \xrightarrow[\text{(Alcoholic)}]{\Delta} C_6H_5-NC + 3KCl + 3H_2O$

Phenylisocyanide

Formation of Schiff base

When aryl amines react with aromatic aldehyde it results in the formation of Schiff base.
These Schiff bases are easily hydrolysed to the free amines and so their formation offers a mean of 'protecting' an amino group (during nitration).

e.g. $C_6H_5-NH_2 + OHC-C_6H_5 \xrightarrow{-H_2O} C_6H_5-N{=}HC-C_6H_5 \longrightarrow C_6H_5-NH_2$

Schiff base — Aniline

Oxidation

Aromatic amines are readily oxidised in presence of $K_2Cr_2O_7 / H_2SO_4$ and Caro's acid

e.g. $C_6H_5-NH_2 \xrightarrow{K_2Cr_2O_7/H_2SO_4} O{=}C_6H_4{=}O$

p-benzoquinone

$p\text{-}NH_2C_6H_4NO_2 \xrightarrow[H_2SO_5]{\text{Caro's acid}} p\text{-}ONC_6H_4NO_2 \xrightarrow{\text{Dil. }HNO_3} p\text{-}O_2NC_6H_4NO_2$

p-nitroaniline — *p*-nitrosonitrobenzene — *p*-dinitrobenzene

(*i*) **Primary aromatic amines** react with HNO_2 under ice cold conditions to form benzene diazonium chloride. The reaction is called **diazotisation reaction.**

e.g. $C_6H_5-NH_2 + HCl \xrightarrow{273\text{ K}} C_6H_5-NH_3^+Cl^- \xrightarrow[-2H_2O]{HNO_2\text{ (ice cold)}} C_6H_5-\overset{+}{N_2}Cl^-$

Benzene diazonium chloride

(*ii*) **Secondary aromatic amine** form a yellow oily layer of nitroso derivatives which is stable at room temperature. However, on reaction with HCl in ether and alcohol –NO group migrates to *para*-position (Fischer-Hepp rearrangement).

$C_6H_5-NH(CH_3) + HNO_2 \longrightarrow C_6H_5-N(CH_3)-NO$

N-nitroso-N-methylaniline (yellow oily liquid)

$\xrightarrow{HCl,\text{ ether}/C_2H_5OH} ON-C_6H_4-NHCH_3$

p-nitroso-N-methyl aniline

(iii) **Tertiary aromatic amines** form *p*-nitroso derivatives

e.g. $C_6H_5N(CH_3)_2 + HNO_2 \xrightarrow{HCl} p\text{-}ON-C_6H_4-N(CH_3)_2 + H_2O$

(Strongly *p*-activating) — *p*-nitroso-N,N-dimethylaniline

Hydrolysis of *p*-nitroso derivative gives 2° amines.

$(H_3C)_2N-C_6H_4-NO + H_2O \xrightarrow{NaOH} ON-C_6H_4-OH + (H_3C)_2NH$

p-nitrosophenol — 2° amine

This provides a path for the preparation of **2° amines.**

Hofmann mustard oil reaction

Primary aromatic amines like aniline when heated with ethanolic solution of CS_2 and solid KOH, gives N,N-diphenylthiourea which reacts with concentrated HCl to give phenyl isothiocyanate.

e.g. $2\,C_6H_5NH_2 + S{=}C{=}S \xrightarrow[\Delta,-K_2S,\;-2H_2O]{2\text{ KOH}(s)} C_6H_5-NH-\overset{S}{\overset{\|}{C}}-NHC_6H_5$

N,N-diphenylthiourea

$\xrightarrow{\text{Conc. HCl}} C_6H_5N{=}C{=}S + C_6H_5\overset{+}{N}H_3\overset{-}{Cl}$

Phenyl isothiocyanate — Aniline hydrochloride

Secondary aliphatic amines do not undergo in reaction.

Electrophilic substitution reactions

—NH_2 group is *o*- and *p*-directing and activating group. The ring undergoes electrophilic substitution easily even under mild conditions.

(a) **For *ortho* product**

(Resonance stabilised σ-complex)

(b) **For *para* product**

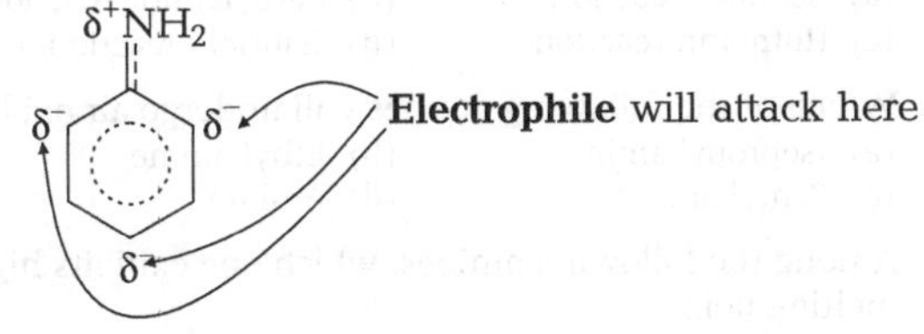

(i) Bromination

On bromination, in aqueous medium, aniline gives a *tri*-substituted product.

$$C_6H_5NH_2 + 3Br_2(aq) \longrightarrow C_6H_2Br_3NH_2 + 3HBr$$

(2, 4, 6-tribromoaniline)

—NH_2 group is greatly activating group hence, reaction takes place rapidly.

However, if ring is deactivated by acetylation then monohaloaniline derivative, i.e. *o*-and *p*-isomers are obtained.

Aniline $\xrightarrow[\text{Acetylation}]{CH_3COCl}$ Acetanilide $\xrightarrow{Br_2}$ *o*-bromo acetanilide (minor) + *p*-bromo acetanilide (major) $\xrightarrow[C_2H_5OH,\ \Delta]{KOH}$ *o*-bromo aniline (Steam volatile) + *p*-bromo aniline

(ii) Nitration

Aniline in the presence of strongly acidic medium gets converted into anilinium ion which being electron deficient, deactivates *o*/*p*-position and thus, a *meta*-nitroderivative is obtained.

$$C_6H_5NH_2 \xrightarrow{H^+} C_6H_5\overset{+}{N}H_3 \xrightarrow[-H_2O,\ -H^+]{HNO_3/H_2SO_4} m\text{-}NO_2C_6H_4NH_2$$

If —NH_2 group is protected by acetylation and then nitrated, *o*/*p*-isomers are formed. Among which *p*-isomer is the major product.

Aniline $\xrightarrow{CH_3COCl}$ Acetanilide $\xrightarrow[H_2SO_4]{Conc.\,HNO_3}$ *p*-nitroacetanilide (major) + *o*-nitroacetanilide (minor)

$\xrightarrow{KOH/C_2C_5OH}$ *o*-nitroaniline (minor) + *p*-nitroaniline (major)

(iii) Sulphonation

Aniline when treated with conc. H_2SO_4 forms anilinium hydrogen sulphate which on heating at 453-473 K gives sulphanilic acid.

$$C_6H_5NH_2 \xrightarrow{\text{Conc. } H_2SO_4,\ \Delta} C_6H_5\overset{+}{N}H_3HSO_4^- \xrightarrow[-H_2O]{453\text{-}473\text{ K }\Delta} C_6H_5NHSO_3H \xrightarrow{\text{Rearrangement}} p\text{-}H_2NC_6H_4SO_3H$$

p-amino benzene sulphonic acid (sulphanilic acid)

Like anthranilic acid, sulphanilic acid also exists as **Zwitter ion** due to internal neutralisation.

$$p\text{-}H_2NC_6H_4SO_3H \rightleftharpoons p\text{-}\overset{\oplus}{N}H_3C_6H_4SO_3^{\ominus}$$

Zwitter ion (dipolar ion)

CHECK POINT 20.4

1. The IUPAC nomenclature for the given compound is

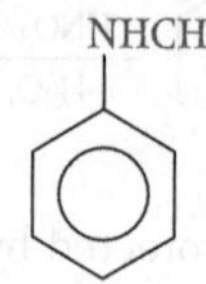

(a) N-methylaniline
(b) N-methylbenzenamine
(c) N-ethyl benzenamine
(d) None of the above

2. Which type of isomerism is shown by aromatic amines?
(a) Position isomerism
(b) Chain isomerism
(c) Geometrical isomerism
(d) Both (a) and (b)

3. Consider the following reaction,

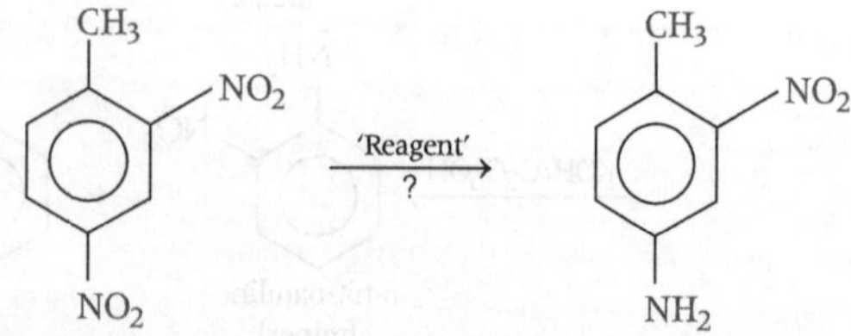

Which of the following reagent is used to convert the above reaction?
(a) NH_4HS (b) $SnCl_2/HCl$
(c) CS_2 (d) All of these

4. Benzaldehyde reacts with Ar_3NH_2 forming
(a) Shiff base (b) Saytzeff base
(c) Hofmann base (d) None of these

5. The method used to convert ArCOOH into $ArNH_2$ by using N_3H and conc. H_2SO_4 is
(a) Schmidt reaction (b) Carbylamine reaction
(c) Hofmann reaction (d) Gabrial's reaction

6. Which of the following amines will undergo air oxidation?
(a) Isopropyl amine (b) Ethyl amine
(c) Pentyl amine (d) Aniline

7. Among the following amines, which one exhibits highest melting point?
(a) *o*-toluidine (b) *p*-toluidine
(c) *m*-toluidine (d) None of these

8. Aniline can be converted to benzene by
(a) $\xrightarrow[273\text{ K}]{HNO_2} \xrightarrow{P_2O_5}$ (b) $\xrightarrow[273\text{ K}]{HNO_2} \xrightarrow{H_3PO_2,\ +\ H_2O}$
(c) $\xrightarrow[273\text{ K}]{HNO_2} \xrightarrow{H_3PO_4}$ (d) $\xrightarrow{H_3PO_2} \xrightarrow[273\text{ K}]{HNO_2}$

9. Consider the following reaction sequence.

$$C_6H_5\text{—}NHCH_3 \xrightarrow{HNO_2} \xrightarrow{HCl,\ ether/C_2H_5OH}$$

The end-product formed in above reaction is
(a) ON—C_6H_4—$NHCH_3$ (b) C_6H_5—N(NO)CH_3
(c) C_6H_4(NO)—$NHCH_3$ (d) H_3C—C_6H_4—NHNO

Distinction and separation of 1°, 2° and 3° amines

Tests for distinction between three types of amines

S.No.	Test	1° amine	2° amine	3° amine
1.	Carbylamine test Amine + $CHCl_3$ + alcoholic KOH $\xrightarrow{\Delta}$	Unpleasant smell of alkyl isocyanide is obtained $RNH_2 + CHCl_3 + 3KOH \xrightarrow{\Delta} RNC + 3KCl + H_2O$	2° amine does not give this reaction	3° amine does not give this reaction
2.	Reaction with HNO_2	N_2 is evolved $R{-}NH_2 + HNO_2 \longrightarrow ROH + N_2\uparrow + H_2O$	Oily nitrosamine is formed $R_2NH + HNO_2 \longrightarrow \underset{\text{(oily)}}{R_2NNO} + H_2O$	Only dissolves, forming $R_3NH^+\ NO_2^-$
3.	Hinsberg's test [with benzene sulphonyl chloride i.e. $C_6H_5SO_2Cl$]	N-alkyl benzene sulphonamide is formed. It is soluble in alkali. $C_6H_5{-}SO_2Cl + RNH_2 \longrightarrow C_6H_5{-}SO_2NHR + HCl$ N-alkyl benzene sulphonamide (soluble in NaOH)	N,N-dialkyl benzene sulphonamide is formed. It is insoluble in alkali $C_6H_5{-}SO_2Cl + R_2NH \longrightarrow C_6H_5{-}SO_2NR_2$ N,N-dialkyl benzene sulphonamide (insoluble in NaOH)	No reaction $R_3N + C_6H_5{-}SO_2Cl$ No reaction
4.	Hofmann mustard oil reaction (reaction with CS_2 followed by $HgCl_2$)	$RNH_2 + CS_2 \longrightarrow S{=}C(NHR)(SH)$ $\xrightarrow{HgCl_2} \underset{\text{Black ppt.}}{HgS} + RNCS + 2HCl$	$R_2NH + CS_2 \longrightarrow S{=}C(NR_2)(SH)$ $\downarrow HgCl_2$ No ppt.	$R_3N + CS_2 \longrightarrow$ No reaction.
5.	Oxidation by $KMnO_4$	Aldehydes or ketones are formed. $CH_3CH_2NH_2 \xrightarrow{[O]} CH_3CH{=}NH \xrightarrow{H^+} CH_3CHO$ $(H_3C)_2CHNH_2 \xrightarrow{[O]} (CH_3)_2C{=}NH \xrightarrow{H^+} (H_3C)_2C{=}O$	Tetra-alkyl hydrazine is formed $2R_2NH \xrightarrow{[O]} R_2N{-}NR_2$	No reaction
6.	Oxidation by Caro's acid	Aldoxime and hydroxamic acid are formed. $CH_3CH_2NH_2 \xrightarrow{O} CH_3CH_2NHOH \longrightarrow$ $\underset{\text{Aldoxime}}{CH_3CH{=}NOH} \longrightarrow \underset{\text{Hydroxamic acid}}{CH_3{-}C(OH){=}N{-}OH}$	Dialkyl hydroxyl amine is formed. $R_2NH \xrightarrow{[O]} R_2NOH$	Tertiary amine oxide is formed. $R_3N \xrightarrow{[O]} (R)_3NO$

Table 20.3 Distinction between aliphatic and aromatic amines

S. No.	Test	$C_6H_5-NH_2$	CH_3NH_2
1.	**Diazotisation** (reaction with HNO_2)	Diazonium salt is formed $C_6H_5-\overset{+}{N_2}\overset{-}{Cl}$ which forms azo dye with phenolic compounds.	$-NH_2$ is replaced by $-OH$ forming alcohol (CH_3OH)
2.	S_E	Can be halogenated, sulphonated, nitrated.	No reaction
3.	Basic nature	Less basic than CH_3NH_2, NH_3	More basic than NH_3 and aniline

Separation of 1°, 2° and 3° amines

There are various preparation methods in which mixture of 1°, 2° and 3° amines are obtained such as Hofmann's ammonolysis of alkyl halide and alcohol. Following are the methods which can be used to separate the mixture of amines into respective classes of amines.

Hofmann method

Mixture of amine is treated with diethyloxalate where primary amine forms a solid dialkyl oxamide, secondary amine forms a liquid oxamic ester while tertiary amines remain unreacted.

$$\underset{\text{Dimethyl oxalate}}{\begin{array}{l}CO-OC_2H_5\\ |\\ CO-OC_2H_5\end{array}} + \underset{1^\circ \text{ amine}}{2H-NHR} \longrightarrow \underset{\text{Oxamide solid}}{\begin{array}{l}CONHR\\ |\\ CONHR\end{array}} + \underset{\text{Ethyl alcohol}}{2C_2H_5OH}$$

$$\begin{array}{l}CO-OC_2H_5\\ |\\ CO-OC_2H_5\end{array} + \underset{2^\circ \text{ amine}}{H-NR_2} \longrightarrow \underset{\text{Oxamic ester (liquid)}}{\begin{array}{l}CONR_2\\ |\\ COOC_2H_5\end{array}} + \underset{\text{Ethyl alcohol}}{2C_2H_5OH}$$

The unreacted tertiary amine is distilled off and the remaining mixture is cooled. The solid oxamide is separated from the liquid oxamic ester by filtration. The two derivatives are then separately treated with strong alkali to regenerate the original amines.

$$\underset{\text{Oxamide solid}}{\begin{array}{l}CO-NHR\\ |\\ CONHR\end{array}} + \begin{array}{l}H\overset{-}{O}\overset{+}{K}\\ H\overset{-}{O}\overset{+}{K}\end{array} \longrightarrow \underset{\text{Oxalate}}{\begin{array}{l}CO\overset{-}{O}\overset{+}{K}\\ |\\ CO\overset{-}{O}\overset{+}{K}\end{array}} + \underset{\text{Primary amine}}{2RNH_2}$$

$$\underset{\text{Oxamic ester}}{\begin{array}{l}CO-NR_2\\ |\\ COOC_2H_5\end{array}} + HOK \xrightarrow{\text{Boil}} \underset{\text{Potassium oxalate}}{\begin{array}{l}CO\overset{-}{O}\overset{+}{K}\\ |\\ CO\overset{-}{O}\overset{+}{K}\end{array}} + \underset{\text{Ethyl alcohol}}{C_2H_2OH} + \underset{2^\circ \text{ amine}}{HNR_2}$$

Hinsberg method

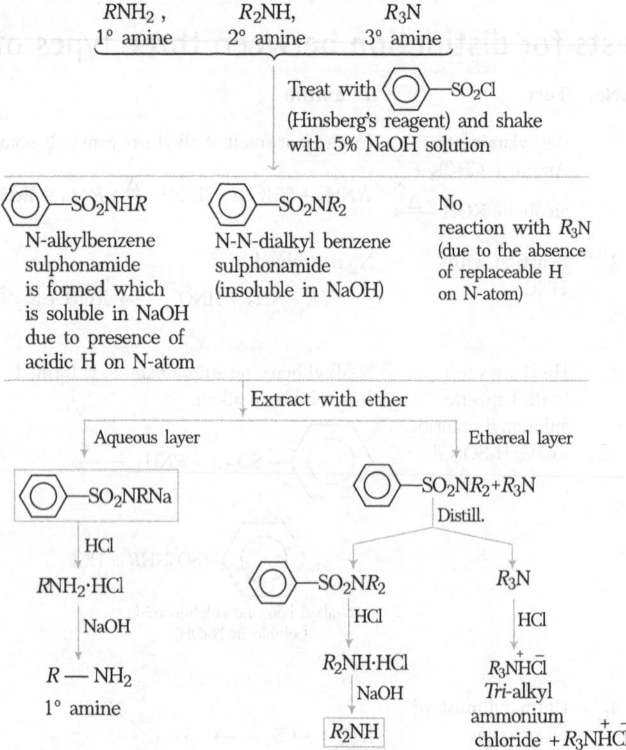

Conversions related to 'NH$_2$' group

(*i*) CH_3NH_2 (methyl amine) into $CH_3CH_2NH_2$ (ethyl amine)

$$\underset{\text{Methyl amine}}{CH_3NH_2} \xrightarrow{HNO_2} CH_3OH \xrightarrow{PCl_5} CH_3Cl \xrightarrow{KCN} CH_3CN \xrightarrow{LiAlH_4} \underset{\text{Ethyl amine}}{CH_3CH_2NH_2}$$

(*ii*) $CH_3CH_2NH_2$ (ethyl amine) into CH_3NH_2 (methyl amine)

$$\underset{\text{Ethyl amine}}{CH_2CH_2NH_2} \xrightarrow{HNO_2} \underset{\text{Ethanol}}{CH_3CH_2OH} \xrightarrow{KMnO_4}$$

$$\underset{\text{Ethanoic acid}}{CH_3COOH} \xrightarrow{NH_3/\Delta} \underset{\text{Ethanamide}}{CH_3CONH_2} \xrightarrow{\Delta/KOH/Br_2} \underset{\text{Methyl amine}}{CH_3NH_2}$$

$$CH_3COOH \xrightarrow[\text{Schmidt reaction, }\Delta]{N_3H + \text{Conc. } H_2SO_4} CH_3NH_2$$

(*iii*) CH_3COOH into CH_3NH_2 (one step)

$$\underset{\text{Acetic acid}}{CH_3COOH} \xrightarrow[\Delta]{N_3H, \text{ conc. } H_2SO_4} \underset{\text{Methyl amine}}{CH_3NH_2}$$

(Schmidt reaction)

(*iv*) Cyclopentanol into cyclopentylamine

Cyclopentanol $\xrightarrow[\text{Oxidation}]{\text{PCC}}$ Cyclopentanone $\xrightarrow{NH_3}$ Cyclopentanimine ($=NH$) $\xrightarrow{H_2,\ Pt}$ Cyclopentylamine ($-NH_2$)

(*v*) Butanamine into butan-2-amine

$$\underset{n\text{-butylamine}}{CH_3CH_2CH_2CH_2NH_2} \xrightarrow{HNO_2} \underset{\text{Butanol}}{CH_3CH_2CH_2CH_2OH} + \underset{\text{Predominant}}{CH_3CH_2CH(OH)CH_3}$$

$$\xrightarrow{Cu,\ \Delta} \underset{\text{Butanone}}{CH_3CH_2COCH_3} \xrightarrow[\text{(ii) } H_2,\ Pt]{\text{(i) } NH_3,\ \Delta} \underset{(sec\text{-butyl amine})}{CH_3CH_2CH(NH_2)CH_3}$$

(*vi*) Acetic acid into glycine

$$\underset{\text{Acetic acid}}{CH_3COOH} \xrightarrow[\Delta]{P+Br_2} CH_2(Br)COOH \xrightarrow{NH_3} CH_2(NH_2)COONH_4 \xrightarrow{H_3O^+} \underset{\text{Glycine}}{CH_2(NH_2)COOH}$$

(*vii*) Ethylcyanide into ethylamine

$$\underset{\text{Ethyl cyanide}}{CH_3CH_2CN} \xrightarrow{H_3O^+} \underset{\text{Propanoic acid}}{CH_3CH_2COOH} \xrightarrow{NH_3,\ \Delta} \underset{\text{Propanamide}}{CH_3CH_2CONH_2} \xrightarrow[\Delta]{KBrO} \underset{\text{Ethyl amine}}{CH_3CH_2NH_2}$$

(*viii*) Benzoic acid into aniline

$$\underset{\text{Benzoic acid}}{C_6H_5COOH} \xrightarrow[\Delta]{N_3H,\ \text{conc. } H_2SO_4} \underset{\text{Aniline}}{C_6H_5NH_2}$$

or

$$\underset{\text{Benzoic acid}}{C_6H_5COOH} \xrightarrow[\Delta]{NH_3} \underset{\text{Benzamide}}{C_6H_5CONH_2} \xrightarrow[\Delta]{KOH/Br_2} \underset{\text{Aniline}}{C_6H_5NH_2}$$

(*ix*) Aniline into benzene and deutero benzene

$$\underset{\text{Aniline}}{C_6H_5NH_2} \xrightarrow[273\ K]{NaNO_2/HCl} C_6H_5\overset{+}{N_2}\overset{-}{Cl} \xrightarrow[\text{reduction}]{H_3PO_2/H_2O} \underset{\text{Benzene}}{C_6H_6}$$

$$C_6H_5\overset{+}{N_2}\overset{-}{Cl} \xrightarrow{H_3PO_2/D_2O} \underset{\text{Deutero benzene}}{C_6H_5D}$$

(*x*) Aniline into N-methyl aniline

$$\underset{\text{Aniline}}{C_6H_5NH_2} \xrightarrow[\Delta]{CHCl_3/KOH} \underset{\text{Phenyl isocyanide}}{C_6H_5NC} \xrightarrow{\text{Reduction}} \underset{\text{N-methyl aniline}}{C_6H_5NHCH_3}$$

(*xi*) Aniline into benzoic acid

$$\underset{\text{Aniline}}{C_6H_5NH_2} \xrightarrow[273\ K]{NaNO_2/HCl} C_6H_5\overset{+}{N_2}\overset{-}{Cl} \xrightarrow{CuCN/HCN} C_6H_5CN \xrightarrow{H_3O^+} \underset{\text{Benzoic acid}}{C_6H_5COOH}$$

(*xii*) Aniline into *p*-nitroaniline

$$C_6H_5NH_2 \xrightarrow{CH_3COCl} \underset{\text{Acetanilide}}{C_6H_5NHCOCH_3} \xrightarrow{HNO_3/H_2SO_4} \underset{p\text{-nitroacetanilide}}{O_2N{-}C_6H_4{-}NHCOCH_3} \xrightarrow{H_3O^+} \underset{p\text{-nitroaniline}}{O_2N{-}C_6H_4{-}NH_2}$$

CHECK POINT 20.5

1. Which of the following amines will respond to carbylamine test?
   (a) N-methyl aniline
   (b) *p*- toluidine
   (c) Phenyl-*n*-butylamine
   (d) N,N-diethyl aniline

2. Consider the following reaction,

   $$A + HNO_2 \longrightarrow B \xrightarrow{H_3O^+} A;$$

   Here, '*A*' is
   (a) primary amine
   (b) secondary amine
   (c) tertiary amine
   (d) quaternary ammonia salt

3. $C_4H_{11}N$ on reaction with HNO_2 forms tertiary alcohol. Thus, $C_4H_{11}N$ is
   (a) primary amine
   (b) secondary amine
   (c) tertiary amine
   (d) quaternary ammonium salt

4. Which of the following is soluble in $NaNO_2$ and dil. HCl mixture forming salt?
   (a) $(CH_3)_3CNH_2$ (b) $(CH_3)_2CHNH_2$
   (c) $(CH_3)_2NH$ (d) $(CH_3)_3N$

5. Secondary amine is best detected by
   (a) Hofmann mustard oil reaction
   (b) Liebermann's nitroso reaction
   (c) Hinsberg's test
   (d) All of the above

6. Distinction between primary and secondary amines is made by
   (a) carbylamine reaction
   (b) Hofmann mustard oil reaction
   (c) Both (a) and (b)
   (d) None of the above

7. An organic compound '*X*' reacts with nitrous acid to form N-methyl-N-nitroso ethanamine. *X* can be obtained by reduction of
   (a) propanenitrile
   (b) acrylonitrile
   (c) methyl isocyanate
   (d) ethyl isocyanide

8. Select the incorrect statement.
   (a) Secondary amines are oxidised to tetra-alkyl hydrazine by $KMnO_4$
   (b) Caro's acid (H_2SO_5) oxidises secondary amine to nitrosamine
   (c) Nitrosamine is hydrolysed to secondary amine
   (d) Nitrosamine is oxidised to nitramine by CF_3CO_3H

9. Which of the following gives black precipitate on reaction with CS_2 followed by addition of $HgCl_2$?
   (a) $(CH_3)_3CNH_2$ (b) $(CH_3)_2NH$
   (c) $(CH_3)_3N$ (d) All of these

10. Which of the following reagents can provide distinction between aliphatic —NH_2 group and aromatic —NH_2 group?
   (a) Benzene diazonium chloride
   (b) Benzene sulphonyl chloride
   (c) $CHCl_3$ /KOH
   (d) CH_3COCl

Chapter Exercises

(A) Topical exercise

I. Classification, nomenclature, isomerism and preparation of aliphatic amines

1 C_3H_9N cannot represent the following
(a) primary amine
(b) secondary amine
(c) tertiary amine
(d) quaternary ammonium salt

2 Melting points are normally the highest for
(a) tertiary amides (b) secondary amides
(c) primary amides (d) quaternary amines

3 The IUPAC nomenclature for the following compound is

$$CH_3CH_2CH_2-\overset{\displaystyle CH_2CH_3}{\overset{|}{N}}-CH_3$$

(a) ethylmethyl propylamine
(b) N-ethyl-N-methyl-1-propanamine
(c) N-methyl-N-ethyl-1-propanamine
(d) None of the above

4 Which of the following category of amine will show metamerism?
(a) Secondary amine (b) Primary amine
(c) Tertiary amine (d) Quaternary amine

5 Match the species given in Column I with their degree (1°, 2°, 3°) given in Column II.

	Column I		Column II
A.	$(CH_3)_3C—NH_2$	1.	Tertiary
B.	$(CH_3)_2NH$	2.	Primary
C.	$(CH_3)_3N$	3.	Secondary
D.	$(CH_3)_4\overset{\oplus}{N}\overset{\ominus}{Cl}$	4.	Quaternary

Choose the correct match.

	A	B	C	D		A	B	C	D
(a)	2	3	1	4	(b)	3	2	4	1
(c)	4	3	2	1	(d)	1	4	2	3

6 Consider the following statements.
Statement I Amines are pyramidal in shape.
Statement II N-atom is sp^3-hybridised in amines.
(a) Both statements I and II are correct
(b) Only statement I is correct
(c) Only statement II is correct
(d) Neither statement I nor statement II is correct

7 Consider the following reaction,

cyclohexyl—Br + NH_3 ⟶

The product formed in above reaction is

(a) cyclohexyl—NH_2 (b) cyclohexyl—N—cyclohexyl
(c) NH_2Br (d) cyclohexene

8 Best method of preparing 1° amines from alkyl halide is by
(a) Hofmann bromamide reaction
(b) reaction with NH_3
(c) Gabriel phthalimide reaction
(d) Sandmeyer reaction

9 A primary amine is formed from an amide by the treatment with bromine and alkali. The primary amine has
(a) 1 carbon atom less than amide
(b) 1 carbon atom more than amide
(c) 1 hydrogen atom less than amide
(d) 1 hydrogen atom more than amide

10 An isocyanide on hydrolysis gives
(a) N-substituted amide
(b) 1°amine and formic acid
(c) carboxylic acid and ammonia
(d) an amide

11 Acid anhydrides on reaction with primary amines give .... **[NCERT Exemplar]**
(a) amide
(b) imide
(c) secondary amine
(d) imine

12 Which of the following is the best method for preparation of only 2° amine?
(a) $CH_3Cl + NH_3 \xrightarrow{3CH_3Cl}$
(b) $CH_3Cl \xrightarrow{KCN} \xrightarrow{Sn/HCl}$
(c) $CH_3NH_2 \xrightarrow[\Delta]{CHCl_3/KOH} \xrightarrow{Sn/HCl}$
(d) Both (b) and (c)

13 Consider the following facts about Curtius and Hofmann rearrangement.

I. are synthetic methods for preparation for primary amines.
II. involve intramolecular alkyl/aryl migration.
III. involve isocyanate as an intermediate.
IV. take place with complete retention of stereochemical configuration in the migrating alkyl group.

Select the correct facts from above and choose the correct option that are given below.
(a) I, II and III
(b) II, III and IV
(c) I, III and IV
(d) All of the above

14 Among the following which one does not act as an intermediate in Hofmann rearrangement?
(a) $RNCO$ (b) $R\overset{O}{\overset{\|}{C}}N$
(c) $RCONHBr$ (d) None of these

15 $C_6H_5CONHCH_3$ can be converted into $C_6H_5CH_2NHCH_3$ by
(a) $NaBH_4$ (b) H_2-Pd/C
(c) $LiAlH_4$ (d) Zn-Hg/HCl

16 Consider the following statements.

Statement I All the three classes of amines are hydrogen-bonding donors, while primary and secondary amines are hydrogen-bonding acceptors.

Statement II Low-molecular-weight amines are soluble in water.
(a) Both statements I and II are correct
(b) Only statement I is correct
(c) Only statement II is correct
(d) Neither statement I nor statement II is correct

17 Consider the following statements.

Statement I Hofmann's bromamide reaction is given by primary amines.

Statement II Primary amines are more basic than secondary amines. **[NCERT Exemplar]**
(a) Both statements I and II are correct
(b) Only statement I is correct
(c) Only statement II is correct
(d) Neither statement I nor statement II is correct

18 The source of nitrogen in Gabriel synthesis of amines is .... **[NCERT Exemplar]**
(a) sodium azide, NaN_3
(b) sodium nitrite, $NaNO_2$
(c) potassium cyanide, KCN
(d) potassium phthalimide, $C_6H_4(CO_2)N^-K^+$

19 CH_3NHCH_3 by reductive amination using CH_2O changes to $(CH_3)_3N$ through which of the following intermediate?
(a) Carbinolamine $CH_3 - \underset{CH_2OH}{\underset{|}{N}} - CH_3$
(b) Imine $CH_3N = CH_2$
(c) Formyldimethylamine $CH_3 - \underset{CHO}{\underset{|}{N}} - CH_3$
(d) None of the above

20 Select the correct statement(s) about the following reaction.

$$R - Br + NaN_3 \xrightarrow{\text{Step I}} RN_3 \xrightarrow{\text{Step II}} R - NH_2$$

(a) Step I is an S_N2 reaction
(b) R—Br should be 1° or 2° alkyl bromide
(c) Step II uses $NaBH_4$ / H_2O or $LiAlH_4$
(d) All of the above are correct statements

21 Select the correct statement.
(a) Methyl cyanide is reduced to ethylamine by $LiAlH_4$.
(b) Methyl isocyanide is hydrolysed to methylamine.
(c) Homologated 1° amine is obtained in the given process.

$$R-Br \xrightarrow{NaCN} \xrightarrow{LiAlH_4}$$

(d) Acetamide is reduced to ethyl amine by $LiAlH_4$.

22 Which is incorrect name for the given reactions?
(a) $RCONH_2 \xrightarrow{KBrO, \Delta} RNH_2$ (Hofmann bromamide)
(b) $RCOOH \xrightarrow{N_3H} RNH_2$ (Schmidt)
(c) $RCOCl \xrightarrow[H_3O^+]{NaN_3} RNH_2$ (Ritter)
(d) $RCl \xrightarrow[H_3O^+]{NH_3, \text{ phthalic acid}} RNH_2$ (Gabriel's phthalimide)

23 Consider the following reaction,

$$RCOOH \xrightarrow{N_3H, \Delta} \text{Product } P$$

Product P and name of reaction are
(a) RNH_2, Schmidt reaction
(b) $RCONH_2$, Schmidt reaction
(c) $RCONH_2$, Hofmann reaction
(d) NRH_2, Hofmann reaction

24 Ethylcyanide (A) can be converted to ethylamine (B) by
(a) $A \xrightarrow{Sn/HCl} B$ (b) $A \xrightarrow{H_3O^+} \xrightarrow[\Delta]{NH_3} \xrightarrow[\Delta]{KBrO} B$
(c) $A \xrightarrow{LiAlH_4} B$ (d) (a) and (c) are correct

25 The reduction of which of the following compounds would yield secondary amine?
(a) Alkyl nitrile
(b) Carbylamine
(c) Primary amine
(d) Secondary nitro compound

26 Select the correct statement (s) about the following reaction ,

$$[(CH_3)_3N^{\oplus}CH_2CH_3]OH^{\ominus} \xrightarrow{\Delta} (CH_3)_3N + CH_2{=}CH_2 + H_2O$$

(a) This reaction is called Hofmann elimination
(b) OH^- ion functions as a base in E2 elimination
(c) Less substituted alkene is the predominant alkene
(d) All of the above are correct statements

27 Best method for preparing primary amines from alkyl halides without changing the number of carbon atoms in the chain is **[NCERT Exemplar]**
(a) Hofmann bromamide reaction
(b) Gabriel phthalimide synthesis
(c) Sandmeyer reaction
(d) reaction with NH_3

28 In order to prepare a 1° amine from an alkyl halide with simultaneous addition of one CH_2 group in the carbon chain, the reagent used as source of nitrogen is .... **[NCERT Exemplar]**
(a) sodium amide, $NaNH_2$
(b) sodium azide, NaN_3
(c) potassium cyanide, KCN
(d) potassium phthalimide $C_6H_4(CO)_2N^-K^+$

29 Amongst the given set of reactants, the most appropriate for preparing 2° amine is .... **[NCERT Exemplar]**
(a) $2^\circ R$—Br + NH_3
(b) $2^\circ R$—Br + NaCN followed by H_2 / Pt
(c) $1^\circ R$—NH_2 + RCHO followed by H_2 / Pt
(d) $1^\circ R$—Br(2 mol) + potassium phthalimide followed by H_3O^+ / heat

30 The best reagent for converting 2-phenylpropanamide into 2-phenylpropanamine is .... **[NCERT Exemplar]**
(a) excess H_2
(b) Br_2 in aqueous NaOH
(c) iodine in the presence of red phosphorus
(d) $LiAlH_4$ in ether

31 The best reagent for converting, 2-phenylpropanamide into 1-phenylethanamine is .... **[NCERT Exemplar]**
(a) excess H_2/Pt (b) NaOH / Br_2
(c) $NaBH_4$/methanol (d) $LiAlH_4$/ether

II. Physical and chemical properties of alkyl amines

32 Dipole moment of

$CH_3CH_2CH_3$ (I), CH_3CH_2OH (II), $CH_3CH_2NH_2$ (III)

will be in order
(a) I < II < III (b) III < II < I
(c) I < III < II (d) II < III < I

33 Which of the following is most volatile?
(a) $CH_3CH_2CH_2NH_2$ (b) $(CH_3)_3N$
(c) CH_3CH_2(CH$_3$)NH (d) CH_3OH

34 Which of the following should be most volatile?
I. $CH_3CH_2CH_2NH_2$ II. $(CH_3)_3N$
III. CH_3CH_2(CH$_3$)NH IV. $CH_3CH_2CH_3$ **[NCERT Exemplar]**
(a) Only II (b) Only IV
(c) Only I (d) Only III

35 Consider the following statements.
Statement I Amines are polar compounds.
Statement II Amines possess electronegativity difference between nitrogen and carbon.
(a) Both statements I and II are correct
(b) Only statement I is correct
(c) Only statement II is correct
(d) Neither statement I nor statement II is correct

36 Amongst the following, the strongest base in aqueous medium is .... **[NCERT Exemplar]**
(a) CH_3NH_2 (b) $NCCH_2NH_2$
(c) $(CH_3)_2NH$ (d) $C_6H_5NHCH_3$

37 The correct decreasing order of basic strength of species H_2O, NH_3, OH^-, NH_2^- is .... . **[NCERT Exemplar]**
(a) $NH_2^- > OH^- > NH_3 > H_2O$
(b) $OH^- > NH_2^- > H_2O > NH_3$
(c) $NH_3 > H_2O > NH_2^- > OH^-$
(d) $H_2O > NH_3 > OH^- > NH_2^-$

38 Among the following the strongest base is
(a) 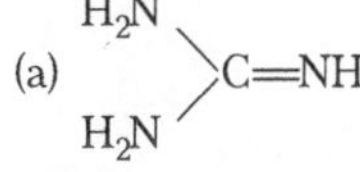 (b) 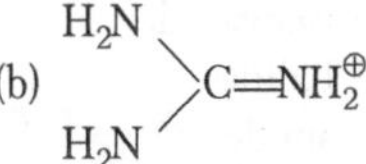
(c) 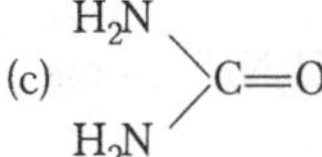 (d) $(H_2N)_2C^{\oplus}$—OH

39 Maximum basic in gas phase is
(a) NH_3 (b) $CH_3CH_2NH_2$
(c) $(CH_3CH_2)_2NH$ (d) $(CH_3CH_2)_3N$

40 Which of the following C_8H_9NO isomers is the weakest base?
(a) *o*-aminoacetophenone
(b) *m*-aminoacetophenone
(c) *p*-aminoacetophenone
(d) Acetanilide

41 Consider the following compound.

I (n-butylamine chain with NH_2) II (branched chain with NH_2) III (branched chain with NH_2) IV (tert-butyl with NH_2)

Correct order of the basic nature of above compounds
(a) I < II < III < IV (b) IV < II < III < I
(c) I < III < IV < II (d) II < IV < III < I

42 $RCH_2CH_2NH_2$ does not undergo E2 elimination. It is attributed to
(a) formation of unstable alkene
(b) $NH_2^{\ominus}$ is a very poor leaving group
(c) $NH_2^{\ominus}$ is a very good leaving group
(d) maximum basic nature

43 RNH_2 can be converted into R_2NH (and no other amine) by
(a) its reaction with RX
(b) its reaction with ($CHCl_3$ / KOH, Δ) followed by reduction
(c) its reaction with ($CHCl_3$ / KOH, Δ) followed by hydrolysis
(d) All of the above

44 Benzylamine may be alkylated as shown in the following equation? **[NCERT Exemplar]**

$$C_6H_5CH_2NH_2 + R - X \longrightarrow C_6H_5CH_2NHR$$

Which of the following alkyl halides is best suited for this reaction through S_N1 mechanism?
(a) CH_3Br (b) C_6H_5Br
(c) $C_6H_5CH_2Br$ (d) C_2H_5Br

45 The compound obtained by heating a mixture of a primary amine and chloroform with ethanolic potassium hydroxide (KOH) is
(a) an isocyanide
(b) an halide
(c) an amide
(d) an amide and nitro compound

46 Exhaustive methylation of $CH_3CH_2NH_2$ forms
(a) $(CH_3CH_2)_4N^{\oplus}OH^{\ominus}$
(b) $(CH_3CH_2)_2\overset{\oplus}{N}(CH_3)_2OH^{\ominus}$
(c) $(CH_3)_3N^{\oplus}(CH_2CH_3)OH^{\ominus}$
(d) None of the above

47 Consider the following statements.
Statement I Amines show basic and nucleophilic character.
Statement II Unshared pair of electrons are present on nitrogen.
(a) Both statements I and II are correct
(b) Only statement I is correct
(c) Only statement II is correct
(d) Neither statement I nor statement II is correct

48 Consider the following reaction,

$$CH_3CH_2CH_2NH_2 \xrightarrow[\text{(ii) } H_2O/H^+]{\text{(i) } HNO_2} A \text{ (mixture)}$$

Here, mixture can contain
(a) $CH_3CH_2CH_2OH$
(b) $CH_3\underset{\displaystyle OH}{\underset{|}{CH}}CH_3$
(c) cyclopropane (CH_2—CH_2, CH_2)
(d) All of the above

49 $\underset{(X)}{C_4H_{11}N} + HNO_2 \longrightarrow C_4H_{10}O$ (3° alcohol) hence, X will give
(a) carbylamine reaction
(b) Hofmann mustard oil reaction
(c) diazonium salt (as the intermediate) with HNO_2
(d) All of the above

50 Consider the following reaction equation ,

1-(aminomethyl)cyclohexan-1-ol (OH, CH_2NH_2) $\xrightarrow{HNO_2}$ A (major product);

Here, 'A' is
(a) cyclohexane with OH and CH_2OH (b) spiro-epoxide (O, CH_2)
(c) cyclohexane with OH and CH_3 (d) cycloheptanone (O)

51 Consider the following reaction ,

$$CH_3\overset{\displaystyle NH_2}{\overset{|}{C}H}\underset{\displaystyle CH_3}{\underset{|}{C}H}CH_3 + O{=}N{-}OH \xrightarrow{H_2O/\bar{O}H}$$

Major product formed in above reaction is

(a) $CH_3CH(N{=}NOH)CH(CH_3)CH_3$ (b) $CH_3C(OH)(CH_3)CH_2CH_3$

(c) $CH_3CH(OH)CH(CH_3)CH_3$ (d) $CH_3CH(CH_3)CH_2CH_3$

52 In the following sequence of reactions ,

$$CH_3CH(NH_2)CH_3 \xrightarrow{HNO_2} A \xrightarrow{\text{Oxidation}} B \xrightarrow[H^+/H_2O]{CH_3MgI} C$$

The compound C formed would be

(a) 1,1-dimethyl ethanol (b) 1-butanol
(c) 2-butanol (d) 2-methyl-propan-2-ol

53 Consider the following reaction sequence ,

$$C_2H_5NH_2 \xrightarrow{HNO_2} A \xrightarrow{PCl_5} B \xrightarrow{NH_3} C.$$

The compound 'C' is identified as

(a) acetamide (b) ethylamine
(c) ethylcyanide (d) methylamine

54 CH_3NH_2 reacts with α, β -unsaturated ketone as shown

$CH_3–NH_2$ + $CH_2{=}CH{-}CO{-}CH_3$ ⟶

Select the correct statements out of I, II and III

$CH_3{-}NH{-}CH_2CH_2{-}CO{-}CH_3$

I. Product is formed by conjugate addition.
II. It is called Michael reaction.
III. Intermediate is Zwitter ion which tautomerises after the proton transfer.

(a) Both I and II (b) Both I and III
(c) Both II and III (d) I, II and III

55 A primary amine on treatment with carbon disulphide and $HgCl_2$ produces black ppt and

(a) alkyl cyanide
(b) alkyl isothiocyanate
(c) alkyl cyanate
(d) carbylamine

56 When $(CH_3)_3CNH_3^{\oplus}Cl^{\ominus}$ is heated, products formed are

(a) $(CH_3)_3CCl, NH_4Cl$
(b) $(CH_3)_2C{=}CH_2, NH_4Cl$
(c) $(CH_3)_3CNH_2$, HC
(d) None of the above

57 Primary amine reacts with carbon disulphide in presence of excess of mercuric chloride and produce isothiocyanate. This reaction is known as

(a) Hofmann bromamide reaction
(b) Hofmann mustard oil reaction
(c) carbylamine reaction
(d) mercuration-demercuration reaction

58 Consider the following reaction,

$$CH_3{-}C(CH_3)_2{-}CH_2NH_2 \xrightarrow{HNO_2} A \text{ (major product)};$$

Here, 'A' is

(a) $CH_3{-}C(CH_3)_2{-}CH_2OH$ (b) $CH_3{-}C(CH_3)(OH){-}CH_2CH_3$

(c) $CH_3{-}C(CH_3)_2{-}CH_2N_2^{\oplus}$ (d) $CH_3{-}C(CH_3)_2{-}CH{=}CH_2$

59 Following 1° amine has chiral carbon as indicated

$H{-}\overset{*}{C}(CH_3)(C_2H_5){-}NH_2$. This on reaction with ($NaNO_2$ + HCl) forms

(a) 1° alcohol with retention of configuration
(b) 2° alcohol with inverted configuration
(c) racemic mixture of 2° alcohol
(d) racemic mixture of 1° alcohol

60 Thermal decomposition of

$$\left[CH_3CH_2{-}N(CH_3)_2{-}C(CH_3)_2\right]^{\oplus} OH^{\ominus} \text{ gives}$$

(a) $CH_3CH{=}CH_2$ (b) $CH_2{=}CH_2$
(c) $CH_3{-}CH_3$ (d) $CH_3CH_2CH_3$

61 Consider the following reaction,

$$CH_3CH(\overset{\oplus}{N}(CH_3)_3OH^{\ominus})CH_2CH_3 \xrightarrow{\Delta}$$

Major alkene formed in above reaction is

(a) $CH_3CH{=}CHCH_3$ (b) $CH_2{=}CHCH_2CH_3$
(c) $(CH_3)_2C{=}CH_2$ (d) $(CH_3)_2C{=}C(CH_3)_2$

III. Nomenclature, isomerism, preparation and properties of aromatic amines

62 In the nitration of benzene using a mixture of conc. H_2SO_4 and conc. HNO_3, the species which initiates the reaction is .... **[NCERT Exemplar]**

(a) NO_2 (b) NO^+
(c) NO_2^+ (d) NO_2^-

63 Which of the following reagents would not be a good choice for reducing an aryl nitro compound to an amine? **[NCERT Exemplar]**

(a) H_2(excess) / Pt (b) $LiAlH_4$ in ether
(c) Fe and HCl (d) Sn and HCl

64 Consider the following statements.

Statement I Chlorobenzene reacts with $NaNH_2$ forming aniline.

Statement II (C—Cl) bond is stabilised due to delocalisation of π-electrons by resonance.

(a) Both statements I and II are correct
(b) Only statement I is correct
(c) Only statement II is correct
(d) Neither statement I nor statement II is correct

65 Consider the following statements.

Statement I *p*-toluidine is solid at room temperature while *o*- and *m*- isomers are liquids.

Statement II Being symmetrical, *p*-isomers have strong intermolecular H-bonding giving solid state of high melting point.

(a) Both statements I and II are correct
(b) Only statement I is correct
(c) Only statement II is correct
(d) Neither statement I nor statement II is correct

66 Consider the following statements.

Statement I Acylation of amines give a monosubstituted product whereas alkylation of amines give a polysubstituted product.

Statement II Acyl group sterically hinders the approach of further acyl groups. **[NCERT Exemplar]**

(a) Both statements I and II are correct
(b) Only statement I is correct
(c) Only statement II is correct
(d) Neither statement I nor statement II is correct

67 Consider the following statements.

Statement I Aromatic 1° amines can be prepared by Gabriel phthalimide synthesis.

Statement II Aryl halides undergo nucleophilic substitution with anion formed by phthalimide. **[NCERT Exemplar]**

(a) Both statements I and II are correct
(b) Only statement I is correct
(c) Only statement II is correct
(d) Neither statement I nor statment II is correct

68 Which of the following is least basic?

(a) Aniline (b) Diethylamine
(c) *p*-sulphanilic acid (d) *o*-nitroaniline

69 Pyridine is less basic than triethylamine because

(a) pyridine has aromatic character
(b) nitrogen in pyridine is sp^2-hybridised
(c) pyridine is a cyclic system
(d) in pyridine, lone pair of nitrogen is delocalised

70 Which of the following is maximum basic in nature?

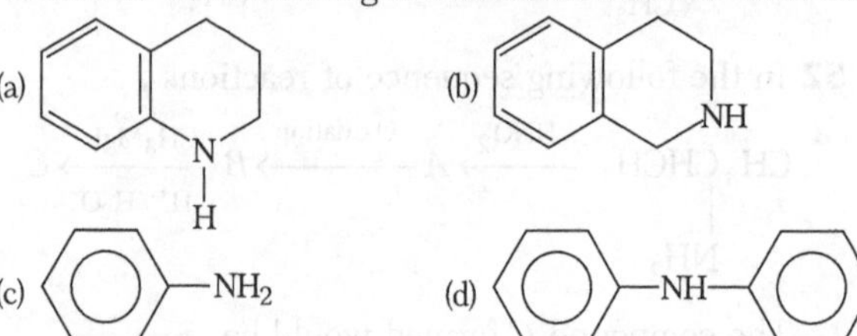

71 Which of the following compounds is the weakest Bronsted base? **[NCERT Exemplar]**

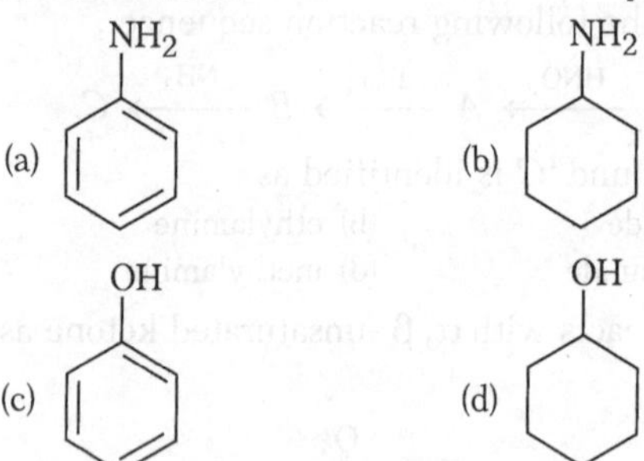

72 Consider the following statements.

Statement I Aromatic amines are generally less basic than alkyl amines.

Statement II The nitrogen lone pair electrons are delocalised by interaction with the aromatic ring π-electron system.

(a) Both statements I and II are correct
(b) Only statement I is correct
(c) Only statement II is correct
(d) Neither statement I nor statement II is correct

73 Consider the following statements.

Statement I Diphenyl amine is less basic than aniline.

Statement II Diphenyl amine is secondary amine.

(a) Both statements I and II are correct
(b) Only statement I is correct
(c) Only statement II is correct
(d) Neither statement I nor statement II is correct

74 Consider the following comparison of basic nature of amines

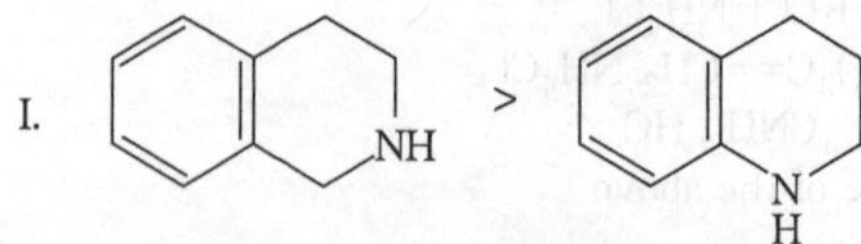

II. 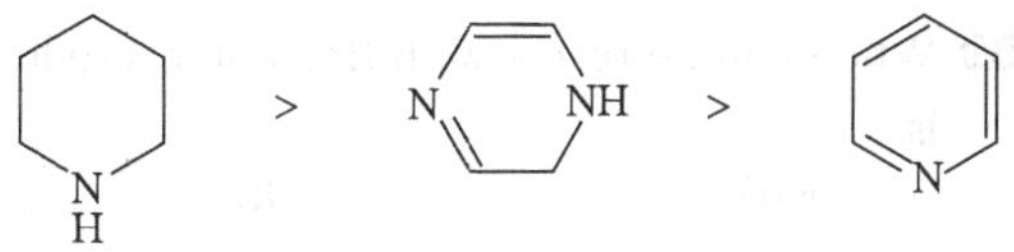 [structure] > [structure] > [structure]

III. 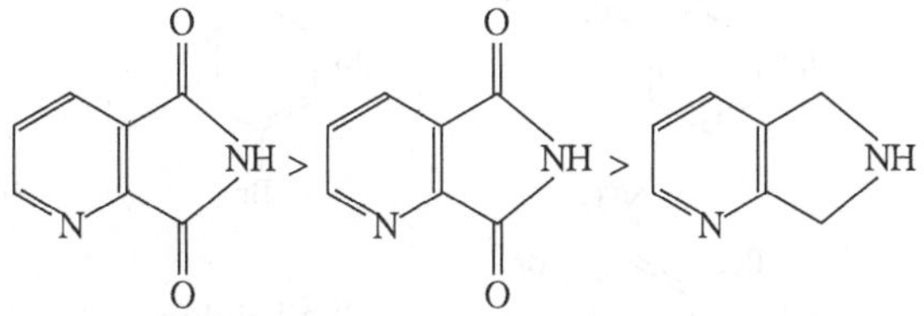 [structure] > [structure] > [structure]

Correct comparisons are

(a) I, II and III (b) Both I and II
(c) Both II and III (d) Both I and III

75 Consider the following statements.

Statement I Pyridine is less basic than aniline.

Statement II It does not use its lone pair of N-atom in delocalisation.

(a) Both statements I and II are correct
(b) Only statement I is correct
(c) Only statement II is correct
(d) Neither statement I nor statement II is correct

76 Identify the strongest and weakest bases among the following and choose the correct code that are given below of these compounds.

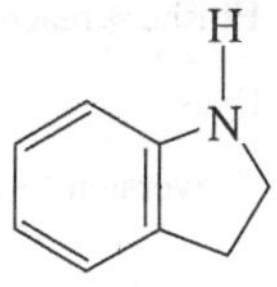

I 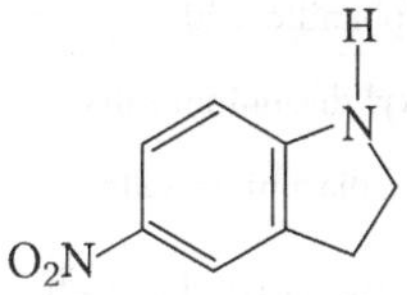 II 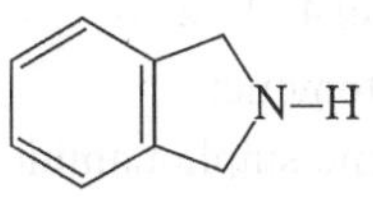 III 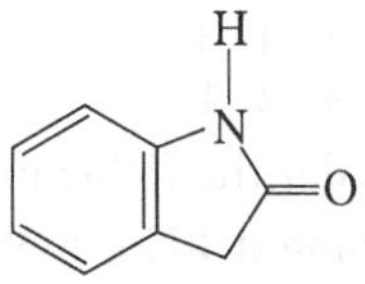 IV

	Strongest	Weakest
(a)	IV	I
(b)	IV	II
(c)	I	IV
(d)	III	IV

77 Among the following compounds, the increasing order of their basic strength is

(I) 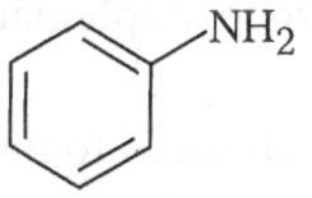 (II) 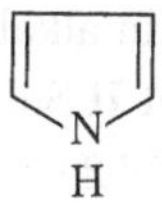 (III) 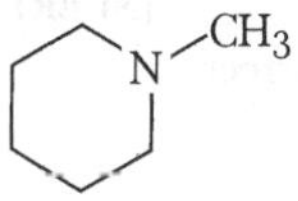 (IV)

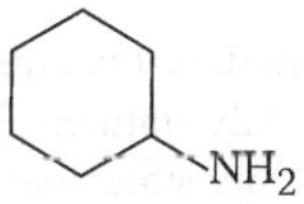

(a) (I) < (II) < (IV) < (III) (b) (I) < (II) < (III) < (IV)
(c) (II) < (I) < (IV) < (III) (d) (II) < (I) < (III) < (IV)

78 Consider the following statements.

Statement I Acetanilide is less basic aniline.

Statement II Acetylation of aniline results in decrease of electron density on nitrogen.

[NCERT Exemplar]

(a) Both statements I and II are correct
(b) Only statement I is correct
(c) Only statement II is correct
(d) Neither statement I nor statement II is correct

79 Aromatic primary amines react with cold HNO_2 to produce

(a) nitrobenzene (b) nitrosamine
(c) diazonium salt (d) benzene

80 Hofmann bromamide degradation reaction is shown by .... **[NCERT Exemplar]**

(a) $ArNH_2$ (b) $ArCONH_2$
(c) $ArNO_2$ (d) $ArCH_2NH_2$

81 Reduction of aromatic nitro compounds using Fe and HCl gives .... **[NCERT Exemplar]**

(a) aromatic oxime
(b) aromatic hydrocarbon
(c) aromatic primary amine
(d) aromatic amide

82 Consider the following statements.

Statement I Anilinium ion on reaction with Br_2 / Fe gives *m*-isomer.

Statement II $-\overset{\oplus}{N}H_3$ group is deactivating and electron-withdrawing group.

(a) Both statements I and II are correct
(b) Only statement I is correct
(c) Only statement II is correct
(d) Neither statment I nor statment II is correct

83 Thermal decomposition of

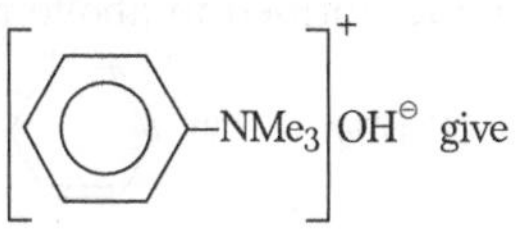

$$\left[C_6H_5-NMe_3\right]^+ OH^{\ominus} \text{ give}$$

(a) 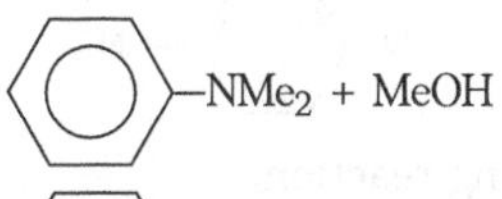 $C_6H_5-NMe_2$ + MeOH
(b) [ring]$=NMe_2$ + MeOH
(c) $C_6H_5-NMe_2 + CH_2$
(d) No reaction

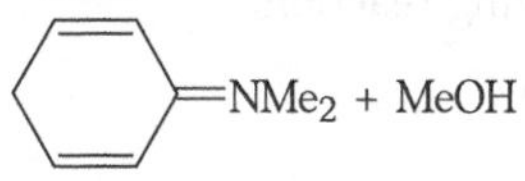

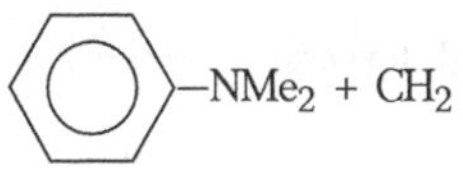

84 Consider the following statements.

Statement I

[Reaction scheme: 4-nitro-2-nitrotoluene (2,4-dinitrotoluene) with NH_4HS gives 2-nitro-4-aminotoluene (CH_3, NO_2, NH_2); with $SnCl_2/HCl$ gives 2-amino-4-nitrotoluene (CH_3, NH_2, NO_2)]

Statement II Action of NH_4HS and $SnCl_2$ / HCl as reducing agent is specific.

(a) Both statement I and II are correct
(b) Only statements I is correct
(c) Only statement II is correct
(d) Neither statement I nor statement II is correct

85 Consider the following statements.

Statement I Nitration of aniline gives *o*-, *p*-nitro aniline.

Statement II group is activating, *o*- and *p*- directing.

(a) Both statements I and II are correct
(b) Only statement I is correct
(c) Only statement II is correct
(d) Neither statement I nor statement II is correct

86 Aniline can be converted into N-methyl aniline by

(a) CH_3Cl
(b) CH_3Cl / $AlCl_3$
(c) $CHCl_3$/KOH, heat followed by reduction
(d) All of the above

87 Consider the following reaction,

$$C_6H_5-NH_2 \xrightarrow[273\text{ K}]{HNO_2} \xrightarrow{CuCN/HCN} \xrightarrow{H_3O^+}$$

The end-product formed in above reaction is

(a) $C_6H_5-CH_2NH_2$ (b) C_6H_5-COOH

(c) C_6H_5-OH (d) C_6H_5-CN

88 Consider the following reaction,

$$\text{C}_6\text{H}_{11}\text{NHCH}_2\text{CH}_3 + CH_3I \text{ (excess)} \longrightarrow ?$$

Final product formed in above reaction is

(a) a primary amine (b) a secondary amine
(c) a tertiary amine (d) a quaternary ammonium salt

89 When aniline reacts with Br_2 water, product formed is

(a) 2-bromoaniline (NH_2, Br) (b) 4-bromoaniline (NH_2, Br)

(c) 2,4,6-tribromoaniline (NH_2, Br, Br, Br) (d) All of these

90 —NH_2 in benzene nucleus is protected before any S_E reaction as —NH_2 group is

(a) greatly activating; *o*, *p*-directing
(b) deactivating; *o*, *p*-directing
(c) activating; *m*-directing
(d) deactivating ; *m*-directing

91 Match the compounds given in Column I with the items given in Column II. **[NCERT Exemplar]**

Column I	Column II
A. Benzene sulphonyl chloride	1. Zwitter ion
B. Sulphanilic acid	2. Hinsberg reagent
C. Alkyl diazonium salts	3. Dyes
D. Aryl diazonium salts	4. Conversion to alcohols

Choose the correct match.

	A	B	C	D		A	B	C	D
(a)	2	1	4	3	(b)	3	4	1	2
(c)	3	4	2	1	(d)	4	1	2	3

92 Consider the following statements

Statement I N-ethylbenzene sulphonamide is soluble in alkali.

Statement II Hydrogen attached to nitrogen in sulphonamide is strongly acidic. **[NCERT Exemplar]**

(a) Both statements I and II are correct
(b) Only statment I is correct
(c) Only statement II is correct
(d) Neither statement I nor statement II is correct

93 Consider the following statements.

Statement I N, N-diethylbenzene sulphonamide is insoluble in alkali.

Statement II Sulphonyl group attached to nitrogen atom is strong electron withdrawing group. **[NCERT Exemplar]**

(a) Both statements I and II are correct
(b) Only statment I is correct
(c) Only statement II is correct
(d) Neither statement I nor statement II is correct

94 Phenyl isocyanide test is used to identify
(a) aromatic primary amine
(b) aromatic secondary amine
(c) aromatic tertiary amine
(d) quaternary ammonium compound

95 Mixture of 1°, 2° and 3° amines can be separated by
(a) Hinsberg's method
(b) Hofmann's method
(c) distillation
(d) All of the above

(B) Miscellaneous questions

1 Strongest base is
(a) $CH_2{=}CHCH_2NH_2$ (b) $CH{\equiv}CCH_2NH_2$
(c) $CH_3CH_2CH_2NH_2$ (d) 

2 Secondary amine is formed by the reduction of
(a) methyl isocyanide (b) methyl cyanide
(c) nitrobenzene (d) nitrotoluene

3 Among the following the weakest base is
(a) $C_6H_5CH_2NH_2$ (b) $C_6H_5CH_2NHCH_3$
(c) $O_2N \cdot CH_2NH_2$ (d) CH_3NHCHO

4 Which of the following is the weakest Bronsted base? **[NCERT Exemplar]**
(a) 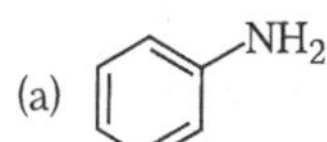 (b) 
(c) 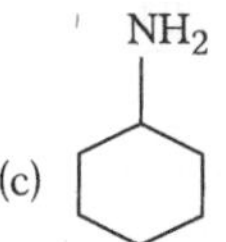 (d) CH_3NH_2

5 The most reactive amine towards dilute hydrochloric acid is .... **[NCERT Exemplar]**
(a) $CH_3—NH_2$ (b) $(CH_3)_2NH$
(c) $(CH_3)_3N$ (d) $C_6H_5NH_2$

6 Among the following amines, the strongest Bronsted base is .... **[NCERT Exemplar]**
(a) 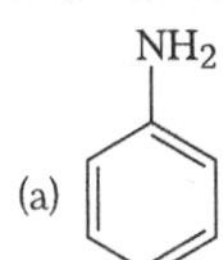 (b) NH_3 (c) 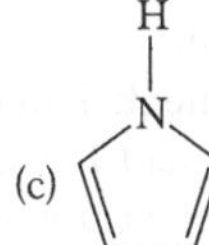 (d) pyrrolidine (N—H)

7 Consider the following compounds.

$$\underset{\text{I}}{CH_3CH_2NH_2}, \quad \underset{\text{II}}{CH_3\overset{O}{\overset{\|}{C}}NH_2}, \quad \underset{\text{III}}{Ph\overset{O}{\overset{\|}{C}}NH_2}$$

Basic nature of above compounds is in the order as
(a) I > II > III
(b) I > III > II
(c) II > III > I
(d) III > II > I

8 Which of the following is Hofmann mustard oil reaction?
(a) Reaction of primary amine with $CHCl_3$
(b) Reaction of primary amine with $CHCl_3$ + KOH
(c) Reaction of primary amine with CS_2 + $HgCl_2$
(d) Reaction of aromatic amine with iodoform

9 Consider the following statements.
Statement I Basicity of $CH_3CH_2NH_2$ (I), NH_3 (II) and $C_6H_5NH_2$ (III) is in order.

I > II > III

Statement II Electron-donating groups (such as alkyl group) increase the basicity of amines and electron-withdrawing groups (such as aryl group) decrease the basicity of amines.
(a) Both statements I and II are correct
(b) Only statement I is correct
(c) Only statement II is correct
(d) Neither statement I nor statement II is correct

10 Which of the following is not the correct reaction?
(a) Cyclohexyl—Cl + NH_3 ⟶ Cyclohexene + NH_4Cl
(b) $(CH_3)_3C$—Cl + $2NH_3$ ⟶ $(CH_3)_3C$—NH_2 + NH_4Cl
(c) $(CH_3)_3C$—Cl + NH_3 ⟶ $(CH_3)_2C{=}CH_2$ + NH_4Cl
(d) $(CH_3)_2CHNH_2$ + HNO_2 $\xrightarrow{273\text{ K}}$ $(CH_3)_2CH$—OH

11 Among the following maximum pK_b value is of
(a) C_6H_5—$NHCH_3$
(b) $(CH_3CH_2)_2NH$
(c) $(CH_3)_2NH$
(d) C_6H_5—NH—C_6H_5

12 Consider the following reaction equation,

(2-oxocyclohexyl)–$CH_2CH_2CH_2NH_2$ $\xrightarrow[\text{Raney Ni}]{H_2}$ A, (A) is

(a) (2-hydroxycyclohexyl)–$CH_2CH_2CH_2NH_2$
(b) octahydroquinoline (N–H)
(c) decahydroquinoline (N–H)
(d) (2-hydroxycyclohexenyl)–$CH_2CH_2CH_2NH_2$

13 Match the reactions given in Column I with the statements given in Column II. **[NCERT Exemplar]**

Column I	Column II
A. Ammonolysis	1. Amine with lesser number of carbon atoms
B. Gabriel phthalimide synthesis	2. Detection test for primary amines
C. Hofmann bromamide reaction	3. Reaction of phthalimide with KOH and R–X
D. Carbylamine reaction	4. Reaction of alkylhalides with NH_3

Choose the correct match.

	A	B	C	D
(a)	2	3	1	4
(b)	3	1	4	2
(c)	3	4	2	1
(d)	4	3	1	2

14 Consider the following statements.

Statement I Sulphanilic acid forms Zwitter ion.

Statement II It is amphiprotic compound.

(a) Both statements I and II are correct
(b) Only statement I is correct
(c) Only statement II is correct
(d) Neither statement I nor statement II is correct

15 Choose the one with the lowest boiling point.

(a) 4-methylpiperidine (N–H)
(b) cyclopentyl–$NHCH_3$
(c) N-methylpiperidine (CH_3–N)
(d) cyclohexyl–NH_2

16 Consider the following compounds .

1. aniline (NH_2)
2. benzylamine (CH_2NH_2)
3. o-nitroaniline (NH_2, NO_2)
4. benzamide ($\overset{O}{\overset{\|}{C}}NH_2$)

Rank the above compounds in order of increasing basicity (weakest → strongest).

(a) 4 < 2 < 1 < 3 (b) 4 < 1 < 3 < 2
(c) 4 < 3 < 1 < 2 (d) 2 < 1 < 3 < 4

17 Consider the following sequence of reactions ,

$(CH_3)_2CHCH_2N(CH_2CH_3)_2 \xrightarrow{CH_3I} \xrightarrow[H_2O]{Ag_2O} \xrightarrow{Heat} ?$

The major end product formed in above reaction are

(a) $(CH_3)_2CHCH_2NH_2 + H_2C{=}CH_2$
(b) $(CH_3)_2NCH_2CH_3 + H_2C{=}C(CH_3)_2$
(c) $(CH_3)_2CHCH_2\underset{}{\overset{CH_3}{\overset{|}{N}}}CH_2CH_3 + H_2C{=}CH_2$
(d) $(CH_3)_3\overset{+}{N}CH_2CH_3I^- + H_2C{=}CH_2$

18 Consider the compounds (I-IV).

I. aniline (NH_2)
II. 3,4-dihydroisoquinolin-1(2H)-one (NH, C=O)
III. CH_3NH_2
IV. $CH_3\overset{O}{\overset{\|}{C}}NH_2$

Choose the option which represents correctly the decreasing order of basicity of the compounds.

(a) II > I > IV > III (b) III > IV > I > II
(c) II > IV > I > III (d) III > I > IV > II

19 Consider the following statements.

Statement I Sulphanilic acid migrates to anode in basic medium and to cathode in acidic medium.

Statement II Sulphanilic acid is a dipolar ion and exists as cation in acidic medium and as an anion in basic medium.

(a) Both statements I and II are correct
(b) Only statement I is correct
(c) Only statement II is correct
(d) Neither statement I nor statement II is correct

20 Consider the following reaction sequence,

$$CH_3\overset{CH_3}{\underset{CH_3}{\overset{|}{\underset{|}{C}}}}CH_2CH_2OH \xrightarrow[H_2O,\ heat]{K_2Cr_2O_7 \cdot H_2SO_4} A \xrightarrow{SOCl_2}$$

$$B \xrightarrow[(2\ mol)]{(CH_3)_2NH} C \xrightarrow[(ii)\ H_2O]{(i)\ LiAlH_4,\ diethyl\ ether} D$$

Identify the product D formed in above reaction ,

(a) $CH_3\overset{CH_3}{\underset{CH_3}{\overset{|}{\underset{|}{C}}}}CH_2C\equiv N$ (b) $CH_3\overset{CH_3}{\underset{CH_3}{\overset{|}{\underset{|}{C}}}}CH_2\overset{N(CH_3)_2}{\overset{|}{C}}HN(CH_3)_2$

(c) $CH_3\overset{CH_3}{\underset{CH_3}{\overset{|}{\underset{|}{C}}}}H_2\overset{O}{\overset{||}{C}}N(CH_3)_2$ (d) $CH_3\overset{CH_3}{\underset{CH_3}{\overset{|}{\underset{|}{C}}}}CH_2CH_2N(CH_3)_2$

21 Consider the following compounds.

I. Methylamine
II. Chloroform/KOH
III. Phosphine
IV. Dimethylamine

Which of the above chemicals are used to manufacture methyl isocyanide that caused "Bhopal tragedy"?

(a) Both I and III
(b) Both III and IV
(c) Both I and II
(d) Both II and IV

22 Amides can be converted into amines by a reaction named after

(a) Perkin (b) Claisen
(c) Hofmann (d) Kekule

23 The major organic product formed from the reaction

$$\text{(3-methylbutan-2-one)} \xrightarrow[\substack{(ii)\ LiAlH_4 \\ (iii)\ H_2O}]{(i)\ CH_3NH_2}$$

is

(a) OH, CH_3 (b) $NHCH_3$

(c) O–$NHCH_3$, OH (d) $NHCH_3$, OH

24 Match the reaction name given in Column I with the reacting substances given in Column II.

Column I	Column II
A. Schmidt reaction	1. $RCONH_2 \xrightarrow[\Delta]{KBrO} RNH_2$
B. Carbylamine reaction	2. $RNH_2 + CS_2 + HgCl_2 \rightarrow$ Black ppt.
C. Hofmann-bromamide degradation reaction	3. $RCOOH + N_3H \xrightarrow{\Delta} RNH_2$
D. Hofmann-mustard oil reaction	4. $RNH_2 + CHCl_3 + KOH \xrightarrow{\Delta} RNC$

Choose the correct match.

	A	B	C	D		A	B	C	D
(a)	2	3	1	4	(b)	3	4	1	2
(c)	2	1	4	2	(d)	4	1	2	3

25 The product formed by the reaction of acetamide with bromine in presence of NaOH is

(a) CH_3CN (b) CH_3CHO (c) CH_3CH_2OH (d) CH_3NH_2

26 Acetamide and ethylamine are distinguished by reacting with

(a) Br_2 water (b) acidic $KMnO_4$
(c) *aq.* NaOH and heat (d) *aq.* HCl and heat

27 Hofmann's elimination product of (*A*) is

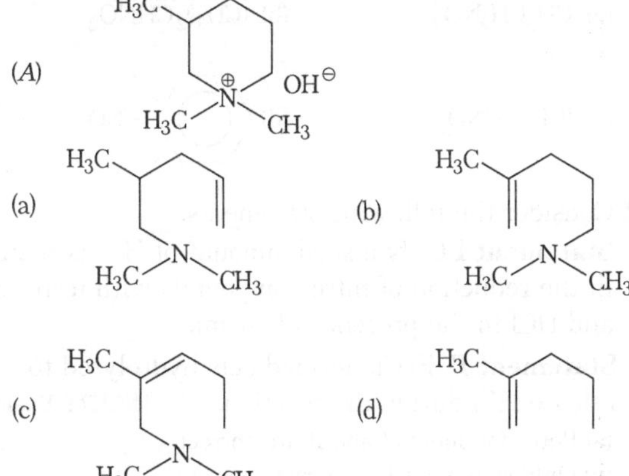

28 Consider the following reaction ,

1. $CH_3CH_2\underset{^+N(CH_3)_3\,^-OH}{\underset{|}{C}}HCH_3$ + Heat $\longrightarrow$ $CH_3\underset{X}{CH=CH}CH_3$ + $CH_3CH_2\underset{Y}{CH=CH_2}$

2. $CH_3CH_2\underset{CH_3}{\underset{|}{C}}HBr + CH_3CH_2\overset{-}{O}\overset{+}{N}a$

What will be the major product of each of the two reactions shown?

(a) 1*X*, 2*X* (b) 1*Y*, 2*X* (c) 1*X*, 2*Y* (d) 1*Y*, 2*Y*

29 Which of the following is the strongest acid?

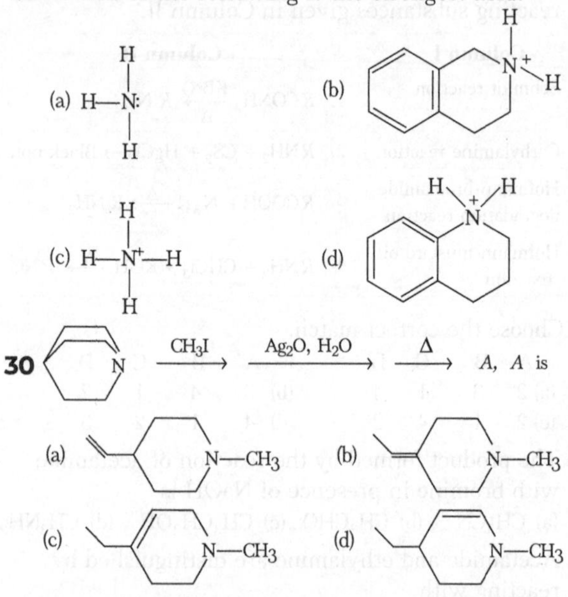

31 Which '*A*' gives red colour in the reactions ?

'*A*' $\xrightarrow[\text{(ii) NaOH}]{\text{(i) HNO}_2}$ red colour

(a) $CH_3CH_2NO_2$ (b) $(CH_3)_2CHNO_2$

(c) $(CH_3)_3CNO_2$ (d) $C_6H_5-NO_2$

32 Consider the following statements.

Statement I Only a small amount of HCl is required in the reduction of nitro compounds with iron scrap and HCl in the presence of steam.

Statement II $FeCl_2$ formed gets hydrolysed to release HCl during the reaction. **[NCERT Exemplar]**

(a) Both statements I and II are correct
(b) Only statement I is correct
(c) Only statement II is correct
(d) Neither statement I nor statement II is correct

(C) Medical entrances' gallery

1 Which of the following amine will give the carbylamine test? **[NEET 2020]**

(a) $C_6H_5NHCH_3$ (b) $C_6H_5N(CH_3)_2$

(c) $C_6H_5NHC_2H_5$ (d) $C_6H_5NH_2$

2 Reaction of propanamide with ethanolic sodium hydroxide and bromine will give **[NEET 2020]**

(a) ethylamine
(b) methylamine
(c) propylamine
(d) aniline

3 The correct order of the basic strength of methyl substituted amines in aqueous solution is **[NEET 2019]**

(a) $(CH_3)_3N > CH_3NH_2 > (CH_3)_2NH$
(b) $(CH_3)_3N > (CH_3)_2NH > CH_3NH_2$
(c) $CH_3NH_2 > (CH_3)_2NH > (CH_3)_3N$
(d) $(CH_3)_2NH > CH_3NH_2 > (CH_3)_3N$

4 The amine that reacts will Hinsberg's reagent to give an alkali insoluble product is **[NEET (Odisha) 2019]**

(a) $CH_3-CH(CH_3)-NH-CH(CH_3)CH_3$

(b) $CH_3-CH_2-N(CH_2CH_3)-CH_2CH_3$

(c) $CH_3-C(NH_2)(CH_3)-CH_2CH_2CH_3$

(d) $CH_3-C(CH_3)(CH_3)-CH(CH_3)-NH_2$

5

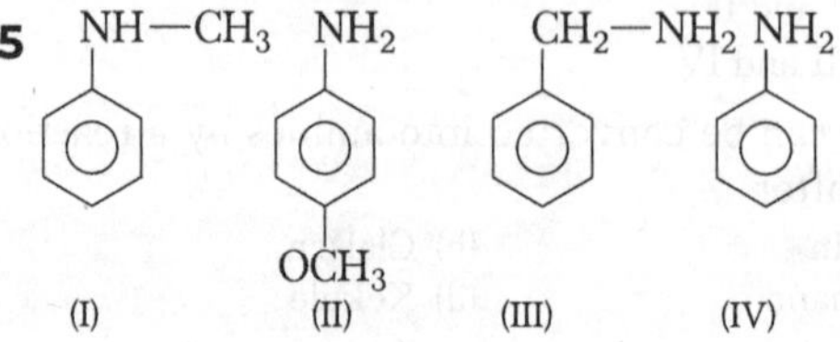

[AIIMS 2019]

Correct order of basic strength is

(a) III > II > IV > I
(b) IV > III > II > I
(c) III > II > I > IV
(d) III > I > II > IV

6 Nitration of aniline in strong acidic medium also gives *m*-nitroaniline because **[NEET 2018]**

(a) in absence of substituents nitro group always goes to *m*-position
(b) in electrophilic substitution reactions amino group is *meta* directive
(c) in spite of substituents, nitro group always goes to only *m*-position
(d) in acidic (strong) medium, aniline is present as anilinium ion

7 The correct increasing order of basic strength for the following compounds is

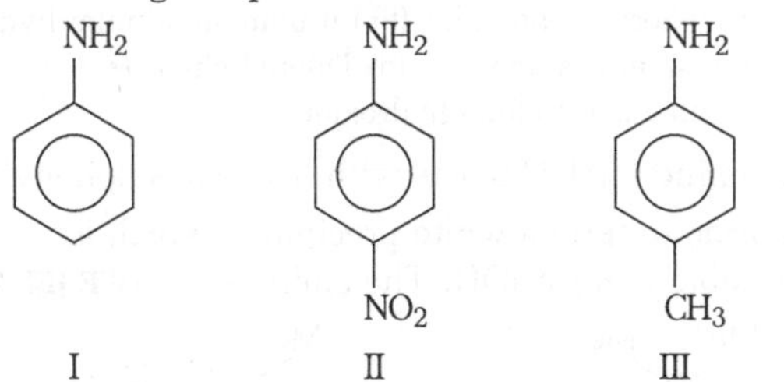

(a) II < I < III (b) II < III < I (c) III < I < II (d) III < II < I

8 Which of the following reactions is appropriate for converting acetamide to methanamine? **[NEET 2017]**

(a) Gabriel's phthalimide synthesis
(b) Carbylamine reaction
(c) Hofmann hypobromamide reaction
(d) Stephen's reaction

9 **Assertion** (A) Aniline on reaction with $NaNO_2$/HCl at 0° C followed by coupling with β-naphthol gives a dark blue coloured precipitate.

Reason (R) The colour of the compound formed in the reaction of aniline with $NaNO_2$/HCl at 0° C followed by coupling with β-naphthol is due to the extended conjugation. **[AIIMS 2017]**

(a) Both A and R are correct and R is the correct explanation of A
(b) Both A and R are correct but R is not the correct explanation of A
(c) A is correct but R is incorrect
(d) Both A and R are incorrect

10 The main product formed in the following reaction is

$$HO-C_6H_4-NH_2 + CH_3COCl$$

[JIPMER 2017]

(a) $HO-C_6H_4-NHCOCH_3$ (b) $CH_3COO-C_6H_4-NH_2$

(c) $C_6H_3(OH)(NH_2)(COCH_3)$ (OH and NH_2 para, $COCH_3$ ortho to NH_2) (d) $C_6H_3(OH)(COCH_3)(NH_2)$ (OH and NH_2 para, $COCH_3$ ortho to OH)

11 The correct order of basic strength of the following are **[JIPMER 2017]**

(I) $C_6H_5-NHCOCH_3$
(II) $O{=}C(-NHCH_3)-C_6H_4-OH$
(III) $O{=}C(-CH_2NH_2)-C_6H_5$
(IV) $C_6H_4(NH_2)(COCH_3)$ (meta)

(a) I > II > III > IV (b) IV > II > III > I
(c) III > IV > II > I (d) III > II > IV > I

12 Consider the following reaction, **[JIPMER 2017]**

$$X \xrightarrow{\text{Bromination}} Y \xrightarrow[\text{+ HCl}]{NaNO_2} Z$$

$$Z \xrightarrow[C_2H_5OH]{\text{Boiling}} \text{Tribromobenzene. } X \text{ is}$$

(a) benzoic acid (b) salicylic acid
(c) phenol (d) aniline

13 Which one of the following nitro-compounds does not react with nitrous acid? **[NEET (II) 2016]**

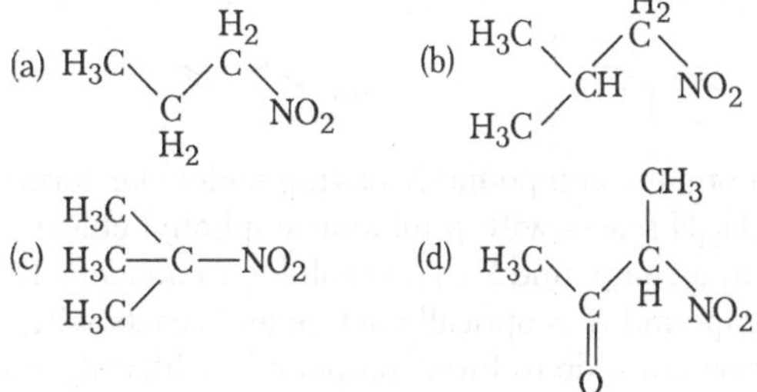

14. In pyrrole **[NEET (II) 2016]**

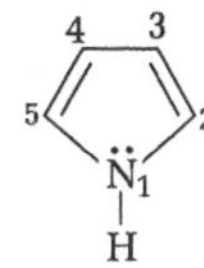

the electron density is maximum on

(a) 2 and 3 (b) 3 and 4
(c) 2 and 4 (d) 2 and 5

15 The product formed by the reaction of an aldehyde with a primary amine is **[NEET (I) 2016]**

(a) ketone (b) carboxylic acid
(c) aromatic acid (d) Schiff's base

16 The electrolytic reduction of nitrobenzene in strongly acidic medium produces **[CBSE AIPMT 2015]**

(a) *p*-aminophenol (b) azoxybenzene
(c) azobenzene (d) aniline

17 Which one of the following amines cannot be prepared by Gabriel phthalimide synthesis? **[Kerala (Medical) 2015]**

(a) Ethylamine
(b) Isopropylamine
(c) *n*-propylamine
(d) Ethylmethylamine
(e) Allylamine

18 Which one of the following amines form a non-acidic and alkali insoluble product with *p*-toluenesulphonyl chloride? **[Kerala (Medical) 2015]**

(a) Tertiary butylamine (b) *n*-butylamine
(c) Isobutylamine (d) Diethylamine
(e) N,N-dimethyl ethylamine

19 Which of the following amide will not undergo Hofmann bromamide reaction? **[KCET 2015]**

(a) CH_3CONH_2 (b) $CH_3CONHCH_3$
(c) $C_6H_5CONH_2$ (d) $CH_3CH_2CONH_2$

20 Identify the alkene that is produced in the following series of reactions. **[MHT CET 2014]**

$\overset{+}{N}(CH_3)_3\ I^-$ $\xrightarrow{\text{Moist}/Ag_2O} X \xrightarrow{\Delta}$ alkene + tert amine + H_2O

(a) (b)

(c) (d)

21 An organic compound X having molecular formula $C_6H_{11}N$ reacts with *p*-toluene sulphonyl chloride to form a compound Y that is soluble in aqueous KOH. Compound X is optically active and reacts with acetyl chloride to form compound Z. Identify the compound Z. **[MHT CET 2014]**

(a) $CH_3CH_2CH_2CH_2NHCOCH_3$
(b) $CH_3CH_2\underset{|}{C}HNHCOCH_3$ with CH_3 on the CH
(c) $CH_3—\underset{}{C}HCH_2NHCOCH_3$ with CH_3 on the CH
(d) $CH_3—C(CH_3)_2—NHCOCH_3$

22 The reagent that is used to distinguish between secondary amine and tertiary amine is **[Kerala CEE 2014]**

(a) *p*-toluenesulphonyl chloride
(b) Lucas reagent
(c) $CHCl_3$ and alc.KOH
(d) Borsch's reagent
(e) Bromine water

23 Which one of the following isomeric amines has the highest boiling point? **[Kerala CEE 2014]**

(a) $CH_3—CH_2—CH_2—NH—CH_3$
(b) $CH_3—CH_2—NH—CH_2—CH_3$
(c) $(CH_3)_2N—CH_2—CH_3$
(d) $CH_3—CH_2—CH_2—CH_2—NH_2$
(e) $(CH_3)_2CH—NH—CH_3$

24 Which one of the following reagents will convert acetamide to ethanamine? **[Kerala CEE 2014]**

(a) Phosphorus pentoxide (b) Lithium aluminium hydride
(c) Potassium cyanide (d) Thionyl chloride
(e) Bromine and sodium hydroxide

25 An amine C_3H_9N reacts with benzene sulphonyl chloride to form a white precipitate which is insoluble in *aq.* NaOH. The amine is **[WE JEE 2014]**

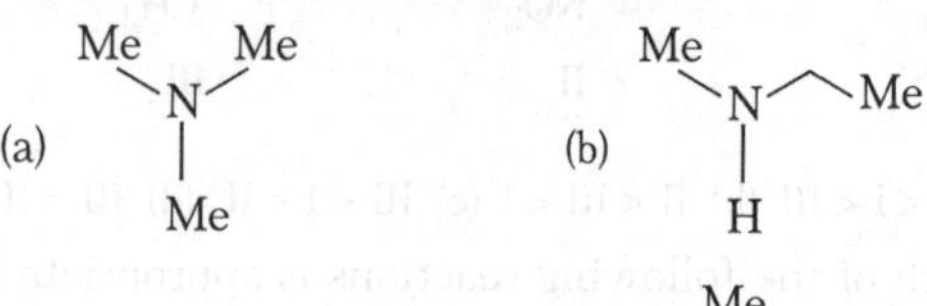

(c) Me$\diagup\!\diagdown$$NH_2$ (d) Me, Me NH_2

26 The reaction of aniline with chloroform under alkaline conditions leads to the formation of **[WB JEE 2014]**

(a) phenyl cyanide (b) phenyl isonitrile
(c) phenyl cyanate (d) phenyl isocyanate

27 The source of nitrogen in Gabriel synthesis of amines is **[UK PMT 2014]**

(a) NaN_3 (b) $NaNO_2$
(c) KCN (d) $C_6H_5(CO)_2N^-K^+$

28 Hofmann bromamide degradation is shown by **[UK PMT 2014]**

(a) RNH_2 (b) $RCONH_2$
(c) RNO_2 (d) RCH_2NH_2

29 An aromatic compound '*A*' (C_7H_9N) on reacting with $NaNO_2$ / HCl at 0° C forms benzyl alcohol and nitrogen gas. The number of isomers possible for the compound '*A*' is **[KCET 2014]**

(a) 5 (b) 7 (c) 3 (d) 6

30 The correct sequence of reaction to be performed to convert benzene into *m*-bromoaniline is **[KCET 2014]**

(a) nitration, reduction, bromination
(b) bromination, nitration, reduction
(c) nitration, bromination, reduction
(d) reduction, nitration, bromination

31 Identify the compounds from the following which form primary amines under suitable reduction conditions. **[EAMCET 2014]**

I. C_2H_5NC II. C_2H_6
III. $C_2H_5CONH_2$ IV. $C_6H_5NO_2$

(a) Both I and IV (b) Both III and IV
(c) I, III and IV (d) II, III and IV

32 Reduction of nitrobenzene with Sn/HCl produces **[AIIMS 2013]**

(a) azobenzene (b) azoxybenzene
(c) nitrobenzene (d) aniline

33. Nitrobenzene gives N-phenylhydroxylamine by [AIIMS 2013]
(a) Sn / HCl
(b) H_2 / Pd - C
(c) Zn/NaOH
(d) Zn / NH_4Cl

34 Nitrobenzene on reaction with conc. HNO_3 / H_2SO_4 at 80-100°C forms which one of the following products? [NEET 2013]
(a) 1, 2-dinitrobenzene
(b) 1, 3-dinitrobenzene
(c) 1, 4-dinitrobenzene
(d) 1, 2, 4-trinitrobenzene

35 The product formed by the reaction of acetamide with bromine in the presence of NaOH is [AIIMS 2013]
(a) CH_3CN
(b) CH_3CHO
(c) CH_3CH_2OH
(d) CH_3NH_2

36 $C_6H_5NH_2 \xrightarrow{NaNO_2/HCl^+} X \xrightarrow[H_2O/H^+]{Cu_2(CN)_2} Y \xrightarrow{H_2O/H^+} Z,$

Z is defined as [Manipal 2013]
(a) $C_6H_5—NH—CH_3$
(b) $C_6H_5—COOH$
(c) $C_6H_5—CH_2NH_2$
(d) $C_6H_5—CH_2—COOH$

37 $RCONH_2 + 4NaOH + Br_2 \longrightarrow RNH_2 + 2NaBr + Na_2CO_3 + 2H_2O$

Above reaction is said to be [WB JEE 2013]
(a) Hofmann bromamide reaction
(b) Schmidt reaction
(c) Curtius reaction
(d) Beckmann reaction

38 Nitration of nitrobenzene at 125° C with mixed acids gives [AFMC 2012]
(a) *meta*-dinitrobenzene
(b) *ortho*-dinitrobenzene
(c) *para*-dinitrobenzene
(d) 1, 3, 5-trinitrobenzene

39 An organic compound (C_3H_9N) (*A*), when treated with nitrous acid, gave an alcohol and N_2 gas was evolved. (*A*) on warming with $CHCl_3$ and caustic potash gave (*C*) which on reduction gave isopropylmethylamine. Predict the structure of (*A*). [CBSE AIPMT 2012]
(a) $(H_3C)_2CH—NH_2$
(b) $CH_3CH_2—NH—CH_3$
(c) $CH_3—N(CH_3)—CH_3$
(d) $CH_3CH_2CH_2—NH_2$

40 Reaction of aniline with acetyl chloride in the presence of NaOH gives [Manipal 2012]
(a) acetanilide
(b) *p*-chloroaniline
(c) a red dye
(d) aniline hydrochloride

41 Aniline + excess of $Br_2 \longrightarrow$? [OJEE 2012]
(a) tribromobenzene
(b) *m*-bromoaniline
(c) 3, 4, 5-tribromoaniline
(d) 2, 4, 6-tribromoaniline

42 The strongest base in aqueous solution among the following amines is [Kerala CEE 2011]
(a) N-N-diethylethanamine
(b) N-ethylethanamine
(c) N-methylmethanamine
(d) ethanamine
(e) phenylmethanamine

43 Aniline is treated with bromine water to give an organic compound '*X*' which when treated with $NaNO_2$ and HCl at 0°C gives a water soluble compound '*Y*'. Compound '*Y*' on treatment with Cu_2Cl_2 and HCl gives compound '*Z*'. Compound '*Z*' is [Kerala CEE 2011]
(a) *o*-bromochlorobenzene
(b) *p*-bromochlorobenzene
(c) 2,4,6-tribromophenol
(d) 2,4,6-tribromochlorobenzene
(e) 2,4-dibromophenol

44 Anilinium hydrogen sulphate on heating with sulphuric acid at 453-473 K produces [Kerala CEE 2011]
(a) benzene sulphonic acid
(b) anthranilic acid
(c) aniline
(d) *m*-aminobenzene sulphonic acid
(e) sulphanilic acid

45 Compare boiling point of isomeric alkyl amines. [Guj. CET 2011]
(a) $1° > 2° > 3°$
(b) $1° > 2° < 3°$
(c) $1° < 2° < 3°$
(d) $1° < 2° > 3°$

ANSWERS

Check point 20.1

1. (a)	2. (a)	3. (d)	4. (b)	5. (b)	6. (c)	7. (a)	8. (c)

Check point 20.2

1. (b)	2. (a)	3. (c)	4. (b)	5. (a)	6. (b)	7. (c)	8. (a)	9. (c)	10. (c)

Check point 20.3

1. (a)	2. (d)	3. (c)	4. (b)	5. (c)	6. (c)	7. (a)	8. (d)	9. (b)	10. (d)

Check point 20.4

1. (b)	2. (a)	3. (a)	4. (a)	5. (a)	6. (d)	7. (b)	8. (b)	9. (a)

Check point 20.5

1. (b)	2. (b)	3. (a)	4. (d)	5. (b)	6. (c)	7. (d)	8. (d)	9. (a)	10. (a)

(A) Topical exercise

1. (d)	2. (c)	3. (b)	4. (a)	5. (a)	6. (a)	7. (d)	8. (c)	9. (a)	10. (b)
11. (a)	12. (c)	13. (d)	14. (b)	15. (c)	16. (a)	17. (d)	18. (d)	19. (a)	20. (d)
21. (a)	22. (c)	23. (a)	24. (b)	25. (b)	26. (d)	27. (b)	28. (c)	29. (c)	30. (d)
31. (b)	32. (c)	33. (b)	34. (b)	35. (a)	36. (c)	37. (a)	38. (a)	39. (d)	40. (d)
41. (a)	42. (b)	43. (b)	44. (c)	45. (a)	46. (c)	47. (a)	48. (d)	49. (d)	50. (d)
51. (b)	52. (d)	53. (b)	54. (d)	55. (b)	56. (b)	57. (b)	58. (b)	59. (c)	60. (b)
61. (b)	62. (c)	63. (b)	64. (c)	65. (a)	66. (b)	67. (d)	68. (d)	69. (b)	70. (b)
71. (c)	72. (a)	73. (a)	74. (b)	75. (c)	76. (d)	77. (a)	78. (a)	79. (c)	80. (b)
81. (c)	82. (a)	83. (a)	84. (a)	85. (c)	86. (c)	87. (b)	88. (d)	89. (c)	90. (a)
91. (a)	92. (a)	93. (a)	94. (a)	95. (d)					

(B) Miscellaneous questions

1. (c)	2. (a)	3. (c)	4. (a)	5. (b)	6. (d)	7. (a)	8. (c)	9. (a)	10. (c)
11. (d)	12. (c)	13. (d)	14. (a)	15. (c)	16. (c)	17. (c)	18. (d)	19. (a)	20. (d)
21. (c)	22. (c)	23. (b)	24. (b)	25. (d)	26. (c)	27. (a)	28. (b)	29. (d)	30. (a)
31. (a)	32. (a)								

(C) Medical entrances' gallery

1. (d)	2. (a)	3. (d)	4. (a)	5. (c)	6. (d)	7. (a)	8. (c)	9. (a)	10. (a)
11. (d)	12. (d)	13. (c)	14. (d)	15. (d)	16. (a)	17. (d)	18. (d)	19. (b)	20. (b)
21. (b)	22. (a)	23. (d)	24. (b)	25. (b)	26. (b)	27. (d)	28. (b)	29. (a)	30. (c)
31. (b)	32. (d)	33. (d)	34. (b)	35. (d)	36. (b)	37. (a)	38. (a)	39. (a)	40. (a)
41. (d)	42. (b)	43. (d)	44. (e)	45. (a)					

Hints & Explanations

Check point 20.3

5 (*c*) $C_6H_5-NH_2$

is the weakest base, as lone pair of N-atom is used in delocalisation of π-electrons. It has minimum K_b, thus, maximum pK_b.

8 (*d*) $CH_3CONH_2 \xrightarrow[\text{(I) Hofmann bromamide reaction}]{KOH/Br_2} CH_3NH_2$

$CH_3CONH_2 \xrightarrow{P_2O_5} CH_3CN$; $CH_3NH_2 \xrightarrow[\text{Carbylamine reaction}]{CHCl_3 / KOH(II)} CH_3NC$

$CH_3CN \xleftrightarrow{\text{Isomers}} CH_3NC$

Thus, (a), (b) and (c) are true.

10 (*d*) $(H_3C)_2CHNH_2 \xrightarrow{KMnO_4} \underset{X}{(H_3C)_2C{=}O}$

(a) is true; $X \xrightarrow{\text{(a) } KOH/I_2\ \Delta} \underset{\text{Yellow}}{CHI_3\downarrow}$

(b) is true; $X \xrightarrow{\text{(b) } LiAlH_4} \underset{2^\circ\text{ alcohol}}{(CH_3)_2CHOH}$

(c) is also true; $X \xrightarrow{\text{(c) } NH_2OH} \underset{\text{Oxime}}{(CH_3)_2\ C{=}NOH}$

Check point 20.4

8 (*b*) $C_6H_5-NH_2 \xrightarrow[273\text{ K}]{HNO_2} C_6H_5-\overset{\oplus}{N}_2$

$(H_3PO_2 + H_2O \longrightarrow H_3PO_3 + 2H)$

$\xrightarrow[2\text{ H}]{H_3PO_2/H_2O} C_6H_6$

Check point 20.5

2 (*b*) *B* is hydrolysed to *A*

$$A + HNO_2 \longrightarrow B \xrightarrow{H_3O^+} A$$

Thus, *B* is nitrosoamine formed when *A* (2° amine) reacts with HNO_2.

$$\underset{(A)}{R_2NH} \xrightarrow{HNO_2} \underset{(B)}{R_2N{-}N{=}O} \xrightarrow{H_3O^+} \underset{(A)}{R_2NH}$$

3 (*a*) $C_4H_{11}N \xrightarrow{HNO_2} 3^\circ$ alcohol

Thus, $C_4H_{11}N$ has $-NH_2$ group. Thus, it is 1° amine. Since, 3° alcohol is produced, hence $C_4H_{11}N$ is 3° butyl amine.

$$\underset{1^\circ\text{ amine}}{(CH_3)_3\ C{-}NH_2} \xrightarrow{HNO_2} \underset{3^\circ\text{ alcohol}}{(CH_3)_3C{-}OH}$$

4 (*d*) $(CH_3)_3N \xrightarrow{HNO_2} \underset{\text{Soluble}}{(CH_3)_3\ NH^{\oplus}NO_2^{\ominus}}$

9 (*a*) 1° amine gives the test.

(A) Topical exercise

1 (*d*) $\underset{1^\circ}{CH_3CH_2CH_2NH_2}$, $\underset{2^\circ}{CH_3CH_2{-}NH{-}CH_3}$, $\underset{3^\circ}{(CH_3)_3N}$

2 (*c*) $R{-}\overset{O}{\overset{\|}{C}}{-}NH_2 \longleftrightarrow R{-}\overset{OH}{\overset{|}{C}}{=}NH$

Intermolecular H-bonding at O-atom and N-atom increases the boiling point.

7 (*d*) $C_6H_{11}{-}Br + NH_3 \longrightarrow C_6H_{10}$ (cyclohexene) $+ NH_4Br$

9 (*a*) $R{-}\overset{O}{\overset{\|}{C}}{-}NH_2 \xrightarrow[\Delta]{KOH/Br_2} \underset{\text{One carbon less}}{RNH_2}$

11 (*a*) Acid anhydride on reaction with primary amine produces amide as

$$R{-}CO{-}O{-}CO{-}R \xrightarrow{\ddot{R}NH_2} R{-}CO{-}O{-}C(O^{\ominus})(\overset{\oplus}{N}H_2R){-}R$$

$$\longrightarrow R{-}\overset{\oplus}{N}H_2{-}CO{-}R + R{-}COO^{\ominus} \longrightarrow \underset{\text{Amide}}{R{-}NH{-}CO{-}R} + RCOOH$$

12 (*c*) (a) $CH_3Cl \xrightarrow{NH_3} CH_3NH_2 \xrightarrow{CH_3Cl} (CH_3)_2NH$

$\xrightarrow{CH_3Cl} (CH_3)_3N \xrightarrow{CH_3Cl} (CH_3)_4N^{\oplus}Cl^{\ominus}$

Mixture is formed

(b) $CH_3Cl \xrightarrow{KCN} CH_3CN \xrightarrow{Sn/HCl} \underset{1^\circ\text{ amine}}{CH_3CH_2NH_2}$

(c) $CH_3NH_2 \xrightarrow[\Delta]{CHCl_3/KOH} CH_3NC \xrightarrow{Sn/HCl} \underset{2^\circ\text{ amine (pure)}}{CH_3NHCH_3}$

14 (*b*) $R\overset{O}{\overset{\|}{C}}NH_2 \longrightarrow R{-}\overset{O}{\overset{\|}{C}}{-}\overset{Br}{\overset{|}{N}}{-}H \longrightarrow R{-}N{=}C{=}O$

17 (*d*) Both statements are incorrect.

Corrected Statement I is Hofmann bromamide reaction is given by amide.

Corrected Statement II is amide on reaction with NaO*X* produces amine having one carbon less than amide.

18 (*d*) Source of nitrogen in Gabriel phthalimide synthesis is potassium phthalimide.

$$\text{Phthalimide} + KOH\ (alc.) \xrightarrow{-H_2O} \text{Potassium phthalimide } (C_6H_4(CO)_2\overset{-}{N}\overset{+}{K}) \xrightarrow[-KI]{RI,\ \Delta}$$

$$C_6H_4(CO)_2N{-}R \xrightarrow[HO^-/H_2O,\ \Delta]{H^+/H_2O,\ \Delta \text{ or}} \underset{\text{Phthalic acid}}{C_6H_4(COOH)_2} + \underset{1^\circ \text{ amine}}{RNH_2}$$

20 (*d*) $\underset{\substack{1^\circ \text{ or } 2^\circ \\ \text{Halide}}}{R{-}X} + NaN_3 \xrightarrow[\text{Step I}]{S_N2} R{-}N{=}\overset{\oplus}{N}{=}\overset{\ominus}{N} \xrightarrow[\text{Step II}]{NaBH_4/H_2O} R{-}NH_2$

21 (*a*) $\underset{\text{Methylcyanide}}{CH_3CN} \xrightarrow{LiAlH_4} \underset{\text{Ethylamine}}{CH_3CH_2NH_2}$

22 (*c*) Ritter reaction

$$(CH_3)_3C{-}OH + R{-}C{\equiv}N \xrightarrow[H_2O]{H_2SO_4} R{-}\overset{O}{\overset{\|}{C}}{-}NH{-}C(CH_3)_3$$

24 (*b*) $CH_3CH_2CN \longrightarrow \underset{\text{One carbon less}}{CH_3CH_2NH_2}$

$\underset{\text{Ethylcyanide } (A)}{CH_3CH_2CN} \xrightarrow{Sn/HCl} \underset{\text{Propyl amine}}{CH_3CH_2CH_2NH_2}$ (incorrect)

$\xrightarrow{H_3O^\oplus} CH_3CH_2\overset{O}{\overset{\|}{C}}OH \xrightarrow[\Delta]{NH_3} CH_3CH_2\overset{O}{\overset{\|}{C}}NH_2 \xrightarrow{\Delta\ \ KBrO} \underset{\text{Ethylamine } (B)}{CH_3CH_2NH_2}$

Thus, option (b) is true.

25 (*b*) (a) $CH_3CN \xrightarrow{[H]} \underset{1^\circ}{CH_3CH_2NH_2}$

(b) $CH_3NC \xrightarrow{[H]} \underset{2^\circ}{CH_3NHCH_3}$

26 (*d*) (a) is true

(b) is true

(c) is also true since (CH_2) cannot be possible, thus $CH_2 = CH_2$ is formed.

27 (*b*) Best method of preparing primary amines from alkyl halides without changing the number of carbon atoms in the chain is Gabriel phthalimide synthesis.

$$\text{Phthalimide (NH)} \xrightarrow{KOH\ (alc.)} C_6H_4(CO)_2\overset{\ominus}{N}\overset{+}{K} \xrightarrow{R^+ X^-} C_6H_4(CO)_2N{-}R \xrightarrow{NaOH/H_2O} \underset{\text{Primary amine}}{R{-}NH_2} + C_6H_4(COOH)_2$$

28 (*c*) In order to prepare 1° amine from an alkyl halide with simultaneous addition of one CH_2 group in the carbon chain. The reagent used as a source of nitrogen is KCN. Chemical transformation can be shown as

$$\underset{\text{Carbon containing haloalkane}}{CH_3CH_2{-}X} \xrightarrow[-KX]{KCN} CH_3CH_2{-}C{\equiv}N \xrightarrow{Na/C_2H_5OH} \underset{1^\circ\text{-amine}}{CH_3CH_2CH_2NH_2}$$

29 (*c*) Chemical transformation can be shown as

$$R{-}NH_2 + RCHO \longrightarrow [R{-}N{=}\underset{H}{\underset{|}{C}}{-}R] \xrightarrow{H_2/Pt} \underset{2^\circ \text{ amine}}{R{-}\underset{H}{\underset{|}{N}}{-}\overset{H}{\overset{|}{\underset{H}{\underset{|}{C}}}}{-}R}$$

While other given set of reactants give primary amine only.

30 (*d*) The best reagent for converting 2-phenylpropanamide into 2- phenylpropanamine is $LiAlH_4$ in ether. Reaction is as given below

$$\underset{\text{2- phenylpropanamide}}{C_6H_5CH(CH_3)CONH_2} \xrightarrow{LiAlH_4 \text{ in ether}} \underset{\text{2- phenylpropanamine}}{C_6H_5CH(CH_3)CH_2NH_2}$$

31 (*b*) The best reagent for converting 2-phenylpropanamide into 1-phenylethanamine is by $NaOH/Br_2$ and chemical transformation can be done as

$$\underset{\text{2-phenylpropanamide}}{C_6H_5CH(CH_3)CONH_2} \xrightarrow{NaOH/Br_2} \underset{\text{1-phenylethanamine}}{C_6H_5\overset{CH_3}{\overset{|}{C}H}{-}NH_2}$$

This occurs due to intramolecular migration of alkyl group. It is an example of Hofmann bromamide reaction.

33 (*b*) $(CH_3)_3N$ has least boiling point, due to lack of H-bonding.

34 (*b*) 1° and 2° amines have higher boiling points due to intermolecular H-bonding but less volatile than 3° amines and hydrocarbons of comparable molecular mass. Further, because of polar C—N bonds, 3° amines are more polar than hydrocarbons which are almost non-polar. Hence, due to weak dipole-dipole interactions, 3° amines have higher boiling point (i.e. less volatile) than hydrocarbons.

In other words, hydrocarbons are more volatile among given compounds as amine are less volatile than hydrocarbon.

36 (*c*)

	Compound	Factors responsible for basic character are
(a)	$CH_3—NH_2$	Inductive effect (+ *I*)
(b)	$NC—CH_2—NH_2$	Inductive effect (– *I*)
(c)	$(CH_3)_2NH$	Inductive effect (+ *I*) and solvation
(d)	$C_6H_5—N(H)CH_3$	– *I*-effect and resonance

37 (*a*) $\overset{\ominus}{N}H_2$ > $\overset{\ominus}{O}H$ > NH_3 > H_2O

Most basic ... Least basic

Basic strength of the above species can be explained on the basis of electronegativity of central atom and its proton accepting tendency. Here, amide ion is most basic among given compounds due to presence of negative charge and two pair of electrons on nitrogen atom.

40 (*d*) (a) o-aminoacetophenone (NH_2, $COCH_3$) (b) m-aminoacetophenone (c) p-aminoacetophenone (d) $C_6H_5—NH\to COCH_3$ (EWG)

In option (d) EWG i.e. $O{=}C—CH_3$ group is directly attached to —N atom which withdraw the electron pair and reduce the basicity an N-atom.

41 (*a*) I (1° alkyl) < II (*iso* alkyl) < III (2° alkyl) < IV (3° alkyl)

43 (*b*) $RNH_2 \xrightarrow[\Delta]{CHCl_3/KOH} RNC \xrightarrow{\text{Reduction}} RNHCH_3$

44 (*c*) S_N1 reaction proceeds through formation of carbocation. Hence, more stable is the carbocation more is the reactivity towards S_N1 mechanism.

	Alkyl halides		Intermediate
(a)	CH_3Br	⟶	$CH_3^{\oplus}$
(b)	C_6H_5Br	⟶	$C_6H_5^{\oplus}$
(c)	$C_6H_5CH_2Br$	⟶	$C_6H_5—CH_2^{\oplus}$ (more stable)
(d)	C_2H_5Br	⟶	$C_2H_5^{\oplus}$

Hence, the reaction will proceed through S_N1 mechanism when, $C_6H_5CH_2Br$ is the substrate. On ionisation, it gives a resonance stabilised carbocation ($C_6H_5\overset{+}{C}H_2$).

46 (*c*) $CH_3CH_2NH_2 \xrightarrow{3CH_3Cl} \xrightarrow{Ag_2O/H_2O} (CH_3)_3N^+(CH_2CH_3)\,OH^-$

48 (*d*) $CH_3CH_2CH_2NH_2 \xrightarrow{HNO_2} CH_3CH_2CH_2N_2^{\oplus} \xrightarrow{-N_2} CH_3CH_2\overset{\oplus}{C}H_2$ (1° carbocation) $\xrightarrow{\text{1, 2–hydride shift}} CH_3\overset{\oplus}{C}HCH_3$ (2° carbocation) $\xrightarrow{H_2O} CH_3CH(OH)CH_3$ (b)

1° carbocation $\xrightarrow{H_2O} CH_3CH_2CH_2OH$ (a); 1° carbocation → (c) cyclopropane

49 (*d*) $\underset{(X)}{C_4H_{11}N} \xrightarrow{HNO_2} C_4H_{10}O$ (3° alcohol)

(*X*) reacts with HNO_2 forming alcohol.

Thus *X* is 1° amine.

$C_4H_{10}O$ is 3° alcohol formed from 1° amine.

3° alcohol can be only, $CH_3—C(CH_3)_2—OH$

Thus, 1° amine (*X*) is $CH_3—C(CH_3)_2—NH_2$

1° amine

50 (*d*) 1-(aminomethyl)cyclopentanol (CH_2NH_2, OH) $\xrightarrow[-H_2O]{HNO_2}$ $CH_2N_2^{\oplus}$ $\xrightarrow{-N_2}$ $\overset{\oplus}{C}H_2$ (1° carbocation) $\xrightarrow[\text{(ring expansion)}]{\text{1,2-alkyl shift}}$ cyclohexyl cation with OH $\xrightarrow{H_2O}$ Enol $\xrightarrow{\text{tautomerise}}$ cyclohexanone

51 (*b*) $CH_3CH(NH_2)CH(CH_3)CH_3 \xrightarrow[-N_2]{HNO_2} CH_3—\overset{\oplus}{C}HCH(CH_3)CH_3$ (2° carbocation) $\xrightarrow{\text{1, 2 - hydride shift}} CH_3\overset{\oplus}{C}(CH_3)CH_2CH_3$ (3° carbocation (more stable than 2° carbocation)) $\xrightarrow{H_2O/OH^-} CH_3C(OH)(CH_3)CH_2CH_3$

52 (*d*) CH_3CHCH_3 (with NH_2) $\xrightarrow{HNO_2}$ CH_3CHCH_3 (with OH) (*A*) $\xrightarrow{Oxidation}$

$CH_3-C(=O)-CH_3$ (*B*) $\xrightarrow[(ii)\ H_3O^+]{(i)\ CH_3MgBr}$ $CH_3-C(CH_3)_2-OH$ (*C*)

53 (*b*) $C_2H_5NH_2 \xrightarrow{HNO_2} C_2H_5OH\ (A) \xrightarrow{PCl_5} C_2H_5Cl\ (B) \xrightarrow{NH_3} C_2H_5NH_2$ Ethylamine (*C*)

54 (*d*) $CH_3\ddot{N}H_2$ + $CH_2=CH-CO-CH_3$ ⟶ $CH_3-\overset{\oplus}{N}H_2-CH_2-CH=C(O^{\ominus})-CH_3 \xrightarrow{H^+}$

$CH_3-NH-CH_2-CH=C(OH)-CH_3 \xrightarrow{Tautomerises} CH_3-NH-CH_2-CH_2-CO-CH_3$

Thus I, II and III are true.

56 (*b*) $CH_3-C(CH_3)_2-\overset{\oplus}{N}H_3\ Cl^- \xrightarrow{\Delta} CH_3-C(=CH_2)-CH_3 + NH_4Cl$

58 (*b*) (due to 1, 2-methyl shift)

$CH_3-C(CH_3)_2-CH_2NH_2 \xrightarrow[-N_2]{HNO_2}$

$CH_3-C(CH_3)_2-\overset{+}{C}H_2$ (1° carbocation) $\xrightarrow{1,2\ methyl}$ $CH_3-\overset{\oplus}{C}(CH_3)-CH_2CH_3$ (3° carbocation)

This 3° carbocation gives option (b) alcohol by using suitable reagent.

59 (*c*) $H-\overset{*}{C}(CH_3)(C_2H_5)-NH_2$ *d*(+) $\xrightarrow{HNO_2}$ $H-C(CH_3)(C_2H_5)-\overset{\oplus}{N}_2$ ⟶ $H-\overset{\oplus}{C}(CH_3)(C_2H_5)$ 2° carbocation

$\xrightarrow{\bar{O}H}$ $H-C(CH_3)(C_2H_5)-OH$ *d*(+) + $HO-C(CH_3)(C_2H_5)-H$ *l*(−)

d(+) and *l*(−) 2° alcohols are formed in equal amounts thus, mixture is racemic.

61 (*b*) Least substituted alkene is formed.
Hence, $CH_2=HCH_2CH_3$ is formed.

62 (*c*) Nitration of benzene using a mixture of conc. H_2SO_4 and conc. HNO_3 proceeds as

$^{\ominus}O-\overset{\oplus}{N}(=O)-O-H + H_2SO_4 \longrightarrow H_3O^{\oplus} + \overset{\oplus}{N}O_2 + 2H_2O + SO_2$

$C_6H_6 \xrightarrow{\overset{\oplus}{N}O_2} C_6H_5NO_2$

This reaction is an example electrophilic substitution reaction.

63 (*b*) Aryl nitro compound can't be converted into amine using $LiAlH_4$ in ether.

$C_6H_5NO_2 \xrightarrow{H_2(excess)/Pt} C_6H_5NH_2$

$C_6H_5NO_2 \xrightarrow{Fe/HCl} C_6H_5NH_2$

$C_6H_5NO_2 \xrightarrow{Sn/HCl} C_6H_5NH_2$

$C_6H_5NO_2 \xrightarrow{LiAlH_4/ether}$ No reaction

Hence, option (b) is the correct choice.

66 (*b*) I is correct statement but II is incorrect statement.
Acylation of amine gives a monosubstituted product whereas alkylation of amine gives polysubstituted product because acylation in amine takes place at N-atom and alkylation takes place at *o*- and *p*- position.

67 (*d*) Both statements are incorrect. Aryl 1° amine can't be prepared by Gabriel phthalimide reaction because aryl halides don't undergo nucleophilic substitution with anion formed by phthalimide.

71 (*c*) Phenol is weakest Bronsted base as phenol after losing H^+ produces least stable conjugate acid among the compounds.
Oxygen is more electronegative than N. So, O—H bond is more polar and it has highest value of acidic character. Since, phenol is more acidic than alcohol, therefore, phenol has the least tendency to accept a proton and hence it is weak Bronsted base. Hence, phenol is least basic among given four choices.

74 (*b*) **I** (*A*) (tetrahydroisoquinoline) 2° amine (aliphatic)

(*B*) (tetrahydroquinoline) 2° amine; lone-pair on N is used in delocalisation of π-electrons

Thus, (*A*) is more basic than (*B*).
Thus, true
II True
III carbonyl group is (EWG), thus electron density is decreased, first is least basic, last one is maximum basic, thus incorrect.
Hence, I and II are true.

76 (*d*) I. 2° aromatic
II. 2° aromatic but less basic than I due to $-NO_2$ (EWG)
III. 2° aliphatic
IV. 2° aromatic but less basic than I and II both due to adjacent carbonyl group.

(EWG) is directly attached to benzene nucleus thus, least basic.

Thus, order is $\underset{\text{Least}}{IV} < II < I < \underset{\text{Maximum}}{III}$

78 (*a*) Both statements are correct.

Acetanilide is less basic than aniline because acetylation of aniline results in decrease of electron density on nitrogen.

80 (*b*) Hofmann bromamide degradation is shown by

$Ar—\overset{O}{\overset{||}{C}}—NH_2$ in which amide is converted into amine *via* undergoing intramolecular migration of phenyl group.

$C_6H_5CONH_2 \xrightarrow{Br_2/NaOH} C_6H_5NH_2 + K_2CO_3 + KBr + 2H_2O$

Aniline

86 (*c*) Aniline $\xrightarrow[\text{carbylamine}]{CHCl_3, KOH, \Delta}$ C_6H_5NC $\xrightarrow{\text{Reduction}}$ $C_6H_5NHCH_3$ (N-methyl aniline)

87 (*b*) $C_6H_5NH_2 \xrightarrow[273\ K]{HNO_2} C_6H_5\overset{\oplus}{N_2} \xrightarrow[\text{Sandmeyer}]{CuCN/HCN}$

$C_6H_5CN \xrightarrow{H_3O^+} C_6H_5COOH$

91 (*a*)

	Compounds	Items
A.	Benzene sulphonyl chloride	Hinsberg reagent $C_6H_5SO_2Cl$
B.	Sulphanilic acid	Zwitter ion (dipolar ion) $\overset{\oplus}{N}H_3—C_6H_4—SO_3^{\ominus}$
C.	Alkyl diazonium salts	Conversion to alcohols $R-N_2^+X^- \xrightarrow{H_2O} ROH$
D.	Aryl diazonium salts	Dyes

92 (*a*) Both statements are correct.

N-ethylbenzene is soluble in alkali because hydrogen attached to nitrogen in sulphonamide is strongly acidic.

93 (*a*) Both are correct statements.

N, N-diethylbenzene sulphonamide is insoluble in alkali due to absence of acidic H attached to nitrogen.

(B) Miscellaneous questions

1 (*c*) Alkyl group increases the electron-density on N-atom hence, basic nature is increased.

$CH_3CH_2CH_2 \rightarrow NH_2$

2 (*a*) $CH_3NC \xrightarrow{4H} CH_3NHCH_3$

3 (*c*) $—NO_2$ is (EWG) that decreases electron-density at N-atom thus, basic nature is decreased.

4 (*a*) Aniline is weakest Bronsted base among the given four compounds. It is due to presence of resonance. Resonance structures of aniline are given below.

Hence, lone pair of nitrogen are less available for donating to the acid.

5 (*b*) Greater the strength of base, greater will be its reactivity towards dilute HCl. Hence, $(CH_3)_2NH$ has highest basic strength as it has highest reactivity.

$(H_3C)_2N—H \xrightarrow{HCl} (H_3C)_2\overset{\oplus}{N}H_2\ \overset{\ominus}{Cl}$

6 (*d*) Aniline is a weaker base than NH_3 due to delocalisation of lone pair of electrons of the N-atom over the benzene ring. Pyrrole is not more basic because the lone pair of electrons on the N-atom is donated towards aromatic sixtet formation. Therefore, pyrrolidine is strongest base as lone pair of nitrogen does not involved in resonance and also due to presence of two alkyl ring residue, basic strength becomes high among the four given compounds.

7 (*a*) $CH_3—\overset{O}{\overset{||}{C}}\leftarrow\ddot{N}H_2$ (II) $\quad C_6H_5—\overset{O}{\overset{||}{C}}\leftarrow\ddot{N}H_2$ (III) $\quad CH_3CH_2\rightarrow NH_2$ (I)

Electron withdrawing, [resonance effect of CH_3CO]	Lone-pair of N-atom is also used in delocalisation of π-electrons, resonance of carbonyl benzene nucleus (least basic)	Electron-donating group. (Maximum basic)

Hence, the decreasing order of basic strength is

I > II > III

10 (*c*) NH_3 (a strong nucleophile) causes elimination from 3° alkyl halide, thus, (c) is incorrect.

12 (c) 2-(3-aminopropyl)cyclohexanone $\xrightarrow{H_2/\text{Raney Ni}}$ amino alcohol (OH, NH_2) $\xrightarrow{-H_2O}$ bicyclic amine (N–H)

13 (d)

Reaction	Statements
A Ammonolysis	Reaction of alkylhalide with NH_3 $R—X \xrightarrow{NH_3} RNH_2 + HCl$
B. Gabriel phthalimide synthesis	Reaction of phthalimide with KOH and $R–X$. Phthalimide (NH) $\xrightarrow{KOH/R—X}$ phthalic acid (OH, OH)
C. Hofmann bromamide reaction	Amine with lesser number of carbon atoms. $RCONH_2 \xrightarrow{NaOX} RNH_2$
D. Carbylamine reaction	Detection test of primary amines.

15 (c) N-methylpiperidine (N—CH_3) lacks H-atom at N-atom (3° amine) hence, there is no H-bonding. Thus, has least boiling point.

16 (c) (1) aromatic amine.

(2) aliphatic —CH_2NH_2, thus maximum basic.

(3) aromatic but $-NO_2$ group is (EWG) decreasing electron density at N-atom.

(4) —C(=O) directly attached of N hence, least basic 4 < 3 < 1 < 2.

17 (c) $(CH_3)_2CHCH_2N(CH_2CH_3)_2 \xrightarrow{CH_3I} \xrightarrow[H_2O]{Ag_2O}$

$(H_3C)_2CHCH_2\overset{\oplus}{N}(CH_3)(CH_2CH_3)_2 + OH^- \xrightarrow{\Delta}$

$(H_3C)_2CHCH_2N(CH_3)—CH_2CH_3 + CH_2=CH_2$

19 (a) $H_2N-C_6H_4-SO_3H \rightleftharpoons H_3\overset{\oplus}{N}-C_6H_4-SO_3^{\ominus}$

Dipolar ion (Zwitter ion)

$H_3\overset{\oplus}{N}-C_6H_4-SO_3^{\ominus} + H^+ \rightleftharpoons H_3\overset{\oplus}{N}-C_6H_4-SO_3H$

Cation in acidic medium thus, migrates to **cathode**.

$\xrightarrow{OH^-} H_2N-C_6H_4-SO_3^{\ominus}$

Anion in basic medium thus, migrates to **anode**.

Thus, both statements are correct.

20 (d) $RCH_2OH \xrightarrow{Cr_2O_7^{2-}/H^+} \underset{(A)}{RCOOH} \xrightarrow{SOCl_2} \underset{(B)}{RCOCl} \xrightarrow{(CH_3)_2NH} \underset{(C)}{RCON(CH_3)_2} \xrightarrow{LiAlH_4} \underset{(D)}{RCH_2N(CH_3)_2}$

Thus, $CH_3—C(CH_3)_2— CH_2CH_2N(CH_3)_2$

21 (c) $\underset{\text{Methylamine}}{CH_3NH_2} \xrightarrow[\Delta]{CHCl_3/KOH} \underset{\text{Methyl isocyanide}}{CH_3NC}$

25 (d) $CH_3CONH_2 \xrightarrow[\Delta]{NaOH/Br_2} CH_3NH_2$

26 (c) $\underset{\text{Acetamide}}{CH_3CONH_2} \xrightarrow{NaOH/\Delta} NH_3\uparrow$

$\underset{\text{Ethylamine}}{CH_3CH_2NH_2} \xrightarrow{NaOH/\Delta}$ No effect

28 (b) In (1) major product is least substituted alkene, i.e. Y and in (2) major product is more substituted alkene, i.e. X. Thus, for reaction (1) major product is X and for reaction (2) major product is Y.

29 (*d*) If amine is least basic, its conjugate acid is strong.

Least basic $\xrightarrow{H^+}$ Maximum acidic

30 (*a*) $\xrightarrow{CH_3I}$ $\overset{\oplus}{N}-CH_3I^{\ominus}$ $\xrightarrow[H_2O]{Ag_2O}$ $\overset{\oplus}{N}CH_3OH$ $\xrightarrow{\Delta}$ $-N-CH_3$

Least substituted product is major product

31 (*a*) $\underset{(A)}{CH_3CH_2NO_2} + O=N-OH \xrightarrow{HNO_2} CH_3\underset{\underset{NOH}{||}}{C}-NO_2$

Nitrolic acid (blue)

$\xrightarrow{NaOH} CH_3\underset{\underset{NONa}{||}}{C}-NO_2$

Red

Thus, 1°

It is also called Victor Meyer test.

32 (*a*) I and II both are correct statements.

Only small amount of HCl is required in the reduction of nitro compounds with iron scrap and HCl in the presence of steam because $FeCl_2$ formed gets hydrolysed to release HCl during the reaction.

(C) Medical entrances' gallery

1 (*d*) Carbylamine test is responded by primary or 1°-amine (aliphatic and aromatic both) only.

$$R-NH_2 \xrightarrow[KOH/\Delta]{CHCl_3} \underset{\text{Carbylamine}}{R-NC} + KCl + H_2O$$

$$Ar-NH_2 \xrightarrow[KOH/\Delta]{CHCl_3} \underset{\text{Carbylamine}}{Ar-NC} + KCl + H_2O$$

Degree of the given options are as follows

(a) Ph—NH—CH_3; 2° amine

(b) Ph—$N(CH_3)_2$; 3° amine

(c) Ph—NH—C_2H_5; 2° amine

(d) Ph—NH_2; 1° amine (aniline)

Option (d) responds to carbylamine test.

2 (*a*)

$$\underset{\substack{\text{Propanamide}\\ \text{(No. of C-atoms = 3)}}}{CH_3CH_2-\overset{O}{\overset{||}{C}}-NH_2} \xrightarrow[Br_2]{NaOH-C_2H_5OH} \underset{\substack{\text{Ethylamine (1°-amine)}\\ \text{(No. of C-atoms = 2)}}}{CH_3CH_2-NH_2} + Na_2CO_3 + NaBr$$

It is Hofmann bromamide reaction through which an acid amide degrades into an one carbon less primary (1°) amine.

Here, the carbon atom of the amide group $\left(-\overset{O}{\overset{||}{C}}-NH_2\right)$ gets decarboxylated in the form of carbonate salt (Na_2CO_3).

3 (*d*) Basic strength of methyl substituted amines in aqueous solution depends upon the ease of formation of cation by accepting a proton from the acid.

$$R-\underset{H}{\overset{H}{N}}: + H^+ \rightleftharpoons R-\underset{H}{\overset{H}{\overset{+}{N}}}-H$$

The basic strength can be decided by both inductive effect and solvation effect of alkyl group. In aqueous phase, the substituted $\overset{+}{N}H_4$ cations get stabilised not only by electron releasing effect of alkyl group (+*I*) but also by solvation with water molecules. Greater the size of ion, lesser will be solvation and less stabilised is the ion. The order is as follows :

Inductive effect $(CH_3)_3N > (CH_3)_2NH > (CH_3)NH_2$

Solvation effect

$H_3C-\overset{+}{N}H(-H\text{----}OH_2)_3$ (1°) > $(H_3C)_2\overset{+}{N}(-H\text{----}OH_2)_2$ (2°) > $(H_3C)_3\overset{+}{N}-H\text{----}OH_2$ (3°)

Strong +*I*-effect and hydrogen bonding favours higher basic strength of 2° amine. Thus, the correct order of basic strength in aqueous media will be $(CH_3)_2NH > CH_3NH_2 > (CH_3)_3N$.

4 (*a*) Secondary amines on reaction with Hinsberg's reagent (benzene sulphonyl chloride) forms N,N-dialkyl benzene sulphonamide. The product formed is not acidic as no H-atom is attached to N thus, it is insoluble in alkali.

The reaction of 2° amine given in option (a) takes place as follows:

$$C_6H_5-\overset{O}{\underset{O}{\overset{||}{\underset{||}{S}}}}-Cl + \underset{\text{2° amine}}{CH_3-\underset{CH_3}{CH}-NH-\underset{CH_3}{CH}-CH_3} \longrightarrow C_6H_5-\overset{O}{\underset{O}{\overset{||}{\underset{||}{S}}}}-\overset{H_3C-CH-CH_3}{N}-\underset{CH_3}{CH}-CH_3$$

On the other hand, 1° amines react with Hinsberg's reagent to yield N-alkyl benzene sulphonamide which is soluble in alkali and 3° amines do not react with $C_6H_5SO_2Cl$. The remaining options contain 1° and 3° amines.

$CH_3CH_2-N(CH_2CH_3)-CH_2CH_3$, 3° amine ; $CH_3-C(NH_2)(CH_3)-CH_2CH_2CH_3$, 1° amine

$CH_3-C(CH_3)_2-CH(CH_3)-NH_2$, 1° amine

Thus, option (a) is correct.

5 (c) Basic nature depends on the ability to donate lone pair of electrons by the atoms present in the species. Structure (III) is most basic as in that, $-NH_2$ group is not directly bonded with the benzene ring.

Structure (II) is 2nd most basic due to presence of $-NH_2$ and $-OCH_3$ groups together.

Structure (I) is next basic species which is less basic than (III) and (II) but more basic than structure (I) due to presence of electron releasing ($-CH_3$) group attached to $-NH$ group.

Structure IV is least basic, because $-NH_2$ group is directly bonded with the benzene ring, thus lone pair of electrons on N - atom is involved in resonance. Hence correct order is,

(III) > (II) > (I) > (IV) and option (c) is the correct.

6 (d) In strongly acidic medium, aniline is protonated to form the anilinium ion.

$$C_6H_5NH_2 \underset{}{\overset{H^+}{\rightleftharpoons}} C_6H_5\overset{+}{N}H_3$$

Aniline ⇌ Anilinium ion

Since, anilinium ion so formed is *meta* directing, thus besides *ortho* and *para* derivatives, significant amount of *meta* derivative is also formed.

$$\underset{\text{Aniline}}{C_6H_5NH_2} \xrightarrow[288K]{HNO_3, H_2SO_4} \underset{p\text{-nitroaniline (51\%)}}{p\text{-}NO_2C_6H_4NH_2} + \underset{m\text{-nitroaniline (47\%)}}{m\text{-}NO_2C_6H_4NH_2} + \underset{o\text{-nitroaniline (2\%)}}{o\text{-}NO_2C_6H_4NH_2}$$

7 (a) $-NO_2$ group decreases basic nature of aniline

$-CH_3$ group increases basic nature of aniline.

8 (c) $CH_3CONH_2 \xrightarrow[\Delta]{\text{(i) } Br_2 \text{ (ii) } KOH} CH_3NH_2$

9 (a) Both A are R are correct and R is the correct explanation for A.

Aniline on reaction with $NaNO_2$ /HCl at 0° C followed by coupling with β-naphthol gives a dark blue coloured precipitate due to extended conjugation.

10 (a) As, $-NH_2$ group is more reactive than $-OH$ group, CH_3COCl will attack on $-NH_2$ group and gives the following product,

$$HO-C_6H_4-NHCOCH_3.$$

Reaction is given as follows

$$HO-C_6H_4-NH_2 + CH_3COCl \text{ (1 equivalent)} \longrightarrow HO-C_6H_4-NHCOCH_3$$

11 (d) Electron releasing group –OH in compound (II) increases the basic strength while electron withdrawing group ($-COCH_3$ in compound I and IV) decreases the basic strength of amines. In compound IV, $-NH_2$ is directly attached to benzene ring, it results in conjugation of unshared pair of electrons with benzene ring and thus making it less available for protonation. Therefore, correct order is III > II > IV > I.

12 (d)

$$\underset{\text{Aniline } (X)}{C_6H_5NH_2} \xrightarrow{\text{Bromination}} \underset{(Y)}{2,4,6\text{-}Br_3C_6H_2NH_2} \xrightarrow[+HCl]{NaNO_2} \underset{(Z)}{2,4,6\text{-}Br_3C_6H_2\overset{+}{N_2}\overset{-}{Cl}} \xrightarrow{\text{Boiling } (C_2H_5OH)} \underset{\text{Tribromobenzene}}{1,3,5\text{-}Br_3C_6H_3}$$

13 (c) 1° and 2° nitro compounds react with HNO_2 while 3°-nitro compound does not.

The reactions of connected compounds are

$$\underset{\text{1°-nitro compound}}{CH_3CH_2CH_2NO_2} \xrightarrow{HON=O} CH_3-CH_2-\underset{\|}{C}(=N-OH)-NO_2$$

$$\underset{\text{1°-nitro compound}}{(H_3C)_2CH-CH_2-NO_2} \xrightarrow{HO-N=O} CH_3-CH(CH_3)-C(=N-OH)-NO_2$$

$$\underset{\text{2°-nitro compound}}{CH_3-C(=O)-CH(CH_3)-NO_2} \xrightarrow{HO-N=O} CH_3-C(=N-OH)-C(CH_3)-NO_2$$

$$\underset{\text{3°-nitro compound}}{(H_3C)_3C-NO_2} \xrightarrow{HO-N=O} \text{No reaction}$$

Thus, option (c) is incorrect.

14 (*d*) The conjugation (delocalisation) of electrons in pyrrole can be visualised as

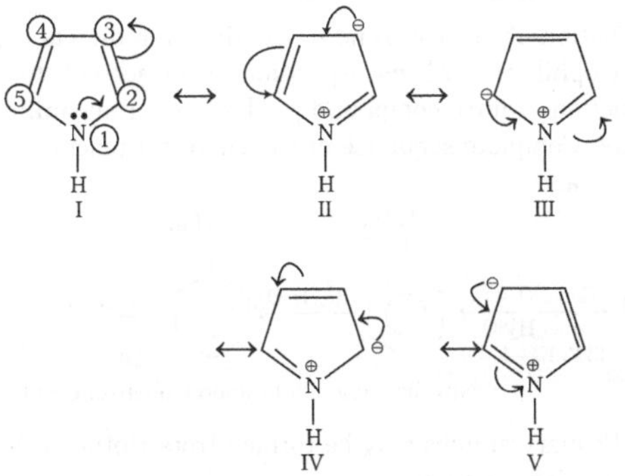

I ⟷ II ⟷ III ⟷ IV ⟷ V

15 (*d*) $RCHO + H_2NR' \longrightarrow RCH{=}NR'$ (1° amine; Schiffs bese)

16 (*a*) Under weakly acidic conditions nitrobenzene on electrolytic reduction gives aniline but under strongly acidic conditions gives *p*-aminophenol.

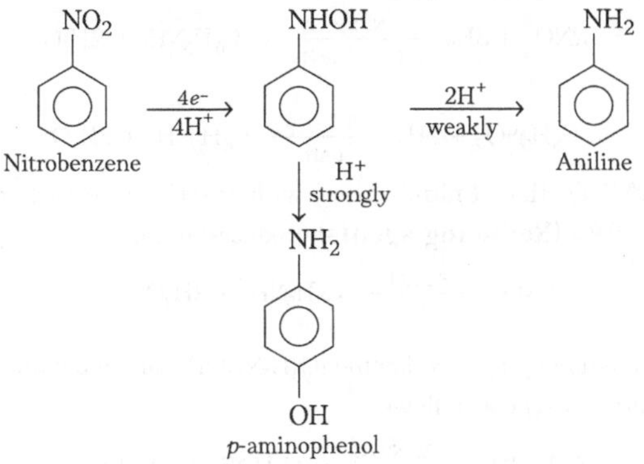

$$C_6H_5NO_2 \text{ (Nitrobenzene)} \xrightarrow[4H^+]{4e^-} C_6H_5NHOH \xrightarrow[\text{weakly}]{2H^+} C_6H_5NH_2 \text{ (Aniline)}$$

$$C_6H_5NHOH \xrightarrow[\text{strongly}]{H^+} p\text{-}NH_2C_6H_4OH \text{ (p-aminophenol)}$$

17 (*d*) Ethylmethylamine is a secondary amine. It cannot be prepared by Gabriel phthalimide synthesis.

18 (*d*) Diethylamine is a secondary amine. It gives an alkali insoluble product with *p*-toluenesulphonyl chloride.

19 (*b*) Hofmann bromamide reaction is

$$R-\overset{O}{\overset{\|}{C}}-NH_2 + Br_2 + 4\,KOH \xrightarrow{\Delta} \underset{\text{Primary amine}}{RNH_2} + K_2CO_3 + 2KBr + 2H_2O$$

This reaction is given by primary amide only.

Option (b) $CH_3CONHCH_3$ is a secondary amide, so it will not undergo Hofmann bromamide reaction.

20 (*b*) Quaternary ammonium salt of alkyl amines, when reacts with silver oxide yield quaternary ammonium hydroxide. On strong heating, it decomposes into alkene, *tert.* amine and water. This reaction is known as Hofmann elimination reaction. Here, elimination of H takes place from β-carbon having more hydrogen atoms.

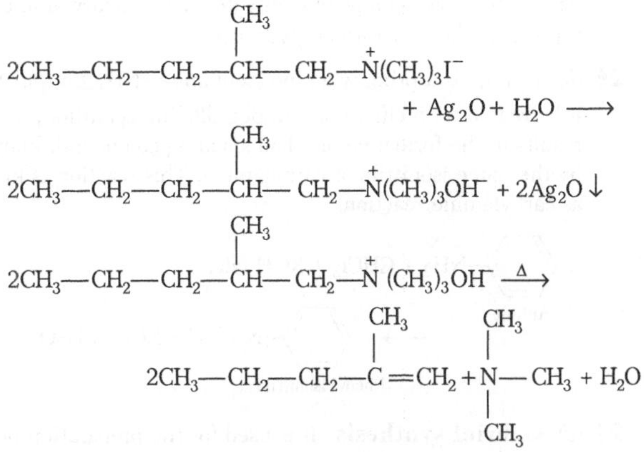

$$2CH_3-CH_2-CH_2-\underset{}{\overset{CH_3}{\overset{|}{C}H}}-CH_2-\overset{+}{N}(CH_3)_3I^- + Ag_2O + H_2O \longrightarrow$$

$$2CH_3-CH_2-CH_2-\overset{CH_3}{\overset{|}{C}H}-CH_2-\overset{+}{N}(CH_3)_3OH^- + 2Ag_2O\downarrow$$

$$2CH_3-CH_2-CH_2-\overset{CH_3}{\overset{|}{C}H}-CH_2-\overset{+}{N}(CH_3)_3OH^- \xrightarrow{\Delta}$$

$$2CH_3-CH_2-CH_2-\overset{CH_3}{\overset{|}{C}}{=}CH_2 + \underset{CH_3}{\underset{|}{\overset{CH_3}{\overset{|}{N}}}}-CH_3 + H_2O$$

Thus, major product is least substituted alkene.

21 (*b*) Since, $X[C_6H_{11}N]$ reacts with *p*-toluene sulphonyl chloride (Hinsberg reagent) to form a compound which is soluble in KOH, therefore X must be a primary amine.

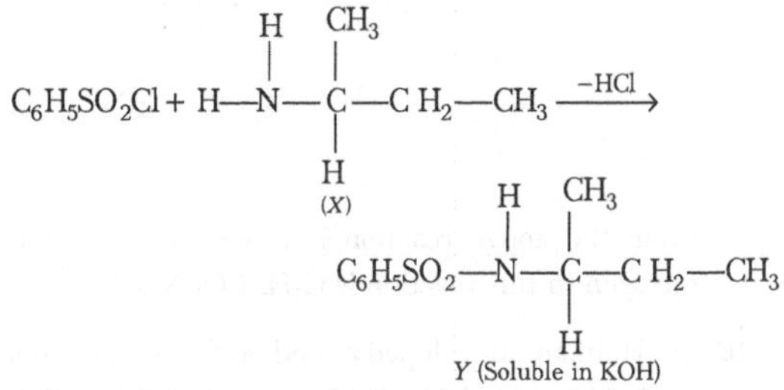

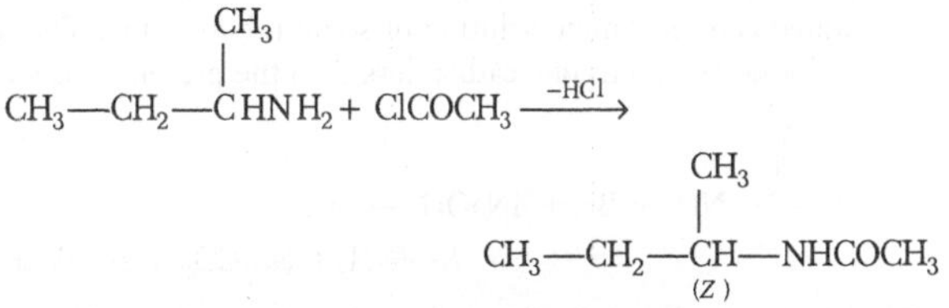

$$C_6H_5SO_2Cl + H-\overset{H}{\overset{|}{N}}-\underset{H}{\underset{|}{\overset{CH_3}{\overset{|}{C}}}}-CH_2-CH_3 \xrightarrow{-HCl} C_6H_5SO_2-\overset{H}{\overset{|}{N}}-\underset{H}{\underset{|}{\overset{CH_3}{\overset{|}{C}}}}-CH_2-CH_3$$

(X) ; Y (Soluble in KOH)

$$CH_3-CH_2-\overset{CH_3}{\overset{|}{C}}HNH_2 + ClCOCH_3 \xrightarrow{-HCl} \underset{(Z)}{CH_3-CH_2-\overset{CH_3}{\overset{|}{C}H}-NHCOCH_3}$$

22 (*a*) Secondary amines form N, N-dialkyl derivative with *p*-toluenesulphonyl chloride which does not dissolve in NaOH whereas tertiary amines do not react with it at all. Hence, secondary and tertiary amines can be distinguished by using *p*-toluene sulphonyl chloride.

23 (*d*) More the tendency to form H-bonding (i.e. more the number of H-atoms attached directly with N atom), higher is the boiling point of amine. Thus, $CH_3CH_2CH_2CH_2NH_2$ has the highest boiling point among the given amines.

24 (*b*) Lithium aluminium hydride ($LiAlH_4$) converts $\rangle C{=}O$ group into CH_2 group thus, it can be used to convert acetamide into ethanamine

$$\underset{\text{Acetamide}}{CH_3CONH_2} + 4[H] \xrightarrow{LiAlH_4} \underset{\text{Ethanamine}}{CH_3CH_2NH_2} + H_2O$$

25 (*b*) Since, the according to given information amine gives white precipitate with benzene sulphonyl chloride which is insoluble in *aq.* NaOH, so it must be a secondary amine (i.e. amine having *R*NH*R* group.)

Among the given compound, only (b) is a secondary amine, so it represents the structure of given amine.

26 (*b*) Aniline is a primary amine (contains —NH_2 group). When it is treated with chloroform under alkaline condition, it results in the formation of a bad smelling compound, known by the name isonitrile or carbylamine. This reaction is known as carbylamine reaction.

$$C_6H_5-NH_2 + CHCl_3 + KOH \text{ (alc.)} \longrightarrow C_6H_5-N \equiv C + KCl + 3H_2O$$

Aniline → Phenyl isonitrile

27 (*d*) **Gabriel synthesis** It is used for the preparation of primary amines as follows

$$C_6H_4(CO)_2NH + KOH \text{ (alc.)} \xrightarrow{-H_2O}$$

From the above reaction it is very clear that the source of nitrogen in this reaction is $C_6H_5(CO)_2N^-K^+$.

28 (*b*) Hofmann developed a method for the preparation of primary amines by treating an amide with bromine in an aqueous or ethanolic solution of sodium hydroxide. The amine so formed contains one carbon less than the present in the amide.

$$R-\overset{O}{\overset{\|}{C}}-NH_2 + Br_2 + 4NaOH \longrightarrow R-NH_2 + Na_2CO_3 + 2NaBr + 2H_2O$$

29 (*a*) Compound *A* (C_7H_9N) is benzyl amine ($C_6H_5CH_2NH_2$). It reacts with $NaNO_2/HCl$ to 0°C to form benzyl alcohol and N_2 gas.

$$\underset{(A)}{C_6H_5CH_2NH_2} \xrightarrow[0°C]{NaNO_2/HCl} \underset{\text{Benzyl alcohol}}{C_6H_5CH_2OH} + N_2\uparrow$$

Benzyl amine shows five isomers.

(I) $C_6H_5CH_2NH_2$; (II) o-toluidine (CH_3, NH_2 ortho); (III) m-toluidine (CH_3, NH_2 meta); (IV) p-toluidine (CH_3, NH_2 para); (V) $C_6H_5NHCH_3$

30 (*c*) Conversion of benzene into *m*-bromoaniline can be done by successive nitration, bromination and reduction using following reagent.

Nitration → conc. HNO_3 | conc. H_2SO_4
Bromination → Br_2 | H_2O
Reduction → Sn | HCl

Formation of nitrobenzene in the first step directs incoming electrophile towards *meta*-position in the second step. Then, reduction of nitro compound produces *m*-bromoaniline very easily. Complete sequence of the chemical reaction can be shown as

$$\underset{\text{Benzene}}{C_6H_6} \xrightarrow[\text{(333 K) }(-H_2O)]{\text{Conc. }HNO_3,\ \text{conc. }H_2SO_4} \underset{\text{Nitrobenzene}}{C_6H_5NO_2} \xrightarrow{Br_2/H_2O} \underset{m\text{-bromonitrobenzene}}{m\text{-}BrC_6H_4NO_2} \xrightarrow{Sn/HCl} \underset{m\text{-bromoaniline}}{m\text{-}BrC_6H_4NH_2}$$

31 (*b*) Primary amines may be formed from Hofmann bromamide degradation reaction.

$$RCONH_2 + Br_2 + 4KOH \xrightarrow{\Delta} RNH_2 + 2KBr + K_2CO_3 + 2H_2O$$

$$C_2H_5CONH_2 + Br_2 + 4KOH \xrightarrow{\Delta} C_2H_5NH_2 + 2KBr + K_2CO_3 + 2H_2O$$

Primary amines may also be formed by reduction of nitroalkanes, nitrobenzene, etc.

$$RNO_2 + 3H_2 \xrightarrow[\text{or }LiAlH_4\text{ or }H_2/Pt\text{ or Metal/ Acid}]{Ni/250°C} C_6H_5NH_2 + 2H_2O$$

$$C_6H_5NO_2 + 2H_2 \xrightarrow[LAH]{Ni/250°C} C_6H_5NH_2 + 2H_2O$$

32 (*d*) Reduction of nitrobenzene with Zn/HCl or Fe/HCl or Sn/HCl **(Reducing agents)** produces aniline.

$$C_6H_5NO_2 \xrightarrow[6[H]]{Sn/HCl} C_6H_5NH_2 + 2H_2O$$

33 (*d*) N-phenylhydroxylamine (C_6H_5NHOH) can be obtained by nitrobenzene as follows

$$C_6H_5NO_2 \xrightarrow{Zn/NH_4Cl} C_6H_5NHOH + H_2O$$

34 (*b*) NO_2 group being electron withdrawing reduces electron density at *o*/*p* positions. Hence, now the *meta*-position becomes electron rich on which the electrophile (nitronium ion) attacks during nitration.

Thus, 1, 3-dinitrobenzene is obtained.

$$HNO_3 + H_2SO_4 \longrightarrow H_2NO_3^+ + HSO_4^-$$

$$H_2NO_3^+ \longrightarrow H_2O + \underset{\text{Electrophile}}{NO_2^+}$$

35 (*d*) This reaction is known as Hofmann bromamide reaction, which is used for the conversion of amides to amines with one carbon atom less than parent amide.

$$\underset{\text{Acetamide}}{CH_3CONH_2} + Br_2 + 4NaOH \longrightarrow \underset{\text{Methyl amine}}{CH_3NH_2} + Na_2CO_3 + 2KBr + 2H_2O$$

36 (*b*) $C_6H_5NH_2 \xrightarrow{NaNO_2/HCl} \underset{(X)}{C_6H_5\overset{+}{N_2}\overset{-}{Cl}} \xrightarrow{Cu_2(CN)_2} \underset{(Y)}{C_6H_5CN} \xrightarrow{H_2O/H^+} \underset{\text{Benzoic acid (Z)}}{C_6H_5-COOH}$

Thus, product *Z* is identified as C_6H_5—COOH .

37 (*a*) $R-\overset{O}{\overset{\|}{C}}-NH_2 + Br_2 \longrightarrow R-\overset{O}{\overset{\|}{C}}-\ddot{N}(H)-Br$

$\xrightarrow[-H^+]{} R-\overset{O}{\overset{\|}{C}}-\ddot{\underset{\ominus}{N}}-Br \longrightarrow R-\ddot{N}=C=O$

$\xrightarrow{KOH/H_2O} RNH_2 + CO_2\uparrow$

The above reaction is known as Hofmann bromamide reaction.

38 (*a*) Nitration of nitrobenzene at 125°C with mixed acids give *meta*-dinitrobenzene.

39 (*a*) On treating with nitrous acid primary amine gives an alcohol and N_2 gas. So, '*A*' should be primary amine.

$C_3H_7NH_2 + HNO_2 \longrightarrow C_3H_7OH + N_2 + H_2O$

$(H_3C)_2CH-NH_2$ ('*A*') $\xrightarrow[\text{(Carbylamine test)}]{CHCl_3 + KOH}$ $(H_3C)_2CH-NC$ ('*C*')

$\xrightarrow{\text{Reduction}}$ $(H_3C)_2CH-NH-CH_3$

iso-propylmethyl amine (1° amine)

40 (*a*) $C_6H_5NH_2 + CH_3COCl \xrightarrow[-HCl]{NaOH} C_6H_5NHCOCH_3 + HCl$

Acetanilide

41 (*d*) Aniline (NH_2) + $3Br_2$ (Excess) $\longrightarrow$ 2, 4, 6-tribromoaniline

42 (*b*) $(C_2H_5)_2NH$, i.e. N-ethylethanamine is the most basic amine among the given compounds.

43 (*d*) Aniline (NH_2) $\xrightarrow{Br_2/H_2O}$ 2,4,6-tribromoaniline (*X*) $\xrightarrow[0°C]{NaNO_2/\ HCl}$ 2,4,6-tribromobenzenediazonium chloride ($\overset{+}{N}_2\overset{-}{Cl}$) (*Y*) (Water soluble compound) $\xrightarrow[HCl]{Cu_2Cl_2}$ 2,4,6-tribromochlorobenzene (*Z*)

44 (*e*) Anilinium hydrogen sulphate ($\overset{+}{N}H_3HSO_4^-$) $\xrightarrow{\Delta,\ Conc.\ H_2SO_4}$ Sulphanilic acid (NH_2, SO_3H)

45 (*a*) Intermolecular association is more in primary amines than in secondary amines as there are two H-atoms available for hydrogen bond formation in it. Tertiary amines do not have intermolecular association due to the absence of H-atom available for hydrogen bond formation. Hence, the order of boiling point of isomeric alkyl amines is as 1° > 2° > 3°.

CHAPTER 21

Diazonium salts, cyanides and isocyanides

In the previous chapter, you have studied the chemistry of amines that belong to the class of organic compounds containing nitrogen atom. Now, you will learn about other important nitrogen containing organic compounds such as **diazonium salts, cyanides** and **isocyanides.** They are constituent of dyes, drugs, fertilisers, proteins etc.

In this chapter, we will study all of them in a sequential manner.

DIAZONIUM SALTS

Diazonium salts are the most reactive derivatives of aromatic primary amines. These are stable for a short period of time in solution at low temperature. Some diazonium salts are given below,

$[Ar - \overset{+}{N} \equiv N:] X^-$; Ar is aryl [General formula]

$C_6H_5-\overset{+}{N} \equiv NCl^-$; Benzenediazonium chloride

$H_3C-C_6H_4-\overset{+}{N} \equiv NCl^-$; *p*-toluenediazonium chloride

$O_2N-C_6H_4-\overset{+}{N} \equiv NBF_4^-$; *p*-nitrobenzenediazonium fluoroborate

They are named by suffixing diazonium to the name of the parent hydrocarbon from which they are formed followed by the name of the anion. The $N_2^{\oplus}$ group is called diazonium group.

Inside

The structure of a diazonium salt is supposed to be derived from ammonium chloride by replacing three hydrogen atoms with a nitrogen atom and the fourth by a phenyl group.

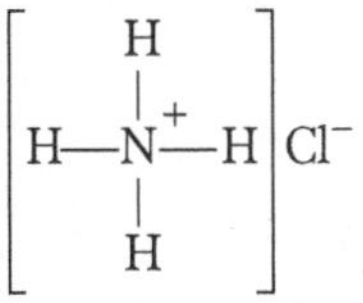

$$\left[H - \overset{H}{\underset{H}{N^+}} - H \right] Cl^- \qquad [C_6H_5 - \overset{+}{N} \equiv N] Cl^-$$

Ammonium chloride — Benzenediazonium chloride

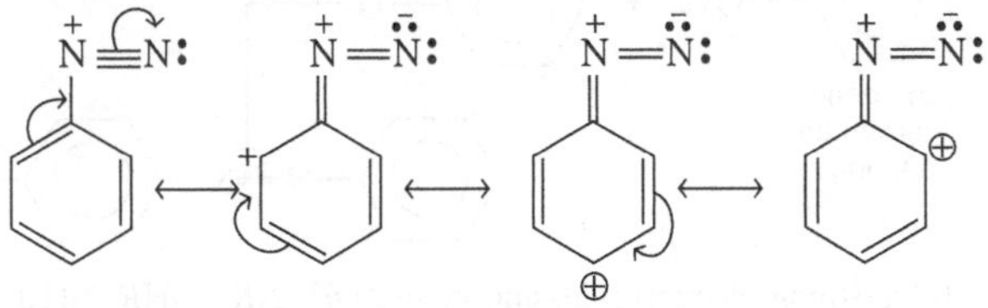

Azo compounds, like alkenes can exist in *cis* and *trans* forms and because of steric strain, the *trans*-isomer is more stable than *cis*-isomer.

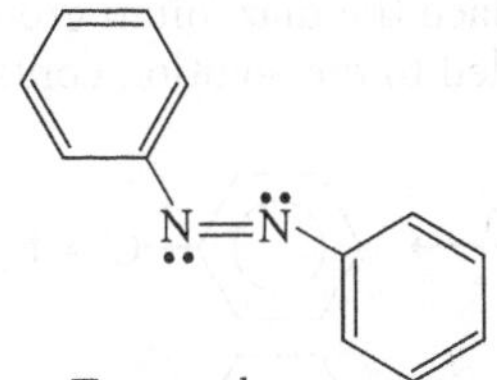

Trans-azobenzene — *Cis*-azobenzene

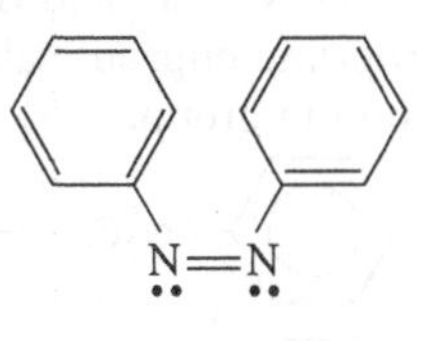

Methods of preparation

Diazonium salts can be prepared by **diazotisation** of aniline. In this reaction, ice-cold solution of HNO_2($NaNO_2$ + HCl or $NaNO_2$ + H_2SO_4) is added to ice-cold solution of primary aromatic amine dissolved in acid to form diazonium salt.

$$NaNO_2 + HCl \longrightarrow HNO_2 + NaCl$$

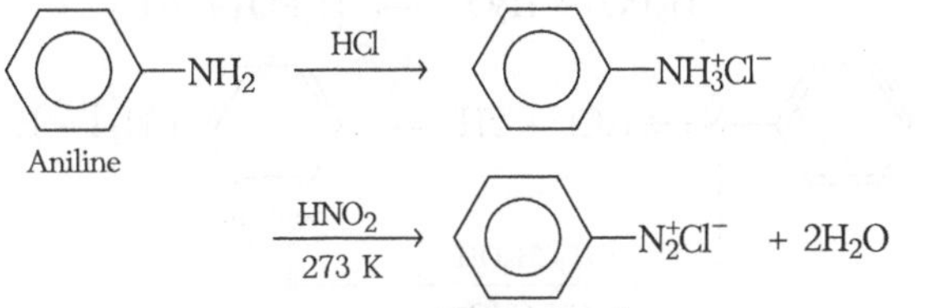

$$C_6H_5NH_2 \text{ (Aniline)} \xrightarrow{HCl} C_6H_5NH_3^+Cl^- \xrightarrow[273\text{ K}]{HNO_2} C_6H_5N_2^+Cl^- \text{ (Diazonium salt)} + 2H_2O$$

Here, some HNO_2 is lost as NO and NO_2. So, excess of $NaNO_2$ should be present which can be tested by KI-starch paper when blue colour appears. (HNO_2 oxidises I^- to I_2 which gives blue colour with starch).

Physical properties

(*i*) Dry diazonium salts are crystalline solids, many of them are explosive in nature. Because of their unstability they are seldom isolated and are used in solution (*in situ*).

(*ii*) They are water soluble due to their ionic nature.

(*iii*) In aqueous solution, they are good conductors of electricity.

(*iv*) Aromatic diazonium salts are more stable than aliphatic diazonium salts (RN_2^+) and the stability is attributed to resonance.

Chemical properties

Decomposition

They slowly decompose even at ice-bath temperature, therefore, the solution is used immediately after preparation (*in situ*).

Alkane diazonium salts are unstable. They decompose to give nitrogen and carbocations.

$$CH_3CH_2 - \overset{\oplus}{N} \equiv N \longrightarrow CH_3CH_2^{\oplus} + N_2 \uparrow$$

The driving force for this reaction is the formation of N_2, an exceptionally stable compound.

Substitution reactions (or replacement reactions)

Due to the better leaving ability of diazonium group, it is substituted by other groups such as Cl^-, Br^-, CN^- and OH^- which displaces nitrogen gas from the reaction. This type of replacement reactions are examples of unimolecular nucleophilic substitution (S_N1) of the diazo group.

Replacement of the diazonium group is shown below.

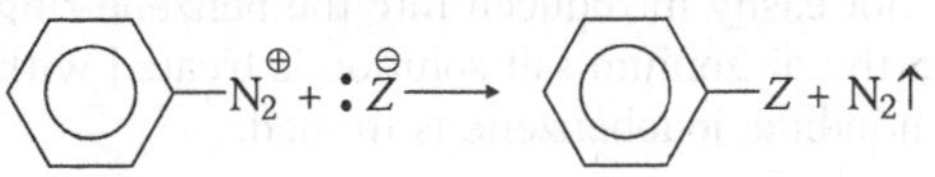

$$C_6H_5 - \overset{\oplus}{N_2} + :\overset{\ominus}{Z} \longrightarrow C_6H_5 - Z + N_2\uparrow$$

1. Hydrolysis

On heating diazonium salts in presence of H_2SO_4, phenol is produced with evolution of nitrogen.

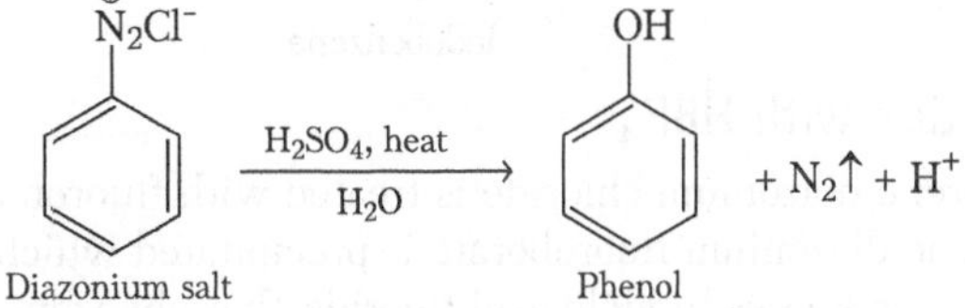

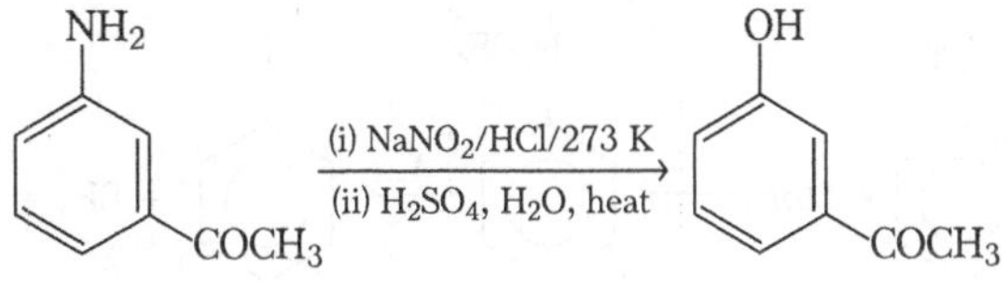

$$C_6H_5\overset{\oplus}{N_2}Cl^- \text{ (Diazonium salt)} \xrightarrow[H_2O]{H_2SO_4,\text{ heat}} C_6H_5OH \text{ (Phenol)} + N_2\uparrow + H^+$$

$$m\text{-}CH_3COC_6H_4NH_2 \xrightarrow[\text{(ii) } H_2SO_4,\ H_2O,\text{ heat}]{\text{(i) } NaNO_2/HCl/273\text{ K}} m\text{-}CH_3COC_6H_4OH$$

2. Sandmeyer reaction

In this reaction, replacement of the diazonium group by chloride, bromide and cyanide takes place. Nucleophiles such as CN^-, Cl^- and Br^- replace the diazonium group, if appropriate cuprous salt is added to the solution containing diazonium group.

$$C_6H_5N_2^+Cl^- \xrightarrow{CuCl,\ HCl} C_6H_5Cl + N_2\uparrow$$

$$C_6H_5N_2^+Cl^- \xrightarrow{CuBr,\ HCN} C_6H_5Br + N_2\uparrow$$

$$C_6H_5N_2^+Cl^- \xrightarrow{CuCN,\ HCN} C_6H_5CN + N_2\uparrow$$

3. Gattermann reaction

Nucleophiles such Cl^- or Br^- can also be introduced in the benzene ring by treating the diazonium salt solution with corresponding halogen acid in the presence of copper powder. This reaction is known as **Gattermann reaction.**

$$CH_3C_6H_4N_2^+Br^- \xrightarrow[HBr]{Cu,\ \Delta} CH_3C_6H_4Br + N_2\uparrow$$

Bromotoluene

$$CH_3C_6H_4N_2^+Cl^- \xrightarrow[HCl]{Cu,\ \Delta} CH_3C_6H_4Cl + N_2\uparrow$$

Chlorotoluene

4. Reaction with KI

Iodine is not easily introduced into the benzene ring directly but when the diazonium salt solution is treated with potassium iodide, iodobenzene is formed.

$$C_6H_5\overset{+}{N}_2\overset{-}{Cl} + KI \longrightarrow C_6H_5I + KCl + N_2\uparrow$$

Iodobenzene

5. Reaction with HBF_4

When arene diazonium chloride is treated with fluoroboric acid, arene diazonium fluoroborate is precipitated which on heating decomposes to yield aryl fluoride/fluorobenzene.

$$C_6H_5\overset{+}{N}_2\overset{-}{Cl} + HBF_4 \longrightarrow C_6H_5\overset{+}{N}_2\overset{-}{BF_4} \xrightarrow{\Delta} C_6H_5F + BF_3 + N_2\uparrow$$

Diazonium salt   Fluorobenzene

This reaction is called **Balz-Schiemann reaction.** The arene diazonium tetrafluoroborate ($C_6H_5\overset{+}{N}_2\overset{-}{BF_4}$) is very reactive.

When it combines with $NaNO_2$ followed by heat then nitrobenzene is formed.

$$C_6H_5\overset{+}{N}_2\overset{-}{BF_4} \xrightarrow{NaNO_2/\Delta} C_6H_5NO_2 + N_2\uparrow + NaBF_4$$

Nitrobenzene

Reduction

Certain mild reducing agents like hypophosphorus acid (phosphinic acid) or ethanol reduces diazonium salts to arenes and themselves get oxidised to phosphorus acid and ethanal, respectively.

$$H_3PO_2 + H_2O \longrightarrow H_3PO_3 + 2H$$

$$C_6H_5\overset{+}{N}{\equiv}NCl^- + 2H \longrightarrow C_6H_6 + N_2\downarrow + HCl$$

$$C_6H_5\overset{+}{N}{\equiv}NCl^- \xrightarrow[-\ CH_3CHO]{CH_3CH_2OH} C_6H_6 \text{ (Benzene)}$$

Benzene diazonium chloride can also undergo reduction with sodium sulphite or $SnCl_2/HCl$ to form phenyl hydrazine hydrochloride.

$$Na_2SO_3 + H_2O \longrightarrow Na_2SO_4 + 2H$$

$$C_6H_5\overset{+}{N}{\equiv}NCl^- + 4H \longrightarrow C_6H_5NH{\cdot}NH_2{\cdot}HCl$$

Phenyl hydrazine hydrochroide

Coupling reaction

Benzene diazonium salts react with highly reactive electron rich compounds such as phenols and amines to form bright coloured azo compounds. This reaction is called **coupling reaction.**

In general,

$$C_6H_5N_2^+Cl^- + H{-}C_6H_4{-}G \longrightarrow C_6H_5{-}N{=}N{-}C_6H_4{-}G$$

Benzene diazonium chloride

(G is strong electron-releasing group: OH, NR_2, NHR, NH_2)

1. Coupling with phenols

Coupling with phenols occurs in the basic medium (pH 9-10) at 273-298 K.

$$C_6H_5-\overset{\oplus}{N}{=}N-\overset{\ominus}{Cl} + C_6H_5-OH \xrightarrow[OH^-]{\text{Weakly alkaline}} C_6H_5-N{=}N-C_6H_4-OH$$

Benzene diazonium salt; Phenol; *p*-hydroxyazobenzene (orange dye)

Coupling with benzene substrates occur preferentially at the less sterically hindered *p*-position to the hydroxy or amino group but if this position is blocked, then *o*-coupling occurs.

$$C_6H_5N_2^+Cl^- + p\text{-}CH_3C_6H_4OH \longrightarrow H_3C-C_6H_3(OH)-N{=}N-C_6H_5$$

Benzene diazoium salt; *p*-hydroxy toluene

At pH = 9 – 10, phenols form phenate ion that readily couples with diazonium salts to form azodyes.

$$C_6H_5-O^- \underset{OH^-}{\overset{H^+}{\rightleftharpoons}} C_6H_5-OH$$

Couples readily; Couples slowly

$$H_2N-C_6H_4-O^- \xleftarrow{OH^-} H_2N-C_6H_4-OH \xrightarrow{H^+} \overset{\oplus}{N}H_3-C_6H_4-OH$$

This position is activated for coupling; *p*-amino phenol; This position is activated for coupling

Various colour dyes can be prepared by this method. Some other examples are given below,

$$O_2N-C_6H_4-\overset{+}{N}{\equiv}N: + C_{10}H_7O^{\ominus} \longrightarrow O_2N-C_6H_4-N{=}N-C_{10}H_6-OH$$

Para-red

$$O_2N-C_6H_4-\overset{+}{N}{\equiv}N: + C_6H_4(O^{\ominus})(COO^{\ominus}) \longrightarrow O_2N-C_6H_4-N{=}N-C_6H_3(OH)(COO^-)$$

Alizarin yellow

2. Coupling with amines

Coupling with amines occur in the acidic medium (pH 4-5) at 273-298 K. e.g.

$$C_6H_5-N{=}\overset{+}{N}\overset{-}{Cl} + C_6H_5N(CH_3)_2 \xrightarrow{H^+} C_6H_5-N{=}N-C_6H_4-N(CH_3)_2$$

In strongly acidic medium, amines get protonated to form a cation, which is considerably deactivated due to the positive charge on nitrogen.

$$C_6H_5\ddot{N}H_2 \overset{H^+}{\rightleftharpoons} C_6H_5\overset{+}{N}H_3$$

(Readily couples) (Does not couple)

Other common example is formation of an indicator called methyl orange.

$$^{\ominus}O-SO_2-C_6H_4-\overset{\oplus}{N}{\equiv}N + C_6H_5-N(CH_3)_2 \longrightarrow {}^{\ominus}O-SO_2-C_6H_4-N{=}N-C_6H_4-N(CH_3)_2$$

Methyl orange (an indicator) (Dye)

CHECK POINT 21.1

1. Which of the following amines will form stable diazonium salt at 273-283 K?
(a) $C_2H_5NH_2$ (b) $C_6H_5NH_2$
(c) $C_6H_5CH_2NH_2$ (d) $C_6H_5N(CH_3)_2$

2. In laboratory, benzene diazonium chloride is not isolated in the crystalline solid state because in solid state it.
(a) explodes (b) is readily oxidised
(c) is readily reduced (d) is polymerised

3. Which of the following statements is incorrect about diazonium salts?
(a) They are crystalline solids
(b) They are water insoluble
(c) They are good conductors of electricity in aqueous solution
(d) Aromatic diazonium salts are more stable than aliphatic diazonium salts

4. Consider the following reaction,

$$C_6H_5\overset{+}{N_2}\overset{-}{Cl} \xrightarrow[H_2O]{H_2SO_4, heat} X$$

What is X in the above reaction?
(a) Benzene (b) Phenol
(c) Aniline (d) Chloro benzene

5. Which of the following reagents are used in Balz-Schiemann reaction?
(a) C_6H_5OH and CO_2
(b) Benzene diazonium chloride and HBF_4
(c) Chlorobenzene and chloral
(d) Benzene diazonium salt

6. Benzene diazonium chloride when reduced with sodium sulphite yields
(a) chloro benzene (b) benzene
(c) phenyl hydrazine chloride (d) phenol

7. Benzene diazonium chloride when reacts with hypophosphorus acid produces.
(a) phenyl phosphate (b) benzene
(c) phenol (d) phenyl isocyanide

8. Reaction of benzene diazonium chloride with alkaline phenol gives an azo dye. This is an example of
(a) electrophilic substitution (b) nucleophilic substitution
(c) oxidative coupling (d) free radical reaction

9. Benzene diazonium chloride forms orange red dye with
(a) phenol (b) cresol
(c) resorcinol (d) All of these

10. During coupling reaction of benzene diazonium chloride with aniline the pH is maintained approximately
(a) 1 − 3 (b) 7−10
(c) 11−14 (d) 4− 5

CYANIDES

Cyanides are the class of organic compounds, which are derived from HCN by replacement of H by an alkyl or aryl group. As HCN is acidic. Therefore, cyanides considered as the acid derivatives.

Both the carbon atom and the nitrogen atom of the cyano group are *sp*-hybridised and the *R*—C≡N bond angle is 180° (linear).

They resemble terminal alkyne, except that the nitrogen atom of the nitrile has a lone pair of electrons in place of acetylenic carbon of the alkyne.

$CH_3—C≡N:$ (180°; C and N are *sp*) $CH_3—C≡CH$ (180°; C is *sp*)

Nomenclature

Commonly, cyanides are named by adding suffix to the alkyl or aryl group. Generally, they are named by replacing the -ic or -oic from the name of the carboxylic acid with the same number of carbon atoms including the nitrile carbon by -o nitrile. In IUPAC system, suffix 'nitrile' is added to the name of the hydrocarbon atoms.

For determining the position of the substituent in a chain, the nitrile carbon is numbered as 1.

Acid	Common name	IUPAC name
$CH_3\overset{O}{\overset{\Vert}{C}}OH$ Acetic acid	$CH_3C≡N$ Acetonitrile	Ethanenitrile
$CH_3CH_2CH_2COOH$ Butyric acid	$CH_3CH_2CH_2CN$ Butyronitrile	Butanenitrile
C_6H_{11}—COOH Cyclohexanecarboxylic acid	C_6H_{11}—CN	Cyclohexanecarbonitrile

When —COOH and —CN both are present, —CN is used in prefix as cyano.

- $CH_3CH(CN)COOH$ 2-cyanopropanoic acid
- $CH_3CH(CN)CN$ 2-methyl-1,3-propanedinitrile
- $CH_2(CN)CH(CN)CH_2(CN)$ Propane-1, 2, 3-tricarbonitrile

$$\begin{array}{l} {}^{4}\overset{5}{CH_2CN} \\ | \\ {}_3CH - CH_2CN \\ | \\ \underset{2}{C}H_2\underset{1}{CN} \end{array}$$

3-cyanomethylpentanedinitrile

Methods of preparation

1. By nucleophilic displacement of the halogen

Alkyl or aryl halides react with inorganic cyanide (i.e. KCN, NaCN) to form alkyl cyanide. In this method, alkyl halides react with potassium cyanide in a solvent that will dissolve both the reactants. Since, cyanide ion is a strong base, it gives elimination reaction with tertiary halides and with secondary halides both the substitution and elimination products are formed.

e.g.
$$R—X \xrightarrow{KCN} RCN + KX$$
$$CH_3Cl + KCN \longrightarrow CH_3CN + KCl$$
$$C_6H_5—Cl + KCN \longrightarrow \text{no reaction}$$

Chlorobenzene

(C — Cl bond is stable due to resonance)

2. By dehydration of primary amide

Nitriles can also be prepared by dehydration of primary amides.

e.g.
$$\underset{\text{Acetamide}}{CH_3\overset{O}{\overset{\|}{C}}NH_2} \xrightarrow[POCl_3]{P_2O_5 \text{ or}} CH_3CN + H_2O$$

$$\underset{\text{Benzamide}}{C_6H_5—\overset{O}{\overset{\|}{C}}NH_2} \xrightarrow[POCl_3]{P_2O_5 \text{ or}} \underset{\text{Benzonitrile}}{C_6H_5—CN} + H_2O$$

3. By reaction with CuCN

Aromatic nitriles are not formed from unreactive aryl halides but are prepared from aryl diazonium salt through the replacement of diazonium group by —CN by allowing the diazonium salt to react with cuprous cyanide.

$$\underset{\text{Diazonium salt}}{C_6H_5—\overset{+}{N}\equiv N\overset{-}{Cl}} + CuCN \longrightarrow \underset{\text{Benzonitrile}}{C_6H_5—CN} + N_2\uparrow + CuX$$

4. By dehydration of aldoximes

Cyanides can be prepared by reaction of aldoximes with acetic anhydride.

$$\underset{\text{Acetaldoxime}}{CH_3—CH = NOH} \xrightarrow{(CH_3CO)_2O} \underset{\text{Methyl cyanide}}{CH_3CN} + H_2O$$

5. By Grignard reagent

Cyanogen chloride reacts with Grignard reagent to give cyanides.

$$CH_3MgBr + CNCl \longrightarrow CH_3CN + Mg\begin{matrix} Br \\ Cl \end{matrix}$$

Physical properties

(*i*) Lower members are colourless liquids but higher members are crystalline solids.

(*ii*) Alkyl cyanides have generally pleasant odour.

(*iii*) Lower alkyl cyanides are soluble in water.

(*iv*) They have higher boiling points than esters and acid chlorides of similar molecular weight. This effect results from a strong dipolar association between adjacent cyano groups.

Different compounds	Molecular weight	Boiling point (K)
$CH_3CH_2CH_2OH$	60	370
CH_3CH_2CN	55	370
$H—\overset{O}{\overset{\|}{C}}—OCH_3$	60	305
$CH_3CH_2CH_2CH_3$	58	273

Due to strong polarisation, cyanides can behave as both electrophile and nucleophile.

$$R—\overset{\delta+}{C} \equiv \overset{\delta-}{N}\!:$$

for electrophilic attack (at N); for nucleophilic attack (at C)

Chemical properties

Some important reactions of cyanides are given below

Hydrolysis

Depending upon the reagents used, hydrolysis can be complete or partial which are as discussed below

1. Complete hydrolysis

Cyanides get hydrolysed to carboxylic acid, when it is refluxed with strong aqueous acid or sodium hydroxide solution. This type of hydrolysis is known as complete hydrolysis. The intermediate in both the reactions is the primary amide.

$$R—C\overset{3OH\ \vdots\ 3H}{\equiv}N \xrightarrow[H_2O/H^+]{H_2O/\bar{O}H \text{ or}} NH_3 + R—C\begin{matrix} OH \\ OH \\ OH \end{matrix} \xrightarrow{-H_2O} R—\overset{}{\underset{O}{\underset{\|}{C}}}—OH$$

An ester can be obtained by using alcohol under acidic conditions.

$$C_6H_5C\equiv N + CH_3OH + HCl(g) \longrightarrow C_6H_5-\underset{\underset{O}{\|}}{C}-OCH_3 + NH_4Cl$$

(Ester)

$$C_6H_5CN + OH^- + H_2O \longrightarrow C_6H_5COO^- + NH_3$$

2. Partial hydrolysis

Alkyl cyanides on shaking with cold conc. HCl or by dissolving it in conc. H_2SO_4 followed by pouring it into water gives partial hydrolysis product.

$$R-C\equiv N \xrightarrow[\text{or conc. } H_2SO_4 \text{ cold}]{\text{Conc.HCl}} \underset{\text{Amide}}{R-\underset{\underset{O}{\|}}{C}-NH_2}$$

Alkyl cyanides can also be reduced to amides by using H_2O_2 and NaOH solution.

$$R-C\equiv N \xrightarrow{H_2O_2 + NaOH} RCONH_2$$

Reduction

Cyanides can be reduced partially as well as completely. These are as follows

1. Partial reduction (Stephen's reduction)

Cyanides are reduced to aldehydes by treating their ethereal solution with stannous chloride and hydrogen chloride gas at room temperature. Imine hydrochloride is initially formed, which upon boiling with water gives aldehyde. This reduction is called **Stephen's reduction.**

$$SnCl_2 + 2HCl(g) \longrightarrow SnCl_4 + 2[H]$$

$$CH_3CN + 2[H] + HCl \xrightarrow{\text{Ether}} \underset{\text{Imine hydrochloride}}{CH_3CH=NH\cdot HCl} \xrightarrow[\text{Boil}]{H_2O} NH_4Cl + \underset{\text{Acetaldehyde}}{CH_3CHO}$$

Partial reduction of cyanides can also be done by following two reagents.

$$R-C\equiv N \xrightarrow[H_3O^+]{\text{DIBAH}} RCHO$$

$$R-C\equiv N \xrightarrow[H_3O^+]{\text{LiAlH}(t\text{-OBu})_3} RCHO$$

2. Complete reduction (Mendius reaction)

The complete reduction can be done by hydrogen in the presence of nickel or platinum and is called **catalytic hydrogenation** to produce primary amines.

$$\underset{\text{Alkylcyanide}}{R-C\equiv N} + 2H_2 \xrightarrow{\text{Ni or Pt}} \underset{\text{Primary amine}}{R-CH_2-NH_2}$$

Here, the complete reduction can also be carried out with $LiAlH_4$ in ether or Na with alcohol (known as **Mendius reaction**).

$$CH_3-C\equiv N + 4[H] \xrightarrow[\text{or Na}/C_2H_5OH]{LiAlH_4/\text{ether}} \underset{\text{Ethylamine}}{CH_3CH_2NH_2}$$

Reaction with Grignard reagent

Cyanides react with Grignard reagents dissolved in anhydrous ether to form addition products, which are decomposed by heating with water in the presence of acid to form ketones.

$$RC\equiv N + R'MgX \xrightarrow{\text{Anhydrous ether}} \underset{\text{Addition product}}{\left[R-\overset{\overset{R'}{|}}{C}=NMgX\right]} \xrightarrow{H_2O/H^+} \underset{\text{Ketone}}{R-\overset{\overset{R'}{|}}{C}=O} + Mg(OH)X$$

Uses

(*i*) Cyanides such as acetonitrile or methyl cyanide are useful solvents.

(*ii*) Acrylonitrile or vinyl cyanide ($CH_2=CHCN$) is used in the manufacture of vinyl polymers.
PAN (poly acrylonitrile) is used to manufacture carpets and blankets.

(*iii*) The cyanide compound **sodium nitroprusside** is used mainly in clinical chemistry to measure urine ketone bodies mainly as a follow up to diabetic patients.

(*iv*) Cyanide is also used in electroplating, where it stabilises metal ions in the electrolyte solution prior to their deposition.

CHECK POINT 21.2

1. The IUPAC name for the compound is

$$\underset{CN}{CH_2}-\underset{CN}{CH}-\underset{CN}{CH_2}$$

(a) 1, 2, 3- tricyanopropane
(b) 3-cyanopentane-1,5- dinitrile
(c) propane-1, 2, 3- tricarbonitrile
(d) 1, 2, 3- propane nitrile

2. Which of the following substances on treatment with P_2O_5 gives ethanenitrile?
(a) Propanamide
(b) Ethanamide
(c) Ethanoic acid
(d) N-methyl ethyl amine

3. Which of the following can behave as both electrophile and nucleophile?
(a) CH_3CN (b) CH_3OH
(c) $CH_2{=}CH{-}CH_3$ (d) CH_3NH_2

4. Boiling of propane nitrile in aqueous alkaline solution will result in the formation of
(a) $CH_3CH_2CH_2NH_2$
(b) $CH_3CH_2COO^- + NH_3$
(c) $CH_3COOH + NH_3$
(d) $CH_3CH_2NH_2$

5. Stephen's reduction converts nitrile into
(a) aldehydes (b) ketones
(c) amines (d) carboxylic acid

6. A mixture of stannous chloride and conc. HCl reduces an alkane nitrile to alkanal. This is known as
(a) Perkin reaction
(b) Nef reaction
(c) Stephen's reaction
(d) Ullmann reaction

7. Which of the following will yield benzaldimine hydrochloride?
(a) Benzene and hydrazine
(b) Hydrazine and HCl
(c) Benzonitrile and $SnCl_2$ / HCl
(d) Nitro benzene and $SnCl_2$ / HCl

8. The product formed by treatment of ethanol and ethane nitrile in the presence of hydrochloric acid is
(a) ethyl acetate
(b) diethyl ether
(c) ethyl methyl ketone
(d) butanol

9. Mendius reaction involves the
(a) reduction of aldehydes to give alcohols
(b) reduction of nitriles with sodium and ethanol
(c) oxidation of nitriles
(d) hydrolysis of cyanides

10. Methyl cyanide on treatment with methyl magnesium bromide followed by hydrolysis gives
(a) propanone (b) ethane
(c) ethanol (d) propanol

ISOCYANIDES

Isocyanide is an isomeric compound of cyanide. They have —NC as a functional group. It has a triple bond but constitutes a dipolar species in which nitrogen has a positive charge and carbon has a negative charge ($R{-}N^+{\equiv}\bar{C}$). This is opposite to the characters of carbon and nitrogen in the cyanide group. It is a resonance hybrid of the following two structures.

$$R-\overset{+}{N}\equiv\overset{-}{C}: \longleftrightarrow R-\ddot{N}=C:$$

Nomenclature

Commonly, the suffix '*carbylamine*' is added to the name of the alkyl or aryl group but in IUPAC system, these are named by adding the suffix '*isocyanide*' to the name of the alkyl or aryl group.

Compound	Common name	IUPAC name
$CH_3{-}N \rightleftarrows C$	Methyl carbylamine	Methyl isocyanide
$C_2H_5{-}N \rightleftarrows C$	Ethyl carbylamine	Ethyl isocyanide
$C_6H_5{-}N \rightleftarrows C$	Phenyl carbylamine	Phenyl isocyanide

Methods of preparation

Various methods for the preparation of isocyanides are as follows

(*i*) **By nucleophilic displacement of the halogen**
Alkyl or aryl halides react with AgCN to form alkyl isocyanide.

$$R-X + AgCN \longrightarrow \underset{\text{Alkyl isocyanide}}{RN \rightleftarrows C} + AgX$$

[where *X*-halogen atom]

(ii) **From primary amines** Primary amines on heating with chloroform and ethanolic solution of potassium hydroxide form isocyanides or carbylamine. This reaction is known as **carbylamine reaction.**

$$R\text{—}NH_2 + CHCl_3 + 3KOH\,(alc.) \xrightarrow[-3H_2O]{-3KCl} \underset{\text{Carbylamine or isocyanide}}{RN \rightleftarrows C}$$

Physical properties

Some important physical properties are as follows

(i) Isocyanides are colourless volatile liquid with extremely unpleasant smell.

(ii) They are poisonous in nature.

(iii) They are almost immiscible in water due to the lack of hydrogen, since the nitrogen atom does not have any lone pair of electron. These are soluble in organic solvents like ether, benzene etc.

(iv) Alkyl isocyanides have lower boiling point than isomeric cyanide. Infact $\text{—}N \rightleftarrows C$ bond is less polar than $\text{—}C \rightleftarrows N$ bond. Therefore, dipolar interactions are also less in this case as compared with alkyl cyanides.

$CH_3\text{—}C \rightleftarrows N$ and $CH_3\text{—}N \rightleftarrows C$ have same molar mass i.e. 41 but have different boiling points. i.e. 355 K and 332 K respectively.

Chemical properties

Various types of chemical reactions shown by isocyanides are as follows

1. Hydrolysis

Isocyanides are hydrolysed by dilute acids (but not by alkali) to form a primary amine and formic acid.

$$R\text{—}N \rightleftarrows C \xrightarrow{H_2O/H^+} \underset{\substack{\text{N-alkyl formamide} \\ \text{(addition intermediate)}}}{\left[R\text{—}\underset{\underset{H}{|}}{N}\text{—}\overset{\overset{O}{\|}}{C}\text{—}H \right]} \xrightarrow{H_2O/H^+} \underset{\text{Amine}}{RNH_2} + \underset{\text{Formic acid}}{HCOOH}$$

2. Reduction

Isocyanides give N-substituted amines i.e. secondary amines on complete reduction.

$$R\text{—}N \rightleftarrows C \xrightarrow{LiAlH_4} R\text{—}\underset{\underset{H}{|}}{N}\text{—}CH_3$$

3. Oxidation

Isocyanides are oxidised to isocyanates on reaction with HgO or with ozone as well as with a halogen and DMSO (dimethyl sulphoxide).

$$RNC \xrightarrow[\text{or } O_2/NiCl_2]{O_3} R\text{—}N{=}C{=}O$$

$$RNC + Cl_2 + CH_3\text{—}\underset{\underset{O}{\|}}{S}\text{—}CH_3 \longrightarrow RNCO + CH_3\text{—}\underset{\underset{Cl}{|}}{\overset{\overset{Cl}{|}}{S}}\text{—}CH_3$$

4. Addition reactions

Isocyanides react with halogens, sulphur, mercuric oxide or ozone to form addition products.

e.g. $R\text{—}\overset{+}{N}{\equiv}\overset{-}{C}\!: + Cl_2 \longrightarrow \underset{\text{Alkyl imino carbonyl chloride}}{R\text{—}N{=}CCl_2}$

$\xrightarrow{S} \underset{\text{Alkyl isothiocyanate}}{R\text{—}N{=}C{=}S}$

$\xrightarrow{O_3} N{=}C{=}O + O_2$

$\xrightarrow{2HgO} \underset{\text{Alkyl isocyanate}}{R\text{—}N{=}C{=}O} + Hg_2O$

Methyl isocyanate is responsible the Bhopal gas tragedy in December 1894.

5. Isomerisation

When isocyanides are heated for a longer duration they rearrange to form more stable cyanides.

$$R\text{—}\overset{+}{N}{\equiv}\overset{-}{C}\!: \xrightarrow{\Delta} R\text{—}C{\equiv}N$$

CHECK POINT 21.3

1. Which of the following represent the correct structure for isocyanide ion?

   (a) $R—N \rightleftharpoons C$

   (b) $R—\ddot{N} = C:$

   (c) $R—\overset{+}{N} \equiv \overset{-}{C}:$

   (d) All of the above

2. Cyanide and isocyanide are the isomers of what type?
   (a) Positional
   (b) Functional
   (c) Tautomer
   (d) All of the above

3. The IUPAC name for C_2H_5NC is
   (a) propyl carbylamine
   (b) propyl isocyanide
   (c) ethyl carbylamine
   (d) ethyl isocyanide

4. Which reagent should be employed to get ethyl carbylamine from ethyl iodide as major product?
   (a) HCN
   (b) AgCN
   (c) KCN
   (d) $AgNO_2$ followed by reduction

5. Reaction of primary amine with chloroform in alkaline solution produces unpleasant smelling compound '*X*'. '*X*' can be
   (a) *R*CN (b) *R*NCO
   (c) *R*NC (d) None of these

6. An organic compound with formula C_3H_5N on hydrolysis forms an acid which reduces Fehling solution. The compound can be
   (a) ethanenitrile (b) isocyanoethane
   (c) ethoxyethane (d) propanenitrile

7. Consider the following reaction

   $$RNC \xrightarrow{'X'} R—\underset{\underset{H}{|}}{\ddot{N}}—CH_3$$

   '*X*' in the above reaction is
   (a) $NaBH_4$ (b) $LiAlH_4$ (c) DIBAl—H (d) H_2O/H^+

8. Consider the following reaction

   $$RNC + Cl_2 + CH_3—\underset{\underset{O}{\|}}{S}—CH_3 \longrightarrow 'X'$$

   What is *X* in the reaction?
   (a) RNH_2 (b) *R*NCO (c) *R*Cl (d) *R*SOCl

9. When isocyanide is heated with chlorine the product formed is
   (a) alkyl imino carboxyl chloride
   (b) alkyl amino carbonyl chloride
   (c) alkyl imino carbonyl chloride
   (d) alkyl amino carboxyl chloride

10. When isocyanides are heated for longer time it forms
    (a) cyanides
    (b) isocyanates
    (c) amines
    (d) None of the above

(A) Topical exercise

I. Diazonium salts

1 Select the correct statement(s) about arene diazonium ions

(a) They act as weak electrophiles in S_E reaction
(b) They react only with strongly activated ring (as phenol)
(c) Both (a) and (b)
(d) None of the above

2 Diazotisation of aniline is done by HNO_2. Excess of HNO_2 can be removed by addition of

(a) urea
(b) ammonium chloride
(c) Both (a) and (b)
(d) $NaNO_2$

3 Consider the following statements.

Statement I Arene diazonium salts react only with strongly activated ring.

Statement II They act as weak electrophiles in S_E reaction.

(a) Both Statements I and II are correct
(b) Only Statement I is correct
(c) Only Statement II is correct
(d) Both Statements I and II are incorrect

4 Consider the following statements.

Statement I Aromatic diazonium salts are more stable than aliphatic diazonium salts.

Statement II Diazonium salts are stabilised by resonance.

(a) Both Statements I and II are correct
(b) Only Statement I is correct
(c) Only Statement II is correct
(d) Both Statements I and II are incorrect

5 Sodium sulphite reacts with benzene diazonium chloride forming

(a) phenyl hydrazine
(b) aniline
(c) benzene
(d) benzonitrile

6 Benzene diazonium chloride reacts with hypophosphorous acid to produce

(a) phenol
(b) benzene
(c) *p*-hydroxyazobenzene
(d) benzonitrile

7 In the following compounds I and II, which positions are activated for coupling reactions?

NH_2 (positions 1, 2, 3, 4) $O^{\ominus}$ (I) $NH_3^{\oplus}$ (positions 1, 2, 3, 4) OH (II)

(a) 2 or 3 in (I) and 1 and 4 in (II)
(b) 1 or 4 in (I) and 2 and 3 in (II)
(c) 1 or 4 in (I) and (II)
(d) 2 or 3 in (I) and (II)

8 Consider the following statements

Statement I Aqueous H_3PO_2 converts benzene diazonium salt into benzene.

Statement II H_3PO_2 is a reducing agent.

(a) Both Statements I and II are correct
(b) Only Statement I is correct
(c) Only Statement II is correct
(d) Both Statements I and II are incorrect

9 Consider the following statements.

Statement I Urea is added in the diazotisation reaction.

Statement II Urea decomposes in presence of excess HNO_2.

(a) Both Statements I and II are correct
(b) Only Statement I is correct
(c) Only Statement II is correct
(d) Both Statements I and II are incorrect

10 Which of the following undergoes coupling reaction at the fastest rate?

(a) C_6H_5–NH_2
(b) C_6H_5–$\overset{\oplus}{N}H_3$
(c) C_6H_5–$O^{\ominus}$
(d) C_6H_5–OH

11 $C_6H_5-\overset{+}{N}{=}N\overset{-}{Cl}$ + (1-naphthol) $\xrightarrow{0°C}$ Orange dye.

It will be

(a) $C_6H_5-N{=}N-$(naphthyl)$-OH$

(b) (naphthyl with OH)$-N{=}N-C_6H_5$

(c) $C_6H_5-N{=}N-$(naphthyl)$-OH$

(d) None of the above

12 Consider the following statements.

Statement I Azo compounds are strongly coloured.

Statement II Benzene rings in conjugation with azo group is called chromophore.

(a) Both Statements I and II are correct
(b) Only Statement I is correct
(c) Only Statement II is correct
(d) Both Statements I and II are incorrect

13 Match the species (in Column I) with the reactions/reagents from which they are obtained given (in Column II) from diazonium salts,

	Column I		Column II
A.	C_6H_6 (benzene)	1.	H_2O
B.	C_6H_5-Cl	2.	Gomberg
C.	$C_6H_5-C_6H_4-NO_2$	3.	Sandmeyer reaction
D.	C_6H_5-OH	4.	H_3PO_2

Codes

	A	B	C	D		A	B	C	D
(a)	1	3	2	4	(b)	4	2	3	1
(c)	1	2	3	4	(d)	4	3	2	1

14 Aromatic diazonium group can be replaced by various groups/atoms. In this context, match Column I with Column II and select the correct answer using the codes given below in the columns

Column I (Reagent reaction types)	Column II
A. H_3PO_2	1. Hydrogen
B. Sandmeyer	2. Hydroxyl group
C. H_2O/Δ	3. Aryl group
D. Gomberg	4. Chlorine

Codes

	A	B	C	D		A	B	C	D
(a)	4	3	2	1	(b)	1	4	2	3
(c)	1	2	3	4	(d)	2	3	4	1

15 Identify Y in the reaction

$$C_6H_5NH_2 \xrightarrow[273\text{ K}]{NaNO_2/HCl} X \xrightarrow[\text{dil. }H_2SO_4]{H_2O} Y$$

(a) $C_6H_5N_2Cl$
(b) C_6H_5OH
(c) C_6H_5NHOH
(d) C_6H_6

16 Aniline when diazotised in cold and then treated with dimethyl aniline gives a coloured product. Its structure would be

(a) $CH_3NH-C_6H_4-N{=}N-C_6H_4-NHCH_3$

(b) $CH_3-C_6H_4-N{=}N-C_6H_4-NH_2$

(c) $(CH_3)_2N-C_6H_4-N{=}N-C_6H_5$

(d) $(CH_3)_2N-C_6H_4-NH-C_6H_5$

17 Which of the following method gives *p*-bromonitrobenzene from aniline?

I. $C_6H_5NH_2 \xrightarrow{HNO_3} \xrightarrow[273\text{ K}]{NaNO_2/HCl} \xrightarrow{CuBr/HBr}$

II. 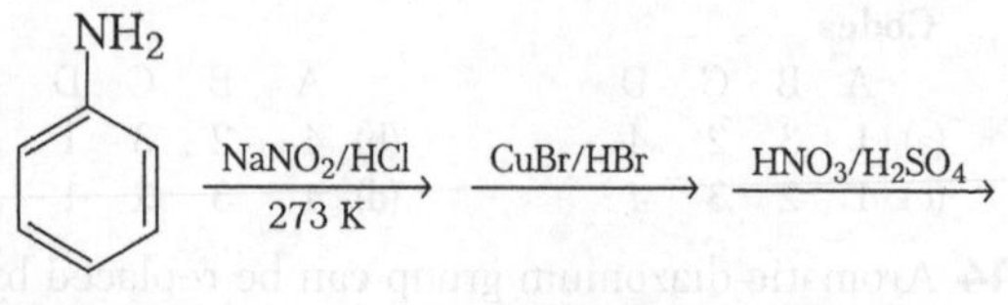 Aniline ($C_6H_5NH_2$) $\xrightarrow[273\text{ K}]{NaNO_2/HCl}$ $\xrightarrow{CuBr/HBr}$ $\xrightarrow{HNO_3/H_2SO_4}$

(a) Only I (b) Only II
(c) Both (a) and (b) (d) None of these

II. Cyanides

18 Consider the following statements.

Statement I IUPAC name of 1,2,3-tricyanopropane is propane-1, 2, 3-tricarbonitrile.

Statement II Cyano at C-2 is not the part of the chain.

(a) Both Statements I and II are correct
(b) Only Statement I is correct
(c) Only Statement II is correct
(d) Both Statements I and II are incorrect

19 Consider the following statements.

Statement I IUPAC name of methyl cyanide is ethanenitrile.

Statement II Cyanides and isocyanides are called nitriles.

(a) Both Statements I and II are correct
(b) Only Statement I is correct
(c) Only Statement II is correct
(d) Both Statements I and II are incorrect

20 For an alkane nitrile
(a) nitrile carbon is *sp*-hybridised
(b) there can be nucleophilic attack and electrophilic attack on nitrile carbon and nitrogen respectively
(c) basic character is less than ammonia or amines
(d) All of the above statements are true

21 Select the incorrect statement(s).
(a) CH_3CN is named as methane nitrile
(b) urea is an amide of carbamic acid
(c) urea is excreted in the urine as the end product of protein metabolism
(d) urea is a diamide

22 Nitriles resemble
(a) terminal alkynes (b) isocyanide
(c) internal alkynes (d) carbon dioxide

23 Lone pair on nitrogen in nitriles makes it
(a) basic (b) acidic
(c) neutral (d) amphoteric

24 $\underset{x}{\overset{\uparrow}{C}}H_3\underset{y}{\overset{\uparrow}{C}}\underset{z}{\overset{\uparrow}{N}}$ hybridisations on *x*, *y* and *z* respectively are
(a) *sp*, *sp*, *sp* (b) sp^2, *sp*, sp^2
(c) sp^3, *sp*, sp^2 (d) sp^3, *sp*, *sp*

25 Aromatic nitriles (ArCN) are not prepared by reaction
(a) ArX + KCN
(b) ArN_2^+ + CuCN
(c) $ArCONH_2 + P_2O_5$
(d) $ArCONH_2 + SOCl_2$

26 Consider the following statements.

Statement I $SOCl_2$ dehydrates $RCONH_2$ into RCN.

Statement II SO_2 is a better leaving group.

(a) Both Statements I and II are correct
(b) Only Statement I is correct
(c) Only Statement II is correct
(d) Both Statements I and II are incorrect

27 Aromatic nitriles (ArCN) are not prepared by reaction
(a) 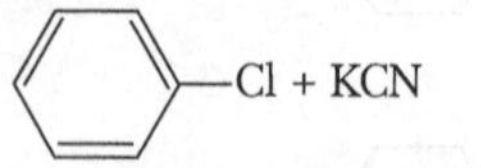 C_6H_5—Cl + KCN
(b) 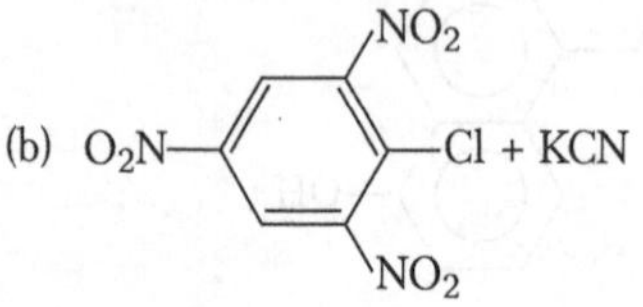 2,4-dinitro: O_2N—$C_6H_2(NO_2)_2$—Cl + KCN (with NO_2 groups ortho to Cl)
(c) 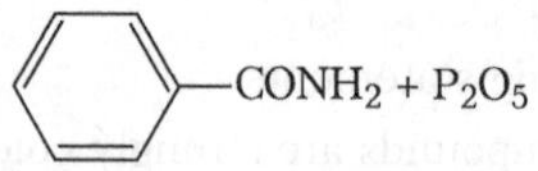 C_6H_5—$CONH_2 + P_2O_5$
(d) 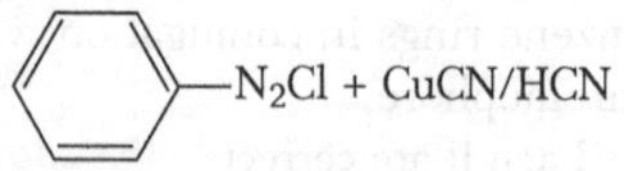 C_6H_5—N_2Cl + CuCN/HCN

28 In the following reaction,

$$CH_3CN \xrightarrow{CH_3Br/NaNH_2} CH_3CH_2NH_2$$

intermediate (s) is/are
(a) $:\overset{\ominus}{C}H_2CN$ (b) $NH_2^{\ominus}$
(c) Both (a) and (b) (d) $\overset{+}{C}H_4CN$

29 $CH_3CH_2CN \xrightarrow{A} CH_3CH_2CHO$

A can be
(a) DIBAH/$H_3O^{\oplus}$ (b) LiAlH (*t*- OBu)$_3$ / $H_3O^{\oplus}$
(c) Both (a) and (b) (d) None of these

30 Methyl cyanide (*A*) can be converted into methylamine (*B*) by
(a) $A \xrightarrow{LiAlH_4} B$
(b) $A \xrightarrow[\text{partial}]{H_3O^+} \xrightarrow[\Delta]{Br_2/NaOH} B$
(c) Both (a) and (b)
(d) $A \xrightarrow{RMgX} \xrightarrow{H_2O/H^+} B$

31 Which of the following best describes the nucleophilic addition step in the acid catalysed hydrolysis of acetonitrile (CH_3CN)?

(a) $H_3C-C{\equiv}N:$ attacked by $H-\ddot{O}:^-$ (b) $H_3C-C{\equiv}N:$ attacked by $H_2\ddot{O}:$

(c) $H_3C-\overset{\oplus}{C}{=}\overset{+}{N}H$ attacked by $H-\ddot{O}:^-$ (d) $H_3C-\overset{\oplus}{C}{=}\overset{+}{N}H$ attacked by $H_2\ddot{O}:$

32 Y in the following reaction is

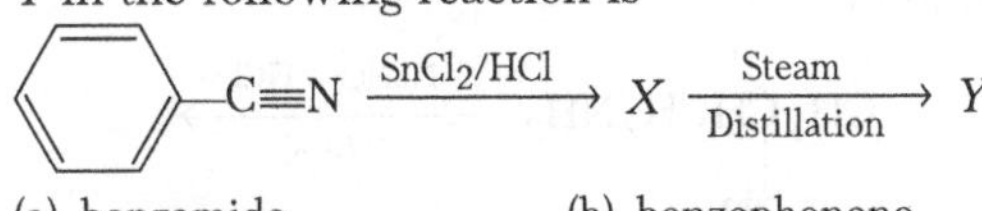

(a) benzamide (b) benzophenone
(c) benzoic acid (d) benzaldehyde

33 Consider the following statements.

Statement I Reduction of methyl cyanide with $LiAlH_4$ gives methyl amine.

Statement II Acetamide is converted to methyl amine by Hofmann bromamide degradation reaction.

(a) Both Statements I and II are correct
(b) Only Statement I is correct
(c) Only Statement II is correct
(d) Both Statements I and II are incorrect

34 Basic hydrolysis of CH_3CN forms

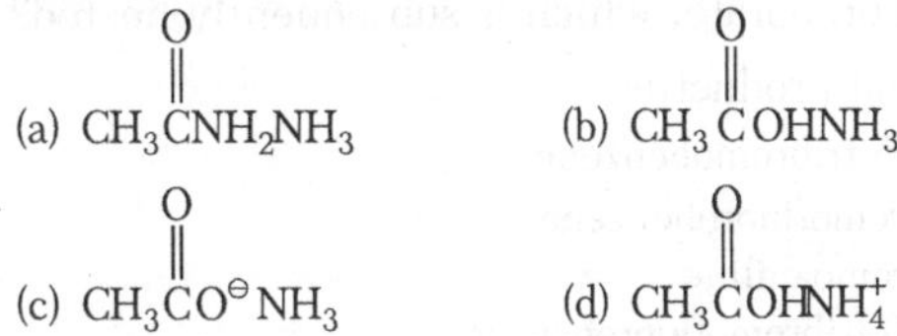

35 $LiAlH_4$/ether reduces methyl cyanide into
(a) methylamine (b) ethylamine
(c) methylimine (d) ethylimine

36 Match the reagents in Column I used with CH_3CN with the products obtained in Column II.

Column I	Column II
A. H_3O^+	1. CH_3CHO
B. $LiAlH_4$	2. CH_3COCH_3
C. DIBAH / H_3O^+	3. CH_3COOH
D. CH_3MgBr / H_3O^+	4. $CH_3CH_2NH_2$

Codes

	A	B	C	D		A	B	C	D
(a)	1	2	3	4	(b)	3	4	1	2
(c)	4	3	2	1	(d)	2	1	3	4

37 Phenyl isocyanides are prepared by which of the following reaction?
(a) Rosenmund's reaction
(b) Carbylamine reaction
(c) Reimer-Tiemann reaction
(d) Wurtz reaction

38 Match the compounds given in Column I with the items given in Column II.

Column I	Column II
A. Benzene sulphonyl chloride	1. Zwitter ion
B. Sulphanilic acid	2. Hinsberg reagent
C. Alkyl diazonium salts	3. Dyes
D. Aryl diazonium salts	4. Conversion to alcohols

Codes

	A	B	C	D		A	B	C	D
(a)	4	3	2	1	(b)	2	3	4	1
(c)	1	4	3	2	(d)	2	1	4	3

III. Isocyanide

39 The IUPAC name of C_2H_5NC is
(a) ethyl isocyanide (b) methyl isocyanide
(d) phenyl isocyanide (d) None of these

40 Consider the following reaction

$R—NH_2 + CHCl_3 + 3KOH(\text{alc.}) \longrightarrow RN\overset{\rightarrow}{=}C$

The name of the reaction is
(a) isocyanide reaction
(b) Stephen's reaction
(c) carbylamine reaction
(d) Heisenberg reaction

41 Which of the following has lowest boiling point?
(a) $CH_3CH_2CH_2CN$ (b) $CH_3\underset{|}{\overset{CH_3}{CH}}—CN$
(c) C_3H_7NC (d) None of these

42 Consider the following statements about isocyanides
I. These are colourless voltatile liquids.
II. They are poisonous in nature.
III. They are miscible with water.
IV. They have lower boiling point than isomeric cyanide.

Which of the given Statement(s) is/are incorrect?
(a) Both I and II (b) Only IV
(c) Only III (d) None of these

43 Which of the following reaction is incorrect?

(a) $R—NC \xrightarrow{H_2O/H^+} RNH_2 + HCOOH$

(b) $R—NC \xrightarrow{LiAlH_4} R—\underset{H}{N}—CH_3$

(c) $RNC \xrightarrow{O_2/NiCl_2} R—N{=}C{=}O$

(d) $R—N{\equiv}C + Cl_2 \longrightarrow RNHCHCl_2$

(B) Miscellaneous questions

1 The gas evolved when methylamine reacts with nitrous acid is

(a) NH_3 (b) N_2
(c) H_2 (d) C_2H_6

2 What is the end-product of the following reaction?

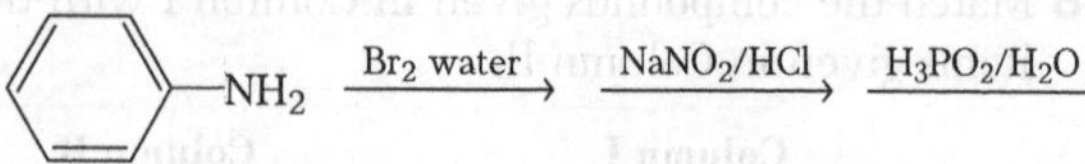

(a) 1,3,5-tribromobenzene (b) Bromobenzene
(c) Bromophosphobenzene (d) Bromonitrobenzene

3 $C_6H_5-N{=}\overset{+}{N}-\overset{-}{Cl}$ + 2-naphthol (OH) $\xrightarrow{273\ K}$ orange dye.

(a) $C_6H_5-N{=}N-$(naphthyl)$-OH$

(b) $C_6H_5-N{=}N-$(naphthyl with OH)

(c) $C_6H_5-N{=}N-$(naphthyl)$-OH$

(d) None of the above

4 In order to prepare a 1° amine from an alkyl halide with simultaneous addition of one CH_2 group in the carbon chain, the reagent used as source of nitrogen is

(a) sodium amide, $NaNH_2$
(b) sodium azide, NaN_3
(c) potassium cyanide, KCN
(d) potassium phthalimide $C_6H_4(CO)_2N^-K^+$

5 What is the end product of the following reaction?

$$CH_3CN \xrightarrow[H_3O^+]{CH_3MgBr/ether} \xrightarrow[H_3O^+]{CH_3MgBr/ether} \xrightarrow{Cu,\ \Delta}$$

(a) $(CH_3)_2CO$ (b) $(CH_3)_3COH$
(c) $(CH_3)_2C{=}CH_2$ (d) $(CH_3)_2CHCHO$

6 What is the end-product of the following reaction?

$$C_6H_5-NH_2 \xrightarrow[273\ K]{NaNO_2/HCl} \xrightarrow{CuCN/HCN} \xrightarrow{Sn/HCl}$$

(a) C_6H_5-CHO (b) C_6H_5-COOH
(c) $C_6H_5-CH_2NH_2$ (d) C_6H_6

7 Most stable intermediate in the following reaction is

$$CH_3CH(CH_3)CH_2NH_2 \xrightarrow[273\ K]{NaNO_2/HCl}$$

(a) $CH_3CH(CH_3)CH_2N_2^{\oplus}$ (b) $CH_3CH(CH_3)CH_2^{\oplus}$
(c) $CH_3\overset{\oplus}{C}(CH_3)CH_3$ (d) $CH_3CH(CH_3)CH_2OH_2^{\oplus}$

8 Aniline is reacted with bromine water and the resulting product is treated with an aqueous solution of sodium nitrite in presence of dilute hydrochloric acid. The compound so formed is converted to a tetrafluoroborate, which is subsequently heated? The final product is

(a) 1,3,5-tribromobenzene
(b) *p*-bromofluorobenzene
(c) *p*-bromoaniline
(d) 2,4,6-tribromofluorobenzene

9 Which of the following can be decompose by HNO_2?

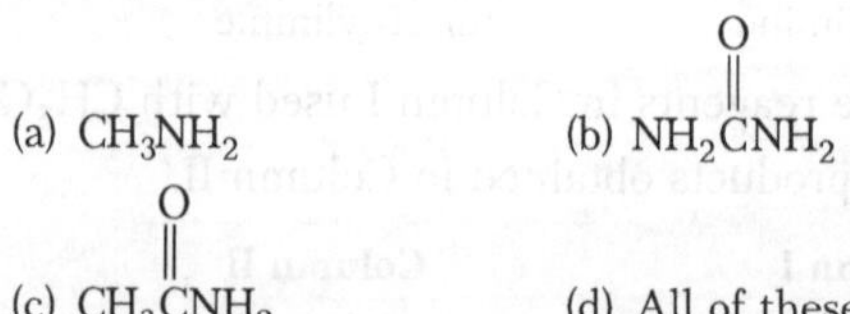

(a) CH_3NH_2 (b) NH_2CONH_2
(c) CH_3CONH_2 (d) All of these

10 $CH_3CH_2CH_2NH_2$ changes to $CH_3CH(OH)CH_3$ on reaction with $NaNO_2$ and HCl which involves

(a) formation of diazonium salt
(b) 1,2-hydride shift
(c) Both (a) and (b)
(d) 1, 3-hydride shift

11 Consider the following substances

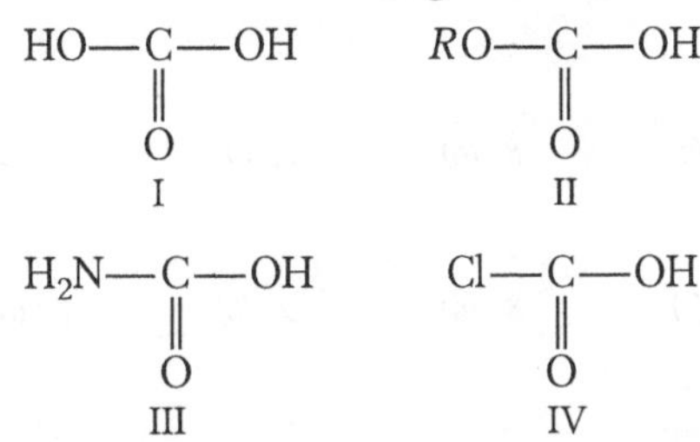

Which of the above can give CO_2 on heating ? Select the correct alternate from the following

(a) Both I and II (b) Both II and III
(c) Only I (d) All of these

12 A compound (*X*) has the molecular formula C_7H_7NO. On treatment with Br_2 and KOH (*X*) gives an amine (*Y*). (*Y*) gives carbylamine test (*Y*) upon diazotisation and coupling with phenol gives an azo dye (*Z*). *X* is

(a) $PhCONH_2$ (b) $PhCONHCOCH_3$
(c) $PhNO_2$ (d) $PhCOONH_4$

13 When $PhNO_2$ is reduced in alkaline medium, the product is

(a) $Ph—\overset{O}{\overset{\uparrow}{N}}=N—Ph$ (azoxybenzene)
(b) Ph—N=N—Ph (azobenzene)
(c) Ph—NH—NH—Ph (hydrazobenzene)
(d) All of the above

14 Toluene is nitrated and resulting product is reduced with tin and hydrochloric acid. The product so obtained is diazotised and then heated with cuprous bromide solution. The reaction mixture so obtained contains

(a) mixture of *o*- and *p*- bromotoluenes
(b) mixture of *o*- and *p*- bromobenzenes
(c) mixture of *o*- and *p*-bromoanilines
(d) mixture of *o*- and *p*- bromotoluenes

15 Consider the following sequence of reactions

$$F—C_6H_4—NO_2 \xrightarrow[\text{DMF, }\Delta]{(CH_3)_2NH} (A) \xrightarrow[0\text{-}5°C]{\text{(i) Fe/HCl, (ii) NaNO}_2\text{/HCl}} (B)$$

The product (*B*) is

(a) $H_2N—C_6H_4—N(CH_3)_2$
(b) $O_2N—C_6H_4—N(CH_3)(CH_3NH_2)$
(c) $H_2N—C_6H_4—NH_2$
(d) $O_2N—C_6H_4—NH_2$

(C) Medical entrances' gallery

1 A given nitrogen containing aromatic compound *A* reacts with Sn/HCl, followed by HNO_2 to give an unstable compound *B*, *B* on treatment with phenol, forms a beautiful coloured compound *C* with the molecular formula $C_{12}H_{10}N_2O$. The structure of compound *A* is **[NEET-II 2016]**

(a) $C_6H_5NH_2$ (b) $C_6H_5NO_2$
(c) C_6H_5CN (d) $C_6H_5CONH_2$

2 Which of the following will be most stable? **[CBSE AIPMT 2014]**

(a) $CH_3N_2^+X$ (b) $C_6H_5N_2^+X^-$
(c) $CH_3CH_2N_2^+X^-$ (d) $C_6H_5CH_2N_2^+X^-$

3 Unpleasant smell of carbylamine is obtained when chloroform and alcoholic KOH are heated with

(a) any aliphatic amine **[KCET 2013]**
(b) any amine
(c) any primary amine
(d) any aromatic amine
(e) any secondary amine

4 In the reaction *A* is **[NEET 2013]**

$$\text{4-}N_2^+Cl^-\text{-3-bromonitrobenzene} \xrightarrow{\Delta} \text{3-bromonitrobenzene}$$

(a) $HgSO_4$ / H_2SO_4 (b) Cu_2Cl
(c) H_3PO_2 and H_2O (d) H^+ / H_2O

5 When CH_3CONH_2 is heated with P_2O_5, the product is **[UP CPMT 2012]**

(a) $CH_3CCl_2NH_2$ (b) CH_3CN
(c) CH_3COCl (d) CCl_3CONH_2

6 Which of the following compound will not undergo azo coupling reaction with benzene diazonium chloride? **[KCET 2012]**

(a) Aniline (b) Phenol
(c) Anisole (d) Nitrobenzene

7 The reaction $Ar\overset{+}{N}_2Cl^- \xrightarrow{Cu/HCl} ArCl + N_2\uparrow + CuCl$ is named as .... **[UP SEE 2012]**

(a) Sandmeyer reaction (b) Gattermann reaction
(c) Claisen reaction (d) Carbylamine reaction

ANSWERS

Check Point 21.1

1. (b)	2. (a)	3. (b)	4. (b)	5. (b)	6. (c)	7. (b)	8. (a)	9. (d)	10. (d)

Check Point 21.2

1. (c)	2. (b)	3. (a)	4. (b)	5. (a)	6. (c)	7. (c)	8. (a)	9. (b)	10. (a)

Check Point 21.3

1. (d)	2. (b)	3. (d)	4. (b)	5. (c)	6. (b)	7. (b)	8. (b)	9. (c)	10. (a)

(A) Topical exercise

1. (c)	2. (c)	3. (a)	4. (a)	5. (a)	6. (b)	7. (b)	8. (a)	9. (a)	10. (c)
11. (a)	12. (a)	13. (d)	14. (b)	15. (b)	16. (c)	17. (b)	18. (b)	19. (b)	20. (d)
21. (a)	22. (a)	23. (a)	24. (d)	25. (a)	26. (a)	27. (a)	28. (a)	29. (c)	30. (b)
31. (d)	32. (d)	33. (c)	34. (c)	35. (b)	36. (b)	37. (b)	38. (d)	39. (a)	40. (c)
41. (c)	42. (c)	43. (d)							

(B) Miscellaneous questions

1. (b)	2. (a)	3. (b)	4. (c)	5. (c)	6. (c)	7. (c)	8. (d)	9. (d)	10. (c)
11. (d)	12. (a)	13. (d)	14. (d)	15. (a)					

(C) Medical entrances' gallery

1. (b)	2. (b)	3. (c)	4. (c)	5. (b)	6. (d)	7. (b)

Hints & Explanations

(A) Topical exercises

2 (*c*) $HNO_2 + NH_4Cl \longrightarrow N_2\uparrow + 2H_2O + HCl$

$2HNO_2 + NH_2CONH_2 \longrightarrow N_2\uparrow + 3H_2O + CO_2$

5 (*a*) $C_6H_5\overset{+}{N_2}\overset{-}{Cl} + Na_2SO_3 \longrightarrow C_6H_5NH{\cdot}NH_2HCl$

Phenyl hydrazine

6 (*b*) $H_3PO_2 + H_2O \longrightarrow H_3PO_3 + 2H$

$C_6H_5N_2Cl + 2H \longrightarrow C_6H_6 + N_2\uparrow + HCl$

10 (*c*) Benzene diazonium salt is a weak electrophile and undergoes coupling reactions with activating group.

15 (*b*) $C_6H_5NH_2 \xrightarrow{NaNO_2/HCl} \underset{(X)}{C_6H_5\overset{+}{N_2}\overset{-}{Cl}} \xrightarrow[\Delta]{H_2O} \underset{(Y)}{C_6H_5OH}$

20 (*d*) $CH_3—\overset{sp}{\overset{\downarrow}{C}}{\equiv}N$, $CH_3—\overset{\oplus}{\underset{\uparrow}{C}}{=}N$ (electrophilic attack)

Nucleophilic attack

Basic character is less than that of ammonia or amines.

22 (*a*) $CH_3—C{\equiv}N$ nitrile

$CH_3—C{\equiv}H$ terminal alkyne

24 (*d*) $\underset{sp^3}{CH_3}—\underset{sp}{C}{\equiv}\underset{sp}{N}$

$x = sp^3$, $y = sp$, $z = sp$

25 (*a*) $C_6H_5—X$ this bond is stable due to resonance, hence S_N reaction is not possible.

30 (*b*) $CH_3CN \xrightarrow{H_3O^+} CH_3CONH_2 \xrightarrow{Br_2/NaOH} CH_3NH_2$

34 (*c*) $CH_3CN + 2H_2O \longrightarrow CH_3COOH + NH_3$

$\downarrow OH^-$

CH_3COO^-

35 (*b*) $\underset{\text{Methyl cyanide}}{CH_3CN} \xrightarrow{LiAlH_4} \underset{\text{Ethylamine}}{CH_3CH_2NH_2}$

37 (*b*) $C_6H_5NH_2 \xrightarrow[\Delta]{CHCl_3/KOH} C_6H_5NC$

Phenylisocyanide

39 (*a*) The IUPAC name of C_2H_5NC is ethyl isocyanide.

40 (*c*) $R—NH_2 + CHCl_3 + 3KOH$ (alc.) $\longrightarrow RN \overset{\rightarrow}{=} C$
Name of the reaction is carbylamine reaction. In this reaction aliphatic and aromatic primary amines on heating with chloroform and eth. KOH form isocyanides.

41 (*c*) C_3H_7NC has lower boiling point than isomeric cyanide.

42 (*c*) Only statement III is incorrect whereas all other statements are correct. Isocyanides are immiscible in water due to the lack hydrogen because the nitrogen atom does not have any lone pair of electron.

43 (*d*) Reaction (d) is incorrect. Corrected reaction is as follows:

$$R—\overset{+}{N} \equiv \overset{-}{C}\text{:} + Cl_2 \longrightarrow \underset{\text{Alkyl imino carbonyl chloride}}{R—N{=}CCl_2}$$

Isocyanides react with halogens, sulphur, mercuric oxide or ozone to form addition products.

(B) Miscellaneous questions

1 (*b*) Chemical reaction takes place during reaction of methylamine with nitrous acid is as follows

$$\underset{\text{Methylamine}}{CH_3—NH_2} \xrightarrow{HNO_2} CH_3—\overset{+}{N}{\equiv}NCl^- \xrightarrow{\Delta\, H_2O} \underset{\text{Methanol}}{CH_3OH} + N_2\uparrow$$

2 (*a*) $C_6H_5NH_2 \xrightarrow{Br_2\ water}$ 2,4,6-tribromoaniline $\xrightarrow[273\ K]{NaNO_2/HCl}$ 2,4,6-tribromobenzenediazonium chloride ($\overset{+}{N_2}\overset{-}{Cl}$) $\xrightarrow[\text{Reduction}]{H_3PO_2/H_2O}$ 1,3,5-tribromobenzene

4 (*c*) In order to prepare 1° amine from an alkyl halide with simultaneous addition of one CH_2 group in the carbon chain. The reagent used as a source of nitrogen is KCN. Chemical transformation can be shown as

$$\underset{\text{2-carbon containing haloalkane}}{CH_3CH_2X} \xrightarrow[-KX]{KCN} CH_3CH_2C{\equiv}N \xrightarrow{Na/C_2H_5OH} \underset{\text{3-carbon containing amine}}{CH_3CH_2CH_2NH_2}$$

5 (*c*) $CH_3CN \xrightarrow[H_3O^+]{CH_3MgBr/\text{ ether}} CH_3COCH_3 \xrightarrow[H_3O^+]{CH_3MgBr/\text{ ether}}$ $CH_3-C(CH_3)_2-OH \xrightarrow[-H_2O]{Cu/\Delta} CH_3-C(CH_3)=CH_2$

6 (*c*) $C_6H_5NH_2 \xrightarrow[273\text{ K}]{NaNO_2/HCl} C_6H_5\overset{+}{N_2}\overset{-}{Cl} \xrightarrow{CuCN/HCN} C_6H_5CN \xrightarrow{Sn/HCl} C_6H_5CH_2NH_2$

7 (*c*) $CH_3CH(CH_3)CH_2NH_2 \xrightarrow{HNO_2} CH_3CH(CH_3)CH_2N_2^{\oplus} \xrightarrow{-N_2} CH_3CH(CH_3)\overset{\oplus}{C}H_2$ (1° carbocation) $\xrightarrow{\text{1,2-hydride shift}} CH_3\overset{\oplus}{C}(CH_3)CH_3$ (3° carbocation (more stable than 1°))

8 (*d*) Aniline $\xrightarrow{Br_2\text{ water}}$ 2,4,6-tribromoaniline $\xrightarrow{NaNO_2/HCl}$ 2,4,6-tribromobenzene diazonium chloride ($\overset{+}{N_2}\overset{-}{Cl}$) $\xrightarrow{NaBF_4}$ $N_2^+BF_4^-$ salt $\xrightarrow{\Delta}$ 2,4,6-tribromo fluorobenzene

9 (*d*) $CH_3NH_2 \xrightarrow{HNO_2} CH_3OH + N_2\uparrow + H_2O$

$NH_2CONH_2 \xrightarrow{HNO_2} CO_2 + N_2\uparrow + H_2O$

$CH_3CONH_2 \xrightarrow{HNO_2} CH_3COOH + N_2\uparrow + H_2O$

10 (*c*) $CH_3CH_2CH_2NH_2 \xrightarrow{HNO_2} CH_3CH_2CH_2N_2^{\oplus} \longrightarrow CH_3CH_2CH_2^{\oplus} + N_2\uparrow$ (1° carbocation) $\xrightarrow{\text{1,2-hydride shift}} CH_3\overset{\oplus}{C}HCH_3$ (More stable 2° carbocation)

14 (*d*) Toluene $\xrightarrow[+\text{ Conc. }H_2SO_4]{\text{Conc. }HNO_3}$ *o*-nitrotoluene (*o*-isomer) + *p*-nitrotoluene (*p*-isomer) $\xrightarrow{\text{Sn/HCl (Reduction of }NO_2)}$ *o*-toluidine + *p*-toluidine $\xrightarrow{NaNO_2\ HCl\text{ diazotisation}}$ *o*- and *p*-methylbenzene diazonium chloride ($\overset{+}{N}\equiv N\overset{-}{Cl}$) $\xrightarrow{\text{CuBr, heat Sandmeyer reaction}}$ *o*- and *p*-bromotoluene $+ 2N_2$

15 (*a*) $F-C_6H_4-NO_2 \xrightarrow[DMF,\ \Delta]{(CH_3)_2NH} (CH_3)_2N-C_6H_4-NO_2$ (*A*) $\xrightarrow[\text{(iii) }H_2/Ni]{\text{(i) Fe/HCl (ii) }NaNO_2/HCl\ 0\text{-}5°C} (H_3C)_2N-C_6H_4-NH_2$ (*B*)

(C) Medical entrances' gallery

1 (*b*) The complete road map of the reaction can be seen as

Nitrobenzene (*A*) $\xrightarrow[\text{Reduction}]{Sn/HCl}$ Aniline $\xrightarrow{HNO_2}$ Benzene diazonium chloride (unstable) (*B*) $\xrightarrow{Ph-OH}$ $C_6H_5-N=N-C_6H_4-OH$ (Red colour dye)

2 (*b*) Diazonium salt containing aryl group directly linked to the nitrogen atom is most stable due to resonance stabilisation between the benzene nucleus and N – atom.

[Resonance structures of benzene diazonium ion: $C_6H_5-\overset{+}{N}\equiv N: \longleftrightarrow \overset{+}{C_6H_5}=N=\overset{-}{N}: \longleftrightarrow \ldots \longleftrightarrow C_6H_5-\overset{+}{N}\equiv N:$]

3 (*c*) Primary amines on heating with chloroform and alcoholic KOH produce foul smelling carbylamine (isocyanide).

$$\underset{\text{Primary amine}}{R\text{—}NH_2} + CHCl_3 + (\text{alc.})\,3KOH \longrightarrow \underset{\text{Foul smelling carbylamine (isocyanide)}}{R\text{—}NC} + 3KCl + 3H_2O$$

4 (*c*) When diazonium salt is treated with H_3PO_2 followed by hydrolysis, it diazonium group is replaced by –H resulting to the formation of hydrocarbon. Thus, *A* must be H_3PO_2 / H_2O.

5 (*b*) $CH_3CONH_2 \xrightarrow[\Delta]{P_2O_5} \underset{\text{Methylcyanide}}{CH_3CN} + H_2O$

6 (*d*) Nitrobenzene will not undergo azo coupling reaction with benzene diazonium chloride, while other three undergo diazo coupling reaction very easily. Diazonium cation is a weak E^+ and hence, reacts with electron rich compounds containing electron donating group i.e., —OH, —NH_2 and —OCH_3 groups and not with compounds containing electron withdrawing group, i.e. NO_2 etc.

$C_6H_5\overset{+}{N}\equiv N\overset{-}{Cl}$ + $C_6H_5NH_2$ ⟶ C_6H_5—N=N—C_6H_4—NH_2

$C_6H_5\overset{+}{N}\equiv N\overset{-}{Cl}$ + $C_6H_5OCH_3$ ⟶ C_6H_5—N=N—C_6H_4—OCH_3

$C_6H_5\overset{+}{N}\equiv N\overset{-}{Cl}$ + C_6H_5OH ⟶ C_6H_5—N=N—C_6H_4—OH

7 (*b*)

$$\underset{\text{Benzene diazonium chloride}}{C_6H_5\overset{+}{N_2}\overset{-}{Cl}} \xrightarrow{Cu/HCl} \underset{\text{Chlorobenzene}}{C_6H_5Cl} + N_2\uparrow + CuCl$$

This reaction is called Gattermann reaction. In this reaction Cl, Br and CN can be introduced into the benzene ring by simply treating diazonium salts with HCl, HBr, KCN, respectively in presence of copper powder instead of using Cu (I) salts.

CHAPTER 22

Biomolecules

Biomolecules are the organic molecules common to all living systems on earth. These molecules perform several physiological functions and act as building blocks of living body. These molecules include carbohydrates, proteins, enzymes, lipids, hormones, vitamins, nucleic acids etc.

CARBOHYDRATES

Carbohydrates (saccharides) occur naturally in the animal and plant kingdom. Sugars, starch and cellulose are the best known members of this class. Chemical analysis of sugar and other carbohydrates shows that they are composed of only the carbon, hydrogen and oxygen and they appear to be the hydrates of carbon. [The general formula is $C_x(H_2O)_y$]

Glucose ($C_6H_{12}O_6$) : $C_6 \cdot 6H_2O$

Sucrose ($C_{12}H_{22}O_{11}$) : $C_{12} \cdot 11H_2O$

But all the compounds which fit into this formula may not be classified as carbohydrates. e.g. acetic acid (CH_3COOH). Similarly, all the carbohydrates do not fit into this formula e.g., rhamnose ($C_6H_{12}O_5$) is a carbohydrate but does not fit in the formula.

Carbohydrates usually contain carbonyl, hydroxyl and ether groups. Hence, chemically, carbohydrates are polyhydroxy aldehydes or ketones or substances which yield such products on hydrolysis. The names of the individual sugars are characterised by the ending-ose.

Classification of carbohydrates

(*i*) On the basis of number of simplest molecules (monomers) produced upon hydrolysis, carbohydrates are classified in three main classes which are as follows

Monosaccharides These are the simplest carbohydrates that cannot be hydrolysed into simpler units (monomers). In other words, it is made up of a single monomer itself, e.g. glucose, fructose, galactose, etc.

These are further classified as **aldose** (functional group is aldehyde) and **ketose** (functional group is ketone) which we discuss later in this chapter.

Oligosaccharides These carbohydrates produce 2-10 monomer units upon hydrolysis. The ones that produce 2 monomer units are called **disaccharides**, the ones that produce 3 monomer units are called **trisaccharides** and so on, e.g. sucrose, lactose, maltose etc.

Inside

Polysaccharides These carbohydrates produce large number of monomer units (at least > 10) upon hydrolysis. e.g. glycogen, starch, cellulose etc.

(*ii*) On the basis of sweetness in taste, carbohydrates are divided into two simple types which are as follows

Sugars They taste sweet on our tongue. This class includes monosaccharides and disaccharides as mentioned above.

Non-sugars They do not taste sweet and include mainly polysaccharides.

(*iii*) On the basis of reducing activity, carbohydrate can be classified as following way

Reducing carbohydrates They contain free functional group (aldehyde or ketone) and reduce Fehling's solution and Tollen's reagent. These include all the monosaccharides (aldose and ketose).

Non-reducing carbohydrates They have bonded functional groups and cannot reduce Fehling's solution or Tollen's reagent. These include oligosaccharides and polysaccharides.

D, L-configuration

'D' or 'L' letter before the name of any compound indicates the relative configuration of particular stereoisomer in relation to two enantiomeric forms of glyceraldehyde.

D-means OH on right

CHO
H — OH
CH_2OH
D-glyceraldehyde

CHO
HO — H
CH_2OH
L-glyceraldehyde

L-means OH on left

The monosaccharides having configuration identical to D-glyceraldehyde are assigned D-configuration whereas those having configuration identical to L-glyceraldehyde are assigned L-configuration.

Monosaccharides

As discussed above, monosaccharides are the simplest carbohydrates and also called **simple sugars**. These cannot be further hydrolysed to simple sugars. On the basis of functional group present, these are further classified as

(*i*) **Aldose** They contain aldehyde ($-\overset{O}{\overset{\|}{C}}-H$) group as main functional group alongwith few hydroxyl (— OH) groups. Thus, they can be termed as polyhydroxy aldehydic monosaccharides. e.g. glucose, galactose.

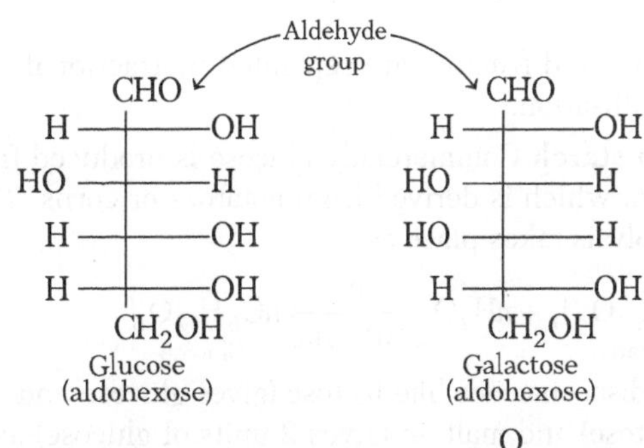

Glucose (aldohexose)   Galactose (aldohexose)

(*ii*) **Ketose** They contain ketone ($-\overset{O}{\overset{\|}{C}}-$) group as main functional group alongwith few hydroxyl (—OH) groups. Thus, they can be termed as polyhydroxy ketonic monosaccharides. e.g. fructose

CH_2OH
Ketone group → C=O
HO — H
H — OH
H — OH
CH_2OH
Fructose (ketohexose)

Note *Depending upon the number of carbon atoms present in a monomer unit, monosaccharide can be called pentose (5-carbon atoms) or hexose (6-carbon atoms) etc.*

Glucose

Glucose having molecular formula $C_6H_{12}O_6$ is an aldohexose. Due to the dextrorotatory optical activity of natural glucose, it is also called **dextrose.** It is white powdery solid, soluble in water and has melting point of 423 K. In nature it is found in sweet fruits, flower nectar etc., in free as well as in combined state. It constitutes about 20% of grapes and for this reason it is also known as **grape sugar.**

In our body, it is essentially present to be consumed up by different organs as an energy source. It is a normal constituent of blood and travels through our blood stream to our cells. Hence, it is also termed as **blood sugar**.

Preparation

It can be prepared by hydrolysis of cane sugar or starch in following ways.

(*i*) **From cane sugar** Cane sugar consists sucrose which when hydrolysed in acidic medium produces glucose and fructose.

$$\underset{\text{Sucrose}}{C_{12}H_{22}O_{11}} + H_2O \xrightarrow{H^+} \underset{\text{Glucose}}{C_6H_{12}O_6} + \underset{\text{Fructose}}{C_6H_{12}O_6}$$

Glucose and fructose are separated by fractional crystallisation.

(*ii*) **From starch** Commercially glucose is produced from starch, which is derived from potatoes or corns. The hydrolysis takes place as

$$\underset{\text{Starch}}{(C_6H_{10}O_5)_n} + nH_2O \xrightarrow[393\text{ K}, 2\text{-}3\text{ bar}]{H^+} \underset{\text{Glucose}}{nC_6H_{12}O_6}$$

Few disaccharides like lactose (gives glucose and galactose) and maltose (gives 2 units of glucose) are also used as sources of glucose.

Structure

Two types of structure, the open chain structure and cyclic structure, have been suggested for glucose molecule which are discussed below

1. Open chain structure

As discussed above, glucose is an aldohexose, having six carbon atoms including one aldehydic group. Its structure can be represented by using Fischer projection as

$$\begin{array}{c} H-\overset{1}{C}=O \\ | \\ H-\overset{2}{C}-OH \\ | \\ HO-\overset{3}{C}-H \\ | \\ H-\overset{4}{C}-OH \\ | \\ H-\overset{5}{C}-OH \\ | \\ \overset{6}{C}H_2OH \end{array}$$

D-glucose
(specific rotation = 53°)

The aldohexoses have four asymmetric carbons (*) and therefore exists as 2^4 (= 16) possible stereoisomers. These can be divided into two enantiomeric sets of eight diastereomers.

$$HOCH_2-\overset{*}{C}H(OH)-\overset{*}{C}H(OH)-\overset{*}{C}H(OH)-\overset{*}{C}H(OH)-CH=O$$

Aldohexoses
four asymmetric carbons (*)
$2^4 = 16$ stereoisomers

This structure suggested the presence of carbonyl ($>C=O$) group, primary alcohol ($-CH_2OH$), secondary alcohol ($>CHOH$) and above all presence of 6-carbon atom. Following chemical reactions give evidences in favour of open chain structure

(*i*) Oxidation of glucose with Br_2/water (mild oxidising agent) gives gluconic acid and with conc. HNO_3, it gives glucaric acid (or saccharic acid).

$$\underset{\substack{\text{Glucaric acid}\\\text{(or saccharic acid)}}}{\begin{array}{c}COOH\\|\\(CHOH)_4\\|\\COOH\end{array}} \xleftarrow[\text{I}]{\text{conc. }HNO_3} \underset{\text{Glucose}}{\begin{array}{c}CHO\\|\\(CHOH)_4\\|\\CH_2OH\end{array}} \xrightarrow[\text{pH 5-6}]{Br_2\text{ water}} \underset{\text{Gluconic acid}}{\begin{array}{c}COOH\\|\\(CHOH)_4\\|\\CH_2OH\end{array}}$$
II

- Reaction I indicates that glucose contains one primary alcoholic group.
- Reaction II indicates the presence of carbonyl group.

(*ii*) Glucose reduces both Tollen's reagent and Fehling's solution (a test of reducing sugars).

$$\underset{\text{Glucose}}{\begin{array}{c}CHO\\|\\(CHOH)_4\\|\\CH_2OH\end{array}} \xrightarrow[\Delta]{\substack{\text{Tollen's reagent,}\\2[Ag(NH_3)_2]OH}} \begin{array}{c}CO\overset{-}{O}\overset{+}{N}H_4\\|\\(CHOH)_4\\|\\CH_2OH\end{array} + \underset{\substack{\text{Silver}\\\text{mirror}}}{2Ag\downarrow} + 3NH_3 + H_2O$$

$$\xrightarrow[\Delta]{\substack{\text{Fehling's solution,}\\2Cu(OH)_2 + NaOH}} \begin{array}{c}COO^-Na^+\\|\\(CHOH)_4\\|\\CH_2OH\end{array} + \underset{\text{Red ppt.}}{Cu_2O\downarrow} + H_2O$$

(*iii*) P/HI reduces glucose into *n*-hexane which shows that all six carbon atoms are linked in a straight chain.

$$\text{Glucose} \xrightarrow{P/HI} n\text{-hexane}$$

(*iv*) Reduction of glucose with $NaBH_4$ or Na(Hg) / H_2O forms D-sorbitol (hexahydric aclohol).

$$\underset{\text{D-glucose}}{\begin{array}{c}CHO\\|\\(CHOH)_4\\|\\CH_2OH\end{array}} + 2[H] \xrightarrow[H_2O]{NaBH_4} \underset{\text{D-sorbitol}}{\begin{array}{c}CH_2OH\\|\\(CHOH)_4\\|\\CH_2OH\end{array}}$$

(*v*) Acetylation with acetic anhydride forms penta-acetate derivative, indicating presence of five —OH groups.

$$\underset{\text{Glucose}}{\begin{array}{c}CHO\\|\\(CHOH)_4\\|\\CH_2OH\end{array}} \xrightarrow[\text{or }5Ac_2O]{5(CH_3CO)_2O} \underset{\substack{\text{Glucose}\\\text{pentaacetate}}}{\begin{array}{c}CHO\\|\\(CHOAc)_4\\|\\CH_2OAc\end{array}} + 5CH_3COOH$$

Here, Ac is $\left(CH_3\overset{O}{\overset{\|}{C}}-\right)$.

(*vi*) Upon fermentation it forms alcohol.

$$\underset{\text{Glucose}}{C_6H_{12}O_6} \xrightarrow{\text{Zymase}} \underset{\text{Ethyl alcohol}}{2C_2H_5OH} + 2CO_2$$

(*vii*) It forms cyanohydrin on reaction with HCN.

$$\underset{\text{Glucose}}{CHO-(CHOH)_4-CH_2OH} + HCN \longrightarrow \underset{\text{Glucose cyanohydrin}}{CH(OH)(CN)-(CHOH)_4-CH_2OH}$$

Thus, one new (C—C) bond is formed. This reaction suggest that glucose contains a carbonyl group.

(*viii*) Glucose reacts with hydroxyl amine to form an oxime. This reaction confirms the presence of $>C=O$ group.

$$\underset{\text{Glucose}}{CHO-(CHOH)_4-CH_2OH} + NH_2OH \longrightarrow \underset{\text{Glucose oxime}}{H-C=NOH-(CHOH)_4-CH_2OH} + H_2O$$

(*ix*) It forms osazone with phenyl hydrazine ($PhNHNH_2$).

$$\underset{\text{Glucose}}{CHO-CHOH-(CHOH)_3-CH_2OH} + H_2N\cdot NHPh \xrightarrow[-H_2O]{} CH=N\cdot NHPh-CHOH-(CHOH)_3-CH_2OH$$

$$\xrightarrow[-PhNH_2]{} \underset{\text{Imino ketone}}{CH=NH-CO-(CHOH)_3-CH_2OH} \xrightarrow[-H_2O]{2PhNHNH_2} \underset{\text{Glucosazone}}{CH=NNHC_6H_5-C=NNHC_6H_5-(CHOH)_3-CH_2OH}$$

(Ph is C_6H_5- (phenyl group))

(*x*) Periodic acid splits glucose into formic acid and formaldehyde.

$$\underset{\text{Glucose}}{CHO-(CHOH)_4-CH_2OH} \xrightarrow{5HIO_4 \text{ or } H_5IO_6} \underset{\text{Formic acid}}{5HCOOH} + \underset{\text{Formaldehyde}}{HCHO}$$

2. Cyclic structure

Glucose, also show following chemical reactions that are against the open chain structure.

(*i*) Despite having aldehydic group, glucose does not give 2, 4-DNP test or Schiff's test and some of the other tests that most aldehydes give.

(*ii*) Glucose penta-acetate does not react with hydroxylamine.

(*iii*) Glucose exists in two anomeric forms, i.e. α - D-glucose and β-D-glucose. Thus, they have different melting points and specific rotations $[\alpha]_D$. In this structure —OH of C - 5 is added to the aldehydic carbon (C-1). Thus, a cyclic hemiacetal structure is formed.

This cyclic structure explains the absence of —CHO group and the existence of α-and β-forms.

anomeric carbon

β-D-(+)-glucose [cyclic structure]		D-glucose [open chain structure]		α-D-(+)-glucose [cyclic structure]
HO—¹C—H		H—¹C=O		H—¹C—OH
H—²C—OH		H—²C—OH		H—²C—OH
HO—³C—H	⇌	HO—³C—H	⇌	HO—³C—H
H—⁴C—OH		H—⁴C—OH		H—⁴C—OH
H—⁵C—O (ring to C-1)		H—⁵C—Ö—H		H—⁵C—O (ring to C-1)
⁶CH₂OH		⁶CH₂OH		⁶CH₂OH

When these cyclic forms are dissolved in water then these forms exist in equilibrium with open chain structure in the solution. This equilibrium solution has a different specific rotation from those of α-and β-forms.

α-D-glucose $\rightleftharpoons$ Equilibrium solution $\rightleftharpoons$ β-D-glucose

$[\alpha]_D = +18.7°$, mp = 423 K; $[\alpha]_D = +52.6°$; $[\alpha]_D = +112.2°$, mp = 419 K

This spontaneous change in specific rotations of solution of an optically active compound is called **mutarotation.** The two cyclic hemiacetal forms of glucose differ only in the configuration of hydroxyl group at C-1 called *anomeric carbon.* Such isomers, i.e. α-forms and β-form are called anomers.

Haworth projection of cyclic glucose

The cyclic structure of glucose is sometimes represented by Haworth projections to get an outlook of spatial arrangement of carbon in 3D. Here, bond as solid wedge represents position above plane of paper.

α-D-glucopyranose β-D-glucopyranose

Cyclic glucose is called glucopyranose as it resembles with pyran.

Fructose

Fructose, like glucose has the molecular formula $C_6H_{12}O_6$ but belongs to the class ketohexose as it consists six carbon atoms and a ketonic (>C=O) group. It is found in most of the fruits, hence, also called **fruit sugar.** As the natural fructose is laevorotatory, its called *laevulose.*

Preparation

As discussed earlier fructose can be produced from cane sugar, alongwith glucose. It can also be produced by the hydrolysis of inulin (a polysaccharide).

$$\underset{\text{Inulin}}{(C_6H_{10}O_5)_n} + nH_2O \xrightarrow[\Delta]{H_2SO_4} \underset{\text{Fructose}}{nC_6H_{12}O_6}$$

Structure

On the basis of the reactions, that are shown by fructose, it was found that fructose have a ketonic functional group at C-2 and six membered chain of carbon atoms. It has two primary alcoholic (— CH_2OH) groups and three secondary alcoholic (>CHOH) groups. The open chain structure of fructose is shown as

CH_2OH / C=O / HO—C—H / H—C—OH / H—C—OH / CH_2OH

D-fructose

Like glucose, fructose too exists in cyclic forms that include α- and β- anomeric forms as represented below, using the **Fischer projections.**

α-D-(−)-fructofuranose β-D-(−)-fructofuranose

The above structures can also be represented using **Haworth's projection** as

α-D-(-)-fructofuranose β-D-(-)-fructofuranose

Note *The cyclic fructose is called fructofuranose as it resembles with furan.*

Properties

(*i*) Fructose is a colourless crystalline compound having melting point of 375 K.

(*ii*) It is readily soluble in water but sparingly soluble in alcohols and insoluble in ether and benzene.

(*iii*) It is sweetest among all known carbohydrates.

(*iv*) It forms a laevorotatory solution and like glucose it also shows mutarotation.

(*v*) Most of the chemical properties of fructose are similar to those of glucose. e.g.

(a) **Oxidation by HIO_4** Fructose gives methanoic acid and acetaldehyde.

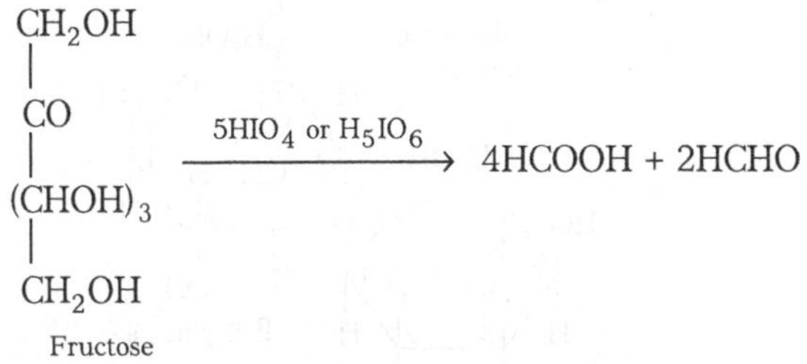

$$\begin{array}{l} CH_2OH \\ | \\ CO \\ | \\ (CHOH)_3 \\ | \\ CH_2OH \end{array} \xrightarrow{5HIO_4 \text{ or } H_5IO_6} 4HCOOH + 2HCHO$$

Fructose

(b) **Oxidation by HNO_3** It gives glycolic acid and tartaric acid.

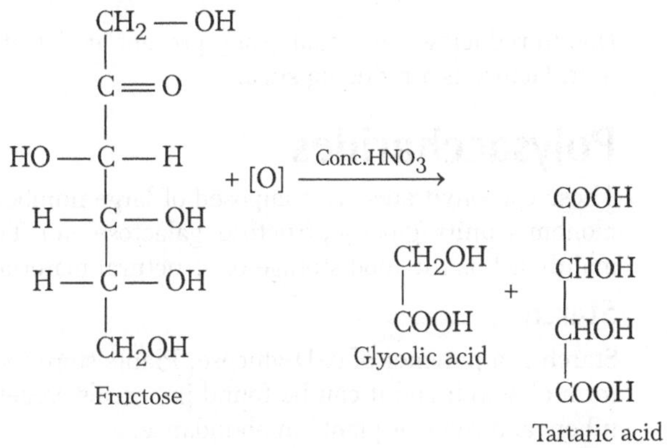

$$\begin{array}{c} CH_2-OH \\ | \\ C=O \\ | \\ HO-C-H \\ | \\ H-C-OH \\ | \\ H-C-OH \\ | \\ CH_2OH \end{array} + [O] \xrightarrow{\text{Conc.}HNO_3} \begin{array}{c} CH_2OH \\ | \\ COOH \end{array} + \begin{array}{c} COOH \\ | \\ CHOH \\ | \\ CHOH \\ | \\ COOH \end{array}$$

Fructose   Glycolic acid   Tartaric acid

(c) **Reduction** by $NaBH_4/H_2O$ Fructose give D-sorbitol and D-mannitol.

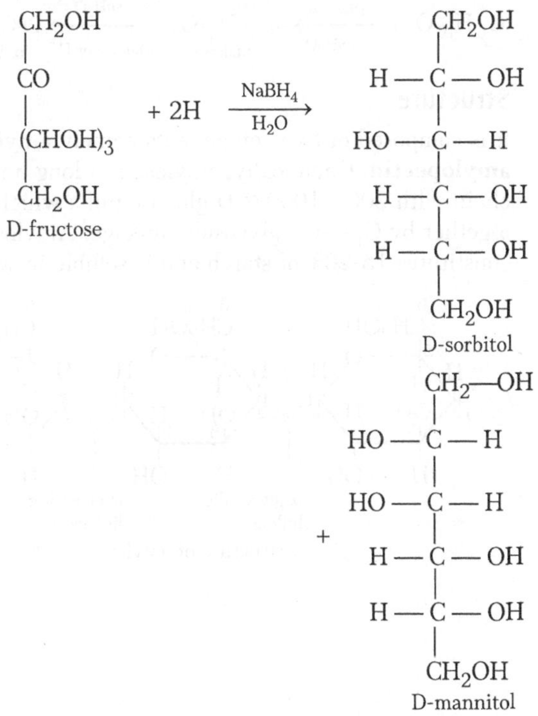

$$\begin{array}{l} CH_2OH \\ | \\ CO \\ | \\ (CHOH)_3 \\ | \\ CH_2OH \end{array} + 2H \xrightarrow[H_2O]{NaBH_4} \begin{array}{c} CH_2OH \\ | \\ H-C-OH \\ | \\ HO-C-H \\ | \\ H-C-OH \\ | \\ H-C-OH \\ | \\ CH_2OH \end{array} + \begin{array}{c} CH_2-OH \\ | \\ HO-C-H \\ | \\ HO-C-H \\ | \\ H-C-OH \\ | \\ H-C-OH \\ | \\ CH_2OH \end{array}$$

D-fructose   D-sorbitol   D-mannitol

Disaccharides

These saccharides on hydrolysis with dilute acids or enzymes gives two molecules of monosaccharides. These monosaccharides may be same or different.

These two monomer units are joined together through **an oxide linkage** which is formed by the loss of a water molecule. Such a linkage between two monosaccharide units through an oxygen atom is called as **glycosidic linkage.**

Some important disaccharides are as follow

1. Sucrose

It is actually the common table sugar we use in our daily lives. It is found in all green plants. It is commercially obtained from sugar cane and sugar beets. The molecular formula of sucrose is $C_{12}H_{22}O_{11}$.

Structure

Sucrose on hydrolysis gives equimolar mixture of D-(+)-glucose and D-(–)-fructose. These two monosaccharides are held together by a **glycosidic linkage** between C-1 of α-glucose and C-2 of β-fructose.

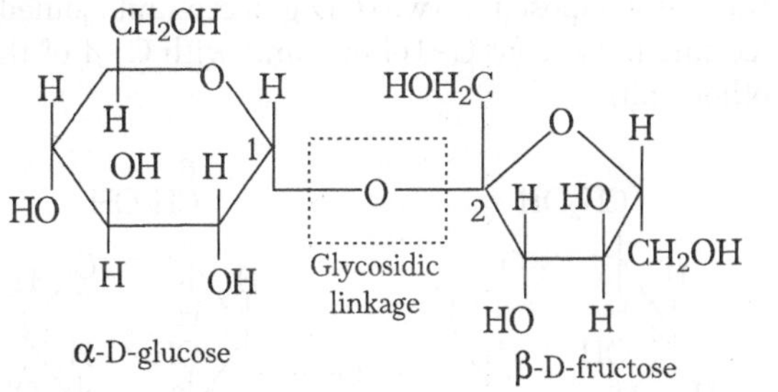

α-D-glucose   Glycosidic linkage   β-D-fructose

Properties

(*i*) It is a colourless, crystalline solid and soluble in water.

(*ii*) It is dextrorotatory but does not show mutarotation.

(*iii*) Since, the reducing groups of glucose and fructose are involved in glycosidic bond formation, sucrose is a **non-reducing sugar.** Thus, it does not reduce Tollen's reagent or Fehling's solution.

(*iv*) Upon strong heating, it forms charred mass of carbon.

$$C_{12}H_{22}O_{11} \xrightarrow{\Delta} 12C + 11H_2O$$

(*v*) On combustion in the presence of air it forms CO_2 gas.

$$C_{12}H_{22}O_{11} \xrightarrow[\text{Combustion}]{12O_2} 12CO_2\uparrow + 11H_2O$$

Inversion of cane sugar

Solution of sucrose is dextrorotatory (specific rotation = + 66.7°). When sucrose is hydrolysed it forms equimolar quantity of D-glucose and D-fructose. The solution of formed products, is found to be laevorotatory (specific rotation = −20.0°).

This change in optical properties of sucrose is called **inversion of cane sugar**. The equimolar product formed (that is D-glucose and D-fructose) is called **invert sugar**. The enzyme involved in this process (produced by a yeast) is called **invertase**.

2. Maltose

It is naturally found in germinating seeds. It can be produced by the partial hydrolysis of starch by the enzymes (amylase or diastase).

$$2(C_6H_{10}O_5)_n + nH_2O \xrightarrow{\text{Enzyme}} \underset{\text{Maltose}}{nC_{12}H_{22}O_{11}}$$

Hydrolysis of maltose by the enzyme maltase produces two units of glucose.

$$\underset{\text{(Maltose)}}{C_{12}H_{22}O_{11}} + H_2O \xrightarrow{\text{Maltase}} \underset{\text{(Glucose)}}{2C_6H_{12}O_6}$$

Structure

Maltose is composed of two α-D-glucose units joined by a glycosidic linkage (at C-1of one unit with C-4 of the another unit).

Glycosidic linkage

α-D-glucose (I) α-D-glucose (II)

It is a reducing sugar as C-1 of the second unit (with functional group) is not involved in the glycosidic linkage.

3. Lactose

It is found commonly in milk and known as **milk sugar**. Upon hydrolysis, it produces one unit of glucose and one unit of galactose.

$$C_{12}H_{22}O_{11} + H_2O \xrightarrow[\text{(enzyme)}]{\text{Lactase}} \underset{\text{Glucose}}{C_6H_{12}O_6} + \underset{\text{Galactose}}{C_6H_{12}O_6}$$

Structure

The molecule of lactose is composed of β-D-glucose unit and a β-D-galactose unit. The units are joined by the glycosidic linkage (at C-4 of glucose and C-1 of galactose).

Glycosidic linkage

β-D-galactose β-D-glucose

Due to reducing functional group present at C-1 of glucose unit, lactose is a reducing sugar.

Polysaccharides

These carbohydrates are composed of large number of monomer units (glucose, fructose, galactose etc). They mainly act as the food storage or structural material.

Starch

Starch is a polymer of α-D-glucose. Plants store food in form of starch and it can be found in cereals, vegetables, tubers and roots of plants in abundance.

It produces maltose on partial hydrolysis and glucose on complete hydrolysis.

$$(C_6H_{10}O_5)_n \xrightarrow[+nH_2O]{\text{Diastase}} \underset{\text{Maltose}}{nC_{12}H_{22}O_{11}} \xrightarrow[\text{Maltase or } H^+]{+nH_2O} \underset{\alpha\text{-D-glucose}}{2nC_6H_{12}O_6}$$

Structure

It is composed of two components namely **amylose** and **amylopectin**. Chemically, amylose is a long unbranched chain with 200 - 1000 α-D-glucose units which are held together by C_1—C_4 glycosidic linkage. Amylose constitutes 15-20% of starch and is soluble in water.

α-glycosidic linkage α-glycosidic linkage

Structure of amylose

On the other hand, amylopectin is branched chain with 25-30 α-D-glucose units, which are held together by $C_1—C_4$ glycosidic linkage.

The branching in chain occurs through $C_1—C_6$ glycosidic linkage.

Amylopectin constitutes 80-85% of starch and is insoluble in water.

Structure of amylopectin

2. Cellulose

It is mainly found in plants and is the main constituent of cell wall of plant cell. It is a polymer of β-D-glucose and is insoluble in water.

Cellulose has a straight chain structure. The β-D-glucose units are linked together by $C_1—C_4$ glycosidic linkage.

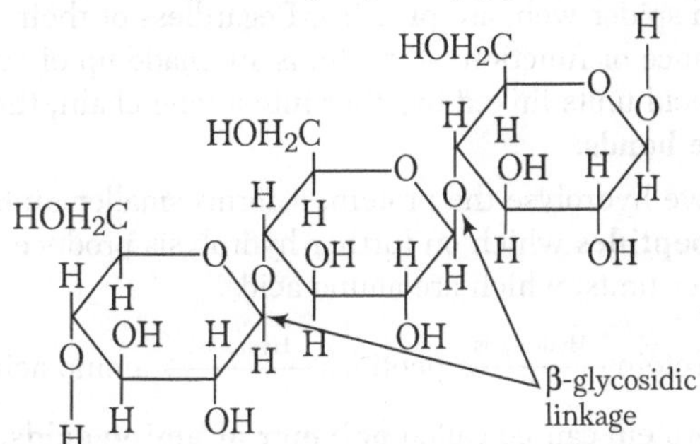

Structure of cellulose

3. Glycogen

It is found in animal body (mainly in liver and muscles) as reserve food and is called *animal starch.* Like starch, it is a polymer of α-D-glucose ($[C_6H_{10}O_5]_n$). When glucose is needed in body it breaks down by the action of enzymes.

Structurily, glycogen is highly branched and resemble with the structure of amylopectin.

CHECK POINT 22.1

1. Carbohydrates are called hydrates of carbon because of their general formula, which is
(a) $C_nH_{2n}O_n$ (b) $C_n(H_2O)_n$
(c) $C_x(H_2O)_y$ (d) $(CH_2O)_n$

2. Which of the following is not a monosaccharide?
(a) Lactose (b) Galactose
(c) Glucose (d) Fructose

3. Reducing carbohydrates can reduce
(a) Fehling's solution
(b) Tollen's reagent
(c) Both (a) and (b)
(d) Neither (a) nor (b)

4. Identify the sugar that belongs to the category of ketohexose.
(a) Glucose (b) Galactose
(c) Maltose (d) Fructose

5. Which of the following reagents is suitable to convert glucose into *n*-hexane?
(a) P/HI (b) $NaBH_4$
(c) Ac_2O (d) HNO_3

6. The cyclic hemiacetal structure of glucose resembles closely with the structure of
(a) benzene (b) pyran
(c) furan (d) cyclohexane

7. Which of the following is the product of fermentation of glucose, in the presence of enzyme zymase?
(a) Acetic acid (b) Ethyl alcohol
(c) Formic acid (d) Methane

8. Which of the following compounds upon reaction with glucose forms glucosazone?
(a) HCN (b) NH_2OH
(c) $NaBH_4$ (d) $C_6H_5NHNH_2$

9. Which of the following biomolecules upon hydrolysis produces only fructose?
(a) Inulin (b) Insulin
(c) Starch (d) Cane sugar

10. Which of the following carbohydrates is found exclusively in plant kingdom?
(a) Cellulose (b) Glucose
(c) Glycogen (d) Starch

PROTEIN

Proteins are large biomolecules that occur in every living organism. They are of many different types and have many different biological functions.

The keratin of skin and fingernails, the fibroin protein of silk and spider web, are proteins. Regardless of their appearance or function, all proteins are made up of several amino acid units linked together into a long chain, through **peptide bonds.**

When we hydrolyse the protein, it forms smaller units called **peptides** which on further hydrolysis produce monomer units, which are amino acids.

$$\text{Protein} \xrightarrow{\text{Hydrolysis}} \text{peptides} \xrightarrow{\text{Hydrolysis}} \text{amino acids}$$

Thus, protein can be called **polymer of amino acids.**

Amino acids

As their name implies, amino acids are bifunctional. They contain an acidic carboxyl group (—COOH) and a basic amino group (—NH_2). Depending upon the relative position of these two groups, amino acids are classified as $\alpha, \beta, \gamma, \delta$ etc.

Hydrolysis of proteins gives only α-amino acid. In this type, both —COOH and —NH_2 groups are attached to the same carbon atom as represented below

α-carbon, H, O, Carboxyl group

R—C—COH

Alkyl chain, NH_2 ← Amino group

This type of structure where both acidic group (acquiring negative charge) and basic group (aquiring positive charge) are present is called **Zwitter ionic structure.**

Zwitter ion is a cation in acidic medium and migrates toward the cathode on passing electric current. It is an anion in basic medium and migrates toward the anode on passing electric current.

$$\underset{\text{Anion in basic medium}}{R\text{—}\underset{NH_2}{\underset{|}{CH}}\text{—}COO^-} \xleftarrow{OH^-} R\text{—}\underset{\overset{\oplus}{NH_3}}{\underset{|}{C}}HCOO^- \xrightarrow{H^+} \underset{\text{Cation in acidic medium}}{R\text{—}\underset{\overset{\oplus}{NH_3}}{\underset{|}{C}}HCOOH}$$

> **Isoelectric point**
>
> (PI) An amino acid bears a positive charge in acidic solution (low pH) and a negative charge in basic solution (high pH). The isoelectric point ((PI)) of the amino acid is the pH at which it has no net charge. It is the pH at which the amount of negative charge on an amino acid exactly balances the amount of positive charge.
>
> PI (isoelectric point) = pH at which there is no net charge

There are about 20 amino acids commonly found in proteins (refer Table 22.1) and are standard amino acids. All are α-amino acids that means, amino group in each amino acid is a substituent on the α-carbon, the one next to the carboxyl carbon. **Most of them have primary amino group** — NH_2. However, **proline is a secondary amine.**

Each of the amino acids in Table 22.1 is referred by a three-letter symbol, e.g. gly for glycine, ala for alanine. In addition, a **one-letter symbol** is also used, as shown in parentheses in the table.

Table 22.1 Amino acids, their symbols and structures

Name	3-letter symbol (1-letter)	Structure	Isoelectric point (PI)
Neutral			
Glycine	Gly (G)	$H_2N—CH_2—COOH$	5.97
Alanine	Ala (A)	$CH_3—\underset{NH_2}{\underset{\vert}{CH}}—COOH$	6.02
Valine*	Val (V)	$CH_3—\underset{CH_3}{\underset{\vert}{CH}}—\underset{NH_2}{\underset{\vert}{CH}}—COOH$	5.97
Leucine*	Leu (L)	$CH_3—\underset{CH_3}{\underset{\vert}{CH}}—CH_2—\underset{NH_2}{\underset{\vert}{CH}}—COOH$	5.98
Isoleucine*	Ile (I)	$CH_3—CH_2—\underset{CH_3}{\underset{\vert}{CH}}—\underset{NH_2}{\underset{\vert}{CH}}—COOH$	6.02
Serine	Ser (S)	$HO—CH_2—\underset{NH_2}{\underset{\vert}{CH}}—COOH$	5.68

Name	3-letter symbol (1-letter)	Structure	Isoelectric point (PI)
Proline	Pro (P)	(pyrrolidine ring; N–H, COOH)	6.30
Phenylalanine*	Phe (F)	$C_6H_5CH_2—CH(NH_2)—COOH$	5.48
Cysteine	Cys (C)	$HS—CH_2—CH(NH_2)—COOH$	5.07
Glutamine	Gln (Q)	$H_2N—C(=O)—CH_2—CH_2—CH(NH_2)—COOH$	5.65
Methionine*	Met (M)	$H_3C—SCH_2CH_2CH(NH_2)—COOH$	5.71
Tryptophan*	Trp (W)	(indole ring, N–H)$—CH_2CH(NH_2)—COOH$	5.92
Threonine*	Thr (T)	$H_2N—CH(COOH)—CH(OH)—CH_3$	5.60
Tyrosine	Tyr (Y)	$HO—C_6H_4—CH_2CH(NH_2)—COOH$	5.64
Threonine*	Thr (T)	$H_2N—CH(COOH)—CH(OH)—CH_3$	5.60
Acidic			
Aspartic acid	Asp (D)	$HOOC—CH_2—CH(NH_2)—COOH$	2.77
Glutamic acid	Glu (E)	$HOOC—CH_2—CH_2—CH(NH_2)—COOH$	3.22
Asparagine	Asn (N)	$H_2NC(=O)CH_2—CH(NH_2)—COOH$	5.43
Basic			
Lysine*	Lys (K)	$H_2N—CH_2—(CH_2)_3—CH(NH_2)—COOH$	9.74
Arginine*	Arg (R)	$HN=C(NH_2)—NH—(CH_2)_3—CH(NH_2)—COOH$	10.76
Histidine*	His (H)	$CH=C—CH_2—CH(NH_2)—COOH$ (imidazole ring: N, NH, CH)	7.59

* *Essential amino acids*

Classification

I. On the basis of relative number of amino and carboxyl groups present, amino acids are classified in following three types.

(*i*) **Acidic amino acids** They contain more number of carboxylic (—COOH) groups than amino (—NH_2) groups, e.g. asp, glu etc.

(*ii*) **Basic amino acids** They contain more number of amino (—NH_2) groups than carboxyl (— COOH) groups, e.g. lys, arg, his etc.

(*iii*) **Neutral amino acids** They contain equal number of amino (—NH_2) groups and carboxyl (—COOH) groups. e.g. gly, ala, val etc.

II. On the basis of requirement in human diet, amino acids are classified into following two types

(*i*) **Essential amino acids** These are the amino acids that are not synthesised in human body and must be obtained from our diet. These are marked with asterisk (*) in Table 22.1, e.g. val, leu, lys etc.

(*ii*) **Non-essential amino acids** These are the amino acids that can be synthesised in our body and hence are not essentially required in our diet. e.g. gly, ala, ser etc.

Physical properties

(*i*) Amino acids are generally colourless, crystalline solids having melting point above 475 K.

(*ii*) They are soluble in water but insoluble in non-polar organic solvents. It is due to formation of Zwitter ion (inner salt).

$$R\text{CH}(\text{NH}_2)\text{COOH} \rightleftharpoons R\text{CH}(\overset{\oplus}{\text{N}}\text{H}_3)\text{COO}^{\ominus}$$

Zwitter ion

(*iii*) With the exception of glycine, all naturally occurring α-amino acids having asymmetric α-carbon atom and are called **chiral molecules.**

(*vi*) These exist in both 'D' and 'L' forms. Most naturally occurring amino acids have L-configuration.

$$\overset{\oplus}{\text{H}_3\text{N}}\text{—C(CO}_2^{\ominus}\text{)(CH}_2\text{OH)—H} \qquad \text{H—C(CO}_2^{\ominus}\text{)(CH}_2\text{OH)—}\overset{\oplus}{\text{N}}\text{H}_3$$

L-serine (*S*) D-serine (*R*)

Fischer projections

Peptides

Amino acids may be joined together by an amide linkage called as **peptide linkage** ($-\overset{O}{\overset{\|}{C}}NH-$). Each amino acid unit in a peptide is called a **residue.** In this linkage, an amino group from one residue forms an amide bond with the carboxyl of a second residue; the amino group of the second, forms an amide bond with the carboxyl of a third, and so on. e.g. alanylserine is a **dipeptide** that results when an amide bond is formed between the carboxyl group of alanine and the amino group of serine. A tripeptide has two -amide linkages fromed between three amino acids.

$$\underset{\text{Alanine (Ala)}}{\text{H}_2\text{N—CH(CH}_3\text{)—CO—OH}} + \underset{\text{Serine (Ser)}}{\text{H—NH—CH(CH}_2\text{OH)—COOH}} \xrightarrow{-\text{H}_2\text{O}} \underset{\text{(Ala-ser) Alanylserine}}{\text{H}_2\text{N—CH(CH}_3\text{)—CO—NH—CH(CH}_2\text{OH)—COOH}}$$

Amide bond

If the alanine amino group reacts with the serine carboxyl, it results in serylalanine. The term oligopeptide is generally used for the peptides containing about four to ten amino acid residues. While the term polypeptide is used if the unit consist of more than 10 amino acids. A polypeptide with more than hundred amino acid residues, having molecular mass more than 10000 is called **protein**.

Polypeptides are amphoteric because of the presence of terminal ammonium and carboxylate ions as well as the ionised side chains of amino acid residues. Most of the toxins (poisonous substances) in animal venoms and in plant sources are polypeptides. Oligopeptides with as few as three modified amino acid residues are effective as hormones. Aspartame (a derivative of dipeptide) is an artificial sugar substitute.

$$\text{H}_2\text{N—CH(CH}_2\text{—COOH)—CO—NH—CH(CH}_2\text{—C}_6\text{H}_5\text{)—COOCH}_3$$

Characteristics of peptide bond

(*i*) Peptide bond is formed by the linkage of carboxyl ($-\overset{O}{\overset{\|}{C}}-OH$) group of one amino acid with amino (—NH_2) group of another amino acid, with removal of a water molecules (as discussed above).

(*ii*) A peptide bond has about 40% double-bond character because of resonance.

(*iii*) Steric hindrance causes the *trans* configuration to be more stable than the *cis*-configuration, so the α-carbons of adjacent amino acids are *trans* to each other.

(*iv*) As a consequence of the partial double-bond character of the peptide bond, free rotation is not possible around this bond.

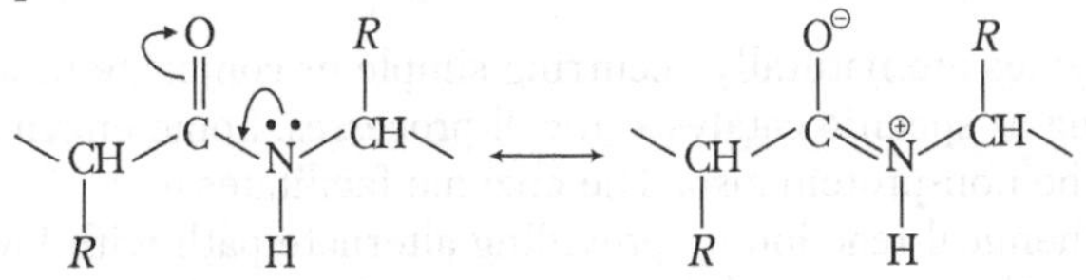

$$\diagdown CH(R)-C(=O)-\ddot{N}(H)-CH(R)\diagup \longleftrightarrow \diagdown CH(R)-C(O^{\ominus})=\overset{\oplus}{N}(H)-CH(R)\diagup$$

Types of protein

Proteins are classified into two types on the basis of their molecular shape

(*i*) **Fibrous proteins** These are elongated thread like molecules held together by hydrogen and disulphide bonds. These proteins are insoluble in water (due hydrophobic part) and are highly resistant to digestive enzymes. They form protective covering of body.
e.g. keratin, collagen.

(*ii*) **Globular proteins** These are spherical, having highly coiled chains and are soluble in water.
e.g. myoglobin, albumin, enzymes, antibodies.

Note *If a protein molecule is combined with a substance other than non-amino acid it is called **conjugated protein,** e.g. nucleoproteins (nucleic acid + protein), phosphoproteins (phosphate group + protein).*
*If a smaller protein is formed by partial hydrolysis of larger protein, it is called **derived protein**, e.g. protease, peptones etc.*

Structure of protein

As we discussed above, protein is a long peptide chain of amino acids. This long chain further acquire different configurations that results into a complex structure. On the basis of different configuration or conformations, there are mainly four types of structure of protein. These are as follows

1. Primary structure

The most basic aspect of protein structure is the primary structure-a specific sequence in which the α-amino acids are joined together by peptide bonds. Primary structure is represented by using the three-letter abbreviations for the α-amino acids.

e.g. if a peptide is formed by three amino acids say glycinc, alanine and proline, then it can be represented as gly-ala-pro. Here, each hyphen represent a peptide bond.

Primary structure gives an idea about the linkage of amino acid. All the amino acids are arranged in single, linear, polypeptide chain. Primary structure of insulin has two chains of insulin held together by two (—S—S—) linkage.

2. Secondary structure

In a primary structure few bonds are free to rotate while others are not. It can give many conformation. Out of these conformations, the stable one, which is held by many hydrogen bonds, gives rise to the secondary structure of protein.

Thus, the structure that describes, how a peptide chain is arranged or oriented in three-dimensional space, is called **secondary structure of protein.**

Secondary structure can exist in following two forms

(*i*) **α-helix** It is the spiral chain of peptide. In this structure, a protein chain is coiled about a central axis. It is held in place by hydrogen bonding that occurs between the hydrogens of —NH groups and oxygens of C=O groups that are oriented toward each other on the inside of the helix. It takes 3.6 amino acids to complete one turn of the helix to enable such hydrogen bonding. Thus, a 13-member ring is formed by H-bonding. α-helix is also known as 3.6_{13} helix. e.g. α-keratin, myosine, actin.

(*ii*) **β-pleated sheet** It resembles the pleated folds of drapery in which two protein chains are lined up parallel to each other running in opposite directions i.e. the N-terminus of one chain is near the C-terminus of the other. This puts the —NH group of one chain opposite the C=O group of the other chain, permitting hydrogen bonding that holds the sheet together.
e.g. fibroin, silk fibre, dragline of silk (spider wave).

3. Tertiary structure

The overall shape of a protein molecule is defined by the tertiary structure. It gives us idea that how the protein chains, whether pleated or helical are folded. The tertiary structure determines whether a protein is elongated (and thus, fibrous) or folded (thus, globular).

The tertiary structure of a protein brings into close proximity of α-amino acids that are very far apart in the primary structure, thereby creating chemical sites. The tertiary structure is held together through a number of chemical interactions as discussed below

(*i*) **Hydrogen bond** Present between oxygen atom of —COOH group of one amino acid and hydrogen atom of $—NH_2$ group of another amino acid, e.g. all types of protein.

(ii) **Electrostatic bond** Present between —COO^- group of one amino acid and —NH_3^+ of another amino acid, e.g. aspartic acid, glutamic acid.

(iii) **Disulphide bond** Formed between the —SH group of sulphur containing amino acid. e.g. cysteine and methionine.

(iv) **Hydrophobic bond** No true bond formed between hydrophobic alkyl chains of amino acids. e.g. all types of protein.

4. Quaternary structure

Certain proteins, usually globular proteins organise into aggregates, or assemblies with spatial orientation. These aggregates are described as the **quaternary structure** of the protein. It is composed of two or more protein. The binding forces of these peptides are hydrogen bond or electrostatic bond or hydrophobic bond (other than covalent bonds).

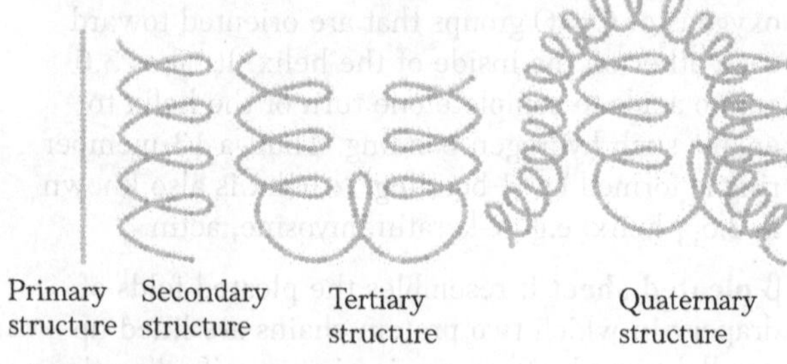

Fig 22.1 Schematic representation of different structure of protein.

Test for proteins

Proteins can be easily characterised by colour reactions. Their presence can be detected by the following tests.

(i) **Ninhydrin test** Protein when treated with a pyridine solution of ninhydrin, it result in formation of intense purple coloured complex.

(ii) **Biuret test** When an alkaline solution of proteins is treated with a drop of aqueous copper sulphate, it gives reddish-violet colour indicating presence of peptide linkage.

(iii) **Nitroprusside test** When sodium nitroprusside is added to proteins containing —SH group, a violet colour is observed.

(iv) **Million's test** When Million's reagent (a solution of mercurous and mercuric nitrates in HNO_3 containing little HNO_2) is added to protein, a white precipitate is obtained. This is characteristic of phenol and of only those proteins having this group, e.g. tryosine.

Denaturation of proteins

The loss in biological activity of proteins due to unfolding of globules and uncoiling of helix is called denaturation of proteins. During the denaturation, primary structure remains intact but secondary and tertiary structures are destroyed.
e.g. the coagulation of egg white on boiling, curding of milk caused by bacteria due to formation of lactic acid etc.

ENZYMES

Enzymes are naturally occurring simple or conjugate proteins acting as specific catalysts in cell processes. Some enzymes can be non-protein also. The enzyme facilitates a biochemical reaction by providing alternate path with lower activation energy and thus increasing the rate of reaction. During reactions, enzymes get inactivated and thus, periodically their replacement is necessary.

Enzymes are grouped into six classes according to the kind of reaction they catalyse (Table 22.2). Hydrolases catalyse hydrolysis reactions; isomerases catalyse isomerisations; ligases catalyse the bonding together of two molecules; lyases catalyse the breaking away of a small molecule such as H_2O from a substrate; oxidoreductases catalyse oxidations and reductions; and transferases catalyse the transfer of a group from one substrate to another.

Table 22.2 Classification of enzymes

S. No.	Main class	Some subclasses	Type of reaction catalysed
1.	Hydrolases	Lipases	Hydrolysis of an ester group
		Nucleases	Hydrolysis of a phosphate group
		Proteases	Hydrolysis of an amide group
2.	Isomerases	Epimerases	Isomerisation of stereogenic centre
3.	Ligases	Carboxylases	Addition of CO_2
		Synthetases	Formation of new bond
4.	Lyases	Decarboxylases	Loss of CO_2
		Dehydrases	Loss of H_2O
5.	Oxidoreductases	Dehydrogenases	Introduction of double bond by removal of H_2
		Oxidases	Oxidation
		Reductases	Reduction
6.	Transferases	Kinases	Transfer of a phosphate group
		Transaminases	Transfer of an amino group

All the 2000 or so known enzymes are globular proteins. In addition to the protein part, most enzymes also have small non-protein parts called **cofactors**. The protein part in such enzymes is called an **apoenzyme**, and the combination of apoenzyme plus cofactor is called a **holoenzyme.**

Only holoenzyme have biological activity, neither cofactor nor apoenzyme catalyses reactions by themselves.

Holoenzyme = Cofactor + apoenzyme

Cofactors can be either inorganic ions, such as Zn^{2+}, or small organic molecules, called **coenzymes.**

The requirement of many enzymes for inorganic cofactors is the main reason for our dietary need of trace minerals. Iron, zinc, copper, manganese, and numerous other metal ions are all essential minerals that act as enzyme cofactors, though the exact biological role is not known in all cases. A variety of organic molecules act as coenzymes.

Many though not all, coenzymes are **vitamins**, small organic molecules that must be obtained in the diet and are required in trace amounts for proper growth.

Optimum conditions for enzymatic activity

The optimum condition of temperature for enzymatic activity is the human body temperature (i.e 37°C or 310 K).

Their activity decreases with decrease or increase in temperature.

The optimum pH to carry out the enzyme-catalysed reactions is 7.4 under the 1 atm of pressure.

CHECK POINT 22.2

1. What do we get up on complete hydrolysis of protein?
   (a) Amino acids
   (b) Dipeptide
   (c) Oligopeptide
   (d) Polypeptide
2. In α-amino acid —NH_2 group is attached to the carbon atom
   (a) of —COOH group
   (b) adjacent to the —COOH group
   (c) Only of (a) or (b)
   (d) Other than (a) and (b)
3. Which of the following amino acids is neutral?
   (a) Arginine
   (b) Valine
   (c) Lysine
   (d) Histidine
4. Which of the following amino acids is not essential to be taken in diet?
   (a) Alanine (b) Valine
   (c) Lysine (d) Tryptophan
5. Which of the following amino acids consists of only secondary amino group?
   (a) Glycine (b) Leusine
   (c) methionine (d) Proline
6. Peptide bond is formed with release of a
   (a) water molecule (b) ethanol molecule
   (c) ammonia molecule (d) carbon dioxide molecule
7. The bond that give rise to secondary structure of protein is
   (a) amide bond (b) ionic bond
   (c) hydrogen bond (d) van der Waals' forces
8. In α-helix structure of protein, approximately how many amino acids take part in formation of turn of the helix?
   (a) 2.6 (b) 3.6
   (c) 6.3 (d) 13.0
9. Which of the following bonds is/are found in tertiary structure of protein?
   (a) H-bond (b) Ionic bond
   (c) Disulphide bond (d) All of these

HORMONES

The chemical substances secreted into blood by ductless glands are called hormones. They participate in different specific biochemical reactions and hence control specific physiological processes. Based on their chemical compositions, hormones are divided into two main classes (i) based on their structure –**steroid hormones** and **non-steroid hormones** (ii) based on activity in the cell. Important hormones, their type, source and functions are listed in Table 22.3.

Table 22.3 Classification and function of hormones

Type	Sub-type	Examples	Source	Functions
Steroids	Sex hormones	(i) Male sex hormones (Androgens)	Testis	**Testosterone** is responsible for male characteristics (deep voice, facial hair, physical constitution) during puberty. Synthetic testosterone analogs are used in medicine to promote muscle and tissue growth.
		(ii) Female sex hormones (Estrogens and progesterone)	Ovary	(a) **Estradiol** is responsible for secondary female characteristics and participates in control of menstrual cycle. (b) **Progesterone** is responsible for preparing the uterus for implantation of the fertilised egg and also plays important role as **birth-control agents.**
	Corticosteroids (Adrenal cortical hormones)	(i) Mineralocorticoids	Adrenal cortex	These regulate NaCl in the body and cause excretion of potassium in urine; are made by different cells in adrenal cortex.
		(ii) Glucocorticoids	Adrenal cortex	They modify certain metabolic reactions and have anti-inflammatory effect; are made by adrenal cortex.
Non-steroids	Peptide hormones	Insulin	Pancreas	Insulin has much effect on carbohydrate metabolism. This decreases glucose concentration in blood and is thus known as **hyperglycemic** factor. It promotes anabolic processes and inhibits catabolic ones. Its deficiency causes **diabetes mellitus.** Sanger (Nobelist) determined the structure of insulin.
		Oxytocin	Posterior pituitary	Causes contraction of smooth muscle, including contraction of uterus during child birth. It helps in milk ejection from mammary gland.
		Vasopressin	Posterior pituitary	It inhibits excretion of water from the body in form of urine.
		Glucogen	Pancreas	It hydrolyses glycogen and elevates blood sugar level.
	Amino acid derivatives	Thyroidal hormones	Thyroid	Thyroxin and triiodothyronine affect the general metabolism independent of their nature and specific activity. Thyroid gland is known as pace-setter of the endocrine system.
		Epinephrine (adrenaline)	Adrenal medulla	It increases blood pressure and heart beat. It increases alterness in body including secretion of sweat and dilation of pupil.

VITAMINS

Vitamins are the organic molecules which function in trace amounts in the body. They are required for the proper functioning of the cell, normal growth and reproduction and their absence causes specific deficiency diseases. Plants can synthesise all vitamins whereas only a few are synthesised in animals.

Vitamins are designated by the alphabets, e.g., A, B, C, D, E etc., in order of their discovery. Further any subgroup of individual vitamins is designated by the number subscript, e.g. B_1, B_2, B_6, B_{12}, etc.

Phsiological functions

The daily dose of any vitamin for any individual is not a fixed quantity and varies according to the size, age and rate of metabolism of the individual. A lack of one or more vitamins leads to characteristic deficiency symptoms in human (refer to Table 22.4). Multiple deficiencies caused by lack of more than one vitamin are common and this condition is known as **a vitaminosis**.

Classification of vitamins

Depending upon their solubility, all the known vitamins can be classified into following two groups.

(*i*) **Water-soluble vitamins** These include vitamins B and C. These vitamins are regularly eliminated from the body in the urine. Fruits and vegetables are a good source of water soluble vitamins. These vitamins are destroyed by prolonged heat (cooking).

(*ii*) **Fat-soluble vitamins** These include vitamins A, D, E and K. These vitamins are non-polar, hydrophobic (water insoluble), fairly stable and are not destroyed by heat. These are stored in liver and adipose tissues.

Table 22.4 Vitamins, their enzyme functions, sources and deficiency symptoms

S. No.	Vitamin	Enzyme function	Source	Deficiency symptom
Water-soluble vitamins				
1.	Ascorbic acid (Vitamin C)	Hydroxylases	Citrus fruits, green vegetables	Bleeding gums, bruising (scurvy)
2.	Thiamine (Vitamin B_1)	Reductases	Yeast, milk, green vegetables	Fatigue, depression, beri-beri (loss of apetite and retorted growth)
3.	Riboflavin (Vitamin B_2)	Reductases	Yeast, milk, vegetables, egg-white, liver, kidney	Cheilosis (cracked lips,) scaly skin, digestive disorders
4.	Pyridoxine (Vitamin B_6)	Aminotransferases	Cereal , grams, molasses, yeast, egg- yolk and meat	Anaemia, irritability, weakness
5.	Niacin (Vitamin B_5)	Reductases	Fresh meat, liver, fish, cereals, milk	Dermatitis, dementia
6.	Cyanocobalamine (Vitamin B_{12})	Isomerases	Liver of ox, sheep, pig, fish	Megaloblastic anaemia, neurodegeneration
7.	Pantothenic acid (Vitamin B_3)	Acyltransferases	Yeast, liver, kidneys, tomatoes	Weight loss, irritability
Fat-soluble vitamins				
8.	Retinol (Vitamin A)	Visual system	Fish oil, rice-polishing, liver, kidney	Night blindness, xerophthalmia, dry skin
9.	Ergocalciferol (Vitamin D)	Calcium metabolism	Sunlight, liver, egg, meat, milk	Rickets, osteomalacia
10.	α-Tocopherol (Vitamin E)	Antioxidant	Wheat germ oil, cotton seed oil and soyabean oil	Hemolysis of red blood cells
11.	Phylloquinone (Vitamin K)	Blood clotting	Cereals, leafy-vegetables	Hemorrhage, delayed blood clotting

NUCLEIC ACIDS

The nucleic acids are the chemical carriers of a cell's genetic information. The chemical composition of nucleic acids acts as coded information. This information when expressed, determines the nature of the cell, controls cell growth (and division) and directs biosynthesis of proteins (including enzymes) required for cellular functions.

Nucleic acids are of two types

(*i*) DNA-Deoxyribonucleic acid (*ii*) RNA-Ribonucleic acid

Chemical composition of nucleic acids

Nucleic acids are biopolymers made of **nucleotides** joined together to form a long chain. Hence, these are also called **polynucleotides.** Each nucleotide is composed of an aldopentose sugar linked to a heterocyclic nitrogenous base and further to a phosphate group. Combined form of sugar and base is called **nucleoside** and that of nucleoside and phosphate group is called **nucleotide.**

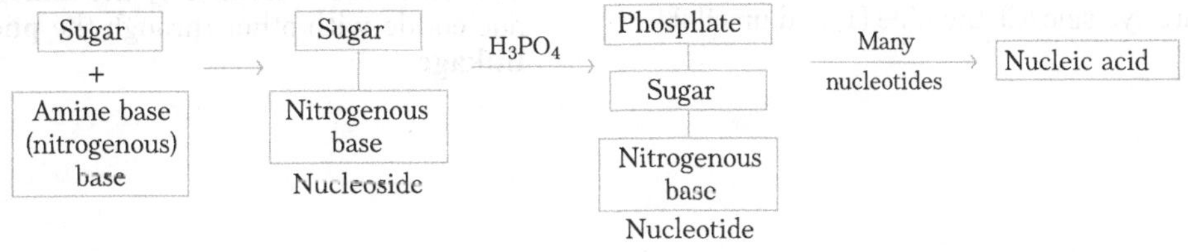

The nucleotides are abbreviated by three capital letters, preceded by '*d*' in case of deoxy series.

AMP – adenosine monophosphate
dAMP – deoxyadenosine monophosphate
ATP – adenosine triphosphate
UTP – uridine diphosphate

Thus, on complete hydrolysis of nucleic acid, three compounds (pentose sugar, phosphoric acid and nitrogenous base) are formed as discussed below.

(*i*) **Pentose sugar** The sugar component in RNA is ribose, and the sugar in DNA is 2′-deoxyribose.

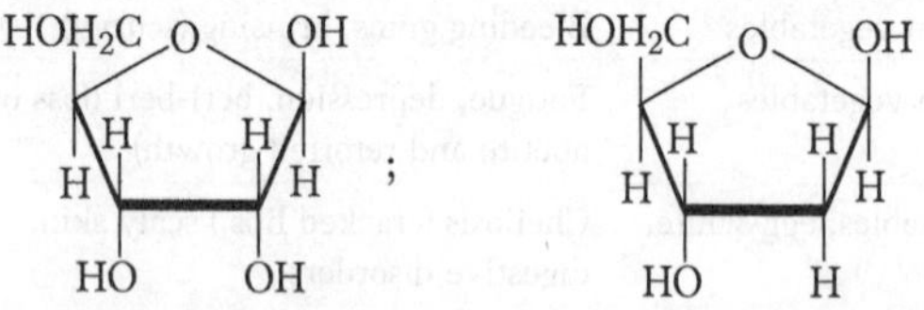

β-D-ribose used in RNA ; β-D-deoxyribose used in DNA

The prefix 2′-deoxy means that oxygen is missing from the 2′ position of ribose; numbers with a prime superscript refer to positions on the sugar component of a nucleotide and the numbers without a prime refer to positions on the heterocyclic amine base.

(*ii*) **Phosphoric acid** It is present as a phosphate (PO_4^{3-}) group in the nucleotide. It plays an important role of linking two nucleotides. The linkage is called **phosphodiester linkage** (discussed later).

(*iii*) **Nitrogenous bases** These are the heterocyclic compounds of nitrogen. These are further divided into two groups.

(a) **Purines** These are bicyclic groups containing N-atoms in each of the cycles. Examples are adenine (A) and guanine (G)

Adenine (A) Guanine (G)

(b) **Pyrimidines** These are single cyclic bases. The examples are cytosine (C), thymine (T) and uracil (U).

Cytosine (C) Thymine (T) Uracil (U)

- Three bases, A, G and C are common to both DNA and RNA. While the base 'T' is exclusive to DNA and the base 'U' is exclusive to RNA.
- So, there are four bases in each DNA and RNA. i.e. DNA = A, G, C, T and base in RNA = A, G, C, U
- Thus, DNA differ from RNA in sugar as well as nitrogenous base.

Structure of nucleic acids

In both DNA and RNA, the heterocyclic amine base is bonded to C-1′ of the sugar, and the phosphoric acid is bonded by a phosphate ester linkage to C-5′ sugar position. Thus, nucleosides and nucleotides have the general structure as shown below

Nucleoside

Nucleotide

Fig. 22.2 Structure of nucleic acid

The structure of nucleic acid can be divided into two following categories (with special reference to DNA).

(*i*) **Primary structure** It gives information about the sequence of nucleotides in the chain of nucleic acid. This structure is formed by the linkage of one nucleotide with other through the **phosphodiester linkage**.

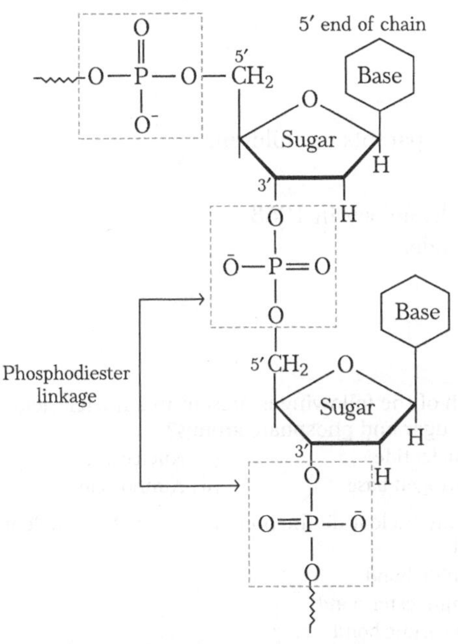

Fig. 22.3 Primary structure of DNA

Phosphodiester linkage, connects a sugar base with other one. The phosphate group in this linkage is attached to C–5′ position of one sugar and the C–3′ position of the other one. One end of the nucleic acid polymer has a free hydroxyl group at C—3′ (the 3′ end and the other end) has a phosphate at C—5′ (the 5′ end)

(*ii*) **Secondary structure** This structure explains the coiling of polynucleotide chains (primary structure). Watson and Crick proposed the secondary structure for DNA. This structure is called **double stranded helical structure**.

Here, two separate polynucleotide chains (or strands) are helically coiled with each other. These two strands are held together by H-bonds. The direction (with reference to sugar base) is opposite to each other (3′ or 5′).

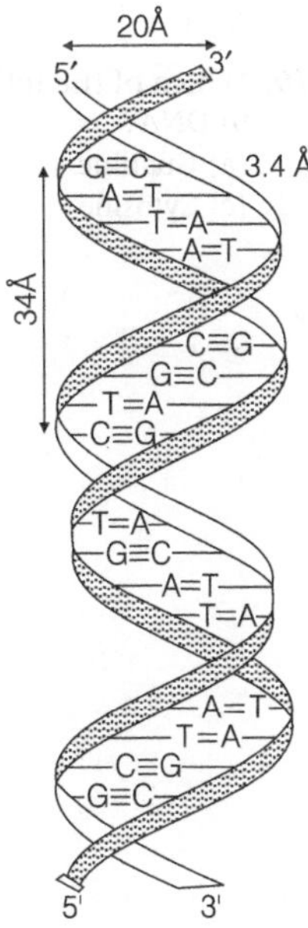

Fig.22.4 Double helix structure of DNA

The two strands are complementary to each other as adenine (A) of one strand is bonded to thymine (T) of the other one, (with 2 H-bonds), while guanine (G) is bonded to cytosine (C) of the other one (with 3 H-bonds) and *vice-versa.*

$$\underset{\text{2 H-bonds}}{A=T} \quad ; \quad \underset{\text{3 H-bond}}{C\equiv G}$$

Nucleic acids and heredity

A DNA molecule is the chemical repository of an organism's genetic information, which is stored as a sequence of deoxyribonucleotides strung together in the DNA chain. For this information to be preserved and passed on to future generations, a mechanism must exist for copying DNA. For the information to be used, a mechanism must exist for decoding the DNA message and for implementing the instructions it contains.

Function of DNA is to store information and pass it on to RNA at the proper time. The function of RNA, in turn, is to read, decode, and use the information received from DNA to make proteins . By decoding the right bit of DNA at the right time in the right place, an organism can use information to synthesise the many thousands of proteins necessary for carrying out its biochemical reactions.

Three fundamental processes take place in the transfer of genetic information.

$$\text{Replication} \circlearrowleft \textbf{DNA} \xrightarrow{\text{Transcription}} \textbf{RNA} \xrightarrow{\text{Translation}} \textbf{Proteins}$$

1. **Replication** is the process by which identical copies of DNA are made so that information can be preserved and handed down to offspring.
2. **Transcription** is the process by which the genetic messages contained in DNA are read and carried out of the nucleus to parts of the cell called ribosomes where protein synthesis occurs.
3. **Translation** is the process by which the genetic messages are decoded and used to build proteins.

DNA fingerprinting

In 1984, a British geneticist, Alec Jeffreys developed the technique of fingerprinting. DNA is the master molecule of all the life forms. It constitute the blue prints of every living organism through which character are passed from one generation to other generation.

Every living organism differs only due to nucleotide sequences of chromosome. The nucleotides form codes; this coded genetic information can be profiled to produce most authentic identity card of any organism. This identification at genetic level is called DNA fingerprinting or DNA profiling.

Uses

- In forensic laboratories for identification of criminals.
- To determine paternity of an individual.
- To identify the dead bodies in any accident by comparing the DNA's of parents or children.
- To identify racial groups and to rewrite biological evolution.

Imp : Har Gobind Khorana shared the Nobel Prize for Medicine and Physiology in 1968 with **Marshall Nirenberg** and **Robert Holley** for cracking the genetic code.

CHECK POINT 22.3

1. Which of the following is not a steroid hormone?
(a) Testosterone
(b) Estradiol
(c) Epinephrine
(d) Progesteron

2. Deficiency of which of the following hormones causes diabetes mellitus?
(a) Glucocorti coids
(b) Insulin
(c) Glucogen
(d) Oxytocin

3. Deficiency of which of the following vitamins causes rickets and osteomalacia?
(a) Ergocalciferol (b) Biotin
(c) Riboflavin (d) Thiamine

4. Most appropriate source of vitamin C, is
(a) yeast (b) milk
(c) citrus fruit (d) cereals

5. Which of the following vitamins does not act as an antioxidant?
(a) Vitamin A (b) Vitamin B_6
(c) Vitamin C (d) Vitamin E

6. Which of the following is present in a nucleic acid along with sugar and phosphate groups?
(a) Nucleotide (b) Nucleoside
(c) Nitrogen base (d) Amino acid

7. In a polynucleotide the bond that joins two nucleotide is called
(a) amide bond
(b) aminoester bond
(c) glycosidic bond
(d) phosphodiester bond

8. Type of linkage present in a nucleoside is
(a) C_1-C_4 glycosidic linkage
(b) N-glycosidic linkage
(c) phosphodiester bond
(d) hydrogen bond

9. In double helical structure of DNA, how many hydrogen bond is involved between cytosine and guanine?
(a) 1 (b) 2
(c) 3 (d) more than 3

10. Which of the following bases is complementary to adenine in DNA?
(a) Guanine (b) Thymine
(c) Cytosine (d) Uracil

Chapter Exercises

(A) Topical exercise

I. Carbohydrates

1 Consider the following statements.

Statement I Deoxyribose, $C_5H_{10}O_4$ is not a carbohydrate. **[NCERT Exemplar]**

Statement II Carbohydrates are hydrates of carbon so compounds which follow $C_x(H_2O)_y$ formula are carbohydrates.

(a) Both statements I and II are true
(b) Only statement I is true
(c) Only statement II is true
(d) Neither statement I nor statement II is true

2 A sugar is classified as a D-isomer if the hydroxyl group

(a) on the chiral carbon nearest to the carbonyl points to the left
(b) on the chiral carbon nearest to the carbonyl points to the right
(c) on the chiral carbon farthest from the carbonyl points to the left
(d) on the chiral carbon farthest from the carbonyl points to the right

3 All of the statements concerning monosaccharides are correct except

(a) the number of stereoisomers possible is 2^n, where n is the number of chiral carbon atoms in the molecule
(b) monosaccharides with 5 or 6 carbon atoms exist in solution in cyclic form
(c) the two different cyclic forms of a particular monosaccharide are called tautomers
(d) a molecule is classified as a D- or L- isomer by the position of the hydroxyl group on the chiral centre farthest from the carbonyl group

4 Which of the following statements about glucose is/are correct?

(a) It is an aldohexose
(b) It is dextrorotatory
(c) It is found in our blood as a source of energy
(d) All of the above

5 Apart from aldehydic group, glucose consists of hydroxyl group. The number of primary and secondary hydroxyl groups respectively are

(a) 1 and 4 (b) 2 and 3 (c) 3 and 2 (d) 4 and 1

6 Which of the following compounds is formed when glucose react with Br_2 water at pH 5-6 ?

(a) Glucaric acid (b) Saccharic acid
(c) Gluconic acid (d) Acetic acid

7 Which of the following tests is/are not given by glucose?

(a) Tollen's test (b) 2, 4-DNP test
(c) Schiff's test (d) Both (b) and (c)

8 Which of the following statement is not true about glucose?

(a) It is an aldohexose
(b) On heating with HI it forms *n*-hexane
(c) It is present in furanose form
(d) It does not give 2, 4- DNP test

9 Common reducing reactions of monosaccharides are due to

(a) their cyclic structures
(b) the presence of at least one hydroxyl group
(c) the presence of more than one hydroxyl group
(d) the presence of a carbonyl group, usually on the C-1 carbon atom

10 Haworth's projection of α-D-glucose is

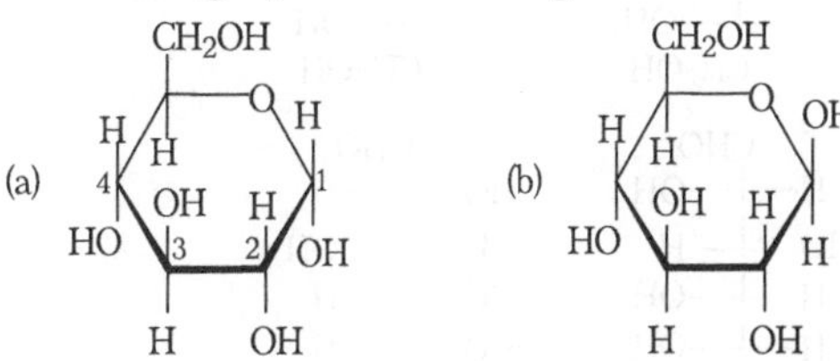

(c) Both (a) and (b) (d) None of these

11 Consider the following statements.

Statement I D-(+)-glucose is dextrorotatory in nature. **[NCERT Exemplar]**

Statement II 'D' represents its dextrorotatory nature.

(a) Both statements I and II are true
(b) only statement I is true
(c) only statement II is true
(d) Neither statement I nor statement II is true

12 There cyclic structures of monosaccharides are given below which of these are anomers.

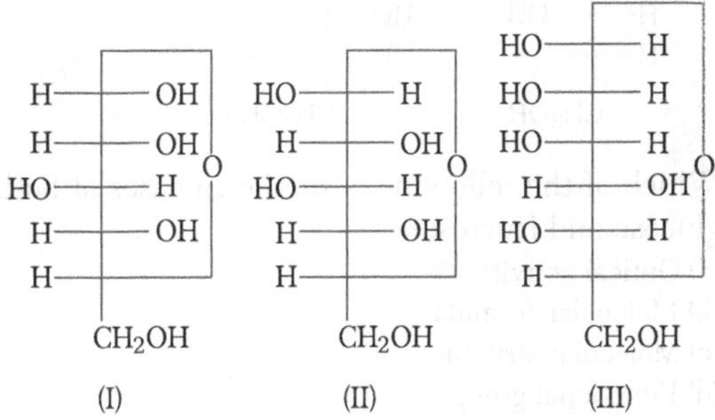

(a) Both I and II
(b) Both II and III
(c) Both I and III
(d) III is anomer of I and II

13 When a monosaccharide forms a ring by interaction of one of its hydroxyl group with its aldehyde group, the bond is referred to as a (an) .... linkage?
(a) ether
(b) ester
(c) acetal
(d) hemiacetal

14 Which of the following reactions of glucose can be explained only by its cyclic structure?
(a) Glucose forms pentaacetate
(b) Glucose reacts with hydroxylamine to form an oxime
(c) Pentaacetate of glucose does not react with hydroxyl amine
(d) Glucose is oxidised by nitric acid to gluconic acid

15 Which of the following pairs represents anomers? **(NCERT Exemplar)**

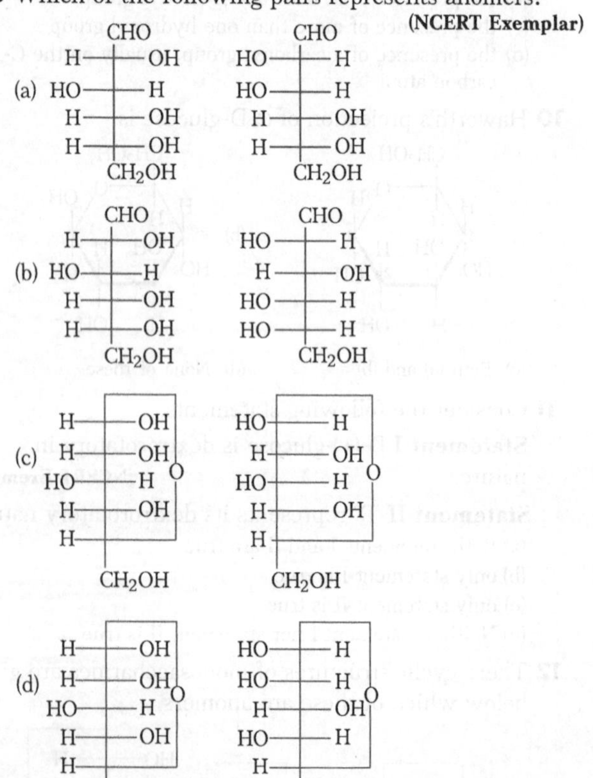

16 Which of the following is similar in cases of both glucose and fructose?
(a) Optical activity
(b) Molecular formula
(c) Molecular structure
(d) Functional group

17 Which of the following is a furanose structure of D-glucose?

(a) (b)

(c) Both (a) and (b) (d) None of these

18 Which of the following constitute(s) invert sugars?
(a) Sucrose
(b) Glucose
(c) Fructose
(d) Both (b) and (c)

19 Optical rotations of some compounds along with their structures are given below which of them have D-configuration.

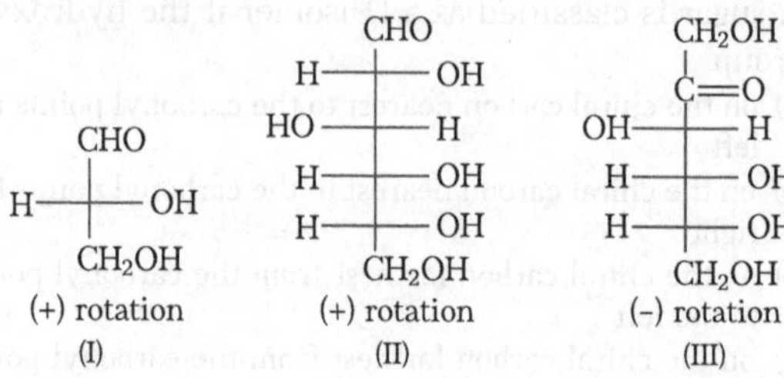

(a) I, II, and III (b) Both II and III
(c) Both I and II (d) Only III

20 A glycosidic bond is
(a) a bond between any two carbohydrate molecules
(b) a bond formed between an anomeric carbon atom and any —*OR* group
(c) any bond that can be broken by reaction with water
(d) a bond between anomeric carbon atoms on two separate monosaccharides

21 Match the following columns.

Column I	Column II
A. D-sorbitol	1. Anomeric carbon
B. L-ascorbic acid	2. A disaccharide
C. Glycoside	3. A sugar lactone
D. C-1 of glucose	4. Sugar alcohol

Choose the correct match.

	A	B	C	D
(a)	4	3	2	1
(b)	3	1	4	2
(c)	2	4	1	3
(d)	3	4	2	1

22 The carbon atoms involved in glycosidic linkage of glucose and fructose in sucrose are respectively.
(a) C_1 and C_2
(b) C_1 and C_4
(c) C_1 and C_6
(d) C_2 and C_4

23 Structure of disaccharide formed by glucose and fructose is given below. Identify anomeric carbon atoms in monosaccharide units. **[NCERT Exemplar]**

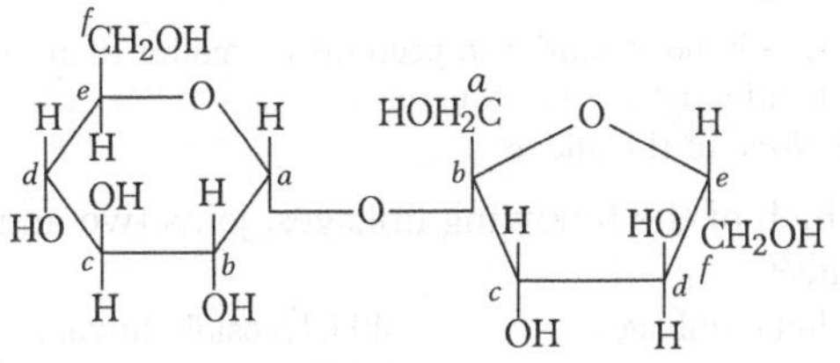

(a) '*a*' carbon of glucose and '*a*' carbon of fructose
(b) '*a*' carbon of glucose and '*e*' carbon of fructose
(c) '*a*' carbon of glucose and '*b*' carbon of fructose
(d) '*f*' carbon of glucose and '*f*' carbon of fructose

24 Which of the following is the correct structure of sucrose?
(a) 1-α-D-glucosyl-4-fructose
(b) 1-β-D-glucosyl-2-fructose
(c) 1-α-D-glucosyl-2-fructose
(d) 1-α-D-fructosyl-4-glucose

25 Sucrose is not a reducing sugar because its
(a) hemiacetal groups cannot be converted to aldehyde groups
(b) disaccharide bond is an α-1,4-linkage
(c) disaccharide bond is a β-1,4-linkage
(d) disaccharide bond is a 1,2-anomeric link

26 Sucrose (cane sugar) is a disaccharide. One molecule of sucrose on hydrolysis gives .... **[NCERT Exemplar]**
(a) 2 molecules of glucose
(b) 2 molecules of glucose + 1 molecule of fructose
(c) 1 molecule of glucose +1 molecule of fructose
(d) 2 molecules of fructose

27 Consider the following statements.
Statement I Sucrose is not a reducing sugar.
Statement II Reducing groups of glucose and fructose are involved in glycosidic bond.
(a) Both statements I and II are true
(b) Only statement I is true
(c) Only statement II is true
(d) Neither statement I nor statement II is true

28 Consider the following statements.
Statement I β-glycosidic linkage is present in maltose. **[NCERT Exemplar]**
Statement II Maltose is composed of two glucose units in which C - 1of one glucose unit is linked to C - 4 of another glucose unit.
(a) Both statements I and II are true
(b) Only statement I is true
(c) Only statement II is true
(d) Neither statement I nor statement II is true

29 All of the following are common monosaccharides of interest in human biochemistry except
(a) glucose (b) fructose
(c) ribose (d) maltose

30 The bond connecting the two monosaccharides in maltose is a(an) .... linkage.
(a) α-1,4-glycosidic (b) β-1,4-glycosidic
(c) 1,2-anomeric (d) α-1,6-glycosidic

31 Out of maltose (I), cellobiose (II), sucrose (III) and lactose (IV), reducing sugars are
(a) I, II and III (b) I, II and IV
(c) II, III and IV (d) All of these

32 The enzyme(s) used to convert starch into maltose is/are
(a) amylase (b) diastase
(c) maltase (d) Both (a) and (b)

33 Which of the following is not a disaccharide?
(a) Sucrose (b) Glycogen (c) Maltose (d) Lactose

34 The structure of glycogen resembles with
(a) starch (b) amylose (c) amylopectin (d) cellulose

35 In disaccharides, if the reducing groups of monosaccharides, i.e. aldehydic or ketonic groups are bonded, these are non-reducing sugars. Which of the following disaccharide is a non-reducing sugar? **[NCERT Exemplar]**

(a)

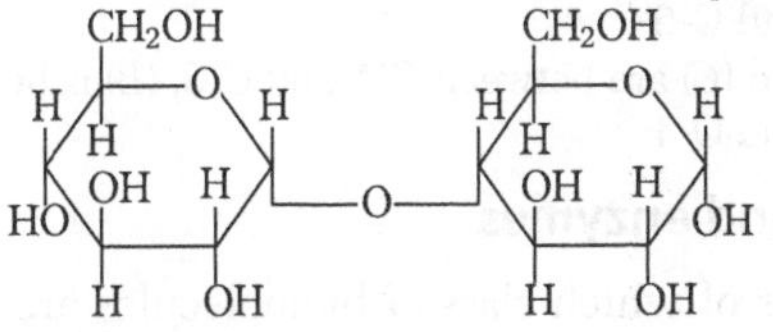

(b)

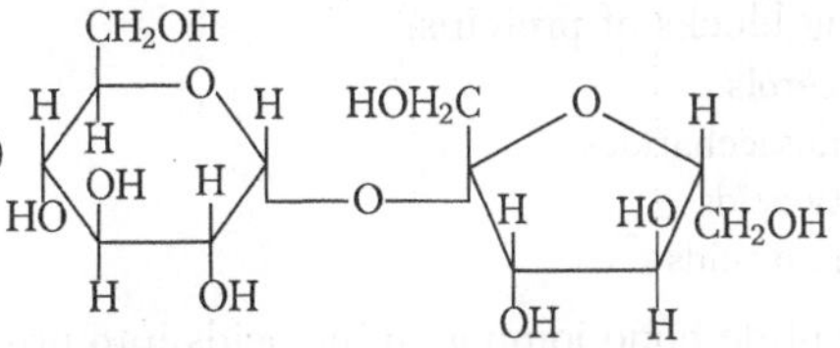

(c)

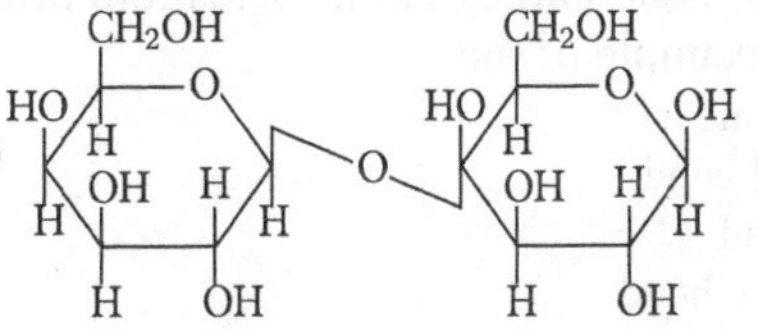

(d)

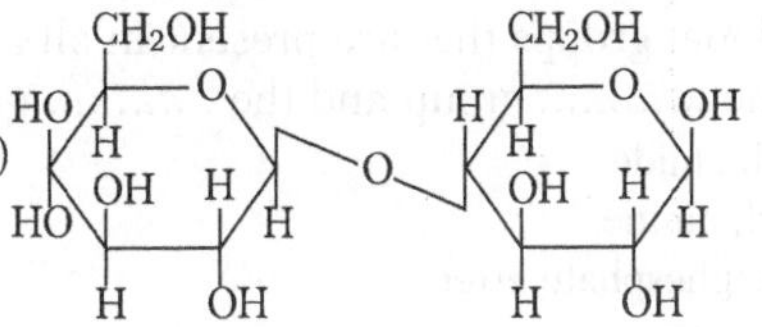

36 Three structures are given below in which two glucose units are linked. Which of these linkages between glucose units are between C-1 and C-4 and which linkages are between C-1and C-6?

(I)

(II)

(III)

(a) (*A*) is between C-1 and C-4, (*B*) are between C-1 and C-6
(b) (*A*) and (*B*) are between C-1 and C-4, (*C*) is between C-1 and C-6
(c) (*A*) and (*C*) are between C-1 and C-4, is between C-1 and C-6
(d) (*A*) and (*C*) are between C-1 and C-6, (*B*) is between C-1 and C-4

II. Proteins and enzymes

37 Members of which class of biomolecules are the building blocks of proteins?
(a) Glycerols
(b) Monosaccharides
(c) Fatty acids
(d) Amino acids

38 The peptide bond joining amino acids into proteins is a specific example of the
(a) amide bond
(b) carbonyl bond
(c) ester bond
(d) glycosidic bond

39 Two functional groups that are present in all amino acids are the .... group and the .... group.
(a) hydroxyl, amide
(b) carboxyl, amine
(c) carboxyl, phosphate ester
(d) acetal, amine

40 Which statement is incorrect about peptide bond $-\overset{\overset{O}{||}}{C}-\ddot{N}H-$?
(a) C—N bond length in proteins is longer than usual bond length of C—H bond
(b) Spectroscopic analysis show planar structure of $-\underset{\underset{O}{||}}{C}-NH-$ group
(c) C – N bond length in proteins is smaller than usual bond length of C–N bond
(d) None of the above

41 Which of the following linkages, joins two amino acids?
(a) Ionic linkages (b) Glycosidic linkages
(c) Amide linkages (d) Hydrogen

42 The isoelectric point of an amino acid is
(a) the pH at which it exists in basic form
(b) the pH at which it exists in acid form
(c) the pH at which it exists in Zwitter ion form
(d) the pH equal to its pK_a

43 Consider the following statements.
Statment I Glycine can be obtained through diet. **[NCERT Exemplar]**
Statement II It is an essential amino acid.
(a) Both statements I and II are true
(b) Only statement I is true
(c) Only statement II is true
(d) Neither statement I nor statement II is true

44 Generally all the amino acids have primary amino function with the exception of
(a) alanine (b) proline
(c) leucine (d) isoleucine

45 Which amino acid is classified as neutral and non-polar?
(a) Aspartic acid (b) Histidine
(c) Phenylalanine (d) Lysine

46 Which of the following is not chiral?
(a) Serine (b) Tryptophan
(c) Glycine (d) Threonine

47 How many minimum number of, amino acids does a normal polypeptide chain have?
(a) 2 (b) 10 (c) 100 (d) 1000

48 Collagen is an example of a(an)
(a) enzyme (b) fibrous protein
(c) globular protein (d) derived protein

49 Serum albumin is an example of a(an)
(a) enzyme (b) structural protein
(c) storage protein (d) transport protein

50 Which of the following is not a globular protein?
(a) Albumin (b) Myoglobin
(c) Keratin (d) Antibodies

51 Out of the following, the example of conjugated protein is/are
(a) lipoprotein (b) nucleoprotein
(c) phosphoprotein (d) None of these

52 The bond present in helical structure of protein is the
(a) dipeptide bonds (b) hydrogen bonds
(c) ether bonds (d) peptide bonds

53 Consider the following statements.
Statement I Covalent linkage that maintains a higher order protein structure is the disulphide bond.
Statement II These interactions are weak and easily disrupted by thermal motion.
(a) Both statements I and II are true
(b) Only statement I is true
(c) Only statement II is true
(d) Neither statement I nor statement II is true

54 All of the following are examples of fibrous proteins except
(a) wool (b) fingernails (c) skin (d) insulin

55 The amino acid sequence of a protein is known as its
(a) primary structure (b) secondary structure
(c) tertiary structure (d) quaternary structure

56 Each polypeptide in a protein has amino acids linked with each other in a specific sequence. This sequence of amino acids is said to be .... **[NCERT Exemplar]**
(a) primary structure of proteins
(b) secondary structure of proteins
(c) tertiary structure of proteins
(d) quaternary structure of proteins

57 The secondary structure of a protein refers to
(a) mainly denatured proteins and structures of prosthetic groups
(b) linear sequence of amino acid residues in the polypeptide chain
(c) regular folding patterns of continuous portions of the polypeptide chain
(d) three-dimensional structure, specially the bond between amino acid residues that are far apart from each other in polypeptide chain

58 Proteins are found to have two different types of secondary structures *viz* α-helix and β-pleated sheet structure. α-helix structure of protein is stabilised by **[NCERT Exemplar]**
(a) peptide bonds
(b) van der Waals' forces
(c) hydrogen bonds
(d) dipole-dipole interactions

59 All of the following can denature proteins without hydrolysis except
(a) heat (b) mechanical stress
(c) enzyme treatment (d) lowering of pH

60 All of the following are examples of denaturing proteins except
(a) souring of milk
(b) a mild sunburn
(c) using a curling iron on your hair
(d) digestion of a meal

61 Match the following columns.

Column I		Column II
A. Globular protein	1.	Protein with the tertiary structure in which it normally occurs in living systems.
B. Zwitter ion	2.	Protein that is usually water soluble having a hydrophilic exterior and hydrophobic interior and an overall rounded shape.
C. Fibrous protein	3.	A protein that is water insoluble, is very tough and has a long shape.
D. Native protein	4.	Overall a neutral molecule but migrates to cathode or anode depending on pH.
E. Simple protein	5.	A protein that produces only amino acids on hydrolysis.

Choose the correct match.

	A	B	C	D	E
(a)	2	4	3	1	5
(b)	3	5	1	2	4
(c)	4	1	5	3	2
(d)	5	3	4	2	1

62 The part of an enzyme that is formed of protein is called
(a) apoenzyme (b) holoenzyme
(c) cofactor (d) coenzyme

63 The enzyme which hydrolyses triglycerides to fatty acids and glycerol is called
(a) maltase (b) lipase (c) zymase (d) pepsin

64 Enzymes having two binding sites are
(a) conjugate enzyme (b) apoenzyme
(c) holoenzyme (d) allosteric enzyme

65 The non-protein portion required by some enzymes for proper functioning is called a(an)
(a) zymogen (b) substrate (c) inhibitor (d) cofactor

66 Enzymes are members of which class of biomolecules?
(a) Carbohydrates (b) Lipids
(c) Nucleic acids (d) Proteins

67 The reactant in an enzyme catalysed reaction is called the
(a) hormone (b) substrate (c) cofactor (d) inhibitor

68 All of the following phrases correctly describes enzymes except
(a) dissolve in water (b) have a globular shape
(c) behave as substrates (d) contain an active site

69 When substrate molecules occupy all of the active sites in the enzyme available for a particular reaction, the enzyme is said to be
(a) denatured (b) hydrolysed (c) activated (d) saturated

70 Poisons such as heavy metal ions or insecticides can disrupt enzyme function by
(a) competititve inhibition
(b) irreversible inhibition
(c) disruption of tertiary structure and covalent modification
(d) All of the above

71 Which of the following is the function of enzymes?
(a) Structure and support
(b) Nutritional reserves
(c) Biochemical catalysts
(d) Communication between cells

72 A competitive inhibitor of an enzyme acts by
(a) modifying one of the amino acid residues of the enzyme
(b) modifying structure of the substrate
(c) competing with the substrate to bind to the active site of the enzyme
(d) Increasing temperature of the reaction mixture

73 Consider the following statements.
Statement I The non-protein part of an enzyme is called a cofactor.
Statement II These cofactors are basically the additional chemical groups, which appear in those enzymes that are conjugated protein molecules.
(a) Both statement I and II are true
(b) Only statement I is true
(c) Only statement II is true
(d) Neither statement I nor statement II is true

74 Consider the following statements.
Statement I Enzymes are highly specific in their action, both with respect to the substrate they act upon and the kind of reaction they catalyse.
Statement II The **"lock and key"** hypothesis cannot explain the specificity of enzyme-substrate interaction.
(a) Both statements I and II are true
(b) Only statement I is true
(c) Only statement II is true
(d) Neither statement I nor statement II is true

75 Consider the following statements.
Statement I In presence of enzyme, substrate molecule can be attacked by the reagent effectively.
Statement II Active sites of enzymes hold the substrate molecule in a suitable position.
Consider the following statements. **[NCERT Exemplar]**
(a) Both statements I and II are true
(b) Only statement I is true
(c) Only statement II is true
(d) Neither statement I nor statement II is true

76 Out of the following enzymes:
I. Urease II. LDH
III. Alcohol dehydrogenase IV. Amylase
hydrolases are
(a) Both I and IV (b) Both I and II
(c) Both II and III (d) Both II and IV

77 Consider the following statements
I. Enzyme urease which catalyses the hydrolysis of urea was shown to be non-protein in nature.
II. The combination of an apoenzyme and a cofactor is known as the holoenzyme.
III. The rates at which different enzymes catalyse their reactions are the same.
Select the correct statements.
(a) Both II and III (b) Both I and II
(c) Both I and III (d) I, II and III

78 Which of the following statements about enzymes are true?
I. Enzymes lack in nucleophilic groups.
II. Pepsin is proteolytic enzyme.
III. Enzymes catalyse chemical reactions by lowering the activation energy.
IV. Enzymes are highly specific both in binding chiral substrates and in catalysing their reactions.
(a) Both I and II (b) Both I and III
(c) Both I and IV (d) II, II and IV

79 Match the following enzymes given in Column I with the reactions they catalyse given in Column II.

Column I (Enzymes)		Column II (Reactions)
A. Invertase	1.	Decomposition of urea into NH_3 and CO_2
B. Maltase	2.	Conversion of glucose into ethyl alcohol
C. Pepsin	3.	Hydrolysis of maltose into glucose
D. Urease	4.	Hydrolysis of cane sugar
E. Zymase	5.	Hydrolysis of proteins into peptides

Choose the correct match.

	A	B	C	D	E
(a)	5	1	4	2	3
(b)	4	3	5	1	2
(c)	3	4	2	5	3
(d)	2	5	1	3	4

III. Hormones

80 Hormones are produced in the .... system.
(a) digestive (b) endocrine
(c) excretory (d) nervous

81 Which type of hormone actually enters the target cell?
(a) Amino acid derivatives (b) Enzymes
(c) Peptides (d) Steroids

82 Steroids are
(a) simple esters of long chain alcohols and fatty acids
(b) esters of glycerol with three predominantly saturated fatty acids
(c) esters of glycerol with three predominantly unsaturated fatty acids
(d) based on a tetracyclic ring system with substituents at various positions

83 Androgens are a specific type of steroid whose function is to
(a) maintain pregnancy
(b) promote development of secondary male sex characteristics
(c) promote development of secondary female sex characteristics
(d) regulate the sodium/potassium balance in cellular fluids

84 The hormone that helps in the conversion of glucose to glycogen is
(a) cortisone (b) bile acids
(c) adrenalin (d) insulin

85 Which of the following hormones is an amino acids derivative?
(a) Adrenaline (b) Vasopressine
(c) Oxytocin (d) Testosterone

86 Which of the following glands secrete triiodothyronine?
(a) Testis (b) Pituitary
(c) Pancreas (d) Thyroid

87 Insulin is secreted by
(a) stomach (b) intestine
(c) pancreas (d) brain

88 Which hormone is an example of a polypeptide?
(a) Epinephrine (b) Thyroxine
(c) Growth hormone (d) Estrogen

89 All of the following types of molecules can be classified as lipids except
(a) triacylglycerols (b) steroids
(c) glycoproteins (d) eicosanoids

90 The hormone which controls the processes like burning of fats, proteins and carbohydrates to liberate energy in the body is
(a) cortisone (b) thyroxine
(c) adrenalin (d) insulin

91 The diseases identified as diabetes are primarily associated with a malfunction of the hormone
(a) glucagon (b) insulin
(c) epinephrine (d) cortisone

92 Which of the following compounds is not an amino acid derivative hormone?
(a) Dopamine
(b) Epinephrine
(c) Thyroxine
(d) Norepinephrine

93 All of the following types of molecules function as chemical messengers in the body except
(a) polypeptide hormones such as insulin
(b) steroid hormones such as progesterone
(c) neurons, including axons and dendrons
(d) amino acid derivatives classified as catecholamines

94 Which of the following statements about hormones is not correct?
(a) Hormones can affect biochemical reactions by inhibiting enzymes
(b) Hormones are transported in the bloodstream
(c) Hormones can affect biochemical reactions by activating enzymes
(d) Hormones are very specific, affecting only chemical reaction

95 Hormones regulate biochemical reactions by
(a) performing the same role as enzymes in certain reactions
(b) acting rapidly on receptor sites close to the point where the hormone was produced
(c) acting to produce indirect effects on reactions occurring at some distance from the site of hormone production
(d) interacting with substrates to prevent their participation in specific reactions

96 Which statement regarding polypeptide hormones is incorrect?
(a) This class of hormones includes all the regulatory hormones produced by the pituitary gland
(b) Vasopressin is an example
(c) Insulin is an example
(d) Sex hormones are included in this category

97 Which of the following description is true for hormones?
I. Produced by the nervous system.
II. Transported in the blood stream.
III. Interact with any cell having the appropriate receptor.
IV. Act within 1 second or less after release.
V. Effects may last several hours or days.
Choose the correct option.
(a) I, II and III
(b) II, III and V
(c) I, III and V
(d) Both I, IV

IV. Vitamins

98 A vitamin is
(a) a nutrient involving trace metals that is necessary for good health
(b) a polypeptide used to manufacture enzymes
(c) a lipid complex which is an important part of all cells
(d) a small organic molecule obtained from the diet and necessary for good health

99 An example of water soluble vitamin is
(a) vitamin A (b) vitamin B
(c) vitamin E (d) Vitamin D

100 Consider the following statements.
Statement I Vitamin C is easily oxidised in air, especially in presence of Cu^{2+} or Fe^{2+} ions.
Statement II Both the forms of vitamin C are biologically active, due to the conversion of dehydroascorbic acid to ascorbic acid and are found in body fluids.
(a) Both statements I and II are true
(b) Only statement I is true
(c) Only statement II is true
(d) Neither statement I nor statement II is true

101 All of the following vitamins are fat soluble except
(a) A (b) K (c) C (d) D

102 Which of the following B group vitamins can be stored in our body?
(a) Vitamin B_1 (b) Vitamin B_2
(c) Vitamin B_6 (d) Vitamin B_{12}

103 Which of the following acids is a vitamin?
(a) Aspartic and (b) Ascorbic acid
(c) Adipic acid (d) Saccharic acid

104 All of the following vitamins B are except
(a) ascorbic acid (b) cobalamine
(c) niacin (d) pantothenic acid

105 The vitamin most closely associated with carbohydrate metabolism is
(a) vitamin B_1 (b) vitamin B_2
(c) vitamin D_2 (d) vitamin B_{12}

106 Consider the following statements.
Statement I Vitamin D can be stored in our body.
Statement II Vitamin D is fat soluble vitamin.
[NCERT Exemplar]
(a) Both statements I and II are true
(b) Only statement I is true
(c) Only statement II is true
(d) Neither statement I nor statement II is true

107 All of the following statements concerning vitamins are true except
(a) fat-soluble vitamins that have a high proportion of polar carbonyl and hydroxyl groups
(b) it is possible to overdose on fat-soluble vitamins because they accumulate in fatty tissues
(c) it is difficult to overdose on water-soluble vitamins because excess amounts can be excreted in the urine
(d) vitamins A, D, E and K are fat-soluble

108 Choose the correct statement about vitamin A, B and D.
(a) Vitamin A and D are fat soluble and thus can accumulate in body fat, where vitamin C is water soluble and the excess will be excreted in urine
(b) Vitamin C is fat soluble and thus, can accumulate in body fat, where vitamins A and D are water soluble and the excess will be excreted in urine
(c) Vitamin C is biologically active as consumed in foods, but vitamins A and D must be converted into biologically active forms before utilisation
(d) Vitamins A and D are biologically active as consumed in food, but vitamin C must be converted into biologically active forms before utilisation

109 Match the vitamins given in Column I with the deficiency disease they cause given in Column II.

	Column I (Vitamins)		Column II (Diseases)
A.	Vitamin A	1.	Pernicious anaemia
B.	Vitamin B_1	2.	Increased blood clotting time
C.	Vitamin B_{12}	3.	Xerophthalmia
D.	Vitamin C	4.	Muscular weakness
E.	Vitamin D	5.	Beri-beri
F.	Vitamin E	6.	Bleeding gums
G.	Vitamin K	7.	Osteomalacia

Choose the correct match from the following options.

	A	B	C	D	E	F	G
(a)	7	5	8	3	2	1	3
(b)	3	7	1	8	7	5	2
(c)	5	4,9	3	2	8,1	6	1
(d)	4,2	8	5,1	9	3	7	6

V. Nucleic acids

110 Which of the following is a major function of nucleic acids?
(a) Storage and transfer of genetic information
(b) Storage and intracellular transfer of energy
(c) Catalysis of virtually all biochemical reactions
(d) Structural support in both plants and animals

111 What is the repeating unit in nucleic acids?
(a) Monosaccharides
(b) Amino acids
(c) Nucleotides
(d) Nucleosides

112 Nucleic acids are the polymers of
(a) nucleosides (b) nucleotides
(c) bases (d) sugars

113 A nucleotide is composed of a ... with a ... added to it.
(a) pentose; phosphate group
(b) pentose; nitrogen base
(c) nucleoside; nitrogen base
(d) nucleoside; phosphate group

114 The sequence in the structure of nucleic acid is
(a) base + phosphate + pentose
(b) phosphate group + pentose + base
(c) pentose + base + phosphate group
(d) All of the above

115 The backbone of a nucleic acid molecule consists of
(a) alternating sugar and phosphate groups linked by phosphate ester bonds
(b) alternating sugar and nitrogen base groups linked by amide bonds
(c) alternating nitrogen bases and phosphate groups linked by amide bonds and strengthened by hydrogen bonds
(d) sugar molecules bonded from the C-3 of one molecule to the C-5 of the other by glycosidic linkages

116 The nucleic acid base having two possible binding sites is
(a) thymine (b) cytosine
(c) guanine (d) All of these

117 What is the complementary strand for a single DNA strand with the sequence TCGA?
(a) TGCA (b) AGCT (c) CTAC (d) GATC

118 In DNA, the complementary bases are
(a) adenine and thymine, guanine and cytosine
(b) uracil and adenine, cytosine and guanine
(c) adenine and guanine thymine and cytosine
(d) adenine and thymine, guanine and uracil

119 In DNA, a DNA sequence complementary to the strand shown below is

C-G-G-T-T-A-G

(a) G-C-C-A-A-T-C (b) C-G-G-T-T-A-G
(c) A-T-T-G-G-C-T (d) G-C-C-U-U-U-C

120 Which of the following is a bicyclic nitrogenous base
(a) adenine (b) uracil
(c) cytosine (d) thymine

121 Presence of which of the following bases, distinguishes RNA from a DNA.
(a) Adenine (b) Guanine (c) Cytosine (d) Uracil

122 Which base is normally found in DNA but not in RNA?
(a) Thymine (b) Adenine
(c) Guanine (d) Uracil

123 A sequence of how many nucleotides in messenger RNA makes a codon for an amino acid?
(a) Three (b) Four
(c) One (d) Two

124 Consider the following statements.
Statement I DNA constitutes the blue prints of every living organism.
Statement II The nucleotides form codes and coded genetic information can be easily profiled.
(a) Both statements I and II are true
(b) Only statement I is true
(c) Only statement II is true
(d) Neither statement I nor statement II is true

125 All of the following are associated with replication of DNA except
(a) DNA polymerase (b) DNA ligase
(c) hydrogen bonds (d) amino acids

126 Which amino acid can form covalent sulphur-sulphur bonds?
(a) Cysteine (b) Glycine
(c) Proline (d) Methionine

127 Dinucleotide is obtained by joining two nucleotides together by phosphodiester linkage. Between which carbon atoms of pentose sugars of nucleotides are these linkages present? **[NCERT Exemplar]**
(a) 5′ and 3′ (b) 1′ and 5′ (c) 5′ and 5′ (d) 3′ and 3′

128 The process of manufacturing an identical copy of a DNA molecule is called
(a) mutation (b) replication
(c) transcription (d) translation

129 DNA multiplication is called
(a) translation (b) transduction
(c) transcription (d) replication

130 In replication of DNA, each new double strand consists of one template strand and one new strand. Therefore, replication is said to be
(a) repetitive (b) semiconservative
(c) polymeric (d) catabolic

131 The process in which information from DNA is used to manufacture RNA is called
(a) mutation (b) replication
(c) transcription (d) translation

132 The process in which information contained in RNA is used to manufacture proteins is called
(a) mutation (b) replication
(c) transcription (d) translation

133 The central dogma of molecular genetics states that
(a) every copy of DNA is an exact replica of the parent molecule
(b) DNA stores genetic information and RNA reads, decodes, and uses that information
(c) in DNA and RNA adenine always pairs with thymine and cytosine always pairs with guanine
(d) each amino acid is coded by a unique set of three base pairs

(B) Miscellaneous questions

1 Which of the following cannot be classified as biomolecules?
(a) Carbohydrates
(b) Proteins
(c) Lipids
(d) All of the above are biomolecules

2 In the tetrapeptide ala–cys–val–leu, the N-terminal amino acid is ...
(a) ala (b) cys (c) val (d) leu

3 Two isomeric sugars that are cyclic and only differ in the position of the –OH group at the hemiacetal group are called
(a) enantiomers (b) mutarotations
(c) anomers (d) epimers

4 Which among the following is α-amino acid?
(a) 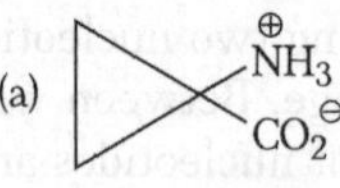

(b) 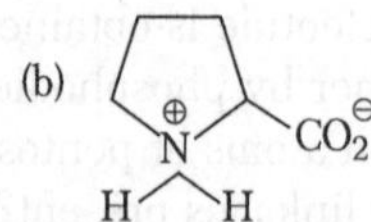

(c) Both (a) and (b) (d) None of these

5 The β-pleated sheet is an example of
(a) primary structure (b) secondary structure
(c) tertiary structure (d) quaternary structure

6 The tertiary structure of most enzymes is
(a) an α-helix (b) a β-pleated sheet
(c) fibrous (d) globular

7 The chemical combination of ribose and one of the five nitrogen bases results in formation of a
(a) nucleotide (b) nucleoside
(c) DNA molecule (d) chromosome

8 Consider the following statements.
Statement I Glycine exists as Zwitter ion.
Statement II It behaves as Lowry-Bronsted acid and base.
(a) Both statements I and II are true
(b) Only statement I is true
(c) Only statement II is true
(d) Neither statement I nor statement II is true

9 Pure form of β-sheets are present in
(a) silk fibroin (b) α-keratin
(c) collagen (d) None of these

10 Globular proteins always contain
(a) some α-helices (b) some β-sheets
(c) some random coil regions (d) α-helix and β-sheets

11 Haemoglobin is evolutionarily related to
(a) haem (b) adenylate kinase
(c) myoglobin (d) None of these

12 Utilisation of calcium and phosphorus in the formation of bones is dependent on
(a) pyridoxal phosphate (b) carboxybiotin
(c) pantothenic acid (d) vitamin D

13 How many different tripeptides can be formed from one molecule each of the amino acids tyrosine, valine and alanine?
(a) 3 (b) 6 (c) 9 (d) 12

14 α-D (+) – glucose and β-D (+) – glucose are
(a) anomers (b) epimers
(c) enantiomers (d) geometrical isomers

15 Which one of the following gives positive Fehling's solution test?
(a) Sucrose (b) Glucose (c) Fats (d) Protein

16 Which of the following polymer is stored in the liver of animals?
(a) Amylose (b) Cellulose
(c) Amylopectin (d) Glycogen

17 Which of the following can destroy antigens?
(a) Insulin (b) Antibodies
(c) Chromoprotein (d) Phosphoprotein

18 Anti-haemorrhagic factor is
(a) vitamin K (b) vitamin D_2
(c) vitamin A_2 (d) vitamin B_{12}

19 The theory of enzyme mechanism that suggests a rigid, inflexible molecule is the ... model.
(a) coenzyme (b) induced-fit
(c) lock and key (d) substrate specific

20 One example of an amino acid derivative that acts as both a hormone and a neurotransmitter is
(a) epinephrine (b) thyroxine
(c) growth hormone (GH) (d) estrogen

21 The drug AZT is used to treat AIDS patient. It has one (or more) of the linkages
(a) amide (b) ether
(c) Both (a) and (b) (d) glycosidic

22 Consider the following statements.
Statement I Zwitter ions are electrically neutral and do not migrate in an electric field.
Statement II They are covalent compounds.
(a) Both statements I and II are true
(b) Only statement I is true
(c) Only statement II is true
(d) Neither statement I nor statement II is true

23 The type of bond that is most important in maintaining secondary structure of a protein is
(a) disulphide bridges
(b) hydrogen bonding within the backbone
(c) hydrogen bonding between *R* groups
(d) salt bridges

24 A compound that is necessary for the synthesis of another compound is called a(an)
(a) allosteric inhibitor (b) cofactor
(c) precursor (d) proenzyme

25 A hemiacetal linkage consists of a
(a) carbon atom bonded to both — OH and —*OR* groups
(b) carbonyl group that is also bonded to a hydroxyl group
(c) carbonyl group that is also bonded to an ether linkage
(d) carbon atom bonded to two —OH groups

26 Consider the following statements.
Statement I The carbon atom next to the carboxyl group is known as α-carbon in an amino acid.
Statement II The biologically significant amino acids are principally, α-amino acids.
(a) Both statements I and II are true
(b) Only statement I is true
(c) Only statement II is true
(d) Neither statement I nor statement II is true

27 When the imidazole ring of histidine is protonated, nitrogen that is protonated is

$CH_2CH(NH_2)COO^{\ominus}$ (imidazole ring: N labelled α, NH labelled β)

(a) at α (b) at β
(c) Both (a) and (b) (d) None of these

28 Which of the following is correct statement?
(a) Starch is polymer of α-glucose
(b) Amylose is a component of cellulose
(c) Proteins are composed of only one type of amino acid
(d) In cyclic structure of fructose, there are four carbons and one oxygen atom

29 When a monosaccharide forms a cyclic hemiacetal, the carbon atom that contained the carbonyl group is identified as the anomeric carbon atom because
(a) the carbonyl group is drawn to the right
(b) the carbonyl group is drawn to the left
(c) its substituents can assume α or β-position
(d) it forms bond to an —*OR* and an —*OR*′

30 Increasing the temperature of an enzymatic reaction is accompanied by
(a) activity remaining constant
(b) activity increasing continuously
(c) activity decreasing continuously
(d) activity first increasing and then decreasing

31 When a new strand of RNA is transcribed from the DNA strand shown, its base sequence will be
Informational strand of DNA : TTACGGAT
Template strand of DNA : AATGCCTA
(a) TTACGGAT (b) AATGCCTA
(c) UUACGGAU (d) AAUGCCUA

32 In a normal biochemical system, the rate of a specific reaction is determined by
(a) pH and enzyme efficiency
(b) temperature and enzyme efficiency
(c) temperature and substrate concentration
(d) enzyme efficiency and substrate concentration

33 Glycogen is a branched chain polymer of α-D-glucose units in which chain is formed by C_1-C_4 glycosidic linkage where as branching occurs by the formation of C_1-C_6 glycosidic linkage. Structure of glycogen is similar to .... **(NCERT-Exemplar)**
(a) amylose (b) amylopectin
(c) cellulose (d) glucose

34 Which statement about cholesterol is not correct?
(a) It is an essential component of cell membranes
(b) It is the precursor for steroid hormones
(c) It is the precursor for bile acids and salts
(d) It cannot be synthesised by the body

35 Enzymes functions as catalysts by
(a) changing the value of the free energy change to a more favourable value
(b) changing the value of the equilibrium constant to a more favourable value
(c) lowering the value of the activation energy
(d) increasing the amount of time needed to reach equilibrium

36 Match the following columns.

	Column I		Column II
A.	Thymine	1.	Pyrimidine base
B.	Thiamine	2.	Enzyme
C.	Insulin	3.	Cell-wall component
D.	Pepsin	4.	Hormone
E.	Phospholipids	5.	Vitamin B_1

Choose the correct match.

	A	B	C	D	E
(a)	1	5	4	2	3
(b)	2	3	1	5	4
(c)	3	4	5	1	2
(d)	4	1	2	5	3

37 Enzymes have an optimum temperature for their catalytic activity. This is best explained by the balance between the ... number of collisions and the ... rate of denaturation of the enzyme as temperature increase.
(a) increased; increased (b) decreased; increased
(c) increased; decreased (d) decreased; decreased

38 Consider the following statements.

Statement I The pH at which amino acids (also proteins) carry no net charge is called its isoelectric point.

Statement II The isoelectric point of glycine is 6.

(a) Both statements I and II are true
(b) Only statement I is true
(c) Only statement II is true
(d) Neither statement I nor statement II is true

39 Biomolecules can be classified as lipids on the basis of

(a) a common structure consisting of long hydrocarbon chains
(b) the presence of many hydroxyl groups and atleast one carbonyl group
(c) the presence of atleast one amine group and one carboxylic acid group on each molecule
(d) the physical property as solubility in non-polar organic solvents

40 When energy from stored triacylglycerols is needed, they are ... and then transported in the bloodstream *via* ... as fatty acids.

(a) hydrolysed; high-density lipoproteins
(b) hydrolysed; albumins
(c) oxidised; chylomicrons
(d) oxidised; albumins

41 Consider the following reagents.

I. Br_2 water II. Tollen's reagent
III. Fehling's solution

Which can be used to make distinction between an aldose and a ketose?

(a) I, II and III (b) Both II and III
(c) Only I (d) Only II

42 Consider the following statements.

Statement I All naturally occurring α-aminoacids except glycine are optically active. **[NCERT Exemplar]**

Statement II Most naturaly occuring amino acids have L-configuration.

(a) Both statements I and II ae true
(b) Only statement I is true
(c) Only statement II is true
(d) Niether statement I nor statement II is true.

43 Which is the correct order for the steps in digestion and absorption of proteins?

I. Hydrolysis by pancreatic lipases
II. Denaturation by stomach acid
III. Hydrolysis by pepsin
IV. Active transport into the bloodstream

(a) I, II, III and IV
(b) IV, II, III and I
(c) II, I, III and IV
(d) II, III, I and IV

44 The following steps are all involved in the action of water soluble hormones. What is the correct order?

I. Receptor interacts with a G protein found in the cell membrane.
II. An enzymes in the cell membrane is activated.
III. Hormone binds to receptor on cell surface.
IV. The desired chemical change occurs inside the cells.
V. The second messenger is produced inside the cell.

(a) III, V, I, II and IV
(b) II, I, III, V and IV
(c) V, III, I, II and IV
(d) V, II, I, IV and III

45 Match the following columns.

Column I		Column II
A. Hydrophilic	1.	A carbon atom bonded to four different groups and therefore able to form enantiomers.
B. Chiral carbon	2.	A biomolecule that produces amino acids and other bimolecules or inorganic substances upon hydrolysis.
C. Hydrophobic	3.	Refers to *R* groups that do not interact readily with water due to their non-polar nature.
D. Conjugated protein	4.	Refers to *R* groups which form H-bonds with water because of their polarity.

Choose the correct match.

	A	B	C	D
(a)	1	4	2	3
(b)	2	3	4	1
(c)	3	4	1	2
(d)	4	1	3	2

(C) Medical entrances' gallery

1 The following metal ion activates many enzymes, participates in the oxidation of glucose to produce ATP and with Na, is responsible for the transmission of nerve signals **[NEET 2020]**
(a) copper (b) calcium (c) potassium (d) iron

2 Which of the following is a basic amino acid? **[NEET 2020]**
(a) Alanine (b) Tyrosine
(c) Lysine (d) Serine

3 Sucrose on hydrolysis gives **[NEET 2020]**
(a) α-D-glucose + β-D-glucose
(b) α-D-glucose + β-D-fructose
(c) α-D-fructose + β-D-fructose
(d) β-D-glucose + α-D-fructose

4 The reaction of concentrated sulphuric acid with carbohydrates ($C_{12}H_{22}O_{11}$) is an example of **[NEET 2020]**
(a) dehydration (b) oxidation
(c) reduction (d) sulphonation

5 Deficiency of which vitamin causes osteomalacia? **[NEET 2020]**
(a) Vitamin A (b) Vitamin D
(c) Vitamin K (d) Vitamin E

6 Which of the following statements is not true about glucose? **[NEET 2020]**
(a) It is an aldohexose.
(b) It contains five hydroxyl groups.
(c) It is a reducing sugar.
(d) It is an aldopentose.

7 The non-essential amino acid among the following is
(a) leucine (b) alanine **[NEET 2019]**
(c) lysine (d) valine

8 Which structure(s) of proteins remains(s) intact during denaturation process? **[NEET (Odisha) 2019]**
(a) Both secondary and tertiary structures
(b) Primary structure only
(c) Secondary structure only
(d) Tertiary structure only

9 **Assertion** Tryptophan is an example of non-essential amino acids.
Reason The amino acids that are not synthesised in human body are non-essential amino acids. **[AIIMS 2019]**
(a) Both A and R are correct and R is the correct explanation of A
(b) Both A and R are correct but R is not the correct explanation of A
(c) A is correct but R is in correct
(d) Both A and R are incorrect

10 **Assertion** Glucose does not give 2, 4- DNP test.
Reason Glucose exists in cyclic hemiacetal form.
(a) Both A and R are correct and R is the correct explanation of A **[AIIMS 2019]**
(b) Both A and R are correct but R is not the correct explanation of A
(c) A is correct but R is incorrect
(d) Both A and R are incorrect

11 **Assertion** Two sugar units are joined by 1, 2-glycosidic bond in sucrose. **[AIIMS 2019]**
Reason It contains C_1-glucose and C_2-fructose glycosidic bond.
(a) Both A and R are correct and R is the correct explanation of A
(b) Both A and R are correct but R is not the correct explanation of A
(c) A is correct but R is incorrect
(d) Both A and R are incorrect

12 Invert sugar is mixture of **[JIPMER 2019]**
(a) D-glucose+D-fructose (b) L-glucose+D-fructose
(c) L-glucose+D-glucose (d) L-glucose+L-glucose

13 Which of the following amino acid is optically inactive? **[JIPMER]**
(a) Glycine (b) Valine (c) Alanine (d) Histidine

14 Which of the following compounds can form a *Zwitter* ion?
(a) Benzoic acid (b) Acetanilide **[NEET 2018]**
(c) Aniline (d) Glycine

15 The difference between amylose and amylopectin is **[NEET 2018]**
(a) amylopectin have (1, 4) α-linkage and 1, 6 β-linkage
(b) amylose have (1, 4) α-linkage and 1, 6 β-linkage
(c) amylopectin have (1, 4) α-linkage and 1, 6 α-linkage
(d) amylose is made up of glucose and galactose

16 Hydrolysis of sucrose with dilute aqueous sulphuric acid yields
(a) 1:2 D-(+)- glucose; D-(–)- fructose **[AIIMS 2018]**
(b) 1:2 D-(–)-glucose; D-(+)-fructose
(c) 1:1 D-(–)- glucose; D-(+)-fructose
(d) 1:1 D-(+)-glucose; D-(–)-fructose

17 Which one of the following represents the correct structure of cellulose? **[JIPMER 2018]**
(a) Branched, α (1, 4) and β (1, 6) - glucose
(b) Unbranched, α (1, 6) - glucose
(c) Branched, α (1, 4) - glucose
(d) Unbranched, β (1, 4) - glucose

18 The central dogma of molecular genetics states that the genetic information flows from **[NEET (II) 2016]**
(a) amino acids → proteins → DNA
(b) DNA → carbohydrates → proteins
(c) DNA → RNA → proteins
(d) DNA → RNA → carbohydrates

19 The correct corresponding order of names of four aldoses with configuration given below **[NEET (II) 2016]**

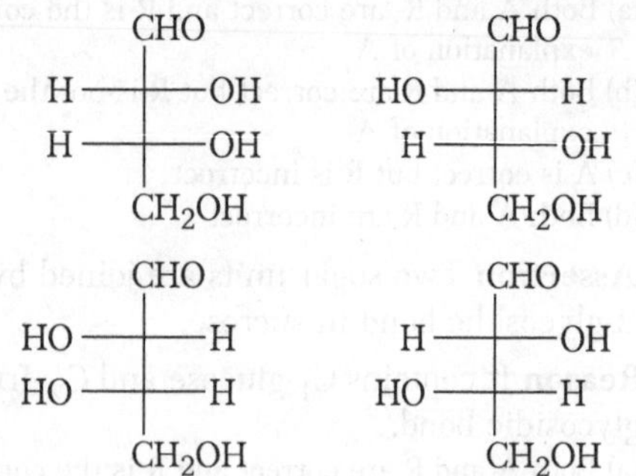

respectively, is
(a) L-erythrose, L-threose, L-erythrose, D-threose
(b) D-threose, D-erythrose, L-threose, L-erythrose
(c) L-erythrose, L-threose, D-erythrose, D-threose
(d) D-erythrose, D-threose, L-erythrose, L-threose

20 In a protein molecule, various amino acids are linked together by **[NEET (I) 2016]**
(a) β-glycosidic bond (b) peptide bond
(c) dative bond (d) α-glycosidic bond

21 The correct statement regarding RNA and DNA respectively is **[NEET (I) 2016]**
(a) the sugar component in RNA is ribose and the sugar component in DNA is 2′ deoxyribose
(b) the sugar component in RNA is arabinose and the sugar component in DNA is ribose
(c) the sugar component in RNA is 2′ deoxyribose and the sugar component in DNA is arabinose
(d) the sugar component in RNA is arabinose and the sugar component in DNA is 2′ deoxyribose

22 Which one given below is a non-reducing sugar? **[NEET 2015]**
(a) Lactose (b) Glucose
(c) Sucrose (d) Maltose

23 Adenosine is an example of **[KCET 2015]**
(a) nucleotide (b) purine base
(c) pyrimidine base (d) nucleoside

24 Glycogen is **[KCET 2015]**
(a) a polymer of β-D-glucose units
(b) a structural polysaccharide
(c) structurally very much similar to amylopectin
(d) structurally similar to amylopectin but extensively branched

25 Cheilosis and digestive disorders are due to the deficiency of **[KCET 2015]**
(a) thiamine (b) ascorbic acid
(c) riboflavin (d) pyridoxine

26 Glucose does not react with **[KCET 2015]**
(a) hydroxylamine (b) conc. HNO_3
(c) acetic anhydride (d) sodium bisulphite
(e) Br_2/H_2O

27 Which of the following is a water soluble vitamin? **[Kerala CEE 2015]**
(a) Vitamin A (b) Vitamin D
(c) Vitamin B_1 (d) Vitamin E
(e) Vitamin K

28 D - (+) - glucose reacts with hydroxyl amine and yields an oxime. The structure of the oxime would be **[CBSE AIPMT 2015]**

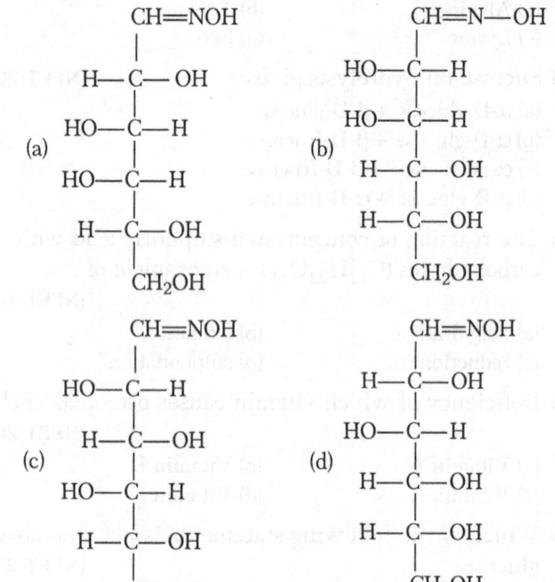

29 One mole of sucrose on hydrolysis yields **[MHT CET 2014]**
(a) 1 mole of glucose +1 mole of fructose
(b) 2 moles of glucose +1 mole of fructose +1 mole of galactose
(c) 1 mole of glucose +2 moles of fructose +1 mole of galactose
(d) moles of glucose +2 moles of fructose

30 Which one of the following is an essential amino acid? **[KCET 2014]**
(a) Methionine (b) Tyrosine
(c) Proline (d) Glycine

31 The one letter code for the amino acid tryptophan is **[KCET 2014]**
(a) G (b) V (c) W (d) H

32 Thymine is held by two H-bonds with the base **[UK PMT 2014]**
(a) guanine (b) cytosine
(c) thymine (d) adenine

33 Metal ion responsible for the minamata disease is **[WB JEE 2014]**
(a) Co^{2+} (b) Hg^{2+}
(c) Cu^{2+} (d) Zn^{2+}

34 Which one of the following sets of vitamins is fat soluble? **[EAMCET 2014]**
(a) C, D, B_6, B_{12} (b) A, D, E, K
(c) A, D, B_1, B_2 (d) D, B_1, B_2, E

35 Sucrose on treatment with conc. HCl produce [AIIMS 2013]
(a) glucose (b) fructose
(c) glucose + fructose (d) laevunilic acid

36 Starch is a polymer of [UP CPMT 2013]
(a) glucose (b) fructose
(c) Both (a) and (b) (d) None of these

37 Maltose on hydrolysis produces [UP CPMT 2013]
(a) glucose (b) fructose (c) mannose (d) galactose

38 The carbohydrate which is an essential constituent of plant cell is [Manipal 2013]
(a) sucrose (b) starch (c) cellulose (d) vitamins

39 Which of the following reactions estabilishes difference between glucose and fructose? [WB JEE 2013]
(a) Tollen's reagent reaction (b) Phenyl hydrazine
(c) P/HI (d) Conc. HNO_3

40 Which of the following is an aldohexose? [KCET 2013]
(a) Cellulose (b) Sucrose
(c) Glucose (d) Raffinose

41 A certain compound gives negative test with ninhydrin and positive test with Benedict's solution, it is [KCET 2013]
(a) an amino acid (b) a monosaccharide
(c) a lipid (d) a protein

42 Subunit present in haemoglobin are [AIIMS 2013]
(a) 2 (b) 3 (c) 4 (d) 5

43 **Assertion** (A) Glycosides are hydrolysed in acidic conditions.
Reason (R) Glycosides are acetals. [AIIMS 2013]
(a) Both A and R are correct and R is the correct explanation of A
(b) Both A and R are correct but R are not the correct explanation of A
(c) A is correct but R is incorrect
(d) Both A and R are incorrect

44 The linkage present in proteins and peptides is [UP CPMT 2013]
(a) $-\overset{O}{\overset{\|}{C}}-$ (b) $-\overset{O}{\overset{\|}{C}}-O-\overset{O}{\overset{\|}{C}}-$
(c) $-NH-$ (d) $-\overset{O}{\overset{\|}{C}}-NH-$

45 The pair of bases in DNA are held together by [Manipal 2013]
(a) ionic bonds (b) H-bond
(c) phosphate group (d) deoxyribose group

46 Interferon is [WB JEE 2013]
(a) the commercial name given to a tonic containing iron
(b) an alloy of iron and chromium used in space vessels
(c) a protein derived from the body against viruses
(d) a trade name of a potent insecticide

47 The human body does not produce [WB JEE 2013]
(a) enzymes (b) vitamins
(c) proteins (d) nucleic acids

48 Glucose and mannose are [AIIMS 2012]
(a) anomers (b) epimers
(c) ketohexoses (d) disaccharides

49 Iodine value is related to [AFMC 2012]
(a) fats and oils (b) alcohols
(c) ethers (d) esters

50 Which one of the following is an example of non-reducing sugar? [UP CPMT 2012]
(a) Sucrose (b) Lactose (c) Maltose (d) Cellobiose

51 Polysaccharides have which of the following linkage? [OJEE 2012]
(a) Glycosidic linkage (b) H-bond
(c) Peptide linkage (d) No linkage

52 Which one of the following statements is incorrect about enzyme catalysts? [CBSE AIPMT 2012]
(a) Enzymes are mostly protein in nature
(b) Enzyme action is specific
(c) Enzymes are denatured by ultraviolet rays and at high temperature
(d) Enzymes are least reactive at optimum temperature

53 A codon has a sequence of *A* and specifies a particular *B* that is to be incorporated into *C*. What are *A*,*B*,*C* [Manipal 2012]

	A	*B*	*C*
(a)	3 bases	amino acid	carbohydrate
(b)	3 acids	carbohydrate	protein
(c)	3 bases	protein	amino acid
(d)	3 bases	amino acid	protein

54 The enzyme which have control site in addition to active site are called [AFMC 2012]
(a) holozymes (b) coenzymes
(c) apoenzymes (d) allosteric enzymes

55 Which one of the following statements is not true regarding (+) lactose? [CBSE AIPMT 2011]
(a) (+)-lactose, $C_{12}H_{22}O_{11}$ contains 8-OH groups
(b) On hydrolysis (+)-lactose gives equal amount of D(+)-glucose and D(+)-galactose
(c) (+)-lactose is a β-glycoside formed by the union of molecule of D(+)-glucose and a molecule of D(+)-galactose.
(d) (+)-lactose is reducing sugar and does not exhibit mutarotation

56 What is nature of glucose-glucose linkage in starch that makes it so susceptible to acid hydrolysis? [DUMET 2011]
(a) Starch is hemiacetal
(b) Starch is acetal
(c) Starch is polymer
(d) Starch contains only few molecules of glucose

57 α-maltose consists of **[KCET 2011]**
(a) one α-D-glucopyranose unit and one β-D-glucopyranose unit with 1, 2-glycosidic linkage
(b) two α-D-glucopyranose units with 1, 2-glycosidic linkage
(c) two β-D-glucopyranose units with 1, 4-glycosidic linkage
(d) two α-D-glucopyranose units with 1, 4-glycosidic linkage

58 Glucose does not react with **[KCET 2011]**
(a) Br_2/H_2O (b) $H_2N—OH$
(c) HI (d) $NaHSO_3$

59 The linkage between the two monosaccharide units in lactose is **[Kerala CEE 2011]**
(a) C-1 of β-D-glucose and C-4 of β-D-galactose
(b) C-1 of β-D-galactose and C-4 of β-D-glucose
(c) C-1 of α-D-galactose and C-4 of β-D-glucose
(d) C-1 of β-D-galactose and C-4 of α-D-glucose
(e) C-1 of α-D-glucose and C-4 of α-D-glucose

60 How glucose is related with fructose? **[Guj. CET 2011]**
(a) Functional group isomerism
(b) Rotamers
(c) Position isomerism
(d) Geometrical isomerism

61 How can you say that glucose is cyclic compound?
(a) Glucose undergoes Tollen's reaction **[Guj. CET 2011]**
(b) Glucose reacts with phenyl hydrazine
(c) Glucose fails to react with sodium hydrogen sulphite
(d) Glucose reacts with nitric acid

62 In an alkaline medium, glycine predominantly exists as/in a/an **[KCET 2011]**
(a) cation (b) anion
(c) Zwitter ion (d) covalent form

63 Which one of the following does not correctly match with each other? **[KCET 2011]**
(a) Silk-polyamide (b) Lipase-enzyme
(c) Butter-fat (d) Oxytocin-enzyme

64 Which is not the correct statement about RNA and DNA? **[DUMET 2011]**
(a) DNA is active in virus while RNA never appears in virus
(b) DNA exists as dimer while RNA is usually single stranded
(c) DNA contains deoxyribose as its sugar and RNA contains ribose
(d) RNA contains uracil in place of thymine (found in DNA) as a base

65 Which of the following does not exhibit the phenomenon of mutarotation? **[CBSE AIPMT 2011]**
(a) (–)-fructose (b) (+)-sucrose
(c) (+)-lactose (d) (+)-maltose

66 Hydrolysis of sucrose is called **[AFMC 2011]**
(a) inversion (b) esterification
(c) hydration (d) saponification

67 Which of the following is a trisaccharide?
(a) Stachyrose (b) Sucrose **[MHT CET 2011]**
(c) Raffinose (d) Ribose

68 Glucose on reaction with Fehling solution's gives **[MHT CET 2011]**
(a) cupric oxide (b) cuprous oxide
(c) saccharic acid (d) Both (b) and (c)

69 Water hydrolysis of sucrose produces **[MP PMT 2011]**
(a) two molecules of glucose
(b) two molecules of fructose
(c) one molecule of each glucose and fructose
(d) one molecule of each glucose and mannose

70 The monomer units of starch are **[RPMT 2011]**
(a) α-D-glucose (b) β-D-glucose
(c) pyranose (d) galactose

71 Glucose reacts with X number of molecules of phenylhydrazine to yield osazone. The value of X is **[RPMT 2011]**
(a) three (b) two (c) one (d) four

72 The letter 'D' in D-glucose signifies **[KCET 2011]**
(a) configuration at all chiral carbons
(b) dextrorotatory
(c) that is a monosaccharide
(d) configuration at a particular chiral carbon

73 The change in optical rotation of freshly prepared solution of cane sugar with time is known as
(a) mutarotation (b) inversion **[Manipal 2011]**
(c) specific rotation (d) rotatory motion

74 Which of the following is the correct order of priority of group in D-glyceraldehyde? **[Guj. CET 2011]**
(a) OH(1), CHO(2), CH_2OH(3) and H(4)
(b) OH(1), CH_2OH(2), CHO(3) and H(4)
(c) CH_2OH(1), CHO(2), OH(3) and H(4)
(d) CHO(1), OH(2), CH_2OH(3) and H(4)

75 Which of the following compounds, when heated at 483 K turns to caramel? **[Guj. CET 2011]**
(a) Glucose (b) Sucrose
(c) Fructose (d) Lactose

76 Which of the following cannot be hydrolysed into aldose and ketose? **[OJEE 2011]**
(a) Disaccharide (b) Monosaccharide
(c) Polysaccharide (d) Oligosaccharide

77 Glucose + Tollen's reagent $\longrightarrow$ Silver mirror the above process shows **[VMMC 2011]**
(a) presence of —COOH group
(b) presence of keto group
(c) presence of —CHO group
(d) presence of $—CONH_2$ group

ANSWERS

Check point 22.1

1. (c)	2. (a)	3. (c)	4. (d)	5. (a)	6. (b)	7. (b)	8. (d)	9. (a)	10. (a)

Check point 22.2

1. (a)	2. (b)	3. (b)	4. (a)	5. (d)	6. (a)	7. (c)	8. (b)	9. (d)

Check point 22.3

1. (c)	2. (b)	3. (a)	4. (c)	5. (b)	6. (c)	7. (d)	8. (b)	9. (c)	10. (b)

(A) Topical exercise

1. (d)	2. (d)	3. (c)	4. (d)	5. (a)	6. (c)	7. (d)	8. (c)	9. (d)	10. (a)
11. (b)	12. (a)	13. (d)	14. (c)	15. (c)	16. (b)	17. (a)	18. (d)	19. (a)	20. (b)
21. (a)	22. (a)	23. (c)	24. (c)	25. (d)	26. (c)	27. (a)	28. (c)	29. (d)	30. (a)
31. (b)	32. (d)	33. (b)	34. (c)	35. (b)	36. (c)	37. (d)	38. (a)	39. (b)	40. (a)
41. (c)	42. (c)	43. (d)	44. (b)	45. (c)	46. (c)	47. (c)	48. (b)	49. (d)	50. (c)
51. (d)	52. (b)	53. (a)	54. (d)	55. (a)	56. (a)	57. (c)	58. (c)	59. (c)	60. (d)
61. (a)	62. (a)	63. (b)	64. (d)	65. (d)	66. (d)	67. (b)	68. (c)	69 (d)	70. (d)
71. (c)	72. (c)	73. (a)	74. (c)	75. (a)	76. (a)	77. (b)	78. (d)	79. (b)	80. (b)
81. (d)	82. (d)	83. (b)	84. (d)	85. (a)	86. (d)	87. (c)	88. (c)	89. (c)	90. (d)
91. (b)	92. (a)	93. (c)	94. (d)	95. (c)	96. (d)	97. (b)	98. (d)	99. (b)	100. (a)
101. (c)	102. (d)	103. (b)	104. (a)	105. (a)	106. (a)	107. (a)	108. (a)	109. (b)	110. (a)
111. (c)	112. (b)	113. (d)	114. (b)	115. (a)	116. (a)	117. (b)	118. (a)	119. (a)	120. (a)
121. (d)	122. (a)	123. (a)	124. (a)	125. (d)	126. (a)	127. (a)	128. (b)	129. (d)	130. (b)
131. (c)	132. (d)	133. (b)							

(B) Miscellaneous questions

1. (d)	2. (a)	3. (c)	4. (c)	5. (b)	6. (d)	7. (b)	8. (a)	9. (a)	10. (c)
11. (c)	12. (d)	13. (b)	14. (a)	15. (b)	16. (d)	17. (b)	18. (a)	19. (c)	20. (a)
21. (c)	22. (b)	23. (b)	24. (c)	25. (b)	26. (a)	27. (a)	28. (a)	29. (c)	30. (d)
31. (c)	32. (d)	33. (b)	34. (d)	35. (c)	36. (a)	37. (a)	38. (a)	39. (d)	40. (b)
41. (c)	42. (a)	43. (d)	44. (c)	45. (d)					

(C) Medical entrances' gallery

1. (c)	2. (c)	3. (b)	4. (a)	5. (b)	6. (d)	7. (b)	8. (b)	9. (d)	10. (a)
11. (a)	12. (a)	13. (a)	14. (d)	15. (c)	16. (d)	17. (d)	18. (c)	19. (d)	20. (b)
21. (a)	22. (c)	23. (d)	24. (d)	25. (c)	26. (d)	27. (c)	28. (d)	29. (a)	30. (a)
31. (c)	32. (d)	33. (b)	34. (b)	35. (d)	36. (a)	37. (a)	38. (c)	39. (d)	40. (c)
41. (b)	42. (c)	43. (d)	44. (d)	45. (b)	46. (c)	47. (b)	48. (b)	49. (a)	50. (a)
51. (a)	52. (d)	53. (d)	54. (b)	55. (d)	56. (b)	57. (d)	58. (d)	59. (b)	60. (a)
61. (c)	62. (b)	63. (d)	64. (a)	65. (b)	66. (a)	67. (c)	68. (b)	69. (c)	70. (a)
71. (a)	72. (d)	73. (b)	74. (a)	75. (b)	76. (b)	77. (c)			

Hints & Explanations

(A) Topical exercise

1 (*d*) Both statements are incorrect. Deoxyribose $C_5H_{10}O_4$ is a carbohydrate because it follows $C_5(H_2O)_2$ formula and exist as polyhydroxy carbonyl compound whose cyclic structure is as shown below

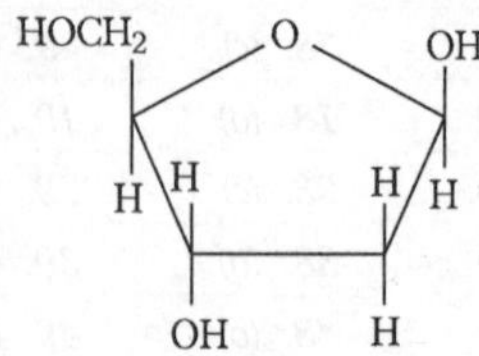

β-2-deoxyribose

8 (*c*) Glucose is a aldohexose having structural formula.

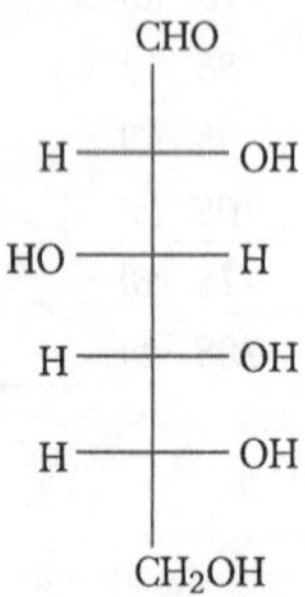

Glucose on heating with P/HI produces *n*-hexane

$$C_6H_{12}O_6 \xrightarrow{P/HI} C_6H_{14}$$

11 (*b*) Statement I is correct but statements II is incorrect D-(+)-glucose is dextrorotatory because it rotates the plane polarised light to right.

Here, D-represents relative configuration of —OH group in glucose present at the bottom chiral centre point.

12 (*a*) Cyclic structures of monosaccharides which differ in structure at C-1 are known as anomers.

Here, I and II are anomer because they differ from each other at C-1 only.

14 (*c*) "Pentaacetate of glucose does not react with hydroxylamine" showing absence of free CHO group. This cannot be explained by open structure of glucose. While all other properties are easily explained by open structure of glucose.

Hence, option (c) is the correct choice.

15 (*c*) Anomers have different configuration at C-1. If OH is present at right side anomeric carbon is known as α- form and if OH is present at left side of anomeric carbon is known as β-form

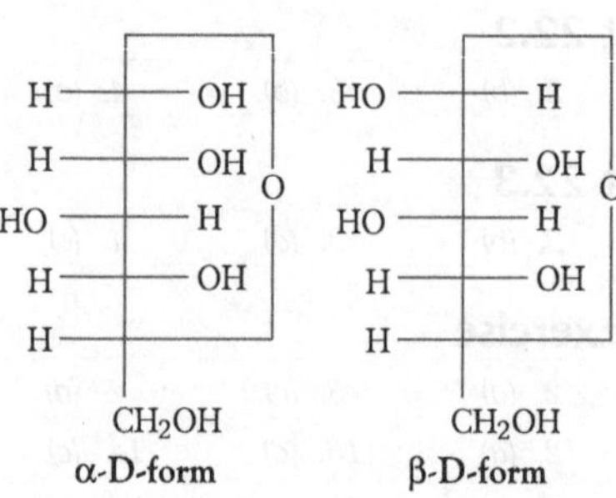

α-D-form β-D-form

19 (*a*) D and L configuration are relative configuration decided by relating structure of given saccharide with D or L-glyceraldehyde.

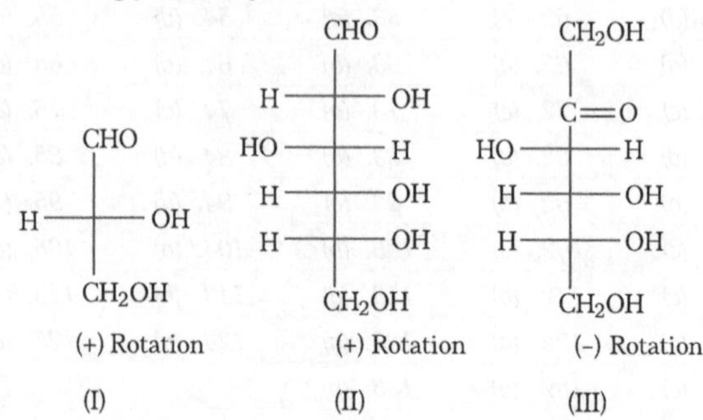

When OH an lowest asymmetric carbon is written at right hand side, it is represented as D-configuration and when OH is written on left hand side, it is represented as L-configuration.

23 (*c*) Carbon adjacent to oxygen atom in the cyclic structure of glucose or fructose is known as anomeric carbon. As the structure shown in question in which '*a*' and '*b*' are present at adjacent to oxygen atom. Both carbons differ in configurations of the hydroxyl group.

26 (*c*) Sucrose (cane sugar) is a disaccharide. One molecule of sucrose on hydrolysis gives one molecule of glucose and one molecule of fructose.

$$\underset{\text{Cane sugar}}{C_{12}H_{22}O_{11}} \xrightarrow[H^+]{H_2O} \underset{\text{D-(+)-glucose}}{C_6H_{12}O_6} + \underset{\text{D-(–)-fructose}}{C_6H_{12}O_6}$$

28 (*c*) Statements I is wrong statement and statements II is correct statement. α-glycosidic linkage is present in maltose

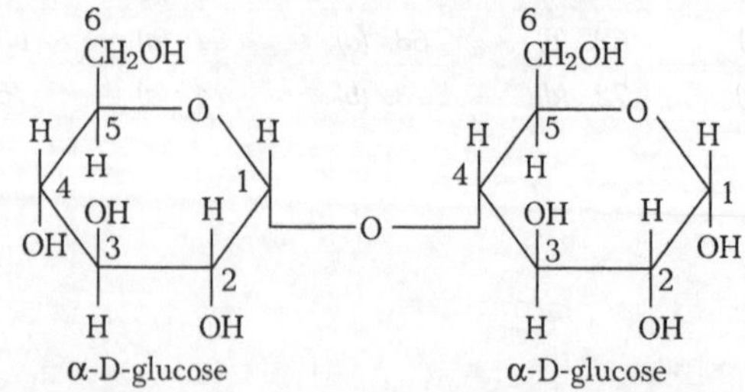

α-D-glucose α-D-glucose

Because maltose is composed of two glucose unit in which C-1 of one glucose unit is linked to C-4 of another glucose unit.

35 (*b*)

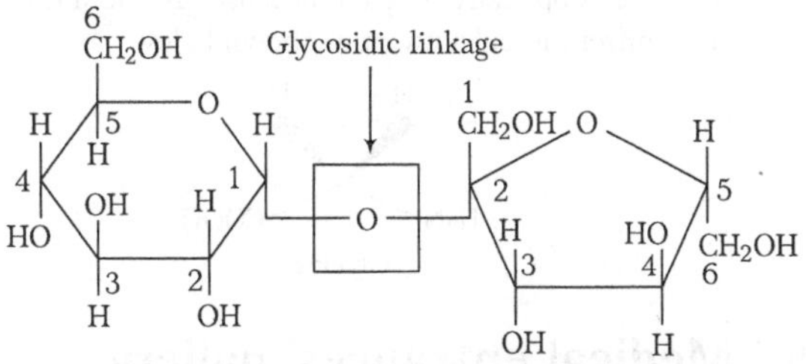

[Structure: α-D-glucose linked to β-D-fructose through glycosidic linkage (–O–)]

α-D-glucose β-D-fructose

This structure represents sucrose in which α-D-glucose and β-D-fructose is attached to each other by C_1—C_2 glycosidic linkage. Since, reducing groups of glucose and fructose are involved in glycosidic bond formation, this is considered as non-reducing sugar.

36 (*c*) Numbering of glucose starts from adjacent carbon of O-atom to the other carbon atom ending at last CH_2OH group as shown below :

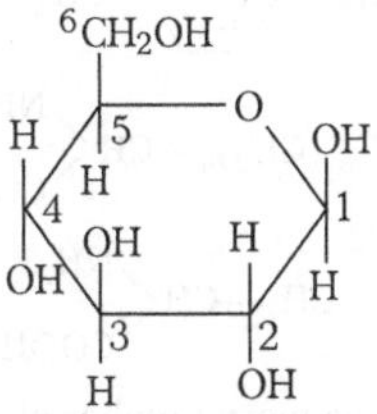

[Structure of glucose with numbered carbons 1–6]

In this way, numbering for the disaccharides can be done as

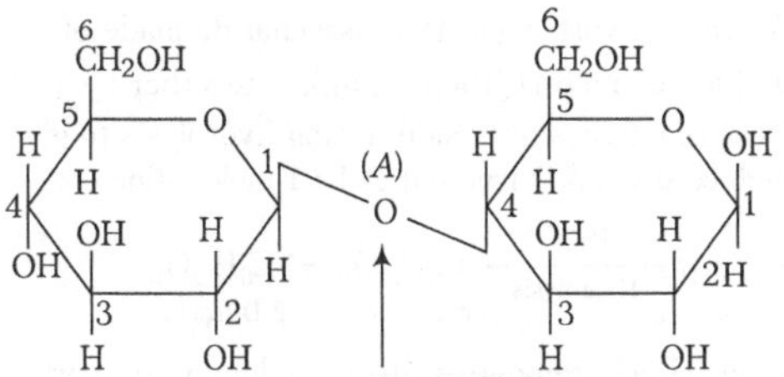

[Structure (I): C_1–C_4 linkage (A)]

[Structure (II): C_1–C_6 linkage (B)]

[Structure (III): C_1–C_4 linkage (C)]

(I) (II) (III)

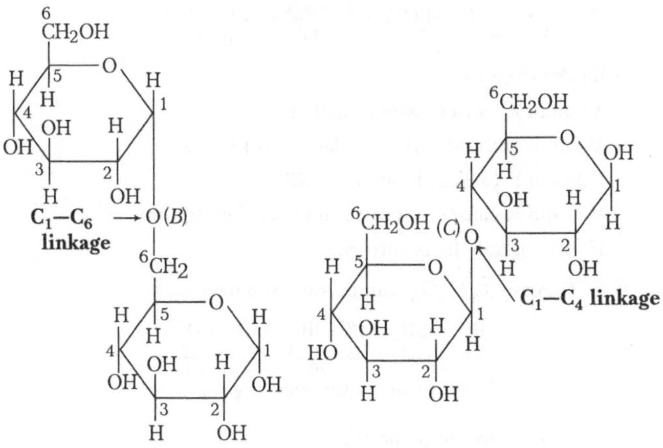

43 (*d*) Both are wrong statements. Correct statements II are glycine must not be taken through diet because it can be synthesised in our body and a non-essential amino acid.

56 (*a*) In primary structure of proteins, each polypeptide in a protein has amino acids linked with each other in a specific sequence. This type of structure is known as primary structure of proteins.

58 (*c*) Secondary structures of protein denotes the shape in which a long polypeptide chain exists. The secondary structure exist in two type of structure α-helix and β-pleated structure.

In α- helix structure, a polypeptide chain forms all possible hydrogen bonds by twisting into a right handed screw with — NH group of each amino acid rest hydrogen bonded to $>C=O$ of adjacent amino acid, which form a helix.

102 (*d*) Vitamin B_{12} can be stored in our body and belongs to B group vitamins, because it is not water soluble.

103 (*b*) Ascorbic acid is the chemical name of vitamin C while others are not vitamins. Aspartic acid is an amino acid. Adipic acid is a dicarboxylic acid having 8 carbon chain. Saccharic acid is a dicarboxylic acid obtained by oxidation of glucose using HNO_3.

106 (*a*) Both are correct statements. Vitamin D can be stored in our body because vitamin D is a fat soluble vitamin.

112 (*b*) Nucleic acids are polymer of nucleotides in which nucleic acids are linked together by phosphodiester linkage.

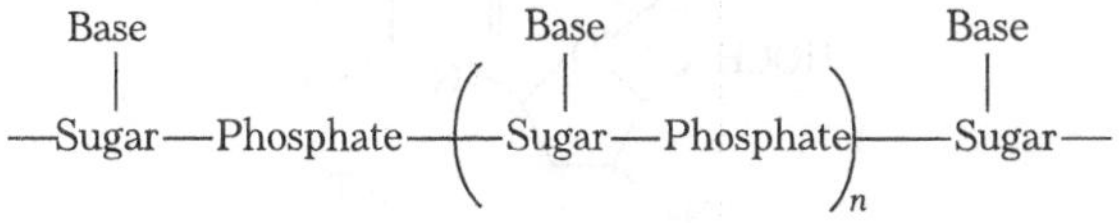

—Sugar—Phosphate—(—Sugar—Phosphate—)$_n$—Sugar— (each Sugar bearing a Base)

127 (*a*) **Nucleoside** Species formed by the attachment of a base to 1′ position of sugar is known as nucleoside. The sugar carbon are numbered as 1′, 2′, 3′, ...to distinguish them from bases.

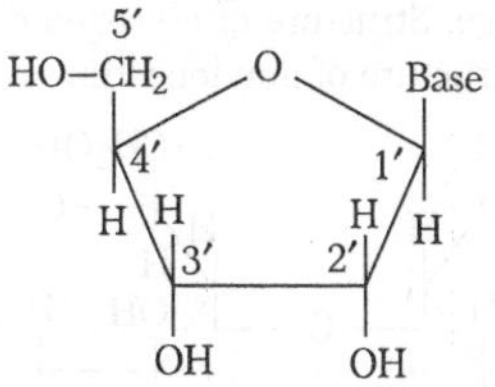

[Structure: HO–CH_2 (5′) on pentose ring with Base at 1′, OH at 2′ and 3′]

Structure of nucleoside

Nucleotide Species formed by attachment of phosphoric acid to nucleoside at 5′ position of sugar nucleotide.

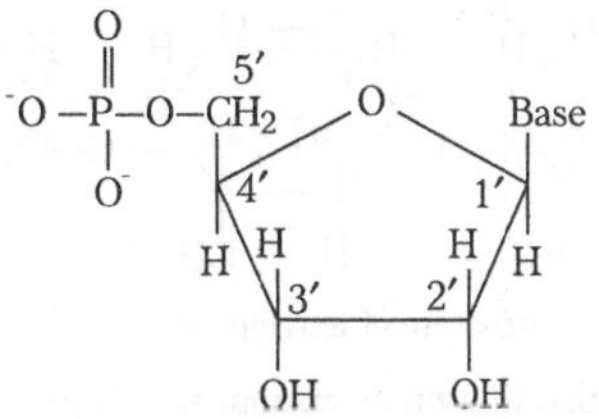

[Structure: ^-O–P(=O)(O^-)–O–CH_2 (5′) on pentose ring with Base at 1′, OH at 2′ and 3′]

Structure of nucleotide

Dinucleotides are formed by phosphodiester linkage between 5′ and 3′ carbon atoms of the pentose sugar.

5′ end of chain

Phosphodiester linkage

3′ end of chain

(B) Miscellaneous questions

2 (*a*) In the given tetrapeptide, the N-terminal amino acid is *ala* and C-terminal amino acid is *leu.*

16 (*d*) Glycogen is a polymer of α-D-glucose stored in the liver, brain and muscles of animals, also known as animal starch.

21 (*c*)

AZT

33 (*b*) Glycogen is a branched chain polymer of α-D-glucose units in which chain is formed by C_1-C_4 glycosidic linkage whereas branching occurs by the formation of C_1-C_6 glycosidic linkage. Structure of glycogen can be shown below similar to the structure of amylopectin.

Structure of amylopectin

Glycogen is also known as animal starch present in liver, muscles and brain.

42 (*a*) Both are correct statements. All naturally occurring α-amino acid except glycine are optically active.

Glycine is optically inactive because glycine does not have all four different substituent as shown below.

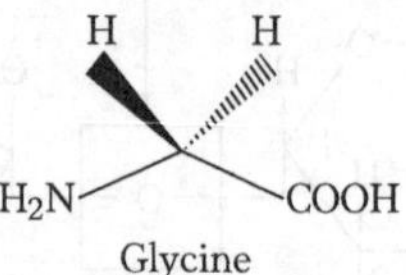

Glycine

(C) Medical entrances' gallery

1 (*c*) The ionic-gradients between two sides of a cell membrane, is operated by Na—K pump and consumes one-third of the ATP used by a resting animal.

Potassium ion (K^+) present in synapse helps transmission of nerve signals from one neuron to another.

2 (*c*)

(a) Alanine ⇒ $CH_3-CH(NH_2)COOH$

(b) Tyrosine ⇒ $HO-C_6H_4-CH_2-CH(NH_2)COOH$

(c) Lysine ⇒ $H_2N-(CH_2)_4-CH(NH_2)COOH$

(d) Serine ⇒ $HO-CH_2-CH(NH_2)COOH$

Since, lysine contains more number of $-NH_2$ groups as compared to $-COOH$ groups, hence it is a basic amino acid.

3 (*b*) Sucrose (invert sugar) is a disaccharide made of α-D-glucose and β-D-fructose, linked together by 1, 2-linkage. Sucrose undergoes inversion during hydrolysis to give α-D-glucose and β-D-fructose in 1 : 1 mole ratio.

$$\underset{\text{Sucrose}}{C_{12}H_{22}O_{11}} \xrightarrow[\text{Hydrolysis}]{H_3O^+} \underset{\alpha\text{-D-glucose}}{C_6H_{12}O_6} + \underset{\beta\text{-D-fructose}}{C_6H_{12}O_6}$$

4 (*a*) Conc. H_2SO_4 removes water, i.e. dehydrates carbohydrates into carbon (black residue) or charred sugar.

$$C_{12}H_{22}O_{11} \text{ or } C_{12}(H_2O)_{11} \xrightarrow[-11\,H_2O]{\text{Conc.}H_2SO_4} \underset{\text{(Black)}}{12\ C}$$

5 (*b*) Deficiency of

Vitamin A causes—Xerophthalmia

Vitamin D causes—Osteomalacia and Rickets

Vitamin E causes—Fragility of RBCs

Vitamin K causes—Increase in blood clotting time.

Hence, option (b) is correct.

6 (*d*) Glucose ($C_6H_{12}O_6$) can be simply shown as,

$$\underset{\text{one } 1°\text{- OH group}}{HO-CH_2}-\underset{\text{Four } 2°\text{- OH group}}{(CHOH)_4}-\underset{\text{aldehyde group}}{CH{=}O}$$

It shows reducing property.

Glucose is an aldohexose (made of six carbon atoms) an contain five —OH groups.

Hence, option (d) is incorrect statement.

7 (*b*) Alanine is a non-essential amino acid while leucine, lysine and valine are essential amino acid. Non-essential amino acids are synthesised in the body.

8 (*b*) The loss in biological activity of proteins due to unfolding of globules and uncoiling of helix is called denaturation of proteins. During the denaturation, primary structure remains intact but secondary and tertiary structures are destroyed. Some examples are the coagulation of egg white on boiling, curdling of milk caused by bacteria due to formation of lactic acid.

9 (*d*) The amino acids that are not synthesised in human body and must be obtained from our diet are essential amino acids. Tryptophan is an example of essential amino acids.

Therefore, both Assertion and Reason are incorrect.

10 (*a*) In nature, glucose exists in cyclic hemiacetal form and has no free —CHO group, thus it does not give 2, 4-DNP test.

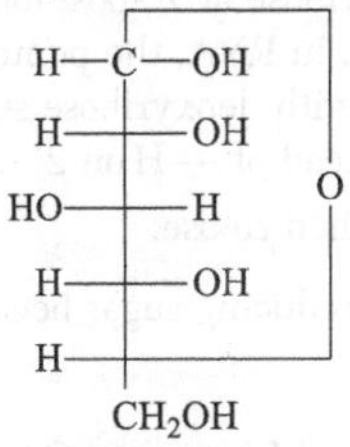

H—C—OH
H—OH
HO—H
H—OH
H—
O
CH_2OH

Cyclic structure

Therefore, Assertion and Reason both are correct and Reason is the correct explanation of the given Assertion.

Hence, option (a) is the correct.

11 (*a*) Sucrose on hydrolysis gives equimolar mixture of D-(+)-glucose and D-(–)-fructose. These two monosaccharides are held together by a glycosidic linkage between C-1 of α-glucose and C-2 of β-fructose. Therefore, both, Assertion and Reason are true and Reason is the correct explanation of the given Assertion.

Hence, option (a) is the correct.

α-(D)-glucose — Glycosidic linkage — β-(D)-fructose

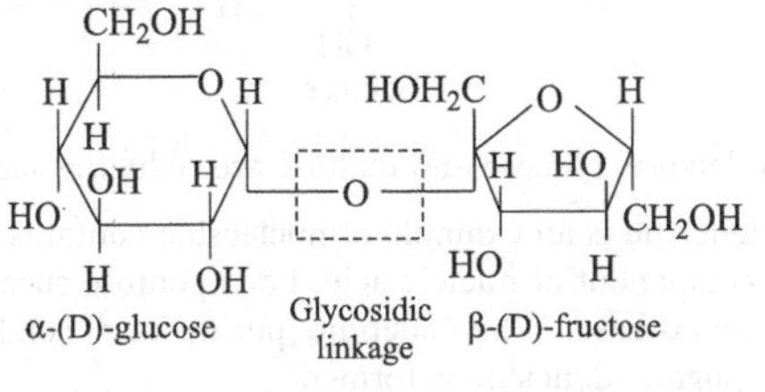

12 (*a*) Invert sugar is a mixture of D-glucose and D-fructose. Solution of sucrose is dextrorotatory. When sucrose is hydrolysed it forms equimolar quantity of D-glucose and D-fructose. The solution of formed products is found to be laevorotatory. This change in optical properties of sucrose is called inversion of cane sugar. The equimolar product formed, i.e D-glucose and D-fructose is called invert sugar.

13 (*a*) Except glycine, all other α-amino acids are optically active as their α-carbon is asymmetric.

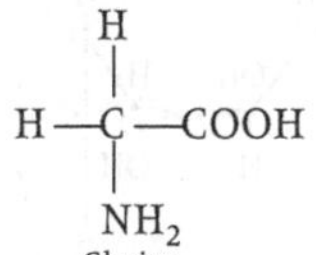

$$H-\underset{NH_2}{\overset{H}{C}}-COOH$$

Glycine

Thus, glycine is optically inactive.

14 (*d*) Ion containing positive as well as negative charge is called Zwitter ion. Among the given options, only glycine ($H_2N—CH_2—COOH$) is an amino acid which contains both acidic (acquiring negative charge) and basic group (acquiring positive charge). Therefore, glycine can form a Zwitter ion. It is because glycine behave like salt rather than simple amines or carboxylic acids. In aqueous solution, the carboxyl group can lose a proton and amino group can accept a proton giving rise to a dipolar ion known as Zwitter ion.

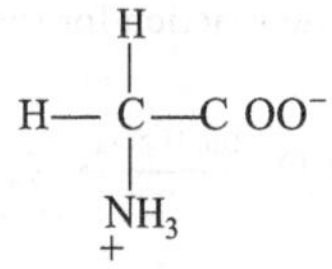

$$H-\underset{\overset{+}{N}H_3}{\overset{H}{C}}-COO^-$$

Zwitter ion

Zwitter ion is behaves as a cation in acidic medium and migrates toward cathode on passing electric current. It is behave as an anion in basic medium and migrates toward anode on passing electric current.

$$\underset{\text{Anion in basic medium}}{H-\underset{NH_2}{CH}-COO^-} \xleftarrow{\overline{O}H} H-\underset{\overset{+}{N}H_3}{CH}-COO^- \xrightarrow{H^+} \underset{\text{Cation in acidic medium}}{H-\underset{\overset{+}{N}H_3}{CH}-COOH}$$

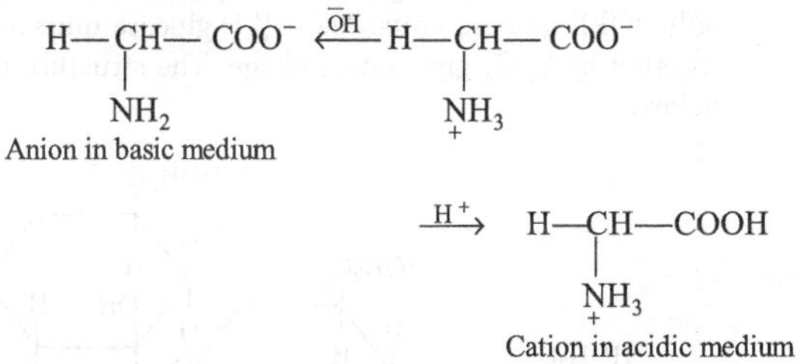

15 (*c*) Starch consists of two components amylose and amylopectin. Chemically, amylose is a long unbranched chain with 200-1000 α-D- (+)-glucose units held by C_1-C_4 glycosidic linkage.

α-link α-link

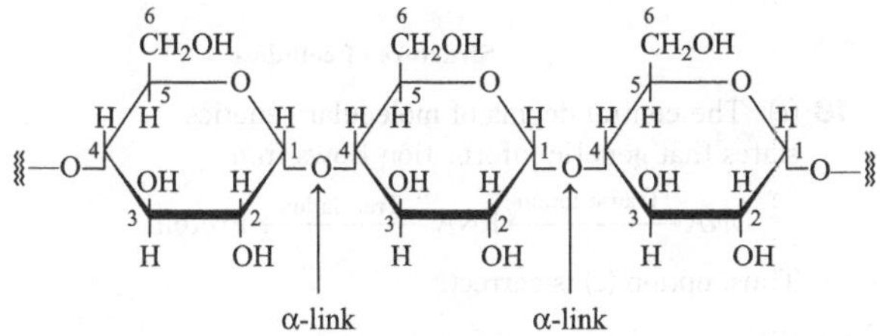

Amylose

Amylopectin is a branched chain polymer of α -D-glucose units in which straight chain is formed by C_1-C_4 glycosidic linkage and branching occurs by C_1-C_6 glycosidic linkage.

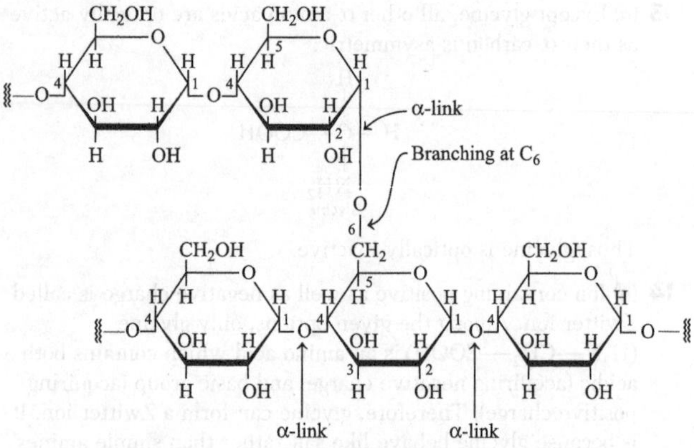

α-link

Branching at C_6

α-link α-link

Amylopectin

16 (*d*) On hydrolysis with dilute aqueous sulphuric acid, sucrose forms an equimolar mixture of D-(+)-glucose and D-(–)-fructose. The reaction for the hydrolysis of sucrose is as follows :

$$\underset{\text{Sucrose}}{C_{12}H_{22}O_{11}} + H_2O \xrightarrow{\text{Dil. } H_2SO_4} \underset{\text{D-(+)-glucose}}{C_6H_{12}O_6} + \underset{\text{D-(–)-fructose}}{C_6H_{12}O_6}$$

Solution of sucrose is dextrorotatory (specific rotation = + 66.5°) but after hydrolysis gives dextrorotatory glucose and laevorotatory fructose. The solution of formed product is found to be laevorotatory (specific rotation = –20.0°)

17 (*d*) Cellulose has a straight chain polysaccharide composed only of β-D glucose units. These β-D glucose units are linked together by C_1-C_4 glycosidic linkage. The structure is shown below:

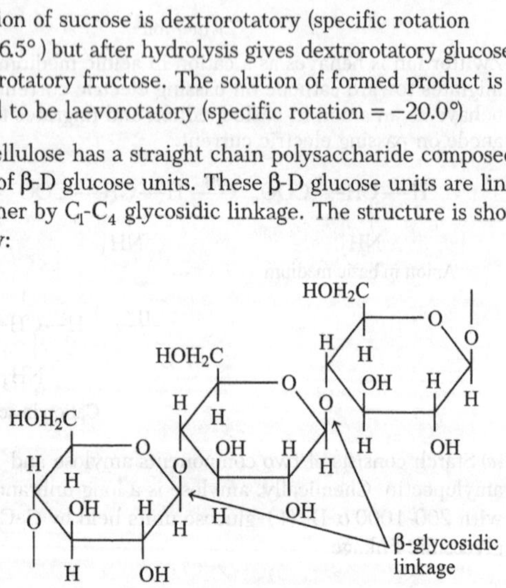

β-glycosidic linkage

Structure of cellulose

18 (*c*) The central dogma of molecular genetics states that genetic information flows from

$$\text{DNA} \xrightarrow{\text{Transcription}} \text{RNA} \xrightarrow{\text{Translation}} \text{Protein}$$

Thus, option (c) is correct.

19 (*d*) Base on glyceraldehydes are D or L

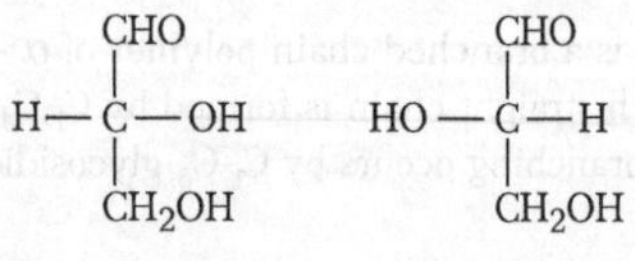

D: CHO, H—C—OH, CH_2OH L: CHO, HO—C—H, CH_2OH

Thus, correct configuration is given in option (d)

Erythrose are pairs of enantiomers with similar groups on same side. Threose are pais of enantiomers having similar groups on opposite side

20 (*b*) Two amino acids in a protein are linked by a peptide bond. e.g. glycylalanine is formed when carboxyl group of glycine combines with the amino group of alanine.

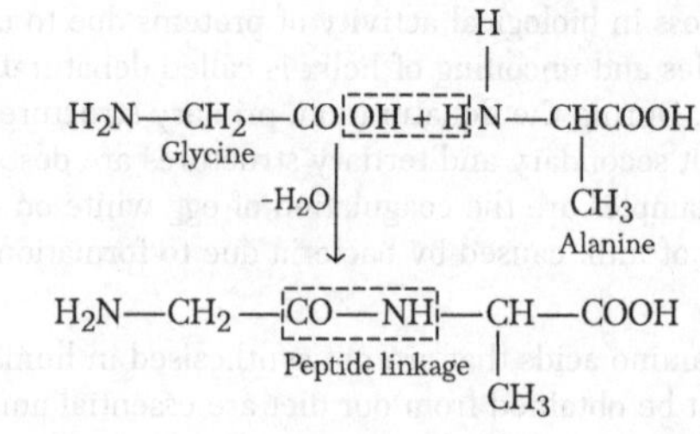

$$\underset{\text{Glycine}}{H_2N—CH_2—CO\,OH} + \underset{\text{Alanine}}{H\,N(H)—CH(CH_3)COOH} \xrightarrow{-H_2O} H_2N—CH_2—\underset{\text{Peptide linkage}}{CO—NH}—CH(CH_3)—COOH$$

Glycylalanine (gly-ala)

21 (*a*) In DNA, two helically twisted strands are connected together by H-bond. Each strand consists of alternating molecules of deoxyribose at 2′-position and phosphate groups. On the other hands, in RNA, the pentose sugar has an identical structure with deoxyribose sugar except that there is an —OH group instead of —H on 2′ carbon atom. Hence, it is only called ribose.

22 (*c*) Sucrose is non-reducing sugar because reducing part of glucose (—C(=O)—H) and fructose (>C= O) are involved in glycosidic linkage.

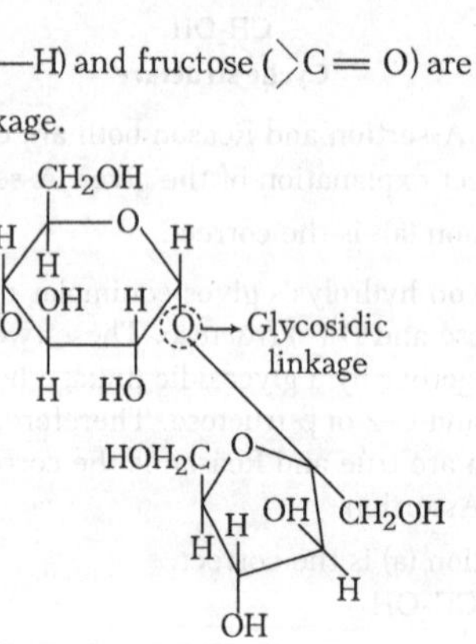

Glycosidic linkage

Sucrose

While lactose, glucose and maltose are reducing sugars.

23 (*d*) Adenosine is an example of nucleoside contains only two basic component of nucleic acid, i.e. a pentose sugar and a nitrogenous base. When adenine [purine base] combines with ribose sugar, adenosine is formed.

A nucleotide contains all the three basic component of nucleic acids, i.e. a phosphoric acid group, a pentose sugar and a nitrogenous base (purines, pyrimidines).

24 (*d*) Glycogen $(C_6H_{10}O_6)_n$ is a condensation polymer of α-D-glucose. Glycogen is also called animal starch because it's structure is similar to that of amylopectin. Amylopectin chains consist of about 20-25 glucose units but glycogen chains are much shorter, i.e. they consist of about 10-14 glucose units. In other words glycogen is more highly branched than amylopectin.

25 (*c*) Cheilosis (cracking or fissuring at corners of mouth and lips), digestive disorders and burning sensation of skin are deficiency diseases, caused by the deficiency of vitamin B_2 (riboflavin)

26 (*d*) Glucose does not react with sodium bisulphite ($NaHSO_3$) even through it has an aldehydic group. This is because in glucose the aldehydic group is not completely free. It is involved in acetal formation with hydroxyl group at C-5.

27 (*c*) Vitamin B_1 is soluble in water.

28 (*d*) D-(+)-glucose contains aldehydic group which reacts with NH_2OH to yield an oxime.

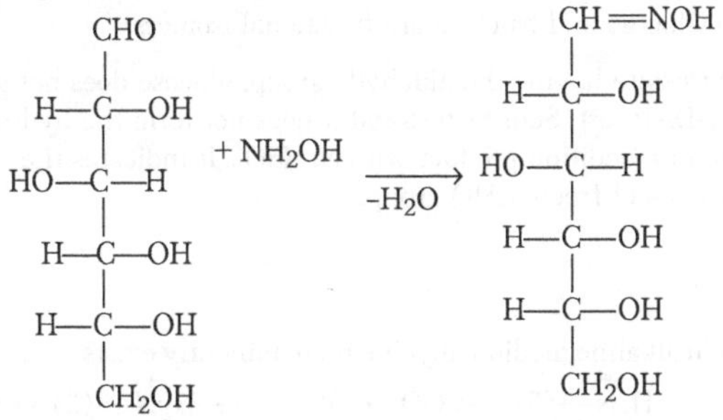

CHO–(H–C–OH)–(HO–C–H)–(H–C–OH)–(H–C–OH)–CH_2OH (D-(+)-glucoside) + NH_2OH $\xrightarrow{-H_2O}$ CH=NOH–(H–C–OH)–(HO–C–H)–(H–C–OH)–(H–C–OH)–CH_2OH (Glycoxime)

29 (*a*) One mole of sucrose on hydrolysis gives 1 mole of glucose and 1 mole of fructose.

30 (*a*) Among the given amino acids, methionine [$CH_3SCH_2CH_2CH(NH_2)COOH$] is an essential amino acid because it cannot be synthesised by our body and hence, its presence is necessary in our diet.

31 (*c*)

Letter code	Amino acid
G	Glycine
V	Valine
W	Tryptophan
H	Histidine

Thus, W is the letter code of tryptophan.

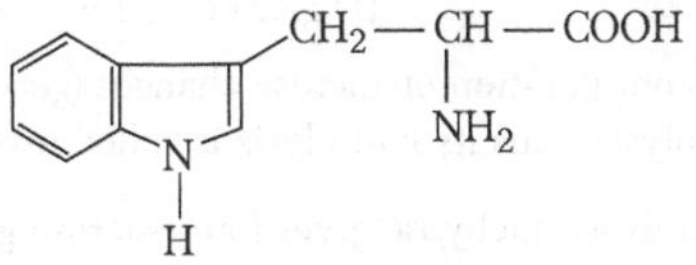

(Indole ring with N–H)–CH_2–CH(NH_2)–COOH

32 (*d*) DNA and RNA are polynucleotides. In secondary structure of nucleic acids, **Watson** and **Crick** gave a double strand helix structure for DNA. Two nucleic acid chains are bound about each other and held together by H-bonds between pair of bases. Adenine forms two H-bonds with thymine, whereas cytosine forms three H-bonds with guanine.

Structures are

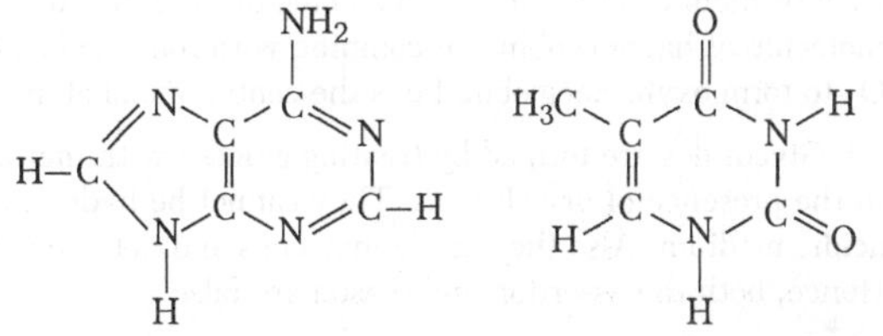

Adenine (A) Thymine (T)

33 (*b*) Eating animal poisoned with mercury (Hg^{2+}) causes deformity, known as minamata (minimata) disease which is characterised by diarrhoea, hemolysis, impairment of various senses, meningitis and death.

34 (*b*) Fat soluble vitamins are A, D, E and K are stored in body for long periods of time. They are found mainly in fatty foods like animal fats.

35 (*d*) Sucrose on reaction with conc. HCl forming laevunilic acid.

$$\underset{\text{Sucrose}}{C_{12}H_{22}O_{11}} \xrightarrow{\text{Conc. HCl}} \underset{\text{Laevunilic acid}}{CH_3COCH_2CH_2COOH}$$

Glucose and fructose are formed by the hydrolysis of sucrose

$$C_{12}H_{22}O_{11} \xrightarrow[\text{Enzymes}]{\text{Inversion}} \underset{\text{Glucose}}{C_6H_{12}O_6} + \underset{\text{Fructose}}{C_6H_{12}O_6}$$

36 (*a*) Starch $(C_6H_{10}O_5)_n$ is a polymer of glucose $(C_6H_{12}O_6)$.

37 (*a*) As we know, disaccharides upon hydrolysis gives two molecules of monosaccharides. So, maltose is a disaccharides which upon hydrolysis gives two molecules of glucose a monosaccharides.

$$\underset{\text{Maltose}}{C_{12}H_{22}O_{11}} + H_2O \xrightarrow[\text{Maltase}]{H_2SO_4} \underset{\text{Glucose}}{2C_6H_{12}O_6}$$

38 (*c*) Cellulose is the most important constituent in the plant give cell.

39 (*d*) Glucose and fructose both give positive Tollen's test, they gives same osazone on reaction with phenyl hydrazine and give *n*-hexane with P/Hg but with conc. HNO_3 they gives different products that are shown below

$$\text{Glucose} \xrightarrow{\text{Conc. } HNO_3} COOH-(CHOH)_4-COOH$$

$$\text{But fructose} \xrightarrow{\text{Conc. } HNO_3} HOOC-(-CHOH-)_2-COOH + HOCH_2-COOH$$

40 (*c*) Glucose is an aldohexose because it contains six C-atoms with $-\overset{O}{\overset{\|}{C}}-H$ group.

41 (*b*) Monosaccharides give positive test with Benedict's solution because they are reducing sugars but they give negative test with ninhydrin because this test is performed by proteins, peptides and amino acids.

42 (*c*) One molecule of haemoglobin have four haem groups and each of them can take up one molecule of oxygen. Thus, one molecule of haemoglobin can combine with four molecules of O_2 to form oxyhaemoglobin. Fe is the central metal atom here.

43 (*d*) Glycosides are formed by treating glucose with methanol in the presence of dry HCl gas. They cannot be hydrolysed in acidic medium. Also they are hemiacetals and not acetals. Hence, both the assertion and reason are false.

44 (*d*) The bond formed between adjacent carboxyl —(COOH) and amino (—NH_2) groups with the elimination of water molecule is called an 'peptide bond'.

All proteins and peptides contain peptide linkage.

$$\left[-\overset{\overset{\displaystyle O}{\|}}{C}-NH- \right]$$

$$\underset{\text{Amino acid}}{H_2NCH(R)—COOH} + \underset{\text{Amino acid}}{H—NH—CH_2COOH} \longrightarrow H_2NCH(R)—\underset{\text{Peptide bond}}{CO—NH}—CH_2COOH$$

45 (*b*) The pair of bases held with each other by H-bond.

46 (*c*) Interferon is a protein which is derived from the body against viruses.

47 (*b*) Vitamins are naturally occurring organic substances which are needed in minute amount to maintain various metabolic reactions. They are not synthesised in animal body.

48 (*b*) Glucose and mannose are isomers differ in configuration at C - 2. The isomers whose configuration is different at any chiral carbon atom are known as 'epimers'.

49 (*a*) Iodine value shows the amount of iodine needed to saturate 100 g oil or fat. In other words, the degree of unsaturation of an oil or fat is expressed in terms of iodine number.

50 (*a*) All polysaccharides including sucrose are called non-reducing sugars because these cannot reduce Fehling's solution and Tollen's reagent.

51 (*a*) In polysaccharides, different monosaccharides units are linked to one another by glycosidic linkages.
e.g. amylose and amylopectin are composed of α - D- glucose units linked by glycosidic linkage.

52 (*d*) Most of the enzymes have proteinous nature. They are highly specific and get denatured at high temperature, or by UV-rays. At optimum temperature, which is generally in between $25° C - 35° C$, enzyme activity is maximum.

53 (*d*) A codon is a specific sequence of three adjacent bases on a strand of DNA or RNA that provides genetic code information for a particular amino acid, that is to be incorporated into protein.

54 (*b*) A coenzyme is a substance that works with an enzyme to initiate or aid the function of enzyme. They act as control site and active sites.

55 (*d*) D (+)-lactose is a reducing sugar and it exhibits mutarotation.

56 (*b*) Starch forms acetal. Monosaccharides are hemiacetal. Acetals and hemiacetals get readily hydrolysed by acids to give alcohol.

57 (*d*) Maltose is composed of two α-D-glucopyranose units which C-1 of one glucose is linked to C-4 of another glucose unit.

58 (*d*) Glucose does not react with $NaHSO_3$.

59 (*b*) In lactose, C-1 of β-D-galactose is joined to C-4 of β-D-glucose.

60 (*a*) Glucose and fructose are functional isomers.

61 (*c*) Despite having the aldehyde group, glucose does not give 2,4-DNP test, Schiff's test and it does not form the hydrogen sulphite addition product with $NaHSO_3$. It indicates the absence of free —CHO group.

62 (*b*) In alkaline medium glycine predominently exists as an anion.

$$H_3\overset{+}{N}—CH_2—COO^- + OH^- \longrightarrow \underset{\text{(Anion)}}{H_2\ddot{N}—CH_2—COO^-} + H_2O$$

63 (*d*) Oxytocin – hormone

64 (*a*) A retrovirus is an RNA virus that is duplicated in a host cell.

65 (*b*) During mutarotation, the ring opens up to give the open chain form which then recloses either in the inverted position or in the original position giving an equilibrium mixture of two anomers with a small amount of the open chain form. Thus, all reducing monosaccharides and disaccharides (except sucrose-due to the absence of hemiacetal linkage) undergo mutarotation in aqueous solutions.

66 (*a*) On hydrolysis, sucrose gives an equimolar mixutre of D-(+)-glucose and D-(–)-fructose.

$$\underset{\text{(+)-sucrose}}{C_{12}H_{22}O_{11}} + H_2O \longrightarrow \underset{\text{(+)-glucose}}{C_6H_{12}O_6} + \underset{\text{(–)-fructose}}{C_6H_{12}O_6}$$

Here, the configuration of sucrose changes (gets inverted) after hydrolysis, thus its hydrolysis is called inversion.

67 (*c*) On hydrolysis stachyose gives four, sucrose gives two, raffinose gives three and ribose gives no molecule of monosaccharides. Thus, only raffinose is a trisaccharide.

$$\underset{\text{Raffinose}}{C_{18}H_{32}O_{16}} + H_2O \longrightarrow 3C_6H_{12}O_6$$

68 (*b*) $$\underset{\text{Glucose}}{C_5H_{11}O_5CHO} + \underset{\text{(from Fehling's solution)}}{CuO} \xrightarrow[\Delta]{OH^-} \underset{\text{Gluconic acid}}{C_5H_{11}O_5COOH} + \underset{\text{Cuprous oxide (red ppt.)}}{Cu_2O}$$

69 (*c*) On hydrolysis with dilute acid or enzyme, sucrose produces an equimolar mixture of D-glucose and D-fructose.

$$\underset{\text{Sucrose}}{C_{12}H_{22}O_{11}} + H_2O \xrightarrow{HCl} \underset{\text{D-glucose}}{C_6H_{12}O_6} + \underset{\text{D-fructose}}{C_6H_{12}O_6}$$

70 (*a*) Starch is a mixture of amylose (15-20%) and amylopectin (80-85%). Both amylose and amylopectin are polymer of α-D-glucose.

71 (*a*)

$$\underset{\text{Glucose}}{\begin{array}{c}CHO\\|\\CHOH\\|\\(CHOH)_3\\|\\CH_2OH\end{array}} \xrightarrow[\substack{2H_2O,\ -NH_3\\ -C_6H_5NH_2}]{3C_6H_5NHNH_2} \underset{\substack{\text{Osazone}\\\text{(glucosazone)}}}{\begin{array}{c}CH{=}NNHC_6H_5\\|\\C{=}NNHC_6H_5\\|\\(CHOH)_3\\|\\CH_2OH\end{array}}$$

Hence, three molecules of phenyl hydrazine is used to yield osazone.

72 (*d*) Letter 'D' before the name of monosaccharide reveals that the —OH group at the bottom chiral centre points to the right, i.e. it only shows the configuration at a particular chiral carbon (in glyceraldehyde second carbon atom).

73 (*b*) Inversion is the change in specific rotations of a cane solution with time.

74 (*a*) The structure of D-glyceraldehyde is as

$$\begin{array}{c}CHO\\|\\H-C-OH\\|\\CH_2OH\end{array}$$

The priority of groups is decided by the following rules

(*i*) Atom having higher atomic number gets higher priority.

(*ii*) If the priority cannot be decided by rule 1 then the next atoms are considered for priority assignment.

(*iii*) Where there is a = bond or ≡ bond both atoms are considered to be duplicated or triplicated. ($-\overset{H}{\overset{|}{C}}=O$ has higher priority than $-CH_2OH$). Hence, the corret order of priority of group in D-glyceraldehyde is as : OH(1), CHO(2), CH_2OH(3) and H(4)

75 (*b*) On heating slowly sucrose melts and if allowed to cool, solidifies to pale-yellow glassy mass called **barley sugar.** At 483 K it loses water and forms a brown amorphous mass called **caramel.**

76 (*b*) Monosaccharides are the simplest carbohydrates which cannot by hydrolysed to smaller molecules.

77 (*c*)

$$\underset{\text{Glucose}}{\begin{array}{c}CHO\\|\\(CHOH)_4\\|\\CH_2OH\end{array}} + \underset{\text{Tollen's reagent}}{2[Ag(NH_3)_2]OH} \longrightarrow \begin{array}{c}CO\bar{O}NH_4^+\\|\\(CHOH)_4\\|\\CH_2OH\end{array} + \underset{\substack{\text{Silver}\\\text{mirror}}}{2Ag\downarrow} + 3NH_3\uparrow + H_2O$$

This reaction indicates the presence of —CHO group in glucose.

CHAPTER 23

Polymers

A polymer is defined as a large molecule of high molecular weight formed by combination of a number of one or more types of molecules. Being molecules of high molecular weight, they are also known as **macromolecules.**

The smallest (or the simplest) molecules, which combine to form a polymer are called **monomers.** The process by which monomers combine to form a polymer is known as **polymerisation.** e.g. polyethylene or polythene is a polymer of ethylene (monomer unit).

$$\underset{\text{Ethylene (monomers)}}{n\,CH_2{=}CH_2} \xrightarrow{\text{Polymerisation}} \underset{\text{Polyethylene (polymer)}}{-\!\!(CH_2-CH_2)\!\!-_n}$$

CLASSIFICATION OF POLYMERS

Polymers are usually classified on the following four basis

1. Source of availability

On the basis of sources of availability, polymers are of following three types.

(i) **Natural polymers** Polymers which are synthesised from plants and animals are called natural polymers. Some of these polymers such as polysaccharide (starch, cellulose), proteins, nucleic acids control various life processes. Therefore, these are called as **biopolymers.**

(ii) **Semi-synthetic polymers** These are mostly derived from naturally occurring polymers by chemical modifications. Cellulose diacetate (used in making threads), vulcanised rubber and cellulose nitrate (gun cotton) are some examples of semi-synthetic polymers.

(iii) **Synthetic polymers** Polymers which can be synthesised in laboratories are called synthetic polymers. These are man-made polymers. They have high tensile strength. The first synthetic fibre was **rayon.**

2. Structure of polymers

On the basis of structure, polymers can be of following three types

(i) **Linear polymers** Here, monomers are joined together to form long straight chains of polymer molecules. The various polymeric chains are stacked over one another to give a well packed structure and therefore, linear polymers have high melting points, high densities, high tensile strength. e.g. high density polythene, polyvinylchloride, polyesters.

Inside

(*ii*) **Branched chain polymers** Here, monomers are joined to form long chains alongwith side chains or branches of different lengths. These are irregularly packed and thus, have low tensile strength, melting points and densities in contrast to linear polymers, e.g. low density polythene.

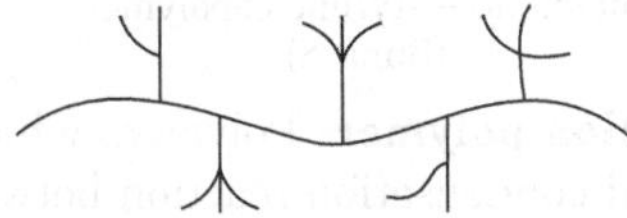

(*iii*) **Cross-linked or network polymers** Here, the initially formed linear polymer chains are cross linked together to form a three-dimensional network structure. These are hard, rigid and brittle due to cross-links. e.g. bakelite, melmac, etc.

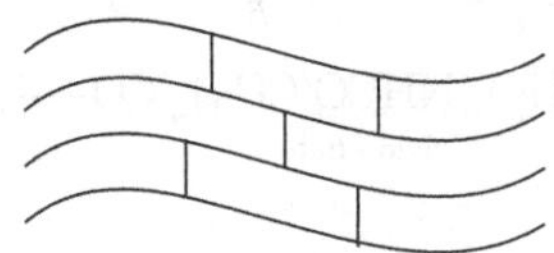

3. Molecular forces

On the basis of molecular forces present, polymers can be classified into following five types

(*i*) **Elastomers** In elastomers, the polymer chains are held together by the weakest intermolecular forces. As a result of these weak forces, polymer can be reverted to its original position after it stretches. It is a randomly oriented amorphous polymer. It must have some cross-linking so that chains do not slip over one another.

When elastomers are stretched, the random chains stretch out but, there are insufficient van der Waals' forces to maintain them in that configuration.

When the stretching force is removed, they go back to their random shapes. e.g. vulcanised rubber, buna-S or SBR, etc.

(*ii*) **Oriented polymers** Polymers that are stronger than steel or that conduct electricity significantly. These kind of polymers can be made by taking the polymer chains obtained by conventional polymerisation, stretching them out and putting them back together in a parallel fashion. Such polymers are called oriented polymers.

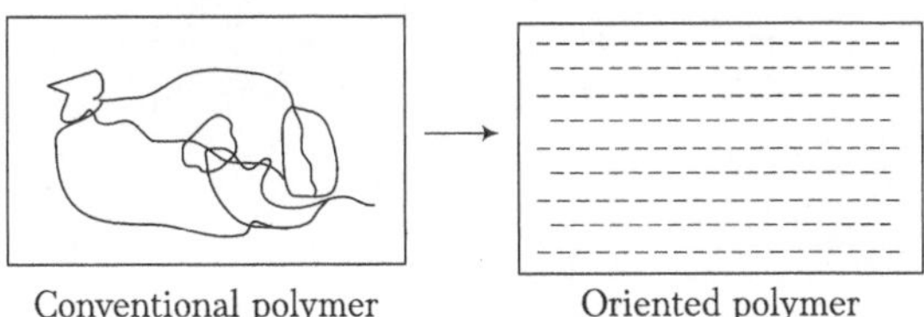

Conventional polymer — Oriented polymer

(*iii*) **Fibres** (crystallines) The polymers that are thread forming solids possess high tensile strength and have high modulus of elasticity.

This can be due to strong intermolecular forces like hydrogen bonding (as in polyamide nylon-66) or powerful dipole-dipole interactions (as in polyesters). These strong forces also lead to close packing of chains and thus, impart crystalline nature and show sharp melting point.

The more crystalline the polymer is, the denser, harder and more resistant to heat it is. If the polymer chains possess substitutes [as in poly (methyl methacrylate)] or have branches that prevent them from packing closely together, the density of the polymer is reduced.

(*iv*) **Thermoplastics** The intermolecular forces of attraction in thermoplastic polymers are intermediate between elastomers and fibres. They have both ordered crystalline regions and amorphous, non-crystalline regions and have no cross-linking between chains.

Such polymers are hard at room temperature but when they are heated, the individual chains can slip past one another and the polymer becomes soft enough to be moulded. **Polyethylene, polystyrene etc. are some common example of thermoplastics**. These are formed by addition polymerisation and usually have linear structure.

(*v*) **Thermosetting plastics** It is a very strong and rigid materials and can be obtained if the polymer chains are cross-linked. The greater degree of cross-linking exists between them, that makes the polymer more rigid. Such cross-linked polymers are known as **thermosetting polymers.**

These are semifluid substances with low molecular masses which when heated, undergo a permanent change in chemical composition to give a hard infusible and insoluble mass. Once they get hardened cannot be remelted by heating because cross-linking reduces the mobility of the polymer chains, causing them to be relatively brittle materials, and thus less widely used. e.g. bakelite and melmac with three-dimensional network.

Plasticisers

Those plastics which do not soften very much on heating can be made soft by the addition of certain organic compounds called plasticisers. Dialkyl phthalates (e.g. di *n*-butylphthalate) or cresyl triphosphates $(CH_3—C_6H_4—O)_3PO$ are the most commonly used plasticisers. They do lower the melting point of a polymer but cannot convert a thermosetting polymer to thermoplastic one.

They just convert a hard and brittle plastic into soft and rubber like by reducing the interparticle forces between polymer molecules. e.g. PVC is extremely stiff and hard even when hot but the addition of di-*n*-butyl phthalate makes it soft and rubber like.

4. Mode of polymerisation

On the basis of mode of polymerisation, polymers are of following two types

(i) **Addition polymers** Polymers which are formed by repeated addition of unsaturated monomers without the elimination of small molecules as by product are called addition polymers. In these polymers, molecular mass is a whole number of the monomer.

These are further classified into following two types

- **Homopolymer** Polymer formed by single monomeric species. e.g. polythene

$$\underset{\text{Ethene}}{n\,CH_2 = CH_2} \longrightarrow \underset{\substack{\text{Polythene}\\\text{(homopolymer)}}}{\lbrack CH_2 - CH_2 \rbrack_n}$$

- **Copolymer** Polymer formed by two or more than two different monomeric species.

e.g. Buna-S is formed from butadiene and styrene.

$$\underset{\text{1, 3-butadiene}}{n\,CH_2{=}CH{-}CH{=}CH_2} + \underset{\text{Styrene}}{n\,CH_2{=}CH{-}C_6H_5} \longrightarrow \underset{\substack{\text{Butadiene – styrene copolymer}\\\text{(Buna-S)}}}{\lbrack CH_2{-}CH{=}CH{-}CH_2{-}\underset{C_6H_5}{\overset{|}{CH}}{-}CH_2 \rbrack_n}$$

(ii) **Condensation polymers** Polymers which are formed by repeated condensation reaction between two different bi-functional or tri-functional monomeric units with the loss of small molecules such as water, alcohol etc. e.g. nylon-6,6 formed by condensation of hexamethylene diamine (*A*) and adipic acid (*B*).

$$n\underset{A}{[NH_2(CH_2)_6NH_2]} + n\underset{B}{[(CH_2)_4(COOH)_2]} \longrightarrow \underset{\text{Nylon-6, 6}}{\lbrack NH(CH_2)_6NHCO(CH_2)_4CO \rbrack_n} + nH_2O$$

CHECK POINT 23.1

1. Molecule of high molecular weight formed by combination of many molecules, is known as
   (a) biomolecule (b) macromolecule
   (c) monomer (d) Both (a) and (b)
2. Which of the following is not a natural polymer?
   (a) Starch (b) Nucleic acid
   (c) Proteins (d) Polythene
3. The first synthetic fibre is
   (a) rayon (b) cotton
   (c) silk (d) None of these
4. Polymers which have high tensile strength and high melting point
   (a) thermoplastic (b) thermosetting
   (c) branched polymer (d) linear polymer
5. Low density polythene is an example of
   (a) linear polymer
   (b) branched polymer
   (c) thermoplastic polymer
   (d) thermosetting polymer
6. Polymers which have molecular forces in between elastomers and fibres
   (a) cross linked polymer (b) thermosetting
   (c) thermoplastic (d) branched
7. Among the following, thermosetting polymer is
   (a) teflon
   (b) melamine-formaldehyde
   (c) bakelite
   (d) Both (b) and (c)
8. Polymers formed without elimination of small molecules are called
   (a) addition polymer (b) condensation polymer
   (c) Both (a) and (b) (d) cross linked polymer
9. Buna-S is an example of
   (a) homopolymer (b) condensation polymer
   (c) copolymer (d) None of these
10. Monomer units of nylon-6, 6 is (are)
   (a) hexamethylene diamine (b) adipic acid
   (c) caprolactam (d) Both (a) and (b)

POLYMERISATION

The process by which monomer units combine to form a large molecule or polymer, is known as polymerisation. Polymerisation can be done by the following two methods.

Addition polymerisation

It is the process in which monomers (same or different) unit combine together to form a polymer without the loss of small molecules. It is also known as **chain growth polymerisation** due to increase in length of the chain. e.g. polystyrene, polyethylene etc.

Following ways are possible in which monomers can combine

(*i*) **Head-to-tail addition** When the head of one monomer combines with the tail of another.

$$\underset{\text{Monomer}}{n\overset{\text{Tail}}{CH_2}{=}\underset{X}{\overset{\text{Head}}{CH}}} \longrightarrow \underset{\text{Polymer}}{\left[CH_2 - \underset{X}{\overset{\text{Head}}{CH}} - \overset{\text{Tail}}{CH_2} - \underset{X}{CH} \right]_n}$$

(*ii*) **Head-to-head and tail-to-tail addition** When the head (or tail) of one monomer combines with the head (or tail) of another.

$$\underset{\text{Monomer}}{n\overset{\text{Tail}}{CH_2}{=}\underset{X}{\overset{\text{Head}}{CH}}} \longrightarrow \underset{\text{Polymer}}{\left[\underset{X}{CH} - \overset{\text{Tail}}{CH_2} - \overset{\text{Tail}}{CH_2} - \underset{X}{\overset{\text{Head}}{CH}} - \underset{X}{\overset{\text{Head}}{CH}} - CH_2 \right]_n}$$

Note *Head-to-tail addition is preferred for steric reasons as propagating site attack the less sterically hindered carbon of the alkene. Groups that stabilise adjacent radicals also favour head-to-tail addition.*

Mechanism

Addition polymerisation can proceed by following mechanisms

Free radical addition polymerisation

In this mechanism, free radicals are generated by **initiators.** Initiator is a compound that readily undergoes homolytic cleavage to form radicals, which are sufficiently energetic to convert an alkene into radical.

e.g.

$$\underset{\text{Hydrogen peroxide}}{HO{-}OH} \longrightarrow 2HO^{\bullet}$$

$$\underset{t\text{-butyl hydrogen peroxide}}{CH_3\underset{CH_3}{\overset{CH_3}{C}}O{-}OH} \longrightarrow CH_3\underset{CH_3}{\overset{CH_3}{C}}O^{\bullet} + {}^{\bullet}OH$$

$$\underset{\text{Potassium perdisulphate}}{KO\underset{O}{\overset{O}{S}}O{-}O\underset{O}{\overset{O}{S}}OK} \longrightarrow 2KO\underset{O}{\overset{O}{S}}O^{\bullet}$$

$$\underset{t\text{-butyl peroxide}}{CH_3\underset{CH_3}{\overset{CH_3}{C}}O{-}O\underset{CH_3}{\overset{CH_3}{C}}CH_3} \longrightarrow 2CH_3\underset{CH_3}{\overset{CH_3}{C}}O^{\bullet}$$

$$\underset{\text{Benzoyl peroxide}}{C_6H_5\overset{O}{\overset{\|}{C}}O{-}O\overset{O}{\overset{\|}{C}}C_6H_5} \longrightarrow 2C_6H_5\overset{O}{\overset{\|}{C}}O^{\bullet}$$

Synthesis of polymers involve following steps

Step I. **Chain initiation**

To generate free radicals, which adds on double bonds to form a large free radical.

$$C_6H_5{-}\overset{O}{\overset{\|}{C}}{-}O{-}O{-}\overset{O}{\overset{\|}{C}}{-}C_6H_5 \longrightarrow 2\,C_6H_5{-}\overset{O}{\overset{\|}{C}}{-}\dot{O} \xrightarrow{-2CO_2} \underset{\text{Phenyl free radical}}{2\,\dot{C}_6H_5}$$

$$\dot{C}_6H_5 + CH_2{=}CH_2 \longrightarrow \underset{\text{(Small free radical)}}{C_6H_5{-}CH_2{-}\dot{C}H_2}$$

Step II. **Chain propagation step**

A large free radical reacts with another molecule containing double bond to form another free radical which carries the reaction forward.

$$C_6H_5{-}CH_2{-}\dot{C}H_2 + CH_2{=}CH_2 \longrightarrow C_6H_5{-}CH_2{-}CH_2{-}CH_2{-}\dot{C}H_2 \xrightarrow{+\dot{C}H_2} \underset{\text{(Big free radical)}}{C_6H_5{-}(CH_2{-}CH_2)_n{-}CH_2{-}\dot{C}H_2}$$

Step III. **Chain termination step**

- By combination of free radicals

$$C_6H_5 \text{+}CH_2-CH_2\text{+}_n \dot{C}H_2-CH_2 + \dot{C}H_2-CH_2\text{+}CH_2-CH_2\text{+}_n C_6H_5$$

$$\xrightarrow{\text{Combination}} C_6H_5 \text{+}CH_2-CH_2\text{+}_n CH_2CH_2CH_2CH_2 \text{+}CH_2-CH_2\text{+}_n C_6H_5$$

Polythene

- By disproportionation of free radicals

$$C_6H_5\text{+}CH_2-CH_2\text{+}_n CH_2-\dot{C}H_2 + C_6H_5\text{+}CH_2-CH_2\text{+}_n CH_2-\dot{C}H_2$$

$$\xrightarrow{\text{Disproportionation}} C_6H_5\text{+}CH_2-CH_2\text{+}_n CH_2CH_3 + C_6H_5-\text{+}CH_2CH_2\text{+}_n CH=CH_2$$

Polythene

Molecular weight of polymer can be controlled by chain transfer process. Reagents like carbon tetrachloride (CCl_4) or CBr_4, when added to the reaction, growing chain gets terminated. These reagents are called chain transfer agents. These agents determine the chain reactions and inhibit further polymerisation and are also called inhibitors.

$$CH_2=CH(C_6H_5) \xrightarrow{\text{Initiator}} \dot{C}H_2-\dot{C}H(C_6H_5) \xrightarrow{CCl_4} CH_2-CH(C_6H_5)-Cl + \dot{C}Cl_3$$

$$CH_2=CH(C_6H_5) + \dot{C}Cl_3 \longrightarrow Cl_3C-CH_2-\dot{C}H(C_6H_5)$$

$$\xrightarrow{\text{Styrene}} \left[Cl_3C-CH_2-CH(C_6H_5)-CH_2-\dot{C}(H)(C_6H_5)\right]$$

In polystyrene, radical is stabilised by resonance of benzene of benzyl group.

$$-CH_2CH(C_6H_5)CH_2\dot{C}H(C_6H_5)$$

Conjugated diene polymerisation

1,3-butadiene, a conjugated diene can be polymerised like a simple alkene but there are **two modes** by which this process can take place.

(i) **1,4-polymerisation** when the polymerisation takes place at C_1 and C_4 of butadiene, an unbranched polymer is formed and can exist as geometrical isomers.

$$\dot{R} + CH_2=CH-CH=CH_2 \longrightarrow$$

$$RCH_2-\dot{C}H-CH=CH_2 \longleftrightarrow R-CH_2-CH=CH-\dot{C}H_2$$

$$\downarrow$$

$$\left[R\text{+}H_2C(H)C=C(H)CH_2\text{~}\right]_n \text{ (Trans)} + \left[R\text{+}H_2C(H)C=C(H)CH_2\text{~}\right]_n \text{ (Cis)}$$

(ii) **1,2-polymerisation** When the polymerisation takes place at C_1 and C_2 of butadiene, product formed is polyvinyl polythene.

$$2nCH_2=CH-CH=CH_2 \xrightarrow{\dot{R}} R\text{+}CH_2-CH(CH=CH_2)-CH_2-CH(CH=CH_2)\text{+}_n$$

Ionic addition polymerisation

This mode of polymerisation results in the generation of ionic species such as cations and anions. Vinylic monomers easily undergo addition polymerisation. It can be classified into following two types.

(i) Cationic addition polymerisation

When the initiator is cationic in nature, intermediate is a carbocation and the process is initiated by an acid. Monomers containing electron-releasing groups favour cationic addition polymerisation.

A cationic initiator is a compound, which readily undergoes heterolytic cleavage to give a proton or cation that is able to convert an alkene into carbocation.

It involves following three steps

Step I **Initiation step** In this step, the proton is generated from either Lewis acid or protonic acid, attacks on monomer unit and a carbocation is formed.

$$H_2SO_4 \rightleftharpoons H^+ + HSO_4^-$$

$$BF_3 + H_2O \rightleftharpoons H^+ + BF_3$$

$$H^+ + CH_2=C(CH_3)_2 \longrightarrow H_3C-C^+(CH_3)_2$$

Step II **Propagation step**

$$H_3C-\overset{CH_3}{\underset{CH_3}{C^+}} + CH_2{=}\overset{CH_3}{\underset{CH_3}{C}} + CH_2{=}\overset{CH_3}{\underset{CH_3}{C}}$$

$$\downarrow$$

$$CH_3-\overset{CH_3}{\underset{CH_3}{C}}-\left[CH_2-\overset{CH_3}{\underset{CH_3}{C}}-\right]_n CH_2-\overset{CH_3}{\underset{CH_3}{C^+}}$$

Butyl rubber

Step III **Termination step** The growing chain is generally terminated by the loss of hydrogen ion

$$H_3C-\overset{CH_3}{\underset{CH_3}{C}}-\left[CH_2-\overset{CH_3}{\underset{CH_3}{C}}\right]_n-CH_2-\overset{CH_3}{\underset{CH_3}{C^+}} \xrightarrow{-H^+}$$

$$H_3C-\overset{CH_3}{\underset{CH_3}{C}}-\left[CH_2-\overset{CH_3}{\underset{CH_3}{C}}\right]_n-CH_2{=}\underset{CH_3}{C}-CH_3$$

e.g. monomers like propylene ($CH_2{=}\underset{CH_3}{CH}$),

isobutylene ($CH_2{=}\underset{CH_3}{C}-CH_3$) and

undergo this type of polymerisation.

(ii) Anionic addition polymerisation

In this mode of polymerisation, intermediate is a carbanion and initiator is an anion. Monomers containing electron-withdrawing groups give this reaction. Such groups tend to stabilise the carbanion hence, addition is preferred.

It involves following three steps

Step I **Initiation step**

$$K^{\oplus}\overset{\ominus}{N}H_2 + CH_2{=}\underset{X}{CH} \xrightarrow[\text{Initiation}]{\text{Step I}} H_2N-CH_2-\underset{X}{\overset{\ominus}{C}H}\,K^{\oplus}$$

(X = electron withdrawing group)

Step II **Propagation step**

$$H_2N-CH_2-\underset{X}{\ddot{C}H} + CH_2{=}\underset{X}{CH} \longrightarrow$$

$$H_2N-\left[CH_2-\underset{X}{CH}\right]_n-CH_2-\underset{X}{\ddot{C}H}$$

Step III **Termination step** The growing chain is terminated by the addition of a proton or Lewis acid

$$H_2N-\left[CH_2-\underset{X}{CH}\right]_n-CH_2-\underset{X}{\overset{\ominus}{\ddot{C}H}} \xrightarrow{-H^{\oplus}}$$

$$H_2N-\left[CH_2-\underset{X}{CH}\right]_n-CH_2-\underset{X}{CH_2}$$

e.g. monomers like vinyl chloride

$\left(CH_2{=}\underset{Cl}{CH}\right)$, acrylonitrile $\left(CH_2{=}\underset{C\equiv N}{CH}\right)$

and

styrene $\left(C_6H_5-CH{=}CH_2\right)$

undergo this type of polymerisation.

Table 23.1 Some examples of addition polymers

Name of polymer	Structure	Monomer	Uses
Polythene	$-[CH_2-CH_2]_n-$	$CH_2=CH_2$ [Ethene]	As insulator, anticorrosive, packing material, household and laboratory wares.
Polystyrene	$-[CH(C_6H_5)-CH_2]_n-$	$CH(C_6H_5)=CH_2$ [Styrene]	As insulator, wrapping material, manufacture of toys and household articles.
Polyvinylchloride (PVC)	$-[CH_2-CH(Cl)]_n-$	$CH_2=CHCl$ [Polyvinylchloride]	In manufacture of raincoats, hand bags, vinyl flooring and leather clothes.
Styrene butadiene rubber (SBR or Buna-S)	$-[CH_2-CH=CH-CH_2-CH(C_6H_5)-CH_2]_n-$	(a) $CH_2=CH-CH=CH_2$ [1,-3 butadiene] (b) $CH(C_6H_5)=CH_2$ [Styrene]	In making automobile tyres and footwear.
Nitrile rubber (Buna-N)	$-[CH_2-CH=CH-CH_2-CH(CN)-CH_2]_n-$	(a) $CH_2=CH-CH=CH_2$; [1, 3-butadiene] (b) $CH(CN)=CH_2$; [Acrylonitrile]	In making oil seals, manufacture of hoses and tank linings.
Neoprene	$-[CH_2-C(Cl)=CH-CH_2]_n-$	$CH_2=C(Cl)-CH=CH_2$ [Chloroprene]	As insulator, making conveyor belts and printing rollers.
Polyethyl acrylate	$-[CH_2-CH(COOC_2H_5)]_n-$	$CH_2=CH-COOC_2H_5$ [Acrylate]	In making films, hose pipes and finishing fabrics
Polytetrafluoroethylene (PTFE) or teflon	$-[CF_2-CF_2]_n-$	$CF_2=CF_2$ [Tetrafluoroethylene]	As lubricant, insulator and for making cooking wares.
Polymethyl methacrylate (PMMA) or Plexi glass	$-[CH_2-C(CH_3)(COOCH_3)]_n-$	$CH_2=C(CH_3)-COOCH_3$ [Methylmethacrylate]	As substitute of glass and making decorative materials.
Polyacrylonitrile or PAN (Orlon)	$-[CH_2-CH(CN)]_n-$	$CH_2=CHCN$ [Acrylonitrile or vinyl cyanide]	In making synthetic fibres and synthetic wool.
Dynel	$-[CH_2-CH(CN)-CH_2-CH(Cl)]_n-$	(a) $H_2C=CH-C\equiv N$ [Acrylonitrile] (b) $HClC=CH_2$ [Vinyl chloride]	Use espacially for marine applications

Condensation polymerisation

When two or more types of reactants undergoing polymerisation reaction, are bi-functional, they undergo a series of condensation reactions in a controlled stepwise manner with the elimination of water to form polymers and the process is called **condensation polymerisation.**

The product of each step is again bi-functional, so that sequence of condensation goes on repeatedly. Hence, this process is also called **step-growth polymerisation.**

$$\underset{\text{Ethylene}}{HOCH_2CH_2OH} + \underset{\text{Terephthalic acid}}{HOOC-C_6H_4-COOH} \longrightarrow \underset{\text{Dacron or Terylene}}{HOOC-C_6H_4-CO\left(OCH_2CH_2OOC-C_6H_4-CO\right)_n OCH_2CH_2OH}$$

Table 23.2 Some examples of condensation polymers

Name of polymer	Structure	Monomer	Uses
Nylon 6	$\left[-NH-(CH_2)_5-\overset{O}{\overset{\Vert}{C}}-\right]_n$	Cyclic amide (H–N–C=O ring) [Caprolactam]	In making fibres, plastics, tyrecords and ropes.
Nylon 6, 6	$-\!\!\left(NH(CH_2)_6NHCO(CH_2)_4CO\right)\!\!-_n$	(a) $HOOC-(CH_2)_4-COOH$ (Adipic acid) (b) $H_2N-(CH_2)_6-NH_2$ (Hexamethylene diamine)	In making brushes, synthetic fibres, parachutes, ropes and carpets.
Bakelite	$\left[-C_6H_2(OH)-CH_2-C_6H_2(OH)-CH_2-\right]_n$	(a) HCHO (Formaldehyde) (b) C_6H_5OH (Phenol)	For making gears, protective coating and electrical fittings.
Terylene (dacron)	$\left[-OOC-C_6H_4-COO-CH_2-CH_2O-\right]_n$	(a) $HOOC-C_6H_4-COOH$ [Terephthalic acid] (b) $HO-CH_2-CH_2-OH$ (Ethylene glycol)	For making fibres, safety belts, tyrecords, tents etc.
Dextron	$\left[-OCH(CH_3)-\overset{O}{\overset{\Vert}{C}}-O-CH_2-\overset{O}{\overset{\Vert}{C}}-O-\right]_n$	(a) $CH_3-CH(OH)-COOH$ [Lactic acid] (b) $CH_2(OH)-COOH$ [Glycolic acid]	Used in packaging, orthopaedic devices in controlling drug release etc.

Copolymerisation

When two or more different monomers combine to form a polymer or copolymer, the process is known as **copolymerisation.** It can proceed through either addition or condensation polymerisation methods. Properties of copolymer is entirely different from the properties of homopolymers.

For instance, polymer styrene butadiene rubber (SBR) is a copolymer of styrene and butadiene which has properties like high abrasion resistance, high load bearing capacity, therefore, used in manufacture of tyres. On the other hand, homopolymer of styrene (called polystyrene) is a good insulator and can be easily moulded into toys, combs etc.

$$n\,C_6H_5CH{=}CH_2 \text{ (Styrene)} + nCH_2{=}CH{-}CH{=}CH_2 \text{ (1, 3-butadiene)} \longrightarrow \left[CH(C_6H_5){-}CH_2CH_2{-}CH{=}CH{-}CH_2 \right]_n$$

SBR copolymer

On the basis of arrangement of monomer units, copolymers are further classified into following two types,

(*i*) **Block copolymers** Copolymers in which different blocks of identical monomer units alternate with each other.

$$\left(A{-}A{-}A{-}B{-}B{-}B{-}A{-}A{-}A{-}B{-}B{-}B \right)$$

(*ii*) **Graft copolymers** Copolymers in which homopolymer branches of one monomer units are grafted on the homopolymer chains of another monomer units.

$$\left(A{-}A{-}A{-}A{-}A{-}A{-}A{-}A \right)_n$$

with $(B{-}B{-}B)_n$ branches attached to the first, fifth and eighth A units.

Table 23.3 Some examples of copolymers and their uses

Monomers	Copolymer name	Uses
$CH_2{=}CHCl$ (Vinyl chloride) + $CH_2{=}CCl_2$ (Vinylidene chloride)	Saran	Film for wrapping food
$CH_2{=}CH{-}C_6H_5$ (Styrene) + $CH_2{=}CH{-}C{\equiv}N$ (Acrylonitrile)	SAN	Dishwasher-safe objects, vacuum cleaner parts
$CH_2{=}CH{-}C{\equiv}N$ (Acrylonitrile) + $CH_2{=}CH{-}CH{=}CH_2$ (1, 3-butadiene) + $CH_2{=}CH{-}C_6H_5$ (Styrene)	ABS	Bumpers, crash helmets, telephones, luggage
$CH_2{=}C(CH_3)CH_3$ (Isobutylene) + $CH_2{=}C(CH_3){-}CH{=}CH_2$ (Isoprene)	Butyl rubber	Inner tubes, balls, inflatable sporting goods.

Coordination polymerisation

This polymerisation proceeds with the formation of coordination compound as intermediate e.g. *Ziegler-Natta* catalyst. *Ziegler* and *Natta* scientist found that structure of a polymer could be controlled if the growing end of the chain and the incoming monomer were coordinated with an aluminium-titanium initiator called *Ziegler-Natta* catalysts. It is also known as *Ziegler-Natta* catalysed polymerisation.

$$\underset{\text{Acetylene}}{CH{\equiv}CH} \xrightarrow[\text{Catalyst}]{\textit{Ziegler-Natta}} \underset{\text{Polyacetylene}}{{-}CH{=}CH{+}CH{=}CH{+}_n\,CH{=}CH{-}}$$

Polyacetylene thus, formed has conjugated double bonds and they cause it **to conduct electricity down** its backbone after several electrons are removed from or added to the backbone; thus it is a **conducting polymer.**

SOME IMPORTANT POLYMERS

Some of the examples of important polymers are given below

Rubber

It is a polymer which can return to its original shape, size or length after being stretched. It is an elastomer. It can be naturally obtained or synthesised chemically.

(i) Natural rubber

It is obtained as latex from rubber trees. Latex is a colloidal suspension of rubber in water. Latex is coagulated with formic or acetic acid. Natural rubber is sticky, gummy, soft mass and is insoluble in water, dilute acids and alkalies but soluble in organic solvents like benzene, chloroform, ether, petrol and CS_2. It has low elasticity and tensile strength.

The monomer of natural rubber is isoprene, whose 11,000 to 20,000 units are linked together in a chain.

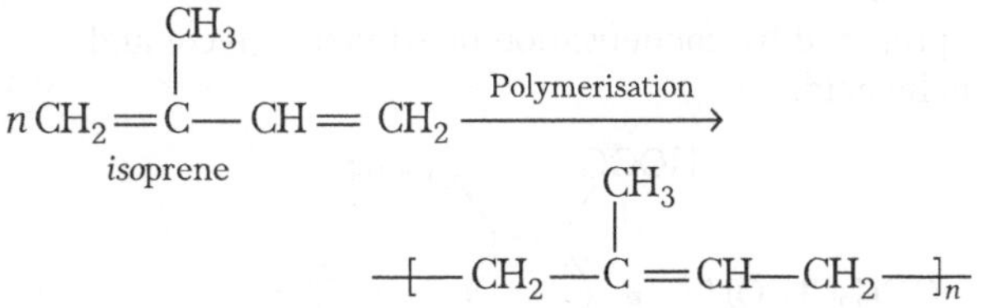

$$\underset{\text{isoprene}}{n\,CH_2{=}\overset{\overset{\large CH_3}{|}}{C}{-}CH{=}CH_2} \xrightarrow{\text{Polymerisation}} \underset{\text{Natural rubber}}{\left[CH_2{-}\overset{\overset{\large CH_3}{|}}{C}{=}CH{-}CH_2 \right]_n}$$

- Weak van der Waals' forces are present in natural rubber due to which it is elastic and non-crystalline in nature.

  Natural rubber is *cis* -1,4-polyisoprene and has all *cis*-configuration about the double bonds.

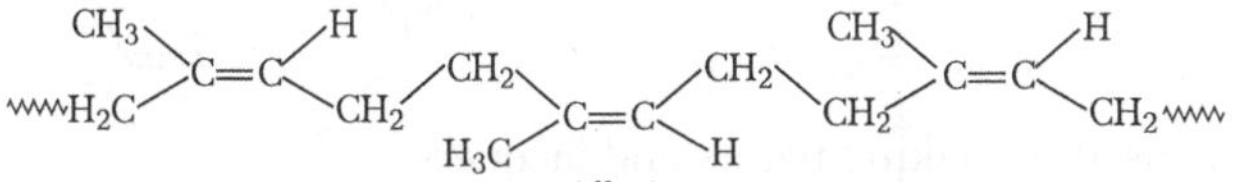

$$\text{〰}H_2C{-}C(CH_3){=}CH{-}CH_2{-}CH_2{-}C(CH_3){=}CH{-}CH_2{-}CH_2{-}C(CH_3){=}CH{-}CH_2\text{〰}$$

All *cis*

- However, its *trans* configuration has highly regular *zig-zag* pattern which fits together well. The *trans* configuration called ***Gutta-percha*** that occurs naturally. It is highly crystalline and non-elastic in nature.

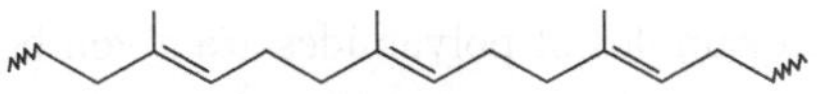

trans configuration in *Gutta-percha*

(ii) Vulcanised Rubber

To improve the properties of natural rubber, vulcanisation is done. Hardening of raw rubber by heating in presence of sulphur at 373-415 K is called **vulcanisation.** It causes **cross-linking** of separate polymer chains through disulphide bonds and thus, vulcanised chains are locked together in one giant molecule. Sulphur cross-links at the sites of double bonds.

Cross-linking prevents the polymer from being torn when it is stretched and the cross-links provide a reference framework for the material to return to when the stretching force is removed.

Vulcanisation process is slow and to accelerate the process, additives like ZnO are added.

The vulcanised rubber has excellent elasticity, low water absorption tendency and is resistant to the action of organic solvents and oxidising agents.

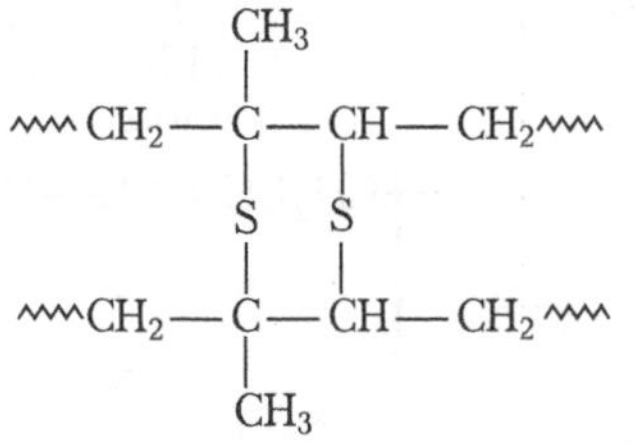

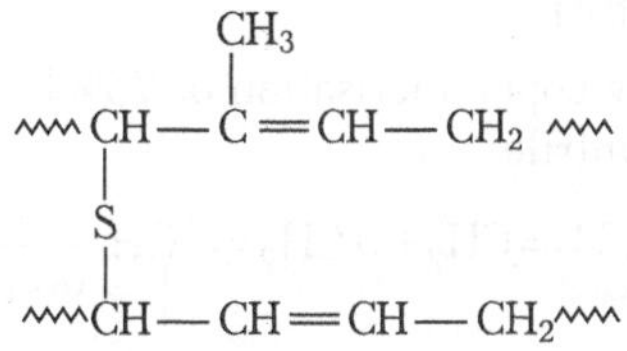

$$\begin{array}{c} \text{〰}CH_2{-}\overset{\overset{\large CH_3}{|}}{C}{-}CH{-}CH_2\text{〰} \\ \quad\;\;\; |\quad\;\; | \\ \quad\;\;\; S\quad\; S \\ \quad\;\;\; |\quad\;\; | \\ \text{〰}CH_2{-}\underset{\underset{\large CH_3}{|}}{C}{-}CH{-}CH_2\text{〰} \end{array}$$

$$\begin{array}{c} \text{〰}CH{-}\overset{\overset{\large CH_3}{|}}{C}{=}CH{-}CH_2\text{〰} \\ | \\ S \\ | \\ \text{〰}CH{-}CH{=}CH{-}CH_2\text{〰} \end{array}$$

The physical properties of rubber can be controlled by regulating the amount of sulphur used in the vulcanisation process. Some of them are explained as follows.

- Rubber made with 1-3% sulphur, is soft and stretchy and is used to make rubber bands.
- Rubber made with 3-10% sulphur, is more rigid and is used in the manufacture of tyres.
- Rubber made with 20-25% sulphur, is used for making ebonite.
- Rubber with 30% sulphur, is used for making battery case.

(iii) Synthetic rubber

It is obtained by either homopolymer of buta-1, 3-diene derivatives or copolymer in which one of the monomer is 1, 3-butadiene. These are more resistant, have high tensile strength. Some of the examples of synthetic rubber are given below.

(a) Neoprene

It is made by polymerising 2-chloro-1,3-butadiene (chloroprene) in the presence of *Ziegler-Natta* catalyst that causes all the double bonds in the polymer to have the ***trans***-configuration.

$$\underset{\text{Chloroprene}}{CH_2=\underset{|}{\overset{Cl}{C}}-CH=CH_2} \xrightarrow[\text{Catalyst}]{\text{Ziegler-Natta}} \underset{\text{Neoprene}}{\left[CH_2-\underset{Cl}{C}=CH-CH_2\right]_n}$$

It is used in preparing gaskets, hoses, conveyor belts etc.

(b) **Buna-S** (styrene-butadiene rubber SBR)

It is prepared by copolymerisation of 75% buta-1,3-diene and 25% styrene.

$$\underset{\text{Butadiene}}{nCH_2=CH-CH=CH_2} + \underset{\text{Styrene}}{nCH_2=\underset{Ph}{CH}} \xrightarrow[\text{copolymerisation}]{\text{Addition}} \underset{\text{Buna-S}}{\left[CH_2CH=CH-CH_2CH_2\underset{Ph}{CH}\right]}$$

(c) **Buna-N** (NBR)

It is prepared by copolymerisation of 75% buta-1, 3-diene and 25% acrylonitrile.

$$\underset{\text{Buta-1, 3-diene}}{nCH_2=CH-CH=CH_2} + \underset{\text{Acrylonitrile}}{nCH_2=\underset{CN}{CH}} \xrightarrow[\text{copolymerisation}]{\text{Addition}} \underset{\text{Buna-N}}{\left[CH_2-CH=CH-CH_2CH_2-\underset{CN}{CH}\right]}$$

(d) **Thiokol**

It is prepared by the condensation copolymerisation of 1, 2-dichloroethane with sodium tetrasulphide in presence of $Mg(OH)_2$

$$\underset{\text{1, 2-dichloroethane}}{nCl-CH_2-CH_2-Cl} + \underset{\text{Sodium tetrasulphide}}{nNa_2S_4} \xrightarrow{Mg(OH)_2} \underset{\text{Thiokol}}{\left[CH_2-CH_2-S-S-S-S\right]_n} + 2nNaCl$$

It is exceptionally resistant to the action of mineral oils solvents, oxygen and ozone.

Polyesters

Polymers prepared by condensation of dialcohols and diacids. These polymers have ester linkage. Some of the examples of polyesters are given below.

(i) Terylene

It is prepared by condensation of ethylene glycol and terephthalic acid at 420-460 K in presence of mixture of zinc acetate and antimony trioxide as catalyst.

$$\underset{\text{Ethylene glycol}}{nHOCH_2CH_2OH} + \underset{\text{Terephthalic acid}}{nHOOC-C_6H_4-COOH} \xrightarrow[-nH_2O]{Zn(CH_3CO_2)_2,\ Sb_2O_3,\ \Delta} \underset{\text{Terylene}}{\left[O-CH_2CH_2-OCO-C_6H_4-CO\right]_n}$$

It is also known as **Dacron.** It is a crease resistant fibre and used in blending with cotton and wool fibres. It is also used in making wash and wear fabrics, safety belts etc.

(ii) Glyptal

It is prepared by condensation of ethylene glycol and phthalic acid.

$$\underset{\text{Ethylene glycol}}{nHOCH_2CH_2OH} + n\ \underset{\text{Phthalic acid}}{C_6H_4(COOH)_2} \xrightarrow[-nH_2O]{\Delta} \underset{\text{Glyptal}}{\left[OCH_2CH_2-O-\overset{O}{\overset{\|}{C}}-C_6H_4-\overset{O}{\overset{\|}{C}}\right]_n}$$

It is used in making paints and lacquers.

(iii) Polyamides

These polymers are prepared by condensation copolymerisation of diamines and dicarboxylic acids. They have amide linkages.

Some of the examples of polyamides are given below.

(a) **Nylon-6**

It is prepared by polymerisation of caprolactam at high temperature.

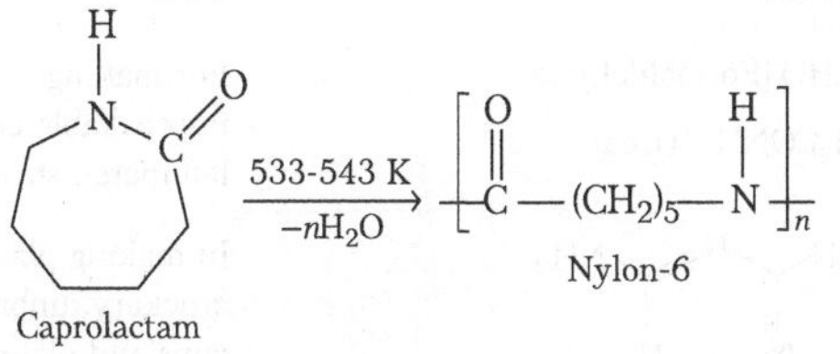

$$\text{Caprolactam} \xrightarrow[-nH_2O]{533\text{-}543\ K} \left[-\overset{O}{\overset{\|}{C}}-(CH_2)_5-\overset{H}{\overset{|}{N}}- \right]_n$$

Caprolactam   Nylon-6

It is used for manufacturing fabrics, tyrecords etc.

(b) **Nylon-6, 6**

It is prepared by condensation copolymerisation of hexamethylenediamine and adipic acid at high temperature and at high pressure.

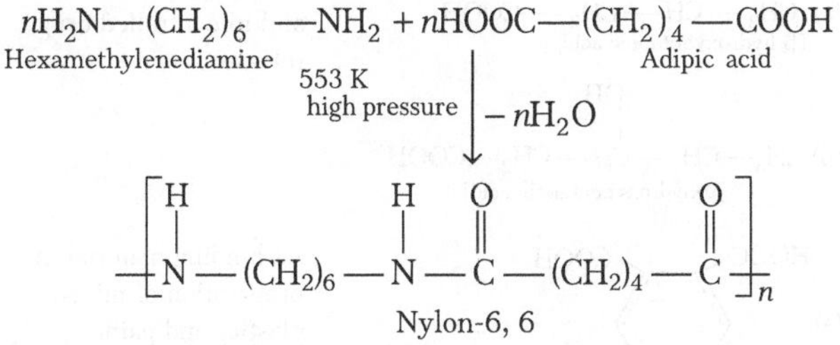

$$nH_2N-(CH_2)_6-NH_2 + nHOOC-(CH_2)_4-COOH$$

Hexamethylenediamine   Adipic acid

$$\xrightarrow[\text{high pressure}]{553\ K} -nH_2O$$

$$\left[-\overset{H}{\overset{|}{N}}-(CH_2)_6-\overset{H}{\overset{|}{N}}-\overset{O}{\overset{\|}{C}}-(CH_2)_4-\overset{O}{\overset{\|}{C}}- \right]_n$$

Nylon-6, 6

It is used in textile industry, in making bristles of brushes etc.

Bakelite

It can be prepared by condensation polymerisation of phenol and formaldehyde in presence of a base, to obtain a mixture of *ortho* and *para*- hydroxy benzyl alcohol.

$$\text{Phenol} + \underset{\text{Formaldehyde}}{HCHO} \xrightarrow{OH^-} o\text{-hydroxy benzyl alcohol} + p\text{-hydroxy benzyl alcohol}$$

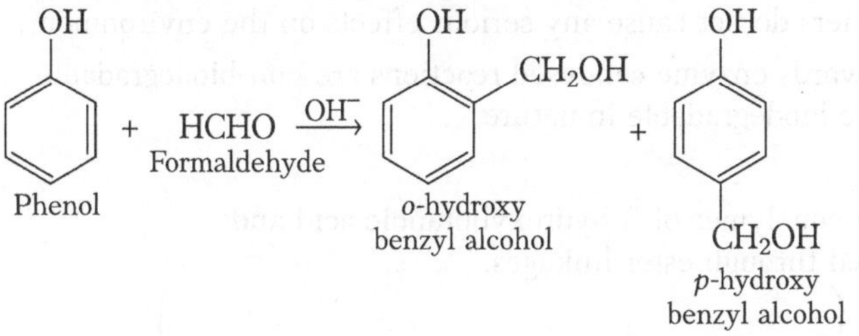

ortho-hydroxy benzyl alcohol undergo self-condensation to give a linear polymer called as **novolac**.

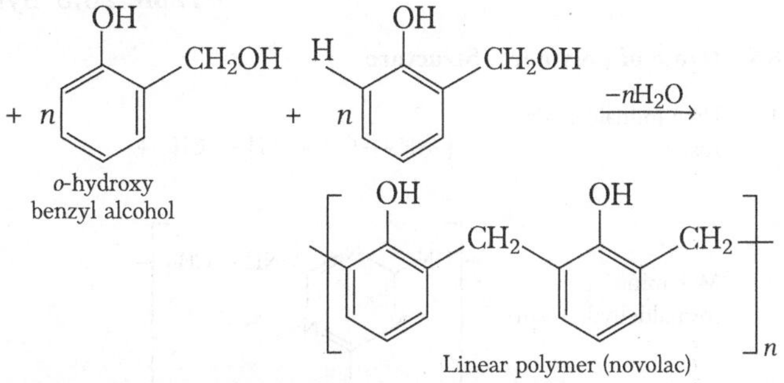

$$n\ (o\text{-hydroxy benzyl alcohol}) + n\ (o\text{-hydroxy benzyl alcohol}) \xrightarrow{-nH_2O} \left[-C_6H_3(OH)-CH_2-C_6H_3(OH)-CH_2- \right]_n$$

Linear polymer (novolac)

When both *o* and *p*-hydroxy benzyl alcohol undergo condensation polymerisation, bakelite is formed. It is a highly cross linked polymer.

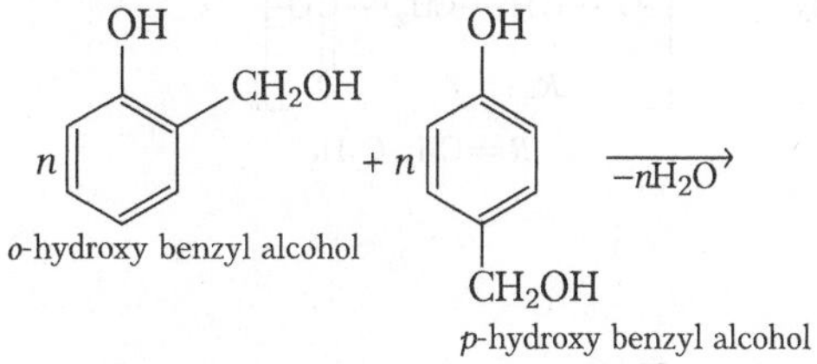

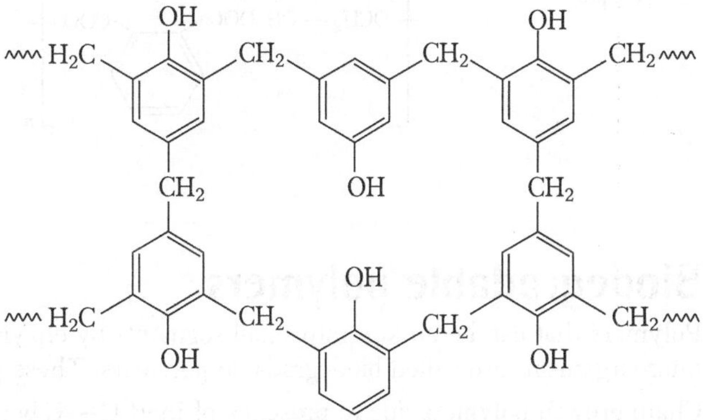

$$n\ (o\text{-hydroxy benzyl alcohol}) + n\ (p\text{-hydroxy benzyl alcohol}) \xrightarrow{-nH_2O}$$

Bakelite (branched polymer)

Uses

(*i*) Bakelite with low degree of polymerisation is used in making glues, in varnishes and lacquers.

(*ii*) It is used in making electrical goods and combs etc.

Structure and uses of some synthetic polymers are shown in the table given below

Table 23.3 Synthetic polymers

SN	Name of polymer	Structure	Monomer	Uses
1.	Urea formaldehyde resin	$\left[NH-CO-NH-CH_2 \right]_n$	(a) HCHO [Formaldehyde] (b) NH_2CONH_2 [Urea]	For making unbreakable cups and laminated sheets.
2.	Melamine formaldehyde resin	$\left[HN-C_3N_3(NH-)-NH-CH_2 \right]_n$ (triazine ring)	(a) H_2N, NH_2, NH_2 on triazine ring [Melamine] (b) HCHO (Formaldehyde)	In making plastic crockery, unbreakable cups and plates.
3.	Poly-β-hydroxy butyrate -co-β-hydroxy valerate (PHBV)	$\left[O-CH(R)-CH_2-C(=O) \right]_n$ $R = CH_3, C_2H_5$	(a) $CH_3-CH(OH)-CH_2-COOH$ [β-hydroxybutanoic acid] (b) $CH_3-CH_2-CH(OH)-CH_2-COOH$ [β-hydroxypentanoic acid]	As packaging, orthopaedic devices and in controlled drug release.
4.	Glyptal	$\left[OCH_2-CH_2OOC-C_6H_4-COO \right]_n$	(a) HOOC–C_6H_4–COOH [Phthalic acid] (b) $HO-CH_2-CH_2-OH$ [Glycer]	As binding material in preparation of mixed plastics and paints.

Biodegradable polymers

Polymers that can be broken into small segments by enzyme-catalysed reactions, using enzymes produced by microorganisms are called biodegradable polymers. These polymers do not cause any serious effects on the environment. Chain growth polymers due to presence of inert C—C bond towards enzyme catalysed reactions are non-biodegradable in nature but step growth polymers having functional groups are biodegradable in nature.

Some example of these polymers are given below

(*i*) **PHBV** (Poly β-hydroxybutyrate-co-β hydroxyvalerate) is a copolymer of 3-hydroxybutanoic acid and 3-hydroxypentanoic acid and monomer units are connected through ester linkages.

$$\underset{\text{3-hydroxybutanoic acid}}{CH_3\underset{|}{CH}(OH)CH_2\overset{O}{\overset{||}{C}}OH} + \underset{\text{3-hydroxy pentanonic acid}}{CH_3CH_2\underset{|}{CH}(OH)CH_2\overset{O}{\overset{||}{C}}OH} \xrightarrow[-nH_2O]{} \underset{\text{[PHBV]}}{\left(OCH(CH_3)-CH_2-\overset{O}{\overset{||}{C}}-O-CH(CH_2CH_3)-CH_2-\overset{O}{\overset{||}{C}} \right)_n}$$

(*ii*) **Nylon-2-nylon-6** is an alternating polyamide copolymer of glycine and amino caproic acid.

$$\underset{\text{Glycine}}{nH_2NCH_2COOH} + \underset{\text{Amino caproic acid}}{nH_2N(CH_2)_5COOH} \xrightarrow{-nH_2O} \underset{\text{Nylon-2-nylon-6}}{\left[HN-CH_2CONH(CH_2)_5\overset{O}{\overset{||}{C}} \right]_n}$$

(*iii*) **Polyglycolic acid** (PGA) is prepared by chain polymerisation of cyclic dimer of glycolic acid.

$$n\text{HO—CH}_2\text{COOH} \xrightarrow{\Delta} \left[\text{—OCH}_2\overset{\overset{\text{O}}{\|}}{\text{C}}\text{—} \right]_n$$

Glycolic acid PGA

(*iv*) **Polylactic acid** (PLA) is prepared by polymerisation of cyclic dimer of lactic acid.

$$\text{HO}\underset{\underset{\text{CH}_3}{|}}{\text{CH}}\text{—}\overset{\overset{\text{O}}{\|}}{\text{C}}\text{—OH} \xrightarrow{\text{Condensation}} \left[\text{—O}\underset{\underset{\text{CH}_3}{|}}{\text{CH}}\overset{\overset{\text{O}}{\|}}{\text{C}}\text{—} \right]_n$$

Lactic acid PLA

Uses

(*i*) Biodegradable polymers are used for medical goods like surgical structure, plasma substitutes.

(*ii*) They are used in agriculture materials like seed coating, fast food wrappers, personal hygiene products.

(*iii*) PGA and PLA are used to make absorbable structure to close a wound and has replaced catgut is a type of cord which is prepared from the natural fiber found in the walls of animal intestive fibre. These are used for making films for packaging and in moulded items.

Molecular mass of polymers

The molecular mass of a polymer is expressed as,

- number average molecular mass $\overline{M}_n$

$$\overline{M}_n = \frac{\Sigma N_i M_i}{\Sigma N_i}$$

- weight average molecular mass $\overline{M}_w$

$$\overline{M}_w = \frac{\Sigma N_i M_i^2}{\Sigma N_i M_i}$$

where, N_i is the number of molecules of molecular mass M_i

Ratio of $\overline{M}_w$ and $\overline{M}_n$ is called **Poly Dispersity Index** (PDI).

(*i*) PDI = 1 for natural polymers which are generally monodispersed and thus $\overline{M}_w = \overline{M}_n$

(*ii*) PDI > 1 for synthetic polymers which are polydispersed and thus, $\overline{M}_w > \overline{M}_n$

(*iii*) Number average molecular mass ($\overline{M}_n$) is determined by methods which depend on the number of molecules present in the polymer sample- known as **osmotic pressure** method based on **colligative property**.

(*iv*) Weight average molecular mass ($\overline{M}_w$) is determined by methods which depend on the mass of individual molecules, known as **light-scattering** and **ultracentrifuge method.**

CHECK POINT 23.2

1. Ethene undergoes polymerisation through
(a) addition polymerisation
(b) condensation polymerisation
(c) homopolymerisation
(d) copolymerisation

2. Among the following, which is not a free radical generator?
(a) Benzoyl peroxide (b) Hydrogen peroxide
(c) *t*-butyl peroxide (d) Benzophenone

3. Electron releasing groups favour
(a) cationic polymerisation
(b) anionic polymerisation
(c) *Zeigler-Natta* catalyst polymerisation
(d) None of the above

4. Aluminium-titanium initiator is used as
(a) plasticiser (b) vulcaniser
(c) telomer (d) *Zeigler-Natta* catalyst

5. Monomer unit of dacron is
(a) acrylonitrile (b) ethylene glycol
(c) vinyl chloride (d) buta-1, 3-diene

6. Monomer unit of natural rubber is
(a) buta-1, 3-diene (b) neoprene
(c) chloroprene (d) isoprene

7. Glyptal is a condensation polymer of
(a) ethylene glycol
(b) terephthalic acid
(c) phthalic acid
(d) Both (a) and (c)

8. Which of the following is not a biodegradable polymer?
(a) PHBV (b) PGA
(c) PLA (d) Nylon-6, 6

9. The monomer of PLA polymer is
(a) glycine (b) glycolic acid
(c) lactic acid (d) None of these

10. Polydispersity index of natural polymers is
(a) greater than 1
(b) less than 1
(c) equal to 1
(d) equal to or greater than 1

Chapter Exercises

(A) Topical exercise

I. Classification of polymers

1 Which of the following is not a semisynthetic polymer? **[NCERT Exemplar]**
(a) *cis*-polyisoprene (b) Cellulose nitrate
(c) Cellulose acetate (d) Vulcanised rubber

2 Consider the following statements.
Statement I Rayon is a semi-synthetic polymer and is taken as a better choice than cotton fabric.
Statement II Mechanical and aesthetic properties of cellulose can be improved by acetylation.
(a) Both statements I and II are correct
(b) Only statement I is correct
(c) Only statement II is correct
(d) Neither statement I nor statement II is correct

3 Consider the following statements.
Statement I Network polymers have high molecular mass.
Statement II Network polymers are thermosetting.
(a) Both statements I and II are correct
(b) Only statement I is correct
(c) Only statement II is correct
(d) Neither statement I nor statement II is correct

4 Which of the following is a fibre?
(a) Teflon (b) Dacron (c) PVC (d) Cellulose

5 Which one among the following is a thermoplastic?
(a) Bakelite (b) Melamine (c) PVC (d) Melmac

6 Which of the following properties is/are true for thermoplastics?
(a) The intermolecular forces of attraction are intermediate between elastomers and fibres
(b) They have cross-linked structure
(c) They cannot be moulded by heating
(d) All of the above

7 Which one of the following is an example of a thermoplastic polymer?

(a) $\left[-C_6H_3(OH)-CH_2-C_6H_3(OH)-CH_2- \right]_n$

(b) $\left[-CH_2-CH_2O-\overset{O}{\overset{\|}{C}}-C_6H_4-\overset{O}{\overset{\|}{C}}- \right]_n$

(c) $\left[-H_2C-\underset{Cl}{\underset{|}{CH}}- \right]_n$

(d) $\left[-CH_2-\underset{Cl}{\underset{|}{CH}}=CH-CH_2- \right]_n$ (with a benzene ring)

8 Which of the following is not classified as thermoplastics?
(a) Polyethylene (b) Polystyrene
(c) Bakelite (d) Neoprene

9 Consider the following statements.
Statement I Teflon has high thermal stability and chemical inertness.
Statement II Teflon is a thermoplastic.
(a) Both statements I and II are correct
(b) Only statement I is correct
(c) Only statement II is correct
(d) Neither statement I nor statement II is correct

10 Consider the following statements.
Statement I Thermoplastics soften on heating.
Statement II Individual chains are held together by weak cohesive forces.
(a) Both statements I and II are correct
(b) Only statement I is correct
(c) Only statement II is correct
(d) Neither statement I nor statement II is correct

11 Which one is not classified as thermosetting plastic?
(a) Bakelite (b) Polyethylene
(c) Melmac (d) Formaldehyde resin

12 $\left[NH(CH_2)NHCO(CH_2)_4CO \right]_n$ is a
(a) thermosetting polymer (b) homopolymer
(c) copolymer (d) addition polymer

II. Types of polymerisation

13 Which of the following is not an example of addition polymerisation?
(a) Polystyrene (b) Polyethene
(c) Polypropene (d) Terylene

14 Which one of the following is a chain-growth polymer?
(a) Starch (b) Nucleic acid
(c) Polystyrene (d) Protein

15 Arrange the following in increasing order of anionic addition polymerisation.

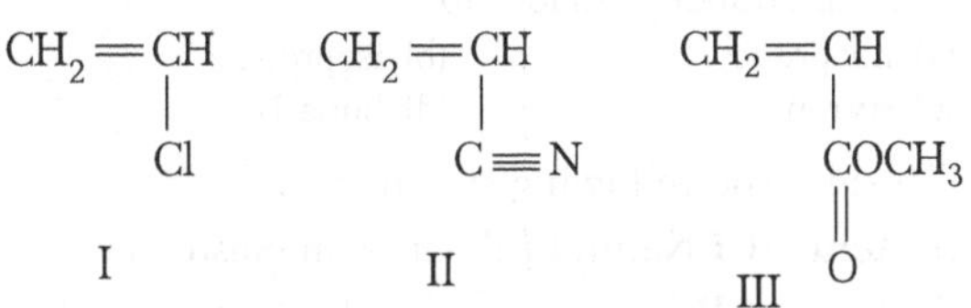

(a) I < II < III (b) II < I < III
(c) III < II < I (d) I < III < II

16 Which carbocation is formed most easily when *A* (given below) is initiated by H^+ in cationic addition polymerisation?

$A : C_6H_5-\underset{CH_3}{CH}CH{=}CH_2 \xrightarrow{H^+}$

(a) $C_6H_5-\underset{CH_3}{CH}\overset{\oplus}{C}HCH_3$ (b) $C_6H_5-\underset{CH_3}{CH}CH_2\overset{\oplus}{C}H_2$

(c) $C_6H_5-\underset{CH_3}{\overset{\oplus}{C}}CH_2CH_3$ (d) $C_6H_5-\overset{\oplus}{C}H\underset{CH_3}{CH}CH_3$

17 Consider the following polymer.
I. Polystyrene II. Dacron III. Orlon
They are classified as

	I	II	III
(a)	chain-growth;	step-growth;	step-growth
(b)	chain-growth;	chain-growth;	step-growth
(c)	chain-growth;	step-growth;	chain-growth
(d)	step-growth;	step-growth;	chain-growth

18 Molecular weight of the polymer can be controlled by
(a) chain-transfer (b) chain-growth
(c) Both (a) and (b) (d) chain initiation

19 Chain-transfer agent is
(a) CCl_4 (b) CBr_4
(c) Both (a) and (b) (d) None of these

20 Consider the following monomers.

$\underset{\text{I}}{CH_2{=}\underset{CH_3}{CH}}$; $\underset{\text{II}}{CH_2{=}\underset{CH_3}{C}CH_3}$; $\underset{\text{III}}{CH_2{=}\underset{\underset{O}{\|}}{\underset{COCH_3}{CH}}}$; $\underset{\text{IV}}{C_6H_5-CH{=}CH_2}$

Which of the above monomer can undergo cationic and anionic addition polymerisation?
(a) Both I and II (b) Both II and IV
(c) Both II and III (d) Both III and IV

21 Arrange the following in increasing order of cationic addition polymerisation initiated by H^+.

$\underset{\text{I}}{CH_3-\underset{CH_3}{C}{=}CH_2}$ $\quad$ $\underset{\text{II}}{CH_3CH{=}CHCH_3}$

$\underset{\text{III}}{C_6H_5-CH{=}CH_2}$, $\underset{\text{IV}}{CH_3-C_6H_4-CH{=}CH_2}$

(a) I < II < III < IV (b) II < I < IV < III
(c) II < I < III < IV (d) III < IV < II < I

22 Head-to-tail addition is not preferred in
(a) $CH_2{=}CHCl$ (b) $CH_2{=}CHCN$
(c) $C_6H_5-CH{=}CH_2$ (d) $HO\overset{O}{\overset{\|}{C}}(CH_2)_4\overset{O}{\overset{\|}{C}}OH$

23 Which of the following polymers describes tail-to-tail addition?
(a) $-CH_2-\underset{X}{CH}-CH_2-\underset{X}{CH}-$
(b) $-\underset{X}{CH}-CH_2-CH_2-\underset{X}{CH}-$
(c) $-CH_2-\underset{X}{CH}-\underset{X}{CH}-CH_2-$
(d) All of the above

24 Cationic addition polymerisation reaction initiated by an acid is maximum in
(a) $CH_3-\underset{CH_3}{\overset{CH_3}{C}}-CH_3$ (b) $CH_3-\overset{CH_3}{C}{=}CH_2$
(c) $CH_3CH{=}CH_2$ (d) $CH_2{=}CH_2$

25 Polyethylene is
(a) random copolymer
(b) homopolymer
(c) alternate copolymer
(d) cross-linked polymer

26 The polymer used in valve packing
(a) teflon (b) terylene
(c) polystyrene (d) polythene

27 Which of the following statements is not true about low density polyethene? **[NCERT Exemplar]**
(a) Tough
(b) Hard
(c) Poor conductor of electricity
(d) Highly branched structure

28 $\left(\!-CH_2-\overset{CH_3}{\underset{CH_3}{\overset{|}{\underset{|}{C}}}}-CH_2-\overset{CH_3}{\underset{CH_3}{\overset{|}{\underset{|}{C}}}}-\!\right)$ is a polymer having monomer units .... **[NCERT Exemplar]**

(a) (b)

(c) H H (d)

29 The monomer of the polymer

$$-CH_2-\overset{CH_3}{\underset{CH_3}{\overset{|}{\underset{|}{C}}}}-CH_2-\overset{CH_3}{\underset{CH_3}{\overset{|}{\underset{|}{C^{\oplus}}}}}$$ is

(a) $CH_2{=}C(CH_3)_2$
(b) $(CH_3)_3C{=}C(CH_3)_2$
(c) $CH_3CH{=}CHCH_3$
(d) $CH_3CH{=}CH_2$

30 The catalyst used in *Ziegler* process for polyethylene manufacture
(a) consists of aluminium chloride and titanium dioxide
(b) is vanadium pentoxide
(c) is finely divided nickel
(d) consists of triethyl aluminium and titanium tetrachloride

31 Consider the following statements.
Statement I Olefinic monomers undergo addition polymerisation.
Statement II Polymerisation of vinyl chloride is initiated by peroxides/persulphates. **[NCERT Exemplar]**
(a) Both statements I and II are correct
(b) Only statement I is correct
(c) Only statement II is correct
(d) Neither statement I nor statement II is correct

32 Match the polymers given in Column I with their types listed in Column II and choose the correct code. **[NCERT Exemplar]**

Column I	Column II
A. Buna-S	1. Addition copolymer
B. Nylon-6	2. Condensation copolymer
C. Nylon-6, 6	3. Addition homopolymer
D. Neoprene	4. Condensation homopolymer

Codes

	A	B	C	D
(a)	1	4	2	3
(b)	4	3	2	1
(c)	1	2	3	4
(d)	3	4	2	1

III. Some important polymers

33 Natural rubber is made up
(a) neoprene (b) isoprene
(c) styrene (d) buna-N

34 Consider the following statements.
Statement I Natural rubber is an elastomer.
Statement II The intramolecular forces of attraction are due to dipole-dipole interactions.
(a) Both statements I and II are correct
(b) Only statement I is correct
(c) Only statement II is correct
(d) Neither statement I nor statement II is correct

35 Consider the following statements.
Statement I 1,3 butadiene is the monomer for natural rubber.
Statement II Natural rubber is formed through radical addition polymerisation.
(a) Both statements I and II are correct
(b) Only statement I is correct
(c) Only statement II is correct
(d) Neither statement I nor statement II is correct

36 Consider the following statements.
Statement I In vulcanisation of rubber, sulphur cross polymers are introduced.
Statement II In vulcanisation, free radical initiates the chain reaction.
(a) Both statements I and II are correct
(b) Only statement I is correct
(c) Only statement II is correct
(d) Neither statement I nor statement II is correct

37 Consider the following polymers.
I. Neoprene; II. Poly (1,3-butadiene)
Stereoisomerism shown by them is

	I	II
(a)	*cis-*	*cis-*
(b)	*cis-*	*trans-*
(c)	*trans-*	*cis-*
(d)	*trans-*	*trans-*

38 Which one of the following is an example of copolymer?
(a) Buna-S (b) PAN
(c) Polythene (d) PTFE

39 Which of the following represents Buna-S polymer?

(a) $\left[-CH_2-\underset{Cl}{\underset{|}{C}}{=}CH-CH_2-\right]_n$

(b) $\left[-CH_2-CH{=}CHCH_2\,\underset{C_6H_5}{\underset{|}{CH}}CH_2-\right]_n$

(c) $\left[\!\!-\underset{\underset{C_6H_5}{|}}{CH}-CH_2-\!\!\right]_n$

(d) $\left[\!\!-\overset{\overset{CH_3}{|}}{CH_2}-CH_2-\!\!\right]_n$

40 Nitrile rubber is
(a) buna-N (b) neoprene
(c) buna-S (d) PMMA

41 Substance having ester linkage is
(a) bakelite (b) polythene
(c) PVC (d) terylene

42 Which one of the following is not an example of addition polymer?
(a) Terylene (b) Polystyrene
(c) PCTFE (d) PVC

43 In which of the following polymers ethylene glycol is one of the monomer units? **[NCERT Exemplar]**

(a) 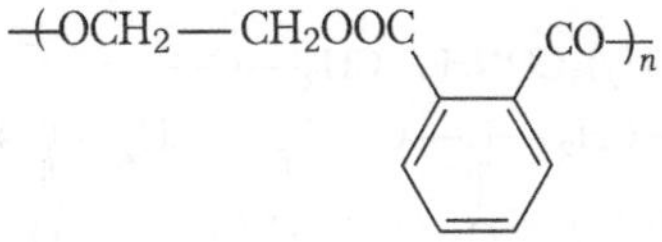 $-\!\!(OCH_2-CH_2OOC-C_6H_4-CO)\!\!-_n$ (ortho-substituted benzene ring)

(b) $-\!\!(CH_2-CH_2)\!\!-_n$

(c) 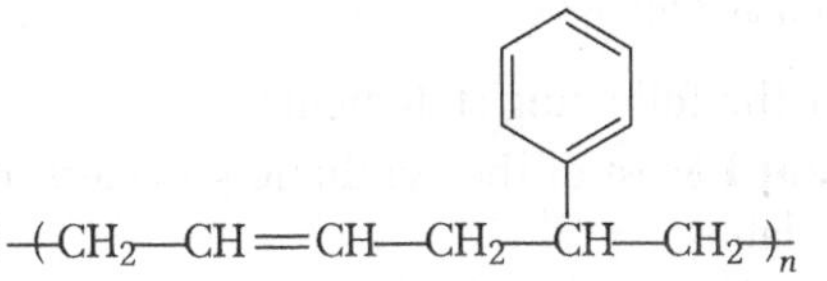 $-\!\!(CH_2-CH=CH-CH_2-\underset{C_6H_5}{CH}-CH_2)\!\!-_n$

(d) $-\!\!(O-\underset{\underset{CH_3}{|}}{CH}-CH_2-\underset{\underset{O}{||}}{C}-O-\underset{\underset{CH_2CH_3}{|}}{CH}-CH_2-\underset{\underset{O}{||}}{C})\!\!-_n$

44 Polymer obtained by condensation polymerisation is
(a) teflon (b) nylon-6
(c) glyptal (d) orlon

45 Terylene is a condensation polymer of ethylene glycol and
(a) benzoic acid (b) phthalic acid
(c) salicyclic acid (d) trephthalic acid

46 $F_2C=CF_2$ is a monomer of
(a) Teflon (b) Glyptal
(c) Nylon-6 (d) Buna-s

47 Which of the following polymer is used in paints?
(a) Terylene (b) Nylon
(c) Glyptal (d) Chloroprevene

48 Glyptal polymer is obtained by the reaction of glycerol with
(a) malonic acid (b) acetic acid
(c) phthalic acid (d) maleic acid

49 Consider the following statements.
Statement I Glyptal is obtained by the condensation polymerisation of ethylene glycol and terephthalic acid.
Statement II Glyptal is used in the manufacture of paints and lacquers.
(a) Both statements I and II are correct
(b) Only statement I is correct
(c) Only statement II is correct
(d) Neither statement I nor statement II is correct

50 Glyptal is a polymer of
(a) ethylene glycol and terephthalic acid
(b) ethylene glycol and phthalic acid
(c) ethylene glycol and adipic acid
(d) ethane-1, 2-diol and tetrafluoro ethylene

51 Glyptals are employed in
(a) toy making (b) surface coating
(c) photofilm making (d) electrical insulators

52 Which of the following is not polyamide?
(a) Nylon-6 (b) Glyptal
(c) Protein (d) Nylon-66

53 Caprolactam is the monomer of
(a) melamine (b) dacron (c) nylon-6 (d) All of these

54 Perlon is
(a) terylene (b) rubber
(c) nylon-6 (d) polyester

55 A raw material used in making nylon is
(a) ethylene (b) adipic acid
(c) butadiene (d) acetylene

56 Which of the following polymer can be formed by using the following monomer unit? **[NCERT Exemplar]**

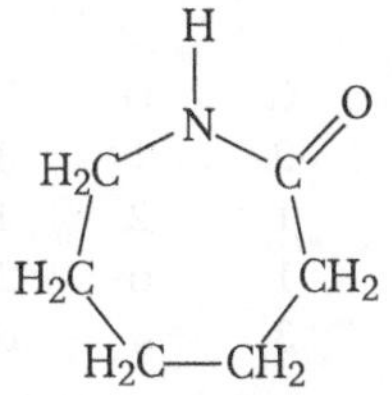

(seven-membered ring: N(H)–C(=O)–CH_2–CH_2–CH_2–CH_2–CH_2 back to N; atoms labelled H, N, O, C, H_2C, H_2C, CH_2, H_2C—CH_2)

(a) Nylon-6, 6 (b) Nylon-2-nylon-6
(c) Melamine polymer (d) Nylon-6

57 The bakelite is made from phenol and formaldehyde. The initial reaction between the two compounds is an example of
(a) aromatic electrophilic substitution
(b) aromatic nucleophilic substitution
(c) free radical reaction
(d) aldol reaction

58 Soft drinks and baby feeding bottles are generally made up of
(a) polyester (b) polyurethane
(c) polyurea (d) polyamide

59 Which one of the following is used to make 'non-stick' cookware?
(a) PVC (b) Polystyrene
(c) Polyethylene terephthalate (d) Polytetrafluoroethylene

60 Which polymer is not correctly matched in terms of physical properties?
(a) Polyethylene–crystalline (b) Rubber–amorphous
(c) Nylon–semicrystalline (d) Bakelite–amorphous

61 The polymer used in orthopaedic devices and in controlled drug release is
(a) orlon (b) PTFE
(c) SBR (d) PHBV

62 Consider the following statements.
Statement I Bakelite is a thermosetting polymer.
Statement II Bakelite can be melted again and again without any change.
(a) Both statements I and II are correct
(b) Only statement I is correct
(c) Only statement II is correct
(d) Neither statement I nor statement II is correct

63 Match the polymers given in Column I with their main applications given in Column II. **[NCERT Exemplar]**

Column I	Column II
A. Bakelite	1. Unbreakable crockery
B. Low density polyethene	2. Non-stick cookwares
C. Melamine-formaldehyde resin	3. Packaging material for shock absorbance
D. Nylon-6	4. Electrical switches
E. Polytetrafluoroethene	5. Squeeze bottles
F. Polystyrene	6. Tyre, cords

Codes

	A	B	C	D	E	F
(a)	6	5	2	1	3	4
(b)	5	3	1	2	4	6
(c)	4	5	1	6	2	3
(d)	1	2	3	4	6	5

64 Match the polymers given in Column I with their commercial names given in Column II. **[NCERT Exemplar]**

Column I	Column II
A. Polyester of glycol and phthalic acid	1. Novolac
B. Copolymer of 1, 3-butadiene and styrene	2. Glyptal
C. Phenol and formaldehyde resin	3. Buna-S
D. Polyester of glycol and terephthalic acid	4. Buna-N
E. Copolymer of 1,3- butadiene and acrylonitrile	5. Dacron

Codes

	A	B	C	D	E
(a)	3	2	1	5	4
(b)	4	5	1	2	3
(c)	3	2	1	4	5
(d)	2	3	1	5	4

65 Which one of the following sets forms the biodegradable polymers?
(a) $OH—CH_2—CH_2OH$ and $COOH(C_6H_5)COOH$
(b) $CH_3—CH_2—\underset{OH}{\underset{|}{C}H}—CH_2COOH$ and $CH_3—CH(OH)CH_2COOH$
(c) $CH_3CH(OH)COOH$ and $H_2N(CH_2)_5COOH$
(d) $HOCH(CH_3)COOH$ and $H_2N(CH_2)_5COOH$

66 Which of the following polymers is biodegradable? **[NCERT Exemplar]**
(a) $\left[CH_2—\underset{Cl}{\underset{|}{C}}=CH—H_2\right]_n$
(b) $\left[HN—(CH_2)_5CONH—CH_2—\overset{O}{\overset{\|}{C}}\right]_n$
(c) $\left[O—\underset{CH_3}{\underset{|}{C}H}—CH_2—\underset{O}{\underset{\|}{C}}—O—\underset{CH_2CH_3}{\underset{|}{C}H}—CH_2—\underset{O}{\underset{\|}{C}}\right]_n$
(d) Both (b) and (c)

67 Consider the following statements.
Statement I Most of the synthetic polymers are not biodegradable.
Statement II Polymerisation process induces toxic character in organic molecules.
(a) Both statements I and II is correct
(b) Only statement I is correct
(c) Only statement II is correct
(d) Neither statement I nor statement II is correct

68 The number average molecular mass and mass average molecular mass of a polymer are respectively 30,000 and 40,000. The polydispersity index of the polymer is
(a) < 1 (b) > 1 (c) 1 (d) 0

(B) Miscellaneous questions

1 The polymer used for making contact lenses for eyes is
(a) polymethylmethacrylate (b) polyethylene
(c) polyethylacrylate (d) nylon

2 'Celanese' silk is basically
(a) cellulose trinitrate (b) cellulose acetate
(c) cellophane (d) pyroxylin

3 The commercial name of polyacrylonitrile is .... **[NCERT Exemplar]**
(a) dacron (b) orlon (acrilan)
(c) PVC (d) bakelite

4 Which is not the copolymer out of the following?
(a) SAN (b) ABS
(c) Saran (d) PVC

5 Thermosetting polymers are except
(a) bakelite (b) PET
(c) epoxy resins (d) melmac

6 The polymer with strong intermolecular forces of attraction is
(a) rubber (b) dacron
(c) teflon (d) PVC

7 Kevlar is used in the following except
(a) bullet proof vests
(b) crash helmets
(c) radial tyres
(d) shatter-resistant eye glasses

8 Synthetic polymer which resembles natural rubber is
(a) neoprene (b) chloroprene
(c) glyptal (d) nylon

9 Which is not the step growth polymer?
(a) Polysulphones
(b) Polycarbonates
(c) Polyvinyl acetate
(d) Polyethylene oxide

10 The turbidity of a polymer solution measures a
(a) light absorbed by the solution
(b) light transmitted by the solution
(c) light scattered by the solution
(d) None of the above

11 Select the incorrect statement.
(a) Nylon is a step growth condensation polymer.
(b) H_2O_2, $(C_6H_5CO)_2O_2$ are radical initiators.
(c) Natural rubber is *trans*-polyisoprene.
(d) *Gutta-percha* is *trans*-polyisoprene.

12 If polymer becomes more crystalline, then
(a) it becomes harder
(b) it becomes denser
(c) it becomes more resistant to heat
(d) All of the above properties are true

13 Which is the best monomer for chain-growth polymer?
(a) Vinyl cyanide (b) Vinyl chloride
(c) Styrene (d) Vinyl acetate

14 IUPAC name for polystyrene is
(a) poly (1-ethenyl benzene)
(b) poly (2-ethenyl benzene)
(c) poly (1-phenylethylene)
(d) poly (2-phenylethylene)

15 Following polymer will be named as

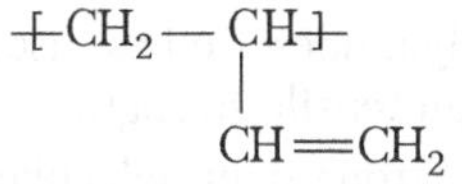

$$\left[CH_2 - \underset{\underset{CH=CH_2}{|}}{CH} \right]$$

(a) 1,4-poly (1,3-butadiene) (b) 1,2-poly (1,3-butadiene)
(c) 1,2-poly (1-butene) (d) 1,2-poly (3-butene)

16 Orlon is a hard, horny and a high melting material. Which of the following represents its structure?

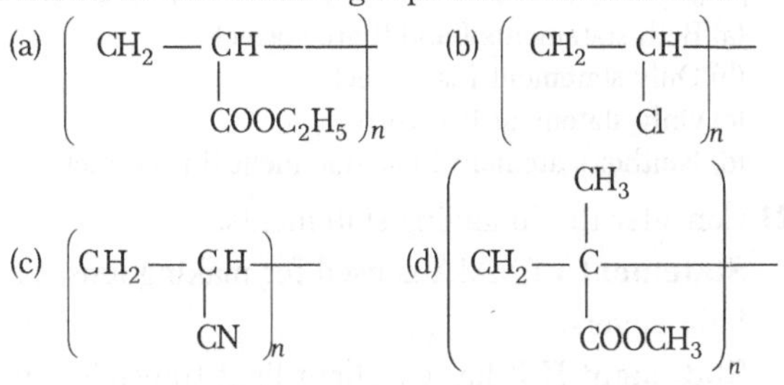

(a) $\left(CH_2 - \underset{\underset{COOC_2H_5}{|}}{CH} \right)_n$ (b) $\left(CH_2 - \underset{\underset{Cl}{|}}{CH} \right)_n$

(c) $\left(CH_2 - \underset{\underset{CN}{|}}{CH} \right)_n$ (d) $\left(CH_2 - \overset{\overset{CH_3}{|}}{\underset{\underset{COOCH_3}{|}}{C}} \right)_n$

17 Synthetic human hair wigs are made from a copolymer of vinyl chloride and acrylonitrile. This polymer is called
(a) orlon (b) dynel (c) thiokol (d) acrilon

18 Which one of the following is not correctly matched?

(a) Neoprene $\left[CH_2\underset{\underset{Cl}{|}}{C}=CHCH_2 \right]_n$

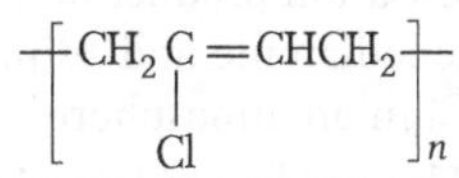

(b) Nylon-6, 6 $\left[NH(CH_2)_6NH\overset{\overset{O}{||}}{C}(CH_2)_4\overset{\overset{O}{||}}{C} \right]_n$

(c) Terylene $\left[OCH_2CH_2\overset{\overset{O}{||}}{C}-C_6H_4-\overset{\overset{O}{||}}{C} \right]_n$

(d) PMMA $\left[CH_2 - \overset{\overset{CH_3}{|}}{\underset{\underset{\underset{O}{||}}{COCH_3}}{C}} \right]_n$

19 Cyclohexanone $+ NH_2OH \longrightarrow A \xrightarrow{H_2SO_4} B$

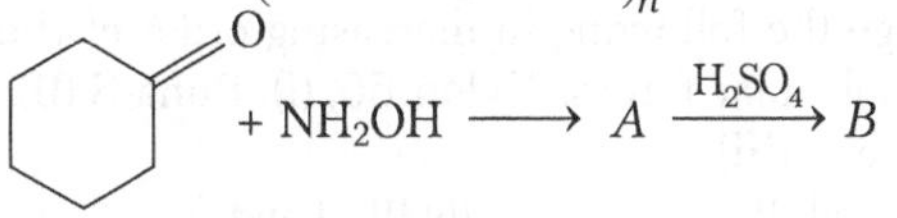

The product B is

(a) 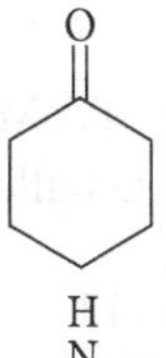 cyclohexanone (b) 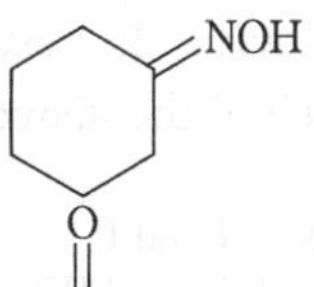 cyclohexanone oxime (=NOH) (c) 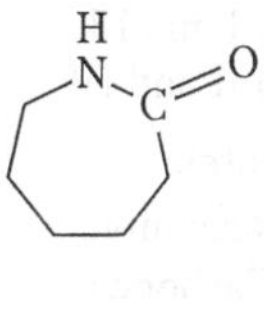 caprolactam (seven-membered ring with N–H and C=O) (d) 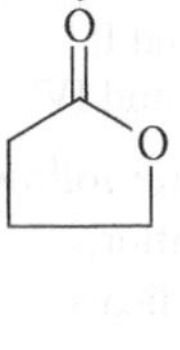 five-membered lactone (ring O with C=O)

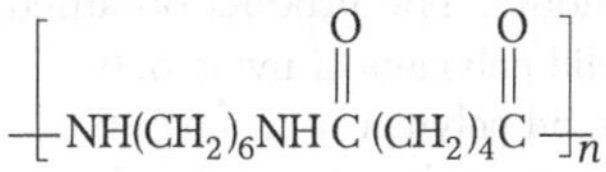

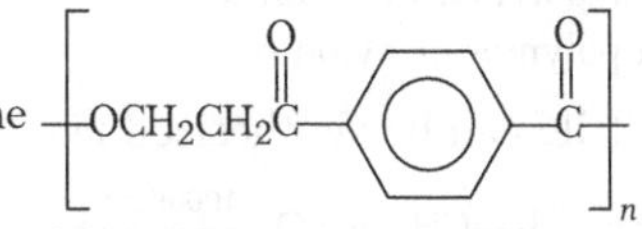

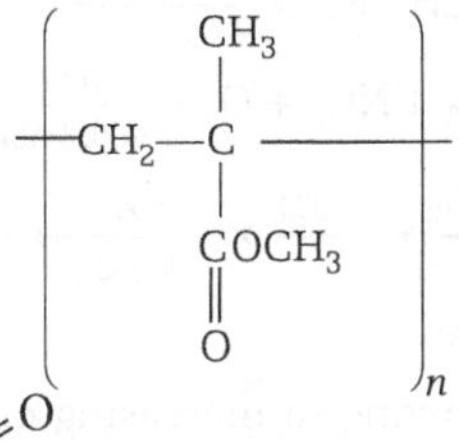

20 Consider the following statements.

Statment I Plyamides are best used as fibres because of high tensile strength.

Statement II Strong intermolecular forces (like hydrogen bonding within polyamides) lead to close packing of chains and increases the crystalline character, hence, provide high tensile strength to polymers. **[NCERT Exemplar]**

(a) Both statements I and II are correct
(b) Only statement I is correct
(c) Only statement II is correct
(d) Neither statement I nor statement II is correct

21 Consider the following statements.

Statement I PMMA is used for making lenses and light covers.

Statement II It has excellent light transmission properties.

(a) Both statements I and II are correct
(b) Only statement I is correct
(c) Only statement II is correct
(d) Neither statement I nor statement II is correct

22 When condensation product of hexamethylenediamine and adipic acid is heated to 553 K (80°C) in an atmosphere of nitrogen for about 4-5 hours. The product obtained is

(a) solid polymers of nylon 6, 6
(b) liquid polymer of nylon 6, 6
(c) gaseous polymer of nylon 6
(d) liquid polymer of nylon 6

23 Acrylonitrile can be synthesised *via*

(a) $CH_2 = CH—CH_3 + NO \xrightarrow{1000K, Ag}$

(b) $CH_2 = CHCH_3 + NH_3 + O_2 \xrightarrow[FeO,\ 3\ Atm]{700K}$

(c) $CH_2 = CH_2 \xrightarrow[Ag]{O_2} \xrightarrow{HCN} \xrightarrow[MgCO_3]{500\ K}$

(d) All of the above

24 Arrange the following in increasing order of their intermolecular forces. Nylon-66, (I), Buna-S (II), Polythene (III)

(a) II, I and III
(b) III, II and I
(c) I, II and III
(d) II, III and I

25 Consider the following polymers.

I. Kevlar II. Lexan III. Polyester IV. Mylar

Which of the above are used to prepare bullet proof vests?

(a) Both I and II
(b) Both I and III
(c) Both III and IV
(d) Both II and IV

26 Consider the following intermediates.

I. Carbocations
II. Carbanions
III. Free radicals
IV. Carbenes

Polymerisation may occur through intermediate formation of

(a) Only I
(b) I and II
(c) III and IV
(d) I, II and III

27 Which of the following sets contains only thermoplastics?

(a) Plyethene, bakelite, nylon 6
(b) Glyptal, melmac, PAN
(c) PVC, PMMA, polystyrene
(d) Polypropylene, urea-formaldehyde, teflon

28 Consider the following statements regarding polymer.

I. In the configuration, all the *R* (substituent) groups may lie on the same side of the plane.
II. Substituent groups regularly alternate from one side of the plane to the other.
III. Substituents have no preferred placements.

The correct term for the statement I, II and III respectively are

(a) isotactic, syndiotactic, atactic types of polymer
(b) isotactic, atactic, syndiotactic types of polymer
(c) syndiotactic, atactic, isotactic types of polymer
(d) atactic, syndiotactic, isotactic types of polymer

29 Match the polymers given in Column I with the preferred mode of polymerisation Column II. **[NCERT Exemplar]**

Column I	Column II
A. Nylon-6,6	1. Free radical polymerisation
B. PVC	2. *Ziegler-Natta* polymerisation or coordination polymerisation
C. HDP	3. Condensation polymerisation

Codes

	A	B	C		A	B	C
(a)	3	2	1	(b)	1	2	3
(c)	2	1	3	(d)	3	1	2

30 Match the polymers given in Column I with their chemical names given in Column II. **[NCERT Exemplar]**

Column I	Column II
A. Nylon-6	1. Polyvinyl chloride
B. PVC	2. Polyacrylonitrile
C. Acrilan	3. Polycaprolactum
D. Natural rubber	4. Low density polythene
E. LDP	5. *cis*-polyisoprene

Codes

	A	B	C	D	E
(a)	3	1	2	5	4
(b)	1	2	3	5	4
(c)	4	3	2	1	5
(d)	2	3	4	1	5

31 Match the copolymers in Column I with their monomers in Column II. **[NCERT Exemplar]**

Column I		Column II
A. Saran	1.	$C_6H_5-CH{=}CH_2, CH_2{=}CH-CH_2{=}CH_2$
B. SAN	2.	$CH_2{=}CH-CH{=}CH_2$, $C_6H_5-CH{=}CH_2$, $CH_2{=}CHCN$
C. ABS	3.	$CH_2{=}CHCl$, $CH_2{=}CHCl_2$
D. SBR	4.	$C_6H_5-CH{=}CH_2$, $CH_2{=}CHCN$

Codes

	A	B	C	D
(a)	4	3	2	1
(b)	3	4	2	1
(c)	1	2	3	4
(d)	3	4	2	1

32 Match the polymers given in Column I with their repeating units given in Column II. **[NCERT Exemplar]**

Column I		Column II
A. Acrilan	1.	$\left[CH_2-CH(C_6H_5)\right]_n$
B. Polystyrene	2.	$\left[CH_2-C(Cl){=}CH-CH_2\right]_n$
C. Neoprene	3.	$\left[CH_2-CH{=}CH-CH_2-CH_2-CH(Cl)\right]_n$
D. Novolac	4.	$\left[CH_2-CH(CN)\right]_n$
E. Buna-N	5.	$\left[-C_6H_2(OH)-CH_2-C_6H_3(OH)-CH_2-\right]_n$

Codes

	A	B	C	D	E
(a)	4	1	2	5	3
(b)	1	4	2	3	5
(c)	3	2	4	1	5
(d)	5	2	3	1	4

(C) Medical entrances' gallery

1 Which of the following is a natural polymer? **[NEET 2020]**
(a) Poly (butadiene-styrene)
(b) Polybutadiene
(c) Poly (butadiene-acrylonitrile)
(d) *cis*-1, 4-polyisoprene

2 Whch of the following statement is correct about bakelite? **[NEET 2020]**
(a) It is a cross linked polymer
(b) It is an addition polymer
(c) It is a branched chain polymer
(d) It is a linear polymer

3 The biodegradable polymer is **[NEET 2019]**
(a) nylon-2-nylon-6
(b) nylon-6
(c) buna-S
(d) nylon-6, 6

4 The polymer that is used as a substitute for wool in making commercial fibres is **[NEET (Odisha) 2019]**
(a) melamine
(b) nylon-6, 6
(c) polyacrylonitrile
(d) buna-N

5 Match the following Columns :

	Polymer		Monomer
(i)	Buna -S	(P)	Styrene
(ii)	Ethylene glycol	(Q)	Terylene
(iii)	Elastomer	(R)	Chloroprene

Correct set of answer: **[AIIMS 2019]**
(a) (i) → (P), (ii) → (Q), (iii) → (R)
(b) (i) → (R), (ii) → (P), (iii) → (Q)
(c) (i) → (Q), (ii) → (R), (iii) → (P)
(d) (i) → (P), (ii) → (R), (iii) → (Q)

6 Regarding cross-linked or network polymers, which of the following statements is incorrect?
(a) Examples are bakelite and melamine. **[NEET 2018]**
(b) They are formed from bi- and tri-functional monomers.
(c) They contain strong covalent bonds between various linear polymer chains.
(d) They contain strong covalent bonds in their polymer chains.

7 Arrange the following polymers in increasing order of their intermolecular forces.
Nylon-6, 6, buna-S, polythene, neoprene **[JIPMER 2018]**
(a) Neoprene < buna-S < nylon-6, 6, < polythene
(b) Neoprene < buna-S < polythene < nylon-6, 6
(c) Polythene < nylon-6, 6 < buna-S < neoprene
(d) Polythene < buna-S < nylon-6, 6 < neoprene

8 Assertion (A) Most of the synthetic polymers are non-biodegradable.

Reason (R) Polymerisation always induces toxic effect in environment. **[AIIMS 2018]**

(a) Both A and R are correct and R is correct explanation of A.
(b) Both A and R are correct but R is not the correct explanation of A.
(c) A is correct but R is incorrect.
(d) Both A and R are incorrect.

9 Polymer formation from monomers starts by **[JIPMER 2017]**

(a) condensation reaction between monomers
(b) coordination reaction between monomers
(c) conversion of one monomer into other monomer
(d) hydrolysis of monomers

10 Which of the following sets contain only addition polymers?

(a) Polyethylene, polypropylene, terylene **[AIIMS 2017]**
(b) Polyethylene, PVC, teflon
(c) Buna-S, nylon, polybutadiene
(d) Bakelite, PVC, polyethylene

11 Which of the following fibres is made of polyamides? **[JIPMER 2017]**

(a) Dacron (b) Orlon
(c) Nylon (d) Rayon

12 Which one of the following structures represents nylon-6, 6 polymer? **[NEET-II 2016]**

(a) $\left(\overset{H_2}{C}-\underset{NH_2}{\overset{H}{C}}-\overset{H_2}{C}-\underset{CH_3}{\overset{H}{C}} \right)_{6,6}$

(b) $\left(\overset{H_2}{C}-\underset{NH_2}{\overset{H}{C}}-\overset{H_2}{C}-\underset{NH_2}{\overset{H}{C}} \right)_{6,6}$

(c) $\left(\overset{H_2}{C}-\underset{NH_2}{\overset{H}{C}}-\overset{H_2}{C}-\underset{Cl}{\overset{H}{C}} \right)_6 \left(\overset{H_2}{C}-\underset{CH_3}{\overset{H}{C}}-\overset{H_2}{C}-\underset{COOH}{\overset{H}{C}} \right)_6$

(d) $\left(\overset{O}{\overset{\|}{C}}-\underset{H_2}{C}-\overset{H_2}{C}-\underset{\underset{O}{\|}}{C}-NH-(CH_2)_6-NH \right)_n$

13 Natural rubber has **[NEET-I 2016]**

(a) All *trans*-configuration
(b) alternate *cis* and *trans*-configuration
(c) random *cis* and *trans*-configuration
(d) All *cis*-configuration

14 The polymer used in the manufacturing of squeeze bottle is **[Kerala 2015]**

(a) polystyrene
(b) teflon
(c) polypropene
(d) high density polythene
(e) low density polythene

15 Biodegradable polymer which can be produced from glycine and aminocapric acid is **[AIPMT 2015]**

(a) nylon-2-nylon 6
(b) PHBV
(c) buna-N
(d) nylon-6, 6

16 Which one of the following is an example of a thermosetting polymer? **[CBSE AIPMT 2014]**

(a) $\left(CH_2-\underset{Cl}{\underset{|}{C}}=CH-CH_2 \right)_n$

(b) $\left(CH_2-\underset{Cl}{\underset{|}{CH}} \right)_n$

(c) $\left(\overset{H}{\overset{|}{N}}-(CH_2)_6-\overset{H}{\overset{|}{N}}-\overset{O}{\overset{\|}{C}}-(CH_2)_4-\overset{O}{\overset{\|}{C}} \right)_n$

(d) $\left(C_6H_3(OH)-CH_2-C_6H_3(OH)-CH_2 \right)_n$

17 Which of the following organic compounds polymerises to form the polyester dacron? **[CBSE AIPMT 2014]**

(a) Propylene and *para* HO $-(C_6H_4)-$ OH
(b) Benzoic acid and ethanol
(c) Terephthalic acid and ethylene glycol
(d) Benzoic acid and *para* HO $-(C_6H_4)-$ OH

18 Which of the following polymer is stored in liver of animals? **[UP CPMT 2014]**

(a) Amylose
(b) Cellulose
(c) Amylopectin
(d) Glycogen

19 The two monomers used in the preparation of dextron are **[MHT CET 2014]**

(a) 3-hydroxybutanoic acid and 3-hydroxy pentanoic acid
(b) E-amino caproic acid and glycine
(c) isobutylene and isoprene
(d) Lactic acid and glycolic acid

20 Which one of the following is an example for biodegradable polyester? **[Kerala CEE 2014]**

(a) PHBV (b) PET
(c) Nylon 6 (d) Bakelite
(e) Glyptal

21 Identify from the following monomers, which will undergo condensation polymerisation. **[EAMCET 2014]**

I. $H_2C = CH—CH=CH_2$

II. (benzene ring with CO_2H at top and bottom)

III. $H_2C=CHCl$

IV. (seven-membered ring with =O and NH)

V. $F_2C=CF_2$

VI. $(H_3C)_2C=CH_2$

(a) Both III and V (b) Both I and III
(c) Both I and VI (d) Both II and IV

22 Terylene is not a **[KCET 2014]**

(a) copolymer (b) polyester fibre
(c) chain growth polymer (d) step growth polymer

23 Which is the monomer of neoprene in the following? **[NEET 2013]**

(a) $CH_2= CH—C\equiv CH$ (b) $CH_2 = C(CH_3)— CH=CH$
(c) $CH_2 = C(Cl)— CH = CH_2$ (d) $CH_2= CH— CH=CH_2$

24 Nylon is an example of **[NEET 2013]**

(a) polyster (b) polysaccharide
(c) polyamide (d) polythene

25 Natural rubber is a polymer of **[Manipal 2013]**

(a) *cis*-isoprene (b) *trans*-isoprene
(c) Both (a) and (b) (d) None of these

26 Acrilan is a hard, horny and a high melting material. Its structure is **[WB JEE 2013]**

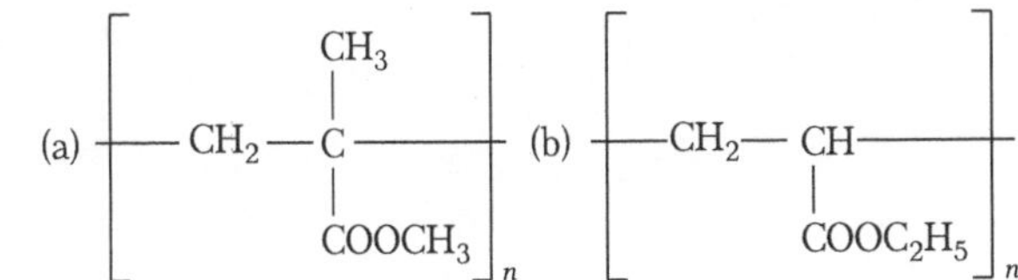

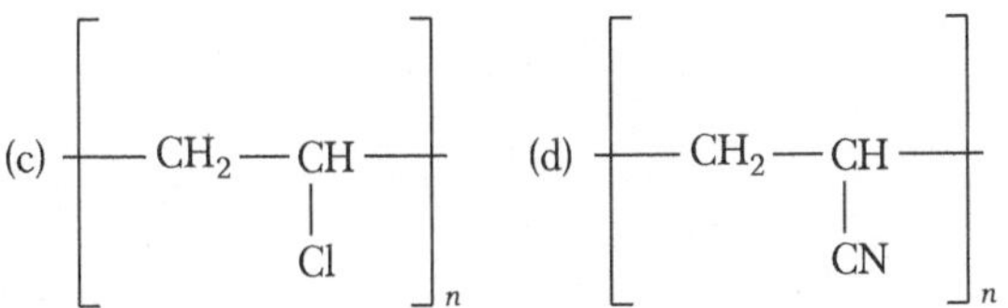

27 A condensation polymer among the following polymers is **[KCET 2013]**

(a) teflon (b) polystyrene
(c) PVC (d) dacron

28 Which one of the following is not a condensation polymer? **[CBSE AIPMT 2012]**

(a) Melamine (b) Glyptal
(c) Dacron (d) Neoprene

29 Which one of the following sets forms the biodegradable polymer? **[CBSE AIPMT 2012]**

(a) $CH_2 = CH—CN$ and $CH_2 = CH — CH = CH_2$
(b) $H_2N — CH_2 — COOH$ and $H_2N — (CH_2)_5 — COOH$
(c) $HO— CH_2CH_2— OH$ and $HOOC—C_6H_4—COOH$
(d) $C_6H_5—CH= CH_2$ and $CH_2 = CH — CH = CH_2$

30 Which of the following is a polymer? **[AFMC 2012]**

(a) Carnuaba wax
(b) Carbo wax
(c) Bees wax
(d) Paraffin wax

31 Which of the following is a biodegradable polymer? **[Manipal 2012]**

(a) Polythene (b) Bakelite
(c) PHBV (d) PVC

32 Neoprene is a polymer of **[OJEE 2012]**

(a) vinyl chloride (b) propene
(c) isoprene (d) chloroprene

33 Of the following which one is classified as polyester polymer? **[CBSE AIPMT 2011]**

(a) Nylon-6, 6
(b) Terylene
(c) Bakelite
(d) Melamine

34 Which type of polymer is bakelite? **[Guj CET 2011]**

(a) Addition polymer
(b) Homopolymer
(c) Condensation polymer
(d) Biopolymer

35 Natural rubber is not used in making footwear for polar regions because **[Guj CET 2011]**

(a) natural rubber becomes soft at temperature lower than 10°C
(b) natural rubber becomes brittle at temperature lower than 10°C
(c) natural rubber melts at temperature lower than 10°C
(d) natural rubber becomes stronger at temperature lower than 10°

ANSWERS

Check point 23.1

1. (b)	2. (d)	3. (a)	4. (d)	5. (b)	6. (c)	7. (d)	8. (a)	9. (c)	10. (d)

Check point 23.2

1. (a)	2. (d)	3. (a)	4. (d)	5. (b)	6. (d)	7. (d)	8. (d)	9. (c)	10. (c)

(A) Topical exercise

1. (a)	2. (b)	3. (b)	4. (b)	5. (b)	6. (a)	7. (c)	8. (c)	9. (a)	10. (a)
11. (b)	12. (c)	13. (d)	14. (c)	15. (d)	16. (c)	17. (c)	18. (a)	19. (c)	20. (d)
21. (c)	22. (d)	23. (b)	24. (b)	25. (b)	26. (a)	27. (c)	28. (a)	29. (a)	30. (d)
31. (a)	32. (a)	33. (b)	34. (b)	35. (c)	36. (c)	37. (d)	38. (a)	39. (b)	40. (a)
41. (d)	42. (a)	43. (a)	44. (c)	45. (a)	46. (a)	47. (c)	48. (c)	49. (c)	50. (b)
51. (b)	52. (b)	53. (c)	54. (c)	55. (b)	56. (d)	57. (a)	58. (a)	59. (d)	60. (d)
61. (d)	62. (b)	63. (c)	64. (d)	65. (b)	66. (d)	67. (b)	68. (b)		

(B) Miscellaneous questions

1. (a)	2. (b)	3. (b)	4. (d)	5. (b)	6. (b)	7. (d)	8. (a)	9. (c)	10. (c)
11. (c)	12. (d)	13. (c)	14. (c)	15. (b)	16. (c)	17. (b)	18. (c)	19. (c)	20. (a)
21. (a)	22. (a)	23. (d).	24. (d)	25. (a)	26. (d)	27. (c)	28. (a)	29. (d)	30. (a)
31. (b)	32. (a)								

(C) Medical entrances' gallery

1. (d)	2. (a)	3. (a)	4. (c)	5. (a)	6. (d)	7. (b)	8. (c)	9. (a)	10. (b)
11. (c)	12. (d)	13. (d)	14. (b)	15. (a)	16. (d)	17. (c)	18. (d)	19. (d)	20. (a)
21. (d)	22. (c)	23. (c)	24. (c)	25. (a)	26. (d)	27. (d)	28. (d)	29. (b)	30. (b)
31. (c)	32. (d)	33. (b)	34. (c)	35. (b)					

Hints & Explanations

(A) Topical exercise

1 (*a*) *cis*-polyisoprene is not a semisynthetic polymer while other three cellulose nitrate, cellulose acetate and vulcanised rubber are semisynthetic polymer made from cellulose, and natural rubber respectively.

3 (*b*) Extensive cross-linking during polymerisation of three dimensional and network solid which is hard, infusible and insoluble.

10 (*a*) Thermoplastics soften on heating because they are held by weak cohesive forces.

27 (*c*) Low density polythene has good electrical property.

28 (*a*) $\text{-}\!\!(CH_2-\underset{CH_3}{\overset{CH_3}{C}}-CH_2-\underset{CH_3}{\overset{CH_3}{C}}\!\!\text{-})_n$ can be obtained by addition polymerisation of $=\!\!<$ as shown below

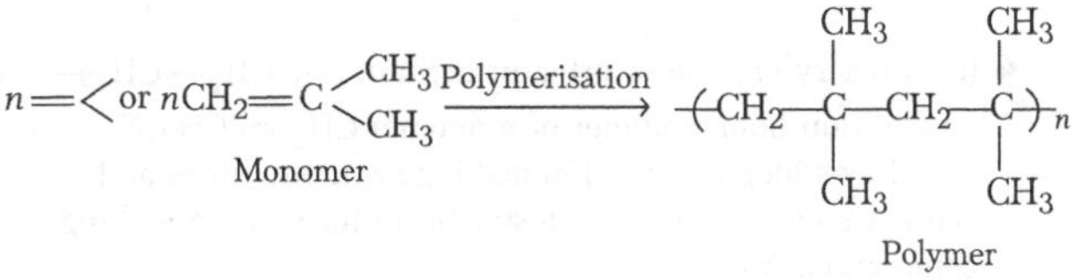

$$n=\!\!< \text{ or } nCH_2=C\begin{matrix}CH_3\\CH_3\end{matrix} \xrightarrow{\text{Polymerisation}} \text{-}\!\!(CH_2-\underset{CH_3}{\overset{CH_3}{C}}-CH_2-\underset{CH_3}{\overset{CH_3}{C}}\!\!\text{-})_n$$

Monomer — Polymer

31 (*a*) Both statements are correct.

Olefin monomers undergo addition polymerisation because it adds on to other monomers of olefin through free radical mechanism due to the presence of peroxide initiator and double bond. It also produces high molecular mass polymers.

34 (*b*) Statement I is correct but statement II is incorrect. Elastomers are held together by the weakest intermolecular forces.

36 (*c*) Vulcanisation is a process of treating natural rubber with sulphur or some compounds of sulphur under heat so as to modify its properties. This cross-linking gives mechanical strength to the rubber.

43 (*a*) Given polymer is glyptal which can be obtained by condensation polymerisation of ethylene glycol and phthalic acid with the elimination of water molecule.

49 (*c*) Glyptal is obtained by condensation polymerisation of ethylene glycol and phthalic acid.

$$nHOCH_2-CH_2OH + n\,C_6H_4(COOH)_2 \xrightarrow[-nH_2O]{420\text{-}460\text{ K},\ Zn(OCOCH_3)_2+Sb_2O_3} \left(\!-O-CH_2-CH_2OOC-C_6H_4-\overset{O}{\overset{\|}{C}}-\!\right)_n$$

Glycol + Phthalic acid → Glyptal

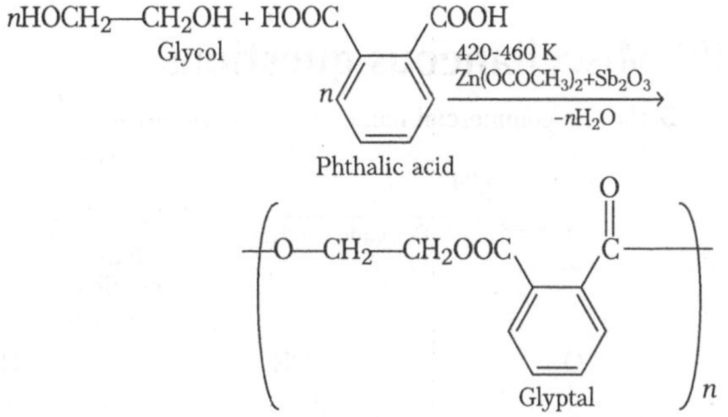

56 (*d*) Given, monomer is the structure of caprolactam which on polymerisation produces nylon-6 polymer.

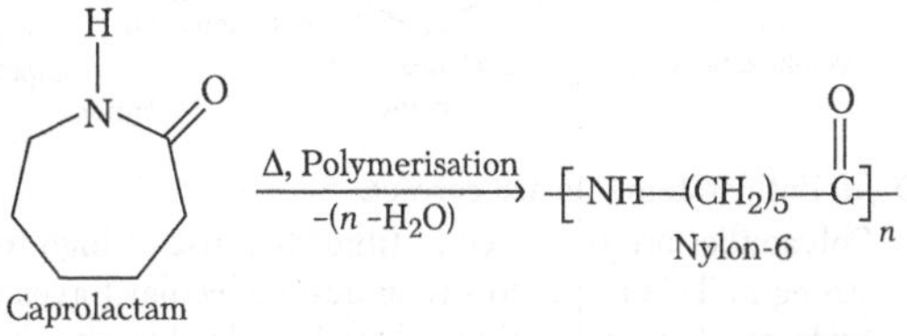

$$\text{Caprolactam} \xrightarrow[-(n\text{ -}H_2O)]{\Delta,\text{ Polymerisation}} \left[\!-NH-(CH_2)_5-\overset{O}{\overset{\|}{C}}-\!\right]_n$$

Nylon-6

57 (*a*) The initial reaction between phenol and formaldehyde is aromatic electrophilic substitution.

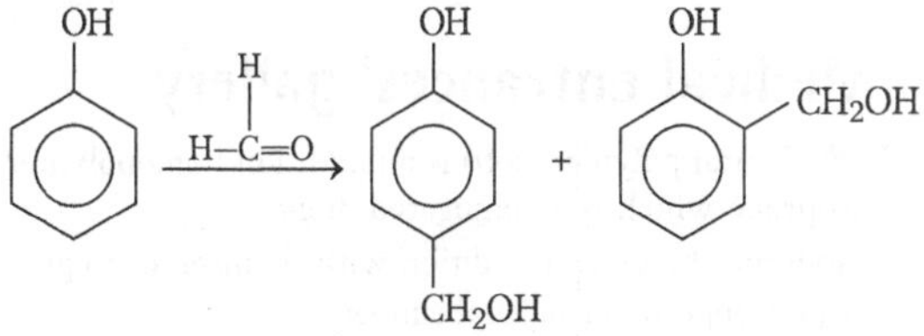

$$C_6H_5OH \xrightarrow{H-\overset{H}{\overset{|}{C}}=O} p\text{-}HOC_6H_4CH_2OH + o\text{-}HOC_6H_4CH_2OH$$

61 (*d*) PHBV (Poly β-hydroxybutyrate-co-β-hydroxy valerate) a copolymer of 3-hydroxybutanoic acid and 3-hydroxypentanoic acid, provides flexibility.

62 (*b*) Bakelite is a thermosetting polymer which can be obtained by condensation polymerisation of phenol with formaldehyde. These polymer once get hardened cannot be remelted by heating because cross-linking reduces the mobility of the polymer chains causing them to be brittle.

66 (*d*) $\left(\!-O-\underset{CH_3}{\underset{|}{CH}}-CH_2-\underset{O}{\underset{\|}{C}}-O-\underset{CH_2CH_3}{\underset{|}{CH}}-CH_2-\underset{O}{\underset{\|}{C}}-\!\right)_n$

Poly β-hydroxybutyrate - co-β-hydroxy valerate is an example of biodegradable polymer. Nylon-2-nylon-6 is also polymer of this type.

67 (*b*) Only statement I is correct. Most of synthetic polymers are not biodegradable because they are not easily broken by soil organisms and hazardous to the environment.

68 (*b*) $\frac{\overline{M}_w}{\overline{M}_n} = \frac{40,000}{30,000} = 1.33$

(B) Miscellaneous questions

3 (*b*) The commercial name of polyacrylonitrile is orlon (acrilan).

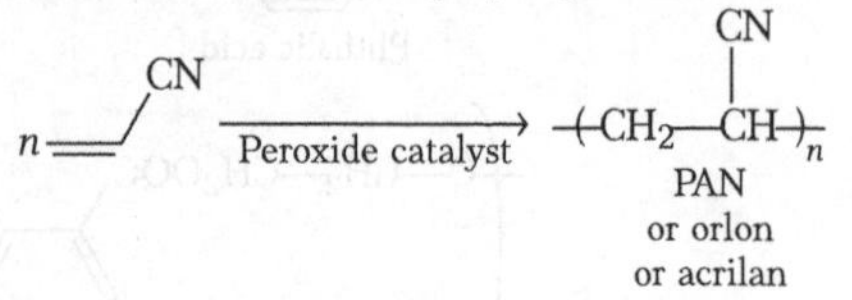

$$n\,CH_2{=}CH(CN) \xrightarrow{\text{Peroxide catalyst}} \left(CH_2-\underset{CN}{CH}\right)_n$$

PAN or orlon or acrilan

19 (*c*)

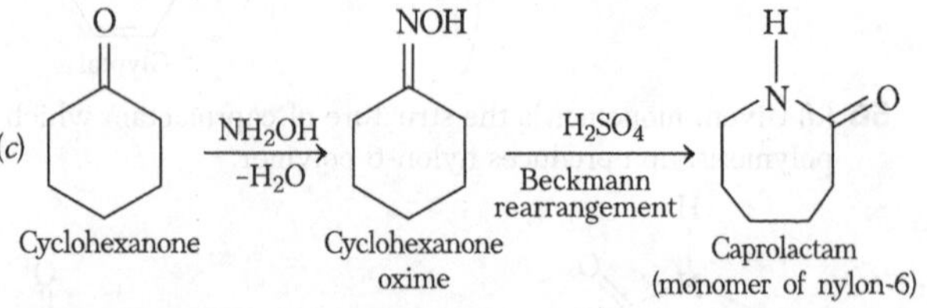

$$\text{Cyclohexanone} \xrightarrow[-H_2O]{NH_2OH} \text{Cyclohexanone oxime} \xrightarrow[\text{Beckmann rearrangement}]{H_2SO_4} \text{Caprolactam (monomer of nylon-6)}$$

20 (*a*) Both statements are correct.

Polyamides are best used as fibres because of high tensile strength. This is due to strong intermolecular forces (like hydrogen bonding within polyamides) lead to close packing of chains and increase the crystalline character. Hence, provide high tensile strength to polymers.

(C) Medical entrances' gallery

1 (*d*) Natural polymer (soft) is an addition homopolymer of isoprene which is a conjugated diene.

Isoprene shows 1, 4-addition with themselves to give *cis*-1, 4-polyisoprene or natural rubber.

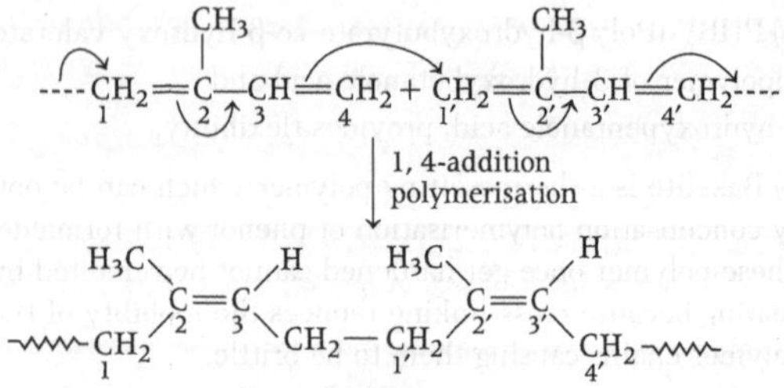

$$\cdots CH_2{=}\underset{}{\overset{CH_3}{C}}-CH{=}CH_2 + CH_2{=}\overset{CH_3}{C}-CH{=}CH_2\cdots \xrightarrow{\text{1, 4-addition polymerisation}} \sim CH_2-\overset{H_3C}{C}{=}\overset{H}{C}-CH_2-CH_2-\overset{H_3C}{C}{=}\overset{H}{C}-CH_2\sim$$

Natural rubber

All polymers mentioned in options (a), (b) and (c) are synthetic 1,4-addition polymers. Where (a) and (c) are buna-S and buna-N respectively.

2 (*a*) Bakelite is a cross linked condensation thermosetting copolymer of phenol and formaldehyde.

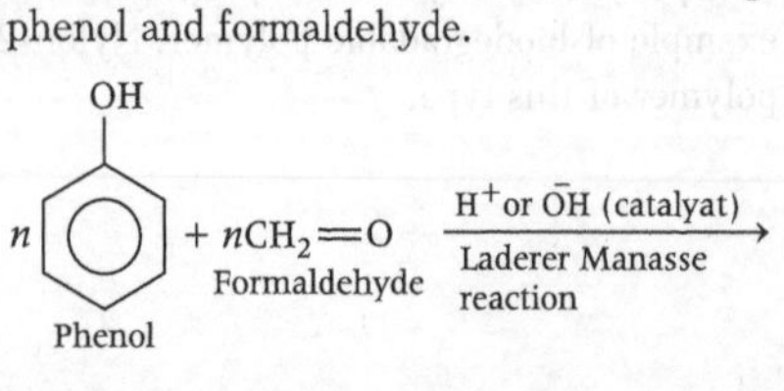

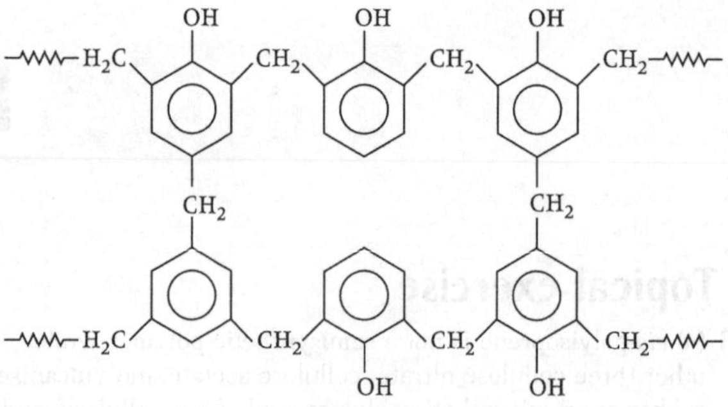

$$n\,C_6H_5OH \;(\text{Phenol}) + nCH_2{=}O \;(\text{Formaldehyde}) \xrightarrow[\text{Laderer Manasse reaction}]{H^+ \text{ or } \bar{O}H \text{ (catalyat)}}$$

Cross linked polymer (bakelite)

3 (*a*) Nylon-2-nylon-6 $\left(HN-CH_2CONH(CH_2)_5\overset{O}{\overset{\|}{C}}\right)_n$ is a biodegradable polymer. It is an alternating polyamide copolymer of glycine (H_2N-CH_2-COOH) and amino caproic acid $[H_2N(CH_2)_5COOH]$.

The remaining polymers, i.e. nylon-6,6, $\left(NH(CH_2)_6NHCO(CH_2)_4CO\right)_n$, nylon-6- $\left(CO(CH_2)_5NH\right)_n$ and buna-S $\left(CH_2-CH{=}CH-CH_2-\underset{}{\overset{C_6H_5}{CH}}\right)_n$ are non-biodegradable polymers. Hence, option (a) is correct.

4 (*c*) Polyacrylonitrile or orlon or acrilan, $\left(CH_2-\overset{CN}{CH}\right)_n$ is an addition homopolymer of monomer $CH_2{=}CH\,CN$ (vinyl cyanide). It is used in making synthetic fibres and synthetic wool. Thus, it is a substitute for wool in making commercial fibres.

5 (*a*) (i) Buna - S → (P) styrene, it is one of the monomer of buna - S, i.e. $[C_6H_5-CH{=}CH_2]$

(ii) Ethylene glycol → (Q).

Terylene is made by the monomers, ethylene glycol $\left[\underset{OH}{CH_2}-\underset{OH}{CH_2}\right]$ and terephthalic acid

(iii) Elastomer → (R), chloroprene is an example of elastomer.

Hence, the correct match is: (i) - P, (ii) - Q, (iii) - R.

6 (*d*) Cross-linked or network polymers are formed from bi-functional and tri-functional monomers and contain strong covalent bonds between various linear polymer chains. These are hard, rigid and brittle due to cross-links, e.g. bakelite, melamine etc. Thus, option (d) is incorrect.

7 (*b*) The increasing order of intermolecular forces in different polymers is elastomers < plastics < fibres. So, in order to solve such problem, first classify the given polymer into the above classes and then arrange them in increasing order of forces. Neoprene (synthetic rubber) is an elastomer, buna-S is also an elastomer, polythene is plastic and nylon-6, 6 is a fibre. So, the increasing order of intermolecular forces of the given polymer is : Neoprene < buna-S < polythene < nylon-6, 6

8 (*c*) Assertion is correct but reason is incorrect. Most of the synthetic polymers are not biodegradable because they are not easily broken by soil organisms (micro-organism) and are hazardous to the environment. Polymerisation does not always causes toxic effect in the environment.

9 (*a*) Polymers are obtained by condensation reaction between monomers. This reaction involves the loss of simple molecules such as water, alcohol etc.

10 (*b*) Polyethylene $+\!\!(CH_2—CH_2)\!\!+_n$, PVC

$+\!\!(CH_2—\underset{\displaystyle Cl}{\underset{|}{CH}})\!\!+_n$ and teflon $+\!\!(CF_2—CF_2)\!\!+_n$ are addition

polymers because all these contain C=C unit in their respective monomers and are formed by the repeated addition of monomer molecules.

11 (*c*) Nylon is a polyamide, e.g. nylon-6,6, nylon-6, etc. The structures are as follows :

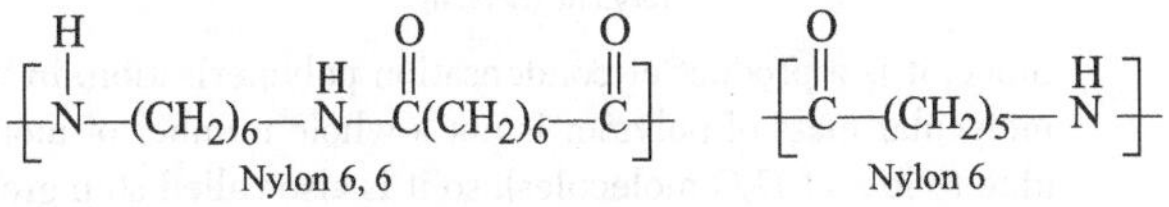

$$\left[\overset{H}{\overset{|}{N}}—(CH_2)_6—\overset{H}{\overset{|}{N}}—\overset{O}{\overset{\|}{C}}(CH_2)_6—\overset{O}{\overset{\|}{C}}\right] \qquad \left[\overset{O}{\overset{\|}{C}}—(CH_2)_5—\overset{H}{\overset{|}{N}}\right]$$

Nylon 6, 6 Nylon 6

12 (*d*) Nylon-6, 6 polymer is formed as

$$\underset{\text{Adipic acid}}{HOOC—(CH_2)_4—COOH} + \underset{\text{Hexamethylene diamine}}{H_2N—(CH_2)_6—NH_2}$$

$$\Big\downarrow \text{Polymerisation}$$

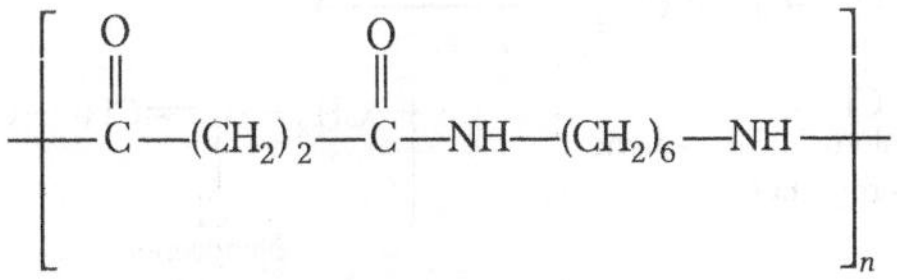

$$\left[\overset{O}{\overset{\|}{C}}—(CH_2)_2—\overset{O}{\overset{\|}{C}}—NH—(CH_2)_6—NH\right]_n$$

Thus, option (d) is correct.

13 (*d*) The repeating unit in natural rubber has the *cis*-configurations with chain extensions on the same side of the ethylene double bond, which is essential for elasticity. If the configuration is *trans*, the polymer is either a hard plastic or a substance like *Gutta-percha.*

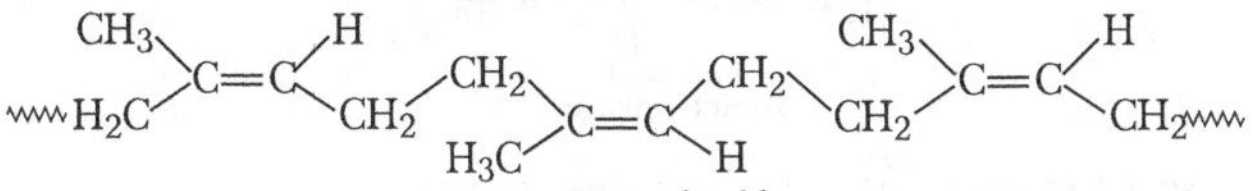

$$\text{~~}H_2C\!\!\overset{CH_3}{\diagdown}\!C{=}C\!\overset{H}{\diagup}\!\!\underset{CH_2}{\diagdown}\!\overset{CH_2}{\diagup}\!\!\diagdown C{=}C \diagup\!\!\overset{CH_2}{}\!\!\diagdown\!\underset{CH_2}{}\!\!\overset{CH_3}{\diagdown}C{=}C\overset{H}{\diagup}\!\!\diagdown CH_2\text{~~}$$

(with H_3C and H on the central C=C)

Natural rubber

All *cis*-configuration

14 (*b*) $+\!\!(CF_2—CF_2)\!\!+_n$ polymer is used in manufacturing of squeeze bottle.

15 (*a*) **Nylon-2-nylon-6** It is an alternating polyamide of glycine (containing two carbon atoms) and amino caproic acid or 6-aminohexanoic acid (containing six carbon atoms).

$$\underset{\text{Glycine}}{nH_2N—CH_2—COOH} + \underset{\text{Amino caproic acid}}{nH_2N—(CH_2)_5—COOH} \longrightarrow$$

$$\left[NH—CH_2—\underset{\underset{\displaystyle O}{\|}}{C}—NH—(CH_2)_5—\underset{\underset{\displaystyle O}{\|}}{C}\right]_n$$

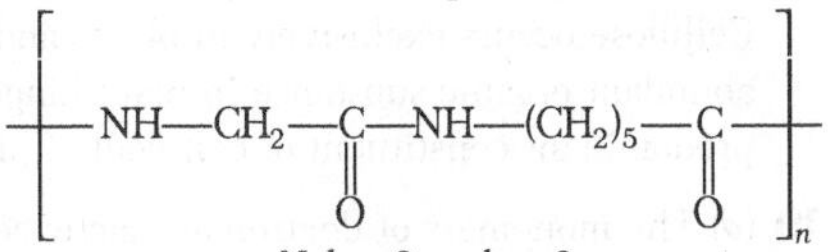

Nylon-2- nylon-6

It is a biodegradable step-growth copolymer.

16 (*d*) (*a*) $+\!\!(CH_2—\underset{\displaystyle Cl}{\underset{|}{C}}=CH—CH_2)\!\!+_n$ is neoprene

(synthetic rubber) which is a thermoplastic polymer.

(b) $+\!\!(CH_2—\underset{\displaystyle Cl}{\underset{|}{CH}})\!\!+_n$ is polyvinyl chloride which is also a

thermoplastic polymer.

(c) 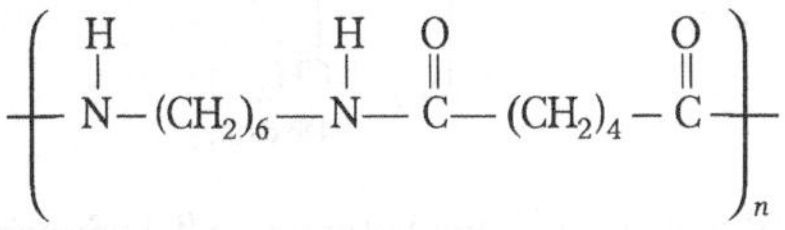 $\left(\overset{H}{\overset{|}{N}}-(CH_2)_6—\overset{H}{\overset{|}{N}}—\overset{O}{\overset{\|}{C}}—(CH_2)_4-\overset{O}{\overset{\|}{C}}\right)_n$ is

nylon 6, 6 a polyamide commonly known as fibre.

(d) 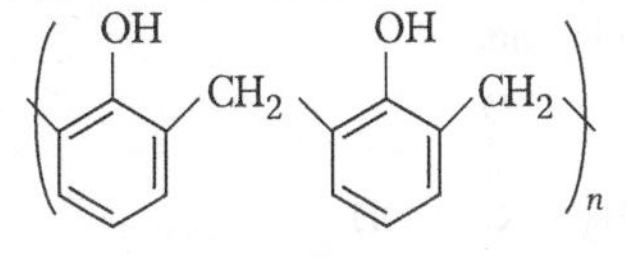 (phenol–CH_2–phenol–CH_2 chain, each ring bearing OH)$_n$

is a linear polymer called novolac. It is a thermosetting polymer formed by the condensation reaction of phenol and formaldehyde.

17 (*c*) Dacron, commonly known as terylene, is obtained by heating mixture of terephthalic acid and ethylene glycol at 420-460 K in the presence of catalyst like zinc acetate and antimony trioxide.

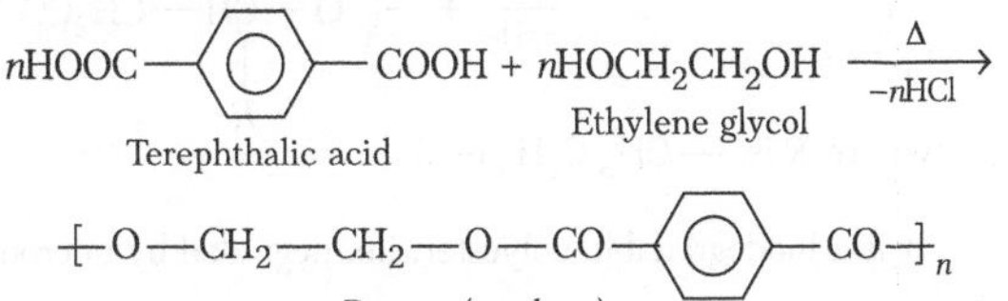

$$\underset{\text{Terephthalic acid}}{nHOOC—C_6H_4—COOH} + \underset{\text{Ethylene glycol}}{nHOCH_2CH_2OH} \xrightarrow[-nHCl]{\Delta}$$

$$\underset{\text{Dacron (terylene)}}{+\!\!O—CH_2—CH_2—O—CO—C_6H_4—CO\!\!+_n}$$

18 (*d*) Glycogen is a multibranched polysaccharide of glucose that serves as a form of energy storage in animals. Glycogen is the analogue of starch, a glucose polymer in plants, having a similar structure of amylopectin, but more extensively branched and compact than starch. Starch is polymer of α-glucose and consist of two components amylose and amylopectin. Amylose is water soluble component which constitutes about 15-20% of starch. Amylose and amylopectin is insoluble in water and constitutes about 80-85% of starch.

Amylose and amylopectin are the main storage polysaccharide of plants.

Cellulose occurs exclusively in plants and it is the most abundant organic substance in plant kingdom. It is a predominant constituent of cell wall of plants cells.

19 (*d*) The monomers of dextron are lactic acid and glycolic acid as it is a polyacetic and polyglycolic acid polymer. It is copolymer and has ester linkage. It is used in lubricants to reduce surface friction.

$$\underset{\text{Lactic acid}}{nCH_3-CH(OH)-COOH} + \underset{\text{Glycolic acid}}{nCH_2(OH)-COOH} \xrightarrow[-nH_2O]{} \underset{\text{Dextron}}{\left[-O-CH(CH_3)-\overset{O}{\overset{\|}{C}}-O-CH_2-\overset{O}{\overset{\|}{C}}-O\right]_n}$$

Ester linkage

20 (*a*) PHBV (poly β -hydroxy butyrate-co-β-hydroxyvalerate) a polyester, is a copolymer of 3-hydroxybutanoic acid and 3-hydroxy pentanoic acid.

$$\underset{\beta\text{-hydroxy butanoic acid}}{nCH_3-CH(OH)-CH_2-\overset{O}{\overset{\|}{C}}-OH} + \underset{\beta\text{-hydroxy pentanoic acid}}{nCH_3-CH_2-CH(OH)-CH_2-\overset{O}{\overset{\|}{C}}-OH}$$

$$\xrightarrow[-nH_2O]{} +O-\underset{R}{\underset{|}{CH}}-CH_2COO+_n$$

(where *R* is $-CH_3, C_2H_5$ etc.)

It is a biodegradable polyester, i.e. degraded by microorganisms.

21 (*d*) In condensation polymerisation, molecules of same or different monomers are combined in a controlled stepwise manner with elimination of small molecules such as water, ammonia, alcohols, etc.

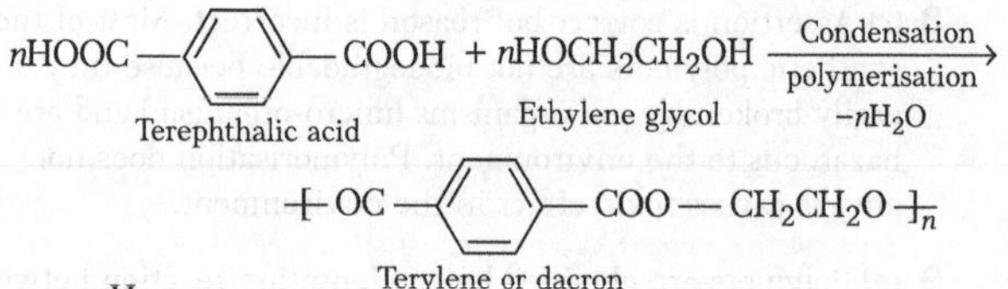

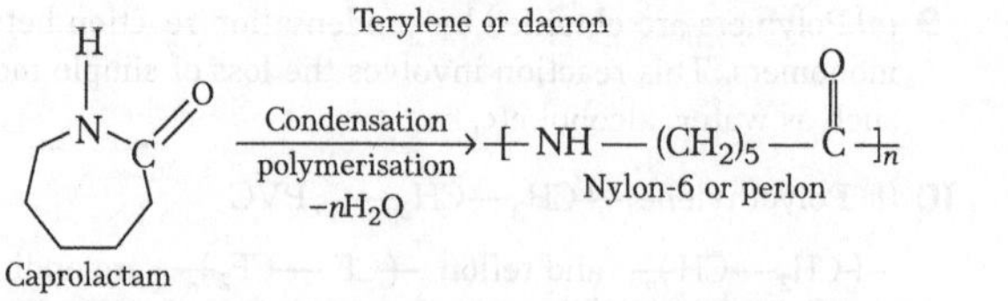

$$\underset{\text{Terephthalic acid}}{nHOOC-C_6H_4-COOH} + \underset{\text{Ethylene glycol}}{nHOCH_2CH_2OH} \xrightarrow[-nH_2O]{\text{Condensation polymerisation}} \underset{\text{Terylene or dacron}}{+OC-C_6H_4-COO-CH_2CH_2O+_n}$$

$$\underset{\text{Caprolactam}}{\text{(cyclic) } -NH-C(=O)-} \xrightarrow[-nH_2O]{\text{Condensation polymerisation}} \underset{\text{Nylon-6 or perlon}}{+NH-(CH_2)_5-\overset{O}{\overset{\|}{C}}+_n}$$

22 (*c*) Terylene (or dacron) is a copolymer of terephthalic acid and ethylene glycol. It is a polyester fibre.

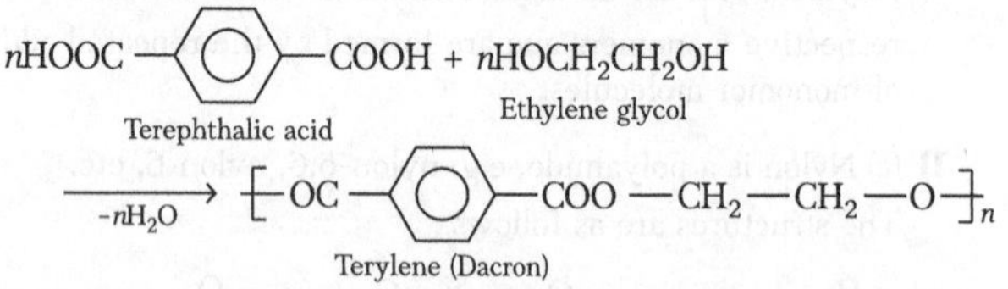

$$\underset{\text{Terephthalic acid}}{nHOOC-C_6H_4-COOH} + \underset{\text{Ethylene glycol}}{nHOCH_2CH_2OH} \xrightarrow[-nH_2O]{} \underset{\text{Terylene (Dacron)}}{\left[OC-C_6H_4-COO-CH_2-CH_2-O\right]_n}$$

Since, it is a product of condensation polymerisation, in which molecular mass of polymer is not a whole number of monomer (due to loss of H_2O molecules), so it is also called step growth polymer.

Note Unsaturated monomers mainly undergo addition polymerisation.

23 (*c*) Neoprene is synthetic rubber and is a homopolymer of chloroprene which is chemically 2-chlorobuta-1, 3-diene. It is obtained by free radical polymerisation of chloroprene.

$$\underset{\substack{\text{2-chloro-1, 3-butadiene}\\ \text{(Chloroprene)}}}{nCH_2=\underset{Cl}{\underset{|}{C}}-CH=CH_2} \xrightarrow{\text{Polymerisaton}} \underset{\text{Neoprene}}{\left[CH_2-\underset{Cl}{\underset{|}{C}}=CH-CH_2\right]_n}$$

24 (*c*) Nylon is a polyamide containing amide linkages $+CONH+$. These are prepared by condensation copolymerisation of diamines with dicarboxylic acids or by condensation polymerisation of lactam.

$$\left[R-\overset{O}{\overset{\|}{C}}-\underset{H}{\underset{|}{N}}-R\right]_n$$

Amide linkage

25 (*a*) Natural rubber is a polymer of *cis*-isoprene, [2-methyl-1, 3-butadiene].

$$CH_2=\overset{CH_3}{\overset{|}{C}}-CH=CH_2$$

26 (*d*) Acrilan is a cyanide compound. It has a structure

$$\left[-CH_2-\underset{\large CN}{\underset{|}{CH}}- \right]_n$$

27 (*d*) The polymer formed by the condensation polymerisation is known as condensation polymer. Dacron (terylene) is formed by the condensation polymerisation of terephthalic acid and ethylene glycol.

28 (*d*) Condensation polymers are obtained by bi-functional molecules (monomers) with the elimination of smaller molecules whereas additional polymers are obtained from multiple bond containing monomers. Neoprene is a polymer of chloroprene ($CH_2{=}C(Cl){-}CH{=}CH_2$), so it is an addition polymer, not a condensation polymer.

29 (*b*) Biodegradable polymers have a tendency to break down and lose their initial integrity.

e.g. $nH_2N-CH_2-COOH + nHNH-(CH_2)_5-COOH \xrightarrow[-nH_2O]{\Delta}$

$$\underset{\text{Nylon–2–nylon–6}}{+HN-CH_2-CO-NH-(CH_2)_5-CO+_n}$$

30 (*b*) Carbowax is a polymer of ethylene oxide whereas carnauba wax and bees wax are esters of higher fatty acids with higher alcohols, and paraffin wax is a mixture of hydrocarbons containing 20 or more C-atoms.

31 (*c*) The poly-β-hydroxy butyrate-co-β -hydroxy valerate (PHBV) is a biodegradable polymer.

32 (*d*) Neoprene is an addition polymer of chloroprene. It gives artificial rubber when heated with ZnO or MgO.

$$\underset{\text{Chloroprene}}{CH_2{=}H{-}\overset{\overset{\large Cl}{|}}{C}{=}CH_2} \xrightarrow{\text{Polymerisation}} \underset{\text{Neoprene}}{+CH_2CH{=}\overset{\overset{\large Cl}{|}}{C}{-}CH_2+_n}$$

33 (*b*) Terylene (dacron) is a polyester polymer because it is formed by the condensation polymerisation of the monomer units terephthalic acid (an acid) and ethylene glycol (an alcohol).

34 (*c*) Bakelite is an example of condensation polymer. It is a obtained by the condensation reaction of phenol with formaldehyde in the presence of either an acid or a base catalyst.

35 (*b*) Natural rubber becomes soft at high temperatures (> 335 K) and brittle at low temperatures (< 283 K). Hence, it is not used in making footwears for polar regions.

CHAPTER 24

Chemistry in Everyday Life

Chemistry play an important role in our life. Their are many chemicals that can cure our sickness and disease. Daily use materials such as soaps, detergents, tooth pastes etc., have made our life very comfortable. That means, all our daily activities are controlled by chemicals.

In this chapter, we will going to study the importance of chemicals and their effects when they are taken in excess.

DRUGS

Drugs are chemicals of low molecular masses (~ 100 to 500 u).

There is interaction of these drugs with macromolecular targets to produce a biological response.

(*i*) When the biological response is therapeutic and useful, these chemicals are called **medicines** and are used in diagnosis, prevention and treatment of diseases.

(*ii*) If taken in doses higher than those recommended, most of the drugs can act as potential poisons.

(*iii*) Use of chemicals for therapeutic effect is called **chemotherapy.**

Classification of drugs

Drugs can be classified on the basis of following criteria

(*i*) **Pharmacological effect** It provides the whole range of drugs available for the treatment of a particular type of problem. e.g. analgesics—pain killing effect, antiseptic—kill or arrest growth of microorganisms.

(*ii*) **Drug action** It explains the action of a drug on a particular biochemical process.

(*iii*) **Chemical structure** Drugs classified in this manner show common structural features and often have similar pharmacological activity. For exampe, all the sulphonamides have common structural feature shown as

$$H_2N-C_6H_4-SO_2-NH-R$$

($H_2N-C_6H_4-SO_2-$: Common; R: Varies)

Inside

(*iv*) **Molecular targets** Drug usually interact with biomolecules such as carbohydrates, lipids, proteins and nucleic acids. These are called **target molecules** or **drug targets.** Drugs with some common structural features may have the same mechanism of action on targets.

Mechanism of drug action

Various functions of the body are performed by the macromolecules of biological origin. Drugs interact with enzymes which are proteins and perform the role of biological catalyst in the body. This interaction leads to formation of chemical compounds that help to cure the ailing body.

Sometimes drugs directly interact with receptors and message them to produce certain chemical compounds necessary to heal body. These receptors are embedded protein molecules that receive drug molecules and perform crucial role in communication system.

Another important class of macromolecules is the carrier proteins, that carry polar molecules through the cell membrane. In this section, we will study enzyme and receptors as drug targets to explain the drug-target interaction.

Enzymes as drug targets

Enzymes perform various activity due to its catalytic action. In their catalytic activity, enzymes perform two major functions as discussed below

(*i*) The enzyme holds the substrate in its active site in such a way that it can be effectively and readily attacked by the reagent. Substrate binds to active site of the enzyme through ionic bonding or hydrogen bonding or van der Waals' interaction or dipole-dipole interactions.

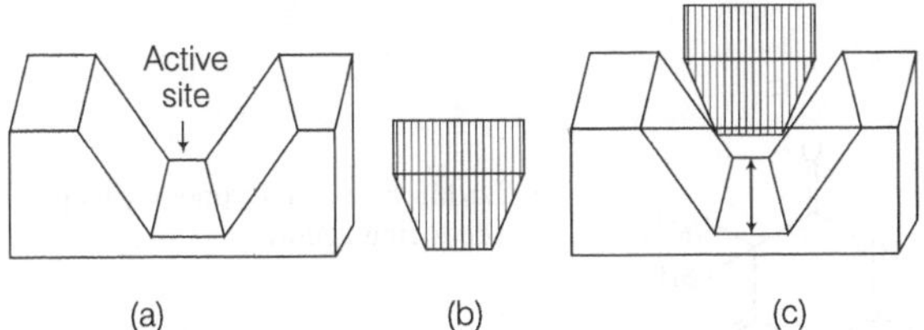

Fig. 24.1 (a) Active site of an enzyme; (b) substrate; (c) substrate held in active site of the enzyme

(*ii*) The enzymes provide the functional groups that will attack the substrate and carry out chemical reaction.

Drug-enzyme interaction

Drugs which inhibits any of the two activities of enzyme discussed above are called **enzyme inhibitors**.

Drugs inhibit the attachment of natural substrate on active site of enzymes in two different ways.

(*i*) Drugs which bind to active site of enzyme and thus, compete with natural substrate for their attachment on the active site are called **competitive inhibitors**.

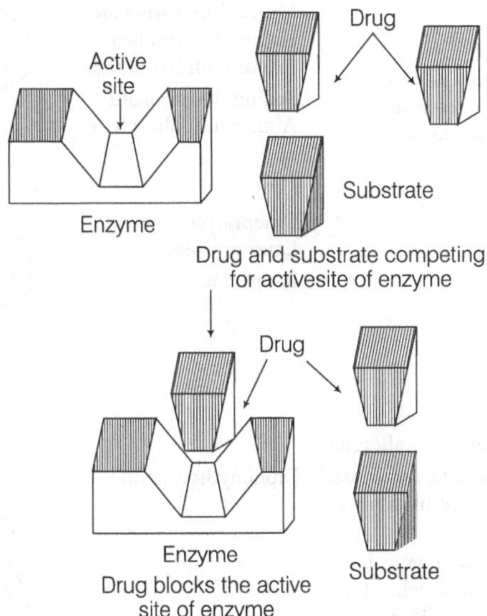

Fig. 24.2 Drug and substrate competing for active site

(*ii*) Some drugs bind to a different site (other than the active site) of enzyme called as **allosteric site**. This binding changes the shape of the active site in such a way that substrate cannot recognise it. Hence activity of enzyme is inhibited. These drugs are called **non-competitive inhibitors.**

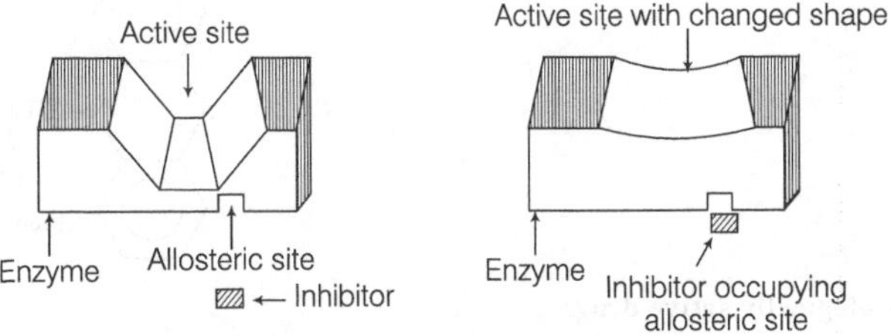

Fig. 24.3 Non-competitive inhibitor changes the active site of enzyme after binding at allosteric site

Receptors as drug targets

Receptors are the proteins which are embedded in cell membrane in such a way that their small part possessing active site project out of the surface of the membrane and opens on the outside region of the cell membrane. These are crucial to body's communication process.

Note ***Chemical messengers*** *are the chemicals received at binding sites of receptor proteins. It fits into the binding site of the receptor protein, passes on its message and then leaves without undergoing any chemical reaction. Thus, these are the chemicals in body through which message between two neurons and that between neurons to muscles is communicated.*

Different classes of drugs, their examples and therapeutic actions

The therapeutic action of a few important classes of drugs are as follows

Classes	Examples	Structures	Functions and uses
Antacids These drugs are used to neutralise the excess acid produced in the stomach.	Magnesium hydroxide Magnesium carbonate Magnesium trisilicate Aluminium hydroxide Sodium bicarbonate Aluminium phosphate	Omeprazole	Remove the excess acid and raise the pH to appropriate level in the stomach. Acid gastritis (caused by excess of HCl in gastric juice) is one of the common ailment associated with digestion that leads to ulcer.
	Omeprazole Lansoprazole Ranitidine	Lansoprazole	**Note** Antacids should not be taken with other drugs particularly with antibiotics as they slow down their absorption.
Antihistamines (anti-allergic) • Allergic reactions are caused by generation of histamine in the body. • Histamine is a chemical released by mast cells of a body during an allergic reaction. • The drugs that interfere with the natural action of histamine by competing with histamine for binding sites of receptor are called antihistamines.	Diphenylhydramine Chloropheniramine (also called chlorophenamine) Promethazine	$CH—O—(CH_2)_2N(CH_3)_2$ Diphenylhydramine $CH—(CH_2)_2N(CH_3)_2$ Chloropheniramine $N—CH_2CHN(CH_3)_2$, CH_3 Promethazine	These are helpful in treating fever i.e., seasonal allergic, nasal allergies like sneezing, itching of eyes, nose, throat etc.
Neurologically active drugs These drugs affect the message transfer mechanism from nerve to receptor. Tranquilizers and analgesics are some examples. (a) **Tranquilisers** These are the chemical compounds that are used for the treatment of strain or stress for mild and severe mental diseases. It is a essential component of sleeping pills and a part of psychotherapeutic drugs.	Barbiturates (Derivatives of barbituric acid *viz.* veronal, amytal, nembutal, luminal and seconal constitutes barbiturates) Mild tranquilizers are chlordiazepoxide and meprobamate.	Veronal (C_2H_5, C_2H_5) Phenobarbital (C_6H_5, C_2H_5)	Barbiturates are hypnotic (sleep producing agent)

Classes	Examples	Structures	Functions and uses
	Equanil	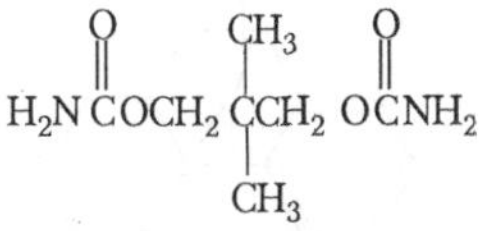 $H_2N\overset{O}{\overset{\Vert}{C}}OCH_2\underset{CH_3}{\overset{CH_3}{C}}CH_2O\overset{O}{\overset{\Vert}{C}}NH_2$ Equanil	Equanil is used in controling depression and hypertension.
	Valium	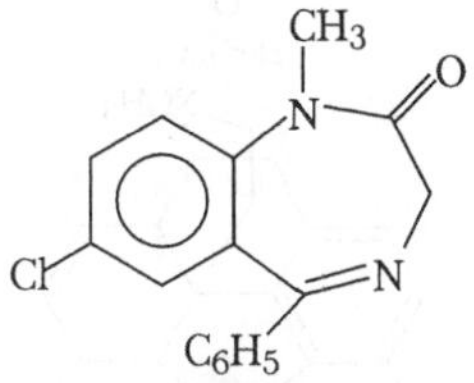 Valium	Noradrenaline (one of the neurotransmitters) plays a major role in mood swings. When the level of this neurotransmitter is low then the message transfer process become slow and the person suffers from depression. In such cases these drugs (also called mood booster and antidepressant drugs) are used.
(b) **Analgesics** (pain relievers) Reduce or abolish pain without causing impairment to consciousness, mental confusion, in coordination or paralysis or some other disturbances of nervous system. These are of two types.		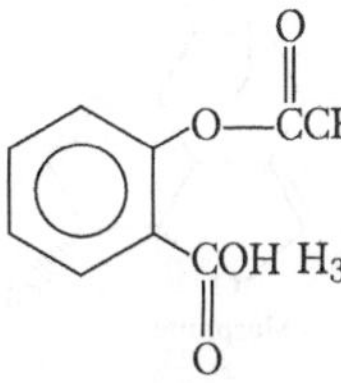 Acetyl salicylic acid or 2-acetoxy benzoic acid (aspirin) 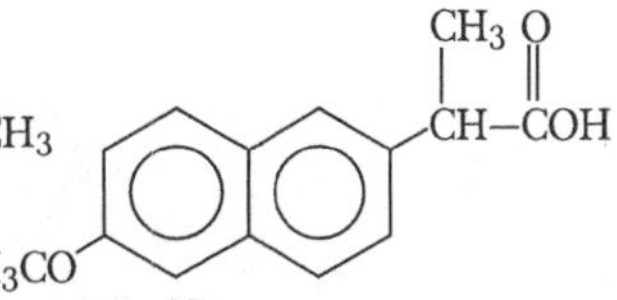 Naproxen	These drugs are antipyretic (temperature lowering properties). Asprin inhibits the synthesis of prostagladins (chemical that stimulate inflamamation in the tissues and cause pain. It is used in the prevention of heart-attacks as it also has anti-blood clotting action, used to check pregnancy related complications, viral inflammation in **AIDS** patients, Alzheimer's disease and cancer. However, it is toxic to liver and causes bleeding from stomach wall and is a gastric irritant. Because of these problems, other analgesics like diclofenac sodium or potassium, ibuprofen and naproxen use as an alternatives.
(i) **Non-narcotic** (non-addictive) These drugs are devoid of any opiate derivative and hence do not produce any hallucinating effect.	**Examples of non-narcotics** Aspirin, naproxen, lbuprofen, dichlorofenac sodium, novalgin, paracetamol etc.	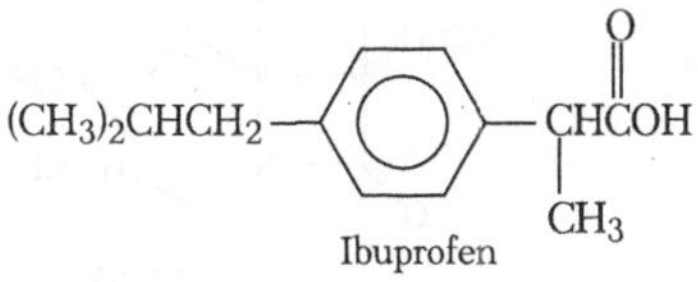 $(CH_3)_2CHCH_2-C_6H_4-\underset{CH_3}{C}H\overset{O}{\overset{\Vert}{C}}OH$ Ibuprofen 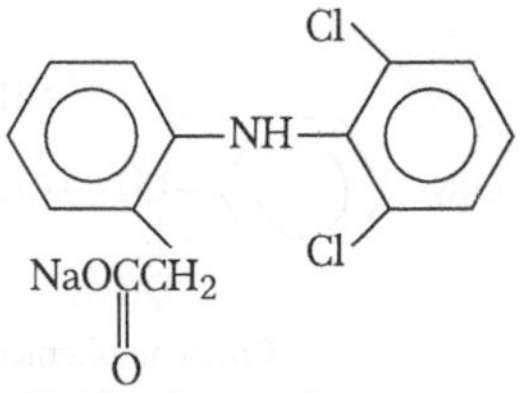 Dichlorofenac sodium	Asprin inhibits the synthesis of prostagladius chemical that stimulate inflammation in the tissue and cause pain.

Classes	Examples	Structures	Functions and uses
(ii) **Narcotics** (addictive) These are used for the relief for postoperative pain, cardiac pain and pains of terminal cancer and in child birth.	Examples of narcotics morphine, codeine, heroine (acetylation of morphine with acetic anhydride gives heroin)	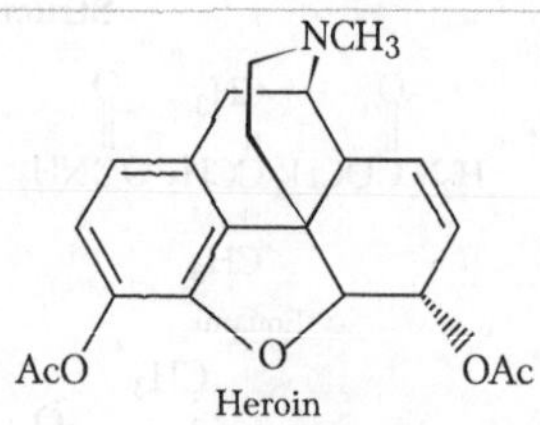 Heroin; 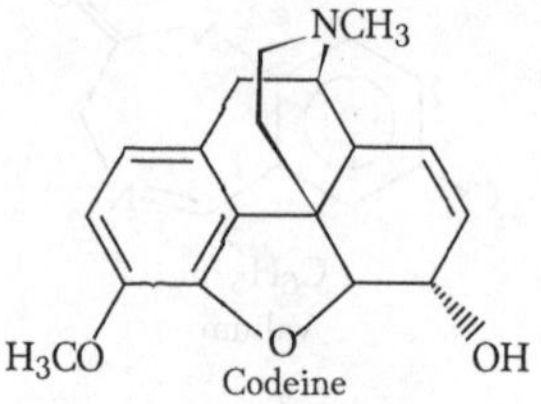 Codeine; 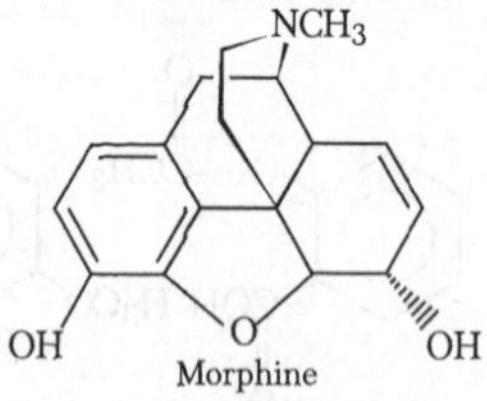 Morphine	They produce sleep and unconciousness. In large doses, they produce stupor, coma, convulsions and ultimately death.
Anti-microbials These drugs prevent the development or inhibit the pathogenic action of microbes such as bacteria (antibacterial drugs) fungi (antifungal drugs), virus (antivirus agents), or other parasites (antiparasite drugs) selectively. Antibiotics, antiseptic and disinfectants are antimicrobial drugs.			Following are the three ways by which the control of microbial diseases can be achieved. • Bactericidal (a drug which kills the organism in the body) • Bacteriostatic (drug which inhibits or arrests the growth of the organism) • By increasing the immunity.
(i) **Antibiotics** It is a chemical substance produced wholly or partly by chemical synthesis, which in low concentrations inhibits the growth or destroys microorganisms by intervening in their metabolic processes.	(a) Bactericidal (killing affect) e.g. penicillin, amino glycosides, ofloxacin.	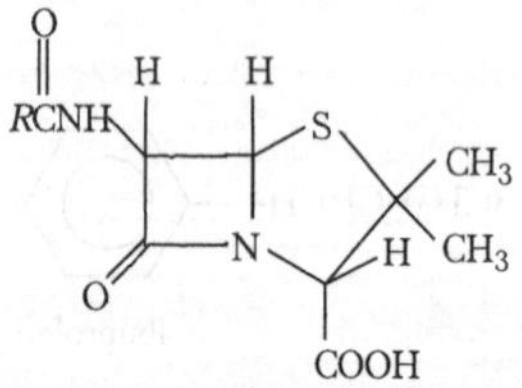 Penicillin	Penicillin is a narrow spectrum antibiotics i.e., these are effective mainly against specific families of bacteria [gram-positive or gram-negative specific families of bacteria]. While amino glycosides and ofloxacin are broad spectrum antibiotic i.e., they kill or inhibit a wide range of gram-positive and gram-negative bacteria.
	(b) Bacteriostatic (inhibiting effect) e.g. erythromycin, tetracycline, chloramphenicol.	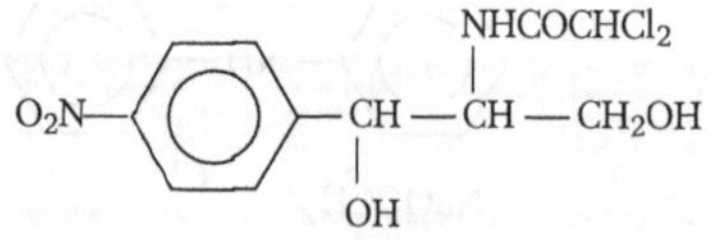 $O_2N-C_6H_4-CH(OH)-CH(NHCOCHCl_2)-CH_2OH$ Chloroamphenicol	Chloramphenicol—a broad spectrum antibiotic is rapidly absorbed from the gastro-intestinal tract, hence can be given orally in case of typhoid, dysentery, acute fever, urinary infections, meningitis and pneumonia.
Sulphadrugs These are another class of antibiotic drugs	Sulphanilamide Sulphadiazine	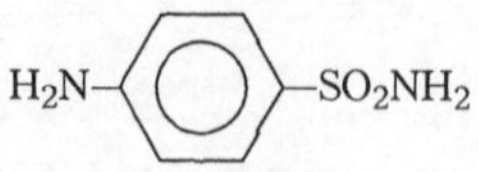 $H_2N-C_6H_4-SO_2NH_2$ Sulphanilamide	
	Sulphaguanidine	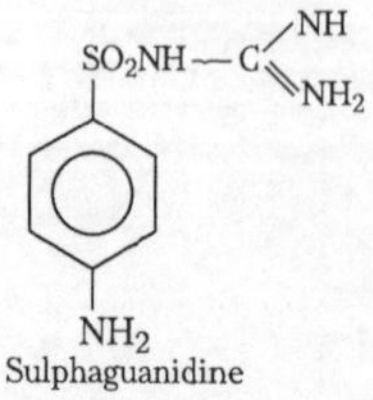 $H_2N-C_6H_4-SO_2NH-C(=NH)NH_2$ Sulphaguanidine	

Classes	Examples	Structures	Functions and uses
(ii) **Antiseptic and disinfectants** (which either kill or prevent the growth of microorganisms) Antiseptics are applied to living tissues like wounds etc, whereas disinfectants are applied to inanimate objects like floors etc.	Chlorine		• 0.2 to 0.4 ppm (parts per million) in aqueous solution of chlorine used as disinfectant for drinking water.
	Phenol		• 0.2 per cent solution of phenol is an antiseptic but 1 per cent solution is a disinfectant.
	Sulphur dioxide		
	Dettol	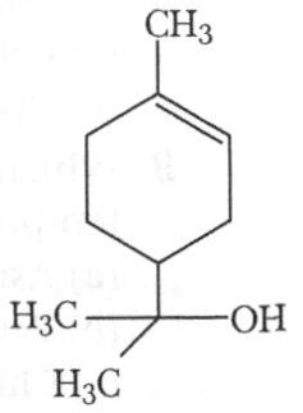 Terpineol; 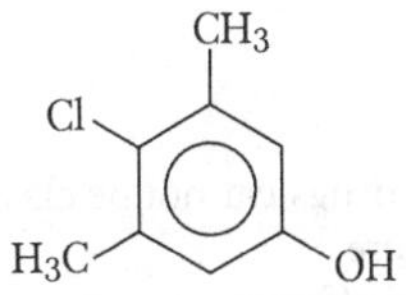 Chloroxylenol	• Sterilising and preservation of squashes. • Mixture of chloroxylenol and terpenol. • Imparts antiseptic property to bathing soap and eliminates odour as a result of bacterial decomposition of organic matter on the skin.
	Bithional	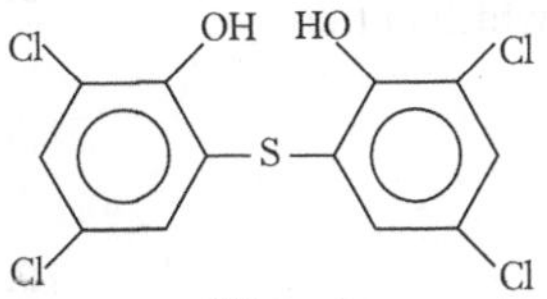 Bithional	
	Iodoform		• 2-3 per cent iodine solution in alcohol-water mixture called *tincture* of *iodine* is an antiseptic solution. It is used on wounds. (Iodoform too has been used for same purpose)
	Boric acid		• Boric acid as a dilute aqueous solution is used as a weak antiseptic for eyes and also in baby talcum powder.
	Hydrogen peroxide		• Non-irritating strong antiseptic.
Antifertility drugs These are the chemical substances which are used to control the pregnancy. They are also known as birth control pills or oral contraceptives.	Steroids (oral contraceptives)		Active ingredients of the pill functioning as antifertility agent, controls the female menstrual cycle ovulation
	Synthetic estrogen and progesterone derivatives e.g. norethindrone, ethynylestradiol	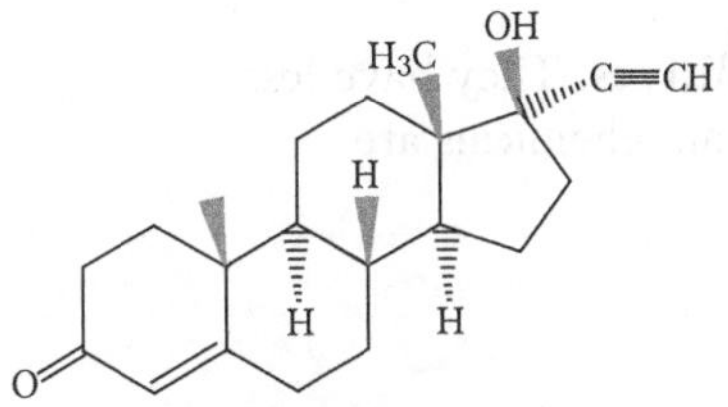 Norethindrone; 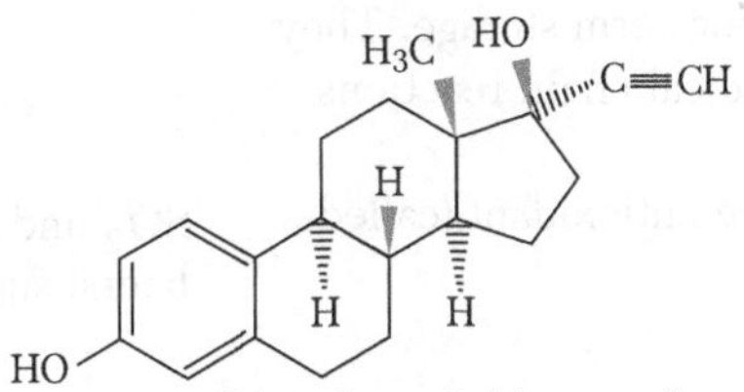 Ethynylestradiol (novestrol)	more potent than the natural hormones. Synthetic steroid that blocks the effect of progesterone and is used as a "Morning after pill" in many countries.

CHECK POINT 24.1

1. The range for the molecular masses of drugs is
   (a) ~ 100 to 200 u
   (b) ~ 100 to 150 u
   (c) ~ 100 to 500 u
   (d) ~ 100 to 600 u
2. Treating a body ailment with drugs is called
   (a) psychotherapy
   (b) chemotherapy
   (c) electrotherapy
   (d) physiotherapy
3. On which of the following basis, drugs cannot be classified?
   (a) On the basis of chemical structure
   (b) On the basis of pharmacological effect
   (c) On the basis of molecular target
   (d) On the basis of colour
4. Substrate can bind to the active site by which of the following force(s) of attraction?
   (a) Ionic bonding
   (b) Hydrogen bonding
   (c) van der Waals' force
   (d) All of the above
5. Which of the following is a common antacid?
   (a) NaOH
   (b) $Ca(OH)_2$
   (c) $Mg(OH)_2$
   (d) KOH
6. Which of the following terms means pain killing agent?
   (a) Antibiotic (b) Analgesic
   (c) Antipyretic (d) Penicillin
7. Pheneacetin is used as a/an
   (a) antipyretic
   (b) antiseptic
   (c) antimalarial
   (d) disinfectant
8. Which of the following medicines brings down the temperature of the body?
   (a) Aspirin
   (b) Penicillin
   (c) Chloroquin
   (d) Quinine
9. Morphine is
   (a) narcotic drug
   (b) non-narcotic drug
   (c) analgesics
   (d) antipyretic
10. ...... substance which can act both as an antiseptic as well as a disinfectant.
    (a) Aspirin
    (b) Chloroxylenol
    (c) Bithional
    (d) Phenol

CHEMICALS IN FOOD

For the preservation of foods and enhancing of their appeal, different chemical compounds like flavouring agents, sweeteners, dyes, antioxidants, emulsifiers, etc. are added.

These chemicals are called **food additives.** They have less nutritive values. Some of the important chemicals are discussed below.

Antioxidants

Antioxidants are added to fat containing foods to prevent autooxidation, spoilage and allow long-term storage. They function by interrupting the free radical chain reactions involved in the oxidation of fat.

Following are the important additive antioxidant (called **radical inhibitors**).

OH, $(CH_3)_3C$, $C(CH_3)_3$, CH_3

BHT (Butylated hydroxytoluene)

OH, $C(CH_3)_3$, OCH_3 — (a) *o*-isomer

OH, $C(CH_3)_3$, OCH_3 — (b) *m*-isomer

BHA (Butylated hydroxyanisole)

SO_2 and Na_2SO_3 are useful antioxidants for wine and beers, sugar-syrups, peeled fruits and vegetables.

Artificial sweeteners

The chemicals which are not sugars (carbohydrates) but give sweetening effect to the food and enhance its odour and flavour are called **artificial sweeteners.** The first widely used synthetic low-calorie sweetener was saccharin (commonly called a sweet '*n*' low). It has proved to be a life-saver for countless diabetics and is of great value to people who need to control intake of calories. It is about 550 times sweeter than sucrose, but leaves a bitter metallic taste afterwards for some users. Some other important artificial sweeteners are given in the Table 24.3.

Table 24.1 Artificial sweeteners

Artificial sweetener	Relative sweetness value (in comparison to sucrose)	Structure
Saccharin (insoluble in H_2O). So, it is available in market as sodium or calcium salt.	550	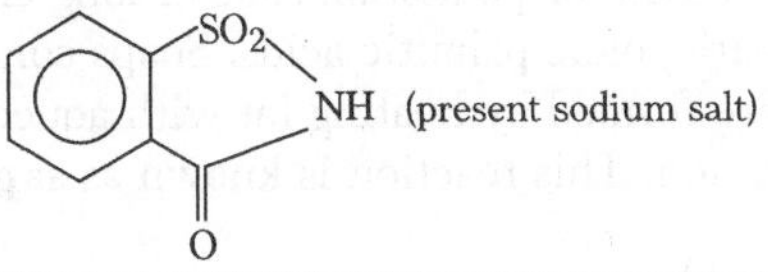 SO_2, NH (present sodium salt), O
Cyclamate (Causes tumour in animals)	7	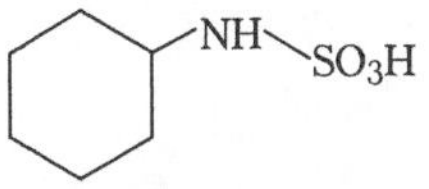 NH, SO_3H
Dulcin (Area sweetener)	25	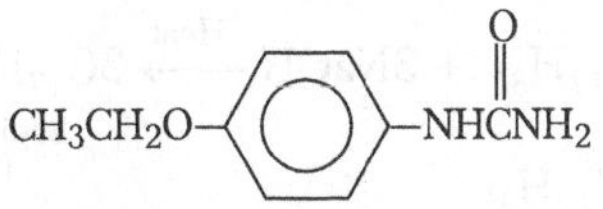 CH_3CH_2O–, –NHCNH$_2$, O
Aspartame (Nutra sweet) (Limited to cold food and soft drinks as it gets decomposed at high temperatures)	100	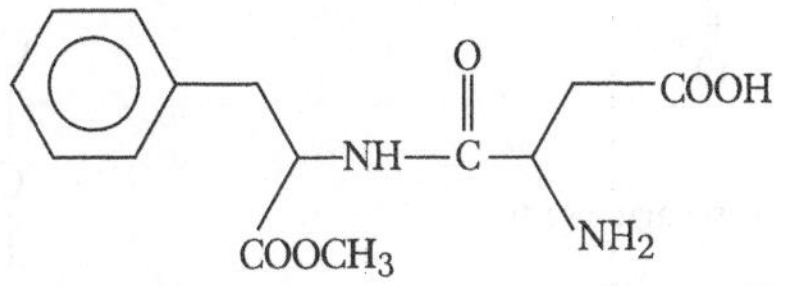 O, COOH, NH—C, $COOCH_3$, NH_2
Sucralose (Stable at high temperature, used for cooking and baking)	600	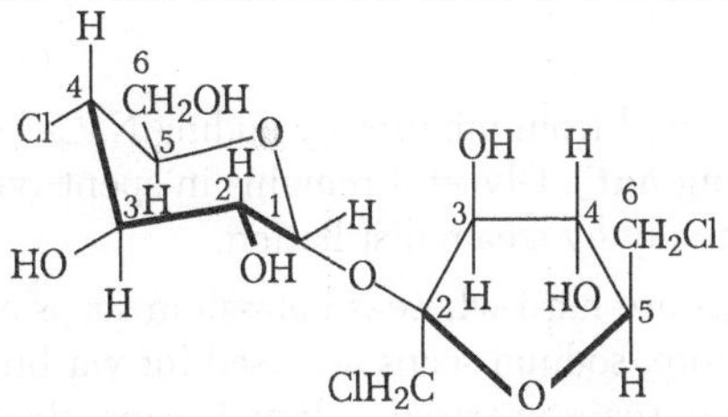 H, 6, 4, CH_2OH, Cl, 5, O, H, OH H, 3H, 2, 1, H, 3, 4, 6, CH_2Cl, HO, OH, O, H, 2 H, HO, 5, ClH_2C, O, H
Alitame (More stable than aspartame)	2000	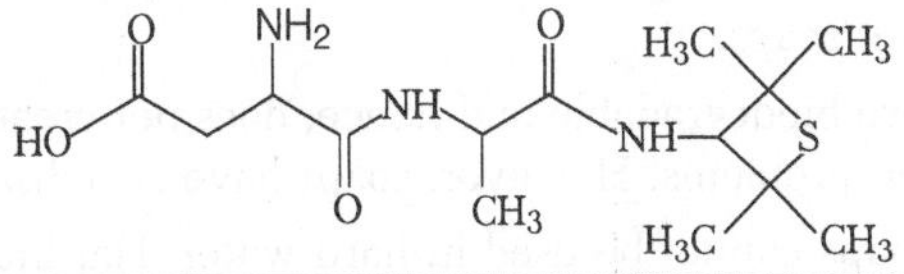 O, NH_2, O, H_3C, CH_3, NH, HO, NH, S, O, CH_3, H_3C, CH_3

Food preservatives

Chemical substances which are used to protect food against bacteria, yeasts and moulds are called **food preservatives.** The preservatives are added to improve the shelf life of processed and packed food materials.

Important examples of food preservatives are listed below.

(*i*) Vitamin E—naturally occurring food preservative found in vegetable oil.

(*ii*) BHA and BHT—synthetic radial inhibitors.

(*iii*) Sodium benzoate (metabolised to hippuric acid and excreted in urine).

(*iv*) Sodium salts of propionic acid and sorbic acid.

Note ***Edible colours*** *Few chemical additives are used to add desired colour to foods. e.g. Tetrazine (now banned), caramel, etc.*

CLEANSING AGENTS

These are the chemical substances that improve the cleansing property of water by reducing the surface tension of the solution. The help in removing the fat/oil by emulsifying them. These are also known as surfactants or surface active agents. Soaps and detergents are the examples of cleansing agents which are discussed below in detail.

Soaps

Soaps are the sodium or potassium salts of long chain fatty acids. e.g., stearic, oleic palmitic acids. Soaps containing sodium salts are formed by heating fat with aqueous sodium hydroxide solution. This reaction is known as **saponification**.

$$\begin{array}{l} CH_2-O-\overset{O}{\overset{\|}{C}}-C_{17}H_{35} \\ | \\ CH-O-\overset{O}{\overset{\|}{C}}-C_{17}H_{35} \\ | \\ CH_2-O-\underset{O}{\underset{\|}{C}}-C_{17}H_{35} \end{array} + 3NaOH \xrightarrow{\text{Heat}} \underset{\substack{\text{Sodium stearate} \\ \text{(Soap)}}}{3C_{17}H_{35}CO\overset{-}{O}\overset{+}{Na}} + \underset{\text{Glycerol}}{\begin{array}{c} CH_2OH \\ | \\ CHOH \\ | \\ CH_2OH \end{array}}$$

(Glyceryl ester of stearic acid)

Soap is separated from mixture by adding NaCl (process is called **'salting out'**). Glycerol remains in spent-lye and is recovered finally by steam distillation.

Sodium soaps are hard whereas potassium soaps are soft to skin. Therefore, sodium soaps are used for washing purposes and potassium soaps are used as liquid soaps, shaving creams and toilet soaps.

Soaps are biodegradable and hence, does not create any pollution problems. However, soaps have two disadvantages.

(*i*) Soaps cannot be used in hard water. Hardness in water is due to dissolved sulphates, chlorides and bicarbonates of calcium and magnesium. When such type of water is used for cleaning clothes with soap (sodium stearate), a curdy precipitate of calcium and magnesium salts are formed.

$$\underbrace{RCO\overset{-}{O}\overset{+}{N}a}_{\text{Soap}} + \underbrace{Ca^{2+} \text{ or } Mg^{2+}}_{\text{in hard water}} \longrightarrow \underset{\text{Curdy precipitate}}{(RCOO)_2 Ca \text{ or } Mg}$$

This precipitate adheres to clothing and blocks the ability of soaps to remove grease from fabrics.

(*ii*) Soaps cannot be used in acidic solution as acids precipitate the insoluble free fatty acids which adhere to the fabrics and reduce the ability of soap to remove oil from fabric.

Detergents

These are the cleaning agents which, unlike soaps, are effective even in hard water.

Synthetic detergents (also called syndets) are alkyl benzene sulphonates. Detergent comes from the Latin word *detergere*, which means "to wipe off".

$$R-C_6H_4-SO_3^- \overset{+}{Na}$$

These compounds are neutral in aqueous solution and do not hydrolyse to create irritating basic or acidic solutions. Also, their metal salts formed in hard water do not precipitate, but remain in solution.

Straight-chain alkyl groups are biodegradable, while branched-chain alkyl groups are not. Non-biodegradable detergents become pollutants in rivers and lakes. Cationic detergents are used widely as fabric softeners and germicides.

Phosphates and more recently silicates, are added as "builders" to form complex with the hard water cations. Other additives including fabric softeners, enzymes (to degrade protein stains), fragrances, antistatic agents and brightners are also mixed.

The hydrocarbon side chain stops bacteria from attacking and breaking the chains. This results in slow degradation of detergent molecules leading to their accumulation. Unbranched chains are more prone to attack by bacteria so the molecules are more easily biodegraded and thus, pollution is prevented.

Types of detergent

The detergents are classified into three types depending on the ionic charge present at the soluble end of their chain

Type	Property	Example
Anionic Thsese are called *anionic detergents* because the anionic part of the molecule is involved in the cleansing property. These are of two types sodium alkyl sulphates and sodium alkyl benzene sulphonates (biodegradability depends on chain of alkyl group)	Polar head of the detergent is negatively charged. Used in toothpastes, household detergent.	$CH_3(CH_2)_{11}-C_6H_4-SO_2-\overset{\ominus}{O}\overset{\oplus}{Na}$ Sodium *p*-*n*-dodecyl benzenesulphonate $CH_3(CH_2)_{11}-O-SO_2-\overset{\ominus}{O}\overset{\oplus}{Na}$ Sodium lauryl sulphate
Cationic The cationic part is involved in the cleansing property.	They are also called as invert detergent. The polar head of these detergent is positively charged. Used as germicides, in hair shampoos and conditioners.	$CH_3(CH_2)_{15}-\overset{\oplus}{N}(CH_3)_2-CH_3Cl^{\ominus}$ Cetyltrimethyl ammonium chloride $C_6H_5-CH_2-\overset{\oplus}{N}(CH_3)_2-(CH_2)_{15}CH_3\overset{\ominus}{Cl}$ Benzylcetyldimethyl ammonium chloride Octahedral ammonium chloride ($C_{18}H_{37}\overset{+}{N}H_3Cl^-$) Cetyl pyridinium chloride, $C_{16}H_{33}-C_5H_4\overset{+}{N}-Cl^-$
Non-ionic (these molcules do not undergo dissociation)	• They have groups at one end of the alkyl chain showing H-bonding, include monoesters of polyhydric alcohols or polyethers derived from ethylene oxide. • Used as liquid dish washing detergent.	$CH_3(CH_2)_{16}-\overset{O}{\overset{\parallel}{C}}-OCH_2-C(CH_2OH)_3$

Advantages of detergent

The advantages of detergent as compared to soap.

(*i*) Synthetic detergents can be used in hard water without any wastage while some of the soap gets wasted in hard water.

(*ii*) Unlike soaps, they can be used in acidic medium.

(*iii*) They are more soluble in water.

(*iv*) They decrease the surface tension of water and have stronger cleansing action.

Disadvantages of detergent

Detergents are highly branched hydrocarbon chains which causes water pollution. If the hydrocarbon chain is highly branched, then bacteria cannot degrade it easily and will leads to accumulation.

Cleansing action of soap and detergents

The cleansing action of soap is due to affinity of soap anions, e.g. stearate ion $C_{17}H_{35}COO^-$. The anionic carboxylic group $-COO^-$ has an affinity for water. It is therefore, hydrophilic and is called '*head*' of the anion. The hydrocarbon chain of fatty acid, ($C_{17}H_{35}$), has an affinity for grease, oil or dirt. It is therefore, hydrophobic and is called '*tail*' of the anion.

This tail part dissolves in the grease (or oil) deposit and thus micelle is formed, which in turn is removed by rinsing with excess of water. Thus, the main function of a soap is to reduce oily and greasy dirt to colloidal particles (an emulsion). Soaps, therefore, are known as emulsifying agents.

This mechanism of cleansing is also valid with any synthetic detergents (such as sodium lauryl sulphate, $CH_3(CH_2)_{11}SO_4^-Na^+$). Repulsion of like charge, $—COO^-$, or $—SO_4^-$ covering each grease micelle prevents them to come together. Thus, the grease remains in suspension and is washed away with water.

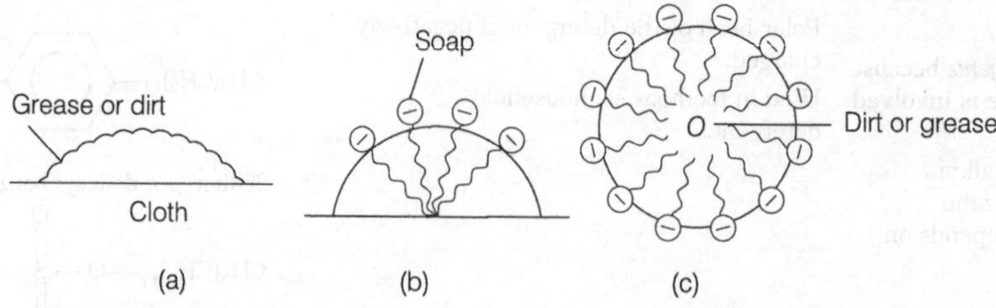

Fig. 24.4 (a) Grease on cloth (b) Stearate ions arranging around the grease droplet and (c) Grease droplet surrounded by stearate ions (micelle formed)

CHECK POINT 24.2

1. Which of the following is used as an antioxidant in food packed itemy?
(a) Aspartame
(b) Sodium benzoate
(c) Tetrazine
(d) Butylated hydroxy toluene

2. Select the incorrect statement about saccharin
(a) It is an artificial sweetener.
(b) It is 650 times sweeter than cane sugar.
(c) It is non-biodegradable.
(d) It has no calorific value.

3. Consider the following structure of compound *A*

$CH_3CH_2—O—C_6H_4—NH\overset{O}{\overset{\|}{C}}NH_2$

(*A*)

The correct name of the given structure (*A*) is
(a) Dulcin
(b) Aspartame
(c) Saccharin
(d) Cyclamate

4. Which of the following is the naturally occurring food preservative found in vegetable oil?
(a) BHA
(b) BHT
(c) Vitamin E
(d) Sodium benzoate

5. Soaps can be formed by which of the following process?
(a) Hydrolysis
(b) Saponification
(c) Neutralisation
(d) Acidification

6. Which of the following represents a synthetic detergent?
(a) $C_{15}H_{31}CO\overset{-}{O}\overset{+}{K}$
(b) $CH_3(CH_2)_6CO\overset{-}{O}\overset{+}{Na}$
(c) $C_{12}H_{25}—C_6H_4—SO_3^-\overset{+}{Na}$
(d) None of the above

7. Structurally, the biodegradable detergent should contain
(a) normal alkyl chain
(b) branched alkyl chain
(c) phenyl side chain
(d) cyclohexyl side chain

8. Compound which is added to soap to impart antiseptic properties is
(a) sodium lauryl sulphate
(b) sodium dodecyl benzene sulphonate
(c) rosin
(d) bithional

9. Which of the following detergent will cause maximum pollution?
(a) $CH_3(CH_2)_{11}—C_6H_4—SO_3^-Na^+$
(b) (zig-zag chain)$SO_3^-Na^+$
(c) Detergents are always pollution free
(d) $CH_3—\underset{C_4H_9}{\underset{|}{CH}}—(CH_2)_2—\underset{C_4H_9}{\underset{|}{CH}}—(CH_2)_3—\underset{C_4H_9}{\underset{|}{CH}}—CH_2CH_2SO_3^-Na^+$

10. Which of the following is a cationic detergent?
(a) Sodium stearate
(b) Dodecyl trimethyl ammonium chloride
(c) Peridinium chloride
(d) Sodium oleate

Chapter Exercises

(A) Topical exercise

I. Drugs

1 Which of the following is not a target molecule for drug function in body? **[NCERT Exemplar]**

(a) Carbohydrates
(b) Lipids
(c) Vitamins
(d) Proteins

2 Consider the following statements.

Statement I Enzymes have active sites that hold substrate molecule for a chemical reaction.

Statement II Drugs compete with natural substrate by attaching covalently to the active site of enzyme. **[NCERT Exemplar]**

(a) Both statements I and II are correct
(b) Only statement I is correct
(c) Only statement II is correct
(d) Neither statement I nor statement II is correct

3 Which of the following statements is incorrect?

(a) Enzyme holds the substrate in its active site.
(b) Enzyme provide the functional groups that will attack the substrate and carry out chemical reaction.
(c) Drugs when bind to the active site then these sites are called allosteric site.
(d) All statements are incorrect.

4 The chemicals in the body through which message between two neurons is communicated is known as

(a) receptors
(b) chemical messengers
(c) enzymes
(d) Both (b) and (c)

5 Consider the following statements.

Statement I Chemical messengers are chemicals that enable communication of message between two neurons or between neurons and muscles.

Staement II Chemicals enter the cell through receptor. **[NCERT Exemplar]**

(a) Both statements I and II are correct
(b) Only statement I is correct
(c) Only statement II is correct
(d) Neither statement I nor statement II is correct

6 Match the class of compounds gives in Column I with their functions given in Column II. **[NCERT Exemplar]**

	Column I		Column II
A.	Antagonists	1.	Communicate message between two neurons and that between neurons to muscles
B.	Agonists	2.	Bind to the receptor site and inhibit its natural function
C.	Chemical messenger	3.	Crucial to body's communication process
D.	Inhibitors	4.	Mimic the natural messenger
E.	Receptors	5.	Inhibit activities of enzymes

	A	B	C	D	E		A	B	C	D	E
(a)	5	3	2	1	4	(b)	2	4	1	5	3
(c)	3	5	4	2	1	(d)	4	1	5	3	2

7 Consider the following statements.

Statement I Competitive inhibitors compete with natural substrate for their attachment on the active sites of enzymes.

Statement II In competitive inhibition, inhibitor binds to the allosteric site of the enzyme.

(a) Both statements I and II are correct **[NCERT Exemplar]**
(b) Only statement I is correct
(c) Only statement II is correct
(d) Neither statement I nor statement II is correct

8 Consider the following statements.

Statement I Non-competitive inhibitor inhibits the catalystic activity of enzyme by binding with its active site.

Statement II Non-competitive inhibitor changes the shape of the active site in such a way that substrate cannot recognise it. **[NCERT Exemplar]**

(a) Both statements I and II are correct
(b) Only statement I is correct
(c) Only statement II is correct
(d) Neither statement I nor statement II is correct

9 Consider the following statements.

Statement I Chemical messenger gives message to the cell without entering the cell.

Statement II Chemical messenger is received at the binding site of receptor proteins. **[NCERT Exemplar]**

(a) Both statements I and II are correct
(b) Only statement I is correct
(c) Only statement II is correct
(d) Neither statement I nor statement II is correct

10 Consider the following statements.

Statement I Receptor proteins show selectivity for one chemical messenger over the other.

Statement II Chemical messenger binds to the receptor site and inhibits its natural function. **[NCERT Exemplar]**

(a) Both statements I and II are correct
(b) Only statement I is correct
(c) Only statement II is correct
(d) Neither statement I nor statement II is correct

11 Chemical compounds used for the treatment of depression and hypertension are called
(a) analgesics (b) tranquilizers
(c) antibiotics (d) pheromones

12 The compound that causes general antidepressant action on the central nervous system belongs to the class of .... **[NCERT Exemplar]**
(a) analgesics (b) tranquilizers
(c) narcotic analgesics (d) antihistamines

13 Aspirin finds use in the
(a) pain relieving drugs
(b) prevention of heart attacks
(c) prevention of alzheimer's disease
(d) All of the above

14 Paracetamol is an
(a) analgesic (b) antimalarial
(c) antipyretic (d) Both (a) and (c)

15 Aspirin is used to cure Alzheimer's disease which is caused by
(a) Al^{3+} (b) Mg^{2+} (c) Fe^{3+} (d) All of these

16 Viral inflammation in AIDS can be checked by use of
(a) aspirin (b) equanil
(c) oil of winter green (d) All of these

17 Which statement about aspirin is not true? **[NCERT Exemplar]**
(a) Aspirin belongs to narcotic analgesics
(b) It is effective in relieving pain
(c) It has antiblood clotting action
(d) It is a neurologically active drug

18 Consider the following statements.

Statement I Naproxen is being used as an analgesic in place of aspirin.

Statement II Aspirin is obtained by acetylation of *o*-hydroxy benzoic acid.

(a) Both statements I and II are true
(b) Only statement I is true
(c) Only statement II is true
(d) Neither statement I nor statement II is true

19 Consider the following statements.

Statement I Aspirin has been used in the prevention of heart attacks.

Statement II Aspirin has blood-clotting action.

(a) Both statements I and II are true
(b) Only statement I is true
(c) Only statement II is true
(d) Neither statement I nor statement II is true

20 Aspirin is chemically
(a) methyl benzoate
(b) ethyl salicylate
(c) acetyl salicylic acid
(d) *o*-hydroxy benzoic acid

21 Which of the following can possibly be used as analgesic without causing addiction and modification?
(a) Morphine
(b) Diazepam
(c) Tetrahydrocational
(d) N-acetyl-*para*-aminophenol

22 Which of the following is a non-narcotic drug?
(a) Aspirin (b) Ibuprofen
(c) Novalgin (d) All of these

23 Which of the following kills the organisms in the body (bactericidal drug)?
(a) Penicillin
(b) Ofloxacin
(c) Both (a) and (b)
(d) Aspirin

24 A narrow spectrum antibiotic is active against .... **[NCERT Exemplar]**
(a) gram positive or gram negative bacteria
(b) gram negative bacteria only
(c) single organism or one disease
(d) Both gram positive and gram negative bacteria

25 Which of the following statements is not true?
(a) Sulphadiazine is a synthetic antibacterial.
(b) Ampicillin is natural antibiotic.
(c) Aspirin is analgesic and antipyretic both.
(d) Some disinfectants can be used as antiseptics at low concentration.

26 Which one of the following is a broad spectrum antibiotic?
(a) Chloramphenicol (b) Plasmoquin
(c) Xylocaine (d) Antiseptic

27 Consider the following statements.

Statement I Antibiotic therapy has been likened to "setting one thief against another".

Statement II Antibiotic themselves are products of microbial growth.

(a) Both statements I and II are correct
(b) Only statement I is correct
(c) Only statement II is correct
(d) Neither statement I nor statement II is correct

28 Chloramphenicol is an

(a) analgesic (b) antipyretic
(c) antiseptic (d) antibiotic

29 Consider the following statements. **[NCERT Exemplar]**

Statement I Penicillin (G) is an antihistamine.

Statement II Penicillin (G) is effective against gram positive as well as gram negative bacteria.

(a) Both statements I and II are correct
(b) Only statement I is correct
(c) Only statement II is correct
(d) Neither statement I nor statement II is correct

30 Which of the following is an antibiotic?

(a) Insulin (b) Ascorbic acid
(c) Penicillin (d) Albumin

31 Which one of the following is a broad-spectrum drug?

(a) Chloroquine (b) Chloroxylenol
(c) Plasmoquine (d) Chloramphenicol

32 Salvarsan is arsenic containing drug which was first used for the treatment of .... **[NCERT Exemplar]**

(a) syphilis (b) typhoid
(c) meningitis (d) dysentry

33 Select the correct statement(s) from the statements given below.

(a) Any organism that causes disease is called a pathogen.
(b) A drug which kills the bacteria is called bactericidal.
(c) Both (a) and (b).
(d) A drug which kills the bacteria is called bacteriostatic.

34 Main component of sulpha drug is

(a) sulphonic acid (b) sulphanilamide
(c) sulphanilic acid (d) None of these

35 Chloramine-T is used as an

(a) analgesic
(b) antiseptic
(c) antipyretic
(d) anti-inflammatory

36 Which of the following set has different class of compounds?

(a) Analgesics : naproxen, morphine, aspirin
(b) Tranquilisers : equanil, heroin, valium
(c) Antiseptics : bithional, dettol, boric acid
(d) Bactericidal : penicillin, aminoglycosiders, ofloxacin

37 Consider the following statements.

Statement I Boric acid as a dilute aqueous solution is used as a weak antiseptic for eyes.

Statement II Boric acid is a monobasic acid.

(a) Both statements I and II are correct
(b) Only statement I is correct
(c) Only statement II is correct
(d) Neither statement I nor statement II is correct

38 Dettol is a mixture of

(a) chloroxylenol and terpeneol
(b) phenol and chlorophenol
(c) phenol and chloroxylenol
(d) chlorophenol and chloroxylenol

39 Which is not the correct matching of medicine with its disease/activity?

(a) Aspirin—pain reliever
(b) Equanil—hypertension
(c) Chloramphenicol—typhoid
(d) 0.2 per cent phenol—disinfectant

40 Match the classes of drugs given in Column I with their action given in Column II. **[NCERT Exemplar]**

Column I	Column II
A. Analgesics	1. Inhibit the growth of microorganisms can be given orally
B. Antiseptics	2. Treatment of stress
C. Antihistamines	3. Applied to inanimate objects
D. Antacids	4. Prevents the interaction of histamine with its receptor
E. Tranquilisers	5. Pain killing effect
F. Antibiotics	6. Applied to diseased skin surfaces
G. Disinfectants	7. Treatment of acidity

	A	B	C	D	E	F	G
(a)	4	5	1	2	3	7	6
(b)	5	6	4	7	2	1	3
(c)	7	4	6	5	1	3	2
(d)	6	4	7	3	4	2	1

41 Match the medicines given in Column I with their uses given in Column II. **[NCERT Exemplar]**

Column I	Column II
A. Ranitidine	1. Tranquiliser
B. Furacine	2. Antibiotic
C. Phenelzine	3. Antihistamine
D. Chloramphenicol	4. Antiseptic
	5. Antifertility drug

	A	B	C	D		A	B	C	D
(a)	2	1	4	3	(b)	3	4	1	2
(c)	4	1	2	3	(d)	4	3	2	1

42 Which is the correct statement about birth control pills? **[NCERT Exemplar]**
(a) Contain estrogen only
(b) Contain progesterone only
(c) Contain a mixture of estrogen and progesterone derivatives
(d) Progesterone enhances ovulation

II. Chemicals in Food

43 Trade name of aspartame is
(a) nutra sweet (b) alitame
(c) saccharin (d) sucralose

44 Consider the following statements.
Statement I All chemicals added to food items are called food preservatives.
Statement II All these chemicals increase the nutritive value of the food. **[NCERT Exemplar]**
(a) Both statements I and II are correct
(b) Only statement I is correct
(c) Only statement II is correct
(d) Neither statement I nor statement II is correct

45 Which of the following will not enhance nutritional value of food? **[NCERT Exemplar]**
(a) Minerals (b) Artificial sweeteners
(c) Vitamins (d) Amino acids

46 Which is not under the class of saccharide?
(a) Glucose (b) Fructose
(c) Saccharin (d) Sucrose

47 Which of the following is an artificial sweetener?
(a) Penicillin (b) Aspartame
(c) Aspirin (d) Chloroquine

48 Consider the following statements.
Statement I Artificial sweeteners are added to the food to control the intake of calories. **[NCERT Exemplar]**
Statement II Most of the artificial sweeteners are inert and do not metabolise in the body.
(a) Both statements I and II are correct
(b) Only statement I is correct
(c) Only statement II is correct
(d) Neither statement I nor statement II is correct

49 Consider the following statements.
Statement I Preservative are added to food items.
Statement II Preservatives inhibit the growth of mircroorganisms. **[NCERT Exemplar]**
(a) Both statements I and II are correct
(b) Only statement I is correct
(c) Only statement II is correct
(d) Neither statement I nor statement II is correct

III. Cleansing agents

50 Surfactant can be
(a) a soap
(b) detergents
(c) syndets
(d) All of the above

51 Which of the following enhances leathering property of soap? **[NCERT Exemplar]**
(a) Sodium carbonate
(b) Sodium rosinate
(c) Sodium stearate
(d) Trisodium phosphate

52 Consider the following statements.
Statement I Bithional is added to soap to impart antiseptic property.
Statement II It eliminates undesirable odours resulting from the bacterial decomposition of organic matter on the skin.
(a) Both statements I and II are correct
(b) Only statement I is correct
(c) Only statement II is correct
(d) Neither statement I nor statement II is correct

53 Consider the following statements.
Statement I Sodium chloride is added to precipitate soap after saponification. **[NCERT Exemplar]**
Statement II Hydrolysis of esters of long chain fatty acids by alkali produces soap in colloidal form.
(a) Both statements I and II are correct
(b) Only statement I is correct
(c) Only statement II is correct
(d) Neither statement I nor statement II is correct

54 With hard water ordinary soap forms curdy precipitate of
(a) $(RCOO)_2Mg$
(b) $(RCOO)_2Ca$
(c) Both (a) and (b) type
(d) $RCOO^-Na^+$

55 Glycerol is added to soap. It functions .... **[NCERT Exemplar]**

(a) as a filler
(b) is to increases leathering
(c) is to prevent rapid drying
(d) is to make soap granules

56 Consider the following statements.

Statement I Transparent soaps are made by dissolving soaps in ethanol.

Statement II Ethanol makes things invisible. **[NCERT Exemplar]**

(a) Both statements I and II are correct
(b) Only statement I is correct
(c) Only statement II is correct
(d) Neither statement I nor statement II is correct

57 Match the soaps given in Column I with items given in Column II. **[NCERT Exemplar]**

Column I	Column II
A. Soap chips	1. Dried miniature soap bubbles
B. Soap granules	2. Small broken pieces of soap formed from melted soaps
C. Soap powder	3. Soap powder + abrasives + builders (Na_2CO_3, Na_3PO_4)
D. Scouring soap	4. Soap powder + builders like Na_2CO_3 and Na_3PO_4

	A	B	C	D
(a)	2	1	4	3
(b)	3	4	1	2
(c)	3	4	2	1
(d)	4	3	2	1

58 Which of the following detergents can be used as germicide also?

(a) A cationic detergent
(b) An anionic detergent
(c) Non-ionic detergent
(d) None of the above

59 Consider the following statements.

Statement I Detergents can be used in hard water too.

Statement II Detergents are sodium salts of alkyl benzene sulphonic acid.

(a) Both statements I and II are correct
(b) Only statement I is correct
(c) Only statement II is correct
(d) Neither statement I nor statement II is correct

60 Which of the following is not detergent?

(a) $PhCH_2-\overset{+}{N}(CH_3)_2-(CH_2)_{15}CH_3Cl^-$

(b) $CH_3(CH_2)_{15}-\overset{+}{N}(CH_3)_2-CH_3Cl^-$

(c) $CH_3(CH_2)_{16}-\overset{O}{\overset{\|}{C}}-\bar{O}\,Na^+$

(d) $RCH_2(O-CH_2-CH_2)_n\,OH$

61 Which of the following detergent can cause maximum pollution?

(a) $SO_3^-Na^+$ (branched chain)

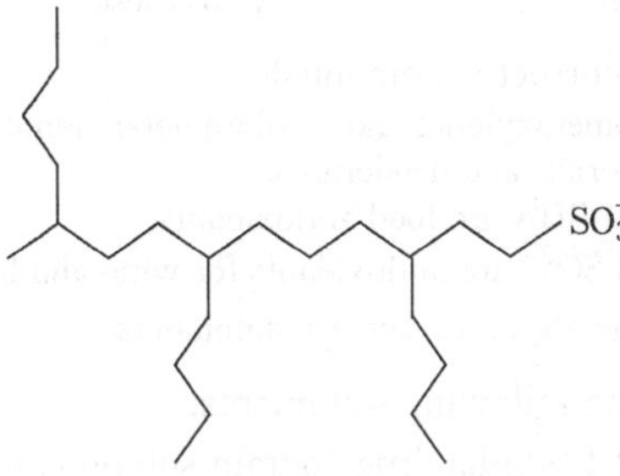

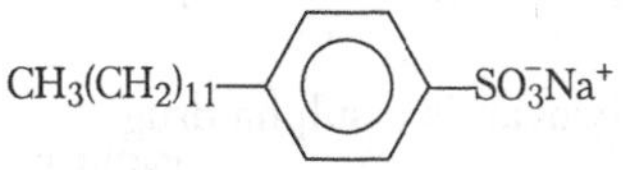

(b) $CH_3(CH_2)_{11}-C_6H_4-SO_3^-Na^+$

(c) $SO_3^-Na^+$ (straight chain)

(d) Detergent is always pollution free

62 Match structures given in Column I with the type of detergents given in Column II. **[NCERT Exemplar]**

Column I	Column II
A. $CH_3(CH_2)_{16}COO(CH_2CH_2O)_n CH_2CH_2OH$	1. Cationic detergent
B. $C_{17}H_{35}COO^-Na^+$	2. Anionic detergent
C. $CH_3-(CH_2)_{10}CH_2SO_3^-Na^+$	3. Non-ionic detergent
D. $\left[CH_3(CH_2)_{15}-N(CH_3)_2-CH_3\right]^+ Br^-$	4. Soap

	A	B	C	D
(a)	3	4	1	2
(b)	3	4	2	1
(c)	4	3	2	1
(d)	2	1	4	3

(B) Miscellaneous questions

1 Equanil is .... **[NCERT Exemplar]**
(a) artificial sweetener
(b) tranquiliser
(c) antihistamine
(d) antifertility drug

2 A newer family of very effective lactains antibiotics is
(a) ampicillin (b) amoxicillin
(c) cephalosporins (d) phenobarbital

3. Which of the following is an ester?
(a) Coconut oil (b) Kerosene oil
(c) Soap (d) Glycerine

4 –COOH is one of the functional groups present in
(a) ascorbic acid (b) aspartame
(c) ibuprofen (d) All of these

5 Select the correct statement(s).
(a) *p*-chlorometaxylenol and dichlorometaxylenol are antibacterials and deodorants.
(b) BHT and BHA are food antioxidants.
(c) SO_2 and SO_3^{2-} are antioxidants for wine and beers.
(d) All of the above are correct statements.

6 Consider the following statements.
Statement I Sulpha drug contain sulphonamide group.
Statement II Salvarsan is a sulpha drug. **[NCERT Exemplar]**
(a) Both statements I and II are true
(b) Only statement I is true
(c) Only statement II is true
(d) Neither statement I nor statement II is true

7 Consider the following statements.
Statement I Receptors are crucial to body's communication process. **[NCERT Exemplar]**
Statement II Receptors are proteins.
(a) Both statements I and II are correct
(b) Only statement I is correct
(c) Only statement II is correct
(d) Neither statement I nor statement II is correct

8 Select the correct statement(s).
(a) Synthetic detergents are alkyl benzene sulphonates.
(b) Straight-chain alkyl groups are biodegradable, while branched-chain alkyl groups are not.
(c) Phosphates and silicates are added in detergents as 'builders' to complex the hard water cations.
(d) All of the above are correct statements.

9 Aspartame is a non-nutritive sweetener. Assuming that both amide and ester bonds are hydrolysed in the stomach, amino acids obtained are
(a) $C_6H_5CH_2CH(NH_2)C(=O)OH$ (benzene ring–$CH_2CHCOOH$ with NH_2)
(b) $H_2N\,CHCH_2\,\overset{O}{\overset{\|}{C}}OH$ with $\underset{O}{\underset{\|}{C}}OH$ on CH
(c) Both (a) and (b)
(d) CH_3CH_2O–C_6H_4–NH–$\overset{O}{\overset{\|}{C}}$–$NH_2$

10 Which of the following statements is correct? **[NCERT Exemplar]**
(a) Some tranquilisers function by inhibiting the enzymes which catalyse the degradation of noradrenaline.
(b) Tranquilisers are narcotic drugs.
(c) Tranquilisers are chemical compounds that do not affect the message transfer from nerve to receptor
(d) Tranquilisers are chemical compounds that can relieve pain and fever

11 Polyethyleneglycols are used in the preparation of which type of detergents? **[NCERT Exemplar]**
(a) Cationic detergents
(b) Anionic detergents
(c) Non-ionic detergents
(d) Soaps

12 Which of the following statement is not correct? **[NCERT Exemplar]**
(a) Some antiseptics can be added to soaps.
(b) Dilute solutions of some disinfectants can be used as antiseptic.
(c) Disinfectants are antimicrobial drugs.
(d) Antiseptic medicines can be ingested.

13 Which of the following statements is not true about enzyme inhibitors? **[NCERT Exemplar]**
(a) Inhibit the catalytic activity of the enzyme.
(b) Prevent the binding of substrate.
(c) Generally a strong covalent bond is formed between an inhibitor and an enzyme.
(d) Can be competitive or non-competitive.

14 Select the incorrect statement.
(a) 0.2 per cent solution of phenol is an antiseptic while 1.0 per cent solution is a disinfectant.
(b) Equanil is used in controlling depression and hypertension.
(c) Mifepristone is a synthetic steroid used as "morning after pill".
(d) A drug which kills the organism in the body is called bacteriostatic.

15 Vitamin E is a natural product used as an antioxidant. Select correct function(s) of vitamin E as antioxidant.
(a) It protects unsaturated fatty acid chains of phosphoacyl glycerols from atmospheric oxidation.
(b) It prevents biological membranes from reacting with oxygen and other radicals.
(c) It retards aging process.
(d) All of the above

16 Sodium benzoate is used as food-preservative. It is
(a) metabolised by conversion to hippuric acid and is excreted in the urine
(b) metabolised to benzoic acid and deposited in the bone
(c) decomposed by gastric juice and escapes as CO_2
(d) decomposed by heat of the digestion process and escapes as CO_2

17 Match the following.

Column I (Compound)		Column II (Type)	
A.	Alizarin	1.	Artificial sweetener
B.	Duclin	2.	Tranquiliser
C.	Aspirin	3.	Dye
D.	Barbiturates	4.	Analgesics

	A	B	C	D		A	B	C	D
(a)	4	3	2	1	(b)	3	1	4	2
(c)	2	4	1	3	(d)	2	1	4	3

18 Following structure

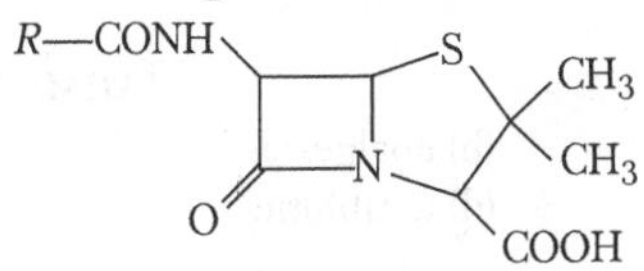

is called penicillin *G* if *R* is
(a) $C_6H_5CH_2$ (b) $C_6H_5OCH_2$
(c) $C_6H_5—CH(NH_2)$ (d) p-$HOC_6H_4CH(NH_2)$—

19 Following medicine was extracted from medicinal plants in early 1960 for treatment of cancer

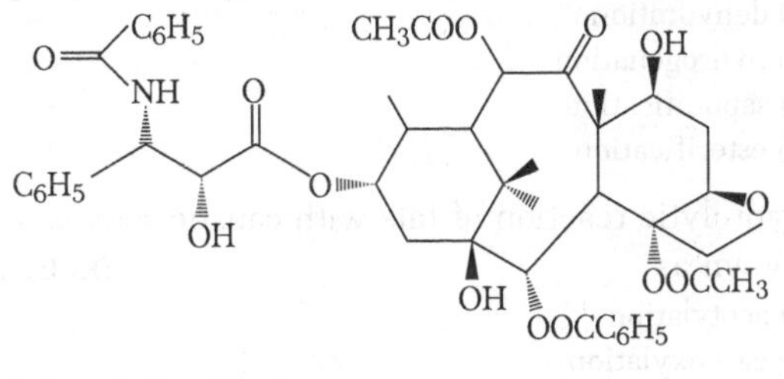

It has following groups
(a) ester, amino, amide, carbonyl, epoxy
(b) ester, amide, carbonyl, epoxy, hydroxy
(c) ether, amide, carbonyl, epoxy, hydroxy
(d) peptide, ether, carbonyl, hydroxy

20 A penicillin is a member of a family of drugs that have a
(a) four-membered cyclic amide fused to a five-mebered thiazole ring
(b) three-membered cyclic amide fused to a five-membered thiazole ring
(c) four-membered cyclic amide fused to a four-membered thiazole ring
(d) five-membered cyclic amide fused to a five-membered thiazole ring

21 Consider the following statements.
I. 0.2 to 0.4 ppm Cl_2 solution is used as disinfectant for drinking water.
II. Many body secretions either kill the microbes or inhibit their growth.
III. Body possesses an efficient natural defence mechanism which operates at all times against potential pathogenic microbes.

Select the correct statements.
(a) I, II and III (b) Both II and III
(c) Both I and III (d) Both I and II

22 Consider the following statements.
I. *R*(-)-enantiomer of the common analgesic ibuprofen is biologically inactive.
II. *R* (-)-enantiomer of the analgesic naproxen is liber toxin.
III. Sodium salt of naproxen is marked as Aleve.
IV. Each of ibuprofen and naproxen has one stereocentre.

Select the correct statements.
(a) Both I and III (b) I, II and III
(c) II, III and IV (d) I, II, III and IV

(C) Medical entrances' gallery

1 Which of the following is a cationic detergent?
(a) Sodium stearate **[NEET 2020]**
(b) Cetyltrimethyl ammonium bromide
(c) Sodium dodecyl benzene sulphonate
(d) Sodium lauryl sulphate

2 Among the following, the narrow spectrum antibiotic is **[NEET 2019]**
(a) ampicillin (b) amoxycillin
(c) chloramphenicol (d) penicillin-G

3 The artificial sweetener stable at cooking temperature and does not provide calories is **[NEET (Odisha) 2019]**
(a) saccharin (b) aspartame (c) sucralose (d) alitame

4 Saccharin an artificial sweeterner is manufactured from **[JIPMER 2019]**
(a) toluene (b) cyclohexane
(c) starch (d) cellulose

5 Consider the statements.

Statement I For liquid dishwashing non-ionic type of detergent are used.

Statement II Remove grease and oil by micelle formation. [AIIMS 2019]

(a) Both statements I and II are correct
(b) Only statement I is correct
(c) Only statement II is correct
(d) Neither statement I nor statement II is correct

6 Food preservatives prevent spoilage of food due to microbial growth. The most commonly used preservatives are [AIIMS 2018]

(a) C_6H_5COONa
(b) table salt, sugar
(c) vegetable oils and sodium benzoate
(d) All of the above

7 Which type of drugs come under anti-microbial drugs? [JIPMER 2018]

(a) Antiseptics, antacids, antihistamines
(b) Tranquilizers, antiseptics, antifertility
(c) Tranquilizers, analgesics, antiseptics
(d) Antiseptics, antibiotics, disinfectants

8 Which one among the following is added to soap to impart antiseptic property? [JIPMER 2018]

(a) Sodium lauryl sulphate
(b) Sodium dodecyl benzene sulphonates
(c) Rosin
(d) Bithional

9 Mixture of chloroxylenol and terpineol acts as [NEET 2017]

(a) antibiotic (b) analgesic
(c) antiseptic (d) antipyretic

10 Which of the following is an analgesic? [NEET(I) 2016]

(a) Penicillin
(b) Streptomycin
(c) Chloromycetin
(d) Novalgin

11 The drug Tagamet is [Kerala CEE 2015]

(a) analgesic (b) anti-depressant
(c) anaesthetic (d) antacid
(e) antibiotic

12 Which one of the following is a cationic detergent? [Kerala CEE 2015]

(a) Sodium lauryl sulphate
(b) Cetyltrimethyl ammonium bromide
(c) Sodium dodecylbenzene sulphonate
(d) Glyceryl palmitate
(e) Glyceryl oleate

13 Which of the following is not true? [KCET 2015]

(a) Erythromycin is a bacteriostatic antibiotic
(b) Ampicillin is not a natural antibiotic
(c) Prontosil is not converted into sulphanilamide in the body
(d) Vancomycin is a broad spectrum antibiotic

14 Which one of the following is a bacteriostatic drug? [Kerala CEE 2014]

(a) Aminoglycosides (b) Penicillin-G
(c) Ofloxacin (d) Ampicillin
(e) Tetracycline

15 Butylated hydroxyltoluene as a food additive acts as [KCET 2014]

(a) antioxidant (b) flavouring agent
(c) colouring agent (d) emulsifier

16 Antiseptics and disinfectants either kill or prevent growth of microorganisms. Identify which of the following is not true? [NEET 2013]

(a) A 0.2% solution of phenol is an antiseptic while 1% solution acts as a disinfectant.
(b) Chlorine and iodine are used as strong disinfectants
(c) Dilute solutions of boric acid and hydrogen peroxide are strong antiseptics.
(d) Disinfectants harm the living tissue.

17 Salol is [AIIMS 2013]

(a) acetyl salicylic acid
(b) phenyl salicylate
(c) methyl salicylate
(d) None of the above

18 Salol is an [AIIMS 2013]

(a) antipyretic (b) analgesics
(c) antiseptic (d) antibiotic

19 Paracetamol is used as an [Manipal 2013]

(a) antibiotic (b) antimalarial
(c) antipyretic (d) arsenical

20 The reaction of fat and sodium hydroxide is known as [Manipal 2013]

(a) dehydration
(b) hydrogenation
(c) saponification
(d) esterification

21 Hydrolytic reaction of fats with caustic soda is known as [KCET 2013]

(a) acetylation
(b) carboxylation
(c) esterification
(d) saponification

22 Chloramphenicol is an [CBSE AIPMT 2012]
(a) antifertility drug
(b) antihistamin
(c) antiseptic and disinfectant
(d) antibiotic broad spectrum

23 The antiseptic present in dettol is [AIIMS 2012]
(a) quaeline (b) chloroxylenol
(c) bithional (d) None of these

24 Parkinson's disease is linket to abnormalities in the level of dopamine in the body. The structure of dopamine is [Manipal 2012]

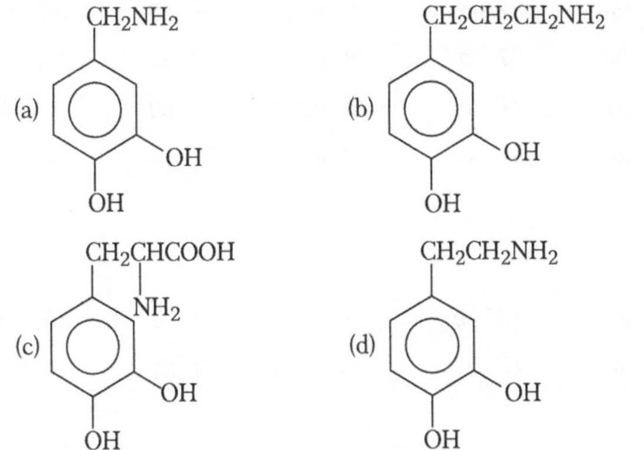

25 Which of the following antibiotics was first discovered? [OJEE 2012]
(a) Ampicillin (b) Amoxycillin
(c) Penicillin (d) Chloramphenicol

26 Saccharin an artificial sweetener, is manufactured from [Kerala CEE 2012]
(a) toluene (b) cyclohexane
(c) starch (d) cellulose
(e) vitamin D

27 Which of the following enzymes is present in animals like cow, buffaloes, etc., to digest compound like paper, cloth, etc.? [Guj. CET 2011]
(a) Ureaze (b) Cellulase
(c) Silicons (d) Sucrase

28 Which one of the following is employed as antihistamine? [CBSE AIPMT 2011]
(a) Omeprazole
(b) Chloramphenicol
(c) Dipenyl hydramine
(d) Norothindrone

29 Which one of the following has antiseptic property? [Kerala CEE 2011]
(a) Dichloromethane (b) Trifluoromethane
(c) Triiodomethane (d) Tetrachloromethane
(e) Freon-12

30 Trefenadine is commonly used as a/an [Kerala CEE 2011]
(a) tranquiliser (b) antihistamine
(c) antimicrobial (d) antibiotic
(e) antifertility drug

31 Which of the following is pheromone? [Guj. CET 2011]
(a) Linalool (b) Disparlure
(c) BHA (d) Alitame

ANSWERS

Check point 24.1

1. (c)	2. (b)	3. (d)	4. (d)	5. (c)	6. (b)	7. (a)	8. (a)	9. (a)	10. (d)

Check point 24.2

1. (d)	2. (b)	3. (a)	4. (c)	5. (b)	6. (c)	7. (a)	8. (d)	9. (d)	10. (b)

(A) Topical exercise

1. (c)	2. (b)	3. (c)	4. (b)	5. (b)	6. (b)	7. (b)	8. (c)	9. (a)	10. (b)
11. (b)	12. (b)	13. (d)	14. (d)	15. (a)	16. (a)	17. (a)	18. (a)	19. (a)	20. (c)
21. (d)	22. (d)	23. (c)	24. (a)	25. (b)	26. (a)	27. (a)	28. (d)	29. (d)	30. (c)
31. (d)	32. (a)	33. (c)	34. (b)	35. (b)	36. (b)	37. (a)	38. (a)	39. (d)	40. (b)
41. (b)	42. (c)	43. (a)	44. (d)	45. (b)	46. (c)	47. (b)	48. (a)	49. (a)	50. (d)
51. (b)	52. (a)	53. (a)	54. (c)	55. (c)	56. (b)	57. (a)	58. (a)	59. (a)	60. (c)
61. (a)	62. (b)								

(B) Miscellaneous questions

1. (b)	2. (a)	3. (c)	4. (d)	5. (d)	6. (b)	7. (a)	8. (d)	9. (c)	10. (d)
11. (c)	12. (d)	13. (c)	14. (d)	15. (d)	16. (a)	17. (b)	18. (a)	19. (b)	20. (a)
21. (a)	22. (d)								

(C) Medical entrances' gallery

1. (b)	2. (d)	3. (c)	4. (a)	5. (a)	6. (d)	7. (d)	8. (d)	9. (c)	10. (d)
11. (d)	12. (b)	13. (c)	14. (e)	15. (a)	16. (c)	17. (b)	18. (c)	19. (c)	20. (c)
21. (d)	22. (d)	23. (b)	24. (d)	25. (c)	26. (a)	27. (b)	28. (c)	29. (c)	30. (b)
31. (b)									

Hints & Explanations

(A) Topical exercise

1 (*c*) Drugs usually interact with biomolecules such as carbohydrates, lipids, proteins and nucleic acids. These are called drug targets. Vitamins are not a target molecule for drug function in body.

2 (*b*) Statement I is correct statement, statement II is incorrect statement. Correct statement is drug compete with natural substrate by attaching by attaching by weak bonds such as hydrogen bond, van der Waals' interaction and ionic bonding etc., to the active site of the enzyme.

5 (*b*) Statement I is correct statement, statatement II is incorrect statement. Correct statement is, chemicals bind to the receptor sites present on the surface of the membrane or the cell. They give message to the cell without entering the cell.

6 (*b*) A. Antagonists drugs are used when blocking of message is required. e.g., dopamine antagonist is a drug which blocks the dopamine receptors by receptor antagonism.

B. Agonists drugs are useful when there is lack of chemical messenger.

C. These chemical messengers are received at the binding sites of receptor. These communicate message between two neurons and that between neurons to muscles.

D. Inhibitors block the binding site of the enzyme and prevent the binding of the substrate, or inhibit the catalytic activity of the enzyme.

E. Receptors are proteins that are crucial to body's communication. They are embedded in the cell membrane in such a way that their small part possessing active site projects out of the surface of the membrane and opens on the outside region of the cell membrane.

7 (*b*) Statement I is correct but statement II is incorrect. The correct statement II is in competitive inhibition, inhibitor binds to the active site of the enzyme.

8 (*c*) Statement I is incorrect and statement II is correct statement. Non-competitive inhibitor inhibits the catalytic activity by binding the enzyme at allosteric site.

9 (*a*) Both statments are correct. Chemical messenger gives message to the cell without entering the cell because it is received at the binding site of receptor proteins.

10 (*b*) Statement I is correct and statement II is incorrect statement. Chemical messenger binds the receptor site and gives the message to the cell without entering the cell.

12 (*b*) Analgesics reduce or abolish pain without causing impairment of consciousness, mental confusion, incordination or paralysis or some other disturbances of nervous system. Analgesics are non-narcotic or narcotic. Narcotic analgesics when administered in small doses relieve pain and produce sleep.

Tranquilisers are a class of chemical compounds used for the treatment of stress and mild or even severe mental disease.

Histamine is a substance which on release produces hypersensitivity to dust, some drugs etc (i.e., allergy). Antihistamines are the drugs which interfere with the natural action of histamine by competing with histamine for binding sites of receptor where histamine exerts its effect. They are anti-allergy drugs.

17 (*a*) Aspirin inhibits the synthesis of compound known as prostaglandins which stimulate inflammation in the tissues and cause pain.

So, it is effective in relieving pain. Beside this, aspirin has many other effects such as reducing fever (antipyretic) and preventing blood platelet coagulation.

Because of this anti-blood clotting action, it is widely used to prevent heart attacks. It does not make a person addictive as it is non-narcotic drug.

29 (*d*) Both statements are incorrect statements.

Correct statements I is penicillin (G) is an antibiotic.

Correct statement II is penicillin is effective against some gram positive bacteria.

32 (*a*) Syphilis is an acute and chronic infectious disease caused by the bacterium Treponema pallidum. It is a sexually transmitted infection. The primary route of transmission is through sexual contact but it may also be transmitted from mother to foetus during pregnancy or at birth.

Paul Ehrlich investigated arsenic based structures in order to produce less toxic substances for the treatment of syphilis as he got Nobel prize for medicine in 1908 for this discovery. Salvarsan was the first effective medicine discovered for syphilis. Although salvarsan is toxic to human beings, its effect on bacteria is much greater than on human being. Structure of salvarsan is

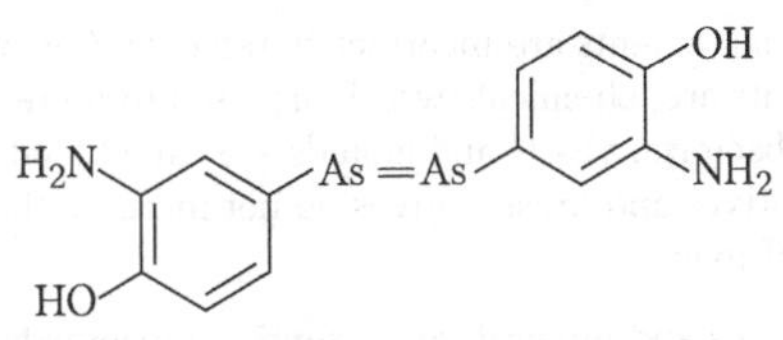

H_2N, HO, As = As, OH, NH_2

40 (*b*) A. Analgesics reduce or abolish pain without causing impairment of consciousness, mental confusion, in coordination or paralysis or some other disturbances of nervous system, e.g., aspirin.

B. Antiseptics are the chemicals which either kill or prevent the growth of microorganisms. They are applied to living tissues such as wounds, cuts etc., tincture of iodine.

C. Antihistamines are anti-allergic drugs. These drugs interfere with the nature action of histamine by competing with histamine for binding sites of receptor where histamine exerts its effect. e.g. seldane.

D. Antacids are used to neutralise excess of acid released in stomach e.g. mixture of $Mg(OH)_2$ and $Al(OH)_3$.

E. Tranquilisers are used for the treatment of stress, and mild or even severe mental diseases. e.g. chloramphenicol.

F. Antibiotics inhibit the growth of microorganisms, can be given orally.

G. Disinfectants are the chemicals which either kill or prevent the growth of microorganisms but they can be applied on non-living objects e.g. 1 per cent solution of phenol.

41 (*b*) A. **Ranitidine** It prevents the interaction of histamine with the receptors present in the stomach wall. Thus, it controls the secretion of HCl and pepsin in stomach.

B. **Furacine** It is an antiseptic. It can be applied to the living tissues to kill or to prevent the growth of microorganisms.

C. **Phenelzine** It is also known as Nardil. Its structure is

$C_6H_5-CH_2-CH_2-NH{\cdot}NH_2$

It is used to treat depression.

D. **Chloramphenicol** It is a broad spectrum antibiotic. It can be given orally in case of typhoid, acute fever, dysentary, certain urinary track infections, meningitis and pneumonia.

42 (*c*) Chemical substances which are used to control the pregnancy in women are called birth control pills. These pills essentially contain a mixture of synthetic estrogen and progesterone derivatives, both of these are sex hormones. It is known that progesterone suppresses ovulatioin and estrogen control the menstrual cycle. Some of the commonly used pills contain a mixture of norethindrone (a progesterone derivative) and novestrol (an estrogen).

44 (*d*) Both statements are incorrect statements. Correct statements are, chemicals which are used to protect food against bacteria, yeasts and moulds are called food preservatives and preservatives do not increase the nutritive values of food.

45 (*b*) Vitamins and minerals are essential nutrients because they perform hundred of roles in the body. They help shore up bones, heal wounds and to bolster the immune system. They also convert food into energy and repair cellular damage.

Artificial sweeteners are non-caloric substitutes for sugar. They are often intensly more sweet than sugar but do not enhance nutritional value of food.

Amino acids are the monomers of proteins. Proteins require for the growth and maintenance of the body. Therefore, amino acids will also enhance the nutritional value of food.

48 (*a*) Both statements are correct. Artificial sweeteners are added to the food to control the intake of calories because most of them are inert and do not metabolise in the body.

49 (*a*) Both statements are correct.

Preservatives are added to food items because it inhibit the growth of microorganisms.

53 (*a*) Both statements are correct. Sodium chloride is added to precipitate soap after saponification because hydrolysis of esters of long chain fatty acids by alkali produces soap in colloidal form.

55 (*c*) Glycerol is added to shaving soap to prevent rapid drying while to enhance the leathering property of soap, a gum called rosin is added to them. It forms sodium rosinate which lathers well. Soap granules are dried miniature soap bubbles.

56 (*b*) Statement I is correct statement. Statement II is incorrect statement. Correct statement ethanol removes air and moisture which scatter light.

57 (*a*) A. Soap chips are made by running a thin sheet of melted soap into a cool cylinder and scraping off the soaps in small broken pieces.

B. Soap granules are dried miniature soap bubbles.

C. Soap powders contain soap powder and builders like sodium carbonate and trisodium phosphate. Builders make the soap act more rapidly.

D. Scouring soaps contain soap powder, a scouring agent (abrasive) such as powdered pumice or finely divided sand and builders.

62 (*b*) Non-ionic detergent

(A) $CH_3(CH_2)_{16}COO(CH_2CH_2O)_n, CH_2CH_2OH$

(B) Soap—$C_{17}H_{35}COO^-Na^+$

(C) Anionic detergent—$CH_3—(CH_2)_{10}CH_2SO_3^-Na^+$

(D) Cationic detergent— $\left[CH_3(CH_2)_{15}-\overset{\displaystyle CH_3}{\underset{\displaystyle CH_3}{\overset{|}{\underset{|}{N}}}}-CH_3\right]^+ Br^-$

(B) Miscellaneous questions

1 (*b*) Equanil is a tranquiliser.

$$H_2N-\overset{O}{\overset{\|}{C}}-O-CH_2-\overset{CH_3}{\underset{CH_3}{\overset{|}{\underset{|}{C}}}}-CH_2-O-\overset{O}{\overset{\|}{C}}-NH_2$$

Structure of equanil

6 (*b*) Statement I is correct and statement II is a incorrect statement. Correct statement II is salvarsan is a antibacterial but does not contain sulphonamide group.

7 (*a*) Both are correct statements. Correct explanation is, Neurotransmitters are small molecules which bind to the receptor for a very short time to transfer message to it and depart quickly unchanged after transferring the message. The receptor then forwards the message inside the cell.

11 (*c*) Polyethyleneglycols are used in the preparation of non-ionic detergents. Non-ionic detergents do not contain any ion in their constitution.

Reaction is as follow

$$\underset{\text{Stearic acid}}{CH_3-(CH_2)_{16}-COOH} + \underset{\text{Polyethyleneglycol}}{HO-(CH_2CH_2O)_n-CH_2-CH_2OH} \xrightarrow{-H_2O} \underset{\text{Non-ionic detergent}}{CH_3(CH_2)_{16}COO(CH_2CH_2)_n\,CH_2CH_2OH}$$

12 (*d*) An antiseptic is an antimicrobial drug. It tends to destroy or prevent development or inhibit the pathogenic action of microbes. Antiseptics are applied to the living tissues such as wounds, cuts, ulcer and diseased skin surfaces e.g., soframicine.

Bithionol is added to soaps to impart antiseptic properties.

Dilute solutions of some disinfectants can be used as antiseptic e.g., 0.2 per cent solution of phenol is an antiseptic while its one per cent solution is disinfectant. But, antiseptic medicines can not be ingested like antibiotics.

(C) Medical entrances' gallery

1 (*b*) (a) Solution stearate, $C_{17}H_{35}COONa$ is a soap.

(b) Cetyltrimethyl ammonium bromide,

$CH_3-(CH_2)_{15}-\overset{\oplus}{N}(CH_3)_3\overset{\ominus}{Br}$ is a cationic detergent.

(c) and (d) are anionic detergents.

Sodium dodecyl benzene sulphonate :

$p\text{-}CH_3(CH_2)_{11}-C_6H_4-S\overset{\ominus}{O_3}\overset{\oplus}{Na}$

Sodium lauryl sulphate:

$CH_3-(CH_2)_{10}-CH_2-OS\overset{\ominus}{O_3}\overset{\oplus}{Na}$

2 (*d*) Antibiotics which are effective mainly against either gram positive or gram negative bacteria are called narrow spectrum antibiotics. Penicillin-G has a narrow spectrum. Ampicillin and amoxycillin are synthetic modifications of penicillin. These have broad spectrum. Chloramphenicol is a broad spectrum antibiotic.
Hence, option (d) is correct.

3 (*c*) Sucralose is a trichloro derivative of sucrose and is about 650 times sweeter than cane sugar. It is a zero calorie sugar and stable at cooking temperature.

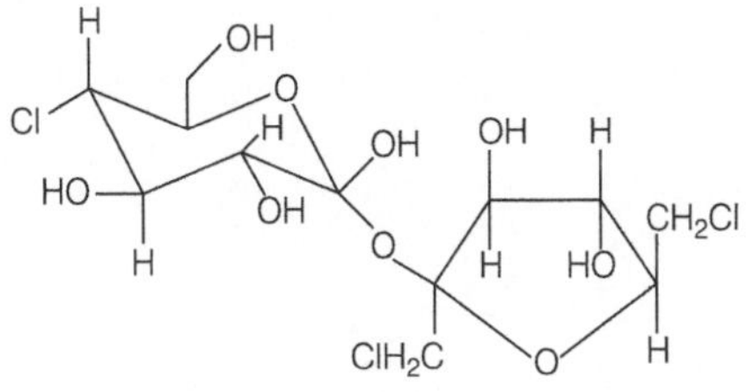

Sucralose

4 (*a*) Saccharin is a low calorie sweetener manufactured from toluene.

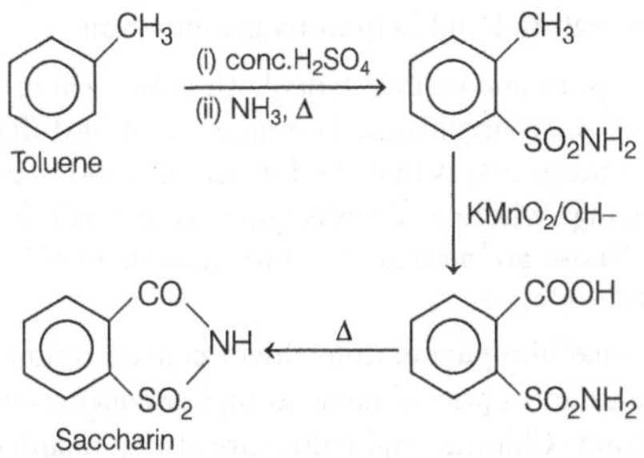

Toluene (CH3) —(i) conc.H_2SO_4 (ii) NH_3, Δ→ o-toluenesulphonamide (CH_3, SO_2NH_2) —$KMnO_2/OH^-$→ ($COOH$, SO_2NH_2) —Δ→ Saccharin (CO, NH, SO_2)

Toluene

Saccharin

5 (*a*) Non-ionic detergents have groups at one end of the alkyl chain showing H-bonding that includes monoesters of polyhydric alcohols or polyethene derived from ethylene oxide. It is used as liquid dishwashing detergent. It removes grease and oil by micelle formation.
Hence, A and R are correct and given R is the correct explanation of the A.

6 (*d*) The chemicals which are used to protect food from microbial action i.e. which inhibit the process of fermentation, acidification and any other decomposition of food are known as *food preservatives.* Table salt, sugar, vegetable oil, vinegar, sodium benzoate (C_6H_5COONa), sodium metabisulphite ($Na_2S_2O_5$), vitamin E. etc. are the common examples of food preservatives.

7 (*d*) Antibiotics, antiseptics and disinfectants are antimicrobial drugs. These type of drug prevents the development or inhibit the pathogenic action of microorganisms such as bacteria (antibacterial drug), virus (antiviral drug), fungi (antifungal drug) or other parasite (antiparasitic drugs selectively).

8 (*d*) Bithional is an antiseptic. It is mixed with soap to impart antiseptic property.

9 (*c*) Mixture of chloroxylenol and terpineol acts as antiseptic.

10 (*d*)
- Novalgin is a non-narcotic analgesic.
- Penicillin is an antibiotic used for curing fever.
- Streptomycin and chloromycetion are also antibiotic drugs.

11 (*d*) Antacid

12 (*b*) Cetyltrimethyl ammonium bromide

$$H_3C(H_2C)_{15}-\underset{CH_3}{\overset{CH_3}{N^+}}-Br^-CH_3$$

13 (*c*) Prontosil is metabolised to sulphanilamide in the body.

14 (*e*) Tetracycline is an antibiotic that fights bacteria in the body. It is used to treat many different bacterial infection in the body like urinary tract infections etc.

15 (*a*) Butylated *p*-hydroxyanisole (BHA), butylated *p*-hydroxy toluene (BHT), esters of gallic acid are examples of antioxidants. These chemicals retard the action of oxygen on the food material and help in its preservation.

16 (*c*) Antiseptics and disinfectants both either kill or prevent the growth of microorganisms. The main point of difference between these two is that the former(antiseptics) are used for living beings where as disinfectants are not safe for living tissues. These are actually used for inanimate objects like floors, tiles etc.

A substance like phenol in its lower concentration (0.2%) behaves as antiseptic whereas in higher concentration (1%) as disinfectant. Chlorine and iodine are strong disinfectants whereas dilute solution of boric acid and hydrogen peroxide are mild antiseptics.

17 (*b*) Salol is phenyl salicylate. It is used as an antiseptic.

$$\text{Salicylic acid } (C_6H_4(OH)COOH) + C_6H_5OH \text{ (Phenol)} \xrightarrow{POCl_3} \text{Phenyl salicylate (Salol) } (C_6H_4(OH)COOC_6H_5) + H_2O$$

18 (*c*) Salol is phenyl salicylate. It is used as an antiseptic.

19 (*c*) Paracetamol is a drug used to decrease body temperature (antipyretic).

20 (*c*) The reaction of NaOH with fat is known as saponification because this is the way of manufacture of soap.

21 (*d*) The hydrolytic reaction of fats with caustic soda is known as saponification.

$$\begin{array}{l} CH_2OCOC_{17}H_{35} \\ | \\ CHOCOC_{17}H_{35} \\ | \\ CH_2OCOC_{17}H_{35} \end{array} + 3NaOH \longrightarrow \begin{array}{l} CH_2OH \\ | \\ CHOH \\ | \\ CH_2OH \end{array} + \underset{\text{(Soap)}}{3C_{17}H_{35}CO\overset{-}{O}\overset{+}{N}a}$$

22 (*d*) Chloramphenicol is an antibiotic broad spectrum used in the treatment of bacterial infections.

23 (*b*) Dettol contains chloroxylenol. It kills bacterial and provides protection against germs causing infection and illness.

24 (*d*) The structure of dopamine is

$$C_6H_3(CH_2CH_2NH_2)(OH)_2 \text{ (3,4-dihydroxy, with } CH_2CH_2NH_2\text{ at position 1)}$$

25 (*c*) The first discovered antibiotic was penicillin. Antibiotics are the compounds produced by bacteria and fungi capable of killing microbial species.

26 (*a*) Saccharin is a low calorie sweetener manufactured from toluene.

$$\underset{\text{Toluene}}{C_6H_5CH_3} \xrightarrow[\text{(ii) } NH_3,\ \Delta]{\text{(i) Conc. } H_2SO_4} C_6H_4(CH_3)SO_2NH_2 \xrightarrow{KMnO_4/OH^-} C_6H_4(COOH)SO_2NH_2 \xrightarrow{\Delta} \underset{\text{Saccharin}}{C_6H_4(CO)(SO_2)NH}$$

27 (*b*) Chemically paper and cloth consists of cellulose. In plant eating animals digestion of cellulose takes place in presence of enzyme cellulase.

28 (*c*) Diphenyl hydramine is employed as antihistamine.

29 (*c*) Triiodomethane or iodoform has antiseptic property and it is applied to wounds.

30 (*b*) Terfenadine is an antihistamine. Antihistamine interfere with the natural action of histamine by competing with histamine for binding sites of receptor where histamine exerts its effect.

31 (*b*) Disparlure is an example of pheromone (sex attractants).

APPENDIX

APPENDIX A

1. Conversion

- **$CH_3CH_2CH_2CH_2I$ to $CH_3CH_2CH_2CH_2Br$**

$$CH_3CH_2CH_2CH_2I \xrightarrow{\text{alc. KOH}} CH_3CH_2CH_2=CH_2 \xrightarrow[\text{peroxide}]{\text{HBr}} CH_3CH_2CH_2CH_2Br$$

- **Cyclobutene to *trans*-1, 2-dihydroxycyclobutane**

$$\text{Cyclobutene} \xrightarrow{HCO_3H} \textit{trans}\text{-1,2-dihydroxycyclobutane (H, OH; H, OH)}$$

- **Cyclohexane to cyclohexene and 1,3-cyclohexadiene**

$$\text{Cyclohexane} \xrightarrow[\Delta]{Br_2} \text{Bromocyclohexane} \xrightarrow{\text{alc. KOH}} \text{Cyclohexene} \xrightarrow{\text{NBS}} \text{3-Bromocyclohexene (Br)} \xrightarrow{\text{alc.KOH}} \text{1,3-Cyclohexadiene}$$

- **2-bromobutane to *cis*- and *trans*-2-butene**

$$CH_3CH_2CH(Br)CH_3 \xrightarrow{\text{alc. KOH}} CH_3CH=CHCH_3 \xrightarrow{Br_2}$$

$$\underset{||}{CH_3\,CH\,CH\,CH_3}\ (\text{Br Br}) \xrightarrow[\Delta]{\text{KOH }(s)} CH_3C\equiv CCH_3 \xrightarrow{Li/NH_3} \textit{trans-2- butene}$$

$$CH_3C\equiv CCH_3 \xrightarrow{H_2,\ \text{Lindlar catalyst}} \textit{cis}\text{-2-butene}$$

- **2-butyne to *meso* and *racemic*-2,3-dideutero butane**

$$CH_3C\equiv CCH_3 \xrightarrow{H_2/\text{Lindlar catalyst}} \underset{cis}{(H_3C)(H)C=C(CH_3)(H)} \xrightarrow[syn-\text{addition}]{D_2/Pt} \underset{meso}{CH_3-C(H)(D)-C(H)(D)-CH_3}$$

$$CH_3C\equiv CCH_3 \xrightarrow{Li\ /NH_3} \underset{trans}{(H_3C)(H)C=C(H)(CH_3)} \xrightarrow[syn-\text{addition}]{D_2/Pt} \underset{racemic\ dl\text{-}\ (\pm)}{CH_3-C(H)(D)-C(D)(H)-CH_3}$$

- **2-butyne to *meso* and *racemic*-2,3-dibromobutane**

$$CH_3C\equiv CCH_3 \xrightarrow{H_2/\text{Lindlar catalyst}} \underset{cis}{2-\text{butene}} \xrightarrow[anti-\text{addition}]{Br_2} \underset{dl\text{-}\ (\pm)}{CH_3-C(H)(Br)-C(Br)(H)-CH_3}$$

$$CH_3C\equiv CCH_3 \xrightarrow{Li/NH_3} \underset{trans}{2-\text{butene}} \xrightarrow[anti-\text{addition}]{Br_2} \underset{meso}{CH_3-C(H)(Br)-C(H)(Br)-CH_3}$$

- ***Cis*-2-butene to *trans*-2-butene and *vice-versa***

cis (H_3C)(H)C=C(H)(CH_3) and *trans* (H_3C)(H)C=C(CH_3)(H) $\xrightarrow{Br_2}$ $CH_3CH(Br)—CH(Br)CH_3$ $\xrightarrow[\Delta]{KOH\ (s)}$ $CH_3C\equiv CCH_3$

$CH_3C\equiv CCH_3$ $\xrightarrow{H_2,\ \text{Lindlar catalyst}}$ *cis*–2–butene

$CH_3C\equiv CCH_3$ $\xrightarrow{Li/NH_3}$ *trans*-2 - butene

- **1-butyne to 1-butanol and 2-butanol**

$CH_3CH_2C\equiv CH \xrightarrow{H_2/\text{Lindlar catalyst}} CH_3CH_2CH{=}CH_2 \xrightarrow[H_2O_2/OH^-]{BH_3/THF} \text{1-butanol}$

$CH_3CH_2C\equiv CH \xrightarrow[H_2SO_4]{HgSO_4} \text{2-butanone} \xrightarrow{LiAlH_4} \text{2-butanol}$

$CH_3CH_2CH{=}CH_2 \xrightarrow{H_3O^+} \text{2-butanol}$

- **1-butene to 1,3-butadiene**

$CH_2{=}CHCH_2CH_3 \xrightarrow{NBS} CH_2{=}CHCH(Br)CH_3 \xrightarrow{\text{alc. KOH}} CH_2{=}CHCH{=}CH_2$

- **CH≡CH to 1,3-butadiene**

$2CH\equiv CH \xrightarrow[NH_4Cl]{CuCl} CH_2{=}CHC\equiv CH \xrightarrow[BaSO_4]{H_2/Pt} CH_2{=}CHCH{=}CH_2$

- **Benzene to *o*, *p* and *m*-chloronitrobenzene**

m-chloronitrobenzene $\xleftarrow{Cl_2/Fe}$ nitrobenzene $\xleftarrow{HNO_3/H_2SO_4}$ benzene $\xrightarrow{Cl_2/Fe}$ chlorobenzene $\xrightarrow[H_2SO_4]{HNO_3}$ p-chloronitrobenzene

If *p*-position is first blocked by sulphonation, —NO_2 will go exclusively in *o*-position and desulphonation will give required product.

benzene $\xrightarrow{Cl_2/Fe}$ chlorobenzene $\xrightarrow[SO_3]{H_2SO_4}$ p-chlorobenzenesulphonic acid (SO_3H) $\xrightarrow[H_2SO_4]{HNO_3}$ (Cl, NO_2 ortho, SO_3H para) $\xrightarrow[\Delta,\ \text{steam}]{H_2SO_4}$ o-chloronitrobenzene

- **Benzene to *p*-, *o*- and *m*- bromobenzoic acid**

benzene $\xrightarrow[AlCl_3]{CH_3Cl}$ toluene $\xrightarrow{Br_2/Fe}$ p-bromotoluene $\xrightarrow{\text{hot } KMnO_4}$ p-bromobenzoic acid (*p*-)

toluene $\xrightarrow{H_2SO_4\ |\ SO_3}$ p-toluenesulphonic acid (CH_3, SO_3H) $\xrightarrow{Br_2/Fe}$ (CH_3, Br ortho, SO_3H para) $\xrightarrow[\Delta,\ \text{steam}]{H_2SO_4}$ o-bromotoluene $\xrightarrow{\text{hot } KMnO_4}$ o-bromobenzoic acid (*o*-)

(as discussed above)

benzene $\xrightarrow[AlCl_3]{CH_3Cl}$ toluene $\xrightarrow{\text{hot } KMnO_4}$ benzoic acid (COOH) $\xrightarrow{Br_2/Fe}$ m-bromobenzoic acid (*m*-)

- **Benzene to *m*-bromotoluene**

$$C_6H_6 \xrightarrow[AlCl_3]{CH_3Cl} C_6H_5CH_3 \xrightarrow[\Delta]{3Cl_2} C_6H_5CCl_3 \xrightarrow{Br_2/Fe} m\text{-}BrC_6H_4CCl_3 \xrightarrow[\text{reduction}]{Zn/HCl} m\text{-}BrC_6H_4CH_3$$

(– CCl_3 group in *m*-directing)

- **Toluene to 3,5-dinitrobenzoic acid**

$$C_6H_5CH_3 \xrightarrow{\text{hot } KMnO_4} C_6H_5COOH \xrightarrow[H_2SO_4]{HNO_3} m\text{-}NO_2C_6H_4COOH \xrightarrow[H_2SO_4,\ \Delta]{\text{fuming } HNO_3} 3,5\text{-}(O_2N)_2C_6H_3COOH$$

(—COOH, NO_2 both are deactivating groups hence, further nitration requires more vigorous conditions)

- **3,3-dimethyl-1-butene to 3,3-dimethyl-1-butanol**

$$CH_3C(CH_3)_2CH{=}CH_2 \xrightarrow[(ii)\ H_2O_2,\ OH^-]{(i)\ BH_3\cdot THF} CH_3C(CH_3)_2CH_2CH_2OH$$

There is again hydration but by **anti-Markownikoff rule.** In such case use BH_3 /THF followed by oxidation with H_2O_2/OH^- .

Remember

- Product (alcohol) is required according to Markownikoff rule without shift –use oxymercuration [by $Hg(OAc)_2$] H_2O followed by reduction with $NaBH_4$.
- Product (alcohol) is required according to anti-Markownikoff rule –use $BH_3 \cdot$ THF followed by oxidation with H_2O_2/OH^- .
- Product (alcohol) is required with shift, use simple H_3O^+

$$\underset{\text{(major)}}{CH_3C(CH_3)(OH)CH_2CH_3} \xleftarrow{H_3O^+} CH_3CH(CH_3)CH{=}CH_2 \xrightarrow[NaBH_4]{Hg(OAc)_2-H_2O} CH_3CH(CH_3)CH(OH)CH_3$$

$$CH_3CH(CH_3)CH{=}CH_2 \xrightarrow[H_2O_2,\ OH^-]{BH_3\cdot THF} CH_3CH(CH_3)CH_2CH_2OH$$

- **CH≡CH to 1-butanol and 2-butanol**

$$CH_3CH_2C{\equiv}CH \xleftarrow[CH_3CH_2Br]{NaNH_2} CH{\equiv}CH \xrightarrow[1\ mol]{H_2} CH_2{=}CH_2 \xrightarrow{HBr} CH_3CH_2Br$$

$$CH_3CH_2C{\equiv}CH \xrightarrow{H_2,\ 1\ mol} CH_3CH_2CH{=}CH_2$$

$$CH_3CH_2CH(OH)CH_3 \xleftarrow{H_3O^+} CH_3CH_2CH{=}CH_2 \xrightarrow[H_2O_2,\ OH^-]{BH_3\cdot THF} CH_3CH_2CH_2CH_2OH$$

- **CH_3OH to CH_3CH_2OH, $(CH_3)_2CHOH$, $(CH_3)_3COH$ in sequence**

$$HCH_2OH \xrightarrow{Cu,\ \Delta} H{-}C(H){=}O \xrightarrow[(ii)\ H_3O^+]{(i)\ CH_3MgBr} CH_3CH_2OH \xrightarrow{Cu,\ \Delta} CH_3CH{=}O$$

$$\xrightarrow[(ii)\ H_3O^+]{(i)\ CH_3MgBr} (CH_3)_2CHOH \xrightarrow{Cu,\ \Delta} (H_3C)_2C{=}O \xrightarrow[(ii)\ H_3O^+]{(i)\ CH_3MgBr} (CH_3)_3COH$$

- **Ethene to 1-propanol**

$$CH_2{=}CH_2 \xrightarrow{C_6H_5CO_3H} \underset{\text{O}}{CH_2{-}CH_2} \xrightarrow[\text{(ii) } H_3O^+]{\text{(i) } CH_3MgBr} CH_3CH_2CH_2OH$$

- **Benzene to styrene**

$$C_6H_6 + CH_2{=}CH_2 \xrightarrow{HF} C_6H_5CH_2CH_3 \xrightarrow{NBS} \underset{\text{(benzylic substitution)}}{C_6H_5CH(Br)CH_3} \xrightarrow{alc.KOH} \underset{\text{styrene}}{C_6H_5CH{=}CH_2}$$

- **$CH{\equiv}CH$ to CH_3COCH_3**

$$CH{\equiv}CH \xrightarrow{HgSO_4/H^+} CH_3CHO \xrightarrow[\text{(ii) } H_3O^+]{\text{(i) } CH_3MgBr} (CH_3)_2CHOH \xrightarrow{Cu,\ \Delta} (CH_3)_2C{=}O$$

$$CH{\equiv}CH \xrightarrow{NaNH_2} CH{\equiv}CNa \xrightarrow{CH_3I} CH{\equiv}CCH_3 \xrightarrow{HgSO_4/H^+} (CH_3)_2C{=}O$$

- **$CH_2{=}CH_2$ to CH_3CHO**

$$CH_2{=}CH_2 \xrightarrow[H_2O]{PdCl_2\cdot CuCl_2} CH_3CHO \text{ (Wacker process)}$$

- **Propene to butanal**

$$CH_3CH{=}CH_2 + CO + H_2 \xrightarrow{Co_2(CO)_8} CH_3CH_2CH_2CHO$$

- **$CH{\equiv}CH$ to *t*-butyl alcohol**

$$CH{\equiv}CH \xrightarrow[CH_3I]{NaNH_2} CH_3C{\equiv}CH \xrightarrow{HgSO_4/H^+} CH_3\overset{O}{\overset{\|}{C}}CH_3 \xrightarrow[\text{(ii) } H_3O^+]{\text{(i) } CH_3MgBr} CH_3{-}\underset{CH_3}{\overset{CH_3}{C}}{-}OH$$

- **$CH{\equiv}CH$ to lactic acid**

$$CH{\equiv}CH \xrightarrow{HgSO_4/H^+} CH_3CHO \xrightarrow{HCN} CH_3\underset{OH}{CH}CN \xrightarrow{H_3O^+} \underset{\text{lactic acid}}{CH_3\underset{OH}{CH}COOH}$$

- **$CH{\equiv}CH$ to** (a) **crotonic aldehyde** (b) **crotonic acid**

$$CH{\equiv}CH \xrightarrow{HgSO_4/H^+} CH_3CHO \xrightarrow[aldol]{NaOH} CH_3\underset{OH}{CH}CH_2CHO \xrightarrow{\Delta} \underset{(a)}{CH_3CH{=}CHCHO} \xrightarrow[\Delta]{\text{Tollen's reagent}} \underset{(b)}{CH_3CH{=}CHCOOH}$$

(a) **$CH_3CH_2NH_2$ to CH_3NH_2 (*A*)** (b) **CH_3CH_2OH to CH_3OH (*B*)** (c) **CH_3COOH to HCOOH (*C*)**

$$\underset{(a)}{CH_3CH_2NH_2} \xrightarrow{HNO_2} \underset{(b)}{CH_3CH_2OH} \xrightarrow[\Delta]{KMnO_4} \underset{(c)}{CH_3COOH} \xrightarrow{NH_3,\ D}$$

$$CH_3CONH_2 \xrightarrow[\Delta]{KBrO} \underset{(A)}{CH_3NH_2} \xrightarrow{HNO_2} \underset{(B)}{CH_3OH} \xrightarrow{oxidn} \underset{(C)}{HCOOH}$$

- **Glycol to succinic acid**

$$\begin{matrix} CH_2OH \\ | \\ CH_2OH \end{matrix} \xrightarrow{PCl_5} \underset{(a)}{\begin{matrix} CH_2Cl \\ | \\ CH_2Cl \end{matrix}} \xrightarrow[H_3O^+]{CN^-} \begin{matrix} CH_2COOH \\ | \\ CH_2COOH \end{matrix}$$

[or by carbonation of (a)]

- **Glycerol to (a) acrolein (b) acrylic acid (c) allyl alcohol**

$$\begin{array}{l} CH_2OH \\ | \\ CHOH \\ | \\ CH_2OH \end{array} \xrightarrow[-2H_2O]{KHSO_4} \begin{array}{l} CH_2 \\ \| \\ CH \\ | \\ CHO \\ (a) \end{array} \xrightarrow[\Delta]{\text{Tollen's reagent}} \begin{array}{l} CH_2 \\ \| \\ CH \\ | \\ COOH \\ (b) \end{array}$$

$$(a) \xrightarrow{LiAlH_4} \begin{array}{l} CH_2 \\ \| \\ CH \\ | \\ CH_2OH \\ (c) \end{array}$$

$$\text{Glycerol} \xrightarrow[D,\ 490\ K]{\text{oxalic acid}} \begin{array}{l} CH_2 \\ \| \\ CH \\ | \\ CH_2OH \\ (c) \end{array}$$

- **Formic acid to oxalic acid**

$$HCOOH \xrightarrow{NaOH} HCOONa \xrightarrow[-H_2]{\Delta} \begin{array}{l} COONa \\ | \\ COONa \end{array} \xrightarrow{H^+} \begin{array}{l} COOH \\ | \\ COOH \end{array}$$

- **1-butanol to 2-butanol**

$$CH_3CH_2CH_2CH_2OH \xrightarrow[\Delta]{\text{conc. } H_2SO_4} \underset{\text{2-butene (major)}}{CH_3CH{=}CHCH_3} \xrightarrow{H_3O^+} CH_3\underset{\displaystyle OH}{\underset{|}{C}H}CH_2CH_3$$

- **Acetic acid to glycolic acid**

$$CH_3COOH \xrightarrow[HVZ]{P/Br_2} \underset{\displaystyle Br}{\underset{|}{C}H_2}COOH \xrightarrow[H^+]{NaOH} \underset{\displaystyle OH}{\underset{|}{C}H_2}COOH$$

- **Propanoic acid to 2-hydroxypropanoic acid** (lactic acid)

$$CH_3CH_2COOH \xrightarrow[HVZ]{P/Br_2} CH_3\underset{\displaystyle Br}{\underset{|}{\overset{\alpha}{C}H}}COOH \xrightarrow[H^+]{NaOH} \underset{\text{lactic acid}}{CH_3\underset{\displaystyle OH}{\underset{|}{C}H}COOH}$$

- **Acetic acid to (a) lactic acid (b) crotonic acid**

$$CH_3COOH \xrightarrow{PCl_5} CH_3\overset{\displaystyle O}{\overset{\|}{C}}Cl \xrightarrow[Pd/BaSO_4]{H_2} CH_3CHO \xrightarrow{HCN} CH_3\underset{\displaystyle OH}{\underset{|}{C}H}CN \xrightarrow{H_3O^+} \underset{(a)}{CH_3\underset{\displaystyle OH}{\underset{|}{C}H}COOH}$$

$$CH_3CHO \xrightarrow{\text{NaOH (Aldol condensation)}} CH_3\underset{\displaystyle OH}{\underset{|}{C}H}CH_2CHO \xrightarrow{\Delta} CH_3CH{=}CHCHO \xrightarrow[\Delta]{\text{Tollen's reagent}} \underset{(b)}{CH_3CH{=}CHCOOH}$$

- **CH_3NH_2 to $CH_3CH_2NH_2$ and $(H_3C)_2NH$**

$$CH_3NH_2 \xrightarrow{HNO_2} CH_3OH \xrightarrow{PCl_5} CH_3Cl \xrightarrow{AgCN} CH_3NC \xrightarrow{LiAlH_4} CH_3NHCH_3$$

$$CH_3NH_2 \xrightarrow{CHCl_3\ +\ KOH,\ \Delta} CH_3NC$$

$$CH_3Cl \xrightarrow{KCN} CH_3CN \xrightarrow{LiAlH_4} CH_3CH_2NH_2$$

- **Aniline to (a) benzoic acid (b) benzyl amine (c) benzyl alcohol (d) benzaldehyde (e) benzene (f) benzonitrile (g) chlorobenzene**

$$C_6H_5NH_2 \xrightarrow[273\ K]{NaNO_2/HCl} C_6H_5N_2Cl \xrightarrow[K_3F_e(CN)_6]{KCN/CuCN} \underset{(f)}{C_6H_5CN} \xrightarrow{Sn/HCl} \underset{(b)}{C_6H_5CH_2NH_2}$$

$$C_6H_5N_2Cl \xrightarrow{CuCl/HCl} \underset{(g)}{C_6H_5Cl}$$

$$C_6H_5N_2Cl \xrightarrow{H_3PO_2} \underset{(e)}{C_6H_6}$$

$$\underset{(f)}{C_6H_5CN} \xrightarrow{H_3O^+} \underset{(a)}{C_6H_5COOH}$$

$$\underset{(b)}{C_6H_5CH_2NH_2} \xrightarrow{HNO_2} \underset{(c)}{C_6H_5CH_2OH}$$

$$\underset{(c)}{C_6H_5CH_2OH} \xrightarrow[KMnO_4]{hot} \underset{(a)}{C_6H_5COOH}$$

$$\underset{(c)}{C_6H_5CH_2OH} \xrightarrow{Cu,\ \Delta} \underset{(d)}{C_6H_5CHO}$$

- **Ethyl cyanide to (a) ethyl amine (b) methyl cyanide and (c) methylamine**

$$CH_3CH_2CN \xrightarrow{H_3O^+} CH_3CH_2COOH \xrightarrow[\Delta]{NH_3} CH_3CH_2CONH_2 \xrightarrow[\Delta]{KBrO} \underset{(a)}{CH_3CH_2NH_2}$$

$$CH_3CH_2COOH \xrightarrow{N_3H,\ \Delta} \underset{(a)}{CH_3CH_2NH_2}$$

$$\xrightarrow{HNO_2} CH_3CH_2OH \xrightarrow[hot]{KMnO_4} CH_3COOH \xrightarrow{NH_3,\ \Delta} CH_3CONH_2 \xrightarrow{KBrO,\ \Delta} \underset{(c)}{CH_3NH_2}$$

$$CH_3CONH_2 \xrightarrow{P_2O_5} \underset{(b)}{CH_3CN}$$

- **Aniline to sulphanilic acid**

$$C_6H_5NH_2 \xrightarrow[473\ K]{\text{fuming } H_2SO_4} C_6H_5NH_3^+HSO_4^- \xrightarrow{\Delta} p\text{-}NH_2C_6H_4SO_3H$$

- **Phenol to (a) salicylic acid and (b) aspirin**

$$C_6H_5OH \xrightarrow[\Delta,\ H^+]{CCl_4,\ NaOH} \underset{(a)}{o\text{-}HOC_6H_4COOH} \xrightarrow[CH_3COOH]{(CH_3CO)_2O} \underset{(b)}{o\text{-}CH_3COOC_6H_4COOH}$$

(product (b): OCOCH$_3$ and COOH on the ring)

- **Benzene to cumene**

$$C_6H_6 \xrightarrow[AlCl_3]{CH_3CH_2CH_2Cl} C_6H_5CH(CH_3)_2$$

$$C_6H_6 \xrightarrow{CH_3CH{=}CH_2} C_6H_5CH(CH_3)_2$$

- **Benzene to benzaldehyde**

$$C_6H_6 \xrightarrow[AlCl_3]{CH_3Cl} C_6H_5CH_3 \xrightarrow{CrO_2Cl_2} C_6H_5CHO$$

- **Aniline to (a) *N*-phenyl ethanamide (b) *p*-nitroaniline**

$$C_6H_5NH_2 \xrightarrow{CH_3COCl} \underset{(a)}{C_6H_5NHCOCH_3} \xrightarrow[H_2SO_4]{HNO_3} O_2N\text{-}C_6H_4\text{-}NHCOCH_3 \xrightarrow[H_2O]{CH_3COOH} \underset{(b)}{O_2N\text{-}C_6H_4\text{-}NH_2}$$

- **Benzene to *m*-bromophenol**

$$C_6H_6 \xrightarrow{HNO_3/H_2SO_4} C_6H_5NO_2 \xrightarrow{Br_2/Fe} m\text{-}BrC_6H_4NO_2 \xrightarrow{Sn/HCl} m\text{-}BrC_6H_4NH_2 \xrightarrow[273\ K]{NaNO_2\text{-}HCl} m\text{-}BrC_6H_4N_2Cl \xrightarrow{H_3O^+,\ \Delta} m\text{-}BrC_6H_4OH$$

- **Benzene to 1,4-cyclohexadiene**

$$C_6H_6 \xrightarrow[EtOH]{Na/NH_3} \text{1,4-cyclohexadiene}$$ (The Birch reduction)

- **CH_3COOH to CH_3NH_2** (in one step)

$$CH_3COOH \xrightarrow[(-\ CO_2,\ -\ N_2)]{N_3H} CH_3NH_2$$ (Schmidt)

(mixture of NaN_3 and conc. H_2SO_4 is used $NaN_3 + H_2SO_4 \longrightarrow N_3H + NaHSO_4$)

- **HCOOH to CH_3COOH**

$$HCOOH \xrightarrow{LiAlH_4} CH_3OH \xrightarrow{PCl_5} CH_3Cl \xrightarrow{KCN} CH_3CN \xrightarrow{H_3O^+} CH_3COOH$$

- **Acetophenone to phenyl acetylene**

$$C_6H_5\overset{O}{\overset{\|}{C}}\text{—}CH_3 \xrightarrow{PCl_5} \xrightarrow[\text{(as above)}]{NaNH_2,\ \Delta} \xrightarrow{H^+} C_6H_5\text{—}C{\equiv}CH$$

- **Propanone to propanal**

$$CH_3\overset{O}{\overset{\|}{C}}CH_3 \xrightarrow{PCl_5} CH_3-CCl_2-CH_3 \xrightarrow[H^+]{NaNH_2,\ D} CH_3C{\equiv}CH \xrightarrow[H_2O_2,\ OH^-]{BH_3{\cdot}THF} CH_3CH_2CHO$$

- **Aniline to 1,3,5-tribromo benzene**

$$C_6H_5NH_2 \xrightarrow{Br_2} \text{2,4,6-}Br_3C_6H_2NH_2 \xrightarrow[273\ K]{NaNO_2\ +\ dil.\ HCl} \text{2,4,6-}Br_3C_6H_2N_2Cl \xrightarrow{H_3PO_2} \text{1,3,5-}Br_3C_6H_3$$

- **Propyne to *trans*-1-phenyl-2-butene**

$$CH_3C\equiv CH \xrightarrow[(ii)\ PHCH_2Cl]{(i)\ NaNH_2} CH_3C{\equiv}CCH_2Ph \xrightarrow{Li/NH_3} \underset{H}{\overset{H_3C}{}}\!\!>C{=}C<\!\!\underset{CH_2Ph}{\overset{H}{}}$$

- **$CH_3CH_2\overset{13}{C}\equiv CH \longrightarrow CH_3CH_2O\overset{13}{C}\overset{O}{\overset{\|}{}}CH_3$** ($^{13}C$ is isotopically labelled carbon)

$$CH_3CH_2\overset{13}{C}{\equiv}CH \xrightarrow{HgSO_4 + H_2SO_4} CH_3CH_2\overset{13}{C}(OH){=}CH_2 \xrightarrow{\text{tautomerises}} CH_3CH_2\overset{13}{C}(=O)CH_3 \xrightarrow[\text{(Baeyer-Villiger oxidation)}]{C_6H_5CO_3H} CH_3CH_2O\overset{13}{C}(=O)CH_3$$

- ***Trans*-2-hexene to *cis*-2-hexene**

$$\underset{\textit{trans}}{\underset{H}{\overset{H_3C}{}}\!\!>C{=}C<\!\!\underset{CH_2CH_2CH_3}{\overset{H}{}}} \xrightarrow[CCl_4]{Br_2} CH_3CHBr\text{—}CHBrCH_2CH_2CH_3 \xrightarrow[470\ K]{\text{alcoholic KOH}} CH_3C{\equiv}CCH_2CH_2CH_3$$

$$\xrightarrow[\text{Lindlar catalyst}]{H_2} \underset{\textit{cis}}{\underset{H}{\overset{H_3C}{}}\!\!>C{=}C<\!\!\underset{H}{\overset{CH_2CH_2CH_3}{}}}$$

APPENDIX B

2. Comparison

Section A

	Acid strength	Order	Factor
1.	$HCOOH$ (I), CH_3COOH (II), C_6H_5COOH (III)	II < III < I	III resonance stabilised; II < I by inductive effect
2.	CH_3COOH (I), CH_3CH_2COOH (II), $CH_3CH_2CH_2COOH$ (III)	III < II < I	(I_+) inductive effect
3.	CH_3COOH (I), $CH_2ClCOOH$ (II), $CHCl_2COOH$ (III)	I < II < III	(I_-) of Cl increases acidic strength
4.	1- (I), 2- (II), 3- (III) chlorobutanoic acid	III < II < I	Farther the (I_-) group (Cl), lesser the acid strength.
5.	1- (I), 2- (II), 3- (III) methyl pentanoic acid	I < II < III	Farther the (I_+) group, greater the acid strength.
6.	1- (I), 2- (II), 3- (III) isomeric butyl alcohol	III < II < I	(I_+) group (methyl) increases electron density hence acidic nature is decreased, and basic nature is decreased.
7.	alkane (I), alkene (II) and terminal alkynes (III)	I < II < III	Acidic nature of C — H bond is $sp > sp^2 > sp^3$ hybridised carbon.
8.	(I) OH; (II) OH, CH_3; (III) OH, NO_2	II < I < III	—CH_3 is electron repelling and —NO_2 is electron withdrawing.
9.	(I) OH, CH_3; (II) OH, CH_3; (III) OH, CH_3	II < I < III	—CH_3 is electron repelling; decreases acidic strength of phenol (Table 18.8).
10.	(I) OH, OCH_3; (II) OH, OCH_3; (III) OH, OCH_3	III < I < II	—do—
11.	(I) OH, NO_2; (II) OH, NO_2; (III) OH, NO_2	I < II < III	—NO_2 is electron withdrawing; III is more resonance stabilised than I and also than II. In I, only inductive effect is operative.
12.	FCH_2COOH (I), $ClCH_2COOH$ (II), $BrCH_2COOH$ (III), ICH_2COOH (IV)	I > II > III > IV	Greater the electronegativity (F > Cl > Br > I) greater the (I_-) effect, hence greater the ionisation giving H^+.

	Acid strength	Order	Factor
13.	$C_6H_5CH_2COOH$ (I), $CH_2{=}CHCH_2COOH$ (II), CH_3CH_2COOH (III)	I > II > III	sp^2-hybridised carbon of I, II are more electronegative hence acid strength is increased. Benzylic ($C_6H_5CH_2$) is more stabilised than allylic ($CH_2{=}CHCH_2$).
14.	$NCCH_2COOH$, CH_3COOH, CH_3CH_2COOH (I, II, III)	I > II > III	—CN is electron withdrawing and $—CH_3$ is electron repelling.
15.	CH_3OH (I), CF_3OH (II), CCl_3OH (III)	I < III < II	Inductive effect (electron-attracting) of F > Cl > H.
16.	C_6H_5OH (I), H_2O (II), CH_3OH (III)	I > II > III	Phenyl group is electron withdrawing and phenoxide ion is resonance stabilised. $—CH_3$ is electron repelling.
17.	oxalic acid (I), succinic acid (II), malonic acid (III), adipic acid (IV) (all dibasic)	I > II > III > IV	Effect of one —COOH on the other decreases as its distance between them increases.
18.	C_6H_5OH (I), C_6H_5COOH (II), CH_3COOH (III)	I < III < II	(I_+) effect in III; resonance effect in benzene nucleus in I, resonance effect in ($—COO^-$) and benzene nucleus in II.
19.	*o*- (I), *m*- (II), *p*- (III) toluic acid	III < II < I	$—CH_3$ is electron repelling.
20	*o*- (I), *m*- (II), *p*- (III) nitrobenzoic acid	II < III < I	$—NO_2$ is electron attracting.
21.	*o*- (I), *m*- (II), *p*- (III) hydroxy benzoic acid	III < II < I	—OH shows electron withdrawing nature at *o*- and *m*- and electron repelling at *p*-; *o*-isomer due to intramolecular bonding in salicylate ion is stronger than *m*-isomer.
22.	*o*- (I), *m*- (II), *p*- (III) methoxy benzoic acid	III < II = I	–do–
23.	*o*- (I), *m*- (II), *p*- (III) amino benzoic acid	I < III < II	$—NH_2$ is electron repelling.

Section B

	Basic strength	Order	Factor
1.	NH_3 (I), CH_3NH_2 (II), $(CH_3)_2NH$ (III)	I < II < III	(I_+) effect of CH_3 group increases electron density at N-atom hence basic nature.
2.	OH^- (I), CH_3COO^- (II), Cl^- (III) (conjugate base of H_2O, CH_3COOH, HCl)	III < II < I	If acid is weak its conjugate base is strong and *vice versa.*
3.	$CH{\equiv}C^-$ (I), $CH_2{=}CH^-$ (II), $CH_3CH_2^-$ (III)	I < II < III	–do–

	Basic strength	Order	Factor
4.	1°, 2°, 3° (I, II, III) Isomeric butyl alcohol $\underset{\text{base}}{ROH} + \underset{\text{acid}}{HCl} \xrightarrow[\text{(anhydrous)}]{ZnCl_2} \underset{\text{white turbidity}}{RCl} + H_2O$ hence, reactivity of alcohol with Lucas reagent ($HCl + ZnCl_2$)	I < II < III	(I_+) effect of alkyl group increases basic nature of alcohol.
5.	NH_3 (I), CH_3NH_2 (II), $C_6H_5NH_2$ (III)	III < I < II	(I_+) effect of methyl increases basic nature. (I_-) effect of C_6H_5 (due to conjugation) decreases electron density hence, the basic nature.
6.	$CH_3CH_2NH_2$; (I) $CH_3\overset{O}{\overset{\parallel}{C}}NH_2$; (II) C_6H_5–C(=O)–NH_2 (III)	I > III > II	(—$\overset{O}{\overset{\parallel}{C}}$—) is electron- withdrawing decreasing basic nature. In III there is resonance stabilisation along with C=O (cross-conjugation) hence III is more basic than II.
7.	Pyridine (I), Pyrrole (N–H) (II), Aniline (NH_2) (III)	I > III > II	Lone pair on N is not used in delocalisation of π-electrons in I. In II lone pair of the ring is itself used in delocalisation while that of outside ring in III.
8.	p-OCH_3 aniline (I), aniline (II), o-OCH_3 aniline (III), m-OCH_3 aniline (IV)	I > II > III > IV	—OCH_3 is strong electron donating group. This is due to *ortho* effect. All the aniline are less basic than *p*-substituted aniline due to **steric hindrance.**
9.	p-CH_3 aniline (I), m-CH_3 aniline (II), aniline (III), o-CH_3 aniline (IV)	I > II > III > IV	I (hyper conjugation and induction) II (induction) IV (*ortho effect*)
10.	$CH_2{=}CHCH_2NH_2$, I $CH_3CH_2CH_2NH_2$ II $CH{\equiv}CCH_2NH_2$ III	II > I > III	Electron-withdrawing order is : propargyl > allyl > propyl and thus, electron density is decreased in same order.
11.	$C_6H_5CH_2NH_2$ (I), cyclohexyl–CH_2NH_2 (II), p-$NO_2C_6H_4CH_2NH_2$ (III)	II > I > III	In II there is sp^3–hybridised C, in I, sp^2. NO_2 is electron withdrawing.
12.	$CH{\equiv}CH$ (I), $CH_2{=}CH_2$ (II), $CH_3{-}CH_3$ (III)	I < II < III	Electron-withdrawing tendency of C, $sp > sp^2 > sp^3$

	Basic strength	Order	Factor
13.	CH_2OH-benzene (I), CH_2OH-benzene with NO_2 (II), CH_2OH-benzene with OCH_3 (III), CH_2OH-benzene with Cl (IV)	III> I> IV > II	The positive charge that develops on the benzylic carbon is most effectively delocalised by $—OCH_3$ (electron repelling group) group, order being OCH_3 > Cl > H > NO_2 ($—NO_2$ is electron withdrawing).
14.	$CH_3CH_2NH_2$, $HOCH_2CH_2NH_2$, $OHCH_2CH_2CH_2NH_2$ I II III	I > II > III	Electron withdrawing inductive effect of the OH group decreases the electron density on N, lowering the basic nature.
15.	NH_2-benzene (I), NHC_6H_5-benzene (II), NH_2-cyclohexane (III)	III > I > II	Lone pair on N is used in delocalisation of π-electrons in aromatic amines while cyclohexyl is electron-repelling (III); in II, lone pair on N is used by two benzene ring.
16.	NH_2-benzene (I), NH_2-benzene with meta NO_2 (II), NH_2-benzene with para NO_2 (III), NH_2-benzene with ortho NO_2 (IV)	I> II>III > IV	NO_2 is electron-withdrawing, thus nitro-anilines are less basic than aniline. IV is less basic than III because $—NO_2$ is closer and exerts a stronger inductive effect.
17.	NH_2-benzene (I), $NHCH_3$-benzene (II), $N(CH_3)_2$-benzene (III)	I < II < III	CH_3 directly attached to N is electron donating hence basic nature is increased.
18.	NH_2-benzene (I), NHC_6H_5-benzene (II), $NHCOCH_3$-benzene (III)	I > II > III	Phenyl and — $COCH_3$ are electron withdrawing and $—C_6H_5$ < $COCH_3$.
19.	NH_2-benzene (I), $NHCH_3$-benzene (II), NHC_2H_5-benzene (III)	I < II < III	Electron donating nature of CH_3 < CH_2CH_3.
20.	NH_2-benzene with ortho Cl (I), NH_2-benzene with meta Cl (II), NH_2-benzene with para Cl (III)	I < II < III	*ortho* effect in I.

	Basic strength	Order	Factor
21.	NH_2 / NH_2 (ortho) (I), NH_2 / NH_2 (meta) (II), NH_2 / NH_2 (para) (III)	I < II < III	*ortho* effect in I
22.	Cl^- (I), $RCOO^-$ (II), OH^- (III), RO^- (IV), NH_2^- (V)	I < II < III < IV < V	If acid (*HX*) is weak its conjugate base (X^-) is strong.
23.	NH_3, CH_3NH_2, $(CH_3)_2NH$, $(CH_3)_3N$ (I, II, III, IV)		
	(a) (in aqueous solution)	I < IV < II < III	(I_+) effect of three methyl groups in 3° amine (IV) alongwith protonation causes steric hindrance hence less basic than 1° and 2° amines.
	(b) (in gas phase)	I < II < III < IV	As no of alkyl group increases basicity increases.

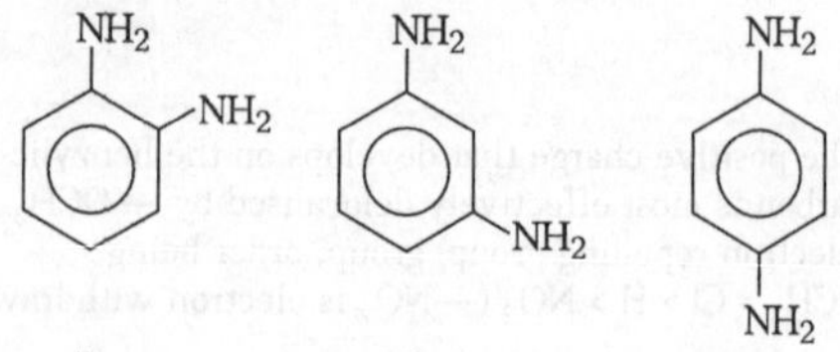

Section C

	Species and its effect	Order	Factor
1.	**Stability of free radical** $CH_3CH_2\dot{C}H_2$ (I), $(CH_3)_2CH\dot{C}H_2$ (II), $(CH_3)_3C\dot{C}H_2$ (III)	I < II < III	Inductive effect is in order 1°< 2°< 3° alkyl group.
2.	**Stability of alkene** *trans* – 2 –butene (I), *cis*–2–butene (II), 1 –butene (III)	I > II > III	In *cis*-isomer steric hindrance decreases stability.
3.	**Heat of hydrogenation of I, II, III** (above)	I < II < III	Greater the stability, smaller the heat of hydrogenation.
4.	**Order of Boiling point** CH_3CH_2OH (I), CH_3OCH_3 (II), CH_3CH_3 (III)	I > II > III	There is intermolecular H-bonding I. III has weak force of attraction and is most volatile.
5.	**Boiling point of *o*, *m*, *p*-nitro phenol**	*o* < *m* < *p*	Intramolecular H-bonding in *o*-isomer makes it most volatile.
6.	**Reactivity of ... with Tollen's reagent** HCHO (I), CH_3CHO (II), CH_3COCH_3 (III), C_6H_5CHO (IV)	I > II > IV > III	—CHO group is easily oxidised compared to keto group.
7.	**Reactivity of ... with Fehling's solution** I, II, III, IV (above)	I > II > IV > III	—do—
8.	**Order for extent of hydration** cyclohexanone (I), 3-bromocyclohexanone (II), cyclohexanecarbaldehyde (CHO) (III), 2-bromocyclohexanecarbaldehyde (CHO, Br) (IV)	I < II < III < IV	Aldehydes are more hydrated than ketones. Halide makes C of carbonyl group more electropositive.
9.	**Stability of ... carbanion** $\overset{\ominus}{C}H_3$ (I), $CH_3\overset{\ominus}{C}H_2$ (II), $(CH_3)_2\overset{\ominus}{C}H$ (III), $(CH_3)_3\overset{\ominus}{C}$ (IV)	I > II > III > IV	(I_+) effect of alkyl group further increases electron density of $C^{\ominus}$.

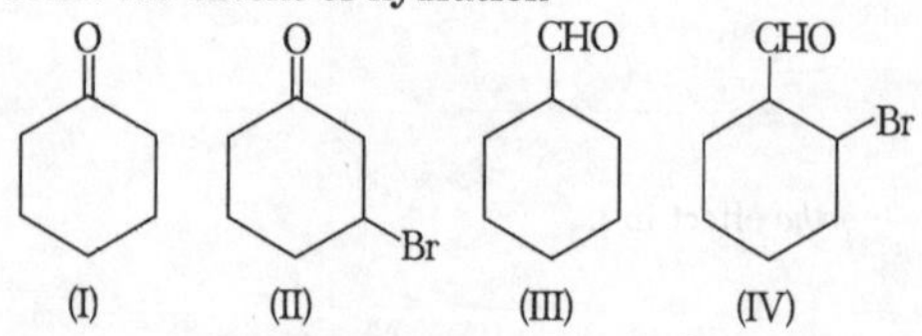

	Species and its effect	Order	Factor
10.	**Reactivity of ... with HBr**		
	p-methoxybenzyl alcohol (CH_2OH, OCH_3) (I); p-nitrobenzyl alcohol (CH_2OH, NO_2) (II); benzyl alcohol (CH_2OH) (III)	II < III < I	OCH_3 is electron-repelling group, increasing basic nature of benzyl alcohol hence reactivity with HBr (acid). —NO_2 is electron withdrawing group hence decreasing basic nature of benzyl alcohol.
11.	**Electrophilic nature of .... for nucleophilic attack.**		
	$H_2\overset{\oplus}{C}$—OH (I); $(H_3C)(H)\overset{\oplus}{C}$—OH (II); $(H_3C)_2\overset{\oplus}{C}$—OH (III)	I > II > III	CH_3 group decreases positive charge on C hence nucleophilic attack.
12.	**Reactivity of isomeric 1°, 2°, 3° butyl halide towards E2 or E1 (elimination)**	1° < 2° < 3°	Due to stability of intermediate carbocation ion. (1° < 2° < 3°)
13.	**Reactivity of .. for S_N1 and S_N2 reaction**		
	MeCl (I); ethyl chloride (II); 2-chloropropane (III); 2-chlorobutane (IV); tert-butyl chloride (V)	S_N1 : I < II < III < IV < V	Due to stability of intermediate carbocation.
		S_N2 : I > II > III > IV > V	Due to steric hindrance in 3° in which attack of nucleophile is retarded.
14.	**Tendency of 1°, 2°, 3° *R–X* for S_N1 and E1 reaction**	Elimination 1° < 2° < 3° Substitution 1° > 2° > 3°	
15.	**Order of dehydration**		
	cyclohexanol (I); cyclohex-2-en-1-ol (II); cyclohexa-2,4-dien-1-ol (III); cyclohex-1-en-1-ol (IV)	IV < I < II < III	Alcohol leading to increase in conjugation due to dehydration is more easily dehydrated. IV is vinylic, hence least.
16.	**Order of stability**		
	$CH_2{=}CH_2$ (I), $CH_3CH{=}CH_2$ (II), $CH_3CH{=}CHCH_3$ (III), $(H_3C)_2C{=}CH_2$ (IV), $(H_3C)_2C{=}CHCH_3$ (V), $(H_3C)_2C{=}C(CH_3)_2$ (VI)	I < II < III < IV < V < VI	Substituted alkenes are more stable.

	Species and its effect	Order	Factor
17.	**Order of stability** (I) (II) (III)	I < III < II	II is more substituted than III.
18.	**Order of stability** (I) (II) (III) (IV)	III > II > I > IV	IV is vinylic while III conjugative, II allylic
19.	**Order of stability** $-CH_2=\dot{C}H$, $\dot{C}H_2CH=CH_2$ I II CH_3 $CH_2\bullet$ III IV	I < IV < II < III	III is 3° allylic and II is 1° allylic.
20.	**Cleavage of ... and reactivity** C—F, C—Cl, C—Br C—I bond I II III IV	I < II < III < IV (IV easiest)	BE of (C—*X*) bond is in order: C–F > C–Cl > C–Br > C–I
21.	**Order of dehydration** 1°, 2°, 3° isomeric butyl alcohol	1° < 2° < 3°	Greater the basic nature of alcohol, greater the tendency of dehydration.
22.	**Order of boiling points** *n* - butyl amine, *n*-butyl alcohol, *n*-pentane I II III	II > I > III	I, II have H-bonding but electronegativity of O > N hence H-bonding in II > I.
23.	**Stability of free radicals** $R_3C\bullet$, $R_2\dot{C}H$, $R\dot{C}H_2$, $\dot{C}H_3$ I II III IV	I > II > III > IV	(I_+) effect of *R*
24.	**Order of formation** $C_6H_5\dot{C}H_2$, $R_3\dot{C}$, $R_2\dot{C}H$, $R\dot{C}H_2$ I II III IV	I > II > III > IV (easiest I)	Greater the stability, easier the formation.
25.	**Order of stability and formation** (a) $C_6H_5\overset{\oplus}{C}H_2$, $(CH_3)_3\overset{\oplus}{C}$, $CH_2=CH\overset{\oplus}{C}H_2$ I II III	I> III > II	I and III are resonance stabilised. (I_+) effect of methyl group decreases positive charge.
	(b) $(CH_3)_2\overset{\oplus}{C}H$, $CH_3\overset{\oplus}{C}H_2$, $\overset{\oplus}{C}H_3$, $CH_2=\overset{\oplus}{C}H$ IV V VI VII	IV >V > VI >VII	
26.	**Stability of free radicals** $CH_3CH_2-\underset{CH_3}{\overset{CH_3}{C}}\bullet$ $CH_3-\underset{CH(CH_3)_2}{\overset{H}{C}}\bullet$ $CH_3CH_2\dot{C}H_2$ I II III	III < II < I	(I_+) effect increases stability 1° < 2° < 3°

	Species and its effect	Order	Factor
27.	**Reactivity of C—H bond** (abstraction of H) $CH_2=CH—H$, vinylic (I); $CH_3—H$, methyl 1° (II); $CH_3CH_2—H$ (III) $(CH_3)_2CH—H$, 2° (IV); $(CH_3)_3C—H$, 3° (V); $CH_2=CHCH_2—H$ (allylic) (VI)	I < II < III < IV < V < VI	Vinyl < methyl 1° < 2° < 3° < allylic
28.	**Leaving nature** (tendency) **of ... in S_N reaction** H^- (I), R^- (II), MeO^- (III), OH^- (IV), CN^- (V), CH_3COO^- (VI), $ROSO_2^-$ (VII), $ArSO_3^-$ (VIII)	I < II < III ~ IV <V < VI < VII < VIII	If acid is strong, its conjugate base is weak and greater the leaving tendency.
29.	**Order of leaving nature** CH_3COO^- (I), CCl_3COO^- (II), PhO^- (III), $PhSO_3^-$ (IV)	III < I < II < IV	—do—
30.	**Order of nitration** 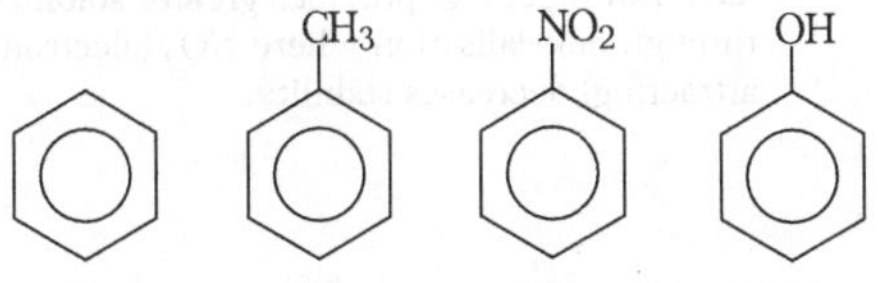 Benzene (I), Toluene (CH_3) (II), Nitrobenzene (NO_2) (III), Phenol (OH) (IV)	III < I < II < IV	—NO_2 is deactivating group, —CH_3 and –OH activating, but due to lone pair on O in phenol, it is more resonance stabilised for S_E reaction.
31.	**Rate of esterification of the following acids with MeOH** $MeCH_2COOH$ (I), $Me_2CHCOOH$ (II), Me_3CCOOH (III), Et_3CCOOH (IV), $(i\text{-Pr})_2CHCOOH$ (V)	I > II > III > IV > V	As the size of the substituents on the α–C increases, the tetrahedrally bonded intermediate becomes more crowded and thus slower the rate.
32.	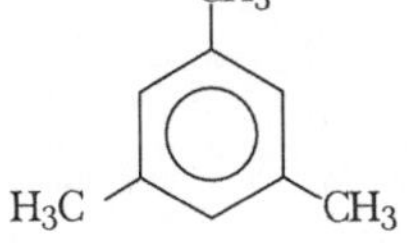 **Relative reactivity of .... with electrophile in S_E reaction** 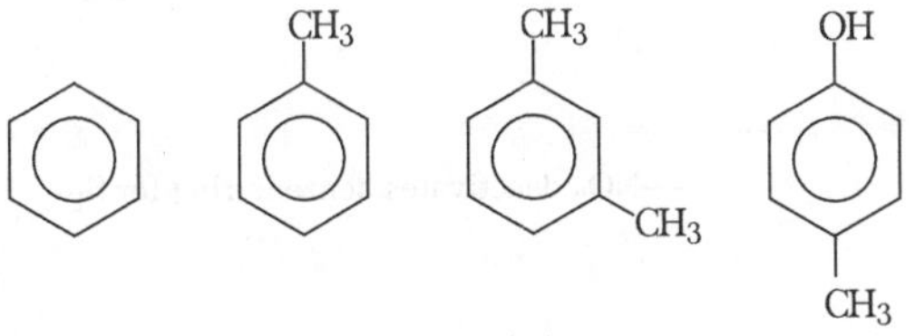 Benzene (I), Toluene (CH_3) (II), m-Xylene (CH_3, CH_3) (III), p-Cresol (OH, CH_3) (IV), Mesitylene (CH_3, H_3C, CH_3) (V)	V > III > IV > II > I	—CH_3 is *o*-, *p*-directing and activating group.
33.	**Relative reactivity of ... with electrophile in S_E reaction** 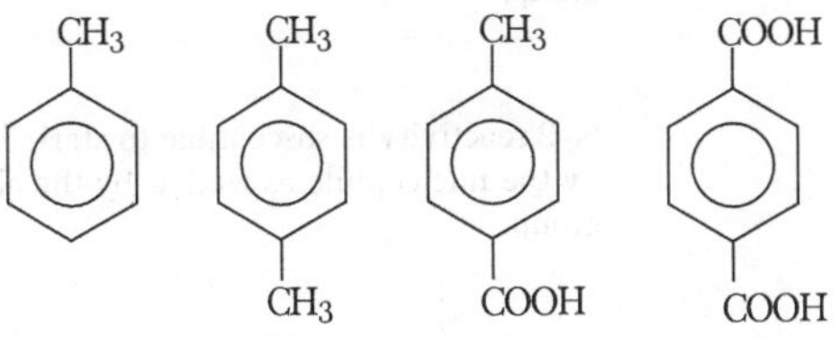 Toluene (CH_3) (I), p-Xylene (CH_3, CH_3) (II), p-Toluic acid (CH_3, COOH) (III), Terephthalic acid (COOH, COOH) (IV)	II > I > III > IV	–CH_3 is *o*-, *p*-directing and activating group while –COOH is *m*- directing and deactivating group.

	Species and its effect	Order	Factor
34.	**Relative reactivity of ... with electrophile in S_E reaction** CH_3 (I), $N(CH_3)_2$ (II), $\overset{\oplus}{N}(CH_3)_3$ (III), $CH_2\overset{\oplus}{N}(CH_3)_3$ (IV) (substituted benzenes)	II > I > IV > III	As the number of sp^3 hybridised C atoms separating the ring from the positively charged substituent increases, deactivating effect decreases.
35.	**Activating effects of the following *o*, *p*- directors.** —OH, I; $—O^{\ominus}$, II; $CH_3\overset{O}{\overset{\parallel}{C}}O—$, III	II > I > III	$—O^{\ominus}$ is best able to donate electrons thereby giving a very stable uncharged intermediate. In $CH_3\overset{O}{\overset{\parallel}{C}}O—$cross conjugation diminished its ability to donate electrons to an arenium ion.
36.	**Relative reactivity of ... towards S_N1 reaction** benzyl chloride, I *p*-methoxy benzyl chloride, II and *p*-nitro benzyl chloride, III	II > I > III	Intermediates are benzylic cations. CH_3O (electron repelling) provides greater stability through delocalisation where NO_2 (electron attracting) decreases stability.
37.	**Relative reactivity of ... towards S_N1 and S_N2 reaction** $PhCH_2Cl$ (I), $PhCHClMe$ (II), $PhCClMe_2$ (III) (Ph stands for phenyl, C_6H_5)	S_N1: III > II > I S_N2: III < II < I	S_N1: 1° < 2° < 3° alkyl halide. (also benzyl carbocation resonance stabilised) S_N2: 3° < 2° < 1° alkyl halide.
38.	**Relative reactivity of ... with E^+ (electrophile) in S_E reaction.** *p*-nitrochlorobenzene (NO_2, Cl) (I), chlorobenzene (Cl) (II), 2,4-dinitrochlorobenzene (NO_2, NO_2, Cl) (III)	II > I > III	$—NO_2$ deactivates benzene ring for S_E.
39.	**Relative reactivity of ... with E^+ in S_E reaction.** $NHCOCH_3$ (I), NH_2 (II), $COCH_3$ (III), benzene (IV)	II > I > IV > III	$—COCH_3$ is deactivating while $—NH_2$ is activating group $—COCH_3$ attached to N is electron withdrawing hence deactivates $—NH_2$ group.
40.	**Order of S_N2 reactivity of alkoxide nucleophiles** Me_3CO^-, I; MeO^-, II; $MeCH_2O^-$, III; Me_2CHO^-, IV; cyclopentyl–$O^{\ominus}$, V	I < IV < V < III < II	S_N2 reactivity is susceptible to steric-hindrance by the nucleophile as well as by the alkyl group.

	Species and its effect	Order	Factor
41.	**Order of S_N2 reactivity** MeX, RCHX, R_2CHX, R_3CX I II III IV	IV < III < II < I	Due to R groups electron density at carbon atom increases hence attack of nucleophile is retarded.
42.	**Order of S_N1 reactivity** I II III IV (above)	I < II < III < IV	C—X bond breaks in slow step (order being 3° > 2° > 1° alkyl halide)
43.	**Order of hyperconjugation effect** $CH_3CH{=}CH_2$ I $CH_3CH_2CH{=}CH_2$ II $(H_3C)_2CHCH{=}CH_2$ III $(CH_3)_3CCH{=}CH_2$ IV	I > II > III > IV	Greater the number of H at α—C *wrt* allylic, greater the hyperconjugation. **(refer sec. 5.6)**
44.	**Reactivity of halogens (X_2) for a given type of abstraction of H (say 1°)**	$F_2 > Cl_2 > Br_2 > I_2$	BE of I_2 is maximum.
45.	**Order of extent of hydration** CH_3COCH_3, CH_3COCH_2Cl, CH_3CHO, $ClCH_2CHO$ I II III IV	I < II < III < IV	Aldehyde is more hydrated than ketone. Cl is electron attracting producing positive charge at C of carbonyl group hence attack of H_2O (hydration) is increased.
46.	**Leavng tendency of ... in S_N reaction** NH_2^-, RO^-, OH^-, $RCOO^-$, Cl^- I II III IV V	I < II ~ III < IV < V	If base is weaker, its leaving tendency is higher.
47.	**Order of hydrolysis tendency** CH_3COCl, CH_3CONH_2, $CH_3COOC_2H_5$, $(CH_3CO)_2O$ I II III IV	II < III < IV < I	Cl^-, NH_2^-, $^-OC_2H_5$, CH_3COO^- are replaced by OH^- in hydrolysis; weaker the base greater the leaving nature.
48.	**Reactivity of ...... NH_3** CH_3COCl, CH_3COOH, $CH_3COOC_2H_5$, $(CH_3CO)_2O$ I II III IV	II < III < IV < I	NH_2^- replaces X^- (Cl^-, OH^-, $^-OC_2H_5$, CH_3COO^-) weaker the X^-, greater the leaving nature.

APPENDIX C

Appendix I. Important Processes

Process	Product
Bosch process	Hydrogen
Castner process	Sodium
Down process	Sodium
Nelson cell	NaOH
Castner-Kellner cell	NaOH
Pidgeon process	Mg
Lowig process	NaOH
Solvay process or Ammonia soda process	Na_2CO_3 and $NaHCO_3$
Leblanc process	K_2CO_3, NaOH
Precht process	K_2CO_3
Mac Arthur Forrest process (Cyanide process)	Ag
Parke's process	Ag
Pattinson's process	Ag
Cupellation process	Ag (Purification)
Mond process	Ni
Baeyer's process	Al
Serpeck's process	Al
Hoop's process	Al (Purification)
Hall Heroult process	Al
Goldschmidt process	Thermite welding
Carter process	Basic lead carbonate (White lead)
Haber's process	NH_3
Brikeland-Eyde process	NO, HNO_3
Ostwald process	NO, HNO_3
Deacon's process	Cl_2
Lead chamber process	H_2SO_4
Contact process	H_2SO_4
Kaldo and L.D. process	Steel
Corey-House	alkane
Oxo process	alcohol
Dow's sea water process	Mg
Cyanamide process	NH_3
Bessemer-Thomas process	Steel
Siemens process	Steel
Frasch process	Sulphur
IMI process	Ti
Wet process	H_3PO_4
Furnace process	H_3PO_4
Salt cake method	Cl_2
Hargreaves process	Cl_2

Appendix II. Some Important Compounds/ Minerals

Name	Formula
Magnesite	$MgCO_3$
Dolomite	$MgCO_3 \cdot CaCO_3$
Kieserite	$MgSO_4 \cdot H_2O$
Epsomite	$MgSO_4 \cdot 7H_2O$
Talc	$Mg_2(Si_2O_5)_2Mg(OH)_2$
Magnesia alba	$(MgCO_3)_x[Mg(OH)_2]_y \cdot zH_2O$
Sorel's cement (magnesia cement)	$MgCl_2 \cdot 5MgO \cdot xH_2O$
Lime stone, marble or chalk	$CaCO_3$
Gypsum	$CaSO_4 \cdot 2H_2O$
Plaster of Paris	$CaSO_4 \cdot \frac{1}{2} H_2O$
Anhydrite	$CaSO_4$
Dolomite	$CaCO_3 \cdot MgCO_3$
Fluorspar	CaF_2
Phosphorite	$Ca_3(PO_4)_2$
Quick lime	CaO
Slaked lime	$Ca(OH)_2$
Soda lime	$Ca(OH)_2$ + NaOH
Chlorapatite	$3Ca_3(PO_4)_2 \cdot CaCl_2$
Apatite (Fluorapatite)	$3Ca_3(PO_4)_2 \cdot CaF_2$
Pearl ash, Potash	K_2CO_3
Sylvite	KCl
Carnallite	$KCl \cdot MgCl_2 \cdot 6H_2O$
Felspar	$KAlSi_3O_8$
Salt petre (Indian salt petre)	KNO_3
Kainite	$KCl \cdot MgSO_4 \cdot 3H_2O$
Schonite	$K_2SO_4 \cdot MgSO_4 \cdot 6H_2O$
Permutit (Zeolite)	$Na_2Al_2SiO_8 \cdot xH_2O$
Calgon (Graham's salt)	$(NaPO_3)_n$, $Na_2[NH_4(PO_3)_6]$
Chile salt petre	$NaNO_3$
Salt cake	Na_2SO_4 (*anhydrous*)
Glauber's salt	$Na_2SO_4 \cdot 10H_2O$
Nitre salt	$NaHSO_4$
Borax	$Na_2B_4O_7 \cdot 10H_2O$
Soda ash (Crystal carbonate)	Na_2CO_3 (*anhydrous*)
Washing soda	$Na_2CO_3 \cdot 10H_2O$
Crystal carbonate	$Na_2CO_3 \cdot H_2O$
Baking soda	$NaHCO_3$

Appendix I. Important Processes

Coinage metals	Cu, Ag and Au
Caro's acid	H_2SO_5
Marshall's acid	$H_2S_2O_8$
Borazine (Inorganic benzene)	$B_3N_3H_6$
Boron nitride (Inorganic graphite)	$(BN)_x$
Sylvinite	(NaCl + KCl)
Bordeaux mixture (fungicides)	$CuSO_4/Ca(OH)_2$
Copper pyrites	$CuFeS_2$ or $Cu_2S \cdot Fe_2S_3$
Cuprite (Ruby copper)	Cu_2O
Copper glance	Cu_2S
Malachite	$Cu(OH)_2 \cdot CuCO_3$
Azurite	$Cu(OH)_2 \cdot 2CuCO_3$
Blue vitriol	$CuSO_4 \cdot 5H_2O$
Argentite (Silver glance)	Ag_2S
Horn silver (Chlorargyrite)	AgCl
Ruby silver (Pyrargyrite)	Ag_3SbS_3
Argentiferrous galena	$PbS + Ag_2S$
Lunar caustic	$AgNO_3$
Zinc blende	ZnS
Calamine	$ZnCO_3$
Zincite (Red ZnO)	ZnO
Philosopher's wool	ZnO
Willemite	Zn_2SiO_4
Bauxite	$Al_2O_3 \cdot 2H_2O$
Corundum (emarld)	Al_2O_3
Diaspore	$Al_2O_3 \cdot H_2O$
Cryolite	Na_3AlF_6
Kaolinite	$Al_2O_3 \cdot 2SiO_2 \cdot 2H_2O$
Mica	$K_2O \cdot 3Al_2O_3 \cdot 6SiO_2 \cdot 2H_2O$
Alum shale	$Al_2O_3 \cdot xSiO_2 + FeS_2$
Alum stone (Alunite)	$K_2SO_4 \cdot Al_2(SO_4)_3 \cdot 4Al(OH)_3$
Cassiterite (Tin stone)	SnO_2
Galena	PbS
Anglesite	$PbSO_4$
Cerussite	$PbCO_3$
Lanakite	$PbO \cdot PbSO_4$
Masscote (Litharge)	PbO
Haematite (Red haematite)	Fe_2O_3
Limonite (Brown haematite)	$Fe_2O_3 \cdot 3H_2O$
Magnetite	Fe_3O_4
Siderite (Spathic iron ore)	$FeCO_3$
Iron pyrites	FeS_2
Mohr's salt	$FeSO_4 \cdot (NH_4)_2SO_4 \cdot 6H_2O$
Green vitriol (copperas)	$FeSO_4 \cdot 7H_2O$
Asbestos	$CaMg_3(SiO_3)_4$

Appendix II. Some Important Compounds/ Minerals

Aqua-regia	Conc. HNO_3 + Conc. HCl (1: 3 part)
Anhdrone	$Mg(ClO_4)_2$
Barytes or barite	$BaSO_4$
Baryta water	$Ba(OH)_2$ solution
Bleaching powder	$CaOCl_2$
Boranes	Hydride of boron
Brine	NaCl solution
Baryta	BaO
Asbestos (Chrysotile)	$Mg_3Si_2O_2O_5(OH)_4$
Spinel	$MgAl_2O_4$
Carborundum	SiC
Cementite	FeC
Chinese white	ZnO
Caliche	$NaNO_3 + NaIO_3$
Caustic potash	KOH
Calomel	Hg_2Cl_2
Corrosive sublimate	$HgCl_2$
Dry ice (Drikold)	solid CO_2
Freon	CF_2Cl_2
Hydrolith	CaH_2
Hypo	$Na_2S_2O_3 \cdot 5H_2O$
Lead pencil	Graphite
Laughing gas	N_2O
Litharge (Masscote)	PbO
Lanakite	$PbO \cdot PbSO_4$
Mortar	Slaked lime + sand (1: 3 in water)
Matte	$Cu_2S + FeS$
Minium	Pb_3O_4
Micro cosmic salt	$NaNH_4 \cdot HPO_4$
Milk of magnesia (Antacid)	Paste of $Mg(OH)_2$ in water
Marsh gas	CH_4
Nitro chalk	$[NH_4NO_3 + (NH_4)_2CO_3]$
Nitrolim	$CaCN_2$ (a fertilizer) + C
Nitrophos	$Ca(H_2PO_4)_2 + 2Ca(NO_3)_2$
Baking powder	$NaHCO_3 + Ca(H_2PO_4)_2$ + Starch
Oil of vitriol	Conc. H_2SO_4
Oleum	$H_2S_2O_7$
Phosgene	$COCl_2$
Pig iron	Impure form of iron
Producer gas	A mixture of $CO + N_2 + H_2$
Quartz (agate)	SiO_2
Quick silver	Hg
Refrigerant	CO_2, NH_3, CF_2Cl_2 etc.
Red lead	Pb_3O_4
Rochelle salt	Sodium-potassium tartrate
Rust	$Fe_2O_3 \cdot xH_2O$

Appendix I. Important Processes

TNT	Trinitro toluene (an explosive)
TNB	Trinitro benzene (an explosive)
Tincal	$Na_2B_4O_7 \cdot 10H_2O$
Water glass	Na_2SiO_3
Water gas	$CO + H_2$
Yperite	Mustard gas
Langbeinite	$K_2Mg(SO_4)_3$
Celestite	$SrSO_4$
Artificial explosive	Dynamite
First Noble Prize of chemistry was given to	vant Hoff
Some commonly used bleaching substances	H_2O_2, SO_2, SO_3, Cl_2
Neutral oxides of non-metals	CO, N_2O, NO, H_2O
Natural explosive	NCl_3

Appendix II. Some Important Compounds/Minerals

Some polymorphic elements	O, S, P
Wrought iron	Pure form of iron
Zinc white	ZnO
Olivine	$(Mg, Fe)_2SiO_4$
Strontianite	$SrCO_3$
Tincture iodine (antiseptic)	I_2 in alcohol
Some commonly used reductants	SO_2, H_2S, Cl_2
	Bleaching powder
Dry bleacher	O_3
Amphoteric oxides	ZnO, PbO, Al_2O_3, SnO, BeO
Some isomorphous substances	$FeSO_4 \cdot 7H_2O$, $MgSO_4 \cdot 7H_2O$, $ZnSO_4 \cdot 7H_2O$, $Na_2S_2O_3 \cdot 5H_2O$, $FeSO_4 \cdot 7H_2O$

Appendix III. Important Reagents

Reagents	Constituents	Uses
Aqua regia,	Conc. HCl and conc. HNO_3 (1: 3 by volume)	To dissolve noble metals Au, Pt
Baeyer's reagent	Cold alkaline $KMnO_4$	To test unsaturation (alkene, alkyne) in organic compounds
Baryta water	5% $Ba(OH)_2$ solution	As a medicine to control digestive system
Barfoed reagent	$(CH_3COO)_2Cu$ and CH_3COOH	To make distinction between monosaccharides and disaccharides
Brown's catalyst	Nickel boride	To prepare stereoisomer from alkene by hydrogenation
Ceric ammonium nitrate (CAN)	CAN in HNO_3	To test alcohols
Candy water	$KMnO_4$ solution	For oxidation
Denige's reagent	Acidic solution of HgO	To test citrate
Etard's reagent	CrO_2Cl_2	Oxidising agent
Fehling's solution	A : $CuSO_4$ solution B : Sodium potassium tartrate	To test –CHO (only aliphatic) and reducing sugars
Fenton's reagent	$FeSO_4$ (or other Fe^{2+} ion) and H_2O_2	To oxidise polyhydric alcohols
Fusion mixture	Na_2CO_3 and K_2CO_3	In salt mixture analysis
Grignard's reagent	$RMgX$ (R = alkyl, aryl X = Cl, Br, I)	To synthesise alcohols, aldehydes, ketones acids, etc.
Hinsberg's reagent	Benzene sulphonyl chloride, $(C_6H_5SO_2Cl)$	To separate and to distinguish 1°, 2° and 3° amines
Lucas reagent	Conc. HCl and anhydrous $ZnCl_2$	To distinguish 1°, 2° and 3° alcohols
Millon's reagent	$Hg(NO_3)_2$, $Hg_2(NO_3)_2$ in HNO_3	To test proteins
Magnesia mixture	$MgCl_2 \cdot 6H_2O$, NH_4Cl, and NH_3	To test arsenate and phosphate
Molisch's reagent	Alcoholic solution of α-naphthol	To test carbohydrates
Nessler's reagent	K_2HgI_4 in NaOH	To test NH_4^+
Schiff's reagent	*p*-rosaniline hydrochloride solution decolourised by SO_2	To test –CHO group
Sodium nitroprusside	$Na_2[Fe(CN)_5NO]$	To test S^{2-} ion (sulphide)
Tollen's reagent	Ammoniacal $AgNO_3$	To test –CHO group and reducing sugars

Appendix IV. Scientists and Their Contributions

Scientist	Contribution
Anderson	Discovery of positron
Aspidin, J	Portland cement
Aston, F. William	Isolation of non-radioactive elements and discovery of mass-spectrograph
Bacon, R	Gun powder
Bohr	Atomic model, Long form of Periodic Table
Becquerrel H.	Radioactivity
Berzelius, J.J.,	Modern system of symbolising elements, relative atomic masses.
Bunsen R	Bunsen burner
Chadwick	Discovery of neutron
Curie I.J.	Radioactive elements
Fermi E.	Nuclear reactor
Goldstein	Discovery of proton
Mendeleef	Periodic Table
Meyer L.	Classification of elements based upon atomic volume
Moseley	Modern Periodic Law
Pauling Linus C.	Pauling scale of electronegativity
Ramsay, Sir William	Discovery of noble gases Ar, Ne, He
Rontgen	Discovery of X-ray
Seaborg G.T.	Discovery of *trans*-uranium elements
Solvay, E	Solvay ammonia soda process for the manufacture of Na_2CO_3
Teller E	Hydrogen bomb
Thomson J.J.	Discovery of electron, *e/m* ratio of electron, photoelectric effect
Yukawa	Discovery of mesons
Wohler F.	Synthesis of first organic compound (urea) from inorganic salt

Appendix V. Facts to Remember

Most reactive solid element : Li

Most reactive liquid element : Cs

Most reactive gaseous element : F

Highest electronegativity : F

Highest ionisation potential : He

Lowest ionisation potential : Cs

Lowest electron affinity : Noble gases (zero)

Highest electron affinity : Cl

Liquid element of radioactive nature : Francium (Fr)

Total number of radioactive elements in Periodic Table : 25

Appendix V. Facts to Remember

Volatile *d*-block elements : Zn, Cd, Hg

Element containing no neutron : ^1_1H

Most abundant element of earth : Oxygen (O)

Rarest element of earth : Astatine (At)

Most abundant metal of earth : Al

Element having maximum tendency for catention : Carbon

Lightest element : H

Heaviest naturally occurring element : ^{238}U

Poorest conductor of current : Pb(metal), S(non-metal)

Amphoteric non-metal : Si

Elements showing diagonal relationship : Li—Mg; Be—Al; B—Si

Metalloids elements : B, Si, Ge, As, Sb, Te

Non-metals having metallic lustre : Iodine, graphite

Element sublime on heating : I

Noble metals : Au, Pt

Amphoteric metals : Zn, Al, Sn, Pb

Non-metal having highest melting point, boiling point : Diamond

Metals showing highest oxidation number. : Ru, Os

Element having highest tensile strength : Boron

Coolant in nuclear reactors : D_2O

Best electricity conductor among metals : Ag

Best conductor among non-metals : Graphite

Most poisonous element : Pu

Liquid non-metal : Br_2

Liquid metals : Hg, Ga, Cs, Fr, Eka

Element kept in water : P

Elements kept in kerosene oil : Na, K, I_2, Cs

Dry ice : CO_2

Most recently elements named by IUPAC : Ds (atomic number = 110)

'All purpose' grease : Lithium stearate

Old name of Astatine : Albamine

Most abundant gas in atmosphere : N_2

Rarest gas in atmosphere : Rn

Lightest gas in atmosphere : H_2

Metal with highest melting point : W

Metal with lowest melting point : Hg

Non-metal with highest refractive index : Diamond

Appendix V. Facts to Remember

Fact	
Lowest refractive index	: Air
Lowest boiling point	: H_2
Group containing highest number of gaseous elements	: Group 18 (Noble gases)
Most electrovalent compound	: CsF
Heaviest solid metal	: Os
Lightest solid metal	: Li
Lightest solid non-metal	: B
Hardest naturally occurring non-metal	: Diamond
Hardest artificial substance good conductor of heat	: B_4C (norbide)
but bad conductor of electricity	: Mica
Strongest basic hydroxide	: CsOH
Strongest basic oxide	: CsO_2 (caesium peroxide)
Most stable metal carbonate	: Cs_2CO_3
Element with highest radioactivity	: Ra
Lighest radioisotope	: Tritium (^3_1H)
Compound with maximum covalent nature	: H_2, N_2, O_2, Cl_2 (diatomic gases)
Strongest reducing agent	: Azide (N_3^-)
Strongest oxidising agent	: OF_2
Largest atomic size	: Cs
Smallest atomic size	: H
Smallest anion	: H^- (hydrides)
First Noble Prize in Chemistry	: van't Hoff
Most abundant noble gas	: Ar
Most abundant element in human body	: Oxygen
Most abundant gas in Sun	: H_2
Most abundant gases in universe	: H_2, He
Element with maximum number isotopes	: Ag (46)
Element with minimum number of isotopes	: H(3)
Element with maximum number of allotropes	: Sn
Softest form of carbon	: Lamp black
Latest allotrope of carbon	: Fullerene or bucky ball
Strongest acid	: $HSO_3F + 90\% SbF_5$ called magic acid
Most reactive form of P	: White
Least reactive form of P	: Red
Purest form of silica	: Quartz
Most ductile metal	: Gold

Appendix VI. Important Alloys

Alloy	Component	Alloy	Component
Aluminium bronze	Cu, Al	Electron	Mg, Zn
Alpax	Si, Al	Ferrochrome	Fe, Cr
Alnico	Al, Ni, Co, Fe, Cu	Ferromanga-nese	Fe, Mn
γ-alloy	Al, Cu, Mg, Ni	Ferrosilicon (silicon steel)	Fe, Si
Amalgam	Metal coated with Hg	German silver (Nickel silver)	Cu, Zn, Ni
Brass	Cu, Zn	Gun metal	Cu, Sn, Zn
Bronze	Cu, Sn	Hydrone	Na, Pb
Bell metal	Cu, Sn	Hastelloy C Nicrome	Ni, Cr
Britannia metal	Sn, Sb, Cu, Zn	Invar	Fe, Ni
Babbit metal	Sn, Sb, Cu	LA 141	Li, Al, Mg
Bearing metal	Cu, Zn, Sb	Magnalium	Mg, Al
Coin alloy (white)	Cu, Ag, Zn, Ni	Manganium	Ni, Cu, Mn
Coin alloy (red)	Cu, Zn, Sn	Manganese steel	Fe, C, Mn
Constantan	Cu, Ni	Monel metal	Cu, Ni, Fe, Mn
Cunico	Cu, Ni, Fe, Co	Nichrome	Ni, Cr, Fe, Mn
Delta metal	Cu, Zn, Fe	Newton metal	Sn, Pb
Dutch metal	Cu, Zn	Nickeloy	Ni, Cu, Al
Dental alloy	Sn, Hg, Ag	Nickel steel	Ni, Fe
Devarda's alloy	Cu, Zn, Al	Nimonic series	Ni, Cr, Co, Al, Ti
Duralumin	Al, Cu, Mg, Zn	Perm alloy	Ni, Fe, C
Phosphor bronze	Cu, Sn, P	Pewter	Pb, Sn
Rolled gold	Cu, Al	Platinite	Fe, Ni
Rose metal	Bi, Zn, Pb, Cd	Type metal	Pb, Sb, Sn
Soft silver (solder)	Sn, Pb	White metal	Cu, Sb, Sn
Stainless steel	Fe, Ni, Cr, C	Wood's metal	Bi, Pb, Sn, Cd

Appendix VII. Alloy Steels

Mechanical and other properties of steel are greatly modified by alloying with different metals and silicon. Some of these alloy steels are :

Type	Composition	Properties	Uses
Silicon steel	3.5% Si and very little carbon	magnetic	electromagnets and transformers.
Stainless steel	12-15% Cr	resists corrosion and action of acids	Stainless steel utensils
Manganese steel	12-13% Mn	resistant to wear	grinding machinery, safe, rock drill
Cobalt steel	up to 35% cobalt	magnetic	permanent magnets
Nickel steel	2-4% Ni	high tensile strength	cables, propeller shafts gears
Invar	up to 36% Ni	low coefficient of thermal expansion	scientific instruments
Chrome-nickel steel	1-4% Ni; 0.5-2% Cr	high tensile strength, great corrosion resistance	motor car and bicycle parts
Platinite	46% Ni	coefficient of thermal expansion same as that of glass	glassmetal-seals, electric bulbs
Chrome-vanadium steel	1-10% Cr, 0.14-0.55% V	high tensile strength, resistant to stress and torsion	automobile axles and springs
Tungsten steel	10-20% W, 3-8% Cr	hard, corrosion-resistant and retains temper at high temperature	high-speed cutting tools
Molybdenum steel	6-7% Mo	retains temper at high temperature	high-speed cutting tools
Misch-metal	70% Ce and traces of rare-earth metals.	produces spark on strike	to produce spark in gas and cigarette lighters.

Appendix VIII. Electronic Configurations of the Atoms of the Elements

Element	Atomic number	Populations of subshells										
		1*s*	**2*s***	**2*p***	**3*s***	**3*p***	**3*d***	**4*s***	**4*p***	**4*d***	**4*f***	**5*s***
H	1	1										
He	2	2										
Li	3	2	1									
Be	4	2	2									
B	5	2	2	1								
C	6	2	2	2								
N	7	2	2	3								
O	8	2	2	4								
F	9	2	2	5								
Ne	10	2	2	6								
Na	11	Neon core			1							
Mg	12				2							
Al	13				2	1						
Si	14				2	2						
P	15				2	3						
S	16				2	4						
Cl	17				2	5						
Ar	18	2	2	6	2	6						
K	19	Argon core						1				
Ca	20							2				
Sc	21						1	2				
Ti	22						2	2				
V	23						3	2				
Cr	24						5	1				
Mn	25						5	2				
Fe	26						6	2				
Co	27						7	2				
Ni	28						8	2				
Cu	29						10	1				
Zn	30						10	2				
Ga	31						10	2	1			
Ge	32						10	2	2			
As	33						10	2	3			
Se	34						10	2	4			
Br	35						10	2	5			
Kr	36	2	2	6	2	6	10	2	6			
Rb	37	Krypton core										1
Sr	38											2
Y	39									1		2
Zr	40									2		2
Nb	41									3		2
Mo	42									5		1
Tc	43									5		2
Ru	44									7		1
Rh	45									8		1
Pd	46									10		
Ag	47									10		1
Cd	48									10		2

Element	Atomic number		Populations of Subshells									
			4*d*	4*f*	5*s*	5*p*	5*d*	5*f*	6*s*	6*p*	6*d*	7*s*
In	49		10		2	1						
Sn	50		10		2	2						
Sb	51		10		2	3						
Te	52		10		2	4						
I	53		10		2	5						
Xe	54		10		2	6						
Cs	55		10		2	6			1			
Ba	56		10		2	6			2			
La	57		10		2	6	1		2			
Ce	58		10	1	2	6	1		2			
Pr	59		10	3	2	6			2			
Nd	60		10	4	2	6			2			
Pm	61		10	5	2	6			2			
Sm	62		10	6	2	6			2			
Eu	63		10	7	2	6			2			
Gd	64		10	7	2	6	1		2			
Tb	65		10	9	2	6			2			
Dy	66		10	10	2	6			2			
Ho	67		10	11	2	6			2			
Er	68		10	12	2	6			2			
Tm	69	Krypton core	10	13	2	6			2			
Yb	70		10	14	2	6			2			
Lu	71		10	14	2	6	1		2			
Hf	72		10	14	2	6	2		2			
Ta	73		10	14	2	6	3		2			
W	74		10	14	2	6	4		2			
Re	75		10	14	2	6	5		2			
Os	76		10	14	2	6	6		2			
Ir	77		10	14	2	6	7		2			
Pt	78		10	14	2	6	9		1			
Au	79		10	14	2	6	10		1			
Hg	80		10	14	2	6	10		2			
Tl	81		10	14	2	6	10		2	1		
Pb	82		10	14	2	6	10		2	2		
Bi	83		10	14	2	6	10		2	3		
Po	84		10	14	2	6	10		2	4		
At	85		10	14	2	6	10		2	5		
Rn	86		10	14	2	6	10		2	6		
Fr	87		10	14	2	6	10		2	6		1
Ra	88		10	14	2	6	10		2	6		2
Ac	89		10	14	2	6	10		2	6	1	2
Th	90		10	14	2	6	10		2	6	2	2
Pa	91		10	14	2	6	10	2	2	6	1	2
U	92		10	14	2	6	10	3	2	6	1	2
Np	93		10	14	2	6	10	4	2	6	1	2
Pu	94		10	14	2	6	10	6	2	6		2
Am	95		10	14	2	6	10	7	2	6		2
Cm	96		10	14	2	6	10	7	2	6	1	2
Bk	97		10	14	2	6	10	9	2	6		2
Cf	98	Krypton core	10	14	2	6	10	10	2	6		2
Es	99		10	14	2	6	10	11	2	6		2
Fm	100		10	14	2	6	10	12	2	6		2
Md	101		10	14	2	6	10	13	2	6		2
No	102		10	14	2	6	10	14	2	6		2
Lr	103		10	14	2	6	10	14	2	6	1	2
Rf	104		10	14	2	6	10	14	2	6	2	2
Db	105		10	14	2	6	10	14	2	6	3	2
Sg	106		10	14	2	6	10	14	2	6	4	2
Bh	107		10	14	2	6	10	14	2	6	5	2
Hs	108		10	14	2	6	10	14	2	6	6	2
Mt	109		10	14	2	6	10	14	2	6	7	2
Ds	110		10	14	2	6	10	14	2	6	9	1

Appendix IX. The Elements : Discoverers, Symbols and their Names

Element	Symbol	Atomic number	Date of discovery	Discoverer and nationality
Actinium	Ac	89	1899	A. Debierne (Fr.)
Aluminium	Al	13	1827	F. Woehler (Ge.)
Americium	Am	95	1944	A. Ghiorso (USA), R.A. James (USA) G.T. Seaborg (USA), S.G. Thompson (USA)
Antimony	**Sb**	**51**	**Ancient**	
Argon	Ar	18	1894	Lord Rayleigh (GB), Sir William Ramsay (GB)
Arsenic	As	33	1250	Albertus Magnus (Ge)
Astatine	At	85	1940	D.R. Corson (USA), K.R. MacKenzie (USA) E. Segre (USA)
Barium	Ba	56	1808	Sir Humphry Davy (GB)
Berkelium	Bk	97	1950	G.T. Seaborg (USA), S.G. Thompson (USA) A. Ghiorso (USA)
Beryllium	Be	4	1828	F. Woehler (Ge.), A.A.B. Bussy (Fr.)
Bismuth	Bi	83	1753	Claude Geoffroy (Fr.)
Bohrium	Bh	107	1976	USA
Boron	B	5	1808	Sir Humphry Davy (GB), J.L. Gay-Lussac (Fr.) L.J. Thenard (Fr.)
Bromine	Br	35	1826	A.J. Balard (Fr.)
Cadmium	Cd	48	1817	Fr. Stromeyer (Ge.)
Calcium	Ca	20	1808	Sir Humphry Davy (GB)
Californium	Cf	98	1950	G.T. Seaborg (USA), S.G. Thompson (USA) A. Ghiorso (USA), K. Street, Jr. (USA)
Carbon	**C**	**6**	**Ancient**	
Cerium	Ce	58	1803	J.J. Berzelius (Swe.), William Hisinger (Swe.) M.H. Klaproth (Ge.)
Cesium	Cs	55	1860	R. Bunsen (Ge.), G.R. Kirchhoff (Ge.)
Chlorine	Cl	17	1774	K.W. Scheele (Swe.)
Chromium	Cr	24	1797	L.N. Vauquelin (Fr.)
Cobalt	Co	27	1735	G. Brandt (Ge.)
Copper	**Cu**	**29**	**Ancient**	
Curium	Cm	96	1944	G.T. Seaborg (USA), R.A. James (USA) A. Ghiorso (USA)
Darmstadtium	**Ds**	**110**	**2003**	**Sigurd Hofmann (Ge)**
Dubnium	Db	105	1970	
Dysprosium	Dy	66	1886	Locoq de Boisbaudran (Fr.)
Einsteinium	Es	99	1952	A. Ghiorso (USA)
Erbium	Er	68	1843	C.G. Mosander (Swe.)
Europium	Eu	63	1896	E. Demarcay (Fr.)
Fermium	Fm	100	1953	A. Ghiorso (USA)
Fluorine	F	9	1886	H. Moissan (Fr)
Francium	Fr	87	1939	Marguerite Perey (Fr.)
Gadolinium	Gd	64	1880	J.C. Marignac (Fr.)
Gallium	Ga	31	1875	Lecoq de Boisbaudran (Fr.)

Element	Symbol	Atomic number	Date of discovery	Discoverer and nationality
Germanium	Ge	32	1886	Clemens Winkler (Ge.)
Gold	**Au**	**79**	**Ancient**	
Hafnium	Hf	72	1923	D. Coster (Du.), G. von Hevesey (H.)
Helium	He	2	1868	P. Janssen (spectr) (Fr.), Sir William Ramsay (isolated) (GB)
Hassium	Hs	108	1984	–
Holmium	Ho	67	1879	P.T. Cleve (Swe.)
Hydrogen	H	1	1766	Sir Henry Cavendish (GB)
Indium	In	49	1863	E. Reich (Ge.), T. Richter (Ge.)
Iodine	I	53	1811	B. Courtois (Fr.)
Iridium	Ir	77	1803	S. Tennant (GB)
Iron	**Fe**	**26**	**Ancient**	
Krypton	Kr	36	1898	Sir William Ramsay (GB), M.W. Travers (GB)
Lanthanum	La	57	1839	C.G. Mosander (Swe.)
Lawrencium	Lr	103	1961	A. Ghiorso (USA), T. Sikkeland (USA) A.E. Larsh (USA), R.M. Latimer (USA)
Lead	**Pb**	**82**	**Ancient**	
Lithium	Li	3	1817	A. Arfvedson (Swe)
Lutetium	Lu	71	1907	G. Urbain (Fr.), C.A. von Welsbach (Au.)
Magnesium	Mg	12	1808	Sir Humphry Davy (GB)
Manganese	Mn	25	1774	J.G. Gahn (Swe.)
Mendelevium	Md	101	1955	A. Ghiorso (USA), G.R. Choppin (USA) G.T. Seaborg (USA), B.G. Harvey (USA) S.G. Thompson (USA)
Meitnerium	Mt	109	1982	–
Mercury	**Hg**	**80**	**Ancient**	
Molybdenum	Mo	42	1778	G.W. Scheele (Swe.)
Neodymium	Nd	60	1885	C.A. von Welsbach (Au.)
Neon	Ne	10	1898	Sir William Ramsay (GB), M.W. Travers (GB)
Neptunium	Np	93	1940	E.M. McMillan (USA), P.H. Abelson (USA)
Nickel	Ni	28	1751	A.E. Cronstedt (Swe.)
Niobium	Nb	41	1801	Charles Hatchett (GB)
Nitrogen	N	7	1772	Daniel Rutherford (GB)
Nobelium	No	102	1958	A. Ghiorso (USA), T. Sikkeland (USA) J.R. Walton (USA), G.T. Seaborg (USA)
Osmium	Os	76	1803	S. Tennant (GB)
Oxygen	O	8	1774	Joseph Priestley (GB), C.W. Scheele (Swe.)
Palladium	Pd	46	1803	W.H. Wollaston (GB)
Phosphorus	P	15	1669	H. Brandt (Ge.)
Platinum	Pt	78	1735	A. De Ullao (Sp.)
			1741	Charles Wood (GB)
Plutonium	Pu	94	1940	G.T. Seaborg (USA), E.M. McMillan (USA) J.W. Kennedy (USA), A.C. Wahl (USA)
Polonium	Po	84	1898	Marie Curie (P.)

Element	Symbol	Atomic number	Date of discovery	Discoverer and nationality
Potassium	K	19	1807	Sir Humphry Davy (GB)
Praseodymium	Pr	59	1885	C.A. von Welsbach (Au.)
Promethium	Pm	61	1945	J.A. Marinsky (USA), L.E. Glendenin (USA) C.D. Coryell (USA)
Protactinium	Pa	91	1917	O. Hahn (Ge.), L. Meitner (Au.)
Radium	Ra	88	1898	Pierre and Marie Curie, (Fr.; P)
Radon	Rn	86	1900	F.E. Dorn (Ge.)
Rhenium	Re	75	1925	W. Noddack (Ge.), I. Tacke (Ge) Otto Berg (Ge)
Rhodium	Rh	45	1804	W.H. Wollaston (GB)
Rubidium	Rb	37	1861	R.W. Bunsen (Ge.), G. Kirchhoff (Ge.)
Ruthenium	Ru	44	1844	K.K. Klaus (R.)
Rutherfordium	Rf	104	1965	USA
Samarium	Sm	62	1879	Lecoq de Boisbaudran (Fr.)
Scandium	Sc	21	1879	L.F. Nilson (Swe.)
Seaborgium	Sg	106	1976	—
Selenium	Se	34	1817	J.J. Berzelius (Swe.)
Silicon	Si	14	1824	J.J. Berzelius (Swe.)
Silver	**Ag**	**47**	**Ancient**	
Sodium	Na	11	1807	Sir Humphry Davy (GB)
Strontium	Sr	38	1808	Sir Humphry Davy (GB)
Sulphur	**S**	**16**	**Ancient**	
Tantalum	Ta	73	1802	A.G. Ekeberg (Swe.)
Technetium	Tc	43	1937	C. Perrier (I.)
Tellurium	Te	52	1782	E.J. Muller (Au.)
Terbium	Tb	65	1843	C.G. Mosander (Swe.)
Thallium	Tl	81	1861	Sir William Crookes (GB)
Thorium	Th	90	1828	J.J. Berzelius (Swe.)
Thulium	Tm	69	1879	P.T. Cleve (Swe.)
Tin	**Sn**	**50**	**Ancient**	
Titanium	Ti	22	1791	W. Gregor (GB)
Tungsten	W	74	1783	J.J. and F. de Elhuyer (Sp.)
Uranium	U	92	1789	M.H. Klaproth (Ge.)
			1841	E.M. Peligot (Fr.)
Vanadium	V	23	1801	A.M. Del Rio (Sp.)
			1830	N.G. Sefstrom (Swe.)
Xenon	Xe	54	1898	Sir William Ramsay (GB), M.W. Travers (GB)
Ytterbium	Yb	70	1907	G. Urbain (Fr.)
Yttrium	Y	39	1843	C.G. Mosander (Swe.)
Zinc	Zn	30	1746	A.S. Marggraf (Ge.)
Zirconium	Zr	40	1789	M.H. Klaproth (Ge.)

Appendix X. Minerals with Metallic Luster

Mineral (Formula)	Colour	Crystal system	Uses and other properties
Graphite (C)	black to grey	hexagonal	pencil lead, lubricants for locks, rods to control some small nuclear reactions, battery poles
Silver (Ag)	silvery white, tarnishes to black	cubic	coins, fillings for teeth, jewellery, silverplate, wires; malleable and ductile
Galena (PbS)	grey	cubic	source of lead, used in pipes, shields for X-rays, fishing equipment sinkers
Gold (Au)	pale to golden yellow	cubic	jewellery, money, gold leaf, fillings for teeth, medicines; does not tarnish
Bornite [Cu_5FeS_4]	bronze, tarnishes to dark blue, purple	tetragonal	source of copper; called "peakcock ore" because of the purple shine when it tarnishes
Copper (Cu)	copper red	cubic	coins, pipes, gutters, wire cooking utensils, jewellery, decorative plaques; malleable and ductile
Chalcopyrite ($CuFeS_2$)	brassy to golden yellow	tetragonal	main ore of copper
Chromite ($FeCr_2O_4$)	black or brown	cubic	ore of chromium, stainless steel, metallurgical bricks
Pyrrhotite (FeS)	bronze	hexagonal	often found with pentlandite, an ore of nickel; may be magnetic
Haematite (specular) (Fe_2O_3)	black or reddish brown	hexagonal	source of iron; roasted in a blast furnace, converted to "pig" iron, made into steel
Magnetite (Fe_3O_4)	black	cubic	source of iron, naturally magnetic, called lodestone
Pyrite (FeS_2)	light, brassy, yellow	cubic	source of iron, "fools's gold," alters to limonite
Fluorite (CaF_2)	colourless, white, blue, green, red, yellow, purple	cubic	used in the manufacture of optical equipment; glows under ultraviolet light
Limonite (hydrous iron oxides)	yellow, brown black	—	source of iron; weathers easily colouring matter of soils
Hornblende ($CaNa(Mg, Al, Fe)_5$ $(Al, Si)_2Si_6O_{22}(OH)_2$)	green to black	monoclinic	will transmit light on thin edges; 6-sided cross section
Feldspar (orthoclase) ($KAlSi_3O_8$)	colourless, white grey green and yellow	monoclinic	insoluble in acids; used in the manufacture of porcelain
Feldspar (plagioclase) ($NaAlSi_3O_8$) ($CaAl_2Si_2O_8$)	grey, green white	triclinic	used in ceramics; striations present on some faces
Augite (Ca, Na) (Mg, Fe, Al) $(Al, Si)_2O_6$	black	monoclinic	square or 8-sided cross-section
Olivine ($(Mg, Fe)_2SiO_4$)	olive green	ortho-rhombic	gemstones, refractory sand
Quartz (SiO_2)	colourless, various colours	hexagonal	used in glass manufacture, electronic equipment, radios, computers, watches, gemstones
Garnet $(Mg, Fe, Ca)_3$ $(Al_2Si_3O_{12})$	deep yellow-red green, black	cubic	used in jewellery, also used as an abrasive
Topaz $(Al_2SiO_4)(F, OH)_2$	white, pink yellow, pale blue, colourless	ortho-rhombic	valuable gemstone
Corundum (Al_2O_3)	colourless, blue, brown green white, pink, red	hexagonal	gemstones; ruby is red, sapphire is blue; industrial abrasive

Appendix XI. Mineral with Non-Metallic Lusture

Mineral (Formula)	Colour	Crystal system	Uses and other properties
Talc ($Mg_3(OH)_2Si_4O_{10}$)	white, greenish	monoclinic	easily cut with fingernail; used for talcum powder; soapstone is used in paper and for table tops
Bauxite (hydrous aluminium compound)	grey, red, white, brown	—	source of aluminium; used in paints, aluminium foil, and airplane parts
Kaolinite ($Al_4Si_2O_5(OH)_4$)	white, red, reddish brown, black	triclinic	clays; used in ceramics and in china dishes; common in most soils; often microscopic-sized particles
Gypsum ($CaSO_4 \cdot 2H_2O$)	colourless, grey, white, brown	monoclinic	used extensively in the preparation of plaster of Paris, alabaster, and dry wall for building construction
Sphalerite (ZnS)	brown	cubic	main ore of zinc; used in paints, dyes, and medicine
Sulphur (S)	yellow	orthorhombic	used in medicine, fungicides for plants, vulcanisation of rubber. Production of sulphuric acid
Muscovite ($KAl_3Si_3O_{10}(OH)_2$)	white, light grey, yellow, rose, green	monoclinic	occurs in large flexible plates, used as an insulator in electrical equipment, lubricant
Biotite ($K(Mg, Fe)_3AlSi_3O_{10}(OH)_2$)	black to dark brown	monoclinic	occurs in large flexible plates
Halite (NaCl)	colourless, red, white blue	cubic	salt; very soluble in water; a preservative
Calcite ($CaCO_3$)	colourless, white pale, blue	hexagonal	fizzes when HCl is added; used in cement and other building materials
Dolomite ($CaMg(CO_3)_2$)	colourless, white, pink, green, grey, black	hexagonal	concrete and cement, used as an ornamental building stone

Appendix XII. Solubility and Colour of Different Salts

Salt/Complex	Formula	Colour	$\left(\frac{\text{Soluble}}{\text{Insoluble}}\right)$ in water
Ammonium carbonates	$(NH_4)_2CO_3$	white	soluble
Ammonium chloride	NH_4Cl	white	soluble
Ammonium nitrate	NH_4NO_3	white	soluble
Ammonium phosphate	$(NH_4)_3PO_4$	white	soluble
Ammonium oxalate	$(NH_4)_2C_2O_4$	white	soluble
Ammonium sulphate	$(NH_4)_2SO_4$	white	soluble
Ammonium dichromate	$(NH_4)_2Cr_2O_7$	orange (due to $Cr_2O_7^{2-}$ ion)	soluble
Ammonium molybdate	$(NH_4)_2MoO_4$	white	soluble in presence of NH_4OH
Arsenic(III) sulphide	As_2S_3	yellow	insoluble
Arsenic(V) sulphide	As_2S_5	yellow	insoluble
Arsenic(III) iodide	AsI_3	red	insoluble
Antimony sulphide	Sb_2S_3	orange	insoluble
Antimonyl chloride	$SbOCl$	orange	insoluble
Antimony iodide	SbI_3	red-yellow	insoluble
Bismuth sulphide	Bi_2S_3	black	insoluble
Bismuth iodide	BiI_3	black	insoluble
Bismuth hydroxide	$Bi(OH)_3$	white	insoluble
Bismuthyl chloride	$BiOCl$	white	insoluble
Bismuth(V) oxide	Bi_2O_5	brown	insoluble
Bismuthyl iodide	$BiOI$	red-orange	insoluble
Barium chloride	$BaCl_2$	white	soluble
Barium nitrate	$Ba(NO_3)_2$	white	soluble
Barium sulphate	$BaSO_4$	white	insoluble
Barium chromate	$BaCrO_4$	yellow (due to CrO_4^{2-})	insoluble (also in CH_3COOH) but soluble in dil. HCl
Chromium(III) oxide	Cr_2O_3	green	insoluble
Chromium sulphate	$Cr_2(SO_4)_3$	green	partially soluble
Chromium chloride	$CrCl_3$	green	soluble
Chromium chloride (mineral)	$CrCl_3 \cdot 6H_2O$	violet	soluble
Chromyl chloride	CrO_2Cl_2	orange red gas	soluble giving H_2CrO_4
Chromium hydroxide	$Cr(OH)_3$	dark green	insoluble
Chromate ion	CrO_4^{2-}	yellow	soluble (salts of alkali and ammonium ion)
Copper(I) sulphide (cupric sulphide)	Cu_2S	black	insoluble
Copper(II) sulphide (cupric sulphide)	CuS	black	insoluble
Hydrated copper (II) sulphate	$CuSO_4 \cdot 5H_2O$	blue	soluble
Anhydrous copper(II) sulphate	$CuSO_4$	white	soluble (colour changes to blue)
Copper(II) hydroxide	$Cu(OH)_2$	sky blue	insoluble
Copper(II) chloride (cupric chloride)	$CuCl_2$	bluish green	soluble
Copper(I) oxide (cuprous oxide)	Cu_2O	red	insoluble

Salt/Complex	Formula	Colour	$\left(\frac{\text{Soluble}}{\text{Insoluble}}\right)$ in water
Cuprous iodide	Cu_2I_2	white	insoluble
Cuprous thiosulphate	$Cu_2S_2O_3$	white	soluble
Tetrammine copper(II) sulphate	$[Cu(NH_3)_4]SO_4$	deep blue	soluble (insoluble in ethanol)
Cupric ferrocyanide	$Cu_2[Fe(CN)_6]$	brown	insoluble
Cupric oxide, (copper(II) oxide)	CuO	black	insoluble
Copper(II) cyanide	$Cu(CN)_2$	yellow	insoluble
Cpper(I) cyanide	CuCN	white	insoluble
Copper(II) thiocyanate	$Cu(SCN)_2$	black	insoluble
Copper(I) thiocyanate	CuSCN	white	insoluble
Cobalt sulphide	CoS	black	insoluble
Potassium cobalti nitrite	$K_3[Co(NO_2)_6]$	yellowish orange	insoluble
Cobalt(II) chloride dihydrate	$CoCl_2 \cdot 2H_2O$	brown	soluble
Cobalt(II) chloride tetrahydrate	$CoCl_2 \cdot 4H_2O$	brown	soluble
Cobalt(II) chloride hexahydrate	$CoCl_2 \cdot 6H_2O$	red	soluble
Cobalt(II) cyanide	$Co(CN)_2$	buff-coloured	insoluble
Cadmium oxide	CdO	brown	insoluble
Cadmium sulphide	CdS	yellow	insoluble
Cadmium carbonate	$CdCO_3$	white	partly soluble
Cadmium hydroxide	$Cd(OH)_2$	white	insoluble
Dichromate ion	$Cr_2O_7^{2-}$	orange	soluble
Zinc sulphide	ZnS	white	insoluble
Zinc hydroxide	$Zn(OH)_2$	white	insoluble, soluble in acid as Zn^{2+}, soluble in alkali as ZnO_2^{2-}
Zinc oxide	ZnO	white	(-do-)
Zinc carbonate	$ZnCO_3$	white	insoluble
Lead sulphite	$PbSO_3$	white	insoluble
Lead sulphate	$PbSO_4$	white	insoluble
Lead chloride	$PbCl_2$	white	soluble in hot water
Lead bromide	$PbBr_2$	white	soluble in hot water
Lead oxide (litharge)	PbO	orange-red	insoluble
Lead double oxide	$2PbO \cdot PbO_2(Pb_3O_4)$	red	insoluble
Lead dioxide	PbO_2	blackish brown	insoluble
Lead carbonate	$PbCO_3$	white	insoluble
Lead chromate (basic)	$PbO \cdot PbCrO_4$	red	insoluble
Lead chromate	$PbCrO_4$	yellow	insoluble (soluble in HNO_3 as $PbCr_2O_7$ and in NaOH as $[Pb(OH)_4^{2-}]$
Lead iodide	PbI_2	yellow	insoluble
Lead sulphide	PbS	black	insoluble
Lead hydroxide	$Pb(OH)_2$	white	insoluble (soluble in NaOH as $[Pb(OH)_4]^{2-}$

Salt/Complex	Formula	Colour	(Soluble/Insoluble) in water
Lead cyanide	$Pb(CN)_2$	white	insoluble
Lead carbonate (basic)	$PbO \cdot PbCO_3$	white	insoluble
Lead phosphate	$Pb_3(PO_4)_2$	white	insoluble
Manganese dioxide	MnO_2	brownish-red	insoluble
Manganese sulphide	MnS	buff coloured	insoluble
Manganese chloride	$MnCl_2$	light pink	soluble
Manganese hydroxide	$Mn(OH)_2$	light pink	insoluble
Permanganate (potassium)	$KMnO_4$	violet (purple)	soluble
Manganate (potassium)	K_2MnO_4	green	soluble
Manganese oxide	MnO	green	insoluble
Manganese carbonate	$MnCO_3$	brown	insoluble
Calcium chloride	$CaCl_2$	white	soluble
Calcium carbonate	$CaCO_3$	white	insoluble
Calcium oxalate	CaC_2O_4	white	insoluble
Calcium nitrate	$Ca(NO_3)_2$	white	soluble
Calcium bicarbonate	$Ca(HCO_3)_2$	white	soluble
Calcium sulphate	$CaSO_4$	white	insoluble
Magnesium chloride	$MgCl_2$	white	soluble
Magnesium ammonium phosphate	$Mg(NH_4)PO_4$	white	insoluble
Iron(III) oxide	Fe_2O_3	red	insoluble
Iron(II) carbonate	$FeCO_3$	green	soluble
Iron(II) chloride	$FeCl_2$	green	soluble
Iron(II) sulphide	FeS	black	insoluble
Iron (I) sulphide	FeS_2	yellow	insoluble
Iron(II) phosphate	$Fe_3(PO_4^2)_2$	brown	insoluble
Iron(III) hexacyanoferrate(II) (Prussian blue) ion	$Fe^{III}[Fe^{II}(CN)_6]^-$	blue	soluble, $Fe_4[Fe(CN)_6]_3$ insoluble
Iron(II) hexacyanoferrate(III) ion (Turnbull's blue)	$Fe^{II}[Fe(CN)_6]^-$	blue	soluble
Complex of S^{2-} with sodium nitroprusside	$Na_2[Fe(CN)_5NOS]$	purple	soluble
Magnetite	Fe_3O_4	red	insoluble
	$[Fe(CNS)]^{2+}$	blood red	soluble
Iron(III) carbonate	$Fe_2(CO_3)_3$	red	insoluble
Iron(III) fluoride	FeF_3	white	insoluble
Iron(II) fluoride	FeF_2	green	insoluble
Iron(III) oxalate	$Fe_2(C_2O_4)_3$	blackish-red	insoluble
Iron(III) hydroxide	$Fe(OH)_3$	reddish-brown	insoluble
Mercury(II) sulphide	HgS	black	insoluble
Mercury(I) oxide	Hg_2O	black	insoluble
Mercury(II) chloride (corrosive sublimate)	$HgCl_2$	white	soluble
Mercury(I) chloride (calomel)	Hg_2Cl_2	white	insoluble

Salt/Complex	Formula	Colour	(Soluble/Insoluble) in water
Mercury(II) bromide	$HgBr_2$	white	
	$HgNH_2Cl \cdot HgO$	black	insoluble
Mercury(II) oxide	HgO	yellow	insoluble
Mercury(II) iodide	HgI_2	reddish-orange	insoluble
Iodide of Millon's base	NH_2–Hg–O–Hg–I	brown	insoluble
Mercury(I) chromate	Hg_2CrO_4	red	insoluble
Mercury(I) carbonate	Hg_2CO_3	yellowish-white	insoluble
Mercury(I) hydrogen phosphate	$HgHPO_4$	white	insoluble
Mercury(II) cyanide	$Hg(CN)_2$	white	soluble
Nickel sulphide	NiS	black	insoluble
Nickel chloride	$NiCl_2$	green	soluble
Nickel hydroxide	$Ni(OH)_2$	green	insoluble
Nickel(III) oxide	Ni_2O_3	black	insoluble
Silver chloride	$AgCl$	white	insoluble (soluble in aq. NH_3 as $[Ag(NH_3)_2]Cl$)
Silver bromide	$AgBr$	yellow	insoluble
Silver iodide	AgI	yellow	insoluble
Silver sulphide	Ag_2S	black	insoluble
Silver oxide	Ag_2O	black/brown	insoluble
Silver thiosulphate	$Ag_2S_2O_3$	white	insoluble
Silver nitrate	$AgNO_3$	white crystals	soluble
Silver phosphate	Ag_3PO_4	yellow	insoluble
Silver carbonate	Ag_2CO_3	yellowish-white	insoluble
Silver arsenate	Ag_3AsO_4	red	insoluble
Silver chromate	Ag_2CrO_4	red	insoluble
Silver sulphite	Ag_2SO_3	white	insoluble
Silver arsenite	Ag_3AsO_3	yellow	insoluble
Silver arsenate	Ag_3AsO_4	brownish-red	insoluble
Tin(IV) sulphide	SnS_2	yellow	insoluble
Tin(II) chloride	$SnCl_2$	white	soluble
Tin(II) sulphide	SnS	brown	insoluble
Tin(II) iodide	SnI_2	red	insoluble
Strontium carbonate	$SrCO_3$	white	insoluble
Strontium sulphate	$SrSO_4$	white	(insoluble in dilute mineral acid)
Strontium sulphite	$SrSO_3$	white	insoluble
Strontium oxalate	SrC_2O_4	white	insoluble in CH_3COOH but soluble in mineral acids